Dinichthys terrelli

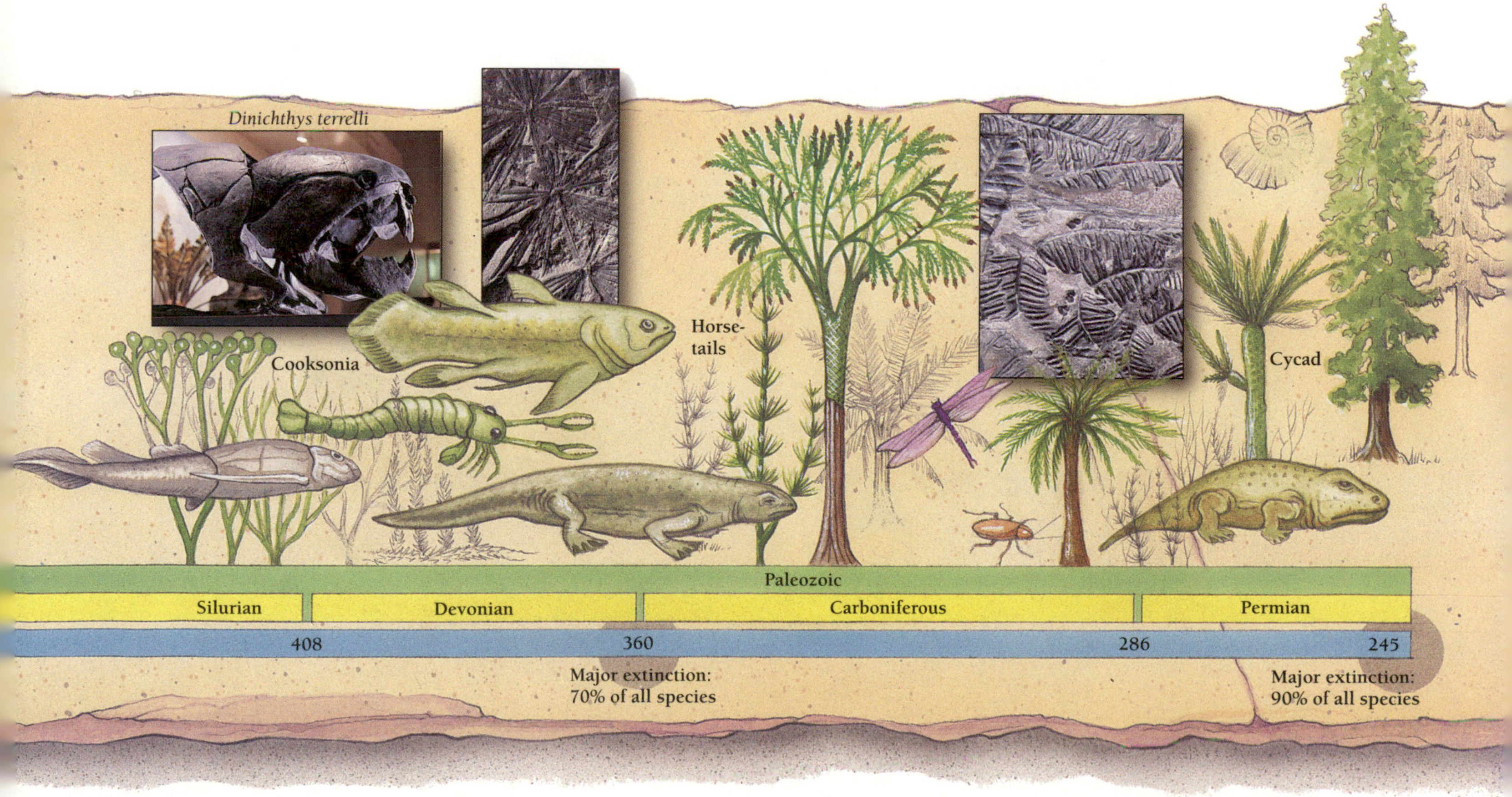

Cooksonia

Horse-
tails

Cycad

		Paleozoic		
Silurian	Devonian	Carboniferous		Permian
408	360		286	245

Major extinction:
70% of all species

Major extinction:
90% of all species

Asking About Life

Second Edition

Allan J. Tobin
*University of California,
Los Angeles*

Jennie Dusheck
*University of California,
Santa Cruz*

HARCOURT COLLEGE PUBLISHERS

Fort Worth Philadelphia San Diego New York Orlando Austin
San Antonio Toronto Montreal London Sydney Tokyo

Publisher: Emily Barrosse
Executive Editor: Nedah Rose
Senior Marketing Strategist: Kathleen Sharp
Senior Developmental Editor: Lee Marcott
Senior Project Editor: Bonnie Boehme
Senior Production Manager: Charlene Catlett Squibb
Senior Art Director: Caroline McGowan
Text Designers: Ruth A. Hoover and Kim Menning

Cover: African elephants near Mount Kilimanjaro, Tanzania *(Renee Lynn/Stone Images)*

Frontispiece: Copyright © 1999 Mauritius/Nawrocki Stock Photo, Inc. All Rights Reserved.

ASKING ABOUT LIFE, Second Edition
ISBN: 0-03-027044-8
Library of Congress Catolog Card Number: 99-068520

Address for domestic orders:
Harcourt College Publishers, 6277 Sea Harbor Drive, Orlando, FL 32887-6777
1-800-782-4479
e-mail collegesales@harcourt.com

Address for international orders:
International Customer Service, Harcourt, Inc.
6277 Sea Harbor Drive, Orlando, FL 32887-6777
(407) 345-3800
Fax (407) 345-4060
e-mail hbintl@harcourt.com

Address for editorial correspondence:
Harcourt College Publishers, Public Ledger Building, Suite 1250, 150 S. Independence Mall West, Philadelphia, PA 19106-3412

Web Site Address:
http://www.harcourtcollege.com

Printed in the United States of America

0123456789   032   10 987654321

Preface

Many good conversations start with a question. What do you think of the new ballpark downtown? Is the new sociology professor a good lecturer? Science is one long conversation, like a thousand-year cocktail party. And, as at any party, the topics of conversation often start with questions. Why are plants green? How do cells replicate?

The history of science reveals that the answers to such questions change over time, but the questions themselves, if they are good ones, remain the same. One good question has been, "How do new species form?" Ever since biologists realized that species do form, we have been trying to find out *how* they do it. But it is a question with many overlapping answers. The answers that seemed correct in 1880 or 1960 continue to be refined, expanded, or even overthrown.

In this new edition of *Asking About Life,* we present some new answers and some new approaches. But our starting point, that science is about curiosity, remains the same. We emphasize *how* and *why* scientists ask questions, how they test hypotheses, and how they reach conclusions. As much as possible, rather than merely presenting dry conclusions, we show our readers how science actually works. For example, we present the many pieces of evidence that convinced scientists that all life evolved from simple organisms. And we present the experimental steps that revealed how plants use sunlight to build sugar molecules, during photosynthesis.

Although our philosophy remains the same, we have improved the book in several important ways. Above all else, we have continued to refine and clarify our written explanations. As we write each sentence, we think first and foremost about how we can help our readers better understand biology.

We have added new metaphors and analogies, new lead-in stories, and new art. Some of our metaphors appear in the text alone, while others appear in figures as art or photos. As in the first edition, we compare chromosomes to socks and the functional groups of molecules to the attachments on a Swiss Army knife. But we have introduced new metaphors, such as the windmill in Chapter 25 that illustrates the idea that organisms recycle materials but not energy. Good visual metaphors help readers remember important concepts.

We regard illustration as a teaching tool in its own right, not just a reinforcement of the text. We have continued to work closely with nationally prominent illustrator Elizabeth Morales, whose skill, attention, and insight have contributed not only to the art manuscript but to the text as well. Morales, the current president of the Guild of Natural Science Illustrators, has been our art developmental editor since the first art meetings for the first edition. She has created a warm, friendly style that perfectly complements the conversational tone of our text. Her clean designs and precise illustrations greatly clarify sometimes difficult material. In addition, we retain many of the beautiful pieces done by natural science illustrator Elizabeth McClelland for the first edition.

We have updated and expanded our coverage of ecology. We think students need to know, for example, that ecology now includes global oceanic and atmospheric studies. Readers can also learn that population biologists study not only questions about how populations regulate their numbers (or fail to), but also the sad story, repeated all too often in modern times, of how species go extinct.

We have introduced a new section on statistics in Chapter 1. We hope this brief overview will be especially useful in

Stephen J. Krasemann/Photo Researchers, Inc.

classes where students encounter the first rudiments of statistics in a lab component of a course. Our purpose, however, is not to teach statistics *per se,* but to help students see why statistics is an essential analytical tool in biology and how it affects experimental design. We hope our brief introduction to the concepts of sample size and variation will prompt students to ask skeptical questions about science stories in the news.

Despite all these new features, the basic vision of *Asking About Life* remains the same. We continue to present biology in the context of the development of biological ideas, never as disconnected facts. We present a relatively few ideas in some depth rather than many ideas superficially.

Asking About Life consistently emphasizes the importance of questions and the process of finding answers. To remind ourselves and our readers of this emphasis, headings and subheadings are more often questions than statements. In addition, at the end of most subsections we summarize the main points covered. The question heading and the summary statement together act as a reality check for both readers and the authors. In our first edition, many readers pointed out that we didn't always answer the questions in the headings. Did we ask the wrong question? Or, did we not answer our own question? Truth to tell, sometimes the former, sometimes the latter. In this edition, we have taken special pains to improve our questions, as well as answers.

Fixing our mistakes has convinced us more than ever that asking explicit questions is the best way to talk about science. We have done our best to improve the question headings. But we welcome sharp-eyed readers to catch us up now and again. Readers of this edition who do not think we have answered the question in the heading should alert us at jennie.dusheck@pobox.com. We look forward to hearing from you.

Refining questions and fixing mistakes is the way science works. What scientist hasn't been brought up short by an insightful comment from a colleague? The way questions, or hypotheses, are phrased determines how they can be answered. Sometimes scientists ask the wrong questions. Sometimes they come up with the wrong answers to the right questions. The process is a delightfully social one, in which everyone can play a key role.

Throughout *Asking About Life,* we have emphasized the passionate engagement of individual scientists. Each chapter begins with a story illustrating how an individual biologist or group of biologists pursued a scientific question—often in the face of intense intellectual and social adversity. We tell the story, for example, of how Barry Marshall convinced first himself and then others that bacteria, not "frustrated personalities," cause ulcers; and how Rosalind Franklin struggled in deep social isolation to elucidate the structure of DNA. Our anecdotes are about real people—their triumphs, their frustrations, their genius, and their persistence. Biology is a story, and, as such, it must be presented as continuously as possible.

In every kind of learning, story and context give shape and meaning to dry facts and ideas. When we are told that someone named Charles Drew invented a way to supply clean blood plasma, a new technology that saved many lives, we may remember this fact briefly. But what if, as happens in Chapter 37, we also learn that Drew was one of only a handful of African-American physicians in the United States in the 1940s? At that time, black physicians did little or no research, but were mostly relegated to quiet country practices in the rural South. Yet Drew ran the "Plasma for Britain" program for the American Red Cross during World War II and invented a way to preserve blood plasma so that it could be shipped across the Atlantic to Great Britain. His work saved the lives of thousands of Londoners injured by Nazi bombs. Midway through the program, however, the American Red Cross decided to refuse blood from any American donor whose skin was black, and Drew resigned in protest. An eminent professor of surgery at Howard University, Drew suffered serious injuries in an auto accident a few years later. Within hours of the accident, he was dead. He bled to death in a small Southern emergency room where neither whole blood nor plasma could save him.

Life's too short not to hear the whole story. Our philosophy of telling the stories of biology sometimes runs counter to current trends. Many textbooks break up information into modules in which each bit of information is presented dictionary-fashion, as if unrelated to the information in the rest of the book. Important ideas may be marooned on illustration islands. The result is a series of disconnected facts. Publishers say that such an approach is necessary because today's students belong to the "visual-information generation" and are incapable of sustained reading or synthesis.

We have more faith in students. We know they enjoy reading, provided the reading is interesting and rewarding. Indeed, student reviewers have raved about the clarity of the writing and the engaging stories that reveal scientists as ordinary human beings.

In the evolution section, for example, we mention the intense frustration Charles Darwin experienced while struggling to distinguish among different species of barnacles. Each species seemed to blend into the next. In the next chapter, on classification, we show readers that biologists have been arguing about the definition of a species for nearly 200 years. In the context of Darwin's difficulties, the highly politicized debate over whether the red wolf is a species (that deserves legal protection) or a hybrid (that doesn't) takes on a different and deeper meaning. At the same time, the red wolf story brings to life what might otherwise be an abstract discussion of classification.

While we stress the continuity of ideas within a discipline, however, we know that every instructor takes a different approach to teaching biology. Therefore, each unit of the book is understandable on its own terms. Nothing prevents instructors from teaching the units in a different order.

The first edition of *Asking About Life* was well received. The Italian publisher Edizioni Bruno Mondadori selected *Asking About Life* for translation into Italian. And, in 1999, the

Text and Academic Authors association conferred on the book an award of excellence for best new textbook in life sciences. But we are most proud of the many instructors and students who have told us they love our book. Thanks to them for their remarkable support.

Who Made This Book?

Our developmental editor, Lee Marcott, continues to be an active participant in the work on *Asking About Life.* In addition to editing the manuscript, and choosing excellent reviewers, Lee has also tactfully managed the two authors—pushing us to meet deadlines, organizing us, and giving us pep talks.

Photo researcher Amy Ellis Dunleavy's novel photo ideas continue to give the book much of its personality. Project editor Bonnie Boehme kept the flow of paper moving smoothly during production and broke up many a logjam with quick-witted solutions and generous contributions of her own time. Art director Caroline McGowan's consistent sense of design gives the book much of its unity and elegance, and her sharp eye caught mistakes in biology as well as the occasional misplaced "leader line" or missing label.

We also thank production manager Charlene Catlett Squibb, manager of art and design Carol Bleistine, and designers Ruth Hoover and Kim Menning. And we deeply appreciate the work of many others who have contributed to this book—including Donald Jackson, Michael Brown, Edward Murphy, Elizabeth Widdicombe, Julie Alexander, and Edith Beard Brady. We thank especially publisher Emily Barrosse for her special combination of toughness and warmth and executive editor Nedah Rose for her cool head and patience. Senior marketing strategist Kathleen Sharp and field editorial specialist Dave Theisen have made creative contributions to the sales and marketing efforts of both the first and second editions.

Special thanks also to Naomi Cappuccino of Carleton University, Mary Kay Cassani of Edison Community College, Kerry S. Kilburn of Old Dominion University, and Jorge Rey of the University of Florida, all of whom helped us with the ecology section in different ways. We also thank science writer Marina Chicurel, who wrote the new story about Stanley Prusiner (Chapter 3); science writer Liese Greensfelder, who revised the discussion of phylloxera in Chapter 31; Marni Fylling, who wrote and illustrated the accompanying lab manual; Harry W. Greene of Cornell and Daniel J. Meinhardt of St. Olaf College, both of whom helped ease us through the subtleties of cladistics; Nancy Segal of Cal State Fullerton, who provided the striking pictures of twins in Chapter 9 and suggested changes to the text; Barney Schlinger and Gordon Fain, of UCLA, Warren Burggren, of the University of North Texas, and Michael Stryker of UCSF for their help in refocusing some of the physiology chapters; John E. Wilkes of the University of California, Santa Cruz, who read and skillfully copyedited many chapters of this edition; Primavera Hernandez, Annie Gallo, April Hosein, and Vanessa Flores, who helped with the book in dozens of ways; and the students in Life Sciences 2 at UCLA, whose comments, suggestions, and enthusiasm were especially valuable in preparing the second edition. Finally, we thank Richard Robinson, Mary K. Miller, Karen Scanlon, and Natalie Peretti for their work on the Web site that accompanies this book.

We thank all the reviewers who took the time to read and comment on this manuscript—correcting our errors, asking thought-provoking questions, and suggesting examples, alternative wordings, or new ways of thinking. Although we have met only a few of our reviewers in person, working with them has been a rewarding intellectual experience. Both the process of writing this book and the resulting book itself would not have been the same without the reviewers.

What Kinds of Pedagogy Does *Asking About Life* Employ?

Asking About Life has a variety of features designed to engage the reader and to aid student learning. Each chapter begins with a story about a piece of research that draws students into the subject of the chapter and also introduces the key questions and ideas that are discussed. The chapter-opening image is described at the end of each chapter in a new feature called **About the Chapter-Opening Image.**

At the start of each chapter, readers will find a list of **Key Concepts,** which are the most basic ideas covered in the chapter. Most headings are posed in the form of questions. These question headings focus the reader's attention on the most significant question to be explored in that section. Following most subsections is a **Summary Statement**—a brief summary of the take-home message. Summary Statements provide students with a reality check. If the student doesn't understand the summary statement, that is a cue to study the preceding material more closely.

Visual metaphors rendered as photos or illustrations drive home key points introduced in the text. Drawings and photographs support concepts covered in the text and help students visualize the structures of objects as diverse as molecules and ecological communities. Photographs of structures that are too small to be seen with the naked eye are accompanied by size bars to give a sense of scale. And **tables** and **graphs** summarize key facts and additional material. Most chapters feature one or more **Boxes,** which introduce science in the news or discuss special topics in greater depth. Figure legends are designed to stand alone, so that a student flipping through the chapter for the first time, glancing at the diagrams and reading the legends, will learn something.

 The second edition of *Asking About Life* introduces our book-specific Web site with additional material in the form of **Web Bits.** An icon in the margin lets the reader know that a trip to the *Asking About Life* Web site offers related information, extra exercises, quizzing, and so on.

Throughout the text, **boldface key terms** help students to locate key terms and their definitions. At the end of each chapter, all of the boldface terms are used again in a highly compressed summary called the **Study Outline with Key Terms.** The Study Outline provides students with another opportunity to check their understanding of the chapter. If they encounter terms they don't remember or ideas that seem unfamiliar, they can return to the main text and illustrations.

Following the Study Outline is a set of **Review and Thought Questions.** Our Thought Questions are especially engaging, frequently bringing ideas in the chapter into the everyday world. At the end of the book, a periodic table (Appendix A), a table of standard weights and measures (Appendix B), and a **Glossary** provide useful reference information for students.

Supplements

To further facilitate learning and teaching, we provide a carefully designed supplements package for students and instructors.

Marni Fylling's **Laboratory Manual to accompany Asking About Life,** Second Edition, covers laboratory topics drawn from six units of the text. Each of the 15 labs accompanies a specific chapter (or chapters) from the book. These interactive labs use familiar materials to give students a hands-on approach to basic biological principles. Author Marni Fylling is an accomplished artist/illustrator, and the manual is fully illustrated with nearly 150 full-color line drawings and photographs.

The **Study Guide,** by Lori Garrett of Danville Area Community College, includes the Key Concepts, an Extended Chapter Outline that gives an overview of the most important topics covered in the chapter, Vocabulary Building exercises, and Chapter Tests. Each Chapter Test has four parts: Multiple Choice; Matching; Short Answer; and Critical Thinking—Using Your Knowledge. All answers are provided, with the exception of the Critical Thinking—Using Your Knowledge questions.

A Guide to Asking About Life for Teachers and TAs by Donald Cronkite of Hope College is an instructor's manual for teaching biology using the inquiry-based approach of the textbook. It includes suggested syllabi for teaching using various themes, such as biodiversity, cellular and molecular approaches, and social issues, as well as one- and two-semester courses. Chapters offer intriguing demonstrations, interactive exercises, and group learning. The Preface provides a listing of where the various types of exercises can be found in the manual. Answers to the Review and Discussion Questions from the textbook also appear in the teaching guide.

The **Test Bank,** by Alma Moon Novotny of the University of St. Thomas, and Frederick Peabody of University of South Dakota, consists of 2950 questions of assorted type (multiple choice, fill-in-the-blank, and short-answer essay questions). These are organized by the main chapter headings. The **Computerized Test Bank** is available for both Windows™ and Macintosh® platforms. The computer program allows the instructor to sort the questions by chapter, section head, and question type.

The *Asking About Life* supplements package also includes a set of 200 **Overhead Transparencies** consisting of drawings from the book and a **Student Art Notebook,** which is a set of 300 unlabeled, full-color line drawings from the text. Our Custom Overhead Transparency Program allows instructors to choose additional four-color images from the text.

An **Instructor's Resource CD-ROM for Biology 2001** features selected photographs and all the line art illustrations from *Asking About Life,* Second Edition, and can be used with PowerPoint™ and Persuasion™ on both Macintosh® and Windows™ platforms. A variety of file formats include Web-ready PDFs, print-ready PDFs, and files for use in creating custom overhead transparencies and custom PowerPoint™ presentations. Eric Rabitoy and Terry Damron of Citrus College have created a **PowerPoint™ Presentation** that is available at the *Asking About Life* Web site for downloading as well as on the IRCD-ROM.

The **Process of Science: Discovering Biology™** CD-ROM reflects the spirit of inquiry that characterizes *Asking About Life,* Second Edition. It allows students to explore the discoveries of some of the most important concepts in biology. In the *Interactive Investigations,* students retrace the steps of scientists' experiments and discoveries using the scientific process as their road map. An *Investigator's Notepad* allows the students to track their progress through each investigation, to pose new questions for themselves to pursue, or to initiate a discussion with the instructor or other students via an Internet connection. *Concept Tutorials* provide students with essential background information in general biology for the course and the investigations.

The CD-ROM set **Introduction to Biology,** created by Archipelago Distributed Learning, consists of four CD-ROM disks with an **Instructor's Guide** and a **Student User Guide.** The course covers all the topics that would be covered in a two-term introduction to general biology and can be used as an ancillary to any general biology text, such as *Asking About Life.* Heavily illustrated and complete with animations and videoclips it can be used by students for self-study or as projection software for the instructor. A demonstration of the course is posted at the *Asking About Life* site.

A **WebCT** course for *Asking About Life,* Second Edition, is in development for the text, the first semester of which will be ready for use in Fall, 2000. A demonstration of Chapter 4 from the WebCT course can be accessed at the book site.

Please visit our *Asking About Life* Web site at http://www.harcourtcollege.com/lifesci/aa2/

Students will find:

- A set of Web links for each chapter, including a bibliography of books, readings, and other media for each unit of the text.

- Sets of quiz questions and answers for self-testing of each chapter.
- On-line glossary of words in *Asking About Life*.
- View-only versions of the PowerPoint™ Presentation.
- **Web Bits**—short articles that expand ideas from certain chapters. (These are indicated by an iconic spider web.)
- Samples of the **Lab Manual, Study Guide,** and **Student Art Notebook** will be posted so that students are aware that these study aids are available to them for purchase.

Instructors will find:

- The **Teacher's Guide** posted chapter-by-chapter.
- Sample syllabi for one- and two-semester courses and courses emphasizing a particular theme or approach.
- Answers to the Review and Thought Questions in the text.
- Downloadable **PowerPoint™ Presentations** for all the line drawings in the text.

- **Test Bank** for the Second Edition.
- Image bank of the line art illustrations from the text.
- WebCT demo of Chapter 4.

Harcourt College Publishers may provide complimentary instructional aids and supplements or supplemental packages to those adopters qualified under our adoption policy. Please contact your sales representative for more information. If, as an adopter or potential user, you receive supplements you do not need, please return them to your sales representative or send them to:

Attn: Returns Department
Troy Warehouse
465 South Lincoln Drive
Troy, MO 63379

Reviewers of the Second Edition

Susan Bandoni, *SUNY Geneseo*

Claudia Barreto, *University of Wisconsin, Milwaukee*

Mark S. Blackmore, *Valdosta State University*

Steve Blumenshine, *Arkansas State University*

Susan L. Bower, *Pasadena City College*

Sara Brenizer, *Shelton State Community College*

Arthur L. Buikema, Jr., *Virginia Polytechnic Institute and State University*

Mary Kay Cassani, *Edison Community College, Lee Campus*

Kerry L. Cheesman, *Capital University*

Harold Cones, *Christopher Newport University*

Walter Conley, *St. Petersburg Junior College*

Patricia B. Cox, *University of Tennessee*

Charles Creutz, *University of Toledo*

Forbes Davidson, *Mesa State College*

Jerry D. Davis, *University of Wisconsin, LaCrosse*

David DeGroote, *St. Cloud State University*

Jean DeSaix, *University of North Carolina, Chapel Hill*

Donald Deters, *Bowling Green State University*

Matthew Douglas, *Grand Rapids Community College* and *the University of Kansas*

Patrick Duffie, *Loyola University, Chicago*

Thomas C. Emmel, *University of Florida*

Kathy McCann Evans, *Reading Area Community College*

Robert Evans, *State University of New Jersey Rutgers, Camden*

Victor Fet, *Marshall University*

Richard F. Firenze, *Broome Community College*

James Fitch, *Jones Junior College*

Jennifer Fritz, *University of Texas, Austin*

Shirley Porteous-Gafford, *Fresno City College*

Wendy Jean Garrison, *University of Mississippi*

Robert George, *University of Wyoming*

Karen F. Greif, *Bryn Mawr College*

Charles J. Grossman, *Xavier University*

Lonnie J. Guralnick, *Western Oregon University*

Betsy Harris, *Appalachian State University*

Patricia Hauslein, *St. Cloud State University*

Chris Haynes, *Shelton State Community College*

James A. Hewlett, *Finger Lakes Community College*

Nan Ho, *Las Positas College*

Margaret Hollyday, *Bryn Mawr College*

Stephen Hudson, *Furman University*

Charles W. Jacobs, *Henry Ford Community College*

John D. Jenkin, *Blinn College, Bryan*

Victoria Johnson, *San Jose State University*

Diane Auer Jones, *Community College of Baltimore County, Catonsville Campus*

Marlene Kayne, *College of New Jersey*

Chris Kellner, *Arkansas Tech University*

Robert M. Kitchin, *University of Wyoming*

Dan E. Krane, *Wright State University*

Keith A. Krapf, *John A. Logan College*

Cheryl Laursen, *Eastern Illinois University*

Siu-Lam Lee, *University of Massachusetts, Lowell*

Thomas Lord, *Indiana University of Pennsylvania*

Jon H. Lowrance, *Lipscomb University*

Craig E. Martin, *University of Kansas*

Timothy D. Metz, *Campbell University*

Alison M. Mostrom, *University of the Sciences in Philadelphia*

Anne-Marie Murray, *Arkansas Tech University*

Ken Nadler, *Michigan State University*

Judy H. Niehaus, *Radford University*

Debra Pearce, *Northern Kentucky University*

Andrew J. Penniman, *Georgia Perimeter College*

Helen K. Pigage, *U.S. Air Force Academy*

Jay Pitocchelli, *Saint Anselm College*

David Polcyn, *California State University, San Bernardino*

Steven M. Pomarico, *Louisiana State University*

Carol Pou, *St. Cloud State University*
Paul A. Rab, *Sinclair Community College*
Eric Rabitoy, *Citrus College*
Lynda Randa, *College of DuPage*
David Ribble, *Trinity University*
Laurel Roberts, *University of Pittsburgh*
Frank A. Romano, III, *Jacksonville State University*
Christopher S. Sacchi, *Kutztown University*
K. Sata Sathasivan, *University of Texas, Austin*
Robert M. Schoch, *Boston University*
Brian W. Schwartz, *Columbus State University*
Angelica P. Seitz, *Auburn University*
David Seigler, *University of Illinois at Urbana-Champaign*
Gail Stratton, *University of Mississippi*
Robert J. Swanson, *North Hennepin Community College*
Sandra J. Turner, *St. Cloud State University*
Linda Tyson, *Santa Fe Community College*
Jack Waber, *West Chester University*
Ken Revis-Wagner, *Clemson University*
Timothy S. Wakefield, *Auburn University*
Tom Weeks, *University of Wisconsin, LaCrosse*
J.D. Wilhide, *Arkansas State University*
Edmund B. Wodehouse, *Skyline College*
Anne E. Zayaitz, *Kutztown University*
Henry H. Ziller, *Southeastern Louisiana University*

The following reviewers graciously shared their expertise with the authors in the effort to maintain currency and accuracy across the many sub-disciplines in biology:

Warren Burggren, *University of North Texas*
Naomi Cappuccino, *Carleton University*
Richard P. Elinson, *University of Toronto*
Carol Erickson, *University of California, Davis*
Gordon Fain, *University of California, Los Angeles*
Robert Full, *University of California, Berkeley*
Harry W. Greene, *Cornell University*
Susan Hughmanick, *University of California, Santa Cruz*
Laura J. Jenski, *Indiana University–Purdue University at Indianapolis*
Peter Kareiva, *University of Washington*
Daniel J. Meinhardt, *St. Olaf College*
Barney Schlinger, *University of California, Los Angeles*
Cynthia V. Sommer, *University of Wisconsin, Milwaukee*
Michael P. Stryker, *University of California, San Francisco*
Chris Tarp, *Contra Costa College*
Robert Thornton, *University of California, Davis*
Robert Turgeon, *Cornell University*
Philip C. Withers, *University of Western Australia*

Reviewers of the First Edition

Juan Aninao, *Dominican College of San Rafael*
Edwin A. Arnfield, *Macomb Community College*
Linda W. Barham, *Meridian Community College*
George W. Barlow, *University of California, Berkeley*
Brenda Blackwelder, *Central Piedmont Community College*
Mildred Brammer, *Ithaca College*
Richard B. Brugam, *Southern Illinois University, Edwardsville*
William Bowen, *Jacksonville State University*
Bradford Boyer, *Suffolk County Community College*
Hara Dracon Charlier, *Miami University*
H. Tak Cheung, *Illinois State University*
Karen Crombie, *Fresno City College*
Donald Cronkite, *Hope College*
Tom Daniel, *University of Washington*
Darleen A. DeMason, *University of California, Riverside*
Leah Devlin, *Pennsylvania State University, Abington College*
Ernest F. DuBrul, *University of Toledo*
Peter Ducey, *State University of New York at Cortland*
Steven H. Everhart, *Campbell University*
Lynn Fancher, *College of DuPage*
Cynthia Fitch, *Seattle Pacific University*
Dietrich Foerstel, *Champlain Regional College & Bishop's University*
Sally Frost-Mason, *University of Kansas*
Jack Gallagher, *William Rainey Harper College*
Lori K. Garrett, *Danville Area Community College*
Ben R. Golden, *Kennesaw State College*

Glenn A. Gorelick, *Citrus College*
Nels H. Granholm, *South Dakota State University*
Herbert H. Grossman, *The Pennsylvania State University*
Lonnie J. Guralnick, *Western Oregon State College*
Ross Hamilton, *Okaloosa-Walton Community College*
Richard Harrison, *Cornell University*
Wiley Henderson, *Alabama A&M University*
Bob Highley, *Bergen Community College*
Kathleen L. Hornberger, *Widener University*
Linda Hsu, *Seton Hall University*
David Inouye, *University of Maryland*
William A. Jensen, *The Ohio State University*
J. Morris Johnson, *West Oregon State University*
Peter Kareiva, *University of Washington*
James Karr, *University of Washington*
Tanseem Khaleel, *Montana State University, Billings*
Robert Kitchin, *University of Wyoming*
Ross Koning, *Eastern Connecticut State University*
Dan Krane, *Wright State College*
James W. Langdon, *University of South Alabama*
Anton Lawson, *Arizona State College*
Charles Leavell, *Fullerton Community College*
Kathleen Lively, *Marquette University*
Melanie Loo, *California State University, Sacramento*
Linda A. Malmgren, *Franklin Pierce College*
Nilo Marin, *Broward Community College*
Theresa Martin, *College of San Mateo*

Dorrie Matthews, *Sage Jr. College of Albany*

Gary F. McCracken, *University of Tennessee, Knoxville*

Robert J. McDonough, *DeKalb College*

Joseph McGrellis, *Atlantic Community College*

John Mertz, *Delaware Valley College*

Debbie Meuler, *Cardinal Stritch College*

Robert Morris, *Widener University*

Alison M. Mostrom, *Philadelphia College of Pharmacy and Science*

David M. Ogilvie, *The University of Western Ontario*

Bruce Parker, *Utah Valley State College*

Lee R. Parker, *California Polytechnic State University, San Luis Obispo*

Frederick Peabody, *University of South Dakota*

Ed Perry, *Faulkner State Community College*

Gary Pettibone, *State College of New York at Buffalo*

Jay Pitocchelli, *St. Anselm College*

David M. Polcyn, *California State University, San Bernadino*

Shirley Porteous-Gafford, *Fresno City College*

Paul F. Ramp, *University of Tennessee, Knoxville*

Franklin Robinson, *Mountain Empire Community College*

Lyndell Robinson, *Lincoln Land Community College*

Earle Rowe, *Walters State Community College*

Donna Rowell, *Holmes Community College*

Andrew Scala, *Dutchess Community College*

Shirley Seagle, *Jacksonville State University*

Phillip R. Shelp, *Brookhaven College*

John Simpson, *Gadsden State College*

Linda Simpson, *University of North Carolina, Charlotte*

Michael E. Smith, *Valdosta State University*

Fred L. Spangler, *University of Wisconsin, Oshkosh*

Steven Spilatro, *Marietta College*

Herbert Stewart, *Florida Atlantic University*

Cindy Stokes, *Kennesaw State College*

Jeffrey Thompson, *California State University, San Bernadino*

Janice Toyoshima, *Bakersfield College*

John Tramontano, *Orange County Community College*

Michael J. Ulrich, *Elon College*

Kristin Vessay, *Bowling Green University*

James A. Winsor, *The Pennsylvania State University, Altoona Campus*

Daniel Wivagg, *Baylor University*

Tom Worcester, *Mt. Hood Community College*

For George Dusheck and Eve Tobin
And in Loving Memory of Nina Dusheck and Maurice Tobin

About the Authors

Allan J. Tobin

Is Director of the UCLA Brain Research Institute. He holds the Eleanor Leslie Chair in Neuroscience at UCLA, where he is both Professor of Neurology and Professor of Physiological Science. Tobin is also Scientific Director of the Hereditary Disease Foundation (HDF), where he helped organize the consortium that identified the gene responsible for Huntington's disease. Both at UCLA and at HDF, he has encouraged close interactions among cell and molecular biologists, geneticists, physiologists, and clinicians, with the goal of bringing basic science discoveries from the laboratory bench to the patient's bedside.

Tobin's undergraduate degree is in literature and biology (MIT, 1963), and his doctoral degree is in biophysics, with an emphasis on physical biochemistry (Harvard, 1969). Tobin did postdoctoral work at the Weizmann Institute of Science, in Israel, and at MIT. At UCLA, his active research laboratory studies the production and action of GABA, the major inhibitory signal in the brain. These studies may eventually lead to new therapeutic approaches to epilepsy, Huntington's disease, and juvenile diabetes. Tobin is the recipient of a Jacob Javits Neuroscience Investigator Award from the National Institute of Neurological Disorders and Stroke.

For more than 35 years, Tobin has taught introductory courses in general biology, cell biology, molecular biology, developmental biology, and neuroscience. He is the recipient of a UCLA Faculty Teaching and Service Award and is regarded as an excellent and highly interactive teacher.

Jennie Dusheck

Is a writer living in Santa Cruz, California. She is a lecturer in science writing and illustration at the University of California, Santa Cruz, and a member of the National Association of Science Writers. Her undergraduate degree is in zoology (U.C. Berkeley, 1978) and her master's degree (by thesis) is in zoology (U.C. Davis, 1983). She also holds a certificate in science writing from U.C. Santa Cruz (1985).

Dusheck has written for *Science News, Science* magazine, and other publications. From 1985 to 1993, she worked as a Principal Editor at U.C. Santa Cruz. She has received several national awards, including the Gold Medal for Best-in-Category from the National Council for the Advancement and Support of Education. Besides *Asking About Life,* she recently coauthored *Life Science,* a middle school text in the Holt Science & Technology Series.

In her previous life as a biologist, Dusheck studied the effects of light intensity on bird song; social behavior in field mice; food preferences of deer, cattle, and skipper butterflies; as well as axis formation in *Xenopus laevis*. While working for the Department of Molecular Biology at U.C. Berkeley, she designed and wrote the protocol for a Space Shuttle experiment that sent live frog embryos into space in the fall of 1992. She has taught university labs in introductory zoology, embryology, and comparative anatomy.

Contents Overview

1 The Unity and Diversity of Life   2

I Chemistry and Cell Biology   21

2 The Chemical Foundations of Life   22
3 Biological Molecules Great and Small   46
4 Why Are All Organisms Made of Cells?   70
5 Directions and Rates of Biochemical Processes   102
6 How Do Organisms Supply Themselves
   with Energy?   118
7 Photosynthesis: How Do Organisms Get Energy
   from the Sun?   140

II Genetics: The Continuity of Life   161

8   Cell Reproduction   162
9   From Meiosis to Mendel   180
10 The Structure, Replication, and Repair of DNA   208
11 How Are Genes Expressed?   234
12 Jumping Genes and Other Unconventional
    Genetic Systems   258
13 Genetic Engineering and Recombinant DNA   276
14 Human Genetics   296

III Evolution   317

15 What Is the Evidence for Evolution?   318
16 Microevolution: How Does a Population Evolve?   344
17 Macroevolution: How Do Species Evolve?   368
18 How Did the First Organisms Evolve?   398

IV Diversity   417

19 Classification: What's in a Name?   418
20 Prokaryotes: Success Stories, Ancient and Modern   432
21 Classifying the Protists and Multicellular Fungi   452
22 How Did Plants Adapt to Dry Land?   476
23 Protostome Animals: Most Animals Form
    Mouth First   496

24 Deuterostome Animals: Echinoderms and
    Chordates   522

V Ecology   543

25 Ecosystems   544
26 Terrestrial and Aquatic Ecosystems   562
27 Communities: How Do Species Interact?   584
28 Populations: Extinctions and Explosions   602
29 The Ecology of Animal Behavior   626

VI Structural and Physiological Adaptations of Flowering Plants   641

30 Structural and Chemical Adaptations of Plants   642
31 What Moves Water and Sugars?   660
32 Growth and Development of Flowering Plants   676
33 How Do Plant Hormones Regulate Growth and
    Development?   694

VII Structural and Physiological Adaptations of Animals   709

34 Form and Function in Animals   710
35 How Do Animals Obtain Nourishment
    from Food?   734
36 How Do Animals Coordinate the Actions of Cells
    and Organs?   754
37 How Do Animals Move Blood Through
    Their Bodies?   770
38 How Do Animals Breathe?   788
39 How Do Animals Manage Water, Salts,
    and Wastes?   808
40 Defense: Inflammation and Immunity   826
41 The Cells of the Nervous System   844
42 The Nervous System and the Sense Organs   862
43 Sexual Reproduction   882
44 How Do Organisms Become Complex?   906

Contents

1 THE UNITY AND DIVERSITY OF LIFE 2

Enough To Give You an Ulcer 2

HOW DO BIOLOGISTS ASK QUESTIONS? 4
Do Scientists Use the Scientific Method? 4
How Do Scientists Design Experiments? 7
Why Do Biologists Study Groups of Organisms? 7
BOX 1.1 Reductionism 8
How Do Biologists Use Statistics To Plan and Evaluate
Experiments? 10
What Is a Theory? 13
HOW ARE ALL ORGANISMS ALIKE? 13
How Do Organisms Self-Regulate? 13
What Kinds of Cells Are Organisms Made Of? 15

HOW ARE ORGANISMS DIFFERENT FROM ONE
ANOTHER? 16
How Do Organisms Become Different from One
Another? 16
Species Differ in Their Adaptations to Distinct
Environments 17
How Do We Know That the Diversity of Life Resulted
from Evolution? 18

Study Outline with Key Terms 20
Review and Thought Questions 20

Flip Nicklin/Minden Pictures

**I Chemistry and Cell
Biology 21**

**2 THE CHEMICAL FOUNDATIONS
OF LIFE 22**

Just Say NO 22

WHAT IS MATTER? 23
What Is Special About the Chemistry of Life? 24
WHAT DETERMINES THE PROPERTIES OF AN
ATOM? 26
What Are Atoms Made Of? 26
What Is the Internal Structure of an Atom? 27
BOX 2.1 The Uses and Dangers of Radioisotopes 29
Where Are the Electrons in an Atom? 30
WHAT HOLDS MOLECULES TOGETHER? 30
Covalent and Ionic Bonds Hold Atoms Together Strongly 30
What Kinds of Forces Hold Separate Molecules
Together? 35
HOW IS WATER ESPECIALLY WELL SUITED
FOR ITS ROLE IN LIFE? 36
Why Does Ice Float? 36
How Does Water Resist Changes in Temperature? 36
Why Do Water Molecules Cling to One Another? 39
Why Do Water Molecules Cling to Other Substances? 39
Why Is Water Such a Powerful Solvent? 39
Water Participates in Many Biochemical Reactions 40
How Do Water Molecules Change in Solution? 40
Why Is pH Important to Organisms? 42
Buffers: How Do Organisms Resist Changes in pH? 43

Study Outline with Key Terms 44
Review and Thought Questions 45

3 BIOLOGICAL MOLECULES GREAT AND SMALL   46

Prions: A Slow-Growing Idea   46

HOW DO ORGANISMS BUILD BIOLOGICAL MOLECULES?   49

How Big Are Biological Molecules?   49

Why Are Biological Structures Made from So Few Building Blocks?   49

What Kinds of Structures Do Carbon Atoms Form?   49

Functional Groups: How Do the Small Molecules Differ from One Another?   50

How Do Cells Build Complex Molecules?   52

What Links the Building Blocks of Life?   52

LIPIDS INCLUDE A VARIETY OF SMALL NONPOLAR COMPOUNDS   53

How Are Fatty Acids Linked Together?   56

BOX 3.1 The Biochemistry of Cholesterol   58

HOW DO SUGARS FORM POLYSACCHARIDES?   59

How Do Organisms Use Sugars To Store Energy?   60

How Else Do Organisms Use Polysaccharides?   60

HOW DO NUCLEOTIDES LINK TOGETHER TO FORM NUCLEIC ACIDS?   61

How Are the Nucleotides of a Nucleic Acid Linked Together?   61

HOW DO AMINO ACIDS LINK TO FORM POLYPEPTIDES?   63

What Distinguishes the 20 Amino Acids from One Another?   64

How Do Amino Acids Join Together?   64

What Do Proteins Look Like?   65

HOW DO CELLS CONSTRUCT PROTEINS?   66

How Does the Folding of a Protein Depend on Interactions Among Its Amino Acids?   66

Are the Interactions Among Amino Acids Enough To Determine Protein Structure?   67

What Factors Determine How Proteins Fold?   68

Study Outline with Key Terms   68

Review and Thought Questions   69

4 WHY ARE ALL ORGANISMS MADE OF CELLS?   70

Very Little Animalcules   70

WHY ARE ALL ORGANISMS MADE OF CELLS?   71

All Organisms Are Made of Cells   71

BOX 4.1 How Do Microscopes Help Biologists Study Cells?   72

Every Cell Consists of a Boundary, a Cell Body, and a Set of Genes   74

How Are Cells Alive?   75

What Are the Advantages of Cellular Organization?   75

WHAT'S IN A CELL?   77

What Role Does the Nucleus Play in the Life of a Cell?   79

The Cytosol Is the Cytoplasm That Lies Outside the Organelles   80

The Endoplasmic Reticulum Is a Folded Membrane   81

The Golgi Complex Directs the Flow of Newly Made Proteins   82

The Lysosomes Function as Digestion Vats   83

Peroxisomes Produce Peroxide and Metabolize Small Organic Molecules   84

Mitochondria Capture the Energy from Small Organic Molecules in the Form of ATP   84

A Plant Cell's Chloroplast Is Just One Kind of Plastid   84

What Does the Cytoskeleton Do?   86

WHAT DO MEMBRANES DO?   87

What Kinds of Molecules Do Membranes Contain?   87

BOX 4.2 The 9 + 2 Structure of Flagella   89

How Does the Structure of a Membrane Establish Its Functional Properties?   90

HOW DO MEMBRANES REGULATE THE SPACES THEY ENCLOSE?   91

Why Does Water Move Across Membranes?   91

What Determines the Movement of Molecules Through a Selectively Permeable Membrane?   93

HOW DO MEMBRANES INTERACT WITH THE EXTERNAL ENVIRONMENT?   96

Membrane Fusion Allows the Import and Export of Particles and Bits of Extracellular Fluid   96

How Do Cells Communicate in Multicellular Organisms?   97

Study Outline with Key Terms   100

Review and Thought Questions   101

5 DIRECTIONS AND RATES OF BIOCHEMICAL PROCESSES   102

Ludwig Boltzmann: Left Behind, or Ahead of His Time?   102

WHAT DETERMINES WHICH WAY A REACTION PROCEEDS?   104

How May Work Be Converted to Kinetic or Potential Energy?   104

How Does Thermodynamics Predict the Direction of a Reaction?   104

How Do Changes in Free Energy Predict the Direction of a Reaction?   107

What Is the Source of the Free Energy Released or Consumed During a Reaction?   108

How Can One Process Provide the Energy for Another?   108

How Do Concentration (and Entropy) Affect Equilibrium?   109

WHAT DETERMINES THE RATE OF A CHEMICAL REACTION?   110

How Does Molecular Motion Help Explain Reaction Rates?   110

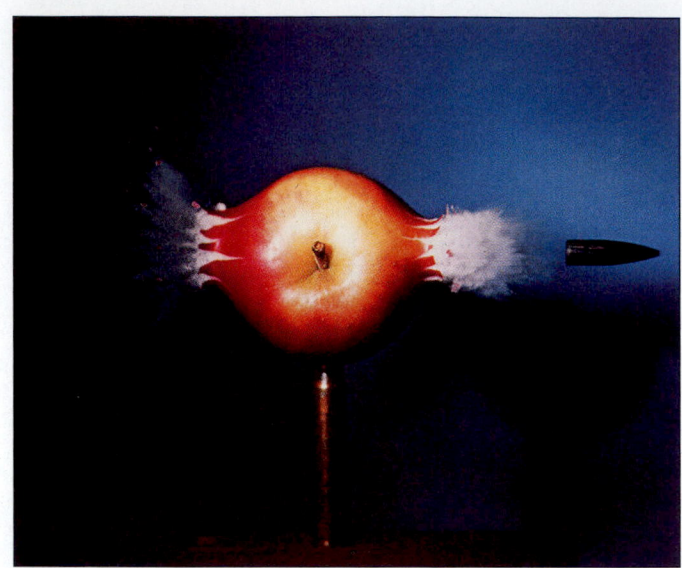

Photograph by Harold E. Edgerton, Harold and Esther Edgerton Foundation, ©1999. Courtesy Palm Press, Inc.

What Stops a Chemical Reaction?   110
What Starts a Chemical Reaction?   110
HOW DO ENZYMES WORK?   112
    How Does an Enzyme Bind to a Reactant?   112
    How Does an Enzyme Lower the Activation Energy
        of a Chemical Reaction?   113
    How Do Environmental Conditions Affect the Rates of
        Enzymatic Reactions?   114
HOW DOES A CELL OR ORGANISM REGULATE
ITS OWN METABOLISM?   114
    Why Do Enzymatic Reactions Often Occur in Small
        Steps?   114
    How Do Organisms Regulate Enzyme Function?   115
    How Do Inhibitors Influence Enzymes?   115

Study Outline with Key Terms   116
Review and Thought Questions   117

6   HOW DO ORGANISMS SUPPLY THEMSELVES WITH ENERGY?   118

Louis Pasteur and Vitalism   *118*

HOW DO ORGANISMS SUPPLY THEMSELVES
WITH ENERGY?   119
    What Is the Common Currency of Energy for
        Organisms?   120
    How Do Heterotrophs Extract Energy from
        Macromolecules?   121
    What Are the Stages of Cellular Respiration?   121
HOW DO CELLS EXTRACT ENERGY FROM
GLUCOSE?   122
    What Is Oxidation?   122
    Glycolysis: How Do Cells Capture Energy in ATP
        and NADH?   123

BOX 6.1 Sprints, Dives, and Marathons   125
    What Happens to Pyruvate When Oxygen Is Present?   127
    How Does a Cell Generate ATP from Acetyl-CoA?   127
ELECTRON TRANSPORT: HOW DOES THE
ENERGY IN GLUCOSE REACH ATP?   130
    How Does the Flow of Electrons from Electron Donors
        to Oxygen Release Energy to the Phosphate Bonds
        in ATP?   130
    Which Molecules Serve as Electron Carriers?   131
    How Do Cells Harvest the Energy of Electron
        Transport?   131
    How Do Mitochondria Generate a Proton Gradient?   133
    What Pumps Protons out of the Mitochondrial
        Matrix?   133
    How Does the Flow of Protons Back into the Matrix
        Cause the Synthesis of ATP?   134
    Are Proton Pumping and ATP Synthesis Really Separate
        Processes?   135
BOX 6.2 The Metabolism of Alcohol   136
    How Much Usable Energy Can a Cell Harvest from a
        Molecule of Glucose?   136

Study Outline with Key Terms   138
Review and Thought Questions   139

7   PHOTOSYNTHESIS: HOW DO ORGANISMS GET ENERGY FROM THE SUN?   140

The Chemical Evangelist   *140*

HOW DO WE KNOW HOW PLANTS OBTAIN
CARBON AND OXYGEN?   142
    What Do Plants and Air Do for Each Other?   142
    Where Do the Atoms Go in Photosynthesis?   143
HOW DO PLANTS COLLECT ENERGY FROM
THE SUN?   145
    What Is Light?   145
    How Can We Tell Which Molecules Help with
        Photosynthesis?   146
    Why Do Plants Use Two Types of Reaction Centers?   148
    How Do the Light Reactions Generate NADPH
        and ATP?   151
HOW DO PLANTS MAKE GLUCOSE?   153
    What Is the Calvin Cycle?   153
    What Happens to Carbon Dioxide in Photosynthesis?   155
BOX 7.1 How Do Some Herbicides Kill Plants?   156
    How Many Photons Does a Chloroplast Need To
        Make One Glucose?   157
HOW DO PLANTS COPE WITH TOO LITTLE
WATER OR CARBON DIOXIDE?   157
    What Factors Limit Productivity?   157
    How Do Plants Prevent Photorespiration?   158

Study Outline with Key Terms   159
Review and Thought Questions   159

II Genetics: The Continuity of Life 161

8 CELL REPRODUCTION 162

The Unfortunate Henrietta Lacks 162

CELL DIVISION 163
How Do Cells Divide? 163
BOX 8.1 Cervical Cancer and the Pap Smear 164
What Did Early Biologists Discover About Chromosomes? 165
How Did Biologists Discover the Function of the Chromosomes? 166
HOW DOES A DIVIDING CELL ENSURE THAT EACH DAUGHTER CELL RECEIVES AN EXACT COPY OF THE PARENT CELL'S DNA? 167
How Do Prokaryotic Cells Divide? 167
How Do Eukaryotic Cells Divide? 168
HOW DOES MITOSIS DISTRIBUTE ONE COPY OF EACH CHROMOSOME TO EACH DAUGHTER CELL? 172
Mitosis Is a Continuous Process, but Biologists Distinguish Four Phases 172
What Propels the Chromosomes During Mitosis? 173
HOW DOES A CELL FIT ALL ITS DNA INTO A NUCLEUS? 173
How Does the DNA Fold Up? 174
How Do DNA, Histones, and Other Proteins Form Such Compact Structures? 175
HOW DOES A CELL DIVIDE ITS CYTOPLASM? 175
HOW DOES A CELL REGULATE PASSAGE THROUGH THE CELL CYCLE? 176
How Do Normal Cells Determine When To Stop Dividing? 176
How Do Normal Cells Determine When It Is Time To Divide? 177
What Triggers the Main Events of Mitosis? 177

Study Outline with Key Terms 178
Review and Thought Questions 179

9 FROM MEIOSIS TO MENDEL 180

Why Is the Yellow Dog Yellow? 180

WHY WAS THE CHROMOSOMAL THEORY OF INHERITANCE SO HARD TO ACCEPT? 182
Blending Inheritance: A Wrong Turn 182
Why Did Biologists Doubt That Chromosomes Could Carry Information? 183
HOW DO ORGANISMS PASS GENETIC INFORMATION TO THEIR OFFSPRING? 183
How Is Phenotype Related to Genotype? 184
The Same Laws of Inheritance Apply to All Sexually Reproducing Organisms 185

HOW DO SEXUALLY REPRODUCING ORGANISMS KEEP THE SAME NUMBER OF CHROMOSOMES FROM GENERATION TO GENERATION? 185
The First Cell of the New Generation Has Two Sets of Chromosomes 186
How Do the Egg and Sperm Find One Another? 187
HOW DOES MEIOSIS DISTRIBUTE CHROMOSOMES TO THE GAMETES? 188
During Prophase I, Homologous Chromosomes Form Pairs 189
During the Rest of Meiosis I, One Chromosome from Each Homologous Pair Goes to Each Daughter Cell 190
Meiosis II Distributes Sister Chromatids to Daughter Cells 191
Disjunction and Nondisjunction 191
WHY SEX? 192
BOX 9.1 What Happens When Meiosis Goes Wrong? 193
What Good Is Genetic Variation? 193
What Are Sex Chromosomes? 194
HOW DO THE NUMBER AND MOVEMENTS OF CHROMOSOMES EXPLAIN THE INHERITANCE OF GENES? 194
What Are the Genotypes and Phenotypes of the F2 Generation? 195
HOW DID GREGOR MENDEL DEMONSTRATE THE PRINCIPLES OF GENETICS? 196
What Did Mendel's F1 Crosses Show? 196
The Principle of Segregation 197
The Principle of Independent Assortment 198
WHAT WAS THE EVIDENCE FOR THE CHROMOSOMAL THEORY OF INHERITANCE? 200
What Were Sutton's Arguments? 200
How Did Thomas Hunt Morgan Bolster the Theory of Chromosomal Inheritance? 201
Do Genes on the Same Chromosome Assort Independently? 203
Why Did Some Flies Have Normal Wings but Purple Eyes? 204
Can All the Genes on a Chromosome Cross Over and Recombine? 205

Study Outline with Key Terms 205
Review and Thought Questions 207

10 THE STRUCTURE, REPLICATION, AND REPAIR OF DNA 208

Rosalind Franklin and the Double Helix 208

WHAT IS THE STRUCTURE OF DNA? 210
How Much Did Franklin Discover About the Structure of DNA? 211
What Were Chargaff's Rules? 215
Franklin Comes Within Two Steps of the Correct Model 216
Modern Science Is a Social Endeavor 217

WHAT IS A GENE?   217
   How Do Genes Affect Biochemical Processes?   217
   One Gene—One Enzyme   218
   One Gene—One Polypeptide   219
HOW DID BIOLOGISTS LEARN WHAT GENES
ARE MADE OF?   221
   How Could Dead Bacteria Change the Genetic
     Properties of Other Bacteria?   221
   What Was the Transforming Factor?   221
   What Is a Bacteriophage?   223
   What Convinced Biologists That Genes Are Made
     of DNA?   223
HOW DOES ONE DNA STRAND DIRECT THE
SYNTHESIS OF ANOTHER?   225
   How Do Dividing Cells Copy the DNA?   225
BOX 10.1 How to Sequence DNA   **226**
   How Does a Cell Begin Assembling a New Strand
     of DNA?   227
   How Does a Cell Copy Both Strands of DNA
     Simultaneously?   227
   How Does DNA Replicate in the 3′ to 5′
     Direction?   228
   How Does DNA Synthesis Begin?   228
WHAT IS THE ULTIMATE SOURCE OF GENETIC
DIVERSITY?   229
   What Causes Mutations?   230
   How Often Do Mutations Occur?   231
   What Kinds of Mutations Are There?   231
HOW DO CELLS HANDLE MISTAKES IN THE
NUCLEOTIDE SEQUENCE OF DNA?   231

Study Outline with Key Terms   232
Review and Thought Questions   233

11   HOW ARE GENES EXPRESSED?   **234**

How Was the Genetic Code Discovered?   *234*

   What Is the Genetic Code Like?   237
GENETIC INFORMATION FLOWS FROM DNA
TO RNA TO POLYPEPTIDES   238
**BOX 11.1 Which Strand of the DNA Is
    Transcribed?**   **240**
RNA POLYMERASE TRANSCRIBES DNA
INTO RNA   241
   How Does RNA Polymerase Begin Transcription?   241
   How Does the Cell Alter the mRNA Transcribed
     from the DNA?   242
   One Gene—Heaven Only Knows How Many
     Polypeptides   243
HOW DOES A CELL TRANSLATE mRNA INTO
A POLYPEPTIDE?   243
   What Do Ribosomes Do?   243
   How Do tRNAs Serve as Adapters Between mRNAs
     and Amino Acids?   244

**BOX 11.2 How Do the Ribosomes Read Different
    Kinds of Mutations?**   **245**
   How Do tRNAs Recognize the Right Amino Acid?   246
   How Do Ribosomes Begin Translation?   246
   How Do Ribosomes Make Peptide Bonds?   247
   Several Ribosomes Can Read a Single Strand of mRNA
     Simultaneously   249
   How Does Polypeptide Synthesis Stop?   249
   How Do Cells Direct Newly Made Proteins to Different
     Destinations?   250
HOW DO CELLS REGULATE GENE
EXPRESSION?   250
BOX 11.3 Antibiotics and Protein Synthesis   **251**
   How Do Prokaryotes Regulate Gene Expression?   251
   How Do Eukaryotes Regulate Gene Expression?   254

Study Outline with Key Terms   255
Review and Thought Questions   257

**12   JUMPING GENES AND OTHER
    UNCONVENTIONAL GENETIC
    SYSTEMS**   **258**

Jumping Genes and Indian Corn   *258*

WHAT IS A VIRUS?   260
   What Is the Form of a Virus?   261
   How Do Viruses Make More Viruses?   262
   How Does a Virus Attach to a Cell?   262
HOW DO VIRUSES AND OTHER MOBILE GENES
REPLICATE THEIR GENES?   264

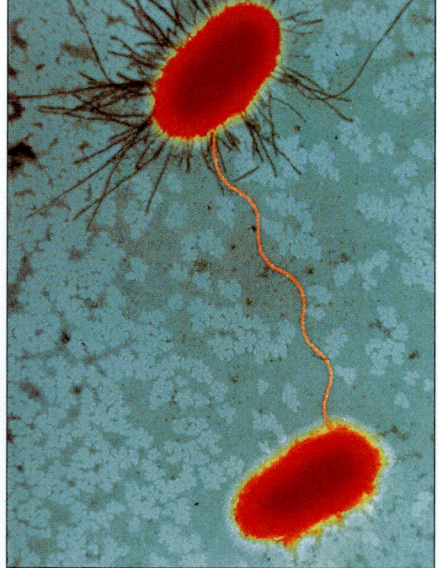

Dennis Kunkel/Phototake NYC

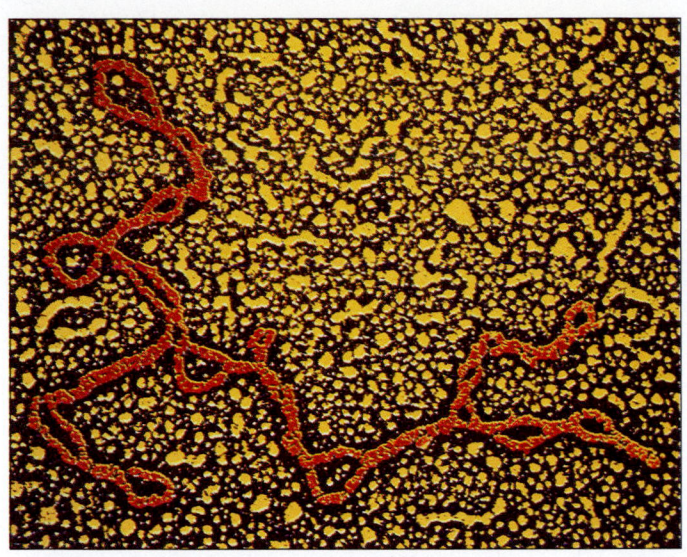

CNRI/Phototake NYC

BOX 12.1 Can Viruses Make You Well? **265**
How Do Plasmids Replicate Autonomously? 265
Many Mobile Genes Can Replicate Only When
Integrated into a Host Cell's DNA 266
How Do RNA Viruses Reproduce and Express Their
Genes? 269
HOW DO MOBILE GENES EVOLVE? 270
Although Viruses Are Not Alive, They Are Still Subject
to Natural Selection 270
BOX 12.2 How Can Researchers Design Drugs That
Will Combat AIDS? **271**
How May Viruses Have Evolved? 272
MITOCHONDRIA AND CHLOROPLASTS
CONTAIN THEIR OWN DNA 272
Studies of the DNA and RNA of Energy Organelles
Support the Theory That They Derive from Ancient
Prokaryotes 273

Study Outline with Key Terms 274
Review and Thought Questions 275

13   GENETIC ENGINEERING AND
RECOMBINANT DNA **276**
The Maverick *276*
WHAT IS GENETIC ENGINEERING? 278
Genetic Engineering Makes Use of Both Natural and
Induced Variation 280
Knowing Biochemical Pathways Helps Molecular
Biologists Design Useful Organisms 280
WHAT IS RECOMBINANT DNA AND HOW IS
IT USEFUL? 281
How Do Molecular Biologists Use Recombinant DNA? 281

How Do Restriction Enzymes Cut Up a Genome? 283
How Do Molecular Biologists Join Restriction
Fragments Together? 283
How Do Molecular Biologists Express Recombinant
DNA in Bacteria and Other Hosts? 283
How Do Researchers Make Multiple Copies of
Recombinant DNA? 285
How Do Biologists Find the Right DNA Sequence in a
Recombinant DNA Library? 286
HOW CAN BIOLOGISTS STUDY THE EXPRESSION
AND STRUCTURE OF MANY INDIVIDUAL
GENES? 287
BOX 13.1 Southern, Northern, and Western
Blotting **288**
RECOMBINANT DNA CAN REPROGRAM CELLS
TO MAKE NEW PRODUCTS 289
Genetically Engineered Bacteria and Eukaryotic Cells
Can Make Useful Proteins 289
RECOMBINANT DNA CAN GENETICALLY
ALTER ANIMALS AND PLANTS 289
How Do Researchers Produce a Transgenic Mammal? 290
The Genetic Engineering of Plants Is Easier Than
That of Animals 291
BOX 13.2 Can Biologists Clone Dinosaurs? **292**
What Are the Risks of Recombinant DNA in
Agriculture? 293
The Application of Recombinant DNA Technology
Poses Moral Questions for Society 294

Study Outline with Key Terms 294
Review and Thought Questions 295

14   HUMAN GENETICS **296**
To Map the Genome or Hunt the
Huntington's Gene? *296*
GENETIC DIVERSITY IN HUMAN BEINGS 299
How Reliable Is DNA Fingerprinting? 299
What Are Human "Races"? 302
THE RELATIONSHIP BETWEEN GENOTYPE
AND PHENOTYPE IS COMPLEX 304
Genes Are Expressed to Different Degrees 304
A Single Gene Can Affect Many Traits 305
A Single Trait Can Be Influenced by Many Genes 306
A Single Gene May Have Multiple Alleles 307
THE HUMAN GENOME PROJECT 307
What Is the Human Genome Project? 308
How Do Researchers Locate Genes for Specific
Diseases? 308
Of What Use Is a Sequenced Gene? 310
PREVENTING GENETIC DISEASE 310
BOX 14.1 Can Human Genes Be Patented? **311**
Genetic Counseling: How Can Parents Decide? 311
Prenatal Testing: Can Parents Tell Before? 311
Gene Therapy: Can We Fix It Later? 313

Study Outline with Key Terms   315
Review and Thought Questions   316

III   Evolution   317

15   WHAT IS THE EVIDENCE FOR EVOLUTION?   318

The Scopes Trial   *318*

WHAT IS EVOLUTION?   320
  Evolution Before Darwin   321
  How Did Charles Darwin Convince Others That
    Species Evolve?   325
WHAT IS THE EVIDENCE FOR EVOLUTION?   327
  The Fossil Record Tells a Story of Evolution   327
BOX 15.1 Radioactive Dating   **329**
  What Does the Fossil Record Show?   329
  Taxonomy   332
  Comparative Anatomy   333
  Comparative Embryology   334
  Comparative Molecular Biology   336
  Biogeography   338

NATURAL SELECTION: DARWIN'S MECHANISM
FOR EVOLUTION   338
BOX 15.2 Continental Drift   **339**
  Artificial Selection   340
  Darwin's Argument for Natural Selection   340
  Natural Selection Analyzed   341

Study Outline with Key Terms   341
Review and Thought Questions   342

16   MICROEVOLUTION: HOW DOES A POPULATION EVOLVE?   344

A Fatal Disagreement   *344*

HOW ARE VARIANTS CREATED AND
MAINTAINED?   346
  Why Did Charles Darwin Accept Blending
    Inheritance?   346
  Why Was the Idea of Blending Inheritance a
    Problem?   347
HOW DID BIOLOGISTS COME TO REJECT THE
THEORY OF ACQUIRED CHARACTERISTICS?   347
  How Did Weismann Explain Variation in Individuals?   348
  How Did the Rediscovery of Mendel's Work Transform
    Biology?   349
  How Did the Modern Synthesis Reconcile Genetics and
    Natural History?   349
WHAT IS THE SOURCE OF THE VARIATION
THAT FUELS EVOLUTION?   350
  What Is Genetic Variation?   350
  Are Traits Determined by Single Genes?   351
  What Determines the Genetic Variability of a
    Population?   352
  How Much Do Populations Vary Genetically?   353
  Is All Genetic Variation Subject to Natural Selection?   353
  Why Is Genetic Variation Essential to Evolution?   353
  How Do Populations Maintain Genetic Equilibrium?   354
WHAT CAUSES ALLELE FREQUENCIES TO
CHANGE?   355
  How Can New Mutations Cause Changes in Allele
    Frequencies?   355
  How Do Chance Events Change Gene Frequencies?   356
  How Does Nonrandom Mating Cause Changes in Allele
    Frequencies?   358
  How Does Breeding Between Populations Change Allele
    Frequencies?   358
  How Does Selection Decrease the Frequency of Certain
    Traits?   358
  How Does Selection Increase the Frequency of Certain
    Traits?   360
**BOX 16.1 Does Selection Act on Preexisting
    Variants?**   **361**
  In What Directions Does Natural Selection Push
    Populations?   362

Tim Flach/Stone

Fred Bruemmer/DRK Photo

Study Outline with Key Terms   366
Review and Thought Questions   367

**17   MACROEVOLUTION: HOW DO
       SPECIES EVOLVE?**   368

*What Darwin Saw at the Galápagos
Islands*   368

WHAT IS A SPECIES?   369
  Why Is It So Hard To Define a Species?   370
  What About Other Taxonomic Categories?   372
HOW DO SPECIES FORM?   373
  What Barriers Reproductively Isolate Populations and
    Species?   373
  How Does Geographic Isolation Initiate Speciation?   374
  How Can Species Form Without Geographic Isolation?   374
  Can Species Form with Limited Gene Flow?   376
HOW DO SPECIES MULTIPLY?   376
  How Does Adaptive Radiation Occur on Chains
    of Islands?   376
  How Do Species Diversify on Continents?   379
THE CAMBRIAN EXPLOSION   380
BOX 17.1 Did It Come from Outer Space?   382
EXTINCTION AND THE RATE OF EVOLUTION   383
  The Mass Extinctions   383
  How Fast Do Species Evolve?   383
**BOX 17.2 Modern Tropical Rain Forests: Mass
            Extinction or Not?**   385
  What Is Phylogeny?   388
HOW DID HUMANS EVOLVE?   388
  How Do We Classify the Apes?   389
  Why Did *Australopithecus* Stand Up?   390
**BOX 17.3 What Is the Nature of Anthropological
            Evidence?**   392

Who Were the Earliest Known Hominids?   393
Where Did We Come From?   394

Study Outline with Key Terms   396
Review and Thought Questions   397

**18   HOW DID THE FIRST ORGANISMS
       EVOLVE?**   398

*Can Organisms Arise Spontaneously from
Nonliving Matter?*   398

WHERE DOES LIFE COME FROM?   399
  When and Where Did Life First Appear?   401
BOX 18.1 Is There Life Elsewhere in the Universe?   403
PREBIOTIC EVOLUTION: HOW COULD
COMPLEX MOLECULES EVOLVE?   404
  What Was the Earth Like When It Was Young?   404
  Is Prebiotic Synthesis Plausible?   405
  Was Prebiotic Synthesis Necessary?   407
  How Can Biological Building Blocks Assemble
    Outside of Cells?   407
  How Can Prebiotic "Cells" Form?   408
EARLY BIOTIC EVOLUTION: HOW DID THE
FIRST GENETIC SYSTEM EVOLVE?   409
  Could RNA Have Served as the Original Genetic
    Material?   409
  How Did Translation Evolve?   411
  How Did Ribosomes Evolve?   411
  How Did tRNAs Evolve?   411
  How Did the First Cells Use Genetic Information?   412
HOW DID MODERN CELLS EVOLVE?   412
  How Did Early Life Affect the Environment?   412
  How Did Mitochondria and Chloroplasts Arise?   413

Study Outline with Key Terms   415
Review and Thought Questions   416

IV   Diversity   417

**19   CLASSIFICATION:
       WHAT'S IN A NAME?**   418

Is the Red Wolf a Species?   418

BOX 19.1 Foucault's Chinese Encyclopedia   421
WHY IS IT DIFFICULT TO NAME AND
CLASSIFY ORGANISMS?   421
  Why Do We Name Things?   422
  What Did Linnaeus's System of Classification
    Accomplish?   422
  Who Names a Species?   423
  How Do We Recognize Species?   423

Gerard Lacz/Peter Arnold, Inc.

BOX 19.2 Are Guinea Pigs Rodents?   **424**
HOW DO BIOLOGISTS GROUP SPECIES?   425
   Classical Evolutionary Taxonomy   426
   How Do Advocates of Phenetics Classify Organisms?   426
   How Do Advocates of Cladistics Classify Organisms?   427
   Do DNA and Protein Sequences Provide "Objective"
     Criteria for Classification?   429
   Why Is Classification So Difficult?   430
HOW MANY GROUPS OF ORGANISMS ARE
THERE?   430

Study Outline with Key Terms   431
Review and Thought Questions   431

**20   PROKARYOTES: SUCCESS STORIES,
     ANCIENT AND MODERN   432**

Tracking Down a Killer   **432**

PROKARYOTES AND PEOPLE   435
   How Do Bacteria Make Us Ill?   435
   How Do Bacteria Acquire Resistance to Antibiotics?   437
HOW DO WE KNOW THAT AN ORGANISM IS
A PROKARYOTE?   438
   Most Bacteria Have Complex Cell Walls   440
   How Do Bacteria Move?   441
   How Do Prokaryotes Grow So Rapidly?   442
   Prokaryotes Are Metabolically Diverse   442
HOW DO BIOLOGISTS CLASSIFY
PROKARYOTES?   443
   How Do Biologists Classify the Archaea?   443
   How Do Biologists Classify the Eubacteria?   445
BOX 20.1 Carl Woese and the Archaea   **447**
   How Has DNA Sequencing Changed the Classification
     of Prokaryotes?   448

Study Outline with Key Terms   450
Review and Thought Questions   451

**21   CLASSIFYING THE PROTISTS AND
     MULTICELLULAR FUNGI   452**

The Fungus Fighters   **452**

WHAT ARE PROTISTS?   454
   Why Are Protists So Diverse?   455
   How Do Protists Reproduce?   456
ALGAE ARE PHOTOSYNTHETIC (PLANTLIKE)
PROTISTS   457
   Dinoflagellates   457
   Chrysophytes (Golden Algae and Diatoms)   458
   Euglenophytes   459
   Phaeophytes (Brown Algae)   460
   Rhodophytes (Red Algae)   460
   Chlorophytes (Green Algae)   461
PROTOZOA ARE HETEROTROPHIC (ANIMAL-
LIKE) PROTISTS   462
   Rhizopoda (Amebas)   462
   Foraminifera   462
   Sporozoans   462
   Ciliophora (Ciliates)   464
   Zoomastigina (Flagellates)   464
FUNGUSLIKE PROTISTS   464
   Acrasiomycota (Cellular Slime Molds)   464
   Myxomycota (Plasmodial Slime Molds)   465
   Oomycota   465
   What Can We Learn from DNA About the Diversity
     and Evolution of Protists?   465
HOW DO TRUE FUNGI DIFFER FROM OTHER
ORGANISMS?   465
   How Can We Characterize the Fungi?   466
   How Do Mycologists Classify Fungi?   467
BOX 21.1 Fungal Infections Today   **468**
   Zygomycetes (Conjugation Fungi)   468
   Ascomycetes (Sac Fungi)   469
   Basidiomycetes (Club Fungi)   469
   Deuteromycetes (Imperfect Fungi)   470
BOX 21.2 Ergotism   **471**
WHAT IMPORTANT ECOLOGICAL ROLES DO
FUNGI PLAY?   472
   What Role Do Fungi Play in Lichens?   472
   What Role Do Mycorrhizae Play in the Lives of
     Plants?   473
   Where Did the Fungi Come From?   474

Study Outline with Key Terms   474
Review and Thought Questions   475

**22   HOW DID PLANTS ADAPT TO
     DRY LAND?   476**

The Loves of Plants   **476**

HOW DID PLANTS ADAPT TO LIFE ON LAND?   478
   What Are the Advantages and Disadvantages of Life on
     Land?   478

What Is Alternation of Generations? 478
How Are Vascular Plants Different from Nonvascular
Plants? 479
How Do We Know About the Evolution of Plants? 480
What Does the Fossil Record Tell Us About the
Evolution of Plants? 481
How Did Angiosperms Evolve? 481

THE NONVASCULAR PLANTS, OR BRYOPHYTES,
INCLUDE MOSSES, LIVERWORTS, AND
HORNWORTS 483
How Do Bryophytes Absorb Water? 483
How Do Bryophytes Reproduce? 483
THE VASCULAR PLANTS, OR TRACHEOPHYTES,
INCLUDE SEED PLANTS AND SEEDLESS
PLANTS 485
Of What Advantage Is a Transport System to a Plant? 485
How Do Vascular Plants Reproduce and Disperse? 485
Four Divisions of Tracheophytes Lack Seeds 486
How Do Seed Plants Reproduce? 487
Gymnosperms: Four Divisions of Seed Plants Without
Flowers 489
THE ANGIOSPERMS ARE SEED PLANTS WITH
FLOWERS AND FRUITS 490
Of What Use Are Flowers? 491
Of What Use Are Fruits? 491
How Do Angiosperms Reproduce? 491
What Are the Parts of a Flower? 492
Why Are There So Many Species of Angiosperms? 493

Study Outline with Key Terms 494
Review and Thought Questions 495

23   PROTOSTOME ANIMALS: MOST
ANIMALS FORM MOUTH FIRST 496
Progress and the Burgess Shale 496

WHAT IS AN ANIMAL? 498
What Features Characterize the Animals? 499
How Do Zoologists Classify the Animals? 499
THE PARAZOA HAVE NO ORGANS 503
RADIALLY SYMMETRICAL ACOELOMATES 504
Cnidarians 504
Ctenophores (Comb Jellies) 505
BILATERALLY SYMMETRICAL ACOELOMATES 506
Platyhelminthes (Flatworms) 506
Nemertea (Ribbon Worms) 508
BOX 23.1 The Origin of Feces 509
ASCHELMINTHS 509
Nematodes (Roundworms) 510
PROTOSTOME COELOMATES 510
How Do Protostomes Differ from Deuterostomes? 511
Mollusks Share a Common Body Plan 511
What Do All Annelids Have in Common? 514
Why Are Arthropods So Successful? 517
How Do Biologists Classify Arthropods? 518

Study Outline with Key Terms 520
Review and Thought Questions 521

24   DEUTEROSTOME ANIMALS:
ECHINODERMS AND
CHORDATES 522
Are We Upside Down? 522

DEUTEROSTOME COELOMATES 524
Echinoderms 524
Arrow Worms and Acorn Worms 527
What Traits Characterize the Chordates? 527
What Have We in Common with Tunicates? 528
What Have We in Common with Cephalochordates? 528
What Distinguishes the Vertebrates from Other
Chordates? 528
Agnatha Have No Jaws 529
Placoderms, Now Extinct, Were Armored Fish
with Jaws 530
Chondrichthyes Are Fish with Cartilaginous Skeletons 530
Osteichthyes Are Fish with Bony Skeletons 531
WHAT ADAPTATIONS HAVE VERTEBRATES
EVOLVED FOR LIFE ON LAND? 532
Amphibians Are Terrestrial Animals That Begin Their
Lives in Water 532
Reptiles Can Live Their Entire Lives on Land 534
BOX 24.1 Why Are Amphibians Disappearing? 535
How Do Terrestrial Vertebrates Regulate Body
Temperature? 538
What Distinguishes the Birds from the Reptiles? 538
What Distinguishes the Mammals? 539

Study Outline with Key Terms 541
Review and Thought Questions 542

V Ecology 543

25   ECOSYSTEMS 544
Gaia: Is the Biosphere an Organic
Entity? 544

WHAT IS ECOLOGY? 546
What Is an Ecosystem? 546
HOW DOES ENERGY FLOW THROUGH
ECOSYSTEMS? 548
How Much Energy Flows Through an Ecosystem? 549
HOW DO ECOSYSTEMS RECYCLE MATERIALS? 552
What Drives the Water Cycle? 552
Where Does Carbon Go? 553
How Have Humans Altered the Carbon Cycle? 555
BOX 25.1 Measuring Carbon Dioxide Levels for
40 years 556

Why Is Nitrogen Both Common and in Short Supply?   557
How Have Humans Altered the Nitrogen Cycle?   559

Study Outline with Key Terms   560
Review and Thought Questions   561

26 TERRESTRIAL AND AQUATIC ECOSYSTEMS   562

How Many Biomes Can You See in a Day?   562

WHAT IS A BIOME?   564
How Do Sunshine and the Earth's Movements Determine Climate?   564
How Does Topography Influence Climate?   566
HOW DO BIOMES DIFFER?   567
Desert Ecosystems Are Surprisingly Delicate   567
How Can a Dry Biome Be So Wet?   569
Why Are There No Trees in Grasslands?   570
Chaparral Always Burns Eventually   571
Savannas Host Some of the Most Spectacular Grazing Animals   572
Taiga, the Spruce–Moose Biome   572
Why Do the Temperate Deciduous Forests Lack Top Predators?   573
BOX 26.1 How Does Acid Rain Alter Biomes?   575
Tropical Rain Forests Are the Most Productive and Diverse Biomes in the World   575
How Are Tropical Rain Forests Being Destroyed?   576
WHAT ARE AQUATIC ECOSYSTEMS LIKE?   578
How Do Oceanographers Divide the Oceans?   578
Doesn't El Niño Do More Than Kill Fish and Birds?   579
Why Are Estuaries and Marshes Valuable?   580
Why Are Lakes, Ponds, Rivers, and Streams Susceptible to Pollution?   580

Study Outline with Key Terms   582
Review and Thought Questions   583

27 COMMUNITIES: HOW DO SPECIES INTERACT?   584

Volcano: An Ecosystem Is Destroyed   584

ARE COMMUNITIES INTEGRATED OR INDIVIDUALISTIC?   585
HOW DO SPECIES IN A COMMUNITY INTERACT?   587
Who Eats Whom?   587
How Do Plants and Animals Defend Themselves?   588
Symbiosis: How Do Organisms Live Together?   590
Do Organisms Coevolve?   590
How Does Competition Affect Organisms?   591
WHAT DETERMINES BIODIVERSITY?   593
How Can We Measure Diversity?   593

How Do Biotic Factors Influence Diversity?   593
How Do Abiotic Factors Influence Diversity?   593
How Do Groups of Populations Interact?   595
COMMUNITIES CHANGE OVER TIME   596
Why Do Communities Change?   596
BOX 27.1 Succession on a Corpse   598
Why Does Succession Occur?   598
What Determines Which Community is the Climax Community?   599
Why Do Communities Stay the Same for Long Periods?   599
Can Ecosystems Bounce Back?   600

Study Outline with Key Terms   600
Review and Thought Questions   601

28 POPULATIONS: EXTINCTIONS AND EXPLOSIONS   602

Fighting for Life   602

HOW DO POPULATIONS GO EXTINCT?   604
How Do Deterministic Factors Drive Species to Extinction?   604
How Do Stochastic Factors Drive Species to Extinction?   606
How Can We Preserve Ecosystems and the Species that Live in Them?   607
HOW DO POPULATIONS GROW?   608
What Is Exponential Growth?   609
What Limits the Size of a Population?   610
What Determines Whether a Population Grows?   611
Do Natural Populations Actually Grow Exponentially?   612
Why Do Some Populations Cycle Up and Down?   612
WHAT FACTORS INFLUENCE THE VALUES OF R AND K?   612
How Do Life Histories Affect Survivorship and Age Distributions?   613
BOX 28.1 Why Are Age Distributions of Males and Females Different?   614
Demography: How Do Scientists Measure Populations?   616
Why Is Population Momentum Important?   616
WHAT IS THE EARTH'S CARRYING CAPACITY FOR HUMANS?   617
How Well Is Our Energy Supply Keeping Pace with Human Energy Consumption?   617
How Well Has Agriculture Met the Human Demand for Food?   619
BOX 28.2 How Much Do We Have and Need?   620
HOW IS HUMAN OVERPOPULATION POLLUTING THE BIOSPHERE?   621
Pollution Has Degraded the Water   621
Pollution Has Also Degraded the Air   622

Study Outline with Key Terms   624
Review and Thought Questions   625

29   THE ECOLOGY OF ANIMAL BEHAVIOR   626

Why Is Sociobiology Controversial?   626

HOW IS BEHAVIOR ACQUIRED?   627
- What Is Innate Behavior?   627
- How Much Can Animals Learn?   630
- How Do Genes Influence Behavior?   631

HOW DOES ECOLOGY INFLUENCE BEHAVIOR?   632
- How Do Animals Make the Best Choices?   632
- How Do Animals Compete for Resources or Mates?   634

SOME ANIMALS LIVE IN SOCIAL GROUPS   634
- How Is Social Behavior Advantageous?   635
- Honeybees Are an Extraordinary Example of Social Behavior   635
- Is Altruism Adaptive?   636

BOX 29.1 Tit for Tat and Prisoner's Dilemma   638

Study Outline with Key Terms   639
Review and Thought Questions   640

VI   Structural and Physiological Adaptations of Flowering Plants   641

30   STRUCTURAL AND CHEMICAL ADAPTATIONS OF PLANTS   642

Should We Eat Our Vegetables?   642

WHAT ARE THE PHYSICAL CONSTRAINTS ON PLANTS?   644
- How Did Terrestrial Plants Evolve?   644
- What Are the Roles of Shoots and Roots?   645

HOW ARE FLOWERING PLANTS STRUCTURED?   646
- How Is the Vascular System Organized?   647
- How Do Monocots and Dicots Differ?   649

HOW DO TERRESTRIAL PLANTS OBTAIN WATER AND MINERALS?   651
- How Do Roots Carry Water?   651

BOX 30.1 The Distinctive Anatomy of C_4 Plants   652
- Stems Carry Water and Minerals and Nutrients Between Roots and Leaves   653
- What Do Leaves Do?   654

HOW DO PLANTS DEFEND THEMSELVES AGAINST HERBIVORES?   656
- Mechanical Defenses   656
- Chemical Defenses   656

Study Outline with Key Terms   658
Review and Thought Questions   659

31   WHAT MOVES WATER AND SUGARS?   660

The Devastating Dry Leaf Creature   660

WHAT DRIVES WATER UP?   662
- What Is the Route of Water from Soil to Air?   662
- Root Pressure: Can the Roots Push Water to the Tops of Tall Trees?   662
- Can Capillary Action Raise Water to the Tops of Tall Trees?   663
- The Driving Force for the Ascent of Most Water Is Transpiration   664
- How Can Transpiration Draw Water to the Tops of Tall Trees?   664
- Is the Flow of Water Through the Stem Coupled to Transpiration?   665
- Is Water Cohesive Enough To Sustain the Transpiration Pull in a Tall Plant?   665
- Is Water in the Xylem Really Under Tension?   666

STOMATA REGULATE TRANSPIRATION BY OPENING AND CLOSING   666
- How Do Stomata Open and Close?   666
- How Can Plants Save Water Without Limiting Photosynthesis?   667

BOX 31.1 Adaptations to Limited Water Supply   668
- Water in the Xylem Carries Minerals from the Soil   669
- Most Plants Obtain Both Water and Essential Minerals from the Soil   669
- Why Do Roots Need Oxygen from the Soil?   670
- Why Do Some Plants Obtain Minerals from Animals?   670

HOW DO PLANTS TRANSPORT THE PRODUCTS OF PHOTOSYNTHESIS?   670
- Which Way Does Sugar Move?   670
- What Is Inside a Sieve Tube?   671
- The Osmotic Flow of Water Drives Translocation in the Phloem   672
- What Draws Water and Sugar into the Phloem?   672

Study Outline with Key Terms   674
Review and Thought Questions   675

32   GROWTH AND DEVELOPMENT OF FLOWERING PLANTS   676

The Origin of Corn   676

HOW DO PLANT EMBRYOS ESTABLISH A BODY PLAN?   678
- How Do Plants Make Sperm?   678
- How Do Plants Make Egg Cells?   679
- How Does Fertilization Occur in Plants?   679
- How Does a Plant Form a Seed?   680
- How Does the Embryo Resume Development?   682
- How Does the Embryo Obtain Nourishment?   683
- What Environmental Cues Trigger Germination?   684

HOW DO PLANTS GENERATE A SHOOT AND A ROOT?   684

How Do Meristem Cells Elongate and Differentiate?   686
Root Tips Grow and Mature in Three Zones   686
Primary Growth of Shoots Is More Complicated than
That of Roots   687
How Does a Flower Derive from a Specialized Bud?   687
How Do Genes Regulate the Development of Flowers?   687
**BOX 32.1 Why Is *Arabidopsis thaliana* Well Suited
to Genetic Analysis?   689**
Secondary Growth Depends on Two Kinds of Lateral
Meristems   689
MERISTEMS MAY PRODUCE WHOLE PLANTS   691

Study Outline with Key Terms   692
Review and Thought Questions   693

33   HOW DO PLANTS REGULATE GROWTH AND DEVELOPMENT?   694

A Life Lived Full   694

HOW DO PLANTS DETECT ENVIRONMENTAL
CHANGES, AND HOW DO THEY COORDINATE
THEIR RESPONSES TO THOSE CHANGES?   697
How Do Plants Respond to the Pull of Gravity?   697
What Pigment Facilitates Phototropism?   699
How Do Plants Detect Changes in Season or Daylight?   699

Stan Osolinski/Dembinsky Photo Associates

HORMONES COORDINATE ACTIVITIES IN
CELLS THROUGHOUT THE PLANT BOTH
DURING DEVELOPMENT AND IN RESPONSE
TO ENVIRONMENTAL CUES   702
Auxin Coordinates Many Aspects of Plant Growth and
Development   702
BOX 33.1 What Is a Bioassay?   703
How Do Gibberellins Act on Target Cells?   704
Cytokinins Stimulate Growth and Differentiation   705
How Did Researchers Identify Ethylene as a Plant
Hormone?   706
Abscisic Acid Promotes Dormancy in Buds and Seeds   707
Are There Other Plant Hormones?   707

Study Outline with Key Terms   708
Review and Thought Questions   708

VII   Structural and Physiological Adaptations of Animals   709

34   FORM AND FUNCTION IN ANIMALS   710

Tyrannosaurus Wrecks: T. Rex *Treks, But Slowly*   710

WHY ARE ANIMALS SHAPED THE WAY THEY
ARE?   711
How Are the Parts Laid Out?   713
What Are Animal Bodies Made of?   713
How Are Size, Surface Area, and Shape Related?   713
Why Does Metabolic Rate Depend on Size?   714
How Do Animals Use Expanded Surface Areas?   714
Bulk Flow   716
How Do Animals Use Countercurrent Systems?   716
HOW DOES STYLE OF LOCOMOTION
INFLUENCE SHAPE?   718
What Kinds of Adaptations Do Swimmers Have?   718
How Do Snakes Crawl?   719
What Kinds of Adaptations Do Flyers and Gliders
Have?   719
How Do Animals Walk and Run?   720
What Kinds of Adaptations Do Runners Have?   720
HOW DO ANIMALS USE SKELETON AND
MUSCLE TO MOVE?   721
How Do Muscles and Skeletons Work Together?   721
Naming the Parts of the Vertebrate Endoskeleton   722
What Forces Act on Bones and Connective Tissue?   724
How Does Bone Respond to Use?   725
How Do Skeletal Muscles Move Bones?   726
How Do Smooth and Cardiac Muscle Differ from
Skeletal Muscle?   726

How Do the Two Kinds of Muscle Fibers Work? 726
How Are Contractile Proteins Arranged? 728
How Do Striated Muscles Contract? 729
INTRODUCTION TO PHYSIOLOGY:
HOMEOSTASIS AND TOLERANCE 730
Homeostasis: How Do Organisms Self-Regulate? 730
How Much Change Can Animals Tolerate? 731

Study Outline with Key Terms 732
Review and Thought Questions 733

35 HOW DO ANIMALS OBTAIN NOURISHMENT FROM FOOD? 734

Another American Shot Heard Round The World 734

WHY MUST ANIMALS EAT? 735
How Much Must an Animal Eat? 736
How Does an Animal Know How Much Food To Eat? 736
What Are the Consequences of Too Much or Too Little Food? 736
What Are the Causes of Malnourishment? 737
HOW DO SINGLE CELLS AND INVERTEBRATES DIGEST FOOD? 739
Harsh Conditions Help Disrupt Tissue Interactions 739
Digestion by Individual Cells 740
Digestion in Cavities with One Opening 741
What Are the Advantages of a Two-Ended Digestive Tract? 741
HOW DOES THE VERTEBRATE DIGESTIVE TRACT FUNCTION AS A "DISASSEMBLY LINE"? 743
How Is the Vertebrate Digestive Tract Organized? 743
The Human Digestive System and Some Vertebrate Variations 743
HOW DOES AN ANIMAL COORDINATE PROCESSES IN INDIVIDUAL PARTS OF THE DIGESTIVE SYSTEM? 751
Hormones and Nerve Cells Coordinate the Actions of the Gut 751
How Does the Gastrointestinal Tract Replenish Itself when Cells Die? 751

Study Outline with Key Terms 752
Review and Thought Questions 753

36 HOW DO ANIMALS COORDINATE THE ACTIONS OF CELLS AND ORGANS? 754

Environmental Effects on Sexual Development 754

HOW DO CHEMICAL SIGNALS COORDINATE RESPONSES IN MANY CELLS? 755
How Do Organisms Send Chemical Signals? 756
BOX 36.1 How Do Researchers Identify Hormones and Glands? 757

HOW DO HORMONES COORDINATE PHYSICALLY DISTANT ORGANS? 758
How Do Hormones Control Insect Metamorphosis? 758
How Do Hormones Help Keep Blood Glucose at a Nearly Constant Level? 758
How Do Hormones Coordinate the Response to Stress? 760
How Do the Pituitary Gland and Hypothalamus Regulate the Production of Hormones by Other Endocrine Organs? 761
HOW DO HORMONES FIND THEIR TARGETS? 762
How Do Target Cells Detect Chemical Signals? 763
BOX 36.2 How Does an Extracellular Chemical Lead to Intracellular Responses in a Target Cell? 764
Hormones May Target More Than One Kind of Receptor 764
How Many Other Endocrine Organs in a Mammal? 766
Thyroid and Parathyroid Glands 766
Testes and Ovaries 766
PARACRINE SIGNALS ACT OVER SHORT DISTANCES 766

Study Outline with Key Terms 768
Review and Thought Questions 769

37 HOW DO ANIMALS MOVE BLOOD THROUGH THEIR BODIES? 770

Charles Drew and the Battle of Britain 770

WHY DO ANIMALS NEED BLOOD? 771
What Is the Connection Between Blood and the Extracellular Fluid? 771
What Does Blood Contain? 772
HOW DID SCIENTISTS DISCOVER THAT THE BLOOD CIRCULATES? 773
Why Do Many Animals Need a Blood Pump? 776
How Does the Heart Pump Blood Through the Body? 776
How Do Birds and Mammals Prevent Oxygen-Rich Blood from Mixing with Oxygen-Poor Blood? 776
HOW DO THE BLOOD AND THE EXTRACELLULAR FLUID EXCHANGE SMALL MOLECULES AND IONS? 778
Arteries, Veins, and Capillaries 779
The Lymphatic System 780
The Structure of Blood Vessels 781
How Does the Blood Acquire Oxygen? 782
WHAT MAKES THE HEART BEAT? 782
Systole and Diastole 782
What Coordinates the Beating of the Heart Muscle? 783
Blood Pressure Depends on the Action of the Heart and the Properties of Blood Vessels 785
NERVES, GLANDS, AND OTHER TISSUES REGULATE THE FLOW OF BLOOD 786
Regulation by Nerves and Hormones 785
How Does the Body Deliver Oxygen Where It Is Needed? 786

Study Outline with Key Terms   786
Review and Thought Questions   787

38   HOW DO ANIMALS BREATHE?   788

Stanton Glantz and the Tobacco Industry 788

BOX 38.1 The Green Ball   790
HOW DO ANIMALS EXCHANGE CARBON
DIOXIDE AND OXYGEN?   791
   Can Animals Take Oxygen Directly From the Air? 791
BOX 38.2 Who Does Smoking Hurt?   792
   How Do Gases Exert Pressure?   792
BOX 38.3 Why Do People Smoke?   793
   How Much Dissolved Gas Will a Solution Hold? 793
   Why Is Surface Area Important in Gas Exchange? 794
HOW DO AQUATIC ORGANISMS OBTAIN
DISSOLVED OXYGEN?   795
   How Do Gills Obtain Enough Oxygen To Power a
   Fast-Swimming Fish?   795
   Why Can't Land Animals Breathe With Gills? 795
HOW DO AIR-BREATHING ORGANISMS
OBTAIN OXYGEN?   796
   What Qualities Enable the Lung to Exchange Gases? 796
   By What Path Does Air Enter the Lungs? 796
   How Do We Breathe?   796
   What Keeps the Airways Moist and Clean? 799
   How Does Tobacco Smoke Damage the Lungs? 799
   How Do Alveoli Overcome the Surface Tension That
   Resists Expansion?   800
   How Do Birds Breathe So Well?   800
   How Do Insects Distribute Oxygen?   801
   How Do the Lungs Foster Gas Exchange? 801
HOW DOES OXYGEN GET TO THE TISSUES?   802
   How Do Red Blood Cells Transport Oxygen? 802
   How Does Hemoglobin Load and Unload Oxygen? 802
   What Determines Hemoglobin's Affinity For Oxygen? 803
   How Does Blood Carry Carbon Dioxide?   804
WHAT FACTORS CONTRIBUTE TO OXYGEN
HOMEOSTASIS?   804
   How Can Blood Deliver Oxygen at High Altitudes? 804
BOX 38.4 How Does a Fetus Obtain Oxygen? 805
   How Do Animals Regulate Breathing in Response to
   Changes in Oxygen and Carbon Dioxide Delivery? 805

Study Outline with Key Terms   806
Review and Thought Questions   806

39   HOW DO ANIMALS MANAGE
WATER, SALTS, AND WASTES?   808

*Will Corn Chips Raise Your Blood
Pressure?*   808

WHY DO ANIMALS NEED TO REGULATE
WATER AND SALT?   810
   Which Way Does Water Flow?   810
   Water and Salt Balance in Aquatic Animals 811

   How Do Saltwater Fish Hold on To Water? 812
   How Do Land Animals Balance Water and Salt? 812
HOW DO ANIMALS DISPOSE OF NITROGEN-
CONTAINING WASTES?   813
HOW DO KIDNEYS WORK?   814
   How Does a Kidney Filter the Blood?   814
   How Do the Tubules Recycle Essential Molecules
   and Ions and Recapture Water?   817
BOX 39.1 The Artificial Kidney   819
   How Does the Kidney Form Concentrated Urine? 819
   How Do the Kidneys of Nonmammalian Vertebrates
   Meet the Challenges of Different Environments? 820
HOW DO ANIMALS ADJUST WATER AND SALT
IN RESPONSE TO CHANGING CONDITIONS?   821
   How Does the Brain Control Fluid Intake? 821
   The Rate of Filtration Depends on Renal Blood Flow 821
   Angiotensin and Renin   821
   Aldosterone   821
   Antidiuretic Hormone (ADH), or Vasopressin 823
   Atrial Natriuretic Factor (ANF)   823
HOW DO INVERTEBRATES SOLVE PROBLEMS
OF WATER, SALTS, AND WASTES?   823
   Flatworms   823
   Earthworms   823
   Marine Invertebrates   824
   Insects   824

Study Outline with Key Terms   824
Review and Thought Questions   825

40   DEFENSE: INFLAMMATION AND
IMMUNITY   826

The Danger Model   826

THE CAST OF CHARACTERS   829
HOW DOES THE SKIN KEEP PATHOGENS
OUT?   829
HOW DOES THE BODY DEFEND ITSELF
WHEN THE SKIN IS BROKEN?   830
   How Does the Body Control Blood Loss and Keep
   Pathogens Out of the Circulatory System? 830
   How Does the Body Prevent Damaging Blood Clots? 831
   How Do Plasma Proteins Contribute to Nonspecific
   Defense?   832
   How Does Inflammation Help the Body Resist
   Infection?   832
HOW DOES THE IMMUNE SYSTEM
RECOGNIZE MOLECULES?   833
   What Distinguishes the Immune System from Other
   Defenses?   833
   What Gives the Immune System its Specificity? 834
   Which Cells Mediate the Immune Response? 835
   How Do Antibodies Recognize Cells and Molecules? 835
HOW DOES THE IMMUNE SYSTEM GENERATE
SO MANY ANTIBODIES?   837

How Can Thousands of Genes Produce Millions of
Antibodies? 837

The Selection Theory Explains Memory and Self-
Nonself Recognition 837

BOX 40.1 Acquired Immunodeficiency Virus 839

HOW DOES THE BODY DEFEND AGAINST
GOOD CELLS GONE BAD? 839

How Do T Cells Recognize Self? 840

How Do T Cells Recognize Altered Self? 841

How Does the Body Eliminate T Cells That Might
Attack Healthy Cells? 841

Individuals Have Distinctive Antigens on Blood Cell
Surfaces 841

Study Outline with Key Terms 842
Review and Thought Questions 843

**41 THE CELLS OF THE
NERVOUS SYSTEM 844**

Rita Levi-Montalcini 844

HOW ARE NERVE CELLS SPECIAL? 846

How Do Nerve Cells Carry Information? 847

How Does a Neuron Generate an Electrical Pulse? 848

How Does a Neuron Maintain Its Resting Potential? 849

How Does a Neuron Generate an Action Potential? 850

How Does a Membrane Change Its Permeability to
Sodium Ions? 851

How Do Voltage-Dependent Sodium Channels Open
and Close? 851

How Does an Action Potential Move Down an Axon? 852

How Do Nerves Speed the Transmission of Action
Potentials? 853

HOW DO NEURONS COMMUNICATE WITH
ONE ANOTHER? 853

How Does a Chemical Synapse Work? 855

What Does the Neurotransmitter Acetylcholine Do? 856

Neurotransmitters May Produce Either Excitatory or
Inhibitory Effects in Postsynaptic Neurons 856

Neurons May Respond to the Same Neurotransmitters
in Different Ways 856

How Are Neurotransmitters Cleared from the Synapse? 857

Many Psychoactive Drugs Act on Chemical Synapses 857

HOW DO NEURAL NETWORKS MEDIATE
BEHAVIOR? 857

**BOX 41.1 What is the Physical Basis of Drug
Action? 858**

How Do Neurons Mediate a Reflex? 858

HOW DOES EXPERIENCE MODIFY NEURAL
NETWORKS? 859

Experience Can Modify the Gill Withdrawal Reflex
of the Sea Hare, *Aplysia* 859

How Does an *Aplysia* Learn? 860

Study Outline with Key Terms 861
Review and Thought Questions 861

**42 THE NERVOUS SYSTEM AND
THE SENSE ORGANS 862**

*Does a Falling Tree Make a Sound When
There Is No One to Hear It? 862*

HOW DOES THE NERVOUS SYSTEM RESPOND
TO ITS ENVIRONMENT? 864

Do Invertebrate Animals Have Brains? 864

How Is the Vertebrate Central Nervous System
Organized? 865

How Do Humans Learn? 868

How Is the Vertebrate Peripheral Nervous System
Organized? 868

How Does Information Move Through the Nervous
System? 869

HOW DO ANIMALS SENSE THEIR
ENVIRONMENT? 870

What Senses Contribute to Sensation and Perception? 870

What Path Does Sensory Information Take to the
Brain? 870

HOW DO ANIMALS HEAR? 871

How Does the Mammalian Ear Detect Frequency and
Amplitude? 871

How Does the Inner Ear Help Maintain Balance? 872

HOW DO ANIMALS SEE? 873

How Does a Vertebrate's Eye Form an Image of the
Outside World? 873

Why Must Many of Us Wear Glasses? 874

How Does the Retina Detect an Image? 874

How Does Information Travel from the Retina to the
Brain? 875

There Is Much More to Vision Than Meets the Eye 876

OTHER SENSORY PATHWAYS 877

Somatic Sensations 877

How Do We Detect Molecules as Smells and Tastes? 878

Study Outline with Key Terms 880
Review and Thought Questions 881

43 SEXUAL REPRODUCTION 882

Why Did Sex Evolve? 882

HOW DO MAMMALS FORM GAMETES? 883

How Do Female Mammals Produce Ova? 883

Where Do Oocytes Go After They Leave the Ovary? 886

How Do Male Mammals Produce Sperm? 886

How Do Sperm Cells Travel from the Testes to the
Penis? 888

SEXUAL INTERCOURSE: HOW DO THE EGG
AND SPERM RENDEZVOUS? 888

When and Where Does Fertilization Occur? 888

How Do the External Genitalia Facilitate Sexual
Intercourse? 889

How Are the Male and Female Genitalia Alike? 890

The Male and Female Sexual Responses Each Consist
of Four Stages 890

How Do the Sperm Reach the Egg and Penetrate Its
Surface?   891
HOW DO HORMONES CONTROL GAMETE
PRODUCTION?   892
How Do Hormones Control Sperm Production?   893
What Are the Main Events of the Menstrual Cycle?   894
How Do Hormones Regulate the Ovarian Cycle?   894
How Do Hormones Regulate the Menstrual Cycle?   896
WHAT HAPPENS DURING PREGNANCY AND
LACTATION?   896
How Can a Woman Tell if She Is Pregnant?   896
What Happens Between Conception and Delivery?   897
Parturition: What Happens on D(elivery) Day?   898
Nursing the Baby   898
HOW DO HUMANS CONTROL
REPRODUCTION?   899
Abstinence   900
Withdrawal   901
Barrier Methods Prevent the Union of Sperm and
Ovum   901
Hormonal Contraceptives and Intrauterine Devices   901
Sterilization Provides Effective But Generally Irreversible
Birth Control   902
Induced Abortions Terminate Pregnancies After
Implantation   903
What Can Couples Do To Overcome Infertility?   903

Study Outline with Key Terms   904
Review and Thought Questions   905

44   HOW DO ORGANISMS
     BECOME COMPLEX?   906

How Do We Become Complex?   906

HOW DOES FERTILIZATION INITIATE
EMBRYONIC DEVELOPMENT?   908
HOW DO CELLS OF THE EMBRYO GIVE RISE
TO CELLS OF THE ADULT?   908
Biologists Separate Vertebrate Development into Eight
Stages   909
Why Does a Zygote Divide?   909
How Do Zygotes Cleave?   910
Germ-Layer Formation: How Does Gastrulation Set
Up the Three-Layered Structure?   911
How Do Amphibians Gastrulate?   912
How Do Mammals Gastrulate?   913
Organ Formation: How Does the Nervous System
Form?   914

Programmed Cell Death Contributes to Normal
Development   914
BOX 44.1 How Does the Eye Form?   916
HOW DOES HUMAN DEVELOPMENT PROCEED
AFTER IMPLANTATION?   916
EMBRYONIC CELLS BECOME INCREASINGLY
DIFFERENTIATED   918
What Kinds of Experiments Distinguish Potency and
Fate?   918
Are Differentiated Animal Cells Totipotent?   918
Can a Differentiated Nucleus Direct Development
from Egg to Adult?   920
How Did the Scottish Researchers Clone Dolly?   920
Did Dolly Really Come from a Differentiated Cell?   921
HOW DOES DEVELOPMENTAL FATE DEPEND
ON CHEMICAL SIGNALING?   921
Can the Extraordinary Actions of the Primary Organizer
in Amphibian Development Be Explained in Terms
of Cells and Molecules?   921
How Do Cells "Know" Where They Are and What To
Do About It?   923
Mammals and Flies Use Many of the Same Transcription
Factors To Establish Patterns During Development   924
POSTEMBRYONIC DEVELOPMENT   924
Metamorphosis Converts a Larva into an Adult   924
Aging Depends on the Action of Both Genes and
Environment   926
STEM CELLS: HOW CAN SOME CELLS RETAIN
THE CAPACITY FOR REGENERATION?   926
AND SO WE SAY GOODBYE   926

Study Outline with Key Terms   927
Review and Thought Questions   928

APPENDIX A   PERIODIC TABLE OF
             THE ELEMENTS   A-1

APPENDIX B   UNITS OF MEASURE:
             THE METRIC
             SYSTEM   A-2

GLOSSARY   G-1

INDEX   I-1

Asking
About
Life

Second Edition

The Unity and Diversity of Life

Enough To Give You an Ulcer

In 1984, an obscure Australian physician named Barry Marshall secretly performed a dangerous experiment that would ultimately deprive some of the world's largest corporations of billions of dollars. Marshall's radical idea was born in 1979, when his colleague and friend J. Robin Warren noticed that samples of stomach tissue taken from ulcer patients were often infected with bacteria.

To find one bacterial infection in the stomach would have been strange; to find dozens was bizarre. The human stomach secretes acid so concentrated that few organisms survive it for more than a few minutes, let alone live and reproduce in it. Yet Warren found bacteria flourishing there.

Warren's discovery suggested an alternative to doctors' long-standing belief that ulcers are caused by excess stomach acid. Ulcers, every medical textbook reported, were caused by the oversecretion of stomach acid in people with overanxious, frustrated personalities. Such personality problems were thought to be aggravated by the stressful pace of modern life. But if a bacterium could infect the stomach, Marshall and Warren realized, maybe it could cause ulcers. Intrigued by this idea, Marshall began ordering biopsies for any patient who had stomach problems (Figure 1-1). He found that nearly every patient with ulcers was infected with the same bacterium.

The most common kind of ulcer is a peptic ulcer, an open wound located where the stomach joins the small intestine, at the bottom of the stomach. The word "peptic" comes from the Greek word *peptein*, to digest. Nearly one in ten adults has a peptic ulcer. Some people with ulcers feel no discomfort. But most feel at least mild pain, and many suffer excruciating pain for weeks at a time throughout their adult lives. In rare cases, blood may pour from the wound so freely that the person bleeds to death. The standard treatment for ulcers had always been a bland diet consisting of eggs, milk, creams, custards, and overcooked cereals. In addition, doctors often prescribed tranquilizers, psychotherapy, and, in severe cases, surgery. Mainly, however, doctors prescribed antacids—lots of antacids.

Until recently, prescription antacids were the biggest-selling prescription drugs in the world. In 1992, Americans alone bought $4.4 billion worth. These drugs are remarkably effective at controlling the secretion of stomach acid, but remarkably ineffective at controlling ulcers. Ninety-five percent of ulcer patients have a new ulcer within 2 years of treatment. That means people who had ulcers took the $100-a-month antacids almost continuously. In a lifetime, an ulcer-sufferer could spend tens of thousands of dollars on antacids.

Yet, if ulcers were caused by a bacterium, as Marshall was suggesting, then a simple 2-week course of antibiotics could cure an ulcer forever. Millions of people could be saved from a lifetime of suffering. If Marshall's hunch was right, he had very good news, although not for the companies selling antacids.

In 1983, Marshall decided to present his hypothesis at a scientific conference in Brussels, Belgium. But he had a problem. None of the other researchers had ever heard of him. He was barely out of medical school and he had no credentials at all as a researcher. His idea seemed to have about as much chance of winning recognition as bacteria had of flourishing in the acidic environment of the stomach.

Predictably, Marshall's presentation was a disaster. He was unknown; he was young, inexperienced, and overexcited. Worst of all, he had what looked like a screwball idea. "He didn't have the de-

- Scientific models suggest testable hypotheses that help answer questions and suggest entirely new questions.

- A good experiment includes one or more controls and a large enough sample to compensate for variation in the data.

- All living organisms reproduce, evolve, respond to their environment, and consist of a single cell or ordered arrangements of many cells.

- All organisms on Earth have descended from a common ancestor through a process called evolution.

meanor of a scientist," recalled Martin Blaser, professor of medicine at Vanderbilt University. "He was strutting around the stage. I thought, 'this guy is nuts'." When Marshall's presentation was over, his audience of eminent medical researchers shifted uneasily in their seats, embarrassed. A few laughed. They couldn't believe he was serious. Most bacteria can barely survive a brief passage through the stomach. How could they flourish there for months or years?

In any case, Marshall had no scientific evidence to back up his claim. Maybe, his audience told him, the bacteria had contaminated the stomach samples after the stomach tissues had been removed. Or maybe the bacteria were harmless and unrelated to the ulcers. Or maybe the bacteria were only able to grow in the stomach because the ulcer was there. The audience peppered him with challenging questions, and Marshall realized he couldn't answer any of them.

The only way to settle all these questions was to study the bacterium in an experimental animal. But to do that, Marshall would need to find an animal whose stomach could be infected with the bacterium. After returning to Australia, he began feeding rats the still-unnamed organism. The bacteria all died in the rats' stomachs without having any effect. He fed the stuff to pigs, with the same result. Now he began to wonder, Could the bacteria really infect a stomach? Maybe the researchers in Brussels had been right to laugh at him.

Desperate to prove that he was no nut, Marshall planned a highly unusual "experiment." He told no one ahead of time—not the medical ethics board at the hospital, not his wife. They wouldn't have approved, he knew. He began with a stomach exam and biopsy to make sure his stomach was healthy. Then he made himself an "ulcer bug" cocktail containing at least a billion bacteria, grown from bacteria he had isolated from biopsies of ulcer patients. Then, in a few swift gulps,

he drank it down. The cocktail was enough, he hoped, to infect his stomach.

Nothing happened. Days passed. Then, 8 days later, nausea woke him early and he vomited. For another week he was tired, irritable, and hoarse. He had headaches and foul breath. A second stomach exam and biopsy showed that his stomach was inflamed and swarming with bacteria.

By the third week, Marshall was lucky enough to have recovered completely. In the April 15, 1985, issue of *The Medical Journal of Australia*, Marshall reported the results of his trial. He had not proved that the bacterium could cause ulcers, or even that it could infect the stomach for years at a time. But he had done something that strongly suggested that the bacterium, still unnamed, could infect a healthy human stomach—one that didn't already have an ulcer.

His disastrous debut in Brussels had ensured that other researchers would remember him. They soon began to take an interest in Marshall's idea. Mainly, they were interested in proving him *wrong*. Yet, by the end of the 1980s, the evidence that the bacterium could infect the stomach was unassailable. The bacterium, they found, has a twisted, helical shape and usually lives in the "pylorus," near the bottom of the stomach. *Helicobacter pylori*, as it was finally named in 1989, twisted down into the mucous lining of the stomach and settled there.

By 1993, medical researchers had found that about 80 percent of all ulcers were caused by *Helicobacter pylori*. Nearly all could be swiftly cured with ordinary antibiotics.

At first, drug companies resisted Marshall's idea. It was true that people treated with $20 worth of antibiotics had a relapse rate of only 5 or 10 percent. It was true that millions of ulcer-sufferers would each spend thousands of dollars less on prescription antacids. But, gradually, they began to see a silver lining in the cloud the Australian had created.

The good news, for drug companies, was that infection by *H. pylori* was one of the most common bacterial infections in the world.

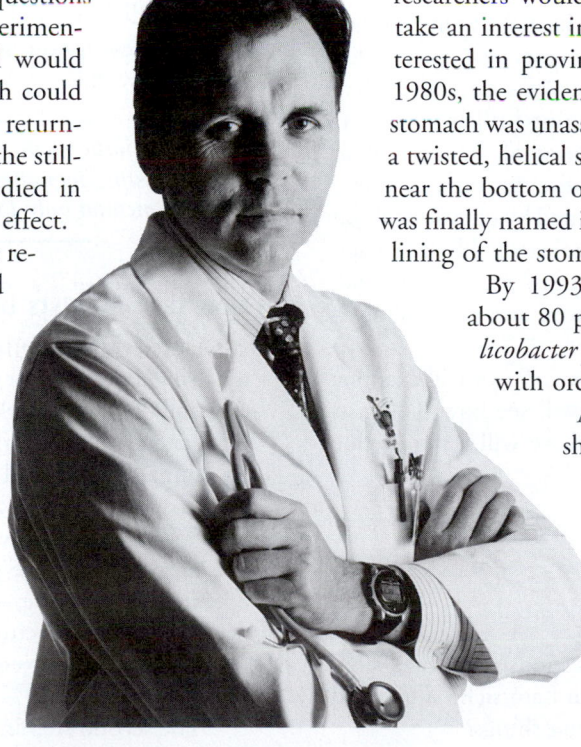

Figure 1-1   An independent thinker. Barry Marshall showed that bacteria, not cranky personalities, cause ulcers. *(Courtesy, Dr. Barry Marshall)*

3

As many as half of all people worldwide harbor the infection. And studies of war veterans showed that people infected by *H. pylori* had six times the rate of stomach cancer as other people. Here was a potential market for antibiotics consisting of half the population of the world, some 3 billion people.

Drug companies hastily developed diagnostic tests for *H. pylori* and new combinations of antibiotics to treat the infection. Biotechnology companies were trying to develop a vaccine, to be given in childhood, that would protect against *H. pylori* infection, ulcers, and maybe even stomach cancer—the world's second leading cause of cancer death after lung cancer.

Early in 1994, the National Institutes of Health declared antibiotics the official treatment for most ulcers. Eleven years after Marshall first presented his hypothesis, doctors could offer their ulcer patients a permanent cure. Even then, change came slowly. As late as 1996, only one-third of doctors in the United States were prescribing antibiotics for ulcer patients. The rest continued to prescribe antacids only. Today, antibiotics are standard treatment for this widespread infection, and every family doctor knows what causes ulcers.

In the end, Marshall won full acceptance as a scientist. Ultimately, he moved to the United States and joined the faculty at the University of Virginia Medical School. Marshall succeeded because he possessed many of the attributes of a good scientist. He had curiosity, intelligence, vision, and the dogged determination to pursue an idea—even when his stubbornness made him appear foolish. Perhaps most importantly, Marshall displayed an unusual independence of thought that allowed him to pursue an idea unimaginable to more dogmatic thinkers.

In this book we will meet many scientists who are as curious, independent, and stubborn as Marshall. Some are impulsive, like Marshall. Others take years to reach their conclusions. Most work long hours for years on end. A very few seem merely to play at science, reaping brilliant discoveries from a few hours' work. We'll see some of them risk their reputations to defend the ideas they believe in. A few even lose their lives. Right or wrong, all are fascinated by questions about what makes living things tick. Good biologists, like other scientists, are people who are intensely engaged with life. And, like all people, they work in intense social and political environments.

In this first chapter of *Asking About Life*, we will examine two important aspects of biology. We will see how biologists try to answer questions about life, and we will try to define life itself.

HOW DO BIOLOGISTS ASK QUESTIONS?

From early childhood, everyone asks questions about life: What makes me alive? What goes wrong when I am sick? Where did I come from? How am I like other living things?

The beginning of all scientific inquiry is curiosity. Something in the world captures our attention, and we begin to ask questions. The complex lives of bees may inspire in us pure wonderment. Or perhaps the slow unfolding of a rose arrests

our eye. Sometimes, a practical problem commands our attention. Maybe we would like to see a dying friend get well or a hungry village grow enough food to feed its children. Whatever engages our minds—whether beauty, complexity, or misfortune—can be the beginning of scientific inquiry (Web Bit 1-1: Dorothy Crowfoot Hodgkin).

The physicist Albert Einstein attributed his scientific achievements to his childlike curiosity. Einstein asked the kinds of questions that most of us ask when we are children. But unlike most people, he continued asking questions when he had the intellectual power to answer them.

Many scientists pride themselves on their childlike curiosity. But scientific inquiry is more than asking questions. Science is curiosity that is controlled and channeled. For example, when Barry Marshall asked whether a bacterium could cause ulcers, he knew it was a question that could be answered. Good scientists try to channel their energy into questions that can be answered.

Do Scientists Use the Scientific Method?

In the first part of this chapter, we will discuss how scientists choose questions to ask and how scientists answer those questions. Because science has been so successful in increasing knowledge, many philosophers and historians have tried to describe how scientists work. One description of the way scientists work is the **scientific method,** a set of formal rules for expressing, testing, and eliminating ideas. In reality, scientists do not think of themselves as following the rules of this or any other single "method." Rather, they pursue knowledge in a variety of individual and creative ways. Nonetheless, we can give a general description of how most scientists go about learning about the world.

> *The "scientific method" is a formal set of rules for forming and testing hypotheses. Most scientists use the scientific method only loosely and often unconsciously.*

How do scientists begin?

A scientist's first step in understanding something is to focus on a single question or small set of questions. Because science, especially biology, works in tiny steps, scientists rarely answer big, general questions all at once. A scientist does not ask, How do animals move? and expect to come up with a simple answer to that question.

When we ask how something works, whether it be a piano or a pancreas, we generally want to know how the parts operate. For example, consider the unbending of an arm. We can ask, What structures change as an arm unbends? What forces cause the movement? These are specific questions that can be answered.

The second step in understanding is *observation*. Looking, hearing, smelling, and touching may all contribute to our observations. For example, during a pushup, one can feel the bulging and hardening of the triceps, the long muscle at the back side of the upper arm.

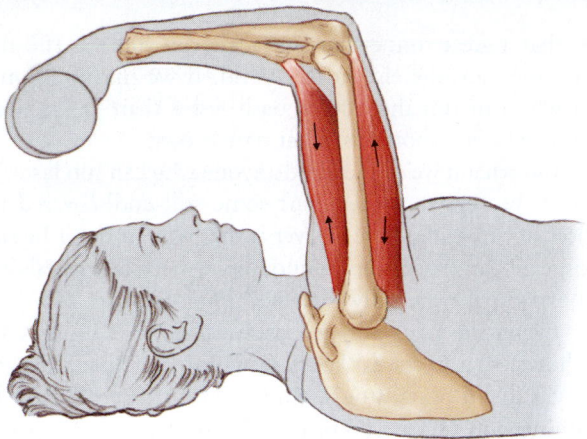

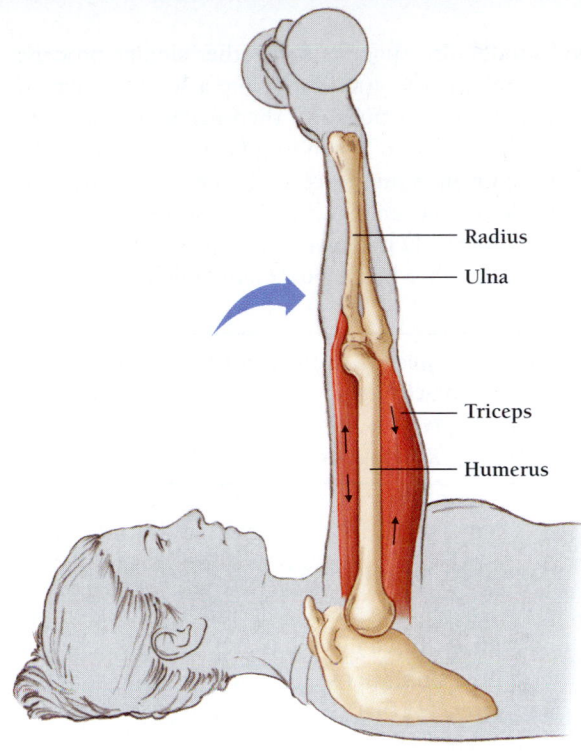

Figure 1-2   Observing the human arm. The upper arm consists of a long bone called the humerus and the lower arm consists of two shorter bones, the ulna and the radius. The ulna pivots on the end of the humerus. The triceps muscle pulls the end of the ulna and swings the lower arm in a wide arc.

Now, let's examine (observe) the internal structure of the arm (Figure 1-2). The skeleton of the upper arm consists of a single, long bone, called the "humerus," between the shoulder and the forearm. The forearm consists of two smaller bones, the radius and the ulna. The ulna pivots, rocking across the elbow end of the humerus. The triceps muscle connects to the elbow end of the ulna and to the shoulder.

Now we are ready for the next step. We need a **model**— a highly simplified (often cartoonlike) view of how the arm works when we unbend it. Let's imagine that the arm is like a seesaw on a children's playground (Figure 1-3). When a child sits on one end of a seesaw, the child's body exerts a force that pushes the seesaw down. As a result, the other end of the seesaw moves up. The ulna pivots on the tip of the humerus in the same way that a seesaw pivots on its base. When one end of the ulna moves up, the other end moves down.

In science, the first step is to ask a sharply focused question about some process. The next step is to observe closely, looking for a familiar pattern of behavior. The next step is to devise a model that suggests how the phenomenon works.

What does a hypothesis do?

A model must be consistent with what we already know. But it must also suggest things we might not have thought of otherwise. A good model should suggest a **hypothesis**—an educated guess about the way a process works that allows us to make predictions. Scientists could not work without hypotheses. Hypotheses are as essential to the daily functioning of science as vitamins are to the normal functioning of our bodies. So where do scientists get hypotheses?

The seesaw model suggests a good hypothesis. If the ulna is like a seesaw, some force must cause it to rock back and forth during a pushup. Maybe when the triceps muscle contracts, it pulls on the end of the ulna. The elbow end would move toward the shoulder, and the wrist end would move away (Figure 1-3).

If that's true, we can predict that if we try to unbend our arm without using our triceps muscle, we won't be able to do it. We could perform an experiment to test this prediction. We could, for example, temporarily immobilize the triceps with an injection of some paralyzing drug.

In general, if the predictions of a hypothesis are right, we begin to guess that we are on the right track. On the other hand, if the predictions of a hypothesis are wrong, then both the hypothesis and the model it is based on are likely wrong.

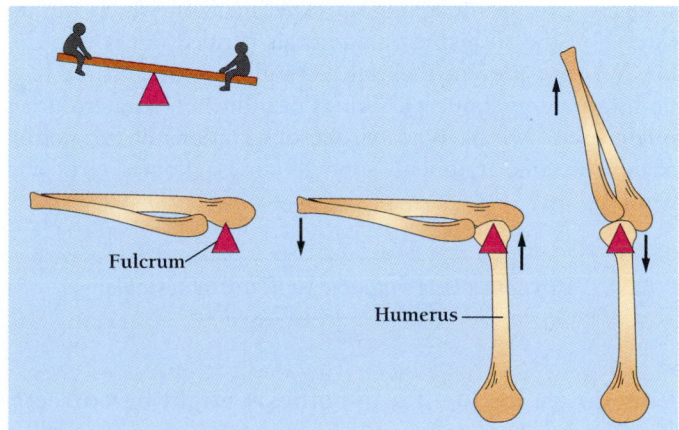

Figure 1-3   It rocks. Like a children's seesaw, the ulna (in the lower arm) pivots on the tip of the humerus (in the upper arm).

A good model also suggests how other similar processes work. For example, if the arm works like a seesaw when the arm is unbending (as in a pushup), then maybe it also works like a seesaw when the arm is bending (as in a chin-up). Perhaps during a chin-up some force pulls the wrist end of the ulna. Our model of the unbending of the arm accomplishes two important things: (1) it allows us to make predictions and (2) it suggests how other body movements might work.

A hypothesis is an informed guess about the way a process works that enables a scientist to predict what will happen in different situations. It is a product of logic, previous knowledge, insight, and creativity.

What is a testable hypothesis?

We can quickly generate lots of models and hypotheses to explain any observation. If we note that ants in the kitchen are crowding around the sink, we might hypothesize: (1) that they are looking for water, (2) that they are looking for food particles, (3) that they are looking for a new nest site, or (4) that they want to clean the dishes for us. Clearly, some hypotheses are better than others. Scientists need ways of deciding which hypotheses to explore.

In general, a hypothesis is valuable only when it is **testable,** meaning that someone can devise an experiment that would disprove the hypothesis if it were incorrect. In the last section we predicted that someone whose triceps were temporarily paralyzed would not be able to do a pushup. This is a test of the hypothesis that the triceps provides the force that powers a pushup.

Whether a hypothesis is testable does not depend on whether it is correct. The hypothesis that bears eat berries is testable. The hypothesis that bears eat rocks is equally testable. Testable (or *disprovable*) means that we can make a prediction that we can test by experiment.

The hypothesis that extraterrestrial beings influence the movement of ants in your kitchen is untestable—but *not* because it is wrong. We have no way to prove that it is wrong. This hypothesis is untestable—and therefore of no interest to science—because designing an experiment in which the influence of the alleged extraterrestrials is ruled out is impossible. We can't prevent their alleged influence because we have no information about what kinds of influence extraterrestrials might exert. We have no reliable observations of extraterrestrials. The same arguments apply to fairies, gnomes, elves, and deities.

A hypothesis is valuable only if it is testable.

How do we decide if a hypothesis might be correct?

We can often prove that a hypothesis is wrong, but we can never prove that it is right. This is because a prediction can turn out right—even when the hypothesis is wrong. We could hypothe-size that a great runner named Jackie can run the 100-m dash faster than anyone else in the world. If we find even one person who can run the 100-m dash faster than Jackie, we have disproved our hypothesis. That part is easy.

But what if we cannot find anyone who can run faster? How can we be sure that there isn't some still-undiscovered person who can run faster? The answer is that we can never be sure. In most cases, an experiment cannot prove that a hypothesis is right. An experiment can only disprove a hypothesis.

Even disproving a hypothesis can be hard. Suppose we hypothesize that bears eat berries. We predict that we will find berry skins in bears' stools. And we test this prediction by examining stools from 100 bears. We find grass, insect parts, and even small plastic bags, but no sign of berries. Does this mean the hypothesis was wrong? Maybe not. Our results might mean only that berries were not in season. Maybe we should repeat our study in the fall. We might also wonder if berry skins are digested completely and leave no trace in the stool. We could test that hypothesis by feeding berries to captive bears. In this way, one experiment leads to another.

Any experiment can fail to give the predicted result for reasons that have nothing to do with the hypothesis being tested. The prediction might have been wrong because of faulty logic, or the experiment might have been poorly designed or executed. It is the scientist's job to figure out why the model's prediction was incorrect and then to come up with either a better model or a better experiment.

When scientists test a hypothesis, they can sometimes prove that the hypothesis is wrong, but they can never prove that it is right.

What makes a good experiment?

An **experiment** is a way of testing a hypothesis or of searching for some unknown effect. An experiment is a procedure that is carried out under "controlled" conditions.

Suppose, for example, that we observe that people who live in countries where cinnamon and other spices are used extensively are shorter than other people. Based on this observation, we hypothesize that cinnamon inhibits growth. We decide to test this hypothesis by measuring cinnamon's effect on the growth of laboratory mice. We choose mice because experimenting on humans would be unethical and because mice are easy to keep.

We inject baby mice with extract of cinnamon to see if the cinnamon slows their growth. We find that their growth is slow compared with that of mice in other studies. Can we conclude that cinnamon slowed their growth? We cannot. Maybe we were not feeding the mice enough food. Maybe the room they are in is too hot. Maybe chemicals in the extract other than the cinnamon are slowing their growth. The outcome of this experiment could depend on many causes besides the cinnamon.

To ensure that the outcome of an experiment depends only on the proposed cause, biologists compare the results of every experiment with the results of a "control" experiment. A **control** is a version of the experiment in which everything is the same except for the one thing being tested. To test our cinnamon hypothesis properly, for example, we need at least two controls. First, we need, in our own lab, a group of mice that receive no injection. These will show us how fast mice normally grow under the conditions in our laboratory, not someone else's. Second, we need a group of mice each of which receives an injection containing everything in the extract except the cinnamon. This second control group will show whether something else in the extract slows the growth of baby mice.

The design of a control is critically important and not always obvious. For example, the trauma of being handled might be enough to slow a mouse's growth. To make sure that is not the case, we might want to include a third control in which researchers merely pretend to give baby mice injections—sticking them with a needle, but not injecting anything. This might sound silly, but we know that both humans and other animals respond in both positive and negative ways to handling. It is good science to test anything that may affect our results. Because an experiment must distinguish causes from irrelevant factors, an ideal experiment allows a researcher to evaluate the effects of one factor at a time.

When Barry Marshall swallowed his mixture of stomach bacteria, he was performing an experiment—but only just barely. It was not a good, controlled experiment. His symptoms—nausea, tiredness, and bad breath—might have been caused by the flu, a sinus infection, or any number of things. Even the inflammation of his stomach shortly after he drank the ulcer bug cocktail might have been a coincidence.

Experiments on a single individual may be strongly suggestive, as Marshall's was. But such experiments do not provide results that mean anything scientifically. To do the experiment properly, Marshall would have had to persuade at least three groups of people to participate. The first group, the "experimental" group, would have gotten the ulcer bug cocktail. The second group would have gotten a different bacterium—one not believed to cause ulcers—to rule out the possibility that any concentrated mixture of bacteria can make someone sick. The third group would have gotten the same mixture except without any bacteria, to rule out the possibility that the cocktail—even without any bacteria—is itself a sickening mixture.

An experiment must include carefully designed controls.

How Do Scientists Design Experiments?

Designing experiments is the greatest challenge to a scientist's ingenuity and the greatest source of excitement. Experimental design takes skill and artistry. Like a custom paint job on a car or the cut of a special dress, great experiments are the product of both mastery and inspiration. One of the most famous stories of inspiration in science comes from the German physiologist Otto Loewi.

Loewi was trying to understand how nerves control muscles. He had been studying the heart, a large muscle that contracts with each beat, and a nerve called the vagus, which controls how fast the heart beats. When Loewi stimulated a frog's vagus nerve with electricity, the frog's heart rate slowed.

But Loewi didn't know if the nerve slowed the heart through direct contact or by means of some chemical. He suspected that nerves make a chemical that stimulates muscles. If that were true, then a muscle should respond to the right chemical even if the nerve weren't there. But Loewi had no idea what kind of chemical might do that. And he didn't see how he could possibly test his hypothesis without knowing.

In fact, 17 years passed before he thought of a way. Loewi himself later described how the experiment finally came to him:

> The night before Easter Sunday of 1920 I awoke, turned on the light, and jotted down a few notes on a tiny slip of thin paper. Then I fell asleep again. It occurred to me at six o'clock in the morning that during the night I had written down something most important, but I was unable to decipher the scrawl. The next night, at three o'clock the idea returned. It was the design of an experiment to determine whether or not the hypothesis of chemical transmission that I had uttered seventeen years ago was correct. I got up immediately, went to the laboratory, and performed a simple experiment on a frog heart according to the nocturnal design.

In hindsight, Loewi's 3 a.m. experiment seems straightforward, almost routine. Loewi isolated two frog hearts, one with the vagus nerve still attached, the other without the nerve. He stimulated the vagus nerve of the first heart, collected some of the fluid surrounding it, and transferred the fluid to the second heart, the one with no nerve. The heart slowed.

The dramatic result showed that the slowing was a response to some substance produced by the vagus nerve. Loewi called this substance "vagus stuff," which we now know as acetylcholine—an important chemical signal that nerves use to communicate. Loewi eventually received a Nobel prize for his work.

Designing experiments to test a hypothesis requires both creativity and logic.

Why Do Biologists Study Groups of Organisms?

We have talked about experiments in groups of mice or other organisms. Why can't scientists just measure growth rate in one mouse with cinnamon and one mouse without? Why do scientists study groups of organisms?

The answer is that every organism is unique. Every fruit fly is unique and every cat is unique (Figure 1-4). So even are identical twins and the individual clones of bacteria or sheep. For example, one pair of identical twin girls differed in height by 7 inches as adults. The difference apparently resulted from prenatal trauma in one of the twins. Other identical twins

BOX 1.1

Reductionism

One way to study a thing is to look at its parts. If we look at the wheels of a bicycle, for example, we can see that they turn and are attached to the bicycle's frame by means of intri-cately designed hubs. Much of biology involves studying the parts of organisms, whether their organs, their cells, or their molecules. For every level of organization, biologists seek to explain structure and function in terms of the next finer level of structure. We can, for example, understand something about the action of arm muscles by studying the anatomy of the bones and muscles. The effort to understand the

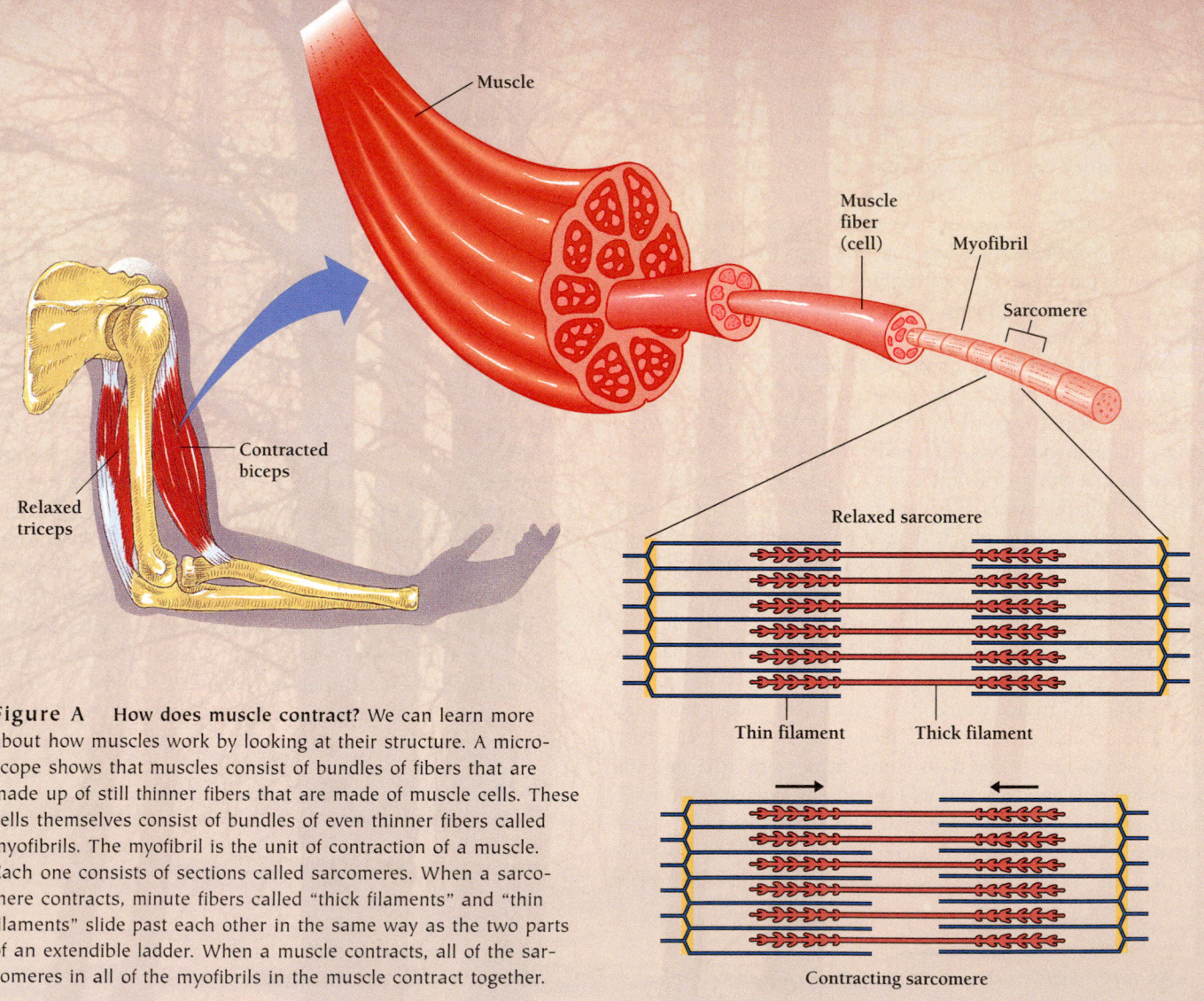

Figure A **How does muscle contract?** We can learn more about how muscles work by looking at their structure. A micro-scope shows that muscles consist of bundles of fibers that are made up of still thinner fibers that are made of muscle cells. These cells themselves consist of bundles of even thinner fibers called myofibrils. The myofibril is the unit of contraction of a muscle. Each one consists of sections called sarcomeres. When a sarco-mere contracts, minute fibers called "thick filaments" and "thin filaments" slide past each other in the same way as the two parts of an extendible ladder. When a muscle contracts, all of the sar-comeres in all of the myofibrils in the muscle contract together.

may have deformities on opposite sides of their faces, heart defects in one but not the other, and so on. Variation among organisms is the rule. Whether we are talking about humans, mice, worms, oak trees, dandelions, mushrooms, or even one-celled organisms, every individual is unique.

Some of this individuality is due to genetic differences. Some of it is due to the impact of the environment. An *H. pylori* bacterium on the wrong end of your stomach may not get the same nutrients as a bacterium in the prime posi-tion down by the pylorus.

whole in terms of the parts is called **reductionism**.

Understanding the parts of a process can help explain larger, more complex biological processes. Many of the questions that we ask in this book concern the relationships of structures and processes. The questions are often the same: What are the involved structures? What are they made of? How do they capture energy? What controls their operation? The answers to our questions may lie at the level of ecosystems, populations, individual organisms, cells, or even individual molecules.

The reductionist approach to science has been immensely successful. The most successful and enthusiastic reductionists are molecular biologists, who seek to answer a wide variety of questions just by looking at the structures of molecules (Figure A). These questions range from the inner workings of cells to the evolution of whole families of organisms over millions of years.

But reductionism has definite limits. All of the properties of an object are not explainable in terms of the object's parts. To understand how the circulatory system works, we need to study the veins, arteries, and heart in whole animals. Looking at the individual molecules of the veins and blood would tell us nothing in the absence of a good understanding of how blood circulates in the body.

A whole is greater than the sum of its parts. The painting in Figure B illustrates this principle simply and beautifully. When we examine the painting under high magnification, we see only an abstract pattern of colored dots. It is only when we step back and look at the whole picture at once that we see a river, trees, and flowered hillsides. The landscape is an **emergent property,** a characteristic that arises only at complex levels of organization. Living organisms are no different.

To discover how bees pollinate flowers, we need to study hives of bees and fields of flowers, not their molecules. Just as a painting is much more than a collection of brush strokes, a group of organisms is more than a collection of individuals. A detailed knowledge of the behavior of solitary humans, for example, would never allow us to predict such bizarre group behaviors as war and Tupperware parties.

Figure B **Art as emergent property.** A landscape painting is an emergent property—a characteristic that arises only at complex levels of organization. When we look at the individual brushstrokes, we see only an abstract pattern of colored dots. When we step back and look at the whole picture at once, we see a landscape— "La Seine à Herblay," by Maximilien Luce (Musée d'Orsay, Paris). *(Erich Lessing/ Art Resource)*

Because each individual organism is unique, each one can react to an experiment differently from every other individual. For example, every mouse in the cinnamon-extract experiment could grow at a different rate. Individual mice, like individual humans, grow at different rates. We'd be amazed if every mouse in the experiment grew at exactly the same rate, cinnamon or no cinnamon. Because individuals vary, biologists always experiment with groups of organisms rather than single individuals. Then they compare the average of the experimental group with that of the control group.

Figure 1-4 Why are individuals unique? When organisms reproduce, the genetic information in the DNA allows the offspring to be both like their parents and unlike their parents. Here, the offspring of a cat show a diversity of phenotypes because each has a unique set of genes inherited from mother and father. Our uniqueness comes not only from our genes, however, but strongly from our environment. For example, even genetically identical twins can differ dramatically if the environment of the womb favors one over the other. *(Francois Gohier/Photo Researchers, Inc.)*

How many individuals should an experiment include?

How many individuals an experiment should include depends on how much individual variation exists. For example, a group of genetically identical mice raised in a laboratory under identical conditions would be expected to show only minimal variation. Imagine that all the mice are genetically identical, that they all receive the same amount of food and water each day, and that the temperature and humidity in their cages was identical. A biologist wouldn't need many mice to show an effect on growth rate—ten in each group might be enough.

Now imagine a group of wild mice living outdoors. Each mouse is genetically unique. Each one lives a slightly different life from the others. Some get more milk from their mother from the start. Some contract a disease that the others don't get. The list of possible sources of variation is endless. The result is mice whose growth rates vary by as much as 50 percent. In this case, detecting a 10 percent difference between the experimental mice and control mice would be difficult. Like wild mice, human populations vary enormously, which is why the best studies of humans often involve thousands of individuals.

The greater the variation, the more individuals we need for an experiment. But we can only find out how many individuals we need by first measuring variation. To measure variation, we have to do a simple version of the experiment. Fortunately, there is a shortcut. A biologist might need hundreds of mice. But thanks to statistics, it's possible to measure a difference between two groups, even when both groups vary enormously.

How Do Biologists Use Statistics To Plan and Evaluate Experiments?

Biologists often plan experiments and evaluate the results of experiments with **statistics**—the mathematics of collecting and analyzing numerical data. Statistics is a powerful tool with two formal uses. (1) Statistics can help us estimate the actual (or "true") value of something. (2) Statistics can help us test a hypothesis. Statistics cannot do experiments for us. But it can measure the probability that what we measure is what it seems to be and not a random effect. The arrangement of data into graphs and pictures can also suggest new questions and new hypotheses.

How do scientists sample a population?

Statistics relies on certain assumptions. But statistics always states its assumptions clearly and then examines those assumptions. If we study the effects of aspirin on heart attack rates in groups of men, we are really interested in the effect of aspirin on *all* men. The men in the experiment are supposed to represent all men. Statisticians call all men the **population** of men. The group selected for the experiment is a **sample** of the population. In statistics, we want to be able to assume that the sample represents the population.

One way of ensuring that the sample represents the population is to take a **random sample.** A sample is random when we ignore any differences among the members of a population while choosing the sample. Suppose we have a population of 100 men, one-third of whom eventually have heart attacks (Figure 1-5). One-fourth of the 100 men agree to volunteer for a study of heart attacks and aspirin.

Figure 1-5A shows the whole population. Figure 1-5B shows what happens if you choose a sample of 25 men randomly. Figure 1-5C shows what happens if you choose a sample based on some difference. In this case, only volunteers, who do not represent the population, were chosen. Apparently, the volunteers are more likely to have heart attacks than the general population. Maybe the volunteers were men who volunteered for the study because they knew they were at risk for heart attacks. Compared to the general population, they may include more smokers, more older men, and more sedentary men than the population at large.

A sample can also fail to represent the population because the sample size is too small. Figure 1-5D shows two samples of only 10 men each. In neither one does the number of heart attack victims represent the rate in the population. This difference is due to random errors.

If we flip a coin 100 times, we will get equal or almost equal numbers of heads and tails. We might get 52:48 or 47:53, but the ratio will be close to 50:50. On the other hand, if we flip the coin only 10 times, we will probably not get a ratio very close to 50:50. We would have a good chance of getting 2, 2, 3, or 4 heads instead of 5. But chances are we wouldn't get exactly 5 heads and 5 tails. Random effects are big in small samples and tiny in big samples.

In a biological experiment, a sample should represent the whole population of organisms. This can happen only if the sample is chosen randomly and the sample is large enough.

Biologists are primarily interested in two statistical measures. The first measure is called the *mean*, which is just the average value of something. The second measure is the *standard deviation,* which is an estimate of the variation in the data. If we were studying heart attack rates in men, we would want to know the mean heart attack rate both in the experimental group and in each of the control groups. We would also need to measure the standard deviation in each sample. Both the mean and the standard deviation are essential to thoughtful evaluation of the experimental results.

How does statistics help biologists estimate value?

The **mean** is the average value, or arithmetic average, that we all learned about in elementary school. Real-life studies have shown that the mean heart attack rate in the population of men over 50 years old who are taking aspirin is lower than that in the population of men over 50 not taking aspirin. Because this is known to be true, millions of older men now take an aspirin every other day to reduce their chance of getting a heart attack.

How can biologists tell if two groups differ?

It might seem as if all we need to know is whether the mean heart attack rate of aspirin-takers is lower than that of the non-aspirin-takers. If it is lower, then all men over 50 should take their aspirin. But it turns out that we need to know more.

Imagine that the sample size is too small and the sample mean is different from the population mean. In that case, the mean for the whole population of aspirin-takers could actually be the *same* as the mean for all non-aspirin-takers. In fact, things could be much worse. The aspirin-takers could have a *higher* rate of heart attack. Meanwhile, the sample means could show the opposite. Indeed, in medical studies of just a few people, this often happens. When you read a newspaper article about the latest health findings, always look to see whether the study included thousands of people or just a few dozen. If the article doesn't say, don't take the conclusions too seriously.

Fortunately, statisticians have devised ways of checking for this problem. In order to find out if two means are truly different, we have to estimate the variation in the data. We have to find out, in other words, how much individual variation

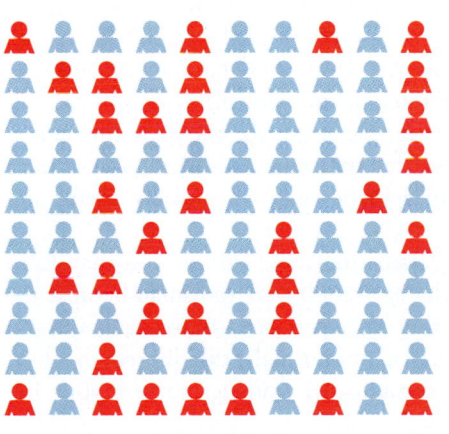

A. 33/100 will have heart attacks. 33%

B. 7/25 will have heart attacks. 28%

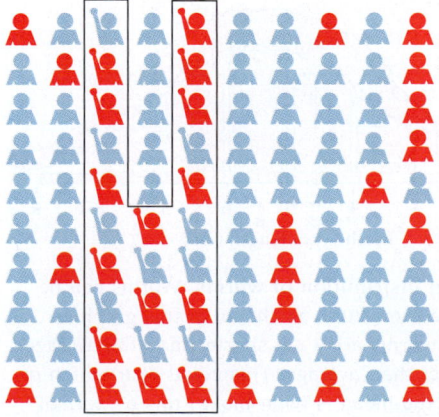

C. 15/25 will have heart attacks. 60%

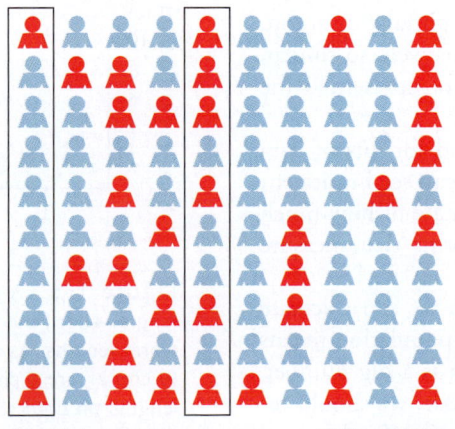

D. 2/10 and 6/10 20% and 60%
will have heart attacks.

Figure 1-5 Sampling a population.
A. A population of 100 men includes 33 who have heart attacks. B. A random sample of 25 men yields 7 who later have heart attacks. This rate, 28 percent, approximately reflects the heart attack rate in the general population, but with some sampling error. C. 25 men volunteer for a study of heart attacks, but the heart attack rate in the volunteers—15 out of 25, or 60 percent—does not reflect the rate in the whole population. The volunteers are not a random sample. D. Two small samples of just 10 men each show a larger sampling error than in B. In one case, the heart attack rate is 2/10, or 20 percent. In the other case, it is 6/10, or 60 percent. Neither is a good measure of the overall heart attack rate in the general population.

there is. Statistics measures variation by measuring how far each data point is from the mean. When those distances are averaged, we get a measure called the **standard deviation,** which summarizes how much the data deviate from the mean.

This sounds more complex than it is. Here's a simple example. Suppose we want to determine the average length of earthworms we have just dug up from our back yard. We start with only three worms whose lengths are 3 cm, 5 cm, and 7 cm. We say we have three "data points." The mean worm length is simply the average of those measurements, or 5 cm. We can then measure the deviation from the mean as follows:

$$7 \text{ cm} - 5 \text{ cm} = 2 \text{ cm}$$
$$5 \text{ cm} - 5 \text{ cm} = 0 \text{ cm}$$
$$5 \text{ cm} - 3 \text{ cm} = 2 \text{ cm}$$

To calculate standard deviation, statisticians square each of these differences, then add them all together, and then divide by the number of data points minus 1. That would be $(2^2 + 0^2 + 2^2)/(3 - 1) = 8/2 = 4$. (We leave out the "cm" here for simplicity.) Then, because they did all that squaring, statisticians bring the number back to reality by taking the square root of 4, which is 2. In this case, the standard deviation is the mean difference from the average.

We express this by saying that the length of the earthworms in our sample is 5 cm "plus or minus" 2 cm.

$$5 \text{ cm} \pm 2 \text{ cm}$$

The standard deviation here is huge, a full 40 percent of the mean. We can see that our sample of earthworms is highly variable. But does the population vary that much? If we sampled 100 worms, would the mean still be 5 cm $\pm$ 2 cm? Not necessarily. We may have accidentally picked the very longest and shortest worms. Indeed, the mean length may not be 5 cm at all. It may be something quite different. If the standard deviation is large, we know that our estimates are unreliable. Only with a much larger sample size could we accurately measure the mean of the population.

Imagine, on the other hand, that the standard deviation had been very small. Suppose the three earthworms were 5.1 cm, 5.0 cm, and 4.9 cm. We would be more confident that our 5-cm sample mean was close to the population mean. And if we had 20 earthworms whose lengths were so similar, we could be extremely confident that 5 cm represented the population mean.

Calculating standard deviations by hand for lots of data is not something anybody does for fun. We can see intuitively, however, that 2 cm is probably a good estimate of how much variation there is in the group of earthworms that measure 3 cm, 5 cm, and 7 cm.

Once we know the standard deviation, we can calculate how big a sample we need to tell if the two population means are truly different. If the standard deviation is a big number, our sample size may be too small. In that case, we should repeat the experiment using the right sample size. In this case, we need to measure many more earthworms. Alternatively, if

the standard deviation is small enough, then we may already have enough data. We can then calculate the probability that the two population means are different. Statisticians do this all the time. They might say, The probability that the aspirin group has a lower heart attack rate than the controls is 95 percent. By this, they mean that there is only a 5 percent chance that the aspirin-takers have either more heart attacks or the same number of heart attacks as the controls.

Can the mean and standard deviation explain all data sets?

Knowing the standard deviation is essential. But it is accurate only in groups of numbers that are distributed around the mean "normally." A **normal distribution** is one in which there are equal numbers of data points on each side of the mean. When graphed, they are shaped something like a bell (Figure 1-6). Such distributions are surprisingly common in biology. In Chapter 16, Figure 16-6 shows a graph of student heights. That's a normal distribution.

When data have a normal distribution, biologists can assume that about 68 percent of the data will fall within one standard deviation of the mean, and about 95 percent will fall within two standard deviations (Figure 1-7). Once we know that, we can test the hypothesis that the two means are different.

We will stop now. But just remember this. If the numbers in a population are normally distributed (shaped like a bell curve) and if you know the mean and standard deviation of the two samples (experimental and control), then you can calculate the probability that the mean for the experimentals is different from the mean for the controls—in the whole population. This aspect of statistics is a fantastically useful and powerful tool.

In the course of calculating these things, you may learn that the sample sizes that you took were too small because the variation is huge. In this case, you need more data. Or you may learn that your data are not normally distributed, in which case you need more complicated statistics to test your hy-

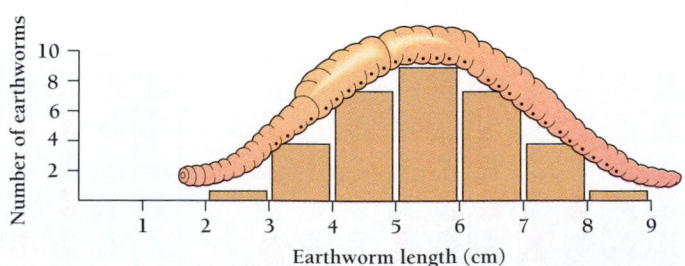

Figure 1-6   A normal distribution of worms. Worms can be long or short, but most are medium, as shown in this graph of worm lengths. As many worms are 1 cm shorter than the average (5 cm) as are 1 cm longer than average. This equal distribution of lengths on both sides of the mean is what makes this a "normal distribution." If most of the worms were either very short or very long and few were average, the distribution would not be normal.

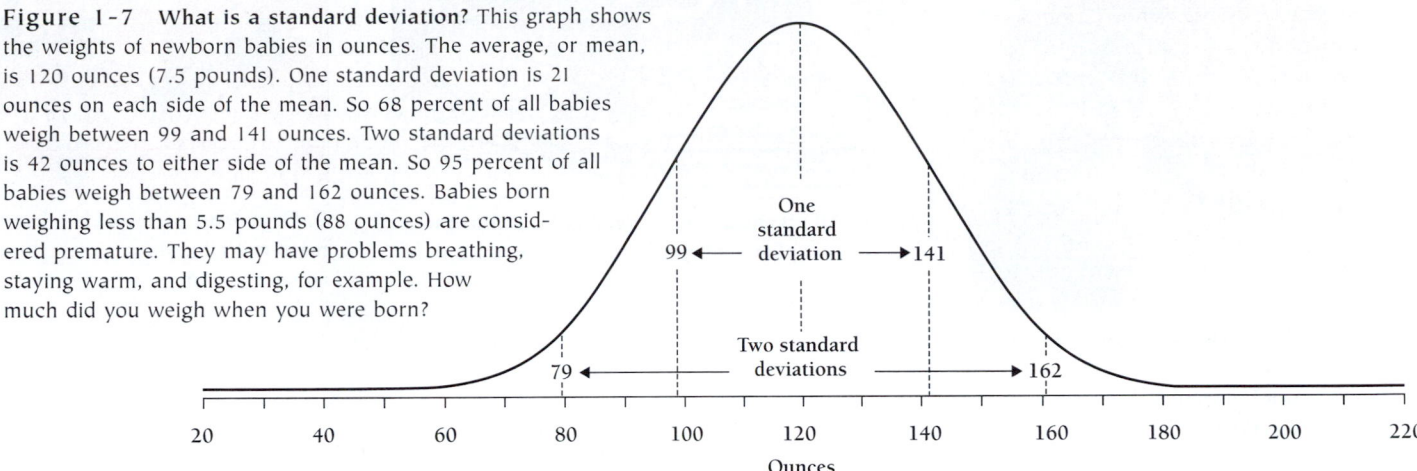

Figure 1-7 What is a standard deviation? This graph shows the weights of newborn babies in ounces. The average, or mean, is 120 ounces (7.5 pounds). One standard deviation is 21 ounces on each side of the mean. So 68 percent of all babies weigh between 99 and 141 ounces. Two standard deviations is 42 ounces to either side of the mean. So 95 percent of all babies weigh between 79 and 162 ounces. Babies born weighing less than 5.5 pounds (88 ounces) are considered premature. They may have problems breathing, staying warm, and digesting, for example. How much did you weigh when you were born?

pothesis. But don't worry. You don't have to calculate anything right now.

What Is a Theory?

When a series of experiments fails to disprove the predictions of a hypothesis, scientists gain confidence in the hypothesis and its underlying model. In addition, the more things a model explains, the more important the model becomes to other scientists.

Once scientists have tested a model for a group of related processes, they may decide that they can accept a more general hypothesis. If we continued to study the movements of muscles and bones, for example, we might conclude that our limbs work mostly by simple mechanical laws. Scientists often call a group of related hypotheses a **theory**—a system of statements and ideas that explains a group of related facts or phenomena. For example, "atomic theory" includes many ideas, all of which are based on the conclusion that matter is made of tiny entities called atoms. "Evolutionary theory" includes many ideas, all of which are based on the conclusion that living organisms are related to one another and have descended from a common ancestor.

Scientists use the word "theory" differently from many nonscientists. A nonscientist might use the word "theory" for any half-baked idea. Joe has a "theory" that if he reads every other paragraph in this book, he'll get an A on the final. To a scientist, that's only a hypothesis, and one that needs testing. It doesn't rate as a formal theory. A scientific theory is a set of interconnected hypotheses that have withstood rigorous experimental testing.

A set of related hypotheses that consistently resist scientists' attempts to disprove them may become recognized as a formal theory.

HOW ARE ALL ORGANISMS ALIKE?

Everywhere on Earth, there is life: from the depths of the Pacific Ocean to the top of Mt. Everest, from the frozen wastes of the Arctic Ocean to the tropical rain forests of the Amazon and the arid Sahara Desert. For each living thing, the same questions arise: How is it put together? How does it work? How did it get here? In answering these questions, biologists have accumulated millions of individual facts (only a tiny fraction of which we will study). Tying these disconnected facts together, however, are a few basic themes. Once we grasp these few generalities, the study of life becomes much easier.

Since **biology** [Greek, *bios* = life + *logos* = word, argument] is the study of life, the first general question we must answer is, What is life? Most of the time we can distinguish the living from the nonliving as easily as we can tell a live bear from a teddy bear. Although agreeing on a definition of life that is both precise and general is harder, we can agree on some common characteristics of all living things. Figure 1-8 lists and illustrates eight features of all living organisms.

In summary, all living things are organized into cells and other parts, perform chemical reactions, obtain energy from their surroundings, change with time, respond to their environments, reproduce, and share a common evolutionary history.

How Do Organisms Self-Regulate?

To survive, organisms need to keep their insides separate from the outside world. The first step to doing this is to create a barrier. In the case of our bodies, it is our skin. In the case of our cells, it is the cell membrane.

The second step is to actively resist change. The process of resisting change is called **homeostasis** [Greek, *homeo* = same + *stasis* = standing still]. Homeostasis always involves two stages: (1) detecting change and (2) counteracting such change. Counteracting change is called *negative feedback*.

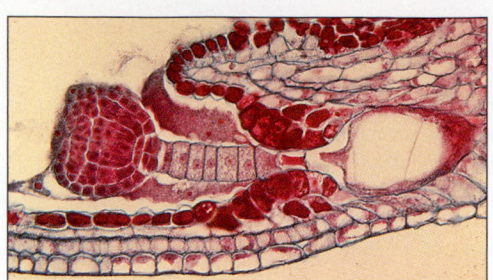

B. All organisms are made of cells, as illustrated by this plant embryo. Just as the atom is a unit of matter, a cell is the unit of life.

C. Organisms perform chemical reactions. Honeybees convert flower nectar into honey.

A. Living things consist of organized parts, as illustrated by this beautiful chambered nautilus. The word "organism" has the same root as "organization."

E. Organisms respond to their environments. The plants in this pot are bending toward a sunlit window.

F. Organisms mature over time. A tadpole maintains its organization while developing into a frog.

D. Living organisms obtain energy from their surroundings. Sunlight is the ultimate source of energy for most organisms. Plants obtain energy directly from the sun. Animals obtain energy from plants, or from animals that have eaten plants.

G. Organisims reproduce. Each organism, including each of these rabbits, comes from another organism or from a pair of organisms.

H. Organisms share a common evolutionary history. All organisms on Earth today are descended from organisms that lived in earlier times. Some have changed dramatically. Others have remained the same. The horseshoe crab of Atlantic beaches shown here is nearly identical to one that lived 200 million years ago.

Figure 1-8 Some characteristics of living organisms. *(A, J.H. Robinson/Animals Animals; B, John D. Cunningham/Visuals Unlimited; C, M.A. Chappell/Animals Animals; D, Otto Willner/OKAPIA/Photo Researchers, Inc.; E, N. Pecnik/Visuals Unlimited; F, Dan Suzio/Photo Researchers, Inc.; G, Jane Burton/Bruce Coleman, Inc.; H, [top] John Cancalosi/Peter Arnold, Inc., [bottom] Alan Desbonnet/Visuals Unlimited)*

One example of the use of negative feedback in homeostasis is temperature regulation. Birds and mammals maintain a remarkably constant internal temperature. A person can be exposed to temperatures as low as 55°F or as high as 140°F and still maintain a body temperature of between 98°F and 99°F.

When our temperature rises, we perspire, drink water, or look for a cool place to rest. When our body temperature drops, we conserve heat by not perspiring and by reducing blood flow to the skin, hands, and feet. In addition, we increase heat production by moving around, shivering, or increasing our metabolic rate. And, of course, we can also put on warm clothes or turn up the heater. All of these responses, whether physiological or behavioral, are homeostatic responses that keep the temperature of the body between 98°F and 99°F.

Using negative feedback, animals regulate such things as body temperature, thirst, hunger, and even sperm production. Plants regulate their intake of water and carbon by means of negative feedback. In all organisms, cells regulate the production of proteins and other molecules using negative feedback.

All organisms resist change through homeostasis and negative-feedback mechanisms.

What Kinds of Cells Are Organisms Made Of?

Whether dandelion or dachshund, *Helicobacter pylori* or bathroom mold, we organisms are all made of cells. A cell is a tiny mass of water, protein, and other molecules, all surrounded by a thin membrane. Indeed, all cells are made from the same few molecular building blocks. But a cell is much more than a bag of protein and water. It can do all of the things listed in Figure 1-8, including reproduce itself.

Cells are wonderfully diverse and specialized. In our own skin, flat "epithelial" cells protect our body, "follicle" cells produce hairs, and nerve cells carry messages. Plant cells are similarly specialized. Tough outer cells protect the plant's body, others transform light energy into sugar, and still others form passages that carry sugar, water, and other materials throughout the plant. There is no easy way to count all the different kinds of cells. Mammals alone have at least 200 different kinds of cells (Figure 1-9).

Nonetheless, we can group cells into just two basic kinds. Our own cells are eukaryotic cells. **Eukaryotic** cells [Greek, *eu* = true + *karyon* = nucleus] contain a **nucleus,** which is a package of genetic material enclosed in a thin membrane. Eukaryotic cells contain about 20 kinds of little membrane-enclosed parts called **organelles** [little organs]. These include the energy-supplying mitochondria and chloroplasts (in plants only). Plants, animals, fungi, and the microscopic protists are all made of eukaryotic cells.

The second kind of cell is the **prokaryotic** cell [Greek, *pro* = before]. Prokaryotic cells contain no nuclei and no membrane-enclosed organelles. They have genetic material, but they are not enclosed in a membrane. All prokaryotic cells are kinds of bacteria and all bacteria are prokaryotic cells. So far, biologists have named only 4800 species of prokaryotes, compared to more than a million species of eukaryotes.

Yet if the tiny prokaryotes are few in kind, they are great in number. More bacterial cells live in our intestines, for example, than we have cells in our bodies. And it is from the prokaryotes that all eukaryotes—including humans—are descended.

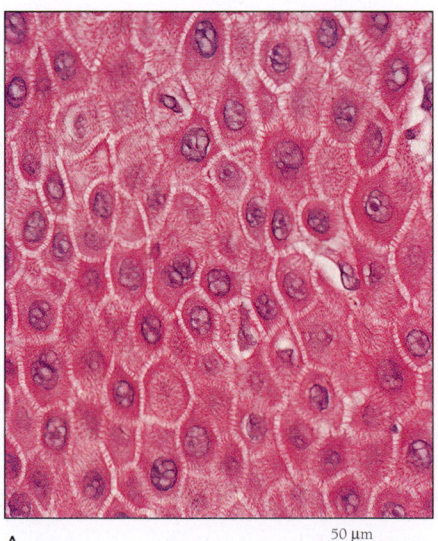 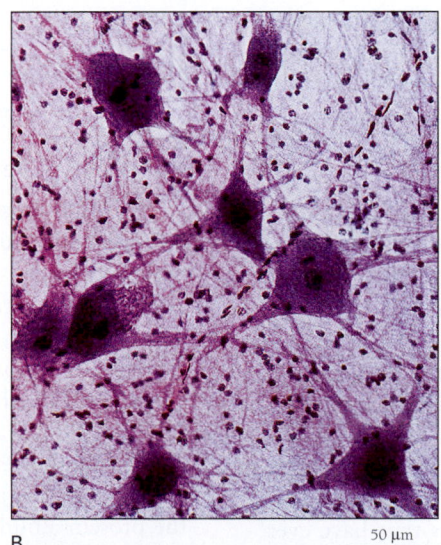 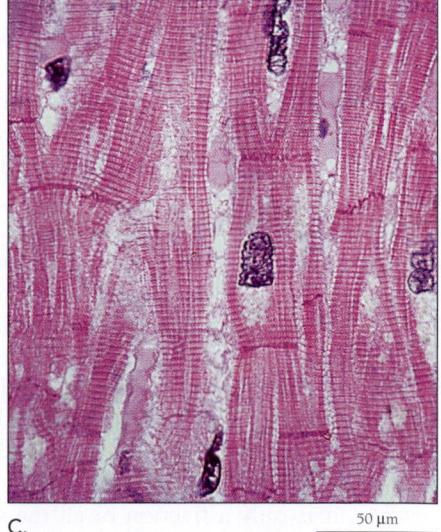

A. 50 µm B. 50 µm C. 50 µm

Figure 1-9 Every organism is composed of many kinds of cells. A. Human skin cells. B. Nerve cells from the spinal cord. C. Cells of heart muscle. *(A, G.W. Willis, M.D./ Biological Photo Service; B, Manfred Kage/Peter Arnold, Inc.; C, Ed Reschke/Peter Arnold, Inc.)*

Eukaryotic cells have a nucleus and other organelles, each surrounded by a thin membrane. Prokaryotic cells lack a nucleus or other organelles.

Life is enormously diverse. About 1.4 million living species have names, but millions more have no names or are already extinct.

HOW ARE ORGANISMS DIFFERENT FROM ONE ANOTHER?

Organisms come in many sizes, shapes, and colors. They range in size from *Chlamydia*, one of the smallest species of bacteria, to California's coastal redwoods, the most massive organisms in the world. Organisms range in shape from streamlined dolphins to prickly cacti, and in color from the softest brown moth to the brightest red poppy. The more we look, the more different forms of life we see. Biologists now divide these many organisms into three large categories, called **domains** (Figure 1-10).

The first two domains, **Archaea** and **Eubacteria,** comprise all the prokaryotes, commonly called bacteria. The great majority of these simple, single-celled organisms are free living. Only a handful of bacteria infect humans or other organisms. Each domain is further divided into one or more **kingdoms** of bacteria.

The third and last domain is **Eukarya,** which consists of four kingdoms. The two biggest kingdoms are the **Plantae** and the **Animalia,** the plants and animals. All plants and animals are multicellular. The 248,000 species of plants range from tiny mosses and ferns to giant redwood and mahogany trees.

Animals are even more diverse, with well over a million known species. Yet, fewer than 1 percent of animal species are what many people consider animals—birds, mammals, fish, reptiles, and amphibians. About 200,000 animal species are kinds of sponges, jellyfish, worms, spiders, snails, or sea urchins. But the vast majority of animals, some 800,000 species, are insects. Beetles alone constitute about 30 percent of all known animals, with some 300,000 to 350,000 species.

The third kingdom in the domain Eukarya is the **Fungi,** with just under 70,000 named species. The funguses include both multicellular organisms (such as mushrooms) and single-celled organisms (such as yeast). The fourth kingdom is the **Protista,** most of which are single celled. These tiny organisms include more than 30,000 species. Protists include different kinds of amebas, such as those that live in pond water; dinoflagellates, some of which cause poisonous red tides in coastal waters; the delicate and lacy foraminifera, and numberless algae, water molds, and slime molds.

Altogether, the named species of organisms within these three domains total about 1.4 million. Biologists believe, however, that most living species have yet to be named. The world may hold as many as 10 million or even 100 million species. No one really knows how many. Whatever the current total, however, it is only a fraction of all the species that have ever lived. Life first appeared on Earth almost 4 billion years ago. Fewer than 1 percent of all species that have ever lived are still living today. The vast majority of species—in all their diversity—went extinct millions of years ago.

How Do Organisms Become Different from One Another?

Even though organisms are enormously diverse, all share the same chemical building blocks and the same cellular organization. Such common characteristics have persuaded biologists that all organisms are genetically related and are descended from a common ancestor. But how have related organisms become so different? How can plants and animals have evolved from the same common ancestor?

Scientists have discovered that during the last 4.5 billion years, our planet has changed from a lifeless sphere of rock and water to a rich, green world alive with fabulously diverse organisms. This transformation is the result of **evolution**—the process by which species have arisen and changed as they descended from common ancestors.

Why do children look like their parents?

Evolutionary change occurs because offspring inherit traits from their parents. A visitor to a museum may be surprised to find that a cat living in ancient Egypt 6000 years ago looked just like a house cat from Houston. Species change very little from one generation to the next. Just as today's children look much like those born 30 years ago, they also closely resemble children born 3000 years ago.

Children look like their parents because every individual inherits a set of **genes** [Greek, *gen* = to produce], which carry detailed information about the molecular parts of an organism. The preservation of similarity depends on the way that parents transmit genes to their offspring.

Genes ensure continuity from generation to generation.

What are genes made of?

As we will discuss in detail later in this book, biologists have learned that organisms store genetic information in long, thin molecules called **deoxyribonucleic acid (DNA).** DNA contains coded instructions for building different kinds of proteins. A cell can read each gene and translate it into a protein. The DNA of humans, for example, contains about 100,000 genes, most of which code for proteins. A set of genes—called a *genome*—is in many ways like the list of ingredients in a recipe. But the genes are a recipe for building an organism. Instead of "two eggs," the genome says, "42,000 molecules of the protein albumin."

Genes are the means by which organisms transmit the information needed to build and maintain new individuals.

Species Differ in Their Adaptations to Distinct Environments

The members of a species share most of the same genes (and proteins). As a result, all the members of a species share characteristic **adaptations,** special structures or behaviors that fit individuals for life in a particular environment. All adaptations are inherited. Adaptations increase the chance that an individual will live and reproduce. In Figure 1-10, the hard, carapacelike upper wings of the spotted lady beetle have evolved to protect the delicate flying wings beneath. Its habit of attacking and eating aphids provides it with a steady supply of nutrients and energy.

Size, form, color, internal structure, and special chemical properties all contribute to an organism's ability to survive in a particular environment and to reproduce more of its own kind. Despite the differences among individuals within a species, all the members of a species are similarly adapted for a particular way of life. Given the opportunity, all cows stand around eating grass, all polar bears hunt seals, all kittens pounce on small objects, and all human children laugh and chase one another.

Life is diverse because different species have different sets of adaptations. The adaptations prompt biologists to ask questions such as, How does a particular structure or behavior help this organism to survive and reproduce? Put more concretely,

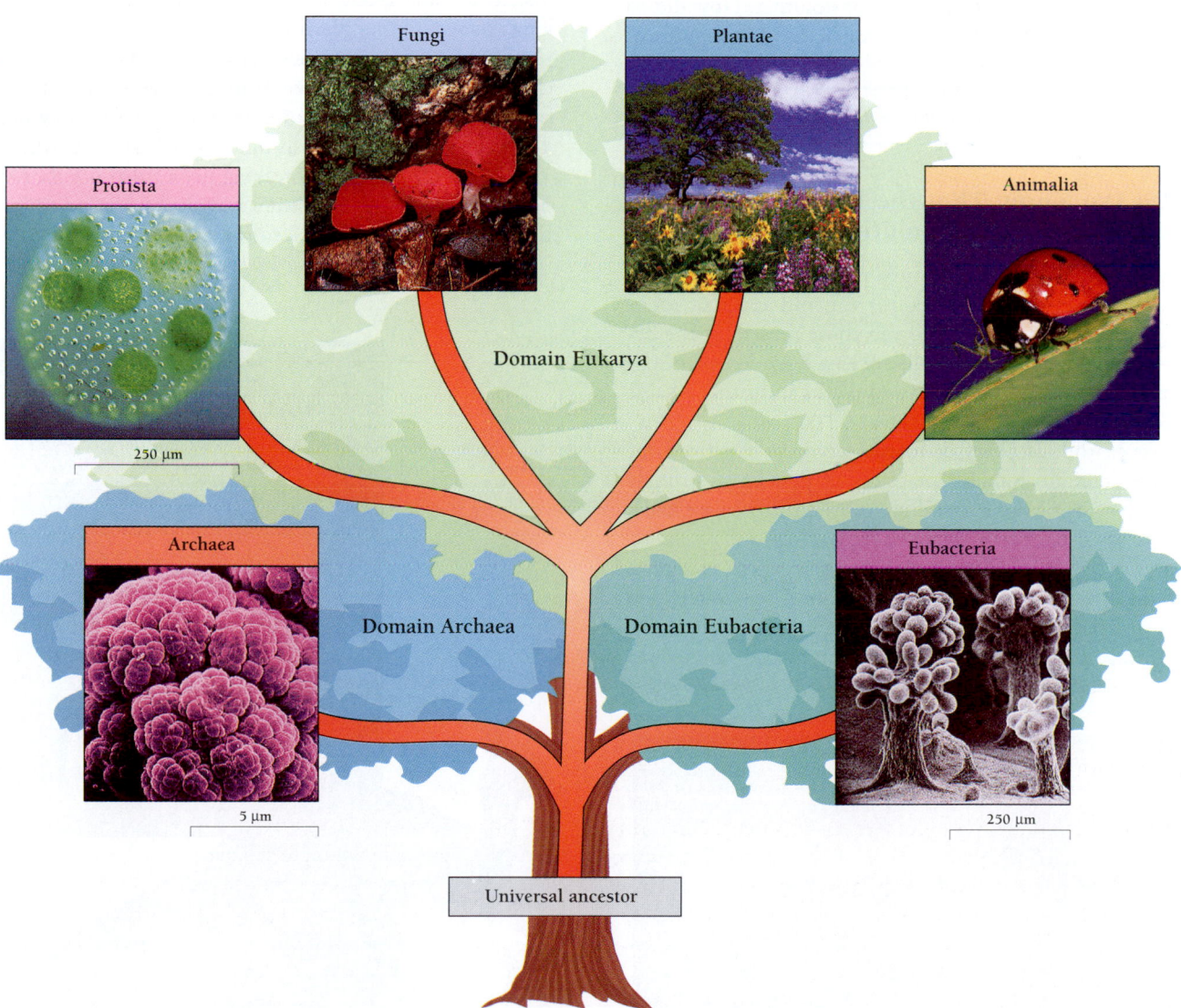

Figure 1-10 Three domains of organisms. At bottom are the first two domains, two very different kinds of bacteria. The first domain, the Archaea, is represented by a colony of methanobacteria, *Methanosarcina.* The second domain, the Eubacteria, is represented here by aggregating myxobacteria. The third domain, the Eukarya, includes four kingdoms. The seven-spotted lady beetle represents the Animalia. An oak tree surrounded by wildflowers (arrowleaf balsam root, paintbrush, and lupine) represents the Plantae. The stalked scarlet cup, *Sarcoscypha occidentalis,* represents the Fungi. A colony of one-celled organisms called *Volvox* represents the Protista. *(Lady beetle, Runk/Schoenberger from Grant Heilman; oak tree, F. Stuart Westmorland/Photo Researchers, Inc.; stalked scarlet cup, R.M. Meadows/Peter Arnold, Inc.; Volvox, Biological Photo Service; myxobacteria, Patricia L. Grilione/Phototake; Methanosarcina, R. Robinson/Visuals Unlimited)*

What good is an elephant's trunk, a mallard's mating dance, or a cactus's thick stems? Of ourselves we might ask, Of what use are a large brain and hairless skin? These are the same general questions that we might ask about the parts of a sewing machine or a car engine: What does it do? How else could it be done?

Because every species has so many adaptations, the same general questions arise again and again. After a while, biology students might begin to wonder how such a variety of adaptations came to be in the first place. As we mentioned earlier, all organisms appear to share a common origin. Biologists must therefore ask how individual species became so different from one another and why there are so many different species. The question, Why are organisms so diverse? becomes, How did so many different species derive from a single kind of organism?

Every species has a unique set of adaptations.

How Do We Know That the Diversity of Life Resulted from Evolution?

Until the middle of the 19th century, most Europeans believed that the Earth was only about 6000 years old and that modern organisms were no different from those present at the beginning of life on Earth. After all, ancient Egyptian paintings and mummies revealed creatures and plants indistinguishable from those of modern times (Figure 1-11).

However, the discovery of **fossils**—objects dug from the ground that contain remains or traces of past life—strongly suggested that organisms very different from the ones we know today once lived on Earth (Figure 1-12).

Early in the 19th century, geologists began to recognize that the Earth was far older than 6000 years. They understood for the first time that over millions of years, huge parts of the Earth's surface had moved about, new mountain ranges had arisen, and wind and water had shifted land from one place

to another. Organisms that had been near the surface became covered over with new layers of mud and other materials that later turned to stone. The successive layers of fossil-rich rock produced by these processes, geologists realized, contained the history of the Earth and its inhabitants (Figure 1-8G).

Fossils represent extinct forms of life that are often different from anything living today.

Charles Darwin provided evidence for evolution

By the early 19th century, many scientists realized two things: first, that some species of plants and animals had disappeared long ago and, second, that new species had appeared to take their places. Many scientists suspected that species of organisms changed over time, *evolving* from one species into another. At first, no one understood how this could possibly have occurred. Then, in 1859, the English biologist Charles Darwin (1809–1882) published a groundbreaking book titled *On the Origin of Species*, in which he suggested a way that evolution could work. Darwin called his theory *evolution through natural selection*.

Figure 1-11 Organisms may remain unchanged for many generations. This hippopotamus mother, carved more than 2000 years ago, closely resembles hippopotami that live today. *(Larry Tackett/Tom Stack & Associates)*

Figure 1-12 Parasaurolophus. The unearthing of fossils in the 18th century suggested that strange organisms had once lived on Earth. Other geologic evidence suggested that the Earth was unimaginably old and that the layers of the Earth's crust contained the history of the Earth and its inhabitants. *(Francois Gohier/Photo Researchers, Inc.)*

Darwin's model for how evolution could work came from farming. For thousands of years, plant and animal breeders have bred new varieties of corn, wheat, dogs, or horses. Their method is simple: a breeder selects certain individuals to breed—the most productive wheat plants, the most intelligent dogs, the fastest horses. Slow race horses and other animals with undesirable characteristics were not allowed to breed.

The power of this strategy is undeniable. Every kind of dog, from chihuahua to mastiff, is the result of selection from a line of animals descended from the ordinary wolf. Similarly, a surprising number of vegetables have been bred from a single plant, the wild sea cabbage of the Mediterranean. Figure 1-13 shows the results of selection in wild sea cabbage to produce broccoli, cabbage, Brussels sprouts, and more. By selecting for leaves, gardeners developed cabbage. By selecting for flower buds, gardeners developed broccoli.

Darwin saw that a similar kind of selection, which he called natural selection, could operate in nature. **Natural selection** is the superior survival and reproduction of some individuals that have certain inherited traits compared to others lacking those traits.

Figure 1-13 **European gardeners developed a great variety of vegetables from a single species of plant,** *Brassica oleracea.* Shown above are cabbage, cauliflower, broccoli, Brussels sprouts, kale, and kohlrabi. In the case of kale and cabbage, farmers selected only those plants with large, edible leaves. To create broccoli and cauliflower, they selected plants with the most flower buds. For Brussels sprouts, they chose those with the largest, most edible leaf buds. In kohlrabi, breeders selected plants with the biggest, most edible stems. What other plant parts could be selected for?

While artificial selection by breeders depends on the effects of human preferences, natural selection depends on the ever-changing demands of an organism's natural environment. For example, organisms whose adaptations improve their ability to escape predators or to compete for food or sunlight usually live longer and produce more offspring than organisms without these adaptations. To the extent that a trait increases an organism's ability to survive and reproduce, that trait appears more often in the next generation. In this way, Darwin suggested, one organism can gradually evolve into another.

Charles Darwin supplied extensive evidence for the idea that species have evolved from common ancestors. He also suggested that species have evolved by means of natural selection.

How are traits passed from one generation to the next?

Darwin understood how evolution might work, but he had no idea how living organisms pass traits to their offspring or why the offspring are different from their parents and different from each other. In short, Darwin did not know why individuals are unique or how they inherit traits from their parents. Darwin knew neither why we look like our parents nor why we look slightly different from our parents. The question of how heredity works puzzled and frustrated him for the rest of his life. Ironically, the answer lay close at hand, although the answer would remain hidden for another 30 years.

In 1866, just 7 years after Darwin published *On the Origin of Species,* the Austrian naturalist monk Gregor Johann Mendel published a paper describing the laws of heredity. Mendel's work laid the groundwork for the modern science of genetics. Mendel's discovery was profound, yet his paper was virtually unknown until 1900, when his work was rediscovered (Chapter 9).

Gregor Mendel discovered the genetic rules that underlie natural variation.

We have now seen how biologists define life and how they try to answer questions about living organisms. We have also seen, briefly, how living organisms evolve. In the next few chapters, we will survey the chemistry of life, view the structure of cells, and see how cells obtain energy to fuel life.

As you read, keep in mind that in the next 20 years, Earth's rapidly growing human population and dwindling resources will force the world's leaders to make many hard decisions. To participate as citizens in these decisions, we need to do more than simply learn the facts of biology. We need to know how to scrutinize those facts as they come to us from television and radio, the Internet, or a newspaper or magazine. We should know how to ask what biologists and other scientists mean when they describe their findings. To do that, we need to know how scientists think and how they themselves ask questions.

Study Outline with Key Terms

The **scientific method** is a formal set of rules intended to describe the way scientists work. Most scientists, however, do not stick to a rigid set of rules. Both thinking up questions and finding ways to answer those questions require inspiration and creativity, as well as hard work, patience, and a methodical approach. In general, a scientist begins by narrowing an area of interest to a specific question, then observing a phenomenon.

A scientist may sometimes visualize the phenomenon by constructing a **model.** A **hypothesis** is a formal guess at the answer to the question, often stimulated by a model. A good hypothesis allows a scientist to make a **testable** prediction about the way the process or structure will act under specific circumstances. A set of hypotheses that consistently resists scientists' attempts to disprove them may become accepted as a **theory.**

An **experiment** is a way of testing the predictions of a hypothesis or, sometimes, of searching for some unknown effect. An experiment is a procedure that is carried out under "controlled" conditions. An experiment can prove a hypothesis wrong but not right. Essential elements of most biology experiments are the **control** experiment and **statistical** analysis.

Meaningful statistics requires a **sample** that represents the whole **population** of organisms. A sample can represent the population only if the sample is a **random** sample and the sample is large enough to overcome random variation. Statisticians can determine if the sample represents the population using the **mean** and **standard deviation.** Statistics is most easily done on data with a **normal distribution,** or bell curve.

Biology is the study of living organisms. All organisms are composed of one or more cells, which are themselves alive, and which all come from other cells. All are organized into parts, obtain energy from their surroundings, perform chemical reactions, mature over time, reproduce, share a common evolutionary history, and respond to their environments. Mechanisms that maintain a constant internal environment are a form of **homeostasis.**

Eukaryotic cells contain a **nucleus** and other **organelles,** each of which is enclosed by a membrane. **Prokaryotic** cells have no membrane-enclosed nucleus. The millions of species of organisms are organized by biologists into three **domains: Archaea, Eubacteria,** and **Eukarya.** Eukarya comprise just four **kingdoms: Animalia, Plantae, Fungi,** and **Protista.**

Genes, which are made of long, thin molecules of **deoxyribonucleic acid (DNA),** ensure continuity from generation to generation. Traits that are inherited in the genes and help a species thrive are **adaptive.** The discovery of **fossils** suggested to 18th- and 19th-century scientists that organisms had changed, or evolved, over long periods. Charles Darwin assembled compelling evidence for the idea of **evolution** and suggested a mechanism, called **natural selection.** Gregor Mendel discovered and articulated the first rules of genetic inheritance.

Review and Thought Questions

Review Questions

1. Explain the differences between a hypothesis and a theory.
2. Explain why a hypothesis can sometimes be proved wrong but can never be proved right.
3. Give an example of an untestable hypothesis. Why is it untestable? What would you need to know or do to test it?
4. Suppose you decide to test the hypothesis that your favorite soft drink contains substances that cause cancer. You plan to feed the drink to 100 mice and then count the number of tumors they develop. Describe an appropriate control experiment.
5. Describe eight distinguishing characteristics of living things. Give an example of each one.
6. Define "homeostasis." List two examples of homeostasis.
7. Explain what is meant by "natural selection." What is "selected"?

Thought Questions

8. The ants in your kitchen are driving you nuts. Using the "scientific method," how would you find out where they were entering your house? How would you determine why they were entering your house? Think of several alternative hypotheses. Now design a series of experiments whose results will either support or eliminate the various alternative explanations.
9. A virus is a bit of DNA surrounded by a protein coat. A virus can attach to the wall of a cell and inject the DNA into the cell. Inside the cell, the viral DNA forces the cell to make new viruses. The cell makes so many new viruses that the cell eventually bursts open, releasing thousands of new viruses, which go on to infect other cells. Is the virus alive? Using the definitions of life given at the beginning of this chapter, explain why a virus is alive or why a virus is not alive.
10. From your own experience, describe a moment of inspiration—intellectual or otherwise. How did it feel? How did it change you?

About the Chapter-Opening Image

The amazing story of how we came to know that the twisted bacterium Helicobacter pylori *causes ulcers is a classic example of how science works.*

 On-line materials relating to this chapter are on the World Wide Web at **http://www.harcourtcollege.com/lifesci/aal2/**

part I

Chemistry and Cell Biology

Chapter 2
The Chemical Foundations of Life

Chapter 3
Biological Molecules Great and Small

Chapter 4
Why Are All Organisms Made of Cells?

Chapter 5
Directions and Rates of Biochemical Processes

Chapter 6
How Do Organisms Supply Themselves with Energy?

Chapter 7
Photosynthesis: How Do Organisms Get Energy from the Sun?

Photo: Scanning electron micrograph of the mucous membrane lining the stomach. *(P. M. Motta, K. R. Porter, and P. M. Andrews/Science Photo Library/Photo Researchers, Inc.)*

2

The Chemical Foundations of Life

Just Say NO

Alfred Nobel, one of the names most associated with scientific discovery, was not a scientist but an inventor who made a fortune from the invention of dynamite. Nobel devoted his wealth to establish the Nobel prizes, now given to recognize extraordinary achievements in peace, chemistry, physiology and medicine, physics, literature, and economics. Nobel wanted to use his fortune—much of it made from the military use of dynamite—to promote peace and peaceful achievement. Strangely, the 1998 Nobel Prize in Physiology and Medicine actually rewarded insights that came from the biological use of Nobel's troubling explosive.

The active chemical in dynamite is *nitroglycerin*, a colorless oily liquid. Nitroglycerin is so unstable that just dropping it causes an explosion. Nobel's invention consisted of a way of stabilizing nitroglycerin by mixing it with porous soil, so that it could be safely transported.

Despite the dangers of pure nitroglycerin, however, physicians have been prescribing it for more than 100 years. In small doses, nitroglycerin reduces blood pressure, and it is especially effective in treating the chest pain associated with heart disease (*angina*). Nobel's own doctor recommended nitroglycerin treatment for his heart condition, but (perhaps understandably) Nobel refused.

Almost 80 years later, in the early 1970s, biologist Ferid Murad, then at the University of Virginia, was trying to learn exactly how nitroglycerin actually helped the heart. In 1977, Murad discovered that nitroglycerin releases nitric oxide, a gas that is a well-known by-product of automobile engines. Nitric oxide, he argued, is the agent that caused blood vessels to dilate, thereby lowering blood pressure. But nitric oxide—abbreviated as NO by chemists—is a major component of smog, a noxious chemical that causes irritation and burning in the lungs. No one believed that it could possibly be an effective biological agent. According to Murad's former professor, "People used to say 'C'mon, Fred [as they called him]. Why don't you work on something important?'"

About the same time, Robert Furchgott, working at the State of New York Health Science Center at Brooklyn, was trying to understand the effect of other drugs on blood vessels. Furchgott's work was extremely frustrating, since the same chemical sometimes caused vessels to contract and other times caused them to dilate. Trying to get to the bottom of these contradictions, Furchgott demonstrated that dilation-causing drugs worked only when the lining of the blood vessels, the *endothelium*, was intact. He concluded that the drugs were somehow inducing the endothelium to release an unknown chemical, which he called "endothelium-derived relaxing factor," or EDRF. In 1986, Furchgott, Louis Ignarro (from the University of California, Los Angeles),

Dennis Galante/FPG

- An atom is the fundamental unit of an element, and a molecule is the fundamental unit of a compound.

- In an ionic bond, electrical attraction holds atoms together. In a covalent bond, atoms share electrons.

- If the atoms in a covalent bond do not share electrons equally, the resulting molecule will be polar.

- Individual molecules may also interact with each other via hydrogen bonds, hydrophobic interactions, or van der Waals interactions.

and Salvador Moncada (from University College London) found that EDRF was, in fact, identical to nitric oxide. It was the same molecule that Murad had already implicated in the action of nitroglycerin.

The effect of these announcements was electrifying. Furchgott, Ignarro, and Moncada were claiming—against all precedent—that gas molecules could carry messages from one cell to another. Although biologists knew of some bacteria that produced NO, skeptics argued that there was no known mechanism by which eukaryotes could produce the toxic substance. Besides, they argued, NO itself was unstable and was destroyed within 10 seconds after it appeared. How could it be a biologically important regulator? The critics dismissed NO as "hot air."

But the researchers persisted. They learned how cells produced NO and how it acted. Pharmaceutical companies began to develop medicines that increase NO production for heart patients. Perhaps most famously, the anti-impotence drug Viagra was shown to act by increasing the action of NO, allowing more blood flow to the penis. The increased blood flow enables many men suffering from impotence to achieve a normal erection.

"'C'mon, Fred. Why don't you work on something important?'"

Nitric oxide is not always beneficial. In people suffering from septic or toxic shock, for example, white blood cells pump out too much NO in response to a bacterial infection. The resulting rapid drop in blood pressure can be fatal. Therefore, many researchers are trying to discover ways of reducing as well as increasing NO production. On the other hand, since white blood cells also use NO to defend against tumor cells, scientists are using new knowledge of NO to develop new anti-cancer drugs. By 1992, the influential journal *Science* had named NO the "molecule of the year," and the 1998 Nobel prize (to Furchgott, Ignarro, and Murad) again underscored its importance.

Nitric oxide is invisible but it still has weight and occupies space. It is a particular form of matter. Understanding the composition and the transformations of matter is the business of chemistry, the subject of this chapter.

WHAT IS MATTER?

Anyone whose car has ever run out of gas knows that when the gauge says empty, the gas is gone. Where does it go? A car's engine burns (or "oxidizes") these chemicals into water and carbon dioxide (the same stuff that makes the bubbles in soft drinks). This chemical transformation happens only when the engine provides a spark (from a spark plug) to get things going. Our bodies perform similar chemical reactions, though without spark plugs. Like a car engine, we convert chemicals into water and carbon dioxide. But we also use the molecules in our food to make thousands of other chemicals, which are the stuff of our bodies.

Chemistry is the science dealing with the properties and the transformations of all forms of matter—whether gasoline or food, people or planets. **Matter** is anything that occupies space and has weight (or "mass"). Matter can exist in different states or phases—solid, liquid, or gas.

Every bit of matter consists of one or more *pure substances,* each of which has its own unique properties. **Chemical reactions** transform pure substances into other substances, without changing the total amount of matter. ("Biochemical reactions" are those chemical reactions that occur in organisms.)

Pure substances that cannot be converted into simpler substances in chemical reactions are called **elements.** Hydrogen, helium, oxygen, and carbon are elements. So are silver and gold, the carbon in a pencil, the copper in a wire, and the mercury in a thermometer. Altogether, scientists know of 92 naturally occurring elements. Elements are composed of tiny particles called **atoms**—the smallest units that still have the properties of an element. Atoms themselves are made of smaller "subatomic" particles, including protons, neutrons, and electrons.

Each element has a standard one- or two-letter abbreviation. These usually correspond to the beginning of the element's English name, though sometimes it's the Latin or Greek name. "C" stands for carbon and "O" for oxygen, while "Na"

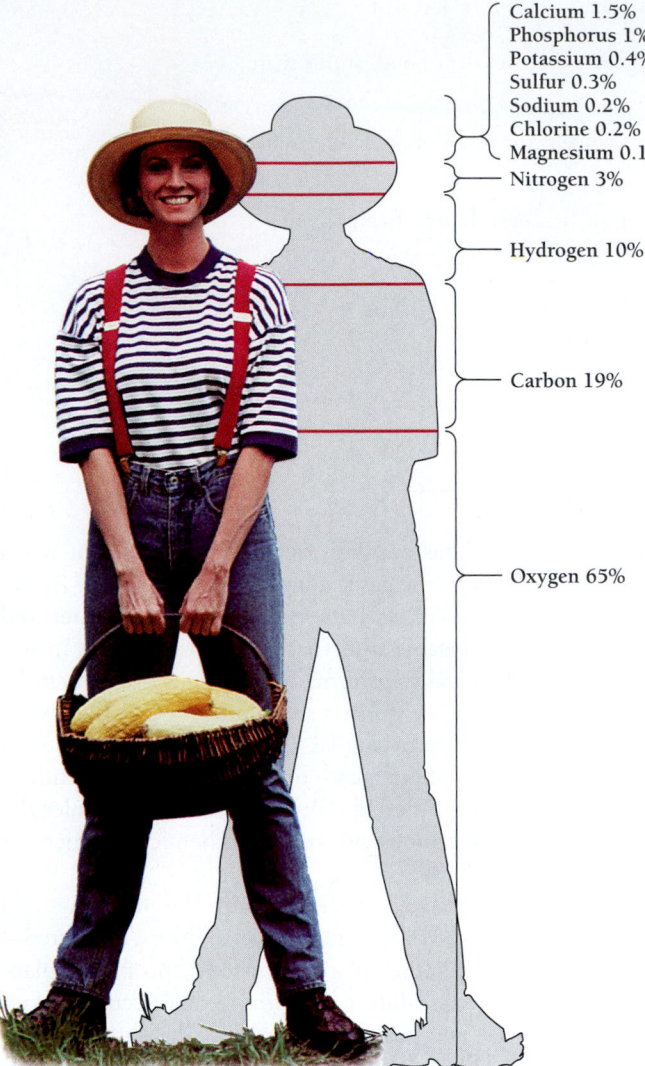

Calcium 1.5%
Phosphorus 1%
Potassium 0.4%
Sulfur 0.3%
Sodium 0.2%
Chlorine 0.2%
Magnesium 0.1%
Nitrogen 3%

Hydrogen 10%

Carbon 19%

Oxygen 65%

Figure 2-1   Humans, like other organisms, are primarily water. The commonest two atoms in our bodies, then, are hydrogen and oxygen (H_2O). Hydrogen is a much lighter atom than oxygen, so most of our mass is made up of oxygen atoms. Next comes carbon, which, like hydrogen and oxygen, is in virtually every other molecule in our bodies. *(Roy Morsch/The Stock Market)*

stands for sodium [Latin, *natrium*]. Four elements—carbon (C), oxygen (O), nitrogen (N), and hydrogen (H)—make up more than 99 percent of the matter in organisms. Figure 2-1 shows the main elements of the human body.

Chemical reactions cannot change or destroy elements. Carbon cannot be converted to oxygen. Silver cannot be converted to gold. However, elements can be created or changed by means of **nuclear reactions,** the high-energy transformations of individual atoms that occur in nuclear reactors or in the bowels of stars such as our sun.

Although oxygen is an element, we do not breathe oxygen in its atomic form. Instead, we breathe "molecular" oxygen, which is two atoms of oxygen bound together by a "chemical bond." A **molecule** is a stable assembly of two or more atoms.

Most molecules are **compounds**—pure substances forged in chemical reactions from two or more *different* elements. Water, for example, is a compound of the elements hydrogen and oxygen (Figure 2-2A). Carbon dioxide is a compound of carbon and oxygen.

A compound is unlike the elements that compose it. The properties of water are different from those of either hydrogen alone or oxygen alone, and the properties of carbon dioxide are different from those of carbon and oxygen.

A compound is made of a single kind of molecule. Chemists designate the composition of a molecule by writing the number of atoms of each type as subscripts to the right of the symbol for each element. Water is always H_2O, carbon dioxide is CO_2, and ordinary table sugar (sucrose) is $C_{12}H_{22}O_{11}$.

In our day-to-day lives, we rarely encounter either compounds or elements in pure form. Most matter is a **mixture** of different compounds (Figure 2-2A). Sand and salt poured together, for example, is a mixture. One component of a mixture does not affect the properties of the other components. If we pour salt into boiling water to make spaghetti, the salt is still salt and the water is still water. If we boil away all of the water, the salt forms crystals at the bottom of the pot, and we can retrieve the same salt we poured into the pot.

Unlike the components of a compound, which are always present in the same fixed ratio, the components of a mixture may be present in any ratio. A molecule of water always contains twice as much hydrogen as oxygen. But you can mix a little sugar into your cup of coffee, or a lot.

Elements are pure substances that cannot be converted into simpler substances. An atom is the fundamental unit of an element, and a molecule is the fundamental unit of a compound.

What Is Special About the Chemistry of Life?

Until the 16th century, chemistry as we know it did not exist. Alchemists combined and recombined various compounds and elements, hoping to convert lead and other cheap metals into silver and gold (Figure 2-3). Alchemists did not yet recognize that the elements could not be changed one into another by ordinary means. Some alchemy was science of a sort, but much of it was simple superstition. None of the alchemists in the Middle Ages possessed a modern understanding of what matter was and why it acted as it did. In their efforts, however, they learned a lot about the properties and transformations of matter (Web Bit 2-1: A Case of Modern-Day Alchemy).

A still older kind of chemistry was *metallurgy*—the study of the properties and transformations of metals. Thousands of years ago, bronze tools replaced stone tools and, later, iron tools replaced bronze tools. Steel tools have, in turn, replaced iron

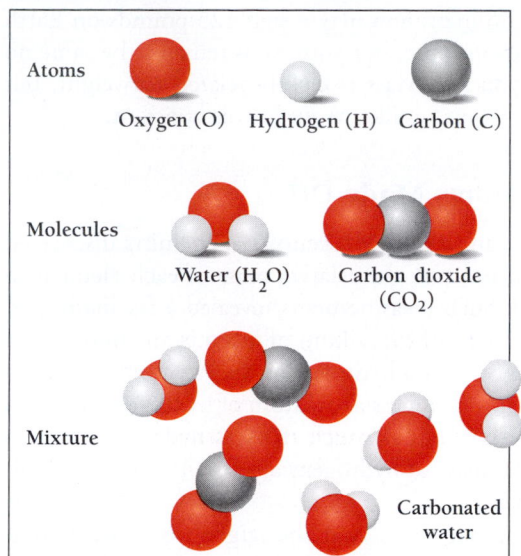

A.

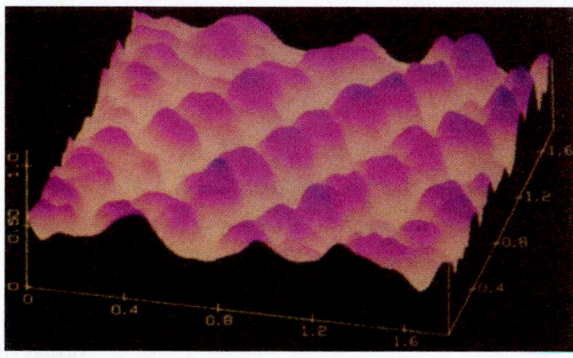

B.

Figure 2-2 **Atom, molecule, compound, and mixture.** A. Two atoms of the same element, such as two oxygen atoms, may form a molecule. But most molecules are compounds forged from two or more different elements. Water is a compound of hydrogen and oxygen, and carbon dioxide is a compound of carbon and oxygen. Most substances are mixtures of different compounds. Sparkling water, for example, is a mixture of carbon dioxide and water. B. An atomic force microscope reveals the arrangement of individual atoms of copper. (B, *Xueping Xu. From Science, vol. 258, p. 788, 1992*)

ones. Metallurgy was, and is, so important that its mastery remains a measure of the progress of a civilization.

In the 1500s, the chemistry of medicines emerged as a serious discipline. Then, bit by bit, Renaissance scientists and philosophers began reinterpreting the chemical facts discovered by the alchemists. Not until the 19th century, however, did scientists begin to understand the chemical nature of matter. Today, at the beginning of the 21st century, chemistry extends beyond metals and medicines to all matter, including the substances that make up organisms. One of the great achievements of 20th-century biology was to explain the chemistry that underlies all cellular processes—from the workings of a gene to the contraction of a muscle.

One of the great questions in biology was whether living matter was in any way different from nonliving matter. Are a horse and a plow made of the same basic stuff? In this chapter, we will see that the molecules of living organisms indeed obey the same chemical laws as the molecules of nonliving matter.

From a chemist's point of view, just two aspects of living matter distinguish it from nonliving matter—the elaborate organization of the molecules and the limited number of kinds of molecules that comprise living matter. All organisms consist of the same few kinds of small *organic* (carbon-containing) molecules, which we may think of as "building blocks." Just as a good cook can make a dazzling array of cookies, cakes,

Figure 2-3 *The Alchemist,* **painted by David Teniers the Younger in 1648.** Alchemists of the medieval period and later attempted to turn lead and other cheap metals into gold and silver. (*Scala/Art Resource, NY*)

and pies from just a few ingredients, so organisms can produce extraordinary structures from a small number of standard molecules.

The chemistry of living matter is the same as that of nonliving matter.

WHAT DETERMINES THE PROPERTIES OF AN ATOM?

By the early 19th century, chemists were able to distinguish compounds—which could be broken down into simpler substances—from elements—which could not. They then began to ask, Does the way in which a substance breaks down reveal something about its basic structure? Are there units that cannot break down further?

To answer these questions, the English chemist John Dalton revived an idea, first developed by the Greek philosopher Democritus in the 5th century B.C., that matter is actually built from small particles. Dalton proposed that each element is composed of tiny particles called "atoms." Only in the past 20 years has it been possible to see atoms, using the *atomic force microscope,* which provides a detailed image of the surfaces of solids. Figure 2-2B shows rows of individual copper atoms aligned on a surface.

The atoms of each element have a characteristic weight. For example, an atom of carbon weighs 12 times as much as an atom of hydrogen. Chemists and physicists prefer to talk of the "mass" of an atom, rather than its weight. **Mass** is the amount of matter in something. Weight results from the ac-

tion of gravity on mass. You may weigh 125 pounds on Earth and zero pounds in space, but your mass remains the same no matter where you are. Mass is closely related to weight, but the mass of an object, unlike its weight, is constant.

What Are Atoms Made Of?

In the late 18th and early 19th centuries, chemists discovered that they could measure the relative mass of each element in any compound. Such measurements revealed a fascinating relationship: The mass of every kind of atom is an almost exact multiple of the mass of a hydrogen atom. An atom of helium has four times as much mass as an atom of hydrogen. An atom of lithium has six times as much mass as hydrogen. Carbon has 12 times the mass of hydrogen; oxygen, 16 times; and nitrogen, 14 times.

This pattern suggested that the larger atoms were composed of groups of hydrogen atoms. We now know that atoms are not actually made of hydrogen atoms but of subatomic particles, each of which has about the same mass as a hydrogen atom. An atom of carbon is composed of 12 particles each about the size of a hydrogen atom, whereas an atom of oxygen is composed of 16 such particles.

The masses of atoms are extremely small. An oxygen atom, for example, weighs less than 3×10^{-23} gram: that's 0.00000000000000000000003 gram. (A gram is about as heavy as one-third of an animal cracker.) For convenience, chemists have defined a measuring unit for the mass of an atom or a molecule that is much smaller than a gram. A dalton, named after the chemist John Dalton, is approximately the mass of a hydrogen atom. An atom of hydrogen has a mass of 1 dalton, and an atom of oxygen has a mass of 16 daltons.

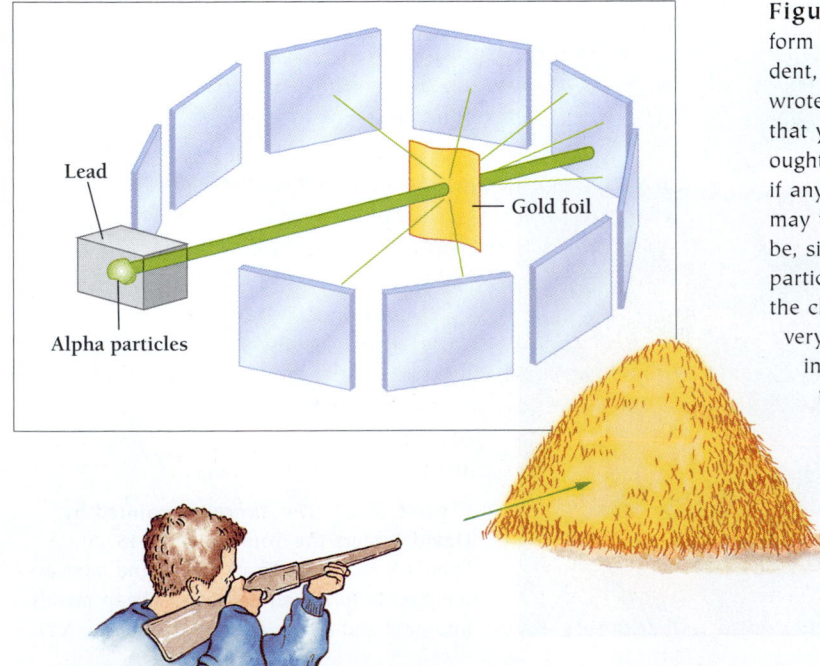

Lead

Gold foil

Alpha particles

Figure 2-4   Rutherford's experiment. Rutherford did not perform his now famous experiment himself. He delegated it to a student, to whom he initially did not even speak. Rutherford later wrote, "One day Geiger came to me and said, 'don't you think that young Marsden, whom I am training in radioactive methods, ought to begin a small research?' . . . I said, 'Why not let him see if any alpha particles can be scattered through a large angle?' I may tell you in confidence that I did not believe that there would be, since we knew that the alpha particle was a very fast massive particle, with a great deal of energy, and you could show that . . . the chance of an alpha particle being scattered backwards was very small. Then I remember two or three days later Geiger coming to me in great excitement and saying, 'We have been able to get some of the alpha particles coming backwards . . . ' It was quite the most incredible event that has ever happened to me in my life. It was almost as incredible as if you fired a 15-inch shell at a piece of tissue paper and it came back and hit you." (Rutherford and the Nature of the Atom, by E.N. da Costa Andrade, Doubleday, Garden City, NY, 1964; quoted by S. Weinberg, The Discovery of Subatomic Particles, Scientific American Library, New York, 1983, p. 124.)

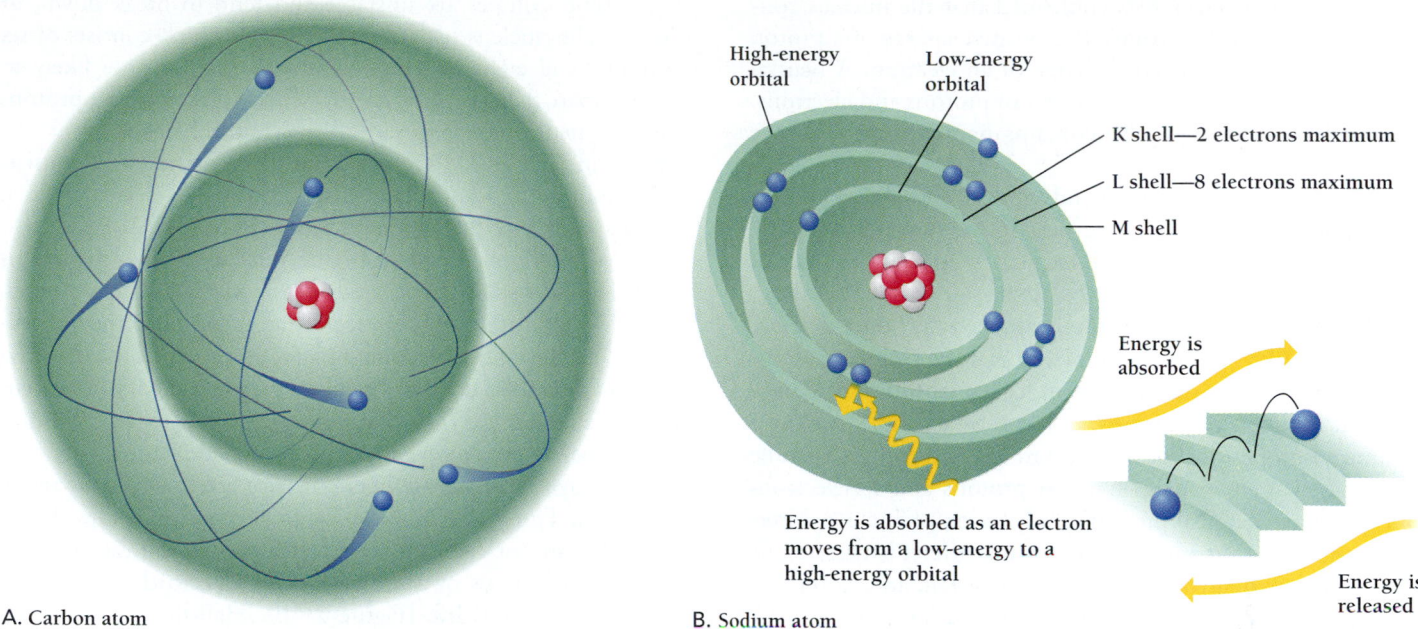

High-energy orbital   Low-energy orbital

K shell—2 electrons maximum

L shell—8 electrons maximum

M shell

Energy is absorbed

Energy is absorbed as an electron moves from a low-energy to a high-energy orbital

Energy is released

A. Carbon atom

B. Sodium atom

Figure 2-5 **The structure of an atom.** A. A carbon atom. By convention, orbitals are drawn as flat rings on the page as in B, but in reality, electron orbitals are diffuse, three-dimensional spheres (or other shapes), as shown here. B. A sodium atom. Orbitals are grouped into shells, each of which has a different energy. Closest to the nucleus is the K shell. It has the lowest energy and holds a single orbital with a capacity for two electrons. Farther out is the L shell, which contains four orbitals with a capacity for eight electrons. Each orbital has a defined shape, which describes the most probable locations of the electrons within them.

The number of daltons in 1 gram is 6×10^{23} and is called "Avogadro's number" [after Amadeo Avogadro, an 18th-century Italian lawyer and physicist]. To get an idea how large this number is, imagine that hydrogen atoms were huge—the size of unpopped popcorn kernels—and that you poured Avogadro's number of kernels out of an airplane window. You would find that Avogadro's number of kernels would cover the whole United States to a depth of 9 miles.

Since a molecule consists of a fixed number of atoms, every molecule has a characteristic molecular mass, usually called its **molecular weight.** A molecule of water (H_2O), for example, has a molecular weight of 18:

$$18 = 16 \text{ (oxygen)} + 1 \text{ (hydrogen)} + 1 \text{ (hydrogen)}$$

Chemists and biologists also say that a water molecule has a mass of 18 daltons.

Atoms are made of subatomic particles similar in size to one hydrogen atom.

What Is the Internal Structure of an Atom?

Once physicists knew that atoms were made up of hydrogen-sized particles, they began to wonder about those particles. What were the hydrogen-sized particles, and how did they fit together to make an atom? The first important finding about atomic structure was that atoms are almost entirely empty space. In 1910, a New Zealand physicist working in England devised an ingenious way to probe the structure of the atom. Ernest Rutherford's approach, diagrammed in Figure 2-4, was like searching for a stone in a haystack by shooting bullets into the haystack. If a bullet goes straight through the haystack, it means that it has failed to hit anything solid. If it comes out at an angle, then it must have bounced off something. The "bullets" Rutherford used were small, positively charged alpha (α) particles, produced from the radioactive decay of radium. The "haystack" was a thin piece of gold foil.

Rutherford found that his α particles passed right through most parts of a gold atom, only bouncing off a tiny, central part of each atom. He proposed that the atom is mostly empty space, with almost all of its mass concentrated in a small heavy core, or **nucleus,** at its center.

The diameter of the nucleus is only 1/10,000 that of the whole atom. Imagine a Ping-Pong ball (the nucleus) at the center of two vacant city blocks (the atom), and you will have an idea of the proportions of an atom. Because the tiny nucleus reflected Rutherford's positively charged particles, he reasoned that the nucleus must itself have a positive charge that repelled his "bullets." The positive charge of the nucleus is balanced by negatively charged particles, called **electrons,** which surround it (Figure 2-5). Rutherford imagined that the negative electrons orbit the positive nucleus in the same way that the planets orbit the sun.

In fact, later researchers confirmed that the nucleus contains particles called **protons.** The positive charge of a proton exactly balances the negative charge of an electron. A neutral (uncharged) atom has equal numbers of protons and electrons. The nucleus of an atom also contains **neutrons,** which have no electric charge.

The mass of an atom depends almost entirely on the total number of protons and neutrons in the nucleus. The tiny electrons have little mass. Each has just 1/2000th the mass of a proton.

The atoms of each element always have the same number of protons. Oxygen, for example, always has eight protons. However, some forms of an element, called **isotopes,** may have different numbers of neutrons. Carbon, for example, has three common isotopes: carbon 12, carbon 13, and carbon 14. The atoms of all three isotopes have six protons and six electrons and the same chemical properties. However, each has a different number of neutrons and, therefore, different masses: carbon 12 has six neutrons, carbon 13 has seven, and carbon 14 has eight. The three isotopes of carbon are also written ^{12}C, ^{13}C, and ^{14}C.

Some isotopes are unstable and tend to break down, or *decay*. The nucleus of ^{14}C, for example, which consists of six protons and eight neutrons, is unstable. It is more likely to come apart than a ^{12}C nucleus, which consists of six protons and six neutrons. Every second, the nuclei of about four ^{14}C atoms in 1 trillion (10^{12}) break up. Each decay event consists of a neutron decomposing into a proton and an electron, which then speeds away from the nucleus (Figure 2-6A).

An electron ejected from the nucleus in this manner is called a "beta (β) particle" and is easily detected and counted. It is distinct from the six electrons that surrounded the ^{14}C nucleus. The nucleus now contains a total of seven neutrons (having lost one). Because it has gained a proton, the carbon nucleus is now a nitrogen (N) nucleus, but the new nitrogen (^{14}N) nucleus does acquire a seventh electron outside its nucleus.

Isotopes that release beta particles are called **radioactive isotopes.** The decay process occurs at a constant rate that is characteristic of each radioactive isotope. The time required for half the atoms of a radioactive isotope to decay is called that isotope's **half-life** (Figure 2-6B). Half-lives for different isotopes range from as little as 0.3 microsecond to as long as

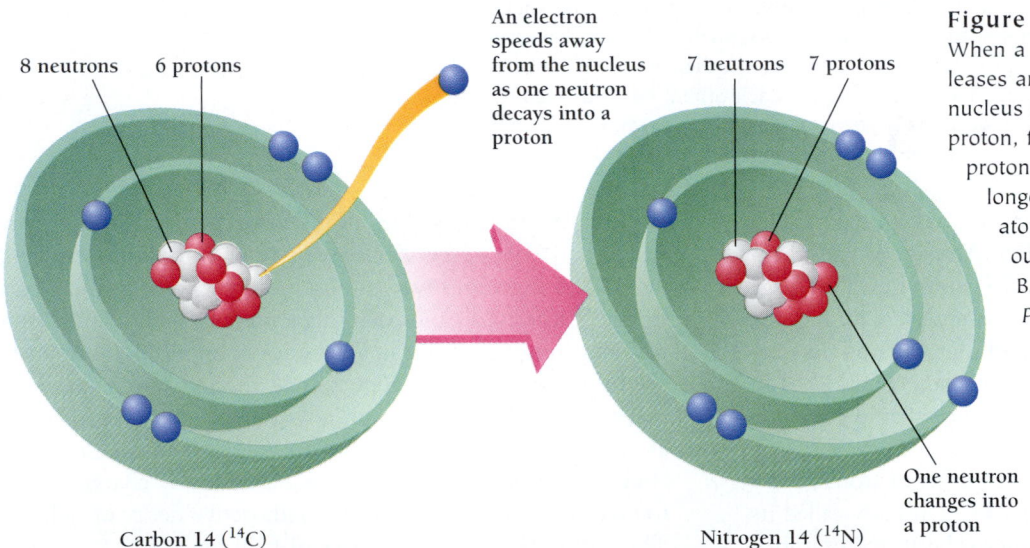

8 neutrons   6 protons

An electron speeds away from the nucleus as one neutron decays into a proton

7 neutrons   7 protons

Carbon 14 (^{14}C)

One neutron changes into a proton

Nitrogen 14 (^{14}N)

A.

Figure 2-6 Half-life. A. Breakdown of ^{14}C. When a neutron decays into a proton, it releases an electron (a beta particle). The altered nucleus has one fewer neutron and one more proton, for a total of seven neutrons and seven protons. A nucleus with seven protons is no longer carbon, however, but nitrogen. The atom then acquires an extra electron in its outer shell and stays electrically neutral. B. Half-life of chocolates. (*B, Paraskevas Photography*)

B.

BOX 2.1

The Uses and Dangers of Radioisotopes

Most of the isotopes found in nature are stable forms. Many of the isotopes of the heaviest elements, however, are unstable, or radioactive. Radioisotopes break down spontaneously into entirely different elements. As they break down, or decay, they release high-energy radiation.

The radiation from radioisotopes can be used to kill cancer cells. Radiation is now a standard therapy for many forms of cancer. Radioisotopes can also be used as tracers. For example, because the radioactive form of carbon (^{14}C) has the same properties as the regular form (^{12}C), organisms process both forms in the same way. Researchers can use sensitive instruments or photographic films to trace the movements of ^{14}C and answer many questions about molecular processes.

The Polish-born French chemist Marie Curie did some of the earliest and best work on radioisotopes. Marie Curie and her husband, Pierre, shared the 1903 Nobel Prize for Chemistry with H. Becquerel for the discovery of radioactivity. After Pierre's death in 1906, Marie Curie assumed his professorship at the Sorbonne and went on to win a second Nobel prize, in 1911, for the isolation of pure radium. Her subsequent work laid the groundwork for the discovery of the neutron and the synthesis of artificial radioactive elements. This latter work was carried out by Curie's daughter Irène Joliot-Curie and her husband, Frèdèric Joliot. In 1934, Marie Curie died of leukemia at age 67, a result of her lifelong exposure to radiation. The following year, Irène and Frèdèric won the Nobel Prize for Chemistry. Irène Curie also died of leukemia, in 1956, at age 59, and her husband Frèdèric assumed her professorship at the University of Paris.

Figure A   Marie Curie and her two daughters, Irène and Eva. *(Culver Pictures)*

Radium became a major weapon in the war on cancer. No one, especially Marie Curie, wanted to believe that radium could also cause harm. She hid the cataracts that clouded her eyes lest they implicate her beloved radium as harmful, and she could not believe that radium had anything to do with the early deaths of four of her colleagues who also worked with radium. Even today, her desk and laboratory notebooks are still highly radioactive, testimony to the massive amounts of radium that she accumulated for study.

Perhaps the saddest result of the discovery of radium was the inadvertent poisoning of young working women during World War I. Glowing radium on watch dials helped the Allied effort by allowing soldiers to read their watches in the dark and to synchronize their attacks. Young women in U.S. factories were painting radium onto watch dials, drawing their brushes to fine points between their lips. In doing so, they were ingesting radium. Because radium chemically resembles calcium, the workers were also incorporating it into their teeth and bones. One worker later recalled that after work, "We'd get the neighborhood kids to watch Alice glow in a dark closet. They'd get a good laugh out of that." Only 10 years later did the women start to get sick, with cancers of their jaws and mouths and the failure to make blood cells. Finally, the dangers of radiation became known.

2,000,000,000,000,000 (2×10^{15}) years. The half-life of ^{14}C is about 5760 years, while that of uranium 235 (^{235}U) is about 704 million years.

The predictable decay of isotopes is extremely useful to researchers. Isotopes can be used, for example, to determine the age of rocks and fossils; to track individual molecules in viruses, cells, organisms, or ecosystems; and even to identify the origins of products ranging from brand-name beverages to the explosive materials used in bombs.

Atoms are mostly empty space. They consist of a tiny central nucleus orbited by even tinier electrons. The nucleus has a positive charge, balanced by the negative charge of the electrons.

What does radioactivity do to living organisms?

Beta particles released by radioactive isotopes are a form of radiation that damages and kills cells. Alpha particles, beta particles, and x rays, for example, are all different forms of *ionizing radiation*. Ionizing radiation causes the water molecules in tissues and cells to break down into *free radicals*. Free radicals are chemically active, oxygen-containing molecules that easily react with and damage other molecules, such as proteins and DNA. Free radicals can kill cells outright or alter their ability to function. The most lasting and serious effects occur, however, when DNA is damaged.

Cells whose DNA has been altered by radiation cannot divide properly. Ionizing radiation has its greatest effect in tissues whose cells need constant replacement—cells that line the mouth, the stomach, and the intestines; skin cells that produce hair; bone marrow cells that produce red and white blood cells. As a result, people and other mammals exposed to intense doses of ionizing radiation—from nuclear weapons used in World War II, from accidents, or as therapy for cancer—suffer nausea, vomiting, diarrhea, hair loss, and anemia.

Because ionizing radiation can damage the DNA in cells and prevent them from dividing, it is commonly used to treat cancers. In some cases, however, DNA damage can itself cause the failure of normal cell regulation, thereby leading to a subsequent bout of cancer. An entire medical discipline—radiation oncology—is devoted to understanding this trade-off in order to maximize the effectiveness of radiation treatment of cancer (Box 2.1).

Where Are the Electrons in an Atom?

Although Rutherford envisioned the electrons neatly orbiting the nucleus like tiny planets, physicists have found that electrons actually move around the nucleus randomly, like a cloud of gnats. Although we still speak of electron **orbitals,** the distributions of electrons more closely resemble clouds. Like clouds, orbitals have fuzzy boundaries but definite shapes (Figure 2-5).

An electron in a particular orbital has a characteristic **energy**—the capacity to perform work. Each orbital has a different amount of energy. When an electron moves from a high-energy orbital to a low-energy orbital, an atom may release energy, often in the form of light. Conversely, when an atom absorbs energy, one or more electrons may move from a low-energy orbital to a high-energy one. We can visualize the energy of electrons as steps on a staircase. A ball on a staircase tends to bounce down the stairs. Just so, electrons tend to fall into orbitals with the lowest energy.

Two rules determine the distribution of electrons in an atom: (1) electrons occupy orbitals with the lowest possible energy, and (2) each orbital may contain only two electrons.

The atoms that are most common in organisms contain fewer than 18 electrons. The electrons of each of these atoms occupy only the nine orbitals with the lowest energies. Chemists group orbitals into three electron **shells**—groups of orbitals whose electrons have nearly equal energy. The first shell has just one orbital containing two places for electrons. The second shell includes four orbitals, with room for eight electrons. The third shell includes 9 orbitals, with room for 18 electrons. Most biologically important atoms (except for some metals), however, use only the four innermost orbitals of the third shell, with a capacity of eight electrons.

The arrangement of electrons in the outermost shell of each atom is the basis for the Periodic Table (Figure 2-7). The elements in each column have the same number of electrons in their outermost shell. These outer electrons are those that interact with other atoms and determine the chemical properties of an atom.

Electrons occupy orbitals with the lowest possible energy, and each orbital may contain only two electrons.

WHAT HOLDS MOLECULES TOGETHER?

An atom is most stable when its outermost shell contains eight electrons (or an "octet" of electrons). (This generalization is called the "octet rule.") The atoms of some elements—such as neon and argon—already contain eight electrons in their outer shells and are chemically unreactive. Most atoms, however, do not have full octets. These atoms can attain an octet by sharing electrons with other atoms, by stealing electrons from other atoms, or by losing electrons to other atoms, as illustrated in Figure 2-8. Such arrangements are the basis of chemical bonds.

A few molecules, however, have atoms that do not obey the octet rule. The most significant rule-breaking molecule of this sort is nitric oxide, discussed at the beginning of this chapter.

Covalent and Ionic Bonds Hold Atoms Together Strongly

Two kinds of bonds—covalent bonds and ionic bonds—hold atoms together especially strongly and are difficult to break. Such "strong" bonds hold together all the major molecules of life—from the carbohydrates in a loaf of bread to the DNA in your cells.

Two atoms that share one or more electrons are said to be held together by a **covalent bond.** An atom that gains or loses electrons becomes an **ion**—an atom or a molecule with a net electrical charge. Ions with opposite charges may be drawn together to form an **ionic bond.**

What holds two atoms together in a covalent bond?

The molecules of life are mainly built from just six kinds of atoms: hydrogen (H), carbon (C), nitrogen (N), oxygen (O), phosphorus (P), and sulfur (S). One way to remember this list is to arrange the symbols for each element into the nonsense word "SPONCH." The outer shells of these six atoms determine their ability to form molecules. Carbon, for example, has only six electrons—two in the first shell and four in the second (Figure 2-8A). To fill its outer shell with eight electrons, it must find other atoms with which to share an additional four electrons. These four shared electrons can form four covalent bonds, as in the case of methane (CH_4), the simplest organic compound (Figure 2-8B).

Two atoms can sometimes share more than one pair of electrons. For example, carbon may share two pairs of electrons with a single other carbon atom in a **double covalent bond.** The molecule ethylene (C_2H_4), important in the ripening of apples, bananas, and other fruits, has a double bond between its two carbon atoms. Some other molecules, including the nitrogen (N_2) in our atmosphere, contain **triple covalent bonds,** in which two nitrogen atoms share three pairs of electrons.

A covalent bond holds two atoms together because they share electrons. Most of the molecules in organisms are made from just six kinds of atoms.

Figure 2-7   The first three periods in the Periodic Table of the elements. The chemical and physical properties of the elements depend on the atomic numbers *(left)* and the atomic mass *(right)* of the nucleus. The atomic number is the number of protons in the nucleus. The atomic mass is the total number of protons plus neutrons. On the far right are elements such as helium and argon, whose electron orbitals are full. These "noble gases" remain aloof from other elements and rarely form chemical compounds. On the far left are hydrogen, sodium, and other highly reactive elements. The atoms of these elements contain a single electron in their outer orbital and are "eager" to bond with other atoms. In the middle, near the top, are carbon, nitrogen, and oxygen, all fairly reactive, but not explosively so. These three elements, together with hydrogen, form the basis of all organic molecules. *(Silicon chip, Alfred Pasleka/Peter Arnold, Inc.; neon sign, Allan Kaye/DRK Photo; blimp, Gianni Tortoli, Photo Researchers, Inc.; aluminum can, Paraskevas Photography; fireworks, AlaskaStock)*

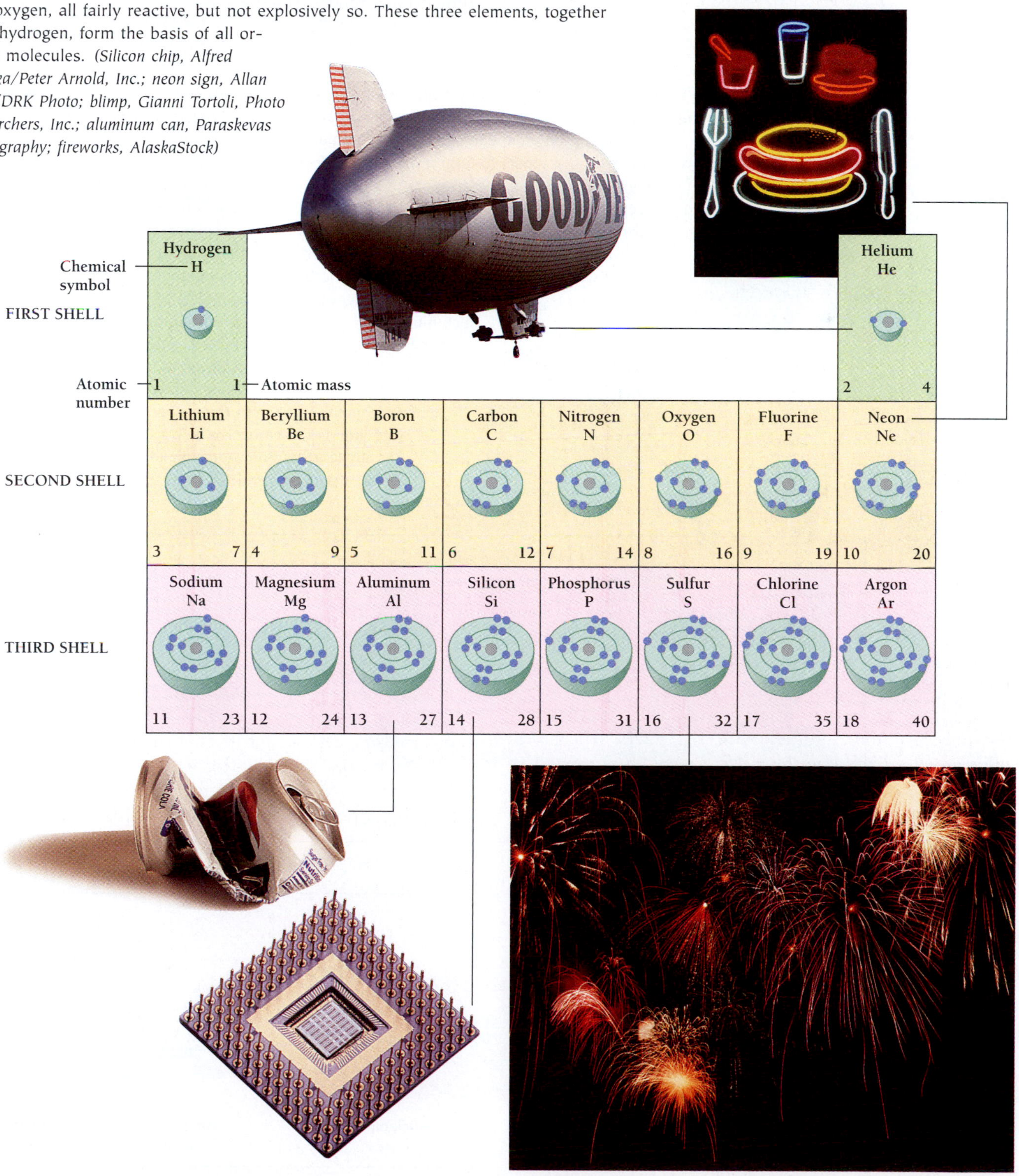

What determines the shape of a molecule?

The covalent bonds within a molecule have fixed directions and lengths, which result from the characteristics of electron orbitals. For example, the distance between a carbon atom and a hydrogen atom is the same whether in a complex molecule (such as DNA) or a simple one (such as methane) (Figure 2-8). Similarly, the angles between bonds are always the same.

Because bond lengths and angles are constant, every molecule has a definite size and shape. A water molecule, for example, always has the same boomerang shape. We can calculate the shape of a molecule by making a three-dimensional model, based on what we know about the positions of the atoms and the bond lengths and angles. Chemists and biologists represent the structures of molecules in a variety of ways—on a two-dimensional page, and in several types of three-dimensional models (Figure 2-8).

Figure 2-8   Covalent bonding usually depends on the sharing of electrons to produce octets in each participating atom. A. Covalent bonding between carbon and hydrogen. Each hydrogen atom can share one electron, and each carbon atom can share four. The four electrons in carbon's outer shell form covalent bonds with the hydrogen atoms to form a methane molecule. To emphasize their capacity to form four covalent bonds, this drawing separates the electrons of the carbon atom's outer shell. B. An exception to the octet rule—nitric oxide. The NO molecule is relatively stable despite the absence of an octet of electrons. C. Molecules may be depicted in several ways. Simple structural formulas (far right) are easy to write, but they do not clearly show the angles at which the atoms connect to one another—the "bond angles." Space-filling models (second from right) suggest the arrangements of the atoms. But ball-and-stick models (third from right) show bond angles most clearly. In methane (CH_4), each bond points to one corner of a tetrahedron. The angle between any two bonds is 109.5°. In the structural and space-filling models, bonds drawn vertically are understood to point away from the viewer toward the back of the page; bonds drawn horizontally point out at the viewer. C. In ethylene, carbon atoms make double covalent bonds by sharing two pairs of electrons.

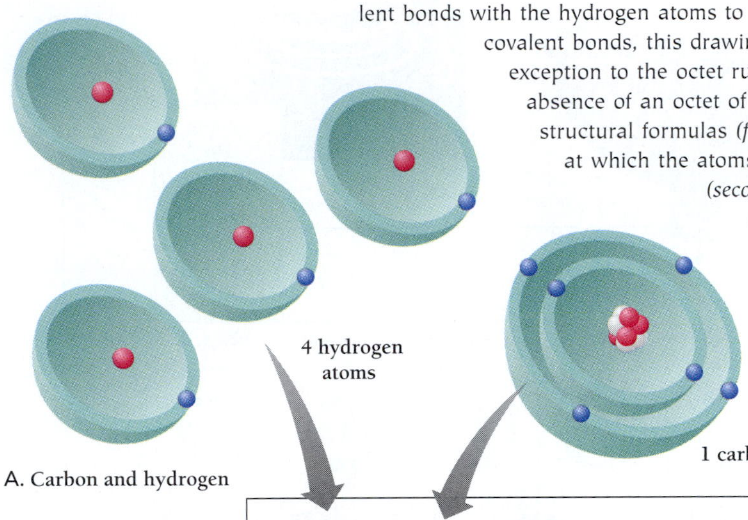

4 hydrogen atoms

1 carbon atom

A. Carbon and hydrogen

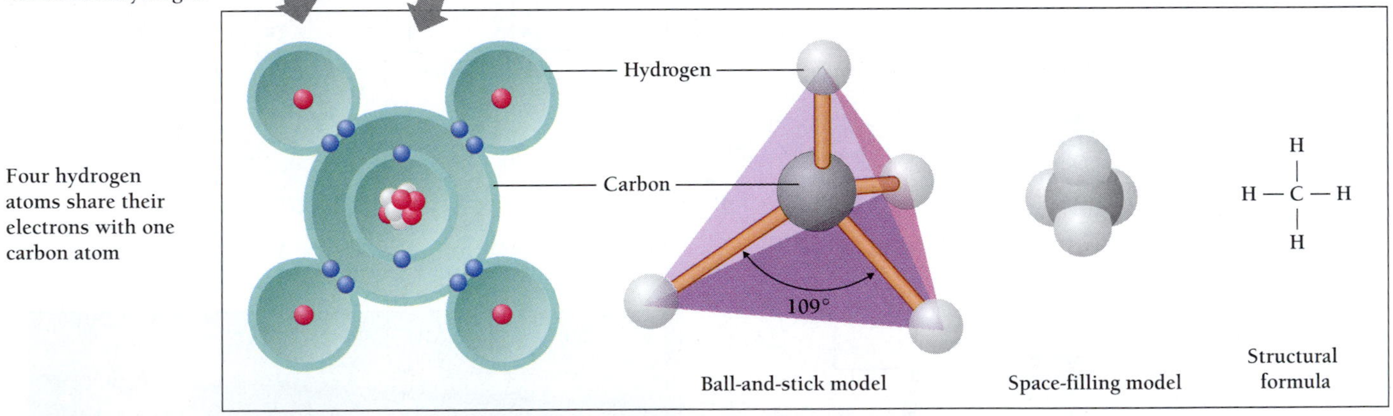

Four hydrogen atoms share their electrons with one carbon atom

Hydrogen

Carbon

109°

Ball-and-stick model

Space-filling model

H
|
H — C — H
|
H

Structural formula

B. Methane (CH_4)

Hydrogen

Carbon

Ball-and-stick model

Space-filling model

Structural formula

C. Ethylene (C_2H_4)

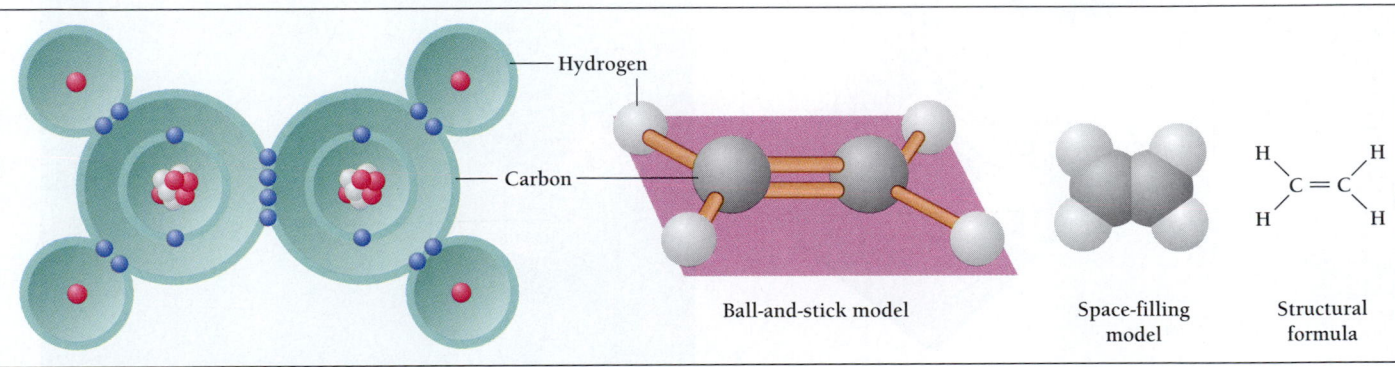

> *Because bond lengths and angles are constant, every molecule has a definite size and shape.*

What holds two atoms together in an ionic bond?

The most familiar ionic bonds are those in salts, such as ordinary table salt—sodium chloride. The sodium atoms and the chlorine atoms in sodium chloride have stable electron con-figurations, with eight electrons in their outer shells (Figure 2-9). Instead of sharing electrons, however, the sodium atom gives away an electron and the chlorine atom takes one. The result is positively and negatively charged ions. The positively charged sodium ions and the negatively charged chloride ions (abbreviated Na$^+$ and Cl$^-$) are held together by simple electrical attraction, in ionic bonds. Electrostatic attraction is the same thing that makes your socks cling to your shirts when they come out of the dryer.

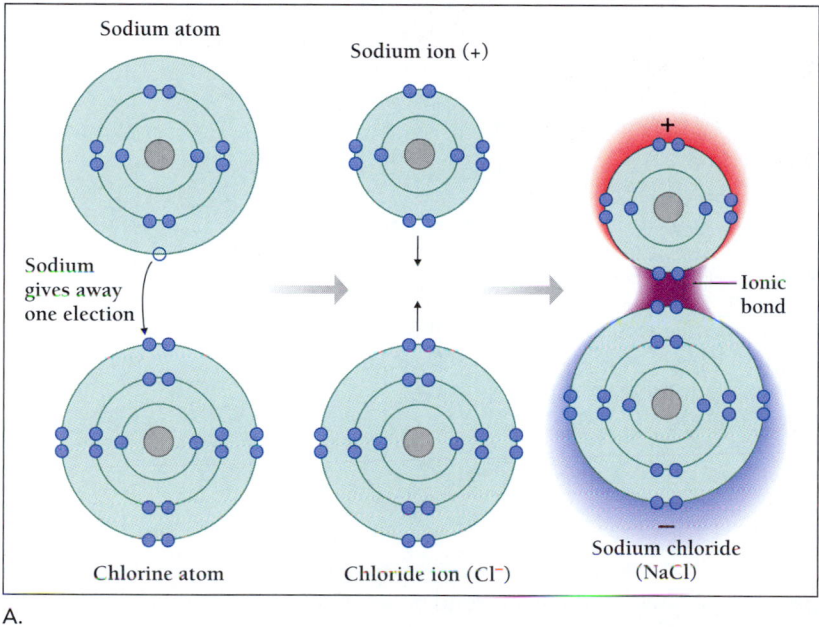

Sodium atom

Sodium ion (+)

Sodium gives away one election

Ionic bond

Chlorine atom

Chloride ion (Cl$^-$)

Sodium chloride (NaCl)

A.

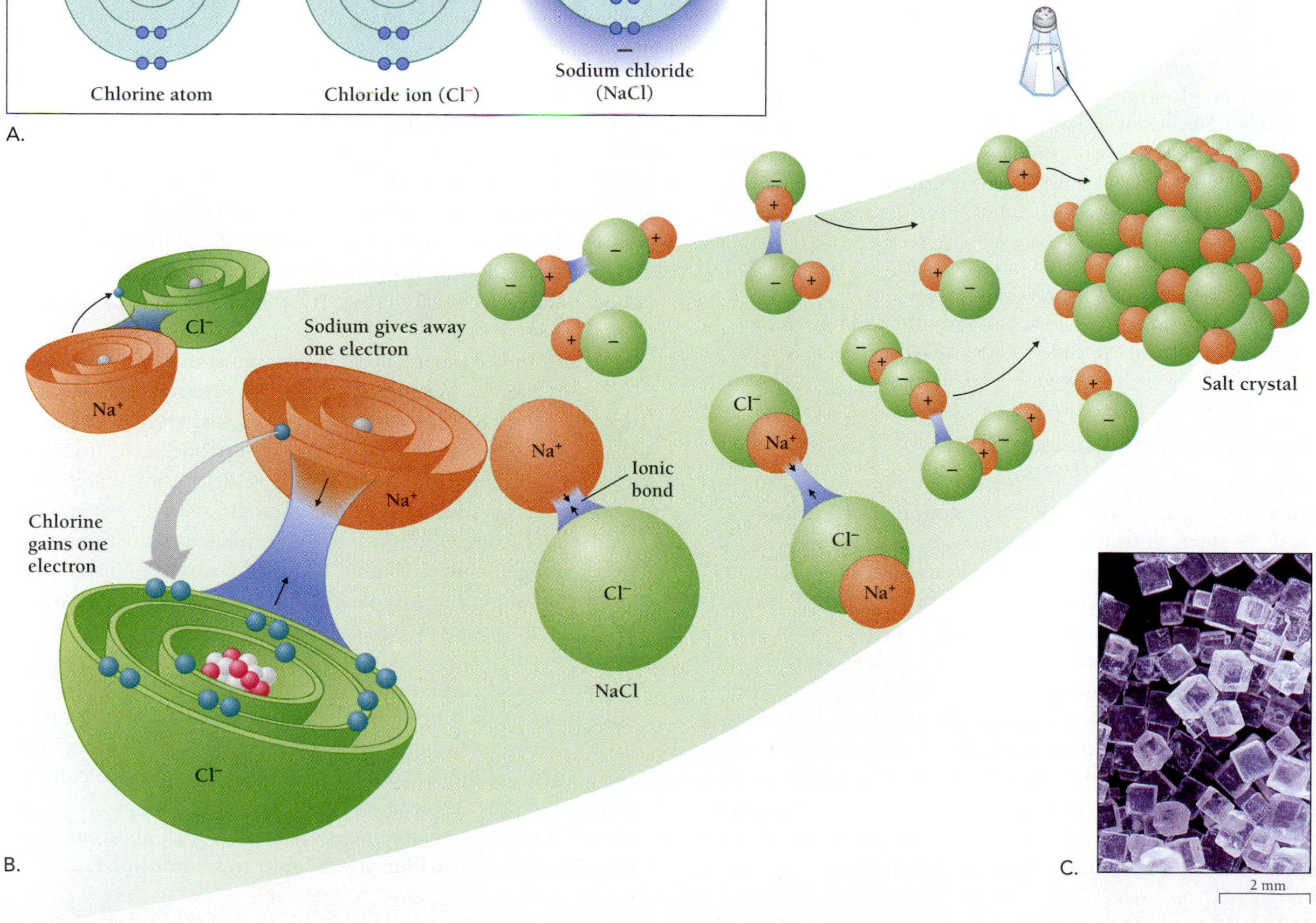

Figure 2-9   Table salt. A. Ionic bonding between sodium and chloride ions in a crystal of sodium chloride. B. Reaction of sodium and chloride to form salt. C. Salt crystals. *(C, Bruce Iverson)*

Cl$^-$

Na$^+$

Sodium gives away one electron

Na$^+$

Na$^+$

Chlorine gains one electron

Ionic bond

Cl$^-$

Cl$^-$

Na$^+$

Cl$^-$

NaCl

Cl$^-$

Na$^+$

Salt crystal

2 mm

B.

C.

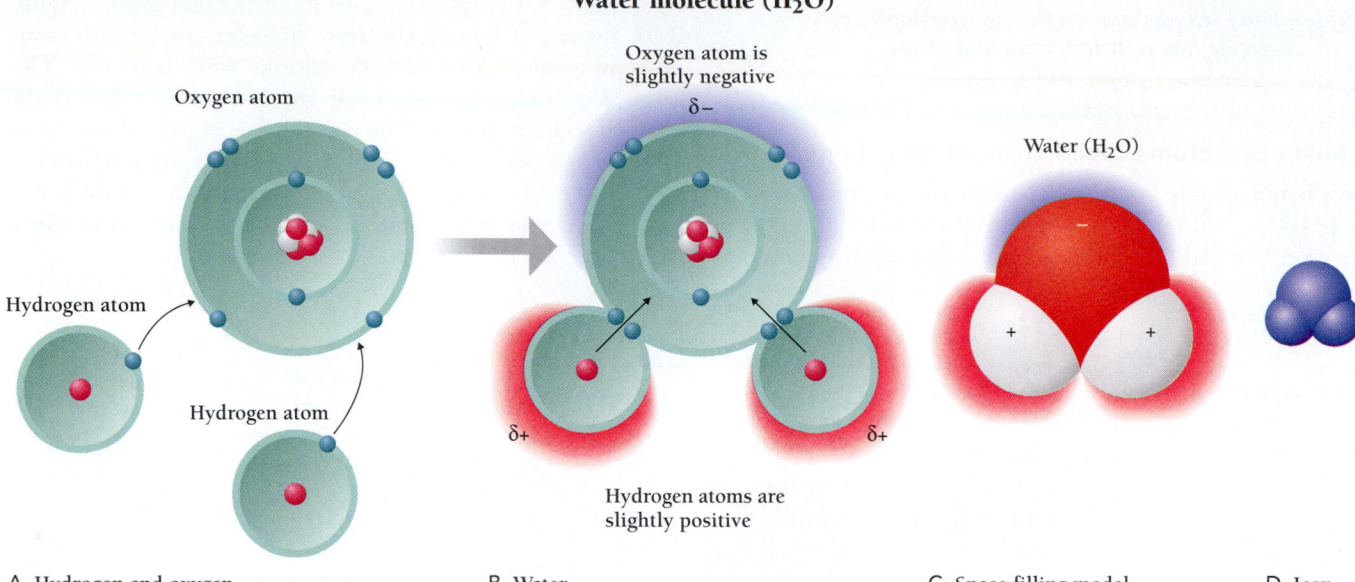

Water molecule (H₂O)

Oxygen atom is slightly negative
$\delta-$

Oxygen atom

Hydrogen atom

Hydrogen atom

$\delta+$ $\delta+$

Hydrogen atoms are slightly positive

Water (H₂O)

A. Hydrogen and oxygen   B. Water   C. Space-filling model   D. Icon

Figure 2-10   The polarity of water. A. Two atoms of hydrogen and one atom of oxygen share electrons in two covalent bonds to form a molecule of water. B. Because the oxygen atom is more electronegative than the hydrogen atoms, the electrons spend more time hovering around the oxygen end of the water molecule. As a result, the oxygen end has a slightly negative charge, while the hydrogens have a slight positive charge. Such a partial charge, less than one full electron, is symbolized by δ. C. Space-filling model of water molecule. D. Icon for water. *(C, Tom Pantages/Phototake)*

Electrical forces hold together ions of opposite charge, forming a complex of positive and negative ions rather than a true molecule. In a salt crystal, for example, each Na⁺ ion is surrounded by Cl⁻ ions, and each Cl⁻ ion, by Na⁺ ions (Figure 2-9). If salt is dissolved in water, however, each ion is surrounded by water molecules, as we discuss later.

An ionic bond is a simple electrical attraction between two oppositely charged ions.

Why are some molecules polar?

A chlorine atom has a high tendency to gain an electron to form a chloride ion, while a sodium atom has a high tendency to lose an electron to form a sodium ion. A carbon atom, on the other hand, tends to share its outer electrons to form covalent bonds. The American chemist Linus Pauling realized that the atoms of each element have a characteristic **electronegativity**—the tendency to gain electrons. Oxygen and chlorine, for example, have a great tendency to gain electrons and are among the most electronegative elements. In contrast, sodium, potassium, and lithium are among the least electronegative elements—they have the greatest tendency to lose electrons. In between are elements such as carbon that have no tendency to lose or gain electrons. (We can think of electronegativity as a measure of electron greediness.)

Knowing the electronegativity of two atoms allows one to predict whether a bond between them will be covalent or ionic. The larger the difference in the electronegativities of two atoms, the more likely they are to form an ionic rather than a covalent bond. Sodium and chlorine, for example, have a large difference in electronegativities and therefore form ionic bonds. Carbon and nitrogen, on the other hand, have similar, moderate electronegativities, and they usually form covalent bonds.

Even in a covalent bond, however, atoms may not share electrons equally. When atoms differ in electronegativity, they do not share electrons equally. Instead, the diffuse clouds of shared electrons tilt toward the more electronegative atoms. In a water molecule, for example, an oxygen atom shares electrons with two hydrogen atoms. But the shared electrons are more concentrated around the oxygen nucleus than around the two hydrogen nuclei (Figure 2-10). As a result, the oxygen atom has a slight negative charge, and the two hydrogen atoms have a slight positive charge.

Molecules that have uneven distributions of electrical charge are said to be **polar,** since they have positive and negative poles in the same way that a magnet has two poles (Figure 2-10). When a polar molecule, such as water, comes close to an ion or to another polar molecule, its negative pole points toward the other molecule's positive pole, and its positive pole toward a nearby negative pole. Molecules with approximately uniform charge distributions are said to be **nonpolar.**

Atoms with great differences in electronegativity tend to form ionic bonds. Atoms with only slight differences in electronegativity tend to form covalent bonds. Molecules with uneven charge distributions are said to be polar.

What Kinds of Forces Hold Separate Molecules Together?

The forces that hold atoms together in covalent bonds are strong. Breaking them requires lots of energy. Electrostatic interactions in salt crystals (ionic bonds) are also strong. Most of the chemistry that concerns biologists, however, happens in or around water. In such an **aqueous** (watery) environment, a variety of weak interactions bind molecules together. Although individually each interaction is weak, the summation of many weak interactions determines the structure of many complex molecules.

Why do nonpolar molecules cling together in an aqueous solution?

When a bartender mixes an alcoholic drink, such as brandy and soda water, the result is a smooth mixture, uniform throughout. Yet, when a cook mixes oil and vinegar to make salad dressing, the oil runs together in globs and floats to the top of the vinegar (Figure 2-11). Why is this? The answer is that polar molecules, such as alcohol, mix well with water and are said to be **hydrophilic** [Greek, *hydro* = water + *phili* = love], or water loving. But because water molecules attract and surround themselves with other polar molecules, nonpolar molecules, such as oil, seem to be repelled by water. We say that nonpolar molecules are **hydrophobic** [Greek, *hydro* = water + *phobos* = fear], or water fearing. In water, hydrophobic molecules are driven together by their mutual tendency to avoid water. The aggregation of nonpolar molecules is called a **hydrophobic interaction** (Figure 2-11).

The properties of polar and nonpolar molecules play a key role in life. The tendency of nonpolar molecules to separate from water underlies the formation of the membranes that separate the inside of a cell from the outside (Chapter 4).

Oils and other hydrophobic molecules cling together when confronted with an aqueous environment.

Van der Waals attractions reinforce hydrophobic interactions

Because electrons move around rapidly, the charges on the different parts of a molecule change constantly. At a given instant, such fluctuations may temporarily produce positive and negative poles, even in a nonpolar molecule. A temporarily polar molecule may, at the instant of polarity, induce a charge in an adjacent molecule, so that it too becomes temporarily polar. For a moment, then, the two molecules are attracted to

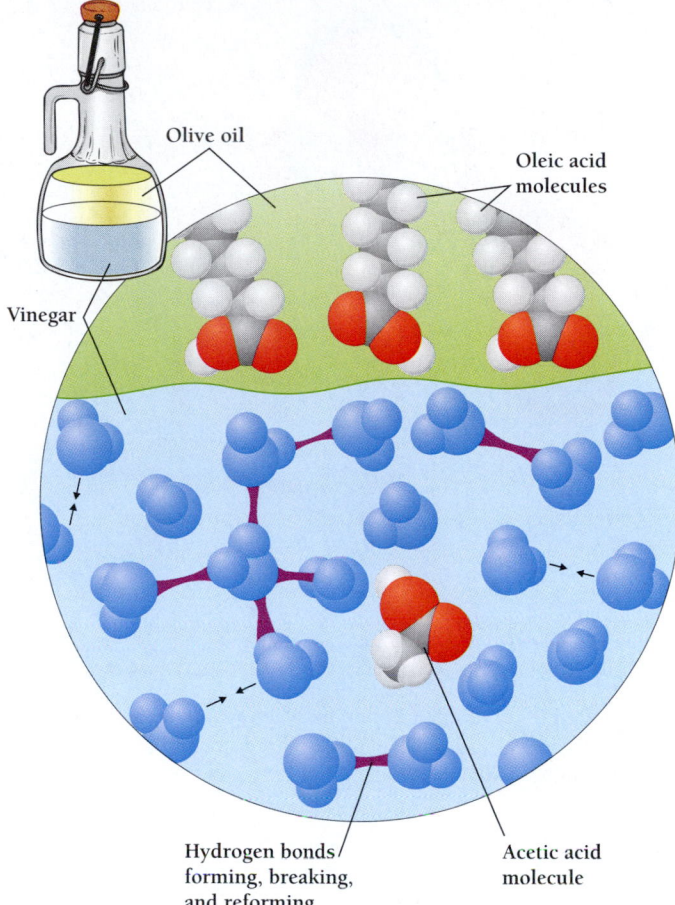

Figure 2-11   Fear of water. Separation of oil and vinegar in a salad dressing. Vinegar, a dilute solution of acetic acid, is mostly water. In hydrophobic interactions, water molecules attract and surround themselves with other polar molecules (including water) but repel nonpolar molecules such as oils. Nonpolar molecules are driven together by their mutual tendency to avoid water.

each other. Such **van der Waals interactions** operate only over very short distances and tend to reinforce the hydrophobic interactions among nonpolar molecules in water. Van der Waals interactions are important in maintaining the three-dimensional arrangements of proteins and other large molecules, such as DNA.

If two molecules are already very close, van der Waals attractions can pull them closer together.

What are hydrogen bonds?

When a hydrogen atom attaches to a highly electronegative atom, such as oxygen or nitrogen, the resulting covalent bond is polar. In this case, the hydrogen atom acquires a slight positive charge. Such a hydrogen atom can then participate in a

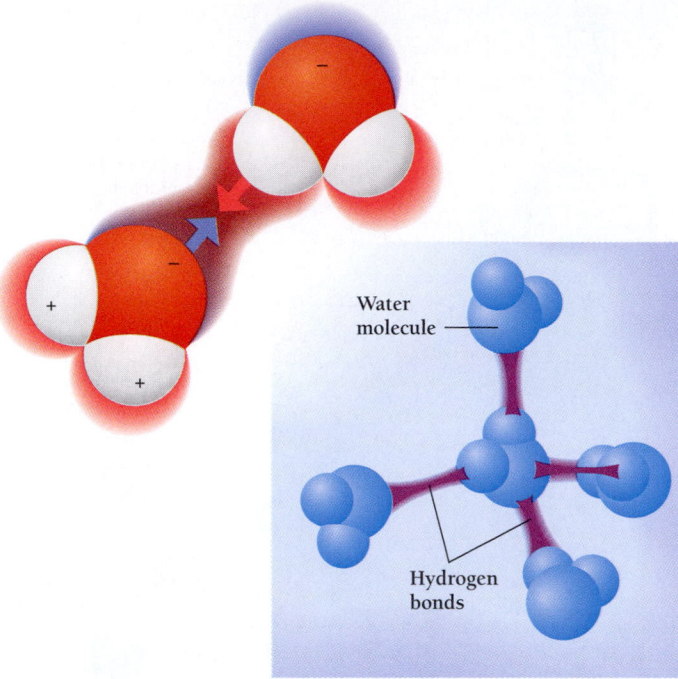

Figure 2-12 Hydrogen bonds between water molecules. The hydrogen atoms of one water molecule are attracted to the oxygen atom of another water molecule.

hydrogen bond—a weak attraction to a negatively charged atom in another molecule (Figure 2-12). The most common hydrogen bonds are those between water molecules, but other hydrogen bonds also play a critical role in the structure of proteins and DNA.

The polarity (or nonpolarity) of molecules underlies their noncovalent interactions.

HOW IS WATER ESPECIALLY WELL SUITED FOR ITS ROLE IN LIFE?

Life and water are intimately connected (Figure 2-13). As we discuss in Chapter 18, life probably originated in water. Wherever life is found, there is water, and wherever liquid water is found, there is life. Water makes up more than 70 percent of the material of living organisms themselves and covers more than 75 percent of the Earth's surface. It is the medium in which most cells are constantly bathed and the major component of cells themselves. Not only do most biochemical reactions occur in water, but water itself participates in many biochemical reactions. Although water is common—at least on Earth—its properties are highly unusual. Water's special attributes make it uniquely fit for its important role in life.

Why Does Ice Float?

Most solids sink into their corresponding liquids, but ice floats on water (Table 2-1A). Because ice floats, lakes and ponds freeze from the top down, instead of from the bottom up. The surface ice of a wintry pond, such as that in Figure 2-14, insulates the liquid water below the surface from the freezing air above. As a result, ponds, lakes, and rivers remain liquid at the bottom, allowing fish and other organisms to survive through long winters.

Ice has a lower density than liquid water because the water molecules in ice form a rigid lattice or crystal that holds them somewhat apart from one another: Each water molecule is hydrogen bonded to four other water molecules. In liquid water, the network of water molecules constantly changes, and there is less open space between the molecules.

Ice floats on water because ice has a lower density than water.

How Does Water Resist Changes in Temperature?

In our daily lives, we know that temperature is a measure of how warm or cold something is. But to a scientist, temperature is a measure of the motion of molecules.

Molecules move constantly. In solids, molecules merely jiggle. In gases, molecules whiz around at mind-boggling speeds, bumping into other molecules. At room temperature, for example, the average speed of an oxygen molecule in the air is about 500 meters per second (more than 1000 mph). The higher the temperature, the more rapidly molecules move about. Heat is energy that increases the random movement of molecules.

For most substances, any heat is converted into the more rapid motion of whole molecules, that is, into increased tem-

Figure 2-13 Water is essential to every organism. Killer whales (*Orcinus orca*) spend their entire lives in the ocean. (*Francois Gohier/Photo Researchers, Inc.*)

perature. In water, however, additional heat is converted not only into the increased motion of individual molecules but also into a more rapid exchange of hydrogen ions between adjacent molecules. As a result, for a given amount of additional heat, water temperature increases much less than in other substances, that is, water has a high **heat capacity.**

One consequence of water's heat capacity is that the oceans heat up and cool down very slowly, moderating the climates of nearby land. In winter, the interiors of the world's continents disappear under a layer of snow, while temperatures in coastal areas remain relatively mild. Conversely, in summer, the continental interiors reach blazing temperatures, while coastal areas remain cool. Every large body of water has a similar influence on the land nearby.

Water not only cools the coasts of continents, it also cools the human body. To keep our body temperature within the narrow range that many biochemical reactions demand, we must rid ourselves of excess heat. Humans do so largely by perspiring. As the water in sweat evaporates, it carries the heat away. Dogs accomplish the same thing by panting (Figure 2-15).

When a substance changes from a liquid to a gas, it absorbs heat. As water evaporates, it absorbs a lot of heat—more than twice as much as the same amount of ethanol, for example. (Ethanol is also called "ethyl alcohol," or just plain "alcohol" when in wine, beer, or spirits.) Water absorbs a lot of heat when it turns into water vapor because each of the hydrogen bonds that hold the water molecules together must be broken.

Figure 2-14 Pond life under winter ice. Because water is denser than ice, the ice floats and covers lakes and ponds in the winter. This layer of ice insulates the water below from freezing air temperatures. Fish, frogs, insects, copepods, protists, and other life forms survive the winter, all because water is denser than ice.

TABLE 2-1   HOW DOES WATER COMPARE WITH ETHANOL AND OLEIC ACID?

	Water	Ethanol	Oleic acid

A. **Density.** In water, the liquid phase is more dense than the solid phase (ice), which is why ice floats. In most substances, including ethanol and oleic acid, the solid phase is denser than the liquid phase.

B. **Cohesion.** In water, the attraction of the water molecules for one another creates a web of molecules that resists the sinking of the spoon. Neither ethanol nor oleic acid is as cohesive as water, and the spoon sinks.

C. **Adhesion.** Water molecules' polarity enables them to form stronger interactions with the molecules in paper, so water rises faster than does ethanol or oleic acid.

D. **Ability to dissolve other substances.** Water dissolves table sugar (sucrose) more quickly than does ethanol or oleic acid.

E. If you were trapped on a desert island which of these would you choose?

(A–D, Charles D. Winters; E, © 1994 Zefa Germany/The Stock Market; © 1997 Paraskevas Photography; © Camerique/The Picture Cube)

Water's enormous heat capacity stabilizes temperatures in oceans, lakes, coastal areas, and even within individuals.

Why Do Water Molecules Cling to One Another?

An attraction between molecules of the same substance is called **cohesion.** Water and chewing gum are both highly cohesive. Under tension, they stretch but do not break. The cohesion of water molecules also creates high **surface tension**—the tendency of a substance to form a smooth round surface. Because water molecules cohere, a water surface resists pressure. For example, as illustrated in Table 2-1B, the bowl of a teaspoon balanced on the edge of a teacup will rest on the surface of the tea as though floating. But a cup of alcohol or olive oil will not support the same spoon.

Water's cohesiveness is caused by hydrogen bonds among its molecules. The attraction among water molecules is so great that water has the highest surface tension of any liquid except for liquid mercury. The surface of water acts like a thin skin. Some insects, such as the water strider, take advantage of water's surface tension and walk easily over the surface of a pond.

Hydrogen bonding makes water molecules cohere, creating surface tension.

Why Do Water Molecules Cling to Other Substances?

An attraction of one substance to another is called **adhesion.** Water clings to and "wets" surfaces, such as glass, that are composed of polar or charged molecules. Water's adhesiveness allows a close interaction between it and many other biological surfaces.

Because water both clings to itself (cohesion) and clings to other surfaces (adhesion), it readily enters and climbs tiny tubes, in a process called **capillary action.** If we place fine glass tubes (capillaries) into beakers of water, ethanol, and olive oil, the water moves upward a fair distance, the ethanol moves less, and the olive oil not at all (Table 2-1C). Such capillary action is important in the movement of water through soil, and up the narrow tubes inside of trees and other plants. Water's cohesiveness and adhesiveness help trees raise water from roots deep in the ground to leaves hundreds of feet in the air (Figure 2-16).

Water's adhesiveness results from the tendency of water molecules to form hydrogen bonds with other polar or charged molecules. Water molecules stick to glass capillaries because glass contains charges on its surface. Similarly, paper towels, which are made of long polar molecules called cellulose, absorb water very well. In contrast, olive oil, which is mostly nonpolar, is relatively poorly absorbed by the cellulose in paper towels. That's why cleaning up water is so much easier than cleaning up olive oil.

Hydrogen bonding allows water to adhere to other compounds.

Why Is Water Such a Powerful Solvent?

Most people think of a solvent as the fluid that the dry cleaner uses to clean a wool coat, or maybe the gasoline you use to get grease off the wheels of your car. In fact, a **solvent** is any fluid in which other substances dissolve. Anyone who has poured salt into boiling water or stirred sugar into a cup of hot coffee will have noticed how rapidly salt and sugar dissolve in water. Water is a powerful solvent. It is an especially good solvent for ions, such as those of table salt, and for polar molecules, such as sugar. Salt and sugar, which dissolve so well in water, dissolve poorly in ethanol and hardly at all in olive oil, which is nonpolar (Table 2-1D).

The substances that dissolve within a solvent such as water are called **solutes.** All salts are composed of ions, and all are extremely soluble in water. When sodium chloride dissolves in water, the highly polar water molecules surround the Na^+

Figure 2-15   Why are dogs' noses wet? Water's enormous heat capacity makes it a useful coolant. As the water in a dog's nose evaporates, it carries heat away, cooling the inside of the nose. On a hot day, overheated blood from the body flows through the cold nose before reaching the brain, so that the dog's brain tissues remain cool. *(John Cancalosi/DRK Photo)*

Figure 2-16 Only because water is so cohesive can trees raise it hundreds of feet off the ground. Individual California coast redwoods (*Sequoia sempervirens*) may grow to heights of 360 feet, taller than a 25-story building. (*Larry Ulrich/DRK Photo*)

and Cl⁻ ions like devoted fans crowding around a sports star (Figure 2-17). This shell of water molecules shields the positive and negative ions from each other, allowing the ions to remain farther away from each other than they can in their solid form (crystals), where they have a regular arrangement.

Sugars and other polar molecules also dissolve well in water, but for a slightly different reason. When water molecules surround a polar molecule such as sugar, hydrogen bonds form between the water molecules and the electronegative oxygen atoms of the sugar. These water molecules isolate the molecules of sugar from one another, breaking the attachments that keep the sugar molecules in a crystal.

The polar properties of water explain why some molecules are hydrophilic and others are hydrophobic. Hydrophilic molecules are polar or charged molecules that interact strongly with water molecules. Nonpolar molecules are hydrophobic and do not interact with water molecules. Oil, a mixture of nonpolar molecules, does not readily mix with water but it mixes easily with nonpolar solvents. We can summarize the tendency of polar molecules to dissolve in water and nonpolar molecules to dissolve in oil with the phrase "like dissolves like."

Many biologically important molecules contain both hydrophilic and hydrophobic regions and are said to be **amphi-**

pathic [Greek, *amphi* = both + *pathos* = feeling]. When an amphipathic molecule is placed in water, its hydrophobic regions avoid the water, and its hydrophilic regions bond with water (Figure 2-18). Mayonnaise, for example, consists of tiny oil droplets suspended in a watery solution.

All **detergents** are made of amphipathic molecules that interact both with water and with hydrophobic molecules such as oil. Detergents remove oil from dirty dishes and clothes by making the oil soluble in water.

Water is the most important polar molecule both within cells and in their external environments. Water molecules interact strongly with each other and with other charged and polar molecules. These interactions, and the lack of interactions with nonpolar molecules, are the most important factors in establishing biological structures.

Water's polar properties make it an excellent solvent for ions and polar molecules.

Water Participates in Many Biochemical Reactions

Besides serving as the medium in which most biochemical reactions occur, water itself participates in many reactions. For example, when we break down the large carbohydrate molecules in spaghetti into the small sugar molecules that we use for fuel, we do so by adding water molecules. The negatively charged oxygen atom of a water molecule can easily approach positively charged atoms in other molecules. The oxygen may then form a covalent bond with the other molecule by simultaneously breaking its covalent bond with one of its hydrogen atoms.

How Do Water Molecules Change in Solution?

Recall that the electrons that form the chemical bonds in a water molecule are closer to the oxygen atom than to the hydrogen atoms. Sometimes the electrons get so close to the oxygen atom that one of the hydrogen atoms has no electron at all and becomes a naked nucleus, or **hydrogen ion** (H⁺). A hydrogen ion without an electron is just a proton, and it can jump back and forth between the oxygen atoms of two different water molecules (Figure 2-19). When that happens, one water molecule can end up with an extra proton and the other water molecule ends up missing a proton. The deprived water molecule is called a **hydroxide ion** (OH⁻) and has a charge of −1. The water molecule with the extra hydrogen ion (H⁺) is called a **hydronium ion** (H₃O⁺) and has a charge of +1. We can represent this exchange of hydrogen ions by the following chemical equation:

$$H_2O + H_2O \longrightarrow OH^- + H_3O^+$$

water + water ⟶ hydroxide ion + hydronium ion

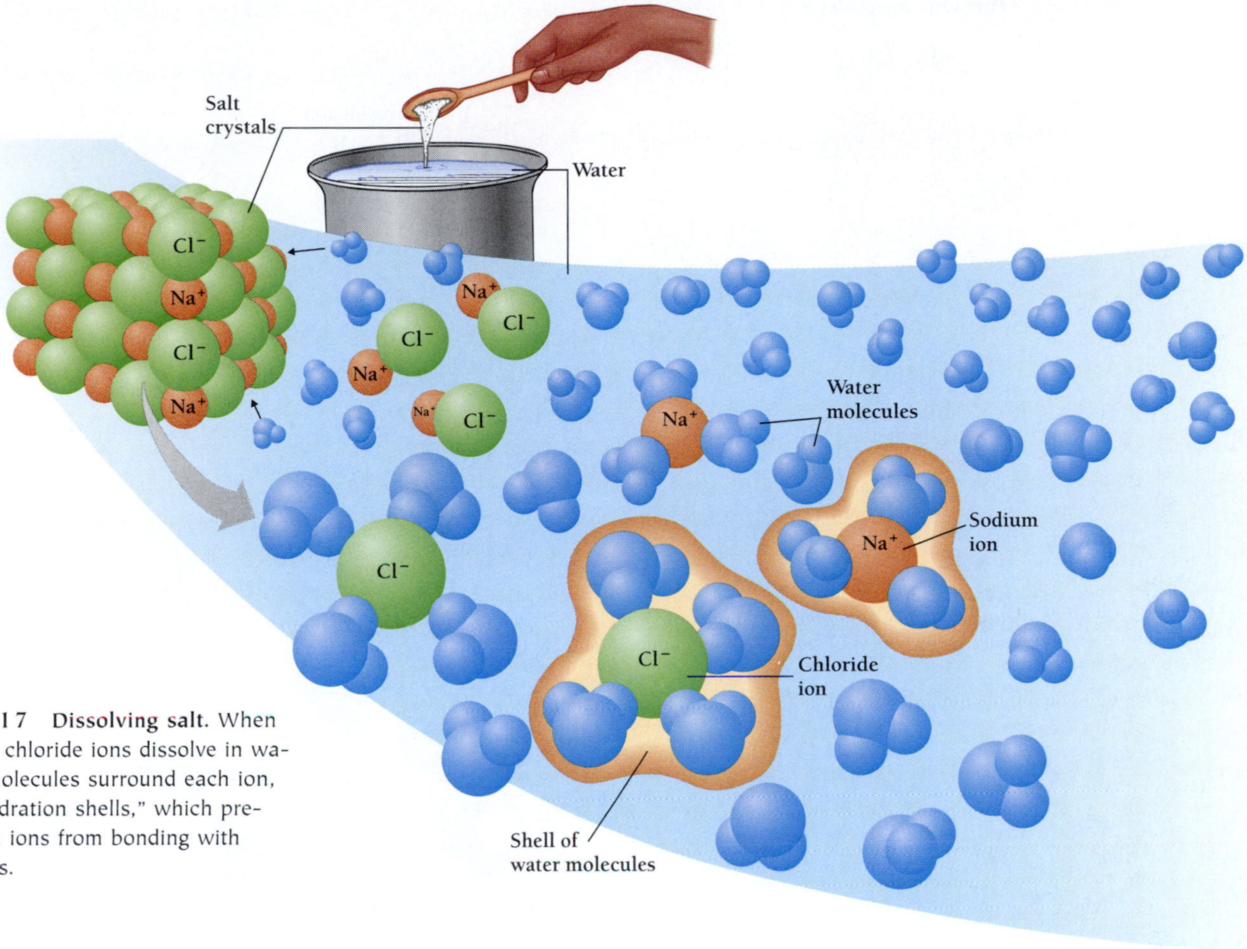

Salt crystals

Water

Cl⁻

Na⁺

Cl⁻

Na⁺

Na⁺

Cl⁻

Cl⁻

Na⁺

Cl⁻

Cl⁻

Na⁺

Water molecules

Na⁺   Sodium ion

Cl⁻

Chloride ion

Shell of water molecules

Figure 2-17 **Dissolving salt.** When sodium and chloride ions dissolve in water, water molecules surround each ion, forming "hydration shells," which prevent sodium ions from bonding with chloride ions.

Notice that each side of the arrow has equal numbers of hydrogen atoms and oxygen atoms and that the charges are balanced. For the sake of brevity, we can leave one of the water molecules out of the equation. We can represent the shuttling of hydrogen ions as the splitting of water into a hydroxide ion and a naked hydrogen ion:

$$H_2O \longleftrightarrow H^+ + OH^-$$

water $\longleftrightarrow$ hydrogen ion + hydroxide ion

The opposite charges of hydrogen ions and hydroxide ions ensure that they do not go far. They quickly recombine with other hydroxide ions and hydrogen ions to form uncharged water molecules again. Water, then, constantly splits into hydrogen ions and hydroxide ions, and the ions constantly recombine to form water.

When the splitting of water into hydrogen ions and hydroxide ions exactly balances the recombining of hydrogen ions and hydroxide ions back into water, the number of ions at any instant is always the same, and we say that the solution is "at equilibrium." Such a balance between forward and reverse reactions is called **chemical equilibrium** [Latin, *aequus* = equal + *libra* = balance].

In pure water at equilibrium, the number of hydrogen ions exactly equals the number of hydroxide ions. However, in most solutions, the number of hydrogen ions differs from the number of hydroxide ions. Vinegar has a million times more hydrogen ions than hydroxide ions. Stomach acid is even stronger. It contains about one trillion times more hydrogen ions than hydroxide ions.

The **concentration** of hydrogen ions—the number of hydrogen ions in a given amount of liquid—varies so widely in different solutions that chemists have defined a scale, called **pH,** to measure the concentration of hydrogen ions in a solu-

Amphipathic molecule

Hydrophobic region   Hydrophilic region

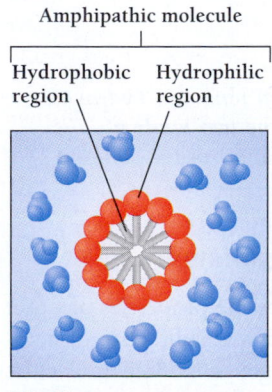

Micelle

Figure 2-18 **Amphipathic molecules contain a hydrophobic tail and hydrophilic head.** A ball of these molecules, called a "micelle," forms when water molecules cluster around the hydrophilic heads and the hydrophobic tails hide together in the middle.

Two water molecules

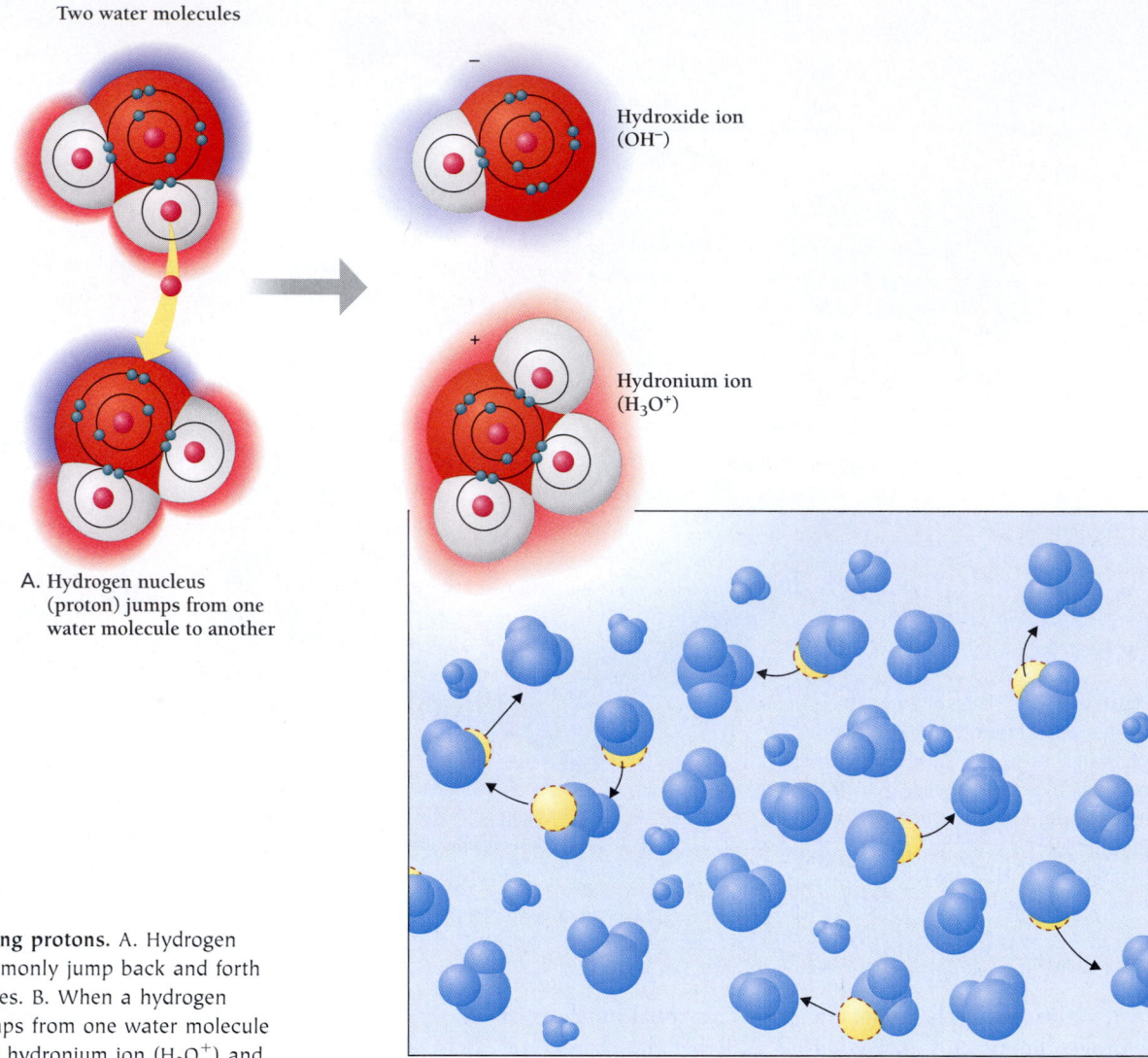

Hydroxide ion
(OH⁻)

Hydronium ion
(H_3O^+)

A. Hydrogen nucleus
(proton) jumps from one
water molecule to another

Figure 2-19 Juggling protons. A. Hydrogen
nuclei, or protons, commonly jump back and forth
between water molecules. B. When a hydrogen
nucleus, or proton, jumps from one water molecule
to another and stays, a hydronium ion (H_3O^+) and
a hydroxide ion (OH⁻) result.

B.

tion (Figure 2-20). The pH scale is based on the logarithm (to the base 10) of the concentration of hydrogen ions. We can think of pH as standing for "power of hydrogen." But the logarithm is actually a negative number. So a pH of 7, for example, represents a concentration of hydrogen ions equal to 10^{-7} moles per liter (a *mole* is Avogadro's number of molecules, so moles per liter is a unit of concentration). The middle of the pH scale is 7, which represents a neutral solution, such as water, in which H^+ and OH^- concentrations are the same. Acid solutions have a lower pH. For example, the pH of vinegar is 4 (10^{-4} moles per liter). The pH of the hydrochloric acid secreted by the stomach is less than 1 (10^{-1} moles per liter), or 1/10 mole per liter—very concentrated. In short, the greater the concentration of hydrogen ions, the lower the pH value. A tenfold increase in concentration means a decrease of one pH unit.

A molecule that can give up a hydrogen ion is called an **acid.** A molecule that can accept a hydrogen ion from an acid

is called a **base.** An acid dissolved in water contributes hydrogen ions to water molecules and raises the concentration of hydrogen ions, which decreases the pH. Conversely, bases dissolved in water result in a decreased concentration of hydrogen ions and increased pH. Bases, then, have pH values higher than 7.

Water molecules split into hydrogen ions and hydroxide ions. The exact balance between the two kinds of ions determines the pH (the relative acidity) of the solution.

Why Is pH Important to Organisms?

Changing the pH of a solution affects the properties of other molecules. For example, the characteristic sour taste of acids results from the interaction of hydrogen ions with molecules on the tongue that report taste to the brain. Similarly, basic

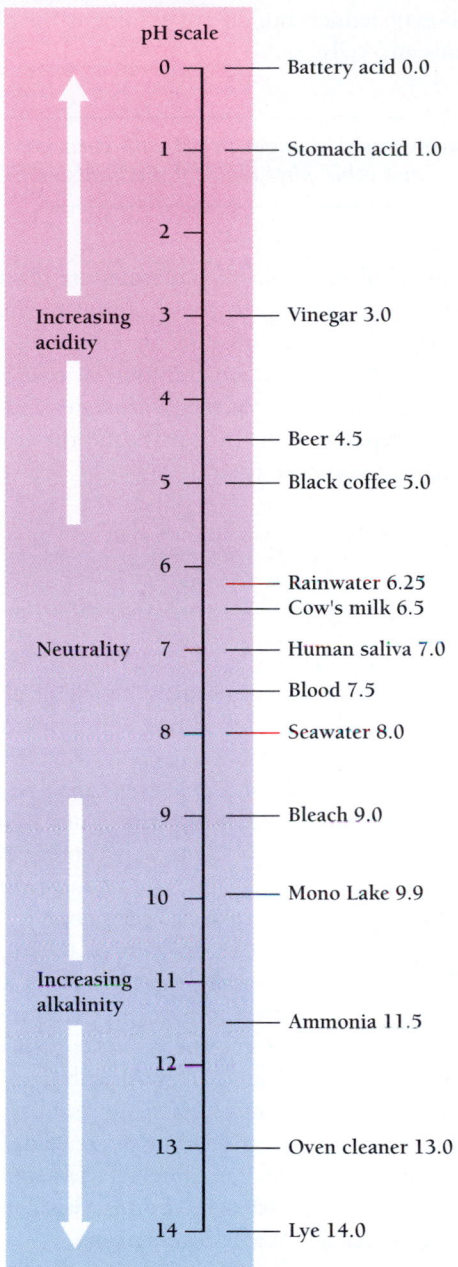

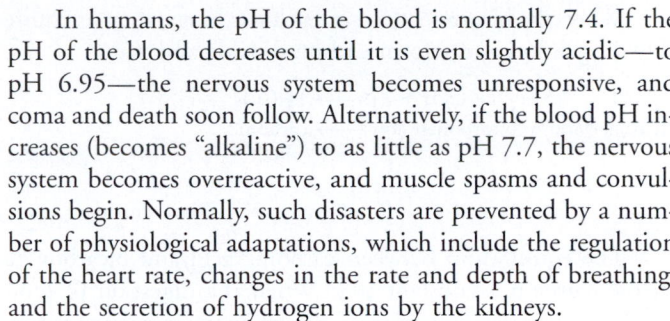

Figure 2-20 The pH scale. The pH scale measures acidity and alkalinity. Pure water, with a pH of 7.0, is considered neutral. Anything above 7.0 is alkaline and anything below 7.0 is acid.

In humans, the pH of the blood is normally 7.4. If the pH of the blood decreases until it is even slightly acidic—to pH 6.95—the nervous system becomes unresponsive, and coma and death soon follow. Alternatively, if the blood pH increases (becomes "alkaline") to as little as pH 7.7, the nervous system becomes overreactive, and muscle spasms and convulsions begin. Normally, such disasters are prevented by a number of physiological adaptations, which include the regulation of the heart rate, changes in the rate and depth of breathing, and the secretion of hydrogen ions by the kidneys.

However, a wide variety of illnesses, injuries, and drugs can cause acid or alkaline blood. One of the most common causes of acid blood, for example, is severe diarrhea. Basic blood can result when a physician gives a patient excessive amounts of diuretics (drugs that cause the body to eliminate large amounts of water and hydrogen ions) or if a person takes too much bicarbonate of soda for an upset stomach.

Changes in pH do not have to be dangerous. Many organisms use pH to control normal processes. For example, a sperm in the testes remains motionless, saving energy for the time when it is near an egg. Yet something must activate the sperm when the moment is right. In sea urchins, that something is a decrease in pH. The pH of the sea urchin's testes and sperm is about 7.2, very slightly basic. When the sea urchin releases its sperm into seawater, which has a pH of about 8.0, the sperm become more basic (about 7.6), which activates the sperm. Off they swim.

Both inside and outside of cells, pH powerfully influences the chemistry of life.

Buffers: How Do Organisms Resist Changes in pH?

Because pH dramatically affects both the structure and the chemical reactivity of most biologically important molecules, cells have mechanisms that maintain nearly constant pH. The interior contents of cells almost always have a pH of about 7.

Organisms use both chemical and physiological devices to maintain constant pH. The chemical strategy is to employ **buffers**—molecules that easily interconvert between acidic and basic forms by donating or accepting hydrogen ions. In human blood, the main buffer is bicarbonate (which is in chemical equilibrium with carbonic acid).

Many chemical reactions in organisms involve exchanges of hydrogen ions. For example, when a person exercises, the muscles produce carbon dioxide (CO_2), which the blood carries to the lungs. Most of the carbon dioxide in the blood, however, combines with water to form carbonic acid (H_2CO_3), which tends to break down into a hydrogen ion and a bicarbonate ion (HCO_3^-):

$$CO_2 + H_2O \longrightarrow H_2CO_3 \longrightarrow$$
carbon dioxide + water $\longrightarrow$ carbonic acid

$$H^+ + HCO_3^-$$
hydrogen ion + bicarbonate ion

solutions feel slippery because they change the characteristics of molecules on the skin.

When a molecule donates or accepts a hydrogen ion, its net charge changes. Such changes alter a molecule's interactions with other molecules and ions. As a result, pH dramatically affects both the structure and the chemical reactivity of most biologically important molecules. (For information on organisms that can withstand extremes of pH, see Web Bit 2–2: Life at Low pH.)

When the bicarbonate ions reach the lungs, they reacquire a hydrogen ion and reform carbon dioxide and water:

$$H^+ \quad + \quad HCO_3^- \quad \longrightarrow \quad H_2CO_3 \longrightarrow$$

hydrogen ion + bicarbonate ion $\longrightarrow$ carbonic acid

$$CO_2 \quad + H_2O$$

carbon dioxide + water

The equilibrium between carbonic acid and bicarbonate helps maintain a constant pH. When the digestion of food produces extra hydrogen ions, bicarbonate ions can absorb them by becoming carbonic acid, and the pH of the blood changes little. Similarly, hydrogen ions provided by carbonic acid can "soak up" hydroxide ions added to the blood. In both cases, the presence of the buffer minimizes changes in the total concentration of hydrogen ions. Buffers such as carbonic acid/bicarbonate reduce, but do not eliminate, the pH changes of organisms and cells.

Organisms resist changes in pH with chemical buffers and other physiological mechanisms.

The biological functions of molecules are determined by their structures. Since water is the universal medium of life, the attraction or avoidance of water by molecules or parts of molecules is the basis of the organization of cells and organisms. The interactions of molecules with water and with each other, in turn, depend on the properties and the arrangements of the atoms that compose them.

Study Outline with Key Terms

Chemistry is the study of the properties and transformations of matter. All **matter** is composed of **atoms.** An atom is the smallest particle of an element that has the properties of that element. **Elements** are forms of matter that cannot be broken down to simpler substances. **Compounds** can be broken down into elements. Most matter consists of **mixtures** of compounds.

Compounds consist of **molecules,** which are fixed arrangements of two or more atoms, with a characteristic **molecular weight.** The atoms of each element have characteristic properties that determine what kinds of molecules can be formed. **Chemical reactions** change the arrangements of atoms in molecules.

Atoms are made of still smaller particles. Almost all of an atom's **mass,** or weight, is contained in its **nucleus.** An atom's nucleus contains both protons and **neutrons. Nuclear reactions** change the number of protons and neutrons in a nucleus.

Typical atoms are electrically neutral; the positive charge of the **protons** in the nucleus is balanced by an equal number of negatively charged **electrons.** Electrons are outside the nucleus and occupy most of the space of the atom.

The atoms of an element all have the same atomic number, but they may have different mass numbers, that is, different numbers of neutrons. The different kinds of atoms of a single element are called **isotopes. Radioactive isotopes** spontaneously disintegrate, with a characteristic **half-life,** giving off radioactive particles that can be detected photographically or electronically.

Electrons are arranged in **orbitals,** each of which can hold a different number of electrons. Groups of orbitals, in turn, form **shells,** which differ in **energy.** The chemical properties of an atom are largely determined by the number of electrons in the outermost shell.

In forming chemical bonds, electrons may be rearranged in two distinct ways: (1) by sharing with other atoms to form **covalent bonds,** or (2) by donating or accepting electrons to form **ionic bonds.** Carbon atoms, unlike most other atoms, can form **double,** or even **triple,** covalent bonds. Individual atoms have different tendencies to gain or to lose electrons. Atoms that are most apt to gain electrons are said to be the most **electronegative.** Differences in the electronegativity of atoms in a molecule determine the extent to which the electrons in a chemical bond are equally distributed. Molecules with uneven distributions of electrons are said to be **polar;** those with uniform charge distributions are said to be **nonpolar.** Individual molecules may interact via several types of relatively weak forces—**hydrogen bonds, hydrophobic interactions,** and **van der Waals interactions.**

Water is the most abundant molecule in all organisms, and its unusual properties are especially suited for its role in life. Among these properties are its **cohesiveness** (which leads to water's strong **surface tension**), its higher density as a liquid than as a solid, its high **heat capacity,** its **adhesiveness** (which leads to water's **capillary action**), its ability to serve as a powerful **solvent,** its ability to absorb heat, and its participation in many biochemical reactions.

The properties of water result from its polarity. The polarity of water permits interaction with **ions** and polar molecules in **aqueous** solutions. **Hydrophilic solutes** readily dissolve in water. Nonpolar molecules do not dissolve readily in water, and they are said to be **hydrophobic.** Many biologically important molecules are **amphipathic,** containing both hydrophilic and hydrophobic regions. **Detergents** are amphipathic molecules that allow hydrophobic molecules to dissolve in water.

The hydrogen nuclei that participate in hydrogen bonds can also jump between water molecules. This ability leads to water's high heat capacity.

A molecule that can donate a **hydrogen ion** to another molecule is called an **acid.** A molecule that can accept a hydrogen ion from an acid is called a **base.** Some of the time,

the jumping hydrogen ions create **hydroxide ions** and **hydronium ions,** which lie in **chemical equilibrium** with water. Water is thus both an acid and a base. The **pH** scale measures the **concentration** of H^+ and OH^- ions in a solution. Changes in pH alter the charge on molecules and affect their chemical and physical properties. The fluids of organisms contain **buffers,** which resist changes in pH. Cells and organisms also have physiological adaptations that maintain nearly constant pH.

Review and Thought Questions

Review Questions

1. Define the words "element," "atom," "compound," and "molecule."
2. What is an isotope? What is meant by the term "radioactive isotope"? Why are only some isotopes radioactive?
3. What is an electron orbital?
4. Explain the difference between a covalent bond and an ionic bond. How is each formed? Name a common compound that exhibits each type of bond.
5. Explain how a covalent bond can be polar or nonpolar. Give examples of each type of bond. Name a polar molecule that is of great importance to organisms and their environments.
6. Describe each of the following: hydrogen bonds, hydrophobic interactions, and van der Waals interactions. Which of these reactions are likely to occur between (1) polar compounds or (2) nonpolar compounds?
7. Define "cohesion" and "surface tension." What is the relationship between these two properties of water?
8. How does adhesion differ from cohesion? What is a capillary?
9. Give examples of polar and ionic compounds that dissolve easily in water as well as nonpolar ones that do not. Which type of molecule would you use to make waterproof bags?

10. What type of bond is formed between adjacent water molecules? Why do they cohere? Is this more similar to a covalent or ionic bond?
11. What types of bonds tend to produce (1) hydrophilic substances or (2) hydrophobic ones?
12. What is a hydrogen ion? If a hydrogen ion is formed from a water molecule, what other ion is also formed? If a hydrogen ion is formed idiot from HCl, what other ion is also formed? How is the pH of water defined?

Thought Questions

13. Explain how the results of Rutherford's experiment caused him to propose that the nucleus of an atom was very small and dense and that most of the atom was nearly empty space.
14. The tendency of living things to remain in a narrow range of temperatures is part of homeostasis—one of the characteristics of living things. How do the properties of water help organisms regulate temperature? Consider organisms that live in water, then organisms, such as humans, that live outside water.

About the Chapter-Opening Image

Alfred Nobel invented a way of packaging highly explosive nitroglycerin into dynamite, but he would not take nitroglycerin as a medicine for his heart condition. We now know that nitroglycerin, as a medicine, releases nitric oxide (NO), which dilates blood vessels.

 On-line materials relating to this chapter are on the World Wide Web at **http://www.harcourtcollege.com/lifesci/aal2/**

Biological Molecules Great and Small

Prions: A Slow-Growing Idea

In July of 1972, a young medical resident at a San Francisco hospital admitted an elderly woman whose memory was so poor that she was having difficulty doing everyday chores. The resident was 30-year-old Stanley Prusiner, who would someday win a Nobel prize for his work on the kind of disease she suffered from. But that was far into the future. Prusiner soon learned there was nothing he could do to help the woman.

She had a rare illness of the brain called Creutzfeldt-Jakob disease. No one knew what caused it or how to cure it. Her body's defenses mounted no attack against it. After only 7 weeks, she died. An autopsy showed that her brain was riddled with holes, like a sponge. Intrigued, Prusiner resolved to read everything he could about the mysterious killer. He soon learned that it was one of a group of fatal brain diseases known as transmissible spongiform encephalopathies (TSEs).

One TSE, called kuru, was killing people in Papua New Guinea who had honored the dead by eating their brains. Another, called scrapie, damaged the brains of goats and sheep. Scrapie, Prusiner learned, caused sheep to lose coordination and scratch themselves so viciously that they scraped off their wool.

What caused these different forms of TSEs? One clue was that tissues from an animal with a TSE could be used to infect another animal. From this, biologists concluded that some infectious agent—perhaps a bacterium or a virus—was causing the disease. Most researchers concluded that TSEs were caused by "slow viruses," which, like the AIDS virus, take a long time to produce symptoms. But no one had been able to isolate the presumed virus. Prusiner soon encountered a radically different hypothesis about TSEs—an idea that slowly took over his life (Figure 3-1).

In 1967, radiobiologist Tikvah Alper had published a scientific paper on scrapie so controversial that scientists around the world buried her in mail. Indeed, the number of letters reached the point that mail room employees at Hammersmith Hospital, in London, refused to deliver any more. They told Alper to come get the bags of mail herself.

Alper's results implied that scrapie couldn't be caused by a virus or any known organism. Her experiments had shown that tissues infected with scrapie could infect animals even after apparently all DNA and RNA had been destroyed by radiation. Viruses, like bacteria and other organisms, rely on the nucleic acids DNA and RNA to carry genetic information. To destroy their DNA or RNA was to destroy their ability to infect cells.

Alper proposed that whatever was causing scrapie had no genes, or didn't need them. But it was heresy to suggest that a virus or any infectious microbe could exist and reproduce without genes. And it was heresy to suggest that anything else besides a microbe or a virus could infect the body and spread.

Alper knew what her experiments had shown, however, and she was not one to shy away from controversy. When at first her paper was rejected for publication, she sent an outraged letter to the editor of the magazine and convinced him to revoke his decision. "Tikvah is remembered by most people for her forceful discussion style, which could be quite alarming," two of Alper's colleagues later wrote in the *International Journal of Radiation Biology*. Indeed, at 79, Alper sent a burglar fleeing from her home after tackling him and grappling for his gun.

John Wilkes/Octopus Photos

Key Concepts

- Organisms build most structures from just four molecular building blocks: lipids, sugars, amino acids, and nucleotides.

- Functional groups of atoms determine the chemical properties of each building block.

- Dehydration reactions, which remove the equivalent of one molecule of water, join small building blocks into macromolecules.

- The three-dimensional structure of a protein depends on interactions among the functional groups of its amino acid building blocks.

But as much as Alper would have liked to pursue her work on scrapie and convince her colleagues of her results, she was unable to do so. The Agricultural Research Council of Great Britain refused to fund additional research. The council thought scrapie was of little financial importance. Disappointed but not defeated, Alper turned her attention to her many other interesting research projects.

Later events showed that the council's decision to cut funding for scrapie research was a mistake. In the 1980s, an

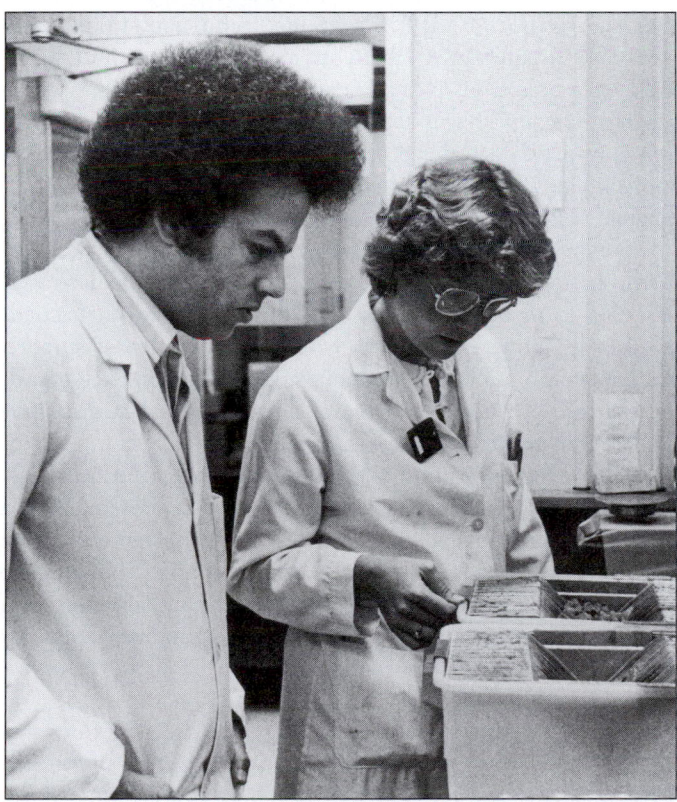

Figure 3-1 Stanley Prusiner. Prusiner's research on the brain disease scrapie has persuaded most scientists that the way proteins fold can carry biological information and even cause disease. Yet despite Prusiner's Nobel prize for this remarkable work, some biologists still doubt that renegade proteins cause scrapie. (*Courtesy of Stanley Prusiner, UCSF*)

outbreak in England of the scrapie-like TSE called "mad cow disease" shut down the English beef industry. Nearly 2 million beef cattle and other farm animals had to be slaughtered at a cost of more than $6 billion. Despite these precautions, more than 40 people have so far died of mad cow disease, apparently from having eaten tainted beef years earlier.

Like scrapie itself, however, Alper's discovery would not go away. There was, after all, a possible explanation for her idea that neither virus nor bacterium caused the disease. Shortly after her article was published, J.S. Griffith, of Bedford College in London, proposed that proteins could sometimes fold into unusual shapes and then coax normal proteins into similar misfoldings. Could scrapie proteins cause disease by forcing the proteins in the brains of sheep to misfold? Almost no one thought so.

In 1974, Stanley Prusiner had just begun working as an assistant professor at the University of California at San Francisco (UCSF). Inspired by his early brush with Creutzfeldt-Jakob disease, he had chosen scrapie as his research subject. Despite Alper's findings, Prusiner assumed that the cause was some sort of virus.

But as Prusiner began purifying and testing extracts from scrapie tissues, the evidence gradually forced him to change his mind. The more he repeated Alper's experiments and refined them, the more he came to believe that she must be right. When he damaged the nucleic acids in scrapie extracts, they continued to infect animals. But when he denatured—or unfolded—scrapie's proteins, the extracts lost their ability to infect. These results could only mean that the infectious agent was a protein. But for something to cause an infection, it must make more of itself. And everyone knew that proteins were incapable of reproducing themselves.

Even as Prusiner became more fascinated by his findings, the agency that had supported his work cut his research grant. Almost simultaneously, the university informed him that he would not be given tenure. Prusiner was devastated. He was losing his job, his lab, and the grant he needed to pay for his research. As he later wrote, "Everything seemed to be going wrong, including the conclusions of my research studies."

Yet the moral support of a few of his closest colleagues carried him through. Skilled at selling his ideas, Prusiner soon

the group determined the order of amino acids in the PrP protein. Prusiner and other colleagues then showed that PrP could fold up in at least two ways. One form was *scrapie PrP,* found in the brains of sick sheep, and the other was *cellular PrP,* a protein made in the cells of healthy animals. When Prusiner and Fred Cohen, also of UCSF, compared the shapes of cellular and scrapie PrP, they discovered an important difference. Where the normal, cellular PrP was twisted into spirals called "α helices" (alpha helices), scrapie PrP was folded into parallel strands called "β pleated sheets" (beta pleated sheets), as shown in Figure 3-2.

Was it possible that scrapie PrP was compelling the body's own PrP to misfold and become scrapie PrP? Prusiner and Cohen hypothesized that scrapie PrP could turn the α helices of cellular PrP into β sheets. Working with short pieces of PrP, they found that β sheets could indeed nudge α helices into becoming β sheets. Amazingly, other biologists were even able to transform full-length cellular PrP into scrapie PrP by mixing cellular PrP and scrapie PrP together.

These astonishing experiments brought to light a completely new phenomenon, the ability of a protein (scrapie PrP) to alter the three-dimensional structure of another protein. The experiments shattered the established view that genes alone determine the three-dimensional structures of proteins and carry biological information. Proteins too could convey information.

On October 6, 1997, a young technician walked into Prusiner's usually quiet lab to start his day's work and stumbled on a boisterous party. "People were shouting and drinking champagne. They practically tackled me," he recalled. After more than two decades of struggling against the scientific establishment, Prusiner had won one of the highest honors in science, a Nobel prize.

For more than 20 years, Prusiner had probed the enigmatic nature of the scrapie agent. Like Alper, Prusiner faced disbelief from his colleagues and difficulties getting funded. Ultimately, however, Prusiner uncovered what most agree is the explanation underlying scrapie's odd nature. In the process, he changed our understanding of how biological information can be transmitted.

Following a winding path of experiments, Prusiner realized that scrapie's secret lay in the folding of a protein. By coercing proteins into adopting altered shapes, scrapie proteins can carry and transmit biological information free of nucleic acids. But even today, many questions remain. Like most scientific adventures, the story keeps evolving and reshaping itself as scientists add new findings to their growing collection of clues.

Some nagging questions still lingered when Prusiner was awarded the Nobel prize for discovering a new class of disease-causing agents and for showing how they could infect an organism. Most importantly, no one had yet infected an animal with lab-made PrP. Some researchers continue to insist that the real cause of TSEs could be viruses that are exceptionally good at protecting and repairing their damaged nucleic acids. Scrapie comes in different forms, and some biologists doubt that different shapes of a single protein can explain the different strains

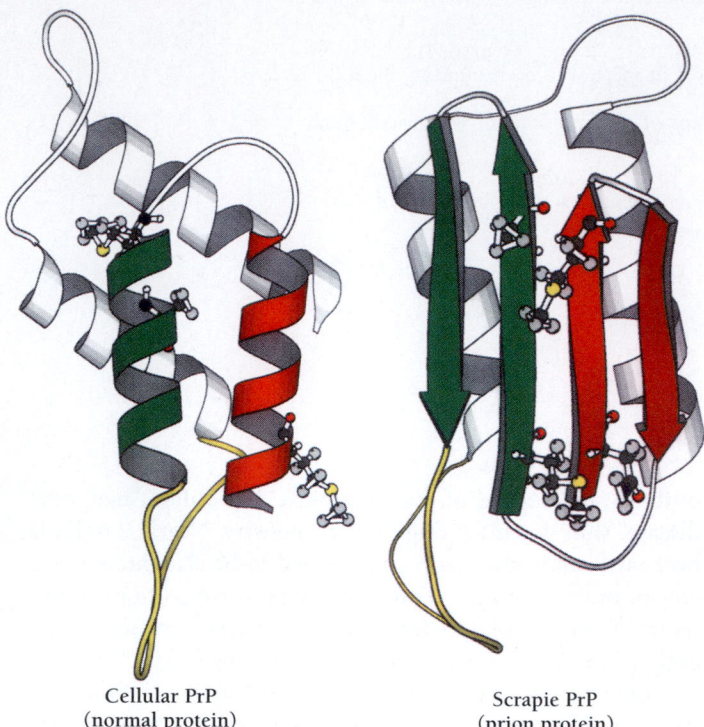

Cellular PrP
(normal protein)

Scrapie PrP
(prion protein)

Figure 3-2   An α helix and a β sheet. In the prion protein PrP, the difference between an α helix and a β sheet can spell the difference between life and death. When regions of the protein that are normally α helices *(left)* fold into β sheets *(right),* the result may be a fatal disease of the brain. *(Courtesy of Stanley Prusiner, UCSF)*

convinced two private foundations to fund his work. And, amazingly, even his tenure decision was reversed. He would be allowed to stay at UCSF.

By 1982, Prusiner still lacked a convincing explanation for his data. Nonetheless, he boldly proposed that scrapie was transmitted by infectious proteins he called *prions* (pronounced "pree-onz"). Many scientists said his claim was premature, even brazen.

Former allies turned against him. Frank Masiarz, a researcher who had worked closely with Prusiner, quit. As Masiarz told *Discover* magazine in 1986, "There's no point in creating a name for something that we don't know even exists yet. . . . [Prusiner] tends to jump to conclusions, to give less credence to facts that would discount his preferred interpretation of the results." Feeling betrayed, Prusiner refused to talk to the press for a decade after the publication of the *Discover* article.

But like Tikvah Alper, Prusiner was determined to change the minds of those around him. His research team succeeded in isolating the prion protein, which they named PrP. The next step was to describe its structure. All proteins are made of long chains of 20 small molecules called amino acids that are strung together like beads on a string. These long chains fold into the complex shapes of functioning proteins.

Working with Leroy Hood of the California Institute of Technology and Charles Weissman of the University of Zurich,

of scrapie. Such biologists claim that Prusiner's Nobel prize was premature at best and an outright error at worst.

Even Tikvah Alper has disagreed with Prusiner. In 1993, Alper began to revisit her early work on scrapie. Only 2 weeks before her death, in 1995, she was still writing to scientific journals. She argued that the scrapie agent was neither a protein nor a virus, but a rogue piece of membrane from a cell. But few people have taken her suggestion seriously.

The study of protein folding is now one of the most exciting fields in biology. Understanding how and why proteins fold the way they do is critical to understanding the physiology and genetics of cells and whole organisms. Yet, to understand what causes a protein to fold, we need to understand the chemistry of the smaller molecules that make up proteins. In this chapter, we look at the way cells build small building blocks into larger molecules (Web Bit 3-1: Mad Cows and Promiscuous Fungi).

HOW DO ORGANISMS BUILD BIOLOGICAL MOLECULES?

If we measure the sizes of all the molecules in a cell, we discover a surprising thing. Cells have many small molecules, with molecular weights less than 300, and many large molecules, with molecular weights greater than 10,000. But cells have very few molecules with intermediate sizes.

How Big Are Biological Molecules?

Biologists call large molecules **macromolecules** [Greek, *macro* = large]. Macromolecules consist of small molecules linked together in long chains of small building blocks.

Nearly all molecules found in living organisms share one important trait. They are made of chains of carbon atoms. In fact, living organisms are associated so closely with carbon that all carbon-and-hydrogen-containing compounds are called **organic** molecules (even when they have nothing to do with organisms). **Organic chemistry** is the study of the structures and reactions of carbon-and-hydrogen compounds. **Biochemistry** is the study of the chemistry of living organisms.

The carbon-based building blocks of life are simple and universal. Even though there are a nearly infinite number of carbon-containing molecules, organisms use only a few. We are all of us made of these few building blocks. In this chapter, we will discuss 35 small molecules that together form almost all biological structures.

The 35 building blocks fall into four classes: sugars, amino acids, nucleotides, and lipids. Sugars join together to form long-chain macromolecules called polysaccharides, like the starch in bread and potatoes. Amino acids form polypeptide chains, which make up proteins.

Nucleotides form long chains called nucleic acids, such as the DNA that our genes are made of. Lipids, which include fats and oils, do not form long chains. However, they do form sheets of thin membrane. Along with other molecules, lipid

membranes form the membranes that surround all cells and regulate their interactions with their environment.

All organisms are made of the same four kinds of molecules: sugars, amino acids, lipids, and nucleotides.

Why Are Biological Structures Made from So Few Building Blocks?

Given the enormous diversity of organic molecules, the relatively small number that are used by organisms comes as a surprise. But why should there be so few building blocks? For example, Why do almost all organisms use the sugar glucose as their principal energy source? And why do all of the millions of species on this planet use the same 20 amino acids to make proteins?

The most likely explanation for this sameness is that we are all related. We all use the same building blocks because all life on Earth has a common origin. The same molecules are used again and again because organisms inherit from their ancestors the tools to use these few building blocks.

All living organisms are made of the same few small molecules.

What Kinds of Structures Do Carbon Atoms Form?

Because all the building blocks of life include carbon, we begin by discussing carbon. Carbon atoms often form long chains of atoms. Each carbon bonds to the next by single or double covalent bonds. Each carbon atom can bond with up to four other atoms.

Along the sides of a simple carbon chain, hydrogen atoms fill the remaining slots. Such long chains of carbon and hydrogen are called **hydrocarbon chains.** One simple hydrocarbon chain is octane, a component of gasoline, which consists of 8 carbon and 18 hydrogen atoms (Fig. 3-3A). Most of the 35 building blocks are based on hydrocarbon chains.

Hydrocarbon chains can also form rings. Because carbon can form covalent bonds only at certain angles, most carbon rings have either five or six atoms. A ring sometimes includes other types of atoms, often oxygen or nitrogen.

When the atoms of a ring are connected by alternating single and double bonds, the rings lie flat, like the benzene ring shown in Figure 3-3B. Such flat rings belong to a special group of organic molecules called **aromatic** compounds.

Aromatics include, for example, acetylsalicylic acid (aspirin), benzene (an industrial solvent), and dimethylpyrazine (a component of chocolate). Some aromatic compounds consist of carbon atoms only, and some include one or two nitrogen atoms in their rings. Many are quite literally aromatic,

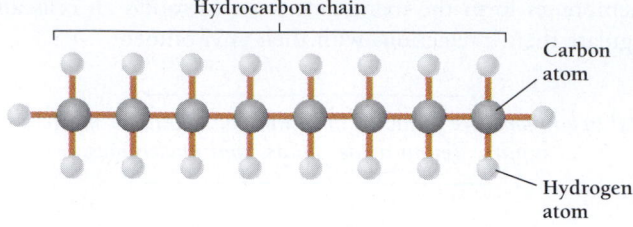

A. Octane (C$_8$H$_{18}$)

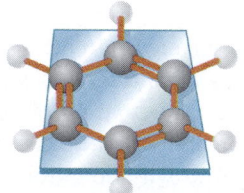

B. Benzene (C$_6$H$_6$)

Figure 3-3   A chain of carbon atoms forms the skeleton of many building blocks. A. Octane, a component of gasoline, has 8 carbons, with 18 hydrogen atoms hanging off the sides. Inside a working automobile engine, carbon-carbon bonds in octane are first broken, using the energy supplied by the spark plug. Following this, new bonds form with oxygen, releasing stored potential energy. B. Many building blocks consist of aromatic rings, like that of benzene. Aromatic rings have alternating double bonds that make them lie flat.

including such strong-smelling compounds as the oil that makes almonds bitter, the quinine in tonic water, and the benzene, toluene, and xylene that together give high-performance gasoline its zing. Because many of the building blocks of life are aromatics, they are also planar. Such aromatics include the nucleotides that make up our DNA.

Not all carbon rings lie flat. When the atoms of a ring are connected by single covalent bonds (instead of double bonds), the ring can bend into a structure that looks a little like a chair or a boat (Figure 3-4).

Carbon atoms form long hydrocarbon chains and five- or six-carbon rings.

Functional Groups: How Do the Small Molecules Differ from One Another?

The structure of a molecule determines its biological function. Just as the parts of a car or a bicycle must fit together for the machine to work, the molecules in a cell must also fit together or interact for the cell to work. Every molecule has a distinctive shape and charge distribution. And every molecule interacts with water and other molecules in unique ways.

The special properties of a molecule depend on the arrangement of its atoms. In analyzing the structures of many organic molecules, chemists have repeatedly found standard small groupings of atoms called **functional groups,** which help predict the properties of a molecule.

Functional groups are comparable to the standardized parts of a Swiss Army knife, which may include, for example, a knife, a corkscrew, and a can opener (Figure 3-5). A Swiss Army knife may have only one or two such attachments or it may have many, in a variety of combinations. In the same way, a biological molecule may have a few functional groups or many, in a variety of combinations. The exact combination of functional groups on a biological molecule profoundly affects the way the molecule behaves.

Figure 3-5 shows the structures of four simple molecules: ethane, ethanol, acetic acid, and aminoethane. Each molecule contains two carbon atoms and different numbers of hydrogen atoms. Of these molecules, ethane has no functional groups, while each of the other three molecules has a different functional group. These are a hydroxyl group, a carboxyl group, and an amino group.

Ethane has the simplest structure: two carbon atoms, each with three hydrogens attached. It is an odorless gas that makes

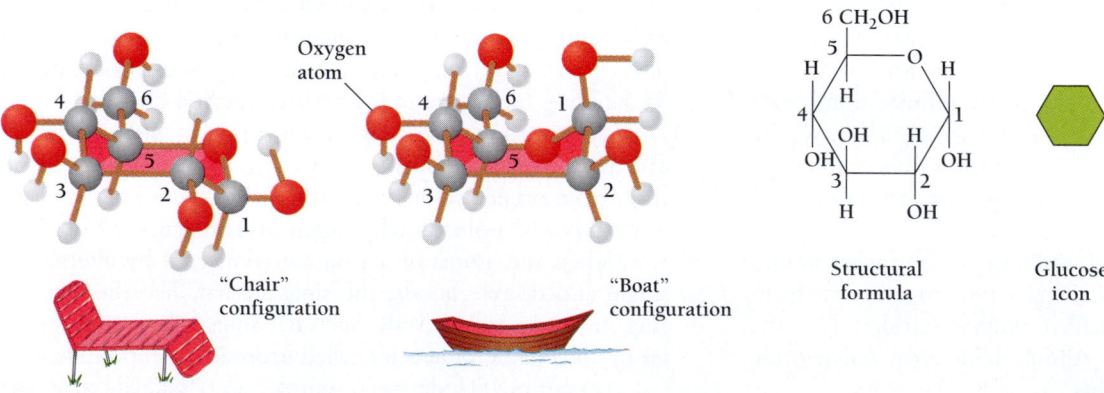

Glucose molecule (C$_6$H$_{12}$O$_6$)

Figure 3-4   Most rings bend. Like benzene, the simple sugar glucose consists of a ring of atoms. But while benzene's double bonds keep it flat, glucose's single bonds easily twist around, allowing the molecule to fold into a "chair" structure *(left)* or a "boat" structure *(right).* The green hexagon is our icon for glucose.

up about 9 percent of natural gas, the kind we use to heat homes (Figure 3-5A). In high concentrations, ethane has a narcotic effect, but it is not particularly toxic.

Ethanol—the kind of alcohol found in beer, wine, and liquor—looks like ethane except that in place of one hydro-

gen atom is a **hydroxyl,** or OH group (Figure 3-5B). That single oxygen atom makes ethanol very different from ethane. Ethanol (or ethyl alcohol) is a sweet-smelling liquid that is comparatively toxic. It is used as an antiseptic to kill both prokaryotic and eukaryotic cells. Ethanol is also used as an industrial solvent, as an additive in gasoline, and, in veterinary medicine, to kill nerve tissue.

Ethanol's most familiar use is in beverages such as beer and liquor. In small amounts, it gives a pleasant high for a short period. In large amounts, ethanol causes nausea, vomiting, mental excitement or depression, loss of coordination, stupor, coma, and even death.

The hydroxyl group in alcohol is an extremely common functional group, which easily forms hydrogen bonds. Because oxygen is more electronegative than hydrogen, electrons in a hydroxyl group spend more time near oxygen than hydrogen. This leaves the hydrogen with a slight positive charge, so it readily forms a hydrogen bond with the slightly negative oxygen atom in a water molecule. As a result, ethanol is extremely hydrophilic and dissolves readily in water. In contrast, ethane, which has no hydroxyl group, does not interact much with water. Ethane bonds are all nonpolar, and ethane molecules do not interact much, even with one another. Most molecules that contain an OH group attached to a carbon atom—including other alcohols and also sugars—have properties similar to ethanol's.

We can take ethane and lop off one carbon with its three hydrogens and substitute a **carboxyl** group (—COOH). The result is acetic acid, the acid that gives vinegar its familiar sour smell and sharp bite (Figure 3-5C). Just as the OH group characterizes alcohols and sugars, the carboxyl group characterizes organic acids.

The carboxyl group is acidic because it gives up its hydrogen easily. Both of the two oxygens in a carboxyl group pull electrons toward them so strongly that the hydrogen atom actually loses the electron that binds it to the carboxyl group. The hydrogen ion breaks away, leaving behind a COO^- ion.

Suppose now that, instead of a hydrogen atom or a hydroxyl or carboxyl group, we substitute an **amino** group, a single atom of nitrogen bonded to two atoms of hydrogen (NH_2). An amino group tends to make molecules basic because it attracts hydrogen ions, forming NH_3^+ ions. Substituting an amino group for one of the hydrogen atoms in ethane creates aminoethane (Figure 3-5D). Like ethanol, aminoethane is a liquid, but it is otherwise very different. Highly alkaline, aminoethane's powerful vapors smell intensely of ammonia and can severely irritate the skin, eyes, and mucous membranes. Aminoethane is 20 times more toxic than ethanol.

Aminoethane belongs to a class of compounds called amines. Amines all have amino functional groups and nearly all smell terrible or at least unpleasant. Two of the most offensive amines are putrescene and cadaverine, which together give memorable odors to rotting fish, aging corpses, urine, semen, and bad breath. Table 3-1 shows the properties and structures of half a dozen of the most important functional groups. Many more functional groups exist that are not shown.

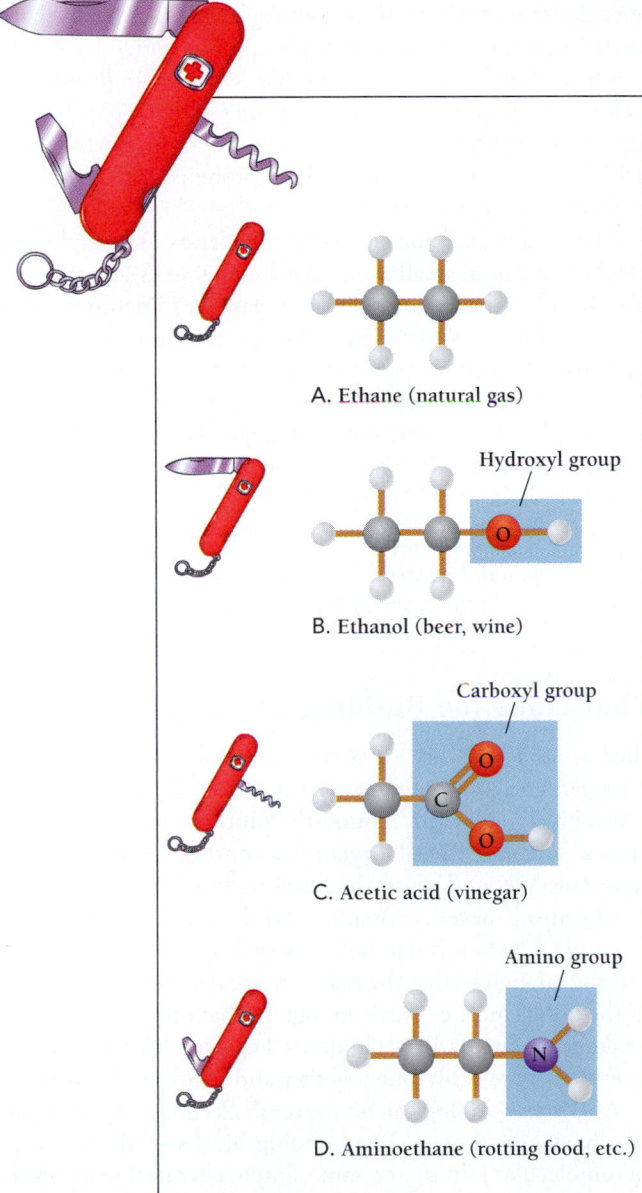

A. Ethane (natural gas)

Hydroxyl group

B. Ethanol (beer, wine)

Carboxyl group

C. Acetic acid (vinegar)

Amino group

D. Aminoethane (rotting food, etc.)

Figure 3-5 **Functional groups are like the different attachments on a Swiss Army knife.** Functional groups determine the structure and function of a molecule. A. In the nontoxic gas ethane, only hydrogen atoms attach to the two-carbon chain. B. In ethanol, a hydroxyl (OH) group transforms the carbon chain into an alcohol, a toxic liquid. C. In acetic acid, a carboxyl (COOH) group transforms the two-carbon chain into an organic acid. D. In aminoethane, an amino group (NH_2) makes the molecule alkaline, or basic. Like many other amines, aminoethane is strong smelling and toxic.

TABLE 3-1   FUNCTIONAL GROUPS

Functional Group	Ball and stick model	Structural formula
Hydroxyl group	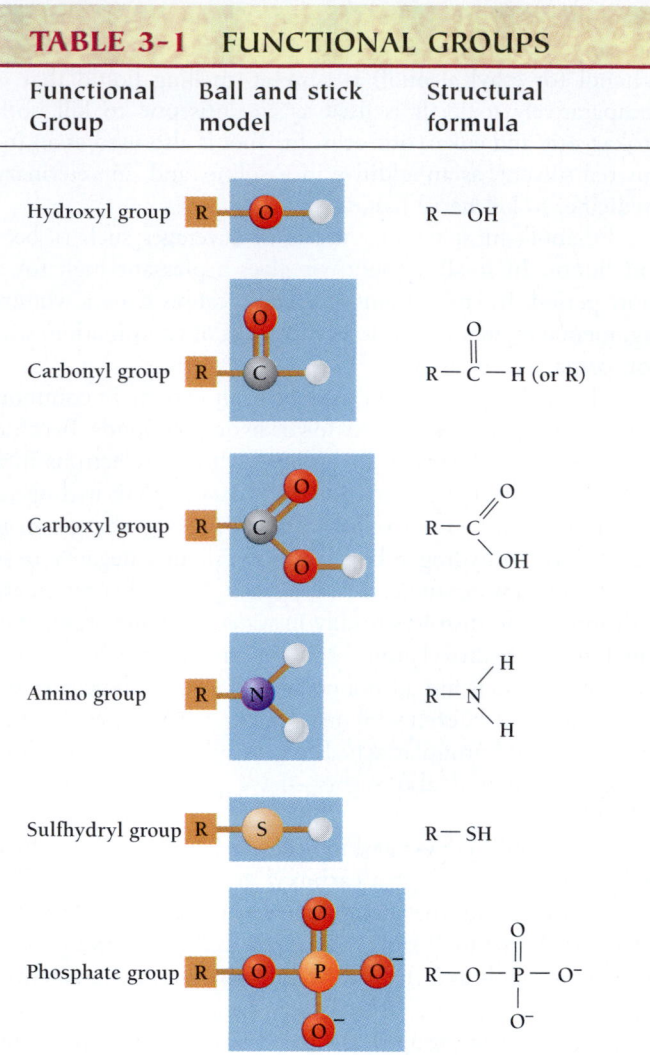	R — OH
Carbonyl group		$\underset{\displaystyle R-\overset{\textstyle O}{\overset{\|}{C}}-H\ (or\ R)}{}$
Carboxyl group		$R-C\overset{\textstyle O}{\underset{\textstyle OH}{}}$
Amino group		$R-N\overset{\textstyle H}{\underset{\textstyle H}{}}$
Sulfhydryl group		R — SH
Phosphate group		$R-O-\overset{\textstyle O}{\underset{\textstyle O^-}{\overset{\|}{P}}}-O^-$

A small number of functional groups determine the properties of an array of biological molecules.

How Do Cells Build Complex Molecules?

We are now in a position to understand the properties of many organic molecules. Each building block has a carbon skeleton to which are attached hydrogen atoms or functional groups. These groups determine the properties of the molecule.

Some of a molecule's most important properties depend on how it interacts with water. Recall from Chapter 2 that whether a molecule is hydrophobic, hydrophilic, or amphipathic depends on the polarity of the molecule's bonds. Polar bonds have a positive charge at one end that can form hydrogen bonds with the negative oxygen in a water molecule. This is why polar molecules dissolve easily in water.

As we mentioned earlier, almost all the structures of a cell are built from combinations of the 35 small-molecule building blocks listed in Table 3-2. These building blocks fall into four classes: (1) **lipids,** fatty or oily nonpolar compounds that tend to dissolve in oils, but not in water; (2) **sugars,** molecules that have two hydrogen and one oxygen atom (water) for every atom of carbon; (3) **amino acids,** molecules that contain both amino and carboxyl groups; and (4) **nucleotides,** molecules that each consist of a nitrogen-containing aromatic ring, a sugar, and a molecule of phosphoric acid.

Of the four kinds of building blocks, three kinds link end to end in long chains, just like a child's plastic "pop beads." Sugars link together to form polysaccharides, nucleotides form nucleic acids, and amino acids form proteins. Only lipids do not form long chains.

In proteins and nucleic acids, the chains are simple, unbranched strings of small molecules (Figure 3-6A). In contrast, polysaccharide chains formed from sugars are sometimes highly branched (Figure 3-6B). Even though proteins and nucleic acids are unbranched, they form much more complex molecules than polysaccharides. Protein chains can coil and fold into complex three-dimensional shapes. As Stanley Prusiner showed with scrapie PrP, how a protein folds is crucial to the way it behaves in a living organism.

Cells build nearly all complex molecules from four types of building blocks.

What Links the Building Blocks of Life?

When we digest a slice of bread, we break its carbohydrates into sugars and its proteins into amino acids. At the same time, we also break down and rebuild the proteins of our own skin, muscles, and bones. All organisms continually break down macromolecules and reuse the building blocks.

Organisms have to assemble and disassemble macromolecules easily. The bonds that hold macromolecules together must be strong enough so that the macromolecules will not fall apart. But the bonds must be weak enough so that organisms can easily take them apart. Like children's Lego bricks, the building blocks of life are easily put together and easily taken apart.

Amazingly, biological building blocks snap together just as easily as Lego bricks. The building blocks of all the major macromolecules join by the same simple chemical reaction. In every case, enzymes remove two hydrogen atoms and one oxygen atom from between pairs of building blocks, forming a bond. Removing two hydrogens and an oxygen—the equivalent of a molecule of water—is called a **dehydration reaction** (Figure 3-7A).

The linking of two building blocks by a dehydration reaction is also sometimes called a "condensation" reaction. Amino acids are "condensed" into polypeptides in the same sense that milk is condensed by the removal of water. In each case, something is made more compact by removing water.

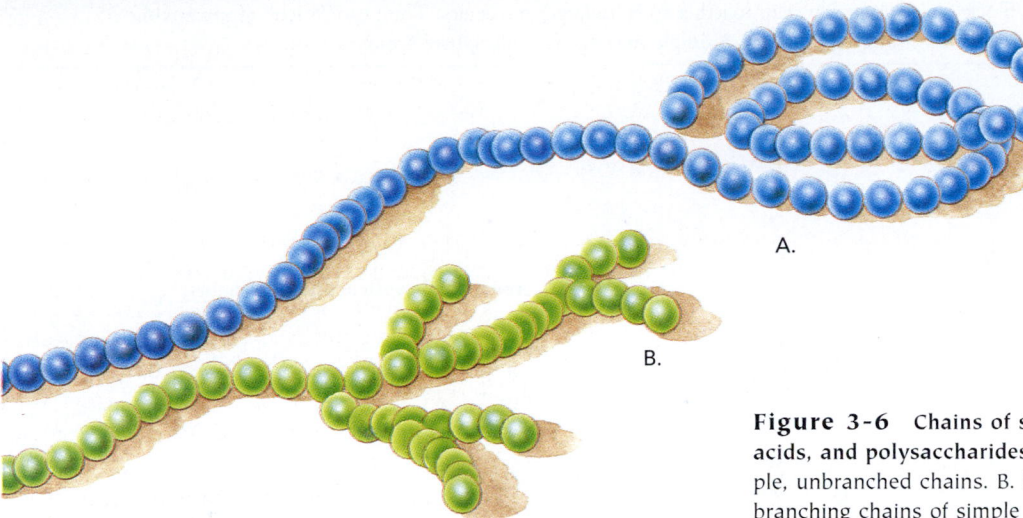

Figure 3-6 Chains of small molecules form proteins, nucleic acids, and polysaccharides. A. Proteins and nucleic acids are simple, unbranched chains. B. Polysaccharides are more complex branching chains of simple sugars.

To take apart macromolecules, cells reverse the dehydration process. Enzymes detach each small molecule from a macromolecule by adding one molecule of water, a process called **hydrolysis** [Greek, *hydro* = water + *lysis* = breaking] (Figure 3-7B).

Although all the building blocks are joined by similar dehydration reactions, the exact bonds that form are different in each case. For example, sugars form *glycosidic bonds,* while amino acids form *peptide bonds.*

Enzymes link two building blocks by taking away the equivalent of one water molecule. Adding a water molecule breaks a building block from a chain.

LIPIDS INCLUDE A VARIETY OF SMALL NONPOLAR COMPOUNDS

Look around your kitchen and you'll find lipids. Oils, fats, and candle wax, to name a few, are all lipids. Because lipids contain more chemical energy per gram than other biological molecules, lipids often serve as energy stores.

One characteristic of lipids is that they dissolve poorly in water but dissolve well in nonpolar solvents such as gasoline or olive oil. The reason that lipids do not dissolve in water is that they have few functional groups that are polar. The limited solubility of lipids in water explains why removing a butter or gravy stain often requires dry cleaning, that is, treatment with nonpolar organic solvents such as benzene.

But not all lipids are hydrophobic. Some are amphipathic: they contain a polar functional group as well as nonpolar chains of carbon and hydrogen atoms. An important class of amphipathic lipids is the **fatty acids**—a major component of the membranes that enclose cells.

Fatty acids differ from one another both in the number of carbon atoms they contain and in the numbers of single and double bonds. The hydrocarbon chain in a fatty acid consists of carbon atoms linked either to one or two other carbon atoms and to hydrogen atoms. When all the bonds between carbon atoms are single bonds, the hydrocarbon chain contains the maximum number of hydrogen atoms and is said to be **saturated**

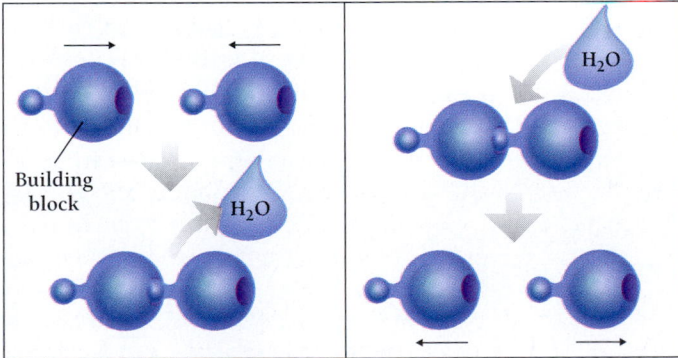

A. Dehydration   B. Hydrolysis

Figure 3-7 Dehydration and hydrolysis. Two simple reactions allow molecular building blocks to snap together or apart. A. In a dehydration reaction, building blocks join together when a hydrogen is lost from one molecule and a hydroxyl is lost from the other. Each time two building blocks join, the equivalent of one molecule of water is lost. B. In a hydrolysis reaction, the addition of one molecule of water breaks the bond between each pair of small molecules.

(Text continued on page 56)

TABLE 3-2   THE BUILDING BLOCKS OF LIFE

Like children's Lego™ bricks, building block molecules easily snap together to form larger molecules. The Lego™ icon represents the 35 building blocks of life—20 amino acids, 6 lipids, 5 nucleotide bases, 3 simple sugars, and a phosphate group.

Amino acid

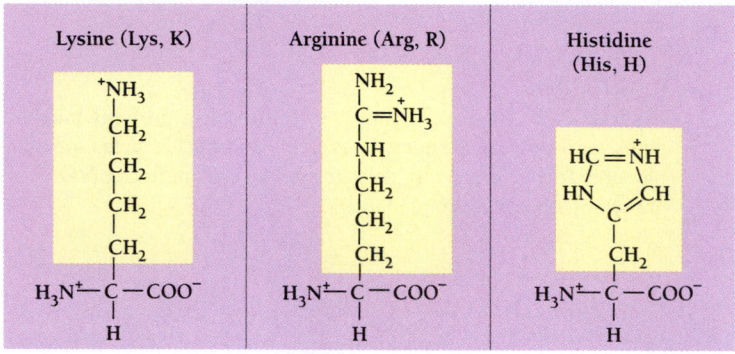

Amino acids with basic side chains

Lysine (Lys, K)   Arginine (Arg, R)   Histidine (His, H)

Amino acids with acidic side chains

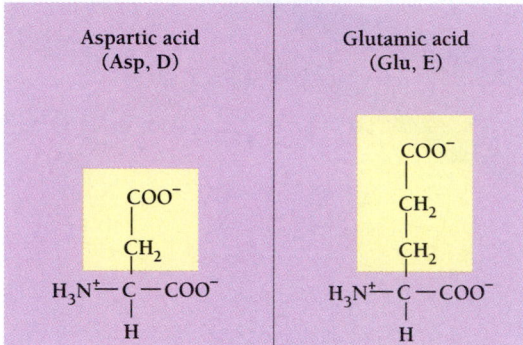

Aspartic acid (Asp, D)   Glutamic acid (Glu, E)

Amino acids with uncharged polar side chains

Asparagine (Asn, N)   Glutamine (Gln, Q)   Serine (Ser, S)   Threonine (Thr, T)   Tyrosine (Tyr, Y)

Amino acids with nonpolar side chains

Glycine (Gly, G)   Alanine (Ala, A)   Valine (Val, V)   Leucine (Leu, L)   Isoleucine (Ile, I)

Phenylalanine (Phe, F)   Methionine (Met, M)   Cysteine (Cys, C)   Proline (Pro, P)   Tryptophan (Trp, W)

Glucose

Ribose

Deoxyribose

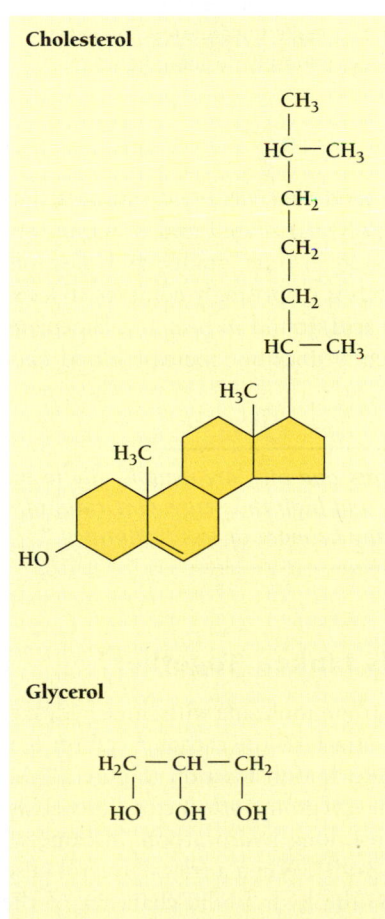

Bases

Pyrimidines (single ring)

Cytosine Uracil Thymine

Purines (two rings)

Adenine Guanine Phosphate group

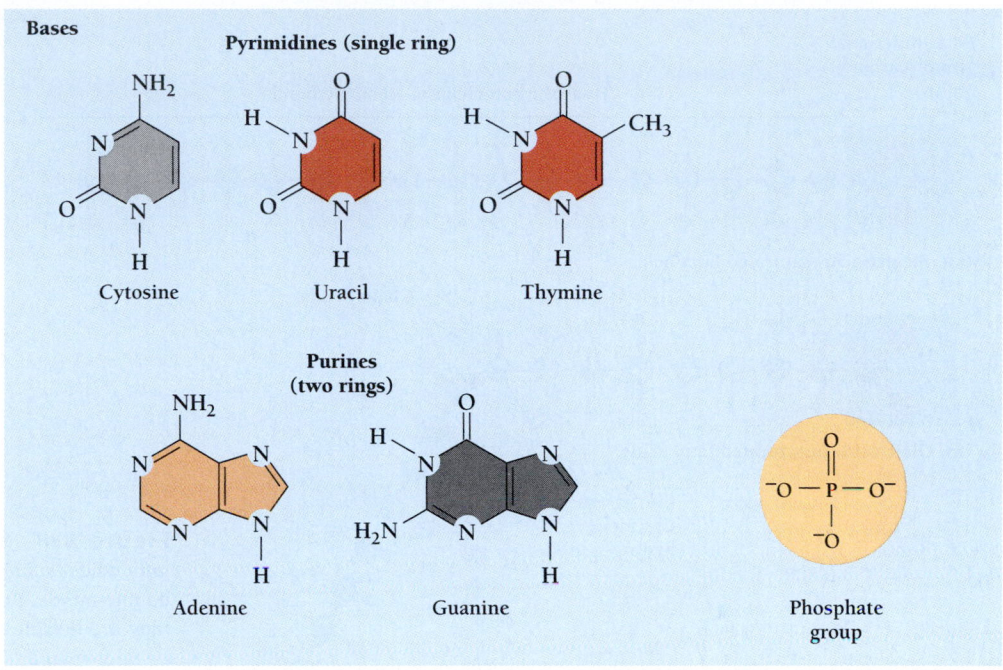

Cholesterol

Glycerol

Stearic acid (saturated) **Oleic acid (unsaturated)**

Phosphatidylcholine

Phosphatidylinositol

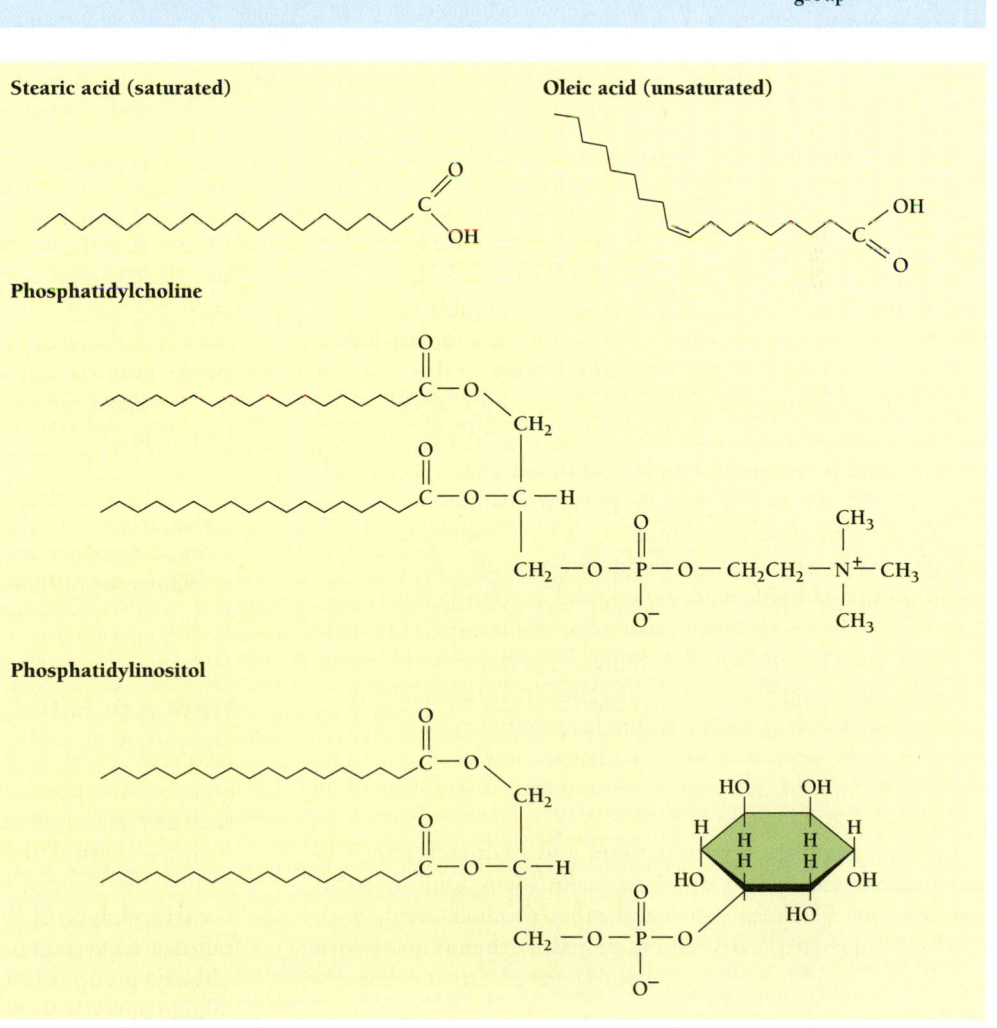

LIPIDS

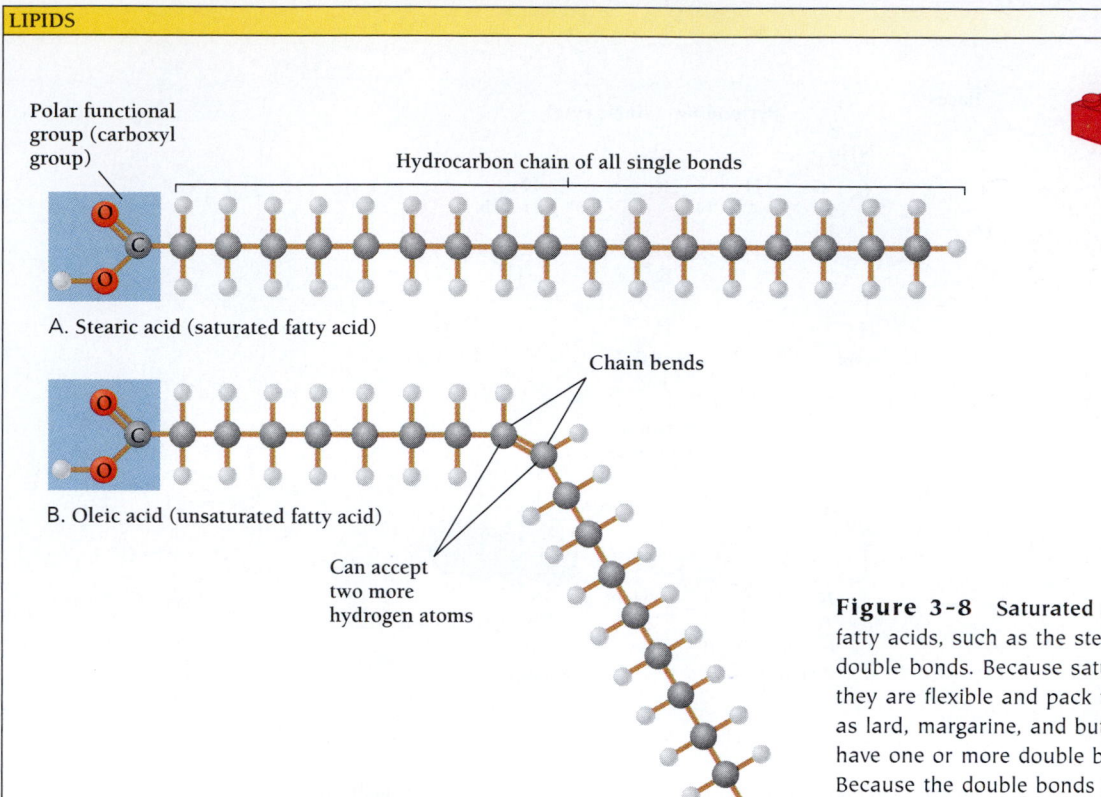

Polar functional
group (carboxyl
group)

Hydrocarbon chain of all single bonds

A. Stearic acid (saturated fatty acid)

B. Oleic acid (unsaturated fatty acid)

Chain bends

Can accept
two more
hydrogen atoms

Figure 3-8 **Saturated and unsaturated.** A. Saturated fatty acids, such as the stearic acid in beef fat, have no double bonds. Because saturated fats lack double bonds, they are flexible and pack tightly together into solids such as lard, margarine, and butter. B. Unsaturated fatty acids have one or more double bonds between carbon atoms. Because the double bonds make unsaturated fats more rigid, unsaturated fats pack loosely, forming liquid oils.

with hydrogen atoms. Stearic acid, an important component in beef fat, is a highly saturated fatty acid (Figure 3-8A).

When a fatty acid contains one or more double bonds between adjacent carbons, and is therefore missing hydrogen atoms, the fatty acid is said to be **unsaturated.** Oleic acid, an important component of vegetable oils, is unsaturated because it has double bonds (Figure 3-8B). A fatty acid with just one double bond is said to be monounsaturated. A fatty acid with many double bonds is said to be **polyunsaturated.**

Fats, waxes, and other lipids that have saturated hydrocarbon chains are solid at room temperature. In contrast, polyunsaturated lipids tend to be liquid at room temperature, and we call them oils. The reason is in the bonds. The single C — C bonds in saturated fats rotate freely, allowing the chains to pack tightly together to form a solid. In contrast, the double C = C bonds in saturated lipids are more rigid and tend to form kinky polycarbon chains that do not pack well. The result is a fluid, or oil. In general, animals contain more fats and plants contain more oils.

Food manufacturers sometimes add hydrogen atoms to polyunsaturated oils in order to make them solid at room temperature. Cookies, chips, and other products made with such "hydrogenated" fats tend to be crisper than if made with oil. Butter would work, but hydrogenated fats are less expensive.

Both naturally and artificially saturated fats seem to increase cholesterol levels and overall risk of heart disease. Nu-

tritionists generally recommend that we avoid saturated fats and stick to diets rich in polyunsaturated and monounsaturated oils, such as those found in nuts, grains, and olive oil. However, this area of research is surprisingly complex. For example, even though stearic acid, found in beef and chocolate, is a saturated fat, it apparently does not increase blood cholesterol levels.

Lipids, which include both fats and oils, are usually oily to the touch, insoluble in water, and high in energy. One class of lipids, the fatty acids, may be more or less "saturated."

How Are Fatty Acids Linked Together?

Glycerol is a simple three-carbon molecule with three hydroxyl groups. Each hydroxyl can attach to the carboxyl carbon of a fatty acid by means of a dehydration reaction (Figure 3-9A). A glycerol with all three hydroxyl groups attached to fatty acids, so that the glycerol has three, long hydrocarbon "streamers," is a **triacylglycerol** (Figure 3-9B). When a triacylglycerol (also called a triglyceride) forms, the hydrophilic character of the carboxyl group is lost. As a result, triacylglycerols are even less soluble in water than fatty acids. Most of the fats that humans and other animals store, as well as the oils of plants, consist of triacylglycerols.

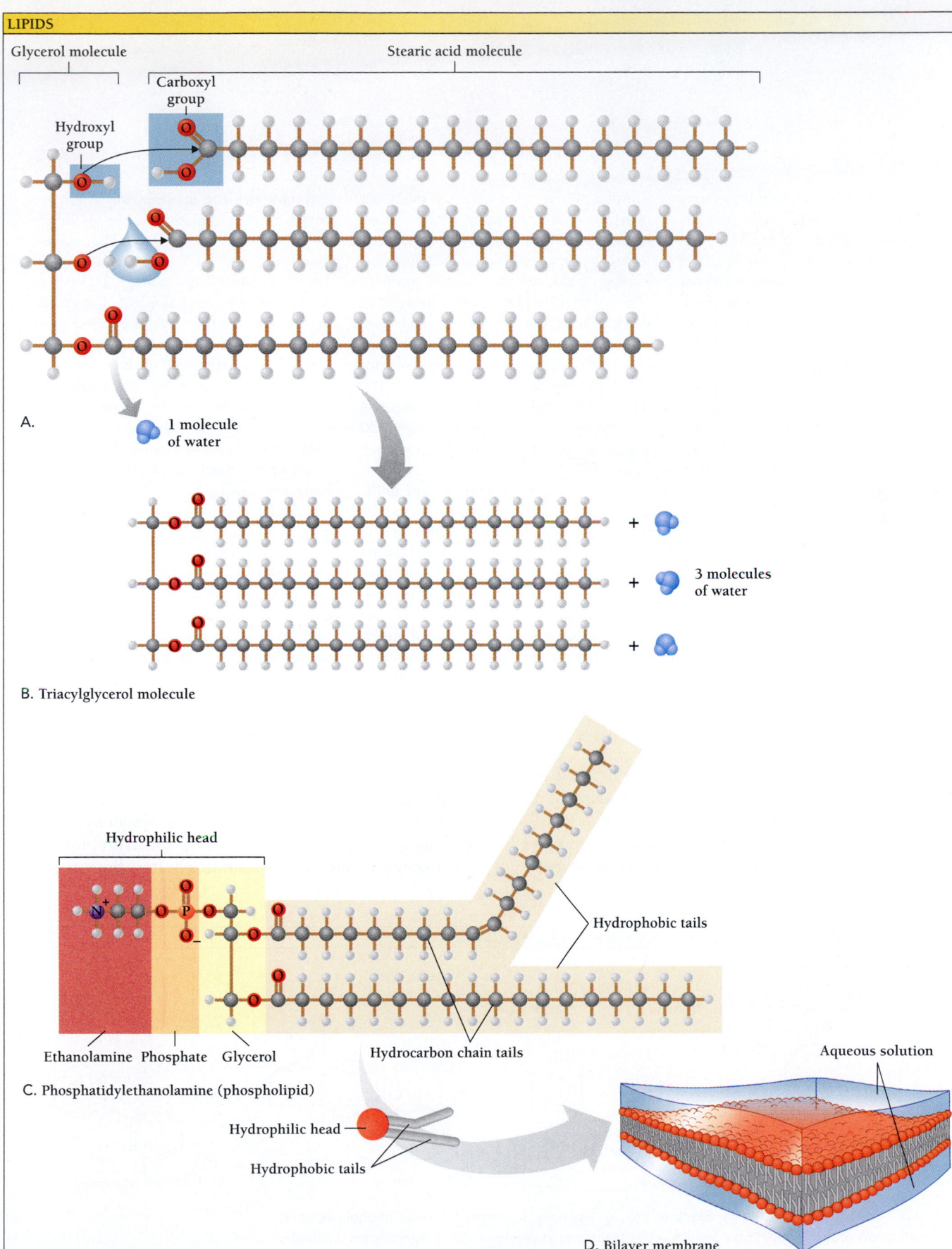

Glycerol molecule | Stearic acid molecule

Hydroxyl group

Carboxyl group

A.

1 molecule of water

B. Triacylglycerol molecule

+

+ 3 molecules of water

+

Hydrophilic head

Hydrophobic tails

Hydrocarbon chain tails

Ethanolamine Phosphate Glycerol

Aqueous solution

C. Phosphatidylethanolamine (phospholipid)

Hydrophilic head

Hydrophobic tails

D. Bilayer membrane

Figure 3-9 **Putting fatty acids to work.** A and B. A triacylglycerol (or triglyceride) forms when three fatty acid chains attach to glycerol. One molecule of water is lost at each attachment. C. A phospholipid forms when just two fatty acids attach to glycerol. D. Phospholipids form membranes in which the hydrophilic heads face out into the aqueous environment and the hydrophobic tails hide together in the middle.

BOX 3.1

The Biochemistry of Cholesterol

A quick scan of any health magazine would leave most people with the impression that cholesterol is always bad for you. In fact, you can't live without it. The outer membranes of your cells are rich in this lipid, having as much as one cholesterol molecule for every phospholipid molecule. Cholesterol's bulky shape contributes to membrane fluidity, in the same way the double bonds in phospholipids' fatty acid chains contribute to the fluidity of vegetable oils. Unlike phospholipids, however, cholesterol is hydrophobic, which means that, by itself, it doesn't dissolve well in blood.

Cholesterol also serves as a chemical precursor to the sex hormones progesterone, testosterone, and estradiol (a form of estrogen). All steroids contain the large, nonpolar ring structure of cholesterol and are fat soluble (hydrophobic).

Cholesterol comes from two sources: the liver and the diet. The liver makes cholesterol as a component of bile, a deter-

gentlike substance secreted into the upper intestine to aid digestion. In addition, most Americans consume about 550 mg of cholesterol a day, all of which comes from butter, eggs, meat, and other animal products. Some of this cholesterol passes through the digestive tract into the feces, and some gets absorbed into the blood. The blood plasma of a typical college-aged American contains about 180 mg/deciliter (1 deciliter = 100 ml) of cholesterol. Total plasma cholesterol rises slowly with age to peak at about 230 mg/dl in men and 250 mg/dl in women by age 55.

Cholesterol doesn't travel alone in the blood. If it did, its hydrophobic nature would cause it to form big clumps, like oil in a bottle of salad dressing. Instead it travels in lipoprotein complexes. The most common are called low-density lipoproteins, or LDLs. These are responsible for delivering cholesterol to cells and tissues where it is needed. High-density lipoprotein, HDLs, remove cholesterol from cells

and transport it to the liver, where it is secreted into the digestive juices, or "bile."

Sometimes people call LDL cholesterol "bad" and HDL cholesterol "good." This is because if not enough HDLs are present to remove cholesterol, the excess builds up along the walls of arteries, contributing to deposits called "plaques." Plaques cause hardening of the arteries, or atherosclerosis (Figure A). Atherosclerosis reduces blood flow to the organs, including the heart, and damages the arteries. Tiny bits of plaque may break off and lodge in an artery, blocking blood flow to the heart and causing a heart attack.

While limiting cholesterol intake is a good idea, one of the best indicators of the chance of developing heart disease from cholesterol deposits in the arteries is not total cholesterol, but the ratio of bad to good cholesterol. The average American has an LDL-to-HDL ratio of 5:1. Reduce that ratio to 3.5:1, by increasing plasma HDL, and you'll cut your risk of heart disease in

A glycerol molecule with a phosphate group attached in place of the third hydroxyl group and with just two fatty acid chains instead of three is a **phospholipid** (Figure 3-9C). Phospholipids are highly amphipathic. Each one has a hydrophobic "tail," which consists of the hydrocarbon chains of two

fatty acids, and a hydrophilic "head," which contains the charged phosphate group.

In water, the hydrophobic tails of phospholipids clump together to form an oily interior, while the hydrophilic heads point outward into the surrounding water. Both the outer

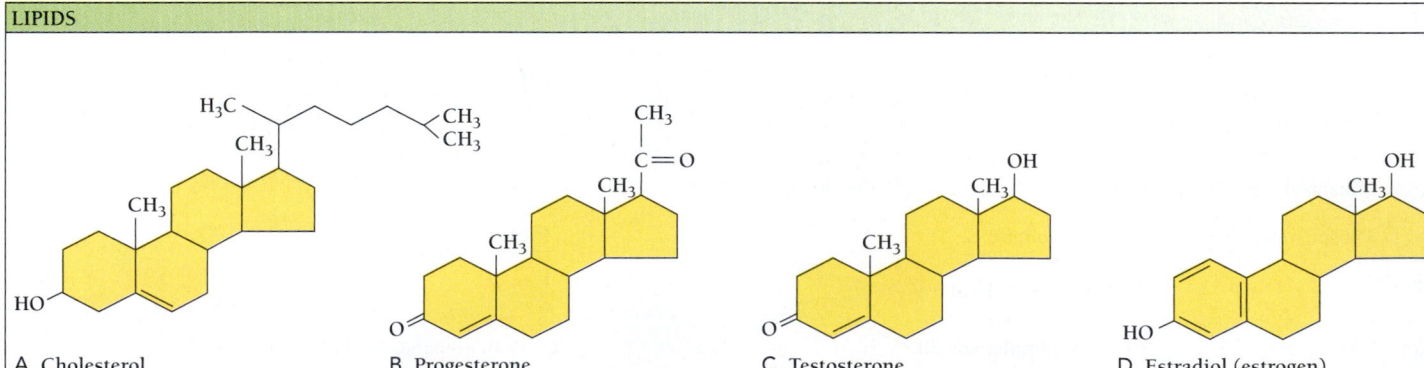

LIPIDS

A. Cholesterol   B. Progesterone   C. Testosterone   D. Estradiol (estrogen)

Figure 3-10   Cholesterol, a building block of life. A. The basic four-ring structure of cholesterol. B. With the substitution of a few functional groups, cells transform cholesterol into progesterone, a female sex hormone. C. Progesterone becomes testosterone, a sex hormone that gives males many of their most striking characteristics. D. A few more changes, and testosterone becomes estradiol, a powerful form of the female hormone estrogen.

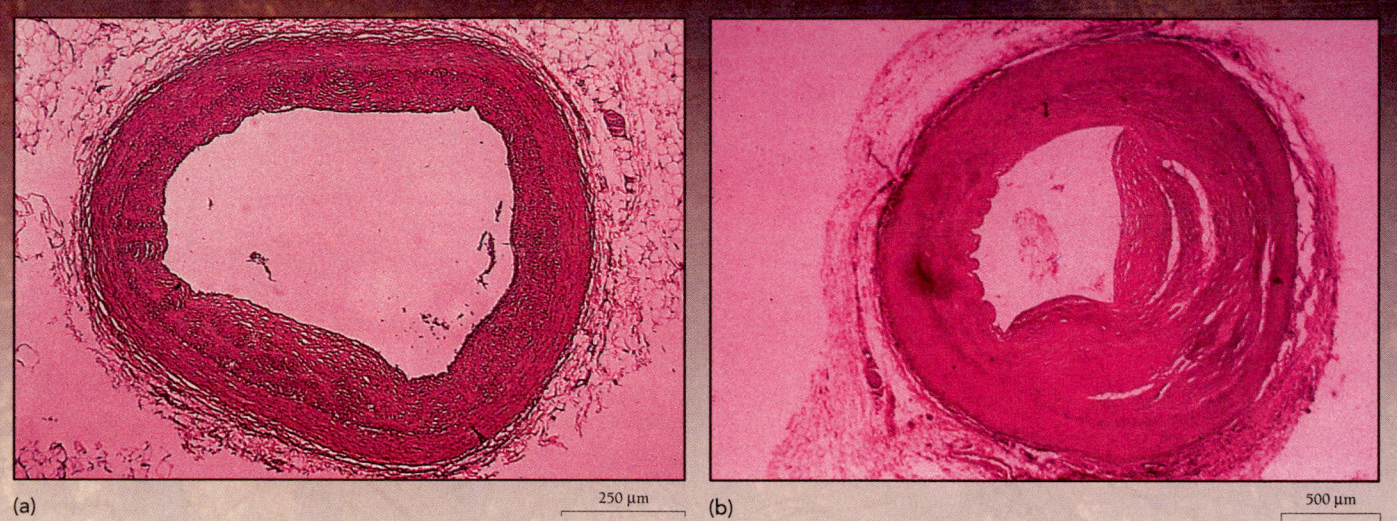

Figure A Atherosclerosis results when plaque deposits narrow and block our arteries. (a) One artery is clear. (b) The other is blocked by plaque. *(a, Cabisco/Visuals Unlimited; b, Ober/Visuals Unlimited)*

half. An even better predictor of heart disease are high levels of IDLs, intermediate-density lipoproteins, which, for now, few doctors measure.

How can people increase HDLs? The following items all tend to raise HDLs: female sex hormones; regular aerobic exercise; monounsaturated fats, such as olive oil (instead of, not in addition to, saturated fats); weight loss; and certain forms of

fiber. On the other hand, tobacco smoke lowers plasma HDL levels and also contributes to heart disease in other ways.

But don't assume science has found all the answers. Research in the area of diet and heart disease raises baffling new questions almost daily. For example, while high-fat diets are clearly bad for us, people on extremely low-fat diets have increased numbers of both LDLs and IDLs. And

low-fat meals can have other effects. In another study, 12 overweight teenagers were given either a high-fat meal or a low-fat meal, of equal calories. Over the next five hours, the adolescents were allowed to eat whatever they liked. Those who had eaten the low-fat meal ate twice as many calories as those who had eaten the high-fat meal. Apparently, the super-low-fat meal made them hungrier.

membranes of cells and their interior membranes are composed primarily of phospholipids. All of these membranes have a hydrophobic interior and a hydrophilic exterior (Figure 3-9D).

Another class of lipids is the **steroids**—hydrocarbon chains with four interconnected rings (Figure 3-10). Many of the hormones responsible for sexual development and reproductive functions in animals are steroids. The starting point for the synthesis of all steroids is **cholesterol**—a hydrocarbon with the same pattern of four interconnected rings (Box 3.1). Small differences in the attached functional groups can make enormous differences in a steroid's biological properties. For example, cells make the female hormone progesterone from cholesterol by replacing an OH group with an oxygen atom with a double bond, along with some other changes (Figure 3-10A and B). Cells then make testosterone from the progesterone and then estrogen from the testosterone, as needed (Figure 3-10C and D). All four lipids have profoundly different effects in the body, yet differ by only a few functional groups.

Most of the fatty acids in organisms are linked to glycerol to form lipids as diverse as triacylglycerols, phospholipids, and steroids.

HOW DO SUGARS FORM POLYSACCHARIDES?

We have all been told to avoid eating too much sugar. But while this may be sound advice, avoiding all sugar is neither possible nor desirable. Sugar is the fundamental energy storage molecule of all living organisms.

The smallest sugars are "simple sugars," or **monosaccharides** [Greek, *mono* = one + *saccharine* = sugar), which can have from three to nine carbon atoms. Two of the most important simple sugars are **glucose** and fructose, which have six carbon atoms each (Figure 3-11). Cells link sugar molecules using a dehydration reaction. The result is a **glycosidic bond,** in which one oxygen atom forms a bridge between carbon atoms on two adjacent sugar molecules. Two sugars can be linked to form a **disaccharide** [Greek, *di* = two + *saccharine* = sugar), such as **sucrose,** or table sugar (Figure 3-11). Several sugars can form short chains, called **oligosaccharides** [Greek, *oligos* = few + *saccharine* = sugar]. The longest chains of sugars, called **polysaccharides** [Greek, *poly* = many + *saccharine* = sugar], may include thousands of simple sugars. Polysaccharides include the starch in bread and the cellulose in wood and paper.

Besides glucose, two other important sugars are **ribose** and **deoxyribose,** simple five-carbon sugars that combine with

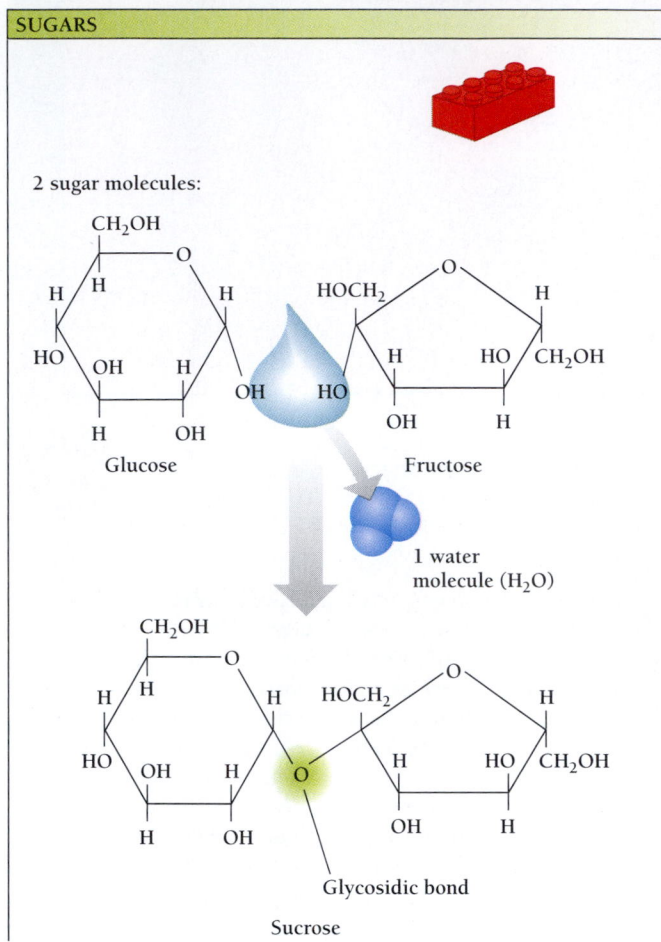

Figure 3-11 **Life is sweet.** A simple dehydration reaction joins the two simple sugars glucose and fructose, forming the disaccharide sucrose—regular table sugar.

other small molecules to form the building blocks for DNA (deoxyribonucleic acid) and RNA (ribonucleic acid), as shown in Table 3-2.

All sugars and polysaccharides are **carbohydrates,** molecules that contain the equivalent of one water molecule for every carbon atom. Most of the carbons in a carbohydrate have a hydrogen atom on one side and a hydroxyl group on the other, giving big carbohydrates hundreds or even thousands of hydroxyl groups. Because these hydroxyl groups easily form hydrogen bonds with water molecules, most carbohydrates dissolve well in water. Stir some sugar into a cup of hot tea and you'll see.

How Do Organisms Use Sugars To Store Energy?

Organisms use monosaccharides and disaccharides to store energy for a few hours, and they use longer polysaccharides, such as starch, to store energy for weeks, months, or years. In the same way, we keep our money in cash if we want to buy something right away, but put it in the bank if we want to save our money for later.

The single most important simple sugar is glucose, what we sometimes call blood sugar. Almost all biochemical reactions are in some way connected to glucose. Synthesizing glucose is a way of storing energy. Breaking down glucose is a way of releasing energy. The energy from glucose drives everything that happens in an organism from cell division to long division. So, linking long chains of glucose molecules is just another way for organisms to store energy.

Animals store energy in a long chain of glucose molecules called **glycogen.** Because glycogen has many short branches, it is easy for enzymes to quickly pick off many glucose molecules at a time. Instead of breaking down the whole molecule to get energy, enzymes harvest glucose from the ends of the branches as needed.

Plants store energy as **starch,** another large polysaccharide made of glucose molecules. Potatoes, pasta, bread, rice, and many other familiar foods contain lots of energy-rich starch, which may be either branched or unbranched. Two of the most common kinds of starch are amylopectin and amylose. Smaller amylose consists of hundreds to thousands of glucose molecules in simple, unbranched chains. Larger amylopectin consists of up to 50,000 glucose molecules in highly branched chains (Figure 3-12).

How Else Do Organisms Use Polysaccharides?

Plants and animals use polysaccharides not only as a way to store energy, but also as a building material. Arthropods such as butterflies and crabs construct skeletons from the polysaccharide **chitin.** Plants build their bodies using **cellulose**—the major component of wood, paper, cotton, and linen.

The structure and properties of polysaccharides depend both on the kinds of sugars from which they are made and on the way in which the sugars are joined. Although cells can use different sugars to make polysaccharides, many of the most important consist entirely of glucose. Starch, cellulose, and glycogen are all made of glucose alone. The differences in structure come from the way the glucose molecules are linked.

For us humans, the most important difference between cellulose and starch is that—due to the arrangement of the glycosidic bonds between the glucose molecules—one polysaccharide is digestible and the other is not. In cellulose, the links between glucose are oriented differently from those in starch. The result is an extremely straight—and strong—chain of glucose molecules that makes an excellent building material for plants. But the digestive enzymes that work well on starches cannot break the bonds of cellulose. Different enzymes are needed to digest cellulose.

Because most animal cells do not have enzymes that break the links in cellulose, they cannot digest it. The few animals that can digest cellulose—cows and termites, for example— do so with the help of microorganisms that have the needed enzymes. For the rest of us, the cellulose in bran flakes and prunes is just "roughage" or fiber—a water-absorbing material

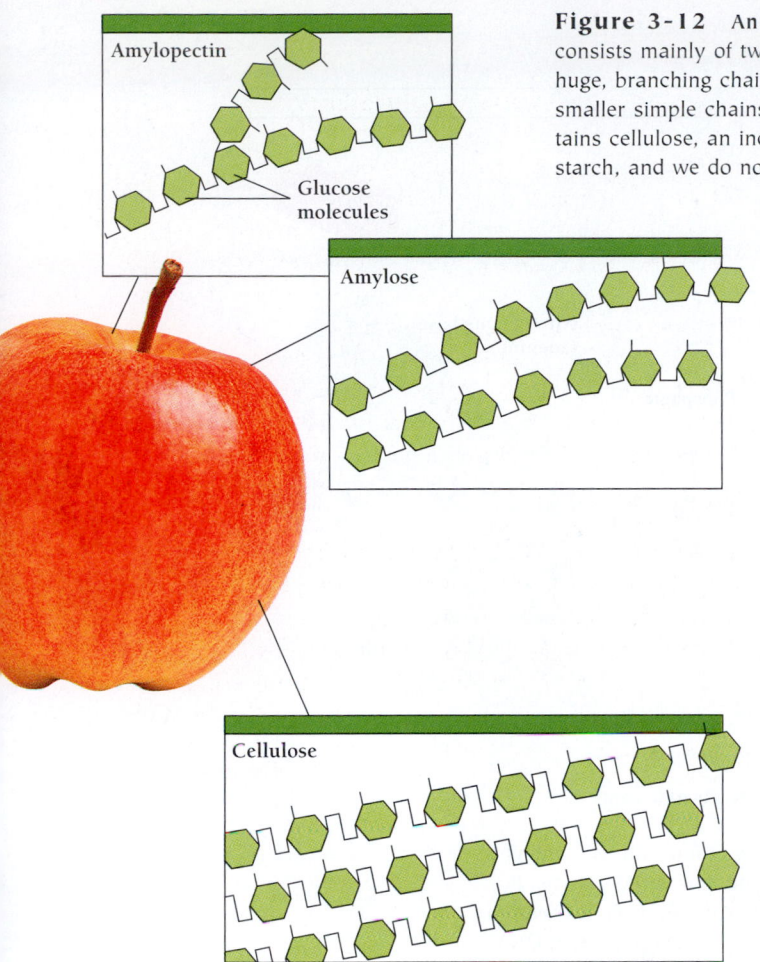

Figure 3-12 An apple contains three important polysaccharides. The starch in most plants consists mainly of two polysaccharides, amylopectin and amylose. Amylopectin molecules are huge, branching chains consisting of up to 50,000 glucose building blocks. Amylose molecules are smaller simple chains of a few hundred or a few thousand glucose molecules. An apple also contains cellulose, an indigestible polysaccharide. The bonds in cellulose are different from those in starch, and we do not have the enzymes necessary to break them. *(Photo, Harcourt Photo Library)*

rings, each of which includes two nitrogens. Attached to these rings are hydrogen atoms and various polar functional groups. Like all bases, nitrogenous bases accept hydrogen ions.

The five bases are divided into two types: the two-ring **purines**—adenine and guanine—and the one-ring **pyrimidines**—cytosine, uracil, and thymine. Uracil and thymine are almost identical, differing by the presence of a methyl group ($-CH_3$) in thymine.

Nucleotides serve many important biological roles. First and foremost, they are the building blocks of RNA and DNA (Figure 3-13B). In addition, nucleotides with one, two, or three phosphate groups play a central role in the production of energy within cells. **ATP (adenosine triphosphate),** which has three phosphates, is the immediate source of energy for all biological processes. Energy from the breakdown of glucose is stored in ATP. Cells then release energy from ATP by hydrolyzing its phosphate-phosphate bonds. The resulting energy drives all the chemical reactions of a cell (Figure 3-14).

A nucleotide consists of a sugar, a phosphate, and a base. Nucleotides are essential to life both as the units of DNA and RNA and for energy storage.

How Are the Nucleotides of a Nucleic Acid Linked Together?

Like sugars, nucleotides can connect one to another by a simple dehydration reaction (Figure 3-15). With the removal of one molecule of water, nucleotides form **phosphodiester bonds,** which connect the nucleotides into long, unbranched chains called **nucleic acids.** Two nucleic acids—DNA and RNA—are so important in the storage and transmission of genetic information that this book devotes eight chapters to them.

DNA serves as a library of information for every cell. Along the thin strands of DNA are genes that carry instructions for building every protein a cell will ever need. The genes of all organisms are made of DNA. The only genes that are not DNA are those of a few viruses—including those responsible for influenza and AIDS—whose genes are made of RNA. But viruses are not organisms, for they are not made of cells. In cells, RNA acts only as an interpreter, carrying information from the DNA to the rest of the cell.

In a chain of nucleotides, the sugar and phosphate parts of each nucleotide link together alternately to form a sugar-phosphate "backbone." RNA consists of a single long backbone. DNA usually consists of two parallel backbones connected to each other by thousands of hydrogen bonds

that keeps things moving through the digestive tract. Indeed, cellulose is one of those no-calorie foods that the food industry uses to thicken milkshakes or to make low-calorie breads. Because we cannot digest it, these foods are said to be "low in calories." But if we were cows, these foods would have plenty of calories.

One of the most important uses for sugars is in short polysaccharides called oligosaccharides. Oligosaccharides are often tacked on to proteins to form **glycoproteins,** which just means "sugar protein." Glycoproteins are commonly found on the outsides of cell membranes. Like big signs, they help cells to recognize one another and communicate. Inside cells, glycoproteins tacked onto newly made proteins act as address labels that tell a cell where to ship the new proteins.

HOW DO NUCLEOTIDES LINK TOGETHER TO FORM NUCLEIC ACIDS?

Nucleotides are the building blocks of RNA (ribonucleic acid) and DNA (deoxyribonucleic acid), the two kinds of molecules that carry genetic information. Each nucleotide building block consists of three parts: a sugar (ribose or deoxyribose), one or more phosphate groups, and a nitrogenous base.

Most cells use just five nitrogenous bases (Figure 3-13A). Each **nitrogenous base** consists of either one or two aromatic

NUCLEOTIDES

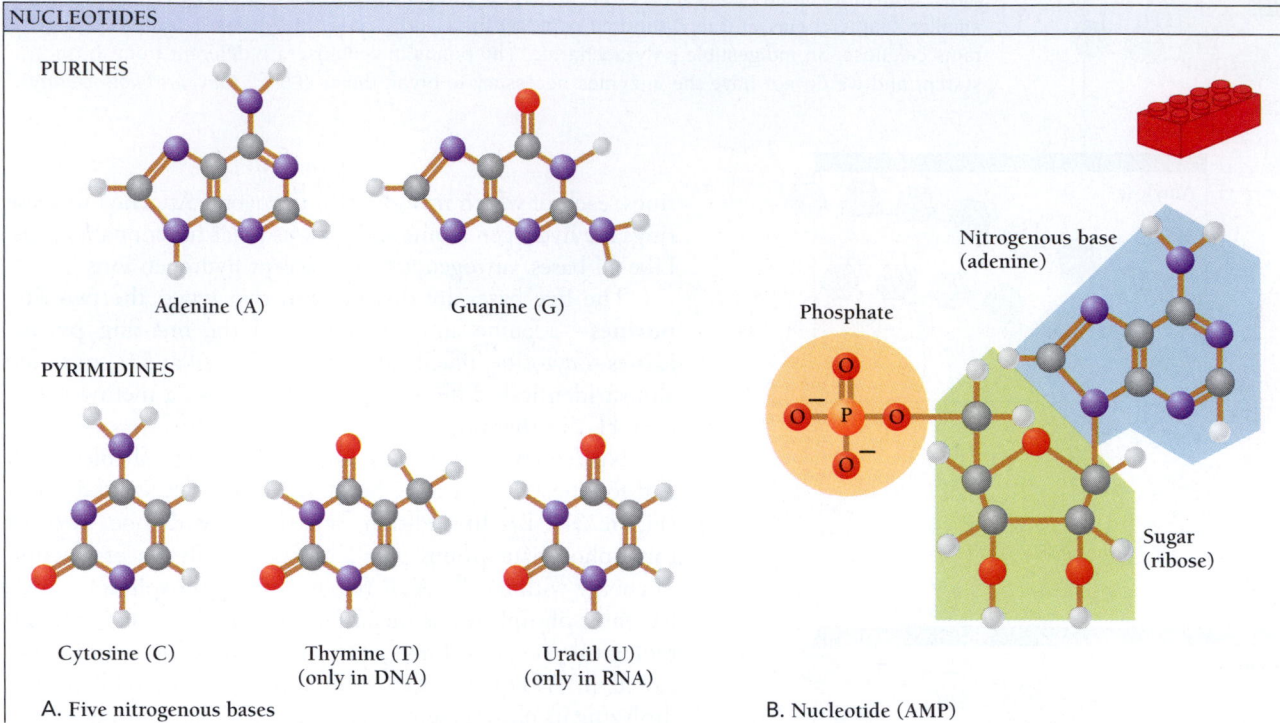

PURINES

Adenine (A)   Guanine (G)

PYRIMIDINES

Cytosine (C)   Thymine (T)
(only in DNA)   Uracil (U)
(only in RNA)

A. Five nitrogenous bases

Phosphate

Nitrogenous base
(adenine)

Sugar
(ribose)

B. Nucleotide (AMP)

Figure 3-13   The five **nitrogenous bases**. A. The purines have double rings of four nitrogen atoms and five carbon atoms. The pyrimidines have one ring of two nitrogen atoms and four carbon atoms. B. Each nitrogenous base can join a sugar and a phosphate group to form a nucleotide, such as adenosine monophosphate (AMP). AMP is one of just four nucleotide building blocks that form long chains of DNA.

NUCLEOTIDES

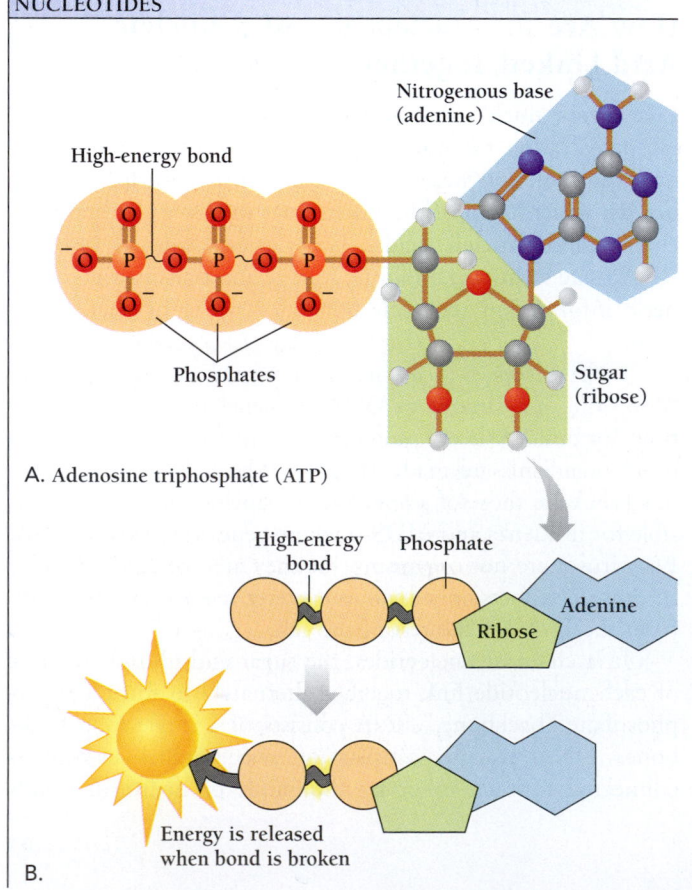

High-energy bond

Nitrogenous base
(adenine)

Phosphates

Sugar
(ribose)

A. Adenosine triphosphate (ATP)

High-energy
bond   Phosphate

Adenine

Ribose

Energy is released
when bond is broken

B.

between adjacent bases. The result is a molecule that resembles a long ladder, with each rung consisting of a pair of nucleotide bases (Figure 3-16). The whole ladder is twisted along its long axis to form a "double helix."

DNA and RNA each consist of just four kinds of nucleotide bases. Three bases—adenine, guanine, and cytosine—occur in both DNA and RNA. But DNA's fourth base is thymine, and RNA's fourth base is uracil. The structures of RNA and DNA are important to the way that they function as information molecules. In Chapters 10 and 11, we discuss the structure and function of DNA and RNA in great detail.

The long chains of nucleotides in the information molecules DNA and RNA are linked—sugar-phosphate-sugar—by phosphodiester bonds.

Figure 3-14   For all organisms, ATP is the common currency of energy. A. The nucleotide ATP consists of the nitrogenous base adenine, the sugar ribose, and three phosphate groups. B. ATP releases energy for use in chemical reactions when one of the phosphate groups breaks off.

NUCLEOTIDES

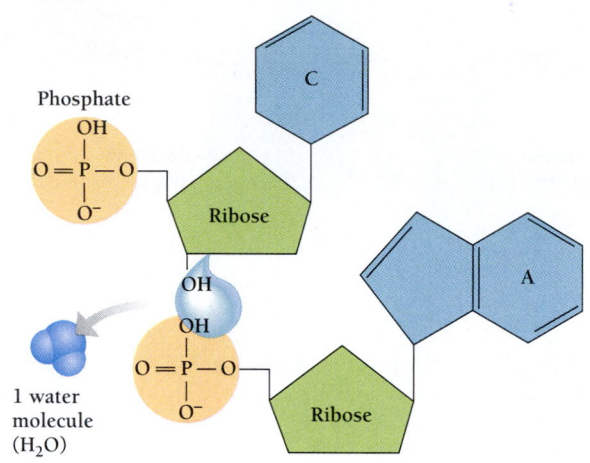

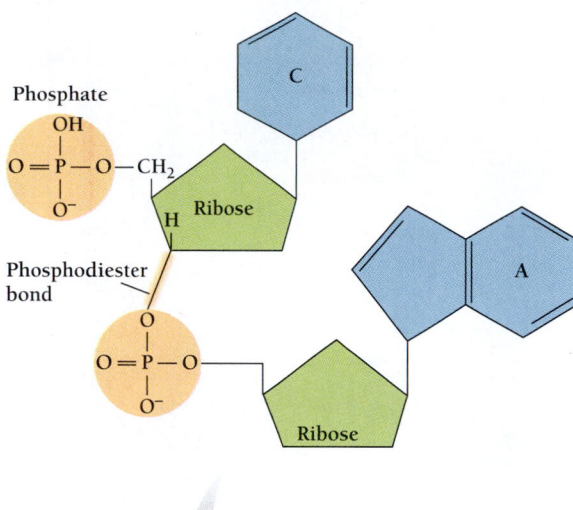

A. Formation of a phosphodiester bond

Figure 3-15 Building a library of information.
A. All organisms store genetic information in long chains of nucleotides—either DNA or RNA. The nucleotide building blocks join by a dehydration reaction to form a phosphodiester bond, losing one water molecule in the process. B. RNA consists of a single chain of nucleotides.

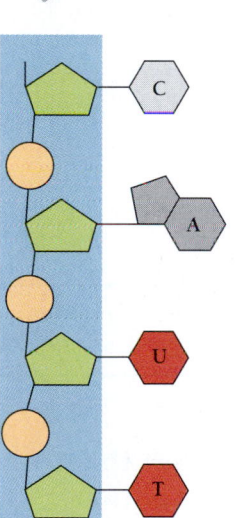

B. RNA

HOW DO AMINO ACIDS LINK TO FORM POLYPEPTIDES?

Proteins perform thousands of jobs. They hold the body together, lend support to bones, transport molecules and ions in and out of cells, and much more. Every organism has hundreds to thousands of different proteins. The human body alone contains at least 80,000 different proteins.

Collagen, the commonest protein in the human body, gives strength to connective tissues in tendons, ligaments, bones, muscle, and skin. Dissolved in sugar water, collagen gives Jell-O its jiggle. Yet, in the body, collagen is as strong as steel. Elastin makes skin stretchy, while stringy keratin toughens hair, horns, nails, and claws.

Some proteins taxi materials from one place to another. Hemoglobin, for example, binds to oxygen in the lungs and transports it, by way of the blood, to the cells in the rest of the body. Other proteins act as hormones, antibodies, or poisons (rattlesnake venom, for example). Still others transport molecules across the membranes of cells. Some proteins—often glycoproteins with attached oligosaccharides—act as identification tags on the surfaces of cells. Other proteins help cells to communicate with one another. Protein "signaling molecules" carry messages from one cell to another, and protein "receptors" on cell surfaces pick up those signals.

One of the most important roles for proteins is as **enzymes**—biological molecules that speed up reactions between other molecules. Enzymes are indispensable to life. Without them, many chemical reactions would occur too slowly to sustain life. Enzymes copy and repair DNA, help cells to extract energy from sugars, digest our food, and much more. Enzymes in the mouth, for example, help break down the polysaccharides in bread into simple sugars. That is why bread gets sweeter and sweeter as you chew it. We discuss enzymes in greater detail in Chapter 5.

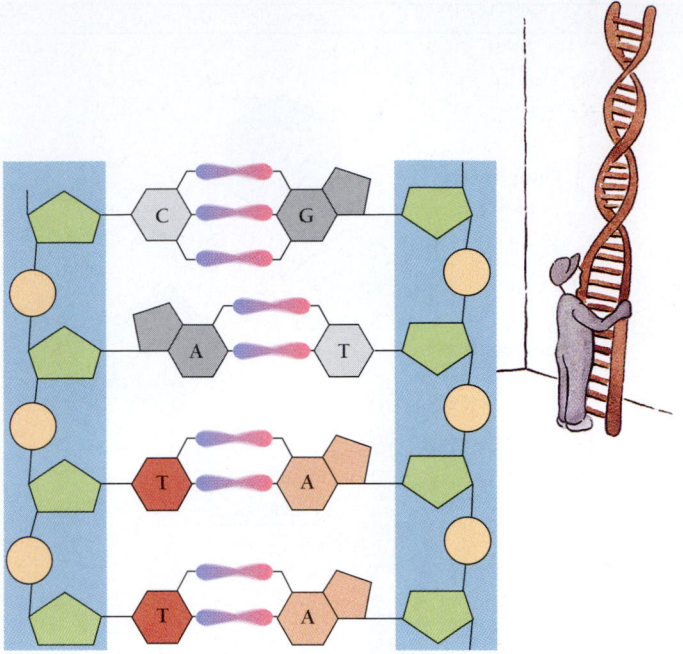

Figure 3-16　The DNA ladder. The DNA of our genes consists of two long chains of nucleotides linked together by bonds between the nitrogenous bases of each nucleotide. The bonds are often compared to the rungs of a ladder. The whole, long double chain is twisted like a phone cord into a helix.

Cells contain thousands of proteins, each of which has a different role to play.

What Distinguishes the 20 Amino Acids from One Another?

Every protein is made of chains of amino acids called **polypeptides.** A protein may be made of one simple chain or of several different chains. A chain may consist of only a few hundred amino acids or 10,000 or more. Regardless of how long the chains are or how many of them there are, cells build proteins one amino acid at a time.

Each of the 20 amino acids has the same basic structure. Each contains a carboxyl group, which makes it an acid, and an amino group, which makes it a base (Figure 3-17). Both the carboxyl group and the amino group are attached to the same carbon atom, called the **alpha (α) carbon.**

Each amino acid also has a characteristic side chain, called an R group, which is also attached to the α carbon atom. *The α carbon, the carboxyl group, and the amino group are exactly the same in 19 of the 20 amino acids.* (The one exception is proline.) But each amino acid has a different R group. Table 3-2 shows the 20 amino acids found in proteins with their 20 different R groups.

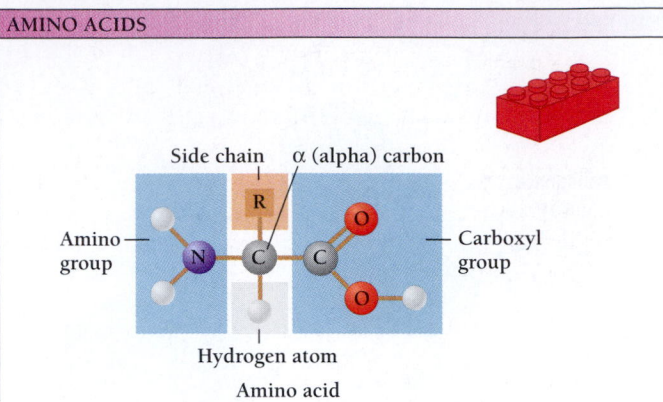

AMINO ACIDS

Side chain　　α (alpha) carbon

Amino group　　Carboxyl group

Hydrogen atom

Amino acid

Figure 3-17　What's in an amino acid? An amino acid includes a carboxyl group, an amino group, a hydrogen atom, and a side chain (R). All four are attached to a single, central carbon atom, called the "α carbon."

R groups vary in length and in their functional groups. Some are hydrophobic, some hydrophilic. Some have a positive charge, some have a negative charge, and some are neutral. The R group side chains establish the distinctive properties of each amino acid and the properties of the resulting polypeptide.

Just as the 26 letters of our alphabet allow a nearly infinite number of unique sentences, the 20 amino acids allow a nearly infinite number of different proteins. A polypeptide just 100 amino acids long can be made in 10^{130} (or 20^{100}) ways by varying the order and selection of amino acids. Recall that 10^{130} is 1 followed by 130 zeros.

Although amino acids usually end up folded inside of massive proteins, such as hemoglobin, a few perform other roles. For example, some amino acids are neurotransmitters—chemicals secreted by nerve cells that convey information to cells.

All amino acids have a carboxyl group, an amino group, and an R group, all attached to the α carbon. In each of the amino acids, all the parts are identical except for the R group.

How Do Amino Acids Join Together?

Like polysaccharides and nucleic acids, polypeptides are composed of chains of building blocks joined by dehydration reactions. The bond that links the amino acids together into polypeptides is called a **peptide bond.** Two amino acids join when the carboxyl group of one joins to the amino group of the next (Figure 3-18).

Each amino acid in the string is oriented in the same way as the others—amino, carboxyl, amino, carboxyl—forming an unvarying backbone of repeating atoms. The R groups hang off the sides of the polypeptide backbone. Uninvolved in the peptide bond, the R side chains may interact with one another or with other molecules in their environment. The polypep-

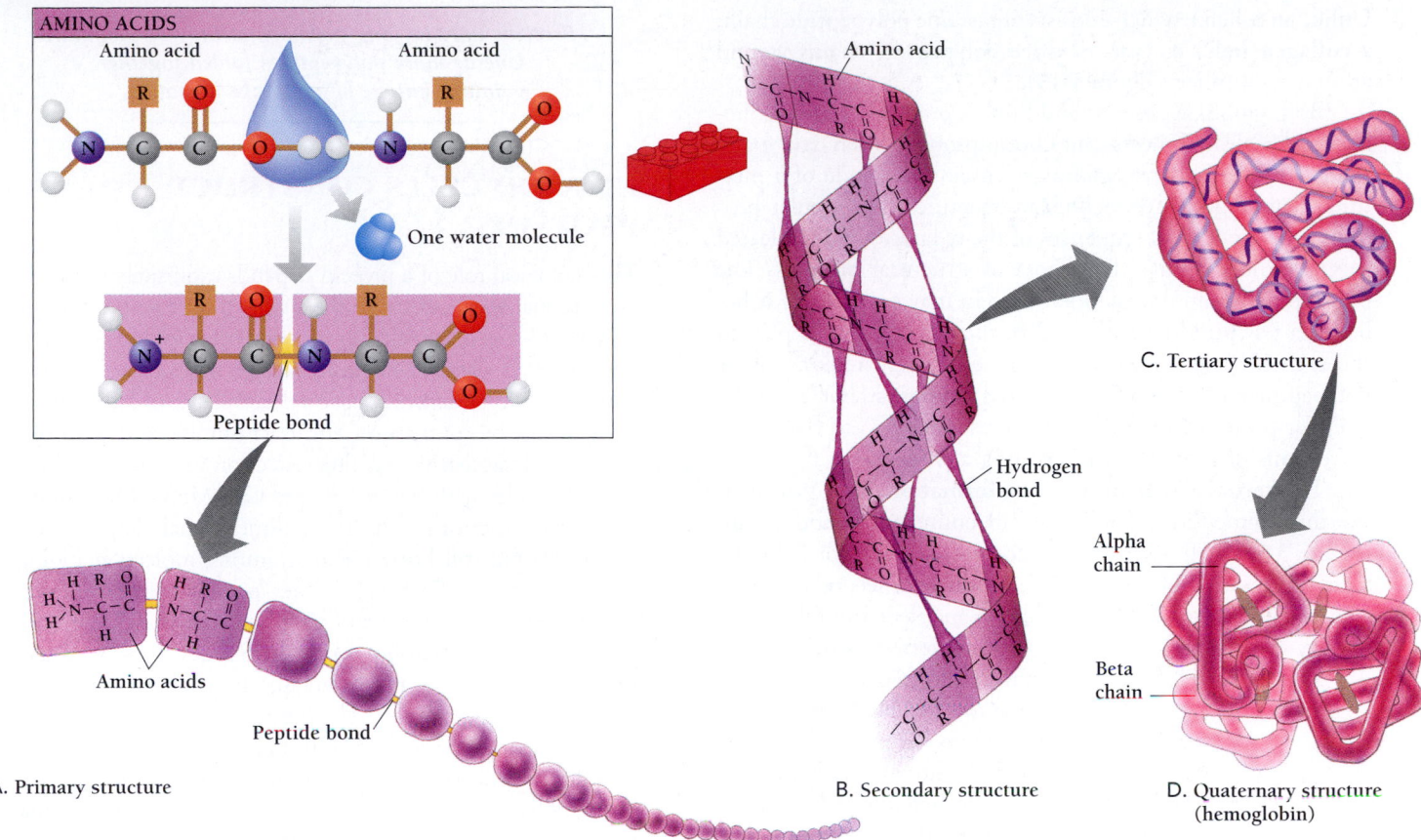

AMINO ACIDS

Amino acid   Amino acid

One water molecule

Peptide bond

Amino acids

Peptide bond

A. Primary structure

Amino acid

Hydrogen bond

B. Secondary structure

C. Tertiary structure

Alpha chain

Beta chain

D. Quaternary structure (hemoglobin)

Figure 3-18 Building a protein. The usual dehydration reaction joins two amino acids by a peptide bond. A. A long chain of amino acids forms a polypeptide chain. The primary structure of a protein is simply the sequence of amino acids in its polypeptide(s). B. The secondary structure is the curling of each polypeptide into a three-dimensional form, such as the α helix shown here. C. Tertiary structure is the folding of the curled polypeptide into a more complex three-dimensional shape, such as that of the myoglobin polypeptide shown here. D. Quaternary structure is the fitting together of two or more folded polypeptides, as in the hemoglobin molecule. Hemoglobin is the major oxygen-carrying protein in the blood of many animals.

tide backbone is identical from one end to the other, except for the very beginning and very end.

Peptide bonds link amino acids together into polypeptide chains, in which the R groups hang off the sides.

What Do Proteins Look Like?

Proteins are extraordinarily diverse. For each of the thousands of jobs that proteins perform, there is a unique protein to do it. They can be as short as 50 amino acids or as long as 10,000. They can be as regular in their structure as a stack of Lego bricks or as irregular as a wad of crumpled paper.

Despite their diversity, biochemists have found ways to classify proteins. Most are dense, roundish molecules called **globular proteins.** A classic example of a globular protein is hemoglobin, the protein that carries oxygen in the blood. Like many proteins, hemoglobin is gigantic compared to the small

molecules of the cells. Hemoglobin is almost 400 times as big as a molecule of glucose and more than 2000 times as big as each molecule of oxygen that it ferries from the lungs.

In addition to the globular proteins are **fibrous proteins**—long, thin molecules such as the keratin in hair or the collagen in skin and bone. Fibrous proteins provide structural support to the bodies of plants, animals, and other organisms.

Proteins have up to four levels of structure. The **primary structure** of a protein is the sequence of amino acids in each chain (Figure 3-18A). This is like the sequence of letters in a sentence. It is simply the order and kind of amino acids. The **secondary structure** is the coiling, bending, or folding of individual sections of the polypeptide chain. For example, polypeptides can coil into an **α helix,** which resembles the coil of a phone cord (Figure 3-18B).

Other secondary structures include the β pleated sheet and the collagen helix, found in collagen. The **β pleated sheet** is a long chain that folds back and forth against itself—like the coils of a snake or the long lines of people at an airport.

Unlike an α helix, which consists of just one polypeptide chain, a **collagen helix** consists of three polypeptide chains wound around each other (Figure 3-19).

Both globular proteins and fibrous proteins contain α helices and β pleated sheets. But fibrous proteins often have many rigid α helices, whereas globular proteins are made of a mixture of many kinds of secondary structures. In globular proteins, the amino acid sequences of the α helices and β pleated sheets are less regular than those of structural proteins. The polypeptide chain of a globular protein may coil into an α helix for a few hundred amino acids, then switch to a β pleated sheet, then back to an α helix. These different regions of the polypeptides of a protein are called **domains.** For example, globular proteins contain β structures, with two to five parallel sections of a single chain forming a β sheet.

The **tertiary structure,** or **conformation,** of a protein is the three-dimensional folding of an entire polypeptide chain (Figure 3-18C). If you curl a ribbon into a spiral (or helix) by scraping it with a blade, you have produced a secondary structure. If you then take the curled ribbon and tie it around a birthday present, you have given the ribbon a tertiary structure.

Many proteins also have **quaternary structure**—the fitting together of two or more folded chains (Figure 3-18D). This is analogous to piecing together several different ribbons into an elaborate bow. Just as the ribbons can be either all the same color or a mixture of different colors, proteins can be made of several identical polypeptide chains or of several different ones. Hemoglobin, for example, is made of four polypeptide chains, two of one kind of chain and two of another (Figure 3-18D).

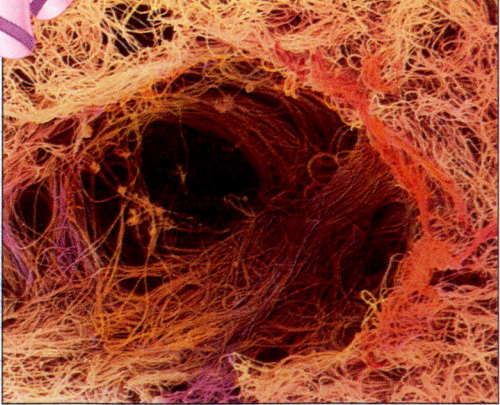

A. Collagen molecule

B. Collagen fibers   5 μm

Figure 3-19   What is skin made of? Collagen is the major constituent of skin, tendons, ligaments, cartilage, and bones. A. Collagen is made of three helical polypeptides wound together into a superhelix. B. Collagen fibers. *(B, Professor P.M. Motta and E. Vizza/Science Photo Library/Photo Researchers, Inc.)*

Proteins may be long or short, stringy or globular. One or more polypeptides folded together into a unique shape make a protein.

HOW DO CELLS CONSTRUCT PROTEINS?

The biological role of a protein depends exquisitely on its three-dimensional shape. Because the shape of a protein is so important, the chains that make up proteins must fold in precisely the right way to form biologically active molecules. The way a polypeptide folds depends first on the sequence of building blocks in its chains. By comparison, the exact number and order of sugars in a polysaccharide has little effect on its function (Web Bit 3-2: Molecules with Similar Shapes Can Mimic One Another).

What determines the three-dimensional shape of a protein is a topic still hotly debated among molecular biologists and biochemists. They call it the "protein-folding problem." Contributions to this worldwide debate come from many sources. But the ongoing work on Stanley Prusiner's scrapie PrP has raised many important questions about protein folding that scientists might not otherwise have asked.

By making different versions of a polypeptide, biochemists have found that tiny changes in amino acid sequence can affect a polypeptide's biochemical behavior. Sometimes changing one amino acid has little effect. For example, a muscle protein from a cow can have a sequence slightly different from that of a human. Yet the difference does not affect the protein's three-dimensional structure or the way it functions in the body. Cow and human proteins differ slightly, but they function the same.

On the other hand, sometimes changing an amino acid can completely alter a protein's structure and abolish its biological activity. The kinds of amino acid changes that make the biggest differences are those that substitute amino acids with very different properties. For example, in the blood protein hemoglobin, the substitution of a hydrophobic amino acid (valine) for a hydrophilic one (glutamate) produces a defective hemoglobin molecule that crystallizes. The result is the disease sickle cell anemia. Sickle cell anemia is caused by a mistake in the nucleotide sequence of the gene for hemoglobin.

A small change in the structure of a polypeptide can enormously alter the way it behaves or have no effect at all.

How Does the Folding of a Protein Depend on Interactions Among Its Amino Acids?

Polypeptides are always linear, never branched like some of the polysaccharides. As we've seen, however, polypeptides can coil or fold into helices, sheets, balls, and other shapes. A protein may consist of a single, folded polypeptide chain or of several

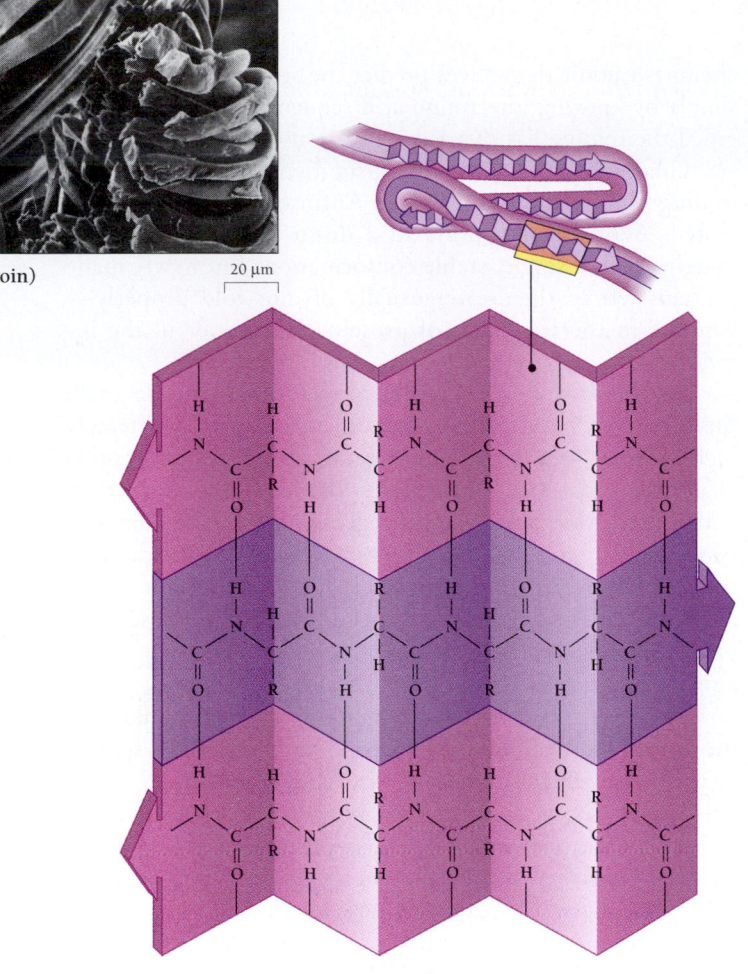

(fibroin)

20 µm

B. β (beta) structure

Figure 3-20 **What is silk made of?** A. The protein fibroin that makes up silk consists of β sheets. B. In a β sheet, polypeptide chains can snake back and forth past each other, held fast by hydrogen bonds between adjacent segments. Biochemists say that the segments that run in one direction are "antiparallel" to those that run the other way. *(A, K. Jakes, The Ohio State University)*

antibodies in the immune system, some act as transport proteins (e.g., hemoglobin), and some act as enzymes.

The tertiary folding of polypeptide chains results from interactions of the side chains on each amino acid. Each side chain has its particular characteristics. Some are polar, some nonpolar. Some can serve as proton donors in hydrogen bonds. Others serve as proton acceptors. Either a hydrogen bond or the attraction between positively and negatively charged side chains can stabilize a particular shape. Conversely, some interactions, such as the repulsion between two like-charged side chains, prevent certain shapes. The one characteristic of a polypeptide that most determines how it folds is the hydrophobic interactions of the nonpolar side chains.

Singly, any of these interactions contributes only slightly to a given molecular structure. Together, however, the various interactions establish a three-dimensional structure that can be extremely stable in certain chemical environments. Because the internal environment of a cell is relatively constant, each protein usually maintains its characteristic conformation. Subtle changes in the chemical environment of the cell, however, can change the shape of a protein and change the way it behaves.

Structural proteins tend to have regular sequences of amino acids and regular secondary structures, such as α helices and β structures. Globular proteins have nonrepeating amino acid sequences and irregular and unique shapes.

Are the Interactions Among Amino Acids Enough To Determine Protein Structure?

Just seconds after it forms, a polypeptide begins folding into exactly the right shape to fulfill its biological role. A polypeptide chain is as flexible as a snake and could easily take any of thousands of shapes. A short polypeptide just 100 amino acids long has 200 movable angles, not counting the R side chains.

In one study of more than 40 proteins, researchers found that about two-thirds of all hydrogen bonds link the hydrogen of one backbone amino group to the oxygen of a backbone carboxyl group. These hydrogen bonds are very specific. For example, in one protein, the 117th amino acid might link to the 245th amino acid every single time. What ensures that the 117th doesn't link to the 247th amino acid, or even the 10,005th? After all, the elements of the backbone are identical.

chains folded together. In proteins with more than one chain, the individual chains are held together tightly, so that, although made of several separate molecules, the protein behaves like a single, huge molecule.

The α helix, the β structure, and the collagen helix all share an important trait. As regular repeating forms, they are strong (Figure 3-20). Their simple shapes allow them to function as standardized building materials, comparable to the wooden beams and bricks with which a house is built. All fibrous proteins, whose role is to provide mechanical support, consist largely of these simple repeating forms.

In structural proteins, the α helix, β structure, and collagen helix each result from an extremely regular sequence of amino acids in the polypeptide chain. In the collagen helix, for example, every other amino acid is glycine. This smallest of the amino acids fits snugly in the interior of collagen's triple helix.

In globular proteins, the α helix, β structure, and other regular structures are composed of nonrepeating sequences of amino acids. As a result, globular proteins are more flexible in the way they fold and often form intricate shapes that allow them to play more diverse roles than the fibrous proteins. Some globular proteins act as hormones, for example, some act as

Despite all the ways that protein folding can go wrong, these remarkable molecules almost always seem to fold almost inevitably into their "native" functional shape. How do they do it?

The protein-folding problem has plagued biochemists for decades. In the 1950s, biochemists wondered if the sequence of amino acids was enough to exactly determine the precise three-dimensional structure of a protein. Might not some outside molecule somehow help direct the formation of a polypeptide? In that case, the shape of a polypeptide would depend not only on its amino acid sequence but also on the presence of some other "instructional" molecule.

In 1957, the biochemist Christian Anfinsen attempted to answer this question with a now classic experiment. He chose to study an enzyme called "ribonuclease," which breaks the bonds in RNA molecules. Anfinsen first exposed ribonuclease to a chemical solution that interfered with hydrogen bonds and hydrophobic interactions. This treatment changed the enzyme's shape and destroyed its ability to function. It could no longer break the bonds in RNA. When Anfinsen removed the ribonuclease from the harsh ("denaturing") solution, however, the ribonuclease spontaneously reverted back to its "native," functional shape. Once again, it could function properly, breaking bonds in RNA.

The enzyme's recovery required help from no other molecule. The amino acid sequence of ribonuclease and the right chemical environment were enough to determine the enzyme's three-dimensional structure and biological activity. Anfinsen's experiment suggested that proteins spontaneously fold into their biologically active forms, a conclusion that satisfied biochemists for many years.

Christian Anfinsen's experiment suggested that in a normal cell environment amino acid sequence alone determines the structure of a protein.

What Factors Determine How Proteins Fold?

For many years, biologists accepted the idea that in a normal cell environment, the amino acid sequence of a polypeptide implied one and only one three-dimensional structure. Yet biochemists cannot themselves predict the structure of a protein simply by knowing the amino acid sequence. What rules governed the folding has puzzled generations of biochemists.

Gradually, molecular biologists discovered that protein folding is more complicated than Anfinsen's experiment suggested. A pure polypeptide in a dilute solution may ultimately adopt a single, stable conformation. But newly made proteins left to themselves usually do not fold properly—whether in the test tubes of protein chemists or in the interiors of cells.

Several obstacles can prevent correct folding. Inside a cell, the concentration of proteins and other molecules is extremely high. At these high protein concentrations, individual peptide chains can easily form bonds with other peptides, rather than with themselves. Molecular biologists call such bonding between separate molecules *promiscuous interactions.* In addition, polypeptide chains are assembled gradually, one amino acid at a time. So one end of a chain is forming among all these other proteins, while the other end does not even exist yet. This is a recipe for promiscuous interactions.

To prevent the parts of polypeptide chain from linking to the wrong molecule or to the wrong part of itself, a staff of enzymes and special proteins, called **chaperones,** bind to the newly emerging polypeptide and prevent promiscuous interactions. Chaperones not only help other proteins fold into the correct shape, they also help proteins maintain that shape or change shape in order to play a new role.

Amino acid sequence alone was once thought to be sufficient to determine the three-dimensional structure of a protein. But enzymes and chaperone molecules guide the correct folding of a protein in a cell.

In this chapter, we have seen how the structure of a molecule—whether at the level of individual functional groups or at the level of amino acid sequences and tertiary structure—profoundly affects the chemistry of the molecule. The structure of a cell determines its behavior and capabilities just as profoundly, as we will see in the next chapter.

Study Outline with Key Terms

Other than water, nearly all molecules found in living organisms contain one or more carbon atoms. In fact, all **organic** molecules contain carbon. **Organic chemistry** is the study of the structures and reactions of carbon compounds, while **biochemistry** is the study of the chemistry of living organisms.

Most organic molecules are made of **hydrocarbon chains.** But the specific properties of small molecules depend on **functional groups,** such as the **hydroxyl, carboxyl,** and **amino** groups. The properties of **macromolecules** depend on the nature and sequence in which the component building blocks assemble.

Some hydrocarbon chains form rings of carbon. Flat, five- or six-carbon rings are called **aromatic** rings. Most macromolecules are linear, however. The small building blocks all assemble into different kinds of long-chain macromolecules, by the same kind of general reaction. This chemical reaction is called a **dehydration reaction** because the equivalent of one molecule of water is eliminated each time two building blocks join together. Conversely, the breaking down of a macromolecule into its constituent building blocks is called **hydrolysis** because the equivalent of one molecule of water is added as each small molecule is removed from the chain.

The small building blocks of cells fall into four classes—**sugars, amino acids, nucleotides,** and **lipids. Glycosidic bonds** link sugars into macromolecules called **polysaccharides. Peptide bonds** link amino acids together into **polypeptides,** and **phosphodiester bonds** link nucleotides into **nucleic acids.**

Although lipids do not form long chains, some, called **fatty acids,** do contain long carbon chains. Fatty acids may be **saturated, unsaturated,** or **polyunsaturated.** A **glycerol** molecule with two fatty acid chains and a phosphoric acid group is a **phospholipid,** an amphipathic molecule that forms cell membranes. A glycerol molecule with three fatty acid chains and a phosphoric acid group is a **triacylglycerol.** Different kinds of triacylglycerols make up the fats (solid at room temperature) and oils (liquid at room temperature) of animals and plants. A third class of lipids consisting of four interconnected rings is the **steroids,** including both **cholesterol** and steroid hormones such as progesterone and testosterone.

Simple sugars, or **monosaccharides,** are the building blocks of **disaccharides, oligosaccharides,** and **polysaccharides. Carbohydrates** include disaccharides, such as **glucose** and **sucrose,** and polysaccharides, such as **starch, glycogen, chitin,** and **cellulose.** For labels and signals, cells use **glycoproteins,** which are oligosaccharides attached to proteins.

The building blocks of nucleic acids are nucleotides. Each nucleotide consists of a **nitrogenous base,** either a **purine** or a **pyrimidine,** connected to a sugar, either **ribose** or **deoxyribose,** and (by way of the sugar) a phosphate group.

Many proteins, including **enzymes** and transport molecules such as hemoglobin, are dense, roundish molecules called **globular proteins.** But many structural proteins are **fibrous proteins**—long, thin molecules such as the keratin in hair or the collagen in skin and bone.

All polypeptides are made of strings of amino acids. Each amino acid has a carboxyl group, an amino group, and an R group, all attached to the **α carbon.** In each amino acid, all the parts are identical except for the R group. Proteins can have up to four levels of structure: **primary structure,** the sequence of amino acids; **secondary structure,** the coiling, bending, or folding of individual sections of the polypeptide chain into forms such as the **α helix,** the **β pleated sheet,** and the **collagen helix; tertiary structure,** or **conformation,** the three-dimensional folding of an entire polypeptide chain; and **quaternary structure,** the fitting together of two or more folded chains. Different regions of the polypeptides of a protein are called **domains.**

The folding of a polypeptide into a precise three-dimensional structure generally requires energy. Some of this energy comes from interactions among the atoms of the polypeptide backbone and those of the amino acid side chains. Additional energy may come from **ATP,** which is used by **chaperones** that help proteins fold.

Review and Thought Questions

Review Questions

1. What evidence persuaded biologists that scrapie could not be caused by a bacterium or virus?
2. Where would you find a large concentration of lipids in a cell? Why?
3. What building blocks make up proteins? How many kinds of these building blocks are there?
4. What three components make up a nucleotide?
5. What small molecule is removed during a dehydration reaction? What is the opposite reaction called? What happens in that case?
6. What is the relationship between polysaccharides, sugars, and carbohydrates? What three roles do carbohydrates play in cells?
7. Define peptide bond, polypeptide, protein, and enzyme.
8. Define nucleic acid.
9. Describe the four levels of structure that occur in protein folding.
10. Distinguish between the α helix, the β structure, and the collagen helix, all of which occur in fibrous proteins.

11. What is the major component of each of these materials? Use C (carbohydrate), P (protein), and L (lipid) to answer.

   ____Cotton T shirt   ____Cooking oil
   ____Chicken nugget   ____Wooden toothpick
   ____Apple juice   ____Cheese
   ____Silk scarf   ____Fat on a piece of steak
   ____Tortilla   ____Newspaper
   ____Rice
   ____Lipstick

Thought Questions

12. What is it about the shape of a protein that determines its properties and therefore its functions? Describe the shape of a globular protein and of a fibrous protein. How are these two shapes usually put to different purposes in a cell?
13. The 35 small molecules listed in Table 3-2 are called building blocks. How can the same building blocks be used to construct billions upon billions of different macromolecules?

About the Chapter-Opening Image

This white lab mouse is constructed with building blocks. These represent the building blocks of life.

 On-line materials relating to this chapter are on the World Wide Web at **http://www.harcourtcollege.com/lifesci/aal2/**

4 Why Are All Organisms Made of Cells?

Very Little Animalcules

More than 300 years ago, an uneducated Dutch cloth merchant named Antonie van Leeuwenhoek (1632–1723) discovered something so interesting that he received visits from the Czar of Russia, King James II of England, and King Frederick II of Prussia. Like many young men, Leeuwenhoek had a consuming interest that distracted him from his work (Figure 4-1). He liked to make glass lenses by grinding and polishing bits of glass and then mounting the lenses between gold or copper plates. He used these simple, single-lens microscopes to examine anything he encountered.

To Leeuwenhoek's delight, his lenses—some no larger than a pinhead—revealed hundreds of tiny living beings never before seen by human eyes. Leeuwenhoek astonished his friends and acquaintances with descriptions of red blood cells in blood, sperm in semen, and "very little animalcules" in pond water. With his lenses, Leeuwenhoek also demonstrated that fleas, ants, weevils, and other pests do not arise spontaneously from dust or wheat, as was commonly believed, but developed from larvae that hatch from tiny eggs laid by the adults.

In 1673, when Leeuwenhoek was 41, he began sending reports of his discoveries to the recently created Royal Society in London, then one of the few scientific organizations in the world. Over the next 50 years, until Leeuwenhoek's death in 1723, the Society received and published 375 letters from him, the last when the Dutch lens grinder was 90.

Over the years, Leeuwenhoek described how everywhere he looked, he found tiny organisms whose existence few people had even suspected. In 1683, he became the first person to see bacteria. Leeuwenhoek's great contribution was to demonstrate that life is not limited to organisms visible to the naked eye.

In England, the Royal Society welcomed Leeuwenhoek's fascinating observations. The Society had already hired Robert Hooke to be its "curator of instruments." Hooke's job was to demonstrate the microscope and other new technologies for the entertainment of the gentlemen at the Royal Society's meetings. In 1665, Hooke published his book *Micrographia*, a series of illustrations and descriptions of his observations using new and improved microscopes. Hooke showed, for example, the structure of feathers, the stinger of a bee, and the foot of a fly.

Hooke also coined the term "cellulae" [Latin, *cella* = small room] to describe the boxlike cavities (cells) he saw when he examined a slice of cork under a microscope. "Cells," in those days, referred to the small rooms of a monastery, nunnery, or prison. Hooke later published detailed drawings of cells in other plant tissues, noting that they contained "juices." Leeuwenhoek, Hooke, and the few other microscopists of the day described a whole new world of tiny living organisms, and their research revealed an unsuspected level of structural complexity in larger organisms. These discoveries might well have stimulated the beginning of a new science. Yet, oddly, these discoveries lay dormant for nearly 200 years.

Neither Leeuwenhoek's nor Hooke's studies led directly to any unifying biological principles. Most scientists dismissed Leeuwenhoek's animalcules as little more than amusing natural curiosities. Hooke's "cellulae" were simply bubbly stuff in plants. No one guessed that cells come from other cells by the process of cell division, that cells are the fundamental unit of life, and that all tissues in all organisms are made of cells. No one guessed that some microscopic organisms cause disease. To most people, Leeuwenhoek's microorganisms and Hooke's

- Every cell has a boundary, a cell body, and a set of genes. Different kinds of cells have characteristic organelles and other structures.

- In eukaryotic cells, internal membranes enclose organelles and help regulate the chemical composition of the spaces inside. Prokaryotic cells have no internal membranes or compartments.

- The cytoskeleton contributes to the organization and internal communication of eukaryotic cells.

cells seemed no more important than the specks of dust in a ray of sunshine.

One reason that the important thinkers took neither Hooke nor Leeuwenhoek seriously was the two men's social class. Leeuwenhoek was an uneducated amateur, and Hooke was an employee of the Royal Society, not a full member. Neither man was, in other words, a "gentleman," so neither commanded respect.

Equally important, 17th-century scientists and philosophers carried a bias, left over from the Middle Ages, that valued theory over experiments and observations. Hooke's and Leeuwenhoek's microscope work was entirely descriptive. Leeuwenhoek described thousands of objects but, probably because of his lack of education, he did not attempt to present any unifying explanation for why things were as they were.

Most scientists saw the microscope as more of a toy than a scientific instrument. By the end of Leeuwenhoek's life, at the beginning of the 18th century, few people were even making lenses anymore. Those simple microscopes already in existence were playthings for ladies, of whose observations there is little record.

> Leeuwenhoek showed that life is not limited to organisms visible with the naked eye.

Another important reason that study of the microscopic details of life languished was the poor quality of 17th- and 18th-century microscopes. Leeuwenhoek's handmade lenses were far better than any others in existence. Some of the 400 lenses he left behind could magnify objects 300 times, to show the largest bacteria. And his mysterious techniques—probably sophisticated lighting—which he never revealed, allowed him to see details that neither Hooke nor any of the other prominent microscopists of the time could match. Still, even Leeuwenhoek's microscopes were crude by today's standards (Box 4.1). His finely polished lenses distorted both the shapes and the colors of the objects under the microscope. It was not until the late 19th century, 100 years after the beginning of the Industrial Revolution, that technology provided microscopes adequate for the study of the insides of cells.

WHY ARE ALL ORGANISMS MADE OF CELLS?

In the 1820s, better microscopes paved the way for rapid advances in cell biology. Within 10 years, biologists discovered two important parts of cells. They found that both animal and plant cells contain a nucleus, a dark mass that we now know contains the genetic material (DNA). Researchers also finally realized that the "juice" observed by Robert Hooke was a living substance, which they named **protoplasm** [Greek, *proto* = first + Latin, *plasma* = a thing molded or formed].

All Organisms Are Made of Cells

In 1838, the German botanist Matthias Schleiden proposed that all plants consist of cells—an idea seconded for animal cells in 1839 by German zoologist Theodor Schwann. Cells, they wrote, are the elementary particles of all living organisms. In addition, Schwann and Schleiden argued that all cells are alive—independent of the organisms to which they belong. In other words, a liver cell is alive in its own right even though it is only a small part of a living organism. Remove the cell from the liver and the cell is still alive.

Recall from Chapter 1 that animals and plants are all multicellular, that the Fungi and Protista include both multicellular organisms and single-celled organisms, and that the Archaea

Figure 4-1 **Antonie van Leeuwenhoek** built primitive microscopes so effective that he saw bacteria 200 years before anyone else. (*Science Photo Library/Photo Researchers, Inc.*)

(Text continued on page 74)

BOX 4.1

How Do Microscopes Help Biologists Study Cells?

An important way to learn about cells is to look at them. Each major improvement in microscopy has led to a burst of new knowledge and a host of new questions. In the 17th century, the development of the microscope made it possible to see cells for the first time. In the 19th century, successive improvements in microscope design revealed that all organisms are made of cells and that their elaborate adaptations allow specific functions. By the mid-20th century, the development of new types of microscopes had further revolutionized the study of cells—revealing the rich internal structure of cytoplasm, nuclei, and membranes.

How can we see such details? We can see an object only when it is large, clear, and stands out from its background. In order to see an object as small as a cell, first we must use lenses to magnify its image. The ratio between the size of the image and the size of the object itself is called the *magnification.*

Even when we can magnify an object, however, we may not be able to see it clearly enough to learn anything about it. When small objects are too close together, we cannot tell whether we are looking at one or several objects. The *resolution* of a microscope or other optical instrument is the minimum distance between two objects that allows them to form separate images. We can say that resolution is a measure of image clarity. Our eyes have a resolution of about 0.1 mm. A microscope resolves images less than 1/100th that size (0.001 mm or 1 μm). Such resolving power is enough to distinguish individual prokaryotic cells and some internal structure in eukaryotic cells.

Visibility depends not only on magnification and resolution but also on *contrast*—how well an object stands out from its background. Contrast depends on an object's tendency to absorb more or less light than its surroundings. A transparent

object is invisible. So is an object that is the same shade as its background.

Microscopists have devised ingenious ways of increasing contrast. Nineteenth-century biologists, for example, found that different dyes would bind to different parts of the cell—one dye to the nucleus, another to the membrane, another to parts of the cytoplasm. Dyes that bind differently to cell components are called stains.

But staining cells usually kills them, so early biologists could study living cells only with difficulty. Beginning in the 1930s, biologists developed techniques for studying living cells. In conventional light microscopy, contrast depends on how different cellular structures absorb light. Most materials in unstained cells absorb little visible light. However, substances that do not absorb or reflect much light can still change the speed at which light passes through. For example, light passes through the cell nucleus more slowly than it does through the watery cytoplasm. The light passing through the nucleus becomes "out of phase" with light waves that pass through the cytoplasm. Special optical arrangements, such as "phase contrast" and "interference contrast" microscopy, can heighten this effect, revealing phase differences among different cell components without the use of stains. These techniques permit the study of living cells.

Figure A(a) shows the workings of an ordinary light microscope. Because a light microscope contains several lenses, it is also called a *compound microscope.* It consists of a tube with lenses at each end, a stage that holds a specimen, and a light source. The light source has its own lens, which allows the user to control the light that passes through the specimen. The lens near the eye is called the *eyepiece,* and the lens at the other end of the tube, near the object being examined, is called the *objective.*

The objective lens magnifies the image of the specimen and projects the im-

age to the eyepiece. The eyepiece magnifies the image again. The final magnification is the product of the magnification of the objective and that of the eyepiece. For example, an objective that magnifies 40 times (40×) and an eyepiece that magnifies 10 times (10×) would produce an image 400 times (400×) as large as the original specimen.

A light microscope often has three objective lenses (for example, with magnifications of 10×, 40×, and 100×) mounted on a rotating turret, so that the viewer can change magnifications. Because eyepieces usually contribute another 10-fold magnification, the final image usually ranges from 100 to 1000 times the size of the specimen.

Even the best lenses, however, have a resolution limited by the nature of light. Light is composed of waves, and lenses cannot resolve structures smaller than the wavelength of light. The wavelength of visible light ranges from about 0.4 mm (for blue light) to about 0.7 mm (for red light). Light microscopes reveal no details of subcellular structures smaller than about 0.5 mm.

One solution is to use wavelengths far smaller than those of visible light. Electron microscopes, which see objects as small as 0.002 mm, depend on the wave properties of moving electrons. In a 100,000-volt electron microscope, the wavelength of an electron is 0.004 nanometers (nm), or 0.000004 mm, less than the diameter of an individual atom.

The lenses of an electron microscope are electromagnets that bend the path of electrons just as a glass lens bends the path of light. In reality, the resolution for looking at cells is only about 2 nm (0.002 μm) but still about 100 times better than the best light microscope (Figure A(b) and (c)).

Biologists commonly use two kinds of electron microscopes—the transmission electron microscope (TEM) and the scan-

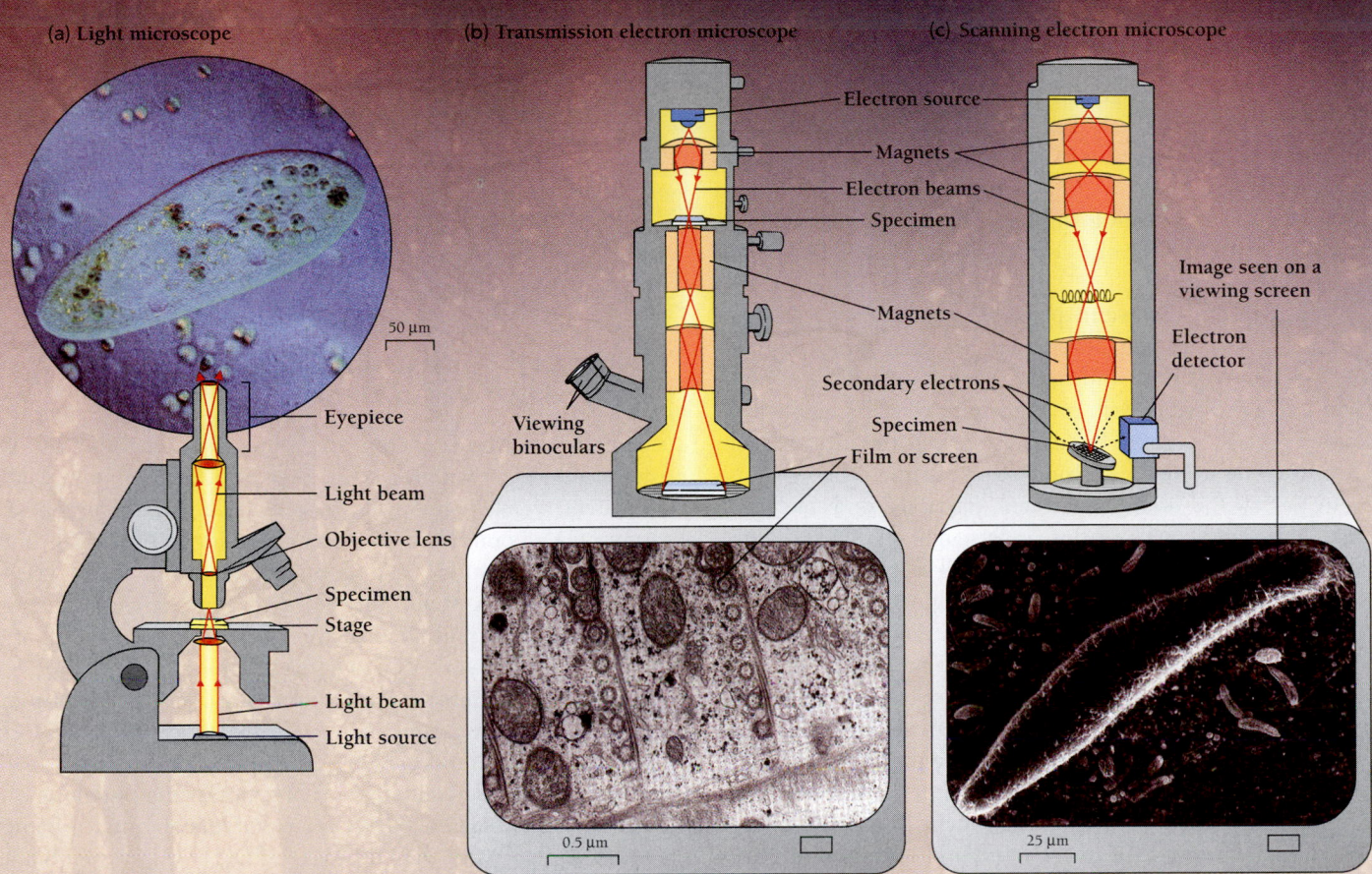

(a) Light microscope

Eyepiece
Light beam
Objective lens
Specimen
Stage
Light beam
Light source

50 μm

(b) Transmission electron microscope

Electron source
Magnets
Electron beams
Specimen
Magnets
Viewing binoculars
Film or screen

0.5 μm

(c) Scanning electron microscope

Electron source
Magnets
Image seen on a viewing screen
Electron detector
Secondary electrons
Specimen

25 μm

Figure A **Views of *Paramecium* with different types of microscopes.** (a) With a light microscope, we see the internal anatomy of this one-celled organism and a fringe of hair-like cilia, which these creatures use to propel themselves through the water. (b) With a transmission electron microscope, we see a cross section through three rows of cilia at much greater magnification and in great detail. (c) With a scanning electron microscope, set at low power, we see the cilia protruding from the surface in greater detail than with the light microscope but nothing of the *Paramecium*'s internal anatomy. (a, Eric Grave/Phototake; b, David Phillips/Visuals Unlimited; c, Dr. Dennis Kunkel/Phototake)

ning electron microscope (SEM). In each case, the electrons form an image on a phosphorescent screen, like that of a television set. Figure A compares a *Paramecium* seen with a light microscope, TEM, and SEM.

In TEM, a beam of electrons passes through the specimen in the same way that a beam of light would in a light microscope. SEM is different. In SEM, a narrow beam of electrons moves back and forth across the surface of the specimen. Wherever the beam hits the surface of the spec-

imen, it causes the surface to emit more electrons, which produce a detailed, almost three-dimensional, picture of the specimen's surface.

Another powerful technique for examining the structure of cells relies on the immune response. One of the ways that mammals resist infection is by making proteins, called antibodies, that form complexes with foreign molecules, such as those of infectious bacteria. Biologists can induce experimental animals to produce antibodies against specific cell components, such

as specific proteins in the cilia of *Paramecium*. These antibodies can form complexes with the specific protein, allowing biologists to visualize its subcellular distribution. Before doing such a study, a researcher may link the antibodies to a fluorescent dye. A special microscope can focus a beam of light from above, and the researcher can look for fluorescence, which shows the distribution of the particular subcellular component.

and Eubacteria are all single-celled organisms. Noncellular organisms do not exist. With the discoveries of Schleiden and Schwann, biologists recognized the importance of cells. Cells, they saw, were everywhere. But where did cells come from? Where did the new cells that heal a wound come from? Where did the cells that build a child's growing body come from? Where did the burgeoning masses of cells in cancers come from? And where did the one-celled organisms living in a pond come from?

Schleiden and Schwann proposed that cells crystallized spontaneously out of shapeless matter. But ever-improving microscopes allowed more and more scientists to see that individual cells can divide, each giving rise to what biologists call "daughter" cells. In time, biologists accepted the idea that all new cells come from the division of preexisting cells.

In 1858, the widely respected physician and biologist Rudolf Virchow formalized this understanding with the Latin phrase, *omnis cellula e cellula,* meaning "all cells from cells." In a book written for physicians, Virchow summarized his theory about the role that cells play in disease. In doing so, he simultaneously revolutionized both biology and medicine. Virchow argued (1) that cells never arise from noncellular material and (2) that diseases result from changes in specific kinds of cells.

As a result of these arguments, biologists came to accept the principle that cells always come from other cells. Cells healing a wound result when preexisting skin cells begin dividing. A child's growth results from the rapid multiplication of cells. Cancers result from the uncontrolled division of an organism's cells. Single-celled organisms in ponds descend from other single-celled organisms.

Today, we summarize the work of Schleiden, Schwann, and Virchow in the **cell theory,** which states that (1) all organisms are composed of one or more cells; (2) cells, themselves alive, are the basic living unit of organization of all organisms; and (3) all cells come from other cells. As we dis-

cussed in Chapter 1, we use the word "theory" to denote a system of statements and ideas that explains a group of related facts or phenomena, *not* to indicate uncertainty.

The "cell theory" says that all organisms are composed of one or more cells, that cells are the basic living unit of organization of all organisms, and that all cells come from other cells.

Every Cell Consists of a Boundary, a Cell Body, and a Set of Genes

The cells of organisms from the three domains of life can differ greatly. Even within a single organism, cells come in wildly different types. Nonetheless, all cells have three common features: (1) a boundary that separates the inside of the cell from the rest of the world, (2) a set of genes, and (3) a cell body.

The **plasma membrane** is a cell's boundary, a highly organized and responsive structure. Like the outer wall of a house, with its windows and doors, the plasma membrane not only defines the limits of a cell, but also helps to regulate the cell's internal environment by selectively admitting and excreting specific molecules.

Each cell also contains a set of genes strung out along one or more molecules of DNA. In eukaryotes (animals, plants, fungi, and protists), the DNA is confined within a membrane-enclosed structure called the **nucleus** (Figure 4-2A). The presence of a true nucleus alone defines the eukaryotes and distinguishes them from the prokaryotes (eubacteria and archaebacteria). In the prokaryotes, the DNA occupies a limited region of the cell, called the **nucleoid,** which has no membrane (Figure 4-2B).

Throughout the 19th century and well into the 20th century, biologists continued to think of the protoplasm as a more

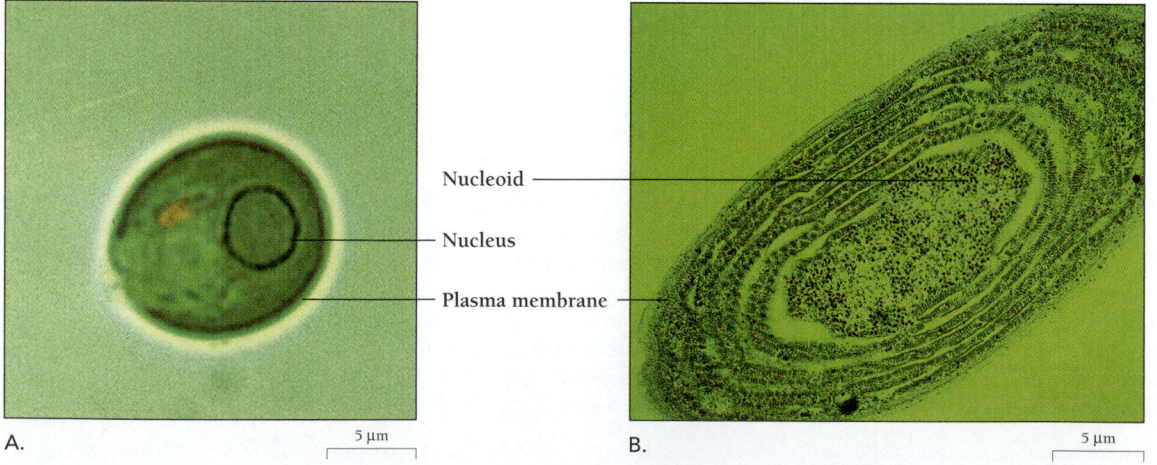

Nucleoid

Nucleus

Plasma membrane

A.   5 µm   B.   5 µm

Figure 4-2   Eukaryotes have a nucleus. A. A eukaryotic cell, such as the green alga *Chlamydomonas,* has both a plasma membrane and a nucleus. B. A prokaryotic cell, such as a cyanobacterium, has a plasma membrane, but no nucleus. *(A, Philip Sze/Visuals Unlimited; B, Elizabeth Gentt/Visuals Unlimited)*

or less homogeneous jellylike substance. Improved techniques for looking at cells, however, revealed more and more tiny structures in eukaryotic cells, though not in prokaryotic cells. We now know that each of these tiny structures, called **organelles,** performs a specialized task.

Biologists finally discarded the word protoplasm and renamed the cell body—that part of the cell outside the nucleus but inside the membrane—the **cytoplasm** [Greek, *cyto* = a hollow container (a cell) + Latin, *plasma* = a thing molded or formed]. The part of the cytoplasm not contained within membrane-enclosed organelles was named the **cytosol.** Most of a cell's biochemical work occurs within the cytosol. Running through the cytosol is an intricate network of protein fibers, called the **cytoskeleton,** which gives the cell its shape, holds organelles in place, and participates in cell movement.

Every cell is enclosed by a plasma membrane and keeps its DNA in a nucleus (eukaryotes) or nucleoid (prokaryotes). Eukaryotic cells contain other organelles within the cytoplasm.

How Are Cells Alive?

Cells—from Leeuwenhoek's one-celled "animalcules" to the individual cells of the liver—are the fundamental living units of life. They have all the characteristics of life that we listed in Figure 1–8. Made up of organized parts, cells perform chemical reactions, obtain energy from their surroundings, respond to their environments, change over time, reproduce, and share an evolutionary history. Cell components work together to achieve homeostasis, maintaining an internal cellular environment that is constant enough to carry out the thousands of chemical reactions that allow cells to live and to reproduce.

The ability of cells to obtain energy from their environment depends on a set of chemical reactions, some of which we discuss in Chapters 6 and 7. Almost every cell can capture energy from glucose, usually by oxidizing it to carbon dioxide and water. In eukaryotic cells, many of the energy-producing reactions take place in specialized organelles, while others take place in the cytosol.

Cells change with time, both chemically and mechanically. For example, individual muscle cells shorten as muscles contract. Other cells, too, change their size and shape. To allow for these changes, almost every eukaryotic cell contains a network of fibers, called the cytoskeleton, that supports the cell's components and brings about change in cell shape. The cytoskeleton also serves as train tracks along which cell components can move, pulled by locomotives called *molecular motors*.

In order to reproduce, every cell must be able to copy its genes for its descendants. All cells have elaborate mechanisms for precisely duplicating DNA.

Cell reproduction also depends on the cell's ability to read and to use the coded information in DNA. In a single-celled organism, cells adjust the readout of their genetic information in response to environmental changes. A bacterial cell growing in a glass of milk, for example, makes different proteins from the same type of cell growing in a glass of sugar water. In a multicelled organism, different types of cells make different sets of proteins, according to their specialized characteristics: a red blood cell, for example, makes lots of hemoglobin, whereas a muscle cell makes lots of contractile proteins.

Once a cell produces the appropriate proteins, however, it must deliver them to the right address within the cell. Cells have elaborate mechanisms for directing molecular traffic to different parts of the cell.

Finally, cells must deal with their own garbage—waste products such as carbon dioxide and ammonia, as well as cell remnants damaged by ordinary wear and tear. Special molecules and mechanisms serve as cellular toilets and garbage disposals.

A cell could conceivably employ a wide variety of solutions to its problems. Different cells might use different strategies to derive energy, to distribute ions, and to organize its organelles. But all cells use the same molecules, share the same biochemical pathways, and contain the same types of organelles. This sharing of solutions reflects the common evolutionary heritage of all cells, which almost certainly descended from a common ancestor that lived about 3.5 billion years ago.

What Are the Advantages of Cellular Organization?

Single-celled organisms are nearly all very small, but there are some exceptions. Plants, animals, and other multicellular organisms can be as large as trees and elephants, but a free-living cell never grows that big. In fact, with some remarkable exceptions, cells are nearly all about the same size. Most eukaryotic cells range from 10 to 100 micrometers (μm), while prokaryotic cells are smaller, ranging from 0.4 to 5 μm. (For comparison, a period on this page is about 100 μm in diameter, the size of a large eukaryotic cell, such as those in an onion skin.) No cells are smaller than 0.4 μm, and almost none are larger than 100 μm. Exceptional cells may be enormous. (See Web Bit 4–1 on *Caulerpa*, the world's largest single-celled organism.) Some plant fiber cells are a meter long (1 million μm), while the nerve cells in the legs of a giraffe may be several meters long.

One reason that cells are nearly all so small is that subdivision into tiny cells offers organisms many advantages. To understand these advantages, we can start by asking a simple question: What problems would you face if you were one very large cell?

A cell's need to regulate its internal environment limits its size

For the biochemical machinery of a cell to function, the cell must maintain a relatively constant internal environment. Otherwise, enzymes and other proteins will not adopt the shapes needed for biological activity. To maintain a constant internal environment, the cell must keep both the concentration of salts and the pH from changing. At the same time, the cell must also be able to take in useful molecules and dispose of waste molecules.

If you were one big cell, you would have to maintain just the right environment for each of the different molecular activities in your body. The highly acidic (low) pH that you use to digest food would have to somehow coexist with the neutral pH needed for protein synthesis.

All the materials that come into a cell or leave a cell must go through the thin membrane that envelops the cell. If you were one big cell, your plasma membrane would have to admit or exclude specific molecules selectively. Somehow you'd have to absorb enough pizza to get you through the day.

Protists, such as the *Didinium* pictured in Figure 4-3, perform all these functions and more. In fact, unlike most protists, the *Didinium* even has a sort of mouth, called a cytostome [Greek, *cyto* = cell + *stoma* = mouth], with which it is eating another protist (a *Paramecium*). Protists, however, are all small. Accomplishing all that a protist does wouldn't be so easy for a one- or two-hundred-pound cell like yourself. The membrane's ability to admit or to excrete molecules is limited by its size. A big cell certainly has more external membrane than a small one. As a cell grows, however, its volume increases faster than its surface area.

Let's think, for a minute, about the relationship between the surface area and the volume of any object. Suppose we have a tiny cardboard cube, one centimeter on each side. If we double the linear dimensions of the cube, so that each side is 2 cm by 2 cm, the surface area of the cube increases by a factor of 4. But its *volume* increases by a factor of 8 (Figure 4-4).

The larger cube has proportionately less surface area than the smaller cube. We can see this if we look at the **surface-to-volume ratio**—the amount of surface area for each bit of volume. For the small cube, the surface-to-volume ratio is 6:1. For each cm^3 of volume, the smaller cube has 6 cm^2 of surface. But the surface-to-volume ratio for the larger cube is 24:8 (or 3:1). For each cm^3 of volume, the larger cube has only 3 cm^2 of surface.

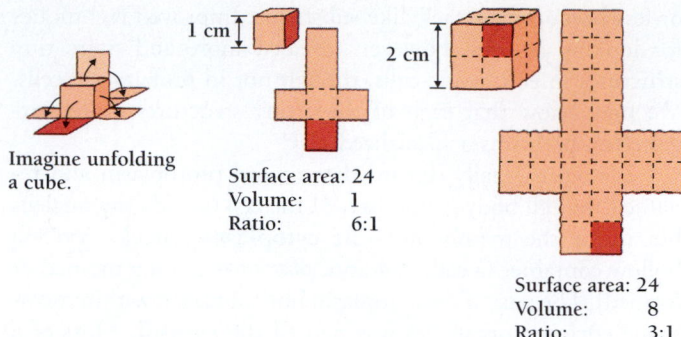

Surface area: 24
Volume: 1
Ratio: 6:1

Surface area: 24
Volume: 8
Ratio: 3:1

Figure 4-4   The relationship between surface area and volume determines many of the properties of cells. In particular, smaller cells have more surface area per volume than do larger cells. Here we can see that a cube 1 cm on a side has a surface area of 1 cm^2 on each of its 6 sides, so that it has a total surface area of 6 cm^2. Its volume is 1 cm^3. When we double the linear dimensions of the cube, we have a cube 2 cm on a side, which gives a surface area of 24 cm^2. Its volume is 8 cm^3. We can see that when the linear dimensions of an object double, surface area increases four times, while volume increases eight times.

Because larger objects have smaller surface-to-volume ratios, larger cells have proportionately less membrane with which to regulate their internal space. As a result, larger cells have a harder time obtaining nutrients, getting rid of wastes, and regulating the internal concentrations of ions and molecules. If you were one big cell, molecules of oxygen and food would take decades to get from your membrane to your interior. You could bury yourself in pizza and you'd still go to bed hungry. You simply would not be able to digest the pizza's macromolecules fast enough to survive.

Knowing how a cell's size affects the regulation of its internal environment, we might wonder how eukaryotic cells, which range from 10 to 100 μm, can be so much larger than prokaryotic cells, which range from 0.4 to 5 μm. The answer is that eukaryotic cells have special adaptations to increase their surface areas. For example, many eukaryotic cells have convoluted surface membranes, and almost all have elaborate internal membrane systems, which we will discuss later in this chapter. Indeed, some eukaryotic cells, such as some white blood cells, have 50 times more membrane inside than outside.

The larger a cell, the smaller is its surface-to-volume ratio. A low surface-to-volume ratio limits a cell's ability to absorb nutrients and expel wastes.

Individual cells may specialize for different tasks

In multicellular organisms, individual cells often perform separate tasks. For example, red blood cells carry oxygen, while muscle cells contract and move. Cells in the roots of a tree absorb nutrients from the soil, while cells in the leaves harvest the energy of sunlight. Just as a division of labor makes hu-

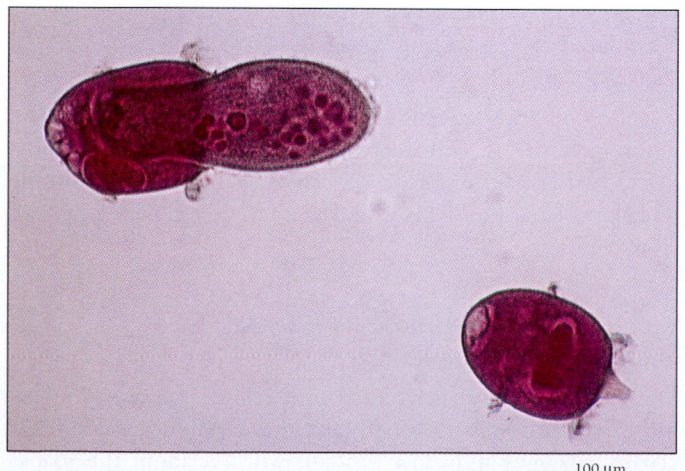

100 μm

Figure 4-3   *Didinium* eating a *Paramecium*. *Didinium* has a mouth with which it can consume other protists. *(Stan W. Elems/Visuals Unlimited)*

TABLE 4-1 THE CELLS OF ALL ORGANISMS HAVE BOTH COMMON AND DISTINCTIVE FEATURES

Cell component	Function	Eubacteria Archaebacteria	Protist	Fungus	Plant	Animal
Cell wall	Protects and supports cell	Yes	Yes	Yes	Yes	None
Plasma membrane	Regulates communication with other cells and movement of materials in and out of cell		Yes	Yes	Yes	Yes
Membrane-enclosed nucleus	Isolates DNA from cytoplasma	None	Yes	Yes	Yes	Yes
DNA	Encodes information for proteins	Single molecule	Many molecules			
RNA	Constructs proteins from information in DNA	Yes	Yes	Yes	Yes	Yes
Nucleolus	Synthesizes ribosomes	None	Yes	Yes	Yes	Yes
Ribosome	Synthesizes proteins	Yes	Yes	Yes	Yes	Yes
Endoplasmic reticulum	Modifies proteins, makes lipids	None	Yes	Yes	Yes	Yes
Golgi complex	Tags proteins and lipids, packages them for export	None	Yes	Yes	Yes	Yes
Lysosome	Contain digestive enzymes	None	Yes	Yes	Yes	Yes
Mitochondrion	Synthesize ATP	None	Yes	Yes	Yes	Yes
Chloroplast	Photosynthesize	None	Some	None	Yes	None
Nonchloroplast plastids	Store food, pigments	None	Some	None	Yes	None
Central vacuole	Provides turgor pressure; contains water and wastes	None	None	Yes	Yes	None
Cytoskeleton	Shapes, strengthens, and moves	None	Yes	Yes	Yes	Yes
Flagellum, cilium	Moves cell through fluid or fluid past cell surface	None	Yes	Yes	Yes	Yes
Bacterial flagellum	Moves cell through fluid or fluid past cell surface	Yes	None	None	None	None

man societies more productive, the specialization of cells allows an organism to function efficiently.

If you were one big cell, organizing your body to perform all its different jobs would be difficult. You wouldn't have a heart or arteries to move oxygen and nutrients around your body. You wouldn't have lungs to extract oxygen from the air. You'd just have 125 pounds (or so) of cytoplasm.

Cellular organization allows organisms to make a division of labor among specialized cells.

A multicellular organism can lose and replace individual cells

Subdividing the body into cells has another advantage. Cells often live and die independently of the whole organism. In fact, the death of some cells is a normal part of development. The fingers of the hand, for example, develop from paddlelike appendages as a result of the death of cells between the fingers. The lenses of the eyes, the red blood cells that carry oxygen to all the tissues, and the outer layer of the skin all develop because of the death of specific cells. Equally important, the cells of the skin, blood, and intestines continuously replace themselves. The advantage of such replacement is that the life of a multicellular organism can extend far beyond the life of an individual cell.

Cellular organization allows organisms to outlive the cells that compose them.

WHAT'S IN A CELL?

Eukaryotic cells contain so many different organelles, each with a special set of tasks, that we can imagine each cell as a tiny walled city (Figure 4-5). Inside the wall of the city are power stations, a central library of genetic information, warehouses for packaging proteins, and much more. Table 4-1 summarizes the characteristics of cells from the six kingdoms of organisms.

When scientists first began using the transmission electron microscope in the 1940s and 1950s, one of the great surprises was the huge amount of internal membrane in

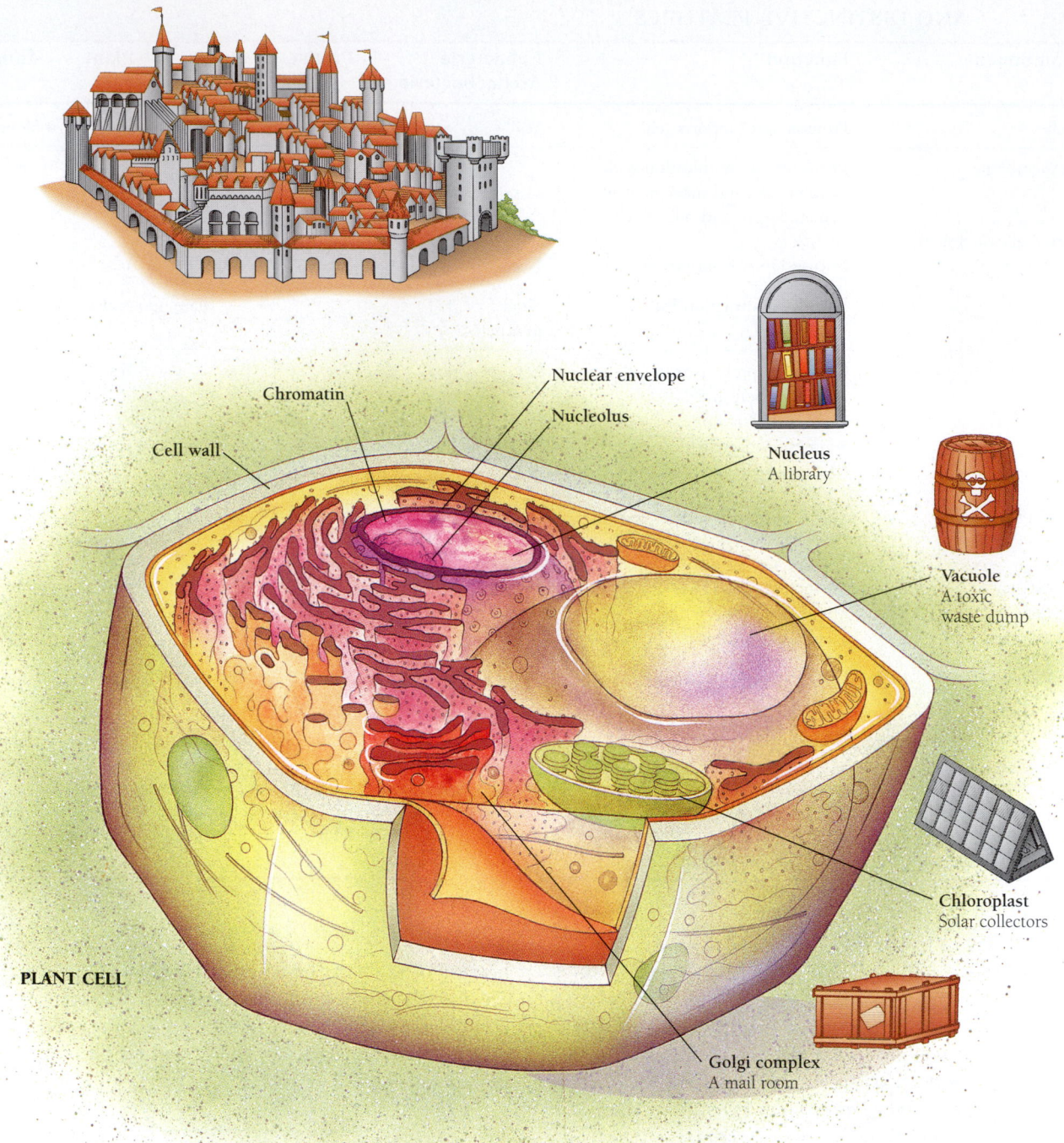

Chromatin

Nuclear envelope

Nucleolus

Cell wall

Nucleus
A library

Vacuole
A toxic
waste dump

Chloroplast
Solar collectors

PLANT CELL

Golgi complex
A mail room

Figure 4-5   **Two eukaryotic cells, a plant cell and an animal cell. A eukaryotic cell, with its plasma membrane and membrane-enclosed organelles, is like a walled medieval city.** The plasma membrane regulates the passage of molecules in and out of the cell, just as the city wall regulates the passage of people. Inside, the mitochondria supply power to the cell in the form of ATP. The nucleus is a library, housing all of the information a cell needs to produce proteins. The nuclear membrane determines which molecules have access to the information inside and which bits of information (books) may be checked out for use in the cytoplasm. The ribosomes act as workbenches for the production of proteins. The vacuole serves as a depository for waste materials, much like a toxic waste dump. The Golgi complex is like a mail room. Here, glycoproteins are packaged and labeled for export or intracellular transport. See if you can think of analogies for some of the other structures in these cells. To what would you compare a chloroplast?

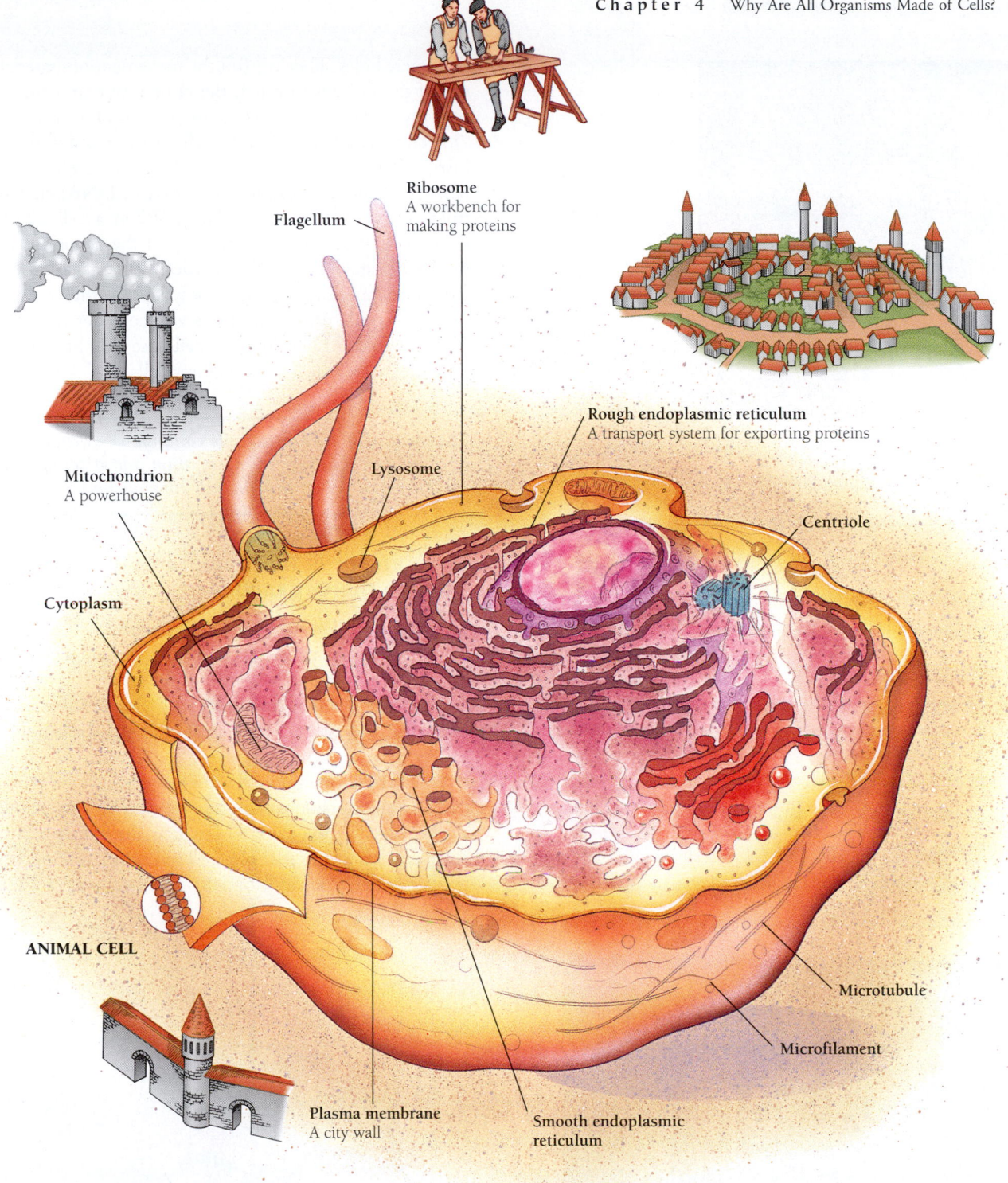

Flagellum

Ribosome
A workbench for
making proteins

Rough endoplasmic reticulum
A transport system for exporting proteins

Lysosome

Centriole

Mitochondrion
A powerhouse

Cytoplasm

Microtubule

Microfilament

ANIMAL CELL

Plasma membrane
A city wall

**Smooth endoplasmic
reticulum**

eukaryotic cells (Box 4.1). These membranes divide cells into at least seven kinds of compartments: nucleus, cytosol, endoplasmic reticulum, Golgi complex, lysosomes, peroxisomes, and mitochondria. In addition, a plant cell contains one more set of compartments—plastids.

Eukaryotic cells also contain a variety of membranous sacs with nothing visible inside. The **vesicles** of plant and animal cells are small, apparently empty, sacs—generally about 100 nanometers (nm) in diameter. Eukaryotic cells (especially plant

cells) also contain **vacuoles**—large sacs that may occupy as much as 95 percent of the volume of the cell.

What Role Does the Nucleus Play in the Life of a Cell?

The nucleus usually makes up 5 to 10 percent of the volume of the cell. In cells that are not dividing, the nucleus stains as a diffuse mass. This pattern results from the binding of stains

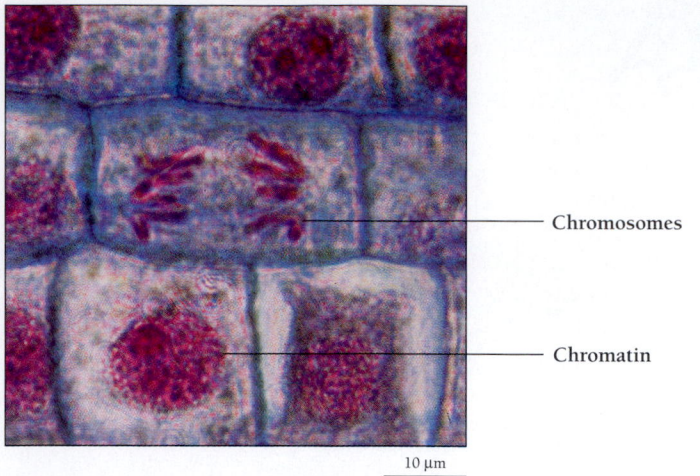

10 μm

Figure 4-6 **Chromatin in the nucleus of an onion cell.** Chromatin condenses in dividing cells, making the chromosomes visible. *(Runk/Schoenberger from Grant Heilman)*

Chromosomes

Chromatin

to **chromatin** [Greek, *chroma* = color], a complex of DNA and protein. In cells that are dividing, the chromatin forms distinct objects, visible in the microscope, called **chromosomes** [Greek, *chroma* = color + *soma* = body] (Figure 4-6).

An electron microscope shows that the nucleus's boundary is a double membrane called the **nuclear envelope** (Figure 4-7). The two membranes are separated by a space about 20 to 40 nm wide. In many places, however, the two membranes are fused to form interruptions called **nuclear pores,** which form channels between the nucleoplasm (the contents of the nucleus) and the cytoplasm. These pores resemble the holes in a tea ball that allow water to enter the tea ball and tea to leak out into the surrounding water. However, nuclear pores play a more active role

than mere holes. They control the movement of materials in and out of the nucleus in much the same way that a turnstile controls the movements of people in and out of a subway.

The nucleus serves as the cell's library, and it also provides instructions for the construction of new cells. The nucleus contains the genetic information (coded in DNA) that is passed from one cell to its descendants. When a cell divides, each daughter cell gets a complete copy of the genetic instructions in the nucleus. The DNA in each cell nucleus includes instructions for building every polypeptide the body will ever use. In Part II of this book, we will consider how DNA carries genetic information, how cells use this information, and how they pass it to their daughter cells.

The nucleus of eukaryotic cells contains DNA and protein. The DNA acts as a library of information and a set of instructions for making cell components.

The Cytosol Is the Cytoplasm That Lies Outside the Organelles

The cytosol makes up about half of a cell's volume. Cell biologists usually isolate the cytosol from the organelles by breaking open cells and then spinning the mixture in a centrifuge. A centrifuge is a machine that spins a set of tubes arranged so that when they are spinning their contents come under a force greater than gravity (Figure 4-8). This force—centrifugal force—is the same force that presses clothes in a washing machine to the outside wall of the tub during the spin cycle. In a centrifuge, a tube spins so fast that the contents sink to the bottom, with bigger particles and molecules sinking faster than smaller ones. When cell biologists centrifuge cytoplasm from

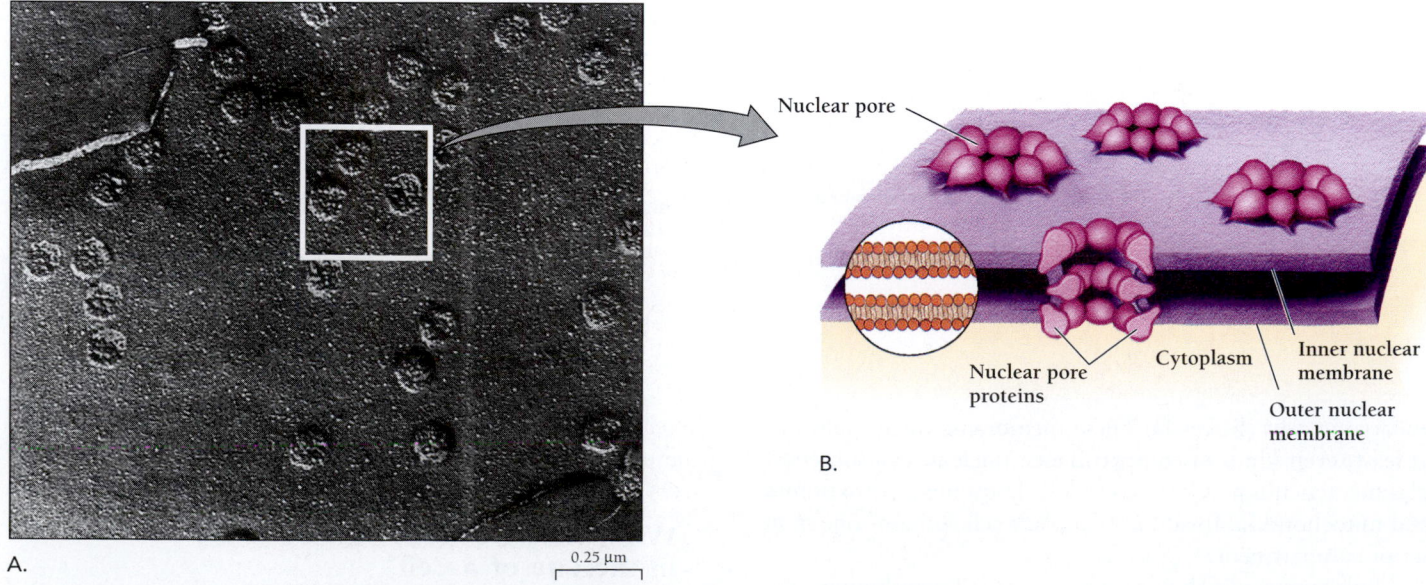

A.

0.25 μm

B.

Nuclear pore

Nuclear pore proteins

Cytoplasm

Inner nuclear membrane

Outer nuclear membrane

Figure 4-7 **Nuclear pores span the nucleus's double membrane.** A technique called freeze fracture reveals their structure by splitting the membrane. *(A, R. Kessel-G. Shih/Visuals Unlimited)*

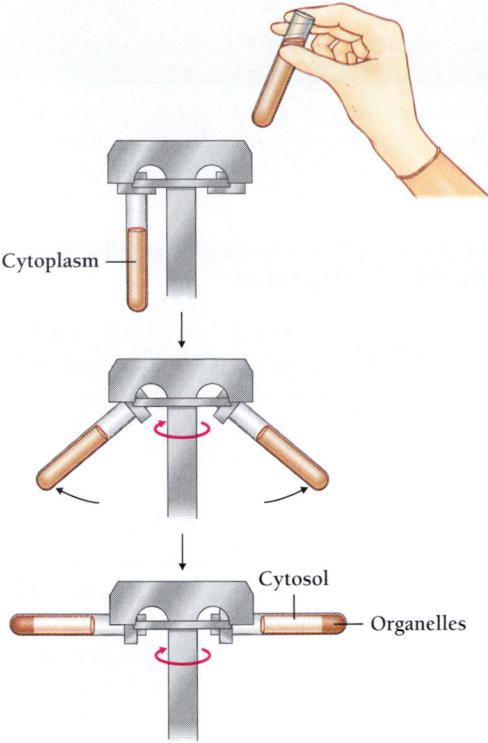

Figure 4-8 **A centrifuge spins a tube of macromolecules at up to 80,000 revolutions per minute.** Such speed places the contents of the tube under forces more than 100,000 times that of gravity. Large molecules travel to the bottom of the tube fastest.

cells, the organelles, which are heavier than the cytosol, drop soonest to the bottom of each tube.

The cytosol contains thousands of different kinds of enzymes responsible for producing building blocks, degrading small molecules to yield energy, and synthesizing proteins. Although the cytosol is aqueous (mostly water), about 20 percent of its weight is protein, giving it the consistency of Jell-O. Embedded in the cytosol are organelles, vesicles, and vacuoles, as well as smaller structures that are not enclosed by mem-

branes. These include granules of energy-rich glycogen, droplets of stored fat, *ribosomes*—tiny, round organelles, about 15 to 30 nm in diameter—and *proteasomes*—barrel-shaped organelles, slightly smaller than ribosomes.

Ribosomes are tiny assemblies of RNA and protein that are indispensable to protein synthesis and therefore to cell maintenance and cell reproduction. Ribosomes [Latin, *soma* = body + "ribo" because they contain ribonucleic acid (RNA)] act as workbenches on which protein molecules are stitched together with peptide bonds. Ribosomes in the cytosol are called "free ribosomes" (Figure 4-9A). Ribosomes associated with a cell's internal membranes are called "membrane-bound ribosomes."

Proteasomes are protein complexes that degrade old proteins in the cytosol and recycle their amino acids. Proteins, like other machines, eventually wear out or become damaged, and the cell must dispose of them. Proteasomes function as molecular garbage disposals: each contains a hollow central chamber, in which denatured polypeptides are broken down, and feeder assemblies at each end, which help push denatured polypeptides into the central chamber (Figure 4-9B).

The jellylike cytosol, which surrounds the cell's organelles, contains ribosomes, glycogen, fat, thousands of enzymes, and much more.

The Endoplasmic Reticulum Is a Folded Membrane

The **endoplasmic reticulum (ER)** [Greek, *endon* = within + *plasmein* = to mold + Latin, *reticulum* = network] is an elaborate structure needed for cell reproduction, since it is important in the synthesis of proteins and lipids, which the cell exports.

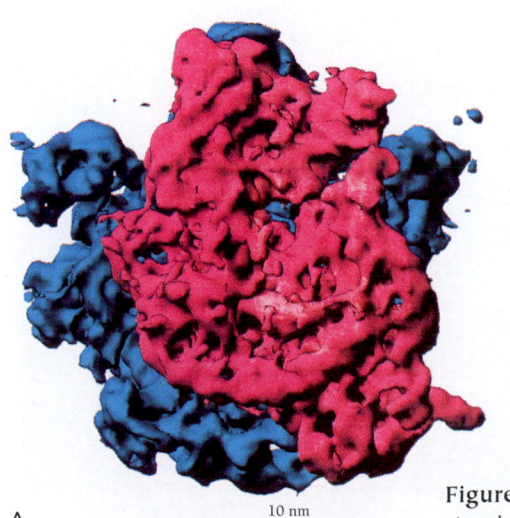

A.

10 nm

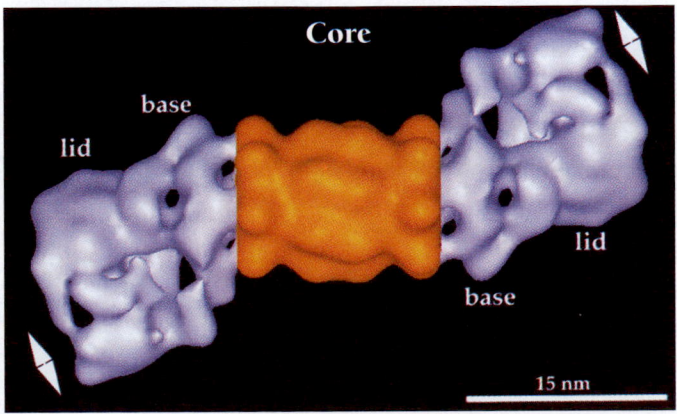

B.

Figure 4-9 **Ribosomes and proteasomes.** A. A computer-generated model of the three-dimensional structure of a single ribosome. B. A proteasome, the cytosolic garbage disposal. Denatured proteins enter through the "lids" and are digested in the central "core." (*A, Harry Noller, University of California, Santa Cruz; B, Wolfgang Baumeister, Max-Planck Institute, Martinsried, Germany*)

The ER resembles the marbling in marble cake. When biologists cut a cell into uniform slices, as we would slice a loaf of marble cake, the ER in each slice, or "section," forms an extensive and convoluted network. Careful study of successive sections, like those in Figure 4-10, show that the ER is actually a single sheet of membrane enclosing a network of interconnected cavities and channels called the **lumen** [Latin, *lumen* = light, an opening]. The ER lumen is like a three-dimensional maze. The ER membrane is continuous with the outer membrane of the nucleus. In many eukaryotic cells, the lumen of the ER makes up about 15 percent of the cell's volume, and the membranes of the ER more than half the total membrane.

The ER consists of two parts that are distinct in both appearance and function. The **rough ER** is studded with ribosomes on the cytosolic side of the membrane and is mostly devoted to modifying recently synthesized proteins, especially those destined for export to organelles or out of the cell (Figure 4-10B). The **smooth ER** has no ribosomes and is mostly devoted to the synthesis and metabolism of lipids (Figure 4-10C). Another important function of the smooth ER is detoxification of substances such as alcohol.

Both rough and smooth ER are present in most eukaryotic cells. But since the two kinds of ER serve different purposes, some specialized cells may have especially large amounts of one or the other. Cells specialized for the production of secreted proteins, such as those in the pancreas, have large amounts of rough ER. Cells specialized for the production of lipids, such as the cells that produce steroid hormones, contain large amounts of smooth ER.

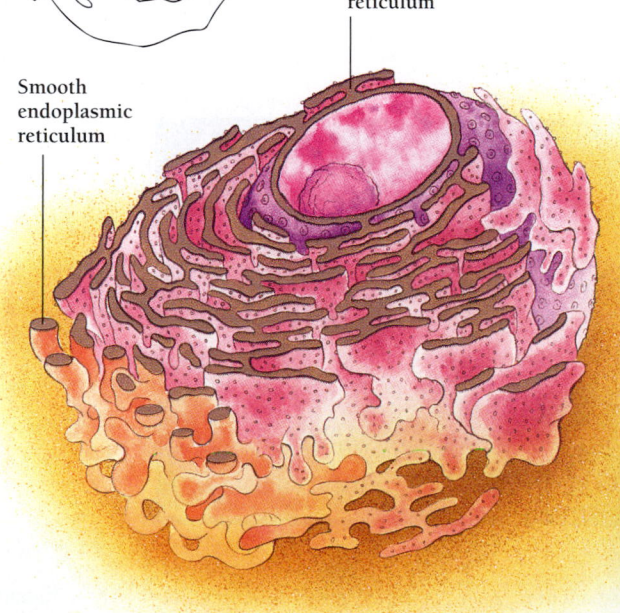

Smooth endoplasmic reticulum

Rough endoplasmic reticulum

The endoplasmic reticulum (ER), with its mazelike internal lumen, consists of rough ER and smooth ER.

The Golgi Complex Directs the Flow of Newly Made Proteins

The **Golgi complex** functions as both a packaging center and a traffic director. During cell reproduction and maintenance, it helps form structures that stay in the cell, such as lysosomes. It also contributes to specialized cell function by helping prepare materials for export outside the cell.

The Golgi complex, usually found near the nucleus, consists of sets of flattened discs, like stacked dinner plates (Figure 4-11). Each stack is about 1 μm in diameter and typically consists of about 6 discs. Surrounding the stacks of flat discs are some smaller, spherical vesicles.

Every eukaryotic cell has a Golgi complex, but the number of stacks varies among different cell types. The cells with the most Golgi stacks are those that secrete large quantities of glycoproteins, which are proteins, such as the albumin in egg white, that have sugars or carbohydrates attached. Mucus is also made of glycoproteins, and the mucus-secreting goblet cells of the intestines have particularly large numbers of Golgi stacks (Figure 4-11). Most of the proteins that are exported to the outside of the cell are glycoproteins. Biologists like to say, "All exported proteins are sugar-coated."

To see what the Golgi complex does, George Palade and Marilyn Farquhar, at Rockefeller University, labeled newly made

---

Figure 4-10   The two parts of the endoplasmic reticulum, the rough ER and the smooth ER, have different functions. A. The rough ER is continuous with the smooth ER. B. The rough ER is studded with ribosomes and is mostly devoted to synthesis and modification of proteins, especially those destined for export to organelles or out of the cell. C. The smooth ER has no ribosomes and is mostly devoted to the synthesis and metabolism of lipids and the detoxification of substances such as alcohol. (*B, Don Fawcett/Science Source/Photo Researchers, Inc.; C, David M. Phillips/Visuals Unlimited*)

Ribosomes on rough ER

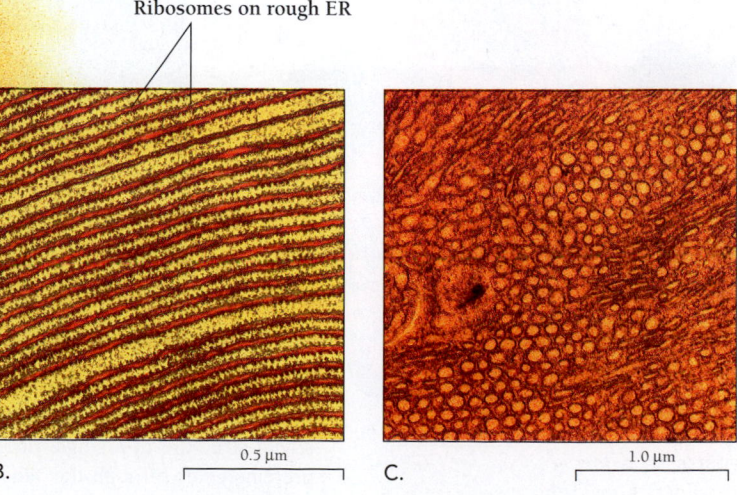

A.   B.   0.5 μm   C.   1.0 μm

glycoproteins with radioactive tracers. Proteins destined for export appeared first in the rough ER and then in the Golgi complex (Figure 4-11). The Golgi complex modifies the glycoproteins, then packages them into membrane-enclosed secretory vesicles. The vesicles eventually fuse with the plasma membrane and discharge their contents to the outside of the cell.

The Golgi complex also manages the flow of proteins to different destinations. The Golgi complex "addresses" glycoproteins by modifying the carbohydrates (which were originally attached to the glycoproteins in the lumen of the ER). Proteins acquire specific "tags" that direct them to specific destinations within the cell—such as lysosomes, or to the cell's export machinery. Other proteins are labeled specifically for export outside of the cell.

The Golgi complex packages and labels glycoproteins for destinations both inside and outside the cell.

The Lysosomes Function as Digestion Vats

Lysosomes, present in all eukaryotic cells, are small vesicles, enclosed by a single membrane. They are the sites of many chemical reactions. In some cells, they are the sites of the first digestion of food.

Lysosomes vary widely in size, but all contain enzymes that break down proteins, nucleic acids, sugars, lipids, and other complex molecules. The large vacuole of a plant cell contains similar enzymes, and we may consider it to be a giant lysosome. Lysosomes are particularly numerous in cells that actively perform **phagocytosis** [Greek, *phagein* = to eat + *kytos* = hollow vessel, or cell], the process by which cells take in and consume large particles of solid food. Free-living amebas and many other single-celled eukaryotes feed primarily by phagocytosis (Figure 4-12). Many of our own white blood cells regularly engulf and digest bacteria, viruses, and assorted cellular debris, for example, and these white cells are packed with lysosomes. Once a cell has engulfed material within a vesicle, the vesicle fuses with a lysosome, whose enzymes then digest the engulfed material.

The membrane of a lysosome keeps its enzymes from digesting the cytosol's own proteins. If the lysosome membrane breaks down, however, the cell begins to digest itself and dies.

Lysosomes are made by other organelles in the cell. Lysosomes first form from membranes derived from the Golgi complex. Their digestive enzymes are made on the ribosomes of the rough ER. They are packaged into lysosomes after passing through the lumen of the ER, where each lysosomal enzyme acquires a special attached sugar. This sugar provides the lysosomal proteins with an address tag, so that the Golgi complex directs them to lysosomes rather than exporting them to the outside of the cell.

Lysosomes contain enzymes that the cell uses to digest food, wastes, and foreign invaders. Lysosome membranes normally isolate these dangerous enzymes from the rest of the cell.

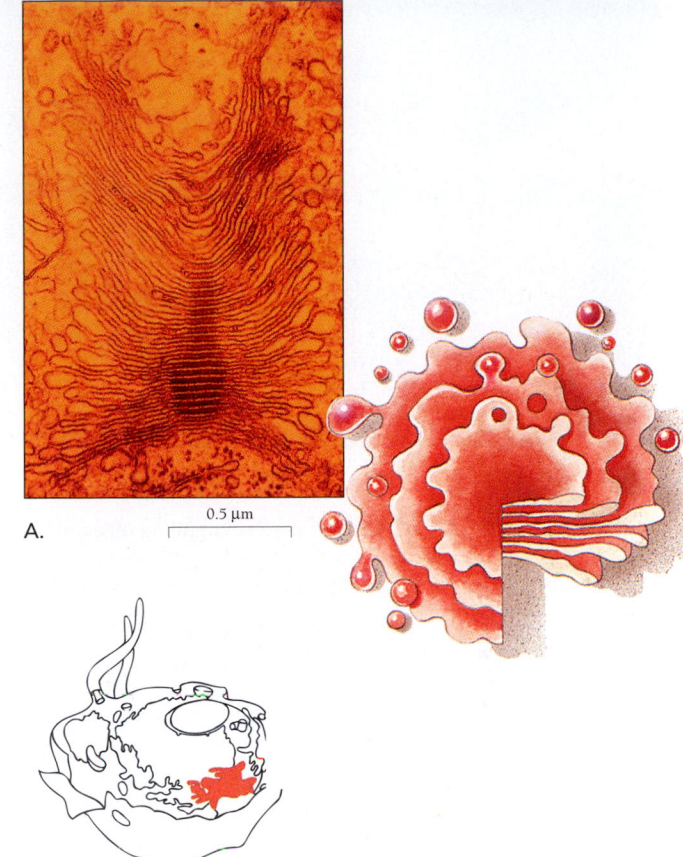

0.5 μm

A.

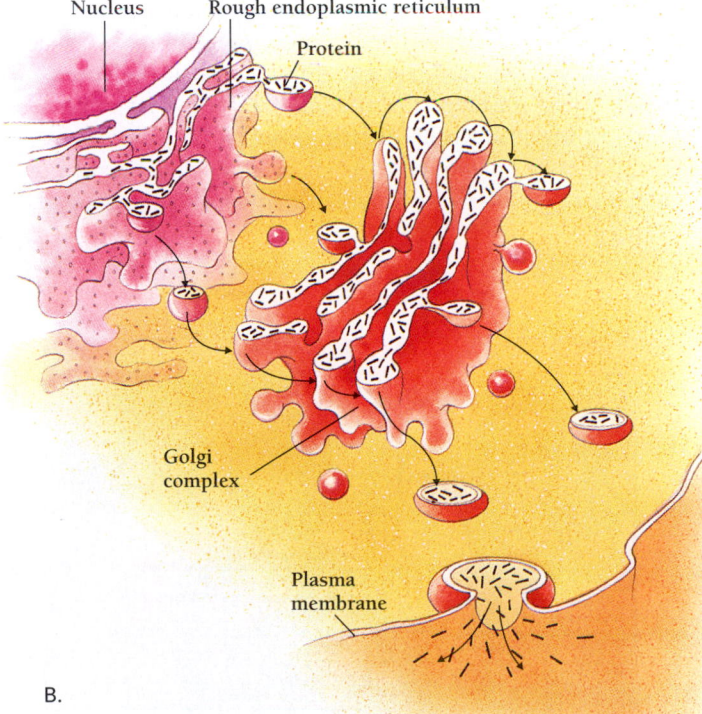

Nucleus   Rough endoplasmic reticulum

Protein

Golgi complex

Plasma membrane

B.

Figure 4-11   Newly made proteins pass through the ER and then the Golgi complex. A. Transmission electron micrograph of a Golgi complex or "stack." B. Vesicles on the surface of the Golgi complex shuttle glycoproteins from the rough ER to the Golgi stack, which "labels" the glycoproteins and packages them into vesicles for secretion. *(A, Cabisco/Visuals Unlimited)*

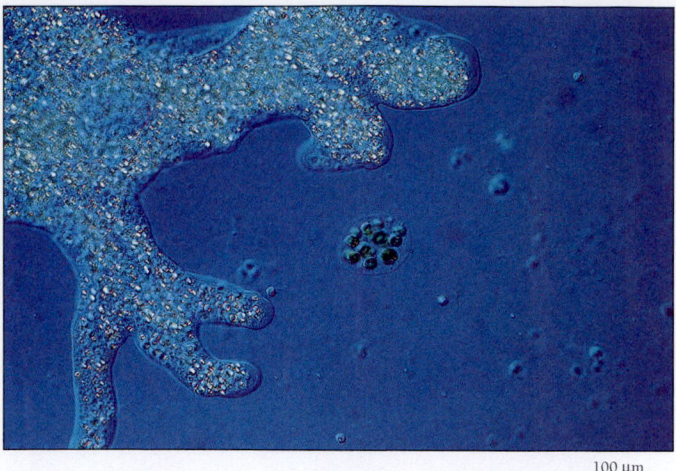

Figure 4-12   **Phagocytosis.** An ameba engulfing algae. (*Michael Abbey/Visuals Unlimited*)

Peroxisomes Produce Peroxide and Metabolize Small Organic Molecules

Peroxisomes are another kind of membrane-enclosed vesicle, widespread in all kinds of eukaryotes (Figure 4-13). **Peroxisomes** contain enzymes that use oxygen to break down various molecules by means of biochemical reactions that produce hydrogen peroxide (H_2O_2), hence their name. Because hydrogen peroxide is highly reactive and can easily damage cells, peroxisomes also contain a protective enzyme, called *catalase*, that breaks down hydrogen peroxide into water and oxygen.

Peroxisomes, whose diameters can range from 0.15 μm to more than 1 μm, can occupy as much as half the volume of a cell. They are especially abundant in cells that synthesize, store, or break down lipids. For example, peroxisomes in the seeds of some plants contain an enzyme that breaks down stored fats and therefore provides energy for the growing plant embryo.

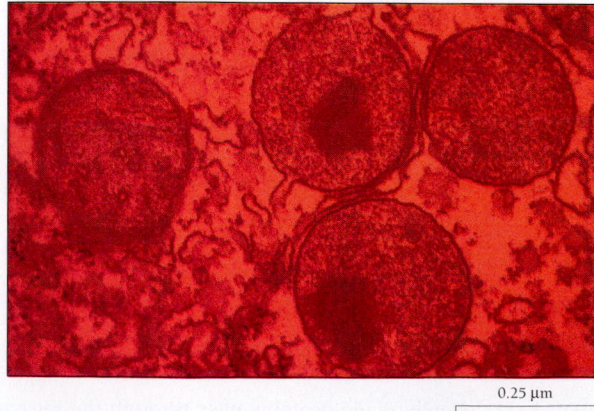

Figure 4-13   **Peroxisomes.** Peroxisomes contain enzymes that degrade molecules in reactions that produce hydrogen peroxide. (*D. Friend and D. Fawcett/Visuals Unlimited*)

Peroxisomes contain a chemical dangerous to the cell itself—hydrogen peroxide. Cells use enzymes that give off hydrogen peroxide to break down fats and oils.

Mitochondria Capture the Energy from Small Organic Molecules in the Form of ATP

Mitochondria [singular, mitochondrion; Greek, *mitos* = thread + *chondrion* = a grain] are the organelles most responsible for the cell's ability to obtain energy from nutrients. In cells that do not directly harvest sunlight, these subcellular power plants produce nearly all of the ATP that powers the chemical reactions of the cell. This ATP production depends on the organization of the internal and external membranes of mitochondria.

Even under the best light microscope, mitochondria, often less than 1 μm in diameter, can barely be seen. However, the transmission electron microscope (TEM) reveals that mitochondria are some of the most numerous organelles in most eukaryotic cells (Figure 4-14). Different kinds of cells have different numbers of mitochondria. A liver cell, for example, may contain several thousand mitochondria, which together make up about a quarter of the cell volume. Cells that use particularly large amounts of energy, such as those of the heart, may contain still more.

The transmission electron microscope also reveals the mitochondrion's characteristic structure. Two membranes, 6 to 10 nm apart, separate a mitochondrion's interior from the cytosol. The outer membrane of a mitochondrion is smooth, but the inner membrane forms elaborate folds, called *cristae*. Mitochondria come in a variety of sizes and shapes. Time-lapse photography with a phase contrast microscope reveals that mitochondria may be constantly in motion, changing shape, fusing, and dividing.

Mitochondria are unusual among organelles not only in having double membranes: they also have their own DNA and make some of their own proteins. These characteristics have led biologists to think that mitochondria arose by the incorporation of captured bacteria, as we discuss in Chapter 18.

Mitochondria synthesize energy-rich ATP in their mazelike interior.

A Plant Cell's Chloroplast Is Just One Kind of Plastid

Plastids are organelles in plant cells that, like mitochondria and nuclei, are surrounded by a double membrane. Plastids are present in nearly all plant cells. The most important plastid in the **chloroplast,** present only in green plant cells and some protists, which uses the energy of sunlight to build energy-rich sugar molecules in the process of **photosynthesis**

Figure 4-14 Mitochondria manufacture ATP. The outer membrane encloses the whole organelle, and the inner membrane encloses the matrix. The inner membrane is folded into cristae, which extend into the matrix. *(A, Don Fawcett/Visuals Unlimited)*

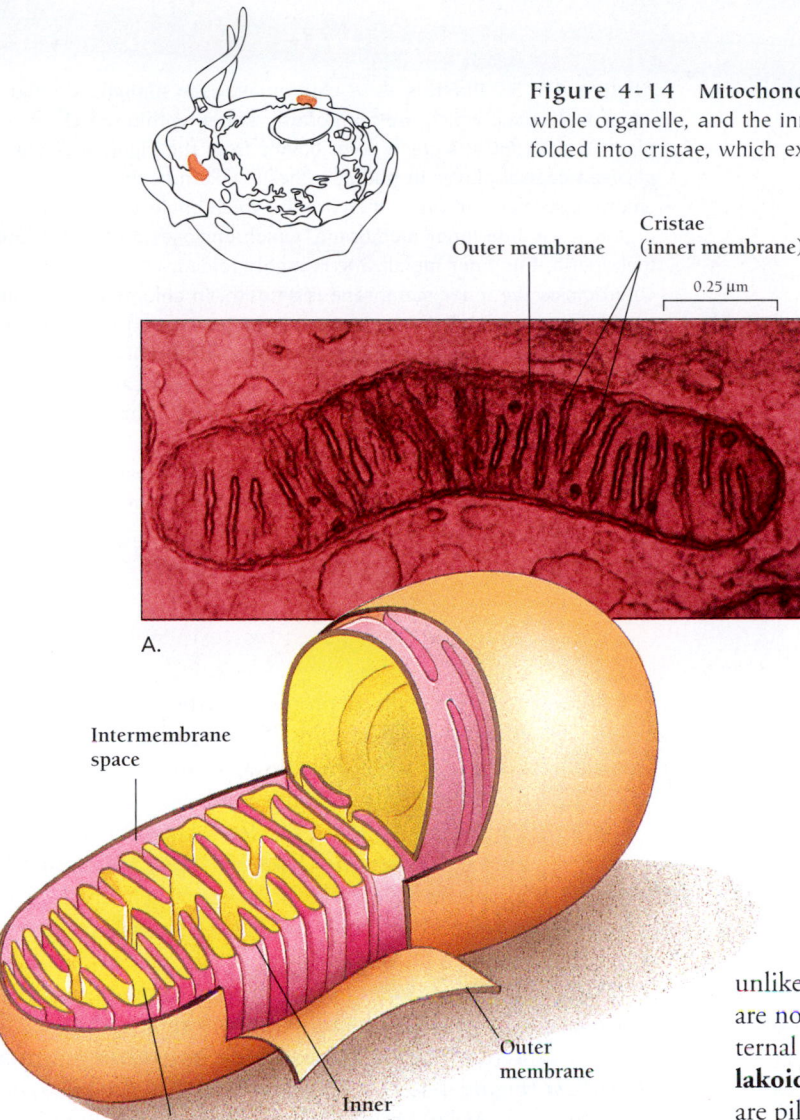

[Greek, *photo* = light + *syntithenai* = to put together]. Most organisms on Earth therefore ultimately depend on chloroplasts for food. Plants also have other kinds of plastids, which serve as storage depots for various kinds of molecules (Figure 4-15).

During photosynthesis, chloroplasts make sugars. The chloroplasts also serve as temporary storage depots for the newly made sugars (in the form of starch grains). The size, shape, and number of chloroplasts in a cell vary from species to species and from cell type to cell type. A leaf cell in a flowering plant, for example, typically contains several dozen chloroplasts. A stem cell contains only a few, and a flower cell may contain none at all.

A typical chloroplast is a large, round, green organelle. The distinctive shade of green comes from the pigment *chlorophyll.* The chloroplast is 5 to 10 μm in diameter, but only 2 to 4 μm high, a thick disc. Like a mitochondrion, a chloroplast has an intricate internal structure of folded membranes. But,

unlike in a mitochondrion, the chloroplast's folded membranes are not continuous with the inner membrane. Instead, the internal membranes lie in flattened, disclike sacs, called **thylakoids** [Greek, *thylakos* = sac + *oides* = like]. The thylakoids are piled like plates (Figure 4-15A). Each stack, of ten or more discs, is called a **granum** [plural, grana; Latin, *granum* = grain] because the stacks resemble the granaries that farmers use to store grain. Each chloroplast contains many grana.

Like mitochondria, chloroplasts have double membranes and DNA, and they perform protein synthesis. Because of these and other clues, biologists have concluded that chloroplasts originated as free-living organisms that were somehow "captured" by early eukaryotic cells.

Two other kinds of plastids are chromoplasts and amyloplasts. **Chromoplasts** contain the pigments that give yellow, orange, and red colors to flowers, fruits, vegetables, and autumn leaves (Figure 4-15B). Chromoplasts arise from chloroplasts through the reshaping of the inner membrane and the breakdown of chlorophyll. **Amyloplasts** do not derive directly from chloroplasts, but from *proplastids,* the common precursors of all plastids. Amyloplasts, which store starches, are abundant in roots such as sweet potatoes and turnips, as well as in wheat, rice, and other seeds (Figure 4-15C).

Chloroplasts turn solar energy into chemical energy in the process of photosynthesis. Other kinds of plastids store pigments or starches.

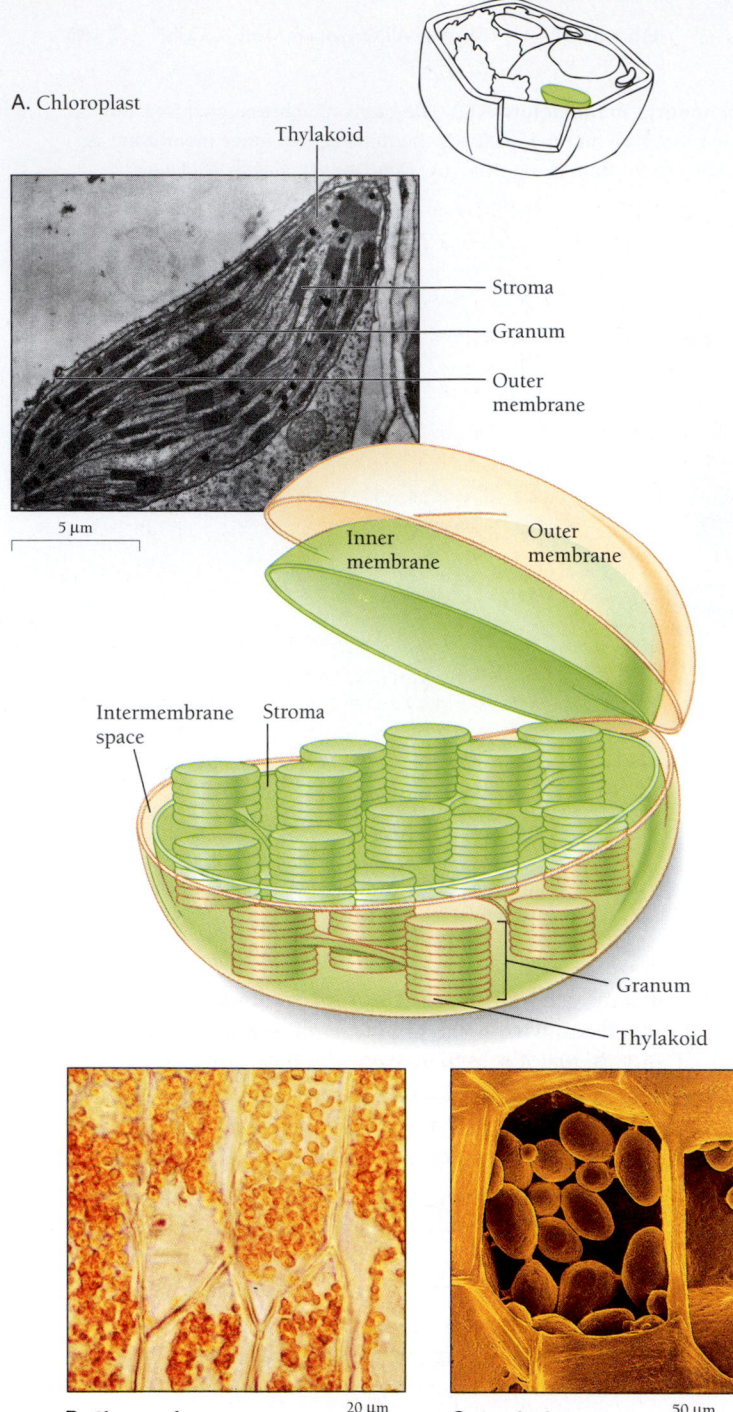

A. Chloroplast

Thylakoid

Stroma

Granum

Outer membrane

5 μm

Inner membrane

Outer membrane

Intermembrane space

Stroma

Granum

Thylakoid

B. Chromoplasts   20 μm

C. Amyloplasts   50 μm

Figure 4-15   Plastids. A. A chloroplast uses sunlight to build sugar molecules, which are temporarily stored within the chloroplast. A chloroplast's green color comes from the pigment *chlorophyll*—a central player in photosynthesis. Like mitochondria, chloroplasts have an outer membrane, which encloses the whole organelle, and an inner membrane, which encloses a matrix. In mitochondria, the inner membrane is highly folded, whereas in chloroplasts the inner membrane is smooth. In chloroplasts, a third membrane, derived from the inner membrane, is even more highly folded than the inner membrane of mitochondria. This membrane forms pockets called thylakoids, arranged in stacks called grana. The intermembrane space is called the stroma. B. Chromoplasts in the cells of a yellow petal (from a tulip). C. Amyloplasts in raw potato. (*A, Grant Heilman Photography; B, Biophoto Associates/Science Source/Photo Researchers, Inc.; C, Jeremy Burgess/Science Photo Library/Photo Researchers, Inc.*)

Some of the specialized proteins of the cytoskeleton are closely related to muscle proteins. While muscle proteins are essentially permanent structures, however, the cytoskeleton is constantly dissolving and reforming itself.

Microtubules are hollow cylinders about 25 nm in diameter, which vary in length (Figure 4-16A). Microtubules play a critical role in cell division. When a cell divides, microtubules distribute the DNA and other materials to the two daughter cells.

Although microtubules appear throughout the cytoplasm, and even in the nucleus, their assembly always seems to originate in structures called **microtubule organizing centers (MTOCs)** (Figure 4-17). In the cytoplasm of most eukaryotic cells, the major MTOCs are near the nucleus, in a zone called the **centrosome**—a region that in many cells also contains a distinct organelle called the **centriole.** The MTOCs of the centrosome play an important role in cell division, as we discuss in Chapter 8.

The cylinders of microtubules consist of two globular protein molecules called **tubulins.** Pure tubulin in a test tube can form microtubular structures all by itself. In living cells, however, up to 50 other proteins may associate with microtubules, influencing their distribution and stability. Some of these *microtubule-associated proteins* (or *MAPs*) bundle microtubules together, allowing them to establish the shapes of cells. In plants, for example, the layout of the cellulose fibers that determine the shapes of plant cells depends on the arrangement of microtubules and MAPs.

Other microtubule-associated proteins function as **motors,** which use the energy of ATP to move along microtubules. The motors can attach to other subcellular components and carry them, as freight, along microtubule tracks. Some motor molecules (called *dyneins*) travel away from the MTOC end of microtubules, while others (called *kinesins*) travel toward the MTOCs, as illustrated in Figure 4-17B.

Actin filaments, also called *microfilaments,* are much finer than microtubules—only about 7 nm in diameter (Figure

What Does the Cytoskeleton Do?

Eukaryotic cells have a complex network of protein filaments, called the cytoskeleton, which is visible with the TEM (Figure 4-16). Three types of filaments make up the cytoskeleton: microtubules, actin filaments, and intermediate filaments (Figure 4-16). Like our own bony skeletons, the cytoskeleton provides structure and support. The name "cytoskeleton" is slightly misleading, however, since the cytoskeleton does more than give mechanical support for the cytoplasm and its component organelles. It also provides the force for most of the movements of cells, including the transport of materials and changes in the overall shape of a cell.

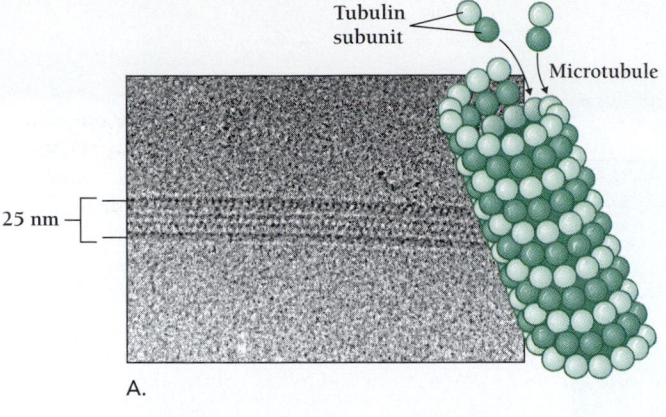

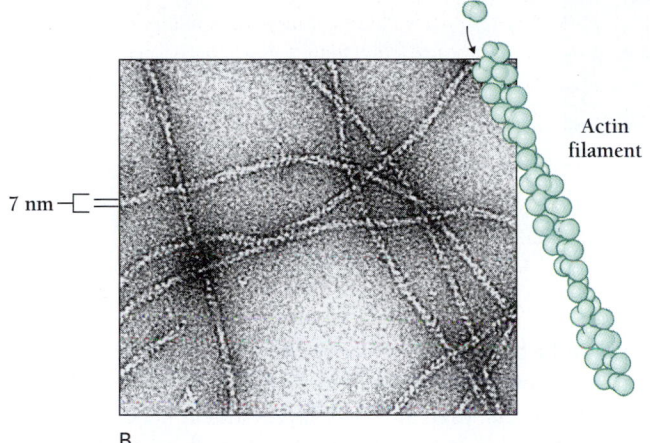

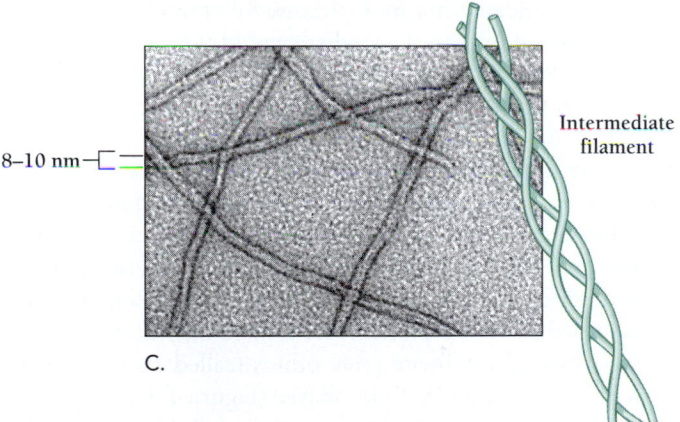

Figure 4-16   The cytoskeleton of a cell. The cytoskeleton is a complex network of three types of protein filaments, each with distinctive appearances and distributions. A. The microtubules emanate from a microtubule organizing center (MTOC) and serve as tracks for freight carried by motor molecules. B. Actin filaments are often closer to the cell surface. C. Intermediate filaments reinforce cells, especially at places of particular stress. (*A, Richard Wade; B, Roger Craig; C, Roy Quinlan*)

4-16B). Actin filaments are often anchored to the cell surface and provide the force for movement and shape changes. Actin filaments consist of a single kind of protein, called actin, which is also present in muscle fibers.

In muscle, actin filaments are permanent structures that interact with other proteins to produce the force leading to contraction. In nonmuscle cells, actin filaments are constantly being built and destroyed. Even in the test tube, actin molecules can self-assemble to form a two-stranded helix apparently identical to the actin filaments in intact cells. In cells, they form cross-linked cables that provide mechanical support. They also associate with other proteins to furnish musclelike contractile forces. Other proteins associate with actin filaments to act as molecular winches that pull on actin cables.

Intermediate filaments, usually 8 to 10 nm in diameter, are fibrous proteins such as keratin (Figure 4-16C). Keratin strengthens wool, hair, fingernails, the outer layers of the skin, and other epithelial sheets. All intermediate filaments seem to be common in parts of cells that are subject to mechanical stress, suggesting that these filaments strengthen cells and tissues (Box 4.2).

The cytoskeleton is both skeleton and muscle system for the cell. Hollow microtubules, along with the microtubule organizing center near the centrosome or centriole, provide the cell with a shape. Microfilaments supply the dynamic contractile force, and intermediate filaments provide strength.

WHAT DO MEMBRANES DO?

We saw earlier in this chapter that cell size is partly limited by the amount of membrane, which controls the transfer of materials in and out of the cell, just as a system of freeways and rail lines carries materials in and out of a city. Membrane, then, is as important to cells as a transportation system is to a city.

But membranes do more than just move materials in and out of the cell. In the past 30 years, cell biologists and biochemists have found four major roles for cellular membranes:

1. Membranes are essential boundaries that separate the inside from the outside.
2. Membranes regulate the contents of the spaces they enclose.
3. Membranes serve as a "workbench" for a variety of biochemical reactions, especially those involving the metabolism of lipids and the secretion of proteins.
4. Membranes participate intimately in energy conversions (as we discuss in Chapters 6 and 7).

What Kinds of Molecules Do Membranes Contain?

The plasma membranes of all cells look the same, appearing in cross section as two dark lines, 7 to 10 nm apart, with a lighter zone between them. The membrane consists, then, of at least two parallel layers separated by a slight space. Our first question concerns the structure of these layers. What molecules make up the layers, and what holds them in the same characteristic structure in all organisms?

The first insight into the molecular structure of membranes came from studies of red blood cells. Mammalian red

Microtubules

B.

Figure 4-17 Microtubules radiating out from the microtubule organizing center. Most MTOCs are near the nucleus in a zone called the centrosome, which plays a central role in cell division. (*B, K.G. Murti/Visuals Unlimited*)

0.5 μm

MTOC

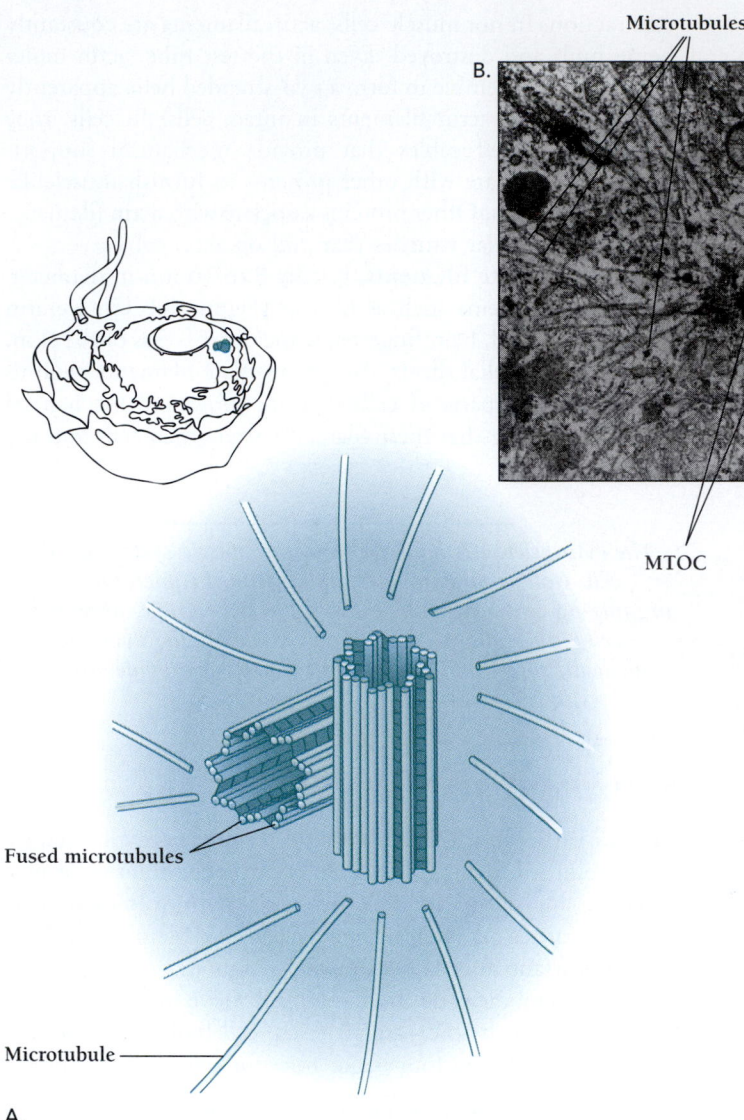

Fused microtubules

Microtubule

A.

blood cells are ideally suited for membrane studies: they have no internal membranes, no organelles, and few types of macromolecules other than hemoglobin, the red oxygen-binding protein. Red blood cells do not even have nuclei. Like empty plastic bags, they can be squashed into many shapes. This flexibility allows them to pass more easily through the tiny vessels of the circulatory system.

Biochemists can obtain essentially pure plasma membrane by breaking open the red blood cells to release the hemoglobin and then using a centrifuge to separate the empty plasma membranes, called red blood cell "ghosts." Biochemical analysis has shown that these membranes consist mostly of lipids and proteins, as well as some carbohydrates. In the 1920s, biochemists calculated that a red cell ghost contained just enough lipid to enclose the cell with a continuous layer two molecules thick. This result suggested that the ghosts might indeed consist of a lipid "bilayer."

The most important lipids in the membrane are phospholipids. Recall from Chapter 3 that phospholipids are highly amphipathic; that is, they have a hydrophilic (water-loving)

head and a hydrophobic (water-fearing) tail. When pure phospholipids are placed in water, they can spontaneously organize into sheets that are two layers thick. The oily tails of the lipids point inward toward one another, while their hydrophilic heads point outward. Once biologists understood this, they wondered if a biological membrane was in fact just a phospholipid bilayer. If so, where would the membrane proteins lie? Since biologists knew that proteins studded both the inner and outer surfaces of the cell membrane, they hypothesized that the membrane proteins were scattered on the inner and outer surfaces of the lipid bilayer, in contact with the polar groups of the phospholipid. But this is not quite correct.

Researchers have now determined that proteins not only occupy the two layers of the plasma membrane, but also occupy the space *in between* the two layers. Some membrane proteins (called *peripheral proteins*) are located only on the outside or the inside of the membrane; others (called *transmembrane proteins*) span the entire lipid bilayer (Figure 4-18).

Membrane proteins, like membrane lipids, are amphipathic, with both hydrophilic and hydrophobic regions. The hydrophilic regions interact with the aqueous solutions of the cytosol and the cell exterior, as well as with the hydrophilic heads of the lipids. The hydrophobic regions interact with the hydrophobic interior of the membrane.

The two lipid layers of the membrane differ slightly from one another (Figure 4-18B). Certain enzymes, for example, can break down the lipids in one layer, but not the other. Another difference is that carbohydrates appear only on the outer face. Some of these carbohydrates are attached to proteins (glycoproteins), and some are attached directly to membrane lipids (glycolipids). The inner and outer layers also differ in the way they interact with membrane proteins. Proteins that span the entire thickness of the membrane, for example, consistently orient themselves in a certain direction (Figure 4-18). We will see later in this chapter that the orientation of a protein is crucial because some membrane proteins work like turnstiles,

BOX 4.2

The 9 + 2 Structure of Flagella

Many prokaryotes, including *Escherichia coli,* have long fibers on their surfaces. Each of these structures is about 10 to 20 nm thick and up to 10 μm long and is called a **flagellum** [Latin, = whip; plural, flagella]. The flagellum's proportions are similar to those of a half-inch rope 40 feet long. Flagella act as propellers, driving bacteria at speeds of tens of micrometers per second. In terms of the bacteria's own dimensions, this would be equivalent to a 15-foot-long car moving at 100 miles per hour.

The flagella of eukaryotes are different from those of prokaryotes. When these structures are long and few in number, they are called **flagella** (Figure A). When they are short and numerous, they are called **cilia** (Figure B). However, both cilia and flagella have the same characteristic cross section. Each consists of a ring of nine sets of microtubules, surrounding two more sets in the center. The microtubule organizing center of a cilium or flagellum also has a nine-membered ring of microtubular structures and is called a basal body. Protists and some cells of animals (including sperm) use flagella or cilia to propel themselves. Cilia on the surfaces of the respiratory and reproductive tracts push fluids along the surface.

 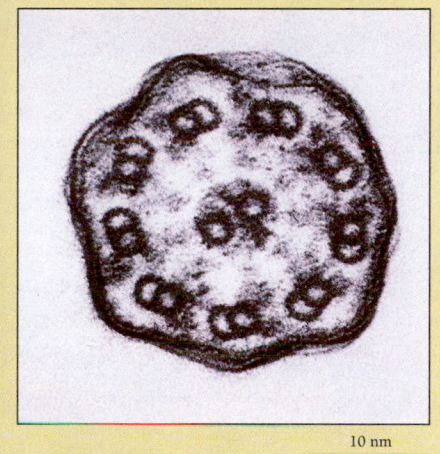

10 μm
10 nm

Figure A Flagella. The protist *Euglena (left)* has a long flagellum, which it whips around like a propeller. An electron micrograph *(right)* reveals the characteristic 9 + 2 organization of the *Euglena* flagellum, which resembles that of the cilia shown in Box 4.1 and the 9-fold symmetry of the microtubule organizing center in Figure 4-17. *(Left, David M. Phillips/Science Source/Photo Researchers Inc.; right, Dr. Gopal Murti/Science Photo Library/Photo Researchers, Inc.)*

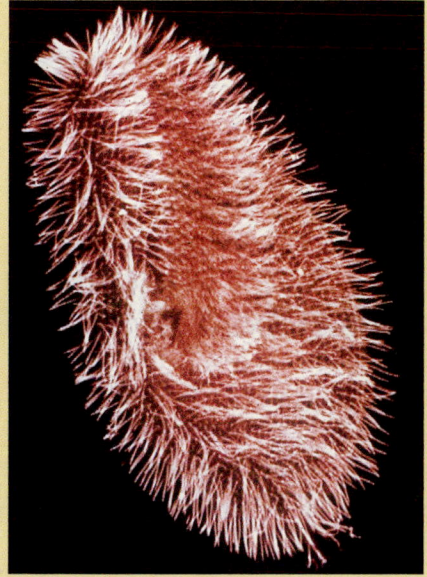

25 μm

Figure B   Cilia. Cilia of a *Paramecium*, as seen with a confocal microscope. *(Anne Fleury, Laboratoire de Biologie Cellulaire 4, Université Paris Sud, CNRS)*

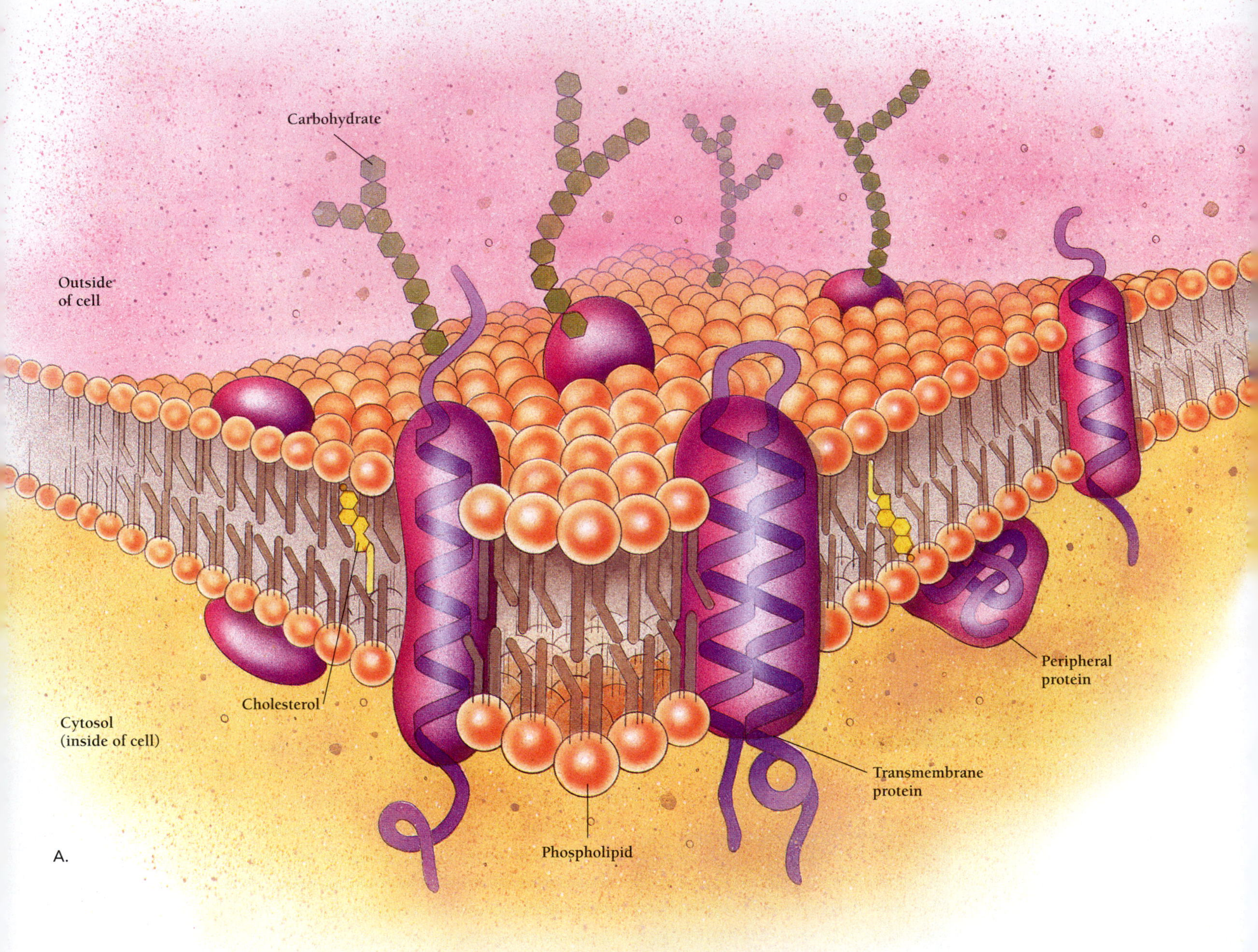

Outside
of cell

Carbohydrate

Cytosol
(inside of cell)

Cholesterol

Phospholipid

Peripheral
protein

Transmembrane
protein

A.

Figure 4-18   A plasma membrane. A. Carbohydrates stud the outer surface of the membrane (as glycolipids and glycoproteins). Transmembrane proteins pass through the lipid bilayer. B. Edge of cell showing plasma membrane. *(Biophoto Associates/Photo Researchers, Inc.)*

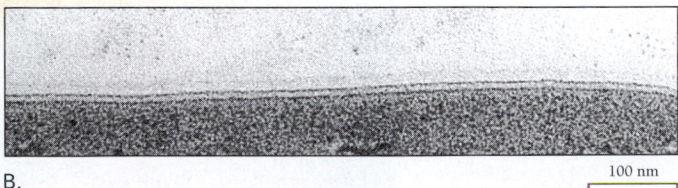

B.   100 nm

letting molecules pass through the membrane in only one direction.

The plasma membrane consists of an asymmetric phospholipid bilayer. Some membrane proteins pass through the interior of this bilayer, and others associate with the inner or outer surface. Carbohydrates occupy the outer surface only.

How Does the Structure of a Membrane Establish Its Functional Properties?

The plasma membrane has a consistency similar to that of thin oil, such as you might use to lubricate a sewing machine or a bicycle. Water can diffuse rapidly from one side of the membrane to the other. Small, nonpolar molecules can diffuse across a membrane and within it. The membrane selectively pumps other molecules from one side to the other.

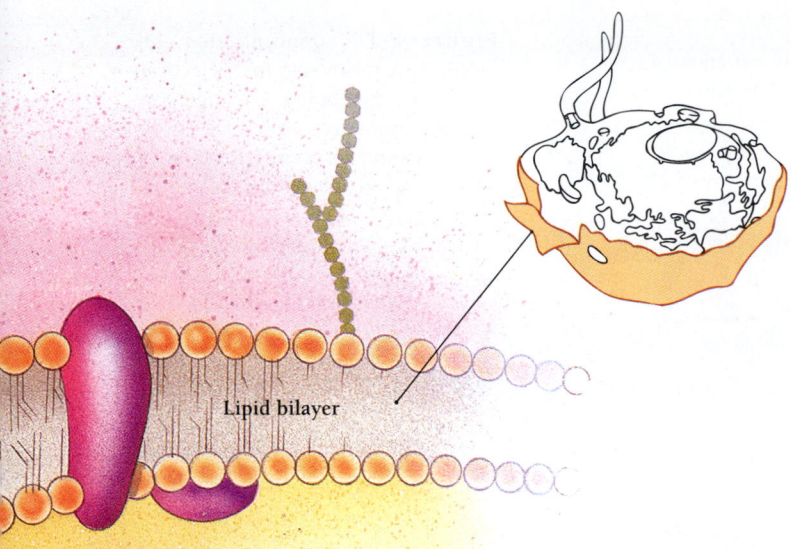

Lipid bilayer

What about the molecules of the membrane itself? Are they also able to move about? To answer this question, biochemists marked certain membrane phospholipids with a chemical group that they could detect with a special method (called electron spin resonance). The biochemists could then study the movements of the marked lipids within the membrane. These experiments showed that membrane lipids move freely within the plane of the membrane but that the lipids in one layer do not usually move to the other layer.

Proteins also move freely within each layer of the membrane. To study the movement of proteins, biochemists marked membrane proteins with a fluorescent (glowing) dye, then used a laser to bleach the dye in a small region of the membrane. They then watched the membrane through a microscope to see if new fluorescent molecules would appear in the bleached area. They found that glowing proteins entered from adjacent areas. These experiments showed that about half of the proteins in a membrane move freely, and about half remain tightly bound in their original position within the membrane.

These experiments showed that both lipid and protein molecules move about with relative ease within each layer of the membrane. The molecules move easily within each layer because the layers are **fluid.** On the basis of this fluidity, two biochemists at the University of California San Diego, Jonathan Singer and Garth Nicholson, proposed a new model of membrane structure, called the **fluid mosaic model.** Singer and Nicholson's model emphasizes that proteins and lipid molecules can usually move freely within the lipid bilayer. Since this model was first advanced in 1972, cell biologists have used it as a basis for understanding all cellular membranes. Today we summarize the fluid mosaic model—somewhat modified since Singer and Nicholson first proposed it—as follows:

1. The basic structure of the membrane is a lipid bilayer, with two sheets of phospholipids arranged tail to tail.
2. Proteins are dispersed through the membrane like the individual pieces of stone or tile in a mosaic. Proteins contribute to membrane structure and function. Some proteins span the membrane, while others are confined to the inner or outer surface.

3. The membrane is fluid; both protein and lipid molecules move freely within the plane of the membrane.
4. The lipid bilayer serves as a hydrophobic barrier, confining hydrophilic molecules to the inside or the outside of a cell (or organelle).
5. Some membrane proteins help transport specific molecules across the membrane.

HOW DO MEMBRANES REGULATE THE SPACES THEY ENCLOSE?

Throw a handful of raisins into a pot of boiling water and the raisins gradually fill with water, swelling until they resemble the grapes from which they were made (Figure 4-19). The raisin skin is a kind of membrane. Water and other small molecules easily pass through the skin, but larger molecules, such as proteins or cellulose, cannot pass through the skin. Large molecules inside the raisin are trapped; those outside are excluded. Such a membrane is said to be **selectively permeable,** meaning that it allows the passage of some molecules (especially of water), but not of others.

Plasma membranes are also selectively permeable. Proteins, for example, cannot directly move through plasma membranes at all. Nor can many ions, such as sodium ions. The regulation of a cell's internal environment—cellular homeostasis—depends on membrane molecules that actively pull specific small molecules and ions through the membrane.

Why Does Water Move Across Membranes?

Before we can discuss how membranes regulate a cell's internal environment, however, we must first talk about the general rules that govern the movement of water and dissolved substances. These rules are the same physical laws that govern the movement of water into and out of raisins.

What is diffusion?

If you pour a teaspoon of salt into a pot of water, most of the salt will at first drop to the bottom of the pot in a little pile. Gradually, the salt will dissolve into sodium ions and chloride ions. At first, the concentration of ions will be very high near the spot where you first poured the salt. The **concentration** is simply the number of molecules in a given volume. However, as time passes, the salt ions will move about randomly in amongst the water molecules until the ions are distributed evenly throughout the pot of water. Even after the ions are distributed evenly, they continue to move about randomly. Since as many ions move out of a given area as into it, however, the net charge is still zero. The random movement of like molecules or ions from an area of high concentration to an area of low concentration is called **simple diffusion** [Latin, *diffundere* = to pour out]. Stirring speeds this process, but diffusion will happen all by itself if you wait long enough.

The presence of an area of high concentration adjacent to an area of low concentration is called a **concentration**

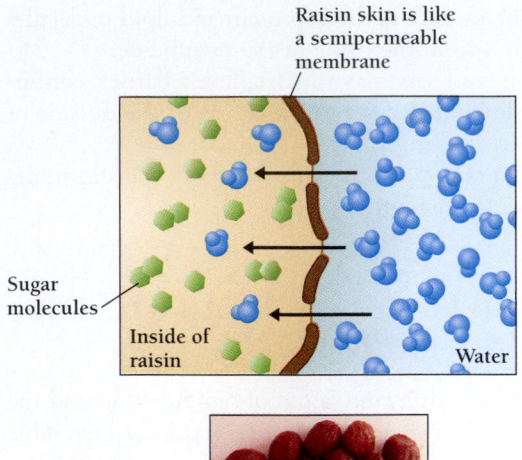

Raisin skin is like a semipermeable membrane

Sugar molecules

Inside of raisin   Water

A.

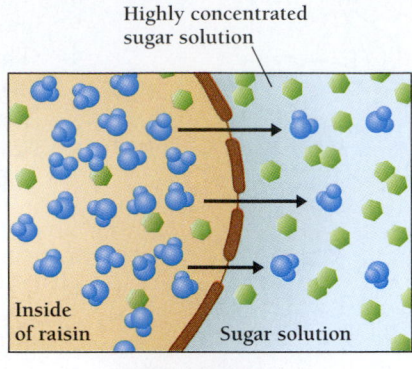

Highly concentrated sugar solution

Inside of raisin   Sugar solution

B.

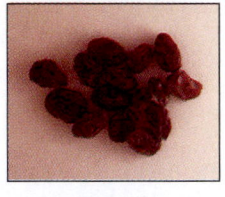

Figure 4-19   Concentration gradient. A. A raisin skin is a selectively permeable membrane. The high concentration of sugar molecules inside the raisin creates a concentration gradient. Water flows inside. B. In a concentrated sugar solution, the water in the swollen raisin flows outward. Because raisins are so sweet, such a sugar solution would have to be very concentrated. *(Photos, courtesy of Amy Dunleavy)*

gradient (Figure 4-19). Concentration gradients are virtually everywhere. Cigarette smoke moving from the smoking section of a restaurant to the nonsmoking section creates a concentration gradient from an area of dirtier air to an area of cleaner air. A teaspoon of sugar poured into a cup of coffee creates a concentration gradient—for a while.

Concentration gradients are always temporary: simple diffusion, ordinary mixing, and turbulence all tend to eliminate concentration differences. How fast diffusion occurs depends on a variety of factors, including temperature, the sizes of the molecules, and the steepness of the gradient. At higher temperatures, molecules move faster and so diffuse faster. Small molecules generally diffuse faster than large ones. Finally, the greater the difference in concentration at the two ends of a gradient, the faster diffusion occurs.

In simple diffusion, solutes move from an area of high concentration to an area of low concentration.

What is osmosis?

Diffusion occurs across membranes as well as through open solutions. For example, raisins immersed in a pot of plain water fill with water because the raisins have a high concentration of sugar and other molecules. The sugar molecules cannot pass easily through the skin of the raisin and out into the water. However, water molecules move in and out of the raisin easily. From the perspective of a water molecule, then, a raisin is an area of low water concentration. The difference between the sugary inside of the raisin and the watery outside constitutes a concentration gradient, along which the water molecules move.

The movement of the water molecules along this concentration gradient into the raisin is called osmosis. **Osmosis**

[Greek, *osmos* = push, thrust] is the movement of water across any selectively permeable membrane in response to a concentration gradient.

As water flows in, it creates a force against the inside surface of the membrane. This **osmotic pressure,** comparable to physical pressure, stretches the raisin's skin until it begins to resist. If the skin is weak, it may break. Otherwise, it pushes back, meeting osmotic pressure with mechanical pressure. The physical pressure exerted by the raisin's skin counterbalances the osmotic pressure.

Osmosis is the movement of water across any selectively permeable membrane in response to a concentration gradient. The pressure exerted by osmosis, called osmotic pressure, is comparable to mechanical pressure.

Which way does water flow through a membrane?

We have seen what happens when a closed membrane (the raisin) with a high concentration of solutes (in this case, various sugars) is dropped into a solution with a low concentration of solutes (plain water). What happens if we take the water-filled raisins and put them into a highly concentrated solution of sugar?

The answer depends on how concentrated the sugar is. Look inside a can of fruit cocktail and you'll see what happens when the concentrations of sugar in a grape and the solution are about the same. The grape neither shrinks nor expands, but retains its normal shape. Put the grape into a highly concentrated sugar solution, however, and you will see it actually shrink as the water molecules inside move out through the skin, following the concentration gradient (Figure 4-19).

In the raisin example, the movement of water into the raisin is greater than the movement of water out of the raisin.

The water in which the raisin sits is said to be **hypotonic** [Greek, *hypo* = under + *tonos* = tension] compared to the inside of the raisin. In ordinary canned fruit cocktail, the water movements in and out of grapes are balanced perfectly. In that case, the sugary solution in which the grape is sitting is said to be **isotonic** [Greek, *isos* = equal] with the inside of the grape. If we put grapes into too heavy a syrup, however, the concentration of the syrup solution is so high that water moves out of the grapes. In that case, the solution outside the grape is said to be **hypertonic** [Greek, *hyper* = above] compared to the inside of the grape.

Cells, it turns out, behave in the same way. A red blood cell immersed in an isotonic solution, such as blood plasma, or a suitable salt solution, remains intact (Figure 4-20A). But if we immerse the red blood cell in a highly concentrated (hypertonic) solution, the cell shrivels (Figure 4-20B). The selectively permeable membrane allows water to pass freely out of the cell. When enough water flows out so that the new concentration of molecules within the cell is again the same as outside the cell, no more water flows.

Conversely, a red blood cell immersed in a hypotonic solution such as plain water will tend to fill with water and burst. When we decrease the solute concentration outside the cell, water then flows into the cell, causing it to expand (Figure 4-20C). In principle, water will flow in until the total solute concentration is equal on both sides of the membrane. In practice, however, there may be too few solute molecules on the outside for this equalization to occur, as, for example, when the cells are suspended in pure water. In this case, the water will continue to flow into the cell until the cell bursts and releases its contents. (This is how cell biologists can prepare red cell ghosts.)

When a selectively permeable membrane separates two solutions that differ in their total solute concentrations, water flows in the direction that tends to equalize the solute concentrations on the two sides of the membrane. A solution may be hypotonic, isotonic, or hypertonic, relative to the concentration of solutes on the other side of the membrane.

How does turgor pressure support plants?

When plant cells, such as those in the onion skin shown in Figure 4-21, are exposed to a hypotonic solution, water moves into the cell but does not accumulate in the cytoplasm. Instead, the inflowing water passes from the cytoplasm to a vacuole. The expansion of the vacuole pushes the surrounding cytoplasm against the **cell wall**—the rigid boxlike structure that encloses all plant cells. Conversely, when an onion cell is placed in a hypertonic solution, water flows out of the vacuole and the cell separates from the cell wall, a phenomenon called **plasmolysis** [Greek, *plasma* = form + *lysis* = loosening].

In a hypotonic solution, the flow of water into the cell creates pressure against the cell membrane. The cell wall pushes back, exerting mechanical pressure that prevents the further flow

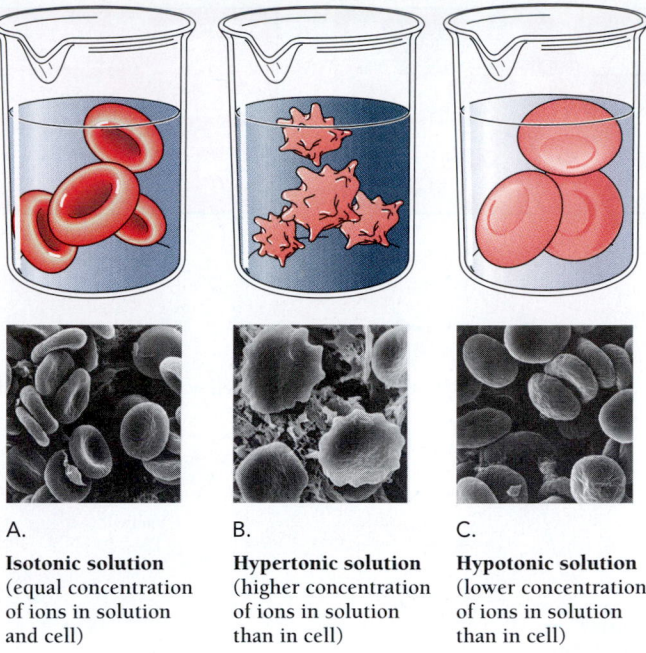

A.
Isotonic solution
(equal concentration of ions in solution and cell)

B.
Hypertonic solution
(higher concentration of ions in solution than in cell)

C.
Hypotonic solution
(lower concentration of ions in solution than in cell)

Figure 4-20   Iso, hyper, hypo. A. A red blood cell in a isotonic salt solution remains intact. B. A red blood cell in a hypertonic salt solution shrivels. C. A red blood cell in a hypotonic salt solution swells with water. In a still more hypotonic solution, such as plain water, the cells in C would fill up and burst open. *(Photos, Dennis Kunkel/Phototake NYC)*

of water into the cell. (Water actually continues to flow equally in both directions through the plasma membrane, but there is no *net* inflow of water.) The mechanical pressure thus balances the osmotic pressure. The pressure of a plant cell's contents against its cell wall is called **turgor pressure** [Latin, *turgere* = to swell].

Turgor pressure supports nonwoody plants in the same way that water supports the shape of a water balloon. This is why plants (and water balloons) begin to droop, or wilt, when they do not have enough water. The same phenomenon accounts for the way lettuce wilts soon after it is covered with salad dressing. Because the dressing is hypertonic, it draws water out of the cells of the lettuce leaves, causing them to wilt.

The flow of water into a plant cell gives mechanical strength to plant tissue.

What Determines the Movement of Molecules Through a Selectively Permeable Membrane?

We can ask two questions about the passage of molecules through a membrane: "Which way?" and "How fast?" In addressing the first question, membrane researchers distinguish between "passive transport" and "active transport." **Passive transport** occurs spontaneously, without the expenditure of energy (Figure 4-22B). The rate of passive transport can vary,

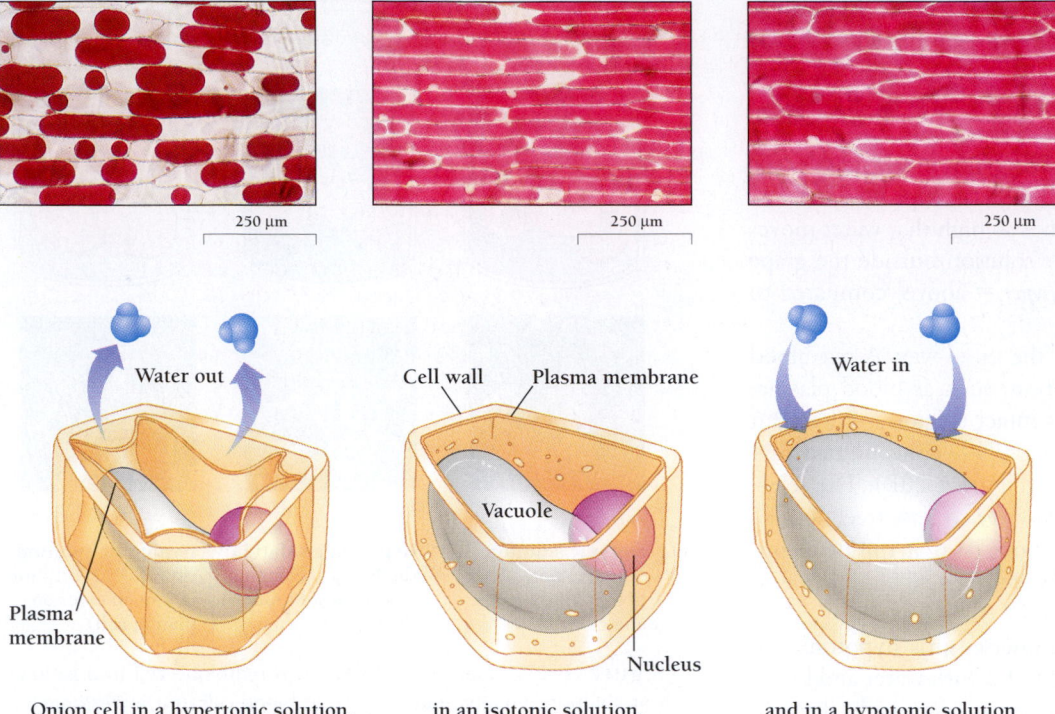

250 μm   250 μm   250 μm

Water out   Cell wall   Plasma membrane   Water in

Vacuole

Plasma
membrane

Nucleus

Onion cell in a hypertonic solution,   in an isotonic solution,   and in a hypotonic solution

Figure 4-21   Why does water move in and out of a plant cell vacuole? Plant cells
like this onion cell behave much like red blood cells (and raisins). The water content of
the vacuole increases or decreases according to whether the cell is in a hypertonic, isotonic, or hypotonic solution. A. Hypertonic: the vacuole has emptied, and the cytoplasm
has concentrated away from the cell wall. B. Isotonic: there is some space between the
cytoplasm and the cell wall. C. Hypotonic: the cytoplasm has pushed against the cell
wall, filling all the available space. *(Photos, Michael Abbey, Photo Researchers, Inc.)*

depending on whether the membrane contains specific carriers that participate in a process called "facilitated diffusion,"
which we discuss shortly. **Active transport** moves molecules
against a concentration gradient and requires energy (Figure
4-22C).

The answer to the question "Which way?" is different
for passive transport and active transport. Passive transport
results in equal concentrations of a molecule on the two sides
of a membrane, at least for uncharged molecules. For charged
molecules or ions, passive transport results in an unequal distribution of ions. This unequal distribution happens because
all cells have an excess of negative charges inside the cell membrane. (This excess results from the unequal pumping of
sodium and potassium ions, which we discuss later.) The result is that the cell interior repels negative ions and attracts
positive ions. Biologists can calculate how unequal the distribution of ions should be, and any deviation from the predicted distribution indicates that active transport has
occurred.

Active transport comes in several forms. In one form, cells
move ions across the membrane so that a net charge accumulates on one side. For example, an important active transporter,
the **sodium-potassium pump,** simultaneously transports

sodium ions (Na^+) out of cells and potassium ions (K^+) into
cells, using the energy of ATP. Cells have more negative charges
inside than outside. The net negative charge results from the
sodium-potassium pump, which moves three sodium ions out
of cells for every two potassium ions that it moves in.

Facilitated diffusion accelerates passive transport

Some molecules move across membranes much faster than
would be expected on the basis of diffusion alone, even though
their final concentrations are consistent with passive transport.
Such **facilitated diffusion**—an increased rate of passive transport—depends on the action of specific molecules within the
membrane that help, or "facilitate," transport. Glucose, for example, usually moves across cell membranes by means of facilitated diffusion.

A transmembrane protein called the glucose transporter is
responsible for moving glucose across the cell membrane, for
example, in red blood cells. To show that this protein causes
the facilitated diffusion of glucose, researchers studied the rates
of entry of radioactive glucose into liposomes—tiny bubbles
of artificial phospholipid bilayers that behave like membranes.
Pure liposomes allow the entry of scarcely any glucose. But

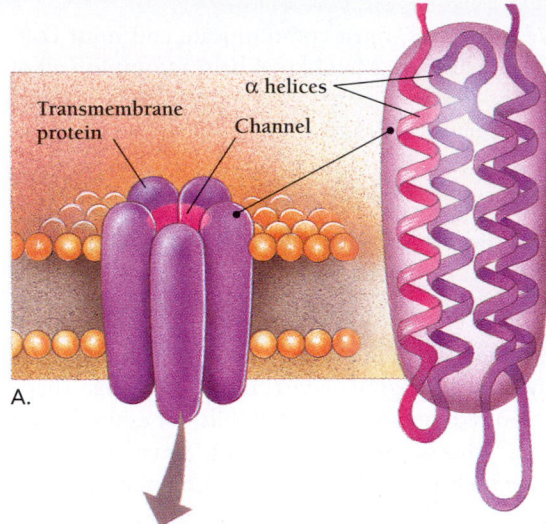

Figure 4-22 Passive and active transport. A. Transmembrane proteins help transport molecules and ions across the cell membrane. B. In passive transport, no energy is required. In channel-mediated transport, ions or small molecules pass through the transmembrane channels without regulation or help. In facilitated diffusion, molecules move faster than they ordinarily would, thanks to the action of transporter proteins that carry molecules across the membrane. Transporters bind only to certain molecules, such as glucose. C. Active transport requires energy, as well as a carrier protein that couples transport with energy expenditure. (1) The sodium-potassium pump binds to ATP and Na$^+$ ions on the *inside* of the plasma membrane and to K$^+$ ions on the *outside* of the membrane. It then transports Na$^+$ and K$^+$ in opposite directions, using the energy of ATP. (2) In cotransport, a cell membrane transports molecules using the energy from a concentration difference across a membrane (instead of using ATP). Cotransporter proteins transport glucose using a concentration difference in Na$^+$ ions, linking the "uphill" transport of one molecule or ion to the "downhill" transport of another.

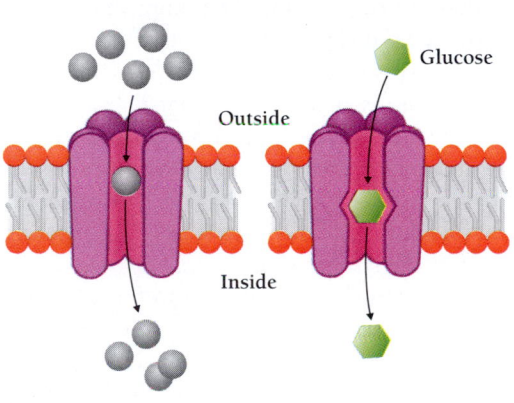

when the glucose transporter is added, the rate of glucose entry increases dramatically, showing that the transporter facilitates the entry of glucose into the liposome.

Transport proteins, such as the glucose transporter, bind to the molecules they transport, then carry the molecules across the membrane (Figure 4-22). Transporters bind only to specific molecules. The glucose transporter, for example, transports only glucose and other similar sugars.

Active transport uses energy to move ions

How can a membrane accomplish active transport? To transport a substance actively across a membrane requires a special "pump" mechanism that couples transport to an energy source (such as ATP). For example, the sodium-potassium pump binds to ATP and Na$^+$ ions on the inside of the plasma membrane and to K$^+$ ions on the outside of the membrane (Figure 4-22). When inserted into liposomes, the pure protein transports Na$^+$ and K$^+$ in opposite directions, using the energy obtained by hydrolyzing ATP.

Somewhat surprisingly, other examples of active transport do not directly depend on ATP. Instead, the cell harnesses energy from an already existing concentration difference across a membrane, usually of Na$^+$ or H$^+$ ions.

The coupled transport of two substances is called **cotransport** and depends on specific transmembrane protein molecules, called cotransporters. Cotransporters couple the "uphill" transport of one molecule or ion to the "downhill" transport of another. This mechanism is comparable to the action of a children's seesaw. As gravity pulls one child down, the child at the other end is raised up against the force of gravity. In a cotransporter used by the cells that line the small intestine, for example, the concentration difference of Na$^+$ ions supplies the energy for the cells to take up glucose (Figure 4-22).

We can see that membranes are important regulators of a cell's contents. They allow water to pass freely, with the net movement dependent on the relative solute concentrations inside and outside the cell. They also admit slightly larger molecules, but selectively. In some cases, they speed the passive transport of specific small molecules such as glucose with transport proteins. In other cases, they harness the energy of ATP or of existing concentration gradients to accumulate or exclude particular molecules or ions.

While we have spoken almost entirely about plasma membranes, the internal membranes of eukaryotic organelles are similarly important in regulating the contents of the spaces that they enclose. In Chapters 6 and 7, we will discover how crucial the membranes of some organelles are to the production of ATP.

Molecules pass through plasma membranes by means of diffusion, facilitated diffusion, and active transport. Active transport mechanisms use energy from ATP or from a concentration gradient of an ion or molecule different from the one being transported.

HOW DO MEMBRANES INTERACT WITH THE EXTERNAL ENVIRONMENT?

The plasma membrane of a cell functions like the outer surface of an organism. In multicellular organisms, the cell itself creates its own immediate environment. Most nonanimal cells, for example, surround themselves with a rigid carbohydrate structure, called the **cell wall** (Figure 4-23A). The cell wall can continue to function even after the death of the cell. In fact, some plant cells function *only* after death. Many of the vessels that conduct fluids in vascular plants, as well as many structural fibers, are composed of the walls of dead cells. Animal cells can similarly contribute to their own extracellular environment, which in many cases consists of a more diffuse network of carbohydrates and proteins, called the **extracellular matrix** (Figure 4-23B).

Cells, like organisms, must communicate and must consume and excrete materials. Membranes help accomplish all of these activities and ensure that the cell's contents are best suited to the cell's immediate challenges.

Membrane Fusion Allows the Import and Export of Particles and Bits of Extracellular Fluid

The tendency of membranes to avoid water, together with their fluidity, means that biological membranes can change shape rapidly. The ruptured membranes of red cells, for example, can reform as intact "ghosts" of cells that do a remarkably good job of regulating their internal environments by pumping ions and molecules.

The reorganization of membranes is an important part of many cellular functions. In cell division, for example, the plasma membrane of the parent cell divides and reseals itself around each daughter cell. In phagocytosis, cells engulf relatively large particles, such as microorganisms or cell debris. First, part of the plasma membrane extends outward and surrounds the particle (Figure 4-12). The membrane then fuses to form a large vesicle, which travels to the interior of the cell. The vesicle's membrane later fuses with the membrane of a lysosome, thereby exposing the vesicle's contents to the lysosome's digestive enzymes.

In contrast, in **endocytosis,** the cell takes in tiny amounts of materials in vesicles that arise by the inward folding (or invagination) of the plasma membrane. The pouchlike vesicles formed in endocytosis are tiny, about one-tenth the size of those formed in phagocytosis.

Cell biologists now recognize two types of endocytosis. In **pinocytosis** [Greek, *pinein* = to drink], the cell takes up any bits of liquid and dissolved molecules. In **receptor-mediated endocytosis,** the cell takes up only specific substances, which

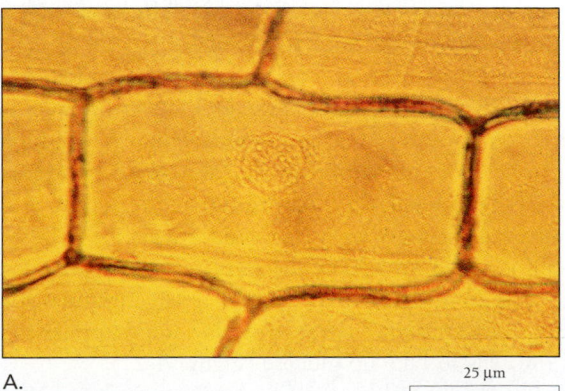
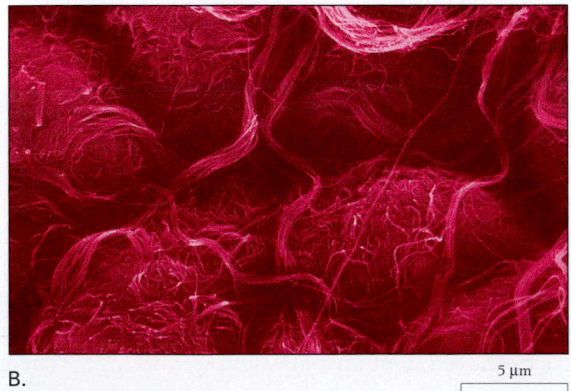

A.   25 µm   B.   5 µm

Figure 4-23   The cell wall. A. A rigid, external cell wall made of carbohydrate surrounds the plasma membranes of plant and fungus cells, most prokaryotes, and many protist cells. B. An extracellular matrix of carbohydrates and proteins encloses many animal cells. *(A, Kevin Collins/Visuals Unlimited; B, Fred Hossler/Visuals Unlimited)*

it recognizes by means of special proteins on the cell's surface, called receptors. Cholesterol, which travels through the blood in complexes with an amphipathic protein called LDL (low-density lipoprotein), enters cells through receptor-mediated endocytosis (Figure 4-24A; Box 3.1).

Exocytosis is the export of molecules from a cell by a process that is approximately the reverse of pinocytosis. Cells accumulate molecules to be exported in vesicles, surrounded by membranes. These vesicles then move to the cell surface, where they fuse with the plasma membrane and release their contents to the outside of the cell (Figure 4-24B). Molecules exported by exocytosis include proteins that function outside of cells, such as the digestive enzymes of the gut, and chemical signals, such as hormones and neurotransmitters.

The regulation of exocytosis is an extremely important part of cellular function. In the nervous system, for example, intercellular signaling depends on the precise timing of neurotransmitter release by exocytosis. Rapid increases in intracellular calcium ions stimulate exocytosis, and dozens of proteins are involved in its coordination. Some of these pro-

teins are part of the vesicle's own membrane. Other proteins are part of the plasma membrane, where the vesicle will fuse, while still other proteins form temporary bridges that allow the two membranes to fuse. Figure 4-24C shows the fusing of a vesicle membrane and a plasma membrane in a nerve cell.

Plasma membranes fuse easily and can spontaneously form vesicles. These abilities allow cells to change shape and to perform phagocytosis, endocytosis, and exocytosis.

How Do Cells Communicate in Multicellular Organisms?

Because the individual cells in all multicellular organisms must cooperate, communication among cells is paramount. In general, cells communicate by means of chemical signals such as hormones. Such signals allow the activities of one cell to influence those of another.

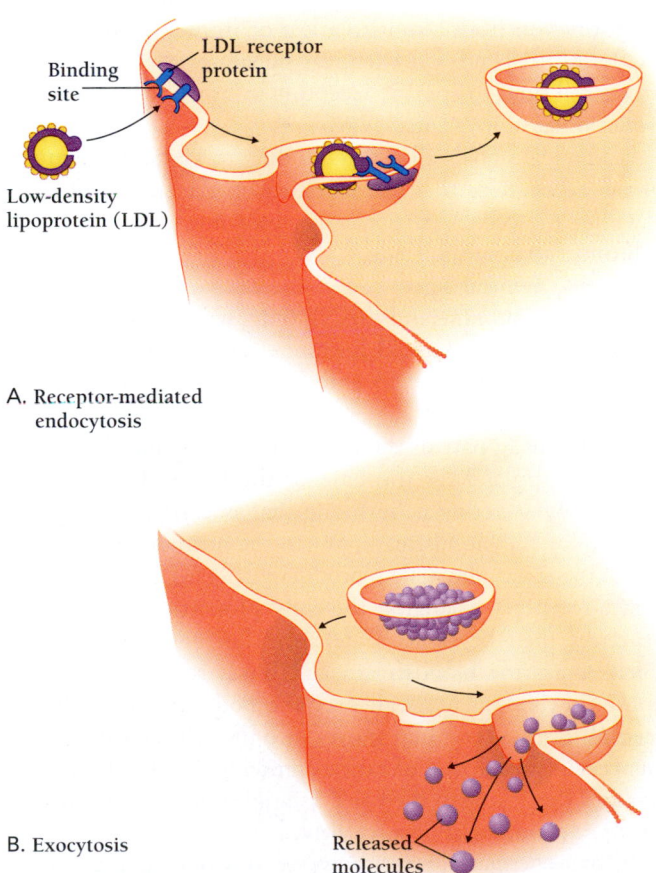

A. Receptor-mediated endocytosis

B. Exocytosis

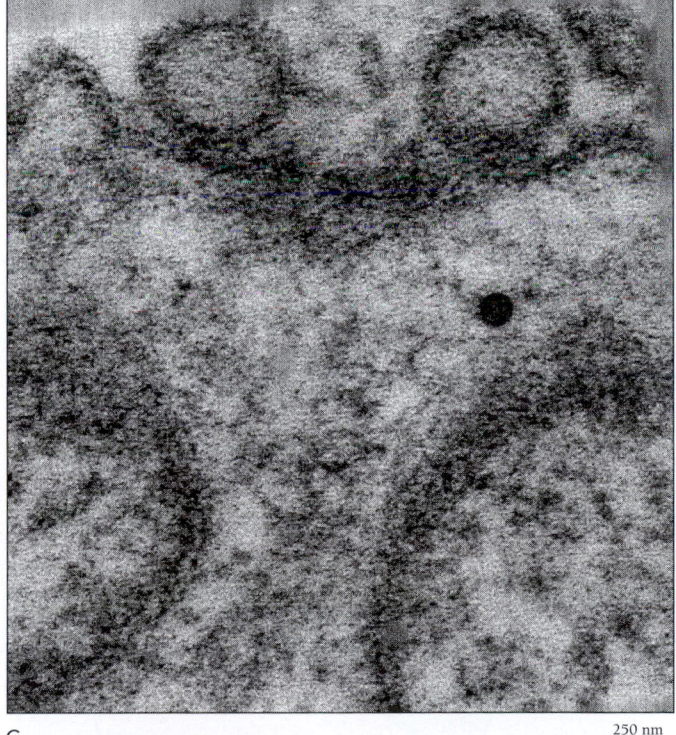

C.   250 nm

Figure 4-24   Endocytosis and exocytosis. A. Endocytosis plays a central role in the regulation of low-density lipoprotein (LDL) and cholesterol levels in the blood, an important predictor of heart disease. In one form of receptor-mediated endocytosis, LDL particles full of cholesterol bind to LDL receptors on the cell membrane. The cell membrane folds around the LDL particle and engulfs it, then the cell digests it. B. In exocytosis, the reverse appears to happen: vesicles full of proteins, hormones, or neurotransmitters made inside the cell fuse with the cell membrane and diffuse into the outside environment of the cell. Exocytosis depends on the coordinated action of proteins on the membranes that will fuse. C. A new technique, called electron microscopic tomography, has allowed researchers to reconstruct a three-dimensional picture of exocytosis. In this photo, we can see exocytosis of vesicles in a nerve cell. The vesicles contain acetylcholine, which (as we discussed in Chapter 1) stimulates muscle contraction. (*C, courtesy of U. J. McMahan, Stanford University*)

How do cells communicate with their immediate neighbors?

Plant cells are both cemented together and separated from each other by rigid cell walls. Plant cells are nonetheless in intimate contact through fine intercellular channels called **plasmodesmata** (singular, plasmodesma) [Greek, *plassein* = to mold + *desmos* = to bond]. Plasmodesmata, which are between 20 and 60 nm in diameter (Figure 4-25), allow molecules to pass directly from cell to cell. Some kinds of cells, such as those in the tips of a plant's roots, have huge numbers of plasmodesmata. The more plasmodesmata between neighboring cells, the more rapidly materials pass from one cell to the next.

Animal cells are not separated from each other by cell walls, and they communicate with each other in a variety of ways. Some animal cells easily pass ions and molecules, even proteins, through specialized regions of their membranes called **gap junctions.** In gap junctions, the membranes of adjacent cells are only 2 to 4 nm apart, instead of the usual 10 to 20 nm (Figure 4-26A). In addition, gap junctions connect cells with a protein called connexin that allows the cells to exchange chemical signals. Gap junctions, for example, connect heart muscle cells and allow them to coordinate their contractions.

Other animal cells communicate using "adhering junctions." **Adhering junctions** help provide strength to resist mechanical stress without affecting the passage of molecules between them. Adhering junctions usually consist of **desmosomes** [Greek, *desmos* = to bond + *soma* = body], buttonlike structures that hold two membranes together (Figure 4-26B). Desmosomes may be localized in spots, like the rivets on a pair of jeans, or they may be spread in bands or belts.

Several types of specialized junctions connect **epithelial cells** [Greek, *epi* = upon + *thele* = nipple], the cells that line the surfaces throughout the body, for example the lining of the gut. Epithelial cells are linked together tightly to form a sheet, called an **epithelium** [plural, epithelia]. Epithelia enclose many fluid-containing cavities, such as kidney tubules and intestines. One task of these epithelia is to keep the fluids from escaping. Impermeable junctions not only weld cells together but also prevent any molecules from leaking between them. In mammals and other vertebrates, the main kind of impermeable junction is called a **tight junction,** in which the membranes of adjacent cells actually fuse (Figure 4-26C).

Plasmodesmata and gap junctions allow materials to pass directly from cell to cell. Adhering junctions strengthen the connections between cells, and tight junctions keep fluids from passing between them.

How do cells communicate with distant cells?

Much intercellular communication, such as the transmission of information from a root to a leaf or from a nerve cell to a muscle cell, depends on chemical signaling. Molecules made by one cell bind to specific receptors in other cells, triggering a series of specific events.

Chemical signals all act by binding to specific **receptors,** specialized proteins that can lie either on the surface of a cell or within the cell. A particular receptor normally only binds with one signaling molecule.

Some chemical signals enter the cell by passing through the plasma membrane and then binding with receptors within the cytoplasm. The steroid hormone testosterone, for example, acts in this way, actually entering cells and altering the expression of individual genes. At puberty, a young man's hypothalamus (a part of the brain) sends a chemical signal to the pituitary gland, at the base of the brain. The pituitary responds by secreting a hormone that is carried throughout the body by the bloodstream. When the hormone reaches the cells of the testes, the male reproductive organs, it binds to cell surface receptors, changing the cells' properties. The cells of the testes then begin to secrete the hormone testosterone, which enters the bloodstream and binds to cytoplasmic receptors in different cells all over the body. In each cell, the receptorhormone complex moves into the cell nucleus and influences the production of various proteins. The results include a gradual deepening of the voice, a rapid spurt in body growth, and the development of a sex drive.

Other chemical signals, however, stay outside the cell, influencing what is going on inside by means of cell surface receptors. Some surface receptors can open channels in the plasma membrane to allow the passage of new materials into or out of the cell. Other receptors can regulate the levels of an influential chemical—called a *second messenger*—inside the cell. Many extracellular signals, for example, trigger an increase

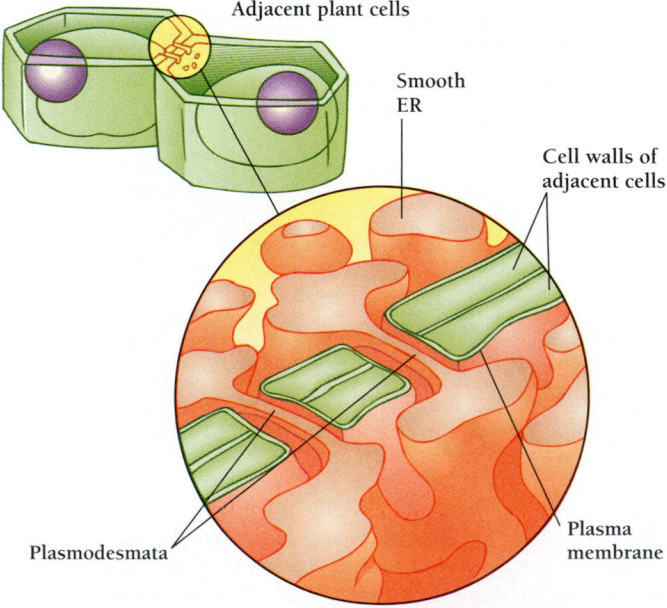

Figure 4-25   Cell to cell. Fine intercellular channels called plasmodesmata connect all the living cells of higher plants.

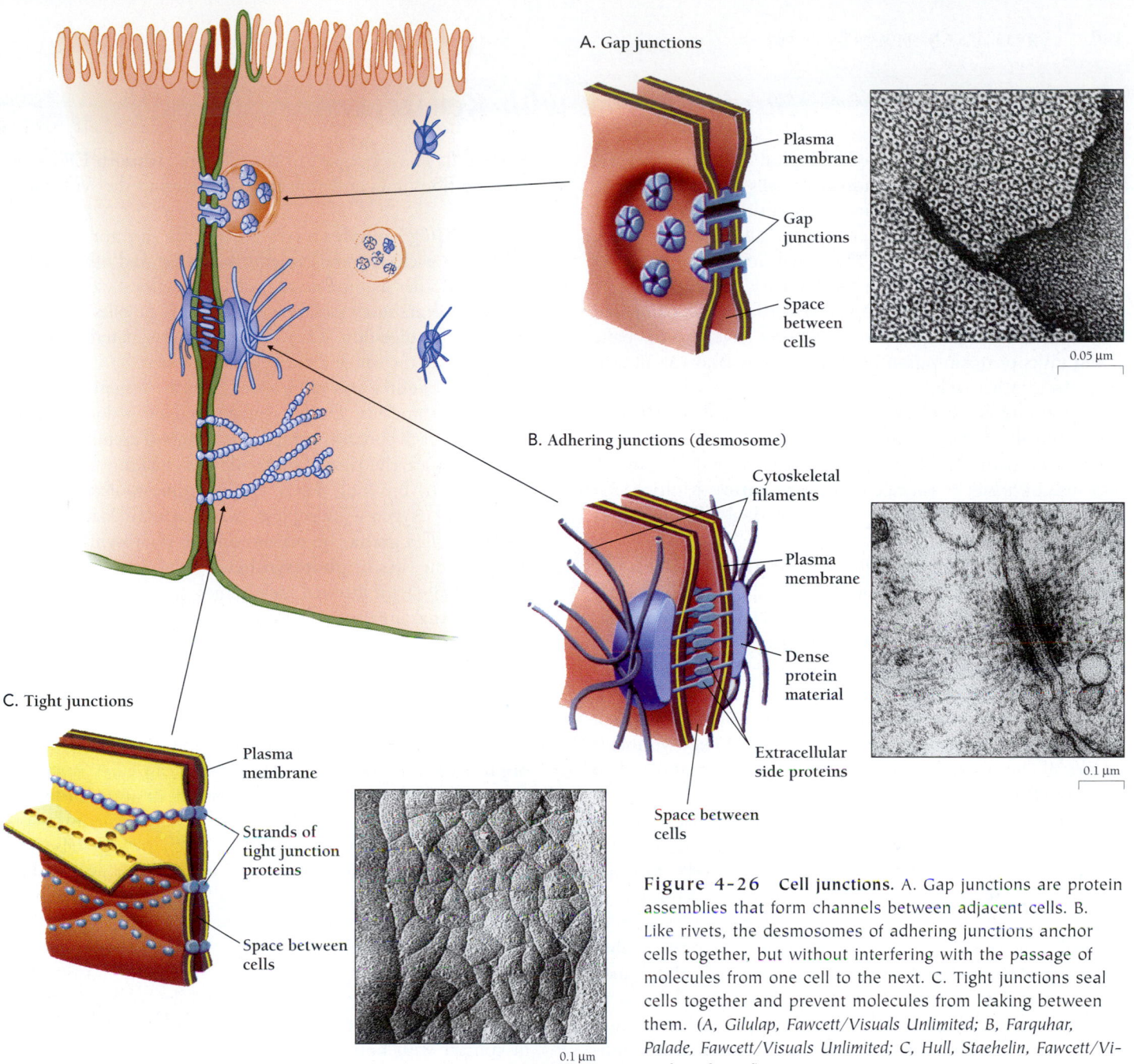

A. Gap junctions

Plasma membrane

Gap junctions

Space between cells

0.05 μm

B. Adhering junctions (desmosome)

Cytoskeletal filaments

Plasma membrane

Dense protein material

Extracellular side proteins

Space between cells

0.1 μm

C. Tight junctions

Plasma membrane

Strands of tight junction proteins

Space between cells

0.1 μm

Figure 4-26 Cell junctions. A. Gap junctions are protein assemblies that form channels between adjacent cells. B. Like rivets, the desmosomes of adhering junctions anchor cells together, but without interfering with the passage of molecules from one cell to the next. C. Tight junctions seal cells together and prevent molecules from leaking between them. (*A, Gilulap, Fawcett/Visuals Unlimited; B, Farquhar, Palade, Fawcett/Visuals Unlimited; C, Hull, Staehelin, Fawcett/Visuals Unlimited*)

in the concentration of calcium ions, which serve as a common second messenger.

Still other surface receptors act directly on the surface itself. When receptors bind to the hormone insulin, for example, they trigger the incorporation of additional glucose transporters into the cell membrane. The result is an increase in the cell's rate of glucose uptake.

In multicellular organisms, cells communicate both physically (through plasmodesmata and gap junctions) and chemically (through hormones and other chemical signals). These routes of communication allow organisms to coordinate their responses to changes in the external environment.

Cells respond to chemical signals by means of specialized proteins called receptors.

In this chapter, we have seen how biologists first recognized the importance of cells as the fundamental living units of all organisms. We have examined the internal structures of cells and membranes and seen how cells must coordinate the activities of their many internal components. In the next three chapters, we will use some of this knowledge to try to understand the biochemical processes that take place inside of cells.

Study Outline with Key Terms

Cells are not only the units of biological structure; they are also the units of function. Individual cells are specialized to perform distinct tasks. Cells are living entities that are highly organized, perform chemical reactions, obtain energy from their surroundings, respond to their environments, change over time, reproduce, and share an evolutionary history.

The **cell theory** states that (1) all organisms are composed of one or more cells; (2) cells, themselves alive, are the basic living units of organization of all organisms; and (3) all cells come from other cells.

The subdivision of organisms into cells and of cells into organelles allows for an efficient division of labor and makes it easier to maintain a constant environment within each cell (because of a higher **surface-to-volume ratio**). Multicellular organisms can survive the deaths of individual cells because of cell replacement.

Cells from all six kingdoms contain three common features: a **plasma membrane** that separates the inside of the cell from the rest of the world, a set of genetic instructions encoded in DNA, and a cell body (or **cytoplasm**). In the cells of all eukaryotes, the genetic material is enclosed in a membrane-enclosed **nucleus**. The nucleus contains the cell's genetic instructions in the form of **chromosomes,** which are made of **chromatin,** a complex of DNA and protein. The two membranes of the **nuclear envelope** are fused in many places to form **nuclear pores.**

The Archaea and Eubacteria, or prokaryotes, lack a membrane-enclosed nucleus. They keep their DNA in a region called the **nucleoid.** The cytoplasm of most eukaryotic cells (formerly referred to as the **protoplasm**) is divided into well-defined compartments, or **organelles,** by internal membranes. The jellylike **cytosol,** which surrounds all the cell's organelles, contains **ribosomes, proteasomes,** glycogen, fat, and thousands of enzymes.

Within the cytoplasm are several well-defined kinds of membrane-enclosed organelles: the endoplasmic reticulum (ER), Golgi complex, **lysosomes, peroxisomes, mitochondria,** and such **plastids** as **chloroplasts, chromoplasts,** and **amyloplasts.** In addition, eukaryotic cells contain **vesicles** and **vacuoles,** which are also enclosed by membranes but whose structures and functions vary widely.

The **endoplasmic reticulum (ER)** is an extensive network of membranes that encloses a space called the ER **lumen.** One part of the endoplasmic reticulum—the **rough endoplasmic reticulum**—is studded with ribosomes and makes proteins mainly destined for export to the outside or the outer surface of the cell. The other part of the ER—the **smooth ER**—contains no ribosomes and appears to function in the synthesis of lipids.

The **Golgi complex** serves as the traffic director in the cell, helping to package some proteins for export and others for inclusion in other membrane-enclosed organelles. Lysosomes and peroxisomes are small membrane-enclosed organelles that are responsible for digestion. They dispose of both material taken in from outside the cell and the cell's own debris.

Mitochondria and chloroplasts are the largest organelles outside the nucleus. Each is surrounded by a double membrane, and each contains complicated internal membranes. Mitochondria are the powerhouses of cells, making most of the cell's ATP by breaking down small, organic molecules in the presence of oxygen. Chloroplasts, which contain stacks of **thylakoids,** called **grana,** are present only in photosynthetic cells. They are the sites of **photosynthesis.**

Together with internal membranes, the **cytoskeleton** provides internal organization for eukaryotic cells. It consists of three sets of filaments: **microtubules** (which are made of two kinds of **tubulin**), **actin filaments,** and **intermediate filaments.** In most cells, the cytoskeleton consists of structures that are assembled and disassembled continually. Their assembly seems always to originate in structures called **microtubule organizing centers (MTOCs). Motors** can carry organelles along microtubule tracks either toward or away from the MTOC. The major MTOCs are near the nucleus in a zone called the **centrosome,** a region that (in many cells) also contains a distinct organelle called the **centriole.** Some cells have permanent structures, such as **cilia** and **flagella,** which are composed of elements of the cytoskeleton.

All cells are surrounded by membranes. Membranes consist mainly of lipids and proteins. The lipids, mostly phospholipids, are arranged in two layers, a "bilayer." Some proteins are located on the external and internal surfaces of the bilayer, in contact with both the hydrophobic interior of the membrane and the hydrophilic cytoplasm or cell exterior. Other proteins pass completely through the membrane. The membrane is **fluid,** both protein and lipid molecules move freely within it. This picture of membrane structure is called the **fluid mosaic model.**

Membranes are **selectively permeable:** they allow some molecules to pass in and out freely and confine others to the inside or the outside. In **simple diffusion,** solutes move from an area of high **concentration** to an area of low concentration. That is, solutes move along a **concentration gradient. Osmosis** is the movement of water across any selectively permeable membrane in response to a concentration gradient. A solution may be **hypotonic, isotonic,** or **hypertonic,** relative to the concentration of solutes on the other side of the membrane. The pressure created by the concentration gradient, called **osmotic pressure,** is comparable to mechanical pressure.

In plants, the central vacuole fills with water, which presses against the **cell wall,** creating **turgor pressure**—the principal mechanical support of nonwoody plants. When a plant cell is immersed in a hypertonic solution, water flows out of its vacuole and the cell separates from the cell wall in **plasmolysis.**

Molecules pass through plasma membranes by means of **passive transport** (which requires no energy), **facilitated diffusion,** and **active transport** (which requires energy from ATP or **cotransport**). The **sodium-potassium pump** is one active transport mechanism. Certain cells engulf large particles by extending their plasma membranes in a process called **phagocytosis.** Most cells are also capable of **endocytosis** and **exocytosis.** Endocytosis may be nonspecific (**pinocytosis**), or it may depend on specific receptors on the membrane (**receptor-mediated endocytosis**).

The outer surfaces of cells are responsible for their interactions with other cells and with other molecules that bind to **receptors** on the cell surface or within the cytoplasm. Cell walls enclose most prokaryotic and many eukaryotic cells (though not animal cells). Animal cells often are surrounded with **extracellular matrix,** a complex of carbohydrates and proteins.

Multicellular organisms depend on coordination of activities among many cells. Plants and animals have a variety of specialized structures by which cells interact. Some of these structures—**plasmodesmata** in plants and **gap junctions** in animals—allow the free passage of some molecules between adjoining cells. Other specialized junctions between cells—**adhering junctions**—rely on **desmosomes** and provide mechanical strength to sheets (**epithelia**) of **epithelial cells. Tight junctions** weld some cells together and prevent the passage of molecules through them.

Review and Thought Questions

Review Questions

1. State the cell theory. Be sure your statement includes the concepts of what a cell is, where cells occur, where they come from, and what they do.
2. Describe three features that all cells have in common.
3. Describe the differences between prokaryotic and eukaryotic cells.
4. Explain how the surface-to-volume ratio of a cell is related to its ability to take in resources and to rid itself of wastes. Why are most cells so small? How can a chicken egg, which is one cell, be so big?
5. Sketch and label a mitochondrion and a chloroplast. What is the principal function of each? Where does each obtain fuel?
6. Describe the fluid mosaic model of membrane structure.
7. A plasma membrane functions to keep the material outside a cell separate from the material inside. What is one property of the membrane that allows this?
8. Explain how membrane-enclosed vesicles move materials into, out of, and within cells.
9. What is the role of the cytoskeleton in eukaryotic cells?
10. Compare the sizes of microtubules, actin filaments, and intermediate filaments. Describe one activity or structure in which each participates.

Thought Questions

11. Between conception and death, most of your cells die while new ones are produced by the division of other cells. What makes you really "you"? What is it that is constant about your structure?
12. Explain how the endoplasmic reticulum might increase the effective surface-to-volume ratio for a cell, even though it is located inside the outer perimeter of the cell. Can it really assist in bringing resources in and getting rid of wastes?
13. Chloroplasts are also called the "powerhouses" of cells. What do they produce to earn this description? What types of cells contain chloroplasts?

About the Chapter-Opening Image

One-celled organisms, such as this diatom, were among the first studied by early microscopists.

 On-line materials relating to this chapter are on the World Wide Web at **http://www.harcourtcollege.com/lifesci/aal2/**

5 Directions and Rates of Biochemical Processes

Ludwig Boltzmann: Left Behind, or Ahead of His Time?

In September 1906, the eminent Austrian physicist Ludwig Boltzmann fell into the last of many deep depressions. Hoping to lift his spirits, he took his wife and daughters on a trip to Duino, Italy, a spectacular resort on the Adriatic Sea.

The mercurial Boltzmann had always been emotional (Figure 5-1). He could be moved to tears of joy by a well-played sonata, the color of the sea, or a piece of poetry. At other times he lapsed into long, grief-stricken silences. He wept at the railroad station whenever he left his wife for any extended period.

Although Boltzmann was unstable, his growing despondency also had roots in the scientific milieu of his time. At 60, the physicist was convinced that much of his life's work had been wasted. For years, he had championed the idea that matter is composed of atoms and that the behavior of atoms accounts for all chemical transformations. Other chemists and physicists discussed atoms extensively, but for them, atoms were just a useful way to think about the formation of

chemical compounds. Almost no one thought atoms were real, and no one before Boltzmann thought of them as the key to understanding energy.

The hypothesis that matter was made of tiny particles called atoms was first popularized by the Greek philosopher Democritus in the fifth century B.C. Not until the 17th century, however, did "atomism" get serious attention. And the same renaissance in thinking that gave birth to atomism also gave birth to another idea that would help drive Boltzmann to despair. According to "positivism," observable facts are the only basis for a true knowledge of the world.

Since nobody could see atoms, 19th-century positivists insisted that there was no reason to believe in atoms. Boltzmann's peers, all eminent scientists and heartfelt positivists, insisted that assuming that atoms were real—actually basing scientific theories on their existence—was no better than mysticism. Indeed, one of Boltzmann's stoutest critics, the physicist Ernst Mach, used to harass Boltzmann as Boltzmann lectured on atoms by calling out from the audience, "Have you ever seen one?" Of course, Boltzmann had not. (Indeed, only with the development of atomic force microscopes in the 1980s could scientists actually "see" atoms.)

But Boltzmann felt certain that atoms were real, and for years he had written scathing critiques of even his closest friends' work. In his personal life, Boltzmann was charming, modest, and earnest, but in debates and in the lecture hall, he was a bulldog, relentlessly attacking arguments he believed were wrong and proclaiming his own ideas.

For almost 40 years, Boltzmann worked to show how atoms accounted for the behavior of all forms of matter. His greatest achievements were to find a way to understand the behavior of matter in terms of large collections of molecules and atoms. He showed that the statistical laws of large numbers could be applied to collections of atoms or molecules.

But Boltzmann's work could be correct only if atoms were real. And, after 40 years of arguing, fewer and fewer people seemed to think that atoms existed. As Boltzmann wrote to a friend, "I am conscious of being only one individual struggling weakly against the stream of time," and later, "I present myself to you therefore as a man left behind."

Key Concepts

- The Second Law of Thermodynamics predicts the direction of any process because energy becomes less available for work as a reaction or process occurs.

- The rate of a chemical reaction depends on the random movements of molecules.

- Enzymes speed up biochemical reactions by lowering their activation energies.

- Cells and organisms regulate their own metabolism by changing the shapes and activities of enzymes.

Boltzmann did not know it, but by 1906, the year he took his family to Italy, the atom was about to take center stage in the world of science. The darkest days for his ideas were over, and bright young scientists were taking a second look at atoms. In 1905, for example, 26-year-old Albert Einstein had published a paper showing that the constant jiggling of particles that scientists noticed when looking down a microscope (called Brownian motion) was most likely due to bombardment by atoms—real, massive objects, not abstract statistical averages. In just a few years, Ernest Rutherford would locate the tiny, central nucleus of the atom. And soon after that, the physicist Niels Bohr would calculate the movements of electrons.

But Boltzmann had given up. On a sunny day in September 1906, while his wife and daughters were swimming, Boltzmann ended his rich and productive life by hanging himself in his hotel room. Had he endured just a few more years of frustration, he would have lived to see the triumph of his own ideas and the humiliation of his intellectual enemies.

Indeed, Boltzmann's intellectual victory was so complete that the material in this chapter—written nearly 100 years after his death—rests heavily on his work. The most important applications of Boltzmann's idea were in the field of **thermodynamics** [Greek, *thermo* = heat + *dynamis* = power, force], the study of the relationships among different forms of energy. For example, as a car converts the chemical energy of gasoline into the energy of motion, the car moves forward. Every event, every reaction, every process is accompanied by a "transformation of energy."

The direction a transformation takes depends on the rules of thermodynamics. A scientist can predict the direction of a reaction without knowing exactly how the reaction takes place. In the same way, we know that water always flows downhill, even if we do not know whether it gets there through a pipe, a stream, or a waterfall.

Figure 5-1 Ludwig Boltzmann (1844–1906). *(Science Photo Library/Photo Researchers, Inc.)*

Before Boltzmann, thermodynamics was understood as a set of abstract mathematical laws. These two laws could predict which way a reaction would go, but they said nothing about the actual process of the reaction.

Boltzmann showed that the reason for thermodynamics was the behavior of large numbers of individual atoms and molecules. He showed how chemical reactions occur, what heat is, and why some chemical reactions produce heat while others absorb heat. Most important, Boltzmann's work paved the way for the theory of **kinetics** [Greek, *kinetikos* = moving], the study of the rates of reactions. Before Boltzmann, chemists could predict only whether a reaction should take place, not how long it would take. Kinetics, which explains why a reaction takes place, can also predict how fast it will happen.

Understanding a biological or chemical process (or any process, for that matter) requires more than noticing changes in structures. For example, we may observe that most of the gasoline in a car's engine is converted to water and carbon dioxide, with a tremendous release of heat and energy of motion. But we cannot understand what is happening unless we know the answers to two general questions: (1) Which way will the reaction occur? and (2) How fast will it occur?

We must ask, Why does the conversion of gasoline to water and carbon dioxide occur in one direction only? Why don't the carbon dioxide and water present in the air we breathe combine spontaneously to form gasoline? And why does the conversion from gasoline to carbon dioxide and water occur in fractions of a second? What if converting a tank of gasoline to water and carbon dioxide took several years? In that case, gasoline would be a poor fuel. Why does the reaction occur as rapidly as it does? What factors determine how fast any reaction goes?

Predicting the speed of a process requires that we know something about the process itself. For example, we know that water plunges quickly downhill in a waterfall and meanders slowly in a gentle stream. The rate at which the water moves depends on the steepness of its descent. Thermodynamics predicts that water will flow downhill. But only kinetics can tell us how fast the water will flow. In this chapter, we will see that thermodynamics tells us which way a reaction will go and that kinetics tells how fast.

WHAT DETERMINES WHICH WAY A REACTION PROCEEDS?

Everyone can tell when a movie runs backward: a waterfall goes up instead of down; a rubber ball bounces up a flight of stairs; a smashed apple spontaneously reassembles (Figure 5-2). We all know the direction of time's arrow. Biological processes also have direction: a child matures into an adult; a bud unfolds into a leaf; dead organisms decay.

The key to predicting direction lies in understanding the conversion of energy. In any process—whether the flow of water, the growth of a tree, or the maturation of a child—energy changes form. The study of thermodynamics has provided rules that govern all energy changes, both biological and nonbiological. These rules establish the direction of time's arrow.

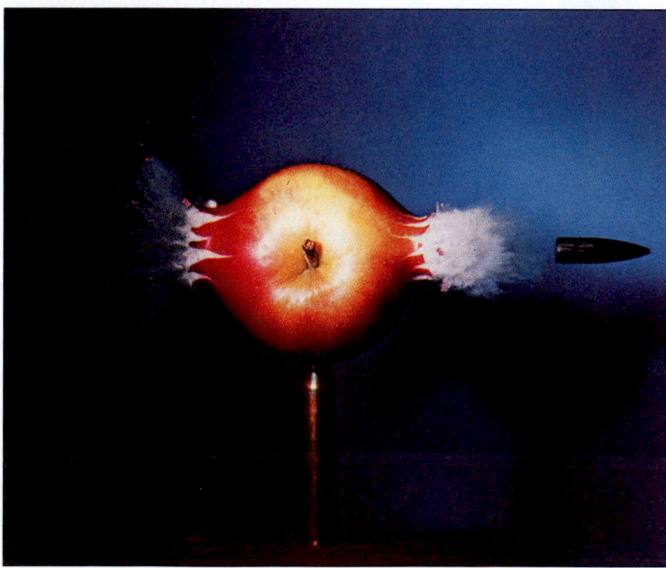

Figure 5-2   We all know the direction of time's arrow. If we saw two photos of an apple taken before and after a bullet passed through it, we would know instantly which photo was taken first. The destruction of an apple is a process that increases entropy. The only way to reassemble such an apple would be to return the pieces of apple to an apple tree, which might construct a new apple in a few months with the aid of energy from the sun. *(Photograph by Dr. Harold E. Edgerton, Harold and Esther Edgerton Foundation, ©1999. Courtesy Palm Press, Inc.)*

How May Work Be Converted to Kinetic or Potential Energy?

Most of us think of work as any process that requires energy. There is, however, a scientific definition of work: **work** is the movement of an object against a force. For example, lifting a sack of concrete against the force of gravity is work. So is winding the spring in a child's toy (Figure 5-3).

Work, in this sense, can be stored for later use, the way energy is stored in a battery. The wound-up spring in a toy beetle can propel the toy by operating its legs. When the spring is wound, we say that the work of winding the spring is stored as **potential energy.** When the spring unwinds and the beetle leaps forward, we say that the toy has **kinetic energy**—the energy of moving objects. A boulder at the edge of a cliff has potential energy. The same boulder falling to the bottom of the cliff has kinetic energy. A moving car, a running athlete, and a contracting muscle all have kinetic energy. Fast-moving molecules also have kinetic energy, which we call heat. Like kinetic energy, potential energy exists in many forms, including gravitational, electrical, and chemical.

Rearranging atoms in a chemical reaction is also a form of work. Recall, for example, that the energy-rich molecule ATP consists of a molecule of adenosine linked to three negatively charged phosphate groups (Figure 5-3). Making a molecule of ATP requires the squeezing together of three negatively charged phosphate groups. This squeezing is work against an electrical force. Once the atoms are squeezed together into a molecule, the work is stored as potential energy. Thus, ATP has potential energy. This energy can later be released to do work. If, for example, the third phosphate group is removed, stored energy is released. That energy can drive another chemical reaction, which can make a muscle move, heat the body, or help us understand a sentence.

> *Work, the movement of an object against a force, can be stored as potential energy. The construction of a molecule of ATP stores potential energy that can be released later as kinetic energy.*

How Does Thermodynamics Predict the Direction of a Reaction?

Scientists and engineers first began to understand the relationship among different forms of energy in the late 18th century. The major stimulus for the development of the rules of thermodynamics was the increasing use of the steam engine, which converted the potential energy of coal into kinetic energy capable of powering a locomotive or a factory. As we shall see, the same rules that apply to steam engines also apply to the energy conversions in organisms.

One of the first and most important theories of the new field of thermodynamics was that, in any process, energy is neither created nor destroyed—it only changes form. The **First Law of Thermodynamics** states that the total amount of energy in any process stays constant (Figure 5-4A). In this view,

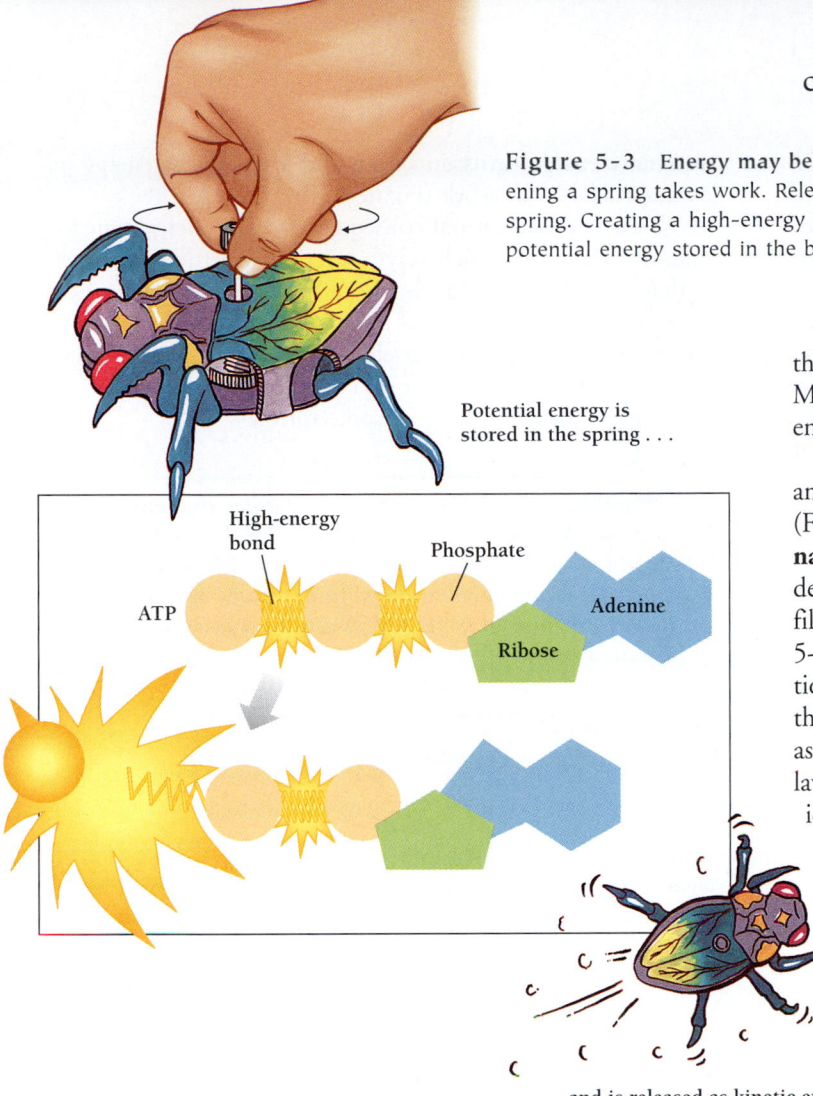

Figure 5-3 **Energy may be stored in a spring or in the high-energy bonds of ATP.** Tightening a spring takes work. Releasing the spring discharges the potential energy stored in the spring. Creating a high-energy phosphate bond takes work. Breaking the bond discharges the potential energy stored in the bond.

Potential energy is stored in the spring . . .

. . . and is released as kinetic energy.

then heat is just the rapid movement of atoms and molecules. Molecules and atoms—though invisible to us—have kinetic energy in the same way that a ball moving through the air has.

The second major theory of thermodynamics was that, in any process, energy becomes less and less available to do work (Figure 5-4B). More briefly, the **Second Law of Thermodynamics** states that in any process the energy available for work decreases. The Second Law explains why we can tell when a film is running backward—why time has a direction (Figure 5-2). At 27, Boltzmann worked out a mathematical description of the behavior of large groups of atoms. He showed that the properties of these groups accounted for the Second Law, as well the first. Oddly, even though everyone accepted the laws of thermodynamics, Boltzmann's peers mostly ignored his ideas about atoms. They all assumed Boltzmann was a genius, but few understood or wanted to understand what he was talking about. In the rest of this book, we will see this pattern again and again. Interesting and important ideas often meet rejection when they are first proposed.

Today, scientists have come to accept that the two laws of thermodynamics together predict the direction of a reaction. For example, we know that if we do a chin-up, our muscles contract and perform work. The energy for this work comes from the chemical energy in ATP. In the muscle cells, ATP molecules give up part of their stored energy by splitting off one of three phosphate groups. The remaining molecule has two phosphate groups and is then called adenosine *di*phosphate, or ADP [Greek, *di* = two]. ADP still has potential energy, but less than ATP.

But, according to the Second Law, muscle contraction can convert only a part of the energy obtained from ATP into useful work. The rest of the energy from splitting ATP becomes

energy may switch from potential energy to kinetic energy and back again, but it is neither created nor destroyed. Any heat generated during a process is merely another form of kinetic energy. When Ludwig Boltzmann was only 22, he showed that the movements of atoms nicely explain the First Law of Thermodynamics. If atoms and molecules actually exist, he argued,

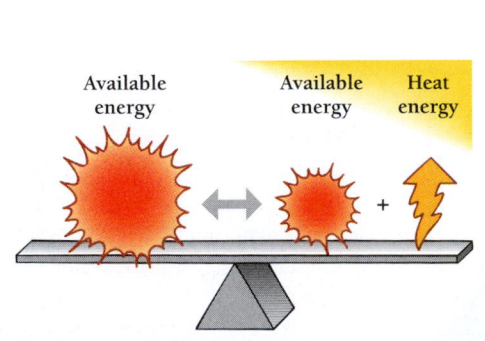

A. First Law of Thermodynamics

B. Second Law of Thermodynamics

Available energy

Available energy

Heat energy

Available energy

Heat energy, which is unavailable for work

Time's arrow

Figure 5-4 **Two laws of thermodynamics.** A. First Law: Energy may change from one form to another, but it is never lost. B. Second Law: The energy available to do work decreases as energy is lost as heat.

heat—the random motion of molecules. From an organism's point of view, heat is mostly wasted energy. Even though some energy is lost as heat, however, no energy is actually destroyed. All of the energy released from splitting ATP goes either to contracting the muscle or to heat.

The First Law implies that all forms of energy—including chemical energy, electrical energy, mechanical energy (work), and heat—can be measured in the same units. The unit of energy that we use in this book is the **calorie**—the amount of energy needed to raise the temperature of 1 gram of water by 1°C. The energy contained in food is often reported in Calories (with a capital C). One Calorie equals 1 kcal ("kilocalorie"), which equals 1000 calories. The daily diet of a college student usually contains from 2000 to 3000 kcal (Calories) of energy.

Every time energy changes form—whether from chemical to electrical or from electrical to light energy—a little more of the energy turns into heat (Figure 5-5). Over time, more and more energy turns into heat, and less and less energy becomes available for work (Figure 5-4B).

Heat loss is a natural consequence of all energy transformations. If we ask which way a reaction will run, the answer is that usually it will run in the direction that releases heat. The formation of a chemical bond between two atoms generally releases energy. Conversely, the breaking of a chemical bond usually consumes energy. The only way to reverse such a reaction is to add energy in some form.

The First Law of Thermodynamics states that energy changes form but is neither created nor destroyed. The Second Law of Thermodynamics states that in any process, the energy available to do work decreases as energy is lost in the form of heat. The loss of heat establishes the direction of the process. Without an input of energy, the process cannot reverse itself.

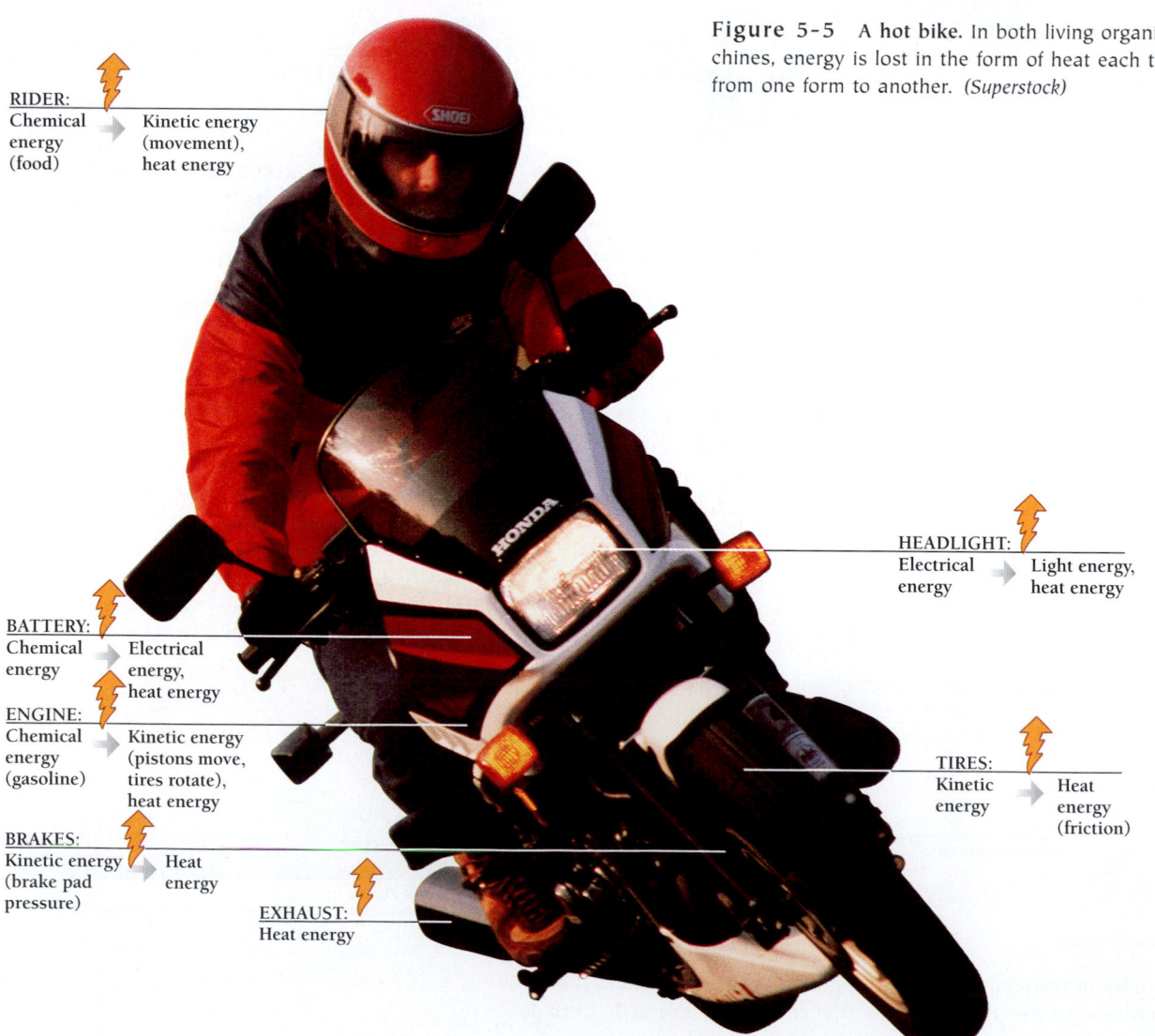

Figure 5-5 A hot bike. In both living organisms and in machines, energy is lost in the form of heat each time energy changes from one form to another. *(Superstock)*

RIDER:
Chemical energy (food) → Kinetic energy (movement), heat energy

HEADLIGHT:
Electrical energy → Light energy, heat energy

BATTERY:
Chemical energy → Electrical energy, heat energy

ENGINE:
Chemical energy (gasoline) → Kinetic energy (pistons move, tires rotate), heat energy

TIRES:
Kinetic energy → Heat energy (friction)

BRAKES:
Kinetic energy (brake pad pressure) → Heat energy

EXHAUST:
Heat energy

How Do Changes in Free Energy Predict the Direction of a Reaction?

To predict the direction of a chemical reaction, we need a measure of how much energy is available. **Free energy** is the energy in a system available for doing work. Free energy is abbreviated **G**, after the American chemist J. Willard Gibbs, who, in the 1870s, single-handedly worked out the theory of chemical thermodynamics.

Every chemical has a certain amount of free energy. And every chemical reaction involves a "change in free energy." When two molecules (or **reactants**) react to form one or more new molecules (or **products**), the change in free energy is equal to the free energy of the products minus the free energy of the reactants. The change in free energy is abbreviated **ΔG**. (Δ, pronounced "delta," is the mathematical symbol for a change in a quantity.) Chemists abbreviate this statement by writing

$$\Delta G = G_{products} - G_{reactants}$$

This equation might look intimidating, but it is no different from saying that the amount of weight you have gained or lost in a week is equal to the amount you weigh today minus the amount you weighed last week:

$$\Delta lbs = lbs_{today} - lbs_{last\ week}$$

Here, Δlbs ("delta pounds") equals the change in your weight, the amount of weight you lost or gained. In fact, without adding energy to your body's "system" by eating, you will inevitably lose weight, and Δlbs will be negative. That's because your body's activities are energy-releasing processes.

All processes that release energy are said to be **exergonic.** The breaking of ATP into ADP and phosphate, for example, is exergonic. Likewise, a coconut falling from a tree releases energy as it falls. Stand underneath it and you will not doubt that it is an exergonic process.

On the other hand, forcing another phosphate onto ADP to make ATP is an **endergonic,** or energy-consuming, process. Carrying a coconut up a tree is work and an endergonic process. It requires an input of energy, but it increases the capacity of a system to do further work (Figure 5-6).

In all exergonic processes, free energy decreases, so ΔG is negative. As a result, exergonic processes, such as ATP changing to ADP, occur spontaneously (given enough time). In endergonic processes, free energy increases and ΔG is positive. Endergonic reactions will not occur spontaneously, but require energy from some other source.

Finally, a change in free energy depends only on the beginning and end of a reaction. The change does not depend on anything that happens in between or on how long the process takes. A ball on a staircase might fall down the whole staircase in one big bounce or in several smaller bounces. The ball might even sit unmoving on the top step for years on end until it is jostled by an earthquake, a passing truck, or a curious child. But in all these cases, the change in free energy is the same.

Thermodynamics cannot predict how long it will take for a ball to fall down the stairs or for a muscle cell to break down ATP into ADP and phosphate. All that thermodynamics tells us is that, if we wait long enough, the ball will roll down the stairs—not up—and that ATP will break down into ADP and phosphate. In both examples, the process goes "downhill," from higher free energy to lower free energy.

If, during a reaction, the change in free energy is negative, the reaction is said to be exergonic and the reaction can occur spontaneously.

ADP + P + Energy → ATP

ATP → ADP + P + Energy

Endergonic

Exergonic

Figure 5-6 Work and play. Carrying a coconut up a tree is an endergonic (energy-consuming) process that scientists call "work." Dropping a coconut is an exergonic (energy-releasing) process. In the same way, adding a phosphate group to ADP consumes energy and removing the same phosphate group releases energy. It's work to make ATP.

What Is the Source of the Free Energy Released or Consumed During a Reaction?

When two isolated atoms form a chemical bond, energy is released. The tighter the bond, the more energy is released. During a chemical reaction, some of the bonds in the reactants break, while other bonds form to create the products. For example, consider the breakdown of ATP to ADP:

$$ATP + H_2O \longrightarrow ADP + phosphate$$

$$(Reactants) \longrightarrow (Products)$$

In this reaction, the oxygen atom in the water molecule separates from one of its hydrogens and replaces the oxygen atom in ATP's third phosphate group (Figure 5-7). Altogether, two bonds break—an O–H bond (in water) and a P–O bond (in ATP). In addition, one new bond forms—a P–O bond (in phosphate). The new P–O bond (in the phosphate) is more stable than the old one. In general, free energy decreases when unstable bonds break and stable ones form. And, overall, the free energy of the products (ADP and phosphate) is less than the free energy of the reactants (ATP and water).

The ATP → ADP reaction favors the breaking of less stable bonds and the forming of more stable bonds. Biochemists

REACTANTS: Water molecule and ATP molecule

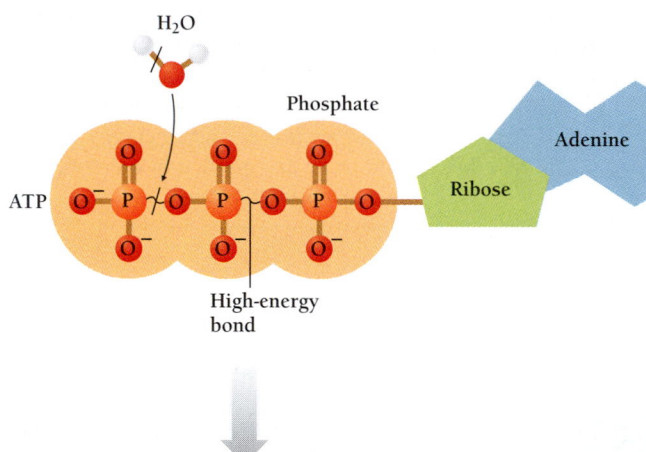

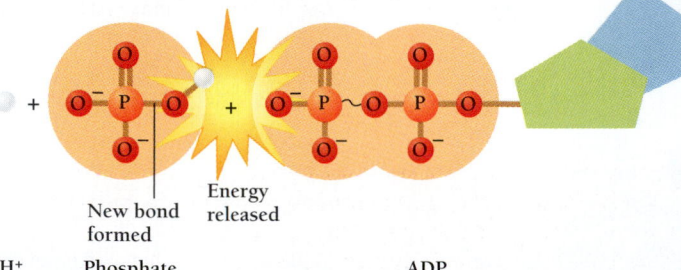

PRODUCTS: Hydrogen ion, phosphate group, ADP molecule

Figure 5-7   The most important exergonic reaction in the body. Splitting a high-energy phosphate bond in ATP releases energy that can be used in thousands of other reactions.

call unstable chemical bonds that give up energy easily **high-energy bonds.** They are sometimes represented as a squiggle (~). The bonds that hold the second and third phosphates of ATP are both high-energy bonds.

Chemists can predict the direction of a chemical reaction by measuring the change in free energy for the making and breaking of chemical bonds. For example, the change in free energy (ΔG) for the breakdown of ATP to ADP and phosphate is −7.3 kcal/mole. (As mentioned in Chapter 2, a mole is a fixed, huge number of molecules—6×10^{23}.) Because ΔG is negative, we can tell that free energy decreases and the reaction is exergonic. It will occur spontaneously.

The bonds between the atoms of a molecule have characteristic energies that change in the course of a reaction. Rearrangements of these chemical bonds may consume or release energy.

How Can One Process Provide the Energy for Another?

We saw earlier that an energy-consuming, or endergonic, reaction needs a source of energy. That energy often comes from an exergonic reaction. The most common source of energy for biological processes is the breakdown of ATP. When ATP breaks down into ADP and phosphate, it releases energy that becomes available to drive other chemical reactions, such as the linking of glucose molecules into a polysaccharide.

When an exergonic reaction drives an endergonic reaction, we say that the two processes are **coupled.** The chimpanzees and coconuts in Figure 5-8 are a coupled system. The falling coconut releases energy that lifts the small chimpanzee. A coupled reaction must be exergonic overall. That is, the exergonic reaction and the endergonic reaction must together release energy. If the picture in Figure 5-8 showed an elephant at the end of the log, we could guess that the coconut could not lift it. In the same way, the energy from splitting a single ATP molecule can drive only small reactions. But most reactions in biology occur in small steps.

How do cells couple endergonic and exergonic reactions? The answer is **"enzymes,"** large molecules, usually proteins, that speed up chemical reactions. Enzymes do not affect the change in free energy of a reaction. That is, enzymes do not change whether a reaction will happen or not, just how fast. You can slide down a hill faster if it's covered with ice than if it's just dirt. But no matter how slippery a hill is, you can't slide up the hill. Enzymes are like the ice on the hill: they speed up a spontaneous reaction but cannot reverse its direction.

Enzymes can also couple endergonic and exergonic reactions. In muscle contraction, for example, an enzyme couples the breakdown of ATP to the contraction of tiny muscle filaments.

Endergonic

Exergonic

Figure 5-8   A coupled reaction. When the adult chimpanzee drops the coconut, the baby chimp is thrown up into the air. The falling coconut is *coupled* to the flying chimp.

Enzymes couple exergonic reactions, such as the splitting of ATP, to endergonic reactions.

How Do Concentration (and Entropy) Affect Equlibrium?

The Second Law of Thermodynamics—that the energy available to perform work always decreases—also implies that uncoordinated motion is more probable than coordinated motion. Our common experience with "time's arrow" is that it is more likely that the pieces of a burst balloon will fly apart than it is that they will spontaneously reassemble. It is more likely that the books in your room will become disorganized than that they will remain neatly shelved in alphabetical or-

der. The Second Law is a formal statement that the amount of disorder in the universe always increases.

Physicists have defined a formal measure of disorder, called **entropy.** Entropy has a high value when objects are disordered or distributed at random, and a low value when they are ordered. Surprisingly, because entropy tends to increase, it can be used to do work (Figure 5-9).

Any process that converts an orderly arrangement to a less orderly one can perform work. For example, when we burn wood (converting well-ordered molecules of cellulose to disordered molecules of carbon dioxide and water), we can use the energy from this exergonic reaction to warm a house, to grill a hamburger, or to power a steam engine.

On the other hand, any process that converts a disorderly arrangement to an orderly one consumes energy. An oak tree uses energy from the sun and carbon dioxide from the air to build cellulose in its massive trunk and branches. Likewise, a cell uses energy to maintain concentrations of molecules inside the cell that are different from those outside the cell. But cells must perform work to keep their internal environments constant. When no energy supply is available to perform this work, the differences between the inside and the outside of the cell disappear, and the cell dies.

The movement of molecules from a more concentrated to a less concentrated state constitutes an increase in entropy, an increase that can be harnessed to do work. The free energy—the energy available to perform work—depends on the concentration of the molecules (Figure 5-9). The greater the difference in concentration between two sides of a cell membrane or other barrier, the more work can be done. In summary, whether a reaction occurs depends not only on the energy in the individual bonds but also on the concentrations of both chemical reactants and products.

Imagine a chemical reaction in which reactant A is converted to a product B. If the reaction is exergonic (energy releasing), then the reaction will proceed, and product B will begin to accumulate. As the concentration of B increases, however, so does B's free energy (the energy available to do work). Gradually, the difference in energy between the reactants and the products decreases. When the free energy of B equals the free energy of A (and $\Delta G = 0$), the reaction is finished. It is

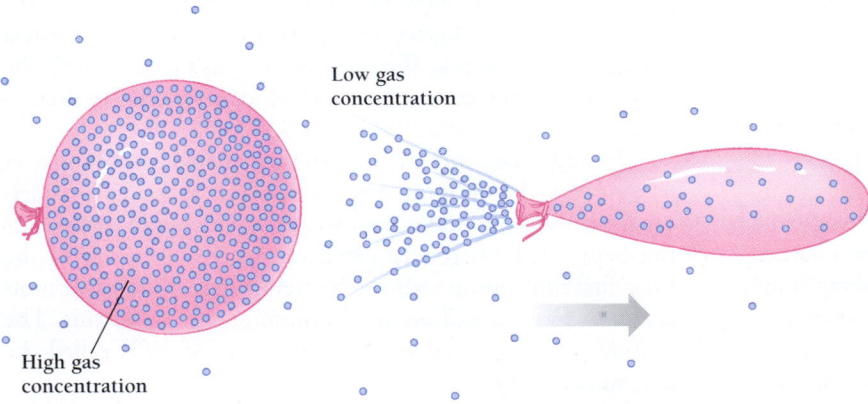

Low gas concentration

High gas concentration

Figure 5-9   The more you blow it up, the faster it flies. Any difference in concentration creates potential energy. The greater the concentration of gas molecules inside the balloon, the faster the balloon flies when we release it. More precisely, the greater the difference between the concentration inside the balloon and outside, the faster it flies.

no longer "downhill." The products and reactants have reached **equilibrium.**

Thermodynamics predicts where the equilibrium lies, but gives no information about how long a reaction will take to get to equilibrium. In the next section, we will find out how the rules of kinetics predict the rates of chemical reactions.

The equilibrium of a reaction depends both on the free energy in the bonds between the atoms of each molecule and on the concentration of the reactants and products.

WHAT DETERMINES THE RATE OF A CHEMICAL REACTION?

In the 19th century, archaeologists opened the tomb of an Egyptian pharaoh and found, among other artifacts, a perfectly preserved breakfast laid out for him. If the pharaoh had lived long enough to eat this meal, the food would have been oxidized in a matter of minutes. Yet, even after sitting for several thousand years in an oxygen-filled room, the pharaoh's breakfast was chemically unchanged. What kept the pharaoh's breakfast from being oxidized? In other words, How can cells initiate and speed up chemical reactions that run slowly or not at all outside the body?

We have seen that thermodynamics predicts both whether a reaction will occur and the direction of a reaction, independent of the precise mechanism. But to find out how *rapidly* a reaction occurs, we must understand its mechanism. We need to know what actually happens—what "path" a process takes from start to finish. For chemical reactions, this means going beyond discussing changes in bond energies and concentrations. Predicting the rates of chemical reactions requires an understanding of how molecules behave.

How Does Molecular Motion Help Explain Reaction Rates?

Molecules and atoms are constantly in motion. When a substance absorbs heat, its temperature increases and the atoms and molecules bounce and jostle about more rapidly. This simple fact is the basis for kinetic theory, or kinetics. Kinetics allows us to understand the rates of chemical reactions and how temperature and other environmental factors influence these rates.

For a moment, picture molecules as tiny frogs constantly jumping about at random (Figure 5-10A). Some of these frogs (molecules) are in an area called "Level A," and some are in an adjacent, but lower area, called "Level B." Some frogs will naturally jump down to Level B. But frogs will have a harder time jumping back up to A. Because more frogs can jump from A to B than can jump back up, frogs will tend to accumulate at B. Only the frogs (molecules) with the highest kinetic energy will be able to jump back to Level A. (Of course, real

frogs do not hop about at random, but in molecules, the kinetic energy of "hopping" is exactly the same as heat energy.)

Kinetic theory, which is based on the random movements of atoms and molecules, helps predict the rates of chemical reactions.

What Stops a Chemical Reaction?

In the 1870s, Ludwig Boltzmann proposed that molecules are constantly in motion and that their behavior explains many of the properties of matter. Boltzmann's ideas inspired the concept that heat is the kinetic energy of moving molecules. The faster molecules move, the higher their temperature. However, temperature measures only the average energy of the molecules. Molecules may have a range of kinetic energies.

In our frog model, we would say that some frogs jump higher than others. At high temperatures, most frogs can jump the barrier (Figure 5-11). Only a few will be stuck at Level B. After the frogs have jumped for a while, they reach equilibrium. At equilibrium, just as many frogs jump from A to B as from B to A, and the number of frogs at each energy level remains the same.

The equilibrium for a reaction depends on the temperature. At equilibrium at a high temperature, lots of frogs will be at Level A. At equilibrium at a low temperature, more frogs will be at Level B.

We can see that this model leads to the same conclusion that thermodynamics does: molecules move in the direction of less free energy (Level B). In addition, the jumping-frog model allows us to see once more that the concentrations of reactants and products influence chemical equilibrium.

A chemical reaction stops when it reaches equilibrium. Equilibrium depends in part on the temperature and concentration of the reactants and products.

What Starts a Chemical Reaction?

A ball sitting at the top of a flight of stairs will not bounce down the stairs until something starts it moving. Just so, every process—no matter how much it is favored thermodynamically—needs to be started in the right direction. In chemical reactions, we must ask, Where does this push come from? The answer is that the energy for the reactions between molecules comes from the random movements of molecules.

To understand the factors that start chemical reactions, let us return to the jumping-frog model. The difficulty in starting a chemical reaction is comparable to the height of the barrier between the frogs. To get from Level A to Level B, the frogs first must jump over the barrier (Figure 5-10). The molecules or frogs must have a minimum energy to do this. The minimum energy needed for a process to occur is called the **activation energy.**

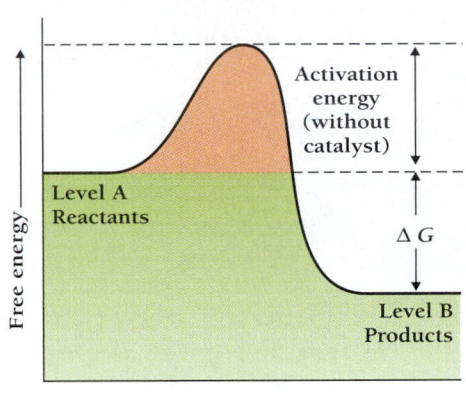

Level A

Level B

High activation energy barrier

A.

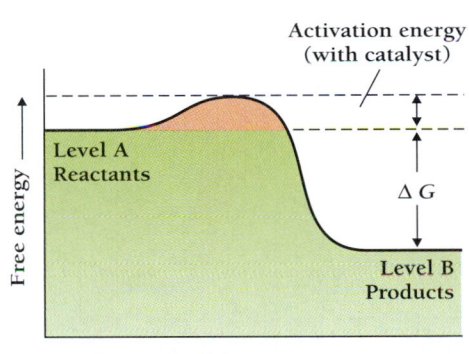

Level A

Level B

Low activation energy barrier

B.

Figure 5-10   Leaping frogs and bouncing molecules. The random bouncing of molecules resembles random leaping in a group of frogs. A. Even if all the frogs begin at Level A, more will end up at Level B because it is harder for them to jump back up. B. With a decreased activation energy barrier, frogs can more easily jump between the two levels, and it will take them less time to reach equilibrium.

The number of the molecules in a space that are able to react depends first on the temperature and second on the height of the activation energy barrier (Figure 5-10). The higher the temperature, the greater the number of molecules with enough energy to jump the barrier. The lower the activation energy, the greater the number of molecules that can jump the barrier and the more rapidly equilibrium will be achieved.

Laboratory and industrial chemists can speed up a specific chemical reaction by using a **catalyst**—a substance that lowers the activation energy of a reaction but is not consumed or changed in the reaction. A catalyst, in essence, lowers the barrier between the two groups of frogs (Figure 5-10B).

An enzyme is a biological catalyst. Almost all enzymes are proteins. Enzymes allow organisms to lower the activation en-

ergy for thousands of specific chemical transformations and so shorten the time required to attain equilibrium. Hastening reactions is essential to life. We all need to be able to digest our breakfast in something less than 5000 years.

Kinetic theory helps us understand why some thermodynamically favored reactions take place while others do not. Spontaneous reactions will actually occur only when most of the molecules have the needed activation energy. The carbohydrates in a piece of bread can be oxidized either by increasing the temperature (who hasn't burned toast?) or by exposing them to enzymes such as those in the human digestive tract.

The presence of an active enzyme allows a cell to use a molecule such as ATP in specific ways. Depending on how particular enzymes couple the splitting of ATP to other reactions, the result may be the contraction of muscle, the

Low temperature

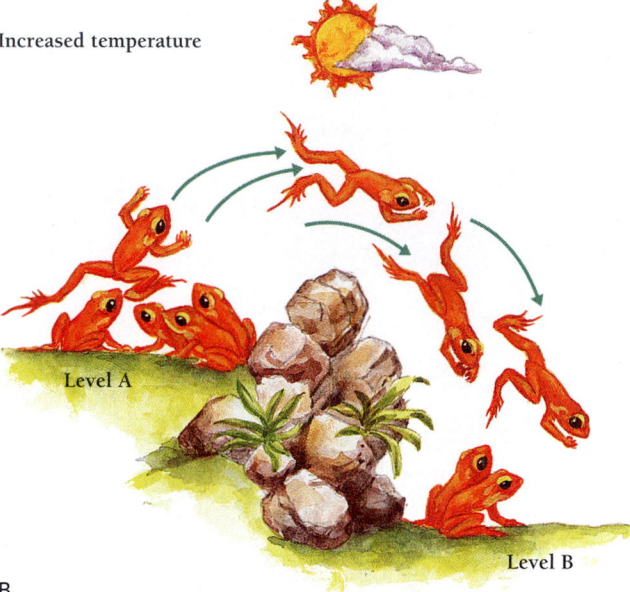

Figure 5-11 Hot frogs, cold frogs. The equilibrium point of a reaction depends on temperature. A. At a low temperature, few, if any, frogs have the activation energy to reach Level B. At equilibrium, all the frogs remain at Level A. B. At a higher temperature, half the frogs have enough energy to get to Level B. C. At an even higher temperature, all the frogs have enough energy to get to Level B, but only some can leap back. The reaction equilibrates with seven frogs at Level B and three at Level A.

A.

Increased temperature

Level A

Level B

B.

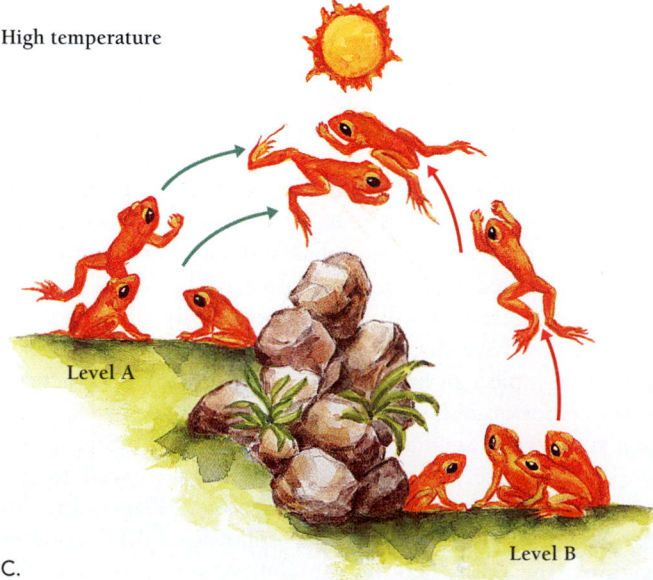

High temperature

Level A

Level B

C.

pumping of ions into or out of cells, or the synthesis of other molecules. Cells control their chemistry by regulating the production and activity of individual enzymes.

The initiation of a chemical reaction depends on the temperature of the molecules (their kinetic energy) and on the activation energy of the reaction. In organisms, enzymes can hasten a chemical reaction by lowering the activation energy.

HOW DO ENZYMES WORK?

In the last section, we saw that enzymes speed up biological reactions by lowering activation energies. In this section, we will see how enzymes accomplish this feat.

How Does an Enzyme Bind to a Reactant?

Enzymes speed up chemical reactions by binding to the reacting molecules, which are called **substrates.** Enzymes have enormous *catalytic power.* The right enzyme can make a chemical reaction go a million times faster than it would without the enzyme. For example, at 0°C, the freezing point of water, a single molecule of an enzyme called catalase can break down 5 million molecules of hydrogen peroxide each minute.

Enzymes are also highly *specific.* Each enzyme usually catalyzes only a single chemical reaction. What makes enzymes specific is their shape and also subtle interactions with the substrate. Most enzymes are large, globular proteins with irregular shapes. Every enzyme has an **active site**—a groove or "cleft" on its surface with which it binds to the small substrate molecule (Figure 5-12A). The binding of the enzyme to the substrate lowers the activation energy for a particular chemical reaction. Different kinds of enzymes may be able to bind to the same substrate, but in most cases *each enzyme lowers the activation energy for only a single kind of reaction.*

The enzyme temporarily binds to the substrate by weak noncovalent bonds—including hydrogen bonds, ionic bonds, hydrophobic interactions, and van der Waals attractions. Sometimes, as the two molecules bind, the shape of the enzyme changes (Figure 5-12A). When this occurs, the interaction of enzyme and substrate is called an **induced fit.**

The binding of the enzyme and the substrate is reversible, and the reactant or product molecules are quickly released back into the solution. Some enzymes are capable of binding, altering, and releasing individual substrate molecules more than 100,000 times a second.

Even so, the interactions between the enzyme and the substrate occur one molecule at a time. So although the rate of any biochemical reaction depends on temperature and the concentration of the reactants (as in the case of the jumping frogs or molecules), it also depends on how many molecules of enzyme are present and how fast the enzyme works.

The active site on an enzyme binds to a substrate by means of weak noncovalent bonds. Each enzyme lowers the activation energy for one reaction.

How Does an Enzyme Lower the Activation Energy of a Chemical Reaction?

The formation of noncovalent bonds between enzyme and substrate lowers the activation energy for a reaction. In all cases, the enzyme itself remains unchanged after the completion of the reaction. So, if the enzyme is unchanged, how does it change the activation energy of its substrate? The answer is that the enzyme briefly changes the shape of the substrate.

First, enzymes bring substrates together in a position that favors a reaction (Figure 5-12B). Without the enzyme, substrate molecules in solution bump into each other at random, but only a few of these collisions lead to a reaction. Most molecules do not react. We say the reaction is occurring very slowly.

Second, when an enzyme binds to a substrate molecule, the enzyme may strain and distort the covalent bonds within the substrate. Bonds under such strain can break more easily. The distorted form of a substrate is called a **transition state**—a molecule intermediate in form between the starting reactant and the final product. The transition state exists only for a brief moment—as little as a billionth of a second.

If we use our fingers to flex a spring clip just before clipping together a sheaf of papers, we are performing a role similar to that of an enzyme (Figure 5-12A). Our hand is the enzyme, the spring clip is one of the substrate molecules, and the sheaf of papers—held firmly in our other hand—is a second substrate. By using our hand, we can briefly alter the shape of the substrate (spring clip) so that it can bond with the second substrate (sheaf of papers). We have created a transition state. At the end of the "reaction," our hand is unchanged, just as an enzyme is unchanged.

An enzyme lowers the activation energy barrier. In some cases, enzymes may contribute directly to the chemical reaction—for example, by lending or temporarily accepting an

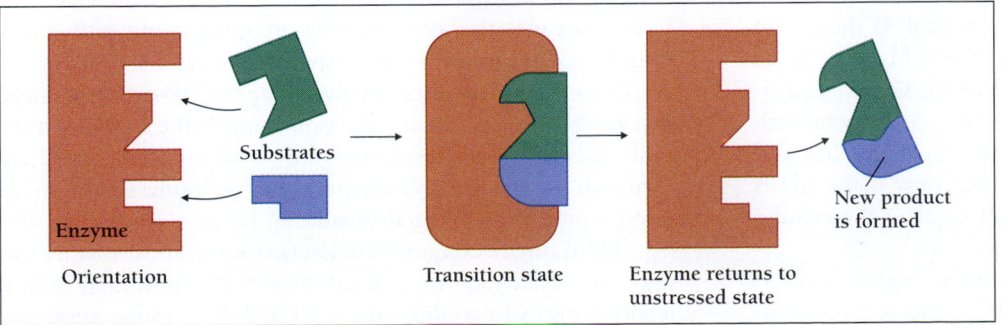

A.

B.

Figure 5-12 How enzymes work. A. Induced fit model of enzyme-substrate interaction. Both the enzyme and the substrate change shape, straining bonds during the reaction. The strained form of a substrate—the transition state—exists for as little as a billionth of a second. B. How does an enzyme catalyze a reaction between two other molecules? The enzyme brings both substrates close together in the proper orientation and also strains the covalent bonds of the substrates, lowering the activation energy of the reaction.

atom or an ion. This change strains the substrate molecule, bringing it to its transition state. Again, the net effect is to lower the activation energy barrier to the reaction.

Enzymes lower the activation energy for a reaction by aligning two or more substrate molecules, by straining the covalent bonds within the substrates, and even by participating directly in the reaction.

How Do Environmental Conditions Affect the Rates of Enzymatic Reactions?

Enzymes work most efficiently at certain temperatures, pH values, and salt concentrations. In mammals, for example, most enzymes work best at body temperature. In our previous discussion of kinetics, we saw that all chemical reactions occur more rapidly at higher temperatures because more molecules possess enough kinetic energy to get over the activation energy barrier.

But increasing the temperature of an enzymatic reaction increases the rate only to a point. Once the temperature reaches a certain point, the bonds that maintain the enzyme's structure begin to loosen and break. When an enzyme loses its shape and activity, the enzyme is **denatured.** When an organism's enzymes denature, it dies. We have all seen denatured proteins in cooked egg white, for example. Because heat denatures enzymes and other protein, most organisms cannot survive temperatures much higher than 40°C. A few unusual organisms live at higher temperatures. Bacteria that live in hot springs, for example, have evolved enzymes that function at high temperatures.

Besides temperature, pH (or acidity) also matters. Most enzymes work best in only a narrow range of pH values. The pH of the environment can, for example, alter the shape of an enzyme and add or remove hydrogen ions from the enzyme, changing the way the enzyme binds to the substrate. Most cellular enzymes work best at a pH of between 6 and 8. (Recall that pure water is neutral, with a pH of 7.) However, pepsin, the protein-digesting enzyme secreted by the stomach, works best at a pH of about 2. The fuller the stomach becomes, the more the stomach acid is diluted and the higher the pH rises, though usually no higher than about 3.5. If the pH of the stomach contents were to rise above 5, the pepsin would no longer digest our food.

Many enzymes work by acting as acids or bases. That is, they donate or accept hydrogen ions to a reaction. For example, when the amino acid histidine is part of the active site of certain enzymes, it can act as a base or an acid. If the pH of the environment is about 6.5 (slightly acid), histidine is an acid. At a pH of about 7.5 (slightly alkaline), however, histidine loses a hydrogen ion and becomes a base. An enzyme with an active histidine side chain can donate a hydrogen ion and help break a particular chemical bond, but only if the pH is right. Such relationships can help a cell regulate the cellular environment and maintain homeostasis.

Most enzymes, however, do not work in highly acid environments. This is why vinegar is such an effective preserva-tive. The acetic acid in vinegar disables any enzymes present either in food itself or in any microorganisms that slip into the food. Without enzymes, reactions take place so slowly that pickled cucumbers, onions, peppers, herring, and cabbage can remain almost unchanged for decades or longer.

The rates of enzymatic reactions increase with temperature, but only up to a point. Most enzymes work in a narrow range of pH values.

HOW DOES A CELL OR ORGANISM REGULATE ITS OWN METABOLISM?

Organisms and cells are enormously flexible in the way they process, or metabolize, food. They function in lean as well as in fat times. They can obtain energy when food is available or break down their own stores when it is not. They can make building blocks not available in their diet, then stop making a particular building block when it becomes available. An organism with a diet rich in protein, for example, need not waste hard-won ATP making its own amino acids.

Organisms do all this by regulating the amounts and activities of enzymes. They both control the total flow of energy and focus that energy on specific tasks.

Why Do Enzymatic Reactions Often Occur in Small Steps?

When organisms and cells transform one molecule into another, they frequently do so by means of a series of small steps or subreactions, each one mediated by a different enzyme. For example, if a chemical reacts to form product B, which reacts to form C, which reacts to form D,

$$A \longrightarrow B \longrightarrow C \longrightarrow D$$

each step is catalyzed by a different enzyme. In humans, separate reactions number in the thousands. For example, ten different reaction steps and ten different enzymes are necessary in the initial extraction of energy from the sugar glucose. Almost all organisms—including all bacteria, all eukaryotes, and most archaebacteria—use the same ten enzymes to extract energy from this sugar (Chapter 6).

The network of biochemical reactions in organisms is complex for at least four reasons. First, some of the chemical transformations that cells perform are quite complex and can only be accomplished in many steps. Second, vast numbers of reactions are endergonic. For these to proceed, each one must be coupled to an exergonic reaction, such as the breakdown of ATP to ADP. Coupled endergonic reactions are especially common when cells synthesize large molecules from small ones, a process called **anabolism.**

Third, many exergonic transformations produce more energy all at once than the cell can spend. If you wanted to buy a soft drink, you couldn't use a $100 bill in most cities (besides Las Vegas). A cell "makes change" by separating a highly exergonic reaction into steps, so that the cell can capture smaller denominations of energy. A series of smaller reactions allow the cell to more easily store the energy for later use and also minimize the buildup of heat. Stepwise transformations of this kind are especially common in the breakdown of food molecules, called **catabolism.**

Fourth, in some stepwise reactions, intermediate products are the starting materials for the synthesis of essential building blocks. For example, some of the intermediate steps in the breakdown of sugars to carbon dioxide and water are precursors for amino acids, the building blocks from which proteins are made.

Many biological reactions occur in series of small steps. Stepwise reactions allow cells to carry out complex transformations, to couple reactions, to capture smaller denominations of energy, and to generate useful intermediate products.

How Do Organisms Regulate Enzyme Function?

The huge variety of molecules in a cell can undergo countless thermodynamically favored chemical reactions. The only reactions that happen rapidly enough to matter, however, are those that are catalyzed by enzymes. These include all the reactions that provide cells with the energy they need for life. Not only do enzymes serve as catalysts for specific reactions, they also regulate the rates of reactions, often in response to environmental changes.

For example, organisms need to regulate the quantities of building blocks they make. For an organism to make unnecessary building blocks would waste energy. Organisms that conserve energy by regulating the activity of enzymes therefore have an advantage over organisms that waste energy by building molecules they do not need. An example of such regulation is the synthesis of the amino acid isoleucine. When isoleucine is not available, cells make it from another amino acid, threonine. Making isoleucine from threonine requires five enzymatic reactions. But this succession of steps is energetically expensive, so the pathway from threonine to isoleucine shuts down when enough isoleucine accumulates. Isoleucine, the end product of the pathway, specifically inhibits the first step of this pathway (Figure 5-13).

This is a classic case of negative feedback, first mentioned in Chapter 1. Other molecules besides end products may also change the activity of enzymes—sometimes inhibiting enzyme action, sometimes increasing enzyme action. Any molecule or ion that changes the activity of an enzyme is called an **effector.** Effectors that inhibit enzymes are called **inhibitors.** But other effectors increase the activity of enzymes.

One way cells regulate enzyme activity is through effectors, which increase or decrease the activity of enzymes.

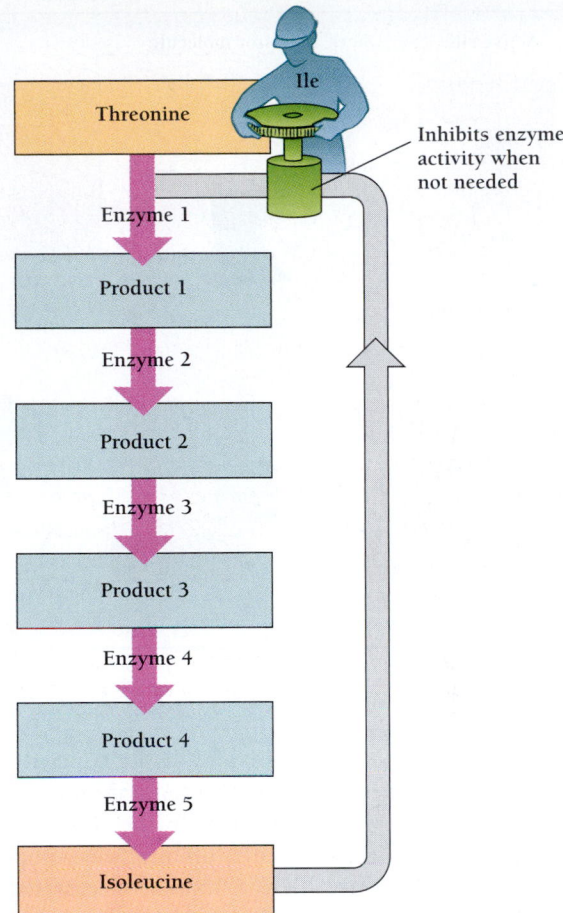

Figure 5-13 How can cells regulate enzyme function? The production of the amino acid isoleucine self-regulates by means of negative feedback. In the absence of isoleucine, a sequence of five enzymatic reactions synthesizes isoleucine from threonine. However, when isoleucine molecules accumulate, they inhibit the enzyme that catalyzes the first of the five reactions, shutting down the isoleucine pathway.

How Do Inhibitors Influence Enzymes?

Because an enzyme recognizes the shape of its substrate, enzymes can be "fooled" into binding to molecules that are similar in size and shape. Molecules that resemble the substrate can block an enzyme's active site and prevent the enzyme from working (Figure 5-14). Such a molecule is called a **steric inhibitor** [Greek, *stereos* = solid, three-dimensional] because it inhibits the work of the enzyme by means of its physical shape.

In carbon monoxide poisoning, for example, carbon monoxide (produced by burning gasoline or other fuel) competes with oxygen for a site on the hemoglobin molecule in our blood. Carbon monoxide binds so tightly to the hemoglobin that the hemoglobin can no longer carry oxygen to the brain and body. Unlike carbon monoxide, however, most steric inhibitors are reversible. They detach from the enzyme, so that a given enzyme sometimes binds to the substrate and sometimes to the inhibitor.

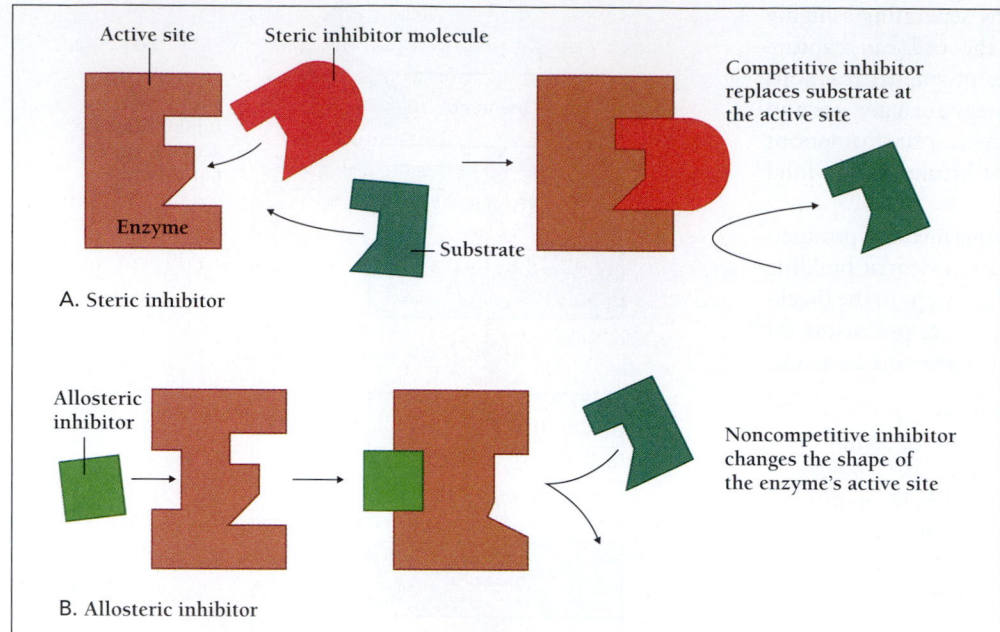

Figure 5-14   What can prevent an enzyme from catalyzing a reaction? A. Steric (competitive) inhibition. A steric inhibitor is a molecule that binds to the enzyme's active site, preventing the substrate from binding. A steric inhibitor competes with the substrate for the enzyme's active site. B. Allosteric (noncompetitive) inhibition. An allosteric inhibitor does not compete with the substrate but binds to another site on the enzyme in such a way that the active site no longer functions.

A steric inhibitor can be overcome. If the concentration of substrate is much greater than the concentration of a steric inhibitor, the enzyme will mostly bind to the substrate, not the inhibitor. The substrate and the inhibitor are competing for access to the enzyme. So the more concentrated the substrate, the more effective it will be at binding enzyme. And the more concentrated the steric inhibitor, the more effective it will be at binding enzyme. Because of such competition, a steric inhibitor is also called a "competitive inhibitor."

Many effectors can act as "noncompetitive inhibitors." Such molecules do not compete directly with the substrate. Lead and other heavy metals, for example, work by blocking specific groups outside the active site, sometimes changing the overall shape and chemistry of an enzyme. An effector that binds to an enzyme at a place other than the active site is called an **allosteric effector** [Greek, *allos* = other, different]. For example, an allosteric inhibitor blocks an enzyme by binding to a site on the enzyme other than the active site.

Some noncompetitive inhibitors act only temporarily. Others act irreversibly by forming a permanent bond with the enzyme and altering its shape. The antibiotic penicillin, for example, irreversibly damages an enzyme that bacteria use to build their cell walls.

Many noncompetitive inhibitors are poisons, like lead. Other inhibitors are essential parts of cells. Cells use allosteric inhibitors, for example, to regulate their own chemical processes. Most enzymes have evolved so that their most active shape is also their most stable shape. When an allosteric inhibitor binds to an enzyme, however, it changes the enzyme's shape to a less active form. The allosteric inhibitor acts as a switch, changing the enzyme from its active shape to its less active shape.

Effector molecules increase or decrease the activity of enzymes—either by competitively binding to the active site or by altering the shape of the enzyme.

Study Outline with Key Terms

All processes—biological and nonbiological—are accompanied by changes in the form of energy, which are subject to the laws of **thermodynamics.** The **First Law of Thermodynamics** states that the total amount of energy does not change. The **Second Law of Thermodynamics** states that the total amount of energy in a closed system becomes less and less available for **work;** more and more of it is converted to heat. In short, **entropy** increases.

The laws of thermodynamics allow scientists to predict whether a reaction will occur spontaneously. The making and breaking of chemical bonds between atoms leads to changes in available energy. A reaction will occur spontaneously if the energy available to do work—the **free energy (G)**—decreases. A reaction is **exergonic** if free energy (ΔG) decreases and **endergonic** if free energy increases.

Organisms can accomplish "uphill reactions" by **coupling** endergonic processes to exergonic reactions with protein **catalysts** called **enzymes.** In many cases, the energy that drives endergonic reactions comes from the **high-energy bonds** of ATP.

During chemical reactions and other processes, **kinetic energy** may be converted to **potential energy.** The free-energy change in a chemical reaction depends on the concentrations of **reactants** and **products,** as well as on the number of chemical bonds that are made and broken. Reactions tend to go in the direction of increased disorder, or entropy, in which the concentrations of molecules become more uniform.

To predict how long a process will take to occur, we need to know how it occurs. The study of **kinetics** allows scientists to predict the rates of chemical reactions from the properties of molecules. The basis of kinetics is that molecules are constantly in motion, with the average energy of a group of molecules determined by the temperature, measured in **calories.**

At chemical **equilibrium,** the direction of a reaction is balanced by the opposite reaction. Enzymes and other catalysts speed up specific chemical reactions but do not affect the free-energy changes of chemical reactions. They increase the speed at which equilibrium is approached, but they do not affect the character of the equilibrium.

Enzymes hasten chemical reactions by lowering the **activation energy** of the reaction. Enzymes accomplish this by binding to the reacting molecules by means of noncovalent interactions. An enzyme may join two or more **substrate** molecules. Sometimes, in an **induced fit,** enzymes strain the bonds of a single substrate molecule, forcing it into a **transition state.** Some enzymes actually participate in the chemical reaction by temporarily giving or receiving atoms or ions. The activity of an enzyme is highly sensitive to environmental conditions that can affect its activity. Temperature, pH, and the presence of other molecules may **denature** an enzyme.

Many chemical transformations performed by organisms in **anabolism** and **catabolism** consist of numerous small reactions. Such small reactions more easily allow the cell to capture the energy for later use.

Cells regulate the activity of many enzymes with **effectors,** including **steric inhibitors** and **allosteric inhibitors.** Steric inhibitors block the enzyme's **active site,** while allosteric inhibitors change the enzyme's shape and activity by binding to a nonactive site.

Review and Thought Questions

Review Questions

1. State the First and Second Laws of Thermodynamics and explain their relationships to an example of work.
2. What is free energy? What is its relationship to the Second Law of Thermodynamics? In what direction must it change in all reactions (or processes of work)? Compare this to the meaning of "entropy."
3. Define endergonic and exergonic in terms of the free-energy change associated with reactions. Which reaction involving ATP is endergonic? Exergonic? Which reaction occurs spontaneously?
4. Which part of downhill skiing is endergonic (work), and which part is exergonic?
5. What is the activation energy for a reaction? How might it influence the speed with which a reaction proceeds?
6. Enzymes are biological catalysts. How do they perform the task of speeding up biological reactions?
7. What is meant by the "active site" of an enzyme? How is it related to the substrate of that enzyme? What is meant by "induced fit?" How does this make the substrate more likely to undergo a reaction?
8. Describe four benefits that cells derive from organizing their reactions into pathways.

9. Decide if these reactions are exergonic or endergonic. Mark "X" for exergonic and "N" for endergonic.
   __Growing hair
   __Digesting pizza
   __Baking soda and acetic acid (vinegar) reacting to give off fizzy carbon dioxide that makes a cake rise
   __Liver cells produce fibrinogen, a blood clotting protein
   __The adrenal glands produce cortisol derived from cholesterol
   __ATP splits during muscle contraction

Thought Questions

10. Where does the energy come from when you perform the work of riding your bicycle? Did *all* of the energy in the food you used for this task actually help turn the wheels of your bicycle? Explain using the First and Second Laws of Thermodynamics.
11. Explain why unraked leaves might form a neat pile in a specific spot in your backyard. Does this mean that the Second Law of Thermodynamics is not working? What is the source of energy for this organization of the leaves?
12. Why do you think that some exergonic reactions will *not* go at all in the absence of a catalyst? If the change in free energy is no obstruction, what is?

About the Chapter-Opening Image

A car rusting in a field is a good example of entropy. The Second Law of Thermodynamics says that as entropy (disorder) increases, less and less energy is available to do work.

 On-line materials relating to this chapter are on the World Wide Web at **http://www.harcourtcollege.com/lifesci/aal2/**

6

How Do Organisms Supply Themselves with Energy?

Louis Pasteur and Vitalism

In 1835, the French scientist Charles Cagniard de la Tour observed that the yeast in fermenting beer consisted of tiny living cellular organisms. Cagniard de la Tour noted that these tiny organisms multiplied by budding, and he suggested that they were somehow linked to the process of alcoholic fermentation—the conversion of sugar to ethanol and carbon dioxide. In the same year, two other scientific papers described similar observations. One was by Theodor Schwann, the German botanist whose work helped establish the cell theory described in Chapter 4.

Schwann also showed that if grape juice were boiled to kill the yeast, the juice would not ferment to wine unless yeast cells were added back to the juice. Schwann correctly concluded that yeast plays an essential role in fermentation. He then showed that fermentation began at the same time that the yeast cells first appeared, progressed with their multiplication, and stopped as soon as the yeast cells stopped multiplying.

Amazingly, the scientific community rejected these important discoveries out of hand, owing to a handful of powerful personalities and the scientific and philosophical temper of the time. The most influential scientists insisted that yeast was not a living organism at all, but a substance, and probably an unimportant one.

This mistaken idea was held by no less than the father of modern chemistry, Antoine Lavoisier (1743–1794). Lavoisier had already argued that fermentation could be understood as a simple chemical reaction. Fermentation, he and others had written, consisted of a reaction in which the six-carbon sugar glucose broke down into two molecules of ethanol and two molecules of carbon dioxide:

$$C_6H_{12}O_6 \longrightarrow 2\ C_2H_5OH + 2\ CO_2$$

$$\text{glucose} \longrightarrow \text{ethanol} + \text{carbon dioxide}$$

This nicely balanced formulation seemed so simple and so right that chemists, riding high on their tremendous successes (such as Lavoisier's own discovery of oxygen) could not imagine why yeast would be necessary. Even if yeast did turn out to be necessary, where could it possibly go in the equation? They accepted that yeast was present during fermentation. But there was no question of yeast actually taking part in the reaction. At most, they said, dead and dying yeast cells might release some substance that could speed up the chemical reaction. Nowadays, we would call such a substance a "catalyst."

Nineteenth-century chemists firmly believed that all biological processes were chemical in nature. To believe otherwise, to insist on some mysterious role for living organisms that was not purely chemical in nature, was condemned as **vitalism**—the belief that living systems have powers beyond those of nonliving systems.

The idea that a simple chemical reaction needed tiny living organisms to occur seemed absurd. In 1839, two of the most prominent early chemists, Friedrich Wöhler and Justus von Liebig, composed and published a lampoon designed to humiliate Schwann and the other yeast researchers. Wöhler and Liebig pretended to give a scientific description of the anatomy and behavior of yeast cells. A yeast cell, they said, looks exactly like a miniature still (the apparatus used to distill whiskey). Yeast cells suck in sugar and excrete ethanol and carbon dioxide, Wöhler and Liebig joked, writing, "Although teeth and eyes are not to be seen, one can distinguish a stomach, intestine, the anus (a rose-pink spot), and the organs of urine secretion. . . . The urine bladder in the full condition is shaped like a champagne bottle."

The lampoon must have worked, for Schwann retired from the debate, and Liebig and Wöhler's mistaken opinions prevailed unquestioned for 20 years. However, in 1854, young Louis Pasteur, a bright new star in the scientific sky, was appointed professor of chemistry at Lille, France.

Key Concepts

- Cellular respiration converts the chemical energy of food molecules into the chemical energy of ATP.

- Almost all cells can use glycolysis to derive energy, but cellular respiration provides much more energy.

- ATP production from cellular respiration depends on oxidative phosphorylation, which couples electron transport to the production of high-energy phosphate bonds in ATP. This process depends on the intact organization of mitochondria.

Lille was a major center for the manufacture of ethanol, through the fermentation of beet juice, and it was not long before one of Pasteur's students appealed to him to help solve problems at the family distillery. Pasteur immediately became intrigued by the process of fermentation. Just as quickly, he realized that fermentation depended entirely on the presence of yeast.

Pasteur was ambitious, and he must have known that proving the eminent Liebig wrong would be a magnificent coup. In September, 1855, Pasteur began a series of studies in fermentation. As his wife, Marie Pasteur, wrote to her father-in-law, "Louis is up to his neck in beet juice. He spends all his days in the distillery. He has probably told you that he teaches only one lecture a week; this leaves him much free time which, I assure you, he uses and abuses."[1]

By 1860, Pasteur had succeeded in growing the yeast in a medium seeded with tiny numbers of yeast cells. The ethanol produced was proportional to the multiplication of the yeast. Normally, vast numbers of yeast cells are present in fermenting solutions. And Liebig had long insisted that some chemical released by the decomposing bodies of yeast cells initiated fermentation. But Pasteur's yeast cells were multiplying, not dying, and the scientific community soon rallied to Pasteur's side. Liebig stubbornly rejected Pasteur's results to the end of his life. Fermentation, he said, was a chemical process, and vitalism was not science.

Ironically, although fermentation normally depends on cells, Liebig was largely right. In fact, fermentation *is* a chemical process—though far more complicated than that envisioned by early chemists. And fermentation can take place outside of living cells—provided the right enzymes are available.

In 1897, Hans and Eduard Büchner decided to make an extract of yeast, called "zymase," to be used as a patent medicine. However, the extract kept rotting, so the brothers added

The scientific community rejected these important discoveries out of hand, thanks to a handful of powerful personalities.

sugar, for the same reason that fruit canners add sugar to canned peaches and berries. In high concentration, sugar is an excellent preservative. To their amazement, the yeast extract began bubbling furiously, as carbon dioxide gas burst from the surface of the extract. Hans Büchner immediately recognized that the sugar was fermenting in the absence of living yeast cells. The Büchners' zymase contained essential enzymes that catalyzed the alcoholic fermentation reaction—just as Liebig had predicted.

Nonetheless, Pasteur also was right. Fermentation normally does not occur outside of living organisms. It is a highly specialized process by which microorganisms extract energy from sugar. And although biologists now all agree with Liebig's idea that life will ultimately be understood in terms of chemistry, organisms are more than mere bags of enzymes. Their organization and complex structures guide and control the chemical reactions that contribute to life. In this chapter, we will see how cells of all kinds extract energy from biological molecules.

HOW DO ORGANISMS SUPPLY THEMSELVES WITH ENERGY?

All organisms need energy, and the ultimate source of energy for most organisms is sunlight. Plants convert light energy from the sun into chemical energy through photosynthesis. Plants and all other organisms that obtain energy and synthesize organic molecules from inorganic material are **autotrophs** [Greek, *auto* = self, same + *trophe* = to nourish]. The vast majority of autotrophs, including most plants and many algae and bacteria, photosynthesize. Nonetheless, not all autotrophs photosynthesize. Autotrophs include a few bacteria (called *chemotrophs*) that obtain energy by oxidizing inorganic substances such as sulfur and ammonia.

Animals, fungi, and other organisms that obtain chemical energy from other organisms are **heterotrophs** [Greek, *hetero* = other + *trophe* = to nourish]. Both heterotrophs and

[1]From Dubos, Rene, *Louis Pasteur, Free Lance of Science*, Little, Brown, 1950.

Figure 6-1 When it comes to sustained aerobic activity, the arctic tern, *Sterna paradisaea*, is the champion. This arctic tern nests each summer near the Arctic Circle in North America, then migrates across the Atlantic Ocean to Europe, then south to South Africa, then across the South Atlantic to Antarctica, a distance of 18,000 km (11,000 miles). In the spring, the bird flies all the way back around the world to the Arctic Circle to nest once more. *(Dwight Kuhn)*

autotrophs obtain energy by breaking down organic molecules in the process called cellular respiration. The chemical reactions of cellular respiration are part of a complex network of biochemical conversions that are collectively called **metabolism** [Greek, *meta* = after + *ballein* = to throw]. The difference between heterotrophs and autotrophs is that autotrophs make these organic molecules themselves, while heterotrophs take them from others—either from autotrophs or other heterotrophs. Ultimately, nearly all the energy that powers living organisms comes from sunlight.

What Is the Common Currency of Energy for Organisms?

If you spend all your time today studying biology, you will use the energy from about 10^{25} molecules of ATP in the next 24 hours—some 40 kg (90 lb) of ATP. If you were to run a marathon, you would use the same amount in just 2 hours. Almost every time one of your cells performs an energy-consuming reaction, it derives the energy it needs by splitting ATP, the universal currency of biological energy. Fortunately, however, cells recycle ATP (so you don't really have to carry around an extra 90 lb).

All cells use ATP. Plants produce ATP during photosynthesis, as we will see in Chapter 7. But both plants and animals can produce ATP in another way—by capturing energy from the breakdown of carbohydrates, lipids, and proteins.

The main way that organisms break down organic molecules is **cellular respiration,** the oxygen-dependent process by which cells extract energy from food molecules. Cellular respiration perfectly illustrates how chemical pathways proceed step by step, how some pathways occur in cycles, and how eukaryotes all share the same metabolic pathways. Animals and other heterotrophs obtain almost all of their energy through cellular respiration. Plants also depend on cellular respiration for ATP at night or at other times when they cannot photosynthesize.

Cellular respiration is distinct from ordinary respiration, or breathing. Breathing—inhaling oxygen and exhaling carbon dioxide—is nonetheless closely related to cellular respiration. Breathing supplies our cells with the oxygen needed for cellular respiration and disposes of the carbon dioxide that is its waste product.

Most eukaryotic cells (and many prokaryotic cells) are **aerobic,** that is, they depend on oxygen for life. All aerobic organisms, from earthworms to arctic terns, produce the bulk of their ATP by means of cellular respiration (Figure 6-1).

Many prokaryotes and a few eukaryotes can live **anaerobically,** without oxygen. The microbe *Clostridium botulinum,* which can poison canned meats and other foods and causes botulism (food poisoning), grows anaerobically (Figure 6-2A). Even our own cells, which are mainly aerobic, can produce energy anaerobically for short periods by a process similar to alcoholic fermentation. Some organisms, like the red-eared turtle shown in Figure 6-2B, can sustain themselves on glycolysis for days at a time.

In all organisms, the common currency of energy is ATP.

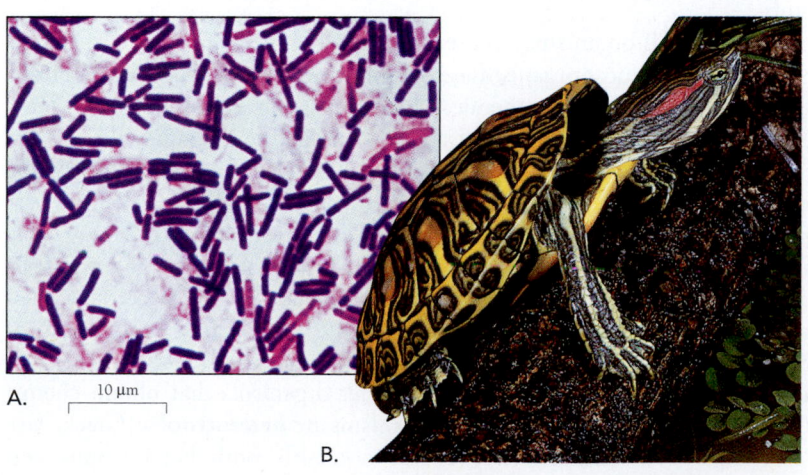

A.   10 µm

B.

Figure 6-2 Life without air. A. The obligate anaerobe *Clostridium botulinum*, which causes a serious form of bacterial food poisoning, cannot reproduce in the presence of oxygen. It is known, however, for its ability to multiply inside of sealed canned goods. B. Among vertebrates, the red-eared turtle, *Chrysemys scripta elegans*, is unusual in its ability to live without oxygen. It can stay under water for 2 weeks at a time, relying on glycolysis for energy production. *(A, CNRI/Phototake; B, Zig Leszczynski/Animals Animals)*

How Do Heterotrophs Extract Energy from Macromolecules?

Cellular respiration begins with small molecules such as glucose—a simple six-carbon sugar. Cells cannot extract energy directly from complex carbohydrates, proteins, or fats. Large molecules must therefore undergo **digestion,** splitting into smaller units—proteins to amino acids, polysaccharides to glucose and other simple sugars, and fats to fatty acids and glycerol. Digestion occurs through the process of hydrolysis—breaking each link in a polymer through the addition of a molecule of water.

Most digestion takes place outside the cytosol of the cell so that cells are less likely to digest themselves. In animals and fungi, most digestion takes place outside the cell through the action of secreted enzymes, in the intestinal tract of animals, or in the space surrounding a fungus (Figure 6-3). Even in protists and in those animal cells that digest inside the cell, digestion occurs inside lysosomes separated from the cytosol. The lysosomal membrane separates digestive enzymes from the rest of the cell (Figure 4-12A).

Digestion does not provide energy, and it does not occur in a fixed order of reactions. Digestion generates small, energy-rich molecules that move across the cell membrane into the cytosol. One of the most useful of these molecules is glucose. (In vertebrates, for example, the blood carries glucose to all the cells of the body.) Inside the cell, enzymes transfer the energy from glucose into the chemical bonds of ATP.

Digestion also breaks proteins, nucleic acids, and lipids into small molecules that can enter cells. These molecules (or molecules derived from them) ultimately enter the same biochemical pathways that extract energy from glucose, as we will discuss later in this chapter.

Heterotrophs prepare for cellular respiration by capturing macromolecules and by digesting them with enzymes into their component building blocks. Digestion itself does not produce energy.

What Are the Stages of Cellular Respiration?

Cells extract energy from glucose by breaking the bonds between each of its six carbon atoms and moving the energy from those bonds to the high-energy bonds of ATP. During cellular respiration, glucose is completely broken down to carbon dioxide and water. The overall reaction is summarized by the following equation:

$$C_6H_{12}O_6 + 6\,O_2 \longrightarrow 6\,CO_2 + 6\,H_2O$$

glucose + oxygen ⟶ carbon dioxide + water

The complete breakdown of glucose to carbon dioxide and water and the packing away of the energy into ATP occurs in four stages. In eukaryotes, the first stage, glycolysis, takes place in the cytosol. In this stage, a six-carbon glucose is broken in half, forming two molecules of pyruvate, each with

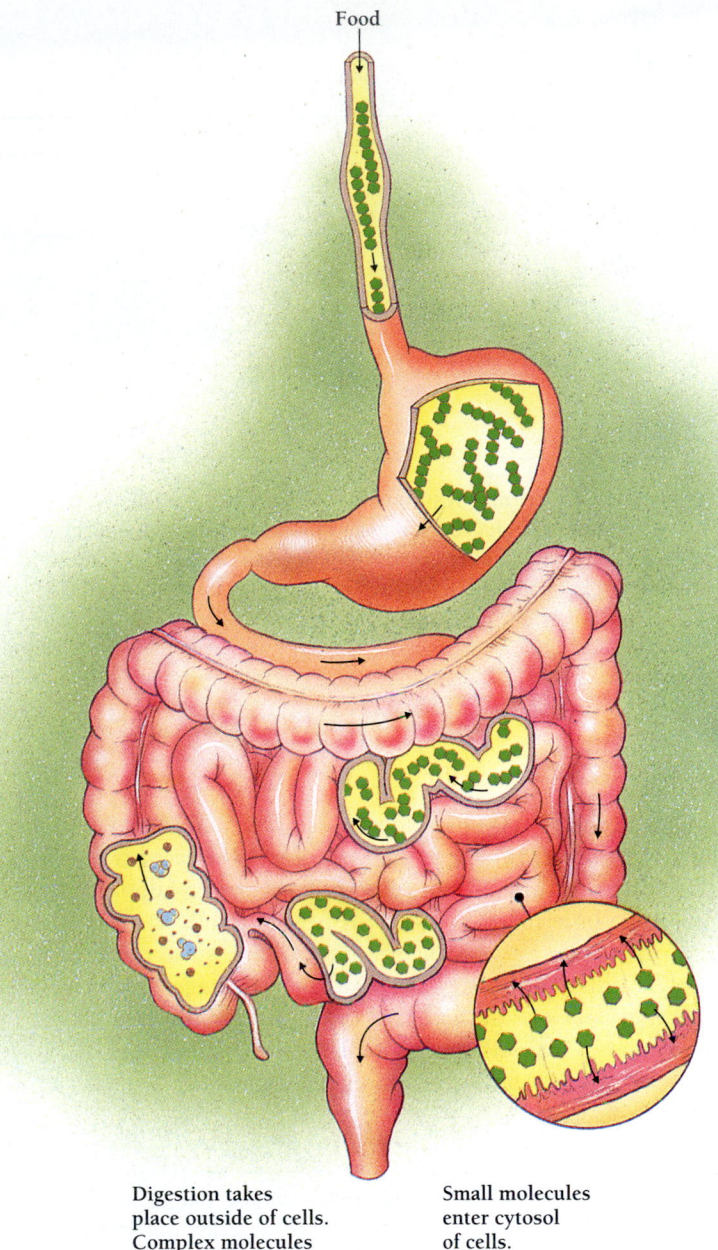

Food

Digestion takes place outside of cells. Complex molecules are broken down.

Small molecules enter cytosol of cells.

Figure 6-3   Digestion nearly always occurs outside of the cytosol. Digestion occurs either outside the cells altogether in some kind of stomachlike cavity, or in lysosomes, separated from the cytosol. Digestion does not produce energy but makes building blocks available to the metabolic pathways.

three carbon atoms (Figure 6-4). In the second stage, the three-carbon pyruvate molecules each lose a carbon atom, which is ultimately released as carbon dioxide. The remaining two-carbon molecules go into the third stage, the citric acid cycle. Here, the last remaining carbon-carbon bonds are broken and the single carbon atoms combine with oxygen to form more carbon dioxide. The second and third stages take place in the mitochondrial matrix.

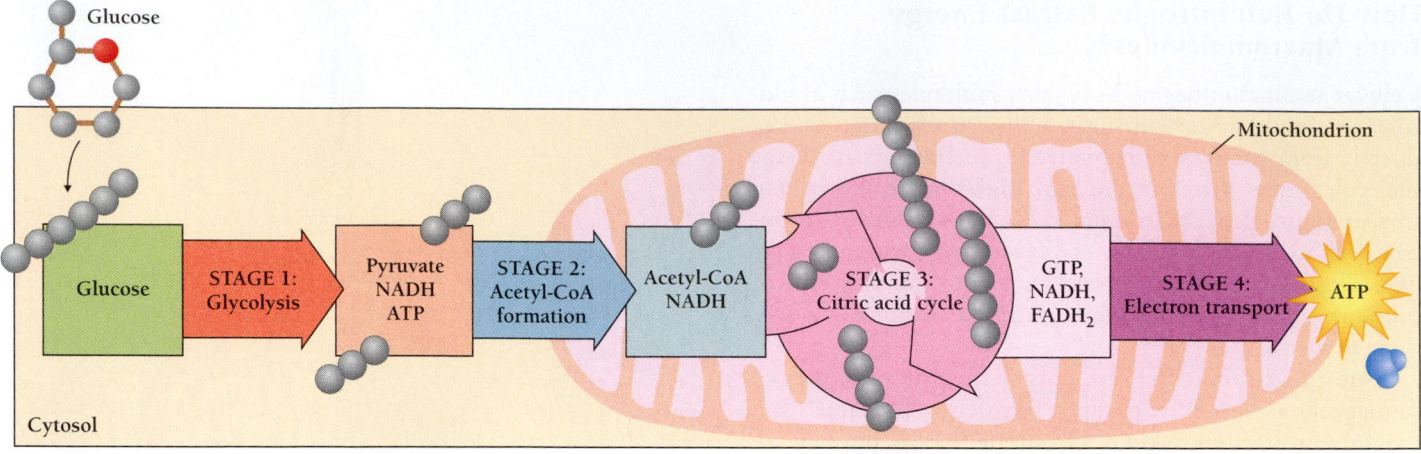

Figure 6-4   Cellular respiration. The complete breakdown of glucose occurs in four stages. (1) In gly-colysis, six-carbon glucose is broken in half, forming two molecules of pyruvate, each with three carbon atoms. Glycolysis produces two ATPs from ADP and two NADHs from NAD^+. (2) In the second stage, each pyruvate molecule loses a carbon atom (as carbon dioxide) to form an acetyl-CoA molecule and one NADH. (3) The citric acid cycle converts two more carbon atoms into carbon dioxide. The cycle generates two high-energy phosphate bonds (in GTP) and uses the eight electrons of each acetyl-CoA molecule to generate six NADHs and two FADH₂s. (4) Oxidative phosphorylation captures energy from NADH and FADH₂ in the high-energy bonds of ATP.

The fourth stage of cellular respiration is **oxidative phosphorylation.** During this stage, the energy released from the breaking of carbon-carbon and carbon-hydrogen bonds in glucose is transferred to phosphorus-oxygen bonds in ATP. The fourth stage is not actually the last in time, since much of it occurs simultaneously with the other three stages. As we will see later, much of the action of the fourth stage occurs within the inner mitochondrial membrane. Of the four stages of cellular respiration, oxidative phosphorylation (stage 4) makes by far the most ATP.

Cellular respiration occurs in four stages, one of which takes place in the cytosol of eukaryotic cells, and three of which take place in mitochondria.

HOW DO CELLS EXTRACT ENERGY FROM GLUCOSE?

The first of the four stages in the metabolism of glucose is **glycolysis** [Greek, *glykys* = sweet (referring to sugar) + *lyein* = to loosen], a set of ten chemical reactions (or steps). Glycolysis converts glucose, a sugar with six carbon atoms, into two identical smaller molecules, called **pyruvate,** with three carbon atoms each (Figure 6-5). The ten reactions also convert some of the energy of glucose into two high-energy ATP bonds, and some into two high-energy molecules of NADH. All ten reactions take place in the cytosol, outside the mitochondria.

The most remarkable fact about glycolysis is that *all* organisms accomplish it in exactly the same way. This univer-sality suggests that the common ancestors of all present-day organisms performed glycolysis in the same way that we do to-day. The enzymes that catalyze the ten reactions must have evolved at least 3.8 billion years ago, at the very dawn of life on Earth.

What Is Oxidation?

Most of the energy in food molecules is stored in covalent bonds between carbon and hydrogen atoms. We obtain energy from these bonds by arranging for the electrons shared by car-bon and hydrogen atoms to be captured by oxygen atoms. Oxygen atoms are extremely electronegative, so the flow of electrons from carbon atoms in food molecules (such as glu-cose) to oxygen is highly exergonic, releasing energy that cells use to make more ATP.

In fact, most of the energy-producing processes in cells in-volve the transfer of electrons from one molecule to another. In cellular respiration, most of the high-energy electrons are first transferred to a molecule called NAD^+, which we will discuss later.

A molecule that accepts an electron is called an electron acceptor, or **oxidizing agent.** A molecule that donates an elec-tron is called an electron donor, or **reducing agent.** Oxida-tion and reduction reactions always go hand in hand. The oxidation of one atom, molecule, or ion provides the electrons for the reduction of another atom, molecule, or ion. Such re-actions are called **reduction-oxidation reactions** or **redox re-actions.**

In cellular respiration, electrons from the chemical bonds of glucose molecules combine with oxygen (and hydrogen ions)

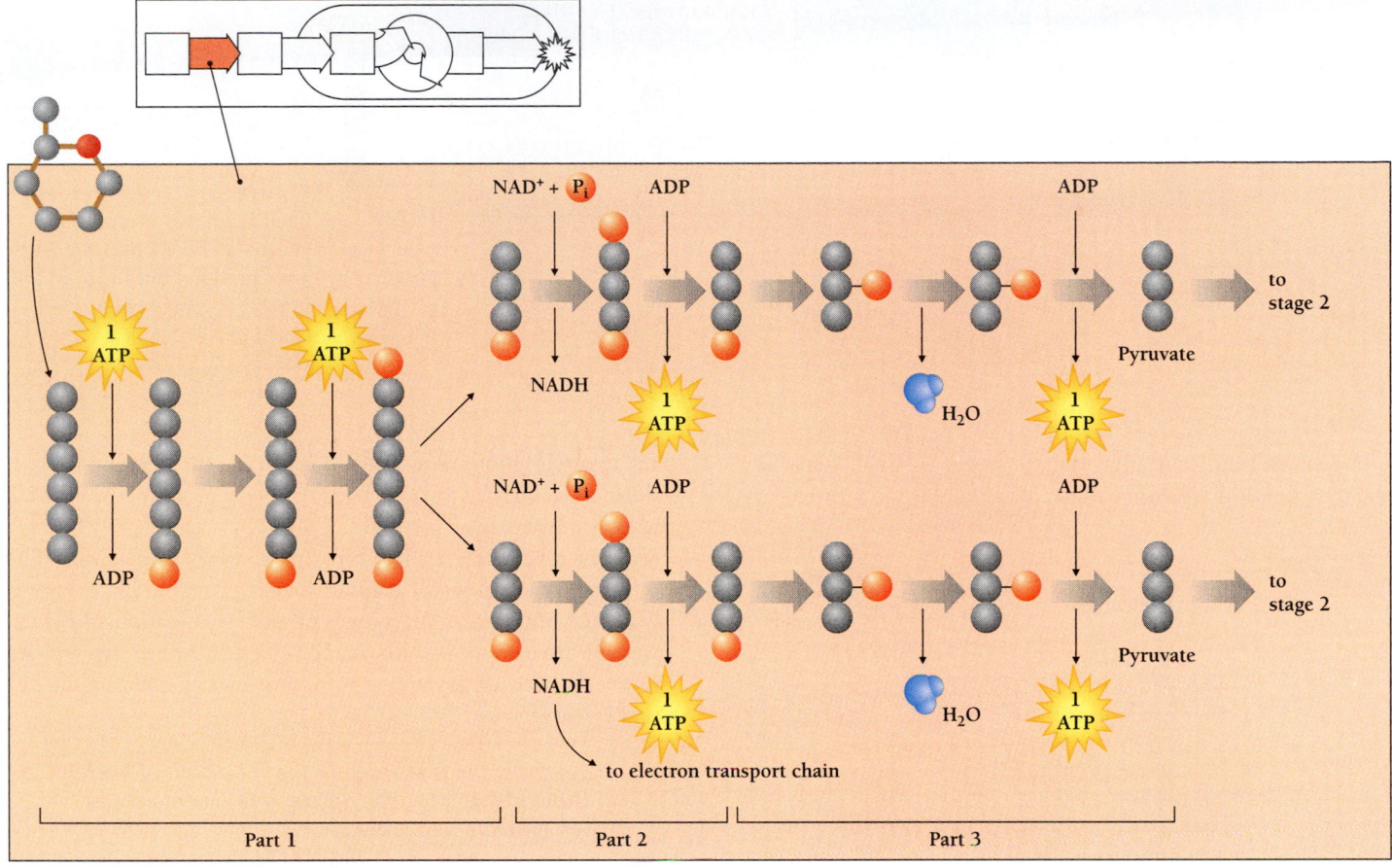

STAGE 1: GLYCOLYSIS

Figure 6-5   Stage 1: the first stage of metabolism is glycolysis. Glycolysis is a set of ten chemical reactions that break each six-carbon glucose into two molecules of three-carbon pyruvate. The ten reactions—nine of which are shown here—also convert some of the energy of glucose into high-energy ATP and NADH. All these reactions take place in the cytosol, outside the mitochondria.

to form water. As we have seen, the six carbons in glucose ultimately break down into six molecules of carbon dioxide. Carbon atoms donate electrons, oxygen atoms ultimately accept electrons, and hydrogen ions join the reduced oxygen to form water.

In Chapter 2, we saw that different atoms have different affinities for electrons. Oxygen is the most electron-hungry atom in the environment. Organisms have evolved mechanisms for using oxygen's enormous electronegativity to obtain energy from reactions in which oxygen is the ultimate electron acceptor. As electrons move from glucose and other energy-rich molecules to oxygen, however, they must pass through many intermediate electron acceptors, as we will discuss later.

The two most important electron acceptors in energy metabolism are **NAD⁺** and **FAD,** which take electrons from glucose and transfer them to oxygen in a series of steps. Each NAD^+ can accept two electrons and a hydrogen ion, which converts it to its reduced form, **NADH.** Each FAD can accept

two electrons (and two hydrogen ions) to form **FADH₂.** Most of the ATP produced by cell respiration comes from energy originally captured in NADH and converted into ATP energy by oxidative phosphorylation.

NAD^+ (and NADH) and FAD (and $FADH_2$) are examples of **coenzymes**—organic molecules that are necessary participants in certain enzyme reactions, often as electron donors or acceptors. Examples of coenzymes are most of the B vitamins, which are found in whole grains. NAD^+ and NADH participate in many enzyme reactions. NADH donates electrons, and NAD^+ accepts electrons.

Glycolysis: How Do Cells Capture Energy in ATP and NADH?

As glycolysis breaks glucose into two molecules of pyruvate, it also converts two molecules of NAD^+ to NADH and two molecules of ADP to ATP. The major question about glycolysis is: How do cells capture energy in ATP and NADH? Like other

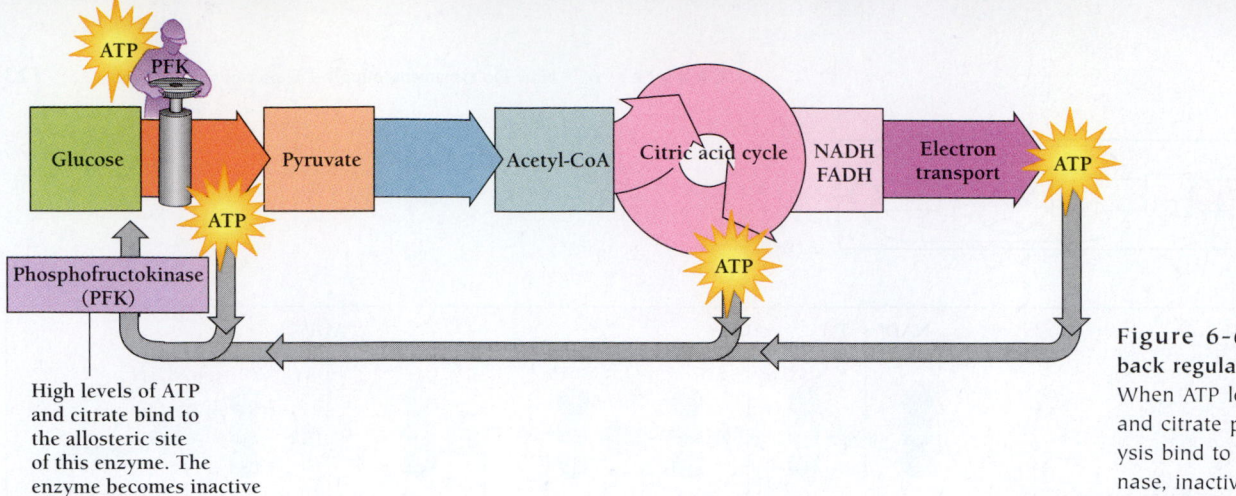

High levels of ATP
and citrate bind to
the allosteric site
of this enzyme. The
enzyme becomes inactive
and glycolysis stops.

Figure 6-6 Negative feedback regulates glycolysis. When ATP levels rise, the ATP and citrate produced by glycolysis bind to phosphofructokinase, inactivating it and slowing glycolysis.

biochemical pathways, glycolysis proceeds in small steps. But the ten chemical reactions of glycolysis fall into just three parts (Figure 6-5):

Part 1. Splitting of the six-carbon glucose into two three-carbon molecules.

Part 2. Oxidation of the three-carbon molecules and reduction of NAD^+.

Part 3. Further oxidation of the new three-carbon molecules to pyruvate.

Here's how glycolysis is organized: Part 1 of glycolysis converts the six-carbon glucose molecule into two three-carbon molecules. Part 1 requires five separate reactions (or steps), each catalyzed by a different enzyme. This process transfers two phosphate groups from two ATP molecules to glucose and splits glucose in half. The energy realized from the splitting of the high-energy phosphates pushes the reactions in the direction of the three-carbon product, glyceraldehyde-3-phosphate (G3P). At the end of part 1, glycolysis has consumed rather than harvested energy.

In part 2 of glycolysis, NAD^+ oxidizes each molecule of G3P to phosphoglycerate. Each molecule of G3P generates one ATP and one NADH. In addition, a phosphate ion from the cytosol is attached first to G3P, then to ADP to form a new molecule of ATP. For each starting glucose molecule, then, part 2 yields two ATPs and two NADHs, thus recovering the ATP investment of part 1.

The reaction steps of part 2 require both phosphate ions and NAD^+. If either is missing, the reaction simply cannot occur. Phosphate and NAD^+ are essential: cell survival requires a constant supply. Phosphate is usually not a problem, since it is present at fairly high concentrations in most cells and is continuously released as ATP is split. In aerobic organisms, the regeneration of NAD^+ generally requires transfer of NADH's electrons to oxygen. In the absence of oxygen (and in anaerobes), however, the cell must regenerate NAD^+ by other means, as we will discuss later.

Part 3 converts phosphoglycerate to pyruvate and transfers its phosphate to ADP, forming one more ATP molecule for each phosphoglycerate (that is, two ATP molecules for each glucose). Since the reactions in part 2 had already recovered

the two ATP molecules invested in part 1, the net gain of ATP during glycolysis is two ATPs per molecule of glucose, as well as two molecules of NADH.

But this is only a small bite out of a very large apple. The pyruvate molecules still contain lots of energy. This energy is extracted only in the later stages of glucose metabolism, which we will discuss later in this chapter. First, however, we will see how cells can use glycolysis to obtain energy even in the absence of oxygen.

Glycolysis, the first stage in the extraction of energy from glucose, breaks glucose into two molecules of pyruvate and produces two molecules each of ATP and NADH.

What regulates the overall rate of glycolysis?

In the cell, the overall process of glycolysis is highly exergonic. This means that glycolysis proceeds spontaneously and irreversibly. Since all the needed enzymes are present in cells, the reactions should also proceed quickly. How, then, does the cell keep all its glucose from being immediately converted to pyruvate?

We can see that glycolysis must be regulated. The question is, Where and how? In general, biochemical pathways are regulated at the first step that is unique to the pathway. This strategy ensures that molecules are not processed needlessly when they could be used for alternate purposes. In glycolysis, regulation takes place in reaction 3, which is catalyzed by the enzyme phosphofructokinase (Figure 6-6).

We would expect glycolysis to proceed rapidly when the cell needs ATP and to proceed slowly or not at all when the cell already has lots of ATP. This is indeed the case. ATP inhibits the action of phosphofructokinase, so when ATP is present, it slows glycolysis at part 3. As a result, glycolysis stops producing ATP when it is not needed.

Because ATP inhibits phosphofructokinase, reaction 3 of glycolysis, cells with plenty of ATP stop extracting energy from glucose. Cells with little ATP can proceed with glycolysis.

BOX 6.1

Sprints, Dives, and Marathons

Respiration yields at least 15 times more ATP than glycolysis. It's no wonder that animals—especially warm-blooded ones—depend on respiration for most of their energy. However, because oxygen is the ultimate electron acceptor in respiration, available energy depends on a steady supply of oxygen. For vertebrates, continuous energy production requires efficient lungs (to take in oxygen-rich air) and hearts (to pump oxygen-rich blood from the lungs to the muscles).

Training can improve the performance of the systems that deliver oxygen, but humans and other animals often need to spend energy more quickly than our oxygen supply will allow. We all need to sprint occasionally, whether we are a pedestrian dashing to avoid an inattentive driver, an outfielder swooping down on a wayward baseball, a cheetah chasing a gazelle, or a gazelle eluding a cheetah. Several energy reservoirs allow such bursts of energy: (1) muscles have stores of ATP and other compounds with similar high-energy phosphate bonds, enough to allow several minutes of work without any new ATP production; (2) muscles can use blood glucose and stored glycogen to produce ATP by anaerobic glycolysis, producing lactate as the end product.

The length of time that an organism can depend on anaerobic glycolysis for ATP production is limited both by training and by genetic capacity. Humans can manage only a few minutes. But some animals, such as whales, seals, and other diving animals, can function by means of glycolysis alone for much longer periods. The freshwater red-eared turtle shown in Figure 6-2B, for example, can stay under water for as long as 2 weeks.

Anaerobic glycolysis provides animals with a way of borrowing energy. But the loan is only short term. After anaerobic exertion, all animals must get rid of the accumulated lactate and restore their supplies of glucose and glycogen. These processes require oxygen. In most animals, the "oxygen debt" must be repaid almost immediately—through heavy breathing and increased circulation. Anyone who has run after a bus or away from an angry sibling has experienced the recovery from an oxygen debt. Animals extract energy from the accumulated lactate by converting it to pyruvate, which can be used in aerobic respiration. In some vertebrates, some of the lactate travels via the blood to the liver, where it can be recycled into glucose. The liver then resupplies the glucose reservoir of the blood.

Anaerobic glycolysis allows untrained schoolchildren to sprint up to 17 miles per hour (27 km/hr) in a 100-yard (or 100-meter) dash. But the necessity for oxygen prevents even champion athletes from averaging more than about 11.5 miles per hour in a 26-mile marathon (Figure A).

Nonmammalian vertebrates (fish, amphibians, and reptiles), which live their lives at a slower metabolic pace than we do, may function without air for long periods. Instead of using oxygen to recycle the accumulated lactate, fish, amphibians, and reptiles may excrete the lactate in their urine. This represents a huge waste of energy, but it allows these animals to feed and function in environments in which they could not otherwise live.

Figure A 1995 World women's marathon. The speed that a marathoner can attain is limited by the need for oxygen. *(David Madison)*

How can glycolysis continue without oxygen?

To sustain glycolysis, a cell must regenerate more NAD^+ from NADH. Just how it does this depends on the species of organism and on whether oxygen is present.

In the presence of oxygen, most cells can oxidize NADH to NAD^+ by the reactions of oxidative phosphorylation. In the absence of oxygen, however, cells must regenerate NAD^+ by using NADH to reduce pyruvate. Their manner of doing so depends on what enzymes they have.

Many microorganisms live by **fermentation** [Latin, *fervere* = to boil]—the anaerobic extraction of energy from organic compounds (Figure 6-7). Yeast cells, for example, ferment the sugars in beer, wine, and bread, converting glucose into carbon dioxide and ethanol. The carbon dioxide makes beer bubble and bread rise. The ethanol gives beer its punch and freshly baked bread its sweet aroma. To perform this feat, yeasts employ all the reactions of glycolysis that we have discussed. In addition, they use the accumulated NADH

to reduce pyruvate to form ethanol. In the process, NADH is oxidized back to NAD^+. Yeast is a *facultative anaerobe*—an organism that can live either anaerobically, by fermentation, or aerobically, by oxidative phosphorylation. Other microorganisms, such as *Clostridium botulinum,* are *obligate anaerobes,* meaning that they can grow only in the absence of oxygen (Figure 6-2A).

Animals, including humans, often suffer a temporary lack of oxygen in some of their cells. For example, during sprinting or other vigorous exercise, muscle cells need more oxygen (to oxidize NADH) than the blood and lungs can deliver (Figure 6-7B). In this case, and for short periods only, the cells can regenerate NAD^+ in the absence of oxygen. Instead of producing carbon dioxide and ethanol (as a yeast cell does), however, muscle cells make lactate (lactic acid, after it has given up a hydrogen ion), a compound with three carbon atoms. During a vigorous sprint, lactate accumulates in the muscles, sometimes causing soreness. Once oxygen becomes available

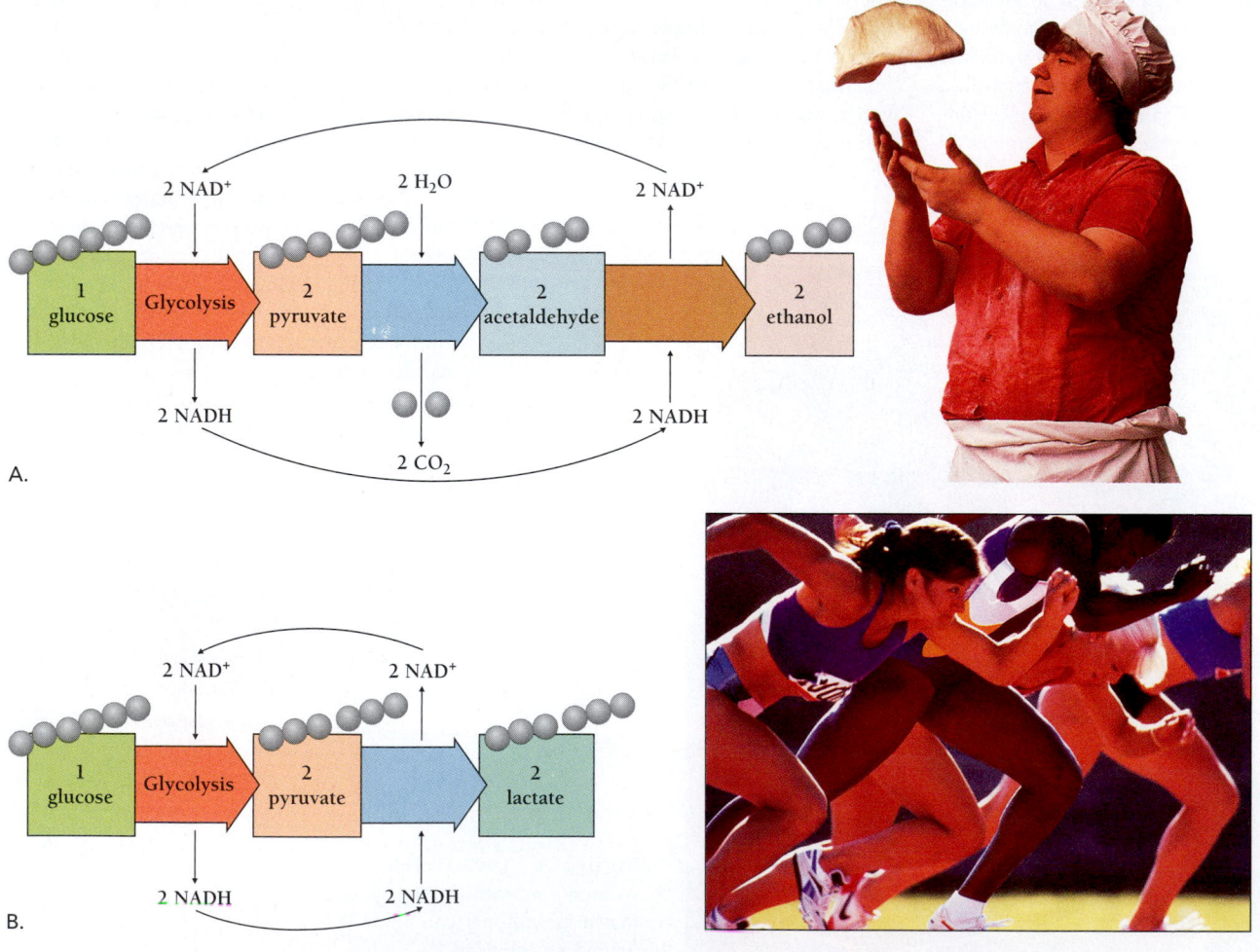

Figure 6-7   Two kinds of fermentation. In each case, cells regenerate NAD^+ for further glycolysis.
A. When yeast cells grow in limited oxygen (as in a fermentation vat or a bread dough), pyruvate forms acetaldehyde and ethanol. B. In muscle cells that have exceeded their oxygen supply, pyruvate forms lactic acid (lactate). (*Pizza maker, Jim Erickson/The Stock Market; sprinters, David Madison*)

again, however, cells convert most of the lactate back to pyruvic acid. Heart muscle is especially efficient at extracting energy from lactate by first converting it to pyruvate (pyruvic acid, after it has given up a hydrogen ion).

Glycolysis can continue without oxygen if cells use pyruvate to regenerate NADH from NAD^+.

What Happens to Pyruvate When Oxygen Is Present?

In eukaryotic cells, the steps of energy extraction after glycolysis take place in the mitochondria. The pyruvate produced by glycolysis must enter the mitochondria for cellular respiration to proceed. After pyruvate arrives in the **mitochondria,** it is oxidized through a series of steps to a compound called acetyl-CoA, whose acetyl group contains two carbons (Figure 6-8).

Acetyl-CoA forms from a molecule of acetate (the same as the acetic acid of vinegar) linked by a high-energy bond to a molecule called **coenzyme A (CoA).** During the formation of acetyl-CoA, the reaction removes one molecule of carbon dioxide from pyruvate, as well as two electrons and two protons. NAD^+ serves as the electron acceptor. This reaction cannot occur if there is no oxygen to drive the fourth stage of respiration, which regenerates NAD^+ from NADH.

Acetyl-CoA has only two principal fates: (1) it can enter the third stage of energy metabolism and generate more ATP, as discussed in the following section, or (2) it can be used to synthesize new lipids, either fatty acids or cholesterol. When a cell's ATP level is high, the cell redirects acetyl-CoA into a set of reactions that make energy-rich fats and oils. This redirection of acetyl-CoA allows the body to store energy for later use. It is the main reason that we accumulate fat when we consume too many energy-rich molecules, whether they are starches in a sandwich or stearic acid in a steak.

When oxygen is present, the pyruvate produced by glycolysis enters a cell's mitochondria, where it is converted to acetyl-CoA. Acetyl-CoA can generate still more ATP.

How Does a Cell Generate ATP from Acetyl-CoA?

Most of the energy in a glucose molecule is still present in the two molecules of acetyl-CoA produced by stages 1 and 2 of cellular respiration. The reactions of stage 3, which we will now discuss, are responsible for capturing most of the high-energy electrons in NADH. Oxidative phosphorylation (stage 4) then harnesses the energy of NADH into the production of ATP.

In nearly all organisms, acetyl-CoA enters directly into a set of reactions that converts the carbon atoms of acetyl-CoA

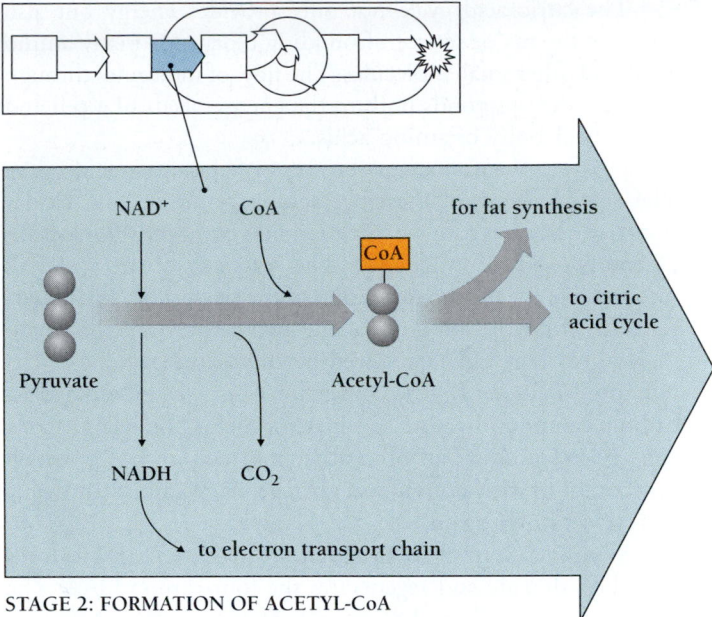

STAGE 2: FORMATION OF ACETYL-CoA

Figure 6-8 Stage 2: pyruvate to acetyl-CoA. After glycolysis, pyruvate enters the mitochondria for the second stage of energy extraction. In the mitochondria, pyruvate is oxidized to form acetyl-CoA. Acetyl-CoA can either enter the third stage of energy metabolism and generate more ATP or it can be used to synthesize body fat.

into carbon dioxide. These reactions are called the **citric acid cycle,** or the **Krebs cycle,** after Hans Krebs, who began working out the details of the cycle in the 1930s.

In eukaryotes, the citric acid cycle takes place within the mitochondrial matrix. In its barest outlines, the citric acid cycle takes a molecule of the two-carbon compound acetyl-CoA, adds it to a four-carbon molecule (oxaloacetate, or OAA) to form a six-carbon molecule (citric acid), removes two carbons as carbon dioxide, and produces another four-carbon OAA, ready to combine with another molecule of acetyl-CoA.

Altogether, the citric acid cycle removes eight electrons from the two carbons of acetyl-CoA. Three NAD^+ molecules accept six of these electrons, and one molecule of FAD accepts the other two. The citric acid cycle also forms one high-energy phosphate bond. Unlike the other reactions that we have discussed, however, the phosphate acceptor in the citric acid cycle is guanosine diphosphate (GDP), rather than ADP. The resulting guanosine triphosphate (GTP) carries the same amount of energy as ATP.

Oxygen is the ultimate acceptor of electrons and is therefore absolutely essential for the citric acid cycle to proceed. In stage 4 of cellular respiration (the electron transport chain), oxygen accepts electrons from NADH and $FADH_2$, allowing the cell to regenerate more NAD^+ and FAD and permitting both the cycle (stage 3) and the oxidation of pyruvate (stage 2) to continue.

The citric acid cycle not only provides energy but also serves as the major source of building blocks for many amino acids and other small molecules. The flow of molecules through the cycle then responds both to the energy needs of a cell and to the availability of amino acids.

Figure 6-9 summarizes the steps of the citric acid cycle. The central events of the citric acid cycle are its four oxidation-reduction reactions, which employ two kinds of electron acceptors—NAD$^+$ and FAD. The reactions of the cycle fall into three parts: (1) the formation of the six-carbon citrate (citric acid) from the addition of the two-carbon acetyl-CoA to the four-carbon OAA, followed by the rearrangement of citrate into isocitrate; (2) the conversion of isocitrate into a four-carbon compound called succinate, which is linked to CoA; and (3) the production of another molecule of OAA, which can combine with acetyl-CoA to regenerate citrate and start the cycle all over again.

With each turn of the cycle, citrate loses two carbon atoms as carbon dioxide and regenerates the four-carbon OAA. The carbon atoms of the carbon dioxide from each turn, however, are not those that entered the cycle as acetyl-CoA. Nor is the OAA regenerated by the cycle the same as the one that started the cycle. In each turn of the cycle, citrate loses a total of eight electrons.

The citric acid cycle converts acetyl-CoA to carbon dioxide, transferring, in the process, eight electrons from acetyl-CoA to NADH and FADH$_2$.

How was the citric acid cycle discovered?

Biochemists tried for years to understand how cells use oxygen to obtain energy from organic molecules. But it was the brilliant Hungarian biochemist Albert Szent-Györgyi, a man obsessed by all oxidative processes, who solved the problem. The first observation that Szent-Györgyi made was that extracts of pigeon breast muscle were strikingly efficient at taking up oxygen for cellular respiration. This is not surprising, since the flight muscles of birds require lots of energy and are packed with mitochondria.

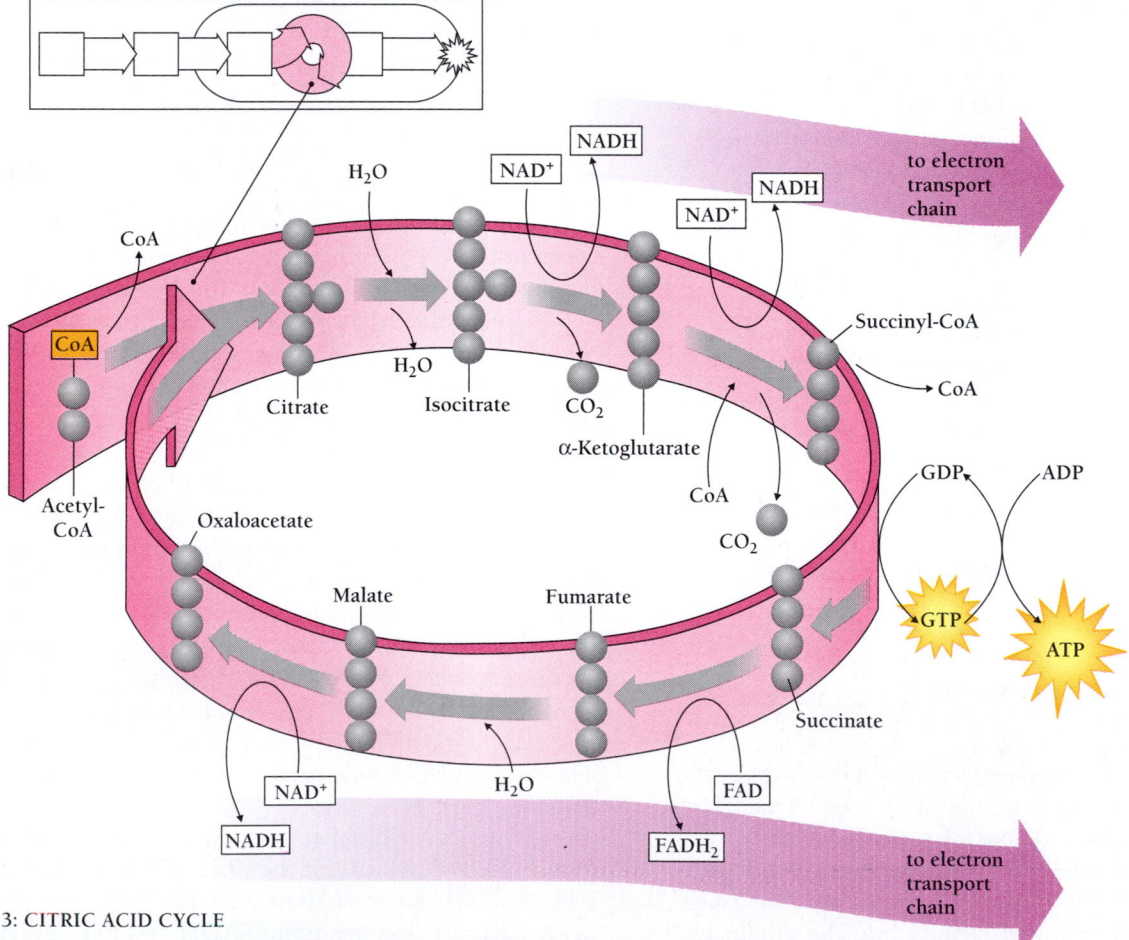

STAGE 3: CITRIC ACID CYCLE

Figure 6-9 Stage 3: the citric acid cycle. During the citric acid cycle, a single molecule of acetyl-CoA generates three NADH, one FADH$_2$, one GTP, and two carbon dioxide (CO$_2$) molecules.

Szent-Györgyi then added various compounds to his extracts to see how these compounds were oxidized. To his surprise, he found that certain four-carbon compounds stimulated the uptake of oxygen far more than he would have expected if each carbon atom were converted to carbon dioxide and each hydrogen became part of a water molecule. He had calculated the amount of oxygen required for the complete oxidation of each added compound, but the extracts were using more oxygen than complete oxidation would require.

The reason for this discrepancy, we now realize, is that the four-carbon compounds that Szent-Györgyi added were intermediates in the citric acid cycle—they increase the rate at which acetyl-CoA enters the cycle. The higher levels of acetyl-CoA production, in turn, increase the production of NADH and FADH$_2$ and therefore the amount of oxygen needed to regenerate NAD$^+$ and FAD (Figure 6-9).

By 1937, biochemists knew that pigeon muscle extracts could convert citric acid to these same four-carbon compounds. In 1938, Hans Krebs, a refugee from Nazi Germany working in England, provided the missing link and the final interpretation. Krebs showed that Szent-Györgyi's muscle extracts could make citric acid from the four-carbon compound OAA and acetyl-CoA.

Krebs's crucial contribution was the idea that the conversion of acetate to carbon dioxide involved a cycle. Krebs showed how the particular four-, five-, and six-carbon compounds fit together in a single cycle. He also showed that the same chemical transformations occurred both in cells and in test tubes.

> *The first clue to the existence of the citric acid cycle was Szent-Györgyi's demonstration that four-carbon OAA stimulated the uptake of more oxygen than was required to oxidize the OAA.*

The citric acid cycle intersects other biochemical pathways

The citric acid cycle lies at the center of the complex network of metabolic reactions. Metabolism is divided into two classes of pathways: **catabolism** [Greek, *cata* = down + *ballein* = to throw]—the breakdown of complex molecules, such as those in food; and **anabolism** [Greek, *ana* = up + *ballein* = to throw]—the synthesis of complex molecules (Figure 6-10). In general, catabolism produces energy and usually involves oxidation—the removal of electrons and the production of NADH. Anabolism, on the other hand, usually consumes energy and involves reduction, so it uses up NADH and other reducing agents.

Figure 6-11 shows that different molecules enter the cellular respiration pathway at different points. Carbohydrates in such foods as bread are broken down into simple sugars that usually take the same path as glucose. Fats and oils are broken down into glycerol, which may enter glycolysis, and into fatty acids, which enter the pathway as acetyl-CoA. Proteins are broken down into component amino acids, which are then con-

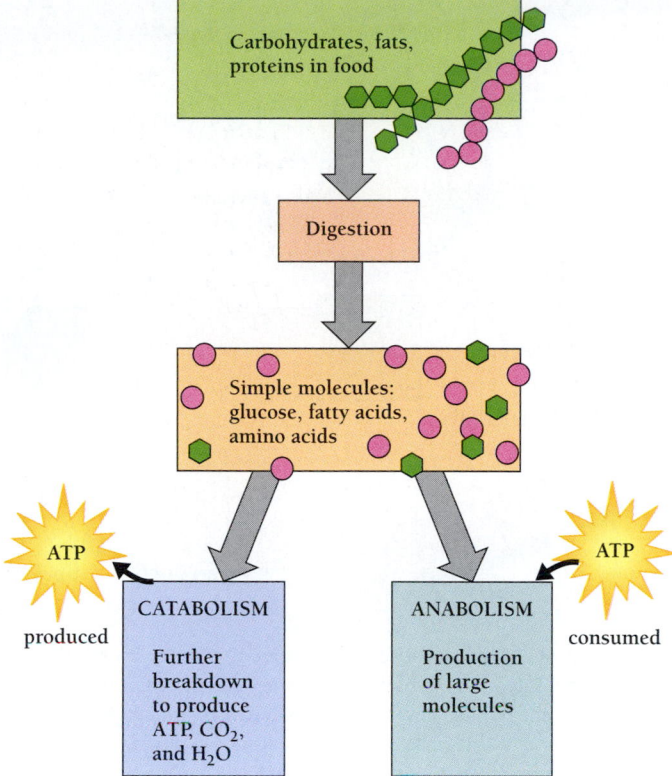

Figure 6-10   Metabolism consists of two pathways: catabolism and anabolism. Catabolism produces energy through the oxidation of molecules. Anabolism usually consumes energy in the process of building large molecules.

verted to pyruvate, acetyl-CoA, or other intermediates of the cellular respiration pathway.

Cells regulate the rates at which pathways or sets of pathways operate. High levels of ATP favor anabolism, and low levels favor catabolism. When ATP levels are high but the products of anabolism are abundant, the energy in food is stored for future use—as carbohydrates, proteins, or fats. Despite the importance of glucose and other carbohydrates in energy metabolism, the preferred form of energy storage in animals is fat.

The end product of the catabolism of fats is acetyl-CoA. Fats are more reduced compounds than sugars; that is, for each carbon atom, fats provide more electrons for oxidative phosphorylation, the most productive stage of cellular respiration. Fats, therefore, contain more free energy—about 9.4 kcal per gram of fat, instead of the 4.1 kcal per gram of carbohydrate and 5.6 kcal per gram of protein. In addition, fats are less polar and less hydrophilic than carbohydrates, which may bind as much as twice their weight in water. As a result of these factors, each gram of stored fat contains about six times as much energy as each gram of stored glycogen. A 50-kg (110-lb) woman, for example, has fuel reserves of about 70,000 kcal in fats, about 18,000 kcal in proteins (mostly in muscle), about 400 kcal in glycogen, and about 30 kcal in glucose. Her fat

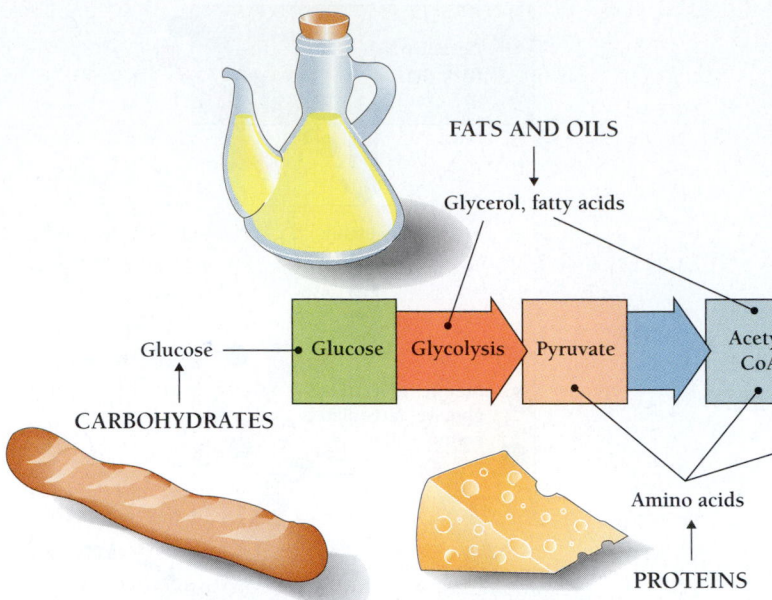

Figure 6-11   How do other food molecules enter metabolism? Fat, carbohydrates, and proteins enter the cellular respiration pathway at different points. The busiest crossroad is at acetyl-CoA, which then enters the citric acid cycle. All fatty acids and many amino acids are converted into acetyl-CoA.

stores, however, weigh only about 8 kg (18 lb). To store the same amount of energy as carbohydrates (with all their bound water), she would need to weigh at least 90 kg (200 lb)!

The citric acid cycle lies at the intersection of the catabolic and anabolic pathways. High levels of ATP favor anabolism, and low levels favor catabolism. When ATP levels are high, the most efficient way to store energy is in the form of fat.

ELECTRON TRANSPORT: HOW DOES THE ENERGY IN GLUCOSE REACH ATP?

A major goal of biologists and biochemists has been to learn how cells convert the energy of sunlight or food into the high-energy bonds of ATP. Until the early 1960s, biochemists believed that they would soon understand how organisms make ATP simply by studying the rearrangements of chemical bonds in reactions catalyzed by enzymes in test tubes. Indeed, decades of ingenious research produced detailed maps of such enzyme-catalyzed transformations, many of which made or used ATP. Yet researchers gradually realized that in living cells most ATP was made in some other way. But how?

Surprisingly, the first clue came not from the properties of enzyme reactions, but from the properties of membranes and subcellular organelles. In producing ATP, cells rely on their internal structure in a way that eluded most biochemists.

It was the biochemists' rejection of vitalism that misled them. Because biochemists (in Liebig's tradition) were more interested in chemistry than in cells, they tended to take a "grind-and-find" approach to ATP production. Ignoring the complex structure of living organisms, they would simply grind up cells to find out what chemicals are in them. This approach has worked very well. Decades of research have shown that the majority of cellular processes depend on simple reactions that can be reproduced in a test tube. In fact, the grind-and-find approach was so successful that, until the 1970s, most biochemists were blind to the importance of intact cellular structures.

As it turns out, however, the steps of cellular respiration that release the most energy occur in functioning, intact mitochondria. Cellular respiration depends so heavily on mitochondria that we can pick out the cells that are most active in cellular respiration simply by counting the number of mitochondria in them.

Biochemists' "grind-and-find" approach to cell biology prevented them from recognizing the crucial role of cellular organization in cell respiration.

How Does the Flow of Electrons from Electron Donors to Oxygen Release Energy to the Phosphate Bonds in ATP?

Altogether, the process of cellular respiration transfers 24 electrons from glucose to oxygen. The first electron acceptor for 20 of the 24 electrons is NAD^+. The other four electrons of glucose are transferred to FAD. Altogether, about 90 percent of the energy stored in the chemical bonds of a glucose molecule is converted to chemical energy in the form of the high-energy compounds, NADH and $FADH_2$.

The two electrons received by each molecule of NADH or $FADH_2$ ultimately go to reduce a single oxygen atom. The reduced atom of oxygen, together with two hydrogen ions from the surrounding water, results in the formation of one molecule of water.

The energy from the oxidation of NADH or $FADH_2$ is used to make high-energy phosphate bonds in ATP by a process that releases the energy in a series of small steps. This stepwise release resembles the breaking of a big waterfall into a gentle cascade.

Electrons move to oxygen as spontaneously as water flows downhill, and for the same reason. The flow of electrons to oxygen is a "downhill"—energy-releasing—process. However, the making of a high-energy phosphate bond to form ATP from ADP is definitely an uphill, energy-consuming process.

One important question, then, is how oxidative phosphorylation couples the downhill flow of electrons to the uphill production of high-energy bonds in ATP. The answer—discovered only in the 1960s—was a complete surprise to the thousands of researchers who were trying to figure it out.

Cellular respiration transfers electrons from glucose to oxygen by way of a series of intermediaries. The first electron acceptor is either NAD^+ or FAD. NADH and $FADH_2$ then release their electrons to the electron transport chain, which transfers electrons to oxygen through a series of small steps. Oxidative phosphorylation couples this transfer to the production of ATP.

Which Molecules Serve as Electron Carriers?

The electrons from NADH and $FADH_2$ make their way to oxygen by way of a series of other molecules, called **electron carriers.** The pathway of the electrons (from one carrier to another) is called the **electron transport chain.** Biochemists like to compare the electron transport chain that carries electrons to oxygen to the "bucket brigade" that old-time firefighters used to carry water to a fire (Figure 6-12D). Each electron carrier passes its electrons to the next carrier in the line. A reduced carrier becomes oxidized as it gives up its electrons, and the next carrier becomes reduced as it receives electrons. In eukaryotes, the transfer occurs in the mitochondrial membrane. In prokaryotes, these reactions occur in association with the cell membrane.

In 1925, the British biochemist David Keilin discovered that cellular respiration depends completely on a set of specific proteins that change color as they accept or donate electrons. These proteins, called **cytochromes** [Greek, *kytos* = hollow vessel, cell + *chroma* = color], have a reddish color that derives from **heme,** the same iron-containing ring structure that gives hemoglobin its color. Heme, like NAD^+ and FAD, may both donate and accept electrons. Keilin and others showed that cytochromes could accept electrons from NADH and also from each other. By comparing the flow of electrons both in isolated cytochromes and in whole cells, biochemists worked out the path of electrons from NADH to oxygen in the electron transport chain.

Iron-containing cytochromes contribute to the transfer of electrons through the electron transport chain.

How Do Cells Harvest the Energy of Electron Transport?

After the discovery of the electron transport chain, the big question still remained: How do cells harvest the energy released by the flow of electrons? Biochemists spent decades looking for intermediate molecules, particularly for small molecules that transferred phosphate groups to ATP in the same way that phosphoglycerate generates ATP in part 3 of glycolysis. (Such a phosphate group transfer to ATP is called "substrate-level phosphorylation.") But no one was able to find such an intermediate. Biochemists (incorrectly) concluded that the intermediate molecules were so unstable that they fell apart before they could be isolated in an experiment.

The situation was this: an electron transport chain in a test tube could pass electrons, but it would generate no ATP. Isolated mitochondria could perform both electron transport and ATP synthesis, but as soon as the mitochondria were broken open, ATP synthesis stopped even though electron transport continued. These experiments clearly showed the importance of intact mitochondrial organization for the production of ATP. But what did the mitochondria do?

The puzzle was solved in the early 1960s by Peter Mitchell, an English biochemist. Mitchell had been studying the way bacteria pump hydrogen ions across their membranes. The inside of a bacterium has a higher pH (a lower hydrogen ion concentration) than its environment. The difference in hydrogen ion concentration across a membrane is called a **proton gradient.**

Mitchell showed that bacteria could use the energy in this proton gradient to transport other substances into the cell. In 1961, he suggested that mitochondria, like bacteria, might be able to make use of the energy stored in a proton gradient. The idea that a proton gradient could be used to make ATP was, at the time, thoroughly unconventional. Mitchell was unorthodox in other respects, as well. For example, he performed his research not in a university laboratory, but in a manor house in the English countryside where he lived.

Mitchell proposed a two-part hypothesis. He suggested first that the flow of electrons through the electron transport chain establishes a proton gradient across the mitochondrial membrane. Second, he argued, some special molecular machinery in the membrane captures the energy stored in the proton gradient by making ATP. Mitchell's hypothesis showed why biochemists had not found the intermediate molecules that supposedly coupled electron flow to ATP production: they simply did not exist. Mitchell argued instead that the formation of ATP was the result of **chemiosmosis** [Greek, *osmos* = to push]—the harnessing of the energy stored in the chemical gradient created by the electron transport chain.

In the 1960s and 1970s, Mitchell, his colleague Jennifer Moyle, and others showed that the coupling of a proton gradient to the synthesis of ATP helps explain not only cellular respiration in mitochondria but also photosynthesis in chloroplasts. As we shall see, Mitchell's radical suggestion, called the **chemiosmotic hypothesis,** has now become the basis for understanding the harvesting of energy both in photosynthesis and cellular respiration.

The English biochemist Peter Mitchell proposed that the energy from the electron transport chain drives ATP production by harnessing the energy in a proton gradient, not by means of intermediate molecules.

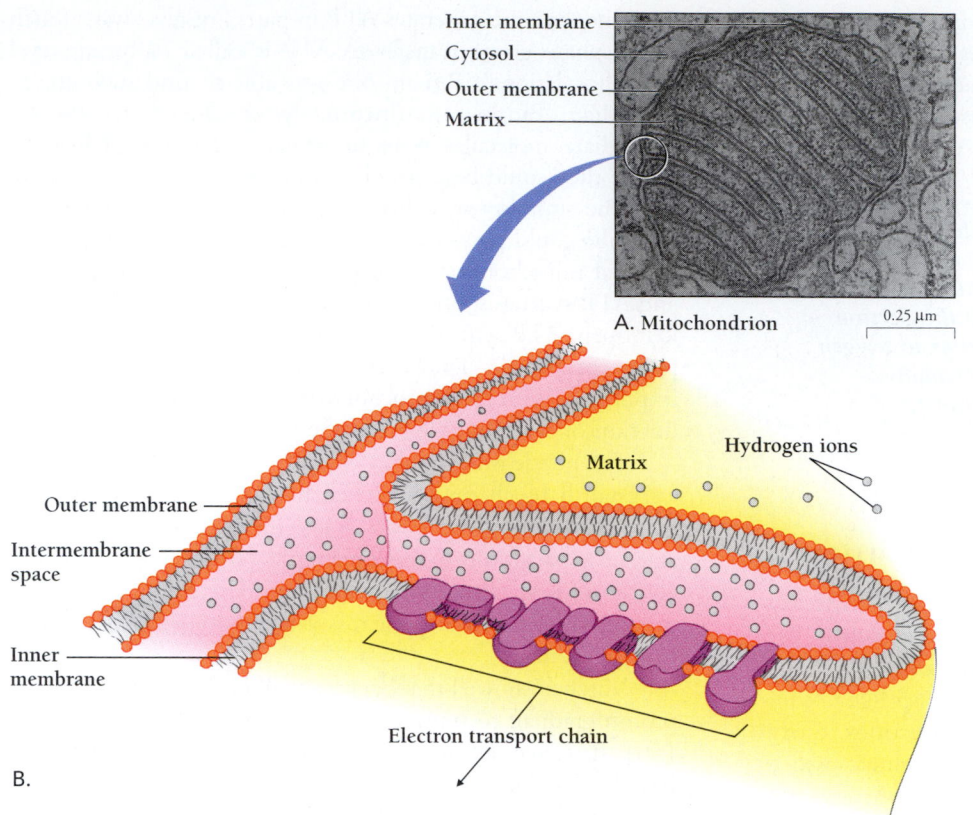

Inner membrane
Cytosol
Outer membrane
Matrix

A. Mitochondrion

0.25 μm

Figure 6-12   The electron transport chain is like a bucket brigade. A. Transmission electron microscopy (TEM) of a single mitochondrion. **B.** The electron transport chain is located in the inner membrane. It consists of four protein complexes and a number of small lipid molecules that carry electrons between the complexes. The intermembrane space, between inner and outer membranes, has a high concentration of hydrogen ions. In contrast, the mitochondrial matrix, inside the inner membrane, has a low concentration of hydrogen ions. **C.** The electron transport chain drives hydrogen ions out of the matrix, creating an electrochemical gradient. **D.** The electron transport chain consists of a series of electron carriers that pass electrons from one to the next like the members of an old-fashioned bucket brigade trying to put a fire out. Instead of passing buckets of water, however, electron carriers pass electrons, derived from the breakdown of glucose. The electrons ultimately go to oxygen molecules. *(A, D. Friend, D. Fawcett/Visuals Unlimited)*

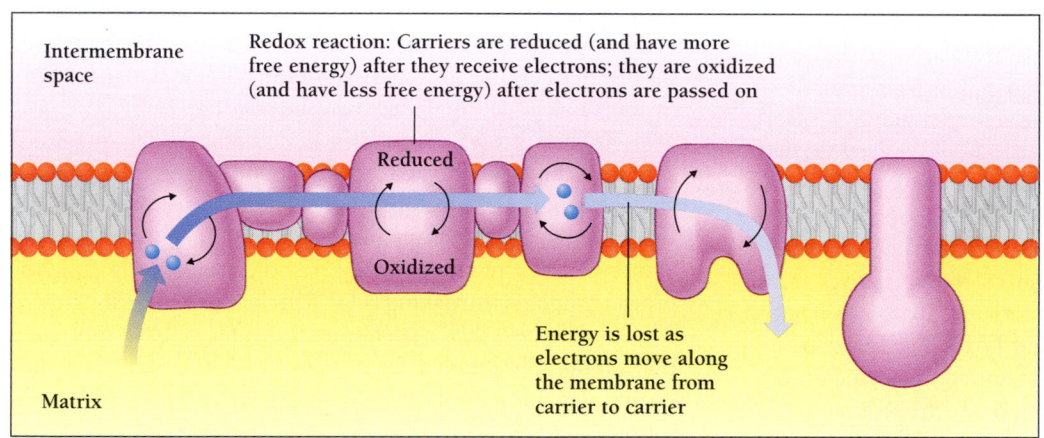

Matrix
Hydrogen ions
Outer membrane
Intermembrane space
Inner membrane
Electron transport chain

B.

Intermembrane space

Redox reaction: Carriers are reduced (and have more free energy) after they receive electrons; they are oxidized (and have less free energy) after electrons are passed on

Reduced

Oxidized

Energy is lost as electrons move along the membrane from carrier to carrier

Matrix

C.

D.

Glucose

How Do Mitochondria Generate a Proton Gradient?

Recall that a mitochondrion has two membranes. A mitochondrion, therefore, provides four separate locations for the different parts of cellular respiration: the outer membrane; the inner membrane; the space between them, called the intermembrane space; and the **mitochondrial matrix,** the space surrounded by the inner mitochondrial membrane (Figure 6-12A and B).

Most of a cell's NADH and $FADH_2$ is produced within the mitochondrial matrix, which makes up about two-thirds of the volume of a typical mitochondrion. The electron transport chain lies within the inner membrane. The electron carriers are oriented within the inner membrane so that, as they pass electrons from one to another, hydrogen ions move from the matrix out to the intermembrane space by crossing the inner mitochondrial membrane (Figure 6-12B). This movement of hydrogen ions ("protons") out of the matrix establishes a proton gradient across the inner membrane, which lies between the matrix and the intermembrane space. Careful measurements with intact mitochondria confirm that a proton gradient really exists across the inner mitochondrial membrane, with the pH lower outside the membrane than inside. The movement of electrons through the electron transport chain in the inner membrane causes protons to move *out* of the matrix.

Experiments have demonstrated that even fragments of inner membranes can pump protons. In these studies, mitochondria were broken open and their membranes allowed to reassemble spontaneously into empty vesicles. As long as electrons flowed, protons moved. When the experimenter stopped electron flow with a poison, however, proton pumping stopped also. These experiments established the ability of the electron transport chain to pump protons.

The electron transport chain within the inner membrane of the mitochondrion pumps protons out of the matrix and generates a proton gradient.

What Pumps Protons out of the Mitochondrial Matrix?

By taking apart the inner membrane still further, biochemists found four distinct electron transport complexes in the inner mitochondrial membrane. Three of these complexes (complexes I, III, and IV) use the energy derived from transporting electrons part of the way down the electron transport chain to pump protons across the inner membrane (Figure 6-13).

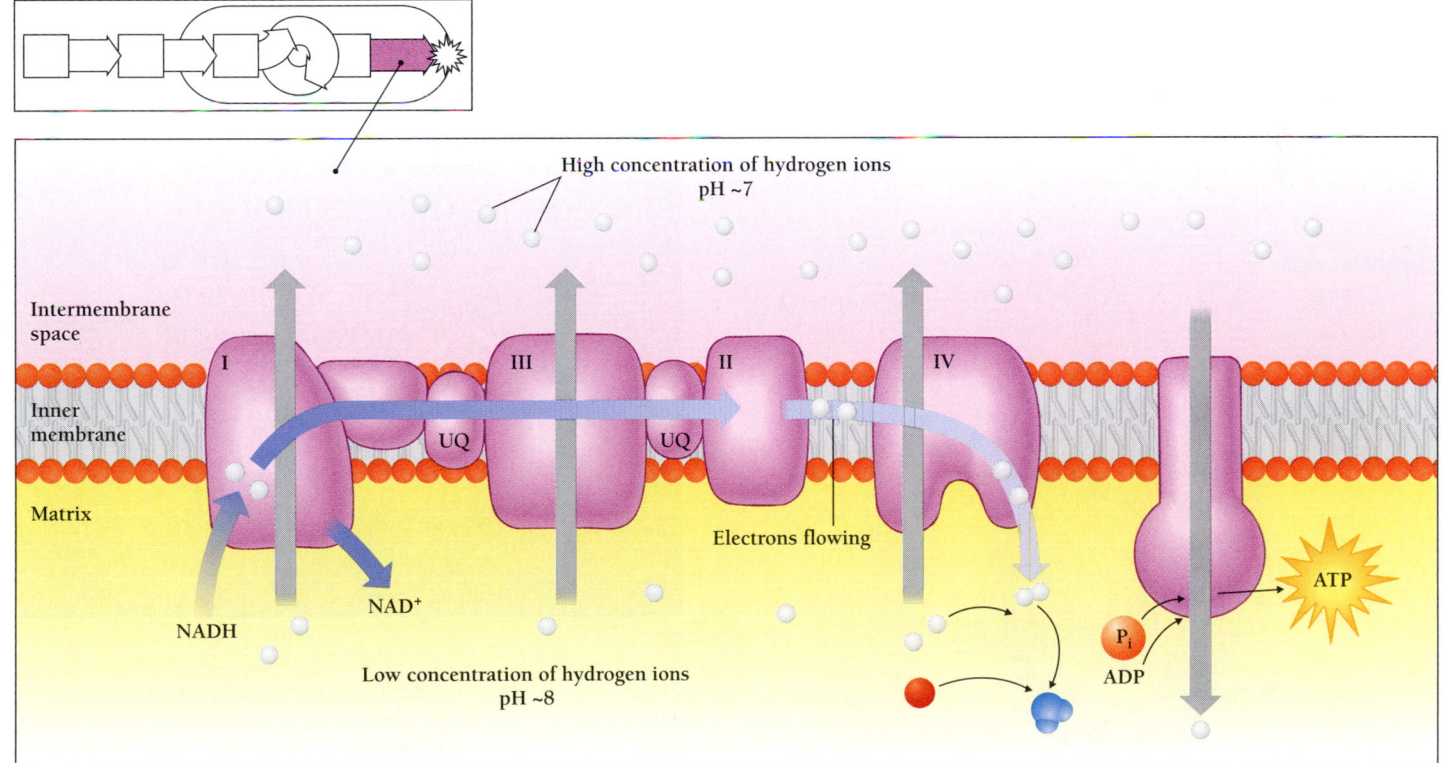

STAGE 4: ELECTRON TRANSPORT CHAIN

Figure 6-13 Stage 4: the electron transport chain. Each electron carrier is alternately reduced and oxidized as it gains and loses each electron passing down the chain. As the electrons pass, they lose some of their energy to proton pumps that push hydrogen ions, or protons, out of the matrix. The electrons ultimately combine with oxygen (and two hydrogen ions) to form water.

(The fourth complex, called complex II, transfers electrons but does not move protons.)

The complexes have fixed orientations in the membrane, so that they pump protons in a single direction only—from the matrix to the intermembrane space. For example, one of the respiratory complexes, called cytochrome oxidase, passes electrons from the last carrier in the electron transport chain (cytochrome *c*) to oxygen. Purified cytochrome oxidase complex can transfer electrons from reduced cytochrome *c* to oxygen and simultaneously pump protons, even in a vesicle made from artificial membrane.

The outer mitochondrial membrane allows the easy passage of small molecules and ions, so the intermembrane space has the same neutral pH as the bulk of the cytosol, about pH 7. In contrast, the pH of the matrix increases to about pH 8 as the electron transport chain pumps protons out of the matrix. The inner membrane is not permeable to protons (except through special channels), so the pH difference between the matrix and the intermembrane space is stable as long as electron flow continues (Figure 6-13).

Proton pumping establishes a charge gradient as well as a pH gradient. This is because pumping the positive protons out of the matrix leaves behind an excess of negative ions. The proton pump generates an **electrochemical gradient,** a double gradient composed of a *chemical* gradient (the difference in hy-drogen ion concentration, or pH) and an *electrical* gradient (the difference in charge). Both components of the electro-chemical gradient favor the return of protons to the matrix.

Three electron transport complexes in the inner membrane of the mitochondrion pump protons out of the matrix, generating an electrochemical gradient that favors the return of protons to the matrix.

How Does the Flow of Protons Back into the Matrix Cause the Synthesis of ATP?

Once hydrogen ions leave the matrix, there is only one way back in—through special channels in the otherwise imperme-able inner membrane. Protons flow "downhill" back into the matrix—from an area of high concentration to an area of low concentration and from an area of net positive charge to one with a net negative charge. As the protons flow, a protein com-plex called **ATP synthase** uses the flow of protons to drive the synthesis of ATP (Figure 6-14). ATP synthase works like a windmill, which converts the kinetic energy of wind to elec-trical energy.

We can actually see ATP synthase protein complexes as components of the inner mitochondrial membrane. Each looks

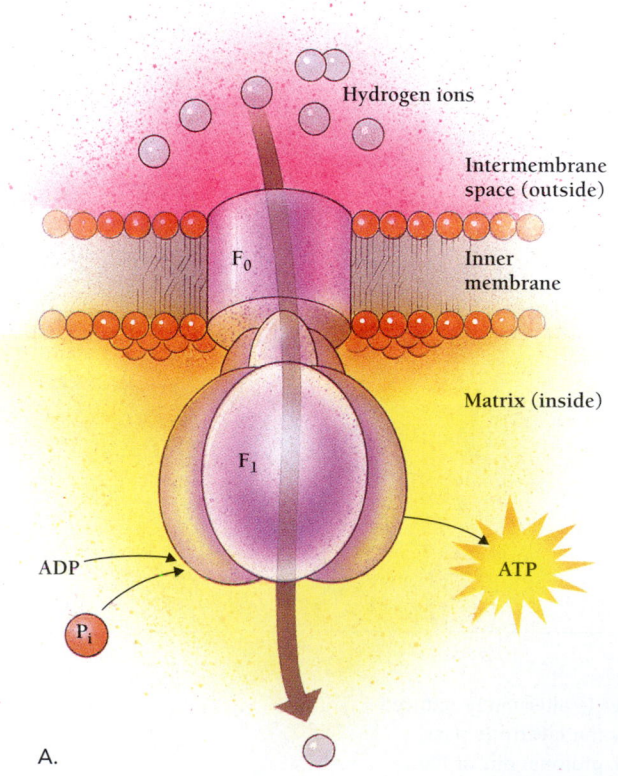

A.

B.

Figure 6-14 ATP synthase, the F_0-F_1 complex. A. The F_0-F_1 complex harnesses energy stored in the proton gradient across the inner mitochondrial membrane. F_0 serves as a channel for protons flowing back into the matrix, and F_1 is an enzyme that makes ATP as the protons pass. B. A windmill's blades spin around as it gen-erates electricity. In the same way, the F_1 blades of the F_0-F_1 com-plex spin around as the complex generates ATP. (*B, Alan Reininger/The Stock Market*)

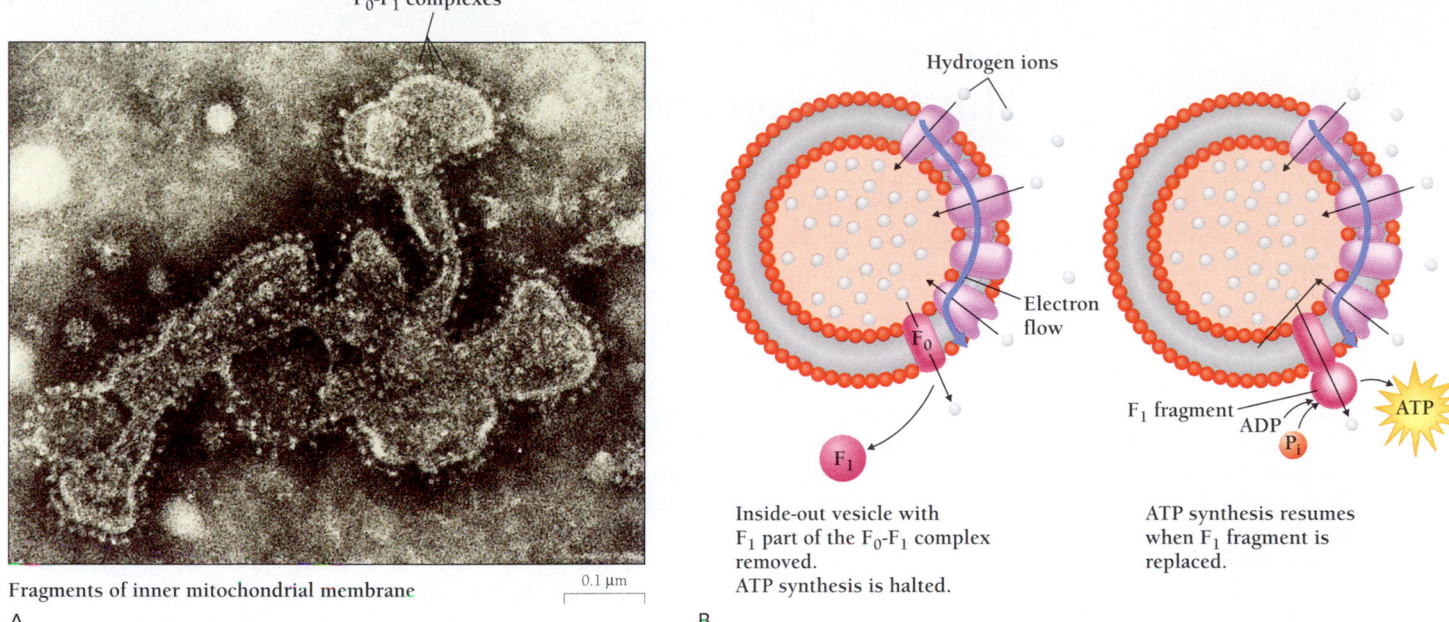

F_0-F_1 complexes

Hydrogen ions

Electron flow

F_0

F_1

Inside-out vesicle with F_1 part of the F_0-F_1 complex removed. ATP synthesis is halted.

0.1 µm

Fragments of inner mitochondrial membrane

A.

F_1 fragment

ADP

P_i

ATP

ATP synthesis resumes when F_1 fragment is replaced.

B.

Figure 6-15   ATP synthase, the F_0-F_1 complex. A. This micrograph shows hundreds of F_0-F_1 complexes dotting the surface of an inside-out inner mitochondrial membrane. F_0 is the channel through which protons can move back into the matrix, while F_1 actually produces ATP from ADP and phosphate. The inside of the vesicle corresponds to the intermembrane space, which is continuous with the cytosol. The outside of the vesicle corresponds to the matrix of an intact mitochondrion. B. When the F_1 fragment (the coupling factor) is replaced, ATP synthesis resumes. (A, R. Bhatnagar/Visuals Unlimited)

like a miniature lollipop protruding from the inner mitochondrial membrane into the mitochondrial matrix (Figure 6-14). Biochemists have found that each contains two major components in a single complex: (1) the stick, which spans the inner membrane, called **F_0,** and (2) the top, called **F_1.** The "F_0F_1 complex" is the same as ATP synthase.

Under certain conditions, fragments of inner mitochondrial membrane form inside-out vesicles from which F_0-F_1 complexes protrude (Figure 6-15). Gentle shaking removes F_1 from the membrane, leaving F_0 behind. After this treatment, the inside-out vesicles still allow protons to flow down an electrochemical gradient, suggesting that F_0 serves as a channel for protons to move back into the matrix. But the removal of F_1 abolishes the ability of the vesicles to make ATP. Replacing the F_1 fragment causes ATP synthesis to resume (Figure 6-15). So F_1 appears to be responsible for making ATP and is called the **coupling factor.** We still do not fully understand, however, exactly how the F_0-F_1 complex harnesses the proton flow to create new high-energy phosphate bonds in ATP. We do know, however, that F_1 spins like a three-bladed windmill as it cranks out ATP.

The F_0-F_1 complex (ATP synthase) channels the flow of protons back into the mitochondrial matrix and uses the free energy from that flow to synthesize ATP.

Are Proton Pumping and ATP Synthesis Really Separate Processes?

Even before the discovery that the F_0-F_1 complex was responsible for the coupling of proton flow to ATP synthesis, biochemists became convinced of the chemiosmotic hypothesis because of a particularly elegant demonstration of the separateness of proton pumping and ATP synthesis. The convincing experiment started with photosynthetic bacteria from the salt flats near San Francisco Bay. In these bacteria, a purple protein called *bacteriorhodopsin* pumps protons across the bacterial membrane in response to light, as part of their process of photosynthesis. Even when researchers put this protein into artificial membrane vesicles, light stimulates the purple protein to pump protons into the vesicles. This proton pumping establishes a pH gradient between the inside and the outside (Figure 6-16).

To these artificial vesicles researchers then added the F_0-F_1 protein complex isolated from the mitochondria of a cow's heart (Figure 6-16). The F_0-F_1 complex allowed protons to flow back out of the vesicles and generate ATP. This experiment demonstrated that the F_0-F_1 complex can harness the energy of proton flow even in an artificial situation, with separate components from two extraordinarily different sources—a rare microscopic bacterium and a familiar large animal.

BOX 6.2

The Metabolism of Alcohol

For thousands of years and in many societies, ethanol has provided a popular means of changing mood and behavior. In heavy drinkers, however, ethanol can cause permanent damage to cells in the liver, nerves, brain, and heart. The metabolic pathways described in this chapter help us to understand how this happens.

In humans, liver cells are practically the only cells that can metabolize ethanol. In these cells, NAD^+ oxidizes ethanol, first to acetaldehyde and then to acetic acid, generating abundant amounts of NADH. The NADH then donates its electrons to the electron transport chain. Oxidative phosphorylation then occurs, but without the participation of the citric acid cycle. Sustained use of large amounts of alcohol leads to strange-looking mitochondria, the result of active oxidative phosphorylation and the shut-off of the citric cycle. In addition, carbohydrates that would ordinarily enter the citric acid cycle are instead converted to fat. The fat is secreted by the liver into the blood in little vesicles (produced by the endoplasmic reticulum).

Alcohol accumulates in the blood because it moves from the intestines into the blood more rapidly than the body can eliminate it. The blood carries the alcohol to cells throughout the body. Nerve and brain cells are particularly vulnerable to damage by alcohol. Heavy alcohol use irreparably damages heart muscle after about 10 years in men, sooner in women.

However, the most common organic disease among heavy drinkers is liver failure. After a few years, the liver cells begin to fill with fat and to cease functioning. Eventually, the liver becomes heavily scarred—a condition called *cirrhosis* (Figure A). Once the damaged liver loses its ability to metabolize alcohol and other toxins, damage to the body accelerates. Cirrhosis of the liver is the seventh leading cause of death in the United States.

Our ability to metabolize alcohol demonstrates how adaptable animals are. Alcohol is just one of hundreds of chemicals from which we can obtain energy. No matter what we eat, biochemical homeostasis keeps the cells supplied with ATP and reducing power. Sources of energy are rarely wasted; they are used or stored. But animals also have evolved within a relatively narrow set of conditions. We cannot tolerate too large a variation of diet for an extended period without damage to the cells of our body.

The chemiosmotic hypothesis explains how cells transfer the energy from the reduced coenzymes NADH and $FADH_2$ to ATP: (1) the electron transport chain creates an electrochemical gradient across the inner mitochondrial membrane, and (2) the F_0-F_1 complex couples the flow of protons down the gradient to the synthesis of ATP.

The main point is that the creation of the proton gradient and the generation of high-energy bonds in ATP are separate processes. The experiments with *bacteriorhodopsin* and cow F_0-F_1 complex support this idea and further suggest that this strategy for ATP synthesis is nearly universal. In Chapter 7, we will see that proton gradients not only produce ATP in mitochondria but also produce ATP in chloroplasts during photosynthesis.

Proton pumping and ATP synthesis are separate processes.

How Much Usable Energy Can a Cell Harvest from a Molecule of Glucose?

Actual measurements in isolated mitochondria show that the oxidation of each molecule of NADH leads to the movement of 10 hydrogen ions across the inner membrane and to the synthesis of 2.5 molecules of ATP. Similar experiments show that each $FADH_2$ can yield 1.5 molecules of ATP. For each molecule of glucose, then, the NADH produced in the mitochondria can yield 20 molecules of ATP, and the $FADH_2$ yields 3 molecules of ATP (Figure 6-17). The actual yield of ATP is somewhat less than this because the mitochondrion also uses the energy of the proton gradient to drive the transport across the mitochondrial membrane of other molecules and ions, including phosphate.

Glycolysis produces two more molecules of NADH. These are made in the cytoplasm and do not readily cross the inner mitochondrial membrane. The two molecules of NADH made during glycolysis produce either three or five molecules of ATP, depending on the tissue.

Finally, two molecules of ATP are produced in glycolysis and two molecules of GTP in the mitochondria. Since the high-energy phosphates of the GTPs readily produce ATP, we have an additional four molecules of ATP per molecule of glucose.

Cellular respiration thus produces 30 or 32 molecules of ATP for each molecule of glucose. (Until recently, biochemists estimated that each molecule of glucose yielded 36 to 38 molecules of ATP because it was believed that each NADH yielded 3 molecules of ATP, rather than 2.5, and each $FADH_2$ yielded 2 molecules of ATP, instead of 1.5.) All but four of the ATP molecules produced from a glucose molecule are produced by oxidative phosphorylation. In the absence of oxygen, however, glycolysis alone can produce only two molecules of ATP.

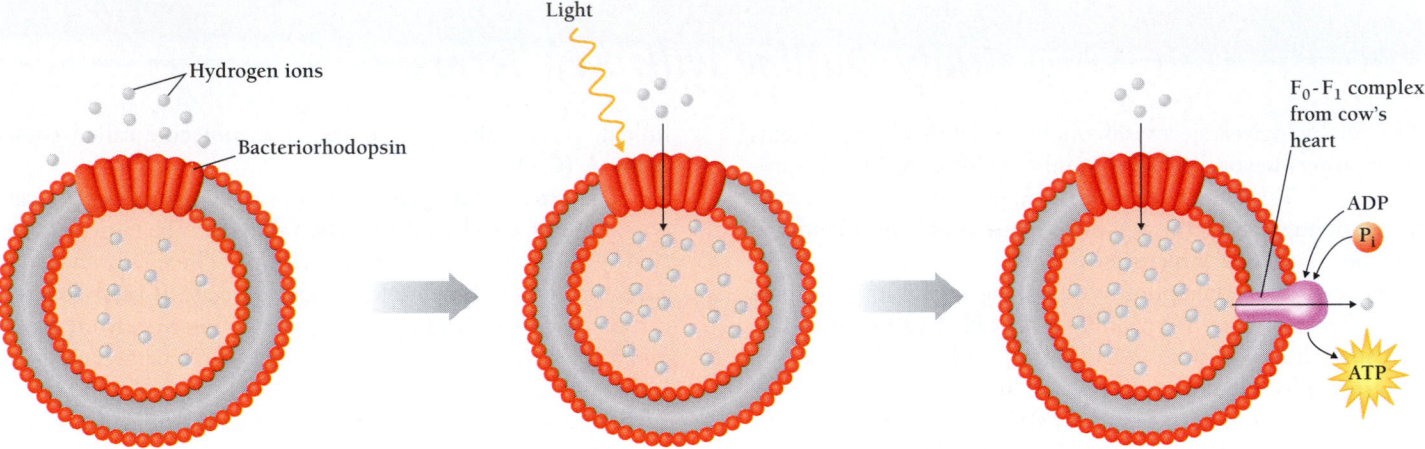

Figure 6-16 Proof that a proton gradient can drive ATP synthesis. Even an isolated F_0-F_1 protein complex can generate ATP, as long as it is embedded in a membrane spanning a proton gradient. In this experiment, researchers took an F_0-F_1 protein complex isolated from the heart muscle of a cow and placed the F_0-F_1 complex into the membrane of a photosynthetic bacterium. A purple protein called *bacteriorhodopsin* pumped protons across the bacterial membrane in response to light. But the F_0-F_1 complex allowed protons to flow back out of the vesicle and generate ATP.

The ATP molecules produced in the mitochondria travel across the mitochondrial membranes to the cytoplasm, where they can be used. Once their high-energy phosphate bond is broken, ADP molecules return to the mitochondria to be reconverted to ATP. Within the mitochondria, the rates of the citric acid cycle and oxidative phosphorylation depend on the relative amounts of ATP and ADP. Because of such regulation, the body catabolizes sugars and fats five to ten times more rapidly during exercise than when at rest.

The process by which energy is extracted from glucose under cellular conditions is highly efficient. A gasoline engine has an efficiency of only 10 to 20 percent. That is, only 10 to 20 percent of the energy in the gasoline is converted to forward motion. In contrast, cells extract energy from glucose with a phenomenal efficiency of 50 percent. It is hard to think of another process with such high efficiency.

Cellular respiration is a highly efficient process for extracting energy—in the form of 30 or 32 molecules of ATP—from glucose.

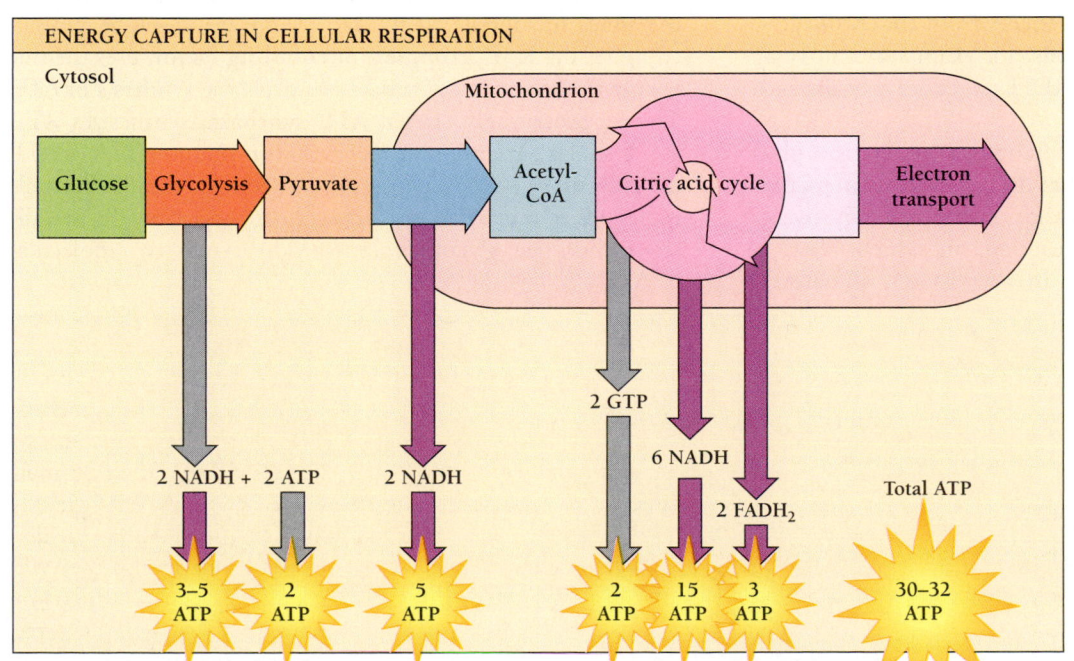

ENERGY CAPTURE IN CELLULAR RESPIRATION

Figure 6-17 Energy capture in cellular respiration. For each glucose molecule, glycolysis produces two ATP and two NADH molecules, and the production of acetyl-CoA yields another two NADHs. The citric acid cycle then yields two GTP, six NADH, and two $FADH_2$ molecules. The electron transport chain (represented by the downward purple arrows at the bottom of the figure) captures the energy of NADH and $FADH_2$ in 26 to 28 ATPs.

Study Outline with Key Terms

Modern biologists reject **vitalism,** the belief that living systems have powers beyond those of nonliving systems. But the production of ATP by the **mitochondria** presents an excellent example of how the chemical transformations of living organisms depend on cellular structures.

Plants and other photosynthesizing organisms that produce their own food from sunlight are called **autotrophs.** Organisms that get their energy from other organisms are **heterotrophs.** Heterotrophs prepare for cellular respiration by capturing macromolecules and **digesting** them into their component building blocks. Digestion itself does not produce energy.

Almost all eukaryotic cells (and many prokaryotic cells as well) are **aerobic,** producing most of their ATP by **cellular respiration.** Most energy-producing reactions in cells are **reduction-oxidation (redox) reactions,** which transfer electrons from the **reducing agent** to the **oxidizing agent.**

Cellular respiration transfers 24 electrons from glucose to oxygen. For 20 of these electrons, the first acceptor is NAD^+, which can accept two electrons and a hydrogen atom to give its reduced form **NADH.** The other four electrons of glucose are transferred to **FAD,** whose reduced form is $FADH_2$. These reduced **coenzymes**—the major energy storage molecules in energy metabolism—ultimately transfer their electrons to oxygen in a series of steps. In eukaryotes, the steps occur within the inner mitochondrial membrane. In prokaryotes, these reactions occur in association with the plasma membrane.

Glycolysis is the first stage in the metabolism of glucose. Glycolysis is a set of ten chemical reactions that breaks glucose into two molecules of **pyruvate.** These ten reactions also convert some of the energy of glucose into two high-energy ATP bonds.

Many microorganisms obtain energy from **anaerobic** glycolysis, or **fermentation.** Yeast cells, for example, can avoid the need for oxygen by using NADH to reduce pyruvate to form ethanol.

In eukaryotic cells, energy metabolism (other than glycolysis) takes place in the mitochondria. Pyruvate enters the mitochondria, where cellular respiration proceeds. Pyruvate then undergoes an oxidation reaction, in which NAD^+ serves as the electron acceptor. In acetyl-CoA, acetate is linked, via a high-energy bond, to a molecule called **coenzyme A (CoA).**

The **citric acid cycle,** or **Krebs cycle,** converts the carbon atoms of acetyl-CoA to carbon dioxide. The citric acid cycle is a nearly universal set of reactions. The citric acid cycle adds the four-carbon molecule oxaloacetate (OAA) to two-carbon acetyl-CoA to form a six-carbon molecule, from which two carbons are removed in succession. OAA then reenters the cycle once more.

The citric acid cycle, which not only provides energy but also serves as the major route for the synthesis of many small molecules, lies at the center of the complex network of biochemical conversions that collectively are called **metabolism.** Metabolism is divided into two classes of pathways: **catabolism** and **anabolism.**

Oxidative phosphorylation couples the oxidation of NADH and $FADH_2$ to the production of high-energy phosphate bonds in ATP. The electrons from NADH and $FADH_2$ reach oxygen, their ultimate acceptor, through the **electron transport chain** (or respiratory chain), which consists of a series of **electron carriers.** Five of the electron carriers are **cytochromes,** whose reddish color derives from **heme.**

Oxidative phosphorylation operates by a mechanism (called **chemiosmosis**) summarized in the **chemiosmotic hypothesis:** (1) the flow of electrons through the electron transport chain pumps protons out of the **mitochondrial matrix,** establishing an **electrochemical gradient** across the mitochondrial membrane; and (2) molecular machinery in the membrane captures the energy, as protons flow back into the mitochondrial matrix.

The orientation of the electron carrier complexes establishes the direction of the **proton gradient.** As the protons flow down the electrochemical gradient through channels in the inner membrane, their free energy decreases. A protein complex, the F_0-F_1 **complex,** or **coupling factor,** uses the free energy from the flow of protons to drive the synthesis of ATP. The F_1 protein, also called **ATP synthase,** synthesizes ATP.

Cellular respiration produces 30 (or 32) molecules of ATP for each molecule of glucose. All but four of these ATP molecules are produced by oxidative phosphorylation. In the absence of oxygen, glycolysis alone produces only two molecules of ATP.

Review and Thought Questions

Review Questions

1. Write down a reaction that describes the following situation: substance A carries a net charge of -1, that is, it has one extra electron, while substance B is electrically neutral. During a reaction, the extra electron spontaneously moves from A to B, so that A becomes neutral and B acquires a negative charge. Indicate each of the following: the oxidized substance, the reduced substance, the oxidizing agent, the reducing agent, the more electronegative substance, and the substance with lower affinity for electrons.

2. Name the process in which most of the energy from glucose is converted to ATP. What role does oxygen play in this process? What gas is produced as a waste product? Where in a cell do these reactions occur?

3. Why is the electron transport pathway divided into so many small reactions?

4. Across what eukaryotic membrane is a gradient established during cellular respiration? What is unequally distributed across these membranes? What caused this unequal distribution?

5. Describe and name three stages that aerobic organisms use to extract energy from glucose.

6. How many carbon atoms does a glucose molecule have? Name the end product molecule that is produced during glycolysis. How many carbon atoms does it have?

7. How many ATPs are produced in the citric acid cycle? How does this number compare with that produced in glycolysis?

8. Define metabolism, anabolism, and catabolism, and give examples of catabolic and anabolic pathways. Which type of metabolism is most favored when a cell (a) has abundant ATP? (b) contains little ATP?

Thought Questions

9. Why might you lose weight (a lot of weight) if your mitochondria suddenly lost the ability to couple electron transport to the production of ATP?

10. When you exercise heavily, you consume oxygen faster than you can take it in through your lungs and circulatory system, so your cells have to rely at least partly on anaerobic processes to extract energy from food molecules. Which of the three stages in energy extraction can your cells not perform under these conditions? How much ATP can you form from food molecules under these conditions?

About the Chapter-Opening Image

Louis Pasteur proved that the fermentation of wine and beer depends on the presence of microorganisms (yeasts). Twentieth-century biochemists figured out just what the yeasts were doing.

 On-line materials relating to this chapter are on the World Wide Web at **http://www.harcourtcollege.com/lifesci/aal2/**

7

Photosynthesis: How Do Organisms Get Energy from the Sun?

The Chemical Evangelist

Early in the 17th century, a Belgian alchemist named Johann Baptista van Helmont (1579–1646) performed a simple and dramatic experiment that scientists have been admiring for nearly 400 years. Although van Helmont dared not publish his results, his studies laid a firm foundation for the modern understanding of **photosynthesis** [Greek, *photos* = light + *syntithenai* = to put together]—the process that plants and other organisms use to transform the energy of light into the chemical bonds of sugars.

Van Helmont was a man who, even in our time, would be considered highly unusual. He was deeply religious, nearly fanatical in his beliefs. Yet he was a born skeptic, always questioning the beliefs of those around him. It was an attitude that would bring upon him the cold fury of the Spanish Inquisition.

As a teenager, van Helmont found his education at the Catholic University of Louvain so disappointing that he refused the title "Master of Arts" because he said he had learned nothing substantial or true. On leaving the university, he then refused a well-paid position as a clergyman because, he said, he opposed "living on the sins of the people." For a while, he tried living the austere life of a Christian Stoic, until the hardship made him ill.

Perhaps illness inspired the young van Helmont, for he then began to study medicine, with particular emphasis on herbal remedies. Again he was disappointed, however. The herbalists, he complained, had discovered nothing new about medicine since ancient Greek times, and argued endlessly over details that had nothing to do with making people well. When van Helmont was 20 years old, he accepted a medical degree from Louvain. Yet he never practiced medicine, believing that he had learned nothing that would actually help anyone who was sick.

His new interest was alchemy, a result, he later wrote, of a call from God. (Some called him the "chemical evangelist.") He had frequently claimed to detest money, but at 28 he married a woman so wealthy that he was freed from ever having to work for a living again. For 15 years, he happily pursued his own researches into alchemy, religious philosophy, medicine, and natural science.

He found, for example, that humans digest food with acid and bile. He recognized gas as a form of matter, defining it as the volatile part of a substance, distinct from the mixture of gases we call air. He recognized that the lung is an organ for gas exchange. By weighing reactants and products, he showed that matter is conserved during chemical reactions.

Van Helmont wrote profusely of his discoveries and investigations, never hesitating to criticize the thinking of others. He especially singled out physicians—for their ignorance and superstition—and the Jesuits, a powerful order of Catholics—for what he viewed as their ignorance and arrogance.

Van Helmont was smart enough not to publish his offensive tirades against physicians and Jesuits. But, eventually, one of his many enemies published one of the worst. The faculty at the Louvain medical school in 1623 denounced the pamphlet as "monstrous." And 2 years later, the Spanish Inquisition declared that 27 statements in van Helmont's manuscript were almost certainly heretical. Between

- Photosynthesis is the process by which plants use energy from the sun to build sugars from carbon dioxide and water.

- Specialized photosynthetic machinery absorbs light energy, transforms it into the chemical energy of NADPH and ATP, and then stores it in the chemical bonds of sugar.

- Photosynthesis entails two light-dependent reactions—photosystem I and photosystem II.

1624 and 1642, the Spanish Inquisition imprisoned and interrogated van Helmont, then locked him up in a convent, and finally confined him to house arrest for many years. During all of this time, van Helmont published nothing. Yet he continued to work and, equally important, to write.

Much of van Helmont's work was directed at undermining the teachings of the Jesuits. The Jesuits based their teachings on the writings of Aristotle, the most influential of the Greek philosophers. Aristotle had argued that plants derive their sustenance from the soil. A plant's roots, Aristotle said, extract materials from soil in much the same way that an animal's stomach extracts materials from its food. This view seemed logical: farmers had long known that even the richest soils eventually lose the ability to support luxuriant plant growth.

But van Helmont knew Aristotle was wrong. Anyone could see that a potted plant didn't use much soil; it used water. Besides, van Helmont believed that all natural bodies come from water. To knock Aristotle from his pedestal and bolster his own theory, van Helmont performed a now famous experiment (Figure 7-1). He planted a 5-lb willow sapling in 200 lb of soil, which he had carefully dried in a furnace, weighed, and then placed in a large pot. He covered the pot with an iron plate to keep dust from falling into it and watered the growing tree as needed. He did not weigh the leaves that fell from the tree each autumn. After 5 years, he again weighed both the tree and the soil. The willow tree now weighed 169 lb, while the soil weighed only 2 oz less than the starting 200 lb. Van Helmont concluded, "164 pounds of wood, bark, and roots arose out of water alone."

Van Helmont's experiment was excellent, but his interpretation was partly mistaken. His experiment did prove that Aristotle (and the Jesuits) were wrong: the tree's mass could not have come from the soil. However, van Helmont himself was also wrong in assuming that his experiment showed that the tree came entirely from water. It was ironic that van Helmont, the discoverer of gas, never guessed that something in the air might also have contributed to the growth of his willow tree. Only years after he died did other scientists discover the role of gas in plant growth.

Young willow tree: 5 lbs

Soil: 200 lbs

Only water is added

5 years later

Tree: 169 lbs

Soil: 199.8 lbs

◀ **Figure 7-1 Plants are made of water and air.** Van Helmont's willow tree experiment showed that the material that makes up plants does not come from the soil. After 5 years, the willow tree gained 164 lb, but the soil lost only a few ounces.

HOW DO WE KNOW HOW PLANTS OBTAIN CARBON AND OXYGEN?

The first suggestion that the atmosphere contributed to the stuff of plants came not from chemists or alchemists, but from microscopists. With the new microscopes developed in the late 17th century, an Englishman and an Italian almost simultaneously discovered minute pores in the leaves of plants, called **stomata** [Greek, *stoma* = mouth]. Stomata appeared to allow air to pass to the interior of the leaves (Figure 7-2). The activities of plants, they speculated, must somehow depend on interactions with air. By the mid-18th century, scientists had concluded that "vegetation is planted in the air as well as in the earth."

What Do Plants and Air Do for Each Other?

The next big clue to the interaction between plants and air came from a series of experiments in which a mouse or a candle was placed in an airtight container. After some time, the mouse would die from lack of oxygen or the candle would go out for the same reason. This was because the mouse, or the burning candle, consumed the oxygen in the container.

In the 1770s, the English clergyman Joseph Priestley performed an experiment that showed that a sprig of mint could restore the air in the jar and keep a mouse alive. Priestley showed that plants put back what living animals (or burning candles) take out (Figure 7-3). That "something" that plants release is the gas oxygen. Around the same time, the French chemist

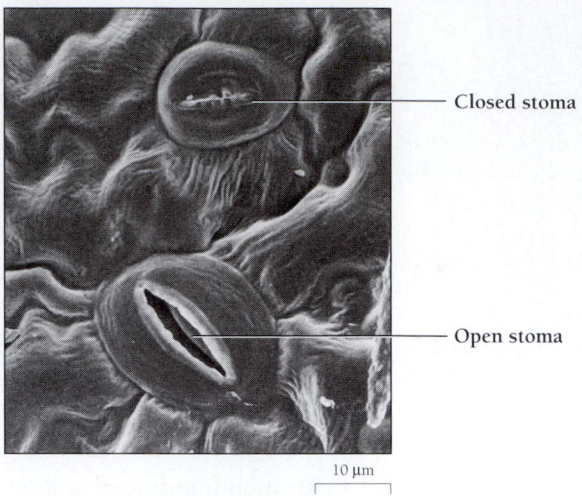

Figure 7-2   Stomata from the underside of a leaf. Tiny pores in the surfaces of leaves called stomata open and close depending on whether plants need to conserve water or collect carbon dioxide. The discovery of stomata in the surfaces of leaves suggested that plants need gases from the air as well as water from the ground. Scientists concluded that "vegetation is planted in the air as well as in the earth." (©*Dwight Kuhn*)

Antoine Lavoisier independently discovered oxygen, so credit for the discovery usually goes to both men (Figure 7-4).

In the dark, plants act just like animals: they consume oxygen and release carbon dioxide (Figure 7-5A). In light, however, plants put back the oxygen and remove something very different—carbon dioxide. During the day, plants remove carbon dioxide from the air and release oxygen as a waste product (Figure 7-5B).

Today we know that during photosynthesis plants take in water and carbon dioxide with which they synthesize the simple sugar glucose. To a large extent, then, plants are made of water and air. However, as we saw in Chapter 3, proteins and most other biological molecules also include atoms of nitrogen, phosphorus, and other elements. These come, in small quantities, from the soil. Indeed, the 2 oz of soil that van Helmont measured most likely included nutrients essential to the growth of the willow tree.

In this chapter, we will ask how cells accomplish photosynthesis. How does photosynthesis capture light energy? How does it store the captured energy? What molecules and subcellular structures participate?

Figure 7-3   What did Priestley's experiment show? Plants put something into air that mice need. A. The mouse in the jar uses up the oxygen and fills the jar with carbon dioxide, which suffocates the mouse. B. A photosynthesizing plant placed in the jar takes up the carbon dioxide released by the mouse and replaces it with oxygen. The mouse is fine.

During photosynthesis, plants take carbon dioxide from the air and water from the soil and release oxygen back into the air.

Figure 7-4 Joseph Priestley. Even as modern chemistry took its first breath, its greatest exponents—Joseph Priestley and Antoine Lavoisier—were caught up in the turmoil surrounding the American and French revolutions. In 1791, Priestley's sympathies for the French Revolution inspired an angry mob to destroy his home, library, and laboratory. Within 3 years, he fled to the United States. In the same year, on the other side of the English Channel, the nobleman Lavoisier, who had served King Louis XVI, was guillotined. His executioners proclaimed, "The Republic has no need for scholars," and threw his body into a common grave. (*The Granger Collection, New York*)

Where Do the Atoms Go in Photosynthesis?

Now that we know the raw materials of photosynthesis, we can summarize the overall chemical transformation in the following equation:

$$6 \, CO_2 + 6 \, H_2O \longrightarrow C_6H_{12}O_6 + 6 \, O_2$$

carbon dioxide + water $\longrightarrow$ glucose + oxygen

Notice that this is the exact reverse of the overall equation for respiration given in Chapter 6 on page 121. Another way of writing the equation for photosynthesis is:

$$CO_2 + H_2O \longrightarrow C(H_2O) + O_2$$

carbon dioxide + water $\longrightarrow$ carbohydrate + oxygen

Here, $C(H_2O)$ is the general formula for a carbohydrate, with each carbon atom associated with the equivalent of one molecule of water.

To early 20th-century chemists, this equation suggested a hypothesis for how photosynthesis might occur: light splits the carbon dioxide into oxygen and carbon. Then oxygen is expelled into the atmosphere, and the carbon combines with water to form carbohydrate. Although this idea is appealing, it is wrong.

The first realization that this view of photosynthesis was wrong came from studies of photosynthetic bacteria. In the early 1930s, a graduate student at Stanford University named Cornelius van Niel was examining photosynthesis in bacteria

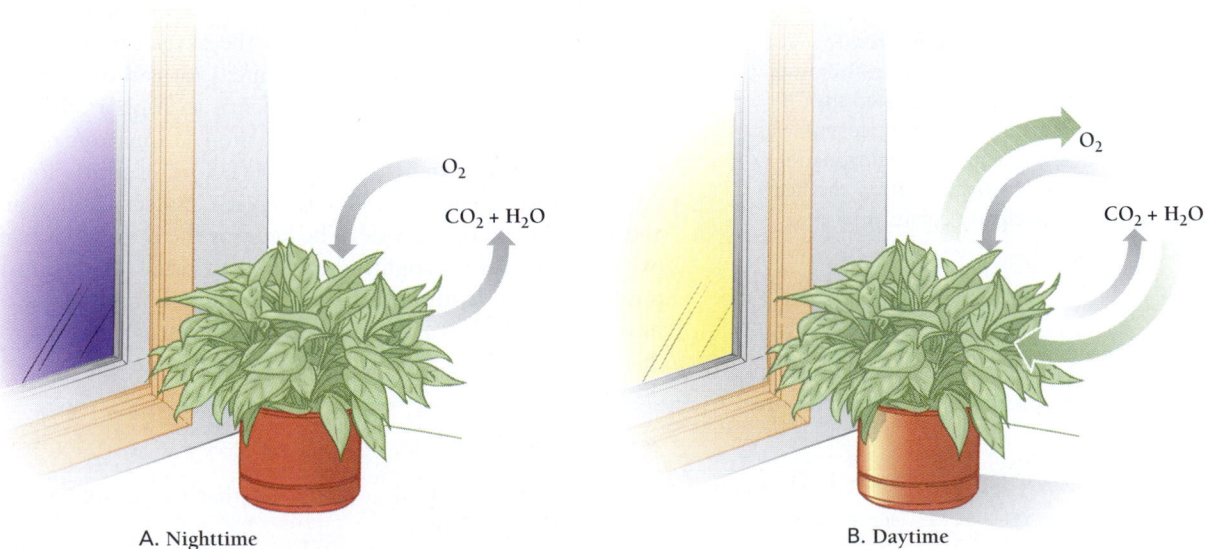

A. Nighttime

B. Daytime

Figure 7-5 Photosynthesis A. In the dark, plants respire only. Like animals, they use oxygen to break down sugars. B. In light, plants also photosynthesize, using solar energy to make sugar from carbon dioxide and water.

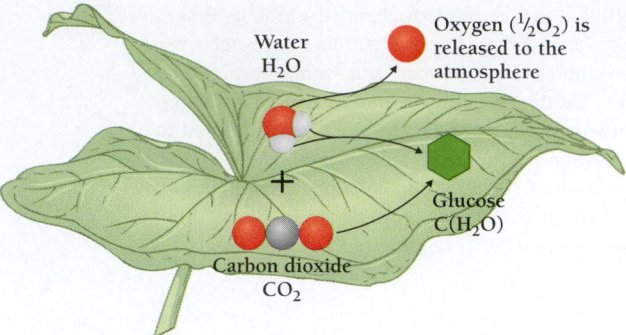

A. Photosynthesis in plants

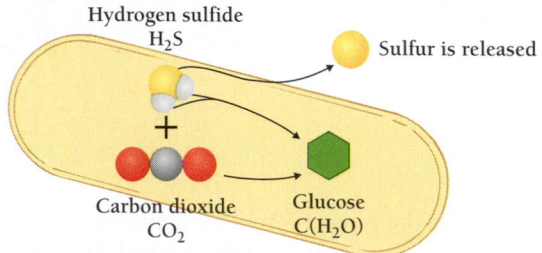

B. Photosynthesis in bacteria

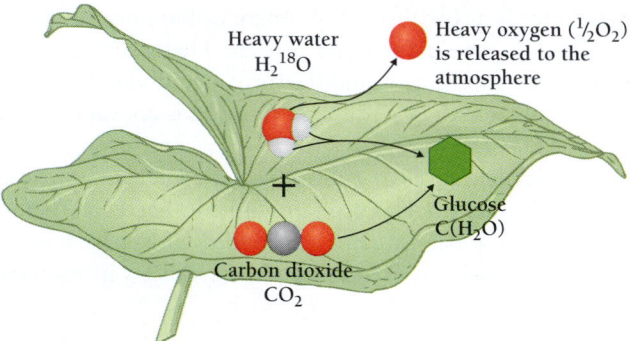

C. Photosynthesis in plants using heavy water

Figure 7-6   Where do the oxygen atoms come from? A. The oxygen released by plants during photosynthesis comes from water, not from carbon dioxide. Two experiments suggested this: B. First, the sulfur released by certain bacteria during photosynthesis comes from hydrogen sulfide, which is chemically analogous to water. C. Second, in plants supplied with "heavy water," the heavy oxygen released during photosynthesis clearly comes from the water.

that used hydrogen sulfide (H_2S) instead of water (H_2O). Instead of producing oxygen as a by-product, these bacteria produce sulfur (Figure 7-6).

$$CO_2 \;+\; 2\,H_2S \;\longrightarrow\; C(H_2O) \;+\; 2\,S \;+\; H_2O$$

carbon dioxide + hydrogen sulfide ⟶ carbohydrate + sulfur + water

Van Niel reasoned that these bacteria probably used the same biochemical pathways as other photosynthesizing organisms. Water (H_2O) and hydrogen sulfide (H_2S) probably served the same biochemical role. Yet, the sulfur produced by

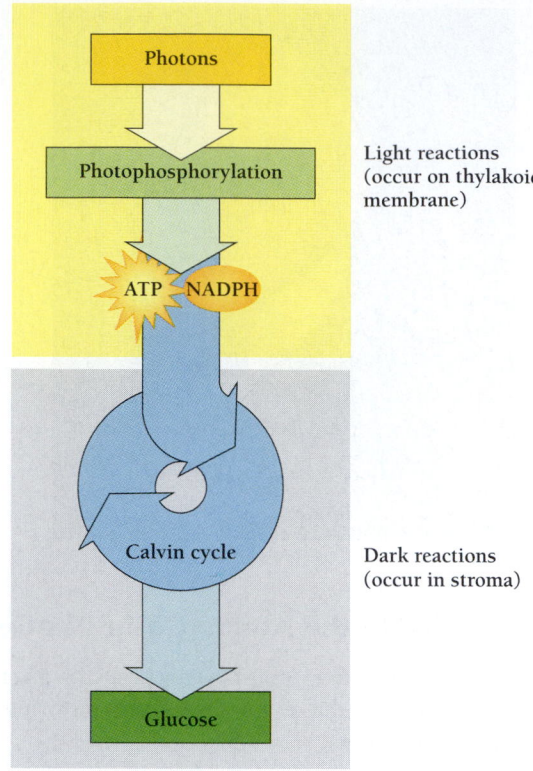

Figure 7-7   The two parts of photosynthesis. The light-dependent reactions, called photophosphorylation, generate energy-transfer molecules such as ATP and occur in the thylakoid membrane of the chloroplast. The light-independent, or "dark," reactions use the energy packaged in the light reactions to make glucose from carbon dioxide. These reactions occur in the stroma, outside the thylakoid membrane.

photosynthetic bacteria clearly comes from splitting hydrogen sulfide. It stood to reason, then, that the oxygen produced by plants might come from splitting water, not carbon dioxide.

However, van Niel had no evidence to support this hypothesis. No one knew which oxygen atoms came from which molecules. More than 10 years passed before scientists found a way to distinguish between the oxygen molecules in water and those in carbon dioxide. The key was the isotope ^{18}O, an oxygen atom that is slightly heavier than usual. Biochemists provided plants with normal carbon dioxide and so-called heavy water, $H_2^{18}O$. The plants split the heavy water and produced heavy oxygen, but the carbohydrates were normal.

$$CO_2 \;+\; 2\,H_2^{18}O \;\longrightarrow\; C(H_2O) \;+\; H_2O \;+\; ^{18}O_2$$

carbon dioxide + water ⟶ carbohydrate + oxygen

This experiment showed conclusively that plants split water, not CO_2 (Figure 7-6C).

The oxygen released by plants comes from water, not from carbon dioxide.

Biochemists now had a clear understanding of what raw materials plants used to synthesize glucose. But two difficult questions still remained: How do plants channel light energy into this chemical reaction? And how do plants use that chemical energy to synthesize sugar?

Photosynthesis consists of two parts. In the first part, energy from sunlight powers the addition of a phosphate group to ADP to form ATP. Adding a phosphate group is called "phosphorylation." Because light comes in packets of energy called "photons," using light energy to phosphorylate something is called **photophosphorylation.** Sometimes light energy is used to reduce $NADP^+$ to NADPH, another energy-storage molecule. Both kinds of reactions are called photophosphorylation, or **light-dependent reactions.**

The second part of photosynthesis uses the energy from the light reactions (ATP and NADPH) to make glucose. Because these reactions use no photons and can occur in the dark, they are sometimes called the "dark reactions." But the dark reactions do not have to occur in darkness; they just do not require an immediate supply of photons. The dark reactions are more correctly termed **light-independent reactions.** In the rest of this chapter, we will review both the light reactions and the light-independent reactions (Figure 7-7).

HOW DO PLANTS COLLECT ENERGY FROM THE SUN?

Julius Mayer, one of the discoverers of the First Law of Thermodynamics, elegantly summarized the business of photosynthesis: "Nature has put itself the problem of how to catch in flight light streaming to the earth and to store the most elusive of all powers in rigid form."

What Is Light?

In the morning, sunlight streams in your window, shining on the curtains, the furniture, and the floor. All of these objects absorb sunlight and transform it into heat. In photosynthesis, plant cells absorb light and transform it into chemical energy. To understand how plants use the energy in light, we need to understand what light is.

Light is the energy of particles, called **photons,** that travel at the speed of light (186,000 miles per second or 3×10^8 meters per second). Like other moving particles, photons travel in straight lines. They can hit a surface and bounce off at an angle (Figure 7-8). Light can also vary in intensity. The intensity of light shining on the pages of this book corresponds to the number of photons hitting the page each second. If you turn on a bright light, or even six dim ones, you will increase the total number of photons hitting the page.

Photons all travel at the same speed, but they nevertheless have different amounts of energy. How much energy depends on their wavelength, for light behaves like a wave as well as a particle. Light has troughs and crests that interfere with

Figure 7-8 **Light is made of particles.** Particles of light called photons move in straight lines and bounce off surfaces in the same way as marbles or billiard balls. *(Leonard Lessin/Peter Arnold, Inc.)*

one another, just like the ripples in a rain puddle (Figure 7-9A and B).

One of the most important properties of light is its **wavelength**—the distance between successive crests (Figure 7-9C). Light energy with a long wavelength has long troughs and crests; light with a short wavelength has short troughs and crests. The shortest wavelengths are measured in nanometers (nm), each of which is a billionth of a meter. Wavelengths vary in length from as short as a few nanometers, which we call

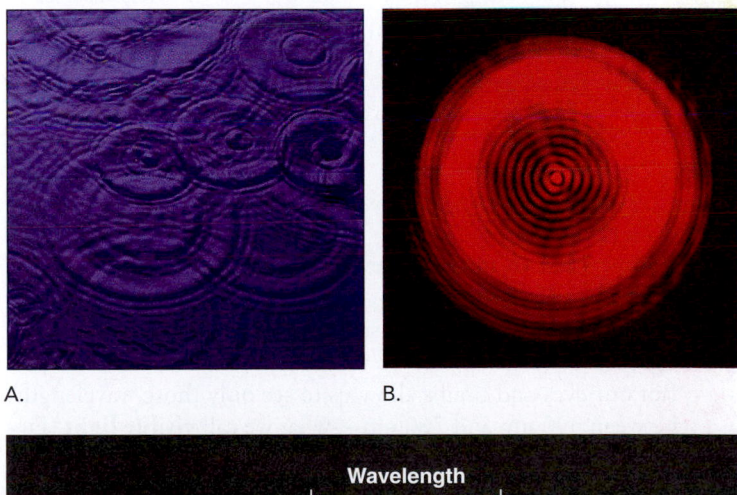

A.   B.

C.

Figure 7-9 **Light is a wave.** A. The circular ripples created by raindrops on a pond make a pattern of troughs and crests. B. Light also forms waves with crests and troughs. C. Wavelength is the distance from the crest of one wave to the next. Light with a long distance from one crest to the next (long wavelength) is red. Light with a short wavelength is blue. *(A, Runk/Schoenberger from Grant Heilman; B, Yoav Levy/Phototake)*

A. Sunlight is a mixture of many wavelengths

Figure 7-10   Sunlight. A. Sunlight is a mixture of many wavelengths of light, from infrared to ultraviolet. B. Irises, tiger lilies, and other living organisms have evolved spectacular pigments that absorb some wavelengths of light and reflect others.

Light

All colors absorbed except purple

All colors absorbed except orange

B.

gamma rays, to meters or even kilometers, which we call radio waves.

Most light, including sunlight and the light from most light bulbs, is a mixture of many wavelengths (Figure 7-10). But our eyes and brains allow us to see only those wavelengths between 390 nm and 760 nm—what we call **visible light** (Figure 7-11). Within this narrow range, we can distinguish all the different wavelengths of the rainbow. The shortest wavelengths of light appear to us as blue or purple, and the longest wavelengths appear red. For example, we see 400-nm light as violet, 500-nm light as blue-green, and 600-nm light as orange-red.

A photon is a package of energy. Unlike an electron or a proton, a photon has no mass. It is pure energy. The amount of energy in a photon depends on its wavelength. Photons of shorter wavelengths (more bluish) have more energy than photons of longer wavelengths (more reddish). Blue light, for example, has about twice as much energy per photon as red light. The energy, wavelength, and color of light are all measures of the same thing. (But light intensity, how bright the light seems to us, is differ-

ent. Light intensity is a measure of numbers of photons, not the energy, or wavelength, of the individual photons.)

Plants photosynthesize by taking advantage of the way light interacts with matter. An atom or a molecule can absorb a photon of light if the energy of the photon matches the energy needed to boost one of its electrons to a higher energy level. Our own vision is the result of visible light exciting electrons in special molecules in our eyes. Visible light can also alter the chemicals in photographic film to create beautiful photographs. A photon can also induce an electrical signal, for example, in a video camera or the photoelectric cell of a solar collector.

In photosynthesis, light excites electrons in special molecules, increasing their free energy. The light and "dark" reactions capture some of this energy and convert it into high-energy chemical bonds.

The wavelength of a photon is a measure of its energy. Short-wavelength photons have high energy, and long-wavelength photons have low energy. Photons with the right amount of energy (the right wavelength) can increase the free energy of molecules by exciting their electrons.

How Can We Tell Which Molecules Help with Photosynthesis?

Much of the light from the sun does not reach the Earth's surface (Figure 7-12). High-energy ultraviolet ("beyond violet") light is absorbed by ozone in the upper atmosphere, and low-energy infrared ("below red") is absorbed by water vapor and carbon dioxide in the atmosphere. Not surprisingly, living organisms have evolved molecules that take full advantage of the energy in visible light—the light that best penetrates our atmosphere. The energy contained in a visible photon stimulates the chemical reactions of both photosynthesis and our own eyes.

Different molecules absorb different wavelengths of light. For example, a red flower absorbs almost all the wavelengths

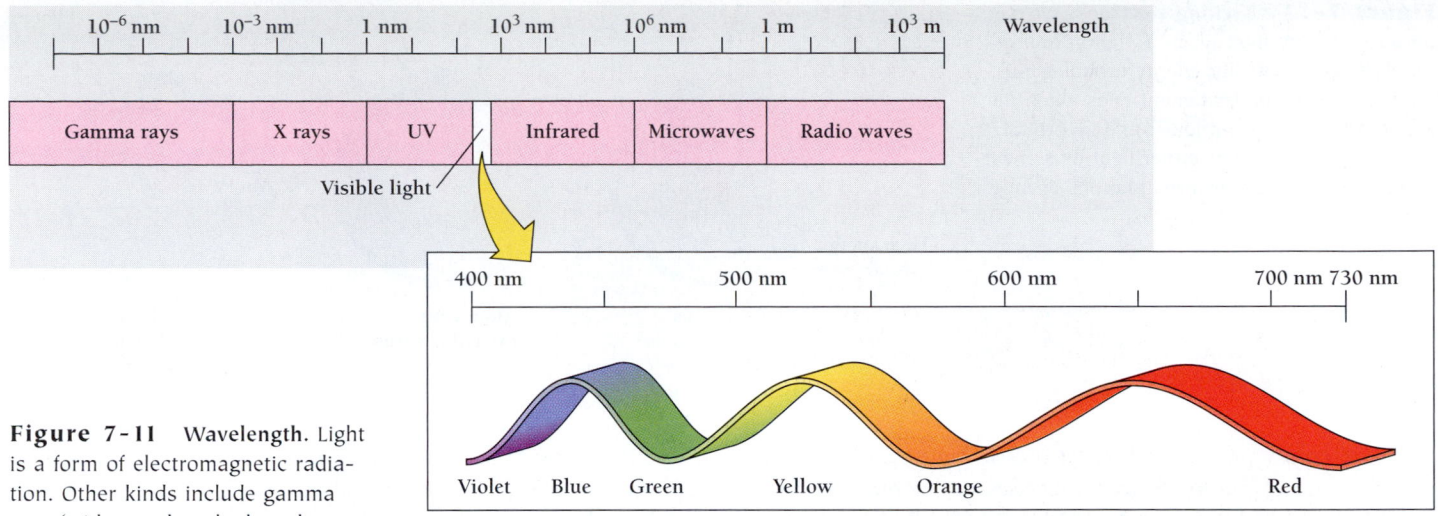

Figure 7-11 Wavelength. Light is a form of electromagnetic radiation. Other kinds include gamma rays (with wavelengths less than 1 nm) to radio waves (with wavelengths measured in meters and even kilometers). Visible light ranges from about 390 nm to about 760 nm. To our eyes, 400 nm light appears violet, 500 nm is blue-green, and 600 nm is orange-red.

in the visible range. The only ones the flower does not absorb are those that we see as red, which the flower's petals reflect into our eyes and elsewhere. Likewise, the green pigment in plants absorbs every color but green.

When any molecule absorbs a photon, the extra energy kicks an electron to a higher-energy orbital. The high-energy electron can return to the lower orbital by losing energy as another photon (fluorescence), or as heat. Alternatively, the high-energy electron may move to another molecule, leaving the

first molecule in an oxidized state. We will see that such a transfer of electrons drives photosynthesis.

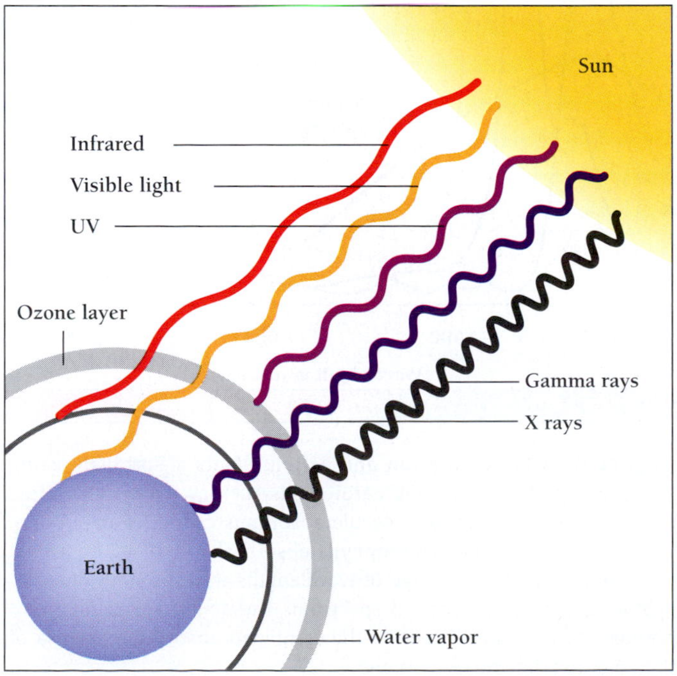

The particular set of wavelengths that a molecule absorbs is called its **absorption spectrum.** A molecule's absorption spectrum tells us how much energy the molecule needs to move an electron from one orbital to another (Figure 7-13).

Most of the molecules in an ordinary cell absorb ultraviolet light only. This is because the energy required to move an electron to a higher-energy orbital is usually quite high. But some molecules, called **pigments,** can absorb low-energy light from the visible spectrum. Any molecule that absorbs visible light is a pigment.

In plants, the most important molecule of photosynthesis is the green pigment **chlorophyll,** which absorbs every color but green and yellow. Plants use two kinds of chlorophyll—chlorophyll *a* and chlorophyll *b*. Each has a slightly different absorption spectrum. A solution of pure chlorophyll absorbs red and blue photons, while green and yellow photons pass through or are reflected. These unabsorbed photons excite pigment molecules in our eyes, and we perceive a green color.

We can look at plant pigments another way. Every light-dependent process has an **action spectrum**—the relative effectiveness of different wavelengths in promoting the process. In plants, for example, we can measure how well lights of

◀ **Figure 7-12 Our protective atmosphere.** Because the Earth's atmosphere absorbs most infrared light and reflects most ultraviolet light, most sunlight that reaches the Earth's surface is visible light. Organisms have evolved biochemical pathways that allow us to use this light to build complex molecules.

Figure 7-13   Exciting electrons. We say an atom *absorbs* light when a photon excites an electron to a higher energy orbital. This temporary state of excitement ends when the electron falls back to a lower energy orbital, releasing the energy as either light or as heat. A pigment molecule absorbs photons of only a certain range of wavelengths.

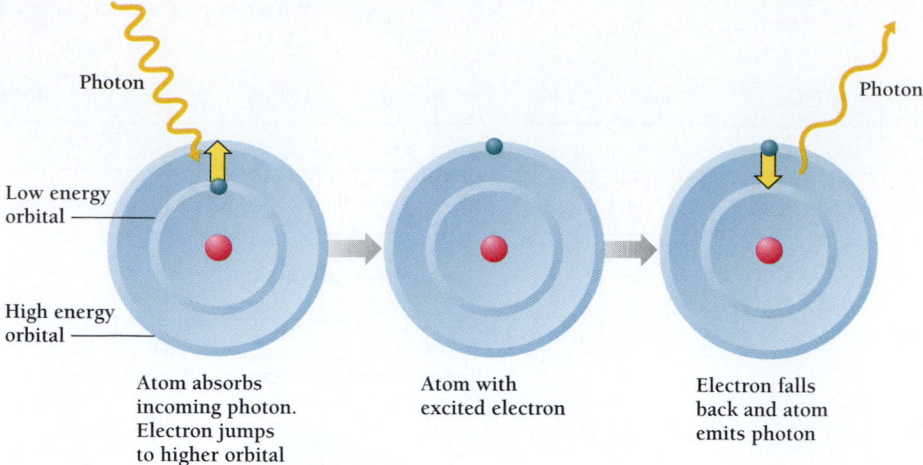

Photon

Low energy orbital

High energy orbital

Photon

Atom absorbs incoming photon. Electron jumps to higher orbital

Atom with excited electron

Electron falls back and atom emits photon

different colors promote photosynthesis. Figure 7-14 shows that the rate of photosynthesis increases under blue and orange-red light and decreases under green and yellow light. The action spectrum of photosynthesis roughly follows the absorption spectrum of chlorophyll. We might quickly conclude that the absorption of light by chlorophyll must be responsible for photosynthesis.

But, surprisingly, the action spectrum of photosynthesis does not correspond exactly to the absorption spectrum of chlorophyll *a*. Chlorophyll *a* does not absorb blue-green light (500 nm), yet blue-green light can stimulate photosynthesis because plants have other pigments that can take advantage of other wavelengths. These pigments include both chlorophyll *b* and the yellow and orange **carotenoids,** such as beta (β)-carotene (Figure 7-15). Carotenoids, such as β-carotene, work by transferring the energy of their excited electrons to chlorophyll.

Most chlorophyll molecules transfer energy the same way. Researchers have found that most of the pigment molecules in a leaf do not actually photosynthesize. Instead, hundreds of chlorophyll molecules associate with a few carotenoids to form a weblike **antenna complex,** which traps light energy. The pigments of the antenna complex then transfer energy to a **photochemical reaction center,** where a small number of chlorophyll molecules actually converted the captured light energy to chemical energy.

The action spectrum of photosynthesis corresponds to the absorption spectrum of chlorophyll and other pigments. Chlorophyll and other pigments trap and transfer light energy to the photochemical reaction center.

Why Do Plants Use Two Types of Reaction Centers?

We can now describe a fairly simple model of photosynthesis: (1) chlorophyll and carotenoid molecules absorb light; (2) these pigments transfer the captured energy to the reaction center; and (3) in the reaction center, the energy splits oxygen from water and forms chemical bonds.

If photosynthesis in plants were this simple, each absorbed photon would contribute equally to photosynthesis, which is indeed the case in photosynthesis by bacteria. Photosynthesis in plants, however, appears to operate by a more complicated mechanism.

Some of the wavelengths that are readily absorbed by plant pigments are inefficient at promoting photosynthesis. As the action spectrum in Figure 7-14 shows, when absorbed light has a wavelength longer than 680 nm, efficiency drops steeply.

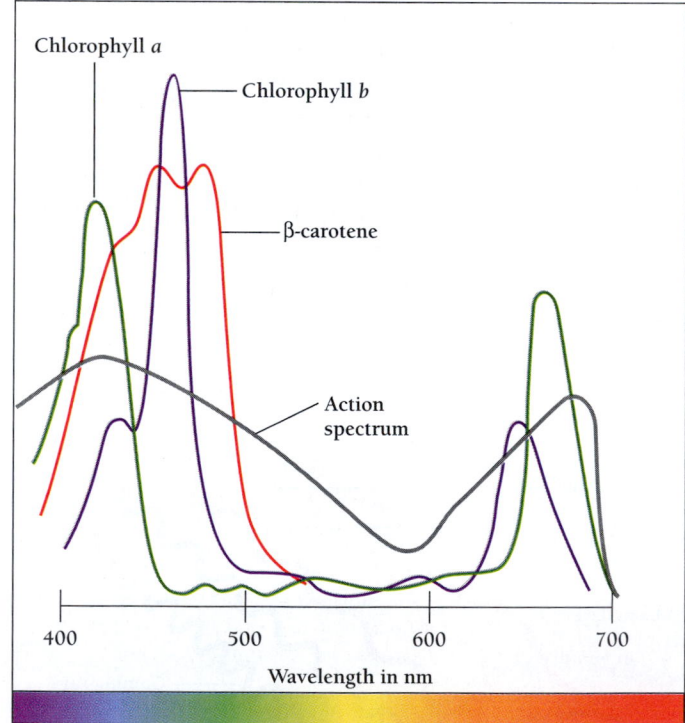

Chlorophyll *a*

Chlorophyll *b*

β-carotene

Action spectrum

400   500   600   700

Wavelength in nm

Figure 7-14   Absorption and action. Plants absorb light using two kinds of chlorophyll and carotenoids such as β-carotene. The set of wavelengths that a molecule absorbs is called its absorption spectrum. For example, chlorophylls absorb blue and red light but reflect green light. The range of wavelengths that promote photosynthesis is called the action spectrum. The action spectrum of photosynthesis roughly follows the combined absorption spectra of the chlorophylls and β-carotene.

Yet deep red light (with a wavelength of 700 nm or longer) will promote photosynthesis if shorter-wavelength light (680 nm) also illuminates the plant. And when both 680-nm (red) and 700-nm (deep red) light shine on a plant, photosynthesis is more efficient than with 680-nm light alone. In other words, photosynthesis works better with light of two wavelengths than with either alone.

Plants' need for more than one wavelength of light suggested to researchers that photosynthesis consists of two distinct sets of reactions. Each set of reactions depends on light of different wavelengths. One set of reactions, called **photosystem I,** best absorbs light with wavelengths of about 700 nm (deep red). The other set, called **photosystem II,** requires light with wavelengths of about 680 nm (red light). (While plants use two photosystems, photosynthetic bacteria use only one.)

The existence of the two photosystems in plants explains why indoor plants grow better when lit with both a bluish fluorescent bulb and a yellowish incandescent bulb than with either kind of bulb alone. Later, we will see how the two photosystems each contribute to harnessing light energy.

Where are these photosystems? Recall from Chapter 4 that inside the chloroplasts of each cell are stacks of thylakoids called grana (Figure 7-16). Each thylakoid is a hollow disc of thylakoid membrane. The two photosystems are distinct particles, but both lie in the thylakoid membranes. Each photosystem particle consists of chlorophyll molecules together with carotenoids, specific proteins, and other molecules and ions, and each photosystem performs its own biochemical reactions (Figure 7-17).

But both photosystems are required for plant photosynthesis. The chlorophyll in photosystem I absorbs light most efficiently at 700 nm and is called P_{700} ("P" stands for "pigment"). The chlorophyll in photosystem II absorbs light best at 680 nm and is called P_{680}.

The excited electrons of photosystem I can take either of two paths. The first path, called "cyclic photophosphorylation," produces ATP. The second, called "noncyclic photophosphorylation," produces NADPH but requires electrons supplied by photosystem II. The excited electrons of photosystem II all take one path—to the electron-hungry P_{700} molecules generated by photosystem I.

The two separate photosystems of plants each absorb a different wavelength of light most efficiently.

What path do the excited electrons take in cyclic photophosphorylation?

When P_{700} absorbs a photon from the antenna complex, its free energy increases, and it becomes a powerful electron donor. An excited P_{700} readily gives up a high-energy electron to one of two electron transport chains (Figure 7-18). When P_{700} loses its excited electron, however, it also loses most of the free en-

Figure 7-15 Autumn leaves along the Kancamagus Highway, in New Hampshire. Some of the beautiful colors of autumn leaves result from the carotenoid pigments that help plants photosynthesize sugar. (*Stephen J. Krasemann/Photo Researchers, Inc.*)

ergy it gained from the photon. Because P_{700} has now lost an electron, it is "oxidized."

One of the two electron transport chains associated with photosystem I leads the excited electron back to oxidized P_{700}. This cycle restores P_{700} to its original state and makes it ready to absorb another photon, a process called **cyclic photophosphorylation** (Figure 7-18).

The electron acceptor in cyclic photophosphorylation is a cytochrome complex that passes the electron on to another acceptor and back to oxidized P_{700}, completing the cycle. As in mitochondria, the passage of electrons through the cytochrome complex results in a proton gradient across a membrane. Protons move from a space just outside the thylakoids, called the **stroma,** into the thylakoid space (Figure 7-19). The stroma corresponds to the mitochondrial matrix. In mitochondria, protons move out of the matrix to the intermembrane space. In chloroplasts, protons move out of the stroma into the thylakoid space.

Like the inner mitochondrial membrane, the thylakoid membrane contains spinning, windmill-like ATP synthase molecules, which produce ATP as the protons flow back down their gradient (Figure 7-18). The ATP is made in the stroma, just as ATP is made in the mitochondrial matrix. The net effect of the electron route just described is the production of ATP from the energy of absorbed light. The ATP made in the stroma is later used to synthesize glucose.

In cyclic photophosphorylation, excited electrons from P_{700} pass through an electron transport chain and back to P_{700}. This cyclic passage of electrons creates a proton gradient across a membrane in the chloroplast capable of generating ATP.

Leaf cross section

Plant cell

Figure 7-16 Anatomy of a chloroplast. A. The cells in an oak leaf contain photosynthetic organelles called chloroplasts. Inside each chloroplast, the inner membrane is folded into pockets called thylakoids, arranged in stacks called grana. The inside of each thylakoid pocket is called the thylakoid space. The space outside the stacks of thylakoids (but inside the chloroplast) is called the stroma. Light-*in*dependent reactions occur in the stroma. Light-*de*pendent reactions occur in the photochemical reaction center in the antenna complex of the thylakoid membrane. B. TEM of a chloroplast. *(B, James Dennis/CNRI/Phototake)*

What path do the excited electrons take in noncyclic electron flow?

Photosystem I also includes an alternate, noncyclic electron path, which does not produce ATP (Figure 7-19). Instead, the energy of the excited electron of P_{700} is used to make NADPH, a derivative of NADH. NADPH is the electron-carrying coenzyme that chloroplasts use (called $NADP^+$ when lacking its electron). Recall from Chapter 6 that the similar molecules NADH and NAD^+ are used in pathways such as glycolysis and respiration (catabolism). But NADPH and $NADP^+$ are different. They are used in the building of high-energy molecules such as glucose, in chloroplasts, in photosynthetic bacteria, and even in nonphotosynthetic organisms.

The noncyclic route of photosystem I supplies most of the reducing power for glucose synthesis. But after P_{700} has given up its excited electron to form NADPH, P_{700} needs another electron before it can again serve as an electron donor. That is, some other molecule with an available electron must oxidize P_{700} before it can function again.

In the noncyclic route of photosystem I, the excited electron from P_{700} goes to help form NADPH. P_{700} is then missing one of its electrons.

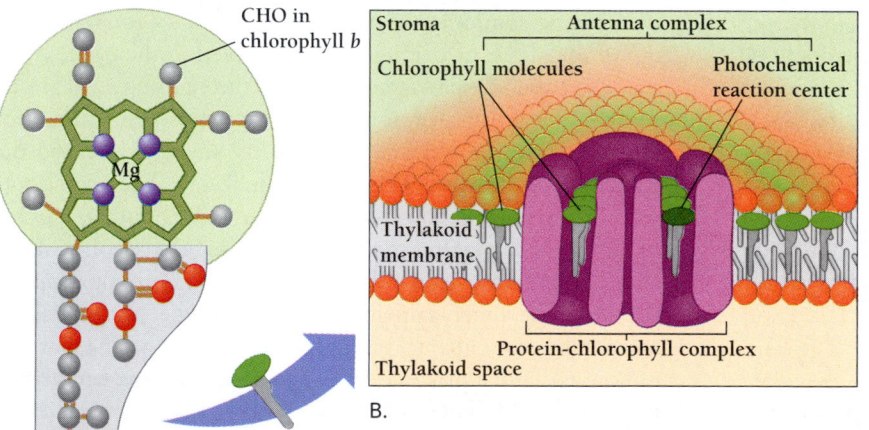

CHO in chlorophyll *b*

Mg

Chlorophyll *a* molecule

A.

Stroma

Chlorophyll molecules

Antenna complex

Photochemical reaction center

Thylakoid membrane

Protein-chlorophyll complex

Thylakoid space

B.

Figure 7-17 The heart of photosynthesis. A. Plants use two kinds of chlorophyll, chlorophyll *a* (shown here) and chlorophyll *b*. B. The photochemical reaction centers are made up of clusters of chlorophyll molecules embedded in the thylakoid membrane. Light excites electrons in the chlorophyll molecules, stimulating electron flow similar to that in the electron transport chain of mitochondria.

How does photosystem II provide electrons to P_{700}?

The chlorophyll molecules of photosystem II absorb light best at 680 nm and are called P_{680}. Just as photon-excited P_{700} is a powerful reducing agent that can donate a high-energy electron, photon-excited P_{680} is an even more powerful electron donor (Figure 7-20). Electrons move from excited P_{680}, through the electron transport chain of photosystem II, to oxidized P_{700} (from photosystem I). By regenerating reduced P_{700}, photosystem II supplies electrons to photosystem I even when there is no cyclic flow of electrons.

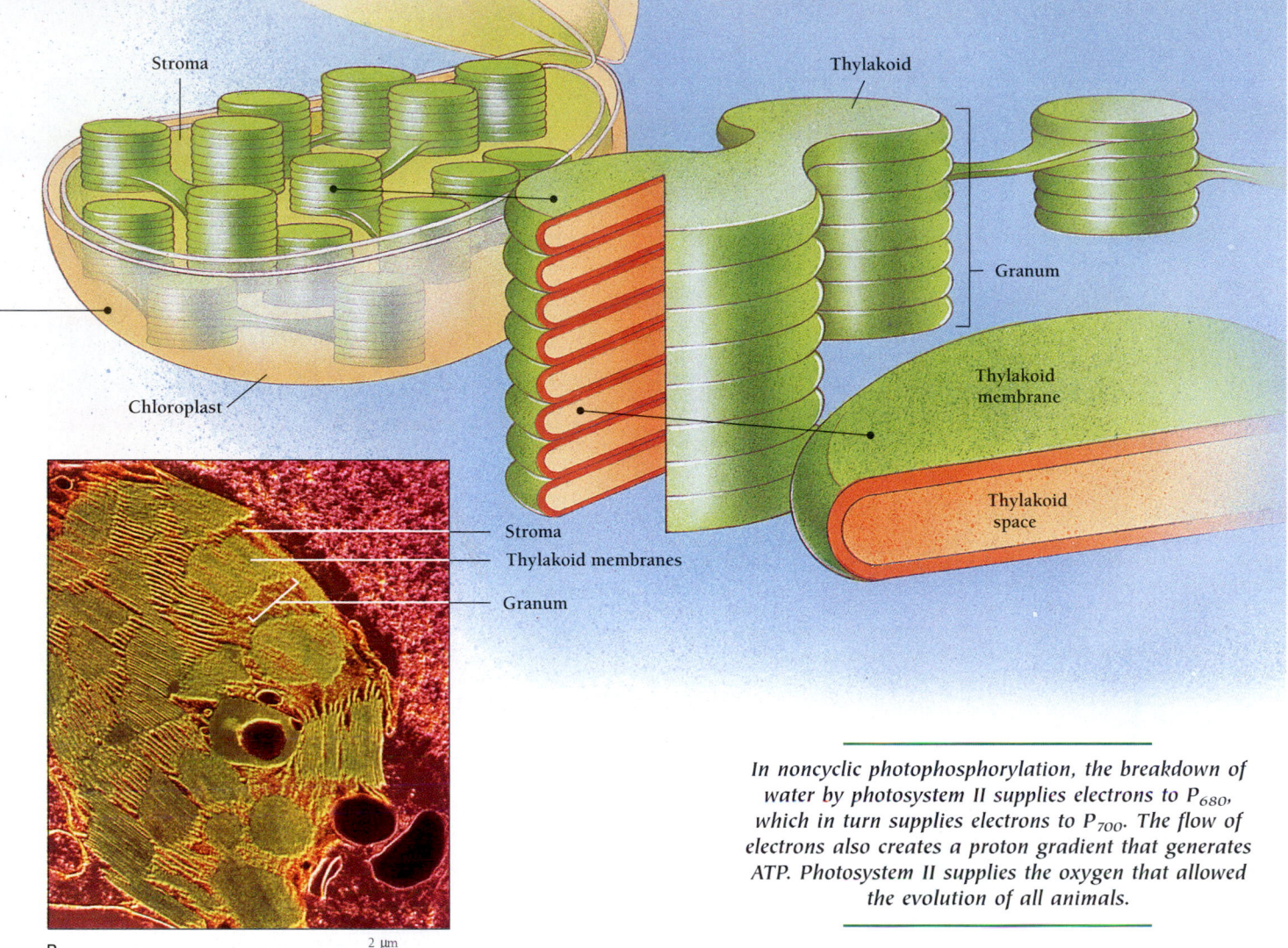

Stroma

Thylakoid

Chloroplast

Granum

Thylakoid membrane

Thylakoid space

Stroma
Thylakoid membranes
Granum

B.

2 μm

In noncyclic photophosphorylation, the breakdown of water by photosystem II supplies electrons to P_{680}, which in turn supplies electrons to P_{700}. The flow of electrons also creates a proton gradient that generates ATP. Photosystem II supplies the oxygen that allowed the evolution of all animals.

Photosystem II also produces ATP as protons flow down the gradient generated by its electron transport chain. This ATP production is called **noncyclic photophosphorylation,** to distinguish it from the ATP production in the cyclic path of photosystem I.

But we have the same question all over again: If the electron from excited P_{680} is transferred to oxidized P_{700}, how is the reduced form of P_{680} regenerated? How does it get an electron back again? The answer is that the electrons for regenerating P_{680} come from water. Recall from the beginning of this chapter that, using heavy oxygen isotopes, biochemists were able to show that plants split water molecules into oxygen and hydrogen. This fact is central to our existence.

In all plants, photosystem II strips electrons from water molecules, separating the hydrogens from the oxygen molecule. The waste product of this reaction is oxygen gas, which we gladly breathe. Indeed, until photosystem II evolved, at least 3 billion years ago, the Earth's atmosphere contained virtually no oxygen. No animals existed; nor could they have existed. Only when photosystem II evolved did oxygen begin to accumulate in our atmosphere and allow the evolution of animals, fungi, and other heterotrophic organisms.

How Do the Light Reactions Generate NADPH and ATP?

Figure 7-20 shows how photosystems I and II together make up the light-dependent reactions of photosynthesis. The absorption of light by the two photosystems causes (1) the splitting of water to oxygen and hydrogen ions, (2) the production of ATP, and (3) the production of NADPH. Oxygen is a waste product released into the atmosphere through the stomata. ATP and NADPH mostly remain in the chloroplast, where they contribute energy to the synthesis of sugar—the light-independent reactions of photosynthesis.

The mid-20th century biochemist Albert Szent-Györgyi, whose work on cellular respiration we discussed in Chapter 6, characterized this efficient process as "a little electric current driven by the sunshine." Each photon of light excites a single electron. Reducing a molecule of $NADP^+$ to NADPH, however, requires two electrons. To move these two excited electrons along the electron path to $NADP^+$, photosystems I and II must each absorb one photon. The production of a molecule of NADPH requires two photons of light.

As each electron flows from photosystem II to photosystem I, it pumps enough protons to produce a molecule of ATP. So the energy of two photons is transformed into the chemical energy of one molecule of NADPH and two molecules of ATP.

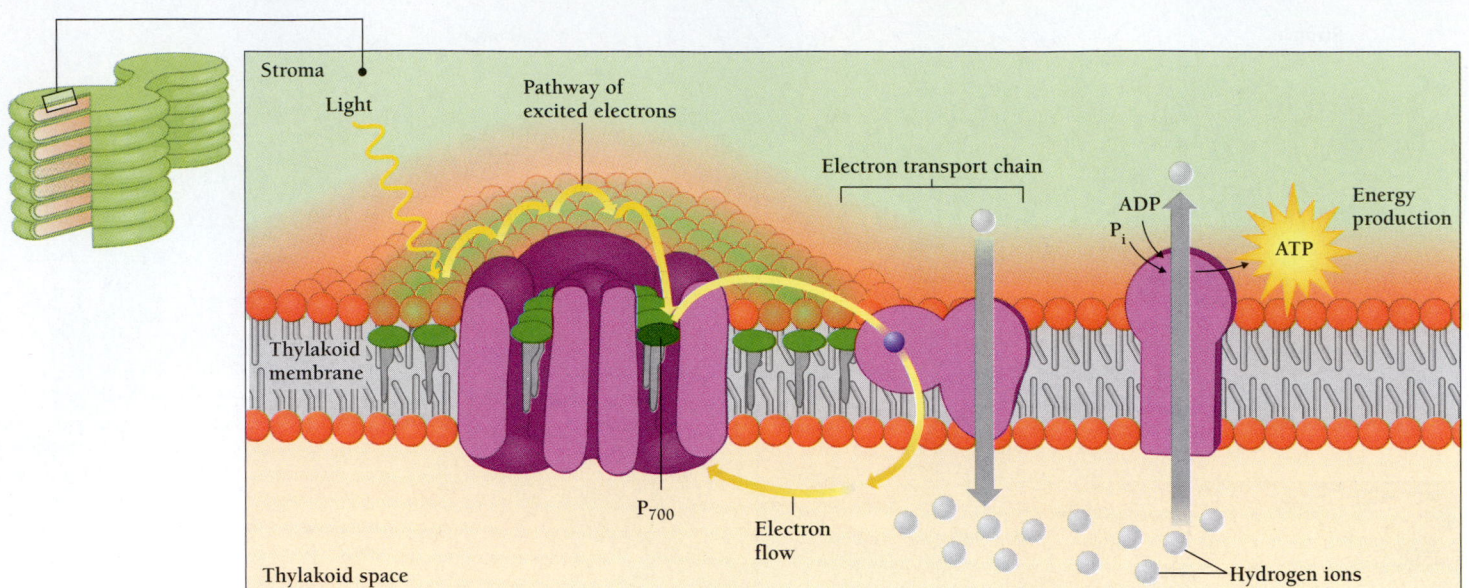

Stroma

Light

Pathway of
excited electrons

Electron transport chain

ADP
P_i

Energy
production

ATP

Thylakoid
membrane

P_{700}

Electron
flow

Thylakoid space

Hydrogen ions

PHOTOSYSTEM I: Cyclic photophosphorylation

▲ **Figure 7-18** Cyclic photophosphorylation in photosystem I. Excited electrons from P_{700} pass through an electron transport chain and back to P_{700}. This cycle of electrons creates a proton gradient across the thylakoid membrane that can generate ATP.

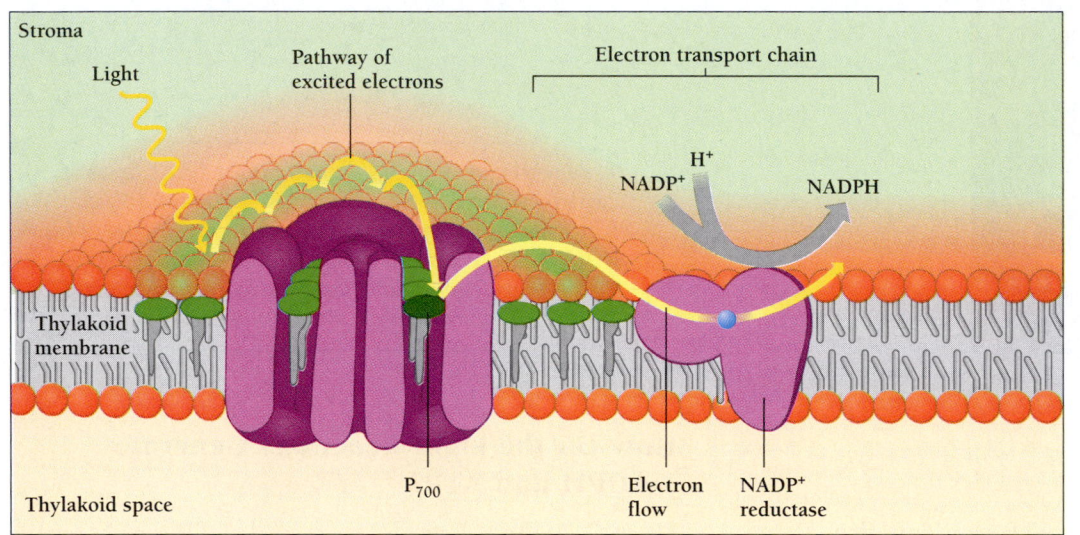

Stroma

Light

Pathway of
excited electrons

Electron transport chain

H^+
$NADP^+$

NADPH

Thylakoid
membrane

P_{700}

Electron
flow

$NADP^+$
reductase

Thylakoid space

PHOTOSYSTEM I: Noncyclic photophosphorylation

◄ **Figure 7-19** Noncyclic photophosphorylation in photosystem I. An excited electron from P_{700} goes to help form NADPH. P_{700} then is missing an electron, giving it a positive charge.

Figure 7-20 ▶
Photosystems I and II. Photosystem II's P_{680} molecule acquires electrons from water, then supplies these electrons to P_{700} by way of an electron transport chain. The flow of electrons generates a proton gradient that allows ATP synthesis. The breakdown of water into oxygen and hydrogen by photosystem II is the ultimate source of the oxygen we breathe. The evolution of photosystem II underlies the evolution of animals and other heterotrophs.

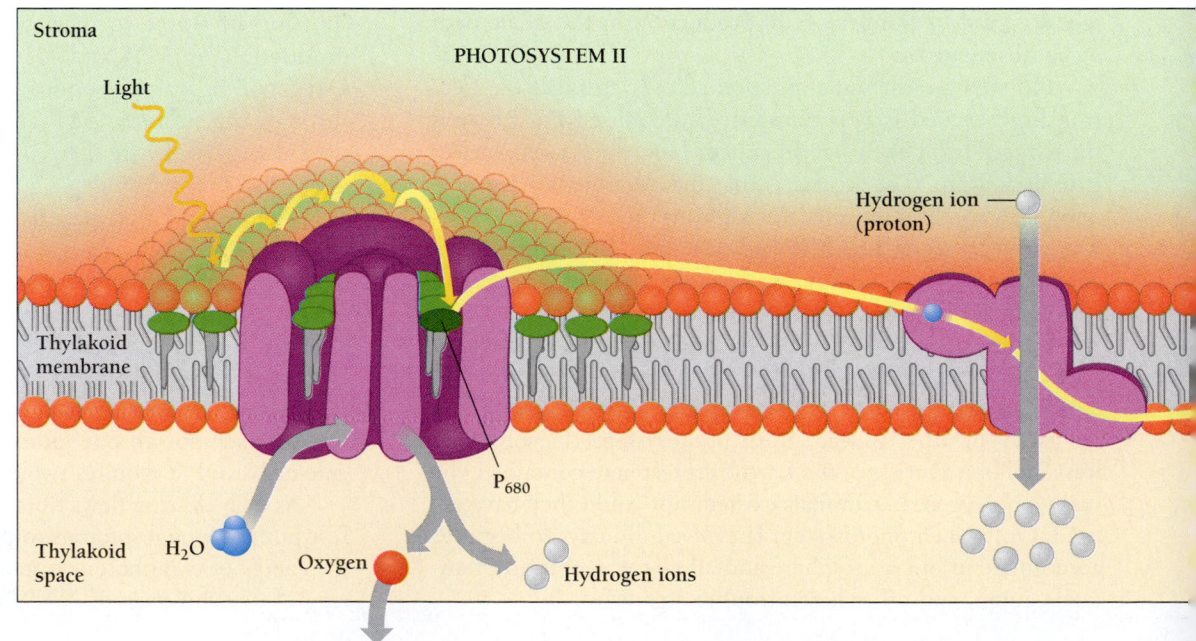

Stroma

PHOTOSYSTEM II

Light

Hydrogen ion
(proton)

Thylakoid
membrane

P_{680}

Thylakoid
space

H_2O

Oxygen

Hydrogen ions

Released to
atmosphere

The energy stored in the chemical bonds of NADPH and ATP is used to make glucose, a more stable form in which to store energy. Plants then convert glucose to sucrose, starch, or some other carbohydrate. Starch forms solid deposits in chloroplasts. In contrast, sucrose, or table sugar, remains in solution and can move throughout the plant. Just as animals transport sugar from cell to cell in the form of glucose, plants transport sugar from cell to cell in the form of sucrose.

The process by which plants transform the energy in ATP and NADPH into more stable forms such as glucose is the subject of the next section. The synthesis of sugar is the second part of photosynthesis.

The light-dependent reactions of photosynthesis split water into oxygen and hydrogen ions and generate a flow of electrons that produces ATP and NADPH. These molecules supply the energy to synthesize sugar.

HOW DO PLANTS MAKE GLUCOSE?

Glucose is plants' most important way to store both energy and the building blocks carbon, hydrogen, and oxygen all in one convenient package. We've seen that the hydrogen for glucose comes from water, and the carbon and oxygen come from carbon dioxide in the atmosphere.

The overall reaction of photosynthesis is the reverse of respiration (though the individual steps are not). So we can guess, from what we learned in Chapter 6, that the synthesis of glucose from carbon dioxide and water must consume a lot of energy:

$$6\ CO_2 + 6\ H_2O \longrightarrow C_6H_{12}O_6 + 6\ O_2$$

carbon dioxide + water $\longrightarrow$ glucose + oxygen

Recall that the individual reactions in the breakdown of glucose divide the huge downhill process into many small steps whose energy can be captured. In the same way, the individual reactions of glucose synthesis allow the energy of ATP and NADPH to be converted to glucose a little bit at a time. The energy "hill" is climbed in a set of small steps.

We know, from the last section, that photosystems I and II split the H_2O in this equation and provide electrons. But we still need to find out what happens to those electrons and what happens to the carbon dioxide. That is, what is the process by which the single carbon atoms in carbon dioxide are combined into six-carbon glucose?

What Is the Calvin Cycle?

The discovery of the radioactive isotope ^{14}C provided a way to answer this question. In 1940, Martin Kamen and Samuel Ruben, then working at the University of California, Berkeley, discovered the isotope ^{14}C and immediately showed that $^{14}CO_2$ (carbon dioxide made from ^{14}C) could be used to study photosynthesis. In 1945, Melvin Calvin, another researcher at Berkeley, began the crucial experiments.

Calvin injected a small amount of $^{14}CO_2$ into a flask of algae that were actively growing in light. Quickly, sometimes within 5 seconds, he killed the algae by adding alcohol. Then he worked out what compounds the algae had formed using the radioactive carbon atoms. To do this, Calvin used a technique called paper chromatography (Figure 7-21). In paper chromatography, the compounds made by the algae were dotted on a piece of

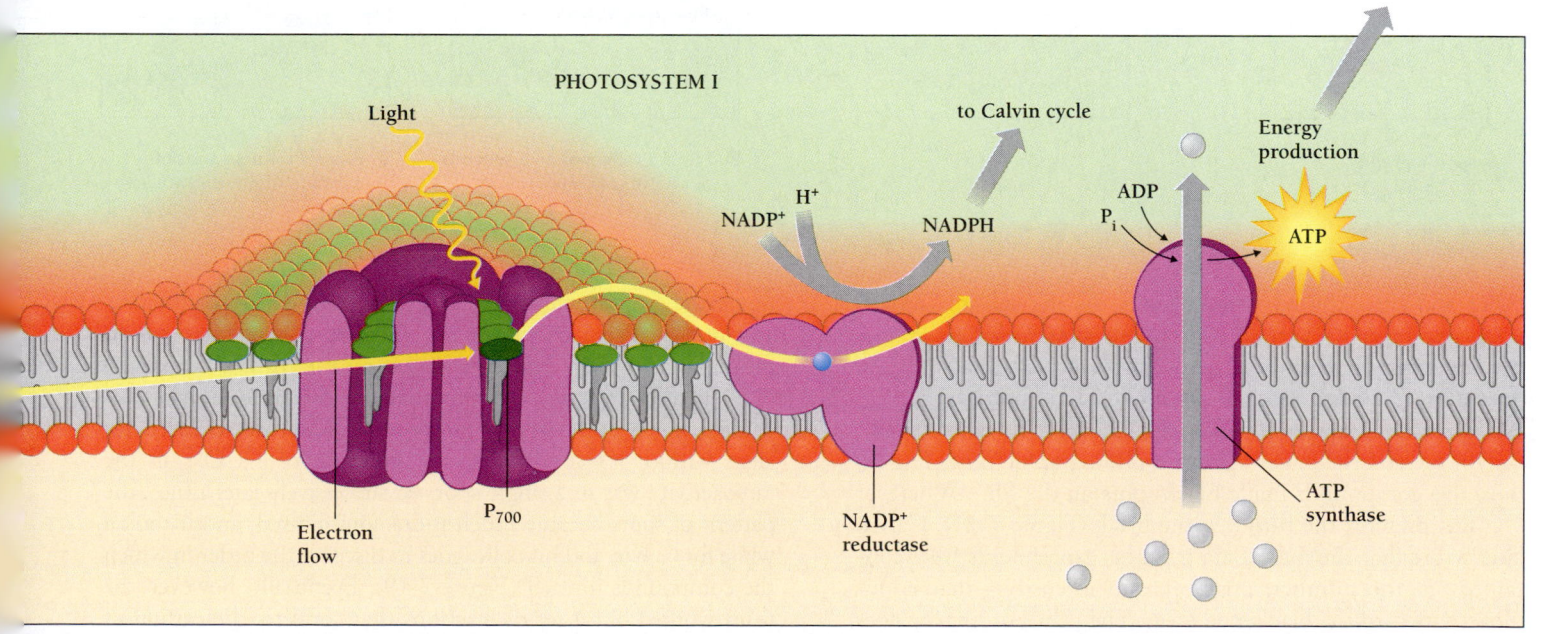

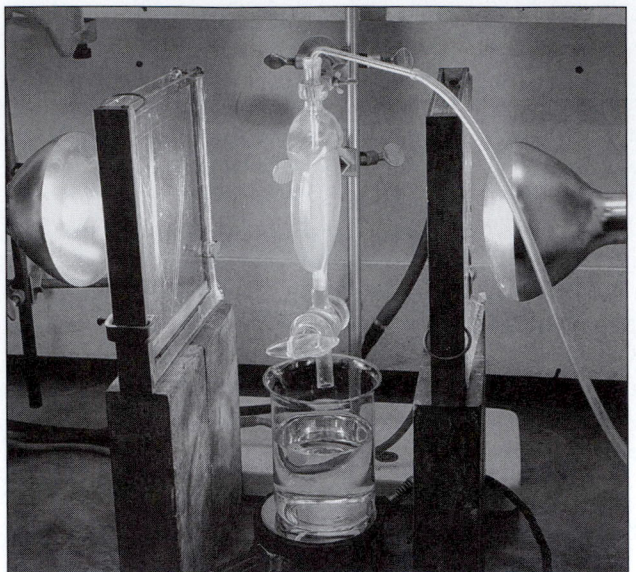

Figure 7-21 **How did Melvin Calvin discover the products of photosynthesis?** Calvin sorted out the individual products of photosynthesis by using paper chromatography. Calvin then suggested a sequence of reactions that explained the order in which the various chemicals appeared—a sequence that came to be known as the Calvin cycle. A. Calvin added radioactive $^{14}CO_2$ to photosynthesizing algae growing under bright lights. He stopped photosynthesis at different times by killing the algae and then extracted whatever molecules were present. B. Calvin separated the many compounds from one another by two-dimensional paper chromatography. C. The radioactive compounds show up as black spots on photographic film, which is sensitive to radiation. D. After 5 seconds of photosynthesis, most of the radioactivity is in a single spot, corresponding to a three-carbon molecule called 3-phosphoglycerate. E. Later, the pattern becomes more complicated, reflecting the subsequent conversion of 3-phosphoglycerate to other compounds. *(A, Melvin Calvin, U. C. Berkeley)*

A. Experimental apparatus of Calvin and Benson

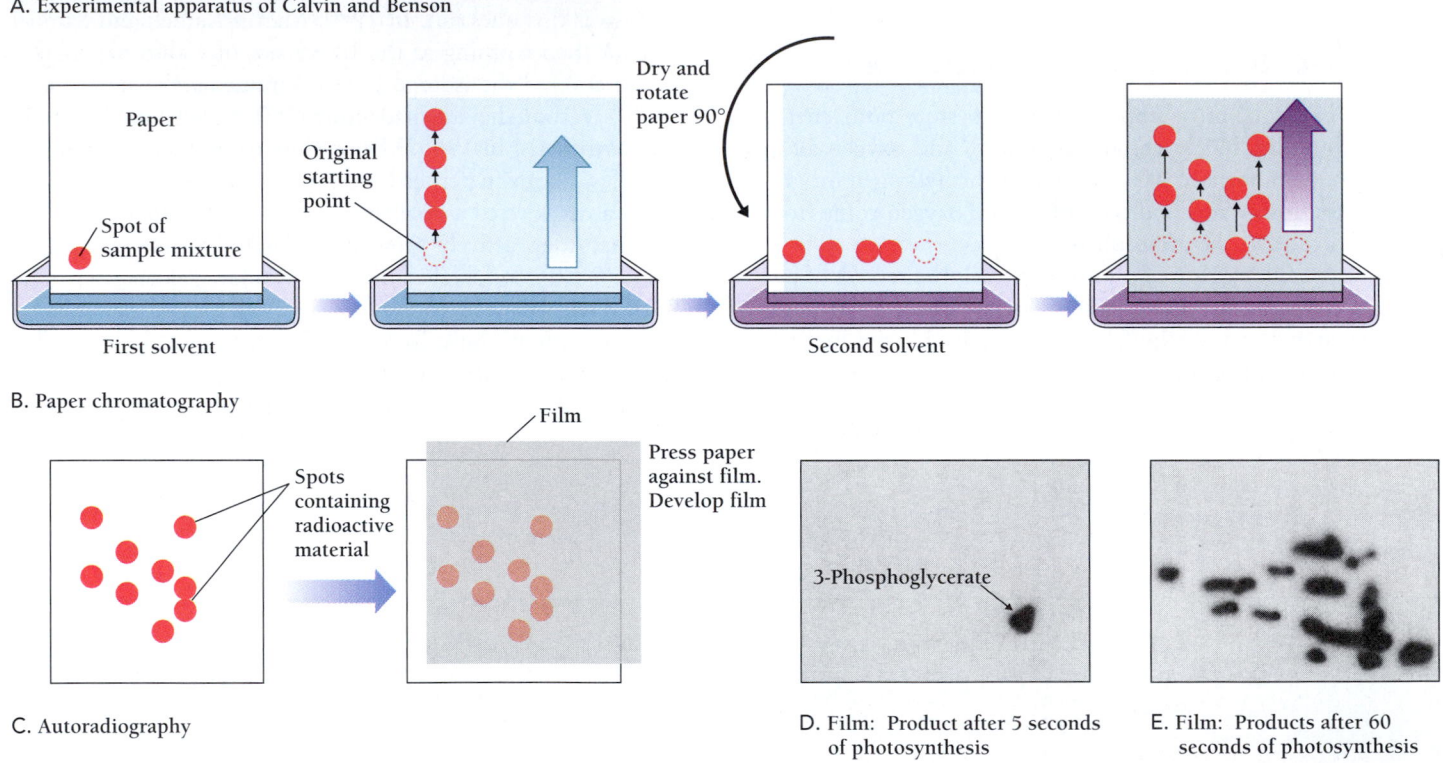

B. Paper chromatography

C. Autoradiography

D. Film: Product after 5 seconds of photosynthesis

E. Film: Products after 60 seconds of photosynthesis

paper and then treated with solvents. As the solvents moved across the paper, the various compounds were carried along, too. But the compounds moved across the paper at different rates, separating from one another like runners in a footrace.

But the compounds were still invisible. To detect them, Calvin pressed the paper against a photographic film. The radioactive compounds made black spots on the film. Wherever ^{14}C was present, the film became black (Figure 7-21). Calvin later wrote that the main data from his experiments were contained "in the number, position, and intensity—that is, radioactivity—of the blackened areas. The [film] ordinarily does

not print out the names of these compounds, unfortunately, and our principal chore for the succeeding ten years was to properly label those blackened areas on the film."

The first detectable product made by the algae, Calvin concluded, was a three-carbon compound. When he collected his algae within 5 seconds after the injection of $^{14}CO_2$, all the radioactivity was in a single spot. At successively later times, the pattern of spots became much more complicated, and it took a while for Calvin and his colleagues to discover the order in which the compounds formed (Figure 7-19). Eventually, however, the team worked out a set of reactions that converts the carbon of carbon dioxide into glucose. The reactions form a cycle called the **Calvin cycle,** after its principal discoverer (Figure 7-22).

Melvin Calvin used paper chromatography to obtain an ordered list of the chemical products of photosynthesis. Calvin was then able to suggest a reaction sequence that explained the order in which the various chemicals appeared in time. This reaction sequence came to be known as the Calvin cycle.

What Happens to Carbon Dioxide in Photosynthesis?

Carbon enters the Calvin cycle as carbon dioxide and exits as sugar. The transformation of carbon dioxide into sugar occurs in three parts:

Part 1. Capturing the carbon. In the first part of the Calvin cycle, carbon dioxide combines with a five-carbon compound called ribulose bisphosphate. The resulting six-carbon molecule immediately breaks down into two molecules of phosphoglycerate.

Part 2. Making sugar. In the second part of the Calvin cycle, each of the two molecules of phosphoglycerate is converted into a true three-carbon sugar. These two sugars may either be converted to glucose or other compounds, or they move into part 3.

Part 3. Regenerating ribulose bisphosphate. In the third part of the Calvin cycle, the three-carbon sugars made in part 2 are used to regenerate ribulose bisphosphate, which, recall, captures carbon dioxide.

What is the first thing that happens to carbon dioxide?

The first reaction of the Calvin cycle is the addition of carbon dioxide to a five-carbon compound called ribulose bisphosphate, to form a six-carbon molecule (Figure 7-22). This six-carbon molecule is unstable, however, and immediately breaks into two molecules of three-carbon phosphoglycerate.

The enzyme that catalyzes this reaction is ribulose bisphosphate carboxylase, but it has a friendlier name—**Rubisco.** Rubisco has the world on its shoulders. This small enzyme is almost solely responsible for recycling carbon dioxide from the atmosphere into plants and, ultimately, for supplying carbon to all living things.

But Rubisco is no superenzyme. Most enzymes can process 1000 molecules per second, and some champions even manage 10,000 per second. But Rubisco is so slow that it barely deserves to be called an enzyme, managing only three molecules per second. To compensate for this enzyme's sluggishness, plants typically produce huge amounts of Rubisco. In some plants, Rubisco makes up more than 50 percent of the total protein in the chloroplasts. Rubisco is probably the most abundant protein on the planet.

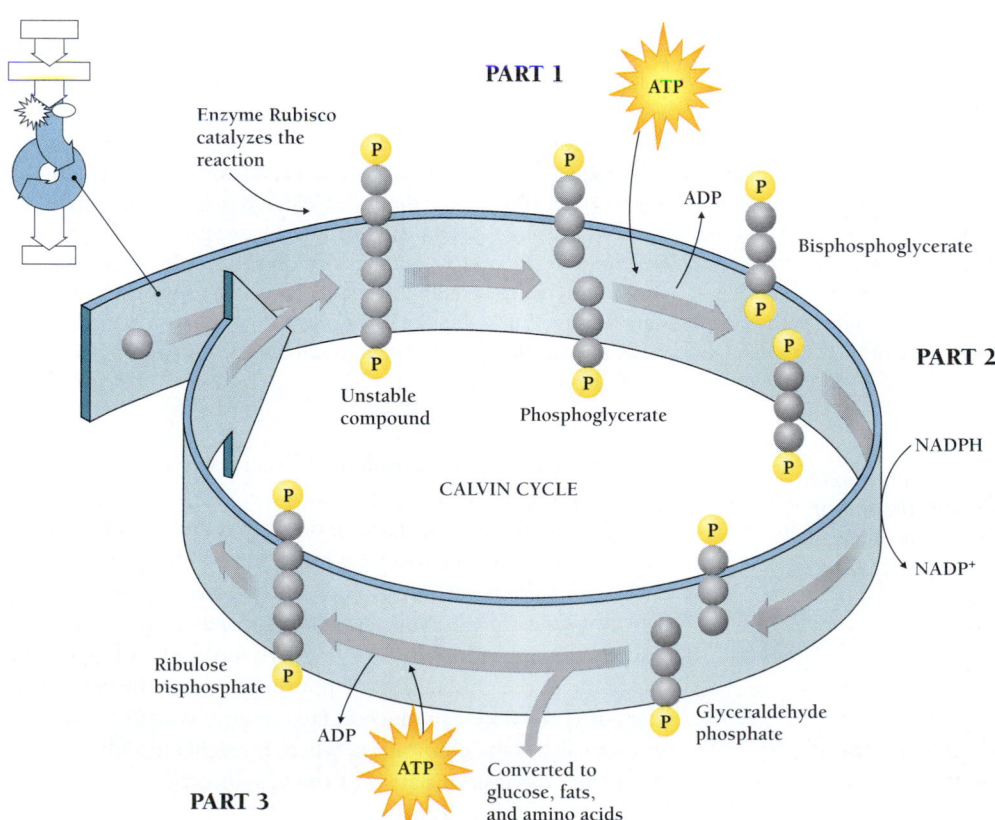

Figure 7-22 The Calvin cycle turns carbon dioxide into sugar. In part 1, CO_2 combines with ribulose bisphosphate to form a three-carbon molecule (phosphoglycerate). In part 2, the three-carbon molecule changes into a true three-carbon sugar (glyceraldehyde phosphate). Part 3 makes more ribulose bisphosphate for part 1.

BOX 7.1

How Do Some Herbicides Kill Plants?

When we think of farmers and gardeners, we imagine them nursing tender seedlings into healthy plants. But growing plants also involves killing other plants, weeds that compete with whatever a farmer or gardener is trying to grow. A weed is simply a plant that is unwelcome for any reason. The easiest way to kill a weed, or any plant, is to spray it with an herbicide, a chemical that kills plants. Farmers began using modern herbicides long before anyone understood how these chemicals hurt plants. In many cases, however, research has revealed precisely the metabolic processes in which herbicides act.

Some herbicides selectively kill different types of plants. For example, 2,4-D—a synthetic compound that mimics a plant hormone—kills C_3 plants more effectively than it kills C_4 plants. Many weeds are C_3 plants, while corn, a major U.S. crop, is a C_4 plant. Thus, farmers can apply 2,4-D to corn fields, killing the weeds, but not the corn.

Most herbicides kill all plants indiscriminately. Such herbicides act by interfering with photosynthesis or other basic processes. Two herbicides, atrazine and diuron, do so by interfering directly with electron transport in photosystem II. Both compounds prevent the transfer of electrons from excited P_{680} to one of the cytochromes. The site of interference is a thylakoid protein, called QB, that helps transfer electrons. Atrazine and diuron bind to QB and prevent electron transfer.

Corn farmers have used atrazine and related compounds heavily, since corn contains enzymes that make it resistant to atrazine. In contrast, diuron is as toxic to corn as it is to weeds. So farmers can only use it when there is no risk of killing their crop. In recent decades, however, many species of weeds have evolved tolerance to atrazine. In other words, these weeds thrive despite being sprayed with massive amounts of atrazine.

Researchers have been able to identify the exact cause of this resistance. Atrazine-resistant weeds have an altered QB protein. In nonresistant weeds, QB becomes disabled when it binds to atrazine. But a single change in one of QB's 300 amino acids can prevent atrazine from binding, rendering the weed safe from atrazine.

Another widely used herbicide that inhibits photosynthesis is paraquat. Paraquat and related compounds accept electrons from photosystem I and transfer them directly to oxygen to produce *free radicals,* highly reactive ions that destroy delicate cellular machinery. These radicals rapidly attack thylakoid membranes, destroying their ability to sustain photophosphorylation and leading to plant death. Because paraquat serves as an electron shuttle—taking electrons from photosystem I and transferring them to oxygen—it continues to do damage for as long as photosystem I is operating. Unfortunately, paraquat is also toxic to humans and must be used with great caution. In humans and other animals, paraquat can potentially damage the digestive tract, the kidneys, and the lungs—thereby causing diarrhea, vomiting, and sometimes death.

Some herbicides act on other basic metabolic pathways that are unique to plants. The widely used herbicide glyphosate (Roundup), for example, interferes with the synthesis of aromatic amino acids (phenylalanine, tyrosine, and tryptophan). Unlike paraquat, however, glyphosate is not particularly toxic to animals. In animals, the synthetic pathways for the aromatic amino acids are completely different from those in plants.

A number of herbicide-producing companies are now using genetic engineering techniques to produce herbicide-resistant crop plants. For example, some crops now carry the atrazine-resistant QB protein. Chemical companies argue that such herbicide-resistant crops allow farmers to destroy weeds selectively by applying a particular herbicide. As a result, farmers do not have to plow their fields to get rid of weeds, and less topsoil is lost to erosion. However, most environmentalists doubt the wisdom of this strategy, since it will ultimately increase herbicide use.

Hundreds of herbicides are used everywhere today: in fields where fruits and vegetables are grown; on crops we feed to animals, themselves bound for our dinner table; along roadsides; and on our lawns and gardens. We find herbicides useful, indeed, hard to live without. Some break down soon after they are applied and are nearly harmless to the environment in the long run. Most do not harm humans. A few are hazardous in high doses, such as those that farm workers might be exposed to. Some herbicides remain in the environment for months or years, continuing to poison plants and, in some cases, animals. All plants regularly exposed to herbicides ultimately evolve resistance, forcing researchers to develop new toxins.

In the first reaction of the Calvin cycle, carbon dioxide combines with ribulose bisphosphate with the aid of the enzyme Rubisco. The resulting six-carbon molecule immediately breaks down into two molecules of three-carbon phosphoglycerate.

What happens to the two molecules of three-carbon phosphoglycerate?

The second part of the Calvin cycle involves two steps that together convert phosphoglycerate into a three-carbon sugar called glyceraldehyde phosphate. Glyceraldehyde phosphate is a true carbohydrate.

These two reactions require one molecule of ATP and one molecule of NADPH for each molecule of phosphoglycerate. Each molecule of carbon dioxide entering the cycle results in two molecules of phosphoglycerate, so each turn of the cycle requires two molecules of ATP and two molecules of NADPH.

Some of the glyceraldehyde phosphate enters the cytoplasm, where it is converted to glucose, fats, or amino acids. And some of it remains in the chloroplast, where it regenerates ribulose bisphosphate, the task of part 3 of the Calvin cycle.

Each of the two molecules of phosphoglycerate is converted into a three-carbon sugar. This sugar is either converted to glucose, fats, or amino acids (in the cytoplasm), or it is used to regenerate the ribulose bisphosphate consumed in part 1 of the Calvin cycle.

How is ribulose bisphosphate regenerated?

The molecule that first binds to carbon dioxide is ribulose bisphosphate. But once it combines with carbon dioxide and splits into two molecules of three-carbon phosphoglycerate, the ribulose bisphosphate is gone. How can a plant cell make more of this important molecule?

The short answer is that a cell uses some of the glyceraldehyde phosphate to make more ribulose bisphosphate. This process is the messy part of the Calvin cycle and involves seven different enzymatic steps. In the last of the seven steps of part 3, ATP contributes a second phosphate group to form more ribulose bisphosphate.

The important molecule ribulose bisphosphate is regenerated in seven steps from glyceraldehyde phosphate.

How Many Photons Does a Chloroplast Need To Make One Glucose?

For each molecule of glucose produced, six molecules of carbon dioxide must enter the Calvin cycle. As the cycle turns six times, it produces 12 molecules of three-carbon glyceraldehyde phosphate. Of these, two form glucose, and the remaining ten regenerate six molecules of five-carbon ribulose bisphosphate.

The cost of making a molecule of glucose is considerable. Each turn of the Calvin cycle requires one molecule of ribulose bisphosphate and two molecules each of ATP and NADPH. The six turns needed to make a single molecule of glucose use 12 molecules of ATP and 12 molecules of NADPH. In addition, the production of the six molecules of ribulose bisphosphate requires an additional six molecules of ATP. The total cost of making a molecule of glucose, then, is 18 molecules of ATP and 12 molecules of NADPH.

All the NADPH and ATP for the Calvin cycle comes from the light-dependent reactions in the thylakoid membrane. To make the 12 molecules of NADPH, a chloroplast must absorb 48 photons of light by means of photosystems I and II. The same 48 photons provide energy for up to 24 ATP molecules—more than enough for the needed glucose synthesis.

To make one molecule of glucose, the chloroplasts need 18 molecules of ATP and 12 molecules of NADPH, or the energy from 48 photons.

HOW DO PLANTS COPE WITH TOO LITTLE WATER OR CARBON DIOXIDE?

The reactions of photosynthesis do not take place in a test tube, but in living, growing plants, subject to storms, drought, and attacks by plant-eating animals. Grasses, small plants, shrubs, and trees all live in complex and demanding ecosystems where many plants die without reproducing. Many factors determine whether a plant survives and even thrives. One of these factors is how well the plant uses sunlight to synthesize sugars.

We have seen that plants use the three-carbon sugar glyceraldehyde phosphate to make glucose, fats, and amino acids. These, in turn, are used to make the carbohydrates and proteins that make up the bulk of the body of the plant. The construction of leaves, stems, trunks, and flowers creates a massive demand for glyceraldehyde phosphate from the Calvin cycle.

If a plant lacks any of the raw materials of photosynthesis—water, carbon dioxide, or photons—meeting this demand becomes impossible. As a result, plants have evolved a variety of adaptations for increasing their productivity, the efficiency with which they collect and use these three raw materials.

The factors that influence plant productivity are not only important to plants but important to humans as well. All the world's major crops—including rice, wheat, and corn—depend on plant breeders' efforts to increase the productivity of those crops. The world food supply and the lives of billions of people depend on these super-productive crops.

What Factors Limit Productivity?

The major factors that affect photosynthesis are the amount and wavelength of incoming photons, the amount of water available, and the amount of carbon dioxide available. All of these can be in short supply at times.

We have already seen how the wavelength of light might influence the efficiency of photosynthesis. Plants cannot absorb light that is the wrong wavelength. Only light in the action spectrum of photosynthesis is of any use at all. Equally important is the *intensity* of light, the number of photons striking a leaf surface per day. Plants cannot produce as much sugar in the shade as they can in full sun.

Many factors influence how much light a plant gets. These include, for example, the length of the day, the length of the growing season, and the amount of cloud cover. In London, for example, pollution and fog reduce by half the total sunlight reaching a plant. Plants also compete with one another for light. In a tropical rain forest, 15 to 20 layers of vegetation may absorb and reflect more than 95 percent of the light before it reaches plants at ground level.

Too little water can also limit photosynthesis. Plants pull water from the soil and transport it through stems and trunks to their leaves. Some of this water is used in photosynthesis, but most passes out of the stomata in the leaves and evaporates into the surrounding air.

When the weather is hot and dry and water is in short supply, plants shut their stomata to avoid losing water. But the stomata need to be open to allow waste oxygen to leave the plant and carbon dioxide to enter. As photosynthesis continues, oxygen levels inside the plant rise, and carbon dioxide levels drop. For desert plants, which must keep their stomata closed as much as possible to avoid drying out, stocking enough carbon dioxide for photosynthesis is a problem.

How Do Plants Prevent Photorespiration?

Plummeting carbon dioxide levels not only slow photosynthesis, they also initiate a wasteful chemical pathway called **photorespiration,** in which plants squander much of their hard-won carbohydrate. The problem occurs because Rubisco requires a lot of carbon dioxide to work well. Normally, in the first step of the Calvin cycle, Rubisco converts ribulose bisphosphate into two molecules of phosphoglycerate. However, when carbon dioxide levels are low, Rubisco converts ribulose bisphosphate into one molecule each of phosphoglycerate and another, almost useless, two-carbon molecule. The result is that during photorespiration plants waste as much as half of the carbohydrate they produce. Photorespiration, unlike cellular respiration, produces no ATP or NADH. Its function, if any, is unknown.

There is only one cure for photorespiration: lots of carbon dioxide. Some plants have evolved a special biochemical pathway that allows them to maintain high levels of carbon dioxide inside their cells, even when their stomata are closed most of the time.

In these plants, the first reaction of carbon dioxide does not produce the usual three-carbon molecule phosphoglycerate. Instead, they make a four-carbon compound called oxaloacetate, or OAA (which is also part of the citric acid cycle of Chapter 6). The cells then convert OAA to another four-carbon compound, which moves into special cells that make carbohydrates. There it breaks down to release carbon dioxide, which immediately enters the Calvin cycle.

Plants that use the three-carbon phosphoglycerate, which includes most plants, are called **C_3 plants.** Plants that use the four-carbon OAA pathway are called **C_4 plants.** C_4 plants, such as corn and sugar cane, have a distinctive anatomy and grow well in hot, dry climates. The C_4 pathway and anatomy are adaptations that allow plants to minimize water loss by keeping the stomata closed, while still providing enough carbon dioxide to supply photosynthesis and prevent photorespiration. C_4 plants waste far less sugar than C_3 plants, but C_4 plants need more ATP (and therefore more light) than C_3 plants.

In hot, dry climates with plenty of light, C_4 plants predominate. In temperate zones with less light and more water, C_3 plants have the advantage. In Kentucky, for example, the humid summers allow plants to leave their stomata open much of the time, and C_3 plants, such as Kentucky bluegrass, thrive. But Kentuckians who move to California are in for a rude surprise. When they plant beautiful Kentucky bluegrass lawns, ugly yellow-green crabgrass gradually takes over. Crabgrass is a C_4 plant that grows better in the bright but dry California summers.

C_4 crops are especially efficient at capturing sunlight, converting as much as 8 percent of the solar energy falling on a field to carbohydrates. As oil and coal reserves decline, some countries have begun to fuel their cars and trucks with alcohol produced from corn and sugar cane.

Desert-adapted plants, such as cacti, use a modified C_4 pathway called crassulacean acid metabolism, or **CAM,** a type of photosynthesis that conserves even more water. CAM plants open their stomata to collect CO_2 at night only, when the air is cool and water loss is minimal. They then store the CO_2 in a four-carbon molecule until the next day, when photosynthesis can proceed. But collecting and storing CO_2 at night requires enormous amounts of ATP. This extra ATP comes from the breakdown of carbohydrates made during the day. The enormous energy cost of CAM plants' nighttime respiration keeps their growth rates extremely low.

C_4 plants minimize water loss and photorespiration and maximize photosynthesis by storing high carbon. C_4 plants are more efficient than C_3 plants, but they need more light than C_3 plants.

In these first few chapters of *Asking About Life*, we have examined some of the basic chemical processes that enable organisms to live. In the next few chapters, we will see how cells multiply and grow. Then we will see how organisms recombine their genetic material in novel arrangements in the process of sexual reproduction. We will learn how the structure of DNA was discovered and find out how genes are expressed. Finally, we will explore the ways that molecular biologists have learned to control the expression of genes and how this dizzying array of new knowledge applies to humans.

Study Outline with Key Terms

In **photosynthesis,** green plants and other organisms use the energy of **photons** to synthesize carbohydrates from carbon dioxide and water. In the **light-dependent reactions (photophosphorylation),** photosynthesis splits water into oxygen and hydrogen atoms, generating oxygen and high-energy electrons. In the **light-independent reactions,** the electrons from the light reactions help link the carbon atoms in carbon dioxide into three-carbon sugars.

Each photon carries a fixed amount of energy, which corresponds to its **wavelength. Visible light** ranges from about 400 nm to about 750 nm. Plants contain **pigments** that absorb light, including the **chlorophylls** and the **carotenoids.** Every molecule has an **absorption spectrum,** and every light-dependent process has an **action spectrum.**

Antenna complexes, consisting of large numbers of chlorophyll and carotenoid molecules, absorb photons and transfer them to one of two kinds of **photochemical reaction centers,** located in the thylakoid membrane of the chloroplast. Each kind of reaction center participates in a separate photosystem.

In **photosystem I,** the main pigment is P_{700}. In **photosystem II,** the main pigment is P_{680}. When P_{700} or P_{680} absorbs a photon, the pigment's energy increases, which excites its electrons. The excited electrons transfer to special electron acceptors.

The excited electrons of photosystem I can take either of two paths. In **cyclic photophosphorylation,** excited electrons from P_{700} pass through an electron transport chain and back to P_{700}. This cyclic passage of electrons creates a proton gradient between the **stroma** and the thylakoid space in the chloroplast that generates ATP. In **noncyclic photophosphorylation,** the excited electrons of photosystem I flow down a different electron transport chain to produce NADPH. But the noncyclic path leaves P_{700} short an electron. Only electrons from photosystem II can make up this deficit.

The breakdown of water by photosystem II supplies electrons to P_{680}, which in turn supplies electrons to P_{700}. The flow of electrons from P_{680} to P_{700} also creates a proton gradient that generates ATP. The breakdown of water by photosystem II supplies the oxygen that allowed the evolution of all animals.

The **Calvin cycle** is the heart of the light-independent reactions of photosynthesis. The Calvin cycle uses the energy of ATP and NADPH (both generated by the light reactions) to make sugar out of carbon dioxide. In the Calvin cycle, the enzyme **Rubisco** synthesizes three-carbon sugars from ordinary carbon dioxide. These small sugars can then be converted to glucose, fats, amino acids, or ribulose bisphosphate. The production of one molecule of glucose from six carbon dioxide molecules requires 48 photons of light.

The efficiency of photosynthesis depends on the amount of light, water, and carbon dioxide. In hot, dry weather, plants close their **stomata** to limit water loss. With the stomata closed, however, waste oxygen builds up, and carbon dioxide levels plummet. Under such conditions, C_3 **plants** may begin **photorespiration,** an unhealthy biochemical pathway that wastes about half of all carbon dioxide. C_4 **plants** and **CAM** plants avoid photorespiration by maintaining high levels of carbon dioxide inside the cells that make glucose.

Review and Thought Questions

Review Questions

1. What is photosynthesis? What material does it synthesize? Where in a cell does this process occur?
2. Write the overall reaction for photosynthesis. How do the reactants get into the leaf and into the cells? How does oxygen get out of the leaf cells and back into the air outside the plant?
3. Which parts of the electromagnetic spectrum do plants use in photosynthesis? Which parts of the spectrum are reflected or absorbed by the Earth's atmosphere?
4. Describe the cyclic photophosphorylation associated with photosystem I. What is the relationship between the absorption of light and the excitation of electrons? How does the transfer of electrons lead to the synthesis of ATP?
5. Describe the reactions that make up the Calvin cycle. What enzyme captures carbon dioxide from the atmosphere? What are the products of this reaction? How is the energy from light introduced in the Calvin cycle? Why must the cycle produce more ribulose bisphosphate?
6. Under what environmental conditions might a plant close its stomata? Compare the rates at which plants can carry out photosynthesis when their stomata are open and when they are closed. Why should the opening of stomata affect the rate?

7. Explain how photorespiration causes C_3 plants to waste some of their photosynthetic efforts under conditions of low carbon dioxide concentration.

8. Describe Priestley's experiment that showed that a plant could restore the air in a bell jar. Make a sketch of a bell jar containing only a mouse, and explain why the mouse dies. Then sketch a bell jar containing both a mouse and a plant, and show with arrows how oxygen is produced and consumed.

9. What advantages do C_4 plants have over C_3 plants? Why don't C_4 plants take over everywhere? What limits them?

10. Put the letter "R" in the blank if the event is part of cell respiration. Use the letter "P" for photosynthesis. Use the letter "B" for both.
    ____Sugar is product
    ____Uses oxygen
    ____Requires light
    ____Uses electron transport chain
    ____Releases carbon dioxide
    ____Produces a proton gradient
    ____Releases oxygen
    ____Uses an ATPase to make ATP

Thought Questions

11. What prevents the roots of a plant from carrying out photosynthesis? By what metabolic process do you suppose roots extract energy from their available resources?

12. Describe in a few sentences the concept of how light energy is used to drive the reaction of glucose synthesis. (Be careful to give the outline of the story, so that a nonbiologist could understand what you mean. Don't just list a lot of details.)

13. Why is the process of photosynthesis important to feeding our entire planet? Can you name any living things that do not depend on plants (directly or indirectly)?

About the Chapter-Opening Image

A large plant in a small pot of soil emphasizes the idea that plants make their bodies primarily from air and water, not from soil.

 On-line materials relating to this chapter are on the World Wide Web at **http://www.harcourtcollege.com/lifesci/aal2/**

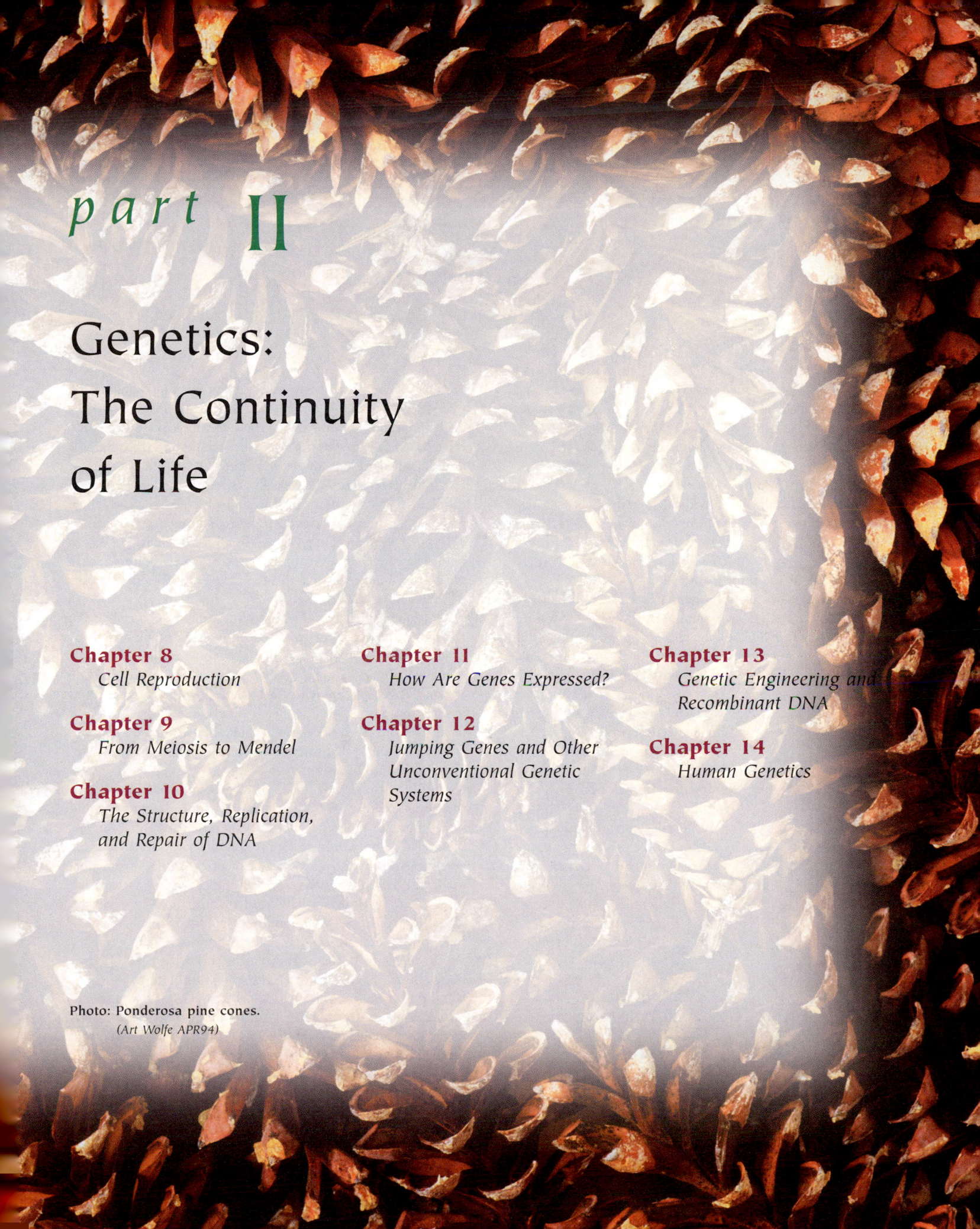

part II

Genetics:
The Continuity
of Life

Chapter 8
Cell Reproduction

Chapter 9
From Meiosis to Mendel

Chapter 10
*The Structure, Replication,
and Repair of DNA*

Chapter 11
How Are Genes Expressed?

Chapter 12
*Jumping Genes and Other
Unconventional Genetic
Systems*

Chapter 13
*Genetic Engineering and
Recombinant DNA*

Chapter 14
Human Genetics

Photo: Ponderosa pine cones.
   (Art Wolfe APR94)

The Unfortunate Henrietta Lacks

Early in 1951, an anxious 31-year-old woman checked into the medical clinic at Johns Hopkins University in Baltimore (Figure 8-1). Her anxiety was confirmed when her doctor found a one-inch purple growth on her cervix, the neck-shaped entrance to the womb. He told her he would have to do a biopsy, and he cut out a small piece of the growth and sent it to a pathologist, a medical specialist who can tell, among other things, whether a human cell is cancerous or not. The report came back positive. Henrietta Lacks's growth was dangerously malignant. Its cells were dividing rapidly, and, without treatment, the tumor would kill her.

The news came as a blow. She and her husband had two children to raise. She was too young, she thought, for so serious a disease, one that does not usually afflict women until their 40s or 50s. Within days, the doctors at Johns Hopkins began treating her with high doses of radiation, in what was to be a futile effort to save her life. In the autumn of 1951, the best part of Henrietta Lacks died. Ironically, her tumor cells survived and live today in laboratories all over the world.

The day of her biopsy, Lacks's doctor took a piece of her tumor to a group of researchers who were struggling to grow human cells in plastic dishes containing a mixture of nutrients and blood products. They tried adding different nutrients or changing the pH of the mixtures. No one had yet succeeded. Most human cells died immediately. A few divided several times, then died. Still, the researchers hoped that any day they would find the right cells or the right conditions for growing human cells in the laboratory.

When Henrietta Lacks's cells arrived in the lab, they were different from anything anyone had ever seen before. They grew without restraint, dividing and redividing every 24 hours. The same virulence that would finally kill Henrietta Lacks made these cells unstoppable in the laboratory. Before Lacks's life was over, her cells took up an independent existence, thanks to the work of medical researchers.

Although Henrietta Lacks's cells, named HeLa cells, were the first human cells to be successfully cultured, hundreds of others soon followed. Unlike her cells, most of these were difficult to culture. In general, human cells die quickly outside the body, even in the rich broths that allow many other types of cells to grow. A few stray cells on the side of a glass flask, or in the fluid left in a pipette, expire almost immediately. Beginning a culture of human cells in a plastic culture dish requires care and attention.

Still, in the 1950s and 1960s, growing human cell cultures was exciting work. For the first time, scientists could study human biology without experimenting on humans themselves. They could study the way human cells divide, the way human cells move, the way cancer cells differ from normal cells. The possibilities seemed endless. Every week a lab somewhere developed a new kind of human cell to culture. These new cell lines were rapidly distributed to other labs—some down the hall, others on the other side of the globe.

In the early 1970s, odd things began to happen. The National Institutes of Health (NIH), in Washington, DC, sent five kinds of human cells, which had come from five different research labs in the Soviet Union, to a lab in Berkeley. There, an alert researcher named Walter Nelson-Rees discovered that the five supposedly different cell lines were all HeLa cells.

Soon HeLa cells began to turn up everywhere. Nelson-Rees began checking every cell line he knew of. He published lists of cell cultures that were not what they were supposed to be. By 1981, he had found 90 different cell lines, each of which was HeLa cells. Researchers studying the biology of breast cells,

Amy Dunleavy

Key Concepts

- Cell reproduction in eukaryotes involves three distinct but interconnected processes: the duplication of DNA in the nucleus, its equal distribution to each of two daughter cells, and the division of the cytoplasm.

- During mitosis, microtubules and other subcellular components bring about chromosome movements that distribute one copy of each chromosome to each daughter cell.

- Eukaryotic cells, ranging from yeast to human cells, use similar molecular mechanisms to regulate passage through the cell cycle. Failure of these mechanisms can result in cancer.

both normal and cancerous, were horrified to find that their years of research had not been done on breast cells—as they had thought—but on Henrietta Lacks's cervical cells. The story was the same for research on prostate cells, kidney cells, liver cells, and more.

The repercussions were enormous. Years of research by scores of researchers had become meaningless. Much of the biology of cancer had to be rethought. For years researchers had believed that all cancer cells have the same nutritional needs and all cancer cells have abnormal chromosomes, the so-called mark of cancer. Then, as one cell line after another proved to be not prostate cells, not kidney cells, not this and not that, but HeLa cells, biologists came to realize that, in fact, cancer cells are not all alike. The only cancer cells that were all alike were HeLa cells. The abnormal chromosomes were typical of HeLa cells, but not of all cancer cells.

But how had it happened? The answer was that Henrietta Lacks's cancer cells were unbelievably vigorous. If so much as a single one of her cells fell by accident onto a culture dish, her cells could take over, no matter what other cells were living in the dish. One group of researchers found that HeLa cells could float through the air inside the tiny bits of spray created when a rubber stopper was pulled from a bottle containing HeLa cells. The floating cells could then drop down into nearby culture dishes. Within a week, a culture of prostate cells was taken over by cervical cells. It was a scenario that scientists had never imagined.

Today, cell biologists have learned to be more cautious. They work with cell cultures whose ancestry they've checked. And, using DNA-based techniques, they can now make sure that they have exactly the cells that they think they have.

But HeLa cells are still used, and Lacks's cells live on. All over the world her cells continue to divide relentlessly day after day. In the 1950s, her cells were used as hosts for growing poliovirus, enabling researchers to develop the vaccine against polio. Today, genetic engineers insert genes for valuable human proteins into HeLa cells, which then synthesize the proteins. Henrietta Lacks died before she could finish raising her own two children, yet her cells have been instrumental in basic research and have helped save the lives of millions of children all over the world.

Figure 8-1 **Young Henrietta Lacks** died of cancer in 1951, but biologists have kept her cancer cells alive ever since. In the 1950s, her cells, called HeLa cells, were used to develop a vaccine against polio. Today, genetic engineers insert genes for valuable human proteins into HeLa cells, which then synthesize the proteins. (*Science Photo Library/Photo Researchers, Inc.*)

CELL DIVISION

Cells are the fundamental living units of all organisms. An organism may consist of a single cell or many cells. Bacteria such as *Lactobacillus,* yeasts such as *Candida albicans,* and protists such as *Paramecium* are all single-celled organisms. In contrast, an adult human contains about 10 trillion (10^{13}) cells. Because all the cells in a multicellular organism originate from a single cell, the continuation of life ultimately depends on cell reproduction.

How Do Cells Divide?

In the five minutes or so that it will take you to read this page, your body will produce a billion (10^9) new cells. Most of these replace battered and dying cells of your skin, intestines, and

BOX 8.1

Cervical Cancer and the Pap Smear

In the 1930s, more women died of cervical cancer in the United States than of any other form of cancer. Today, the rate of cervical cancer is less than one-tenth that of breast cancer and declining rapidly. Even more encouraging, the overall survival rate (more than two-thirds) is one of the highest of all cancers, thanks to a diagnostic test invented by George N. Papanicolaou in 1943.

In the past 50 years, the widespread use of the Pap test, or Pap smear (named after its inventor), has transformed a diagnosis of cervical cancer from a virtual death sentence to a highly survivable disease. The survival rate for cervical cancer would be even higher if more women were tested regularly.

The test consists of a simple office procedure in which a doctor or nurse scrapes a few living cells from the surface of the cervix. The scraping may hurt very slightly, but the life-saving test is well worth the second or two of discomfort. Once the cells have been collected from the cervix, they are sent to a laboratory pathologist for analysis.

In most forms of cervical cancer, cells undergo changes in appearance years before they become cancerous. Such "precancerous" cells have distinctive characteristics that a good pathologist can easily recognize (Figure A). The cervical cells are usually classified into five grades: normal, atypical, low-grade precancerous lesion,

high-grade precancerous lesion, and invasive cancer. In the great majority of cases, all the cells collected will be normal. Occasionally, some of the cells will be "atypical," usually because of some temporary infection of the cervix or vagina.

Women with low-grade or high-grade lesions are immediately tested for any sort of growth. Any precancerous growths that are found are removed from the cervix, using cryosurgery (freezing) or lasers. After treatment, a woman is normally cancer free and able to bear children if she wishes.

If the Pap smear reveals cancerous cells, as opposed to precancerous ones, the recommended treatment is a hysterectomy (the removal of the uterus). If the cancer has spread, the surgeon will also remove the upper part of the vagina. If the cancer is caught early, when it is still localized on the cervix, the survival rate is 91 percent. If the cancer is diagnosed late, after it has spread beyond the reproductive organs, the survival rate is less than 9 percent.

Cervical cancer strikes women of all ages and ethnic groups. African-American women, however, are twice as likely as white women to die of the disease. Researchers attribute the difference to lack of good medical care, especially regular Pap tests. Fortunately, increasing numbers of African-American women are receiving the Pap test,

and the difference in survival is therefore decreasing.

Because most cervical cancer develops slowly, it is quite easy to diagnose the disease while it is still in one of the precancerous stages. Public health experts recommend that all women have annual Pap smears starting at age 18 or at the onset of sexual activity, whichever occurs first. After three or more negative tests, a woman is safe in having an exam every other year. Women who are at risk for cervical cancer should continue to have the exam every year.

The main risk factor for cervical cancer is infection with human papilloma virus (HPV), also known as genital warts. In 93 percent of all cases of both cervical cancer and precancerous lesions, HPV DNA is present. Indeed, HPV incorporates two genes into the genomes of previously healthy cervical cells. These two genes encode proteins that cause the cells to form tumors. In Chapter 12, we will learn more about how viral DNA infects cells and how tumor viruses cause these cells to form tumors.

Because HPV is a sexually transmitted disease (STD), the best ways to avoid infection are the same as for AIDS and other STDs: abstinence or else barrier methods of birth control (condom or diaphragm) and fewer than five sexual partners over a lifetime.

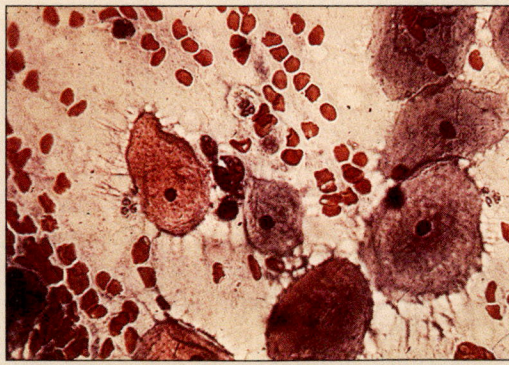
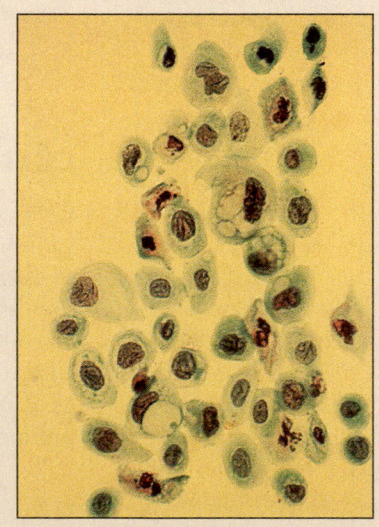
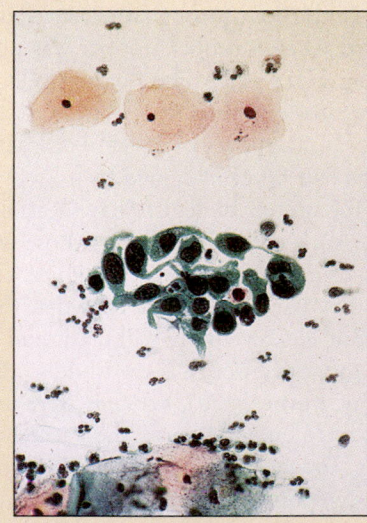

Figure A Normal cervical cells *(left)* are distinctly different from precancerous *(middle)* and cancerous *(right)* cervical cells. Note the sperm-shaped cells that characterize invasive cancer cells. *(©Biology Media/ Photo Researchers, Inc.; Science Photo Library/Photo Researchers, Inc.; M. Rotker/Photo Researchers, Inc.)*

blood. Each newly made cell contains both a copy of the genetic information you inherited from your parents and the molecular machinery and materials needed to interpret that information, including membranes, organelles, macromolecules, and small molecules.

Cell reproduction is so common that it is easy to forget how extraordinary it is. Our wonder at this copying process may be heightened by our everyday experience with nonbiological copying—the recording of music, the filming and taping of movies and television programs, and the printing of books. In each case, complicated machinery carries out a complex but ordered process. But cell reproduction is even more remarkable than these other processes, since the cell is both the thing being copied and its own copying machine.

Cell reproduction has different results in single-celled and multicelled organisms. Single-celled organisms divide to create new individuals. Cell division in yeast, for example, leads directly to an increase in the number of individuals. In a multicellular organism such as a frog the same kind of cell reproduction allows one individual to grow from a single cell into a whole organism. Cell reproduction also allows a multicellular organism to replace dying cells. A dandelion root can regenerate its leaves and flowers after they are cut away by a lawnmower. In the same way, children quickly replace the skin cells that they scrape and tear off in their inevitable bangs and falls. Before cells divide, they must first duplicate their contents, including their DNA. Eukaryotic cells must also duplicate their internal membranes and organelles.

An altogether different kind of cell division also occurs in sexually reproducing organisms. Plants, animals, and other sexually reproducing organisms make sperm and eggs (or their equivalents) through an elaborate series of chromosomal divisions and fusions. We discuss this process, called *meiosis,* in Chapter 9. In this chapter, we examine simple cell division, the kind that bacteria use to reproduce. Humans and other multicelled organisms use the same process to grow and to replace the cells of tissues such as leaves, bark, muscle, and hair. In simple cell division, the cell duplicates its DNA, then divides it in a process called mitosis.

Both single-celled organisms and multicellular organisms make more cells by dividing. Just before they divide, cells double their DNA. In eukaryotic cells, they then distribute identical copies of their DNA in a process called mitosis.

What Did Early Biologists Discover About Chromosomes?

In the 19th century, the cell theory established that all organisms are made of cells and that all cells come from other cells. But the cell theory suggested a problem. How does a cell know how to make another cell just like itself? How does the cell pass on the knowledge of what kind of cell it is? How, for example, does a cell from Koko the gorilla know how to make another

Koko cell, a cell unlike that found in any other gorilla and unlike that found in other primates such as your classmates?

The answer, we now know, is in the **chromosomes** [Greek, *chroma* = color + *soma* = body]—the threads of DNA and protein that carry the genes. Yet, 19th-century microscopists saw chromosomes and examined them in great detail without recognizing that they were the means by which organisms inherit traits. Early biologists held the key to the problem of inheritance without knowing it.

In the late 19th century, improvements in the lenses of microscopes for the first time allowed microscopists to see the tiny structures within cells. Some of the most important advances, however, came from the use of natural dyes developed for the burgeoning 19th-century textile industry. The two most useful dyes, still in use today, were the purple-and-blue dye hematoxylin (which comes from the logwood tree of Central America) and the synthetic red dye eosin.

Thanks to these dyes, microscopists of the 1870s were able to describe in detail the threadlike chromosomes in a variety of animal and plant cells. In fact, during one 4-year period, researchers published over 200 papers on the complex dance of the chromosomes during cell division. By the 1880s, biologists had discovered that cells reproduce themselves in a process called **cell division,** in which a parent cell gives rise to two **daughter cells.** Cells stained with hematoxylin and eosin showed the cell's nucleus to be full of a grainy material called **chromatin** [Greek, *chroma* = color] (Figure 8-2). Biologists also knew that when a cell divides, its nucleus divides, providing one-half of its material to each daughter nucleus.

The German physician and microscopist Walther Flemming pieced together a detailed description of the process by

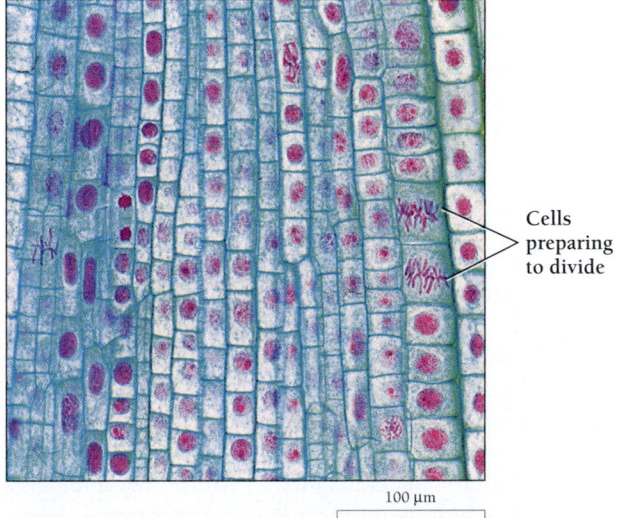

Cells preparing to divide

100 μm

Figure 8-2 Onion cells reproduce by dividing in two. Cells stained with a dye that binds to DNA show the DNA in the cell's nucleus. In some cells, the DNA stains as a diffuse grainy material called chromatin. In dividing cells, the chromatin condenses into distinct chromosomes. Chromatin is present in nearly all eukaryotic cells, while chromosomes are visible only in dividing cells. (*Dwight Kuhn*)

which the chromosomes are equally divided and distributed to the two new cells during cell division. Flemming named this elaborate division of the chromosomes **mitosis** [Greek, *mitos* = thread].

One of the first events that Flemming noted was the gathering of the chromatin into visible threads (later called chromosomes). He also observed in salamander cells the duplication of an object that he called an **aster** [Latin, *aster* = star]—the center of a finer set of threads just outside the nucleus. During mitosis, the two asters migrate to opposite ends of the cell. While the asters are migrating, the chromosomes line up in a row in the middle of the cell, and then each one splits into two. Each of the two half-chromosomes then migrates toward one of the asters. Finally, the chromosomes disappear, and the nuclei regain their grainy appearance.

Late 19th-century microscopy revealed all the major details of cell division in eukaryotes without explaining the role of the nucleus or chromosomes in the life of the cell.

How Did Biologists Discover the Function of the Chromosomes?

But what was the point of this careful division? For the moment, no one knew. The nucleus of the cell had been recognized as a standard component of cells as early as the 1830s, but its function was elusive. During cell division, the nuclear membrane breaks down, and the nucleus seems to disappear. Because the nucleus disappears, biologists did not at first consider that it might play an important role in cell division. Many biologists assumed that the cell constructs a new nucleus after each cell division.

The first hint that the nucleus might be important came with studies of *fertilization,* the process by which a sperm and egg fuse to form a *zygote.* In the 1870s, several microscopists noticed that the sperm is little more than a nucleus with a tail to propel it. Further studies showed that a zygote forms from exactly one sperm and one egg and that the sperm nucleus and the egg nucleus fuse to form a single nucleus. In time, the zygote develops into an embryo by dividing into more and more cells.

By 1880, biologists knew that all the nuclei of a developing organism are descended from a single nucleus created by the fusion of the nuclei of the egg and the sperm. This fact did not, however, suggest to biologists that the nucleus contained the hereditary material. First, 19th-century biologists were not necessarily looking for a hereditary material. Many biologists of the time believed that the process of fertilization did not transmit a material substance, but merely provided some form of energy or force that triggered the development of the embryo.

In addition, biologists of the time were unaware that the material of the chromosomes carries information. Biologists assumed that the individual chromosomes in a cell are all alike, no more different from one another than are individual pretzels. Although they might look different from one another, dividing them would be similar to dividing a bag of pretzels between two people. It wouldn't matter who got which pretzel, as long as each person got half.

Flemming and others had demonstrated the presence of individual chromosomes. For example, we now know that every human cell has the same set of 23 pairs of chromosomes. Each member of a pair has a distinctive and unique shape and size. Flemming saw the uniqueness of chromosomes in other organisms and described how each chromosome splits down its length, one-half "predestined for one new cell and [one-half] for the other." If we divided a bag of pretzels in half, however, we would not cut each pretzel down its whole length. We would just give each person half the pretzels in the bag.

It was the great embryologist Wilhelm Roux who first observed that the complex way the chromosomes split during mitosis strongly suggested that the chromosomes were not uniform at all. Why, he asked, would the cell use such a complex process to divide the nucleus, dividing each chromosome neatly in half, if simply cutting the nucleus in two would work? The chromosomes in the nucleus, argued Roux, must contain material of different kinds. The cell must be carefully dividing the material so that each daughter cell receives some of each.

In 1889, August Weismann insisted that the nucleus contained the hereditary material, which he called the *germ plasm.* "Heredity," Weismann wrote, "is brought about by the transference from one generation to another, of a substance with a definite . . . molecular constitution." But Weismann was known for his radical ideas, and most biologists of the time dismissed Weismann's claim.

In the same year, the German cell biologist Theodor Boveri performed an unusual experiment that lent firm support to the idea that the nucleus can determine the characteristics of an organism. Boveri discovered that, by vigorously shaking the eggs of a sea urchin until they broke open, he could knock the nucleus right out of the eggs. He then fertilized these damaged "enucleated" eggs with sperm from a different species of sea urchin. Unlike normal zygotes, these had no genetic material from the mother (Figure 8-3).

Miraculously, these zygotes developed into normal sea urchins. But the resulting adult sea urchins did not look like the species from which the eggs came. Instead they looked like the species from which the sperm came. In a control experiment, Boveri used sperm that belonged to the same species as the eggs. The resulting sea urchins looked like the species from which the eggs and sperm both came. Since the sperm contributes virtually nothing but a nucleus, here was dramatic evidence that the nucleus by itself can determine the characteristics of the organism. In 1891, Boveri wrote, "in all cells derived . . . from the fertilized egg, one half of the chromosomes are of strictly paternal origin, the other half of maternal."

But Boveri's experiment left hardly a ripple in the scientific community. As we will see in the next chapter, evidence that was even more persuasive failed to persuade many biologists that the chromosomes are the hereditary material. And,

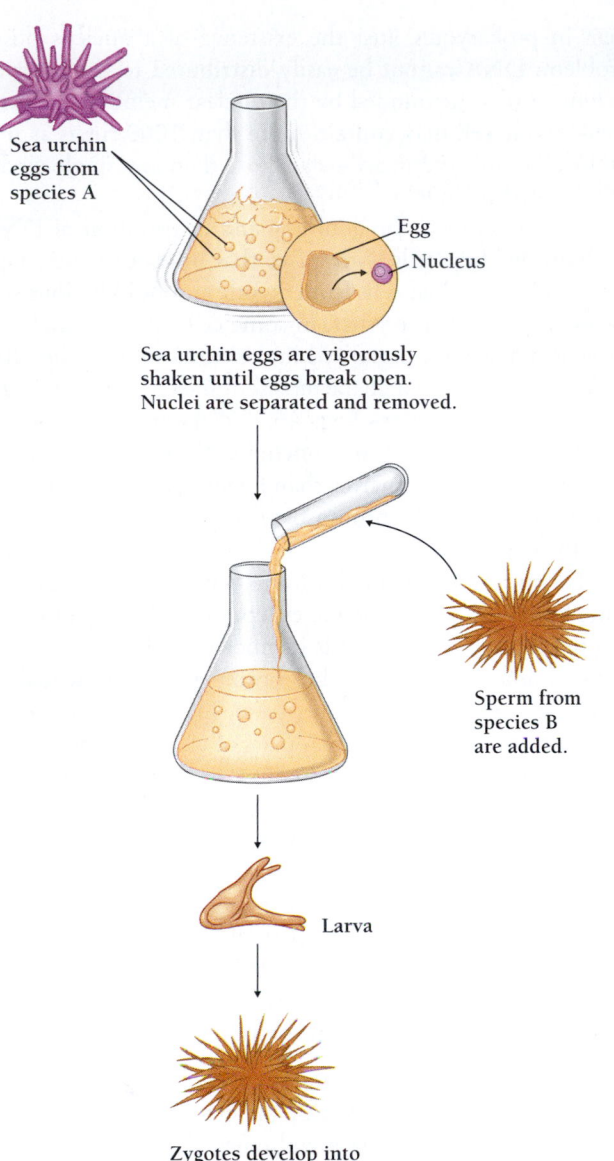

Sea urchin eggs from species A

Egg

Nucleus

Sea urchin eggs are vigorously shaken until eggs break open. Nuclei are separated and removed.

Sperm from species B are added.

Larva

Zygotes develop into normal sea urchins resembling species B.

Figure 8-3 The nucleus determines the characteristics of an organism. In an experiment in 1889, Theodor Boveri shook sea urchin eggs with such force that the nuclei fell out. "Enucleated" eggs that were fertilized with sperm from a different species of sea urchin developed into individuals like the species from which the sperm came. Boveri concluded that a nucleus (in this case, the sperm's) determines the characteristics of the organism.

as recently as the 1930s, a few respected biologists still rejected the notion that the chromosome is the vehicle of inheritance.

Today we have much more information about both chromosomes and the nucleus. In later chapters, we will see how biologists learned more about chromosomes, nuclei, and heredity. We will see that each daughter cell carries genetic information inherited from the parent cell and that the genetic material determines how the organism interacts with its environment. In all organisms (and all cells) the genetic material is DNA.

Despite detailed descriptions of the division of chromosomes during mitosis, 19th-century biologists did not realize that the chromosomes carry hereditary information.

HOW DOES A DIVIDING CELL ENSURE THAT EACH DAUGHTER CELL RECEIVES AN EXACT COPY OF THE PARENT CELL'S DNA?

One of the main tasks of cell reproduction is to **replicate,** or copy, the cell's DNA (Chapter 10). After the DNA replicates, the cell's machinery distributes one copy to each of the two daughter cells. When a single molecule of DNA is involved, as in most bacteria, equal distribution is relatively simple. In eukaryotes, where several DNA molecules are involved, the distribution process is more complex. Cell division also splits the rest of the cell's contents between the two daughter cells. Cell reproduction, then, accomplishes three tasks: the replication of DNA, the equal distribution of the DNA to the two daughter cells, and the division of the other cell components.

We can describe the process of cell reproduction in terms of three M's—materials (the small molecules that carry energy or serve as building blocks), machinery (the organelles and macromolecular structures needed to carry on cellular processes), and memory (the information, contained in DNA, for building cellular machinery from the building blocks). Before a cell divides, it usually doubles its materials, machinery, and memory. In most cases, each daughter cell then receives an equal share of each.

Each of the three M's is essential to cell reproduction. DNA, for example, contains the information needed to produce the next generation of cells and organisms. Without the memory encoded in DNA, the cell's materials and machinery are useless. On the other hand, without the cell's materials and machinery, the DNA has no environment in which to express itself.

The information in the DNA comes from both parents. But the materials and machinery come exclusively from the mother. The precise doubling of DNA employs some of the cell's most elaborate molecular mechanisms. We discuss these mechanisms at length in Chapter 10. In this chapter, we see how dividing cells coordinate the duplication and distribution of DNA and cytoplasm.

Each DNA molecule consists of two polynucleotide chains. After the DNA replicates, each daughter DNA molecule contains a new chain and one of the parent chains. The cell's machinery then distributes one copy to each of the two daughter cells.

How Do Prokaryotic Cells Divide?

Cell division in prokaryotes is fairly straightforward. Prokaryotes divide by **binary fission,** in which a cell pinches in two, distributing its materials and molecular machinery more or less evenly to the two daughter cells. Since prokaryotes have no membrane-enclosed nucleus, binary fission is simpler than mitosis. In most prokaryotes, genetic information is stored in a single molecule of DNA, which usually forms a closed loop (Figure 8-4A). This loop is referred to as a "circle" of DNA, even though it looks like a true circle only in a diagram. Before the prokaryotic cell divides, it duplicates its DNA. Each of the two resulting daughter DNA molecules then attaches to the membrane fold at which binary fission begins (Figure 8-4B and C). As the cell divides, the two DNA molecules are pulled apart by their attachments to the plasma membrane, so that each daughter cell receives a single molecule of DNA.

Bacteria divide by binary fission in a process that does not involve mitosis.

How Do Eukaryotic Cells Divide?

The internal membranes and compartments of eukaryotic cells make the process of cell reproduction far more complicated than in prokaryotes. Just the existence of a nucleus poses a problem: DNA cannot be easily distributed to daughter cells as long as it is surrounded by the nuclear membrane. Further, a eukaryotic cell may contain more than 1000 times as much DNA as a prokaryotic cell—far more than any single circle of DNA could include.

We can observe the doubling and distribution of DNA if we stain dividing cells with a dye that binds to DNA. Figure 8-2 shows individual cells from an onion root in various stages of division. We can see that, in some cells, the material of the nucleus has become organized into chromosomes. The cells of each species contain a characteristic number of chromosomes: onions have 16, humans have 46, and the fruit fly *Drosophila melanogaster* has 8. (Don't conclude, however, that 46 chromosomes make you smarter than a fruit fly or an onion. Your baked potato had 48 chromosomes in each cell before it went into the oven.)

Nearly every cell in the human body has copies of the same 46 chromosomes. (One exception is the red blood cell; as a red cell develops, it sheds its nucleus and ends up with no chromosomes at all.) The chromosomes of most sexually reproducing eukaryotes come in pairs, like socks (Figure 8-5). Humans, for example, have 23 pairs of chromosomes, whereas fruit flies have 4 pairs. Each member of the pair is similar but not identical to its mate. Pairs of matching chromosomes are called **homologous chromosomes** [Latin, *homo* = same], and each member of the pair is called a **homolog.**

Everything that happens to a cell from the time that it first forms until it divides is part of the **cell cycle**—the orderly sequence of events that accomplish cell reproduction. The time from one division to the next is one cycle. Biologists distinguish three major stages in the cell cycle:

1. **Mitosis,** the division of the nucleus.
2. **Cytokinesis,** the division of the cytoplasm and formation of two separate plasma membranes.

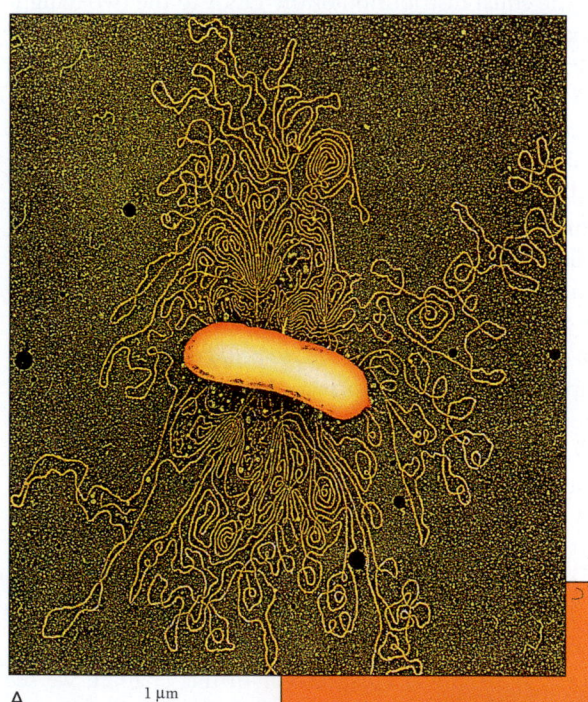

A.   1 μm

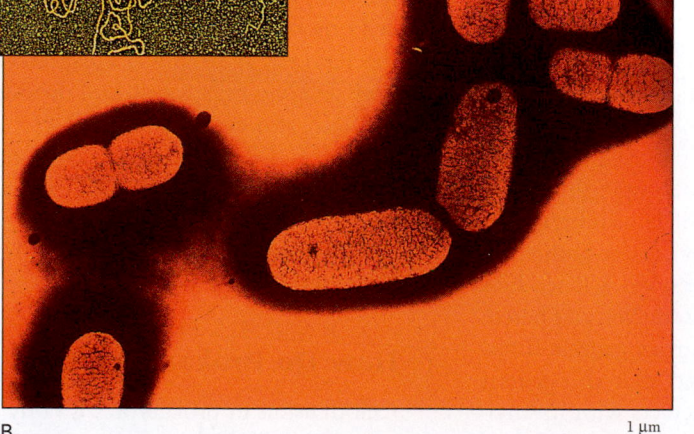

B.   1 μm

Figure 8-4 Prokaryote cells divide by pinching in two. A. In *Escherichia coli* and most other prokaryotes, genetic information is stored in a circular loop of DNA. Here, circular DNA is spilling out in loops from a broken *E. coli* bacterium. B. As the cell divides, the two DNA molecules are pulled apart by their attachments to the membrane. (*A, Dr. Gopal Murti/Science Photo Library/Photo Researchers, Inc.; B, K.G. Murti/Visuals Unlimited*)

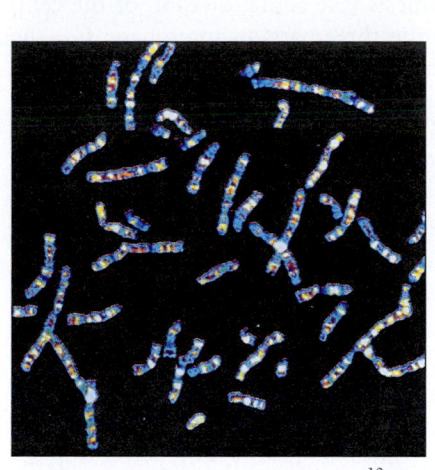

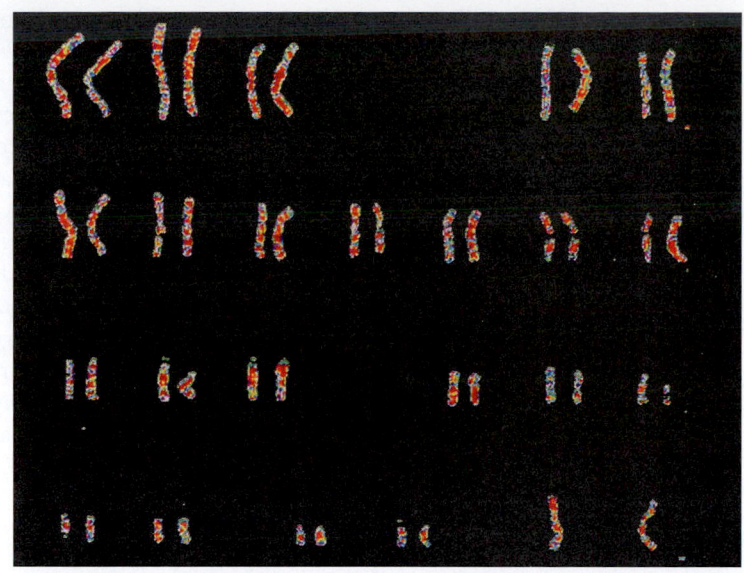

10 μm

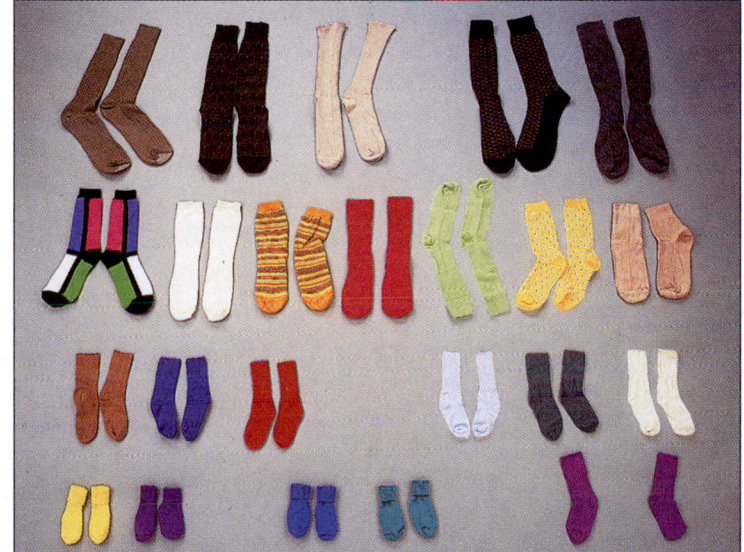

A.

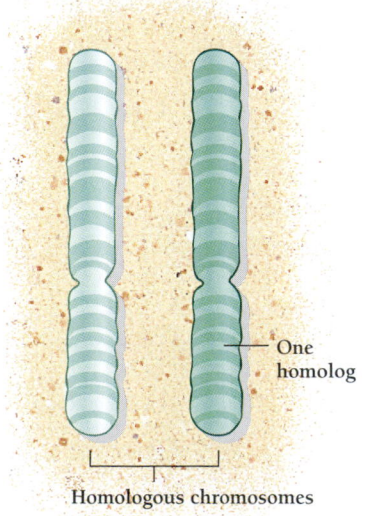

One
homolog

Homologous chromosomes

B.

Figure 8-5   The 46 chromosomes in the human karyotype come in pairs, like socks. A. If we take a photograph of a set of chromosomes, then cut out the 46 individual chromosome images and arrange them in pairs of decreasing size, we create a tidy picture called a karyotype. In this micrograph, we can see chromosome bands that provide landmarks for the location of a hemoglobin gene in human chromosomes. B. Each member of a pair is similar, or homologous, but not identical to its mate. In each chromosome, the DNA loops and clusters in characteristic ways to form a pattern of bands that becomes visible under a microscope when the bands are stained. Biologists can detect abnormal chromosomes by comparing the banding patterns of abnormal chromosomes with those of normal chromosomes. (*A, top, Custom Medical Stock; bottom, Paraskevas Photography*)

3. **Interphase,** the time when the DNA replicates and when the chromosomes are not condensed and the cytoplasm is not dividing.

The amount of time that a single cell cycle takes depends on the organism and on its circumstances. Most plant and animal cells are capable of dividing in about 1 day, with mitosis and cytokinesis occupying 1 or 2 hours. But every cell is different. For example, a neuron spends most of its life in interphase, while rapidly dividing cancer cells spend very little time in interphase. The replication of DNA during interphase is invisible, and nothing appears to be happening in the cell. Only when biologists recognized the importance of DNA replication did they realize that what happens during interphase is just as important as what happens during mitosis and cytokinesis.

The amount of DNA in a cell provides a useful way of dividing the cell cycle into different phases (Figure 8-6). The most visually dramatic phase is **M,** for *mi*tosis, which includes both mitosis and cytokinesis (the actual division of the cell). After cytokinesis come the three parts of interphase: **G₁,** the *g*ap, or growth phase, between the completion of M and the beginning of DNA synthesis; **S,** the period of DNA *s*ynthesis, when a cell doubles its DNA content; and **G₂,** the *g*ap between the completion of DNA synthesis and the beginning of M (of the next cell cycle). During G₁, a cell accumulates the materials and machinery needed for the DNA synthesis that will occur during S. In G₂, the cell prepares for mitosis and cytokinesis, assembling the molecular machinery needed to sort the chromosomes and divide the cell.

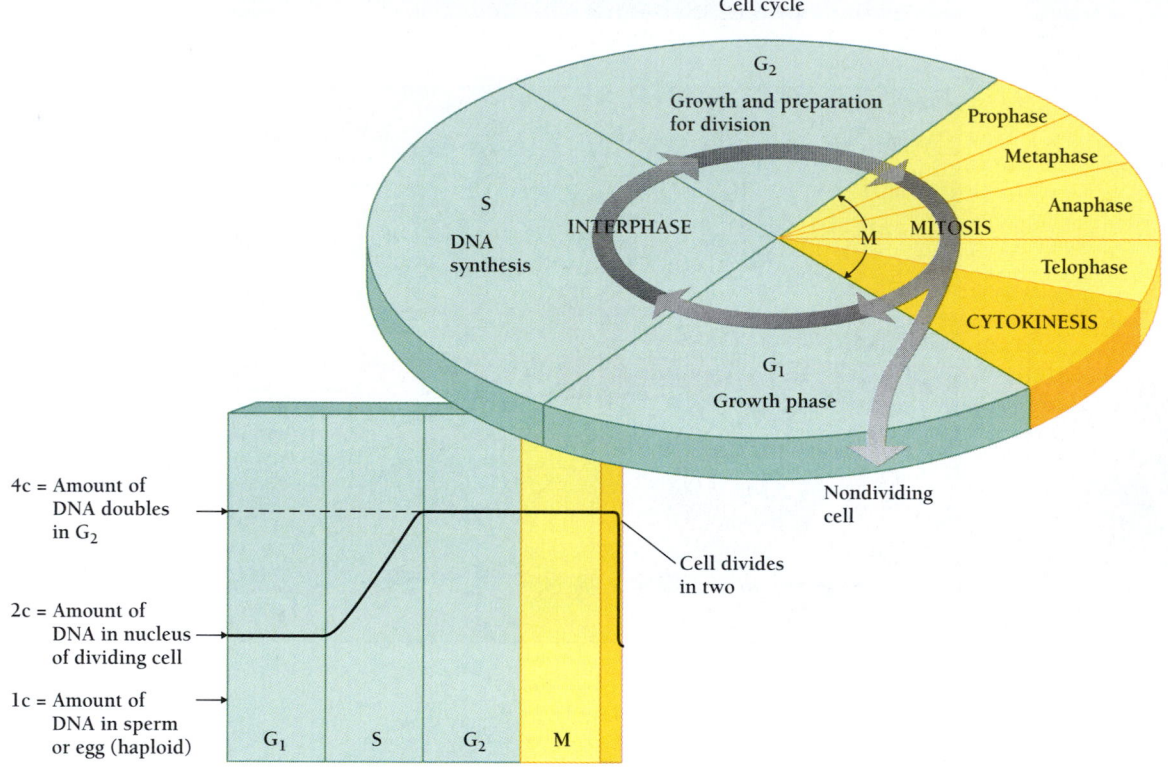

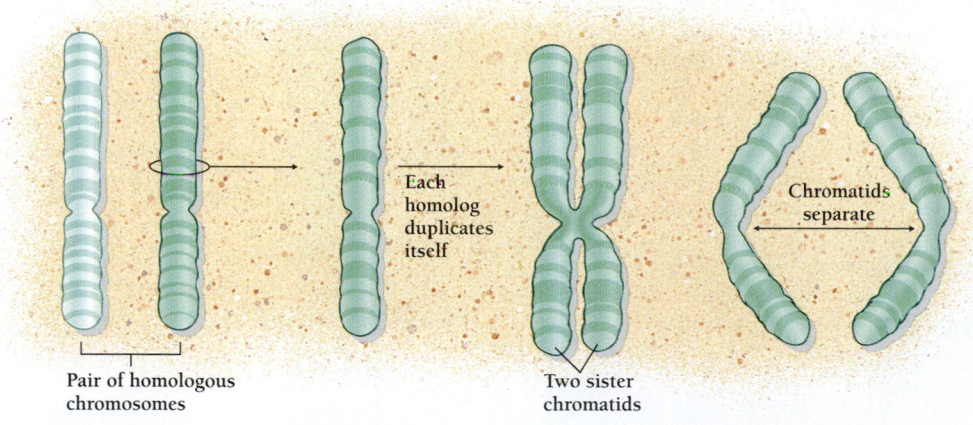

Figure 8-6 The cell cycle consists of G₁, S, G₂, mitosis, and cytokinesis. After the first growth phase (G₁), the DNA doubles (S). The cell then increases materials (G₂) in preparation for the separation of identical sister chromatids (mitosis) and the separation of the two daughter cells (cytokinesis).

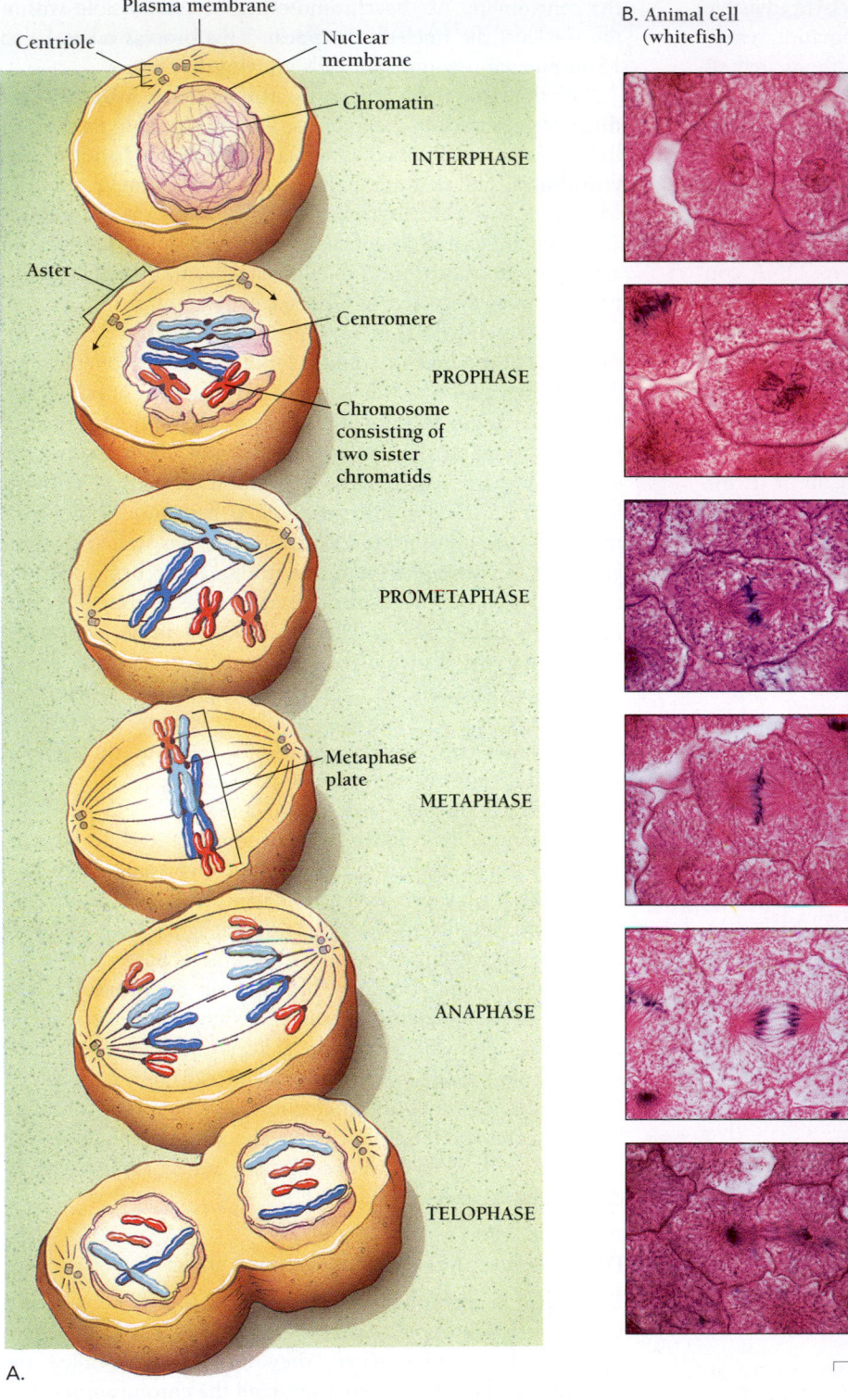

Plasma membrane
Centriole
Nuclear membrane
Chromatin
INTERPHASE

Aster
Centromere
PROPHASE
Chromosome consisting of two sister chromatids

PROMETAPHASE

Metaphase plate
METAPHASE

ANAPHASE

TELOPHASE

A.

B. Animal cell (whitefish)

25 μm

Figure 8-7 The stages of mitosis appear clearly in microscopic images of cells whose chromosomes have been stained. Look for the mitotic spindle, sister chromatids, centriole, kinetochores, centromere, kinetochore microtubules, polar microtubules, metaphase plate, and asters. *(B, M. Abbey/Photo Researchers, Inc.)*

With each new phase of the cell cycle, the characteristics of the chromosomes change. At the beginning of mitosis, when the chromosomes first become visible, each chromosome consists of two separate but connected bodies, called **sister chromatids** (Figure 8-7). The two chromatids are joined at a single point called the **centromere.** Both the chromatids and the centromere consist of molecules of DNA joined to various proteins. The two sister chromatids that make up a single chromosome are duplicate copies with exactly the same genetic information. In contrast, homologs are similar but different. During mitosis, the sister chromatids within each chromosome separate and go to opposite poles of the cell.

During cytokinesis each chromosome consists of a single chromatid, which gradually unfolds into chromatin and disappears. All the chromosomes remain unchanged throughout G_1, and we cannot see them under a microscope. During S, each chromosome duplicates, a process that takes 8 to 10 hours in most eukaryotic cells. The amount of DNA also doubles, so that each chromosome again consists of two (still invisible) sister chromatids. During G_2, the paired chromatids remain invisible. Only at the beginning of mitosis do the paired sister chromatids again become visible.

Researchers cannot tell whether a cell is in S, G_1, or G_2 by looking. They can sometimes tell what phase a cell is in, however, by measuring the amount of DNA in each cell. One way to measure the DNA content of individual cells is to treat cells with a fluorescent dye that binds to DNA. The amount of fluorescence from each cell is a measure of its DNA content and so of its phase in the cell cycle (Figure 8-7). To see if a cell is in S, researchers supply radioactively labeled precursors called nucleosides. Cells in S take up these radioactive nucleosides (to make new DNA), while cells in G_1 or G_2 do not.

During G_1, the cell and the DNA appear to do nothing. Yet this is the time when the cell does most of the work of reproduction. During G_1, the cell doubles nearly all of its materials and machinery (other than DNA). If the cell runs out of nutrients, it stops the cycle in G_1.

The cell cycle consists of the three parts of interphase (G_1, S, and G_2), mitosis, and cytokinesis. During interphase, a cell doubles all of its materials, including the DNA. During mitosis, the sister chromatids separate. During cytokinesis, the cell divides into two daughter cells.

HOW DOES MITOSIS DISTRIBUTE ONE COPY OF EACH CHROMOSOME TO EACH DAUGHTER CELL?

The major task of mitosis is to distribute one chromatid from each chromosome to each daughter cell. Mitosis normally ensures that each offspring cell inherits two sets of chromosomes. Remarkably, all eukaryotic cells accomplish mitosis with the same chromosomal ballet that Flemming first described more than 100 years ago. Modern optical techniques, however, allow biologists to watch the whole process in living cells far more conveniently than could Walther Flemming.

Mitosis Is a Continuous Process, but Biologists Distinguish Four Phases

Cell biologists divide mitosis into four phases. These are prophase, metaphase, anaphase, and telophase. The boundaries between these phases are arbitrary, since mitosis is a continuous process. Nonetheless, the events described in these four phases always occur in exactly the same order (Figure 8-7).

During prophase, chromosomes condense and the mitotic spindle forms

During the first phase of mitosis, **prophase** [Greek, *pro* = before], the diffuse chromatin condenses into discrete chromosomes, each consisting of two chromatids, joined together at the centromere. As the chromosomes become visible within the nucleus, the nucleoli disappear. This process takes 10 to 15 minutes in mammalian cells growing in culture.

At the same time, a new structure, called a **mitotic spindle,** develops outside the nucleus. The mitotic spindle, shaped like an American football, consists of prominent bands of **microtubules,** hollow tubes constructed of the protein tubulin. Microtubules form the machinery that moves the chromatids apart. The microtubules of the mitotic spindle are identical in appearance to those found in cilia, flagella, and the cellular cytoskeleton, as discussed in Chapter 4.

In animal cells (and in some other eukaryotes as well), each pole of the mitotic spindle contains a pair of small cylindrical **centrioles,** which lie at right angles to one another. Microtubules originate from each pole of the spindle. Some of the microtubules, the "polar microtubules," run toward the equator. Others, called "astral microtubules," extend outward from each centriole to form an aster, as first noticed by Flemming. By the end of prophase, the spindle has started to elongate, and the two poles (with or without asters) begin to move apart. (The cells of most plants usually have neither centrioles nor asters, and mitosis in these cells is said to be "anastral.")

During the transition from prophase to metaphase, so much happens that researchers give this period its own name, "prometaphase." The nuclear membrane disappears, and the chromosomes are free to attach to the mitotic spindle. Throughout mitosis, the homologs in each pair of chromosomes behave independently of one another.

In mammalian cells, the chromosomes take 10 to 20 minutes to attach to the mitotic spindle. Each chromatid develops a **kinetochore,** a specialized disc-shaped structure that attaches the mitotic spindle to the centromere (Figure 8-8). Some of the polar microtubules attach to the kinetochores (and so to chromosomes) and become "kinetochore microtubules." By the end of prometaphase, animal cells have three sets of microtubules, all of which originate at a spindle pole: kinetochore microtubules, polar microtubules, and astral microtubules. Most plant cells contain kinetochore and polar microtubules but no astral microtubules. After the spindle apparatus has formed, the attached chromosomes are moved along the spindle toward the cell's equator.

During metaphase, chromosomes align

By the end of **metaphase,** the second and longest stage of mitosis, the chromosomes have moved halfway between the two poles of the spindle, where they form a disc, called the **metaphase plate.** For about 1 hour, all the chromosomes then lie in a single plane at the equator (at right angles to the spindle fibers). At this stage, the individual chromosomes look different from one another, in size and in the positions of the centromeres.

During anaphase, chromatids separate

The most dramatic stage of mitosis is **anaphase,** the time of chromatid separation. The centromere of each chromosome splits so that each chromatid contains its own centromere con-

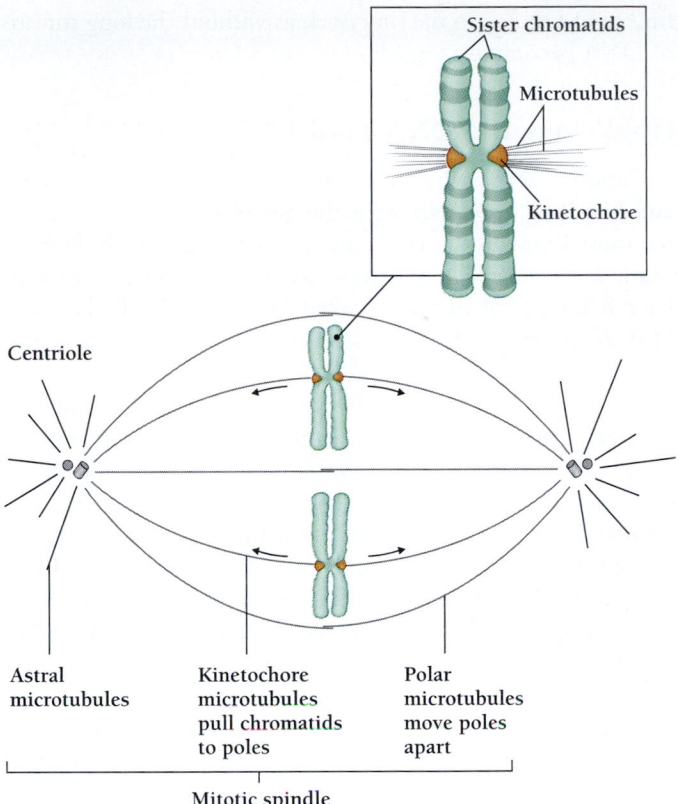

Sister chromatids

Microtubules

Kinetochore

Centriole

Astral microtubules

Kinetochore microtubules pull chromatids to poles

Polar microtubules move poles apart

Mitotic spindle

Figure 8-8 Each chromatid develops a disc-shaped kineto-chore that attaches the mitotic spindle to the chromosome's centromere. Polar microtubules attach to the kinetochores (and so to chromosomes) and become "kinetochore microtubules." By the end of prometaphase, animal cells have three sets of microtubules originating at a spindle pole: kinetochore microtubules, polar microtubules, and astral microtubules.

nected to the kinetochore microtubules. Then, all at once, the two chromatids that make up each chromosome begin to move away from the metaphase plate. The paired chromatids separate and move in opposite directions. For 5 to 10 minutes, all the chromatids move at the same speed (about 1 μm [millionths of a meter] per minute) toward the two poles of the spindle. Each centromere leads the way, with the rest of the chromatid dragging behind, as if the kinetochore microtubules are pulling the chromatid toward the poles.

Even as the chromosomes move toward the spindle poles, the poles themselves are moving further apart, often doubling the distance between them. By the end of anaphase, the two processes, acting simultaneously, have separated the two sets of chromatids.

During telophase, the mitotic apparatus disappears

During **telophase** [Greek, *telos* = end], the last phase of mitosis, the mitotic apparatus (including kinetochore, polar, and astral microtubules) disperses. The chromosomes once more assume the diffuse appearance of chromatin, losing their distinct identities, and the nucleoli again become visible. The nu-

clear membranes form and enclose the two new sets of daughter chromosomes. Each daughter nucleus has the same number of chromosomes as the parent nucleus, but now each chromosome consists of a single chromatid. (Each chromosome will replicate during the next S phase.) After the completion of telophase, the cell usually completes cytokinesis, which we will discuss later.

What Propels the Chromosomes During Mitosis?

We have described three sets of movements whose molecular mechanisms we would like to understand: (1) the movement of the chromosomes toward the equator to form the metaphase plate, (2) the movement of the separated chromatids toward the poles, and (3) the movement of the spindle poles away from each other during prophase and again during anaphase. The chromosomes themselves do not play active roles in their own movement. Rather, the major players are the microtubules. These seem to push the chromosomes around by lengthening and shortening, as necessary.

Experiments have demonstrated the importance of microtubules in mitosis. In one experiment, a researcher used a laser to cut the kinetochore microtubules that attach one of the two sister chromatids to one pole at metaphase. The loose chromatid followed its sister to the pole opposite the one to which it normally would have gone. This experiment showed that the kinetochore microtubules normally pull the two chromatids in opposite directions. Cell biologists still do not fully understand the mechanism of mitosis, however, and the identities and roles of the components of the mitotic machinery are the subject of intense research.

During mitosis, microtubules move the chromosomes by attaching to the chromatids and growing or shortening.

HOW DOES A CELL FIT ALL ITS DNA INTO A NUCLEUS?

The DNA in a cell carries information that the cell uses to construct cell machinery and to carry out cell processes. When a cell divides, the cell must duplicate this information and pass it on to its two daughter cells.

Each molecule of DNA contains two polynucleotide strands, wrapped around each other in a double helix. This ropelike structure is so fine that a DNA molecule long enough to circle the Earth at the equator would weigh less than a grain of fine sand. In contrast, the same length of fine cotton thread would weigh about 275 kg (over 600 lb). The DNA in a human cell weighs about 6×10^{-12} g (6 trillionths of a gram) and has a total length of about 2 m. Yet a chromosome is only a millionth of a meter long, and the nucleus is only 5 μm in diameter. A cell must therefore solve a major problem: how to

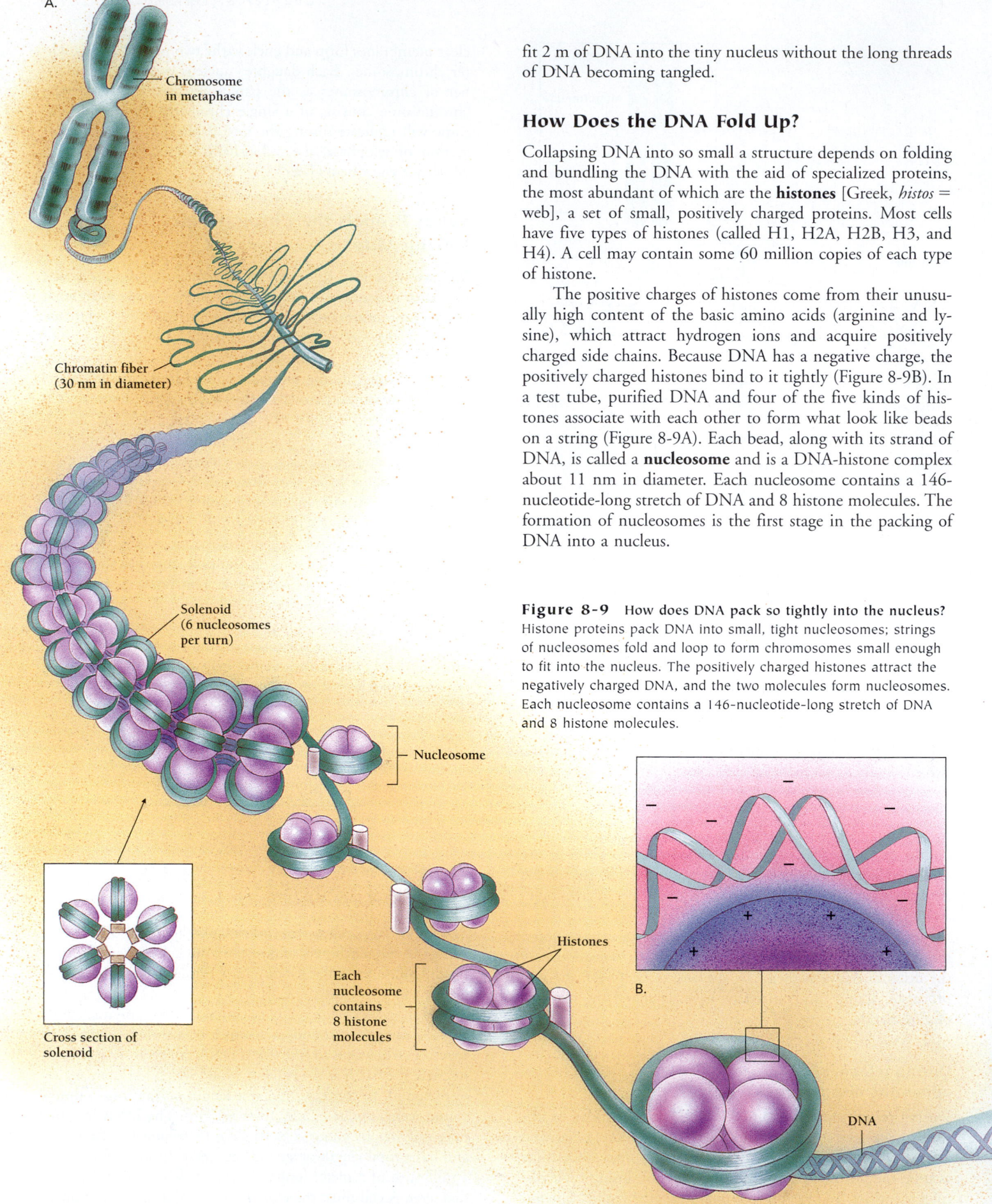

A.

Chromosome in metaphase

Chromatin fiber (30 nm in diameter)

Solenoid (6 nucleosomes per turn)

Nucleosome

Cross section of solenoid

Each nucleosome contains 8 histone molecules

Histones

B.

DNA

fit 2 m of DNA into the tiny nucleus without the long threads of DNA becoming tangled.

How Does the DNA Fold Up?

Collapsing DNA into so small a structure depends on folding and bundling the DNA with the aid of specialized proteins, the most abundant of which are the **histones** [Greek, *histos* = web], a set of small, positively charged proteins. Most cells have five types of histones (called H1, H2A, H2B, H3, and H4). A cell may contain some 60 million copies of each type of histone.

The positive charges of histones come from their unusually high content of the basic amino acids (arginine and lysine), which attract hydrogen ions and acquire positively charged side chains. Because DNA has a negative charge, the positively charged histones bind to it tightly (Figure 8-9B). In a test tube, purified DNA and four of the five kinds of histones associate with each other to form what look like beads on a string (Figure 8-9A). Each bead, along with its strand of DNA, is called a **nucleosome** and is a DNA-histone complex about 11 nm in diameter. Each nucleosome contains a 146-nucleotide-long stretch of DNA and 8 histone molecules. The formation of nucleosomes is the first stage in the packing of DNA into a nucleus.

Figure 8-9 **How does DNA pack so tightly into the nucleus?** Histone proteins pack DNA into small, tight nucleosomes; strings of nucleosomes fold and loop to form chromosomes small enough to fit into the nucleus. The positively charged histones attract the negatively charged DNA, and the two molecules form nucleosomes. Each nucleosome contains a 146-nucleotide-long stretch of DNA and 8 histone molecules.

Histones from all eukaryotes resemble one another. For example, histone H4 of cows differs by only 2 amino acids (out of 102) from that of peas. This extraordinary similarity suggests that the structure of histone H4 has remained almost unchanged for more than a billion years. We can conclude that this method of packing DNA evolved long ago.

Proteins called histones pack DNA into small, tight bundles called nucleosomes that resemble beads on a string.

How Do DNA, Histones, and Other Proteins Form Such Compact Structures?

Histone H1 appears to be responsible for holding together groups of nucleosomes to form a fiber that is 30 nm in diameter. As a result of this packing, the DNA of a typical human chromosome could be condensed from an average length of 5 cm to about 1 mm—a terrific achievement, but still not small enough for it to fit into a 5-μm nucleus. We still do not know exactly how DNA is made more compact. In some specialized cells, microscopists can see loops extending from chromosomes (Figure 8-10). Many biologists now think that such loops are common to the chromatin of all cells.

In chromosomes, the DNA is even more condensed than in the diffuse chromatin. Still more loops form, and these loops cluster in characteristic ways in each chromosome. The specific pattern of DNA clustering results in **chromosome banding,** the pattern (visible under a microscope) that results from the selective binding of certain dyes. Biologists can detect abnormal chromosomes by comparing the banding patterns of abnormal chromosomes with those of normal chromosomes. Using DNA technology, it is now possible to determine the location of individual genes on specific chromosomes, using chromosome bands as landmarks (Figure 8-5A).

Strings of nucleosomes fold and loop to form the chromosomes, the highly compact arrangement of chromatin that fits into the nucleus. No one knows how this occurs.

HOW DOES A CELL DIVIDE ITS CYTOPLASM?

So far we have discussed how a cell distributes the two copies of its genetic information. But we still need to describe cytokinesis—the process by which a dividing cell partitions its cytoplasm.

Cytokinesis and mitosis are separate processes, and they depend on different molecular machinery. In some organisms, for example, mitosis sometimes occurs without cytokinesis, leading to two or more nuclei within a single cell, as in the early fruit fly embryo and in the specialized tissues of a plant seed. This shows that cytokinesis and mitosis are truly separate processes. Nonetheless, cytokinesis almost always accompanies mitosis, usually beginning during anaphase and finishing shortly after the end of telophase.

During cytokinesis, the plane of cell division is always perpendicular to the axis of the mitotic spindle (Figure 8-11). In early anaphase, dividing animal cells form a beltlike **contractile ring,** a bundle of actin filaments that surrounds the dividing cell. The contractile ring works like a purse string, pinching the cell into two parts through the spindle's equator. The force for this movement comes from a mechanism similar to that which causes muscle contraction. In contrast to the permanent contractile structures of muscle cells, however, the contractile ring is a temporary structure like the mitotic spindle.

As the beltlike contractile ring tightens, the cell membrane is pulled into a deepening groove, called the **cleavage furrow** (Figure 8-11). As telophase proceeds, the two daughter cells

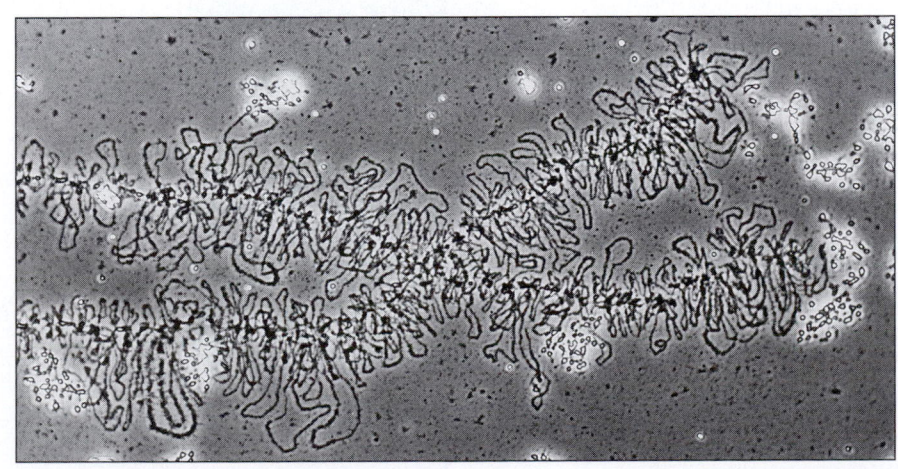

25 μm

Figure 8-10   In some cells, the DNA forms loops, obvious in this micrograph. *(Joseph Gall, Carnegie Institution)*

become almost, but not quite, completely separated. A thin connection between the daughter nuclei, called the midbody, persists until the end of mitosis. The midbody is packed with microtubules from the spindle apparatus. Finally, at the end of telophase, the microtubules disassemble, and the midbody breaks, leaving the daughter cells completely separated.

As the two daughter cells separate, they must enlarge their cell membranes (since two cells have more surface area than a single cell of the same volume). The additional membrane material comes from a supply of extra membrane made by the parent cell during interphase.

A plant cell, which has a rigid cell wall, cannot simply pinch itself in two the way an animal cell does. Cytokinesis in a plant cell requires the building of a new cell wall between the daughter cells. The new wall begins in telophase as a small, flattened disc, called the early **cell plate,** in the space between the two daughter nuclei (Figure 8-12). The disc grows to become a cell plate, which, except for connecting plasmodesmata, completely seals off the two daughters from one another.

During cytokinesis in animal cells, actin filaments form a contractile ring that pinches the cell in two. In plant cells, telophase includes the building of a new cell wall between the two daughter cells.

HOW DOES A CELL REGULATE PASSAGE THROUGH THE CELL CYCLE?

In this chapter, we have seen how normal cells accomplish the equal distribution of chromosomes by mitosis and the division of cytoplasmic components by cytokinesis. But how do cells

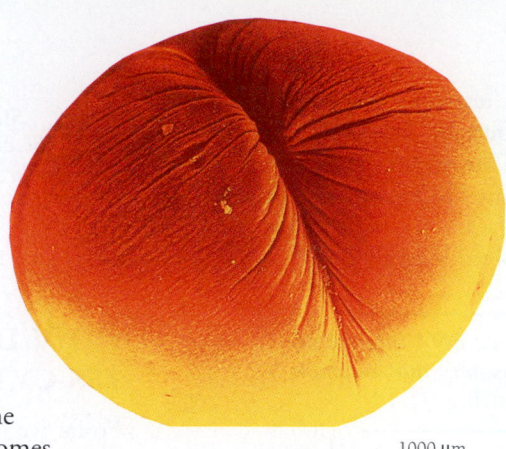

Figure 8-11   During cytokinesis, actin filaments inside the cell form a contractile ring, or cleavage furrow, and pinch the dividing cell into two. *(David M. Phillips/Visuals Unlimited)*

1000 µm

know when to begin the cell cycle? How do they coordinate its separate parts? And how do they know when to stop dividing?

If cell division continued without stopping, single-celled organisms would soon run out of space and food. For example, after just 4 days of undisturbed growth, a single yeast cell could have divided 48 times and produced 2^{48} (about 10^{15}, or 10 million billion) descendants, as many cells as are in 100 people.

In multicellular organisms, cells normally subordinate such growth and division to the needs of the body's tissues and organs. Cells divide rapidly during periods of growth and more slowly (or not at all) in mature organisms. The only exception to this rule is the uncontrolled growth and division of cancer cells, such as those of Henrietta Lacks. These multiply at the expense of all the other cells in the body.

How Do Normal Cells Determine When To Stop Dividing?

What normally prevents such runaway cell division? Two general mechanisms appear to operate—cell senescence (aging) and growth control. **Cell senescence** limits the number of

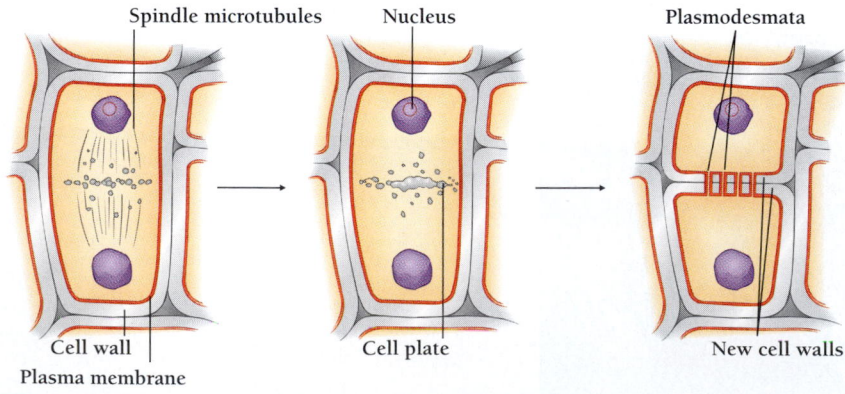

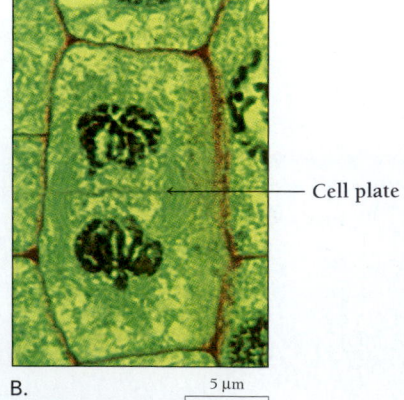

Figure 8-12   In plants, a new cell wall forms between daughter cells. The cell wall begins during telophase as a small, flattened disc, called the early cell plate, then grows into a full-sized cell wall, complete with interconnecting plasmodesmata. *(B, R. Calentine/Visuals Unlimited)*

times a cell can divide: the more times a cell divides (at least under conditions of laboratory culture), the more likely it is to withdraw from the cell cycle. An average cell taken from a newborn baby, for example, will divide about 50 times in a standard culture medium. But cells taken from an 80-year-old stop cycling after about 30 divisions. Despite the general occurrence of cell senescence, however, even a mature mammal contains cells that are capable of unlimited cell divisions in the presence of the right "growth factors."

A second set of regulatory mechanisms—those involved in **growth control**—prevent uncontrolled division by allowing the cell cycle to proceed under some conditions and to stop under others. As an example of growth control, think about what happens when you cut yourself. The skin cells on the edges of the cut begin to divide and fill the space left by the wound. The cells divide until the two edges again touch. When the cells on each side of the cut meet, and the wound is healed, cell division ceases.

In a culture dish, cells regulate their division in much the same way (Figure 8-13). Cells attach to the bottom of a plastic dish and divide until they form a single, continuous layer of cells that occupies the whole surface. Then they stop dividing. If we now make a "wound" in this monolayer by scraping away a swath of cells, the cells on the margins move into the open space and begin to divide. Division stops when the space is filled.

We can summarize the growth control of normal cells in culture with the following rules: (1) cells stop dividing during G1 when they run out of free space on which to spread—that is, when neighboring cells all touch each other; and (2) cells that have proceeded beyond G1 begin to divide when contact with their neighbors ceases. This kind of growth control is called **contact inhibition** of cell division.

Cancer cells provide an excellent example of what happens when contact inhibition fails. Cancer cells do not exhibit growth control in cell culture (Figure 8-13). Instead of forming a continuous monolayer, they pile up on top of one another and grow to much higher densities than normal cells.

Growth control depends on the actions of several types of proteins: (1) growth factors, which stimulate cell division or promote cell survival, depending on the type of cell; (2) proteins that bind to growth factors and contribute to the sequence of events that they trigger; (3) proteins that regulate the expression of genes into RNA and protein; and (4) "tumor-suppressor" proteins, which regulate the passage of cells through the cell cycle. Changes in just one kind of tumor-suppressor protein, called *p53*, occur in more than 50% of human cancers.

Growth control mechanisms such as contact inhibition prevent normal cells from growing out of control the way cancer cells do.

How Do Normal Cells Determine When It Is Time To Divide?

Most cells divide only after they have first doubled their mass. Otherwise, daughter cells would get smaller with every gener-

ation. Somehow cells sense when they have reached a critical size and have enough nutrients to supply two daughter cells. We do not know how eukaryotic cells accomplish this sensing, but once they do, they become irreversibly committed to cell division. Once a cell proceeds beyond a certain point in G_1, the cell proceeds through the rest of the cycle, including mitosis and cytokinesis. This "point of no return" is called **Start.** Arrival at Start depends heavily on the environment of the cell—including the availability of nutrients and signals from other cells. Once the cell has gone through Start, however, the rest of the cycle proceeds, independent of the extracellular environment.

One important exception to the general rule that cells double their mass before dividing are egg cells, which grow to enormous sizes without dividing. For example, the ostrich egg is a single cell. Another important exception are the early embryos of many animal species. Embryos derived from large eggs begin life after fertilization with a series of rapid cleavage divisions, which divide the embryo into many cells without increasing its mass. A frog zygote, for example, is an enormous single cell about 1 mm in diameter. Cleavage divisions rapidly divide it into cells that are more nearly the size of cells in the adult, about 10 to 20 μm in diameter. Even a tiny human zygote divides into 50 or 60 cells in the week before it connects to its mother's blood supply. Only after the human zygote connects can it increase the size of its cells.

The cells of mature organisms only divide when they have doubled their mass, unless they are egg cells or the cells of an early embryo.

What Triggers the Main Events of Mitosis?

Cell reproduction requires the coordination of three cycles:

1. The duplication and packaging of DNA, which we discuss in Chapter 10
2. The duplication of the centrosomes and the operation of the mitotic spindle apparatus
3. Cytokinesis, which depends on the contractile ring

Recent research has shown that all eukaryotes, from yeasts to humans, use almost identical mechanisms to coordinate and initiate the events of these three processes. One way researchers study cell cycles is to create **synchronous cell populations,** groups of cells that are all at the same stage of the cell cycle. One source of such synchronous populations are the rapidly dividing cells of early embryos from marine invertebrates; a researcher can start thousands of embryos dividing simultaneously just by mixing sperm and eggs, collected separately. Another way of producing synchronous populations is to block DNA synthesis with a chemical inhibitor that stops the cell cycle just before S. When the inhibitor is removed, all the cells begin S at the same time and continue in lock step through the rest of the cell cycle.

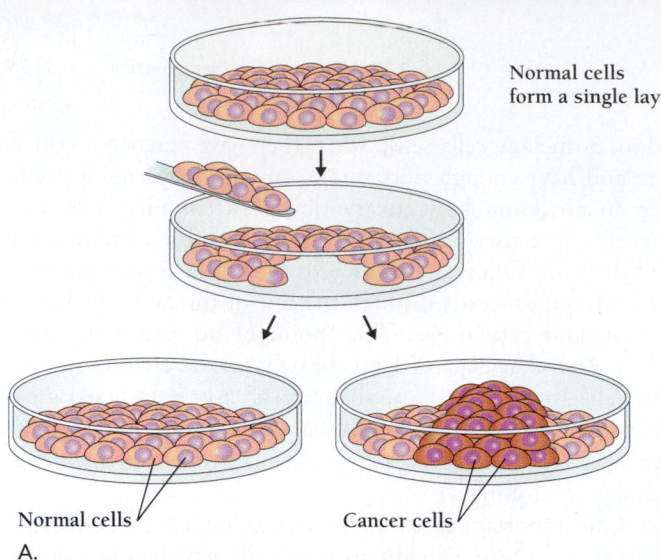

Normal cells form a single layer

When cells are scraped away, normal cells divide until they form a single layer. Cancer cells continue to divide and form a mound of cells

Normal cells Cancer cells

A.

Figure 8-13 Cells regulate their division by means of contact inhibition. A. Cancer cells fail to regulate their growth in tissue culture. They pile up on top of one another, unlike normal cells. B. Normal cells stop dividing when they come into contact with other cells. Cancer cells, in contrast, continue to divide even after they come into contact with other cells. In cell cultures, normal cells divide until they form a single, continuous layer of cells, then stop. If we scrape a "wound" in this monolayer of cells, cells from the edge move into the open space and divide until the wound is filled. (*B, courtesy of G. Steven Martin*)

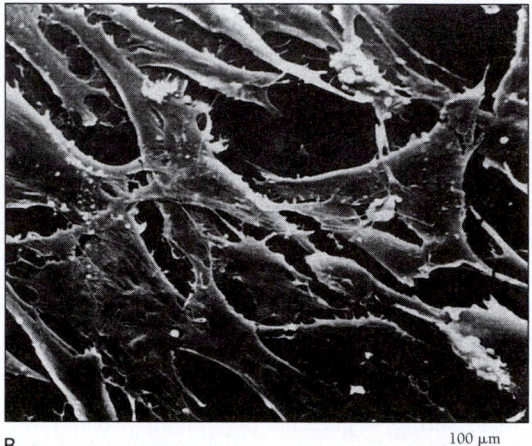

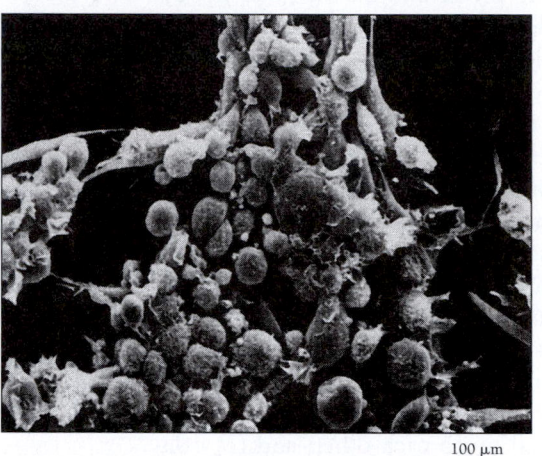

B.   100 μm   100 μm

Studies with all three kinds of synchronized cells have led to the discovery of specific proteins, called **cyclins,** whose concentrations change through the cell cycle. Each type of cyclin accumulates during a particular phase of the cell cycle and stimulates the passage of the cell into the next phase of the cycle.

Each type of cyclin stimulates a specific enzyme, called a cyclin-dependent kinase, which alters proteins of the cell (by adding phosphates from ATP). Among the targets of these enzymes are components of the mitotic spindle. These protein modifications are the "ticks" of the cell cycle clock, propelling the cell to enter the next phase. As the cell moves into the next phase of the cycle, the cell's protein-digesting machines (the proteasomes) digest each type of cyclin, causing the oscillations that give the cyclins their names.

Various events regulate a cell's passage through the cell cycle, including the completion of DNA synthesis and fluctuations in cyclins.

In this chapter we have seen how eukaryotic cells with two sets of chromosomes divide their genetic material to create two daughter cells, each with two sets of chromosomes. Eukaryotes, such as ourselves, employ such mitotic division to create genetically identical copies of cells. In the next chapter, we will see how eukaryotic cells divide a cell with two sets of chromosomes to create gametes (for example sperm and eggs)—cells with just one set of chromosomes. This process, called meiosis, creates cells that are genetically different from the parent cells and also genetically different from one another. Along the way, we will see that sex and meiosis are two sides of the same coin.

Study Outline with Key Terms

Cells reproduce themselves in a process called **cell division,** in which a parent cell gives rise to two **daughter cells,** each with the same genetic information as the parent cell. Dividing cells replicate DNA and distribute it during an orderly sequence of events called the **cell cycle.**

Prokaryotes divide by a simple process called **binary fission,** which distributes one copy of the parent cell's DNA to each daughter cell. Cell reproduction is more complicated in eukaryotic cells, whose DNA is enclosed in a nucleus. Eukaryotes employ a spindle apparatus, but prokaryotes do not.

Chromosomes are highly condensed and visible in dividing eukaryotic cells and more diffuse (**chromatin**) in nondividing cells. Chromosomes come in nonidentical pairs called **homologous chromosomes.** Each member of a pair is a **homolog.** When the DNA in a cell has doubled, each chromosome consists of two identical **sister chromatids.**

Cell biologists divide the cell cycle into five phases: mitosis, cytokinesis, G_1, S, and G_2. The process of evenly distributing chromosomes to daughter cells is called **mitosis.** The process of dividing a cell's cytoplasm is called **cytokinesis.** Mitosis and cytokinesis together make up **M.** During G_1, a cell accumulates the materials and machinery necessary to **replicate** its DNA; during **S,** it synthesizes DNA, making two copies of the parent cell's DNA; during G_2, it prepares for mitosis and cytokinesis. G_1, S, and G_2 together are called **interphase.** The point of no return is **Start.**

Just before a cell begins mitosis, each chromosome consists of two identical chromatids. Mitosis is a continuous process, but biologists distinguish four major phases: prophase, metaphase, anaphase, and telophase. During **prophase,** chromosomes condense and the **mitotic spindle** forms. As the **centrioles** of animal cells move apart, **microtubules** radiating outward from the centrioles form **asters.** Between prophase and metaphase, the nuclear membrane disappears and the chromosomes attach to the spindle fibers. During **metaphase,** the chromosomes align in a disc called a **metaphase plate,** then bind to the **kinetochores** that form on the **centromere** of each chromatid. During **anaphase,** the two chromatids that make up each chromosome separate. During **telophase,** the mitotic apparatus disappears.

During interphase, the cell duplicates its single centriole. During prophase, a mitotic spindle assembles, with microtubules emerging from each centriole to form a mitotic spindle. Microtubules connect to each chromatid in a structure called a kinetochore. The mitotic apparatus is responsible for chromosome movements during mitosis.

The DNA of eukaryotic cells is highly folded. Protein molecules called **histones** bind to DNA and form particles called **nucleosomes** that resemble beads on a string. DNA, histones, and other proteins fold to form still more compact structures. The orderly packing of DNA produces a characteristic pattern of **chromosome banding** in each metaphase chromosome.

During cytokinesis, a **contractile ring** pulls the membrane inward to form a **cleavage furrow,** which deepens and separates the two daughter cells. Cytokinesis in plant cells requires a special mechanism for building a new cell wall. The new wall begins in telophase as a small, flattened disc that grows to become a **cell plate.**

Cells actively regulate passage through the cell cycle. Regulated cell division in multicellular organisms depends on **cell senescence** and **growth control.** Cell senescence limits the number of times a cell can divide. Growth control regulates cell reproduction according to external conditions. For example, many cells exhibit **contact inhibition** of cell division, in which cells divide only when they are not in contact with their neighbors. Growth control may also involve growth factors—extracellular proteins that stimulate cell division. Cancer cells fail to show normal growth control. Studies of **synchronous cell populations** show that changes in specific proteins, called **cyclins,** control a cell's passage through the cell cycle.

Review and Thought Questions

Review Questions

1. Why must a cell double its materials, machinery, and memory before it can divide?
2. Why did the discovery that DNA was the genetic material change the way that cell biologists regarded interphase?
3. Describe a method for measuring the amount of DNA in a cell. How does the amount of DNA in a cell change during the cell cycle?
4. Describe and draw the events of the four stages of mitosis. When does the actual division of the chromosomes occur?
5. Using diagrams, compare the events in prophase to those in telophase. Why do you think they are so nearly opposite?
6. Compare the role of the centriole in mitosis with that of a kinetochore.

7. What factors regulate a cell's ability to proceed to the next round of cell division?
8. Describe the role of histones in the structure of eukaryotic chromosomes. How do histones help DNA fit into a nucleus?
9. How does cytokinesis in a plant cell differ from that in an animal cell?

Thought Questions

10. How would prevention of DNA synthesis affect mitosis?
11. What is the relationship between the materials accumulated in G_1 and G_2 and the processes that follow?
12. How might cancer be associated with improper regulation of the cell cycle?

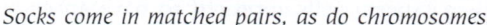

About the Chapter-Opening Image

Socks come in matched pairs, as do chromosomes.

On-line materials relating to this chapter are on the World Wide Web at **http://www.harcourtcollege.com/lifesci/aal2/**

9 From Meiosis to Mendel

Why Is the Yellow Dog Yellow?

In the spring of 1902, a young graduate student at Columbia University, in New York City, approached his professor with barely suppressed excitement and declared that he had solved the problem of heredity. The student, Walter S. Sutton (1877–1916), was a 6-foot-tall, 215-pound Kansas farm boy who towered over everyone around him (Figure 9-1). Gleefully, he announced, "I know why the yellow dog is yellow." At just 25, Sutton had discovered that tiny bars in cells, called chromosomes, must carry the hereditary material.

Chromosomes, realized Sutton, were the mechanism by which organisms inherit traits. They were thus the key to the then-infant science of genetics. But disappointingly, Sutton's professor only nodded indulgently. Neither he nor anyone else took Sutton seriously.

It was disappointing, especially because Sutton's professor was E.B. Wilson (1856–1939), the most distinguished cell biologist in the United States. But when summer came, Wilson invited Sutton to the seaside at Beaufort, North Carolina—to study marine animals, to sail, to swim, and to talk. For the usually preoccupied Wilson, it was also a slow time when he would really listen to young Sutton. "It was only then," Wilson later wrote, "that I first saw the full sweep and the fundamental significance of his discovery."

Sutton had come to biology partly by chance, and he hadn't yet decided if he would stay. In 1896, he had begun studying engineering at the University of Kansas. When he returned home for the summer in 1897, he took with him a case of typhoid fever that felled his entire family. When it was over, Walter Sutton's beloved 17-year-old brother John was dead. In his grief, Sutton decided to abandon engineering for medicine. That fall, he returned to the University of Kansas at Lawrence and began to study biology.

Before long, Sutton became friends with a young professor of zoology named Clarence McClung. On vacations at home, Sutton collected giant grasshoppers from the wheat fields of his family's farm to send to McClung, who admired the grasshoppers' enormous cells (Figure 9-2). The cells of *Brachystola magna,* the lubber grasshopper, in fact, made perfect subjects for the study of the fine structure of the cell.

McClung was especially interested in the fact that some of the grasshopper sperm cells appeared to have an extra chromosome. At the time, no one

Key Concepts

- Because genes lie on chromosomes, the inheritance of genes parallels the inheritance of chromosomes.

- Sexually reproducing organisms have *pairs* of homologous chromosomes, which meiosis distributes—one chromosome from each pair—to each gamete (egg or sperm).

- At fertilization, two gametes unite to form a zygote, which has paired chromosomes. Each gamete contributes one chromosome to each pair.

- The phenotype of an individual results from the two-way interaction between its genotype and the environment.

Figure 9-1 **Walter Sutton.** At age 25, Sutton discovered the location of the genes in a cell. Genes, he said, had to be in the nucleus on the chromosomes. (*University of Kansas Archives*)

knew what chromosomes were, but McClung called this chromosome the "accessory chromosome" and hypothesized that it determined sex. As it turned out, he was essentially correct. McClung's idea that a chromosome could give an individual grasshopper a specific trait—its sex—suggested to Sutton that the other chromosomes might do the same thing for other traits.

In April 1900, Sutton published a paper describing in great detail the sex cells of the lubber grasshopper, including 41 drawings and 10 photographs. A year later, Sutton received his master's degree based on his work on cell division in the sperm cells of the grasshopper. In the fall of 1901, encouraged by

McClung, he transferred to Columbia University to work with the eminent cell biologist E.B. Wilson.

Within 6 months, Sutton was hanging on Wilson's sleeve trying to persuade him that he had solved the mystery of heredity. That fall, the famous British biologist William Bateson came to Columbia to talk about the new science of genetics. Two years before, Bateson told his audience, three botanists had rediscovered the laws of inheritance first described and published by the Austrian monk Gregor Mendel in 1866. These laws laid a foundation for all future work on inheritance. The years between 1900 and 1910 would be a breathtaking period in biology, and Bateson's inspired talk about inheritance electrified his American audience.

Bateson banished any doubts Sutton may have had about the importance of his idea. Everything Bateson said made perfect sense in the light of Sutton's knowledge of chromosomes. Within one month, Sutton had published a brief paper describing his idea that the chromosomes are the physical basis for heredity. Four months later, he had published a full-length paper describing all of the theoretical implications of his idea.

Research in the 100 years since has confirmed nearly every detail of Sutton's hypothesis. His ideas are so well accepted today that it is hard to imagine why more-experienced biologists

Figure 9-2 **Lubber grasshopper.** Walter Sutton's interest in biology began early with his research on the chromosomes of the lubber grasshopper, *Brachystola magna*, which he collected on his parents' farm in Kansas. (*Skip Moody/Dembinsky Photo Associates*)

in his time did not notice the role that chromosomes play in inheritance. Yet, as is so often the case in science, older biologists not only did not see what now seems obvious to us, but some continued to resist Sutton's idea for nearly 40 years.

Maybe it was partly the cool response of his elders. Maybe it was Sutton's idea that he could do more immediate good in the world as a doctor. In any event, after doing work so important that would forever influence biology, Sutton dropped out of graduate school, abandoning biology forever. Two years passed before, pressured by his father, he returned to Columbia, where he resumed his long-delayed medical training. He was as outstanding a medical student as he had been a biologist, and in 1907 he received his medical degree. Returning to Kansas in 1909, he practiced as a surgeon for the rest of his short life. Still a bachelor, he died of a burst appendix at 39.

Nevertheless, his research in biology would change the world. At 25, he discovered a fundamental fact that had eluded older scientists for decades. Sutton revealed the basic mechanism by which organisms inherit traits—why, as he put it, "the yellow dog is yellow." His work is the foundation for all of modern genetics and molecular biology. In the next section, we will see why it took a young biologist such as Sutton to recognize what now seems so obvious. In the rest of this chapter, and in succeeding chapters, we will try to find out exactly what the chromosome does.

WHY WAS THE CHROMOSOMAL THEORY OF INHERITANCE SO HARD TO ACCEPT?

Historically, biologists had no trouble posing theories to explain how organisms stay the same between generations. Before the 17th century, a favorite theory was that miniature versions of future organisms were already "preformed" in sperm or eggs, so that the properties of all future generations were already determined. Many philosophers believed that the egg contributed the materials for an individual and that the sperm contributed some life force that determined the form or shape of the individual. Philosophers compared the creative action of the sperm to the divine creation of the universe from formless matter.

Blending Inheritance: A Wrong Turn

In contrast, plant breeders had long known that when two varieties of plants are crossed, both parent plants contribute equally to the form of the next generation. No matter which kind of plant provided the sperm (pollen) and which kind provided the egg, the resulting cross was the same. In the face of such evidence, most biologists reasoned that the different types of offspring somehow resulted from the "blending" of the genes of the parents. We are not surprised, for example, when a black mare and a white stallion produce a gray foal or when a red

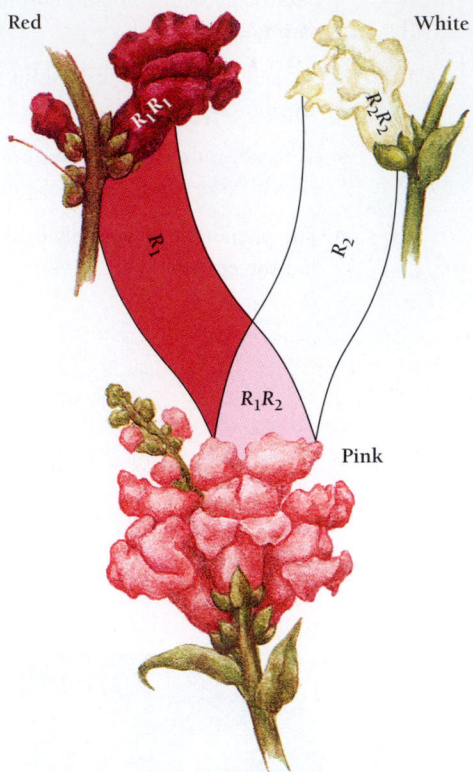

Figure 9-3   Blending inheritance? The offspring of red and white snapdragons are pink. Blending inheritance seemed reasonable to early biologists. But organisms inherit color and other traits by way of discrete units called genes. Here the genes for flower color are represented by R_1 (*red*) and R_2 (*white*).

snapdragon crossed with a white snapdragon produces a plant with pink flowers (Figure 9-3). The color of the offspring appears to be the result of blending, as if two cans of paint have been poured together and mixed.

The **blending** model of inheritance worked well for traits that varied smoothly, such as height, weight, and even, sometimes, color. But the blending model did not always make sense. Pink flowers can give rise to red or white offspring, and brown-eyed parents often have blue-eyed children. If traits blend, how can unblended characteristics reappear in later generations?

The blending model also created a serious problem in understanding the mechanism of evolution (Chapter 15). If traits blended, variation would eventually disappear, and natural selection would have no inherited variation on which to act. For 19th-century biologists, the problem of blending inheritance was so intriguing that, in time, nearly every important biologist of the second half of the 19th century had a different hypothesis about how inheritance worked.

Most 19th-century biologists mistakenly accepted blending inheritance.

Why Did Biologists Doubt That Chromosomes Could Carry Information?

Recall from the last chapter that a wealth of clues suggested that the chromosomes carry the hereditary material. Walther Flemming had shown that during mitosis, the chromosomes divide down the middle, so that half of each chromosome goes to each cell. Wilhelm Roux had argued that this complex way of dividing the chromosomes at mitosis suggested that each chromosome was unique. Theodor Boveri had shown that a nucleus, which contains the chromosomes, can determine the characteristics of an organism. Boveri also showed that in each and every cell of the body, half the chromosomes come from the father and half come from the mother.

But the idea that the chromosomes were responsible for heredity was still no more than a possibility, and most 19th-century biologists did not take the idea seriously. One exception was the German biologist August Weismann. Weismann hypothesized that organisms inherit traits by means of some information-carrying molecule in the nucleus of the cell. Moreover, he argued, since the hereditary material from the mother and father fuse at fertilization, the egg and sperm must each contain only half the normal amount of hereditary material. Otherwise, said Weismann, the hereditary material would double every generation.

Within 3 years, cell biologists confirmed Weismann's prediction. First, they showed that sperm and eggs contain only half the usual number of chromosomes. Second, they observed the special process of **meiosis,** by which cells produce daughter cells (eggs and sperm) with only half the normal number of chromosomes. By the late 19th century, most cell biologists accepted that the chromosomes were central to the process of both mitosis and meiosis and probably somehow essential for the normal development of an embryo into an adult. But they did not know what the chromosomes did. No one yet knew that the chromosomes were the hereditary material.

In 1900, however, an intellectual earthquake rocked the scientific landscape. In that year, three scientists independently rediscovered the work of an obscure but brilliant scientist named Gregor Mendel. In 1866, Mendel had discovered several important principles of heredity while studying inheritance in pea plants. In particular, Mendel had shown that every organism has two sets of genes.

The rediscovery of Mendel's work gave birth, in just a few months, to an entire new field of science—genetics. Biologists began to suspect that the behavior of chromosomes during the creation of sperm and egg cells (meiosis) might explain Mendel's principles. Chromosomes suddenly claimed the intense attention of a great many biologists.

But it was one thing to suspect that chromosomes *could* carry genes, quite another to show that they *did.* Two important problems stood in the way. First, chromosomes disappear between cell divisions, and it seemed possible, even likely, that cells create new chromosomes after each cell division. In short,

biologists had no idea if the chromosomes passed intact from generation to generation. If not, biologists reasoned, chromosomes could not carry information from parent cell to daughter cell.

Second, biologists had no evidence that the individual chromosomes in a cell were different from one another. The chromosomes seemed to come in a variety of shapes, but they all behaved in the same way, and they all seemed to be made of the same material. Like pretzels, they might come in slightly different shapes, but they were all the same stuff. There was nothing to suggest that each one might be carrying unique information.

Weismann's argument that each chromosome always maintains its individuality and integrity was only a guess. Young Walter Sutton, the Columbia University graduate student, was the first person to state explicitly that Mendel's genes were on the chromosomes and to give good reasons why this must be so.

Nineteenth-century biologists doubted that the chromosomes were the genetic material because the chromosomes disappeared during cell division and because biologists had no reason to think that each chromosome carried unique information.

HOW DO ORGANISMS PASS GENETIC INFORMATION TO THEIR OFFSPRING?

Genetics, the study of inheritance, today faces the same contradictory challenges it did 100 years ago. Genetics must explain both how like begets like—why children resemble their parents—and also the origin and maintenance of variation—why children differ from their parents. In this chapter, we will discuss **transmission genetics,** the study of how variation is passed from one generation to the next. We will see that the behavior of chromosomes during meiosis and fertilization does much to explain the inheritance of variation.

Distinct from transmission genetics is **molecular genetics,** the study of how DNA carries genetic instructions and how cells carry out these instructions. Molecular genetics helps explain how a single cell divides and eventually develops into liver cells, skin cells, nerve cells, and more. Molecular genetics also explains how the DNA in these different kinds of cells helps direct day-to-day operations in the cell. We discuss molecular genetics in Chapters 10 through 14. In this chapter, we begin our discussion of transmission genetics by briefly defining how the zygote (the fused egg and sperm) uses the genetic information it receives from its parents to become a unique individual. No two organisms are identical, and the unique collection of traits that define an individual is called the "phenotype."

How Is Phenotype Related to Genotype?

Everything about an organism, from its size and chemistry to its behavior, constitutes its **phenotype** [Greek, *phenein* = to show]. Phenotype encompasses both physical and behavioral characteristics. In humans, eye color, height, and skin color are obvious *phenotypic traits,* the individual aspects of phenotype. In contrast, the **genotype** is the particular collection of genes of a cell or organism. Biologists use the word "genotype" to refer both to all the genes in an organism, its **genome,** or, alternatively, to the subset of genes (or even one gene) that influences a particular trait, such as eye color. A genome, the set of all the genes, is a relatively passive entity within the cell. Like a cookbook, a genome provides lists of ingredients (proteins) and directions for how much of each protein to use.

Phenotype results from the interaction of the genes with each other and with their environment. A **gene** is a region of the DNA that does two things: it specifies a particular protein and it regulates "expression." A gene is *expressed* when the cell makes that protein. How much protein is made, and when, is *regulation.* Each gene responds to signals within a cell that tell the cell when to make more and when to make less of its particular protein.

An organism's phenotype reflects both its genotype (what genes it has) and its environment. In the case of plants and animals, the phenotype reflects the environment in which an individual has developed. "Environment" includes many effects that most people don't associate with the word. For example, an egg cell contains proteins made by the mother that influence the development of an embryo. But which proteins and how much of each depends on the external environment of the mother. In zebra finches, for example, red stripes painted on the father bird's legs induce the mother bird to load up her eggs with an extra dose of the hormone testosterone. The extra testosterone influences the expression of genes and makes the baby birds more aggressive.

Testosterone directly affects the expression of genes throughout the body, but its own production is influenced by the social environment of the individual. In humans and other mammals, hormones and nutrients in the blood supply from the mother similarly affect the development of an individual. After birth, nutrition and many other factors continue to influence the expression of genes. In men, winning a chess game, watching the home team win a basketball game, and picking up a paycheck can all increase the production of testosterone. Losing a game or enduring a chewing-out from the boss can reduce testosterone levels.

Figure 9-4   More than genes. Even identical twins, with exactly the same genes, often look and behave differently. Identical twins occur when a single-celled embryo divides into two separate cells, which then develop into two separate people. Identical twins have identical genetic information. As the two individuals develop in the womb, however, subtle differences in environment—such as a different blood supply—can lead to distinct differences in phenotype. The two 8-year-old twins shown here are genetically identical. Yet the facial bones of one twin are more robust, those of the other twin, more delicate. Prenatal differences in the womb account for the striking differences in facial contours. *(Dr. Nancy L. Segal, Entwined Lives: Twins and What They Tell Us About Human Behavior, 1999, New York, Dutton)*

Studies of genetically identical twins beautifully illustrate that both genes and environment shape individual differences during development. (The relative contribution from genes and environment varies from trait to trait.) Identical twins occur when a zygote divides into two separate cells, which then develop separately. Identical twins thus have identical genetic information. Yet, they can have different phenotypes owing to environmental influence (Figure 9-4). (Fraternal twins, which develop from two eggs fertilized by two sperm, are no more related than ordinary sisters or brothers.) Most of the differences between identical twins result from differences in their environments, in the womb and outside it as they grow to adulthood.

One of the most difficult tasks in genetics is to distinguish between environmental and genetic influences on phenotype. Genes and environment affect the phenotype separately but also interactively. Genes influence the environment of an individual. To take a common case, people treat tall men differently than short men, on average. This different treatment affects their behavior. Similarly, environment influences the way genes are expressed. A dominant male baboon can inhibit the manufacture of testosterone in subordinate males, altering the way the subordinates look and behave.

Although most geneticists study the inheritance of the genes only, the phenotype of an organism is ultimately more interesting to us, first, because we are fellow organisms, and second, because the phenotype determines survival and reproduction. It is the phenotype that is subject to natural selection. The genotype plays an essential but indirect role. The crucial biological questions—How well does a given individual compete with other organisms? and How much will it contribute to the next generation?—depend on the phenotype.

Phenotype includes all of the physical characteristics of an organism. The word "genotype" is used to mean either all the genes an individual has—the genome— or, alternatively, a particular gene or group of genes that produce a trait such as blue eyes.

The Same Laws of Inheritance Apply to All Sexually Reproducing Organisms

All sexually reproducing eukaryotes—from the most primitive algae to the most complex primates—follow the genetic rules that we describe in this chapter. There are some notable exceptions to these rules, including the inheritance of genes carried in the DNA of mitochondria and chloroplasts, which we will discuss in Chapter 12. These rules also do not apply to bacteria and archaebacteria, which do not reproduce sexually. But the rules of genetics are otherwise nearly universal. The reasons for this unity are:

1. All organisms use DNA as the genetic material.
2. The DNA of all eukaryotic organisms is organized into chromosomes.

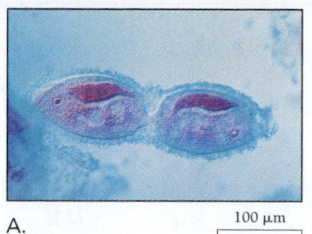 

100 µm

A.   B.

Figure 9-5 Asexual reproduction. A. The single-celled protozoan *Paramecium* can reproduce by simply doubling its DNA and dividing into two cells, as in mitosis. B. Many plants also reproduce "vegetatively," making genetically identical "clones" through mitosis. Shown here is a head of garlic. Each clove can take root and form a separate individual that is genetically identical to the parent plant. (A, *Biophoto Associates/Photo Researchers, Inc.;* B, *Kevin Schafer/Peter Arnold, Inc.*)

3. Almost all chromosomes exist in pairs at some time during a sexual life cycle.
4. These pairs of chromosomes behave in the same ways during meiosis and at fertilization in all eukaryotes.

HOW DO SEXUALLY REPRODUCING ORGANISMS KEEP THE SAME NUMBER OF CHROMOSOMES FROM GENERATION TO GENERATION?

In Chapter 8, we saw how cells distribute two identical sets of chromosomes to their daughter cells. In this chapter, we will see how sexually reproducing organisms provide single sets of chromosomes to their sperm and egg cells. Later, when a sperm and egg fuse, the resulting individual has a full, double set of chromosomes. This is the essence of sexual reproduction.

For single-celled eukaryotes, such as the common freshwater protozoan *Paramecium,* the simplest way to reproduce is to divide in two through mitosis (Figure 9-5A). Even multicelled eukaryotes can reproduce through mitosis, by splitting off a single cell or a group of cells, which then develops into a whole individual. Plants, for example, often reproduce "vegetatively," by sprouting a new individual from a root or branch (Figure 9-5B). Orchardists can create hundreds of genetically identical trees by taking cuttings of branches from a single individual. Indeed, this method has been in use for thousands of years and is called "cloning" [Greek, *klon* = twig].

But all reproduction through mitosis, called **asexual reproduction,** produces offspring with genes from just one parent. All the offspring have the same genes, those of their one parent. We say the offspring that result from asexual reproduction constitute a **clone,** a set of genetically identical individuals.

In contrast, **sexual reproduction** produces offspring that inherit genetic information from two parents. Sexual repro-

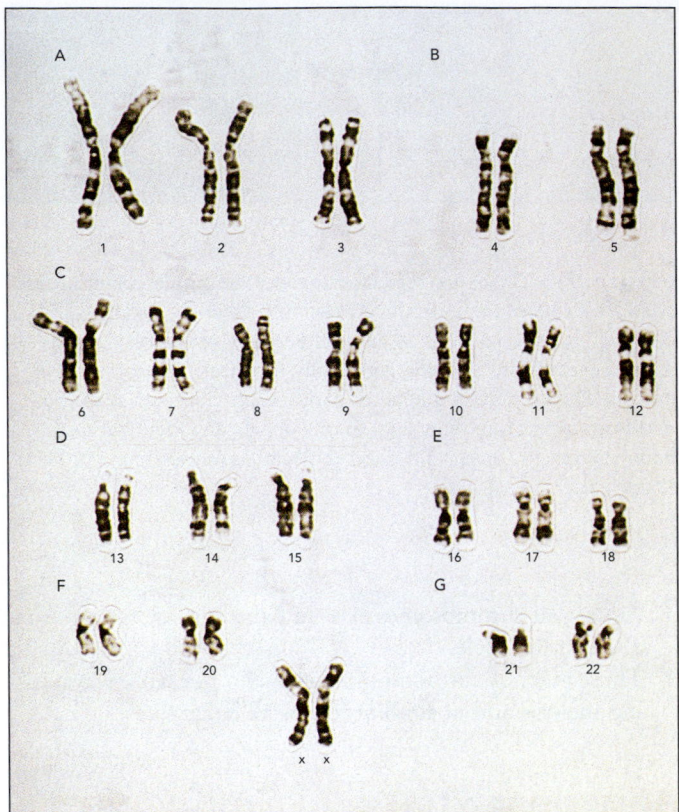

Figure 9-6 **Like socks, chromosomes come in matched pairs.** These are the 23 pairs of chromosomes of a human. Is it a male or a female? How can you tell? *(Leonard Lessin/Peter Arnold, Inc.)*

duction combines two sets of genes in new ways, so that, in every generation, a new set of unique individuals is created. Sexual reproduction creates enormous diversity. But sexual reproduction creates a problem: How can each new organism receive chromosomes from both parents and still maintain the same total number of chromosomes?

The answer depends on the fact that chromosomes, like socks, come in pairs. Figure 9-6 shows the 46 chromosomes of a human being. Notice that for each chromosome with a characteristic size and banding pattern there is another that looks just like it. So the 46 chromosomes in a human cell are actually 23 *pairs* of chromosomes. Cells that contain paired sets of chromosomes are said to be **diploid** [Greek, *di* = double + *ploion* = vessel]. Pairs of matching chromosomes are called **homologous chromosomes,** and each member of the pair is called a **homolog.** Like the individual socks in a pair, homologous chromosomes are basically alike, but they are not perfect copies. One may have a slight defect at one end; the other may have a different color in one spot near the middle.

Organisms that reproduce sexually ensure that their offspring have the right number of chromosomes by giving the offspring only half of the needed chromosomes. Each parent provides one of the chromosomes in a pair. In humans, for ex-

ample, each parent contributes 23 chromosomes. This is analogous to receiving for your birthday 23 unmatched socks from your mother and 23 unmatched socks from your father. You put them together and discover that, miraculously, you have 23 matching pairs.

An organism that reproduces sexually passes the half-set of chromosomes to its offspring in specialized reproductive cells called **gametes** [Greek, *gamos* = marriage]. Each gamete is **haploid** [Greek, *haploos* = single], meaning that it carries a single set of chromosomes—in humans, 23. (Sometimes the word "haploid" is confusing, since it sounds like "half." But haploid means single, not half.)

Nearly all sexually reproducing organisms produce two kinds of gametes. One kind, called the **egg,** or **ovum** [Latin, = egg; plural, **ova**], is large and usually cannot move itself. The other, called the **sperm,** or **spermatozoon** [Greek, *sperma* = seed + *zoos* = living], is small and able to move under its own power. Males and females of other species may be completely unlike humans and other mammals. But no matter how small or strange an organism seems to us, if it produces sperm, we say it is male. Likewise, anything that produces eggs is female. (A few odd species produce more than two kinds of gametes— up to 13, any two of which can fuse to form a new individual. We say such organisms have 13 sexes, instead of the usual 2.)

Gametes are the genetic link between generations. But where do these haploid cells come from? The answer is that haploid gametes arise from diploid cells by **meiosis,** a process that allots one haploid set of chromosomes to each of four daughter cells. Gametes and the special cells from which they arise are called **germ cells** or the **germ line** (Figure 9-7A). Gametes can come only from germ cells. In adult animals, the germ cells are found in special gamete-producing organs, called **gonads** [Greek, *gonos* = seed]. The gonads are called **ovaries** in females and **testes** in males.

All of the rest of the cells in the body of a multicelled organism are called **somatic cells** [Greek, *soma* = body]. Mature somatic cells may divide by means of mitosis, but not by means of meiosis. In normal sexual reproduction, genetic instructions pass exclusively through the germ line.

The germ cells of sexually reproducing organisms undergo meiosis to produce haploid gametes, which fuse to form a diploid individual.

The First Cell of the New Generation Has Two Sets of Chromosomes

Every new generation begins with **syngamy,** or **fertilization,** the union of the two haploid gametes (one from each parent) to form a single diploid cell called a **zygote** (Figure 9-7B). In humans, syngamy and fertilization are often called "conception." Many biologists prefer the term "syngamy" to "fertilization," because "fertilization" wrongly suggests that the sperm

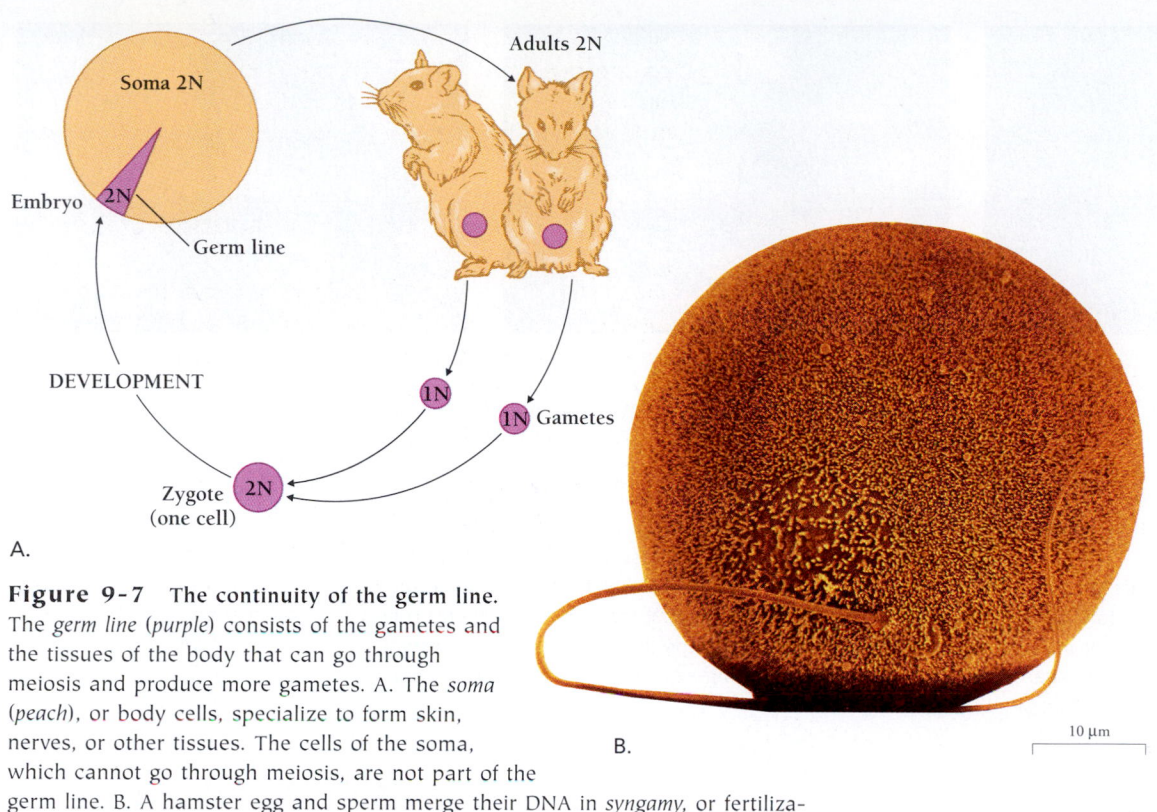

Figure 9-7   The continuity of the germ line.
The *germ line* (*purple*) consists of the gametes and the tissues of the body that can go through meiosis and produce more gametes. A. The *soma* (*peach*), or body cells, specialize to form skin, nerves, or other tissues. The cells of the soma, which cannot go through meiosis, are not part of the germ line. B. A hamster egg and sperm merge their DNA in *syngamy*, or fertilization to create a zygote. (*B, David Phillips/Visuals Unlimited*)

makes the egg fertile. In fact, the egg and sperm are both equally fertile before they fuse to form the zygote. Nonetheless, in this book we will use the older and still commonly used term, "fertilization."

Fertilization depends heavily on adaptations of the gametes. For example, the spermatozoon's small size, whiplike tail, and ATP-packed mitochondria specialize it for rapid movement. The egg cell, in turn, is a virtual supermarket of useful molecules and organelles that are essential for the early development and growth of the embryonic organism. In addition, the egg has special adaptations for moving the sperm nucleus and its own nucleus together, for fusing the two haploid sets of chromosomes into one diploid set, and for dividing rapidly after fertilization.

How Do the Egg and Sperm Find One Another?

Organisms bring the egg and sperm together in many ways. Some aquatic animals produce millions of eggs and sperm, which they "broadcast" into the water. Only a tiny fraction of these eggs and sperm meet and fuse, however. Most of the rest are filtered from the water by filter-feeding animals such as clams and sponges. Many plants broadcast their sperm (but not their eggs) in the form of grains of pollen blown by the wind. Other kinds of plants entice insects, bats, birds, and other animals to carry pollen from plant to plant. Females of

many species of fishes and amphibians lay eggs externally while a male sheds his sperm over them.

In many animals, fertilization is "internal," meaning that it takes place inside the female's reproductive system and therefore requires copulation. Mammals, birds, reptiles, insects, and many other animals that practice internal fertilization often rely on elaborate behavioral and structural adaptations to accomplish fertilization (Figure 9-8). Putting the egg and sperm in the same place is only the beginning, of course. Many factors determine whether fertilization occurs, including pH, temperature, and immune responses.

Despite the many different ways that species achieve fertilization, it always serves the same purpose: to combine two haploid sets of chromosomes into one diploid set of chromosomes (Figure 9-9). Meiosis, in turn, always serves to produce the gametes with haploid sets of chromosomes. All sexually reproducing eukaryotes accomplish meiosis in essentially the same way, in an elaborate chromosomal ballet that both resembles and differs from mitosis.

Every species of organism has behavioral and structural adaptations that bring eggs and sperm together. In animals, fertilization may be either external or internal. In all sexually reproducing eukaryotes, meiosis creates haploid gametes, which combine to form diploid individuals at fertilization, or syngamy.

Figure 9-8 **Adaptations to internal fertilization.** Internal fertilization, or copulation, requires special behavioral and physical adaptations. A. Here a pair of black-tailed rattlesnakes (*Crotalus molossus*) copulate, lying together under a juniper tree in Arizona with their tails crossed (*center of photo*). The male is at the left, and the female, at the right. The most striking physical adaptation to internal fertilization is the penis. Male snakes and lizards have two penises, called "hemipenes," but use only one at a time. In A, the male rattlesnake's left hemipenis is inserted into the female's bulging cloaca. B. A close-up shows the female's cloaca and the base of the male's purple-red left hemipenis. C. The two everted hemipenes of a male rattlesnake under anesthesia. The forking groove along the left organ is a channel through which sperm travels into the female. The circle of downward-pointing spines most likely keeps the two snakes from coming apart during copulation; it's not unusual for the female to begin crawling away, literally dragging the male with her. (*Harry W. Greene/Cornell University*)

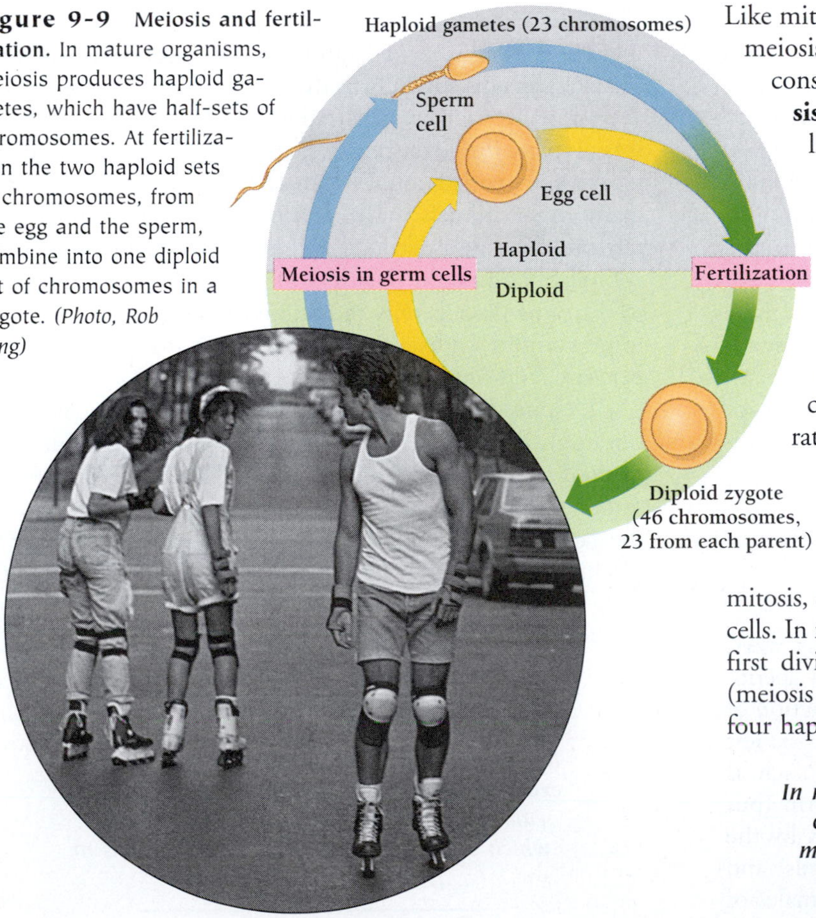

HOW DOES MEIOSIS DISTRIBUTE CHROMOSOMES TO THE GAMETES?

▶ **Figure 9-9** **Meiosis and fertilization.** In mature organisms, meiosis produces haploid gametes, which have half-sets of chromosomes. At fertilization the two haploid sets of chromosomes, from the egg and the sperm, combine into one diploid set of chromosomes in a zygote. (*Photo, Rob Lang*)

Haploid gametes (23 chromosomes)

Sperm cell

Egg cell

Haploid

Meiosis in germ cells   Fertilization

Diploid

Diploid zygote (46 chromosomes, 23 from each parent)

Like mitosis, meiosis is a continuous process. Biologists divide meiosis into separate stages for convenience only. Meiosis consists of two cell divisions, called **meiosis I** and **meiosis II.** These divisions occur only in cells of the germ line during the production of gametes. Biologists divide each of these two meiotic divisions into four phases (Figure 9-10). These are prophase, metaphase, anaphase, and telophase.

In general outline, each phase of meiosis resembles the corresponding stage of mitosis. During each **prophase,** chromosomes condense, and the nuclear membrane breaks down. During each **metaphase,** the chromosomes move to the equator of the spindle apparatus. During each **anaphase,** spindle fibers move the chromosomes toward the poles. And during each **telophase,** the nuclear membrane reforms.

Despite the resemblance to the stages of mitosis, the end result of meiosis is crucially different. In mitosis, each and every division produces two diploid daughter cells. In meiosis, by contrast, DNA replication occurs before the first division (meiosis I), but *not* before the second division (meiosis II). As a result, the two divisions of meiosis produce four haploid daughter cells.

In mitosis, a cell replicates its DNA once and divides once, resulting in two diploid daughter cells. In meiosis, a cell replicates its DNA once but divides twice, resulting in four haploid daughter cells.

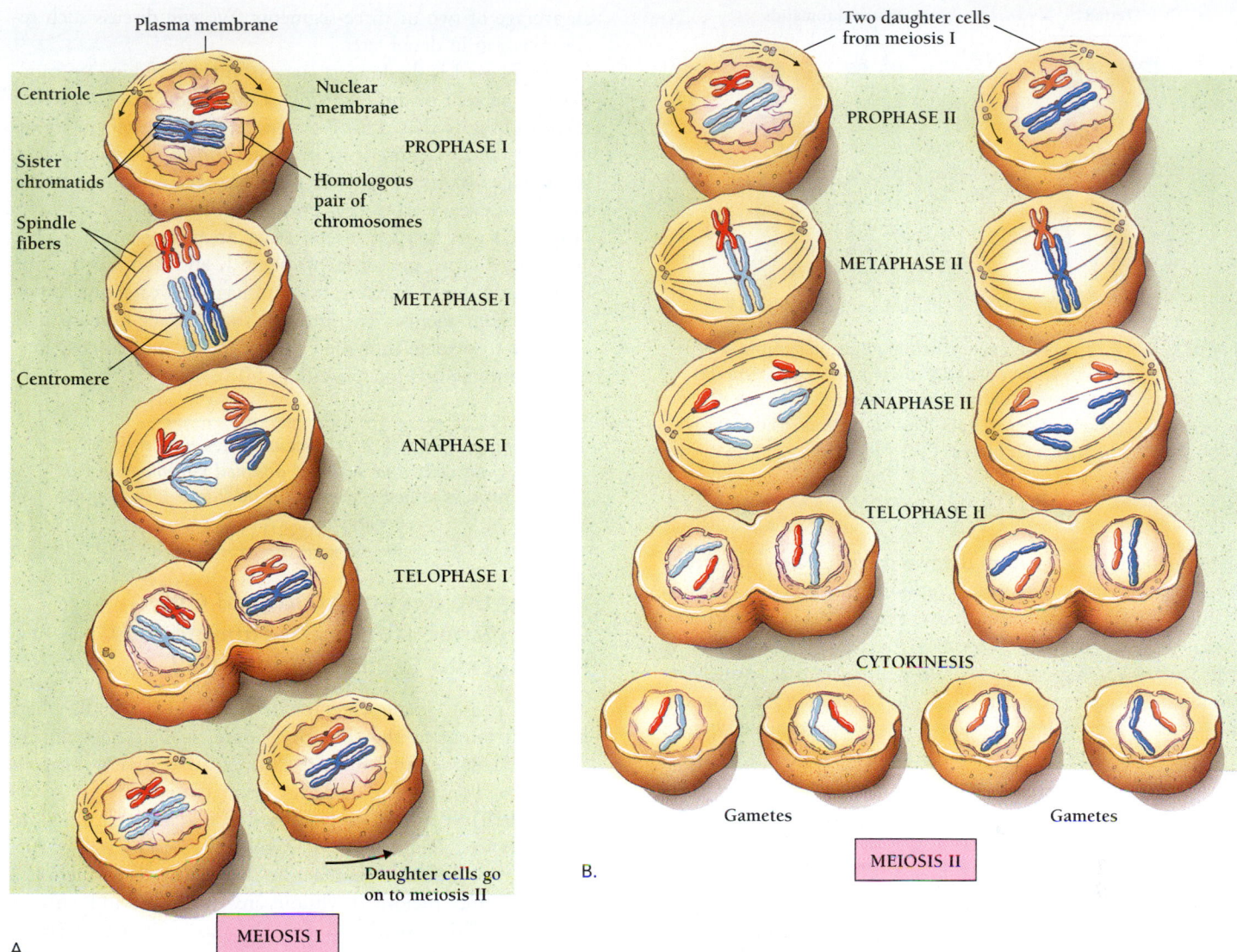

Figure 9-10 **Meiosis.** In meiosis, a cell replicates its DNA once but divides twice, producing four haploid daughter cells. Each of the four daughter cells (*bottom, right*) receives two of the eight homologous chromosomes in the original cell (*top, left*)—one from each tetrad. In the first division, meiosis I, one chromosome from each homologous pair goes to each daughter cell. Meiosis II, the second division, resembles mitosis, but in a haploid cell.

During Prophase I, Homologous Chromosomes Form Pairs

During interphase, before prophase I, the cell copies, or replicates, the chromosomes. As in mitotic prophase, the (doubled) chromosomes condense, the nucleoli and the nuclear membrane disappear, and the spindle apparatus begins to form. But meiotic prophase differs from mitotic prophase in one essential way: after the chromosomes begin to condense, the homologous pairs come together in the process of **synapsis** [Greek, = union], in which homologous chromosomes align exactly. Each chromosome itself consists of two sister **chromatids,** so that the four

chromatids form a **tetrad** (Figure 9-10). Synapsis resembles the process of pairing socks before putting them away. From a disorganized pile of 46 socks, we can put together 23 pairs.

Synapsis is the central event of meiosis. It is the process by which homologous chromosomes assort into two haploid sets. Following synapsis, the homologous chromosomes go their separate ways, while the chromatids of each chromosome stay together. (In both mitosis and in meiosis II, it is the *sister chromatids* that separate, not the homologous chromosomes.)

Another important event that occurs during synapsis is **crossing over,** in which homologous chromosomes break and exchange equivalent pieces. Crossing over results in new

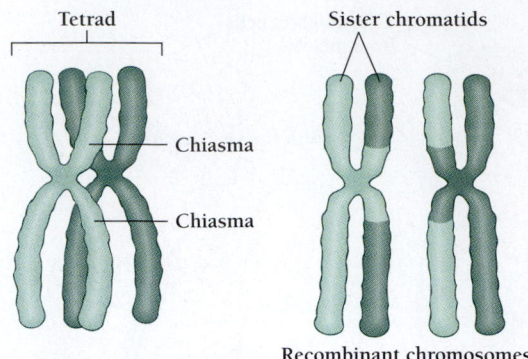

Figure 9-11   Crossing over. When homologous chromatids pair up during synapsis (prophase I), they may break at special places called chiasma and exchange equivalent pieces. The result is new combinations of genes from the two parents. These chromatids are exchanging two pieces.

combinations of genes from the two parents. The visible result of crossing over is a crosslike configuration called a **chiasma** [Greek, = cross; plural, **chiasmata**] (Figure 9-11). At each chiasma, homologous chromosomes may exchange chromatid segments. Crossing over is an important way that sexually reproducing organisms recombine genes and increase variation. In humans, each pair of homologous chromosomes exchanges

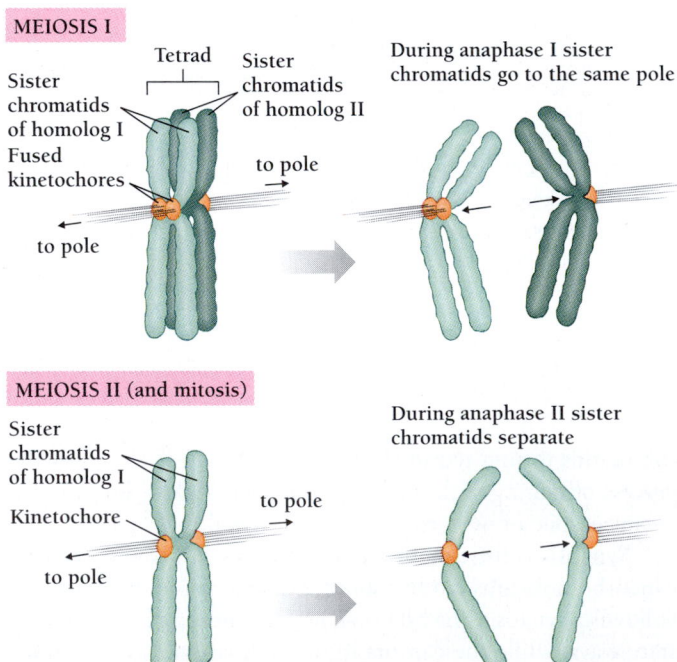

Figure 9-12   Separation of the chromatids. During meiosis I, sister chromatids go to the same pole. During meiosis II, sister chromatids go to opposite poles (as in mitosis).

an average of two or three segments. We will discuss such recombination in detail later.

Prophase I is the longest stage of meiosis, taking up more than 90 percent of the total time. One reason that prophase I takes so long is that during this time, the egg cell builds up materials and machinery to be passed on to the gametes. In the females of some species, meiosis may seem to stop for months or even years during prophase I. In frogs, for example, prophase I may last for several years. In humans, meiosis begins before birth, pauses in prophase I, and does not resume until after puberty. Starting at around age 12 years, one or two eggs complete meiosis each month. Since female humans may continue to ovulate into their 50s, some individual cells may not resume meiosis for more than 50 years.

During prophase I, a cell accumulates materials for its daughter cells and sorts its doubled chromosomes in the process called synapsis. In addition, homologous chromosomes cross over.

During the Rest of Meiosis I, One Chromosome from Each Homologous Pair Goes to Each Daughter Cell

After prophase I comes metaphase I, in which spindle fibers pull each tetrad to the equator (Figure 9-12). Then, during anaphase I, something dramatic happens that is very different from what occurs in mitosis: the sister chromatids go to the *same* pole. (In contrast, in mitosis, the two chromatids go to *opposite* poles.) Telophase I and cytokinesis rapidly follow. Another critical difference is that the daughter cells of meiosis I move right into meiosis II without any interphase and without any further DNA synthesis. The chromosomes may even remain partly condensed.

An important question is how the tetrads sort out during metaphase I. Do all the chromosomes from the mother go to one pole, and all the chromosomes from the father to the other? If so, meiosis would recreate the same sets of chromosome generation after generation. Instead, the tetrads sort themselves randomly. One pair of homologs aligns one way, another pair aligns another way. Biologists say that each pair aligns "independently" of all the other pairs.

Because chromosome pairs sort themselves independently, meiosis is a major source of genetic diversity. Sutton showed that just two pairs of chromosomes can form 4 (2^2) kinds of gametes (Figure 9-13). Similarly, three pairs of chromosomes can form 8 (2^3) kinds of gametes. A single human germ cell, with 23 pairs of chromosomes, can therefore form 2^{23} (about 8 million) kinds of gametes, each one unique. Together, a woman and a man could produce more than 64 trillion genetically unique offspring (8 million kinds of eggs × 8 million kinds of sperm). In addition, crossing over increases the actual diversity of possible gametes and zygotes far beyond 64 trillion. All this diversity comes

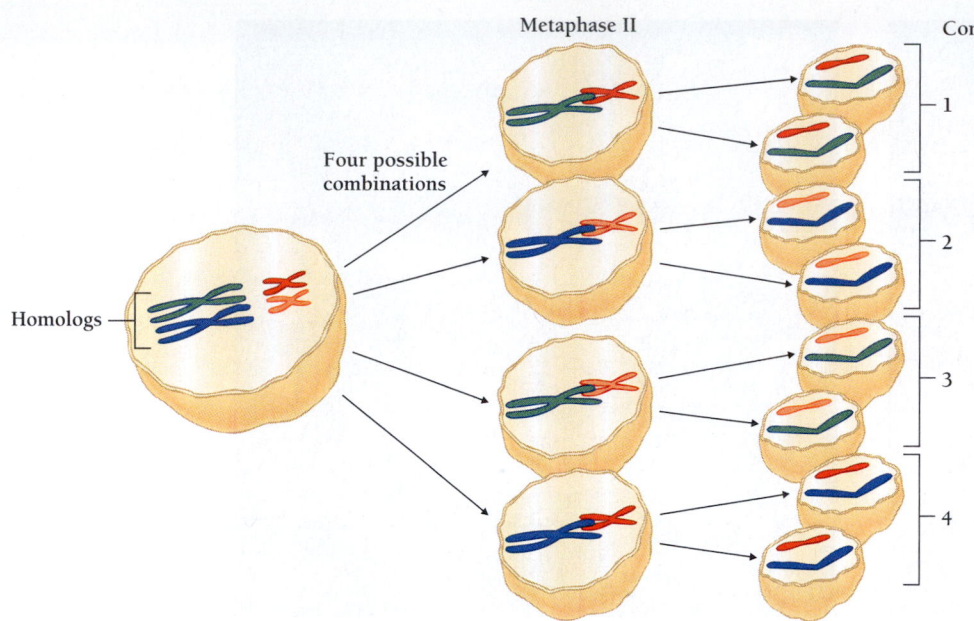

Metaphase II

Combination

Four possible combinations

Homologs

1

2

3

4

Figure 9-13 **Just two pairs of chromosomes can form four kinds of gametes.** If the chromosomes arrange themselves as in the first cell in this figure, the resulting gametes have chromosome combinations 1 and 2. If, in other germ cells, the big chromosomes on the left swap places (which occurs at random), then the resulting gametes have combinations 3 and 4.

from (1) crossing over and (2) the random alignment and subsequent separation of tetrads in metaphase I. Given this random recombination, it is no wonder that all sexually reproducing species (including our own) are so wonderfully diverse.

During anaphase I of meiosis, sister chromatids go to the same pole in the cell, but homologous chromosomes go to opposite poles. Telophase I and cytokinesis I resemble their counterparts in mitosis.

Meiosis II Distributes Sister Chromatids to Daughter Cells

Meiosis II looks just like mitosis would look in a haploid cell. At the end of meiosis I, the cell is haploid. After it divides again in meiosis II, it is still haploid. Meiosis II simply separates the sister chromatids, so that each chromosome of the daughter cell contains a single chromatid.

Prophase II (unlike prophase I) is brief, since the chromosomes are still mostly condensed from meiosis I. If a nuclear membrane has reappeared during telophase I, it breaks down during prophase II. The chromosomes attach to newly assembled spindle fibers. Then, during metaphase II, the chromosomes line up across the equator of the spindle apparatus, with each centromere connected to spindle fibers from both poles. During anaphase II, the chromosomes split—one chromatid moves to each pole. In telophase II, the nuclear membrane forms once

more. The daughter cells each have a complete haploid set of chromosomes, each consisting of a single chromatid.

Meiosis II resembles mitosis except that each resulting cell is haploid, rather than diploid.

Disjunction and Nondisjunction

The separation of homologous chromosomes in anaphase I and of sister chromatids in anaphase II is called **disjunction.** Occasionally, separation fails to proceed normally, an event called **nondisjunction.** This results in a gamete having too many or too few chromosomes. In humans, for example, a sperm or egg might have 22 or 24 chromosomes instead of 23. After fertilization, the resulting zygote likewise has the wrong number of chromosomes—45 or 47. A cell or individual with the correct number of chromosomes is said to be **euploid** [Greek, *eu* = good, true + *ploion* = vessel], while one with an abnormal number of chromosomes is said to be **aneuploid** [Greek, *an* = not + euploid].

In humans, one of the most common results of nondisjunction (and therefore aneuploidy) is **Down syndrome,** a disorder that leads to mental retardation and the abnormal development of the face, heart, and other parts of the body (Figure 9-14A). Down syndrome is almost always associated with a chromosomal abnormality called **trisomy 21** [Greek, *tri* = three + *soma* = body], the presence of three, rather than two, copies of chromosome 21 (Figure 9-14B).

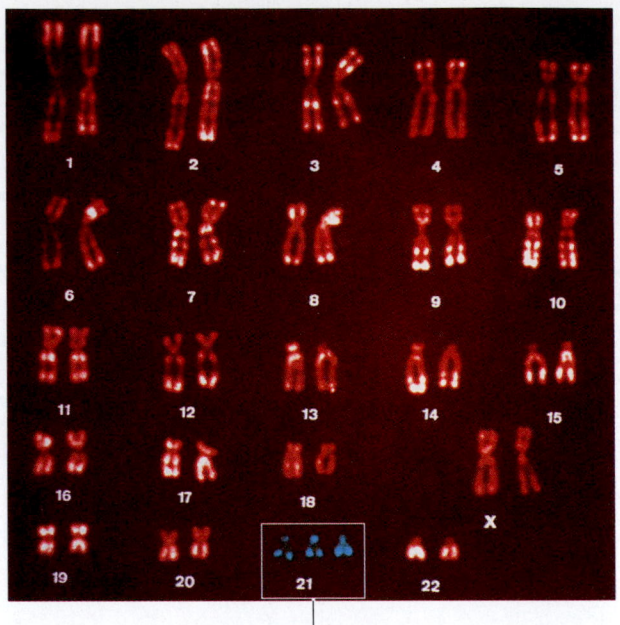

A.

B.   3 copies of chromosome 21

Figure 9-14   Children with Down syndrome have distinctive characteristics. These include mental retardation, heart defects, a flat facial profile, excess skin on the neck, double-jointedness, an unusual crease in the palm of each hand. B. Down syndrome usually results from the presence of three copies of chromosome 21, a condition called "trisomy 21." (*A, Hattie Young/Science Photo Library/Photo Researchers, Inc.; B, Kunkel/Phototake NYC*)

Trisomy 21 usually results from nondisjunction during meiosis I. If, during anaphase I, the two chromosomes 21 fail to separate, then two of the four possible gametes produced during meiosis will each have two copies instead of one, and the other two will have no copies of chromosome 21 (Figure 9-15). When these gametes unite with normal gametes at fertilization, they produce abnormal zygotes—some with three copies of chromosome 21, and some with only one copy (Box 9-1).

If, during anaphase, homologous chromosomes fail to separate, the gametes will have too many or too few chromosomes.

WHY SEX?

We saw earlier that many organisms can reproduce either sexually or asexually. For example, plants can reproduce sexually, as when pollen finds its way to a flower, or asexually, as when the roots of an aspen tree sprout new trees. Even some animals, such as the many-armed hydra, can reproduce asexually, by budding.

Asexual reproduction is the quickest and easiest way to reproduce. Sexual reproduction is enormously expensive. To reproduce sexually, organisms must make hundreds, thousands, or even millions of gametes to ensure that a few meet for fertilization. A female codfish produces some 9 million eggs per year, of which only a small fraction are fertilized. A typical male human may produce 20 to 30 billion sperm over the course of the weeks or months it takes him to successfully fertilize a single egg. Even a female human is born with 2 million egg cells (oocytes). In addition, humans and other animals invest huge amounts of energy into finding, courting, and mating with other individuals. Many plants construct elaborate and showy

MEIOSIS I

Normal disjunction of homologs

Metaphase I

Figure 9-15   How does trisomy occur? In nondisjunction, homologous chromosomes fail to separate during anaphase. The resulting gametes have either two copies of a chromosome or none, but not the usual one. Later, when the gamete combines with another, normal gamete, the resulting zygote has one homolog or three, but not two.

Nondisjunction of one pair of homologs

Anaphase I

What Happens When Meiosis Goes Wrong?

Among human embryos with Down syndrome, or trisomy 21, about 80 percent die before birth in **spontaneous abortions,** natural abortions that occur because the embryo cannot live. Only about 20 percent of trisomy 21 embryos survive until birth, although those that survive may live for many years. For comparison, about 70 percent of all human embryos normally survive to birth. Although the death rate among embryos with trisomy 21 is high, the death rate is even higher—almost 100 percent—among those with only one copy of chromosome 21.

About 1 baby in every 700 has Down syndrome. Although the extra chromosome usually comes from the mother, about one-fifth of the time, the extra chromosome 21 comes from the father. Older women, whose eggs are older, are much more likely than younger women to produce eggs with an extra chromosome 21.

As a rule, physicians test the embryos of women older than 35 years for Down syndrome, but there is nothing magic about age 35. The likelihood of giving birth to a child with Down syndrome increases gradually (Table A).

Cell biologists believe that the chance of nondisjunction (at least for chromosome 21) increases with maternal age because of the pattern of meiosis in humans. The cells that form human eggs begin meiosis while the mother is herself still an embryo in the womb. But meiosis stops in the middle of prophase I, well before birth, and does not resume until after the egg is released from the ovaries, anywhere from 13 to 55 years later. Nondisjunction of chromosome 21 may be the result of interrupting the complicated process of meiosis for so long a time.

TABLE A   RISK OF TRISOMY 21

Age of mother (y)	Risk of giving birth to child with trisomy 21
Under 25	1/1400
Under 30	1/1000
At age 35	1/350
Over 40	1/100

flowers stocked with sweet nectar, adaptations for attracting insects and other animals that carry pollen from one plant to another. What makes sexual reproduction worth a great expense?

One answer is genetic variation in one's offspring. We have seen that, because of crossing over and because of the independent assortment of chromosomes during metaphase I, a single pair of humans can produce a seemingly infinite number of unique zygotes. In contrast, asexual reproduction produces offspring that are genetically identical to the parent.

What Good Is Genetic Variation?

In an unchanging environment, a single, perfectly adapted genotype might persist unchanged for millions of years. Most environments, however, change constantly. A polar bear with short hair might survive fine in a period of global warming. But as temperatures drop, its siblings with longer fur do better. Parents that produce a diversity of offspring are more likely to have some of them survive and reproduce than parents that produce offspring that are all identical. Indeed, evolutionary biologists generally find that the less predictable an environment is, the more likely that the organisms living in it will have evolved methods for increasing diversity.

Some species actually change their mode of reproduction depending on how predictable their environment is. For example, a female water flea reproduces asexually when food and space are plentiful, giving birth to huge numbers of genetically identical offspring, but reproduces sexually when food is limited (Figure 9-16). The resulting diversity increases the chance that at least some individuals will survive.

▶ **Figure 9-16   Where's Dad?** When food is abundant, the water flea, *Daphnia*, reproduces asexually. Each female is born pregnant with more female water fleas, each one a clone of its mother. A drop in water temperature or a decrease in food supply results in the birth of males and sexual reproduction. (*Oxford Scientific Films/Animals Animals*)

500 μm

 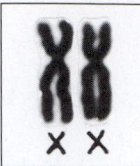

Male   Female

Figure 9-17   Sex chromosomes in humans. In certain organisms, one sex has a pair of unmatched sex chromosomes. For example, a female human has 23 pairs of matched chromosomes, including two X chromosomes, which are called sex chromosomes. A male human has 22 pairs of matched chromosomes, plus two sex chromosomes called X and Y. *(Left, Don Kelly/Grant Heilman Photography; right, Photo Researchers, Inc.)*

Many biologists think that the need for genetic variation alone is not a good enough reason for sex. After all, many organisms, including all bacteria, reproduce asexually. Bacteria have survived quite well for billions of years. In Chapter 43, we discuss one alternative hypothesis for why so many eukaryotes reproduce sexually. The reasons for sex remain controversial.

Sexual reproduction is energy expensive, and biologists wonder why organisms bother with it. One reason may be that sex increases genetic diversity, and genetic diversity increases the chances that some individuals will survive, no matter how the environment changes.

What Are Sex Chromosomes?

In mammals, birds, and some other organisms, either the male or the female may have a pair of unmatched chromosomes. In mammals (including humans), for example, the female has two homologous "X" chromosomes, while the male has one "X" chromosome and one "Y" chromosome (Figure 9-17). In birds, the males have two matched sex chromosomes ("WW"), and the females have one "W" chromosome and one "Z." In many insects, including the lubber grasshoppers that Sutton collected, the female has two homologous X chromosomes and the male has only one X chromosome. The **X** and **Y** chromosomes (as well as the W and Z chromosomes) are called **sex chromosomes.** The rest of the chromosomes are called the **autosomes.** A human cell has 22 pairs of autosomes and 1 pair of sex chromosomes.

Protists, fungi, plants, and many animals lack sex chromosomes. In most organisms, males and females both have complete sets of homologous (matching) chromosomes. In these organisms, what sex an individual becomes depends on different factors, including, for example, plant hormones in plants or temperature during development in some animals.

In humans and other mammals, the X chromosome in a male is the same as the X chromosome in a female. The Y chromosome, however, is much smaller and bears far fewer genes than the X chromosome. In both females and males, the two sex chromosomes (XX in females and XY in males) pair and segregate during meiosis in the same way as homologous pairs of autosomes.

How can X and Y pair off during synapsis if they differ so much in size and form? The answer is that they share a region that is homologous. Despite their striking differences in size and form, then, they still behave as homologs during meiosis.

Sex chromosomes are chromosomes involved in sex determination. In mammals, the female has two homologous X chromosomes, and the male has one X chromosome and one Y chromosome, which are not homologous.

HOW DO THE NUMBER AND MOVEMENTS OF CHROMOSOMES EXPLAIN THE INHERITANCE OF GENES?

If a cell has two copies of each chromosome, then the cell must also have two copies of every gene. This is the key to understanding genetics in higher organisms.

The copy of a gene on one chromosome may differ slightly from its counterpart on the other chromosome. Alternate versions of the same gene are called **alleles.** For example, in humans, both chromosomes in a pair may have a certain gene that influences eye color, but one chromosome might carry an allele for brown eyes, while the other chromosome carries an allele for blue eyes.

Most genes are simply instructions for the production of a particular protein. Alleles are variations in those instructions. Just as your aunt June makes chocolate brownies with extra chocolate and your cousin Ralph makes them with a teaspoon of vanilla, different alleles specify slightly different proteins. Sometimes the differences are subtle and the resulting protein is the same either way; sometimes an allele specifies a protein so different from normal that the protein cannot function at all.

For a given gene, an individual can carry two identical alleles or two different alleles. In snapdragon flowers, for example, one allele of a particular gene specifies white flowers and the other allele specifies red flowers. If the flower has two red-flower alleles, the flowers are red. If the snapdragon has two white-flower alleles, the flowers are white. On the other hand, if the snapdragon has one red-flower allele and one white-flower allele, the flowers are pink (Figure 9-3). If an individual has two copies of the same allele, we say it is **homozygous** for that allele. If an individual has two different alleles of a gene, we say it is **heterozygous.**

Biologists name the alleles of a gene with letters or short abbreviations. In snapdragons, for example, we can call the red-flower allele R_1 and the white-flower allele R_2. We can imagine that the R_1 allele allows the plant to make a red protein pigment, while the R_2 allele allows the plant to make no functioning pigment at all. The genotype of the homozygous red snapdragon would be "R_1R_1," and that of the homozygous white snapdragon would be "R_2R_2." The genotype of the pink heterozygote would be "R_1R_2."

Plant breeders knew about homozygous plants for hundreds of years before the first geneticists. But, knowing nothing about genes, plant breeders called plants that were homozygous for flower color "true breeding" for flower color. In **true-breeding** plants, all the offspring of the red-flowered variety produce red flowers only, generation after generation; and all the offspring of the white-flowered variety produce white flowers only, generation after generation. A line of plants (or other organisms) can be true breeding for other traits besides flower color. In all true-breeding varieties, both chromosomes carry the same allele for a certain trait. (We say they are homozygous.)

Snapdragons normally "self-fertilize." Each snapdragon flower produces both pollen (which produces sperm) and egg cells. The sperm from the pollen fertilizes eggs within the same flower. So, on a red snapdragon, gametes carrying red-flower alleles always pair up with gametes carrying red-flower alleles. The same is true for white-flowered snapdragons.

A plant breeder can cross a true-breeding red-flowered plant with a true-breeding white-flowered plant only by preventing self-fertilization. To cross the two homozygous plants, the breeder must prevent each plant from self-pollinating. This is done by snipping off the pollen-carrying structures and then artificially pollinating the flower with pollen from another plant. This method of breeding two genetically distinct organisms is called **cross breeding** (or **crossing**). The offspring of such a cross are called **hybrids.** In snapdragons, crossing a homozygous red snapdragon with a homozygous white snapdragon results in a heterozygous hybrid that has pink flowers (Figure 9-3).

Geneticists call the original parents in such a cross the **parental,** or **P,** generation, and the offspring the "first filial" [Latin, *filia* = daughter, *filius* = son], or **F1,** generation. The **F2** generation are the offspring of the F1 generation. In other words, the F2 generation are the grandchildren of the P generation. In the snapdragon cross, then, the parents are R_1R_1 and R_2R_2 homozygotes, and the F1 generation are R_1R_2 heterozygous hybrids. The genotypes of the heterozygous hybrids are all R_1R_2, and their phenotypes are pink flowers.

Eukaryotes that have two sets of chromosomes (diploid) have two copies of each gene. An individual may have two identical alleles, or two different alleles.

What Are the Genotypes and Phenotypes of the F2 Generation?

Both a plant breeder and a geneticist would want to know what color flowers result from crossing two hybrid snapdragons. We

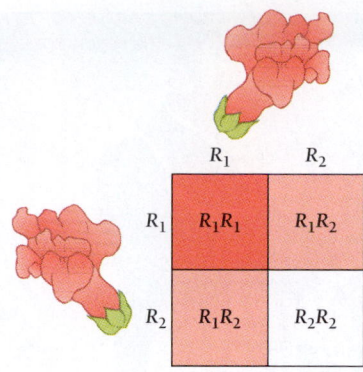

Figure 9-18 Inheritance of flower color in snapdragons. A cross between two R_1R_2 heterozygotes, shown here in a Punnett square, produces three genotypes: snapdragons homozygous for R_1, snapdragons homozygous for R_2, and R_1R_2 heterozygous snapdragons.

can easily figure this out before we even cross the plants. We know that meiosis produces haploid gametes with just one of each chromosome. The F1 hybrids each have one R_1 allele and one R_2 allele, each on a separate chromosome. So each snapdragon gamete receives either R_1 or R_2. Half of all the gametes get the R_1 allele, and half get the R_2 allele. An R_1 can pair up either with another R_1 or with an R_2, and an R_2 can pair up with either another R_2 or with an R_1. So the F2 generation can have R_1R_1 (red flowers), R_2R_2 (white flowers), or R_1R_2 (pink flowers). But how many of each kind will we get?

A convenient way to find out is to use a checkerboard of numbers invented by an early 20th-century British geneticist named Reginald Punnett (Figure 9-18). In a **Punnett square,** we show each kind of gamete made by one parent along the top of the square and each kind of gamete made by the other parent along the left side of the square. Within the small squares, we write the genotypes that would be produced by each combination. Usually, the R_1R_2 genotype is the same as the R_2R_1 genotype. The flowers are pink no matter which allele comes from which parent. But we end up with twice as many pink flowers as white or red.

The Punnett square shows that the three possible genotypes of the F2 generation of our snapdragon experiment should be in the ratio 1:2:1. If we actually cross the flowers, we can directly count the number of red, white, and pink flowers. Because the pink phenotype of the heterozygote is different from the phenotypes of either homozygote (red or white), we can directly count the ratios of the genotypes in the F2 generation. Experiments of this kind always produce a ratio fairly close to the expected 1:2:1.

If we cross two R_1R_2 heterozygotes, one-quarter of the offspring are homozygous for R_1, one-quarter are homozygous for gene R_2, and one-half are heterozygous R_1R_2.

Figure 9-19   Gregor Mendel in his monastery garden. *(The Bettmann Archive)*

HOW DID GREGOR MENDEL DEMONSTRATE THE PRINCIPLES OF GENETICS?

Today, the inheritance of flower color in snapdragons makes sense to us because we know about chromosomes and how they assort during meiosis. But 19th-century biologists knew little about chromosomes and nothing about genes. For them, inheritance was mystifying. Our knowledge of the behavior of chromosomes and genes gives us a great advantage.

The first person to penetrate the mystery of the ratios of different phenotypes in genetic crosses was Gregor Mendel (Figure 9-19). In his monastery's garden, in what is now the Czech Republic, Mendel studied seven different pairs of traits in pea plants (Figure 9-20): round versus wrinkled seeds, yellow versus green seeds, purple versus white flowers, inflated versus constricted pods, green versus yellow pods, axial versus terminal flowers, and tall versus dwarf stems. Each of these pairs of traits showed the same pattern of inheritance. Mendel chose pairs of traits in which each variant was distinct (in other words, tall could not be confused with short).

In each case, Mendel showed that the parental stocks were true breeding. For example, self-pollination of plants with round seeds always produced offspring with round seeds.

Mendel then cross-pollinated flowers with different traits, a purple-flowered pea with a white-flowered pea, for example (Figure 9-21).

But F1 hybrids do not always have an intermediate phenotype like the pink flowers of hybrid snapdragons. Instead, a heterozygous hybrid may have the same phenotype as one of the parents. For example, when Mendel crossed true-breeding, purple-flowered peas with true-breeding, white-flowered peas, all the hybrids had purple flowers (Figure 9-22A). The F1 hybrids are all heterozygotes, but only the allele for purple flowers is expressed. We say that an allele is **dominant** when it alone determines the phenotype of a heterozygote, as in the case of the purple allele in peas. An allele is **recessive** when it contributes nothing to the phenotype of a heterozygote, as is the case of the white allele of peas. (The names of dominant alleles begin with a capital letter, and those of recessive alleles with a small letter.)

Today we know that a dominant allele usually specifies a functional protein, and a recessive allele usually specifies a protein that doesn't function at all. Recessive alleles that are harmless, such as the one that causes type O blood, may be extremely common. Others, such as the allele for the genetic disease Tay-Sachs, which kills children at 3 or 4 years of age, are quite rare. Lethal recessives such as Tay-Sachs are rare because people who are homozygous for this allele never pass the gene on to any offspring.

When two alleles each contribute to the phenotype, as in the case of the R_1 and R_2 alleles of snapdragons (which together gave a color different from either homozygote), the alleles are said to lack dominance. In the snapdragons, for example, both alleles are expressed and we say they each have **partial dominance** or "codominance." But blending inheritance cannot explain the reappearance of red and white snapdragons after crossing the hybrids.

In each of Mendel's seven crosses, all of the F1 hybrids had the same phenotype. Round pea stocks crossed with wrinkled pea stocks gave F1 hybrids with round peas, tall pea stocks crossed with short pea stocks gave tall F1 hybrids, and so on. We say that the purple flower allele is dominant to the white flower allele, the round allele is dominant to the wrinkled allele, and the tall allele is dominant to the short allele.

> *Mendel began his experiments with true-breeding (homozygous) pea plant stocks. He chose pairs of traits in which the two alternate forms were distinct. Each pair of traits was determined by just two alleles, and each allele in a pair was either dominant or recessive.*

What Did Mendel's F1 Crosses Show?

Mendel allowed the F1 hybrids to self-pollinate, then counted the F2 offspring, or *progeny*. In every case, the dominant allele phenotype made up about three-fourths of the progeny, and the recessive allele phenotype made up the remaining one-fourth of the progeny.

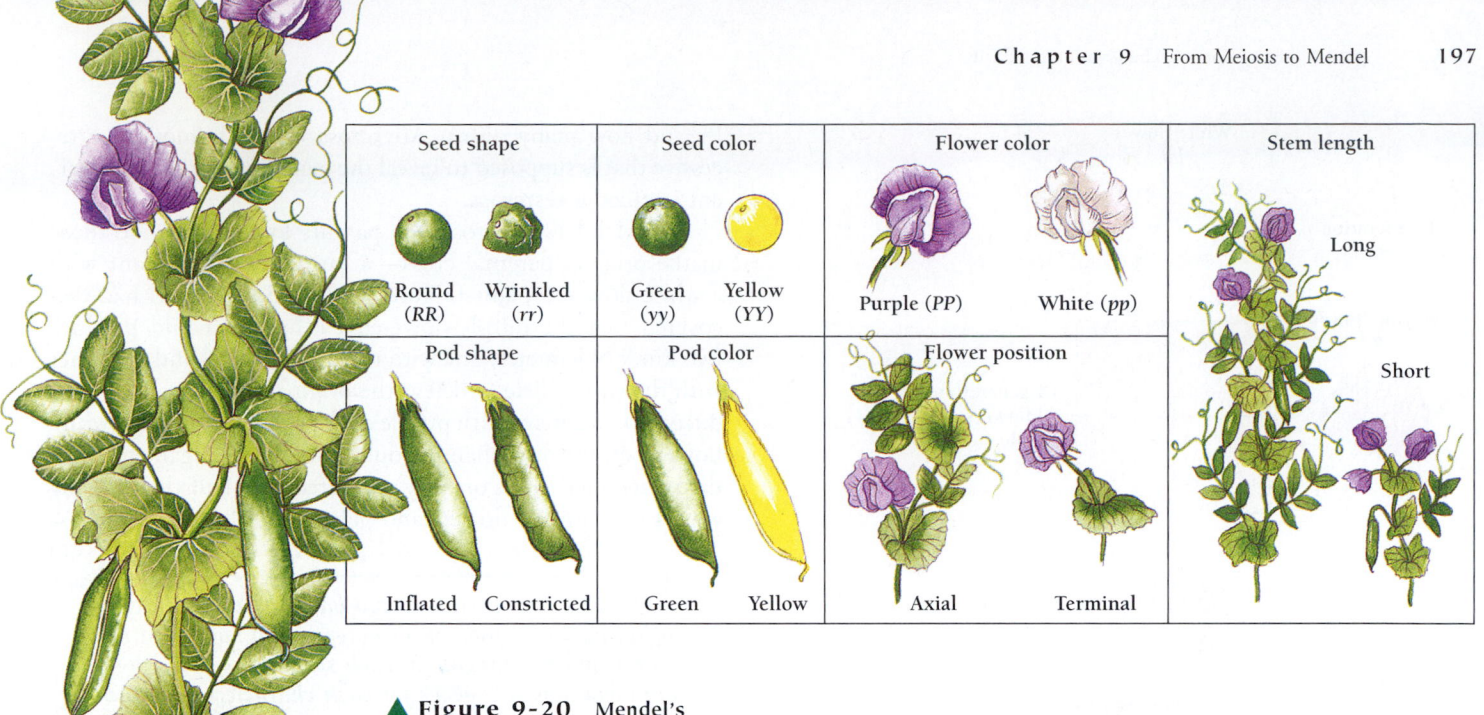

▲ **Figure 9-20** Mendel's seven traits in pea plants. Mendel chose seven traits, each with two forms that corresponded to two alternate alleles of the gene.

This is the same result we got with the snapdragons, except that the heterozygotes (which make up half of the offspring) look the same as the dominant homozygotes (which make up a quarter of the offspring). This was because, in each of Mendel's crosses, one allele was dominant and one was recessive.

In the case of the round versus wrinkled cross, for example, $^1/_4$ were homozygous for the round allele, $^1/_4$ were homozygous for the wrinkled allele, and $^1/_2$ were heterozygous. Since the round allele is dominant, the heterozygotes had the same phenotype as the homozygote for the round allele. So $^3/_4$ ($^1/_4 + ^1/_2$) were round, and $^1/_4$ were wrinkled.

The Principle of Segregation

Mendel had to make sense of his results without the knowledge of genes and chromosomes that we now have. Amazingly, he was able to do it just by studying the ratios of phenotypes in his breeding experiments.

Mendel's first conclusion was the **principle of segregation:** *Each sexually reproducing organism has two genes for each characteristic; these two genes segregate (or separate) during the production of gametes.* (Of course, Mendel didn't use the word "gene," which is a modern term. He called genes "determinants." But his meaning was the same.)

To test the principle of segregation, Mendel performed a different kind of cross, called a **backcross.** Instead of allowing the F1 heterozygotes to self-pollinate, he crossed them with

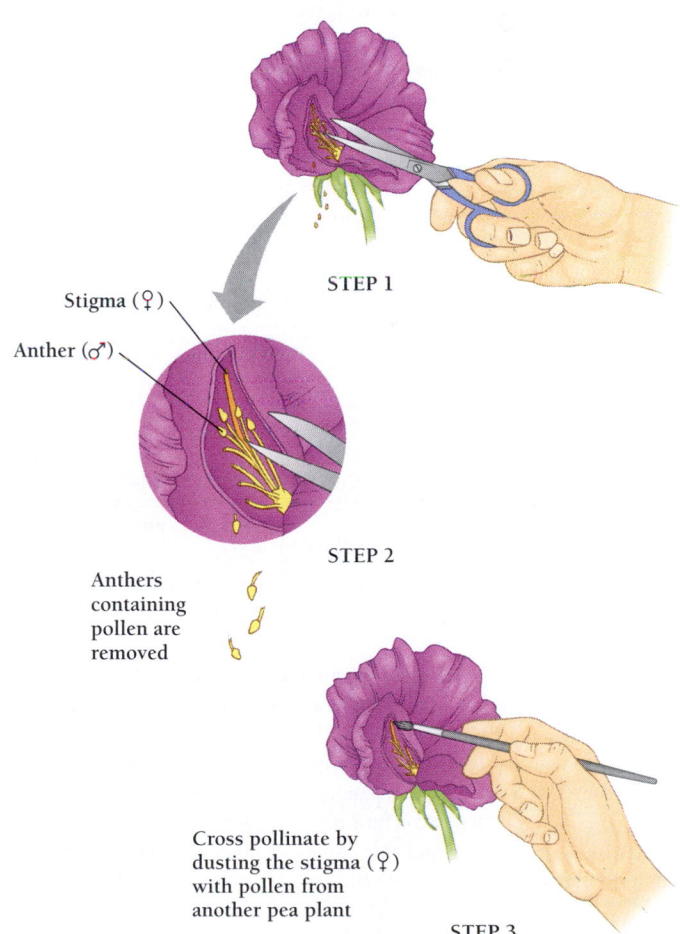

Figure 9-21   Cross-pollinating peas. Pea flowers carry both male and female parts, which allows them to self-pollinate. But Mendel's studies of inheritance required him to prevent the peas from self-pollinating so he could cross different lines. He snipped the male, pollen-producing anthers from each pea flower and then sprinkled the female stigma with pollen from another flower.

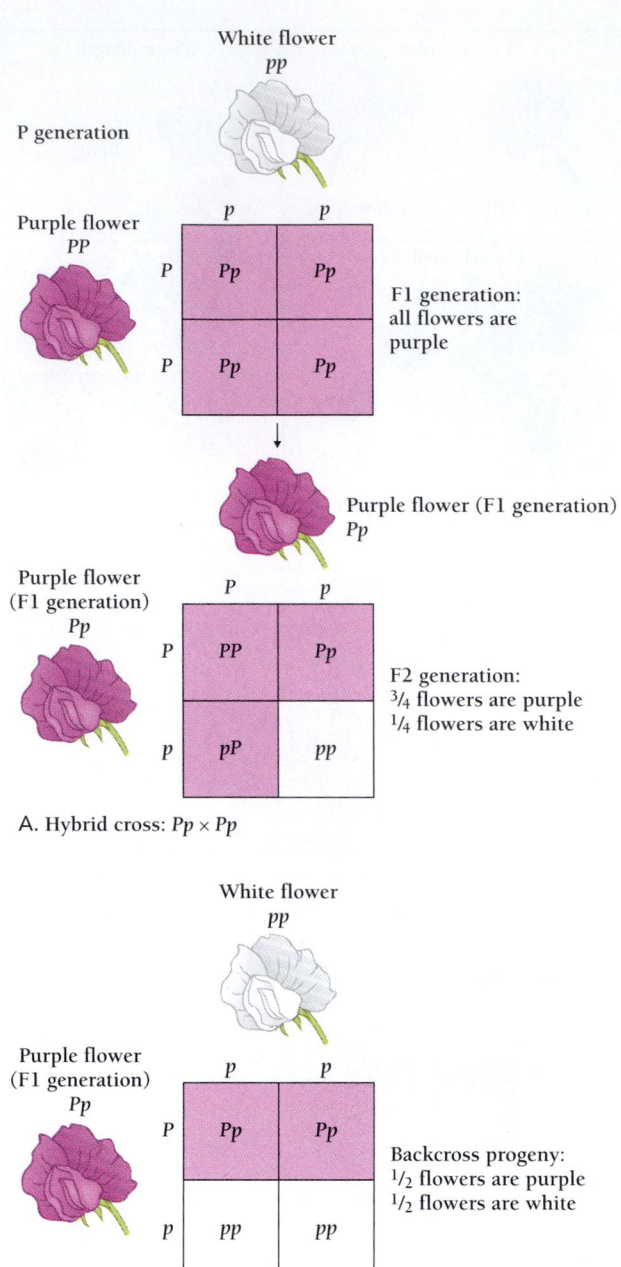

Figure 9-22   A backcross of purple- and white-flowered peas. A. First, a homozygous white flower is crossed with a homozygous purple flower to produce a 1:2:1 mix of homozygotes and heterozygotes. B. Then, in the backcross, a white homozygote is crossed with a purple heterozygote. Half the offspring are purple heterozygotes and half are white homozygotes. (By a convention begun by Mendel, we begin the names of dominant alleles with a capital letter and those of recessive alleles with a small letter.)

the homozygous recessive parental (P) stock—those which contained only the recessive allele (Figure 9-22B). By crossing the purple F1 plants to a white homozygous recessive, Mendel could find out how many of the F1 plants had recessive alle-

les and how many didn't. Any cross with a homozygous recessive that is supposed to reveal the genotype of the other parent is called a **testcross.**

In Mendel's testcross, the parents looked just like those in the original parental cross—a purple-flowered plant with a white-flowered plant (Figure 9-22A). But Mendel had discovered that the purple-flowered F1 hybrids could produce two kinds of gametes, one with the purple allele, and the other with the white allele (whereas the original purple parents produced only gametes with purple alleles). The plants with white flowers, on the other hand, would only produce gametes with the white allele. He correctly predicted that half the progeny would have purple flowers and half, white flowers.

Mendel's hybrid crosses and backcrosses produced offspring with ratios of phenotypes that suggested the principle of segregation; each sexually reproducing organism has two genes for each characteristic, which segregate during the production of gametes.

The Principle of Independent Assortment

Mendel next wanted to know if different traits were inherited together. In other words, were traits "linked?" To find out, he took two true-breeding stocks that differed in both the color and the form of the peas they produced. One parent produced round, yellow peas; the other, wrinkled, green peas. From previous crosses, he knew that round was dominant to wrinkled and yellow was dominant to green. As he expected, all the F1 peas were yellow and round. He then determined how these traits were inherited in the F2 generation. The allele specifying round peas is *R*, and the one for wrinkled peas is *r*; the allele specifying yellow peas is *Y*, and the one for green peas is *y* (Figure 9-23). The genotypes of the parents are therefore *RRYY* and *rryy*.

Homozygous parents can produce only a single kind of gamete: *RRYY* plants produce only *RY* gametes, and *rryy* plants produce only *ry* gametes. The genotype of the F1 plants therefore must be *RrYy*.

But what happens in the F2 generation? Mendel allowed the F1 to self-pollinate and then counted 556 peas in the F2. Of these, 315 were yellow and round (representing the action of the dominant alleles), and only 32 were green and wrinkled (Figure 9-23). The rest represented new phenotypes, unlike either of the parental stocks: 101 were yellow and wrinkled, and 108 were green and round. The two crosses had produced new combinations of genes, illustrating again how sexual reproduction provides diversity in a population. But how did Mendel make sense of the numbers?

First, Mendel noted that the traits associated with each gene were in a ratio of about 3:1, just as in his experiments with only one trait. In the F2 generation there were three times as many yellow as green peas (416:140 = 3:1), and three

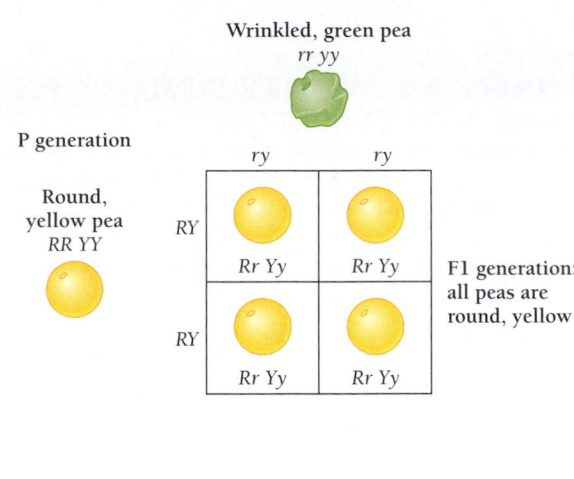

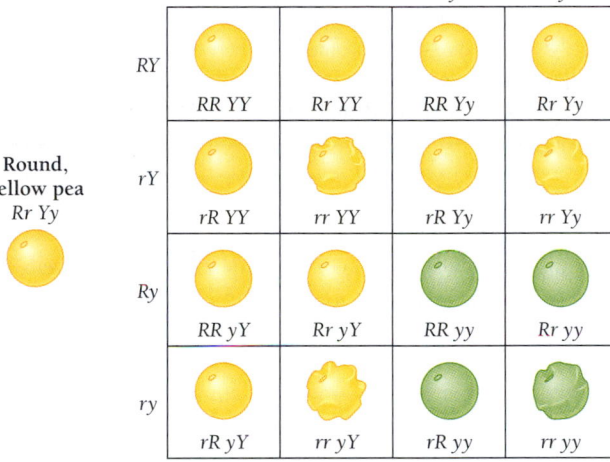

Figure 9-23 Testing two traits at once. In this two-factor cross, Mendel crossed round, yellow peas with wrinkled, green peas. In the first (F1) generation, all the peas were round, yellow heterozygotes. When he crossed the F1 generation, he got an assortment of different peas.

TWO-FACTOR BACKCROSS

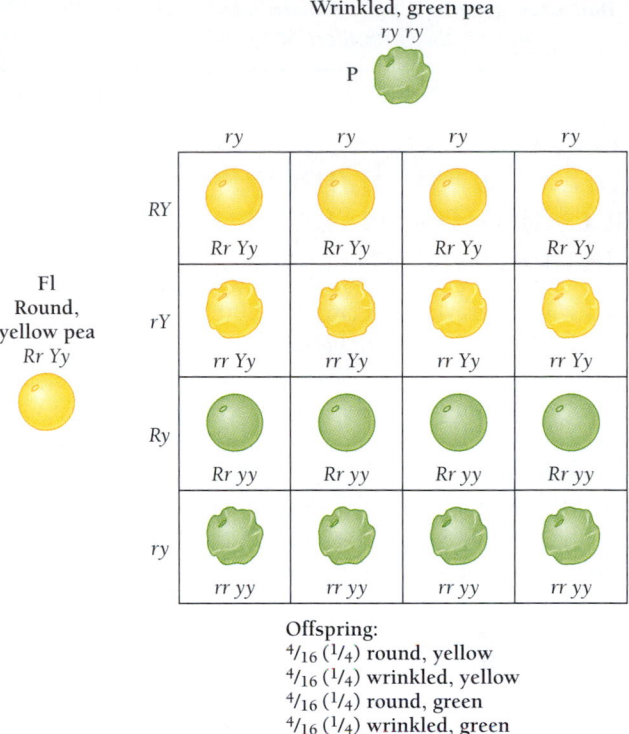

Figure 9-24 Independent assortment. By backcrossing the heterozygous F1 peas with the homozygous recessives (green, wrinkled), Mendel was able to show that each gene was distributed independently of the others.

times as many round as wrinkled ($423:133 = 3:2:1$). Mendel immediately saw that the two traits were behaving independently; whether a pea was yellow or green had no effect on whether it was wrinkled or round.

Mendel could explain the data if he assumed that each F1 plant produces four kinds of gametes: *ry*, *Ry*, *rY*, and *RY*. The Punnett square of Figure 9-23 predicts the distribution of genotypes and phenotypes of the F2 generation of this cross. It shows that 9 out of the 16 combinations of gametes produce plants with round, yellow peas, but only one combination produces plants with wrinkled, green peas. The overall ratio of phenotypes predicted by this analysis is $9:3:3:1$.

The correspondence of Mendel's experiments to this prediction is called the **principle of independent assortment.** This principle states that each pair of genes is distributed independently from every other pair during the formation of the gametes. The independent assortment of homologous chromosomes during meiosis causes the independent assortment of genes that are on different chromosomes.

Mendel tested his understanding of the genotypes of the gametes produced by the heterozygous F1 hybrids by doing another testcross with plants that produce green, wrinkled peas (Figure 9-24). Because both *r* (for wrinkled peas) and *y* (for green peas) are recessive, the genotype of the plants used in the backcross must be *rryy*. The gametes produced by these plants all must be *ry*.

If the genotype of the F1 plants is really *RrYy* and the chromosomes bearing the two genes assort independently, then the four kinds of gametes (*RY*, *rY*, *Ry*, and *ry*) should be made in equal numbers (Figure 9-24). When these gametes form zygotes with *ry* gametes, we expect equal numbers of the four phenotypes, in contrast to the $9:3:3:1$ ratio expected in the case of self-fertilization of the F1 plants. Again, Mendel obtained the expected ratios, confirming his understanding of the process.

Mendel's principle of independent assortment states that each gene in a pair is distributed independently during the formation of the gametes.

WHAT WAS THE EVIDENCE FOR THE CHROMOSOMAL THEORY OF INHERITANCE?

The rediscovery of Mendel's laws in 1900 hit British biologist William Bateson like a bolt of lightning. Within hours of reading the news, Bateson began devoting himself to spreading the word. As we saw earlier in this chapter, Bateson's talk at Columbia inspired Walter Sutton to quickly publish his own theory of inheritance.

In 1902, Walter Sutton and the German biologist Theodor Boveri independently published papers that showed that the behavior of chromosomes during meiosis perfectly explained Mendel's principles of segregation and independent assortment. Both biologists concluded that the genes had to be on the chromosomes.

What Were Sutton's Arguments?

Sutton's argument consisted of six separate points that together formed an almost fortresslike argument for what came to be known as the **Sutton-Boveri chromosomal theory of inheritance.** Sutton's first three points merely confirm the ideas of others; the second three are his own new ideas.

First, said Sutton, chromosomes come in pairs, one of which comes from the mother and one from the father. Second, synapsis consists of the pairing of homologous maternal and paternal chromosomes. Because the chromosomes of the lubber grasshopper had distinctive sizes and shapes, Sutton was able to see that the chromosomes occurred in pairs and that they separate during meiosis.

Third, the chromosomes retain their distinctive shapes throughout the cell cycle. Sutton argued that the constant differences in size and shape of the chromosomes suggested that the chromosomes are not recreated every cell cycle. The best evidence for this idea came from Boveri, who had studied sea urchin cells with reduced numbers of chromosomes. Only cells with the correct number of chromosomes developed normally. Boveri concluded that since every chromosome had to be present for normal development, each chromosome must be unique and must make a unique contribution to development.

Fourth, said Sutton, the chromosomes contain Mendel's genes. Fifth, meiosis creates new combinations of genes in each generation. During meiosis, when the diploid cell is reduced to two haploid cells, each homolog in a pair orients randomly with respect to the two poles of the dividing cell.

1. Chromosomes come in pairs. One chromosome comes from the mother and one from the father.
2. Synapsis is the pairing of homologous maternal and paternal chromosomes. These pairs separate during meiosis.
3. The chromosomes are not broken down and created anew during each cell cycle.
4. The chromosomes carry Mendel's genes.
5. Meiosis creates new combinations of genes in each generation. During meiosis, either homolog may end up in either new cell, regardless of the way all the other homologous chromosomes divide. This accounts for Mendel's principle of segregation.
6. Each chromosome carries a different set of genes, and all of the genes on one chromosome are inherited together. That different chromosomes are inherited independently accounts for Mendel's principle of independent assortment.

As a result, either homolog may end up in either new cell, regardless of the way all the other chromosomes divide. This, said Sutton, accounted for Mendel's principle of independent assortment.

In addition, because the maternal and paternal homologs of each chromosome migrate to each new cell independently of the other homologs, each parent can produce 2^n kinds of gametes, where n is the number of chromosome pairs. As we saw earlier, and as Sutton himself argued, an organism with two pairs of chromosomes, for example, would have 2^2, or 4,

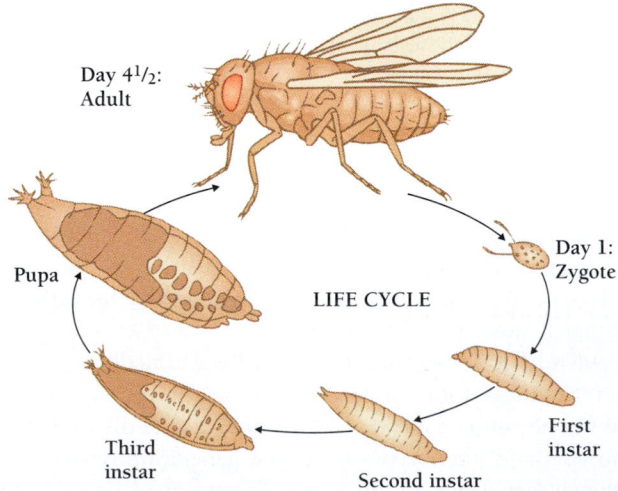

Figure 9-25   Growing up fast. Fruit flies (*Drosophila melanogaster*) mature from egg to adult in just 2 weeks, making them an ideal organism for studies of inheritance.

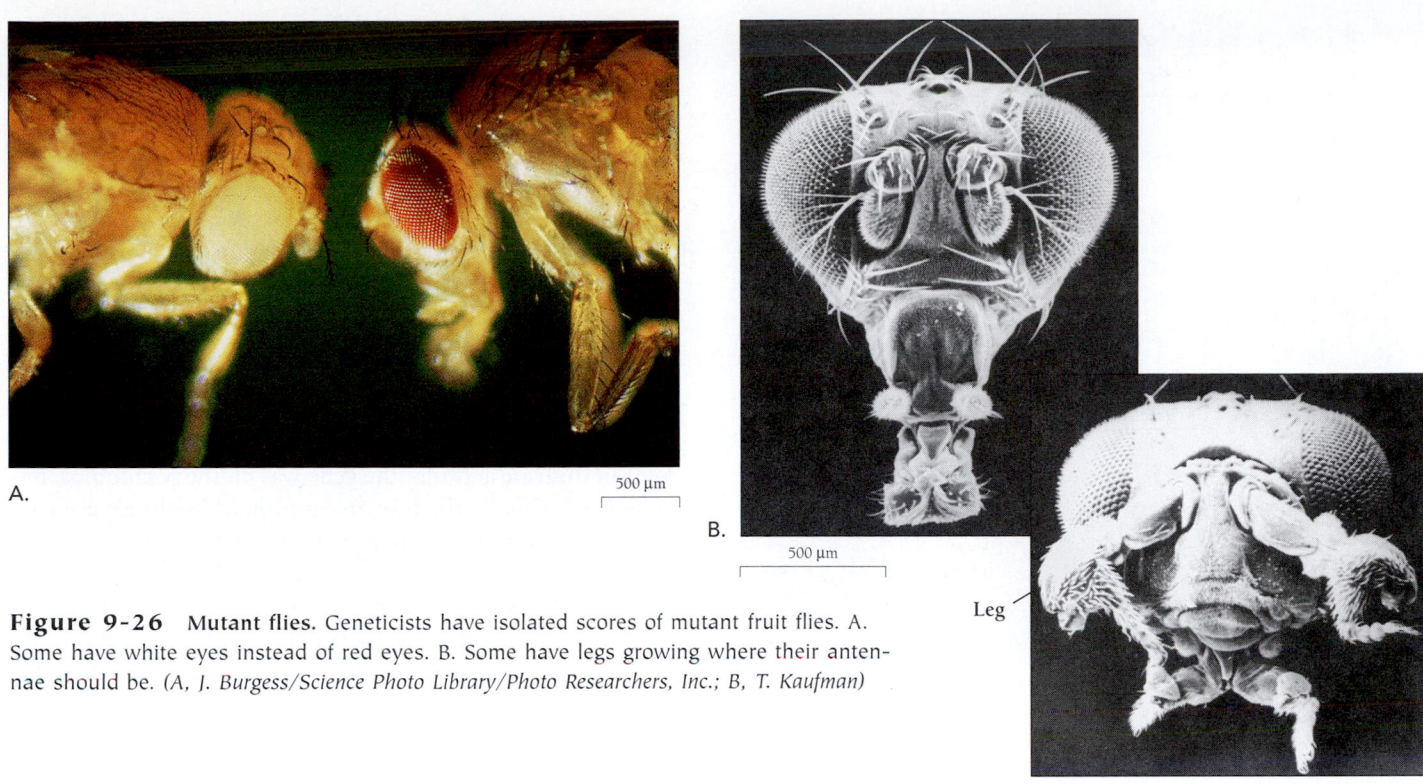

Figure 9-26 **Mutant flies.** Geneticists have isolated scores of mutant fruit flies. A. Some have white eyes instead of red eyes. B. Some have legs growing where their antennae should be. (*A, J. Burgess/Science Photo Library/Photo Researchers, Inc.; B, T. Kaufman*)

types of gametes, while a human could produce 2^{23}, or more than 8 million kinds of gametes.

Sixth, argued Sutton, each chromosome carries a different set of genes, and all of the genes on one chromosome are inherited together. That different chromosomes are inherited independently accounts for Mendel's principle of independent assortment. We now see Sutton's insights as the basis for understanding the chromosomal basis of the laws of inheritance. Table 9-1 summarizes the rules of inheritance that we can take both from Sutton's work and from later work. Because of Sutton's synthesis and because of vast amounts of subsequent work in genetics, we are in a better position to understand genetics than were the 19th-century biologists who first discovered the physical basis of heredity.

How Did Thomas Hunt Morgan Bolster the Theory of Chromosomal Inheritance?

With the publication of Sutton and Boveri's papers, the field of genetics took off. One of the most important contributors to the new field was Thomas Hunt Morgan, who began breeding tiny fruit flies called *Drosophila melanogaster* in his laboratory at Columbia University, in New York City, in 1909. Fruit flies have many advantages for genetic research. Among them are: (1) they breed rapidly, growing from eggs to sexually mature adults in less than 2 weeks (Figure 9-25); (2) they

are prolific, with each female laying hundreds of eggs; (3) they are small (about 3 mm long) and easy to keep; (4) they are intricately and precisely built organisms, with dozens of easily identifiable traits under genetic control; and (5) they have only four pairs of chromosomes.

Morgan first collected his flies by leaving a piece of ripe pineapple on his windowsill. He kept the flies captive in small milk bottles "borrowed" from his home milk delivery man. Before long he was raising thousands of the tiny flies, and his laboratory, called "the fly room" by other biologists, became the center of genetic research in the United States for more than 25 years. Since then, thousands of geneticists and hundreds of thousands of undergraduate biology students have studied inheritance in fruit flies, demonstrating over and over that traits such as eye color, body color, and wing shape are all inherited according to Mendel's laws.

One of Morgan's first discoveries concerned a gene for eye color, which, he showed, lies on the X chromosome. Most wild fruit flies have red eyes. One day, however, Calvin Bridges, an undergraduate hired to wash the hundreds of dirty milk bottles in Morgan's laboratory, noticed a single fly with white eyes (Figure 9-26). Morgan crossed the white-eyed male with several normal, red-eyed females. All the F1 progeny had red eyes, which showed that the white-eye allele must be recessive and the corresponding red-eye allele, dominant.

Morgan then crossed the F1 males and females and obtained an overall ratio of 3:1 of red-eyed flies to white-eyed

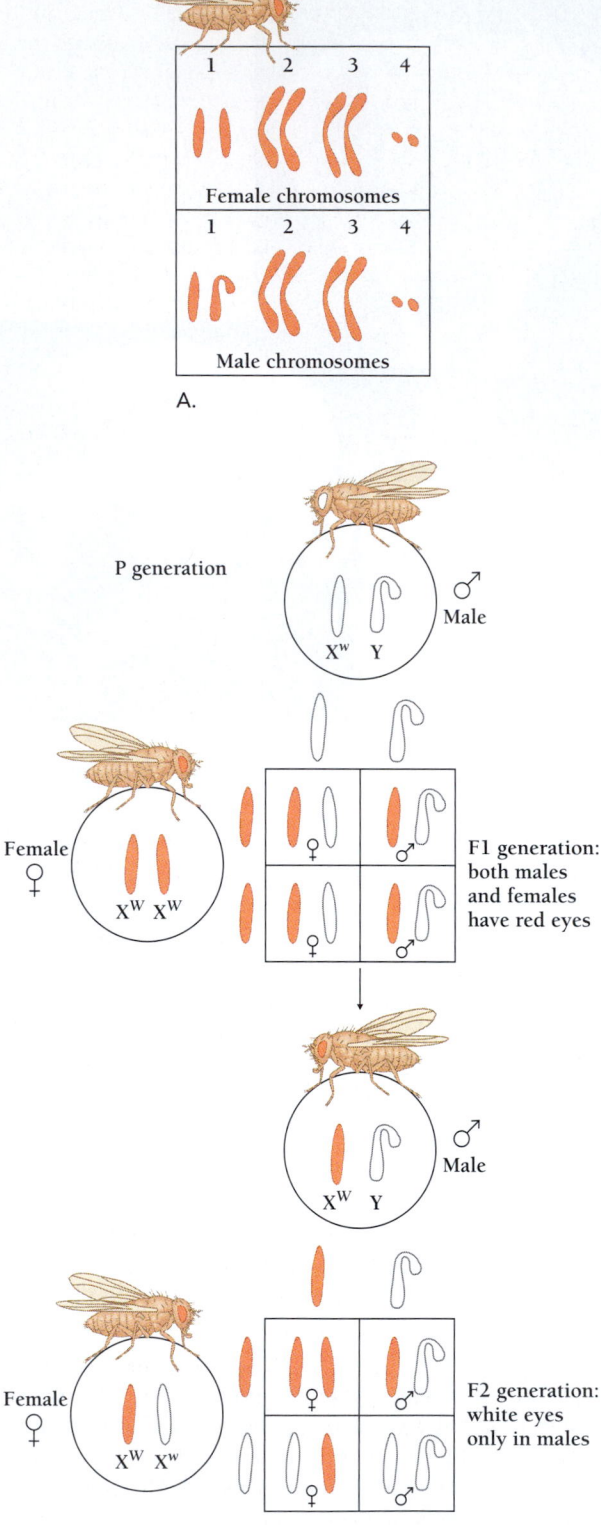

A.

P generation

♂ Male
X^w Y

Female ♀
X^W X^W

F1 generation: both males and females have red eyes

♂ Male
X^W Y

Female ♀
X^W X^w

F2 generation: white eyes only in males

B.

Figure 9-27   Sex-linked genes. A. Fruit flies have four pairs of chromosomes. As in humans, the male has an unmatched set of sex chromosomes. B. The white gene is on the sex chromosome, and we can trace its descent through two crosses by looking at the way the sex chromosomes assort.

flies. But all of the white-eyed flies were male! None of the females had white eyes. Half of the F2 males had red eyes, but all of the females had red eyes. What was going on?

Drosophila have only four pairs of chromosomes (Figure 9-27A). Males have three pairs of autosomes plus one pair of unmatched X and Y sex chromosomes, while females have three pairs of autosomes plus one pair of X chromosomes. As in humans, some genes are on both the X and Y chromosomes. But the Y chromosome is smaller than the X chromosome, so many genes on the X chromosome are not on the Y chromosome. Many genes on the X chromosome are present in two copies in females but in only one copy in males. Biologists say the males are *hemizygous* for genes on the X chromosome.

Morgan hypothesized that a single gene determined eye color in this case and that the gene was on the X chromosome. In Figure 9-27B, Punnett squares explain Morgan's experiment, showing the dominant red allele as X^W (because it is on the X chromosome) and the recessive white allele as X^w (because it too is on the X chromosome). The *W* stands for *white,* the name of the eye-color gene. Geneticists frequently name a gene after a mutant allele that functions badly or not all, rather than after the normal allele. So, ironically, the normal *white* allele (*W*) specifies red eye color!

The males and females in the F1 generation have different genotypes. The females are all heterozygous, with the white-eye allele (*w*) on one X chromosome and the dominant red-eye allele (*W*) on the other X chromosome. So all the females have red eyes. The males are all hemizygous, with the red-eye allele on their single X chromosome and no eye-color allele at all on their Y chromosome. So all the males have red eyes, too, which they got from their mothers.

In the F2 generation, the females receive one X chromosome from each parent. All the females receive the dominant X^W from their fathers and have red eyes. Of these, however, half are homozygotes that received the dominant X^W from their mothers and half are heterozygotes that received the recessive X^w from their fathers. The males receive their Y chromosomes from their fathers and their X chromosomes from their mothers. Because the Y chromosome has no eye-color allele, they receive no allele from their fathers. From their mothers, half of the male F2 flies receive X^W and have red eyes, and half receive X^w and have white eyes.

To double-check this conclusion, Morgan performed a backcross, mating what he thought were heterozygous F1 females with a hemizygous white-eyed male. His results were exactly as he predicted. Half of the males and half of the females had white eyes. This result established that the *white* eye-color gene indeed lay on the X chromosome.

The gene studied in this experiment is one of almost 1000 genes in *Drosophila* that are sex linked. In flies and other organisms with unmatched sex chromosomes, **sex-linked genes** have different patterns of inheritance in males and females because they lie on chromosomes that also determine the sex of the offspring. In mammals and flies, for example, the X chromosome contains many more genes than the Y chromosome,

A. ORIGINAL CROSS

Wild type parent:
red eyes, normal wings

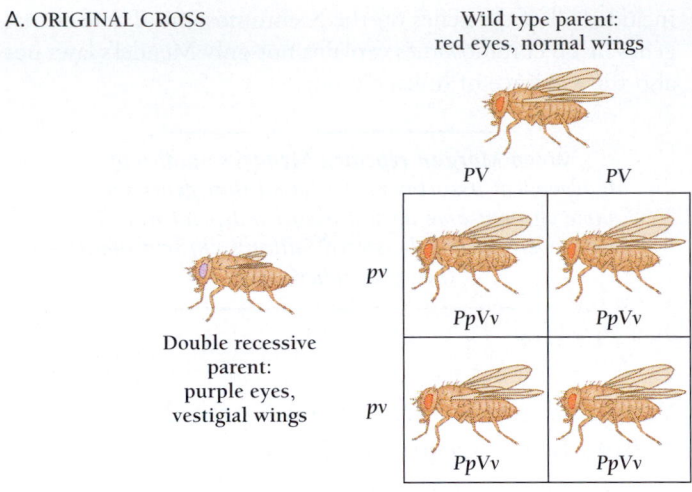

Double recessive
parent:
purple eyes,
vestigial wings

Results: F1 all *PpVv* (wild type)

B. BACK CROSS

Heterozygous F1 parent:

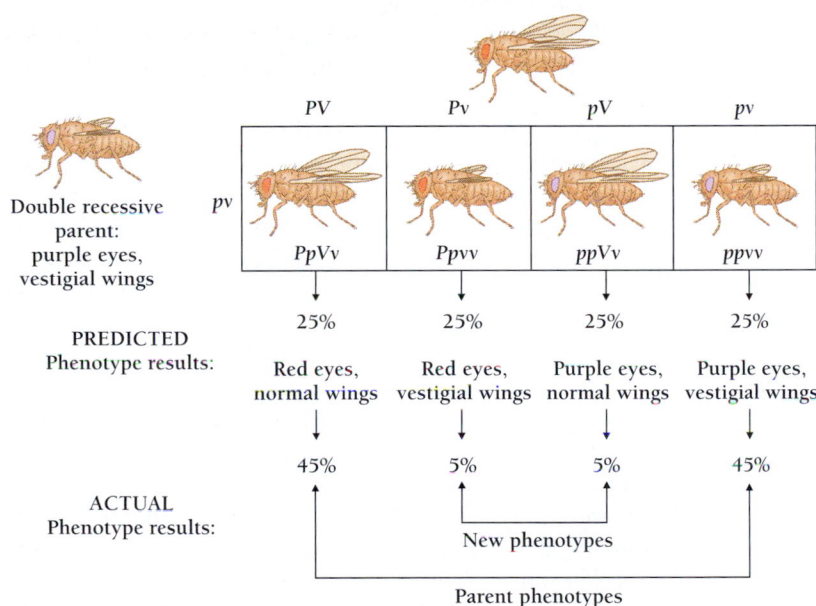

Double recessive
parent:
purple eyes,
vestigial wings

Figure 9-28 Independent assortment? No, genetic linkage. *P* and *V* indicate the dominant wild-type alleles, red eyes and normal wings. Small *p* and *v* indicate the recessive alleles, purple eyes and vestigial wings. A. In a cross between a red-eyed parent with normal wings and a purple-eyed parent with vestigial wings, all the offspring are heterozygotes with red eyes and normal wings ("wild-type"). B. If the two gene assorted independently, a back-cross between the heterozygous and homozygous flies should produce a ratio of 1:1:1:1. Instead, nearly all the flies had either red eyes and wings or purple eyes and vestigial wings. The two genes do not assort independently because they are on the same chromosome.

and sex linkage therefore nearly always refers to genes on the X chromosome. This experiment and others performed by Morgan and his students helped establish that genes were indeed on chromosomes, just as Sutton had proposed a decade earlier (Web Bit 9-1: Sex Linkage in Humans).

> *Thomas Hunt Morgan's experiments with sex-linked genes confirmed Sutton's theory that the genes lie on the chromosomes.*

Do Genes on the Same Chromosome Assort Independently?

Mendel's paper did not attempt to explain what happened when two genes were on the same chromosome, or linked. But Mor-

gan, with his new knowledge of genes and chromosomes, was able to figure it out. The principle of independent assortment depends on the independent assortment of chromosomes, but it cannot hold for genes that are close together on the same chromosome. That is, if Mendel had chosen an eighth trait, it would not have assorted independently. Sutton recognized this problem in 1903, but Morgan and his students were the first geneticists to design experiments around this question. In so doing, they further established the chromosomal basis of inheritance.

Morgan crossed two types of flies. One kind had wings of a more or less standard length and red eyes—characteristics usually found in wild populations. These flies were called *wild type*. The second kind of fly had short, stubby "vestigial" wings and purple eyes (Figure 9-28). Morgan had previously shown that purple-eyed flies with vestigial wings bred true, and he concluded that the purple, vestigial phenotype results from the ho-

Chromosome 9
of corn

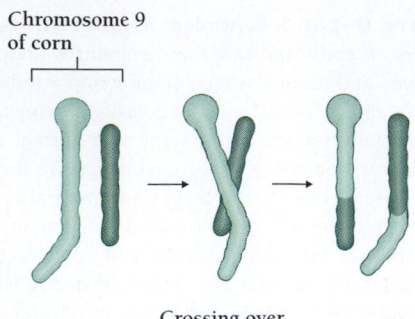

Crossing over

Figure 9-29   Crossing over made easy. In corn, one homolog of chromosome 9 has a knob at one end and a tail at the other. When homologous chromotids cross over, one gets the knob and one gets the tail. In 1931, Harriet Creighton and Barbara McClintock easily showed that what looked like crossing over in chromosome 9 corresponded to the pattern of genetic linkage of genes on chromosome 9.

mozygous condition of two genes. Morgan named the purple allele *pr* and the vestigial allele *vg*. For simplicity we call them *p* and *v* and the corresponding wild-type alleles *P* and *V*.

Purple vestigial flies crossed with true-breeding, wild-type flies produced normal (wild-type) F1 flies (Figure 9-28A). This showed that *p* and *v* were recessive alleles and that *P* and *V* were dominant. Since *p* and *v* were recessive, the genotype of the purple vestigial flies was *ppvv*.

To find out whether the *p* and *v* genes would assort independently, like Mendel's pea genes, Morgan performed a backcross. He mated the heterozygous F1 flies with purple vestigial flies. If the two genes assorted independently, this cross should have yielded four possible phenotypes in equal proportions. Morgan expected: $^1/_4$ purple eyes, vestigial wings; $^1/_4$ purple eyes, normal wings; $^1/_4$ red eyes, vestigial wings; and $^1/_4$ red eyes, normal wings (Figure 9-28B).

Instead, almost 90 percent of the flies had the same phenotypes as the two original parents—either purple eyes and vestigial wings or red eyes and normal wings. The remaining 10 percent of the flies were divided into flies with purple eyes and normal wings and flies with red eyes and vestigial wings.

Morgan concluded that the two genes did not assort independently. The *v* and *p* alleles almost always stayed together, and the *V* and *P* alleles almost always stayed together. Morgan guessed that these alleles stayed together because they were on the same chromosome and, more generally, that genes assort independently only when they are on separate chromosomes.

The tendency of genes on the same chromosome to stay together is called **genetic linkage.** Morgan and his students found that all of the fly alleles they studied fell into just four distinct **linkage groups,** sets of genes that do not assort independently. The four linkage groups, Morgan realized, corresponded to the four pairs of chromosomes in *Drosophila*. One linkage group consisted of sex-linked traits, indicating that it

included all of the genes on the X chromosome. The fact that genes lie on chromosomes explains not only Mendel's laws but also the existence of linkage groups.

When Morgan repeated Mendel's studies of independent assortment, he found that genes on the same chromosome do not assort independently. This discovery further bolstered Sutton's chromosomal theory of inheritance.

Why Did Some Flies Have Normal Wings but Purple Eyes?

Genetic linkage explains why 90 percent of the flies in the cross shown in Figure 9-28B resemble one or the other parental phenotype. But why do some flies in the backcross have purple eyes and normal wings?

The answer is that homologous chromosomes frequently exchange material with each other during meiosis. Earlier in this chapter, we mentioned that crossing over occurs when homologous chromosomes pair up during meiosis I. Single chromatids in homologous chromosomes break, exchange material, and rejoin (Figure 9-11).

After crossing over, a new chromosome contains some genes from each of the two homologs. These new combinations of genes then stay together through the rest of meiosis. In this way, the chromosomes of the resulting gametes can differ dramatically from the chromosomes of the parents.

The physical basis of crossing over was first demonstrated in 1931 by graduate student Harriet Creighton and her mentor, geneticist Barbara McClintock, at Cornell University. McClintock, an expert on the genetics of corn, had noticed a pair of homologous chromosomes that differed in appearance. One homolog had a knob at one end and a long tail at the other, while the other homolog was relatively plain (Figure 9-29). McClintock suggested that Creighton find out whether homologous chromosomes literally traded parts by comparing the inheritance of alleles on the odd pair of chromosomes with their appearance after meiosis. Creighton, not realizing that the work was important, dawdled, and McClintock had to press her to finish the work. In time, young Creighton was able to show that crossing over wasn't just an abstract concept but that the chromosomes actually swapped long sections, so that the new homologs had either a knob but no tail or a tail but no knob (Figure 9-29).

In 1931, Thomas Hunt Morgan came to visit Cornell. When he heard what Creighton had done, he insisted that she write up her results and publish them immediately. Although he didn't mention it, Morgan knew that another famous geneticist was finishing up the same work in *Drosophila* and would soon publish. Morgan sat down on the spot and wrote a letter to the editor of a scientific journal, telling him to expect Creighton and McClintock's paper in 2 weeks. Because *Drosophila* reproduce quickly, each experiment lasts just a couple of weeks, while the same experiment with corn takes a whole summer. As Morgan later admitted, "I thought it was about time to give corn a chance to beat *Drosophila!*"

Crossing over recombines alleles in new ways. Harriet Creighton and Barbara McClintock showed that crossing over is a physical exchange of material between homologous chromosomes.

Can All the Genes on a Chromosome Cross Over and Recombine?

Once Sutton and Boveri had persuaded biologists that genes were on chromosomes, the door was open for a new kind of question: Just *where* on a chromosome did a particular gene lie? At least 1000 genes lie on the *Drosophila* X chromosome, and at least 10,000 or more lie on the human X chromosome. How could they map these genes?

The answer came from Alfred Sturtevant, another undergraduate in Morgan's laboratory. Sturtevant discovered that the farther apart two genes are on a chromosome, the greater the

likelihood that they will cross over. Sturtevant expressed the "genetic distance" between two genes as the chance that recombination would occur between them. He defined a **map unit** as the distance between two genes that would produce 1 percent recombinant gametes (and 99 percent parental gametes). He then constructed a **genetic map** (or **linkage map**) that summarized the distances between many different genes. Geneticists refer to the position of a gene on a chromosome as its **locus** [Latin, = place; plural, **loci**]. The distance between the *v* locus and the *p* locus, for example, is about ten map units.

The chance that any two genes on a chromosome will recombine is proportional to the physical distance between them.

In this chapter, we have seen how cells divide to create haploid gametes by means of meiosis and how the twin processes of syngamy and meiosis complement each other. We have also seen that the behavior of the chromosomes during meiosis explains the patterns in which genes are inherited. In the next chapter, we learn what a gene does and what it is made of. We will also see that the structure of DNA explains how it can copy itself during the S phase of the cell cycle.

Study Outline with Key Terms

Genetics is divided into **transmission genetics,** the study of how variation is passed from one generation to the next, and **molecular genetics,** the study of how DNA carries genetic instructions and how cells carry out these instructions. For many years, biologists thought that the characteristics of offspring represented a **blending** of the characteristics of their parents. But most genes are present in two copies, one from each parent, and—most of the time—these genes do not change as they pass from generation to generation. The **Sutton-Boveri chromosomal theory of inheritance** states that the units of heredity are genes carried on chromosomes.

Phenotype encompasses all physical and behavioral characteristics of an organism. **Genotype**—the genetic constitution of a cell or organism—is a collection of genes. The word "genotype" can refer to the entire **genome** or to a subset of genes. A **gene** is a region of DNA that specifies the amino acid sequence of a polypeptide and its pattern of expression.

Asexual reproduction, through mitosis, produces offspring with genes from just one parent, a **clone. Sexual reproduction** produces offspring that inherit genetic information from two parents. This genetic information is contained in two sets of chromosomes, one set from each parent.

The **somatic cells** of an organism all have two copies of each chromosome and are said to be **diploid.** The **sperm,** or **spermatozoon,** and **ovum (egg),** or **gametes,** each carry only one copy of each chromosome and are said to be **haploid.** Gametes arise from the **germ line** or **germ cells** in the **gonads—ovaries** in females and **testes** in males.

Male and female gametes come together at fertilization (or syngamy) to form a **zygote. Fertilization** (or **syngamy**) restores the diploid chromosome number. The genetic information in the zygote, however, is a mixture of information from the parents. Adult organisms produce new haploid gametes by means of **meiosis,** a process that allots one haploid set of chromosomes to each of four daughter cells. Because the distribution of chromosomes occurs randomly, meiosis generates combinations of genes that are different from those present in the parents and thus accounts for much of the genetic diversity of individuals within a species.

Meiosis consists of two cell divisions (**meiosis I** and **meiosis II**), each of which consists of **prophase, metaphase, anaphase,** and **telophase.** During prophase I, homologous chromosomes form precisely aligned pairs in a process called **synapsis.** Each chromosome consists of a **tetrad.** During the rest of meiosis I, one chromosome from each pair (with two **chromatids**) goes to each daughter cell. Meiosis II distributes the sister chromatids of each chromosome to daughter cells.

Each member of a matching pair of chromosomes, or **homologous chromosomes,** is called a **homolog.** Homologous chromosomes pair and **cross over** during synapsis, forming **chiasmata. Disjunction,** the separation of homologous chromosomes (resulting in **euploid** gametes), sometimes fails. **Nondisjunction** leads to the production of gametes with an abnormal number of chromosomes, or **aneuploidy.** One example is **Down syndrome,** or **trisomy 21.** Most aneuploid embryos die long before birth in **spontaneous abortions.**

Alternative versions of the same gene are called **alleles.** An organism with two copies of the same allele is **homozygous,** one with two different alleles is **heterozygous,** and one with only a single allele is hemizygous.

An allele is **dominant** when it alone determines the phenotype of a heterozygote and **recessive** when it contributes nothing to the phenotype of a heterozygote. In **partial dominance,** both alleles are expressed.

Mendel made use of **true-breeding** varieties of peas to show his **principles of segregation** and **independent assortment.** Breeding two genetically distinct organisms is called **cross breeding** (or **crossing**), and the progeny of such a cross are called **hybrids.** To test the principle of segregation, Mendel performed a new kind of cross, called a **backcross.** A backcross to a homozygous recessive individual that is intended to reveal the genotype of the other parent is called a **testcross.** The original parents in a cross are called the **parental,** or **P,**

generation, and the progeny are the "first filial," or **F1** generation, **F2** generation, and so on. All kinds of crosses can be represented in a **Punnett square,** with each kind of gamete produced by one parent along the top of the square and each kind of gamete produced by the other parent along the left side of the square.

The position of a gene on a chromosome is its **locus.** The tendency of genes on the same chromosome to stay together is called **genetic linkage.** Geneticists can construct a **genetic map** (or **linkage map**), which summarizes the distances between genes in **map units. Linkage groups** are sets of genes that do not assort independently. Compelling evidence that genes are on chromosomes came from the analysis of genetic linkage in *Drosophila.* Genetic maps of *Drosophila* and other organisms identified the location of the genes on the chromosomes and explained deviations from the principle of independent assortment.

In mammals and fruit flies, males have two unmatched **sex chromosomes,** called **X** and **Y,** as well as homologous **autosomes.** Females have the same autosomes plus a pair of matched X chromosomes. The X and Y chromosomes have homologous regions and pair during synapsis. Because males have only one X chromosome, they are susceptible to **sex-linked** genetic defects such as hemophilia and color-blindness that result from being hemizygous.

Harriet Creighton and Barbara McClintock showed that crossing over is a physical exchange of material between homologous chromosomes. Their work was immediately accepted as confirming the chromosomal theory of inheritance.

Review and Thought Questions

Review Questions

1. How are homologous chromosomes similar to each other? How are they different?

2. Describe and draw the steps of meiosis I and explain what they accomplish. Then do the same for meiosis II. Which part of meiosis actually accomplishes the reduction in chromosome number?

3. Explain how crossing over of chromatids in prophase I leads to genetic recombination, so that the gametes produced from one parent are virtually all unique.

4. Describe the main ways in which meiosis differs from mitosis.

5. Define the principles of independent assortment and segregation. What aspects of meiosis explain these principles?

6. Explain the difference between autosomes and sex chromosomes. How many of each do you have? How many of each does a potato have?

7. What causes Down syndrome and other forms of aneuploidy? Why should the frequency of aneuploid eggs increase as a mother ages?

Thought Questions

8. In what way is fertilization the opposite of meiosis? Why is it important for a complete sexual life cycle to contain both processes?

9. If you found a new species in which the sexes look different, what criteria would you use to decide which to call "female" and which to call "male"?

10. Textbooks make a point of contrasting the genotype and the phenotype as mutually exclusive. If the phenotype is every measurable aspect of an organism, including its structure, chemical makeup and behavior, could an organism's genotype be considered a part of the phenotype? Why or why not?

About the Chapter-Opening Image

When young Walter Sutton solved the mystery of heredity in 1902, he cried, "I know why the yellow dog is yellow!"

 On-line materials relating to this chapter are on the World Wide Web at **http://www.harcourtcollege.com/lifesci/aal2/**

10 The Structure, Replication, and Repair of DNA

Rosalind Franklin and the Double Helix

On February 28, 1953, James D. Watson and Francis Crick discovered the structure of DNA, the hereditary material of the genes. Their breathtaking discovery laid a foundation for all of the most spectacular advances in biology until the present time. It was "the secret of life," as they announced at the time. The two men were helped in this momentous accomplishment by conversations with many other scientists. Yet, a young scientist named Rosalind Franklin, working in near-total isolation, almost beat them to the punch (Figure 10-1).

In 1951, Jim Watson was a 23-year-old former radio Quiz Kid, cherished by his elders as an up-and-coming genius. He had just received his Ph.D., studying the genetics of bacteria and the viruses that infect them, and his professors had arranged for him to go to Copenhagen to study the chemistry of nucleic acids. But the chemistry of nucleic acids bored him, and he began looking about for something else to work on.

While on a trip to Naples, Italy, Watson found that something—a question so fundamental that its answer seemed guaranteed to bring a Nobel prize. At a lecture by an English physicist turned biochemist named Maurice Wilkins, Watson saw a slide of an x-ray diffraction pattern from what Wilkins said was a crystalline form of DNA. Wilkins argued that his photos, taken with the help of graduate student Raymond Gosling, were the first step in deducing the structure of DNA. The molecular structure of DNA would, of course, shed light on how genes work.

Watson was transfixed. He understood that if DNA could form crystals, it must therefore have a regular and relatively simple shape. Then, within just a few days, he read a sensational paper by famed chemist Linus Pauling on the alpha-helical structures within proteins. Watson longed to do something equally spectacular. With DNA and helices swirling in his head, Watson persuaded his former professors back in the United States to get him out of Copenhagen and into a lab in England where he could study DNA.

In a few months, Watson arrived at the Cavendish Laboratories in Cambridge, in the same build-

ing as Max Perutz and John Kendrew, pioneers in the study of macromolecules using x-ray diffraction techniques. Also in the lab was Francis Crick, a 31-year-old graduate student working on his Ph.D. project. Working under Perutz, Crick was studying the x-ray diffraction of polypeptides and proteins, especially the blood protein hemoglobin.

Crick immediately liked Watson, and the two began to do an enormous amount of talking, so much so, in fact, that Perutz and Kendrew finally banished the two younger scientists to a room of their own. Watson knew nothing about x-ray crystallography except that he needed it to solve the structure of DNA. He nagged Crick relentlessly to help him work on DNA.

Crick liked to talk about DNA, but he wasn't about to start working on its structure. With a wife and daughter to support, he had to finish his Ph.D. and get a job. Besides, etiquette forbade Crick from working on the DNA problem. Crick and Wilkins were old friends going back to their joint work for the British Admiralty during World War II. Just as important, the Cavendish and the King's College Laboratory in London, where Wilkins worked, were sister facilities— and the DNA problem formally belonged to King's. At King's Col-

- DNA consists of two polynucleotide strands, which wind together to form a double helix.

- The two strands of DNA contain complementary versions of the same information. During DNA replication, each strand directs the synthesis of a complementary strand.

- In all organisms, genes are made of DNA, and each allele of a gene usually specifies the structure of a single polypeptide.

lege, Wilkins and Rosalind Franklin were already working on the structure of DNA, and they hadn't asked for help.

Still, it didn't hurt to talk. The first step to solving the three-dimensional structure of DNA was to study its density, size, and other properties using x-ray crystallography. Because DNA is a long, thin molecule, it would be impossible to obtain true DNA crystals until the 1960s, when researchers learned to produce crystals from short pieces of DNA synthesized in the laboratory (Figure 10-2). The DNA Wilkins had acquired was not crystals, but *fibers,* in which hundreds of millions of DNA molecules lay parallel to one another.

In 1951, the best x-ray diffraction photos of DNA fibers were those taken by Rosalind Franklin at the King's Laboratory. In November 1951, Watson took the train down to London to visit Maurice Wilkins and to hear Franklin give a lecture on the structure of DNA.

Franklin had arrived at King's at the beginning of 1951. A talented and recognized authority in industrial physical chemistry, she had just completed 4 years of work on the structure of coals at a lab in Paris. Her work laid the foundation for modern industrial carbon-fiber technology. Today, carbon-fiber technology is used in making strong, lightweight (and expensive) bicycle frames and sailboats, for example. For her successful analysis of coals, Franklin learned and improved on known x-ray diffraction techniques, developed new mathematical techniques for interpreting the resulting photos, and constructed scale models like those that Linus Pauling had used in his work. She returned to her native England eager to apply x-ray diffraction techniques to an important biological molecule.

John Randall, the head of King's Laboratory, invited Franklin to set up and head an x-ray crystallography lab, with all the newest, most powerful equipment available. Once the lab was set up, her assigned task was to work out the structure of DNA. He assigned a graduate student to assist Franklin. It was Raymond Gosling, the same young man who had taken pictures for Wilkins.

What Randall did not explain to Franklin was that Maurice Wilkins, just down the hall, had already begun work on DNA and considered the DNA problem his own. In fact, even

Figure 10-1 Rosalind Franklin on a walking tour of France in about 1950. Franklin produced most of the data needed to discover the structure of DNA. (*Vittorio Luzzati*)

in his letter offering her the job, Randall had never mentioned Wilkins. Worse, at her first meeting with Randall and Gosling to discuss the project, Wilkins was not present. With that peculiar oversight, Randall set the stage for one of the most disastrous personality clashes in 20th-century science.

Wilkins was equally in the dark about Franklin. He viewed her more as a technician than a scientist, and he appears to have believed that she was hired specifically to do his x-ray crystallography work, in which he had no expertise. He handed over his best DNA fibers, hoping she would produce for him the photos that would reveal their structure. But Franklin displayed none of the deference that Wilkins expected. Instead, she treated Wilkins as a colleague, vigorously arguing with his ideas about DNA in the spirited style she had picked up in Paris. It was her way of trying to make friends. Offended by what he perceived as combativeness, he struggled to put her in her place. Franklin found him unaccountably touchy; he was given to turning away in the middle of conversations he didn't like.

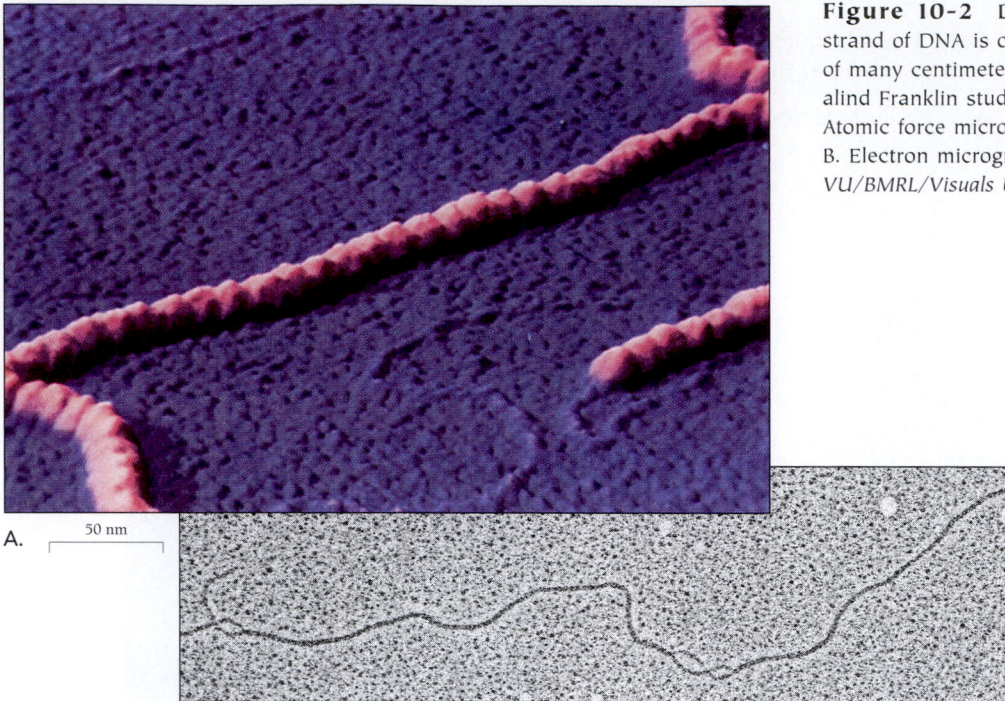

Figure 10-2   DNA is a long, thin molecule. A double strand of DNA is only 2 nm thick, but many reach a length of many centimeters. DNA fibers such as those that Rosalind Franklin studied consist of many parallel molecules. A. Atomic force micrograph showing a close-up view of DNA. B. Electron micrograph of a strand of DNA. *(A, Science VU/BMRL/Visuals Unlimited; B, K.G. Murti/Visuals Unlimited)*

A.   50 nm

B.   1 µm

Wilkins spitefully attacked her behind her back, undermining her relationships with other scientists at the lab. This was all too easy, for Franklin was never around when the other scientists gathered to talk at lunchtime or teatime: women were not allowed in the King's dining room.

Over the next 6 months, Franklin set up the x-ray crystallography lab and then, with Gosling's assistance, began working out the structure of DNA. Though she liked her work, the social atmosphere was poisonous, unlike anything she had experienced in other labs. Franklin was used to making warm friendships with her colleagues. But between Wilkins' hostility and her banishment from the dining room, this seemed impossible. She longed for her happy days in Paris. Unable to do the x-ray crystallography himself, Wilkins resentfully moved on to other projects. Randall, who might have reconciled the two researchers, did nothing.

WHAT IS THE STRUCTURE OF DNA?

When Franklin began her work in 1951, the chemical makeup of DNA—but not its three-dimensional structure—had been known for about 30 years. Franklin knew that DNA consisted of long chains of nucleotides linked together into polynucleotides. She knew that each nucleotide consisted of a sugar, a phosphate, and a base, and that the sugar and phosphate units were linked together alternately (sugar-phosphate-sugar-phosphate, etc.) into a sugar-phosphate "backbone" (Figure 10-3A).

Biochemists also knew that from the sugar-phosphate backbone hung four kinds of nucleotide bases. Two of the four bases

are **pyrimidines,** which consist of rings made up of four carbon atoms and two nitrogen atoms. The other two bases are **purines,** which contain a double ring of six carbon atoms and four nitrogen atoms (Figure 10-3B). The purine bases are **adenine** and **guanine,** and the pyrimidine bases are **cytosine** and **thymine.**

Each base has a long, correct name: deoxyadenosine monophosphate, or dAMP, for adenine; deoxyguanosine monophosphate, or dGMP, for guanine; deoxycytidine monophosphate, or dCMP, for cytosine; and thymidine monophosphate, or TMP, for thymine. But biologists rarely use these names. In fact, they generally refer to each nucleotide by a single letter abbreviation: A, G, C, or T. This convention not only saves space, it also emphasizes that the nucleotides are indeed the four letters in the DNA alphabet.

What was *not* known by anyone at the time that Franklin set to work was that a molecule of DNA consists of exactly two polynucleotide chains that are twisted together into a long helix, like a ladder twisted about its long axis (Figure 10-4). The two legs of the ladder are the two sugar-phosphate chains. Each rung of the ladder consists of a pair of bases (A always goes with T, and G always goes with C). Each twist of the ladder contains ten rungs, which are 0.34 nanometer (nm) apart. A single molecule of DNA may be either short or long. In humans, a molecule of DNA may be very long—up to 9 cm (3½ inches) long—and may consist of millions of nucleotide pairs. The width of the ladder, however, is the same from one end to the other—2 nm.

Importantly, each sugar-phosphate chain has a direction, and the two chains—the legs of the ladder—run in opposite directions (Figure 10-5). In each sugar-phosphate backbone,

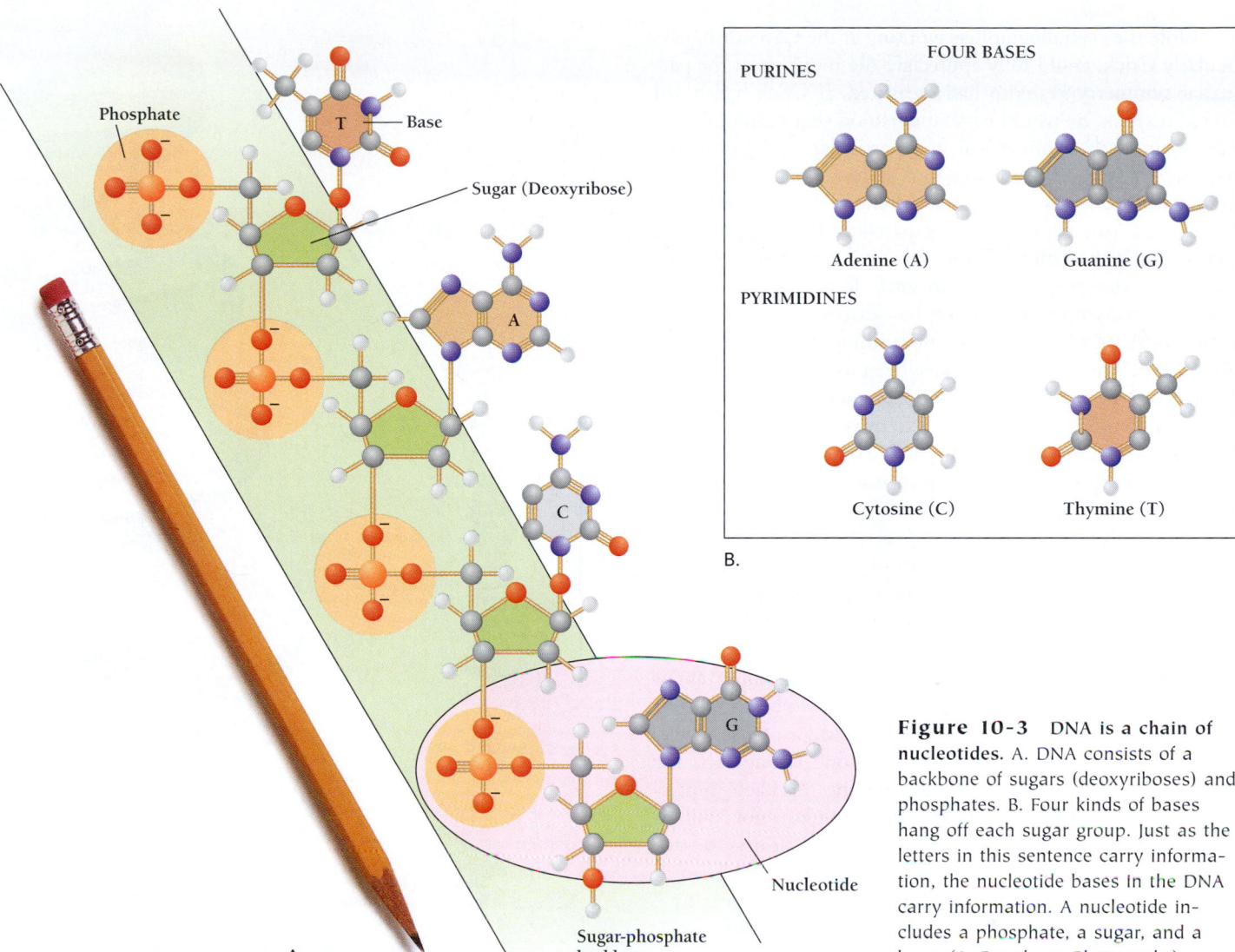

Figure 10-3 DNA is a chain of nucleotides. A. DNA consists of a backbone of sugars (deoxyriboses) and phosphates. B. Four kinds of bases hang off each sugar group. Just as the letters in this sentence carry information, the nucleotide bases in the DNA carry information. A nucleotide includes a phosphate, a sugar, and a base. (*A, Paraskevas Photography*)

the third carbon of one sugar is attached by a phosphate group to the fifth carbon of the next sugar. The third carbon is called the 3′ carbon, pronounced "three-prime carbon," while the fifth carbon is called the 5′ carbon, or "five-prime carbon." One end of each polynucleotide has a free 5′ carbon, called the **5′ end,** while the other end has a free 3′ carbon, called the **3′ end.** If we draw an arrow from the 5′ end to the 3′ end of each strand in Figure 10-5, one arrow points up, the other, down.

> *Each nucleotide in DNA consists of a sugar, a phosphate, and a base. The sugar and phosphate units are linked together alternately (sugar-phosphate-sugar-phosphate, etc.) into a sugar-phosphate "backbone." The two parallel backbones (running in opposite directions) resemble a ladder twisted around its long axis, with each rung consisting of a pair of bases (A-T or G-C).*

How Much Did Franklin Discover About the Structure of DNA?

By the time Franklin gave her talk in November 1951, her 5 months of work had already revealed some essential facts about the structure of DNA. She believed that DNA was probably a big helix with two, three, or four parallel chains, with the phosphates wound around the outside of the helix, like the uprights of a ladder. She had accurately measured the density of DNA and the number of water molecules associated with each nucleotide, and discovered that DNA fibers have two slightly different structures, depending on whether they are wet or dry. Finally, she had identified the "symmetry" of DNA. Crystallographers classify crystals into 230 types according to their symmetry. Which of these symmetries DNA had might seem like an obscure point, but it would be crucial to Watson and Crick's solution to the structure of DNA. DNA's symmetry was the clue that the two strands run in opposite directions.

Only the crystallographers working at the Cavendish, particularly Crick, could fully appreciate the meaning of the particular symmetry Franklin had measured. If Crick had heard Franklin speak, he would have understood that Franklin's parallel chains were symmetrical in a particular way. Just as the two ends of a sharpened pencil look different, a chain that ran in one direction only would look different when rotated 180°. In contrast, two pencils running parallel but in opposite directions look the same at both ends, and a pair of symmetrical chains running opposite to each look the same when rotated. The symmetry Franklin had discovered implied that DNA consisted of two chains, one running up, the other running down. The chains could have been four (two up and two down), but Franklin's density measurements ruled that out.

Franklin did not have the extensive training of a full-fledged crystallographer, and unlike Crick, she had never before worked with a large biological molecule, so she couldn't yet deduce the shape of the DNA molecule. Nonetheless, she was on her way.

Watson, who knew almost nothing about crystallography, watched Franklin talk, but he understood little of what she said and took no notes. After the talk, he visited Wilkins, who said that Franklin didn't know what she was doing. Back at Cambridge, Watson could tell Crick almost nothing about Franklin's talk. Crick was annoyed, but it would be 14 months before he learned how much he had missed.

What Watson did bring back was one wrong fact, the amount of water in the molecule, and the mistaken impression, gained from Wilkins, that neither Franklin nor Wilkins was capable of solving the structure of DNA. Clearly, the two at King's were not working together, and Wilkins seemed to be going nowhere by himself. Watson does not appear to have considered that Franklin might solve the structure. It seemed to him ridiculous to allow Franklin to bumble away, not solving this important problem, when he and Crick could be deciphering DNA themselves.

Back in Cambridge, Watson finally persuaded Crick to help him attack the problem. Within days, the two produced a model of DNA consisting of three intertwined helices, sugar-phosphate backbones on the inside, and far too little water. They called Wilkins and Franklin to come up to Cambridge the next day and look over their model. Franklin glanced at the model, knew at once that it was completely wrong, and said so. Crick was humiliated, for he knew instantly that she was right.

Watson had missed everything important that she had said. Worse, Watson and Crick had worked on a problem that belonged to King's, and they had

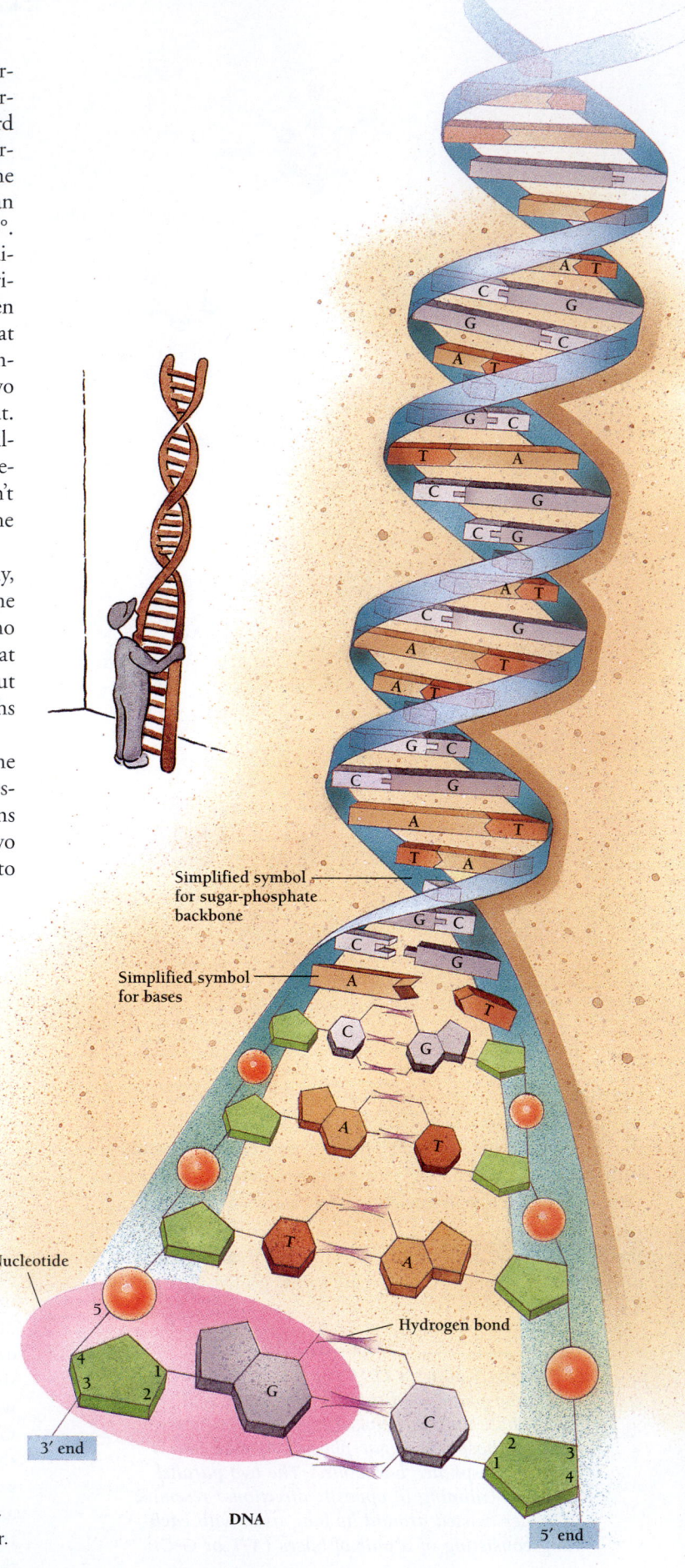

Simplified symbol for sugar-phosphate backbone

Simplified symbol for bases

Nucleotide

Hydrogen bond

3′ end

5

4
3   1
    2

G

C

DNA

5′ end

2   3
1   4

Figure 10-4   DNA is like a twisted ladder. The two sugar-phosphate backbones run in opposite directions. The bases hanging off the sugars connect in the middle, like the rungs of a ladder.

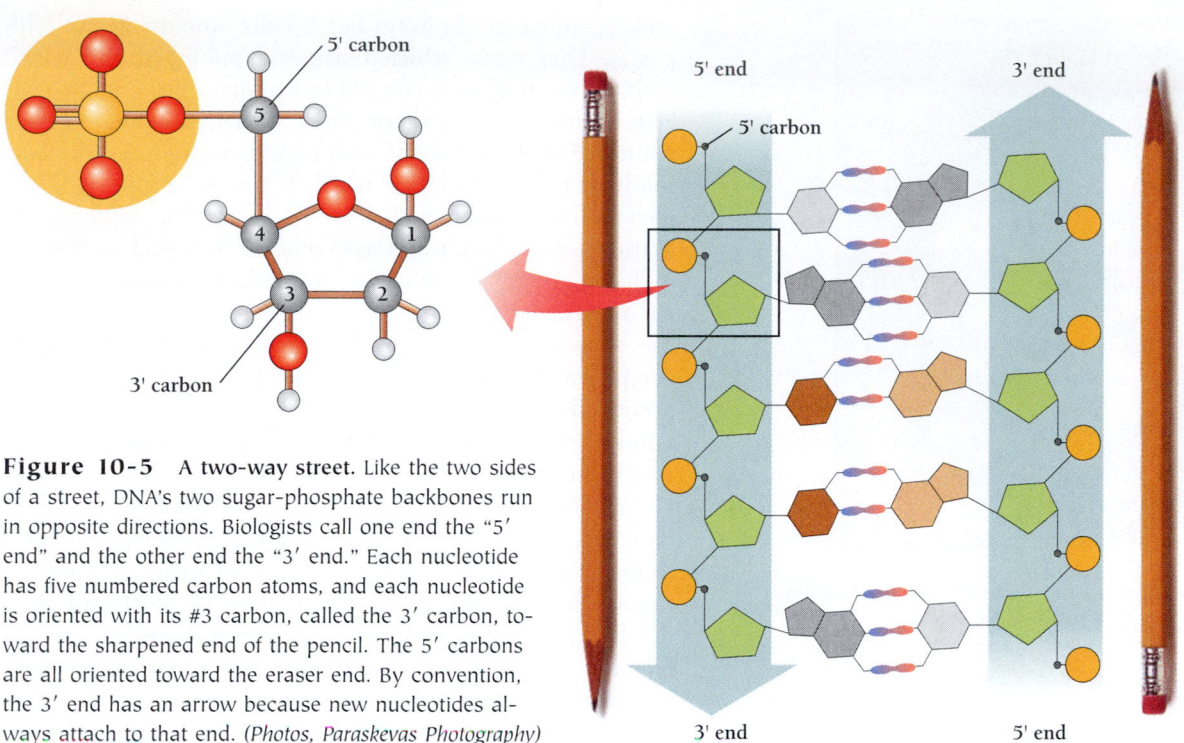

5' carbon

5' end

3' end

5' carbon

3' carbon

3' end

5' end

Figure 10-5 A two-way street. Like the two sides of a street, DNA's two sugar-phosphate backbones run in opposite directions. Biologists call one end the "5′ end" and the other end the "3′ end." Each nucleotide has five numbered carbon atoms, and each nucleotide is oriented with its #3 carbon, called the 3′ carbon, toward the sharpened end of the pencil. The 5′ carbons are all oriented toward the eraser end. By convention, the 3′ end has an arrow because new nucleotides always attach to that end. (*Photos, Paraskevas Photography*)

been wrong. The head of the Cavendish Lab, Sir William Lawrence Bragg, sternly reminded them that the DNA problem was not theirs to work on. Reluctantly, they agreed to stop trying to model its structure, but they continued to talk about DNA. They couldn't help it. They loved to talk (Figure 10-6).

For another year, Franklin studied DNA in near solitude, talking only to Gosling. For a while she thought that maybe the structure couldn't be a helix after all. The dry form of the DNA fiber gave pictures that seemed to suggest DNA was not a helix. There was, in fact, no certain evidence for a DNA helix early on. Pauling had argued that a repeating chain of any kind, whether a molecule or a telephone cord, will tend to twist about its axis into a helix. A helix seemed likely. But the pictures of the dry form did not lend support to a helix.

Franklin played with other possibilities, including repeating figure eights. In the summer she took an excellent picture of wet DNA fibers, showing a pattern that fairly shouted "helix." She labeled it photo number 51. She liked it a lot, and spent several days thinking about it (Figure 10-7). She had been working on the dry form for so long, however, that she soon returned to that. If she had had a collaborator, someone to argue with, she might have given more thought to photo number 51. But the atmosphere at King's was as bad as ever. She was already making plans to go to another lab.

Back at the Cavendish, Watson frittered away his time, playing tennis, going to parties, and fretting vaguely about DNA. He was, in his own words, "totally underemployed." Then, in the fall of 1952, two young researchers arrived at the

Figure 10-6 Francis Crick and James D. Watson share a cup of tea in the early 1950s. (*A, Barrington Brown/Science Source/Photo Researchers, Inc.*)

Cavendish—Linus Pauling's son Peter and one of Pauling's graduate students. In mid-December, Peter announced that his father had discovered the structure of DNA. It was more than Watson could stand. In January, he begged Peter to get his father to send the manuscript describing his structure for DNA.

Franklin, who had also heard the news, wrote to Pauling's lab with the same request. Meanwhile, she began calculating bond lengths and angles for the different building blocks of DNA. Her months of work had yielded the main dimensions

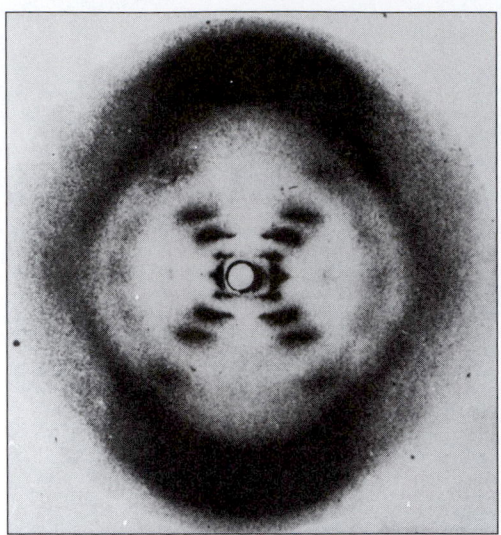

Figure 10-7 An x-ray diffraction photo of DNA by Rosalind Franklin. (*Cold Spring Harbor Laboratory Archives*)

of the molecule. She suspected that DNA was a double helix. She knew for certain that it was 2 nm in diameter and that if it was a helix, each complete turn of the helix would be 3.4 nm long. Now, with these and other hard-won clues at hand, she quietly prepared to build a model of DNA, the same approach that had worked so well for Pauling in deducing the α helix of proteins.

On January 30, 1953, Watson went to King's College and stopped to visit Wilkins. Finding Wilkins busy, Watson dropped in on Franklin to taunt her with his copy of Pauling's manuscript. When she looked at the manuscript she could see as well as Watson that Pauling's three-chain helix was hopelessly wrong.

She also discovered that Pauling and his co-author, Robert Corey, had snubbed her. The previous spring, Corey had come to King's and asked to see her pictures, and she had graciously shown him her work. Wilkins and Gosling had taken earlier pictures of DNA also, but these were vastly inferior to Franklin's. Now Pauling and Corey had not only ignored her letter requesting their manuscript, they had also cited Wilkins' inferior x-ray diffraction pictures of DNA—which they had never seen—and not her own, which they had. Watson later recalled that he then began to goad her again, implying that she was incompetent to interpret her own x-ray diffraction photos. Although few people believe him, Watson claimed that the petite Franklin was so angry that she chased him from her lab.

According to Watson, he then found Wilkins, and the two men took turns complaining about Franklin. Wilkins reminded Watson of Franklin's ideas on DNA, the same ideas she had presented 14 months earlier, then showed Watson her striking picture of the wet form of DNA, number 51. Crick and another scientist had already worked out a mathematical

understanding of the way that a helix would interact with x rays. That work, which Crick had published and which Franklin was fully aware of, had shown that a helix should generate a cross-shaped pattern. Now, with that knowledge and without Franklin's scores of other pictures to confuse him, Watson leapt to the conclusion that DNA must be a helix. It was obvious to everyone who had seen it—Franklin, Wilkins, and Watson—that this photo was of a helix. Nonetheless, Wilkins told Watson that Franklin stubbornly insisted that DNA was *not* a helix.

Watson could only conclude that Franklin really was incompetent. He was tired of the British laboratory etiquette that prevented him from working on DNA himself. He would build a model of the wet form of DNA, whose picture looked so simple. Instinctively, he decided to try two backbones. He had a feeling that for DNA to duplicate itself most easily, it should have two chains. Four chains seemed needlessly complex, and three chains seemed illogical. He was still convinced, however, that the sugar-phosphate backbones must be on the inside of the helix, with the bases hanging off the outside, despite Franklin's assertion that the sugar-phosphate backbones were on the outside. Watson rushed back to the Cavendish and at last succeeded in persuading Bragg to let him work on DNA. Feverishly, Watson began working to build a model. Crick, still writing his Ph.D. thesis, watched from across the room with amusement and growing interest.

Watson made little progress until, on February 5, 1953, Crick finally persuaded him to try putting the two sugar-phosphate backbones on the outside, as Franklin had suggested. For the first time, Watson had the beginnings of a model. The following weekend, Wilkins came up to Cambridge and reluctantly gave Watson and Crick permission to work on the DNA problem. Wilkins promised Crick in a letter to "tell you all I can remember & scribble down from Rosie."

No one asked Franklin's permission. Unaware that Wilkins and Bragg had now given away her research problem, she restarted her work on a structure for the easier, wet form of DNA. She was trying to decide if it was one chain or two.

For the first time, Crick felt free to work on DNA. The best source of information about DNA was Franklin, but Crick had no more intention of talking to her than had Watson or Wilkins. Crick soon learned, however, that Franklin and Gosling had summarized their unpublished work in an annual report on research at King's. Max Perutz had a copy of the report, and when Crick asked for it, Perutz handed it over.

Now Crick knew almost everything that Franklin and Gosling had so laboriously worked out. For the first time, he began to take the whole enterprise rather seriously. When he read Franklin's report describing DNA—its density, its water content, and, critically, its symmetry—it gradually dawned on Crick that DNA had to have two chains and that they had to run in opposite directions.

Franklin and Gosling also reported that the distance along the wet form of the molecule through one complete turn was 3.4 nm. Previous work had shown that the distance between the bases was 0.34 nm, so it was easy to see that there were probably ten bases per 360° turn. Wilkins and Gosling had earlier shown that the diameter of DNA was about 2 nm, and Franklin and Gosling had confirmed this. Things were beginning to come together.

Crick tried to persuade Watson to build a model with the chains running in opposite directions. But Watson couldn't figure out how to make it come out right. So, while Watson was off playing tennis, Crick abandoned his Ph.D. dissertation for an afternoon and took over the model. He built a more open double helix than Watson's, with sugar-phosphate chains running in opposite directions as required by the symmetry, and 10 bases per turn of the helix. The building blocks, with their bond angles and lengths, all fit nicely. Each complete turn was 3.4 nm, as Franklin had measured, and the distance from one base to the next was 0.34 nm. The only problem now was to find a way to fit the bases together in the middle.

It wasn't immediately obvious how this could be done. The four bases were of very different sizes and shapes. Some pair combinations would be wider than 2 nm and would make the molecule bulge and strain. Other combinations would have been too narrow and would have left a gap in the middle or pinched the two backbones in. Franklin's x-ray data showed, however, that the helix had a constant diameter. It neither bulged nor pinched in.

Watson began reading all the biochemistry of bases that he could find. If DNA were to carry genetic information, it should be able to accommodate any sequence of nucleotides, just as each line of print on this page must be able to accommodate any sequence of letters. Two weeks after he first put the sugar-phosphate backbones on the outside, Watson guessed that like bases paired (guanine with guanine and so on). But within days, Jerry Donohue, the young chemist from Pauling's lab, completely demolished Watson's hypothesis, explaining that the molecular forms that Watson had used, from standard textbooks, were all wrong. Besides, added Crick, Watson's like-with-like base pairing did not square with Chargaff's rules.

What Were Chargaff's Rules?

In the 1920s and 1930s, chemists believed that DNA contained almost equal amounts of each base. Then, in the 1940s, Erwin Chargaff developed a new method for analyzing the base composition of DNA. Chargaff found that the proportions of the bases could vary widely from species to species. But the amount of A always equaled the amount of T, and the amount of G always equaled the amount of C. Neither Chargaff nor anyone else at the time understood the significance of his findings, summarized in what came to be called **Chargaff's rules.**

In the spring of 1952, 9 months before Watson and Crick discovered the correct structure of DNA, Chargaff had come to Cambridge. Watson and Crick were introduced to Chargaff, and the older scientist took an immediate dislike to the two young men. "It struck me as a typical British intellectual atmosphere, little work and lots of talk," Chargaff later complained to the writer Horace Judson. Crick knew almost nothing about Chargaff's work, but the older man brusquely set him straight.

At the time, Crick saw that Chargaff's rules might imply complementary base pairing. That is, A always pairs with T, and G always pairs with C. Such a relationship could, he reasoned, provide a means for DNA to copy, or replicate, itself. Nine months later, however, Crick had forgotten his own insight.

By Saturday, February 28th, Watson had constructed cardboard cutouts representing the shapes of the four bases. He fiddled with them, fitting them together this way and that. He struggled to come up with pair combinations that would have the same symmetry and diameter, and still allow any order of bases on the backbone. Finally, he tried pairing A with T and G with C. The fit was unbelievably good. An A-T pair was the same width as a G-C pair (Figure 10-8).

Donohue quickly assured Watson that the two sets of bases could in fact fit together that way. The arrangement of atoms in the bases is such that A is perfectly arranged to make two hydrogen bonds with T, while G can form three hydrogen bonds with C. The chemistry worked, and for the first time, Watson and Crick understood the entire structure. It would be another week before they had the materials to build a complete model, but their work was done. Euphoric, they announced to everyone they met that they had "discovered the secret of life."

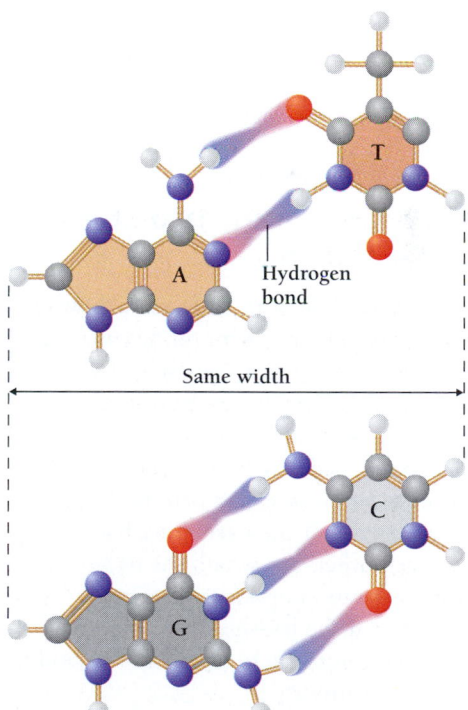

Figure 10-8   Base pairing. In DNA the bases connect in the middle of the twisted ladder by means of hydrogen bonds. The nucleotides A and T can form two stable bonds with each other, while G and C can form three. When matched in this way, the width of a G-C pair is the same as the width of an A-T pair, as predicted by Franklin's x-ray diffraction data.

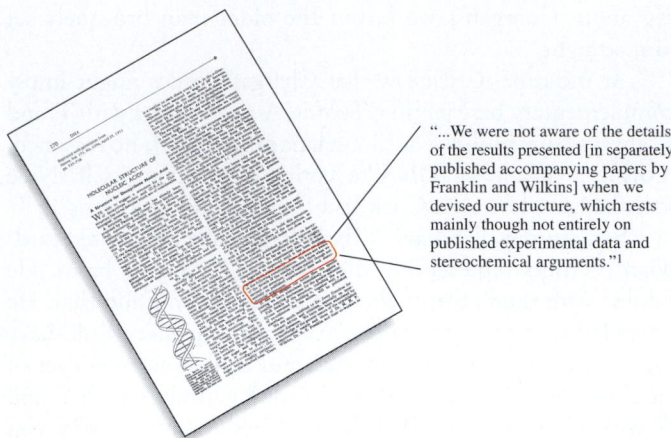

"...We were not aware of the details of the results presented [in separately published accompanying papers by Franklin and Wilkins] when we devised our structure, which rests mainly though not entirely on published experimental data and stereochemical arguments."[1]

On Saturday, March 7, Watson and Crick eagerly took model bases that had been made for them in the Cavendish machine shop and built an accurate model of the DNA double helix. On the same day, Wilkins, unaware of the model, wrote a note to Crick, including the following:

> I think you will be interested to know that our dark lady leaves us next week and much of the . . . data is already in our hands. I . . . have started up a general offensive on Nature's secret strongholds on all fronts: models, a theoretical chemistry and interpretation of data, crystalline and comparative. At last the decks are clear and we can put all hands to the pumps!

> It won't be long now.
> Regards to all,
> Yours ever,
> M.

Franklin Comes Within Two Steps of the Correct Model

But the race was over. Wilkins little knew how close Franklin had come to solving the problem on her own, or how late he was in restarting his own work. Nearly 2 weeks earlier, on February 23, Franklin had begun remeasuring photo number 51. By the end of the next day, she had correctly concluded that DNA was a double helix, with a diameter of 2 nm. She was now certain that the sugar-phosphate backbones were on the outside. She guessed that the backbones had some peculiar relationship to each other, but what? She hadn't yet realized that the two backbones ran in opposite directions. She knew about Chargaff's rules, and she had begun playing with the problem of how the bases would fit together in the middle. She was two steps from the answer.

But Franklin's work on DNA was at an end. Randall had told her she could leave King's for another lab if she wanted, but she could not take DNA with her. Accordingly, she left King's and, with Gosling, wrote up the results of their 18 months of work on the structure of DNA. Their work was not complete, but it was substantial. The potent combination of Franklin's experimental work, Crick's theoretical work, and

Watson's infectious enthusiasm had provided everything needed to solve the problem. Yet, amazingly, when Watson and Crick wrote up their results for the prestigious British journal *Nature,* Franklin was the one person whose work was not formally cited. This deficiency was apparently no oversight. According to writer Horace Judson, Wilkins asked Watson and Crick to delete a sentence from their manuscript that said, "It is known that there is much unpublished experimental material." Instead, Watson and Crick explicitly asserted that they were unaware of Franklin's work, writing that "We were not aware of the details of the results presented [in separately published accompanying papers by Franklin and Wilkins] when we devised our structure, which rests mainly though not entirely on published experimental data and stereochemical arguments."[1] In fact, according to later accounts, Watson and Crick's structure rested heavily on the unpublished data of Rosalind Franklin and Raymond Gosling, details of which Watson and Crick were well aware.

Not until March 18, when Wilkins first received Watson and Crick's manuscript, did Franklin learn of their achievement. Eagerly, she took the train up to Cambridge to see their model. Watson was astonished at how quickly and graciously she agreed that the model was correct. He had been sure that she would find some grounds on which to criticize it, for he little appreciated how much she knew and understood. She, for her part, had no reason not to be gracious, for she little suspected how much Watson and Crick had depended on her own work. The elegant beauty of the structure delighted her. She was filled with admiration for Crick's genius.

Franklin was glad, in any case, to be hard at work in her new lab. She was now studying the structure of tobacco mosaic virus. Over the next 5 years, Franklin directed a highly productive research team. Later, she collaborated with researchers at Berkeley and Yale in the United States, and at Tübingen in Germany. When she discovered she had cancer, in 1956, she continued to work cheerfully and productively until the very end, so devoted was she to the research she loved. She died in 1958, only 37 years old, without ever learning of her own crucial contribution to the discovery of the structure of DNA.

When, a couple of years later, it appeared that Watson and Crick might get a Nobel prize for their work, Sir Lawrence Bragg, who had headed the Cavendish, made sure that Wilkins received the prize with them. Recounting the story to writer Horace Judson, Bragg said that "when it came to the Nobel Prize, I put every ounce of weight I could behind Wilkins getting it along with them. It was just frightfully bad luck, really." Wilkins was lucky enough, however, to receive in 1963 a third share of a Nobel prize for the discovery of the structure of DNA. Franklin herself was ineligible, for only the living are honored.

In 1968, 10 years after Franklin's death, James Watson wrote a personal account of the discovery of the structure of

[1]From Watson, J.D., and Crick, F.H.C., "Molecular Structure of Nucleic Acids: A Structure for Deoxyribose Nucleic Acid," *Nature,* April 25, 1953, vol. 171, pages 737–738.

DNA. In *The Double Helix,* Watson for the first time made clear how essential Franklin's experimental results had been to their discovery. He also falsely depicted her as Wilkins' rebellious and shrewish assistant, who seemed to deserve any bit of ill treatment that came her way. Watson's book was attacked and reviled by nearly everyone who had been present during the discovery. It has never been considered an accurate account. Yet, as a work of literature and as a window into the mind of a successful molecular biologist, *The Double Helix* is a classic.

Modern Science Is a Social Endeavor

The discovery of DNA's structure is the best-documented story in the history of science. Watson, Crick, and others have all written extensively and passionately about both the science and the scientists. It is clear that Franklin's failure to discover the structure of DNA before Watson and Crick was a direct result of her social isolation. Watson himself said in 1993, "[Franklin] would be famous for having found DNA if she'd just talked to Francis [Crick] for an hour."

Watson and Crick had each other to talk to. They had Max Perutz, John Kendrew, Jerry Donohue, Peter Pauling, and even, grudgingly, Erwin Chargaff. All told, at least a half dozen scientists besides Rosalind Franklin herself provided essential clues that guided Watson and Crick toward the correct answer. Franklin, isolated by her feud with Wilkins, as well as her exclusion from the King's dining room, had no one to talk to but Gosling, who had little useful knowledge.

Watson and Crick's discovery of the structure of DNA was a borrowed victory that rested heavily on Franklin's discoveries. Their main contribution to biology was their early recognition that DNA's structure explains how it works. In fact, Watson and Crick's insight into the implications of base-pairing stands as one of the most influential advances in the history of biology. It has provided the sole basis for the twin fields of molecular genetics and genetic engineering, and transformed every other area of biology.

What was so important about their model? We've seen that the structure was like a twisted ladder with the uprights running in opposite directions. Each rung of the ladder was a pair of bases held together by hydrogen bonds. Watson and Crick's rules of complementary base pairing meant that the sequence of one polynucleotide strand exactly specified the sequence of the second strand. That is, one strand could give directions for building the other. This explained how a chromosome could replicate itself during cell division. Further, it was clear that the bases could be arranged in a nearly infinite variety of different sequences. That meant that the sequence of bases in DNA, like the sequence of letters on a page, could contain coded information.

In the rest of this chapter, we will examine the structure of DNA in more detail. Along the way, we will find out how biologists found out that the genetic material was DNA, what a gene is, and how DNA replicates and repairs itself. In the next chapter, we will see how cells actually use the information encoded in the DNA.

WHAT IS A GENE?

Defining a gene remains a problem not only to students taking biology examinations but also to scientists searching for biological molecules. To Mendel in 1865, a gene was an abstraction—some form of information that determined the traits of individuals. To Sutton and Morgan in the early years of the 20th century, a gene was a concrete physical structure that was part of a chromosome.

But biologists still did not know what genes actually were or how cells used and copied the information that genes contain. That knowledge came only when geneticists tried to understand the function and the structure of genes in chemical terms. In the rest of this chapter, we will see how researchers established (1) what genes do, (2) what they are made of, and (3) how they replicate. We will learn that a gene spells out the sequence of amino acids in a polypeptide, that a gene is made of DNA, and that the structure of DNA explains its ability to replicate and repair itself.

In this and later chapters we will also see that genes are not only physical entities or places on the chromosomes, but pure information. In this sense, genes are abstractions, as Mendel conceived them. In the same way that information about biology, for example, can be communicated as printed pictures or text, as spoken words in a lecture, or as electrons dispersed on the Internet, genetic information can be relayed in many forms. The physical expression of that information is separate from the information itself (Figure 10-9).

How Do Genes Affect Biochemical Processes?

The first hint of the biochemical importance of genes came in 1902 from the work of an English physician, Archibald Garrod. Among Garrod's patients at the Hospital for Sick Children in London was a baby whose diapers developed black urine stains. The baby's disease, alkaptonuria, had first been described in 1649 and named for dark "alkapton bodies" that formed after the urine was exposed to air. The precursor of the alkapton bodies was identified as homogentisic acid (HA), which oxidized to form the alkapton bodies.

Garrod, working just 2 years after the rediscovery of Mendel's paper, suspected that alkaptonuria might be an inherited disease because the baby's parents were first cousins (Figure 10-10). After studying the way the disease was inherited, he realized that alkaptonuria behaved as if it were caused by a recessive allele, much like the white flowers of Mendel's peas. Each parent had contributed the same recessive allele to the baby, who therefore had two copies of the same allele. We say the baby was homozygous for that allele.

Like the amino acids phenylalanine and tyrosine, HA contains a benzene ring. Garrod hypothesized that HA was an intermediate in the metabolism of phenylalanine and tyrosine. Most people have an enzyme called *HA oxidase,* which helps oxidize HA so that it can be excreted as carbon dioxide and water (Figure 10-11). Alkaptonuria patients, however, cannot

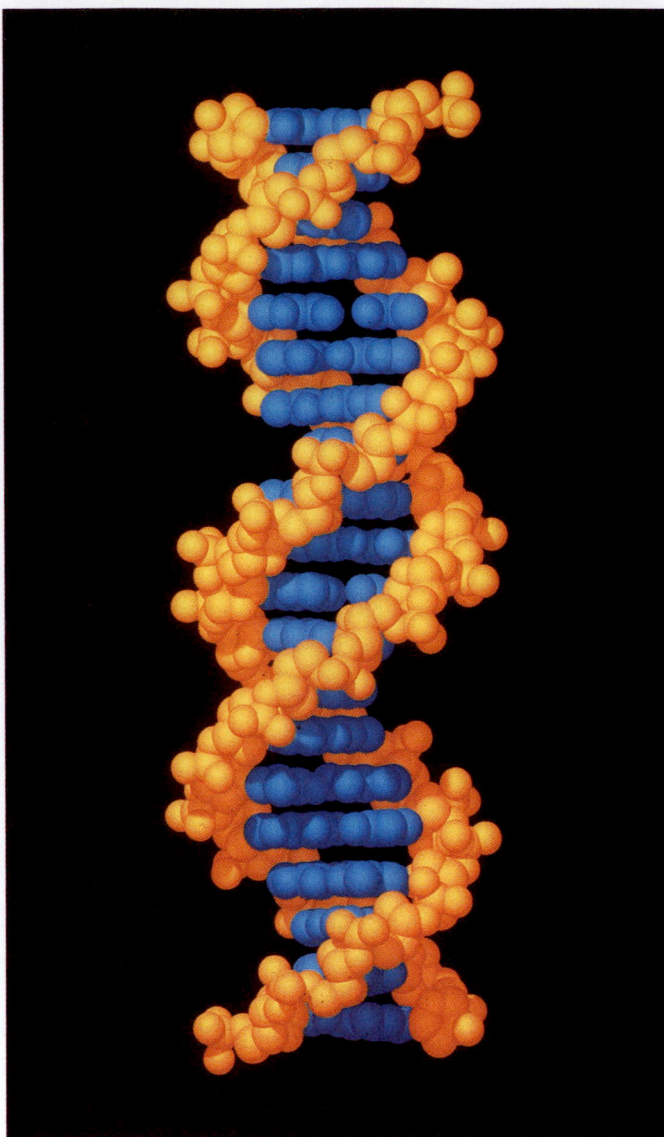

Figure 10-9 **Computer-generated model of DNA.** The structure of DNA is now so well known that most of us have seen the familiar double helix somewhere. *(Ken Eward/Biografx/Science Source/Photo Researchers, Inc.)*

oxidize HA, and the HA accumulates and turns to alkapton bodies in the urine. Garrod realized that the absence of HA oxidase might cause alkaptonuria. He hypothesized that the recessive allele whose inheritance he had charted could explain the missing enzyme. If an alkaptonuria patient had two copies of a defective allele of the gene for HA oxidase, he or she would not be able to make the enzyme.

Archibald Garrod's study of the genetic disease alkaptonuria suggested that genes affect biochemical processes by specifying enzymes.

One Gene—One Enzyme

Garrod's insight, that genes control the activity of enzymes, at first seemed to apply to only a few rare diseases—alkaptonuria and albinism, for example. In these conditions, the failure to make an enzyme explained an obvious abnormality. But do all genes specify enzymes?

The answer did not come until 1941, when two Stanford University geneticists, George Beadle and Edward Tatum, began looking for other genetic defects in metabolism. Beadle and Tatum chose to study a simple fungus, the pink bread mold, *Neurospora.* Beadle and Tatum induced mutations in *Neurospora* with x rays and then looked for biochemical defects (Figure 10-12). Such biochemical defects were easy to detect because *Neurospora* is a haploid organism—it has only a single copy of each gene. When that single copy is altered, the change immediately shows up in the phenotype. (In diploid organisms such as humans, a normal allele on the homologous chromosome can compensate for a recessive allele.)

Wild-type *Neurospora* can grow on an extremely simple medium (called a minimal medium) that contains glucose, ammonia, and salts. But Beadle and Tatum identified a large number of mutant *Neurospora* molds that could not grow on such a minimal medium. These mutants required the addition of various compounds that the molds were incapable of synthesizing themselves. One mutant strain, for instance, would grow

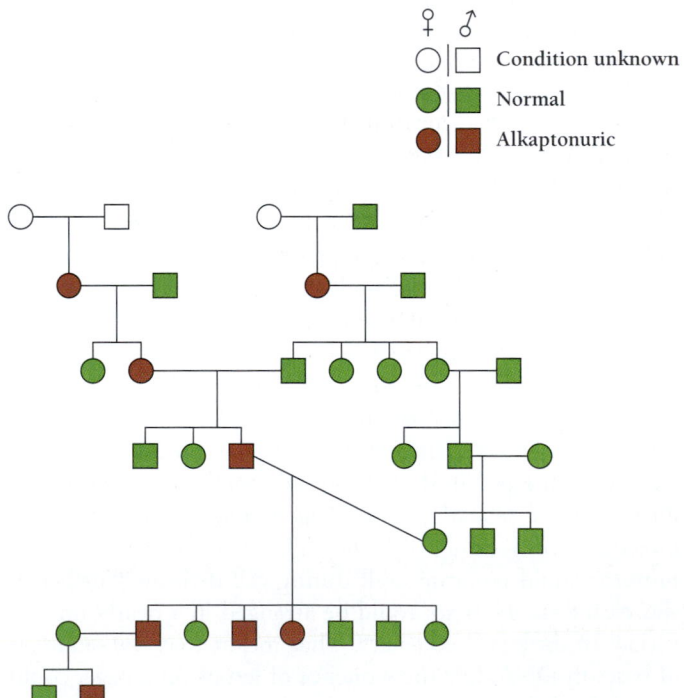

Figure 10-10 **How do people inherit alkaptonuria?** A pedigree of alkaptonuria from a Lebanese family. The marriage of cousins and other closely related family members increases the incidence of homozygous recessive individuals for both alkaptonuria and other genetic diseases.

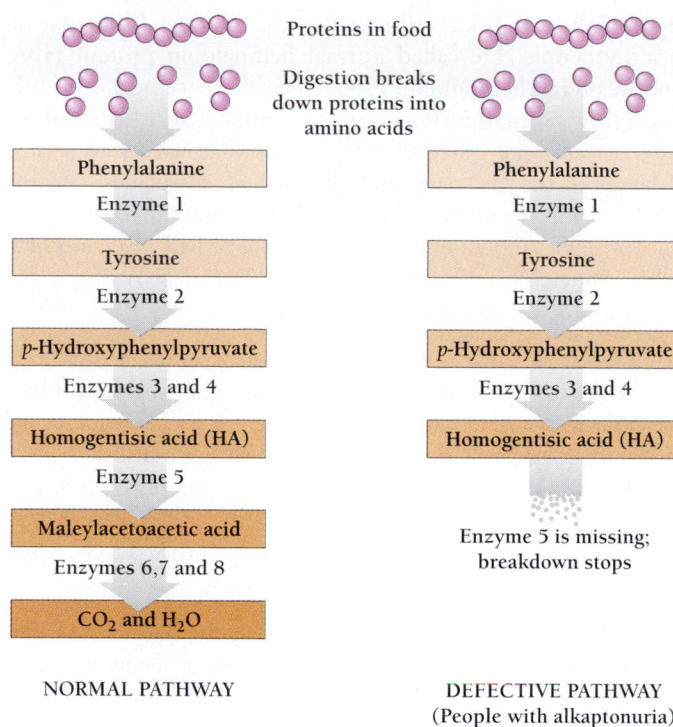

Proteins in food

Digestion breaks
down proteins into
amino acids

NORMAL PATHWAY

| Phenylalanine |
| Enzyme 1 |
| Tyrosine |
| Enzyme 2 |
| *p*-Hydroxyphenylpyruvate |
| Enzymes 3 and 4 |
| Homogentisic acid (HA) |
| Enzyme 5 |
| Maleylacetoacetic acid |
| Enzymes 6,7 and 8 |
| CO₂ and H₂O |

DEFECTIVE PATHWAY
(People with alkaptonuria)

| Phenylalanine |
| Enzyme 1 |
| Tyrosine |
| Enzyme 2 |
| *p*-Hydroxyphenylpyruvate |
| Enzymes 3 and 4 |
| Homogentisic acid (HA) |

Enzyme 5 is missing;
breakdown stops

Figure 10-11 What causes alkaptonuria? Cells can break down phenylalanine into homogentisic acid (HA), but without a crucial enzyme (#5) that normally breaks down HA to carbon dioxide and water, HA accumulates in the body. After it is excreted in the urine, the HA becomes oxidized and turns to dark alkapton bodies. Alkaptonuria is a relatively harmless genetic anomaly.

only if the minimal medium was enriched with vitamin B₆. Another mutant strain needed the amino acid arginine (Figure 10-12).

Beadle and Tatum then mapped the chromosomal location of each of their mutants by studying genetic linkage, as described in the last chapter. The mutants that needed arginine, for example, fell into three groups. Beadle and Tatum established that the three groups corresponded to mutations in three separate genes, on three separate chromosomes. Each gene coded for a separate enzyme needed to make arginine.

One group of mutants (*arg*-1) would grow when arginine was added and also in the presence of two related compounds, ornithine and citrulline. The second group of mutants (*arg*-2) would grow in the presence of arginine or citrulline, but not ornithine alone. The third group (*arg*-3) required arginine; neither citrulline nor ornithine would do.

Beadle and Tatum realized they could explain their results with the following model of arginine synthesis:

enzyme 1   enzyme 2   enzyme 3

precursor ⟶ ornithine ⟶ citrulline ⟶ arginine

Each step in the synthesis of arginine, they reasoned, is catalyzed by a different enzyme. The *arg*-1 mutations, they argued, must lie in the gene for enzyme 1. The absence of enzyme 1, then, could be overcome by ornithine, citrulline, or

arginine. Similarly, defects in the gene for enzyme 2 can be overcome by the addition of citrulline or arginine, but not of ornithine, since the organism cannot process it. Finally, defects in the gene for enzyme 3 can only be overcome by the addition of arginine itself (Figure 10-13). Beadle and Tatum's work helped unravel previously unknown biochemical pathways, since the mutations blocked the pathways at different steps and allowed biochemical analysis. Still more importantly, however, their work suggested that every gene somehow affects the function of a single enzyme. Beadle and Tatum captured this concept in a single phrase: **"One gene—one enzyme."**

Beadle and Tatum showed that one gene could specify one enzyme.

One Gene—One Polypeptide

Recall from Chapter 3 that all enzymes are proteins, but cells contain thousands of proteins that are not enzymes. Do genes contain information for making enzymes only? Are enzymes the only link between genotype and phenotype? Or can genes code for other proteins besides enzymes?

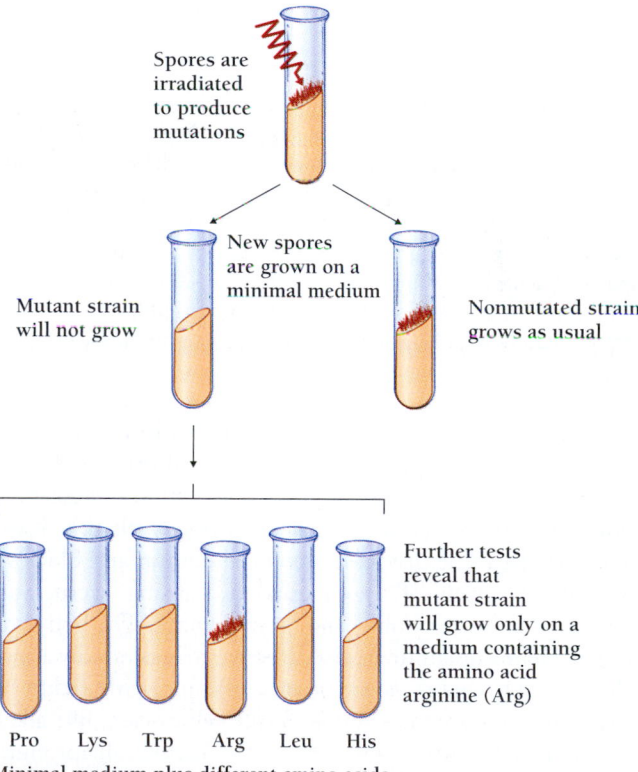

Spores are irradiated to produce mutations

New spores are grown on a minimal medium

Mutant strain will not grow

Nonmutated strain grows as usual

Further tests reveal that mutant strain will grow only on a medium containing the amino acid arginine (Arg)

Pro Lys Trp Arg Leu His

Minimal medium plus different amino acids

Figure 10-12 Mutant *arg* genes in the bread mold *Neurospora crassa*. Irradiation causes various mutations in bread mold spores. One set of mutant strains of *Neurospora* could not make the amino acid arginine in the normal way and needed either arginine or other amino acid supplements to survive.

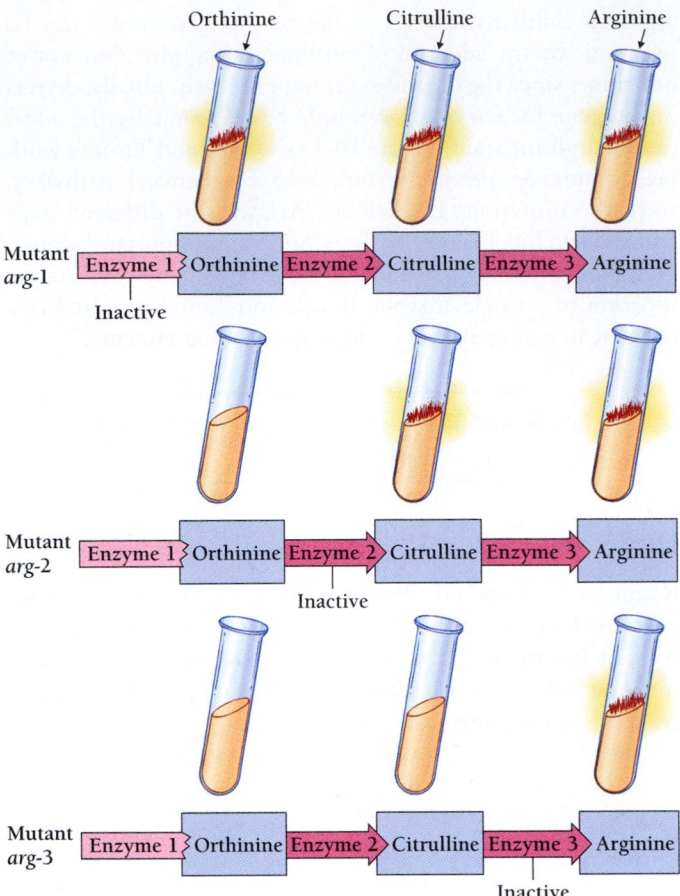

Figure 10-13   One gene—one enzyme. In the bread mold *Neurospora crassa*, the three normal *arg* genes (*arg*-1, *arg*-2, and *arg*-3) each specify one enzyme. Together these three enzymes allow the synthesis of arginine. A mutation in any of the *arg* genes that makes the specified enzyme defective disrupts the chain of reactions. *Arg*-1 mutants grow in the presence of arginine or two related compounds, ornithine and citrulline. *Arg*-2 mutants grow in the presence of arginine or citrulline, but not ornithine alone. *Arg*-3 mutants require arginine; neither citrulline nor ornithine will do.

The answers to these questions came with the unraveling of the molecular basis of another genetic disease, sickle cell anemia. The name "sickle cell" comes from the crescent or sickle shape of the red blood cells of people with this disease (Figure 10-14). The distorted blood cells clump and block the small blood vessels of the circulation, causing excruciating pain, anemia, heart trouble, and brain damage. The body's defense mechanisms eliminate the defective cells as rapidly as possible, but this causes severe anemia. People with sickle cell disease almost always die young. Sickle cell disease, like alkaptonuria, results from a recessive allele. Unlike alkaptonuria, however, sickle cell anemia is both serious and common—1 in 13 African Americans carry the allele, and 1 in 625 have two copies of the allele and have the disease.

In 1949, the biochemist Linus Pauling and his collaborators showed that the red, oxygen-binding blood protein

hemoglobin of sickle cell patients was different from that of healthy people. He called normal hemoglobin protein HbA and sickle cell hemoglobin HbS. The difference between HbS and HbA is a matter of a couple of amino acids. But Pauling used a technique called electrophoresis that reveals even tiny differences in the sizes and charges of molecules (Figure 10-15). Pauling and his colleagues showed that the healthy parents of sickle cell patients have a mixture of half HbS and half HbA. Their children, however, lack HbA entirely. A simple explanation of the disease emerged:

1. A recessive allele codes for the abnormal protein, HbS.
2. Heterozygotes such as the parents have genes for both HbA and HbS.
3. Heterozygotes with both HbA and HbS are healthy.
4. Homozygotes for the recessive allele make only HbS and are sick.
5. In the absence of the normal protein, HbA, HbS causes sickle cell disease.

In sickle cell disease, differences in the gene had caused differences in the hemoglobin protein, not in an enzyme. Pauling's work suggested that genes controlled other proteins besides enzymes. Pauling extended "one gene—one enzyme" to "one gene—one protein."

Exactly how does an abnormal gene change a protein? To answer this question, Vernon Ingram, a young biochemist working at Cambridge University in 1956, used new methods of protein chemistry to analyze the structural differences between HbS and HbA.

Normal hemoglobin is made of four polypeptides folded together—two of one kind, α-globin, and two of another, β-globin. Ingram showed that the α-globins in HbS were normal, but the β-globins contained a tiny "point" mutation. In the β chain's entire sequence of 146 amino acids, a single amino acid change produced an altered protein that could cause a deadly disease.

In the β-globin of HbS, amino acid number 6, normally glutamic acid, has been changed to valine. But while glutamic acid has a negative charge, valine is neutral. This single amino

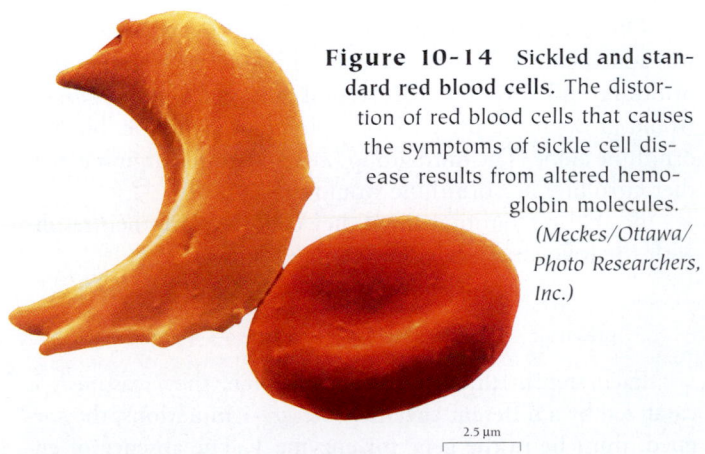

Figure 10-14   Sickled and standard red blood cells. The distortion of red blood cells that causes the symptoms of sickle cell disease results from altered hemoglobin molecules. *(Meckes/Ottawa/ Photo Researchers, Inc.)*

2.5 μm

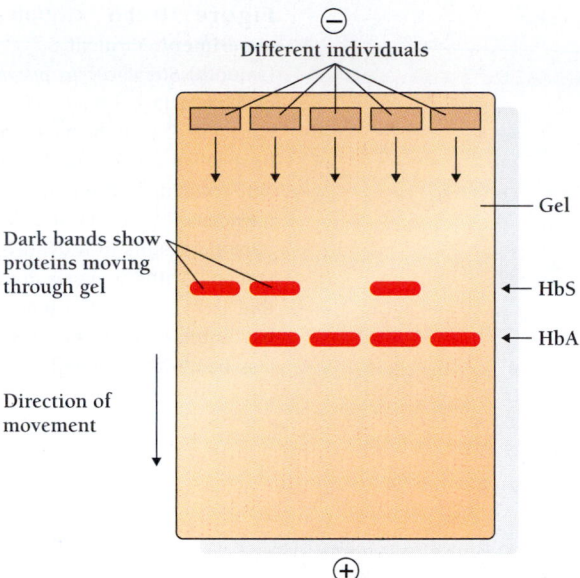

Figure 10-15 Electrophoresis. Biologists can separate macromolecules according to charge, shape, and size. In electrophoresis, an electric current pulls DNA, RNA, or proteins through a gel. Molecules move through the gel at different rates depending on their size and charge. Smaller ones move faster than larger ones, and those with a greater negative charge move faster than those that are less negative. The sickle cell hemoglobin polypeptide, HbS, is less negative than the normal hemoglobin polypeptide, HbA. So HbS moves more slowly through the gel than HbA.

acid substitution alters the way that hemoglobin molecules pack into the crowded red cell. When oxygen concentration drops—due to heavy exercise, for example—the HbS forms a fibrous precipitate that distorts the cell. This same charge difference is what makes HbS move more slowly than HbA during electrophoresis.

Ingram found that the sickle cell gene produced abnormalities only in the β-globin polypeptide, never in α-globin. Biologists concluded that each gene is responsible for the production not of an entire protein, but of a single polypeptide chain. "One gene—one protein" became **"one gene—one polypeptide."** But even this rule has important exceptions. As we'll see later on, some genes specify no polypeptide at all. And in special immune system proteins, several genes determine the sequence of a single polypeptide.

Nonetheless, we can say that, generally, genes influence phenotype by carrying the information necessary for the synthesis of polypeptides. But exactly how does a gene code for a polypeptide? To answer this question, we must look more closely at what a gene is.

Genes carry information necessary for the synthesis of polypeptides.

HOW DID BIOLOGISTS LEARN WHAT GENES ARE MADE OF?

Research on an amazing variety of organisms—including humans, peas, flies, and bread mold—has contributed to our understanding of what genes do. To find out what genes are made of, biologists turned to less complex, smaller organisms: bacteria and the viruses that infect them.

How Could Dead Bacteria Change the Genetic Properties of Other Bacteria?

Before the development of antibiotics during World War II, many people died of bacterial pneumonia. Intense research directed at developing a pneumonia vaccine led, by chance, to the discovery of the nature of genes.

In 1928, Frederick Griffith, an English bacteriologist, was trying to stimulate the immune defenses of mice against the pneumonia bacterium *Streptococcus pneumoniae*, also called pneumococcus. Griffith worked with two strains of pneumococcus. One caused pneumonia when injected into the mice, and one was harmless. When Griffith grew the two strains in dishes of gelatin, the harmless pneumococci formed rough (R) colonies, while the disease-causing strain formed smooth (S), glistening colonies. Today, we know that the virulent S bacteria form smooth colonies because they produce a polysaccharide capsule that protects them from attack by the body's immune cells. The R bacteria are mutants—they lack an enzyme needed to produce the polysaccharide capsule, which is why they cannot cause disease.

Griffith tried to produce immunity to pneumonia in mice by injecting them with different combinations of live and heat-killed pneumococci. As he expected, the mice injected with live S strain bacteria got pneumonia and died, while the mice injected with either live R strain bacteria or with heat-killed S bacteria survived. Griffith was amazed, however, by the results of injecting a mouse with a mixture of heat-killed S strain bacteria and live R strain bacteria (Figure 10-16). The mice died! Equally astonishing, the blood of the sick mice contained live S bacteria. Something in the dead S bacteria had transformed the harmless R bacteria into virulent S bacteria.

Frederick Griffith discovered that dead virulent bacteria could transform harmless bacteria into virulent ones.

What Was the Transforming Factor?

Other scientists soon found that they could convert R strains into S strains in a test tube as well as in a mouse. Extracts of S strains, dead or alive, could give R strains the ability to make the protective polysaccharide coat and become virulent. Miraculously, once the R strains were transformed into S strains, their descendants inherited the properties of S strains. How could the

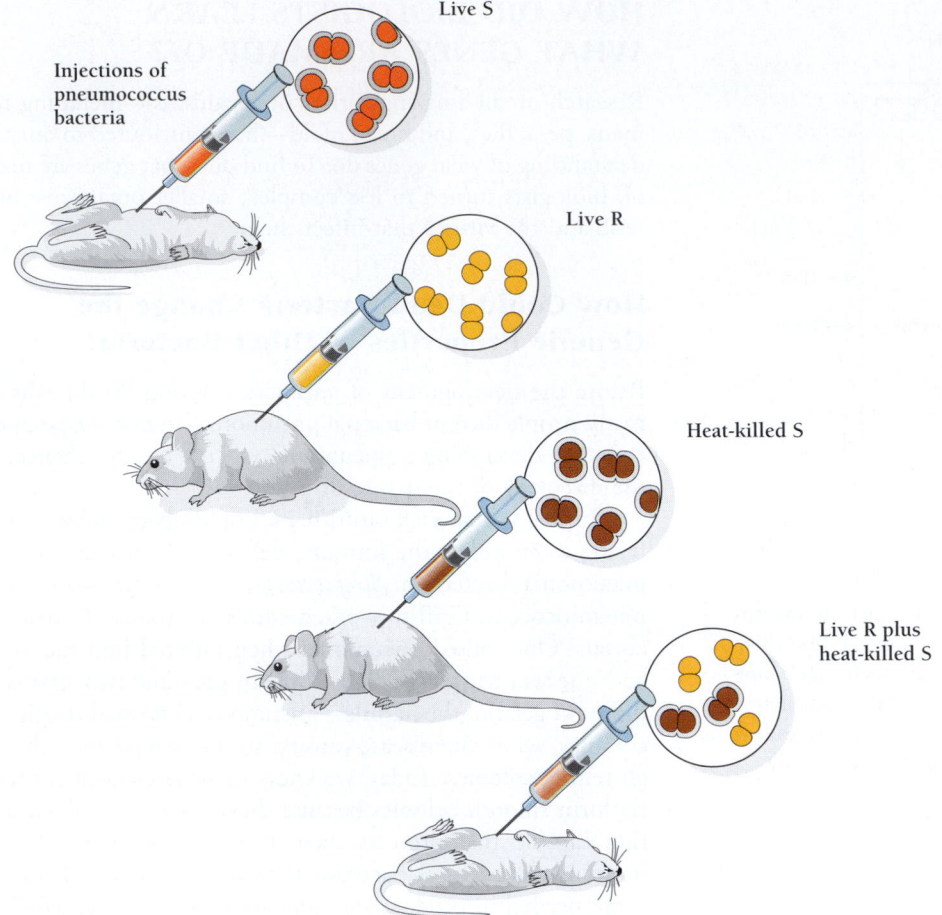

Injections of
pneumococcus
bacteria

Live S

Live R

Heat-killed S

Live R plus
heat-killed S

R bacteria "inherit" traits from individuals they were not descended from? What was *transforming* them into S bacteria?

It looked as if whatever was transforming the bacteria was some sort of genetic material. In the early 1930s, Oswald Avery at the Rockefeller Institute in New York set out to determine what that material was. With his coworkers Colin MacLeod and Maclyn McCarty, Avery carefully isolated different kinds of molecules from heat-killed S bacteria and tested to see which ones transformed the R bacteria.

At the time, almost everyone expected that the transforming factor would turn out to be made of protein, since proteins were already known to perform so many other cellular functions. Yet transformation had occurred even when the bacteria had been heat killed. Because heat denatures, or unfolds, proteins, it destroys their ability to function.

Avery and his colleagues worked for almost 10 years, isolating substance after substance from heat-killed S bacteria and then testing each substance to see if it could transform R bacteria. They treated each isolated substance with a variety of enzymes that would break down proteins into amino acids, polysaccharides into sugars, and nucleic acids into nucleotides. Finally, in 1944, they were forced to conclude that the transforming factor was DNA. First, only DNA from the S bacteria transformed the R bacteria. Second, the extract's ability to

transform was destroyed only by an enzyme (DNAase) known to break down DNA but not proteins.

Avery's work strongly suggested that genes are also made of DNA. Despite the care with which Avery and his colleagues had carried out their work, most biologists rejected Avery's conclusion. Biologists had two important reasons for being skeptical. First, if DNA were the genetic material, the DNA of bacteria should differ from the DNA of butterflies and bananas. Yet, all DNA seemed to be chemically identical. Second, most biochemists accepted an idea proposed in the 1920s that DNA was a "tetranucleotide," the same sequence of four nucleotides repeated over and over. In that case, DNA would be unable to carry information. So, as late as 1944, most biologists continued to believe that DNA was a simple repeating molecule, no more capable of carrying information than a string of identical beads. DNA was regarded as a structural component of the chromosomes, no more.

By 1944, Oswald Avery and his colleagues had shown that bacteria are transformed by DNA, not proteins. Yet most biologists rejected the idea that DNA could be the genetic material.

What Is a Bacteriophage?

In 1915, long before Griffith's work on pneumococcus, a young French microbiologist named Felix d'Herelle discovered that bacteria themselves could be killed by a tiny infectious agent called a **bacteriophage** [Greek, *bakterion* = little rod + *phagein* = to eat], or simply **phage.**

D'Herelle first became aware of bacteriophages while trying to stop a plague of locusts that was then sweeping the Yucatan, in Mexico. The young biologist noticed that some of the locusts were dying from some disease that gave them diarrhea. He collected bacteria from the diarrhea of dying locusts and cultured it. Then, as hoards of the locusts advanced across the Mexican countryside, d'Herelle dusted the vegetation before them with his bacterium. As the locusts ate, they sickened and died. D'Herelle's plan worked magnificently.

D'Herelle was puzzled, however, by the appearance of clear spots in his cultures, places where the bacteria themselves seemed to have died. What was killing his bacteria? At first, he had no idea. The answer did not come until a moment of inspiration some 5 years later. Back in France, d'Herelle was studying an epidemic of dysentery in a squadron of French cavalry stationed near Paris. The soldiers had Shiga dysentery, and d'Herelle had been culturing the Shiga bacteria when, once more, the bacteriophages made their presence known.

> The next morning . . . I experienced one of those rare moments of intense emotion which reward the research worker for all his pains: at the first glance I saw that the broth culture, which the night before had been very turbid [cloudy], was perfectly clear: all the bacteria had vanished, they had dissolved away like sugar in water . . . in a flash I . . . understood: what caused my clear spots was in fact an invisible microbe, . . . a virus parasitic on bacteria. Another thought came to me also: "If this is true, the same thing has probably occurred during the night in the sick man, who yesterday was in a serious condition. In his intestine, as in my test tube, the dysentery bacilli will have dissolved away under the action of [the virus]. He should now be cured."

And, indeed, d'Herelle dashed to the hospital to find the sick man greatly recovered.

D'Herelle saw that bacteriophages were small enough to pass through the filters he used to remove bacteria. Also, these "filterable agents" could reproduce themselves within bacteria and, therefore, seemed to be alive.

Biologists do not consider bacteriophages alive, however, because they are not made of cells. Bacteriophages are **viruses,** molecular assemblies of protein and either DNA or RNA. Viruses are not cells. They cannot reproduce outside of a host cell, and they themselves cannot obtain energy from their environments. A single virus can, however, infect a cell and subvert the cell's molecular machinery to make more viruses. A virus, then, is a molecular assemblage of nucleic acid and protein that parasitizes cells. Viruses are not confined to infecting bacteria. They may infect plant cells, animal cells, and other kinds of eukaryotic cells.

Bacteriophages are a kind of virus—a parasitic assembly of protein and DNA or RNA capable of forcing host cells to make more viruses.

What Convinced Biologists That Genes Are Made of DNA?

In the 1940s, Max Delbruck and Salvador Luria began to study the genetics of the bacteriophages that infect the common intestinal bacterium (Figure 10-17A). Because bacteriophages reproduce so rapidly, they were, Delbruck and Luria decided, the ideal material for genetic studies. A bacterium such as *E. coli* infected with a single bacteriophage breaks open after about 20 minutes and releases about 200 new phage particles (Figure 10-17B). Each of these new phages can then infect another bacterium. If the bacteria are growing on a solid culture medium, the phages create a clear spot, or "plaque," on the otherwise continuous bacterial "lawn." As the phages multiply, their descendants may eventually consume almost the entire lawn.

Delbruck, Luria, and other members of the "phage group" catalogued many inherited variants of the bacteriophage. The researchers produced many mutant phages with distinct phenotypes. Some phages produced clear plaques, and others produced cloudy plaques. Others infected one strain of *E. coli* but not another.

The different kinds of phages passed their traits to their descendants. Phage particles isolated from cloudy plaques produced viruses that made cloudy plaques. Bacteriophages, then, like organisms, have genes, and alleles of these genes specify alternative phenotypes such as clear or cloudy plaques. The phage researchers knew that bacteriophages consisted of complexes of proteins surrounding a molecule of nucleic acid, usually DNA. Their genes therefore had to be in either the protein or the DNA.

Which one was settled in 1952 by Alfred Hershey and Martha Chase (Figure 10-18A). They performed an experiment specifically designed to discover if DNA was also the hereditary material in bacteriophages. Hershey and Chase knew from microscope studies that during a phage infection only part of the virus enters the host cell. The rest remains outside (Figure 10-18B). They reasoned that the material that actually enters the cell must carry the virus's genetic information. So the question was, What was that material?

To find out, Hershey and Chase used radioisotopes to label the protein and DNA. The work on the atomic bomb during World War II had greatly increased the availability of radioisotopes, so Hershey and Chase grew bacteria ^{32}P-labeled phosphate and ^{35}S-labeled sulfate. Because some of the amino acids that make up proteins contain sulfur but no phosphorus and because nucleic acids have phosphorus but no sulfur, the bacteria therefore had ^{35}S incorporated in their proteins and ^{32}P in their DNA. Hershey and Chase then allowed phages to infect these radioactive bacteria. The newly made phages picked up the same radioactive labels, so their DNA contained ^{32}P and their protein contained ^{35}S.

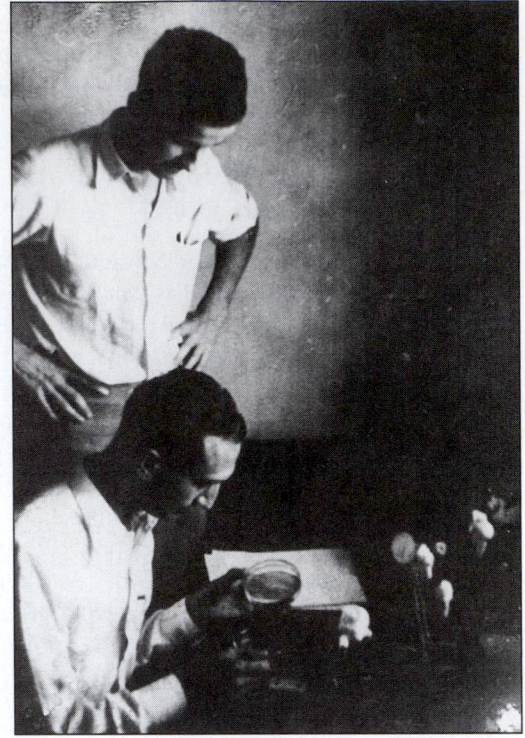

A.

Figure 10-17   The life of a phage. A. Max Delbruck and Salvador Luria examining a petri dish containing bacteria and phages. B. Life cycle of a bacteriophage. (*A, Cold Spring Harbor Laboratory Archives*)

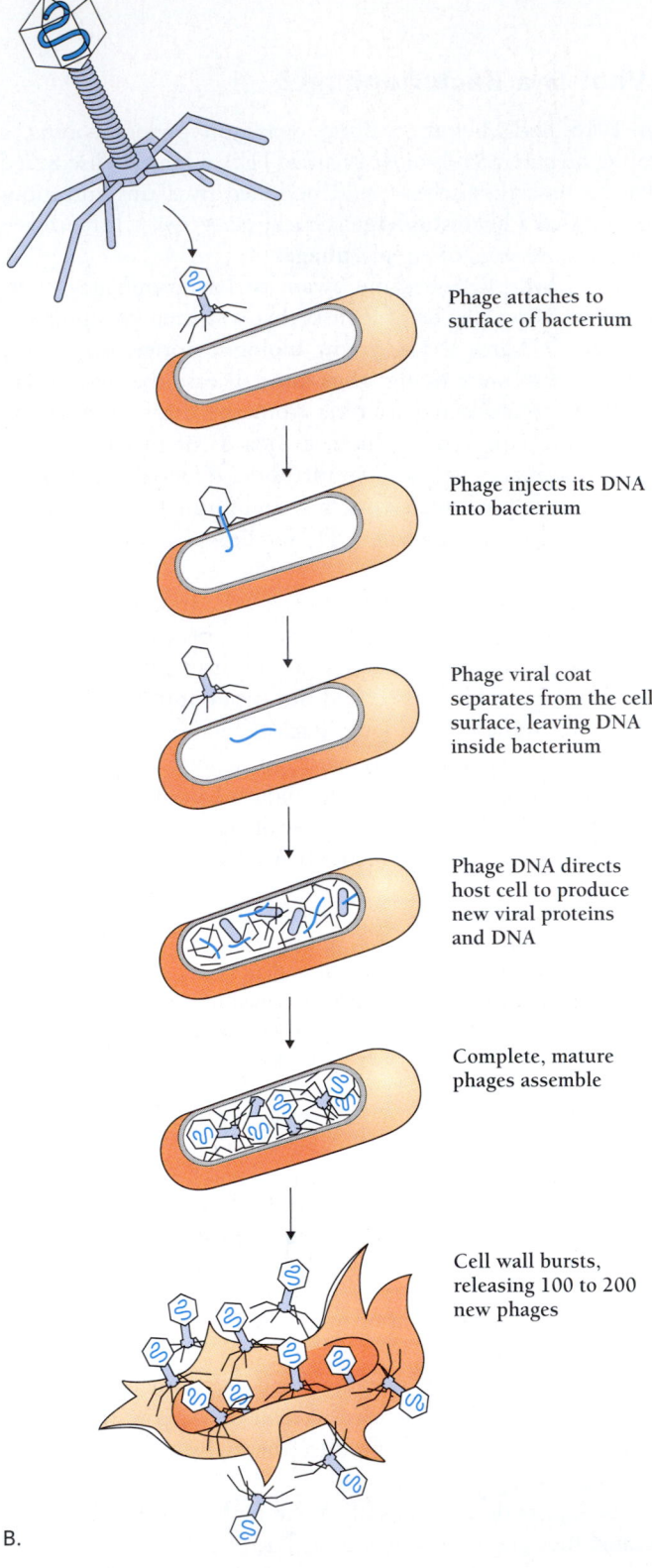

Phage attaches to surface of bacterium

Phage injects its DNA into bacterium

Phage viral coat separates from the cell surface, leaving DNA inside bacterium

Phage DNA directs host cell to produce new viral proteins and DNA

Complete, mature phages assemble

Cell wall bursts, releasing 100 to 200 new phages

B.

Hershey and Chase then used the radioactive phage to infect new, nonradioactive bacteria. They waited just long enough for the infection to begin and then interrupted the process by dumping the bacteria into a kitchen blender. The force of the blender's whirling blades knocked the empty phage particles off the outside of the bacteria and separated these particles from the bacteria. Hershey and Chase then examined the infected bacteria to see what part of the phage had entered the bacteria. Was it the DNA or the protein?

It was the DNA. The ^{32}P-labeled phage DNA was inside of the bacteria, and the ^{35}S-labeled phage protein was outside. Hershey and Chase also showed that the phage DNA alone could direct the synthesis of new phages and that these new phages—made and released by the infected cell—contained ^{32}P, but never ^{35}S. Hershey and Chase therefore concluded that the genes of the bacteriophage were made of DNA, not protein.

Why was Hershey and Chase's experiment more persuasive than Avery's? When Avery's paper came out in 1944, most molecular biologists still believed that DNA was a simple repeating structural molecule. But by 1950, Erwin Chargaff had shown that all DNA was not the same—that different species had different proportions of bases. Without equal proportions

of each base, DNA could not possibly be a regular repeating tetranucleotide.

In addition, Delbruck and Luria's phage group had convinced many biologists that the genes of bacteriophages must be similar to the genes of organisms. Hershey and Chase's 1952 experiment therefore immediately drew the attention of a large

A.

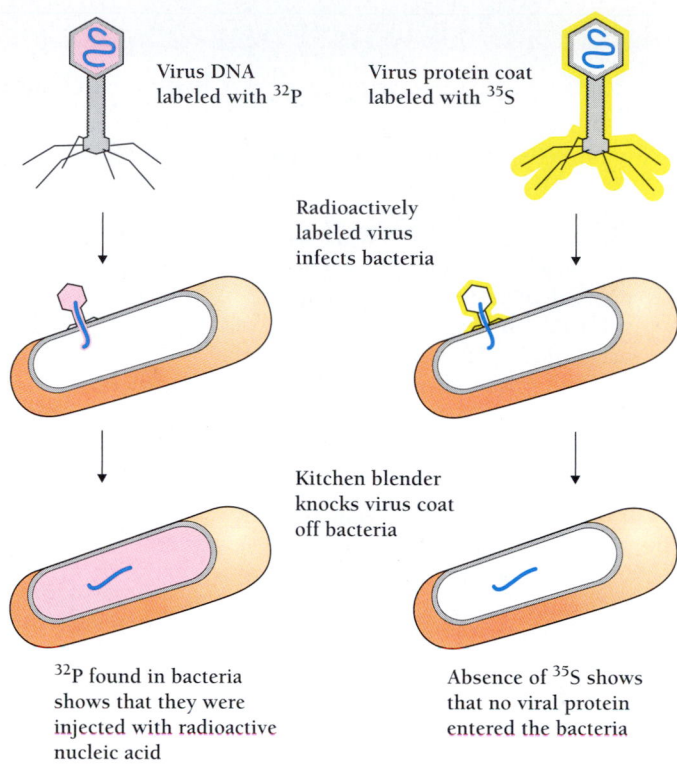

Virus DNA labeled with ^{32}P

Virus protein coat labeled with ^{35}S

Radioactively labeled virus infects bacteria

Kitchen blender knocks virus coat off bacteria

^{32}P found in bacteria shows that they were injected with radioactive nucleic acid

Absence of ^{35}S shows that no viral protein entered the bacteria

B.

Figure 10-18   What convinced biologists that the genetic material was DNA? A. Martha Chase and Albert Hershey showed that when a phage infects a bacterium, it is the viral DNA, not protein, that enters and infects the bacterium. B. In the Hershey-Chase experiment, radioactive DNA from a phage shows up inside the infected bacterium. Radioactive protein stays outside. This 1952 experiment finally convinced biologists that DNA, not protein, was the genetic material. (Later work showed that a few viruses use RNA as their genetic material.) (*A, Cold Spring Harbor Laboratory Archives*)

group of influential biologists in a way that Avery's paper, 8 years earlier, had not. Finally, Watson and Crick's 1953 paper on how the structure of DNA might allow it to carry information and replicate silenced any remaining skeptics. After 1953, nearly all biologists accepted the idea that DNA was the genetic material.

It took from 1903 to 1952 to persuade biologists that DNA was the genetic material. The next questions were, What is the structure of DNA? and How does it work? As we have seen, Watson and Crick substantially answered both questions within a month of Hershey and Chase's paper.

> *Hershey and Chase's experiments with radioactively labeled phages showed that when a phage infects a bacterium, it is the DNA that enters and infects the bacterium, not the protein. This experiment convinced biologists that DNA is the genetic material.*

HOW DOES ONE DNA STRAND DIRECT THE SYNTHESIS OF ANOTHER?

Watson and Crick's complementary base-pairing rule established the basis of DNA's ability to replicate itself. Since a base on one strand can match with only one other base, the se-

quence of bases on each strand implies the sequence of bases in the complementary strand. In their first paper describing their model, the two coyly commented on this property: "It has not escaped our notice that the specific pairing we have postulated immediately suggests a possible copying mechanism for the genetic material." (Years later, they confessed that they wanted to claim the idea without saying so much that they sounded foolish if they turned out to be wrong.) In the same way that a photographer can make a positive print from a negative, or a negative from a positive print, each strand of DNA has the capacity to direct the synthesis of the other.

This complementarity is the basis for understanding both how a chromosome replicates itself during cell division and how genes actually work. The structure of DNA explains both how cells copy their DNA and how genes specify proteins (Chapter 11).

How Do Dividing Cells Copy the DNA?

Each of the four bases in DNA can pair with only one other base—A with T and G with C. The rules of base pairing in DNA mean that the sequence of one polynucleotide strand specifies the sequence of the other strand. The two strands of DNA are complementary, meaning that they fit together to make a unified whole—the double helix.

The basis for DNA replication is that the sequence of bases in one chain specifies the sequence of bases in the other. That

BOX 10.1

How To Sequence DNA

One of the most important questions a biologist can ask about a gene or any stretch of DNA is, What is its sequence of nucleotides? With that information, medical researchers can begin to understand complex genetic diseases such as sickle cell anemia, Tay-Sachs disease, and many others. The most widely used method for DNA sequencing was developed in England in the 1960s by Fred Sanger. Sanger's method depends on DNA polymerase and five "rules" of DNA synthesis:

1. DNA synthesis is semiconservative.
2. A newly made DNA strand grows in a single direction, adding one nucleotide at a time to the 3′ end.
3. The nucleotide sequence of the template strand determines the sequence of nucleotides in the growing strand.
4. DNA polymerase requires a primer polynucleotide.
5. The copying of a single strand into a new complementary strand is catalyzed by a single enzyme, DNA polymerase.

In addition, DNA sequencing requires that a researcher be able to separate DNA fragments of different lengths.

Sanger began by heating the DNA until it separated into two single strands. He then added a specific complementary primer, DNA polymerase, and plenty of A, C, G, and T. The DNA polymerase automatically set to work extending the primer until it produced a complete copy of the template strand.

Sanger then devised a way of stopping the polymerase reaction at different places. He could stop the reaction at A, C, G, or T by replacing each kind of base with one that would prevent further polymerization. For example, Sanger replaced A with a similar molecule that lacked a hydroxyl group at the 3′ end. Once this modified A "base" was added to the growing strand, DNA polymerase could add no more bases. The result was a set of fragments of different lengths, each ending in A. Sanger did the same for C, G, and T.

The next problem was to put all these fragments in order, shortest to longest. Fortunately, electrophoresis does just that. If we run a small electric current through a flat sheet of gel containing the fragments, the smallest fragments speed across the gel, while the longest ones move more slowly as they work their way through the complex structure of the gel. Using electrophoresis, Sanger could put the fragments in order, shortest to longest—even ones that differ in length by no more than one nucleotide (Figure A). Each fragment showed up on the gel as a dark spot. Each column of spots represents the different fragments ending in A, G, C, or T (Figure B). From these four sets of bands, researchers can work out the complete sequence of any newly made DNA.

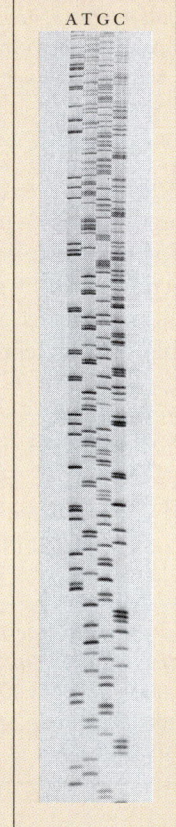

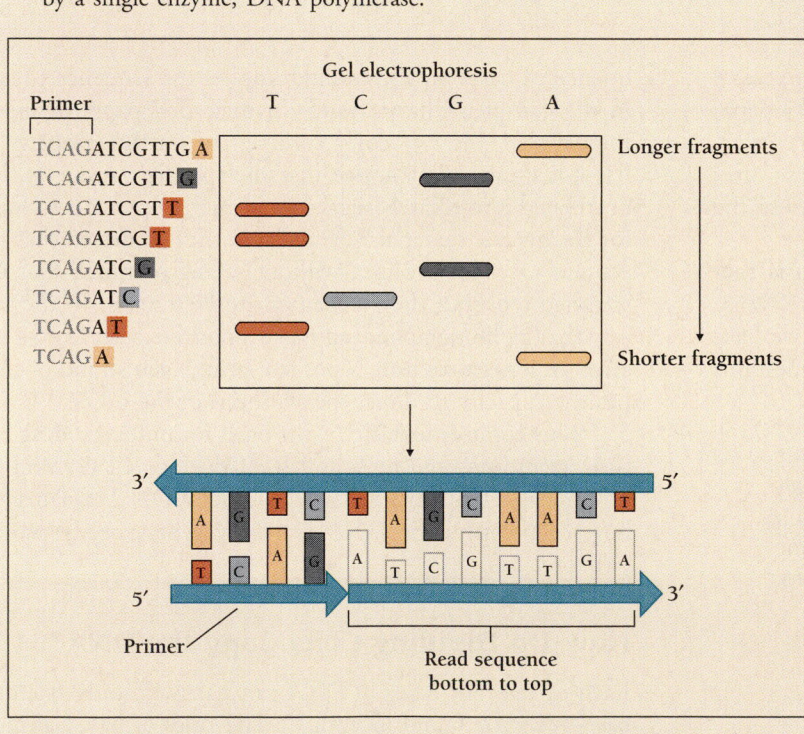

Figure A In electrophoresis, an electric current causes fragments of DNA to spread across a gel according to the speed with which they move through the gel. The biggest fragments are the slowest. Biologists then "read" the last letter of each fragment, starting with the shortest fragment. Reading from shortest to longest, the nucleotide letters are A-T-C-G-T-T-G-A.

Figure B Photo of an actual DNA sequencing gel. Reading from the 5′ end (bottom of the gel), the sequence is TATCCCGTTGGAAGGTCGTCTGCTCCCTG-GAAGTAG. (*James D. Colandene, University of Virginia*)

is, the two strands of the DNA double helix contain the same information in two different forms. Picture a molecule of DNA unzipping to form two complementary single strands. If the bases and the proper cellular machinery are available, each strand can act as a template for the manufacture of a new complementary strand. When each strand has created its one complement, the result is two new double-stranded DNA molecules, each a replica of the parent DNA.

According to this model, DNA replication is **semiconservative.** That is, half of each parent molecule is present in each daughter molecule (Figure 10-19). Each new DNA molecule, then, is really half old and half new.

DNA replication is semiconservative: When DNA replicates, each half of the double helix acquires a new mate.

Our discussion has so far focused on how DNA passes information to daughter DNA molecules. But what is the mechanism? How do cells actually produce new DNA? As in the case of other biochemical reactions, DNA replication depends on enzymes. The full process of DNA replication requires at least a dozen enzymes. Even now, more than 40 years after molecular biologists first synthesized DNA in a test tube,

they still do not know all the details of DNA replication. The most basic steps of the process, however, are well understood.

How Does a Cell Begin Assembling a New Strand of DNA?

The assembly of DNA involves linking many small molecules (nucleotides) into a long chain, a process called **polymerization** [Greek, *polys* = many + *meros* = part]. The enzyme that strings together the nucleotides is called **DNA polymerase.** In each strand, the nucleotides must polymerize in a fixed order, guided by the order of bases in the complementary strand (Figure 10-20).

The Watson-Crick model for DNA replication suggests how each strand of DNA serves as a **template,** or guide, for the assembly of a complementary sequence. Our understanding of how the template works and how the new chains are assembled depends heavily on the work of Arthur Kornberg and his colleagues. By 1957, Kornberg had demonstrated that extracts of bacterial enzymes could catalyze DNA synthesis. He and his colleagues then isolated DNA polymerase and studied the reactions it catalyzed.

Kornberg found that DNA polymerase can add nucleotides only to an already existing chain of nucleotides called a **primer** (Figure 10-21). DNA polymerase adds nucleotides to the 3′ end of the primer, so that the polynucleotide grows in a single direction—from its 5′ end toward its 3′ end.

The particular nucleotide added at each step depends upon the base sequence of the template strand. Each added nucleotide forms hydrogen bonds with the next base on the template strand. So each growing strand extends in a single direction, starting with the sequence closest to the 5′ end of the primer and extending from its 3′ end.

Once a free 3′ end exists, we can see how the process continues. But if DNA can only add nucleotides to the 3′ end of existing chains of nucleotides, how can the replication process ever begin? The resolution of this problem is that DNA replication does not begin with a DNA primer, but with a temporary RNA primer.

The RNA primer arises from the action of **primase,** an RNA polymerase that makes short stretches (fewer than ten nucleotides) of RNA using DNA as a template. Primase does not itself require a primer. With the RNA primer in place, DNA polymerase can then begin synthesizing the new strand of DNA, adding nucleotides step by step to the RNA primer's 3′ end.

An RNA polymerase called primase creates an RNA primer that jump-starts DNA polymerase, enabling it to begin making a copy of each strand of DNA.

How Does a Cell Copy Both Strands of DNA Simultaneously?

Researchers can actually see cellular DNA in the process of replication with the electron microscope. The DNA in a dividing tissue culture cell looks like a long, thin molecule

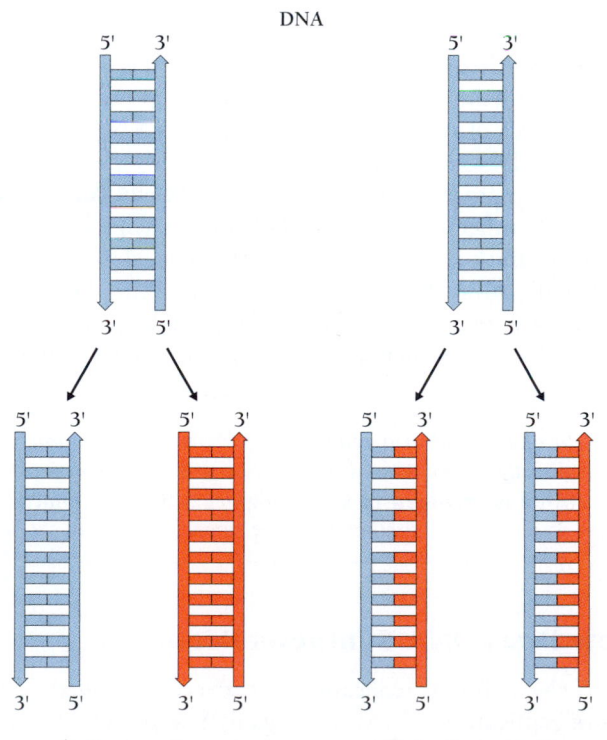

DNA

A. Conservative   B. Semiconservative

Figure 10-19 Semiconservative replication. A. At one time biologists hypothesized that DNA replication was conservative. Both halves of the old molecule stayed together, they thought incorrectly, while a whole new molecule was created. B. In fact, when DNA replicates, each half of the old double helix acquires a new mate.

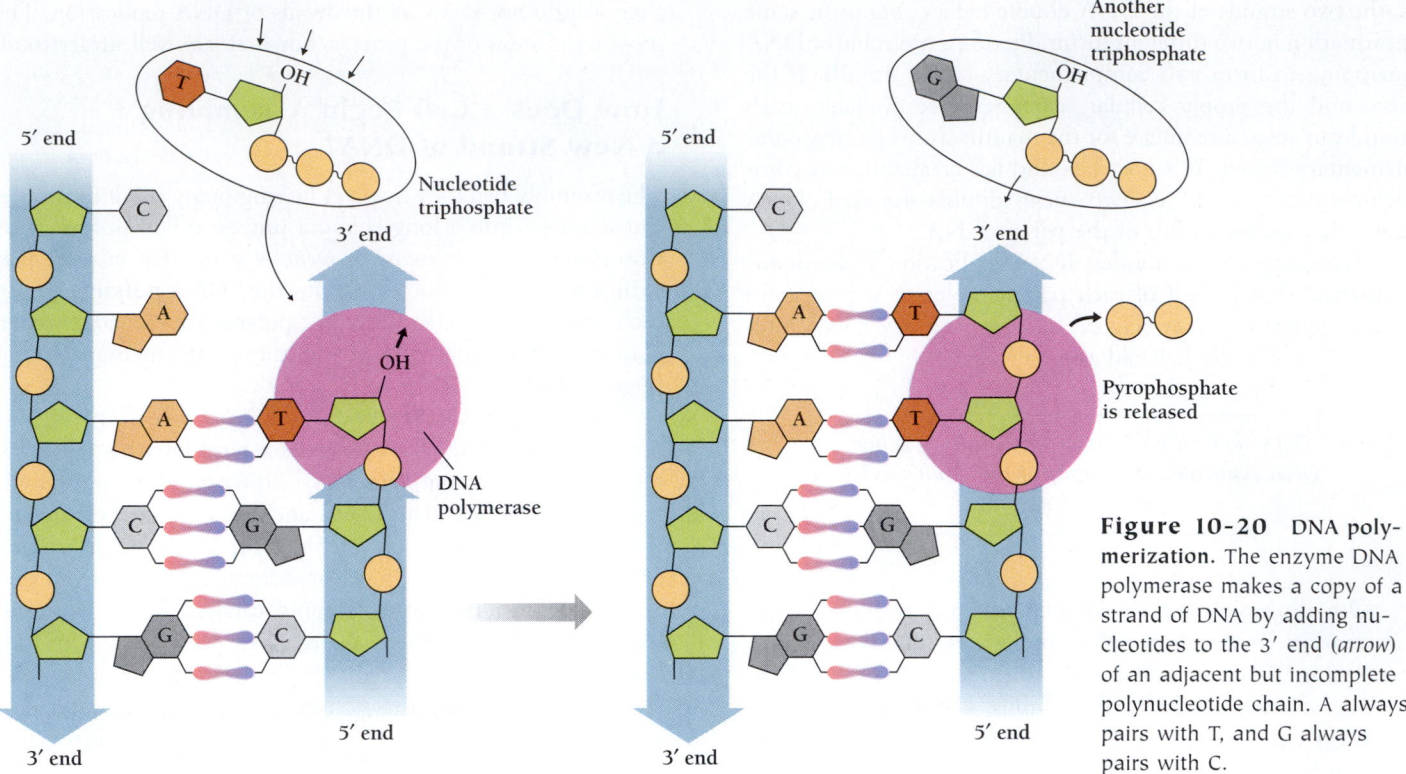

Figure 10-20 DNA polymerization. The enzyme DNA polymerase makes a copy of a strand of DNA by adding nucleotides to the 3′ end (*arrow*) of an adjacent but incomplete polynucleotide chain. A always pairs with T, and G always pairs with C.

punctuated by bubbles. As DNA replicates, these bubbles expand. Beyond each bubble are the two strands of parental DNA. Within each bubble are the four strands of replicated DNA—two parental strands and two newly made strands.

Each bubble consists of two **replication forks,** Y-shaped regions of DNA where the two strands of the helix have come unzipped (Figure 10-22A). At the replication fork, each strand begins to direct the assembly of a new complementary strand. As replication proceeds, the forks move away from each other. In bacteria, the replication forks move away from each other at about 500 nucleotides per second, and, in mammals, at about 50 nucleotides per second.

DNA replication occurs in replication forks, Y-shaped regions of DNA where the two strands of the helix have come apart.

How Does DNA Replicate in the 3′ to 5′ Direction?

As a replication fork moves along both strands of DNA, it moves from the 5′ end to the 3′ end along one strand, but from the 3′ end to the 5′ end on the other strand. We have seen how simply and continuously the first strand is replicated. But the second strand cannot be replicated in this way because nucleotides must be added to its 5′ end, a task that DNA polymerase cannot perform. How is this strand made?

The answer is that the second strand is produced in bits and pieces. After the moving fork has exposed about 1000 nucleotides of template for the lagging strand, primase produces a short piece of complementary RNA. This RNA serves as a primer for DNA polymerase, which attaches nucleotides to its 3′ end. This process produces fragments, called **Okazaki fragments,** after their discoverer Reiji Okazaki. Each Okazaki fragment contains a short stretch of RNA connected to about 1000 nucleotides of a DNA strand (Figure 10-22B). DNA polymerase then removes the RNA primer, and yet another enzyme, called **DNA ligase,** stitches the fragments together.

In one strand of a replication fork, nucleotides add directly to the 3′ end of an RNA primer. The other strand is produced in short fragments that are joined together by DNA ligase.

How Does DNA Synthesis Begin?

DNA always begins replication at a special site called an **origin of replication.** *E. coli,* the organism whose DNA synthesis has been most studied, has a single piece of circular DNA with a single origin of replication. Starting from this single site, two replication forks move in opposite directions away from the origin of replication, and DNA synthesis stops when the fork that is moving clockwise meets the counterclockwise fork (Figure 10-23).

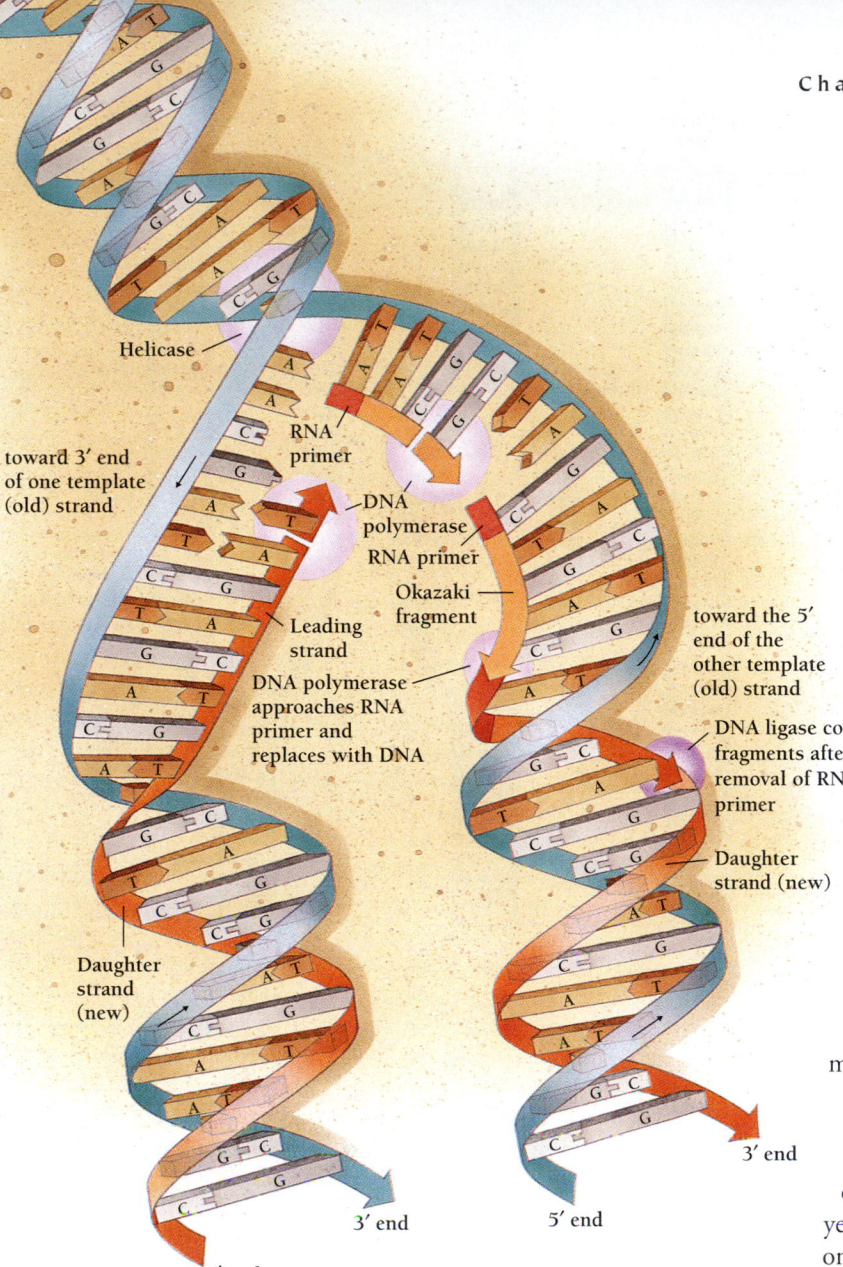

Figure 10-21   How do enzymes replicate DNA? The blue strand of DNA acts as a template for the creation of a new complementary strand of DNA (rust). The enzyme *helicase* unwinds the strands. On the left, DNA polymerase adds nucleotides in a continuous string, working from the 3′ end of the template to the 5′ end. On the right (*center*), DNA polymerase works in the opposite direction, starting with RNA *primers* (also rust) to create Okazaki fragments, which are later linked together by *DNA ligase*.

Helicase

toward 3′ end
of one template
(old) strand

RNA
primer

DNA
polymerase

RNA primer

Okazaki
fragment

Leading
strand

DNA polymerase
approaches RNA
primer and
replaces with DNA

toward the 5′
end of the
other template
(old) strand

DNA ligase connects
fragments after
removal of RNA
primer

Daughter
strand (new)

Daughter
strand
(new)

3′ end

3′ end   5′ end

5′ end

WHAT IS THE ULTIMATE SOURCE OF GENETIC DIVERSITY?

Every living organism shares the same chemistry of life. Your DNA is made of the same stuff and follows the same rules as the DNA of the roses nodding in your neighbor's yard. Yet you are different from a rose. If, for example, you want lunch, you must get up and make yourself a sandwich, while the rose simply waits passively for the sun to emerge from behind a cloud. What makes us so different from a rose?

Each species has its own characteristic DNA. But life most likely began on Earth just once. If that is the case, then at one time all organisms were of the one kind and all of their genetic material was the same. How, over billions of years of evolution, did we organisms become so different from one another? The short answer is **mutations,** permanent changes in the sequence of DNA. The different alleles of a gene are created by changes in the sequence of the original gene. The original sequence itself is a mutation of some other gene. All genetic differences ultimately depend on the accumulation of mutations in the genome.

Although the evolution of organisms depends on the accumulation of changes in the DNA, the vast majority of mutations are either harmful or meaningless. If we opened up the back of a television set and switched two wires at random, the chances that this would improve the picture or the general performance of the television are almost zero. The same is true of mutations in highly evolved genomes, such as those of butterflies or mammals. Most gene mutations are a disaster for the phenotype. Still, every now and then, a new mutation is useful under certain circumstances. In that case, natural selection may increase its representation in a population of organisms.

We have already seen in Chapter 9 how the recombination (crossing over) and the mixing of gametes during sexual reproduction create unique combinations of alleles. But, ultimately, all new alleles result from mutations. Our focus in the

Several proteins participate in DNA replication. An enzyme called *helicase* uses energy from ATP to untwist the helix, while another enzyme called DNA gyrase prevents the accumulation of kinks in the stems of the replication forks. Still another protein binds to the unwound DNA and keeps the jaws of the replication fork open.

In eukaryotic cells, DNA replication is more complex. Each cell contains thousands of origins of replication. Groups of 20 to 50 origins, called **replication units,** form replication forks at the same time. Within a replication unit, the forks move in opposite directions from each origin until they encounter a fork from an adjacent origin.

DNA synthesis always begins at special sequences in the DNA called origins of replication. Proteins bind to these origins of replication and open and untwist the double helix.

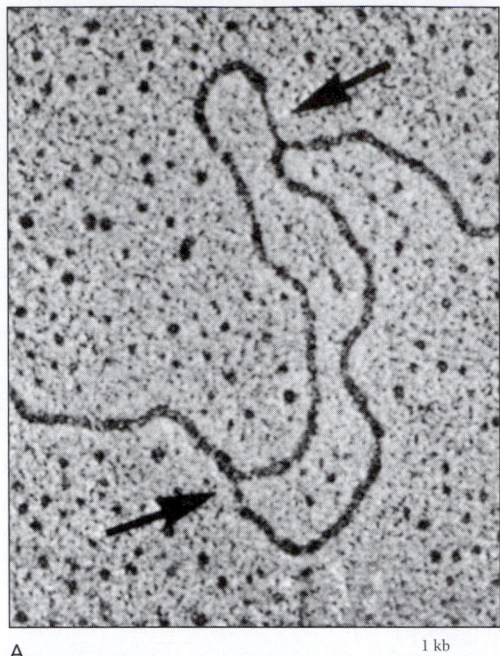

A.

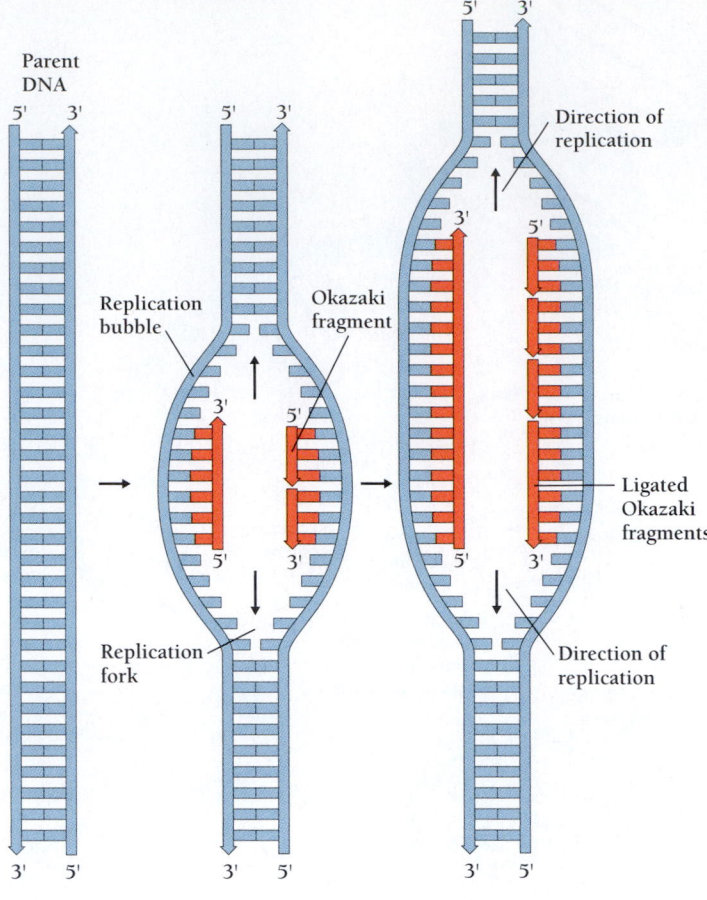

B.

Figure 10-22 **Replicating DNA.** A. TEM of a replication bubble, with a replication fork at each end. B. A simplified view of a DNA replication bubble. Replication begins in the middle of the bubble and moves away from the middle in two directions at once. (A, D. Hogness and H. Kriegstein, Stanford University, Proceedings of the National Academy of Science, 71, 135–139, 1974)

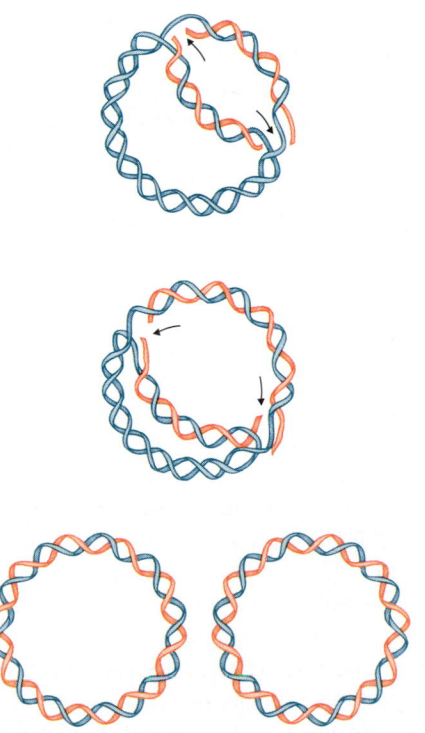

Figure 10-23 **Replication of bacterial DNA.** Replication begins at the origin of replication (*arrows*), then works in both directions around the circle.

last part of this chapter will be on the origin and consequences of mutations: How do mutations arise? How do they influence the phenotype of the organism?

Mutations are the ultimate source of all genetic diversity.

What Causes Mutations?

Mutations occur spontaneously as random events in the cell. The structure of any molecule—including DNA—can change as a result of a random collision with another molecule or ion. For example, the bonds between the A and G (purine) bases in DNA and the backbone sugar are slightly unstable. As a result, each cell in the human body loses some 5000 to 10,000 A and G nucleotides every day. (Of course, most of this damage is repaired, as we'll see later.)

Occasionally, mutations occur because DNA polymerase places the wrong nucleotide in a sequence during the synthesis (S) phase of the cell cycle (Chapter 8). In addition, certain agents, called **mutagens,** increase the rate of mutation. Ultraviolet light, high-energy radiation such as x rays, and many chemicals can all act as mutagens. Each mutagen has a characteristic spectrum of effects. Ultraviolet light, for example, may link together adjacent Ts in the same DNA strand. X rays are particularly effective producers of chromosomal mutations, causing DNA to break and rejoin in strange new arrangements. Chemical mutagens, such as alcohol and dioxin, for example, also increase mutation rates.

Mutations occur spontaneously. But they occur more frequently when DNA is exposed to ultraviolet light, x rays, chemicals, or other mutagens.

How Often Do Mutations Occur?

Changes in DNA sequence occur spontaneously all the time. Mutations can initially result from random collisions between molecules, from mistakes in replication (for example, from a failure to form a correct base pair), or from changes in the structure of a nucleotide.

The **mutation rate** is how often mutations occur. More specifically, the mutation rate is the number of changes per nucleotide per generation. The mutation rate depends both on how often a sequence mutates and on how efficiently cells repair these mutations. Some sequences within a gene, called **mutation hot spots,** are more likely to mutate than others. In bacteria, for example, some sequences mutate 25 times more rapidly than others. Even the highest rates of mutation, however, are far less than the error levels to which we are accustomed in everyday life. In the days before computers and word processors, for example, an outstanding typist could type this whole chapter without going back to correct anything and with only one mistake. Yet, the average rate of mutation in bacteria during ordinary DNA replication is only 1 mutation in 10 million base pairs each generation. DNA replication is at least 1000 times more accurate than the best human typist.

Mutations in bacterial DNA occur at the rate of about 1 mutation in 10 million base pairs each generation.

What Kinds of Mutations Are There?

We can categorize mutations on the basis of the sizes of the changes in DNA. Geneticists often distinguish between **point mutations,** those which change one or several nucleotide pairs, and **chromosomal mutations,** those which change relatively large regions of chromosomes. A point mutation may be (1) a **base substitution**—the replacement of one base (nucleotide) by another, (2) an **insertion**—the addition of one or more nucleotides, or (3) a **deletion**—the removal of one or more nucleotides.

Chromosomal mutations may affect large regions of chromosomes or even whole chromosomes. Chromosomal mutations include (1) **deficiencies**—deletions that are larger than a few nucleotides, (2) **translocations**—in which part of one chromosome is moved to another chromosome, (3) **inversions**—in which a segment of a chromosome is flipped over

180° from its normal orientation, and (4) **duplications**—in which part of a chromosome appears twice. **Aneuploidy,** abnormal chromosome number, is regarded as another type of chromosomal mutation. For example, **trisomy** is the presence of three copies of a certain chromosome, and **monosomy** is the presence of only a single copy in a cell that normally has pairs of chromosomes.

A mutation is any change in the sequence or number of nucleotides, ranging from small changes of one or a few nucleotides to chromosomal inversions or aneuploidy.

HOW DO CELLS HANDLE MISTAKES IN THE NUCLEOTIDE SEQUENCE OF DNA?

Because DNA mutates, cells must be able to repair damage to DNA if they are to maintain the continuity of life. This is especially important for the germ line cells that pass information to the next generation by way of eggs and sperm. But mutations in somatic (body) cells are dangerous as well. In skin cancer, for example, a cell with a defective gene for growth control multiplies so fast that immature daughter cells take over the area. More than half of all forms of cancer arise from cells with one or more mistakes in just a few genes. The normal forms of these genes specify proteins called "tumor suppressors."

All cells have sophisticated mechanisms for repairing damaged DNA and correcting mistakes in replication. These mechanisms depend on the redundancy of information in DNA's structure and on special enzymes that recognize common mistakes.

To correct a mistake, an enzyme detects something wrong in one strand of the DNA and removes the error. For example, cytosine may be accidentally converted to uracil, a base that is ordinarily not present in DNA. A special enzyme called uracil-DNA glycosidase can recognize the strange base and remove it. In another example, two T nucleotides may bond to one another instead of to their respective A nucleotides (Figure 10-24).

Once correction enzymes remove the offending nucleotides, DNA polymerase goes to work. It copies the information in the intact second strand and creates a new stretch of DNA that is properly matched. DNA ligase then seals the gap, and the original sequence is restored. Over 50 enzymes "proofread" DNA for errors such as unusual bases, gaps, and bulges.

Errors still can slip through this elaborate system. For example, the genes for error-correction enzymes may themselves be damaged. In that case, large numbers of mutations can accumulate (Figure 10-25). If these uncorrected mutations occur in genes that help control cell division, the cells become cancerous.

Figure 10-24 Healthy cells are quick to repair damaged DNA. Ultraviolet light can damage DNA by linking together adjacent Ts in the same strand. Repair enzymes excise the abnormal nucleotides and restore the original sequence, then DNA polymerase and DNA ligase repair the gap as in normal replication.

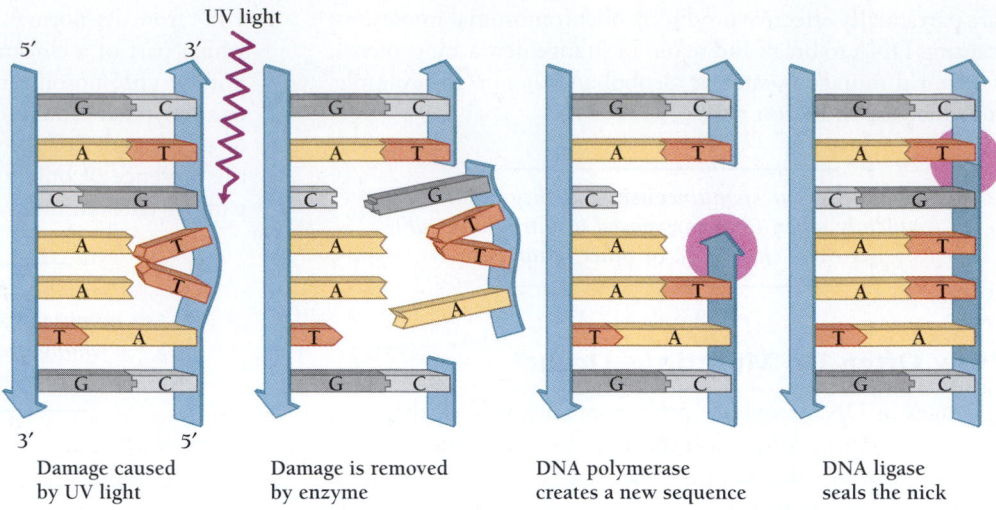

Damage caused by UV light   Damage is removed by enzyme   DNA polymerase creates a new sequence   DNA ligase seals the nick

Cells have sophisticated mechanisms for repairing damage and for correcting mistakes made during DNA replication.

In this chapter, we have seen that DNA has a special structure that enables it to replicate. We have also seen how we know that genes are made of DNA and that genes are physical regions in the DNA. In the next chapter, we will see how DNA's structure allows the "expression" of genes as polypeptides and how cells control the expression of genes—so that a given polypeptide is synthesized at the right time and in the right amount. We will also further explore how mutations in the structure of DNA change the information in genes.

Figure 10-25 Sunburn damages DNA and leads to skin cancer. The ultraviolet light in sunlight can damage DNA repair mechanisms. In the absence of DNA repair, many genes become damaged, including those that control the cell cycle. In early rounds of sunburn, most cells with damaged DNA self-destruct, and the skin peels. One theory of skin cancer holds that a few cells with damaged DNA remain but do not reproduce because they are surrounded by healthy cells. During subsequent sunburns, however, more cells with damaged DNA self-destruct, clearing the field for cancerous cells to begin growing. (© 1990 Peter Menzel/Stock, Boston)

Study Outline with Key Terms

Biologists first hypothesized that genes specified the structures of enzymes. When an organism cannot make an enzyme, it cannot carry out the biochemical conversions catalyzed by the enzyme. In alkaptonuria, for example, the absence of a specific enzyme causes the accumulation of a compound that turns the urine black. Studies of similar interruptions in biochemical pathways in the bread mold *Neurospora* in the 1940s suggested that every gene specifies the structure of a single enzyme. Researchers Beadle and Tatum coined the phrase, **One gene—one enzyme.** But within just a few years, other research showed that genes could specify the structures of proteins other than enzymes. Genes could specify single polypeptides only, however, not necessarily whole proteins. A more accurate statement of gene function is, **One gene—one polypeptide.**

Different alleles of a gene produce different versions of the same polypeptide. Thus, an alteration in the gene coding for the β-globin polypeptide leads to a **hemoglobin** with a structure different from normal hemoglobin. Sickle cell disease results when an individual inherits two copies of an allele that alters the structure of β-globin at a single amino acid.

The idea that genes are made of DNA came from work on bacteria and bacterial **viruses,** called **bacteriophages,** or simply **phages.** A virulent strain of bacteria can transfer the genetic instructions for virulence to a nonvirulent strain by means of DNA.

DNA consists of nucleotides held together by bonds between the third (**3′**) carbon of the sugar of one nucleotide and the fifth (**5′**) carbon of the sugar of the next nucleotide. Each polynucleotide chain has a direction with a 5′ end and a 3′ end. Hanging off the sugar-phosphate backbone of DNA are four kinds of nitrogen-containing bases. The bases are two **purines** (**adenine** and **guanine,** abbreviated A and G) and two **pyrimidines** (**cytosine** and **thymine,** abbreviated C and T). According to **Chargaff's rules,** the amount of C always equals the amount of G, and the amount of A equals the amount of T in the DNA of a given species.

In cells, DNA consists of two polynucleotide strands twisted around each other in a double helix. In the early 1950s, Rosalind Franklin's research showed that DNA was a helix. She was also the first to describe its density, water content, symmetry, and the number of bases (10) per turn of the helix. Previous work had shown that DNA was 2 nm wide, which Franklin confirmed, further reporting that the molecule was consistently 2 nm along its whole length.

Using Franklin's results, James Watson and Francis Crick built a model of DNA. They found that with the two strands of DNA running in opposite directions, hydrogen bonds could link the base pairs, G with C and A with T. These base pairings, argued Watson and Crick, explain the equal concentrations of G and C, and of A and T. The base pairings also meant that the information contained in the sequence of nucleotides of one strand was equivalent to that in the other strand. Each strand was a template for the other, and each strand could direct the synthesis of a complementary strand during DNA replication.

DNA replication is **semiconservative:** When DNA replicates, each strand of the double helix acquires a new mate. The enzyme **DNA polymerase** assembles nucleotides into a sequence specified by one strand of DNA, the **template** strand, a process called **polymerization.** DNA polymerase can add nucleotides only to a preexisting RNA **primer** made by the RNA polymerase **primase.** DNA polymerase can add nucleotides only to the 3′ end of the primer and continues to add nucleotides to the 3′ end of the growing polynucleotide.

DNA replicates in Y-shaped regions of the DNA called **replication forks.** Nucleotides add directly to the 3′ end of one strand in the fork, called the leading strand. The other strand is made discontinuously in **Okazaki fragments** of about 1000 nucleotides. Each fragment starts with a short RNA primer. **DNA ligase** links the fragments into a continuous strand. Special sequences in DNA serve as **origins of replication,** where DNA synthesis always begins. The sequences bind to proteins that open and untwist the double helix. Groups of 20 to 50 origins are called **replication units.**

DNA molecules suffer inevitable damage in the form of mutations. A **mutation** is any change in the sequence or number of nucleotides. **Point mutations** include multiple **base substitutions, insertions,** and **deletions. Chromosomal mutations** include **deficiencies, translocations, inversions,** and **duplications,** as well as **aneuploidies** such as **trisomy** and **monosomy.** Mutations are the ultimate source of all genetic diversity.

Mutation rates are usually low—occurring in bacterial DNA, for example, at the rate of about 1 mutation in 10 million base pairs each generation. **Mutation hot spots** mutate more often than other parts of the genome. Mutations can occur spontaneously, but they are hastened by ultraviolet light, x rays, certain chemicals, and other **mutagens.** Cells have sophisticated mechanisms to repair damage to DNA and to correct mistakes made during DNA replication. One result of uncorrected mistakes is cancer.

Review and Thought Questions

Review Questions

1. What facts about the structure of DNA did Franklin contribute? What did Watson and Crick contribute?
2. How does Watson and Crick's theory of base pairing explain how DNA replicates itself?
3. Do genes specify enzymes, proteins, or polypeptides? Why did biologists worry about this distinction?
4. What does it mean for a bacterial cell to be *transformed*? What does the transforming?
5. How is semiconservative replication different from conservative replication?
6. What serves as a template for the replication of a single strand of DNA?
7. What enzyme polymerizes nucleotides into a new strand of DNA?
8. Why does the cell need an RNA primer for DNA replication to begin? Where on the DNA does DNA polymerase begin replication?
9. What is the ultimate source of all genetic variation?
10. Why do cells need mechanisms for repairing DNA?

Thought Questions

11. D'Herelle and others thought that bacteriophages might provide a way to kill disease-causing bacteria—an idea that was especially exciting in the days before the discovery of penicillin and other antibiotics. Yet, until recently, no one found a way to use bacteriophages to destroy bacteria that have already infected the body. What problems might prevent medical researchers from making bacteriophages kill bacteria in the body?
12. Why does the dentist place a lead shield over your lap before x-raying your teeth? What might happen if the dentist did not take this precaution?

About the Chapter-Opening Image

Like these two pencils, the two strands of a molecule of DNA are parallel but run in opposite directions.

On-line materials relating to this chapter are on the World Wide Web at **http://www.harcourtcollege.com/lifesci/aal2/**

How Are Genes Expressed?

How Was the Genetic Code Discovered?

As soon as Watson and Crick discovered the structure of DNA, scientists all over the world wanted to become involved in the obvious next phase of the work, solving one of life's oldest secrets—the genetic code. Especially attracted were mathematicians and physicists, who believed they could crack the genetic code through logic alone. Yet, after years of effort, the code was ultimately cracked, not by the theoreticians, but by the rigorous experimental work of two young and unknown molecular biologists.

The first chink in the armor of the coding problem came in 1955, when Marianne Grunberg-Manago, working at New York University, discovered an enzyme that could link nucleotides together into long polymers. Grunberg-Manago took a year to isolate the enzyme. Once she had the enzyme, however, she began making artificial RNA polymers. She could, for example, link together long chains of A (adenine) nucleotides, which she called "poly A," or long chains of C (cytosine) nucleotides, called "poly C." In addition, she could make poly U, poly G, poly AU, and other combinations.

For the moment, however, no one knew what combination of nucleotides could actually be translated by a cell into a polypeptide. It was 6 more years before the next big break came. Then, in 1960, 31-year-old Johann Heinrich Matthei arrived at the National Institutes of Health (NIH) near Washington, DC. Matthei had come from Germany earlier that year looking for interesting scientific work. He had not found anything satisfactory yet, but he was especially interested in working on protein synthesis. At NIH, he got the names of three researchers who were trying to synthesize proteins artificially outside of a cell. Thirty-three-year-old Marshall Nirenberg stood out from the other two (Figure 11-1). His interests and instincts matched Matthei's own, and in November 1960, Matthei and Nirenberg embarked on a productive, if sometimes strained, collaboration.

Although, technically, Nirenberg was Matthei's boss, Matthei had come to NIH for collaboration, not direction. He was fully able to work on his own. Nirenberg's goal was to synthesize the polypeptides that make up proteins. He wanted to make them in a test tube. To do this, he planned to add ATP (for energy) and free amino acids to a mixture of ribosomes, nucleic acids, and enzymes extracted from cells. Other researchers had used this technique and had shown that the amino acids were incorporated into polypeptides. No one, however, knew what substances in the cell extracts were mak-

Paraskevas Photography

Key Concepts

- Protein synthesis in test tubes, using cell extracts, revealed how information flows from messenger RNA (mRNA) to polypeptide.

- DNA and RNA carry information in a 4-letter alphabet, while proteins carry information in a 20-letter alphabet.

- Translation of messenger RNA requires correct initiation, elongation, and termination of polypeptide synthesis.

- In both prokaryotes and eukaryotes, many proteins cooperate to accomplish and influence the expression of a single gene.

ing the polypeptides, and no one knew what kinds of polypeptides were being synthesized.

The obvious next step was to try to make a particular polypeptide by putting genetic information into the test tube. Several labs in the United States were hotly pursuing this idea. Before Matthei arrived, Nirenberg had tried to synthesize penicillinase, the enzyme that antibiotic-resistant bacteria use to break down the antibiotic penicillin. Nirenberg added bacterial DNA and RNA to his test-tube system. In experiment after experiment, however, he got no penicillinase. But without knowing how the code worked, no one would ever be able to synthesize any polypeptide, let alone a particularly valuable one such as penicillinase. Nirenberg needed to ask a simpler question.

After Matthei arrived, the two researchers asked a basic question: What kinds of RNA stimulate the synthesis of polypeptides? To answer this question, they took an extremely methodical approach. Before trying to synthesize any protein, they would make sure that their test-tube system could respond to RNA, any RNA. The two spent months perfecting their system. In the end, they could detect even minute amounts of protein synthesis. Their test-tube system was now highly sensitive to added RNA.

Nirenberg and Matthei knew that protein synthesis required ribosomes and that ribosomes contained RNA. Like most other scientists of the period, they believed that RNA must carry a sort of message from the DNA in the nucleus to the ribosomes in the cytoplasm, where protein synthesis normally happened. But although Nirenberg and Matthei's test tubes were packed with ribosomal RNA, they

Figure 11-1 Marshall Nirenberg. Nirenberg and Johann Matthei cracked the genetic code in 1960. (UPI/Corbis-Bettmann)

were seeing very little protein synthesis. Instinctively, they felt that, besides ribosomal RNA, there must be another kind of RNA—a kind that carried information that could specify a polypeptide and would make protein in their test tubes. But where was it? What was it?

They drew up a list of some 200 kinds of RNA to test. One of the first possibilities was RNA from tobacco mosaic virus. (In some viruses, RNA carries the genetic information.) The results were spectacular. Nirenberg and Matthei's test-tube system incorporated huge amounts of amino acids into some mystery protein.

Nirenberg instantly recognized a golden opportunity. Another molecular biologist, Heinz Fraenkel-Conrat, at the University of California, Berkeley, had recently worked out the sequence of the tobacco mosaic virus protein. Nirenberg called Fraenkel-Conrat, forged an instant collaboration, and in May 1961 left to spend a month in California to try to synthesize tobacco mosaic virus proteins. If he succeeded, he would gain scientific celebrity.

Matthei stayed behind at NIH, patiently testing the other possibilities on the list of 200 different kinds of RNA. High on the list were the artificial RNAs made using the methods invented by Marianne Grunberg-Manago. A week after Nirenberg left, Matthei set up the test tubes, the different enzymes, the ribosomes, the ATP, and the 16 amino acids he had on hand. Then to the different test tubes he added poly U, poly A, and poly AU. He knew that if he got any synthesis at all, poly U should make a polypeptide consisting entirely of one kind of amino acid. Poly A should too, but it should be a different amino acid.

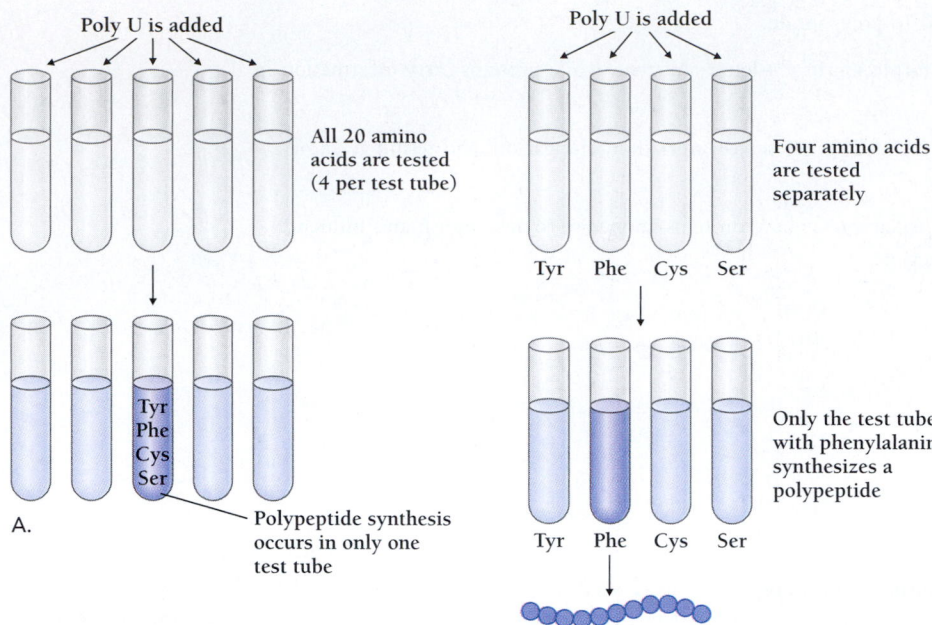

Poly U is added

All 20 amino acids are tested (4 per test tube)

Tyr
Phe
Cys
Ser

A.   Polypeptide synthesis occurs in only one test tube

Poly U is added

Four amino acids are tested separately

Tyr   Phe   Cys   Ser

Only the test tube with phenylalanine synthesizes a polypeptide

Tyr   Phe   Cys   Ser

B.

Figure 11-2 A simplified view of Nirenberg and Matthei's experiments. A. Matthei added four different radioactively labeled amino acids to each of five test tubes containing enzymes, ribosomes, and ATP, as well as poly U. One test tube synthesized large amounts of an unknown polypeptide. B. To find out which amino acid had been joined into polypeptides, Matthei added just one amino acid to each of five more test tubes. The test tube that synthesized protein was the one that contained the amino acid phenylalanine. The synthesized polypeptide consisted entirely of phenylalanine.

The results again were dramatic (Figure 11-2). Poly A and poly AU did almost nothing, but poly U stimulated 12 times the protein synthesis Matthei and Nirenberg got with nothing else added. The only question now was, What polypeptide had been made? In other words, which of the amino acids had been polymerized? Matthei attacked the problem aggressively. For 5 days, he worked almost round the clock, testing one amino acid after another.

On Friday night, Matthei stayed up all night again. Early on Saturday morning, bleary eyed but elated, Matthei had his answer. Poly U coded for a polypeptide made exclusively of the amino acid phenylalanine. He now knew that U nucleotides coded for phenylalanine, but how many nucleotides it took—whether one, two, three, or even more nucleotides—was another question. The number of nucleotides required to specify one amino acid was still unknown. Nonetheless, as Matthei stood in the lab, he realized that he alone in all the world knew the first word of the genetic code. He told Nirenberg's boss at NIH, but Matthei did not call Berkeley.

Nirenberg did not hear the news until he returned from California nearly 3 weeks later. His experiments at Berkeley had been inconclusive. The protein that tobacco mosaic virus RNA seemed to be synthesizing could not be identified. He must have heard Matthei's news with a mixture of emotions—delight at Matthei's success, disappointment that he had not been there to share in it.

Everyone at NIH soon knew what Matthei and Nirenberg had done. But the rest of the world either had not heard or did not believe it. Nirenberg was quiet, soft spoken. As one

well-placed molecular biologist later said apologetically in reference to Nirenberg, "either a person . . . is someone who's in the club and you know him, or else his results are unlikely to be correct, because he [isn't] in the club." Nirenberg was definitely not in the club. And, as far as the fast-moving world of molecular biology was concerned, Matthei might as well have not existed.

Together Nirenberg and Matthei wrote up the results of all their work so far and sent it to a scientific journal. A week later, NIH sent Nirenberg to Moscow for the Fifth International Congress of Biochemistry. In a small, nearly empty room, Nirenberg presented the results of his and Matthei's months of work. Of the handful of scientists present, only molecular biologist Matthew Meselson really listened.

As Meselson later recalled, "I was bowled over by the results, and I went and chased down Francis [Crick], and told him that he must have a private talk with the man." Crick talked to Nirenberg, then arranged for him to give his talk once more in front of an audience of hundreds of biochemists and molecular biologists. Nirenberg's scientific fortune was made. He might not belong to the club, but he had been heard by nearly everyone who did.

While Nirenberg was in Moscow, Matthei nailed down still another "word" in the genetic code. Poly C, he found, made a polypeptide composed entirely of the amino acid proline.

Nirenberg returned to NIH in triumph. Yet within days of his announcement in Moscow, another research lab began using Nirenberg and Matthei's refined technique to crack the

First letter (5′ end)	Second letter				Third letter (3′ end)
	U	**C**	**A**	**G**	
U	UUU Phe UUC UUA Leu UUG	UCU UCC Ser UCA UCG	UAU Tyr UAC UAA Stop UAG Stop	UGU Cys UGC UGA Stop UGG Trp	U C A G
C	CUU CUC Leu CUA CUG	CCU CCC Pro CCA CCG	CAU His CAC CAA Gln CAG	CGU CGC Arg CGA CGG	U C A G
A	AUU AUC Ile AUA AUG Met	ACU ACC Thr ACA ACG	AAU Asn AAC AAA Lys AAG	AGU Ser AGC AGA Arg AGG	U C A G
G	GUU GUC Val GUA GUG	GCU GCC Ala GCA GCG	GAU Asp GAC GAA Glu GAG	GGU GGC Gly GGA GGG	U C A G

Nucleotide base   Codon   Amino acid

A.

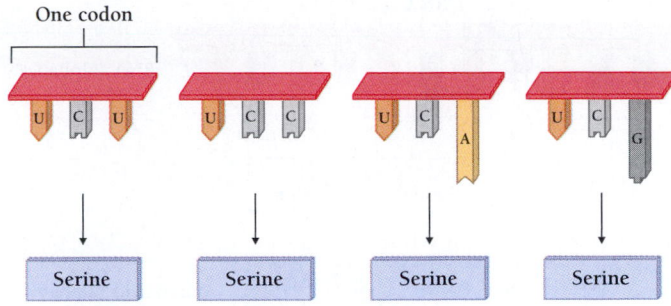

B.

Four different codons code for the same amino acid. Note difference is in last base.

Figure 11-3   The genetic code. A. This table shows which amino acid corresponds to each RNA triplet. All the DNA Ts are replaced by Us, since this is RNA. B. Some amino acids have more than one codon. Usually the first two bases are the same, but the third base is different.

rest of the genetic code. Although initially taken aback by the aggressiveness of this move, Nirenberg soon threw himself into the competition, determined to be first.

In his excitement, Nirenberg began making all the decisions, and Matthei began to feel like a fifth wheel. His friendship and collaboration with Nirenberg soured, and 18 months after they began to work together, Matthei was back in Germany working by himself once more.

Nirenberg never missed a step. He began working with another collaborator, Philip Leder, and the two soon showed that a length of RNA just three nucleotides long could specify an amino acid. They were able to define a **codon** as a group of three nucleotides that specifies a single amino acid within a polypeptide. (Web Bit 11-1: How Do We Know That a Codon Is Three Nucleotides Long?)

Then another researcher stepped into the picture. Organic chemist Gobind Khorana found a way to link nucleotides together in any order he wanted. In one experiment, he put together long chains of poly AC, which contained the sequence ACACACACAC and so on. Khorana found that poly AC RNA stimulated the production of a polypeptide consisting of two alternating amino acids—threonine and histidine.

Khorana showed that ACA specified threonine and CAC specified histidine. Thus ACACACACACAC coded for

threonine-histidine-threonine-histidine. From a series of such experiments, Khorana and his colleagues established a dictionary of codons. It took 5 years, but by 1966, Nirenberg, Khorana, and others had written a complete dictionary for the genetic code (Figure 11-3). There were 64 different codons. Sydney Brenner and Francis Crick showed that 3 of the 64 possible codons coded for no amino acids at all. They were "nonsense" codons whose role was to signal the end of a polypeptide chain, just like the period at the end of a sentence.

In 1968, Nirenberg and Khorana won a Nobel prize for their elucidation of the genetic code. In the same year, James Watson's best-selling book, *The Double Helix,* was published. In the world of biology, the first era in the history of molecular biology had come to an end.

The genetic code was deciphered by means of careful laboratory experiments.

What Is the Genetic Code Like?

The nucleotide sequences of DNA and RNA are written in a different language from the amino acid sequences of polypep-

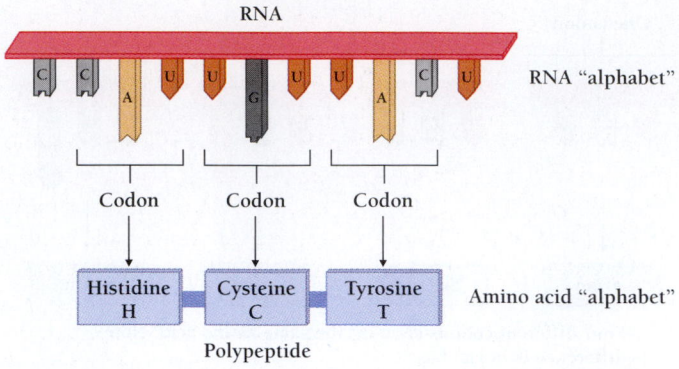

Figure 11-4   Three nucleotides code for one amino acid. The DNA/RNA language is written with an alphabet of nucleotides, while the polypeptide language is written with an alphabet of amino acids.

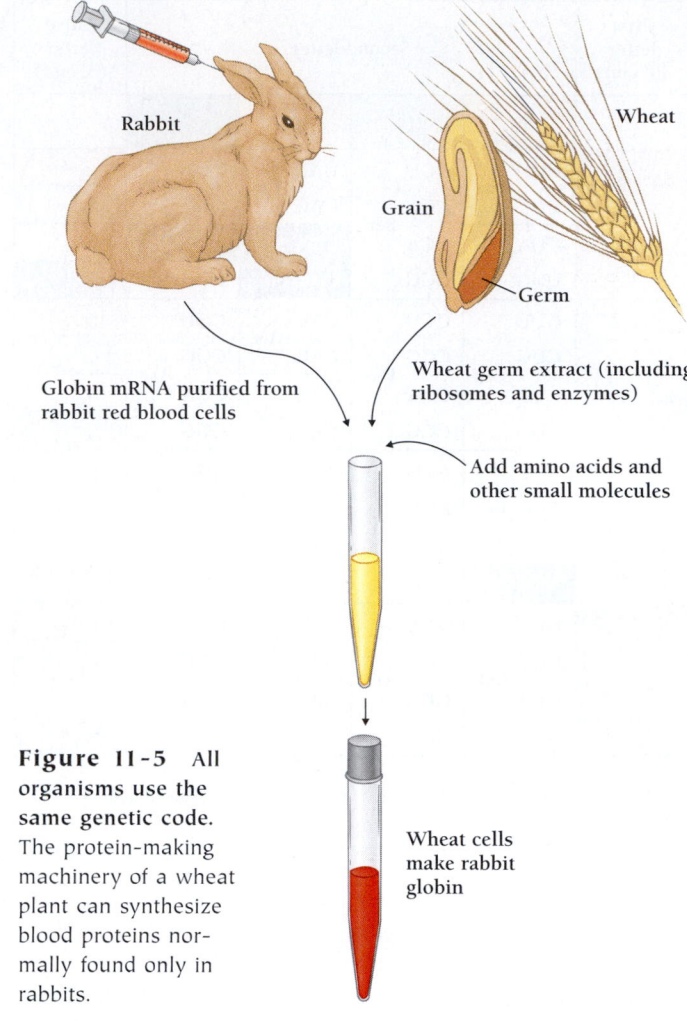

Figure 11-5   All organisms use the same genetic code. The protein-making machinery of a wheat plant can synthesize blood proteins normally found only in rabbits.

tides. The DNA/RNA language is written with an alphabet of just 4 nucleotides, while the polypeptide alphabet has 20 amino acid letters (Figure 11-4).

Although the DNA and protein languages are different, the relationship between them is simple. This is because the **genetic code** clearly specifies which nucleotide sequences correspond to which amino acid sequences. Each set of three nucleotides, called a codon, specifies an amino acid.

Since different combinations of any 3 nucleotides can code for 64 different amino acids, and there are only 20 amino acids, what do the extra 44 codons do? The answer is that most of these extra codons also code for the same 20 amino acids. The genetic code has several ways of specifying each amino acid. Because several codons may have the same meaning, we say that the genetic code is "redundant."

Of the 64 triplets, only 61 specify amino acids. The other three codons (UAA, UAG, and UGA) are called **nonsense,** or **stop, codons** because rather than specifying an amino acid, they normally signal the end of a polypeptide chain.

Because the same code is used by all species, the genetic code is said to be **universal.** One demonstration of this universality comes from studies of hemoglobin, the oxygen-carrying molecule in red blood cells. Before red blood cells enter the bloodstream, they synthesize large amounts of hemoglobin from two polypeptides, called globins, made from two kinds of RNA. When this RNA is purified from rabbit red blood cells and then added to wheat germ cells, the wheat cells make rabbit globin (Figure 11-5). This demonstrates that wheat cells and rabbit cells use the same code to translate RNA into protein. Indeed, all eukaryotes not only use the same code but they also use interchangeable molecular machinery.

Nonetheless, the genetic code is not completely universal. Mitochondria, whose DNA is passed on separately from the DNA in the nucleus, use a slightly different code from the one in Figure 11-3. And yeast mitochondria use a different genetic code from human mitochondria.

We have seen how molecular biologists solved the genetic code, which helps explain how genetic information flows from

DNA to RNA to proteins. But we must also ask about the mechanism by which this information flows. What happens, step by step?

The genetic code translates the language of DNA and RNA into the language of polypeptides. Three nucleotides—a codon—specify one amino acid. The genetic code is redundant and nearly universal, and 3 of the 64 codons are nonsense, or stop, codons.

GENETIC INFORMATION FLOWS FROM DNA TO RNA TO POLYPEPTIDES

By the mid-1950s, biologists knew that genes were sequences of nucleotides that specify the sequences of amino acids in polypeptides. The next question was, How does a sequence of nucleotides specify a sequence of amino acids? Molecular biologists such as Nirenberg and Khorana confirmed what Watson and others had long suspected: RNA is always an intermediate in the production of proteins.

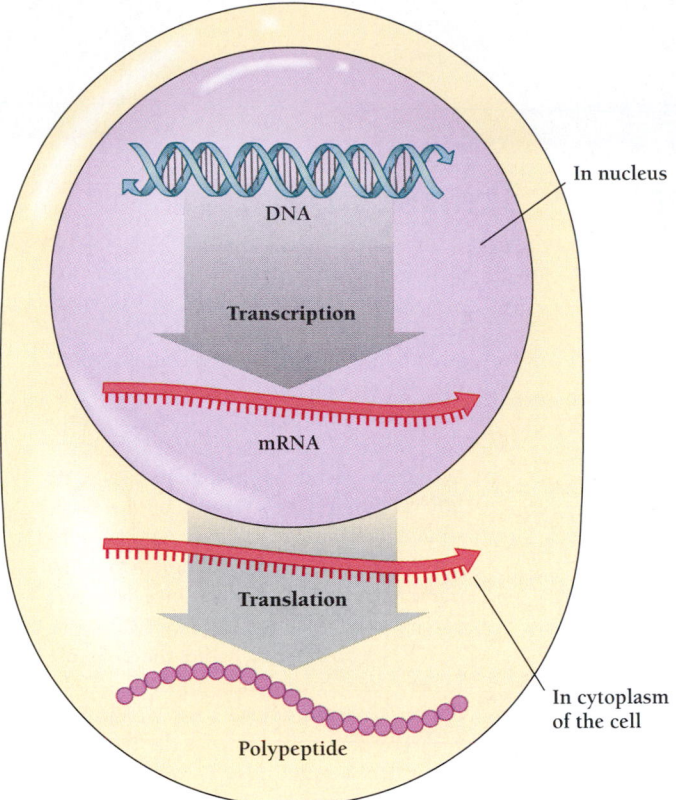

Figure 11-6 **The central dogma.** DNA is transcribed into RNA, which is translated into protein. In eukaryotes, transcription occurs in the nucleus and translation occurs outside the nucleus, in the cytoplasm. In prokaryotes, which lack a nucleus, both transcription and translation occur in the cytoplasm.

Francis Crick summarized cellular information flow in a statement he called the **central dogma** of molecular biology (Figure 11-6): "DNA specifies RNA, which specifies proteins." According to Crick's central dogma, information can flow *only* from DNA to RNA to proteins. It cannot flow from proteins to RNA or DNA. Crick put this idea another way, writing, "Once 'information' has passed into protein, *it cannot get out again.*" Proteins cannot change the information in the genes.

Crick used the word dogma ironically because "dogma" means an opinion, belief, or a tenet of faith. It is not the word to use to describe scientific facts or well-supported theories. At the time Crick formulated it, molecular biologists had no good evidence supporting the central dogma. Crick and others just thought it must be true. Crick's joke was lost on most other researchers, however, and the "central dogma" has now been presented in textbooks *as dogma* for decades. As we will see in the next chapter, Crick had good reason for his twinge of doubt. The central dogma is mostly true, but not completely: some viruses—like the AIDS virus—carry information in RNA, which infected cells can incorporate into their own DNA. In addition, information in RNA and proteins influences the expression of the DNA's information.

Nevertheless, the idea that DNA specifies RNA, which specifies proteins, is a good starting point for understanding how genes are expressed through the two processes of transcription and translation. The process by which DNA transmits its information to RNA is called **transcription** (Figure 11-7A). During transcription, the information in the DNA is

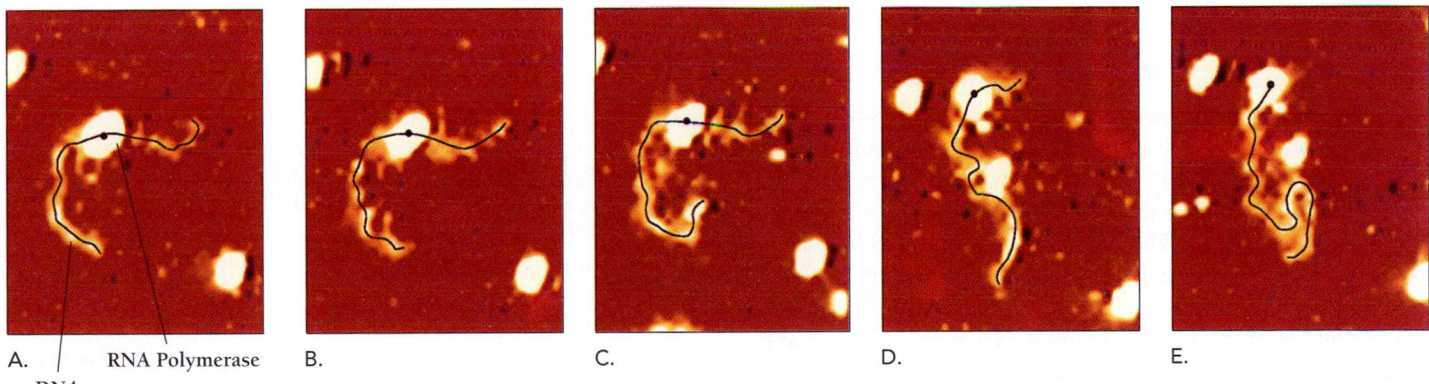

A.   B.   C.   D.   E.

RNA Polymerase

DNA

Figure 11-7 **Transcription and translation.** A–E. In this sequence of micrographs, RNA polymerase transcribes a strand of DNA into mRNA (which is not visible). RNA polymerase, the large, round blob, moves from the middle of the strand (A) to the end (E) in less than 5 minutes. F. In prokaryotes, which lack a nuclear membrane, transcription and translation can occur simultaneously. In this electron micrograph, the DNA runs diagonally from the upper left to the lower right. Transcription began at the upper left end of the DNA, so the longest mRNAs (with the most ribosomes) are at the lower right. Clusters of up to 13 ribosomes read individual strands of mRNA, even as they are being transcribed from the DNA. (*A–E, M. Guthold, X. Zhu, and C. Bustamante, University of Oregon; F, Dr. Barbara Hamkalo, University of California, Irvine*)

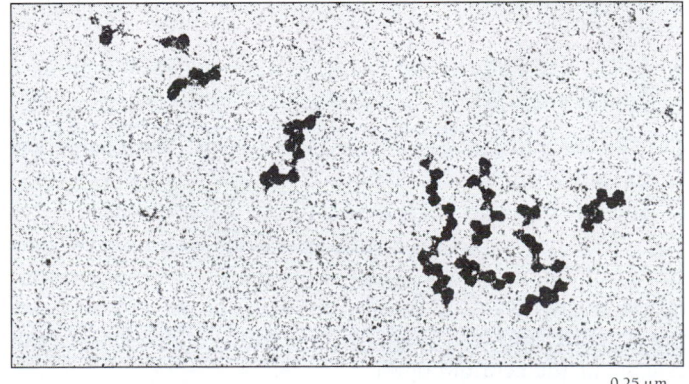

F.   0.25 μm

BOX 11.1

Which Strand of the DNA Is Transcribed?

At the beginning of this chapter, we saw that protein synthesis in a test tube helped solve the genetic code. Such test-tube protein synthesis also allowed researchers to discover how RNA polymerase transcribes the DNA.

Just as DNA polymerase assembles new molecules of DNA one nucleotide at a time, RNA polymerase assembles RNA by adding one nucleotide at a time. However, DNA polymerase copies both strands of the DNA double helix, while RNA polymerase copies only one of the two strands of DNA. The strand that is copied is called the *template strand*. Although the two strands are chemically the same, the information in the template strand is completely different from that in the nontemplate strand of DNA. The template strand, also called the *antisense strand*, is a reverse copy of its mate, the *sense strand*. If this sentence were the sense strand, we might think the template strand would look something like this: :siht ekil gnihtemos

kool dluow dnarts etalpmet eht kniht thgim ew ,dnarts esnes eht erew ecnetnes siht fI

Actually, however, the template strand is not only a reverse copy of the sense strand. It is also composed of the Watson-Crick base pairs, as if we had substituted different letters into our reversed sentence. For example, if the sense strand of DNA read CATTAG, then the antisense strand, read forward by RNA polymerase, would be CTAATG.

$$\text{sense strand}$$
$$\overrightarrow{\text{CATTAG}}$$
$$\underset{\longleftarrow}{\text{GTAATC}}$$
$$\text{antisense strand}$$

The most important result of this reversal and substitution is that the antisense strand cannot code for a protein. The antisense strand is a chaotic sequence of codons that includes so many stop codons

that any strange polypeptides made following its instructions could not be more than a few amino acids long. The sense strand, in contrast, has stop codons only at the end of each gene.

The antisense strand is useful, however, because the mRNA created from the antisense strand is not nonsense. The mRNA is an RNA version of the sense strand, the reverse of a reverse. The antisense strand at left, for example, would be transcribed as CAUUAG, since wherever DNA has a T, the mRNA has a U.

RNA polymerase copies the antisense DNA strand by holding in place an RNA nucleoside triphosphate base, which pairs with the next nucleotide in the template. The mRNA strand grows from its 5′ end to its 3′ end, as it copies the template DNA strand from its 3′ end to its 5′ end. The new mRNA molecule has the same sequence of bases as the sense strand of DNA, except that wherever DNA has a T the RNA has a U.

---

rewritten, or transcribed, to RNA. It is that information in RNA that cells use to make polypeptides. Using RNA as a secondary information source is analogous to working from a photocopy of an important document, while keeping the original in a safe place, in this case the nucleus.

The RNA copy of the information from DNA is composed of **messenger RNA (mRNA).** Molecules of mRNA have the same information as the DNA original. RNA, like DNA, is a polynucleotide. It differs from DNA in just three ways:

1. The backbone sugar is ribose instead of deoxyribose.
2. Uracil (U) replaces thymine (T) as one of the pyrimidine bases.
3. RNA is usually single stranded, whereas DNA is usually double stranded.

Despite these differences, a message in the RNA alphabet (A, G, C, and U) is just the same as in the DNA alphabet (A, G, C, and T), but with U replacing T.

The conversion of the message carried by the RNA into strings of amino acids—actual polypeptides—is called **translation** (Figure 11-7F). Translation takes the mRNA text and translates it into the language of amino acids. This is the essence of the central dogma.

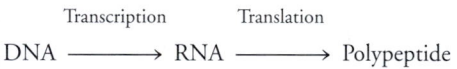

$$\text{DNA} \xrightarrow{\text{Transcription}} \text{RNA} \xrightarrow{\text{Translation}} \text{Polypeptide}$$

In eukaryotes, transcription occurs in the nucleus and translation occurs in the cytoplasm (outside the nucleus). Messenger RNA passes through the nuclear membrane out to the cytoplasm. There, ribosomes, acting like molecular sewing machines, stitch together amino acids into polypeptides.

In the next part of this chapter, we will discuss the details of the molecular machinery responsible for the flow of information from DNA to RNA to polypeptide. The process of building a cell from the information in DNA results from chemical reactions that biologists can now describe in detail. These reactions are subject to the same thermodynamic laws as other chemical processes. The same kinds of forces drive them, and no special force is needed to build the machinery of life.

The message in the DNA is transcribed into mRNA, which is translated into polypeptides. The central dogma says that information flows from DNA to RNA to protein only. Information in protein is not supposed to be able to influence the information in the DNA.

RNA POLYMERASE TRANSCRIBES DNA INTO RNA

When a cell makes a polypeptide specified by a particular gene, we say that the cell is "expressing" that gene. Gene expression begins when the enzyme **RNA polymerase** transcribes the DNA gene into mRNA. Actually, there are several kinds of RNA polymerases, but since they all make RNA in the same way, we will discuss them as if they were a single enzyme.

How Does RNA Polymerase Begin Transcription?

In the last chapter, we saw that DNA polymerase can build DNA in two directions (extending the 3′ end of each DNA strand) and that the enzyme begins by adding nucleotides to an existing polynucleotide called a primer. In contrast, RNA polymerase transcribes DNA into RNA in only one direction

and does not require a preexisting RNA primer (Figure 11-8). The starting signal for mRNA synthesis is a special sequence of DNA called a **promoter.** RNA polymerase (working with other proteins) recognizes the promoter, unwinds and separates the two strands of DNA near the promoter site, and then attaches to the promoter and begins to "read" the sequence of bases in the gene. Wherever the template strand has a G, RNA polymerase will place a C; wherever the template has an A, it places a U.

In bacteria, the promoter consists of two characteristic sequences, each about 6 nucleotides long, with about 25 other nucleotides between them. RNA polymerase binds to the promoter and begins to make an RNA molecule that is complementary to the template strand.

In eukaryotes, the production of RNA involves an extra step. Unlike prokaryotic RNA polymerases, eukaryotic RNA polymerases do not themselves recognize the promoter. Instead, they depend on the assistance of other proteins, called

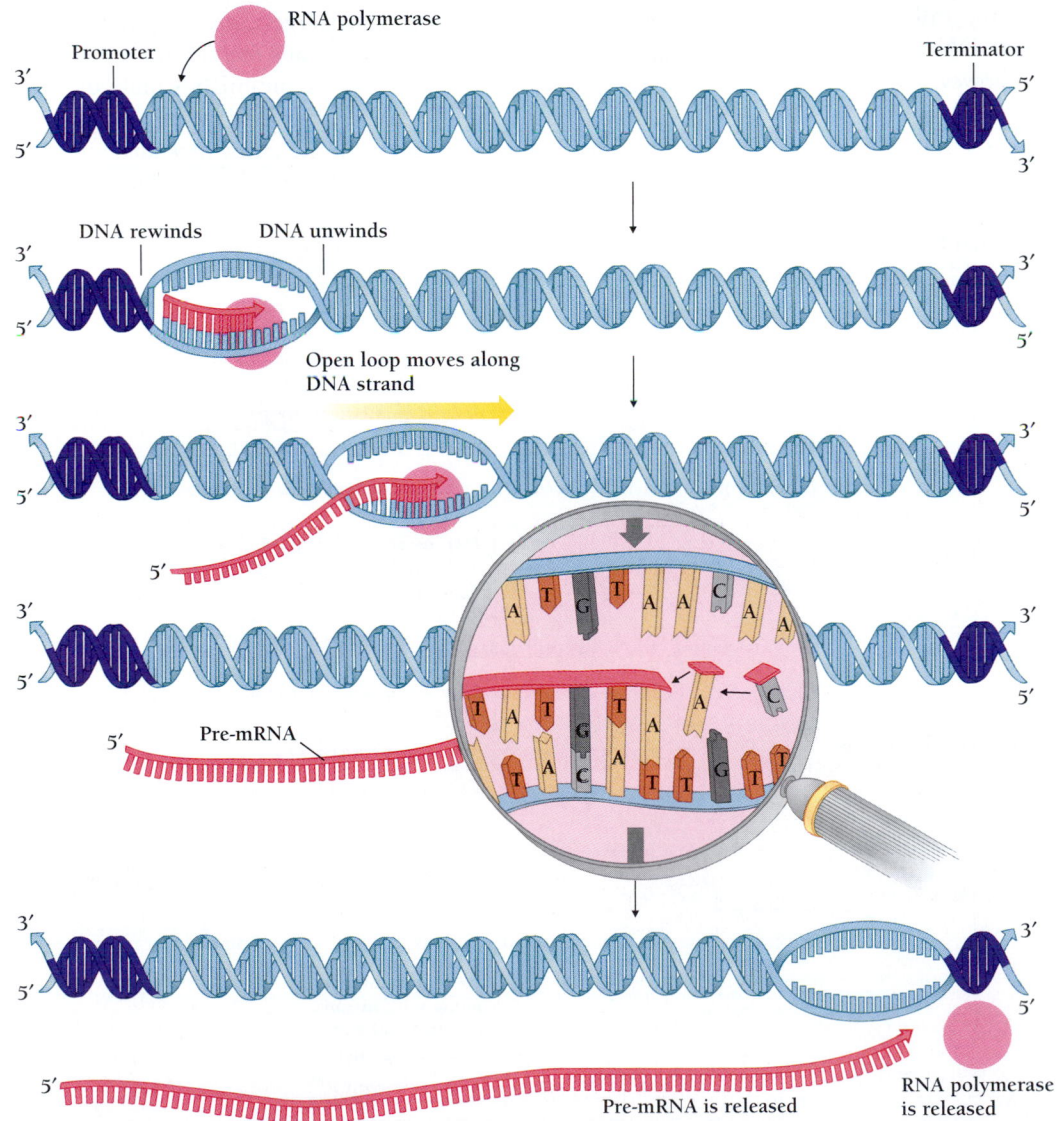

Figure 11-8 How does RNA polymerase begin transcription? Starting at the DNA promoter, RNA polymerase transcribes DNA into mRNA until it reaches the terminator.

activators, or **transcription factors,** which lead the RNA polymerase and its accompanying proteins to the promoter. One set of activators, for example, guides RNA polymerase to a DNA sequence called a "TATA box," since it usually contains the sequence TATA. Other activators guide RNA polymerase to other specific DNA sequences.

Both eukaryotic and prokaryotic RNA polymerases stop transcription at special sequences, called **terminators,** which stop the enzyme from transcribing any more of the DNA.

With the help of activators, RNA polymerase produces molecules of RNA by copying one strand of DNA starting at a promoter sequence.

How Does the Cell Alter the mRNA Transcribed from the DNA?

In prokaryotes, the mRNA binds to the ribosomes and begins directing protein synthesis as soon as the mRNA is made. Sometimes translation begins even as the DNA is being transcribed (Figure 11-7F). In eukaryotes, however, transcription and translation occur separately. Transcription occurs in the nucleus, and translation occurs—minutes, hours, or even days later—in the cytoplasm.

Transcription and translation in eukaryotes are separated by much more than just the nuclear membrane. After the

mRNA is transcribed from the DNA, in the nucleus, it is called pre-mRNA. While still in the nucleus, pre-mRNA undergoes a number of modifications that transform it into a molecule of "mature" mRNA that can be translated into a polypeptide. These modifications include: (1) capping the 5′ end of the RNA with a GTP (a relative of ATP), (2) adding a "tail" of 150 to 200 As to the 3′ end of most mRNAs, and (3) removing large pieces of the RNA and splicing the remaining pieces together. In addition, using a process called RNA editing (only recently discovered by molecular biologists), cells can add, remove, or change a few selected nucleotides in some regions of the mRNA (Figure 11-9A).

The cutting and splicing of pre-mRNAs was a surprising discovery, since it demonstrated that genes are interrupted by nucleotide sequences that do not code for proteins. These interruptions in pre-mRNA (and the corresponding sequences in DNA) are called **introns,** for intervening sequences. The RNA sequences that directly code for proteins are spliced together to form the mature mRNA molecule. These sequences (and the corresponding sequences in DNA) are called **exons,** for expressed sequences. In a typical eukaryotic gene, most of the sequence consists of introns, with exons making up only a tiny fraction of the total length. A single pre-mRNA molecule may contain as many as 70 introns, each one ranging in size from 80 to 10,000 nucleotides. Yet all the exons in a single mRNA are, together, rarely more than 3000 nucleotides long.

Most of the DNA in a gene, then, consists of introns and does not code for a functioning polypeptide. In fact, more than

Figure 11-9 RNA splicing. A. In this example, introns are snipped from the mRNA, the five exons are spliced together, and each end of the spliced mRNA is punctuated. Like a capital letter, the GTP cap indicates the beginning of the gene sentence. Like a period, the poly A tail indicates the end. B. The same pre-mRNA can be snipped and spliced in different ways to produce mRNAs that code for different polypeptides.

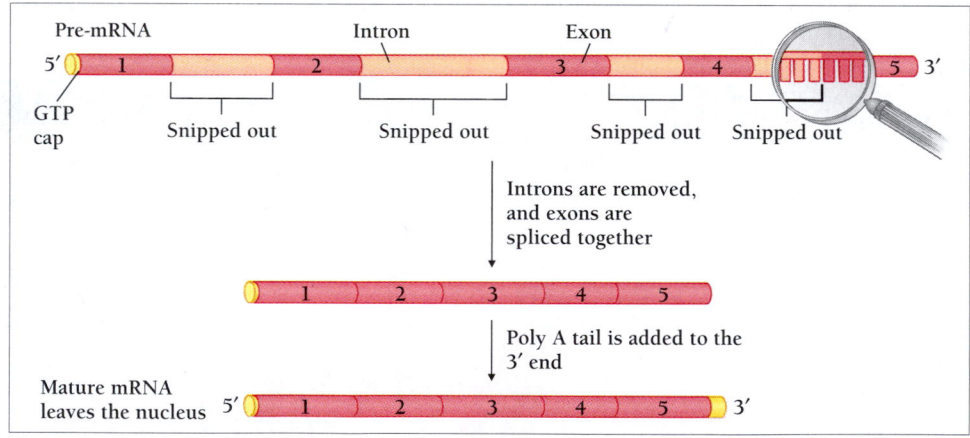

98 percent of human DNA does not directly specify polypeptide sequences, and a large part of this apparently useless DNA is introns. No one knows what introns do. Some scientists believe that the DNA in the introns is basically "junk DNA" that has accumulated in the genome in the same way that old keys and odd parts of things accumulate in the back of a kitchen drawer. Others believe that all DNA must be in the genome for a purpose, even if that purpose is not yet apparent.

Pre-mRNA includes long stretches of noncoding DNA sequences, called introns, that must be removed. The remaining exons are spliced together to form mature mRNA.

One Gene—Heaven Only Knows How Many Polypeptides

Most eukaryotic genes consist of exons interspersed with introns. Transcription, we have said, then yields a single kind of pre-mRNA containing all the introns and exons. Molecular machinery within the nucleus cuts and splices the pre-mRNA to produce mature mRNAs without introns.

As it turns out, transcription can be far more complicated than this simple picture suggests. In some cases, cells can actually transcribe many different mRNAs from the same gene.

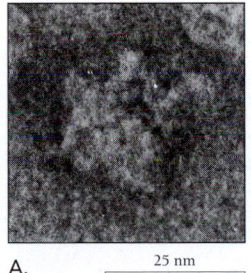

A.   25 nm

For example, the mRNA transcript may start or end at different sites on the same gene. Or a single pre-mRNA may be broken up and spliced together in several different ways, so that one mRNA's exon becomes another mRNA's intron (Figure 11-9B). As a result, a single gene may encode several polypeptides. What is more, the cell may express all of these various polypeptides at once.

Geneticists spent decades arriving at our definition from Chapter 10, "one gene—one polypeptide." But studies of gene expression beginning in the late 1970s have conclusively shown that one gene can actually encode several polypeptide chains, occasionally with very different functions. Most molecular biologists now say that a gene is a DNA sequence that is transcribed as a single unit and encodes either a single polypeptide or a set of related polypeptides.

A single gene can encode several different polypeptides.

HOW DOES A CELL TRANSLATE mRNA INTO A POLYPEPTIDE?

The synthesis of proteins requires a multitude of molecular tools. Among them are three distinct kinds of RNA: (1) messenger RNA carries genetic information from DNA that specifies the amino acid sequence of a polypeptide; (2) transfer RNAs carry each amino acid to a codon in the mRNA; (3) ribosomal RNA, complexed with proteins to form ribosomes, serves as the site for polypeptide synthesis. In addition, an array of enzymes and other proteins assist in the synthesis and folding of new proteins.

What Do Ribosomes Do?

Ribosomes are like subcellular sewing machines that link amino acids into polypeptides. All cells contain ribosomes. Each ribosome consists of a small and a large subunit (Figure 11-10). Each subunit contains both RNA and proteins. The

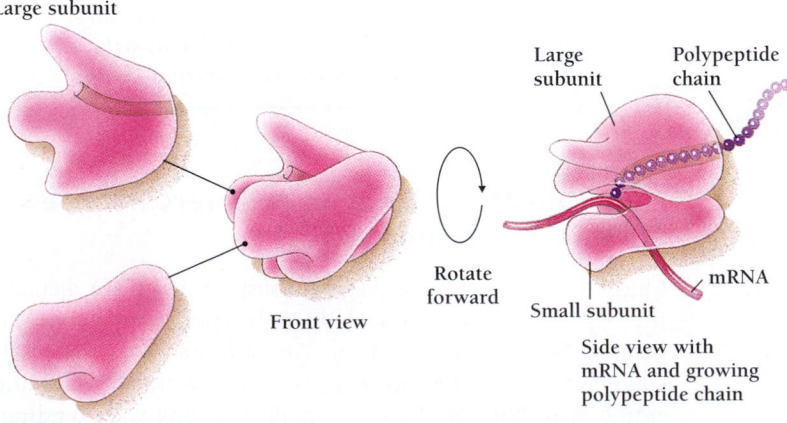

Large subunit

Small subunit

B.

Front view

Rotate forward

Large subunit

Polypeptide chain

mRNA

Small subunit

Side view with mRNA and growing polypeptide chain

Figure 11-10 How do ribosomes work? A. Electron micrograph of whole ribosome. B. The two parts of a ribosome fit together around the mRNA molecule and the growing polypeptide chain. (*A, courtesy of James Lake*)

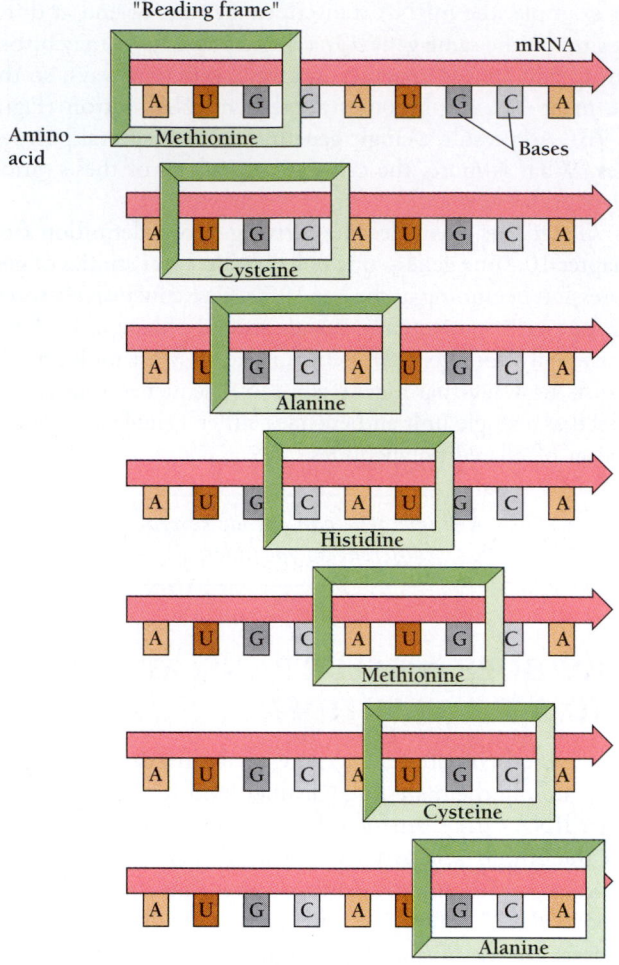

"Reading frame"

mRNA

Amino acid — Methionine

Bases

Cysteine

Alanine

Histidine

Methionine

Cysteine

Alanine

Figure 11-11   The reading frame. The exact base at which the ribosome begins translating the mRNA determines how it interprets all of the three-base codons from there onward.

E. coli ribosome, for example, consists of 3 molecules of RNA, called **ribosomal RNAs,** or **rRNAs,** and 54 kinds of protein molecules. Eukaryotic ribosomes are even more complicated, with 4 molecules of rRNA and some 82 kinds of protein.

Ribosomal RNA is very different from messenger RNA. The sequence and length of messenger RNA varies, depending on the kind of polypeptide it encodes. But the RNA that makes up the ribosome is always the same, no matter what kind of protein the ribosome is producing.

Ribosomal RNAs from organisms of all kingdoms are highly similar to one another. Such similarity suggests that the exact shape of the ribosome is important to the way it works. We now know that ribosomal RNAs play an important role in holding ribosomal proteins together in just the right orientation. Some researchers argue that ribosomal RNA may also act as a "ribozyme," helping to catalyze the joining of amino acids into polypeptides. We can guess that any individual with

even minor changes in the standard ribosome design would be unable to make proteins and would therefore die immediately.

In 1999, after nearly 40 years of work, a number of research groups obtained the first detailed information about the three-dimensional structure of ribosomes. Knowledge of ribosome structure should soon allow biologists to understand the precise roles of individual components of the ribosome, including those of the rRNAs.

Ribosomes help catalyze the formation of peptide bonds between amino acids, in an order determined by mRNA. Ribosomes' main tasks are to assemble amino acids into polypeptides in the order specified by mRNA. In addition, ribosomes must recognize where to start and where to stop reading the mRNA for the production of each polypeptide.

For example, consider the translation of the mRNA sequence AUGCAUGCA. If a ribosome started reading at the first nucleotide, it would assemble the peptide methionine-histidine-arginine. If, however, the ribosome started with the second nucleotide, it would assemble cysteine-methionine. If the ribosomes started with the third nucleotide, it would assemble alanine-cysteine. The grouping of nucleotide triplets for translation is called the **reading frame** (Figure 11-11).

Because starting at the right place is crucial to reading mRNA correctly, mRNAs must contain signals that say "start" and "stop." Start and stop signals are like punctuation that tells where a sentence begins and ends. The smaller ribosomal subunit finds the right place to start, using slightly different mechanisms in prokaryotes and in eukaryotes.

Once a ribosome has found the start site, prokaryotic and eukaryotic ribosomes work in much the same way. Each ribosome contains two grooves. On the small subunit is a groove for reading the mRNA. On the large subunit is a groove for the growing polypeptide chain. Protein synthesis requires the participation of both subunits. The small subunit is responsible for starting synthesis at the right place, while the large subunit contains the machinery for making the peptide bonds that hold together the polypeptide.

Ribosomes, which are made of rRNA and proteins, link amino acids together into polypeptides. The small ribosomal subunit reads the mRNA, and the large ribosomal subunit makes the peptide bonds between each pair of amino acids in the growing chain.

How Do tRNAs Serve as Adapters Between mRNAs and Amino Acids?

Once molecular biologists began to suspect that RNA dictated the sequences of amino acids in polypeptides, they tried to imagine how this could occur. An early idea was that each codon in an mRNA molecule bound directly to a specific amino acid. No one, however, could find any such binding. Besides, it was hard to imagine how a sequence of just three

BOX 11.2

How Do the Ribosomes Read Different Kinds of Mutations?

A base substitution in a gene may result in a *missense mutation,* which changes the codon for one amino acid into the codon for another amino acid. The resulting polypeptide may be either nonfunctional or altered in the way it functions. For example, some altered polypeptides result in *conditional mutations,* in which the resulting protein may be able to function under some conditions, but not under others. In one *Drosophila* mutant, for example, the fly functions normally at 20°C but becomes paralyzed at 30°C. This is due to a protein that regulates ion transport across nerve cell membranes, which fails at 30°C. Such temperature sensitivity results when an amino acid substitution reduces a protein's stability at higher temperatures, so that it denatures, or changes shape, more readily than the normal protein.

A base substitution can also result in a *nonsense mutation,* which changes the codon for an amino acid into a nonsense or stop codon, leading to a shortened polypeptide (Figure A). Such a shortened polypeptide almost certainly will not function at all in its former role.

Mutations that have no effect on phenotype are called *silent mutations.* Some are simply single-base substitutions that result in a codon that codes for the same amino acid (because of the redundancy of the genetic code). In that case, the polypeptide is unchanged. Most silent mutations, however, lie in DNA sequences that do not actually code for a polypeptide. Other supposedly silent mutations may merely produce changes in the phenotype so subtle that researchers cannot tell the difference.

Insertions and deletions in structural genes produce *frameshift mutations,* which alter the groupings of nucleotides into codons. The addition or subtraction of a single base can wreak havoc in the ribosomes' reading of mRNA downstream from the mutation. The result is often a string of missense changes, followed by a stop codon.

Sometimes point mutations can reverse themselves in a process called *reversion.* Reversion mutations, sometimes called *suppressor mutations,* consist of a second mutation that reverses the effects of the first mutation. A reversion might be an exact restoration of the original nucleotide sequence. Alternatively, the DNA might have a nucleotide sequence that is not identical to the original but that specifies the same amino acid—a silent mutation.

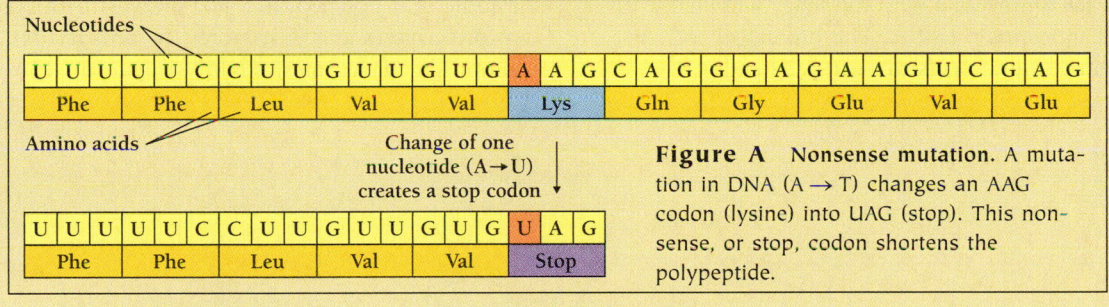

Figure A Nonsense mutation. A mutation in DNA (A → T) changes an AAG codon (lysine) into UAG (stop). This nonsense, or stop, codon shortens the polypeptide.

nucleotides could distinguish among amino acids as similar, for example, as leucine, isoleucine, and valine.

The actual explanation, proposed by Francis Crick, is that the amino acids are first linked to special adapter molecules, called **transfer RNAs,** or **tRNAs.** The mRNA codons bind to the tRNA, not to the amino acid itself. Each tRNA consists of 75 to 85 nucleotides and folds to form a secondary structure shaped like a cloverleaf (Figure 11-12). The "stems" of the tRNAs are double strands of RNA held together by base pairing. The "leaves" are loops of RNA with unpaired bases. X-ray diffraction studies of tRNA's tertiary structure show that the whole cloverleaf bends into a shape resembling the letter **L.**

One loop of each tRNA attaches to the amino acid corresponding to that codon (Figure 11-12). The opposite side of each tRNA molecule binds to a specific codon in the mRNA. In one of the loops is a sequence of three nucleotides, called the **anticodon,** that forms base pairs with the matching codon sequence in the mRNA (Figure 11-12). Each tRNA molecule has only one kind of anticodon and can pick up only one kind of amino acid.

We may wonder how many tRNAs a cell needs—one for each amino acid or one for each codon? A cell must have at least one tRNA for each of the 20 amino acids. Yet we might expect 61—one for every codon except the 3 codons that do not specify any amino acid. In fact, different species have different numbers of tRNAs. However, every species has more than 20 and fewer than 61 kinds of tRNA. Organisms can have fewer than 61 tRNAs because some tRNA anticodons recognize more than 1 codon.

How do we know that mRNA recognizes the tRNA, rather than the amino acid itself? To find out, researchers "tricked" the mRNA. They took the tRNA for cysteine, attached the cysteine, then chemically converted the attached cysteine into alanine so that the tRNA for cysteine actually carried an alanine. Finally, researchers added this incorrectly loaded tRNA to a cell extract that was synthesizing protein. The extract made incorrect polypeptides. Wherever the mRNA specified cysteine, the tRNA inserted alanine instead. This result showed that the mRNA recognizes the tRNA, not the amino acid itself.

Transfer RNA helps assemble amino acids into polypeptides by binding first to an amino acid and then to the correct codon in the mRNA.

How Do tRNAs Recognize the Right Amino Acid?

We have seen how the tRNA anticodon recognizes the correct codon. But how does tRNA recognize and load the correct amino acid? The task is to match the unique shape and charge distribution of each amino acid with the shape and charge distribution of the appropriate tRNA. The linking of tRNAs to their corresponding amino acids to form "charged" tRNAs—

tRNAs joined to their amino acids by high-energy bonds—depends upon a set of enzymes. The tRNAs and amino acids recognize each other because the different tRNAs have different three-dimensional structures recognized by the enzymes. Each charging enzyme has two binding sites: one site holds a particular tRNA, and the other site holds the corresponding amino acid (Figure 11-13).

Special charging enzymes help tRNAs hook up with the correct amino acid.

How Do Ribosomes Begin Translation?

We have seen how each molecule of mRNA contains signals that tell the ribosomes where to start and where to stop. We have also seen that starting translation at the right place is crucial, since a mistake of only a single nucleotide would lead to a complete misreading of the information.

The first step of translation is **initiation,** which begins with the attachment of the ribosome's small subunit to the mRNA. The actual initiation site, or **start site,** in mRNA is always the codon AUG, which specifies methionine. This means that *all* recently translated polypeptides begin with methionine. In most cases, however, a special enzyme later clips away the methionine.

But how can the ribosome distinguish between the AUG meaning "start" and the AUGs that specify methionines in the middle of a polypeptide? The answer to this question differs in prokaryotes and eukaryotes. In prokaryotes, a sequence of about six nucleotides, slightly upstream from the initiation

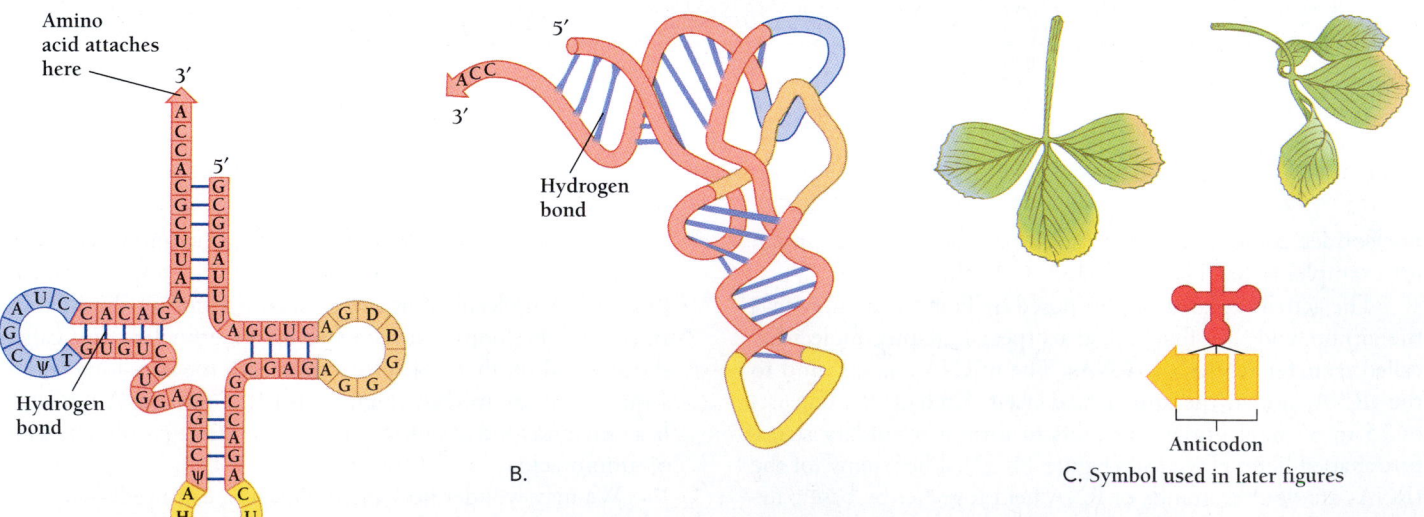

Figure 11-12 Cloverleaf-shaped tRNAs translate codons into amino acids. A. The anticodon is at the tip of the middle "leaflet." B. Like many macromolecules, tRNAs have a complex folded structure. Like so many hairpins, hydrogen bonds hold this complex shape together. C. The anticodon binds to the codon.

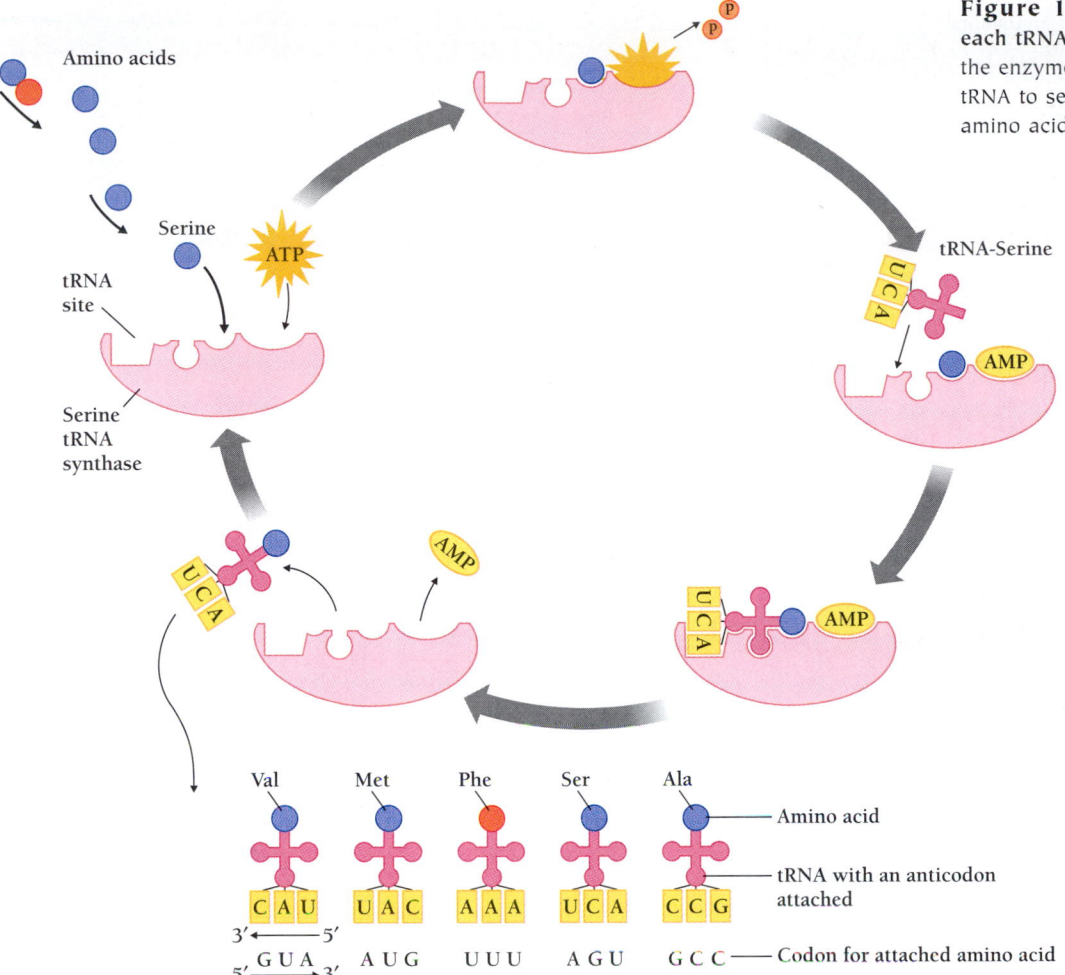

Figure 11-13 Special enzymes charge each tRNA with the right amino acid. Here, the enzyme serine tRNA synthase joins a tRNA to serine. Different tRNAs join different amino acids.

AUG, distinguishes between initiation sites and internal methionine. In eukaryotes, each ribosome simply attaches to the 5′ end of the mRNA and moves along until it encounters the first AUG, where translation then begins (Figure 11-14).

The initiation process aligns the mRNA for proper reading and moves the first amino acid into place. After the first tRNA and its attached methionine are in place, elongation of the polypeptide chain can begin.

In the first step of translation—initiation—the ribosome attaches to the start site—the codon AUG. In prokaryotes, a sequence of six nucleotides distinguishes the start AUG from the AUG codon for methionine. In eukaryotes, the start AUG is simply the first one from the 5′ end of the mRNA.

How Do Ribosomes Make Peptide Bonds?

The process of polypeptide-chain **elongation** consists of three steps (Figure 11-14): (1) putting the next amino acid into position, (2) forming a peptide bond between the growing polypeptide and the next amino acid, and (3) moving the ribosome to the next codon. Every amino acid is joined to its neighbor in exactly the same way.

Elongation, unlike initiation, requires the participation of the large ribosomal subunit (which binds to the initiator tRNA) as well as the mRNA and the small ribosomal subunit. The large subunit contains three special pockets for tRNAs, called the P site [for *p*eptide], the A site [for *a*mino acid], and the E site [for *e*xit]. The P and A sites bind two tRNAs to adjacent codons in the mRNA; the E site binds each tRNA as it exits from the ribosome. The binding of the two charged tRNAs to the large subunit is the first step of elongation.

After the P and A sites are filled, the ribosome links the attached amino acids to form a peptide bond. This linkage is the second step of elongation. The tRNA in the P site then falls away from the ribosome via the E site. Left behind is the tRNA in the A site and its attached amino acid, which is now linked to the entire polypeptide chain. In the third step, the ribosome moves forward so that the tRNA in the A site with its attached polypeptide chain ends up in the P site. The

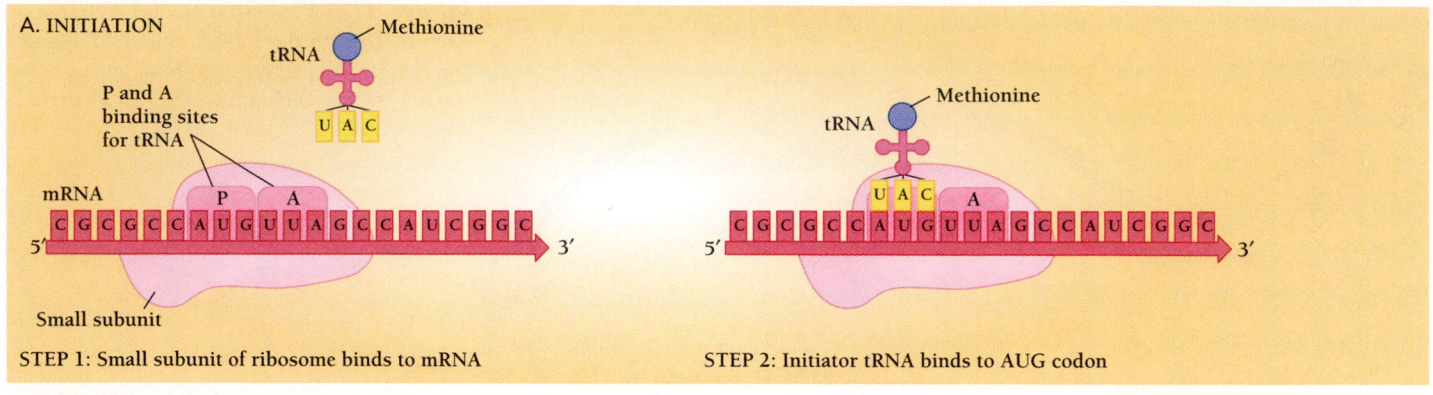

A. INITIATION

STEP 1: Small subunit of ribosome binds to mRNA

STEP 2: Initiator tRNA binds to AUG codon

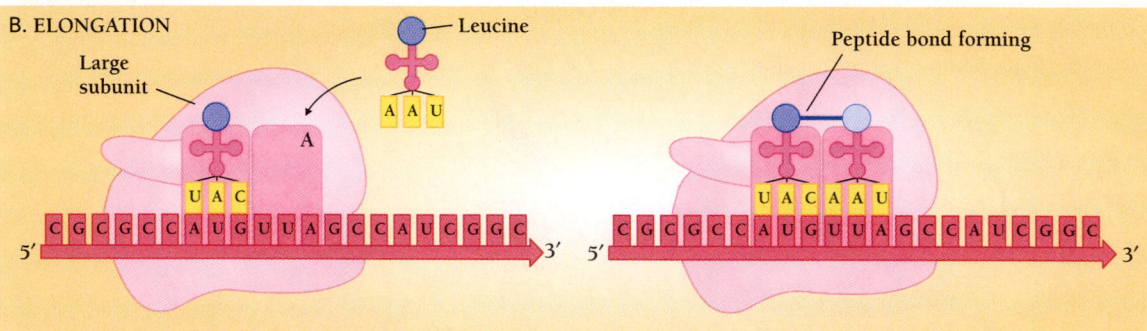

B. ELONGATION

STEP 1: Large subunit of ribosome binds to mRNA; next tRNA with amino acid leucine binds to UUA codon

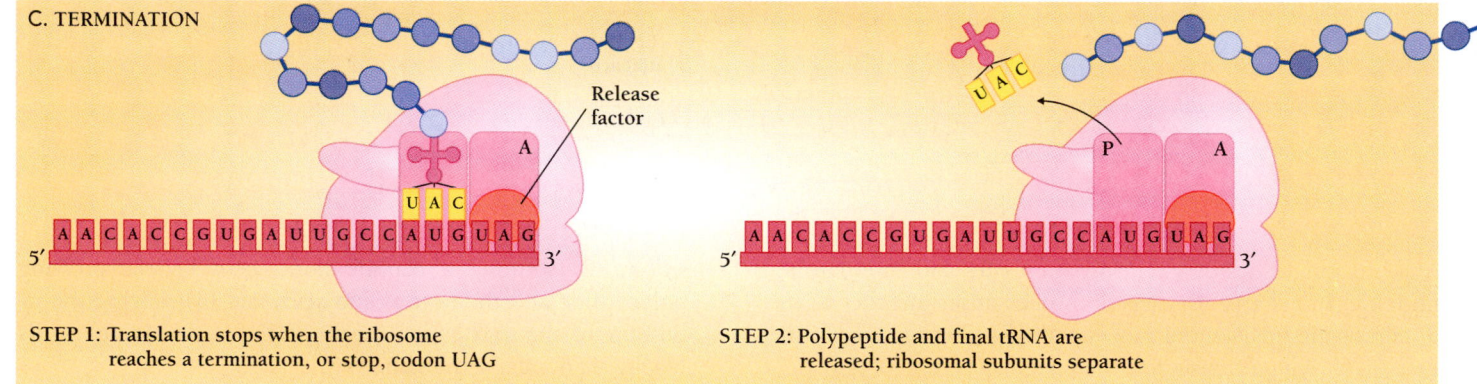

C. TERMINATION

STEP 1: Translation stops when the ribosome reaches a termination, or stop, codon UAG

STEP 2: Polypeptide and final tRNA are released; ribosomal subunits separate

▲
Figure 11-14 Translation: initiation, elongation, and termination. A. During initiation, the small subunit of the ribosome binds to AUG (the start site) of the mRNA and moves the amino acid methionine into place. B. During elongation, a second charged tRNA arrives and the large subunit of the ribosome—holding the two charged tRNAs side by side in the P and A sites—binds the two amino acids together. The ribosome and its attached polypeptide chain then move forward on the mRNA, and another charged tRNA moves into place. The tRNA released from the P site binds temporarily to the E site as it exits from the ribosome. (For clarity, the drawing does not show the E site.) C. When the ribosome reaches a stop codon, a protein called release factor binds to the stop codon, and an enzyme cuts the newly made polypeptide chain from the last tRNA. The ribosomal subunits then separate from each other and become available for another round of initiation and elongation.

repositioning of the tRNA and the attached polypeptide to the P site completes the three steps of the elongation cycle.

The cycle now repeats over and over until all the codons in the mRNA are read. In a bacterium, each elongation cycle (the addition of one amino acid) takes about 1/20th of a second, so the synthesis of an average-sized polypeptide (including 400 to 1200 amino acids) takes 20 to 60 seconds.

During elongation, amino acids are joined one after the other in exactly the same way. A tRNA first moves an amino acid into position next to the last amino acid. Once two amino acids are in place, an enzyme in the ribosome joins them together with a peptide bond. The ribosome then moves forward on the mRNA, repositioning the tRNA and its attached polypeptide chain.

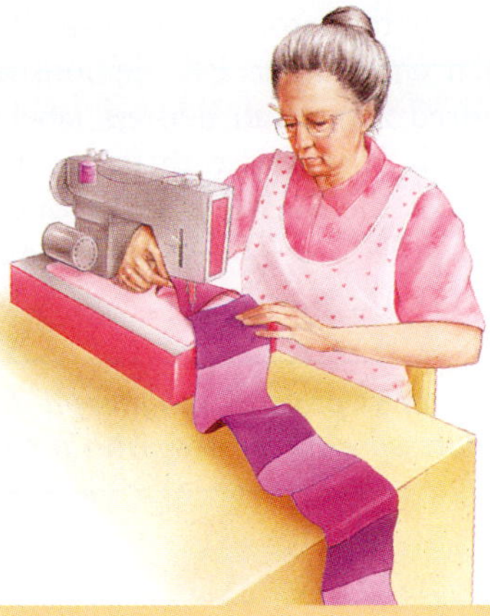

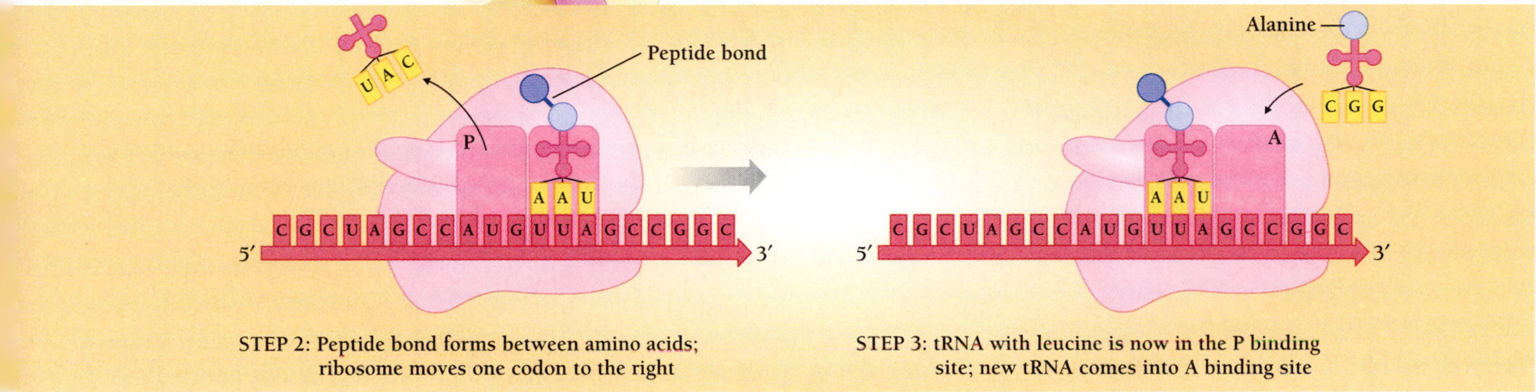

Peptide bond
U A C
P
A A U
C G C U A G C C A U G U U A G C C G G C
5′
3′
STEP 2: Peptide bond forms between amino acids; ribosome moves one codon to the right
Alanine
C G G
A
A A U
C G C U A G C C A U G U U A G C C G G C
5′
3′
STEP 3: tRNA with leucine is now in the P binding site; new tRNA comes into A binding site

Several Ribosomes Can Read a Single Strand of mRNA Simultaneously

As soon as a ribosome has moved far enough away from the AUG start site, the start site is free for another ribosome to attach and begin translation. As a result, several ribosomes, spaced as little as 80 nucleotides apart, may be attached to a single mRNA at one time, each one overseeing the production of identical polypeptides. This complex is called a **polysome,** short for polyribosome (Figure 11-15). In this way, the cell can synthesize many copies of a polypeptide at once, greatly speeding the mass production of proteins. We will see later in this chapter, for example, that an *E. coli* bacterium can begin producing enzymes to break down milk sugar (lactose) as soon as the bacterium detects the milk sugar.

How Does Polypeptide Synthesis Stop?

The appearance of a stop codon in the mRNA signals the end of a polypeptide chain, a process called **termination.** When the ribosome reaches one of the three stop codons (UAA, UAG, or UGA), there will be no corresponding tRNA. Instead a protein called a **release factor** binds to the stop codon (Figure 11-14). The enzyme that makes peptide bonds during elongation now cuts the polypeptide from the last tRNA, releasing the completed polypeptide into the cytoplasm. The ribosomal subunits then separate from each other and become available for another round of initiation and elongation.

All of the steps of protein synthesis, including transcription and translation, are summarized in Figure 11-16. We can

0.1 μm

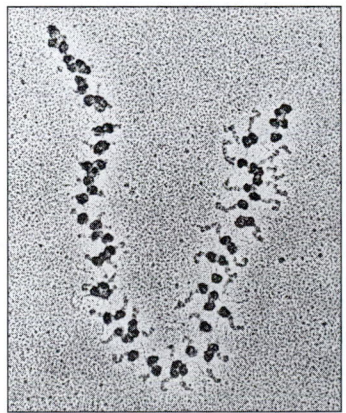

Figure 11-15 **A large polysome.** Individual ribosomes move along a strand of mRNA, translating as they go. The ribosomes on the far left are just beginning translation and the polypeptide strands are not yet visible. The ribosomes on the right and in the middle have produced visible polypeptide strands. *(Oscar L. Miller, University of Virginia)*

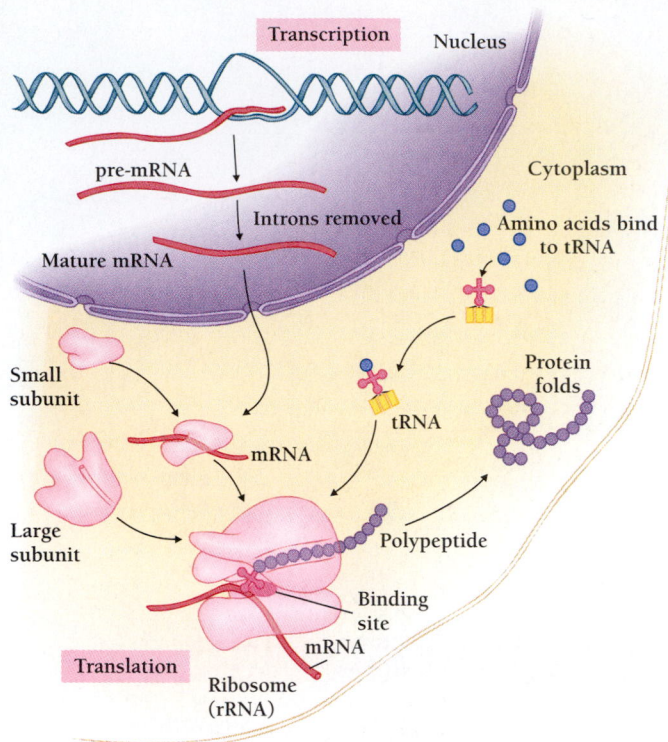

Figure 11-16   Transcription and translation in a eukaryotic cell.

see that once a polypeptide chain is assembled, it must then fold into a functioning protein.

When the ribosome reaches a stop codon, a protein called release factor binds to the stop codon and an enzyme cuts the newly made polypeptide chain from the last rRNA.

How Do Cells Direct Newly Made Proteins to Different Destinations?

Most of the proteins produced on the ribosomes stay in the cytosol, but many must find their ways to various cytoplasmic organelles, to the plasma membrane, or to the outsides of cells. But how do newly made proteins find their way to the right destinations?

The answer is that many newly made polypeptides have "address tags," additional blocks of amino acids that specify where the protein should go in the cell. In proteins destined for the plasma membrane or for secretion outside of the cell, for example, a **signal peptide** of 4 to 12 amino acids directs the newly made polypeptide into the space (or "lumen") that is surrounded by the endoplasmic reticulum (ER) during protein synthesis. In one experiment, molecular biologists attached

a signal peptide to a protein not normally sent to the ER. The protein then entered the ER, demonstrating that the signal peptide indeed acts as an address label. Other polypeptides contain signals that direct them to the mitochondria, the chloroplasts, or the nucleus. In the case of secreted proteins, the signal peptide is removed. In other cases, the address tag remains part of the protein.

Cells direct proteins to different destinations by addressing the protein with a signal peptide.

HOW DO CELLS REGULATE GENE EXPRESSION?

We have seen now how cells express the information in the DNA. But cells contain thousands of genes. Some of these are expressed as polypeptides all the time, some only occasionally, and many are never expressed. Prokaryotes transcribe almost all of their genes into RNA. But eukaryotes regulate gene expression more closely. Cells, especially eukaryotic cells, do not make all of their polypeptides at once for the same reason that we do not leave all of the water taps in our homes flowing at once. Just as we turn on hot or cold water according to our needs, a eukaryotic cell produces different proteins according to its needs. Further, just as a water tap can be turned on full blast or at a trickle, a cell can turn out thousands of copies of a polypeptide, or only two or three copies.

Eukaryotes, most of which are multicellular, face an additional problem. The dramatic cell specialization that characterizes most multicellular organisms requires the differential expression of genes in cells that have identical genomes. A human, for example, consists of more than 200 distinguishable types of cells. Nearly every one of these cells contains the same set of about 100,000 genes. Yet each kind of cell expresses only a fraction of its 100,000 genes.

A typical eukaryotic cell transcribes only about 20 percent of its DNA into RNA. In a human skin cell, for example, this 20 percent would include information for basic life-support proteins that all human cells need—such as membrane proteins, ribosomal proteins, and metabolic enzymes—as well as information for specialized proteins, such as the stretchy collagen that makes skin so tough. In a skin cell, the 80 percent of the DNA not transcribed would include, for example, the genes for α- and β-globin, various muscle proteins, and a host of digestive enzymes.

Cells can regulate the expression of genes at many levels. The most efficient and most usual way is so-called **transcriptional control,** in which cells increase or decrease the amount of mRNA transcribed from the DNA. Every cell in the body has the gene for the globin polypeptides, for example, yet only the developing red blood cells transcribe them. In every other kind of cell in the body, the globin genes sit unused.

BOX 11.3

Antibiotics and Protein Synthesis

Antibiotics are usually complex organic molecules that allow bacteria and fungi to dispose of (and often to eat) their competitors. The medical uses of antibiotics began at least 2500 years ago, when the Chinese began to use the moldy curd of soybeans to treat boils and similar infections. In the 20th century, the pursuit of antibiotics as a treatment for human disease started with Alexander Fleming. In 1928 Fleming discovered that the growth of bacteria in one of his experiments had been stopped by the presence of a green mold.

Fleming identified the mold as a species of *Penicillium*. Later, other scientists isolated the active substance and named it penicillin. Penicillin and its many derivatives have remained the most frequently used antibiotics. But dozens of other antibiotics have been isolated from molds or bacteria.

Penicillin acts by interfering with cell wall synthesis in many bacteria. Other antibiotics act by interfering with protein synthesis. This characteristic has made them extremely important tools for dissecting the steps of protein synthesis. Each of the steps of elongation, for example, is subject to inhibition by a different set of antibiotics: in prokaryotic cells (bacteria), tetracycline interferes with positioning of the ribosome, chloramphenicol with peptide bond formation, and erythromycin with the movement of the ribosome on the mRNA (Figure A).

Some of the most useful antibiotics interfere with prokaryotic protein synthesis without affecting that in eukaryotic (human) cells. For example, while eryth-romycin blocks translocation of the growing peptide chain in bacteria, it does not inhibit this step during protein synthesis on eukaryotic ribosomes. This specificity reflects differences in the proteins of prokaryotic and eukaryotic ribosomes.

Antibiotics have dramatically decreased the number of deaths caused by infectious disease. They can not only cure an initial infection but also prevent long-term consequences such as rheumatic fever, which, until this century, often resulted from strep throat. In addition, killing the bacteria that make one person sick prevents the spread of an infection. Finally, the ability to deal with infections has made surgery safer and allowed the development of operations ranging from the removal of cancers to the replacement of vital organs.

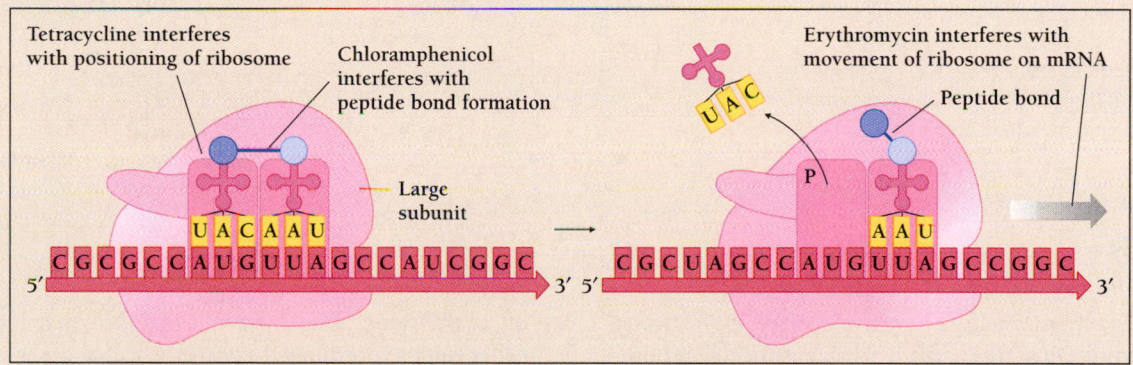

Figure A **Each of the steps of elongation is inhibited by a different set of antibiotics.** In prokaryotic cells, tetracycline interferes with positioning, chloramphenicol with peptide bond formation, and erythromycin with the movement of the ribosome along the mRNA.

Cells can regulate the expression of genes in other ways. A cell may process pre-mRNA in alternative ways—either by altering the patter of exon splicing or by changing the efficiency of export to the cytoplasm (**posttranscriptional control**). Or a cell may vary the rate at which it translates individual types of mRNAs (**translational control**). Finally a cell can modify the structure of a protein and change the rate at which it turns over (**posttranslational control**). The addition of phosphate groups to particular sites in proteins is an especially important form of posttranslational control, as we will discuss in Chapter 36.

Cells regulate the expression of genes in many ways, but transcriptional control is the most efficient and most common way.

How Do Prokaryotes Regulate Gene Expression?

Bacteria provide the best-understood examples of gene regulation. Most of the time, bacteria regulate gene expression by controlling transcription. Like other organisms, bacteria must

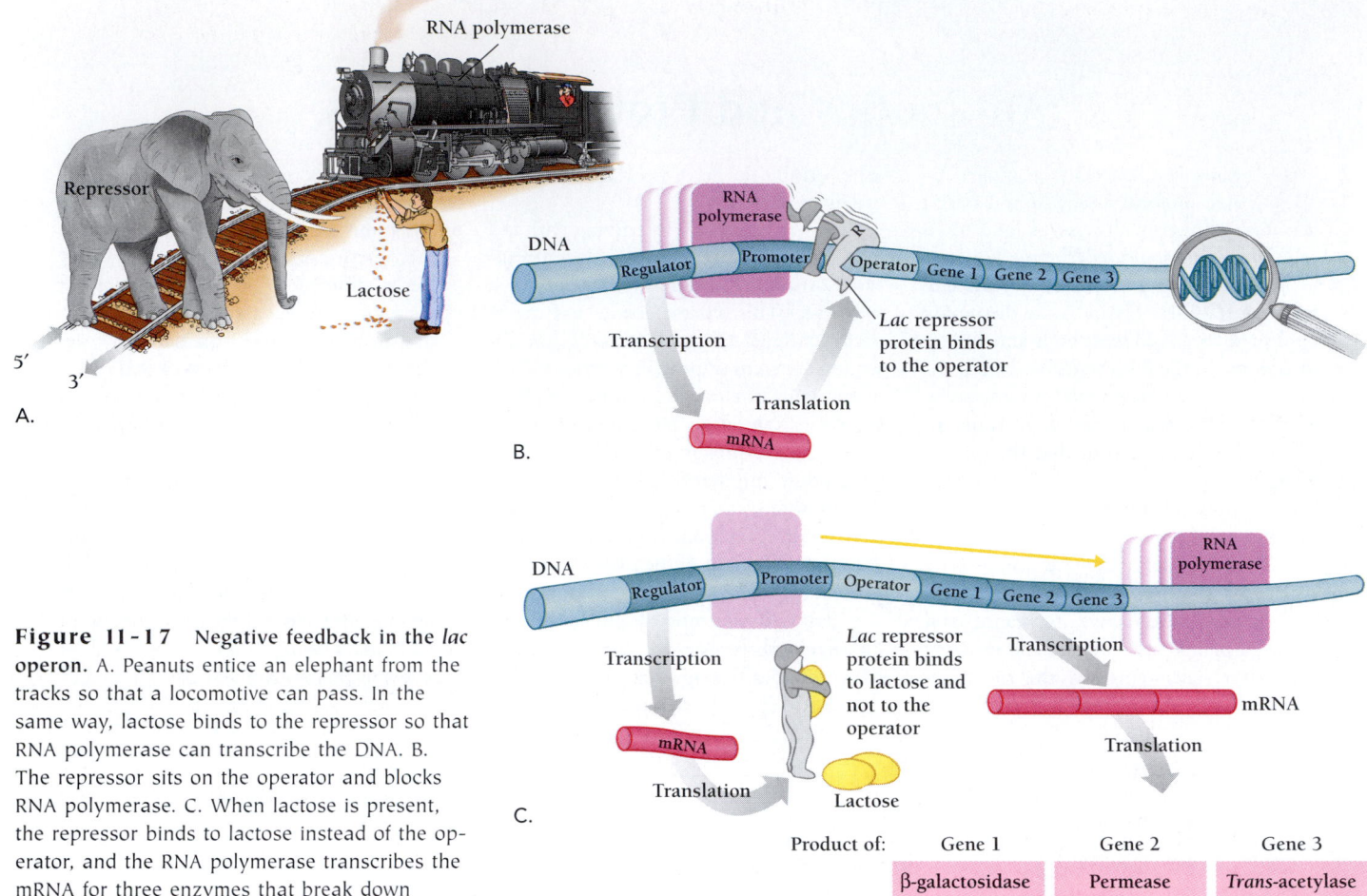

Figure 11-17   Negative feedback in the *lac* operon. A. Peanuts entice an elephant from the tracks so that a locomotive can pass. In the same way, lactose binds to the repressor so that RNA polymerase can transcribe the DNA. B. The repressor sits on the operator and blocks RNA polymerase. C. When lactose is present, the repressor binds to lactose instead of the operator, and the RNA polymerase transcribes the mRNA for three enzymes that break down lactose.

respond to their environment. They have evolved sophisticated and sensitive molecular mechanisms that ensure the production of enzymes when they need them. When a bacterium's environment contains high concentrations of amino acids, it devotes most of its energy to rapid division and does not squander energy making more amino acids. If no amino acids are available, however, the bacterium rapidly synthesizes amino acids from simple molecules, usually taking carbon atoms from sugar and nitrogen atoms from ammonia, for example.

To do all this, the bacteria must make the right enzymes at the right time. For example, the synthesis of just one amino acid, histidine, from simple carbon- and nitrogen-containing compounds requires the participation of seven different enzymes. Conveniently, the genes for these seven enzymes lie right next to each other on the bacteria's DNA.

Molecular biologists were surprised to find that the cell transcribes the seven genes into a single huge molecule of mRNA instead of into seven separate lengths of mRNA. Transcribing the seven genes together ensures that all the enzymes needed to make the final product (histidine) are turned on and off at the same time. A stretch of DNA that includes several genes under coordinated control is called an "operon."

Negative regulation in prokaryotes

The best-studied operon in prokaryotes regulates the use of lactose (the main sugar in milk) by *E. coli,* a common bacterium in the intestines of humans and other mammals. This elegant system enables bacteria to produce costly enzymes only when they are needed. As long as glucose is available in the intestines, *E. coli* will not digest the milk sugar lactose. When glucose is absent, however, *and* the host organism (that's you) drinks milk, then *E. coli* begins synthesizing three proteins that allow it to digest lactose. One of these is the enzyme β-galactosidase (which splits lactose into the sugars galactose and glucose).

The three genes encoding these three proteins lie next to each other on the DNA of *E. coli* and are part of the lactose operon, or ***lac* operon.** *E. coli* transcribes the DNA of the *lac* operon into an mRNA that encodes all three proteins.

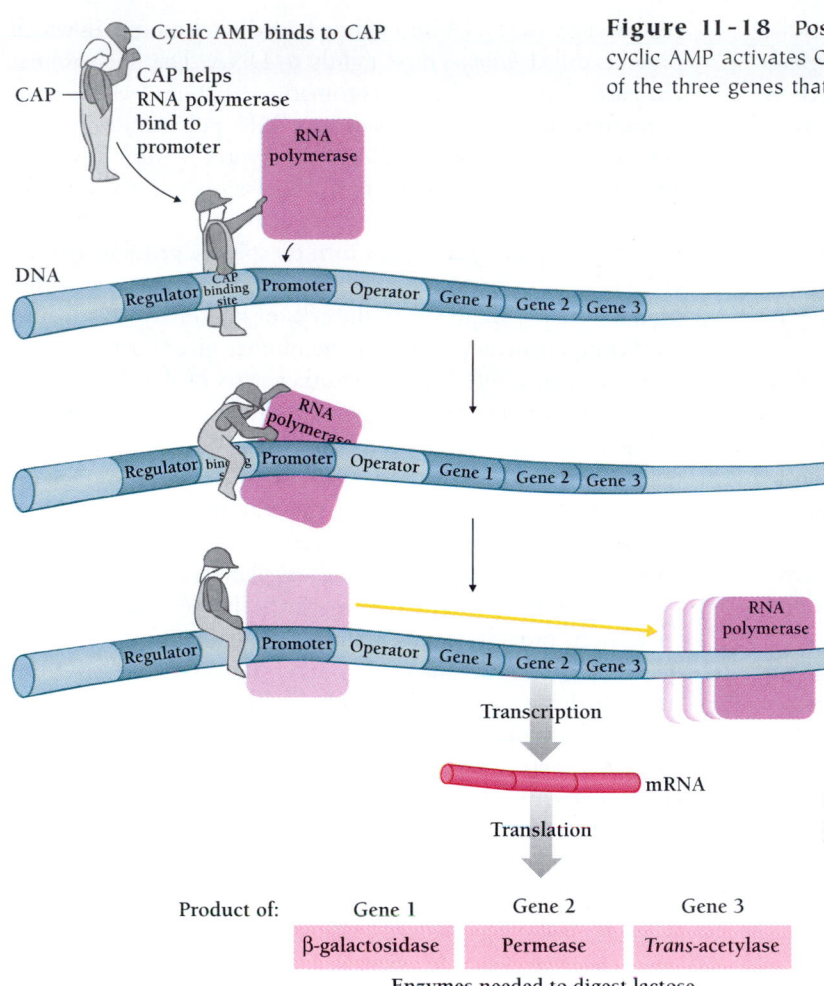

Figure 11-18 **Figure 11-18**   Positive feedback in the *lac* operon. In the presence of glucose, cyclic AMP activates CAP, which binds to the promoter and stimulates transcription of the three genes that code for lactose-digesting enzymes.

The operator overlaps with the operon's *promoter,* which, recall, is the DNA sequence that signals the start of a gene. When the *lac* repressor binds to the operator, it prevents RNA polymerase from binding to the promoter and transcribing the three enzyme genes in the operon (Figure 11-17B). The *lac* repressor was the first example of a protein that regulates gene expression. The operator was the first example of a regulatory site.

How do the repressor and operator allow lactose to control gene expression in the *lac* operon?

The key to understanding the induction of the *lac* operon is the realization that the *lac* repressor has two binding sites: one site binds to the operator (a DNA sequence), and the other binds to lactose (a sugar). But the repressor cannot bind to both at the same time. When the repressor binds to lactose, it can no longer bind to the operator and prevent transcription (Figure 11-17C). RNA polymerase can then bind to the *lac* operon's promoter and transcribe the RNA that specifies the three proteins needed to digest lactose. The *lac* repressor is an example of a **negative regulator,** a molecule that inhibits transcription.

When lactose binds to the lac *repressor (a protein), the* lac *operon's promoter becomes unblocked, and RNA polymerase can transcribe the three genes that encode proteins needed to digest lactose.*

Positive regulation in prokaryotes

Bacteria also use **positive regulators,** or activators, molecules that increase the rate of transcription. One such activator is the **catabolite activator protein,** or **CAP** (Figure 11-18). Like the *lac* repressor, CAP has two binding sites. One site binds

In the absence of lactose, each bacterium contains only about three molecules of β-galactosidase. However, when glucose is absent and lactose becomes available, the lactose stimulates, or **induces,** the expression of the *lac* operon, and the bacterium synthesizes about 3000 molecules of β-galactosidase. Lactose is called the "inducer" because it induces gene expression.

In 1960, two French molecular biologists—François Jacob and Jacques Monod—discovered how *E. coli* bacteria control these dramatic changes. Jacob and Monod discovered two genes that regulate the expression of the *lac* operon. One gene, which lies a short distance from the gene for β-galactosidase, is the gene that encodes the ***lac* repressor,** a protein that prevents the expression of the *lac* operon (Figure 11-17A). The other gene, a sequence of 21 nucleotides called the **operator,** lies immediately next to the gene for β-galactosidase and is part of the operon. The operator, we now know, is not actually a separate gene because it does not encode a protein (or RNA). It is a regulatory site to which the *lac* repressor can bind.

to a particular sequence in DNA, and the other binds cyclic AMP, a nucleotide derived from ATP. Cyclic AMP is a *signaling molecule,* a molecule that conveys signals from one cell to another. In bacteria, cyclic AMP is a "hunger" signal: high concentrations of cyclic AMP tell the cell that glucose is unavailable and that the cell needs to get energy from another source, such as lactose.

In the presence of lactose, CAP binds to the *lac* promoter and stimulates the transcription of the *lac* operon. When the cell has lots of glucose and little cyclic AMP, however, the *lac* operon is not active even in the presence of lactose.

The combination of negative control (by the *lac* repressor) and positive control (by the CAP activator) allows bacteria to respond quickly to changes in their environment. When lactose is present but glucose is not, *E. coli* quickly begin to make the enzyme β-galactosidase. Similarly, when glucose becomes available or when lactose is exhausted, *E. coli* almost immediately stops production of the enzyme. This system saves the bacteria from expending energy making polypeptides they do not need.

CAP activates transcription not only of the *lac* operon but also of several other genes. All of these genes code for enzymes needed to obtain energy from molecules other than glucose. CAP provides an example of how a single activator can affect the transcription of genes that are physically separated but functionally related.

The activator protein CAP increases the transcription rate for certain mRNAs.

How Do Eukaryotes Regulate Gene Expression?

Until the late 1980s, molecular biologists knew far more about regulation in prokaryotes than in eukaryotes. More recent work, however, has resulted in an explosion of information about gene regulation in eukaryotic cells. We now know that eukaryotes control gene expression at many more levels than prokaryotes. Nonetheless, as in prokaryotes, most regulation occurs at the level of transcription.

The major differences between regulation in prokaryotes and eukaryotes result from the presence, in eukaryotes, of a nuclear membrane and of a much larger amount of DNA. The presence of the nuclear membrane has several important consequences. One is that only proteins in the nucleus can contribute to regulation. Another consequence of the nuclear membrane is that the pre-mRNA can be modified before it moves into the cytoplasm. Posttranscriptional regulation may occur in any of the steps involved in modifying the pre-mRNA.

Eukaryotes also have much more DNA than prokaryotes. Each human cell, for example, contains about 1000 times as much DNA as is in an *E. coli* bacterium. Recall from Chapter 10 that to fit the DNA of a eukaryotic cell into its nucleus, the cell packages the DNA into tightly folded nucleosomes in which proteins called *histones* bind tightly to DNA. The nucleosomes are part of more elaborate chromatin structures, which limit transcription by restricting access to RNA polymerase. In general, the highly packaged DNA of eukaryotes is much less available for transcription than that of prokaryotes. As a result, eukaryotic transcription requires an elaborate set of activators. In eukaryotes, these activators turn on specific genes in specific cell types at specific times. In mammals, for example, genes that promote milk production in the cells of mammary glands turn on during pregnancy and after the mother gives birth, but not at other times of life and not in other parts of the body.

Eukaryotes also have many intracellular compartments and elaborate mechanisms for the distribution of newly made proteins. Sending the right proteins to the nucleus, for example, is especially important for building the chromatin and regulating gene expression. Varying the delivery of these regulatory proteins is another way by which cells regulate gene expression.

The protein tubulin, from which the cell builds its cytoskeleton, provides a nice example of regulation at the level of translation. The microtubules of the cytoskeleton form through the assembly of α- and β-tubulin polypeptides into αβ-tubulin "dimers" [Greek, *dimeres* = have two parts]. When enough dimers accumulate in the cytoplasm, they activate an enzyme that blocks the mRNA on ribosomes that are producing tubulin polypeptides. Once the ribosomes stop producing α- and β-tubulin polypeptides, the concentration of dimers in the cytoplasm drops, and the enzyme stops interfering with the production of α- and β-tubulin.

Eukaryotes can regulate gene expression in more ways than prokaryotes because of the presence of the nuclear membrane, because of the way the DNA is folded and packaged, and because of the many compartments in eukaryotic cells.

How complex is the problem of gene regulation in eukaryotes?

Just as in prokaryotes, the best-studied cases of transcriptional regulation in eukaryotes depend on the interaction of proteins with regulatory sites on the DNA. Unlike prokaryotic DNA, however, most eukaryotic DNA is largely "repressed" because of its tight association with histones. Most eukaryotic regulation therefore depends on specific activators of transcription.

Regulatory proteins may affect transcription in two ways. The first class of regulators change the rate of transcription: some of these resemble the *lac* repressor in decreasing initiation by RNA polymerase; others resemble the CAP protein in increasing initiation. The second class of regulators (sometimes called "chromatin-remodeling machines") produce permanent changes in chromatin that make a particular gene more or less accessible to the transcriptional machinery. Such a change in chromatin structure can persist long after the regulator itself has disappeared.

Many of the proteins that contribute to transcriptional regulation bind directly to specific sequences in DNA. Among these sequences are (1) *promoters,* which specify the starting point and direction of transcription; (2) *response elements,* also called *enhancers* because they usually stimulate the transcription of neighboring DNA without themselves serving as promoters; (3) *silencers,* which help inhibit transcription; and (4) *termination signals.*

Some regulatory proteins bind not to DNA but to other proteins, forming large assemblies of proteins. Such a protein assembly can interact with activator proteins around a set of enhancer elements in a specific gene.

The regulation of the 100,000 or so genes in the human genome is an enormously complex problem. In *E. coli,* researchers have estimated that as many as 10 percent of the approximately 1000 genes contribute to transcriptional regulation. Gene regulation in the mammalian nervous system is certain to be far more complex. Biologists estimate that mammals have at least 5000 transcriptional activators and a corresponding number of response elements (silencers and enhancers). Each transcriptional activator may interact with more than one response element, and each response element can respond to more than one activator. The number of possible interactions at the transcription level alone is huge. Yet the sets of 10,000 to 100,000 kinds of protein molecules in each of the trillion or so nerve cells in the human brain probably require even more complex regulation than that.

In eukaryotes, the huge numbers of regulatory proteins and their corresponding binding sites in DNA allow nearly unlimited varieties of interactions. The problem of gene expression is enormously complex.

A single transcription factor may affect many genes

The hormone testosterone, which works by binding to a transcriptional activator, shows the diverse effects of a single protein. In human and other mammalian embryos, testosterone is one of a group of hormones called androgens that stimulate the formation of the male genitalia (penis, scrotum, and the internal ducts that carry sperm). Later, at puberty, androgens stimulate the development of male secondary sexual characteristics—male distribution of fat, muscle, and body hair; deepening of the voice; and other patterns of growth and behavior.

We know that androgens accomplish all this by binding to an **androgen receptor.** This androgen receptor binds both to testosterone and to response elements in the DNA, thereby influencing the expression of genes related to the secondary sexual characteristics.

In a condition called **testicular feminization,** a chromosomal (XY) male produces testosterone but no androgen receptors. Despite the presence of the Y chromosome and plenty of testosterone, such an individual is insensitive to androgen (testosterone) and develops into a woman, with breasts and female external genitalia. However, she has no uterus, or only a rudimentary one, and no ovaries. She has testes, but these are high in the body cavity, rather than descended. This XY individual is female in nearly every other respect. Testicular feminization shows how the actions of testosterone and its receptor increase the rate of transcription of a host of important genes.

The absence of a single transcription factor—the androgen receptor, for example—can greatly alter gene expression and the resulting phenotype.

In this chapter, we have seen how the coded information in the gene flows from DNA to RNA to polypeptides and how cells regulate that flow. In the next chapter, we will see that genetic information (in the form of individual genes) is not confined to the DNA in the nucleus but can move about within the cell or from cell to cell. The consequences of this genetic mobility are enormous.

Study Outline with Key Terms

DNA and RNA carry information in the sequences of nucleotides. Protein synthesis translates this information into a sequence of amino acids. The **genetic code** is the **universal** dictionary that specifies the relationship of the nucleotide sequence to the amino acid sequence. A **codon** is a group of three nucleotides that specifies a single amino acid. There are 64 codons. Of these, 61 specify the 20 amino acids and

3—**nonsense,** or **stop, codons**—signal the end of a polypeptide.

According to Francis Crick's **central dogma:** "DNA specifies RNA, which specifies proteins." Cells **transcribe** information from DNA into RNA and **translate** information from **messenger RNA (mRNA)** into protein. **RNA polymerase** copies information from one strand of DNA into RNA. The starting point for transcription is called a **promoter.** In eukaryotes, **transcription factors** lead the RNA polymerase to the right promoter. Eukaryotic and prokaryotic RNA polymerases stop transcription at **terminators.**

In prokaryotes, RNA polymerase produces mature RNAs directly. In eukaryotes, cells modify the initially transcribed pre-mRNAs into functional mRNAs. Such modifications include cutting out **introns** and splicing together the remaining **exons** to form a patchwork mRNA.

Ribosomes, made of **ribosomal RNAs (rRNAs)** and proteins, are the molecular factories that assemble amino acids into polypeptides. Several ribosomes may associate with a single mRNA molecule to form a **polysome.** Before ribosomes can arrange the sequence of amino acids according to the information encoded in mRNA, however, the amino acids are first linked to adapter molecules called **transfer RNAs (tRNAs).** Each tRNA contains an **anticodon,** a specific nucleotide sequence that forms base pairs with a codon in mRNA. The pairing of specific tRNAs with their appropriate amino acids is the task of enzymes that recognize one amino acid and the corresponding tRNA and link them together.

The translation of information from mRNA into the sequence of amino acids in a polypeptide occurs in three distinct steps, called initiation, elongation, and termination. **Initiation** positions the mRNA in the correct **reading frame.** Initiation always starts with AUG, which is both a signal for methionine and the **start site. Elongation** consists of three steps: (1) putting the next amino acid in place, (2) forming a peptide bond, and (3) moving the ribosome to read the next codon. **Termination** occurs when the ribosome encounters a stop codon to which a protein called a **release factor** has bound.

Eukaryotic cells sort newly made polypeptides to different addresses such as the cytosol, organelles, membranes, or locations outside of the cell. The destination of a polypeptide depends on the presence of **signal peptides.**

Most molecular biologists now say that a gene is a DNA sequence that is transcribed as a single unit and encodes either a single polypeptide or a set of related polypeptides. Cells regulate the expression of genes in many ways, including **posttranscriptional control, translational control, and posttranslational control.** But **transcriptional control** is the most efficient and the most common way.

Two types of transcriptional control in prokaryotes are positive regulation and negative regulation. The *lac* **operon** is an example of **negative regulation.** When lactose binds to the *lac* **repressor** (a protein), the *lac* operon's promoter becomes unblocked, and RNA polymerase can transcribe the three genes that encode enzymes for digesting lactose. We say lactose **induces** the expression of the *lac* operon. When lactose is absent, the *lac* repressor binds to a regulatory site called the **operator** and prevents transcription.

CAP is an example of an **activator,** or **positive regulator.** CAP increases the transcription rate for certain mRNAs, provided that other regulatory mechanisms will allow these mRNAs to be transcribed at all.

Eukaryotes can regulate gene expression in more ways than prokaryotes because of the presence of the nuclear membrane, because of the way the DNA is folded and packaged, and because of the many compartments in eukaryotic cells. In eukaryotes, the huge numbers of transcription factors and regulatory sites allow nearly unlimited varieties of interactions. The problem of gene expression is enormously complex. In **testicular feminization,** the absence of a single transcription factor—the **androgen receptor,** for example—can greatly alter gene expression and the resulting phenotype.

Review and Thought Questions

Review Questions

1. What is the central dogma?
2. What is the difference between the template, or antisense, strand of DNA and its mate, the sense strand? Which one is transcribed into RNA? What would happen if the sense strand were transcribed?
3. What kinds of proteins help RNA polymerase recognize the right promoter? What is a promoter?
4. How do tRNAs mediate between the mRNA and the amino acids? What does "t" stand for? What did Francis Crick call tRNAs before they were discovered?
5. What is the difference between positive regulation and negative regulation of gene expression? Give an example of each.
6. Why do bacteria need to regulate the expression of their genes?
7. What is a regulatory site? Give an example of one.
8. Why do eukaryotes need to regulate the expression of their genes? Why do multicellular organisms need to regulate gene expression?

Thought Questions

9. Why is it hard to clone humans and other mammals? If you could clone someone, whom would you choose to clone? How many copies of the person would you make? Would you expect them to be completely identical in every respect? What would you hope to accomplish by this? Would you send them all to college?
10. Suppose that you spent half your girlhood and all of your teenage years training to be a sprinter. You have qualified for the Olympic team and will compete in the 2000 Olympics. The night before the first heat, your coach tells you that your chromosome test has revealed that you have a Y chromosome and you will not be allowed to compete. You have never thought of yourself as anything except a girl, now a young woman. How do you feel? How will this change your life?
11. If you are a heterosexual man, what would you do if you discovered that your wife of 5 years had a Y chromosome?

About the Chapter-Opening Image

The sewing machine is a metaphor for the ribosome, a tiny molecular machine that stitches amino acids together into polypeptides.

 On-line materials relating to this chapter are on the World Wide Web at **http://www.harcourtcollege.com/lifesci/aal2/**

12 Jumping Genes and Other Unconventional Genetic Systems

Jumping Genes and Indian Corn

In the 1940s, a geneticist named Barbara McClintock made a discovery that would eventually revolutionize molecular biology, a science that did not even exist yet (Figure 12-1). Like Gregor Mendel, McClintock was so much ahead of her time that her research made no impact on the world of biology for 30 years.

McClintock began her career in 1918 at Cornell University, where she was first an undergraduate student, then a graduate student, and finally a researcher. In the 1920s and 1930s, geneticists there noticed that the normal purple color of corn kernels switched to pale yellow or white as a result of a mutation in a pigment gene. Multicolored "Indian" corn is normal corn: the pale corn we buy at the store has been bred to prevent the expression of these purple pigment genes. Oddly, however, the mutation that turns purple kernels yellow or white can reverse itself, or partly reverse itself, from generation to generation.

Each kernel of corn on a cob contains a separate, unique individual—the result of a separate pairing of sperm and egg, a separate pairing of maternal and paternal chromosomes. As a result, some kernels on a cob might have the mutation and some might have it reversed: some kernels are white and some are purple. Still others may be streaked or spotted—the result of partial expression of the pigment gene. When McClintock was at Cornell, no one understood how these "unstable mutations" could flip back and forth so easily.

In 1941, McClintock moved her own research on corn to Cold Spring Harbor Laboratory on Long Island, New York. She was especially interested in broken chromosomes. She had noticed a corn plant in which all of the chromosomes always broke in the same place—an oddity, since chromosomes usually break randomly. McClintock called the place where the chromosome always broke the *Ds* gene. *Ds* was short for "dissociator," the gene that took apart (dissociated) the chromosome. She soon discovered that the *Ds* gene could move around on the chromosome, or "transpose," from generation to generation. Then came an even more startling revelation. McClintock noticed that the same corn plants in which *Ds* had moved also acquired the unstable pigment mutation.

That connection suggested to McClintock that the unstable pigment mutation might be caused by the movement of the *Ds* gene into the middle of the pigment gene. Taking another step, she saw that if the *Ds* gene jumped back out of the pigment gene, the pigment gene would function again, coloring the kernel purple. This would explain how the kernels seem to gain and lose the mutation, switching back and forth between a normal phenotype and an abnormal one, from generation to generation. McClintock's hypothesis not only explained how unstable mutations worked, but introduced a completely new idea to the science of genetics: the idea that genes could move. She coined the word **transposon** to describe these genes that jumped around the genome. Others would come to call them "jumping genes."

In time, McClintock found more complex examples of transposons. She also realized that just as the *Ds* gene could influence the expression of the pigment gene, regulatory genes could finely regulate the expression of structural genes. McClintock was the first person to recognize that one gene could regulate the expression of another. It was McClintock who first discovered promoter and

Gene Kelly: The Kobal Collection; corn: Runk/Schoenberger from Grant Heilman

suppressor genes. Unfortunately, she discovered them in corn, an organism far more complex than the bacteria in which Jacob and Monod discovered the *lac* operon. The system of regulation that McClintock discovered was far more difficult—both to understand and to explain—than the *lac* operon, and most geneticists did not know what to make of her results.

McClintock was nonetheless a highly respected geneticist. In 1931, recall, she and Harriet Creighton had demonstrated that chromosomes cross over, exchanging genes during meiosis (Chapter 9). The two women had provided the conclusive evidence for the theory of chromosomal inheritance. Yet, in 1951, when McClintock presented her work on transposons to a meeting of geneticists specifically interested in mutations, her talk "met with stony silence," as McClintock's biographer Evelyn Fox Keller wrote in her biography of McClintock, *A Feeling for the Organism.* "With one or two exceptions, no one understood. Afterward, there was mumbling—even some snickering—and outright complaints. . . . It was impossible to understand." McClintock knew her work was complex, but she had not anticipated such rejection. "It was such a surprise that I couldn't communicate," she told Keller. "It was a surprise that I was being ridiculed, or being told I was really mad."

The idea that genes can move from one part of the genome to another is not, in fact, extraordinarily difficult to understand. But it was extraordinarily difficult for geneticists to accept. McClintock's colleagues just assumed that—except for crossing over and other changes that occur during meiosis—the genome is a rigid, stable structure. McClintock might as well have told them that the streets they drove every day moved from place to place. She was viewed as something of a mad genius. Her explanations were extraordinarily complicated; her writing lacked the spare elegance of Jacob and Monod's discussions of the *lac* operon. For 30 years,

Figure 12-1 Barbara McClintock in the early 1920s, at her parents' home in Brooklyn, New York. (*Cold Spring Harbor Laboratory Archives. Permission of Marjorie M. Bhavnani*)

most geneticists did not even try to understand McClintock's work.

Not until the early 1970s, when bacterial geneticists began to discover the same things that McClintock had discovered in corn in the 1940s and 1950s, did biologists begin to recognize the importance of her work. They realized that McClintock's transposons, or jumping genes, are just one example of a whole class of **mobile genes**—genes that move around.

Finally, biologists began to see how broadly McClintock's ideas applied to all sorts of mobile genes, not just jumping genes and not just in corn, but to mobile genes in all living organisms.

By the 1980s, young molecular biologists had begun to use mobile genes as tools for the study of how genomes work. And, as we will see in the next chapter, mobile genes also play a pivotal role in the development of a great range of genetic engineering techniques. In 1983, when McClintock was 81, her great contributions to biology were finally recognized with a Nobel prize.

Since the beginning of the 20th century, geneticists have carefully studied both the information in a gene and the gene's place in the chromosome. While the information in a gene is important, it is a gene's location that determines how the gene will be inherited. We have already seen, for example, that genes that are far apart on the chromosome are more likely to recombine than genes that lie next to one another. Barbara McClintock's work showed that individual genes on the chromosome can actually move from place to place, with profound consequences.

In this chapter we will also examine an odd collection of **unconventional genetic systems.** Unconventional genes either move about within the genome or are not in the nuclear (or nucleoid) genome at all. Mitochondria and chloroplasts have their own DNA. This DNA does not move, but it constitutes an unconventional genetic system because it is passed down from organelle to organelle separately from the DNA in the nucleus and without any kind of sexual recombination.

The other unconventional genes we will discuss are all mobile genes. Jumping genes, or transposons, are the simplest mobile genes. Transposons replicate as part of a host cell's DNA. In contrast, **plasmids** usually exist as simple loops of DNA separate from the cell's own DNA. Plasmids, which exist in prokaryotic cells (and yeast cells), can pass from cell to cell and may sometimes integrate into the cell's DNA. Viruses, familiar to us as the purveyors of colds and flu, are another variety of mobile gene. Viruses move genetic material from cell to cell and can insert genetic information into the genomes of both prokaryotic cells and eukaryotic cells.

As we survey this odd collection of unconventional genes, however, we will see that the replication and expression of all genes, unconventional or conventional, can occur only within a cell. The study of unconventional genes has fundamentally changed cell biology. It has given biologists new insights into diseases ranging from the common cold to cancer and AIDS. It has suggested new views about the origin of eukaryotic cells and of life itself. And it has led directly to the extraordinary growth of the biotechnology industry.

WHAT IS A VIRUS?

Viruses are examples of unconventional genetic systems. A virus is an assemblage of nucleic acid (DNA or RNA) and proteins (and occasionally other components, such as lipids or carbohydrates). Viruses have many properties of life: they are highly organized, they reproduce, they respond to their environments, and they evolve. But viruses cannot obtain energy from their surroundings, and they cannot form the building blocks needed for the synthesis of their components. Because

a virus cannot perform metabolic reactions, it can replicate its genetic material and build its proteins only by using the biochemical machinery of a host cell. In short, a virus is not alive.

Most viruses that infect bacteria, plants, and animals do not cause disease. Many are relatively harmless, and some even introduce useful genes into the genomes of other organisms. In fact, in the next chapter, we will see how molecular biologists have used this attribute of viruses to create new strains of bacteria, plants, and animals.

Nonetheless, the study of viruses has, historically, been the study of disease-causing viruses. In the 19th century, the word "virus" [Latin, slimy liquid or poison] came to mean any material associated with death and disease. By the late 19th century, Louis Pasteur and others had shown that bacteria and other small cells cause many infectious diseases and also spoil food. Pasteur's colleague Charles Chamberland developed a way of removing even the smallest cells from liquids by forcing them through a fine filter (Figure 12-2A). But biologists soon discovered that some infectious agents managed to get through the bacterial filters. These mysterious entities were called "filterable viruses." One by one the bacteria specifically responsible for many diseases were identified and named. The only infectious agents left unnamed were the mysterious filterable viruses, which microbiologists simply call "viruses."

In the 1890s, just a few years after Chamberland's filter became available, two men—the Russian pathologist Dmitri Iwanowski and the Dutch microbiologist Martinus Beijerinck—independently demonstrated that a filterable virus was responsible for a disease of tobacco leaves (Figure 12-2B). Because the diseased leaves have a patchy, or mosaic, appearance, the disease was called *tobacco mosaic disease,* and the virus that causes it is now called *tobacco mosaic virus.*

Other microbiologists soon discovered that viruses cause diseases in animals as well, including yellow fever (a deadly human disease carried by mosquitoes) and hoof-and-mouth dis-

Figure 12-2 Viruses are much smaller than bacteria. A. Fine filters remove the bacteria that cause milk to sour. B. Fine filters do not remove the virus that causes tobacco mosaic disease.

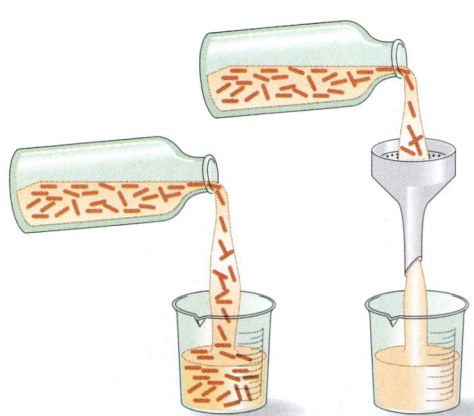

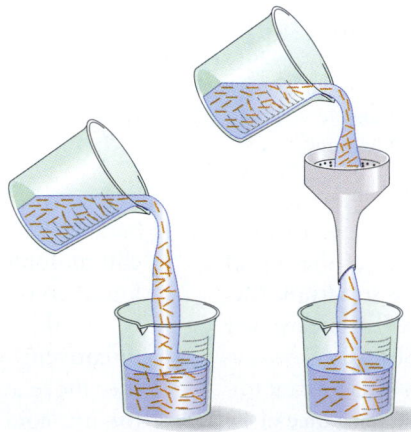

A. Porcelain filter removes bacteria from sour milk.   B. Tobacco mosaic virus passes through filter.

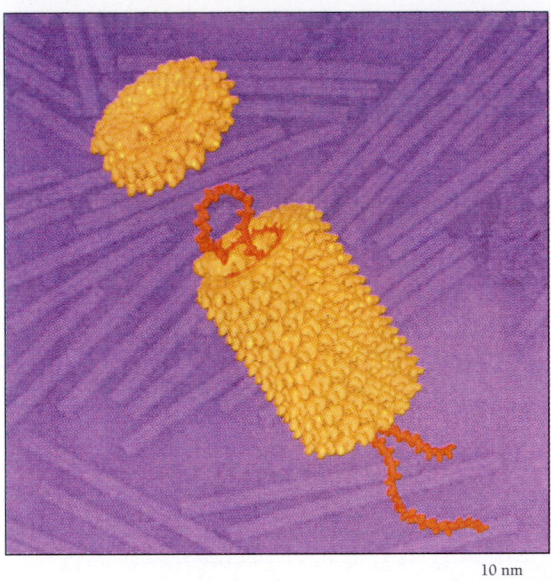

10 nm

Figure 12-3 Three-dimensional model of tobacco mosaic virus. *(Drs. Gerald Stubbs and Hong Wang, Vanderbilt University)*

ease (a disease of cattle). In 1915, the English bacteriologist Frederick Twort discovered that viruses even infect bacteria. Twort called his bacteria-infecting virus a **bacteriophage** [Greek, *phagein* = to eat].

Discoveries such as these demonstrated that disease-causing agents could be smaller than the 0.5-μm pores in a Chamberland bacterial filter. No one, however, knew what viruses actually were. Most people thought that they were just cells too small to be stopped by a filter and too small to be seen through a microscope.

In 1935, this view fell apart, however, when Wendel Stanley, a biochemist working at the Rockefeller Institute, produced crystals of tobacco mosaic virus (Figure 12-3). This was confusing indeed. A crystal forms only when identical molecules arrange themselves in a regular lattice (as in the case of the familiar salt and sugar crystals on the breakfast table). How could an infectious agent—something that was assumed to be alive—form the precisely ordered arrays of a crystal? One certainly couldn't imagine a crystal of bacteria or of mice.

Stanley's discovery seemed to imply that a virus was just a large molecule and that viral diseases were similar to chemical poisoning. But a poisonous substance would be diluted and broken down if it were passed from individual to individual, while tobacco mosaic virus from a dissolved crystal could be passed indefinitely from plant to plant. Clearly, tobacco mosaic virus was replicating itself—something ordinary molecules are not known to do.

We now understand that viruses are relatively large, regular molecular complexes, some of which can form crystals. By analyzing such crystals by x-ray diffraction, researchers have

been able to determine the structure of some viruses in great detail. Most viruses consist of a nucleic acid core (either DNA or RNA) that carries the virus's genes, all surrounded by a protein coat. (A few viruses direct the host cell to make a viral coat out of the host cell's own plasma membrane. This trick helps an infecting virus conceal its identity.) Some of a virus's genes direct the synthesis of proteins and nucleic acids that make up the virus, while other genes specify proteins that regulate the host cell's DNA. Such regulatory genes force the host cell to make new virus particles.

A virus is a large complex of nucleic acid and protein capable of infecting a cell. Infection may be helpful, harmless, or fatal to the cell.

What Is the Form of a Virus?

With the development of the electron microscope in the 1940s, scientists could actually see viruses for the first time. They saw that viruses are large particles, much smaller than cells but larger than single protein molecules. The biggest viruses are the pox viruses such as smallpox virus, which range from 250 to 400 nm in diameter. The smallest ones, including, for example, poliovirus, are 20 to 30 nm in diameter (Figure 12-4).

Viruses have many shapes, but most have one of two forms: a long helix, such as tobacco mosaic virus (Figure 12-4A); or a compact, nearly spherical particle, such as poliovirus (Figure 12-4B). Poliovirus is not actually spherical. It is a 20-sided object, called an **icosahedron** [Greek, *eikosi* = twenty + *hedra* = base], a little like a soccer ball. Unlike the sides of a soccer ball, however, each facet is shaped like a triangle. Other viruses have more complex forms. The bacteriophage T4 is among the most complex of viruses (Figure 12-4C). With an icosahedral head and an intricate tail structure, it looks and functions like a tiny medical syringe, injecting the bacteriophage's DNA into its bacterial host.

The outside surface of a virus is either a coat of protein—a **capsid**—or a **membrane envelope**—a membrane like that which surrounds a cell. Many viruses that infect animals, such as influenza virus and human immunodeficiency virus (HIV, the AIDS virus), direct the host cell to make a viral membrane from the host's own cell membrane, together with proteins specified by the virus's genes (Figure 12-4D). Otherwise, viruses consist almost entirely of proteins and nucleic acid. The nucleic acid may be either DNA or RNA, depending on the virus. Just as in humans and other living organisms, the nucleic acid serves as the virus's genetic material and contains the information for most, if not all, of the virus's proteins.

Viruses have regular shapes, such as an icosahedron or a helix. They consist of a core of genetic material—DNA or RNA—enclosed in a protein capsid or a membrane envelope.

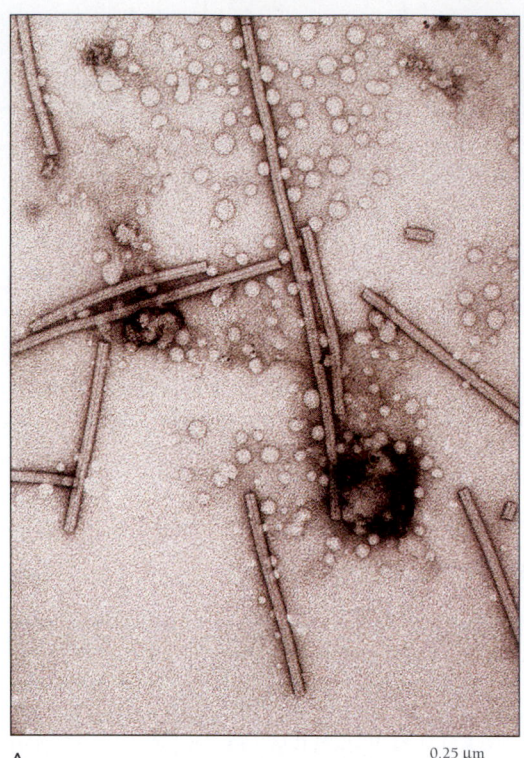

A.   0.25 μm

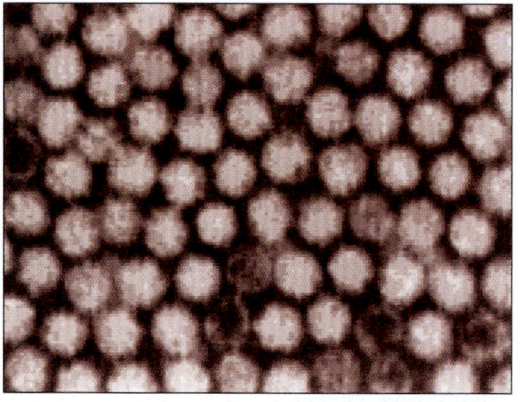

B.   50 nm

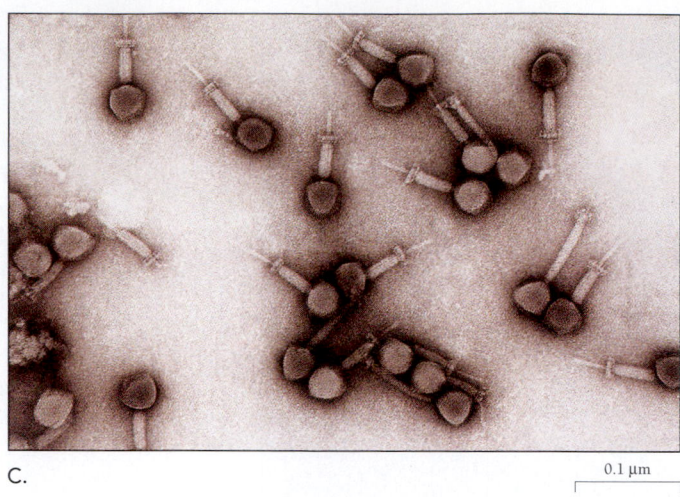

C.   0.1 μm

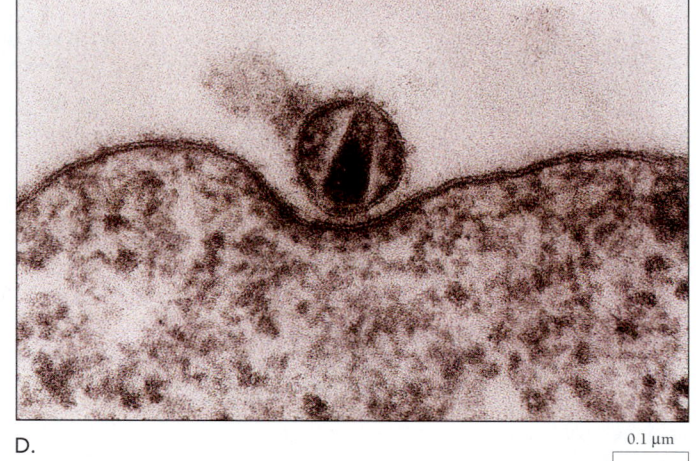

D.   0.1 μm

Figure 12-4 Viruses come in many sizes and shapes. A. Tobacco mosaic virus. B. Like the poliovirus, tomato bushy stunt virus is a 20-sided icosahedron. C. T4 bacteriophage. D. HIV, which causes AIDS. *(A, Jack D. Griffith, University of North Carolina; B, Andrew O. Jackson, University of California, Berkeley; C, Thomas Broker/Phototake NYC; D, Hans Gelderblom/Visuals Unlimited)*

How Do Viruses Make More Viruses?

Virologists' studies of how viruses infect cells in the laboratory have revealed much about how viruses replicate. Although viruses can make more viruses in many different ways, their histories always include the following stages: (1) the virus attaches to the host cell; (2) the viral nucleic acid enters the cell; (3) the cell synthesizes proteins specified by the virus's genes; (4) the cell replicates the virus's DNA or RNA; (5) the new viral protein and DNA or RNA made by the host cell assembles into new viruses, also called virus particles; and (6) the new viruses are released from the cell (Figure 12-5).

Each stage in a virus's multiplication cycle depends on the interaction between the host cell and the virus. A given kind of virus can infect only certain kinds of cells. We say that viruses are adapted to specific hosts. Most cold viruses and the HIV virus, for example, are specific to humans. In this section, we will briefly discuss four of these six stages of viral infection. After that, we will discuss the replication of the viral DNA or RNA in greater detail.

How Does a Virus Attach to a Cell?

The attachment of the virus to a cell depends on an interaction between a viral protein and receptor molecules on the host

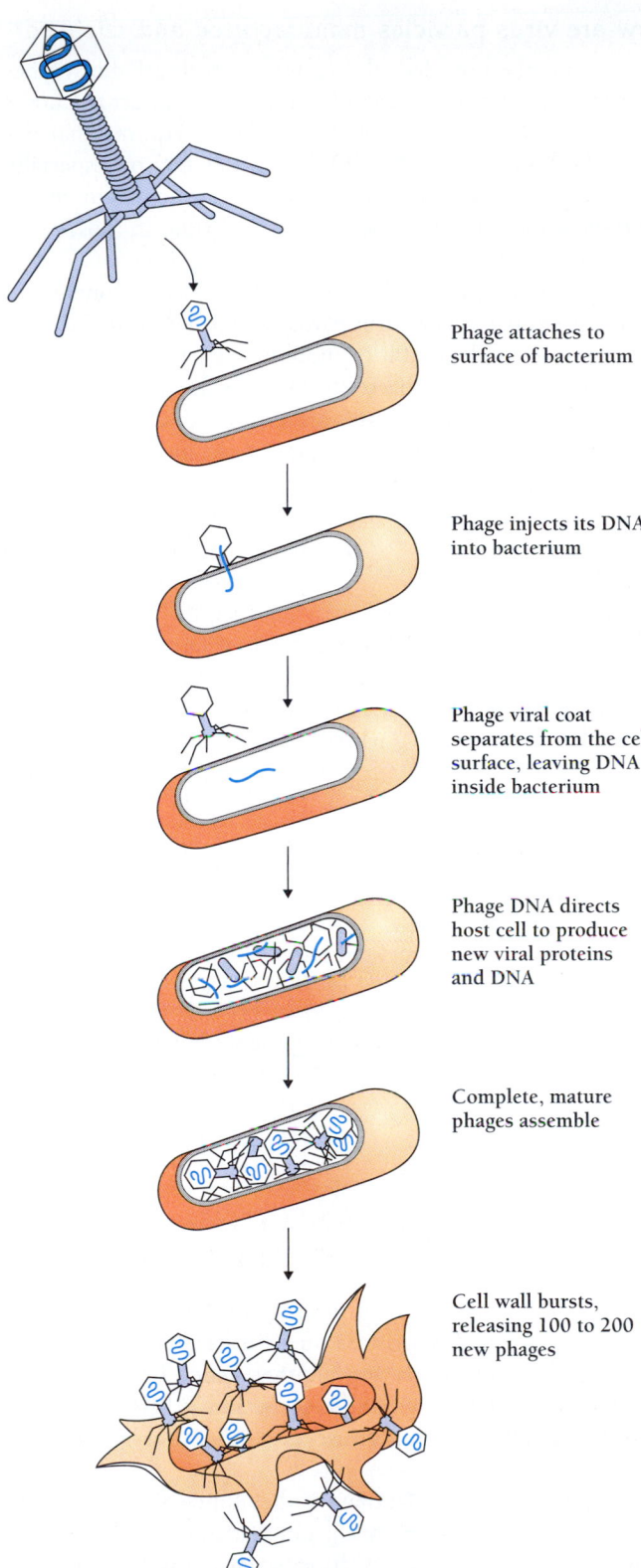

Phage attaches to surface of bacterium

Phage injects its DNA into bacterium

Phage viral coat separates from the cell surface, leaving DNA inside bacterium

Phage DNA directs host cell to produce new viral proteins and DNA

Complete, mature phages assemble

Cell wall bursts, releasing 100 to 200 new phages

Figure 12-5   The reproductive cycle of a bacteriophage.

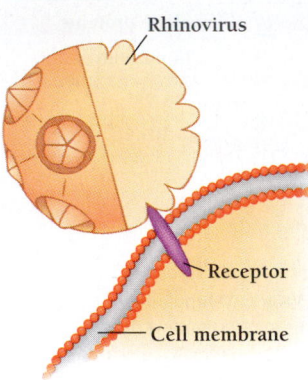

Rhinovirus

Receptor

Cell membrane

Figure 12-6   Viruses latch onto cell surface receptors.

cell's membrane (Figure 12-6). Specific proteins on the surface of the virus bind to molecules on the surface of the host cell. The attachment site may be on all of the host's cells or just on a special set of cells. Human cold and flu viruses, for example, specifically infect cells of the respiratory tract. Others, such as chicken pox virus and some types of herpes viruses, infect skin and nerve cells. Poliovirus initially attacks cells in the digestive tract, then moves into the bloodstream, where it can gain access to all the cells in the body. In 1 percent of polio cases, poliovirus attaches to and destroys nerve cells, causing paralysis or even death. In comparison, HIV primarily infects cells of the immune system and the brain.

Viruses attach to protein receptors on the surfaces of cells.

How does the viral DNA or RNA enter the host cell?

The mechanism by which the viral nucleic acid enters the host cell depends on the kind of virus and the kind of host cell. Bacteriophage T4, for example, contains an enzyme that digests away part of a bacterial cell wall, allowing the phage to inject its DNA into its host. Other viruses, such as the flu virus, are taken into cells by endocytosis, the process by which the cell's membrane enfolds particles (Figure 12-7). Still other viruses, including HIV, are enclosed in bits of the cell's own membrane, which the virus particles capture as they leave an infected cell. Such viruses easily enter their next host cell, when the virus's stolen membrane envelope fuses with the host cell's membrane (Figure 12-7).

Viruses enter cells either by cutting a hole in the cell membrane, or by inducing the cell to engulf the virus particle, or by fusing with the cell's membrane.

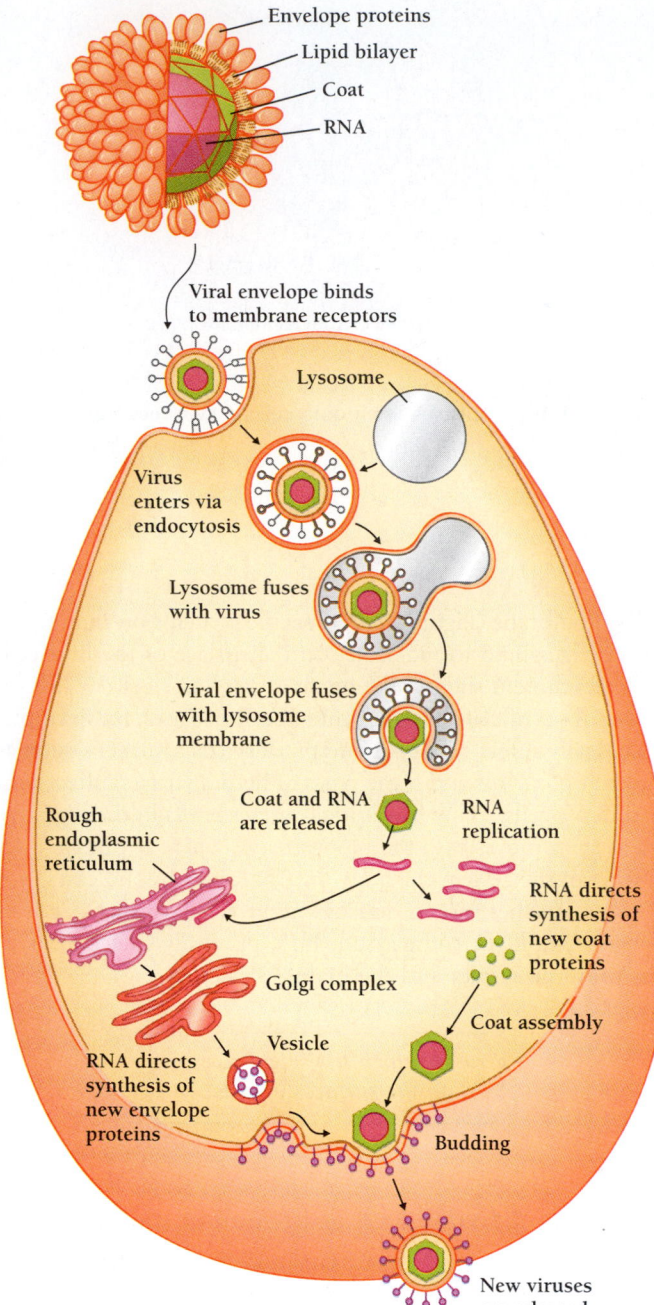

Envelope proteins
Lipid bilayer
Coat
RNA

Viral envelope binds
to membrane receptors

Lysosome

Virus
enters via
endocytosis

Lysosome fuses
with virus

Viral envelope fuses
with lysosome
membrane

Coat and RNA
are released

RNA
replication

Rough
endoplasmic
reticulum

RNA directs
synthesis of
new coat
proteins

Golgi complex

Coat assembly

Vesicle

RNA directs
synthesis of
new envelope
proteins

Budding

New viruses
are released

Figure 12-7   Like the Trojan horse, an influenza virus tricks a cell into taking up the virus. The virus attaches to cell surface receptors, which initiate endocytosis of the virus. Once inside, the virus releases RNA, which the cell duplicates and packs in viral protein coats for release to the outside.

How are virus particles manufactured and released?

Once inside the host cell, the virus's genes direct the synthesis of viral proteins in a variety of ways. Some direct the host's protein-synthesizing machinery to make an enzyme that destroys the host's own DNA. Other viruses rely on especially powerful promoters and other regulatory sequences to ensure rapid replication, transcription, and translation. In most cases, viral protein completely dominates protein synthesis.

For a bacteriophage, the whole cycle, from attachment and entry to assembly of new virus particles, may take only 20 minutes and may produce 100 or more new bacteriophages from each original invading bacteriophage inside the cell (Figure 12-5). After these new viruses are produced, a viral enzyme causes the host cell membrane to break open in a process called **lysis** [Greek, lysis = loosening]. The ruptured cell then releases the newly made viruses, which are ready to infect more host cells. Viruses that destroy their host cells in this way are called **lytic viruses**.

Not all viruses cause cell lysis, however. Killing the host—whether the host is a single cell or a large multicellular organism—is not the only strategy by which a virus can successfully replicate. After all, a host that a virus kills may be the last one it encounters for a while, so killing the host could spell the end of the line for a virus. Some viruses have a multiplication cycle similar to that of a lytic virus, but they exit the cell without destroying the host. Other viruses stay inside the cell for long periods, as we will see.

Viruses force the host cell to make viral proteins—sometimes by destroying the host's DNA, sometimes through subtler forms of gene regulation. Many viruses can escape from their hosts by breaking open the cell membrane.

HOW DO VIRUSES AND OTHER MOBILE GENES REPLICATE THEIR GENES?

"A hen is only an egg's way of making another egg," wrote the English novelist and essayist Samuel Butler. Just so, a virus is only a gene's way of making another gene. This statement might sound extreme, but the idea behind it helps us to understand the strategies that viruses and other mobile genes use to replicate.

A virus is the most elaborate of the mobile genes, but there are many other kinds of mobile genes that consist only of DNA or RNA. Such mobile genes dispense with any kind of exterior coat or membrane. Like viruses, however, they include sequences that induce the host cell to make copies of their genes.

Mobile genes, including viruses, plasmids, and transposons, have two reproductive strategies. In **autonomous replication,** the genes are separate from the DNA of the host cell and can replicate even when the host DNA is not replicating. In **integrated replication,** the genes become integrated

BOX 12.1

Can Viruses Make You Well?

Imagine one day going to the doctor with a case of strep throat and getting a spoonful of virus to cure it. As more and more bacterial infections become resistant to antibiotic treatments, that day may not be so far off. In some parts of the world, bacteriophages, the viruses that infect bacteria, are already in use to combat disease.

Many infections that were once easily cured have become resistant to all but a few antibiotics. The list includes some kinds of pneumonia, ear infections, strep throat, urinary tract infections, cholera, tuberculosis, and others. The heavy use of antibiotics during and after surgery has made hospital recovery rooms into breeding grounds for antibiotic-resistant strains of bacteria. In recent years, some hospitals have reported outbreaks of hospital-acquired pneumonia that are resistant to every known antibiotic.

Bacteriophages are natural parasites of bacteria. Once a phage invades a bacterial cell, it reproduces rapidly, and within an hour the cell bursts open to release millions of phage progeny, which then hunt down more bacteria.

Phages have several advantages over antibiotics. Because antibiotics are toxic, they often cause side effects, such as intestinal disorders and hearing problems. Some people are violently allergic to antibiotics. And many people don't take the complete course of antibiotic. In such cases, all of the bacteria may not be eliminated—either by the antibiotic or by the immune system—and the surviving bacteria in such a lingering infection tend to be those that are resistant to the antibiotic. Phages have none of these problems. They function regardless of whether the bacteria are antibiotic resistant or not. They produce no side effects and tend not to be allergenic. And, since they multiply themselves in the process of slaughtering the bacteria, forgetting to take a dose or stopping treatment too early is never a problem.

But phages are no panacea. Each strain is so specific for a particular type of bacteria that doctors have to know exactly what kind of bacterium is causing an infection. That could mean a delay of several days while the species of bacterium is identified. Also, some harmful bacteria have no known phage predators. And, unfortu-nately, bacteria may evolve resistance to infection by phage, just as they do to antibiotics. Finally, little is known about the efficacy of phage therapy.

In the 1930s and 1970s, researchers made some attempts to treat infections with phages, and many hoped that they would be the ultimate treatment for bacterial infections. In Sinclair Lewis's popular novel *Arrowsmith,* written in 1924, the hero tried to combat a devastating epidemic with a bacteriophage treatment. Nonetheless, the method was all but abandoned with the advent of antibiotics, but only in the West.

In parts of the former Soviet Union, however, phages have been used routinely to treat everything from burn infections to food poisoning. Doctors at a hospital in Tbilisi, Georgia, report using phages successfully to disinfect operating rooms and surgical instruments, as well as to reduce the spread of antibiotic-resistant pneumonia and strep throat. While bacteriophage therapy may be low tech, a few researchers hope that this therapy will be the wave of the future.

---

into the host's DNA and are replicated along with those of the host. Integrated genes can replicate *only* when the host DNA replicates. Some mobile genes can switch back and forth from autonomous to integrated replication depending on circumstances.

In the next few sections, we will discuss plasmids, mobile genes that usually replicate only autonomously; transposons, mobile genes that can replicate only when integrated into the host's DNA or into a plasmid; and **episomes,** viruses and plasmids that can switch back and forth between integrated and autonomous replication.

How Do Plasmids Replicate Autonomously?

A plasmid is a small piece of circular DNA or RNA in a bacterial or yeast cell that includes from 3 to 300 genes. Most plasmids do not integrate themselves into the host cell's DNA but exist as independent entities. Usually, plasmids replicate only when the DNA of their hosts does. However, many plasmids can replicate even when host DNA synthesis has stopped.

Such plasmids are said to replicate autonomously. A plasmid can do this because it contains an origin of replication, the DNA sequence where DNA replication begins. As a result of this autonomous replication, a cell may contain up to 1000 copies of each plasmid. And the cell may carry many different kinds of plasmids.

Plasmids, unlike viruses, do not produce protein coats that would allow them to move easily from one cell to another. However, plasmids have another way to get around. Bacteria sometimes engage in a sexlike process called **conjugation,** during which two bacterial cells temporarily attach to one another and exchange genetic material (Figure 12-8). During conjugation, plasmids can move from one bacterium to another.

Plasmids often contain genes that bring their hosts some advantage. For example, some plasmids—called **resistance factors,** or **R factors**—contain genes for enzymes that make the cell resistant to antibiotics (Figure 12-9). Other plasmids contain genes for resistance to heavy metal poisoning, for the metabolism of unusual compounds (such as those found in oil spills), or for the production of toxins that kill other bacteria.

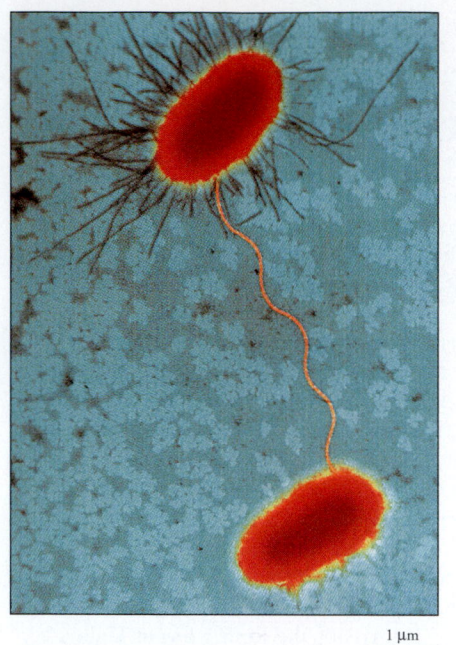

1 μm

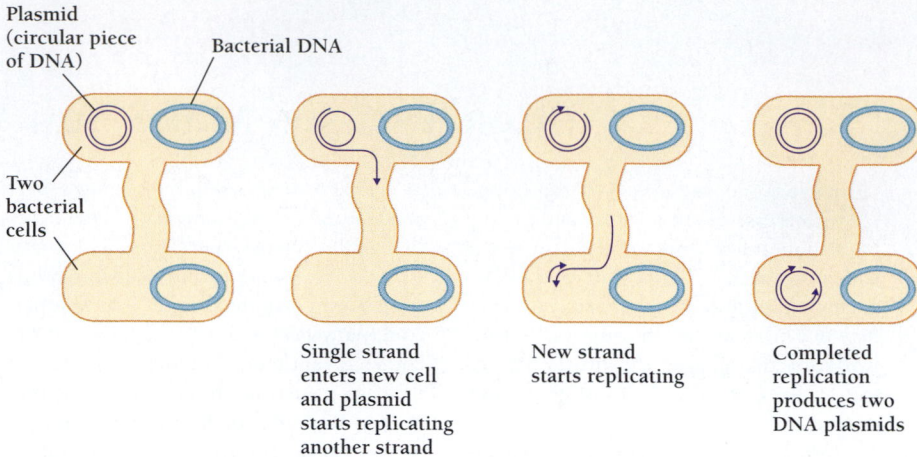

Plasmid (circular piece of DNA)   Bacterial DNA

Two bacterial cells

Single strand enters new cell and plasmid starts replicating another strand

New strand starts replicating

Completed replication produces two DNA plasmids

Figure 12-8   Conjugation in bacteria. During a brief encounter, bacteria exchange genes carried on the bacterial DNA or on plasmids. The bacteria may, for example, transfer genes on plasmids that confer antibiotic resistance. By means of conjugation, bacteria can spread genes for antibiotic resistance among different strains and even different species of bacteria. *(Photo, Dennis Kunkel/Phototake NYC)*

In addition, molecular biologists have synthesized thousands of **recombinant plasmids,** plasmids that contain combinations of genes never found in nature. We will discuss the production and uses of such recombinant DNA in Chapter 13.

A plasmid is a circular piece of DNA or RNA capable of autonomous replication that carries genes, such as those for antibiotic resistance, from one bacterial cell to another.

Many Mobile Genes Can Replicate Only When Integrated into a Host Cell's DNA

Unlike a plasmid, a transposon can replicate only when it is part of the host cell's DNA. A transposon, or jumping gene, is a DNA sequence that moves, either by changing its own

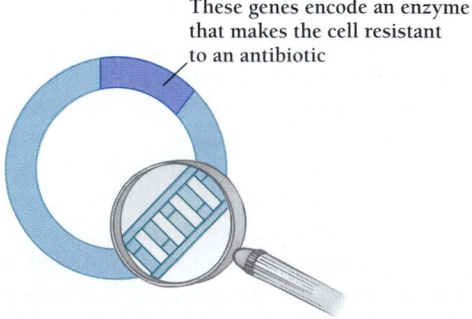

These genes encode an enzyme that makes the cell resistant to an antibiotic

Figure 12-9   A plasmid carrying a gene for antibiotic resistance.

chromosomal location or by duplicating itself for repeated insertions elsewhere. A transposon always includes a gene for an enzyme called **transposase,** which catalyzes insertion into new sites (Figure 12-10). While plasmids occur mostly in bacteria and in some yeast cells, transposons are common in all cells, including McClintock's corn.

Eukaryotic transposons mostly reproduce only in the integrated form. They are surprisingly mobile, however. In eukaryotic organisms from corn to humans, they can jump from place to place within the genome, duplicating themselves, disrupting other genes, deleting and repeatedly duplicating whole stretches of DNA, breaking chromosomes, and sometimes even attaching part of one chromosome to another. The mapping of eukaryotic genomes has revealed that virtually all genomes are littered with DNA sequences that appear to be relics of ancient transpositions of jumping genes. Corn is especially rich in such transposons.

A transposon may be copied many times and each copy moved about the host's genome separately. For example, at least six different transposons are present in the genome of *E. coli.* Several of these transposons have been copied five to ten times within each *E. coli* genome.

A **simple transposon** specifies transposase and nothing else. As far as biologists know, a simple transposon's only function is to make more simple transposons. Still, simple transposons may help bacteria in an indirect way. Because transposons rearrange the genome and interrupt genes, they increase genetic variability in their hosts. Greater genetic variability could allow species to adapt to new conditions more quickly and more effectively.

When two simple transposons lie near each other on the host genome, they can jump together, carrying with them any

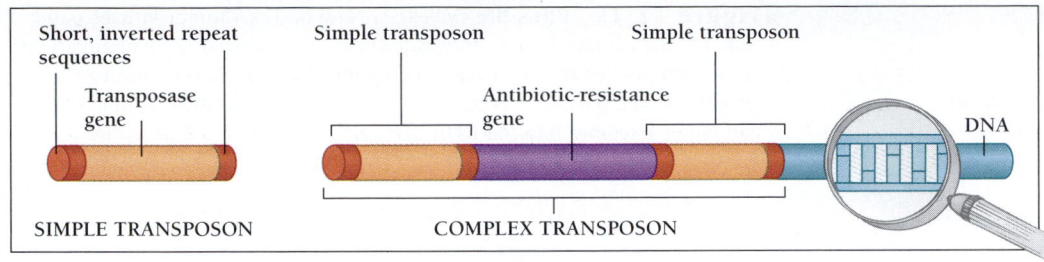

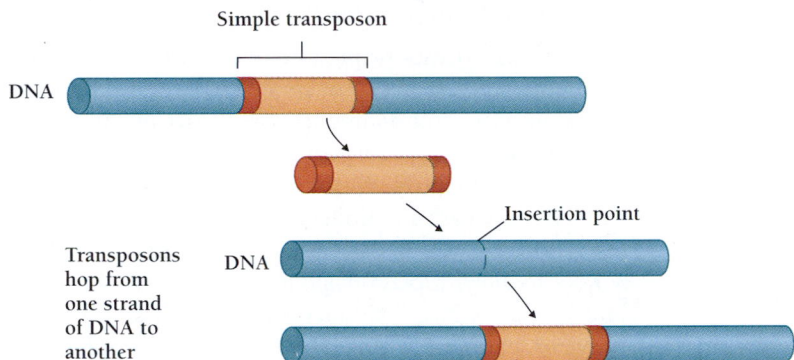

part of the host's genome that lies between them. Such combinations are called **complex transposons.** Some complex transposons carry antibiotic resistance genes, for example (Figure 12-10). Transposons can move these genes into plasmids or into the host cell's chromosomal DNA. In this way, transposons can help a plasmid accumulate genes for resistance to multiple antibiotics (tetracycline and penicillin, for example). The plasmid can then carry this resistance from one bacterium to another, spreading antibiotic resistance even to different species of bacteria.

As we will see in Chapter 20, transposons' ability to move genes ultimately explains how antibiotic resistance spreads so rapidly among bacteria. Some strains of the bacteria that cause tuberculosis (TB) are now resistant to every known antibiotic. The medical treatment for people infected with such bacteria is essentially the same as that used 50 years ago, before the availability of antibiotics. TB patients with multiple-antibiotic-resistant TB are kept isolated to prevent them from passing this dangerous form of TB to others. But there is no specific medical treatment that will help them get well.

A simple transposon codes for nothing but transposase, which is all the transposon needs to insert itself into a new site in the host's DNA. Complex transposons carry other genes as well.

Tumor viruses also insert themselves into chromosomal DNA

Many viruses replicate autonomously, lyse the cell, and escape to infect more cells. But viruses that cause tumors are a spe-

cial case. Rather than multiplying as rapidly as possible and killing their host cells, tumor viruses multiply only when their eukaryotic hosts do (Figure 12-11A).

Tumor viruses exist independently of the host's DNA only during a brief, independent, mobile phase when the tumor virus finds the eukaryotic cell and infects it. After infection, the genes of most tumor viruses become integrated into the DNA of their hosts and stay there. The virus uses the host's enzymes to express one or more viral genes. The resulting viral proteins ruin the host's ability to control cell division, forcing the cell to divide rapidly, which causes a tumor to form. With each round of cell division, a new generation of viral genes is created.

The viral genes whose expressed proteins interfere with the host cell's ability to control cell division are called **oncogenes** [Greek, *onkos* = mass, swelling], a word that means "genes that cause tumors." Most oncogenes evolved from normal genes in humans or other animals. The original versions of oncogenes, called **proto-oncogenes,** normally control growth and differentiation. Tumor viruses have somehow captured these genes, which have evolved further into **viral oncogenes.** Most oncogenes of tumor viruses provide an overdose of normal cellular proteins that ordinarily stimulate normal cell division.

Tumor viruses integrate their DNA into that of a eukaryotic host, then force the host to replicate both viral and genomic DNA more often than normal. The result is many copies of the viral DNA.

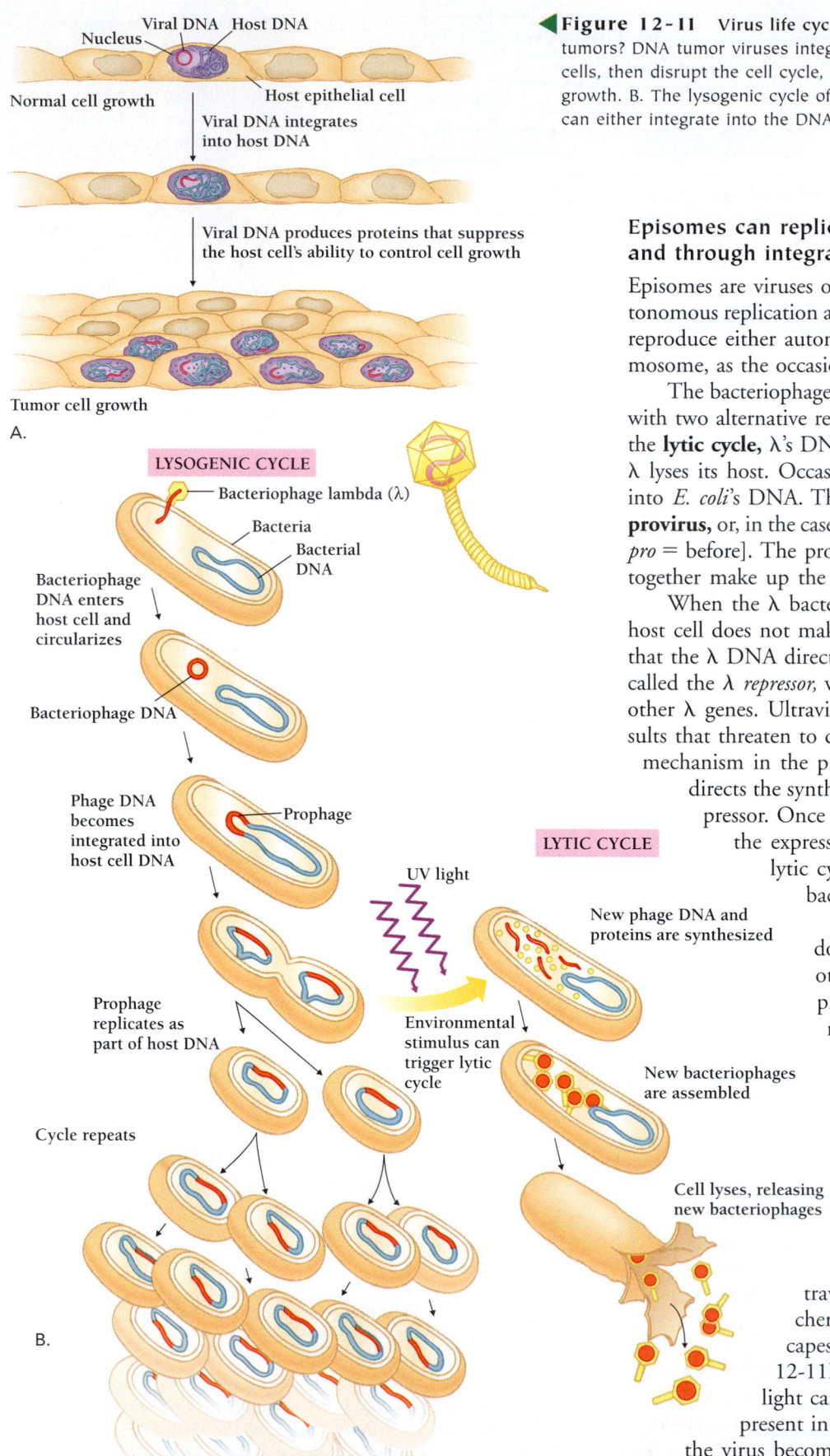

Figure 12-11 Virus life cycles. A. How do DNA tumor viruses cause tumors? DNA tumor viruses integrate their DNA into the DNA of eukaryotic cells, then disrupt the cell cycle, forcing infected cells into excessive growth. B. The lysogenic cycle of the bacteriophage lambda. Lambda (λ) can either integrate into the DNA of its host or initiate a lytic cycle.

A.

Viral DNA   Host DNA
Nucleus
Normal cell growth
Host epithelial cell
Viral DNA integrates into host DNA

Viral DNA produces proteins that suppress the host cell's ability to control cell growth

Tumor cell growth

LYSOGENIC CYCLE
Bacteriophage lambda (λ)
Bacteria
Bacterial DNA
Bacteriophage DNA enters host cell and circularizes
Bacteriophage DNA
Phage DNA becomes integrated into host cell DNA
Prophage
Prophage replicates as part of host DNA
Cycle repeats

LYTIC CYCLE
UV light
New phage DNA and proteins are synthesized
Environmental stimulus can trigger lytic cycle
New bacteriophages are assembled
Cell lyses, releasing new bacteriophages

B.

Episomes can replicate both autonomously and through integration

Episomes are viruses or plasmids that switch hit between autonomous replication and integrated replication. Episomes can reproduce either autonomously or as part of the host's chromosome, as the occasion requires.

The bacteriophage lambda (λ), for example, is an episome with two alternative reproductive cycles in its *E. coli* hosts. In the **lytic cycle,** λ's DNA is separate from that of its host, and λ lyses its host. Occasionally, however, λ integrates its DNA into *E. coli*'s DNA. The integrated form of a virus is called a **provirus,** or, in the case of a bacteriophage, a **prophage** [Greek, *pro* = before]. The provirus (or prophage) and the lytic cycle together make up the viral **lysogenic cycle** (Figure 12-11B).

When the λ bacteriophage is in the prophage stage, the host cell does not make viral proteins. The reason for this is that the λ DNA directs the synthesis of a powerful repressor, called the λ *repressor,* which prevents the transcription of any other λ genes. Ultraviolet light and other environmental insults that threaten to destroy the host cell trigger an "escape" mechanism in the prophage. In escape mode, the λ DNA directs the synthesis of a protein that represses the λ repressor. Once the λ repressor is no longer preventing the expression of the λ genes, the virus enters a lytic cycle, destroys the host cell, and new λ bacteriophages escape the dying cell.

Because a virus in the provirus stage does not lyse its host cell, it cannot infect other cells. Although a provirus or prophage does not force the cell to make new virus particles, the nucleic acid of the provirus is copied along with the cell's own DNA each time the cell divides or otherwise copies its DNA. As a result, all of the daughter cells of an infected cell are also infected with proviruses, and an infected eukaryotic organism cannot rid itself of the virus.

If the host cell is exposed to ultraviolet light or certain DNA-damaging chemicals, however, the viral DNA then escapes and begins a lytic cycle (Figure 12-11B). This may explain why intense sunlight can activate the herpes virus that may be present in the cells of the mouth and lips. When the virus becomes active, it enters a lytic cycle, lysing cells and causing a painful cold sore.

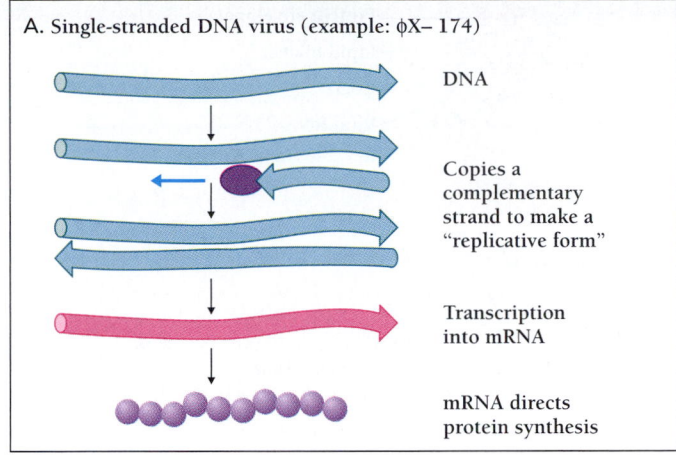

A. Single-stranded DNA virus (example: φX– 174)

DNA

Copies a complementary strand to make a "replicative form"

Transcription into mRNA

mRNA directs protein synthesis

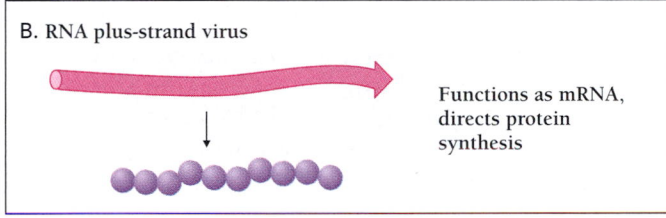

B. RNA plus-strand virus

Functions as mRNA, directs protein synthesis

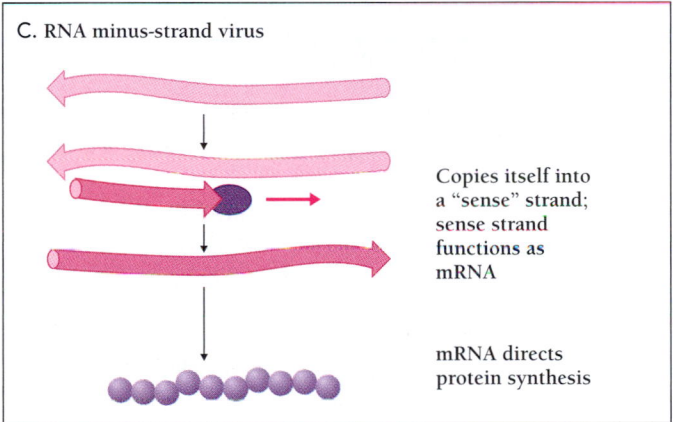

C. RNA minus-strand virus

Copies itself into a "sense" strand; sense strand functions as mRNA

mRNA directs protein synthesis

Figure 12-12 How do viruses express their genes? A. Single-stranded DNA virus. B. RNA plus-strand virus. C. RNA minus-strand virus.

Episomes are viruses or plasmids that switch back and forth between integrated and autonomous replication as the occasion demands.

How Do RNA Viruses Reproduce and Express Their Genes?

Many mobile genes use genetic material other than double-stranded DNA. These genetic entities must use special tricks to replicate themselves. For example, the virus ΦX-174 uses single-stranded DNA. This virus simply copies the single strand into a complementary strand of DNA to make a double-stranded "replicative form" (Figure 12-12A).

Other viruses and plasmids have RNA genes. These include the viruses that cause polio, influenza, AIDS, and many animal tumors. How do these genes function? The short answer is that they use special enzymes, encoded in the virus's genome. The particular function of the enzyme varies with the virus, but RNA viruses all depend on standard base-pairing to make a complementary strand.

Figure 12-12 shows several different ways that RNA viruses express their genes. Some viruses carry their genes on a strand of RNA that, like messenger RNA, actually codes for various viral proteins. Such RNA is expressed directly by the host cell's machinery as soon as the virus enters the cell. Such viruses are called "plus-strand" RNA viruses (Figure 12-12B). Other RNA viruses, called minus-strand viruses, use an "antisense" RNA strand. Such viruses contain a viral enzyme that first copies the RNA into a complementary (sense) strand that functions as mRNA and codes for proteins (Figure 12-12C).

Plus-strand viral RNA works like mRNA. Minus-strand viral RNA is antisense RNA, which is transcribed into RNA that works like mRNA.

RNA tumor viruses present special problems

By the mid-1960s, virologists had discovered many of the ways that RNA viruses could reproduce. But the reproductive cycle of RNA tumor viruses, such as Rous sarcoma virus, mystified them. DNA tumor viruses inserted their genes directly into the DNA of their hosts, permanently changing their genetic makeup. But where could an RNA tumor virus leave a permanent copy of its genes? Certainly single-stranded RNA could not insert directly into chromosomal DNA.

The answer was first suggested in the 1960s by Howard Temin, a virologist at the University of Wisconsin. Perhaps, Temin argued, the RNA is recopied into DNA, and the DNA is then inserted into the host cell's genome. Many scientists found this suggestion heretical, since it violated Francis Crick's central dogma: "DNA specifies RNA, which specifies proteins."

In 1970, however, Temin and David Baltimore, a molecular biologist then at MIT, independently discovered that RNA tumor viruses contain an enzyme, called **reverse transcriptase,** which indeed copies RNA into DNA. The enzyme received its name because the direction in which information flows is backward from that found in normal transcription (where information goes from DNA to RNA). As we will see in the next chapter, reverse transcriptase has become an essential tool for molecular biologists. Viruses that use reverse transcriptase in this way are called **retroviruses.** The discovery of reverse transcriptase supported Temin's hypothesis.

Subsequent work showed that Temin was exactly right. In most viruses, only the virus's nucleic acid actually enters the host. In retroviruses, however, the reverse transcriptase enzyme is actually packaged *with* the viral RNA, so the enzyme (not just the gene for the enzyme) enters the cell along with the virus's RNA (Figure 12-13). The reverse transcriptase then copies the RNA first into single-stranded and then into double-stranded DNA. The double-stranded DNA then integrates into the host cell's DNA, so that the virus's genes are replicated along with those of the host. Like DNA-containing tumor viruses, many (but not all) retroviruses contain an oncogene, whose expression leads to tumor formation. Retroviruses are known to cause cancer in birds, mice, monkeys, and humans. HIV is a retrovirus that infects cells of the human immune system but does not contain an oncogene.

> Retroviruses use reverse transcriptase to transfer information from RNA to DNA, in the opposite direction to that specified by the central dogma.

HOW DO MOBILE GENES EVOLVE?

Viruses and other mobile genes live so closely with cells that we might not realize at first that they are influenced by evolutionary forces independent of those that influence their hosts. Even though mobile genes are not alive, they are still independent entities whose replication is subject to natural selection—the mechanism by which Darwin explained the evolution of organisms.

Although Viruses Are Not Alive, They Are Still Subject to Natural Selection

Because a virus uses energy-expensive materials from its host cell, the cell's ability to grow and reproduce is impaired. As a result, hosts that evolve ways to resist viral infection are more likely to survive and reproduce. This is simply natural selection. Cells have therefore evolved general mechanisms for resisting most kinds of viruses and also resistance mechanisms aimed at specific viruses. Many strains of *E. coli,* for example, guard against viruses by producing **restriction enzymes,** enzymes that cut foreign DNA into little pieces. Most restriction enzymes do not cut the DNA at random, but at special "target sites"—particular sequences of nucleotides that occur commonly in all genomes. Although a restriction enzyme can protect *E. coli* (or another bacterial strain) from infection by a bacteriophage, such an enzyme would be self-destructive if it also digested *E. coli*'s own DNA. To prevent such molecular suicide, *E. coli* modifies its DNA by adding methyl groups that protect the bacterial DNA from digestion.

As hosts evolve mechanisms for resisting a virus, the virus, in turn, evolves mechanisms for getting past the cell's defenses. The kinds of viruses that are able to infect *E. coli,* for example, all have DNA that is missing the sequence targeted by *E. coli* restriction enzymes. Because the viral DNA lacks this target sequence, the cell's restriction enzymes cannot destroy the viral DNA.

Viral multiplication cycles—from infection to release—have evolved to become highly efficient. Cycles are generally

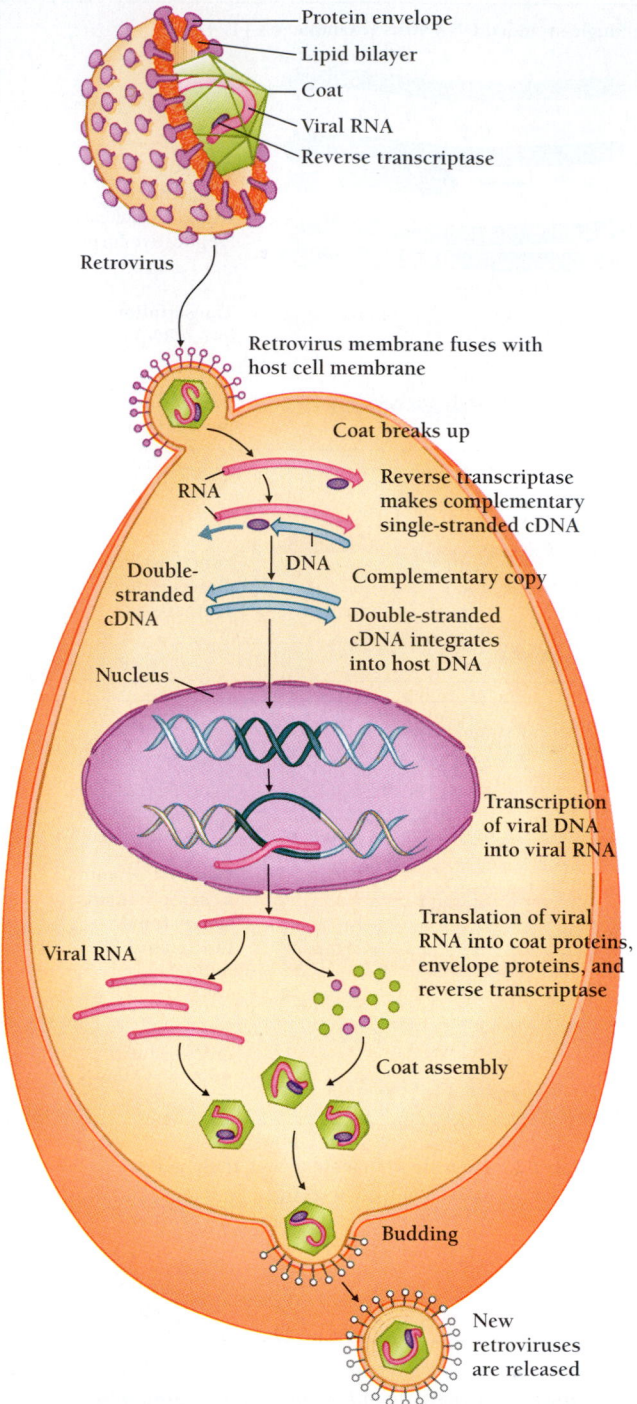

Figure 12-13   How do HIV and other retroviruses reproduce?

BOX 12.2

How Can Researchers Design Drugs That Will Combat AIDS?

The greatest epidemic in recent history is acquired immune deficiency syndrome (AIDS). Most biologists now agree that AIDS is caused by one of two types of human immunodeficiency virus (HIV1 and HIV2). HIV infects specialized white blood cells, called T lymphocytes, that normally help find and destroy invading pathogens. Over several years, the virus reduces the number and effectiveness of these crucial immune cells, making the body susceptible to infections and cancers that a healthy immune system normally fends off. These secondary illnesses, rather than the HIV virus itself, are what kill most people infected with AIDS.

Because HIV grows and replicates in the body for years before producing any symptoms, a person can test positive for HIV antibodies, indicating the presence of the virus, long before the onset of any illness. During this early stage of infection, bodily fluids such as blood and semen harbor active viruses that can transmit the infection to a new host. Most people contract the virus by sharing contaminated hypodermic needles or by having unprotected sex with an infected partner.

Like all retroviruses, HIV carries its genetic material in a single RNA strand accompanied by a molecule of reverse transcriptase. Once inside the cell, the reverse transcriptase generates a DNA strand copied from the RNA template. A second round of synthesis produces a double-stranded DNA copy of the viral RNA, which inserts itself into a chromosome in the host cell's nucleus. Without reverse transcriptase, HIV could still get into the cell, but it would never be able to replicate.

Since human cells do not depend on reverse transcriptase, the virus's enzyme makes a perfect target for anti-AIDS drugs. One type of antiviral therapy deceives the enzyme with synthetic molecules that resemble ordinary nucleotides, the raw materials for DNA synthesis. Reverse transcriptase incorporates these nucleotide impostors into the growing strand of viral DNA. But since the false nucleotides are subtly different from real ones, replication stops. AZT and 3TC, two drugs now commonly used to slow viral replication in patients with AIDS, are examples of false nucleotides. Human DNA polymerase is more picky than reverse transcriptase in selecting nucleotides and rarely incorporates AZT or 3TC. As a result, human cells exposed to the drugs continue to grow and divide normally. Unfortunately, HIV mutates rapidly, and strains that no longer use the false nucleotides quickly emerge. These resistant strains thrive in patients treated with these drugs, rendering the drugs useless.

Another class of HIV drugs works by interfering with a different viral enzyme—one that cuts a viral protein. After viral DNA integrates into the host's chromosome, host proteins transcribe the viral genes. Rather than producing a separate mRNA for each viral protein, however, the messages for several proteins form a single, long mRNA molecule. The resulting mRNA gets translated into a long polypeptide that must be cut into separate proteins before they will function. One of the proteins in the viral polypeptide chain is the enzyme that does the cutting. This enzyme, called a *protease,* begins cutting even while it is still part of the "polyprotein." Without the protease, the polyprotein remains unclipped and inactive, and the virus cannot go on to infect more cells.

In an attempt to stop HIV by knocking out the protease, researchers devised tiny molecules that nestle into the active site of the protease enzyme and prevent it from doing its job. The first of these protease inhibitors, called *saquinavir,* was released in December 1995—with dramatic results. A drug cocktail containing saquinavir, AZT, and 3TC can reduce the amount of virus in a patient's body to undetectable levels. For the first time since the discovery of HIV, this "triple therapy" has transformed AIDS from a death sentence to a potentially survivable disease.

But triple therapy has its problems. The cost of these AIDS drugs, particularly protease inhibitors, is still extremely high, placing an economic barrier between many sufferers of AIDS and good treatment. In the United States, not all insurance covers AIDS therapies, and in Africa, where most of the world's AIDS cases occur, the price is out of reach of almost everyone who has the disease. Also, triple therapy doesn't work well for everyone, and it will be several years before researchers know whether the drug cocktail can permanently clear the virus out of the body. Indeed, researchers worry that HIV may evolve resistance to protease inhibitors, as it has to other drugs. In that case, it could still return in full force, insensitive once more to all drug treatments.

To help stave off that possibility, research is underway on still another class of drugs that may close the door on the virus, preventing it from getting into T cells in the first place. To enter T cells, an HIV virus must bind to a pair of proteins on the cell surface, one of which is the recently identified chemokine receptor CCR5. Chemokines are chemical signals that induce inflammation at a site of injury or infection. Researchers noticed in 1995 that chemokines block HIV from infecting cultured cells. Not long after, more work revealed that CCR5 helps HIV enter cells and that the chemokine was interfering with that process. Drug companies are now racing to develop a drug that binds to CCR5 without setting off the inflammatory pathway. One reason to believe that this approach may work is that studies of people with a defective CCR5 gene have shown that the defect causes no obvious health problems yet provides genetic resistance to many strains of HIV.

The battle against the AIDS epidemic may finally be succeeding. In 1996, the number of patients dying of AIDS in the United States declined for the first time. While this is a significant landmark, the epidemic is far from over. As has been true of many other infectious diseases, the AIDS virus will continue to evolve, and researchers may always be just one step ahead.

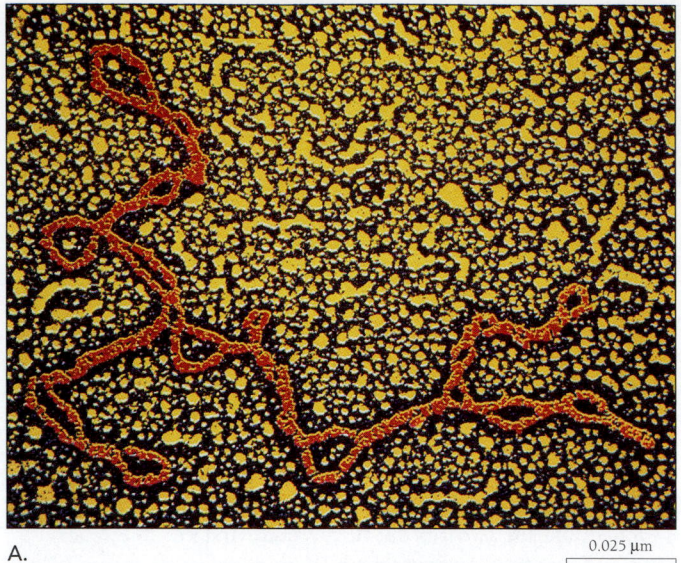

A.   0.025 µm

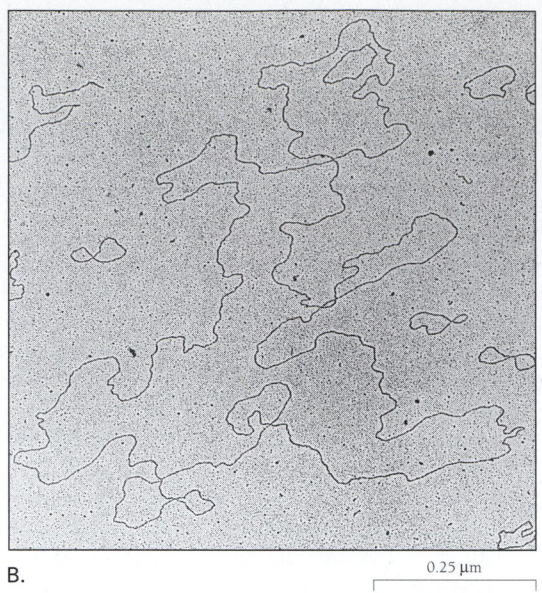

B.   0.25 µm

Figure 12-14   Circular DNA in energy organelles. A. From a mitochondrion. B. From a chloroplast. Chloroplast DNA is ten times longer than that in animal mitochondria. The smaller circular DNA shown is from a virus. (*A, CNRI/Phototake NYC; B, Richard Kolodner, Ludwig Institute for Cancer Research*)

rapid, with lots of new viruses produced in each infected cell. Viral regulatory genes are so powerful that the host's transcription and translation machinery recognizes them in preference to the host's own genes.

One way that viruses speed up their cycle times is to keep their genetic material to a minimum. Unlike eukaryotic genomes, which are loaded with "junk DNA," every gene in a virus matters. One kind of cold virus, for example, specifies several different proteins from a single gene merely by arranging to have the mRNA spliced in different ways.

How May Viruses Have Evolved?

Some scientists believe that viruses are descended from free-living cells that became parasites on other cells. According to this view, the viruses progressively (over eons) lost most of their cellular components, letting the host cells do all the work.

Most biologists, however, believe that viruses may have evolved from plasmids, which evolved from transposons. Knowing what we now know about mobile genes, we can imagine that mobile genes first appeared as simple transposons that could move from place to place in the genome. Transposons that also included genes that benefited the host cell—such as genes that allow the host to break down poisons or antibiotics—would have allowed the host cells that had them to reproduce faster than cells without them. As a result, transposons with beneficial genes would have replicated much faster than those without beneficial genes. The differential survival of the rapidly replicating viruses is simply natural selection.

Later, beneficial genes could have acquired the ability to replicate independently as plasmids do. Finally, some autonomously reproducing plasmids could have acquired genes for protein coats that allowed them to move efficiently from cell to cell. Such a protein-wrapped plasmid would look just like a virus.

Over evolutionary time, transposons may have acquired the ability to replicate autonomously, making them plasmids. Some plasmids may then have acquired the genes for making a protein coat, allowing them to evolve into a modern virus.

MITOCHONDRIA AND CHLOROPLASTS CONTAIN THEIR OWN DNA

Chloroplasts and mitochondria contain their own DNA. Since all eukaryotes have mitochondria and many have chloroplasts as well, almost every eukaryotic cell therefore carries at least one genetic system independent of the nucleus. In a mammalian cell, each mitochondrion contains five to ten identical DNA molecules, each about 16,000 nucleotides long. Since a single cell may have hundreds of mitochondria, up to 1 percent of the total DNA in the cell may consist of mitochondrial DNA. Each piece of mitochondrial DNA is circular, meaning that it forms a continuous loop, like the DNA of bacteria and many viruses (Figure 12-14A).

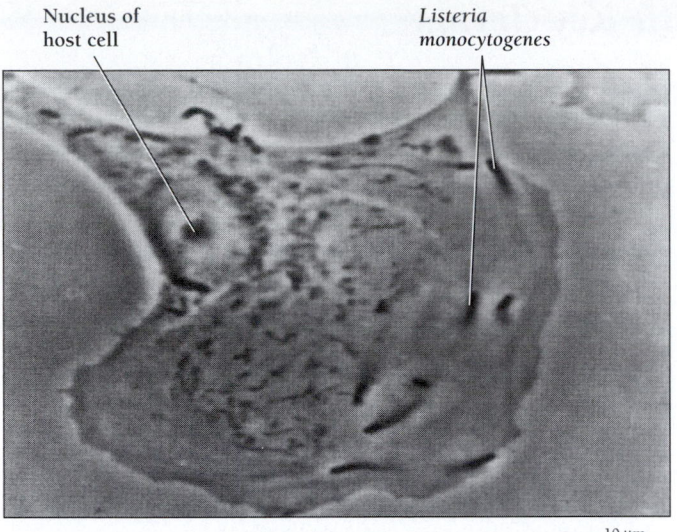

Nucleus of host cell

Listeria monocytogenes

10 µm

Figure 12-15 **An intracellular parasite.** *Listeria monocytogenes* is a bacterium that infects human cells, including cells in the nervous system and the genital area and the white blood cells of the immune system. It most frequently causes disease in pregnant women and their babies. *Listeria monocytogenes* is remarkable for its ability to move about inside cells with the aid of one to four actin filaments, which work like flagella. *(J.A. Theriot)*

Plant mitochondria also contain circular DNA molecules, but usually of several different kinds. The total length and the total information content of mitochondrial DNA in plants can be more than 20 times greater than in animal mitochondria. Chloroplasts contain DNA as well (Figure 12-14B). Each species has a single kind of chloroplast DNA. The chloroplast DNA is circular and usually about 150,000 nucleotides long. In higher plants, chloroplast DNA may constitute up to 15 percent of the total DNA in a cell.

The reproduction of mitochondria and chloroplasts depends on cellular machinery that is specified both by the DNA in the nucleus and by the DNA in the organelle itself. The replication and expression of mitochondrial or chloroplast DNA, for example, requires about 90 proteins—far more than the organelle's DNA can usually encode. These proteins are specified by the DNA in the nucleus and made on cytoplasmic ribosomes. The DNA of mitochondria and chloroplasts nonetheless contains important information. For example, both mitochondria and chloroplasts contain the proton-pumping complex F_0-F_1 (ATP synthase) that harnesses the energy of proton gradients to make ATP. Each complex contains nine different polypeptides, only some of which are specified by the DNA in the nucleus. The other genes are in the mitochondria or chloroplasts.

Mitochondria and chloroplasts contain complete systems for transcription and translation, including not only DNA but also ribosomes, transfer RNAs (tRNAs), and an RNA polymerase similar to that of bacteria. Some components of the machinery of protein synthesis are made in the cytoplasm, and other components are made within the organelles.

Mitochondria and chloroplasts rely on both their own genetic information and that in the nucleus.

Studies of the DNA and RNA of Energy Organelles Support the Theory That They Derive from Ancient Prokaryotes

Like prokaryotic DNA, the DNA of most mitochondria and chloroplasts is circular. The ribosomes of mitochondria and chloroplasts are also more like bacterial ribosomes than like the ribosomes in the cytoplasm of eukaryotic cells. These and many other similarities between the energy organelles and prokaryotes support the **endosymbiotic theory** [Greek, *endon* = within + *syn* = together + *bios* = life]. This theory holds that present-day mitochondria and chloroplasts are descended from prokaryotes that came to live inside ancient eukaryotic cells. In support of this view, biologists point to many modern examples of cells that host endosymbionts (Figure 12-15). Recent studies of DNA sequences have revealed that *all* mitochondrial DNA shares many common features. Researchers now suspect that all mitochondria have evolved from a single type of mitochondrion in an ancient eukaryote.

The genetic code used in mitochondria is slightly different from that used in prokaryotes or in eukaryotic nuclei, with four codon assignments that are different. If mitochondria are descended from free-living prokaryotes, their different genetic code suggests an origin so ancient that it predates the establishment of a universal genetic code. In Chapter 18, we will discuss the origin of mitochondria and chloroplasts in more detail.

The genetic systems of mitochondria and chloroplasts differ from those in both eukaryotic nuclear DNA and in prokaryotes, suggesting that these organelles have an ancient and separate origin.

In this chapter, we have seen that not all genes reside in the chromosomes, nor do all genes pass neatly from generation to generation by means of meiotic divisions. Genes exist in mitochondria and chloroplasts, suggesting that these organelles were once independent organisms. Even more intriguing, however, genes also exist in mobile forms, such as viruses, plasmids, and jumping genes. In the next chapter, we will see that mobile genes are the power behind genetic engineering, a new field that may have profound effects on society in the next century.

Study Outline with Key Terms

Barbara McClintock showed that some genes can move from place to place. She showed that jumping genes, or transposons, cause the curious dappled colors of Indian corn. **Unconventional genetic systems** are DNA or RNA sequences that can replicate and function apart from chromosomal DNA, including the DNA of mitochondria and chloroplasts and mobile genes. **Mobile genes,** which include **viruses, plasmids,** and **transposons,** can move from cell to cell or from place to place within a chromosome. All of these unconventional genes depend on cellular machinery specified by chromosomal DNA.

Viruses, including **bacteriophages,** are assemblages of nucleic acid and proteins that can reproduce within a living cell. Whether shaped like an **icosahedron** or a long helix, viruses consist of an inner core of nucleic acid surrounded by a protein **capsid** or **membrane envelope.** Viruses were first discovered as disease-causing agents that could pass through filters fine enough to remove bacteria. Viruses have many properties of life: they reproduce, they respond to their environments, and they evolve. However, viruses cannot obtain energy from their surroundings and cannot form molecular building blocks.

Although viruses are not alive, they evolve through natural selection. Viruses are able to take over the machinery of their host cells and to produce more viruses. Viral multiplication cycles consist of six steps: (1) attachment to the host cell, (2) entry of viral nucleic acid into the cell, (3) synthesis of viral proteins, (4) replication of viral nucleic acid, (5) assembly of new virus particles, and (6) release of new viruses.

Many viruses kill their hosts in the process of replicating themselves. These viruses, which break open the host membrane in a process called **lysis,** are called **lytic viruses.** Sometimes, viral genes can integrate into the chromosomal DNA of their hosts. The integrated virus, called a **provirus,** does not immediately produce more virus. However, when the host cell replicates, it replicates the viral genes as well, a process called **integrated replication.** Other viruses make new viruses through **autonomous replication.** Actual virus particles are produced only when something induces the provirus to enter a lytic cycle. In a **lysogenic cycle,** a virus or bacteriophage switches back and forth between a **lytic cycle** and an integrated provirus, or **prophage,** stage.

RNA serves as the genetic material in some viruses. In plus-strand viruses, the viral RNA directs polypeptide synthesis. In minus-strand viruses, an enzyme copies the viral RNA into a complementary strand that directs polypeptide synthesis. In AIDS and other **retroviruses, reverse transcriptase** copies viral RNA into DNA, which then integrates into the host cell's chromosomal DNA.

Mobile genes include plasmids, which can replicate only autonomously; transposons, which can replicate only in integrated form; and **episomes,** which can replicate either autonomously or in integrated form. **Simple transposons** carry nothing but the gene for **transposase,** while **complex transposons** carry genes that benefit their bacterial hosts such as those for antibiotic resistance. Plasmids artificially constructed in the lab are called **recombinant plasmids.**

Plasmids called **resistance factors,** or **R factors,** contain genes for enzymes that make the cell resistant to antibiotics. Bacteria engaging in **conjugation** exchange genetic material, including both bacterial DNA and plasmids. In eukaryotic cells, some transposons, or jumping genes, move within the chromosomal DNA, disrupting the genome. Bacteria defend themselves against bacteriophages using **restriction enzymes,** which cut DNA at special "target sites."

Oncogenes specify proteins that help regulate cell division. The **viral oncogenes** of a tumor virus, which stimulate cell division, resemble, and almost certainly evolved from **proto-oncogenes** of eukaryotic host cells.

Energy organelles—chloroplasts and mitochondria—have their own DNA. The circular form of these DNA molecules and the genes they contain suggest that they derived from ancient prokaryotes that lived within ancient eukaryotic cells, a theory called the **endosymbiotic theory.**

Review and Thought Questions

Review Questions

1. What characteristics of a virus makes it nonliving?
2. How can a virus or other nonliving genetic entity evolve through natural selection? Give two examples.
3. How does a virus attach to a cell and transfer its genetic material into the cell?
4. What advantages does integrated replication have over autonomous replication?
5. What is a retrovirus? How does it differ from other RNA viruses?
6. What is the difference between a plasmid and a transposon?
7. What evidence suggests that mitochondria and chloroplasts were once prokaryotes that lived independently of other cells?

Thought Questions

8. If you take antibiotics frequently, you will select for *E. coli* that carry genes for antibiotic resistance. Suppose that you contracted tuberculosis (TB) from your roommate. Could plasmids then transmit antibiotic resistance from the *E. coli* in your intestines to the TB bacteria? Why or why not?
9. Many biologists suspect that transposons move genes among different species of eukaryotes. For example, some evidence suggests that within the past 100 years, a parasitic mite has carried transposons called P elements from one species of fruit fly to another. What evolutionary consequences might result from such transfers?

About the Chapter-Opening Image

Gene Kelly (1912–1996), dancing star of Broadway shows and Hollywood films, jumping over Indian corn, Barbara McClintock's experimental organism. The varied colors of the corn kernels result from the movement of "jumping genes," or transposons.

 On-line materials relating to this chapter are on the World Wide Web at **http://www.harcourtcollege.com/lifesci/aal2/**

13

Genetic Engineering and Recombinant DNA

The Maverick

Early in the morning of October 13, 1993, reporters and a photographer showed up at the door of 50-year-old unemployed surfer and molecular biologist Kary Mullis (1944—). The Nobel prize winners for 1993 had been announced in the middle of the night, and all the winners were being interviewed. The reporters gathered around Mullis's beachfront apartment in southern California and caught him on his way out for a morning of surfing (Figure 13-1). They asked him, Did he know he had won? Was he happy? Surprised? How did he feel when the phone call came from Stockholm, Sweden? Mullis knew—he had already gotten his call in the small hours of the morning. Naturally, he was happy. "I realized I did something significant, the kind of thing you get the Nobel prize for," he told reporters, "but never in my wildest dreams did I ever think I'd win."

He wasn't the only one surprised. Explained one colleague to *Esquire* magazine, "He doesn't fit the normal mold of a Nobel-prize winner. . . . He's a wild man. There's a certain amount of politicking involved, and I was not sure his personality and his lifestyle would have allowed him to win." Mullis is indeed a maverick. Eventually, he offends nearly everyone, and he seems to enjoy doing it. At one scientific conference, he began laughing when someone else was speaking. When the speaker chided Mullis for laughing at a serious matter, he responded, "I'm not laughing at that. I'm laughing at you." Mullis is a far cry from the serious, dedicated researcher who still goes to the lab 7 days a week long after retirement. In fact, he hasn't worked in a lab in years.

Mullis, whose personality is equal parts boyish charm and astonishing boorishness, grew up in South Carolina, where, like many a Southern boy, he learned about life by roaming the woods and building rockets from drugstore supplies. At Georgia Tech, he earned a degree and got married for the first time.

In 1966, he arrived at the University of California, Berkeley, with a young wife and baby daughter, to work on a Ph.D. in biochemistry. Berkeley was all he had hoped it would be, and then some. "Six years in the biochemistry department didn't change my mind about DNA," he told UC Berkeley's alumni magazine, "but six years of Berkeley changed my mind about almost everything else." Besides psychedelics such as LSD, Mullis also discovered free love and divorce. "It wasn't a good place to be married."

By 1981, he was on his third marriage, with two young sons. For a while he worked as the night manager at the Buttercup Bakery in Berkeley. When he took a job as a chemist for one of the earliest biotechnology companies, Cetus Corporation, it was as a low-level technician. At Cetus he fell in love with another biochemist and, at 36, ended his third marriage.

One Friday night in the spring of 1983, the two biochemists headed for Mullis's weekend cabin in the coastal mountains of northern California. Late that night, as Mullis drove along a twisting mountain road through vineyards and redwoods, he filled the silence by thinking about DNA. Specifically, he was interested in making his job at Cetus—building short pieces of synthetic DNA—a little easier.

As the winding road unfolded before him, he thought about the way the two strands of DNA untwist and separate so that they can be copied. He designed experiments in his head, wondering how he might single out one particular nucleotide in the DNA and make copies of it. He saw ways his imaginary experiments might go wrong, and twisting his thoughts this way and that, fixed them. His idea seemed foolproof—he couldn't find anything wrong with it. In theory, he should be able to take a test tube filled with thousands of genes, single out a stretch of nucleotides, and make millions

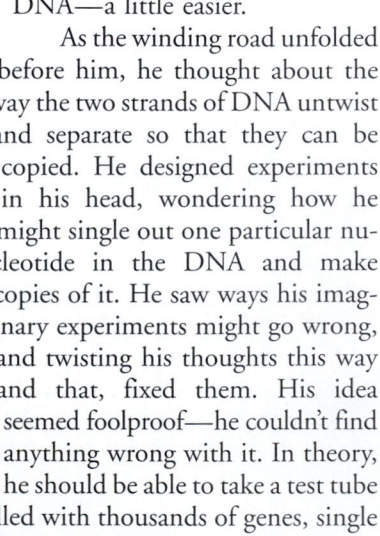

Paraskevas Photography

Key Concepts

- Genetic engineering started with the selective breeding of organisms with high levels of natural or induced variation.

- Recombinant DNA technology allows biologists to create genes that do not exist in nature.

- Using recombinant DNA, molecular biologists can engineer both prokaryotic and eukaryotic cells to make specific proteins in large amounts.

- Biologists can also engineer plants and animals to produce new proteins and to have new biological properties.

of copies within a day. For a molecular biologist, it was undreamed-of power.

He realized with a shock that his idea was almost good enough to win him a Nobel prize. There had to be something wrong with his idea, he thought. It was too obvious, too slick, too easy. The idea was this: Mullis could make an oligonucleotide, a short piece of DNA, that binds to a complementary sequence in a far longer strand of DNA (Figure 13-2). Once the oligonucleotide was bound to the DNA, it could serve as a primer for DNA polymerase. The DNA polymerase would copy only the DNA immediately adjacent to the bound oligonucleotide. In the next step, the newly made DNA could itself be used as a template to make still another strand (as we discuss later in this chapter). The result would be a chain reaction: one copy of a specific DNA sequence yielding 2 copies, then 4, then 8, then 16, and so on.

Back at Cetus, Mullis's boss told him to take all the time he needed to develop his idea. Colleagues pitched in and helped him make the idea work. By mid-December, Mullis had worked out all the bugs in a new technique called the **polymerase chain reaction (PCR)**. PCR specifically and repetitively copies any segment of DNA between any two defined nucleotide sequences. Today, specialized machines about the size of a toaster perform this process automatically, so that a single copy of a chosen sequence can be multiplied into millions or billions of copies in less than a day.

Finding a short sequence of nucleotides in a cell's DNA is like searching for a needle in a haystack. PCR finds the right sequence, then copies it repeatedly until, after 30 rounds of replication, the original haystack is dwarfed by a haystack of needles. Not only does PCR help find particular sequences, it also makes them in usable amounts. For molecular biologists to study a particular nucleotide sequence, they need more than one copy of a sequence; they need thousands of identical

Figure 13-1 Kary Mullis, inventor of PCR.
(©1997 Darryl Estrine Photography)

copies. Mullis's idea was a singularly valuable technique that would allow thousands of other molecular biologists to make important discoveries.

The uses for PCR seem limitless. PCR enables crime experts to use DNA fingerprinting on mere specks of blood and other tissues to identify who has been present at the scene of a crime. PCR allows researchers to pick out trace amounts of DNA in blood from HIV and other viruses, as well as pick up early signs of cancer and genetic defects in human eggs and embryos. PCR allows archaeologists and paleontologists to study minute amounts of DNA from ancient animals and plants—from the 130-year-old tissues of Abraham Lincoln to an 18-million-year-old magnolia leaf preserved in a peat bog.

Mullis's discovery might have launched his career, even made him happy. But it did not. Almost as soon as Mullis had perfected the technique, other researchers at Cetus began using the technique to do experiments. Mullis thought his name would be on those papers—as a coauthor—but his colleagues didn't agree. A simple acknowledgment of his invention near the end of the paper seemed like enough to them. Tempers flared, and Mullis was asked to leave if he couldn't get along with his colleagues. In the end, Mullis left Cetus, taking with him a $10,000 bonus for his invention. It was little enough. Cetus made millions of dollars from his invention and ultimately sold the rights to PCR for a mind-boggling $300 million. But even that was a bargain. Today, PCR technology is a $1-billion-a-year industry.

Except for a brief stint as a consultant, Mullis has mostly abandoned conventional science. Friends and colleagues suggest that his talents are being wasted. Modern molecular biology, however, is as much a social endeavor as an intellectual one, and Mullis's intellectual talents are of little value in an environment where getting along with people is paramount. Doing science, especially doing science that others notice, means

Figure 13-2 PCR is a simple, powerful technique for multiplying specific sequences of DNA. A. When DNA is heated, the two strands uncoil. They are then cooled and replicated. The cycle of heating, cooling, replicating, and then heating again is repeated until millions or billions of copies of the sequence are obtained. B. Short segments of single-stranded DNA called oligonucleotides act as primers and allow researchers to replicate a particular sequence, not just any DNA. The 20 or so bases of the oligonucleotide pair with the correct segment of the DNA and initiate replication.

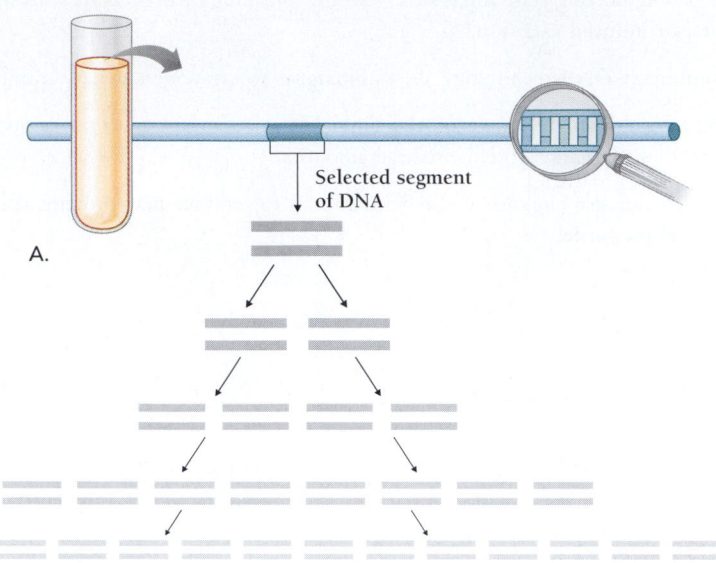

fitting in, belonging to one or another "club." A few researchers—loners like Barbara McClintock—do not entirely mind if no one appreciates their discoveries. They enjoy their work so much that they continue in a virtual social vacuum.

Mullis will be starved for neither money nor attention. He eventually won $450,000 from the prestigious Japan Prize and a $412,500 Nobel prize for his invention of PCR. And although he may not have a desk at a university or his name on a door at a biotechnology company, his Nobel prize has indirectly provided him with many opportunities for fame and fortune.

WHAT IS GENETIC ENGINEERING?

Thousands of years ago, before cities or farms existed, nomadic hunter-gatherers collected grain and seeds from the plant varieties that yielded the most grain. As they worked with the grain, carrying it or storing it, they spilled some of the grain. In the spring it sprouted and people had only to harvest their crop. In time, they learned to deliberately plant some of their seed in places where it would grow well. They became farmers. Populations that had been nomadic settled in one place to farm.

In time, these first farmers learned to observe closely the individual plants in their fields. Grains from the plants that produced the most seed, the largest seeds, or the best-tasting seeds were saved for the next year's crop. Ten thousand years ago, the first farmers began practicing artificial selection, breeding both plants and animals.

The success of agriculture allowed human populations in the Middle East, northwestern China, and southern Mexico to flourish. As these populations became larger and denser,

they built villages and cities. By 5000 years ago, the peoples of the Middle East had mastered enough microbiology to ferment grain into beer and to culture milk into cheese and yogurt.

The early domestication of plants and animals depended primarily on selecting which individuals to breed. Once breeders had chosen desirable varieties of plants or animals, the trick was to preserve and enhance their genetic traits. They did this by allowing individuals with the desired qualities to breed only with each other. Sheep with the thickest wool, for example, were bred to one another. This ancient practice is depicted in the Bible, where Jacob keeps his spotted goats apart from the nonspotted goats of his father-in-law. Brewers and cheese makers likewise learned to select the right strains of microorganisms by starting each batch with the right "starter."

Beer brewing, cheese making, and agriculture itself are early examples of **biotechnology**—the use of living organisms for practical purposes. Modern biotechnology depends as heavily on the selection of organisms for their desirable genetic qualities as the first biotechnology did 10,000 years ago. Genetic engineers can alter the genetic traits of any organism, whether sheep, goats, wheat, corn, yeast, or mold (Figure 13-3). The greatest difference between ancient methods and modern ones is the precision and rapidity with which organisms can be altered. But modern genetic engineering also transgresses the natural boundaries among species. Genetic engineers now happily move genes among all kinds of organisms, including humans, mice, tomatoes, yeasts, and bacteria.

Biotechnology first arose 10,000 years ago when humans began selecting and breeding useful plants, animals, fungi, and microorganisms.

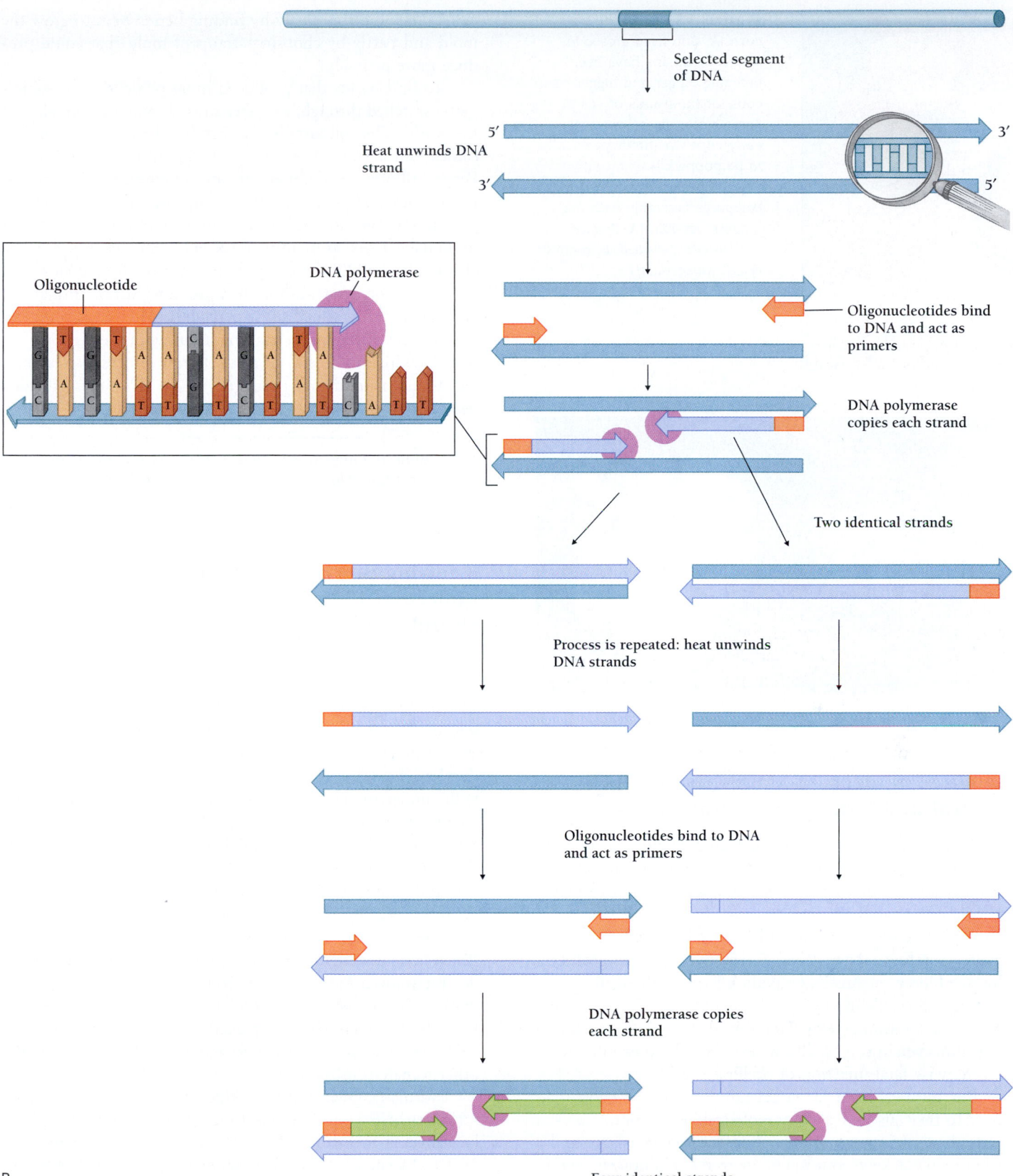

Selected segment of DNA

Heat unwinds DNA strand

5′   3′

3′   5′

Oligonucleotide

DNA polymerase

Oligonucleotides bind to DNA and act as primers

DNA polymerase copies each strand

Two identical strands

Process is repeated: heat unwinds DNA strands

Oligonucleotides bind to DNA and act as primers

DNA polymerase copies each strand

Four identical strands

B.

A.

Figure 13-3 Prehistoric corn. A. Cob from a cave in Mexico. Humans have been breeding bigger and bigger corn cobs for thousands of years. Some biologists believe that early corn was bred specifically to be popped. B. A recently popped prehistoric corn kernel compares favorably with the modern version. (*A, Science VU/Visuals Unlimited; B, courtesy of C.P. Mangelsdorf*)

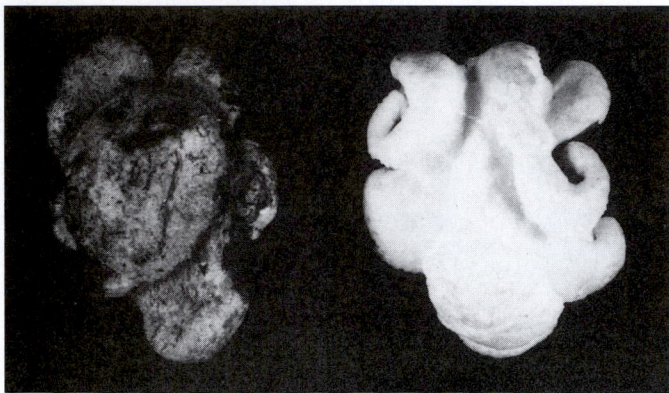

B.

Genetic Engineering Makes Use of Both Natural and Induced Variation

In 1929, Alexander Fleming discovered penicillin as the result of a chance contamination of one of his bacterial cultures. Fleming isolated the mold responsible for killing his bacteria and grew the mold in pure cultures. In 1938, 9 years after Fleming's discovery, two Oxford University scientists—Ernst Chain, a refugee from Nazi Germany, and his professor, Howard Florey, an Australian pathologist—realized that penicillin might be able to play a role in controlling diseases caused by bacteria. Within 2 years, Florey and Chain announced that penicillin, even in very small amounts, could protect mice from an otherwise fatal injection of bacteria.

But humans need far larger amounts of penicillin than mice. To treat just one adult for a bacterial infection, a doctor needs about 7.5 grams of penicillin (the weight of two and a half animal crackers). But to obtain even a millionth of a gram of penicillin, Florey's team had to process enough mold to fill nearly 30 bathtubs. Penicillin was so rare that, in some cases, the precious drug was extracted from the urine of patients and given to them once again in subsequent doses. It was clear that to make penicillin in the amounts needed to control human disease, scientists had to find a way to make much more peni-

cillin. They did this partly by finding better ways to grow the mold and partly by choosing strains of mold that could produce more penicillin.

To find molds that produced more penicillin, microbiologists searched through, or screened, soil samples from all over the world. The soil sample that produced the most penicillin came from a U.S. Department of Agriculture Laboratory in Peoria, Illinois. Microbiologists in government laboratories, universities, and pharmaceutical companies then increased the genetic variation in the Peoria strain of mold by subjecting it to mutagens—x rays, ultraviolet light, and chemicals. The changes in the DNA were random and most were useless, but because of the great number of mutations, a few turned out to increase the production of penicillin. After each round of mutagenesis, researchers chose the strains with the highest yield of penicillin. By the end of this selection process, researchers had found a mold that produced nearly 100 times as much penicillin as the original Peoria strain.

Biotechnology selects useful traits from a range of variation. This variation can be natural or induced by means of mutagens.

Knowing Biochemical Pathways Helps Molecular Biologists Design Useful Organisms

To find the strain of mold that could produce the most penicillin, Florey and Chain and the other researchers depended on the same general approach that breeders have always used, screening large numbers of individuals for those with the best expression of a particular trait—in this case penicillin production. They improved their chances by treating the molds with mutagens. However, they had no idea why the Peoria strain produced more penicillin than other strains. In the 1940s, researchers could not predict what mutations would increase the yield of penicillin.

Today, however, molecular biologists know so much about the biochemical pathways in some organisms that they can predict what kind of mutation will produce a desired trait. The genetic engineering of tomatoes provides one example. In order for tomato farmers to get tomatoes safely to the supermarket, they must pick the tomatoes when they are green and hard. If the tomatoes were picked when they were ripe and soft, they would be crushed to a pulp by the weight of the other tomatoes in the box.

Normally, green tomatoes ripen into sweet, flavorful red tomatoes before we eat them. Once the tomatoes are ready to be delivered to a store, they are treated with the plant hormone ethylene gas, which triggers their ripening (Chapter 33). The tomatoes also produce ethylene themselves, and, regardless of whether the gas is natural or factory made, the tomatoes ripen just the same. Unfortunately, about 40 percent of tomatoes are picked too soon and can never ripen properly. These are the tomatoes we all know so well that turn red but taste like wet cardboard. Only tomatoes that have developed different "flavor components," chemicals that make them taste good, are actually ready to be picked. But it is impossible for the grower to tell the green tomatoes that are ready to be picked from those that are not. All green tomatoes look the same.

One biotechnology company solved this problem by deliberately damaging a gene that controls the production of ethylene in the tomato. Such tomatoes can be left on the vine until they turn a pinkish orange, which signals that they have developed their flavor components. These tomatoes will never soften (or turn red and sweet) on their own, for they cannot make ethylene. Only after the tomato grower has harvested the tomatoes and the tomato distributor has treated the tomatoes with ethylene will they ripen. Since the tomatoes do not need to be picked until they have both color and flavor components, growers won't pick them too soon.

Biotech companies have also used specific knowledge of biochemical pathways to select hundreds of commercially important strains of bacteria and fungi. Some of the molecules produced by these strains include amino acids such as glutamic acid (used as the flavoring "monosodium glutamate," or MSG), vitamins such as riboflavin, and even industrial chemicals such as ethanol and acetone. Microorganisms also produce commercially important enzymes, such as those used to convert starch into fructose for use as a sweetener and the protein-digesting enzymes used in laundry detergents.

Molecular biologists can design useful organisms by inserting or destroying genes that code for proteins involved in specific biochemical pathways.

WHAT IS RECOMBINANT DNA AND HOW IS IT USEFUL?

The discovery, in the early 1950s, that genes are nothing more than chains of nucleotides led directly to our present ability to both alter individual genes and move them around wherever we want them. By the early 1970s, molecular biologists had learned to cut and paste pieces of DNA from different organisms (as well as from various mobile genes), and by the early 1980s, they had learned how to alter the individual nucleotides in a gene. Earlier work had already revealed how viruses and plasmids shuttle genes around the genomes of bacteria and other host cells. Molecular biologists found it easy to use these tiny molecular parasites to carry along and reproduce pieces of DNA that had been patched together in a test tube.

How Do Molecular Biologists Use Recombinant DNA?

Recombinant DNA is any DNA molecule consisting of two or more DNA segments that are not found together in nature. For example, researchers can join a gene for antibiotic resistance with a plasmid, then infect a cell with the plasmid, giving the cell a gene for antibiotic resistance. Because molecular biologists can now design and make recombinant DNA that fulfills particular requirements—scientific, medical, agricultural, or industrial—the resulting recombinant DNA molecules are sometimes called "designer genes."

Recombinant DNA technology is so powerful that it is difficult to summarize all the ways in which it has proved useful. Recombinant DNA has provided scientists with an irreplaceable tool for studying the structure, regulation, and function of individual genes; for unraveling the molecular bases of genetic diseases; for understanding why individuals differ in their responses to drugs and medicines; and for turning plants, animals, and microorganisms into chemical factories that can churn out vast quantities of proteins and other substances that these organisms would never make on their own (Figure 13-4).

Recombinant DNA promises purer vaccines (which have fewer risks), new drugs, gene therapy for treating genetic

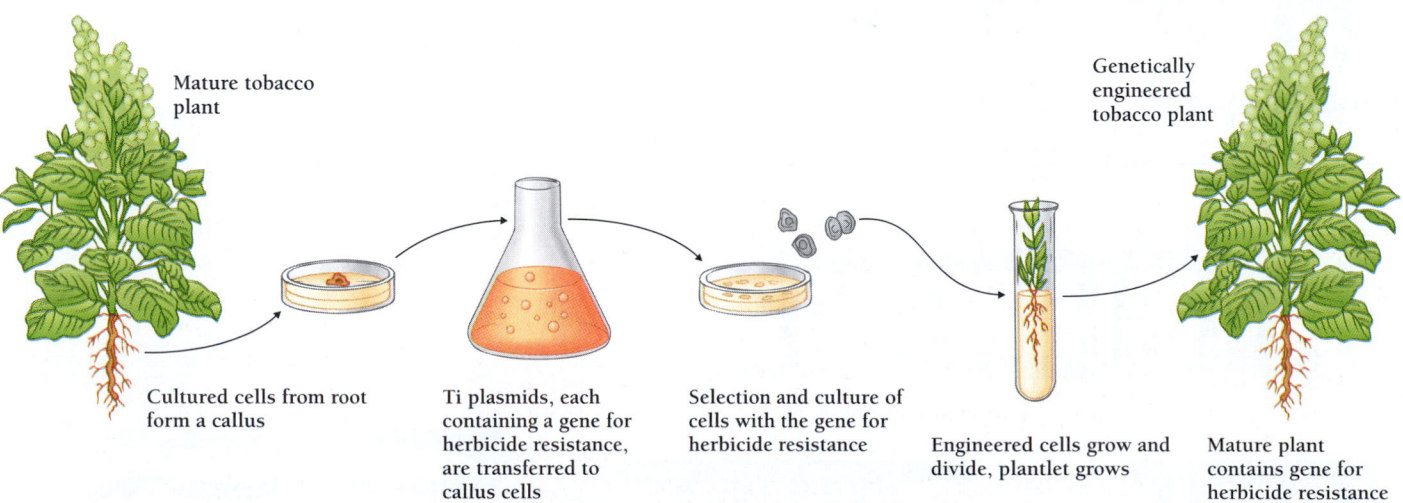

Figure 13-4 Recombinant DNA technology in action. Cells from a tobacco plant are grown in culture, then infected with a Ti plasmid carrying a gene for herbicide resistance and other genes. The growing cells are treated with herbicide, and only the cells expressing the new genes survive. These herbicide-resistant cells can then be grown into mature plants, which will bear seeds carrying the new genes.

Mature tobacco plant

Cultured cells from root form a callus

Ti plasmids, each containing a gene for herbicide resistance, are transferred to callus cells

Selection and culture of cells with the gene for herbicide resistance

Engineered cells grow and divide, plantlet grows

Genetically engineered tobacco plant

Mature plant contains gene for herbicide resistance in all cells

diseases such as cystic fibrosis, and genetic tests for hereditary diseases such as Huntington's disease or partly hereditary diseases such as breast cancer. Recombinant DNA means brand new proteins, oils, and carbohydrates—designed by molecular biologists and manufactured by genetically engineered organisms. Finally, recombinant DNA presents a thousand moral and ethical questions that cannot be answered solely by the scientists who raise them. These questions must be understood and answered by all educated citizens.

The development of recombinant DNA technology depended on the union of many different lines of research, in-

cluding work on (1) restriction enzymes in bacteria, (2) DNA replication and repair, (3) replication of viruses and plasmids, and (4) the chemical synthesis of specific nucleotide sequences. We will see how information and techniques from these different fields all contributed to the ability to make recombinant DNA.

Recombinant DNA techniques join together pieces of DNA in combinations that do not occur in nature.

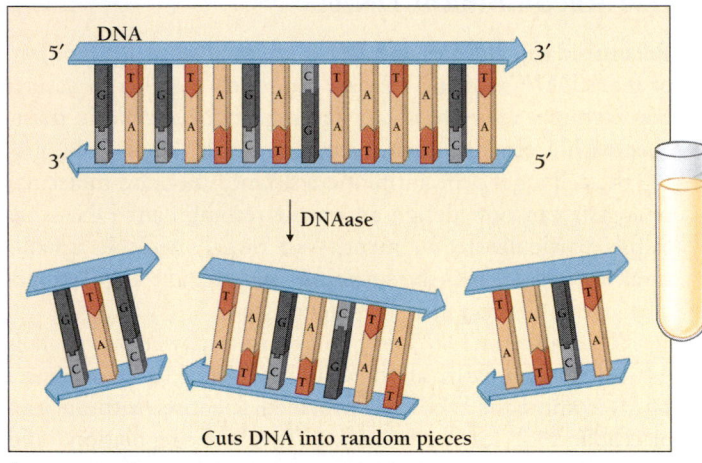

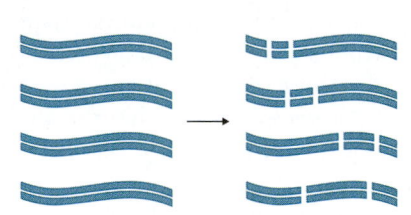

Random cleavage: every strand is different

A.

Figure 13-5   Cutting up DNA. A. DNAases cut DNA at random. Each copy of DNA is cut differently. B. Restriction enzymes cut each copy of DNA at the same places. Matching fragments have the same length and sequence. Restriction enzymes can cut the DNA with blunt ends, as in B, or with sticky ends, as in C.

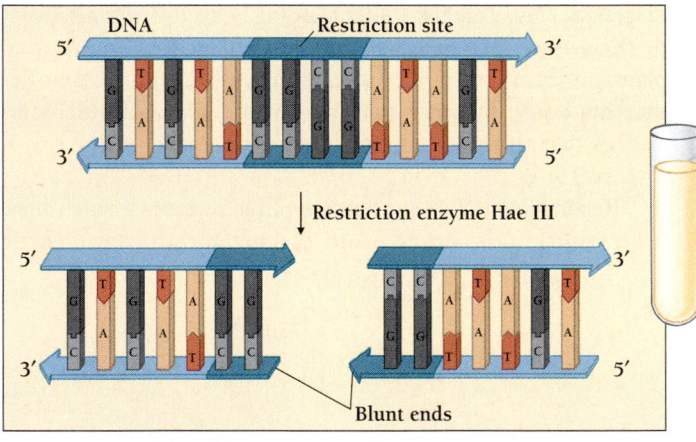

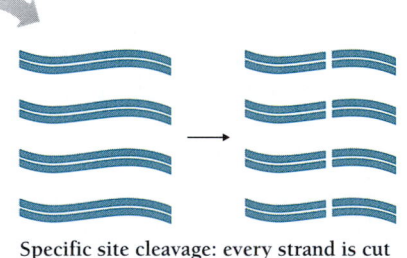

Specific site cleavage: every strand is cut in the same place

B.

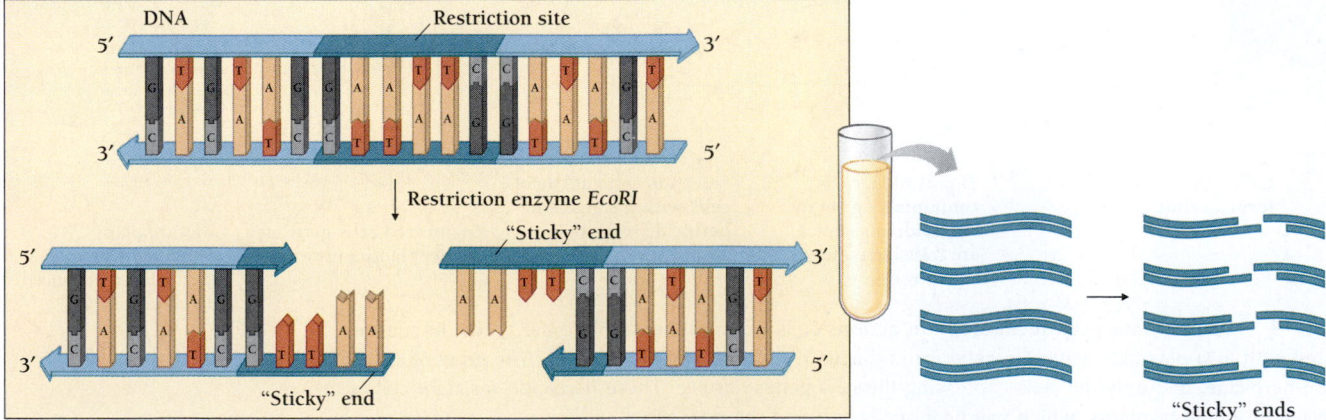

C.

How Do Restriction Enzymes Cut Up a Genome?

DNAase and RNAase are enzymes that cut the links between the nucleotides in DNA or RNA, respectively. DNAase usually cuts DNA into short fragments at random (Figure 13-5A). In a test tube filled with DNA, each of the many copies of a genome will be cut differently. The resulting fragments are all cut into different sizes and the pieces from one genome overlap those in another. In such a mess, there is little hope of finding a specific sequence. To isolate a specific piece of DNA, a researcher must instead use **restriction enzymes,** special DNAases that cut DNA only at particular sequences (Figure 13-5B).

Happily, such enzymes exist. Bacteria use restriction enzymes to defend themselves against viruses and other mobile genes. When a bit of alien DNA enters a bacterium, the restriction enzymes cut it to bits. Bacteria have many kinds of restriction enzymes, each of which recognizes a different nucleotide sequence. The sequence recognized by a given restriction enzyme is called a **restriction site** and typically consists of four or six nucleotides, though some restriction sites are longer. Some restriction enzymes cut each of the two strands of DNA in the same place, so that the resulting fragment of DNA has "blunt" ends (Figure 13-5B). Other restriction enzymes make staggered cuts, which result in single-stranded, "sticky" ends that easily join with other such fragments that have complementary sequences (Figure 13-5C).

Researchers have prepared restriction enzymes from over 200 different bacterial strains. Each enzyme recognizes a specific sequence, with over 90 such restriction sites now described. One strain of *E. coli,* for example, uses a restriction enzyme, called *EcoRI,* that binds to and cuts DNA wherever it finds the sequence GAATTC (Figure 13-5C). A restriction enzyme such as *EcoRI* will cut multiple copies of DNA everywhere a particular sequence appears. The result is a matching set of **restriction fragments**—pieces of DNA that begin and end with a restriction site.

By comparing the sizes of restriction fragments, researchers can establish a **restriction map,** which shows how the restriction sites are placed within a piece of DNA. More importantly, genetic engineers can join these fragments together in novel ways. A restriction enzyme that makes fragments with sticky ends can be used to make human and mouse fragments that stick together (Figure 13-6A). They can, for example, join a restriction fragment from a human being to a fragment from a mouse or a bacterium.

Restriction enzymes, which bacteria use to defend themselves against viruses, cut DNA molecules at specific sequences. Restriction enzymes allow the preparation of DNA fragments with defined lengths and sequences.

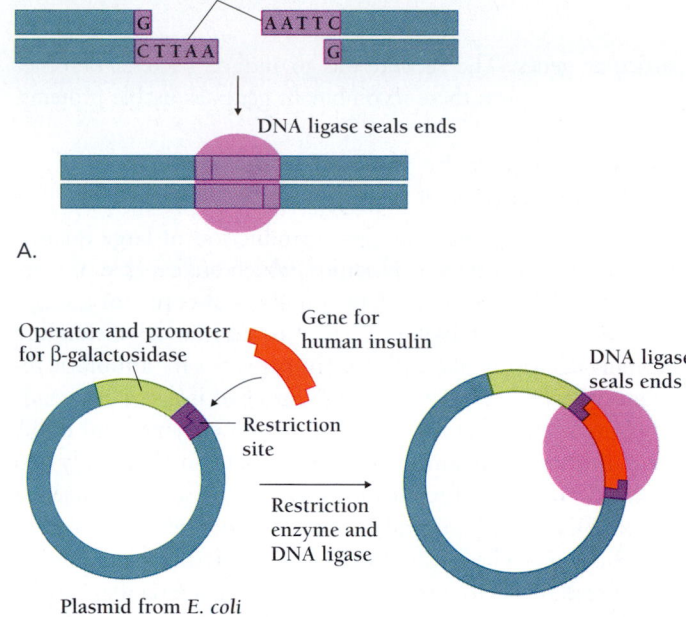

Figure 13-6 Genetic engineering. A. Sticky ends enable researchers to join pieces of unrelated DNA. B. DNA from a mouse and human can be joined, as long as they have been cut at the same sequence by the same restriction enzyme.

How Do Molecular Biologists Join Restriction Fragments Together?

Recall from Chapter 10 that during ordinary DNA replication, the enzyme **DNA ligase** links together the separate pieces of DNA into one continuous strand. Molecular biologists depend on DNA ligase to link restriction fragments together. For example, to make the hormone insulin, researchers initially inserted a sequence coding for human insulin next to the operator and promoter for β-galactosidase from *E. coli* (Figure 13-6B). The manufactured recombinant DNA then contained a piece of a bacterial DNA that regulated the expression of an adjacent mammalian gene. This recombinant DNA, put back into *E. coli,* produces insulin when the bacteria are exposed to the milk sugar lactose. Ordinarily, the presence of lactose would signal the bacteria to prepare to digest lactose. But bacteria carrying the recombinant DNA respond to lactose by producing human insulin.

DNA ligase can link together two pieces of DNA from different sources to produce recombinant DNA.

How Do Molecular Biologists Express Recombinant DNA in Bacteria and Other Hosts?

Once scientists taught themselves to make recombinant DNA using DNA ligase and restriction fragments, they faced two serious challenges. The first challenge was to produce large quantities of particular genes. The second was to induce bacteria or other host cells to express these recombinant genes as usable proteins.

How can bacteria be induced to make great numbers of copies of a gene?

The answer to the first problem (production of large quantities of specific genes) was plasmids, which are capable of forcing a bacterial cell to make large numbers of copies of a single gene. Recall from Chapter 12 that R (resistance) factors are naturally occurring plasmids containing genes for antibiotic resistance. By the early 1970s, researchers had isolated and studied many such R factors. Using restriction enzymes and DNA ligase, they made streamlined plasmids that contained only two parts: (1) an origin of replication—the DNA sequence needed to start DNA synthesis and (2) one or more genes for antibiotic resistance. The antibiotic resistance acted as a marker, allowing researchers to distinguish and select bacteria that carried the plasmid from those without the plasmid.

Using DNA ligase, researchers can hook any gene to such a plasmid. A host bacterium will copy the plasmid DNA as if it were its own, enabling the recombinant plasmid to replicate many times. Researchers then grow the bacteria in an antibiotic. Because the plasmid's antibiotic resistance gene protects the host bacteria against being killed by the antibiotic, only the genetically engineered bacteria survive.

Microbiologists have long used the word **vector** [Latin; carrier, rider] for any organism or virus that carries disease from organism to organism. Molecular biologists therefore use the word "vector" to describe anything that carries genes from organism to organism. Plasmids are not the only vectors. Researchers have also used specially modified bacteriophages, animal and plant viruses, transposable elements, and even fragments of chromosomes.

Vectors are DNA sequences from viruses, plasmids, and other mobile genes that can carry recombinant DNA into cells.

How can bacteria be induced to make eukaryotic genes?

If a researcher wants to express a gene from a virus or a bacterium, pulling out a particular gene for cloning is not too hard. Often enough, the entire sequence is already known. Extracting genes from the chromosomes of plants and animals, however, is not only more difficult, it is often useless. Recall that a eukaryotic gene is interrupted by long sequences of introns. Bacteria, which must express the gene, do not have any machinery for recognizing and removing introns from the resulting pre-messenger RNA. (Prokaryotes, recall, have no introns.) If the bacteria cannot remove the introns, they cannot make a translatable mRNA. So how can researchers acquire a clean, intronless copy of a eukaryotic gene that can be cloned using bacteria?

The solution comes from retroviruses, the viruses that use RNA as their nucleic acid, yet insert themselves into the DNA genomes of eukaryotes. Recall (from Chapter 12) that retroviruses use the enzyme **reverse transcriptase** to copy RNA into DNA. Reverse transcriptase copies almost any RNA into DNA. Researchers who want a particular gene can start with the mature mRNA for that gene, which has already had all the introns removed. (The mature mRNA comes from eukaryotic tissues that express the desired gene.) Reverse transcriptase then copies that mRNA back into DNA. The result is a DNA molecule, called **complementary DNA,** or **cDNA,** that is complementary to the mRNA from which it was copied. Unlike the DNA in the genome, cDNA has no introns, as it is a copy of the mature mRNA (Figure 13-7).

Another enzyme copies the single-stranded cDNA into double-stranded cDNA, the second strand of which is identical in sequence to the original mRNA. The cDNA "gene" is now ready to be joined to a vector and propagated.

Actual genes from eukaryotes cannot be expressed in bacteria. Molecular biologists must make a DNA copy (called cDNA) of a eukaryote's mRNA using reverse transcriptase. The cDNA can then be propagated using standard methods.

Can host cells be induced to express polypeptides in a usable form?

Unfortunately, not all eukaryotic genes can be expressed in bacteria. As we saw in Chapter 11, many genes require extensive modification of the polypeptide itself in order to function. Bacteria cannot perform these modifications. Most membrane proteins, for example, require modifications that can only be made in eukaryotic hosts. Recombinant *E. coli* designed to produce insulin, for example, cannot produce insulin that will function properly in the human body. Nonetheless, hundreds of functional proteins have been expressed from recombinant DNA molecules in bacteria.

The production of other proteins requires that the recombinant DNA be expressed in eukaryotic cells, which are able to modify proteins after translation. One particularly useful cell line comes from a moth that is susceptible to infection by a virus called baculovirus. After infection, these cells devote almost all their protein synthesis to the production of the virus's coat protein. By substituting another protein-coding DNA for the coat protein, molecular biologists can produce large amounts of a desired protein. Other useful cells, which can be grown in large vats under the proper conditions, include yeast, human kidney cells, and hamster ovary cells. Among the commercially important proteins made by eukaryotic cells from recombinant DNAs are insulin (used to treat diabetes), tissue plasminogen activator (TPA, used to treat heart attacks), and erythropoietin (used to stimulate red blood cell production, especially after kidney transplants).

Bacteria and various eukaryotic cells can produce large amounts of specific proteins that would otherwise be available only in small amounts.

How Do Researchers Make Multiple Copies of Recombinant DNA?

Throughout the 1970s and early 1980s, the only practical way to make enough copies of a particular DNA sequence to study it was to introduce a single recombinant DNA molecule (gene plus plasmid vector) into a bacterial host cell. The plasmid can induce the host cell to make up to a thousand copies of the gene it carries. In addition, researchers induce the bacterial host cell to divide rapidly. As the bacteria multiply, so does the recombinant DNA. Each colony of bacteria may grow into 10^{12} or more bacteria. If each one has 1000 (10^3) copies of the plasmid, such a colony may contain 10^{15} identical copies of the recombinant DNA. Researchers call this process "cloning the recombinant DNA."

The word "clone" comes from gardening. Gardeners and fruit growers often grow whole new plants from live twigs cut from mature trees and bushes. All of the plants derived in this way are genetically identical and constitute a **clone** [Greek, *klon* = twig]—a group of genetically identical cells or organisms. In humans, identical twins are a natural clone. Any group of genetically identical individuals constitutes a clone. Similarly, the offspring of a single bacterial cell are called a clone, and the many copies of a plasmid are called a clone. By extension, molecular biologists also refer to the many copies of recombinant DNA as a clone.

Cloning recombinant DNA depends on the DNA replication machinery of bacteria or other cells to copy recombinant DNA. Since Kary Mullis's invention of PCR, however, researchers have been able to make millions, or even billions, of copies of DNA sequences in a test tube, using a DNA polymerase purified from a bacteria. PCR can make huge numbers of individual DNA sequences very rapidly. The sequences must be short, usually no more than 20,000 nucleotide pairs, but the process is much faster than cloning.

PCR depends on the ability to make two specific polynucleotides that correspond to sequences at each end of the se-

quence to be copied. These two synthetic polynucleotides are typically each about 20 nucleotides long and are examples of **oligonucleotides** [Greek, *oligos* = few]. Each oligonucleotide serves as a primer for DNA polymerase. No enzyme is needed to unwind the DNA. Instead, in PCR, the DNA is simply heated until the two strands come apart. The DNA is cooled, and the two primers then bind to the complementary sequences—one on each strand (Figure 13-2). DNA polymerase then copies each strand until the experimenter (or the machine) stops the reaction by again raising the temperature.

The increased temperature also separates the template strand and the newly made complementary copy. Lowering the temperature then allows DNA polymerase to go back to work, again starting with the oligonucleotide primers. The only region that will be copied—again and again with further cycles—is the segment flanked by the two primers. The result is the rapid and exclusive multiplication of one specific segment of DNA.

Cloning and PCR technologies allow researchers to make thousands or millions of copies of individual genes.

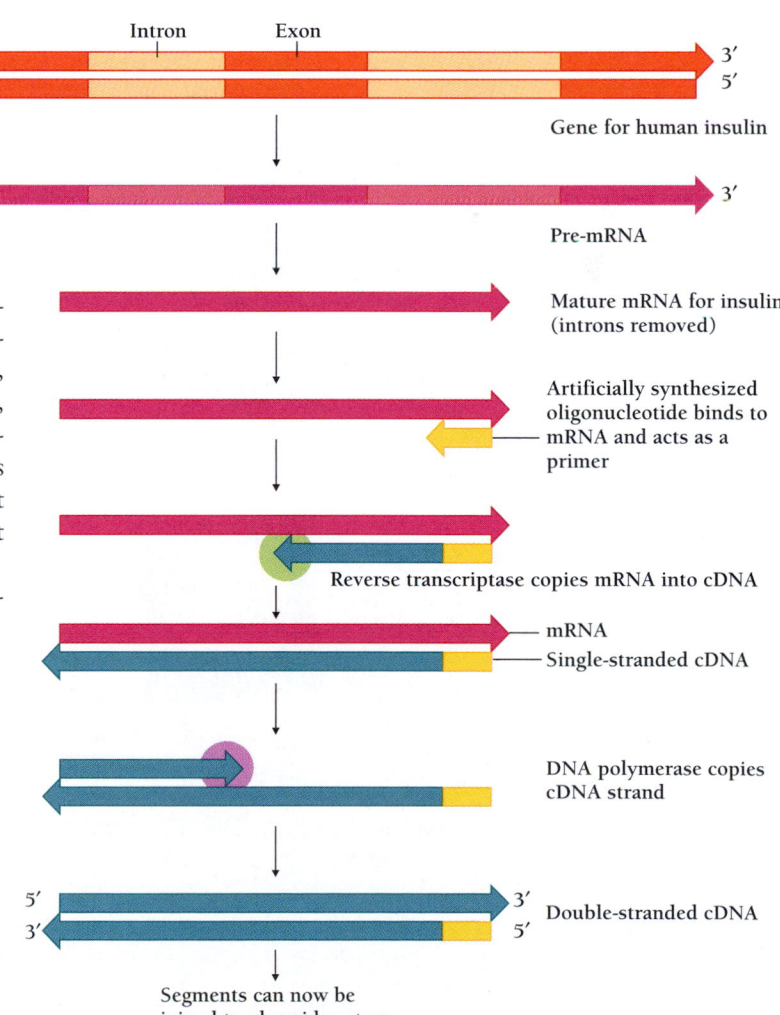

Figure 13-7 **How do researchers derive human genes without introns?** In a human cell, the gene for human insulin is transcribed into pre-mRNA. The cell edits out the introns to form mature mRNA. Researchers then isolate the mature insulin mRNA and copy it into DNA, which can be joined to a plasmid vector and expressed by a bacterium.

How Do Biologists Find the Right DNA Sequence in a Recombinant DNA Library?

The genome of any organism—whether bacterium, plant, or animal—can be cut into pieces with restriction enzymes. The collection of thousands or millions of restriction fragments from a single genome is called a **gene library.** These fragments can be combined with plasmid vectors and introduced into populations of bacteria, which can then replicate each fragment in the library indefinitely. If the library contains at least one representative of every sequence in a genome, it is called "a complete genomic library." Alternatively, a recombinant DNA library may be constructed from mRNA alone (using reverse transcriptase), in which case it is called "a cDNA library." In either case, a gene library is useful only if researchers can find the gene they want. Until recently, isolating the right gene was like trying to find a needle in a haystack.

Molecular geneticists have developed techniques for finding and cataloging the tens of thousands of sequences in a gene library. Just as a magnet can help find a needle even in a haystack, molecular biologists can use two kinds of molecular "magnets" to find particular pieces of DNA. One kind of magnet is a **probe,** a short bit of single-stranded DNA whose sequence is complementary to a small part of the sequence in the gene researchers are looking for, much like the DNA primers used in PCR. The other magnets are **antibodies,** proteins from the immune system that recognize and bind to other proteins. As we will see, antibodies do not detect the recombinant DNA itself, but only the proteins specified by the DNA.

A probe, also called a **hybridization probe,** can hybridize by base-pairing with DNA in the library to form **hybrid DNA**—DNA whose two complementary strands come from different sources (Figure 13-8A). Such a probe, dropped into

Figure 13-8 **Two techniques for locating a gene.** A. A hybridization probe locates a specific DNA sequence. B. Antibodies locate the protein product of the same sequence.

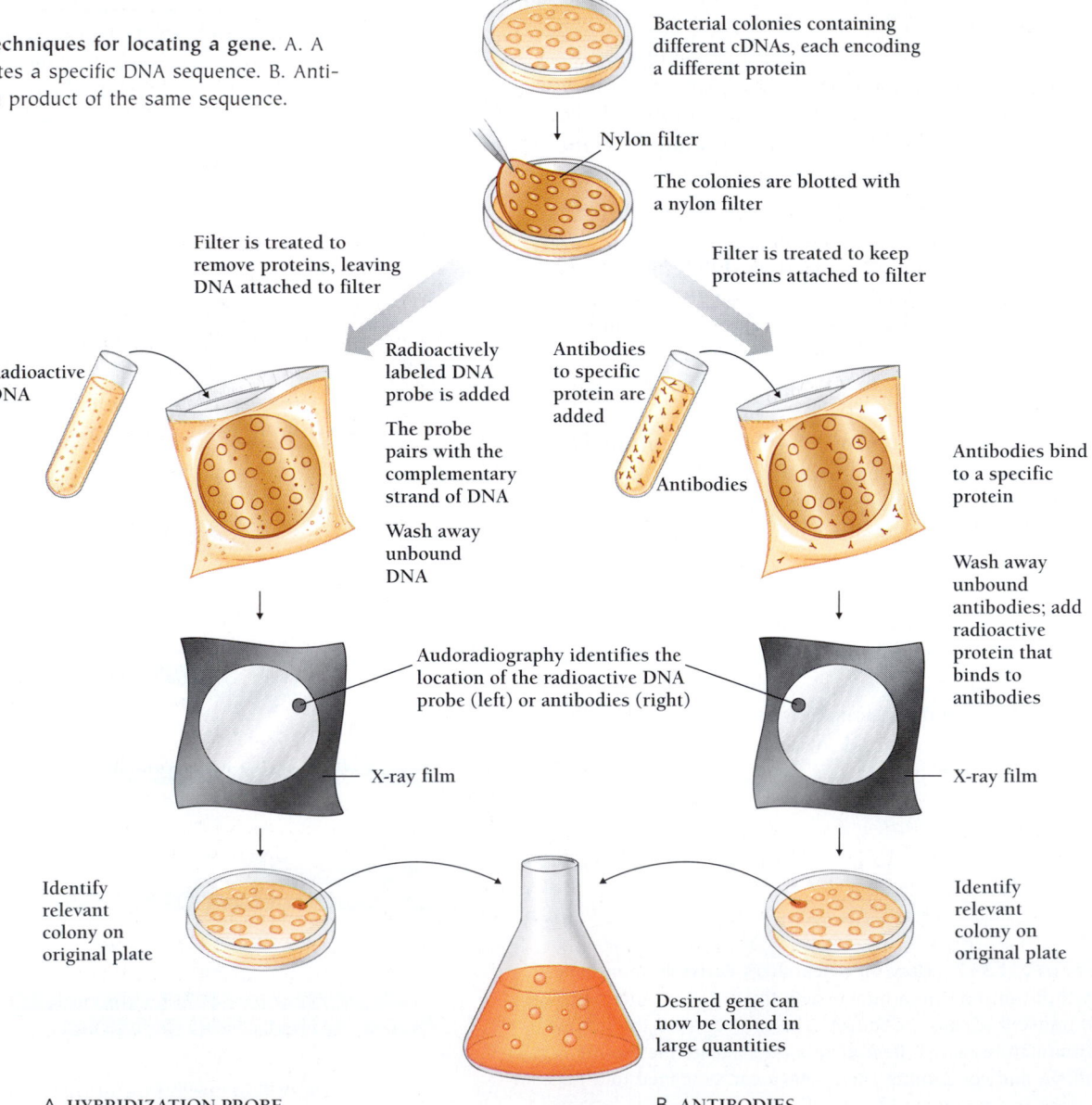

Bacterial colonies containing different cDNAs, each encoding a different protein

Nylon filter

The colonies are blotted with a nylon filter

Filter is treated to remove proteins, leaving DNA attached to filter

Filter is treated to keep proteins attached to filter

Radioactive DNA

Radioactively labeled DNA probe is added

The probe pairs with the complementary strand of DNA

Wash away unbound DNA

Antibodies to specific protein are added

Antibodies

Antibodies bind to a specific protein

Wash away unbound antibodies; add radioactive protein that binds to antibodies

Autoradiography identifies the location of the radioactive DNA probe (left) or antibodies (right)

X-ray film

X-ray film

Identify relevant colony on original plate

Identify relevant colony on original plate

Desired gene can now be cloned in large quantities

A. HYBRIDIZATION PROBE

B. ANTIBODIES

a test tube filled with thousands of restriction fragments, can locate and pair with the one fragment whose sequence complements its own as easily as a bloodhound can track down a lost child. Once the probe binds to the right piece of DNA, it is easy to locate the probe if it has been chemically or radioactively labeled. The technique of hybridizing a labeled DNA probe in order to find the right piece of recombinant DNA in a gene library is called molecular hybridization.

Researchers who know what protein they are looking for can also use antibodies to find the gene for that protein (Figure 13-8B). Antibodies are proteins with surfaces that recognize the shapes of foreign molecules, thereby giving animals an important defense against infection. When an animal is exposed to a foreign protein, it responds by making antibodies that bind specifically to that protein. Researchers start with a complete cDNA library laid out in a half-million colonies of bacteria, each colony carrying and expressing a different cDNA "gene" from the human genome. Over a few days, the 500,000 colonies can be treated with labeled antibody, which binds to and identifies the colony that is synthesizing the protein and therefore carries the gene the researchers are looking for. From the bacterial colony carrying the correct gene, researchers can clone the gene for study.

Two kinds of molecular probes allow biologists to locate specific genes. Hybridization probes detect genes in recombinant DNA clones, in cell extracts, and in cells. Antibodies detect the synthesis of specific proteins in colonies of bacteria containing recombinant DNA.

HOW CAN BIOLOGISTS STUDY THE EXPRESSION AND STRUCTURE OF MANY INDIVIDUAL GENES?

A DNA probe for a specific sequence can hybridize back to the corresponding gene in cellular DNA and to a recombinant DNA that contains that sequence. The same probe can also hybridize to any mRNA transcribed from that gene. Researchers can therefore use a cDNA or oligonucleotide probe to measure the amount of a particular mRNA in a tissue. Until the early 1990s, researchers performed such hybridizations one gene (one mRNA) at a time, as described in Box 13.1 (northern blots). These experiments showed that different cell types are specialized for the production of certain mRNAs: muscle cells thus have high levels of the mRNA myosin (the contractile protein of the thick filaments), but not of the mRNAs for hemoglobin.

In the last few years, new methods have allowed biologists to measure the levels of thousands of types of mRNA all at the same time. Researchers will soon be able to ask about the expression—at the level of mRNA—of *every* gene in a particular cell type. The simultaneous measurement of so many kinds

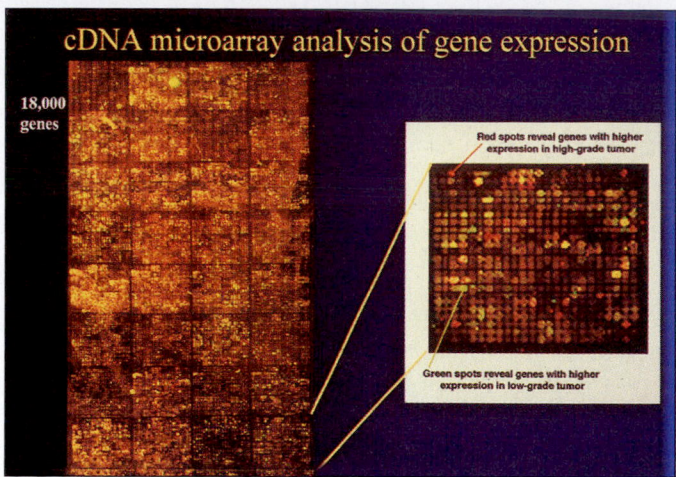

Figure 13-9   Patterns of gene expression in a benign versus a malignant brain tumor. DNA arrays allow the simultaneous monitoring of the levels of thousands of individual mRNAs. Differences in the expression of specific genes (coded here by green and red spots) give researchers clues about the classification and the treatment of different types of tumors. Red spots indicate individual genes whose expression is higher in a high-grade, malignant brain tumor than in a low-grade, less malignant brain tumor. Green spots indicate genes whose expression is higher in the low-grade tumor. *(Courtesy of Linda Liau, UCLA Department of Neurosurgery)*

of RNA depends on the use of thin, dime-sized wafers, containing thousands of tiny DNA spots in a fixed arrangement. Each DNA spot is put in place by a robot or (in some cases) by a lithographic process derived from integrated circuit technology. The whole arrangement of spotted DNA is called a **DNA array** or (in the case of the lithographic wafers) a *DNA chip*. Each spot in the array contains a DNA or an oligonucleotide corresponding to just one gene.

To determine the level of individual mRNAs, a researcher first isolates RNA from a tissue (for example, from a muscle or a tumor) and copies the RNA into cDNA, tagging each of the resulting cDNA molecules with a fluorescent label. Each kind of cDNA molecule now hybridizes to the corresponding DNA spot in the array. The higher the level of a particular mRNA, the greater is the fluorescence from that spot. At the end of the hybridization, the pattern of fluorescent spots tells which genes are expressed in that tissue.

A major problem in this kind of analysis is that there are so many spots. For example, researchers now think that mammals express a total of 80,000 to 100,000 different mRNAs, and they are developing techniques that can determine the levels of each of these at the same time. But no one is yet sure how best to analyze such huge amounts of data. Researchers are developing computer methods that will allow meaningful comparisons of individual spots and of expression patterns in types of tissues.

One of the first applications of this new technique was the study of different kinds of cancers (Figure 13-9). Cancer

BOX 13.1

Southern, Northern, and Western Blotting

Stop by a molecular biologist's lab bench today and chances are you will find a plastic box filled with gel and wired up to run an electric current through the slab of clear gel. Using a technique called **gel electrophoresis,** researchers can separate large molecules on the basis of their size. Gel electrophoresis is one of the most useful techniques available for studying DNA, RNA, and proteins.

In a gel containing sequences of DNA, the current draws the negatively charged DNA through the gel—a porous matrix that slows down the larger molecules as they snake and bump their way toward the positive electrode. A special stain reveals bands of DNA molecules, which group at different positions in the gel according to their size. The smaller fragments have moved farther, and the larger fragments remain closer to the rectangular well in the gel where they were first "loaded."

Sometimes, however, the bands all blur together and it can be hard to tell one fragment from another. For such complex samples, molecular biologist Ed Southern in 1975 developed a technique that lets the experimenter choose which fragments to look at. Usually, only a particular part of the genome is of interest—perhaps a specific gene or a polymorphic region used for DNA fingerprinting. In **Southern blotting** and **hybridization,** the DNA is separated by means of electrophoresis (Figure A). Then, for more convenient handling, the DNA is transferred from the thick, watery gel to special paper. The paper, placed between the gel and a stack of paper towels, wicks the DNA and all the liquid out of the gel. The result is a Southern blot.

Next, the researcher applies a radioactive DNA probe that base pairs, or hybridizes, with the region of interest. (One or more of the phosphorus atoms in the DNA probe is radioactive.) Since the DNA is normally already double stranded, heat is applied to melt the strands apart, allowing the probe to find its target. When the temperature is again lowered, some of the single-stranded probe molecules hybridize with DNA fragments in the blot.

The result of this Southern hybridization is a sheet of paper the size of the original gel with radioactive bands of DNA wherever there were fragments of the gene of interest. The Southern blot is then exposed to a sheet of film, which reveals only the bands of DNA lit up by the probe. The DNA sequences of interest light up, while irrelevant bits of DNA remain invisible.

Several years after Southern developed his blotting technique for DNA, scientists began studying complex samples of RNA with a similar procedure and dubbed the technique **northern blotting.** Other scientists blotted proteins out of protein gels and called it **western blotting.** There is no such thing as an eastern blot, but the other three points of the compass come up in laboratory discussions every day.

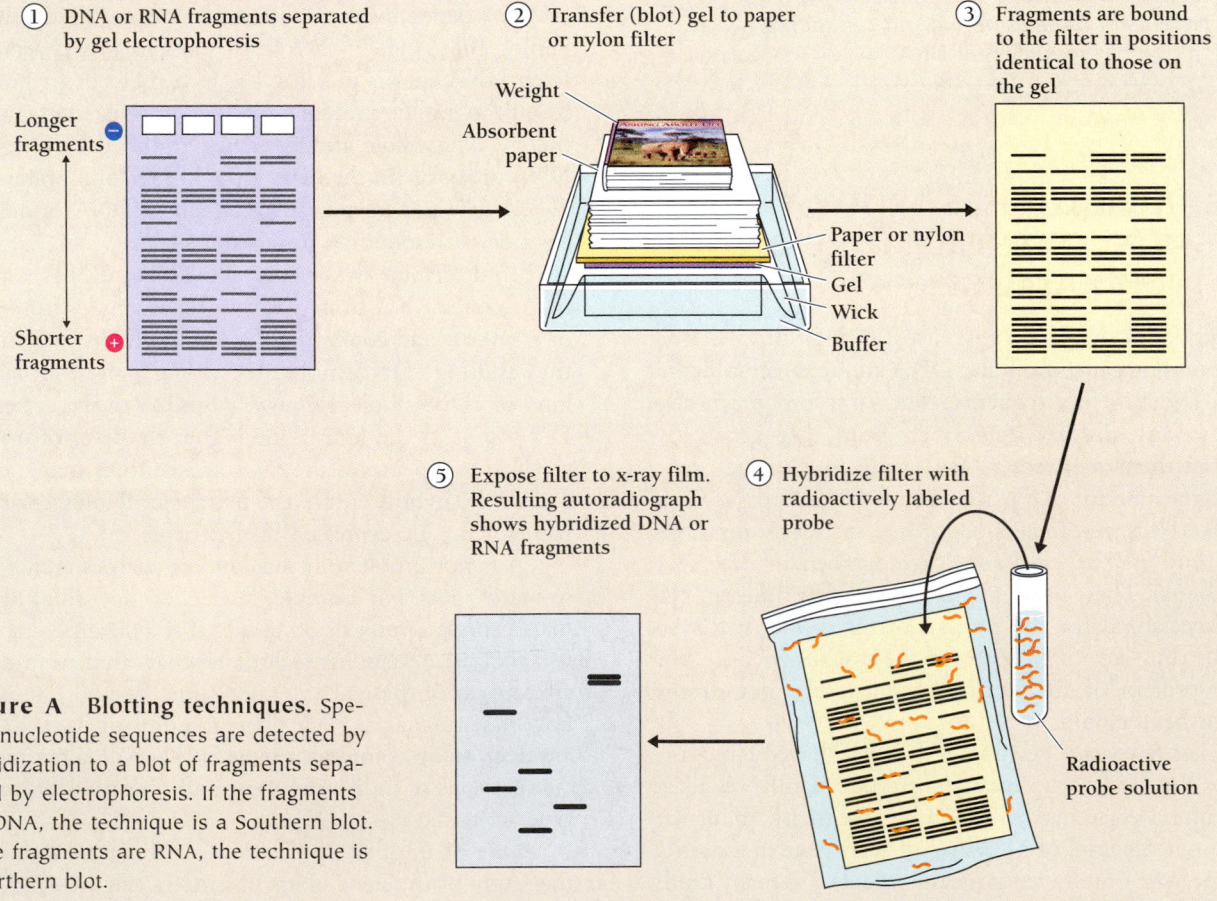

① DNA or RNA fragments separated by gel electrophoresis

Longer fragments
Shorter fragments

② Transfer (blot) gel to paper or nylon filter

Weight
Absorbent paper
Paper or nylon filter
Gel
Wick
Buffer

③ Fragments are bound to the filter in positions identical to those on the gel

⑤ Expose filter to x-ray film. Resulting autoradiograph shows hybridized DNA or RNA fragments

④ Hybridize filter with radioactively labeled probe

Radioactive probe solution

Figure A   Blotting techniques. Specific nucleotide sequences are detected by hybridization to a blot of fragments separated by electrophoresis. If the fragments are DNA, the technique is a Southern blot. If the fragments are RNA, the technique is a northern blot.

researchers have found that different types of tumors have different patterns of mRNAs. Researchers hope that such differences in mRNA patterns will ultimately help physicians predict how a particular tumor will grow and how best to stop it. DNA arrays also provide a powerful way to do human genetic mapping (as we will discuss in Chapter 14).

> *DNA arrays allow researchers to study the expression and the structure of thousands of individual genes at the same time.*

RECOMBINANT DNA CAN REPROGRAM CELLS TO MAKE NEW PRODUCTS

The ability to manipulate genes has already had important practical consequences. Hundreds of biotechnology, chemical, and pharmaceutical companies have begun to use recombinant DNA to make proteins useful to medicine, agriculture, and the chemical industry. Recent products include newly engineered forms of well-known proteins (such as an improved version of an enzyme used in laundry detergents), as well as previously unknown proteins—such as the protein missing in the lungs of cystic fibrosis patients or the protein growth factors that may prevent nerve death in individuals with neurological diseases. In addition, genetic engineers are breeding plants and animals with new characteristics—plants that are resistant to insects or to particular herbicides, sheep that produce a clotting factor in their milk, and pigs that produce human hemoglobin. Such gene products are then purified and used in medicine.

Figure 13-10   Effects of human growth hormone (hGH). hGH cures certain forms of dwarfism. Formerly hGH had to be isolated from human cadavers, and recipients of the growth hormone too often contacted hepatitis and other viral diseases. Today, hGH is produced by genetically engineering bacteria and is free of viruses that infect humans. (*© Jerry Cooke/Photo Researchers, Inc.*)

Genetically Engineered Bacteria and Eukaryotic Cells Can Make Useful Proteins

Recombinant DNA technology allows the reprogramming of cells to make an extraordinary number of products. These products include insulin, growth hormone, and other protein hormones, as well as ingredients for processed foods and enzymes needed to produce valuable small molecules or to destroy pollutants (Figure 13-10).

Vaccines are another protein product that can now be produced by means of recombinant DNA technology. Recombinant DNA techniques allow the preparation of uncontaminated vaccines because the immunizing molecules are made from viral DNA rather than from active viruses. Instead of growing the virus and then inactivating it, virologists isolate a piece of viral DNA that encodes a single viral protein. A single protein cannot cause a viral disease, but it can provoke an immune response that will guard against infection by the actual virus. Recombinant DNA technology has produced safer vaccines for many kinds of disease-causing viruses. Researchers hope that one day such techniques may also produce vaccines for diseases such as herpes and AIDS, for which no effective vaccines now exist.

Some of the protein products of recombinant DNA are so rare and so hard to isolate that little was known of them before they could be produced from recombinant DNA. *Growth factors*, for example, are rare proteins that keep specific kinds of animal cells alive, stimulate cell division, or trigger particular types of cell specialization. *Cytokines* are other rare proteins that coordinate the immune response. Until recently, researchers knew little about the structure of growth factors or how they work. Recombinant DNA techniques, however, have produced relatively large amounts of pure growth factors and cytokines, allowing researchers to study their structure and mode of action. Many academic and commercial laboratories are now investigating the possibility that growth factors may be able to prevent nerve degeneration in such devastating diseases as Alzheimer's disease, Parkinson's disease, amyotrophic lateral sclerosis (ALS, also called Lou Gehrig's disease), and Huntington's disease.

> *Genetically engineered bacteria and eukaryotic cells can make useful proteins, including enzymes, vaccines, drugs, and human proteins for treating genetic diseases.*

RECOMBINANT DNA CAN GENETICALLY ALTER ANIMALS AND PLANTS

Recombinant DNA can be inserted into the cells of whole plants and animals. Organisms that carry recombinant DNA in their genomes are called **transgenic organisms.** The added DNA is called a **transgene.**

To be expressed in a higher organism, a transgene must contain the appropriate control signals. To identify such regulatory regions, researchers usually join together a segment of DNA that is suspected of being regulatory with a protein-coding gene whose expression can be easily determined.

How Do Researchers Produce a Transgenic Mammal?

For a gene to be expressed in all the appropriate cells of an animal, researchers must put the transgene into the zygote before the beginning of embryonic development. Then all the cells of the organism will contain the engineered DNA.

In the early 1980s, several research groups succeeded in producing transgenic mice. Perhaps the most dramatic of these early experiments produced mice carrying the gene for human growth hormone (hGH). At birth, such transgenic mice were already twice the size of their littermates (Figure 13-11A). In the past 20 years, researchers have produced thousands of kinds of transgenic mice. Because some of the diseases in these mice accurately mimic those in humans (including cystic fibrosis, muscular dystrophy, Alzheimer's disease, and AIDS), these mice provide an opportunity to understand the basis of these diseases and to develop and test new therapies.

To produce a transgenic mammal, researchers must surgically remove eggs from the ovaries in the abdomen of the female. As a result, few eggs are available at one time. These few eggs are put into a dish and fertilized with sperm from a male. The resulting zygote is then injected with the engineered gene. This technique has produced transgenic pigs, goats, and sheep, and it seems likely to yield similar results with other species (Figure 13-11B). So far, however, only one transgenic animal is born for every 100 injected zygotes. Despite tabloid claims to the contrary, no one is known to have made any transgenic human embryos, though the basic steps are no more difficult in humans than in mice or goats.

The engineering of transgenic animals faces serious obstacles. First, few eggs actually incorporate the recombinant DNA. Even genes that initially seem to have been incorporated turn out to be not in the chromosomes, but in some mobile DNA outside the nucleus. Such genes can be lost in subsequent cell division. Perhaps more troubling, current genetic engineering techniques put the recombinant gene into the animal's chromosomes at random, both adding to the host's own genes and, often, disrupting other genes. The disruption of key genes creates mutant animals that die young, which is

A.

Figure 13-11 Transgenic animals carry foreign genes that they can pass on to their own offspring. A. The first transgenic mammal (*right*) was a mouse carrying the gene for human growth hormone. B. Today, transgenic goats secrete valuable proteins in their milk. The proteins are isolated from the milk for use as pharmaceuticals. This goat carries the gene for tissue plasminogen activator (tPA), a protein that activates the anticlotting system and is used to save the lives of heart attack patients. (*A, R.L. Brinster/Peter Arnold, Inc.*)

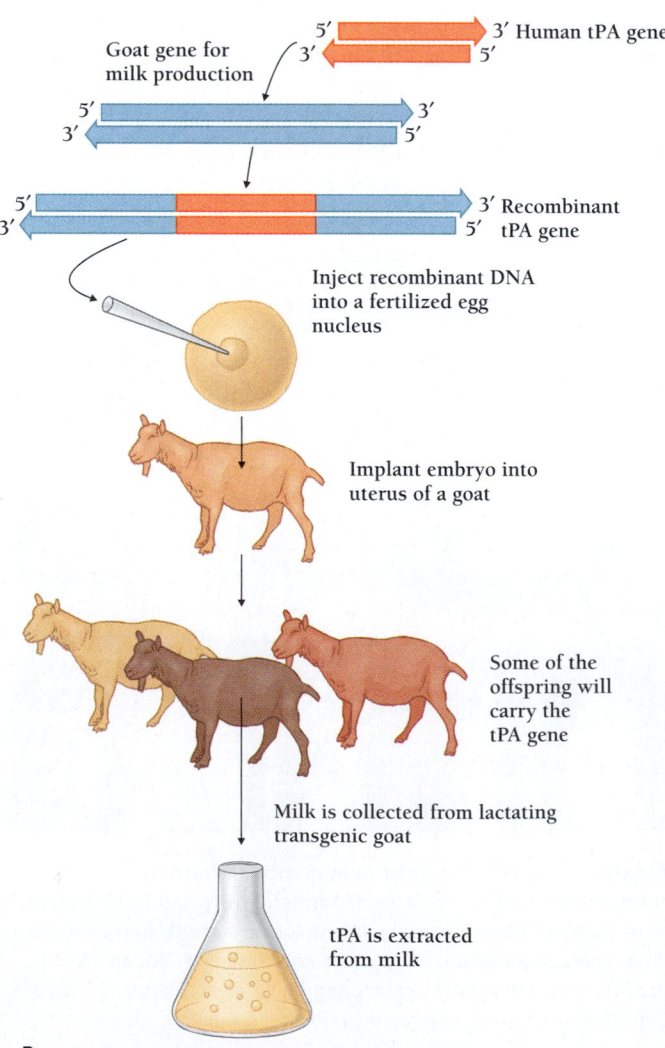

B.

one reason why it would be unethical to attempt this technique in human zygotes.

Researchers would like to replace the animal's own genes with the recombinant version instead of merely adding to them. This is relatively easy in prokaryotes and yeast, but more difficult in animal cells. One area of success is the production of genetic "knockouts"—animals in which a particular gene has been inactivated. Knockouts are a useful way of creating research animals with specific genetic defects, such as those found in humans. Usually, knockouts are used as a way of studying what role a gene plays in determining phenotype. The phenotypes that result from knockouts can surprise biologists. Figure 13-12, for example, shows a mouse in which researchers knocked out a gene specifying a growth factor believed to influence early embryonic development. The researchers expected the knockouts to stop developing long before birth. Instead, normal-looking baby mice were born, but with hair that grew so long that the knockouts looked a bit like Persian cats.

Recently, researchers have succeeded in creating "knockin" mice, in which an allele has been *replaced* rather than merely added. Like knockouts, knockins are difficult to make. In a knockin, however, the allele goes exactly where it is supposed to and the old allele is eliminated.

The knockout approach has several drawbacks. Quite often, all of the individual knockouts die as early embryos, and little can be learned about the adults or even the developing embryos. Even more discouraging for researchers, many knockouts show no effects at all from the loss of a functioning gene. Other genes apparently compensate when a particular protein is absent. In one recent example, researchers created knockout mice unable to make oxytocin, a hormone produced during the birth process and commonly given to women to induce labor. Yet, knockout mice unable to produce oxytocin delivered their pups with no problems.

Interpreting the phenotypic results of knockout organisms is also problematic. Knocking out genes to see what they do is like trying to find out what the different objects in a house do by removing them one at a time. If we remove the washing machine, the laundry no longer gets washed. But the same thing occurs if we remove the laundry detergent, or the clothes, or the electrical outlet into which the washing machine is normally plugged. Each of them is essential to getting the clothes done, but no single one of them is solely responsible for clean clothes. Similarly, when molecular biologists claim to have found a "master gene" for a given trait, the gene could equally well be described as the "weak link" in the intricate chain of events and materials that make up the phenotype. Teasing apart that chain of events will continue to be a major challenge.

New techniques now allow researchers to knock out a gene in an adult mouse. Using such "conditional-knockout" mice, researchers can determine what a gene product does in adult life, rather than during development. A related technique allows researchers to knock out a gene just in a particular organ or set of cells. While these experiments are still in their infancy, researchers are using them to unravel the role of individual genes and proteins in a whole living organism.

Figure 13-12 Knockout mouse. Researcher Gail Martin decided to knock out a growth factor gene, called FGF5, believed to play a role in the very early formation of an embryo. Martin and her colleagues expected that none of the homozygous embryos would finish developing. Instead, the knockouts all developed normally, but with hair nearly 50 percent longer than normal. Although FGF5 is normally expressed in muscle cells, brain cells, and motor neurons, all of these tissues were normal. (*Gail Martin, University of California, San Francisco*)

Transgenic mammals present many new opportunities and problems. The opportunities include researchers' new ability to study the function of specific proteins, including some responsible for human diseases. The problems include the insertion of DNA at random in the genome, which means that the genes may not function as researchers hope and may also disrupt other genes. Genetic knockouts and knockins can nonetheless provide clues about how previously mysterious proteins function in the body.

The Genetic Engineering of Plants Is Easier Than That of Animals

Perhaps the most rapidly expanding area of genetic engineering is in agriculture. Plant cells are generally easier to clone than animal cells. Equally important, plants can be grown in vast fields, so that a human protein, such as hemoglobin, produced by a plant, can be grown in hundreds of pounds per acre, instead of in grams per tub of *E. coli*. Tobacco plants, for example, can synthesize mouse antibodies that seem to behave just like mouse-made antibodies. Tomato plants can synthesize human serum albumin, a protein used to treat burn victims. Biotechnology companies have produced crops (sometimes called "nutriceuticals") that can mass-produce otherwise expensive compounds for use as medicines.

Genetic engineering of plants also has the potential to be enormously lucrative. If one of the engineered tomatoes mentioned earlier in this chapter becomes popular, for example, the inventors stand to gain a virtual monopoly in the tomato

BOX 13.2

Can Biologists Clone Dinosaurs?

Scary as *Tyrannosaurus rex* must have been, few people would pass up a chance to see the great flesh-eating dinosaur alive. Is it possible that DNA technology will someday bring dinosaurs back from extinction to populate theme parks everywhere? Can we scrape together bits of prehistoric DNA from fossils, stuff them in a big egg, and watch a baby brontosaurus poke through the shell?

The answer is no. Researchers cloned frog embryos in the early 1950s, and cloning adult animals became reality in 1997 with the birth of Dolly, a lamb whose genetic makeup was taken from the mammary tissue of an adult ewe. But the creation of a living organism from scraps of ancient DNA is an entirely different problem.

The first challenge for would-be dinosaur cloners is to find the genetic material of a race of animals that died out 65 million years ago. Little more remains of the ancient beasts than fossilized bones, in which minerals have replaced the organic material, essentially turning the bone into rock. Fossils therefore contain little or no DNA. Blood from ancient mosquitoes preserved in amber may be a more likely source of Jurassic genes. But tapping into that resource raises other questions: What animal provided the insect's last meal? Did mosquitoes even bite dinosaurs? And if

they feasted on several different species, which DNA in the insect's belly came from whom?

Even if scientists found a reliable source of dino DNA, a complete dinosaur genome would be nearly impossible to recover. For one thing, many scientists believe that DNA is not likely to survive intact for so long. One study has shown that DNA in a water solution will break down almost entirely into individual nucleotides after about 50,000 years. Still, under the right conditions, DNA can last longer. In 1990, geneticists successfully extracted DNA from a 17-million-year-old fossilized magnolia leaf dug up in Idaho. Two years later, researchers used PCR to amplify short gene fragments from a termite and a stingless bee, both encased in amber for 30 million years.

Attempts to recover dinosaur genes have been less successful. In 1993, scientists in Montana found blood cells in a *T. rex* bone that had only partly fossilized, thus protecting the central cavity from the elements. Unfortunately, the only sequences isolated from the sample later turned out to be human DNA contaminating the specimen. Other labs have had similar contamination problems in trying to squeeze dino DNA from ancient bones. Nevertheless, some researchers remain confident that PCR will eventually produce

authentic dinosaur gene fragments. But even then, those few hundred bases will be a far cry from the millions and possibly billions of bases of sequence that once resided in dinosaur chromosomes.

The lack of an appropriate egg is yet another nail in *T. rex*'s coffin. Even in the unlikely event that scientists discovered a cache of perfectly preserved dinosaur chromosomes, they would need to place them in an egg from some suitable modern animal. Whose egg would they try?

If dinosaurs lie on the evolutionary pathway to modern birds, as many biologists theorize, then the ostrich egg might not be a bad one to start off with. But an egg is more than just a protective coating for DNA. Eggs contain information about how cell division should proceed, or how the embryo will be oriented (which side is up, for example). It seems improbable that an ostrich egg, or any modern egg, would resemble a dinosaur egg closely enough to support the growth of a baby dinosaur, even with the proper genetic instructions.

Cloning living animals is at least technologically feasible. And now that successful clones of sheep have been made, successes in other animals, including humans, may not be far off. But for better or worse, the great dinosaurs are gone for good.

---

market. (Americans eat more tomatoes than any other vegetable, by far.) Already, however, farmers are already planting huge amounts of genetically engineered (and patented) soybeans, corn, and cotton. Some of these seeds, available since 1996, produce a bacterial protein (from an insect pathogen called *Bacillus thuringiensis*) that kills major pests, such as the European corn borer. (In 1998, American farmers planted 2 million acres of this engineered corn.) In 1999, farmers began to plant soybeans engineered to resist a fungus (*Phytophthora*) that causes root rot. Still other seeds, patented by Monsanto Corporation and marketed as "Roundup-Ready," carry a gene that allows crops to resist the herbicide glyphosate (Roundup), also made by Monsanto. These engineered crops allow farmers to spray an entire field with herbicide and kill the weeds but not the crops.

One way that researchers carry genes into plant cells involves a species of soil bacterium called *Agrobacterium tume-*

faciens. A. tumefaciens helps cause plant tumors, called crown galls. The actual tumor-causing agent is not *A. tumefaciens*, however, but a plasmid called the **Ti plasmid** (for *T*umor *in*ducing), which resides within the bacterium. A tumor results when a piece of plasmid DNA moves into the genome of the host plant. Researchers can engineer the Ti plasmid to carry specific genes into the genome of a plant. As in other forms of genetic engineering, the recombinant DNA usually includes an antibiotic marker that allows researchers to quickly distinguish the cells that have taken up the recombinant DNA from those that have not.

In another technique, sometimes called the **gene gun,** tiny particles of gold or tungsten are coated with fragments of recombinant DNA, then fired into a young plant with a miniature "shotgun" powered by gunpowder or compressed gas. The young plant is then chopped up and its cells grown on a cul-

Figure 13-13   Transgenic cotton. Both the normal cotton plant (*left*) and the transgenic cotton plant (*right*) have been sprayed with a potent herbicide. Only the transgenic plant flourishes, however, for it alone possesses a gene for herbicide resistance. (*Calgene*)

ture containing an antibiotic. The antibiotic kills all of the cells except those that have taken up the recombinant DNA introduced with the gene gun.

Genetic engineering and conventional breeding programs have produced new varieties of flowers and remarkable crops (Figure 13-13). Many crops have been engineered or bred for higher yields. Now, however, agronomists and molecular geneticists have begun to collaborate to produce crops engineered to produce well on a specific plot of land, according to the particular farm's microclimate and to the availability of minerals.

Other programs have focused on the production of "value-enhanced" crops, whose value will not come just from higher yields but from specially engineered traits. For example, in 1998, farmers planted more than a million acres of "high-oil" corn, with double the amount of oil in conventional corn. This crop can reduce the cost of feeding livestock and of producing cooking oils. Seed manufacturers are also working on many other crops—soybeans with high levels of lysine and methionine (normally in short supply in grains), canola and sunflowers with high contents of specific fatty acids with desirable properties in cooking or in the manufacture of soap, and colored and wrinkle-resistant cotton, which would reduce the need for chemical dyes and treatments. The production of such specialized crops will require new ways of tracking particular crops from field to market. Such production will also require special testing procedures

to ensure that each crop has the properties that it was supposed to have.

Using either the Ti plasmid or the gene gun, molecular biologists can genetically engineer plants that can, for example, synthesize animal or plant proteins, resist herbicides, or resist infection by plant viruses.

What Are the Risks of Recombinant DNA in Agriculture?

The incredible power of recombinant DNA technology lies partly in how precisely and quickly changes can be made and partly in the ability of molecular biologists to transfer genes from one organism to another. The long-term consequences of moving genes from one organism to another are unknown. Some critics argue that the ecological effects, if any, can be global and irreversible. Genetically engineered plants can, for example, exchange their new genes with wild cousins, either through conventional interbreeding or by way of plasmids and other vectors. Will genes inserted into a crop plant insert themselves into unrelated pest plants, creating, for example, herbicide-resistant weeds? Can a wild plant acquire traits that allow it to take over a local ecology or become a serious agricultural pest? Will engineered plants and animals interact with their environments in unforeseen, potentially destructive ways? Biologists and others can only speculate. No one knows if any of these scenarios are real threats.

Many experimental products are also controversial for other reasons. For example, in one laboratory experiment, milkweed deliberately dusted with pollen from insect-resistant corn led to the death of caterpillars of Monarch butterflies, which ordinarily feed on milkweed. Environmentalists worry that genetically engineered crops could inadvertently kill useful (and beautiful) as well as harmful insects.

Furthermore, because genetically engineered seeds are patented, farmers cannot save seed from one year to plant the next (as is common for many crops, such as wheat). Many critics are concerned that the increasing use of patented seeds will make farmers economically dependent on just a few seed producers, who may be more concerned with multinational profits than with the welfare of farmers in any one country.

Criticism of genetically modified crops has been especially loud in the European Union. Part of this opposition derives from ecological concerns, part from economic concerns (especially about U.S.-based corporations), and part from a general suspicion of genetic engineering. In June 1999, the European Union placed a moratorium on importing genetically modified foods.

Critics of genetically modified crops are concerned with the spread of genes that may be ecologically harmful as well as with increasing dependence on a few seed producers.

The Application of Recombinant DNA Technology Poses Moral Questions for Society

At the other end of the spectrum are the moral and ethical issues that come from recombinant DNA's precision. The diagnosis of genetic diseases, for example, is far in advance of treatment. Thus, in the next century, people may know that they are at higher-than-average risk for certain forms of cancer or mental illness, for example, but they will not necessarily be able to do anything about their fate. This is already true for genes that predispose people to certain forms of cancer of the breast, colon, and prostate. People at risk for Huntington's disease can now find out if they carry the disease-causing allele. Those with the Huntington's allele have no way, however, of avoiding the disease, so the test is of very limited value. Yet, potential employers and insurance companies want to know what people's genetic flaws are so that they can avoid hiring or insuring people likely to get sick.

Many people worry about applying genetic engineering to humans. Researchers are already working to treat genetic diseases by inserting genes in human cells. These techniques, called gene therapy, will be discussed in Chapter 14. Some researchers may try to eliminate certain diseases altogether by replacing defective genes even before conception. Such a heritable genetic cure has already been achieved in mice for a neurological disease characterized by abnormal brain development.

However, while few people object to curing deadly diseases, what constitutes a "genetic disease" may become controversial. Societies have often designated people with certain characteristics as "undesirable," and readily embraced arguments that these characteristics are genetic. If biologists develop the capacity to replace genes in humans, will genetic engineers also try to modify genes that affect characteristics other than those responsible for disease? Will future societies attempt to produce more (or less) intelligent citizens? Or less (or more) aggressive citizens? Already, many physicians treat shorter-than-average children with growth hormone, so that the children will "fit in." There is no reason to think that either parents or physicians would hesitate to "treat" such children with gene therapy.

While many aspects of physical appearance and behavior are strongly genetic, it seems unlikely in principle that gene replacement could ever produce a uniform society. Possibly, the wealthiest members of society could afford to have their children engineered for such traits as good teeth and other aspects of health and appearance. Meanwhile, the rest of society would go on as usual. Conventional medicine already suffers from this defect: the wealthiest members of society get better medical care than the poorest. It is clear that all the people in a society need to discuss the issues raised by the future possibilities of genetic engineering. As in the case of any new advance (such as nuclear weapons and nuclear power), the use of technology is a concern for the society as a whole, not just for scientists.

We do not yet know all the limitations of the new technology. Society must carefully evaluate the risks, benefits, and moral implications that arise from the ability to produce new or previously unavailable products, especially engineered plants and animals. In the next chapter, the last chapter of this section on genetics, we will explore some of these issues in more detail.

Study Outline with Key Terms

Until the development of **recombinant DNA** techniques, **biotechnology** consisted mostly of the selection of varieties of organisms with desired traits.

A molecule of recombinant DNA consists of two or more DNA segments that are not found together in nature. For example, by combining DNA pieces from humans and *E. coli,* researchers made a recombinant DNA molecule with the regulatory region of β-galactosidase and the structural gene for insulin. Bacteria that contain this DNA make insulin in response to the presence of lactose instead of β-galactosidase.

Making recombinant DNA depends on the use of enzymes that can cut, link, and modify DNA. **Restriction enzymes** cut DNA at **restriction sites,** where specific sequences occur, allowing molecular biologists to isolate individual DNA fragments, called **restriction fragments. DNA ligase,** an enzyme used in DNA replication, links pieces of DNA from different sources. A **restriction map** shows the positions of the restriction sites in a piece of DNA.

DNA molecules derived from viruses and plasmids can serve as **vectors** for the propagation of recombinant DNAs within host cells. Some of the most widely used vectors derive from R factors, plasmids that carry antibiotic resistance genes. Other vectors include bacteriophages, as well as plant and animal viruses.

Polymerase chain reaction (PCR) is now a basic tool for finding and multiplying, or **cloning,** fragments of DNA. **Oligonucleotides,** short pieces of single-stranded DNA, act as primers for DNA polymerase. Alternating heating and cooling do much of the rest.

The starting materials for recombinant DNA may come from the genomes of viruses or organisms or from DNA copies of mRNA. Most plant and animal genes contain introns, which would interfere with their expression in bacteria. Many recombinant DNAs coding for plant and animal proteins are therefore derived from **complementary DNAs (cDNAs),** copied from mRNA by the **reverse transcriptase** of a retrovirus.

Collections of recombinant DNAs, called **gene libraries,** may contain thousands or millions of different DNA sequences. Finding a desired DNA sequence usually depends on the use of a **probe.** A **hybridization probe**—often a piece of DNA made in the laboratory—bonds with one strand of the DNA to form **hybrid DNA.** By labeling the probe with radioactive nucleotides or a chemical marker, the experimenter can identify recombinant DNA. **Antibodies,** which have surfaces that recognize the shapes of specific molecules, can help identify bacteria that are producing a particular protein.

In **western blots,** antibodies are used to detect specific proteins after **gel electrophoresis.** Hybridization probes are used to identify specific DNA sequences in **Southern blots** and to determine levels of individual mRNAs in **northern blots** or in **DNA arrays.**

In many cases, researchers do not know the function of a particular protein until recombinant DNA techniques make enough of the protein available to study. Recombinant DNAs

are likely to be useful in the preparation of safe vaccines for use in disease prevention.

Recombinant DNAs can reprogram organisms to make new products. Products such as insulin and **growth factor** that are made by genetically engineered bacteria are already commercially important as pharmaceuticals. Recombinant DNA can be used to alter animals and plants genetically. **Transgenic organisms** carry deliberately added **transgenes** in their genomes. The current method for making transgenic mammals involves the injection of an engineered gene into a zygote produced by in vitro fertilization. Molecular biologists can also damage selected genes to make genetic knockouts.

Making transgenic plants is easier than making transgenic animals because experimenters can easily add or replace genes in single cells in culture. The engineered cells can then be grown into whole plants. Genes are introduced either by means of a tumor-inducing (or **Ti**) **plasmid,** which normally resides in bacteria that cause plant tumors, or by means of a **gene gun.** The agricultural applications of transgenic plants have been widespread and controversial.

Review and Thought Questions

Review Questions

1. How does modern genetic engineering using recombinant DNA differ from breeding techniques practiced over the last 10,000 years? How are these two kinds of genetic engineering the same?
2. How can knowing a biochemical pathway help molecular engineers design useful organisms?
3. What is recombinant DNA?
4. What two enzymes are basic to assembling recombinant DNA? What do the two enzymes do, and what is their role in nature?
5. Why do molecular biologists need multiple copies of a gene?
6. What two techniques allow molecular biologists to make many copies of a gene? What are the strengths and weaknesses of each technique?
7. Why do molecular biologists want to engineer cells to express large amounts of the gene product, or protein? How do they accomplish this?
8. What prevents bacteria from expressing a gene cut directly from a human genome? How do molecular biologists get around this problem?

9. How do molecular biologists find the right DNA sequence in a recombinant DNA library?

Thought Questions

10. What is your opinion about creating transgenic humans? Suppose that you knew that you carried a gene that would predispose your children to a mental illness. Would you opt to conceive by means of *in vitro* fertilization so that your zygotes could be screened and the gene replaced? Suppose large numbers of other people chose not to use this service. Some of their children would continue to develop severe mental illness, eventually becoming rather expensive wards of the state. Would you feel that everyone should be required to screen his or her zygotes? What if you knew that the families of people carrying this gene tend to be more intelligent and more creative than average? Would that make a difference in your opinion? Why or why not?
11. Would you mind eating a genetically engineered tomato? Why or why not?

About the Chapter-Opening Image

Finding a particular DNA sequence is like finding a needle in a haystack. PCR simplifies the process by making many copies of the needle.

 On-line materials relating to this chapter are on the World Wide Web at **http://www.harcourtcollege.com/lifesci/aal2/**

14

Human Genetics

To Map the Genome or Hunt the Huntington's Gene?

David Botstein, burly and bombastic, jumped up from the table and paced the small conference room, loudly lecturing the other 12 scientists still seated at the table. It was October 1979, and the 13 scientists had met at the National Institutes of Health near Washington, DC, to discuss the impact of molecular biology on inherited diseases, especially Huntington's disease. But the discussion had come to a critical point of disagreement, and Botstein, a brilliant and temperamental molecular biologist from the Massachusetts Institute of Technology (MIT), could no longer contain himself (Figure 14-1).

Beginning a search for the Huntington's gene now, argued Botstein, was premature and a waste of time. New technology would soon revolutionize human genetics, allowing scientists to map and identify every human gene and find the molecular basis of every genetic disease. Once a general map was established, it would be child's play to find the molecular defects that cause Huntington's disease, cystic fibrosis, muscular dystrophy, and any of the other 3000 inherited diseases.

At the table, still seated, was David Housman, another molecular biologist from MIT. Housman watched Botstein impassively, then quietly disagreed. Yes, eventually, all would be known, he said, but wasn't it better to start first—right now with a single disease? And why not Huntington's disease? Researchers need not wait for a complete genetic map to find the Huntington's gene. There was no reason not to jump ahead, applying knowledge and techniques that were already available. Housman himself had already shown how the abnormal regulation of hemoglobin genes could cause certain inherited anemias called thalassemias.

Botstein scoffed. It was short sighted to jump into problems that might be personally pressing but were not yet scientifically "ripe." Better, said Botstein, for scientists to wait for basic knowledge. Indeed, despite the efforts of dozens of researchers, the Huntington's gene was not sequenced for 14 more years. The early mapping and sequencing of individual disease genes moved excruciatingly slowly. Only when basic research by Botstein and others provided new techniques for faster advances did the mapping of defective alleles begin to proceed quickly.

Yet it was Housman who prevailed at the 1979 conference. The conference had been sponsored by a small foundation called the Hereditary Disease Foundation (HDF), which is based in Los Angeles and dedicated to promoting research on Huntington's disease. The HDF had been started by Milton Wexler, a prominent Beverly Hills psychoanalyst. Wexler's two daughters, Alice and Nancy, were at risk for developing

Key Concepts

- Even though humans are more than 99 percent genetically identical to one another, all humans, except identical twins, are genetically unique.

- Phenotype is a product of a complex interaction between genotype and environment.

- Many aspects of human genetics, including, for example, diagnostic testing and gene therapy, raise serious ethical, legal, and social issues.

Huntington's; in 1968 their mother, Leonore, had been diagnosed with the fatal disease.

Huntington's disease is a devastating neurological disease that causes jerky movements, mood swings, personality changes, and premature death. The most obvious early symptom of the disease is the constant motion of the arms and legs, giving the disease its previous name, "Huntington's chorea" [Greek, *chorea* = dance, as in "choreography"]. Huntington's usually begins in middle age. Over the course of 10 to 20 years, patients gradually lose control of movements, memory, and mood. The bodies of Huntington's patients gradually waste away, and for the last few months of life patients stop moving altogether.

Huntington's disease is a classic genetic disease. An affected parent passes the disease to half of his or her offspring, indicating that the disease is caused by a single dominant allele of some gene. Nearly everyone who inherits the Huntington's disease allele eventually develops the symptoms. This high level of expressivity makes the disease particularly suitable for genetic analysis. In 1979, however, researchers and physicians knew nothing about the gene and almost nothing about the damage it caused at a biochemical and cellular level.

In 1969, Milton Wexler had begun engaging young scientists to brainstorm new approaches to understanding Huntington's disease. He attracted these biologists both by inviting them to parties that included some of the most famous artists and entertainers in the Los Angeles area and by encouraging these young biologists to think freely and creatively. Few young biologists could resist these brushes with Hollywood glamour. But the real purpose of the gatherings was the scientific workshops, in which Wexler brought together small groups of biologists who did not ordinarily meet—researchers in fields ranging from genetics to neurology. These informal workshops were intellectual incubators for new ideas not only about Huntington's disease but about the application of new techniques to the study of all inherited diseases, especially disorders of the brain. Among the young scientists stimulated by these workshops was one of this book's authors, Allan Tobin. In 1979, Wexler appointed Tobin Scientific Director of the Hereditary Disease Foundation. It was Tobin who organized the 1979 workshop on molecular genetics, with help from his former Cambridge neighbor, David Housman.

Botstein's fears that attempting to map the Huntington's gene would be difficult turned out to be well founded. To find a particular gene among the 100,000 genes on the 46 chromosomes, geneticists had to look for markers, obvious phenotypic characteristics, such as eye color, that are always inherited with the defective allele and are therefore near the defective gene. Geneticists had constructed linkage maps in *Drosophila* through the painstaking analysis of the selective mating of flies with identifiable phenotypes—such as white eyes or notched wings. But geneticists cannot set up matings of humans; human genetics instead depends on the analysis of matings that, to a geneticist, seem arbitrary and senseless. A few markers in humans, such as those for blood proteins, work well, but none of these were inherited with the Huntington's allele.

Figure 14-1 David Botstein. *(Courtesy of Stanford University Medical Center)*

The picture changed, however, when molecular biologists began to use recombinant DNA techniques to clone and sequence specific genes. Researchers then discovered numerous small differences among individuals at the DNA level. Even though two people may make exactly the same hemoglobin molecules, for example, their hemoglobin genes are likely to differ from one another. These "silent" differences lie in DNA sequences that are not transcribed and translated into protein—for example in introns or in flanking sequences. By the late 1970s, researchers were convinced that (except for identical twins) every person is genetically distinct from every other. At the DNA level, the homologous genes inherited from mother and father differ at about 1 nucleotide in 500.

The technology that so excited David Botstein was a method for identifying such differences, which he called **restriction fragment length polymorphisms,** or **RFLPs**

297

(pronounced "RIFFlips"), and which we discuss below. Within a few years, differences in DNA sequences would not only allow geneticists to produce a map of the human genome, it would also allow police and military laboratories to match blood and tissue samples with a previously unimaginable accuracy—changing forever the ability to identify criminal suspects, as well as victims of crimes, disasters, and battles. By 1980, Botstein, his collaborators, and his competitors had established the feasibility of making a map of the human genome based on differences in DNA sequence, rather than on differences in phenotypic traits. By 1994, researchers had succeeded in making a relatively complete RFLP map of the human genome, allowing researchers to locate genes within about 1 million base pairs of DNA. More recently, geneticists have begun a much finer map, with a resolution of a few thousand base pairs of DNA, using variations in single bases. These variations are called *single-nucleotide polymorphisms,* or *SNPs* (pronounced "snips").

Until the late 1980s, enthusiasm for making a map of the human genome was not universal, even among molecular biologists. The research required was repetitive and dull. Indeed, the excitement for learning about human genes came from the public's interest in finding the alleles that underlie diseases ranging from Huntington's to cancer. The first success in the application of the new molecular genetics to human disease was, in fact, Huntington's disease.

Huntington's disease provided the key and the momentum for several reasons. First, its inheritance was particularly clear. Second, motivated families, especially those of Leonore Wexler and of the great folksinger Woody Guthrie, who also died of Huntington's disease, encouraged young researchers both psychologically and financially through their support of several foundations concerned with Huntington's disease (Figure 14-2). In particular, Nancy Wexler, the younger daughter of Milton and Leonore, has devoted her professional life to ad-

vancing the search for the gene for Huntington's disease. As a researcher herself, she documented the inheritance of Huntington's disease within an extended family that lives in three villages on the shores of Lake Maracaibo in Venezuela.

Within this group of about 10,000 people, several hundred have already developed Huntington's disease, and hundreds more carry the gene. By tracking the simultaneous inheritance of Huntington's disease and of DNA markers within this family, James Gusella (a former student of David Housman), Nancy Wexler, and their colleagues were able—in 1983, years before the construction of a complete human genetic map—to determine the approximate chromosomal location of the Huntington's disease gene. This accomplishment depended directly on the RFLP method advocated by Botstein. The Huntington's allele was the first disease gene ever found with that method.

It took an international consortium, organized by Allan Tobin and Nancy Wexler for the Hereditary Disease Foundation, another 10 years to identify the actual gene for Huntington's disease. By then (1993), recombinant DNA techniques had already led to the precise identification of many other disease-causing alleles, including that for cystic fibrosis, the most common genetic disease among Caucasians and a good target for the development of gene-based therapies. But the Hereditary Disease Foundation's first-ever gene search—undertaken despite Botstein's fierce arguments for patience—dramatically demonstrated the power of molecular genetics to help understand human disease.

Because of the efforts of the HDF, dozens of researchers focused on applying known techniques to the specific problem of Huntington's disease. Choosing a single disease on which to focus their efforts gave biologists a sense of mission and a single, clear-cut goal that was emotionally appealing. They could imagine that their day-to-day research might one day save the Wexler daughters and other sufferers of Hunt-

Figure 14-2 The fight against Huntington's disease. The families of the famous singer Woody Guthrie and biology teacher Leonore Wexler used their considerable influence to initiate and support the search for the Huntington's gene. *(Guthrie, Culver Pictures; Wexlers, courtesy of Alice Wexler)*

ington's disease. In addition, the HDF's intimate workshops promoted many fruitful scientific friendships and collaborations. And finally, Gusella's location of the Huntington's gene in 1983 provided the "proof of principle" that may well have convinced both scientists and scientific administrators that mapping the human genome was feasible. In this chapter we will see, however, that it was the basic research carried out by Botstein and others that invested the science of molecular genetics with the promise it now holds.

The genetics of humans is no different from the genetics of other organisms. For this reason, many biologists object to the very idea of having a chapter on human genetics. The *study* of human genetics, however, does differ in one very significant way from the study of the genetics of fruit flies, pea plants, or corn. Humans cannot be crossed at the whim of geneticists. As a result, the study of genetics in humans (as well as in other undomesticated animals) is much harder than in laboratory organisms in which any two individuals can be crossed at will.

Another difference between the genetics of humans and that of other organisms is that our own traits are, to us at least, infinitely more interesting and subtle. The degree to which a trait is the product of genetic influences or the product of environmental influences rouses people to intense political debate that makes objective discussion nearly impossible. In this chapter we will try to present enough information for readers to make intelligent judgments about what they read and hear. We will learn about genetic diversity in humans, the complex relationship between genotype and phenotype, the Human Genome Project, and, finally, both the promise and limitations of applying knowledge about human genetics.

GENETIC DIVERSITY IN HUMAN BEINGS

On a Saturday evening in the winter of 1983, 5-year-old Alice Brown was abducted just steps from her own front door. She and her mother, Sarah Brown, were about to sit down to dinner with a neighbor in their 200-unit apartment complex in northern California when Alice's 3-year-old brother, Brad, begged for his boots. Alice took Brad home, just two doors down, but when Brad returned moments later he was alone. Alice's mother never saw her daughter alive again.[1]

When Alice's body was found a week later, police discovered that she had been raped and then suffocated. Sarah said that Alice would never have gone with any man except one: a neighbor in the same apartment complex with whom Sarah had gone out several times. Police considered Harry Gordon, then 31, a prime suspect, but they had no concrete evidence to link him to the case. They had to have "probable cause."

All of the evidence was circumstantial. He'd given contradictory stories about his whereabouts on the night of the murder. Police had arrested him in 1981 for exposing himself to a little girl and again, in 1984, on suspicion of molesting a third girl. He was eventually convicted of a misdemeanor for

exposing himself while drunk. Gordon's ex-wife said he was a likely suspect, and police pursued dozens of leads. But Gordon denied any involvement in Alice's death, and nothing gave the police the evidence they needed to arrest him.

Sarah and Brad moved away and changed their names. Thirteen years passed. Then, in the spring of 1996, when Alice would have been 18 years old, police finally arrested Gordon for the rape and murder of Alice Brown. It turned out that police had had the concrete evidence they needed all along. Subtle changes in California law and advances in molecular biology had finally made it possible for prosecutors to use that evidence to link Gordon to the crime and to know that they could probably convict him. (The same evidence had cleared a previous suspect.)

Just as every individual has a unique set of fingerprints, so every individual also has a unique set of gene fragments called a **DNA fingerprint.** DNA for a DNA fingerprint can be extracted from any cell that has a nucleus. Such cells can be found in blood stains, in saliva on an envelope, in tears on a cheek, in skin cells, under fingernails, in hair, or in semen. Thanks to the invention of PCR, researchers can now analyze DNA from the barest traces of cells. When police found Alice's body in 1983, they had been able to collect samples of semen from her body and clothes. As evidence, these samples had been carefully stored. But in 1983 they were of little value, for DNA fingerprinting had not yet been invented.

By the early 1990s, prosecutors knew that the DNA in the semen could tell them who had murdered Alice Brown. A legal problem, however, prevented prosecutors from acting. Unlike most states, California had not issued a ruling on the admissibility of DNA fingerprinting as evidence. As a result, lower courts had issued conflicting rulings, and a conviction based on DNA fingerprinting stood a good chance of being overturned on appeal. In 1992, a northern California judge had ruled that DNA fingerprinting was not as reliable as a real fingerprint. If prosecutors tried to convict Gordon and the court rejected the DNA fingerprinting, prosecutors knew they would have lost their only chance to convict Gordon, and at great cost. Just arguing the admissibility of DNA fingerprinting in court could cost $50,000—the price of obtaining testimony from expert witnesses.

How Reliable Is DNA Fingerprinting?

The haploid human genome contains some 3 billion base pairs of DNA. Only 5 percent of this DNA (150 million base pairs) codes for polypeptides. The polypeptide-coding regions are almost identical from one individual to the next because natural selection eliminates most mutations.

The 95 percent of the DNA that does not code for polypeptides is much less subject to natural selection. As far as biologists can tell, such noncoding DNA has no central role in the phenotype, and so it can mutate without impairing an individual's ability to survive and reproduce. Noncoding DNA accumulates large numbers of mutations and contains far more variation than coding DNA. Estimates suggest that each pair of chromosomes differ from each other by about 1 base

[1]We have changed the names of those involved.

in 1000, which on a single chromosome could add up to some 250,000 base-pair differences. As a result, except for identical twins, every individual is genetically unique, and DNA sequencing can reliably distinguish every individual in a population.

Researchers create a DNA fingerprint by using restriction enzymes to break up an individual's DNA into restriction fragments (Figure 14-3). Fragments from some regions of the DNA vary in length from individual to individual. The length of a fragment depends on whether the sequence includes a restriction site for the restriction enzyme in use. (Other parts of the DNA are consistently the same length from one person to another.) Fragments that vary in length are said to be polymorphic and are called restriction fragment length polymorphisms, or RFLPs. This long name merely means that for a given RFLP sequence, the presence or absence of a restriction site determines whether the fragment is long or short. Many of the most useful RFLPs include a repeated sequence, with the number of repeats varying greatly among individuals.

In the next step, electrophoresis, the fragments of DNA are exposed to an electric current that moves them through a jellylike medium, called a "gel." The DNA fragments, which are negatively charged, move through the gel toward the positive electrode. Some move quickly, some slowly, depending on their size. The shortest fragments travel through the gel rapidly, while the longer fragments are left behind. After a while, the fragments are spread out across the gel like runners in a race. When hybridized to DNA probes in a Southern blot, the fragments produce a pattern of labeled bands (Box 13.1).

Every genome yields a unique pattern of bands. Since every individual has a different DNA sequence, every individual (other than an identical twin) has a unique pattern of bands. Fragments taken from the whole genome in different individuals would reveal a DNA fingerprint as unique as a real fin-

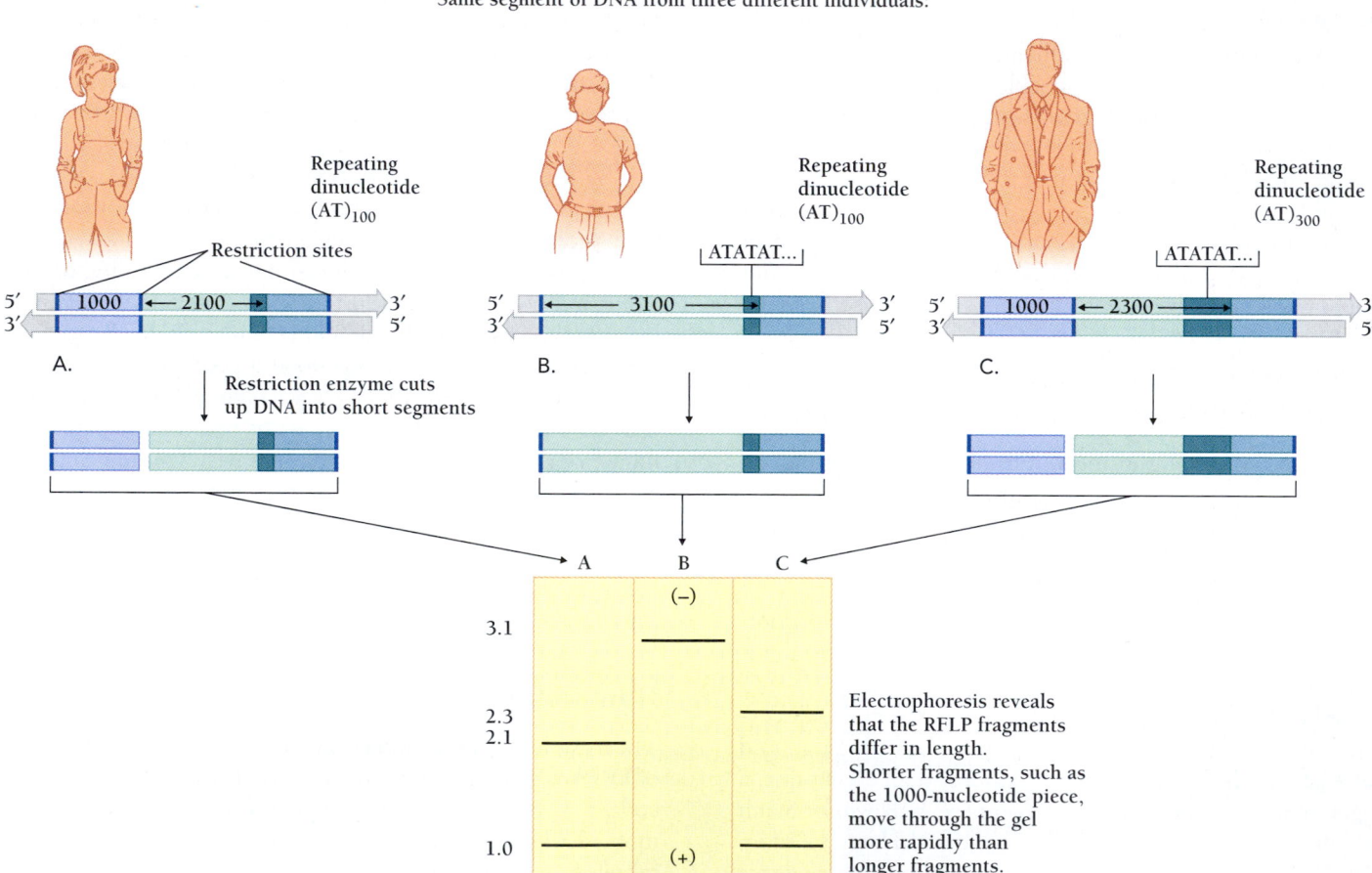

Figure 14-3   **RFLPs (restriction fragment length polymorphisms) in three individuals.** Restriction enzymes break the DNA of each individual into fragments. For individual A, two fragments result that are 1 kilobase (1000 bases) and 2.1 kilobases in length, respectively. The shorter one moves across the electrophoretic gel faster than the longer one. In B, the middle restriction site is absent and a single long fragment results that is 3.1 kilobases long (1 + 2.1 = 3.1). This long fragment moves across the gel very slowly. In C, the 1-kilobase fragment is the same as the one in A. But a repeating sequence has made the 2.1-kilobase fragment in A 200 bases longer. C's longer fragment moves more slowly than the one in A but faster than the one in B.

gerprint. In practice, however, fingerprinting laboratories examine only a small number of DNA fragments, those known to have many variants. Researchers then calculate the probability that someone else in the general population has that exact set of base changes. Researchers can come up with a probability of uniqueness ranging from 1 in 500 to 1 in several billion.

Two factors affect this number. The first is the number of different markers examined. The greater the number of markers surveyed, the more specific the DNA fingerprint will be, and the more other people in the world can be excluded from consideration. The second factor is how "the general population" is defined. As we will see later, different populations are more or less likely to have certain genotypes. The chances of someone having the same DNA fingerprint as Harry Gordon might be one number for the population of San Francisco Bay Area and another number for the population of all of California. In any case, geneticists do not necessarily know what those probabilities are for different populations. They have estimates for large populations that have been studied, such as those of the United States, Argentina, or Spain, but not for individual cities such as San Francisco or Los Angeles.

Why did police finally arrest Gordon?

In the fall of 1995, prosecutors in the Harry Gordon case began to sense that DNA fingerprinting was coming to be accepted in California. They no longer believed that if they took Gordon (or any other suspect) to court that they would have to prove that a DNA fingerprint is unique. In the fall of 1995, police obtained a court-ordered search warrant and compelled Gordon to give a blood sample. They sent the blood sample and the sample of the semen from Alice Brown's body to a laboratory in Maryland. There, technicians determined that the two samples matched. In six highly variable RFLPs, the pattern of bands was the same.

These six RFLPs vary so much from person to person that it was highly unlikely that a perfect match between the recovered semen DNA and Gordon's DNA could have arisen by chance. The testing laboratory estimated that the DNA recovered from Alice's body could be matched to only 1 in 5.7 billion male Caucasians, some ten times the number of male Caucasians in the world. Prosecutors finally had the tangible evidence for which they had waited so long. In the spring of 1996, they arrested Gordon and charged him with Alice's murder. Many people are hoping that his arrest will bring to an end a string of similar unsolved murders in the area.

What might Gordon's defense attorneys argue?

The use of DNA-based evidence has been much debated in U.S. courts and in the pages of scientific journals. Even if the probability of someone else in the world having Gordon's DNA fingerprint is virtually zero, his defense attorneys can still argue that the evidence is inconclusive. They are likely to bring four kinds of arguments: (1) that the DNA samples (like *any* evidence) could have been mixed up (intentionally or unin-

tentionally), so that the perfect match came from duplicate samples of Gordon's own DNA; (2) that the semen samples are so old that the evidence is unreliable; (3) that another person besides Gordon could have the same RFLP pattern; and (4) that the RFLP analysis was itself faulty. Which of these is arguments is likely to be correct?

Questions about the proper handling of evidence apply to any kind of physical evidence (hairs, threads, or fingerprints, for example). The prosecutor must demonstrate—beyond any reasonable doubt—that the police have carefully collected the biological samples, preserved them appropriately, and not mixed them up.

In some cases, defense attorneys have argued that the tissue samples from the site of a murder have been contaminated by DNA from other sources. Because PCR-based DNA fingerprinting is so sensitive, a few cells from a different person could change the results, they argue. In reality, unless the contaminating cells came from the suspect, they would not match the suspect's DNA. The only way for the suspect's cells to be added or substituted would be if someone deliberately set out to frame the suspect. This is possible, but no more likely with DNA fingerprinting than with other forms of evidence.

Gordon's attorneys are also likely to argue that the semen samples are very old and were improperly stored. But how does improper storage affect the results? The answer is that if the DNA were to break down, for example, as a result of exposure to heat, sunlight, or chemicals, the changes in sequence would be random and extremely unlikely to exactly match Gordon's own DNA. In other words, damaged DNA will not convict an innocent person. Can two people have the same RFLPs by chance? Geneticists are unanimous in the conclusion that everyone (other than identical twins) has a unique genome. So if a laboratory sequenced all the DNA in the blood or semen from two individuals, they would certainly find differences. But, as of the year 2000, no complete human genome has been sequenced, and the cost of total sequencing is likely to remain prohibitive. So DNA fingerprinting depends on a limited sampling of the genome, concentrating on sequences (such as RFLPs) that are highly variable within the population. Prosecutors can calculate the chance that each of these DNA variants (alleles) would appear in a large population (such as that of the United States). In general, however, no one knows the frequency of each variant in a local population. For the RFLPs used for DNA fingerprinting, however, the probability of obtaining six identical DNA patterns by chance is typically 10^{-10}, that is, 1 in 10 billion. And, of course, so far the Earth has only 6 billion people.

Finally, there is the question of faulty laboratory analysis. In fact, some of the analyses presented in evidence in the early 1990s were so poor that expert witnesses for the prosecution ended up arguing for acquittal. Recently, however, new standards have been adopted for the certification of laboratories and technicians, and purely technical errors are much less likely than previously. Still, DNA evidence (like other evidence) must be scrutinized for technical artifacts and misinterpretations.

DNA fingerprinting has changed the way that the police investigate crimes by providing a powerful new kind of physical evidence. DNA analysis has not only led to many convictions but has resulted in the release of suspects and even prisoners unjustly accused. DNA evidence, for example, led to the release of a 32-year-old Baltimore man after 9 years on death row.

DNA fingerprinting also led to the identification and return of kidnapped children. One of the most egregious crimes of the Argentine military dictatorship (1975–1983) was the abduction and killing of some 15,000 people, called "the disappeared." Among these "disappeared" were pregnant women, whose babies were delivered and given to military families. After the fall of the dictatorship, a group of women whose daughters or granddaughters had "disappeared" contacted Professor Mary Claire King, a human geneticist then at the University of California Berkeley. King was using RFLP analysis to track a gene variant responsible for breast cancer. By applying her fingerprinting methods to the DNA of the stolen children and their surviving relatives, she was able to help reunite over 50 children with their families.

DNA fingerprinting identifies individuals on the basis of variations in DNA sequences.

What Are Human "Races"?

We all know that the human species varies in appearance and in physiology geographically. People of Indian descent, for example, are recognizably different from those of Chinese descent. We can look at the shape of the nose, the eyes, and the ears, for example, and we see differences (Figure 14-4). We can look at the distribution of individual alleles and see differences. Fifty-one percent of Nigerians have type O blood, compared to only 30 percent of Japanese. Twenty percent of Russians have type B blood, while the Amerindians of Lima, Peru, have no detectable levels of type B blood at all.

Nineteenth-century anthropologists struggled to classify human groups into a few major races. Some systems identified only 12 races, while other systems listed 30 or more. One problem was that no matter how anthropologists classified humans, there always seemed to be tribes or nations that would not fit into any known group. The Basques, who live in the Pyrenees mountains between France and Spain, for example, appear European. Yet their language and culture are unlike any other in the world, and some researchers once argued that they are direct descendants of Stone Age Europeans. Similarly, the Bushmen are unique among African groups in both appearance and physiology (Figure 14-5).

A more serious problem with the grouping of humans into races is that most groups do *not* stand out from those around them; they blend. Because groups of humans inevitably mix, through migration, warfare, and trade, human "races" are never pure. Both the Japanese and British, for example, take pride

A.   B.

Figure 14-4 People from different geographic regions vary in such traits as height, blood type, skin color, and facial features. A. An Indian woman. B. An ethnic minority woman from China. (*A, Superstock; B, Paul Grebliunas/Tony Stone Images*)

in the purity of their island races. Yet the Japanese are a graded mixture of Korean and Ainu north islanders (a people of possibly European descent). This mix shows up in the distribution of blood types from one end of Japan to the other (Figure 14-6).

The British are even more of a melting pot than the Japanese. The Bronze Age Beaker Folk mixed with the Indo-European Celts in the first thousand years B.C. In the next thousand years, the Angles, the Saxons, the Jutes, and the Picts arrived, followed by the Vikings and their descendants, the

A.   B.

Figure 14-5 Classifying humans into a few discrete races has been unsuccessful. Named races frequently include markedly distinct peoples. A. A Bushman from Namibia. B. A rubber plantation foreman from the Ivory Coast. (*A, M.P. Kahl/Photo Researchers, Inc.; B, Charles O. Cecil/Visuals Unlimited*)

races—in the biological sense—do not exist. This is because, in humans, genetic variation within populations is greater than that between nations or races. Of all human genetic variation, 85 percent is variation among the individuals within a country or a continent. Another 6 percent is variation among populations from the same continent. Only about 9 percent of all DNA reflects genetic differences among peoples ("races") from different continents. The greatest genetic variation among humans is found in Africa. If a disaster killed everyone on Earth except for those living in Africa, the human species would still retain at least 91 percent of its genetic diversity. Less than 9 percent of all genetic diversity would be lost.

In practice, we focus on the small numbers of traits (for example, skin color and eye shape) that vary geographically—a subset of the 9 percent—and we use just those few traits to identify groups of people who differ from us more in their cultural practices than in their genetic constitutions.

Recent work, for example, shows that many differences in skin and hair color reflect variations in a single gene. That gene specifies the structure of a protein, called MC1R. MC1R affects our response to a hormone that regulates the balance of different types of skin pigments (melanins). Caucasians who tan well, for example, have the same form of MC1R that is present in Africans.

Researchers suggest that variations in skin color represent adaptations to different amounts of sunlight. A mere handful of variants are probably responsible for the most obvious traits that have historically been used to distinguish human "races." As geneticists Mary Claire King and Kelly Owens state, "The myth of major genetic differences across "races" is . . . worth dismissing with genetic evidence."

Geneticists and anthropologists have also begun to combine DNA-based differences among current populations with fossil evidence to study early human history. Current data suggest, for example, that modern humans originated in Africa. By 100,000 years ago, humans were anatomically identical to modern humans, but it was not until 50,000 years ago that substantial numbers migrated out of Africa through the Arabian Peninsula into Asia.

Modern humans appear to have reached Europe only about 40,000 years ago. By analyzing genetic variations among European populations, geneticists have concluded that the modern human colonization of Europe occurred in five distinct waves. The first two of these were from the Middle East, first westward across the Mediterranean Sea to Greece, Italy, and Spain, and then northwest to northern Germany and Britain. Subsequent migrations gave rise to other distinctive populations—first, Finns, Lapps, and other northern European peoples; then, Celtic populations of Britain, Ireland, and Brittany; and, later, the Basques of Spain and France. In each case, the distinctive genetic characteristics of each population reflect either the particular characteristics of the original immigrants or the results of natural selection in distinctive environments.

A surprising conclusion of these genetic-historical studies is that women and men had distinctly different patterns of

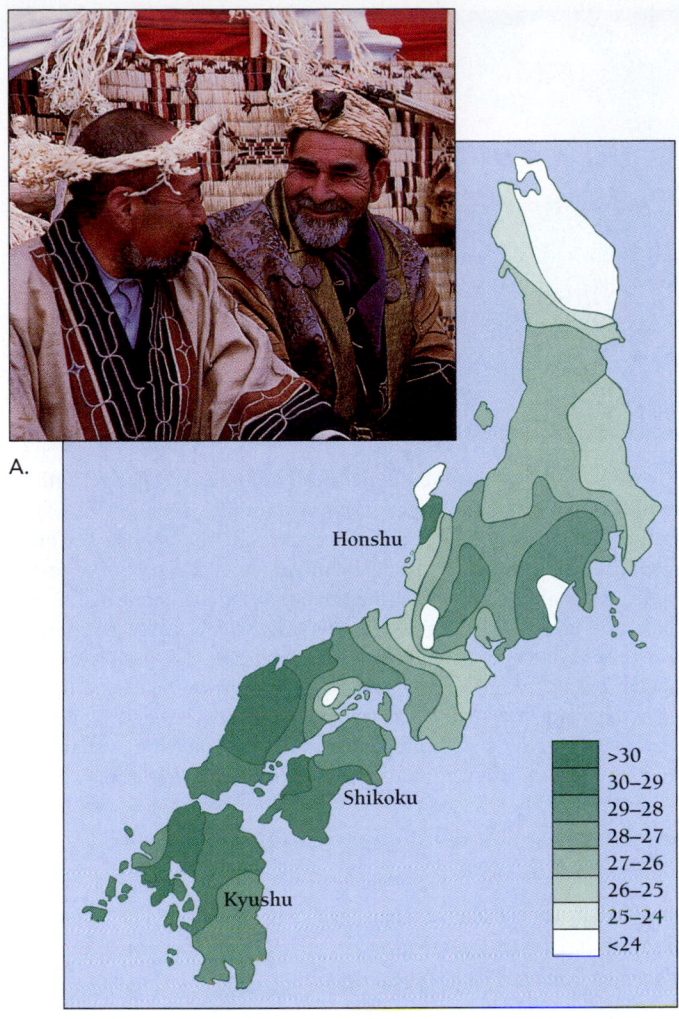

Figure 14-6 Variation in blood group A in modern Japan. A. Even the Japanese, a relatively homogeneous Island race, show variation that demonstrates intermarriage with other groups. Populations closest to mainland Korea have the greatest concentration of A alleles. Populations closest to Hokkaido have the lowest concentrations of A alleles. B. In Hokkaido lives a remnant population of Ainu, sometimes called the "hairy Ainu." The Ainu, who were once a white minority in Japan, are considered primitive by other Japanese. The Ainu had white skin, abundant body hair, and round eyes. Descendants of Causasoid peoples from northern Asia, the Ainu once lived on all four of Japan's major islands. Today they live only in the north. Intermarriage and cultural assimilation by the Japanese have made the traditional Ainu virtually extinct. (A, Superstock; B, from Human Diversity by Lewontin © 1982 by Scientific American Library. Used with permission of W.H. Freeman and Company)

French Norman invaders. In the last century, Africans, Indians, Pakistanis, and others have lent spice to this already heady mix.

If the Japanese and the British, seemingly isolated by geography, are well mixed, most mainland groups represent a true continuum of traits. In fact, studies of the distributions of different alleles have convinced researchers that human

migration. The traditional view was that women were more likely to stay in one place, while men moved away from their homes in hunts and campaigns. Now, however, it appears that men have been more likely to stay in one place, while women have moved from place to place.

The evidence for this idea comes from studies of the sequence variations in mitochondrial DNA and in the Y chromosome. All of the mitochondria in an egg come from the mother, since the sperm delivers chromosomes only. Because mitochondria have their own DNA, a woman's mitochondria come in a direct line from her mother, her grandmother, and so on back in time.

On the other hand, a man's Y chromosome contains DNA that can come only from the father, who got it from his father, and so on. So every woman has a marker for her maternal line and every man has a marker for his paternal line. (Men also have mitochondria, too, of course. So men also have a marker for their maternal lineage.)

The remarkable results of these studies of mitochondrial and Y chromosome DNA are that Y chromosome variations are more likely to be local, while mitochondrial DNA variations are more likely to be common to several populations.

Human races blur into one another, so that defining races is impossible. Almost all human variation is present in an individual population.

Figure 14-7   The phenotype of an animal can change with its social environment. The scalefinned anthias, shown here, lives in the Red Sea. It is one of dozens of fish that can change sex in response to a change in its social environment. If a school of females loses a breeding male, the dominant female changes into a male and begins breeding. Over a period of a few days, a female's ovaries convert into testes, while "her" coloring changes from golden orange to a pale background with magenta head and fins. Four days after the fish's transformation, he begins courting females. His anatomy, coloring, and behavior are indistinguishable from those of any other male's. *(David Hall/Photo Researchers, Inc.)*

THE RELATIONSHIP BETWEEN GENOTYPE AND PHENOTYPE IS COMPLEX

Genes provide the continuity between the generations, but they do not by themselves define individuals. Every individual—as expressed in the phenotype—results from complex interactions between genes and environment. The environment includes both the physical and chemical environment of the developing organism and the ongoing environment of the mature organism. In humans, environmental influences include the quality of the cytoplasm in the egg, hormonal and nutritional influences in the womb, upbringing, and the physical and social atmosphere of the home or workplace.

Studies of other animals illustrate the influence of social environment on gene expression and even brain structure (Figure 14-7). The tiny Japanese goby *Trimma okinawae,* for example, can change sex repeatedly, from female to male and back again. Schools of female fish breed with a single male. If a local breeding male is replaced by a larger male, a juvenile male in the school becomes female and remains to breed with the new male. If the breeding male leaves, however, one of the females becomes male and takes his place. These fish can change the structure of the gonads, genitals, and brain, as well as their behavior, in as little as 4 days. Their anatomy and behavior are the same as that of fish that have always been one

sex. In short, the same genotype can impart more than one phenotype. In this section, we will explore the complex relationship between genotype and phenotype.

Genes Are Expressed to Different Degrees

Genes vary enormously in their **expressivity**—the range of phenotypes associated with a given genotype. Two individuals with the same allele, for example, may express it very differently. The human disease neurofibromatosis, for example, is caused by a dominant allele whose expressivity varies so much that in some people it causes blindness or horribly disfiguring growths, whereas in others the only symptom is a coffee-colored spot (Figure 14-8). Still others with the same neurofibromatosis allele have no symptoms at all. Huntington's disease varies, too, in how strongly it is expressed and when. Although most people show no symptoms until middle age or after, a few children may exhibit severe symptoms as early as 2 years of age.

Genes also vary in their **penetrance.** While expressivity is a measure of how strongly a trait is expressed, penetrance describes the likelihood that an individual with a dominant allele will show the phenotype usually associated with that genotype. A dominant allele, such as the allele for purple flowers in Mendel's peas, has "complete penetrance," meaning that it will always be expressed.

A dominant allele with "incomplete penetrance" will, in certain physical or genetic environments, not be expressed. In

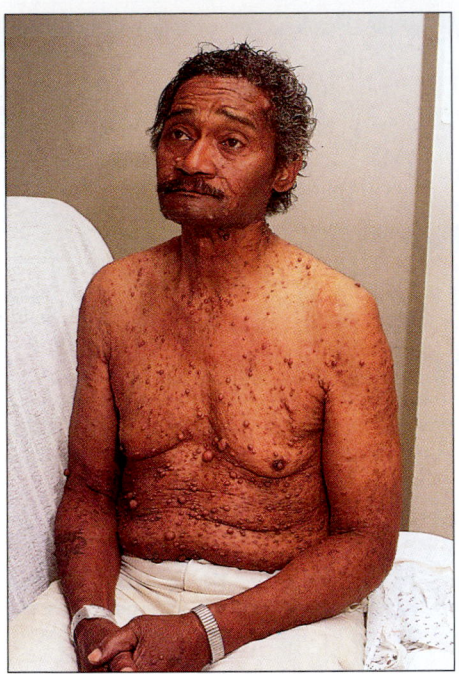

Figure 14-8 The symptoms of neurofibromatosis are distinctive. (*L. Steinmark/Custom Medical Stock Photo*)

frequently sterile, as the *vas deferens*, the tube that carries sperm from the testes, may also be blocked.

Cystic fibrosis results from the failure to produce a polypeptide called cystic fibrosis transmembrane conductance regulator, or CFTR. The disease-causing alleles are recessive, since one normal allele will provide functioning CFTR. In about 70 percent of cases, all of the symptoms of cystic fibrosis are caused by the deletion of just three base pairs in the CFTR gene. CFTR is 1480 amino acids long. The deletion of three base pairs results in the deletion of the 508th amino acid (phenylalanine) from the CFTR protein (Figure 14-10).

The CFTR polypeptide is a plasma membrane protein in *epithelial cells* that pumps chloride ions out of the cell. Epithelial cells are the cells that line the inner and outer surfaces of cavities. They can be found in every kind of duct—in the intestines, the lungs, the nose, and the mouth. Skin cells are also epithelial cells. The altered version of the CFTR polypeptide functions poorly, and epithelial cell fluids do not go where they are supposed to. Thus, one mutation can have wideranging effects on the phenotype.

Not all mutated genes produce polypeptides that function as poorly as the altered CFTR polypeptide. Some altered proteins may be able to function reasonably well (and the phenotype will be normal, or nearly so). In between are cases such

flowers, this would simply mean that only a certain proportion (say, 10 percent or 75 percent) of individuals with a given genotype would show the phenotype. But penetrance can be a slippery concept. The allele for Huntington's disease, for example, is said to be completely penetrant. Theoretically, those who have the allele always get the disease. Because the allele may be expressed late in life, however, some people with the allele die of other causes before they experience any symptoms.

Why a gene varies in its expressivity and penetrance is usually not known. Both the environment (temperature and nutrition, for example) and interactions with other genes play crucial roles in determining the expression of individual genes. But the possession of a particular allele usually does not by itself predict what an individual's phenotype will be.

Because alleles vary in their expressivity and penetrance, the genotype does not define the phenotype.

A Single Gene Can Affect Many Traits

When a baby is born with the genetic disease cystic fibrosis (CF), the baby's doctors know that the baby is destined for a life with chronic respiratory infections. Cystic fibrosis sufferers tend to have unusually salty perspiration and thick, sticky mucus that clogs the lungs and leads to repeated bouts of pneumonia, bronchitis, and other respiratory infections (Figure 14-9). In addition, many have blocked pancreatic ducts, which prevents the normal secretion of digestive enzymes. Males are

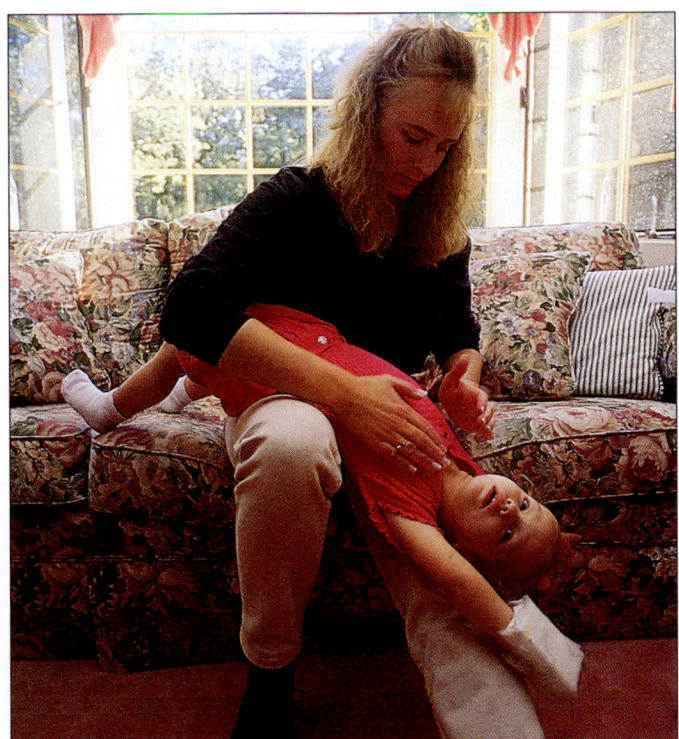

Figure 14-9 A child who has cystic fibrosis. Children who have cystic fibrosis must be thumped on the chest several times a day to loosen accumulated mucus in the lungs. (*©1997 Abraham Menashe*)

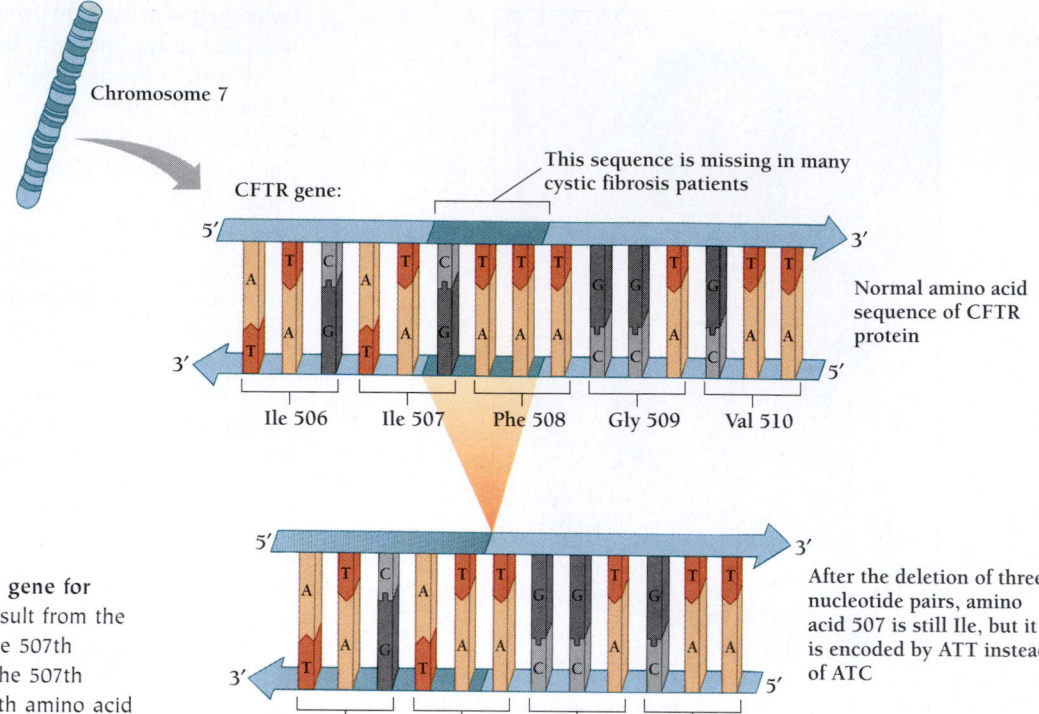

Figure 14-10 Translation of the gene for CFTR. Most cases of cystic fibrosis result from the deletion of three base pairs, one in the 507th codon and two in the 508th codon. The 507th amino acid is unaffected, but the 508th amino acid is missing and CFTR functions poorly.

as sickle cell anemia, in which the mutation produces a protein that functions much of the time, but not always.

The allele that causes cystic fibrosis affects many aspects of phenotype. The capacity of one gene to have diverse effects is called **pleiotropy** [Greek, *pleios* = more + *trope* = turning]. Every gene that has been studied—in humans and other organisms—has been found to be pleiotropic to some degree.

A small mutation in a single gene, such as the one for the CFTR polypeptide, can affect many aspects of phenotype, a phenomenon called pleiotropy. All genes are pleiotropic.

A Single Trait Can Be Influenced by Many Genes

Phenotype depends on the interactions of many genes, as well as the effects of environment. Most traits are **polygenic,** meaning that they are influenced by more than one gene. Usually, we know little about the identities and effects of each of the genes. In a few cases, however, we know a great deal. In mice, for example, fur color depends on the action (and interaction) of at least five genes: one gene determines the distribution of pigment within individual hairs, a second gene determines the color of the pigment, a third gene allows or prevents the production of the pigment molecules, a fourth gene determines the intensity of the pigmentation, and a fifth gene controls the presence or absence of spots (Figure 14-11). The gene that reg-

ulates the intensity of pigmentation is called a modifier gene because it regulates the action or "expression" of another, separate gene.

Many genes are now thought to be back-up systems for other genes. That is, the genome has several ways of accomplishing the same thing. More and more researchers are finding that knocking out seemingly important genes may have no effect on phenotype.

The long-haired mice mentioned in the last chapter are an example of such redundancy in genetic systems. In 1993, Gail Martin, at the University of California, San Francisco, found that knocking out a growth factor gene (*FGF5*) believed to influence early development in mice actually had no effect on development. The researchers were quite surprised and disappointed to see healthy mice born, but astonished when the mice all grew hair that was nearly 50 percent longer than normal (Figure 13-12). *FGF5* normally is expressed in early embryo tissues, in muscle cells, in the hippocampus of the brain, and in motor neurons. Yet, only the hair follicles were affected by the absence of the gene. Every other aspect of phenotype appeared normal.

Martin and other developmental biologists now believe that nerve, muscle, and brain function survive knocking out the *FGF5* gene, not because *FGF5* has no function in these tissues, but because other genes provide backup systems. If this is true, it is reasonable to ask why other genes do not seem to provide backup for hair length. For the moment, no one really knows.

In humans, as in mice and other organisms, the vast majority of traits are polygenic and probably redundant as well.

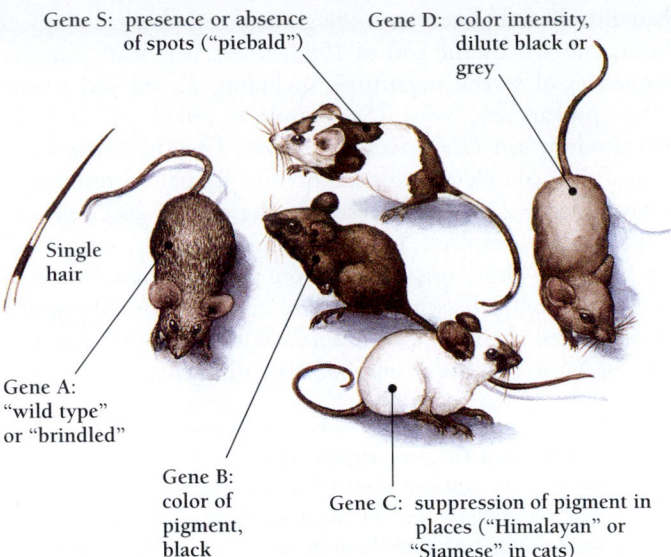

Gene S: presence or absence of spots ("piebald")

Gene D: color intensity, dilute black or grey

Single hair

Gene A: "wild type" or "brindled"

Gene B: color of pigment, black

Gene C: suppression of pigment in places ("Himalayan" or "Siamese" in cats)

Figure 14-11 **Several interacting genes determine coat color in mice.** Gene A determines where on the hair the pigments go. In wild mice, a band of yellow in the middle of the hair creates an "agouti" or "mousy" coat color. One allele confers a narrow yellow band, which gives a dark coat. Another allele confers a broader yellow band, which results in a lighter coat. In mice with a coat of a solid color, the yellow band is not present at all. The normal allele of gene B encodes a black pigment. Gene C suppresses the expression of pigment in the warmest parts of the body. Only the cooler ears, nose, tail, and paws are dark. Gene D regulates how much pigment each hair receives. Gene S determines the presence or absence of pigment in patches.

Eye color, for example, is controlled by at least two genes. Studying polygenic traits is extremely difficult, since with each added set of alleles, the number of possible genotypes increases geometrically. If these are difficult to study in research animals or plants, they are nearly impossible to study in humans. As a result, almost nothing is known about polygenic traits (most traits) in humans.

The expression of each trait is the result of many genes acting together.

A Single Gene May Have Multiple Alleles

Most of the time we have spoken as if each gene can have only two alleles—a dominant allele and a recessive allele. But a gene may exist in many alternate forms, each of which contains a distinct sequence of nucleotides that specifies a distinct polypeptide. Some of the polypeptides cannot function at all, while others may function differently.

One familiar example of a gene locus with multiple alleles is the one responsible for the ABO blood types. This locus has three alleles, called I^A, I^B, and i, which specify different versions of a surface marker on red blood cells. (Although three alleles exist in most human populations, each individual has only two of the three alleles, one on each homologous chromosome.) Alleles I^A and I^B each specify an enzyme that helps to synthesize a glycoprotein marker on the surface of red blood cells. These two glycoproteins are called, respectively, glycoproteins A and B. Allele i specifies no enzyme and so neither cell surface glycoprotein is synthesized. An individual human may have one of six genotypes (Table 14-1): $I^A I^A$, $I^B I^B$, $I^A I^B$, $I^A i$, $I^B i$, or ii. Because alleles I^A and I^B are dominant to i, individuals with $I^A I^A$ or $I^A i$ genotypes have type A blood, individuals with $I^B I^B$ or $I^B i$ have type B blood, and individuals with ii genes lack these surface glycoproteins and are said to have type O blood. Alleles I^A and I^B are codominant, meaning that in individuals with both alleles, both alleles are fully expressed. $I^A I^B$ individuals have type AB blood.

A single gene may have many functioning alleles. While some amino acid changes result in proteins that do not function at all, other amino acid changes alter the activity of the resulting protein. When both alleles are expressed, they are said to be codominant.

THE HUMAN GENOME PROJECT

If each of the 6 billion human beings on Earth is genetically unique, we can conclude that human beings harbor enormous genetic diversity. At the same time, we must remember that we are still more than 99 percent identical. For this reason, it

TABLE 14-1   HUMAN ABO BLOOD GROUPS

Blood genotype	Red blood type	Plasma cell surface	Antibodies
$I^A I^A$	A	A glycoprotein	anti-B
$I^A i$	A	A glycoprotein	anti-B
$I^B I^B$	B	B glycoprotein	anti-A
$I^B i$	B	B glycoprotein	anti-A
$I^A I^B$	AB	A and B glycoproteins	no antibodies
ii	O	no glycoproteins	anti-A and anti-B

makes sense to most people to study "the human genome" as if it were a definable entity.

What Is the Human Genome Project?

In 1988, the National Institutes of Health (NIH), the federal agency that oversees and funds research in the health sciences, launched the largest and most expensive single project in the history of biology. The Human Genome Project is a 15-year, $3 billion project jointly funded by the Department of Energy and by NIH's National Center for Human Genome Research. The project is designed to map the human genome in four major ways:

1. The project has already made linkage maps of more than 50,000 genes, restriction enzyme cut sites, or other markers. The resulting map allows geneticists to determine the chromosomal position of any gene. (Recall from Chapter 9 that the closer two genes are on a chromosome, the less likely they are to cross over during meiosis and assort independently.)
2. The project has also produced a physical map of each human chromosome by first cutting the chromosomes into restriction fragments, then identifying unique sequences that can act as landmarks, and determining the distances to the next landmark.
3. The project also aims to sequence all 3 billion base pairs on one set (23 pairs) of chromosomes for several anonymous volunteers, then load this information onto a computer to create an electronic version of the genome.
4. By reducing the cost of DNA sequencing, the project will enable scientists to study variations in the human genome. It will also determine the complete sequence of the DNA of several other species.
5. The project will also develop new methods for studying the functions of newly identified gene products. These studies will depend on studies of the effects of gene expression in cultured cells and in living organisms.

In 1990, work on the project was begun in laboratories all over the world under the direction of Nobel Laureate James Watson, the codiscoverer, with Francis Crick, of the structure of DNA. In spite of the large number of researchers involved, the task of sequencing the entire human genome seemed insurmountable. At the time, researchers had sequenced fewer than 5 percent of all genes.

Then, just as the project was getting under way, a molecular biologist at NIH named Craig Venter found a way to simplify one part of the job. Since only 3 to 5 percent of the 3 billion bases in the human genome actually code for proteins, he would extract mRNA from human tissues, copy it into cDNA with reverse transcriptase, clone the cDNA, then sequence it. That way, he would sequence only the 150 million nucleotides (100,000 genes × 1500 nucleotides) that actually code for proteins, leaving out all the introns and other junk DNA.

Thousands of researchers all over the world are working on the rest of the Human Genome Project's goals. Technical advances speeded the work, and by the end of 1998, researchers had already completed both the genetic and physical maps of human DNA. By the end of 1999, researchers had complete sequences of several organisms, including *E. coli* and several other prokaryotes, yeast (*Saccharomyces cerevisiae*), and the nematode worm *Caenorhabditis elegans.* They had also determined the complete sequence of one human chromosome (chromosome 22, the smallest). By the end of 2003, geneticists expect a complete, "polished" sequence, with no gaps and no mistakes. Other organisms whose genomes have been targeted for complete sequencing are the fruit fly *Drosophila melanogaster,* the mouse *Mus musculis,* the domestic dog *Canis familiaris,* and the plant *Arabidopsis thaliana.*

The Human Genome Project offers to sequence an entire human genome, as well as the genomes of a few other organisms, and to construct linkage maps and physical maps of each human chromosome. By 2003, scientists will probably have sequenced the entire human genome, as well as the genomes of at least a half-dozen other organisms.

How Do Researchers Locate Genes for Specific Diseases?

The drive to sequence the human genome has rapidly provided a gene library of actual gene fragments stored in numbered test tubes, as well as a computer database listing the nucleotide sequence of each human chromosome. Researchers express many of these genes (from mRNA) to find out what sorts of proteins they make. But these libraries have no catalogue to help researchers identify and locate individual genes. The contents of a gene library are therefore still something of a mystery. What genes does the library contain? How do they function in the body? Which chromosomes contain which genes? How can researchers find individual genes among the 80,000 to 100,000 human genes in the genome?

The time-honored way to locate a gene in the genome is by means of experimental crosses. Recall from Chapter 9 that experimental crosses reveal how frequently genes cross over, providing a measure of how far apart they are on a chromosome or whether they are on separate chromosomes. Although researchers cannot arrange experimental crosses in humans, they can obtain helpful information from many families about physical traits or genetic diseases extending back for generations.

Such records, arranged in the form of family pedigrees, show patterns of inheritance. Family pedigrees are useful for analyzing the inheritance of obvious traits that are passed in a very simple manner (Figure 14-12). Traits such as fatal diseases and whether the earlobe is "attached" are easy to identify.

A more sophisticated approach is to look for genetic markers that seem to travel with a disease from generation to generation, a technique called **pedigree analysis.** Researchers first collect blood samples from individuals in families that carry a given disease, then use DNA fingerprinting to look for RFLP alleles that are consistently associated with the disease. In the search for the Huntington's disease allele, for example, James

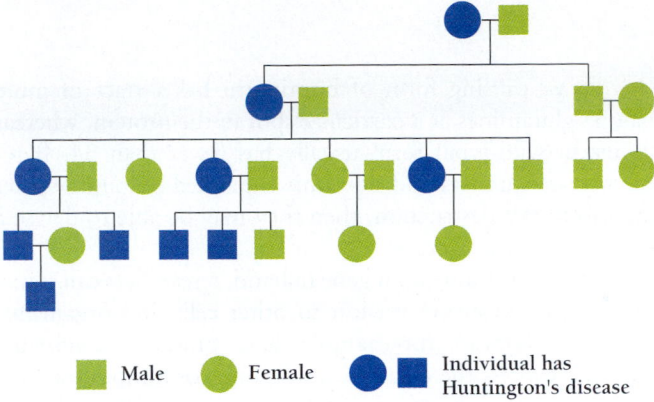

Male Female Individual has Huntington's disease

Figure 14-12 Pedigree for a family with Huntington's disease. The marriages and offspring of unaffected individuals have been omitted.

Gusella, at Massachusetts General Hospital, began looking for a particular RFLP allele that traveled with the Huntington's disease allele in some families. Because the human genome is enormous, and because no one knew even on what chromosome the Huntington's gene might lie, the task appeared Herculean. Yet, miraculously, Gusella immediately happened on a polymorphic marker that, in certain families, including the huge family at Lake Maracaibo, in Venezuela, was always inherited with Huntington's disease. This RFLP site came from the short arm of chromosome 4. In 1983, for the first time, the approximate location of the Huntington's gene was known.

Pedigree analysis allows researchers to find a genetic marker associated with a particular trait or disease that shows approximately where on the genome a gene lies.

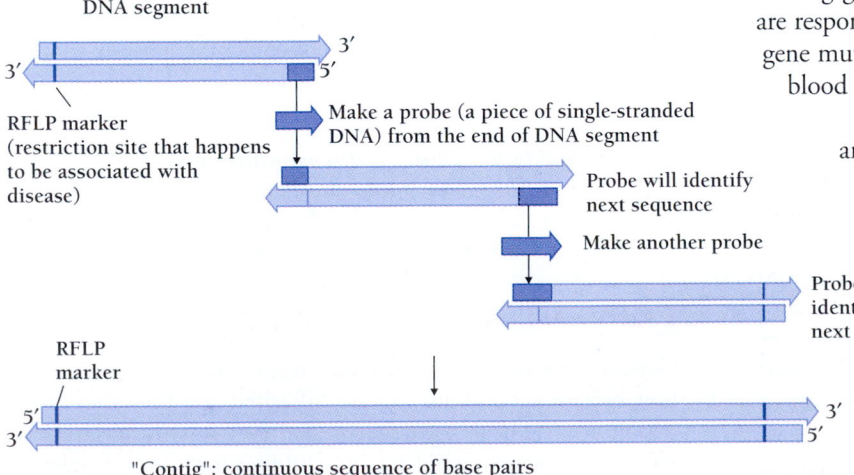

DNA segment

RFLP marker (restriction site that happens to be associated with disease)

Make a probe (a piece of single-stranded DNA) from the end of DNA segment

Probe will identify next sequence

Make another probe

Probe will identify next sequence

RFLP marker

"Contig": continuous sequence of base pairs from one marker to the next

Figure 14-13 Assembly of a contig. To pinpoint a gene, researchers assemble a single long sequence of DNA between two RFLP markers on either side of the gene.

Positional cloning

Once the approximate position of a gene has been located between two RFLP markers some million base pairs apart, researchers can use **positional cloning** to pinpoint the gene. The first step in positional cloning is the assembly of all the clones that cover the region between the two markers on either side of the gene (Figure 14-13). Researchers clone numerous pieces of DNA that are known to overlap, then string them together into a continuous string that extends from one marker to the other, a so-called "contig."

Next, researchers narrow their search down to DNA sequences that are likely to encode a protein. One way of finding such sequences is to hybridize human DNA with labeled DNA from rats, mice, or even flies and worms. DNA from a mouse, for example, can hybridize with human DNA to form a double-stranded molecule, one strand mouse and one strand human. Only sequences that have stayed more or less the same since mice and humans diverged millions of years ago, however, will hybridize. The hybridized sequences mark genes whose exact sequences are so important that they have been conserved in the course of evolution. Most of these highly conserved sequences encode proteins. Mutations in these protein-coding genes are far more likely to cause death or disease than mutations in noncoding DNA.

In the next step, molecular biologists look for mRNA in the tissues where the gene is expressed. A diseased gene that affects the brain, for example, would presumably be expressed in brain tissue. Having identified one or more genes that might cause the disease, researchers then directly compare the sequences from sick individuals with the corresponding sequences from healthy individuals. The purpose of these comparisons is to show that the sequence changes are always associated with disease.

Positional cloning is a time-consuming, tedious procedure, taking anywhere from 10 to 100 person-years to locate one gene. Nonetheless, geneticists have identified hundreds of disease-causing genes and determined exactly which sequence changes are responsible for diseases. The diseases caused by such single-gene mutations can affect every system in the body—from the blood (leukemia) to the brain (Alzheimer's disease).

Positional cloning is becoming much easier. For example, geneticists no longer have to start from scratch to make "contigs" in a region of interest. Because it will soon be possible to find DNA markers (SNPs) within a few thousand bases of any gene, researchers will be able to determine the possible positions of disease-causing genes much more quickly. Instead of examining the pattern of inheritance of a disease (or any other trait) within single families, researchers will be able to find *all* the DNA sequences shared by any group of people who have a particular disease. Using rapid sequencing, they will then be able to determine which DNA changes go along with a particular disease, even in a group of unrelated individuals.

Geneticists have come to realize that only a small fraction of familial diseases result from alterations in single genes. In addition to finding disease-causing mutations, researchers hope to find the genetic basis of more complex (polygenic) inherited traits. Among the genes that geneticists hope to find are those that affect the susceptibility to diseases, such as schizophrenia, that often run in families but that have no clear pattern of inheritance.

Another set of questions focuses on why people respond differently to particular medicines. Some people, for example, have variant forms of enzymes that break down medicines that would otherwise help them. Researchers, physicians, and pharmaceutical companies hope that this new field of study, called *pharmacogenomics,* will allow physicians to choose more effective treatments more rapidly.

Positional cloning allows researchers to locate a particular gene precisely on a chromosome.

Of What Use Is a Sequenced Gene?

Once a gene has been sequenced, geneticists can test individuals to see if they have a normal allele or a disease-causing allele of the gene. In practice, this may be neither easy nor necessarily useful. For example, in recent years, researchers cloned two genes, called *BRCA1* and *BRCA2,* some alleles of which predispose women to breast cancer. Some of the alleles that cause cancer give a woman an 85 percent chance of developing breast cancer. The disease alleles of these two genes, however, can result from any of hundreds of different mutations, each of which may carry a different risk of causing breast cancer. In addition, other genes, both known and unknown, can alter that risk as well.

In any case, the vast majority of women who get breast cancer do not have an abnormal allele of either *BRCA1* or *BRCA2.* It only makes sense to test women who come from families where breast cancer is strongly inherited and common. Even for that minority of women, however, the test is of limited value. Molecular biologists' knowledge of the sequences of *BRCA1* and *BRCA2* does not enable them to either prevent or cure breast cancer. Finally, once tests reveal that a person has a gene that predisposes him or her to a particular disease, insurance companies will not want to provide health insurance to that person for that disease. Most geneticists argue that no genetic information should be made available to insurance companies. But insurance companies have a powerful economic incentive to overturn or sidestep any rules limiting their access to genetic information. For the average person, then, a genetic test can potentially do more harm than good.

On the other hand, researchers who know the sequence of a disease allele are in a good position to begin to understand how the gene does its damage. Huntington's disease provides a good example of this. In 1993, researchers found that the gene involved in Huntington's disease codes for a polypeptide called huntingtin that is expressed in every cell in the body.

The disease-causing form of huntingtin has a tract of more than 37 glutamines at a particular spot in the protein, whereas the usual (wild-type) form usually has fewer than 34. If researchers can understand how this expanded glutamine tract can lead to cell destruction, then they may be able to devise a treatment for this disease.

Having the huntingtin gene in hand, researchers can transfer the disease-causing version to other cells and organisms. Several laboratories, for example, have produced transgenic mice that express the disease-causing form of huntingtin. The characteristics of these genetically engineered cells and mice have led researchers to unexpected ideas about the causes of cell death not only in Huntington's disease but also in other diseases in which nerve cells progressively die or fail to function properly. For example, the huntingtin-expressing transgenic mice develop normally but begin to show abnormal behavior within the first 2 months of life. When researchers examined the brains of the sick mice, they found abnormal deposits of protein in the nucleus of many brain cells. When other researchers looked in the brains of people who had died from Huntington's disease, they found similar deposits. In each case, the protein deposits contain the protein specified by the disease-causing gene.

Many researchers have hypothesized that these protein deposits are the direct cause of cell death. Other researchers, however, have suggested that the deposits are a cell's way of getting rid of a potentially toxic protein. The best way to decide between these alternatives would be to find a chemical that can prevent the deposits without otherwise damaging the cell.

By using robots that can add individual compounds to hundreds of cell cultures at the same time, researchers hope to screen large sets of chemical compounds. This approach, called *high-throughput screening,* takes advantage of the huge "libraries" of compounds that have been produced by organic chemists, especially within pharmaceutical companies. Compounds that are effective in such cell-based assays could then be used to treat transgenic mice to see if they prevent protein deposits and abnormal behavior. If one or more compounds are successful in mice, researchers could then begin to plan clinical trials on people who have Huntington's disease.

The ability to transfer a disease-causing gene to cells and transgenic mice can allow researchers to develop new therapies.

PREVENTING GENETIC DISEASE

Genetic research has greatly advanced our understanding of how certain genetic diseases are inherited and how they make people ill. Medical researchers have used this information to try to help people avoid having children with serious defects, or, in a few rare cases, to actually treat people with genetic diseases through "gene therapy," or "genetic medicine." In this section we will see how this is done.

Only about 3 percent of all human diseases are caused by defects in a single gene. The most common genetic disease in

BOX 14.1

Can Human Genes Be Patented?

In 1991, Craig Venter suggested that the NIH patent the protein-encoding cDNA, so that any commercial uses for the genes would earn royalties for the federal agency. In June 1991, NIH filed applications on 315 gene fragments. But the scientific community protested. The human genome was viewed as a common heritage, not something that could be patented so that one person or one business could make money from it. Even religious coalitions have made formal objections to the idea of patenting human genes, and whether the patents will be upheld if they are challenged legally is still in doubt.

Ultimately, the applications were withdrawn, but Venter was determined. He started a nonprofit organization called The Institute for Genomic Research (TIGR) that would continue his work at NIH on a massive scale but would be funded by private companies. By 1995, TIGR had sequenced pieces of nearly 90 percent of the genes expressed in human tissues, that is, the cDNA version of all the different kinds of mRNA. A second, for-profit company, Human Genome Sciences, Inc. (HGS), patents and sells whatever interesting genes TIGR clones. Pharmaceutical companies have lined up to give TIGR millions of dollars in exchange for a stake in its discoveries. Drug companies are gambling that human proteins made from cDNA will turn out to be useful and lucrative drugs that will, for example, boost the immune system or dull the appetite. Meanwhile, researchers at many universities and several pharmaceutical companies are freely publishing the full sequences of as many genes as they can, so that at least some human gene sequences will be public property and not patentable. Other universities are themselves patenting DNA sequences.

---

the United States is cystic fibrosis, which affects about 1 person in 10,000. None of these single-gene diseases—several of which we have discussed in this chapter—are the ones that kill most adults. For comparison, cardiovascular disease kills more than one person in three. About half that number die of cancer. People fall prey to one disease or another—lung cancer versus stroke, for example—because of lifelong behaviors and experiences and because a constellation of different genes slightly predisposes them to one weakness or another. Every person carries at least five to ten genes that could make them seriously ill under certain circumstances. As geneticist Michael Kaback puts it, "We are all mutants. Everybody is genetically defective."

Genetic Counseling: How Can Parents Decide?

The first kind of genetic testing is simply to test couples who are considering having a baby to see if they carry alleles for genetic diseases, such as sickle cell anemia, Tay-Sachs disease, cystic fibrosis, or Huntington's disease. Geneticists can now test for more than 100 different genetic defects. Usually, however, geneticists test only for the alleles that are relatively common or for those that the parents seem most likely to carry. A genetic counselor can calculate the likelihood that a child will have a certain disease and help the couple decide whether to risk having children or not.

Tay-Sachs, for example, is a genetic disease that results from a single defective gene for an enzyme normally synthesized by nerve cells. In healthy children, this enzyme breaks down lipids. But in children homozygous for the Tay-Sachs allele, lipids accumulate in the brain, causing mental retardation, blindness, and failure to develop control of the muscles.

All Tay-Sachs children die in early childhood, usually by age 4 years. Tay-Sachs has a simple pattern of inheritance because all children homozygous for the single Tay-Sachs allele become ill. Heterozygous individuals, who have one defective allele and one normal allele, are called "carriers." Carriers can be identified by their low levels of the lipid-digesting enzyme, but they are otherwise healthy.

The Tay-Sachs allele is most common among eastern European Jews and their descendants. If both parents are of eastern European Jewish descent, a genetic counselor will likely recommend that they be tested for the allele. If only one parent is Jewish, however, the counselor will probably not recommend the test, because a baby can only become ill if he or she receives the allele from both parents. This population of people has taken such care to avoid having Tay-Sachs children, however, that Tay-Sachs is now quite rare and, at least in America, the disease may now be more common in the population at large than among those descended from eastern European Jews. This decrease in the prevalence of Tay-Sachs is one of the great successes of genetic testing.

Genetic counselors can tell couples whether they carry certain alleles and the likelihood that their offspring will express certain traits.

Prenatal Testing: Can Parents Tell Before?

Once a woman becomes pregnant, physicians can obtain cells from the fetus in order to search for genetic defects. Besides looking for chromosomal abnormalities by examining the chromosomes, genetics clinics can now detect genetic diseases in the cells of an early embryo. Once the parents know that their baby has a genetic disease, however, they have only two options. They may decide to have the mother carry the fetus to term (birth) and cope as best they can with the baby's condition, or they can decide to abort the fetus.

Amniocentesis

By examining chromosomes from a fetus early in pregnancy, geneticists can tell if the parents have produced a zygote with a chromosomal abnormality. The most common chromosomal abnormality found in humans is trisomy 21, or Down syndrome. Geneticists usually obtain the fetal chromosomes by taking a sample of fetal cells from the **amniotic fluid** [Greek, *amnion* = membrane around a fetus], the watery fluid that surrounds the fetus. This procedure, called **amniocentesis** [Greek, *centes* = puncture], is usually recommended to women over 35 years old and to women with a family history of chromosomal abnormalities or other diseases (Figure 14-14A). Amniocentesis can detect more than 100 genetic abnormalities besides Down syndrome.

At 15 or 16 weeks after conception, a physician inserts a needle through the mother's abdominal wall, just below her navel, and into her uterus. The needle is used to remove from the womb amniotic fluid that contains cells from the fetus.

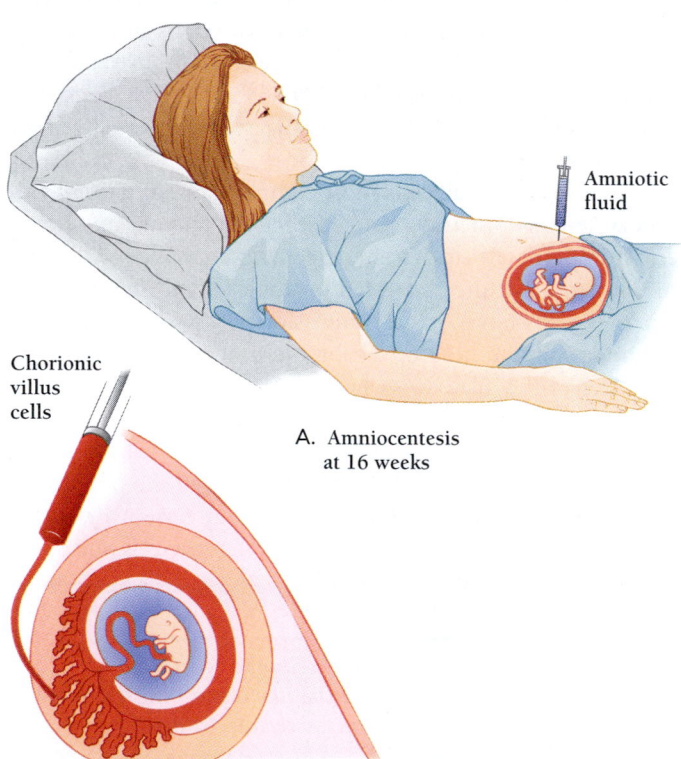

A. Amniocentesis at 16 weeks

Amniotic fluid

Chorionic villus cells

B. Chorionic villus sampling at 6–12 weeks

Figure 14-14   Prenatal testing. A. In amniocentesis, a physician extracts amniotic fluid through a needle inserted through the mother's abdominal wall. B. In chorionic villi sampling, a physician extracts cells from the placenta through the mother's vagina and cervix.

This procedure is virtually painless, but can be emotionally upsetting for some women. Amniocentesis is very safe, but not perfectly so. Approximately 0.5 percent of fetuses exposed to amniocentesis later die because of spontaneous abortion, commonly known as miscarriage. (Some 10 to 15 percent of known pregnancies abort spontaneously before the 20th week, usually because of chromosomal or other abnormalities in the fetus. After 20 weeks, an early delivery is called a "preterm birth.")

To prepare the chromosomes for examination takes 3 to 4 weeks, during which a geneticist grows the fetal cells in culture. During this time, the fetus continues to mature and, indeed, approaches the age when it becomes viable outside of the womb, which is about 20 weeks after conception. (A baby is full term at 30 to 38 weeks after conception. Physicians add 2 weeks to these times, counting from the date of the last menstrual period.) Parents meanwhile think about what they will decide if the fetus turns out to have Down syndrome or some other chromosomal defect.

Amniocentesis allows a geneticist to look for chromosomal or other genetic abnormalities in a 4-month-old fetus.

Chorionic villus sampling

A second procedure, called **chorionic villus sampling (CVS)**, uses fetal cells present in the placenta—the tissue that carries nutrients and oxygen from the mother's blood to the fetus and that carries wastes from the fetus back to the mother (Figure 14-14B). CVS may be done as early as 6 to 12 weeks after conception, much earlier in pregnancy than amniocentesis. Because the results of the test become available so much sooner, a decision to abort becomes more acceptable to some people. Alternatively, news that the fetus is healthy means a shorter period of worry for the parents. The increased risk of spontaneous abortion due to CVS is about the same as in amniocentesis, about 0.5 percent.

Chorionic villus sampling allows geneticists to test for the same genetic defects as amniocentesis, but earlier in pregnancy.

Aborting a fetus

In the next section, we see that researchers are working hard to develop techniques for actually curing genetic diseases, or, at least, mitigating their effects. For now, however, prenatal testing implies only one decision: to abort or not. For many people, abortion is unacceptable under any circumstances. For them, prenatal testing provides time to plan for the care of a child with unusual needs.

For most people in the United States, however, abortion is acceptable under certain circumstances. Each family's circumstances are unique, and what is acceptable varies enormously from individual to individual and from situation to situation. For example, many people find it more acceptable to abort a fetus with Tay-Sachs than a fetus with Down syndrome or sickle cell anemia. A child with Tay-Sachs always dies young and cruelly. The effects of Down syndrome, however, vary enormously, and many of its symptoms can be treated. A child with Down syndrome can experience a happy childhood and may live well into adulthood (Figure 14-15).

Likewise, a person with sickle cell anemia may have few or no symptoms much of the time and live well into middle age. New treatments may prevent any significant ill effects. Even now, when a sickle cell crisis does occur, and such a crisis can be both life threatening and agonizingly painful, it usually can be controlled with prompt medical attention. Many symptoms of cystic fibrosis can also be effectively treated with drugs.

Many parents might decide to abort a fetus with Down syndrome, sickle cell disease, or cystic fibrosis, but many others would choose to let the fetus develop into a baby. In neither case is the decision a light one. Parents must weigh their feelings about abortion against their family's ability to marshal both the emotional and the financial resources necessary to provide adequate care for the child (as well as the adult the child becomes) who has special needs. The ethical values of the family, the developmental age of the fetus, the financial status of the family, and the stability of the parents' relationship may all be considerations.

Test-tube babies from selected gametes

Recently, medical researchers have come up with a way for parents to avoid having a baby with a specific genetic defect without confronting the question of abortion. Very simply, a doctor takes unfertilized eggs from a mother who is heterozygous for a particular genetic defect and selects the eggs that lack the allele in question.

The techniques for doing this are rather complex. In one example, the woman must first be treated with hormones that induce her to ovulate several eggs at once. A surgeon then removes these freshly ovulated eggs from the mother's abdomen and takes them to a laboratory. Freshly ovulated eggs have barely completed meiosis, and they are still attached to a sister cell. Because the two daughter cells do not divide evenly, however, the egg cell is very large, while the other cell, called the *polar body,* is very tiny.

The polar body is normally sloughed off, but geneticists have found a use for it. Because homologous chromosomes separate during meiosis, only one of the two cells gets the defective gene. If researchers find that the polar body has the defective gene, then they know that the egg has the normal copy. The normal eggs can then be combined with sperm and implanted in the mother's uterus.

The number of genes that researchers can examine by means of this method will increase ten times in the next decade, since the Human Genome Project is producing potential new genetic tests at the rate of about one a week. Because this method of making babies is rather expensive, only the wealthiest and most motivated couples—such as those carrying alleles for Tay-Sachs—are likely to take advantage of it.

Medical researchers have invented ways to select human eggs that are free of certain genetic defects, but this technology is cumbersome and expensive.

Gene Therapy: Can We Fix It Later?

Gene therapy is a developing technology whose purpose is to deliver genes as a way of treating disease or injury. Researchers have attempted to treat 13 different genetic diseases, as well as AIDS, heart disease, and cancer. They have used dozens of different methods. So far, however, not one variety of gene therapy has resulted in a complete cure of even one person. Neither the genotype nor the phenotype has been "fixed." In the best cases, researchers can effect a partial, temporary cure that must be renewed every few months, often in combination with conventional drug therapy (Figure 14-16).

All human gene therapy has focused on somatic cells rather than germ cells. Human germ-cell therapy would require that researchers apply to humans the techniques now used to make transgenic and knockout mice. Although current technology might make such experiments possible, ethical and legal (as well as scientific) restrictions have prevented human germ-line manipulations. For the foreseeable future, gene therapy research will deal with somatic cells only.

Much gene therapy research has concerned finding vectors to carry DNA into human cells. Almost all of the current approaches depend on modifying vectors that could otherwise cause diseases ranging from colds to cancers to AIDS. Some approaches, however, depend on the ability of tiny lipid globules ("liposomes") to fuse with cells and deliver DNA. The major obstacle to gene therapy is the lack of vectors that carry DNA efficiently into human tissue.

Among the applications of gene therapy to genetic diseases, cystic fibrosis is among the most attractive targets. Cystic fibrosis is the most common serious genetic defect in the United States, afflicting some 30,000 people, or about 0.01

Figure 14-15 A child who has Down syndrome (*at right*). (*Gary Parker/Science Photo Library/Custom Medical Stock Photo*)

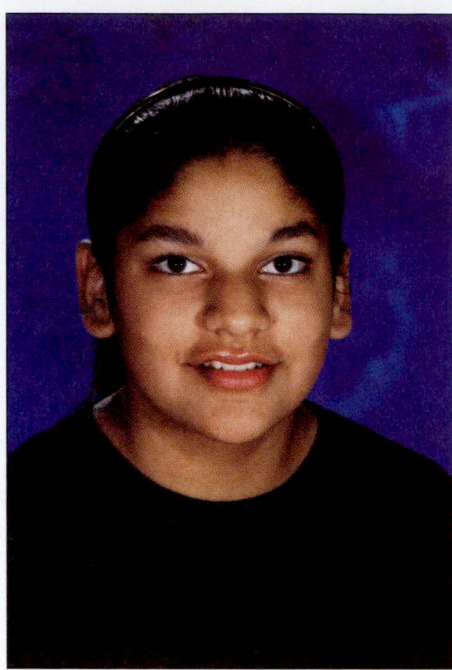

Figure 14-16   The most successful example of gene therapy. In 1990, researchers at the National Institutes of Health treated Ashanthi DeSilva for the genetic disease SCID (severe combined immunodeficiency disease), which almost completely disables the immune system. Researchers removed immune cells from the 4-year-old DeSilva, added a functioning allele for the enzyme adenosine deaminase, then injected the genetically engineered cells back into her body. With repeated treatments, DeSilva has been transformed from a chronically sick child with a short expected life span to an evidently healthy young woman. (*Courtesy of Van DeSilva*)

percent of the population. Recall that the many symptoms of cystic fibrosis, from chronic respiratory infections to blocked pancreatic ducts, all seem to result from a defective ion transport molecule in the cell membrane.

In many genetic diseases, the body's cells normally need to express large amounts of the missing protein. In cystic fibrosis, however, even very low levels of expression of the CFTR gene—say 10 to 100 molecules per cell—reduce symptoms dramatically. This fact makes cystic fibrosis more amenable to treatment through gene therapy than other diseases.

One partly successful approach to treating cystic fibrosis through gene therapy has been to spray genetically engineered cold viruses into the noses of patients who have the disease. The engineered *adenovirus* carries a normal version of the CFTR gene into the cells of the inner surface of the nose. The virus is deliberately damaged so that it cannot replicate in the body and make the patient ill. It can only deliver its cargo— the CFTR gene—into the infected cells.

The treated cells express the normal CFTR gene, and the patient's symptoms are alleviated, but only partly. Treating the back of the nose does not, for example, unblock the pancreatic ducts or help alleviate symptoms in any other part of the body. Further, the cells that the virus infects live only a few weeks before they are shed and replaced by new cells. These new cells

must be infected with another treatment of virus. Patients must therefore return again and again for more treatments.

Why isn't the gene passed on to daughter cells? A gene can integrate into the genome (into one of the chromosomes) or it can stay unintegrated within the cell's nucleus (along with some regulatory DNA that ensures that it will be transcribed). A gene randomly inserted by a viral vector can affect nearby genes; it could, for example, turn off an adjacent tumor suppressor gene and trigger cancerous growth. So far, however, no such tumors have been observed. But no one has yet succeeded in inserting a gene into dividing progenitor cells in a human.

Many gene therapy researchers are now focusing their attention on how best to get genes into stem cells, dividing progenitor cells that can develop into many alternative cell types. The stem cells in bone marrow, for example, give rise to both red and white blood cells. Other stem cells give rise to muscle and even to brain cells. If researchers could deliver genes to stem cells in the body, those cells could provide a reservoir of health replacements for many types of cells. In addition, because researchers are now able to grow stem cells in the laboratory, they are also able to deliver genes to stem cells in culture. The engineered stem cells can themselves be implanted in the tissues most affected by a genetic disease.

Although many researchers are excited about the prospects of stem cell therapies, the potential use of human stem cells raises serious ethical problems, since most stem cell populations come from embryos. People opposed to abortion are concerned that researchers and physicians will directly or indirectly encourage abortions in order to harvest stem cells. On the other hand, some types of stem cells are available from adult humans, and their use does not raise the same ethical problems. In 1999, after much discussion and debate, the National Institutes of Health proposed guidelines for stem cell research. These guidelines establish strict procedures for obtaining approval for any such research.

Until researchers are able to insert genes into dividing cells, especially into stem cells, gene therapy will be only a temporary cure that can last only as long as the engineered cells live. Such genetic medicines will need to be repeated every few weeks or months. And an individual with a genetic disease will always harbor the defective allele in the germ cells and can continue to pass it on to offspring. For now, gene replacement therapy remains an area of basic research, not of practical results.

Gene therapy is still in its infancy despite expenditures of more than $400 million per year and more than 100 clinical trials in humans. Many of these trials, said the report, were poorly designed, and none has demonstrated even one technique that actually cures genetic disease. Several critical reports have accused researchers of "overselling" gene therapy, of understating its dangers, and of neglecting basic questions about gene regulation, vectors, stem cell function, and other areas. In short, people with genetic diseases should not expect effective gene therapy any time soon.

On the brighter side, many gene therapy centers are turning their attention to conditions where effective therapy should not depend on long-term gene expression or on the continuing proliferation of engineered cells. Most gene therapy trials deliver genes whose products can help mobilize the immune system to combat cancer or AIDS. A particularly promising experimental treatment (called a "biobypass") promotes the growth of new blood vessels in conjunction with bypass heart surgery.

Using new gene therapy techniques, researchers can cause human cells to express a missing or otherwise desired gene. Expression is limited, so far, to the cells that originally take up the engineered DNA. Because the cells usually cannot pass the gene on to new daughter cells in the body, treatments must be repeated. Also, the treated patient cannot pass on a "good" allele acquired in this way to any offspring.

In this chapter on genetics, we have used the knowledge gained from previous chapters to examine genetic diversity in humans, the complex relationship between genotype and phenotype, the Human Genome Project, and the promise and limitations of applied human genetics. Now that we understand inheritance, we are ready to tackle evolution, the most important idea that biology has ever produced.

Study Outline with Key Terms

DNA fingerprinting identifies variations in DNA, called **RFLPs (restriction fragment length polymorphisms),** derived from blood or other tissues. It provides an estimate of how common that particular pattern of variation is in the population. Humans are 99 percent genetically identical. Yet every individual who is not an identical twin has a unique genotype. Human races blur into one another, so that defining consistent races is essentially impossible. The differences among the members of one race are greater than the average differences between races.

A small mutation in a single gene, such as the one that causes most cases of cystic fibrosis, can affect many aspects of phenotype. But the relationship between genotype and phenotype is generally extremely loose. One reason is that alleles vary in their **expressivity** and **penetrance.** In addition, each gene affects many aspects of phenotype—a phenomenon called **pleiotropy.** The expression of each trait is **polygenic,** the result of many genes acting together. Finally, a single gene may have many functioning alleles. Because of all this complexity, the expression of any gene is heavily influenced by both the physical and developmental environment and also by the genomic environment of the gene.

The Human Genome Project offers to sequence an entire human genome, as well as the genomes of a few other organisms, and to construct linkage maps and physical maps of each human chromosome. By 2003, the Human Genome Project will probably have sequenced the entire human genome. Meanwhile, businesses and universities are patenting DNA fragments that encode human proteins.

Pedigree analysis allows researchers to find a genetic marker associated with a particular trait or disease that shows approximately where on the genome a gene lies. **Positional cloning** allows researchers to obtain a particular gene on the basis of its precise location.

Genetic counselors can tell couples both whether they carry certain alleles and the likelihood that their offspring will express certain traits. In **amniocentesis,** a geneticist samples a 4-month-old fetus's **amniotic fluid** and looks for chromosomal or other genetic abnormalities. **Chorionic villus sampling (CVS)** allows geneticists to test for the same genetic defects as amniocentesis, but earlier in pregnancy.

Using new **gene therapy** techniques, researchers can induce human cells to express a gene that the cells were previously unable to express. Expression is presently limited, however, to relatively few molecules of the desired protein. Because the cells cannot pass the gene on to new daughter cells in the body, treatments must be repeated. Also, the treated patient cannot pass on the "good" allele acquired in this way to any offspring.

Review and Thought Questions

Review Questions

1. What factors ensure that the relationship between genotype and phenotype is a loose one? What is pleiotropy?
2. What are the limitations and strengths of DNA fingerprinting? What is a RFLP?
3. What is the purpose of the Human Genome Project?
4. How do researchers locate genes for specific diseases such as Huntington's disease or cystic fibrosis?
5. How is prenatal testing different from genetic counseling?
6. If a prenatal test reveals a genetic defect in a fetus, what choices do the parents have?

Thought Questions

7. If you and your spouse both carried the gene for cystic fibrosis, would you get prenatal testing? If you decided to have a test, would you prefer amniocentesis or chorionic villus sampling? Why?
8. How would your current knowledge about research on gene therapy for cystic fibrosis influence your decision?
9. If you knew you had an allele that slightly predisposed you to not get heart disease, would that influence your eating and exercise habits? If you had an allele that slightly predisposed you to get liver cancer, would you care if your insurance company and your employer knew?

About the Chapter-Opening Image

Mapping the human genome is symbolized by this woman, wrapped in a map of the world.

 On-line materials relating to this chapter are on the World Wide Web at **http://www.harcourtcollege.com/lifesci/aal2/**

part III

Evolution

Chapter 15
*What Is the Evidence
for Evolution?*

Chapter 16
*Microevolution: How
Does a Population
Evolve?*

Chapter 17
*Macroevolution: How
Do Species Evolve?*

Chapter 18
*How Did the First
Organisms Evolve?*

Photo: Ammonite fossils display the multiple-chambered shell of the animal.
(*Mauritius/Nawrocki Stock Photo, Inc.*)

What Is the Evidence for Evolution?

The Scopes Trial

I t was a hot July morning in Dayton, Tennessee, when high-school teacher John T. Scopes went on trial for teaching evolution in the public schools. In the stuffy courtroom, spectators mopped their brows and swatted at flies. A dozen journalists crowded together stood on top of a table at the back of the room, trying to get a better view. Midway through the interminable testimony the table collapsed, dumping the men on the floor. Cheers and boos from the rowdy audience outside came through the open windows and drowned out the judge's calls for order. Also outside were a score of evangelists. Their trumpeting voices could be heard throughout days of testimony.

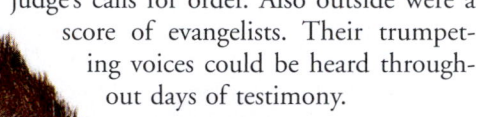

The town of Dayton was bedecked as if for a carnival, with banners and posters of monkeys. Newly constructed stands selling hot dogs, lemonade, and sandwiches lined the sidewalks, and a circus man ran a sideshow with two chimpanzees. The animals had been brought to town to "testify for the prosecution," the man said. Two hundred twenty-five newspapermen had arrived from New York, Baltimore, Philadelphia, and points beyond. It was 1925, before television and even radio. These newsmen would put the tiny southern town of Dayton on the map forever.

The law prohibiting the teaching of evolution—passed by the Tennessee legislature in a moment of boredom—was never intended to be enforced. It nonetheless attracted the attention of the American Civil Liberties Union (ACLU), which believed the law violated the First Amendment right to free speech. Eager to test the law in a federal court, the ACLU advertised in the *Chattanooga News* for a teacher willing to be arrested. Dayton boosters jumped at the chance to liven up their sleepy town with some national attention.

John Scopes agreed to be arrested. An affable high-school physics teacher and athletic coach, Scopes had never actually taught evolution, but he'd once substituted for a biology teacher who was ill and assigned the textbook pages that covered evolution. For this crime he was arrested.

The trial—loudly and forcefully argued by two nationally famous lawyers, William Jennings Bryan and Clarence Darrow, and covered by every major newspaper in the country—was, nevertheless, a rather dull affair. The attorneys representing each side wanted the same outcome—Scopes's conviction. The ACLU wanted a clean conviction so they could appeal the case in a higher court. Their goal was to have the antievolution law overturned.

But it was not to be. In an ironic twist, the ACLU won the battle it was trying to lose and so lost the war. Scopes was initially convicted, but then acquitted on a technicality. The judge technically violated the law when he mistakenly levied Scopes's $100 fine himself, instead of letting the jury do it. Because Scopes was acquitted, the ACLU never got to appeal the case, and the Tennessee antievolution law stayed on the books until the U.S. Supreme Court declared such laws unconstitutional in 1968.

Daniel I. Cox/Natural Selection

Key Concepts

- Evolution is heritable change in populations of organisms, which Darwin called "descent with modification."

- Darwin's theory of evolution described evolution itself, natural selection (a mechanism for how evolution occurs), and the time scales over which evolution occurs.

- The evidence for evolution includes a clear history of evolution in the form of fossils in the ordered layers of the Earth's crust; biogeography; obvious similarities among groups of related organisms; and similarities in structure of all parts of organisms from limbs and organs to individual molecules.

- Natural selection is the differential reproductive success of individuals that leads to the accumulation of genetically inherited traits.

No one was again arrested for teaching evolution. But American textbook publishers—ever shy of controversy—nevertheless took the hint, reducing or even eliminating discussions of evolution in most high-school textbooks. By 1942, fewer than half of all high-school science teachers covered evolution at all. Only with the Soviet Union's 1957 launching of the first satellite, Sputnik, did the United States begin emphasizing better science education and textbooks. But the increasing appearance of evolution in textbooks sparked a new reaction among antievolutionists, who realized that their fight against science was far from won.

In 1973, antievolutionists in Arkansas, Tennessee, and Louisiana passed identical bills calling for "equal time" for teaching evolution and **creationism,** the biblical myth that the universe was created by the Judeo-Christian God in 7 days. But a court ruled that the "equal-time" bill was unconstitutional on the grounds that it violated the separation of church and state. (The United States Constitution outlaws teaching religion in public schools because not everyone has the same, or any, religion; and the United States was founded on the idea of religious freedom.)

Creationists argued that the account given in the Bible's book of Genesis was the basis for "creation science." But creation "science" is not science. In 1982, Federal Judge William R. Overton overturned the Arkansas law, writing, " . . . the evidence is overwhelming that both the purpose and effect of Act 590 is the advancement of religion in the public schools." About creation science, he said, "While anybody is free to approach a scientific inquiry in any fashion they choose, they cannot properly describe the method used as scientific if they start with a conclusion and refuse to change it regardless of the evidence . . . "

On the basis of Overton's decision, another judge struck down the Louisiana law as well. And in June 1987, the United States Supreme Court affirmed this decision in a 7 to 2 vote, effectively ending a dispute that had begun 128 years before.

Eager to test the law in a federal court, the ACLU advertised in the Chattanooga News for a teacher willing to be arrested.

Creationists of the 1990s have found another way to fight evolution, however. Rather than try to persuade public schools to teach the biblical version of creation, they now simply ask schools not to teach evolution. But it is unconstitutional to outlaw the teaching of ordinary science. So some school boards simply say students cannot be tested on any origin science, including the origin and age of the universe (about 14 billion years), the origin and age of the Earth (4.6 billion years), and the origin of species of organisms.

In December 1999, for example, the Kansas State Board of Education adopted a curriculum stipulating that Kansas schoolchildren not be tested on evolution in any biology or other science class. Students would be expected to learn about natural selection, Darwin's mechanism for evolution, but not that species can evolve one into another or that all organisms on Earth are related.

American scientists expressed bewilderment. And the presidents of Kansas's six public universities wrote a letter stating that the new standards would "set Kansas back a century" and drive away science teachers. "The argument that teaching evolution will destroy a student's faith in God," the university presidents argued, "is no more true today than it was during the Scopes trial in 1925."

Modern debate about evolution began in 1859, when Charles Darwin published his arguments for the evolution of species in *On the Origin of Species by Means of Natural Selection.* In that book, Darwin rejected the idea that each species had been specially created. The book disturbed a great many people. Three of Darwin's arguments offended most often: (1) that the Earth was very old, (2) that one species could change into another, and (3) that humans and apes were related (Figure 15-1).

In this chapter we will discuss the evidence that has persuaded biologists that organisms are all descended from a common ancestor. At the end of the chapter, we will introduce the mechanism by which most evolution probably occurs—natural selection—which Charles Darwin first proposed more than 150 years ago.

PRIMATE ANCESTOR

Capuchin

NEW WORLD MONKEYS

Baboon

OLD WORLD MONKEYS

Ring-tailed lemur

PROSIMIANS

Gibbon

Orangutan

Gorilla

Chimpanzee

Human

HOMINOIDEA

Figure 15-1   Our family tree. Biologists classify humans as primates not only because we look like other primates, but because we are descended from the same stock as they are. We are most closely related to chimpanzees, gorillas, and orangutans. *(Capuchin, Richard Gibson/Natural Selection; baboon, Stephen G. Maka/DRK Photo; lemur, Kevin Schafer; gibbon, Joe McDonald/Bruce Coleman, Inc.; orangutan, Daniel J. Cox/Natural Selection; gorilla, Mimi Forsyth/Bruce Coleman, Inc.; chimpanzee, Frans Lanting/Minden Pictures; Jane Goodall, Jack Dykinga, Bruce Coleman, Inc.)*

WHAT IS EVOLUTION?

Biological evolution is the process that has taken life on Earth from perhaps a few mere flecks of RNA to the breathtaking variety of more than a million species. What is popularly understood to be Darwin's theory of evolution is really a group of several ideas that can be roughly divided into two parts.

The first part is **evolution** itself, the idea that species change over time. Darwin called evolution **"descent with modification."** A concise definition of evolution says it is heritable change in organisms over time. But the theory of evolution also says that all life on Earth is descended from a few types of cells that came into being billions of years ago. Just

as the members of a large family that includes hundreds of cousins can trace their family history back to a few great, great grandparents, all species on Earth can trace their lineages back to common origins. Dogs and foxes truly are cousins, of a sort, as are humans and orangutans.

The second part of Darwin's theory is (1) his specific mechanism for evolution, which he called natural selection, and (2) a description of the way he thought evolution occurred, gradualism. **Gradualism** is the idea that species evolve gradually and continuously through the steady accumulation of changes, rather than through sudden changes.

But for Darwin, natural selection was the heart of his theory. It was a mechanism that set his theory of evolution apart from similar theories. Briefly, **natural selection** is the differential reproductive success of individuals due to genetically inherited traits.

As we saw in the genetics section of this book, every individual has unique combinations of genetically influenced traits. Some of these combinations are **adaptive,** which means that they give the individual an advantage in the struggle for survival and reproduction and so increase its chances of perpetuating these useful combinations of genes in future generations.

If, for example, the average beak size of a population of finches increases, the finches have evolved. Whether any trait, such as increased beak size, is adaptive depends on the environment of the organism. When many large, hard seeds are present, a large beak is adaptive. When few large, hard seeds are available but plenty of small soft seeds are available, the large beak may not be adaptive. Birds with smaller beaks may be better at eating the smaller seeds. In that case, smaller beak sizes may slowly return.

Darwin compared natural selection to **artificial selection,** the process by which animal breeders and farmers create new lines of dogs, pigeons, corn, and other organisms. In artificial selection, human beings select the most desirable individuals for breeding, generation after generation. In natural selection, the individuals best adapted to their environment survive and reproduce best, leaving more offspring. At the end of this chapter, we will discuss Darwin's theory of natural selection. In later chapters, we will encounter other mechanisms for evolution. Some evolutionary change can occur through random processes. We will also discuss the rate at which evolution occurs (Darwin's gradualism). But in this chapter, we will focus on the evidence for evolution itself.

Darwin's theory of descent with modification through natural selection rested on four ideas. First, he recognized that the world is ever changing and very old (rather than constant and recently created). Second, he recognized that species change. (As we shall see, neither of these ideas was original with Darwin.) Third, Darwin argued that species are composed of populations of individuals, not identical organisms. Every individual is unique. And fourth, he argued that every species and group of species is descended from a common ancestral species. Ultimately, Darwin said, all organisms—including humans—may derive from a single origin of life on Earth.

These ideas conflicted with the biblical account of creation and with a major idea in Western philosophy, called **essentialism.** According to the Greek philosopher Plato (427–347 B.C.), individuals, whether chairs, daisies, or people, are only distorted shadows of an ideal, or essential, form. Only the ideal form was real. The variations that make every daisy unique, said Plato, are distortions and flaws.

Because of essentialism, most scientists in Darwin's time viewed species in terms of their ideal forms. Like so many Lego bricks, every member of a species was supposed to be identical to every other. Species were supposed to be perfect, unchanging, sharply defined "natural kinds." In contrast, Darwin viewed every individual as an individual, not a distortion of some ideal form. As evolutionary biologist Ernst Mayr has written, "It was Darwin's genius to see that this uniqueness of each individual is not limited to the human species but is equally true for every sexually reproducing species of animal and plant."

Darwin asserted (1) that the world is ever changing and very old; (2) that species are made up of individuals; (3) that species change; and (4) that all organisms are related.

Evolution Before Darwin

Although Darwin was the first to persuade the world that plants and animals had evolved, he was hardly the first to hypothesize that organisms evolve. His own grandfather, Erasmus Darwin (1731–1802), argued in 1794 that all species originated from the same primitive "filament." However, Erasmus Darwin propounded many ideas that seemed outlandish in the 18th century, and no scientist took his ideas on evolution seriously. In fact, in the senior Darwin's time, to be accused of "Darwinizing"—engaging in wild speculation—was an acute embarrassment.

How old is the idea of evolution?

Evolution as an idea actually predates even Plato. In the 6th century B.C., 100 years before Plato, the Greek philosopher Anaximander argued that the world was not created abruptly as most ancient religions said. He believed, instead, that animals evolved, and that humans, like all other vertebrate animals, were descended from fishes. Although we now know that Anaximander was mainly correct, it was 2500 years before his ideas found scientific support and general acceptance.

In large part, no one took evolution seriously because Christian beliefs incorporated Plato's essentialism. And essentialism, along with creationism, established a nearly unbreachable intellectual barrier to the idea of the evolution of species. Another serious barrier to the idea of evolution was the notion—accepted by nearly everyone in 17th- and 18th-century Europe—that the Earth had been created in 4004 B.C. This estimate had been worked out in 1650 by Archbishop James Ussher, who added together the life spans of all the patriarchs named in the Bible, beginning with Adam. Most Christians believed that the Earth was less than 6000 years old.

By the end of the 18th century, however, the biblical account of creation and the notion that species do not change were undermined by three separate lines of evidence. First was the discovery of hundreds of thousands of species not mentioned in the biblical story of creation. Explorers returning to Europe from Africa, the Americas, and the Pacific brought back plants and animals that no European had ever seen or heard of before (Figure 15-2A). Equally disturbing was the great number of fossil plants and animals that were turning up in Europe itself. These fossils were stone remnants of species as alien as any from the far side of the world (Figure 15-2B). Yet, they appeared to have lived in Europe in some previous era. But when? Finally, by the middle of the 19th century, geologists found compelling evidence that the Earth was far older than Ussher's estimated 6000 years.

The enormous difficulty of reconciling these new facts with the story of creation in the Judeo-Christian Bible is apparent in the thinking of five of the greatest scientists of the 18th and early 19th centuries. All of them discovered evidence for evolution, yet because of creationism and essentialism, all of them interpreted the evidence in nonevolutionary terms.

How did Linnaeus's ideas lay the groundwork for evolution?

The first of the five scientists is Carolus Linnaeus (1707–1778), a Swedish physician whose most familiar contribution to biology was the two-part Latin names—genus and species—that biologists assign to all species. The scientific name for a human being, for example, is *Homo sapiens*. *Homo* refers to the **genus** to which we belong, and *sapiens* ("wise") refers to our **species.** Linnaeus named more than 4000 plants and animals, and biologists since Linnaeus have used his system to name some 1.4 million more species. Biologists estimate that 10 to 100 million more remain to be discovered and named. We discuss Linnaeus in more detail at the beginning of Chapter 22.

Even more important than Linnaeus's system of naming organisms was his system of grouping, or classifying, them. For example, the genus *Homo* belongs to the **family** Hominidae. And the family Hominidae belongs to the **order** Primates, one of 16 orders in the **class** Mammalia. Thus, we group species into genera, genera into families, families into orders, and orders into classes (Chapter 19). Each of these groupings is a **taxon** [Greek, *taxis* = arrangement; plural, **taxa**]. The science of classifying organisms is called **taxonomy.**

Linnaeus firmly believed that each species was individually created by God, yet he arranged species into groups that reflected their similarities to one another. He could see, for example, that all the species of mice were more closely related to rats than to horses, and his formal system for recognizing such natural relations was fundamental to Darwin's arguments 100 years later.

Linnaeus avoided asking *why* organisms fall so naturally into families, piously writing, *"Deus creavit, Linnaeus disposuit"*—God creates, Linnaeus arranges. But one of his contemporaries, the great French biologist George-Louis Leclerc, or Count Buffon (1707–1788), couldn't leave the question alone.

Linnaeus introduced a modern system of classifying organisms.

Why was Count Buffon afraid to discuss evolution?

Buffon was an accomplished mathematician as well as biologist. He was also a science popularizer, loved by the public but resented by fellow scientists. In his early manhood, Buffon boldly stated that the Earth must be far older than 6000 years. But when Church elders threatened him with excommunication for this heresy, he quickly took back his arguments.

A.

B.

Figure 15-2   Another world. A. Western grey kangaroo and joey. European explorers and naturalists brought back strange animals and plants no one had ever heard of, let alone seen. These organisms were not mentioned in the Bible. B. Even at home in Europe, paleontologists dug up fossil animals as strange as this fossil pterosaur, a flying reptile of the Jurassic Period discovered in Bavaria. None of these had been named in the Bible nor classified by European scientists. (*A, Tim Flach/Stone; B, Kevin Schafer/Peter Arnold, Inc.*)

In later years, he carefully recanted each heresy as he committed it. In a 36-volume natural history, in which he interspersed descriptions of plants and animals with discussion of natural philosophy, Buffon noted the remarkable similarity of the horse and the ass, or donkey. He wrote, in 1753, "But if we once admit that there are families of plants and animals, so that the ass may be of the family of the horse, and that the one may only differ from the other through degeneration from a common ancestor, we might be driven to admit that the ape is of the family of man." And, as if that were not bad enough, he added, " . . . then there is no further limit to be set to the power of nature, and we should not be wrong in supposing that with sufficient time she could have evolved all other organized forms from one primordial type."

To placate the church, however, he quickly contradicted this insight: "It is certain through Revelation that all animals have shared equally in the grace of creation and that each emerged from the hands of the creator as it appears today."

Buffon's theory of evolution resembled Darwin's, but without natural selection. Fearing excommunication, however, he denied that he believed his own arguments.

What made Hutton think that the Earth was older than 6000 years?

Beginning in the mid-18th century, the Industrial Revolution gave rise to an insatiable demand for coal and metals that turned geology into a practical and hugely popular science. For the first time, scientists began to seriously explore the surface of the Earth. They learned that soil, rock, and mountains were eroded by wind, frost, and running water. In some places, whole mountain ranges seemed to have eroded completely away. But could all this have happened, geologists asked, in just 6000 years?

In 1788, the Scottish geologist James Hutton (1726–1797) suggested that it could not. The Earth, he said, was eternal. The perpetual erosion of mountains was repaired by the uplift of sediments and the expulsion of magma from the Earth's depths. At the time, most geologists doubted continents and mountains could rise from the Earth's surface. But Hutton insisted that the Earth was not a static, passive body but a "beautiful machine," powered by the heat of its interior.

He proposed that endless cycles of erosion, sedimentation, and uplift molded the Earth's surface. The Earth, he said, might be immeasurably old. In his now famous conclusion, he wrote, "The result, therefore, of our present inquiry is, that we find no vestige of a beginning—no prospect of an end."

It was heresy. But if Hutton was a thorn in the side of theologians, Baron Georges Cuvier (1769–1832) was their savior. In 1812, the great French anatomist and paleontologist found an ingenious way to reconcile geology and Genesis—a way to reconcile the great age of the Earth with theological doctrine.

Catastrophism: How did Cuvier try to reconcile Genesis and geology?

Cuvier was a gifted anatomist and the first to study fossils systematically. He invented both **comparative anatomy,** the detailed comparison of the anatomies of different species, and **paleontology** [Greek, *palaios* = ancient + *on* = being + *logos* = discourse], the study of fossils. **Fossils** [Latin, *fossilis* = dug up] are the remains or imprints of past life. The most impressive fossils are the intact skeletons of whole animals, such as 70-million-year-old dinosaurs. Such spectacular finds are rare, however, compared to the hundreds of thousands of small fossilized shells and tiny bone fragments paleontologists have unearthed. Fossils include preserved bones and teeth, petrified trees, bits of fossilized dung, footprints, impressions of the surfaces of ferns, as well as ants and other insects preserved in hardened tree resin (amber).

Fossils had puzzled people for thousands of years. Fossils look like plants and animals or parts of plants and animals. But they are made of stone, and they rarely look like *familiar* plants and animals. Some people thought fossils were God's first experiments with creation or maybe the remains of plants and animals that had literally missed the boat. Failing to board Noah's ark, they had perished in the great flood described in the Bible (Figure 15-3A).

But Cuvier knew that all the different organisms represented by fossils could not have died in the same flood. The different kinds of fossil organisms had lived and died out at different times in the Earth's history. By 1760, geologists had already shown that rocks could be arranged in a series of layers, or strata, according to their age. The oldest layers were usually deepest (Figure 15-3B). The newest layers were on top.

William "Strata" Smith, a geologist and canal engineer, compared these layers to "slices of bread and butter on a breakfast plate." Smith showed that different fossils appeared and disappeared from layer to layer. Older fossils appeared in the deepest layers, then disappeared and were replaced by new fossils—on up to the topmost layers.

Starting with the assumption that species could not evolve, Cuvier concluded that the fossils in the oldest layers had nothing to do with modern life. To account for the different kinds of life that appear in the fossil layers, Cuvier hypothesized that the organisms from the different layers had perished in a series of catastrophes. Noah's flood was merely the most recent and dramatic catastrophe. (The previous catastrophes were not mentioned in Genesis, Cuvier argued, because God didn't think Moses needed to know about them.)

After each catastrophe, argued Cuvier, new species appeared in another creation. **Catastrophism,** as his theory came to be called, allowed for the great age of the Earth and the fossil record without implying that Genesis was wrong. This idea was appealing to a great many pious Christians, and catastrophism remained popular for decades. In fact, the 1981 Arkansas equal-time bill called for a revival of this very theory.

But, although paleontologists have identified several mass extinctions, the evidence does not support catastrophism. First,

A.

B.

Figure 15-3 Catastrophism and uniformitarianism. A. Noah's flood. The deluge as portrayed in the first edition of the Luther Bible, published in 1534. B. The Grand Canyon. James Hutton wrote, "Thus . . . from the top of the mountain to the shore of the sea . . . everything is in a state of change; the rock and solid strata slowly dissolving, breaking and decomposing, for the purpose of becoming soil; . . . without those operations which wear and waste the solid land, the surface of the earth would become sterile." *(B, Jack Dermid/Bruce Coleman, Inc.)*

although Cuvier had originally proposed only four or five catastrophes before Noah's flood, fossil hunters continued to find so many distinct layers of fossils that more and more catastrophes had to be invented to account for them all. The Swiss-American biologist Louis Agassiz (1807–1873) maintained, for example, that the fossil record revealed 50 to 80 separate catastrophes, each one requiring a separate creation.

And while some organisms appeared once in the fossil record and then disappeared forever, others kept reappearing virtually unchanged in "creation" after "creation" (Figure 15-4). Lastly, archaeologists found human bones and other artifacts that suggested that humans predated not only the biblical flood, but the hypothetical creation of the world in 4004 B.C. Catastrophism didn't seem to fit the facts.

Uniformitarianism: Lyell rejects catastrophism

Catastrophism received its most damaging blow in 1830, when the Scottish geologist Charles Lyell (1797–1875) published the first volume of his three-volume *Principles of Geology.* Lyell rejected catastrophism. Instead, expanding on Hutton's ideas, Lyell argued that the slow processes that now mold the Earth's surface—erosion, sedimentation, and upheaval—are the same ones that have always molded it. Geologic change, he said, is slow, gradual, and steady—not catastrophic. Given enough time, mountains could erode down to mere hillocks, sea beds rise to great heights, and rivers cut deep canyons (Figure 15-3B). Gradual processes alone could create the world we know.

Lyell, who had been trained as a lawyer, provided huge amounts of evidence in his book to back up this hypothesis, later named **uniformitarianism.** Darwin was so influenced by Lyell that he later wrote, "I have always thought that the great merit of the *Principles* was that it altered the whole tone of one's mind, and therefore that, when seeing a thing never seen by Lyell one yet saw it partially through his eyes." But even though Lyell's *Principles* provided fertile soil for Darwin's ideas, Lyell himself believed that species could not evolve. It was Darwin's own book *The Origin of Species,* published almost 30 years after the *Principles,* that changed Lyell's mind.

The Industrial Revolution's demand for fossil fuels created the science of geology. The fossil record discovered by geologists showed the Earth was extremely old.

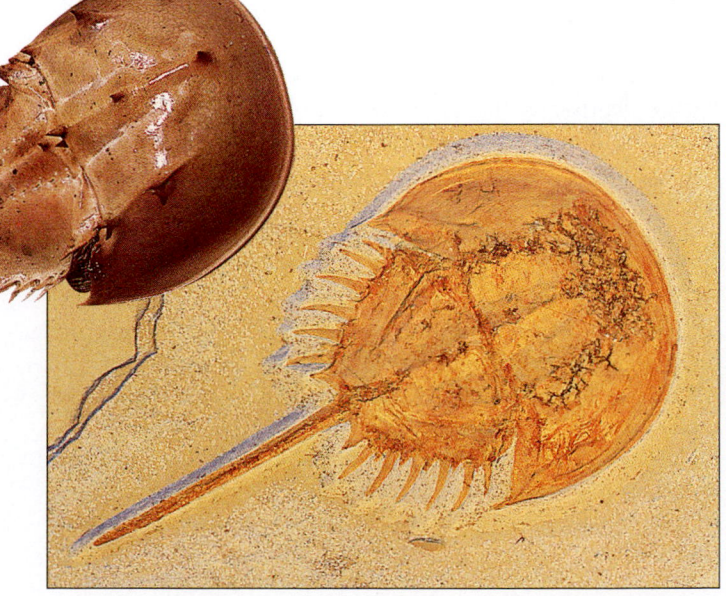

Figure 15-4 Horseshoe crab. Many groups of organisms have changed hardly at all over millions of years. The modern horseshoe crab remarkably resembles its Jurassic ancestor, which lived nearly 200 million years ago. *(Left, Alan Desbonnet/Visuals Unlimited; right, John Cancalosi/Peter Arnold, Inc.)*

Why was Lamarck ignored?

Even together, the fossil record and the great age of the Earth were not enough to convince most scientists that species evolve. Among reputable scientists, evolution was at best a dubious and controversial hypothesis. Count Buffon and Erasmus Darwin had proposed that species evolve, but neither man was taken seriously by fellow scientists. In 1809, the great French biologist Jean Baptiste Lamarck (1744–1829) again proposed the idea that species evolve over time.

Lamarck came from an impoverished noble family and served as a tutor to Count Buffon's son. He studied medicine and botany in Paris and published two works on botany before being appointed professor of zoology at the Jardin des Plantes in 1793. A renowned taxonomist, Lamarck named hundreds of species, wrote more than a dozen volumes on fossil "invertebrates" (a word he invented), and argued that plants and animals are all related. To express this union, he invented the word "biology."

In his studies of fossils, Lamarck saw a clear line of descent—from older fossils to more recent fossils to modern species. *The oldest fossils looked the least like modern organisms. Later groups of fossils included more and more familiar organisms.*

Lamarck hypothesized that the accumulated changes were adaptive—that is, the changes made organisms better able to survive under new conditions. However, Lamarck argued that the changes resulted from a drive toward perfection and complexity.

A giraffe, for example, reaching ever higher in the trees for leaves to eat, would constantly stretch its neck and thus acquire a permanently longer neck. The giraffe's offspring would then inherit longer necks. Lamarck hypothesized that adaptive characteristics acquired by the parent could be inherited by the offspring.

In Lamarck's day, most scientists accepted his mechanism for evolution, the theory of the **inheritance of acquired characteristics.** But they vehemently rejected evolution itself, and he was vilified by the entire scientific community. The fiercest opposition came from Cuvier. As both an eminent scientist and a French cabinet minister, Cuvier's opinion carried enormous weight, and Cuvier criticized Lamarck relentlessly. In 1832, 3 years after Lamarck had died—poor and almost forgotten—Cuvier delivered an "elegy" to Lamarck at the prestigious French Academy, in which he once more dismissed not only Lamarck's theory of evolution but also his reputation as a biologist.

So it was not surprising that when Darwin began writing his theory of evolution, he did everything he could to distance himself from Lamarck (and his own grandfather Erasmus) and to anticipate every possible objection to his own ideas.

How Did Charles Darwin Convince Others That Species Evolve?

By the middle of the 19th century, natural science had become enormously popular with the lay public. Every sector of English society was obsessed with the exotic and the unusual in nature. Even as explorers brought back exotic animals and plants from Africa, Australia, and Asia, every middle-class English family acquired an aquarium, a fern case, a butterfly cabinet, or a shell collection. Mothers routinely taught their children the names of ferns and fungi. Aristocrats turned their estates into parks for exotic animals such as elands, beavers, and kangaroos. Books on natural history were so popular that one—*Common Objects of the Country*—sold 100,000 copies in a week, a total that even today would earn the book a place high on *The New York Times* best-sellers list.

Darwin's education

Charles Robert Darwin (1809–1882) came of age in the middle of this national obsession and was himself entirely caught up by it. An upper-class English boy whose father was a doctor, Charles spent his boyhood tramping through the woods and fields of Shrewsbury, hunting, fishing, and collecting insects. As a young man, his interests narrowed to hunting, and his father despaired of Charles ever making anything of himself. Darwin senior once wrote to his son, "You care for nothing but shooting, dogs and rat-catching, and you will be a disgrace to yourself and all your family."

Darwin's utter lack of ambition forced his father to pick a profession for him, and in 1825 Robert Darwin sent young Charles to Edinburgh University to study medicine. In the course of his studies, Darwin attended lectures on geology and zoology and thought they were the most boring thing he'd ever had to endure in his life. His only interest, outside of shooting birds, was the meetings of a natural history club. As for medicine, if he had ever considered becoming a doctor, he changed his mind after watching two horrifying surgical operations done without anesthesia, one on a child. His resistance to studying medicine, or anything else, only hardened when he realized that his wealthy father would probably support him for the rest of his life, whatever he chose to do (or not to do).

Dr. Darwin then decided that Charles should study for the clergy and in 1828 sent him to Cambridge University. Fortunately for modern biology, the clergy was then a haven for naturalists, and 19-year-old Charles Darwin soon shook off his perpetual boredom and rediscovered his boyhood passion for nature.

In 3 years, he managed to complete his B.A., and when he graduated from Cambridge in 1831, one of his professors recommended him for a position as ship naturalist on board the H.M.S. *Beagle*. The *Beagle* was a survey ship about to embark on a trip around the world, its 5-year mission to map the coast of South America. Darwin's duties were to collect specimens of the plants, animals, and rocks from this unexplored new world. In December 1831, just before the *Beagle* sailed, a professor friend sent Darwin the first volume of Lyell's *Principles of Geology*. His course was set.

During the long trip across the Atlantic, Darwin immersed himself in Lyell's arguments about the slow processes that formed the Earth. He thought of nothing but geology for

A.

B.

Figure 15-5 **Darwin and Wallace.** A. Charles Darwin and Emma Wedgwood Darwin. Even in his most productive years, Charles Darwin was sickly and worked only for short periods. B. Alfred Russel Wallace's ideas about evolution generally matched Darwin's. However, although Wallace believed that humans had evolved from ancestral primates, he did not believe that our capacity for thought and spiritual passion could have evolved in the same way. *(A, Bridgeman Art Library; B, The Granger Collection, New York)*

months. Lyell formalized two important ideas: first, that the Earth is very old, and second, that geologic processes occur very slowly and gradually. Darwin's theory of evolution, as he would later conceive it, likewise stated that life is very old (millions of years) and change (evolutionary, in this case) has been slow and gradual.

As Darwin explored South America over the next 5 years, he began to suspect that all animals and plants were descended from a common ancestor. He wondered if Count Buffon, Lamarck, and even his own grandfather had been right about evolution. He took notes; he collected specimens. He noticed what animals ate, where they lived, what kind of soil this plant or that seemed to prefer.

Almost as soon as he returned to England in 1836, Darwin began to classify his specimens and to organize his notes. But it was 6 more years before he "allowed himself the satisfaction" of even outlining his theory. In the meantime he read the work of the second great intellectual influence in his life after Lyell—Thomas Malthus.

Malthus: Too many children

In 1798, the English economist and clergyman Thomas Robert Malthus (1766–1834) wrote a highly controversial paper, *An Essay on the Principle of Population.* Malthus argued that human populations tend to increase geometrically, while food supplies and other resources increase only arithmetically, if at all. For example, if a couple had three children, and each of their children had three children, and the nine grandchildren each had three children, the family farm could not possibly support the 27 great-grandchildren. The family may be able to increase the amount of food they can produce on their farm, but they cannot possibly triple production every generation. Poverty, famine, overcrowding, disease, and war are all the inevitable consequences of this excessive population growth, argued Malthus. He predicted that without some check on population

growth, human populations would face a continuing, ferocious struggle for existence in the face of limited resources.

When Darwin read Malthus in 1838, he saw for the first time how evolution could work. Malthus's arguments applied to all species, reasoned Darwin, not just humans. Many more individuals of each species are born than can possibly survive, and, consequently, there is a constant struggle for existence.

"It follows," Darwin later wrote, "that any being, if it vary however slightly in any manner profitable to itself, under the complex and sometimes varying conditions of life, will have a better chance of surviving, and thus be *naturally selected.*" His theory of evolution through natural selection was beginning to take shape.

Wallace scoops Darwin

But Darwin delayed publishing. Month after month, year after year, he added new facts to bolster his theory. Lyell, who was by then a friend, still didn't accept evolution, but he urged Darwin to publish his idea before someone else thought of the idea and published first. Then, in 1858, Alfred Russel Wallace (1823–1913), a land surveyor and naturalist working in the Malay Archipelago (in what is now Malaysia), independently hit upon natural selection as a mechanism for evolution.

Wallace had written to Darwin the year before asking for information about selective breeding. Darwin wrote back with the information and added tactfully that he had been working on similar ideas for some 20 years (Figure 15-5). Again Wallace wrote, this time outlining his theory of evolution— but without natural selection. Darwin responded politely, but hinted that he knew more. Wallace was obviously hot on his heels, yet Darwin did nothing.

Then, a few months later, in 1858, Wallace wrote once more. In the interim, he had read Malthus and experienced the same flash of insight Darwin had. Eagerly, Wallace wrote to Darwin, outlining his theory of evolution by natural selection.

Darwin, utterly crushed, was ready to give all of the credit to Wallace. In a letter to Lyell, he wrote:

> "I never saw a more striking coincidence; if Wallace had my [manuscript] sketch written out in 1842, he could not have made a better abstract! Even his terms now stand as heads of my chapters. Please return me the [manuscript], which he does not say he wishes me to publish, but I shall, of course, at once write and offer to send it to any journal. So all my originality, whatever it may amount to, will be smashed."

Then worse misfortune struck. Scarlet fever swept Darwin's family, killing his 19-month-old son Charles in a matter of days. Grief stricken, Darwin threw up his hands and turned everything over to Lyell, who gladly rescued Darwin. Lyell presented both Wallace's paper and an extract from Darwin's unpublished outline to the Linnaean Society, giving credit to both men. The following year, 1859, Darwin finally published his theory in a book, *On the Origin of Species by Means of Natural Selection.*

The Origin, as biologists call it affectionately, is the single most influential scientific book ever written. When it appeared in 1859, all 1250 copies of the first printing sold out in a single day. Its arguments persuaded even those, such as Lyell, who had previously rejected the idea that species evolve. What was so convincing?

WHAT IS THE EVIDENCE FOR EVOLUTION?

Evolution is heritable change in populations of organisms, which Darwin called "descent with modification." Evolution is also "speciation," the division of one species into two or more. Such "diversification" is the means by which a few kinds of simple cells living 3.8 billion years ago gave rise to the millions of complex organisms that live on Earth today.

The Origin's great strength was Darwin's gluttony for evidence. Darwin provided evidence for evolution from the fossil record, from the distribution of plants and animals (biogeography), from taxonomy, from comparative anatomy, from comparative embryology, and from domestic breeding. Since Darwin's time, other biologists have further buttressed his arguments with evidence from fields as diverse as comparative molecular biology, classical genetics, population ecology, developmental biology, and animal behavior. In the rest of this chapter we will briefly examine some of this evidence.

The Fossil Record Tells a Story of Evolution

The sequence of organisms in the fossil record tells a story with an obvious meaning. At the bottom, where sediments are oldest, lie the primitive prokaryotes and eukaryotes (Figures 15-3 and 15-6). Above them come simple multicellular organisms, then trilobites and other invertebrates, and then the first vertebrates, the fishes. Many more fish follow. Then, in rapid succession, amphibians and reptiles arrive. Newer, and higher in the sedimentary strata, are the dinosaurs, crocodiles, birds, and mammals.

Three facts stand out. First, fossils are distributed consistently. Rocks of the same age contain approximately the same groups of organisms. Cambrian rocks, which generally contain trilobites, never contain dinosaurs or horses, only trilobites. Recent Pleistocene rocks contain no trilobites. Such a

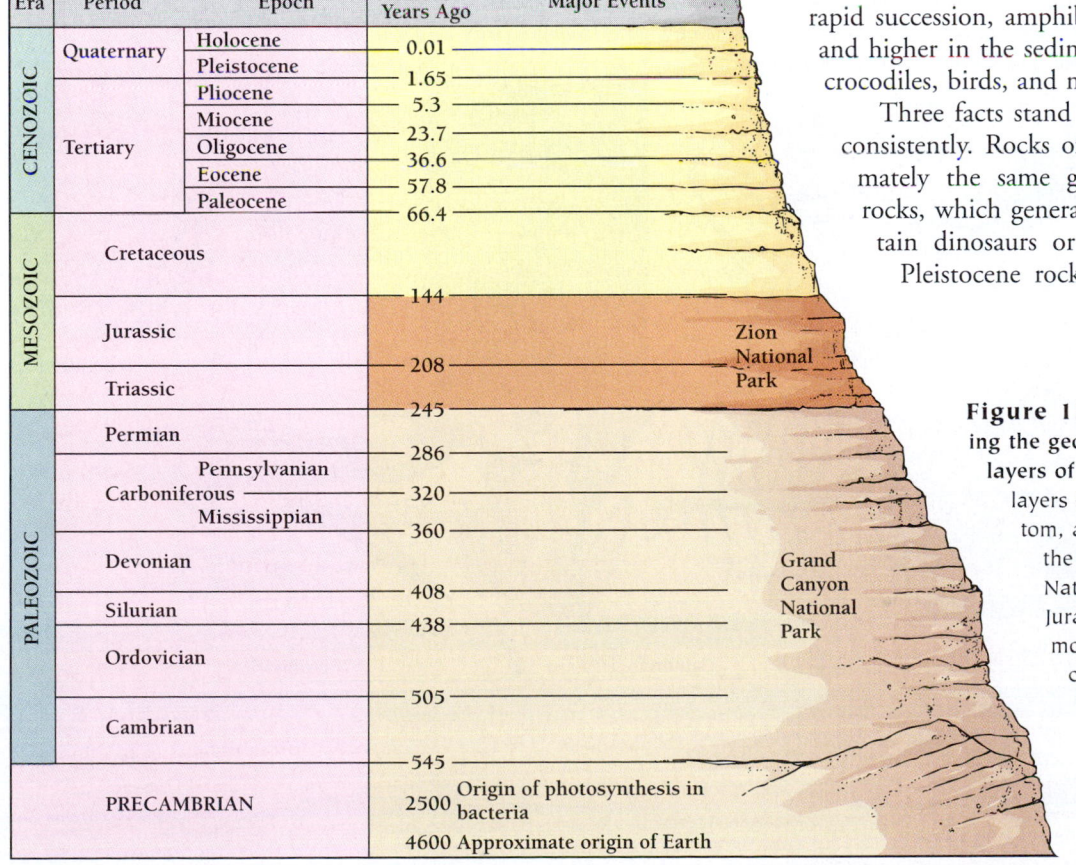

Era	Period	Epoch	Millions of Years Ago	Major Events
CENOZOIC	Quaternary	Holocene	0.01	
		Pleistocene	1.65	
	Tertiary	Pliocene	5.3	
		Miocene	23.7	
		Oligocene	36.6	
		Eocene	57.8	
		Paleocene	66.4	
MESOZOIC	Cretaceous		144	
	Jurassic		208	
	Triassic		245	
PALEOZOIC	Permian		286	
	Carboniferous	Pennsylvanian	320	
		Mississippian	360	
	Devonian		408	
	Silurian		438	
	Ordovician		505	
	Cambrian		545	
	PRECAMBRIAN		2500	Origin of photosynthesis in bacteria
			4600	Approximate origin of Earth

Zion National Park

Grand Canyon National Park

Figure 15-6 A composite canyon showing the geologic eras and periods as discrete layers of rock. The Grand Canyon includes layers from the Precambrian, at the bottom, all the way up to the Permian, near the rim of the canyon. Nearby Zion National Park includes Triassic and Jurassic layers. At both canyons, the most recent layers, from the Cretaceous onward, have all been washed away. But these layers can be found intact in other places.

If you wanted to look for the bones of *Apatosaurus,* would you go to the Grand Canyon or to Zion?

pattern is consistent with the idea that once organisms go extinct, they stay extinct. Extinct species do not suddenly reappear millions of years later as a new creation.

Second, the order in which organisms are laid down in the fossil record suggests a sequence of evolution that is, as we shall see, independently confirmed by other fields of biology. For example, the first true mammals do not appear in the fossil record until *after* the appearance of the mammal-like reptiles, whose anatomy is intermediate between early reptiles and mammals. This pattern is consistent with the idea that mammals evolved from these mammal-like reptiles. Similar patterns appear over and over again in the fossil record.

Finally, recent fossils look most like modern organisms, while the oldest fossils generally look least like modern organisms, a pattern that suggests the accumulation of change—or evolution (Figure 15-7).

The order in which organisms appear in the fossil record is consistent with the theory of evolution.

How do fossils form?

An animal or plant becomes a fossil only if it dies under the right conditions. The vast majority of dead plants and animals are consumed by other organisms and leave no trace of their existence. Dead things leave a trace only if their bodies or imprints are rapidly buried in a bog or at the bottom of a sea or lake. There, protected from scavengers, erosion, and decay, organisms may in time become more deeply buried as successive layers of mud and sand settle over them.

Over time, pressure from all these layers of sediment turns the deepest layers of mud and sand into "sedimentary rock." After millions of years, geologic forces raise the rock up into new mountains. Rivers flow down the sides of the new mountains, eroding canyons through the layers of old sediment, revealing the ancient—and to us, amazing—fossils.

Because organisms fossilize most often when buried in sediment at the bottom of a sea or lake, most fossils are marine or freshwater organisms. Terrestrial plants and animals are usually fossilized only if their bodies happen to fall into a lake or sea. Furthermore, only the hard parts of animals—shells, teeth, and bones—and the woody parts of plants are likely to survive long enough to form fossils. The probability is remote that a soft, terrestrial organism such as a slug will be fossilized.

The chanciness of fossilization means that the fossil record is incomplete. For example, most insects are too small and delicate to be preserved, and so the fossil record for insects is poor. However, where insects are preserved, they often appear in great numbers. Large numbers of insects may appear in one layer, disappear—leaving a gap—then reappear millions of years later. Even vertebrates, some of the best preserved of all

Figure 15-7 (part 1) Distinctive fossils from the geologic past. (*Photos from left to right: Dickinsonia, Biological Photo Service; fossil jellyfish, William E. Ferguson; trilobites, James L. Amos/Photo Researchers, Inc.; Devonian fish Dinichthys terrelli, James L. Amos/Photo Researchers, Inc.; calamites, Ted Clutter/Photo Researchers, Inc.; fern fossil, J.C. Carton/Bruce Coleman, Inc.*)

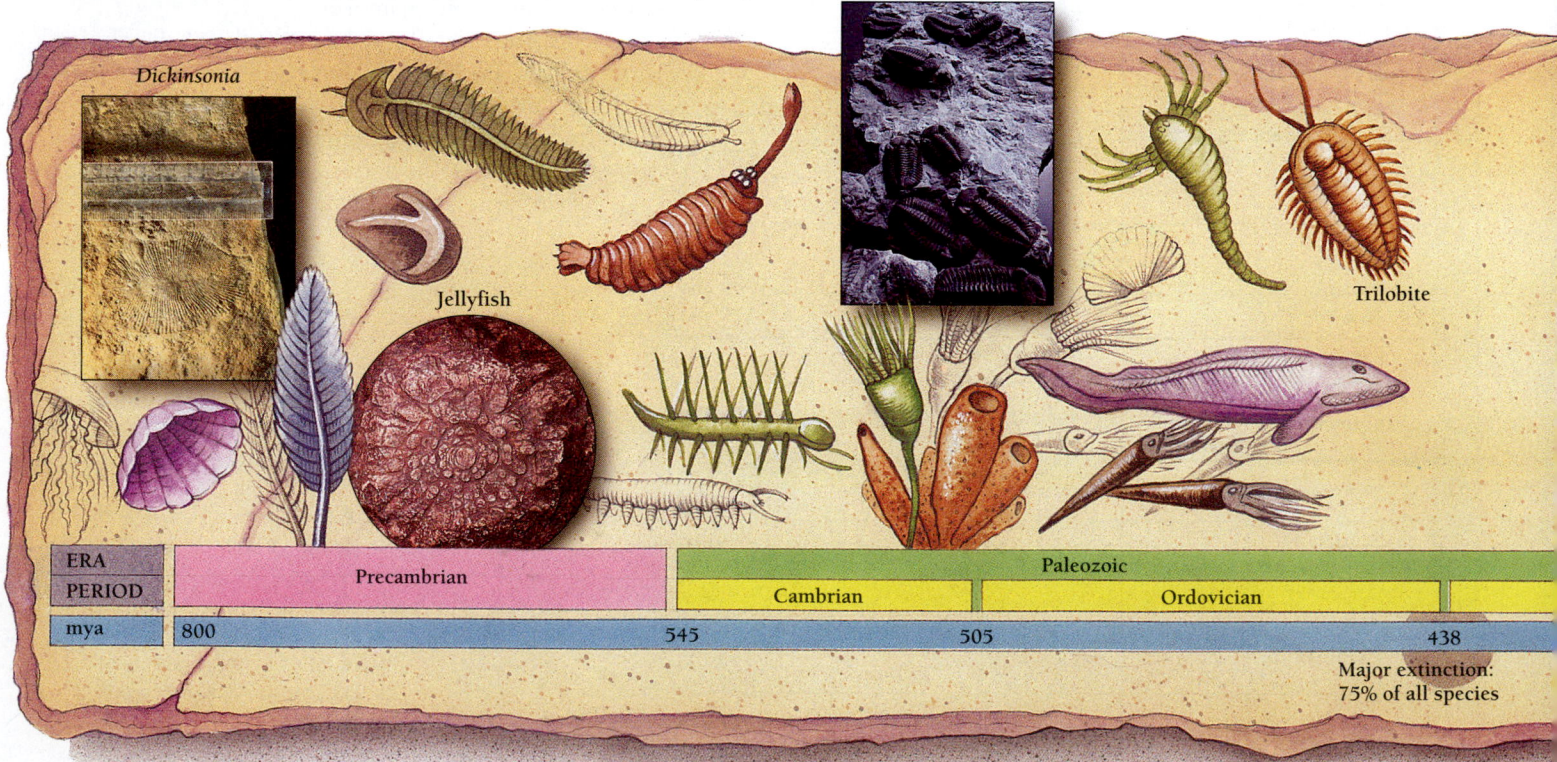

Radioactive Dating

The fossil record provides a useful chronicle of the sequence of organisms that have lived on Earth. Fossils also provide an accurate way to determine the relative ages of different layers. Still, even if scientists know the relative ages of every single layer, they still want to know the **absolute age**—the age in years—of each of the layers. Scientists had no way of estimating the absolute ages of rocks until the discovery of radioactivity by Marie Sklodowska Curie and Pierre Curie in 1898.

As we discussed in Chapter 2, radioactive atoms, such as uranium, break down into other materials by giving off alpha, beta, and gamma rays. The breakdown, or **decay,** occurs at a steady rate that is unaffected by temperature, pressure, or other environmental variables. Every form of an atom, called an **isotope,** breaks down into a series of other isotopes. Half of the isotope carbon-14, for example, decays to nitrogen-14 every 5600 years. We say that the **half-life** of carbon-14 is 5600 years.

By comparing the relative proportions of different isotopes in a rock sample, scientists can tell the age of the rock. Different radioactive isotopes have different half-lives. Carbon-14, denoted as ^{14}C, is useful for dating objects that are only a few thousand years old. In contrast, rubidium-87 has a half-life of about 5×10^{10} years (50 billion years), so it is useful for dating rocks that originated at the same time as the solar system (about 5 billion years ago) or even earlier.

organisms (because of their hard bones), show a discontinuous record. A mammal may disappear from the record, then reappear, slightly altered, ten thousand years later. Despite its limitations, however, the fossil record clearly shows that life has evolved according to the sequence summarized in the next section.

What Does the Fossil Record Show?

In the 18th and 19th centuries, scientists recognized that the fossils in the top layers of sediment were the most recent, while the fossils in the deepest layers were the oldest. For the first time, scientists were able to read the layers of rock as if they

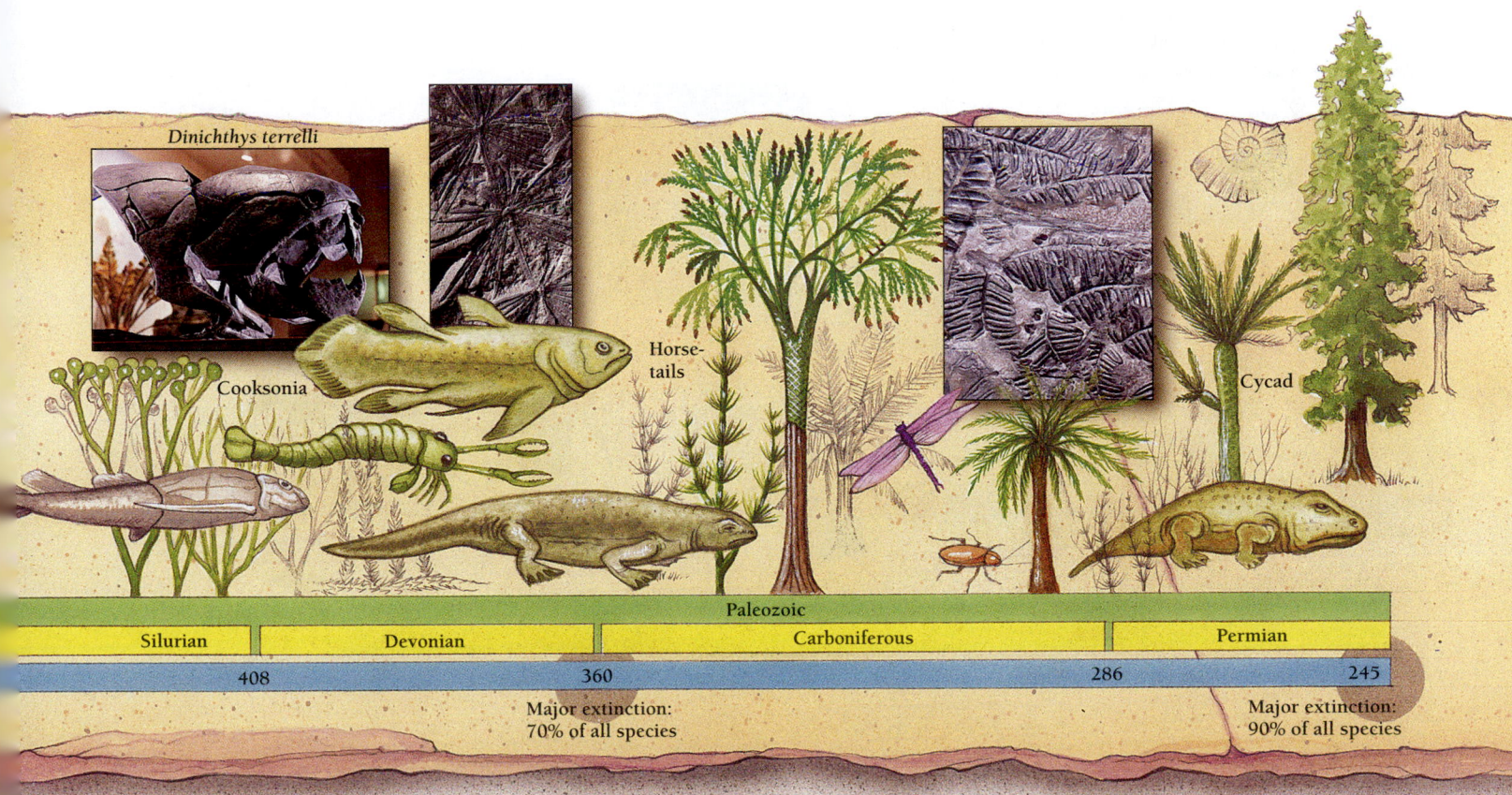

Dinichthys terrelli

Cooksonia

Horse-
tails

Cycad

Paleozoic			
Silurian	Devonian	Carboniferous	Permian

408   360   286   245

Major extinction:
70% of all species

Major extinction:
90% of all species

were the pages of a history book. They discovered a startling history of the last 570 million years. Twentieth-century paleontologists have now extended this history back to the appearance of the first cells on Earth nearly 4 billion years ago.

Geologists divided this history into chapters. Each chapter, or **period,** is 30 to 75 million years long and contains distinctive forms of life in its fossil record (Figure 15-6). Periods are grouped into four long **eras,** of from 65 million to several billion years each. The most recent chapters are subdivided into relatively short **epochs.** The same distinctive layers appear in rock formations around the world.

Nineteenth-century geologists could not give exact ages to each layer. However, they could tell the order in which the layers had been deposited. And they estimated the ages of the various periods and eras recorded by these fossil layers remarkably well. The modern development of radioactive dating techniques, in the 1940s and 1950s, allowed paleontologists to assign actual dates to evolutionary events and confirm many of the old estimates.

The **Precambrian Era** comprises most of the history of the Earth, beginning with the planet's formation 4.6 billion years ago. Before the discovery of the first Precambrian fossils in the 1950s, paleontologists viewed the Precambrian as one long lifeless era. With the discovery of ancient bacteria and algae, paleontologists now subdivide the Precambrian into six

eras. The first fossils, some as old as 3.8 billion years, resemble modern prokaryotes, or bacteria. These ancient fossils vividly illustrate the breathtaking antiquity of life on Earth. Fossils of the first eukaryotes (cells with organelles) appear in rocks about 2 billion years old. And new evidence—in the form of decayed cholesterol molecules in ancient rocks in northwestern Australia—suggests that eukaryotes may have appeared as early as 2.7 billion years ago. In Chapter 18, we discuss the formation of the Earth and how life might have originated.

The Cambrian explosion

The first multicelled organisms appear in the fossil record near the end of the Precambrian, about 800 million years ago. And the first multicelled animals—which resemble jellyfish, worms, and tiny shelled creatures—do not appear until about 640 million years ago. But it is in the Cambrian Period [Latin, *Cambria* = Wales], at the beginning of the **Paleozoic Era** [Greek, *palaios* = ancient + *zoos* = life], that multicelled life first appeared in all its staggering diversity. By 543 million years ago, at the time of the "Cambrian explosion," the Earth was home to trilobites, jellyfishes, corals, segmented worms, fungi, algae, and organisms unlike anything alive today (Figure 15-7). Indeed, by the end of the Cambrian, 11 of the 35 phyla that live today had already evolved, including the first chordates, an-

Figure 15-7 (part 2)   Distinctive fossils from the geologic past. *(Photos from left to right: Archaeopteryx fossil, James L. Amos/Photo Researchers, Inc.; fossil dinosaur eggs (Protoceratops andrewsi), Ken Lucas/Visuals Unlimited; ichthyosaur fossil, E.R. Degginger/Bruce Coleman, Inc.; fossil flower, William E. Ferguson; plesiosaur fossil, E.R. Degginger/Earth Scenes; orangutan and baby, Tim Davis/Photo Researchers, Inc.)*

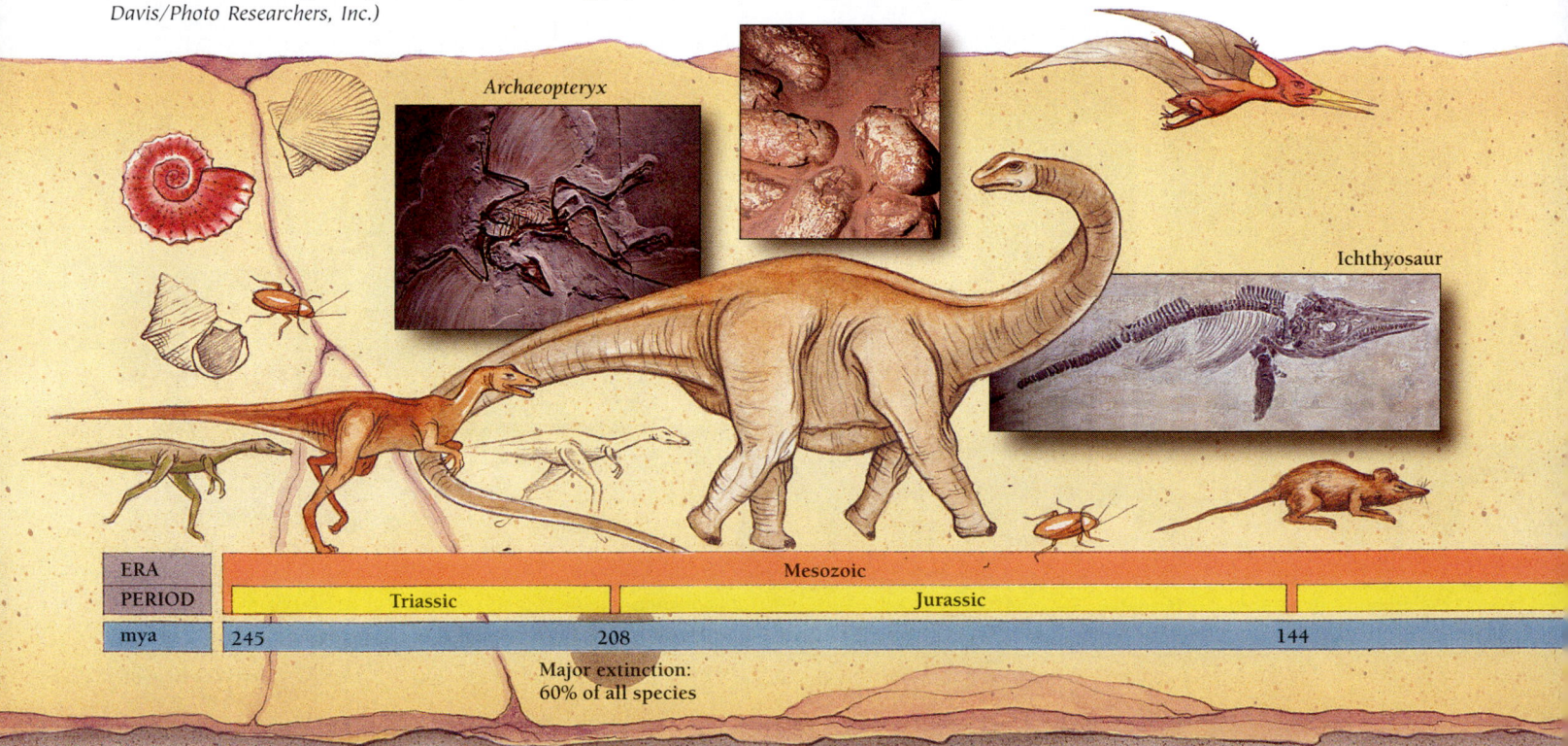

cestors of all vertebrates—animals such as fishes and mammals that have backbones (Chapter 23). Geologic studies suggest this burst of evolution occurred within about 30 million years and, possibly, in as little as 5 million years. By the end of the Cambrian, some 500 million years ago, a mass extinction wiped out whole classes and families of organisms. But the first vertebrates—the armor-plated fishes—survived.

Besides the Cambrian, the Paleozoic Era includes five other periods. During the next two periods of the Paleozoic Era, the Ordovician and the Silurian, the first fossil plants, great numbers of starfishes, nautiluslike animals, and fishes with bony skeletons appeared.

During the Devonian Period, the oceans receded, leaving vast new tracts of dry land. On land, fungi, mosses, and other bryophytes evolved, as well as the first "vascular" plants, those with a system of tubes for transporting water and nutrients. Arthropods such as scorpions, centipedes, millipedes, and springtails appeared as well. In the seas and lakes, so many kinds of fishes evolved that the Devonian is called the Age of Fishes.

During the Carboniferous, the seas rose and fell repeatedly, and the Amphibia, the first vertebrates to live permanently on land, evolved. The Carboniferous Period, together with the next period, the Permian, is sometimes called the Age of Amphibia. But it was also the age of the first great forests. The first reptiles appeared, as well as flying insects, such as cockroaches and dragonflies, some with wingspans of 75 cm (2.5 ft). Swamps filled with ferns and horsetails, and cone-bearing trees covered huge areas of the Earth. These fossilized swamps persist today as the coal deposits from which we get coal, oil, and natural gas.

The extraordinary explosion of plants and animals that occurred during the Carboniferous ended with another mass extinction in the Permian. The continents drifted together to form a single supercontinent, called Pangaea; the seas dropped to their lowest level ever; and the climate became extremely dry. About 90 percent of all marine species went extinct within 5 to 8 million years.

The **Mesozoic Era** [Greek, *mesos* = middle + *zoos* = life], often called the Age of Reptiles, began about 248 million years ago and ended 65 million years ago. The Mesozoic comprised three periods—the Triassic, Jurassic, and Cretaceous.

During the Triassic Period, more kinds of reptiles lived than ever before or since. Soon after the reptiles evolved, they split into three groups. The first group were the turtles. The second group were the ancestors of the first mammals. These were small, shrewlike animals that did not diversify much, but, lucky for us, they hung on for 200 million years until the dinosaurs were gone.

The third group were the ancestors of all the other reptiles—including the lizards, snakes, dinosaurs, and birds. In the seas, long-necked plesiosaurs and ichthyosaurs fed on Mesozoic fish. In the air, pterosaurs flapped across the sky, some with wingspans of as much as 10 meters (30 feet), others as small as sparrows. Late in the Triassic, the first dinosaurs appeared, diversified throughout the Jurassic, then disappeared in another mass extinction at the end of the Cretaceous (Figure 15-7).

Flowers

Early mammals

Toxodon

Mesozoic

Cenozoic

Cretaceous

Tertiary

Quaternary →

66.4

1.65

Major extinction:
75% of all species

Many organisms survived the mass extinction at the end of the Cretaceous. These included the insects, flowering plants, birds, mammals, and many reptiles. The insects and flowering plants, in particular, diversified enormously during the Cretaceous. But the birds and mammals awaited the arrival of the Cenozoic Era.

The shortest era of geologic time, the **Cenozoic Era** [Greek, *cainos* = recent], extends from about 65 million years ago to the present. The Cenozoic has seen the diversification of the insects and flowering plants, the first grasslands, together with grassland animals, and all of the other modern mammals and birds.

Because so little time has elapsed since the beginning of the Cenozoic—a mere 65 million years—fossils from this era are common and well preserved. Fossils of mammals are especially abundant, and paleontologists have been able to piece together nearly continuous evolutionary histories for many modern species, including those of humans and horses.

Taxonomy

The taxonomist Carolus Linnaeus did not believe that species evolved, yet his classifications of organisms provided Darwin with persuasive evidence for evolution. Linnaeus did not intend for his hierarchical classification to be interpreted literally as a family tree. We now understand, however, that the careful anatomical comparisons that Linnaeus and his successors made usually reflect actual lines of descent (Figure 15-8).

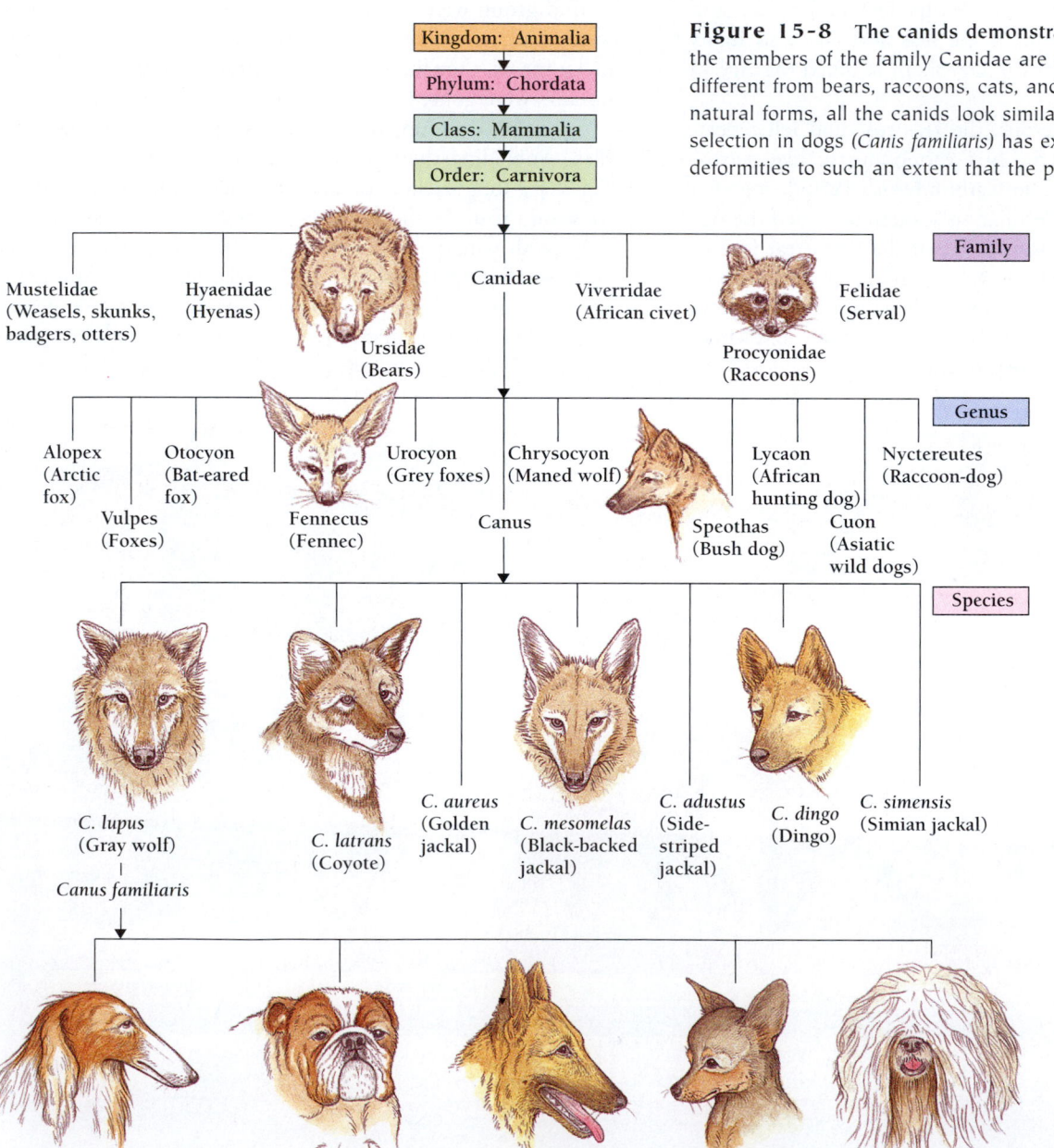

Figure 15-8   The canids demonstrate hidden variation. All the members of the family Canidae are distinctive—recognizably different from bears, raccoons, cats, and other carnivores. In their natural forms, all the canids look similarly doglike. Yet, artificial selection in dogs (*Canis familiaris*) has exaggerated certain traits and deformities to such an extent that the physical variation expressed by just five breeds surpasses that shown by all the other species in the genus *Canis* and even in the whole family Canidae. There are limits to the variation present in a single species, however. No dog could be mistaken for a bear, a cat, a hyena, or any of the other noncanid families in the order Carnivora.

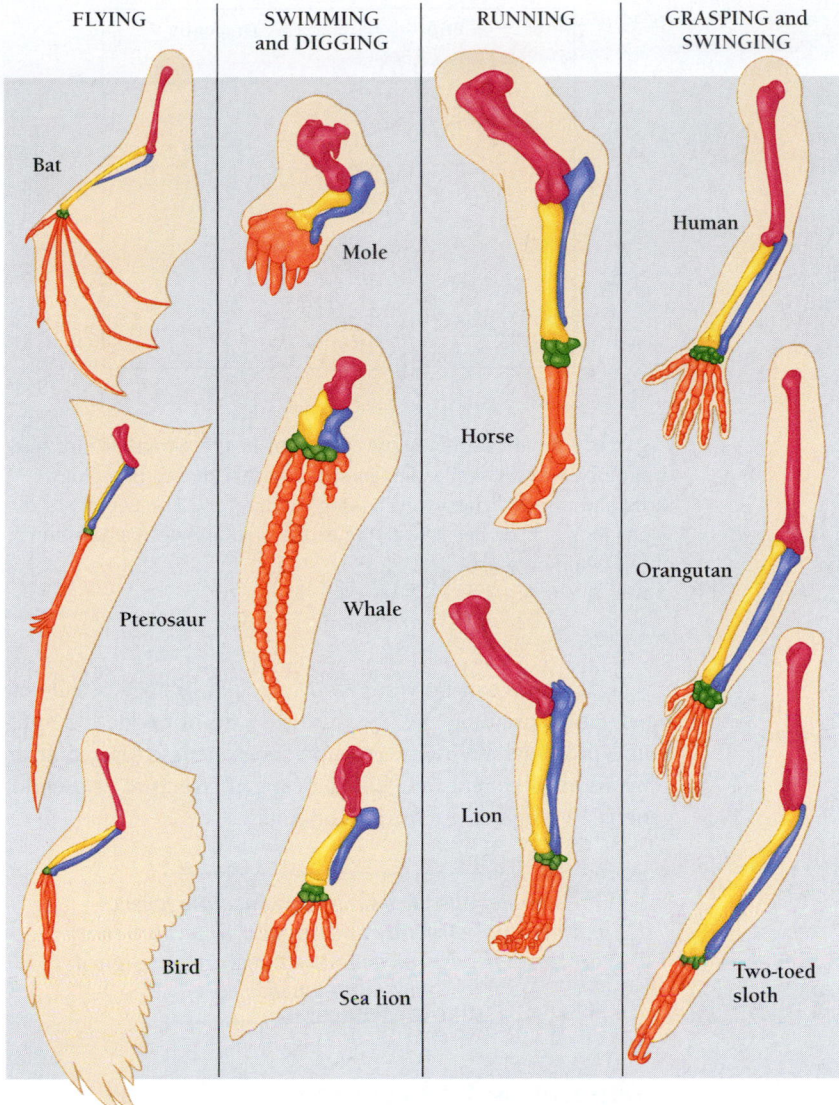

FLYING   SWIMMING and DIGGING   RUNNING   GRASPING and SWINGING

Bat

Pterosaur

Bird

Mole

Whale

Sea lion

Horse

Lion

Human

Orangutan

Two-toed sloth

Figure 15-9 **Homology in vertebrate limbs.** The limbs of bats, moles, horses, and humans contain the same set of bones, modified, respectively, for flying, digging, running, and throwing. This limb homology extends to birds, reptiles, and amphibians.

Comparative Anatomy

Species descended from a common ancestor may evolve in quite different directions and yet retain many of the same characteristics. For example, consider the forelimbs of humans, horses, bats, moles, and whales (Figure 15-9). Each of the different forelimbs possesses the same bones arranged in the same way. But each animal uses its forelimb differently. Humans write, horses stride, bats fly, moles dig, and whales swim. Even though these various bones serve different functions today, they all arose from the same structures in a common ancestor. Similar structures, despite differences in function, imply common ancestry and, therefore, evolution.

Similar structures in two or more species are called **homologous structures.** Homologous structures don't have to perform different functions. For example, the bird, the bat, and the pterosaur evolved different arrangements of the same bones to perform the same function—flying. All these wings evolved separately from homologous structures, and each in its own way allows flight.

And some homologous structures are hidden unless you know what you are looking for. For example, the four tiny bones in the mammalian middle ear are homologous with much larger bones from the jaws of our ancestors (Figure 15-10).

The opposite of "homology" is "convergence." The wings of birds and of insects, for example, evolved from unrelated structures, yet they serve the same function and have a similar shape. All wings must be light and have broad surfaces. The similar wings of a dragonfly and a sandpiper are an example of *convergent evolution* (Figure 15-11).

Evolution is opportunistic. Just as a handyman can use a bit of wire here, a bit of glue there, to fix a bicycle or a light fixture, so evolution may employ a humerus bone here, a radius bone there, to allow one organism to gallop, another to fly.

Paleontologists can trace changes in homologous structures through the fossil record. For example, the first recognizable horse lived about 50 million years ago. The size of a large dog, *Hyracotherium* had four toes on its forefeet and three on its hind feet (Figure 15-12). Its teeth were too weak for it to chew the tough grasses that modern horses live on, so paleontologists believe that these ancestral horses lived on leaves, which are softer and more easily digested.

Subsequent horse fossils show a progressive lengthening of the legs, strengthening of the teeth, and reduction in the number of toes. In the foreleg of a modern horse, the lower bones

We can group species into hierarchies for the same reason that we can group individual humans into families and other hierarchies of relatedness. All organisms are related. We cannot do the same with elements or minerals, which are not related to one another (Chapter 19).

In Figure 15-8, we see that all the dogs, wolves, coyotes, and other species in the genus *Canis* look alike, just as members of the same family should. They are like siblings. The members of related genera—*Canis* and the foxes *Vulpes* and *Fennecus,* for example—are also similar, but not as much. They are like cousins. All the dogs, foxes, and wolves belong to the same family, the Canidae, for the members of the dog family are more closely related to one another than they are to the bears, cats, and other families in the order Carnivora.

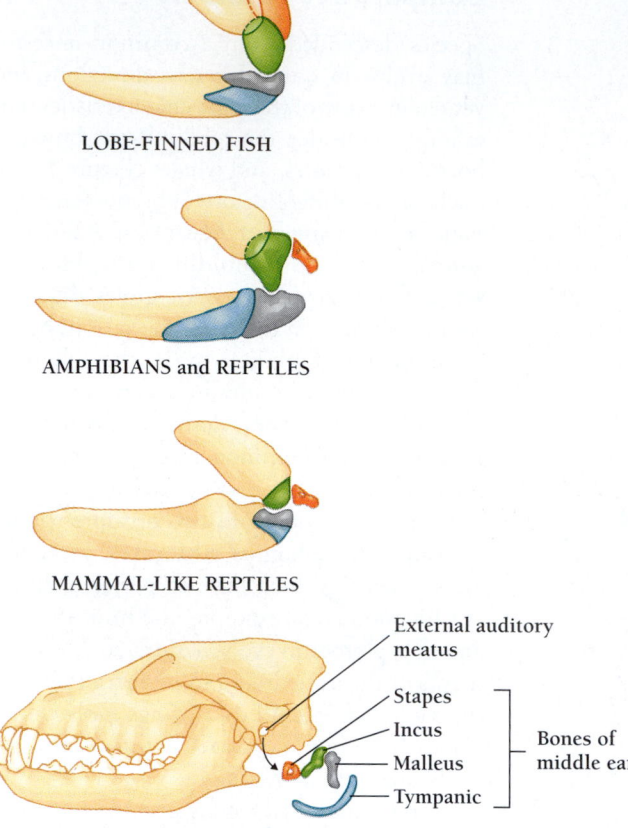

LOBE-FINNED FISH

AMPHIBIANS and REPTILES

MAMMAL-LIKE REPTILES

External auditory meatus

Stapes
Incus
Malleus
Tympanic

Bones of middle ear

MAMMAL

Figure 15-10 Evolution is opportunistic. The four tiny bones of the mammalian middle ear evolved from the jaw bones of early reptiles. The jaw bones evolved, in turn, from the branchial arches (structures that carry gills) of our jawless ancestors.

(radius and ulna) are fused and long. The last segment of a horse's leg is actually a long middle finger, the only one remaining of the five digits found in primitive ancestral mammals. These successive changes in the homologous bones give the modern horse longer legs and extraordinary fleetness. The many fossils that grade from the tiny, horselike animal *eohippus* to modern horse are called **intermediate forms.**

Some homologous structures have no apparent use. The modern horse has a single, useless side toe, called a splint. The splint is an example of a **vestigial structure** [Latin, *vestigium* = footprint], a part of an organism with little or no function but which had a function in an ancestral species. A vestigial structure is a remnant. Our own tailbone, invisible without an x-ray image, is a vestige of another way of life. Similarly, some snakes and whales have vestigial pelvic and leg bones, left over from ancestors that walked (Figure 15-13). Vestigial organs, like other homologous structures, are indicators of biological history and thus offer further evidence for evolution.

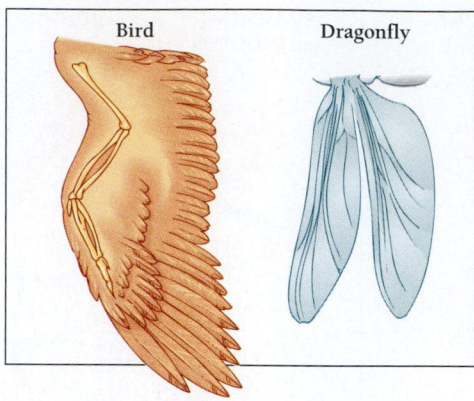

Figure 15-11 Convergent evolution in the wings of bird and dragonfly. Not all similarities in structure are due to homology. Birds and insects have evolved wings separately. The two kinds of wings look similar because flying exerts similar selective pressures on all flying organisms.

Even genes can be vestigial. Molecular biologists call vestigial genes "pseudogenes" because they do not code for functional proteins. Yet these corrupt genes persist, passed from generation to generation. Like a vestigial structure, a pseudogene is a vestige of a common ancestry.

Homologous structures, intermediate forms, and vestigial structures all suggest an evolutionary process whereby ancient structures are redesigned or abandoned.

Comparative Embryology

The early embryos of vertebrates are amazingly alike (Figure 15-14). For example, all vertebrate embryos, including humans, have tails and gill-like *branchial arches.* In fishes and amphibians, the tail develops into a swimming structure. Other groups develop the tail differently. Our own tail is vestigial.

Likewise, the branchial arches of fishes and amphibian embryos develop into gills for breathing under water. Amphibians that change into air-breathing adults lose their gills. In reptiles, birds, and mammals, the branchial arches develop into other structures in the adult ears, mouth, and respiratory tract. These include the bones of the ears and throat, some of the muscles of the face and neck, the parathyroid and thymus glands, as well as a major blood vessel of the lungs (Figure 15-15).

In embryos, the branchial groove or opening between the branchial arches may open to form a gill slit (where water exits from the gills). These are obvious in sharks and tadpoles, for example. Amazingly, the opening between the first and second arches forms, in humans, a hole that forms the Eustachian tube leading from the outside of the ear inward to the eardrum (Figure 15-15). The outer hole, called the "external auditory

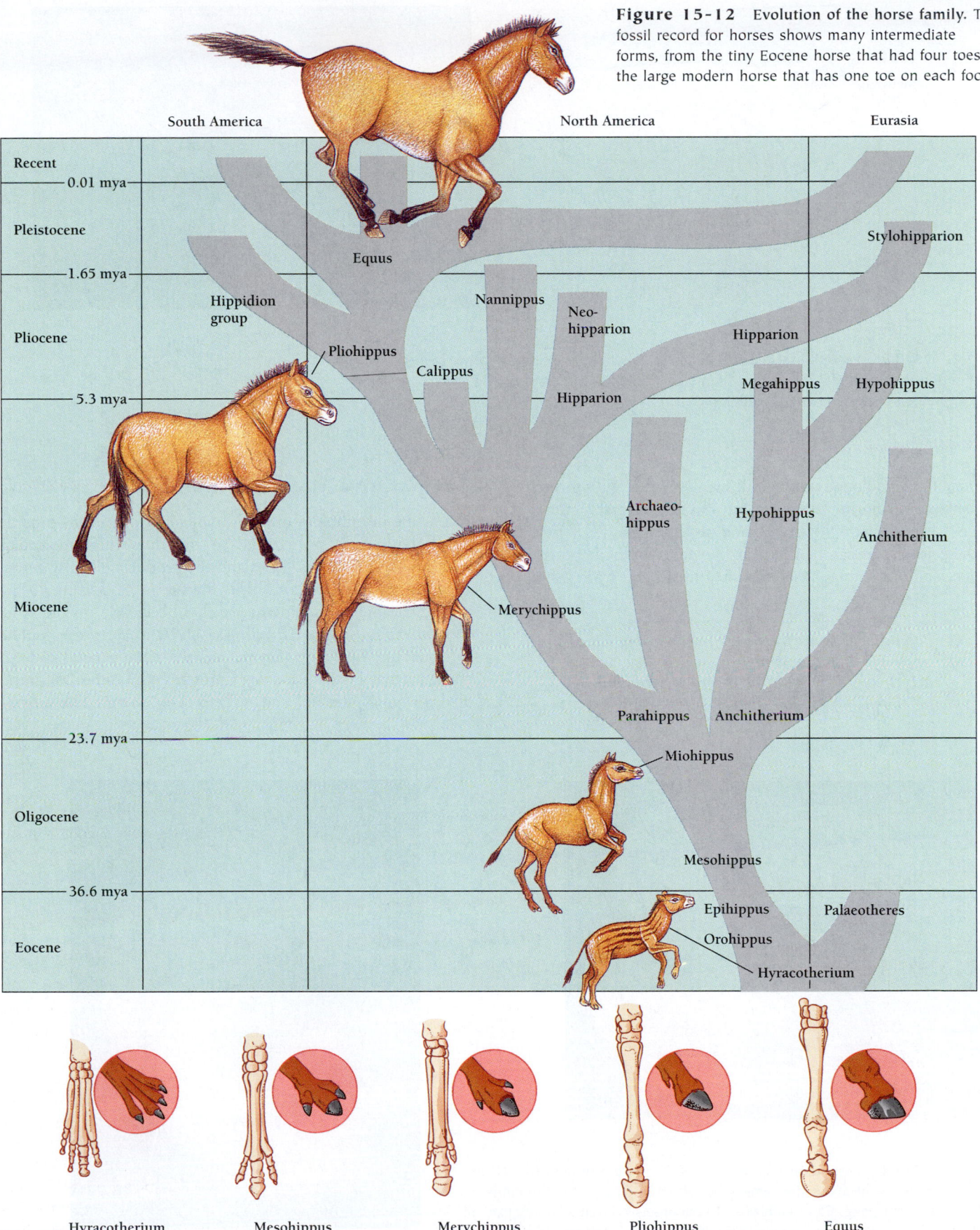

Figure 15-12   Evolution of the horse family. The fossil record for horses shows many intermediate forms, from the tiny Eocene horse that had four toes to the large modern horse that has one toe on each foot.

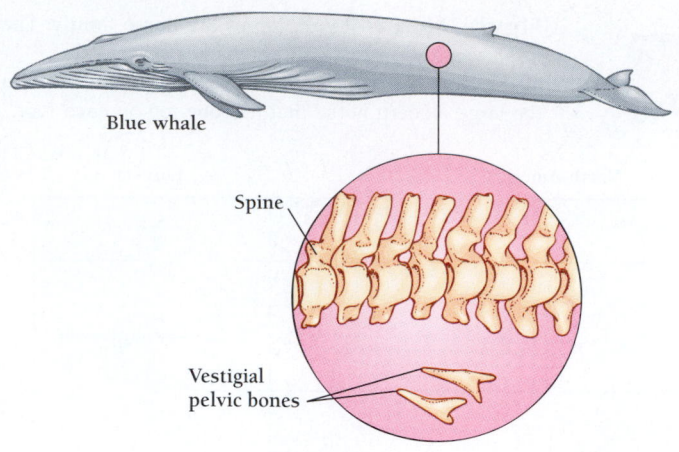

Blue whale

Spine

Vestigial
pelvic bones

A.

B.

Figure 15-13   Vestigial structures. A. The blue whale has vestigial pelvic bones, remnants from the whale's ancestors' life as land animals with legs. B. The South African python's short claw is a vestigial leg. (B, Anthony Bannister Photo Library)

meatus," is homologous with a gill slit. The other gill slits normally close up during development, but in rare cases, human babies are born with gill slits in their necks (Figure 15-15). All of these structures can be traced to the embryonic pharyngeal pouches and unambiguously indicate our descent from fishes.

Developing organisms frequently pass through stages that resemble the organisms from which they evolved, a fact consistent with the theory of evolution.

Comparative Molecular Biology

Just as the morphology and embryology of organisms provide clues about their relatedness, so too does their molecular makeup. An organism's genes and the products of those genes, proteins, are a clear record of its lineage.

At the most fundamental level, all cells rely on the same molecular machinery. All cells use DNA to carry the genetic information; RNA, ribosomes, and the same genetic code to translate that information into proteins; the same 20 amino acids to build proteins; and ATP to carry energy. The univer-

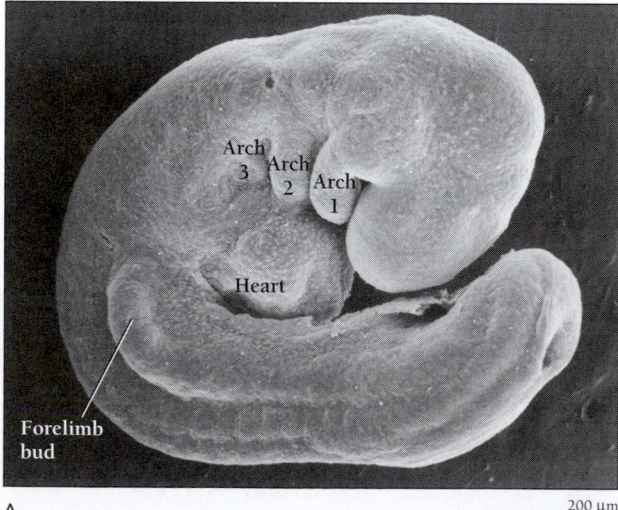

Arch 3
Arch 2
Arch 1

Heart

Forelimb
bud

A.   200 μm

Figure 15-14   Early embryos of a mouse (A) and a human (B). These two mammals are at about the same stage of development. The human is only about 4 mm long. Early embryos of vertebrates are remarkably alike, all displaying branchial arches, paddlelike hands and feet, and tails. (A, courtesy of K. Sulik; B, Professor Hideo Nishimura)

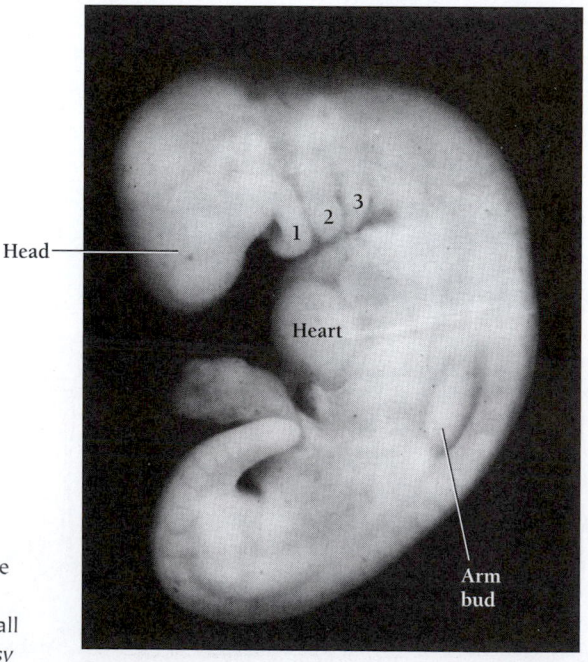

Head

1 2 3

Heart

Arm
bud

B.   1000 μm

Figure 15-15 Gills in adult humans?
Not quite. In some rare individuals, however, the second branchial groove (red) remains open even in adulthood. This opening from the throat to the outside of the neck appears in B as a vestigial gill slit. In all adult humans, the opening into the outer ear and the eustachian tube derive from the first gill slit (blue). In addition, the bones of the inner ear derive from the first and second branchial arches of the gills, as shown in A and B. Bones in the throat derive from the 3rd and 4th branchial arches.

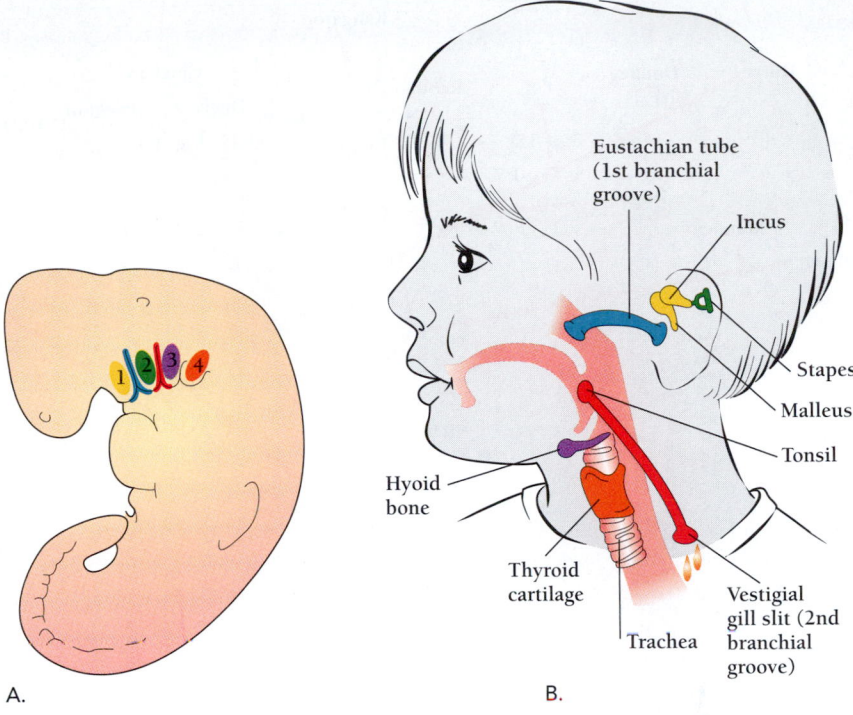

A.   B.

sal homology of this system suggests that it is a common heritage, passed down from some ancient common ancestor.

Other cell proteins are nearly as universal. Organisms as different as yeasts and humans often use similar proteins. By comparing proteins, or the genes that code for the proteins, biologists can construct hierarchies of relatedness much like those derived from comparative anatomy. In general, the more closely related two species are, the greater the similarity between the amino acid sequences of all their proteins.

Today, biologists most often construct phylogenetic trees by comparing genes themselves, rather than proteins. But the results are similar. In fact, biologists can compare not only genes, but also the organization of chromosomes. Such comparisons show that less-related species have more chromosomal rearrangements. A cat genome, for example, differs from a human genome by about 13 rearrangements. These rearrangements occur rarely, and geneticists estimate that in mammals, one rearrangement occurs about every 10 million years.

One of the earliest proteins that biochemists analyzed was cytochrome *c*, a protein in the electron transport chain of the mitochondria (Chapter 6). Cytochrome *c* is a single polypeptide chain of 104 amino acids, whose exact sequence is known in dozens of species. In all of the species examined, 35 of the 104 amino acids are always identical. The remaining 69 amino acids can vary from species to species.

Not surprisingly, similar species have similar cytochrome *c*, while very different species have more changes. For example, humans and chimpanzees have exactly the same sequence of amino acids in cytochrome *c*, while humans and tuna differ by more than a hundred. Evolutionary biologists now routinely use data from such comparisons to construct **phylogenetic trees** (Figure 15-16).

The evolutionary history suggested by amino acid sequences usually agrees rather closely with the history implied by the fossil record. For example, cytochrome *c* sequences suggest that mammals and reptiles diverged from each other long before horses and donkeys did. And the fossil record shows that the first mammal-like reptiles emerged at least 180 million years ago, while the first horselike fossils didn't appear until about 36 million years ago. This is logical: groups of mammals, such as horses, can't diverge from each other if mammals haven't evolved yet. The fossil record, anatomical comparisons, and molecular comparisons are internally logical and independently confirm one another.

Comparative molecular biology has revealed that genes tend to change, or mutate, at a constant rate within certain groups of organisms. Molecular changes seem to accumulate in each type of protein, or in the corresponding piece of DNA, steadily, like the ticks of a *molecular clock*. Molecular clocks help researchers estimate when two species diverged even when no common ancestor exists in the fossil record. By comparing many genes, biologists can construct reasonably reliable phylogenetic trees.

But molecular clocks are imperfect, for individual genes evolve at different rates. And the same gene may evolve faster in one lineage than in another—faster in rodents, say, than in primates. Because genes evolve at different rates in different lineages, molecular clocks have limited value for measuring actual evolutionary time. Nonetheless, molecular clocks can tell us which groups diverged first, second, or third, if not exactly when.

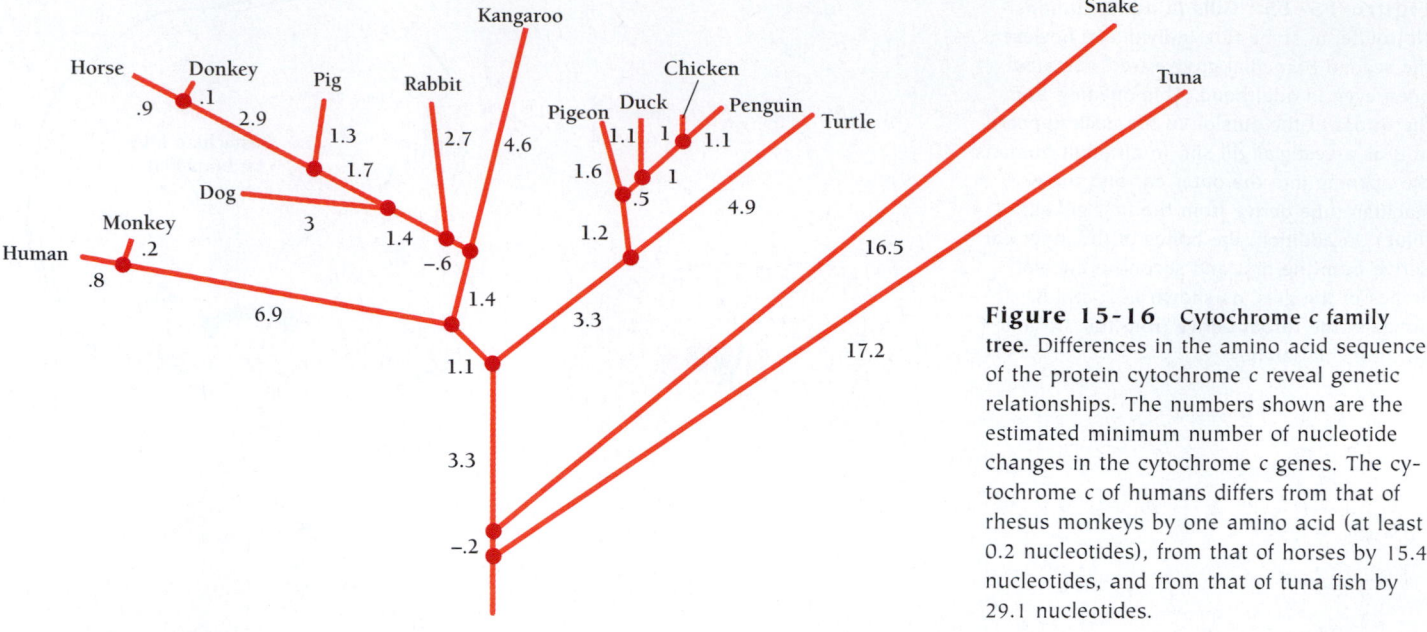

Figure 15-16 Cytochrome *c* family tree. Differences in the amino acid sequence of the protein cytochrome *c* reveal genetic relationships. The numbers shown are the estimated minimum number of nucleotide changes in the cytochrome *c* genes. The cytochrome *c* of humans differs from that of rhesus monkeys by one amino acid (at least 0.2 nucleotides), from that of horses by 15.4 nucleotides, and from that of tuna fish by 29.1 nucleotides.

The molecules of closely related organisms are more similar than the molecules of distantly related organisms. Comparative molecular biology confirms and elaborates on the story told by comparative anatomy and the fossil record.

Biogeography

When Darwin wrote *The Origin of Species,* most Europeans believed that every species had been individually created—a rhinoceros for this continent, a horse for that continent. If there were no tigers in North America, it was because God had not put any there. But Darwin saw that the distribution of many groups of species made no sense unless they had arisen from common ancestors.

From his studies of the distributions of species, called **biogeography,** Darwin concluded that each species has a definite region of origin, becomes distributed by migrating, and evolves descendant species in the regions to which it migrates. Darwin's view is so well accepted that today, if we ask someone why are there no tigers in North America, we will be told it's because tigers have no easy way to get from India and China across the Pacific Ocean. We all take it for granted that species move from place to place. But 19th-century biologists did not.

One of Darwin's strongest arguments for evolution was the close resemblance of island species to those on nearby continents. A 19th-century biologist would have expected that the closest relative of a finch from the Galápagos Islands, off the coast of South America, would be found in a similar group of islands, matched for climate and geology. But Darwin found that the closest relatives of island species were not on nearly

identical islands halfway around the world, but right next door, usually on the nearest continent.

Species all over the world are most closely related to those that live nearby. Darwin's theory of common descent accounts for this pattern.

NATURAL SELECTION: DARWIN'S MECHANISM FOR EVOLUTION

Darwin made a strong case for the theory of evolution. And the evidence that scientists have accumulated since then has strengthened this theory. Indeed, the biologist Theodosius Dobzhansky summed up biologists' feelings best, saying, "Nothing in biology makes sense except in the light of evolution."

We can accept the idea that organisms have changed over time, and we can accept the idea that modern organisms are descended from ancient ones. But what of Darwin's proposed mechanism, natural selection? Darwin believed that natural selection, acting gradually over immense spans of time, created not only new species, but new genera, new families, and all the higher taxa. Why did he think that was possible?

Once Darwin conceived the idea of evolution by natural selection, he developed an obsessive interest in the work of gardeners, farmers, and stock breeders. It was perfectly apparent to him that many of the different breeds of dogs and pigeons, for example, would, if found in nature, be identified as separate species.

BOX 15.2

Continental Drift

Studies in biogeography reveal that species resemble nearby species more than species that live far away. However, similar organisms can occur in places that are remote from one another. For example, rheas, ostriches, and emus, all large flightless birds, occur in South America, Africa, and Australia, respectively. We may wonder if these birds are related to one another and, if so, how they came to be so widely dispersed.

The fossil record shows many examples of a similar pattern of dispersal, with closely related species in the southern continents. The distinctive fossil plant *Glossopteris,* for example, appears in similar rock formations from the late Paleozoic Era in both Brazil and South Africa. How could organisms that are so closely related live so far from one another?

In 1915, the German geologist Alfred Wegener, observing the jigsaw-puzzle fit of the African and South American coastlines, proposed the theory of "continental drift." Wegener argued that Africa and South America, as well as Australia, Antarctica, and India, were once a single supercontinent, which broke up into separate continents that drifted apart. Geologists rejected Wegener's theory, however, and biologists and paleontologists continued to puzzle over the similarities between the species on the southern continents.

Then, in the early 1960s, geophysicists discovered that the Earth's surface consists of large blocks, or **plates,** which move with respect to one another at the rate of a few centimeters a year. The discovery that both the continents and the sea floor were moving about on the surface of the Earth immediately revived Wegener's theory.

Geologists now understand that this movement comes from the continuous formation of new crust under the ocean. As lava emerges from below the surface, the new rock pushes apart the plates on either side. Powered by this pressure, plates can collide, scrape past one another, or dive beneath one another, pushing up mountains and precipitating earthquakes and volcanic activity. The Himalayas, for example, rise a little higher each year as the Indian-Australian plate crashes—ever so slowly—into the Eurasian plate. Similarly, the Pacific plate, beneath the Pacific Ocean, scrapes northward past the North American plate, creating California's earthquake-prone San Andreas Fault.

Geologists have been able to puzzle out many of the past movements of the continents by studying the magnetic orientation of rocks. This history explains many features of the fossil record and the distribution of modern species as well. About 250 to 200 million years ago, all of the continents were united into a supercontinent called Pangaea. Pangaea was surrounded by one huge ocean, Panthalassa (Figure A). In time, Pangaea began to break apart, allowing independent evolution to occur on the two resulting continents—*Laurasia* (Eurasia, Greenland, and North America) and *Gondwanaland* (India, Africa, South America, Madagascar, Antarctica, and Australia).

By the beginning of the Cretaceous, 135 million years ago, Gondwanaland had also begun to break into the separate continents. By about 65 million years ago, South America had split from Africa and Australia had separated from Antarctica, after which Australian species evolved in isolation. The land connection between North and South America formed only about 3 million years ago.

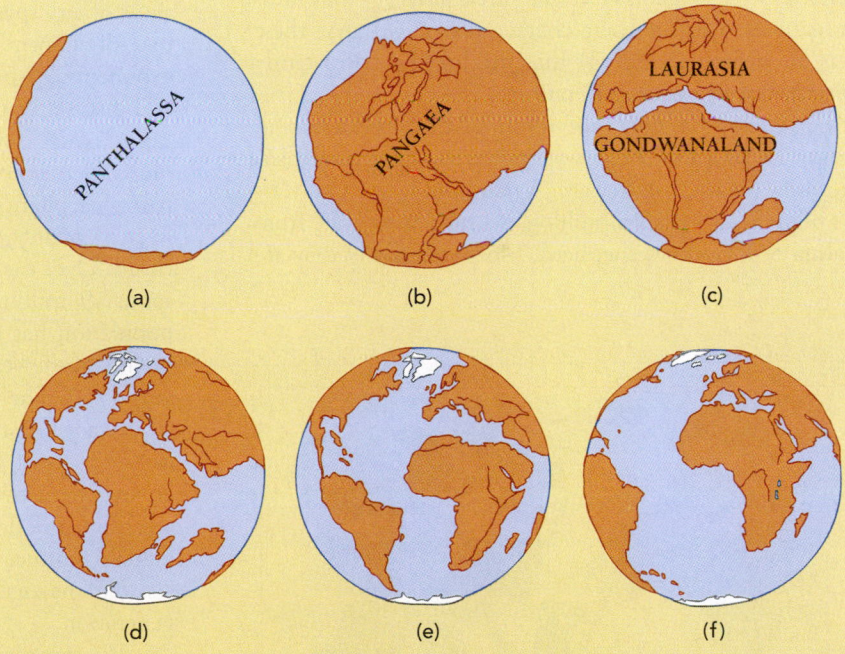

Figure A **The history of the continents.** (a) and (b) Alfred Wegener's view of the Earth as it existed 200 million years ago. At that time, there was one ocean (Panthalassa) and one continent (Pangaea). (c) By about 20 million years later, the supercontinent had begun to split into northern Laurasia and southern Gondwanaland. (d) Separation continued, and (e) looked this way about 65 million years ago. (f) Further widening of the Atlantic and northward migration of India brings the Earth to its present state.

Artificial Selection

Breeders derive such varied forms from a single species by carefully selecting and breeding only individuals that show certain desired traits most strongly. Under this selective regimen, each generation will show the selected trait more clearly than the last.

Dramatic examples of such intense selection come from all kinds of farm animals and crops. Most amazing are the differences among dogs (Figure 15-8). The English bulldog, for example, is the result of selection for a developmental deformity in which the upper jaw fails to develop at the same rate as the lower jaw. The lesson here is that nearly all species possess tremendous untapped genetic variation.

Darwin's particular interest was the work of Victorian pigeon fanciers. He collected many varieties of pigeons, studied the differences in their skeletons, and bombarded fanciers with questions (Figure 15-17). Ultimately, he concluded that all breeds of pigeons were descended from a single species—the European rock pigeon. Yet, the more than 20 different breeds were so different, Darwin argued, that a zoologist who discovered them in the wild would certainly classify them into a dozen or more species and quite possibly into half a dozen genera, as well.

Clearly, artificial selection by breeders could transform the characteristics of a breed. Success required only (1) that individual organisms vary in their characteristics, (2) that these variations be heritable, and (3) that the breeder consistently select certain traits in each generation.

Artificial selection does not result in the production of new species. At least new breeds are not recognized as species, no matter how distinctive they are. A mammologist from Mars wouldn't hesitate to classify a bulldog as a species separate from a chihuahua or a German shepherd. Morphologically they are

distinct. But we know their history and that they can interbreed, so we classify them all as one species. Artificial selection demonstrates how much genetic variation exists in populations of organisms and how much change can occur in just a few years. More importantly, however, artificial selection hints at the much greater degree of change that might occur under the constant pressure of natural selection over millions of years.

Artificial selection provides a model for natural selection, a possible mechanism for evolution.

Darwin's Argument for Natural Selection

How can nature make selections in the same way as human breeders, generation after generation? In Malthus, Darwin found the answer. Malthus, recall, had argued that the production of food and other resources can never keep up with the enormous increases that human populations are capable of, that early death is the fate of a large proportion of individuals born in every generation.

Darwin saw that Malthus's arguments applied to populations of any species. Populations of organisms, although capable of enormous increases, do not continue to grow exponentially, but instead are limited by resources, predation, and disease. Natural selection must occur, Darwin argued, because nature cannot sustain unlimited exponential growth. To illustrate this point, he considered the descendants of a single pair of elephants, among the slowest breeders of all organisms. Even elephants, however, have the potential to breed geometrically. Over the course of 750 years, two elephants could have some 19 million descendants. Yet, historically, the elephant population has been fairly stable. So it is likely that two elephants in the year 1000 had only two descendants alive in the year 1750. The question Darwin posed was, Which two?

According to Darwin's theory, the survivors would be the descendants of whichever elephants were best adapted to the environment over the course of those 750 years. In artificial selection, the decisions of the breeder determine which individuals produce the next generation. In nature, the individuals that manage to survive and reproduce produce the next generation.

We can easily see how this argument might apply to the evolution of horses, for example. Individuals with long legs and teeth capable of chewing the tough grasses so common on the plains apparently survived and reproduced, whereas those with short legs and small teeth died out.

Darwin argued that if large changes can occur in domestic plants and animals over just a few generations, then even larger changes can certainly occur over the millions of years available for evolution.

The rate at which changes accumulate depends both on the intensity of selection and on the extent of inherited vari-

Figure 15-17 Artificial selection in pigeons. Charles Darwin collected different breeds of pigeons to study the variation possible within a single species. Shown here are a pair of baldheads, a carrier cock, two white fantails, two white pouters, and several other varieties. (*The Illustrated London News Picture Library*)

ation. For example, if only a few young horses from each generation survive, because most cannot escape a fast predator or cannot digest grass well enough, selection is strong and advantageous traits would accumulate rapidly. However, if most of the young horses from each generation survive and reproduce, then selection is weak and change occurs slowly or not at all.

The rate of change also depends on the extent of variation within the population. To take an extreme case, if all the individual horses of a generation were identical, then the next generation of horses could not differ from their parents. But, in fact, individuals vary enormously.

Natural Selection Analyzed

Modern evolutionary biologists have broken down natural selection into a series of basic facts and inferences.

1. Fact: **Superfecundity.** All species of organisms are so fertile that their populations would increase exponentially if all offspring survived.
2. Fact: **Not all offspring reproduce.** Populations of organisms do not attain these staggering proportions because many offspring succumb to disease, predation, and other fates before they reproduce. Those that do reach reproductive age find that the world does not contain enough food, territory, and other resources for all of them to reproduce.

   Inference: **Struggle for existence.** Whenever extinction does not occur, some individuals will escape death and disease and avail themselves of the limited resources more successfully than others. In the "struggle for existence," as Darwin called it, some individuals will survive to reproduce, while others will not.
3. Fact: **Individual variation.** Among sexually reproducing species, the individuals of a population differ. Most individuals are unique.
4. Fact: **Heredity.** Many of the differences between individuals are heritable—that is, individual differences can be passed on to offspring. Darwin knew nothing about genetics, but he could see, for example, that the short legs of the dachshund were inherited from its parents.

   Inference: **Adaptive traits may be passed on differentially.** Darwin argued that in the struggle for existence,

individuals whose traits best suited them to their environment survive and reproduce more successfully than those less well suited to the environment. Adaptive traits that are heritable are more likely to be passed on to the next generation. For example, in the desert, rain comes infrequently. The faster a seed germinates when rain comes, the more water it can absorb as it forms a shoot and root. Seeds that germinate rapidly will survive and leave more offspring than seeds that germinate more slowly.

Inference: **Adaptive traits accumulate within a population.** As a result of natural selection, various traits become more or less common in a population, and the overall character of the individuals within that population changes. For example, if the speed with which a certain desert plant can germinate is both heritable and adaptive, then, over time, more and more plants will germinate quickly. In this example, the genetic traits that allow a plant to germinate faster are *adaptive* traits. They are traits that allow it to survive long enough to reproduce.

Biologists accept the reality of both evolution and natural selection. The evidence that natural selection operates in natural populations is now irrefutable, yet some modern biologists question whether natural selection is the primary means by which new species and higher taxa such as genera and families are created. Over the years, biologists have proposed a host of alternatives to natural selection. A few biologists have argued that mechanisms other than natural selection may be responsible for macroevolution, the evolution of species, genera, and higher taxa of organisms. In the next chapter, we will examine natural selection more closely and also consider alternative mechanisms for evolution.

In addition, we will explore the genetic basis of variation. We can see that for natural selection to work, organisms must vary, and that the differences that distinguish individuals must be heritable. But Darwin himself knew nothing about *how* variation was inherited. The details of inheritance would not become clear until after 1900, with the rediscovery of Mendelian genetics and the advent of population genetics. Only in the 20th century did biologists begin to understand how genes, carrying the hereditary information, provide the basis for natural selection.

Study Outline with Key Terms

When Darwin published his book *On the Origin of Species* in 1859, he changed the way we view the world. The theory of **evolution** or **"descent with modification,"** contradicted **creationism,** the Bible's version of the story of creation, the philosophy of **essentialism,** and the Christian doctrine that the Earth was only 6000 years old.

By the middle of the 19th century, Hutton's theory of gradual geologic change and Lyell's **uniformitarianism** had persuaded the scientific community that the Earth was very old, and that change occurs slowly and gradually, not catastrophically. Examination of the fossil record by Cuvier and others suggested a very long history of life on Earth—a history far longer than the 6000 years implied by the Bible. In a theory that came to be called **catastrophism,** Cuvier suggested that fossil species were organisms that had perished in periodic catastrophes, each followed by a creation.

Linnaeus had created a system of naming organisms and a hierarchy that resembled a family tree. Biologists such as Buffon and Lamarck believed in evolution, and Lamarck formally introduced the idea. But Lamarck thought that species evolved in response to "felt needs" and proposed that evolution depended on the **inheritance of acquired characteristics.** Many European scientists accepted the inheritance of acquired characteristics, but rejected evolution.

In 1859, Charles Darwin published the influential *On the Origin of Species,* which defined evolution, and described a mechanism by which evolution could work, **natural selection.** Natural selection is the differential survival and reproduction of individuals with inherited traits that are **adaptive.** Darwin also believed that evolution occurs gradually through the slow accumulation of changes, an idea called **gradualism.**

Darwin argued that natural selection worked like **artificial selection.** In artificial selection, the breeder selects the varieties that will produce the next generation. In natural selection, the individuals who best survive and reproduce are those that will produce the next generation. Natural selection, like artificial selection, favors some forms over others.

Artificial selection suggested a mechanism for evolution, but other lines of evidence suggested evolution itself. One of the most persuasive was the fossil record. **Fossils** are the remains or imprints of past life. Most come from layers of sedimentary rocks, which form from the compression of settling mud, sand, and debris. The order of the layers (or strata) of sedimentary rock establishes the sequences in which ancient life evolved. The simplest organisms lie at the bottom of the record and through **intermediate forms** become both increasingly complex and increasingly like modern organisms, a pattern that suggests evolution. **Paleontologists,** scientists who

study fossils, usually divide the history of life into four long eras—the **Precambrian,** the **Paleozoic,** the **Mesozoic,** and the **Cenozoic** eras. Each era in turn consists of several shorter divisions called **periods,** and, more recently, **epochs.**

Another line of evidence for evolution comes from **biogeography.** Species tend to resemble neighboring species in different habitats more than they resemble species that are in similar habitats but far away. This pattern of geographical distribution suggests that species evolved from one another.

Linnaeus's **taxonomy,** a hierarchical grouping of organisms by **species, genus, family, order, class,** and other **taxa,** suggests patterns of relatedness among different species, not separate creations. The relationships implied by taxonomy—so-called **phylogenies**—are derived from studies in **comparative anatomy.** Species that closely resemble one another are considered to be more closely related than species that do not resemble one another. **Homologous structures,** such as limb bones, provide additional evidence for evolution. **Vestigial structures,** such as the horse's splint, indicate evolutionary history.

Embryology provides another highly persuasive line of evidence for evolution. Embryos repeat the development of their ancestors to a limited but nonetheless remarkable degree. The early embryos of fish, amphibians, reptiles, birds, and mammals all have the tails and pharyngeal pouches. Comparative molecular biology confirms the lines of descent suggested by both comparative anatomy and the fossil record.

Natural selection, Darwin's mechanism for evolution, rests on several facts and inferences. All species are **superfecund,** capable of producing far more offspring than the world can support. Not all of these offspring survive long enough to reproduce. In each generation, the offspring of sexually reproducing organisms vary. That is, each individual is unique, and this uniqueness has a distinct heritable component. Those individuals that survive to reproduce determine which traits will predominate in the next generation. Such individuals can be said to be naturally selected.

Review and Thought Questions

Review Questions

1. What three ideas kept most scientists from seriously considering the idea of evolution until the 19th century?
2. Which of these three ideas did Cuvier's theory of catastrophism address? Describe the theory of catastrophism.
3. Describe the theory of uniformitarianism. In what two ways did it influence Darwin?
4. Who wrote the essay that inspired both Darwin and Wallace to invent the theory of natural selection? What did the essay say?
5. Explain why the fossil record is not complete. Describe two possible interpretations for the gaps in the fossil record.
6. During which era did mammals flourish? Were dinosaurs present at the same time?
7. How did Darwin's observations on the distribution of species suggest that species have evolved?

8. If you were a bat, which bones (that you have as a human) would you use for wings?

9. If you were a fish, instead of a human, how would you use the hole that in your human form lets sound into your ear?

10. Describe the molecular clock. What is its use?

11. Darwin's theory of natural selection can be divided into four facts and three inferences. What are they?

Thought Questions

12. The world population is currently about 6 billion. If every couple in the world had 20 surviving children, what would the world population be in two generations?

13. If the individuals of a species did not vary but were all identical genetically, what would be the consequences of natural selection?

About the Chapter-Opening Image

The orangutan represents our close relations with other primates, an idea that disturbs many nonbiologists.

 On-line materials relating to this chapter are on the World Wide Web at **http://www.harcourtcollege.com/lifesci/aal2/**

16 Microevolution: How Does a Population Evolve?

A Fatal Disagreement

On January 26, 1943, an internationally famous and much beloved plant geneticist died of starvation in a Soviet prison hospital. He must have wondered at the irony. Nikolai Vavilov (1887–1943), director of the prestigious Institute of Plant Breeding, in Leningrad, had devoted his life to the study and improvement of wheat, corn, and other cereal crops, in order to eliminate hunger in the vast, new Union of Soviet Socialist Republics (USSR). His "crime" had been to become involved in a long-running dispute about how evolution works. He contradicted the party line and paid for it with his life.

A cheerful, good-natured scholar, Vavilov traveled all over the world in the 1920s collecting seeds from different kinds of crops. The world's expert on the biogeography of wheat, he was a friend to eminent biologists in every part of the world. In the United States, he picked up the expression "Keep smiling" and liked to use it on anxious Soviet colleagues.

Vavilov's own troubles began in the mid-1920s when he befriended a hardworking and charismatic young scientist named Trofim Lysenko (1898–1976). Lysenko, the son of a peasant, worked from sunup until sundown, talking to his experimental plants as if they were people. He continually imagined new ways of "training" crops to yield more grain or grow in different climates. Lysenko passionately wanted to improve the lot of the Soviet people. However, he had little education in biology and disdained what he did not know. He especially disdained genetics, which he considered to be "harmful nonsense." Even his friends in those days joked, "Lysenko is sure that it is possible to produce a camel from a cotton seed."

Vavilov tolerated the young worker's ignorance, and, although Lysenko was not well received by other biologists, Vavilov began inviting him to scholarly meetings, hoping to win him over to genetics. Vavilov had reason to promote the young man. The Soviet government was pressuring Soviet scientists to abandon basic science and come up with "practical" results. Moreover, the government preferred uneducated workers of peasant stock to educated bourgeois elitists. Peasant scientists were especially hard to come by, and Vavilov leaped at the chance to bring along one who also worked entirely in the pursuit of practical results.

In the early 1930s, Lysenko joined forces with Isai Prezent, a philosophy student. Prezent could see that Lysenko needed a theoretical framework to make his shoot-from-the-hip approach to biology seem more scholarly. Prezent knew no more about genetics than did Lysenko, but he had studied Lamarck's theory of the inheritance of acquired characteristics, finally in its death throes in the West. The theory perfectly supported Lysenko's approach to plant breeding, stating as it did that traits developed during the life of an organism in response to

Frans Lanting/Minden Pictures

the environment—long roots grown during a dry season, for example—could be passed on to the offspring. By then, modern genetics had supplanted Lamarckism, and few biologists in the world believed that acquired characteristics could be inherited. Yet, surprisingly, no one had actually disproved Lamarckism.

For this reason, and because Lysenko's peasant background and pragmatic philosophy were above criticism, government officials granted him more and more privileges and opportunities to carry out his research. Lysenko promised spectacular results with his techniques, claiming, for example, that he would double or triple the wheat crop by "training" wheat seedlings to grow in the far North. Grain would be superabundant, Lysenko promised. Soviet bureaucrats responded by meeting his every demand, and newspaper reporters fell all over themselves to interview the young genius.

No one seemed to notice that Lysenko's techniques consistently failed to increase harvest yields. When one idea failed, he came up with another. Each one's promise eclipsed the failures of the last. In the late 1930s, Lysenko was put in charge of all of Soviet agriculture. The whole nation became his "experimental garden," and every peasant farmer became a scientist. Once, Lysenko ordered tens of thousands of collective farms to test a theory about cross-pollinating different varieties of wheat. He requisitioned 800,000 pairs of tweezers for transferring pollen, one pair for each of the peasant farmers who would implement his "experiment."

Joseph Stalin, the totalitarian premier of the Soviet Union, was entranced by the enormous scale of Lysenko's work. Imagine! Eight hundred thousand tweezers! But Vavilov and his fellow biologists were horrified. They knew that cross-pollinating most of the wheat crop in the Soviet Union would eliminate many of the country's best varieties of wheat. And it did.

Despite Lysenko's failures, Lysenko and Prezent became more outspoken than ever in their disdain for genetics. Under their influence, Stalin agreed that genetics would no longer be taught at the universities, medical schools, or high schools. Increasingly, seminars at Vavilov's Institute of Plant Breeding, a world-class research center for genetics, deteriorated into confrontations between geneticists and Lysenko's student supporters. Lysenko came to regard Vavilov's persistent advocacy of genetics as a personal affront. In an era when millions of Soviet citizens were executed for far less than irritating a friend of Stalin's, Vavilov's defense of genetics was courageous but fatal.

Imagine! Eight hundred thousand tweezers!

In the summer of 1940, Vavilov was arrested for "wrecking" Soviet agriculture and sentenced to be shot. That sentence was later commuted to life in prison. But in the two and a half years he spent in prison, he was never moved to a work camp. Instead, along with thousands of other Soviet intellectuals, he died a slow and humiliating death in prison. In his last year, starving and wasted by dysentery, Vavilov and two other imprisoned professors passed the time by giving a series of lectures on history, biology, and forestry. Vavilov delivered more than a hundred hours of lectures. But his efforts to bring a bit of civilization to the dirty, crowded cells could not stave off the hungry prisoners who stole his bread and the slow death that finally took him.

Vavilov knew that by continually contradicting Lysenko he had signed his own death warrant. Yet he never renounced genetics. Why did the truth not prevail? If Lysenko was wrong in his belief in Lamarckian inheritance, why couldn't Vavilov and his fellow geneticists prove that inheritance was genetic?

The short answer to this question is that although geneticists of the 1920s and 1930s knew that genes were somehow passed from generation to generation, they could not prove beyond doubt that genes were real physical objects rather than just theoretical constructs. Nor could they prove Lamarck and Lysenko wrong in their belief that acquired characteristics somehow influenced the genes.

The proof that genes have a physical reality in DNA did not come until the 1950s—too late for Vavilov. Later work demonstrated that information goes from the genes to the body, not the other way around. In general, an individual's behavior—whether reaching for higher leaves, lifting weights, or tanning in the sun—cannot alter the genes it receives from its own parents and passes to its offspring. Nonetheless, actual disproof of Lamarck's theory of acquired characteristics has remained surprisingly difficult to pin down.

Lysenko's influence continued only a few years after Stalin died. In the early 1960s, disastrous crop failures forced the Soviet Union to buy wheat from the United States and Canada, a humiliation that irreparably weakened then Soviet premier Nikita Khrushchev's political power. In 1964, Soviet biologists rejected Lysenko's ideas and resumed the research in genetics that had been suspended for 30 years. Lysenko himself, however, retained his academic titles and his freedom until his death in 1976.

HOW ARE VARIANTS CREATED AND MAINTAINED?

We saw in Chapter 15 that evolution as an idea has a long and venerable history, dating back to the ancient Greeks. Yet even when biologists finally accepted that evolution must have occurred, they couldn't agree on how it could work. Darwin had proposed natural selection. But without an understanding of inheritance, Darwin's mechanism was incomplete. We know now that Darwin's theory of natural selection was, in fact, largely correct. But it took biologists more than 70 years of patient research and bitter dispute to appreciate Darwin's ingenious mechanism for evolution—natural selection.

In this chapter we will see how biologists gradually came to understand how organisms inherit traits. We will study variation in populations, the basis for Darwin's theory of natural selection, the sources of variation, and the means by which that variation is passed from generation to generation. We will examine, as well, some of the theories that have been proposed to account for the evolution of populations.

Why Did Charles Darwin Accept Blending Inheritance?

Darwin wrote the *Origin of Species* primarily to show that evolution had occurred and to discredit the idea that species were specially and individually created by God. His theory of natural selection was a separate idea—advanced to account for how evolution might have happened. Darwin thoroughly understood selection—the differential survival and reproduction of genetic variants. But he did not know where the variation came from.

The gene, as concept or fact, was completely unknown in 1859. Most scientists, including Darwin, believed that inheritance was "blended"—much as yellow and blue watercolors blend to make green (Figure 16-1). Biologists could see that when two races or species were crossed, the resulting offspring generally appeared to possess a smooth blending of the characters of each of

Blending inheritance   Particulate inheritance

Figure 16-1 Blending inheritance or particulate inheritance? Nineteenth-century biologists believed that the hereditary material from each parent blended together in the offspring much as watercolors blend. Darwin suspected the truth: each parent's genetic contribution remains separate.

the parents. For example, a cross between a horse and a zebra produces an intermediate animal, called a "zebroid," not simply a horse with stripes or a zebra without stripes (Figure 16-2). Many 19th-century scientists assumed that the hereditary material, whatever it was, must blend smoothly also.

Yet Darwin intuitively suspected the true particulate nature of inheritance. As he wrote to Thomas Huxley in 1856, "I have lately been inclined to speculate, very crudely and in-

Figure 16-2 Zebroid. A zebroid is a hybrid of a horse and a zebra with some of the traits of each. How are its stripes not quite like those of a zebra? (*Left, Carl Purcell/Photo Researchers, Inc.; right, John Gerlach/Visuals Unlimited*)

distinctly, that propagation by true fertilisation will turn out to be a sort of mixture, and not true fusion, of two distinct individuals, or rather of innumerable individuals, as each parent has its parents and ancestors." However, Darwin had no evidence for this and did not dispute the theory of **blending inheritance** publicly.

In 1866, just a few years after Darwin's publication of the *Origin of Species,* Gregor Mendel published his groundbreaking research, which showed, among other things, the particulate nature of inheritance (Chapter 9). For evolutionary theory, Mendel's most important conclusion was that the individual units of inheritance, which we now call genes, remained intact and did not blend or fuse with one another. But neither Darwin nor any of his contemporaries ever read Mendel's paper.

Darwin's public acceptance of blending inheritance was unfortunate, for it left him open to a criticism for which he had no answer and led him into a second fundamental misunderstanding about the nature of inheritance.

Why Was the Idea of Blending Inheritance a Problem?

In 1867, Fleeming Jenkin, an enterprising Scottish engineer, announced that natural selection could never work in sexually reproducing organisms. Favorable variants, Jenkin argued, would not be able to pass on their advantages to the next generation. Instead, interbreeding would quickly "dilute" any new traits. His idea was that when a fast cheetah bred with a slow one, for example, the offspring would be only half as much faster as the fast parent. And *their* offspring would be only a quarter as much faster as the fast grandparent. Eventually, the increased speed would disappear altogether. How, he asked, could natural selection operate on variants that faded away as soon as they appeared? Assuming blending inheritance, he had a point.

Thanks to Mendel's work, however, we know that inheritance is particulate, not blending. Sexual reproduction does not necessarily dilute traits, but instead creates an assortment of different traits. In the cheetah example, the offspring would not all be exactly half intermediate in speed between the faster parent and the slower parent. Each young cheetah would be different, ranging from fast to slow, and each one's ability to reproduce would be correspondingly different.

Jenkin's logic was fine, but his grasp of biology was weak, and he was mistaken in nearly everything he said. Darwin could easily have demolished the engineer's argument on a variety of grounds. Yet Darwin was so uncertain about the nature of inheritance that, rather than counter the particulars of Jenkin's arguments, or question the basic assumption behind it—blending inheritance—Darwin made a second mistake. He took refuge in the inheritance of acquired characteristics, an idea that had been universally accepted for hundreds of years.

Darwin knew very well that individuals are unique. He knew that natural populations contain enormous reservoirs of

variability on which natural selection can act. If Jenkin was right that variability was constantly blending away, where were new variants coming from? The **inheritance of acquired characteristics,** which Darwin accepted in principle, seemed to provide the only way for new variants to be continuously generated and passed on. According to the theory:

1. Changes in the environment create changes in the needs of organisms.
2. Changes in needs occasion changes in behavior to satisfy these needs.
3. Changes in behavior result in the increased use and development of certain parts, or the use and development of new parts.
4. These changes are passed on to the offspring.

After Jenkin's critique, Darwin began to concede a greater role to the inheritance of acquired characteristics. Darwin died in 1882, one year short of seeing the problem of inheritance solved by one of his most ardent followers.

Fleeming Jenkin proposed that blending inheritance and evolution by natural selection were incompatible. Darwin's acceptance of blending inheritance, as opposed to particulate inheritance, forced him to accept the theory of acquired characteristics.

HOW DID BIOLOGISTS COME TO REJECT THE THEORY OF ACQUIRED CHARACTERISTICS?

Of all the world's biologists, the German biologists were the most enthusiastic about Darwinism. Among them was a man named August Weismann (1834–1914), who became one of the first biologists to unequivocally reject the inheritance of acquired characteristics and the first to suggest a coherent theory of molecular genetics.

As a boy, Weismann happily collected butterflies, beetles, and plants. Like Darwin, he studied medicine, but then returned to natural science, devoting himself to the embryology of insects. Ultimately, an eye disease forced him to give up these studies, which required long hours at the microscope, and he turned increasingly to theoretical work.

Weismann soon realized—as Darwin apparently had not—that the controversy about natural selection could not be resolved until inheritance was fully understood. Like other biologists of the time, Weismann at first accepted the inheritance of acquired characteristics. But in the 1870s he began testing the theory in a series of experiments. By the time he was 49 years old he had completely changed his mind, and, in 1883, he published a paper rejecting Lamarckism entirely.

It was a radical proposal, and biologists of the day nearly universally repudiated it. Only Wallace, Darwin's codiscoverer of evolution through natural selection, publicly supported

Weismann. Wallace and Weismann alone believed that natural selection—without the inheritance of acquired characteristics—was the major mechanism for evolution. Other biologists had no reason to abandon the inheritance of acquired characteristics. Lamarckism was not, in fact, inconsistent with natural selection; it had not been disproved; and it constituted a likely source of variation. On the other hand, as Weismann saw it, no one had any idea how the inheritance of acquired characteristics could work, and Lamarckism was not necessary for natural selection to work.

Furthermore, Weismann had a better idea—another mechanism for inheritance. In 1882, a German cell biologist had discovered the cell's nucleus and noticed the way the chromosomes divide at mitosis. Weismann theorized that the genetic material was all in the nucleus. His theory, he said, was "founded upon the idea that heredity is brought about by the transmission from one generation to another of a substance with a definite chemical, and above all, molecular constitution." He was, of course, exactly right. But no one knew that, and not even he could be sure.

Weismann went on to argue that the hereditary material, once it is formed, is essentially isolated from the rest of the body. Embryologists had already noted that in vertebrate animals, at least, the tissue that eventually forms the ova or testes—the **germ plasm**—separates from the rest of the body early in development. In humans, for example, the cells that will later form the ova or testes appear early in the fourth week of development (Figure 16-3). The ovaries and testes themselves become distinctly formed at about four and a half months.

Weismann concluded that the germ plasm is separate from the rest of the body from the very beginning, and virtually nothing in the development or environment of the body can influence the germ cells or the hereditary material. The genetic material, he said, is passed from generation to generation intact, unblended, and unchanged. This theory came to be known as the "continuity of the germ plasm." Nothing could influence the germ plasm, the location of the hereditary material. And only the germ plasm was passed on to the offspring. Therefore, the inheritance of acquired characteristics must be impossible.

Modern biology has mainly confirmed Weismann's conclusion. A parent who works outside all day and develops a deep tan cannot pass this trait to his or her children. Tanning does not influence the molecular structure of the DNA in the mother's ova, which formed months before she was born, or even in the DNA of the father's sperm. Genetic information flows almost entirely from the DNA to RNA to protein.

August Weismann proposed that the hereditary material was in the nucleus and that it was molecular. He argued that the hereditary material was not susceptible to influence by developments in the rest of the organism and that, therefore, the inheritance of acquired characteristics was impossible.

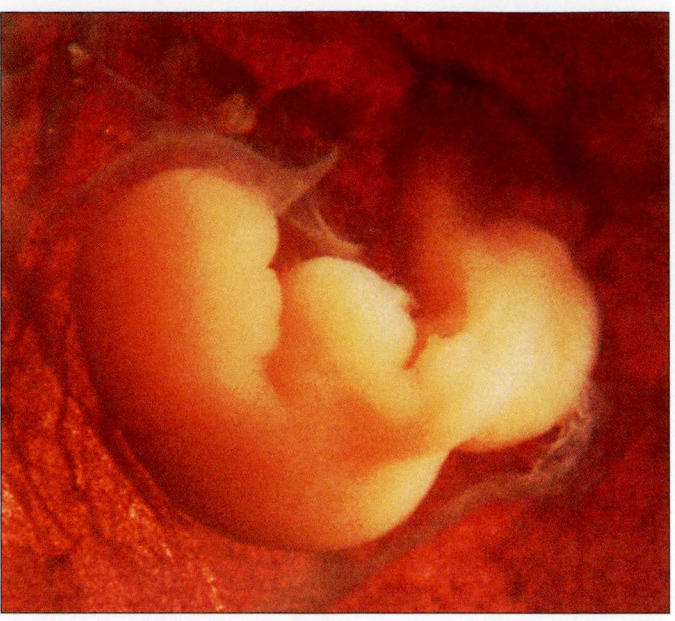

Figure 16-3   Four-week-old human embryo. At four weeks, the germ plasm, from which the ova or testes will form, is already separate from the other tissues of the developing body. *(Petit Format/Nestle/Science Source/Photo Researchers, Inc.)*

How Did Weismann Explain Variation in Individuals?

Weismann's theory of inheritance ruled out Lamarckism. However, he still needed to explain the origin of the variation on which natural selection acts. Remarkably, he had an answer that was partly correct—sexual reproduction.

By 1883, cell biologists had observed that maternal and paternal chromosomes do not fuse during fertilization, but merely restore the two sets of chromosomes normally found in most cells. From this observation, Weismann leapt to the conclusion that the effect of sexual reproduction was not to blend parental characters (as blending inheritance implied) but to recombine separate "hereditary tendencies."

Weismann rightly concluded that sexual reproduction, far from homogenizing parental traits, as Jenkin had argued, brings together old traits into new combinations. Even though Mendel's work was still buried in obscurity and the word "gene" had yet to be coined, Weismann had, to a remarkable degree, solved the problem of heredity.

Weismann's view remained a minority opinion throughout the last part of the 19th century and into the beginning of the 20th century. But he and Wallace kept alive the idea of natural selection without Lamarckism. Without these two scientists, the idea might have been forgotten. Not until the 1940s did biologists appreciate Weismann's enormous contribution to evolutionary theory.

In the meantime, one major piece of the heredity puzzle remained to be discovered. While sexual reproduction results in an almost unlimited supply of new variants, the mixing of parental genes into new combinations does not account for the origin of new traits. With the rediscovery of Mendelian genetics in 1900, another piece of the puzzle fell into place.

Weismann showed how sexual reproduction, by creating new combinations of traits, could act as a source of new variants for natural selection.

How Did the Rediscovery of Mendel's Work Transform Biology?

In 1900, three biologists working in three different countries independently rediscovered Mendel's laws. In the Netherlands, Hugo Marie de Vries (1848–1935) published a paper showing that if he crossed even numbers of hairy and smooth primroses he got 3 : 1 ratios of hairy and smooth plants. From this he concluded that the alternative traits—smooth and hairy—were controlled by a single pair of what we now call genes. Each parent, he explained, passes on only one of each of its pairs.

As soon as de Vries's paper came out, he immediately received copies of papers from two younger biologists—Karl Franz Joseph Correns (1864–1935), in Germany, and Erich Tschermak (1871–1962), in Austria—each claiming that they too had discovered the 3 : 1 ratios, each offering the same genetic explanation as de Vries, and each pointing out to him that Gregor Mendel had anticipated them all by 35 years.

Historians believe that both Correns and Tschermak probably read Mendel's paper before they had their respective "flashes of insight." De Vries himself received a copy of Mendel's paper from a fourth biologist sometime before he published his first paper. Whether there were actually any "independent discoveries" besides Mendel's we may never know. One thing is clear: Mendel's ideas made a splash.

In 1900, the British biologist William Bateson (1861–1926) was traveling to London to deliver a paper to the British Horticultural Society. As the train trundled through the English countryside, Bateson passed the time reading. When he got to de Vries's paper, he was thunderstruck. In London, he cast aside his prepared notes and proclaimed the death of Darwinism (by which he meant gradual evolution) and "the birth of a new science" (later called "genetics").

Bateson, de Vries, and others, who came to be known as the Mendelians, argued that evolution occurred not gradually, through natural selection, but in leaps and bounds, or saltations, by means of genetic mutations. These earliest geneticists firmly rejected the idea that evolution occurred through the slow selection of variants in a population. The Mendelians rejected natural selection and gradualism and any suggestion that evolution occurred among groups of individuals. They believed that evolution occurred suddenly, from one generation to the next, when genetic mutations transformed a single individual into a radically new type. They argued that genetic mutations in an individual were the cause of new species, not to men-

tion new genera, families, and other higher groupings of plants and animals.

The Mendelians contended that every evolutionary change, large or small, was the result of a separate individual mutation, and that natural selection, individual variation, and sexual reproduction played no important role in evolution. Normal continuous individual variation, such as adult height in humans—the very foundation for Darwin's theory of natural selection—was not even heritable, argued the Mendelians.

Naturalists of the era were horrified. They knew, from their close observations of wild populations of plants and animals, that most variation was continuous and that continuous variation was indeed heritable. But they, in turn, overreacted. The naturalists claimed that although the rules of Mendelian inheritance might apply to discontinuously inherited traits such as smooth or hairy plants or blue or brown eyes, such traits were unusual and irrelevant to evolution. Continuous variation, the stuff of evolution, they argued, was not governed by the rules of Mendelian inheritance. Rejecting all of genetics, they clung to a variety of other theories, including Lamarckism. The polarized debate between the Mendelians and the naturalists raged for nearly 40 years.

The rediscovery of Mendelian genetics precipitated an intense debate. Early geneticists believed that evolution occurred from one generation to the next as a result of radical mutations in individuals. Biologists who studied natural populations of organisms insisted that Mendelian genetics did not apply to the kind of variation that was the basis of evolution. Both sides were wrong.

How Did the Modern Synthesis Reconcile Genetics and Natural History?

In the late 1930s, a remarkable marriage of genetics and evolutionary theory occurred, which, for the first time, elucidated the genetic basis of variation and natural selection. This marriage, called the **modern synthesis,** did much to end the international feud between geneticists and naturalists. The modern synthesis fully embraced Weismann's theories and added to them all that was known about genetics. Much of what we will consider in the rest of this chapter comes from a field of genetics that contributed substantially to the modern synthesis—population genetics. Population genetics combined Mendelian genetics with the recognition that evolution can occur at the population level.

Population genetics explains the processes by which variation is generated and passed on within populations of organisms in precise mathematical terms. These processes are sometimes called **microevolution.** Population geneticists define microevolution narrowly as *changes in the frequencies of alleles of genes in a population.* For example, consider a wildflower population that includes some individuals with a gene for red flowers and other individuals with a gene for white flowers. If the relative proportions of red and white flowers change over the course of several generations, the population of flowers is said to evolve. Microevolution is distinct from **macroevolution**—the processes by which species and higher groupings (taxa) of organisms originate, change, and go extinct, a topic we will discuss in the next chapter.

Population genetics explains variation at the population level and provides a basis for natural selection and other evolutionary forces. In the next section, we will discuss what genetic variation is, where it comes from, and how it spreads and persists in populations of organisms.

WHAT IS THE SOURCE OF THE VARIATION THAT FUELS EVOLUTION?

Until the 1930s, few biologists believed that genes were real entities. No one had ever seen a gene, and there was no other evidence for their existence. Geneticists carefully stated that genes were purely theoretical constructs. Gradually, however, geneticists began to notice more and more cases where one genetic trait, such as eye color, was consistently linked to some other genetic trait, such as size. It began to look as if the genes themselves were physically linked together when they were passed from parent to offspring.

As we saw in Chapter 10, genes are very real, consisting of strings of nucleotides in the DNA. Each set of three nucleotides codes for one amino acid. Strings of amino acids, constructed on the ribosomes, fold into proteins.

Recall also that most sexually reproducing organisms have two of each chromosome. The two chromosomes in a pair are said to be **homologous.** One homologous chromosome possesses the same genes in the same positions as its homologue (unless they are unmatched sex chromosomes). Eggs and sperm have only one of each chromosome. When the egg and sperm fuse into a zygote, the two sets of chromosomes pair up. Thus each parent supplies one of the chromosomes in each pair of chromosomes.

Genetic variation originates when a gene mutates. The different forms of a single gene are called **alleles.** If each chromosome in a pair has the same allele at a particular gene position, the organism is said to be **homozygous** for that gene. If each chromosome has a different allele at a particular gene position, the organism is said to be **heterozygous** for that gene (Chapter 9).

What Is Genetic Variation?

Most genetic variability comes from gene mutations. Such mutations change only one or a few nucleotides in a gene and, thus, alter only a single protein. Chromosomal mutations, which change the number of chromosomes or the number or arrangement of genes on a chromosome, can affect a large number of proteins at once. Most chromosomal mutations are lethal or extremely damaging in animals, though less so in plants. But chromosomal rearrangements can, on occasion, benefit the organism. For example, the duplication of a chromosome segment, if harmless, can be passed on. In time, mutations in the new copy of the gene can allow it to take on new functions. As a result, the species has more genes than it did formerly.

In a population at large, a given gene may have just one form, or allele. In that case, every member of the population would have to be homozygous for that gene. On the other hand, a given gene may have two or even hundreds of alleles. If a gene had several alleles, members of the population could be either homozygous or heterozygous in one of several different ways. For example, members of a population possessing a gene with three alleles *A, a,* and *a'* could be homozygous in three ways—*AA, aa,* or *a' a'*—or heterozygous in three ways—*Aa, Aa',* or *aa'*. If a population has two or more alleles of a given gene, it is said to be **polymorphic** [Greek, *poly* = many + *morph* = form]. The word "polymorphic" also has a second, related meaning. If the members of a population come in two or more forms, the population is also said to be polymorphic. Humans, for example, are polymorphic for eye color (Figure 16-4).

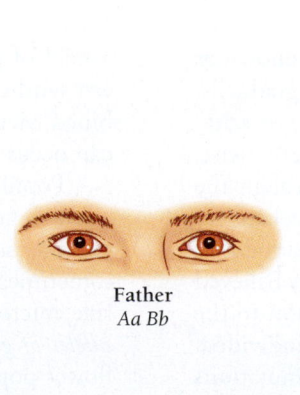
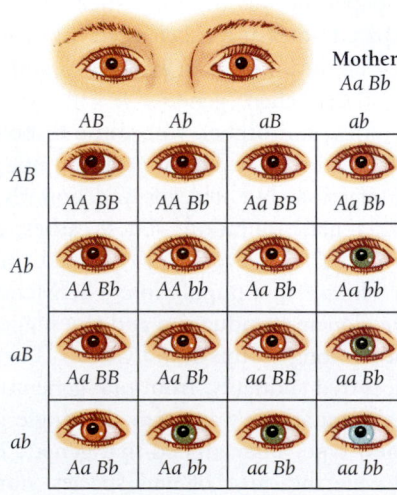
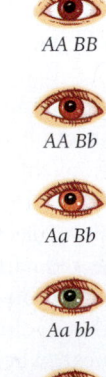
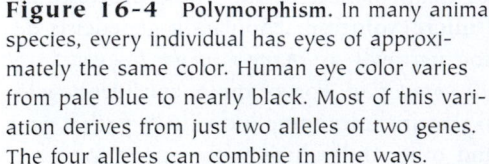

Figure 16-4   Polymorphism. In many animal species, every individual has eyes of approximately the same color. Human eye color varies from pale blue to nearly black. Most of this variation derives from just two alleles of two genes. The four alleles can combine in nine ways.

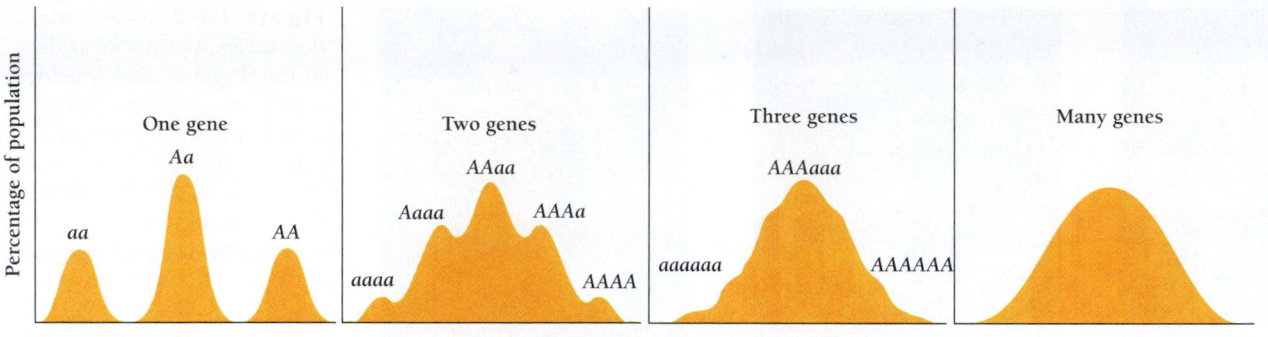

Figure 16-5 **Most traits are polygenic.** The more genes involved that influence a trait, the more continuously the trait will vary in a population. If only one gene with two alleles is involved, then no more than three basic types will appear. Two genes with two alleles each can result in five phenotypes (as in eye color). However, even those individuals with the same genotype differ from one another because of environmental influences. Most traits are influenced by many genes and display a smooth bell-shaped curve.

The original source of variation is gene mutations or chromosomal mutations. A population is polymorphic for a given gene if that gene has more than one allele.

Are Traits Determined by Single Genes?

The Human Genome Project's attempt to map the entire human genome has focused the public's attention on single genes that, when mutated, cause certain genetic diseases. A single-gene trait with just two alleles creates a pattern of variation like that in the left-hand panel of Figure 16-5. But most genes work in concert with other genes and under the influence of the environment to create complex phenotypes (Chapters 9 and 14). The same figure shows that the more genes are involved, the more a trait can vary.

In a research lab, most genetic research deals with alleles that specify discrete characters such as the presence or absence of a specific enzyme. In the natural world, however, most traits are **polygenic,** influenced by many genes. Everything about individual organisms—height and weight, bone structure, ability to withstand stress or infection, hormone levels, personality, and intelligence—is influenced by multitudes of genes. The constellation of traits we call personality, for example, is probably influenced by thousands of genes, as well as by environment. Height, weight, and skin color vary smoothly and continuously within a population, often forming a bell curve (Figure 16-6).

In Chapter 1, we discussed bell curves in more detail. But two aspects of such curves deserve special attention—the average value, or mean, and the breadth of the curve. The breadth of the curve is a measure of the variability of the trait. We all

Figure 16-6 **Height is a polygenic trait.** Because adult height in humans is controlled by many genes, the distribution of heights in adults is a smooth curve, as illustrated by this photo of students. The students are arranged according to height, with the shortest on the left and the tallest on the right. Most students are of average height, with very few at the extremes. (*Photo, courtesy of Joiner Associates, Inc.*)

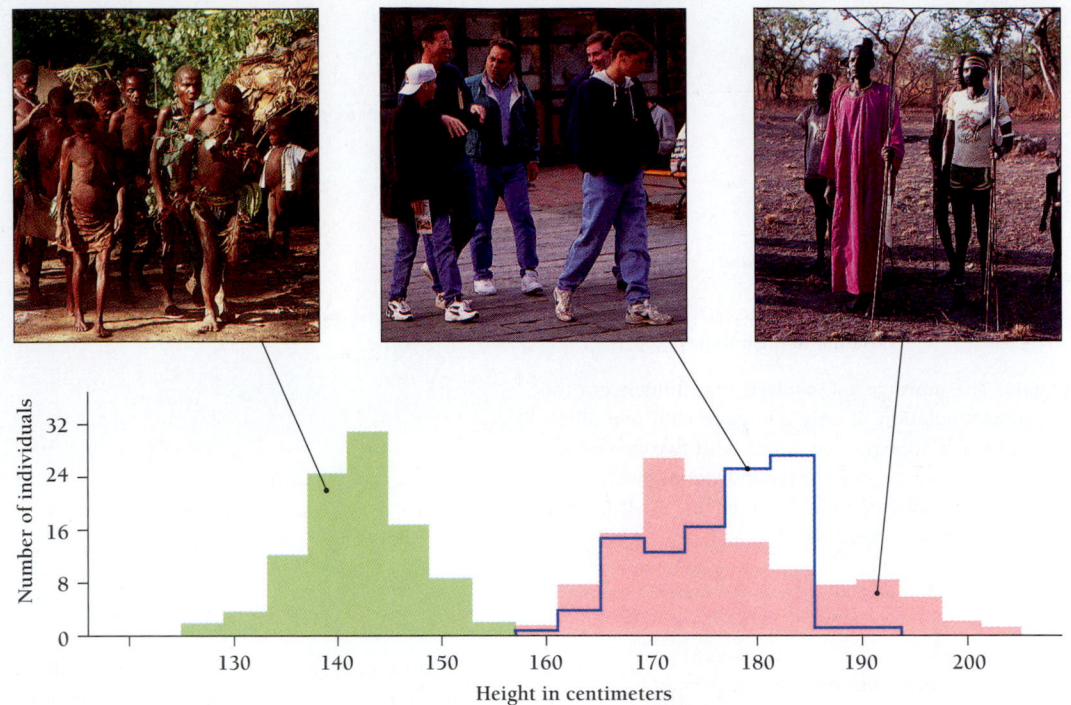

Figure 16-7 **Even variation varies.** Heights for male African Pygmies, male American Bostonians, and male African Dinkas. Nearly all Pygmies are shorter than Bostonians and Dinkas. Dinkas are, on average, taller than Bostonians, but they have a great range of heights, so that many Dinkas are actually shorter than many Bostonians. Which group varies most? (Pygmies, Klaus Payson/Peter Arnold, Inc.; Bostonians, Frank Siteman/Stock, Boston; Dinkas, F. Jackson/Bruce Coleman, Inc.)

know that the average value of a trait may vary from population to population. African Pygmies are, on average, shorter than people living in Boston (Figure 16-7). But populations may also vary in the variability of a trait. The population of African Dinkas pictured in Figure 16-7, for example, shows a greater range of heights than that of either the Pygmies or the Bostonians. An adult Dinka can be taller than almost any Bostonian, or as short as the tallest Pygmies. Genetic variation, the fuel of evolution, varies from population to population. We should not forget, however, that we cannot tell from the graph if the Dinkas' great variability is all due to genetic variation or also to environmental effects such as diet.

Most traits are influenced by many genes.

What Determines the Genetic Variability of a Population?

Populations that contain many alleles vary more than populations that have few alleles per gene. Highly variable populations can evolve more rapidly than more uniform populations because natural selection has more to select from. Three factors determine genetic variability:

1. How fast mutations accumulate in the DNA
2. How fast such mutations spread through a population
3. How fast selection eliminates mutations from a population, which we discuss later in this chapter

How often do mutations occur?

Mutations cause random changes in protein structure and, thus, most mutations damage the organism. Occasionally, however, random changes may cause no harm or may even give an advantage to the organism that bears them. Such rare advantageous mutations are the fuel of evolution. Without a constant supply, evolution would stop. Instead, mistakes and mutations persist and occasionally provide every species with adaptive variants. For example, on average, each new human baby carries one or two new mutations.

Mutations are not the result of directed molecular surgery but arise constantly as the result of random processes. Because they occur constantly and randomly in all genes, geneticists have been able to estimate the probability that a given nucleotide will mutate in each generation. The allele for eyelessness in the fruit fly *Drosophila melanogaster* occurs in 6 of every 100,000 fly gametes. Because mutation occurs constantly, and because the total number of genes is great (about 100,000 for humans, for example), the total number of variations within a species can be enormous.

How do mutations spread through a population?

The rate at which a population evolves depends on how rapidly new mutations spread through the population. In fact, although evolution could not long continue without new mutations, the spread of those mutations, rather than their original occurrence, is the most important factor in increasing variability in a population. The main way for mutations to spread is through sexual reproduction. Recombination, which occurs during sexual reproduction, mixes mutations that have arisen independently into endless new combinations. Differential reproduction determines which mutated alleles increase in frequency and which decrease.

Genetic variation originates when genes or chromosomes mutate. Genetic variation spreads when sexual reproduction and recombination mix these changes into endless new combinations.

How Much Do Populations Vary Genetically?

Since variability is the raw material for selection, one of the first questions population geneticists ask is How much variation is there in single genes within a population?

To examine the extent of variation of single genes, we can look at the polypeptides and proteins that are the products of genes or at the genes (DNA) themselves. One way of estimating the level of variation within proteins is electrophoresis, a technique for separating molecules according to their charge, size, and shape (Chapter 10). Using electrophoresis, researchers estimate the degree of genetic polymorphism in humans and other species.

Electrophoresis shows that in populations of the fruit fly *Drosophila,* for example, about 53 percent of genes that code for proteins are polymorphic. In a survey of invertebrate animals, nearly half of all proteins examined had more than one form in the population at large.

Another way of expressing the level of genetic variation in a population is to estimate the likelihood that any individual will be heterozygous for a given gene. Any two fruit flies, for example, are likely to differ at about 25 percent of their gene loci. The cheetah, of Africa, has some of the lowest variability found in any animal, with an average heterozygosity of only 0.07 percent. Cheetahs have most likely descended from a small number of individuals, called a "bottleneck." For comparison, invertebrates average 13 percent, and plants are even more diverse (17 percent).

Humans fall somewhere in the middle. In us, about 7 percent of genes that code for proteins are heterozygous. Interestingly, humans appear to possess much less genetic diversity than our closest relatives, the chimpanzees. A 1999 study of a noncoding region of the X chromosome revealed that common chimpanzees possess four times as much variability in this region as do humans. The researchers suggested that humans, like cheetahs, may have passed through a bottleneck at some time in their evolutionary history.

Most natural populations vary enormously.

Is All Genetic Variation Subject to Natural Selection?

The only variants important for natural selection are those that contribute to changes in the whole individual in ways that affect the individual's ability to survive and reproduce. The expressed traits of a whole individual organism are called the **phenotype** [Greek, *phainein* = show]. The phenotype, which includes the appearance, the physiology, and the biochemical makeup, is a result of both the genetic makeup of the individual—its **genotype**—and the effects of the environment on the development of the individual.

Selection acts on the phenotype. For example, an individual homozygous for some lethal, recessive trait, such as cystic fibrosis, expresses the trait and dies. The gene has expressed itself in the phenotype, and selection has occurred. Another individual, heterozygous for the trait, will have only one copy of the deadly allele and will have a phenotype that is perfectly healthy. In the heterozygote, the allele is not expressed in the phenotype, and so selection against the allele cannot operate even though the allele is present in the genotype. Most lethal mutations are recessive and are usually not subject to selection unless they are homozygous.

With the advent of high-speed DNA sequencing machines, evolutionary biologists have begun to accumulate more information on the extent of variation in DNA. On average, about 1 nucleotide in every 500 is different in the two homologous chromosomes. Most of these differences, however, lie in nucleotides that do not code for proteins. They contribute to high variability at the DNA level but do not alter the structure of a protein or change the way the organism functions. Such mutations do not affect phenotype and are therefore not subject to natural selection.

Mutations that do not affect the phenotype are not subject to selection.

Why Is Genetic Variation Essential to Evolution?

If mutations are the fuel of evolution and recombination is its motor, then the driver—the guiding force—of evolution is natural selection. Because some variants survive and reproduce more successfully than others, the mutations that best equip their possessors to survive and reproduce will increase in frequency in each generation.

When biologists study genetic variation within a species, they almost always find lots. In some exceptional cases, however, genetic variation is limited. For example, most thoroughbred horses are direct descendants of a few dozen horses that lived in the late 18th century. Although breeders have allowed some outbreeding—breeding with nonthoroughbreds—the relative lack of genetic variability has led to stagnation. Despite the best efforts of highly skilled breeders, the winning times for the Kentucky Derby have not changed significantly in 50 years. The lack of change in winning Derby times could be considered an example of microevolution that has come to a standstill. Later in this chapter we will discuss some examples of natural populations that lack genetic diversity.

Evolution by natural selection depends on the continued existence of variation. Yet selection tends to reduce variation by eliminating organisms less able to survive and reproduce. We might expect, then, that species would evolve so that all individuals will have the same "best" genotype. If this were to happen, evolution of that species could not continue. (However, environments change constantly, and so do selection pressures. What is "selected" changes over time.)

In the absence of genetic variation,
evolution would cease.

In the rest of this chapter, we will see how populations maintain variability, even as their gene distributions change during evolution. Such variability allows populations to adapt to continuing changes in their environment.

How Do Populations Maintain Genetic Equilibrium?

In 1908, G. Hardy, an English mathematician, and G. Weinberg, a German physician, independently reconsidered Fleeming Jenkin's idea that sexual reproduction eliminates variation in light of the fresh knowledge that heredity is nonblending. Recall that Mendel's Principle of Segregation says that alleles do not blend with other alleles; they stay intact. Hardy and Weinberg each reached the same conclusion, now called the **Hardy-Weinberg Principle:** *Sexual reproduction by itself does not change the frequencies of alleles within a population.* Mendelian genetics, argued Hardy and Weinberg, overturned Jenkin's case against natural selection.

To see what this means, we will need to define some terms. We speak of all the genes of all the individuals in a population as the **gene pool.** And we calculate the *frequency* of a given allele within the gene pool by dividing the number of times the allele is present in a population by the number of times it could be present if every organism in the population had the allele on both of the paired homologous chromosomes. In that case, the allele would be present on every single chromosome on which the gene appears. *The allele frequency is the number of times the allele is present in the population divided by the total number of chromosomes on which the gene appears.*

To understand the Hardy-Weinberg Principle, let us consider a gene that has two alleles, *A* and *a*. For example, the color of mussels on the north coast of California depends on a single gene with two alleles. One allele, *A*, gives the mussels a blue color and is *dominant*. The other allele, *a*, makes the mussels brown and is *recessive*. That is, whether the mussel is *AA* or *Aa*, it will be blue. Only when the mussel is *aa* will it be brown.

The fraction of sperm and ova in the population that carries the *A* allele we call "*p*," and the fraction that carries the *a* allele we call "*q*." The fraction of the blue allele (*A*) in a particular mussel population is $p = 0.6$, while the fraction of the brown allele (*a*) is $q = 0.4$. Since there are only two alleles for this gene, $p + q = 1$. Geneticists call *p* and *q* allele "frequencies." Thus the *A* allele has a frequency of 0.6 and the *a* allele has a frequency of 0.4.

But the frequencies of the alleles are not the same as the frequencies of the genotypes. Suppose the mussels release their eggs and sperm into the seawater. We know that 60 percent will carry the *A* allele and 40 percent will carry the *a* allele. Now the sperm and eggs begin joining up at random and fusing into zygotes with genotypes *AA*, *Aa*, and *aa*. What are the frequencies of those genotypes? We can calculate the frequencies of the three genotypes among the offspring, or F1 generation, using either a Punnett square, as described in Chapter 9, or simple math.

If the frequency with which a sperm or egg carries the *A* allele is 0.6, then the frequency of an *AA* homozygote will be the probability that an *A* sperm will fertilize an *A* egg. That probability is $p \times p = p^2 = 0.36$. Similarly, the frequency of an *aa* homozygote is $q^2 = 0.16$. The frequency of a heterozygote is the probability that an *A* sperm will fertilize an *a* egg plus the probability that an *a* sperm will fertilize an *A* egg, that is:

$$(pq) + (qp) = 2pq = 2(0.4)(0.6) = 0.48$$

The frequencies of *AA*, *aa*, and *Aa* are 0.36, 0.16, and 0.48, respectively. Notice that the sum of the frequencies is $p^2 + 2pq + q^2 = (p + q)^2 = 1^2 = 1$. Likewise, $0.36 + 0.48 + 0.16 = 1$. In other words, we have accounted for 100 percent of the population.

From the genotype frequencies we can derive the allele frequencies in the F2 generation and see if they are the same as in the previous generation, in short, $p = 0.6$ and $q = 0.4$.

First, let's assume that we have 100 individuals, each with 2 alleles, for a total of 200 alleles in the gene pool. What fraction of these 200 are the *A* allele? Each of the *AA* homozygotes carries two copies of *A*, so they contribute $2p^2$ of the *A* alleles. That's $2(0.36) = 0.72$. So, among the 100 mussel offspring are 72 *A* alleles from homozygotes. In addition, each of the *Aa* heterozygotes also carries a copy of an *A* allele, so they contribute $2pq$, or 0.48, *A* alleles. That's 48 more *A* alleles. In Table 16-1, notice that we can derive the frequency for the *a* allele in much the same way. We see that despite sexual reproduction, the allele frequencies have not changed. In successive generations, the frequencies for the genotypes *AA*, *Aa*, and *aa*—$p^2 + 2pq + q^2$—will also remain the same.

TABLE 16-1   THE ESSENCE OF HARDY–WEINBERG

The frequency of the A allele remains p and the frequency of the a allele remains q. Generally, then, sexual reproduction does not change allele frequencies. The total number of A alleles is $2p^2 + 2pq$. Because there are 2 chromosomes, the *frequency* of the A alleles is this number divided by two, or $(2p^2 + 2pq)/2 = p^2 + pq = p(p + q)$. Remember, however, that $p + q = 1$, so $p(p + q) = p$. So if a population starts with gene frequencies of $p = 0.6$ and $q = 0.4$, the next generation will have the same frequencies.

Genotype	AA	Aa	aa
Frequency	p^2	$2pq$	q^2
	0.36	0.48	0.16
Number of individuals with that genotype in the population	36	48	16
Total alleles	72A	48A 48a	32a

A alleles = 72 + 48 = 120
a alleles = 32 + 48 = 80
Number of individuals is 100.
Total alleles is 200.
So A alleles have a frequency of 120/200 = 0.6
And a alleles have a frequency of 80/200 = 0.4

So far, we have considered only two alleles for each gene, but we know that mutations are possible at every nucleotide of DNA. The number of possible alleles for each gene is large indeed. We could easily accommodate more alleles in our discussion. We might, for example, designate the frequencies of three alleles (A, a, and a') as p, q, and r, so that $p + q + r = 1$. We would find that the Hardy-Weinberg Principle remains the same. The frequencies of the individual alleles in a population do not change as a result of sexual reproduction.

The frequencies of each allele therefore stay constant in successive generations, and sexual reproduction does not lead to the dilution of variation, as Fleeming Jenkin suggested. Equally important, genotype frequencies stay at the same **Hardy-Weinberg equilibrium,** a stable distribution of genotype frequencies maintained by a population from generation to generation.

Such an equilibrium occurs, however, only when the population meets all of the following five conditions:

1. Random mating
2. Large population size
3. No mutations
4. No breeding with other populations
5. No selection

In reality, these conditions are rarely met. Consequently, gene frequencies change and evolution occurs. If a population is not at equilibrium—that is, if the genotype frequencies are changing from generation to generation—then we know that one or more of the five conditions are not being met.

The Hardy-Weinberg Principle shows that sexual reproduction does not lead to the dilution of variation but maintains a stable distribution of genotype frequencies from generation to generation, called the Hardy-Weinberg equilibrium. The Hardy-Weinberg equilibrium is a baseline against which the evolution of populations can be measured and is the foundation for the genetic theory of evolution.

WHAT CAUSES ALLELE FREQUENCIES TO CHANGE?

Of all the causes of changing allele and genotype frequencies—nonrandom mating, small population size, mutations, breeding between populations (gene flow), and natural selection—only natural selection consistently leads to *adaptive* changes in allele frequencies (Figure 16-8). The others tend to cause changes in allele frequencies that are not directly related to the ecology of the population and may or may not be adaptive. These effects are said to be "nonadaptive."

How Can New Mutations Cause Changes in Allele Frequencies?

Slight changes in allele frequencies also occur as the result of mutation. As we saw in the last section, mutation can generate every possible change in the amino acid sequence of each

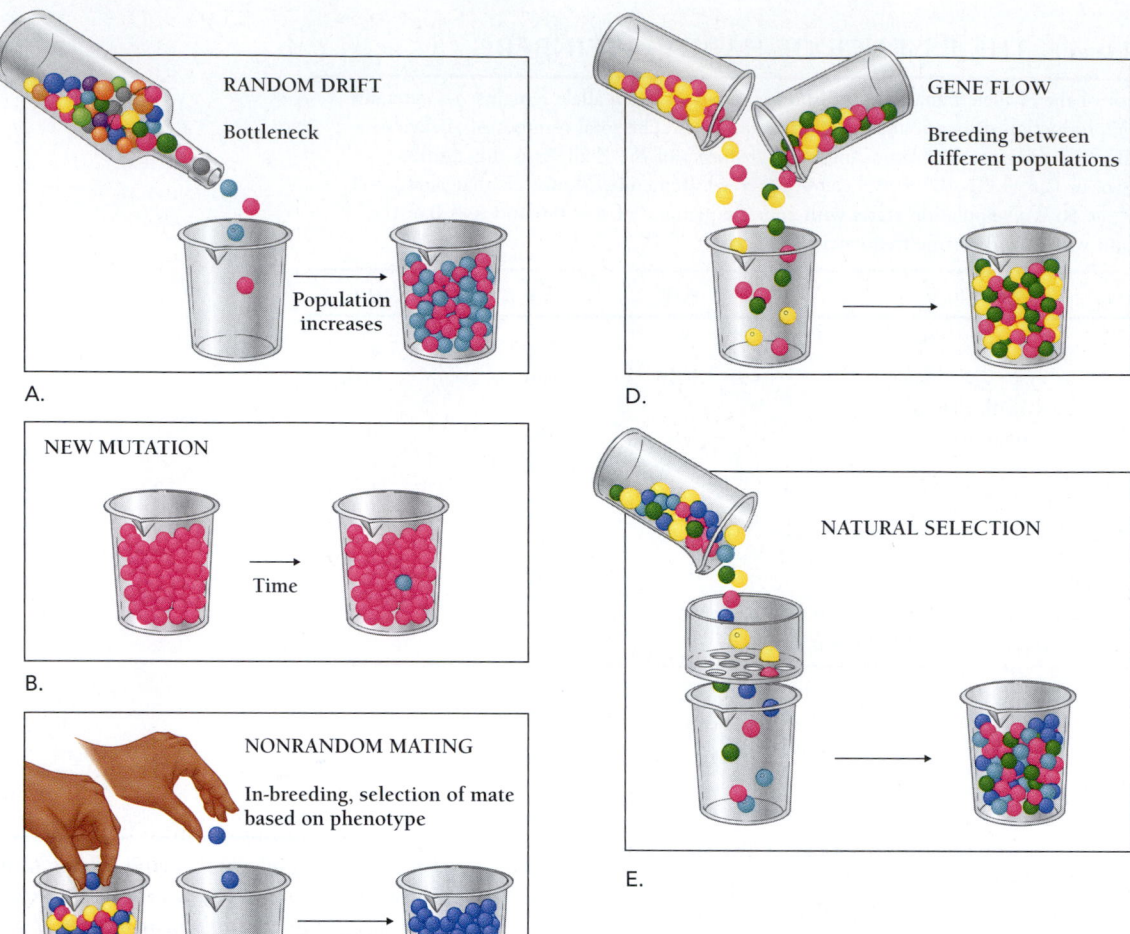

Figure 16-8   **What causes changes in gene frequency?** A. Random drift. B. New mutation. C. Nonrandom mating. D. Gene flow. E. Natural selection.

protein during the lifetime of a species. Indeed, mutations provide the raw material of evolution. How do such mutations affect the genetic equilibrium of a population?

The answer is, very little. Random drift—given enough time—can lead to either the disappearance of a new mutation or its spread through the population. However, the number of new mutations in each generation does not much change allele frequencies. Although the rate of mutation is high enough that each new human baby carries one or two new mutations, the change in the frequency of any single allele is low.

---

*Mutation barely alters the overall frequency of
any given allele.*

How Do Chance Events Change Gene Frequencies?

Flip a coin 1000 times and your chance of getting all heads is virtually zero. You can reasonably expect that the number of heads will be close to 500, the "expected" number. However, if you flip the coin only four times, your chance of getting all heads is rather high (1 in 16), and you're not likely to get heads exactly one-half of the time. In fact, your chance of getting two heads is less than 40 percent. Such chance fluctuations, which affect small samples much more than large ones, are known as "sampling errors." In small populations of organisms, sampling errors can cause allele frequencies to change randomly from generation to generation.

Figure 16-9   A classic bottleneck. In the 19th century, heavy hunting reduced the number of elephant seals to as few as 20 individuals. Legally protected since 1884, the huge seals are now flourishing, but they have become highly inbred. (*Frank S. Balthis Photography*)

For example, in a small population of mussels, the actual proportion of eggs and sperm with the *A* allele may differ widely from 0.6. The smaller the population, the greater the likelihood that the frequency of various alleles will deviate from what they are "supposed" to be. The effects of chance in small populations can lead to **random drift**—changes in gene frequency not due to selection, mutation, or immigration, but due to random events. Such random events can lead to changes in gene frequency within a small population.

The founder effect

In small populations, random drift can have pronounced effects. In one kind of random drift, called the **founder effect,** a whole population descends from a small number of individuals. For example, in the early 19th century, elephant seals flourished along the California coast from Point Reyes, just north of San Francisco Bay, south to Baja California. By 1890, hunters had virtually eliminated the entire population (Figure 16-9). Best estimates are that the population was reduced to between 20 and 100 individuals. Under federal protection, the elephant seal population now numbers some 100,000 animals. All of these animals have descended from the same few individuals. When, as in this case, a large population is reduced to a few individuals by some disaster, population biologists say the population has passed through a **bottleneck.**

The founder effect produces a population whose gene pool is a tiny sample of the original. In this new population, the frequency of an allele is likely to be quite different from that in the original population. Many alleles will not be represented at all. Since the allele frequencies have changed, the population can be said to have evolved in the absence of natural selection.

Founder effects in the human population—together with the inbreeding and increased homozygosity that occur in small populations—are thought to be responsible for the high incidence of genetic diseases in certain human populations. For example, the founder effect is responsible for the high frequency of otherwise rare alleles among Afrikaaners of South Africa (descended from about 30 Dutch families in the 17th century), the Old Order Amish in the United States (descended from only a few individuals in the 18th century), and Swedes in Lapland (descended from three families in the 18th century).

How can random drift spread new mutations?

Random drift can play an important role in the spread of new mutations. Mathematicians have shown that, even in the absence of selection, a new mutation will either disappear completely or spread through the population. The smaller the population, the less time it takes for either to happen. If a new mutation enters the gene pool, it will at first exist in only a few individuals and will have a significant chance of being eliminated. For example, imagine a population of 25 elephants, 3 of which have a new allele for hairy ears. A volcano that erupts every 10,000 years happens to eliminate the three hairy-eared elephants. Their elimination is not selection, since smooth ears do not protect elephants from volcanoes. It's just chance. Yet, the result is the same as if natural selection had eliminated the hairy-eared elephants—a population of smooth-eared elephants.

Does random drift affect large populations?

The role of random drift in speciation in large populations is not so clear cut. However, even in large populations, random drift will *tend* to cause alleles to disappear over long periods of time. In the absence of natural selection, new mutations, recombination, and migration, random drift would cause the frequencies of all alleles in even large populations to go to either 0 percent or 100 percent. It would happen slowly, but it would happen eventually, leading, ultimately, to a population that was completely homozygous. In such a population, evolution would cease. We can see that new mutations "restock" genetic variation, and the constant mixing of such mutations, within populations and between populations, is key to evolution.

In small, isolated populations, random drift can cause significant changes in gene frequencies. In large populations, drift operates so slowly that its importance may be small compared to the effects of mutation, gene flow, and selection.

How Does Nonrandom Mating Cause Changes in Allele Frequencies?

In deriving the Hardy–Weinberg Principle, we assumed that individuals of one genotype do not mate preferentially, either with their own genotype or with others. Our calculation would be invalid, for example, if blue mussels mated only with other blue mussels and not with brown mussels. In fact, mussels mate by releasing into seawater gametes that then come together essentially at random. The same is true of most plants and many other animals.

But mammals and birds usually choose their mates very carefully. Such nonrandom mating, called **assortative mating,** occurs when individuals choose mates on the basis of the mate's genotype (as reflected in the phenotype). Such choices may be made independent of the individual's own genotype. For example, regardless of their own appearance, humans generally show a preference for mates with average facial structure. Other mammals and birds also prefer mates with a high degree of symmetry.

Individuals may also choose mates on the basis of similarities (or dissimilarities) to the individual's own genotype. Humans show positive assortative mating with respect to skin color and height. That is, on average, they mate preferentially with those of similar height and skin color. Positive assortative mating tends to increase the proportion of homozygotes.

An extreme form of positive assortative mating is **inbreeding,** mating among close relatives, which greatly increases the proportion of homozygotes. For example, Ellis–van Creveld syndrome is a genetic defect frequent among the Amish because of 200 years of inbreeding. Recessive genes may have

negative effects that are expressed only in homozygotes. In sexual species, at least, inbred individuals who are homozygous for many genes are, on average, less healthy than individuals with greater heterozygosity.

Neither assortative mating nor inbreeding changes overall allele frequencies. However, any increase in the number of homozygotes facilitates change by natural selection by exposing alleles to selection.

Most species are **outbreeders,** with complex physical or behavioral adaptations that promote crossbreeding with usually not-too-related individuals. That is, individuals choose mates somewhat unlike themselves, a process called negative assortative mating.

How Does Breeding Between Populations Change Allele Frequencies?

Hardy-Weinberg equilibrium requires that a population not interbreed with other populations. But few populations are completely isolated from all others of the same species. Animals immigrate to nearby populations, taking new alleles with them. Plant pollen, carried into a population by wind or insects, introduces new alleles. In either case, the resulting **gene flow** causes changes in allele frequencies. In our own species, we have many obvious examples of gene flow as people move to new countries. Despite a general tendency to find a mate within one's own group, gene flow among different groups is common (Figure 16-10).

Gene flow has two major effects on allele frequencies. Within a given local population, gene flow from outside populations increases genetic variation. The same gene flow makes adjacent populations more like one another.

Gene flow increases variation within a local population but makes adjacent populations more alike.

How Does Selection Decrease the Frequency of Certain Traits?

A group inept
Might better opt
To be adept
And adopt
Ways more apt
To wit, adapt.
—John M. Burns

A large, isolated population whose members mated randomly and whose mutation rate was low would maintain constant allele frequencies—except for one thing: **natural selection.** Because natural selection reduces the frequency of some alleles and increases the frequency of others, it is a powerful way to change gene frequencies (Figure 16-11).

Figure 16-10 **Gene flow.** Humans travel singly or in groups, spreading their customs and genes. In November 1996, hundreds of thousands of Rwandan refugees began returning home after living in Zaire for months. *(Christopher Morris/Black Star)*

A simple example occurs when an individual carries two copies of a lethal allele, such as that for Tay-Sachs disease. Children with Tay-Sachs die before they are 4 or 5 years old. Having Tay-Sachs therefore eliminates the possibility of reproducing. Although the allele can survive in heterozygotes, homozygotes *never* pass the allele to offspring. As a result, the allele survives only at low frequencies.

Lethal recessive alleles such as Tay-Sachs do not disappear entirely because these alleles "hide" from natural selection in the form of heterozygotes, who are perfectly healthy. (In contrast, natural selection eliminates dominant alleles that kill before the age of reproduction.)

How long does it take to eliminate one allele? To estimate the rate at which selection eliminates recessive lethals, imagine that all *aa* homozygotes die before they reproduce (as in Tay-Sachs). Recall that, according to the Hardy-Weinberg Principle, the frequencies of the *AA, Aa,* and *aa* genotypes are p^2, $2pq$, and q^2, respectively. We can say, then, that the fre-

quency of the *a* allele in the population will decrease by q^2 in each generation. When the lethal *a* allele is present at relatively high frequencies, many individuals are eliminated in each generation. When the allele is present at low frequencies, however, selection eliminates very few individuals. For example, if the frequency (q) of *a* is 0.5, then 25 percent of the population will die young each generation. But if $q = 0.01$, then only 1 infant in 10,000 will die from having the *aa* genotype, and only $2(0.01)(0.99) = 2$ percent of the population will carry the gene. Thus, while selection can dramatically change allele frequencies, even the most lethal recessive alleles can still persist indefinitely as heterozygotes.

Nevertheless, even a small difference in the probability of reproduction can lead, over time, to the effective elimination of a deleterious allele. Suppose, for example, that individuals with the homozygous *aa* genotype are slightly less well adapted than individuals with either *AA* or *Aa* genotypes. Say that each *aa* individual leaves only 999 progeny for every 1000 of the *AA* or *Aa* individuals. Even if we start with a gene frequency for the *A* allele of only 0.00001, we can predict a steady increase in the frequency of the *A* allele, until after 23,400 generations its frequency becomes 0.99. For the fruit fly *Drosophila,* this could occur in less than 500 years. For species with generation times as long as those of humans, it would

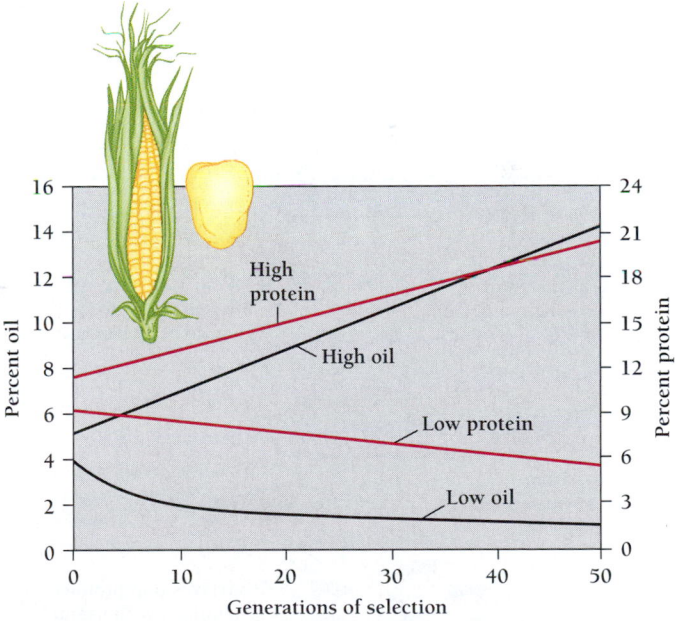

Figure 16-11 **Selection is a powerful way to change gene frequencies.** In four long-running experiments, researchers select for corn with more protein or less protein and for corn with more oil or less oil. After 50 generations, researchers have continued to produce plants with greater and greater amounts of protein and oil. However, in the experiments selecting for less protein or less oil, the amounts of oil and protein reach a minimum and microevolution stops. Apparently, plants need a minimum amount of oil and protein to survive.

take about half a million years—certainly a long time, but only a tick in the history of life. Even subtle selection pressures thus have large effects on allele frequencies.

Natural selection cannot, by itself, eliminate most alleles, but natural selection is, nevertheless, a powerful way to change allele frequencies.

How Does Selection Increase the Frequency of Certain Traits?

Just as selection can eliminate or reduce the incidence of one trait, it can increase others. One example is the development of resistance to toxins by natural populations of "pests." Most people are now aware that populations of bacteria, parasites, and pests have developed "resistance" as a result of humans waging chemical warfare against them.

For example, in many parts of the United States, household cockroaches have even evolved an aversion to the glucose (sugar) used to bait roach traps (Figure 16-12). And, after

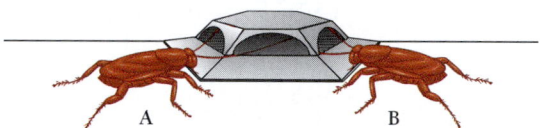

Two apparently identical cockroaches touch/taste glucose-baited poison with their antennae.

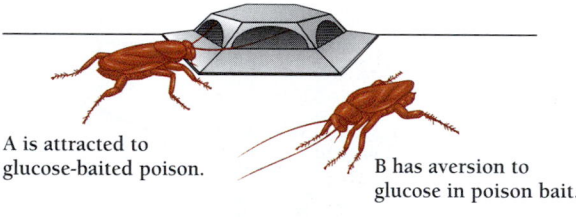

A is attracted to glucose-baited poison.

B has aversion to glucose in poison bait.

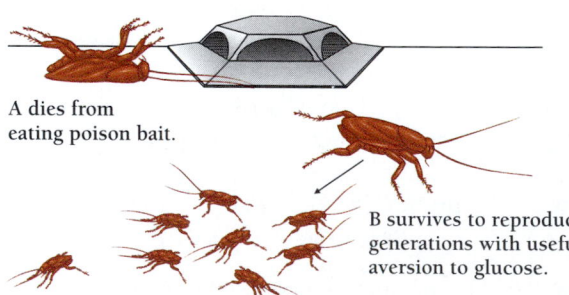

A dies from eating poison bait.

B survives to reproduce generations with useful aversion to glucose.

Figure 16-12   Natural selection in action. Because cockroach traps and poisons are commonly baited with the sugar glucose, intense selection pressure favors cockroaches that dislike the flavor of glucose. Researchers have discovered populations of German cockroaches from Florida to California that no longer take such bait because selection has allowed only the glucose-haters to survive.

World War II, insects developed resistance to the pesticide DDT. In the 1940s, countries around the world began spraying massive amounts of DDT to kill disease-carrying insects such as flies and mosquitoes. At first, the heavy doses of DDT killed all or most insects. But each time the DDT was applied, a few individual flies would survive. These survivors happened to carry alleles that made them resistant to DDT, and they founded a new population of resistant flies. In time, DDT would no longer kill certain insects in some areas. Because DDT is toxic to many forms of life, its use is now banned in the United States, although not elsewhere.

Of more pressing concern is the evolution of pathogenic bacteria, more and more of which carry genes for resistance to antibiotics. It is now all too common to find patients in big-city hospitals infected with bacteria that have genes for resistance to all known antibiotics. In this case, the antibiotics themselves select for resistant bacteria, killing off all that are not resistant. We discuss this in more detail in Chapter 20.

A classic example of selection in nature concerns the peppered moth *(Biston betularia)*. Because of a long British tradition of collecting moths and butterflies, biologists in Great Britain have been able to trace a change in the moth population over more than 100 years. In the early 19th century, all the peppered moths collected in Britain were pale and mottled. Collectors netted the pale moths as they fluttered through the air. Occasionally, a collector found a dark variant of the peppered moth. These were rare and therefore highly prized (Figure 16-13).

The industrial revolution dramatically changed this scene, however. In the 19th century, thousands of coal furnaces filled the air with dark soot, which blackened the landscape. The

Figure 16-13   Directional selection. The English peppered moth *Biston betularia* comes in two forms, light and dark. The light ones, whose colors blend well with the pale lichens found on trees, formerly predominated throughout England. The dark form predominates in areas where industrial pollution has blackened the landscape. *(Michael Tweedie/Photo Researchers, Inc.)*

BOX 16.1

Does Selection Act on Preexisting Variants?

Many 19th-century biologists believed, as Lamarck did, that variation arose in response to the needs of the organism. Thus even after mutations and genetics were understood, an important question arose: Do mutations arise spontaneously without reference to the needs of the organism or selective pressure, or do they arise in response to selective pressure? In short, are the variants that selection acts on preexisting ones, or not?

In the early 1950s, the American geneticists Joshua and Esther Lederberg decided to test the idea that variants preexist. They used bacteria, because they grow so fast, and an elegant technique they had devised, called replica plating. Replica plating allowed them to study the descendants of thousands of individual cells in different environments.

After growing bacteria in a nutritious broth, the Lederbergs spread the bacteria onto a petri dish containing the Jell-O–like medium agar. Each single bacterium on the agar divided to produce a visible colony composed of thousands of cells. The Lederbergs then touched the surface of the agar to a piece of velvet. The velvet picked up a sample of each colony. By touching the velvet to another agar plate, they made a replica of the original plate. Every colony of the replica had the same position as the "parent" colony on the original plate. The corresponding colonies all derived from the same cell in the liquid broth and so had the same genotypes.

The Lederbergs used their replica-plating technique to study antibiotic-resistant variants of the human gut bacterium *E. coli*. In plates containing the antibiotic streptomycin, most of the bacteria died, but a few survived and grew into colonies. All of the cells in each of the colonies themselves produced streptomycin-resistant progeny, showing that the resistance to the drug is a genetic property.

But were individuals with genetic resistance present in the original population, or did they arise only as a result of exposure to the drug? Replica plating provided an easy way to answer this question. The Lederbergs simply examined the replicate colonies that had not yet been exposed to streptomycin. Because they knew the identity of each colony from its position on the plate, they could pick out the ones whose brethren, on other plates, had been able to flourish on streptomycin-spiked agar. Here were colonies of the same genotype that had not yet been exposed to streptomycin. Were they resistant to the drug, too? The answer was yes. The Lederbergs proved that the variation preexisted in the original bacterial population, even though the bacteria had no need for such resistance. This result supports the idea that adaptation to changed environment can result from selection among preexisting variants.

---

light peppered moth became rare, and collectors found more and more of the dark variants. While lighter moths continued to prevail in unpolluted rural areas, the frequency of dark moths around factory cities increased to about 98 percent.

Looking at the historical data (from old collections of moths and butterflies), the British evolutionary biologist J.B.S. Haldane calculated that, in industrial areas, the darker variant was twice as likely to survive and reproduce as its lighter counterpart. A reasonable explanation for this was that, against the now dark background of trees and rocks, the dark variant had a selective advantage because it was less likely to be seen and eaten by birds. A fine theory, but the naturalists of the time said that they had never seen a bird eating a peppered moth. Indeed, no one had actually seen a peppered moth resting on a tree or rock. The moths were always caught while flying. The question remained: Was the change in frequency of dark moths the result of selection by birds?

To answer this question, H.B.D. Kettlewell, a biologist at Oxford University, undertook experiments that now provide among the best evidence concerning the operation of natural selection. In the 1950s, Kettlewell captured several hundred dark and light moths and marked them with little spots of paint. He released some of the dark and light marked moths in a smoky industrial area near Birmingham and the rest in an unpolluted woodland near Dorset. He then recaptured as many moths as he could and calculated the percentage of recovered dark and light moths (Figure 16-13). Near blackened Birmingham, the fraction of dark moths recovered was twice that of the light. Near rural Dorset the situation was reversed; the percentage of light survivors was twice that of the dark. This experiment supported the idea that the change in frequency was related to the color of the tree trunks.

But were birds the agents of selection? Kettlewell showed that the moths really were being eaten by birds and that the color of the moth does, indeed, change the chance of being eaten. After placing moths on tree trunks, Kettlewell recorded the menu decisions of the birds in the area with hidden cameras. As expected, the birds preferentially chose the dark moths on the lighter trees of Dorset and the light moths on the darker trees near Birmingham.

Kettlewell's work seemed to demonstrate both the operation and the actual agents of natural selection. The darkening of moths and butterflies to give better protective coloration in sooty environments is now called "industrial melanism." It has occurred in dozens of species not only in Britain, but also near coal-burning industries in the United States and Germany. Beginning in 1956, clean-air laws returned the forests to their normal greens and browns. Gradually, the fraction of dark moths decreased. In Michigan, black peppered moths increased throughout the 1950s, and then, after the United States instituted its first clean-air laws in 1965, the white form came to predominate.

But there is a flaw in this story. At one site where the moths were studied, the pale moths began to return long before the lichens began coming back. And, despite intensive searching, no researcher has ever found a peppered moth resting on a lichen-covered tree (unless the researcher put it there). Some researchers now postulate that the moths roost in the topmost branches of trees and that selection occurs there.

Selection increases the frequencies of certain traits by eliminating or reducing the number of individuals without those traits. Kettlewell showed that in a blackened environment, dark forms of the peppered moth have a selective advantage over lighter forms. Researchers do not yet know, however, whether selection by birds accounts for such color changes in response to pollution.

In What Directions Does Natural Selection Push Populations?

We think of natural selection as pushing populations in some adaptive direction. And, indeed, in many populations that is true. For Galápagos finches, for example, selection enlarges the beaks of birds that have only big seeds to choose from. But selection can have other effects as well. Evolutionary biologists name several kinds. Here we briefly discuss directional selection, stabilizing selection, disruptive selection, and sexual selection.

How does directional selection influence traits?

Directional selection shifts the frequency of one or more traits in a particular direction (Figure 16-14). Industrial melanism is a classic example of directional selection. Selection for resistance, whether to a pesticide or an antibiotic, is also a form of directional selection. Likewise, agricultural breeders have long practiced directional artificial selection to produce such things as chickens that lay more eggs and cows that give more milk.

In nature, directional selection is typical of a changing environment and is a likely explanation for the directional changes that occur during the evolution of new species. The fossil record shows, for example, that the descendants of early humans and horses have become progressively larger. These changes presumably reflect the action of various selective pressures. In the case of horses, paleontologists suggest that the major selective force was the presence of predators. As horses

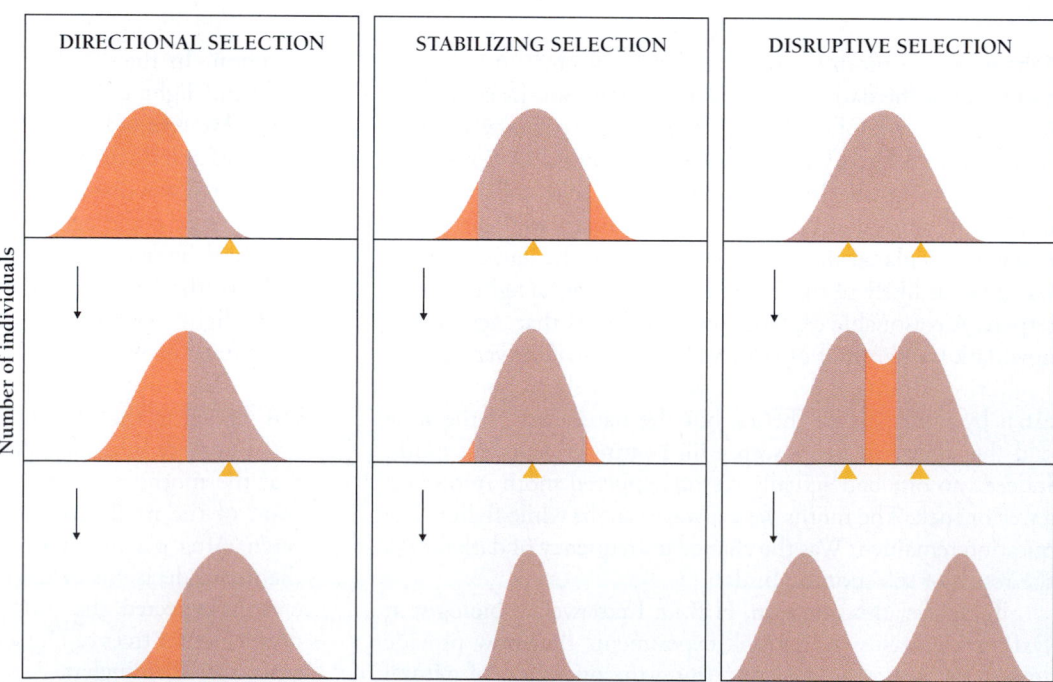

Figure 16-14 Directional, stabilizing, and disruptive selection.

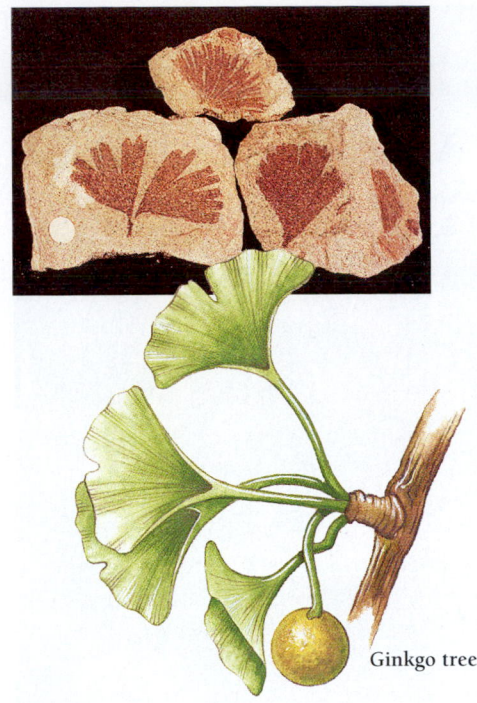

Ginkgo tree

Figure 16-15   They haven't changed a bit. Stabilizing selection probably accounts for organisms that hardly change over millions of years of evolution. Shown here are the ginkgo tree (modern and 250 million years old) and the coelacanth (modern and 350 million years old). *(Fossil ginkgo, Kjell B. Sandved/Photo Researchers, Inc.; fossil coelacanth, John Cancalosi/DRK Photo; modern coelacanth, Tom McHugh/Photo Researchers, Inc.)*

became larger, they could run faster and were better able to avoid being caught and eaten.

It is important to understand, however, that selective pressures can change over time. Traits that are adaptive in one decade or millennium may not be selected for in the next. Thus directional selection can push and pull an organism in many different directions over time.

How does stabilizing selection influence traits?

Stabilizing selection tends to act against extremes in the phenotype, so that the average is favored (Figure 16-14). For example, variability in wasps drops dramatically during the winter, when their mortality is high. An average wasp stands a better chance of surviving a harsh winter than those with unusual phenotypes, which die in great numbers.

Stabilizing selection is characteristic of stable environments because such environments consistently favor the aver-

age and rarely or never favor extreme variants. Stabilizing selection is one explanation for the almost unchanging morphological characteristics of such "living fossils" as the horseshoe crab, the coelacanth, and the ginkgo tree (Figure 16-15). If the environment for such organisms were sufficiently stable, the same phenotype could be adaptive for hundreds of millions of years.

How does disruptive selection influence traits?

Disruptive selection is the opposite of stabilizing selection; it increases the frequency of extreme types in a population, at the expense of intermediate forms (Figure 16-14). Like directional selection, it is a likely cause of the evolution of new species. A nice example of disruptive selection is the African swallowtail butterfly *(Papilio dardanus),* which comes in three distinct variants, each of which lives in a different part of the butterfly's range (Figure 16-16). All three variants appear

Figure 16-16   Disruptive selection. Female African swallowtail butterflies, *Papilio dardanus* (top row), taste good to birds. The butterflies avoid bird predators by mimicking one of three foul-tasting species of butterflies that also live in the area (bottom row). Intermediate forms of the African swallowtail are snapped up by birds, and only the swallowtails that most closely resemble the bad-tasting species survive and reproduce. *(Hope Entomology Collection, Oxford/Biological Photo Service)*

unappetizing to their predators because of their resemblance to another particularly foul-tasting butterfly. However, each variant resembles a different foul-tasting butterfly that lives in the same area. Predators in each area avoid the three extreme forms, but eat any intermediate variants because they resemble none of the three foul-tasting butterflies.

How does sexual selection influence traits?

Physical or behavioral differences between the sexes are extremely common. Among birds, for example, the male is often showy, with brilliant plumage, a long tail or crest, and a lively song, while the female is drab and inconspicuous. Among mammals, the male tends to be larger and more aggressive, as in humans. Many male mammals have horns, antlers, or other rather theatrical weapons. These are used in displays or fights with other males in competitions for territory or access to females. Such differences between the sexes, called **sexual dimorphism,** are sometimes the result of **sexual selection,** the differential ability of individuals with different genotypes to acquire mates (Chapter 29).

Biologists generally recognize two kinds of sexual selection, both of which revolve around sexual reproduction. These are **female choice** (or, rarely, male choice) and **male competition.** In male competition, a male competes with other males for territory or access to females (Figure 16-17). He is more likely to leave numerous descendants if he can win the contest. Anything that gives him an edge in a contest with other

Figure 16-18 **Female choice, in humans.** Among the Wodaabe people of Niger, the men dress up in elaborate costumes and makeup for the Geerewal dance, where women observe and judge, selecting the most beautiful men. In America and in many other cultures, it is women who compete in beauty contests and men who do the choosing. (© *Carol Beckwith/Robert Estall Photo Library*)

Figure 16-17 **Male competition.** Male gemsbok antelopes sparring. Male competition leads to selection for traits that increase the reproductive success of males but are of no value to females. In most animals that have horns, for example, only the males have elaborate horns. (*Fred Bruemmer/DRKPhoto*)

males, whether larger horns or greater aggressiveness, will give him a selective advantage.

The second variety of sexual selection, female choice, involves courtship behavior. The purpose of courtship is to allow the female to assess the male (Figure 16-18). In most cases, females have fewer opportunities to breed than males. Females also invest more time and energy into each of their offspring, on average, than males do. It is important, therefore, for females to be more selective in choosing a mate than males are. In general, females try to mix their own genes with the best set of genes they find. As a result, most species have evolved complex rituals that enable the female to decide if her prospective mate is genetically "good enough" and choose accordingly.

For example, males of outbred and inbred populations of the fruit fly *Drosophila subobscura* differ in their ability to court female flies. To gain a female's favor, a male must perform a series of complex maneuvers, including tapping the female on the head with his front legs and moving around to approach her head on while extending his proboscis, the tonguelike structure with which insects feed and taste. The female quickly sidesteps back and forth, forcing the male to keep adjusting his position. This she does, apparently, to prolong the courtship. The longer the courtship, the more easily she can assess his vigor and skill. Outbred males, which show far greater athletic ability in performing these fly dances, are more successful in their courtship. They have a distinct selective advantage over inbred males.

How can natural selection promote genetic variation?

Given the power of natural selection to change allele frequencies, we may well wonder why significant polymorphism persists at all. The maintenance of genetic heterogeneity in a population is not particularly surprising if different alleles of the same gene are equally useful. For example, within the range of a single interbreeding population, local environments may select for different alleles. The climate of a mountainside may select for short trees that are less likely to be toppled by winter winds, while a milder valley climate may favor larger trees that produce more seeds. Or one allele may be better adapted at one time of year than another. Among the members of one species of ladybug, for example, black ladybird beetles are more frequent in the fall, while red ladybird beetles predominate in the spring. Such a situation creates a balance of different alleles in a population, called a **balanced polymorphism.** The most famous case of a balanced polymorphism in nonhumans is, of course, the current balance between dark and light peppered moths.

Populations may maintain high frequencies of alleles that are deleterious or even lethal when homozygous. In such cases,

polymorphism results from the selective superiority of the heterozygote. Sickle cell anemia is an excellent example of a case in which heterozygote superiority creates a balance of different alleles.

As we discussed in Chapter 10, people with two copies of the "β^S-globin" ("Beta S globin") allele suffer from sickle cell anemia. Recall that hemoglobin consists of four polypeptide chains, two βs ("betas") and two αs ("alphas"). In people of African and Middle Eastern descent, the normal allele for the β chain is sometimes replaced by an allele known as β^S. In people with two copies of the β^S allele, low oxygen levels cause the abnormal hemoglobin to precipitate, leading to distortion and stiffening of the red blood cells. The sickled cells are so stiff that they can block capillaries and so fragile that they break open in the strong currents of larger arteries. In the United States, victims of sickle cell disease rarely live beyond age 40, dying of infections, kidney failure, or the blockage of blood vessels in vital areas such as the lungs. Most suffer periodic bouts of pain, increasing weakness, and constant infections.

In Central and West Africa, where health care is worse, homozygotes often die in their teens and 20s and rarely reproduce. Nevertheless, the β^S allele persists in the population, for it has strong selective value in β^A/β^S heterozygotes. Heterozygotes, whose blood contains a mix of normal and abnormal hemoglobins, are not anemic, yet they are resistant to infection by the parasite that causes the deadly disease malaria. The heterozygotes therefore have a distinct selective advantage over people who are homozygous for the "normal" copies of these genes, who are susceptible to malaria.

In some regions of Central and West Africa, the frequency of the β^S allele is about 20 percent (Figure 16-19). Using the Hardy-Weinberg Principle, we can calculate that the frequency of heterozygotes ($2pq$) is then $2(0.2)(0.8) = 0.32$, or 32 percent. The expected frequency of homozygotes would be

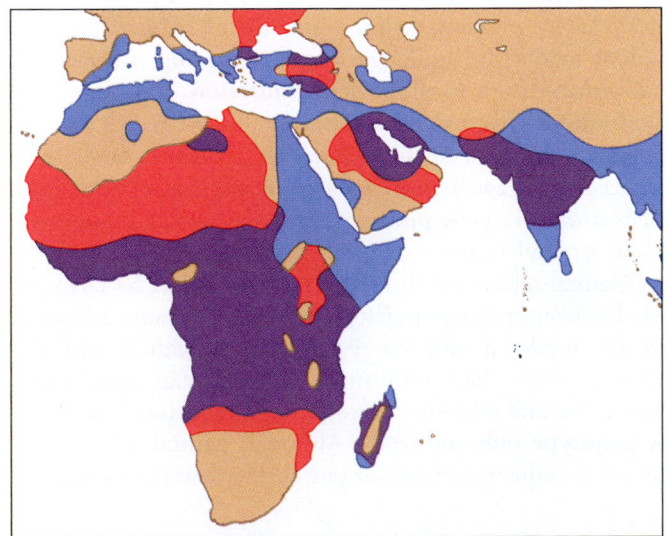

A.

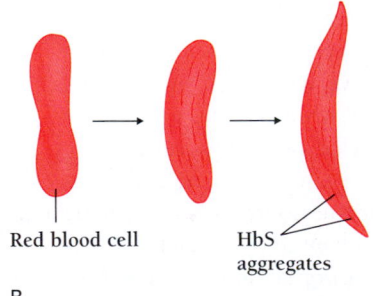

Red blood cell   HbS aggregates

B.

Figure 16-19   A balanced polymorphism. A. The distribution of the sickle cell allele (β^S) closely parallels that of malaria. Blue shows areas of southern Europe, Africa, and the Middle East where malaria is or was once common. Red illustrates areas where sickle cell anemia occurs. Considerable overlap (*purple*) suggests selection for the sickle cell allele (β^S) in areas with malaria. B. The allele for sickle cell alters the shape of a polypeptide in hemoglobin. The altered hemoglobin changes the shape of individual red blood cells.

$(0.2)^2 = 0.04$, or 4 percent. The homozygotes usually die before reproducing, and these deaths represent the "cost" of maintaining the β^S allele in the population. There is a trade-off between the deaths caused by the homozygous condition and the lives saved by the heterozygote condition.

In the United States, where malaria is nearly nonexistent, the β^S allele probably gives people no selective advantage, either in the homozygote or the heterozygote individual. In the United States, the frequency of homozygotes among African Americans is less than one-hundredth the frequency in Central and West Africa, while the frequency of heterozygotes is only about one-third that in Africa.

The reduction in the frequency of the β^S allele is due not only to its selective disadvantage but also to the effects of gene flow. On average, one-third of the genes of contemporary African Americans come from their European ancestry, so the frequency of the β^S allele would be reduced by dilution as well as by selection.

A balanced polymorphism may result from heterozygote superiority or from different selective pressures in time and space within the same population.

In this chapter we have seen where variation comes from, how it is inherited, and some of the ways that allele frequencies can change. Changes in allele frequency in a population, called microevolution, can occur through natural selection or, in small, isolated populations, through random drift. However, explanations for the larger patterns of evolution—the multiplication of species and higher taxa, as well as patterns of extinction, for example—remain controversial. In the next chapter we will consider why it is so hard to define a species, how new species form, and why species multiply and go extinct.

Study Outline with Key Terms

Natural selection was once controversial because heredity was not understood. Darwin believed in **blending inheritance,** which led him to wrongly embrace the theory of **the inheritance of acquired characteristics.**

Weismann knew nothing of genes or mutations, so he could not know the ultimate source of genetic variation. Nevertheless, he was able to correctly guess that inheritance is not blending but particulate, that the genetic material is in the **germ plasm** and is unaffected by the life of the parent, and that sexual reproduction is a means of creating new variants by mixing and recombining old traits.

The **modern synthesis** wedded natural selection and genetics into an essentially modern theory of evolution. Population geneticists attempt to understand evolution in terms of the changes in the frequency of alleles within populations and species. **Microevolution** is shifts in gene frequencies in a population. **Macroevolution,** which we will study in the next chapter, deals with the processes by which species and higher groupings of organisms originate, change, and go extinct.

Genetic variation is the fuel of evolution. Gene and chromosome mutations are the ultimate source of all genetic variation. Genetic recombination, enhanced by sexual reproduction, spreads new **alleles** through a population by recombining different alleles into new genotypes.

Most organisms have two sets of **homologous** chromosomes. When the same allele is on each chromosome in a pair, the organism is said to be **homozygous** for that gene. When a different allele is on each chromosome in a pair, the organism is said to be **heterozygous** for that gene. Traits determined by two or more genes are said to be **polygenic.** Organisms displaying two or more alternative forms are said to be **polymorphic.**

The **Hardy-Weinberg Principle** shows that under certain conditions, a single generation of random mating establishes a distribution of genotype frequencies called the **Hardy-Weinberg equilibrium** ($p^2 + 2pq + q^2$) that will not change in future generations. The Hardy-Weinberg Principle disproves Fleeming Jenkin's assertion that sexual reproduction by itself would change the genetic characteristics of a population. The Hardy-Weinberg Principle applies only if the following conditions are met: random mating, large population size, no mutations, no gene flow between populations, and no selection. **Natural selection** is only one of several forces that can change the frequency of alleles within a population. Others are nonrandom mating (including **assortative mating, inbreeding,** and **outbreeding**), **random drift, mutation,** and **gene flow.** The **founder effect** occurs when a population passes through a **bottleneck,** or a small population founds a larger one. Chance affects gene frequencies in small populations more than in larger ones. A **gene pool** is all the alleles of all the individuals in a population.

Natural selection can bring about evolution by changing gene frequencies in a population. Natural selection can lead to changes in allele frequencies even when individuals with different genotypes have only slight differences in reproductive success. Natural selection acts on the **phenotype** directly, and the **genotype** only indirectly. Although natural selection reduces the frequency of deleterious alleles, deleterious recessive alleles—even when they are lethal in homozygotes—persist at low frequencies for many generations. Natural selection can also act to decrease the genetic variability in a population (**stabilizing selection**), to change the quantitative characteristics of a population in a particular direction (**directional selection**), to increase the frequency of extreme types in a popula- tion (**disruptive selection**), or to cause **sexual dimorphism** (**sexual selection**). Biologists believe that sexual selection re- sults most often from **female choice** and **male competition**. Selection for heterozygotes and simultaneous selection for dif- ferent alleles maintains genetic variation in a population (**balanced polymorphism**).

Review and Thought Questions

Review Questions

1. Darwin understood how natural selection works in the presence of a population with many variant types. What does that varia- tion consist of? How does the variation arise? How does it spread through the population?
2. How much genetic variation do human populations exhibit?
3. Do the eye colors exhibited by your classmates fall neatly into the five colors shown in Figure 16-4? If not, why might they be different?
4. Why doesn't sexual reproduction, by itself, change allele fre- quencies in a population?
5. Why do the frequencies of the alleles in a gene pool add up to 1?
6. What conditions must be met for a population to be at a Hardy- Weinberg equilibrium?
7. How does chance lead to changes in allele frequencies?
8. Why are certain genetic diseases more common in certain hu- man groups than in others?
9. Give an example of gene flow in a human population and in an- other animal population.

10. Why would a dominant lethal allele die out in one generation, while a recessive lethal might persist for hundreds of generations?
11. Why is the lethal recessive allele for sickle cell anemia so com- mon in some human populations?

Thought Questions

12. Why doesn't natural selection eliminate variation in a popula- tion?
13. Why did Darwin think that sexual reproduction must eliminate variation in a population? What was his mistake?
14. What facts tended to suggest that the theory of the inheritance of acquired characteristics was wrong? What facts were needed to prove that it was wrong?
15. Is it possible that the apparent changes in the proportions of light and dark peppered moths are due to the moths being invisible to collectors, rather than to birds? Discuss your rea- soning.

About the Chapter-Opening Image

The cheetah has some of the lowest genetic variability known in a wild organism. Biologists believe that cheetahs passed through a bottleneck within the past few thousand years, in which the entire population was reduced to a few individuals.

 On-line materials relating to this chapter are on the World Wide Web at **http://www.harcourtcollege.com/lifesci/aal2/**

17 Macroevolution: How Do Species Evolve?

What Darwin Saw at the Galápagos Islands

When Charles Darwin arrived at the Galápagos Islands in the summer of 1835, he found a group of dreary volcanic islands consisting of hundreds of craters, the largest some 3000 or 4000 ft high. Five main islands dominated the little group of 14 islands, 600 miles off the coast of Ecuador. Despite their position on the equator, the Galápagos were not the lush, tropical islands typical of other parts of the Pacific, but barren desert islands populated by a small number of cacti, reptiles, and small, brown birds. It was the dry season.

"Nothing could be less inviting than the first appearance," Darwin wrote in his diary. "A broken field of black basaltic lava, thrown into the most rugged waves, and crossed by great fissures, is everywhere covered by stunted, sunburnt brushwood, which shows little signs of life. The dry and parched surface, being heated by the noonday sun, gave to the air a close and sultry feeling, like that from a stove: we fancied even that the bushes smelt unpleasantly." The most striking animals were the reptiles, large marine iguanas, and immense Galápagos tortoises (*Geochelone elephantopus*), some of them 5.5 ft tall and weighing 500 lb (Figure 17-1).

Darwin spent a month in the islands, exploring and collecting rocks, plants, and animals. He discovered that almost all of the terrestrial animals and at least half of the plants were endemic species, that is, found nowhere else in the world. Yet, all of the Galápagan species resembled species found on the nearby South American mainland—all, that is, but a European rat, probably brought by ship to the islands.

The island-bound plants and animals were related to those on the mainland. But how were they related, and why were they different?

Darwin's observation that so many Galápagan species were unique to the islands did not at first strike him as meaningful. Even when the vice-governor of the islands casually mentioned that tortoises from different islands were so distinctive that he could easily tell which island a particular tortoise had come from, Darwin paid no attention. Then, just days before the HMS *Beagle* was to set sail for Tahiti, the vice-governor's remark struck Darwin with sudden force.

Maybe, Darwin thought, each island supported a different set of organisms. Frantically, he went back through his specimens to see if different species came from different islands. But he had labeled most of them "Galápagos." He had hopelessly mixed up the different kinds of tortoises and the finches. There was no way to tell which islands they had come from.

But one group of birds, the mocking thrushes, "thoroughly aroused" his attention. Of three species, one was unique to Charles Island, one to Albemarle, and one to James and Chatham Islands. Darwin's plant collection, more carefully labeled, was a gold mine. For example, of 71 species found on James Island, 38 were endemic to the Galápagos and 30 of those were endemic to James Island alone. Every island had the same story to tell.

"One is astonished," he wrote in his diary, "at the amount of creative force, if such an expression may be used, displayed on these small, barren, and rocky islands; and still more so, at its diverse yet analogous action on points so near each other."

Tui de Roy/Minden Pictures

Key Concepts

- A species is a population of organisms whose members can interbreed to produce viable offspring but that usually do not interbreed with members of other such populations.

- A population may form a separate species when it becomes reproductively isolated by any of a variety of means.

- Adaptive radiation is the generation of diverse new species or other taxa from one or more ancestral species.

- Species probably evolve at different rates at different times.

Two years after his visit to the Galápagos, Darwin began a notebook on the "transmutation of Species." It would be years before he realized the full implications of all that he had seen in the Galápagos. It was no accident, he ultimately realized, that Galápagos species were so closely related to each other and to those on the South American mainland. Each endemic species had evolved on one of the islands, having descended from migrants from other islands or the mainland.

Nowadays, every child knows that different sorts of animals and plants populate each of the different continents. Africa has its lions and camels, South America its jaguars and llamas. Lions and jaguars, although descended from the same ancestral cat, have evolved independently for millions of years. The same is true of camels and llamas.

In Darwin's day, however, the presence of similar animals in very different habitats, or of completely different assemblages of animals in similar habitats, would have been surprising. According to the 19th-century theological concept of special creation, similar habitats should support identical species. Every dark forest should have its bear and every clear stream its trout. And, since each species was thought to have been created by the Judeo-Christian God less than 6000 years ago, species were not supposed to be related to one another. Most biologists, like Linnaeus, could see that animals and plants group naturally into rough hierarchies that resemble family trees. Yet they carefully ignored the suggestion of genealogy suggested by these hierarchies.

But Darwin—scion of a distinguished family—knew the importance of genealogy. He realized that 5 million years ago, the volcanic Galápagos Islands had burst steaming and erupting from the sea, devoid of life. Gradually, plants and animals began to colonize the islands. Seabirds roosted on the barren rocks, leaving seeds that had stuck to their wings or passed through their digestive tracts. Some of these seeds took root and managed to survive. Ocean currents brought more plant seeds and an occasional reptile. Winds also brought seeds, as well as finches and sparrows and other land birds. Some plants and animals survived and reproduced; others did not. Some moved from island to island, spreading and interbreeding with other populations; others remained confined to one island. These slowly evolved away from their cousins on the other islands and the parent population on the mainland. In this way, one species became many.

In this chapter we will study **macroevolution,** the origin and multiplication of species. We will see what a species is, and explore how it comes into being. We will ask what the patterns in the fossil record tell us about evolution. And we will discover why, after over 150 years, evolutionary biologists are still arguing about the questions raised by Darwin after his visit to the Galápagos Islands. Near the end of the chapter, we will briefly discuss the evolution of humans.

> *Frantically, he went back through his specimens to see if different species came from different islands.*

WHAT IS A SPECIES?

In the next section of this book, we will explore the enormous biological diversity of life on Earth. For now, we will merely note that biologists have so far identified about 1.4 million species of living organisms—give or take a hundred thousand. Newly discovered species pour into the museums monthly, so this number is probably less than one-tenth of the total number of species currently living. The actual number could be anywhere from 10 million to 100 million. Of the known species, more than half are insects (Figure 17-2). Nearly half of all insects—290,000 species—are beetles, some 20 percent of all species of organisms combined. The biologist J.B.S. Haldane once ironically remarked that God "must have had an inordinate fondness for beetles."

Figure 17-1   The voyage of the *Beagle*. In 1831, Charles Darwin left England on board the British surveying ship the H.M.S. *Beagle* for a 5-year trip around the world. Darwin and the *Beagle* spent nearly 4 years in South America. After Darwin left the Galápagos Islands, he was struck by the fact that each island had its own kind of tortoise. On each island, the tortoises evolved separately from those on the other islands and so differ with respect to the shape of the shell, the length of the neck, and other traits. The first written account of the differences among Galápagos tortoises came from the captain of the United States frigate *Essex*, which visited the Galápagos during the War of 1812 and, by itself, nearly destroyed the British whaling fleet stationed there. Before the *Essex* was finally captured by the British in 1814, Captain David Porter found time to describe the tortoises and various lava lizards. *(Darwin, The Bridgeman Art Library; H.M.S.* Beagle, *The Granger Collection, New York; marine iguana, Michio Hoshino/Minden Pictures)*

Why Is It So Hard To Define a Species?

The average person, faced with groups of familiar organisms such as mammals or birds, easily recognizes one species from another. Even when presented with a set of unfamiliar mammals, the average person can generally distinguish most species. In 1927, for example, a young biologist named Ernst Mayr, who later became a leading figure in evolutionary biology, led an expedition into the Arafak Mountains of New Guinea to catalogue new species of animals. After he had identified 138 species of birds, he was surprised to discover that the local Papua natives themselves recognized 137. Two of the species were so similar that the Papuans considered them a single species.

Unfortunately for biologists, such similar-looking species are common. Two species may even look identical. Some

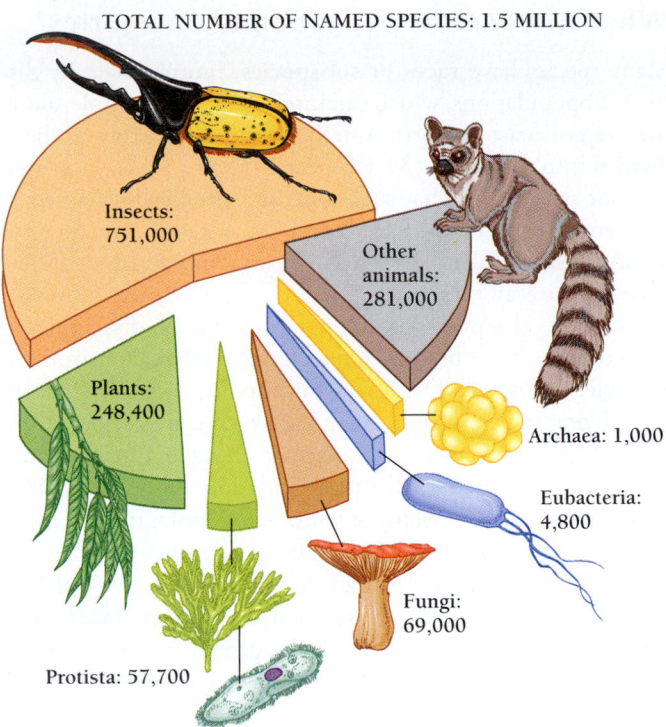

TOTAL NUMBER OF NAMED SPECIES: 1.5 MILLION

Insects: 751,000
Other animals: 281,000
Plants: 248,400
Archaea: 1,000
Eubacteria: 4,800
Fungi: 69,000
Protista: 57,700

Figure 17-2   The diversity of life. At least 1.5 million species now inhabit the Earth.

species of frogs, for example, are distinguishable only by their calls. Adult forms of the malaria-carrying mosquito *Anopheles maculipennis* cannot be distinguished from those of several other species that do not carry malaria. A species that is extremely similar to, but reproductively isolated from, another is called a **sibling species.** We can conclude that appearance alone fails to define a species.

Defining a species is awkward. When biologists believed that organisms were all variants of "ideal types," as 19th-century essentialists believed, the problem was much easier. The definition of a species was a description of the physical features of the ideal type, a sometimes arbitrarily chosen "type specimen." Type specimens are still essential for naming a new species and for identifying other individuals of the same species that are collected later. But we now realize that every individual in a species is unique—ideal types do not exist. Even an individual that happened to be completely average in every respect would not represent its species better than any other individual in the population.

The definition of species must take account of the enormous variation among individuals and recognize that two species can be distinct even if they are very much alike. Darwin found himself exasperated by the problem of classifying species. Writing to his friend J.D. Hooker in 1853, Darwin noted, "After describing a set of forms as distinct species; tearing up my [manuscript], & making them one species; tearing that up & making them separate, & then making them one again (which has happened to me) I have gnashed my teeth, cursed species, & asked what sin I committed to be so punished."

Darwin concluded that since species gradually evolve one from another, the term "species" is a convenient but arbitrary designation that has no biological meaning. He found this a great "relief," for he no longer felt he had to worry about whether organisms were truly separate species or not.

Modern biology takes a different view from Darwin. We now think that species are, for the most part, discrete and very real entities. Further, biologists have found that in order to discuss evolution, they need to define what a species is. In the 1930s and 1940s, the makers of the modern synthesis produced the **biological species concept,** one modern definition of a species. In 1942, Ernst Mayr defined a **biological species** as follows: *"Species are groups of actually or potentially interbreeding populations, which are reproductively isolated from other such groups."* A biological species, then, is the largest group of similar organisms in which gene flow is possible.

Most species conform nicely both to the biological species concept and to our sense that species should look different from one another. Bullfrogs *(Rana catesbeiana)* and wood frogs *(Rana sylvatica)*, for example, look different from each other and do not interbreed (Figure 17-3). They are members of different species.

But the biological species concept is a tricky definition. First, it does not apply to asexual organisms such as prokaryotes, certain fungi, and even some plants.

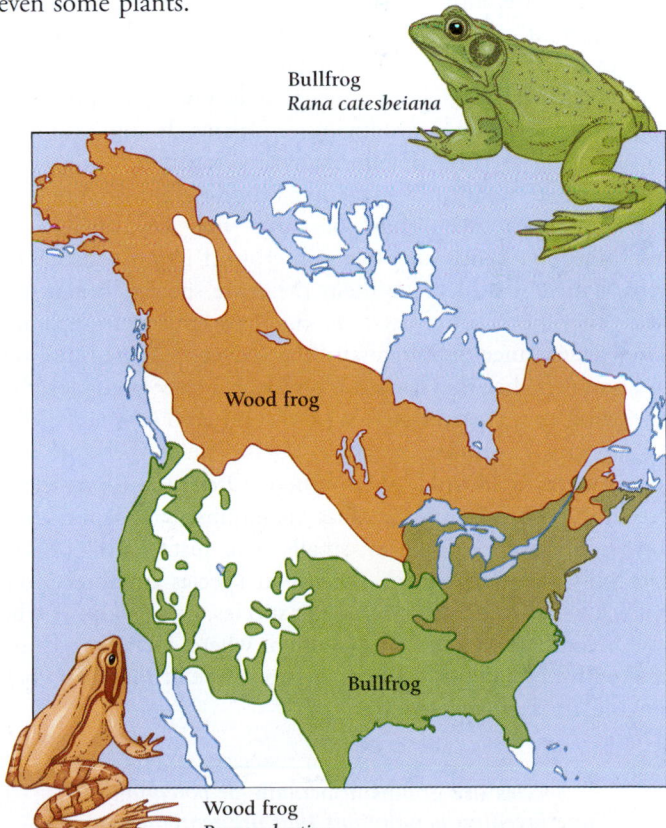

Bullfrog *Rana catesbeiana*
Wood frog
Bullfrog
Wood frog *Rana sylvatica*

Figure 17-3   Separate species. The wood frog *Rana sylvatica* and the bullfrog *Rana catesbeiana* look different and occupy different, though overlapping, habitats. They probably do not interbreed; they are true species. *(After R.C. Stebbins, 1985)*

For example, *E. coli,* the distinctive bacterium that inhabits the human digestive tract, is almost always asexual. Even though it does not satisfy the definition of the biological species, taxonomists have nonetheless assigned it to a genus and species.

Second, the biological species concept does not apply easily to extinct organisms. Paleontologists naming fossils have no way of knowing if two so-called species interbred. Many fossil species gradually give way to increasingly different forms. Paleontologists arbitrarily give an organism a new species name when it looks sufficiently different from its ancestors. But no one knows whether the modern horse *(Equus),* for example, could interbreed with its predecessor *Hippidium* (Figure 15-12).

Furthermore, paleontologists must give taxonomic species names to fossil organisms on the basis of morphological differences in the hard parts alone. Muscles and other soft parts, as well as clues about physiology and behavior, are almost never fossilized. The two mosquito species mentioned earlier, for example, would never be recognized as two species if they appeared only as fossils. And some paleontologists have begun to suspect that several "species" of dinosaurs may turn out to be different ages and sexes of the same species.

Paleontological species are therefore subjective designations. A group of fossils may be lumped together as a single species or divided into several. But these same organisms might be judged differently if they were alive today. This point will be important to keep in mind when we discuss the evolution of humans at the end of this chapter.

The biological species concept does not even apply well to all living sexually reproducing organisms. It applies poorly to organisms that do not normally live together. Geographically isolated organisms—whether very different or very similar in appearance—may or may not be capable of interbreeding. Lions and tigers interbreed if put together in zoos, which should make them the same species. But in the wild, they never interbreed, so they are effectively separate. Then again, since their habitats do not overlap, there is no way to know whether they would interbreed if they lived closer to one another. In garter snakes of the genus *Thamnophis,* the two species *T. hydrophila* and *T. elegans* live in the same area in northern California and southern Oregon without interbreeding. They seem to be what taxonomists call "good" biological species. Yet, in a small area just south of the Oregon-California border, these two species interbreed. Finally, for the vast majority of living species, no one knows who interbreeds with whom, how often, or where. Detailing the reproductive biology of wild organisms is one of the most challenging areas of biology.

Species are groups of actually or potentially interbreeding populations that are reproductively isolated from other such groups. The best definition of species that we have is the biological species concept, but this definition, good as it is, is often hard to apply.

What About Other Taxonomic Categories?

Many species have **races** or **subspecies,** morphologically distinct subpopulations, which can interbreed. For example, along the western coast of North America, the song sparrow has been divided into as many as 34 subspecies (Figure 17-4). However, it is not always clear that such races are specifically adapted to different environments. For example, six races of carpenter bee inhabit a single island in Indonesia, with no apparent differences in the way they live.

In fact, almost any subpopulation can be distinguished from others on the basis of some cluster of characteristics. For example, the people living in Minneapolis might differ from those living in St. Paul in having, on average, longer noses and a slightly higher frequency of ingrown toenails. But we would not necessarily conclude that the populations of the Twin Cities were separate races. Many evolutionary biologists therefore consider the concept of race or subspecies to be just a normal manifestation of natural variation and without any real biological meaning. The differences among human "races" constitute only about 9 percent of the genetic diversity in the human species. The other 85 percent is individual variation (Chapter 14).

Nevertheless, even though most races or subspecies of organisms interbreed easily, a few do not. Subspecies that do not interbreed may be in the process of separating into species and are sometimes called **incipient species.** Distinguishing ordinary subspecies from incipient species is often a judgment call that only a biologist who has thoroughly studied the population in question can make.

As we will see in Chapter 19, genera and higher taxonomic categories are still harder to define. The hierarchical system of

Central California
(M. m. heermani)

Desert form
(M. m. saltonis)

San Francisco Bay
(M. m. samuelis)

Alaskan form
(M. m. inexpectata)

Subspecies of the song sparrow, *Melospiza melodia*

Figure 17-4 **Subspecies.** Here are just 4 of the 34 known subspecies of Western song sparrow.

classification that taxonomists use to organize biological species into higher groupings generally represents degrees of anatomical difference and so, to some extent, evolutionary distance.

A higher taxon is usually defined by a few characters that consistently distinguish its members from the species in other taxa. Mammals, for example, are distinguished by hair, two sets of teeth, and mammary glands. If we find an animal with these characters, we can be sure it is a mammal.

However, the division of a group of species into separate genera, families, or higher taxa can be very subjective. For example, in 1947, a botanist decided that a recently discovered daisylike plant whose flowers had no petals was so novel that it should be put into its own genus. Only later did botanists discover that the plant was a simple variant of a well-known species that normally had petals. The "new genus" was not even a new species.

Species exist as biological entities independent of human definitions. Higher taxa, as well as subspecies and races, are somewhat subjective groupings.

HOW DO SPECIES FORM?

We have seen that for two populations to become separate species, they must become **reproductively isolated,** unable to interbreed. Often, the process by which one species becomes two begins when one population is geographically isolated from the rest of the species. If that happens, the physically isolated population has a chance of becoming reproductively isolated from the parent species and, with that, a chance of becoming a separate species. Some populations merely happen to become reproductively isolated and then evolve apart.

What Barriers Reproductively Isolate Populations and Species?

Other populations, however, may first evolve separate adaptations and then develop reproductive barriers that allow them to retain their respective adaptations. Many of the most remarkable features of organisms function to maintain reproductive isolation. Such features often include the diverse structures of flowers and the elaborate rituals of animal mating.

The larkspur *(Delphinium)* provides one example of instant reproductive isolation. Most species of larkspur have blue or purple flowers (Figure 17-5). Bees visit blue larkspur (and other blue and purple flowers), carrying pollen from flower to flower. But one northern California larkspur has a mutation that gives it scarlet flowers—a color bees cannot see. Fortunately, hummingbirds see red flowers very well, and they pollinate the red larkspurs. The bees and hummingbirds do not intrude on one another's flowers, and the blue and scarlet larkspurs are reproductively isolated.

For populations to become reproductively isolated, gene flow between them must cease. Barriers to gene flow fall into two broad categories: (1) **prezygotic barriers,** which prevent syngamy, the fusion of the sperm and egg to form a zygote, and (2) **postzygotic barriers,** which make a zygote either inviable (certain to die) or sterile. The red and blue larkspurs are an example of a prezygotic barrier, since the flowers' different colors prevent cross fertilization. Prezygotic barriers also include differences in the ways that species live **(ecological isolation),** in the times at which they reproduce **(temporal isolation),** in their mating behaviors **(behavioral isolation),** in the complementarity of male and female reproductive organs **(mechanical isolation),** and in the compatibility of their gametes **(gametic isolation).**

Postzygotic barriers operate when fertilization has already occurred, but the resulting zygote either dies or fails to reproduce successfully. An individual that results from cross fertil-

Figure 17-5 Reproductive isolation. Some time in the past, the scarlet larkspur, shown here, became isolated from the rest of its species. A mutation produced a generation with red flowers instead of the usual blue. But red flowers tend to attract hummingbirds instead of the bees and other insects that pollinate the blue flowers. Because the hummingbirds do not carry pollen between the red and blue flowers, the scarlet larkspur has become reproductively isolated from its blue relatives.

ization between two different species, or, sometimes, between two distinct populations, is called a hybrid. For example, a cross between a horse and a donkey results in a mule, a hybrid that is nearly always sterile. The mule is an example of **hybrid sterility.** However, hybrids between related species often die during development or early in life because of chromosomal and genetic incompatibilities, a barrier referred to as **hybrid inviability.** In some cases, the hybrids of two species may be viable and fertile, but *their* offspring are weak or sterile **(hybrid breakdown).**

In plants, changes in chromosome number can simultaneously create new variants and cause their immediate reproductive isolation. For example, if homologous chromosomes fail to separate during meiosis, then each gamete will have the diploid number ($2n$) of chromosomes instead of the haploid number (n). Fertilization can then produce a tetraploid (or $4n$) zygote, with four rather than two sets of homologous chromosomes (Figure 17-6). Such tetraploid plants will produce gametes with $2n$ chromosomes. These gametes cannot form zygotes with gametes of the parental stock, since such zygotes would contain $3n$ chromosomes, which could not pair at meiosis. Tetraploid plants thus become immediately isolated from the main population. Other multiples of the haploid number of chromosomes also occur. Plants that contain more than two complete sets of chromosomes are said to be **polyploid.** Polyploidy has played an important role in the evolution of grasses, ferns, and other vascular plants.

Another common postzygotic barrier in plants results when the zygote combines the chromosome sets of two species. Organisms formed from the gametes of two different species are ordinarily sterile, if not inviable. Such hybrids rarely produce gametes because the maternal and paternal chromosomes are not homologous, and so they cannot pair during meiosis. However, pairing can occur if the hybrid doubles its chromosomes. Such a doubling forms an **allopolyploid.** Each set of chromosomes is present twice and can engage in proper meiotic pairing and gamete formation. But the gametes of the hybrid can no longer form viable zygotes with the gametes of either parental species, so they become reproductively isolated from both parental populations. Since many plant species can self-fertilize, however, a single plant can, potentially, establish a new species.

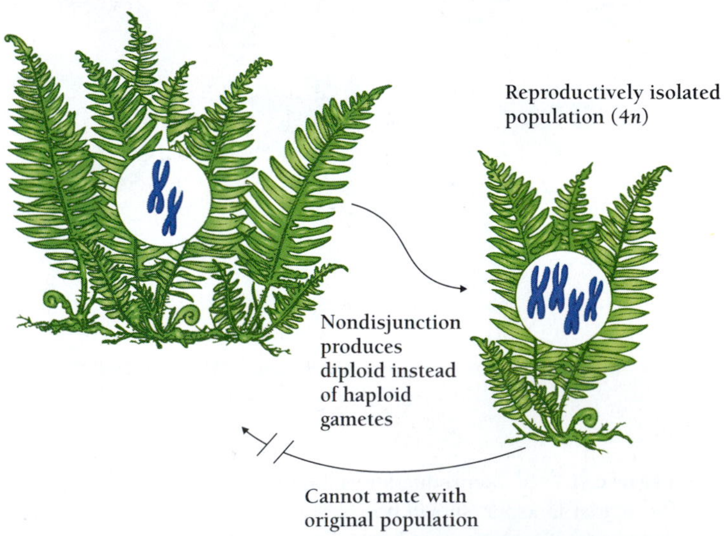

Populations become reproductively isolated either by prezygotic or by postzygotic barriers.

How Does Geographic Isolation Initiate Speciation?

In most cases, populations must be geographically isolated from the parent population before other barriers to reproduction arise. Oceans, mountain ranges, and rivers, for example, are all capable of sharply reducing or eliminating interbreeding with a parent population. Once a population is isolated, it can follow its own evolutionary course (Figure 17-7). Its gene frequencies can change independently, according to the forces of selection and genetic drift (Chapter 16). In time, the isolated population may acquire new chromosomal arrangements and reproductive strategies that prevent interbreeding with the parent population if the two should again come into contact. These isolating mechanisms preserve the isolated population's differences and allow it to **speciate**—form a new species. Species formation by geographic isolation is called **allopatric speciation** [Greek, *allos* = other + *patra* = country].

Allopatric speciation is speciation that proceeds from geographic isolation. Once two populations are geographically isolated, independent selective pressures and genetic drift can further differentiate the two populations.

How Can Species Form Without Geographic Isolation?

The formation of separate species without geographic isolation is called **sympatric speciation** [Greek, *syn* = together + *patra* = country]. For a single population to spontaneously break into two reproductively isolated subpopulations, something must inhibit gene flow between the two subpopulations. Some barriers to gene flow, such as polyploidy in plants, can isolate subpopulations from their parent populations in one generation. However, most barriers to gene flow would have to evolve more gradually.

Original population ($2n$)

Reproductively isolated population ($4n$)

Nondisjunction produces diploid instead of haploid gametes

Cannot mate with original population

Figure 17-6   Reproductive isolation by polyploidy. An error during meiosis produces diploid gametes, which form tetraploid offspring. The offspring can breed only with one another, however, because if the diploid gametes fuse with normal haploid gametes, they form triploid zygotes, which cannot survive.

All grassland.
One species of mouse.

Formation of a river separates the mice. Later, differences in habitat cause the two species to diverge.

Figure 17-7 **Allopatric speciation.** When a single population of mice becomes divided—by a river, for example—the two subpopulations evolve separately and can form separate species.

For example, the ensatina salamander has spread down the two sides of a large valley (Figure 17-8). Increasing geographic distance correlates with both increasing genetic differences and increasing differences in appearance. The populations on either side of the valley are separated from one another and mostly do not interbreed. One exception are the yellow-eyed salamanders on the coast, which can move up a river system to the mountains in the east, where they hybridize with the Sierra Nevada salamanders. Near the bottom, the Monterey salamander and the yellow-blotched salamander, which do not interbreed, are different enough to be considered different species, yet they are linked by a continuous chain of subspecies that can and do interbreed.

Most models of sympatric speciation invoke disruptive selection as a mechanism. The hawthorn fly may present a good example of the first steps in sympatric speciation by disruptive

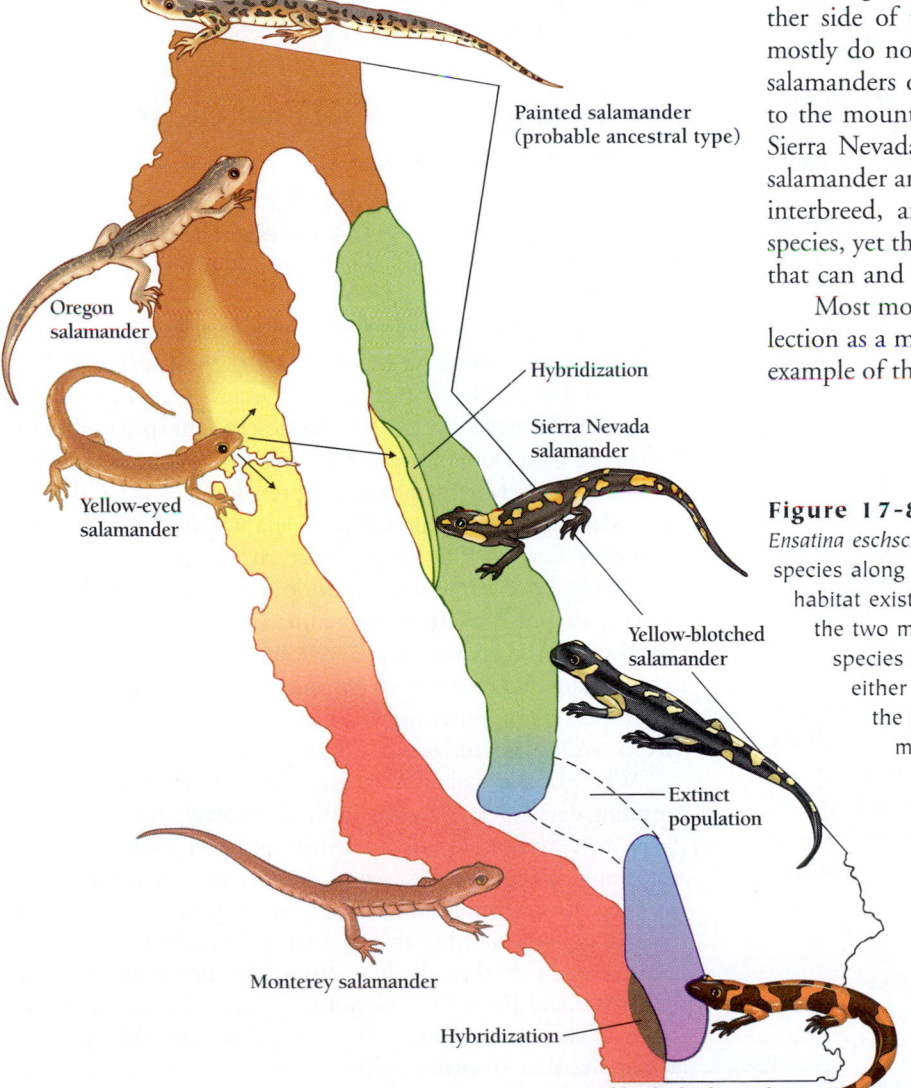

Painted salamander (probable ancestral type)

Oregon salamander

Yellow-eyed salamander

Hybridization

Sierra Nevada salamander

Yellow-blotched salamander

Extinct population

Monterey salamander

Hybridization

Large-blotched salamander

Figure 17-8 **Sympatric speciation.** The ensatina salamander *Ensatina eschscholtzii* has diverged into a series of overlapping subspecies along the Coast Range and the Sierra Nevada. Little or no habitat exists in the Great Central Valley, which runs between the two mountain ranges. Within the mountains, adjacent subspecies of the salamander interbreed. The two subspecies at either end of the salamander's horseshoe-shaped range—the Monterey salamander and the yellow-blotched salamander—do not themselves interbreed. But they are linked by a continuum of subspecies. In the south, the yellow-blotched salamander and the large-blotched salamander are phenotypically and genetically similar, and biologists believe that these two species were once linked by a population that has most likely gone extinct. In 1976, one ensatina salamander was seen in this area—the only one since biologists began searching in 1949. (*Excerpted from* Life on the Edge, *copyright 1994 by Bio Systems Analysis, Inc., with permission from Ten Speed Press, P.O. Box 7123, Berkeley, CA 94707*)

selection. In 1864, farmers in New York State noticed that the small hawthorn fly *Rhagoletis pomonella,* which normally fed on hawthorn fruits, had begun to feast on apples as well. Because the apples at once became both food source and mating site for some of the hawthorn flies, it has been suggested that the apple-eating flies rapidly became reproductively isolated from the rest of the population. In 1960, some of the flies switched to cherries, possibly creating yet another isolated, but sympatric, race.

The flies move back and forth from one fruit to another, however, and researchers are not certain how different or how well isolated these three populations are. The three subpopulations have divided the same environment into three potentially separate habitats, like three islands of habitat. But so far, biologists still recognize only a single species. Whether these are incipient species remains to be seen.

Sympatric speciation, the formation of species without geographic isolation, is controversial.

Can Species Form with Limited Gene Flow?

A third type of speciation, called **parapatric speciation,** occurs when a species splits into two populations that are geographically separated, but still have some contact. Parapatric speciation is thus intermediate between allopatric and sympatric speciation. Reproductive barriers can isolate parapatric populations if the environments of the two populations impose different selective pressures on each, and if gene flow between the two populations is small (lest it overwhelm the divergent changes).

The growth of grasses around the entrance to a Welsh lead mine provides a good example of parapatric isolation and the probable beginning of parapatric speciation. The soil immediately around the mine has a high concentration of lead—high enough to kill most of the species of grass that flourish in the adjacent pasture. But a variety of one species of grass, *Agrostis tenuis,* tolerates lead and grows well on the mine tailings. This variety has become reproductively isolated from the neighboring parental pasture grass because it happens to flower at a different time.

If we could be sure that the separation were complete—that gene flow had completely ceased—then we would call the two populations separate species. At this point, however, we can only note the different selective pressures of tailings and pasture and the limited gene flow between them. Such a situation could easily lead to true reproductive isolation and speciation.

Parapatric speciation may occur if gene flow between adjacent, nonoverlapping populations is sufficiently limited.

HOW DO SPECIES MULTIPLY?

The separation of one species of organism into two (or more) species is also called **divergent evolution. Convergent evolution,** the independent development of similar features in separate groups of organisms, is not the opposite of divergent evolution, but a separate, unrelated idea. Recall from Chapter 15, for example, the convergent evolution of wings in the unrelated birds, bats, and insects. In some sense, evolution tends to be divergent. From a few simple forms, millions of species have evolved. Species are constantly splitting into two, three, or more new species. These separate species diverge even more to form separate families and orders.

Species often diverge and multiply rapidly when a new place to live becomes available. Biologists call such an event an **adaptive radiation,** the generation of many new species with widely varying adaptations. For example, nearly all the major groups of modern mammals—the carnivores, whales, bats, rodents, and primates, for example—evolved during a 10-to 15-million-year period in the Cenozoic. DNA sequence analysis suggests that these diverse groups of mammals split from one another almost simultaneously from a group of closely related species.

How Does Adaptive Radiation Occur on Chains of Islands?

Adaptive radiation is the key to understanding why there are so many species. We will first discuss examples of adaptive radiation in chains of islands and then try to understand how diversity increases worldwide.

The most dramatic cases of adaptive radiation occur on chains of islands (or archipelagos), such as the Galápagos. We have seen that the formation of new species generally requires either reproductive isolation (so that descendant species do not interbreed with one another) and one or both of two other conditions: either distinct selective pressures (so that different descendant species acquire different adaptations) or else small population size (so that genetic drift can operate).

Archipelagos sometimes provide all of these conditions. Geographic isolation from the mainland ensures reproductive isolation from mainland species. Of course, islands don't have to be surrounded by water to encourage speciation. A series of isolated lakes, mountaintops, or even forests are, for many species, equivalent to islands (Figure 17-9).

Whether the isolation is caused by water or by a surrounding desert, relatively few individuals arrive at each island. Plants may arrive as seeds carried by wind or birds. Animals arrive only if they can fly, swim, or float on a piece of driftwood or other natural raft. In short, founder populations tend to be so small in number that founder effects alone cause gene frequencies in the island populations to differ from those in their mainland parent populations. In addition, the environments of different islands differ enough to exert different selective pressures (Figure 17-10).

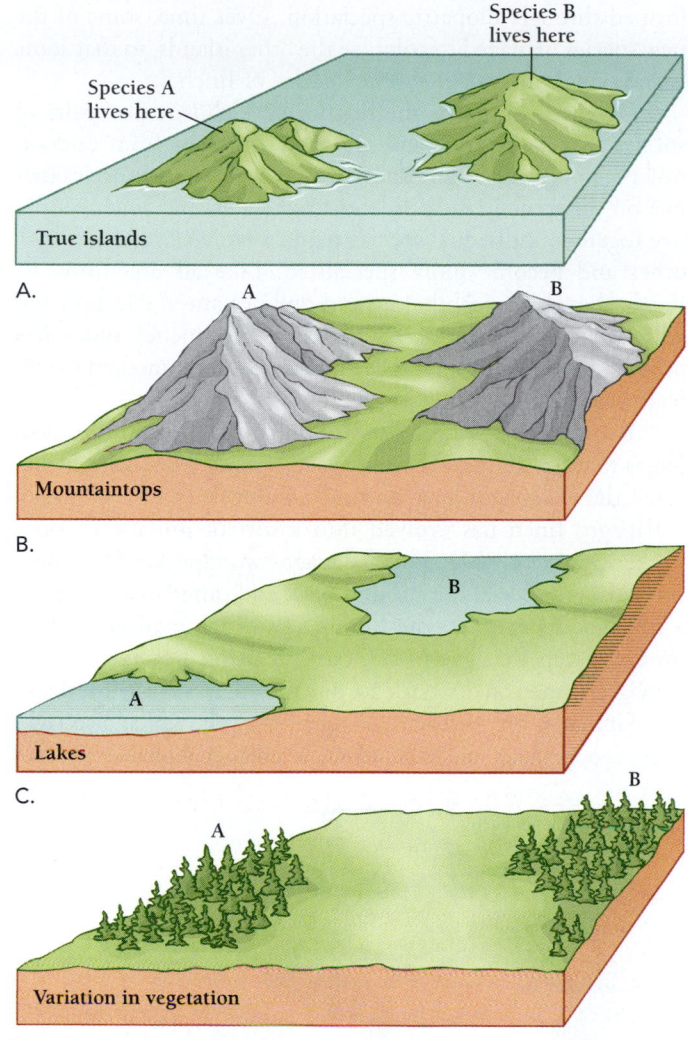

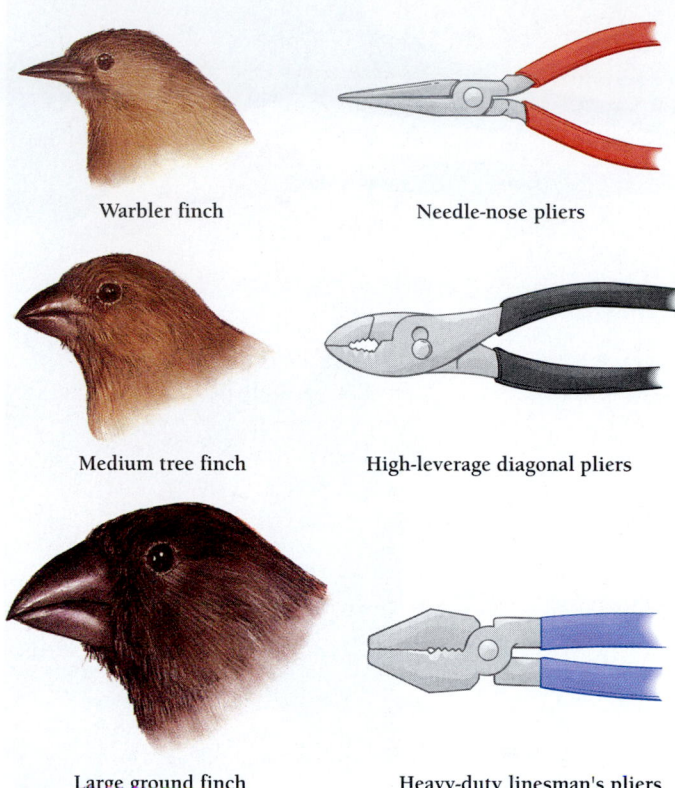

Figure 17-10 Adaptive radiation. Less than 5 million years ago, the Galápagos Islands began erupting from the floor of the Pacific Ocean. Sometime after the islands had been colonized by plants, a single species of finch colonized the islands. This finch then radiated into 13 separate species, which taxonomists have grouped into 4 genera. Here are just three, representing three genera. Each finch is specialized for a different way of life. The finch with the largest beak, for example, breaks open large, hard seeds, impossible for the other two finches to crack.

Figure 17-9 Geographic isolation comes in many shapes. Although oceanic archipelagos feature many classic cases of adaptive radiation, other kinds of "islands" also facilitate adaptive radiations. To a species of fish, a series of lakes or ponds is an archipelago. To a mouse, the individual tops of mountains or plateaus can function as islands. A. True islands. B. Mountaintops. C. Lakes, rivers, and oceans. D. Variation in vegetation.

When Darwin visited the Galápagos Islands in the Pacific and the Cape Verde Islands in the Atlantic, he noted three surprising facts (Figure 17-1):

1. The species in each chain of islands resembled those on the closest mainland, rather than those of other islands. Despite the similar climate and terrain of the Cape Verde Islands and the Galápagos Islands, the birds on the Cape Verde Islands were similar to those of Africa and those of the Galápagos to those of South America. This suggested that the islands were colonized by migrants from the mainland.

2. The archipelagos contained a much smaller number of species than the mainland. The islands' biological poverty supported Darwin's idea that only a small number of ancestral organisms had colonized the islands.

3. The individual islands in a chain had distinct species. Tortoises from the drier islands had longer necks and peaked shells that allowed them to reach the cactus and tree branches, while those from wetter islands had shorter necks, better adapted to feeding on ground vegetation.

Darwin recognized that the distribution of species in chains of islands provides a record of the arrival of immigrants from the mainland. For example, of the hundreds of bird species in South America, only eight succeeded in reaching and colonizing the Galápagos. Yet 1 of those 8 gave rise to no fewer than 14 species of finches. The Galápagos finches offer the most famous example of adaptive radiation (Figure 17-11).

The Galápagos finches are all distinct from any on the continent of South America. Yet they were enough alike to

Figure 17-11 **Vampire finch.** Galápagos finches seem to have evolved to take advantage of any opportunity. This sharp-billed finch (*Geospiza difficilis*), or vampire finch, feeds on the blood of larger birds such as this masked booby. (*Tui de Roy/Minden Pictures*)

suggest to Darwin that they all stemmed from a common ancestor. Contemplating the finches, Darwin wrote his first statement on adaptive radiation: "One might really fancy that . . . one species had been taken and modified for different ends."

The probable mechanism of adaptive radiation on islands such as the Galápagos is relatively easy to understand. Over the course of the 5 million years since the islands first appeared—and possibly as recently as a half-million years ago—a few finches (at least one male and one female) managed to fly the 1000 km (600 miles) from the South American mainland—no easy task, since finches are poor long-distance flyers. The founders were probably ground birds, with short bills useful for crushing seeds. Descendants of these birds managed to colonize the 13 islands in the Galápagos chain, as well as Cocos Island, about 1000 km away. However, because of the finches' limited flying abilities, flights among the islands must have been rare. So the finches on each island mated only with other finches from the same island. In short, the finches on each island became reproductively isolated.

Because each island differs slightly in geology, terrain, vegetation, and insect life, the finch population on each island experienced different selective pressures. Acted upon by both natural selection and genetic drift, the populations came to differ in many ways: size and form of beaks, nesting patterns, songs, plumage, and courtship behavior. Separate species

formed through allopatric speciation. Over time, some of the new species managed to colonize the other islands, so that some islands came to sustain several species of finches.

Detailed studies of the finches have shown that pairs of species that share an island are more different from one another—in beak size and other measures—than are species that live on different islands. In short, where several finch species live together, individual species tend to evolve away from each other and become more specialized. Like all organisms on Earth, these birds feel the tug and pull of natural selection and will evolve (or go extinct). Only in these finches and a few other organisms, however, have researchers documented the effects of natural selection so clearly.

Yet, it is the absence of more specialized birds in the Galápagos that has allowed the finches to radiate. For example, because the Galápagos have no true woodpeckers, one species of Galápagos finch has evolved into a sort of imitation woodpecker (Figure 17-12). The Galápagos woodpecker finch does not have the special tongue and beak and other adaptations of a true woodpecker. Yet the woodpecker finch manages to feed on the insects that live in tree bark by using a cactus spine or a twig clamped in its beak to dig the insects from the bark.

Similarly, on the island chain of Hawaii, which also lacks true woodpeckers, a species of honeycreeper digs insects from

North American hairy woodpecker

Galápagos woodpecker finch

Hawaiian honeycreeper (akiapolaau)

Figure 17-12 **Opportunity knocks.** In the absence of mainland woodpeckers, other kinds of island birds can evolve adaptations for taking grubs from tree bark.

tree bark by chiseling the bark with its lower jaw and picking out insects with its curved upper bill. Neither the Hawaiian honeycreeper nor the Galápagos finch seems as well adapted as a true woodpecker, but each manages to flourish in the "woodpecker" niche in the absence of direct competition from a true mainland woodpecker.

Plants also undergo adaptive radiation. The sunflower family (Asteraceae) provides two especially dramatic examples. In the Hawaiian Islands, 28 species of the most spectacular plants in the islands have all derived from a single small and unremarkable California tarweed. The amazing Hawaiian tarweeds, some of which are shown in Figure 17-13, include the Haleakala silversword, which grows within the cone of a volcano on the island of Maui, as well as the yuccalike iliau, which grows among the scrub vegetation of Kauai. Most of these plants, like their relatives in California, grow in dry areas, but a few have adapted to wet forests and bogs.

An equally dramatic radiation within the sunflower family has occurred on the remote Atlantic island of St. Helena, halfway between Africa and South America. The usually low-growing members of this family have evolved into large "trees." Reproductive isolation and the absence of competitors have evidently allowed the evolution of these distinctive species, which are found nowhere else on Earth.

Adaptive radiation, the evolutionary divergence of a single line of organisms, is the key to biological diversity.

How Do Species Diversify on Continents?

The multiplication of species is not confined to archipelagos. On continents, two ways that species diversity increases are **dispersal,** the spread of a taxon through a large area, and **vicariance,** the fragmentation of an already dispersed species or group of species.

Birds and bats and even insects are, of course, very good at dispersing all over the world. So are plants, whose tough seeds hitch rides on animals and winds. Ants, horses, and other terrestrial organisms may not be able to cross oceans, but they can disperse enormous distances by simply walking a little farther, generation after generation.

In vicariance, a widely dispersed population can become fragmented either through the extinction of populations in between or through the creation of geographic barriers. As we saw in Chapter 15, tectonic plates move around, alternately drifting apart and bumping into one another over millions of years. Such movement not only separates and joins whole continents, it also raises mountain ranges and lifts old sea beds that are then cut by massive canyons. Once populations are isolated by such geologic changes, they can speciate allopatrically.

The fossil record makes sense in terms of the movements of the continents. For example, the giant ground birds of Australia, Africa, and South America are all related and were sep-

arated only by the Mesozoic breakup of Gondwanaland. In contrast, the camel family evolved after the continents were already separated. The Camelidae originated in North America during the Eocene. They later dispersed into Eurasia by way of the Bering Land Bridge (long since disappeared) that connected Alaska and Siberia and into South America by way of the Central American isthmus. Today the Camelidae are extinct everywhere except Asia, North Africa, and South America.

Figure 17-13 Adaptive radiation in Hawaii. In the isolation of the Hawaiian Islands, an unremarkable plant, represented in California by tarweeds in the genus *Madia*, has radiated into 28 species. A. The California tarweed *Madia madiodes*. B. Hawaiian iliau *Wilkesia gymnoxiphium*. C. The Haleakala silversword *Argyroxiphium sandwicense*. (A, N.H. [Dan] Cheatham/Photo Researchers, Inc.; B, Gerald D. Carr; C, C.K. Lorenz/Photo Researchers, Inc.)

Likewise, the modern horse evolved in North America, then migrated to Eurasia and to South America in periods when those continents were connected with North America. Because the current land connection between North and South America was formed only about 3 million years ago, no horse fossils in South America date back before that time.

The Bering and Central American land bridges provided a route for the dispersal of many species between Eurasia and North and South America—including, for example, horses, camels, and humans (Figure 17-14). But there were no such bridges to Australia. Australia became separated from all the other continents, near the end of the Cretaceous Period, some 65 million years ago. After that, Australian species evolved in isolation. At the time Australia became separated, no horses or other grass-eaters had yet evolved anywhere on Earth. Because no horses or similar herbivores reached Australia after its separation, grass-eaters evolved in Australia and populated the plains—the kangaroos.

Kangaroos, like almost all other Australian mammals, are "marsupials"—mammals that bear young as fetuses and carry them in a pouch until the young are able to fend for themselves. In contrast, humans and other "placental mammals" bear more-developed young. Placental mammal mothers nurture their young in the uterus with the aid of a specialized tissue, called the *placenta* [Latin, *placenta* = flat cake], which passes nutrients, oxygen, and wastes between the mother and the developing embryo.

Marsupials and placental mammals diverged from each other in the early Cretaceous, and the marsupials happened to reach Australia first, in the late Cretaceous. With the exception of bats, which evolved elsewhere and later flew there, and various species of rats, which must have rafted from island to island in the East Indies, all the native mammals of Australia are marsupials.

The marsupials of Australia are an extraordinary example of adaptive radiation (Figure 17-15). All appear to have evolved from a primitive, opossumlike marsupial of the Cretaceous. The evolution of the marsupials in Australia parallels that of the placental mammals elsewhere, in what is an equally extraordinary display of large-scale convergent evolution. The present-day marsupials include animals that resemble the placental mammals of other continents, both in appearance and in habit. Although kangaroos do not look much like deer or goats, their teeth are similar, for they live a similar life. Australian marsupials include species that look like rabbits, mice, moles, anteaters, cats, wolves, squirrels, and woodchucks. Distinctly wolflike creatures have separately evolved from both a marsupial and a placental ancestor.

The parallel radiations of marsupials and placental mammals show the opportunism of evolution and help account for some of the enormous diversity we see around us.

Continental diversity increases through the dispersal of organisms to new parts of the world and through vicariance, the fragmentation of dispersed groups into populations that can evolve separately. Australian marsupials illustrate both adaptive radiation and convergent evolution.

THE CAMBRIAN EXPLOSION

Up until the Cambrian Period, almost 600 million years ago, life on Earth seems to have been limited to a few soft, one-celled organisms, such as algae and bacteria. Until quite late in the Precambrian, the only complex organisms that paleontologists find are a few jellyfishlike creatures.

Then, at the beginning of the Cambrian, life bloomed. In a burst of diversification unmatched in the history of the world, all of the modern animal phyla that have fossilizable skeletons appeared, though not necessarily in their current forms. For example, the arthropods, which today include the insects, spiders, crustaceans, and barnacles, were represented by the trilobites, crustaceans, and other marine animals. Also present in the Cambrian were the brachiopods (lamp shells), mollusks (snails, clams, and squid), sponges, echinoderms (starfish and sea urchins), and cnidarians (jellyfish and corals), not to mention the first chordates (today including amphibians and mammals).

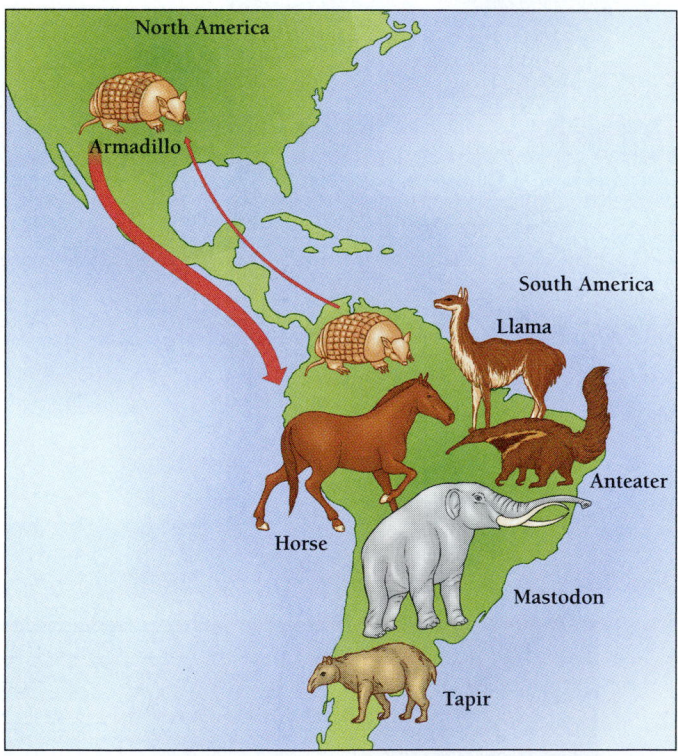

Figure 17-14 Land bridge. The North and South American continents have been isolated and connected at different times in geologic history, depending on the level of the oceans. Land bridges allow many organisms to migrate to new habitats. When sea levels rise and again isolate the continents, separate species arise through allopatric speciation.

MARSUPIALS   PLACENTALS   MARSUPIALS   PLACENTALS

Koala

Tree sloth

Wombat

Woodchuck

Marsupial mole

Common mole

Dunnart

Shrew

Banded anteater

Giant anteater

Sugar glider

Flying squirrel

Tiger cat (Quoll)

Ocelot

Tasmanian wolf

Gray wolf

Figure 17-15 Adaptive radiations in parallel. The marsupial mammals of Australia belong to an ancient and very distinct lineage of mammals. Yet, marsupial mammals (*left in each pair*) and placental mammals (*right*) have separately evolved many of the same body types. The sugar glider and the flying squirrel, for example, both have long, bushy tails and webbed limbs—and for the same reason. Both animals quickly cover great distances by sailing from the tops of tall trees down to the ground.

Taxonomists have based the classification of animal phyla on basic body plans—the two-layered radial symmetry of the jellyfish; the three-layered bilateral symmetry of flatworms; the three-layered bilateral symmetry, plus notochord, of chordates, such as ourselves, to name a few. Although enormous evolutionary change has occurred since the early Cambrian and the level of diversity is arguably as great today as it ever was, almost no new body plans have evolved in the half-billion years since these basic forms first appeared.

Evolutionary biologists refer to this one-time evolutionary radiation as the **Cambrian explosion.** Paleontologists remain divided over the exact time span during which this explosion occurred. Some evidence suggests that the major animal phyla may have diverged as many as 900 million years ago, hundreds of millions of years before the Cambrian. Other evidence suggests that all the phyla may have appeared at once at the very end of the Precambrian. Either way, evolutionary biologists are left with a major puzzle. What has prevented major new body plans from evolving since the Cambrian?

One argument is that evolution has somehow slowed since then, that changes simply do not occur as rapidly as they did in the Precambrian and the Cambrian. However, modern populations are capable of evolving quite rapidly. For example, both bacteria and insects evolve resistance to antibiotics and

BOX 17.1

Did It Come from Outer Space?

In 1979, geologist Walter Alvarez discovered a thin layer of mineral that told him that the Age of Dinosaurs ended with a bang, not a whimper. Alvarez was studying rocks in Gubbio, Italy, at what geologists call the K/T boundary. Here, rocks laid down at the end of the Cretaceous (K), 65 million years ago, are overlaid by rocks from the Tertiary (T).

At Gubbio, the boundary is clearly marked by a 2- to 3-cm layer of clay (Figure A). To find out know how many years of Earth's history was represented by that inch-deep clay layer, Alvarez measured its iridium content. The metal iridium is rare on the Earth's surface, but a constant hail of it from space lightly dusts the Earth's surface. Like the sand in an hourglass, the buildup of iridium can be used to determine how much time has passed.

Alvarez expected to find less than 1 part per billion (ppb) of iridium, and, indeed, both the Cretaceous rock below the clay and the Tertiary rock above had the expected amount—1 ppb or less. To his surprise, however, he found a whopping 10 ppb in the clay layer itself. What had caused the spike in iridium in the clay?

Collaborating with his father, Nobel prize-winning physicist Luis Alvarez, and geochemist colleagues Frank Asaro and Helen Michel, Alvarez found similar iridium spikes at two other K/T sites—in Denmark and New Zealand. In 1980, the team published a groundbreaking paper, "Extraterrestrial Cause for the Cretaceous-Tertiary Extinction," in the journal *Science*. In it, the researchers argued that the iridium spike, or *iridium anomaly*, as it is often called, was the result of a gigantic 10-km asteroid that slammed into the Earth, exploded, and drove to extinction all of the dinosaurs and many other forms of terrestrial and marine life.

Could this really happen? Paleontologists initially scoffed dismissively. But the Alvarez team argued back, and the controversy spread, so that soon even nonscientists began taking sides.

The excitement stimulated other scientists to investigate, and to date they have found iridium anomalies at more than 100 different K/T boundary sites, both on land and in ocean sediments. Even more intriguing is evidence that the asteroid may have fallen into the Gulf of Mexico.

If an asteroid 10 km across smashed into the Earth, the pressure generated would liquefy the meteorite and the rock it smashed into, spewing molten meteorite and rock into the air—just as water droplets fly into the air when a boulder is heaved into a pond. As the molten meteorite droplets cooled, they would solidify into smooth round pebbles called *microtektites* and fall back to Earth. Such microtektites, as well as *shocked quartz*—deformed pieces of the mineral quartz—have been found in association with known meteor sites, such as Meteor Crater in northern Arizona.

But the most dramatic evidence would be a huge hole in the ground. Did it exist? Until the late 1980s, no one could find the 150- to 300-km crater that scientists predicted such an asteroid would make. If the crater was deep in the ocean and now filled with sediment, it would not be easy to find. Off the Gulf Coast of Texas, however, researchers found that, just at the K/T boundary, huge boulders had been moved around as if by a tidal wave. Then another researcher reported a subterranean bowl-shaped structure, 180 km across, located in the northwestern corner of the Yucatan Peninsula near the town of Chicxulub (pronounced CHICKS-uh-loob). In 1991, other researchers reported that samples from the Chicxulub bowl contained shocked quartz. The layer of melted rock underneath the huge bowl has now been dated at 65 million years old—the age of the Cretaceous-Tertiary (K/T) boundary.

Few people now doubt that the structure at Chicxulub is the crater from an enormous asteroid that slammed into Earth 65 million years ago. That this impact caused global catastrophe is also indisputable. The impact would have incinerated plants and animals across the North American continent and spewed debris into the atmosphere, blotting out the sun worldwide for months or even years.

The only debate is over the exact consequences of that global catastrophe. Some scientists still insist that the asteroid alone did not cause the mass extinction at the end of the Cretaceous. For one thing, whole lineages of organisms disappeared long before the asteroid hit. This fact suggests that there was a more gradual increase in extinctions during the late Cretaceous, not a sudden catastrophic reduction in species.

But why else might they have gone extinct? The best alternative is massive volcanic activity in India and the Pacific Rim that released enormous amounts of volatile gases and dust into the atmosphere. Such gases could have altered the global climate, caused acid rain, and damaged the ozone layer, and caused a large-scale climatic catastrophe.

For now, paleontologists still disagree about whether the late Cretaceous extinctions were rapid or gradual. If the extinctions were sudden, the asteroid is the most likely culprit. If the extinctions were more gradual, the asteroid may have acted merely as a *coup de grâce*.

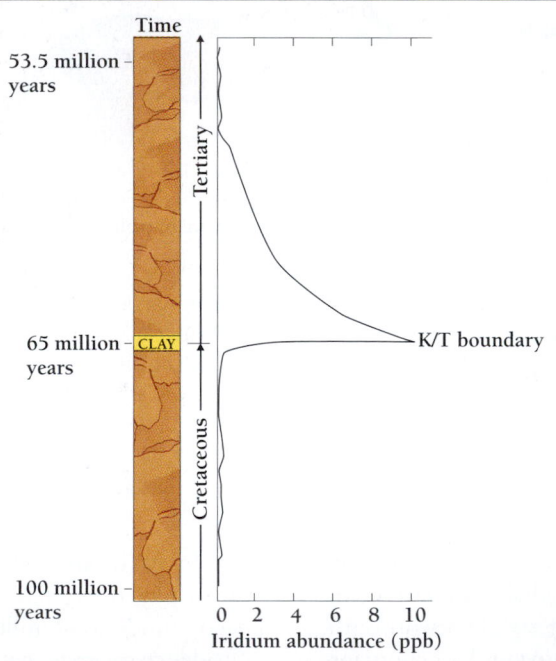

Figure A   An asteroid impact? At the end of the Cretaceous, the amount of iridium present in deposits increases dramatically from 1 part per billion (ppb) to 10 ppb, then gradually decreases again over a period of about 10 million years. Iridium is a metal rare on Earth but common in asteroids and meteors.

pesticides within a few years or even months. Vertebrates, such as fish and birds, are also capable of rapid changes. Research on the finches in the Galápagos has shown that natural selection can change their average beak size, according to what size seeds are available, within a matter of months. All of the Galápagos finches probably evolved from a single species within the last half-million years, a short period in geologic time.

Another argument is that once the major body plans formed, **developmental constraints**—the rules of embryological development that determine the general form of an organism—prevented new body plans from evolving. Many biologists believe that in most organisms, development has become so finely tuned since the Cambrian explosion that any dramatic change results in organisms too defective to flourish. The idea is that modern organisms are so specialized that an organism with an entirely new form could not compete with modern organisms.

But the ancient body plans are not necessarily optimal designs that could never be improved on. Indeed, very likely they are just accidents of history, like the homologous vertebrate limbs illustrated in Chapter 15.

Animal body plans, such as the particular arrangement of bones in the limbs of horses, whales, and humans, may simply be a case of organisms "making do" with what they have. The first airplanes were made of wood, not because this is a better material than light alloy metals, but because that was the best material available at the time. Modern airplanes can be completely redesigned, with new shapes and new materials. But organisms tend to reuse the same parts, reshaping them for new purposes.

Evolution at the species level occurs as rapidly as ever, but the evolution of new phyla seems to have ceased. The Cambrian explosion may have been a one-time phenomenon.

EXTINCTION AND THE RATE OF EVOLUTION

Evolutionary biologists estimate that more than 99 percent of all the species that have ever lived are now extinct and that the vast majority of these went extinct millions of years ago. But why species go extinct is, in most instances, something of a mystery. By looking at species that have gone extinct in historical times, we can say that some species go extinct because of competition from other species, some because their habitat is drastically changed, and some because of a new predator. Human beings, for example, have driven a variety of species to extinction—the American passenger pigeon, for example—through heavy hunting. But in most extinctions, probably no single factor determines whether a species will go extinct. Some species survive such onslaughts, while others do not.

The Mass Extinctions

The fossil record shows that, in the long run, species go extinct at a more or less regular rate. However, superimposed on this ongoing extinction rate are five intriguing mass extinctions, including the death of the dinosaurs (Figure 17-16). Near the end of five geologic periods—the Ordovician, the Devonian, the Permian, the Triassic, and the Cretaceous—the number of extinctions increased by as much as 2.5 times the background extinction rate.

Why these mass extinctions occurred is one of the most controversial questions in paleontology. Some researchers deny that the mass extinctions are qualitatively different from the background extinction rate. They argue that these mass extinctions are just random fluctuations in the usual loss of species over time. Other researchers argue that the five extinctions are the consequence of major worldwide catastrophes. The exact nature of the catastrophes is hotly debated.

The most controversial mass extinction is the one at the end of the Cretaceous. Sixty-five million years ago, all the dinosaurs and half of all species of plants and animals went extinct. Every conceivable catastrophe has been suggested to explain this extinction. But the majority of researchers now seem to believe that this most recent extinction occurred when a giant asteroid hit the Earth with enough force to blast a 180-km crater and plunge the world into almost total darkness for months.

We are now in the midst of a sixth mass extinction. The full extent of this extinction is somewhat controversial. No one knows precisely how many families of organisms have gone extinct, nor how many will go extinct before this extinction event is over. But the cause of this mass extinction is not at all controversial. It is entirely the result of human activity.

After each major extinction in the past, diversity increased dramatically (Figure 17-16). For example, the mammals radiated spectacularly after the Cretaceous extinction of the dinosaurs and other reptiles 65 million years ago. We might guess that the only circumstances under which really new organisms might flourish are where a mass extinction has eliminated much of the competition.

All species go extinct eventually. The rate of extinctions since the Cambrian has been relatively constant except for five mass extinctions. Evolutionary biologists have yet to agree on what caused these mass extinctions.

How Fast Do Species Evolve?

In 1859, Charles Darwin was nearly alone in his belief that evolution was a gradual process. He held that not only individual species but also genera, orders, and other higher taxa all arose through the slow transformation of species. His closest friends tried to persuade him that evolution might very well

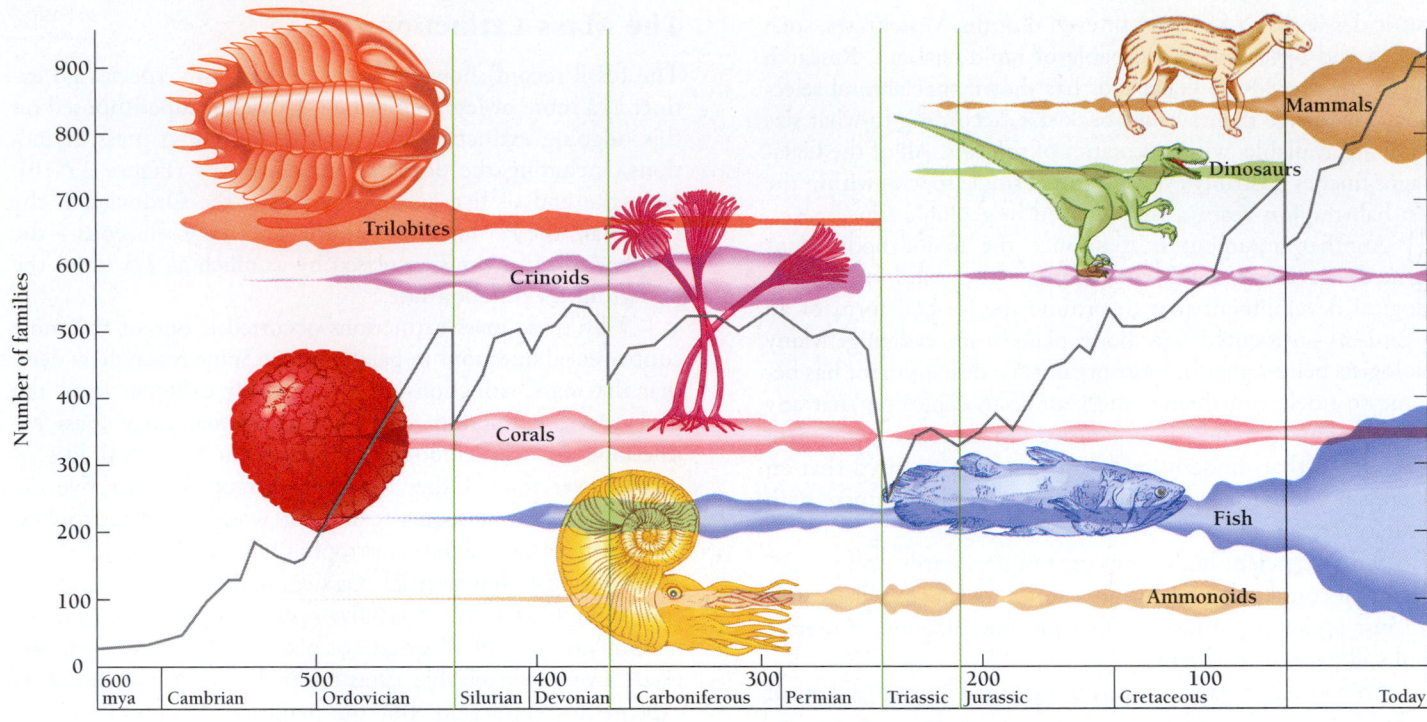

Figure 17-16 Mass extinctions and increasing diversity. This selection of animals shows that numbers of species and families tend to increase over time. Each mass extinction is followed by an enormous adaptive radiation. Recovery from a mass extinction like the one the Earth is experiencing today, however, takes 5 million years or longer. Many groups, such as the dinosaurs and the trilobites, never recover.

occur in jumps, or **saltations.** On the eve of the publication of the *Origin of Species,* in 1859, Darwin's friend Thomas H. Huxley wrote to him, "You have loaded yourself with an unnecessary difficulty in adopting *natura non facit saltum* [nature makes no jumps] so unreservedly."

For Darwin, the conflict was between what he saw everywhere around him in nature and what the fossil record showed. As we noted in Chapter 15, gaps in the fossil record pose a problem in understanding the evolution of species. To demonstrate that a species evolved gradually, we would like to see a fossil representing each small step in an evolutionary line. But, often, the fossil record does not fulfill this wish. Transitional forms between distinct species are disappointingly unusual.

Paleontologists are working with a random sample of organisms through time. They may find a well-preserved fossil of species A in one layer, then no more of that species until 50,000 years later (in the fossil beds). If the next fossil they find is different enough from the first, they may say it is a second species, B, but they won't know how B is related to A. Is B descended directly from A? Are A and B instead descended from a common ancestor C? If so, is C much older than A and B, or just a little older? Did B (or C) gradually turn into A, leaving no unevolved remnants? Or could A, B, and C have all lived at the same time?

The lack of intermediate forms may merely represent the incomplete preservation of ancient organisms. Just as most in-

dividual human beings are not named in the pages of recorded history, so most species are not preserved in the layers of rock that make up the fossil record. However, another interpretation for the gaps in the fossil record is that species change suddenly, evolving in short bursts, or saltations. The intermediate forms are not preserved, either because they never existed or because they existed only momentarily in geologic time (less than 10,000 years).

Darwin firmly rejected this latter line of reasoning. He argued that the gaps in the fossil record were the result of the incomplete preservation of ancient organisms.

Darwin had no evidence, and no reason to believe, that the hundreds of species he had studied all his life could change suddenly. First, species were so subtly different that it was hard to decide if they were true species. Three Galápagos Island mockingbird species, each of which came from a different island, were so alike that Darwin assumed they were merely races of a South American species until he took his specimens back to England. There, however, an ornithologist pronounced each a separate species. Later in his life, Darwin spent eight long years struggling to work out the taxonomy of barnacles. Each barnacle species graded so smoothly into the others that he complained bitterly about how hard it was to tell one from another. In short, everywhere Darwin looked, species seemed to grade into one another. Linnaeus also had commented on the difficulty of telling one species from another.

BOX 17.2

Modern Tropical Rain Forests: Mass Extinction or Not?

The fossil record shows five natural mass extinctions during the past 600 million years. Many biologists believe that a sixth, human-induced mass extinction is now under way.

The most species-rich regions of the world are tropical environments, such as the Amazon Basin. For any group of organisms—whether birds, insects, trees, or fungi—the number of species is greatest in the tropics. Although tropical rain forests cover less than 7 percent of the world's land area, one-half to two-thirds of the Earth's 10 million to 100 million species inhabit these rich forests.

Few of these species are known to science—only 1.5 million of the world's species have been named. A disproportionate number of these are European and North American species, where most trained taxonomists have lived and worked. Researchers are struggling to keep ahead of the chainsaws and forest fires that daily destroy the habitats of thousands of tropical forest species, half of which are insects. Tropical rain forests are dense with species. A 1-hectare (2.5-acre) plot of Brazilian forest may contain 200 species of trees, whereas the same size plot in Canada would support fewer than 10. And each species of tree may support five to ten species of insects that specialize only on that one species of tree. Eliminating that tree species causes all the insect species to vanish also. Such cascades of extinctions are not limited to plant-insect interactions. When peccaries (a piglike mammal) were eliminated in one Brazilian forest, so were three species of frogs. All the frogs, it seemed, lived in the mud wallows the peccaries created.

As of today, fewer than one-half of the original tropical forests remain. And logging, agriculture, cattle ranching, and firewood harvesting continue to degrade and destroy tropical forests (Figure A). In some cases, regions of burning forests are so vast that the fires are visible from space. Harvard biologist E.O. Wilson estimates that if tropical forests continue to be destroyed at current rates, fewer than 10 percent of the original forests will remain by the year 2050, and between 2.5 and 25 million species will go extinct.

How does this loss of species compare with the five prehistoric mass extinctions? By examining the fossil record, scientists estimate that the normal, or background, extinction rate may be as high as one extinction per year. In a mass extinction, the number of extinctions increases to at least 2.5 times the background rate. Since 1600, more than 1000 species of plants and animals have gone extinct—yielding a minimum extinction rate of more than 2.5 species per year (Figure B).

However, there were probably at least ten times as many extinctions during that time period. Recall that scientists have, at best, named only about one-tenth of the 10 million to 100 million species thought to exist. Just as the named species are a small fraction of the total species, the tally of documented extinctions includes a similarly small fraction of the total number of extinctions. For the 400 years since 1600, the minimum extinction rate may well approach 25 species per year. In addition, as the remaining tropical forests grow smaller, species are lost at a faster rate. The current loss of tropical forest species is comparable in magnitude to the prehistoric mass extinctions.

Although evolution will continue, the world will not recover its former biological diversity in our lifetime—or in our great-great-great-grandchildren's lifetime. If we can look to the fossil record for an answer, it may be 5 million years before the Earth once more supports the diversity of life that existed when Columbus sailed to the New World.

Figure A Slash-and-burn agriculture. In Brazil, farmers burn small areas of tropical rain forest. They farm for a while and then move on to burn another plot of land as young trees grow up in the last plot. As human population size increases, however, rain forests are being burned faster than they are growing back. (*Earth Scenes/© 1996 Dr. Nigel Smith*)

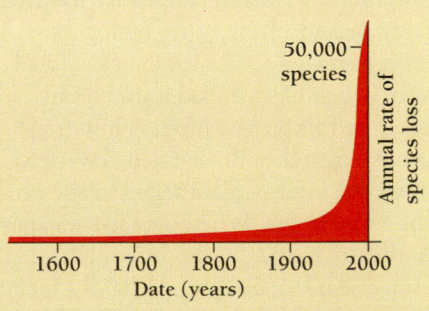

Figure B The rate of species loss has increased precipitously in the last 100 years.

Darwin's studies of artificial selection demonstrated that selection was clearly capable of producing enormous changes gradually. The dramatically different races of domestic pigeons, dogs, and sheep, all without question the result of gradual selection, would be considered separate species if found in nature (Figure 15-17).

Nevertheless, the discontinuous nature of the fossil record convinced many 19th-century biologists that evolution must occur by fits and starts, a view known as **saltationism.** After the rediscovery of Mendel's work in 1900, Mendelians argued strenuously that species arose abruptly from one generation to the next because of so-called macromutations—mutations that caused large changes in individual organisms. By the 1940s, however, Darwin's gradualism was finally in vogue and saltationism was out.

By the 1960s, most biologists subscribed to **gradualism**—the view that most species evolved gradually and continuously in direct response to selective pressures from the environment. Paleontologists could object that the fossil record did not support this idea, but geneticists dominated biology at that time, and few of them were interested in accounting for the discrepancies in the fossil record.

Nonetheless, a few renegades held out for saltationism. The most famous of these was the German-American biologist Richard Goldschmidt (1878–1958). He argued that even though macromutations, such as chromosome rearrangements, usually result in wildly deformed organisms, some of these monstrosities have potential if they happen to arise in the right environment. He called these potential species **hopeful monsters.** "A fish undergoing a mutation which made for a distortion of the skull carrying both eyes to one side of the body is a monster," he argued. "The same mutant in a much compressed form of fish living near the bottom of the sea produced a hopeful monster, as it enabled the species to take to the life upon the sandy bottom of the ocean, as exemplified by the flounders" (Figure 17-17). Goldschmidt's term "hopeful monsters" was both memorable and unfortunate. Evolutionary biologist Ernst Mayr's equally memorable term of derision for Goldschmidt's idea—"hopeless monsters"—was quickly picked up by other biologists.

Nevertheless, by 1950, the fossil record, which Darwin had found so imperfect, had improved little. A hundred years of research still revealed few cases with all the intermediate stages in the evolution of a lineage. When new species appeared, they did so suddenly, the intermediate stages leaving no trace. And many groups hardly seemed to change at all over long periods of geologic time. For example, garfishes and sturgeons show only minimal morphological change over 80 million years. Horseshoe crabs have barely changed at all in 230 million years. Such stability over long periods is called **stasis.**

The periods of apparent stasis in the fossil record may not be what they seem, however. Only hard tissues such as bones and shells are preserved. Even extensive changes in soft anatomy, physiology, and behavior of a species would not appear in the fossil record.

Ironically, paleontologists often ignore intermediate forms. Faced with a series of fossils, paleontologists tend to classify individual fossils in a layer as one species or another, not as "something in between." Like Darwin before them, they may have trouble making distinctions between two organisms, but they have to call them something. And classifiers normally say that two fossils are either the same species or two different species.

An intensive study of 15,000 trilobite fossils from Wales provided a good example of what happens when a scientist looks instead for the in-between species. The trilobites, a massively successful group of arthropods that flourished through the Permian, showed so many intermediate forms that classifying individual fossils into species became an exercise in arbitrariness.

Nonetheless, even if we accept that species may not actually evolve quite as abruptly as paleontology suggests, we can still try to explain why speciation should seem sudden at all. Species' abrupt appearances suggest two alternatives. In one scenario, new species appear almost instantaneously, as, for example, in a single generation through saltations. In the other scenario, new species evolve quite rapidly—by geologic standards—over just millions or even mere thousands of years. In either case, the intermediate forms would not appear in the fossil record. The Mendelians and Richard Goldschmidt had hypothesized that new species appeared almost instantaneously.

But the second idea—rapid, but not instantaneous, speciation—is entirely con-

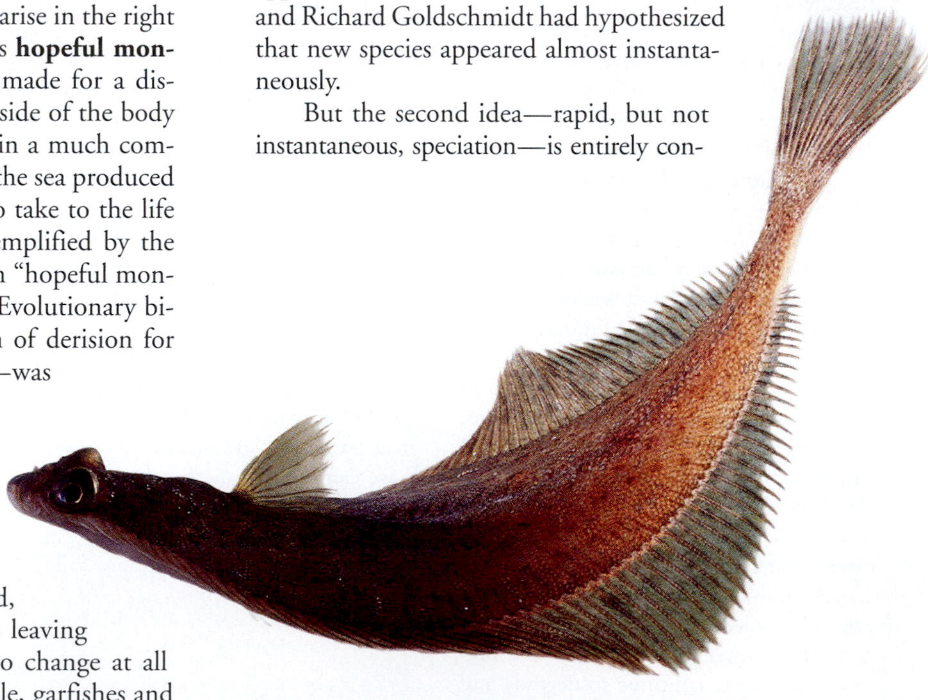

Figure 17-17 Hopeful monster? The flounder lies on the bottom of the sea on its side. To compensate for this position, it has evolved so that its eyes are both on one side of its body (the top side). The flounder begins its life like any other fish, with one eye on each side of its head. As the embryonic fish develops, however, one eye migrates over the top of the head to join the other eye. As an adult, the fish lies on its side in the sand and the mud at the bottom of the ocean and looks up. (*Heather Angel/Biofotos*)

sistent with what population genetics had suggested about speciation in small, isolated populations. The founder effect, allopatric speciation, and even sympatric speciation are all consistent with the idea of rapid speciation in small, isolated populations. Ernst Mayr himself had suggested this idea in 1942. In the case of rapid speciation, intermediate forms would exist in such small numbers and for such a short time geologically that the chances of their being preserved in the fossil record would be almost nil.

Until the 1970s, biologists paid little attention to the connection between rapid speciation and the gaps in the fossil record. Most biologists were interested in microevolution and population studies, not macroevolution. And paleontologists, interested primarily in macroevolution, wondered how biologists could go on believing in gradualism when there were so few examples of it in the fossil record.

In the 1970s, two paleontologists—Niles Eldredge at the University of California, Berkeley, and Stephen J. Gould at Harvard—proposed that the fossil record is exactly what we would expect it to be if the formation of new species was very rapid compared to the accumulation of changes within a species.

This view, called **punctuated equilibrium,** proposes two main ideas. First, *species change very little most of the time* (stasis). Second, *most anatomical or other evolutionary change in individual species occurs during a geologically brief period at the time of speciation.* In this view, speciation events punctuate an otherwise stable equilibrium, and the most important events in evolution are those that lead to reproductive isolation. Discontinuities (even catastrophes) are paramount, and gradual changes within species are mere fine-tuning.

But punctuated equilibrium does not contradict the findings of population geneticists or Darwin's view that evolutionary change occurs gradually. The brief periods during which populations can probably speciate—5000 to 50,000 years—are invisible in the fossil record (and therefore "instantaneous"), but these brief periods in evolutionary history provide ample time for *gradual* change to occur.

How fast can species form? At the extreme end, plants can speciate in one generation through polyploidy. Several species of fish have formed in a lake near Lake Victoria, in Africa, in just 4000 years. And Lake Victoria itself gave rise to 300 species of fish in only 200,000 years. One group of researchers has even suggested that the lake had dried out 12,000 years ago, which would mean that the fish evolved since then, a blink of the eye in geologic time.

Punctuated equilibrium and gradualism are extreme ends of a continuum of possible scenarios (Figure 17-18). The two ideas are not truly alternative theories that can be tested against one another. It's just a question of how much change accumulates how fast. No biologist believes that *all* evolutionary change occurs only gradually over millions of years. And few biologists would argue that all speciation occurs suddenly in just a few generations. Real species can most likely evolve rapidly or slowly, depending on circumstances such as how much genetic variation they have and what their environment is like. Whether *most* species have formed in relatively brief time spans, of 5000 to 10,000 years, or in longer time spans, say 75,000 to 100,000 years, is a question we may never answer. The fossil record is mute on this point.

The gaps in the fossil record between an ancestral species and its descendant species represent periods of rapid evolution that are invisible in the fossil record. However, such rapid evolution—on the order of 5000 to 50,000 years—is consistent with Darwin's ideas about gradual evolution.

Punctuated
equilibrium

Gradualism

Figure 17-18 Is evolution gradual or abrupt? The answer probably depends on the organism. Punctuated equilibrium and gradualism are the ends of a continuum.

Figure 17-19 **Phylogeny of primates.**
By measuring base-pairing between matching sequences of DNA from different species, biologists can determine relative differences and estimate the time in years since two species diverged. This phylogenetic tree, for example, suggests that humans diverged from bonobos 5 to 10 million years ago.

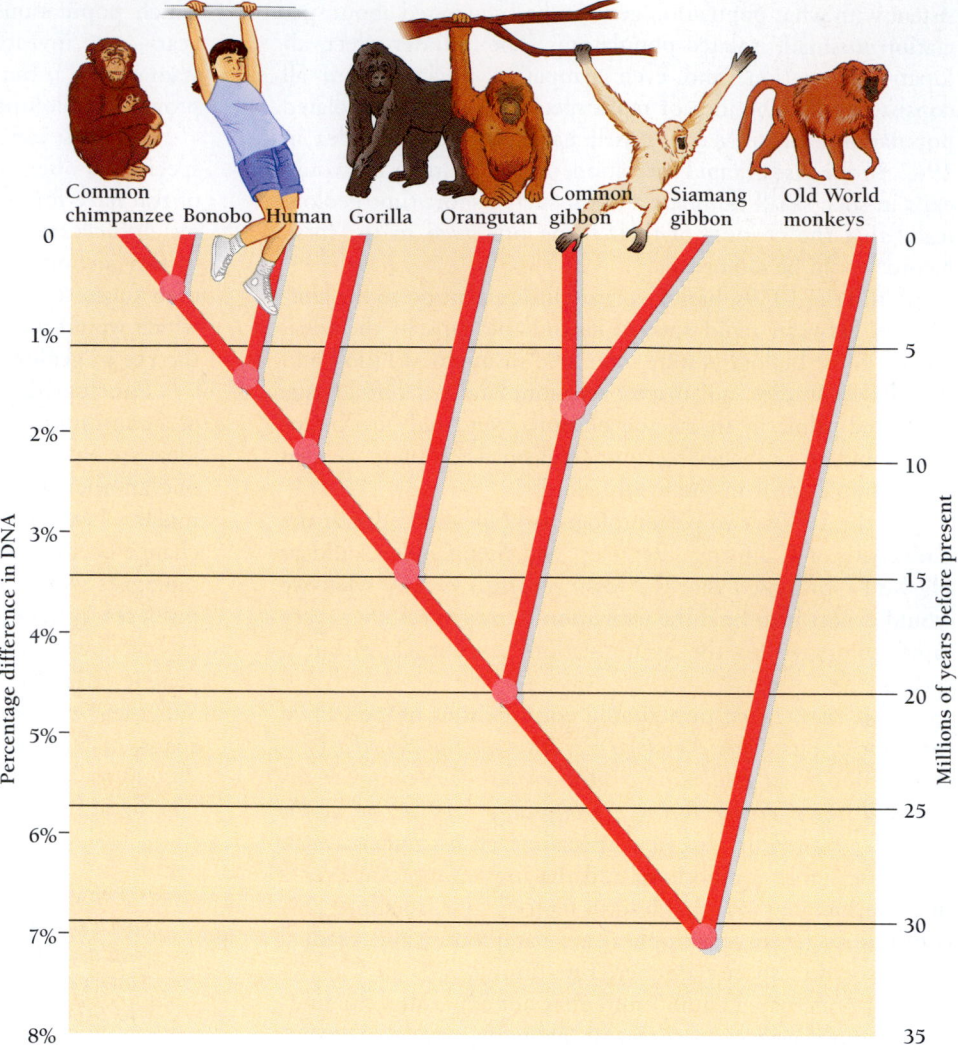

What Is Phylogeny?

Phylogeny is the history of descent of groups of organisms from their common ancestors. Phylogeny includes the order in which different groups branch off and the time in history when branching occurred. Phylogeny is very like what we humans called "genealogy," except that phylogeny applies to all living organisms instead of just one human family (Figure 17-19).

Biologists classify species according to their phylogeny, how they are related. Groups of related species are put into a genus, groups of related genera are put into families, and so on. (In the same way, we may classify our cousins by which side of the family they're from.)

But species don't always leave a clear record of who's descended from whom, which makes biologists' job difficult. When the fossil record is poor, dozens of large groups of species may appear to emerge all at once. And DNA studies suggest that rapid speciation leaves an equally muddy record in the genes.

Yet, we know that these species did evolve and that there is one true phylogeny. We just don't always know what it is. And sometimes we know what it is, but it's not what we used

to think. As we'll see in Chapter 19, biologists now know that crocodiles are more closely related to birds than to lizards. Despite this knowledge, however, birds and reptiles are two separate classes. And the crocodiles are tucked in with the lizards. Even though we still classify crocodiles this way, we know better.

HOW DID HUMANS EVOLVE?

When we visit the zoo, we are often drawn to the animals that most resemble ourselves—the monkeys and apes. Taxonomists say that apes, monkeys, and lemurs all belong to the same order of mammals. This group, called the **primates,** are remarkably unspecialized. Like the earliest mammals, we have five digits on each hand or foot and unspecialized teeth. All primates lack, for example, the sharp cutting teeth and broad grinding surfaces that characterize the teeth of horses, rabbits, and other herbivores. We also lack the long canines and shearing teeth of carnivores such as dogs.

Nonetheless, despite our lack of specialization, we primates have a set of common characteristics that together dis-

A.

B.

Figure 17-20 Platyrrhines and catarrhines. A. Howler monkey. The platyrrhine monkeys (from South and Central America) have nostrils that point sideways. B. Green monkey. The catarrhines—the Old World monkeys and apes—diverged from the platyrrhine monkeys from 17 to 20 million years ago. The catarrhines' nostrils point downward. *(A, Tom & Pat Leeson/Photo Researchers, Inc.; B, Glenn Vanstrum/Animals Animals)*

tinguish us from other mammals. These include large brains and binocular vision, flexible shoulder joints, and hands and feet with five grasping digits, a grasping thumb, and flat fingernails rather than claws.

Most of the 166 species of living primates are tree dwellers. Our flexible shoulder joints and grasping hands and feet enable us to climb trees and swing from one branch to another. As a group, primates are basically herbivores that live on leaves and fruits. Nonetheless, many primates are opportunistic omnivores who will occasionally eat grains, grubs, and even flesh. Most 20th-century humans eat grains and flesh regularly.

How Do We Classify the Apes?

Primates are one of the oldest orders of mammals, dating back some 65 million years. Early primates, which more closely resemble modern lemurs than monkeys or apes, had smaller bodies and longer snouts than modern monkeys and apes (Figure 17-19).

Monkeys and apes consist of two large groups. The more primitive group are the **platyrrhines** [Greek, *platyus* = flat + *rine* = nose], monkeys whose flat noses have widely separated nostrils that point sideways (Figure 17-20A). Many platyrrhines also have prehensile (grasping) tails. All platyrrhines live in South and Central America. Based on fossil and anatomical evidence, zoologists have concluded that they first evolved about 26 million years ago.

The second group, called **catarrhines** [Greek, *kata* = down + *rina* = nose], evolved in Africa and Asia, somewhat later. The catarrhines have downward-pointing nostrils and no prehensile tail (Figure 17-20B). Gibbons and other primitive catarrhines first appeared in East Africa 20 to 17 million years ago (Figure 17-21A). All of the catarrhines except for the humans still live only in Africa or Asia.

A.

B.

Figure 17-21 Two hominoids. The hominoids include the primitive gibbon and the more evolved bonobos and humans. A. White-handed gibbon. The gibbons are the most primitive of the hominoids. B. Bonobo and her baby. *(A, Gerard Lacz/Peter Arnold, Inc.; B, M. Long/Visuals Unlimited)*

Taxonomists group the catarrhines into the catarrhine monkeys, on one hand, and the **hominoids** [Latin, *homo* = human], on the other. Hominoids have large skulls and long arms, and they tend to walk at least partially erect (Figure 17-21B). The hominoids include the gibbons, which are relatively primitive, as well as gorillas, baboons, orangutans, humans, and chimpanzees.

Exactly how all the different apes are related to each other and to humans is a subject of great debate among all kinds of biologists. Molecular biologists compare protein or DNA sequences and draw one conclusion. Zoologists and **physical anthropologists** [Greek, *anthropo* = human], researchers who study humans, compare anatomy and behavior and draw slightly different conclusions. Paleontologists focus on the fossil record and see a subtly different picture. All these researchers agree, however, that humans share a common ancestry with other apes.

Were it not for our tendency to distance ourselves from other animals, humans would probably be lumped into the same family as the gorillas, chimpanzees, and orangutans—together called the **Pongidae.** We are closely related to both gorillas and chimpanzees. Our own DNA differs from that of chimpanzees and their cousins the bonobos ("pygmy chimps"), for example, by less than 1 percent.

Despite the similarities between humans and pongids, we humans have given ourselves a separate family, the **Hominidae.** Fossil and molecular data suggest that the common ancestors of orangutans diverged from those of the other pongids and the hominids about 12 million years ago. Hominids diverged from the chimpanzees and bonobos only 5 to 6 million years ago.

Homo sapiens is not the only hominid. We count as relatives a number of different species of hominids, some of which lived more than 4 million years ago. Two characteristics distinguish all hominids from other apes: (1) hominids are **bipedal,** that is, they consistently walk on two rather than four feet, and (2) hominids have rounded rather than rectangular jaws (Figure 17-22).

All hominid species except ours are now extinct, but anthropologists have discovered the buried bones of hominids throughout Africa, Asia, and Europe. In 1996, researchers digging near the shores of Lake Turkana, in East Africa, discovered the oldest known hominid. *Australopithecus anamensis,* as its discoverers named it, is 4.2 million years old and the earliest example yet of an **australopithecine** [Latin, *australis* = south + Greek, *pithekos* = ape]. *A. anamensis* had a small brain, a squarish jaw, and long apelike arms. But it walked upright.

How are we different from *Australopithecus?* Later hominids such as ourselves have large brain cases. Our backs arch, and our thigh bones attach differently to the pelvis and angle inward, so that we are "knock-kneed" compared to the pongids (Figure 17-23A). We have lost the opposable toe that characterizes other primates. Instead, we have the reduced toes and long feet seen in other running animals, such as horses and dogs (Figure 17-23B).

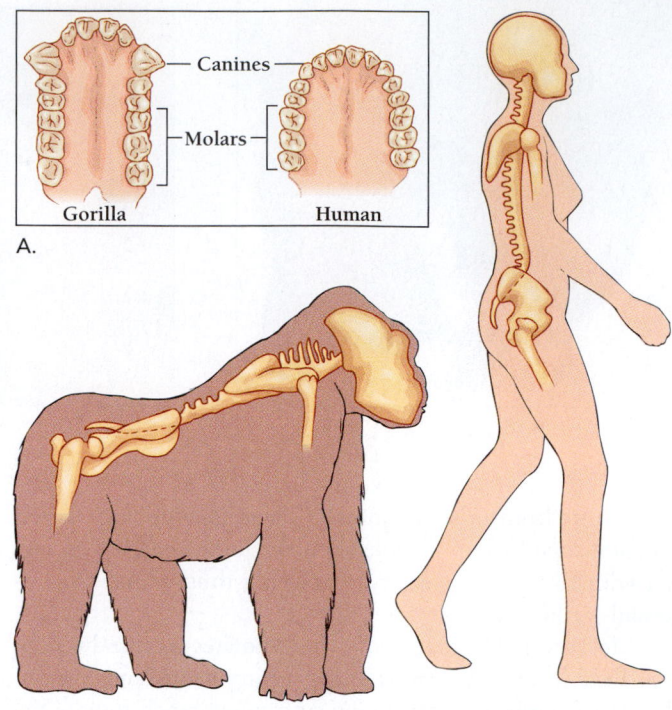

Figure 17-22 Hominid characteristics. Several characteristics distinguish *hominids* from other *hominoids*. A. Hominids tend to have more rounded jaws than chimps and gorillas. B. In addition, because hominids walk upright, we have arched rather than straight spines.

In addition, *Homo sapiens* combines extremely sparse body hair (compared to other apes) with an extravagant growth of hair on the head and, in the case of males, on the face as well. When illustrating hominids, artists are forced to decide whether to draw our ancestors naked like us, or hairy like pongids. Most artists compromise and draw our ancestors as semihairy, but with short hair on the head. In reality, the degree to which other species of *Homo* shared our strange distribution of hair will probably never be known, as hair is rarely fossilized.

Primates are divided into the lemurs, the platyrrhine monkeys, and the catarrhine monkeys and apes. The catarrhine apes, including every ape from gibbons to humans, are called hominoids. Humans and their relatives are hominids, while chimpanzees, gorillas, and orangutans are pongids.

Why Did *Australopithecus* Stand Up?

The question of how walking hominids evolved from a more primitive ancestor has haunted anthropologists for more than a century. Some researchers have argued that the evolution of

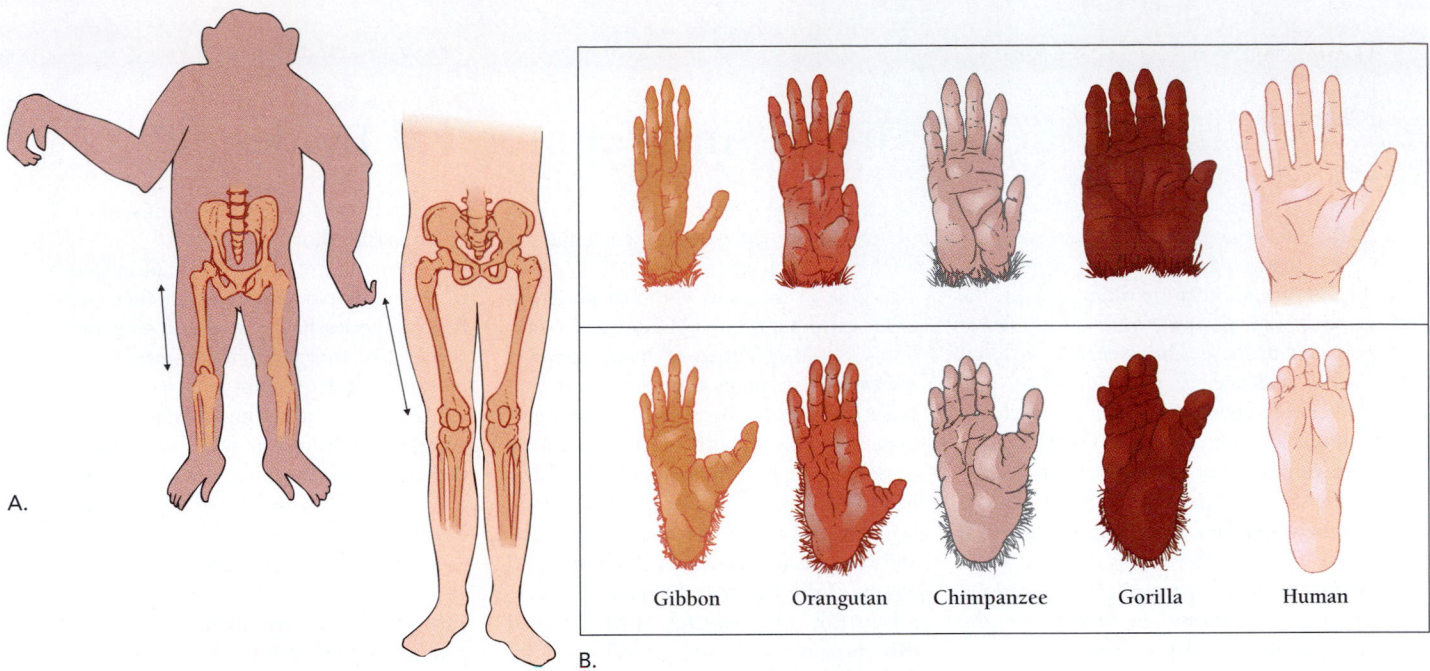

A.

B.

Gibbon   Orangutan   Chimpanzee   Gorilla   Human

Figure 17-23 Tools for walking. A. Humans and other hominids have evolved the long lower limbs that characterize dogs, cheetahs, horses, and other animals that walk or run. Our legs attach to our hips differently from those of chimps, making us relatively "knock-kneed." B. Our feet are proportioned differently. Our big toe is no longer very good at grasping, and all our toes are much reduced compared to those of other apes. In this respect, we resemble horses, which also have reduced digits.

an upright stance and a hairless body was an adaptation for remaining cool on a hot African savanna millions of years ago. The idea is that a person (or ape) standing up exposes less surface area to the sun. Other researchers have suggested that the first hominids stood up because they could run faster that way—either from predators, or, alternatively, in pursuit of prey. Still others have argued that hominids stood up so that they could carry and wield weapons and other tools. Some now argue that hominids became walking apes when they began scavenging over large territories.

Most of these ideas lack specific evidence in their favor. The only certainty is that an ancestral ape did evolve an upright stance and the long legs of a long-distance walker.

Nearly all of the hypotheses about why hominids evolved an upright stance depend on the idea that hominids left the forest to live in an open grassland, or savanna. One of the most persistent theories is that our upright stance evolved in association with a global cooling 2.5 to 2.0 million years ago. At that time, the world was a colder and drier place. So much water was frozen into glaciers in the far north and south that ocean levels worldwide were much lower than today.

Researchers have theorized for many years that the dry climate caused African forests to turn into savanna. Our herbivorous, tree-dwelling forebears then abandoned both the remaining forest and their vegetarian habits and began hunting other animals out on the savanna, or so goes this widely accepted story. Out on the savanna, hominids evolved an erect

posture, stone tools, large brains, and hairless skin. This story has been repeated so many times and with such assurance that most nonscientists are under the impression that it is an accepted scientific fact.

Unfortunately, it is just a story. Recent studies of the remains of plants in 2-million-year-old soils in East Africa suggest no dramatic change from forest to grassland between 2.5 and 2.0 million years ago. Researchers have concluded that hominids of the area most likely never encountered an open grassland. Other research, on the dust in ocean sediments, has shown only a single *dramatic* change in climate in Africa in the last 5 million years. This dry period came about 2.8 million years ago, more than a million years too late to account for the evolution of the first known bipedal hominid, 4.2 million years ago *(A. anamensis)*. This arid period, therefore, could not have been responsible for the evolution of bipedalism.

Later hominids did not necessarily live on the savanna, either. The 3.2-million-year-old hominid *Australopithecus afarensis,* until recently one of the oldest hominids known, was found at two sites in East Africa. One site was a dry, open habitat, as anthropologists had expected, but the other site was a woodland divided by ancient rivers.

Anthropologists trying to save the "savanna hypothesis" argue that bipedalism evolved in a dry, open habitat. It does not matter, they say, where other, later hominids lived or what the climate was like worldwide. They are pinning their hopes for validation on the idea that *A. anamensis,* the oldest known

BOX 17.3

What Is the Nature of Anthropological Evidence?

The kinds of evidence that molecular biologists use when studying living organisms and their relationships to one another are different from the evidence that paleontologists and anthropologists must use. The essence of most scientific evidence is reproducibility, and studies of living organisms tend to yield reproducible results, provided the studies are done carefully. For example, if a molecular biologist finds that chimpanzees share a certain proportion of the same alleles with humans on Tuesday in Los Angeles, a different researcher, using the same material and methods, should be able to get the same result on Friday in Boston.

Anthropologists, however, have limited material with which to work, and much of their evidence is not reproducible. For example, the number of fossils of early apes and humans is minuscule. A species of ape may be represented by a single jaw bone or a few teeth.

One of the most complete skeletons of an early hominid is "Lucy," a 3.2-million-year-old specimen of *Australopithecus afarensis*. Researchers strongly suspect that Lucy was female. Based on that assumption, they have hypothesized what a male of the same species would look like. They have further assumed that the male would be larger by a certain percentage than Lucy, and some textbooks report the height and weight of this theoretical male as if it were a fact. But other researchers have argued that there is no certainty that Lucy is female. Maybe "she" was a male. Further, without hundreds of complete skeletons of the same species, there is no way to know if Lucy was in any way typical of males *or* females. Lucy may have been unusually tall or unusually short.

Nonetheless, the scarcity of evidence in no way detracts from the fact that a long line of species has existed and that those species are morphologically intermediate between modern apes and modern humans. We may never know either the true average dimensions of skulls and limbs of each species or whether males and females were mostly the same size or different sizes. Nor may we ever know with certainty how each ancient hominid species is related to any other. But we do know that we are closely related genetically to modern chimpanzees and other apes. And we do know that the 3.2-million-year-old fossil Lucy walked upright, just as we do. There may be only one Lucy, but Lucy lived.

---

bipedal hominid, lived in an arid, open habitat. Other researchers argue that huge rivers fed Lake Turkana, where *A. anamensis* lived. These rivers and the lake itself, these researchers assert, must have been rimmed with forests. Anthropologists are now vigorously debating whether the habitat around Lake Turkana was arid or wet 4.2 million years ago.

The hypothesis that the first hominids stood up as a result of moving from the dark forest to a bright, open savanna carries great appeal. So far, however, no one has provided much evidence for it. Other related theories are similarly weak. The idea that standing up and losing the fur were adaptations to the heat of the savanna has little support, as most other animals on the African savanna walk and run on four legs and retain their fur. The ostrich, which is bipedal and runs well, has a broad, horizontal, black back, well suited to absorbing heat. It seems unlikely that the ostrich's bipedal gait is an adaptation for staying cool. Excluding naked mole rats, which live underground, the few African mammals that lack fur are large herbivores—elephants, rhinoceroses, and hippopotamuses, for example.

Did the first hominids stand up in order to carry something? Possibly, but that something was emphatically not stone weapons. The earliest stone tools, which experts say were most likely used to chop and mash vegetables (not to kill zebras), do not appear in the fossil record until about 2.5 million years ago, long after our ancestors stood up. Absent stone tools, it's difficult to imagine our australopithecine ancestors fashioning any other tools or objects important enough to confer a selective advantage on an upright posture. Some possibilities are wooden tools or woven baskets, which usually rot away and so do not appear in the paleontological record.

Another object an ape or australopithecine might have found worth carrying around is a baby. The babies of other monkeys and apes cling to their mother's fur, however. Human babies are born relatively undeveloped compared to other primate infants and are unable to cling. Biologists hypothesize that this is because our heads are so big that if babies waited any longer in the womb, they could not pass through the mother's pelvic bones. But our large heads apparently evolved *after* our upright stance. So researchers would have to hypothesize that hominids lost their fur before they evolved an upright posture, an idea for which there is no evidence.

With luck, researchers may someday agree about what selective pressures caused hominids to evolve an upright stance. For the moment, we can only say with certainty that they did. From the earliest australopithecines to the later ones, the skeleton increasingly became specialized for walking and running while upright.

No one knows exactly why hominids evolved the back, hips, and legs of a bipedal walker and runner. Most anthropologists assume that our ancestors left a life in the trees to live in an open, probably grassy, habitat. Researchers assume that the first hominids had to walk long distances and that this new lifestyle created selective pressures for an upright stance. But we may never know exactly where our ancestors were walking or why.

Who Were the Earliest Known Hominids?

The australopithecines fall into three to six distinct species, which lived at slightly different but overlapping times (Figure 17-24). All of them come from Africa, south of the Sahara desert. In fact, all hominids older than about 2 million years come from Africa, and most anthropologists agree that the first hominids evolved in Africa. The different species of australopithecines appear to have lived between 4.2 and 2.5 million years ago. For example, *A. afarensis* has turned up in Chad and in East Africa, where a nearly complete skeleton was unearthed (Figure 17-25).

This specimen of *A. afarensis,* named "Lucy" by her discoverer, was small, only about 1.1 meters (3 feet 7 inches) tall and weighing less than 30 kg (about 60 pounds). Lucy's adult body was the size of a 9-year-old *Homo sapiens,* but her brain was far smaller. Brain size for most normal adult humans ranges from 1000 to 1600 cubic centimeters (cm^3) (although the full range is 900 to 2000 cm^3). In contrast, the brain of *A. afarensis* was about 400 cm^3, about the size of a chimpanzee's brain (300 to 500 cm^3).

Other australopithecine species, taller and with larger brains, appeared during the next 2 million years. But the first fossils to be recognized as members of the same genus as ourselves—*Homo*—date from 2 to 3 million years ago. These hominids all used crude stone tools, which is one measure that sets them apart from the australopithecines. Members of the

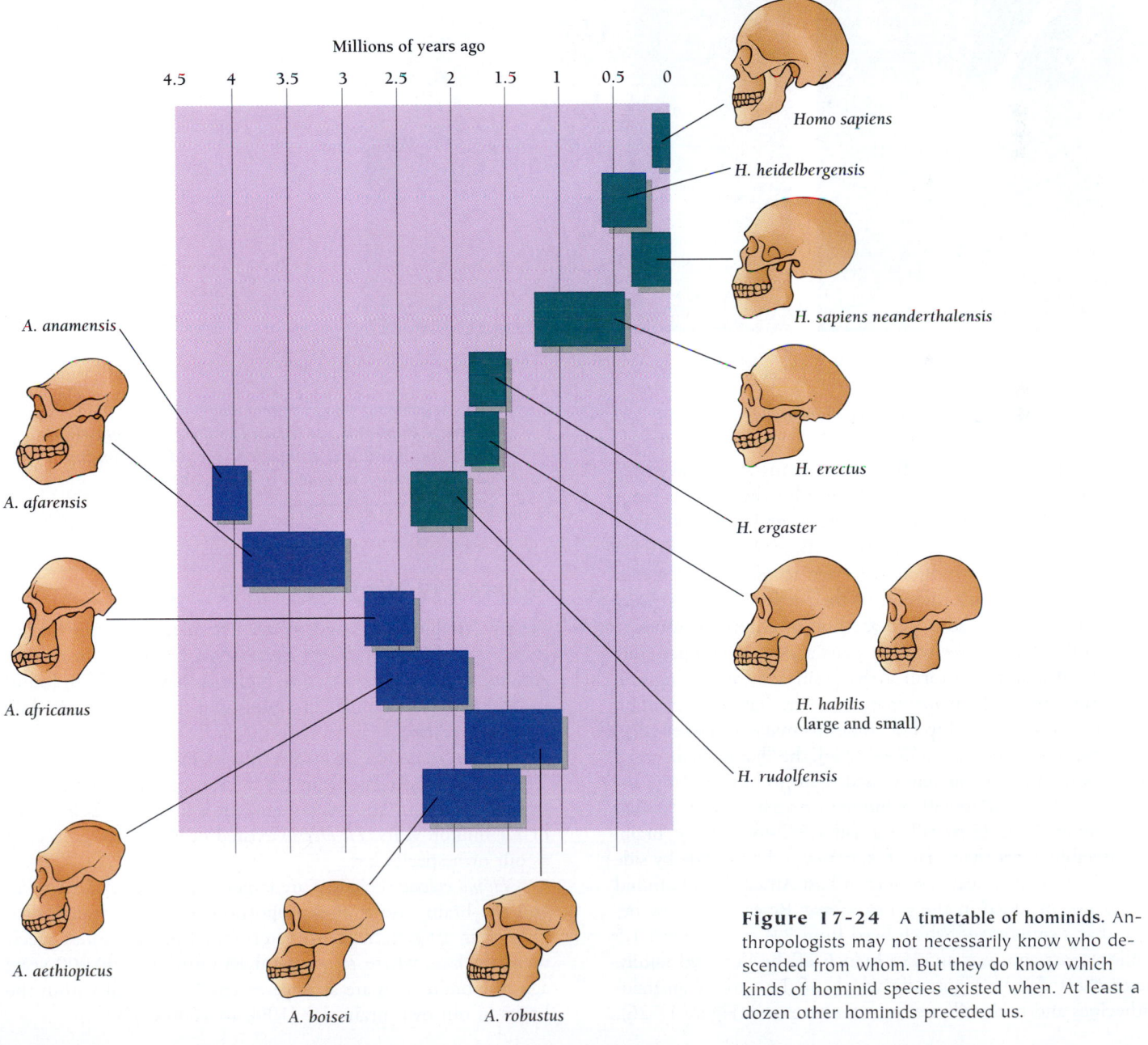

Figure 17-24   A timetable of hominids. Anthropologists may not necessarily know who descended from whom. But they do know which kinds of hominid species existed when. At least a dozen other hominids preceded us.

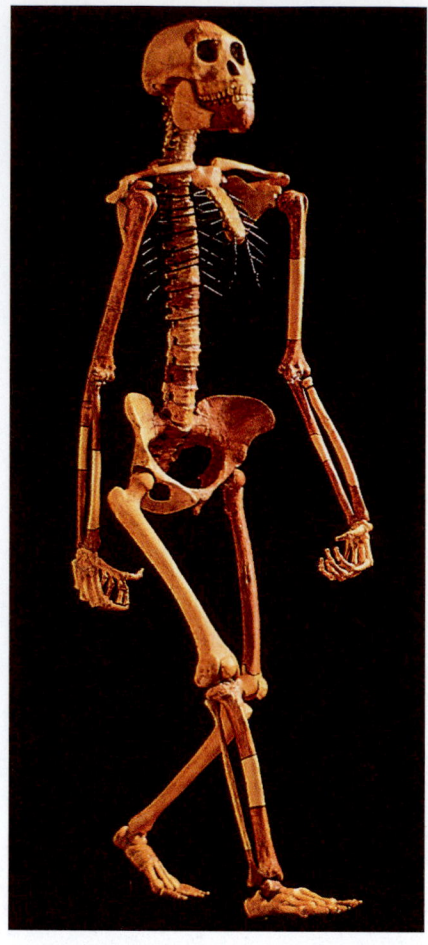

A.

B.

Figure 17-25   *Australopithecus afarensis.* One of the earliest australopithecines, *A. afarensis* is represented both in the nearly complete skeleton of "Lucy"—discovered in Ethiopia in the 1970s—and in other, more fragmented, fossils from Tanzania. Evidence that this australopithecine was bipedal comes not only from the pelvis, but from the extraordinary footprints found by Mary Leakey and her colleagues. A. Reconstruction of the *A. Afarensis* fossil "Lucy." B. Footprints of *A. afarensis*. A, *Dr. Owen Lovejoy and students, Kent State University/Photo © 1985 David L. Brill; B, John Reader/Science Photo Library/Photo Researchers, Inc.)*

genus *Homo* include *H. ergaster, H. habilis, H. erectus, H. heidelbergensis,* and *H. sapiens.*

The earliest fossils classified in the genus *Homo* have been discovered in Ethiopia. Estimated dates for these fossils range from 2.7 to 2.2 million years. No one is yet sure exactly how old they are. *Homo ergaster* appears in different parts of Africa from as early as 2.35 million years ago to as late as 1.4 million years ago. Toward the end of their existence, *H. ergaster* peoples used a more advanced method of chipping stones to form tools. Two or three million years ago, they chipped only one side of a stone to form a sharp surface for cutting and chopping—probably plant material. After about a million years, they learned to chip both sides to make a sharper edge.

Homo habilis [Latin, *habilis* = able], the "handy man," used stone tools and also apparently used them to butcher large animals, which *H. habilis* either hunted or scavenged. *H. habilis* stood about 1.7 m (5 ft) tall and had a 590-to 700-cm³ brain, considerably larger than that of *A. afarensis*. Living side by side with *H. habilis* in at least two sites in East Africa was a hominid that is not included in the genus *Homo*. Researchers now believe that *Paranthropus,* which lived from 2.0 million until 1.5 million years ago, was another branch of the hominid family. Researchers dispute vigorously which of the various australopithecines and hominids are our true ancestors (Figure 17-26).

The early hominids, or australopithecines, were short and small brained compared to modern humans and other members of the genus Homo.

Where Did We Come From?

Nearly 2 million years ago, *Homo* left Africa and migrated to Asia. Asian fossils of *Homo erectus* exist from Java in eastern Asia to northern China. *H. erectus* fossils have also been found in Africa, so it's reasonable to guess that the *H. erectus* in Asia came from Africa.

Homo erectus was by far the longest-lived species of *Homo*. The earliest fossils date from 1.8 million years ago and the last date from just 100,000 years ago, for a total species life span of 1.7 million years. *H. erectus* existed nearly 17 times as long as our own species has.

Homo erectus skeletons are larger than those of *H. habilis*, and the brain case is also proportionally larger, ranging to more than 1000 cubic centimeters (cm³) in the earliest specimens. In Java, where *H. erectus* lived until just 100,000 years ago, the brain cases are as large as 1300 cm³, well within the limits of our own brain size (1000 to 1600 cm³).

Millions of years ago

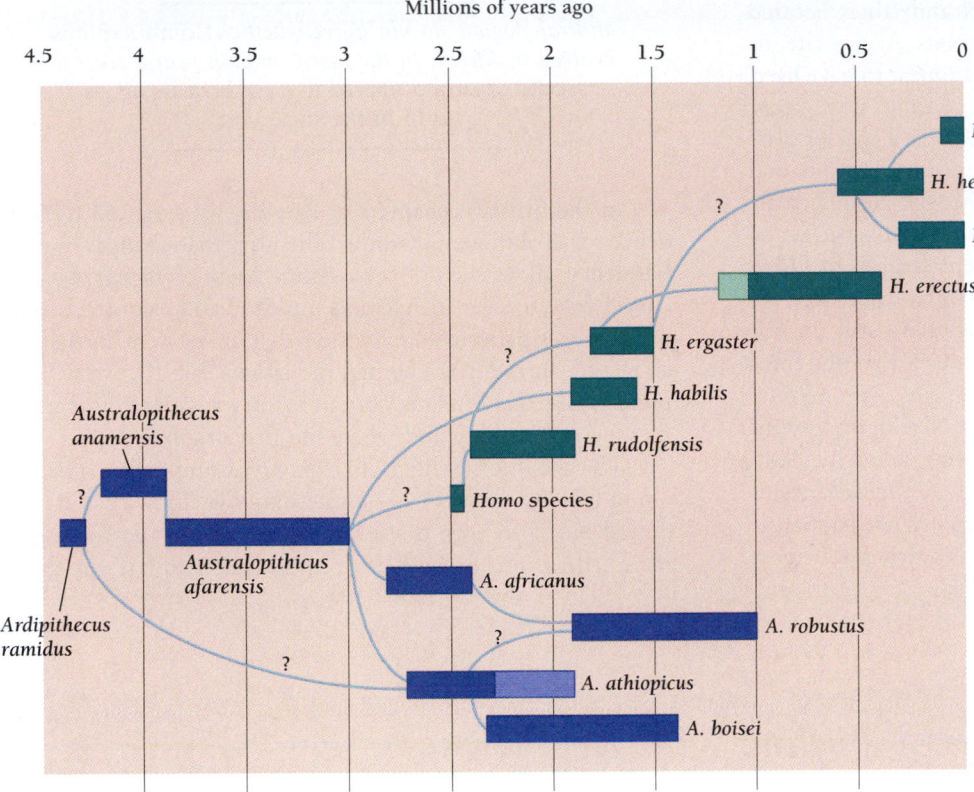

Figure 17-26 Branching evolution of the hominids. How are the various hominids related? Anthropologists do not agree. Anthropologists are increasingly recognizing that human evolution has not consisted of a single line of hominids, with one species neatly evolving into the next to culminate in *H. sapiens.* Between 2 and 3 million years ago, as many as six different hominids may have coexisted. *(Reprinted with permission from Johanson and Edgar: From Lucy to Language, Simon & Schuster Editions)*

Some anthropologists and paleontologists have suggested that *H. erectus* migrated from Africa, starting about 1.5 million years ago, with subsequent waves of migration evolving separately into varieties, or races, that differ from one another. Some of these, they argue, evolved into *H. sapiens.*

Others argue that *H. erectus* living in different parts of the world were replaced first by *Homo heidelbergensis* and later by *H. sapiens,* each of which migrated separately from Africa. In this view, *H. erectus* was a side branch in the evolution of humans, and we are descended from another line of *Homo* that left Africa after *H. erectus.* The most likely candidate seems to be *H. heidelbergensis,* found in Ethiopia and dated at about 130,000 years old. Several skulls that seem transitional between *H. heidelbergensis* and *H. sapiens* have been found in Ethiopia.

Alternatively, *H. sapiens* may have evolved in Africa 200,000 to 100,000 years ago, separately and at the same time as *H. heidelbergensis,* then emigrated to Asia and Europe. The earliest known representatives of *H. sapiens* were found at two sites in Israel and are dated at about 90,000 years old.

The early *H. sapiens* are commonly divided into two kinds, the **Cro-Magnons** and the **Neanderthals.** The Neanderthals, who looked much like us, but with a heavier build, occupied parts of Europe until just 35,000 years ago. Their thick arms and legs and heavy brows provide cartoonists with the classic "caveman" image; and, because of differences in Neanderthal throat bones, some researchers doubt that they spoke as well as we do, if at all. Nonetheless, Neanderthals, who first evolved between 200,000 and 100,000 years ago, probably had a fair

amount of mental horsepower. Their brains were as large or larger than those of the Cro-Magnons, and some Neanderthals may have played musical instruments (Figure 17-27).

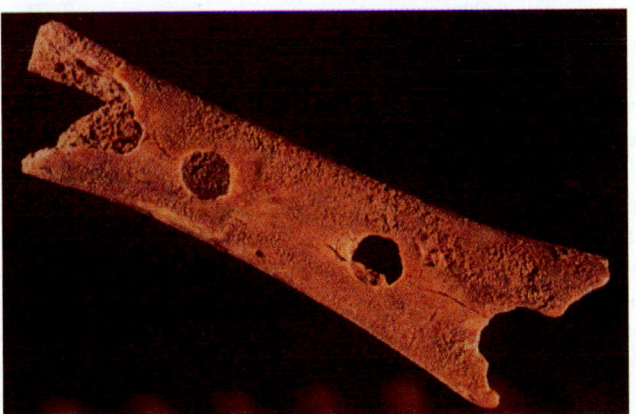

Figure 17-27 Neanderthal flute? In 1996, a team of researchers discovered what appeared to be a bone flute in a 43,000-year-old Neanderthal cave in Slovenia, in eastern Europe. Apparently fashioned from the thigh bone of a young bear, the hollow bone has four holes spaced as if for a diatonic scale (do, re, mi . . .). Similar flutes have been found in newer, Cro-Magnon caves, and researchers around the world reacted with delight to the idea that Neanderthals might have played music. Some researchers, however, flatly reject the idea. They argue that the holes were made by wolves' canine teeth and that the spacing is purely accidental. *(Blackwell, Queens College)*

Researchers do not know why the Neanderthals became extinct or if they interbred with Cro-Magnons. At one site, in Israel, the Cro-Magnons and Neanderthals appear to have lived within a few miles of each other at the same time. If we knew that they interbred, we could apply the biological species concept and conclude that they were one species. But no one knows, and researchers do not agree whether the Neanderthals were a separate species *(H. neanderthalensis)*, a subspecies, or just a now-extinct race of *H. sapiens*. Recent analysis of DNA extracted from a Neanderthal bone suggests that they were genetically distinct from modern humans. But we still do not know if they were equally different from our Cro-Magnon forebears.

Even if we ultimately conclude that the Neanderthals were *H. sapiens,* we probably had other *Homo* cousins in the early days of our existence (Figure 17-24). Both *H. heidelbergensis* and *H. erectus* lived up to 100,000 years ago. What happened to them? No one knows, but anthropologists strongly suspect that *H. sapiens* exterminated them.

Anthropologists do not agree whether Homo sapiens *evolved in Africa. In the last 2 million years, several species of* Homo *appear to have been living on Earth at the same time.*

In the first two chapters of Part III, we reviewed the evidence for evolution and studied the ways that changes in gene frequency allow evolution to occur. In this chapter we discussed how species change and multiply and examined some of the larger patterns of evolution that are evident in the fossil record. In the next chapter, the last in Part III, we will explore a topic that is related, but in many ways quite different—the question of how and where life first originated.

The evolution of life is, like gravity, a simple fact. The origin of life, however, is truly mysterious. We know that life appeared on Earth within a few million years of the cooling of the Earth's crust, but we can only surmise how it got there.

Study Outline with Key Terms

Macroevolution is the origin and multiplication of species. According to the commonly accepted **biological species concept,** a **biological species** is a group of organisms that actually or potentially interbreed to produce viable offspring, but cannot interbreed with members of other such populations. Populations that do not interbreed with other populations are said to be **reproductively isolated. Sibling species** are species that are extremely similar, yet reproductively isolated. **Races** or **subspecies** are subpopulations of a species that are distinct, yet not reproductively isolated. In other words, races can interbreed with other races of the same species. Subspecies that interbreed only rarely are sometimes called **incipient species.**

Populations become reproductively isolated from one another through two kinds of barriers to gene flow: **prezygotic barriers,** which prevent fertilization, and **postzygotic barriers,** which either make the zygote inviable or make the resulting adult sterile. Prezygotic barriers include **ecological, temporal, behavioral, mechanical,** and **gametic isolation.** Once fertilization occurs, postzygotic barriers can come into play. These include chromosomal and genetic incompatibilities that cause the hybrid zygote to die young (**hybrid inviability**); or, if the individual reaches maturity, to produce no gametes (**hybrid sterility**); or, in some cases, if the hybrid produces offspring, to have *its* offspring be weak or sterile (**hybrid breakdown**).

Two other postzygotic barriers, peculiar mostly to plants, reproductively isolate hybrid offspring from the parent population in just one generation. These include the multiplication of chromosomes, or **polyploidy;** and the combining and doubling of nonhomologous chromosomes from different species, or **allopolyploidy.**

Geography frequently isolates populations from one another well before other isolating mechanisms develop. Populations can become isolated through **dispersal** and through **vicariance.** When reproductively isolated populations diverge in morphology or behavior, they are said to **speciate.** Geographic speciation follows several patterns. In **allopatric speciation,** geographic isolation halts nearly all gene flow. Genetic drift or different selective pressures complete the speciation process. In **parapatric speciation,** adjacent populations with limited gene flow manage to speciate. In **sympatric speciation,** populations that occupy the same geographic area become reproductively isolated through barriers to gene flow—temporal, ecological, or other.

Phylogeny is the history of descent of lineages of organisms from a common ancestor. **Divergent evolution** is the separation of a lineage into two or more species or lineages. **Convergent evolution** is a separate idea, referring to the acquisition of similar characters in unrelated lineages. An important kind of divergent evolution is **adaptive radiation,** the generation of diverse new species from a single ancestral species. Because no new body plans evolved after the **Cambrian explosion,** some biologists have suggested that **developmental constraints** limit evolutionary pathways.

Despite the gaps in the fossil record, Darwin believed in **gradualism,** the view that evolution was a gradual process. Early biologists rejected Darwin's gradualism in favor of **saltationism,** the view that evolution occurred in sudden leaps, or **saltations.** The last well-known saltationist suggested that genetic monstrosities, which he called **hopeful monsters,** began new evolutionary lineages.

Although gradualism gained favor from the 1930s through the 1960s, the fossil record seemed to show that many species remained unchanged for millions of years (**stasis**) and then abruptly gave rise to new species. A full record of intermediate stages was usually lacking. Thus, evolution, according to the fossil record, was neither gradual nor continuous. It seemed to happen in fits and starts.

In the 1970s, two paleontologists, Niles Eldredge and Stephen J. Gould, argued that the formation of new species is rapid compared to the rate of accumulation of changes within a species. This view, called **punctuated equilibrium,** rests on two main ideas. The first is that species change very little most of the time. The second is that most anatomical or other evolutionary change in individual species occurs during a geologically brief period at the time of speciation.

Anatomical and molecular evidence suggests that humans are **primates** most closely related to chimpanzees, gorillas, and other pongids. The **Pongidae** are a family of **hominoids,** which are the ape members of the catarrhines. The **catarrhines,** which include Old World monkeys and apes, are a more recently evolved group of primates than the **platyrrhines,** which are tailed monkeys that live in South and Central America.

Humans and their ancestors all belong in the family **Hominidae.**

Hominids are **bipedal** and have rounded jaws. **Physical anthropologists** call the earliest hominids **australopithecines.** Later hominids also have large brains, arched backs, and reduced toes. Two genera of hominids succeed the australopithecines: *Paranthropus* and *Homo.* All species of *Paranthropus* and *Australopithecus* are now extinct, and only one species of *Homo* remains—*H. sapiens.* No one knows exactly why hominids evolved a bipedal gait. Most anthropologists assume that the first hominids left a life in the trees to live in an open, probably grassy, habitat where they had to walk long distances.

Anthropologists do not agree whether *H. sapiens* evolved in Africa, as did all or most other species of *Homo,* or if *H. sapiens* evolved from an intermediate ancestor, which evolved in Africa but then dispersed to Europe and Asia. In the last 2 million years, several species of *Homo* appear to have been living on Earth at the same time. The longest-lived species was *H. erectus,* which lived in parts of Africa and Asia for nearly 2 million years and became extinct only about 100,000 years ago. Early *H. sapiens* are divided into the heavy-boned **Neanderthals** and the more delicate **Cro-Magnons.** Modern humans more closely resemble the Cro-Magnons.

Review and Thought Questions

Review Questions

1. Define the word "species"; then explain why that definition is tricky. What are sibling species? What is a subspecies? Why is a subspecies not a separate species?
2. What conditions are required for speciation?
3. Distinguish between prezygotic and postzygotic barriers to gene flow. Give several examples of each. Is polyploidy a prezygotic barrier or a postzygotic barrier?
4. Distinguish among sympatric, allopatric, and parapatric speciation.
5. What clues from the Galápagos Islands suggested to Darwin the idea of adaptive radiation?
6. Describe the Cambrian explosion. What happened 600 to 900 million years ago that has not happened since? Why hasn't it happened again?
7. Name a postzygotic isolating barrier in plants that is consistent with the saltationist view of evolution.
8. Define gradualism and punctuated equilibrium. What characteristics of the fossil record favor the theory of punctuated equilibrium? Explain why these two ideas are or are not in conflict with each other.
9. What are some weaknesses in the argument for stasis?

Thought Questions

10. Consider an oceanic archipelago consisting of only ten islands. Describe a pattern of colonizations whereby 20 or more species could arise from a colonization by a single mainland species.
11. With Goldschmidt's "hopeful monsters" in mind, make up a science-fiction scenario in which a human baby born with two hearts would have some selective advantage.
12. The current mass extinction will likely result in an adaptive radiation over several million years. What organisms do you think will survive and what niches will they evolve to fill 5 million years from now?

About the Chapter-Opening Image

The Galápagos tortoises are a classic example of the adaptive radiations that lead to speciation, the evolution of new species. The tortoise shown here is a giant tortoise (Geochelone elephantopus vandenburghi).

 On-line materials relating to this chapter are on the World Wide Web at **http://www.harcourtcollege.com/lifesci/aal2/**

18

How Did the First Organisms Evolve?

Can Organisms Arise Spontaneously from Nonliving Matter?

Until the last few hundred years, most people believed that new life arose spontaneously and daily from non-living materials. Anyone could see that flies and maggots arose from rotting meat, frogs from swamps, mice from moist grain. Nearly everyone believed that spontaneous generation of life was common among the simple—so-called "lower"—plants and animals. Some people even argued that, under carefully controlled conditions, it might be possible to generate a man from a corpse.

Of course, every farmer knew that wheat and many other plants came from seeds, and that the cows and sheep were born of their mothers. Here and there, a few skeptics also insisted that insects and other creatures come from the eggs or sperm of parent organisms. Not until 1668, however, was this idea put to the test.

The skeptic was Francesco Redi (1626–1697), an Italian physician, poet, and naturalist, and one of the first modern experimental biologists. Heavily influenced by Galileo's belief that the natural world could be understood best through the senses, Redi strove to perform controlled experiments to test his ideas. Among these was a series of experiments challenging the idea of spontaneous generation.

In his own account of these experiments, Redi first states his hypotheses:

> I shall express my belief that the Earth, after having brought forth the first plants and animals at the beginning by order of the . . . Creator, has never since produced any kinds of plants or animals . . . and everything which we know in past or present times that she has produced came solely from the true seeds of the plants and animals themselves, which thus . . . preserve their species. And, although it be a matter of daily observation that infinite numbers of worms are produced in dead bodies and decayed plants, I feel . . . inclined to believe that . . . the putrefied matter in which they are found has no other office than that of serving as a place . . . where animals deposit their eggs at the breeding season, and in which they also find nourishment; otherwise, I assert that nothing is ever generated therein.

Redi then describes how, after preliminary studies of the development of different kinds of maggots into various species of flies, he put several kinds of fresh meat into two sets of jars. One set of jars was open, the other closed. Those that were open all developed maggots and, later, flies. Those that were closed developed no maggots, though, Redi notes, "Outside on the paper cover there was now and then a deposit, or a maggot that eagerly sought some crevice by which to enter and obtain nourishment."

Finally, Redi wanted to prove that the absence of air was not keeping the maggots from developing on the meat. He placed fresh meat into large vases that were either open or covered with very fine mesh. To prevent maggots from crawling through the holes in the netting, he put the covered vases inside a large net-covered frame and carefully removed any maggots that managed to penetrate the outer netting (Figure 18-1). The results were the same. "I never saw any worms in the meat, though many were to be seen moving

James Cotier/Tony Stone Images

Key Concepts

- Life appeared just a few hundred million years after the Earth's crust hardened.

- Scientists can construct a reasonable scenario for the evolution of the first biological molecules and for the generation of abiotic cell-like structures.

- The first cells required a means of coupling metabolism and protein synthesis to genetic instructions. These cells may have used RNA as both genetic instructions and enzymes.

- Eukaryotic cells probably arose through a symbiotic association of prokaryotic cells.

about on the net-covered frame." The meat in the open vases was, of course, riddled with maggots, which, Redi observed, eventually turned into various kinds of flies. This experiment showed that maggots did not appear spontaneously.

Redi's 1668 experiments convinced most scientists that the common conception of spontaneous generation was incorrect. However, within 10 years, van Leeuwenhoek used his new microscope to find little "animalcules" (microorganisms) everywhere. Most people felt that even if maggots and flies could not arise by spontaneous generation, very likely these microorganisms could.

It was not until a hundred years later that anyone showed that they could not. In the 18th century, Lazzaro Spallanzani (1729–1799), an Italian physicist, microscopist, and physiologist, demonstrated that no organisms grew in a rich broth if it were first heated and allowed to cool in a stoppered flask (Figure 18-2). Other experimenters less careful than Spallanzani, however, contaminated their broth and so obtained the opposite result. Moreover, some scientists argued that, by heating the broth and air, Spallanzani may have destroyed some substance needed for spontaneous generation.

For another hundred years, these objections stood. Finally, in 1860, the controversy over spontaneous generation became so intense that the prestigious French Academy of Sciences offered a

prize to anyone who could resolve the matter. In 1864, Louis Pasteur, the founder of microbiology, designed a simple but irrefutable experiment that won the prize.

Pasteur repeated Spallanzani's experiment, but with one essential difference. Instead of stoppering the flask to prevent organisms or spores from falling into the broth, Pasteur used a flask with a long neck like that of a swan (Figure 18-3). Air could pass freely in and out of such a flask, but microbes could not get through the long, curved neck. Only when Pasteur broke the neck of the flask was the broth "seeded" by bacteria and fungi falling from the air. Some of Pasteur's open flasks—still sterile after more than 130 years—are on display in Paris.

"All life from life," Pasteur emphatically concluded. "Never," he said, "will the doctrine of spontaneous generation recover from the mortal blow of this simple experiment."

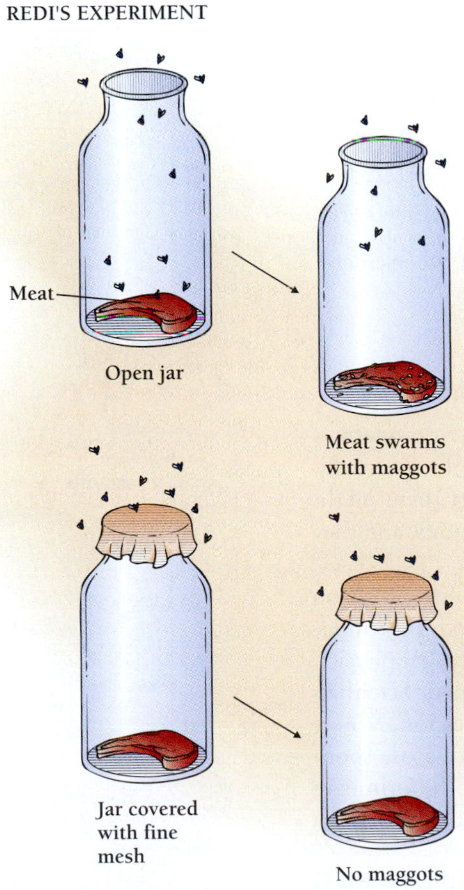

WHERE DOES LIFE COME FROM?

Since Darwin, biologists have known that all modern species have evolved from previous ones. But we must ask, From what did the millions of species on Earth evolve? If Pasteur was correct in saying that all life comes from other life, we must account for the presence of evolved life on a planet that itself came into being a finite number of years ago. Where did life first come from?

REDI'S EXPERIMENT

Figure 18-1 Francesco Redi's experiment. Do flies arise spontaneously from rotten meat? Or do they develop from eggs laid on the meat by adult flies?

SPALLANZANI'S EXPERIMENT

Figure 18-2 Lazzaro Spallanzani's experiment. Do microorganisms arise spontaneously in a rich broth?

Every culture on Earth—ancient, primitive, or modern—has its own answer to this question. Many invoke an intelligent, creative force. However appealing many of these myths may be, they do not constitute scientific explanations, and they tell us little more than that "life happened."

Scientists cannot say with certainty how life arose on Earth, either. However, they can speculate, based on a modern understanding of astronomy, planetary science, geophysics, and biology, what might reasonably have happened. Scientists see essentially two alternative answers to the question, Where did life on Earth come from? Either life on Earth arose spontaneously from nonliving organic molecules in the primitive environment or else it arrived from outer space by means of a meteor, comet, or asteroid.

In this chapter, we will see why neither of these ideas is quite as outlandish as it might at first seem. The bulk of our discussion will be devoted to exploring how life might have originated on Earth. As one prominent biologist put it, this discussion resembles a mystery novel with too many suspects.

Evolutionary biologists have inferred part of the story from the shared characteristics of modern organisms. But much of the story must be guesswork. Because the origin of life is not an ongoing, testable process, as evolution is, biologists can only surmise how life might have arisen. The point is to show that life could have arisen by some series of steps, each of which is reasonable in terms of our current understanding of the laws

PASTEUR'S EXPERIMENT

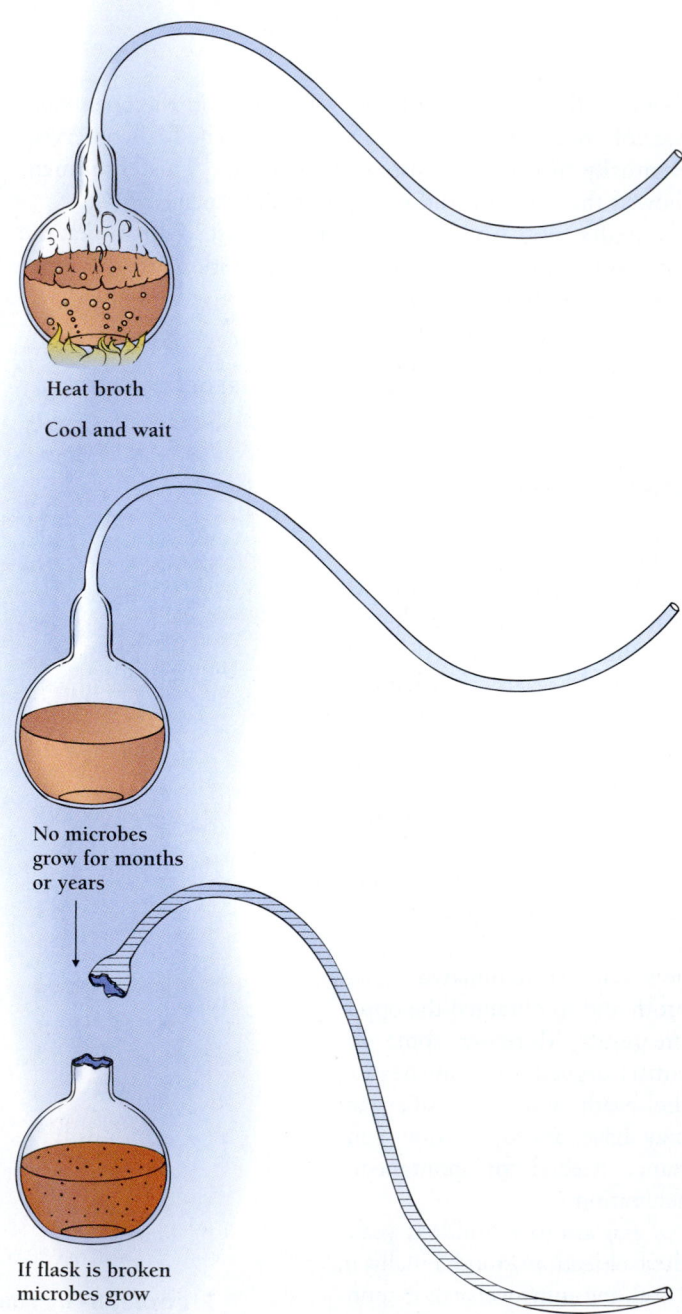

Figure 18-3 Louis Pasteur's experiment. Do microorganisms arise spontaneously in a rich broth provided that air is available?

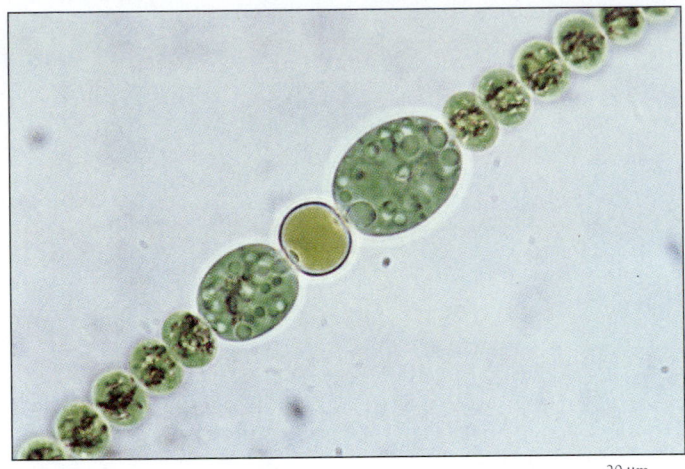

A.   20 μm

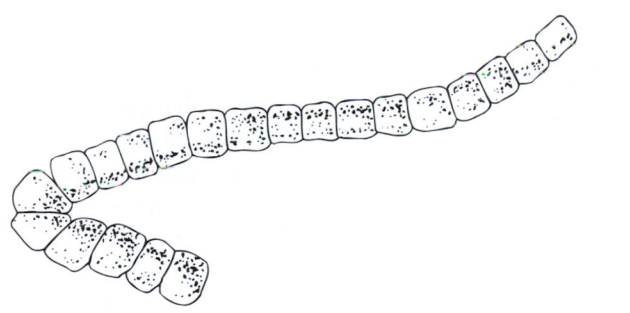

10 μm

B.

Figure 18-4 Filamentous cyanobacteria. Some filamentous fossils closely resemble their modern counterparts. A. Modern filamentous cyanobacterium. B. A 3.5-billion-year-old filamentous cyanobacterium from the Warrawoona rocks of western Australia. (*A, Biological Photo Service; B, Stanley Awramik/Biological Photo Service*)

of physics, chemistry, and biology. By showing that they can tell such a reasonable story, evolutionary biologists have attempted to refute the idea that there is an uncrossable gulf between living and nonliving matter. We will also see why, despite all this, Pasteur was after all correct: spontaneous generation could not happen today.

When and Where Did Life First Appear?

Before 1950, no one had seen a fossil older than about 570 million years (the beginning of the Cambrian Era). Older fossils went unnoticed, it turns out, because they are extremely rare, usually quite small, and found in a kind of rock different from that of later fossils.

Cambrian and later fossils are usually hard bones and shells preserved in shales, sandstones, limestones, and other sedimentary rocks. But most Precambrian organisms had few or no hard parts, and their soft remains usually dissolved or decayed. Their soft, tiny bodies persisted only when caught in an extremely fine-grained rock called chert. Such rocks are now generally far below the Earth's surface, covered by later deposits. Even cherts that have returned to the surface may be hopelessly deformed by heat and pressure. In only a few places on Earth are the cherts that bear Precambrian fossils both accessible and intact.

But for scientists interested in Precambrian life, finding fossil-bearing cherts was only half the battle. In order to see the fossils, paleontologists had to slice the chert into thin sections, polish them until they were transparent, and then examine the sections under a microscope.

The oldest and most primitive fossils known resemble modern prokaryotes—simple rods, spheres, and filaments with no nucleus (Figure 18-4). These ancient fossils, embedded in cherts from southern Africa and western Australia, are about 3.5 billion years old. More recently, researchers have dated fossil organisms that are even older, some 3.8 billion years.

Another important group of ancient fossils are the stromatolites, banded columns of limestone or chert, as old as 3 billion years (Figure 18-5). For many years, scientists did not realize that stromatolites were the remains of living organisms.

Figure 18-5 Living stromatolites like these once filled vast shallow seas. These cushions of cyanobacteria at Shark Bay, western Australia, resemble 3.5-billion-year-old fossil stromatolites. (*John Reader/Science Photo Library/Photo Researchers, Inc.*)

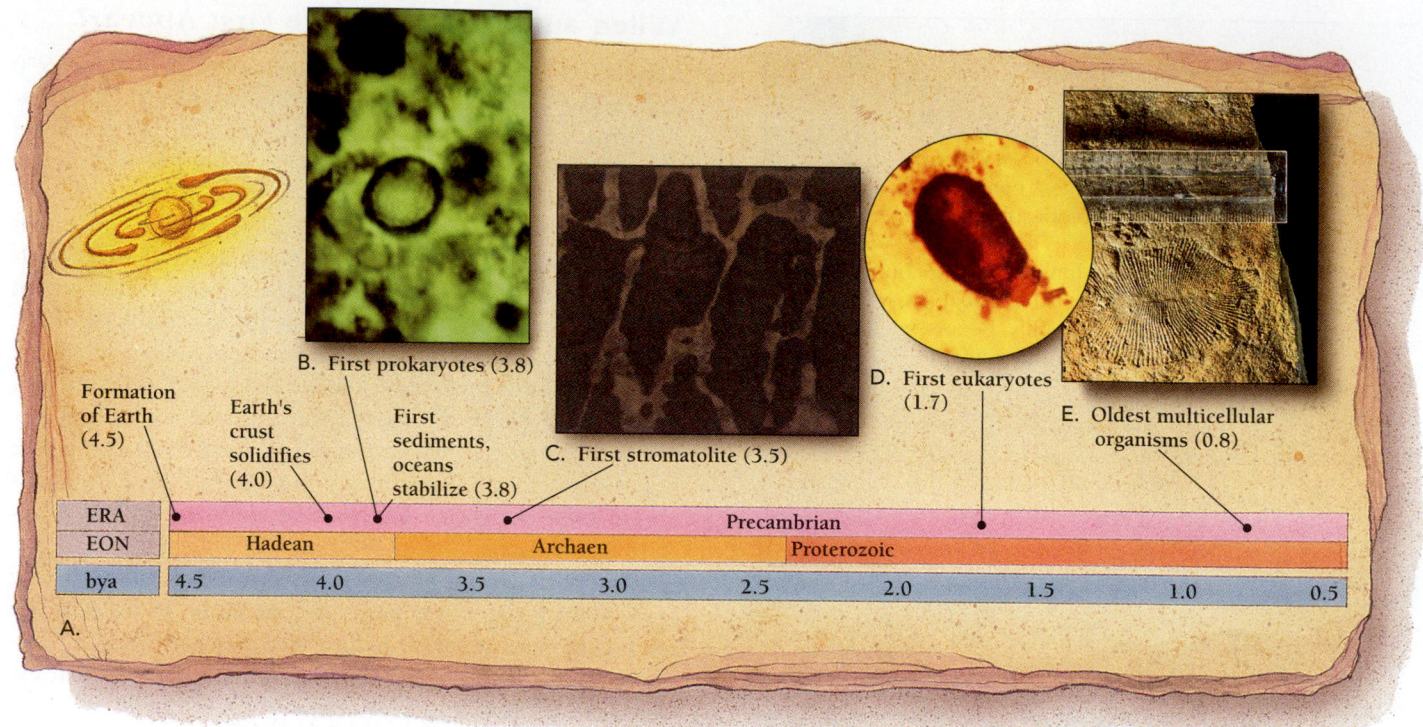

Figure 18-6 A calendar of Earth's early history. Life appeared on Earth almost as soon as conditions would allow it. The first stromatolites (photosynthetic cyanobacteria) appeared 3.5 billion years ago. Eukaryotes appeared nearly 2 billion years after the first prokaryotes. *(B, Science VU-USM/Visuals Unlimited; C, William E. Ferguson; D, Andrew H. Knoll, Knoll and Calder, Palaeontology 26:467, 1983; E, Biological Photo Service)*

This failure was because they consist of an unusually orderly arrangement of sediments, with no trace of the original organisms. But living stromatolites, discovered in a few remote intertidal areas, solved the mystery. Stromatolites are formed when huge, mounded colonies of cyanobacteria and other microorganisms trap fine sediment. The sediment settles in thin layers within each mound. The cyanobacteria in modern stromatolites are photosynthetic. Paleontologists assume that fossil stromatolites were also photosynthetic. In other words, photosynthesis probably evolved at least 3 billion years ago (Figure 18-6). Later in this chapter, we will see how the photosynthetic abilities of early cyanobacteria profoundly changed the world, allowing the evolution of all higher organisms.

As we have noted, there are two general explanations for the relatively sudden appearance of prokaryotic organisms soon after the Earth had cooled: (1) life arose spontaneously from nonliving organic molecules in the primitive environment, or (2) life arrived from elsewhere in the universe. It is almost impossible to imagine how we could prove or disprove either of these explanations. We can, however, examine the reasons for thinking either of them reasonable.

The second hypothesis at first appears to be the simpler explanation for the appearance of life. According to this view, the present diversity of life all derives from a colonization of Earth followed by a huge adaptive radiation. An appealing feature of this hypothesis is that the universe is thought to have originated at least 12 to 16 billion years ago, 8 to 12 billion years before the cooling of the Earth. This far longer history would give life more time to develop. Furthermore, an extraterrestrial origin of life explains some otherwise unexpected chemical properties of organisms. It would explain, for example, the requirement for molybdenum—an extremely rare element on this planet—in key reactions of photosynthesis and nitrogen fixation.

Until recently, no evidence existed for life on the other planets in the solar system. In August of 1996, however, researchers discovered signs of life in a meteorite composed of 3.6-billion-year-old Martian rock, apparently dislodged from the Martian surface by an asteroid collision about 15 million years ago. The rock, which crashed into Antarctica 13,000 years ago, contains what look like tiny fossils and also complex organic molecules that, on Earth, are produced when organisms decay or burn. These molecules are most concentrated in the interior of the rock, suggesting to some researchers that they did not infiltrate the rock from the outside after it landed on Earth.

No one knows if these fossils represent life from Mars. Even if Mars did once harbor living organisms, there is no evidence that life on Earth was seeded by Martian organisms. Earth may have seeded Mars. Or life may have arisen independently on each planet. Another possibility is that life arrived on both planets from some source outside the solar system.

BOX 18.1

Is There Life Elsewhere in the Universe?

In 1959, a young astronomer assigned to the Greenbank Observatory in rural West Virginia began using the big observatory radio telescope to listen for signals from extraterrestrial beings. Frank Drake and his colleagues tried to keep it a secret, but word got out quickly. Newspaper reporters besieged the observatory with calls, but Drake could not honestly report any definite communications from outer space.

Yet, for nearly 40 years, Drake hasn't stopped listening (Figure A). In 1974, he used Cornell University's Arecibo radio telescope, a giant, 300-m telescope in Puerto Rico, to send the first deliberate message from Earth. The message described Earth's position in the Milky Way galaxy, what humans look like, and how many of us there were.

Today, Drake and others are still listening for signs of intelligent life in the universe. This project, called the Search for Extraterrestrial Intelligence (SETI) asks, Is anyone out there? (SETI's work is depicted melodramatically in the recent Hollywood film "Contact.") Biologists tend to be conservative about such things. Many believe that the emergence of life on Earth is so remarkable that it would be astonishing if it occurred elsewhere in the universe. Indeed, the biologist and writer Jared Diamond has argued that the development of the radio on Earth was extremely unlikely. (He adds that, given the way humans treat less advanced civilizations and less intelligent creatures on our own planet, we will be fortunate indeed if we are alone in the universe.) Biologists are also eager to distance themselves from people who claim encounters with intelligent extraterrestrials on Earth.

Astronomers tend to take a different view. They argue that to insist that life could have arisen only on Earth is arrogant and excessively geocentric, or Earth-centered. Of course, they dismiss flying saucers—so-called UFOs (unidentified flying objects). Even the closest stars are too far away for spaceships to get here, let alone home again. And if intelligent life exists, the chances are that it is far away, indeed. Any signals that SETI picks up will be thousands or millions of years old.

Figure A **Are we alone?** Probably not. Frank Drake, president of the SETI Institute and professor emeritus of astronomy at the University of California, Santa Cruz, first began searching for intelligent extraterrestrials in 1959. (*Bill Lovejoy*)

But we are almost certainly not alone. Drake has suggested that the number of civilizations in the galaxy can be calculated using the following equation, called the Drake Equation:

$$N = R f_p\, n_e f_l f_i f_c f_L$$

In English, this equation says, the number of technological (detectable) civilizations in the galaxy (N) is equal to the rate (R) of star formation, times the fraction of stars with planets (f_p), times the number of planets ecologically suitable for life (n_e), times the fraction of those planets where life actually evolves (f_l), times the fraction that develops intelligent life (f_i), times the fraction that communicates (f_c), times the fraction of the planets' life occupied by the communicating civilization (f_L).

Clearly, the answer depends on what values are plugged into this equation. However, many of these values can be estimated with some confidence. For example, astronomers estimate that the Milky Way contains about 400 billion stars. Conservative estimates for most of the rest of the variables give predictions such as 4 billion planets with some form of life and 4000 existing technological civilizations in the Milky Way. And that is in our galaxy only. From astronomers' perspective, the evolution of life was practically inevitable.

But if extraterrestrial life is out there, we haven't found it yet. Soil samples from other planets have revealed no life, and SETI has so far turned up nothing concrete. The only tantalizing hints so far are some complex organic compounds from a meteorite believed to have come from Mars.

Organic molecules seem to abound in the universe. Astronomers have detected the presence of a number of carbon- and nitrogen-containing compounds in interstellar gas clouds. These compounds include those that we think were present in the Earth's primitive atmosphere—hydrogen, hydrogen cyanide, carbon monoxide, and ammonia.

Soil samples brought back from the moon contained several amino acids. Meteors, such as the one that fell in Murchison, Australia, in 1969 (Figure 18-10), contain the amino acids found in organisms as well as other amino acids and more complex molecules. The fragments from the Murchison meteorite contained 1 or 2 percent organic material. If such an endowment is typical of comets, asteroids, and meteors in the rest of the galaxy, it is hard to imagine how life could not have evolved elsewhere.

For life to colonize Earth (and Mars), it would have to have come from at least as far away as the nearest star, some 4 light years away—38 trillion km (23 trillion miles). Living organisms would have had to survive extremely violent events, for nothing else would have propelled them into interstellar space. There they would have had to survive intense radiation for untold millions of years. Recent research has suggested, however, that bacteria might survive the intense radiation if they were encased in ice.

If such a trip were possible, we would still want to understand how (and where) those organisms first arose. The problem of the origin of life would be virtually insurmountable, since we would not know where in the universe such organisms might have come from.

For now, the colonization-from-space hypothesis is only remotely likely. Perhaps more important, the hypothesis is so inaccessible to inquiry that most scientists interested in such things have focused their attention on the possible origin of life on Earth. Most biologists now believe that life must have developed from nonliving molecules on the early Earth's surface. They have accepted the challenge of showing that the development of organisms from nonliving matter—called **prebiotic evolution**—is possible.

Most biologists believe that life originated on Earth. No one knows enough about conditions in the rest of the universe to guess how extraterrestrial life might have evolved or safely arrived on Earth.

PREBIOTIC EVOLUTION: HOW COULD COMPLEX MOLECULES EVOLVE?

By the 1930s biochemists realized that all organisms are made from the same building blocks—the sugars, lipids, amino acids, and nucleotides that make up the "biochemical alphabets" described in Chapter 3. Two influential scientists—Alexander Ivanovich Oparin, a Russian biochemist, and J.B.S. Haldane, the English evolutionary biologist—proposed that these building blocks had been formed in the primitive environment. The original synthesis of these materials, they argued, was **prebiotic** and therefore **abiotic**—before life and therefore without life.

Important differences separate spontaneous generation from theories of prebiotic evolution. First, while spontaneous generation had to occur nearly instantly, the origin of life may have taken as long as 300 million years to unfold. Second, the conditions during which life must have first appeared were dramatically different from those that exist today. One of the most important differences was that the atmosphere contained virtually no free oxygen. Had oxygen been present in any quantity, it would certainly have oxidized the starting materials for life. A second important difference was the absence of life itself. Today, life could not arise on Earth because a plethora of hungry organisms find and consume organic molecules.

Darwin himself grasped this distinction, writing to a friend in 1871:

> It has often been said that all of the conditions for the first production of a living organism are now present which could ever have been present. But if (and oh! what a big if!) we could conceive in some warm little pond, with all sorts of ammonia and phosphoric salts, light, heat, electricity, etc., present, that a protein compound was chemically formed ready to undergo still more complex changes, at the present day such matter would be instantly devoured or absorbed, which would not have been the case before living creatures were formed.

Was Darwin right? In the absence of devouring organisms and destructive oxygen, can the building blocks of life manage to assemble themselves into living organisms? To decide, we must ask whether we can imagine a reasonable scenario for the "spontaneous" production of the building blocks of life and whether laboratory experiments demonstrate that the steps in the presumed process can actually occur. To discuss these questions, however, we must first understand the conditions on Earth shortly after it formed.

The conditions available for the origin of life differ from those for spontaneous generation in three dramatic ways—the long period of time available, the absence of life, and the absence of oxygen.

What Was the Earth Like When It Was Young?

The Earth formed about 4.6 billion years ago, when a spinning cloud of rocks, dust, and gas formed the sun and its planets. Astronomers believe that the spinning cloud gradually spread out and flattened as it rotated, like a lump of spinning pizza dough (Figure 18-7). The bulk of the material, at the center, condensed into an enormous compact mass that became the sun. The rest of the spinning cloud gradually formed the planets, comets, meteors, and other objects that make up the solar system.

The debris orbiting closest to the protosun [Greek, *protos* = primitive] moved fastest, and there the friction of constant collisions generated the most heat. Close to this hot center, pieces of silicon, metals, metal oxides, and other materials that remain solid even at high temperatures fell together, drawn to one another by their own gravities. These hot lumps of solid material gradually formed the inner protoplanets—Mercury, Venus, Earth, and Mars. Water, methane, and other compounds that solidify only at low temperatures formed the cold, outer protoplanets, such as Jupiter, far from the center of the spinning disk. The huge, outer planets attracted and held cold atmospheres of hydrogen, helium, and other gases. But the tiny, inner planets had gravities too weak to hold the hot gases, which escaped into space.

The early protosun was much cooler than today's sun. But Earth was far hotter. Gravity compacted the Earth, and huge

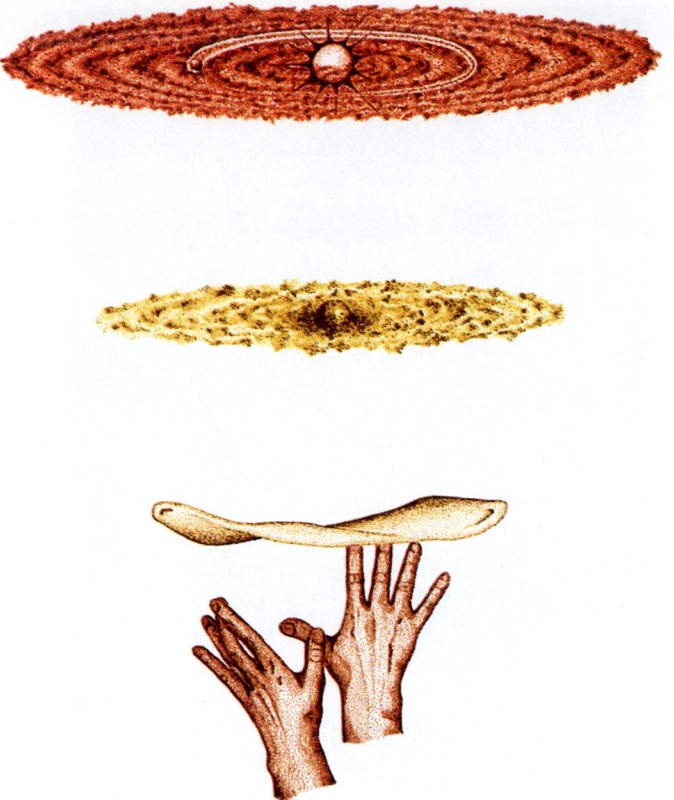

Figure 18–7 **Formation of the solar system.** Like a spinning lump of pizza dough, a cloud of dust and rocks flattened and then condensed into a central mass that became the sun. Material further out condensed into rotating balls of rock and gas that we now know as the planets. *(Source: Kristin Levier, Science Notes, University of California, Santa Cruz)*

meteors, many of them hundreds of miles in diameter, bombarded the planet. (The impact of one of these collisions was so great that it dislodged a huge chunk of Earth, which became the moon.) The compaction and the bombardment kept the molten Earth from solidifying for another half a billion years. The heaviest elements sank to the center of the Earth, while the lightest ones floated to the top. Iron and nickel sank to the core, the lightest silicates (granite, for example) rose to the surface, while denser silicates of iron and magnesium floated in between.

After 500 to 600 million years, the surface cooled enough to form a thin crust. Dense gases escaped from the interior through cracks and volcanoes in the new crust to form a primitive atmosphere. Among these gases was water vapor, immense quantities of which then condensed into clouds. Torrential rains pummeled the still hot Earth for millions of years, creating the oceans and eroding the Earth's rocky surface. Although the oceans formed relatively early, constant bombardment by meteors, large and small, may have repeatedly vaporized the oceans until about 3.8 billion years ago. About that time, the first sediments and the first life appeared.

The evolution of the atmosphere

The exact composition of the gases in the primitive atmosphere has been a matter of much debate. The modern atmosphere is 78 percent nitrogen, 21 percent oxygen, and about 1 percent argon, with varying amounts of water vapor close to the Earth. Another 0.1 percent of the atmosphere consists of traces of carbon dioxide, carbon monoxide, sulfur oxides, nitric oxides, and other gases.

Scientists long assumed that the early atmosphere contained hydrogen (H_2), ammonia (NH_3), methane (CH_4), and water (H_2O), with traces of carbon dioxide (CO_2) and hydrogen sulfide (H_2S). Recent research has cast doubt on these assumptions, however, suggesting that there was no hydrogen and much more CO_2.

About one important gas there has been little argument. The early atmosphere had almost no oxygen. The strongest evidence for this is the near-total absence of oxidized iron (rust) in the earliest sediments. Had oxygen been present in any quantity, it would certainly have oxidized the starting materials for life. Oxygen, we will see later in this chapter, came much later, probably as a by-product of photosynthesis.

The absence of oxygen had other important consequences. Chief among these was the absence of an ozone layer. The ozone (O_3) layer, about 40 km (25 miles) up, today shields the Earth from intense ultraviolet (UV) light from the sun. Modern organisms need that protection because UV light, extremely good at making and breaking chemical bonds, interferes with normal life functions.

However, 4 billion years ago, any chemical synthesis would have required considerable energy. Recall from Chapter 5 that synthetic reactions are almost all thermodynamically uphill. So UV radiation from the sun would have helped to promote novel chemical reactions. Heat from the hot interior of the Earth provided more energy. But the most dramatic energy source was the lightning accompanying the storms that filled the oceans.

The Earth's crust solidified about 4 billion years ago, but the oceans were not stable until about 3.8 billion years ago. Since the first fossil organisms are about 3.8 billion years old, we can conclude that life on Earth originated relatively quickly—within a 300-million-year window.

Is Prebiotic Synthesis Plausible?

By the early 1950s, scientists had some idea of what the early Earth was like: they believed it had the raw materials for organic synthesis—water, methane, ammonia, and hydrogen—and it had sources of energy—UV radiation, heat, and lightning. But scientists did not know whether the conditions of the Earth's early history would allow the abiotic synthesis of biologically important molecules.

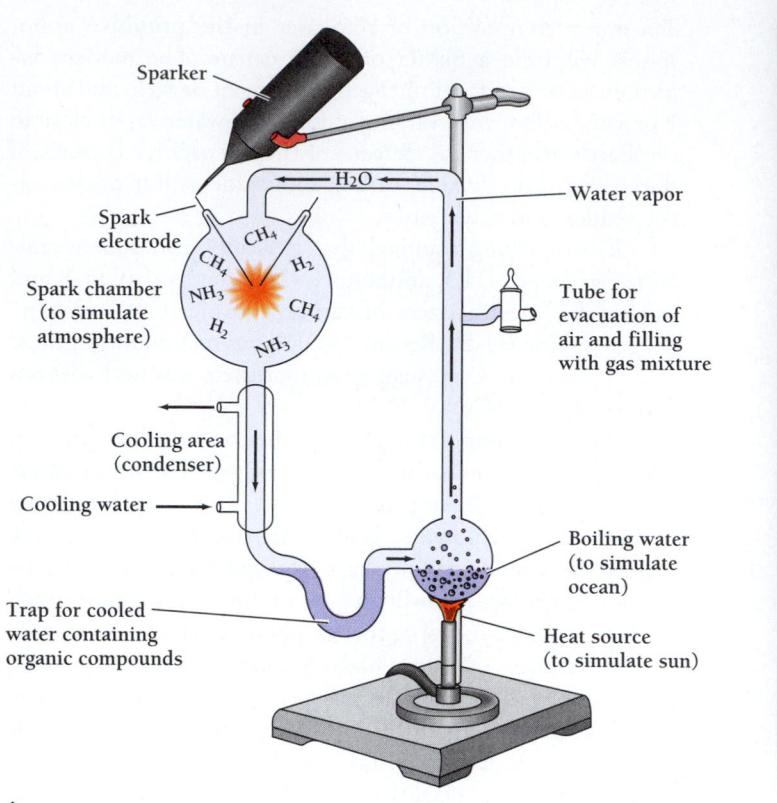

A.

B.

Figure 18-8 Stanley Miller's experiment. Can conditions similar to those on Earth 4 billion years ago give rise to the building blocks of life? A. Miller treated a mixture of water and gases with flashes of artificial "lightning." Among the complex molecules produced after 1 week were five amino acids. B. Miller and his apparatus. *(B, © 1988 Roger Ressmeyer/ Corbis)*

Then, in 1953, a 23-year-old graduate student working in the laboratory of Harold Urey, at the University of Chicago, undertook to simulate the primitive conditions of the Earth. Stanley Miller simulated the early Earth inside a simple glass apparatus built from the standard equipment found in a sophomore organic chemistry laboratory (Figure 18-8). The simulated Earth consisted of water in a boiler (the ocean) and a chamber fitted with electrodes (the atmosphere, with lightning). In one experiment, Miller mixed together an atmosphere of hydrogen (H_2), methane (CH_4), and ammonia (NH_3). Miller let the gases circulate past the flashing electrodes for a week and then analyzed the contents.

He found that more than 10 percent of the carbon from the methane was in organic molecules. These included at least five amino acids, as well as compounds that are precursors of other amino acids and of the bases of nucleic acids. Among these precursors was hydrogen cyanide (HCN), five molecules of which can directly form adenine (Figure 18-9). The mix-

ture was unexpectedly rich in some (but not all) of the same basic building blocks of life, which we listed in Chapter 3.

Miller and others repeated this basic experiment, with different atmospheres and different sources of energy. As long as free oxygen (O_2) was absent, the same building blocks appeared. Other sources of energy also work—most notably UV light, the principal energy source available on the Earth's surface before the Earth developed its protective ozone layer.

As a result of these experiments, it seemed likely that—within a few million years—the primitive oceans loaded up with organic molecules capable of becoming the building blocks of life. This primordial soup may have contained as much as 1 percent organic molecules in some places, about the same as a watery chicken broth. Remarkably, the amino acids most easily produced in these abiotic experiments are the ones most common in proteins today. And the most easily synthesized base—adenine—is biologically the most important of all the nitrogenous bases.

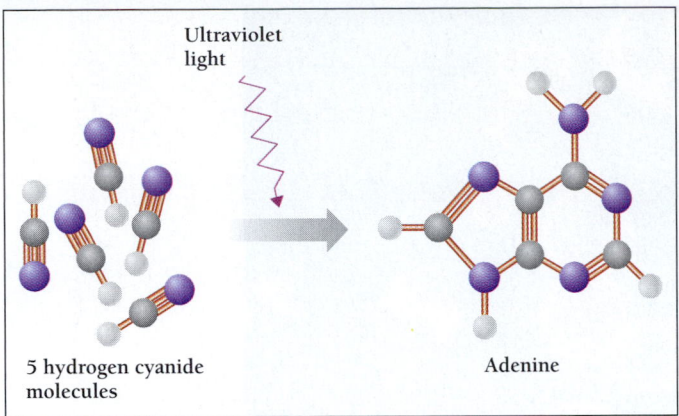

Ultraviolet
light

5 hydrogen cyanide
molecules

Adenine

Figure 18-9   **First steps in the formation of life?** Miller's apparatus also produced hydrogen cyanide. In the presence of ultraviolet light, five molecules of hydrogen cyanide form adenine, the nitrogenous base in ATP.

A distinct problem with this scenario, however, is that recent research by geoscientists suggests that the early atmosphere contained neither methane (CH_4) nor ammonia (NH_3). Heavy doses of UV light from the sun would have destroyed such hydrogen-based molecules, and even free hydrogen is so light that it would have escaped into space. Geoscientists say that the major components of the atmosphere were more likely carbon dioxide and nitrogen, with water and traces of carbon monoxide and free hydrogen, the same gases that are ejected from modern volcanoes.

Such an atmosphere would not have allowed the kind of amino acid synthesis suggested by Miller's experiment. Formaldehyde (H_2CO), essential to the formation of sugars, would have formed easily enough. But a likely biochemical pathway for the synthesis of hydrogen cyanide, the building block for adenine and other amino acids, has yet to be found.

Miller, now a professor of chemistry at the University of California, San Diego, defends his early study, saying that smoke and clouds might have shielded the fragile methane and ammonia molecules from UV light. And several Japanese researchers, supporting Miller, argue that solar particles and cosmic rays would have countered the effect of UV light by releasing free hydrogen from water molecules and so promoting the synthesis of methane and ammonia.

Other researchers have suggested another possible source of the first building blocks—the seas surrounding thermal vents in the ocean. These hot springs lie deep in the oceans, at the crests of underwater ridges. Seawater circulates through the vents, where its temperature rises to 350°C and then abruptly plunges to 2°C as it mixes with cold ocean waters. Modern-day vents are the sites of lots of underwater life, sustained by unusual metabolic reactions that do not depend on solar energy.

Some biologists suggest that the abundant hydrogen sulfide and iron sulfide in the superheated water could have provided the perfect environment for the synthesis of primordial building blocks. In this scenario, the building blocks accumulated on the surfaces of underwater rock instead of dissolving in the ocean. Perhaps, these biologists say, there was no primitive soup.

Prebiotic synthesis of biological building blocks may have taken place in the ocean or on underwater rocks.

Was Prebiotic Synthesis Necessary?

Some astronomers have suggested that the debate over the exact composition of the early atmosphere may be moot. They believe that another source of simple organic molecules is possible. Researchers have discovered that asteroids, meteors, and comets are rich in complex organic compounds created during the formation of the solar system (Figure 18-10). Further, based on the calculated number and size of comets and meteors likely to have entered the Earth's atmosphere, astronomers estimate that as much as 10 million kg of complex organic molecules might have showered the Earth each year. At the end of a billion years, they calculate, the Earth could have accumulated as much as a million billion (10^{15}) kg of organic molecules, more than the total mass of all life on Earth today.

The first organic molecules could have come from asteroids, meteors, or comets.

How Can Biological Building Blocks Assemble Outside of Cells?

If we assume that the Earth's early environment either allowed the production of amino acids and other building blocks or received them passively from space, we must still account for the assembly of those building blocks into other amino acids, nucleic acids, ATP, and other macromolecules. Can this process be simulated in the laboratory?

In the Miller atmosphere, adenine easily forms from hydrogen cyanide (Figure 18-9). The other nucleic acid bases require more complex reactions, so adenine could have been the first nitrogenous base to form. If so, it would not be surprising for its activated product, ATP, to have become the universal energy currency of life.

ATP itself may have formed rather easily. Laboratory experiments show that mixtures of adenine, ribose, and phosphate can form ATP when exposed to UV light. Other triphosphates may have arisen in the same way.

Even prebiotic replication of nucleic acids is not at all far fetched. Biochemists have discovered that nucleotides can combine in the test tube to form short RNAs. These RNAs, more-

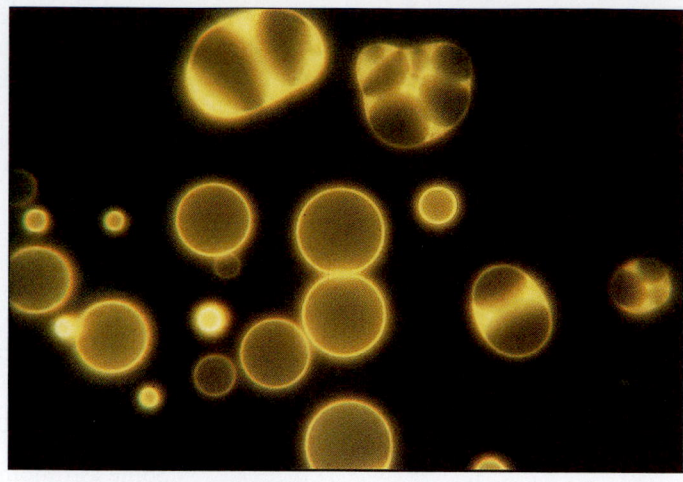

A.

B.

Figure 18-10 **Remains of extraterrestrial life?** A. This fragment of a meteorite hurtled through the atmosphere and landed in Murchison, Australia, in 1969. Scattered through the rock are particles of organic compounds. B. When extracted, these compounds self-assembled into vesicles. The yellow-green color is the fluorescence of complex organic compounds called polycyclic aromatic hydrocarbons. (*D.W. Deamer, University of California, Santa Cruz*)

over, form double-stranded molecules that follow the Watson-Crick rules for base pairing.

Simulating the polymerization of amino acids into polypeptides is more of a challenge. Many researchers, however, think that ATP in the primordial broth may have somehow contributed to the synthesis of activated amino acids. Solutions of such activated amino acids can spontaneously give polypeptide chains up to 50 amino acids long. Polypeptides also form when researchers heat dry mixtures of amino acids. Sidney Fox, a biochemist at Southern Illinois University, has suggested that such polypeptides could have formed on volcanic cinder cones and washed into the sea.

The earliest polypeptides almost certainly would not have had specific sequences of amino acids. In fact, Fox has dubbed the early polypeptides **proteinoids,** to distinguish them from the true proteins, which are the products of living organisms and whose sequences are defined by the genes in DNA.

The abiotic synthesis of RNA, ATP, and even polypeptides seems plausible.

Some variant of the chemical reactions described earlier may have dissolved in the open seas or may have accumulated in small puddles or on the surfaces of rocks. In any case, most biologists think that the building blocks of life became available for prebiotic evolution, perhaps as long ago as 4 billion years.

But the materials of life are not the same as life. Two important steps would have to have occurred—the evolution of

reproduction and the evolution of metabolism. The evolution of reproductive ability depended on establishing a highly specific relationship between the self-replicating nucleic acids and the sequence of amino acids. This meant the development of both the genetic code and a mechanism of translation.

Before this could happen, however, polypeptides and other molecules probably began to come together into organized (and concentrated) associations. As we will see, such primitive cells, or **protobionts,** could concentrate organic molecules and maintain an internal environment different from that of their surroundings. They could, in short, begin to develop the first suggestion of metabolism.

Life required the evolution of reproduction and metabolism.

How Can Prebiotic "Cells" Form?

Polypeptides, nucleic acids, and polysaccharides in solution will spontaneously concentrate themselves into discrete tiny, balloonlike droplets, which the Russian biochemist Oparin called **coacervates.** Coacervates have a discrete inside and outside separated by a membranelike boundary.

If enzymes are present in the solution, these will be bound in the coacervates, which then act like little cells. They absorb molecules from the solution and release the products of the reactions catalyzed by the enzymes. For example, Oparin showed that coacervates containing the enzyme phosphorylase would take up glucose-1-phosphate from the surrounding medium,

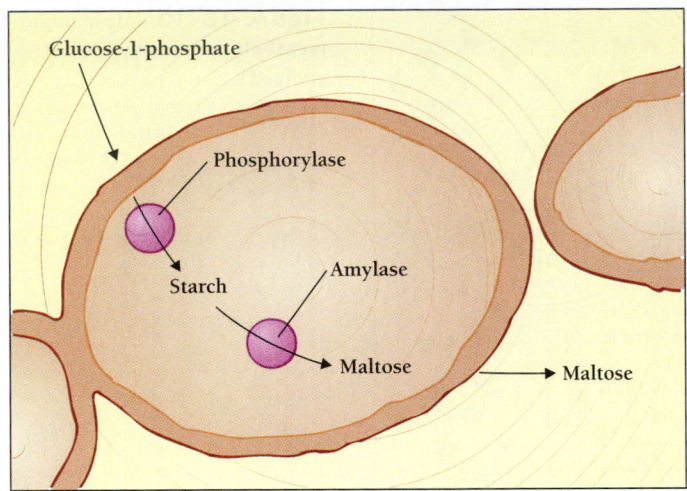

Figure 18-11 Prebiotic metabolism? Macromolecules in solution can form droplets, which Alexander Oparin called coacervates. In one experiment, coacervates that contained the enzymes phosphorylase and amylase could convert glucose-1-phosphate to starch and then to maltose.

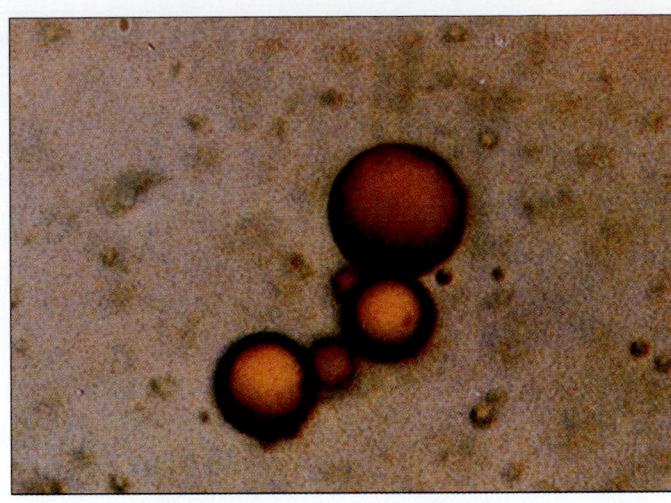

Figure 18-12 Proteinoid microspheres. These nonliving polymer bubbles can reproduce themselves by growing and budding. (*Steven Brooke and Richard LeDuc*)

convert it to starch, and concentrate it inside the droplet (Figure 18-11).

Other mixtures can also give rise to isolated, cell-like structures. For example, Sidney Fox found that the proteinoids formed from dry amino acids would form tiny spheres. These structures, called **proteinoid microspheres,** could absorb more polymers from solution, forming buds or even a new generation of spheres (Figure 18-12). Microspheres dropped in water are differentially permeable. They swell or shrink according to the concentration of salts in the water. Some will even store energy, in the form of a differential electric charge across the membrane. Such microspheres can release a charge in much the same way a nerve cell does. A mixture of phospholipids and proteins can form yet another kind of enclosed structure, called a **liposome.** Some liposome membranes have two layers, much like those in modern cells.

None of these self-organizing structures are cells. Yet they look so much like cells that experienced biologists have occasionally mistaken them for bacteria and even tried to classify them. And protobionts show some of the characteristics of life, including growth, metabolism, and responsiveness to their environment. The existence of protobionts shows that cell-like structures could have formed spontaneously in the primordial soup.

Further, we can envision the evolution of primitive metabolic pathways. As coacervates used up a given compound in the environment, those coacervates able to make the compound from another component of the soup would persist. Once the precursor was exhausted, further growth of protobionts would depend on the ability to produce the needed

product from still another precursor. According to this scheme, a number of pathways would evolve—one step at a time (Figure 18-13).

EARLY BIOTIC EVOLUTION: HOW DID THE FIRST GENETIC SYSTEM EVOLVE?

Even with the evolution of primitive metabolic pathways, protobiont "life" would have been severely limited. Despite protobionts' ability to grow and to bud, they could not reproduce. Even if they had accumulated catalysts for specific chemical reactions, protobionts could not make more of these specific catalysts. As unique molecular catalysts were passed on to increasing numbers of "offspring" protobionts, the catalysts became increasingly diluted. To invest the new protobionts with equal concentrations of functional molecules required the invention of a way of "teaching" the next generation to make its own catalysts. Life would need a way of storing, replicating, and translating the information for the building of enzyme catalysts—in short, a genetic system.

Could RNA Have Served as the Original Genetic Material?

A modern cell stores its genetic information as DNA, transcribes the information into RNA, and then translates the message into specific polypeptides. When the cell divides into two cells, it gives a copy of the DNA to each new cell, and so passes on the recipe for each polypeptide.

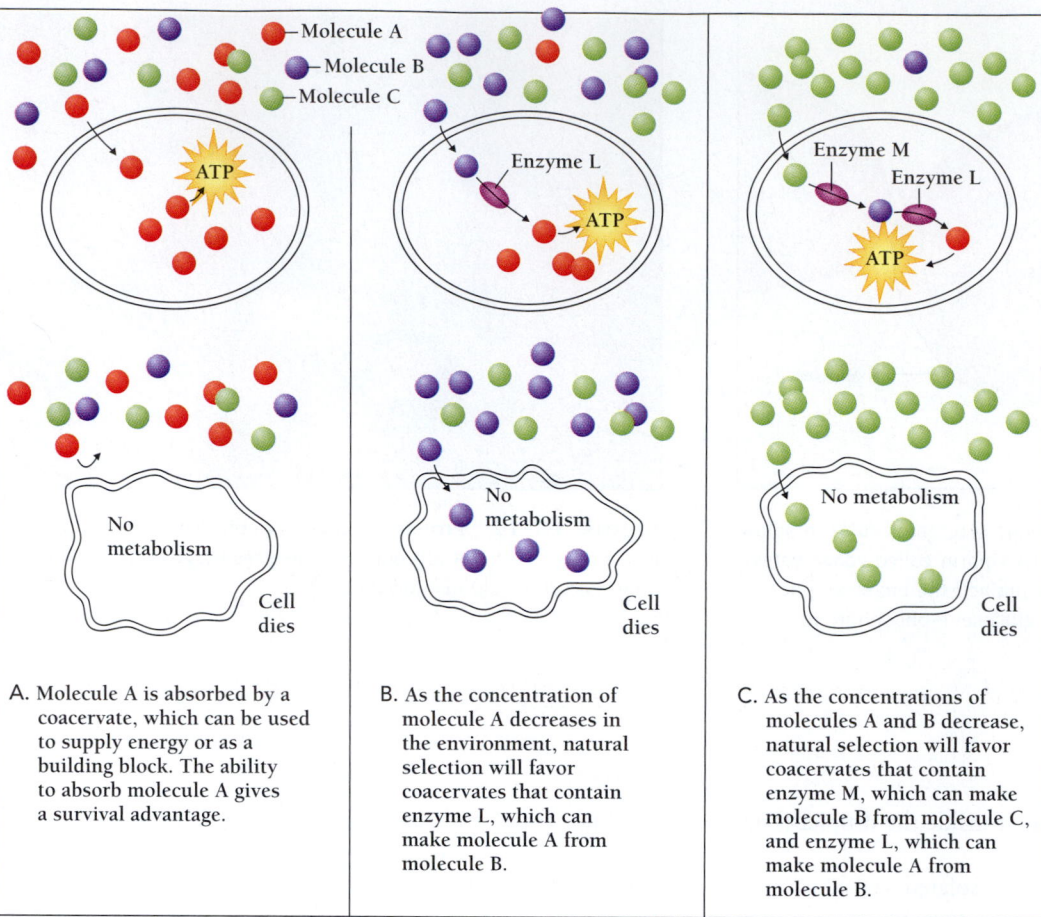

Molecule A
Molecule B
Molecule C

Enzyme L

Enzyme M
Enzyme L

No metabolism

No metabolism

No metabolism

Cell dies

Cell dies

Cell dies

A. Molecule A is absorbed by a coacervate, which can be used to supply energy or as a building block. The ability to absorb molecule A gives a survival advantage.

B. As the concentration of molecule A decreases in the environment, natural selection will favor coacervates that contain enzyme L, which can make molecule A from molecule B.

C. As the concentrations of molecules A and B decrease, natural selection will favor coacervates that contain enzyme M, which can make molecule B from molecule C, and enzyme L, which can make molecule A from molecule B.

Figure 18-13 How might metabolic pathways have evolved?

The DNA-to-RNA-to-polypeptide scheme is too complicated to have arisen all at once in the first cells. So biologists have tried to guess which of these three kinds of molecules could have first stored information, directed the synthesis of proteins, and also replicated itself. But DNA cannot form without catalytic proteins, and proteins cannot form without DNA.

That leaves RNA. In modern cells, RNA plays a central role in translation, supporting the idea that RNA could have been the primitive genetic material. Still, if RNA acted as the original genetic material, some sort of mechanism had to exist for aligning amino acids along sequenced strands of genetic RNA. Such a mechanism would have to allow RNA to store information, direct the synthesis of proteins, and also replicate itself. Considerable evidence now supports the idea that RNA can do this.

First, scientists have found that some RNA can spontaneously create itself and replicate itself in a test tube. Short strands of nonordered abiotic RNA can self-assemble from individual ribonucleotides. And, in the presence of an RNA template, ordered sequences of five to ten nucleotides will polymerize according to the standard base-pairing rules.

In the early 1980s, molecular biologists Thomas Cech and Sidney Altman showed that certain kinds of RNA could act as enzymes, snipping off a bit at an end or cutting out a bit and splicing the remaining pieces together. It was a dramatic discovery that earned the two researchers a Nobel prize in 1989 and suggested that perhaps RNA could replicate itself without the help of protein enzymes (polymerases).

Indeed, in 1989, Jennifer A. Doudna and Jack Szostak created an artificial RNA enzyme, or **ribozyme,** that could make new RNA molecules. However, although the molecule can splice preassembled strands of RNA along a template, it cannot add individual bases to a growing chain. These discoveries suggested to many researchers that the first cells evolved in an **RNA world,** in which RNA molecules functioned both as enzymes and as carriers of genetic information. The hypothesized RNA world is not without problems, however. RNA is notoriously unstable, and it is unlikely that RNA enzymes could have survived outside of a living cell. Many researchers therefore think that the first enzymes were neither RNA nor protein, but a different type of molecule, perhaps a nucleic acid with a sugar other than ribose.

Cells based on such a system allowed the evolution of RNA-based cells. RNA-based cells in turn allowed the evolution of cells that used proteins as enzymes and DNA as the genetic material. Reverse transcriptase, so important in many tumor viruses and in HIV, may have played an important role in this last process (Chapter 12).

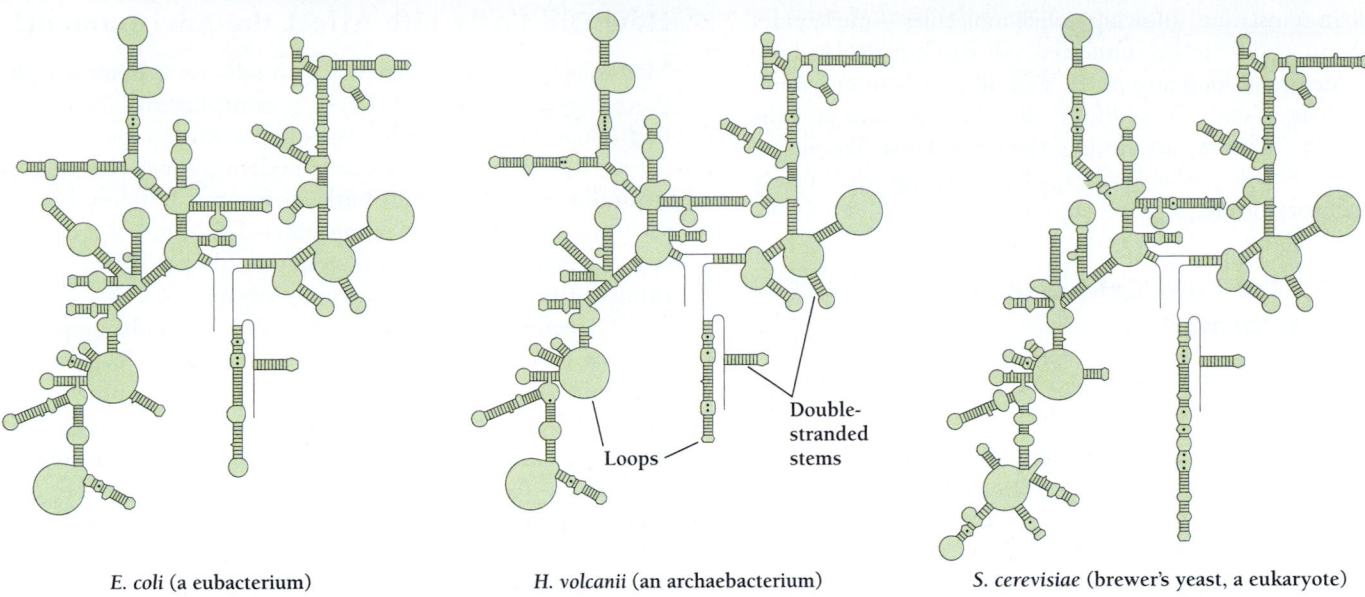

E. coli (a eubacterium)   H. volcanii (an archaebacterium)   S. cerevisiae (brewer's yeast, a eukaryote)

Figure 18-14   Ribosomal RNA. All modern ribosomal RNAs have a similar "stem-and-loop" structure. Notice that the pattern of stems and loops in eukaryote (yeast) RNA is more like the RNA of an archaebacterium than like that of a eubacterium. This supports the theory that we are descended from archaebacteria rather than eubacteria.

How Did Translation Evolve?

We know that RNA is capable of storing information, and it might, perhaps, be able to replicate itself. But the hardest question to answer is how modern translation—of RNA into protein—could have evolved. How a correlation between nucleotide sequences and amino acid sequences could have arisen remains a mystery.

Since the genetic code plays such a pivotal role in translation, we may look to it for clues about the evolution of translation. The first clue is the fact that all organisms use the same triplet genetic code. This suggests that the system originated with life itself. Second, most of the information for each codon is in the first two nucleotides; for 7 of the 20 amino acids, the third "letter" is irrelevant. This does not mean that the code was originally a doublet, however. If the code had started out as a doublet, then all previously evolved genes would have become useless when the code became a triplet. Instead, the present characteristics of the code were almost certainly fixed after its initial success.

As molecular biologist Francis Crick put it, the code is a "frozen accident." Information in mRNA was probably read three bases at a time from the beginning. But initially, the first two letters of each codon contained all the meaning.

Because the code is universal among both prokaryotes and eukaryotes, it must have completed its refinement before the time of the common origin of all modern organisms.

How Did Ribosomes Evolve?

Modern ribosomes contain more than 50 different proteins as well as ribosomal RNAs (rRNAs). Since each protein has a defined amino acid sequence, ribosomal proteins could not have been part of the original translational machinery. It seems more likely that the primitive translational machinery depended primarily on RNA. Indeed, the mRNAs of many prokaryotic and eukaryotic bacteria can form specific base pairs with part of the smaller rRNA.

The anatomy of rRNAs also supports the idea that rRNAs directly participated in protein synthesis. For example, researchers have determined the nucleotide sequences of the smaller rRNAs of many widely divergent organisms, ranging from bacteria to vertebrates, as well as of ribosomes from mitochondria and chloroplasts. For each rRNA, researchers predicted the pattern of stems and loops that could be formed by the formation of duplexes of complementary sequences. While the sequences of the rRNAs vary widely, the patterns of stems and loops are much the same (Figure 18-14). Such conservation suggests an essential function for the pattern of stems and loops, probably as a catalyst for directed protein synthesis.

How Did tRNAs Evolve?

All modern cells attach amino acids to transfer RNAs (tRNAs) before assembling a polypeptide. Transfer RNAs play a critical role in translation, allowing unlike molecules—nucleotides and amino acids—to be connected. All modern tRNAs have similar stem-and-loop structures, and all attach to an amino acid at their 3′ ends. In addition, all tRNAs contain an anticodon with common surrounding elements. Thus, like the genetic code, tRNAs probably diverged from a single tRNA early in the history of life.

How Did the First Cells Use Genetic Information?

The first true cell would have had to have the following properties:

1. A boundary, such as that in artificial protobionts
2. Enzymes for extracting energy from chemicals in the environment
3. Energy storage in ATP
4. RNA that specified the structure of specific enzymes
5. RNA replication, using the Watson-Crick rules of base pairing
6. Specification of each amino acid by a triplet codon, with most of the information conveyed by one or two bases
7. Primitive tRNAs to connect amino acids to nucleotides

The first enzymes probably arose by the random assembly of amino acids into short polypeptide segments. Such primitive enzymes were probably neither as active nor as stable as modern enzymes. But if a polypeptide segment could form the active site of a catalyst, then its possessor would have enormous advantages over its competitors.

Similarly, the first genetically useful RNAs probably had only short stretches of information. Once the first genetic system had evolved, however, natural selection would begin to work. Natural selection would favor cells with increased capacity to reproduce, as well as cells better able to extract energy from chemicals in the environment. Reproduction, protein synthesis, and metabolism must therefore have evolved together. The most successful organisms would be those that best coupled their genetic and metabolic systems—those best able to pass on useful catalysts to their descendants.

HOW DID MODERN CELLS EVOLVE?

The first fossils that are clearly eukaryotic are about 1.7 billion years old. More than 2 billion years elapsed between the first appearance of life 3.8 billion years ago and the first eukaryotes. What, we may ask, happened during that period to allow more complex forms to evolve? What biochemical pathways did the earliest organisms develop, and, equally important, how did these developments fundamentally change the Earth's environment?

How Did Early Life Affect the Environment?

Most biologists think that the first cells were **heterotrophs** [Greek, *heteros* = other + *trophe* = nourishment], organisms that obtain energy only by degrading existing organic molecules. Human beings and other modern animals are all heterotrophs, too. We obtain both energy and building blocks, such as amino acids, by eating other organisms. We can extract energy in the process of respiration by oxidizing preexisting organic molecules from other organisms.

The earliest organisms probably obtained both energy and building blocks from preexisting organic molecules in a rich primordial soup. Like all heterotrophs, they consumed what was in the soup but contributed no energy. But unlike animals, they could break down molecules only by means of glycolysis, or fermentation. They were incapable of true respiration, since there was no oxygen. And they were incapable of obtaining energy from the sun by photosynthesis or from inorganic molecules, as plants and other modern **autotrophs** [Greek, *auto* = self + *trophe* = nourishment] do.

Of the three ways to extract energy from molecules—photosynthesis, respiration, and fermentation—only fermentation is common to all bacteria and eukaryotes. It seems likely, then, that fermentation is the oldest metabolic pathway and that it had already evolved in the earliest cells—at least those whose descendants have survived.

The suspected sources of nourishment for early heterotrophs—the organic molecules in the primordial soup—were in limited supply. Early organisms would have gradually exhausted them of their nutritive value, since no autotrophs existed to renew them. Early organisms probably forestalled the exhaustion of the Earth's first resources by developing elaborate chemical pathways for exploiting more and more of the molecules in the soup. Eventually, though, all the original molecules, and all the energy in them, would have been exhausted. One alternative was to begin extracting energy from inorganic chemicals by oxidizing them and to use the energy to build new organic molecules from carbon dioxide. The present-day *Sulfolobus* bacterium, for example, gets its energy by converting sulfur to sulfur dioxide.

The evolution of photosynthesis

The most abundant energy source was not inorganic molecules, however, but sunlight. We may guess at how photosynthesis gradually developed, at least 2.7 billion years ago, by examining the variety of ways in which modern organisms harness solar energy.

The first step in the direct harnessing of solar energy by organisms was probably the development of a light-driven proton pump, such as that in the modern purple bacterium. This bacterium pumps protons across its cell membrane, which allows the bacterium to store energy in ATP, as well as actively transport other substances across the membrane.

More advanced modern bacteria use light to generate reducing power (hydrogen atoms or electrons) to synthesize carbohydrates from carbon dioxide. Some photosynthetic bacteria obtain the needed electrons from such donors as hydrogen gas or hydrogen sulfide.

The largest potential source of electrons, however, is water. Water is difficult to break down, but it is the electron source for nearly all modern photosynthetic organisms. The first ancient organism to synthesize organic molecules from sunlight, water, and carbon dioxide won for its descendants a freedom from want that has lasted more than 2 billion years. Cells that could photosynthesize would no longer have to compete for energy-bearing organic molecules. They could simply make their own from the abundant water, carbon dioxide, and sunlight.

The first global toxic waste problem

However, breaking down water created a serious global toxic waste problem. The very steps needed to separate the hydrogen atoms from water also released huge quantities of molecular oxygen. Oxygen is a powerful oxidant that easily rips apart many molecules. Sensitive organic molecules would have been destroyed by oxygen. Many of the original heterotrophs, which depended on organic molecules, probably starved, their food supply destroyed by the wastes of their own autotrophic cousins.

The development of photosynthesis drastically altered the Earth's early environment. Major contributors to this change were the cyanobacteria making up the stromatolites, mentioned at the beginning of this chapter. The wide distribution of stromatolite fossils suggests that they covered vast areas of the world's shallow waters from about 3 billion years ago until the end of the Precambrian. In the absence of modern grazing sea animals, photosynthetic cyanobacteria flourished nearly unchecked for millions upon millions of years. For perhaps as long as 2.5 billion years, they pumped oxygen into the oceans and the atmosphere. The cyanobacteria began to decline about a billion years ago. However, single-celled algae began to appear throughout the open oceans, their photosynthetic activity further boosting the oxygen content of the seas and the atmosphere.

The availability of oxygen allowed the evolution of a new class of heterotrophic organisms that could use oxygen to break down organic molecules. The complete oxidation of organic compounds to water and carbon supplies more energy than glycolysis. The advantage of exploiting oxygen was so great that today anaerobic organisms persist in only a few isolated environments.

The vast flats of cyanobacteria and oceans of algae must have presented another undeniable opportunity. Some of the new organisms that had developed the ability to breathe oxygen now developed the ability to eat the very organisms that had created this caustic waste. The first autotrophs thus provided the basis of a food chain that still survives today. Most animals that live in the ocean eat smaller animals that eat planktonic animals that eat single-celled algae.

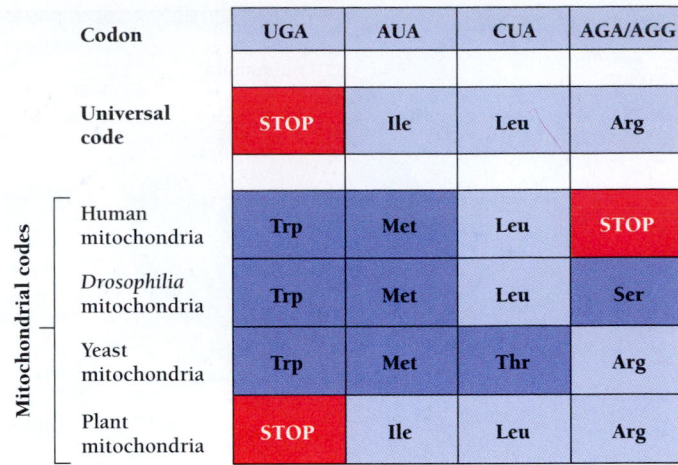

Codon	UGA	AUA	CUA	AGA/AGG
Universal code	STOP	Ile	Leu	Arg
Human mitochondria	Trp	Met	Leu	STOP
Drosophilia mitochondria	Trp	Met	Leu	Ser
Yeast mitochondria	Trp	Met	Thr	Arg
Plant mitochondria	STOP	Ile	Leu	Arg

Mitochondrial codes

Figure 18-15   Genetic code in mitochondria. Mitochondrial DNA encodes polypeptides using a slightly different code than that used by nuclear DNA. The top codons translate to stop, isoleucine, leucine, and arginine, respectively, in the universal code. The mitochondria of humans, fruit flies, and yeast each translate one or more of these codons differently. In contrast, plant mitochondria use the universal code. What does that suggest about plant mitochondria?

How Did Mitochondria and Chloroplasts Arise?

Almost all eukaryotes—protists, fungi, plants, and animals—have mitochondria, the organelles that produce most of a cell's energy. Only plants and some protists, however, have chloroplasts. Biologists conclude from this that the first eukaryotes probably had mitochondria but no chloroplasts.

Because mitochondria depend on oxygen to generate useful energy, they could have evolved only as the photosynthetic prokaryotes filled the world with oxygen. Mitochondria are probably extremely ancient, however, as they use a genetic code that differs from the universal code used by all modern eukaryotes and most prokaryotes. In fact, the mitochondria of different organisms use slightly different codes (Figure 18-15).

Mitochondria are probably older than chloroplasts.

Are mitochondria and chloroplasts endosymbionts?

In both size and internal organization, chloroplasts resemble cyanobacteria (Figure 18-16). These similarities have long sug-

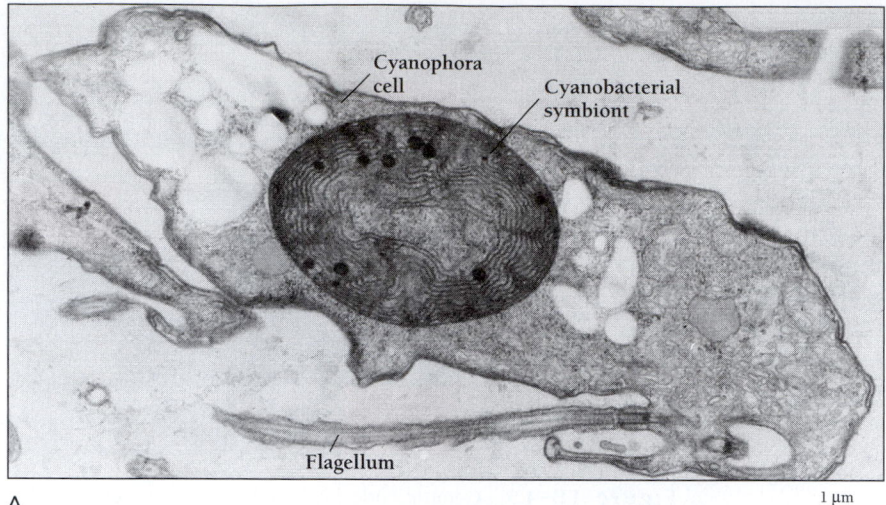

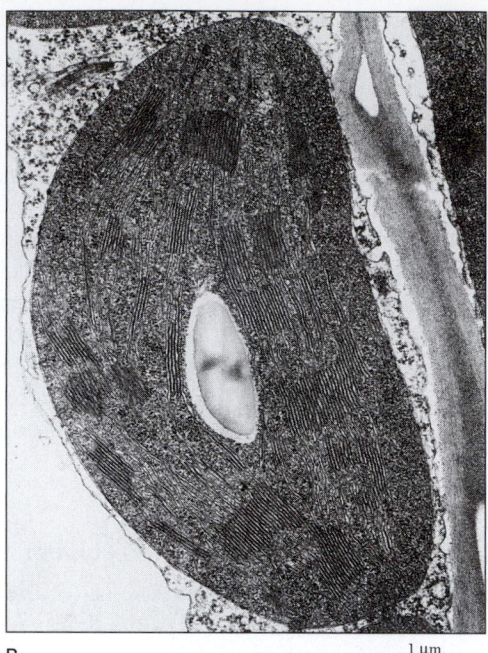

Figure 18-16 Endosymbiosis. A. Living inside the photosynthetic protist *Cyanophora* is a photosynthetic cyanobacterium. B. A chloroplast resembles a cyanobacterium. (*A, Biophoto Associates; B, Omikron/Photo Researchers, Inc.*)

gested that chloroplasts derived from free-living organisms. Indeed, there are many examples today of photosynthetic organisms living within animals or protists. Such a close association of two organisms, one of which lives inside the other, is called **endosymbiosis** [Greek, *endon* = within + *syn* = together + *bios* = life]. Endosymbiosis is mutually beneficial, since it provides an internal source of food for the host and mobility and protection for the inner organism.

Most biologists now favor the view that both chloroplasts and mitochondria are the endosymbiotic descendants of free-living prokaryotes (Figure 18-17). One of the most forceful proponents of this view has been Lynn Margulis, an evolutionary biologist at the University of Massachusetts.

Some biologists still consider the endosymbiosis theory too speculative. Many feel that chloroplasts may well have originated as free-living cyanobacteria but that the case for mitochondria is much weaker. Recent studies of mitochondrial DNA, however, suggest all mitochondria may have evolved from a single kind of ancient bacteria.

Advocates of the endosymbiotic theory are still not agreed about the origin of eukaryotic nuclei, though a strong candidate has recently emerged. Most prokaryotes lack the proteins histone and actin. But the archaebacterium *Thermoplasma acidophilum,* a heat-loving prokaryote, does make histones and actin. The ancestors of this organism, which lacks a cell wall, are a popular best guess for the origin of the nucleus.

The actual origin of eukaryotic cells can never be proven, but Margulis's arguments—together with recent DNA sequence data—have aroused considerable interest in this subject.

Did eukaryotes arise through the proliferation of internal membranes?

Opponents of the endosymbiotic theory argue that eukaryotes arose instead by the proliferation of internal membranes to form new cellular compartments. Lysosomes and the endoplasmic reticulum, for example, form from cellular membranes—not from free-living forms. Perhaps mitochondria and chloroplasts formed when membranes surrounded pieces of mobile DNA (plasmids or transposons) that detached from nuclear genes.

The endosymbiotic theory produces a more complicated evolutionary tree than the traditional view. With this complexity, however, comes a consistency with current views of prebiotic evolution. For the moment, at least, most biologists seem to find endosymbiosis the most convincing general explanation for the origin of eukaryotic cells.

Mitochondria and chloroplasts, which resemble prokaryotes, probably began their association with eukaryotic cells as endosymbionts.

In this chapter, we have explored scientists' hypotheses about how life might have evolved on Earth billions of years ago. Our discussion of such long-ago events has been, necessarily, speculative. In the next section of *Asking About Life*, we return to the present, with a review of the great diversity of modern life. We will examine how biologists classify organisms and then survey the three domains of life.

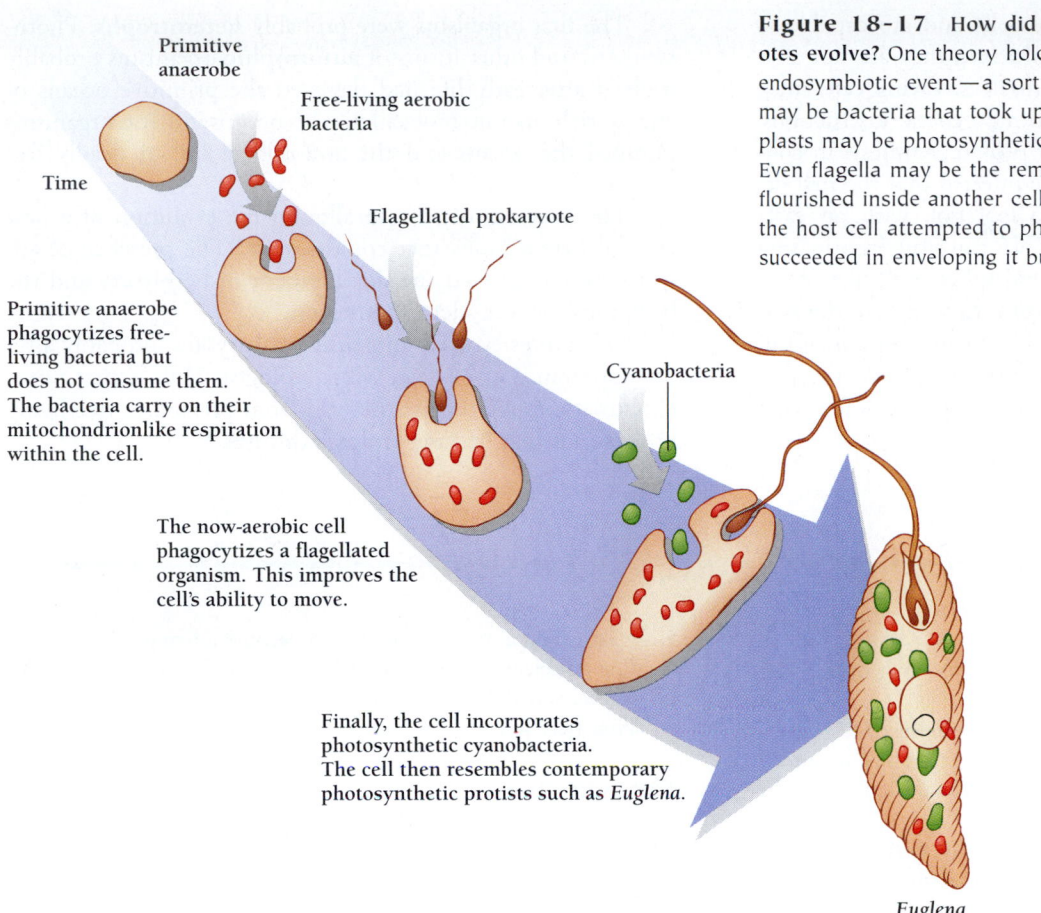

Figure 18-17   How did the subcellular organelles of eukaryotes evolve? One theory holds that each organelle represents an endosymbiotic event—a sort of housing cooperative. Mitochondria may be bacteria that took up residence in another cell. Chloroplasts may be photosynthetic cyanobacteria that likewise moved in. Even flagella may be the remnants of flagellated prokaryotes that flourished inside another cell. In each case, biologists speculate, the host cell attempted to phagocytose a smaller bacterium and succeeded in enveloping it but did not digest it.

Primitive
anaerobe

Free-living aerobic
bacteria

Time

Flagellated prokaryote

Primitive anaerobe
phagocytizes free-
living bacteria but
does not consume them.
The bacteria carry on their
mitochondrionlike respiration
within the cell.

Cyanobacteria

The now-aerobic cell
phagocytizes a flagellated
organism. This improves the
cell's ability to move.

Finally, the cell incorporates
photosynthetic cyanobacteria.
The cell then resembles contemporary
photosynthetic protists such as *Euglena*.

Euglena

Study Outline with Key Terms

By the middle of the 19th century, the work of Redi, Spallanzani, and Pasteur had convinced scientists that life could not arise spontaneously. Yet all life must have some ultimate origin. While some scientists have argued that life may have arrived from elsewhere in the universe, most researchers assume that life evolved from nonliving, or **abiotic,** organic molecules here on Earth.

The theory of the origin of life on Earth is different from the theory of spontaneous generation. Life could not evolve on Earth today because oxygen would oxidize the components and other organisms would eat the components before enough time elapsed for anything to happen. Ideas about the origin of life rest on the demonstration that plausible **prebiotic** chemical reactions could have produced molecules and organized structures like those in modern organisms.

After the Earth formed some 4.6 billion years ago and its crust had hardened half a billion years later, torrential rains began creating the first sediments. Precambrian rocks contain fossils as old as 3.8 billion years, only 300 million years after the formation of the oldest known sedimentary rocks. The Earth's early atmosphere contained no free oxygen. Consequently, no ozone layer protected the atmosphere from ultraviolet (UV) light. UV light and lightning probably contributed the intense energy needed for the formation of the first organic molecules. The early atmosphere may have contained hydrogen, ammonia, and methane, as well as other gases and water. In the laboratory, such a mixture, exposed to electrical discharges, produces many organic molecules, including amino acids and precursors of nucleotides. However, recent research by geoscientists suggests that the early atmosphere was more likely composed of carbon dioxide and nitrogen. Such an atmosphere could have easily given rise to sugars, but to amino acids only with difficulty.

Biologists assume that the primitive oceans became loaded with organic molecules. These organic molecules can form polymers, called **proteinoids.** Mixtures of different kinds of molecules can assemble into organized cell-like structures, called **protobionts.** Polypeptides, nucleic acids, and polysaccharides form **coacervates.** Proteinoids will form tiny spheres called **proteinoid microspheres.** A mixture of phospholipids and proteins form **liposomes,** which have a bilayered membrane similar to that of a true cell.

Although protobionts resemble cells and are capable of cell-like behavior and **prebiotic evolution,** they are not alive because they cannot reproduce themselves. Living cells must be able to store, translate, and replicate genetic instructions. By comparing the characteristics of protein synthesis in contemporary organisms, scientists have inferred that the first organisms used RNA as the genetic material, with an early version of the triplet genetic code. RNA can store information and replicate itself. It can also snip and splice itself like an enzyme. RNA catalysts, called **ribozymes,** can catalyze the synthesis of tRNA, rRNA, and mRNA, the major components of translation. Most evolutionary biologists think that the first cells functioned in an **RNA world,** using RNA both for genes and enzymes.

The first organisms were probably **heterotrophs.** Photosynthetic and other forms of **autotrophic** organisms probably evolved after early life had depleted the primitive oceans of energy-rich organic molecules. The photosynthetic organisms pumped the oceans and the atmosphere full of deadly free oxygen.

The presence of oxygen allowed the evolution of a new class of heterotrophs that could respire. The presence of autotrophs also allowed the development of herbivores and the beginnings of a modern food chain.

Eukaryotes evolved long after prokaryotes, probably after the appearance of oxygen. Most biologists believe that mitochondria and chloroplasts derived from previously free-living prokaryotes that became **endosymbionts.**

Review and Thought Questions

Review Questions

1. Give two reasons why it took paleontologists so long to find Precambrian fossils.

2. Describe the two possible alternative compositions of the Earth's primitive atmosphere, including the disagreements scientists have about them. List two major sources of prebiotic organic molecules.

3. Describe Stanley Miller's landmark experiment to test the idea of prebiotic synthesis. What is the "primordial soup?"

4. What is a coacervate made of, how is it formed, and what can it do? What is a proteinoid microsphere made of, how is it formed, and what can it do? What is a liposome made of, how is it formed, and what can it do?

5. What would a protobiont need to become a living cell?

6. Why do scientists think that RNA was the first genetic material?

7. In what way did the early heterotrophs change the environment? What were some of the consequences of this change?

Thought Questions

8. Can natural selection operate on nonliving structures such as coacervates? Explain.

9. Pasteur's sterile broths contain no organisms that could consume any life that might arise spontaneously. Why have these flasks not given rise to new life in 130 years?

About the Chapter-Opening Image

Until Francesco Redi proved otherwise in the mid-17th century, most people believed that houseflies arose spontaneously.

 On-line materials relating to this chapter are on the World Wide Web at **http://www.harcourtcollege.com/lifesci/aal2/**

part IV

Diversity

Chapter 19
Classification: What's in a Name?

Chapter 20
Prokaryotes: Success Stories, Ancient and Modern

Chapter 21
Classifying the Protists and Multicellular Fungi

Chapter 22
How Did Plants Adapt to Dry Land?

Chapter 23
Protostome Animals: Most Animals Form Mouth First

Chapter 24
Deuterostome Animals: Echinoderms and Chordates

Photo: Tail of Meller's chameleon *(Chamaeleo melleri).*
(Tim Davis/Stone Images)

19

Classification: What's in a Name?

Is the Red Wolf a Species?

In the 18th century, when European naturalists first began mapping and naming the animals and plants of North America, the destruction of the eastern forests had already driven countless species from the landscape. One of the first to go was the gray wolf, which had, before the time of Columbus, occupied nearly the whole of North America. By the 1970s, the gray wolf had been driven out of virtually all of the lower 48 states, its range reduced to Alaska, Canada, and tiny bits of northern Michigan and Minnesota (Figure 19-1). In the gray wolf's wake were left a rapidly shrinking population of red wolves, which had once lived throughout the American South, from Florida to Texas; and a rapidly expanding population of coyotes, which gradually spread from the Rockies and the Great Plains—north, east, and west—into the once-forested areas vacated by the gray and red wolves.

In 1967, the federal government designated the red wolf an endangered species, and biologists working for the U.S. Fish and Wildlife Service began surveying the South for red wolves (Figure 19-2). The red wolf, they discovered, was all but extinct. By 1972, the last few red wolves clung to the southern edge of North America, where Texas and Louisiana drop into the Gulf of Mexico. These last wolves were actively breeding with coyotes.

Government biologists realized with a shock that the only way to save the red wolf from complete extinction was to capture all the remaining wolves and breed them in captivity. From 1974 to 1980, government biologists trapped more than 400 animals. In 1980, in a marsh edging an industrial area of Galveston, Texas, the last wild red wolves in America walked into traps. Of the 400 captured animals, only 43 actually looked like wolves, however, and only 14 produced descendants that looked like wolves rather than coyotes.

Under the watchful eyes of biologists, this group of 14 animals has produced hundreds of offspring. Many of these captive-bred wolves have been released into places such as the Great Smoky Mountains of Tennessee and the Alligator River National Wildlife Refuge, in North Carolina. The rest of the captive wolves continue to produce offspring to be released in years to come. The Red Wolf Reintroduction Program, as this project is called, is the most successful captive-breeding program in existence. By the early 1990s, the red wolf had been brought back from the very brink of extinction. Today, 70 to 80 red wolves live in the wild in the Great Smoky Mountains and at Alligator River.

Not everyone has applauded the program's success, however. Initially, rural residents worried about the safety of their children and pets. Others questioned the price of saving the red wolf. The red wolf reintroduction program costs about $1 million a year. Yet the program's most vocal opponents are not taxpayers' groups, but sheep and cattle ranchers. For them, the reintroduction of wolves to areas of the country now occupied by cattle and sheep might realize the romantic notions of city-dwelling environmentalists, but at the cost of valuable livestock.

Into this heated conflict stepped David Mech, one of the world's leading wolf experts. In 1989, at an Atlanta meeting of experts on wolf biology, Mech challenged his fellow researchers to tell him how they could justify spending so much money rescuing the red wolf when it might not even be a species.

Since the 18th century, biologists had argued over the status of

- Naming organisms allows biologists to distinguish one species from another and to group related species.

- Biologists classify organisms into species and also into larger groups such as genera, families, and orders.

- Natural selection shapes organisms in ways that sometimes obscure patterns of relatedness.

the red wolf. In 1791, naturalists designated the red wolf a subspecies of the gray wolf. But later biologists decided that the red wolf was a separate species. In 1979, U.S. Fish and Wildlife Service biologist Ronald Nowak carefully compared the skulls of gray wolves, red wolves, and coyotes and noticed that the size and shape of the red wolf skull fell midway between that of the coyote and the gray wolf. Nowak's interpretation of the fossil record further suggested to him that intermediate skulls like that of the red wolf skull first appeared in North America more than a million years ago, well before the first wolves or coyotes. Nowak concluded that the red wolf was not only a unique species but also the ancient ancestor of both the gray wolf and the coyote.

Additional paleontological evidence suggested that the gray wolves of Europe and Asia originated in North America. Therefore, argued Nowak, the red wolf must be a direct descendant of the first wolf. The rescue of the red wolf from the brink of extinction, then, could be seen as the salvation of a southern species of nearly royal lineage. It was a compelling idea, one that persisted almost unchallenged for 10 years, throughout the early years of the Red Wolf Reintroduction Program.

But David Mech had a different theory about red wolves. In a 1970 book, Mech had proposed that the red wolf was neither species nor subspecies, but a hybrid produced by interbreeding between the gray wolf and the coyote.

Mech's hypothesis accounted for Nowak's data showing that the size and shape of the red wolf skull was midway between that of the gray wolf and the coy-

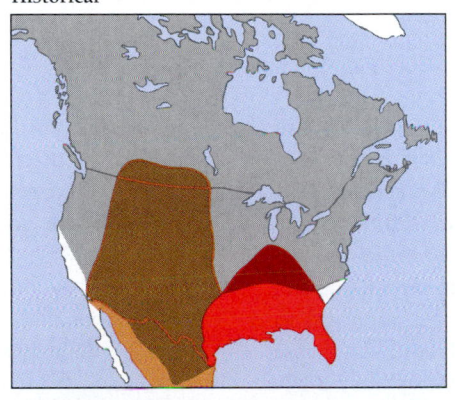

Historical

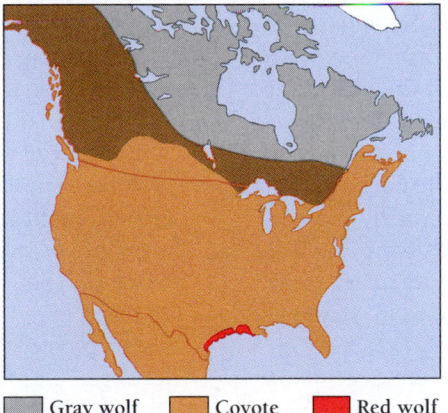

Modern

| ■ Gray wolf | ■ Coyote | ■ Red wolf |

Figure 19-1 The disappearance of the gray wolf. As settlers moved across North America in the 1900s, they used guns, traps, and poison to virtually eliminate the gray wolf from the United States. But the coyote, which has been subjected to similar organized massacres, has flourished. (*Michael Goodman*)

ote. Furthermore, northern gray wolves were known to interbreed with coyotes in some areas, and the offspring had skull measurements similar to those of red wolves. Nowak's skull measurements could be interpreted in two ways, Mech argued: either the red wolf species had split over evolutionary time into three species, the red wolf, the gray wolf, and the coyote; or red wolves were hybrids of gray wolves and coyotes.

In 1989, two University of California biologists, Robert Wayne (of UCLA) and Susan Jenks (of UC Berkeley), approached the U.S. Fish and Wildlife Service and offered to settle the matter once and for all. Like Nowak, Wayne was an expert on the morphology and taxonomy of wolves and other canids (members of the dog family). But unlike Nowak, Wayne had added molecular genetics to his repertoire of research techniques. Wayne and Jenks offered to search the red wolf genome for some unique stretch of DNA that neither the coyote nor the gray wolf had, something that would clearly identify the red wolf as a separate species.

The government agreed to fund the study, and the two biologists began examining DNA from red wolves, gray wolves, and coyotes. Wayne and Jenks began with mitochondrial DNA because, in general, it accumulates mutations more rapidly than DNA from the nucleus. This rapid evolution makes mitochondrial DNA particularly useful for comparing groups of closely related organisms. The base sequence of the mitochondrial DNA of gray wolves (*Canis lupus*) and coyotes (*Canis latrans*), for example, differs by about 4 percent.

The mitochondrial DNA of the gray wolf, Wayne and Jenks found, included

Figure 19-2   The red wolf (*center*) is intermediate between the gray wolf (*left*) and the coyote (*right*) in size, conformation, and behavior. (*Gray wolf, Tom & Pat Leeson/DRK Photo; red wolf, Jane McAlonan/Visuals Unlimited; coyote, C.K. Lorenz/Photo Researchers, Inc.*)

sequences that the coyote did not have. Similarly, the DNA of the coyote included sequences that the gray wolf did not have. The gray wolf and the coyote were clearly separate species. But, to Wayne and Jenks's surprise, the mitochondrial DNA of the red wolf (*Canis rufus*) included *no sequences unique to red wolves.* Red wolf mitochondrial DNA only included sequences identical to those of gray wolves and sequences identical to those of coyotes. The two biologists tentatively and somewhat reluctantly concluded that the red wolf was most likely a hybrid of the gray wolf and the coyote.

Nowak and the other biologists at the U.S. Fish and Wildlife Service could not believe what they were being told. Maybe, argued the government biologists, Wayne and Jenks had simply missed the DNA sequences that distinguished the red wolf. Maybe they had not looked at enough DNA.

To put to rest any lingering doubts, Wayne and other colleagues turned to special repetitive regions of the nuclear DNA called microsatellites. Wayne checked ten microsatellite regions in blood samples taken from hundreds of living animals and skin samples from 16 pre-1930s skins stored in a vast collection of furs at the Smithsonian Institution, in Washington, DC.

The results were the same. Neither the samples of blood from living red wolves nor the samples from the skins of pre-1930s red wolves showed any unique sequences. By 1994, Wayne had found no evidence that the red wolf had ever been reproductively isolated from either gray wolves or coyotes. The red wolf had to be a hybrid of the gray wolf and the coyote.

Wayne's genetic data proved to be an embarrassment to the U.S. Fish and Wildlife Service, which had poured millions of dollars into the reintroduction program in the belief that the red wolf was a unique and endangered species. Yet, the

agency had acted in good faith. Until Wayne and his colleagues finished their research, the U.S. Fish and Wildlife Service had no way of knowing that the red wolf was not a species.

Now the government agency was faced with a terrible dilemma. Wayne's results threatened to discredit the reintroduction program, strip the red wolf of its endangered status, and further undermine the increasingly battered public image of the federal Endangered Species Act. The act had been under attack from its very inception by real-estate developers, ranchers, miners, and other industries that find species preservation inconvenient and expensive.

For now, the red wolf hybrid has not lost protection under the Endangered Species Act, but its hybrid status certainly weakened its claim to protection. The American Sheep Industry Association, for example, formally petitioned the U.S. Department of the Interior to take the animal off the endangered list. The Department of the Interior turned down this petition in 1992, arguing that "limited genetic exchange" between groups does not undermine species status. In short, the agency did not acknowledge that the red wolf was probably not a species.

To protect the red wolf, the U.S. Fish and Wildlife Service began pressuring Wayne to avoid the word "hybrid" in his research papers and to substitute the term "intergrade species" and other similar phrases. By 1994, it was clear, however, that the red wolf was not a species engaging in "limited genetic exchange" with other species.

In 1995, the U.S. Department of the Interior issued a legal opinion that said that hybrids would be protected under the Endangered Species Act if *morphological* evidence showed that the hybrids were similar to the endangered "pure" form. In essence, if they looked like red wolves, they would be pro-

Foucault's Chinese Encyclopedia

"This book first arose out of a passage in Borges, out of the laughter that shattered, as I read the passage, all the familiar landmarks of my thought—our thought, the thought that bears the stamp of our age and our geography—breaking up all the ordered surfaces and all the planes with which we are accustomed to tame the wild profusion of existing things, and continuing long afterwards to disturb and threaten with collapse our age-old distinction between the Same and the Other.

"This passage quotes a 'certain Chinese encyclopaedia' in which it is written that 'animals are divided into: (a) belonging to the Emperor, (b) embalmed, (c) tame, (d) sucking pigs, (e) sirens, (f) fabulous, (g) stray dogs, (h) included in the present classification, (i) frenzied, (j) innumerable, (k) drawn with a very fine camel-hair brush, (l) et cetera, (m) having just broken the water pitcher, (n) that from a long way off look like flies.'

"In the wonderment of this taxonomy, the thing we apprehend in one great leap, the thing that, by means of the fable, is demonstrated as the exotic charm of another system of thought, is the limitation of our own, the stark impossibility of thinking that."

—From the introduction to *The Order of Things, An Archaeology of the Human Sciences,* by Michel Foucault, Vintage Books, New York, 1970.

Figure A A Chinese hanging scroll depicting hens and roosters in a barnyard. *(Werner Forman Archive, National Gallery, Prague/Art Resource)*

---

tected. But the genetic data did not support the idea that a "pure" form of the red wolf had ever existed, certainly not in the last 100 years. In issuing this opinion, the agency excluded all the genetic evidence regarding the red wolf's species status. The only question was whether the red wolf looked different from the coyote and the gray wolf. It did, and, therefore, until such time as the government acknowledges the genetic data, the red wolf will be considered a species.

From the point of view of a taxonomist, the agency's decision might seem illogical. But from a political point of view, the decision made sense. The Endangered Species Act is the keystone law for environmental protection. Anything that weakens it, weakens all environmental protection in the United States. Furthermore, the red wolf is enormously popular. One study by researchers at Cornell University found that nearly 80 percent of people surveyed in eight states surrounding the red wolf reintroduction area favored the wolves. Researchers estimated that tourists attracted by the wolves bring $37 million per year to eastern North Carolina and about $132 million per year to the national park area. In addition, the researchers estimated that people in the eight states surveyed are willing to pay about $3 million per year for the red wolf restoration program, and nationally, the public is willing to pay $18 million per year, figures that far exceed the cost of the program.

In any case, biologists themselves continue to disagree about both the status of the red wolf and the political meaning of that status. In a 1998 issue of the journal *Conservation Biology,* Nowak and another biologist wrote, "Studies support the biological hypothesis that the red wolf is systematically valid. . . . Hybridization has been advanced by individuals and groups who oppose reintroduction of the red wolf as a conservation strategy," while in the same issue, Wayne and his colleagues insisted, "Enough scientific evidence exists to suggest that the red wolf is a hybrid of the gray wolf and the coyote. . . . Conservation efforts directed solely at the red wolf may therefore compromise preservation efforts for other wolf species."

WHY IS IT DIFFICULT TO NAME AND CLASSIFY ORGANISMS?

The story of the red wolf illustrates that deciding how to classify organisms into tidy groups is no easy task. Surprisingly, recognizing species is usually the *least* controversial task facing a **taxonomist,** a biologist who classifies organisms. In contrast, determining whether a species should be subdivided into subspecies (and, if so, how many subspecies) nearly always raises an argument. Likewise, grouping species into higher taxa—such as families, orders, or classes—immediately stirs up a host of controversies, which may be scientific, personal, philosophical, or political. In this chapter, we will briefly discuss how biologists classify organisms. Along the way, we will see why classification is nearly always a controversial topic.

Why Do We Name Things?

"What's in a name?" asks Shakespeare's Juliet, "That which we call a rose by any other name would smell as sweet."

Why do we name things? By naming the rose, we distinguish it from other flowers—both in our own minds and in the minds of others with whom we speak. The name allows us to talk with others about the rose's significance—romantic, evolutionary, or otherwise. To the extent that everyone agrees on a name, we can each talk of roses with confidence that we will be understood. Naming things also allows us to communicate what we know about something, whether the something is an individual object or organism or a class of objects or organisms. We can distinguish poisonous plants from palatable ones, and dangerous animals from docile ones.

Further, by naming things, we force ourselves to make distinctions that we might not otherwise have noticed. In thinking about a rose, most people picture a red flower, yet we would not confuse a red rose with a red poppy or red carnation. Nor would we mistake a yellow rose for a buttercup or a dandelion. Naming an object forces us to know it.

Another reason for naming an object is to classify it, or assign it to a group. A couple might name their children David and Nina simply to distinguish them from others, for example. But parents also give their children a last name to identify the children as part of a particular group, in this case a family.

Naming objects allows us to both distinguish one object from another and assign objects to groups of similar objects.

What Did Linnaeus's System of Classification Accomplish?

To classify objects, we usually establish different categories, or levels, of classification and place these categories in a **hierarchy,** an arrangement in which larger groups contain smaller groups, which contain still smaller groups. We can illustrate a hierarchy of classification with a familiar example. Your address probably falls into each of the following categories: apartment number, street number, street, zip code, city, county, state, country, continent, planet, etc. Notice that the categories form a hierarchy, with each category often (but not necessarily) including more than one example of the lower category (Figure 19-3). Each state, for example, includes many counties, and most counties include several cities and towns.

Recall from Chapter 15 that biologists use a hierarchical taxonomy established by the 18th-century Swedish physician and biologist Carolus Linnaeus. Before Linnaeus's work, biologists had already used a category called a **genus** [plural, **genera;** Greek, *genos* = to be born], a group of species that share many morphological characteristics. Linnaeus and his later followers established higher categories, so that today we traditionally recognize the following hierarchy of classification: species are grouped into genera, genera into families, families into orders, orders into classes, classes into phyla, and phyla

Figure 19-3 "Hierarchical classification" means that each level is included within the levels above.

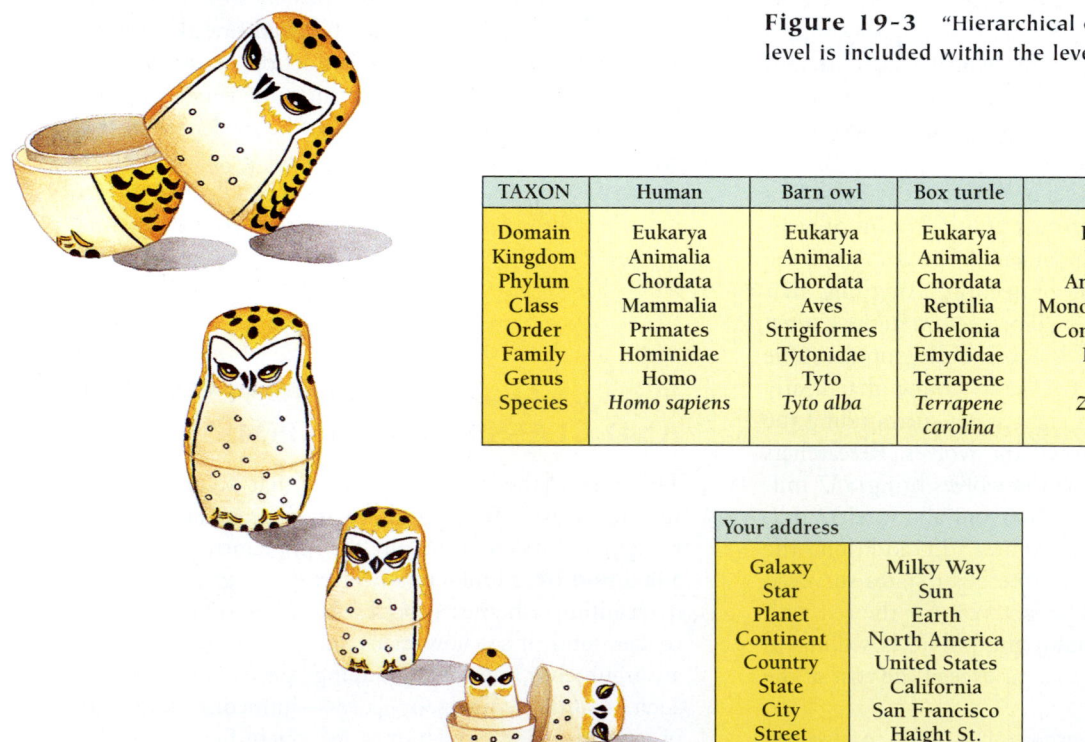

TAXON	Human	Barn owl	Box turtle	Maize
Domain	Eukarya	Eukarya	Eukarya	Eukarya
Kingdom	Animalia	Animalia	Animalia	Plantae
Phylum	Chordata	Chordata	Chordata	Anthophyta
Class	Mammalia	Aves	Reptilia	Monocotyledones
Order	Primates	Strigiformes	Chelonia	Commelinales
Family	Hominidae	Tytonidae	Emydidae	Poaceae
Genus	Homo	Tyto	Terrapene	Zea
Species	*Homo sapiens*	*Tyto alba*	*Terrapene carolina*	*Zea mays*

Your address	
Galaxy	Milky Way
Star	Sun
Planet	Earth
Continent	North America
Country	United States
State	California
City	San Francisco
Street	Haight St.
Number	555

into kingdoms. (Some people remember these categories—in the reverse order—using the mnemonic, "*Kings Play Chess On Fine-Grained Sand.*" But you can invent your own mnemonic.) Recently, biologists have further arranged the kingdoms into three domains: the Archaea, the Eubacteria, and the Eukarya. Figure 19-3 shows the hierarchical classification of several species, including our own. We are members of the **domain** Eukarya, the **kingdom** Animalia, the **phylum** Chordata, the **class** Mammalia, the **order** Primates, the **family** Hominidae, the genus *Homo,* and the species *Homo sapiens.*

Biologists use the word **taxon** [plural, taxa; Greek, *taxis* = arrangement] to mean a group of organisms at any level of the classification hierarchy. So we may speak of all the primates as one taxon and all the bats as another. At the same time, we may speak of all the individuals of a species such as *Homo sapiens* as another taxon.

Linnaeus not only introduced a system of classification but also standardized the naming of organisms. The official name of a species is a **binomial,** a two-part name that contains both the genus (*Homo,* for example) and species (*sapiens,* for example). It is not correct to leave out the genus name or the species name when referring to a species. The fruit fly (or, more properly, the vinegar fly) most often used in genetics experiments is *Drosophila melanogaster.* We may, after writing the name out fully once, abbreviate the genus to its initial letter, as in *D. melanogaster.* But scientists and laypersons alike often refer to *Drosophila melanogaster* as simply *Drosophila.* This is informal and technically incorrect. North America alone has nearly 200 different species that all belong to the genus *Drosophila.* We cannot call all of them *Drosophila* and know what we are talking about.

Binomial names are always Latin or Greek (or at least they are meant to be). Latin was, until this century, regarded as the "universal language" by educated Europeans. The use of Latin names eliminates uncertainty in translation from one language to another (as the Latin name need never be translated). Latin also prevents confusion that results when people in different regions give the same common name to different species. For example, the English robin (*Erithacus rubecula*) bears only passing resemblance and relation to the American robin (*Turdus migratorius*) we know in the United States (Figure 19-4).

Linnaeus standardized both the naming of organisms (with a Latin binomial) and the classification of organisms (into a strict hierarchy).

Who Names a Species?

The first person to describe a new species assigns it its name. Sometimes biologists discover that two people have described and named the same species. The name given first is always considered the correct name and has "priority." This rule is a way of ensuring that species names are stable, that they stay the same from one decade to the next. That way, biologists always know exactly which species is meant.

Figure 19-4 Although unrelated, both the English robin and its American namesake are called "robins." Scientific names are intended to minimize such confusion.

Still, sometimes this is too bad. One hundred years ago, the Yale paleontologist Charles Marsh assigned the charming name eohippus ("dawn horse") to the 55-million-year-old ancestor of the modern horse. Marsh did not know that the tiny fossil horse had already been described and named by someone who thought it was a hyrax (an animal that looks like a guinea pig, only bigger). Eohippus, which is both poetic and accurate, lost out to *Hyracotherium,* a mouthful that means hyraxlike mammal. Because of the rules of priority in naming, we are stuck with this unfortunate name—rules are rules.

Taxonomists use their naming privileges for different purposes. Some name species after colleagues they admire. Some try to describe the species: *Rosa alba,* for example, is a white rose (alba, as in albion or albumen, means white), while *Rosa chinensis* is a Chinese rose. Other biologists just want to have fun: the names for some genera of stink bugs (family Pentatomidae), for example, end with the suffix -chisme, pronounced "kiss me."

Species are named by the first person to recognize the species as new.

How Do We Recognize Species?

Long ago, the Greek philosopher Plato described how we make ordinary classifications. Plato argued that the individual objects of the world are all variations of universal "types," forms

BOX 19.2

Are Guinea Pigs Rodents?

One group of animals whose classification has come under close scrutiny recently is the guinea pigs and their relatives. One of the defining characteristics of the order Rodentia is a single pair of razor-sharp incisors with which they can gnaw through the husks of nuts and seeds. In fact, the word "rodent" means "to gnaw." Because guinea pigs possess these distinctive gnawing incisors, as well as other adaptations for gnawing, taxonomists have traditionally classified them in one of three suborders of rodents.

The three suborders—the squirrels, the mice and rats, and the guinea pigs and their relatives—are classified according to the anatomy of their gnawing muscles. The principal jaw muscle is the masseter muscle, which closes the jaws and also pulls the lower jaw forward, a movement critical to gnawing. In mice and rats, the two sections of the masseter—the lateral masseter and the deep masseter—both attach well forward on the jaw, and these animals' ability to gnaw is unparalleled. In squirrels, only the lateral masseter extends forward. In guinea pigs, only the deep masseter extends forward. Consequently, neither squirrels nor guinea pigs gnaw as well as mice and rats.

The suborder to which the guinea pigs belong, called the caviomorphs, differs from other rodents in numerous ways. Caviomorphs are more diverse in their morphology and habits than either the squirrels or rats and mice. They include porcupines, chinchillas, various guinea pigs (called cavys), capybara, pacas, and agoutis. The capybara, for example, is a large, social herbivore. Individuals may reach 150 lb. Caviomorphs all have proportionately larger heads than other rodents and they reproduce differently (Figure A). While a rat may bear litters of 11 every 3 weeks, a guinea pig bears 2 or 3 young every few months.

Like mice and rats, guinea pigs are frequent subjects of scientific experiment. As a result, quite a bit is known about guinea pig physiology and genetics. In 1991, researchers suggested that the protein sequences of guinea pigs are sufficiently different from those of rats and mice to justify giving the guinea pigs and their relatives a separate order.

A series of studies in the early 1990s on both protein and nucleotide sequences gave contradictory answers, however, and the guinea pigs remained classified as rodents. Then, in 1997, a group of Italian and Swedish researchers carefully analyzed differences in the mitochondrial DNA (mtDNA) of 16 animals, whose complete mtDNA sequences were known.

The results were dramatic. Three different methods of classification all put the guinea pigs in a separate taxon from the mice and rats. Indeed, the guinea pigs appeared to be more closely related to rabbits and primates than to rodents. Although the five researchers called for further research—"the phylogeny will be clearer when a larger number of species can be considered in phylogenetic analysis"—they titled their research report in the journal *Nature* decisively "The guinea pig is not a rodent."

Figure A Guinea pigs. Rodents or not? (*Gerard Lacz/Peter Arnold, Inc.*)

that share an ideal underlying plan (Chapter 15). A chair, for example, always has a seat large enough to support a sitter and, usually, has four legs and a back.

Aristotle extended Plato's concept of types to organisms. In this view, called the **typological species concept,** each species consists of individuals that are variants of a fixed underlying plan. The key word here is "fixed." A typological species cannot evolve and become something different. Aristotle would have argued that a robin and a cat each have a fixed underlying plan.

Eighteenth- and nineteenth-century biologists applied the philosophy of types in the following way. When they first recognized a new species, they killed and collected several examples of the organism. One individual was then selected as the "type specimen" for each species and kept in a museum for reference. All other organisms believed to belong to that species were then compared to the type specimen to verify that they were the same. Because of Darwin, however, biologists have recognized that the typological species concept has two important limitations:

1. All species change over time.
2. The individual members of a species vary.

Twentieth-century biologists have mostly defined **biological species** as groups of actually or potentially interbreeding populations that are reproductively isolated from other such groups (Chapter 17). For example, the cardinals in a Virginia suburb can easily breed with one another, as well as with other cardinals in nearby Maryland. But they cannot breed with mockingbirds. In this view, called the **biological species concept,** individuals are members of a species only in relation to other species. In the same way, "brother" is not a category that can be applied to a single individual. Someone is a brother only in relation to someone else.

To decide whether a group of organisms constitutes a separate species, most biologists like to use the biological species concept. For most organisms, however, the biological species concept is impossible to apply. Certainly we can obtain no information about breeding populations for fossil organisms or for organisms that reproduce asexually, including great numbers of plants, fungi, protists, and bacteria.

Even living, well-studied, sexually reproducing organisms pose problems. No one was certain, for example, how frequently gray wolves and coyotes interbred until biologists tested DNA in blood samples drawn from hundreds of wild wolves and coyotes. Such efforts are expensive and unusual. Many sexually reproducing plants and animals readily form hybrids with groups traditionally classified as separate species.

We can see two objections, then, to the biological species concept: (1) the frequent lack of information about whether two groups can interbreed; and (2) the fact that many recognized species can and do interbreed.

Biologists cope with the practical problems presented by the biological species concept by recognizing that a biological species is usually a reproductive unit, a genetic unit, and an ecological unit. A species is a **reproductive unit** in the sense that the species is the group of potential mates. Among organisms that do not reproduce sexually, that definition of a species does not apply. A species—whether sexually reproducing or not—is also a **genetic unit,** because a species can be defined as a gene pool (usually isolated from other "species" by reproductive barriers).

Finally, a species is an **ecological unit:** within a given habitat, a species interacts in a predictable fashion with other species in the environment. Gray wolves, for example, hunt in packs of 7 to 20 if large animals such as deer, moose, and caribou are available. In contrast, coyotes do not hunt in packs and prefer to prey on smaller animals such as mice and hares. Because of their size and behavior, coyotes interact with their environment differently from wolves. Biologists now know, for example, that although under certain circumstances wolves and coyotes interbreed, they also know that the two species generally remain distinct, genetically and ecologically.

Even these rules are not enough to help classify all organisms. As we will see in the next chapter, bacteria frequently exchange genes with other, unrelated species. Thus bacteria are not generally reproductive units or even genetic units, but only ecological ones, and the way that bacteria interact with their environment can change radically with the sudden acquisition of a single gene.

In reality, biologists usually classify species by the way they look, relying on morphology—the form or visual appearance—of organisms for the provisional definition of a species. As we saw at the beginning of this chapter, biologists name even living, sexually reproducing species such as the red wolf on the basis of their morphology, simply because it is easier.

Fortunately, the morphologically defined species usually corresponds exactly to the biologically defined species. The gray wolf and the coyote, for example, look different and behave differently. In large part, they occupy different habitats. In short, they look and act like true species. Genetic data confirm that the two species have unique sequences of DNA, showing that they have been reproductively isolated from each other. That these two species can interbreed extensively does not make them the same species. In general, the morphologically defined species corresponds to the biologically defined species because reproductive isolation tends to lead to the evolution of divergent morphological characters. Despite most biologists' noisy objections to typological species, then, they frequently begin defining a species on a typological basis.

Biologists use both biological and typological concepts to name and classify species.

HOW DO BIOLOGISTS GROUP SPECIES?

Although biologists with different agendas argue whether particular groups deserve the species label, all modern biologists consider species to be real entities, not abstract or arbitrary classifications. Because species are natural entities, recognizing what constitutes a species is obvious most of the time. Biologists disagree surprisingly rarely on the boundaries that separate individual species. The same researchers may bicker constantly, however, about how to subdivide species into subspecies and how to place species into larger groups such as genera, families, orders, phyla, kingdoms, and now domains.

Naming and classification of species is the main business of **systematics,** the scientific study of the kinds and diversity of organisms and of any and all relationships among them. In contrast, **taxonomy** is the theoretical study of classification, including the study of the principles, procedures, and rules of classification. These two words have different meanings, but, for simplicity, we use the word "taxonomy" only.

Biologists would like to classify organisms according to *relatedness,* but they must do so from measurements of *similarity.* In classifying any kinds of objects—organisms, books, or buildings—we naturally group things according to their similarities. We can understand the problems this presents if we imagine attending a big family reunion of 200 people and trying to decide who is related to whom. A group of redheads might appear to be related. We might find, however, that one of them is simply married to someone else in the family and is not related to anyone else at the reunion. Similarity by itself does not always signify relatedness, and which similarities are important ones is often a matter of much dispute.

For example, in Papua New Guinea, the giant tropical island just north of Australia, biologists have compared their own classification of plants and animals with that used by the tribespeople who live there. The biologists and the tribespeople agree nearly exactly on the number and the identity of species. They disagree, however, on the higher classifications. The biologists group organisms according to their biological similarities—lumping all the birds together in the class Aves, for example. The tribespeople classify according to use—lumping together all the species that are edible, all that are poisonous, or the ones that provide materials for clothing. Both systems of classification serve legitimate purposes.

Taxonomy invites dissent. In fact, biologists now use three methods for classifying organisms. The first is classical evolutionary taxonomy, in use for the last 150 years. In addition, two groups of biologists have each proposed a new way to classify organisms. The first of these new approaches—phenetics—focuses on the practical problem of taxonomy and proposes a seemingly objective step-by-step method for classifying organisms. The second new approach—cladistics—ambitiously proposes to classify organisms according to their evolutionary history.

Classical Evolutionary Taxonomy

Classification by means of classical evolutionary taxonomy relies explicitly on the judgment and experience of the taxonomist. In fact, in one definition of a species, biologists like to say, "A species is what a good taxonomist says it is. And a good taxonomist is one whose work has stood the test of time." A good taxonomist examines the similarities and differences in as many traits as possible.

The most commonly used traits in living animals are morphological ones—including external form, internal anatomy, tissue types, and chromosomal morphology. Mammals, for example, are classified into three broad groups according to their skeletons. Adult monotremes such as the duck-billed platypus and the spiny anteaters have no teeth and lack certain bones of the skull that other mammals have. Their limbs stick out to the sides in a fashion that resembles that of lizards and other reptiles. Because of these and other traits, monotremes are considered primitive mammals.

Classical taxonomists also rely on embryological evidence. For example, the early embryos of all mammals, including humans, include a stage with gill-like pharyngeal slits and pouches (Figure 15-15). This fact, with others, suggests that we are descended from aquatic organisms that needed gills to extract oxygen from water. Finally, taxonomists may also classify on the basis of behavioral traits and, increasingly, on the basis of molecular characteristics, either protein sequences or DNA sequences.

Not all similarities between organisms mean that they are related. Bats and birds have both evolved wings, for example, but bats are no more related to birds than are humans. We say that the wings of birds and bats are "homoplasious." **Homoplasy** [Greek, *homos* = same + *plasis* = mold, form] is the possession by two species of a similar trait that is not derived from a common ancestor. ("Homoplasy" has replaced the older word "analogy," which requires that two similar traits perform the same function—a fact sometimes difficult to determine.) In contrast, similarities that are inherited from a common ancestor are said to be **homologous.** The legs of insects are homoplasious with those of humans and other vertebrates. They are jointed and muscled in a similar manner because they perform the same job, not because they were inherited from a common ancestor. In contrast, the leg of a horse and the leg of a human are homologous. They share similar bones and muscles because they evolved from a common ancestor. Supporting evidence comes from embryology: in humans and horses, the bones and muscles of the leg develop from the same embryological tissues.

When classifying organisms, a classical taxonomist must take into account which characteristics are likely to be homologous and which are homoplasious, giving weight to homologous characters and disregarding homoplasious ones. Although the wings of bats and birds seem obviously different to us, homoplasious characters in closely related species are sometimes hard to distinguish from homologous ones.

Homologous characters indicate relatedness, while homoplasious characters indicate similar adaptations but not necessarily relatedness. Taxonomists try hard to distinguish homologous characters from homoplasious ones.

How Do Advocates of Phenetics Classify Organisms?

Phenetics is a method of classification that avoids subjective choices by making none. **Phenetics** classifies species using *all* observable characteristics. The word "phenetics" comes from the same Greek root as "phenotype," *phainein* = to show.

The goal of phenetics is to list as many traits as possible for each organism and then to use computers to sort organisms according to their degree of similarity. Pheneticists, who are also called "numerical taxonomists," argue that the resulting groupings are objective measures of the similarities among organisms. Phenetic classification requires no knowledge of organisms' breeding behavior or ancestry.

A phenetic taxonomist describes the phenotype of an organism with a series of yes-or-no questions. Each of these questions defines a *unit character*—a characteristic of an organism that cannot be split into finer descriptions. For example, some unit characters for an animal are the answers to questions such as: (1) Does it have five digits on its front limb? (2) Does it have four digits on its front limb? (3) Does it have a backbone? The pheneticist records the answers to these questions in a binary (1 or 0) code: 1 for yes, 0 for no.

In phenetics, each unit character is as important as any other. That is, each of hundreds or thousands of characters is considered to be equally important. Whether an animal has spotted fur would be considered as important as whether it has a backbone.

The problem with phenetics, however, is that while it accurately groups organisms according to their similarities and differences, it doesn't necessarily group them according to lines of descent. Unrelated species often evolve similar characteristics (convergent evolution). For example, a marsupial mole looks very like a placental mole, but these two species are not closely related (Figure 17-15).

Because phenetics views all traits as equivalent, it doesn't distinguish between homologous and homoplasious traits. Homologous traits can tell volumes about ancestry, but homoplasious traits (due to convergent evolution) only confuse us. We know birds and bats are related because both groups have backbones, skulls, and other traits that are unique to vertebrates. But their wings and other adaptations for flight might suggest that they are more closely related than they are.

Phenetics is an objective method of classification, but the results may be meaningless biologically. What biologists really want to know is how species are related to one another. At that big family reunion, we want to know which individuals are siblings and which ones are cousins. And sometimes cousins look more alike than siblings, so grouping only by similarity wouldn't necessarily give us the right answers.

Finally, phenetics could never reveal whether the red wolf was a hybrid of a gray wolf and a coyote or a common ancestor of both. Phenetics can only say that red wolves are intermediate in morphology, not *why* they are intermediate.

Pheneticists classify species and higher groupings by assigning equal value to all characters.

How Do Advocates of Cladistics Classify Organisms?

By contrast, **cladistics** [Greek, *clados* = branch] groups organisms according to their evolutionary history, or **phylogeny** [Greek, *phylon* = race or tribe + *geneia* = birth or origin]. Cladistics starts with the assumption that there is one true family tree and that, if we can discover it, we can produce a nat-

ural grouping or taxonomy of organisms. In contrast to phenetics, which classifies according to all traits, cladistics constructs family trees using only certain traits. These certain traits are called **shared derived characters.** For example, all mammals share fur and mammary glands, both of which are shared characters. In the context of other vertebrates—amphibians and reptiles—fur is a shared derived character in mammals. But context is everything. In primates, fur is a **primitive character,** one that is shared with all other mammals.

Cladists use shared derived characters to construct a family tree, but it is the tree itself, called a **cladogram,** that is the basis for classification, not the characters. We can construct a cladogram for the primates. The cladogram in Figure 19-5 shows just five primates: a ring-tail lemur, a spider monkey, a rhesus monkey, a chimpanzee, and a human. Constructing the cladogram requires that we note characters in which these species differ. Four of these species (all except the lemur) possess a shared derived character—specialized area in the center of the eye's retina that allows more acute vision. Three species (rhesus monkey, chimp, and human) have nostrils that point downward. Humans and chimps differ from the other three species in that they lack tails.

The cladogram in Figure 19-5 is a classification hierarchy that represents a series of hypotheses about the evolutionary history of the primates. Biologists can test these hypotheses by examining other features of these five organisms and by comparing these organisms with the fossil record. They can also compare the sequences of specific proteins and genes.

According to cladistics, every taxon above the species level—whether it is a genus, a family, or a higher category— should be **monophyletic,** meaning that each taxon includes the ancestral species and all its descendants. Birds, for example, form a monophyletic taxon: all birds are descended from a common ancestor that was itself a bird, and all the descendants of this ancestor are birds.

Cladists reject any taxon that is **polyphyletic**—one that includes descendants of more than one ancestor. As we will see in the following chapters, a number of higher taxa appear to be polyphyletic. For example, protists have undoubtedly evolved from many different separate ancestors. They do not constitute a single lineage. For this and other reasons, cladists reject the traditional grouping of protists into the kingdom Protista.

Another group that are not monophyletic are the fishes, the class Pisces. Traditional taxonomists group the lobe-finned lungfishes with all other fishes. But the descendants of the ancestors of today's lungfishes include not only lungfishes but also amphibians, reptiles, birds, and mammals. Because some of the descendants of lungfish ancestors are not fishes, cladists say the group "fishes," which includes bony fishes and lobe-finned fishes, is not monophyletic but "paraphyletic." A **paraphyletic** group contains some but not all of the descendants of a common ancestor. A strict cladistic interpretation would therefore say that the class of animals we call fish—including lungfishes but not mammals—does not exist. In a cladistic sense, we humans are a kind of fish.

The cladists' argument that phylogeny must be the basis of classification also leads them to argue that the Reptilia is not a meaningful classification unless it includes birds. At first glance, reptiles appear to be a monophyletic group. All reptiles appear to have descended from a common ancestor that was itself a reptile. But, in fact, reptiles are paraphyletic. Some of the descendants of reptiles (dinosaurs) have evolved into birds, organisms not usually classified as reptiles.

From the fossil record, we know that the first organisms with "amniotic eggs" (eggs with layers of surrounding tissues such as are found in reptiles, mammals, and birds) arose about 300 million years ago. These "amniotes" split into several lines: the reptilelike mammals and their descendants, the mammals; the turtles; and the reptiles, including the lizards and snakes and the archosaurs. The archosaurs, in turn, split into various dinosaurs, primitive birds, and crocodiles.

The cladogram in Figure 19-6 shows a nested set of taxa for birds, mammals, and reptiles. According to this scheme, reptiles (understood as turtles, lizards and crocodiles, but not birds) are not a legitimate taxon because it is paraphyletic if we leave out the birds and mammals.

Traditional biologists tend to dislike the idea of including birds and mammals in the reptiles. In part, this is because birds and mammals are "warm-blooded" homeotherms, while the other amniotes are mostly "cold-blooded" heterotherms. But homeothermy is a derived character that has evolved separately in birds and mammals. It does not indicate relatedness. That tradition plays a role in such preferences is supported by the fact that taxonomists have adopted no new paraphyletic groups in recent decades. All paraphyletic groupings are left over from previous centuries.

If birds and crocodiles are more closely related to each other than either is to a lizard, why don't they look more alike? One answer is that birds have evolved faster than either crocodiles or lizards. So birds have effectively left the crocodiles and lizards behind. Cladistics raises such questions and offers insights into the ties among seemingly unrelated organisms. For example, knowing that dinosaurs, birds, and crocodiles are closely related, we can guess that the similar nesting behavior found in all three—including parental care and vocalizations (at least in crocodiles and birds)—is probably another sign of their close family ties. We can conclude that, most likely, the nesting behavior of these dissimilar animals is a homologous trait, not a result of convergent evolution.

Still, the cladistic approach has some limitations. Some groups cannot be classified using cladistics at all. In bacteria, for example, deciding which characters reveal ancestry is nearly impossible. Bacteria exchange genes and traits with such freedom

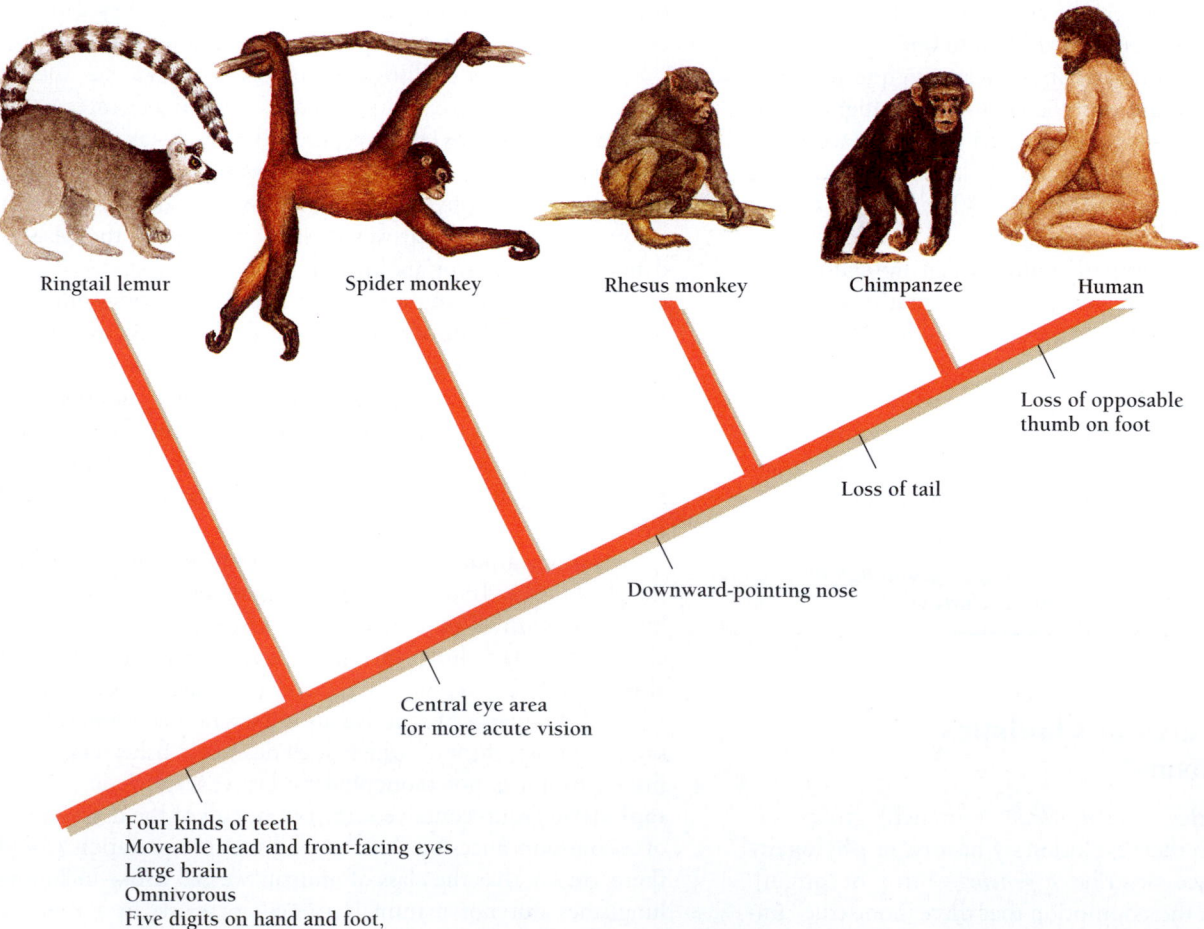

Figure 19-5 A cladogram for five primates.

Cladogram

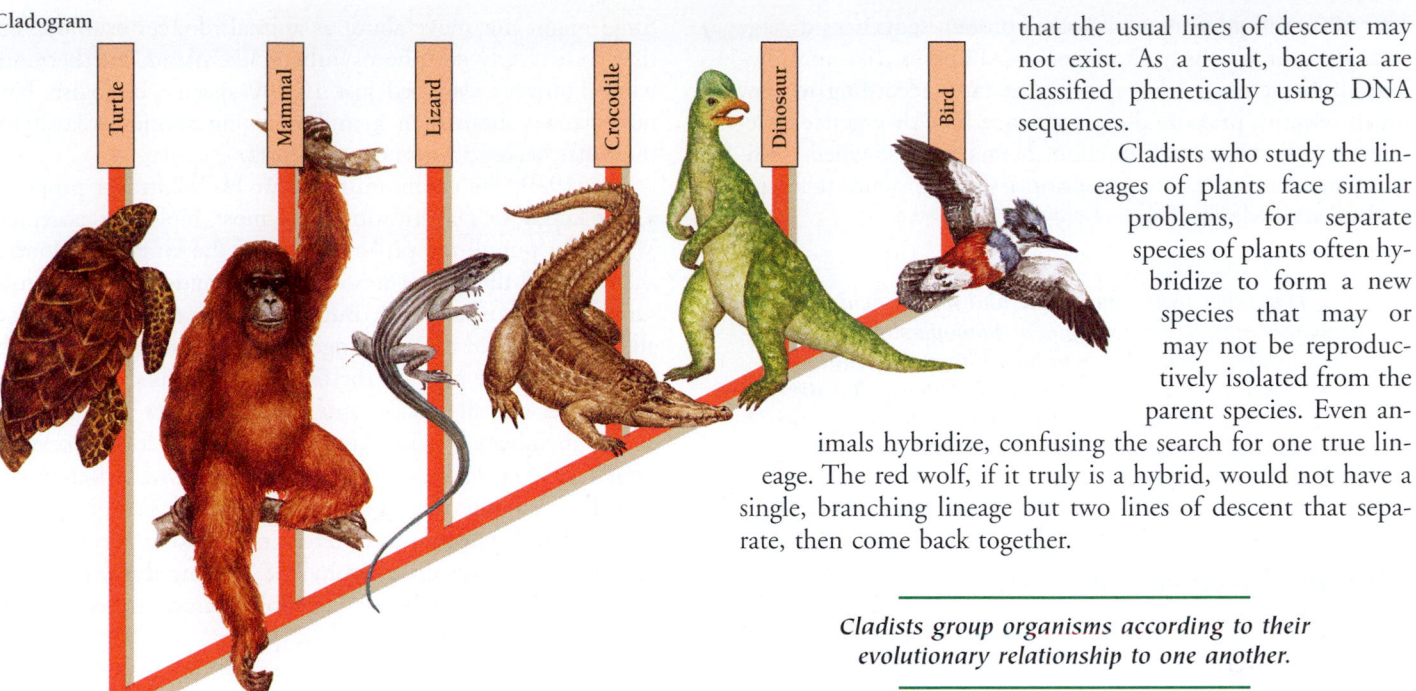

Cladistic Hierarchy

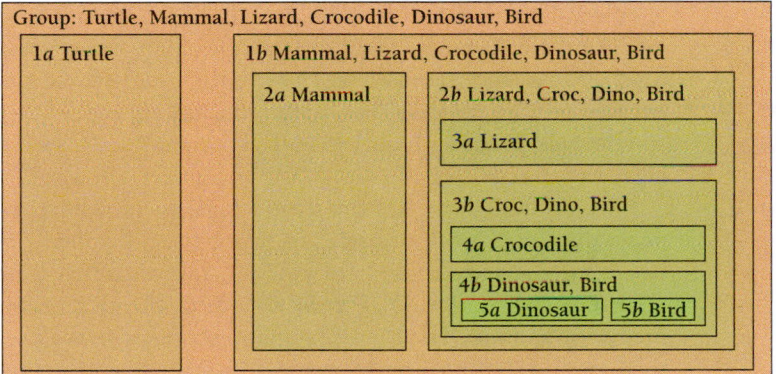

Phylogenetic Tree

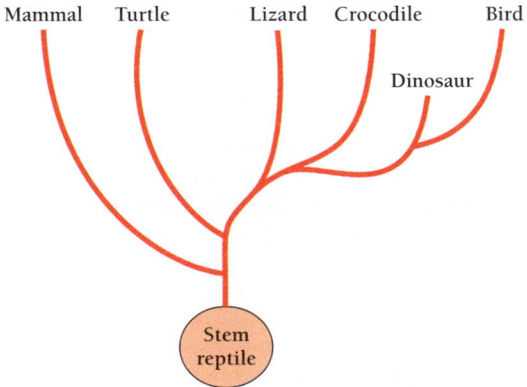

Figure 19-6   Cladistic classification of birds, mammals, and reptiles. In the cladistic view, reptiles are a legitimate taxon only if they include birds and mammals.

that the usual lines of descent may not exist. As a result, bacteria are classified phenetically using DNA sequences.

Cladists who study the lineages of plants face similar problems, for separate species of plants often hybridize to form a new species that may or may not be reproductively isolated from the parent species. Even animals hybridize, confusing the search for one true lineage. The red wolf, if it truly is a hybrid, would not have a single, branching lineage but two lines of descent that separate, then come back together.

Cladists group organisms according to their evolutionary relationship to one another.

Do DNA and Protein Sequences Provide "Objective" Criteria for Classification?

As we saw in Chapter 15, the sequence of a protein evolves in much the same way as a morphological feature. For this reason, comparisons of amino acid sequences (or nucleotide sequences) suggest relationships between species. For example, all other things being equal, two species with identical genes for hemoglobin are more likely to be closely related than two species whose genes for hemoglobin differ by several bases.

Because base sequences are linear and unambiguous, it is easy to imagine ranking all genetic differences equally, as pheneticists suggest we do with morphological differences. Indeed, evolutionary biologists have embraced the molecular approach enthusiastically, partly for this reason. The gene sequences that specify proteins, however, evolve in the same way as larger-scale characters, and distinguishing between homologous and homoplasious sequences may be even harder than with morphological traits. Base sequences, then, are not more objective measures of relatedness than are morphological characters.

In addition, just like morphological traits, different sequences evolve at different rates. We can easily see that a structure such as the backbone may persist for 200 million years, while fur color may change dramatically in just a few generations. It makes more sense to group animals according to whether they have backbones than according to their external coloring.

Genes, and the polypeptides they specify, also evolve at different rates. Histone 4, a protein that binds tightly to nuclear DNA, appears to have remained nearly unchanged for over 500 million years, while other protein sequences diverge strongly over relatively short periods. Proteins, like morphological characters, evolve at different rates according to how much selective pressure they experience. The divergence of sequences reflects natural selection. It makes sense when classifying organisms, then, to give priority to the proteins that have evolved most slowly.

Like other traits, nucleotide and amino acid sequences may be homologous or homoplasious, stable or unstable. Molecular approaches to classification have opened up a rich, new vein of evidence, but the sequences themselves are not more "objective" measures of relatedness than morphological and behavioral traits.

Why Is Classification So Difficult?

If all genes and morphological features changed at constant rates and if homoplasy did not exist, then maybe cladists, pheneticists, and traditionalist taxonomists would always agree on how to classify organisms. But evolution does not conform to the desires of biologists. Instead, the characters of organisms change in response to the chance appearance of new mutations, the changing geography of our planet, and the many forces of natural selection.

In the following chapters, we follow a traditional (and flawed) classification. The views taken in this book are that (1) biologists and students must be aware of the difficulties of classification and (2) we must be able to communicate with one another. We therefore will follow the most widely accepted conventions for classification, while occasionally noting the difficulties of conventional wisdom.

Classification is made difficult by the fact that the forces of natural selection shape organisms and their genes in ways that sometimes obscure patterns of relatedness.

HOW MANY GROUPS OF ORGANISMS ARE THERE?

Linnaeus divided all organisms into two kingdoms, the plants and the animals. As biologists learned more about life on Earth, however, they became dissatisfied with this grouping. The fungi might not move about as animals do, for example, but they were clearly not photosynthetic like plants. Furthermore, while Linnaeus classified just 10,000 species, biologists have now named more than a million living species. Classifying them has become a monumental task.

In 1969, the taxonomist Robert H. Whittaker proposed a five-kingdom classification that most biologists accepted. Whittaker put all the prokaryotes into the kingdom Monera, then divided the eukaryotes into four kingdoms: the protists, fungi, plants, and animals. But in recent years, biologists have all but abandoned the five-kingdom classification. Cladists object to it because none of the five kingdoms is monophyletic. Eukaryotes are all descendants of prokaryotes and therefore clearly members of other kingdoms, and the four eukaryotic kingdoms each include lines of organisms with different origins. For example, among the protists, the green algae and the protozoa each arose from different prokaryotic ancestors. We say that the protists are polyphyletic. Among the animals, the descendants of the sponges probably include other animals. Sponges are paraphyletic.

Researchers have suggested several other classifications, including one with eight kingdoms. But most now agree that all the kingdoms, whatever their number, can be assigned to just three domains—the Archaea, the Eubacteria, and the Eukarya.

The classification that we use in this book distinguishes among organisms in two ways: cellular organization and source of energy. First we can distinguish the cells of prokaryotes from those of the eukaryotes. The Eukarya—including the protists, fungi, plants, and animals—all have nuclei and membrane-enclosed organelles. In contrast, prokaryotic cells of the Archaea and the Eubacteria lack a nucleus and other membrane-enclosed organelles.

We distinguish among the three kingdoms of plants, animals, and fungi by the manner in which their members derive energy—plants from photosynthesis, animals from eating other organisms, and fungi from absorbing chemicals from their surroundings. The fourth kingdom of Eukarya, the protists, are mostly single-celled eukaryotes that do not fit into any of the other three kingdoms of eukaryotes.

In the following five chapters, we will survey some of the major groups within each domain and kingdom. In doing so, we will examine the diverse ways by which species survive and flourish.

Using a traditional taxonomy, we group organisms in terms of differences and similarities in cellular organization and lifestyle, rather than strict evolutionary relationships.

Study Outline with Key Terms

A **taxonomist** is a biologist who classifies organisms. Each classification of organisms is a **taxon.** Many biologists now use the word **systematics** for the naming and classification of organisms and **taxonomy** for the study of *how* organisms should be classified. Until this century, biologists usually viewed the members of a species as variants from a fixed underlying plan, in what is now called the **typological species concept.** Today, biologists usually define a species according to the **biological species concept,** which says that a species is a group of actually or potentially interbreeding organisms that do not interbreed with members of other such groups. Besides being **reproductive units, biological species** are also **genetic units** and **ecological units.**

Most of the time, morphologically defined species correspond to the biologically defined species. Classification of species into larger groups, however, may be difficult. Biologists use a **hierarchical** scheme, in which species are grouped into **genera,** genera into **families,** families into **orders,** orders into **classes,** classes into **phyla** (or divisions), and phyla into **kingdoms.** Kingdoms are further arranged into three **domains**—the Archaea, the Bacteria, and the Eukarya. Each species has a two-word designation, called a **binomial,** which names both the **genus** and the species.

The biological species concept is not always useful. Classifiers may know nothing about breeding patterns, and many species interbreed with other species. **Phenetics** gives equal weight to every character, grouping organisms with similar clusters of characteristics. But not all common characters indicate relatedness. Similarities that are inherited from a common ancestor are said to be **homologous** (and are given more weight). Similar traits that evolved separately are said to be **homoplasious.**

Cladistics insists that the only scientifically meaningful classification scheme is one that reflects **phylogeny,** or evolutionary history. The goal of cladistics is to construct evolutionary trees, or **cladograms,** that show the relationship of groups of species. Cladograms are constructed using **shared derived characters** rather than **primitive characters.** Cladists argue that every recognized group should be **monophyletic,** that is, each group should include an ancestral species and all its descendants. Many traditionally accepted groupings are **paraphyletic** or **polyphyletic** and therefore unacceptable to cladists. The classification used in this book reflects differences and similarities in cellular organization and lifestyle, rather than strict evolutionary relationships.

Review and Thought Questions

Review Questions

1. Define a species according to the biological species concept. What two kinds of problems often make this definition impractical?
2. What two purposes are accomplished by Linnaeus' system of classification? Name the seven levels of taxa commonly used by taxonomists.
3. Explain how a numerical taxonomist would decide whether a group of organisms constituted a species. Define phenetics.
4. What is cladistics? Why do cladists require that a lineage be monophyletic to be considered part of a single taxon?
5. Why do cladists insist that the reptiles are not a legitimate taxon?
6. Cladists consider the birds, class Aves, a legitimate taxon because birds are monophyletic. On the other hand, the very first bird was also most likely a dinosaur. Would a cladist argue that the group we call dinosaurs should include the birds? Explain why or why not.
7. Define "paraphyletic."
8. Does DNA sequencing provide a more objective method for classifying organisms? Why or why not?

Thought Questions

9. Do you think that the red wolf should be saved from extinction? Do you think that the decision to save it should be influenced by whether it is a species, a subspecies, or a hybrid?
10. If you found a new species of bird, what would you name the bird? If you were in charge of classifying this bird, what system of classification would you use? Why? Now describe the process, step-by-step, by which you would classify your bird. You may make up characters.

About the Chapter-Opening Image

The red wolf illustrates the confusion that often arises in classifying organisms. How a species is classified can have important consequences. For example, the red wolf's status could profoundly affect the strength of the Endangered Species Act.

 On-line materials relating to this chapter are on the World Wide Web at **http://www.harcourtcollege.com/lifesci/aal2/**

20 Prokaryotes: Success Stories, Ancient and Modern

Tracking Down a Killer

In 1983, a young scientist named Scott A. Holmberg showed for the first time that feeding antibiotics to farm animals can sicken and even kill people. Most of us have experienced food poisoning, the severe nausea and diarrhea that result from consuming meat, milk, or eggs contaminated with salmonella bacteria. Farm animals—whether cattle, pigs, or chickens—can carry salmonella, but animals fed antibiotics, Holmberg showed, carry especially deadly strains of the bacteria.

In February 1983, Holmberg was working in Atlanta, at the Centers for Disease Control (CDC), the nation's premier center for the study of all public health problems, from AIDS to smoking among teenagers. His boss, Mitchell Cohen, had received a report that 11 Minnesotans had been hospitalized with salmonella poisoning. Eleven people was hardly an epidemic: after all, 40,000 Americans are hospitalized with salmonella every year, and another 5 million suffer the severe diarrhea and vomiting associated with salmonella poisoning but don't need hospitalization. What was both disturbing and exciting about the Minnesota cases was, first, that most of the patients had been taking antibiotics when they came down with food poisoning and, second, that each patient was extremely sick. One man had died. Cohen strongly suspected that antibiotics were somehow causing or helping to cause the food poisoning. But how?

Cohen called Holmberg on a Saturday, and by the next morning, Holmberg was on a flight to Minnesota to find out. In Minnesota, Holmberg met with other **epidemiologists** [Greek, *epidemia* = how widespread a disease is]—researchers who study the incidence and transmission of diseases in populations. By that afternoon, Holmberg and state epidemiologist Michael Osterholm had two facts in hand.

First, all of the victims were infected with the same strain of *Salmonella newport*. This strain of bacteria, resistant to several common antibiotics, including penicillin and tetracycline, was unusual. The fact that all the victims had been made ill by the same rare form of *S. newport* suggested that they had all acquired the infection from a common source. In short, they must have all eaten the same thing. But what?

Paraskevas Photography

Key Concepts

- Prokaryotes are successful because of their rapid growth rates and their metabolic versatility.
- The metabolic abilities of prokaryotes provide a basis for their classification, but DNA-based classification reveals still more diversity.
- Plasmids can transfer genes to one another via plasmids and viruses.

Second, 8 of the 11 victims had been taking antibiotics manufactured by the same drug company. At first glance, it looked like the antibiotics themselves were contaminated with salmonella bacteria. However, because 3 of the 11 patients had not been taking antibiotics at all, Holmberg and Osterholm ruled out that possibility.

Nonetheless, another clue suggested that antibiotics played an important role. One woman had been hospitalized with salmonella poisoning the day after she had begun taking amoxicillin for a sore throat. Her husband then developed a sore throat, too, and took some of his wife's amoxicillin. Within 48 hours, he was in the hospital with salmonella poisoning.

Holmberg and Osterholm hypothesized that the antibiotic had actually triggered the couple's salmonella poisoning (Figure 20-1). The couple must have been infected with antibiotic-resistant salmonella that suddenly flourished when the amoxicillin killed off all other competing bacteria. Without competition, the salmonella bacteria multiplied from a few thousand cells into a full-scale infection of millions upon millions of cells in just a few hours. But why were the couple harboring antibiotic-resistant salmonella bacteria in the first place? Where had they come from?

Holmberg spent the following 2 weeks searching for the common source of the food poisoning in the homes of salmonella victims. He looked through refrigerators and cupboards, searching for some sort of food they had all eaten—some common source for the antibiotic-resistant bacteria.

Finally, with still no answer, he returned to Atlanta frustrated and discouraged. He and his boss, Mitchell Cohen, knew that the most likely scenario was that all of the patients had consumed meat, milk, or eggs infected with resistant *S. newport*. In the United States, large numbers of farm animals are fed low doses of antibiotics to increase their growth rate. This practice kills bacteria that are sensitive to antibiotics, encouraging the

Cohen strongly suspected that antibiotics were somehow causing or helping to cause the food poisoning. But how?

proliferation of antibiotic-resistant bacteria. By 1983, Cohen strongly suspected that farm animals raised on antibiotics were making people sick, but no one had yet been able to prove it.

Cohen knew of a new technique, however, that might help Holmberg make a direct link between individual farm animals and individual salmonella victims. Molecular biologists had recently developed a way to "fingerprint" bacteria by characterizing their plasmids, simple loops of DNA that sometimes live inside bacteria (Chapter 12). With this new tool, every strain of bacteria could be distinguished from all others according to which plasmids it carried.

Cohen suggested that Holmberg find out where else in the United States *S. newport* existed. Holmberg wrote to epidemiologists all over the United States, and, within a month, he had a clue that would solve the case. The state epidemiologist in South Dakota wrote to say that antibiotic-resistant *S. newport* had recently sickened four people.

From Atlanta, Holmberg interviewed the South Dakota victims by phone. One of the victims, a dairy farmer, told Holmberg that his dairy cows had all had diarrhea the previous November, and that one dead calf had been autopsied and found to have died of salmonella. Researchers at the state agricultural laboratory that had analyzed the dead calf's salmonella told Holmberg that it had the same plasmid profile as the salmonella bacteria in the Minnesota patients. Holmberg soon discovered that three of the four South Dakota victims had eaten beef from cattle raised on a farm adjacent to the dairy farm. The beef farmer said that his cattle had been perfectly healthy, adding that he regularly laced their feed with the antibiotic tetracycline, a handful to every ton of grain. Even if the beef cattle had shown no symptoms of salmonella, however, they could have been carrying salmonella just the same.

"I knew what the story was," Holmberg later told *Science* magazine. "I knew that between this point here in South Dakota

Figure 20-1 How antibiotics can trigger an infection.

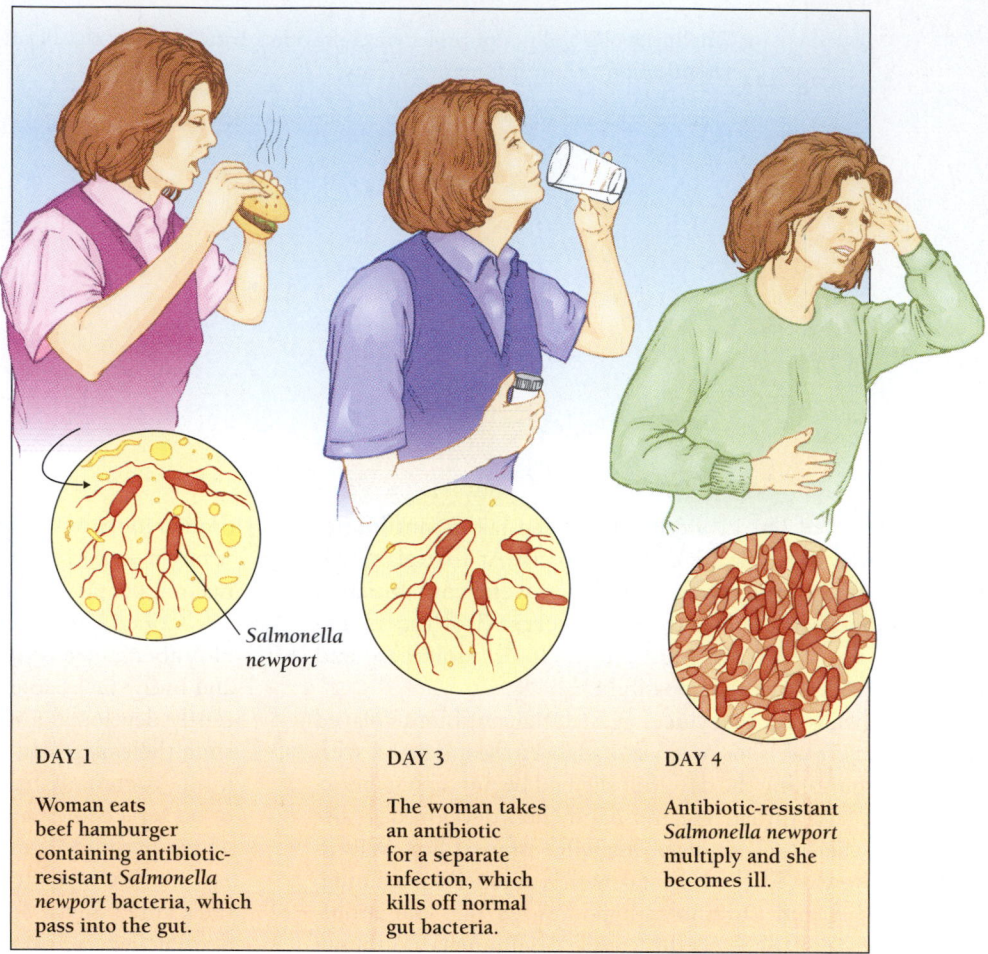

Salmonella
newport

DAY 1

Woman eats
beef hamburger
containing antibiotic-
resistant *Salmonella
newport* bacteria, which
pass into the gut.

DAY 3

The woman takes
an antibiotic
for a separate
infection, which
kills off normal
gut bacteria.

DAY 4

Antibiotic-resistant
Salmonella newport
multiply and she
becomes ill.

and that point there in Minnesota, there was a connection. Nature does not hand you coincidences like this. I was pretty sure that the beef herd was the source [of the *S. newport*]. . . . Now all I wanted to do was connect the two points. . . ."

Connecting the points, it turned out, was not so easy. In mid-January, just days before the salmonella outbreak in Minnesota, the beef farmer had sent all 105 head of cattle to slaughter. Now the beef cattle were gone, and Holmberg had no way to test the cattle for salmonella.

Nonetheless, Holmberg was able to follow the path the tainted meat had traveled to supermarkets in Minnesota and Iowa, where ten of the salmonella victims had shopped. In addition, Holmberg was able to show that the strain of *S. newport* from the dead dairy calf was truly distinctive. He examined 91 samples of different strains of *S. newport* from all over the United States, yet only the bacteria from the dead calf had the same plasmid profile as the salmonella in the sick Minnesotans. Holmberg no longer doubted that the beef cattle in South Dakota were the source of the antibiotic-resistant salmonella in Minnesota.

The plasmid fingerprinting technique had given epidemiologists their first clear evidence that feeding animals antibiotics can make the humans who eat them sick. Antibiotic-resistant bacteria are, it turns out, unusually dangerous. Holmberg showed that people are 21 times more likely to die from an infection by antibiotic-resistant salmonella than from an infection by nonresistant salmonella.

Amazingly, the practice of feeding antibiotics to farm animals continues today. Despite the clear hazards to public health, no laws regulate the practice. As a result of public pressure, only 20 percent of poultry growers now use antibiotics, the poultry industry reports. The beef industry recommends that its members use only antibiotics not used in humans. The pork industry continues to use human antibiotics without restraint.

How bacteria develop resistance to antibiotics, and how they pass that resistance among themselves, is a topic that reveals much about the lives of bacteria. In the rest of this chapter, we will learn a little about how bacteria are different from the rest of us and what kinds of bacteria populate the Earth.

Along the way, we will see why bacteria are so good at becoming resistant to antibiotics.

PROKARYOTES AND PEOPLE

The number of bacteria in your mouth at this moment exceeds the total number of people that have ever lived. You carry many more bacteria on your skin and in your gut than you have cells in your body. From a bacterium's point of view, your body is just another island to be colonized. Still, no human was even aware of their existence until Leeuwenhoek's observations some 300 years ago. And no one thought that they were at all important until about 100 years ago, when Pasteur and others found that they could cause diseases.

Prokaryotes—single-celled organisms with no nuclei—include both eubacteria ("ordinary" bacteria) and archaea, as we mentioned in Chapter 1. Most of the known life (by total mass) of Earth consists of prokaryotes, and some biologists think that there are huge numbers of prokaryotic organisms yet to be discovered in the depths of the oceans and even within the Earth's crust. Despite the incredible diversity of prokaryotic species and environments, most of the prokaryotes we know best are those that interact directly with humans.

But, though we know most about bacteria that live in us, for better or worse, bacteria are not limited to living in or on people or other animals. They live in the bodies of every kind of organism, even inside other cells. They live in soil, ice, and water. A pinch of ordinary soil, for example, contains some 10 billion bacteria. And bacteria can live almost anywhere on Earth, from the ice fields of Antarctica to the steaming cauldron of a Yellowstone hot spring, from the bottom of the ocean to the driest desert soils. In the rest of this chapter, we will see that bacteria are the most numerous, the most diverse, and the most ancient of organisms. We will also see how studies of DNA sequences from prokaryotes have revolutionized our views of the origins of life.

How Do Bacteria Make Us Ill?

Most people think of bacteria as lowly pests—carriers of disease and spoilers of food. In fact, many bacteria are useful to humans in one way or another. The trillions of bacteria in a handful of soil, for example, help recycle carbon and other nutrients. All the organisms on Earth depend on bacteria to pull nitrogen from the air so that we can use it to build proteins. Even the bacteria inside our bodies are useful. Some help us digest various kinds of foods. Others produce vitamins that we would otherwise have to get from special foods. In short, bacteria are a normal and mostly benign part of our lives.

But bacteria can also be **pathogens,** agents that cause disease. To cause a disease, a bacterium usually must accomplish two things:

1. It must invade the host, attaching and multiplying on or in the host's tissues and cells.

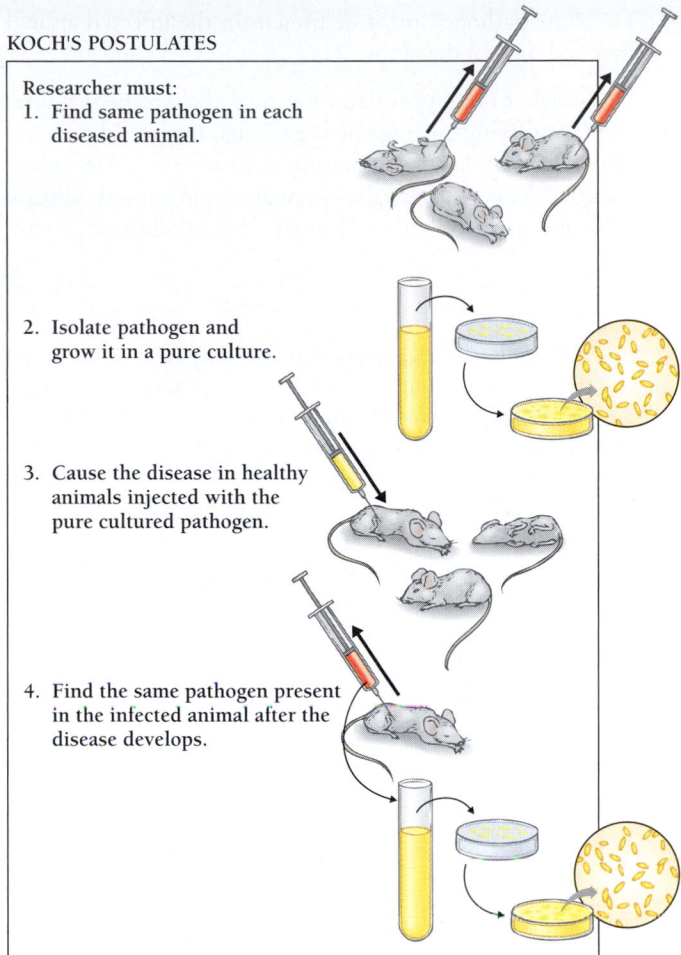

KOCH'S POSTULATES

Researcher must:
1. Find same pathogen in each diseased animal.
2. Isolate pathogen and grow it in a pure culture.
3. Cause the disease in healthy animals injected with the pure cultured pathogen.
4. Find the same pathogen present in the infected animal after the disease develops.

Figure 20-2   Koch's postulates. Pathogens such as viruses that can only be grown inside of cells cannot be isolated in a pure culture.

2. It must produce a toxin—a chemical that destroys or interferes with the host's normal processes.

In some cases, disease results just from the excessive growth of pathogenic bacteria. But in most cases, disease results from toxins made by the pathogen. These toxins may be **endotoxins,** toxic substances that are components of the bacterial cells themselves, or **exotoxins,** molecules that are secreted by the bacteria.

In the late 19th century, Robert Koch, a German physician, established rigorous criteria for identifying the pathogen for a given disease. The four rules, now known as **Koch's postulates** (Figure 20-2), are:

1. The same pathogen must be present in every person with the disease.
2. The pathogen must be capable of being grown in a pure culture—one uncontaminated with other organisms.
3. The pure cultured pathogen, introduced into an experimental animal, must cause the disease.

4. The same pathogen must be present in the infected animal after the disease develops.

Although Koch's postulates are now the accepted guidelines for identifying pathogenic organisms, the postulates cannot always be fulfilled. For example, the bacterium *Treponema pallidum,* which causes the sexually transmitted disease syphilis, has yet to be cultured in the lab. In addition, bacteria that live inside other cells (such as the sexually transmitted bacterium *Chlamydia*) cannot grow in pure cultures because these bacteria always require the presence of host cells. Likewise, all viruses must grow inside other cells (Chapter 12), so pathogenic viruses such as HIV (human immunodeficiency virus) and influenza cannot be grown in a "pure culture."

Not all pathogenic bacteria are strangers. Many are normal residents of our bodies that are out of place or have mul-

Figure 20-3 Microorganisms that normally inhabit the human body. All of the microorganisms listed here live—usually harmoniously—on the surfaces and in the interiors of human bodies. *Candida albicans,* commonly known as yeast, is a fungus that lives on the skin and in the mouth and vagina. *Candida* is, of course, a eukaryote, not a prokaryote. *(Erich Lessing/Art Resource)*

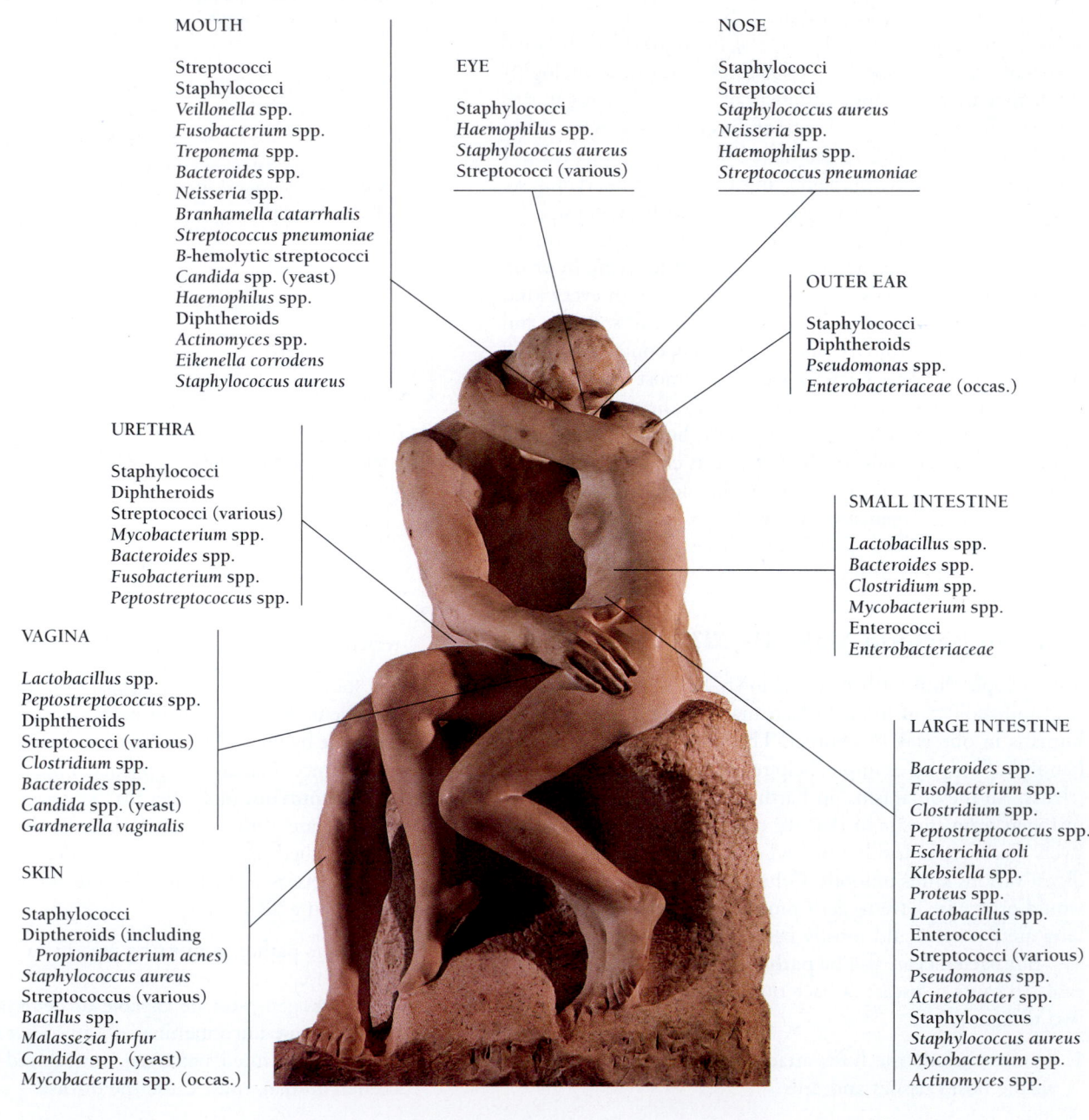

MOUTH

Streptococci
Staphylococci
Veillonella spp.
Fusobacterium spp.
Treponema spp.
Bacteroides spp.
Neisseria spp.
Branhamella catarrhalis
Streptococcus pneumoniae
B-hemolytic streptococci
Candida spp. (yeast)
Haemophilus spp.
Diphtheroids
Actinomyces spp.
Eikenella corrodens
Staphylococcus aureus

EYE

Staphylococci
Haemophilus spp.
Staphylococcus aureus
Streptococci (various)

NOSE

Staphylococci
Streptococci
Staphylococcus aureus
Neisseria spp.
Haemophilus spp.
Streptococcus pneumoniae

OUTER EAR

Staphylococci
Diphtheroids
Pseudomonas spp.
Enterobacteriaceae (occas.)

URETHRA

Staphylococci
Diphtheroids
Streptococci (various)
Mycobacterium spp.
Bacteroides spp.
Fusobacterium spp.
Peptostreptococcus spp.

SMALL INTESTINE

Lactobacillus spp.
Bacteroides spp.
Clostridium spp.
Mycobacterium spp.
Enterococci
Enterobacteriaceae

VAGINA

Lactobacillus spp.
Peptostreptococcus spp.
Diphtheroids
Streptococci (various)
Clostridium spp.
Bacteroides spp.
Candida spp. (yeast)
Gardnerella vaginalis

LARGE INTESTINE

Bacteroides spp.
Fusobacterium spp.
Clostridium spp.
Peptostreptococcus spp.
Escherichia coli
Klebsiella spp.
Proteus spp.
Lactobacillus spp.
Enterococci
Streptococci (various)
Pseudomonas spp.
Acinetobacter spp.
Staphylococcus
Staphylococcus aureus
Mycobacterium spp.
Actinomyces spp.

SKIN

Staphylococci
Diptheroids (including
  Propionibacterium acnes)
Staphylococcus aureus
Streptococcus (various)
Bacillus spp.
Malassezia furfur
Candida spp. (yeast)
Mycobacterium spp. (occas.)

Figure 20-4 A tuberculosis isolation ward. In the 1920s, sick children were isolated on this ferryboat in New York's harbor. With the advent of antibiotics, hospitals no longer needed isolation wards for patients with tuberculosis. Today, because of strains of tuberculosis that are resistant to all known antibiotics, such wards have returned. *(UPI/Corbis-Bettmann)*

tiplied when the body's defenses weaken (Figure 20-3). *Escherichia coli* in our intestines, for example, are normal and useful members of the community of organisms in the body. But the very same organism in the bloodstream can kill in a matter of hours.

> *Most pathogenic bacteria make us ill by releasing toxic chemicals that kill our own cells. Koch's postulates are guidelines for determining whether a microorganism causes a specific disease.*

How Do Bacteria Acquire Resistance to Antibiotics?

Whenever we are sick from a bacterial infection, we and our doctors want to eliminate the pathogenic bacteria as quickly as possible. The easiest way to do that is with an antibiotic. **Antibiotics** are chemicals that kill bacterial cells without harming our own eukaryotic cells. Since the first antibiotics came into use in the 1940s, medical researchers have discovered or synthesized more than 150 different antibiotics.

Amazingly, pathogenic bacteria now exist that are resistant to all antibiotics. In New York City, patients with certain strains of highly resistant tuberculosis, for example, are now treated the same way tuberculosis patients were treated 50 years

ago—by isolating them. There is no cure for such people, and health workers try to prevent them from passing their resistant strains of tuberculosis bacteria to other patients or to hospital staff (Figure 20-4).

Where do bacteria acquire resistance to antibiotics? First we must understand how antibiotics work and where that resistance comes from. Most antibiotics are naturally synthesized by soil bacteria or fungi. In fact, soil bacteria produce so many chemicals that are toxic to one another that pathogenic bacteria usually die when exposed to ordinary dirt.

The genes for resistance to these antibiotics come from the very organisms—mostly soil organisms—that make the antibiotics. A soil bacterium needs to be immune to its own toxins, after all, and they have evolved a variety of tools to fend off specific antibiotics.

What are these tools? That depends on the antibiotic. Some antibiotics, such as penicillin, kill bacteria by preventing the bacteria from synthesizing a cell wall. Other antibiotics prevent bacteria from expressing their genes, interfere with DNA folding (which can disrupt gene expression and DNA replication), inhibit the synthesis of folic acid (important in nucleotide synthesis), or block protein synthesis.

Resistance to antibiotics can be classified into two main varieties (Figure 20-5A). Some resistance genes merely enable the bacteria to survive exposure to an antibiotic. Such genes may alter the shapes of bacterial molecules in such a way that the antibiotics can no longer bind to them, pump the antibiotic out of the cell, or destroy the antibiotic as it enters the cell.

Other resistance genes enable the bacteria to release enzymes that actually destroy antibiotics in the environment (Figure 20-5B). This kind of resistance is the most dangerous. If, for example, a college student regularly took the antibiotic tetracycline to treat acne, he could gradually accumulate intestinal and other bacteria able to destroy tetracycline. If he then became infected by nonresistant *Chlamydia trachomatis,* for example, and his doctor prescribed tetracycline, his intestinal bacteria would destroy the tetracycline before it could kill the invading *Chlamydia* bacteria.

Resistance accumulates in populations of bacteria in several ways. In all populations of bacteria, a few individuals carry genes for resistance to antibiotics. Even baboons living in the wild, far from any humans, carry small numbers of intestinal bacteria that are resistant to some antibiotics. Humans have far higher numbers of resistant bacteria. One study of resistance in a group of students showed that one in ten students carried large numbers of *E. coli* that were resistant to a combination of two antibiotics first introduced in the 1970s.

When humans or other animals are exposed to antibiotics, the resistant bacteria flourish in the absence of competing bacteria, multiplying until they are the majority. This is simply natural selection. The longer the antibiotic is present, the more likely the antibiotic-resistant bacteria are to increase in number. Thus, beef cattle regularly fed tetracycline gradually accumulate populations of bacteria that are resistant to tetracycline. Unfortunately, the story does not end there.

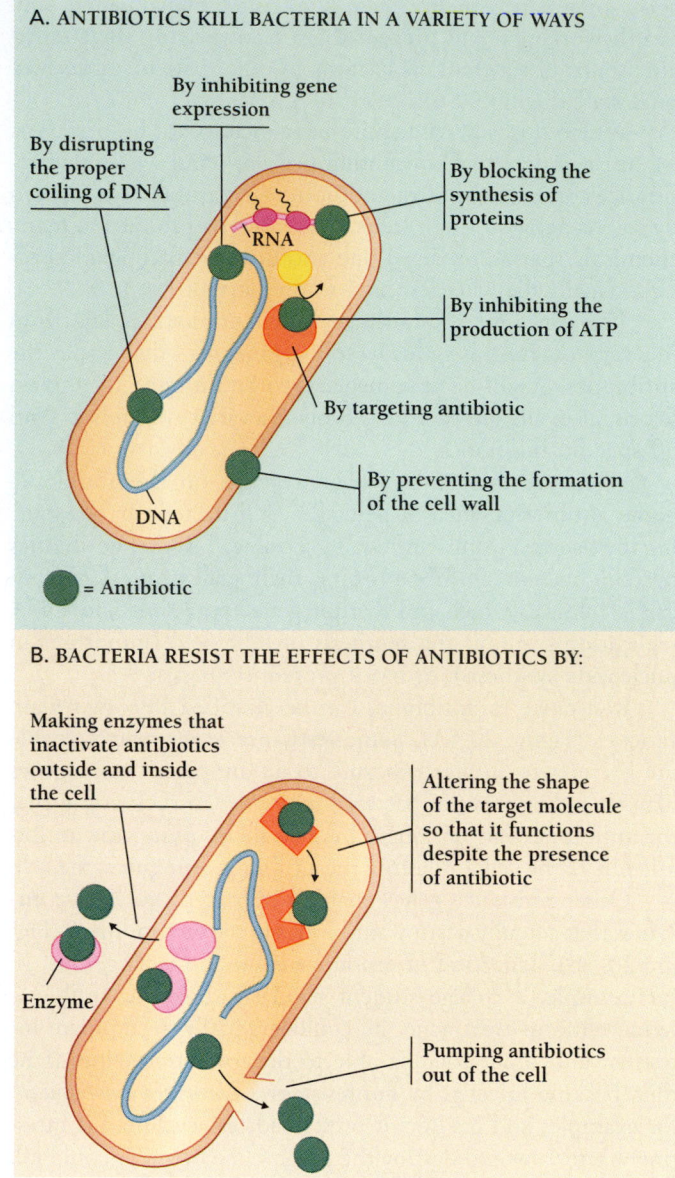

A. ANTIBIOTICS KILL BACTERIA IN A VARIETY OF WAYS

By inhibiting gene expression

By disrupting the proper coiling of DNA

By blocking the synthesis of proteins

RNA

By inhibiting the production of ATP

By targeting antibiotic

By preventing the formation of the cell wall

DNA

● = Antibiotic

B. BACTERIA RESIST THE EFFECTS OF ANTIBIOTICS BY:

Making enzymes that inactivate antibiotics outside and inside the cell

Altering the shape of the target molecule so that it functions despite the presence of antibiotic

Enzyme

Pumping antibiotics out of the cell

Figure 20-5 How does it work? A. Every antibiotic has a distinctive way of preventing bacteria from reproducing. B. Bacteria have different ways of resisting the effects of antibiotics.

gation, in which one bacterium can pass plasmids and other DNA to another bacterium (Figure 12-8). Plasmids, simple loops of DNA carried separately from the bacterium's DNA, generally carry genes for traits such as antibiotic resistance that are beneficial to the bacteria that they inhabit. Because plasmids themselves pass genes around to one another by means of jumping genes, plasmids accumulate genes for resistance the way a snowball accumulates snow as it rolls downhill. Plasmids can carry resistance to as many as eight classes of antibiotics. Bacteria then pass such plasmids, through conjugation, not only to other bacteria of the same species but also to bacteria that are unrelated.

Unlike most species of organisms, bacteria are not discrete genetic entities, but a vast, integrated microbial world. As a result, multiple antibiotic resistance is the rule, rather than the exception. In 1995, for example, hospital workers in Atlanta found that 25 percent of white children fighting ordinary ear and respiratory infections by *Streptococcus pneumoniae* were infected with strains resistant to two or more antibiotics.

The tendency of plasmids to carry genes for resistance to several antibiotics has important consequences. Cattle treated with tetracycline, for example, can acquire bacteria that are resistant not only to tetracycline but to several other antibiotics as well. This is why the South Dakota beef cattle, which had been treated with tetracycline, carried *Salmonella* resistant to both amoxicillin and tetracycline. The plasmid that protected the bacteria from tetracycline also happened to carry genes for resistance to other antibiotics.

Resistance may allow a bacterium simply to survive exposure to an antibiotic or, alternatively, actively to destroy any antibiotic in its environment. Resistance genes accumulate in populations of bacteria, first, through natural selection, and second, by moving from one bacterial strain to another by means of plasmids and other mobile genes.

If antibiotic-resistant bacteria accumulated only in the presence of antibiotics, through selection, then society's problems with antibiotic resistance would be relatively minor. Strains of bacteria without resistance to a specific antibiotic could never acquire resistance except through chance mutations. Unfortunately, bacteria can pass fully functioning antibiotic resistance genes to completely different kinds of bacteria, so that resistance spreads rapidly. Earlier in this book, we mentioned that bacteria engage in a process called conju-

HOW DO WE KNOW THAT AN ORGANISM IS A PROKARYOTE?

The defining characteristic of prokaryotes that sets them apart from all other organisms is the absence of a true nucleus (Chapter 4). Prokaryotes lack not only a nucleus but also chloroplasts and mitochondria and all other membrane-bounded organelles. Some prokaryotes do have extensively folded plasma membranes, however, and use these large surfaces for respiration, photosynthesis, and DNA replication.

Instead of neatly packing its DNA in a membrane-enclosed nucleus (as a eukaryote does), a prokaryote tucks its DNA into a concentrated mass called a **nucleoid.** The DNA

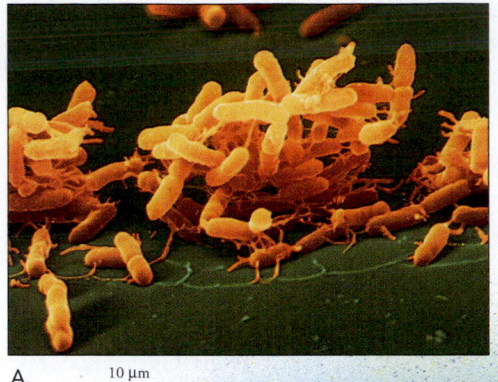

Figure 20-6   The familiar shape of the bacillus typifies both *Salmonella* and *E. coli.* A. *Salmonella sp.* B. *E. coli.* (A, Meckes/ Ottawa/Photo Researchers, Inc.)

A.   10 µm

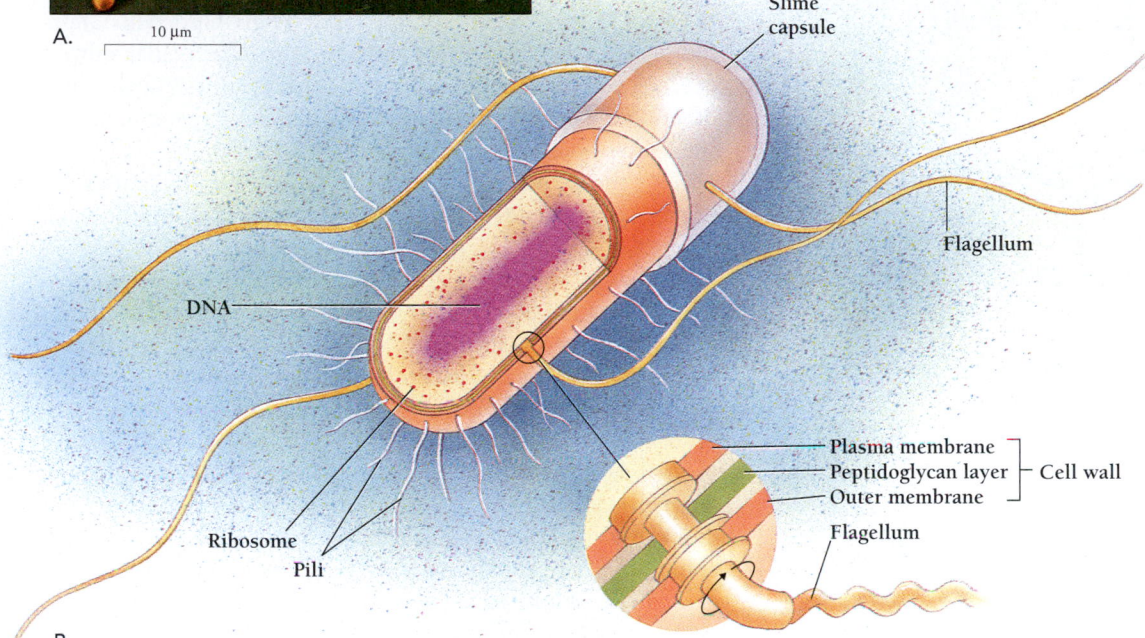

Slime capsule

DNA

Flagellum

Plasma membrane — ⌐ Cell wall
Peptidoglycan layer — |
Outer membrane — ⌐

Flagellum

Ribosome
Pili

B.

in the nucleoid consists of a single molecule of double-stranded DNA that forms a closed loop. A prokaryote may also contain one (or more) small loops of DNA called plasmids (Chapter 12). Because a prokaryote's DNA is circular, it can replicate continuously, unlike eukaryotic DNA, which must replicate in pieces.

Also unlike the eukaryotes, prokaryotes undergo neither mitosis nor meiosis. Prokaryotes do not require sex to reproduce. However, as we have seen, they exchange genetic information so freely that reproductive isolation in bacteria is the exception rather than the rule.

Unlike other species of organisms, then, a "species" of bacterium is neither a reproductive unit nor a genetic unit, and classifying bacteria into species has been something of an art. As we will see later in this chapter, recent studies of the complete DNA sequences of prokaryotic genomes have led to new views of the relationships among groups of prokaryotes. Many

biologists, for example, now argue that we should not lump eubacteria and archaea together in a single group, even one as broadly defined as the prokaryotes.

Prokaryotic cells come in a variety of shapes (Figure 20-6). The most common are the rod-shaped **bacilli** [singular, bacillus; Latin, = little rod]. In addition are the spiral-shaped **spirilla** [singular, spirillum; Latin, *spira* = coil] and the spherical **cocci** [singular, coccus; Greek, *kokkos* = berry]. Most bacteria are single-celled organisms, though some aggregate to form colonies of unspecialized cells (Figure 20-7).

The defining characteristic of prokaryotes is the absence of a nucleus or other membrane-enclosed organelles. The DNA of prokaryotes forms a single circle of double-stranded DNA.

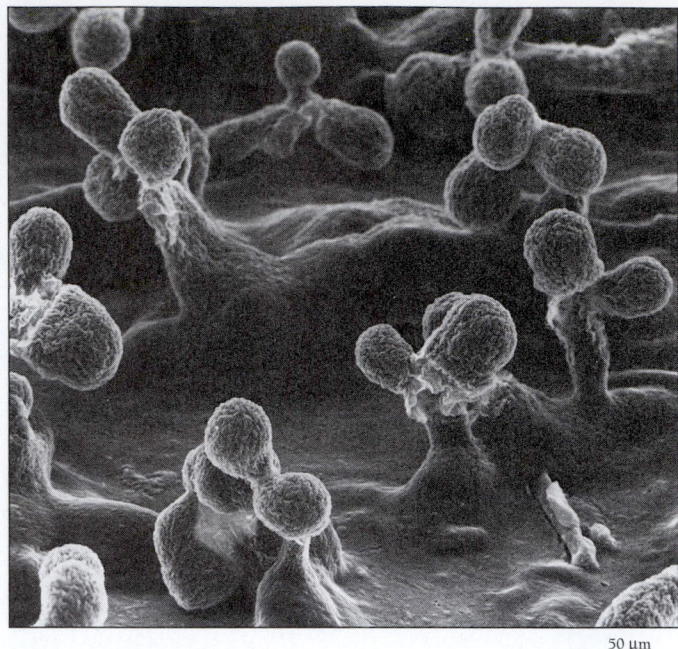

Figure 20-7 Some bacteria, such as this myxobacterium, are colonial. Cells of this species form a "fruiting body" to reproduce. *(Karen Stephens/Biological Photo Service)*

Most Bacteria Have Complex Cell Walls

Like plant cells, most bacterial cells have rigid cell walls outside their plasma membranes. Like the wall of a plant cell, the wall of a bacterial cell gives the cell a shape and protects it

from injury. However, while plant cell walls are made of cellulose, bacterial cell walls are made of a mixture of polysaccharides and polypeptides called **peptidoglycan.** The cell wall is a single, giant, bag-shaped macromolecule composed of a network of cross-linked peptidoglycan. The peptidoglycan cell wall is important to biologists for two reasons. It helps distinguish two major groups of bacteria, and it makes some bacteria susceptible to antibiotics. (Archaea, like eukaryotes, do not produce peptidoglycan.)

Biologists divide all of the bacteria into two major groups according to the way their cell walls soak up various dyes. In the 1880s, the Danish bacteriologist Hans Christian Gram found that when he treated bacteria with a particular purple dye, some kinds of bacteria became permanently stained, while others could be washed clean. Those that became permanently stained he called "stain positive." The others were "stain negative." Today bacteriologists call these two categories of bacteria **gram positive** and **gram negative,** after Gram's stain (Figure 20-8).

What sets gram-negative and gram-positive bacteria apart from one another is the structure of the cell wall. Gram-negative bacteria—such as *E. coli* or the gonococcus bacteria that cause the sexually transmitted disease gonorrhea—have a thin, three-layered cell wall, whose outermost layer does not bind the Gram stain (Figure 20-8C). In contrast, gram-positive bacteria—such as the streptococci (which can cause strep throat) and the staphylococci (which can cause boils and other infections)—have a thick, single-layered cell wall containing at least 20 times as much peptidoglycan as the gram-negative bacteria. The peptidoglycan, itself arranged in up to 40 layers, binds the Gram stain so that it cannot be washed away.

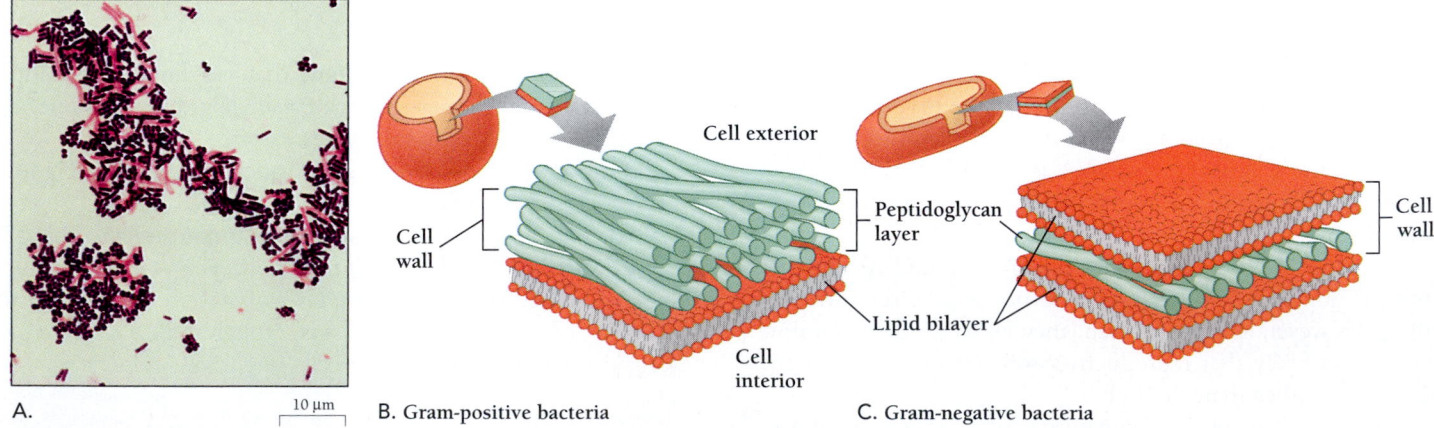

Figure 20-8 It's all in the wall. Bacteria absorb stain or not depending on what kind of cell wall they have. A. A mixture of gram-negative (*light*) and gram-positive (*dark*) bacteria. B. The gram-positive coccus bacterium has a two-layered, peptidoglycan-rich cell wall. C. The gram-negative bacillus has a three-layered, peptidoglycan-poor cell wall. (*A, G.W. Willis/Biological Photo Service*)

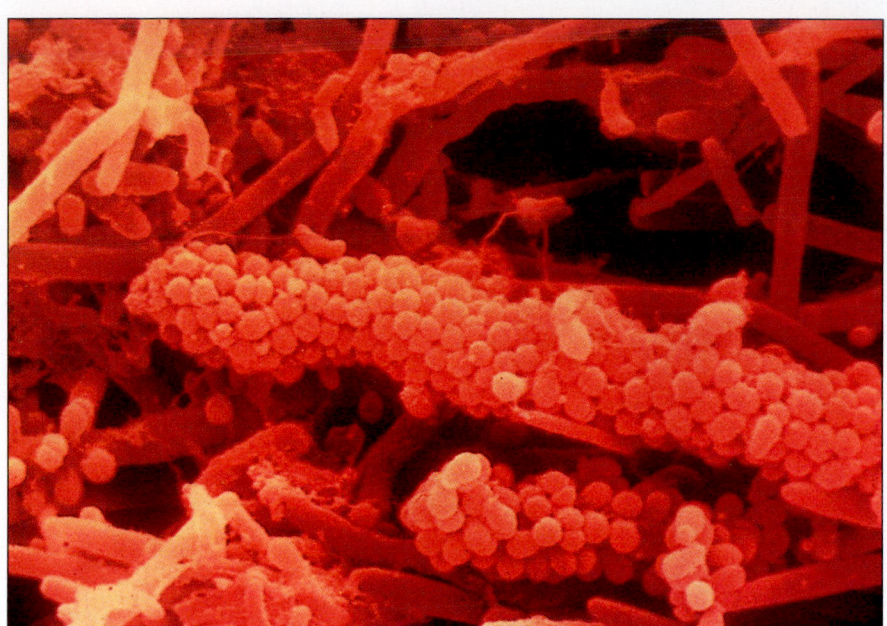

5 µm

In gram-negative bacteria, the same outer layer of the cell wall that prevents the Gram stain from penetrating also prevents antibiotics such as penicillin from penetrating. As a result, penicillin and other "narrow-spectrum" antibiotics kill gram-positive bacteria only. "Broad-spectrum" antibiotics such as tetracycline kill both gram-negative bacteria and many gram-positive bacteria.

Outside of the cell wall, some prokaryotes have other materials. The outer surfaces of some prokaryotes possess *pili* [singular, pilus; Latin, = hair], hairlike structures that attach the bacteria to surfaces and to other cells. The F-pilus, for example, provides a bridge for the transfer of DNA during bacterial conjugation.

The bacterium that causes pneumonia is enclosed in a **capsule,** a gelatinous layer of polysaccharides and proteins outside its cell wall. Capsules protect disease-causing bacteria from attack by the body's white blood cells, enabling the bacteria to overwhelm the body's immune system. Such bacteria are more deadly, or virulent.

Other bacteria possess a **slime layer,** a tangled web of polysaccharides that, like a capsule, lies outside the cell wall. A sticky slime layer protects the bacterial cell against dehydration, helps keep food from diffusing away, and also allows cells to attach to and move about on the surfaces on which they grow. Dental plaque is a slime layer made by the bacteria that inhabit the mouth. Plaque helps bacteria grow smoothly over the surfaces of our teeth (Figure 20-9).

The presence of large amounts of peptidoglycan distinguishes the bacterial cell wall from the cell walls of archaea and eukaryotes. Gram-positive bacteria have a thick, peptidoglycan-rich cell wall that leaves them susceptible to penicillin and other narrow-spectrum antibiotics.

How Do Bacteria Move?

Some bacteria move by secreting slime and gliding on it. Others use a propeller—the bacterial **flagellum** (Chapter 4). Bacterial and eukaryotic flagella are entirely different. The eukaryotic flagellum is an extension of the cytoplasm surrounded by membrane that beats back and forth. In contrast, the prokaryotic flagellum is a thin, rotating protein filament attached to the outside of the cell.

Many bacteria use their flagella for directed movement, or **taxis** [Greek, = to put in order]. (A taxicab performs directed movements.) The direction of movement depends on the operation of the flagella. For example, when the flagella of an *E. coli* rotate in a counterclockwise direction, they twist around each other and propel the bacterium forward. When the flagella rotate in a clockwise direction, however, they separate and become uncoordinated. As a result, the bacterium tumbles aimlessly. When the flagella resume a counterclockwise spin, the bacterium moves in a new direction. By controlling the

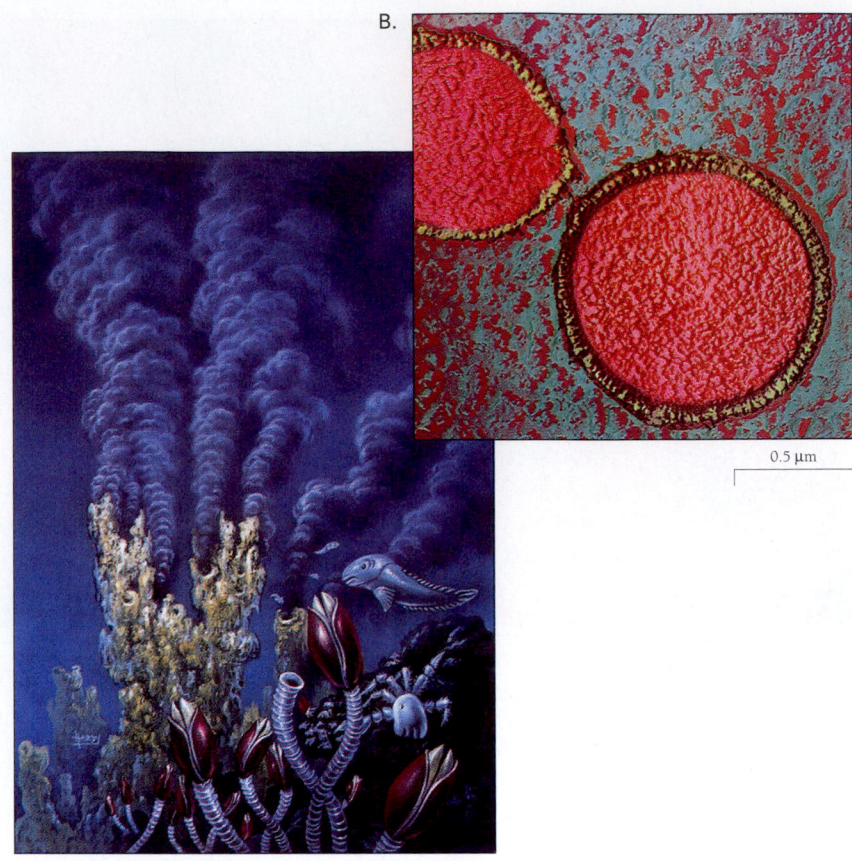

0.5 μm

Figure 20-10 Profiting from prokaryotic adaptations. These tubeworms, *Riftia pachyptila* (A), live near hydrothermal vents, in symbiosis with bacteria, *Staphylothermus marinus* (B), that get energy from inorganic sulfide compounds. (*A, David A. Hardy/SPL/ Photo Researchers, Inc.; B, Wolfgang Baumeister/ SPL/Photo Researchers, Inc.*)

A.

timing of clockwise and counterclockwise rotations, bacteria can move toward or away from food, oxygen, heat, light, and other stimuli. Some bacteria even use the Earth's magnetic field to guide them to the lower depths of the ocean. **Chemotaxis** is taxis in response to varying concentrations of specific chemicals.

Bacteria move in directed fashion, by gliding on slime or by rotating propellerlike flagella.

How Do Prokaryotes Grow So Rapidly?

Because bacteria keep their DNA in a nucleoid instead of a nucleus, bacteria can replicate their DNA continuously and divide very rapidly. In fact, the cells in well-nourished colonies of *E. coli* can divide every 20 minutes. Every time the bacteria divide, the colony doubles in size. That is, after one generation, one cell becomes two. After another generation the 2 cells become 4, then 8, 16, and so forth. After 20 such doublings (in less than 7 hours), a single cell would yield 2^{20}, or about a million, new cells; and after 30 doublings (10 hours),

2^{30}, or about a billion, cells. Biologists describe such growth as exponential or logarithmic (Chapter 28).

If a single *E. coli* bacterium could be supplied with enough food to divide every 20 minutes continuously, the combined mass of the bacterium and all its descendants would exceed that of the entire planet after only 2 days. We can see that, in reality, exponential growth can last for only short times. Nonetheless, the rapid proliferation of prokaryotes means that a favorable new mutation such as antibiotic resistance can spread quickly to a huge number of progeny.

A streamlined metabolism and a simple means of replicating DNA together allow bacteria to grow and divide more rapidly than eukaryotes.

Prokaryotes Are Metabolically Diverse

We eukaryotes are monotonous from a biochemical point of view. Nearly all of us use glycolysis and respiration to produce energy. Nearly all of us need oxygen to live. Likewise, all photosynthetic eukaryotes photosynthesize in the same way. In fact, the relentless sameness of metabolic pathways in all eukaryotes strongly supports the idea that we are all descended from a common ancestor.

In contrast, prokaryotes use a great variety of biochemical pathways. Some, for example, perform glycolysis, but with end products other than ethanol or lactate (Chapter 6). Many of these end products contribute to the pleasant (or not-so-pleasant) odors of wines and cheeses. Some prokaryotes do not use glycolysis at all. Instead, they produce ATP by capturing the chemical energy of molecules in their environment. Methane-producing prokaryotes, for example, capture energy released when carbon dioxide (CO_2) reacts with hydrogen gas (H_2) to produce methane (CH_4). Some archaea capture the energy from the reaction of sulfur-containing compounds with hydrogen gas. Hydrogen gas, unstable in the atmosphere, nonetheless accumulates to high levels under the oceans and within the earth's mantle.

Because prokaryotes are so diverse, they flourish in places where eukaryotes could never survive. Over billions of years, prokaryotes have evolved an amazing variety of adaptations that contribute to their continuing success. Some of these adaptations are (1) spore formation, which allows bacteria to survive, multiply, and disperse no matter how hostile the environment; (2) the production of antibiotics that can kill competitors; (3) the use of resistance transfer factors and plasmids to spread resistance to antibiotics from one bacterium to another; and (4) the ability to obtain energy from the waste products of other species (for example, carbon dioxide and methane).

The adaptations of prokaryotes in turn provide new opportunities for eukaryotic life. A 2-meter-long tubeworm, called *Riftia pachyptila,* lives near hydrothermal vents on the ocean floor. *Riftia* gets energy-rich compounds from symbiotic bacteria that use the energy of inorganic sulfur compounds from the vents (Figure 20-10).

The enormous success of prokaryotes depends on their metabolic diversity and their ability to transfer genes to one another.

HOW DO BIOLOGISTS CLASSIFY PROKARYOTES?

Because prokaryotes do not (usually) reproduce sexually, taxonomists cannot use the biological species concept, which defines a species as an interbreeding population, to classify prokaryotes (Chapter 19). Reconstructing the phylogeny of bacteria is equally difficult, as the fossil record for prokaryotes is nearly nonexistent.

Until recently, taxonomists have relied on morphological and metabolic comparisons. This approach is fraught with difficulties, however, since two strains of a single species may look very different. For example, one may possess a distinctive feature such as an external capsule, while the other lacks the capsule. Similarly, metabolic and biochemical differences may not define species, since genes carrying distinctive traits such as antibiotic resistance can spread rapidly—through viruses, plasmids, or transfer factors—among many unrelated bacteria. So far, some 1700 species of prokaryotes have been named, although estimates of the number of species vary widely—from 5000 up into the millions.

How Do Biologists Classify the Archaea?

Although some biologists still put all of the prokaryotes into a single kingdom, the consensus has changed dramatically in the last 5 years. Most biologists now recognize that prokaryotes include two large domains—the Archaea and the Eubacteria—and that each of these domains is enormously diverse. The archaea consist of at least 2 kingdoms and the eubacteria of at least 25 kingdoms (Figure 20-11).

The **Archaea** [Greek, *archaio* = ancient] differ from eubacteria and from eukaryotes in the composition of their ribosomes and in the kinds of lipids in their cell membranes. The Archaea entirely lack peptidoglycan in their cell walls. Different types of archaea derive energy from hydrogen gas in one of two ways—by converting carbon dioxide (CO_2) to methane (CH_4) or by converting sulfur compounds to sulfur. These two groups form the two recognized kingdoms of the archaea.

The first kingdom (called the Euryarcheota) includes the **methanogens,** shown in Figure 20-12, and the **halophiles,** which live in extremely salty environments (Figure 20-13). The second kingdom (called the Crenarchaeota) includes the **thermoacidophiles,** which inhabit hot sulfur springs such as those in Yellowstone National Park (Figure 20-14).

Methanogens, all obligate anaerobes, live principally in two environments: (1) swamps and sewage treatment plants, and (2) the guts of animals, especially of cows, termites, and other animals that digest cellulose. Even the human gut harbors methanogens that produce methane, which, when released, is called flatulence. Methanogens play an essential role in the ecology of the Earth. Where methanogens flourish in swamps made anaerobic by the growth of other bacteria, they release methane, or "marsh gas," into the atmosphere. There, the methane is quickly oxidized to carbon dioxide, which is then available for photosynthesis. If methanogens did not recycle carbon from organic molecules back into the atmosphere, other organisms would gradually run out of carbon. Natural gas, which we use to heat our homes and cook our meals, is about 98 percent methane—all provided by the methanogens.

The Archaea consist of at least two kingdoms. The most widespread and ecologically important archaea are the methanogens.

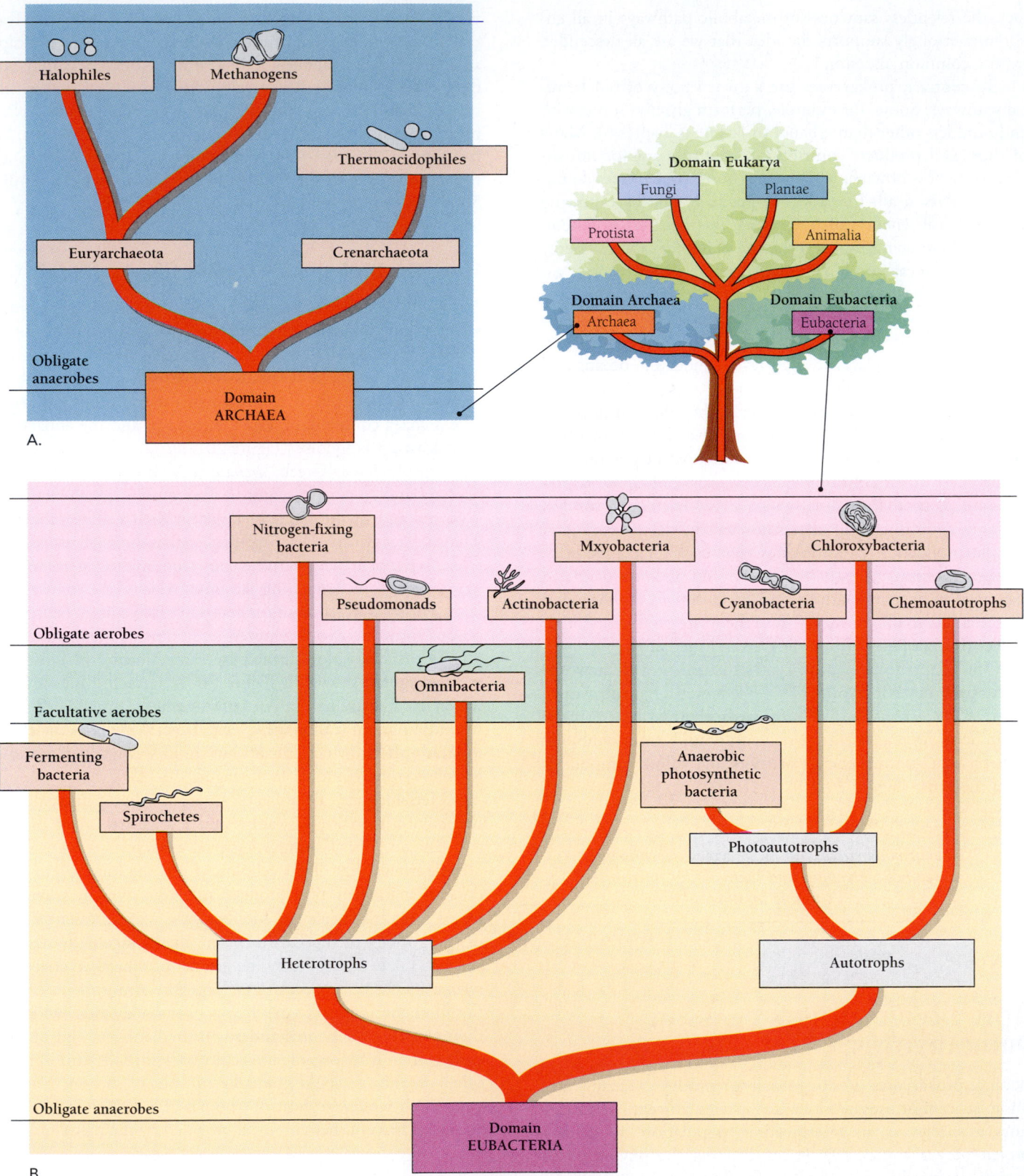

Figure 20-11   Classification of prokaryotes. The domain Archaea includes three distinct groups, which may be kingdoms or phyla. The domain Eubacteria includes at least 25 groups, now given kingdom status, of which a dozen are shown here. The Eubacteria are divided into heterotrophs and autotrophs. The autotrophs are further divided into photoautotrophs and chemoautotrophs.

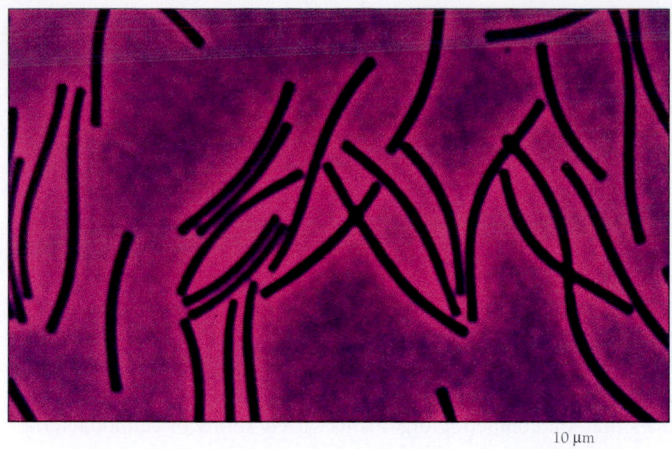

10 μm

Figure 20-12   Methanogens. Methanogens, which *generate* the gas *methane*, include bacteria of all three standard shapes—bacilli, spirilla, and cocci. (*F. Widdel/Visuals Unlimited*)

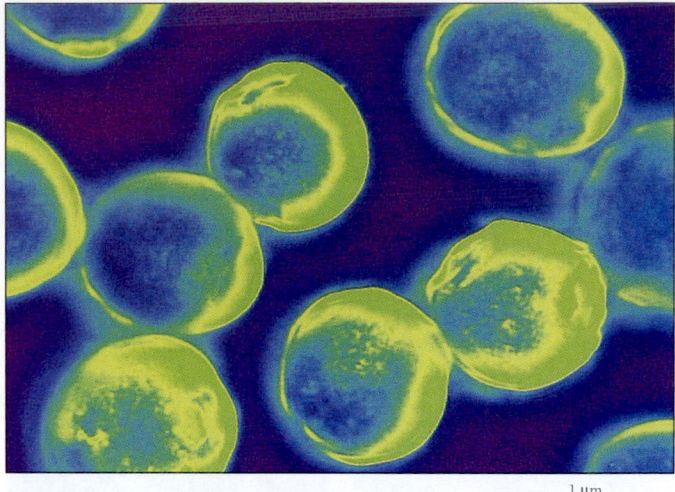

1 μm

Figure 20-13   Halophiles. Most bacteria cannot grow in high concentrations of salt. Because salt prevents the growth of most bacteria, it acts as a preservative in ham, beef jerky, and other salty foods. Unlike any other bacteria, halophiles such as these *Halococcus* bacteria flourish in briny places the world over—from the Dead Sea to Utah's Great Salt Lake and California's Mono Lake. Halophiles, usually flagellated rods or cocci (like these), form a lavender-pink scum over their salty homes. The pink color comes from a purple pigment in the plasma membrane that captures light energy. (*Alfred Pasteka/Science Photo Library/Photo Researchers, Inc.*)

How Do Biologists Classify the Eubacteria?

Despite the difficulties of classifying prokaryotes, biologists do agree on some broad groupings. In one classification, in 1987, taxonomists divided the **Eubacteria** into 12 kingdoms; a more recent list included more than 25 kingdoms. DNA sequences from each of these kingdoms differ from those of the others more than the DNA from the plant kingdom differs from that of the animal kingdom.

Just as zoologists classify some animals according to whether they eat plants or other animals, microbiologists have traditionally classified Eubacteria into broad categories according to the way they derive food. **Autotrophs** [Greek, *autos* = self + *trophos* = feeder] obtain carbon atoms directly from carbon dioxide, while **heterotrophs** [Greek, *heteros* =

Figure 20-14   Thermoacidophiles. Thermoacidophiles may be chemoautotrophs (reaping energy from sulfur compounds or methane) or heterotrophs (reaping energy and carbon from small organic compounds made by other organisms). The best-known genus, *Sulfolobus*, lives in the hot sulfur springs of Yellowstone National Park. *Sulfolobus* can survive a temperature as high as 88°C (190°F) and an acidity as low as pH 1. Although *Sulfolobus* thrives in heat, it dies of cold at temperatures below 55°C (130°F). (*Richard Thom/Visuals Unlimited*)

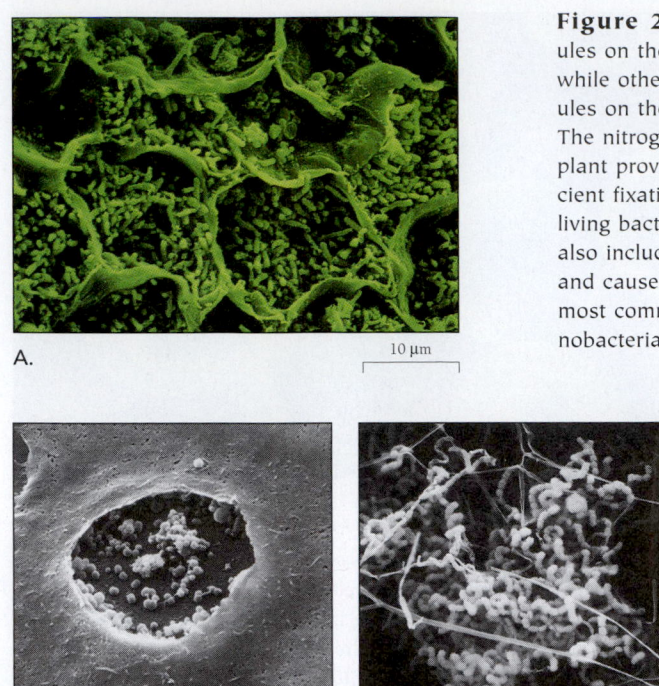

A.   10 μm

B.   5 μm   C.   10 μm

Figure 20-15   **Aerobic heterotrophs.** A. Nitrogen-fixing bacteria growing in nodules on the roots of a plant. Some nitrogen-fixing bacteria live freely in soil and water, while others associate intimately with specific plants, living, for example, in tiny nodules on the roots of peas, beans, lupines, locusts, and other members of the pea family. The nitrogen-fixing bacteria provide the plant with a usable form of nitrogen, and the plant provides carbohydrates for energy. This relationship allows extraordinarily efficient fixation of nitrogen, as much as 100 times faster than that performed by free-living bacteria. B. *Chlamydia trachomatis.* Besides the enteric bacteria, the omnibacteria also include rickettsia and chlamydia, which live only as parasites inside animal cells and cause a variety of diseases. Shown here is *Chlamydia trachomatis,* which causes the most common sexually transmitted disease in the United States. C. *Streptomyces.* Actinobacteria form branching filaments like those of fungi. They are widely distributed in soils, where they decompose organic matter. *Streptomyces,* for example, produces the valuable antibiotic streptomycin and other substances that kill competing organisms. A few forms of actinobacteria cause human diseases such as tuberculosis and leprosy. *(A, C.P. Vance/Visuals Unlimited; B, David M. Phillips/Science Source/Photo Researchers, Inc.; C, Frederick Mertz/Visuals Unlimited)*

other + *trophos* = feeder], which include the great majority of prokaryotic species, derive carbon atoms (and also energy) from organic molecules such as glucose. Autotrophs may be divided into **photoautotrophs,** which use light energy to drive the synthesis of organic compounds from CO_2, and **chemoautotrophs,** which derive energy by oxidizing such inorganic substances as hydrogen sulfide (H_2S) or ammonia (NH_3).

Heterotrophs fall into two groups. **Photoheterotrophs** use light energy but also require organic compounds. **Chemoheterotrophs** use no light, relying exclusively on organic molecules for both energy and carbon atoms. Heterotrophic prokaryotes may be **decomposers,** which absorb nutrients from dead organisms, or **symbionts** [Greek, *syn* = together + *bios* = life], which absorb nutrients from living organisms. Symbionts that cause illness, such as *Salmonella newport,* are called pathogens. Others, such as *E. coli,* are usually harmless, or even helpful. Prokaryotic decomposers, like fungal decomposers, recycle the materials that were formerly part of other organisms.

Prokaryotes also vary in the way they use oxygen. Heterotrophs that require oxygen for respiration are called **obligate aerobes.** Those that can grow either with or without oxygen are called **facultative aerobes.** Still others, for which oxygen is a deadly poison, are called **obligate anaerobes.**

The Eubacteria include aerobic heterotrophs, anaerobic heterotrophs, and autotrophs. We will briefly describe examples of each of these three groups and the incredible diversity of environments in which these simple organisms thrive.

The aerobic heterotrophs

The aerobic heterotrophs have tremendous biochemical diversity: some break down all kinds of organic compounds, while others can use nitrate —NO_3^-—instead of oxygen—as the ultimate electron acceptor, and still others make powerful antibiotics that kill competitors. A few aerobic heterotrophs are the nitrogen-fixing bacteria, pseudomonads, omnibacteria, actinobacteria, and myxobacteria (Figure 20-15).

From the perspective of most other organisms on Earth, the most important prokaryotes may be the **nitrogen-fixing aerobic bacteria** (Figure 20-15A). Nitrogen-fixing bacteria provide nitrogen in forms such as ammonia and amino acids from which other organisms can make proteins. Although nitrogen gas (N_2) constitutes some 80 percent of our atmosphere, it is so chemically stable that most organisms are incapable of breaking the bond between the two nitrogen atoms to form complex organic compounds. As a result, all organisms rely on nitrogen-fixing bacteria to provide nitrogen in

BOX 20.1

Carl Woese and the Archaea

In the summer of 1996, a researcher working on an obscure microbe at The Institute for Genomic Research in Rockville, Maryland, announced a finding that immediately hit the television nightly news and the front pages of newspapers all over the United States. In sequencing the microbe's genome, the researcher, Carol Bult, had provided irrefutable evidence that all organisms, historically classified as either eukaryotes or bacteria, actually fall into three basic categories.

The microbe, named *Methanococcus jannaschii,* was only the fourth organism for which researchers had completed such a sequence, but the implications were profound. Bult's work not only supported the theory that the archaea are distinct from both bacteria and eukaryotes, it also provided independent support for two other ideas: (1) that we eukaryotes are more closely related to the archaea than to the bacteria and (2) that both eukaryotes and bacteria are descended from ancient organisms that resembled archaea.

Yet, despite the drama with which these important results were revealed, it was an oddly anticlimactic end to a very long story. Thirty years before, Carl Woese, a young molecular biologist at the University of Illinois, Urbana, began the painstaking task of working out the evolutionary relationships of the bacteria. Earlier researchers had made the attempt and failed. "The ultimate scientific goal of biological classification cannot be achieved in the case of bacteria," wrote the eminent microbiologist Roger Stanier.

Woese admired Stanier's own considerable efforts to classify the bacteria but did not accept that classifying the bacteria was impossible. Indeed, he had an ingenious plan for discovering their relations to one another. He knew that ribosomal RNAs are among the most conserved parts of all organisms. Because the work of translating RNA into proteins is fundamental, organisms with mutations in their ribosomal RNA nearly always die and ribosomes change hardly at all over millions of years.

Ribosomes are, in short, some of the best sources of information about long-term evolutionary relationships among organisms.

For 10 years Woese (pronounced woes) catalogued short strings of nucleotides that he cut from the ribosomal RNAs of different microorganisms. It was mind-numbing, labor-intensive work that left him, he later said, "just completely dulled down." Yet it was fruitful work, for he gradually unraveled the relationships among some 60 different kinds of bacteria.

Then, in 1976, Woese's colleague Ralph Wolfe sent him some methanogens on which to work his magic. Methanogens were curious bacteria that came in a variety of shapes but all produced methane as a by-product of their metabolism. No one knew where they fit in with the other bacteria. Woese set to work to find out. To his surprise, however, the methanogens lacked the characteristic sequences of ribosomal RNA that all other bacteria share. As he told Wolfe, the methanogens didn't seem to be bacteria at all.

It was dramatic news, for it meant that in addition to the eukaryotes, and the bacterial prokaryotes, we humans were sharing the planet with a completely different domain of organisms—the archaea—never before recognized. Yet, although the story hit the front page of *The New York Times,* the news faded away immediately. For most people, the news was abstract and meaningless. And most of Woese's own colleagues ignored his discovery. They simply didn't believe him. Although Woese provided the answers to many of the questions that most intrigued the eminent microbiologist Roger Stanier, Stanier, who has since died, never acknowledged Woese. For years, Woese says, he waited in vain for some friendly acknowledgment from Stanier.

R.G.E. Murray, the author of the leading microbiology textbook, didn't even include the archaea until 1986, and then only as a subgroup within the prokaryotes—even though Woese's work had demonstrated that the archaea were distinct from other bacteria. A handful of other researchers stood behind Woese, including Wolfe, arguing with their colleagues on his behalf. Yet, Woese's work was otherwise almost universally snubbed.

Woese merely retreated back to his laboratory, continuing his work, gathering new and better data, and publishing a series of landmark papers on the evolution and classification of bacteria. Today Woese is highly acclaimed. He has been elected to the prestigious National Academy of Sciences and awarded both a MacArthur Foundation "genius" grant and microbiology's top honor—the Leeuwenhoek medal.

Great discoveries in science frequently meet initial rejection. As Wolfe told *Science* magazine in 1997, "We can say, 'Oh, that's just the way science works.' But it was a personal experience for him. He's the one who had to live through it."

Figure A Carl Woese discovered the archaea in 1977. The scientific community did not fully accept his results until 1997. (*Courtesy of University of Illinois at Urbana-Champaign News Bureau*)

a usable form. The bacteria perform this service, called **nitrogen fixation,** at great cost, however, for the task requires a lot of energy. (Conversely, processes that yield nitrogen gas as a product release lots of energy. For this reason, many nitrogen-containing compounds, including commercial fertilizers, are explosives.)

The **pseudomonads,** known as "the weeds of the bacterial world," are a large and aggressive group of bacteria that seem to live everywhere—in soil, ponds, infected wounds, hot tubs, and even medicine bottles. Pseudomonads are typically straight or curved gram-negative rods with flagella. They require oxygen and produce CO_2. The extraordinary success of pseudomonads comes from their metabolic diversity. They can survive on carbohydrates, petroleum products, tough polysaccharides in natural fibers, and even antibiotics and pesticides.

Omnibacteria are the most common organisms on Earth (Figure 20-15B). These bacteria can all perform aerobic respiration and also have the extraordinary ability to use nitrate (NO_3^-) as an electron acceptor instead of oxygen. The most prolific and well known are the common enteric bacteria [Greek, *enteron* = intestine], such as *E. coli,* which inhabit the intestines of animals. Omnibacteria may be rod shaped, as is *E. coli,* or comma shaped, as is the species that causes cholera.

The anaerobic heterotrophs

The anaerobic heterotrophs are mostly obligate anaerobes. They include, for example, the **fermenting bacteria,** which

1 µm

Figure 20-16 Swimming syphilis spirochete. The most famous spirochete is *Treponema pallidum,* which causes the sexually transmitted disease syphilis. Spirochetes contain internal flagella enclosed between layers of the cell wall, by means of which they move differently from any other bacteria. Although some live freely and some are parasites, all spirochetes move by undulating quickly, even through the most viscous liquids. Most spirochetes are finicky about their diets and *T. pallidum* is no exception. Its requirements are so specific, in fact, that although it grows well inside the human body, biologists have been unable to grow virulent strains in the laboratory. *T. pallidum* is transmitted from one person to another by means of direct contact between moist mucous membranes. Until the introduction of penicillin in the 1940s, syphilis was incurable, progressive, and fatal. *(CDC/Science Source/Photo Researchers, Inc.)*

derive energy and carbon atoms from a variety of organic compounds in the absence of oxygen; and the **spirochetes,** which have a distinctive corkscrew shape (Figure 20-16).

The autotrophs

The autotrophs consist of four phyla: the chemoautotrophs, which photosynthesize *without* oxygen, and three phyla of photoautotrophs, which photosynthesize *with* oxygen. The photoautotrophs consist of the **anaerobic photosynthetic bacteria,** the **cyanobacteria,** and the **chloroxybacteria.**

Biologists have traditionally classified prokaryotes according to how they acquire energy and building blocks, whether they photosynthesize, and whether they use or tolerate oxygen. Biologists divide the Eubacteria into three main groups: aerobic heterotrophs, anaerobic heterotrophs, and autotrophs. The most ecologically important Eubacteria are (1) the cyanobacteria, the main photosynthetic prokaryotes, which fix carbon; and (2) the nitrogen-fixing bacteria, which supply the world's organisms with fixed nitrogen for making proteins.

How Has DNA Sequencing Changed the Classification of Prokaryotes?

Two kinds of DNA sequence information have dramatically changed the way that biologists view the prokaryotes. The first type of information initially came from the pioneering work of Carl Woese on the genes for ribosomal RNA (Box 20.1). This work led to the originally revolutionary idea that the archaea represent a separate domain of life, as different from the eubacteria as from eukaryotes. The second type of information started with the determination, in 1997, of the complete genomic sequences of two archaea species, one a methanogen and the other a sulfur-metabolizer.

The complete DNA sequences revealed that, in some important respects, archaea resemble eukaryotes more than eubacteria. The most startling resemblances were in the enzymes responsible for handling of genetic information (DNA replication and RNA synthesis).

Analysis of other archaeal genes, however, suggests that they are more closely related to eubacterial than to eukaryotic genes. Researchers continue to disagree about the family tree of prokaryotes. A major difficulty in interpreting the DNA sequence data is that researchers increasingly think that there has been lots of gene transfer among the archaea.

Biologists are beginning to conclude that the task is far more complicated than originally expected (Figure 20-17). As researchers sequence more prokaryotic genomes, however, we

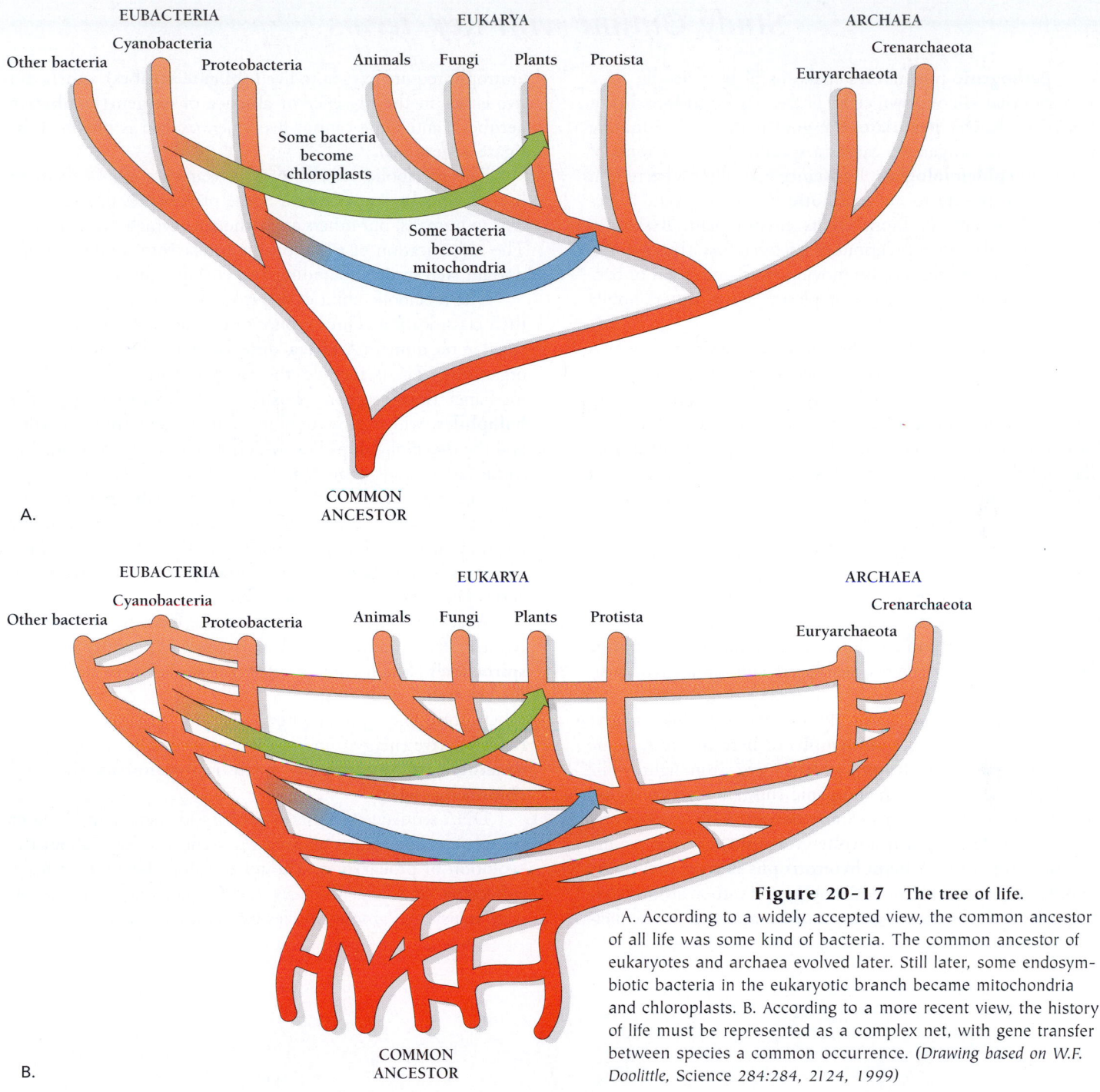

Figure 20-17 The tree of life.
A. According to a widely accepted view, the common ancestor of all life was some kind of bacteria. The common ancestor of eukaryotes and archaea evolved later. Still later, some endosymbiotic bacteria in the eukaryotic branch became mitochondria and chloroplasts. B. According to a more recent view, the history of life must be represented as a complex net, with gene transfer between species a common occurrence. (*Drawing based on W.F. Doolittle*, Science *284:284, 2124, 1999*)

can expect to know more about the evolutionary history of individual genes. Many biologists suspect that the history of prokaryotes will not be one of simple descent from a common ancestor, but will include many instances of gene transfer between species.

In this chapter we have briefly surveyed the two domains of prokaryotes. Just as important, we learned how the biology of bacteria helps them acquire resistance to antibiotics. In the next chapter, we survey the first two kingdoms of eukaryotes, the Protista and the Fungi.

Study Outline with Key Terms

Most **pathogenic** prokaryotes make us ill by releasing toxic chemicals that kill our own cells. These may be **endotoxins** or **exotoxins. Koch's postulates** are guidelines for determining whether a microorganism causes a specific disease, a topic of interest to **epidemiologists.** Resistance may allow a bacterium to survive exposure to an **antibiotic** or, alternatively, to actively destroy any antibiotic in its environment. Resistance genes accumulate in populations of bacteria, first, through natural selection, and second, by moving from one strain of bacterium to another by means of plasmids and other mobile genes.

Prokaryotes tuck their DNA into a **nucleoid,** rather than a true nucleus. They also lack other membrane-bounded organelles. Most are single-celled organisms, though some aggregate to form colonies. They come in many shapes, but three forms dominate: the rod-shaped **bacilli,** the spiral-shaped **spirilla,** and the spherical **cocci.** Most have cell walls made of a **peptidoglycan.** Bacteriologists classify bacteria according to whether they are **gram positive** or **gram negative.** Some bacteria have specialized outer surfaces that form a **capsule,** a **slime layer,** or hairlike structures called pili.

Many prokaryotes are capable of directed movement, or **taxis** (including **chemotaxis**), and many have **flagella,** whose structure and operation are different from those of eukaryotic flagella. Prokaryotes grow continuously and rapidly, and usually reproduce asexually.

Prokaryotes are biochemically more diverse than eukaryotes. They may be either **autotrophs** or **heterotrophs.** Some autotrophs (**photoautotrophs**) obtain energy from light, while others obtain energy from inorganic substances, such as hydrogen sulfide or ammonia. Some heterotrophic prokaryotes derive energy from light (**photoheterotrophs**) but also require organic compounds. **Chemoheterotrophs** depend exclusively on organic compounds for both energy and carbon atoms. Heterotrophs may be **symbionts** or **decomposers.** Some heterotrophs require oxygen to live (**obligate aerobes),** others can live either in the presence or absence of oxygen (**facultative aerobes),** and some cannot live when oxygen is present (**obligate anaerobes).**

The metabolic versatility of prokaryotes allows them to live where eukaryotes cannot. Some prokaryotes provide benefits to humans, but others are pathogens, which cause disease. The identification of a pathogenic prokaryote is often an important step in understanding and curing a disease.

The metabolic abilities of prokaryotes provide a basis for their classification. Three groups of prokaryotes, together classified in the domain **Archaea,** differ dramatically from all other organisms. These include the **methanogens,** which make methane from carbon dioxide and hydrogen gas, the **halophiles,** which grow only in high concentrations of salts, and the **thermoacidophiles,** which derive energy by oxidizing sulfur compounds or methane.

The **Eubacteria** include about ten phyla. Aerobic heterotrophs are biochemically diverse but use the same metabolic pathways that are found in eukaryotes. These bacteria include **nitrogen-fixing bacteria, pseudomonads, omnibacteria, actinobacteria,** and **myxobacteria.**

Anaerobic heterotrophs include the **fermenting bacteria** (which often cannot live when oxygen is present) and the **spirochetes** (which contain internal flagella between layers of the cell wall).

Autotrophic prokaryotes include **chemoautotrophs** (which derive energy from a variety of inorganic compounds), **anaerobic photosynthetic bacteria, cyanobacteria,** and **chloroxybacteria.**

DNA sequence analysis of prokaryotic genes, and even of complete genomes, has allowed researchers to learn about the evolution of prokaryotes. The relationships between archaea, eubacteria, and eukaryotes are not simple, since many genes were transferred among species by plasmids and viruses.

Review and Thought Questions

Review Questions

1. What features of eukaryotes do prokaryotes lack?
2. What molecule makes up much of the cell wall of members of the Eubacteria but not of Archaea?
3. What is the difference between gram-positive and gram-negative bacteria? Which kind is susceptible to penicillin?
4. Describe two ways in which antibiotics work against bacteria. Why do these mechanisms not affect eukaryotic cells?
5. Describe the two major ways that antibiotic resistance is spread through populations of bacteria.
6. What features allow prokaryotes to reproduce so much more rapidly than eukaryotes? What advantages does this give bacteria?

7. What criteria do biologists use to classify bacteria?

Thought Questions

8. Are antibiotics helpful against a viral infection such as a cold or flu? Why or why not?
9. Animal breeders began giving animals antibiotics because studies showed that animals raised on antibiotics grew faster. Today, this effect is not as strong as it was 30 or 40 years ago. Why might antibiotics have this effect? Why might antibiotics no longer be so good at increasing growth rates?

About the Chapter-Opening Image

Salmonella bacteria that infect hamburger typify our problems with antibiotic-resistant strains of bacteria, a theme in this chapter about bacteria.

 On-line materials relating to this chapter are on the World Wide Web at **http://www.harcourtcollege.com/lifesci/aal2/**

21

Classifying the Protists and Multicellular Fungi

The Fungus Fighters

In 1948, two government researchers set out to find a drug that would ultimately save thousands of human lives. In just two years, Elizabeth Lee Hazen and Rachel Brown succeeded in producing a drug that could safely cure deadly fungal infections in humans—a drug that is still a medical mainstay today. Both women had overcome great obstacles to obtain their scientific training. That they were highly educated, practicing scientists at all was something of a miracle. That they were able to make a major contribution to medicine was a testimony to their lifetime dedication to science and the respect they inspired in those around them.

Elizabeth Lee Hazen, born in August 1885, to a Mississippi cotton farmer and his wife, was an orphan by the time she was two years old. Raised by relatives, Hazen grew up in a large family in Lula, Mississippi. After high school, she enrolled in the Mississippi Industrial Institute and College, the first state-supported college for women in the United States. Without such an institution, she might never have gone to college, might never have studied physiology, anatomy, botany, zoology, and physics. Her grades were unremarkable. Like many people with great contributions to make, her strengths would not become obvious until she was much older.

After graduating from college, Hazen taught high school for six years. When she was 31, she enrolled as a graduate student at Columbia University, in New York, where one interviewer sized her up and then told her that she was so ignorant he doubted she had ever been to college. Nonetheless, she soon proved herself, and the following year Columbia awarded her a Master of Arts degree in biology. During World War I, she worked for the Defense Department diagnosing bacterial infections. After the war, she returned to Columbia, and, in 1927, at the age of 42, she received a Ph.D., becoming one of a small number of women with doctorates in the United States. She stayed at Columbia until 1931, when the New York State Department of Public Health lured her away to supervise its large medical pathology lab. But she maintained a connection with Columbia. In 1944, when the health department needed an expert on fungal infections, she returned to study mycology, at the same time continuing her full-time work in the pathology lab.

For most of us, fungi are no more than a mild curiosity. Mushrooms are a nice addition to a pizza. Mold is an annoyance in the shower. Some of us may have had unpleasant experiences with a fungus, a bout of athlete's foot, a yeast infection, nothing that a simple antifungal medication could not treat. In most people, the immune system's white blood cells quickly engulf and digest any fungal spores that make it into the body, and fungi are not a problem.

Under the right circumstances, however, a fungus can destroy a living human being as thoroughly as it can rot a dead stump in the forest. Diabetics, for example, have tissues that are unusually high in sugar. The warm, sugary environment of a diabetic's lung apparently creates a perfect environment for mucormycosis, the lush growth of a fungus in the order Mucorales. A common, usually harmless food mold, some species of Mucorales can grow through the lungs into the heart. Alternatively, this fungus may infect the sinuses, then run hyphae, long strands of the fungal body, throughout a person's head, destroying nerves and brain function. People with damaged immune systems—burn victims, transplant patients being treated with immunosuppressive drugs, and those with immune diseases such as AIDS—are all susceptible to fungal infections. For such people, fungi are a potentially life-threatening problem.

In the 1940s, when the first antibiotics came into use, medical experts discovered that patients treated with antibiotics became highly susceptible to fungal infections. It seemed that many of the bacteria killed by the antibiotics were important in keeping fungi at bay. Yet no antifungal drugs existed, and doctors were helpless to stop the spread of the often fatal infections.

Elizabeth Hazen was determined to find a drug that would help. By 1948, she had established an immense collection of pathogenic (disease-causing) fungi with which she and her team of technicians could diagnose fungal infections from people all over the eastern United States. Even more important, Hazen had for years been testing soil samples for bacteria that could kill infectious fungi. Now she had found two different bacteria that produced substances toxic to fungi. But what were the substances? Hazen needed the help of a talented chemist to isolate the active ingredient. One day in 1948, Hazen took a train from New York City to the health department's main office in Albany, New York, 150 miles away. There she had an appointment with senior researcher Gilbert Dalldorf, himself a talented bacteriologist and physician. Dalldorf listened to Hazen's proposal, then took her to see chemist Rachel Brown.

Like Hazen, Brown had succeeded in science against long odds. She was born in 1898 and raised in Springfield, Massachusetts, in a small, middle-class family. In 1912, however, when Rachel Brown was 14 years old, her father abandoned his family. Rachel's mother Annie then supported herself, her two children, and both her parents on a secretary's salary. Despite her poverty, Annie Brown was determined that her two children should go to the best schools she could afford. By Rachel's senior year in high school, however, her heart was set on Mount Holyoke College, a small, private college for

> *The third substance was only mildly toxic and killed virtually any fungus it came into contact with. It looked like a miracle drug, another penicillin.*

women. But Rachel was turned down for a scholarship, and her mother could not possibly afford the tuition.

Miraculously, a wealthy friend of Rachel's grandmother offered to pay all of Rachel's expenses for a full four-year college education. Brown earned a B.A. in chemistry from Mount Holyoke and then put herself through graduate school at the University of Chicago. In 1926, she completed a Ph.D. dissertation in chemistry and shortly after took a job in Albany with the New York State Department of Public Health, an institution that hired an unusual number of women scientists. There, she worked steadily for almost a quarter of a century, isolating proteins and other molecules from infectious bacteria and protists that could be used for diagnostic tests or for vaccines. When Elizabeth Hazen and Gilbert Dalldorf stuck their heads in the door of her lab, it was the beginning of a fruitful collaboration and friendship that would last for more than 25 years.

Hazen explained that she had two strains of bacteria that could kill fungi, but that she needed to isolate the antifungal substances for further testing. In the months that followed, Brown and Hazen sent bulky package after bulky package between Albany and New York City. Hazen would send samples of bacteria, then Brown would extract different compounds for further testing by Hazen. The two researchers quickly isolated three fungicidal molecules from two kinds of bacteria. Two of these molecules were highly toxic to mice and therefore not useful as drugs. The third substance, however, was only mildly toxic and killed virtually any fungus it came into contact with; they called it fungicidin. It looked like a miracle drug, another penicillin.

At a meeting of the National Academy of Sciences in October 1950, Hazen and Brown presented their results. A reporter from *The New York Times* jumped all over the story, and within a day or two, the new life-saving drug was big news. Pharmaceutical companies began calling the state lab, demanding to be let in on the project.

Dalldorf and Brown nearly panicked. Fungicidin had been tested in animals but never in humans. For clinical trials, huge quantities of the drug had to be manufactured cheaply and quickly. Only a pharmaceutical company could do that. But Dalldorf and Brown knew, from previous experience with another drug, that they needed to patent fungicidin quickly before someone else did. Without a patent, to make clear who owned the rights to the drug, no pharmaceutical company would develop it. And Hazen, Brown, and Dalldorf wanted the drug to be used to save lives, the sooner the better.

By February, the researchers had filed a patent application with the United States Patent Office; a month later, they licensed the patent to Squibb Pharmaceuticals. Fungicidin, it turned out, was a name already in use for another chemical, so Hazen and Brown renamed it nystatin (pronounced nye-stat-in), after New York State. Within five years, Squibb had found a way to produce nystatin cheaply and had successfully tested the drug in humans. As the first treatment for fungal infections in humans, nystatin was tremendously useful.

Squibb sold so much nystatin that between 1955 and 1979 alone, royalties from sales of the drug came to $13.4 million. Half of this went to a nonprofit foundation that funded scientific research. The other half went to the nonprofit Brown–Hazen Fund, which supported research and training in the medical and biological sciences and awarded grants to women scientists. Neither Hazen nor Brown ever accepted any of the enormous wealth that resulted from their discovery. Brown apparently helped put several young women through college, as she herself had been helped. Both Hazen and Brown continued to work into their 70s, content in their achievements, honored with medals, honorary degrees, and elections to prestigious societies.

In this chapter we will quickly survey two kingdoms of organisms: the fungi and the protists. Biologists used to define a fungus as a eukaryotic organism that is neither plant nor animal. However, that definition applies to the protists as well. One important difference between the fungi and the protists is their organization. Most fungi are multicellular, while most protists are single celled. A few kinds of protists form colonies, but these are primitive collections of nearly identical cells. In contrast, fungi resemble plants and animals in having complex internal structures and specialized cells that serve different purposes.

Another difference between fungi and protists is how easily fungi can be classified. Most of the fungi are remarkably alike and make an easily recognized kingdom of organisms. But the protists are so diverse that many biologists believe that they should be divided into several kingdoms. Indeed, analysis of DNA sequences has led some biologists to suggest that protists form at least 12 kingdoms, each of which is at least as different from one another as plants and animals.

Protists, because they are mostly one celled, are considered more primitive than the fungi. Certainly, the first protists lived in the sea long before the first fungi evolved. However, protists are anything but simple, for they show extraordinary and elaborate specializations at the cellular level.

If we look at a drop of pond water under a microscope, we discover an amazing world of tiny, wriggling creatures

Figure 21-1 **How many different protists live in a pond?** A variety of protists live in both fresh water and salt water, including algae, dinoflagellates, and diatoms. Some protists are motile; others drift or anchor themselves to some object. Different protists move by waving flagella or cilia or by ameboid movements.

whose existence no one even suspected until Leeuwenhoek's first microscopic observations. Many of these microscopic creatures are members of the kingdom Protista. The protists range from the simple amebas to the most complicated cells on our planet (Figure 21-1). Their variety raises many questions: Why are protists so diverse? How are the various protists related to one another? How are they different from larger, more familiar organisms—such as mushrooms, mosses, or mammals? How are they different from prokaryotes? And how do they derive energy?

WHAT ARE PROTISTS?

Protists [Greek, *protos* = first] include the algae (which photosynthesize like plants), the protozoa (which act more or less like animals), and some funguslike forms. Protists flourish wherever there is moisture. In some marine ecosystems, the brown algae and the red algae are responsible for trapping nearly all of the solar energy that eventually reaches other organisms. The abundance of protists in the fossil record suggests that protists have been successful for a long time.

Protists are eukaryotic. Protist cells have nuclei and other membrane-bounded organelles, and almost all have mitochondria. Most (but not all) also have chromosomes and undergo mitosis. All can reproduce asexually by fission, and some can perform meiosis and reproduce sexually.

At a biochemical level, protists are more or less identical to plants, animals, and fungi: all derive energy from glycolysis and respiration, and the photosynthetic protists all use water as a source of electrons. Protists lack the biochemical diversity of the prokaryotes, but they are the champions of cellular diversity. No other kingdom includes so many different kinds of cells.

Why Are Protists So Diverse?

Some protists are autotrophic and have chloroplasts. Others are heterotrophic and possess vesicles and lysosomes. Some are shapeless bags, while others have elaborately sculpted tests, or shells. Some are free living, some parasitic. Some produce spores, others do not. Some are single celled and others are multicellular like plants, animals, and fungi.

Some protists are **plankton** [Greek, *planktos* = wandering], organisms that float in the water, carried by currents (Figure 21-2). Others are **motile,** meaning that they move actively in search of food. Of the motile forms, some propel themselves with one or two long flagella (Figure 21-2). Some use hundreds of cilia, and others use protrusions called **pseudopodia** [Greek, *pseudos* = false + *pous* = foot], temporary extensions of the cytoplasm.

With so much diversity, we may ask, What binds the protists together? The answer is, surprisingly little. What protists have in common is that they do not fit in with plants, animals, or fungi. When biologists regarded all organisms as either plants or animals, they classified all the multicelled protists as plants and all the single-celled protists as either **algae** (which, like plants, photosynthesize) or protozoa (which, like animals, move under their own power).

The common pond-water protists of the genus *Euglena* illustrate why this classification scheme never worked very well (Figure 21-2). In the light, most species of *Euglena* develop chloroplasts and photosynthesize. In the dark, however, *Euglena* lives a heterotrophic life, absorbing nutrients from its surroundings, and may even lose its chloroplasts. *Euglena,* then, is neither plant nor animal, but it is a fine protist.

The green alga *Chlamydomonas,* like *Euglena,* is a single-celled organism capable of both photosynthesis and directed movement. But *Chlamydomonas*'s close relative *Volvox* consists of colonies of hundreds to thousands of cells, each of which looks just like a free-living *Chlamydomonas.* Because the two algae are otherwise identical, it would be unreasonable to put one into the plant kingdom and one into the animal kingdom. As a result, biologists now put both groups into the Protista.

Protists are eukaryotic organisms that are defined not by what they have in common, but by not belonging to any of the other three eukaryotic kingdoms (Figure 21-3). In contrast, we define the other eukaryotic kingdoms, fungi, plants, and animals, partly by the way their cells function and develop. Plants and animals are all multicellular organisms that develop from embryos, organized collections of tissue that, through a highly predictable sequence of steps, develop into more-mature forms. Fungi are organisms, nearly always multicellular, that absorb nutrient molecules directly from their environments. The fungi do not photosynthesize and do not form embryos.

The diversity of the protists is both a joy and a headache. How can we classify them? Biologists do not agree on the number of phyla, or even on whether all the protists should be in the same kingdom: one protist researcher has even suggested that they really comprise some 20 separate kingdoms! Many biologists divide them, according to their cellular organization and mode of nutrition, into plantlike, animal-like, and funguslike organisms. We will follow the classification scheme of Margulis and Schwartz, who divide the protists into 27 phyla, mostly on the basis of their cellular structures and life cycles. We will survey 16 of these phyla. For convenience, we group these into plantlike phyla (the algae), animal-like phyla (the protozoa), and funguslike phyla.

Protists are metabolically more uniform than prokaryotes. However, partly because protists are a grab-bag kingdom (including every eukaryote that is not animal, plant, or fungus), they show enormous diversity at the cellular level.

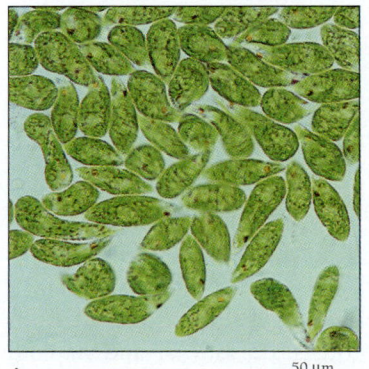
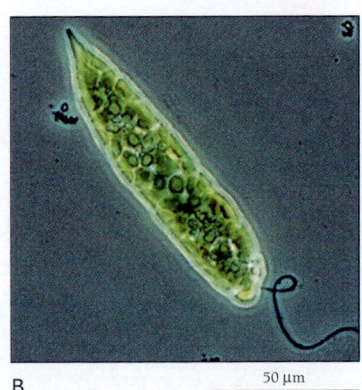

A.   50 μm   B.   50 μm

Figure 21-2 **Plant or animal?** A. Euglena in sunlight are green with chloroplasts. B. Euglena in a dark environment use their flagella to go looking for light. (*A, Dwight Kuhn; B, Biological Photo Service*)

Figure 21-3 The diverse kingdom Protista.

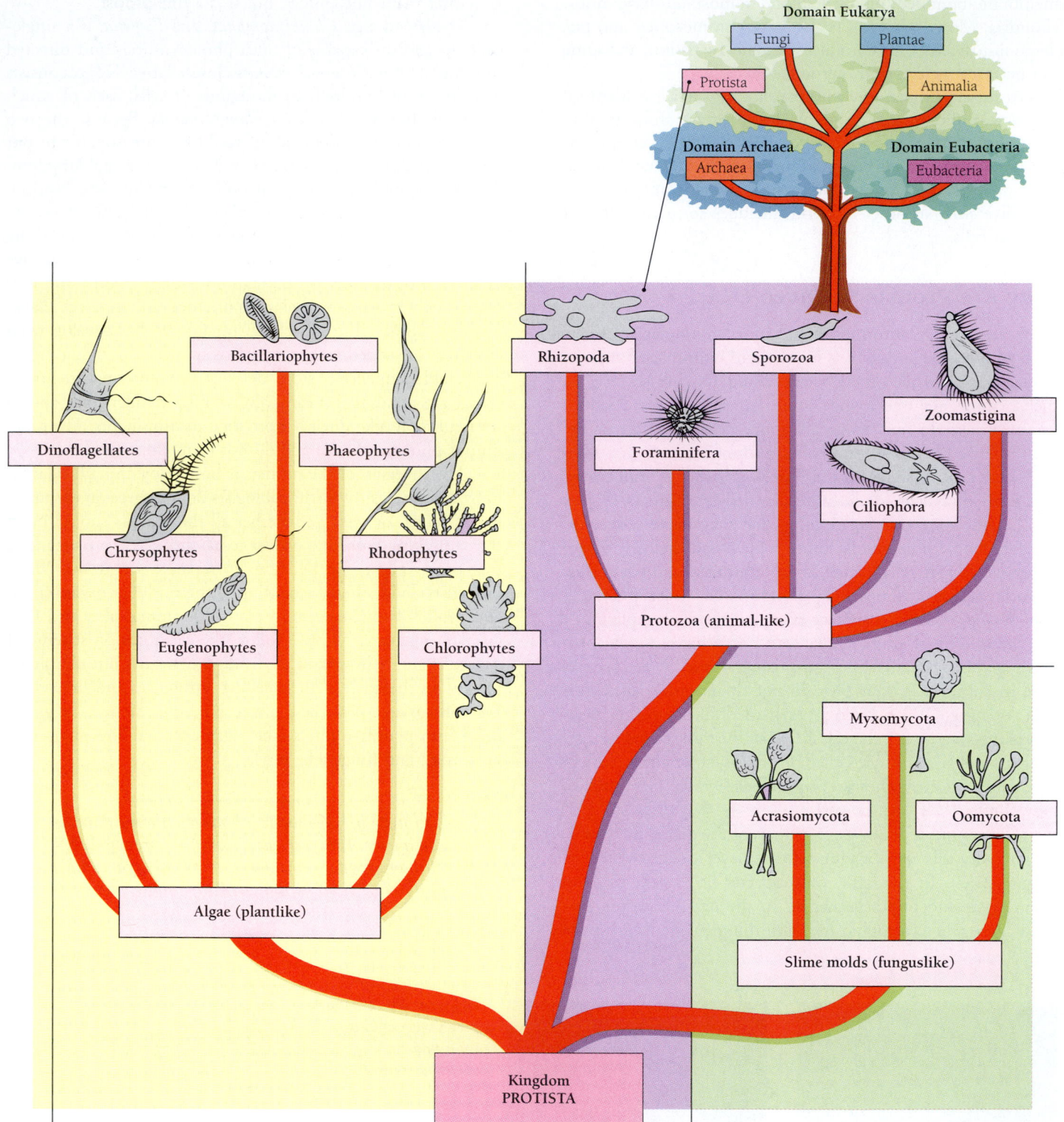

How Do Protists Reproduce?

Every protist species goes through two phases in its life cycle—reproduction and dispersal. Dispersal may take many forms. Some free-living, motile protists swim or crawl. Others scatter seedlike **cysts,** enclosed structures that contain reproduc-tive cells. Still others infect a host that moves, such as an insect, bird, or mammal.

Reproduction may be either asexual or sexual, depending on the species. Some protists are asexual and haploid throughout their life cycles and never undergo meiosis. Other

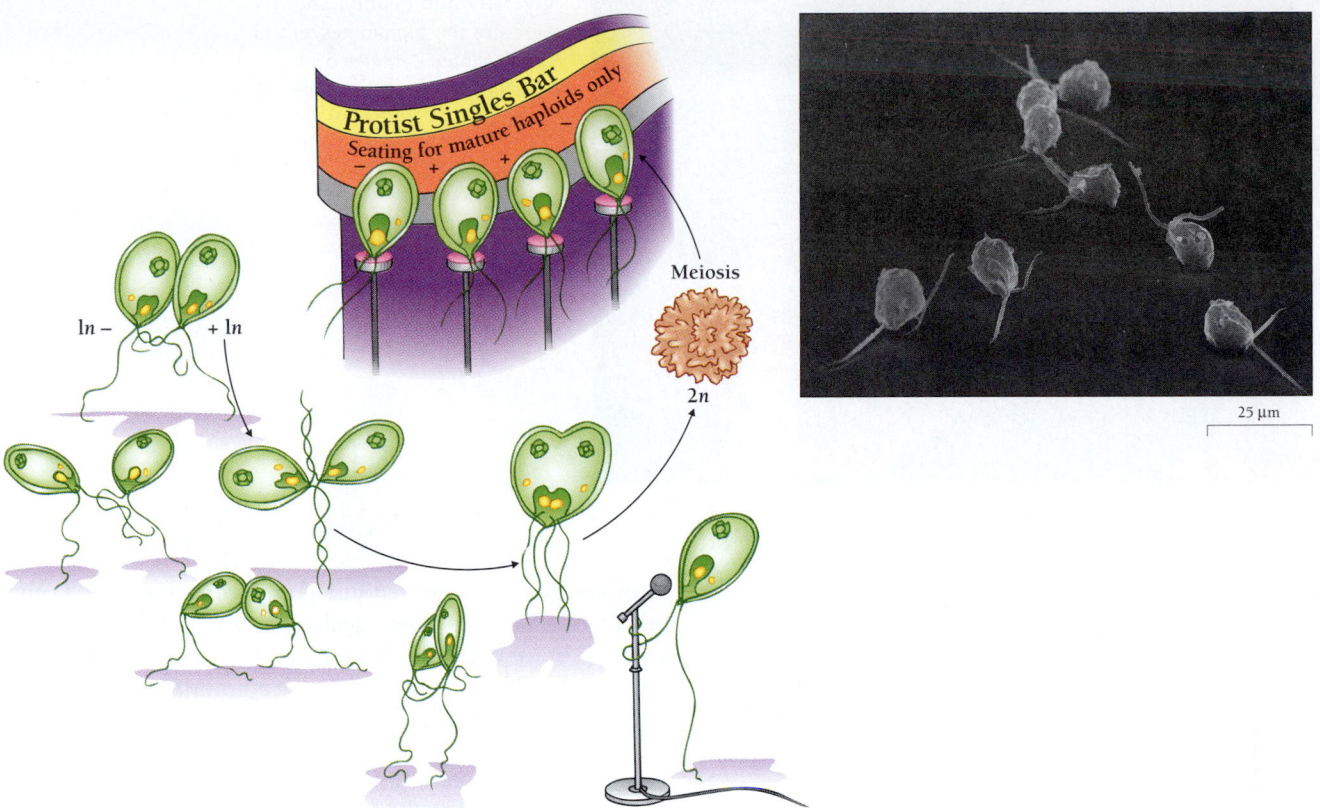

25 µm

Figure 21-4 **Sexuality in protists.** *Chlamydomonas* is a well-studied green alga that grows in ditches, pools, fresh water, and soil. The zygote is the only diploid stage. The mature cell and all other reproductive stages are haploid. Researchers suspect that the first sexually reproducing organism had a life cycle like that of *Chlamydomonas.* (Photo, David M. Phillips/Science Source/Photo Researchers, Inc.)

protists are diploid for most of their life cycle and produce haploid gametes (Figure 21-4). Still other protists are haploid for almost all of their life cycles but produce a short-lived diploid zygote. Finally, some protists undergo **alternation of generations,** passing through both haploid and diploid phases.

ALGAE ARE PHOTOSYNTHETIC (PLANTLIKE) PROTISTS

Algae are relatively simple, plantlike organisms that live in moist environments everywhere: in open water, ice, soil, as well as in puddles or pockets of water in trees or rocks. Most algae are *photoautotrophs,* organisms that supply themselves with energy by means of photosynthesis. Some, such as *Euglena,* can also live as heterotrophs. Just as plants do, algae use chlorophyll and other pigments to capture light for photosynthesis. However, whereas true plants photosynthesize with both chlorophyll *a* and chlorophyll *b,* most protists do not have any chlorophyll *b.* Algal protists also differ from true plants in the structure of their cell walls, and the arrangement of their flagella. In this section we will briefly survey 6 of the 12 phyla

of algae: dinoflagellates, chrysophytes, euglenophytes, phaeophytes, rhodophytes, and chlorophytes.

Dinoflagellates

Most **dinoflagellates** [Greek, *dinos* = whirling + *flagellum* = whip] are single-celled organisms that float freely as plankton in warm oceans (Figure 21-5). A few can form colonies, and some live symbiotically with corals, sea anemones, and clams. The ultimate source of food in coral reef and other communities, in fact, is photosynthetic dinoflagellates. Dinoflagellates make up a major part of the phytoplankton, the microorganisms on which most aquatic ecosystems depend.

Of the thousands of species of dinoflagellates, the best known are those that grow rapidly to cause red tides, in which population explosions, or blooms, of dinoflagellates color the ocean red, brown, or other colors, depending on the species (Figure 21-6). Other red tides release deadly toxins. Because many dinoflagellates are bioluminescent, that is, they glow in the dark, they create dramatic nighttime displays.

Almost all dinoflagellates have a rigid wall, or **test,** made of cellulose and coated with silica. Around its circumference, the test contains two characteristic grooves, in which the two flagella are embedded at right angles to each other. Together,

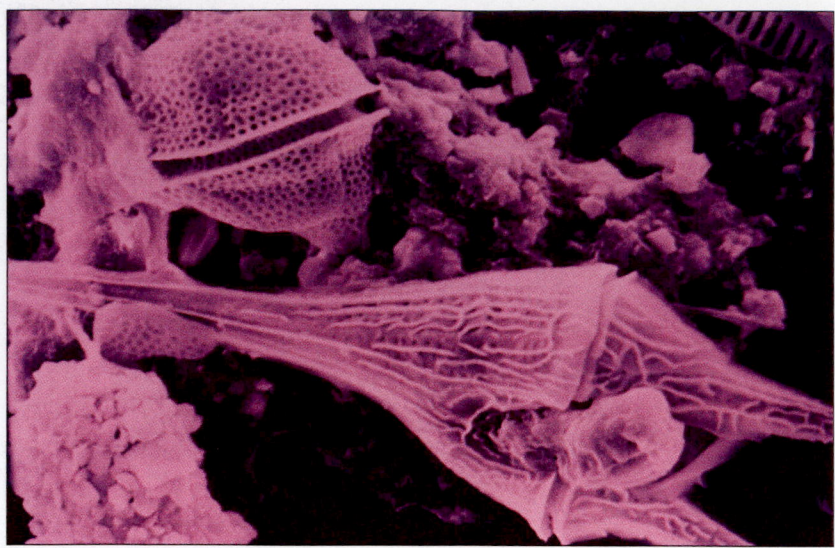

Figure 21-5 Dinoflagellates are renowned for their
diversity and beauty. Shown
here are the bioluminescent dinoflagellates *Ceratium* and
Peridineum. (*Terry Hazen/Visuals
Unlimited*)

25 μm

the two flagella create the spinning that gives these organisms their name.

Dinoflagellates are very old. Because their hard tests are well preserved in the fossil record, we know that they have existed at least since the beginning of the Cambrian Period, some 570 million years ago. They are so ancient, in fact, that their chromosomes are organized differently from those of all other eukaryotes. When a dinoflagellate divides, its chromosomes segregate by attaching to the nuclear membrane rather than to a mitotic spindle. Dinoflagellates have no mitosis, no meiosis, and no sexual reproduction.

Dinoflagellates are among the oldest and most primitive of the eukaryotes.

Chrysophytes (Golden Algae and Diatoms)

Golden-yellow pigments give the golden algae, or **chryso- phytes** [Greek, *chrysos* = golden + *phyta* = plant], their name. As dinoflagellates do, many chrysophytes form hard tests containing silica and other minerals that help preserve them in the fossil record as far back as 500 million years ago. Each chrysophyte family contains both single-celled and colonial forms. The colonial forms may be spherical or loosely branched. Most chrysophytes are plankton, widespread in temperate lakes and ponds. Only one group lives in the oceans.

Chrysophytes reproduce asexually in one of two ways. A single cell, called a swarmer cell, may swim away from a colony to found another colony. Alternatively, the hundreds or thousands of cells of a mature colony may split into two groups, which then float away from each other to form two new colonies. Chrysophytes can also form cysts that resist cold temperatures and desiccation.

The *diatoms* are some of the most numerous and beautiful of the chrysophytes. Most of the 10,000 species of diatoms are single cells, although some form simple filaments or colonies. Each species has a unique and elaborate silica test (Figure 21-7). Most diatoms live as free-floating plankton in fresh or salt water. Some, however, can move on slime that they extrude from a slit between the two parts of the test. The tests of diatoms are well preserved in the fossil record, dating from the Cretaceous Period, between 65 and 135 million years ago.

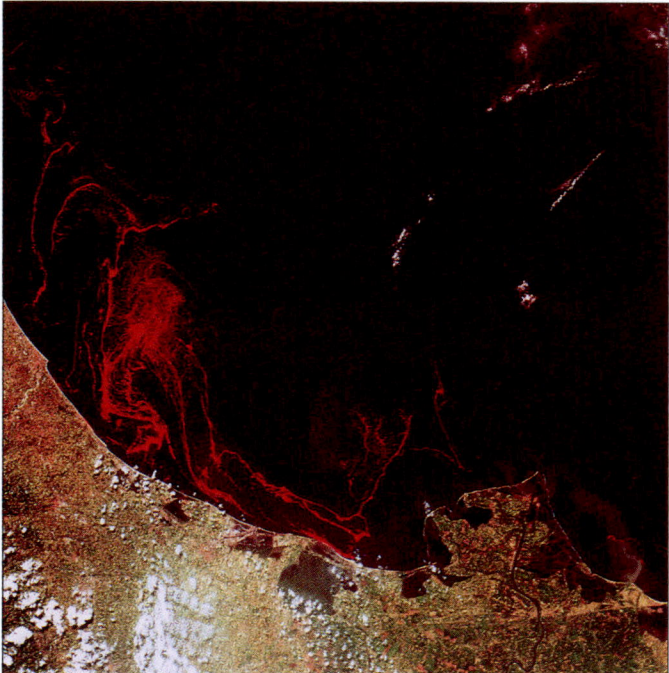

Figure 21-6 Satellite photo of a red tide off the east coast of Italy. Dinoflagellate populations explode into "blooms," or "red tides," when conditions are good. Some species of dinoflagellates make poisons that clams, mussels, and other shellfish accumulate in their bodies, making these ordinarily edible shellfish deadly. Some dinoflagellates secrete toxins that kill whole schools of fish, whose flesh the dinoflagellates then feed on. During a bloom of such dinoflagellates, toxic vapors over the water can make boaters temporarily ill. (*Geospace/Science Photo Library/Photo Researchers, Inc.*)

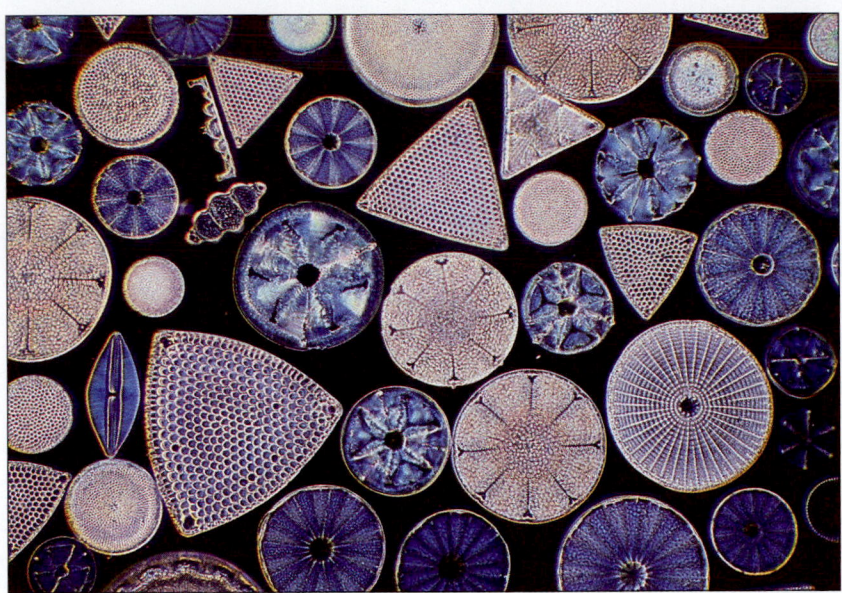

Figure 21-7 Diatoms may be even more beautiful than dinoflagellates. *(Robert Brons/Biological Photo Service)*

50 μm

Diatoms can reproduce sexually, forming gametes by meiosis. Usually, however, they reproduce asexually, dividing their chromosomes through ordinary mitosis. At cell division, each half of the test stays with one of the daughter cells, which then assembles a second half.

Chrysophytes (golden algae) have both single-celled and colonial species. Diatoms have elaborate tests and can reproduce sexually as well as asexually.

Euglenophytes

The **euglenophytes** [Greek, *eu* = true + *glene* = eyeball] derive their name from their light-detecting eye spots, which enable them to live a double life. In the dark, they live heterotrophically, by breaking down dissolved organic matter.

When light becomes available, however, euglenophytes use their eyespots, and a single flagellum, to swim into the sunshine. Euglenophytes photosynthesize with both chlorophyll *b* (a pigment also used by higher plants), which gives them their grass-green color, and chlorophyll *a*. Except for the green algae, no other algae or any other protists make chlorophyll *b* (Table 21-1). Unlike plants and green algae, however, euglenophytes have no cell wall and store energy in a polysaccharide called paramylon, instead of in starch. Euglenophytes all reproduce asexually, using a primitive form of mitosis, with no clear mitotic spindle and no well-defined chromosomes. They also do not go through the stages of metaphase or anaphase.

One of the biggest surprises from analysis of protist DNA was the discovery of the close relationship between the euglenophytes and **trypanosomes,** the flagellated protozoa that cause sleeping sickness in Africa and Chagas disease in South

TABLE 21-1	WHICH PROTISTS USE THE SAME CHLOROPHYLLS AS PLANTS?			
Chlorophyll:	*a*	*b*	*c*	Storage material
Plants	*a*	*b*		Starch
Chlorophytes	*a*	*b*		Starch
Euglenophytes	*a*	*b*		No Starch
Dinoflagellates	*a*		*c*	Starch or oil
Phaeophytes	*a*		*c*	No starch
Chrysophytes	*a*		*c*	No starch
Rhodophytes	*a*			Starch

America. Trypanosomes, which have complex life histories, are transmitted to people and animals via insect bites (the tsetse fly in Africa and the "kissing bug" in South America).

Euglenophytes can live as either autotrophs or heterotrophs.

Phaeophytes (Brown Algae)

The **phaeophytes** [Greek, *phaios* = dusky brown], or brown algae, together with the rhodophytes (red algae), which we will discuss next, make up most of the organisms that we call seaweeds. Phaeophytes are all multicellular, and most live in the cool waters off temperate coasts. Phaeophytes include the biggest protists in the world—the giant kelps, which in deep water may be up to 100 meters tall (Figure 21-8).

Most phaeophytes reproduce sexually. However, meiosis in phaeophytes produces **spores** rather than gametes. The spores germinate into new multicellular organisms in which all the cells are haploid. This haploid form is called a **game-**

tophyte because it then produces gametes. The diploid form is called a **sporophyte** because it produces spores. Such alternation of generations also occurs in green algae, red algae, fungi, and, of course, the true plants.

While the size and complexity of the phaeophytes far exceed those of most other protists, the individual cells of phaeophytes resemble those of other protists. In particular, the flagellated sperm of phaeophytes look just like the cells of chrysophytes. Phaeophytes make chlorophylls *a* and *c* and produce other pigments like those of chrysophytes (Table 21-1). But, unlike higher plants, phaeophytes never produce chlorophyll *b* and do not synthesize starch.

Phaeophytes (brown algae), which are related to the chrysophytes, are the largest and most complex of the protists.

Rhodophytes (Red Algae)

Although the brown algae include the most spectacular seaweeds, most of the world's seaweeds are **rhodophytes** [Greek, *rodon* = rose], or red algae. Nearly all of the 4000 species of rhodophytes are marine, and most live in tropical seas, attached to rocks or other algae. Among the most beautiful of the rhodophytes are the coralline algae. These algae deposit calcium carbonate in their cell walls and contribute to the building of coral reefs. Calcium-depositing rhodophytes have been preserved in the fossil record as far back as 500 million years.

The cell walls of red algae occasionally contain cellulose but more often contain another polysaccharide, whose attached sulfate groups give the rhodophytes their characteristic slippery feel. Two such polysaccharides—agar and carrageenan—are useful to humans. Agar helps gel instant desserts, jellies, and baked goods, and is also the most widely used culture medium for growing bacteria in the laboratory. Carrageenan is used to stabilize oil and water emulsions in foods such as mayonnaise, as well as in paints and cosmetics.

All rhodophytes reproduce sexually and have complex life cycles. Meiosis can take place right after fertilization, with no sporophyte stage. Alternatively, the zygote can develop into a diploid organism, which later produces haploid spores that grow into male and female gametophytes. The female gametophyte produces a large ovum, while the male produces a much smaller sperm.

◄ **Figure 21-8   Kelp forests are home to a variety of other organisms.** In shallow coastal waters, the giant kelp *Macrocystus* may grow at enormous rates, producing masses of brown algae that provide both food and shelter for fish, sea urchins, sea otters, and other organisms. Although the giant kelps are simpler than true plants, they have a few specialized structures. Floating photosynthetic blades capture light for photosynthesis. The blades are kept near the surface by gas-filled bladders and anchored to the sea floor by holdfasts. (*Flip Nicklin/Minden Pictures*)

*Rhodophytes (red algae) have distinctive pigments
and complex sexual cycles.*

Chlorophytes (Green Algae)

Chlorophytes [Greek, *chloros* = green], or green algae, are
more like plants than any other group of protists. In fact, plants
probably arose from a chlorophyte ancestor. Chlorophytes are
unlike most of the other photosynthetic protists in having both
chlorophylls *a* and *b* and in making starch, which they store
within their chloroplasts, as do plants. Among the rest of the
protists, only the euglenophytes have both chlorophylls *a* and
b, and only the dinoflagellates make starch. Although modern
chlorophytes include several multicellular groups, biologists do
not believe that any of them was the ancestor of the multicellular plants (Figure 21-9).

Chlorophytes come in a great array of colonial and multicellular forms. They range in size and complexity from the
single-celled *Chlorella,* about 25 μm in diameter, to the sea-
weedlike *Codium magnum,* which may be 8 m long. Chlorophytes are mostly free living and aquatic, but some species live
on the surface of snow, on tree branches, in the soil, or in close
association with other organisms, including fungi, hydras, and
other protists (Figure 21-10).

The simplest chlorophyte is *Chlorella,* which lacks flagella
and is smaller than *Chlamydomonas. Chlorella* is among the
most widespread of the green algae, living in fresh water, salt

water, and soil. *Chlorella*-like species live within protozoa as
well as within many other invertebrate animals, including
sponges, hydras, and flatworms. A close association between
two organisms is called **symbiosis** [Greek, *sumbios* = living to-
gether]. Organisms living in a symbiotic relationship with one
another are called *symbionts. Chlorella* provide their partners
with energy-rich compounds, produced by photosynthesis.

*Chlorophytes (green algae) are more like plants than
other protists and also the most diverse of the
plantlike protists.*

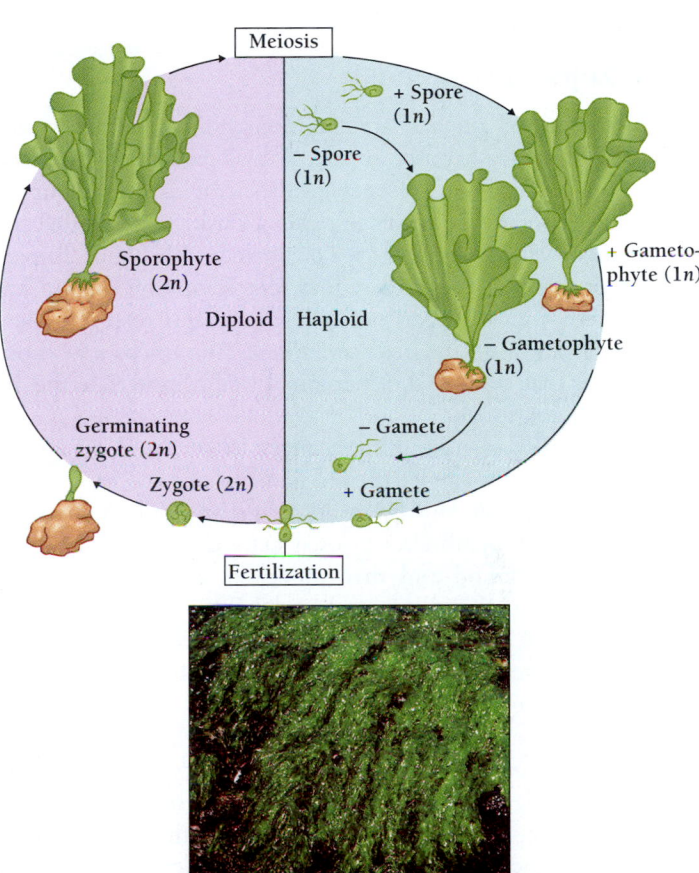

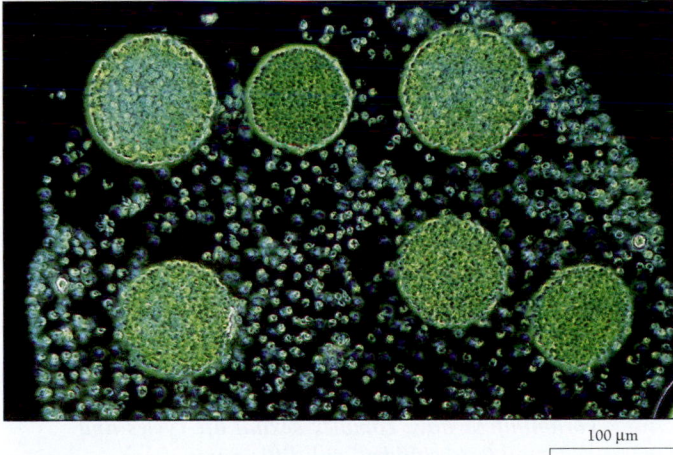

**Figure 21-9 Colonial green alga—blobs of identical cells or
multicellular individuals?** Members of the colonial genus *Volvox*
may contain 500 to 60,000 cells arranged in a single-layered, hollow sphere. The cells in most colonies of green algae are identical.
In *Volvox* and some other genera, however, there is some division
of labor, suggesting a rudimentary form of multicellularity. For example, only some cells are able to reproduce. In addition, the cells
may move smoothly through the water by coordinating the beating
of their flagella, much as our heart cells coordinate their contractions. (*Alfred Owczarzak/Biological Photo Service*)

Figure 21-10 The life cycle of a sea lettuce. The marine Ul-
vophyceae take their name from *Ulva,* or sea lettuce, which grows
in shallow seas throughout the temperate zones. Each ulva has a
flat, leaflike thallus, which may be a meter or more in length, and
a rootlike holdfast, which attaches the thallus to a rock or other
object. (*Photo, Brian Parker/Tom Stack & Associates*)

PROTOZOA ARE HETEROTROPHIC (ANIMAL-LIKE) PROTISTS

Even though we no longer classify protists as plants or animals, we still use the old term **protozoa** [Greek, *protos* = first + *zoe* = life] to describe the protists that most resemble little animals. The protozoa include a wide variety of heterotrophs, most of which obtain their food by phagocytosis, ingesting particles or other cells.

Most protozoa are single celled and motile. They are extraordinarily varied and include eight phyla, of which we will survey just five: rhizopoda, foraminifera, sporozoa, ciliophora, and zoomastigina. Three of these—the rhizopoda, the ciliophora, and the zoomastigina—we distinguish by the way they move. Each of the other two has characteristic traits. We identify the sporozoa by their characteristic life cycles and the foraminifera by their distinctive shells.

Rhizopoda (Amebas)

Rhizopoda, or amebas, are common throughout the world, in oceans, in fresh water, in soil, and as parasites of animals. Hundreds of millions of people in the tropics (and about 10 million in the United States) harbor parasitic amebas that infect the lower intestinal tract. Fortunately, only about 20 percent of such infections produce the symptoms of amebic dysentery, which range from mild discomfort to severe diarrhea and death. In the most severe cases, the ameba may cause abscesses that rupture the abdominal wall or penetrate the diaphragm and lungs.

An ameba continuously changes its shape, in an apparently chaotic way (hence the old name of one species—*Chaos chaos*). Amebas do this by means of pseudopodia, temporary extensions of membrane-enclosed cytoplasm. An ameba extends a pseudopod and then moves its cytoplasm into the pseudopod, so that the pseudopod becomes the body. The movement of amebas only appears chaotic, for they actually move in a directed manner toward sources of food. They then envelop the food in their pseudopodia.

All amebas reproduce through simple cell division. Some, such as the ones that cause amebic dysentery, can form cysts immune to desiccation or digestion by their hosts. A few construct distinctive shells (tests) by gluing together tiny grains of sand and other inorganic bits. Similar tests exist in the fossil record, with some dating back to the Precambrian, more than half a billion years ago.

Rhizopoda (amebas) use pseudopodia to move and ingest food.

Foraminifera

The **foraminifera** [Latin, *foramen* = little hole + *ferre* = to bear], or **forams,** are marine organisms whose tests are full of tiny holes (Figure 21-11). The tests of forams are made of or-

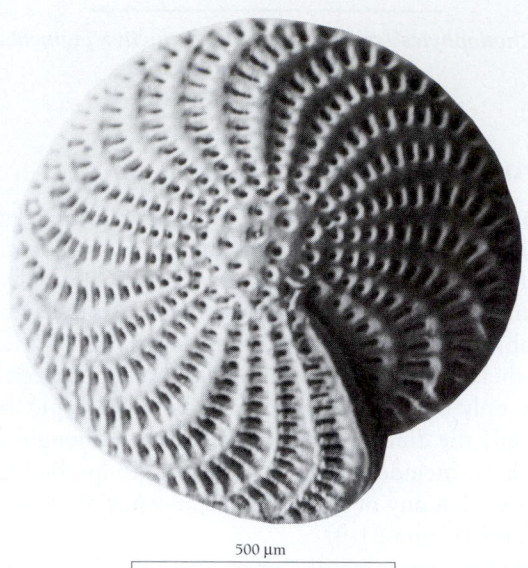

500 μm

Figure 21-11   The tests of foraminifera are rich in calcium. As these organisms die and drop to the sea floor, their tests form thick, white deposits. (*Biophoto Associates*)

ganic materials reinforced with grains of sand or minerals and range in size from about 20 μm to several centimeters. Most are tiny and live in the sand or attached to other organisms or stones. Two large groups are free-floating plankton and serve as food for clams, mussels, and other invertebrates that filter food from water.

Forams have complex sexual life cycles. Unlike most other protozoans, they have true alternation of generations, with extended haploid and diploid phases—just like plants. Some forams harbor within them photosynthetic protists—dinoflagellates, chrysophytes, and diatoms. Although forams are unicellular, they are subdivided into multichambered shells connected by cytoplasm.

The tests of forams are abundant in the fossil record, especially in the last 230 million years. Forams make up much of the chalky sediment on the bottoms of the seas. The famous white cliffs of Dover, England, are just such foramen sediments, uplifted by geologic processes.

Foraminifera have complex sexual life cycles and tests studded with little pores.

Sporozoans

Sporozoans are all parasites. They reproduce by a combination of asexual and sexual processes, including the alternation of generations. Sporozoans have especially complex life cycles, usually involving two or more different hosts. Sporozoans are a scourge to their many animal hosts but a delight for puzzle-solving parasitologists. The members of one genus, *Plasmod-*

ium, cause malaria, a disease that every year infects some 300 to 500 million people and kills 2 to 3 million, mostly in Africa. *Plasmodium* spreads among humans with the help of its other set of hosts—mosquitoes of the genus *Anopheles* (Figure 21-12). The complex life cycles of sporozoans ensure continued success: at any moment, different phases of each species are present in several forms in several places in two or more hosts. They are virtually impossible to eradicate.

Because of the importance of malaria to worldwide human health, researchers have identified *Plasmodium falciparum* as a priority for complete DNA sequencing. A complete DNA sequence may help researchers to understand *Plasmodium*'s metabolism and life cycle more completely, with the hope of de-

signing vaccines and specifically targeted therapies. The sequence of one of the 14 chromosomes, completed in 1998, revealed that *Plasmodium* diverged from other eukaryotes unexpectedly early in evolution. Another surprise was the presence of organelle DNA that, researchers conclude, could only have come from an ancient plastid. As in the case of the euglenophytes, it is almost impossible to make a clear distinction between "plantlike" and "animal-like" protists.

> ***Sporozoans are nonmotile, spore-forming protozoa, which live as parasites in animals. They are known for their complex life cycles.***

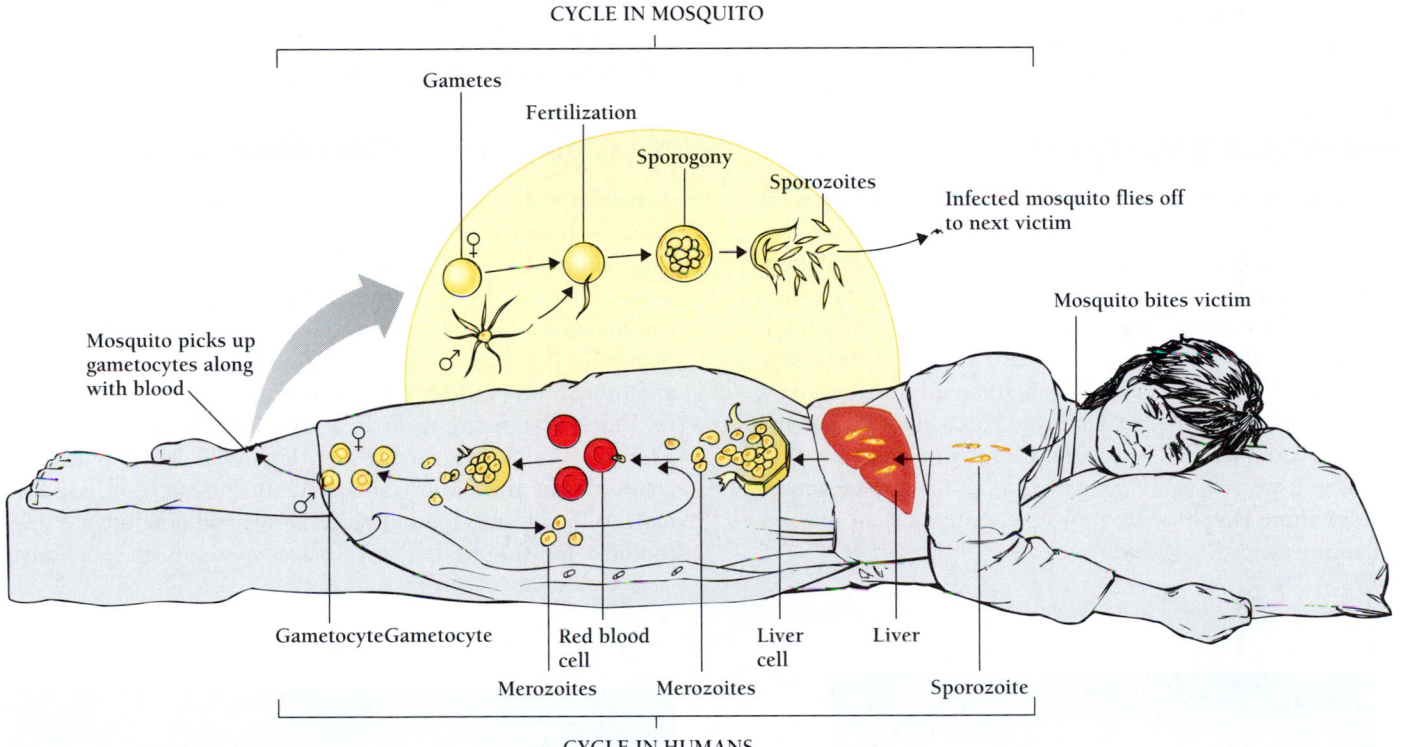

Figure 21-12 How *Plasmodium* causes malaria. When a mosquito sucks blood from a human, it first injects a bit of saliva. The saliva contains an anticoagulant, a chemical that keeps the blood flowing freely and allows the mosquito to drink its fill. A mosquito infected with *Plasmodium* harbors a diploid form of the parasite—called a *sporozoite*—in the wall of its gut and in its salivary glands. When the mosquito injects its saliva, some of these sporozoites enter the human host's blood. In the human bloodstream, the sporozoites travel to the liver, where they divide several times by mitosis and develop into a new diploid phase, called a *merozoite*. The merozoites take up residence inside of red blood cells and divide rapidly. Eventually, the red cells rupture and release merozoites and toxins, which cause the awful fever and chills of malaria. The merozoites infect more red cells or develop into *gametocytes*. The gametocytes, if sucked up by a second hungry mosquito, produce gametes, which form diploid zygotes, which embed themselves in the mosquito's intestinal wall. These develop into sturdy *oocysts*, which undergo mitosis to form thousands of sporozoites, some of which migrate to the salivary gland, where they can begin the cycle again. The simultaneous presence of *Plasmodium* in the gut and salivary glands of mosquitoes and in the blood and liver of humans has made it impossible to eradicate malaria. The draining of swamps and the use of insecticides such as DDT have destroyed breeding grounds for the mosquitoes and reduced the incidence of malaria in some areas. Yet, worldwide, the number of new cases of malaria has more than doubled since the mid-1970s, for the mosquito has evolved resistance to insecticides such as DDT, and *Plasmodium* has evolved resistance to antimalarial drugs.

Ciliophora (Ciliates)

Ciliophora, or the **ciliates,** include about 8000 species of free-living, single-celled heterotrophs that live both in fresh water and in salt water. Their name [Latin, *cilium* = eyelash] reflects the presence, on their surface, of thousands of cilia, shorter and more numerous than eukaryotic flagella. The cilia, which may be arranged in bundles or in sheets, help ciliates move and feed (Figure 21-13). The best-known ciliate is probably *Paramecium,* a common resident of pond water, long used in teaching and research. Most ciliates lack shells, although a few marine ciliates form shells of sand held together by organic cements. The fossil record shows that such shell-forming ciliates have existed for about 100 million years.

Paramecia and other ciliophora (ciliates) are single-celled, free-living heterotrophs coated with cilia.

Zoomastigina (Flagellates)

The **zoomastigina** [Greek, *mastix* = whip], or **flagellates,** are a diverse group of single-celled heterotrophs. All zoomastigotes have at least one long flagellum, and some have thousands. Some are free living in fresh or salt water, while others are parasites in animals. Reproduction of zoomastigotes is mostly asexual, though some species have a sexual cycle as well. Some convert from a flagellated to an ameboid form and back again, depending on the availability of food. DNA sequence studies suggest that the flagellates are far more diverse than previously suspected, as we discussed earlier for trypanosomes, which are more closely related to euglenophytes than to other flagellates.

Zoomastigina (flagellates) are single celled and use flagella to move.

FUNGUSLIKE PROTISTS

We have seen that some protists resemble plants and some resemble animals. The last group of protist phyla most resemble fungi. The names of these phyla reflect the fact that they were initially identified as fungi: all include the Greek root *mukes* [= fungus]. We will describe three of the seven funguslike protist phyla: acrasiomycota, the cellular slime molds; myxomycota, the plasmodial slime molds; and oomycota, which include the water molds, mildews, and potato blight. While these organisms resemble fungi in their general appearance, they more closely resemble other protists in their cellular organization and modes of reproduction.

Acrasiomycota (Cellular Slime Molds)

Acrasiomycota [Greek, *acrasia* = confusing two things + *mukes* = fungus], or cellular slime molds, are peculiar organisms, animal-like at some stages of their lives and plantlike at others. They live in damp soil, in fresh water, on decaying vegetation, and especially on rotting logs. When nutrients are plentiful, they flourish as amebalike cells, consuming bacteria and other food particles by phagocytosis. When food is scarce, the amebas aggregate to form a multicellular stage that looks a bit like a miniature garden slug (Figure 21-14). The slug secretes a slime track and wanders about in search of food and light. When it finds food, it settles down and develops a mushroomlike **fruiting body** that produces spores. Each spore turns

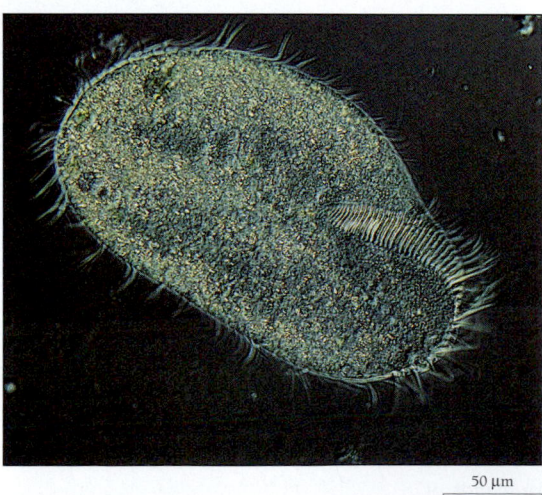

50 μm

Figure 21-13   **The cilia of a ciliate.** The cilia surround the oral groove and form oarlike arrangements for movement. *(M. Abbey/Photo Researchers, Inc.)*

500 μm

Figure 21-14   **The slug of a cellular slime mold.** *(Cabisco/Visuals Unlimited)*

into an ameba. Reproduction is asexual; both the amebas and the slug are haploid.

Acrasiomycota (cellular slime molds) live as amebas and as multicellular slugs.

Myxomycota (Plasmodial Slime Molds)

Myxomycota [Greek, *muxa* = mucus + *mukes* = fungus]—the plasmodial, or acellular, slime molds—closely resemble the cellular slime molds in appearance and in life cycle. The migrating form of a plasmodial slime mold, however, differs greatly from that of a cellular slime mold. Instead of developing from an aggregation of individual hungry amebas, a plasmodial slime mold develops a **plasmodium**—a mass of cytoplasm with many nuclei but no boundaries between cells.

The plasmodial slime molds also differ from the cellular slime molds in having a sexual cycle. The amebas are haploid, and two of them can form a diploid zygote. The nucleus of the zygote then divides repeatedly, but the cell merely grows larger and larger without dividing. The result is a plasmodium with many diploid nuclei. Unlike the cellular slime mold's slug, the plasmodium can both feed and move about. It propels itself by differential growth, growing more in one direction than another. When food or water becomes scarce, the slug forms a fruiting body, undergoes meiosis, and releases haploid spores.

Myxomycota (plasmodial slime molds) live as amebas or as a multinucleated plasmodium.

Oomycota

The **oomycota** most nearly resemble true fungi. They are either parasitic, deriving nourishment from living organisms, or *saprophytic,* deriving nourishment from dead organisms. Like true fungi, the oomycota extend threads into the tissues of their hosts, release digestive enzymes, and absorb the resulting organic molecules. The oomycota differ from the fungi, however, in having flagella on their motile spores. They reproduce sexually, and their life cycles include both haploid and diploid phases. The haploid gametes differ greatly in size; the large ovum is the source of this phylum's name, oomycetes [Greek, *oion* = egg + *mukes* = fungus], meaning "egg fungi."

Among the most common oomycetes are the water molds, which grow in fuzzy masses on the remains of algae or animals. In freshwater ecosystems, water molds decompose and recycle nutrients from dead organisms.

Some oomycotes cause diseases in animals or plants. Historically, the most important of these is *Phytophthora infestans,* which causes blight in potatoes. In 1845 and 1849, this oomycote infected and destroyed the Irish potato crop. A million Irish died of hunger, and another 1.5 million were forced to emigrate to the United States. A similar blight on the German potato crop late in World War I helped force an end to the war.

Oomycota resemble fungi, but their spores have flagella.

What Can We Learn from DNA About the Diversity and Evolution of Protists?

Studies of protist DNA are still in their infancy, and no complete protist genomes have been sequenced. Limited information, from the analysis of a few genes, has nonetheless changed biologists' thinking about protists in two fundamental ways: (1) protists are more diverse than previously thought and (2) endosymbiosis has been a major factor in protist evolution.

Comparisons of the sequences of ribosomal RNAs (and a few other genes) shows that different kinds of protists are much more different from one another than are plants, animals, and fungi. As mentioned earlier, many biologists favor the reclassification of protists into at least 12 kingdoms.

While researchers now think that all mitochondria—in protists and other eukaryotes—had a common origin, protist mitochondria had distinct histories. Most surprisingly, even protists that lack mitochondria have DNA in their nuclei that could only have come from ancient mitochondria. For example, researchers recognize that *Giardia* (a diarrhea-causing protist now common in streams and lakes) has no mitochondria, but it does have genes that, in other eukaryotes, lie on mitochondrial DNA. Researchers have concluded that ancestors of *Giardia* did have mitochondria but that some of the mitochondrial DNA moved to the nucleus before the mitochondria were lost for good.

The other big surprise about organelle DNA was the discovery of genes, for example, in trypanosomes that could only have come from plastids, such as chloroplasts. This result suggests that a trypanosome ancestor was once photosynthetic, and that, during evolution, there was no clear segregation of ("plantlike") algae and ("animal-like") protozoans.

HOW DO TRUE FUNGI DIFFER FROM OTHER ORGANISMS?

We love fungi. We put fungi in our salads and fungi in our sauces. We use fungi to ferment wines and cheeses and to leaven bread. What is more, we need fungi. A walk through a wet wood demonstrates the enormous ecological importance of fungi: they are prodigious decomposers, recycling the nutrients of dead plants and animals. Without fungi, dead trees would fall and accumulate, and the soil, a little bit of which washes away in every storm, would never be replaced.

But fungi also plague us in countless ways. *Cryptococcus neoformans,* for example, is one of many unpleasant fungi that torment humans and other animals. Distributed worldwide, it infects first the lungs, then the membranes surrounding the brain, and sometimes the kidneys, bones, and skin. The first sign of infection is a cough. Usually, the symptoms that bring *C. neoformans's* victims to a doctor, however, are headache, blurred vision, confusion, depression, or inappropriate speech or dress. Another infectious fungus attacks the sinuses, then the brain, producing intolerable headaches and ultimately consuming its victim alive.

Still other fungi destroy entire harvests of wheat, potatoes, and other crops. Fungi destroy in small ways as well as large. In the kitchen, fungi ruin bread, potatoes, jams, and sauces. In the bathroom, they attack our feet, producing athlete's foot, and grow on the walls of the shower and the porcelain surfaces of the toilet.

Fungi grow wherever there is sufficient warmth, moisture, and energy-rich organic molecules. Sometimes fungi find nourishment in surprising and bothersome places—the lining of a camera lens, clothing, shoes, or the wooden planks of a ship. During the American Revolution, the British lost more ships to dry rot than to the fledgling American Navy. In World War II, fungal infections of the skin took more men out of action in the South Pacific than did battle wounds.

Fungi are a little like plants and a little like animals. As long as biologists classified all organisms as either plants or animals, fungi posed a problem. Like animals, fungi are heterotrophic—they derive their energy from other organisms. But fungi enclose their cells in rigid walls as plants do. Fungi are distinct from both plants and animals, however: they take in food differently from animals, their cell walls are chemically different from those of plants, and they lack the embryonic stages of plants and animals. Everyone now agrees that the fungi deserve their own kingdom.

How Can We Characterize the Fungi?

With the exception of the single-celled yeasts and a few others, most fungi are **molds** composed of masses of threadlike filaments called **hyphae** [Greek, *hyphe* = web; singular, hypha]. The hyphae contain many nuclei (Figure 21-15). In some fungi, the nuclei all lie in a common cytoplasm, rather than in separate membrane-enclosed cells. Such fungi are said to be **coenocytic** [Greek, *koinos* = shared + *kytos* = hollow vessel]. In **dikaryotic** fungi (which have two nuclei per cell), walls called **septa** separate the nuclei within a filament. The septa, however, are perforated by large pores that allow the cytoplasm to flow along the hyphae. Hyphae grow, branch, and intertwine to form a mass, called a **mycelium** [Greek, *myketos* = fungus]. Growth may be so rapid that in the time it takes you to read this page a single mycelium can add meters of hyphae.

Fungi's ability to grow rapidly helps compensate for another odd characteristic of their cells—lack of mobility. All fungal cells, even the gametes, lack flagella and cilia. Fungi can

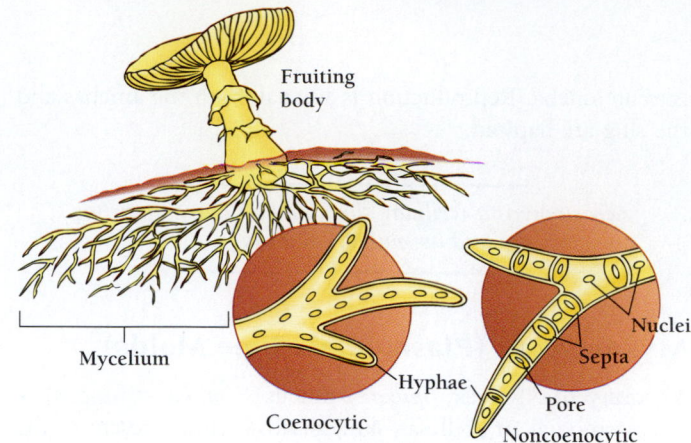

Figure 21-15   Two types of fungal hyphae. The nuclei within hyphae may all share the same cytoplasm (coenocytic), or they may be separated by septa.

move in two ways only. The mycelium can move from place to place by growing, or the mycelium can produce spores that the wind scatters.

Another distinguishing characteristic of the fungi is their cell walls, which are made of a tough polysaccharide called **chitin,** the same substance in the hard shells of insects, spiders, and crustaceans. In plants, the major component of the cell wall is cellulose.

Recall that all fungi are heterotrophs, organisms whose food is organic molecules produced by other organisms. A fungus's food may be either dead or alive. Saprophytic fungi feed on the remains of dead organisms, while parasitic fungi feed on living organisms. Fungi obtain food, whether dead or alive, by infiltrating the bodies of organisms and other food sources with long, thin hyphae, usually just a few micrometers in diameter. The fungi extend their hyphae into the food, secrete digestive enzymes that break down the food, and absorb, rather than ingest, the resulting nutrients.

Fungi may reproduce asexually or sexually. Each cell (or nucleus) within a mycelium is usually capable of generating a new mycelium. Most fungi reproduce vegetatively (asexually). A single mycelium can produce separate individuals by means of simple cell division, by the breaking up of existing hyphae, or by budding. Fungi also reproduce by making and dispersing hardy spores capable of resisting heat, cold, drought, and lack of food. Spores are so small that the wind can scatter them all over the planet.

Spores, which are always haploid, may be made in a variety of intricate structures, called **sporophores.** Ordinary mushrooms are the sporophores of a large mycelium, which may be many meters in diameter (Figure 21-16). In general, asexual spores form when food is plentiful and the environment is otherwise welcoming. Fungi reproduce sexually when food is scarce or the environment has become unfavorable. Recall that sexual reproduction creates a greater diversity of genotypes in the next generation. This diversity increases the chances that a few individuals will survive a hostile environment. Each hap-

Figure 21-16 **A fairy ring.** An underground mycelium grows outward over hundreds of years. When the mycelium "fruits," it forms mushrooms at its outer edge, which release spores. The grass within the circle of mushrooms is often stunted and pale. For this reason, people once believed that dancing fairies caused these rings. (*G. Carleton Ray/Photo Researchers, Inc.*)

loid spore can germinate and produce a new haploid mycelium. Fungi differ from both plants and animals in that they do not go through any embryonic stages. Spores form mycelia directly.

> *Fungi are filamentous, lack flagella, have distinctive cell walls, absorb food, and reproduce by means of spores.*

How Do Mycologists Classify Fungi?

Many mycologists still classify the funguslike protists as fungi. In this book, however, we consider the funguslike protists to be distinct from fungi (Figure 21-17). Fungi lack motility, and

their cell walls differ from those of protists. Fungi also lack embryos and acquire nutrients differently from plants and animals. We will discuss examples of four phyla of fungi: zygomycetes, ascomycetes, basidiomycetes, and deuteromycetes.

Mycologists group fungi into these four phyla according to whether the hyphae have septa and how the fungi produce sexual spores. Most fungi are capable of reproducing both sexually and asexually. Asexual reproduction may occur vegetatively (as when parts of the mycelium divide to form new individuals) or through the production of spores by mitosis. Sexual reproduction always occurs by means of spores. The manner in which fungi produce their sexual spores is one of the two main characteristics that mycologists use to classify fungi.

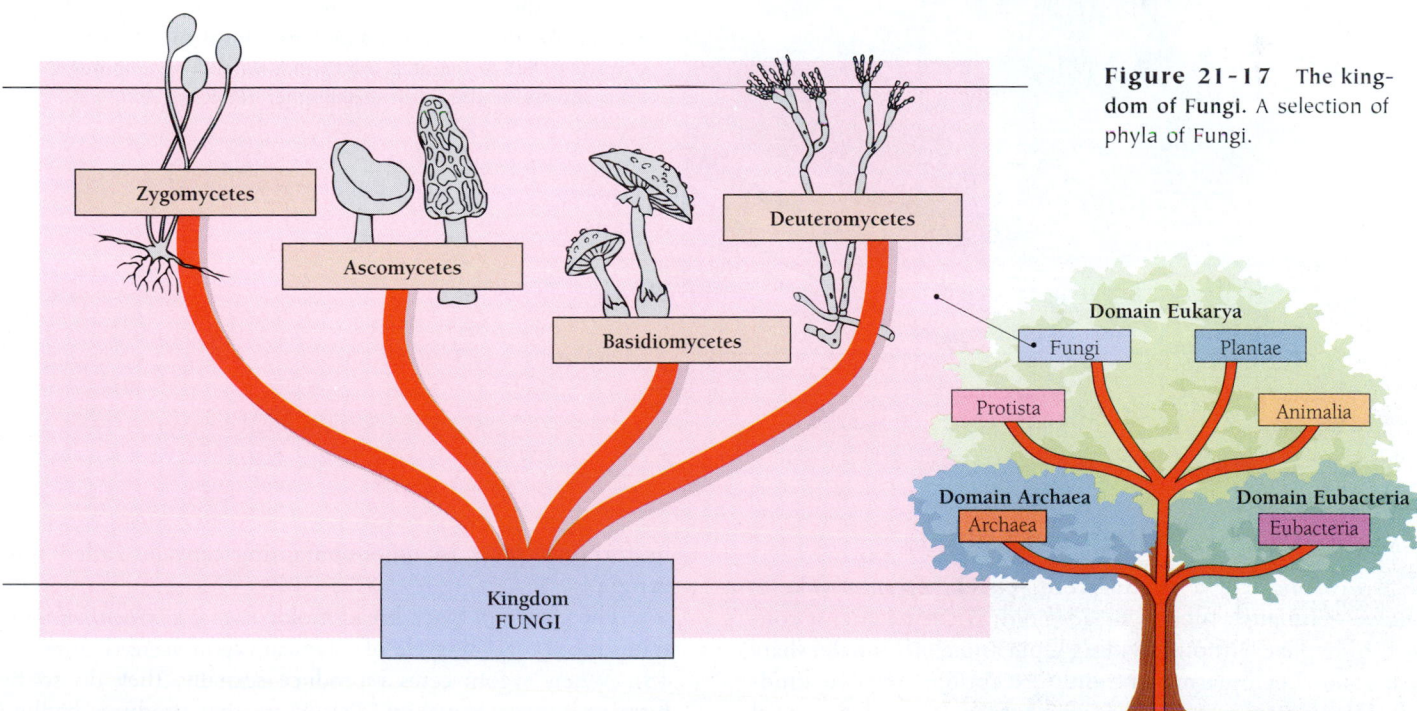

Figure 21-17 **The kingdom of Fungi.** A selection of phyla of Fungi.

BOX 21.1

Fungal Infections Today

In recent years, fungal infections have become increasingly common. Between 1991 and 1994, for example, fungal infections in one county in southern California increased from 400 per year to 4500 per year. In fact, today nearly 40 percent of all deaths from hospital-acquired infections are due not to bacteria or viruses but to a handful of fungi.

Most hospital infections are due to *Candida albicans,* a virulent relative of baker's yeast. *Candida* infects 70 percent of AIDS patients. Even if science could eliminate *Candida,* however, other fungi would quickly take its place. *Cryptococcus, Histoplasma,* and *Aspergillus* are three of the next most common fungi that infect patients with impaired immune systems. Even ordinary baker's yeast, a normal inhabitant of the human mouth and the stuff we use to make bread rise, can grow out of control in individuals who are already very sick.

Healthy people also fall victim to some seemingly ordinary fungi. In California's San Joaquin Valley, for example, many people become infected by the soil fungus *Coccidioides,* which causes symptoms ranging from mild flulike illnesses to attacks in which the fungus spreads throughout the body (Figure A).

Against this onslaught of rot, health care workers have only a few broadly effective drugs. Nystatin and two other drugs derived from it are still the most effective way to treat fungal infections. One of the most powerful antifungals is amphotericin B, a drug with side effects so nasty that patients sometimes call it "amphoterrible." It can cause fever, chills, low blood pressure, headaches, vomiting, inflammation of blood vessels, and kidney damage.

Safe antifungal drugs with few side effects are much harder to find than safe antibiotics. This is because antibiotics, which kill prokaryotes but not our own eukaryotic cells, work by exploiting any of a variety of differences in cell chemistry between prokaryotes and eukaryotes. But fungi are eukaryotes, like us. Their cells work like our own cells, and chemicals that are bad for fungi are usually bad for us, too.

Nonetheless, the chemistry of fungal cells does differ from that of our own, and any differences can be exploited by researchers who design drugs. For example, the three drugs most commonly used all work by attacking ergosterol, a lipidlike solid that helps form the cell membranes of fungi. Our own cell membranes are made of cholesterol. Fungal cells also differ from ours in having a cell wall. Researchers' latest hope is to find a drug that attacks the cell walls of fungal cells without harming our own cells.

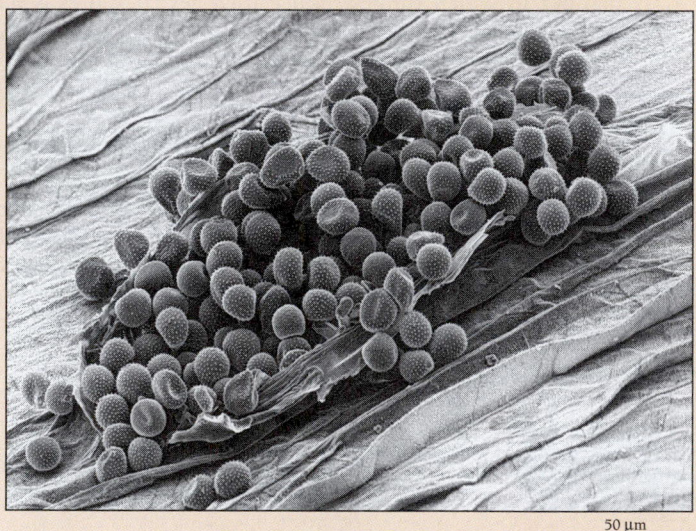

50 μm

Figure A   Fungal spores are everywhere. The tiny haploid particles by which mushrooms and other fungi reproduce float in and out of our nostrils with each breath, land on bread and jams in our kitchens, and grow into dark streaks in our bathrooms. Outside our homes, a rain of fungal spores dusts trees and shrubs, lawns, dead leaves, and soil. Creeks, rivers, and lakes disperse fungal spores from one watershed to the next. And winds carry fungal spores from one end of the Earth to the other. *(Biophoto Associates)*

Zygomycetes (Conjugation Fungi)

The mycelium of a **zygomycete** [Greek, *zygon* = yoke + *mukes* = fungus], such as bread mold, is coenocytic: it consists of hyphae without dividing septa and all the nuclei share a common cytoplasm. The only exceptions are two kinds of spore-forming structures, which form asexual or sexual spores. Most of the time, zygomycetes reproduce asexually by forming spores in spore-containing capsules called **sporangia.**

The common black bread mold, *Rhizopus stolonifer,* illustrates the sexual life cycle of a typical zygomycete (Figure 21-18). When zygomycetes reproduce sexually, they do so by forming a *zygosporangium,* a structure that produces haploid spores from a diploid zygote.

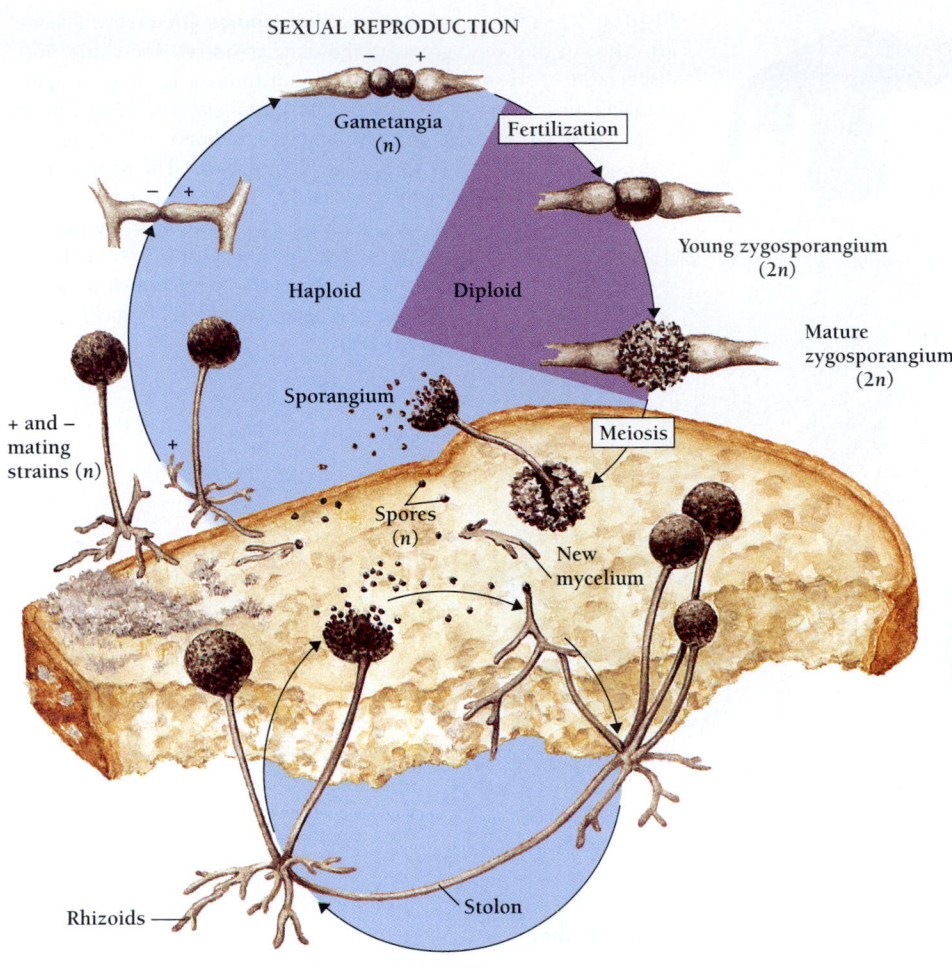

SEXUAL REPRODUCTION

Figure 21-18 Life cycle of black bread mold *Rhizopus stolonifer.*

Gametangia
(n)

Fertilization

Haploid   Diploid

Young zygosporangium
(2n)

Mature
zygosporangium
(2n)

Sporangium

Meiosis

+ and –
mating
strains (n)

Spores
(n)

New
mycelium

Rhizoids   Stolon

ASEXUAL REPRODUCTION

Whether zygomycetes reproduce sexually or asexually, they always use spores. Both kinds of spores travel on air currents and can, in the right environment, divide to produce a new mycelium. Because spores drift randomly, their arrival on a hospitable host (a piece of bread, for example) is entirely a matter of chance. A few zygomycetes, however, have evolved a means of actively launching themselves into likely new environments (Figure 21-19).

Bread molds and other zygomycetes are coenocytic and usually reproduce asexually by means of spores.

Ascomycetes (Sac Fungi)

Ascomycetes are a large and diverse group, with some 30,000 species (Figure 21-20). They include most yeasts, the powdery mildews, the molds, the cup fungi, the bread mold *Neurospora* (so useful to geneticists), and the truffles, prized by chefs for their earthy flavor.

Ascomycetes [Greek, *askos* = wineskin + *myketos* = fungus], or sac fungi, are so named because their sexual spores are always produced in little sacs, called **asci** (singular, ascus). They have no sporangia for making asexual spores. Their hyphae usually have septa separating the nuclei. These septa, however, are perforated. As a result, molecules and even organelles can circulate throughout the cytoplasm.

The truffles, yeasts, mildews, and other ascomycetes have septa dividing the nuclei of their hyphae and no sporangia. They produce sexual spores in little sacs called asci.

Basidiomycetes (Club Fungi)

The 25,000 species of **basidiomycetes,** or club fungi, include many of the most familiar fungi: the mushrooms, the bracket fungi, and the puffballs (Figure 21-21). These fungi take their name from the tiny, club-shaped **basidium** [Latin, = little pedestal; plural, basidia] that they use to produce spores. The basidia develop on the underside of a mushroom's cap, on the sides of the thin gills that radiate from the center like the pages of an open book (Figure 21-22). The mushroom itself, from

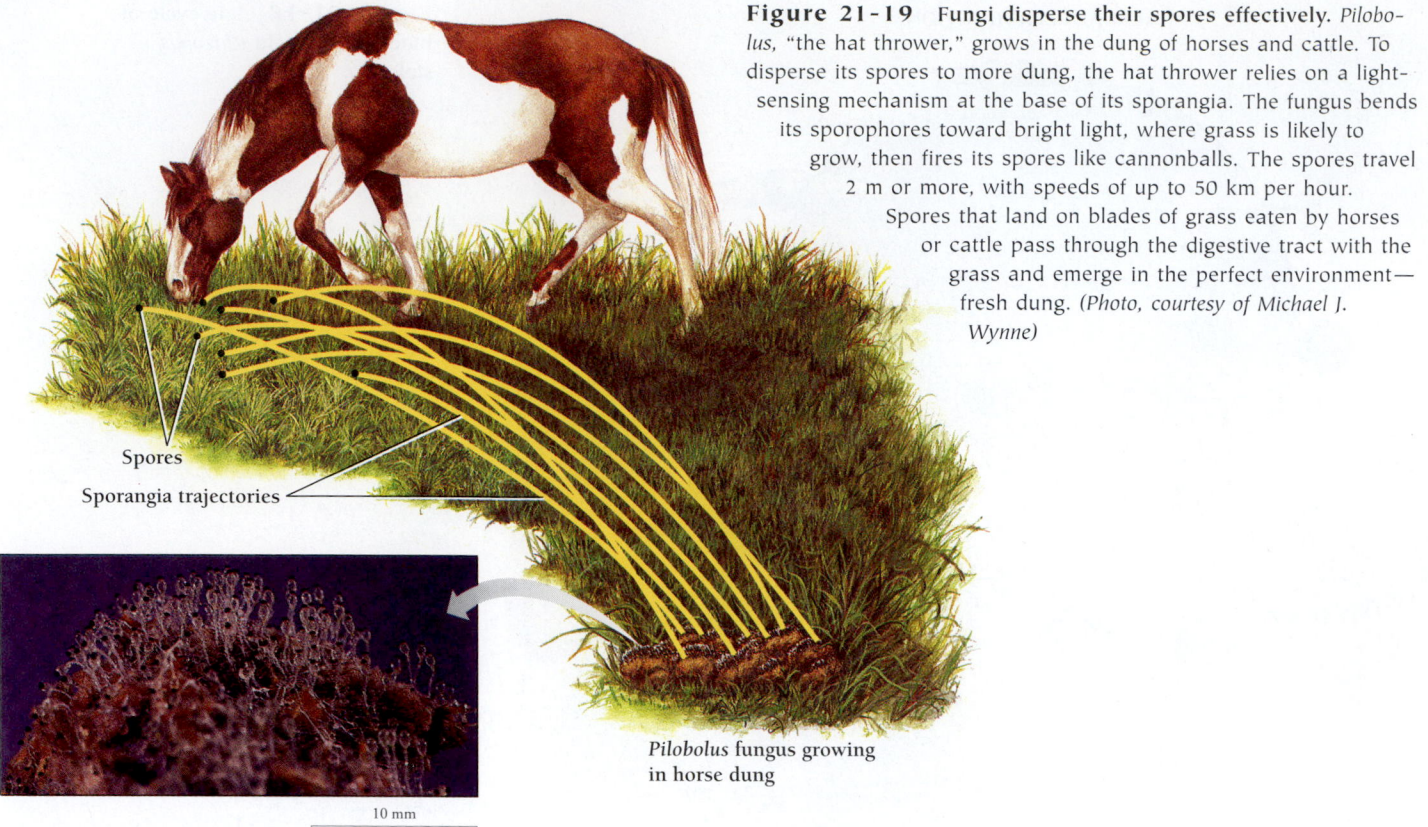

Spores

Sporangia trajectories

10 mm

Pilobolus fungus growing
in horse dung

Figure 21-19 Fungi disperse their spores effectively. *Pilobolus*, "the hat thrower," grows in the dung of horses and cattle. To disperse its spores to more dung, the hat thrower relies on a light-sensing mechanism at the base of its sporangia. The fungus bends its sporophores toward bright light, where grass is likely to grow, then fires its spores like cannonballs. The spores travel 2 m or more, with speeds of up to 50 km per hour. Spores that land on blades of grass eaten by horses or cattle pass through the digestive tract with the grass and emerge in the perfect environment— fresh dung. (*Photo, courtesy of Michael J. Wynne*)

which the basidia develop, is the equivalent of a fruit and biologists call it a **basidiocarp** [Latin, *carpus* = fruit].

In the common field mushroom, the mycelium spreads underground and may be up to 30 m in diameter (Figure 21-16). The basidiocarps (mushrooms) develop at the periphery of the mycelium in "fairy rings"—often overnight. Folk legends once held that mushroom fairy rings appeared at the bidding of fairies who were tired of dancing. In reality, mushrooms are able to grow so quickly because a huge underground mycelium can quickly mobilize vast amounts of material. As the secondary mycelium exhausts the nutrients in the center of the ring, however, the ring grows outward in search of more nutrients. Because the mycelium typically

grows at about 30 cm per year, we can estimate that a fairy ring 30 m in diameter is a century old.

Another group of basidiomycetes, less charming than mushrooms but at least as important, are the rusts and smuts. Rusts and smuts ruin billions of dollars worth of wheat and other grains in the United States and elsewhere every year. As distinct from other basidiomycetes, however, rusts and smuts do not form basidiocarps. Instead, they are parasites with complex life cycles, usually involving hosts of two species.

Mushrooms, rusts, and other basidiomycetes are named for the club-shaped basidia that produce spores. They have septa and distinctive life cycles.

Deuteromycetes (Imperfect Fungi)

Mycologists classify fungi as zygomycetes, ascomycetes, or basidiomycetes, largely according to their mode of sexual spore formation. For some 25,000 named species of fungi, however, mycologists have so far discovered no sexual cycle at all. These are said to be "imperfect." Mycologists disagree, however, about whether the imperfect fungi form a separate phylum (or division).

◀ **Figure 21-20** The familiar cup fungi are ascomycetes. Bird's nest fungi. (*Michael Fogden/DRK Photo*)

Ergotism

After *Penicillium* mold, the most famous fungus may be the ascomycete *Claviceps purpurea,* or ergot, which infects the flowers of rye and other grains. This peculiar sac fungus produces tiny spikes, called sclerotia, or ergots, in the grain. The small, black or brown ergots—compact masses of hardened mycelium—replace a few of the seeds that would normally form in the seed head (Figure A).

Each ergot contains a witch's brew of deadly substances that can poison humans and other animals that eat the infected grain. In extreme cases, the symptoms of this poisoning, called ergotism, consist of hallucinations and fatal convulsions. In milder or chronic poisoning, sufferers may have headaches, muscle pains, numbness, and cold fingers and toes—all signs of constricted blood vessels that can result in gangrene, a usually fatal infection when not treated with antibiotics.

During the Middle Ages, ergotism was common in northern Europe and other areas where rye bread was popular. Also called St. Anthony's fire, ergotism could fell thousands of people in a season. In the year 994, for example, 40,000 people died from eating ergot-infected grain. Modern grains are cleaned of the dark masses of mycelium, and today ergotism is rare.

Ergot in small doses is medically useful. The main ingredient is an alkaloid—ergotamine tartrate. Ergotamine stimulates the smooth muscle of the peripheral and cranial blood vessels and also inhibits the neurotransmitter serotonin. As a result, ergotamine causes small arteries and smooth muscle fibers to contract. It is used to induce labor, to control bleeding from the uterus following childbirth, to treat high blood pressure, and—in conjunction with caffeine—to relieve severe migraine headaches. Two other ergot alkaloids, bromocriptine and pergolide, are used in the treatment of Parkinson's disease. These two alkaloids help relieve the symptoms of Parkinson's disease by activating receptors for the neurotransmitter dopamine.

Ergot was also the original source of the psychoactive drug lysergic acid diethylamide (LSD). LSD induces mood changes (euphoria, anxiety, and depression), impaired judgment, and distorted perceptions of many sensations, including time, space, or self-image. Despite LSD's reputation as a powerful hallucinogen, true hallucinations are apparently rare.

Figure A Ergot. At left, the small black ergot (or sclerotium) in the seed head of a rye plant. At right, a normal, uninfected seed head. (*Dennis Drenner*)

Many imperfect fungi, for example, produce conidia (powdery asexual spores) and are probably ascomycetes. Mycologists therefore recognize the **deuteromycetes** as an artificial category. As the mode of reproduction of each species becomes known, mycologists are reclassifying them into one of the other three divisions of fungi.

The deuteromycetes [Greek, *deuteros* = second + *mukes* = fungus] include many species that are important to humans. Some are parasites on the skin of humans and other mammals, causing ringworm, athlete's foot, and other skin diseases. These fungi are able to displace the bacteria that are normally abundant on skin because they are better able to derive nourishment from skin with their powerful digestive enzymes that break down tough skin proteins. These also compete with bacteria by inhibiting the growth of the bacteria with antibiotics. Recall that the first antibiotic used by humans came from a strain of the common deuteromycete mold *Penicillium*.

Some deuteromycetes are predaceous. Using sticky substances on their surface, these fungi trap protozoa and small animals. One species even snares its prey with quickly tightening hoops (Figure 21-23).

Penicillium and other deuteromycetes always reproduce asexually, which makes them hard to classify.

Figure 21-21 Basidiomycetes.
Bracket fungi and puffballs have the
same sexual life cycle as supermarket
mushrooms. *(Left, J. Robert Stottle-
myer/Biological Photo Service; right,
Michael Fogden/DRK Photo)*

WHAT IMPORTANT ECOLOGICAL ROLES DO FUNGI PLAY?

Fungi have two important ecological roles, both of which depend on the power of the enzymes that fungi secrete. These extraordinarily potent enzymes can reduce substances as different as Wonder Bread and solid rock into small molecules that can be absorbed by fungi or other organisms. Using these enzymes, fungi decompose organic matter and make nutrients available to other organisms. Fungi, together with bacteria, recycle much of the world's organic matter from dead plants and animals, not to mention feces, leaves, and logs. Fungi also participate in close, mutually beneficial symbiotic associations with other organisms. We will discuss two kinds of important fungal associations, lichens and mycorrhizae.

What Role Do Fungi Play in Lichens?

Each of the 25,000 recognized species of lichens looks like a moss or simple plant growing on a rock or a tree trunk.

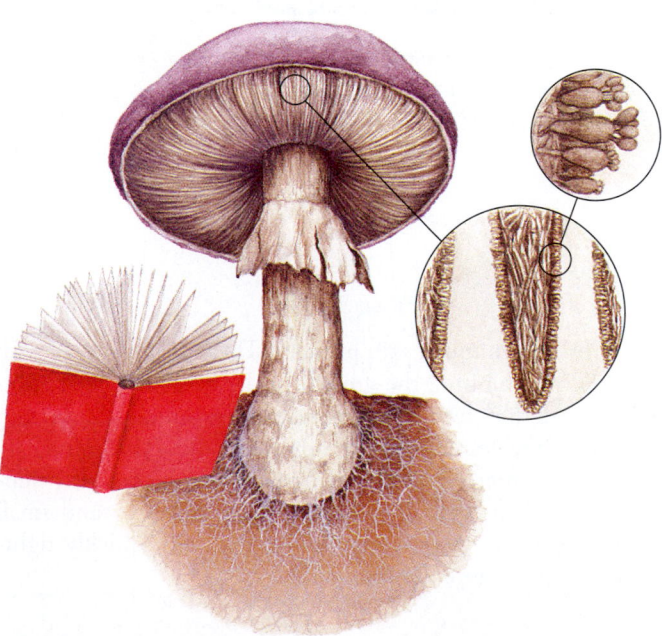

Figure 21-22 Like the pages of a book. In basidiomycetes, the spore-producing basidia develop on the paper-thin gills, which radiate from a central stalk.

25 µm

Figure 21-23 A deuteromycete fungus lassoes a passing nematode worm. *(Ken Barker)*

Lichens may be leafy, shrubby, or crusty. They come in a variety of colors, including black, white, red, orange, yellow, green, and brown. Lichens are not plants at all, however, but associations of fungi with photosynthetic partners (Figure 21-24). In all but about 20 lichen species, the fungal partner is an ascomycete. The photosynthetic partner in a lichen is either a green alga (a protist) or a cyanobacterium (a prokaryote).

The association between a fungus and its photosynthetic partner is not entirely mutual. The various fungi cannot live without the help of photosynthetic partners. The photosynthetic partners, however, are all capable of living without fungi. Most of the diversity of lichens comes from the fungi, for 90 percent of all photosynthetic partners in lichen are the same few organisms (from one of just three genera of green alga or of one genus of cyanobacteria). In contrast, the fungal partner is unique to each lichen. Nonetheless, the symbiotic association between fungi and their photosynthetic partners helps both parties.

The symbiotic association between lichens and their partners makes them especially tough. Fungi extract nutrients from rock and other harsh microenvironments. By doing so, fungi allow the algae or cyanobacteria to grow where they otherwise could not, while the photosynthetic algae provide a continuous source of energy to their fungal partners. Lichens therefore grow where no plant could survive—bare rock, mountaintops, the Arctic tundra, even the seemingly lifeless valleys of Antarctica (Figure 21-25).

Lichens usually reproduce asexually, either by fragmentation or by sporelike starters called soredia, which are clumps of algae enmeshed in a bit of mycelium. The fungal and photosynthetic components can also reproduce independently, either sexually or asexually.

Lichens are mutually beneficial symbiotic associations of fungi and photosynthetic algae or cyanobacteria. In contrast to their photosynthetic partners, lichen fungi cannot live alone.

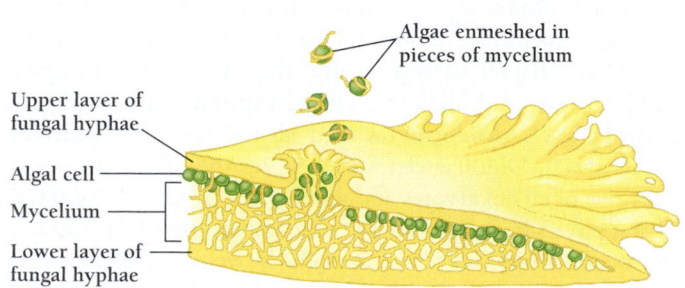

Algae enmeshed in pieces of mycelium

Upper layer of fungal hyphae

Algal cell

Mycelium

Lower layer of fungal hyphae

Figure 21-24   Lichen. Like a pie, a lichen is constructed with a protective crust of fungal hyphae above and below and a filling made of mycelium and algal cells.

Figure 21-25   Lichens are diverse. They are also able to absorb exquisitely tiny amounts of minerals and nutrients dissolved in rain and dew. While this ability is crucial to their survival in harsh settings, it makes lichens especially vulnerable to pollution. The reduced growth of lichens is a valuable indicator of deteriorating air quality. *(Larry Ulrich/DRK Photo)*

What Role Do Mycorrhizae Play in the Lives of Plants?

Fungi form close associations not only with photosynthetic bacteria and photosynthetic protists, but also with true plants. **Mycorrhizae** [Greek, *mukes* = fungus + *rhiza* = root] are symbiotic associations between fungi (usually a zygomycete or a basidiomycete) and the roots of plants (Figure 21-26). At least 80 percent of all plant species have mycorrhizae. Some biologists now think that even more plants depend on mycorrhizae. In such associations, the fungus's hyphae penetrate the plant roots and also extend into the surrounding soil.

What are these fungi doing? Although some fungi obtain sugar from the plant's roots, the mycorrhizae are not parasites, for they help the plant flourish by supplying phosphate and metal ions. Indeed, what the mycorrhizae do for plants is more obvious than what the plants do for the mycorrhizae. In one experiment, plants whose mycorrhizae had been removed grew

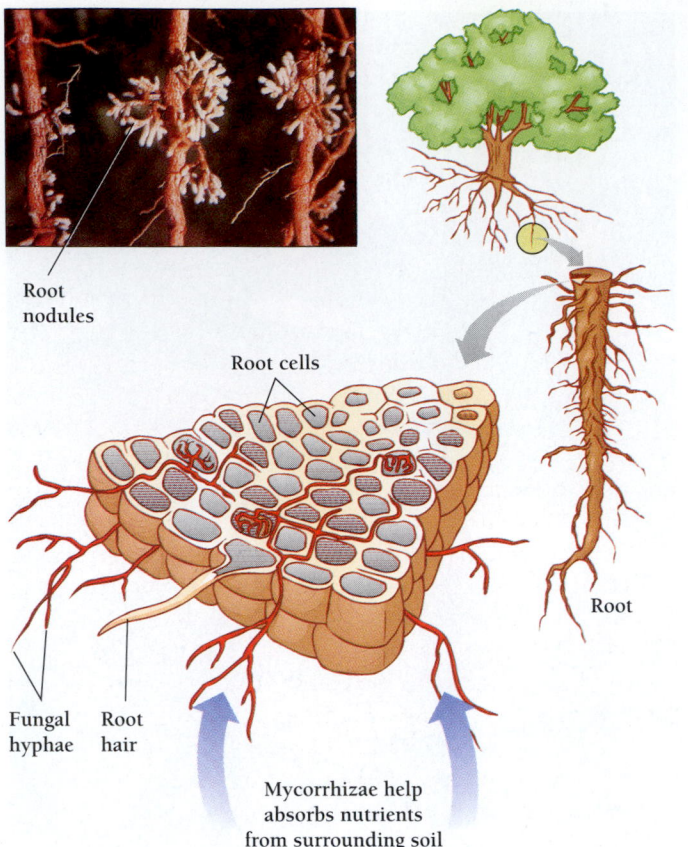

Root
nodules

Root cells

Fungal   Root
hyphae   hair

Mycorrhizae help
absorbs nutrients
from surrounding soil

Root

**Figure 21-26   Mycorrhizae grow inside of the roots of trees
and other plants.** *(Photo, John D. Cunningham/Visuals Unlimited)*

*Mycorrhizae are fungi on the roots of plants
that supply minerals to the plant.*

Where Did the Fungi Come From?

Because fungi have no hard parts, the fossil record of fungi is
not rich. Nonetheless, fossil fungi can be found at least back
to the Ordovician Period, 430 to 500 million years ago. The
earliest fossils of land plants, which are about 400 million years
old, contain petrified mycorrhizae, and some biologists now
believe that plants could not have colonized the land without
the aid of the fungi. Most mycologists think that the first fungi
were zygomycetes, which have the simplest life cycle. Basidiomycetes probably derive from ascomycetes, which are in turn
descendants of the more primitive zygomycetes. Most imperfect fungi probably arose from ascomycetes that lost the sexual stages of their life cycles.

*Fungi evolved from protists at least 430 million years
ago. The fungi may have been responsible for the
colonization of land by plants.*

In this chapter we have surveyed the simplest eukaryotes,
the Protista, as well as the Fungi, the first kingdom that consists entirely of multicellular organisms. In the next chapter,
we will survey the plants, which, along with the fungi, are essential for humans and all other animals. On land, plants provide all the energy and simple organic materials that animals
and fungi need to live. Plants and fungi are thus inextricably
dependent on one another. Animals likewise absolutely depend
on plants and fungi.

more slowly than plants that had mycorrhizae. Plants with no
mycorrhizae were then provided with fertilizer, however, and
they grew as well as the plants with mycorrhizae. This experiment suggested that the mycorrhizae help plants pull nutrients from the soil.

Study Outline with Key Terms

Protists are eukaryotic organisms that are not fungi, plants, or
animals. Protists include organisms that were once called **protozoa** (which move) and **algae** (which perform photosynthesis), as well as some organisms that resemble fungi. Most
protists are single-celled organisms, but many are multicellular. Some protists are **plankton,** which drift. Others are self-
propelled, or **motile.**

Protist life cycles vary greatly from species to species. Each
species goes through two phases: reproduction and dispersal.
Some are haploid throughout their life cycles and never undergo meiosis. Other species undergo **alternation of generations,** that is, they pass through both haploid and diploid
phases. The diploid phase is the **sporophyte,** which produces

spores (sometimes encased in **cysts**), and the haploid phase is
the **gametophyte,** which produces the gametes.

Algae (plantlike protists) include 12 distinct phyla, of
which this chapter surveys 6: **dinoflagellates, chrysophytes**
(golden algae and diatoms), **euglenophytes, phaeophytes**
(brown algae), **rhodophytes** (red algae), and **chlorophytes**
(green algae). Dinoflagellates are among the oldest and most
primitive of all eukaryotes and are marked by distinctive **tests.**
Chrysophytes have both single-celled and multicellular species,
and diatoms can reproduce sexually as well as asexually. Euglenophytes can live as either autotrophs or heterotrophs and
are closely related to **trypanosomes,** flagellated protozoa that
cause some human diseases. Phaeophytes are the largest and

most complex of the protists. Rhodophytes have distinctive pigments, but no flagella. Chlorophytes are the most diverse of the algae.

Protozoa are heterotrophic protists, most of which are motile. They include eight phyla, of which we survey five: **rhizopoda, foraminifera, sporozoa, ciliophora,** and **zoomastigina.** Rhizopods (amebas) move and ingest food with **pseudopodia.** Foraminifera (or **forams**) have tests (shells) studded with little pores. Sporozoans are nonmotile, spore-forming protozoa, which live as parasites in animals. Ciliophora (**ciliates**) are single cells, coated with cilia. Zoomastigina (**flagellates**) are single celled and use flagella to move.

Many protists resemble fungi in appearance but other protists in cellular organization. These protists include seven phyla, of which we described three: **acrasiomycota, myxomycota,** and **oomycota.** Acrasiomycota (cellular slime molds) live as amebas and as multicellular slugs, which can produce **fruiting bodies.** Myxomycota (plasmodial slime molds) live as amebas or as a multinucleated **plasmodium.** Oomycota resemble fungi and are saprophytic. Unlike fungi, however, their cells have flagella.

Despite the tradition of classifying protists into plantlike algae, animal-like protozoans, and fungal-like protists, DNA sequence analysis suggests that evolutionary relationships among the protists are much more complicated. Some DNA-containing organelles persisted and others were lost after some of their genes became part of nuclear DNA.

Fungi are heterotrophic (as are animals) and enclose their cells in rigid walls (as do plants). In contrast to both plants and animals, however, they do not undergo embryonic development. Fungi live mostly on land and, with a few exceptions, consist of a **mycelium,** a mass of intertwined filaments. Their cell walls are made from **chitin,** the same material used to build the external shells of insects. Fungi absorb rather than ingest nutrients after predigesting their food with secreted enzymes.

Most fungi are **molds.** They may reproduce asexually or sexually. They also may reproduce and disperse as spores that resist harsh environmental conditions. Mycologists classify fungi by the way they produce spores (in **sporophores**) and by the appearance of their mycelia. In this chapter we surveyed four phyla of fungi: the zygomycetes, the ascomycetes, the basidiomycetes, and the deuteromycetes. **Zygomycetes** (conjugation fungi) are **coenocytic,** lacking **septa** between the nuclei in their **hyphae. Ascomycetes** (sac fungi) have septa (**dikaryotic**) but no **sporangia.** They produce spores in sacs called **asci. Basidiomycetes** (club fungi) have septa and distinctive life cycles. Each fruiting body of a basidiomycetes mushroom is called a **basidiocarp.** On the thin gills beneath the cap lie the **basidia,** which produce spores. **Deuteromycetes** (imperfect fungi) reproduce asexually.

Fungi play important ecological roles, both as decomposers and as participants in **symbiotic** relationships. **Lichens** are mutually beneficial associations of fungi and photosynthetic algae or cyanobacteria. **Mycorrhizae** are mutually beneficial associations between fungi and the roots of plants. Fungi evolved from protists at least 430 million years ago. Some biologists believe that plants could not have colonized the land without the aid of mycorrhizae.

Review and Thought Questions

Review Questions

1. Why is the kingdom Protista so much more diverse than the plants or animals?
2. How do the algae differ from the protozoa?
3. Which algae contain both chlorophylls *a* and *b*?
4. What features distinguish the slug of the plasmodial slime molds from that of the cellular slime molds?
5. What five features characterize the fungi?
6. To what phylum of fungi do common grocery store mushrooms belong? How is it that mushrooms can grow so quickly after a rain?
7. What are the two partners that compose a lichen? Which partner is more independent?
8. What do mycorrhizae do for plants?

Thought Questions

9. If you wanted to invent a new fungicide that would kill fungal cells but not human cells, what differences between the biology of fungal cells and biology of human cells might you exploit for this purpose?
10. Some paleontologists have speculated that vascular plants could not have evolved to the extent they have without the symbiotic relations with fungi they now depend on. How would you either test such a hypothesis or rephrase it in a way that would make it testable?

About the Chapter-Opening Image

This mushroom represents one of the most familiar and benign fungi.

 On-line materials relating to this chapter are on the World Wide Web at **http://www.harcourtcollege.com/lifesci/aal2/**

22

How Did Plants Adapt to Dry Land?

The Loves of Plants

How to classify organisms is often a matter of dispute. Prokaryotes are commonly classified according to their metabolic pathways; protists, according to their morphology; and fungi, according to their style of reproduction. When 18th-century biologists first began to classify plants, they could not agree on what traits to use. By 1799, naturalists had suggested at least 52 different systems for classifying plants. One system grouped plants according to the form of the flower, seed, and seed coat. Other systems gave priority to the shape of the fruit.

It was Carolus Linnaeus's system that ultimately came into use, but not without heated arguments from both scientists and clergy. Linnaeus based his classification solely on the reproductive structures of flowers. Indeed, he was fascinated by the "marriages of plants," as he called them (Figure 22-1). The son of a country parson, Linnaeus equated sex with marriage, and his descriptions of "vegetable coitus," as some called it, are simultaneously explicit and romantic.

As the historian of science Londa Scheibinger wrote, "Linnaeus emphasized the 'nuptials' of living plants as much as their sexual relations. Before their 'lawful marriage,' trees and shrubs donned 'wedding gowns.' Flower petals opened as 'bridal beds' for a verdant groom and his cherished bride, while the curtain of the corolla lent privacy to the amorous newlyweds."

Linnaeus divided plants first according to whether their marriages were public or private. Public marriages were those in which the flowers were visible to the naked eye; private marriages were those in which the flowers were too small to be seen. (The division corresponded to the public and private marriages then recognized in most of Europe. In England, the 1753 Marriage Act eliminated clandestine marriages by requiring a public proclamation before a marriage could be considered legal.)

Linnaeus further divided plants according to whether the male and female parts were together on the same flower, "in one bed," as he put it. Otherwise, if male and female parts were in separate flowers, they were in "two beds." Those in one bed, also called "hermaphrodites," were divided according to how many male parts (stamens) were present. These Linnaeus termed "husbands" to the single female pistil, or "bride." Flowers with one stamen were monandrian—having one man. (Linnaeus called his beloved wife Sara "my monandrian lily.")

As most flowers, however, have more than one stamen, indeed several, many naturalists and clergymen were horrified by Linnaeus's terminology. One St. Petersburg professor asked what God Almighty would introduce such "loathsome harlotry" into the plant kingdom. The bishop of Carlisle wrote to the Linnaean Society in 1808 that "a literal translation of the first principles of Linnaean botany is enough to shock female modesty."

- A plant is a multicellular organism that photosynthesizes and develops from an embryo.

- Evidence for plant evolution comes from fossils and from comparisons of living species.

- The nonvascular plants, or bryophytes (the mosses, liverworts, and hornworts), lack specialized elements for transporting water and nutrients.

- The development of specialized vascular elements enabled plants to transport water and nutrients, to grow large, and to diversify. Life on land offers both advantages and disadvantages to photosynthetic organisms. Seeds allow plants to withstand dry environments. Flowers and fruits of angiosperms have allowed their rapid diversification.

The Englishman William Smellie, who edited the first edition of the *Encyclopedia Britannica* (1771), rejected the "alluring seductions" of plant sexuality. He was especially repulsed by Linnaeus's description of plants that had reproductive structures of both sexes as "hermaphrodites." Smellie compared Linnaeus's work to that of the most "obscene romance-writer."

Most naturalists avoided Linnaeus's more vivid imagery. One exception was Erasmus Darwin, Charles Darwin's maverick grandfather. His *Loves of the Plants* (1789) is a collection of poems depicting flowers as practitioners of free love and even incest.

Linnaeus's taxonomy had other drawbacks. For example, in his system the number of stamens determined the class to which a plant species belonged. But in some groups of plants, the number of stamens can vary from individual to individual. He also decided that the number of stamens (male parts) would determine class, while the number of pistils (female parts) would determine order, the next taxonomic group down. That is, he deemed the stamens more important than the pistils. This system neither corresponded to the phylogeny (relatedness) of plants, nor was the system arbitrary. It was simply Linnaeus's way of importing into science

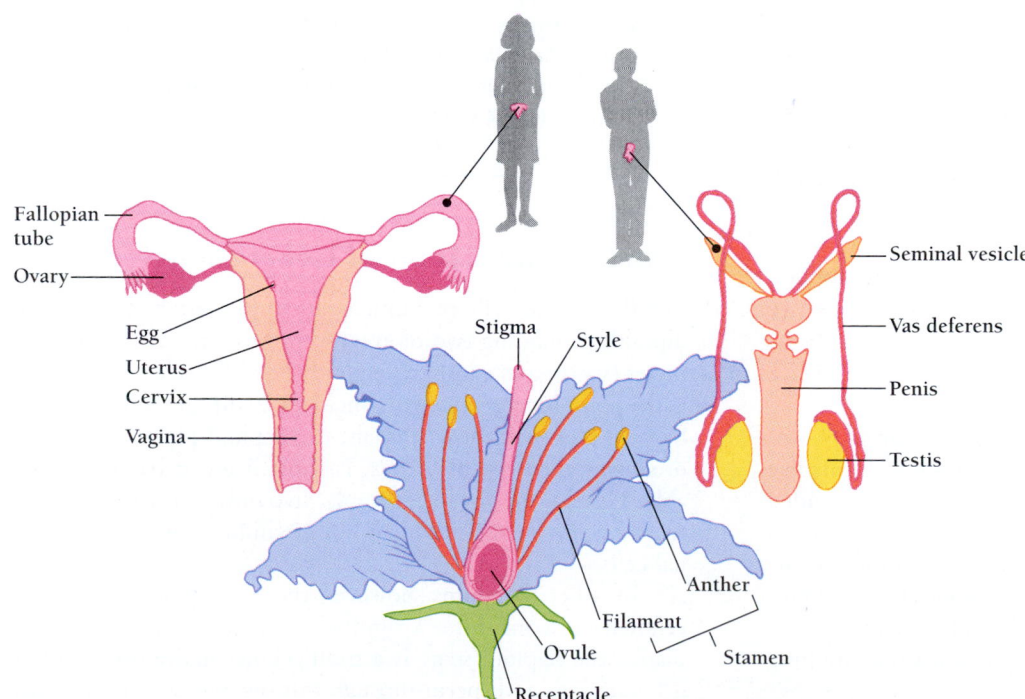

Figure 22-1 **Linnaeus equated flower parts with the sexual parts of animals.** He argued that the anthers are the testes; the filaments, the vas deferens; and the pollen, the seminal fluid. In the female parts, the stigma is the vulva; the style is the fallopian tube; the ovule is the ovary; and the seeds are the zygotes.

477

traditional views about the relative positions of males and females.

Although many of Linnaeus's species designations—genus and species—remain in use today, his classification of groups above the genus level have been abandoned in favor of systems that more accurately reflect evolutionary relatedness. As botanists have learned more about the lives of plants, they have acquired a better understanding of how plants live and how they have evolved.

In the rest of this chapter, we will see that plants are surprisingly diverse. Although the animal kingdom includes four times as many species as the plant kingdom, it is plants that first define a habitat—by their shape, size, and behavior. To a large extent, the animal inhabitants of grasslands differ from those of tropical forests because grasses differ structurally and chemically from trees. We can argue, then, that the plants lay the foundation for more habitat diversity than any other kingdom of organisms.

HOW DID PLANTS ADAPT TO LIFE ON LAND?

> The tree is an aerial garden, a botanical migration from the sea, from those earlier plants, the seaweeds; it has a purchase on crumbled rock, on ground. The human, standing, is only a different upsweep and articulation of cells. How treelike we are, how human the tree.
>
> —Gretel Ehrlich

The beauty of trees and flowers has inspired poets and gardeners alike. Plants delight the eye, tell us the season and even the time of day. They provide us with delicious foods, beautiful clothes, and strong building materials. Botanists like to say that plants provide "food, fiber, feed, fuel, and pharmaceuticals." But even this catchy list doesn't convey the importance of plants.

Through photosynthesis, plants (together with bacteria and protists) provide all of the energy we derive from food, whether that food is itself animal or plant. Plants living millions of years ago captured the energy that we now collect as coal, oil, or gas to heat our homes, to fuel our automobiles, and to drive our power plants. Plants provide the oxygen we breathe. Plants support the entire web of terrestrial life. Without plants, none of the other kinds of organisms could live on land.

A **plant** is a multicellular organism that performs photosynthesis and develops from an embryo. Almost all plants live on the land, and even those that live in water are descended from terrestrial plants. All plants, however, are descended from protists, and plants capture energy from the sun in much the same way as many of the marine protists. Nonetheless, the invasion of the land by plants nearly half a billion years ago depended on the evolution of structures different from anything found among plants' marine ancestors.

What Are the Advantages and Disadvantages of Life on Land?

The land offers three major advantages for photosynthetic organisms: more light (unfiltered by water), more carbon dioxide (which is more concentrated in air than in water), and a more reliable supply of minerals and other nutrients (which are more concentrated in soil than at the surface of a lake or sea).

Plants could not leave the water without solving a host of serious problems. To begin with, the seas provide a continuous supply of water. Water drives photosynthesis and provides a medium in which gametes and spores can disperse or meet for fertilization. Water also provides buoyancy for organisms. Even a 30-meter-tall brown alga can float. So aquatic plants, unlike terrestrial ones, require no skeleton or other structural support system.

To colonize the land, then, plants had to be able to obtain, conserve, and distribute water; to accomplish fertilization with little or no water; and to support their own weight. The evolution of plants depended principally on four adaptations:

1. A waxy **cuticle,** which coats the plant's exposed surfaces and reduces water loss through evaporation
2. The ability to absorb water from dew, rainwater, or groundwater (using roots, for example)
3. Enclosed reproductive organs called **gametangia** in which gametes form
4. Enclosed **sporangia** in which spores form

All plants use some version of a waxy cuticle, an absorption system, and gametangia.

> *Land offers plants more light, carbon dioxide, and nutrients than the sea. But plants living on land need adaptations for absorbing and retaining water, for fertilization with little or no freestanding water, and, in the case of larger plants, some form of structural support.*

What Is Alternation of Generations?

Recall that the cells of humans and most other animals are diploid, containing two of every chromosome, while their gametes (sperm and eggs) are haploid. The life cycle of every sexually reproducing organism includes a diploid phase and a haploid phase. In most animals, the haploid phase, consisting of single-celled sperm or ova, is short lived or barely noticeable. Human sperm, for example, live only a few weeks. Human ova survive much longer but are hidden away as individual cells.

In other kingdoms, however, the haploid stage is multicellular and constitutes a significant part of the life cycle. In plants, the haploid stage is a multicellular **gametophyte** that generates ova and sperm through mitosis, rather than meiosis.

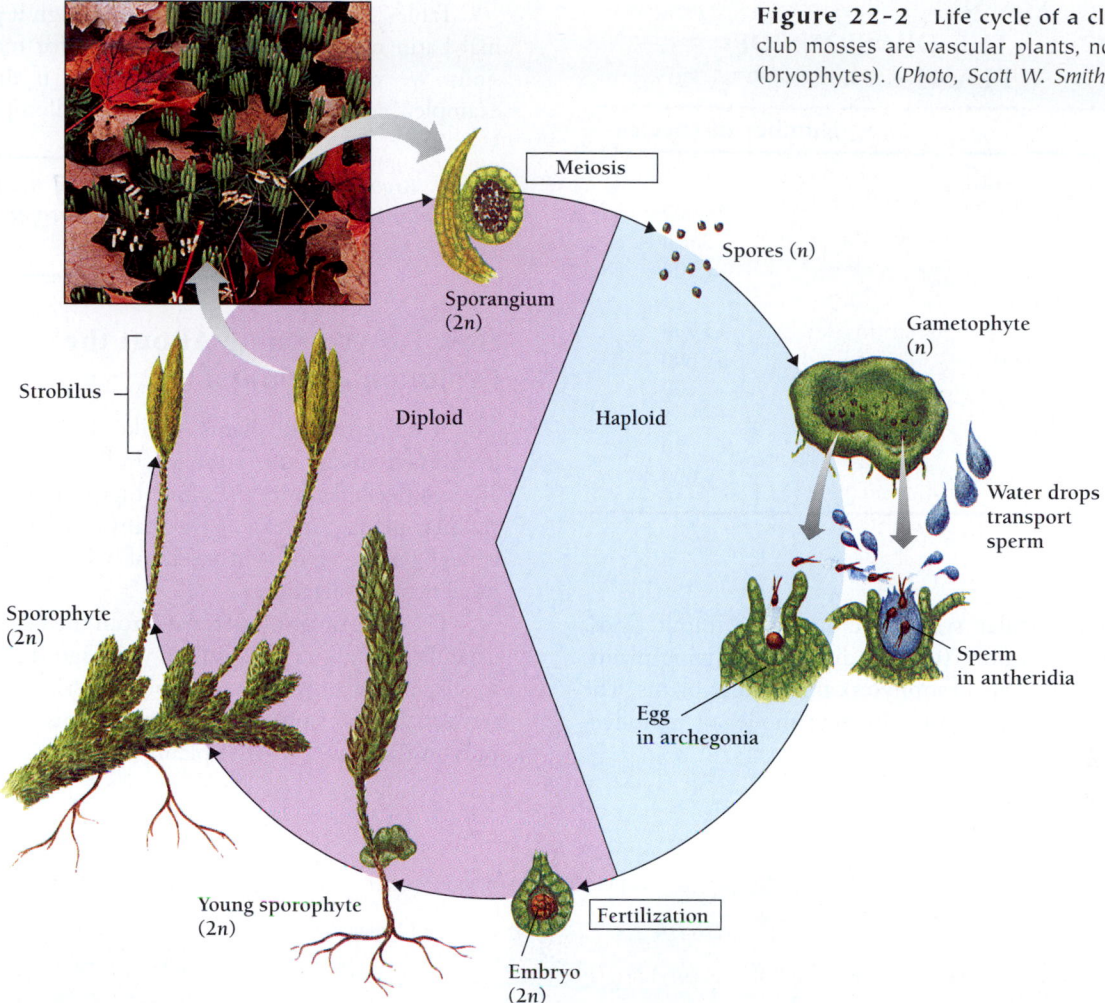

Figure 22-2 **Life cycle of a club moss.** The club mosses are vascular plants, not true mosses (bryophytes). *(Photo, Scott W. Smith/Earth Scenes)*

Meiosis

Spores (*n*)

Sporangium
(2*n*)

Gametophyte
(*n*)

Strobilus

Diploid

Haploid

Water drops
transport
sperm

Sporophyte
(2*n*)

Sperm
in antheridia

Egg
in archegonia

Young sporophyte
(2*n*)

Fertilization

Embryo
(2*n*)

Plants, like many protists and fungi, are said to exhibit **alternation of generations,** a sexual life cycle in which the alternating haploid and diploid phases are both multicellular (Figure 22-2).

> *Plants exhibit alternation of generations, a sexual life cycle in which haploid and diploid phases are both multicellular.*

How Are Vascular Plants Different from Nonvascular Plants?

Despite important similarities among all plants, plants may be divided into two groups—the tracheophytes (vascular plants) and the bryophytes (nonvascular plants). Each group has different solutions for the problems of terrestrial life.

The **tracheophytes** [Latin, *trachea* = windpipe] include all the most familiar living plants, from ferns to fir trees and flowering fuchsias. All tracheophytes have pipelike tissues that conduct water and nutrients from one part of the plant to another. Because this system of conducting tissues resembles the vascular system of humans and other vertebrates, the tracheophytes are also referred to as the **vascular** plants. The tracheophytes' vascular system allows these plants to grow far larger than nonvascular plants, sometimes reaching enormous sizes.

The only true plants that lack such a vascular system are the **bryophytes** [Greek, *bryon* = moss], the mosses and their relatives. Because bryophytes lack a vascular system, they are much smaller than the tracheophytes. The bryophytes are also much less diverse, totaling only about 24,000 species, far fewer than the (approximately) 290,000 named species of tracheophytes.

Most botanists refer to the major groups of plants as divisions, equivalent to the phyla of animals that we will survey in the next chapter. However, the term phylum is also becoming acceptable. Each division, like each animal phylum, contains classes, orders, families, genera, and species.

Some botanists divide the plants into only two divisions—Bryophyta and Tracheophyta. That is, they recognize plants

TABLE 22-1   THE DIVISIONS OF PLANTS

Division	Number of species
Bryophyta (mosses, liverworts, and hornworts)	24,000
Psilotophyta	Small
Lycopodophyta	1,100
Equisetophyta (horsetails)	15
Pteridophyta (ferns)	13,000
Coniferophyta (conifers)	600
Cycadophyta (cycads)	100
Ginkgophyta (ginkgo)	1
Gnetophyta	70
Anthophyta (flowering plants)	275,000

with specialized vascular systems and those without. Most botanists, however, count 10 or 12 divisions of living plants, of which all but one (the bryophytes) are vascular plants. The divisions of vascular plants vary in their modes of reproduction and dispersal.

Table 22-1 lists ten divisions. Although it gives both a formal Latin name and a common name for many of the divisions, we will use the common name if there is one (for example, "conifers" rather than the Coniferophyta).

Bryophyta is the single division of nonvascular plants. The nine divisions of Tracheophyta are all vascular plants.

How Do We Know About the Evolution of Plants?

The fossil record for plants, unlike that for protists, is excellent. Plant fossils are easier to find not only because plants are multicellular and therefore bigger but also because the vascular plants, at least, have many hard parts such as bark, cones, pollen, and veins—all of which nicely survive the process of fossilization.

The oldest plant fossils date from the beginning of the Silurian Period, about 430 million years ago (Figure 22-3). These fossils are not only the oldest plant fossils but also the oldest terrestrial organisms. Until then, organisms apparently lived only in the seas. The first plants appeared on the land, prob-

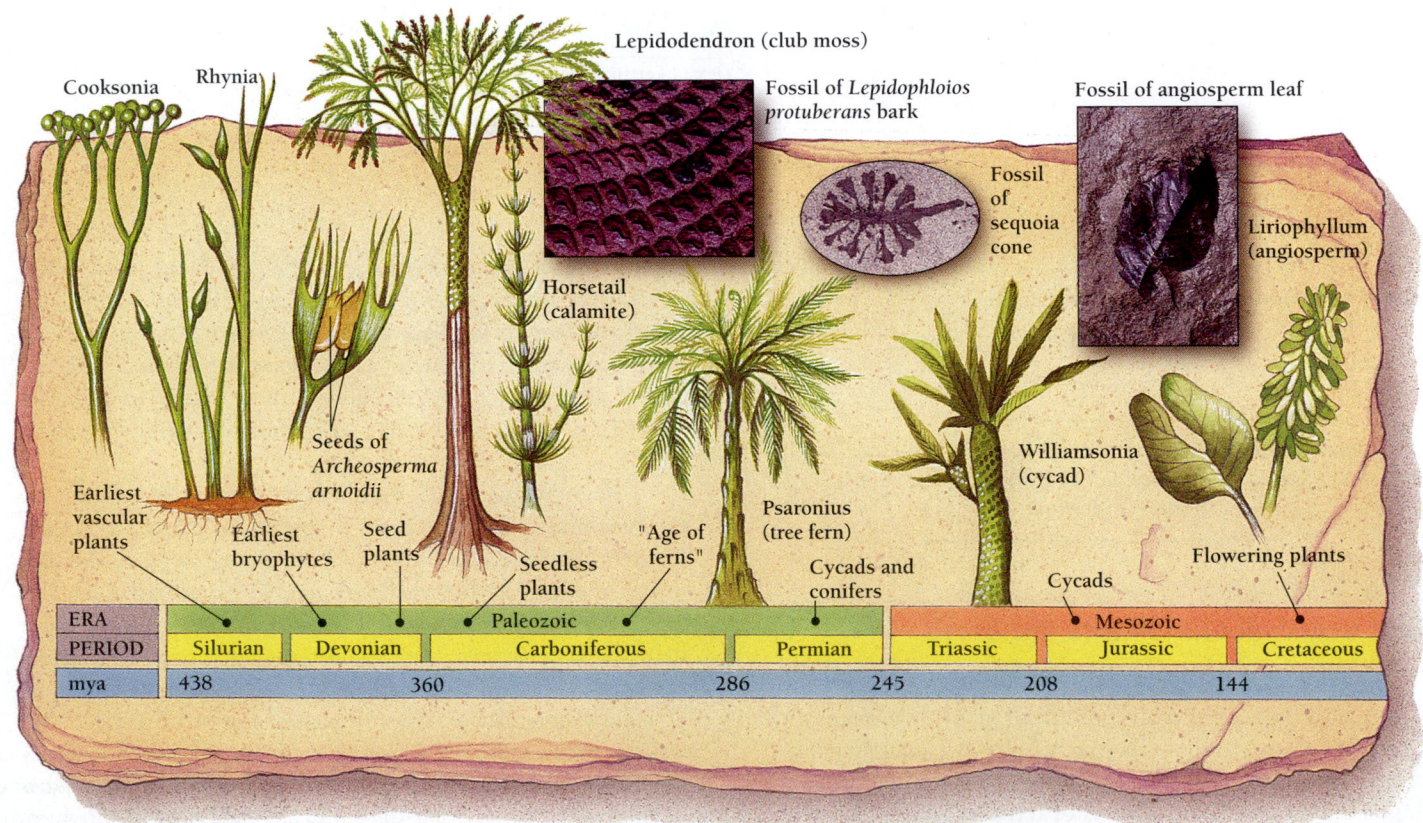

Figure 22-3 The evolution of plants. Notice the first appearance of vascular plants, bryophytes, seed plants, seedless plants, cycads and conifers, and flowering plants. *(Louise K. Broman/Photo Researchers, Inc.; William E. Ferguson; McCutcheon/Visuals Unlimited)*

ably in associations with fungi similar to present-day mycorrhizae (Chapter 21). The successful invasion of the land by plants and fungi allowed the later evolution of terrestrial animals. Among these animals, the most populous and diverse were and continue to be the insects. As we will see, the evolution of the flowering plants (angiosperms) went hand in hand with the evolution of insects and fungi.

We can deduce some of the characteristics of the first plants by noticing what all living plants have in common. All show alternation of haploid and diploid generations, all contain both chlorophylls *a* and *b*, all have cell walls made of cellulose, and all are multicellular. Almost all plants store energy as starch. All develop from embryos that are surrounded by nonreproductive ("sterile") tissue. And, during cell division, all plant cells use a microtubular structure, the phragmoplast, to build a new cell wall between the daughter cells. Because plants share these striking characteristics with some green algae, most botanists believe that both nonvascular and vascular plants descended from the green algae.

True plants, however, have characteristics that set them apart from the green algae. In addition to a waxy cuticle, specialized structures for absorbing water, and specialized enclosed reproductive organs, plants also have **stomata** [Greek, *stoma* = mouth], tiny mouthlike pores that open and close to regulate the flow of carbon dioxide and other gases in and out of the plant.

Evidence for plant evolution comes from fossils and from comparisons of livinf species. The first plants probably evolved from a common ancestor that resembled a green alga.

What Does the Fossil Record Tell Us About the Evolution of Plants?

Although it is tempting to think of nonvascular plants as more primitive than vascular plants, the fossil record of vascular plants is actually older. The earliest bryophyte fossils date from the Devonian Period, about 400 million years ago, while the earliest vascular plants appear at the beginning of the Silurian Period, 430 million years ago (Figure 22-3).

By the time bryophytes first appeared in the early Devonian Period, several types of simple vascular plants were widespread, some of which are strikingly similar to modern species such as ferns and club mosses. Botanists divide the vascular plants into those that produce seeds, such as the flowering plants and the conifers, and those that do not. To the seedless plants, the ferns and their allies, we owe our warm houses and speedy cars. These plants, buried deep beneath layers of sediment, have turned into the famous fossil fuels, mainly oil and coal, on which we are now so dependent. Seedless plants began flourishing in the Carboniferous Period, from 345 to 280 million years ago, when the world's climate was warm and wet. Shallow seas and dense swamps covered most of the land.

The growth of the earliest seedless plants, including ferns, horsetails, and lycophytes, was so luxuriant that a large proportion of these plants did not decay completely before being buried under more dead plants and other sediments. Over millennia, these buried, compressed plants became the major component of all fossil fuels—coal, oil, and natural gas. The large amounts of carbon in these fossils give the Carboniferous Period [Latin, *carbon* = coal + *ferre* = to bear] its name.

During the Carboniferous Period, plants grew taller and taller in an adaptive "light war." The tallest plants had access to the most sunlight and shaded the plants below. The tallest were 40-meter lycophyte trees, as tall as large oaks. Far below this canopy of lycophytes spread vast stands of 18-meter horsetails and 8-meter ferns. Ferns, in fact, became so successful that the late Carboniferous Period is often called the "Age of Ferns."

Although a few Carboniferous plants had seeds, none had flowers. Seed plants, which first appeared in the late Devonian Period, were the *progymnosperms* [Greek, *pro* = first + *gymnos* = naked + *sperma* = seed], the ancestors of the modern gymnosperms. **Gymnosperms** are modern seed plants that do not have fruits or flowers. Gymnosperms include pine trees and other conifers, as well as palmlike cycads.

Besides having seeds, the progymnosperms had more elaborate branching and more complex vascular systems than the earlier vascular plants. The Carboniferous swamps contained conifers and palmlike cycads, as well as lycophytes, horsetails, and ferns. As the climate became drier, the giant lycophytes, horsetails, and ferns began to disappear, and the conifers and the cycads began to dominate the landscape.

At the end of the Paleozoic Era, about 225 million years ago, the Earth's climate changed dramatically, and for 100 million years the dominant plants were the conifers and great palmlike cycads. Then, about 130 million years ago, a new kind of plant appeared—one with flowers.

Flowering plants, which have both seeds and fruits, became the most abundant plants on Earth. Today we live in the Age of Flowers. Of the present 315,000 species of plants, including all bryophytes and all tracheophytes, an astounding 275,000 are flowering plants. All of the flowering plants fall into two groups. These are the **dicots,** or **broad-leafed plants,** such as roses, poppies, oaks, and others; and the **monocots,** such as grasses, lilies, onions, daffodils, and cattails. Of the once dominant gymnosperms, only 529 species remain today.

Vascular plants predate nonvascular plants. To the first seedless plants of the humid Carboniferous Period we owe our current supplies of oil and gas. As the climate dried, first ferns, then conifers and cycads dominated the landscape.

How Did Angiosperms Evolve?

The flowering plants, or **angiosperms** [Greek, *angion* = vessel + *sperma* = seed], make up by far the largest division of

Figure 22-4   The major divisions of the kingdom Plantae.

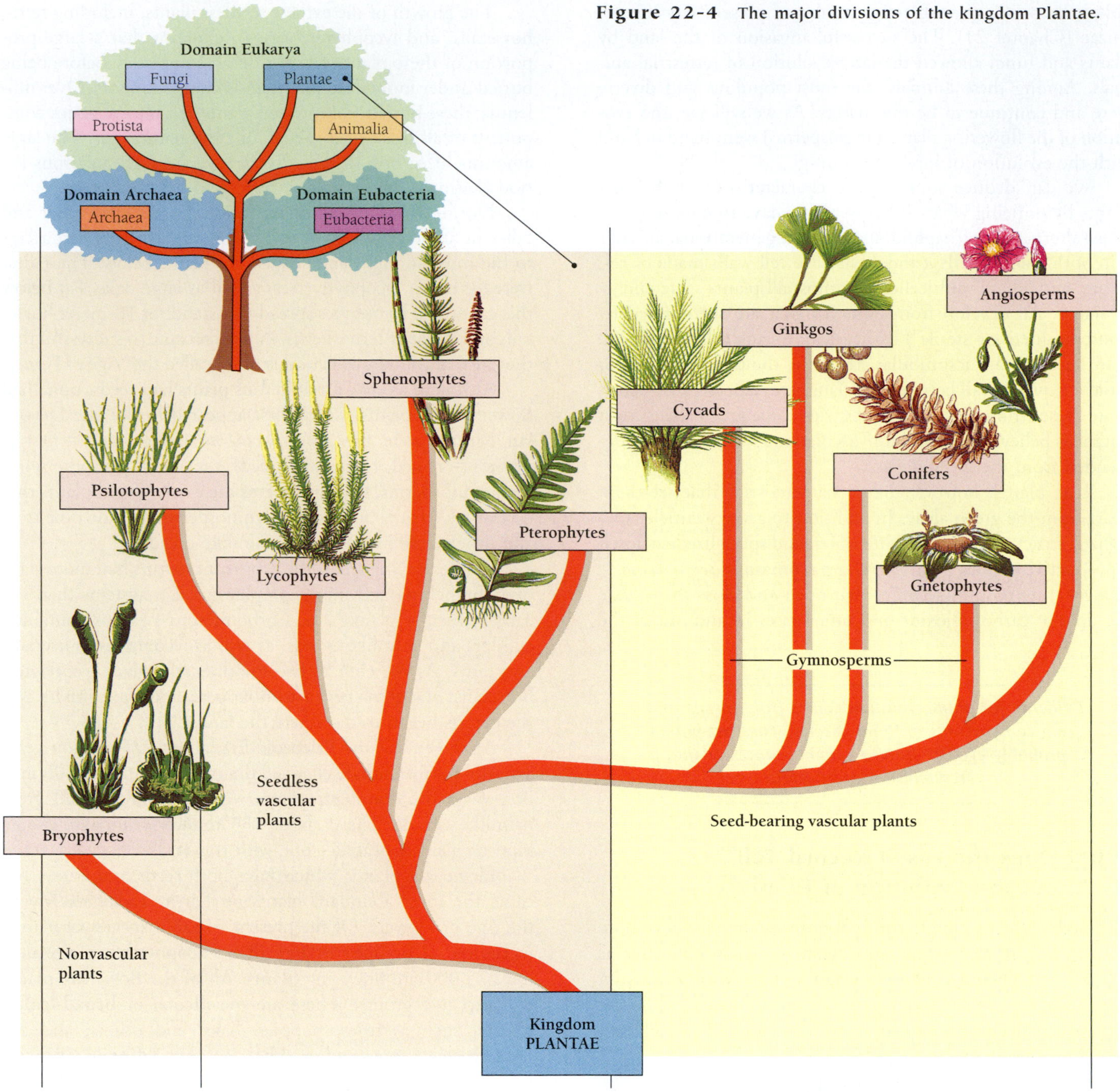

plants (Figure 22-4). The angiosperms enclose their seeds in a vessel that we know as a fruit. A fresh peach is a fruit, and its pit is an enormous seed encased in a hard shell.

The first fossils of flowering plants date from the Cretaceous Period, about 125 million years ago. These early angiosperm fossils are leaves and pollen grains, rather than flowers.

In the absence of fossil flowers, evolutionary biologists who would like to know how angiosperms evolved from the gymnosperms have had to depend on comparisons among living plants. The magnolia, for example, is considered a very primitive plant with few advanced specializations. The mag-

nolia has simple leaves, woody stems, and insect-pollinated flowers, features that probably characterized the first angiosperms.

In the 125 million years since the angiosperms first appeared, they have diversified both morphologically and biochemically. They have improved their transport system and developed *deciduous* leaves, which fall off seasonally, so that angiosperms are better able to withstand periods of drought than gymnosperms.

In addition, many modern angiosperm species have adaptations that prevent self-fertilization. Such species are more genetically diverse than gymnosperms and therefore better able

Figure 22-5 Sphagnum moss surrounding a skunk cabbage in the Quinalt Rain Forest, Olympic National Park, Washington. Most mosses grow in habitats that are always wet or damp. *(Pat O'Hara/DRK Photo)*

to survive environmental changes. Angiosperm diversity and their dominance of the land attest to the success of their many evolutionary experiments.

Thanks to improved transport and deciduous leaves, many angiosperms have been able to colonize ever drier habitats. Adaptations preventing self-fertilization have contributed to angiosperms' enormous diversity and success.

In the rest of this chapter, we will review the ten divisions of plants, beginning with the ancient bryophytes and working our way up to the modern angiosperms, the flowering plants.

THE NONVASCULAR PLANTS, OR BRYOPHYTES, INCLUDE MOSSES, LIVERWORTS, AND HORNWORTS

The nonvascular plants, or bryophytes, include the mosses, which give the division its name, and two less common classes—the liverworts and the hornworts (Figure 22-5). Mosses and other bryophytes generally live in wet environments, some tropical and some temperate.

Because all bryophytes lack specialized vascular systems, they depend on free-standing water for both photosynthesis and fertilization by free-swimming sperm. Although bryophytes can live only where free-standing water occurs, they nonetheless disperse well and grow all over the world. One species of moss even grows high on an Antarctic mountain just a thousand miles from the South Pole.

The bryophytes play important but little understood roles in the world's ecosystems, recycling carbon and absorbing nutrients and toxic metals. Peat moss alone covers up to 2 percent of all land on Earth, an area the size of the United States. In some areas, bryophytes are the principal harvesters of solar energy, and all other organisms depend on them.

Bryophytes are also remarkable because the most obvious part of the bryophyte is not the diploid phase, but rather the haploid phase. When we look at a moss-covered stump, we are looking at the haploid phase, in which each cell has only one set of chromosomes.

How Do Bryophytes Absorb Water?

Although some bryophytes have elongated cells that conduct water, they lack the specialized transporting tissues of the vascular plants. They anchor themselves to surfaces by specialized structures called *rhizoids* [Greek, *rhiza* = root]. Unlike the roots of vascular plants, rhizoids are not specialized for absorbing water and nutrients. Instead, all parts of a bryophyte are absorbent, and the bryophytes consequently have a spongy feel. Bryophytes can grow on rocks, tree trunks, and other hard surfaces that resist the roots of vascular plants.

Bryophytes are small, usually less than 2 cm high and less than 20 cm long. In a moist environment, however, many plants may crowd together in a large mat. The closeness of the short plants allows them to retain water in the tiny spaces within the tangle. Because direct sunlight dries them out, bryophytes usually grow in the shade.

Although bryophytes lack a specialized transport system, all parts of their bodies are adapted to absorb water, which gives them a spongy feel.

How Do Bryophytes Reproduce?

Like the vascular plants, the fungi, and many protists, the bryophytes alternate generations. In contrast to the life cycles of the flowering plants, however, the most prominent phase of a bryophyte's life is the haploid gametophyte, which produces the gametes. Because the gametophyte is already haploid, it produces gametes—sperm or eggs—through mitosis, not meiosis. The gametes come together in the female

Figure 22-6   Life cycle of a moss.

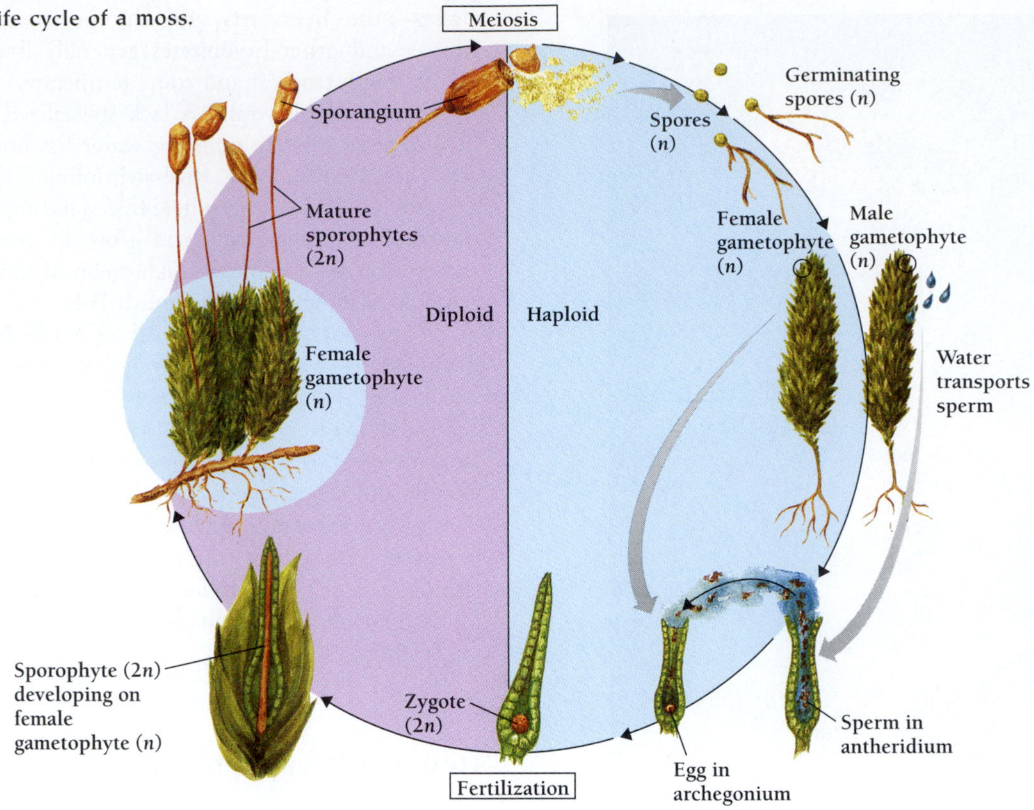

gametophyte to form the diploid phase, called the **sporophyte** (Figure 22-6). The sporophyte produces haploid **spores** by means of meiosis, just as animals and flowering plants do. But these haploid spores are not gametes. They are unlike any phase in our own life cycle, for once the haploid spore finds itself in a welcoming environment, its cells enter mitosis and the bryophyte grows directly into a leafy, green gametophyte.

The haploid gametophytes may be male or female. Each sex develops *gametangia,* structures in which the male or female gametes develop (Figure 22-6). The male gametangium, called an *antheridium,* produces sperm with two flagella. The female gametangium, called an *archegonium,* produces and houses the ova and is generally shaped like a flask with a long neck. Some bryophytes bear archegonia and antheridia on the same plant.

To reach the ovum in the archegonium, the sperm must either swim through water adhering to the spongy mat or be splashed into the vicinity of the neck of the archegonium (Figure 22-6). Sometimes insects carry moss sperm, suspended in water, from male to female plant. Once sperm are near an archegonium, chemical attractants draw them into the neck, and the sperm swim through a fluid in the archegonium's neck to reach the ovum at the center of the swollen base.

Fertilization occurs within the archegonium. Immediately after fertilization, the diploid zygote begins to divide by mitosis to produce the sporophyte. The sporophyte, which is usually not photosynthetic, consists of a stalk, as tall as 15 to 20 cm, at the top of which develops a fruitlike **sporangium** [Greek, *spora* = seed + *angeion* = vessel]. Cells inside the sporangium undergo meiosis to produce the haploid spores, which can develop into a new mossy gametophyte.

Because bryophyte spores need to germinate in an environment that will sustain them, dispersal is essential to bryophyte reproduction. Some bryophytes propel their spores from exploding capsules. Some depend on splashing rainwater. A few bryophyte species live on dung, and flies carry the spores to fresh dung. Once in a hospitable environment, the haploid spores germinate, divide by mitosis, and develop into mature, leafy, male or female gametophytes.

Bryophytes' life cycles differ from those of vascular plants significantly and from those of animals profoundly: in a bryophyte's life cycle, the haploid gametophyte phase dominates.

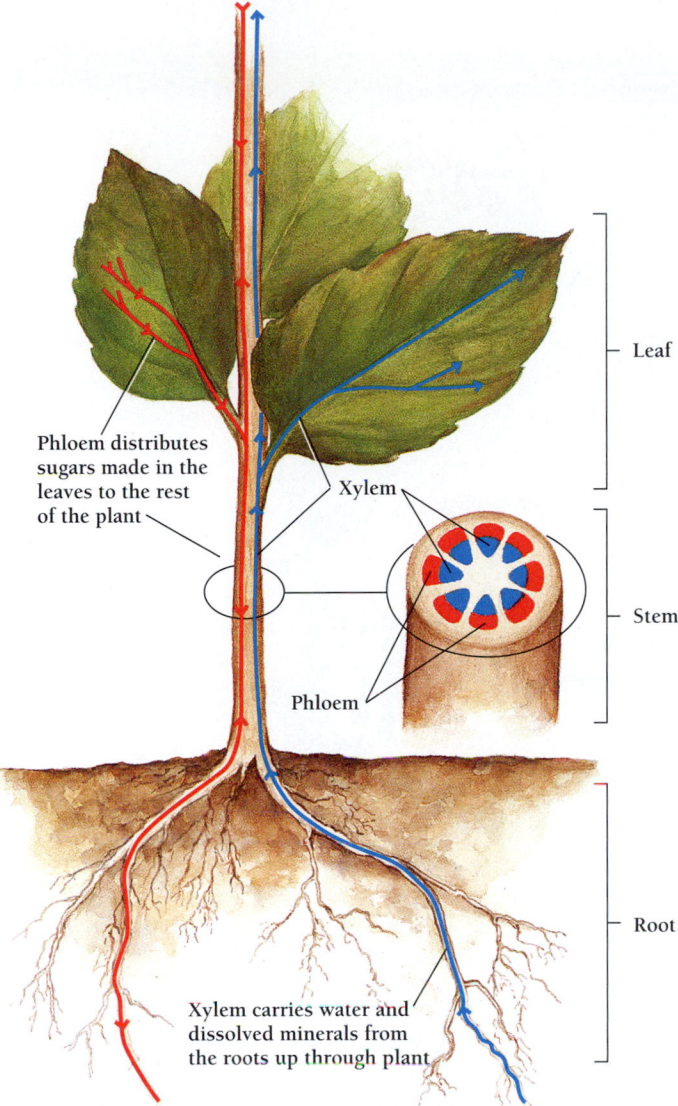

Phloem distributes sugars made in the leaves to the rest of the plant

Xylem

Phloem

Xylem carries water and dissolved minerals from the roots up through plant

Leaf

Stem

Root

Figure 22-7 **Xylem and phloem.** Xylem conducts water and dissolved minerals from the roots to the leaves. Phloem conducts sugar from the leaves to the rest of the plant.

THE VASCULAR PLANTS, OR TRACHEOPHYTES, INCLUDE SEED PLANTS AND SEEDLESS PLANTS

The tracheophytes (vascular plants) are both more diverse and more numerous than the bryophytes. Vascular plants include more than 290,000 named species, but some taxonomists say that the number of vascular plant species runs into the millions. We need only look around us to appreciate the success of the vascular plants—wherever we see green, we nearly always encounter a vascular plant.

Of What Advantage Is a Transport System to a Plant?

The ability of vascular plants to transport water and nutrients allows them to grow to huge sizes and to develop specialized tissues. Vascular plants also have more division of labor than the bryophytes, with distinctive tissues and organs that include roots, leaves, and stems. Their transport systems contain two distinct kinds of conducting tubes: **xylem** [Greek, *xylon* = wood], which carries water and minerals from roots to the photosynthesizing leaves, and **phloem,** which distributes the sugars and other organic molecules made in the leaves to the rest of the plant (Figure 22-7). Phloem consists of elongated, living cells. Xylem consists of the stiff walls of dead cells. Vascular plants also produce a hard, supporting material called *lignin,* which usually lines the walls of the xylem. Lignin, a major component of wood, provides the extra support that allows trees to attain such enormous sizes.

Plants' ability to transport water and nutrients has allowed them to grow larger and develop specialized tissues.

How Do Vascular Plants Reproduce and Disperse?

Recall that in the bryophytes, the haploid gametophyte—the leafy, green part we call moss—is the more prominent part of the plant life cycle. In vascular plants, the diploid sporophytes are more prominent. The seed of a peach tree, for example, develops, in the right environment, into a mature sporophyte—a peach tree. In the spring, the peach tree blooms, producing flowers, which are reproductive structures. Parts of the flowers undergo meiosis to produce short-lived male and female gametophytes, which produce pollen and ova.

Vascular plants that produce seeds have at least one great advantage over the bryophytes. In colonizing new habitats, seed plants do not rely on the chance that a dispersed seed will fall into a benign and nourishing environment. Instead, the seed comes encapsulated in a tough coat that protects the embryo inside.

Within the safety of the plant's female reproductive organ, the plant embryo stops developing, and the mother plant packages the embryo, with a reserve of food, inside a protective seed coat. A seed is able to withstand long periods of cold and drought. When the seed finally encounters water and warmth, it germinates, and the embryo uses the stored food to finish developing into a photosynthetic seedling.

In vascular plants, the diploid phase dominates the life of the plant. The protective coat of a seed, and the food stored inside, allow a seed to wait indefinitely for suitable conditions for germination and growth.

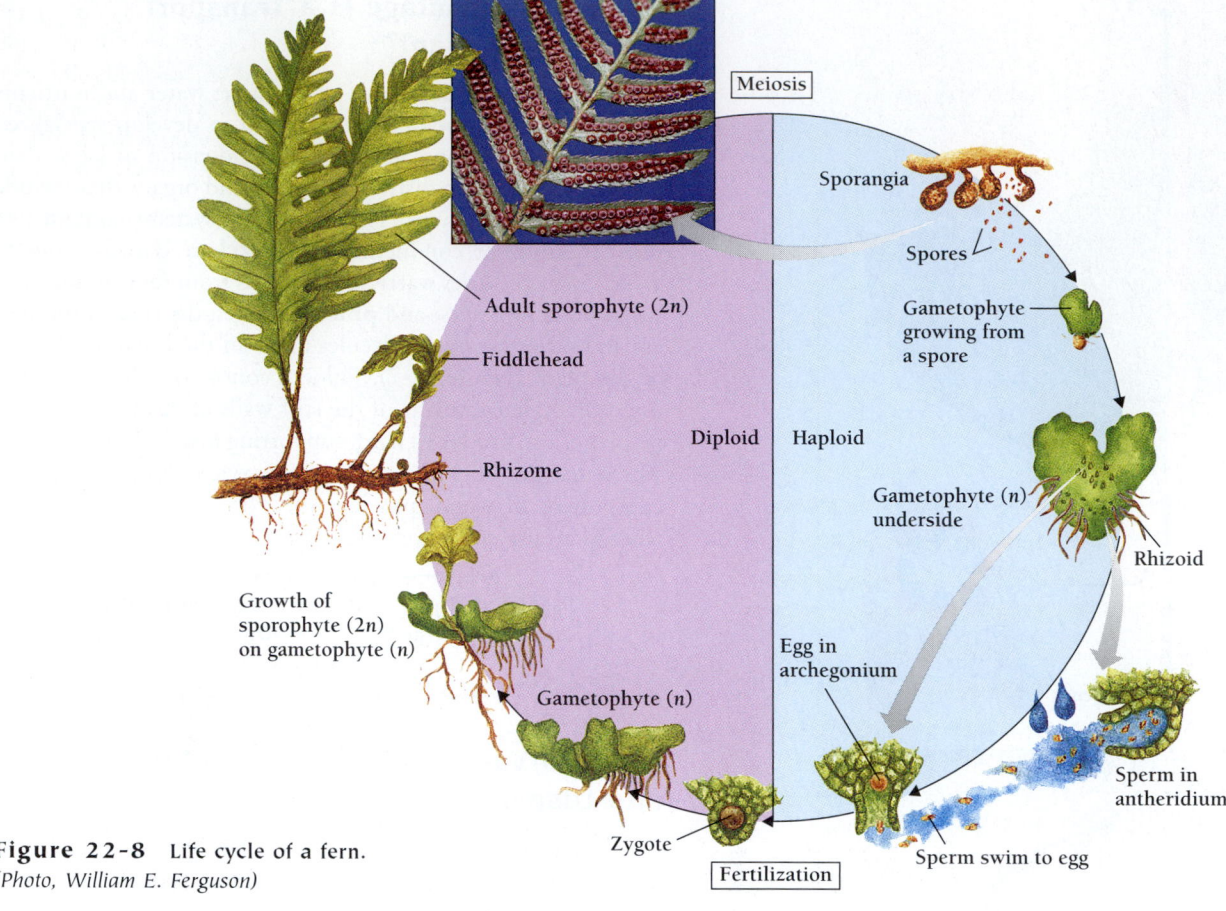

Figure 22-8 Life cycle of a fern.
(Photo, William E. Ferguson)

Four Divisions of Tracheophytes Lack Seeds

Four divisions of living vascular plants lack seeds: the ptero-phytes (ferns), psilotophytes, lycophytes, and equisetophytes. (Another four divisions of seedless plants exist only in the fos-sil record.) We will discuss the ferns first because they are the most familiar.

Pterophytes (ferns)

Some two-thirds of the 10,000 species of ferns live in the trop-ics, and the rest inhabit the temperate zones. Ferns range in size from a few centimeters to giant 30-meter tree ferns, as tall as oak trees. Ferns are common in moist parts of shady forests. Their leaves (called *fronds*) often have a lacy appearance and consist of many *pinnae* [singular, pinna; Latin, = feather], small leaflets that extend from a central stalk (Figure 22-8).

In the large ferns of the tropics, fronds grow from a cen-tral stem. In most ferns, however, fronds grow from a *rhizome,* a horizontal stem that spreads on or below the ground. Each frond develops from a characteristic coil called a fiddlehead. Most other seedless plants have **microphylls**—leaves with just one conducting vein. In contrast, fern fronds are **mega-phylls**—large leaves with several veins that form complex net-works of transport tubes.

Reproduction in ferns and other seedless plants resembles that in the bryophytes (Figure 22-8). Fertilization can occur only if the released sperm can swim to the mature archegonia. Seedless plants therefore require a wet environment to provide at least a film of water in which the sperm can make their way. This need for water, together with the fragility of the game-tophyte, limits the environments in which ferns and other seed-less tracheophytes can grow.

As in other vascular plants, however, the sporophyte (which we recognize as the fern plant) is much more promi-nent than the gametophyte. The fern sporophyte has leaves, called **sporophylls,** which are specialized for reproduction. Clustered on the underside of the sporophylls are groups of sporangia called *sori* [singular sorus; Greek, *soros* = heap] (Fig-ure 22-8). The diploid sporangia produce haploid spores by means of meiosis. Some ferns catapult their ripe spores several meters into the air, where breezes carry them aloft. Such air-borne spores may disperse over enormous distances.

On landing, each spore germinates to form a tiny nonvas-cular gametophyte, which is typically flat, thin, heart shaped, and less than 1 cm wide. This gametophyte is green, photo-

Figure 22-9 Horsetails. Equisetophytes contain bits of silica that give them a rough feel. *(William E. Ferguson)*

synthetic, independent, and fragile. It usually bears both male and female reproductive structures.

Ferns have megaphylls, leaves with networks of transport tubes, and independent gametophytes.

The psilotophytes

The psilotophytes contain just two genera, *Psilotum* and *Tmesipteris*. Both have photosynthetic leaflike scales (which lack conducting tissues) instead of true leaves. The common greenhouse weed *Psilotum*, or "whiskfern," is widespread in the tropics and subtropics. Unlike other vascular plants, *Psilotum* lacks true roots as well as leaves. Underground, it has a rhizome, with tiny branches called rhizoids. Above ground, its vertical stem forks several times and bears sporangia and photosynthetic scales.

Psilotophytes, the simplest of the living vascular plants, lack true leaves.

Lycophytes

Lycophytes (also called lycopods) have true roots, stems, and leaves, as well as a simple arrangement of xylem and phloem. Lycophyte leaves are microphylls (with a single conducting vein). As in ferns, some of the leaves are sporophylls, with spo-

rangia on their upper surfaces. The sporophylls may be green and look like ordinary leaves, or they may be distinct and grouped into spore-producing cones, called **strobili** [singular, strobilus; Greek, *strobilos* = cone]. The most common lycophytes are the club mosses, which form evergreen, mosslike mats beneath temperate forests. Their strobili resemble little clubs.

Lycophytes have true roots, stems, and simple leaves.

Equisetophytes

The living equisetophytes are all members of the single genus *Equisetum*, or horsetails (Figure 22-9). These plants grow in moist places along the banks of streams or at the edges of woods. Like the lycophytes, the equisetophytes have true roots, stems, and leaves. Their jointed stems contain many discrete strands of xylem and phloem. The scalelike leaves that grow from the joints are thought to be megaphylls (with branching vessels). The outer cell layer of the plant contains silica (the same material from which glass is made), which gives the plants a rough texture useful to campers and settlers for scouring pots. Indeed, another name for these plants is "scouring rush."

Equisetophytes have true roots, stems, and complex leaves. Their stems are jointed, and their outer cell walls are reinforced with silica.

How Do Seed Plants Reproduce?

Seeds are of great value in a dry, terrestrial environment because they liberate a plant from the need for free-standing water. **Seed plants** manage fertilization without water in the same way we do, through a form of internal fertilization. The gametophytes have no independent lives but develop within the tissues of the sporophytes. The gametophytes of seed plants are tiny—even less prominent than in the seedless vascular plants.

Many seedless plants are *homosporous,* producing just one kind of spore and one kind of gametophyte. In contrast, all the seed plants are *heterosporous*—producing two kinds of gametophytes (male and female), from two kinds of spores on two kinds of sporangia. The sporophytes produce haploid microspores (male) and megaspores (female) in separate sporangia. In some species, the male and female sporangia occur together on the same plant, while in others, the two kinds of sporangia occur on male and female plants. In either case, the microspores and megaspores develop into distinct male and female gametophytes.

The female gametophyte stays within the megasporangium, which lies safe within the main body of the adult plant (the sporophyte), awaiting the arrival of the sperm. The megasporangium, also called a **nucellus** [Latin, *nucella* = a small nut], is in turn covered by one or two additional layers of tissue, called integuments, which will later develop into the protective seed coat. This combined structure—nucellus and integuments—is the **ovule.**

The small, male gametophytes are grains of pollen, released from the microsporangium and carried, usually by wind or insect, to the female gametophyte. The male gametophyte then develops a slender pollen tube, which carries a sperm nucleus to the ovum produced by the female gametophyte. At fertilization, the sperm and ovum fuse to form a diploid zygote. The resulting zygote, still enclosed within the female gametophyte, begins to develop into an embryo.

After fertilization, the ovule and its contents become a **seed.** The seed includes (1) the new sporophyte embryo; (2) a female tissue that nourishes the developing embryo; and (3) the nucellus and integuments (now the seed coat) from the previous sporophyte generation. Inside the seed coat are the embryo and a source of food for its later development. The seed

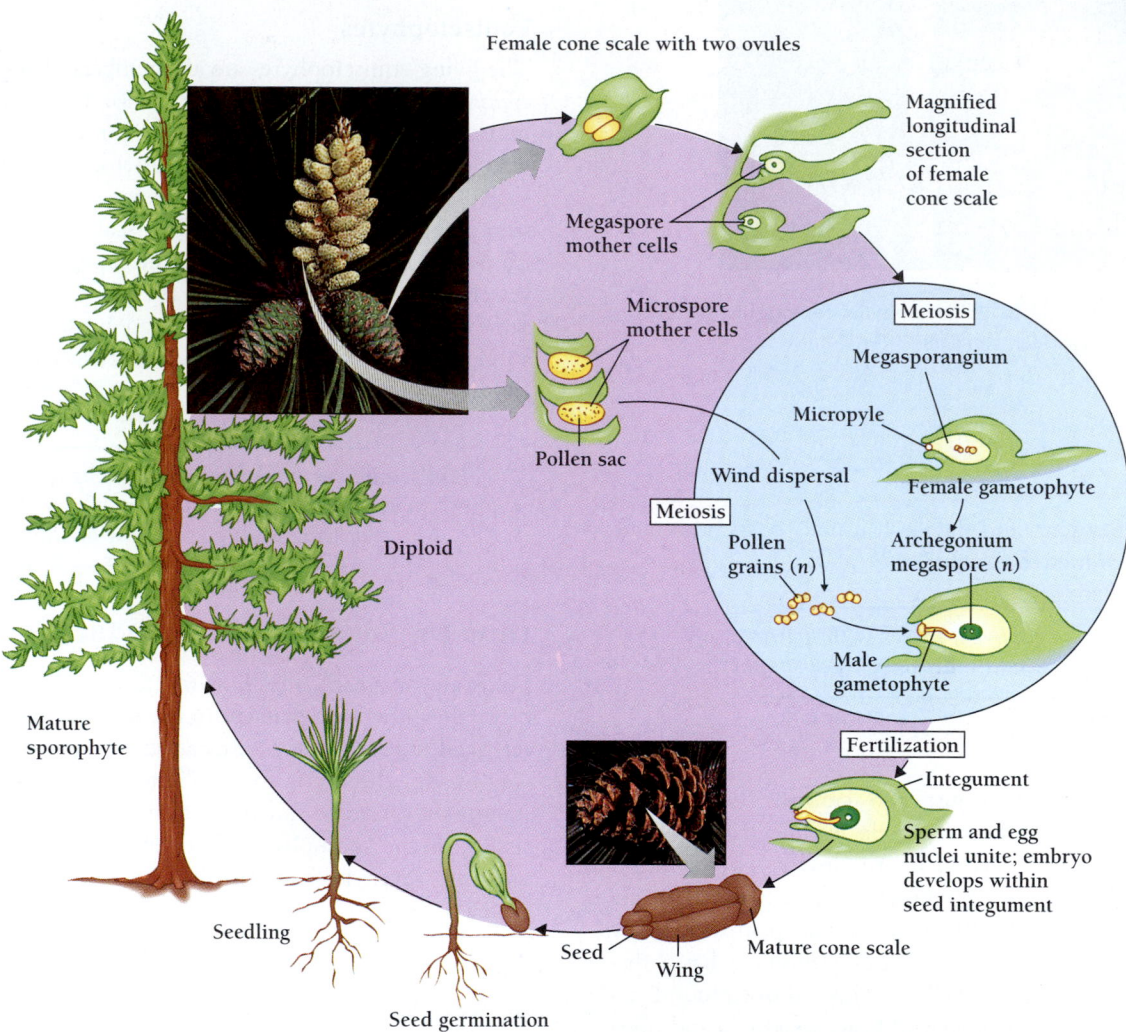

Figure 22-10   **Life cycle of a pine.** Like other conifers, pines have two kinds of cones: pollen cones and seed cones. In the seed cones, which are the familiar "pine cones," each scale bears two ovules. Each ovule in turn contains a female sporangium, surrounded by a protective integument. A cell within the sporangium undergoes meiosis to produce four haploid cells, one of which undergoes mitosis to produce the female gametophyte. The gametophyte remains within the ovule and develops two or three archegonia, each of which in turn produces a single giant ovum, loaded with carbohydrates and proteins. In the integument surrounding the ovule is an opening for the entrance of pollen. In spring, a sticky fluid in the opening captures pollen grains from the wind. After a few months, the pollen gametophyte develops into a pollen tube, which makes its way to one of the archegonia in the ovule. One of the cells of the pollen tube divides to form two sperm nuclei, which then enter the archegonium. As much as a year after pollination, one of these nuclei finally fuses with the ovum to form a zygote, which develops within the ovule. When the new embryo has a root and several leaves, the seed is ready for dispersal. *(Photos: top, Michael Giannechini/Photo Researchers, Inc.; bottom, R.J. Erwin/Photo Researchers, Inc.)*

coat protects the young embryo from drying out or from being eaten. Once the seed germinates, the embryo resumes its development and growth.

In the following sections, we will discuss the five divisions of living seed plants. The 771 named species of the first four divisions are gymnosperms, which produce seeds that are not enclosed in fruits. In contrast, the 275,000 species of angiosperms produce seeds that are enclosed in a fruit.

Seeds, which enable plants to colonize relatively dry environments, consist of a diploid zygote and a source of food encased in a seed coat. Seed plants are heterosporous.

Gymnosperms: Four Divisions of Seed Plants Without Flowers

The four divisions of gymnosperms—conifers, cycads, ginkgos, and gnetae—are all vascular plants with seeds, but without flowers and fruits. They are not closely related to one another. We first discuss the most familiar division, the conifers.

Conifers

The conifers [Greek, *konos* = cone + *phero* = carry] take their name from their characteristic cones, which are male and female strobili. The conifers include many of the most common trees in the Northern Hemisphere—the pines, firs, spruces, and cedars—as well as the magnificent redwoods of the California and Oregon coasts. Conifers grow well in the short growing seasons of northern latitudes and high altitudes. They are mostly evergreen, keeping their leaves throughout the year.

Individual leaves may last through two or three winters. (Bristlecone pines, some of which live to be more than 5000 years old, retain their needles for up to 45 years.) Conifer forests extend almost all the way across North America, Europe, and Asia.

Many conifers, including the familiar pines, have needle-like leaves that are well adapted for dry conditions, including, for example, northern winters, when the air is dry and the groundwater is frozen. A needle's small surface area and heavy cuticle minimize evaporation. Despite its simple shape, however, the needle is a megaphyll, usually with two vascular bundles. The stems of conifers are heavily reinforced with lignin, which gives trees the structural support to reach enormous sizes. The lignin in wood is also what makes it strong enough to support even the largest buildings.

Most conifers bear male and female cones on the same tree. The male, pollen-producing cones are generally small (1 to 2 cm long) and consist of hundreds of sporophylls. Wind carries the pollen grains, often in huge yellow clouds, to the female cones. The female, or ovulate, cone consists of tough scales arranged in a spiral. Figure 22-10 shows the life cycle of a typical conifer, a pine tree.

Conifers produce male and female gametophytes in cone-shaped strobili.

Cycads

The cycads are large-leafed plants that look like palms (Figure 22-11). Some cycads, called "sago palms," are used as ornamental plants in homes and gardens. But true palm trees are flowering plants, while cycads have neither flowers nor fruits: they bear naked seeds on the scales of cones near the top of each plant. Individual plants bear either male or female cones, but not both. Some people eat the seeds and underground

Figure 22-11 *Zamia pumila* is the only species of cycad native to the United States. (*Walter H. Hodge/Peter Arnold, Inc.*)

Figure 22-12 The gnetophyte *Ephedra viridas* (mormon tea) resembles an angiosperm. (*Stephen Ingram/Earth Scenes*)

stems of cycads, but cycads contain toxins that can cause gastrointestinal problems, degenerative brain disease, and even paralysis.

The small male cones of cycads consist of microsporophyll scales, each of which contains many pollen-producing microsporangia. Female cones are much larger. The cone of one Australian cycad may measure 1 m in length and weigh 40 kg (90 pounds). As in conifers, pollen travels by wind to the ovule, enters through a micropyle, germinates to produce a pollen tube and sperm, and fertilizes the ovum.

Cycads produce male and female gametophytes in strobili that lie on separate plants.

Ginkgos

The ginkgo is the last living species in the Division Ginkgophyta, the rest of whose members went extinct as long as 280 million years ago. The ginkgo, or maidenhair tree, is a living fossil that has hardly changed in the last 80 million years. Yet this beautiful tree, which decorates many gardens and parks, is itself now extinct in the wild.

Like the cycads, ginkgo trees are either male or female. The male trees have cones (strobili), while the female carries fleshy (and smelly) seeds at the ends of short stalks. The seeds of the female tree have an odor so rancid that most people prefer to plant male trees only.

Ginkgos resemble cycads in their life cycle and conifers in their growth patterns.

Gnetae have characteristics of both gymnosperms and angiosperms

The Gnetophyta, a small but diverse group of cone-bearing desert plants, consist of three distinct genera—*Gnetum*, *Ephedra*, and *Welwitschia* (Figure 22-12). Of all the gymnosperms, the gnetae most resemble angiosperms. For example, some of their strobili look like flowers, and the vessels of their leaves and stems resemble those of angiosperms. But they have naked seeds and are certainly gymnosperms. Recent DNA studies have led to the rejection of the long-favored hypothesis that angiosperms evolved directly from a gnetophyte. So the origin of angiosperm remains an unsolved, and much debated, problem.

THE ANGIOSPERMS ARE SEED PLANTS WITH FLOWERS AND FRUITS

The diversity of the angiosperms dwarfs that of all other divisions of plants. Angiosperms, also called **Anthophyta** [Greek, *antho* = flower], range in size from the tiny duckweed *Wolffia*, less than 1 mm long, to *Eucalyptus* trees, taller than 100 m. With some 275,000 known species, angiosperms are the most widespread and familiar of plants. We see them daily in our lawns, gardens, parks, fields, and forests. They provide us with furniture, clothes, medicines, and dyes. We stock our kitchens with their seeds, fruits, leaves, stems, and roots. Angiosperms provide us with wine, beer, coffee, and tea. We use the more brilliant angiosperms to brighten our lives and enliven our loves.

Despite their diversity, all angiosperms share two important adaptations: flowers, which promote fertilization, usually by insects or other animals; and fruits, which promote the survival and dispersal of seeds.

Of What Use Are Flowers?

Flowers are reproductive structures that ensure the distribution of pollen, either by animals or by wind. Some flowering plants rely on wind alone to distribute pollen. Grasses, rushes, sedges, plantain, birch, poplar, and oaks, for example, have small inconspicuous flowers and no scent. Because wind pollination is so inefficient, wind-pollinated flowers release mind-boggling amounts of pollen. Individual plants of the weed *Mercurialis annua* may release over a billion grains of pollen.

Other flowers attract insects (especially bees, beetles, butterflies, and moths), birds (especially hummingbirds), and even bats and other small mammals. All these animals' visits are rewarded with a sugary fluid called nectar, a share of the high-protein pollen, or nutritious petals or other flower parts. Animals learn to associate these delicious foods with the seductive perfumes and bright colors of flowers. Not all of a flower's attractions are necessarily attractive to humans: the eastern skunk cabbage (*Symplocarpus foetidus*), for example, attracts pollinating flies by emitting an odor like that of dead fish.

Most flowers are specialized to ensure pollination by just one group of animals. Flowers pollinated by moths, for example, are usually large, white, and overwhelmingly fragrant at night, when moths fly. In contrast, flowers pollinated by butterflies are small and brightly colored, since butterflies fly during the day and rely on sight more than smell. One orchid species produces flowers that look so much like female bees that male bees come to mate with the flowers. As the male bee attempts to copulate with the flower, he is dusted with pollen, which he then carries to another flower in another futile attempt to mate.

Insects are as well adapted to take advantage of flowers as flowers are adapted to take advantage of insects. One South African fly has a tongue nearly 7.5 cm (3 in) long, which allows it to take nectar that is out of reach for other insects. Insects and flowers provide a striking example of **coevolution,** the mutual adaptation of two separate evolutionary lines. Coevolution occurs when two species are significant factors in the lives of one another. In the case of insects and flowers, coevolution ensures a food supply for the insects and pollination for the plants.

Flowers attract animals that distribute pollen.

Of What Use Are Fruits?

A fruit is a mature ovary that encloses and protects seeds. Many fruits promote seed dispersal by animals or wind. The most familiar fruits are those that attract by appearance and taste, for example, peaches, apples, and berries. An animal usually digests the fleshy part of such fruits, but the seeds pass unharmed through the digestive system. (In fact, some seeds cannot germinate until they pass through the digestive tract of an animal.) Seeds excreted by the animal can end up far from the plant that produced them, together with a nourishing supply of fertilizer.

Other fruits use different stratagems to transport their seeds. Thistles and dandelion fruits, for example, soar like kites in the slightest breeze, and maple fruits whirl through the forest like free propellers. Burrs stick to the fur (or socks) of passing animals, while coconuts have spread to countless tropical islands by floating across the ocean. Touch-me-nots (*Impatiens*) propel their seeds from fruits that explode on touch.

A fruit is a ripened ovary that encloses and protects the seeds and usually enhances their dispersal.

How Do Angiosperms Reproduce?

In angiosperms, as in all the other seed plants, the diploid sporophyte generation is more prominent than the haploid gametophyte. In fact, the male and the female gametophytes of angiosperms are even tinier than those of the gymnosperms, and they live their whole lives inside the sporophyte's flowers.

The flowers produce haploid microspores (male) and megaspores (female) in separate sporangia, from which the gametophytes derive. After meiosis, the microspores divide once to give rise to two haploid nuclei—the tube nucleus and the generative nucleus (Figure 22-13). The male gametophyte resumes development only after it arrives on the **carpel,** the flower part that contains the ovule. The pollen tube now grows, and the generative nucleus divides into two sperm nuclei. By the time the pollen tube reaches the ovule, it contains two sperm nuclei.

Meanwhile, within the ovule, the female gametophyte has developed. Meiosis yields four haploid megaspores, only one of which survives. That megaspore divides three times by mitosis to produce the embryo sac, the mature female gametophyte. The embryo sac has eight nuclei and seven cells. One cell (the central cell) contains two haploid nuclei (called the polar nuclei). Another cell becomes the haploid ovum.

Fertilization in angiosperms is different from that in any other organism. As we will discuss in more detail in Chapter 32, there are actually two separate fertilizations, together called a **double fertilization:** (1) one sperm nucleus fuses with the nucleus of the ovum to form the zygote, and (2) the other sperm nucleus fuses with the two nuclei of the central cell to form a triploid nucleus called the primary endosperm nucleus. The zygote then divides by mitosis and develops into an embryo. The primary endosperm nucleus divides to form an **endosperm,** which will provide food for the developing embryo and, in some cases, for the germinating seedling.

Figure 22-13
Life cycle of an angiosperm.

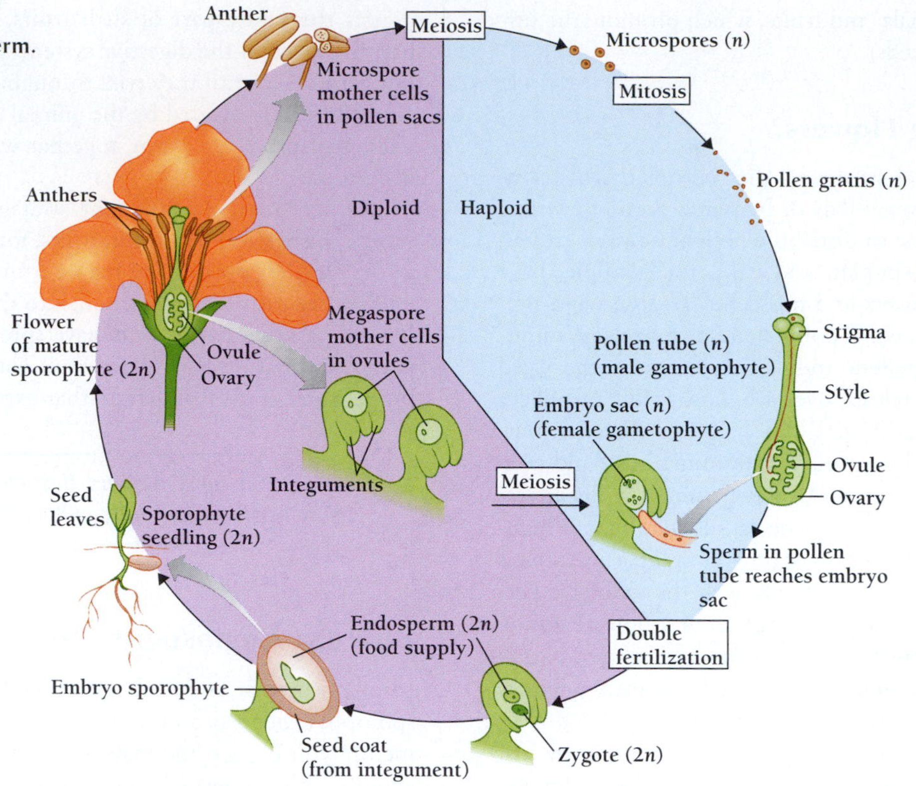

Flowers reproduce by means of double fertilization. Two sperm nuclei from the pollen grain fertilize two ova from the ovary, resulting in a diploid zygote and a triploid cell that forms the nutritious endosperm.

What Are the Parts of a Flower?

Like the gymnosperms, angiosperms are heterosporous. Separate sporangia produce male spores (microspores) and female spores (megaspores). As in the case of the gymnosperms, the sporangia lie on modified leaves, or sporophylls. In angiosperms, however, both male and female sporophylls are flower parts instead of cones. The male sporophylls are **stamens,** and the female sporophylls are carpels.

A flower is a short piece of stem that usually ends with four whorls, or circles, of modified leaves. These are the carpels, the stamens, the petals, and the sepals (Figure 22-13). The innermost parts are the carpels. Each carpel (or pistil) encloses one or more female sporangia (ovules) in a swollen base called the **ovary.** Pollen does not reach the ovules directly but first falls on the stigma, the carpel's receptive area. The style, a slender tube through which the pollen tube enters the ovary, ex-

tends from the stigma to the top of the ovary. After fertilization and development, the ovule becomes the seed, and the ovary becomes the fruit.

Just outside the carpels lies the next whorl of modified leaves, namely the stamens, or male sporophylls. Each stamen consists of an **anther,** a thick, pollen-bearing structure, and a **filament,** a thin stalk that connects the anther to the base of the flower. Within the anther are four sporangia, called pollen sacs, where the pollen grains develop (Figure 22-13).

Outside the stamens is a whorl, or **corolla,** of **petals.** In flowers pollinated by insects or other animals, the petals are usually large and brightly colored. In flowers pollinated by wind, however, the petals are reduced or missing altogether. Sometimes the petals fuse, so that the corolla is an uninterrupted tube.

The fourth and outermost whorl is the **calyx,** a whorl of leaflike parts called **sepals,** which enclose the rest of the flower and protect it before it opens. The sepals are usually small and green, but in some flowers they may be large and colorful.

Flower structures vary widely. Those that lack one or more of the four whorls are said to be incomplete. Those that lack either stamens or carpels are said to be imperfect (as well as incomplete). Flowers may have radial symmetry, so that any cut along the axis of the flower will produce two identical halves. Or they may have bilateral symmetry, so that only one cut through the flower's axis will produce identical (or mirrored) halves.

Fruits are as diverse as the flowers that produce them. Some develop from a single carpel (with one or many seeds) or from several fused carpels with a single ovary. Others, such as raspberries or strawberries, derive from many separate carpels. And still others, like the pineapple, come from many flowers.

The male sporophyll is the stamen—including anther and filament—and the female sporophyll is the carpel (or pistil)—including style and ovary. A corolla of petals and a calyx of sepals generally surround the stamens and carpels.

Why Are There So Many Species of Angiosperms?

Botanists recognize more than 275,000 species of angiosperms, which live in almost every imaginable habitat, from deserts to tropical rain forests, from river valleys to mountaintops. Starting with Linnaeus, botanists have classified flowering plants according to their reproductive organs (flowers), but angiosperms also show bewildering variations in size, woodiness, water transport, seeds, and fruits.

This huge diversity of species depends on mechanisms, discussed in Chapter 17, that maintain barriers to interbreeding. Chromosome incompatibility is usually an important barrier to interbreeding. Chromosomes from different species usually cannot form homologous pairs during meiosis, and so hybrids between two species cannot produce gametes for the next generation.

Many plants, however, have been remarkably adept in jumping this particular barrier. In some cases, plants double the chromosomes of the zygote to form a set of chromosomes that contains a diploid (rather than a haploid) set from each parent. The resulting offspring can then perform meiosis and produce gametes, because each chromosome belongs to a homologous pair.

The coming together of the doubled gametes at fertilization produces a hybrid that is *polyploid* [Greek, *poly* = many + *ploion* = vessel] because it contains several (usually two or four) sets of genes. Polyploid plants, with the combined genetic repertoire of two or more species, are often much better at adapting to new environments than ordinary diploids.

Polyploidy has been a common source of new angiosperm species during evolution. Farmers have also produced many agriculturally important plants by accidentally or deliberately inducing polyploidy. Modern wheats, for example, derive from ancient hybrids of wild ancestors. Studies of angiosperm DNA also suggest that the genomes of many angiosperm species have also expanded during evolution via "jumping genes," (transposons; Chapter 12).

Despite large differences in genome size, diverse plants use genes whose sequences are surprisingly similar, not only among angiosperms but also between angiosperms and gymnosperms. Comparisons of DNA sequences now complement morphological comparisons (especially of flowers) in suggesting the long, obscure origins of modern plants (Figure 22-14).

The terrestrial environment is far more diverse than the marine environment. Extremes of temperature and rainfall, for example, demand a great range of adaptations. As a result, terrestrial organisms are themselves more diverse than their marine cousins. Animals, which we survey in the next two chapters, are especially diverse. But most animal species are insects, and the evolution of insects is closely tied to that of plants. Plants' colonization of the land 430 million years ago may have set the stage for the evolution of insects and other terrestrial animals.

Figure 22-14 *Amborella trichopoda,* **a living fossil, grows only on the island of New Caledonia in the South Pacific.** Botanists often classify *Amborella* as closely related to avocados and sassafras. Analysis of DNA from hundreds of species suggests that the first flowering plants resembled *Amborella* and not a gnetophyte. (*William E. Ferguson*)

Study Outline with Key Terms

Plants are multicellular organisms that perform photosynthesis. They originated on the land, and most are terrestrial. The land offers three major advantages for photosynthetic organisms: more light, more carbon dioxide, and a more reliable supply of minerals and other nutrients. However, for plants to invade the land, they had to be able to obtain and conserve water.

The first plants probably evolved from green algae at least 430 million years ago. The first seed plants appeared by the late Devonian Period. Fossils of the first flowering plants date from the Cretaceous Period, about 125 million years ago.

The evolution of plants depended principally on three adaptations: (1) a waxy **cuticle,** which coats the plant's exposed surfaces and reduces evaporation; (2) the ability to absorb water and minerals from dew, rainwater, or groundwater; and (3) jacketed reproductive organs. All plants also have **stomata** and life cycles characterized by **alternation of generations** between **gametophytes** and **sporophytes.** The **gametangia** of the gametophytes produce gametes, and the **sporangium** of the sporophytes produces **spores.**

The **bryophytes** have no specialized conduction system for distributing water and nutrients and are therefore limited in the sizes to which they can grow. Bryophytes include mosses and two rarer classes, the liverworts and hornworts. The haploid (gametophyte) phase of bryophytes is more prominent than the diploid (sporophyte) phase.

Vascular plants are far more diverse and more numerous than the bryophytes. Vascular plants contain two distinct kinds of conducting tubes, the **xylem** (which carries water and minerals from roots to leaves) and the **phloem** (which distributes sugars made in the leaves to the rest of the plant). In vascular plants, the diploid phase of the life cycle is more prominent than the haploid phase.

Vascular plants disperse as diploid sporophytes. Most kinds of vascular plants disperse as **seeds,** which contain a diploid embryo, a food reserve, and a protective coat. Many **seed plants** have flowers and fruits, which aid in fertilization and in the dispersal of seeds.

Four divisions of vascular plants, or **tracheophytes,** lack seeds: pterophytes, psilotophytes, lycophytes, and sphenophytes. Pterophytes (ferns) have **megaphylls,** leaves with networks of transport tubes. Other tracheophytes have **microphylls,** leaves with just one vein. Psilotophytes are the simplest of the living vascular plants and do not have true leaves. Lycophytes are more complex than psilotophytes, but less so than ferns. Sphenophytes have jointed stems that are reinforced with silica.

Five divisions of tracheophytes bear seeds, but four of these, collectively called the **gymnosperms,** do not have flowers. A seed contains the embryo, the female gametophyte, the organ that produces the female gametophyte (the **nucellus**), and layers of tissue called the integuments from the previous sporophyte generation. The nucellus and integuments form the **ovule.**

Gymnosperms produce male and female gametophytes in **strobili.** We discussed the four divisions of gymnosperms: conifers, cycads, ginkgos, and gnetae. In conifers the strobili are separate cone-shaped structures on the same tree. Cycads produce male and female gametophytes in strobili that lie on separate plants. Ginkgos resemble cycads in their life cycle and conifers in their growth patterns. Gnetae have characteristics of both gymnosperms and flowering plants.

Anthophyta, or **angiosperms,** are the most diverse group of plants. The gametophytes of angiosperms are even more reduced than those of the gymnosperms. Both male and female gametophytes are part of the flowers of the **sporophyte,** as are the male and female **sporophylls.**

A flower is a short piece of stem that usually ends with four circles of modified leaves: the **carpels, stamens, petals,** and **sepals.** Each carpel—including stigma, style, and ovary—encloses the female sporangia. The stamens are the male reproductive structures. Each stamen consists of an **anther** and a **filament.** A **corolla** of petals, often brightly colored and scented, attracts insects that carry pollen grains (male gametophytes) from flower to flower. A **calyx** of sepals encloses and protects the rest of the flower before it opens.

Fertilization in angiosperms is different from that in any other organism and consists of two separate fertilizations—a **double fertilization.** The zygote then divides and develops into an embryo. The primary **endosperm** nucleus divides to form an endosperm, which will provide food for the developing embryo and, in some cases, for the germinating seedling. Flowering plants are divided into **monocots** and **dicots** (or **broad-leafed plants**), according to the form the embryo takes.

A fruit is a ripened, or mature, **ovary** that encloses and protects the seeds. Flowers and fruits allow angiosperms to interact with animals. Flowering plants have **coevolved** with insects.

Review and Thought Questions

Review Questions

1. Describe six traits that together set plants apart from members of the other kingdoms of organisms.
2. Name several advantages and disadvantages to life on land.
3. What traits enable plants to live on land?
4. Compare alternation of generations with the life cycle of an animal, such as ourselves.
5. How long ago do paleontologists find the first association between plants and mycorrhizae? What does this imply about the coevolution of these two groups of organisms?
6. What group of plants that flourished 300 million years ago provided the fossil fuels oil, coal, and natural gas? What is the name of that geologic period?
7. How do bryophytes absorb water? What important roles do mosses play in the world's ecosystems?

8. What is peculiar about the life cycle of a bryophyte?
9. What are xylem and phloem and what functions do they perform in vascular plants?
10. What is the sporophyte phase of a vascular plant?
11. What is the difference between a megaphyll and a microphyll?
12. What are the advantages of seeds to terrestrial plants?
13. What are the advantages of flowers and fruits to terrestrial plants? Name the parts of a flower.

Thought Question

14. Why did plant taxonomists abandon Linnaeus's system of classification? What characteristics of plants reflect their evolutionary relatedness? Why do these characteristics represent relatedness more than the number of stamens or the number of pistils?

About the Chapter-Opening Image

Artist Georgia O'Keeffe is known for her depictions of flowers as sexual organs. Linnaeus had the same idea almost three hundred years ago.

 On-line materials relating to this chapter are on the World Wide Web at **http://www.harcourtcollege.com/lifesci/aal2/**

23

Protostome Animals: Most Animals Form Mouth First

Progress and the Burgess Shale

One day about 530 million years ago, a mud slide the size of a city block suddenly broke loose and buried tens of thousands of small marine animals living in the ooze at the bottom of a shallow sea. Acres of mud sealed off the animals so abruptly and completely that the creatures never had a chance to decay. Over millions of years, the mud slowly turned to rock, perfectly preserving the animals within. Geologic processes slowly lifted the rock thousands of feet high, and today the ancient mud slide, and the creatures entombed within it, sits high in the Canadian Rockies. It is called the Burgess Shale.

When the fossil remains of the tiny animals were finally discovered half a billion years later, they unleashed a bitter debate about the nature of evolution. Within the solid mass of the Burgess Shale, the perfectly preserved creatures revealed a remarkable diversity of animals, a diversity that suggested different things to different people.

The Burgess animals' first discoverer was paleontologist Charles Doolittle Walcott and his family. In 1909, Walcott, together with his wife, Helena Walcott, and several of their children were collecting fossils in the Canadian Rockies when they came across an isolated slab of rock crammed with the small, perfectly preserved organisms. Walcott quickly found the source of this slab, higher up the slope of the mountain, and the following summer, Charles, Helena, and their half-grown children returned and dug more than 80,000 fossils from the mountain.

Walcott was a distinguished leader of the early 20th-century American scientific establishment (Figure 23-1). However, his role as top administrator at the Smithsonian Institution, as well as his positions on the boards of other scientific organizations, left him little time for deep thought or serious research. As discoverer of the Burgess fossils, he was entitled to name and classify the new organisms. In his haste, however, he classified all of the Burgess animals as either trilobites (already well-known ancient organisms) or primitive members of still living groups (such as crustaceans, jellyfish, and polychaete worms). In short, Walcott mistakenly assumed that all his animals were members of already well known phyla.

Walcott well knew how little time he had given the Burgess animals. For the rest of his life, he dreamed of studying them in detail. In 1926, he announced his plans to retire so that he could spend his last years with the Cambrian animals. But he died just 3 months short of his planned retirement, and even though he was one of the leading scientists of his day, he never fully appreciated his own remarkable discovery.

Finally, in the 1970s, three British paleontologists—Harry Whittington, Derek Briggs, and Simon Conway Morris—undertook to reexamine the Burgess fossils. The fossils were not merely flat impressions, but fully formed stone animals whose ancient bodies could be dissected and studied. Using dental drills, Whittington, Briggs, and Conway Morris carefully chipped away layers of

Key Concepts

- An animal is a multicellular organism that eats other organisms and develops from an embryo.

- Zoologists divide animals into two subkingdoms—the Parazoa (sponges) and Eumetazoa (all other animals). The members of the Eumetazoa are grouped according to their symmetry (radial or bilateral) and internal body cavities (coelomate, pseudocoelomate, or acoelomate).

- Arthropods include the greatest number of species and the greatest numbers of individuals of all animal phyla.

stone tissues, revealing the anatomies of some of the most bizarre animals ever found (Figure 23-2). Altogether, argued Whittington, Briggs, and Conway Morris, the Burgess animals represented no fewer than 25 basic body plans, of which only four survive in modern organisms.

In 1989, the Harvard paleontologist and popular writer Stephen Jay Gould wrote a book, *Wonderful Life,* describing the animals of the Burgess Shale and chronicling the work of Walcott, Whittington, Briggs, and Conway Morris. Gould argued that the diversity of life forms in the Cambrian dwarfed anything we know today. The range of anatomical designs among the animals that died in a block of mud 530 million years ago far exceeded that of all the creatures in today's oceans. The Burgess animals, wrote Gould, should have challenged the ablest classifiers of the time. Yet Walcott had classified these bizarre organisms into well-known groups of organisms such as jellyfish and shrimps.

Each of the body plans represented in the Burgess Shale, argued Gould, could have been assigned to a new phylum. Only because many of these types are represented by just a single species had taxonomists hesitated to give each one a whole phylum to itself. Charles Walcott had failed to recognize the true diversity of these creatures, Gould said, for two reasons. Walcott had assumed that the creatures he had found would fit into categories he already knew; and he assumed that evolution, by its nature, inevitably progresses from a few simple forms to a diversity of complex and modern forms (culminating in human beings). This idea—that evolution is progressive—is controversial, however, and depends on how the word "progressive" is defined.

Is evolution progressive? Does diversity increase? Do organisms become more complex? Gould argued that the diversity of the Burgess Shale provided evidence that evolution was *not* a progression from

Figure 23-1 Paleontologist Charles Doolittle Walcott and his wife, Helena Walcott, discovered a rich deposit of ancient organisms in the Canadian Rockies. *(Brown Brothers)*

simple to complex, from few forms to many. He argued that life is less diverse today than during the Cambrian explosion. The Burgess Shale, he said, illustrated that multicellular life began in great variety all at once. Since then, for reasons unknown, most of these forms became extinct. The organisms that survive today, he said, have been merely lucky survivors of various catastrophes in the history of the Earth.

But Gould's arguments were not whole-heartedly accepted by other paleontologists and evolutionary biologists. Soon after the appearance of *Wonderful Life,* further studies of the Burgess fossils by Conway Morris and others suggested that the Burgess animals were not as novel as Conway Morris had initially believed. The animals could be comfortably grouped into smaller numbers of known phyla, although not necessarily the phyla that Walcott had selected.

Were the animals that lived 530 million years ago more diverse than those living today, as Gould had argued? Probably not. While paleontologists have recognized about 11 phyla of animals from the Cambrian, biologists have grouped living animals into 35 phyla. As far as is known, no whole phylum of animals has ever gone extinct. The dinosaurs, for example, were merely a group within the class Reptilia in the phylum Chordata, the phylum to which we humans belong. And the Chordata are represented in the Burgess Shale animals.

Certainly, the history of life suggests that diversity and complexity have increased since the first appearance of life 3.5 billion years ago. All living organisms, in all their current diversity, have descended from the same few simple ancestors. The complex creatures, such as ourselves, that now populate the world have evolved, one generation at a time, from simpler organisms.

In 1995, the philosopher Daniel Dennett attacked Gould, dissecting Gould's arguments about diversity and progress as aggressively as Gould had

Figure 23-2 A few of the animals of the Burgess Shale. Floating in the upper left is a harp-shaped creature named *Marrella*. The large, green creature in the center is *Anomalocaris*, which grew up to a meter long. Just below is *Opabinia*, smaller, but with five eyes and a front-facing "nozzle." The small worm struggling in *Opabinia*'s jaws is a *Burgessochaeta* worm. *Pikaia*, shown lower right, may be the oldest known Chordate—if not our direct ancestor, at least a relative of one. *Odontogriphus* (pink, ribbonlike animal) was a flat, swimming creature. *Hallucigenia* (the green creature walking on the sea floor) had seven pairs of spines and seven tentacles. (Because of an initial misunderstanding by paleontologists, we have depicted *Hallucigenia* upside down.) The three reddish, flame-shaped creatures are *Wiwaxia*, which probably crawled along the sea floor. What look like green plants on stems are actually *Dinomischus*, a sessile animal. *Vauxia*, shown behind *Anomalocaris*, was a sponge. At the lower left is *Aysheaia*, similar to a modern-day onychophoran. Shown in its burrow underneath the sea floor is *Ottoia*, which looked and behaved much like the priapid worms that still live along our coasts today.

scrutinized Walcott's taxonomy of the Burgess animals. Dennett insisted that evolution is, in fact, progressive, in that evolution tends to increase complexity.

As organisms evolve adaptations in response to their environment and to other organisms, complexity increases. For example, the grinding teeth of a horse are an adaptation to tough grasses, which evolved toughness and lack of digestibility in response to grazing by plant-eating animals. In the pre-Cambrian, no land plants existed, so a horse could not have existed.

Increasing complexity may be described as progressive, even though no goal is involved. Such progressive change does not, however, imply that evolution is *designed* to produce increasingly complex organisms. Nor does evolution *necessarily* produce increasing complexity. After all, we are surrounded by relatively simple organisms—including bacteria, amebas, ferns, and flatworms—survivors, like us, of the ravages of major catastrophes, like those that wiped out the dinosaurs, and the constant, local, minor ones that constitute natural selection. All these relatively simple organisms are descendants, as we all

are, of a simple ancestor that lived more than 3.5 billion years ago. All organisms, both ancient and modern, are the product of the same evolutionary processes that have continued for billions of years.

WHAT IS AN ANIMAL?

As animals ourselves, we tend to see our own kingdom as the most interesting. Many of us have spent hours at zoos, observing the beautiful and the bizarre. Even the best zoos, however, barely hint at the variety of animals that exist on Earth. A coral reef is a much better place to observe the tremendous diversity of animals. A coral reef swarms with fishes, shelled animals, worms, sponges, corals, sea anemones, and many other unfamiliar organisms. Most land dwellers are awed by the variety of shapes, domes, branches, fans, tubes, and stars, as well as by the range of colors, from dark red to luminous blue.

What Features Characterize the Animals?

An animal is a multicellular, heterotrophic organism that develops from an embryo. All animals are eukaryotes, and most reproduce sexually. A few, such as the corals, can reproduce by asexual processes, such as budding. Unlike plants, fungi, and protists, however, animals never show alternation of generations. With rare exceptions, the only haploid cells are the gametes (sperm and eggs).

Nearly all animals ingest their food—usually other organisms or their remains. Like fungi, a few parasitic animals absorb nutrients directly from their hosts. But, unlike fungi, these parasites develop from embryos, as other animals do. A few animals—including many corals—appear to perform photosynthesis. Careful examination reveals, however, that these animals harbor symbiotic algae that are actually the ones performing photosynthesis.

Like plants and fungi, animals are highly structured. Almost all animals contain many different kinds of specialized cells. Groups of similar cells, along with the materials they secrete, called **matrix,** form **tissues,** units of structure and function. Two or more kinds of tissue in turn can form an **organ,** a structural unit with a distinctive function. For example, one class of cells can form an **epithelium,** a tissue that lines a surface (such as the outside of the body), while another kind of cell, along with its matrix, can form **connective tissue,** a network of loosely connected cells that may help support other body parts. A kidney, which contains both epithelial and connective tissues, is an example of an organ (Figure 23-3). The specialization of cells and the grouping of cells into tissues and organs allow animals to function efficiently.

Unlike plants, most animals can move about in search of food or mates. Some animals, such as the corals, are stationary but can move body parts to capture prey. Many have stages during development when they disperse. Almost all animals coordinate their movements in patterns that we can describe as behavior. For example, animals capture food, avoid predators, and breed. For all but the sponges, the coordination of such behavior depends on the functioning of a network of nerve cells (neurons).

An animal is a multicellular, heterotrophic organism that develops from an embryo. All animals are eukaryotes, and most reproduce sexually.

How Do Zoologists Classify the Animals?

Zoologists (pronounced zo-OL-ogists), biologists who study animals, estimate that living animal species number at least 4 million, and some estimates run as high as 30 million. Zoologists divide these millions of living animals into about 35 phyla (Figure 23-4). The 40,000 species of vertebrates (animals with backbones)—including fish, frogs, snakes, birds, and mammals—are just a subphylum of one of these 35 phyla. Vertebrates include a mere 1 percent of all animal species. Most animals have no backbones. The vast majority are insects (mostly beetles, in fact), snails, jellyfish, and worms.

Zoologists divide all these creatures into two subkingdoms. The subkingdom **Eumetazoa** [Greek, *eu* = true + *meta* = middle + *zoa* = animals] includes almost all of the 4 million named species of living animals. Excluded are the 5000 species of sponges and one other species, which make up the small subkingdom **Parazoa.** The sponges are the simplest animals, for they have neither regular symmetry nor organs.

How do biologists use developmental patterns to classify animals?

The Eumetazoa are divided into major groups of phyla according to how their embryos develop. At fertilization, or syngamy, the egg and sperm fuse to form a **zygote.** An animal zygote divides by mitosis to produce a hollow ball of cells called a **blastula** (Figure 23-5). The blastula folds in on itself to form a **gastrula,** three cell layers surrounding a simple cavity called the **archenteron** [Greek, *arche* = beginning + *enteron* = gut] (Figure 23-6). The three cell layers of the gastrula develop into distinctly different tissues in the adult. The innermost layer, which surrounds the archenteron, is called the **endoderm** [Greek, *endon* = within + *derma* = skin] and gives rise to the intestines and other digestive organs. The outermost layer, called the **ectoderm** [Greek, *ecto* = outside], gives rise to skin, sense organs, and nervous system. The middle layer, the **mesoderm** [Greek, *mesos* = middle], gives rise to muscle, skeleton, and connective tissue. That all animals develop in the same

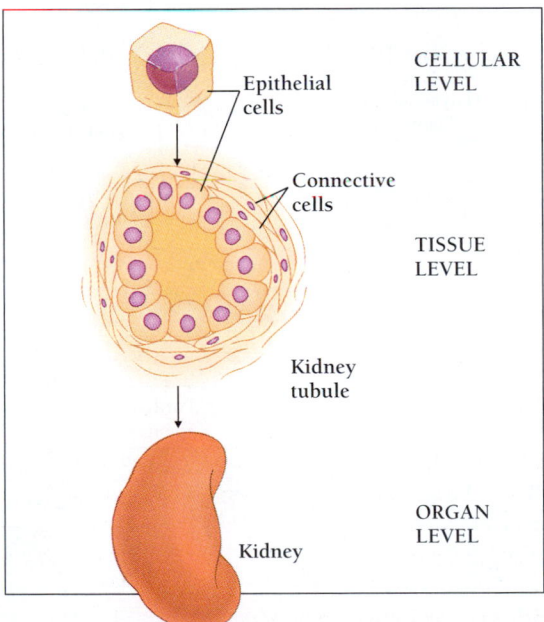

Figure 23-3 The structure of animals. A kidney is an organ, which is made up of two or more tissues. The two tissues in the kidney tubule, for example, are connective tissue and epithelial tissue, which are made of connective cells and epithelial cells, respectively.

CELLULAR LEVEL — Epithelial cells

Connective cells — TISSUE LEVEL

Kidney tubule

ORGAN LEVEL — Kidney

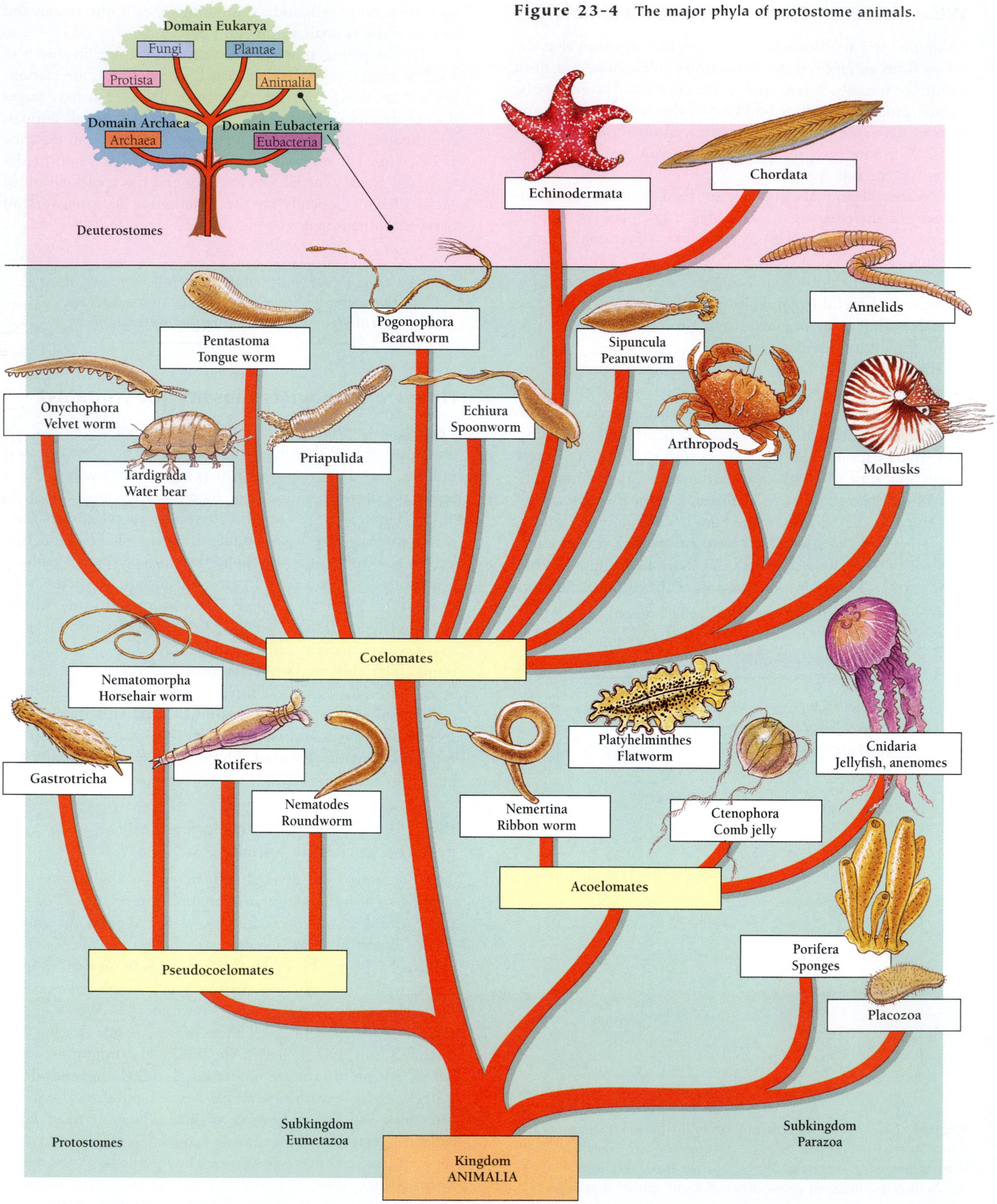

Figure 23-4   The major phyla of protostome animals.

Domain Eukarya

Fungi   Plantae

Protista   Animalia

Domain Archaea   Domain Eubacteria

Archaea   Eubacteria

Deuterostomes

Echinodermata

Chordata

Pentastoma
Tongue worm

Pogonophora
Beardworm

Sipuncula
Peanutworm

Annelids

Onychophora
Velvet worm

Echiura
Spoonworm

Arthropods

Mollusks

Priapulida

Tardigrada
Water bear

Coelomates

Nematomorpha
Horsehair worm

Platyhelminthes
Flatworm

Cnidaria
Jellyfish, anenomes

Gastrotricha

Rotifers

Nematodes
Roundworm

Nemertina
Ribbon worm

Ctenophora
Comb jelly

Acoelomates

Porifera
Sponges

Pseudocoelomates

Placozoa

Protostomes

Subkingdom
Eumetazoa

Subkingdom
Parazoa

Kingdom
ANIMALIA

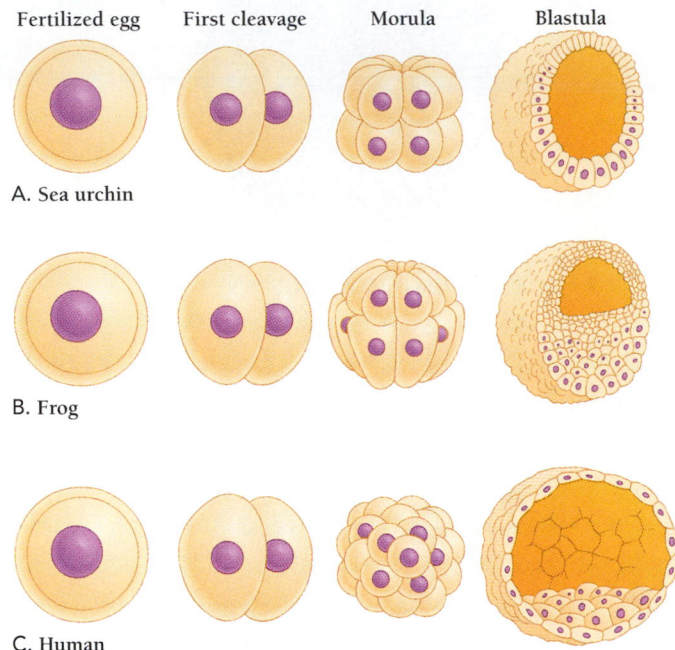

Fertilized egg   First cleavage   Morula   Blastula

A. Sea urchin

B. Frog

C. Human

Figure 23-5 **All eumetazoan embryos begin development in the same fashion.** Whether sea urchin, frog, or human, all true animals begin life by dividing to form a hollow ball of cells called a blastula. Although eggs and their zygotes may differ in size from 1 mm for frogs to 0.1 mm for humans and sea urchins, all share a similar pattern of early development.

general way provides further evidence that we share common ancestors.

Even in the most complex animals, a gastrula has the same organization, or "body plan," as the adult animal. We may view this body plan as a tube within a tube. The inner tube is the digestive tract, with a mouth at one end and an anus at the other. The outer tube is the body wall. Between the two tubes (of most animals) is a fluid-filled space, within which

the internal organs hang. In the most complex animals (including ourselves), a layer of mesoderm cells lines this space, and the lined cavity is called a **coelom** [Greek, *koilos* = hollow]. Animals with a coelom are called **coelomates.** These include, for example, snails, earthworms, insects, and all of the vertebrates.

The simplest animals, including the jellyfish and flatworms, have no cavity and are called **acoelomates** [Greek, *a* = without]. In between are the **pseudocoelomates** [Greek, *pseudo* = false], which have an organ-containing cavity, but without the mesodermal lining of a true coelom (Figure 23-7). A pseudocoel lies between the endoderm and the mesoderm and derives directly from the cavity in the blastula (Figure 23-5). Almost all the pseudocoelomates are tiny worms called nematodes.

The coelomates are further divided into two major groups according to another peculiarity of the early gastrula (Figure 23-6). At one end of the archenteron is an opening called the **blastopore.** In some animals, the blastopore eventually develops into the adult mouth. In other groups of animals, it develops into the anus, and the mouth develops later. Those in which the mouth develops first, from the blastopore, are called the **protostomes** [Greek, *protos* = first + *stoma* = mouth]. These include, for example, snails, earthworms, and all the arthropods—the spiders, insects, and their evolutionary relatives. In contrast, those in which the anus forms first, from the blastopore, and the mouth forms secondarily are called **deuterostomes** [Greek, *deuteros* = second + *stoma* = mouth]. Most coelomates are protostomes. We vertebrates and our cousins the echinoderms (sea stars and sea urchins, for example) are deuterostomes.

Zoologists divide the Eumetazoa according to the presence or absence of a coelom or pseudocoel and according to whether the blastopore develops into a mouth (protostomes) or an anus (deuterostomes).

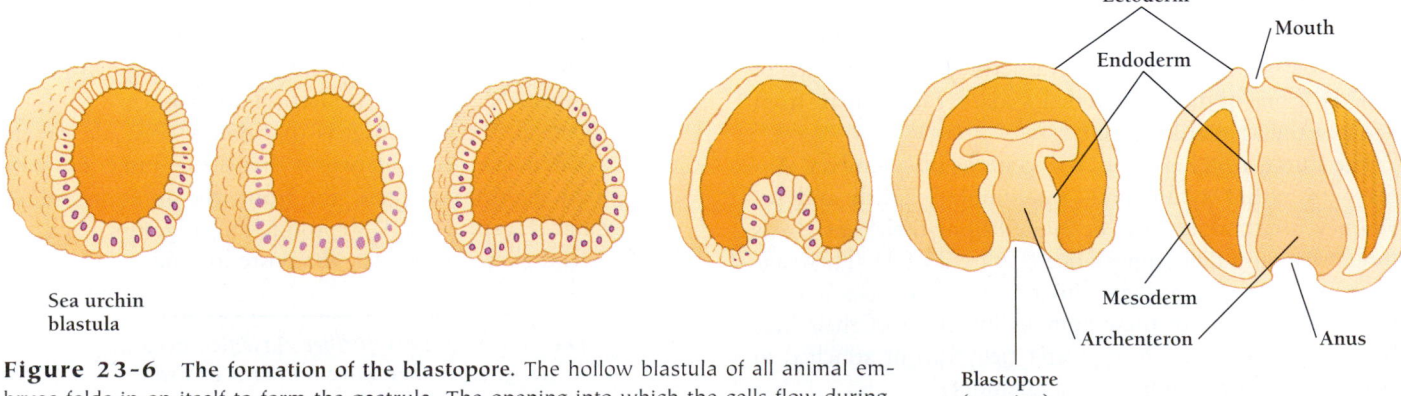

Ectoderm

Mouth

Endoderm

Sea urchin blastula

Mesoderm

Archenteron

Blastopore (opening)

Anus

Figure 23-6 **The formation of the blastopore.** The hollow blastula of all animal embryos folds in on itself to form the gastrula. The opening into which the cells flow during this process is called the blastopore. In some animals (protostomes), the blastopore becomes the mouth; in others (deuterostomes) it becomes the anus. In both kinds of animals, the hollow center (the archenteron) becomes "the tube within a tube," which later forms the gut.

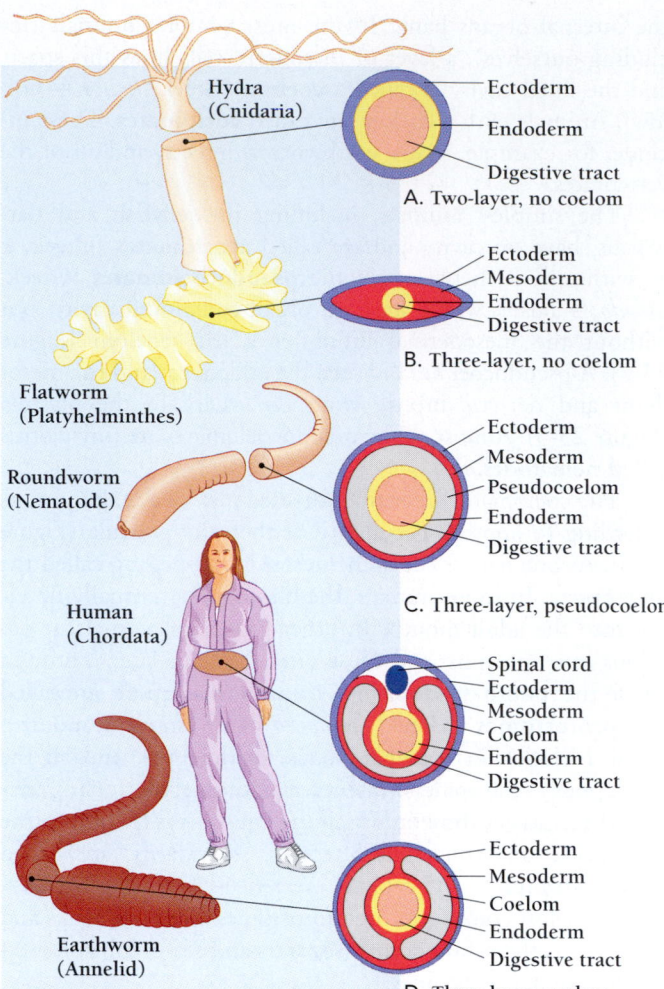

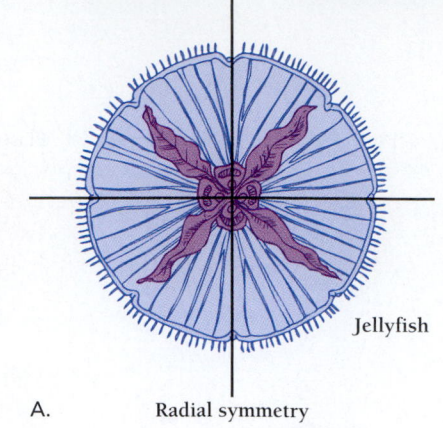

A.   Radial symmetry

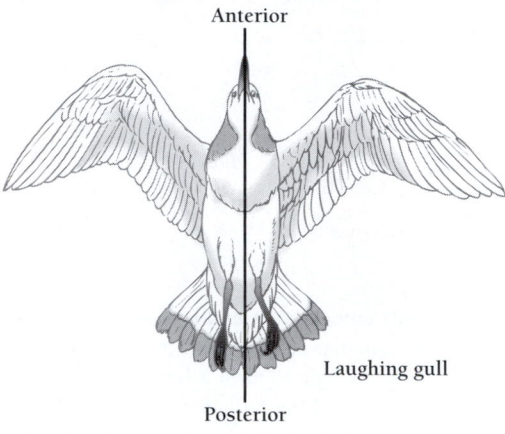

B.   Bilateral symmetry

Figure 23-8   Symmetry. A. Radial symmetry in a jellyfish. B. Bilateral symmetry in a laughing gull and a snake. All animals with bilateral symmetry have anterior and posterior ends and dorsal and ventral surfaces. Shown here are the ventral (under) surface of a jellyfish and a gull and the dorsal surface of a snake.

Figure 23-7   Tissue layers. A. A cnidarian has just two tissue layers and no coelom. B. The flatworms have three layers, but still no coelom. C. The roundworms have three layers and a pseudocoel. D. The true coelomates, including both earthworms and humans, have three layers and a true coelom, in which the gut is suspended.

How do biologists use symmetry to classify animals?

Biologists also classify animals according to their symmetry. The Parazoa (sponges) have no symmetry at all. Their lack of symmetry is one of the characteristics that set them apart from the rest of the animal kingdom.

The Eumetazoa have one of two kinds of symmetry. Animals with **radial symmetry**—the simple jellyfish and sea anemones, for example—can be rotated along their central axis without changing their appearance (Figure 23-8A). These animals have a top and bottom, but no front and back and no left and right. Many of these animals live most of their lives either floating passively in the water (jellyfish) or attached to a rock or some other support (sea anemones).

Most animals are **bilaterally symmetrical,** meaning that their left and right halves are (approximately) mirror images of one another (Figure 23-8B). Such animals have a defined

anterior (front) and **posterior** (back end). The *dorsal* surface is the back, or top, which, in most animals, faces the sky. The *ventral* surface is the part facing the Earth.

Bilateral symmetry implies some sort of head and, ordinarily, a preferred direction of movement—usually "head" first. Directed movement, in turn, demands a highly integrated system of muscles, nerves, and sense organs. For these reasons and others, bilaterally symmetrical animals are usually more complex than the radially symmetrical animals.

In general, the sense organs that detect prey and predators, as well as the neural networks that direct and coordinate movement, lie primarily near the front, the anterior end, of the animal (Figure 23-8B). Reproduction, digestion, and excretion tend to lie in the rear, the posterior end.

The Eumetazoa are further classified according to whether they have bilateral or radial symmetry. Bilaterally symmetrical animals, such as ourselves, are usually more complex than radially symmetrical animals.

In this chapter we will begin with the simplest animals, the sponges; move to the acoelomates, first those with radial symmetry, then those with bilateral symmetry; and then move to the pseudocoelomates. We will finish with an examination of the protostome coelomates. In the next chapter, we will survey the deuterostome coelomates.

THE PARAZOA HAVE NO ORGANS

The Parazoa contain two phyla, the Placozoa and the Porifera (sponges). The Parazoa differ greatly from all other animals, in showing no symmetry and minimal organization and cell specialization. The Parazoa have only simple connective tissues and no organs.

The Placozoa consist of a single species, *Trichoplax adhaerens,* which looks like a large, multicellular ameba. Unlike any protist, however, *T. adhaerens* produces gametes, develops as an embryo, and shows cell specialization. So it is a true animal.

The 5000 species of sponges (most of which are marine) come in many sizes, shapes, and colors, and live at all depths of the ocean. Almost all sponges are **sessile,** permanently anchored to rocks, logs, or coral. They range in size from a few millimeters to two meters. Some are shapeless blobs, while others resemble fans, cups, crusts, and tubes (Figure 23-9A). Like the synthetic "sponges" we use at the kitchen sink, the members of Porifera [Latin, *porifera* = hole-bearer] are full of holes.

A sponge's cells lie in three layers surrounding a central cavity, like the gastrula of all other animals. A sponge's outer, epithelial layer of cells is perforated with tiny holes, through which water enters the sponge. Each sponge pumps an enormous volume of water through its simple body. A sponge the size of a fingertip may pump 20 liters of water a day. This water, loaded with nutritious microorganisms and organic matter, enters through the pores, flows into the cavity, and exits through the sponge's large opening (or openings). The driving force for this flow comes from a layer of flagellated cells lining the sponge's central cavity (Figure 23-9B).

Between the outer epidermal layer and the flagellated cells lining the cavity is a layer of **mesenchyme** [Greek, *mesos* = middle + *enchyma* = infusion], loosely attached cells embedded within a jellylike substance. Within this simple tissue, amebalike cells capture food and shuttle it from the inside cells to the outer epithelial cells. These same amebalike cells also give rise to either sperm or eggs. Most sponges are hermaphrodites: a single organism is both male [like the Greek god Hermes] and female [like the goddess Aphrodite].

Although sponges are multicellular, they are extremely primitive. No nerve cells coordinate a sponge's responses, and each cell functions as an independent unit. Nonetheless, single sponge cells, separated from others by a sponge being forced through a piece of cheesecloth, will spontaneously reorganize themselves into a functioning sponge.

Sponges, whose fossils date from the early Cambrian Period, 530 million years ago, almost certainly evolved from flagellated protistan ancestors and diverged from all other animals early on.

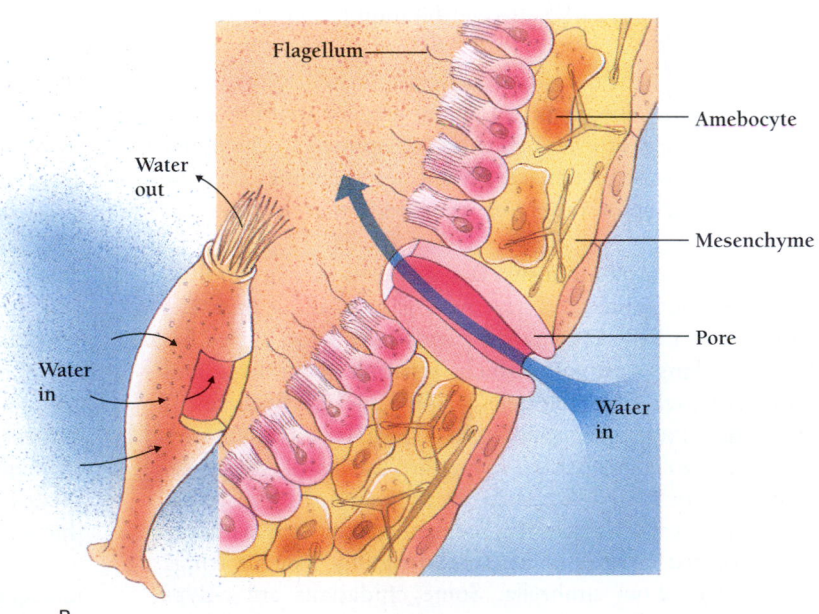

A.   B.

Figure 23-9   **Sponges.** A. An azure sponge from the Caribbean. B. Flagellated cells inside the sponge pump water through the pores of the sponge. Other specialized cells digest microorganisms and organic matter filtered from the flowing water. (*A, Brian Parker/Tom Stack & Associates*)

The individual cells of parazoans resemble protists. But unlike protists, parazoans develop from embryos and produce sperm and eggs.

RADIALLY SYMMETRICAL ACOELOMATES

The 9000 species of radially symmetrical acoelomates fall into two phyla, Cnidaria (pronounced ni-DAR-ee-uh) and Ctenophora (pronounced ten-OFF-or-uh). The cnidarians, which are simpler and far more numerous, are among the oldest animals in the fossil record, present well before the Cambrian explosion.

Cnidarians

Zoologists recognize four classes of **cnidarians:** Hydrozoa (hydroids), Scyphozoa (jellyfish), Anthozoa (corals and sea anemones), and the jellyfishlike Cubozoans. Almost all are marine (with the exception of *Hydra* and a few other freshwater hydrozoans). By far the largest class is Anthozoa, with about 6200 species. Hydrozoa include about 2700 species, and Scyphozoa only about 200.

Anthozoans [Greek, *anthos* = flower] take their name from their flowerlike appearance. They include sea anemones, sea pansies, sea fans, and sea whips. Each anthozoan has a cylindrical body with a crown of tentacles. Symbiotic algae enhance their plantlike appearance and also contribute to their nutrition, often allowing them to grow in water that is poor in nutrients. Especially impressive are the corals, whose secreted skeletons of calcium carbonate are the foundation of gigantic coral reefs.

The anthozoans have the most complex behaviors of the cnidarians. Some species of sea anemones, for example, feed on crabs, mussels, or even fish. A sea anemone can catch and consume prey with its tentacles. Some sea anemones even attack the tentacles of other sea anemones that intrude too closely.

All cnidarians contain two layers of cells that function as true tissues. But cnidarians have no organs, making them the least complex of the Eumetazoa. Cnidarians come in two basic body plans—**polyps,** which resemble cylinders, and **medusae,** which resemble bells (Figure 23-10). Both forms have a single mouthlike opening to a central cavity, and both use tentacles to capture food. Polyps are usually partly sessile, attached to rocks or other surfaces, with mouths and tentacles pointed upward. Medusae generally float free with the mouth pointed down and tentacles dangling, like the fringe on the edge of an umbrella. Some cnidarians are polyps throughout their life cycles, some are only medusae, and some cycle from one form to the other (Figure 23-11). Most cnidarians start out life as free-swimming larvae and become sessile as adults.

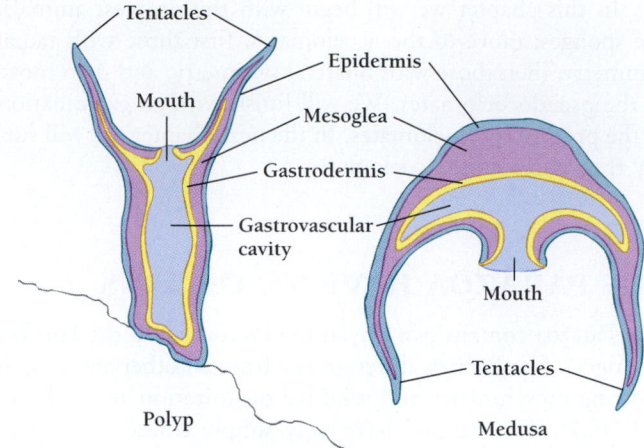

Figure 23-10 **The two body plans of the cnidarians.** Jellyfish and other cnidarians have just two body plans: the polyp and the medusa. The polyp is the sessile form exemplified by sea anemones and hydras. The medusa, exemplified by the jellyfish, is basically a polyp turned over.

Both polyps and medusae have an internal digestive cavity, which is surrounded by an outer layer of epidermis and an inner layer of gastrodermis. Like higher animals, cnidarians perform most of their digestion extracellularly—that is, within the digestive cavity. Digestion in the cavity, however, is not complete, and the cells lining the cavity engulf fragments of the small animals that compose the cnidarian's diet.

Between the two cellular layers is the **mesoglea** [Greek, *mesos* = middle + *glia* = glue], a jellylike material that contains few if any cells. In some medusae, such as the jellyfish, the mesoglea may be quite thick and gelatinous. In polyps, the mesoglea is usually thin, and in some cases (such as in the sea anemones) consists entirely of cells.

Cnidarians [Greek, *cnide* = nettle] take their name from specialized stinging cells, called **cnidocytes,** that lie on their tentacles (Figure 23-12). Each cnidocyte can use water pressure, at some 140 times the pressure of the atmosphere, to fire a tiny barbed spear called a **nematocyst.** As whalers use barbed harpoons threaded with rope to impale whales and drag the animals back to the whaling boat, cnidarians use nematocysts and their attached threads to capture their prey.

Cnidarians work much more quickly than whalers, however. Only a few milliseconds elapse between the detection of the prey by the tickling of the cnidocyte's flagellumlike cnidocil and the firing of the nematocyst. After it has pierced its prey, the nematocyst can discharge a poisonous protein. The nematocysts of one cnidarian—the Portuguese man-of-war—produces a neurotoxin potent enough to kill a human swimmer.

Once the nematocysts have attached their tiny tethers, the tentacles draw the prey mouthward. Pulling and pushing the prey requires coordination among the tentacles, which is controlled by simple muscle cells and nerve cells. A cnidarian's nerve cells are unlike those in more complex organisms, for they transmit impulses in both directions, rather than just one.

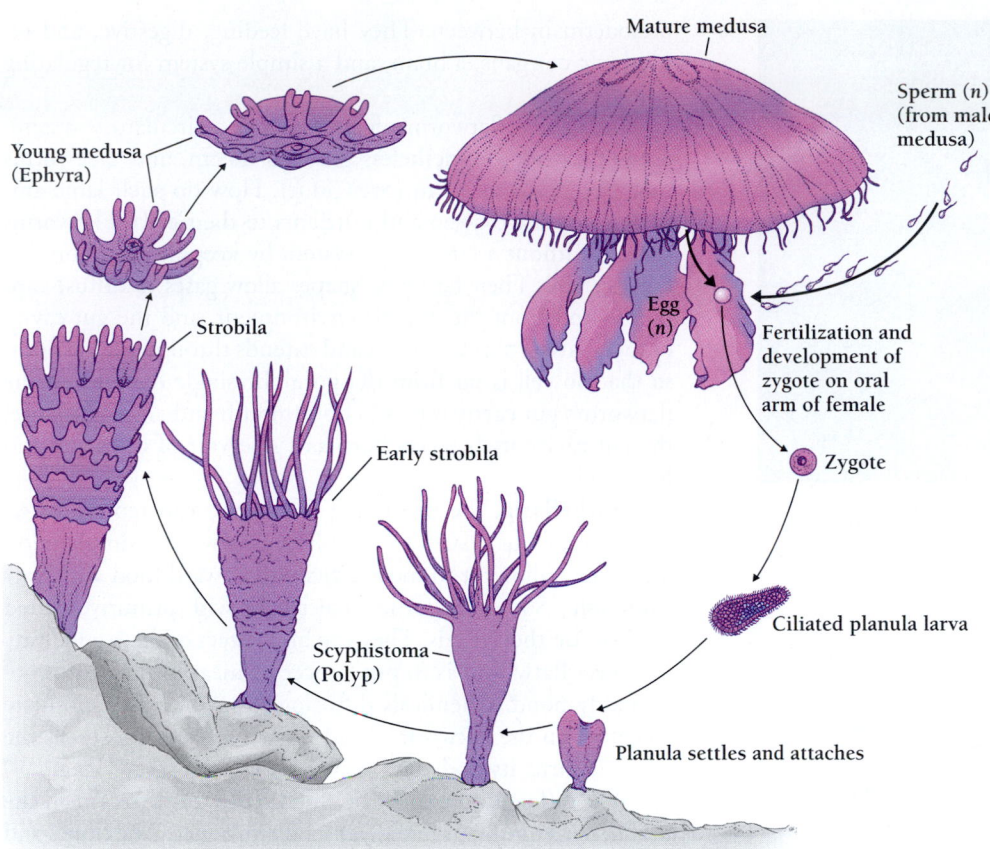

Sperm (n) (from male medusa)

Mature medusa

Young medusa (Ephyra)

Strobila

Early strobila

Scyphistoma (Polyp)

Egg (n)

Fertilization and development of zygote on oral arm of female

Zygote

Ciliated planula larva

Planula settles and attaches

Figure 23-11 The life cycle of a hydrozoan. The class Hydrozoa includes both *Hydra*, a simple polyp often studied in biology classes, and the formidable Portuguese man-of-war (*Physalia*), capable of stinging to death an adult human. Most hydrozoans are jellyfish (also called jellies) that have both polyp and medusa stages. In the polyp stage, they are colonial, with many polyps budding around a branched central cavity. *Hydra* is an exceptional hydrozoan, for it lives as a solitary polyp, with no medusa stage. All hydrozoans shed sperm and eggs from testes and ovaries that project from the body wall.

Cnidarians are radially symmetrical acoelomates with true tissues but no organs. Named for their specialized stinging cells, the cnidarians include the hydras, jellyfish, corals, and sea anemones.

Ctenophores (Comb Jellies)

The **ctenophores** [Greek, *ktenos* = comb + *phora* = motion], or comb jellies, are a small phylum of about 90 living species. Their name refers to the rows of comblike cilia that these beautiful animals carry in bands along their short bodies (Figure 23-13). Ctenophores seem to waltz through the water. By means of the coordinated action of their cilia, they move forward, mouth first, capturing prey with their sticky tentacles, while slowly rotating. They are hermaphrodites (simultaneously male and female) and usually shed both sperm and eggs into the open sea.

Because the combs are arranged symmetrically and because one species of ctenophore has cnidocytes, biologists once classified the ctenophores as cnidarians. But ctenophores' paired tentacles lack the radial symmetry of true cnidarians, and modern zoologists have given the comb jellies their own phylum. The fossil record shows that the ctenophores have existed separate from the cnidarians for at least 400 million years.

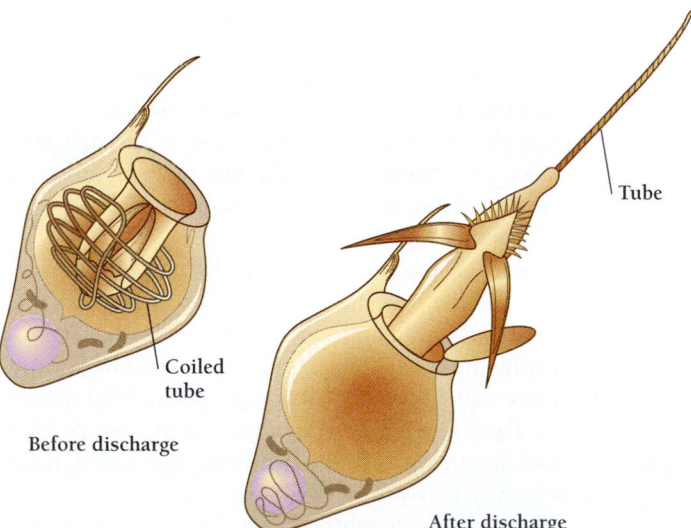

Tube

Coiled tube

Before discharge

After discharge

Figure 23-12 Cnidarians are named for stinging cells, or cnidocytes. Coiled inside each cnidocyte is a tiny barbed harpoon, covered by a lid. When the cnidocyte fires, the lid opens and the harpoon ejects, turning inside out as it unfurls. If the barbs strike the flesh of another animal, they catch and inject a paralyzing toxin. When the victim stops moving, the cnidarian pulls its prey to its mouth and begins to digest it.

Figure 23-13   A ctenophore, or comb jelly. These graceful animals are noted for their beauty. (*James R. McCullagh/Visuals Unlimited*)

Unlike the cnidarians, the ctenophores (comb jellies) are not radially symmetrical, for they possess paired tentacles. In addition, ctenophores move through the water by means of rows of cilia (the "combs").

BILATERALLY SYMMETRICAL ACOELOMATES

The cnidarians and ctenophores are radially symmetrical, or nearly so, but acoelomates may also have bilateral symmetry, as we do. Two phyla, Platyhelminthes (the flatworms) and Nemertea (the ribbon worms), are especially interesting. Flatworms include a host of parasites, which infect hundreds of millions of people, as well as the planarian *Dugesia,* used in biology laboratories all over the world. The flatworms differ from more advanced animals in having only one opening to their digestive cavities. Mouth and anus are the same.

The ribbon worms, in contrast, are the simplest animals with two openings to their digestive tract. Unlike all other acoelomates, ribbon worms have a separate mouth and anus.

Platyhelminthes (Flatworms)

The 12,000 species of **Platyhelminthes** (flatworms) live in a wide range of environments. Most free-living species are marine, although some (such as *Planaria*) thrive in fresh water, and some are terrestrial. One species is specialized for living in bat guano. Zoologists divide the flatworms into three classes: Turbellaria (free-living flatworms), Trematoda (flukes), and Cestoda (tapeworms). All three classes share the same acoelomate body plan.

All flatworms have three body layers—an endoderm that lines the digestive cavity, an ectoderm on the outside, and a mesoderm in between. They have feeding, digestive, and reproductive organs, a brain, and a simple system for regulating water balance.

One thing flatworms do not have is a circulatory system. Some flatworms nonetheless grow to 60 cm, and tapeworms may be as long as 10 m (over 30 ft). How do such large animals distribute oxygen and nutrients to their cells? Flatworms survive without a circulatory system by keeping diffusion distances small. Their flattened shapes allow gases to diffuse rapidly to and from the outside environment, and the gut cavity of each flatworm is branched and extends throughout the body, so that no cell is far from the gut. The single opening to the flatworm's gut cavity serves as both mouth and anus. Likewise, the gut cavity itself serves both as a reservoir of nutrients and for wastes.

Turbellaria, the free-living flatworms, can reproduce either asexually or sexually. A free-living flatworm's simple nervous system allows it to move efficiently toward food and away from light. Some flatworms detect light with primitive paired eyespots on their heads. They can also detect odor. A good way to collect flatworms is to put a piece of meat in the bottom of a muddy pond. Chemicals diffusing from the meat stimulate receptors on the flatworm's head. When a flatworm smells the meat, it turns its body and moves toward its future meal.

The differences among the three classes of flatworms relate to differences in the way they eat and reproduce. The flukes and tapeworms are parasites. Their outer layer of cells resists the digestive enzymes of their hosts, and they lack eyespots and chemical receptors for detecting food. Flukes take in food through a specialized mouth, while tapeworms lack a digestive system altogether and simply absorb digested food from their hosts.

Parasites are specialized for rapid and prolific reproduction in their hosts, and they often evolve stripped-down bodies, lacking many ancestral characteristics that are not useful for their specialized lives. Flukes and tapeworms illustrate many common adaptations of parasites: (1) rapid reproduction, (2) distinct stages that allow passage and dispersal through more than one host, (3) organs for attachment to their hosts, (4) specialized digestion (in the case of tapeworms, direct absorption), and (5) reduced sense organs.

As a result, the more complex free-living flatworms are probably more similar to the original flatworm ancestors than are the streamlined flukes and tapeworms.

Trematoda (flukes) range in size from less than a millimeter to more than 8 cm. All flukes (more than 8000 species) are parasites. Each attaches to the outside or the inside of its host by a hook or sucker. Like the free-living flatworms, flukes have a digestive system with a single opening.

Many flukes are hermaphrodites and generally reproduce by mutual copulation. They are far more prolific than free-living flatworms, producing 10,000 to 100,000 times as many eggs. Some flukes live their whole lives on a single host. Most, however, have more complicated lives, involving two or more hosts. The human liver fluke *Clonorchis sinensis,* for example, uses three distinct hosts to complete its life cycle—a mammal, a snail, and a fish (Figure 23-14).

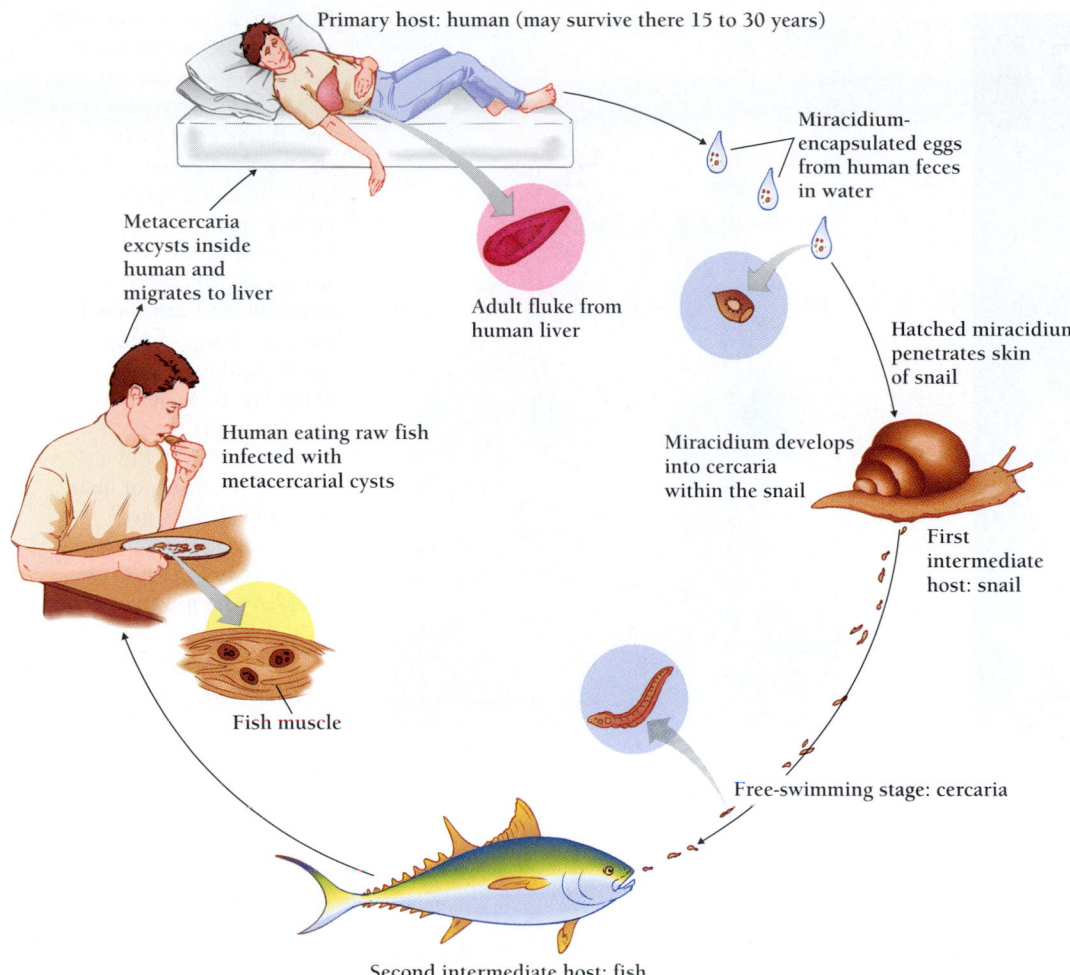

Primary host: human (may survive there 15 to 30 years)

Miracidium-encapsulated eggs from human feces in water

Metacercaria excysts inside human and migrates to liver

Adult fluke from human liver

Hatched miracidium penetrates skin of snail

Human eating raw fish infected with metacercarial cysts

Miracidium develops into cercaria within the snail

First intermediate host: snail

Fish muscle

Free-swimming stage: cercaria

Second intermediate host: fish

Figure 23-14   **The elaborate life cycle of the human liver fluke,** *Clonorchis sinensis.* The liver fluke lives its adult life as a 1- to 2-cm worm in the livers of humans, pigs, cats, and dogs. Each worm mates and reproduces, then deposits thousands of fertilized eggs. These develop into larvae, called miracidia, which leave the mammalian host in the feces. Development of the flukes then stops; the cycle continues only if the feces enter water and the larvae are eaten by a snail. Within the snail's digestive tract, each miracidium develops into a swimming larva, called a cercarium, which then reenters the water. The swimming larvae must burrow into the muscles of a fish within two days or they will die. In the fish's flesh, they form themselves into protected cysts, called metacercaria. Development resumes when a human (or other mammal) eats the raw infected fish.

Cestoda (tapeworms) are even more specialized for reproduction than are the flukes. An adult tapeworm has no digestive system at all. It lives in its host's gut, and its cells directly absorb passing nutrients.

A tapeworm maintains its easy life with a specialized attachment organ, called a **scolex,** at its anterior end. Behind the scolex lie a set of repeated segments, called **proglottids.** In the beef tapeworm, the proglottid chain may be as long as 10 meters (Figure 23-15). Each proglottid is a complete reproductive unit, with both male and female organs. If two or more tapeworms live together, they fertilize one another. Otherwise, a tapeworm self-fertilizes. In either case, the proglottids toward the rear end of the tapeworm are filled with embryos, each within a separate shell, which pass, with the feces, to the out-

side world. If cattle eat vegetation carrying these embryos, they acquire tapeworms. In the United States, about 1 percent of all cattle have tapeworms, an important reason for eating only inspected beef.

Platyhelminthes (flatworms, flukes, and tapeworms) are the simplest animals with heads. They have three body layers, true organs and tissues, a brain, and an organ for regulating water balance. They do not have a one-way digestive system: they take in food and excrete wastes through the same structures. The parasitic members of this phylum possess fewer specializations than the free-living flatworms.

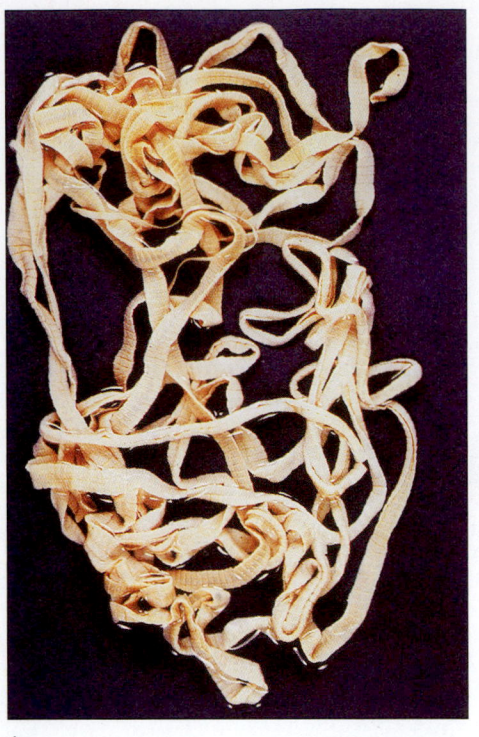

A.

B.

250 μm

Figure 23-15 Most tapeworms have two or more hosts. A. The broad fish tapeworm, *Diphyllobothrium latum*, infects tiny crustaceans called copepods, copepod-eating fish, and fish-eating mammals, including humans and bears. The broad fish tapeworm occurs worldwide, can reach 20 meters in length, and can shed a million eggs a day. A large tapeworm can live for its host's entire life. B. At the front end of a tapeworm is the scolex, which attaches to the inside of the host by means of hooks (top) and suckers (two holes). *(A, Robert Calentine/Visuals Unlimited; B, Oliver Meckes/Photo Researchers, Inc.)*

Nemertea (Ribbon Worms)

Ribbon worms, or **Nemerteans,** are a group of about 750 species of free-living acoelomate animals, most of which live inconspicuously in the sea. A few species live in fresh water or on land in the tropics. Their flat, velvety bodies may be less than a millimeter long or as long as 30 meters (Figure 23-16). Each has a characteristic snout, or **proboscis**—a long, sensitive, retractable, and sometimes venomous, tube. A ribbon worm uses its proboscis to explore its environment, defend itself, and capture prey. Ribbon worms are abundant in tidal mudflats, where they spend their days hidden in the mud, emerging only at night to feed.

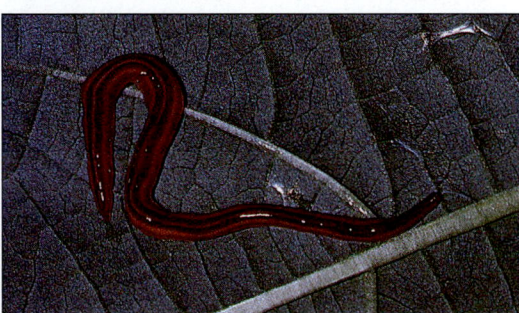

Figure 23-16 Some ribbon worms grow up to 30 m (100 ft) long. All have a two-ended digestive tract and circulatory system. *(Raymond Mendez/Animals Animals)*

Ribbon worms are closely related to free-living flatworms, with similar nervous and reproductive systems (though each ribbon worm is usually either male or female, rather than hermaphroditic). Two important advances distinguish ribbon worms from flatworms: (1) a digestive tract with both a mouth and an anus and (2) a circulatory system. The two-ended gut allows continuous eating, uninterrupted by the need to use the same opening for excretion. The second opening also permits more efficient extraction of nutrients from food, as it progresses in a single direction. The gut is like an assembly line, performing different operations at each stage. The anterior end breaks down the food, the posterior processes wastes.

The ribbon worm's circulatory system is relatively simple; in fact, some species have only two vessels. The blood is usually colorless, but may also be yellow, red, orange, or green. There is no heart to pump the blood in a particular direction: instead the vessels and the body wall contract irregularly, sloshing the blood through the circulatory system. These primitive adaptations make ribbon worms the most complex of the acoelomates.

Ribbon worms (Nemertea) have what flatworms have (nervous system and organs, for example). In addition, ribbon worms have a two-ended digestive system, a circulatory system, and a proboscis.

BOX 23.1

The Origin of Feces

Did the evolution of the one-way digestive tract fuel the Cambrian explosion? In a controversial new theory, geochemists have suggested that the development of one-way digestive systems by metazoans provoked drastic changes in the deep ocean, changes that allowed metazoans to colonize the bottom of the sea in an unparalleled adaptive radiation.

Until the Cambrian explosion, some 540 million years ago (mya), life consisted almost exclusively of single-celled bacteria and algae. As photosynthetic algae proliferated, they pumped increasing amounts of oxygen into the atmosphere. To see how this change in the composition of the atmosphere affected ancient life, a group of geochemists led by Graham Logan measured the chemical signature of fossilized organic matter in ancient ocean deposits.

To their surprise, the researchers found that at the beginning of the Cambrian, a dramatic change in the proportion of different carbon isotopes occurred. Before 590 mya, the sea floor deposits were relatively high in ^{13}C, a heavy isotope of carbon. Sometime between 590 and 540 mya, however, the deposits became lower in ^{13}C and higher in ^{12}C. This switch in "isotope signature" coincided closely with the flowering of multicellular life known as the Cambrian explosion. The researchers could not help asking, Were the two events related?

It seemed possible. Organisms tend to have higher ratios of ^{12}C:^{13}C^{12} than the atmosphere. During photosynthesis, plants incorporate into sugars and other new compounds a lighter proportion of ^{12}C than ^{13}C. (This is because ^{12}C is lighter than ^{13}C and diffuses more quickly.) The result is that ^{12}C accumulates in photosynthetic organisms and all the organisms that eat them. Organic matter therefore has more ^{12}C than the atmosphere.

But nearly all organisms respire—a process that breaks down organic compounds and releases carbon dioxide. Respiration puts the ^{12}C back into the atmosphere. If photosynthesis and respiration are balanced, the ratio of ^{12}C to ^{13}C is also balanced. In that case, researchers would not see an excess accumulation of ^{12}C in ocean floor sediments.

But Logan and his colleagues had an idea that would account for the change in ratios. Before 590 mya, the researchers hypothesized, bacteria in the upper layers of the ocean decomposed nearly all organic waste, releasing organic ^{12}C back into the atmosphere. Early deep-sea sediments, then, were relatively high in ^{13}C.

Huge numbers of bacteria would have filled the upper layers of the ocean, and their respiration would have depleted ocean waters of oxygen. The surface would have remained oxygenated, but the water near the bottom would have contained very little oxygen. Few organisms could have survived there.

But what if the bacteria were suddenly prevented from decomposing organic matter? Logan and his colleagues suggested that the evolution of a one-way digestive tract made all the difference. Animals with a one-way gut can package wastes in concentrated packets (feces), which fall to the ocean floor before surface-dwelling decomposers can absorb and digest them. With less decomposition occurring in the upper layers of the ocean, the overall oxygen content of the water would have risen. Ultimately, argue the researchers, oxygen would have diffused into the deeper waters of the ocean—allowing colonization by oxygen-requiring metazoans. That, the researchers say, paved the way for the Cambrian explosion.

Did the first fecal pellets fuel the Cambrian explosion? Many researchers remain skeptical of this tentative train of hypotheses. But the idea that the origin of all the major animal phyla depended on the evolution of feces has a certain appeal.

ASCHELMINTHS

Zoologists sometimes jokingly call the **Aschelminths** [Greek, *askos* = bag + *helminthos* = worm] the ash-can phylum, since it includes an assortment of unrelated animals that do not fit anywhere else. Like the kingdom Protista, the Aschelminths are united more by what they do not have than by what they do. Because the members of this group are so different, the Aschelminths are not considered a single phylum, but an assemblage of eight or nine distinct phyla.

All of the bilateral animals we have seen so far have been acoelomates, which have solid bodies with no internal cavity (other than the gut). The Aschelminths also lack a coelom. Some Aschelminths, however, have a cavity called a **pseudocoel** [Greek, *pseudo* = false + *koilos* = hollow]—"false" because it lacks the mesodermal lining of a true coelom. The pseudocoel plays two important roles: (1) it is a distribution route for nutrients, gases, and wastes; and (2) it serves as a hydrostatic skeleton, a support for the body that depends on the rigidity of enclosed fluid under pressure.

Aschelminths share a few traits. Most are tiny, less than a few millimeters long, and covered by a cuticle. The animals all have sticky glands for attaching themselves to some surface; the body cavity, if any, is a pseudocoel; and the sexes are usually separate. Aschelminths also lack true circulatory systems. Finally, Aschelminths have a one-way digestive system, with both a mouth and an anus.

The Aschelminths are a diverse group, and most of the phyla have relatively few species each. However, one phylum, the nematode worms, dominates all the rest. Zoologists have so far named 12,000 nematodes, and many researchers suspect that a million more remain to be identified.

Nematodes (Roundworms)

Nematodes—also called roundworms—are slender, cylindrical, and unsegmented. They range in length from a few millimeters to nearly a meter. Most are free living, however, and quite small. Free-living nematodes in the soil are actually aquatic animals, living in the film of water that surrounds each soil particle. They help to aerate the soil and circulate minerals and organic matter. Nematodes are among the most numerous of animals. A rotting apple in an orchard may contain 100,000 nematodes. An acre of soil contains billions of the little worms.

Many nematodes are parasites. They infect virtually all plants and animals. In fact, it has been said that if all organisms except the nematodes suddenly disappeared, a ghostly image of the world's former occupants would remain in the form of billions upon billions of nematode worms.

A nematode has a large pseudocoel between the body wall muscles and the central tubular gut. Because nematodes lack a circulatory system, fluids move about the body within the pseudocoel. These fluids push against the walls of the pseudocoel and thereby stiffen the whole body in the same way that water stiffens a hose. The nematode's muscles work against this support in the same way that the muscles of a vertebrate pull against a bony skeleton. The hydrostatic skeleton allows the worm's body to return to its original shape after muscle contraction has stopped.

The nervous system of a nematode is more complex than that of an acoelomate. A free-living nematode has sense organs that detect movement and chemicals, nerve cords that run the length of its body, and a primitive brain made up of a ring of nerves just behind its mouth.

Nematodes reproduce sexually, with male and female usually separate. In most species, fertilization is internal and therefore requires copulation. Some species are hermaphroditic, and some can even fertilize themselves. Still others are parthenogenic, meaning that females reproduce without males. The unfertilized eggs develop into normal embryos and adults, just as a diploid zygote would. Nematodes are prolific breeders: a single female can produce more than 25 million eggs in her lifetime.

Nematodes (roundworms) are tiny, cylindrical worms that live everywhere in great numbers. They have a fluid-filled pseudocoel that acts as a hydrostatic skeleton and circulatory system, and they have sense organs and a primitive brain.

The other seven phyla of Aschelminths include the *rotifers* (Figure 23-17), tiny aquatic animals with saclike bodies; the *nematomorpha,* or horsehair worms; and the *gastrotricha,* microscopic, bottle-shaped animals that live in freshwater or marine environments.

Despite the great advantages of the pseudocoel, many more possibilities for animal evolution came with the evolu-

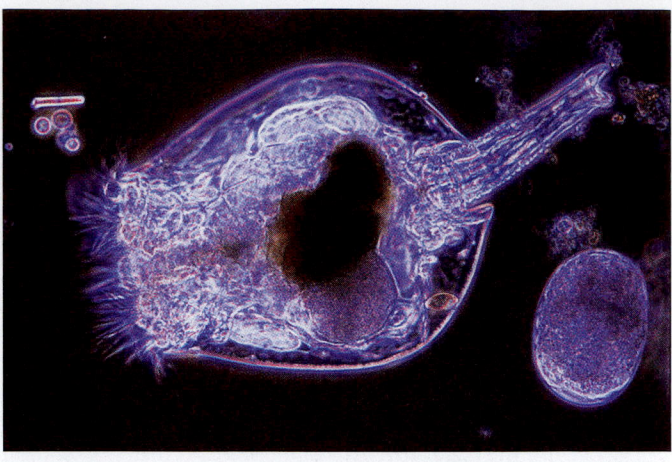

100 μm

Figure 23-17   Brachionus female with eggs. The 2000 species of rotifers are among the smallest of all animals, ranging in size from 40 micrometers (μm) to 2 mm. These common, freshwater animals are often shaped like tiny trumpets, with a crown of cilia around the open end. The cilia direct water containing bacteria, protists, and small animals into the mouth. The whirling cilia make the crown look like a rotating wheel, giving the phylum its name [Latin, *rota* = wheel + *ferre* = to bear]. Rotifers also have tiny but effective jaws, which grind food after it enters the mouth. The rotifers' success depends on the rapidity of their reproduction: some species can double their population every 2 days. And females commonly reproduce without males, through parthenogenesis. Rotifers are a major source of food for other animals that live in fresh water. *(M.I. Walker/Photo Researchers, Inc.)*

tion of the coelom. In the next section, we will see that the coelom allows animals of much greater size, complexity, and mobility.

PROTOSTOME COELOMATES

All of the more complex animals have internal body cavities outside their intestinal tracts. An additional internal space (whether coelom or pseudocoel) brings several advantages. For example: (1) it prevents muscle movement from interfering with the operation of the digestive and circulatory systems, (2) it can provide a hydrostatic skeleton against which muscles can work, and (3) it provides a protected space for the production of sperm and ova.

An internal cavity allows organs to develop and to work far more freely than in acoelomate animals. Your lungs, for example, could not fill and empty so easily if each breath meant struggling against surrounding muscles instead of just filling an empty space within the chest cavity. Similarly, the coelom isolates the working of the intestinal tract from the movement of the surrounding muscles.

The major difference between a true coelom and a pseudocoel lies in the tissues that surround the digestive tract and the internal cavity. A pseudocoel arises from the space between endoderm and mesoderm, so that the gut is lined with a single layer of tissue—endoderm only. In contrast, a true coelom arises within the mesoderm, so that the gut is lined with endoderm surrounded by mesoderm. This allows the gut to develop more complex digestive organs (such as the vertebrate pancreas or liver). The coelom within the mesoderm also allows the mesoderm itself to develop more efficient organs for circulation and excretion.

How Do Protostomes Differ from Deuterostomes?

Nearly all coelomates have digestive tracts with both a mouth and an anus. Recall that coelomates are divided into two groups according to whether the mouth or the anus develops from the blastopore. Those in which the mouth develops first, from the blastopore of the early embryo, are called protostomes; those in which the blastopore becomes the anus and the mouth forms later are called deuterostomes.

The two groups of coelomates develop in other distinctive ways. In protostomes, for example, the cell divisions immediately after fertilization are usually spiral, while the deuterostomes undergo any of several other patterns of cell divisions. In addition, deuterostome coelomates often form the coelom differently from protostomes. None of the other developmental differences, however, are as clear cut as the developmental fate of the blastopore.

In the rest of this chapter, we briefly survey three large invertebrate phyla of protostomes—the mollusks (which include snails, slugs, clams, oysters, squids, and octopods), the annelids (which include earthworms, segmented marine worms, and leeches), and the arthropods (which include horseshoe crabs, spiders, lobsters, and insects). The arthropods, most of which are insects—some three-quarters of a million species—include more named species than any other phylum of organisms.

The annelids, mollusks, and arthropods constitute a group of relatively closely related organisms. The larvae of mollusks, for example, are similar to those of annelid worms. In addition, the segmented parts of earthworms and other annelids recall the repeated body parts found in certain mollusks. The arthropods also resemble annelid worms in certain respects. Like the annelids, arthropods are segmented—think of insects (especially caterpillars), centipedes, and millipedes. Arthropod nervous systems and circulatory systems also resemble those of annelids. Molecular techniques have contributed much to current understanding of how these three phyla are related.

In protostomes, the blastopore develops into a mouth. In deuterostomes, the blastopore develops into an anus.

Mollusks Share a Common Body Plan

If we take the diversity of species as a measure of a phylum's evolutionary success, then, among the animals, mollusks are second only to the arthropods. Over 100,000 species of mollusks are now alive, in virtually every environment where animal life is possible. Mollusks are abundant in the seas, in fresh water, and even on land. The land species, slugs and snails, alone outnumber all species of reptiles, birds, and mammals combined. Mollusks may be as small as a grain of sand (about 1 mm long) or as long as a whale. (The 20-m giant squid is no match for a big whale, however, as it is not nearly as heavy as a whale.) Most mollusks, whether snails or clams, live in protective shells. Indeed, one of mollusks' distinguishing characteristics is their shell. The name **mollusk** [Latin, *molluscus* = soft] refers to the soft body of this invertebrate.

Mollusks are important to humans: oysters, scallops, squid, and other mollusks are among the great delicacies of the culinary world. Pearls and "mother of pearl" (from shells) have long been used to decorate people and their possessions, and the secretion of the marine snail *Murex* long served as the source of the dye royal purple, coveted by kings. Mollusks are also pests: some destroy crops and flowers, others damage docks and boats, and still others are intermediate hosts for such devastating parasites as schistosomes and other flukes.

All mollusks share a common body plan (Figure 23-18A). The intestinal tract, as well as the excretory and reproductive organs, lie within the main body of the mollusk, which is called the **visceral mass** [Latin, *viscera* = internal organs]. Attached to the visceral mass is the **foot,** a muscular extension of the body that the mollusk uses for sensing, grabbing, creeping, digging, and just holding on. The dorsal surface (top) of the visceral mass forms a **mantle,** specialized tissue that, in most mollusks, secretes a shell. Folds of the mantle also enclose a space, called the **mantle cavity,** which contains the mollusk's breathing organs. Both the mouth and the anus usually open into the mantle cavity. Snails and other shelled mollusks can withdraw into their mantle cavities, in some cases pulling behind them a protective flap called the **operculum,** which resembles a trapdoor.

In octopods, squid, sea slugs, and other aquatic mollusks, the mantle cavity contains gills—thin organs that transfer dissolved oxygen to the circulatory system. Air-breathing snails and slugs use the mantle cavity as a lung.

Most mollusks have a **radula,** a rasping tongue covered with teeth made from chitin. Mollusks may use the radula to protect themselves, to capture prey, to scrape vegetation, or to tear food into tiny bits. A snail in an aquarium, for example, uses its radula to scrape algae from the aquarium's glass wall.

All mollusks have a specialized digestive tract, beginning with a mouth and ending with an anus. All have one or two kidneylike **nephridia,** tubular excretory organs that remove nitrogen-containing wastes from the coelomic fluid and regulate water and salt concentration.

All mollusks also have a circulatory system with a heart that receives oxygen-carrying blood from the gills and pumps it to other body tissues. Unlike humans, most mollusks have open circulatory systems, in which the heart pumps blood through spaces between the tissues, rather than exclusively through blood vessels. The only exceptions are the

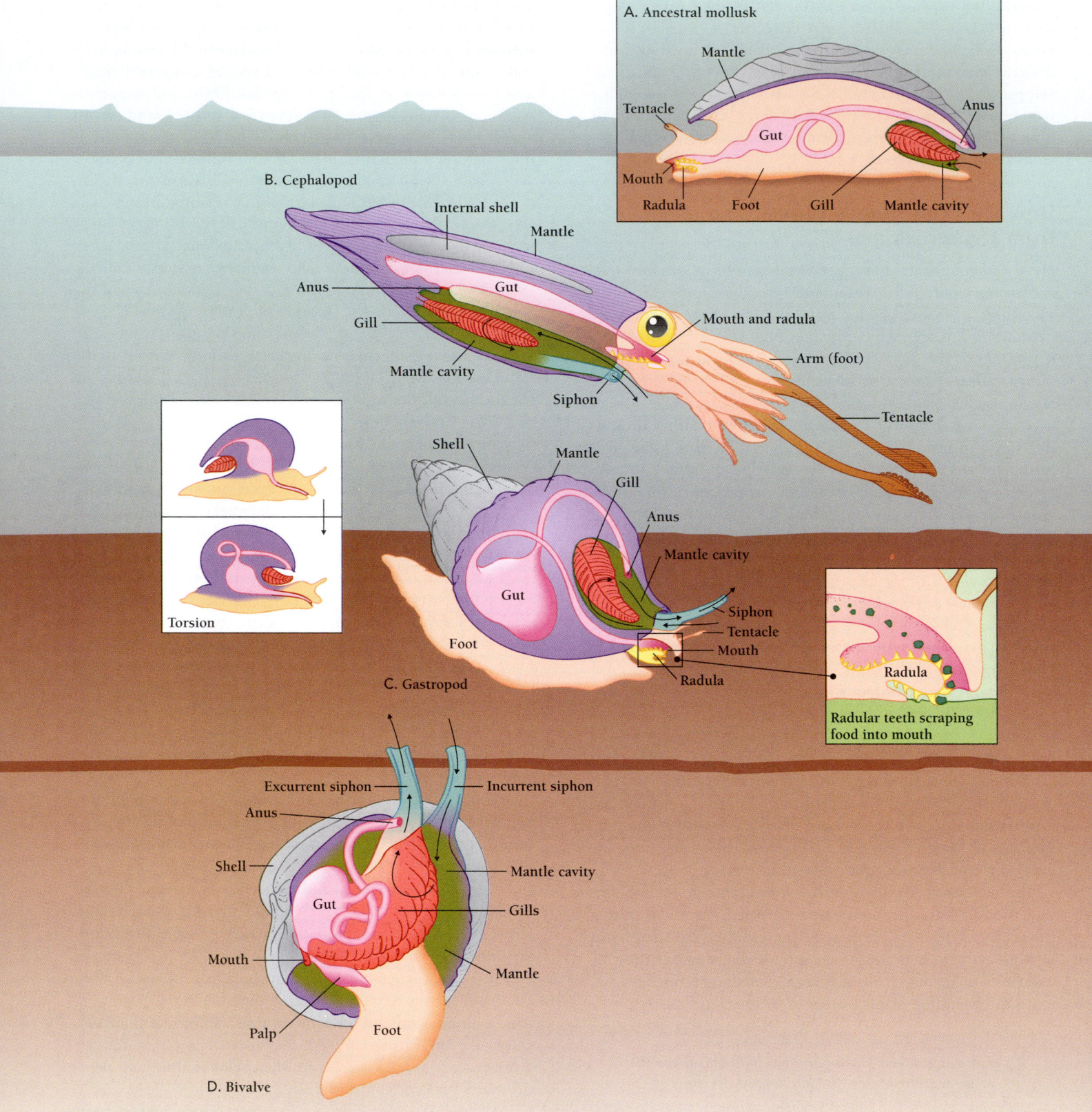

A. Ancestral mollusk

Mantle
Tentacle
Anus
Gut
Mouth
Radula
Foot
Gill
Mantle cavity

B. Cephalopod

Internal shell
Mantle
Anus
Gut
Gill
Mouth and radula
Mantle cavity
Arm (foot)
Siphon
Tentacle

Torsion

C. Gastropod

Shell
Mantle
Gill
Anus
Mantle cavity
Gut
Siphon
Tentacle
Foot
Mouth
Radula

Radula
Radular teeth scraping food into mouth

D. Bivalve

Excurrent siphon
Incurrent siphon
Anus
Shell
Mantle cavity
Gut
Gills
Mouth
Mantle
Palp
Foot

Figure 23-18 **The body of a mollusk.** A. The basic mollusk has a radula, a foot for locomotion, and gills in a mantle cavity. Cilia on the gills drive water through the mantle cavity, from which the gills can absorb oxygen. The passing water also carries away carbon dioxide and other wastes. Land snails and slugs have no gills; instead, oxygen diffuses directly from the mantle cavity into the surrounding circulatory system. B. Cephalopods such as octopods have specializations not shared by other mollusks, including a siphon, which can be pivoted to force a water stream in different directions, and suckered tentacles. Some cephalopod species are quite large. The largest known fossil cephalopods had straight conical shells up to 5 m (16 ft) long. C. Gastropods such as snails depart from other mollusks in having a twisted body and a distinct head with many sense organs, including touch receptors, chemical receptors, and eyes. D. Bivalves have the foot, mantle, shell, cavity, and coelom of other mollusks, but they also have siphons for directing a current of water through the mantle cavity.

cephalopods, which have a closed, well-developed circulatory system.

Mollusks reproduce sexually. Usually, male and female gonads are in separate individuals, but some (such as land snails) are hermaphroditic (both male and female). Individuals of some species, including sea slugs and oysters, change sexes from male to female and back several times during a mating season.

Cross-fertilization is the rule, even among the hermaphrodites. Self-fertilization, however, does occur—an important adaptation for animals that sometimes move too slowly to find a mate. Among the land snails and slugs, fertilization is internal. In other mollusks, gametes are shed externally.

Mollusks are soft-bodied coelomates, with sophisticated organs for respiration, circulation, digestion, and excretion. All the mollusks share a similar body plan, including a foot, a mantle cavity, and a visceral mass. Most have shells and a rasping tongue called a radula. They reproduce sexually, and their embryos develop into a characteristic larval stage.

Zoologists divide the mollusks into seven or eight classes, of which we will briefly discuss three: the bivalves (which include clams, oysters, and mussels), the gastropods (which include snails, slugs, and abalones), and the cephalopods (which include squid, octopods, and the chambered nautilus).

In addition, the mollusks include *aplacophora* [Greek, *a* = without + *plak* = flat plate], shell-less, wormlike animals that live in the deep seas; *monoplacophora* [Greek, *mono* = one], which have a single dorsal shell; the chitons, or *polyplacophora* [Greek, *poly* = many], which have eight plates on their backs; and the tusk or tooth shells, which have tubular shells open at both ends.

Bivalves

Bivalves [Latin, *bi* = two + *valva* = part of a folding door] include clams, oysters, scallops, mussels, and other mollusks

with two shells—a right shell and a left shell. A ligament hinges the two shells (or valves) and a muscular "foot" protrudes from between them (Figure 23-18D). Two large muscles, called adductor muscles, pull the shells together. Although most bivalves lead sedentary lives attached to rocks, some move about a little more. A clam, for example, uses its foot to burrow into the sandy or muddy bottom of the sea or river. The most active bivalves are the scallops, which "swim" by using their adductor muscles to clap their shells together, simulating a human swimmer's "frog kick." A scallop's enormous adductor muscles, tender and delicately flavored, are familiar to seafood lovers (although some restaurants misleadingly give the name "scallops" to circular plugs of shark fin).

Bivalves such as clams, oysters, scallops, and mussels have two shells and usually lead sedentary lives.

Gastropods

The **gastropods** are the most diverse class of mollusks, with some 80,000 named species, including snails, whelks, conches, limpets, abalone, and slugs. They have a distinctive twisted body and, except for the shell-less slugs, asymmetrical spiral shells.

Gastropods differ from all other mollusks in having a mysterious 180-degree counterclockwise twist of the body (viewed from above) that occurs during embryonic development. One result is that both the anus and the mantle cavity end up at the front end (anterior end) of the body, not far from the mouth (Figure 23-18C). Torsion may occur very rapidly, sometimes over the course of just a few minutes. Biologists have many hypotheses about what advantages torsion might confer on gastropods, but they have not yet agreed on an answer.

The shells of gastropods show great variation in form and color. Most gastropods are marine, but many inhabit fresh water or land. Unlike the quiet bivalves, gastropods are active creatures, equipped with a well-developed head and a sensitive nervous system. Gastropods are free living and consume plants, small animals, or decaying organic matter. A few species live as parasites.

Gastropods such as slugs, snails, and abalone are active animals with twisted bodies, single shells (or no shell), and well-developed heads.

Cephalopods

The **cephalopods,** which include nautiluses, squids, cuttlefish, and octopods, are the most active and intelligent mollusks (Figure 23-18B). Although thousands of extinct cephalopods possessed well-developed shells, among the 600 living species only the *Nautilus* possesses an external shell (Figure 23-19). Surrounding the mouth are a set of tentacles that have evolved from the foot. Each species has a characteristic number of tentacles (or arms)—a squid has 10, an octopus 8 (hence its

Figure 23-19   The chambered nautilus. The nautilus shell consists of a series of chambers arranged in a spiral. The nautilus occupies only the outermost chamber, and as the animal grows, it builds a new, larger chamber and moves in. The nautilus uses the remaining chambers to float or sink at will. In the same way that a balloonist makes a balloon rise by adding more hot air to the balloon, the nautilus injects gas into its shell to float higher in the water. To sink, the animal siphons off the gas. Today, only a few species of nautilus exist, all living in the western Pacific Ocean. But relatives of the nautilus, called ammonites, flourished for hundreds of millions of years, from the Devonian Period to the end of the Cretaceous Period. *(Douglas Faulkner/Photo Researchers, Inc.)*

name), and a nautilus 80 or 90 tentacles. Cephalopods, which are all carnivorous, use their tentacles to capture prey, to pull themselves along the sea bottom, and to steer themselves in open water.

Considering the intellectual limitations of their relatives, the bivalves and the gastropods, cephalopods are amazingly intelligent. In a laboratory aquarium, an octopus (plural, octopods) can learn to distinguish objects of different shapes and even to find its way through a maze. Octopods are very good at escaping, and they think nothing of disassembling an aquarium to do so. Presented with stones or small bricks, an octopus can build a protected enclosure, hide behind it, and bob up and down to look around. An octopus also shows much planning and patience in stalking and luring its prey. Cephalopod mating rituals are charmingly complicated (Figure 23-20).

An octopus's interesting behavior is controlled by the most sophisticated brain of any invertebrate. Information about the outside world comes from touch receptors on the tentacles, as well as from large and complex eyes. When we consider that octopus eyes evolved independently of vertebrate eyes, it is remarkable how similar the eye of an octopus is to

Figure 23-20   A male squid caresses his mate with his many arms. *(Randy Morse/Tom Stack & Associates)*

our own (Figure 23-21). This similarity is a result of convergent evolution.

Cephalopods respond to their environments not only with directed movements toward prey and away from predators, but also with a complex set of color changes. Most cephalopods have specialized skin cells, usually with three different pigments. The brain controls these cells in immediate response to the environment. An alarmed octopus, for example, may suddenly acquire dark stripes or spots, which presumably confuse potential predators. Yet another escape trick is to squirt a trailing cloud of dark ink. Some cephalopods, especially squids, can move amazingly fast by jet propulsion, squirting water from their mantle cavities.

Cephalopods are mollusks in which the foot has evolved into tentacles. They are predatory, intelligent, and active.

What Do All Annelids Have in Common?

Earthworms and other annelids are common animals, although not nearly as diverse as the mollusks. The 12,000 species of annelids live as both predators and scavengers in salt water, in fresh water, and on land. They range in size from less than 1 mm to over 3 m. They may be red, pink, green, brown, or purple; plain, striped, or spotted.

Annelids have long, segmented bodies consisting of identical (or nearly identical) sections called **metameres.** Segmentation (which also occurs in arthropods and vertebrates) allows animals to achieve larger sizes by repeating an already successful organizational plan. Injuries to individual metameres are less likely to be lethal, since other segments perform the same functions. In addition, the segments can move independently, which gives the animal more flexibility.

Like mollusks, annelids have specialized excretory organs called *nephridia.* Like us, annelids have a closed circulatory system, with blood contained entirely within vessels. Several vessels carry blood from one end of the body to the other, with many fine capillaries within each segment. Some of the larger vessels serve as hearts and pump the blood along its complicated circuits. Annelids have no gills or lungs, however. Gases move directly through the skin to the blood.

Zoologists divide annelids into three classes: (1) polychaetes are free-living bristle worms, most of which are marine; (2) oligochaetes, which include the earthworms and related forms, live in fresh water or salt water or on land; and (3) hirudinea, or leeches, are mainly freshwater parasites that suck blood.

Annelids have segmented bodies, consisting of identical or nearly identical sections, and closed circulatory systems.

Polychaete worms

The exotic and beautiful polychaete worms include many unusual and arresting species, including clamworms, plumed worms, scale worms, peacock worms, and sea mice (Figure 23-22). Most zoologists think that the earliest annelids were marine polychaetes. Each polychaete segment contains a pair of leglike paddles, called **parapodia,** which the worm uses to swim, crawl, or burrow, as well as to respire. On each parapodium is a bundle of bristles called **setae,** which give the polychaetes their name [Greek, *polukhaites* = many hairs, or bristle]. A polychaete worm's head is well developed and contains a variety of sense organs, including two to four pairs of eyes. Unlike the oligochaetes (earthworms), polychaetes have separate male and female sexes.

Polychaetes are primitive marine annelids. Each of a polychaete's segments has a pair of leglike parapodia with setae.

Oligochaetes

Oligochaetes include the common earthworms, the most familiar of the annelids. Each oligochaete segment, in contrast to that of the polychaetes, lacks parapodia altogether and contains just four pairs of bundled setae, leading to the class's name [Greek, *oligo* = few].

Oligochaetes are masterpieces of segmentation. An earthworm, for example, consists of 100 to 175 nearly identical segments, with a few altered segments at the front and the rear. The forward segments contain specialized structures for eating, coordination, circulation, and reproduction, and the most posterior segment contains the anus (Figure 23-23).

Oligochaetes are mostly hermaphrodites and, typically, two worms will mutually cross-fertilize, with their heads point-

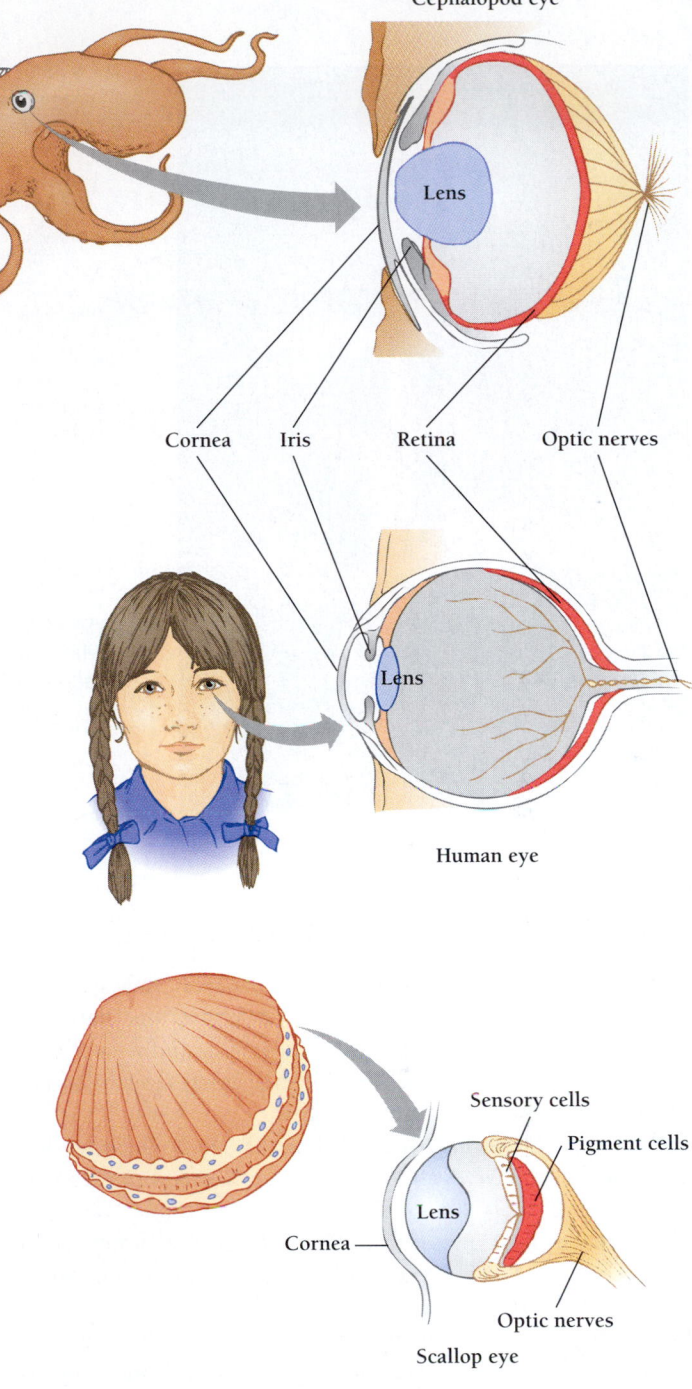

Figure 23-21 The cephalopod eye has long astounded biologists for its resemblance to the vertebrate eye. Cephalopod eyes can be enormous. In 1933, a squid 21 m (67 feet) long washed up on a New Zealand beach; its eyes measured 40 cm (16 in) across! Scallops have simple eyes that cannot focus. However, they have many of these eyes, which are probably useful for detecting the shadows of starfish and other predators.

ing in opposite directions and their undersides touching. Each transfers sperm to specialized pores in its mate.

Oligochaete segments lack parapodia and have many fewer setae than polychaete segments.

515

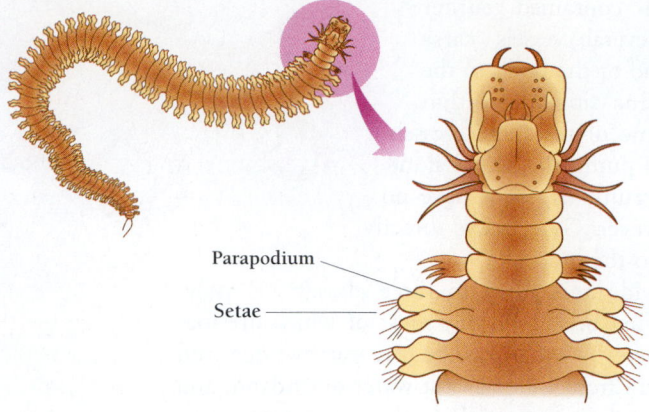

Figure 23-22   Tube worm. Many species of polychaetes—the tube worms—live in hidden places on the shallow sea bottom—under rocks, in burrows, or inside sponges, mollusks, or other animals. Some use their own secretions to build tubes where they live their entire lives. Polychaete feet consist of pairs of leglike paddles called parapodia and tufts of bristles called setae. *(Joyce Burek/Animals Animals)*

Hirudinea

Because most Hirudinea, or leeches, are parasites, they are more specialized than other annelids. Their bodies are flattened rather than cylindrical, and they lack segments and setae. Many species have a sucker at each end of their bodies. Such leeches "walk" like inch worms, attaching first one sucker to a surface and then the other. Like the oligochaetes, from which leeches are thought to have evolved, leeches are hermaphrodites that cross-fertilize.

Leeches range in size from about 1 cm to as much as 30 cm. Most are carnivorous and live in fresh water. Some species—including the medicinal leech, *Hirudo medicinalis*—live on the blood of humans and other mammals. Until the 19th century, leeches were used to treat such conditions as asthma, rheumatism, and drunkenness. Physicians (or barbers, who traditionally performed leech therapy as well as cutting hair) applied leeches, who sucked up to three pints of blood. When the patient fainted from blood loss, the physician re-

▶ **Figure 23-23   A single square meter of meadow soil may contain thousands of earthworms.** As an earthworm moves through soil, it swallows particles of soil. The soil passes through the digestive tract, where any organic matter is ground finely and absorbed. Undigested clay, sand, and other matter pass through the digestive tract and out the anus as small cylindrical "castings." A single worm eats its own weight in soil every day. Because earthworms are so common, they serve to keep soil mixed, which greatly benefits other soil animals and plants.

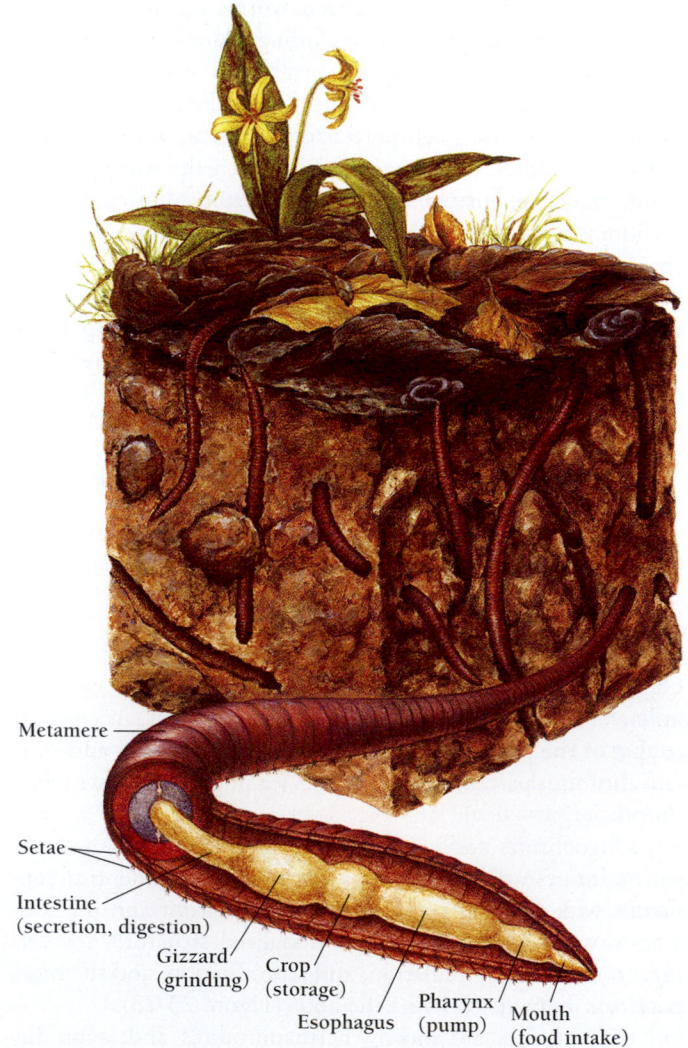

moved the leeches. Today, physicians use leeches, in a more restrained manner, to drain excess blood from surgical wounds.

A leech's impressive capacity to consume blood depends on powerful muscles of the pharynx (the anterior end of the digestive tract), on sharp teeth, and on a chemical that prevents blood clotting, called an anticoagulant. Physicians use leech anticoagulant to treat heart attack, stroke, and cancer.

Why Are Arthropods So Successful?

Arthropods, which include the insects, spiders, scorpions, shrimps, and many others, are the most successful phylum of animals on the planet. Arthropod species number at least a million, and zoologists estimate that between 2 and 30 million species remain to be described and named. The greatest number of species live in the tropics, but a student could probably collect hundreds (or even thousands) of species on any college campus.

Arthropods are more than just diverse, however. Zoologists estimate that the world today contains some 10^{18} (a billion billion) individual arthropods, hundreds of millions for every human being (a figure that might not startle a mosquito-bitten hiker). In a temperate climate, one square kilometer may contain 20 million arthropods. In the tropics, there are even more.

The enormous success of arthropods depends on both external and internal adaptations. All arthropods share two major external adaptations—an exoskeleton secreted by the cells of the epidermis and jointed limbs.

Jointed limbs, which give the arthropods their name [Greek, *arthron* = joint + *podus* = foot], allow them to move agilely, despite the rigidity of their exoskeletons. Like the annelids, arthropods have segmented bodies. Indeed, both phyla are probably descended from the same segmented ancestor. Some, like the centipedes and millipedes, have legs on nearly every segment. But many of the legs on an arthropod have become adapted for other functions. Over millions of years of evolution, legs have changed into mouthparts, antennae, gills, pincers, claws, and egg depositors.

These specialized appendages often develop from fused segments, called **tagmata** [singular, tagma]. For example, the anterior segments of all arthropods are fused into a highly organized head. Other tagmata include the middle portion of an arthropod, called the **thorax,** and the rear portion, called the **abdomen.**

The **exoskeleton,** a hard external supporting framework, consists of layers of chitin and protein. It helps to protect the animals from predators and parasites and provides mechanical support. The exoskeleton also slows water loss, a feature that has allowed the arthropods to flourish away from water and dominate animal life on land since the end of the Devonian Period, more than 350 million years ago.

An exoskeleton has drawbacks, however. Because a hard exoskeleton prevents an animal from growing, many arthropods (spiders, for example) periodically must shed their exoskeletons and secrete new, larger ones in a process called **molting.** The process of molting is dangerous. After an arthropod sheds its old armor and the new one is still hardening, the animal is highly vulnerable to predators. (Soft-shelled crab is a delicacy in many an East Coast restaurant.)

One solution is for an arthropod to do all its growing as a soft larva (as a caterpillar, for example) that lacks an exoskeleton. Once the animal has reached its adult size, it develops into the adult form in a process called **metamorphosis.** Some insects undergo gradual metamorphosis in stages that are not that different from one another (Figure 23-24). Others, such as flies, butterflies, beetles, and bees, undergo a complete transformation from a grub, caterpillar, or maggot to a flying insect. Metamorphosis and molting divide the life cycle of an insect into many distinct stages, each of which may have different food and lifestyle. In butterflies, for example, the caterpillar generally eats plants, while the adult butterflies drink nectar from flowers. In some arthropods, adults do not feed at all; their only role is to mate and die.

The internal anatomy of an arthropod is similar to that of an annelid. Both arthropods and annelids have a tubular gut that stretches from mouth to anus. The arthropod coelom is less prominent than that of an annelid and consists mostly of a cavity that encloses the reproductive organs.

The circulatory systems of arthropods are open. The heart pumps blood through the spaces of the body, and the blood returns through a series of one-way valves. The respiratory systems of arthropods vary according to taxonomic class and source of oxygen. Aquatic arthropods may have gills, with which they extract oxygen from the water in which they live. Spiders and some other air-breathing arthropods use stacks of modified gills, collectively called a **book lung,** to pull oxygen directly from the air.

Most terrestrial arthropods, including the insects, however, have neither lungs nor gills. Instead, they take in air through regulated openings in the body wall called **spiracles.** Air passes from the spiracles to special ducts called **tracheae.** Each trachea branches into tiny tracheoles, which deliver oxygen to individual cells throughout the body.

Reproduction in arthropods is almost always sexual. Male and female are usually separate animals, though some species are hermaphroditic. For terrestrial arthropods, fertilization must be internal, since there is no external water through which sperm can swim. Millipedes, for example, accomplish fertilization in a multilegged embrace. Elaborate courtship rituals help males and females of the same species recognize one another. Male fruit flies for example, do a little dance, during which the female decides whether she wants to mate with him. In an experiment where fruit flies were kept in the dark for several generations, however, the males abandoned the dance—which the females could not see anyway.

Because arthropods that are predaceous do not always distinguish between mates and meals, copulation among predaceous arthropods can be a dangerous matter. Such animals may have especially protracted courtship rituals, which not only help them to recognize one another but also may serve to prevent one partner from eating the other.

Excretion in most terrestrial arthropods depends on unique organs called **Malpighian tubules** (Chapter 39).

| A. | Eggs | Larva | Pupa (inside cocoon) | Adult |

| B. | Eggs | Young nymph | Nymph | Adult |

Figure 23-24   Complete and incomplete metamorphosis. A. During complete metamorphosis an insect, such as this isabella moth, hatches from an egg and develops as a legless larva that looks more like a worm than a mature insect. During development, a larva may go through many successive stages, called *instars*, before settling into a usually inactive stage called a *pupa*, or *chrysalis*. The pupa is usually encased in a protective cover, called a *cocoon*. The metamorphosis from larva to pupa and pupa to adult often involves dramatic changes in both external and internal structures. B. In incomplete metamorphosis, these changes occur more gradually. In the harlequin bug (shown here) early stages of development, called nymphs, look much like the adult, and molting functions mostly to increase size rather than to change form. Even in these metamorphoses, however, the juvenile forms may lack wings.

These tubules absorb fluid from the blood and convert the nitrogen-containing waste molecules into insoluble crystals of uric acid or guanine. The Malpighian tubules also reabsorb and recycle salts and water, so that little water is wasted.

The nervous systems of arthropods are often complex. Many arthropods can move, run, and fly quickly, and many have complicated sexual and social behaviors. The nervous system consists of a double chain of ganglia that runs along the lower surface of the body. At the head, the chain curls upward to form a brain—three pairs of fused ganglia. The sense organs—especially the eyes—of arthropods are more sophisticated than those of annelids. While some arthropods (such as the spiders) have simple eyes, most arthropod species have **compound eyes,** made up of numerous simple light-detecting units, called **ommatidia** [singular, ommatidium].

Arthropods include the greatest number of species of all animal phyla. External adaptations of the arthropods include the exoskeleton and jointed limbs. Internal adaptations of arthropods include circulatory, respiratory, reproductive, excretory, and nervous systems.

How Do Biologists Classify Arthropods?

The specialization of appendages, especially in the first few segments, is the basis of classification of living arthropods into three subphyla. (Extinct arthropods such as the trilobites may have their own subphyla.) In one subphylum—the **Chelicerates**—the first pair of appendages are mouthparts called chelicerae. The chelicerae serve as pincers or fangs, often associated with poison glands (Figure 23-25A).

In the other two subphyla, **Uniramia** and **Crustacea,** the first pair (or the first few pairs) of appendages are antennae, and the next pair are jaws, or mandibles (Figure 23-25B and C). These jaws differ from ours, which move up and down. Instead, an arthropod's mandibles crush and grind as they move from side to side. The Crustacea, which include water fleas, shrimps, lobsters, and crabs, are almost entirely aquatic, while the Uniramia, which include the insects, the centipedes, and the millipedes, are mostly terrestrial. The crustaceans all have branched (or biramous) appendages, while the Uniramia all have unbranched (or uniramous) appendages.

Biologists classify living arthropods into three subphyla according to the specialization of the appendages, especially the first two.

Spiders, mites, and other arachnids have six pairs of appendages (four pairs of legs, plus chelicerae and pedipalps, but no antennae). Like most other arachnids, spiders are predatory, capturing and eating insects and other small animals. Some spiders hunt and pursue their prey. Others trap their prey in elaborate silk webs. One spider genus, *Mastophora*, "fishes" for its prey with a silk line, tipped with a bit of glue. Spiders also use silk to make balloons (for making long trips), droplines (for making a quick exit), egg cases (for protecting their offspring), shrouds (for storing dead prey), and gifts from males (for luring attractive female spiders).

Spiders and other arachnids have no jaws and cannot chew. They digest only liquefied food. Once a spider has trapped its prey, it injects a paralyzing poison through its chelicerae, tears and grinds a bit of its prey with chelicerae and pedipalps, then regurgitates a potion of digestive enzymes into the body of its prey, which liquefies the prey. The spider then sucks its meal into its gut, where it completes digestion.

The scorpions are the oldest order of arachnids, extending back at least to the Silurian Period, about 425 million years ago. Scorpions differ from most chelicerates in having clearly segmented abdomens. This allows the scorpion to curl its abdomen, holding its stinger aloft. Other arachnids include the distinctive daddy longlegs (or harvestmen) and the mites, of which there are some 30,000 named species and perhaps a million more as yet unnamed. Mites, including the ticks, are usually less than 1 mm long, although the largest are 2 cm long. They live almost everywhere, eating plants, fungi, and animals.

Chelicerates—such as spiders, mites, and horseshoe crabs—have no antennae and no jaws.

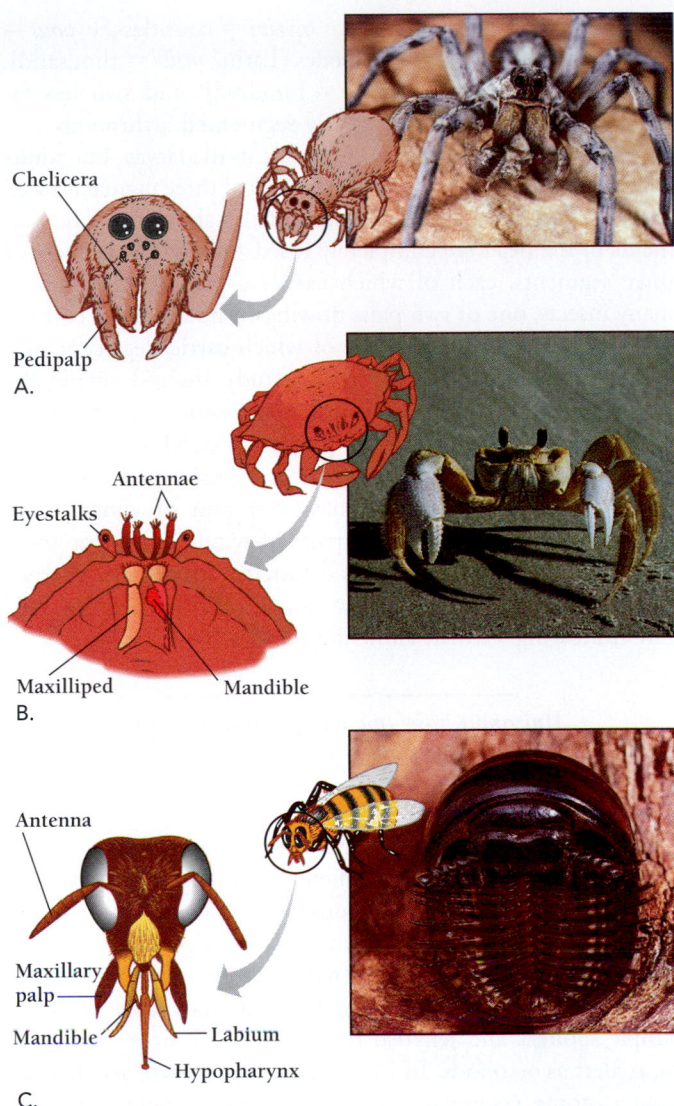

Figure 23-25 The mouthparts of arthropods. A. Chelicerae and pedipalps characterize chelicerates such as spiders and scorpions. B. Crustaceans such as this ghost crab have antennae, mandibles, and biramous appendages. C. The uniramia, including insects, millipedes, and centipedes, have antennae, mandibles, and uniramous appendages. (*A, John Cancalosi/Tom Stack & Associates; B, S.E. Georgia/Animals Animals; C, Tom McHugh/Photo Researchers, Inc.*)

Chelicerates

The chelicerates are the only arthropods without antennae or jaws. The most anterior pair of appendages are the pincerlike chelicerae, and the second are the pedipalps, which perform different functions in the different classes (Figure 23-25A). The most familiar chelicerates are the horseshoe crabs and the spiders, of which there are some 35,000 named species. The spiders are just 1 of 11 orders of arachnids; arachnids are, in turn, just one of three classes of the chelicerate subphylum.

Crustaceans

The crustaceans include some 35,000 species of lobsters, crabs, shrimps, barnacles, and crayfish, as well as less conspicuous species such as water fleas, fairy shrimp, pillbugs, and copepods (Figure 23-26). Crustaceans usually have three pairs of chewing appendages (including the mandibles), many pairs of legs, and, unlike other arthropods, two pairs of antennae. Their legs are biramous, or split into two parts at the tip.

The most numerous crustaceans are the copepods. Indeed, next to the nematodes, the copepods are the most abundant animals on Earth. These tiny crustaceans, each a few millimeters long, are a major component of plankton, the microscopic plants and animals that float near the surface of the oceans. Copepods eat marine algae and are a major food in the diet of whales, nurse sharks, and other filter-feeding marine animals.

The largest crustaceans are the decapods, which include lobsters and crabs. Some lobsters may grow to be 60 cm long, and the Japanese spider crab is more than 3 meters across. The decapods [Greek, *deka* = ten + *pous* = foot] have five pairs of legs.

The oddest of the crustaceans are the barnacles. One wit described these creatures as "nothing more than a little shrimp-like animal, standing on its head in a limestone house and

Figure 23-26 Krill. Small crustaceans such as these copepods are the major component of krill, the marine soup on which the great whales feed. *(William E. Ferguson)*

kicking food in its mouth." Unlike other crustaceans, which are active, a barnacle lives its adult life in a single place, submerged in salt water and attached (by its head) to a submerged rock, piling, boat bottom, or even a whale. Barnacles use their appendages to direct passing food particles into their mouths. They secrete calcium-containing plates, which are firmly cemented to the underlying surface.

Crustaceans—such as lobsters, crabs, and shrimps— have jaws, biramous appendages, and, usually, two pair of antennae.

Uniramia

The Uniramia are the most diverse subphylum of arthropods, including 750,000 insects, 10,000 species of plant-eating millipedes, and 2500 species of carnivorous centipedes. Almost all Uniramia take in air through tracheae, use Malpighian tubules to rid themselves of nitrogenous waste, and have unbranched (uniramous) appendages.

The myriapods [Greek, *myrioi* = countless + *pous* = foot], which include the millipedes [Latin, *mille* = thousand], the centipedes [Latin, *centum* = hundred], and two less familiar classes, are the most clearly segmented arthropods.

In insects, segmentation is clearest in the larvae, but adults are segmented as well. The body consists of three tagmata (fused segments): the head, the thorax, and the abdomen. The segments of the head are completely fused. The thorax consists of three segments, each of which carries a pair of legs, and, in many insects, one or two pairs of wings. The abdomen consists of 12 or fewer segments, none of which carries legs or wings.

Entomologists (scientists who study insects) divide the class Insecta into about 30 orders. These orders differ both in adult morphology and in their pattern of development. Some insects (for example, lice, fleas, and silverfish) have no wings, some (flies and mosquitoes) have one pair of wings, some (dragonflies) have two similar pairs of wings, and some (beetles) have two pairs, each with a distinctive structure. The front wings of beetles are hard shells that serve to protect the delicate hind wings, which alone are responsible for flight.

Uniramia have one pair of antennae and one pair of mandibles.

Zoologists divide the animal kingdom into the Parazoa (mostly sponges) and the Eumetazoa (all the rest). The eumetazoans fall into three groups of phyla: the acoelomates (with no coelom), the aschelminths (with a pseudocoel or no coelom), and the coelomates (with a true coelom). We have briefly surveyed all four groups. These animals range from the simple sponges and jellyfish to animals as complex as insects or as alert as octopods. In the next chapter we will see that the deuterostome coelomates are also diverse, ranging in complexity and intelligence from sand dollars to humans.

Study Outline with Key Terms

An animal is a multicellular, heterotrophic, eukaryotic organism that develops from an embryo. Most animals reproduce sexually, producing a single-celled **zygote** from the fusion of an egg and a sperm. The developing embryo goes through a stage called a **blastula,** which is a hollow sphere made up of identical cells. The blastula folds into a three-layered **gastrula** surrounding the **archenteron.** The three cell layers of the gastrula are the **endoderm, ectoderm,** and **mesoderm.** These three layers develop into different types of **tissues** and **organs.** Tissues such as the **epithelial** and **connective** tissues are composed of specialized cells and their **matrix.** Organs are composed of two or more types of tissues.

Zoologists divide animals into two subkingdoms, the **Parazoa** and **Eumetazoa.** The Parazoa, which have no organs, include two phyla, the Placozoa and the Porifera. There is only one known species of Placozoa, but there are about 10,000 species of Porifera, or sponges. Sponges are usually **sessile.**

Their cells lie in three layers (epidermal cells, **mesenchyme** cells, and flagellated cells) surrounding a central cavity.

Zoologists divided the eumetazoans according to the presence or absence of a **coelom** or **pseudocoel** and also according to whether they have **radial symmetry** or **bilateral symmetry.** Animals with bilateral symmetry generally have distinctive **anterior** and **posterior** ends. Animals with a coelom are called **coelomates.** Those with no coelom are called **acoelomates.** In between are the **pseudocoelomates.**

Almost all coelomate animals have digestive tracts with both a mouth and anus. The coelomates are divided into the **protostomes** and the **deuterostomes,** according to whether the embryonic **blastopore** develops into the mouth (protostomes) or the anus (deuterostomes). The majority of coelomates are protostomes, including the annelid worms, mollusks, and arthropods. However, two major phyla of coelomates are deuterostomes—the echinoderms and the chordates.

Cnidarians and **ctenophores** are radially symmetrical acoelomates. Cnidarians, which include jellyfish, hydras, corals, and sea anemones, have tissues but no organs. They have two layers of tissues enclosing a layer of **mesoglea.** They may have two body plans: **polyps** (which resemble cylinders) and **medusae** (which resemble bells). Both types have specialized stinging cells, called **cnidocytes,** which fire tiny barbs called **nematocysts.** Ctenophores (comb jellies) resemble the cnidarians but also have distinctive features, including bands of cilia.

Platyhelminthes and **Nemertea** are bilaterally symmetrical acoelomates. Platyhelminthes (flatworms) are the simplest animals that have heads and organs. They consist of three layers of cells. They have no circulatory systems, but their flat shapes allow gases and molecules to diffuse rapidly. There are three classes of flatworms: the **Turbellaria, Trematoda,** and **Cestoda.** The first class consists of the free-living flatworms, while the other two classes include only parasites. Parasites are specialized for rapid and prolific reproduction in their hosts. The tapeworm consists of a **scolex** and many **proglottids.** Nemertea (ribbon worms) have a two-ended digestive system, a circulatory system, and a **proboscis.**

The **Aschelminths** are a group of eight phyla, all bilaterally symmetrical animals with either no coelom or a pseudocoel. **Nematodes** (roundworms) are cylindrical and unsegmented, and have a one-way digestive system with both a mouth and an anus. The prominent pseudocoel of nematodes acts as a distribution route for nutrients, gases, and wastes; it also serves as a hydrostatic skeleton.

Mollusks (which include snails, slugs, clams, oysters, squids, and octopods) are unsegmented animals that typically have shells. Mollusks all have a **visceral mass,** a muscular **foot,** a **mantle,** a **mantle cavity,** a **radula,** and one or two kidney-like **nephridia.** Many shelled mollusks have an **operculum.**

There are seven classes of mollusks, of which we discussed three: **gastropods, bivalves,** and **cephalopods.** Gastropods have single shells, well-developed heads, and twisted bodies. Bivalves have two shells and most lead sedentary lives. Cephalopods have forward-pointing tentacles instead of a foot.

Annelids (which include earthworms, segmented marine worms, and leeches) have segmented bodies that consist of identical or nearly identical **metameres.** Annelids have closed circulatory systems, with blood contained entirely within vessels. Annelids are divided into three classes: polychaetes, oligochaetes, and hirudinea. Each segment of a polychaete (or bristle worm) contains a pair of paddlelike projections, called **parapodia,** each with a bundle of bristles, or **setae.** Oligochaete segments lack parapodia. Hirudinea (leeches) are more specialized than other annelids and lack obvious segmentation.

Arthropods include the greatest number of species and the greatest numbers of individuals of all animal phyla. External adaptations of the arthropods include an **exoskeleton** and jointed limbs. Arthropods have many specialized appendages that develop from **tagmata,** fused segments that form the head, the **thorax,** and the **abdomen.** The specialization of appendages, especially in the first few segments, is the basis of classification of arthropods into three subphyla: **chelicerates, crustaceans,** and **uniramia.**

Because an exoskeleton prevents an animal from growing, many arthropods periodically **molt.** Other arthropods do all their growing in larval form, then change into adults during **metamorphosis.**

Internal adaptations include circulatory, respiratory, reproductive, excretory, and nervous systems. Arthropods breathe by means of a **book lung,** or **spiracles** and **tracheae** and regulate excretion by means of **Malpighian tubules.** Some arthropods have simple eyes, but most have **compound eyes,** made up of numerous **ommatidia.**

Review and Thought Questions

Review Questions

1. How is an animal different from a plant or a protist? Why are sponges grouped with animals? What animal features do sponges and cnidarians lack?
2. What criteria do zoologists use to classify the Eumetazoa?
3. What is a coelom and how is it different from a pseudocoel?
4. What specializations set parasites apart from other members of their phyla?
5. What is the difference between a protostome and a deuterostome?
6. Members of which class of mollusks would make the most interesting pets? Why?
7. Which phylum of protostomes includes the greatest number of species of any phylum of organisms?

8. What specific characteristics of a grasshopper tell you that it belongs in the class Insecta?
9. How do millipedes differ from centipedes? Which class includes more species?

Thought Questions

10. What is progress? In what sense is evolution progressive?
11. Zoologists use the following traits (among others) to classify the animals: number of embryonic cell layers, presence or absence of a true coelom, and symmetry. Do you think that such characteristics truly reflect relatedness between different groups of organisms? If you were going to classify the animals, what criteria would you consider using?

About the Chapter-Opening Image

The succulent katydid of Peru is a classic and beautiful invertebrate.

 On-line materials relating to this chapter are on the World Wide Web at **http://www.harcourtcollege.com/lifesci/aal2/**

24 Deuterostome Animals: Echinoderms and Chordates

Are We Upside Down?

In 1830, two French anatomists staged a public debate before the prestigious French Academy of Sciences on a topic so fundamental that historians of science have returned to it again and again. How many basic body types do animals have?

On one side, the politically powerful French anatomist and paleontologist Baron Georges Cuvier (1769–1832) argued that all animals should be divided into four distinct types. On the other side was Étienne Geoffroy Saint-Hilaire (1772–1844), a renowned comparative anatomist, who argued that animals were of just one type, not four.

Geoffroy was not a man to be intimidated even by a figure as powerful as Cuvier. During the French Revolution, Geoffroy had risked his life to save teachers and colleagues from execution. And when in 1807 Napoleon ordered Geoffroy to obtain collections from Portugal's museums by any means possible, Geoffroy slyly traded specimens from French museums instead of seizing the collections by force.

Geoffroy had long maintained that all vertebrates—mammals, birds, reptiles, and amphibians—had the same basic "body plan," a description of their symmetry and the arrangement of their tissues. In all vertebrates, for example, the backbone is at the back (dorsal) and the heart lies along the chest (ventral). The unity of the vertebrates was an idea that Cuvier and his followers had accepted.

But in 1830, Geoffroy Saint-Hilaire took his argument a huge step further, proposing that chordates (vertebrates and their relatives) have the same body plan as arthropods. In essence, he said, an insect and a mammal are similar. The single main difference, he said, was that we vertebrates are turned upside down: where a grasshopper's nerve cord runs along its belly, a human's nerve cord runs along its back (Figure 24-1). And where a human heart beats against the chest wall, a grasshopper's heart lies up along its back. As one modern biologist has put it, "If you lay down on your back and waved your arms, you would be doing what insects do when they walk."

Geoffroy's hypothesis, which he called *unité de plan* (singleness of plan), was provocative. Not only was it in direct opposition to Cuvier's theory of separate, unrelated groups, but it also implied evolution. Cuvier firmly believed in the fixity of species and had publicly ridiculed Lamarck and other proponents of evolution. According to Geoffroy's hypothesis, however, humans and all other vertebrates were each a simple variation on a basic invertebrate body plan. Geoffroy's idea could be interpreted to mean that vertebrates had evolved from invertebrates.

For Geoffroy, *unité de plan* was a serious scheme, based on his dissections of thousands of animals. But during the debate, the pugnacious Cuvier made *unité de plan* look like a joke, listing, one by one, all of the differences between a duck and a squid.

Paraskevas Photography

- Deuterostomes include both invertebrates and vertebrates.

- Chordates are deuterostomes that, at some time during their development, have a notochord, a dorsal hollow nerve cord, pharyngeal gill slits, and a tail.

- Vertebrates have a segmented spinal column and a distinct head with a skull and a brain.

- Terrestrial vertebrates have special adaptations for life on land, one of which is the amniotic egg.

Historians of science still argue over who won the debate on technical points. But the reality was that Cuvier's scientific and political influence was so profound that for 164 years no biologist resurrected *unité de plan* for serious discussion. In fact, a researcher risked his or her reputation even to introduce the discredited idea.

So the matter might well have rested. In the early 1990s, however, biologists working separately on frogs and fruit flies began to untangle the matrix of genes that underlies the development of the overall body plan. To their astonishment, they discovered that the frog genes and the fly genes were the same genes. Unwittingly, they had resurrected Geoffroy Saint-Hilaire's idea of a single body plan.

In fruit flies, a gene called *dpp* codes for a signaling protein that somehow activates other genes necessary for the formation of **dorsal** structures—all the body parts that lie along the back of an animal. In addition, *dpp* seems to suppress the development of **ventral** structures—all the body parts that lie along the abdominal side of an animal. *Dpp* also suppresses the development of nerve cells. A second fruit fly gene, *sog*, counteracts the effects of *dpp*. *Sog*, which is expressed only in the ventral regions of a developing fruit fly, overrides *dpp* in the ventral parts, allowing nerve cells and other ventral structures to develop.

When researchers looked for similar genes in frogs, they found *bmp–4*. Like *dpp*, *bmp–4* codes for a signaling protein. But, oddly, it operated in reverse. *Bmp–4* activates genes that help form the ventral structures in the frog embryo and suppresses the formation of nerve cells in the dorsal regions (Figure 24-1).

No one made an explicit connection between the three genes, however, until September 1994, when two German researchers pointed out that *dpp* and *bmp–4* do the same thing, only acting in opposite areas of frog and fly. Detlev Arendt and Katharina Nübler-Jung suggested that "the longitudinal nerve cords of insects and vertebrates derive from one and the

"If you lay down on your back and waved your arms, you would be doing what insects do when they walk."

same centralized nervous system in their common ancestor." After 164 years, they had revived Geoffroy Saint-Hilaire's *unité de plan*.

They were not ridiculed, however. Instead, researchers in California immediately uncovered a fourth gene. Edward De Robertis (at UCLA) and his colleagues found that the gene *chordin* stimulates the organized development of nerve cells in the dorsal region of a frog embryo, overriding *bmp–4*, just as *sog* overrides *dpp* in a fly.

If Geoffroy Saint-Hilaire and his modern supporters were right, then *dpp* and *bmp–4* were essentially the same gene, just expressed in different parts of the embryo. Likewise, *sog* and *chordin* had to be the same gene, separated by half a billion years of evolution.

In a series of dramatic experiments, De Robertis and several other researchers in California and Wisconsin demonstrated that the fly genes (*dpp* and *sog*) work in frogs, and the frog genes (*bmp–4* and *chordin*) work in flies. *Sog* mRNA injected into frog embryos, for example, promoted the development of dorsal structures. Likewise, *chordin* mRNA injected into flies induced ventral development, as *sog* normally does. Further, when researchers sequenced *chordin* and *sog*, they discovered that the base sequence of *chordin* is 47 percent identical to that of *sog*. And biologists have since uncovered several other developmental genes that are the same in both frogs and flies.

Has molecular biology redeemed *unité de plan*? Was Geoffroy Saint-Hilaire right, Cuvier wrong? Many biologists still hesitate to pronounce the matter settled. But nearly all are intrigued. If organisms as different as lobsters and linebackers share genes that help determine basic body plan, these organisms surely share a common ancestor.

De Robertis, for one, is eagerly seeking more genes shared by vertebrates and invertebrates that might shed light on the structure of that common ancestor. Until he and his colleagues succeed, however, even modern molecular biologists will hesitate a bit, reluctant to embrace an idea so long ridiculed.

Figure 24-1 **Opposites?**
Cross sections through embryos show that vertebrates and invertebrates share similar body plans. The vertebrate body plan features a dorsal nerve and a ventral heart—an upside down version of the arthropod body plan, which has a ventral nerve cord and a dorsal heart. These similarities go deep. Similar genes operate in both kinds of embryos.

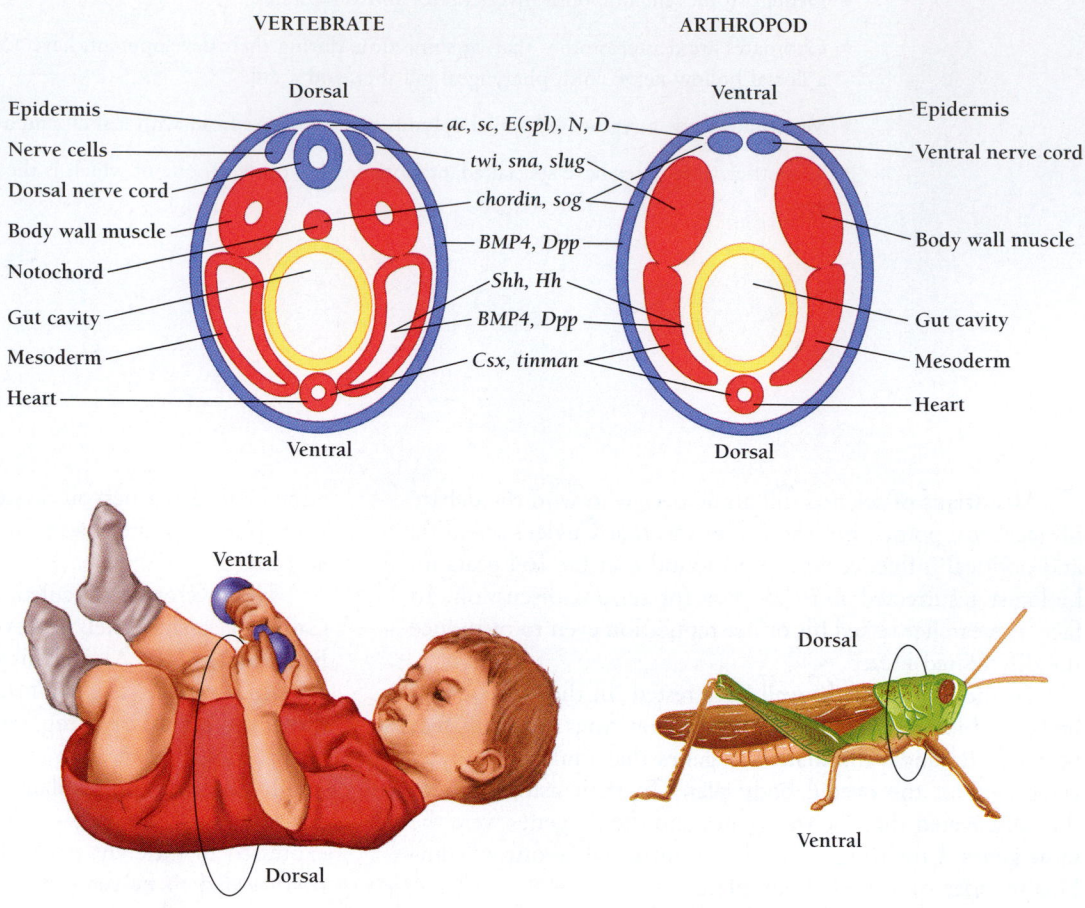

DEUTEROSTOME COELOMATES

As we discussed in the last chapter, the early development of the deuterostome coelomates differs from that of the protostomes in several ways. In all deuterostomes, including ourselves, the tiny hole in the embryo called the **blastopore** develops into an anus instead of a mouth. The mouth develops later (or second), hence the term deuterostome [Greek, *deuteros* = second + *stoma* = mouth].

The deuterostomes consist of four phyla (Figure 24-2): **Echinodermata** ("spiny-skinned" animals such as sea stars), **Chaetognatha** (arrow worms), **Hemichordata** (acorn worms), and **Chordata.** The Chordata include the vertebrates—the fish, the amphibians, the reptiles, the birds, and the mammals—as well as the less well-known sea squirts and lancelets. While sea urchins and humans differ radically as adults, their embryos share many basic features.

Echinoderms

The 6000 species of echinoderms fall into six classes of marine animals. Among these are two of the most familiar inhabitants of tidepools—sea stars and sea urchins. The other four living classes of echinoderms are the sea lilies and feather stars, the brittle stars, the sea daisies, and the sea cucumbers. Another 20 classes of echinoderms are now extinct. The echinoderms are an ancient lineage. Fossil sea stars, sea cucumbers, and sea lilies from the Burgess Shale suggest that the echinoderms had diverged from the protostomes some 530 million years ago.

Echinoderms are plentiful on the bottoms of intertidal zones and deep oceans. The largest is a sea star one meter across. The smallest are just a few millimeters. Most echinoderms have calcium-rich spines that project from their skin, giving the phylum its name [Greek, *echinos* = hedgehog, a small European mammal covered with spines, like a porcupine].

As adults, echinoderms are radially symmetrical, often with five nearly identical parts arranged around a central axis. Like other deuterostomes, however, the larvae of echinoderms have bilateral symmetry. Biologists believe that the echinoderms are descended from bilateral ancestors (Figure 24-3). Neither larvae nor adults show any sign of segmentation.

The sexes of echinoderms are separate, with two sets of gonads (testes or ovaries) in each arm. Males shed sperm and females shed eggs into sea water, where fertilization takes place. In the open ocean, the chances that an individual egg will meet

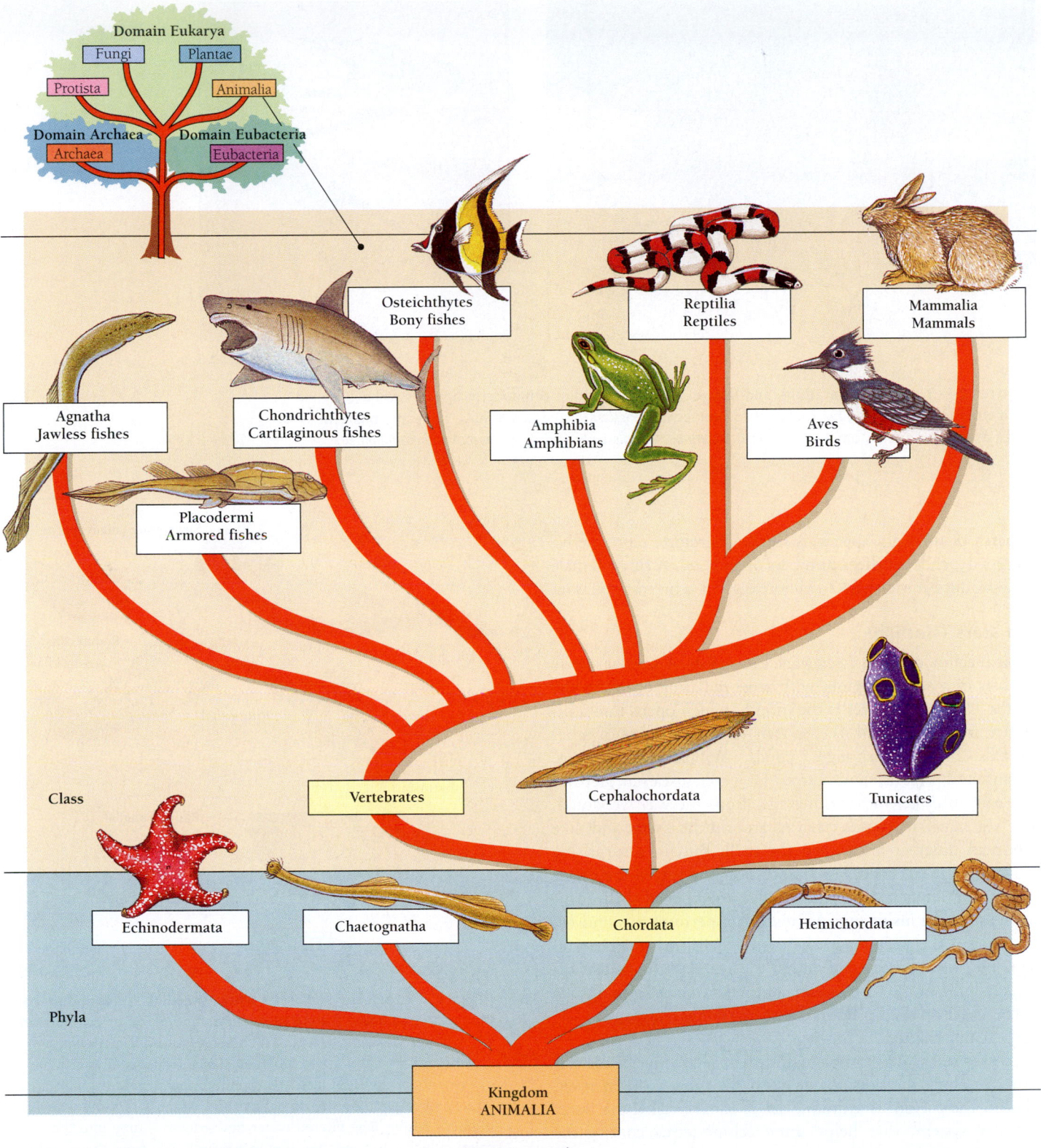

Figure 24-2 The deuterostome phyla of the kingdom Animalia.

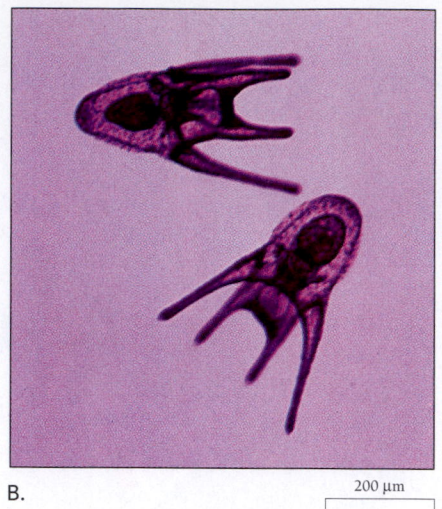

A.

B.

200 μm

Figure 24-3   Sea urchin adults and larvae. Although an adult echinoderm appears to have radial symmetry, a trait regarded by zoologists as somewhat backward, the larvae have the same bilateral symmetry as other deuterostomes. *(A, Dave B. Fleetham/Visuals Unlimited; B, Cabisco/Visuals Unlimited)*

a sperm of the right species is small. Therefore, reproductive success depends on sheer numbers of gametes. A single female may produce more than 2 million eggs in a breeding season.

Sea stars (starfish)

Typical echinoderms are sea stars, or starfish. A sea star's body consists of a central region, with arms radiating out from the center. Under the center is the mouth, which opens into a digestive tract (Figure 24-4). Sea stars feed on all sorts of other animals, including sponges, corals, mollusks, crustaceans, worms, and even fish and oysters. However, a sea star prepares and eats an oyster rather differently from the way we would. The sea star first spreads its arms around the oyster and, like a human diner, it pries open the shell. But, in a surprising move, the sea star places its mouth near the gap in the oyster shell and extends its own gut through the opened shell, turning its stomach inside out. A sea star can perform this unusual feat even if the opening in the oyster shell is as small as 100 μm. During the next few hours, the sea star completely digests its prey. Some sea stars spread their stomachs over a piece of coral reef on the sea floor, digesting sponges, algae, and settled organic matter.

A sea star's arms can bend and twist, allowing the sea star to move along the bottom, to grasp prey, and to right itself. Within the coelom is a unique set of canals, called the water vascular system, that helps some echinoderms control the movement of their arms. A primitive skeleton made of calcium-rich plates reinforces the whole body. Over this supporting structure is a delicate skin, with thousands of cells that are exquisitely sensitive to touch. Tiny, suction cup–like tube feet cover the underside of a starfish.

Starfish and some other echinoderms can regenerate arms that are damaged or broken away. Some species of sea stars can

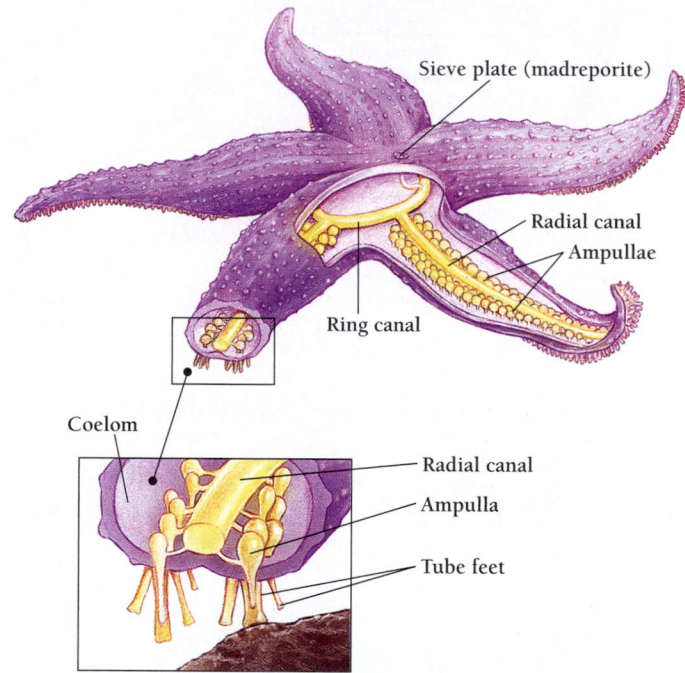

Figure 24-4   The sea star's water vascular system. Perhaps the most remarkable feature of sea stars and all other echinoderms is the water vascular system, which serves as both a circulatory system and an elaborate hydrostatic skeleton that allows complex movements of the tube feet. Water enters the system through a perforated sieve plate on the dorsal surface, which filters the water as it enters a tube. The filtered water flows down a tube into the ring canal, which surrounds the mouth (ventral) and from there into the radial canals that extend into each of the arms. In the arms, the radial canals branch to connect to long rows of tube feet. Each tube foot is also connected to a muscular fluid-filled sac, called an ampulla. In sea stars and other echinoderms, each tube foot ends in a sucker. The ampulla regulates fluid pressure, so that each tube foot may extend or retract, hang on or let go. Using the tube feet, a starfish can slowly pull itself along the bottom of the sea.

regenerate a whole new individual from a single arm. In such asexual reproduction, the starfish breaks into two parts, and each regenerates the missing arms. Usually, however, echinoderms reproduce sexually, with males and females releasing gametes into the sea.

Echinoderms are ancient, spiny-skinned animals that are radially symmetrical as adults and bilaterally symmetrical as larvae.

Arrow Worms and Acorn Worms

The 50 to 70 species of arrow worms are common marine organisms that range in length from 5 mm to 15 cm. Arrow worms feed on a variety of small marine animals, which they grasp with moveable hooks. Fossils of arrow worms date from late Cambrian times. Unlike echinoderm larvae or acorn worms, the body of an adult arrow worm extends beyond the anus, forming a tail. The only other deuterostomes with true tails are chordates, the phylum that includes the vertebrates. Because arrow worms have other properties in common with vertebrates, many evolutionary biologists think vertebrates are descended from ancestors that resembled arrow worms.

Still more similar—though not at first glance—to animals of our own phylum are the acorn worms. Most of the 65 species of acorn worms live in U-shaped burrows on the bottom of the sea. They range in length from 2.5 cm to 2.5 m (nearly 8 ft). Zoologists classify these soft-bodied marine animals as hemichordates ("half chordates") because they resemble chordates in two respects: (1) they have **gill slits,** holes that directly connect the throat to the outside; and (2) in addition to a ventral nerve cord, they have the beginnings of a dorsal nerve cord (like our own spinal cord). As we will discuss later, both of these features are fundamental characteristics of chordates.

Arrow worms and acorn worms may resemble the ancestors of the vertebrates because they possess a tail. The acorn worms, in addition, have gill slits and a dorsal nerve cord like that of the chordates.

What Traits Characterize the Chordates?

The chordates include members of three subphyla. Members of two subphyla have no backbones: the tunicates and the cephalochordates. Members of the third subphylum, the vertebrates, have backbones. Members of all three subphyla exhibit four hallmark characteristics at least at some time in their lives: (1) a flexible rod running along the back (dorsal surface), called the **notochord,** which gives the phylum its name; (2) a **dorsal hollow nerve cord,** running between the notochord and the surface of the back; (3)

pharyngeal slits (sometimes called gill slits), holes in the sides of the body that run from the inside of the gut to the outside surface of the animal; and (4) a segmented body and a "postanal" tail, a tail that extends beyond the anus (Figure 24-5).

The notochord is a long column of fluid-filled cells, with the character of a stiff, but flexible, sausage. In the tunicates and cephalochordates, as well as in some fish and larval amphibians, the notochord provides back support for the adult animal. In many vertebrates, however, the notochord disappears during embryonic development to be replaced by a bony or cartilaginous backbone. Cartilaginous fish have both a notochord and a surrounding vertebral column, while adult reptiles, birds, and mammals have only a vertebral column.

The dorsal nerve cord of embryos and its adult counterparts (the spinal cord and the brain) are partly hollow, and their inner spaces contain fluid. In contrast, the ventral nerve cords of annelids and arthropods are solid. The pharyngeal slits are present in the embryos of all chordates (including humans), but not in air-breathing adults. In early chordates, the slits may

Tunicate larva (urochordate)

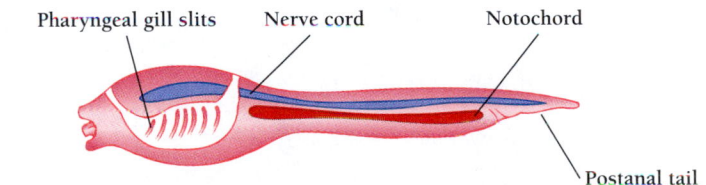

Pharyngeal gill slits   Nerve cord   Notochord

Postanal tail

Amphioxus (cephalochordate)

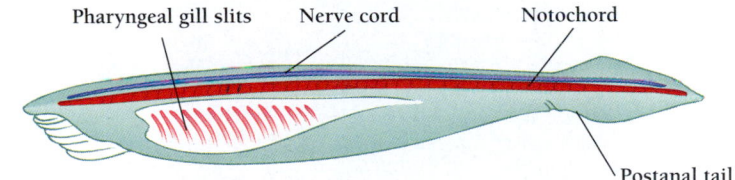

Pharyngeal gill slits   Nerve cord   Notochord

Postanal tail

Lizard (vertebrate)

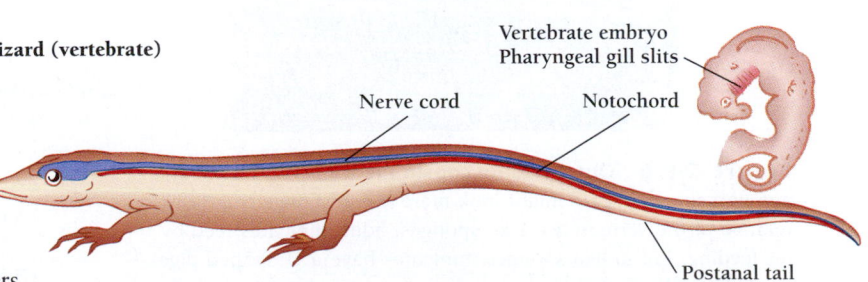

Vertebrate embryo
Pharyngeal gill slits

Nerve cord   Notochord

Postanal tail

Figure 24-5 The four features of chordates. The three chordate subphyla—the tunicates (sea squirts), the cephalochordates (amphioxus), and the vertebrates—all share four distinguishing features. These are pharyngeal gill slits, a dorsal nerve cord, a notochord, and a postanal tail.

have served to allow filter feeding (as they do in tunicates), but fish, tadpoles, and other aquatic chordates use them primarily for respiration.

All chordates have a notochord, a dorsal hollow nerve cord, pharyngeal slits, and a tail at some time in their lives.

What Have We in Common with Tunicates?

Adult tunicates do not look at all like other chordates (Figure 24-6). They are sessile marine animals that attach to rocks, boats, or the ocean floor. Their name derives from a tough outer coating, or tunic, made, amazingly, of cellulose. (Cellulose is common in the cell walls of plants, but rare in animals.) While adult tunicates have pharyngeal gill slits, they have little else to recommend them as chordates. A tunicate larva, however, has a notochord, a dorsal hollow nerve cord, pharyngeal gill slits, and a postanal tail, most of which are lost when it metamorphoses into an adult. It seems likely that the common ancestor of the chordates resembled a tunicate larva.

Figure 24-6   Our chordate cousin the tunicate. To the average person, a tunicate might look more like a sponge than a close relative of the vertebrates. Like sponges, adult tunicates feed by filter feeding. But unlike sponges, tunicates have a U-shaped digestive tract. Cilia in the pharynx pull water through an incurrent siphon into the mouth, past the pharyngeal slits, and out through an excurrent siphon. Sticky mucus traps food particles and carries them down into the digestive tract. The anus empties into the excurrent siphon. *(Dave Fleetham/Tom Stack & Associates)*

As larvae, tunicates have a notochord, a dorsal hollow nerve cord, pharyngeal slits, and a tail, all characteristics of chordates.

What Have We in Common with Cephalochordates?

Cephalochordates [Greek, *cephalo* = head], or lancelets, include about 45 species of small, segmented, fishlike animals. Most belong to the single genus *Branchiostoma,* whose common name is amphioxus. In these animals, the notochord, which is present in adults as well as in embryos, extends all the way through the head, hence their name. Sets of chevron-shaped (<<<) muscles pull against this internal skeleton. The segmentation of a lancelet is superficially similar to that of annelid worms and achieves the same advantages. Cephalochordates have all four chordate characteristics.

Cephalochordates have notochords both as larvae and as adults.

What Distinguishes the Vertebrates from Other Chordates?

Vertebrates have many features that distinguish them from other chordates. Two of the most important are (1) a segmented vertebral column made of units called vertebrae and (2) a distinct head, with a cranium (or skull) and a brain. Arches of the vertebrae encircle and protect the dorsal nerve cord, which becomes the adult vertebrate's spinal cord (Figure 24-7).

Vertebrates also have closed circulatory systems and characteristic internal organs—livers, kidneys, and hormone-secreting (endocrine) organs. The heart pumps blood through the circulatory system, through a system of fine capillaries that helps transport gases and nutrients to nearly every cell of the body. The same system carries metabolic wastes away from cells. Liver, kidneys, and other organs closely regulate the chemical composition of the blood, ensuring a nearly constant internal environment for all cells. (Part VII of this book deals with these organs and their operation in some detail.)

Most modern vertebrates have bony skeletons. **Bone** consists of fibers of the protein collagen, embedded primarily with crystals of calcium phosphate. Within this extracellular material are bone-making cells, as well as blood vessels and nerves. Bone is strong enough to support huge vertebrate bodies. Unlike the exoskeletons of arthropods, bone can grow, allowing vertebrates to reach sizes far greater than those of the arthropods.

The embryos of all vertebrates have skeletons made of **cartilage,** also built from collagen, but softer and more elastic than bone, without calcium phosphate. Some parts of the adult skeleton retain cartilage, such as our ears and noses. But, except in the sharks and other cartilaginous fishes, bone mostly replaces cartilage in adults.

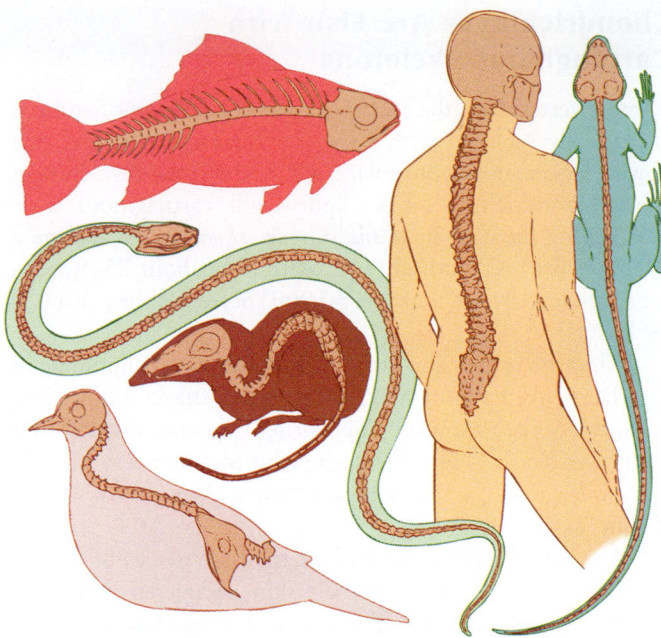

Figure 24-7 **Which of these vertebrates has the strongest back?** All vertebrates share a vertebral column composed of many vertebrae, as well as a thick skull that protects the brain. In most vertebrates, the vertebral column, skull, and skeleton consist largely of bone. In a few, such as the sharks and rays, these elements are flexible cartilage instead. The hero shrew (*center*), of Africa, is reputed to have the strongest backbone of any vertebrate. According to legend, a man can stand on this tiny insectivore's back without injuring the animal.

Vertebrate bone is so distinctive and so well preserved in the fossil record that biologists know more about the evolution of the vertebrates than about any other group. In this chapter, we survey eight classes of vertebrates. Four of these classes are groups of fishes and four are four-footed animals, or **tetrapods** [Greek, *tetra* = four + *pous* = foot]. The four classes of fishes in-clude Agnatha (the jawless fishes), Placodermi (a class of extinct armored fishes), Chondrichthyes (fishes with skeletons of cartilage rather than bone), and Osteichthyes (bony fish). The tetrapods consist of the Amphibia (amphibians), Reptilia (reptiles), Aves (birds), and Mammalia (mammals).

Vertebrates have vertebral columns, skulls, and a well-developed brain. Most vertebrate skeletons are made of bone.

Agnatha Have No Jaws

Most jawless fishes are extinct. These extinct fishes, comprising some 60 families, are called **ostracoderms** [Greek, *ostrakon* = shell + *derma* = skin] because their skin contained a shell-like skeleton of bony plates. Ostracoderm fossils have been found in Cambrian deposits, showing that vertebrates have been around as long as the invertebrates (Figure 24-8). Although, traditionally, the agnathes have been lumped together, ongoing research suggests that they represent several separate lines (or clades).

The living jawless fishes include only about 60 species of **cyclostomes** [Greek, *kyklos* = circle + *stoma* = mouth], soft-bodied fish with round, jawless mouths. The cyclostomes, none of which have bony plates, include the lampreys and the hagfish. Lampreys and hagfish are hardly glamorous animals. Hagfish survive by eating polychaete worms and pieces of dead fish, scavenged from the muddy sea bottom. Lampreys are nearly all parasitic. A lamprey lives by attaching itself to another fish with its circular jawless mouth, then tearing a hole in its living victim with its rasping tongue. The lamprey then sucks whatever fluids flow from the wound.

A lamprey larva, like a lancelet, has a notochord and segmented muscle groups. It differs from a lancelet, however, in possessing

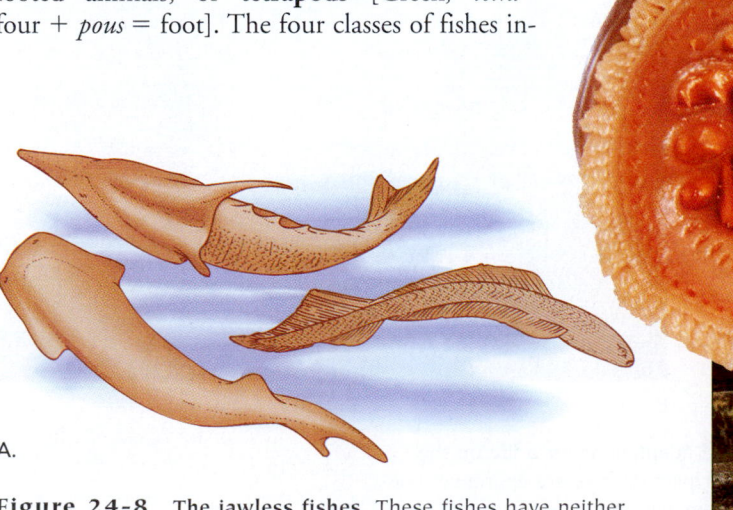

Figure 24-8 **The jawless fishes.** These fishes have neither working jaws nor paired fins. A. Extinct ostracoderms. B. A modern lamprey. *(B, top: Courtesy of Dr. Kiyoko Uehara; bottom, Berthoule-Scott/Jacana/Photo Researchers, Inc.)*

A.

B.

a brain, a liver, and a kidney. Most zoologists think that the first vertebrate resembled a lamprey larva.

Lampreys and hagfish (the living members of the Agnatha) are the only vertebrates that lack jaws.

Placoderms, Now Extinct, Were Armored Fish with Jaws

At the beginning of the Devonian Period, some 405 million years ago, freshwater fish developed the first vertebrate jaws. These fish rapidly spread into the oceans, diversified, and multiplied. Paleontologists sometimes call the Devonian the "Age of Fishes." Among the prominent jawed fishes in Devonian seas were a class of armored fishes called **placoderms** [Greek, *plak* = plate + *derma* = skin]. The placoderms had jaws, paired fins, and much less armor than the ostracoderms.

Today, all vertebrates except the Agnatha have hinged jaws, with which they can seize, bite, and sometimes chew. Jaws permit a much more diverse diet than that of lampreys or hagfish. Jaws probably enabled placoderms and other jawed fishes to replace the ancient ostracoderms. By the end of the Devonian Period, armored placoderms had disappeared, replaced by the Chondrichthyes and Osteichthyes. These two classes of fishes still dominate the aquatic regions of the Earth's surface.

Placoderms were the first vertebrates with jaws.

Chondrichthyes Are Fish with Cartilaginous Skeletons

In most vertebrates, the soft cartilaginous skeleton of embryos is gradually replaced with bone. In sharks, skates, and rays, however, bone does not replace cartilage during development. Adult Chondrichthyes have lightweight cartilaginous skeletons, giving the class its name [Greek, *chondros* = cartilage + *ichthys* = fish]. Chondrichthyes, with only about 750 species, are much less diverse than the 30,000 species of bony fish (Osteichthyes).

Chondrichthyes also lack a swim bladder, a balloonlike organ that helps the bony fish float. Instead, sharks and their allies increase their buoyancy by storing large amounts of oil in their huge livers. Nonetheless, the Chondrichthyes are heavier than water and sink to the bottom when they are not moving (Figure 24-9).

Although sharks and rays lack the bony scales of the placoderms and ostracoderms, their skin is covered with small toothlike scales (denticles) that give it the feel of sandpaper. In addition, they have teeth (derived from their scales), which they use, along with their jaws and their ability to swim rapidly, to live as successful predators (Figure 24-9).

Sharks have particularly well-developed senses that help them find their prey. Among these senses are keen vision and smell. The lateral line system, a row of tiny sense organs along each side of the body, detects changes in water pressure. Other receptors, originally derived from the lateral line system, detect tiny electrical currents, including those generated by muscular contractions of potential prey.

A.

B.

Figure 24-9   The Chondrichthyes. A. Rays and skates have evolved adaptations for a life on the ocean floor, where they feed mostly on mollusks and crustaceans. Their pectoral fins are enormous, giving these fish their characteristic appearance. B. Sharks are mostly fast-moving predators with streamlined bodies, tapered at both ends. Lacking a buoyant swim bladder, a shark keeps from sinking by swimming constantly. (Some sharks take breaks on the sea bottom or in a sea cave.) Paired fins (pectoral and pelvic) on the underside help keep the moving shark afloat. (*A, Jeff Rotman/Tony Stone Images; B, Doug Perrine/DRK Photo*)

The Chondrichthyes have cartilaginous skeletons and no swim bladder.

Osteichthyes Are Fish with Bony Skeletons

The bony fish, Osteichthyes [Greek, *osteon* = bone + *ichthys* = fish], are the most diverse of all the vertebrate classes. More than 30,000 species fill oceans and seas, lakes and streams all over the world. They range in size (as adults) from about 1 cm to more than 6 m (20 feet). The bony fish comprise more than 95 percent of all fish and half of all species of vertebrates.

Like the cartilaginous fishes, the bony fishes first appeared in the Devonian Period, but in fresh water rather than the oceans. Bony fish differ from the Chondrichthyes in having bony skeletons and thin, bony scales. Most bony fish also have a **swim bladder,** an air-filled sac that helps them control their buoyancy (Figure 24-10).

Zoologists divided the living Osteichthyes into two sub-classes: the ray-finned fishes, Actinopterygii [Greek, *aktin* = ray + *pterygon* = wing], and the lobe-finned fishes, Sarcopterygii [Greek, *sarkodes* = fleshy]. Nearly all modern fishes are ray-finned fishes: they have paired fins supported by thin, bony rays, originally derived from bony scales. Ray-finned fishes are some of the best swimmers in the world (Figure 24-11A and B). They are extraordinarily diverse and live virtually wherever there is water.

Unlike the fins of ray-finned fishes, the fins of lobe-finned fishes contain muscle as well as bone. Although once abundant, lobe-finned fishes are now unusual, including only four genera of modern fishes (Figure 24-11C). The lungfishes can use their sturdy fins to walk on land, and biologists believe that the first amphibians must have resembled lobe-finned fishes whose fins gradually evolved into limbs. Modern lobe-finned fishes, such as lungfish, live in stagnant freshwater ponds, using their lungs to extract oxygen they get by gulping air from the surface. Lungfishes can radically alter their metabolism to survive long periods of drought by burying themselves in the mud at the bottom of a pond.

Bony fishes (osteichthyes) have bony skeletons, swim bladders, and paired fins.

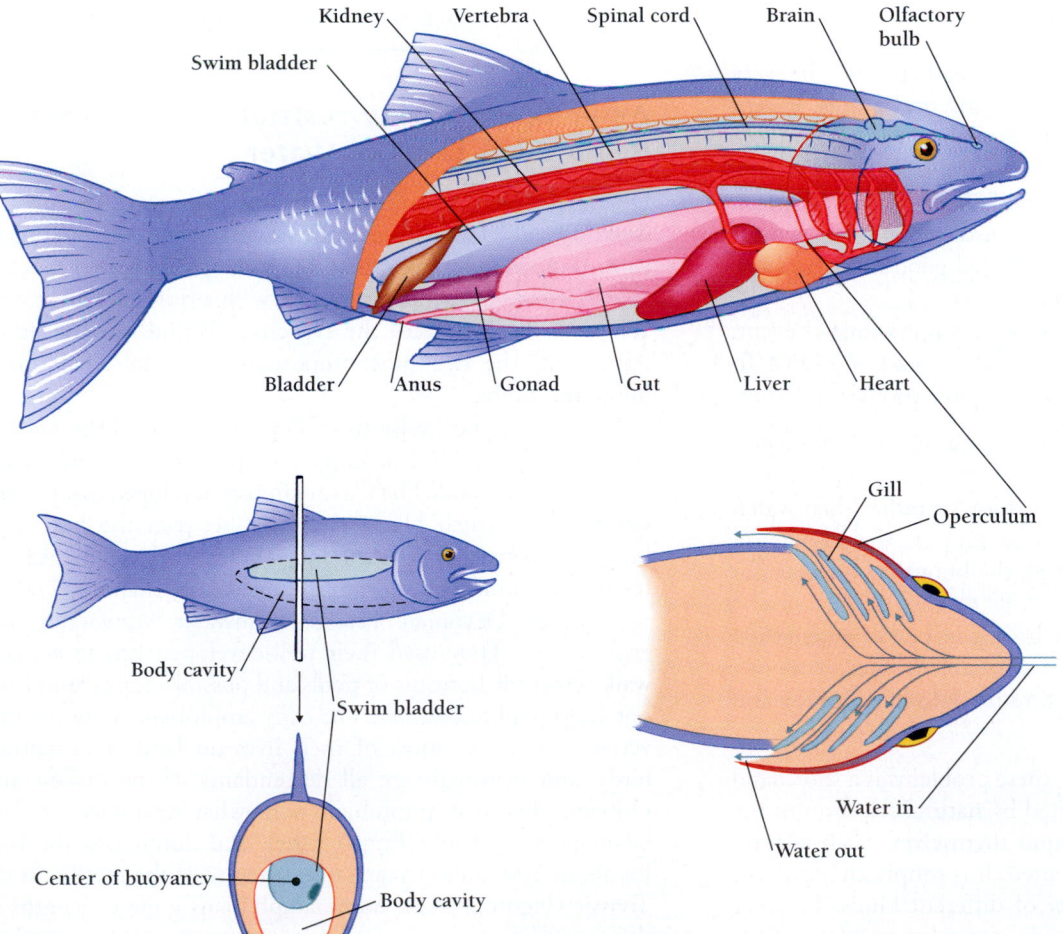

Figure 24-10 The body plan of a bony fish. Fish share many features with terrestrial vertebrates, including a brain, a liver, and true kidneys. In addition, a bony fish has a gas-filled swim bladder from which evolved the terrestrial lung. By regulating gases in the swim bladder, a fish can adjust its buoyancy, allowing it to float at the level it wants without exertion. Bony fishes also possess a bony operculum over the gills that allows them to pump water across the gills, facilitating the exchange of oxygen and carbon dioxide.

A.

B.

C.

Figure 24-11   **The two subclasses of bony fishes:** the ray-finned fishes and the lobe-finned fishes. A. The fast-swimming barracuda typifies the highly evolved ray-finned fishes. B. A pike can achieve accelerations of 12 to 24 g (times gravity), reaching top speeds of 6 meters per second (equal to a 4-minute mile). To find out how fish are able to swim so fast, researchers at MIT have built a robotic pike (*Esox lucius*), named "Robopike," which swims by itself. C. The 350-million-year-old marine coelacanth *Latimeria* is a lobe-finned fish that was thought to be extinct—until museum curator Marjorie Latimer found one in a fisherman's catch in 1938. (A, *Rondi/Tami Church/Photo Researchers, Inc.*; B, © *Sam Ogden*; C, *Sally Bensusen/Science Photo Library/Photo Researchers, Inc.*)

WHAT ADAPTATIONS HAVE VERTEBRATES EVOLVED FOR LIFE ON LAND?

By the end of the Devonian Period, fish had appeared in every conceivable aquatic niche in seas, streams, and ponds. Except for a few lungfish adventurers, however, vertebrates had not found a way to live on the land. Plants and insects had already done so, however, and the land stood all-inviting, with great, lush forests and a diversity of tasty insects and other arthropods. The terrestrial environment offered good meals and few predators.

The movement from water to land was a dramatic event, accompanied by immense changes in vertebrate anatomy. To colonize the land, vertebrates had to solve several problems:

1. They needed to obtain and conserve water in a comparatively dry environment.
2. They needed to extract oxygen from air rather than water.
3. They needed strong skeletons to support their bodies—they could no longer depend on the buoyancy of water for support.
4. They needed to control fluctuations in body temperature, which can be much greater in air than in water.
5. They still needed a watery environment for fertilization and early development.

To say that animals "solved" these problems is a shorthand way of saying that animals adapted by natural selection to the environments in which they found themselves. Such adaptation is not deliberate or goal oriented. It is simply an inevitable result of the differential survival of different kinds. In other words, while individual animals do not solve problems, lineages of animals "solve" certain problems, and they do so by evolving and adapting.

The terrestrial vertebrates—amphibians, reptiles, birds, and mammals—are the survivors that evolved adaptations that solved the problems of life on land.

Amphibians Are Terrestrial Animals That Begin Their Lives in Water

During the Devonian Period, the freshwater lakes and streams dried up and flooded regularly, leaving many aquatic organisms stranded in foul ponds and muddy stream beds. Only fish able to extract oxygen from the air, with a primitive lung, were able to survive. Indeed, the Devonian Period saw the development of the two most important terrestrial adaptations: lungs and limbs.

Nearly all the freshwater fishes that survived the Devonian Period had a kind of lung, a simple sac enhanced with a rich supply of blood. The Devonian fish developed **double circulation,** in which blood circulates between the heart and lungs and between the heart and the rest of the body. All terrestrial vertebrates have double circulation (Chapter 37).

In the Devonian Period, freshwater vertebrates also evolved legs. They used their well-developed fins to actually walk across the bottoms of pools and possibly across land from one deep pool to another. The early amphibians were the first vertebrates to live most of their lives on land. The reptiles, birds, and mammals are all descendants of the earliest amphibians. The first amphibians somewhat resembled modern salamanders and lobe-finned fishes, and dominated the land for about 100 million years, until the rise of the reptiles in the Triassic (Figure 24-12). Some amphibians grew to lengths of three or four meters and were more terrestrial than modern amphibians. Modern salamanders, for example, have evolved several adaptations for moving about in shallow water,

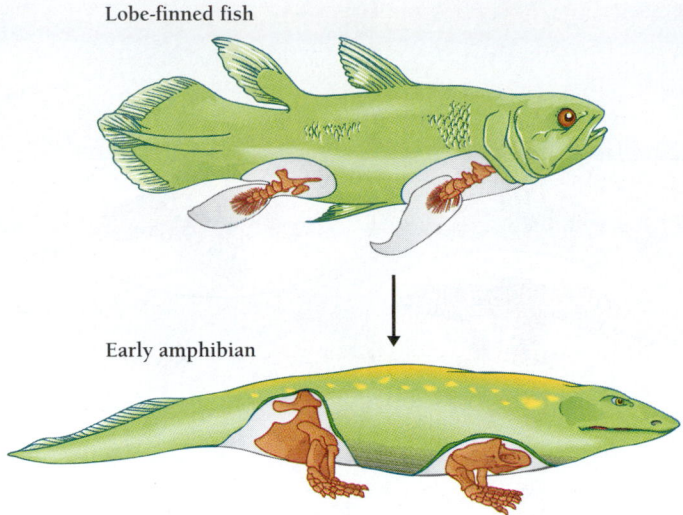

Lobe-finned fish

Early amphibian

Figure 24-12 The first amphibians. The first amphibians evolved from lobe-finned fishes at the end of the Devonian Period.

including limbs, strong tails, and a flat shape, that earlier amphibians did not possess.

Today, three orders of Amphibia exist (Figure 24-13): Urodela (salamanders), Anura (frogs and toads), and Gymnophiona (caecilians). Anurans lack tails; urodeles have them. The names Anura [Greek, *an* = without + *oura* = tail] and Urodela [Greek, *oura* = tail + *delos* = visible] describe this simple distinction. The anurans are by far the most diverse amphibian order, including about 90 percent of the 3000 amphibian species. The Gymnophiona [Greek, *gymnos* = naked + *ophioneos* = snakelike], which lack feet and resemble snakes or earthworms, are the least familiar amphibians. They are tropical amphibians that live underground and live like moles, pushing through the soil and searching for small invertebrates to eat. About 160 species are known.

Adult toads and frogs have powerful leg muscles for jumping and well-developed ears and vocal cords for communicating through air. They have extendible, sticky tongues with which they can expertly capture flying morsels. Amphibians can gulp air into their lungs, but they also absorb oxygen through their skins, which are moist and thin. In fact, some salamanders breathe entirely through their skins and the linings of their mouths.

An amphibian's life cycle generally consists of two stages, the first of which is aquatic and the second terrestrial—hence their name [Greek, *amphi* = double + *bios* = life]. Amphibians are only partly adapted to life on land. Their need for water dominates their lives. Unlike us, however, amphibians do not drink water. Instead, they must absorb it through their thin, damp skins. Such skin is not good at retaining water, and amphibians must confine themselves to very moist environments or dry out.

Most amphibians start life in water, when the male and female shed sperm and eggs together (Figure 24-14). Because amphibian eggs lack shells and special membranes to prevent water loss, their embryos almost always must develop in a very wet environment.

The fertilized egg (zygote) develops into an aquatic larva, or tadpole, which obtains oxygen through gills and uses its long tail to swim. Later, the tadpole becomes a terrestrial animal by losing its gills and developing lungs and limbs, in a process called **metamorphosis** (Figure 24-14). If it is an anuran, it also loses its tail. The adult amphibian is now capable of surviving on land.

A few species of frogs, salamanders, and caecilians have evolved direct development. The zygote develops directly into the terrestrial form, bypassing the larval form. Other species avoid the need for freestanding water for their larvae in other ways. They lay their eggs in moist places on land, in pockets of water in leaves, or in wet mossy places. Several species carry the developing eggs in special pouches on their backs or in the

A.

B.

C.

Figure 24-13 Three orders of amphibians. A. Jordan's salamander, from North Carolina. B. A river frog from Florida. Frogs differ from toads in the quality of their skins: frog skin is smooth and moist, while toad skin is bumpy and dry. C. Gray caecilian. The caecilians are legless amphibians that live in the tropics. Like earthworms, they move by means of a hydrostatic skeleton—even though they have a backbone. (*A, David M. Dennis/Tom Stack & Associates; B, Joe McDonald/DRK Photo; C, Michael Fogden/DRK Photo*)

| Day 0 | Day 2 | Day 14 | Day 28 | Day 38 |

Figure 24-14 **How tadpoles turn into frogs.** Most amphibians lay their eggs in water. The eggs develop into tadpoles, which eat and grow under water. When the time is right, the tadpoles metamorphose into adult amphibians.

mouth (Figure 24-15). The South American frog (*Rhino-derma*), for example, lays its eggs on moist ground, after which groups of adults guard the developing embryos. When an embryo begins to move, a frog quickly takes it into its mouth. When the embryo finishes developing, the frog yawns, and a tiny froglet jumps out of his mouth.

A few salamander species do not complete metamorphosis. The mudpuppy and the axolotl, for example, retain their gills and continue to live exclusively in water even after they are capable of reproducing. If conditions are right, the axolotl (*Amystoma*) can mature into an adult form with lungs instead of gills. But the mudpuppy (*Necturus*) never loses its gills or develops lungs.

Figure 24-15 **Some amphibians care for their young.** Here the reticulated poison dart frog carries two tadpoles to water. (*Michael Fogden/DRK Photo*)

Amphibians live a dual life. The larvae are usually aquatic and the adults are usually terrestrial.

Reptiles Can Live Their Entire Lives on Land

Reptiles [Latin, *reptare* = to crawl], unlike amphibians, are entirely adapted to life on land. Although some reptiles, such as the crocodiles and sea turtles, are adapted to aquatic life, they are descended from terrestrial reptiles and must breath air. They do not have gills.

Most of the 6700 species of living reptiles fall into three orders (Figure 24-16A-D): Chelonia (250 species of turtles and tortoises), Crocodilia (22 species of crocodiles and alligators), and Squamata (lizards and snakes). Of these, the Squamata are by far the most successful, with about 3800 species of lizards and 2700 species of snakes. The Squamata also include the Amphisbaenia, which include some 140 species of legless reptiles that resemble the amphibian caecelians. Like moles, the amphisbaenians have lost the use of their eyes and burrow through soil hunting invertebrates. A fourth order of living reptiles includes only a single species, the tuatara, a lizardlike animal that survives only on a few islands off the New Zealand coast (Figure 24-16E).

Some of the most fascinating reptiles are now extinct. Although many of us think of dinosaurs when we think of extinct reptiles, only a fraction of the animals living in the Mesozoic Era, the Age of Reptiles, were actually dinosaurs. Others were swimming ichthyosaurs, and some were flying pterosaurs, for example (Figure 24-17). The long and glorious history of the reptiles began some 300 million years ago in the

Why Are Amphibians Disappearing?

Costa Rica's spectacular golden toad (*Bufo periglenes*) has not been seen since 1989, Australia's gastric brooding frog (*Rheobatrachus silus*) was last seen in 1979, populations of the common toad (*Bufo bufo*) have declined on Norwegian coastal islands, and many species of frogs in California have vanished from most of their historic ranges (Figure A). Throughout the world, amphibian species are declining in number and, in some cases, going extinct locally.

Is some unknown phenomenon exterminating toads and frogs?

The question first arose at the First International Herpetological Congress, held in 1989 in Canterbury, England. At that meeting, herpetologists recounted what each thought was a personal sorrow: the disappearance of an amphibian species from a study site. So many scientists reported the same kinds of losses, however, that David Wake, a leading herpetologist in the United States, organized a 1990 meeting in Irvine, California, to discuss whether there was a global decline in amphibian populations. At that meeting and at a follow-up symposium, the herpetological community shared information about declines in amphibian populations in Australia, Canada, the western and southeastern United States, Central America, the Amazon Basin, and the Andes.

For most amphibian populations, scientists do not have long-term data on population cycles. Without information about natural population cycles, it is impossible to determine whether, for any given species, a dip in numbers is natural variation or a decrease induced by human activities. Many frog species have boom-and-bust population cycles—populations explode one year and drop to almost nothing in another year. Such natural variation makes it difficult to interpret declines in some species.

Nonetheless, herpetologists agree that an ongoing, worldwide decrease in amphibian populations is occurring. They further agree that the primary cause is the fragmentation and destruction of habitat, as is unfortunately true for many other organisms.

Herpetologists do not yet agree what other reasons might account for the worldwide disappearance of amphibians. Some researchers say amphibian declines are just part of the worldwide loss of biodiversity caused by destruction of forests, wetlands, and other habitats. The disappearance of amphibian species, they argue, is no different from the extinctions of thousands of other species.

Other researchers insist that amphibians—with their permeable skin, permeable eggs, and complex life cycles—are particularly vulnerable to certain environmental changes. In particular, amphibians seem to be especially sensitive to increases in UV radiation and chemical pollutants. Decreases in numbers of the North American tiger salamander (*Ambystoma*

Figure A   Golden toads. *(Michael Fogden/DRK Photo)*

tigrinum) have been linked to pollution, especially acid rain. The Tarahumara frog (*Rana tarahumarae*) is extinct in the United States, and researchers have implicated high levels of the toxic metal cadmium. The massive die-offs of fertilized eggs of the Cascades frog (*Rana cascadae*) and the Western toad (*Bufo boreas*) may be due to increases in UV radiation stemming from the destruction of ozone in the stratosphere; the eggs of both of these latter species are low in *photolyase,* an enzyme that helps repair UV-damaged DNA.

A host of nonnative predators have been accused of decimating amphibian populations. In the western United States, bullfrogs (*Rana catesbeiana*), nonnative trout, bass, and sunfish, all predators of amphibians, have been introduced into countless waterways. Losses of the Yavapai leopard frog (*Rana yavapaiensis*), the mountain yellow-legged frog (*Rana muscosa*), and the Chiricahua leopard frog (*Rana chiricahuaensis*) have all been attributed to such introduced predators.

Some researchers argue that an accumulation of environmental stresses, rather than a single cause, may reduce or destroy a population. For example, a population of frogs may become fragmented and reduced by the loss of wetland habitat. If a subsequent drought kills many more individuals, the population may never recover. Or stress from increased UV exposure and pollution may make individuals more susceptible to disease—one explanation for the extinction of 11 distinct populations of boreal toads (*Bufo boreas*) in western Colorado.

Frogs are showing signs of going the way of the dinosaurs. Species in habitats as different as Costa Rican cloud forest, Sonoran desert, and montane pools in the Sierra Nevada are succumbing. Unfortunately, beyond the obvious effects of habitat destruction on all species worldwide—destruction that proceeds as rapidly as ever—there is probably no one-size-fits-all explanation for the loss of amphibian species.

Figure 24-16 The four orders of living reptiles. A. Giant tortoises wallow in a pond on one of the Galápagos Islands. The Chelonia comprise the turtles and tortoises. B. Nile crocodile washing and releasing baby. The Crocodilia comprise the crocodiles, alligators, and caimans. C. Crested dragon lizard, Malaysia. The Squamata comprise the lizards and snakes. D. Two male Panamint rattlesnakes in ritual combat, White Mountains, California. E. The lone species in the order Sphenodonta is the tuatara of New Zealand. (A, *Tui de Roy/Minden Pictures*; B, *Roger de la Harpe/Animals Animals*; C, *Mike Bacon/ Tom Stack & Associates*; D, *Gordon Wiltsie/ Peter Arnold, Inc.*; E, *John Cancalosi/DRK Photo*)

late Carboniferous Period. During the Permian and Triassic periods, the reptiles radiated into about 17 orders.

Reptiles were the first vertebrates to free themselves of the water. As a result, reptiles share with their descendants, the birds and mammals, some common adaptations to life on land. Their dry, watertight skin helps conserve water, serving the same function as the cuticles of insects and plants and our own dry skin. Because reptiles cannot breathe through such tough skins, they must depend entirely on their lungs for oxygen.

Reptiles and their descendants have also evolved more efficient kidneys for conserving water and legs better suited to walking on land. Among reptiles' many adaptations to land life, however, the most dramatic advance was the evolution of the **amniotic egg** (Figure 24-18). The amniotic egg is so well adapted to a dry environment that even crocodiles and turtles, which live primarily in the water, return to the land to lay their eggs. The **amnion** is a membrane that encloses the developing embryo in its own little pond, eliminating the need for a separate aquatic stage. The amnion, in turn, lies within a porous shell that allows the exchange of gases with the sur-

rounding air. Just beneath the shell is another membrane, called the **chorion** [Greek, = skin]. Also within the shell is the **yolk,** a rich food supply surrounded by a membrane called the **yolk sac.** A fourth membrane, the **allantois,** functions in both respiration and excretion.

All reptiles and their descendants have (or had) an amniotic egg. These include all birds and mammals. The amniotic egg is one of the characteristics that distinguishes all of these groups from other animals and shows that we are related.

In both reptiles and birds, a shell forms around the amniotic egg before it emerges from the female, and so fertilization must occur before the shell forms. That is, fertilization must be internal. All male reptiles except the tuatara accomplish fertilization with a penislike copulatory organ. Crocodilians and turtles have a single penis, while snakes and lizards have a pair of "hemipenes." Although "hemi" means "half," snakes and lizards actually have two penises, and in some species the males alternate using one and then the other (Figure 9-8).

In all reptiles, both sexes possess a **cloaca,** a common entrance and exit chamber for the digestive, urinary, and

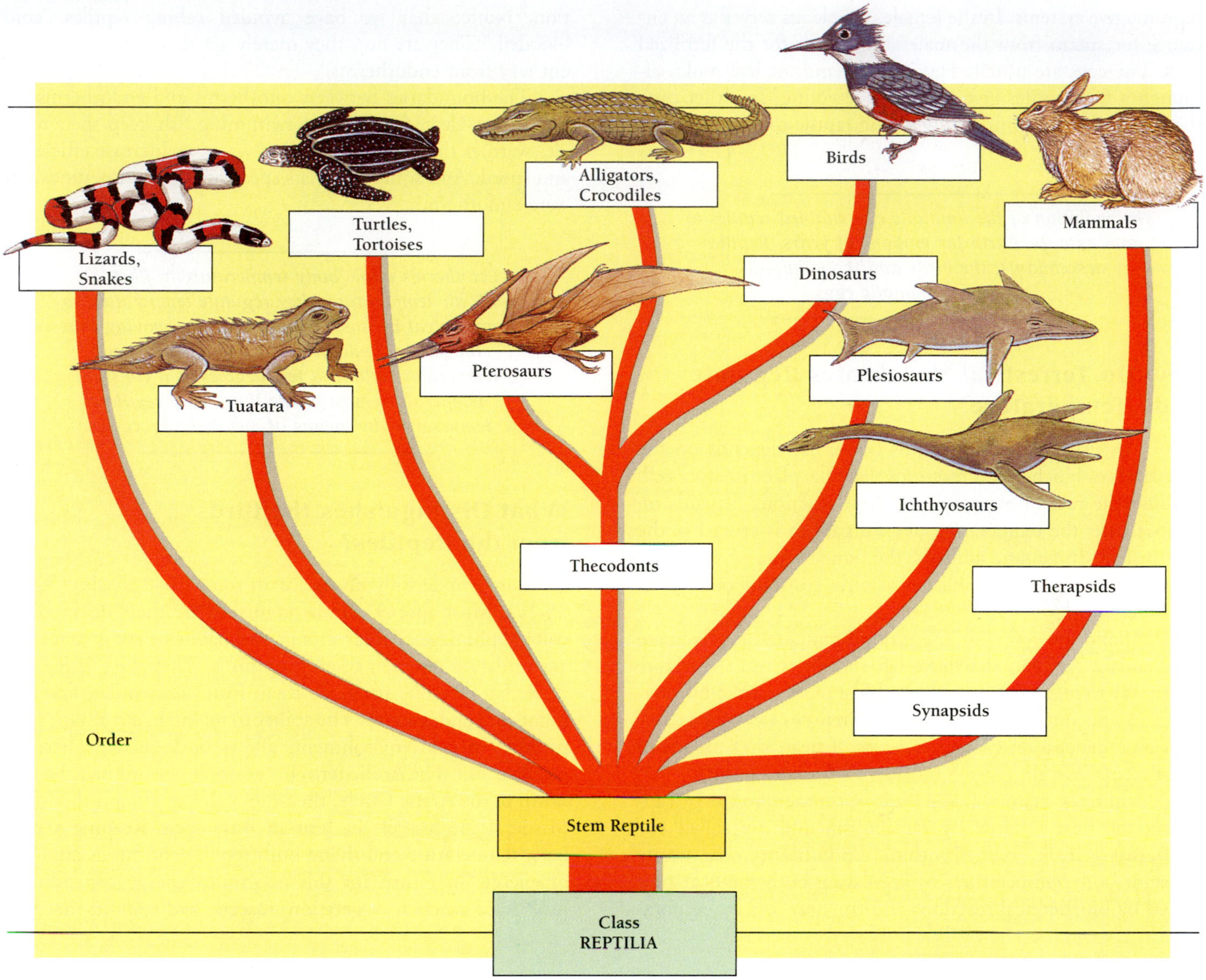

Lizards, Snakes

Turtles, Tortoises

Alligators, Crocodiles

Birds

Mammals

Tuatara

Dinosaurs

Pterosaurs

Plesiosaurs

Ichthyosaurs

Thecodonts

Therapsids

Synapsids

Order

Stem Reptile

Class
REPTILIA

▲ **Figure 24-17**   **Reptiles all.** The familiar lizards, snakes, turtles, and crocodiles are not the only animals in the reptile family album.

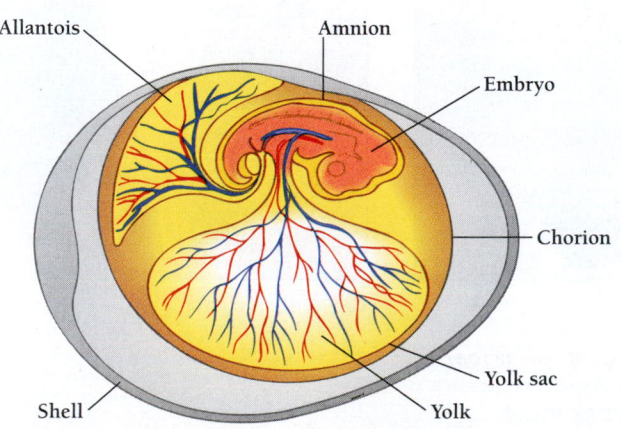

Allantois

Amnion

Embryo

Chorion

Yolk sac

Shell

Yolk

◀ **Figure 24-18**   **The amniotic egg.** The most obvious feature of an egg is the yolk sac, which serves as a pantry, a storage area for fats, proteins, and other nutrients for the developing animal. But fish eggs and amphibian eggs also have a yolk sac. What distinguishes the amniotic egg are three other membranes: the amnion encloses the embryo in water; the allantois serves as a sort of outhouse, a separate area for wastes; and the chorion, the outermost membrane, encloses embryo, yolk, and allantois. The chorion helps regulate the exchange of oxygen and carbon dioxide between the embryo inside and the world outside.

reproductive systems. In the female, the cloaca serves as an entrance for sperm from the male and an exit for the fertilized eggs. The eggs are usually laid in soil, sand, or leaf mold, although a few reptiles give birth to live young. In either case, the zygote develops into a tiny young reptile that fully resembles its parents.

The evolution of the amniotic egg enabled reptiles to dominate the Earth for millions of years. Reptiles' descendants, the birds and the mammals, also have amniotic eggs.

How Do Terrestrial Vertebrates Regulate Body Temperature?

As we discussed in Part I of this book, life depends on the thousands of biochemical reactions that take place in every cell. Because the rate of every one of these reactions depends on temperature, the temperature of an organism is crucial to the way it lives. In aquatic animals, the temperature of the water around them usually establishes a fairly constant body temperature.

On land, however, air and ground temperatures fluctuate enormously. Animals that have left the water must therefore find a way to regulate temperature. Most reptiles do so by moving into and out of the sun. Because reptiles (as well as amphibians) depend on external sources of heat, they are called **ectotherms** [Greek, *ektos* = outside + *thermos* = heat].

In contrast, mammals and birds warm their bodies by capturing the heat released by metabolism and are called **endotherms** [Greek, *endo* = within]. Endotherms keep nearly constant body temperatures by regulating both the heat produced by biochemical reactions and the heat lost by evapora-

tion. Notice that we have avoided calling reptiles "cold blooded." They are not; they merely get their heat in a different way from endotherms.

The boundaries between ectotherms and endotherms are not always clear. Certain fast-swimming fish keep their muscles warmer than the surrounding water, to increase efficiency and speed. And some scientists speculate that many dinosaurs were endothermic.

Ectotherms allow body temperature to fluctuate with air temperature; they regulate temperature to some extent by basking in the sun or moving into the shade. Endotherms maintain their body temperature within a few degrees of a constant temperature most of the time. They regulate temperature by means of metabolic processes.

What Distinguishes the Birds from the Reptiles?

Birds are wonderfully diverse, with some 8700 species (Figure 24-19). Most species are distinguished by their distinctively colored plumage or their strange and haunting songs and calls. Yet birds are also remarkably uniform. They have three features that clearly distinguish them from the reptiles: feathers, flight, and endothermy. The ability to maintain a constant temperature in cold environments allows birds to live where no reptile could. The Arctic tern, for example, spends half the year north of the Arctic Circle, then flies halfway around the world to spend the rest of the year in Antarctica. Keeping a high body temperature and flying both require enormous amounts of energy. In return for this enormous energy expenditure, birds have access to vegetation, insects, and seafood that rep-

A.   B.   C.

Figure 24-19 Birds are both distinctive and diverse. Shown here are a great blue heron, a blue-throated hummingbird, and an emperor penguin and its young. (*A, Jim Zipp/Photo Researchers, Inc.; B, Russel C. Hansen/Peter Arnold, Inc.; C, Barbara Cushman Rowell/DRK Photo*)

tiles can never reach. Flight also provides terrific protection from most predators.

Most birds find food and shelter on land, though many birds depend on streams, ponds, and seas for their food. Like the reptiles, birds produce amniotic eggs and do not need water for fertilization or early development.

Fertilization is always internal, but unlike the reptiles, only a few species possess a penis. Ducks and flightless birds such as ostriches, herons, flamingos, chickens, and turkeys all have rudimentary penises. In most other birds, both males and females have a cloaca. Birds accomplish fertilization by bringing together their cloacas. As in the tuatara, the sperm from the male's cloaca passes into the female's cloaca. Such intimacy often requires a long courtship, and many bird species have elaborate mating rituals.

Feathers are made of keratin, the same protein that forms a lizard's scales and our own fingernails and hair. Flight feathers are an amazing adaptation to flight, but down, which insulates the bodies of birds, is also important. Many biologists now believe that feathers were originally an adaptation for warmth. Flight came later, as did a host of adaptations that lightened the body and made birds still better flyers. These included the loss of teeth (which lessens the weight of the head), the hollowing of the bones (which lightens the whole skele-

ton), and the reshaping of the breastbone (which forms a keel to which flight muscles attach) (Figure 24-20).

Feathers, flight, and endothermy distinguish birds from reptiles.

What Distinguishes the Mammals?

Mammals [Latin, *mammae* = breasts] take their name from their **mammary glands,** milk-producing organs in the female that characterize the mammals. No other vertebrates have mammary glands. Mother mammals nurse their young with warm milk, a rich mixture of fats, sugars, proteins, minerals, and vitamins, as well as antibodies that help defend newborns against infections.

Two other traits distinguish the mammals from other animals. Only mammals have hair and two sets of teeth ("baby" teeth and adult teeth). A mammal's hair, like a bird's feathers, helps conserve heat, a trait that is useful for an endotherm. Mammalian teeth, far more specialized than those of other vertebrates, help mammals get the tremendous amounts of energy they need to maintain a constant body temperature.

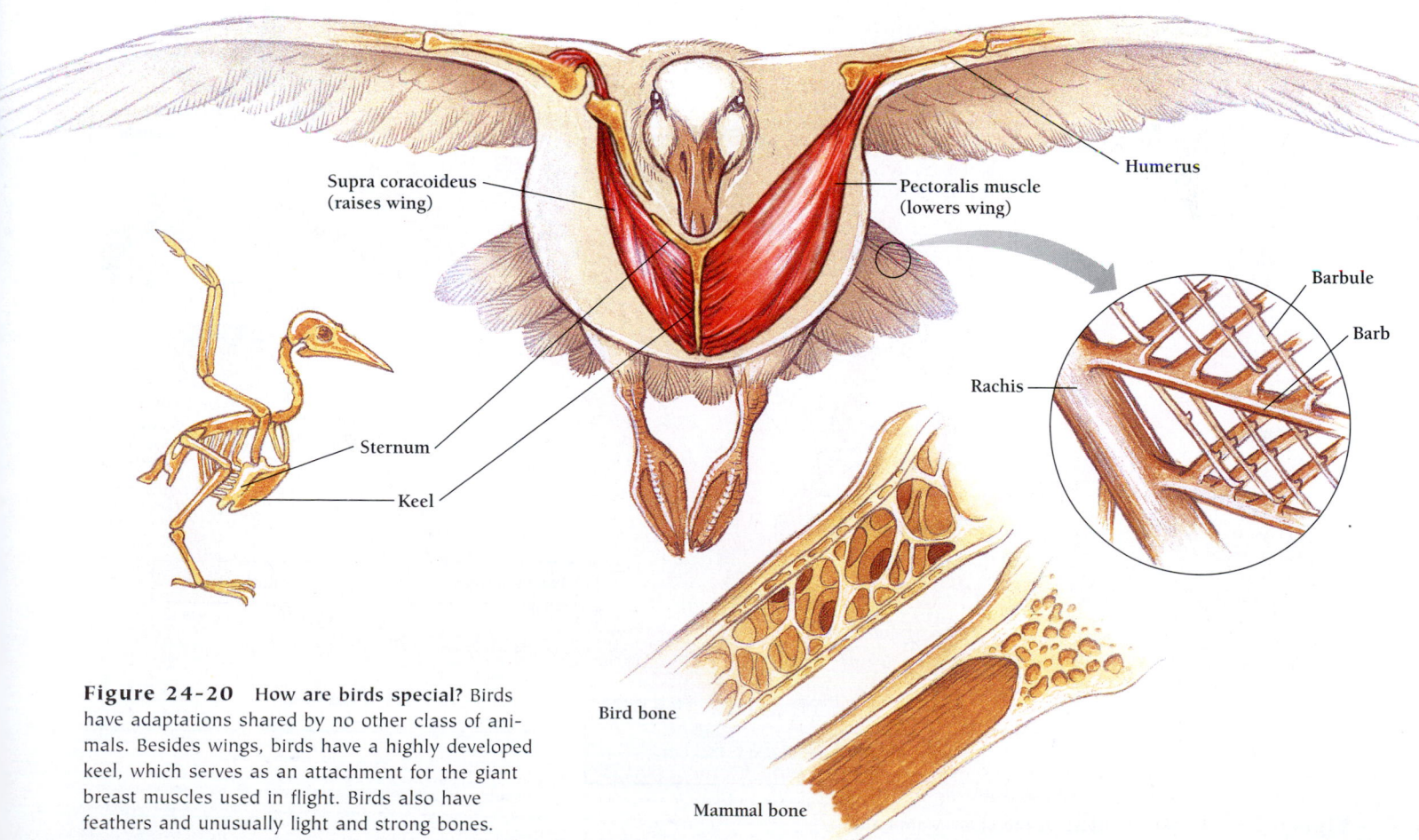

Figure 24-20 How are birds special? Birds have adaptations shared by no other class of animals. Besides wings, birds have a highly developed keel, which serves as an attachment for the giant breast muscles used in flight. Birds also have feathers and unusually light and strong bones.

Mammalian teeth may be specialized, for example, for grinding plants (horses and other herbivores), for shearing flesh (lions and wolves), for breaking bones (hyenas), or for seizing fish (dolphins and whales).

Mammalian teeth are extremely well preserved in the fossil record. Since a single tooth tells volumes about its owner, paleontologists know more about the evolution of mammals than is possible for other classes. In fact, some paleontologists now believe that the efficiency of the mammalian molar is at least partly responsible for the evolutionary success of the mammals since the end of the Cretaceous Period. When the ruling reptiles died off, mammals radiated into about 30 orders, of which about two-thirds still have living representatives (Figure 24-21).

Mammalogists divide the living mammals into three subclasses—prototherians, metatherians, and eutherians. These subclasses have distinctive skulls and teeth that show that they first diverged from each other during the Mesozoic Era, even before the dinosaurs disappeared. These three subclasses differ markedly in the ways they reproduce.

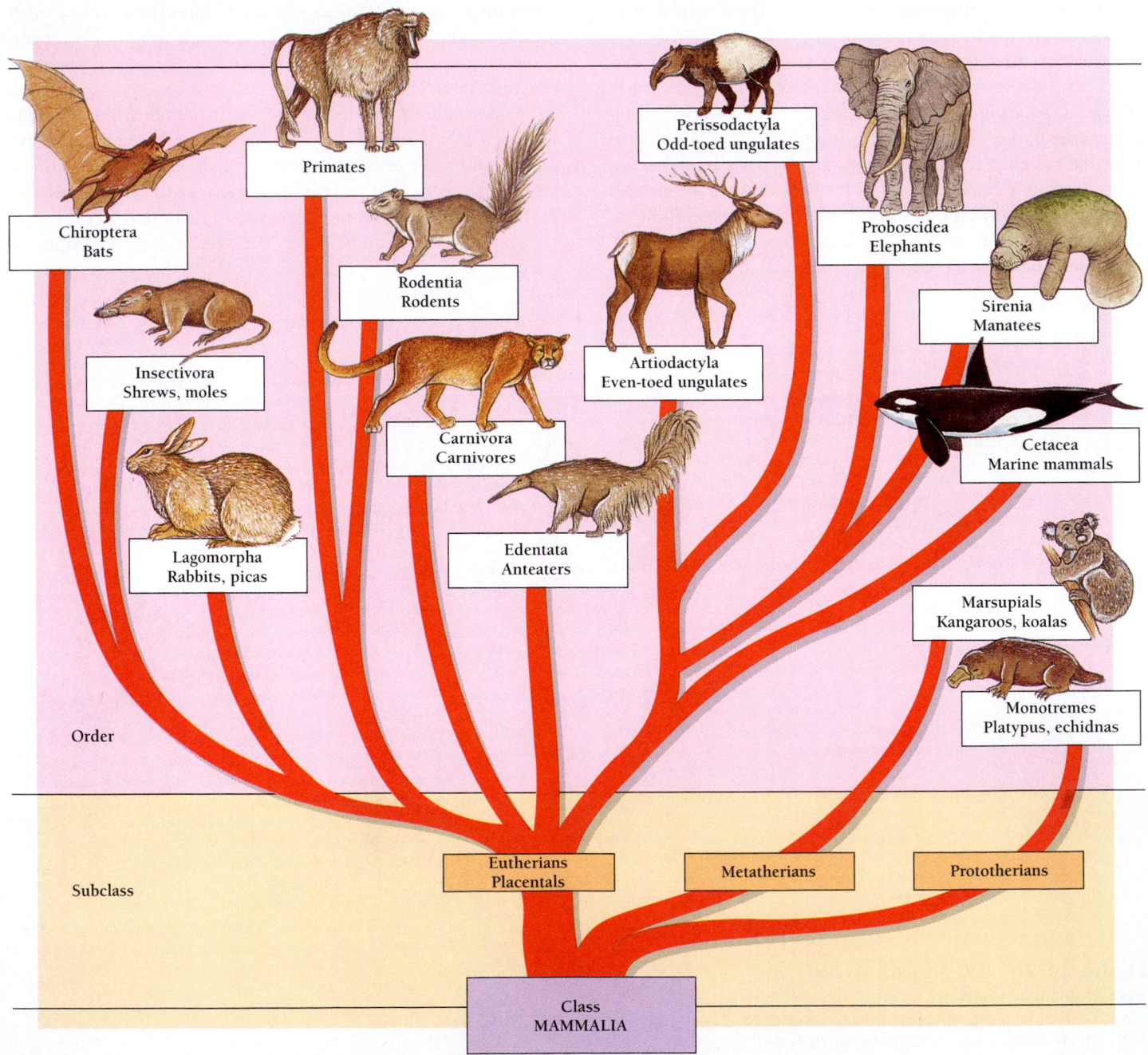

Figure 24-21   Some orders of the class Mammalia.

Most mammals are **eutherians** [Greek, *therion* = wild beast] or **placental** mammals. In humans, elephants, rats, rabbits, and every other kind of eutherian mammal, embryonic development takes place within the uterus of the mother. In fact, the word "placental" refers to the **placenta** [Latin, = flat cake], the flat structure that joins the lining of the uterus with the membranes of the fetus.

Metatherians, or **marsupials** [Latin, *marsupium* = pouch], such as the kangaroo and koala, reproduce differently from eutherians. Newborn marsupials are not babies at all, but undeveloped fetuses. A newborn red kangaroo, for example, is no bigger than a bumblebee (Figure 24-22). Emerging from the uterus only 33 days after fertilization, the tiny creature painstakingly pulls itself along its mother's fur until it finds a nipple. The mother's nipples are in a furry pouch, where the young kangaroo sleeps and nurses for about 6 months. Even after it begins to wander a bit on its own, the young kangaroo regularly returns to the pouch until it is almost 8 months old. Meanwhile, the mother may have another, younger baby in the pouch, as well as an embryo in her uterus.

The **prototherians,** or **monotremes,** are the strangest mammals (Figure 24-23). They are endotherms with hair and

Figure 24-23 Duck-billed platypus. Unlike other mammals, monotremes lay eggs. They are considered the most primitive of all the living mammals. *(Alan Root/Okapia/Photo Researchers, Inc.)*

mammary glands, like other mammals. Like the reptiles from which they and all other mammals are descended, however, monotremes lay eggs that are similar to those of reptiles and birds. In addition, their digestive and reproductive systems empty into a cloaca, hence their name [Greek, *monos* = single + *trema* = hole]. The only living monotremes are the platypus and the two spiny anteaters (or echidnas), which live in Australia and New Guinea. They are believed to be primitive mammals, but exactly how they are related to other mammals is still unknown. Mammals, both modern and ancient, are generally classified according to their tooth structure, and adult monotremes have no teeth.

Three characteristics set mammals apart from other vertebrates—mammary glands, hair, and two sets of teeth.

In this chapter we have seen how creatures as seemingly different as sea urchins and rabbits can share traits that show they are related. This chapter concludes our survey of the six kingdoms of organisms. In the next section, we see how ecological forces shape these many living organisms.

Figure 24-22 A red kangaroo with its offspring. All marsupials carry their young in a special pouch that contains milk-secreting nipples. A female kangaroo may have a youngster at her side, another in the pouch, and a third in her womb. *(Dave Watts/Tom Stack & Associates)*

Study Outline with Key Terms

Research in molecular genetics supports the hypothesis, first advanced in the 19th century, that chordates have an inverted version of the same body plan as most protostome coelomates. Structures that are **dorsal** in frogs and other vertebrates are **ventral** in arthropods. Deuterostomes, however, differ from proto-

stomes in several ways, including, for example, the fate of the embryo's **blastopore.** Deuterostomes include four phyla: **Echinodermata, Chaetognatha, Hemichordata,** and **Chordata.**

Echinoderms, such as starfish, are radially symmetrical as adults and bilaterally symmetrical as larvae. Echinoderms diverged from protostomes more than 500 million years ago. Chaetognatha (arrow worms) and Hemichordata (acorn worms) may resemble the ancestors of the chordates. Hemichordates have pharyngeal slits and a dorsal nerve cord, both of which are also chordate characteristics. They do not have, however, a notochord or a postanal tail, as do the true chordates.

Chordates share four main features at least sometime during their lives: a stiff flexible rod, called a **notochord,** running along the back; a **dorsal hollow nerve cord; pharyngeal slits** (or **gill slits**); and a postanal tail Chordates include the vertebrates and two other subphyla, the tunicates and the cephalochordates. Tunicates resemble other chordates as larvae, but not as adults. Cephalochordates have notochords both as larvae and as adults.

Vertebrates have heads and segmented spinal columns. The embryos of all vertebrates have skeletons made of **cartilage.** Most vertebrates replace the cartilage with **bone** during development. Vertebrates have closed circulatory systems and characteristic internal organs.

Four classes of vertebrates are fishes—Agnatha, Placodermi, Chondrichthyes, and Osteichthyes, and four classes are **tetrapods,** so-called four-footed animals—amphibians, reptiles, birds, and mammals. Agnatha (the extinct **ostracoderms** and the living **cyclostomes**) are the only vertebrates that lack jaws. **Placoderms,** which are now all extinct, were the first vertebrates with jaws. In the Chondrichthyes (sharks and rays),

bone never replaces cartilage. The Osteichthyes (bony fish) have bony skeletons and a **swim bladder.** One type of bony fish, with lobe fins and primitive lungs, resembles the probable ancestors of the tetrapods.

Tetrapods have many special adaptations for land life. They have some or all of the following: **double circulation** and lungs, efficient kidneys, water-resistant skin, strong skeletons and jointed limbs, **ectothermy** and **endothermy,** and the **amniotic egg.**

Amphibians are semiterrestrial animals that always begin their lives in water but later develop lungs and other adaptations to live on land, in the process of **metamorphosis.** Reptiles and their descendants, the birds and mammals, can all reproduce away from water as a result of the development of the amniotic egg. The different parts of the amniotic egg— the **amnion,** the **chorion,** the **yolk** and **yolk sac,** as well as the **allantois**—all contribute to the egg's ability to regulate water balance on dry land. Reptiles also developed watertight skins and strong skeletons. Reptiles and birds have a common exit for digestive, urinary, and reproductive systems, called the **cloaca.** Birds added feathers, flight, and endothermy, while mammals developed hair, two sets of teeth, and **mammary glands** for nursing their young. The mammals are divided into the **eutherians,** or **placental** mammals; the **metatherians,** or **marsupials;** and the **prototherians,** or **monotremes.**

Review and Thought Questions

Review Questions

1. Describe three characteristics that distinguish the deuterostomes from the protostomes.
2. What characteristics do humans share with starfish (sea stars)?
3. What three characteristics define the chordates? Can you locate all of these features in your own body? For each feature, explain where it is or where it once was.
4. Rank the following organisms according to how closely related they are to humans: tunicates, lamprey larvae, and arrow worms.
5. Is the vertebral column, or backbone, actually made of bone in all vertebrates? What is bone?
6. Of what value are jaws?
7. Why must amphibians return to water to reproduce?
8. Name the layers of the amniotic egg. What classes of animals have amniotic eggs?

9. What is thermoregulation? Why is it important? In what two ways do vertebrates accomplish thermoregulation?
10. What three mammalian characteristics distinguish a human from a turkey?
11. What is the function of a placenta?

Thought Questions

12. Compare the vertebrate body plan with the arthropod body plan. What advantage(s) might there be to flipping over the arthropod body plan? How would you test your hypothesis?
13. How would the day-to-day lives of humans be different if we were marsupials instead of placental mammals?

About the Chapter-Opening Image

This child illustrates the idea that the vertebrate body plan is the reverse of that of invertebrates such as lobsters and insects. As one biologist put it, "If you lay down on your back and waved your arms, you would be doing what insects do when they walk."

 On-line materials relating to this chapter are on the World Wide Web at **http://www.harcourtcollege.com/lifesci/aal2/**

part V

Ecology

Chapter 25
Ecosystems

Chapter 26
Terrestrial and Aquatic Ecosystems

Chapter 27
Communities: How Do Species Interact?

Chapter 28
Populations: Extinctions and Explosions

Chapter 29
The Ecology of Animal Behavior

Photo: Mass of starfish, Rialto Beach, Olympic National Park, Washington. *(© Darrell Gulin/Natural Selection)*

Ecosystems

Gaia: Is The Biosphere An Organic Entity?

In the early 1960s, the National Aeronautics and Space Administration (NASA) began planning its first explorations of the solar system. The space agency was especially keen to examine Martian soil for signs of life or conditions that could support life. But analyzing the Martian soil would require new instruments specially designed for the project.

For these, NASA turned to James E. Lovelock, an eccentric English chemist who had invented an ingenious device that detected tiny amounts of chemicals. Lovelock's device was widely used to measure trace amounts of pollutants such as the insecticide DDT.

In the early 1960s, NASA invited Lovelock to the Jet Propulsion Laboratory, in California, to help design instruments for detecting life on Mars. Lovelock's fascination with the search for extraterrestrial life and his insatiable curiosity soon led him beyond this technical chore to reconsider the much greater question of life on Mars. What if Martian or other life didn't look or act anything like terrestrial life? Could it still be detected? If so, how? How could life be described so that scientists could recognize it even in an alien form?

Lovelock approached these questions by considering the concept of "entropy." *Entropy* is a measure of disorder in "systems." A **system** is any collection of interacting parts or objects. If any system—whether a solar system or a dorm room—is left alone, its disorder, or entropy, will increase.

Organisms, by contrast, create order and thus *decrease* local entropy. Plants, for example, convert carbon dioxide and water into highly structured leaves, stems, and flowers. And animals convert amino acids and simple fats into muscle and brain tissue. This decrease in local entropy exacts a cost, however. It requires energy from the sun. And the decrease in entropy is only local. The universe as a whole is running down; its entropy is increasing.

A sure sign of life on Mars, then, Lovelock realized, would be a reduction in local entropy. But what in the world, he wondered, does a "reduction in entropy" look like? Lovelock couldn't stop thinking about the problem. He wondered how an alien being would recognize life on Earth. And it dawned on him that an alien would notice the peculiar chemical composition of Earth's atmosphere.

Earth's atmosphere is loaded with unlikely combinations of chemicals. For example, the atmosphere contains both methane and oxygen, which reacts with methane. Methane is a simple one-carbon hydrocarbon made by bacteria that gives swamp gas its smell. Methane and oxygen readily react with one another to form carbon dioxide and water.

$$CH_4 + 2\,O_2 \longrightarrow CO_2 + 2\,H_2O$$

This reaction constantly destroys methane. If no living organisms were present, methane would soon disappear from Earth's atmosphere. But bacteria in marshes, rice paddies, and cows' stomachs constantly make more methane, so the gas persists.

Lovelock realized that the Earth's chemically unlikely atmosphere could be seen as an extension of our planet's **biosphere**—the system of living things that covers the Earth. Just as a tortoise's shell is part of its body, the Earth's atmosphere is part of the biosphere. Lovelock envisioned the atmosphere, the oceans and soils, as constituting a single, immense living individual. He further hypothesized that this immense organism acts to preserve itself. Lovelock named this entity Gaia after the Greek Earth goddess.

NASA

Key Concepts

- An ecosystem is an interacting group of species, together with nonliving components such as sunlight, soil, water, and air. All of the world's ecosystems, together with the atmosphere and the oceans, make up the biosphere.

- Plants and other autotrophs capture energy from sunlight, but only a small fraction of that energy is available to other organisms.

- Although the biosphere constantly loses energy, it recycles materials such as carbon and nitrogen.

- Without the effect of greenhouse gases, Earth would be a lifeless planet. But too much carbon dioxide and other greenhouse gases can damage the biosphere.

Lovelock reasoned that the biosphere regulates the temperature and chemistry of the atmosphere, just as many animals regulate their blood temperature and chemistry. Such self-regulation is called homeostasis (Chapters 1 and 34).

The Gaia hypothesis, first published in the 1960s, vastly entertained the public. TV anchors glibly announced that the Earth was one big organism. Environmentalists seized on Gaia as a cultural metaphor, stressing the need to consider the biosphere and its environment—the atmosphere, the oceans, and the soils—as a whole. NASA's first breathtaking photographs of Earth as a cloud-swathed blue ball floating in space intensified Gaia's already enormous popular appeal.

Yet while biologists found the idea of Gaia amusing, most also found it embarrassing. Despite Lovelock's impressive credentials as an inventor, many scientists viewed him as a crank. Academic ecologists of the 1960s didn't want their careful scientific work confused with the free-wheeling political activity of environmental groups. But Lovelock's credentials as a scientist seemed to blur the line between ecology and the environmental movement. Evolutionary biologists, for their part, dismissed Gaia, arguing that the biosphere cannot be considered an organism because it neither reproduces nor evolves.

Lovelock now denies that he intended Gaia to be understood as an actual organism. He says Gaia is more like a **superorganism,** a group of individuals that cooperate almost like the cells of the body (Figure 25-1). Yet at times Lovelock has described Gaia in almost human terms, as though

Figure 25-1 Do superorganisms really exist? This Portuguese man-of-war consists of a tightly integrated colony of individuals. Each individual in the colony is a kind of polyp (related to the jellyfish) that cooperates with its comrades to form a complex superorganism. A bright blue, gas-filled float doubles as a sail. Wind drives the colony along the surface of the ocean. Beneath hang tangles of polyps bearing stinging tentacles that can reach 20 meters in length. The polyps quickly immobilize and digest fish or other creatures that come in contact with the deadly tentacles. (*Kelvin Aitken/Peter Arnold, Inc.*)

"she" had intentions and goals. He has referred, for example, to "Gaia's intervention" and "Gaian impatience."

The line between the Gaia metaphor and formal science blurred further when a biologist of national standing stood up for Lovelock. In 1974, Lynn Margulis, a distinguished professor of biology at the Amherst campus of the University of Massachusetts, began enthusiastically promoting a version of the Gaia concept—to the horror of her scientific colleagues.

In Margulis's view, Gaia is not a superorganism but a giant ecosystem, the highest level of organization of life. Margulis, now a member of the prestigious National Academy of Sciences, also proposed a related idea on the cellular level that is now widely accepted (Figure 25-2A). She persuaded her fellow biologists that the mitochondria and chloroplasts of eukaryotic cells were once free-living bacteria that came to live permanently inside other prokaryotic bacteria. In other words, we eukaryotes are a combination of prokaryotes that began cooperating millions of years ago. Such a cooperative arrangement is an example of **symbiosis** [Greek, *syn* = together + *bios* = life], any association between two organisms. Margulis asserts that the different organisms of our biosphere cooperate in similar ways to create a functioning ecosystem. The view that organisms cooperate as well as compete is fundamental to the view that our biosphere is a tightly integrated whole (Figure 25-2B).

Indeed, although most biologists still reject Lovelock's idea that our biosphere is a superorganism, nearly all scientists now accept Margulis's view that the biosphere is a giant, tightly integrated ecosystem. Just don't call it Gaia.

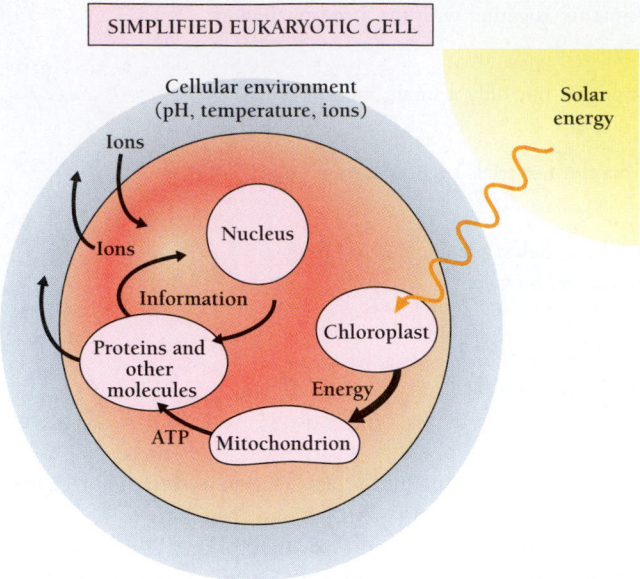

A.

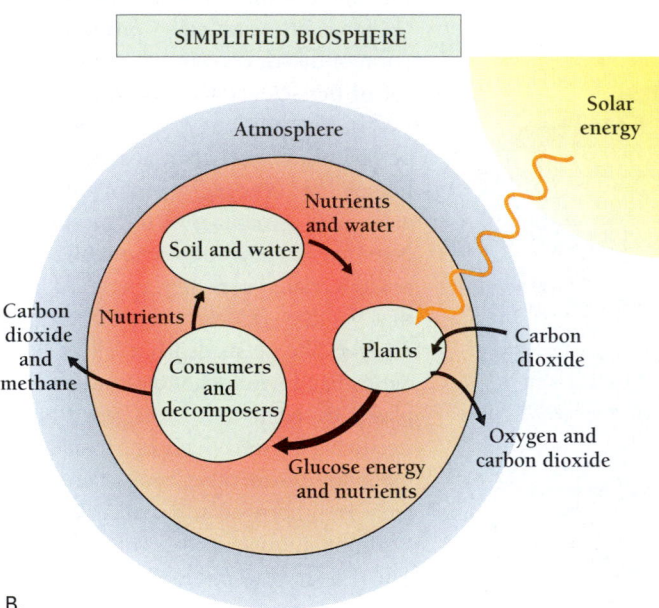

B.

Figure 25-2 Living organisms cooperate. A. Most biologists now agree that the chloroplasts and mitochondria that generate ATP and other energy-rich molecules were once prokaryotic cells that came to live inside our own cells. The idea that eukaryotic cells are symbiotic associations of prokaryotic cells is the brain-child of biologist Lynn Margulis. B. Margulis also asserts that the different organisms of our biosphere cooperate in similar ways to create a functioning ecosystem. Shown here is a simplified biosphere.

WHAT IS ECOLOGY?

The word "ecology" is popularly associated with recycling soft-drink cans, planting saplings in city parks, and campaigning for stricter air and water pollution standards. Yet the science of ecology includes none of these activities. **Ecology** [Greek,

oikos = home] is the study of the interactions of organisms with one another and with their physical environment, or "home." It is also the study of how those interactions determine the abundance and distribution of organisms.

An individual organism is a member of a number of systems. For example, an individual is a member of a **population,** an interbreeding group of individuals; a **community,** an interacting group of many species that inhabit the same area; and an **ecosystem,** a community of organisms together with the nonliving parts of their environment (Figure 25-3).

In addition, ecologists categorize terrestrial communities into various types of *biomes,* such as desert or tropical rain forest, based on the main kinds of plants that grow there. In Chapter 26 we discuss some of the main aquatic communities and the main biomes (types of terrestrial communities).

What Is an Ecosystem?

An ecosystem consists of parts that are living, or **biotic,** and those that are nonliving, or **abiotic.** The abiotic parts of an ecosystem include forces such as wind and gravity; conditions such as light intensity, temperature, salinity, or humidity; and physical substances such as soil, air, and water. These substances can include both inorganic compounds and organic compounds from living things. Soil, for example, usually contains inorganic nitrogen, minerals, and water. But it also contains organic material such as wood particles from trees and shrubs, chitin from the exoskeletons of dead insects, and disintegrating proteins from all kinds of organisms.

An ecosystem has many of the properties of living organisms that we listed in Chapter 1. An ecosystem obtains energy from its environment, transforms chemicals, changes with time, and responds to environmental changes. And, like individual organisms, ecosystems use energy to maintain a stable state. However, ecosystems differ from organisms in important ways. For example, unlike organisms, ecosystems do not reproduce themselves. Further, an ecosystem does something that no organism does. It recycles. No organism supports its own growth by eating its own waste.

The living and nonliving components of an ecosystem interact with one another as a unit. The atmosphere, the water, and the soil both permit and limit life within an ecosystem. A freshwater lake, for example, provides all of the conditions necessary for certain fish and aquatic plants to flourish. Yet, the same lake would be inhospitable to plants and animals adapted to an estuary or a lake with different physical attributes, such as temperature, pH, or nutrient levels.

Just as the physical environment affects organisms, organisms affect their physical environment. Trees, for example, block sunlight, change the characteristics of soil, and release oxygen into the atmosphere. Animals consume plants, altering the physical structure of the environment. Elephants may uproot whole stands of trees in order to eat their leaves, and rabbits can nibble grasses right down to the ground.

An ecosystem's boundaries may seem well defined: a pond is enclosed by its shores, a grassy meadow by the surrounding

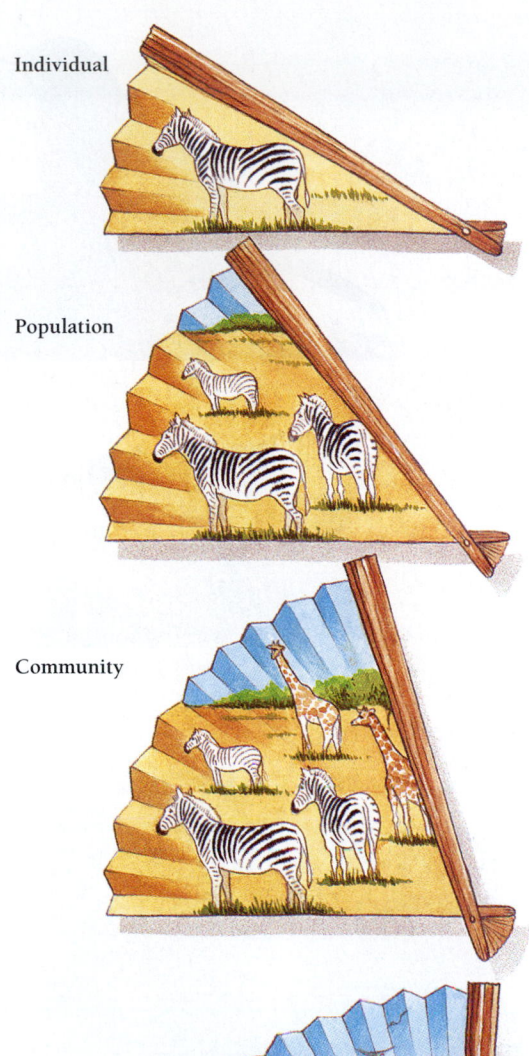

Individual

Population

Community

Ecosystem

forest. But an ecosystem's boundaries are fuzzy: the pond, for example, blends little by little into marsh, and then into a mixture of brush and open meadow. A stream brings nutrients and organisms from other areas into the pond and carries materials to other ecosystems.

Each ecosystem, community, or population is part of a larger one. A pond ecosystem may be part of a meadow ecosystem, the meadow part of a forest, and the forest part of a two-thousand-mile drainage system such as the Mississippi River Valley. We may therefore speak of the frog population of the pond, of the meadow, of the forest, or of the local drainage of streams. How a population is defined depends on the mobility of the organism. Populations of birds may cover vast areas, while populations of frogs may be limited to a few square miles. To some degree, then, biologists choose the boundaries of the populations and ecosystems that they study, always keeping in mind the larger picture.

Most ecosystems interact with other ecosystems. Seeds and spores disperse, animals migrate, and flowing water and air carry organisms—and their products and remains—from one place to another.

All ecosystems taken together make up a much larger ecosystem, the biosphere. The biosphere differs from other ecosystems in having fixed boundaries. Extending over the whole of the Earth's surface, the biosphere begins two miles (3 km) underground and extends into the highest reaches of the atmosphere. It ends where the atmosphere ends and space begins.

In this chapter, we discuss ecosystems as entities, whose parts interact in complex ways. In succeeding chapters of this section on ecology, we will see how ecosystems and communities change over time, how human intervention affects ecosystems, how individual populations behave, and, finally, how individual organisms behave.

Ecology is the study of the interactions of organisms with each other and with their physical environment. An ecosystem is a functional unit consisting of both biotic parts and abiotic parts.

Biosphere

Figure 25-3 Levels of organization. An individual is a member of a population, a community, an ecosystem, and the biosphere.

HOW DOES ENERGY FLOW THROUGH ECOSYSTEMS?

The biotic part of an ecosystem consists of two great groups of organisms, the autotrophs and the heterotrophs. **Autotrophs** [Greek, *auto* = self + *trophe* = to nourish], or **producers,** are organisms that produce their own food. The vast majority of autotrophs are either green plants or photosynthesizing bacteria, both of which obtain energy from the sun.

Heterotrophs [Greek, *hetero* = other + *trophe* = to nourish] are organisms that consume other organisms. Most are animals, fungi, or protists. Biologists divide the heterotrophs into two groups: consumers, such as animals; and decomposers, such as bacteria and fungi. **Consumers** (mostly animals) obtain energy by eating other organisms. Biologists divide consumers according to what they eat. Deer, quail, and caterpillars, for example, eat plants (producers). Such plant eaters are called **herbivores.** Mountain lions, hawks, wasps, and other **carnivores** eat other animals.

Some animals feed from the whole carcasses of dead animals or from large pieces of dead plants. Such **scavengers** include millipedes, sow bugs, earthworms, and termites, which eat dead plant material; as well as flies, lobsters, starfish, clams, and catfish, which often eat animal remains. Many animals, including starfish, lions, and humans, may change from carnivore to scavenger and back, depending on what is available.

But scavengers do not consume all dead material. Most dead organisms are decomposed by **saprophytes** [Greek, *sapro* = putrid + *phyte* = plant]—fungi, bacteria, and even some plants—that live off the energy stored in particles of dead organisms.

Many levels of carnivorous consumers can exist: hawks eat snakes, which eat ground squirrels, which eat lizards, which eat predaceous insects, which eat caterpillars, which eat plants. Such a linear sequence is called a **food chain.** Most food chains have only four or five links (Figure 25-4).

Every ecosystem has numerous food chains, with different numbers of trophic levels. The collection of all the interconnected food chains of an ecosystem is called a **food web** (Figure 25-5). In reality, no food web is static. Instead they are dynamic entities that change with circumstance. In one year, a population explosion of oak moths means that insect predators focus on oak moth caterpillars. In another year, oak moths are rare, and predators eat something else, perhaps mostly one thing, perhaps a diversity of insect prey. Most predators are "generalists," animals that will eat a variety of different foods.

Ecologists assign the organisms in a food web to different "trophic levels," depending on where they get their energy. For example, plants, which get their energy directly from the sun, are in the first trophic level; caterpillars, which get their energy from plants, are in the second; birds that eat caterpillars are in the third. Another way of designating trophic level is to call herbivores "primary consumers," to call carnivores that eat herbivores "secondary consumers," to call animals that eat secondary consumers "tertiary consumers," and so on. **Trophic**

Figure 25-4 A food chain. In a food chain, the productivity of plants and other primary consumers goes to herbivores such as caterpillars, which are eaten in turn by predators such as robins. *(American robin, Gary Meszaros/Dembinsky Photo Associates; orange-tipped oakworm, PSU Entomology/Photo Researchers, Inc.; 500-year-old white oak tree, Jack Rosen/Photo Researchers, Inc.)*

level [Greek, *trophe* = to nourish], then, is an organism's position in the food chain.

The different levels of a food chain are often approximations only. For example, the giant panda normally eats almost nothing but bamboo shoots, which makes it a primary consumer or herbivore. But now and then, this bear will kill and eat a rodent, which makes it a carnivore. Many animals, including humans, bears, pigs, and crows, consistently eat from several levels of a food chain at once. Such animals are called **omnivores** [Latin, *omnis* = all + *vorus* = devouring] because they eat plants, herbivores, and other carnivores. Many ani-

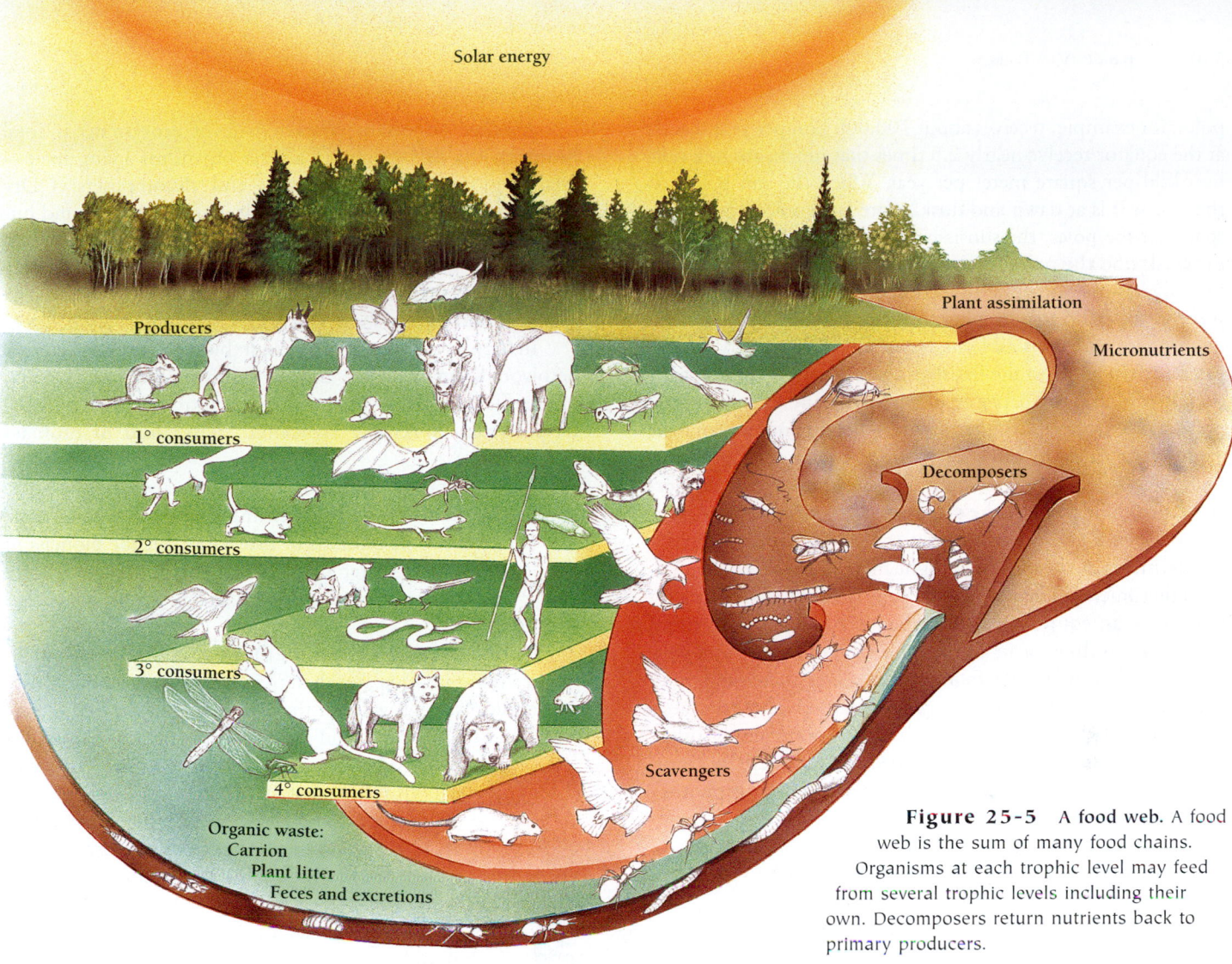

Solar energy

Producers

1° consumers

2° consumers

3° consumers

4° consumers

Organic waste:
Carrion
Plant litter
Feces and excretions

Plant assimilation

Micronutrients

Decomposers

Scavengers

Figure 25-5 A food web. A food web is the sum of many food chains. Organisms at each trophic level may feed from several trophic levels including their own. Decomposers return nutrients back to primary producers.

mals eat one thing as larvae and another as adults. Frogs, for example, are herbivores as tadpoles, but carnivores as adults. Omnivores do not easily fit into any particular trophic level. Unfortunately, for biologists trying to understand food webs, omnivores, far from being the exception, are quite common. One study of a desert food web found that nearly 80 percent of species were omnivorous.

Finally, many species eat each other. The adult fish of species A may eat the young of species B, while the adults of species B eat the young of species A. Which species is at a higher trophic level? Clearly neither. In many food webs, then, trophic level beyond the primary producers is almost meaningless. Still, the *idea* of trophic level is a useful way to think about how energy moves through ecosystems.

Every ecosystem includes one or more food chains that together form a food web. Autotrophs transform energy from the sun into energy-rich molecules. Heterotrophs obtain energy from autotrophs by eating organisms, whether alive or dead.

How Much Energy Flows Through an Ecosystem?

Every ecosystem is unique, whether it is a tiny pond or the vast Serengeti Plain of Africa. Yet, similar ecosystems share fundamental characteristics, including climate, productivity, total mass of living organisms, and numbers of species. We'll discuss many of these characteristics in coming chapters. Here we can look at two important measures of how much energy flows through an ecosystem, *biomass* (the total mass of all living organisms) and *productivity*.

Productivity is a measure of the energy captured in the chemical bonds of new molecules on a square meter of land per unit time, usually per year. Almost all energy for ecosystems comes from sunlight. (The rare exceptions are deep ocean ecosystems fueled by inorganic molecules.) Yet, primary consumers can use only one percent of the energy that comes from the sun. The atmosphere reflects 30 percent back into space and absorbs 20 percent. Most of the rest reaches Earth but is absorbed by rocks, soil, and water.

Different parts of the Earth receive different amounts of sunlight. Each square meter of land near the North and South poles, for example, receives about 700,000 kcal per year. Lands at the equator receive nearly 2.5 times that much, about 2 million kcal per square meter per year. When the sun is low in the sky, as it is at dawn and dusk, more light reflects back into space. At the poles, the sun is never overhead, but always low in the sky. So the poles receive far less sunlight than the equator. You can demonstrate this effect for yourself by holding a piece of paper up to a flashlight. Imagine that the paper is the atmosphere and the flashlight is the sun. If you hold the paper at a right angle to the flashlight, you see more intense light than if you hold the paper at an angle. Equally important, the poles and temperate latitudes receive different amounts of light at different times of year. In contrast, equatorial regions receive the same amount of light all year around. Such differences profoundly affect the ecologies of these regions.

Biologists can easily measure the sunlight that reaches a square meter of forest using a light meter like the ones in automatic cameras. The amount of light is about 1 to 2 million kcal per year, equivalent to the chemical energy in 30 to 60 gallons of gasoline, or in all the food each of us eats in a year.

We can estimate how much of this energy plants and other producers capture by measuring their rate of photosynthesis and their rate of respiration and subtracting one from the other. The result is the "net" energy stored, or productivity. In a pond, for example, we can estimate the total rate of photosynthesis and respiration for all the microorganisms, using the techniques described in Web Bit 25-1: Measuring Photosynthesis in a Pond. The productivity of the producers is called the **primary productivity.**

Primary productivity varies enormously from ecosystem to ecosystem. The most productive ecosystems are marshes. A marsh is twice as productive as a temperate forest, 4 times as productive as a wheat field, and 35 times as productive as a desert. In Chapter 26 we discuss some of the reasons that ecosystems differ in their productivity.

Another important characteristic of ecosystems is total **biomass,** the dry weight of all the organisms living in it—including producers, consumers, and decomposers. ("Dry weight" excludes water, so the dry weight of a 120-pound human being, for example, would be about 35 pounds.) Rain forests have more organisms per square meter and therefore more total biomass than other ecosystems, more even than the more-productive marshes.

In terrestrial ecosystems, plant biomass usually exceeds herbivore biomass, which exceeds carnivore biomass. A graph of the total biomass of organisms at each trophic level of a forest ecosystem looks like a pyramid, a **pyramid of biomass** (Figure 25-6).

But a "pyramid" of biomass does not always have a pyramidal shape. In a pond, for example, the biomass of algae may be less than the biomass of fish. How can the biomass of the consumers be greater than that of the producers (Figure 25-7)? The answer lies in the flow of energy from one trophic level to the next.

During the course of a year, the increase in biomass of algae in a pond is far greater than the increase in biomass of fish. The algae always grow faster than the fish. But the fish eat the algae as fast as it grows, and they convert only a fraction of it into fish. The rest is used to power movement and metabolism or is lost as heat.

Once we compare the energy flowing through the first trophic level with the energy flowing through the succeeding levels, it becomes clear that the producers always process more energy than the consumers. In fact, this relationship holds in all ecosystems.

Plants and other producers use the energy of sunlight to build their bodies and to drive life processes. However, every chemical and mechanical process releases energy in the form of heat. So producers can convert only part of the energy from the sun into chemical en-

Figure 25-6   A pyramid of biomass. In a grassland such as this National Bison Range, in Montana, the biomass of the primary producers is greater than that of herbivores and carnivores—eastern meadow voles (*Microtus pennsylvanicus*) and bobcat (*Lynx rufus*). (*Grassland, Richard J. Green/Photo Researchers, Inc.; voles, Gary Meszaros/Visuals Unlimited; bobcat, R. Lindholm/Visuals Unlimited*)

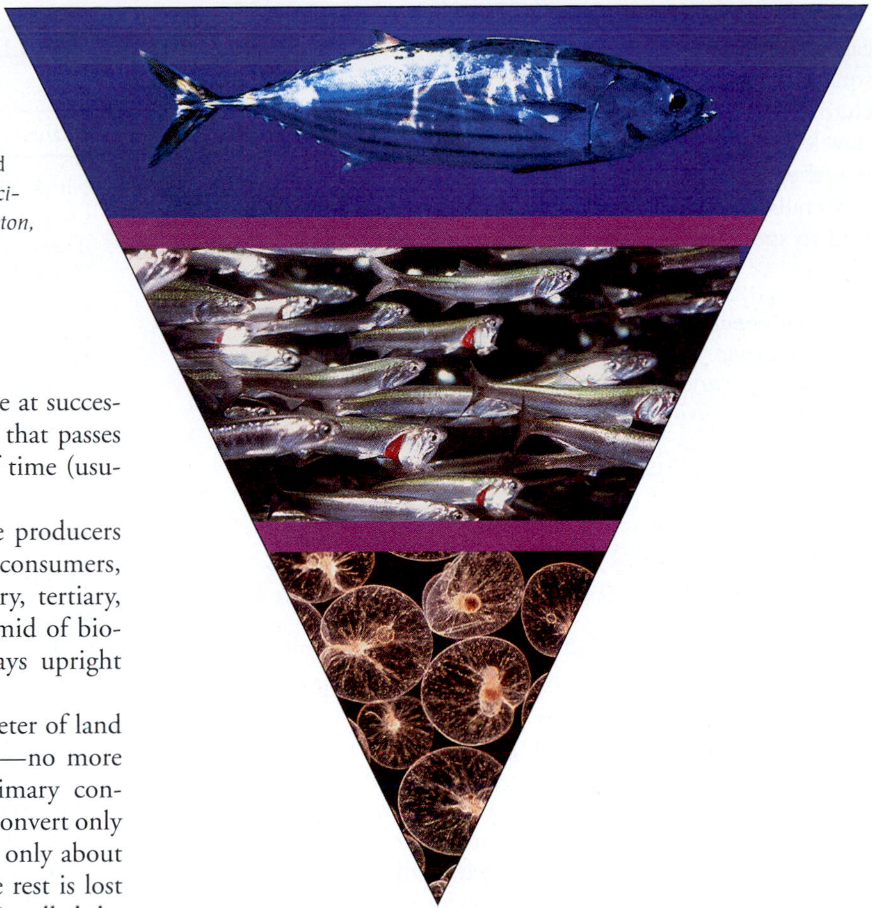

▶ **Figure 25-7** Inverted pyramid of biomass in an aquatic community. In aquatic communities, the primary producers may be eaten as fast as they multiply. So, while the accumulated biomass of the consumers is large, that of the producers may be small. Here schools of tuna feed on northern anchovies, which feed on plankton. (*Tuna, David B. Fleetham/Tom Stack & Associates; anchovies, Norbert Wu/Peter Arnold, Inc.; plankton, D.P. Wilson/Science Source/Photo Researchers, Inc.*)

ergy. Because energy becomes less and less available at successively higher trophic levels, the amount of energy that passes through each trophic level in a certain amount of time (usually a year) looks like a pyramid.

A **pyramid of energy** is always upright. The producers always process more energy than the primary consumers, which always process more energy than secondary, tertiary, and higher-level consumers. Even when the pyramid of biomass is inverted, the pyramid of energy is always upright (Figure 25-8).

Of the 1.5 million kcal that strike a square meter of land in the United States each year, only one percent—no more than 15,000 kcal/yr—is actually available to primary consumers such as deer and cattle. Of this, herbivores convert only 10 percent to tissues. Each trophic level passes on only about 10 percent of the energy of the one below it. The rest is lost as heat in metabolic processes. This generalization is called the **ten percent law.** The ten percent law explains why a pyramid of energy is always upright.

The ten percent law also explains why ecosystems have so few trophic levels and so few individuals at the highest trophic level. If on a square meter of land, primary consumers store 15,000 kcal/yr, secondary consumers (herbivores) will have only about 1500 kcal/yr to live on, and tertiary consumers (herbivore-eating carnivores) will have only 150 kcal/yr to live on, about as many calories as are in a cup of spaghetti.

Worldwide, humans take a huge amount of the Earth's primary productivity. We use 40 percent of terrestrial primary productivity. This includes not only the food we eat and the wood we use to build our houses, but also the areas we use to grow lawns and golf courses and the areas formerly covered by forests and marshes and now covered by asphalt or shopping malls and factories.

▶ **Figure 25-8** Pyramid of energy. The pyramid of energy is always upright, no matter whether for a forest or a pond. Here the grassland of the Masai Mara National Reserve, in Kenya, fixes carbon and energy on which zebras feed. These spotted hyenas, from Kruger National Park, South Africa, are the top predators in a similar ecosystem. (*Masai Mara National Reserve, Renee Lynn/PRI; zebras, Jeremy Woodhouse/DRK Photo; hyena, Gil Lopez-Espina/Visuals Unlimited*)

This is why top carnivores must find food over wide areas. Mountain lions, for example, roam over hundreds of square miles. All top carnivores are highly mobile animals, including, for example, sharks, killer whales, and eagles. Their travels force them to expend even more energy, which means that they need even more food. Top carnivores also tend to be "generalists," animals that eat a variety of foods. They can't afford to specialize too much.

Two important measures of energy flow in ecosystems are primary productivity and biomass. A pyramid of biomass may not always be upright. But a pyramid of energy is always upright, because energy dissipates as it moves from one trophic level to the next.

TABLE 25-1   MAJOR COMPONENTS OF EARTH'S ATMOSPHERE

Molecule	Percent in atmosphere
Nitrogen (N_2)	78.0
Oxygen (O_2)	21.0
Argon (Ar)	0.934
Water (H_2O)	0.10–1.0
Carbon dioxide (CO_2)	0.035
Neon (Ne)	Trace
Helium (He)	Trace
Methane (CH_4)	Trace

HOW DO ECOSYSTEMS RECYCLE MATERIALS?

We can see that energy constantly escapes from the biosphere. In contrast, the biosphere recycles water, carbon, and other materials (Figure 25-9). As nutrients move from one trophic level to another, they may change form. But, unlike energy, nutrients rarely escape from the biosphere entirely. Sometimes these elements are part of organisms; sometimes they are part of the atmosphere (Table 25-1). A single carbon atom in your fingernail, for example, may have been, at different times, part of an apple, part of a bicarbonate ion in the ocean, and part of a lump of coal. Carbon and other materials pass through many forms—both biotic and abiotic—in a system called a **biogeochemical cycle.**

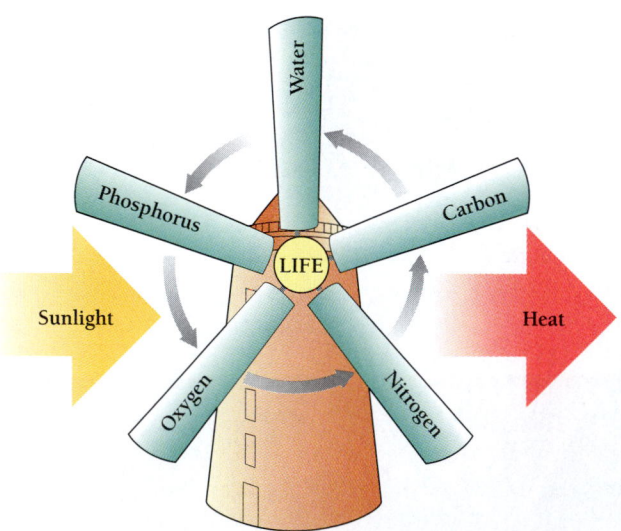

Figure 25-9   Ecosystems recycle material but lose energy forever. Sunlight drives ecosystems the way wind drives windmills. What's left is heat that passes into the atmosphere and, ultimately, out into space. In contrast, carbon, nitrogen, water, phosphorus, and oxygen cycle endlessly through the biosphere, like the vane of a windmill.

Oxygen, nitrogen, phosphorus, and other materials pass through biogeochemical cycles as well. Carbon dioxide (a compound of carbon and oxygen) moves from plants to consumers to water and air and back to plants and other producers. Plants take up carbon dioxide, remove the carbon, and release oxygen. Plants, animals and other organisms then take the same oxygen from the atmosphere, use it to burn carbon compounds, then release carbon dioxide.

The biogeochemical cycles of materials such as carbon and oxygen involve the whole biosphere. A carbon atom may travel all over the world on currents of wind or ocean. In fact, atoms of oxygen and nitrogen become so widely dispersed that each time one of us takes a breath, we breathe in a few atoms from Julius Caesar's dying breath, in 44 B.C. Other materials, such as calcium and phosphorus, tend to stay put and cycle within smaller, more local ecosystems. We will discuss three biogeochemical cycles: those of water, carbon, and nitrogen.

Organisms reuse materials such as water, carbon, oxygen, and nitrogen.

What Drives the Water Cycle?

All life is intimately connected to water. Organisms need a moist environment for the cells of their bodies and plenty of water for the insides of their cells. Vertebrates are about 70 percent water. And plants contain even more water. Plants use some 500 grams of water for every gram of biomass they produce. Some of this water is broken down to help form sugars during photosynthesis. But most of the water passes right through a plant unchanged. Plants pull water from the soil and release it as vapor through tiny pores (stomata) in their leaves, a process called **transpiration** (Chapter 31). On a hot summer day, an average maple tree transpires more than 50 gallons of water—enough to fill a large bathtub—every hour.

Flowing water also refreshes and flushes ecosystems. Running water brings in minerals, organic substances, and even

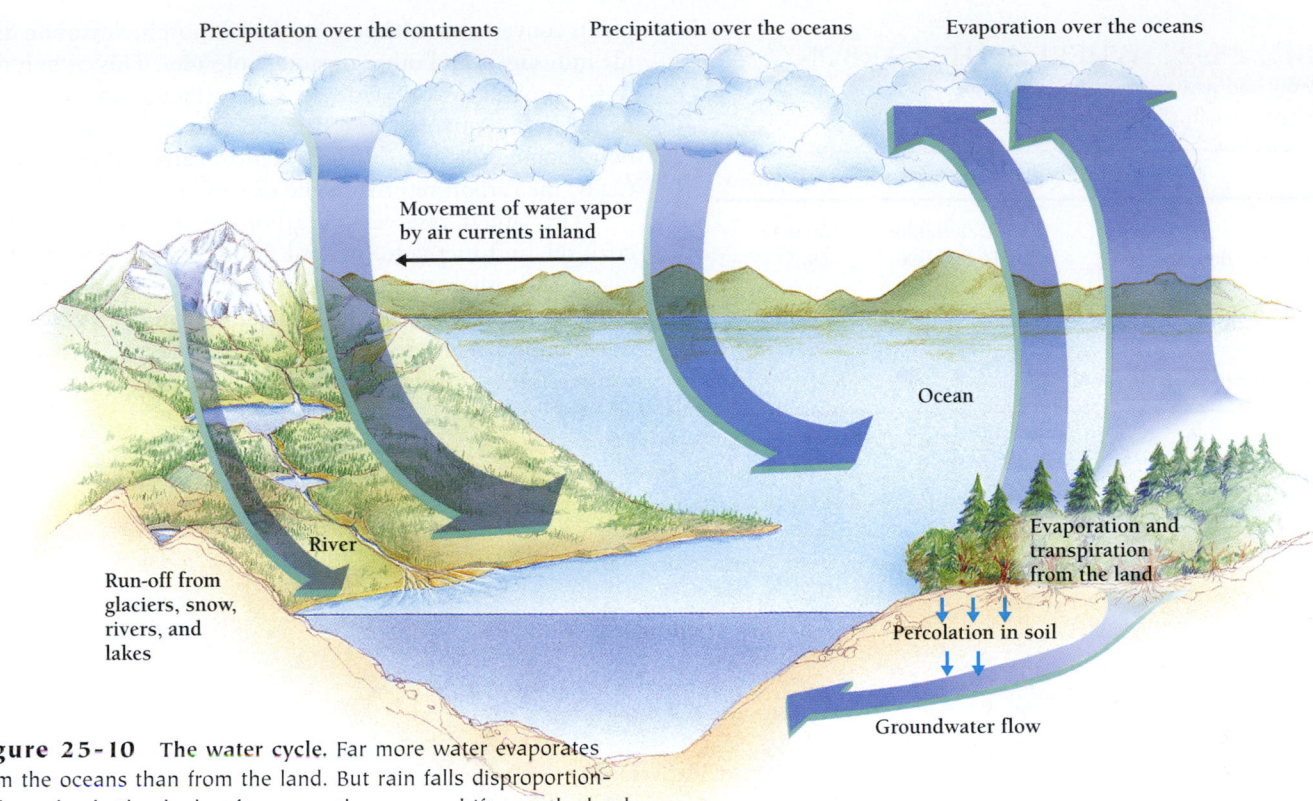

Precipitation over the continents   Precipitation over the oceans   Evaporation over the oceans

Movement of water vapor by air currents inland

Ocean

Evaporation and transpiration from the land

River

Run-off from glaciers, snow, rivers, and lakes

Percolation in soil

Groundwater flow

Figure 25-10 The water cycle. Far more water evaporates from the oceans than from the land. But rain falls disproportionately on land. Clouds that form over the oceans drift over the land, rise up into the mountain, and drop rain and snow that forms streams, rivers, lakes, and underground water flows.

organisms and carries away wastes. Water cycles through both biotic and abiotic parts of the biosphere (Figure 25-10). More than 97 percent of the water on our planet is in the oceans, which cover 70 percent of the Earth's surface.

Sunlight evaporates large quantities of water from the oceans and the land. Of the water that evaporates from the land, plants transpire nearly 90 percent (Chapter 31). Despite all this "evapotranspiration" (evaporation plus transpiration), our huge atmosphere is only a tenth to one percent water vapor. What water vapor there is, however, easily condenses into clouds that drop rain or snow on to the Earth's surface. Day after day, water evaporates, forms clouds, and rains in a rapid cycle that drops nearly a meter of rain over the entire surface of the Earth each year.

Overall, water moves from the oceans to the continents. Of the water that vaporizes into the atmosphere, just 17 percent comes from the land. Yet 25 percent of all rain falls on the land, so more rain falls on the land than comes from the land.

Why does water move from the oceans to the land? Clouds blowing in off the ocean often encounter tall mountains, which push the clouds up higher into the atmosphere. Up high, temperatures tend to be colder, so the water condenses into rain or snow, which drops onto the mountains, filling lakes, rivers, and underground aquifers. The water then returns to the oceans as runoff in streams and rivers.

Sunlight drives the evaporation of water from the oceans, the largest reservoir within the water cycle.

Where Does Carbon Go?

Plants remove carbon dioxide from the atmosphere and transform it into sugars and other hydrocarbons. These forms of carbon eventually pass to herbivores, carnivores, scavengers, and decomposers, all of which release carbon dioxide back into the air.

Our atmosphere contains only trace amounts of carbon. Still, this adds up to about 635 billion tons, most of it in the form of carbon dioxide. But most carbon is not in the atmosphere or even in living organisms. The oceans contain nearly 500 times as much carbon as the atmosphere, mostly in the form of bicarbonate ions (HCO_3^-). Bicarbonate ions tend to settle to the bottoms of oceans, where they form thick sediments. Over millions of years, these immense layers have turned into sedimentary rocks, which today hold 16 million billion tons of carbon (Table 25-2).

By comparison, we organisms are a mere smear of carbon on the surface of the Earth. Living organisms contain only about 2700 billion tons of carbon, one-tenth of what is dissolved in all of the oceans of the world. Each year, terrestrial

TABLE 25-2   WHERE'S THE CARBON?

Coal, oil, and natural gas that are still in the ground contain more carbon than all other sources combined.

Where	Billions of tons	Percent
Fossil fuels	25,000,000	60.0
Sedimentary deposits	16,000,000	39.0
Oceans	30,000	0.07
Living organisms	3,000	0.007
Atmosphere	635	0.0015
Human carbon released/decade	80	

plants convert about 12 percent of all atmospheric carbon dioxide into sugars and other organic molecules. This carbon then flows through a worldwide food web (Figure 25-11). Plants, herbivores, carnivores, and decomposers break up these carbon compounds by oxidizing them to carbon dioxide. Eventually, the carbon returns to the oceans and atmosphere.

The largest reserves of carbon are in the "fossil fuels" coal, oil, and methane (natural gas). Fossil fuels come from plants and other organisms that lived millions of years ago. Humans have discovered that burning these materials releases huge amounts of energy that can be used to run factories, warm houses, and power vehicles of every description (Figure 25-12).

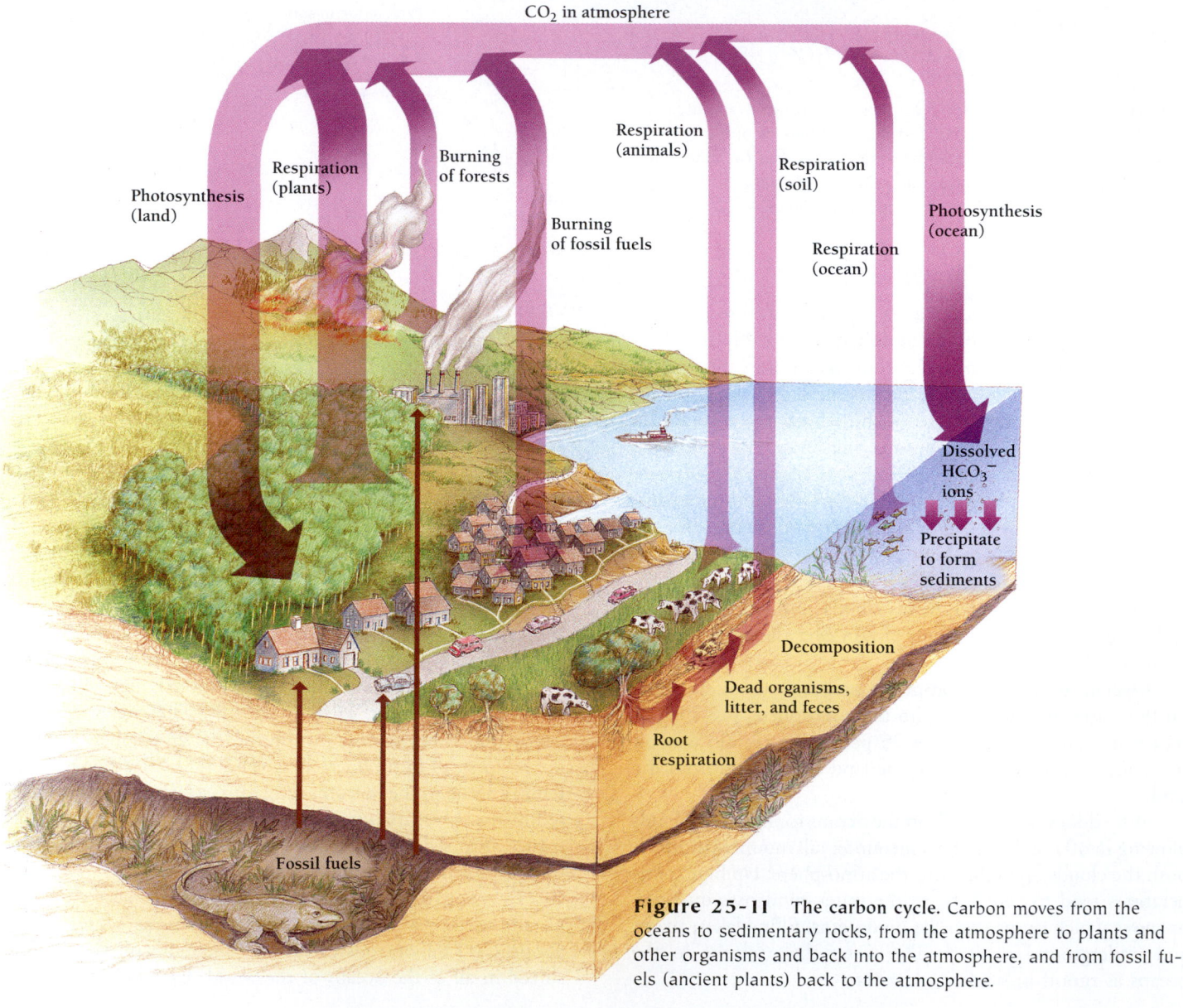

Figure 25-11   The carbon cycle. Carbon moves from the oceans to sedimentary rocks, from the atmosphere to plants and other organisms and back into the atmosphere, and from fossil fuels (ancient plants) back to the atmosphere.

Figure 25-12 The 19-foot-long, 7-foot-tall Ford Excursion weighs more than 7000 pounds. Sport utility vehicles (SUVs) now account for 40 percent of new-vehicle sales. Because these gigantic vehicles are technically trucks, they are subject to looser rules than passenger cars. Brakes and other safety features are less strictly regulated than in cars, and SUVs can legally burn more fuel and emit more pollutants per mile than cars can. The Ford Excursion itself, with the same length and heft as a killer whale, is so big that the law requires no mileage tests. (*Ron Kimball Photography, Inc.*)

Carbon atoms cycle through oceans and atmosphere, with large reservoirs in living organisms and in fossil fuels.

How Have Humans Altered the Carbon Cycle?

Less than 200 years ago, humans began burning large amounts of coal, oil, and gas. In the process, we began returning fossil carbon to the carbon cycle. By also burning vast stretches of the world's forests, human societies have increased the Earth's annual carbon dioxide production by 7.8 billion tons per year. As a result, total carbon dioxide in our atmosphere has increased by 30 percent since 1750.

To measure this change, scientists take ice samples from deep in the polar ice caps. The air trapped inside these frozen samples dates back hundreds of years. And the amount of carbon dioxide in this ancient air can be measured. In about 1750, before the Industrial Revolution, the concentration of atmospheric carbon dioxide was about 280 parts per million (ppm). Now, 250 years later, the average concentration of carbon dioxide in the atmosphere is about 365 ppm (Box 25-1). Most of the increase, however, has occurred in the last 50 years.

Because carbon dioxide traps solar energy, increasing concentrations of carbon dioxide can increase the temperature of the atmosphere. Indeed, global temperatures have increased by about 1°C in the last 100 years (Figure 25-13). Is it really warmer than it was in our grandparents' day? Yes. Of the ten hottest years on record, nine were between 1987 and 1997. A

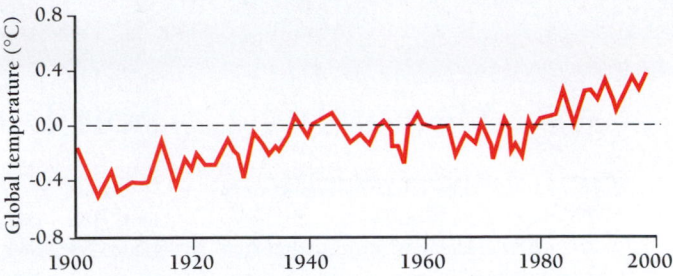

Figure 25-13 **Is the temperature rising?** The average of global temperatures rose about 1 degree C between 1961 and 1990. (*National Oceanic and Atmospheric Administration*)

few scientists still argue that we can't be sure that the increase in carbon dioxide has caused the increase in global temperatures, but most scientists no longer doubt it.

The tendency of carbon dioxide, methane, and other gases to trap solar radiation is called the "greenhouse effect." The glass walls of a gardener's greenhouse allow light in. Once the light is inside, however, it degrades to heat. The heat does not pass easily back out through the glass. Some of the heat is trapped inside the glass, where it warms the air and the plants inside the greenhouse.

For a gardener this is wonderful, but warming the whole Earth is causing major changes in our biosphere. Scientists have predicted that unless we reduce carbon dioxide emissions, the average temperature of the Earth's surface may increase by 3° or 4°C over the next 75 years (Figure 25-14). Such an increase would melt the polar ice caps and raise sea levels worldwide, inundating coastal cities. Glaciers all over the world are already melting, and the polar ice caps have begun breaking up at the edges. Global sea levels have risen about 2 mm per year for the last 50 years.

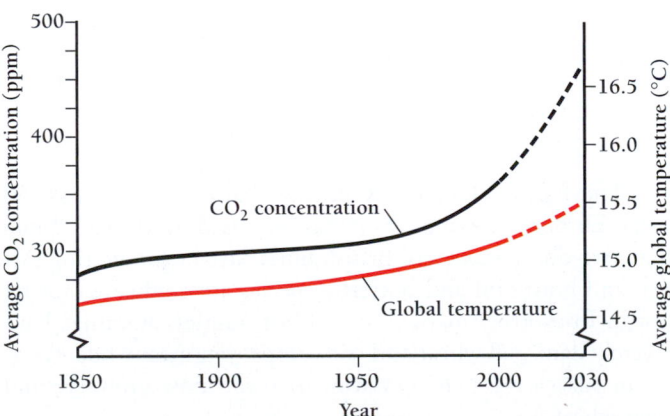

Figure 25-14 **Atmospheric carbon dioxide and temperature.** Scientists cannot prove that excess carbon dioxide in the atmosphere is causing global warming. Historic increases and long-range forecasts like the ones shown here, however, give most scientists little reason for doubt.

BOX 25.1

Measuring Carbon Dioxide Levels for 40 Years

The first hint that carbon dioxide might be a problem came in 1896, when a scientist noticed that high concentrations of "carbonic acid" in the air increased the temperature of the ground. By the 1930s, scientists had begun to believe that atmospheric carbon dioxide concentrations were increasing. Accurate measuring techniques designed in the 1950s removed all doubt.

In 1958, C.D. Keeling set up a system for measuring carbon dioxide concentrations in the air at Mauna Loa, a 4169-meter active volcano on the island of Hawaii. The measurements were taken from towers erected on a barren lava field 3400 meters above sea level. On the rocky slope, the local influences of vegetation and people were small, and Mauna Loa was and is one of the best places in the world for measuring undisturbed air. (When the volcano itself emits carbon dioxide, the researchers leave out those measurements.)

Keeling was not content to make his measurements for just a few years. He and his colleagues have made continuous measurements for more than 40 years, compiling the longest continuous record of atmospheric CO_2 concentrations anywhere (Figure A). Their precise records at Mauna Loa show a 15 percent increase in the annual concentration of CO_2, from about 315 parts per million in 1960 to about 365 ppm in 2000.

Amazingly, their methods and equipment have remained almost unchanged since 1959. The researchers collect air samples from air intakes at the tops of four 7-meter towers and one 27-meter tower. Four air samples are collected every hour and an infrared gas analyzer registers the concentration of CO_2 in the stream of air.

To ensure that the measurements are accurate, the flow is replaced by a stream of gas of known CO_2 concentration every 20 minutes. The gases used to calibrate the instruments are themselves calibrated against standardized gases whose CO_2 concentrations are measured even more accurately.

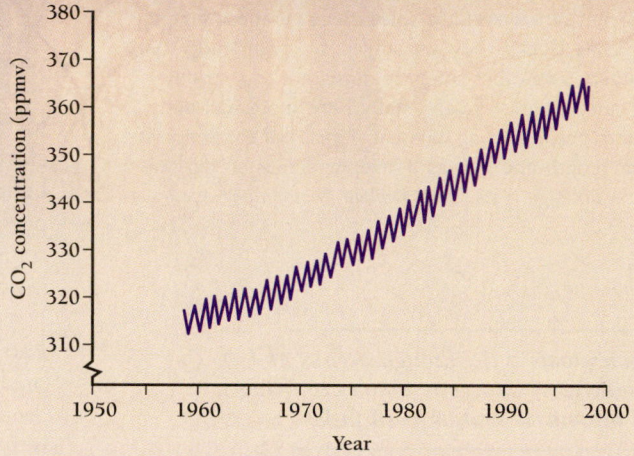

Figure A Increasing carbon dioxide concentrations on Mauna Loa, Hawaii. Carbon dioxide levels rise smoothly from 1960 to 1999. The up-and-down squiggles represent seasonal variation in photosynthesis. In summer, plants photosynthesize more than they respire, taking up carbon dioxide and releasing oxygen. In winter, plants respire more, breaking down carbon-rich molecules and releasing carbon dioxide. Overall, carbon dioxide levels decrease during the growing season and increase during the rest of the year. (Source: Dave Keeling and Tim Whorf, Scripps Institute of Oceanography)

The Earth's climate has already changed. In the past 30 years, Europe's growing season has increased by 10 days. More than two dozen species of British birds have begun nesting earlier, and both bird and butterfly species are gradually moving north. In North America, several hot, rainless summers have severely damaged wheat and corn crops, yet, meanwhile, farmers in Alaska say it is so warm, they can now grow rye and even wheat.

Nations around the world have leaped to halt the trend. In December 1997, 174 nations signed a treaty in Kyoto, Japan, agreeing to reduce overall carbon dioxide output. The "Kyoto Protocol" requires industrialized nations to cut their greenhouse gas emissions by 5 percent by the year 2010. Nearly all industrialized nations have promised to do so. Only Canada, Australia, and the United States, the largest emitter of these gases, have so far refused.

The simplest way to reduce the production of greenhouse gases is to reduce the burning of fossil fuels and forests. This could be accomplished through a combination of energy conservation and the use of alternative sources of energy such as solar power, wind power, or nuclear power. But for complex economic and political reasons, these sources of power remain underused.

How can we get rid of carbon dioxide?

Most solutions to the problem of global warming do not include the obvious solution, which is to burn less fuel. Because most nations want to maintain or even increase fuel consumption, researchers are looking for ways to remove carbon dioxide from the atmosphere.

Hide it in the trees

One way is to plant trees that can mop up excess carbon dioxide through photosynthesis. In Canada, for example, plans are underway to plant billions of poplar trees over millions of acres of land. As these trees grow, they will take up carbon from the atmosphere and store it as wood. In Bolivia, a coalition of three American power companies, a U.S. environmental organization, and the Bolivian government are spending $10 million to protect tropical forests there.

Bury it

Meanwhile, a Norwegian oil company has begun pumping carbon dioxide into the ground under the North Sea at the rate of a million tons a year. And on Hawaii's Kona Coast, the Japanese, Norwegian, and U.S. governments together have built a 1.5-mile pipe that can pump liquid carbon dioxide 1000 m below the surface of the ocean.

Fertilize the ocean

Perhaps the most dramatic plan is to stimulate the growth of ocean plankton by fertilizing the ocean with iron. To test this idea, oceanographer John H. Martin experimented with fertilizing the ocean with iron in the early 1990s. The result was dramatic blooms of photosynthetic plankton.

In one study, iron fertilizer caused a 60 percent drop in carbon dioxide released into the atmosphere. Researchers hypothesize that the extra carbon goes into plankton. These either become part of the ocean food chain or die and sink to the bottom of the ocean, carrying their carbon with them.

Expanding on this idea, environmental engineer Michael Markel founded a private company in Virginia that has obtained the rights to 800,000 square miles of ocean around the Marshall Islands, in the eastern Pacific Ocean. Markel's company, Ocean Farming, Inc., plans to seed this area with iron, stimulating the growth of plankton, and drawing carbon dioxide from the atmosphere. Markel estimates that a 100,000-square-mile area could soak up at least one-third of the carbon dioxide released by the United States. The result, he says, would be to stimulate a huge new fishery and cool our fevered planet. As researcher John Martin once wrote in reference to global warming, "Give me a half-tanker of iron, and I'll give you an ice-age."

Will fertilizing the ocean really cool the planet? Markel says yes. He claims that his patented technique can send all excess carbon dioxide generated by burning fossil fuels to the bottom of the ocean for 1000 to 10,000 years. To do this, he wants to continuously fertilize 6 million square miles of ocean with iron. He says his technique is cheap, harmless, and could be implemented immediately.

But many academic scientists worry that this venture is dangerous and pointless. To make a dent in global levels of CO_2, huge amounts of iron would have to be seeded, and no one knows what kind of plankton will grow. Some plankton are extremely toxic, and a 100,000-square-mile bloom of toxic plankton would be an ecological disaster, they say. Oceanographers worry that the blooms may irreversibly damage Marshall Islands coral reefs, which are the foundation for these ocean ecosystems. Once dead, the reefs might not come back.

For now, biologists can only cross their fingers. Markel's plans to seed the oceans near the Marshall Islands are on hold because of political upheavals there. But he hopes to prove his idea by seeding 5000 square miles of ocean somewhere else.

Humans have increased carbon dioxide concentrations in Earth's atmosphere by about 30 percent by burning fossil fuels and cutting down forests. The apparent result is an increase in global temperatures that may dangerously alter the global climate. The Kyoto Accord is an agreement by 174 nations to reduce carbon dioxide in the atmosphere.

Why Is Nitrogen Both Common and in Short Supply?

Molecular nitrogen (N_2) is the most abundant element in the atmosphere, comprising 78 percent of the air we breathe (Figure 25-15). Nitrogen is also an indispensable part of the proteins and nucleic acids that all organisms need. Yet, ironically, usable nitrogen has been in short supply for millions of years. The huge quantities of atmospheric nitrogen are not directly available to animals, plants, and most other organisms.

Atmospheric nitrogen is useless to most organisms because the two atoms of a nitrogen molecule are held together by a powerful, almost unbreakable triple bond. Breaking this bond requires special enzymes that only a few bacteria possess. It should not surprise us, then, that most organisms cannot perform this reaction. Plants, animals, and most other organisms can absorb only **fixed nitrogen** such as **ammonium ions** (NH_4^+) or nitrate ions (NO_3^-). Both plants and the animals that eat them depend on a steady supply of ammonia and nitrates.

Where does fixed nitrogen come from? One important source is lightning. During storms, the high temperatures generated by lightning—greater than the surface temperature of the sun—convert atmospheric nitrogen to NH_4^+ and NO_3^-. Rain washes the fixed nitrogen to Earth, supplying 5 to 10 percent of all usable nitrogen.

Amazingly, however, most nitrogen is fixed by bacteria. These tiny organisms have evolved powerful enzymes that can

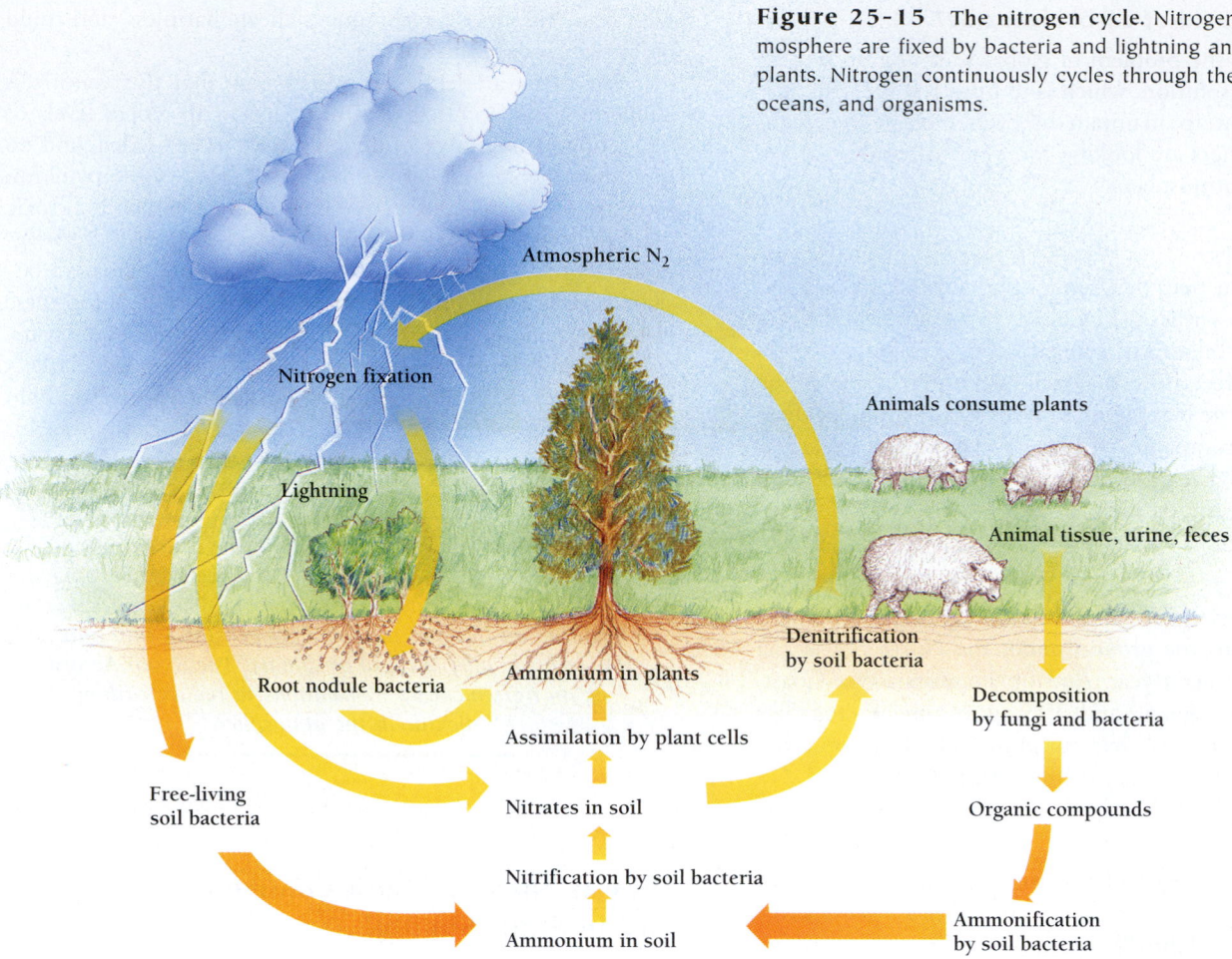

Figure 25-15 **The nitrogen cycle.** Nitrogen atoms in the atmosphere are fixed by bacteria and lightning and assimilated by plants. Nitrogen continuously cycles through the atmosphere, soils, oceans, and organisms.

break the triple bond in N_2 and convert it to ammonia. Some of the **nitrifying bacteria** are free-living bacteria (such as cyanobacteria) that live in soil or water and convert nitrogen to ammonia. Others live in mutualistic relationships with certain plants. For example, plants in the pea family (peas and beans) play host to *Rhizobium* bacteria, which live inside tiny "root nodules" growing on the plants' roots. The *Rhizobium* bacteria convert atmospheric nitrogen into ammonia that the plants use. The plants, in turn, provide the bacteria with energy-rich sugars, an example of symbiosis.

Plants can use ammonium ion directly in the synthesis of amino acids and proteins. Plants that do not harbor nitrifying bacteria, however, must make do with what is in the soil. Unfortunately, bacteria in the soil break down most of the free ammonium to nitrite (NO_2) and then nitrate (NO_3) in a process called **nitrification.** Other bacteria convert nitrate back to N_2, a process called **denitrification**, and the nitrogen returns to the atmosphere. Plants get whatever nitrate is left.

Nitrate is highly soluble, and heavy rains can wash it out of soils and into lakes and streams. Eventually, it ends up in the oceans. Because nitrogen is often in short supply, plants soak up nitrate very easily and conserve it by removing it from dying leaves, stems, and roots. However, such conservation measures are often thwarted. For example, when squirrels and caterpillars cut living leaves, the nitrogen-filled leaves fall to the forest floor, where their nitrates may wash from the ecosystem.

Animals and other consumers also lose nitrogen. When animals break down proteins during respiration, they convert them to ammonia. Because ammonia is toxic when concentrated, aquatic organisms such as fish dilute the ammonia with large quantities of water and flush it from the body. Terrestrial organisms, with less water available to them, first convert ammonia into something less harmful: **urea** in mammals such as ourselves, **uric acid** in birds, reptiles, and insects. Decomposers obtain energy from urea and uric acid by converting them back into ammonia, a process called **ammonification.** The ammonia (or nitrite or nitrate) can then be used once again by plants, completing the cycle of fixed nitrogen among plants, animals, and decomposers.

In natural ecosystems, most fixed nitrogen moves locally from producers to consumers and decomposers and back to producers. Only small amounts of nitrogen are added to the system through nitrification or lost through denitrification.

Fixed nitrogen cycles locally among all the organisms of an ecosystem.

How Have Humans Altered the Nitrogen Cycle?

The world's organisms have evolved over millions of years in an environment with limited amounts of available nitrogen. Yet, in the last few decades, human activity has doubled the amount of fixed nitrogen entering the biosphere and accelerated the movement of fixed nitrogen through ecosystems. In some areas, such as Antarctica, virtually no human-generated nitrogen enters ecosystems. But in highly developed areas such as northern Europe, the extra nitrogen is ruining ecosystems.

The extra fixed nitrogen comes from several sources. The largest single source is industrially manufactured fertilizers. The United States alone manufactures about 40 billion pounds a year for use on crops. But virtually every country in the world manufactures this cheap fertilizer in order to increase the growth of crops (Figure 25-16).

Another source of extra nitrogen is the crops themselves. On nearly one-third of the Earth's surface, crops have displaced natural vegetation. Many of these crops are soybeans, peas, alfalfa, and other members of the pea family that harbor nitrifying bacteria, which fix atmospheric nitrogen.

Burning fossil fuels in automobiles, factories, and power plants also releases nitrogen in the form of nitrous oxide and other gases. Humans also release large amounts of nitrogen

from trees and grasses when we burn forests and grasslands or drain marshes.

The extra fixed nitrogen affects the atmosphere, the carbon cycle, and the functioning of ecosystems. All of the various forms of fixed nitrogen—including nitrous oxide, nitric oxide, and ammonia—are greenhouse gases that are helping to raise global temperatures. In addition, when nitric oxide dissolves in water, it becomes nitric acid, the major component of acid rain. Acid rain kills ponds, forests, and other habitats.

The most profound changes to ecosystems result from excess nitrogen fertilization. In terrestrial ecosystems, nitrogen saturation can leach nutrients from the soil. As ammonium accumulates in the soil, fungal decomposers lose out in competition with bacterial decomposers. The bacteria convert it to nitrate and release hydrogen ions, which acidify the soil. Because the nitrates are water soluble, they leach away into streams and lakes. But the nitrates have negative charges, so when the nitrates leach away, they carry with them a host of positively charged minerals, including calcium, magnesium, and potassium. Thus extra nitrogen can decrease soil fertility.

As calcium leaches away and soil acidity rises, increasing amounts of aluminum dissolve in soil water. The aluminum poisons the roots of trees, and, washed into nearby streams, it also kills fish.

In the long run, soils become less and less fertile, and trees growing in such soils grow very slowly or not at all. In parts of northern Europe, the Front Range in Colorado, and the forests surrounding the Los Angeles Basin, excess nitrogen is so serious that forests are slowly dying off. In many ecosystems, the amount of nitrogen available controls the nature and diversity of plant and animal life and the cycling of carbon and minerals. Excess nitrogen decreases the diversity of species in an ecosystem, as a few species grow wildly out of control and others are crowded out.

Worldwide, 20 percent of all forms of human waste, including sewage and fertilizer, ends up in rivers. In aquatic ecosystems, excess nitrogen, two-thirds of which comes from farms, washes down rivers and accumulates in estuaries, creating "dead zones" in coastal areas such as the Gulf of Mexico and the Adriatic Sea. In a dead zone, massive blooms of algae draw millions of tiny grazing plankton. As the algae die and drop to the bottom, along with the grazers' fecal pellets, staggering numbers of bacteria bloom at the bottom and quickly consume all the oxygen in the water. Fish, shrimp, and crabs swim away from the deoxygenated water, while bottom creatures such as clams, snails, and worms slowly suffocate. In extreme cases, every living organism either swims away or dies.

Dead zones can be enormous. In the summer of 1999, researchers at Louisiana State University mapped a dead zone at the mouth of the Mississippi River that covered 20,000 square kilometers, an area the size of Massachusetts (Figure 25-17).

In this chapter, we have discussed the flow of energy and the cycle of materials within ecosystems. In the next chapter, we will survey a selection of ecosystems. Along the way, we will see how humans are damaging some of these ecosystems.

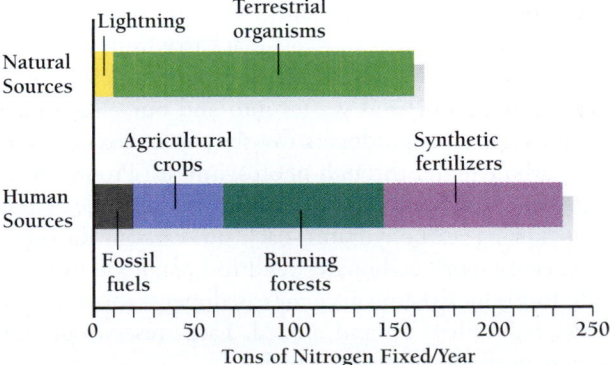

Figure 25-16 **Sources of fixed nitrogen.** Humans produce half again as much fixed nitrogen as all natural sources combined. As a result, available nitrogen in the biosphere has more than doubled. Excess nitrogen damages soils and kills forests and fish.

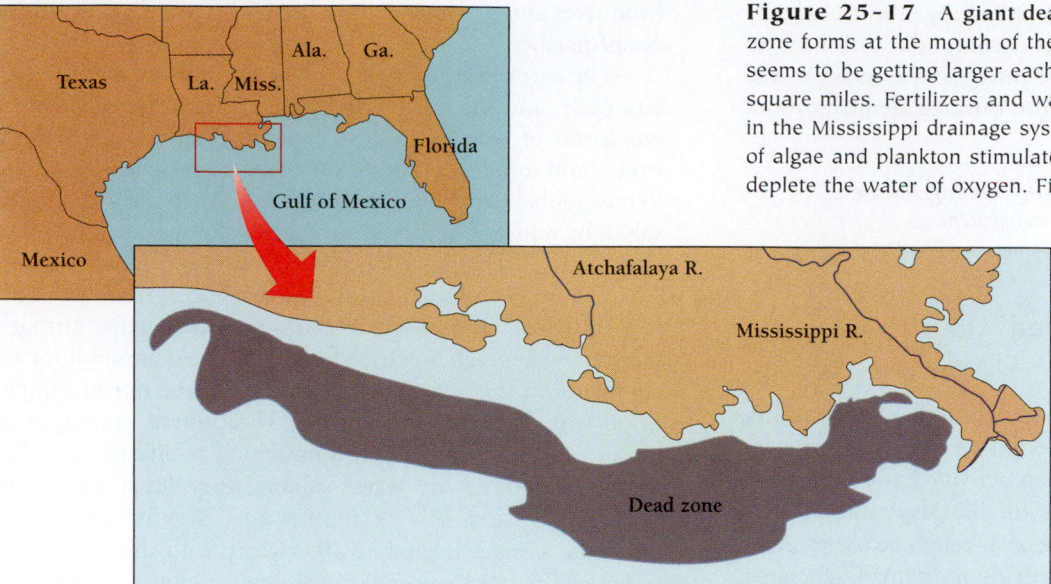

Figure 25-17 A giant dead zone. Each summer, a huge dead zone forms at the mouth of the Mississippi River. This dead zone seems to be getting larger each year. In 1999, it covered 20,000 square miles. Fertilizers and wastes flow from all the major rivers in the Mississippi drainage system into the Gulf of Mexico. Blooms of algae and plankton stimulate blooms of bottom bacteria, which deplete the water of oxygen. Fish leave and other organisms die.

Study Outline with Key Terms

In the 1960s James E. Lovelock proposed that the Earth's atmosphere, oceans, and soils be considered integral with all living organisms, in a single, immense, self-regulating homeostatic superorganism called Gaia. The idea has not been accepted by most scientists, but remnants of his idea persist in the current understanding of what the biosphere is and how it works.

Ecology is the science concerned with understanding the relations among organisms and their environment. Ecologists study **populations, communities,** and **ecosystems. A system** is an assemblage of interacting parts or objects. An ecosystem is the living **(biotic)** community together with the nonliving **(abiotic)** components of the community's environment, including soil, water, air, and weather. The **biosphere** is the sum of all ecosystems. An ecosystem shares some properties of an organism, but not others. Like a **superorganism,** an ecosystem is a functional unit with interacting, or **symbiotic,** parts.

The biotic part of an ecosystem consists of **autotrophs,** or **producers,** and **heterotrophs.** Heterotrophs include **consumers** such as **herbivores, carnivores, scavengers,** and **saprophytes.** Autotrophs belong to the first **trophic level** in every **food chain.** Heterotrophs occupy different trophic levels, from the second level up, depending on what they eat. Animals that eat from several levels of a food chain, including humans, bears, pigs, and crows, are called **omnivores.** Decomposers such as saprophytes and scavengers break down whole organisms and detritus.

Food chains are usually interwoven into a complex **food web.** One way of thinking about communities is in terms of

a **pyramid of biomass** (dry weight) or a **pyramid of energy.** Only the pyramid of energy is always an upright pyramid.

Almost all the energy that drives ecosystems comes from the sun. Producers capture this sunlight energy, and organisms at higher trophic levels obtain what they can. Each trophic level has access to only about 10 percent of the **primary productivity** available to the one beneath it. This is called the **ten percent law** and is another way of describing the upright energy pyramid.

Ecosystems recycle water, carbon, nitrogen, phosphorus, and many other materials in **biogeochemical cycles.** Water enters the atmosphere from the oceans through evaporation and from the continents and islands through evaporation and **transpiration** from plants. Atmospheric water condenses into clouds, then precipitates as rain, sleet, or snow. Rain and snowmelt form streams and rivers that carry fresh water to the oceans. Along the way, fresh water carries minerals, organic materials, organisms, and wastes into and out of ecosystems.

Plants and other producers absorb carbon dioxide and turn it into carbohydrates through photosynthesis. Producers, consumers, and decomposers all produce carbon dioxide by respiring—oxidizing carbohydrates back to carbon dioxide. In aquatic ecosystems, carbonates tend to precipitate to the bottom to form thick sediments. These sediments form rocks that are eventually lifted up and eroded. Large reserves of carbon also exist in the form of fossil fuels.

The Earth's atmosphere is 78 percent nitrogen. But this nitrogen is not available to most organisms until lightning or **nitrifying bacteria** transform it into **ammonium ions** or nitrate, which plants can use to make proteins. Soil bacteria break down ammonium to nitrite (NO_2) and then nitrate (NO_3) in a process called **nitrification.** Other bacteria convert nitrate back to N_2, a process called **denitrification**, and the nitrogen returns to the atmosphere.

Consumers and decomposers get such **fixed nitrogen** by eating plant proteins, or each other. Most aquatic organisms eliminate waste nitrogen as ammonia. Terrestrial organisms eliminate it in the form of **urea** or **uric acid.** Decomposers obtain energy from urea and uric acid by converting them back into ammonia, a process called **ammonification.** Ecosystems conserve nitrogen, but some is lost when rain washes it from fallen organic matter and carries it to the sea.

Review and Thought Questions

Review Questions

1. What is ecology?
2. What distinguishes an ecosystem from a community?
3. What properties do ecosystems share with organisms? How are ecosystems different from organisms?
4. Distinguish between autotrophs and heterotrophs. Give an example of each.
5. Distinguish between a food chain and a food web.
6. Explain why the pyramid of energy is always narrowest at the top, but the pyramid of biomass is not necessarily narrowest at the top.
7. Explain how the biomass of aquatic producers can be smaller than the biomass of the consumers.
8. Where does the carbon dioxide in the atmosphere come from?
9. In what three forms do organisms eliminate broken-down proteins? What form do aquatic organisms use, and why don't terrestrial organisms use the same form?
10. Humans are omnivores and eat from many levels of the food chain. Make a list of primary producers, herbivores, scavengers, and decomposers that humans commonly eat. Include a plant, an animal, a fungus, a protist, and a bacterium.

Thought Questions

11. Describe the steps of the nitrogen cycle. How is nitrogen lost from ecosystems? How is it replaced? (Mention two ways.) Why does it need to be "fixed"?
12. Suppose a strawberry farmer treated his fields with a fumigant (poison gas) that killed all of the bacteria in the soil. Describe two different things the farmer could do to ensure that the strawberries received an adequate supply of nitrogen.
13. Suppose that all of the primary producers were removed cleanly and instantaneously from the biosphere. List ten immediate consequences (e.g., declines in oxygen levels) and ten indirect consequences.
14. A gallon of gasoline weighs only 6 pounds. Yet each gallon of gasoline burned releases almost 20 pounds of carbon dioxide into the atmosphere. Assuming gasoline is all octane, examine the reaction below and explain why the carbon dioxide released weighs more than the gasoline.

$$2\,C_8H_{18} + 25\,O_2 \longrightarrow 16\,CO_2 + 18\,H_2O$$

About the Chapter-Opening Image

NASA's first breathtaking photographs of Earth as a cloud-swathed blue ball floating in space reminded Earth's human population that Earth has definite boundaries and that there is only one Earth.

 On-line materials relating to this chapter are on the World Wide Web at **http://www.harcourtcollege.com/lifesci/aal2/**

26 Terrestrial and Aquatic Ecosystems

How Many Biomes Can You See in a Day?

One hundred fifty years ago, pioneers in covered wagons struggled to cross a flat, sunbaked desert, the last barrier before reaching fertile California. In the salt-encrusted flats of one 140-mile-long valley, not a single plant grew. Even on the slopes above the valley, plants grew only sparsely and water was rare. The oxen pulling the wagons died one by one, and the pioneers who made it to the other side called the place Death Valley.

In the summertime, temperatures in Death Valley, California, can reach an unbearable 134°F. Average yearly rainfall is a minuscule 1.66 inches. The long valley lies sunken along a geologic fault line, surrounded by rocky mountains nearly as dry as the desert itself. It is the lowest and hottest place in North or South America.

Lowest of all is Badwater, 282 feet below sea level. Badwater is the last place most people would want to go in the summer. Yet this salty pool is the starting line for a footrace that makes the Boston Marathon look like a jog around the block. Some 146 miles to the west and more than 15,000 feet higher lies the top of Mt. Whitney, the highest point in the contiguous United States, and the unofficial finish line for the grueling race. Each summer, a dozen runners attempt the 26-hour race from Badwater to Mt. Whitney (Figure 26-1).

Along the way, the runners pass through one of the most abruptly changing series of biological communities in the world, including most of the communities discussed in this chapter. The runners begin the race at 6 P.M. and run the first 50 miles at night. The flat floor of Death Valley itself supports little more vegetation than the moon. Daytime ground temperatures reach nearly 200°F. Nighttime air temperatures sometimes drop no lower than 95°F. Even in the dead of night the runners must stop for water every mile.

During the night, the runners climb out of Death Valley into the foothills and mountains of the Panamint Range, passing miles of widely dispersed shrubs. Desert salt bush has small, leathery leaves as salty as the saltiest cornchip. In the driest place, creosote bushes grow, each one 10 or 20 feet from the last. As in most **deserts,** only 10 percent of the ground has any vegetation whatever. The runners climb several thousand feet, and temperatures drop slightly.

- Temperature and rainfall largely determine the character of an ecosystem.

- Latitude and mountain ranges largely determine climate.

- Areas with heavier rainfall support forests. Areas with less rainfall support grasslands or deserts.

The creosote bush gives way to desert sage, a species that tolerates the higher altitude's cold winters. By the time the first runners reach the top of Townes Pass (5000 feet), it is daylight. A blistering sun rises at their backs. Gratefully, the runners descend into the cool but temporary shadow of the Panamints. As the morning wears on, they cross a small desert valley and another range of mountains.

At noon the lead runners enter the broad Owens Valley. A hundred years ago, rich grasslands and orchards surrounded the 15-mile-long Owens Lake. Towns dotted the edge of the lake and a steamboat ferried silver bullion to the south side of the lake. Today, the lake bottom is dust; all but a film of water at the center has been diverted to the millions of people living in the cities of southern California. Owens "Lake" is so

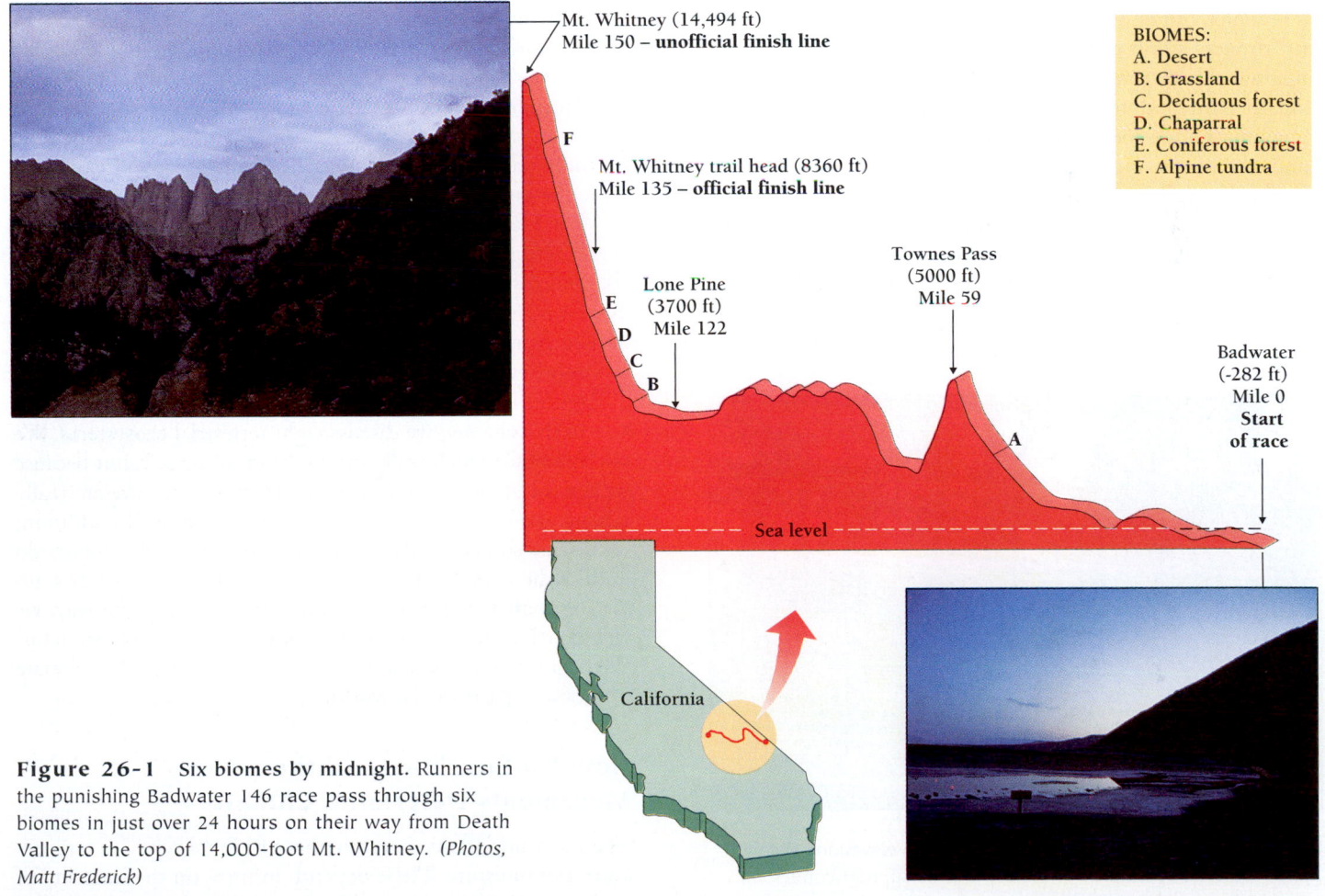

Mt. Whitney (14,494 ft)
Mile 150 – **unofficial finish line**

F

Mt. Whitney trail head (8360 ft)
Mile 135 – **official finish line**

Townes Pass
(5000 ft)
Mile 59

Lone Pine
(3700 ft)
Mile 122

E
D
C
B

Badwater
(-282 ft)
Mile 0
**Start
of race**

A

Sea level

California

BIOMES:
A. Desert
B. Grassland
C. Deciduous forest
D. Chaparral
E. Coniferous forest
F. Alpine tundra

Figure 26-1 **Six biomes by midnight.** Runners in the punishing Badwater 146 race pass through six biomes in just over 24 hours on their way from Death Valley to the top of 14,000-foot Mt. Whitney. *(Photos, Matt Frederick)*

dry that its dust storms are the single largest source of airborne particulate pollution in the United States.

The little band of runners push past the dead lake and north to the town of Lone Pine, where they turn west. After crossing a narrow strip of fertile **grassland,** they begin to climb into the deep, narrow canyons of the eastern slope of the towering Sierra Nevada. There, the runners catch their first whiff of fresh, tumbling mountain water. Now they climb up through the last of the desert and into the rich bottomland at the base of the Sierra Nevada. Here a mountain stream winds through small meadows and dense groves of deciduous cottonwoods and aspens, its banks overgrown with thickets of willow, its bottom lined with red metamorphic rocks and glistening fool's gold. The runners find relief in the odor of lush vegetation and cool water. But the little **deciduous forest** swarms with mosquitoes, which hum in the runners' ears and cluster on their arms and legs, driving them faster through the close, humid air of the canyon.

The runners have just 12 more miles to go. They have been running almost 24 hours, and it is evening again. Before them, rising above the lush valley, is a 6000-foot wall of red metamorphic rock, cut here and there by narrow, shrubby canyons. As the runners climb higher, they pass through dense thickets of prickly shrubs, or mountain **chaparral.**

Gradually, as they gain altitude, the chaparral gives way to open **coniferous forests** of pine and fir (Figure 26-2). The air is cooler, but also thinner. Breathing becomes difficult. The lead runners climb out of the canyon now and begin following a switchback trail up the side of a granite mountain. A few stop to pant and to look back. In the gathering dusk the Panamint Range looms darkly above Death Valley, and beyond, the purple Funeral Mountains.

As the runners continue to climb, the fir trees thin out, replaced by stunted white bark pines. Finally, above the timber line, the sparse, stunted trees give way to alpine **tundra,** composed of grasses, nonwoody plants (herbs), and knee-high

willows and huckleberries. Still higher, even these thin out, and the temperature drops rapidly. It is dark and cold now. An occasional low shrub or wildflower trembles in the wind on the bare slopes of the mountain. But in the last, cold, dark mile, the runners hear only the crunch and chink of running shoes on gravel and rocks.

But even the summit of Mt. Whitney, 14,494 feet high, is not quite bare rock. The summit is too high, too cold, and too exposed to wind and weather to support trees, shrubs, or herbs. But here and there, lichens—symbiotic associations of fungi and photosynthetic algae—grip the cold rocks.

The first runners to make it to the top collapse on the rocks, euphoric with accomplishment and oxygen deprivation. The rest of the runners are scattered across six distinct **biomes**—desert, temperate grassland, temperate deciduous forest, chaparral, coniferous forest, and tundra.

WHAT IS A BIOME?

In every region of the world, ecological succession leads to a characteristic community, or biome. A biome is a geographic region characterized by a distinctive landscape, climate, and community of plants and animals.

Regions with similar climates tend to have similar biomes. A desert biome is the same in the Sahara and in the Mojave. The weather is dry and the plants and animals are adapted to extreme "aridity," or dryness. The deserts of the southwestern United States and northwestern Mexico, central Asia, Africa, and Australia all have plants with spiny or succulent leaves and burrowing animals that emerge only at night. Likewise, the northern parts of Canada, Alaska, Scandinavia, and Russia all share the same kind of forest, called taiga (Figure 26-3).

But biomes differ in the particular species that they include. The grasslands of North America are dominated by deer, elk, and bison, while the grasslands of Africa are dominated by zebras, gazelles, elephants, and giraffes. Differences in the individual species in a biome are partly the result of the unique evolutionary histories of each region and partly the result of differences in geography.

In this chapter, we discuss eight terrestrial ecosystems. We say "biomes" (which really means "communities"), but because we discuss abiotic elements such as temperature, we are really talking about ecosystems (thus the chapter title). In addition, we discuss several kinds of aquatic ecosystems. (Ecologists do not use the word "biome" to refer to aquatic communities, reserving that term for land communities.) Along the way, we will see what effects humans have on each of these ecosystems. We begin with a discussion of what factors determine climate in different parts of the world.

How Do Sunshine and the Earth's Movements Determine Climate?

The most important determinants of biome type are temperature and moisture. These depend, in turn, on the intensity of sunlight and patterns of wind, rain, and ocean currents. All of

Figure 26-2   Coniferous forest. At higher elevations and latitudes, cooler temperatures and moister soils support coniferous forests. (*Superstock*)

A.

B.

Figure 26-3 **Taiga here, taiga there.** Whether in Canada, Scandinavia, or Russia, taiga is characterized by sparse, stunted coniferous trees. A. Siberian taiga. B. Spruce taiga in central Alaska. *(A, © Roland Seitre/Peter Arnold, Inc.; B, Tony Dawson/Words & Pictures/Tony Stone Images)*

these depend on the Earth's movements and the distribution of mountains and valleys.

The single most critical factor is sunlight. Of some 10^{21} kilocalories of solar energy that enters the upper atmosphere each year, about half reaches the ground. The rest is reflected back into space or absorbed by the atmosphere. Plants capture less than 1 percent of the light that strikes the Earth's surface. The energy absorbed by the ground and by the atmosphere raises the temperature of the environment and helps determine a region's climate.

The Earth's daily rotation constantly mixes the air, so that all the air along each latitude has a similar temperature. But different latitudes have different climates. This is because the latitudes closest to the poles receive less solar energy and experience more-extreme seasons than those near the equator

(Figure 26-4). As a result, the climate near the equator is both warmer and less variable than the temperate or polar climates.

Differences in air temperature cause air to move in currents that transfer heat away from the equator and toward the poles. Three rules determine the direction of these currents: (1) hot air rises and cold air falls, (2) hot air holds more moisture than cold air, and (3) the rotation of the Earth twists the moving air.

Figure 26-5 shows the pattern of air currents. Right at the equator, not much wind blows, and old-time sailors called this calm area the *equatorial doldrums* [Old English, *dol* = dull]. There, warm, moist air rises, which creates a region of low air pressure. The low air pressure pulls **tradewinds** toward the equator from about 30° latitude (north and south). Because of the Earth's spin, the tradewinds blow westward. The tradewinds were named when sailing ships used them to sail westward.

As the warm air from the equator rises, it cools. Because cool air can hold less moisture than warm air, the moisture condenses into clouds and falls as rain over the equatorial regions, creating the warm, rainy climate of the **tropics.** The equatorial air spreads north and south, raining and drying out as it goes. By the time this air reaches about 30° latitude, the air is quite dry and the lands beneath are nearly all deserts.

As the air loses water, it cools. The cool, dry air drops down toward the Earth's surface, and the Earth's spin gives the air a twist that makes it blow from west to east. These **westerlies** blow at latitudes between 30° and 60° in both the north and south. As this air moves over the Earth's surface, it warms and picks up moisture. At about 60° latitude, the air again rises, cools, and rains (or snows). At this cool but wet latitude are most of the great coniferous forests. Once the air is dry again, it drops down to create a dry, cold region of Arctic tundra. Here, the winds, called the **polar easterlies,** come from the east (Figure 26-5).

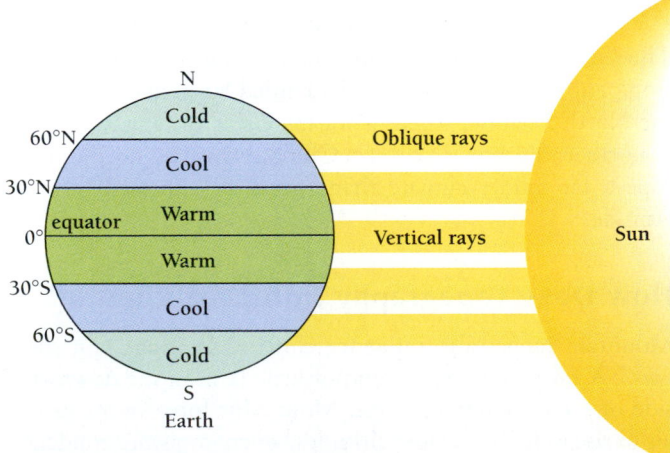

Angle of sun's rays at different latitudes

Figure 26-4 **Temperature depends on latitude.** Temperature and moisture are the two most important determinants of biome type. The latitudes closest to the equator receive the most sun. The latitudes closest to the poles receive less solar energy than those near the equator because the sun's rays arrive at an oblique angle.

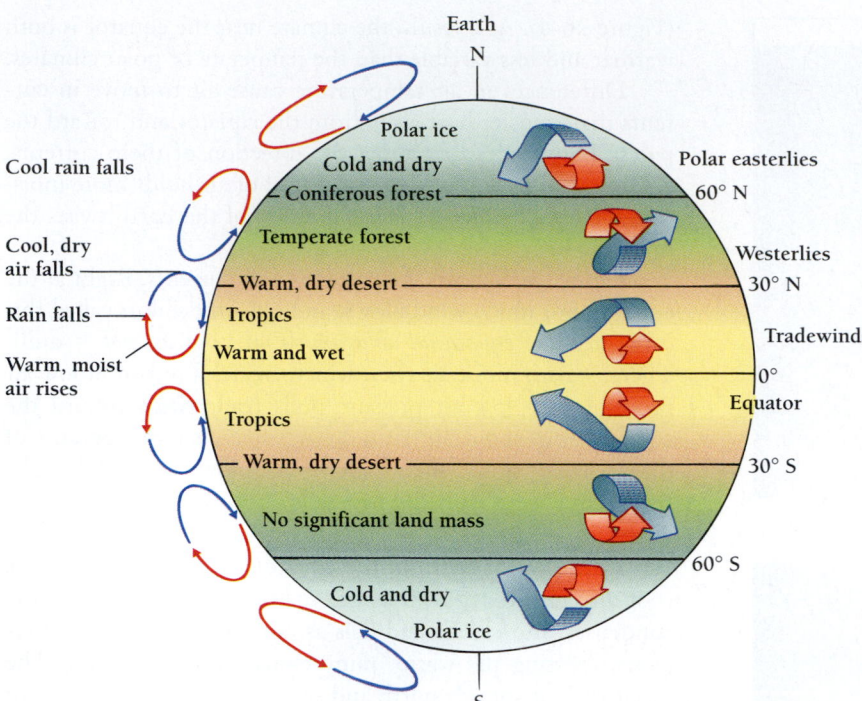

Earth
N

Cool rain falls

Cool, dry
air falls

Rain falls

Warm, moist
air rises

Polar ice
Cold and dry
Coniferous forest
Temperate forest
Warm, dry desert
Tropics
Warm and wet

Tropics
Warm, dry desert

No significant land mass

Cold and dry
Polar ice

S

Polar easterlies
60° N

Westerlies
30° N

Tradewinds
0°

Equator

30° S

60° S

Figure 26-5 **Air currents.** A pattern of rising and falling air, combined with the Earth's rotation, causes air currents to flow east near the north and south poles, to flow west in the middle latitudes, and to flow east along the equator.

The winds at each latitude depend on the overall movement of warm air from the equator toward the poles. The rotation of the Earth also changes the direction of the air movements, causing clockwise air currents in the Northern Hemisphere and counterclockwise currents in the Southern Hemisphere. The effect of rotation on movement, called the **Coriolis effect,** causes Northern Hemisphere winds (including hurricanes, typhoons, and tornadoes) to twist clockwise and Southern Hemisphere winds to twist counterclockwise. The Coriolis effect also induces ocean currents to turn in great circles—clockwise in the north and counterclockwise in the south.

Ocean currents often move along the edges of continents. The Gulf Stream, for example, carries warm water from the tropics up the eastern coast of the United States and then across the North Atlantic, where it warms Great Britain, and much of northern Europe. The warm Gulf Stream gives northern Europe a far milder climate than would be expected from its latitude.

How Does Topography Influence Climate?

Mountain ranges also influence climate, creating "rain shadows." A **rain shadow** is a land of little rain on the downwind side of a tall mountain range. Mountains force warm surface air to rise to higher, colder altitudes, where moisture condenses into clouds and falls as rain or snow. The prevailing westerlies along the California coast, for example, drive warm, moist ocean air up and over two mountain ranges. When warm, wet air coming in off the Pacific Ocean hits the low Coast Range, the clouds rise, cool, and then dump up to 100 inches of rain

a year on the coastal mountains (Figure 26-6A). As the air moves east, down the farther side of the Coast Range, it warms and absorbs moisture, drying the land. The same westerlies then hit the Sierra Nevada, a huge mountain range with some of the tallest peaks in the continental United States. At 3,000 to 14,000 feet, the ocean air cools and drops enough rain or snow to supply water for most of the state's agriculture and most of the state's 35 million residents. By the time the ocean air reaches the far side of the Sierra Nevada, however, it has lost nearly all its water and become very dry. East of the Sierra Nevada, all of the lands in the Owens Valley, Death Valley, most of Nevada, and parts of Oregon, Idaho, Utah, and New Mexico exist in a huge rain shadow, called the Great Basin desert (Figure 26-6B).

Tall mountains have other effects on climate as well. As we climb a mountain such as Mt. Whitney, the climate changes in the same way it does as when we travel from low latitudes to higher ones (Figure 26-7). The Appalachian Mountains, the Sierra Nevada, and the Rocky Mountains all contain extensive coniferous forests, which turn to alpine tundra above the "tree line," the upper limit at which subalpine trees can grow. Even in the tropics, snow covers the highest mountains.

All of these geographic patterns of temperature and rainfall create just a few major types of climate. Within each climate, similar adaptations are useful. For example, the species of a desert biome are recognizably similar, whether the desert is in Asia, North America, Australia, or Africa. A marsupial "wolf" from the forests of Australia has the same long, running legs and powerful jaws as the gray wolf of North America (Figure 17-15).

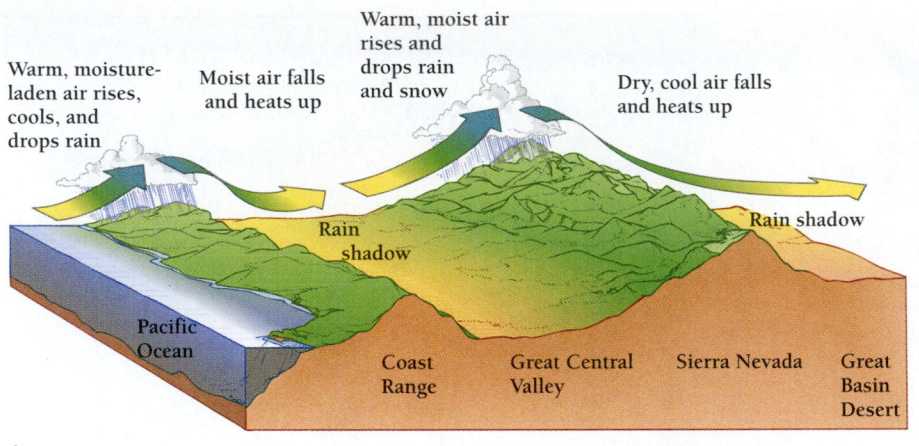

A.

B.

Figure 26-6 A rain shadow. Rain shadows occur in all parts of the world. One of the most dramatic examples is in the western United States. A. Rain clouds driven inland from the Pacific Ocean hit California's Coast Range and ride up the slope, like skateboarders on a ramp. As the clouds rise, they cool and rain precipitates out, soaking the mountains. The clouds drop down the other side, warm up, and hold their remaining moisture. Over the Sierra Nevada, the clouds drop quantities of rain and snow. B. By the time the clouds reach Nevada's Great Basin, the moving air has lost nearly all its moisture and little rain falls in the desert beneath. (*B, Charlie Ott/Photo Researchers, Inc.*)

Latitude and the Earth's spin influence intensity and duration of solar radiation, wind, rain, and ocean currents. These factors, as well as terrain, help create distinct climates that strongly determine what biome develops in an area.

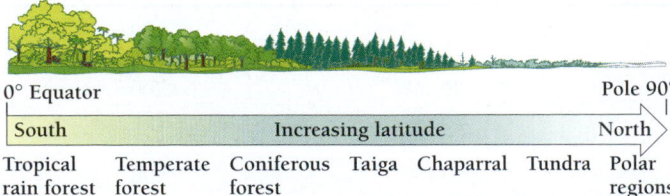

Figure 26-7 Latitude and altitude have similar effects on biomes.

HOW DO BIOMES DIFFER?

Biologists divide the world's communities into a great many sets of biomes. But in this book, we have chosen just eight major biomes that we think illustrate both life's diversity and some unifying themes (Figure 26-8). We will begin with the driest biomes, the warm, dry deserts and the cold, dry tundra, and then move through progressively wetter biomes, including chaparral, temperate grassland, and savanna; taiga, coniferous forest, and deciduous forest; and finally, wettest and warmest of all, the tropical rain forest (Table 26-1).

Desert Ecosystems Are Surprisingly Delicate

Deserts have the lowest net primary productivity of all the biomes. These barren regions produce only about 90 grams dry weight of biomass per square meter per year, about the weight of 12 quarters. Deserts may be hot or cold, but all are dry, receiving less than 25 centimeters of rain per year. In the southwestern United States and northeastern Mexico, desert extends east from the Sierra Nevada to the Rocky Mountains and south to the Sierra Madre. Africa, Arabia, central Asia, central Australia, and parts of South America all have deserts.

What plants dominate a desert depends on the temperature. Sagebrush dominates the cold deserts of Nevada and Utah, while cacti and oily creosote bushes dominate the lower, warmer deserts of Arizona and Mexico (Figure 26-9).

Desert plants have special adaptations to conserve water. Sagebrush, for example, has small leaves that minimize evaporation, while succulents (including the cacti) store water in fleshy tissue for future use. Cacti and other desert plants grow prickly spines that protect them from hungry or thirsty browsers.

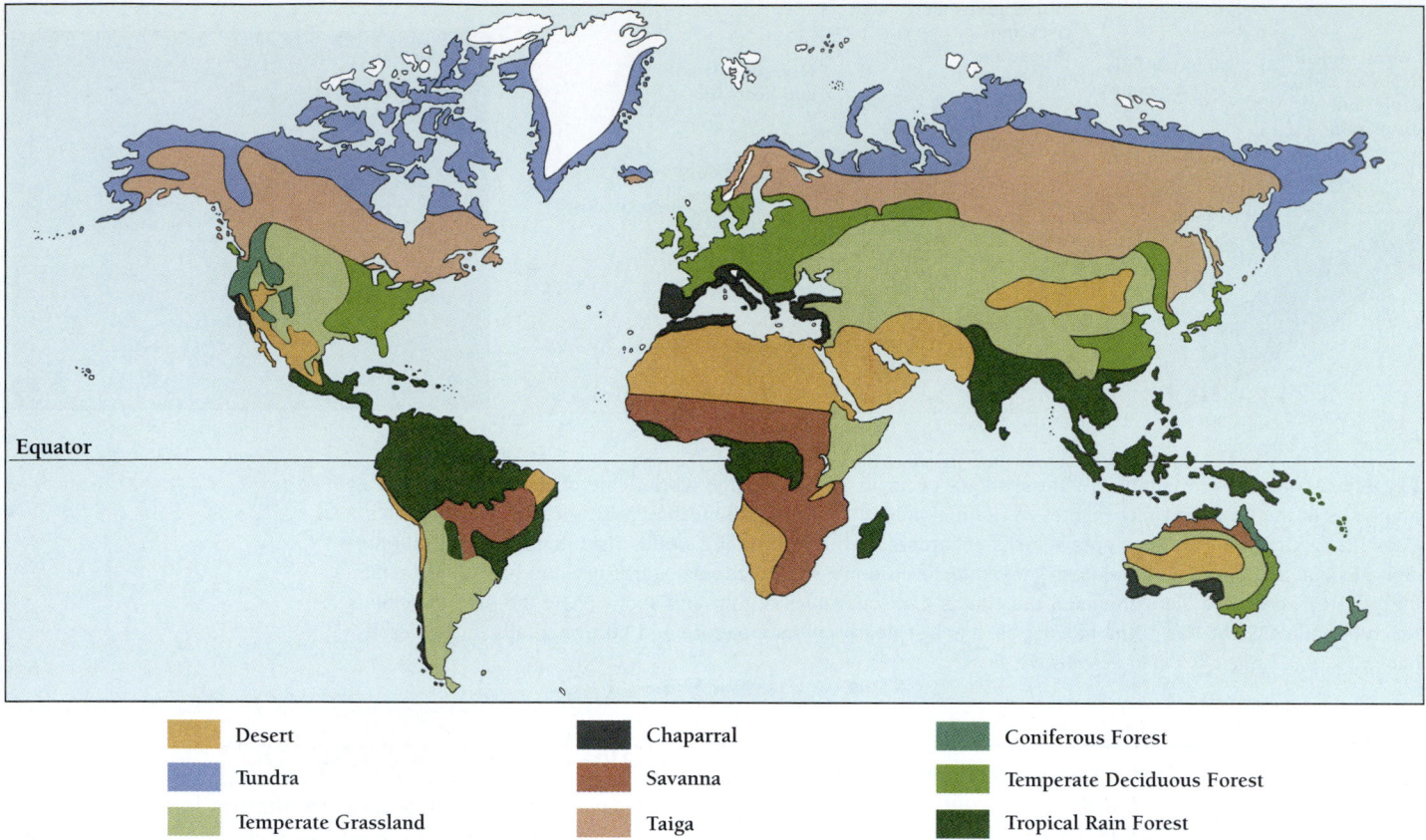

Desert	Chaparral	Coniferous Forest
Tundra	Savanna	Temperate Deciduous Forest
Temperate Grassland	Taiga	Tropical Rain Forest

Figure 26-8   World distribution of biomes. The sharp boundaries in this diagram are misleading, as biomes frequently grade together smoothly. In addition, each solid area actually represents an average of many vegetation types. Botanists divide California alone into 24 "floristic provinces."

Like desert plants, desert animals must also be able to survive extreme heat and dryness. In the summer, many are active only during the cool hours of dawn and dusk. Lizards and snakes thrive, as do birds such as cactus wrens and roadrunners, and rodents such as kangaroo rats and mice. Many rodents **estivate,** retiring to underground burrows and entering a sleeplike state called *torpor* during the hottest and driest months of the summer. Round-tailed ground squirrels (*Ammospermophilus tereticaudus*) and several species of pocket mice estivate, as well as the cactus mouse (*Peromyscus eremicus*). Laboratory experiments suggest that lack of food and lack of water can each trigger estivation in the cactus mouse.

TABLE 26-1   TEMPERATURE, PRECIPITATION, AND PRODUCTIVITY OF MAJOR BIOMES

Biomes with the least seasonality and the most rain are also the most productive. Productivity is measured as the dry weight of new plant material that accumulates on a square meter of land in one year.

Biome	Temperature (winter low/summer high)	Precipitation (cm/y)	Productivity (g/m²/y)
Desert	warm/hot	< 25	90
Tundra	cold/cold	< 25	140
Chaparral	mild/mild	20–40	700
Grassland	cold/warm	25–75	600
Taiga	cold/cool	50	800
Savanna	warm/hot	90–150	900
Temperate deciduous forest	cold/hot	75–125	1200
Tropical rain forest	warm/hot	200–450	2200

Figure 26-9 Deserts. Creosote bushes punctuate the many deserts of the southwestern United States. *(John M. Roberts/The Stock Market)*

Although deserts seem tough, they are delicate. The wheels of off-road recreational vehicles destroy fragile desert soils, as well as plants, lizards, snakes, and tortoises. An even greater threat to deserts is the conversion of desert habitat into suburban developments and farms. Excessive irrigation can destroy

any soil, but desert soils are especially vulnerable. Irrigation water from rivers and wells is loaded with dissolved minerals and salts. As it evaporates from the hot desert soil, the minerals and salts are left behind as a deposit. In time, the soil becomes so caked with minerals and salts that farming is impossible. By then, even desert plants cannot grow in the poisoned soil.

Because deserts lack water, they are the least productive of all biomes. Desert plants and animals are adapted to conserve water.

How Can a Dry Biome Be So Wet?

At the North Pole lies an ice desert utterly devoid of plant life. But just a few hundred miles south lies the cold **tundra** [Finnish, *tunturi* = arctic hill], a vast, open land of bogs, ponds, and lakes. Among the lakes are short grasses, shrubs, mosses, and lichens, nearly all less than 20 centimeters tall. The few small shrubs and trees grow on the banks of lakes or streams. Tundra sweeps round the North Pole, covering more than one-fifth of the Earth's land surface, including northern Canada, Alaska, Scandinavia, and Siberia (Figure 26-10). In addition, alpine tundra cloaks the upper slopes of high mountains and plateaus in temperate regions.

The tundra is cold. Freezing temperatures occur in every season, and **permafrost**—permanently frozen ground less than a meter below the surface—persists throughout even the warmest summers. Still, a one- or two-month summer growing season allows topsoil to thaw and plants to grow. As soon as the brief summer warmth and rains arrive, the plants' extensive underground roots allow them to quickly produce leaves, flowers, and fruits.

Figure 26-10 Alaskan **tundra**. Lupine and ferns on the Pribilof Islands, Alaska. *(John Shaw/Tom Stack & Associates)*

Tundra receives no more rain or snow than a desert. Each year, less than 25 centimeters falls, mostly as summer and autumn rains. In winter, the snow pack rarely exceeds 10 to 20 centimeters. Despite this low precipitation, the soil is perpetually wet, for the cold air prevents evaporation and the icy permafrost keeps water from draining away.

Mammals in the tundra include caribou (or reindeer), polar bears, weasels, foxes, and small rodent herbivores called lemmings. In the summer, large numbers of birds arrive from the south to nest and raise young. Swarms of flies and mosquitoes fill the air. In the summer, the tundra supports an enormous amount of life, but the diversity of species is low.

So far, humans have had only limited impact on the tundra. Inuits of Alaska, Greenland, and Canada live on the tundra hunting wild caribou, and the Lapps (Sami) of northern Scandinavia herd domesticated caribou (reindeer). In the past 30 years, however, oil development, oil pipelines, and roads have begun to impinge on life in even this remote biome.

Tundra has a cold, dry climate and a short growing season.

Why Are There No Trees in Grasslands?

Temperate grasslands have well-defined seasons, with hot summers and cold winters (Figure 26-11). Annual rainfall is low, usually 25 to 75 centimeters per year, enough to keep grasslands from turning to deserts but not enough to sustain forests. In addition, grasslands usually have a dry period, during which fires are common. Wildfires and the lack of rain, together with heavy grazing by animals, keep trees from taking hold in this biome.

Grasslands extend over much of the Earth's surface, mostly in the interiors of the continents. They include, for example, the steppes of Russia, the veld of South Africa, the pampas of Argentina, the puszta of Hungary, and the prairies of the central United States and Canada. Before European settlement, more than a million square miles of grassland covered central North America.

The types of grasses vary with the amount of rain and with season. In the eastern American prairie, for example, tall grasses, especially big bluestem, once dominated; in the drier western plains—in the rain shadow of the Rocky Mountains—short grasses dominated.

Grasslands can support large numbers of mammals. For millions of years, African grasslands supported huge herds of wild gazelles and zebras, while the Asian steppes have supported sheep and horses. All of these herbivores in turn supported populations of wolves, lions, humans, and other predators. When Lewis and Clark crossed the American prairie in 1804, some 60 million bison and 40 million pronghorn antelope grazed in vast herds. The explorers pushed through these herds only with difficulty. Yet 19th-century fur traders and sport hunters drove these populations to the brink of extinction in less than 75 years.

Today, in place of the bison and antelope live some 20 million humans, who have virtually eliminated the temperate grassland biome. In place of hundreds of thousands of square miles of American and Canadian prairie, we have created the world's "breadbasket." Corn has replaced tall bluestem, and wheat has replaced the short grass. Mass-produced pigs, cows, and beef cattle are the chief herbivores; humans, the top carnivores. A few thousand bison live in parks and reserves, and a few hundred thousand pronghorn antelope live in the plains adjacent to the Rocky Mountains. Their natural predators,

Figure 26-11   Temperate grassland. This wild prairie at the Living Prairie Museum in Winnipeg, Manitoba, has never been plowed. *(Tom McHugh/ Photo Researchers, Inc.)*

however, the Great Plains wolf and the grizzly bear, are gone—shot, poisoned, and trapped to extinction.

Temperate grasslands have hot summers and cold winters. In North America, wheat fields have replaced most temperate grasslands.

Chaparral Always Burns Eventually

Chaparral [Basque, *chabarro* = dwarf evergreen] dominates five widely separated temperate regions—California, central Chile, the shores of the Mediterranean Sea, southwestern Africa, and southwestern Australia. These regions share a "Mediterranean" climate, with mild, rainy winters and hot, dry summers. In the summer, prevailing westerly winds blow cool, moist sea air over the warm land. But the warmed air holds the moisture, so no rain falls.

Chaparral is characterized by dense shrubs and scattered, broad-leafed trees (Figure 26-12). The shrubs are typically spiny, thick, and dense. Each of the five chaparral regions has its own species, but all chaparral species share common adaptations to summer drought and to periodic wildfires. Wildfires clear the way for fresh growth and kill trees that would shade out the chaparral. The seeds of some herbs and shrubs actually require exposure to the intense heat of a brushfire before they can germinate. Perennial shrubs such as chemise and coyote brush are well adapted to fire. When fire burns away their dry branches, these species resprout from still-intact roots.

One of the greatest dangers to chaparral communities is overzealous fire prevention by humans who build houses in them. By putting out small fires, humans ensure the accumu-

Figure 26-13 Fire. Brush fires in chaparral communities destroy houses in minutes. Here a 20,000-acre fire sweeps through a suburban neighborhood in southern California. (*Reuters/Corbis-Bettmann*)

lation of dry wood. Inevitably, a fire so big and hot that firefighters cannot put it out sweeps thousands of acres, incinerating the roots of the chaparral vegetation and destroying hundreds of suburban houses (Figure 26-13). Such hot fires also kill seeds and sterilize the soil, so that until the land is recolonized, nothing grows back. In addition, because of the way such superhot fires affect the soil, disastrous erosion may follow (Figure 26-14).

Chaparral biomes are characterized by hot, dry summers and mild, wet winters. Many chaparral species have special adaptations to fire.

Figure 26-12 Chaparral. Dense shrubs and scattered trees characterize Mediterranean climates. (*Tom McHugh/ Photo Researchers, Inc.*)

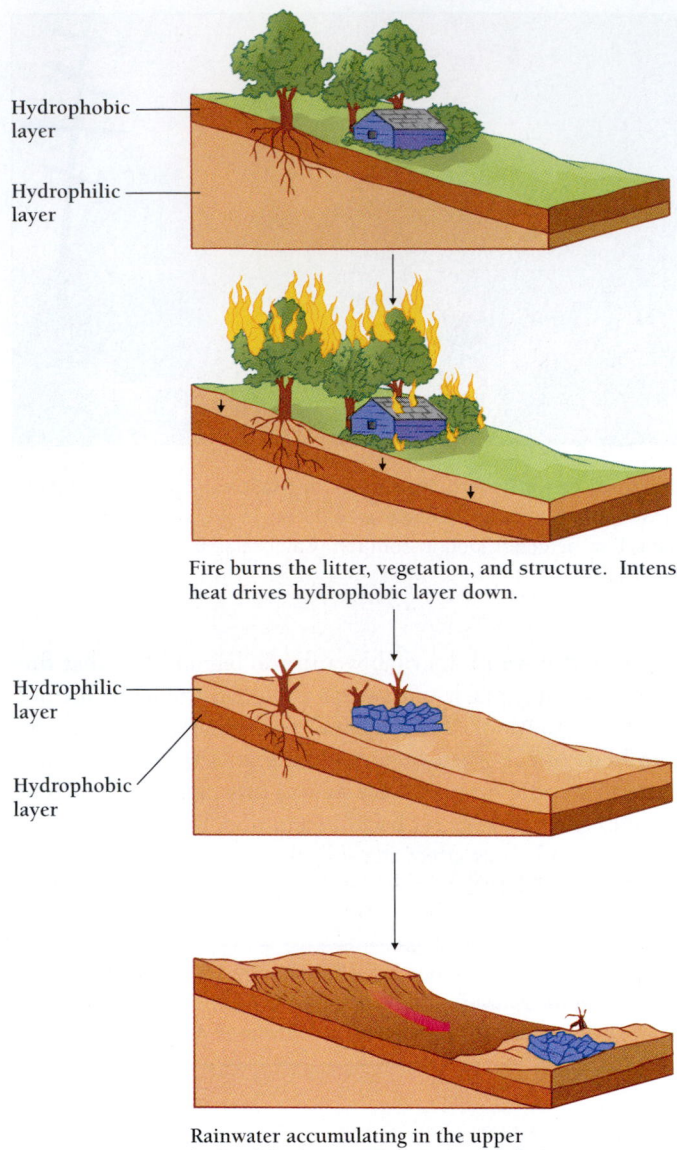

Hydrophobic layer

Hydrophilic layer

Fire burns the litter, vegetation, and structure. Intense heat drives hydrophobic layer down.

Hydrophilic layer

Hydrophobic layer

Rainwater accumulating in the upper hydrophilic layer may cause landslides.

Figure 26-14   Houses in fire-adapted communities. In chaparral communities, the regular control of brush fires leads to a buildup of dead wood. When a fire finally burns out of control, it burns unusually hot. Super-hot fires not only destroy houses, trees, and shrubs but also damage soil. Chaparral soil contains water-resistant plant compounds that prevent water from penetrating the soil. Heat from a hot fire forces this "hydrophobic layer" deeper into the soil. The uppermost layer no longer sheds water but absorbs it instead. When winter rains drench the upper layer of soil, turning it into mud, the top layer slides away.

Savannas Host Some of the Most Spectacular Grazing Animals

Savannas are grasslands with scattered trees or small clumps of trees (Figure 26-15). The trees are usually deciduous, dropping their leaves in the dry season. Like temperate grasslands, savannas have a regular dry season and periodic wildfires that limit the spread of trees. But savannas generally have more rain (usually 90 to 150 centimeters per year) and hotter climates than temperate grasslands. Savannas are often transitional biomes between temperate deciduous forest and prairies or between evergreen tropical rain forests and deserts. Savannas cover much of south and east central Africa, much of Australia, and parts of the Americas, and Southeast Asia.

Savanna grasses are often tall—2 meters or more in Africa, 1.5 meters in South America—and the trees short—usually less than 10 to 15 meters tall. Deep, dense underground networks of roots allow grasses to obtain water where trees cannot and so survive the dry seasons.

Animal life in savannas varies greatly from continent to continent. The best-known animals are those of the African plains. There, herds of zebras, wildebeest, and gazelles graze while elephants and giraffes browse on the trees. Among these herbivores, lions, cheetahs, hyenas, wild dogs, and other carnivores hunt and scavenge. Savannas also support large, flightless, grazing birds—the ostrich in Africa, the emu in Australia, and the rhea in South America. Termites consume up to one-third of all plant litter and serve as a major food source for birds and mammals. In some areas, termite mounds cover the savanna. These mounds may reach 9 meters in height and number 10 to 150 per hectare.

In most savannas, hunters and habitat destruction are driving many species to the brink of extinction. In East Africa, the seemingly infinite herds of grazing animals are becoming food for rapidly growing human populations. Elephants have become particular targets for poachers, not because of their food value, but because of their valuable ivory tusks. Their slaughter continues despite worldwide efforts to outlaw trading in ivory. Nonetheless, the real threat to African savanna is the continuing conversion of savanna to farms and ranches that support the burgeoning human population. While wildlife sanctuaries protect some of the African and South American savannas, the end of the savanna biome seems inevitable.

Tropical savannas are grasslands with scattered deciduous trees that support vast herds of grazing animals.

Taiga, the Spruce–Moose Biome

South of the tundra, most northern lands are both slightly warmer and wetter. Here sparse conifer forests called **taiga** grow monotonously across thousands of miles of Siberia, Scandinavia, Canada, and Alaska (Figure 26-3). Winters are bitterly cold, and summers are short. The one or two species of trees (often spruce and pine) grow only 5 to 10 meters high. Ferns, mosses, and lichens grow sparsely on the forest floor. Shrubs are rare, except for blueberries and gooseberries that flourish in open areas where trees have temporarily died off.

Moose, mice, squirrels, porcupines, and snowshoe hares are the main vertebrate herbivores. Predators include wolves, grizzly bears, lynxes, and wolverines. Birds migrate to the taiga

Figure 26-15 **Savanna.** Grasslands and scattered trees in the Serengeti, in Tanzania. (© Art Wolfe)

in summer, but leave before the first snows fall. Except for garter snakes, reptiles cannot survive the long, harsh winters. Mosquitoes, black flies, and other insects, however, abound.

The taiga is so inhospitable that human populations are sparse. In some areas, logging has eliminated the original forest, and deer have moved in, replacing the moose. A common parasite of the deer—the nematode brain worm—is fatal to moose and, in many areas, is accelerating the disappearance of the moose.

One of the greatest threats to the taiga comes from the air. Factories, power plants, and automobiles produce particulate pollutants and oxides of sulfur and nitrogen that rise and travel thousands of miles to the taiga. The oxides dissolve in the clouds, dropping acid rain and snow on lakes and rivers. The result has been the wholesale death of fish and forests in many regions of the Northern Hemisphere, including Scandinavia and eastern Canada.

Lichens, in particular, accumulate sulfur dioxide, heavy metals, and other toxins. Ecologists use both the health and chemical composition of taiga lichens to monitor global air pollution.

Taiga has cold winters and short, mild summers.

Why Do the Temperate Deciduous Forests Lack Top Predators?

Temperate deciduous forests typically have warm, rainy summers and cold winters. They receive from 80 to 140 centimeters of rain and snow each year. However, temperature and

annual precipitation vary more than in the tropics or the polar regions. As a result, temperate deciduous forests may look rather different from one another. In the northern United States, for example, the dominant trees are maple, oak, hemlock, and birch, while in the southeastern states the dominant trees are pine, oak, and hickory.

Deciduous trees, which shed their leaves each autumn, characterize the temperate forest biome (Figure 26-16). Deciduous plants shed their leaves as a way to minimize water and nutrient loss at times when photosynthesis is minimal. The plants of a temperate forest occupy four vertical layers: trees, shrubs, herbs (nonwoody plants), and ground cover. The trees occupy the highest layer, their tops touching to form a nearly continuous canopy. In the shade beneath the trees are shrubs and bushes whose branches lie closer to the ground. The herb layer may be especially rich and beautiful in the spring. Finally, the ground layer consists of mosses and liverworts, often living among the accumulated leaf litter.

Animal life is abundant in all the plant layers and in the soil beneath the litter (Figure 26-17). Trees and shrubs provide food and nest sites for many species of birds. Woodpeckers and chickadees, for example, nest in cavities in the trees, while warblers and thrushes build nests on their branches. Mammals such as chipmunks and squirrels; reptiles such as snakes, lizards, and wood turtles; and amphibians such as salamanders and tree frogs also find food and homes in the forest.

The eastern third of the United States, as well as western Europe, Japan, Chile, and eastern China, was once dominated by temperate deciduous forests. On every continent, humans have cut down much of the forest for lumber or firewood and to make way for farms, towns, and cities. Before European colonization 400 years ago, forests covered 46 percent of the

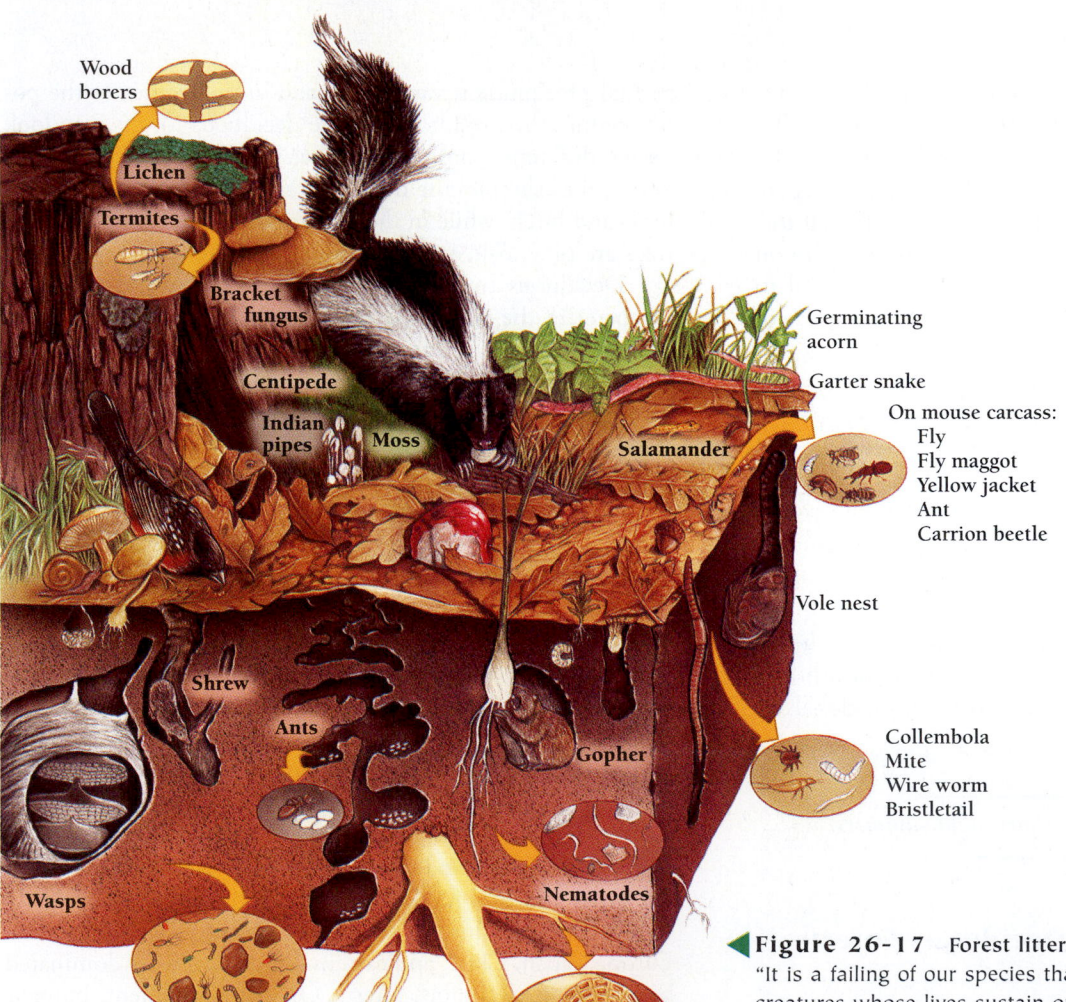

Wood borers

Lichen

Termites

Bracket fungus

Centipede

Indian pipes   Moss

Germinating acorn

Garter snake

On mouse carcass:
Fly
Fly maggot
Yellow jacket
Ant
Carrion beetle

Salamander

Vole nest

Shrew

Ants

Gopher

Collembola
Mite
Wire worm
Bristletail

Wasps

Nematodes

Bacteria

Mycorrhizae

◀ **Figure 26-16** **Deciduous forest.** In autumn, the leaves of deciduous trees turn color, as chlorophyll is pulled from the leaves. North County National Scenic Trail, Itasca State Park, Minnesota. (*G. Alan Nelson/ Dembinsky Photo Associates*)

◀ **Figure 26-17** **Forest litter.** Biologist E.O. Wilson has written, "It is a failing of our species that we ignore and even despise the creatures whose lives sustain our own." The rich leaf litter of a temperate deciduous forest supports myriad organisms from bacteria, fungi, ferns, and worms to skunks, squirrels, foxes, and mountain lions.

How Does Acid Rain Alter Biomes?

Most people now know that severe acid rain can destroy lakes and whole forests (Web Bit 2-2: Life at Low pH). In the worst cases, fish, frogs, plants, and other organisms have disappeared from formerly productive lakes and streams. Yet, increasingly, ecologists are finding that acid rain also has the capacity to transform ecosystems in subtler ways.

Researchers have long known that acid rain increases the nitrogen available to plants. Indeed, some economists have argued that acid rain is a cure for global warming. They say that because nitrates from acid rain can stimulate plants to grow faster, plants in acid environments can take up more carbon dioxide, compensating for the carbon dioxide released when we burn fossil fuels.

Although this is a reassuring thought, ecological research does not support this hypothesis. In one study, researchers at the University of Minnesota applied nitrogen to plots of native Minnesota tallgrass prairie for 12 years to imitate the excess nitrates that result from both acid rain and fertilizer applications by farmers. Because native grasses use nitrogen efficiently, they flourish in places where the nitrogen content of the soil is low. In the nitrogen-treated plots, however, exotic weeds flourished, outcompeting and displacing native Minnesota grasses.

But that was only half of the bad news. As prairie grasses die, their tissues break down very slowly, and over time much of their carbon, nitrogen, and other constituents remain stored in the soil. Such long-term storage is, in fact, the key to the famous, rich soils of the American midwest.

But in the Minnesota study, the weeds behaved differently from grasses. As the weeds died, their huge stores of nitrogen stimulated the growth of nitrogen-hungry soil microbes, which rapidly dismantled the exotic weeds into basic building blocks. The microbes released more nitrates into the groundwater and more carbon dioxide back into the air. The net effect was more carbon dioxide in the air—not less—and more nitrates in the soil, which stimulated the growth of more weeds.

In the spruce forests of Germany's Fichtelgebirge mountains, the excess nitrogen in acid rain supplies the trees with more nitrogen than they can use. Instead of stimulating growth, the acid rain destroys soil. Negatively charged nitrates raining down on forest soils bind with calcium, magnesium, and other positively charged cations in the soil. As the nitrates flow into lakes and streams, they take the calcium and magnesium with them. In some areas, such acid leaching has removed all of the calcium deposited in the soil for the last 500 years.

And, as we saw in the last chapter, as calcium leaches away and soil acidity rises, increasing amounts of aluminum dissolve in soil water. The aluminum poisons the roots of trees and kills fish.

In some areas, all the trees in a forest are dying. In others, ecologists predict, trees deprived of calcium and magnesium will simply habituate and grow quite slowly, like the tiny bonsai trees that gardeners grow in pots. The result may be pygmy forests.

United States, and most of this forest was ancient and deciduous. Although in the eastern United States, forests have returned, they still cover only 32 percent of the land and consist largely of young trees, all the same age and species. More important, these young forests have not brought back the local species of plants, animals, and fungi that went extinct 200 years ago.

Much of the new forest is in small, isolated woodlots, often of 16 hectares (40 acres) or less. These tiny patches tend to hold far fewer species than the extensive virgin forests they replaced (Chapter 28). Such small areas do not provide the large ranges needed to support carnivores at the top of the food chain. A single mountain lion, for example, needs over 75,000 hectares (300 square miles). Even a single family of gray foxes needs some 260 hectares (1 square mile). Gone with these vast areas are nearly all the top carnivores in the eastern United States. Gone, or nearly gone, are the red wolf, the bobcat, and the black bear.

Temperate deciduous forests have warm, rainy summers and cold winters.

Tropical Rain Forests Are the Most Productive and Diverse Biomes in the World

Most **tropical rain forests** are within 10° of the equator—in the Amazon basin of South America; in central and west Africa; and in Malaysia, Indonesia, and New Guinea. The weather in the tropical rain forests is more or less constant throughout the year, with average temperatures of about 27°C (80°F) every month of the year. Rainfall is heavy, ranging from 2 to 4.5 meters per year.

Warmth and moisture are so plentiful in a tropical rain forest that plants compete for light rather than water. Tropical plants are so successful at capturing light that the forest floor is usually rather dim, with little vegetation. Visitors often compare a tropical forest to a cathedral, with its open floor, high vaulted ceilings, and soft filtered light.

Nutrients cycle rapidly through rain forest ecosystems. The soils of tropical rain forests are surprisingly poor in minerals and other nutrients. One reason is the heavy rainfall, which leaches nutrients from the soil. Another is the year-round work of ants, termites, fungi, and bacteria, which rapidly break down leaf litter to simple molecules. In most biomes, a dry season or a long, cold winter slows the breakdown of leaf litter and other organic matter, and soils tend to build up to depths of up to several feet. In the rain forest, only a thin layer of nutrients exists, just beneath the leaf litter at the very surface.

Trees rely on an extensive network of extremely shallow roots to sweep up whatever nutrients are available. The bulk of nutrients and energy in a tropical ecosystem are in the tops of trees and in the other living organisms, rather than in soils or water.

Despite their thin soils, tropical rain forests are the second most productive ecosystem in the world, surpassed only by estuaries and marshes. And tropical rain forests are first in biodiversity. At least half, and perhaps as many as 90 percent, of all the Earth's species live in tropical rain forests. In a hectare (2.5 acres) of temperate forest, we might find two or three dominant species of trees with a scattering of another 10 to 20 species. In contrast, a hectare of tropical forest is home to 40 to 100 different species of trees. More species of butterflies live in one rain forest in Costa Rica, for example, than live in all of the rest of North America.

The diversity of species in a tropical rain forest can be staggering. One ecologist found 163 species of beetles living in just one tree in the Panamanian rain forest. Another ecologist found 445 species of trees on an area of Brazilian rain forest the size of a city block. Fifteen of those trees species were new to science.

More than two-thirds of the plants in a tropical forest are trees. Most are flowering, broad-leaved evergreens (Figure 26-18). Other plants common in tropical forests but rare in temperate forests are **lianas,** vines that are rooted in soil but climb into the tree tops, and **epiphytes,** plants such as orchids that grow entirely on other plants.

As in a temperate forest, the vegetation forms discrete layers. Most of the trees form a continuous layer, called the **canopy,** 30 to 40 meters above the ground (Figure 26-19). Rising above the canopy are occasional, tall, isolated **emergents,** trees with umbrella-shaped crowns extending to a height of 50 meters or more. Below the canopy, a third layer of shorter trees 10 to 25 meters tall catch whatever light they can from gaps in the canopy. Finally, a short shrub layer consists of dwarf palms and giant herbs with especially large leaves, adapted to capture the faint light that filters through upper stories.

All these flowering plants provide a year-round supply of fruits and nectars for an array of birds, insects, snakes, bats, monkeys, and squirrels. Some live permanently in the canopy. Others live in the rivers or on the forest floor. Some eat underground plant parts, others, fruits and nectar. Ants and termites dominate this biome. But anteaters and armadillos, pigs and their relatives, as well as poisonous and constricting snakes, all hold their own in the complex food webs of the tropical rain forest. The top carnivores of the forest floors used to be such large cats as the jaguar in South America and the tiger in Asia. However, these magnificent animals have all but suc-

Figure 26-18   Tropical rain forest. Terra-firma Amazonian rain forest with lianas, Peru. *(David M. Schleser/Nature's Images, Inc./Photo Researchers, Inc.)*

cumbed to trophy and fur hunters and to the destruction of their habitat by humans.

How Are Tropical Rain Forests Being Destroyed?

Until recently, Earth contained 25 million square kilometers of tropical rain forests, an area greater than all of North America, Central America, the Caribbean, and Hawaii combined. Today no more than 8 million square kilometers remain. Estimates of the rate of rain forest destruction show that an area of tropical rain forest the size of the state of Washington is logged or burned every year. And even this estimate may be too low.

In 1999, field surveys of wood mills and forest burning in the Amazon showed that the true rate of destruction may be twice as high as previous estimates. As rain forest disappears, watersheds are destroyed, the local climate becomes drier, and remaining forests dry and become susceptible to

Emergent layer
to 50 m

Canopy layer
30–40 m

Understory
15–25 m

Shrub layer
3–15 m

Ground layer
0–3 m

Figure 26-19 **The layers of a tropical forest.** Tropical rain forests are noted for their extensive layering. Shown here are the protruding emergent trees, the canopy, the understory, the shrubs, and the ground-level plants. Because their roots are so shallow, tropical trees often grow huge, supportive buttresses.

fires. In short, rain forest destruction is an accelerating process. How is it happening?

Like other human populations, those in the tropics are doubling every few decades. To raise cash for these expanding populations, governments support the logging and burning of large areas of tropical rain forest. Tropical nations export timber and plywood from rain forest trees, metals mined from rain forest lands, and beef from cattle raised on clear-cut rain forest.

Once cleared, former rain forest soils do not last long. Recall that the nutrients in tropical rain forest are in the vegeta-

tion, which is carried away or burned. The average cattle ranch in Central America lasts only 6 to 10 years, after which the soil is too depleted of minerals to support grass for grazing cattle. Each hamburger made from the cattle raised on these ranches costs 5 square meters of tropical rain forest, including about a half ton of trees, flowers, birds, mammals, snakes, lizards, frogs, butterflies, beetles, ants, and termites.

One of the greatest causes of rain forest destruction is small-scale farming. Individual farmers cut and burn a small area of land and raise a few crops. Once the soil is depleted, the farmers move on to another patch of land. This kind of farming is called "slash and burn" farming or "shifting agriculture." Because the individual fires set by farmers are so small, the damage is often invisible in satellite imagery until vast areas have been destroyed by thousands of farmers. In Chapter 28, we discuss how isolating small plots of forest creates "edge effects" that cause many species to go extinct.

Future textbooks may list the tropical rain forest as a biome that has disappeared, for human activity threatens to eliminate all but a few parks within the next 20 to 30 years. Accompanying the destruction of the rain forest will be the disappearance of millions of species of plants, insects, and other animals found nowhere else. One-quarter of the world's 250,000 plant species and one-fifth of the 20,000 named butterflies will likely go. Yet most of the species that disappear will never even have been named.

As if that weren't bad enough, recall from Chapter 25 that the burning of forests releases tons of extra carbon dioxide into the atmosphere, increasing global warming, hastening changes in climate and the melting of the polar ice caps.

Tropical rain forests have the greatest biodiversity of any biome in the world. Unless major changes occur, however, 90 percent of it will be gone within 20 to 30 years.

WHAT ARE AQUATIC ECOSYSTEMS LIKE?

Water covers 72 percent of the Earth's surface—70 percent saltwater oceans and 2 percent freshwater lakes, rivers, and streams. Like terrestrial ecosystems, aquatic ecosystems fall into just a few distinct types. Ocean ecosystems differ from one another according to temperature, depth, and distance from shore.

How Do Oceanographers Divide the Oceans?

Although ocean covers more than two-thirds of the Earth's surface, the oceans contain only about 10 percent of the planet's species—about 160,000 species. Because the oceans are all interconnected, they lack habitat islands that promote geographic isolation and speciation (Chapter 17). Oceans provide fewer opportunities for adaptive radiation than land habitats.

Nonetheless, the oldest, largest, and most stable ecosystems on Earth are those of the oceans. The oceans and their connecting seas extend over nearly all of the Earth's latitudes.

In the last few decades, **oceanography,** the study of the seas and their ecosystems, has revealed the shape of the seas. The ocean bottom, for example, contains vast mountain ranges and rifts, as well as a shallow continental shelf bordering each continent. The average depth of the oceans is about 3 kilometers. But the lowest spot on Earth, the Marianas Trench of the western Pacific Ocean, dips to 11 kilometers (7 miles) below sea level—deeper than Mt. Everest is high.

Biologists distinguish **ocean ecosystems** according to ocean depth and distance from shore. The **benthic division** consists of all organisms that live on the ocean floor. These include those that live in the **intertidal zone,** which lies between high tide and low tide, and those that live at the bottoms of deep canyons in the **abyssal zone** (Figure 26-20). The **pelagic division** consists of all organisms that live in open water, above the bottom.

Organisms living in the intertidal zone are submerged at high tide and exposed to air and pounding waves at low tide. They must be able to withstand drying, hot sun, and crashing waves. Intertidal organisms anchor themselves to rocks or sand, and close shells and other structures tightly to keep from drying out when the tide goes out. Tidepools reveal the rich diversity of the intertidal species (Figure 26-21).

In the pelagic division, microorganisms called **plankton** drift in the water, moving only where currents take them. Free-swimming organisms, such as fish and whales, feed on plankton or on each other. The pelagic division is further divided into the **neritic zone,** which is farther from shore, over continental shelf, and the **oceanic zone,** which is out beyond the continental shelf over the deepest water.

The neritic zone consists of shallow coastal areas near islands and continents. In the neritic zone, algae are the dominant primary producers. While the total area of the neritic zone is much smaller than that of the oceanic zone, it contains many more ecological niches and species.

The oceanic zone comprises most of the world's oceans. Here photosynthetic cyanobacteria, diatoms, and dinoflagellates are the primary producers. Their output is extremely low, however, comparable to that of a terrestrial desert. The oceanic zone lacks mineral nutrients, especially phosphorus and iron. Oceanic ecosystems recycle nutrients poorly, depending on a steady supply from terrestrial biomes. As organisms die, their remains fall from the surface layers to the ocean bottom. Once these nutrients drop to the bottom of the ocean, they remain there.

A few exceptions exist. Off the coast of Peru, the Humboldt Current draws nutrients up from the bottom in a process called **upwelling.** Upwelling creates a rich soup that supports a rate of photosynthesis six times that found in other parts of the open ocean.

Where such currents bring nutrients, grazing protozoa and small animals feed on blooms of photosynthetic microorganisms. Small fish eat the tiny invertebrates, and larger fish eat the smaller fish. A marine food chain typically contains four or five trophic levels. (In contrast, terrestrial food chains seldom have more than three trophic levels.) This enormous productivity supports, in turn, huge populations of fish-eating birds, as well as a large human fishing industry.

Every year, around Christmas, the Humboldt Current reverses itself. Warm water flows down the coast, preventing the nutrient-rich cold waters from rising to the surface. Usually these

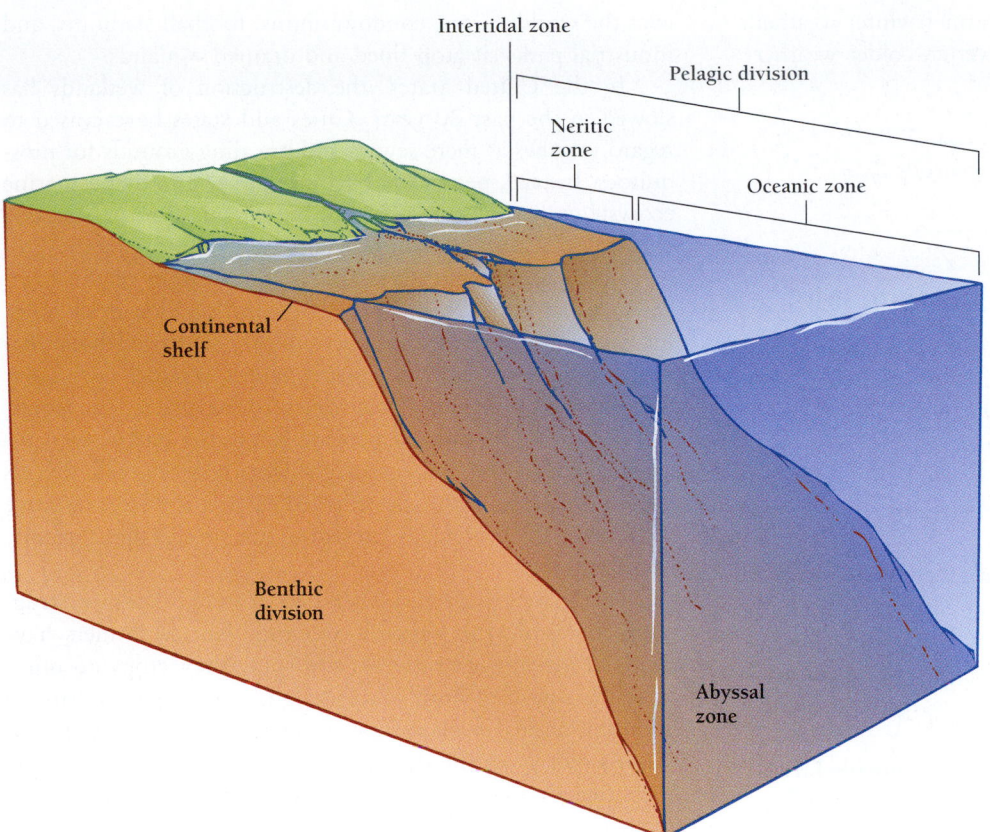

Intertidal zone

Pelagic division

Neritic zone

Oceanic zone

Continental shelf

Benthic division

Abyssal zone

Figure 26-20 **Dividing the ocean into zones.** Oceanographers divide the ocean into divisions and zones according to the depth of the water and the distance from shore. The benthic division includes intertidal organisms and others that live on the bottom. The pelagic division includes the neritic zone (over the continental shelf) and the oceanic zone (deep water).

Figure 26-21 **The intertidal zone.** The intertidal zone is characterized by organisms that cling to rocks as the waves crash around them. Shown here is a Pacific Coast tidepool. (*Nancy Sefton/Photo Researchers, Inc.*)

changes, which local fishermen call "El Niño" [Spanish, = the Christ child], are temporary and harmless. However, every 2 to 7 years, the current reversal lasts long enough to produce catastrophic effects. In 1982 and 1983, for example, the failure of the Humboldt Current to provide its usual supply of nutrients led to a failure of the sardine industry and the wholesale starvation of fish-eating seabirds in the South Pacific.

Doesn't El Niño Do More Than Kill Fish and Birds?

A big El Niño affects weather around the world. In the winter of 1997–1998, an area in the central and eastern equatorial Pacific Ocean the size of the continental United States increased in temperature by 3.5°C. The warm ocean triggered giant rain storms that battered the entire Pacific Coast, from the Pacific Northwest in the United States and Canada to the Andes of Equador and Peru. Meanwhile, eastward, in Indonesia and Malaysia, a severe drought worsened huge wildfires, which burned desiccated tropical rain forests for months.

El Niño has a sibling weather pattern, called La Niña, that has the opposite effect. During La Niña, the waters of the equatorial Pacific Ocean turn cold, and warm thunderstorms return to Indonesia, carrying heavy rains. Over North America, the cold Pacific jet stream weakens and undulates, sweeping across Alaska one month and down across the central United

States the next. La Niña brings drier, warmer winter weather to the southeastern United States and wetter, colder weather to the Northwest.

Ocean ecosystems are more homogeneous than terrestrial ecosystems.

Why Are Estuaries and Marshes Valuable?

Estuaries are partly enclosed waters where freshwater streams or rivers meet the ocean. The water is less salty than the oceans, and the salinity is constantly changing. The flow of salt water and fresh water depends on the tides, which gives the estuaries their name [Latin, *aestus* = tide]. In addition, the flow of fresh water from rivers and streams varies with season.

Estuaries are among the most productive of all natural ecosystems, rivaling that of a tropical rain forest. Tides and rivers provide a constant flow of water and nutrients (Figure 26-22). The primary producers in estuaries include plankton, algae, and large plants such as eelgrasses, sea grasses, marsh grasses, and mangrove trees.

Because of this high productivity, estuaries provide breeding and nursery grounds for many species of fish and shellfish. Countries that destroy their coastal wetlands ultimately destroy their fisheries. Estuaries also provide food and breeding grounds for many species of birds, amphibians, reptiles, and mammals.

Estuaries are especially vulnerable to damage by humans. They are the first to receive pollution from streams and rivers, and they are prime targets for oceanfront development. All over the world, farms, condominiums, football stadiums, and industrial parks sit atop filled and drained wetlands.

In the United States, the destruction of wetlands has slowed in the past 20 years. Cities and states have ceased to regard marshes as mere sewers and breeding grounds for mosquitoes. Instead, people are beginning to realize that estuarine ecosystems are not only places of beauty but also irreplaceable contributors to the world food supply.

In the 1940s, Florida drained large sections of the Everglades in an effort to control floods and provide fresh water for a growing population. These efforts destroyed large sections of the Everglades. In places, the Everglades are so dry that wildfires burn out of control. In other places, fish populations are so low that 90 percent of wading birds have vanished, for there is not enough for them to eat. Excess phosphorus from farms has allowed cattails to take over large sections of the Everglades, crowding out species that formerly flourished there.

Fortunately, state and federal governments are now struggling to restore this valuable wetland. Florida farmers have changed the way they irrigate and spray their crops, to minimize the runoff of both pesticides and excess phosphorus. In addition, animal feeds can be formulated with less phosphorus. As one agricultural researcher put it, cows "can't excrete what they don't eat." In addition, giant artificial marshes now remove up to 80 percent of the phosphorus before it reaches the Everglades. And the U.S. Army Corps of Engineers has proposed to Congress an $8 billion plan that would allow water to flow more freely throughout the Everglades.

Estuaries are the most productive ecosystems in the world. They support a diversity of species.

Figure 26-22 Estuary. Salt marsh at Assateague Island National Seashore, Maryland. *(Michael P. Gadomski/Dembinsky Photo Associates)*

Why Are Lakes, Ponds, Rivers, and Streams Susceptible to Pollution?

Limnology [Greek, *limne* = pool, marshy lake] is the study of freshwater lakes, ponds, rivers, and streams. As rivers and streams flow into lakes, they bring both water and dissolved nutrients. This material, both organic and inorganic, derives from the surrounding land, so the character of lakes and ponds varies from region to region. At the fringes of a lake, marshes and swamps often form transition zones to the adjacent terrestrial ecosystems.

As in ocean ecosystems, primary production in a lake depends on the amount of sunlight penetrating the lake's surface to photosynthesizing microorganisms and water plants. Heterotrophic protists and small animals such as rotifers and crustaceans graze on the photosynthesizing microorganisms, while birds, fish, and invertebrates feed on the larger plants. Fish also eat one another. Despite the efficient aquatic food web in lakes, organic matter falls to the bottom as detritus, where it is consumed by bacteria and in-

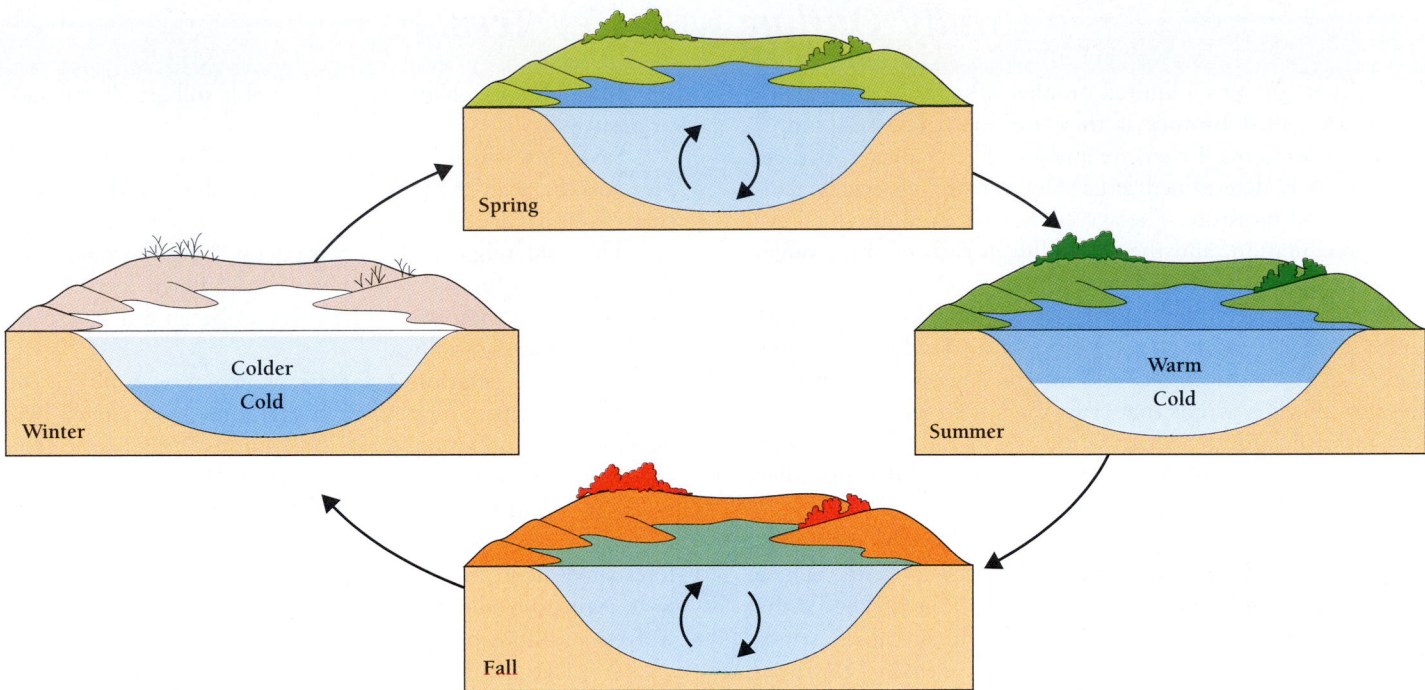

Figure 26-23 Turnover. Deep temperate lakes turn over twice a year. Both turnovers bring nutrients to the surface of the lake, resulting in blooms of algae. Water's highest density occurs at 4°C. In spring, when sun-warmed ice (0°C) melts and reaches 4°C, it sinks below the colder layers, carrying oxygen downward and driving nutrients upward.

vertebrates. Far more energy flows through the detritus food chains than through the chains involving grazing herbivores and carnivorous fish.

In temperate regions, a lake's surface temperature varies greatly from season to season. In the summer, the warm air heats a layer of warm water as much as 20 meters thick (Figure 26-23). As winter approaches, the temperature of the surface layer falls until it is the same as the deeper layer. At that point, the waters of the two layers mix—an event called the *fall turnover.*

A similar turnover occurs in the spring, as the winter ice melts. The two turnovers bring dissolved nutrients from the bottom sediments into the upper waters, where photosynthesis occurs. The result is a seasonal growth of algae. The fall and spring turnovers also move oxygen from the surface down to the bottom-dwelling heterotrophs, which allows them to metabolize settled organic material.

A **eutrophic** lake is one that has abundant minerals and organic matter. The surfaces of eutrophic lakes tend to be clogged with algae and cyanobacteria. As these organisms die and drop to the bottom, they stimulate the lake-bottom decomposers to multiply. During the summer, the decomposers so deplete the deeper layer of oxygen that the lake bottom can no longer support fish and other animals. As in the "dead zone"

mentioned in the last chapter, few organisms can live in a eutrophic lake. In contrast, **oligotrophic** lakes have a limited supply of nutrients. These clear lakes have no permanent algal blooms, and they therefore contain more oxygen and support a more diverse community of organisms, including fish.

Eutrophication happens naturally over thousands of years, but humans often accelerate the process by dumping nutrient-rich sewage and other wastes into streams and lakes. Productivity increases at first, but diversity declines, and eventually the lake dies.

Because rain washes both nutrients and poisons into rivers and streams, freshwater ecosystems easily accumulate pollutants.

We have seen how ecosystems move energy and materials and what kinds of ecosystems exist. But ecosystems are made up of individual species. So we should also consider the characteristic roles each species plays in the flow of energy and the use of materials. In the next chapter, we will see how species interact and how ecosystems change in time. We will also examine the role of evolutionary history and current competition in establishing a community's structure.

Study Outline with Key Terms

Our planet contains a limited number of types of ecosystems. These are called **biomes** if they are terrestrial, and simply aquatic ecosystems if they are aquatic. The character of these ecosystems is determined largely by climate, especially temperature and moisture. The average temperature, the total annual precipitation, and the variability in each of these values help determine the character of a biome.

At the equator, hot, moist air rises and flows north and south over the **tropics,** dropping rain. **Tradewinds** blow cooler air from the tropics toward the equator. The **westerlies** are cool, dry winds that cross the Earth's surface between 30° and 60° latitude. These winds warm and pick up moisture as they go. In the Arctic, cold, dry **polar easterlies** blow from the east.

The climate of a region is determined by the total solar energy absorbed and by wind and ocean currents. These depend on latitude and the rotation of the Earth. The Earth's rotation creates the **Coriolis effect,** which causes winds moving north or south to rotate, clockwise in the Northern Hemisphere and counterclockwise in the Southern Hemisphere.

Mountain ranges create **rain shadows** by forcing warm, moist air to drop most of its moisture on one side of the range. Biomes at high elevations resemble those at high latitudes. At high altitudes, **coniferous forest** gives way to **tundra** at the tree line.

Deserts, which receive less than 25 centimeters of precipitation each year, are characterized by widely spaced drought-tolerant plants, such as sage brush (in cold regions) and cacti (in hot regions). Some animals **estivate** during the hot, dry summers. The greatest human threats to deserts are off-road vehicles and agriculture.

The cold, dry tundra, which receives less than 25 centimeters of precipitation each year, is characterized by short, open grassland, punctuated by bogs, ponds, and lakes. **Permafrost** underlies the soil all year long.

Temperate grasslands, which receive 25 to 75 centimeters of precipitation each year, are characterized by tall grasses in wet areas and short grasses in drier areas. Fire, insufficient water, and grazing animals all keep trees from becoming established in this biome. Much of the world's grassland, midwestern North America, for example, has been converted to agriculture, thus eliminating entire food chains.

Chaparral regions, which have mild, wet winters and hot, dry summers, are characterized by dense shrubs and scattered, broad-leafed trees, mostly drought and fire tolerant. Humans build homes in the chaparral, then suppress natural brush fires so long that when fire finally comes, fuel loads create a firestorm. Houses, chaparral, and valuable soil are all damaged or destroyed.

Savannas, which receive 90 to 150 centimeters of rain each year, are characterized by tall grasses and short, widely spaced trees. Savannas are hotter and wetter than temperate grasslands.

The cold **taiga** is characterized by short, homogeneous pine or spruce forests. This biome has suffered less human-caused damage than most other biomes because it has so little to offer economically.

Temperate **deciduous forests,** which receive 80 to 140 centimeters of precipitation each year, are characterized by a heavy cover of deciduous trees and a limited understory. Temperate deciduous forests, found in eastern North America, eastern China, and Europe, are commonly associated with dense populations of humans. These populations have cleared these forests repeatedly and extensively, and hunted to extinction or near extinction all of the major predators, including bears, wolves, and lions or tigers.

Tropical rain forests, which receive 200 to 450 centimeters of precipitation each year, are characterized by a dense **canopy** with a tall understory and little vegetation on the dim forest floor. **Lianas** and **epiphytes** grow on the trees, and **emergents** rise above the canopy. Tropical countries are cutting their rain forests down at a rate that will eliminate this biome within a few decades.

Oceanography has revealed the depth and character of the oceans and the organisms that live in them. **Ocean ecosystems** can be divided into **benthic** and **pelagic divisions.** The benthic division includes the **intertidal** and the **abyssal zones.** The pelagic division includes the **neritic** and **oceanic zones.** **Plankton** float freely and travel passively only. Fish and other large animals propel themselves. Ocean food chains can have more trophic levels than terrestrial food chains, but the total number of species living in the oceans is far less than the number that live on land. Primary productivity is low, except in areas of **upwelling.**

Estuaries are marshes with partly fresh, partly salt waters. Estuaries are extremely productive and provide breeding grounds for many oceanic species. Humans have a long history of draining marshes and estuaries for farms or developments.

The study of freshwater lakes, ponds, rivers, and streams is called **limnology,** one of the oldest branches of ecology. Lakes that are rich in organic matter are **eutrophic.** Those lower in nutrients support a more diverse community and are said to be **oligotrophic.** Humans frequently pollute fresh water with sewage and industrial wastes, as well as the acid rain that results from burning coal and oil.

Review and Thought Questions

Review Questions

1. Explain why the polar regions experience more extreme seasons than the equatorial regions.
2. What kind of biome exists in a rain shadow? Explain why.
3. How might the division of a forest or other habitat into small sections interfere with top predators?
4. What keeps grasslands from changing to forests?
5. Explain the difference between a liana and an epiphyte. Why would tropical plants make good house plants?
6. What percentage of the Earth's surface is covered by water?
7. Why are estuaries ecologically and economically important?
8. What organisms are the primary producers in the oceanic zone?

What organisms are the primary producers in the neritic zone?

9. What causes fall turnover, and what benefits does turnover have for freshwater organisms?
10. What causes a lake or pond to become eutrophic?

Thought Questions

11. Explain why the soils of tropical rain forests are poor, even though tropical rain forests are so productive.
12. Why do tropical rain forests have so many species?
13. Why are the oceans species poor?

About the Chapter-Opening Image

Nothing says "desert" more clearly than an organ pipe cactus. Every biome includes characteristic species. Desert biomes, for example, often host species of plants that store water and use spines to defend themselves against thirsty herbivores.

 On-line materials relating to this chapter are on the World Wide Web at **http://www.harcourtcollege.com/lifesci/aal2/**

27 Communities: How Do Species Interact?

Volcano: An Ecosystem Is Destroyed

At 8:32 a.m. on May 18, 1980, a medium-sized earthquake shook Mount St. Helens in Washington State. Twenty seconds later, the volcano erupted, and the entire north side of the mountain slid away in a thundering roar. It was the largest landslide in recorded history. Explosions from the volcanic eruption tore through the 2.5 cubic kilometers (0.6 cubic miles) of sliding debris, spewing rocks, ash, gas, and steam across adjacent valleys at velocities approaching the speed of sound. The series of blasts, 500 times as powerful as the bomb that flattened Hiroshima, destroyed an area the size of the city of Chicago—a forest with enough timber to build 300,000 houses. Every tree within 25 kilometers (15 miles) was either blown to bits, burnt to a crisp by 600°C gases, buried under tons of volcanic ash, or blown down. From the north face of the mountain poured a river of melted snow and ice, mud, boulders, and ash. In less than 10 minutes, this colossal river of rocks filled a valley the size of Manhattan to an average depth of 45 meters (deep enough to bury a 15-story building). A black plume of pumice and ash rose 25 kilometers into the air, and 520 million tons of ash began drifting eastward across the United States.

The death toll included an estimated 11 million fish, 27,000 grouse, 11,000 hares, 6000 black-tailed deer, 5200 elk, 1400 coyotes, 300 bobcats, 200 black bears, 57 human beings, and 15 mountain lions. Three more explosions in May, June, and July dropped thick blankets of pumice and ash. In the end, every exposed slope within 10 kilometers (6 miles) of the crater was stripped of trees and other vegetation and buried under 2 meters of ash and rock. In places, the mud, ash, and rock piled up 200 meters (an eighth of a mile) thick. The scene was as gray as a moonscape. For another 16 kilometers (10 miles) from the blast zone, thousands of full-grown fir and hemlock trunks lay flattened in the ash. It was hard to imagine how anything could have survived.

Yet, an enormous amount of life did survive (Figure 27-1). Along snow-covered ridges that faced away from the blast, saplings still bent by the weight of 1 or 2 meters of winter snow, as well as seedlings and shrubs buried under the snow, came through unscathed. Shrubs that were obliterated at the surface, but whose roots survived underground, sprouted and grew anew. From underground, burrowing animals such as pocket gophers and a variety of insects and other invertebrates emerged. Within weeks, plants began growing on the naked slopes of the mountain, wherever rain had washed away the sterile ash to expose the old soil beneath.

Within 10 years, the slopes of Mount St. Helens were covered with dogwood, elder, huckleberry, and other shrubs, as well as seedlings of fir, hemlock, and other conifer trees. Ecologists say that by 2030, the area will begin to look like a young coniferous forest. In 200 years, the area should resemble an old-growth forest. The Mount St. Helens ecosystem—almost completely destroyed—will rebuild itself.

- Over time, communities undergo predictable changes in species composition, a process called succession.

- Communities of species are both integrated and individual.

- Relationships among species, such as symbiosis, competition, and predation, coevolve over time.

- No two species can occupy the exact same niche and habitat indefinitely.

ARE COMMUNITIES INTEGRATED OR INDIVIDUALISTIC?

The process by which Mount St. Helens renews itself is called **succession,** which is the predictable change in numbers and kinds of organisms in a community over time. How and why succession occurs has been one of the burning questions in ecology for more than a hundred years.

In Chapter 25 we discussed Lynn Margulis's view that our biosphere is a tightly integrated whole that is composed of organisms that cooperate as well as compete. Surprisingly, the philosophical differences that fueled the debate over Gaia are old issues in the science of ecology. Do species cooperate to create a cohesive, functioning environment? Or are species free agents that operate independently of other species? For nearly a hundred years, ecologists have been arguing about whether different species form "integrated communities" or whether each species is "individualistic."

It all began in 1898, when one of America's first ecologists, Henry C. Cowles (1869–1933), published a groundbreaking paper on plant succession. Cowles was a warm, funny man of infectious high spirits who inspired generations of young ecologists. As a graduate student, he studied the succession of plants on sand dunes beside Lake Michigan.

Cowles saw that, over thousands of years, the water level of Lake Michigan has gradually dropped, leaving behind a broad, sandy shore.

Fierce winds blow the sand from the former lake bottom into giant dunes. Grasses cover the dunes, growing long roots that stabilize the sand. Soon bearberry, juniper, and other shrubs grow over the dune, displacing the grass. Later on, Cowles found, trees such as jack pine grow among the shrubs. The pines create a moist, shady habitat that kills the shrubs and encourages the growth of black oak trees. Finally, in some places, the black oak forest gives way to beech and maple forests.

This sequence of communities started right by the lake, where the newest sand dunes lay, and ended far up into the woods, where the oldest forests had grown up (Figure 27-2). In every place around the lake, the same series of changes was taking place. Cowles showed that natural communities can change predictably over time. But like the shrubs on the Michigan dunes, Cowles was soon overshadowed by a stronger personality.

Succession's most important booster was Frederick Edward Clements (1874–1945). Clements was Cowles's opposite—arrogant, distant, and ascetic. But he was a prolific researcher and writer all of his life. In one of the earliest automobiles, and before the construction of paved highways, Clements and his wife drove thousands of miles across the roadless western United States, camping and studying. During his life, Clements studied and wrote about every major ecosystem in North America. In 1916, he published the definitive book on plant succession. Clements argued that in every geographic area, as defined by climate and soil, succession always created the same predictable community of plants, the **climax community.**

Clements, sounding rather like James Lovelock of Chapter 25, further argued that

Figure 27-1   Life returns. Mt. St. Helens with Spirit Lake in the foreground before the eruption in 1980 (top photo). Mt. St. Helens in 1994 with Spirit Lake (bottom photo). By 1994, plants had begun to clothe even some of the most devastated areas. Ecologists predict that Mt. St. Helens' ecosystem will return to mature coniferous forest in about 200 years. *(Top, Tom and Pat Leeson/Photo Researchers, Inc.; bottom, John Marshall/Tony Stone Images)*

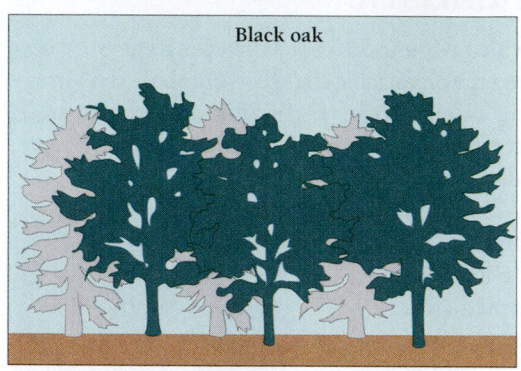

Figure 27-2   Sand dune succession. Henry C. Cowles showed that as the water level in Lake Michigan drops, the sand dunes left behind are gradually colonized by a succession of different plant communities. Beach grasses stabilize sand dunes, allowing the growth of shrubs and the formation of soil. Jack pines grow up through the shrubs, shading the soil. In the cool moist soil, oak trees germinate and grow, eventually shading out the jack pines. Oak forest may, in places, give way to beech and maple forest.

the climax community was an organism. Succession, he said, was a developmental process, like the embryonic development of an organism. Clements wrote, "As an organism, the [climax] formation arises, grows, matures, and dies . . . each climax formation is able to reproduce itself, repeating . . . the stages of its development."

Clements's description is of an **integrated** community, one that consists of characteristic species that always interact with each other in predictable ways. In this view, all maple tree forests, for example, would have approximately the same species of plants and animals, and all these species would interact in similar ways.

Other ecologists found the integrated view compelling. Anyone can see that deserts usually contain cactuses, small mammals, lizards and snakes, while ponds usually contain water plants, frogs, fish, and mosquitoes. Clements's view—that a community is an integrated collection of interdependent species—is one of the most important philosophical themes in ecology. But its simplicity is also its weakness. It invites criticism.

In 1926, a less-famous American ecologist named Henry Allan Gleason published a paper frankly challenging Clements. Gleason argued that communities are not integrated, but composed of separate populations that merely happen to occupy the same habitat. Provocatively, Gleason wrote in part, "In conclusion, it may be said that every species of plant is a law unto itself. . . . [A plant] association is not an organism, scarcely even a vegetational unit, but merely a coincidence." In Gleason's view, communities were **individualistic.**

Gleason's impolite attack on the much-revered Clements so embarrassed other ecologists that most refused to even discuss Gleason's heretical ideas. As Gleason later recalled, "Not one believed my ideas; not one would even argue the matter. . . . For ten years, . . . I was an ecological outlaw."

In time, though, ecologists began to take Gleason's ideas seriously. A healthy and productive scientific debate ensued between Clements's holistic ecologists and Gleason's individualistic ones. Gleason and other individualists argued that every species has an independent distribution, and that, in effect, every community is unique.

The individualistic view garners support from the many careful studies of species distribution patterns that show no obvious clusterings of species. A survey of Wisconsin forests, for example, showed that no two tree species consistently share the same range (Figure 27-3A). Maple trees may be associated with beech trees in one place, elms in another. In the individualistic view, a beech-maple forest is not an organic entity or community, but two species that just happen to share the same moisture requirements. Today, few ecologists accept Clements's idea of a climax community; most research shows that the species composition of forests and other communities varies continuously.

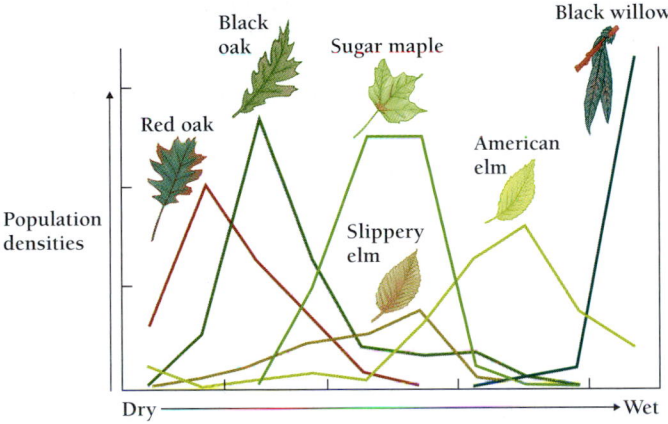

A.

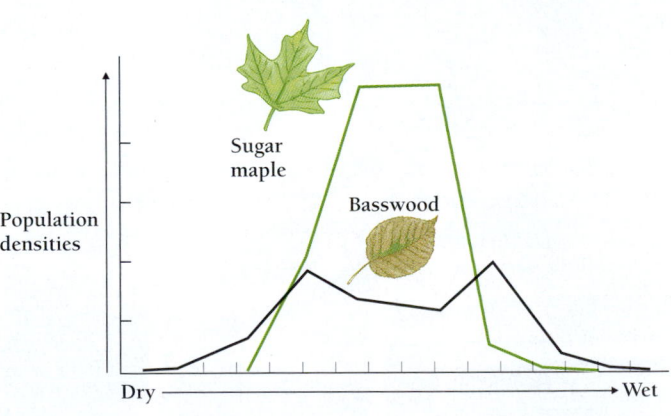

B.

Figure 27-3   Are communities individualistic or integrated? Studies of trees in Wisconsin support both ideas. A. Oaks, maples, and other trees are distributed according to the dampness of the soil, independently of other trees. B. But basswood grows best where sugar maple is absent, even though it is capable of growing in the same soils.

On the other hand, the integrated view of ecological communities rests firmly on thousands of studies that demonstrate interactions among species. To take an obvious example, most woodpeckers and termites could not easily survive in a treeless grassland. Nor could certain plants survive without the fungal mycorrhizae that help them draw nutrients from the soil (Chapter 21). But more subtle interactions also determine who lives where. For example, the presence of sugar maples appears to limit the distribution of basswood trees, perhaps by competition for water or shading (Figure 27-3B). So, even though the unity of a community may not be as well defined as Clements envisioned, ecologists still think of communities as functioning units whose member species all interact.

The integrated and the individualistic views are not mutually exclusive ideas, but the extreme ends of a continuum. Just as the members of a family sometimes behave as individuals and sometimes behave as an integrated whole, the species that make up a community can be both individualistic and integrated. A community is both an integrated set of interacting species and a group of organisms that occur together because their ranges happen to overlap. In the next section, we will examine a few of the ways in which different organisms in a community interact.

The individualistic and integrationist views of ecosystems represent extreme ends of a continuum of thought.

HOW DO SPECIES IN A COMMUNITY INTERACT?

The general nature of a biome depends on abiotic factors such as temperature, moisture, and sunlight, as we saw in Chapter 26. Annual rainfall, for example, largely determines whether an area is a forest or a desert. Abiotic factors alone do not define a community, however. Within each community, the interactions of organisms with one another help to establish the character of a community.

These interactions include those between the consumers and the consumed and between competitors. Interactions between the consumers and the consumed include predator-prey interactions, herbivore-plant relations, and parasitism. The wolf hunts the caribou, and the caribou eats the lichen. Species may also compete—for territory, prey, or other resources—or they may form intimate associations, in which one species benefits from another or where two species help each other. Lichen, for example, turns out to be not one organism, but two, a cooperative association between a fungus and an alga.

Who Eats Whom?

We usually think of consumers as being larger and stronger than the consumed. Hawks, for example, generally prey on mice, squirrels, and other animals smaller than themselves.

Most predators are indeed larger than their prey, although carnivores that hunt in groups, such as wolves, regularly bring down prey far larger than themselves. But herbivores are often smaller than the individuals they consume. Hordes of tiny oak moth caterpillars can strip all the leaves from a 60-foot oak tree. And parasitic consumers, such as ticks, fleas, and tapeworms, are generally smaller than their hosts.

Consumers, then, come in many forms. The main consumers besides the decomposers and scavengers are herbivores, predators, and parasites. A **parasite,** by definition, consumes only part of the blood or tissues of its host and does not necessarily kill the host. In contrast, a **predator** usually (but not always) kills its prey and consumes most of its prey's body.

In some cases, the differences between predators and parasites may blur. For example, parasites do occasionally kill their host outright, and parasites frequently weaken the host so severely that the host dies of other causes. And some predators consume only part of their prey. For example, some predaceous fish bite chunks of flesh from other fish, without necessarily killing the victim.

Herbivores at first seem to fall outside these definitions, but they actually fit rather nicely. Leaf-eating caterpillars and deer, for example, browse on living plants, just like parasites. And seed-eating birds, which kill and consume an entire plant embryo, may be considered predators. In reality, ecologists rarely stress the parasite-predator distinction in herbivores.

How Do Plants and Animals Defend Themselves?

With notable exceptions, consumers' adaptations for finding and eating their favorite food exhibit a certain sameness: in the case of predators, swiftness, intelligence, acute senses, and sharp teeth; in the case of herbivores, patience and a good digestive system. In contrast, organisms have a diverse arsenal of defenses against potential consumers.

Camouflage

The subtlest defense against being eaten is to remain invisible. The rabbit, for example, simply dives down a hole or freezes, so that it blends into a background of shrubs or grass. But many organisms go to far greater lengths to **camouflage** themselves by mimicking twigs, leaves, stones, bark, and other materials in their environment (Figure 27-4). A walking stick insect, for example, looks just like a twig. Some species even sway slightly, as though an errant breeze were blowing.

Chemical defense

Plants defend themselves against consumers in a variety of ways. Trees protect themselves from termites and other wood-eating creatures with thick layers of bark. Roses and cactuses fend off herbivores with thorns. The single most important defense that plants use, however, is toxic chemicals. The unforgettable wallop of hot chili pepper, the sharp tang of raw

Figure 27-4   Camouflage. Many animals avoid being eaten by blending in with their background. This harlequin crab lives on the underside of a sea cucumber, camouflaged against the background of its host. (*Denise Tackett/Tom Stack & Associates*)

broccoli, and the deadly toxin in poison hemlock all result from certain kinds of **secondary plant compounds**—chemicals that are not essential to a plant's normal metabolism but that often serve a defensive purpose.

Chemical defense is not limited to plants. Bees, ants, and wasps inject a powerful acid into attackers. The bombardier beetle and the skunk both spray would-be predators with noxious chemicals (Figure 27-5). The famous poison-dart frog (*Phyllobates*) of South America and the skin and feathers of the pitohui bird of New Guinea contain the same deadly toxin (Figure 27-6).

Figure 27-5   Chemical warfare. Everyone knows that skunks fight back with noxious chemicals. So do many plants and insects, such as this bombardier beetle. (*Thomas Eisner and Daniel Aneshansley/Cornell University*)

Figure 27-6 **Warning coloration in a pitohui bird?** In 1992, researchers discovered three species of birds in the genus *Pitohui*, whose skin and plumage contained the same toxin as that found in poison-dart frogs of South America. Like wasps, skunks, and poison-dart frogs, pitohui birds bear striking colors that may function to warn predators. Research has shown that inconspicuously colored birds taste better than flashy ones. *(W. Peckover/VIREO)*

Warning coloration

If a skunk looked like a rabbit, it would be attacked over and over again, in spite of its potent defense. But because a skunk has a striking and memorable black-and-white pattern, predators quickly learn to avoid the animal. Among all animals, chemical deterrents work best if accompanied by a bright, memorable design called **warning coloration.** Warning coloration, such as the yellow-and-black stripes of the aggressive, stinging yellow jacket, helps predators remember which prey to avoid. Predators will avoid any animal with the colors and patterns they associate with pain, illness, or other unpleasant experiences. Most warning colors are vivid shades of red, yellow, orange, and white, combined with a contrasting shade of blue or black.

Because predators avoid animals with warning coloration, some species copy the warning coloration of toxic or dangerous species. A situation in which one animal *mimics* the color pattern of another is called **mimicry.** Some perfectly harmless and quite edible flies, for example, mimic the black-and-yellow stripes of yellow jackets and bees. A situation in which a harmless animal imitates a dangerous one is called **Batesian mimicry.**

In a second form of mimicry, called **Müllerian mimicry,** two or more dangerous species evolve similar colors. The similar colors signal a similar hazard to any predators the two mimics share in common. For example, the monarch butterfly is toxic to birds. The monarch caterpillar feeds on milkweed, a plant that contains a toxic secondary plant compound. This toxin accumulates in both the monarch caterpillar and butterfly and makes it toxic to birds. A bird that eats a monarch

quickly becomes ill (Figure 27-7A). After a few such experiences, the bird avoids monarchs, and the monarch's striking warning colors help the bird remember.

A second butterfly, the viceroy, is unrelated to the monarch but also contains a toxin that makes birds sick. The viceroy looks remarkably like a monarch, and biologists believe these two butterflies have evolved to resemble one another (Figure 27-7B). The advantage to each butterfly is that birds that have learned to avoid the monarch also avoid the viceroy and birds that avoid the viceroy also avoid the monarch.

Müllerian mimicry helps predators learn the same lesson from a variety of prey species. Bees and wasps both use a similar pattern of yellow and black stripes. Anyone who has been stung by a bee also learns to avoid wasps and hornets. Such encounters lead to greater safety for bees, wasps, and hornets. The more species that use these kinds of patterns and colors, the more this universal signal is reinforced.

A.

B.

Figure 27-7 **Müllerian mimicry.** A. The monarch butterfly is so noxious that birds such as this jay quickly cough them up. B. Viceroy butterflies are unrelated but also toxic. Monarchs (*left*) and viceroys (*right*) have evolved amazingly similar color patterns. Each butterfly's message reinforces the other's: "Don't touch!" *(A, Lincoln Brower; B, Thomas C. Emmel)*

TABLE 27-1   INTERACTIONS AMONG ORGANISMS

Interaction	Effect on A	Effect on B	Example
Competition between A and B	Harmful	Harmful	Otter and mink
Predation by A on B	Beneficial	Harmful	Lion and gazelle
Symbiosis			
Parasitism by A on B	Beneficial	Harmful	Flea and dog
Commensalism of A with B	Beneficial	Harmless	Fish and anemone
Mutualism between A and B	Beneficial	Beneficial	Lichen

Symbiosis: How Do Organisms Live Together?

A species may live in an intimate association with another species, an arrangement called **symbiosis,** meaning literally "living together." Symbiosis may be cooperative or antagonistic. Table 27-1 summarizes some of the possible interactions between any two species. Each species may affect the other positively, negatively, or not at all. In **parasitism,** one species benefits at another's expense. Tiny parasitic wasps, for example, lay their eggs inside of caterpillars. The wasp larvae literally eat the caterpillars alive as both animals mature.

Mutualism describes a symbiotic relationship between two species that mutually benefits each. For example, the bull's horn acacia (*Acacia cornigera*) of Central America has special enlarged thorns that house colonies of ants (*Pseudomyrmex ferruginea*). The acacia's nectar attracts the ants, which use it as their major food source. In return, the ants protect the acacia from being eaten by caterpillars and other herbivorous insects. This arrangement benefits both species. One study showed that acacias grown without ants for 10 months weighed less than one-tenth of those with intact ant colonies.

Another kind of symbiosis is **commensalism,** an intimate relationship between two species that helps one but neither helps nor harms the other (Figure 27-8). For example, the anemone fish lives safely among the poisonous tentacles of sea anemones. The fish, which benefits from the protection of its host, is a commensal. For reasons unknown, the sea anemone, which catches and eats other species of fish, leaves the anemone fish alone.

Species interact as competitors, as predator and prey, and as symbionts.

Do Organisms Coevolve?

The mutualistic relationship between ants and bull's horn acacias demands special adaptations in each species. The acacia has evolved enlarged thorns with a tough, woody exterior and a soft, pithy center, easy for the ants to excavate for nests. The acacia also has "nectaries" at the bases and tips of its leaves to supply the ants with nectar. The ants exhibit special adapta-

tions as well. While most species of ants are active only at night or only during the day, members of the species *Pseudomyrmex ferruginea* remain active and protect the acacia day and night.

These distinctive adaptations in the ant and the acacia appear to result from a long and mutual evolutionary history in which the needs of each species exert selective pressure on the other. The interdependent evolution of two or more species whose adaptations appear to be selected by mutual ecological interactions is called **coevolution.**

Biologists love to suggest likely examples of coevolution. For example, many pollinating insects specialize on flowers that seem, in turn, to be specially constructed to be pollinated by just one kind of insect. Mutualisms such as these present the most likely examples of coevolution. But demonstrating that coevolution has occurred in individual cases is difficult.

Further, all species that have ecological interactions may exert some selective pressure on one another. And virtually every species in a community interacts (however indirectly) with every other species. For example, the bacteria that break down organic matter to nutrients that plants can absorb through their roots are as essential to the herbivores and carnivores as to the plants. Coevolved relationships can be either very specific, as in the case of the acacias and the ants, or very

Figure 27-8 Commensalism. The cleaner shrimp lives safely near the mouth of this predaceous moray eel. (*Mike Severns/Tony Stone Images*)

diffuse. The concept of "diffuse coevolution" is another way of acknowledging that all organisms in a community affect one another.

Interactions between two species influence the evolution of both.

How Does Competition Affect Organisms?

Some species interact as consumers and consumed, and others as commensals or mutualists. Still others compete. In **competition,** one organism uses a resource in a way that limits the availability of that resource to others. For example, if two species of deer both eat the same herbs and live in the same area, and the herbs are in limited supply, the deer are in competition with each other.

Can two similar species share the same habitat?

Measuring competition in natural communities takes careful planning and months or years of research. But laboratory studies can quickly show how competition occurs in simple situations. The Russian ecologist G.F. Gause studied populations of two similar species of *Paramecium*—*P. caudatum* and *P. aurelia.* Each *Paramecium* species, cultured by itself, flourished on the bacterial food that Gause provided. But when Gause mixed the two species, *P. caudatum* disappeared from the culture within 2 weeks. Gause provided only one environment and one kind of food for the two species, and *P. aurelia* was evidently better adapted to Gause's conditions than was *P. caudatum.* One species thus replaced the other.

These and similar experiments have led to a generalization called the **competitive exclusion principle,** which says that when two species compete directly for exactly a resource that is in short supply, one species can eliminate the other. A resource in short supply is a **limiting resource.** For example, territories might be a limiting resource for a pride of lions, but air would not be. Similarly, water is often a limiting resource for terrestrial plants but not for aquatic plants.

The competitive exclusion principle predicts that one species always wins. But the same species does not win under all conditions. For example, in a 40-year series of experiments at the University of Chicago, Thomas Park and his colleagues studied the effects of different environments on competition between two species of grain beetles (*Tribolium castaneum* and *Tribolium confusum*). These beetles were allowed to grow in containers of flour, the favorite food of the larvae. When grown separately, both species did best in warm, moist conditions. When grown together in warm, moist containers, *T. castaneum* drove *T. confusum* to extinction. In a cold, dry container, however, *T. confusum* prevailed.

According to the competitive exclusion principle, no two species can simultaneously exploit the same resources in the same place. The place in which an organism lives is called its **habitat.** In contrast, the way an organism uses its environment is called its **niche.** The habitat of the desert tortoise is the desert. The niche that the tortoise occupies is that of burrowing herbivore.

The competitive exclusion principle states that no two species can indefinitely occupy the exact same niche in the same habitat.

How do competing species coexist?

In the laboratory, species can be made to compete for the same limiting resources. But in nature, two species occupying the same habitat can, in practice, split up the niche in an infinite number of ways. For example, the ecologist Robert H. MacArthur (1930–1972) tested Gause's competitive exclusion principle in field studies of five species of North American warblers. All five bird species hunt for insects and nest in the same kind of spruce tree. The niches of these five birds seem nearly identical, but MacArthur's studies showed that the five warblers had split up the spruce tree niche extremely finely. He showed that each species of bird hunts insects in a different part of the tree and nests at a separate time from the other warblers. As a result, the warblers' niches overlap without being identical. Ecologists call splitting a niche **resource partitioning.**

Another study, on barnacles, shows how two species go about splitting a niche. In tidepools on the coast of Scotland live two species of barnacles. One species, *Balanus balanoides,* lives only on the lowest intertidal rocks, which are wet even during low tide. The other species, *Chthamalus stellatus,* lives higher up, in between the low-tide and high-tide lines.

To find out if the two species were competing, the ecologist Joseph H. Connell removed *Balanus* from some tidepools and *Chthamalus* from others (Figure 27-9). Connell found that where *Balanus* was removed, *Chthamalus* moved down to occupy the newly opened territory and survived there perfectly well. Likewise, where *Chthamalus* was removed, *Balanus* was able to survive on the higher, drier rocks.

Balanus dominated the lower levels because its heavy shell grows under the shell of *Chthamalus* and pries it up off of the rocks, preventing *Chthamalus* from growing on the lowest rocks. Higher up, however, *Chthamalus's* higher tolerance for dry conditions allows it to prevail over the physically stronger *Balanus.*

Connell's two barnacle species illustrate the difference between a species's **fundamental niche,** its potential ability to utilize resources, and its **realized niche,** the resources that it actually uses in a particular community. The fundamental niche of *Chthamalus* included the lower rocks, but its realized niche did not. The fundamental niche of *Balanus* included the higher rocks, but its realized niche did not.

Similar experiments on rat parasites also illustrate competitive exclusion. Although the tapeworm and the spiny-headed worm both prefer the upper intestine of the rat, when the two worms are together, the spiny-headed worm takes the

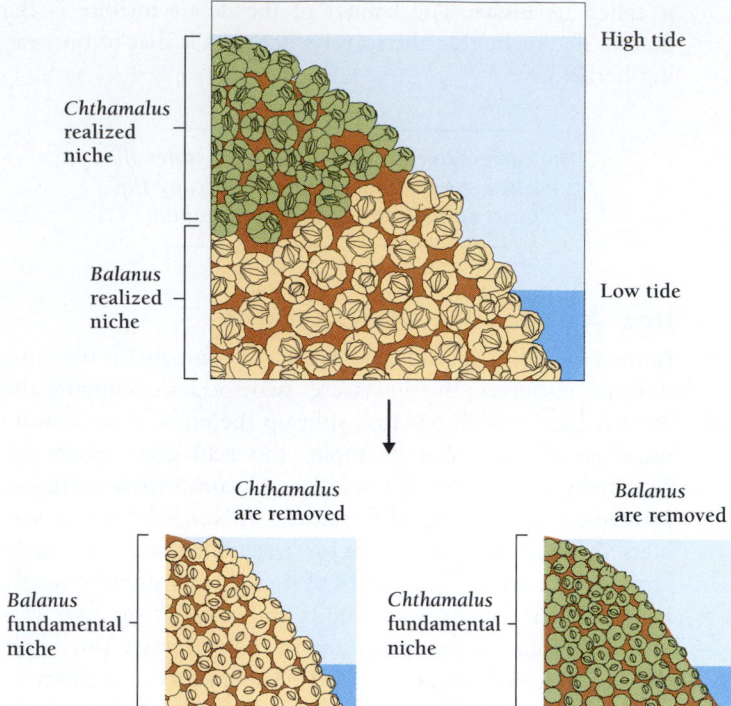

Figure 27-9   Niche splitting in barnacles. In a classic experiment, the ecologist Joseph Connell showed that the two barnacles *Balanus* and *Chthamalus* are capable of growing on the same rocks if the other species is removed. Otherwise they split the tidal niche, *Balanus* below and *Chthamalus* above.

food-rich upper tract and displaces the tapeworm to the lower end of the intestine. All these experiments, and many more like them, suggest that competition in natural populations determines species distributions.

Competing organisms in natural environments may partition resources so that both species persist in the same habitat.

Character displacement: Does competition influence evolution?

We have seen how competition between two species can lead to niche splitting. Species may evolve differences that make niche-splitting strategies permanent. For example, two species of caterpillars that both eat members of the rose family (which includes roses, apples, plums, and other species) might reduce competition by specializing, one on apple trees and one on roses. Any change in morphology, life history, or behavior that results from competition is called **character displacement.**

The finches of the Galápagos Islands provide some of the best-documented examples of character displacement. In the case of the Galápagos finches, character displacement has led to speciation. For example, the small ground finch *Geospiza fuliginosa* and the medium ground finch *Geospiza fortis* occur

together on some islands, while on other islands only one of each species lives. On islands where these birds occur separately, both species's beaks are medium sized, and they eat similar-sized seeds. But on islands where the two birds occur together, *G. fortis* individuals have larger beaks and specialize on larger seeds than do *G. fuliginosa* individuals (Figure 27-10). In other words, when the two species occur together,

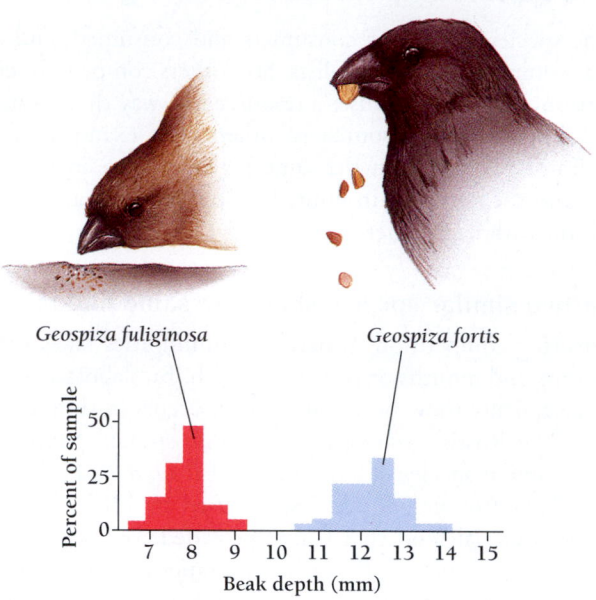

A. Two species together

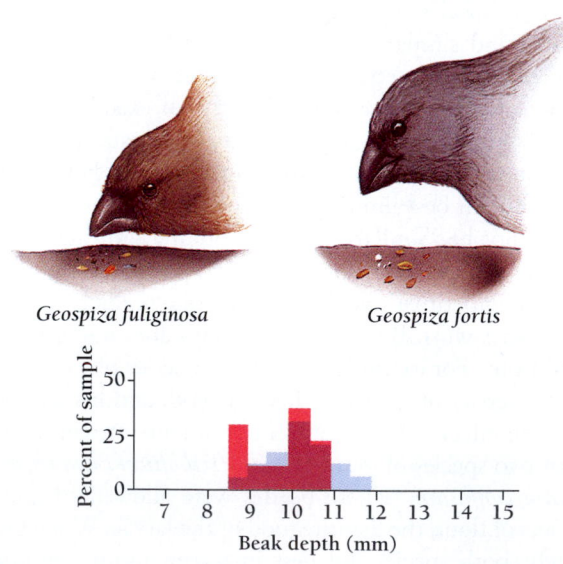

B. Two species separate

Figure 27-10   Character displacement in Galápagos finches. A. On islands where the two species *Geospiza fuliginosa* and *Geospiza fortis* occur together, members of *G. fortis* have much larger beaks and eat larger seeds than do *G. fuliginosa*. B. On islands where the birds occur separately, individuals of each species have beaks of similar size.

they split the niche and specialize on either big seeds or small ones. When they occur separately, they both take the middle ground.

Recent studies have shown that the medium-sized beak is the best option under some circumstances. On islands where the two species occur together, they often hybridize, producing offspring with intermediate-sized beaks. When food is scarce (in times of drought), the hybrids reproduce poorly. When seeds are plentiful, however, the hybrids reproduce *better* than their parent species.

Competition and adversity appear to drive these two species apart, while lack of competition and easy pickings drive them together. Currently, they are separate species. Their distinctive adaptations almost certainly have resulted from previous competition. As Joseph Connell put it, character displacement is "the ghost of competition past."

Competition, over evolutionary time, causes character displacement through natural selection. Strong selection pressure can bring about speciation.

WHAT DETERMINES BIODIVERSITY?

We have seen that for two competing species to coexist, they must exploit different niches. Both resource partitioning and character displacement increase the number of species within an ecosystem.

Each species occupies a unique niche, and each niche has a unique set of properties, such as food supply, sunlight, and other resources. All the unique but overlapping niches fit together to form a kind of multidimensional jigsaw puzzle—the ecosystem. The number of niches is a measure of the richness and complexity of a community.

How Can We Measure Diversity?

The number of species in an ecosystem is called **species richness.** For example, a single river in Amazonia may contain more species of fish than all the rivers of Europe. But species richness is only a crude measure of diversity, since some species may occur in great numbers while others are rare. To take an extreme example, the city of San Diego has an unusually large and well-stocked zoo, and so the city's species richness is high. The vast majority of organisms living in San Diego, however, consist of only a few species of animals—mostly humans, rats, mice, cockroaches, and ants.

A better measure of diversity is **relative abundance,** which takes into account how common each species is. Ecologists use various mathematical summaries of relative abundance. Often, each species is assigned an "importance value" based on the number of individuals in an area, on biomass, or on primary productivity. The relative abundance of species in a tropical rain forest is extremely high. If we find 200 species of trees, most of them will be equally common. In contrast, a temperate forest would be dominated by just a few species. In one West Virginia forest, for example, more than 83 percent of all trees were either yellow poplars or sassafras.

Careful study of a particular ecosystem can reveal the number and kinds of species that compose it. But what determines the particular species makeup of a community? Do latitude, rainfall, and other abiotic components determine the diversity of a biotic community? Or does the biological makeup of each community depend on its unique history—that is, on which plants and animals happen to arrive in what order? Are there any general rules that determine the diversity within a community?

How Do Biotic Factors Influence Diversity?

We have seen how competition and the resulting character displacement can increase the number of species in a community. Predators can also maintain diversity by preventing competitive exclusion. Some predators prevent common species from overwhelming less common ones. When one prey species becomes common, a predator will begin to prey preferentially on that species, switching from prey to prey, according to which is most common. Or the predator may simply keep the population of the dominant competitor at low enough densities that other species can survive. A predator that promotes diversity is called a **keystone species.**

Some ecologists have hypothesized that species diversity increases with productivity and stability. For example, a tropical rain forest has high diversity, enormous productivity, and a stable, predictable climate. On the other hand, some grasslands and deserts are unproductive and unstable, yet they are highly diverse. Overall, there is little correlation between productivity and diversity.

How Do Abiotic Factors Influence Diversity?

The relative importance of biotic and abiotic factors in ecosystem diversity is still a subject of ecological research. That abiotic factors play an important role, however, is certain. Species diversity generally increases from the poles to the equator, and from high elevations to low elevations (Figure 27-11). Physically varied areas (with hills, valleys, and streams) allow more unique niches than flat, featureless landscapes. Finally, the larger the size of an ecosystem, the more species it is likely to contain.

One way that ecologists have studied diversity is by studying the species diversity on islands. They have found that large islands, which have more habitat diversity and more niches, support more species than small islands.

Furthermore, the number of species on an island may remain stable—even as individual species appear and disappear. For example, in 1968, the Channel Islands, off the coast of southern California, had about the same number of bird species as they had had in 1917. But nearly one-third of the species were new.

In the 1960s, Robert MacArthur, and another American ecologist, Edward O. Wilson, argued that the diversity of

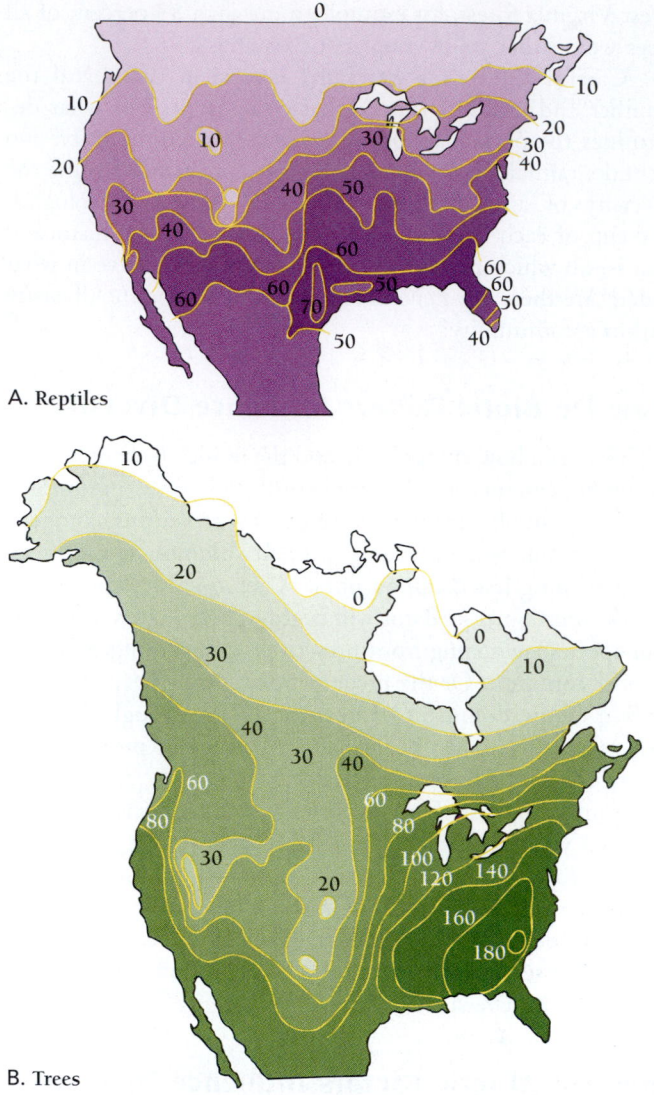

A. Reptiles

B. Trees

Figure 27-11 How does latitude affect species density? In North and Central America, the number of bird species increases in habitats closer to the equator. *(Source: R. L. Smith)*

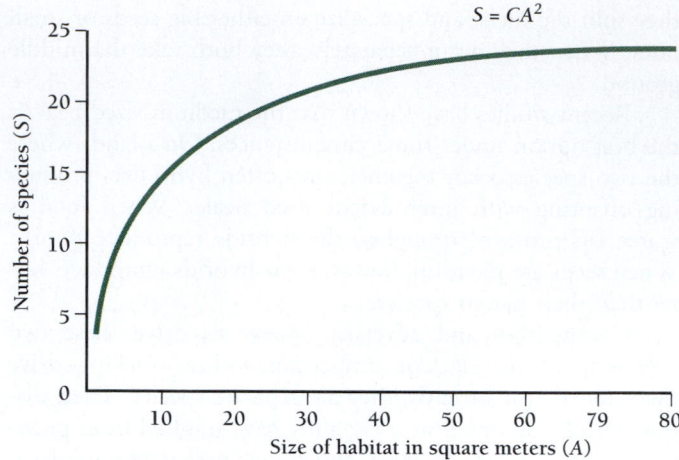

Figure 27-12 Larger habitats support more species. A graph of bird species on islands around New Guinea shows that larger ecosystems support more species than smaller ones. We say that species number (S) is proportional to area (A).

son said the number of species (*S*) could be predicted by the following equation:

$$S = CA^z$$

where *C* and *z* are constants and *A* is the area of the ecosystem or island (Figure 27-12).

In 1969, Wilson and ecologist Daniel Simberloff devised an experimental test of the MacArthur-Wilson theory. They counted all the arthropod species they could find on several small islands (10 to 20 meters in diameter) in Florida, exterminated all the arthropods (mainly insects) with a chemical pesticide, and then monitored the gradual recolonization of the islands. Within 6 months, the islands had recovered their former arthropod diversity (Table 27-2).

island species depends on a balance between the rates of immigration of new species and extinction of old species. According to the MacArthur-Wilson **theory of island biogeography,** a new island, devoid of already established species, should experience a high rate of immigration and a low rate of extinction, since there are fewer species to disappear. As more species arrive, the theory went, newer immigrants would have a harder time establishing themselves because of competition, and the rate of extinction would increase. MacArthur and Wilson also predicted that islands nearer the mainland would be easier for new immigrants to get to and should therefore have a greater rate of immigration and, therefore, more species than more distant islands.

Finally, MacArthur and Wilson formalized what ecologists as far back as Gleason had known—that the number of species in any habitat is proportional to its area. MacArthur and Wil-

TABLE 27-2   DIVERSITY ON ISLANDS BEFORE AND AFTER DEFAUNATION

	Number of species	
Trophic level	**Before**	**After**
Herbivores	55	55
Scavengers	7	5
Detrivores	13	8
Wood borers	8	6
Ants	32	23
Predator	36	31
Parasite	12	9
Unknown	1	3
Total	**164**	**140**

Each island had approximately the same number of species as before it was fumigated. However, the kinds of species that arrived were often different from what had been there before. This study suggested that which species colonize an island may be a historical accident, but the *number* of species depends on the abiotic characteristics of the island. Of course, in this experiment, all of the trees, shrubs, and other plants on the island remained intact. These too probably played a critical role in creating a fixed number of niches for arthropods.

In 1984, Jorge R. Rey repeated this experiment on islands in Oyster Bay, Florida. Rey fumigated islands and then carefully counted the number of insects, spiders, and other arthropod species as they recolonized the islands (Figure 27-13). Rey's work confirmed what Wilson and Simberloff had shown— that an equilibrium number of species exists for each island and that the number of species depends on the area of the island. In addition, Rey showed that, as more species became established on an island, fewer species successfully immigrated to an island and more species on the island went extinct, thus establishing an equilibrium.

How Do Groups of Populations Interact?

Understanding the factors that govern species diversity on islands helps us understand species diversity on the continents. A forest surrounded by croplands or housing subdivisions stands as a kind of island, as does a meadow surrounded by forests or a mountain peak surrounded by valleys. Ecologists call these **habitat islands.** For many species, however, a small island of habitat cannot support a population of sufficient size for plants and animals to find mates or for animals at higher trophic levels to find sufficient prey. Sometimes chance events eliminate all the individuals in such small populations. Eventually, one or more species in these habitat islands goes "locally" extinct.

In the face of such local extinctions, the persistence of a population depends on migration from other habitat islands. A group of local populations that alternately go extinct and recolonize vacant habitat islands are called a **metapopulation** (Figure 27-14). Just as with real islands, the sizes of islands, the number of islands, and their distance from one another all influence the rate of recolonization. In addition, the space between the habitat islands can also influence how easily and quickly species can migrate (Figure 27-15). We discuss extinction and population dynamics in more detail in Chapter 28.

A.

Figure 27-13 Does every island support a certain number of species? A. To find out, Jorge Rey first eliminated all of the invertebrate animals on islands in Apalachee Bay, Florida, by covering the islands with giant tents and filling the tents with pesticides. After all the insects and other invertebrates were dead, he removed the tents and monitored the return of species to each island. B. By about 20 weeks, each island was recolonized by a characteristic number of species. Bigger islands (*circles in top curve*) accumulated more species (from the mainland) than did smaller islands (*triangles in lower curve*). Rey found that the species that colonized an island might not be the same ones that lived on the island before fumigation. The *number* of species, however, was about the same. (*A, Jorge Rey, University of Florida*)

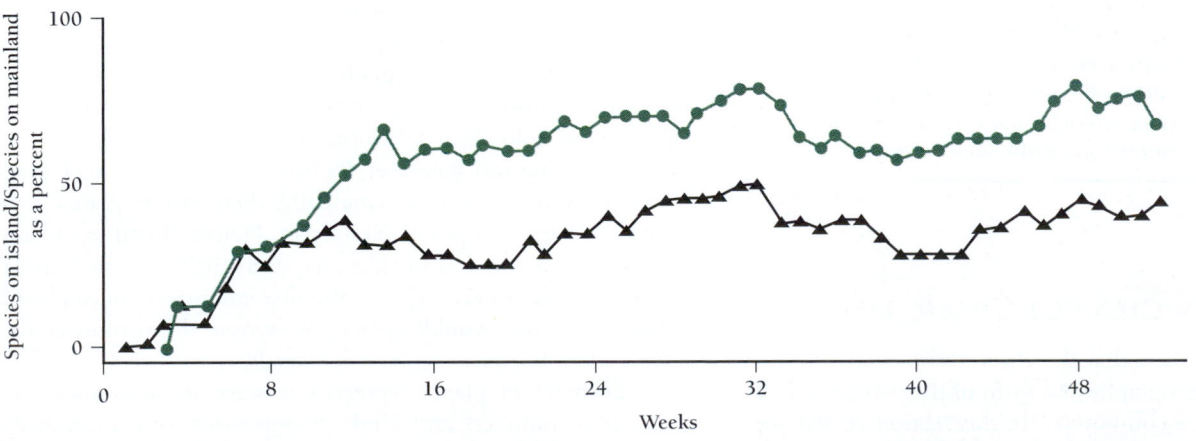

B.

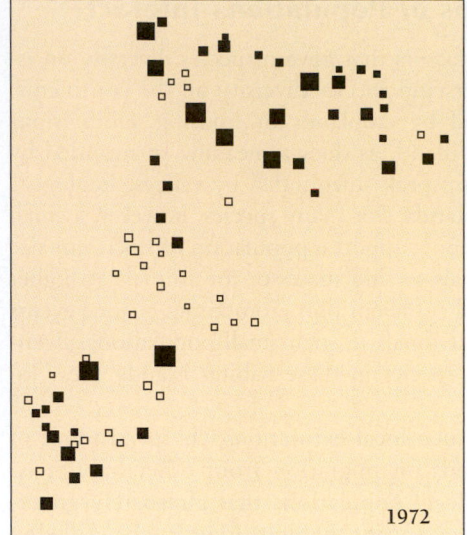
1972

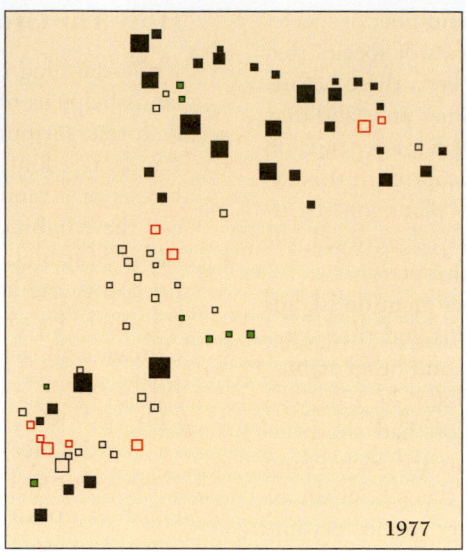
1977

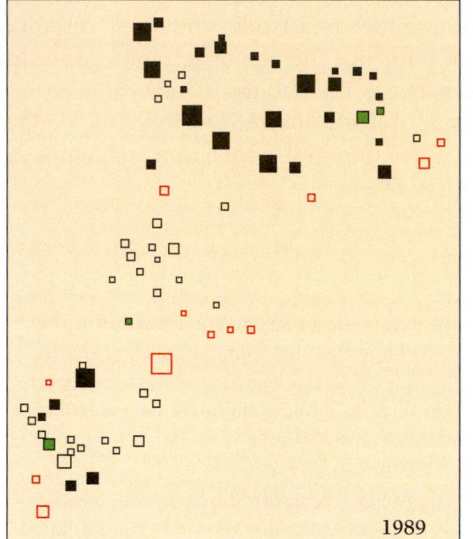
1989

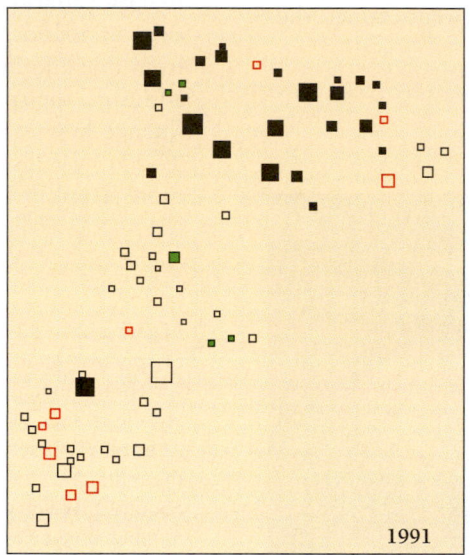
1991

Figure 27-14 Extinction and recolonization in a ghost town. In the desert between Nevada and California is an old gold-mining town called Bodie. Nearby, live small subpopulations of pikas, tiny rabbitlike animals that live in the rocky diggings left over from the old mines. Solid boxes are populated areas, and empty boxes represent extinct subpopulations. Between 1972 and 1991, some subpopulations went extinct (*empty red squares*) and then came back (*green-filled squares*), as pikas from other subpopulations recolonized the empty habitats. Together, all the interacting subpopulations are called a metapopulation. Overall, the number of subpopulations decreased by what percent? (*After Smith and Gilpin; photo, William J. Weber/Visuals Unlimited*).

Species diversity is influenced by both biotic and abiotic factors. Character displacement, keystone predators, and geography all affect diversity. Larger habitat islands support more species than smaller habitat islands. Habitat islands close to a mainland, or other rich source of species, support more species than habitat islands far from such sources.

COMMUNITIES CHANGE OVER TIME

In the last section, we saw that the characteristics of communities depend on the geography and form of the habitat as well as on the history of the community. In this section we will see how communities change over time.

Why Do Communities Change?

After Mount St. Helens blew up in 1980, the near-total destruction of the community of plants and animals that lived on its slopes initiated the gradual development of a new community. Windblown seeds from fireweed, grasses, and other plants first colonized the slopes of the volcano, rooting into the remaining soil wherever the thick volcanic ash had been washed away by erosion. Gradually, these weedy plants covered the barren slopes of Mount St. Helens. In time, shrubs such as dogwood and huckleberry, and small trees such as fir and hemlock, grew in among the first colonizers and replaced them. The trees would replace, or *succeed*, the shrubs, as the shrubs had begun to replace the weeds.

Each set of plants represents a stage in succession, the change in numbers and kinds of organisms in a community over time. Ecologists distinguish between **primary succession,**

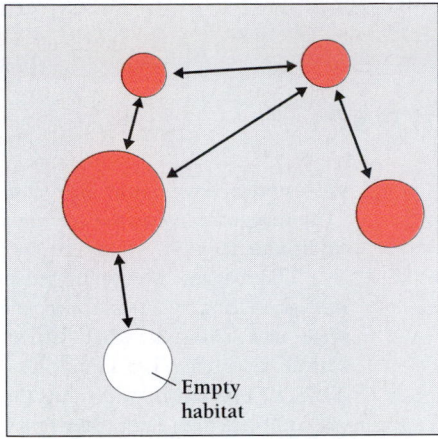

A.

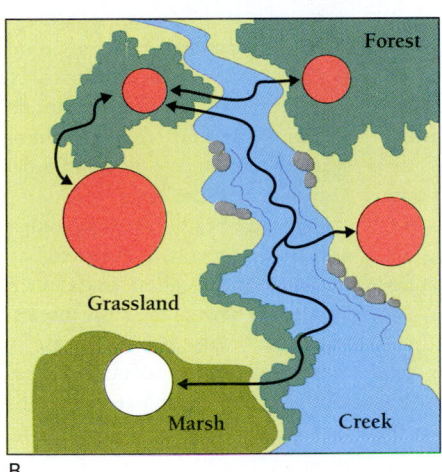

B.

Figure 27-15 **It's what's in between that counts.** A. The islands in MacArthur and Wilson's model of island biogeography were separated by uniform expanses of water. Similarly, the habitat islands of the metapopulation of pikas in Bodie, California, are surrounded by a nearly uniform expanse of uninhabitable desert. Individual pikas must cross this "matrix" to recolonize another habitat island. The greater the distance between habitat islands, the less likely that individuals will be able to cross. B. In nature, most habitat islands are surrounded by habitat that varies and grades from one kind of habitat to another. Depending on what habitat lies between, it may be easier for individuals to cross between islands that are far apart than between those that are closer together. Imagine a frog moving between ponds. Which would be the easier to cross, two miles of stream bed or a quarter mile on a dry, rocky hillside?

the invasion of a completely new environment such as a sandbar or new volcanic island, and **secondary succession,** the sequence of stages in a badly disturbed community. Examples of secondary succession include the reestablishment of a community in a burnt or logged over forest and the return of natural communities to abandoned croplands or vacant lots. In secondary succession, some kind of preexisting community plays a role in the succession.

After the eruption of Mount St. Helens, some areas were entirely sterile. These areas were ready for primary succession to begin. Other areas retained fertile soil with seeds. These seeds sprouted and contributed to the first stages of secondary succession.

The first community in a succession is called a **pioneer community,** while later stages are called **successional communities.** According to the theory of succession, the mix of species in a community changes over time in a way that is cumulative and directional. Climax communities, as they are now understood by ecologists, themselves change over time. Thus, "climax community" has become a relative term.

Pioneer and later communities differ in a variety of other ways, summarized in Table 27-3. For example, pioneer communities usually consist of species that flourish in disturbed areas and breed rapidly. At the other extreme, climax communities often consist of species that breed slowly but gradually take over in undisturbed areas. In general, the short-lived weedy species of pioneer communities (thistles, for example) are gradually replaced, or succeeded, by longer-lived organisms (fir trees, for example).

Most descriptions of succession refer to changes in vegetation because succession is a concept introduced by plant ecologists. However, other kinds of succession exist. For example, a deer's carcass, a rotting tree, or a pile of dung can support a succession of insects, bacteria, and fungi over periods of 2 months or less (Box 27.1). In such cases, there is no climax community, and succession occurs on a small scale in both time and space.

Succession also occurs in patches. If a tree falls in a forest, it leaves a hole in the canopy. Bright shafts of sunlight fall to the forest floor, allowing weedy, light-adapted plant species to crowd out shade-adapted plants. Gradually, the shrubs and trees may take over, and the patch becomes shady forest again. But meanwhile other trees have fallen elsewhere in the forest. The result is a forest with patches of habitat at many different stages of succession and many more species than if the forest were uniform.

TABLE 27-3 HOW PLANT COMMUNITY TRAITS CHANGE DURING SUCCESSION

Trait	Early stages	Later stages
Biomass	Small	Large
Physical structure	Simple	Complex
Role of detritus	Minor	Major
Net primary production	High	Low
Stability (absence of change)	Low	High
Plant species diversity	Low	High
Seed dispersal by	Wind	Animal
Seed longevity	Long	Short

After Barbour, Burke, and Pitts 1980.

BOX 27.1

Succession on a Corpse

"The worms crawl in, the worms crawl out" goes the childhood ditty. But which "worms," and exactly when they crawl in and out, can tell special criminal investigators how long ago, and often where, a person died. By studying the crawling creatures collected from a corpse, forensic entomologists, scientists who apply their knowledge of insects to legal matters, can solve otherwise baffling cases.

The father of American forensic entomology, Bernard Greenberg of the University of Illinois at Chicago, likes to recount the case of a missing 9-year-old girl. After her partly skeletonized body was discovered in an abandoned apartment building, Greenberg collected fly larvae and pupae from the body and took them back to the laboratory. There he raised them at temperatures like those in the apartment building over the days preceding the girl's discovery. He soon learned that the flies took 21 days to mature under those conditions, which meant that the fly eggs had to have been laid 21 days before the girl was found. The last time the little girl had been seen alive was 21 days earlier, near the apartment building in the company of a man in his 20s. As a result of Greenberg's testimony, this man was subsequently convicted of her murder.

Beetles, flies, and other arthropods colonize corpses in a predictable succession. "Flies are the best investigators of corpses," says Carl Olson, an entomologist at the University of Arizona. Metallic blue or green blowflies, drawn by odors that include a compound called cadaverine, can find an uncovered corpse within 10 minutes of death. Blowflies and flesh flies lay eggs in wounds and moist natural body openings, such as eyes, ears, nose, and mouth. Inside the body, bacteria go to work decomposing the body, generating various gases in the process.

As the masses of maggots feed, their activity and that of the bacteria raise the temperature inside the corpse. "It's like when you walk into a room full of people—notice how hot it is?" Olson asks. "The maggots on a corpse create their own environment."

The maggots eventually puncture the skin, initiating what is called the decay stage and releasing what University of Hawaii researcher Lee Goff calls "strong, distinctly unpleasant odors." As the fly larvae complete their feast, they hop or crawl off to pupate in the soil. By that time, the latter part of the decay stage, the fly larvae have created a hospitable environment for the next wave of insects, primarily hide and rove beetles. It is a classic example of ecological succession. Any remaining larvae become food for the visiting beetles, which also finish off the flesh.

By the postdecay stage, only the skin and bones remain. A different set of beetles now arrive. These feed on dried skin and hair. They are the same insects that destroy fur coats and cashmere sweaters. When the body is reduced to bone alone, all of the carrion-feeding arthropods depart.

Any kind of physical disturbance, whether a fallen tree or a collapsed bank along a stream, can create a patch in which secondary succession can begin. Even something as minor as a gopher mound in a meadow creates a disturbed area where a patch of different species can live. A disturbed area can be colonized by seeds already in the soil (a "seed bank") or by organisms from nearby areas.

Succession usually implies a progression through a series of transitional stages to a climax community. Succession can occur in disturbed areas, creating a patchwork of different stages of succession.

Why Does Succession Occur?

Clements argued that succession occurs because organisms degrade their own environment. For example, following a forest fire, the first plants to grow are those that are adapted to live in open sun. But the shade created by these sun-loving plants prevents their own seedlings from growing and favors shade-tolerant plants.

While many of Clements's scientific heirs still contend that each stage in a succession prepares the way for the next, other ecologists argue that this is not always true. Some of the changes in a succession, they say, merely reflect differing rates of growth. Shrubs, for example, appear to come before trees even when both types of seedlings start at the same time because the shrubs take less time to reach maturity.

In reality, **pioneer** species can facilitate succession, tolerate succession, or inhibit succession. In primary successions, however, pioneer organisms always facilitate succession. This is because most organisms cannot colonize bare rock or sand. The formation of soil from bare rock depends on the action of pioneer lichens and plants adapted to thin or nonexistent soils. Once these pioneer species have created a thin layer of soil, other species can grow.

Pioneer or successional species change their environment in such a way as to allow colonization by succeeding species.

What Determines Which Community Is the Climax Community?

The climax community is, by definition, the one that persists the longest without changing. But what factors determine which community will persist? As we saw in Chapter 26, the character of a community depends heavily on such abiotic factors as climate, soil, and terrain. The dry, frozen tundra could never support a tropical rain forest. Early ecologists theorized that an area has only one possible climax community.

Communities assemble themselves flexibly, however, and their particular structure also depends on the specific history of the area. For example, giant sequoia forests in the Sierra Nevada include trees as old as 2000 years. These forests appear to be a classic example of a long-lived climax community. In fact, however, these ancient forests persist only if periodic fires sweep through and kill young white fir and incense cedars growing beneath the mammoth sequoias (Figure 27-16). When humans regularly put out fires, the white fir and incense cedars grow up among the sequoias. Eventually, when a fire too big to put out arises, the smaller trees pass the flames up into the crowns of the giant sequoias, destroying the whole forest. Once that happens, a white fir and incense cedar forest springs up in its place.

Figure 27-16   A grove of giant sequoias. Such rare groves depend on frequent forest fires to prevent succession to white fir and incense cedar. *(Barbara Gerlach/DRK Photo)*

The character of a climax community depends on climate, terrain, and history.

Why Do Communities Stay the Same for Long Periods?

An ecologist who studies a pond today may well find it relatively unchanged a year from today. Individual fish may be replaced, but the number of fish will tend to be the same from one year to the next. We can say that the properties of an ecosystem are more **stable** than the individual organisms that compose the ecosystem.

At one time, ecologists theorized that species diversity made ecosystems stable. They thought that the greater the diversity, the more stable the ecosystem. Support for this idea came from the observation that long-lasting climax communities usually have more complex food webs and more species diversity than pioneer communities. Ecologists concluded that the apparent stability of climax ecosystems depended on their diversity and complexity. To take an extreme example, farmlands dominated by a single crop are so unstable that one year of bad weather or the invasion of a single pest species can destroy the entire crop. In contrast, the many species that make up a native grassland will tolerate considerable damage from weather or pests.

However, research has provided little support for this idea. Indeed, ecologists do not even all agree on what "stability" means. Stability can be defined as simply lack of change. In that case, the climax community would be considered the most stable, since, by definition, it changes the least over time. But, alternatively, stability can also be defined as the speed with which an ecosystem returns to a particular form following a major disturbance, such as a fire. This kind of stability is also called **resilience.** In that case, climax communities would be the most fragile and the least stable, since they can require hundreds of years to return to their former state. As we saw in Chapter 26, tropical rain forests have the greatest biodiversity of any biome and some of the highest primary productivity. Yet once destroyed, they do not quickly return.

Even simple lack-of-change stability is not always associated with maximum diversity. Maximum diversity in some areas is found not in the climax community, but in midsuccessional stages. Once a redwood forest matures, for example, the kinds of species and the numbers of individuals growing on the forest floor are reduced. In general, diversity, by itself, does not ensure stability.

Mathematical models of ecosystems likewise suggest that diversity does not guarantee ecosystem stability—just the opposite, in fact. A more complex system is more likely to break down than a simple one. A computer loaded with conflicting software, for example, is more likely to break down than a simple hand calculator.

Can Ecosystems Bounce Back?

Because ecosystems all over the world are being severely disturbed or eliminated by human activities, ecologists would like to know what factors contribute to the resilience of ecosystems. The severe destruction caused by a volcanic eruption such as Mount St. Helens is nothing in comparison to the huge areas of temperate and tropical rain forests that we humans have destroyed. We need to know what aspects of an ecosystem are most important to its resistance to damage, as well as its recovery.

Many ecologists now think that resilience comes not from diversity of species in one place but from the patchiness of the environment. An environment that varies from place to place supports more kinds of organisms than an environment that is uniform. In an environment with abundant species, a local population that goes extinct is quickly replaced by immigrants from an adjacent community. Even if the new population is of a different species, it can approximately fill the niche vacated by the extinct population and keep the food web intact.

Other factors that contribute to an ecosystem's resilience include resistance to erosion and increased primary productivity. As long as the soil is intact, seeds will quickly sprout, and shrubs, ferns, herbs, and grass will hold the soil in place, preventing erosion. Partly damaged ecosystems often support greater plant growth than mature ones, since younger trees grow more vigorously than trees in a mature climax forest.

Some of the factors that contribute to ecosystem resilience are patchiness of the physical environment, primary productivity, and resistance to erosion.

In this chapter we have seen how communities change over time and what factors, biotic and abiotic, influence the character of a community. In the next chapter, we will examine what factors affect populations and what happens when populations grow too big.

Study Outline with Key Terms

Henry Cowles did the first major American work on **succession,** and Frederic Clements published the first textbook describing succession. Clements also introduced the idea of a **climax community,** a single community of plants that, he said, characterizes a particular area and climate. Clements believed that communities were **integrated**—real organic entities, with species playing the same roles in different communities. In contrast, the ecologist Henry Gleason argued that communities were **individualistic**—mere associations of species that happen to occupy the same area.

The species in communities interact in many ways. They may compete or they may relate as **predator** and prey or as host and **parasite.** Organisms defend themselves from predators and herbivores by using **warning coloration, mimicry** (**Batesian** or **Müllerian**), **camouflage,** and **secondary plant compounds.** Species may share a **symbiotic** relationship, such as **parasitism, mutualism,** or **commensalism.** In evolutionary time, two or more species may **coevolve** mutualisms and other special relations with one another.

Species may also **compete** for a **limiting resource.** The kind of community an organism occupies is its **habitat.** The way the organism uses the resources in its habitat is called its **niche.** The **competitive exclusion principle** says that if two species occupy the exact same niche, one species will eliminate the other through competition. Competition from other species may reduce the **fundamental niche** of an organism to a smaller **realized niche.** In **resource partitioning,** species break up a large niche into several smaller ones. Competition between species can cause them to evolve apart in a process called **character displacement.**

Communities vary in the number of species they contain, **species richness,** and in the **relative abundance** of each kind

of species. Ecologists are still trying to discover what determines the diversity of a community. **Keystone species,** productivity, harshness of the environment, latitude, size of ecosystem, and historical accident all seem to play roles.

In the 1960s, Robert MacArthur and Edward O. Wilson proposed a **theory of island biogeography** to account for species diversity on islands of habitat. Tests of this theory suggest that the number of species on an island is a predictable consequence of the characteristics of the island, and not just a historical accident. **Habitat islands** do not have to be conventional oceanic islands. They can be mountaintops or ponds in a forest, for example. A group of local populations that alternately go extinct and recolonize vacant habitat islands are called a **metapopulation.**

Primary succession occurs when organisms colonize a previously unoccupied area, such as a volcanic island or a sand dune. **Secondary succession** occurs when organisms recolonize an area that has suffered some catastrophe, but which still has its soil intact. In secondary succession, the first **pioneer** plant species in a **pioneer community** are annuals, followed by perennials, shrubs, and trees. The final stage in development is called the climax community, but biologists no longer believe that the idea of a final stage makes sense. All stages can change, depending on circumstances. Intermediate stages are called **successional communities.**

Ecosystems are either more **stable** than pioneer or successional ecosystems or less stable, depending on the definition of stability that is used. Patchiness of the physical environment, primary productivity, and resistance to erosion are some factors that seem to contribute to ecosystem **resilience,** the ability of an ecosystem to recover from fires or other cataclysms.

Review and Thought Questions

Review Questions

1. Contrast the integrationist view of communities with the individualistic one. Describe a piece of research that supports one view or the other.
2. Name a commensal and a parasite to humans.
3. Name a mammal, bird, reptile, amphibian, and insect that have warning coloration.
4. Define the competitive exclusion principle.
5. Give an example of a limiting resource for humans.
6. Define the terms "niche" and "habitat." Give an example of each for the same organism.
7. What factors determine the number of species living on an island?
8. Give two definitions of "ecosystem stability." What factors determine the long-term stability of an ecosystem?

Thought Questions

9. Describe a climax community somewhere near where you live. What criteria did you use to determine whether it was a climax community? How would you go about testing your theory that it is a climax community?
10. Describe a series of small habitat islands that exist within 5 or 10 miles of your house and which might be subject to MacArthur and Wilson's theory of island biogeography.
11. You have discovered two butterflies that look nearly identical, but, from other evidence, clearly belong to different families. Assuming that one is mimicking the other, think of an experiment that would show which one was the mimic and which one was the "original."
12. If succession occurs because pioneer or successional species improve conditions for subsequent species, what can we say about climax species?

About the Chapter-Opening Image

Destruction of an ecosystem. When Mount St. Helens erupted in 1980, it destroyed a 230-square-mile ecosystem, about the size of the city of Chicago. What was destroyed?

 On-line materials relating to this chapter are on the World Wide Web at **http://www.harcourtcollege.com/lifesci/aal2/**

28 Populations: Extinctions and Explosions

Fighting for Life

In 1998, an eminent biologist gave up his tenured job at the University of California. It was the second time Michael Soulé had left the security of a tenured faculty position—one of the strongest guarantees of lifetime employment in the United States. As before, he did so to follow his heart.

The academic life was, for Soulé, too much thought and too little action. He felt impelled to act, to literally save the world. The problem, as he saw it, was that everything was disappearing. All of the animals and special habitats he had studied and grown to love during his 40 years as a biologist were vanishing. His gardens of discovery were winking out one by one.

"Every field biologist has had this experience," he told writer Robert Coontz in 1990. "You return to a study site you knew as a graduate student, or as an assistant professor, or where as a child you first became interested in biology, and it's been devastated."

He began his career in the normal way of biologists. In his boyhood, he roamed the hills and canyons outside San Diego, catch-

ing lizards and snakes, watching birds, rabbits, and other wildlife. In college he studied zoology, then went to graduate school at Stanford. His mentor there was population biologist Paul Ehrlich, whose books on human overpopulation have made Ehrlich far more famous than his decades of research on butterflies.

In 1967, Soulé returned to his boyhood stomping grounds, taking a job as a professor at the University of California at San Diego. He taught biology and studied symmetry in lizards. All animals, including humans, become a little asymmetrical as they develop from embryos to adults. A high degree of symmetry, however, is an indicator of good health. In fact, many animals choose their mates partly on the basis of how symmetrical they are. We could say that symmetry is the biological definition of beauty. Asymmetry, on the other hand, often indicates poor health. Asymmetry, Soulé found, was also an indicator of pollution and other environmental stresses, and he was finding more asymmetry in his animals than he wanted to find.

Before his eyes, the canyons of San Diego

Painting by Charles Knight. Field Museum of Chicago photo # Geo-CK-30Tc by Ron Testa.

- Populations may go extinct when habitat is damaged or destroyed. Chance events, such as the disappearance of alleles from the gene pool, can hasten extinction.

- Natural populations rarely grow exponentially for very long. Usually, growing populations reach a plateau, fluctuate, or crash.

- Both density-dependent and density-independent factors affect the population growth of a species.

- The growth of human populations can interfere with the growth and survival of other species and even whole ecosystems.

County were filling with houses. The fragments of land that survived between housing developments were too small to support many individuals of each species. Not only in San Diego, but around the world, species were going extinct at an unprecedented rate.

These shrinking habitats drew Soulé's interest to the problem of inbreeding. He could see that once a population got small enough, the remaining individuals would have only a few others with which to breed. He theorized that inbreeding would bring out genetic defects in the offspring. Fewer would reach maturity, and those that did would be less fit—less able to survive and reproduce themselves. Small populations spiraled downward toward an extinction from which they'd never be able to escape. Soulé called this downward spiral the **extinction vortex** (Figure 28-1).

It was all interesting material for a researcher. Soulé published his experimental work in journals and presented his ideas at scientific meetings. The university promoted him from assistant professor to tenured associate professor and, finally, to full professor.

Yet something was missing. It was one thing to document the slow destruction of study sites and beloved childhood haunts. It was quite another to do something about it. Real biologists were not supposed to act on the grief they felt when yet another study site disappeared beneath a sewage treatment plant or shopping mall.

In the mid-1970s, Soulé's fellow biologists accepted the wall that then divided biology from the environmental movement. For biologists to affiliate themselves with "ecofreaks" who chained themselves to redwood trees was to abandon any pretense of objectivity. At that time, a biologist who expressed sympathy with down-in-the-trenches environmentalists, let alone became one, could lose credibility as a serious biologist.

Nonetheless, Soulé took action. In 1978, he helped organize the First International Conference on Conservation Biology. It was the first time that field biologists acknowledged that they had both a responsibility and a special ability to do more than just document the demise of the habitats they knew so well. Perhaps their research could provide information that would help save some of these habitats.

By then, Soulé was in his early forties and had been on the UC San Diego faculty for 12 years. It's common for middle-aged people to look for change, but most professors would content themselves with a new line of research, a sabbatical, or even a sports car. For Soulé, that wasn't enough. He quit

his tenured job at UC San Diego and joined a Zen center in Los Angeles.

Although he stayed at the Zen center for 5 years, he never abandoned biology. Instead, he was reconceiving what it should mean to be a biologist. He helped put together a collection of essays on conservation biology. The two volumes, titled *Conservation Biology: An Evolutionary Ecological Perspective,* galvanized field biologists everywhere. Their 1980 publication marked the informal beginning of **conservation biology** as a field separate from traditional ecology.

For the first time, the majority of biologists began to acknowledge openly that human overpopulation was creating a tidal wave of extinctions unprecedented in the history of humankind. In 1987, a group of research biologists formed the

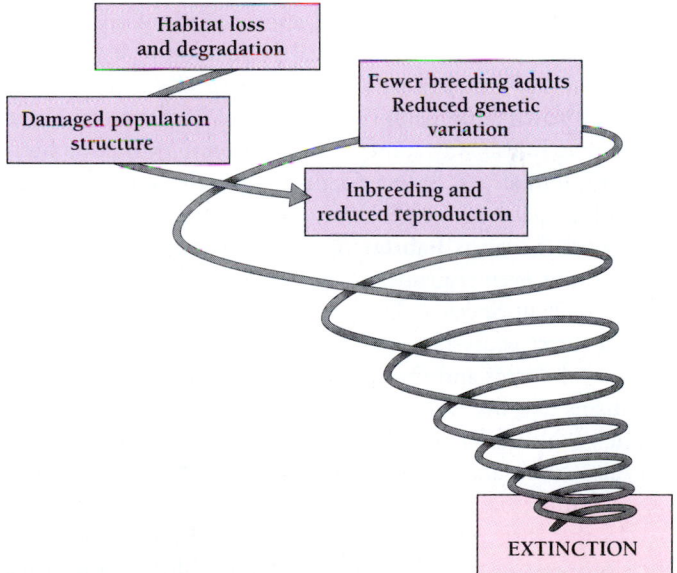

Figure 28-1 The extinction vortex. Biologist Michael Soulé hypothesized that when a population shrinks because of hunting or loss of habitat, the reduced population size—as well as disproportionate numbers of males and females or young and old individuals—causes inbreeding. Inbreeding, in turn, can cause ill health and poor reproduction in each new generation, accelerating the extinction process in a downward spiral. In addition, extinction of one species degrades the environment for other species in the habitat. The feedback loop thus involves more and more species as time passes.

Society for Conservation Biology and began publishing a journal devoted to the science of conservation. Soulé returned to the University of California, this time to the Santa Cruz campus, to lead a department of environmental studies.

HOW DO POPULATIONS GO EXTINCT?

Many factors cause species to go extinct. Biologists today lump all these factors into two categories. One category includes **deterministic** events, such as habitat destruction or deliberate extermination through hunting or trapping. The other category includes **stochastic,** or chance, events such as unusual storms.

How Do Deterministic Factors Drive Species to Extinction?

One deterministic factor that can drive species to extinction is competition from other species. In nature, one kind of tree may shade a smaller species, driving the local population to extinction. The larger tree has successfully competed with the smaller one for light. Usually, the result is not extinction of a whole species, however, but only of a particular population. As we saw in the last chapter, individuals from another population can reestablish a population when opportunity arises.

Humans also drive local populations to extinction. But our large numbers and organized approach to taking over resources allow us to drive *every* local population of a species to extinction. How do humans drive other species extinct?

One way is simply hunting and killing as many organisms as possible. In this way, humans have driven to extinction hundreds of animals, including California grizzly bears, Tasmanian wolves, passenger pigeons, dodos, and great auks.

 Another way is **habitat destruction.** Sometimes we do this without even realizing we are doing it (Web Bit 28.1 Introduced Plant Species). If we remove all the trees and shrubs from a forest ecosystem and then plant wheat, we eliminate not only the trees and shrubs, but also habitat for all the other organisms that live in the forest. If we repeat this pattern from Maine to Nebraska, we eliminate not only trees and habitat but also thousands of subpopulations of insects, spiders, worms, fungi, bacteria, and more.

In the 18th century, pioneer families cleared millions of acres of forest in the eastern United States. Most of the farms they built and tended are now long abandoned, and much of that forest has returned. But while many trees and shrubs are back in great numbers, hundreds of other species are gone forever. For many of these species, not a single population is left to recolonize the new forest. For others, the remaining subpopulations, in Canada or the far West, are simply too far away to recolonize the new forest.

Humans drive other species to extinction by destroying their habitats.

How does fragmentation contribute to habitat destruction?

A subtler process that drives other species to extinction is **fragmentation** of habitat. Natural landscapes are a mosaic of different habitats. Forest blends gradually into savanna, chaparral, or grassland. Riparian (river) forests may run through drier habitats, connecting wetter forest ecosystems together. An organism such as a bird or insect can often travel easily from one patch of forest to another. If a population disappears from one habitat patch, individuals from another subpopulation soon arrive to recolonize.

For organisms that cannot fly, however, recolonizing is usually more difficult. Humans create sharply bounded patches of habitat in an uncrossable sea of pavement, houses, and crops. Many species cannot easily move from one patch to another. Even a simple road can permanently divide a habitat for turtles, amphibians, and other slow-moving animals.

Fragmentation also reduces the total area of a habitat. We saw in Chapter 27 that numbers of species depend on the area of the habitat. According to MacArthur and Wilson's species-area curve, a 90 percent loss of habitat (area) results in a 50 percent loss in species. Most species simply cannot survive in a habitat that is too small. For example, numerous studies have shown that scarlet tanagers, pileated woodpeckers, ovenbirds, Acadian flycatchers, and many other birds rarely live in forests that are smaller than 40 acres. Large animals at the top of the food chain, such as hawks, need large areas to supply food.

Some studies suggest that in Brazilian rain forests, 500,000 acres is the minimum size needed to preserve local biodiversity. That is a little more than one-tenth the area of New Jersey. But rain forests in Peru, Venezuela, and other areas would have different sets of species and so would each need a separate 500,000-acre preserve.

Another important consequence of habitat fragmentation is **edge effects.** The smaller the pieces into which a habitat is cut, the more edge the habitat has—relative to interior. At the edges of a habitat, many interesting things happen. At the edge of a forest, for example, more light enters and different kinds of plants grow. Sometimes safe nesting areas are missing. Predators and other species from the adjacent habitat visit to take birds' eggs, kill young animals, or browse on young trees. In areas heavily populated with humans, these visitors may include dogs, cats, and, of course, humans themselves.

Because the edges of a forest or marsh are often dangerous, many species can survive only in the very center of a habitat patch, far from the edge. For some species, many habitat patches are so small that no habitat exists that is far enough away from the edge (Figure 28-2).

One way humans destroy habitat is by breaking it up into small fragments with lots of edge and little central core.

Draining the gene pool

Recall that Michael Soulé's extinction vortex hypothesis suggested that inbreeding accelerates extinction. One source of

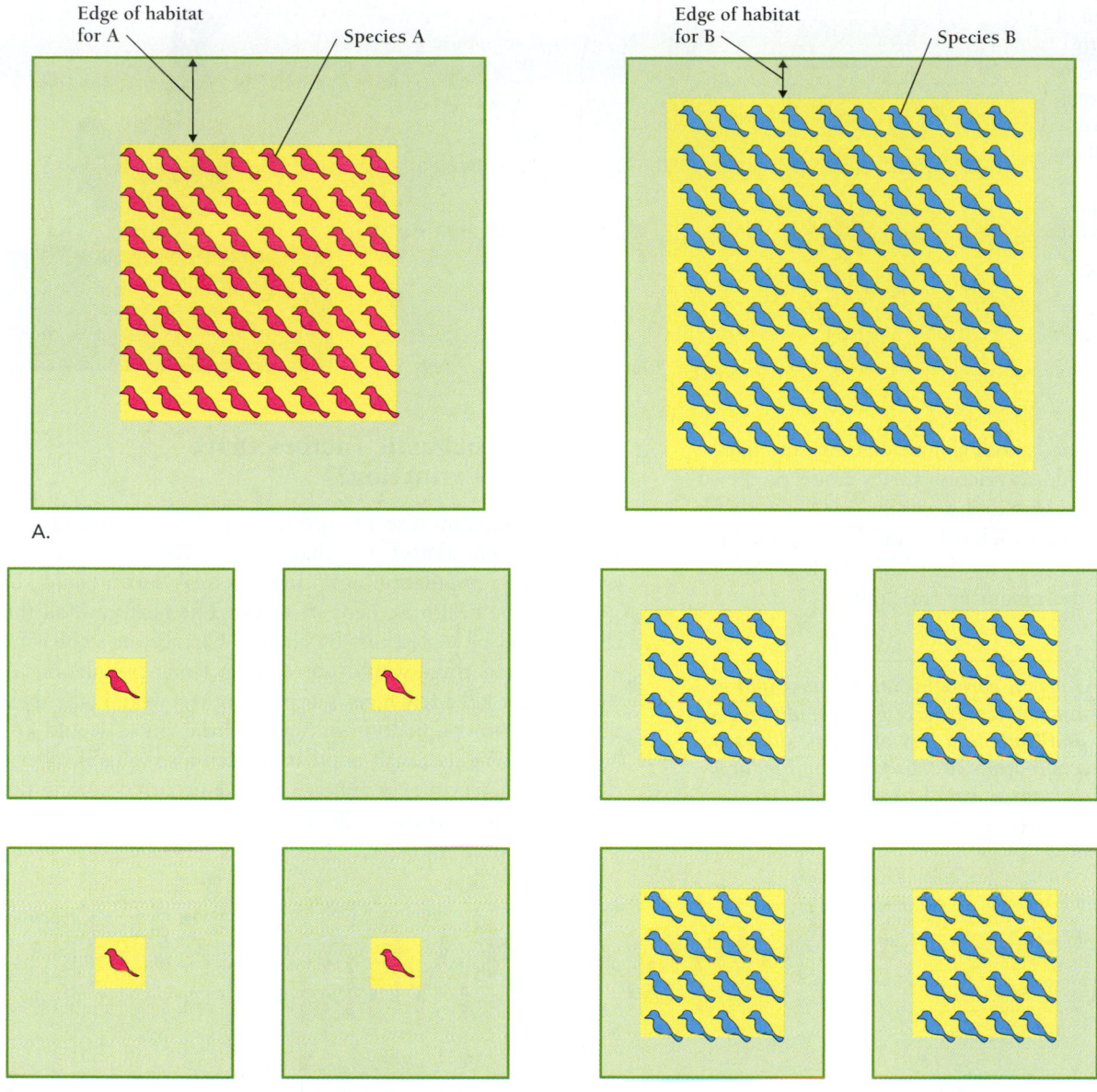

Figure 28-2 Habitat fragmentation and the edge effect. Many organisms cannot survive close to the edge of a habitat. As large habitats are fragmented by roads and other human development, the proportion of edge increases. As a result, a habitat patch that seems large to us is too small to allow many species to survive. Different species have different tolerances for habitat edges. A. An unfragmented habitat patch. B. The same patch broken up into four quadrants is now nearly all uninhabitable edge for species A, but is still habitable for species B.

support for this hypothesis is a 1998 report in the journal *Science* on a long-term study of greater prairie chickens. Beginning in the early 19th century, extensive habitat destruction pushed this native Midwestern bird toward the brink of extinction. In Illinois, the greater prairie chicken is almost gone (Figure 28-3). During a 35-year study of these birds, the population size dropped from 2000 birds in 1962 to 50 birds in 1994. Researchers also recorded a decline in hatching success (hatched eggs per nest) from a high of nearly 100 percent in

1960s to less than 40 percent in 1990. Measures of genetic diversity showed that the Illinois population had less genetic diversity than larger populations in nearby states and that alleles present in the population before the 1970s later disappeared. The data suggest that as the population shrank, so did its genetic diversity.

Biologists hypothesized that genetic defects due to inbreeding were keeping many chicks from hatching (Figure 28-4). When researchers brought in prairie chickens from

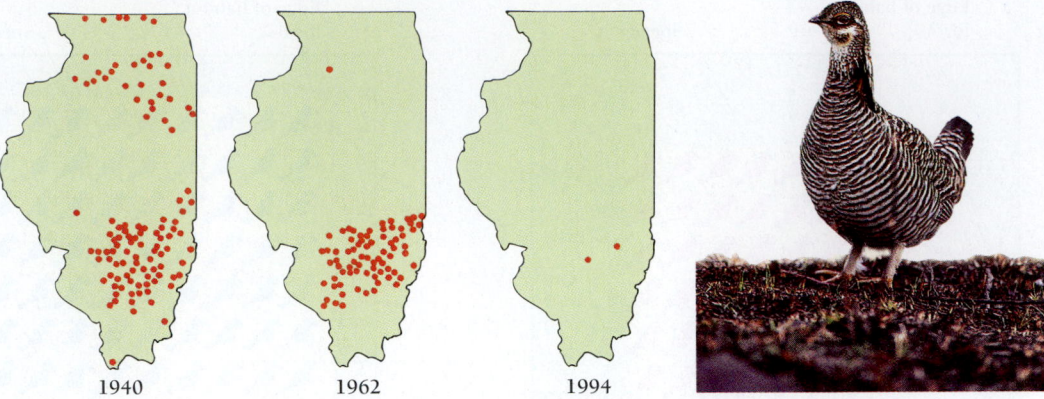

Figure 28-3 The end of the Illinois greater prairie chicken? One hundred years ago, greater prairie chickens (*Tympanuchus cupido*) lived abundantly throughout Illinois. Today, two tiny populations survive in the southern part of the state. *(Photo, Steve & Dave Maslowski/Photo Researchers, Inc.)*

larger populations elsewhere, both genetic diversity and hatching success increased. Soulé's extinction vortex model predicted that declining population size and declining genetic variability would mutually reinforce each other and increase the likelihood of local extinction. Sadly for the prairie chickens, Soulé's prediction seems to be borne out by this study.

According to the extinction vortex hypothesis, loss and degradation of habitat reduces the size of a population, changes its structure, and leads to a loss of genetic diversity. The loss of genetic diversity leads to inbreeding and reduced reproduction, which leads to smaller population size.

How Do Stochastic Factors Drive Species to Extinction?

Once a population is as ravaged as that of the Illinois greater prairie chicken, almost any chance event can drive it to extinction. In a population of 25 birds, a freak storm could kill several of the healthiest hens in 1 year. The next year, all the chicks could be males, just by chance. Chance, or stochastic, events such as these would not affect a larger population. A storm might kill a few hens, but many others would take their place. And in a population of 2000, all the chicks would not be one sex. Finally, small populations can lose valuable alleles through genetic drift, as appears to have occurred among the greater prairie chickens of Illinois.

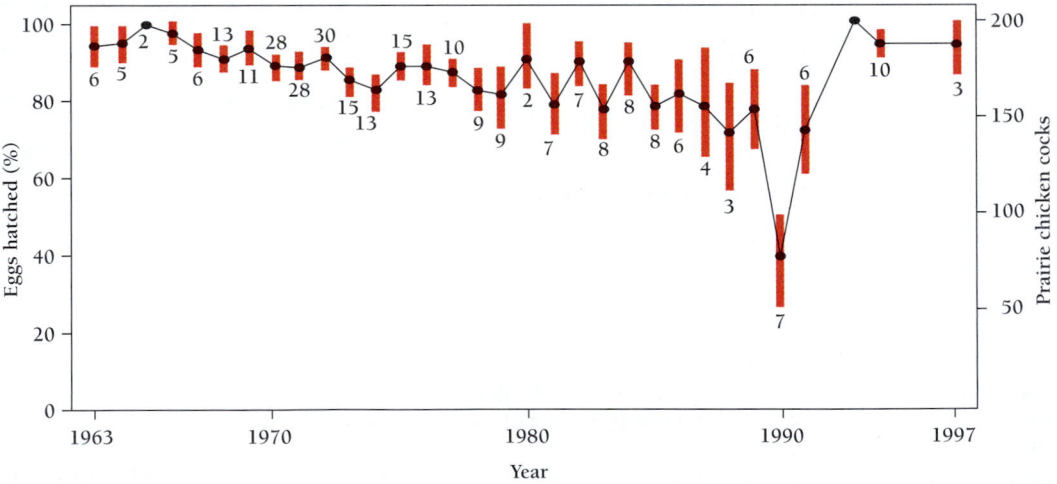

Figure 28-4 **Inbreeding among greater prairie chickens?** As the population of prairie chickens shrank, the percentage of eggs in a nest that successfully hatched dropped from 100 percent in 1963 to less than 40 percent in 1990. Biologists strongly suspect that inbreeding and the expression of deleterious genes may account for this decline. Each data point shown here is the average of hatching success at all the nests in the population. The vertical bars indicate variation in hatching success. Long bars mean more variation. Short bars mean little variation. Increased variation in any trait can be a sign of ill health in a population. Notice that the bars become longer over time. In the late 1990s, biologists introduced prairie chickens from large, genetically diverse populations in other states. The result was a dramatic increase in hatch success in the Illinois prairie chickens. Biologists believe this was likely due to an increase in genetic diversity. Notice that the bar lengths showing variation in hatch success became shorter.

Chance events are more likely to lead to extinction in very small populations.

How Can We Preserve Ecosystems and the Species That Live in Them?

The work of conservation biologists has shown that, unless something dramatic is done soon, approximately half of the world's species will go extinct in the next decade or two. Humankind has now converted one-third of the Earth's land surface to agriculture and covered more with cities, suburbs, roads, and factories. The remaining land includes some of the harshest, least productive areas of the planet, including deserts, jagged mountain ranges, and arctic tundra and taiga. The rate of destruction is increasing. In tropical areas, 29 out of 63 nations have already cut or burned nearly 80 percent of their forests.

As of 1998, 20 tropical nations had committed to saving 10 percent of their land in parks and other preserves by the year 2000. But because far more species live in the tropics than elsewhere, saving only 10 percent of all tropical rain forests means a likely loss of more than half of all the species on Earth.

In the spring of 1998, Michael Soulé expressed his disappointment at this trend in an article in the journal *Science*. He wrote, "Campaigns with targets in this range [10 percent] can create the unintended and false impression that such a paltry tithe to nature is enough to prevent a mass extinction of species."

Not long after the publication of this statement, Soulé left the university once and for all to join the Wildlands Project in Colorado. The ambitious goal of the Wildlands Project is to persuade people and their governments to set aside 50 percent of the North American continent for the preservation of biological diversity (Web Bit 28-2: The Wildlands Project). "Our generation is the key," Soulé said in 1990. "Whatever we save in the next few decades is all future humanity will have to use and to appreciate. If we don't do it, it will be too late."

Conservation biologists have discovered several rules that allow preservation projects to save the greatest numbers of species. The best preserves are large areas, whose size minimizes area effects and edge effects in sensitive species. In addition, many conservation biologists recommend **corridors,** narrow strips of protected habitat that allow species to travel from one preserve to another (Figure 28-5). The most likely beneficiaries of corridors are small, slow-moving animals, such as frogs

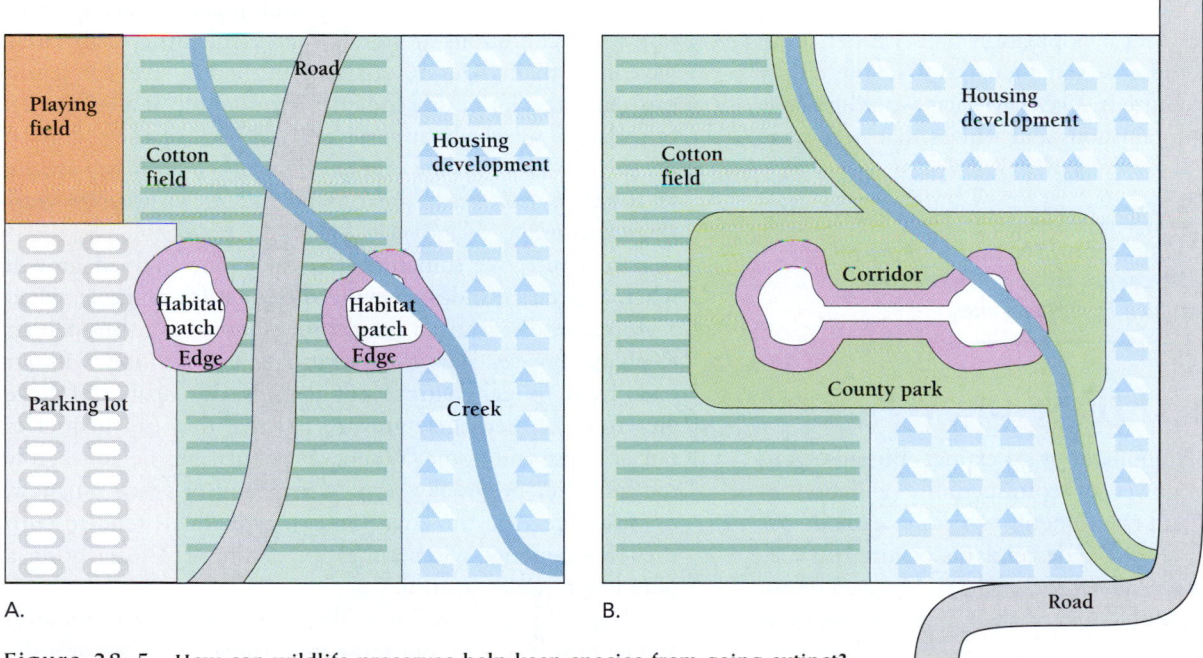

A.

B.

Figure 28-5 How can wildlife preserves help keep species from going extinct?
A. Two small wildlife preserves are bordered by houses, a playing field, and a parking lot. A road and a cotton field separate the preserves. Pesticides from the cotton field and people's backyards flow directly into the stream and then into the right preserve. Dogs, cats, humans, and other animals can enter and destroy plants, animals, nests, and dens. Edge effects are huge. Recolonization would be difficult or impossible for species that cannot easily cross the road and the cotton field. B. In the redesigned preserve, a corridor connects the two habitat patches. The road has been rerouted around the two patches, and the cotton field has been moved to one side. In this arrangement, if a species goes extinct in one habitat patch, individuals can recolonize from the other habitat patch. An inner buffer zone (purple) acts as a wide edge surrounding the corridor and both patches. In this buffer area, human use is restricted to a minimum. An outer buffer zone (green) allows more general human use but still no houses, crops, or industry.

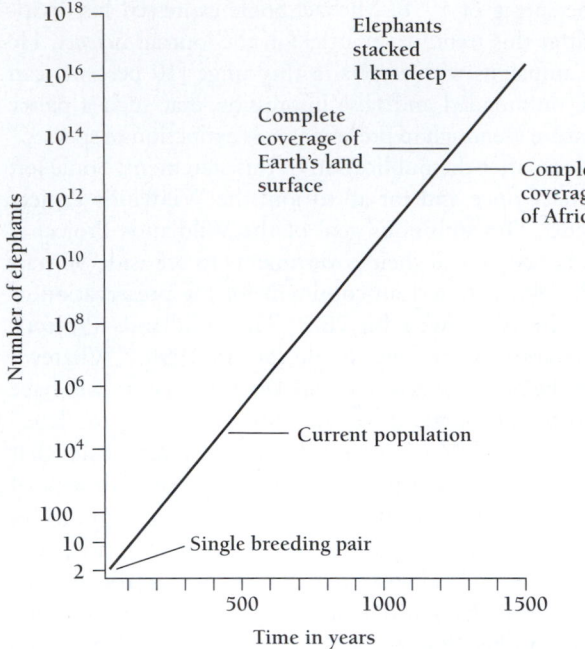

Figure 28-6   Unrestrained growth in elephants. Elephants, like the ones at this water hole, have the lowest r_{max}, or intrinsic rate of natural increase, of almost any species. Even so, one pair of elephants could (hypothetically) produce enough progeny to cover the Earth's surface in only 1000 years. Fortunately, elephants rarely reproduce at the maximum rate. (*Photo, Kennan Ward/The Stock Market*)

and salamanders that need to migrate overland to breeding ponds and streams. If a local population goes extinct, individuals from another preserve can then travel to the empty habitat patch and recolonize it. Other winners would be large predators, such as mountain lions and wolves that need large ranges.

We can preserve many more species if we set aside large preserves, connected by corridors and surrounded by buffer zones.

HOW DO POPULATIONS GROW?

We have discussed how populations go extinct. But all organisms have phenomenal powers of reproduction. Populations of bacteria, cockroaches, and rats sometimes seem to explode out of nowhere. Indeed, if all of the descendants of just 100 California sea stars survived and reproduced for 15 generations, the number of sea stars would exceed the number of electrons in the visible universe.

All organisms are capable of multiplying amazingly rapidly when conditions are ideal. Even elephants, which rarely have more than about 10 offspring in their 60-year life spans, can generate a lot of descendants. For example, if all of the 10 offspring of an elephant survived, and each of them had 10 offspring, and so on, the original elephant could have as many as 100,000 great, great, great grandchildren in just 300 years (Figure 28-6).

In nature, we sometimes see such unrestrained growth when a species invades a new territory, or, for example, when a protozoan discovers an uninhabited human gut. But neither elephants nor sea stars nor protozoans can multiply without end. Even though natural populations are capable of huge growth rates, they rarely grow so fast that they eat themselves out of house and home. Whenever populations begin to grow too fast, they eventually stop. When food or space runs low, or wastes degrade the environment, individuals tend to curb their reproduction. Sometimes, predators or disease bring a population under control.

Natural populations may fluctuate from season to season and from year to year, but most are surprisingly stable over the long run. Two important questions in population ecology are, How do populations grow? and What factors limit population growth?

What Is Exponential Growth?

Under the most favorable conditions, the number of *Escherichia coli* in a nutrient-rich broth can double every 20 minutes. After the first 20 minutes, one bacterium forms two; after another 20 minutes, the two have formed four; after another 20 minutes, there are eight; and so on. We say that the bacteria have a **doubling time** of 20 minutes.

We can calculate the number of bacteria (N) from the number of times the population has doubled since the time at the beginning of the experiment (t) using the following equation:

$$N = 2^t$$

For example, if 60 minutes had elapsed, the population of *E. coli* would have doubled three times. So t would equal 3, and

$$N = 2^3 = 2 \times 2 \times 2 = 8$$

In this equation, t is an "exponent," hence we say the bacteria's growth is **exponential growth,** a growth pattern in which a population doubles in some constant period of time.

Exponential growth means that the larger the population, the faster it increases. For example, we know that 16 *E. coli* cells could double in 20 minutes to 32 cells. The population would increase by 16 cells. But a population of 16 million cells would also double in 20 minutes. That population would increase by 16 *million* cells over exactly the same time period. The large population and the small one would have the same doubling time, but the increase in actual numbers would be much larger in the second case.

When bacteria increase exponentially, a graph of their numbers looks like the letter J. At the bottom of this **J-shaped curve,** the population size (N) increases rather slowly. Then it accelerates. When there are no limits to population increase, N increases faster and faster.

We can describe this ever-increasing growth as follows: During unrestrained exponential growth, the rate of change in the number of organisms is proportional to the number of organisms present in the population. That is, the bigger the population, the faster it grows. Human populations are a good example of a natural population with a J-shaped growth curve (Figure 28-7).

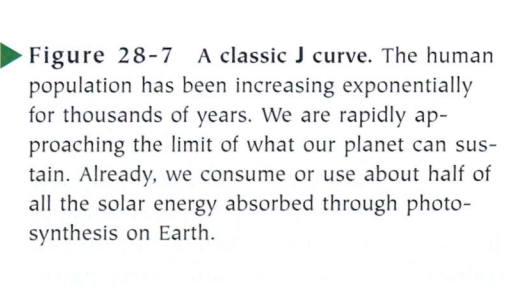

▶ **Figure 28-7** A classic **J curve**. The human population has been increasing exponentially for thousands of years. We are rapidly approaching the limit of what our planet can sustain. Already, we consume or use about half of all the solar energy absorbed through photosynthesis on Earth.

The Plague

Population (billions)

6

5

4

3

2

1

1000 B.C.   0   A.D. 1000   A.D. 2000

The growth rate of a population, called *r*, is the difference between the average birth rate and the average death rate. **Birth rate** is the number of individuals born each year per 1000 individuals. **Death rate** is the number of individuals who die each year per 1000 individuals. The population will grow only if the birth rate exceeds the death rate. When the birth rate of a population, *b*, is at its maximum, and the death rate, *d*, is at its minimum, *r* is as large as it can be. This is called r_{max}, or the **intrinsic rate of natural increase.** For example, if *b* is 10 per 1000 and *d* is 2 per 1000, then *r* (which equals $b - d$) would be 8 per 1000. In populations that are declining in numbers, *d* is greater than *b*, and *r* is a negative number.

The change in total population size per unit time, sometimes called *G*, is simply *r* times *N*, or *rN*.

$$G = rN$$

This is just another way of saying that the larger a population is, the faster it grows.

All populations are capable of exponential growth.
The larger they are, the faster they grow.

What Limits the Size of a Population?

Even in the laboratory, exponential growth cannot continue for long. We may see a J-shaped population curve when a bacterium has first colonized a glass container of a nutrient-rich medium. But even *E. coli* growing under nearly ideal laboratory conditions soon find limits to their growth. The bacteria run out of food and space, and their own wastes begin to poison them. The concentration of bacteria gradually reaches a plateau. If we look at the whole growth curve, we see that it consists of three parts—an initial "J phase," in which the growth rate is accelerating; then a deceleration phase; and finally a gradual leveling off. The whole curve now looks more like a tilted and stretched out "S," a classic **S-shaped curve** (Figure 28-8).

Ecologists modify the growth equation to describe the decelerating growth in natural populations by adding a multiplier. The new equation must express the slowing growth rate in terms of population size. This is because the larger the population size (*N*), the faster growth slows down. Ecologists first define the plateau value of population size as the **carrying capacity** of the environment for a particular species. The carrying capacity, also called **K,** is the maximum population density that the environment can support.

When the environment is not yet filled to its capacity, there remains a fraction of *K* that is left to be filled. For example, if the carrying capacity for houseflies in Australia were

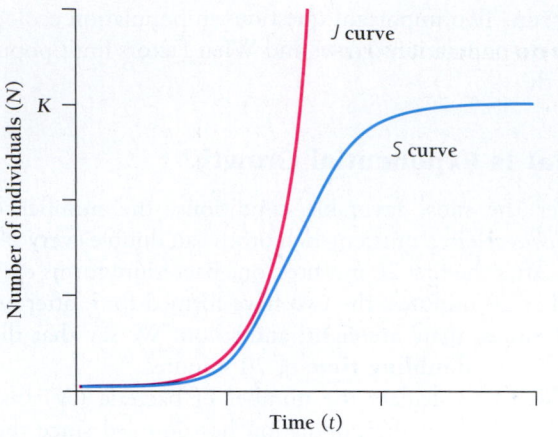

Figure 28-8   How populations grow. All animals are capable of exponential growth for short periods, as represented by the J curve. In natural populations, growth declines as population size increases and approaches the carrying capacity (*K*) of the environment, as represented by the S curve.

100 flies per square meter, and the population density *N* were 70 flies per square meter, then the fraction of *K* left would be $(100 - 70)/100 = 30$ percent.

Population ecologists take that simple fraction of *K* and multiply it by *rN* to get a measure of the change in the population growth over time. The equation they use is called the **logistic growth equation:**

$$G = rN(K - N)/K$$

But it merely describes the **S**-shaped curve. *G* expresses how much population size (*N*) changes as over time (*t*). (*G* is often expressed, more formally, as *dN/dt*, but this means something only if you have studied calculus.)

If the equation is plotted as a curve, it starts out like the J-shaped curve, with an ever-increasing rate of population growth. That is, when the population (*N*) is very small, the unfilled fraction of the carrying capacity—$(K - N)/K$—is about 100 percent, or 1. As the population increases, however, the fraction of unfilled *K* decreases and the growth rate slows. When population size reaches the carrying capacity, the environment cannot support any more growth. Then $(K-N)/K = (K - K)/K = 0$, the fraction of unfilled *K* equals 0, and the population stops growing. The population has reached **zero population growth.**

The carrying capacity of the environment limits
the size of a population.

What Determines Whether a Population Grows?

The carrying capacity of the environment for a particular species depends on the needs of the species and, therefore, on a multitude of factors—food supply, territories, predators, and competitors, to name just a few. Bees, for example, need nectar, pollen, and a safe place for a hive. Even if a given habitat has plenty of flowers, the carrying capacity for bees may be low if there are only a few good places for a hive within flying distance of the flowers.

Changes in a population size ultimately result from just four processes: birth, death, **emigration** (movement out of the population), and **immigration** (movement into the population). The net change in population size is the sum of births and immigration, minus deaths and emigration.

Anything that influences any of these four factors will cause an increase or decrease in carrying capacity. Competition, disease, number of prey species, number of nesting sites, rainfall, soil fertility, and hundreds of other factors can all affect the value of K—the number of individuals of one species that a given habitat can support.

Density-dependent factors

The rates of birth, death, immigration, and emigration all depend on the population density. As a population's density approaches the carrying capacity, its members must increasingly compete with one another for limited resources. For example, one study of a population of 25 pairs of tawny owls at carrying capacity suggested intense competition. The 50 owls, capable of producing as many as 100 offspring per year, successfully fledged only 18 owlets. Because of limitations in the food supply—in this case, rodents—some owls were un-

able to feed their offspring, others did not bother to incubate their eggs, and still others did not even breed. Thus competition for food depressed the birth rate. But even 18 owls was too many. Only 11 territories opened up, as a result of deaths among the adults. So only 11 of the 18 fledglings could find territories on which to breed. The other seven young owls had to emigrate.

We can see how a population at carrying capacity can have a low birth rate. Animals run out of food, nesting sites, or territories. Plants competing with one another for water or light cannot get enough. High population density can also encourage emigration and discourage immigration (Figure 28-9). In addition, high population density can increase the death rate from predation, infectious diseases, and parasites.

Predation, parasitism, disease, competition (both intraspecific and interspecific), and emigration all limit growth in proportion to the density of the population. The denser a population, the more slowly it grows. Such factors are said to be **density dependent.**

Density-independent factors

Some populations can also be limited by **density-independent** factors, including fire, drought, storms, tornadoes, volcanic eruptions, and other natural disasters. For example, severe storms can all but eliminate a population of butterflies long before the caterpillar larvae run out of plants to eat. Changes in population density that result from storms and other density-independent limitations do not conform to the logistic growth equation.

Both density-dependent and density-independent factors limit population growth.

Figure 28-9 A plague of locusts. When the population density of certain species of locusts begins to rise, the individuals develop into longer-winged, more gregarious types that fly together. When the proportion of these gregarious types reaches a certain threshold, they rise in great masses and emigrate to a new area, consuming everything in their path—a classic plague of locusts. *(Gianni Tortoli/Photo Researchers, Inc.)*

Do Natural Populations Actually Grow Exponentially?

The growth of bacteria in the laboratory follows the logistic curve closely. Wild populations also follow the logistic curve but usually fluctuate more. Often, population density will temporarily **overshoot** the carrying capacity (K) of the environment. For example, a population of mountain sheep may wildly exceed the carrying capacity, then drop far below the carrying capacity, and then cycle back up to K again.

When a population exceeds the carrying capacity of its environment, it must decrease. The population may then increase again, fluctuating up and down indefinitely. Many populations fluctuate up and down for a while, then gradually level off at the carrying capacity (K) (Figure 28-10). Whether the fluctuations continue depends partly on the extent of the original overshoot. The overshoot depends, in turn, on r_{max}. A species capable of growing rapidly, such as a sea star, is much more likely to overshoot its carrying capacity than a species such as an elephant, whose intrinsic rate of growth is low.

Natural populations do not usually follow the logistic curve exactly.

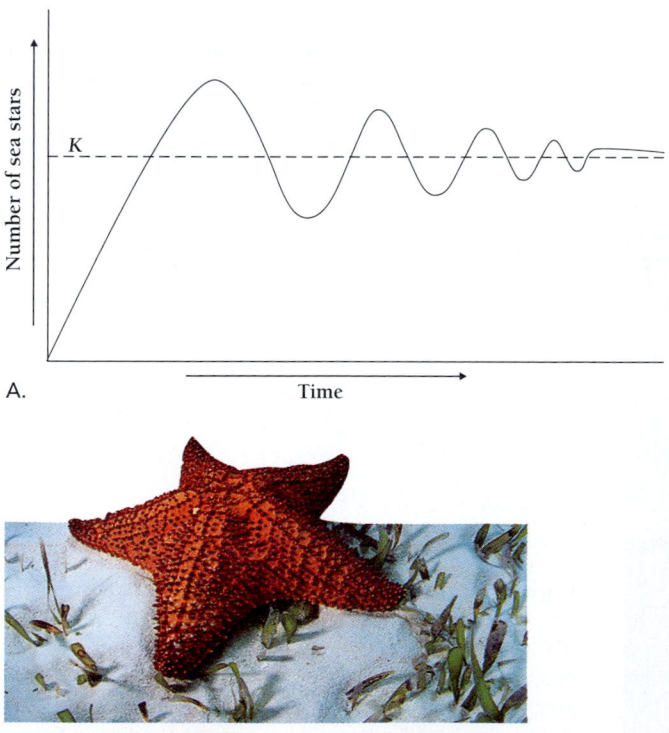

A.

B.

Figure 28-10   Sea stars approach K. This starfish population grew rapidly, then fluctuated around K. (B, Tom Stack/Tom Stack & Associates)

Why Do Some Populations Cycle Up and Down?

If populations rarely show uncontrolled exponential growth, what limits population growth? The answer seems to vary with the species and the circumstances. Predators sometimes control herbivore populations. For example, the Australian ladybird beetle introduced to California orange groves in 1888 successfully controlled an outbreak of scale insects.

Other factors often play roles, as well. The supply of plant food can limit the growth of herbivores. And plants can limit the growth of the herbivores that eat them. For example, many plants produce toxins in response to damage to their leaves. The toxins may interfere with the digestion, growth, or reproduction of insect herbivores, limiting their population growth.

Biologists have long been fascinated by populations that show regular cycles of population density. In Alaska, for example, populations of rodents called lemmings cycle up and down every 3 to 6 years. These animals can start from a density of less than one lemming per hectare (2.5 acres) and reach levels of more than 25 per hectare. At peak densities, lemmings destroy their food supply and attract throngs of predators. With little food, the lemmings stop reproducing and their population plummets. The cycle then repeats itself. Many other animals, including snowshoe hares, arctic foxes, and ptarmigans, show similarly regular cycles.

Exactly why populations cycle, however, remains a mystery. Biologists have hypothesized that cycling results from fluctuations in the food supply. But that may not always be true. In two groups of meadow voles—relatives of lemmings—researchers found that even though one population had ten times as much food as the other population, it cycled as much as the population with less food.

Population cycles probably result from interactions between species. So far, however, biologists have not discovered exactly how or why cycles occur.

WHAT FACTORS INFLUENCE THE VALUES OF *R* AND *K*?

Species differ greatly in their intrinsic rates of increase, r_{max}. Birds that lay six eggs per clutch, for example, have higher values of r_{max} than those that lay only two. Humans that begin reproduction at age 15 have a greater r_{max} than those who delay until age 30. And a tree that makes seeds for 200 years has greater r_{max} than one that reproduces for only 10 years.

Number of offspring, age of first reproduction, and life span are all traits collectively called **life history traits.** Every species has its own life history. But some species possess all of the traits that tend to maximize r_{max}. Such species are often called *r-selected species*. Examples of classic *r*-selected species are insects and other invertebrates. Such species tend to have short life spans, early first reproduction, and many offspring. Other species have just the opposite traits—long life spans, late first reproduction, and few offspring. Such species are called *K-selected species.* Most mammals and birds are *K* selected compared to most insects.

In the 1960s and 1970s, biologists theorized that natural selection generated species with "*K*-selected" and "*r*-selected traits." Unstable environments, such as those with unpredictable storms or volcanic eruptions, were supposed to produce "*r*-selected species." Stable, predictable environments were supposed to produce "*K*-selected species."

But years of experiments have produced no evidence to support this hypothesis. Species fall along a continuum for these traits, with about half of all species showing no tendency in either direction. Increasingly, biologists suspect that the names "*r*-selected" and "*K*-selected" just describe groups of traits that are associated historically. By analogy, we could divide all animals into those with backbones and those without. Vertebrates all have bony skulls and tails, for example. And, except for a few renegades like turtles and armadillos, none have an external skeleton. But we can't conclude that a certain environment selects for vertebrates and another selects for invertebrates. Both kinds of animals live in just about every habitat on Earth. The same is true of "*r*-selected" species and "*K*-selected" species.

How Do Life Histories Affect Survivorship and Age Distributions?

Elephants invest energy in the production and care of a few, large offspring. As a result, the offspring have a good chance of living long enough to reproduce, usually several times. In contrast, when sea stars reproduce, most of the millions of larvae are filtered from the water by larger animals. Only a tiny fraction of the sea star larvae grow up to be sea stars that have offspring of their own.

One way of representing these two contrasting life histories is a **survivorship curve,** a graph that shows the fraction of a population that is alive at successive ages. The opposite of survivorship is **mortality,** the fraction of a population that dies at a given age. Figure 28-11A shows three kinds of survivorship curves. In a **convex** (or **type I**) survivorship curve, survivorship starts out high and decreases slowly with age. Then, at some point, survivorship begins to decline more quickly. Organisms with convex survivorship curves have a good chance of surviving until they have finished reproducing and caring for their young. Then the chance of dying in a given year increases dramatically. Elephants, dall sheep, adult fruit flies, and humans tend to show convex survivorship curves (Figure 28-11B).

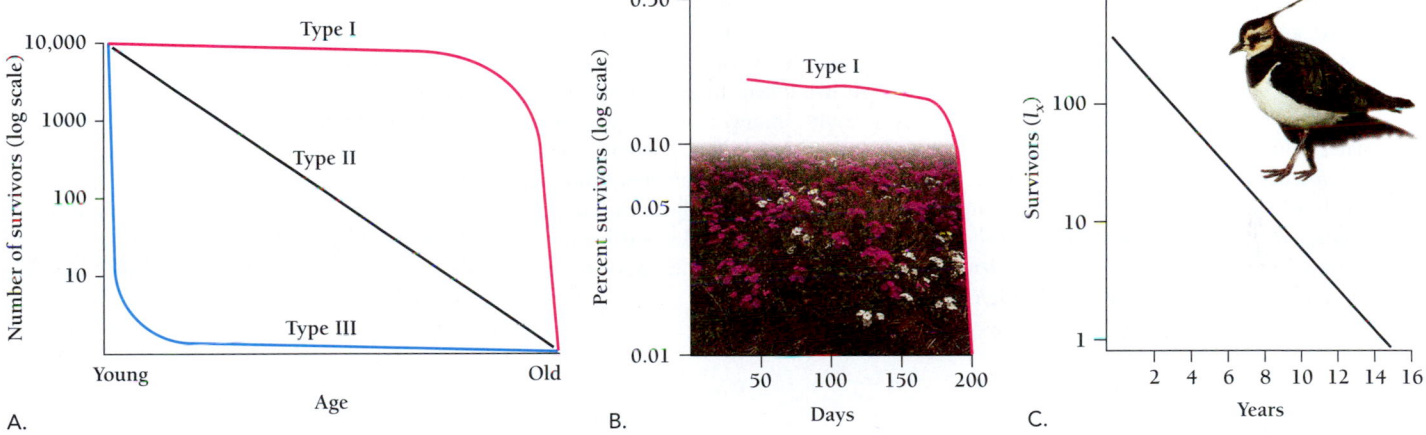

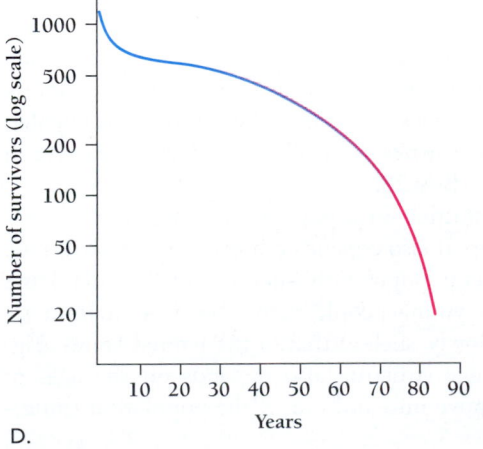

Figure 28-11 Life and death. We all die eventually. But some species are more likely to die young, and some are more likely to die old. Survivorship curves are a way of showing different patterns of mortality and survivorship. A. The members of a type I species survive until they are old and then rapidly die off. The members of a type II species are equally likely to die at every age. Most members of a type III species die soon after they are born, or hatch. Those few that survive live a long time. B. The flower Drummond's phlox shows a classic type I survivorship curve, starting with germination. C. The European lapwing shows a classic straight type II survivorship curve. D. Most natural populations show a combination of shapes, with, for example, high early mortality, a period of steady mortality, and then increased mortality late in life. The survivorship curve shown here is for humans living in the city of Breslau, in Poland in the 17th century. High mortality among infants and children is typical of most human populations in the world today. (*B, Leroy Simon/Visuals Unlimited; C, Roger Wilmshurst/Photo Researchers, Inc.*)

BOX 28.1

Why Are Age Distributions of Males and Females Different?

Age distributions such as those in Figure 28-12 traditionally show males and females separately. We usually think of males and females as being represented in about equal proportions, but in many sexually reproducing organisms, large differences exist in the ratio of males to females.

In certain species of wasps, for example, most offspring are female. All fertilized eggs become females and all unfertilized eggs become males. Consequently, a shortage of males means that many eggs will go unfertilized and more males will hatch. On the other hand, an excess of males means that most of the eggs will be fertilized and will therefore be female. Other organisms use still more complicated negative feedback mechanisms to maintain their sex ratios.

Ants, known for their cooperative behavior, are less cooperative when the subject is sex ratios. A queen ant is equally related to her female and male offspring and bears equal numbers of each. But the males, which develop from an unfertilized egg, are haploid, and the female workers, which develop from fertilized eggs, are diploid. As a result, female workers who share the same father are more related to one another than they are to their brothers.

When Finnish researchers studied the sex ratios of ants from 59 colonies of wood ants, they discovered that in colonies where all the workers share the same father, the males die off disproportionately. Even though the queen lays equal numbers of male and female eggs (the researchers determined the sex of 3000 ant eggs), adult females far outnumber the adult males. What happens to the males? The researchers suspected that female workers either neglect the male larvae or kill them outright. Indeed, in the laboratory, eggs of males added to a nest of closely related females disappeared without a trace.

The situation was very different in colonies in which the queen had mated with two or more males. In such colonies, the females are less related to one another, and the ratio of male and female adult ants remains about equal.

What about humans? Are our sex ratios 50:50? Most people know that women live longer than men, on average. As a result, among older people, women outnumber men. Not everyone knows, however, that in all countries, more baby boys than baby girls are born each day. The difference is not great. In North America, for example, 105 boys are born for every 100 girls.

By the time each cohort reaches reproductive age in the early twenties, however, the sex ratio is nearly exactly 50:50. This adjustment occurs gradually. Baby boys are slightly more likely than baby girls to die of infections and other health problems. Older boys are slightly more likely than girls to die in accidents.

Finally, adolescent boys are dramatically more likely to die in all kinds of accidents than adolescent girls. For example, the leading cause of death for all Americans between 5 and 27 years of age is motor vehicle accidents. But a teenage boy driving a car is more than three times as likely to die in a car accident as a teenage girl driving a car. Among 21- to 24-year-olds, the ratio approaches 4 to 1. It might seem amazing to think that normal 20th-century behavior plays a role in natural selection. But while the kinds of risky behavior that teenage boys engage in may change from century to century, the result—a high death rate—remains the same.

The physiological mechanisms that cause a sex ratio of 105 to 100 are not yet known. We can hypothesize, however, that this skewed ratio has some adaptive value. How would you test such a hypothesis in humans or in some other organism?

Some organisms are as likely to die in midlife as when young or old. When survivorship is plotted logarithmically, birds and many other such organisms show a **diagonal** (or **type II**) survivorship curve—a straight, declining line (Figure 28-11C). Finally, many species are most likely to die early in life. Their survivorship curves are called **concave** (or **type III**).

Few organisms have survivorship patterns that fit any of these curves exactly. The commonest natural pattern is a high mortality rate among the youngest and oldest individuals and a low mortality among those in midlife (Figure 28-11D).

The **age structure** of a population is the fraction of individuals of various ages. We can represent the age structure of a population by a graph such as those in Figure 28-12, which show the number of people in each 5-year age group in Swe-

den, the United States, and India in 1997. Each such group forms a **cohort,** the set of individuals that enter the population, or are born, at the same time. In this case, we show males and females separately, since their age distributions are slightly different. In all countries, more males are born than females. But in developed countries especially, more females survive in every age category (Box 28.1).

The age distribution in a population is not just a function of survivorship; it also depends critically on the birth rate. A rapidly reproducing population, such as that of India, has a greater fraction of young people than does a population reproducing more slowly, such as that of the United States (Figure 28-12). The age structure also depends on the ages of individuals who move into and out of the population (immi-

India: 1997

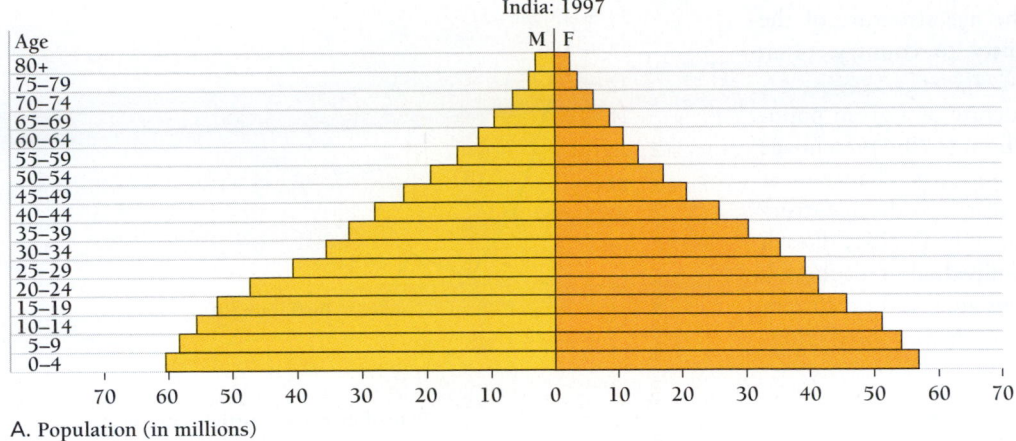

A. Population (in millions)

United States: 1997

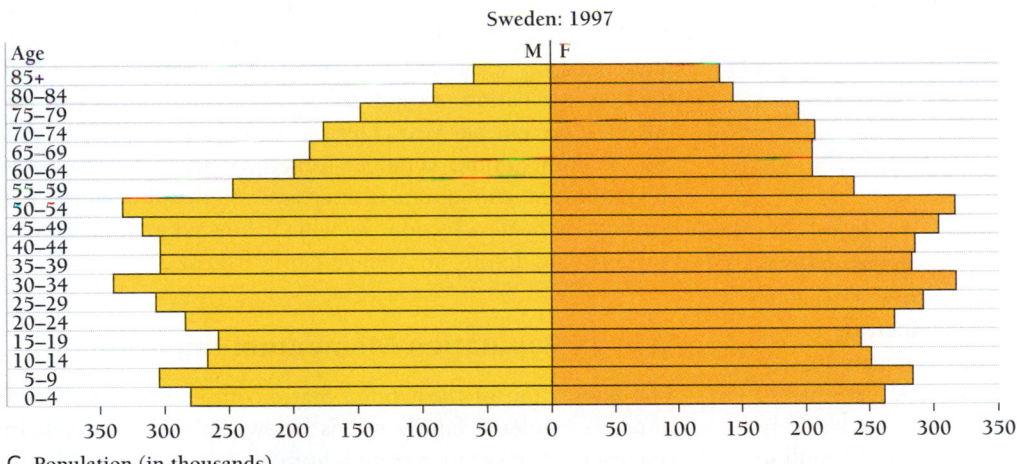

B. Population (in millions)

Sweden: 1997

C. Population (in thousands)

Figure 28-12 Age distributions. A. India, which has a high birth rate, exhibits the classic pyramidal age structure of an expanding population. B. In the United States, a slowly declining birth rate and an increasing life span are gradually transforming the age structure from a pyramid into a rectangle. Generation Y is in green. C. In Sweden, an almost zero growth rate results in stable population size, with a rectangular age structure. Sweden has as many middle-aged people as young people. If you turn a stable age structure on its side, you see a survivorship curve. What type of survivorship curve does Sweden have? (*Source: U. S. Census Bureau, International Data Base*)

grants and emigrants). As a result, the age structure of the human population differs from country to country, often dramatically.

We can predict the future age structure of a given population if we have the following information: (1) the present age structure, (2) the mortality of each cohort (the chances that individuals of a given age will die each year), (3) the age structure of immigrants and emigrants, and (4) the **fertility** of each cohort (the number of offspring each individual in a cohort is likely to produce). Most such predictions are based on current conditions. When mortality or fertility change, predictions may be wrong.

Differences in survivorship curves lead to differences in age distributions.

Demography: How Do Scientists Measure Populations?

Studying the growth and changes in the structure of human populations is the concern of **demography** [Greek, *demos* = people + *graphos* = measurement], the statistical study of populations. Demography is the population biology of humans. It is important for economists, health planners, and insurance companies, as well as for biologists. Demographers use a variety of statistical measures to describe populations. Mortality is expressed in terms of the death rate, the number of individuals per 1000 who die each year. Fertility is expressed in terms of the birth rate, the number of individuals per 1000 who are born each year. For example, in the United States in 1989, 3.9 million babies were born in a population of nearly 250 million, for a birth rate of 15.6 per 1000. In the same year, 2.1 million Americans died (8.4/1000). Net immigration was 700,000. So the total increase in the population was 2.5 million.

$$3.9 \text{ million} - 2.1 \text{ million} + 0.7 \text{ million} = 2.5 \text{ million}$$

Demographers use several measures to express population growth. One is the **annual rate of increase,** the actual percentage by which a population increases each year (Figure 28-13). For example, the 2.5-million-person increase in the 1989 population of the United States came from a population of 250 million, for an annual rate of increase of 1 percent (10 per 1000). The annual rate of increase for the entire world in 1989 was about 1.66 percent (16.6 per 1000), or 87 million more people (Web Bit 28.3 The Human Population is Growing Exponentially).

Another measure of population growth is **completed family size,** the average number of children born to each family that reach reproductive age. For example, the average completed family size in the United States in 1992 was 2.0 children. **Replacement reproduction** is the family size at which each couple is replaced by just two descendants. At the replacement

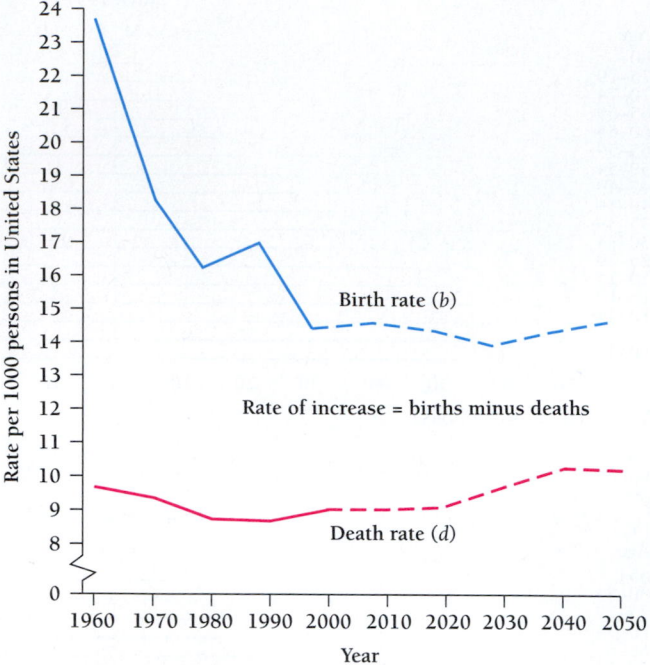

Figure 28-13   What is the natural rate of increase? The birth rate minus the death rate is the natural rate of increase. In the United States, the death rate is increasing as baby boomers age and die, while the birth rate is leveling off. Based on this graph, what was *r* for the year 1980? What is *r* today?

level, a population neither grows nor shrinks. Replacement level in the United States is 2.1 children per family. The additional 0.1 makes up for the small number of children who die young or otherwise have no children themselves.

We can see that the population of the United States is not quite replacing itself each generation. But we can also see that the total population of the United States is still growing, at the rate of 1 percent per year. This paradoxical growth is not due only to immigration, as people sometimes argue. Without immigration, the annual rate of increase would still be about 0.7 percent. In the next section, we will see how population growth can continue despite an average family size of 2.0 or less.

Why Is Population Momentum Important?

The population of the United States continues to grow even though completed family size is below replacement level. In fact, even if tomorrow morning human couples the world over began having no more than two children each, the world population would continue to grow for another 50 years, doubling to 10 billion by about 2050. This paradox is called **population momentum** (Figure 28-14).

The reason for population momentum is simple: in a rapidly growing population, a large proportion of individuals will be young. In fact, about 20 percent of the people in the world today are younger than 15 years old. Even if they only replace

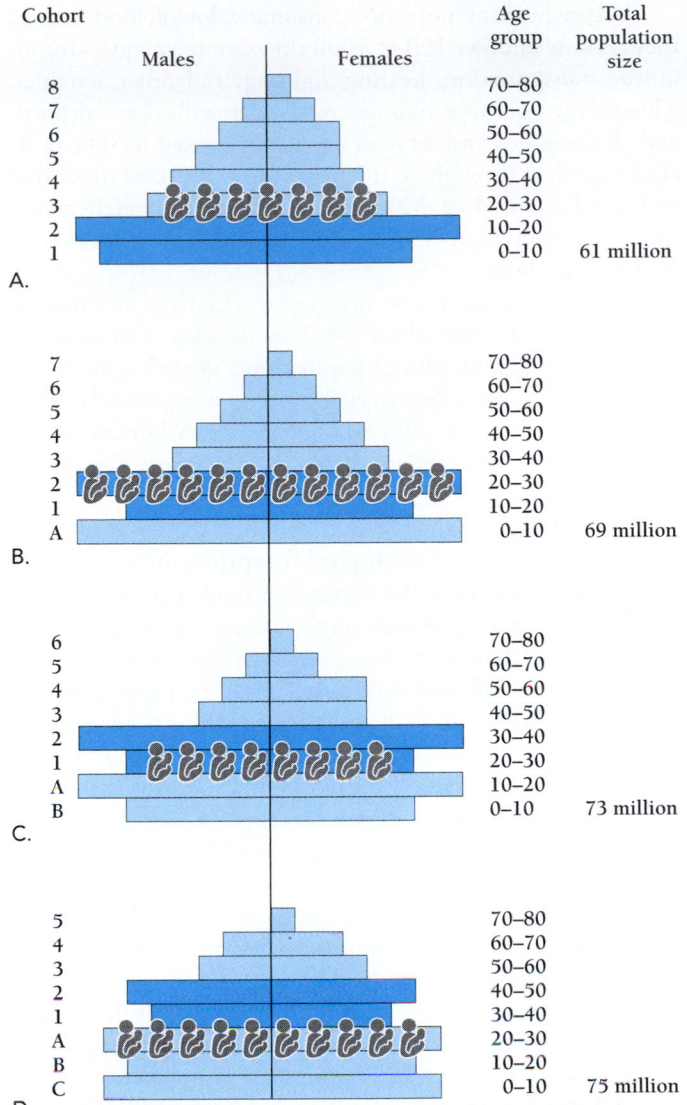

Cohort			Age group	Total population size
	Males	Females		

A.

themselves, they will increase the world population by 20 percent (more than a billion people). And they will meanwhile continue to live alongside their billion-plus children and their billion-plus grandchildren, and they will not begin to die until they are in their 60s, some 50 years from now.

Even if human couples today began having no more than two children each, the world's population would continue to grow for another 50 years.

WHAT IS THE EARTH'S CARRYING CAPACITY FOR HUMANS?

Demographers predict that the world population will increase by 3.3 billion people in the next 50 years. Where will population increase end and how? Just as sea stars cannot go on multiplying exponentially, human beings cannot continue to increase without limit (Figure 28-15). Our growing population raises many questions. Some questions concern our impact on other organisms. Some concern our impact on ourselves. How large, for example, can the human population become before it destroys the biosphere that supports us? If we decrease our birth rate, can we make our population level off or decline to a population size that the Earth can sustain? Or are we headed for a catastrophic population crash caused by famine, disease, or war?

Many people argue about how many people our planet can support. But the Earth's carrying capacity for humans depends partly on what else we want on Earth with us. If all the sunlight that strikes the Earth could be used to feed human beings, we might turn Earth into a giant feedlot for 40 billion or more humans.

Figure 28-14 Population momentum. A population can keep growing even after individuals have no more children than they need to replace themselves. Growth continues when a population has been growing rapidly because the population includes large numbers of children and adolescents. When these cohorts grow up, they replace themselves, each cohort doubling again and again. For simplicity, we have assumed that all reproduction occurs when a cohort is between 20 and 30 years old. A. Rapidly growing population. The members of Cohorts 3 and 4 have more than replaced themselves. Total population size is 61 million. B. Starting here, new parents will have only two children, enough to replace the mother and father. Yet, because Cohorts 1 and 2 are so large, their reproduction causes the total population to increase to 69 million. C. Replacement is still 2.0 children. Yet, as Cohort 1 moves into its reproductive years, the population continues to grow, to 73 million. D. Replacement is still 2.0 children per family. But the many grandchildren of the enormous Cohort 2 are now coming into the world. In addition, more individuals are living into their 70s and 80s than did members of previous cohorts. The population, now 75 million, is still growing.

Whether we have yet exceeded the Earth's carrying capacity for humans is a matter of debate.

How Well Is Our Energy Supply Keeping Pace with Human Energy Consumption?

The vast majority of humans live in cities and suburbs, where little if any food is grown. Yet each city dweller consumes about a million kilocalories per year. The principal source of this food is farms and ranches that cover one-third of the land. Agriculture provides additional energy to, or **subsidizes,** cities and suburbs.

Agriculture is itself subsidized. The productivity of a cornfield depends not only on sunlight, but also on the fossil fuels needed to run tractors, to pump irrigation water, and to make chemical fertilizers. A farm subsidized with fossil fuel produces about ten times what an unsubsidized, natural ecosystem does (Figure 28-16). All of the extra energy is called an **energy subsidy.**

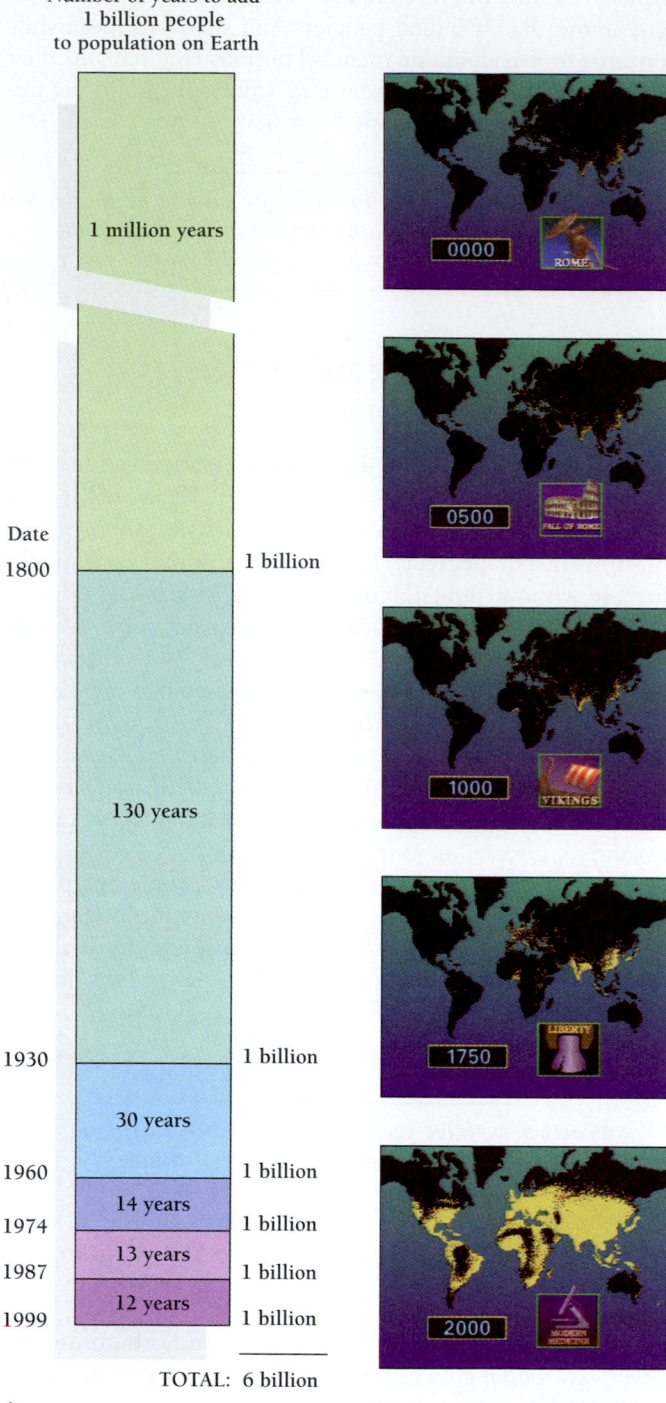

Number of years to add
1 billion people
to population on Earth

1 million years

130 years

30 years

14 years

13 years

12 years

Date
1800 — 1 billion
1930 — 1 billion
1960 — 1 billion
1974 — 1 billion
1987 — 1 billion
1999 — 1 billion

TOTAL: 6 billion

A.

B.

Figure 28-15   Human population growth. A. The number of years it takes to add 1 billion new people to our population has decreased from 130 years in the 19th century to a low of 12 years today. B. Population estimates based on historical records show that large population increases began in India and China centuries earlier than in other countries. In just 250 years, the United States has gone from one of the least populous areas on Earth to the third most populous country in the world. (*B, Reprinted with permission from Zero Population Growth, Inc.*)

Urban humans not only consume a lot of food energy, they also use another 100 to 1000 times more energy—for industry, transportation, heating, lighting, and other activities. This energy goes to buildings, roads, dams, bridges, airports, and all the goods and services we've grown used to. The overall energy flow through an urban area may be 1000 times that of a pond or meadow. Much of this energy ultimately comes from fossil fuels pumped from wells in Alaska, the Middle East, or the ocean floor.

Because humans derive much of their energy and materials from areas far from where they live, the idea of carrying capacity, as applied to natural populations, doesn't quite work. Ecologists talk about the carrying capacity of a particular habitat patch or ecosystem. But technology allows humans to exceed such limits, for humans derive resources from all over the world.

Some ecologists suggest that we think of the human use of resources in terms of **ecological footprint,** the area of land needed to provide all of the resources a single person uses. For example, the average person in the United States uses more than ten times the energy that an average human in the world as a whole does. We use almost ten times the paper, generate five times as much carbon dioxide, and use three times as much fresh water. To fulfill these demands, an average American needs about 5.1 hectares—our ecological footprint. In India, the ecological footprint of the average person is just 0.4 hectares, less than one-twelfth ours (Box 28.2).

When people talk about overpopulation, they often mention the nations with the largest populations—China, India, and the United States—or those with the fastest-growing populations, in Africa and the Middle East. For example, Saudi Arabia's rate of natural increase is 5.5 times that of the United States. But Saudi Arabia has only 20 million people, to our 270 million. So while their population is increasing by 100,000 people per year, ours is increasing by 400,000 people per year.

When we consider total impact on the environment, however, we should look at both population size and ecological footprint together. India's 1.0 billion people each need about 0.4 hectares, or 0.4 billion hectares for the whole country (0.4 × 1.0 billion). But America's 270 million people each need 5.1 hectares, or 1.35 billion hectares for the whole country (5.1 × 0.27 billion).

The area of the United States is only 0.91 billion hectares, however, so we use half again as much land to supply our ever-increasing demand for cars, clothes, and other goods. Even though we are far fewer in number, our impact on the world is more than three times greater than that of India.

Cities depend on energy subsidies and agriculture, which also depend on energy subsidies. Countries with the highest standards of living make the greatest demands on the world's resources.

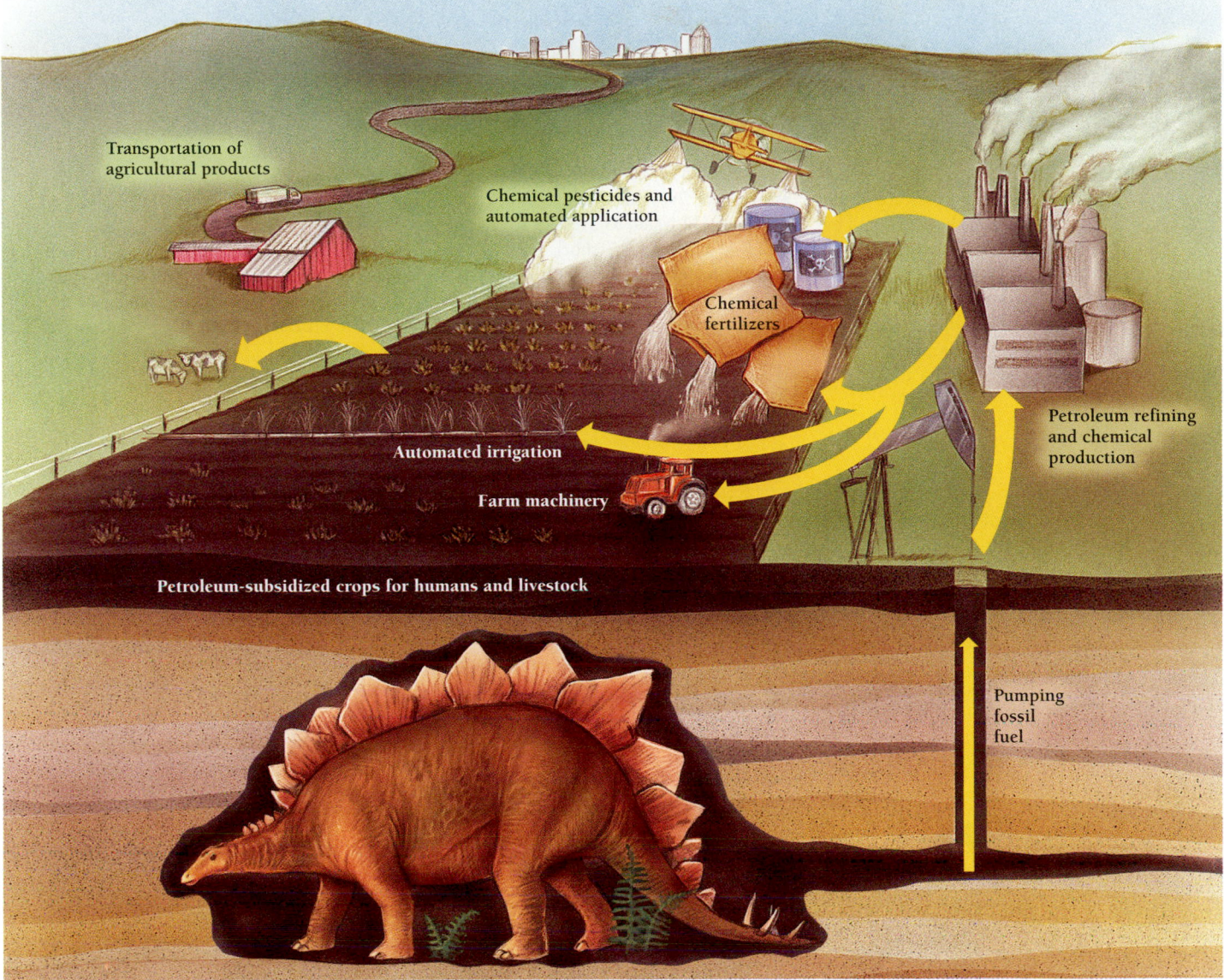

Figure 28-16   Subsidized agriculture. Agricultural subsidies and fossil fuel subsidies are the lifeblood of modern cities and suburbs. Modern agriculture, in turn, depends on energy subsidies from ancient stores of fossil fuels—carbon compounds whose energy came from the sun 65 million years ago. Oil-subsidized agriculture can have up to ten times the productivity of a natural ecosystem. However, the supply of fossil fuels is finite.

How Well Has Agriculture Met the Human Demand for Food?

In the late 1960s, agricultural researchers recognized that many countries were outgrowing their food supplies. Every day, hundreds of millions of people were literally starving. To solve the hunger problem, researchers bred rice, wheat, and other crops that could produce up to ten times as much food as older varieties. The effort, called the *Green Revolution,* was a boon to countries such as India, China, and Mexico, where crop increases temporarily outstripped human population growth.

Unfortunately, many of the Green Revolution's "miracle crops" depend on the use of chemical fertilizers, irrigation, and fossil-fuel-powered farm equipment. As American ecologist

H.T. Odum wrote, the increased food supply is "partly made of oil." The world's poorest farmers cannot afford to buy the fertilizer, machinery, and fuel they need to grow the newer varieties of crops. But because few farmers now use the old crop varieties—bred over hundreds of years to grow well in particular regions—these old standby varieties are becoming unavailable.

Hunger continues to be a serious problem. Most experts agree that the world's farms currently produce enough to feed the world population. Delivering that food to all who need it, however, has proved to be a staggering problem. Each year, about 10 million children as well as 10 to 15 million adults starve to death—more than the number of people in the whole

BOX 28.2

How Much Do We Have and Need?

"In the 21st century, we are going to pass through a bottleneck. The land required to maintain one American's standard of living, on average, is 12 acres; the amount required to sustain the current standard of living of someone in one of the developing countries, where 80 percent of the world's population lives, is one acre. Now here's the problem: The whole world wants to live like Americans. Thus we have the aftershock following the population explosion: not only too many people, but people attempting legitimately, understandably, to increase the quality of their lives. It's been estimated that to bring the whole world up to America's standard of living would require two more planet Earths. The results, politically, economically, and environmentally, of the rush to achieve that impossibility in the next century will be our greatest problem." —E.O. Wilson, Professor of Biology, Harvard University

Figure A Two families, in India (*left*) and Germany (*right*), with their possessions. *(Peter Ginter/Material World)*

state of New York. To put it another way, of the 150,000 people who die each day, 70,000 starve to death. At least another half a billion people are profoundly malnourished, with still more living on inadequate diets that leave them susceptible to disease and premature death. In Bangladesh, life expectancy is 48 years, compared to 75 years in the United States and Canada.

Experts doubt the situation will improve. Between 1950 and 1984, global grain production increased enough to stay ahead of population increases. But by 1986, the rate of increase had begun to decline (Figure 28-17). In 1988, severe droughts and crop failures in North America, the former Soviet Union, and China reduced world grain production by 5 percent. Grain stores from previous good years supplied most of the world population during the drought. But the steadily rising demand for food—due to increasing population size—is making it increasingly difficult to stockpile reserves against future crop failures. Meanwhile, global warming has made weather, and therefore crop yields, less predictable than before. Local crop failures are becoming more and more likely.

The Green Revolution increased the world food supply and so Earth's carrying capacity for humans. Today, however, population growth is outstripping the increased food supply.

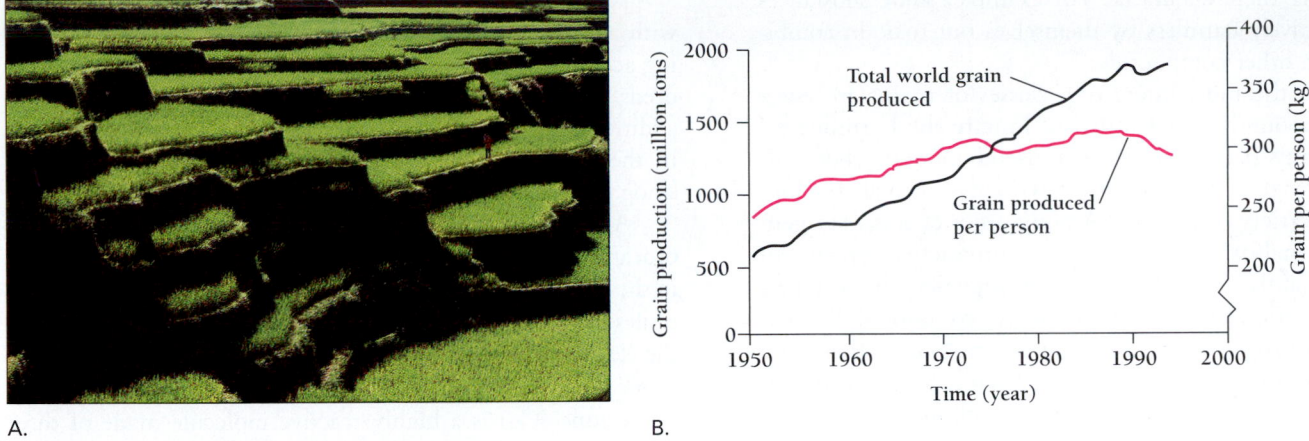

Figure 28-17   Grain production. A. Classic, terraced Asian rice paddy. B. While world grain production is leveling off, the human population continues to grow. As a result, the amount of grain produced per person is declining. (*A, R. Ian Lloyd/The Stock Market*)

HOW IS HUMAN OVERPOPULATION POLLUTING THE BIOSPHERE?

Humans are not only using up limited resources. We are also damaging air, water, and other renewable resources by polluting them with industrial waste, garbage, and sewage. The more of us there are, the more pollution we generate. In addition, by multiplying, we force more and more people to live in dangerous places. Studies show that as we become more numerous, natural disasters such as earthquakes and floods kill more of us and destroy more buildings and other property.

Pollution Has Degraded the Water

Pollution is most obvious in freshwater communities, although marine ecosystems suffer from pollution as well. Historically, most people have disposed of sewage and other wastes by dumping them into the nearest body of water, to be carried away downstream to the ocean. But such polluted water is dangerous to human health, so government agencies have focused on cleaning up rivers and lakes. The worst damage has not been to people, however, but to other organisms. Several kinds of pollution damage ecosystems, including chemical and thermal pollution and sediments.

Thermal pollution occurs when factories dump hot or warm water into rivers and lakes. For example, when a power plant pumps cool river water through a hot nuclear reactor, the resulting hot water flows back into the river, killing some species and causing others to multiply too fast. Warm water can destroy aquatic ecosystems.

Sediments are particles suspended in water that settle to the bottoms of lakes, streams, rivers, and other bodies of water. They damage aquatic ecosystems by blocking light, burying organisms that live on the bottom, and filling in bodies of water. Lack of light prevents photosynthesis, limiting the pro-

ductivity of ecosystems. In coastal waters, sediments can inundate coral reefs and shellfish beds.

Sediments result from the natural process of erosion. Indeed, natural erosion working over millions of years cut the Grand Canyon a mile into the Earth. And erosion created the fertile flood plains of the Mississippi and Nile rivers. But virtually all human activities—including the construction of houses and roads, agriculture, logging, and strip mining—enormously accelerate erosion. The result can be too much sediment building up too fast.

Chemical pollutants include a variety of molecules, not all of which are toxic. For example, *biodegradable pollutants* include carbon dioxide, sewage, fertilizers, and other wastes that organisms can consume. Like thermal pollution, these pollutants promote the growth of some organisms and inhibit the growth of others. Nitrogen and phosphorus in fertilizers used in agriculture and home gardening wash into rivers and lakes and promote "eutrophication" (Chapter 26).

Nondegradable pollutants include solid wastes, such as aluminum cans, plastics, glass, and hundreds of other human products that now accumulate in dumps. Even newspaper, technically biodegradable, does not degrade in most landfills. Up to 40 percent of all landfill is newspaper.

At sea, most waste is dumped overboard as a matter of course. Floating objects may stay on the surface for years. Plastic six-pack loops strangle birds and other animals. Lost fishing nets go on catching fish and other animals long after the nets' owners have retired.

The hardest pollutants to deal with are outright poisons. These include heavy metals, such as mercury and lead; reactive gases from smog; radioactive waste; pesticides; and an unknown but growing number of toxic chemicals used in industry and agriculture. Studies have documented the effects of hundreds of these chemicals, but little is known about the effects of thousands of others. Still less is known about the interactions among these chemicals. For example, some substances can be relatively harmless by themselves but toxic in combination with other compounds.

One of the most interesting classes of toxins are *estrogenics,* compounds that happen to imitate the hormone estrogen. Many plastics and other useful modern chemicals contain estrogenics that interfere with the normal development of a variety of animals. Among other effects, estrogenics cause malformations in the reproductive organs of animals, mainly in males. Some estrogenics are softeners found in plastics, such as those in freezer containers. Another group of estrogenics are *surfactants,* molecules that help spread other molecules over a thin surface. Some dishwashers have a dispenser for surfactants, which keep water from beading up on glasses and leaving spots. Surfactants are also added to pesticides to help disperse, or spread, the pesticides when they are sprayed from airplanes onto forests and crops. Recent research has shown that the surfactants wash into streams over many years and interfere with the normal development of salmon. Lab studies show that when salmon get ready to return to the ocean, 20 to 30 percent of those exposed to surfactants die.

Many toxic chemicals become concentrated as they move up the food chain, a phenomenon called **biomagnification** (Figure 28-18). Plants concentrate a compound to many times that in the groundwater. Herbivores that eat the plants further concentrate the toxin in their flesh, and carnivores concentrate it even more. The highest concentrations are found in top carnivores, such as large freshwater fish, fish-eating birds such as pelicans and eagles, and, of course, humans. Trout in the Great Lakes, for example, have concentrations of pesticides and mercury that range from 1000 to 14 million times that in the water in which they swim. The highest concentrations of the pesticide DDT ever found in humans came from men who regularly caught and ate catfish.

Water and soil pollution come in the form of thermal pollution, biodegradable pollutants, nondegradable pollutants, and poisons.

Pollution Has Also Degraded the Air

Industrialization often means a darkening of the air and surrounding landscape, as particles from incompletely burned coal and oil spewed from factories and power plants. Tiny particles of smoke and soot cause lung disease in humans. But while soot may be the most visible and easily removed product of burning fossil fuels, the invisible chemical products of combustion have the most wide-ranging effects.

As we discussed in Chapter 25, burning fossil fuels has doubled the production of carbon dioxide in the atmosphere, contributing to global warming. Burning fossil fuels also generates many other chemicals that alter the chemistry of our atmosphere.

Nitrogen and sulfur oxides may travel thousands of miles with air currents before dissolving in falling rain. The resulting acid rain may destroy forests and aquatic ecosystems hundreds of miles away from the source of the pollution. In addition, the energy of sunlight can convert nitrogen oxides in the atmosphere to the still more reactive compounds of smog.

Another group of dangerous chemicals are the chlorofluorocarbons, chemicals that have eaten a hole in the ozone layer in the upper atmosphere. **Chlorofluorocarbons** are small molecules used as coolants in refrigerators and air conditioners, in the electronics industry as solvents to clean circuit boards, and as propellants in spray cans.

Ozone (O_3) is a highly reactive molecule made of three oxygen atoms, instead of the two in the oxygen (O_2) we breathe. At ground level, ozone destroys living tissue. But 20 to 50 kilometers above the Earth, a layer of ozone called the **ozone layer** absorbs most of the ultraviolet light from the sun, shielding living things from potentially dangerous amounts of ultraviolet light (Figure 28-19). Ultraviolet light that reaches the Earth's surface hurts organisms by increasing mutations in DNA, destroying protein molecules, and increasing the number of chemically reactive molecules in the lower atmosphere. Some biologists suspect that excess ultraviolet light may have driven some species of frogs to extinction. For humans, the most obvious effect of increased ultraviolet light is higher rates of skin cancer.

Almost 20 years ago, atmospheric scientists began to document a thinning of the ozone layer, particularly over Antarctica. They linked the disappearing ozone to the presence of chlorofluorocarbons in the upper atmosphere. By 1988, the evidence for the connection between increased chlorofluorocarbons and decreased ozone had become so clear that chlorofluorocarbon manufacturers agreed to replace these compounds with others.

Scientists initially expected that the hole in the ozone would soon return to normal. Yet despite dramatic reductions in the use of ozone-destroying chemicals, the ozone hole over Antarctica reached a record size of 27 million square kilometers in 1998—larger than all of North America. In the thinnest places the ozone is only 70 percent of normal. Even over the United States, the ozone layer has thinned by 2 or 3 percent. Computer models by atmospheric scientists at NASA now suggest that the ozone layer will continue to thin for another 10 or 15 years.

Air pollution includes visible particles of soot, invisible products of combustion, and chlorofluorocarbons, which damage the ozone layer.

Human population growth has had enormous effects on the biosphere. Our numbers have crowded out whole ecosystems, fragmented ecosystems, driven thousands of species to

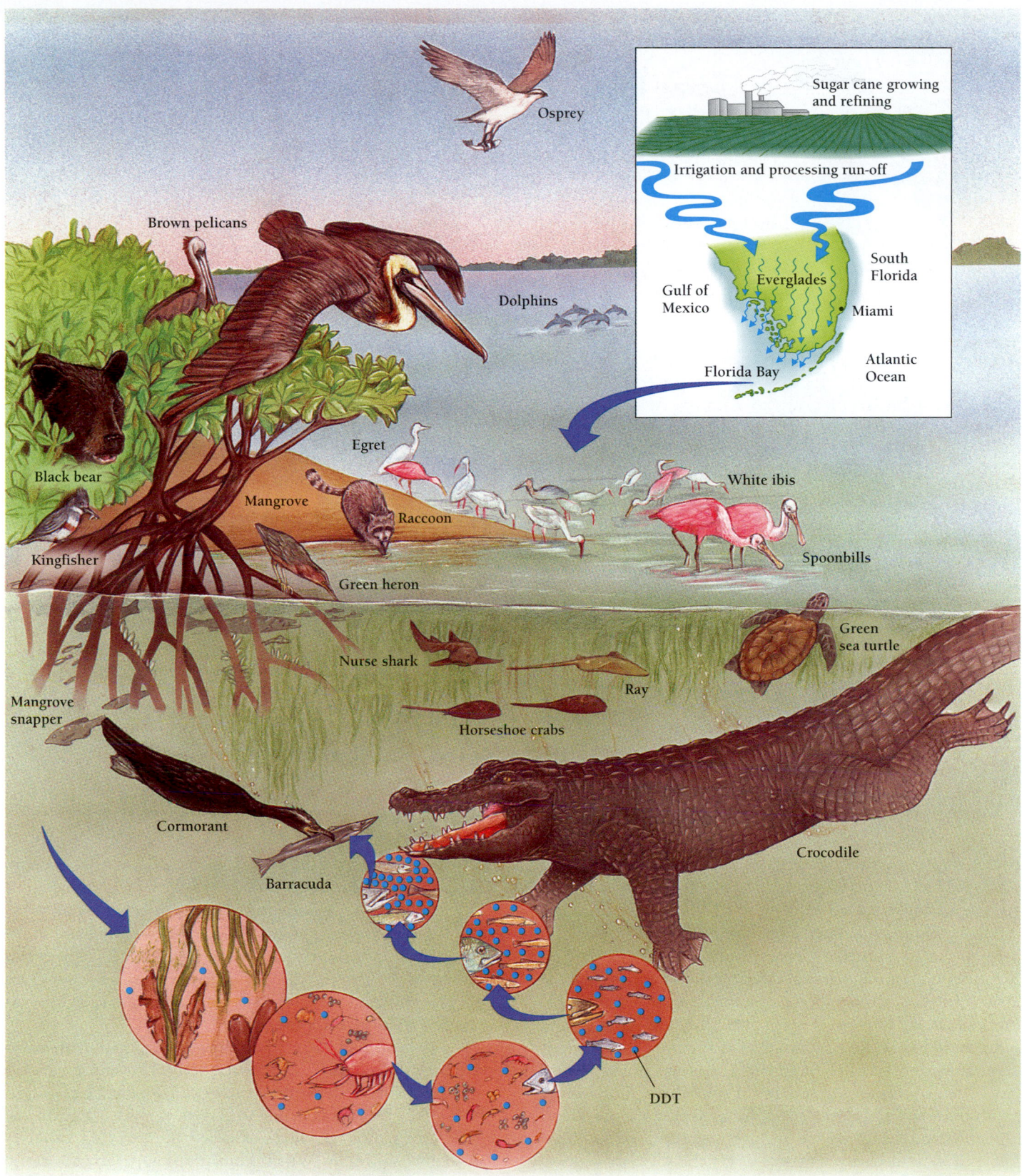

Figure 28-18   Biomagnification. As organisms consume food, they consume whatever toxins are in the food. Toxins that are not broken down or excreted accumulate in the body. At each level of a food chain, the toxin becomes more concentrated—as illustrated here by the increasing concentration of blue dots.

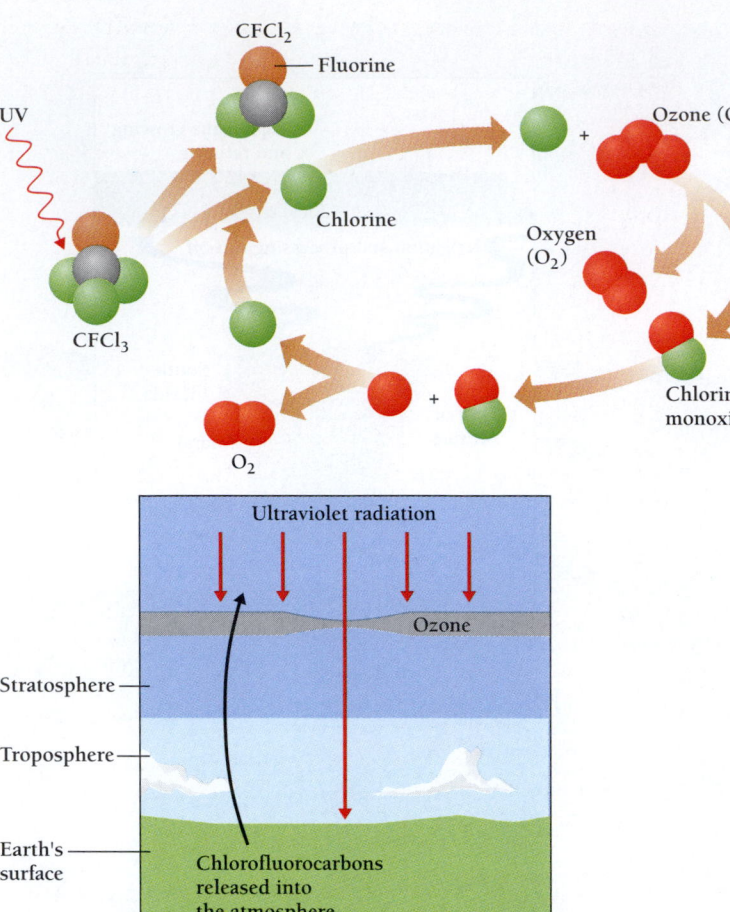

A.

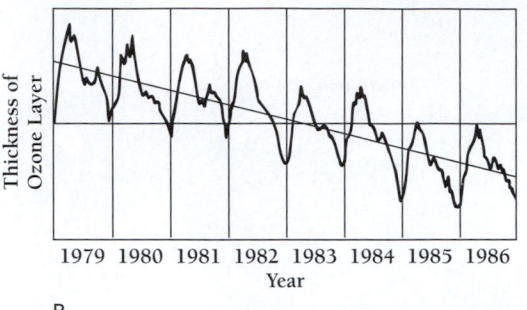

B.

Figure 28-19 **How do chlorofluorocarbons break down the ozone layer?** Ultraviolet light from the sun breaks down CFC_{l3} into CFC_{l2} and chlorine. The chlorine reacts with ozone to form oxygen and chlorine monoxide, thus depleting ozone in the upper atmosphere. As the ozone layer thins, more ultraviolet light penetrates deeper into the atmosphere.

extinction, polluted the air and water, and altered the global climate.

Many researchers still optimistically believe that we can clean up the mess. We can remove pollutants in water and air, we can reduce the greenhouse effect, and we can close the hole in the ozone layer. We can even stop multiplying so fast. Because of population momentum, however, it may take us as much as a hundred years to halt our growth. One thing we cannot do is bring back the species we have driven to extinction. Once they are gone, they are gone for good. Biologists estimate that the evolution of replacement species will take 5 to 10 million years.

Study Outline with Key Terms

Although a conservation movement has existed in the United States for more than 100 years, 1980 marked the informal beginning of **conservation biology** as a field separate from traditional ecology. Both **deterministic** and **stochastic** factors can drive species to extinction. The way humans drive other species to extinction is through **habitat destruction,** hunting, and the introduction of exotic species. Humans destroy habitat by breaking it up into small fragments with lots of edge and little central core. Roads, suburban developments, and agricultural fields all tend to **fragment** remaining habitat and increase **edge effects.** According to the **extinction vortex** hypothesis, loss and degradation of habitat reduce the size of a population, change its age and sex structure, and lead to a loss of genetic diversity. The loss of genetic diversity leads to inbreeding, reduced reproduction, and smaller population size. In a small population, chance, or stochastic, events are more likely to lead to extinction. We can preserve many more species if we set aside large preserves, connected by **corridors** and surrounded by buffer zones.

All living organisms are capable of **exponential growth.** Bacteria have a **doubling time** of as little as 20 minutes. The human population is currently doubling about every 60 years. Unrestrained exponential growth results in a rapid and limitless increase in total population size over time. A graph of this increase is sometimes called a **J-shaped curve.** The maximum possible rate of increase for a species is r_{max}, also called the **intrinsic rate of natural increase.**

The growth of natural populations is limited by **density-dependent factors,** such as space, resources, and predators, and by **density-independent factors,** such as weather and catastrophes. Density-dependent factors impose a limit to the growth of natural populations called the **carrying capacity** of the habitat, or **K.** When populations approach the carrying capacity of their habitats, the size of the population levels off, forming an **S-shaped curve.** Such restricted growth can be described by the **logistic growth equation.** When a population stops growing, it is said to have **zero population growth.**

Populations increase or decrease as a function of **birth rate, death rate, immigration,** and **emigration.** When populations **overshoot** their carrying capacity, they may crash, then rise and fall until the carrying capacity is reached or surpassed once more. Some populations **cycle** regularly and somewhat mysteriously. Ecologists used to think that predators drove such cycles. However, the population of prey may cycle as a result of fluctuations in their food supply, and the population of predators may cycle as a result of cycling in the prey population.

Ecologists classify species by **life history traits,** such as life span and number of offspring. One such measure is the **survivorship curve: convex,** or **type I; diagonal,** or **type II;** and **concave,** or **type III.**

If we know the **age structure** of a population, which represents the size of each age cohort in a population, the age structure of immigrants and emigrants, and the fertility and mortality of each **cohort,** we can predict the future age structure of the population. **Demography,** the study of human populations, expresses **mortality** as death rate and **fertility** as birth rate, **replacement reproduction,** or **completed family size.** Demographers express the rate at which a human population increases as the **annual rate of increase.**

The global population of humans has been increasing exponentially for thousands of years. **Population momentum** means that even if humans immediately and permanently reduced their annual rate of increase to zero, the world population would continue to grow for another 50 years. Human agriculture is heavily **subsidized** with energy from fossil fuels, which increase crop yields. Despite the **energy subsidy,** global food production is beginning to fall behind increases in global population size.

Human overpopulation increases water and air pollution, with attendant problems such as **biomagnification,** global warming, and the destruction of the **ozone layer** by **chlorofluorocarbons.** The impact of humans can be expressed in terms of **ecological footprint.**

Review and Thought Questions

Review Questions

1. The world population of humans is about 6 billion, and the doubling time is now around 60 years. Using the equation $N \times 2^t$, calculate the world population after 30 years, 60 years, and 90 years. Graph your results, plotting N against time.

2. Using the information in the table (below), calculate r for Sweden. Show your work.

	Sweden	United States	India
b (births per 1000)	12	14	26
d (deaths per 1000)	11	9	9
r (rate of natural increase [%])	—	0.6	1.7
Annual rate of growth (%)	0.3	0.9	1.7
Life expectancy (years)	79.2	76.1	62.9
Infant deaths per 1000 live births	4	6	63
Total fertility rate (per woman)	1.8	2.1	3.2
N (population in millions)	9	270	1000

3. What factors make India's r value so much higher than that of the United States?

4. Why does Sweden have a higher death rate than the United States?

5. In your garage live a pair of healthy, mature Norway rats. Assume these animals can produce 10 young every 3 weeks (actually, Norway rats can produce up to 11). Assume also that each litter consists of five females and five males. Females are able to bear young at 12 weeks of age. If none of the adults or offspring die, and if conditions in your garage are so pleasant that none emigrate, how many rats will occupy your garage at the end of 1 year?

6. What is the intrinsic rate of natural increase of this pair of rats, expressed as the rate of increase per year?

7. What factors limit the carrying capacity of a habitat?

8. Give three examples of density-dependent limits to human population growth and three examples of density-independent limits.

9. What is replacement reproduction, and why is replacement reproduction for humans 2.1 instead of just 2.0?

10. If the 1990 annual rate of increase had persisted, what would the global population of humans have been in the year 2000?

11. Why is subsidizing urban populations necessary?

12. Approximately how many people starve to death each year?

Thought Questions

13. If a way could be found for the Earth to support 30 billion people, five times the current density, what other organisms and ecosystems would probably have to go to make room for these people? What could be retained? How would the world be better or worse than it is with a population of 6 billion?

14. Make a list of factors, both density dependent and density independent, that might limit the population of rats in your garage.

About the Chapter-Opening Image

The woolly mammoth—which went extinct 11,000 years ago—led a parade of modern extinctions that now includes millions of species.

 On-line materials relating to this chapter are on the World Wide Web at **http://www.harcourtcollege.com/lifesci/aal2/**

29 The Ecology of Animal Behavior

Why Is Sociobiology Controversial?

In 1979, a shy and bookish expert on the biology of ants was delivering a scientific talk in Washington, DC, when a group of angry students charged the podium and doused him with water. E.O. Wilson, an eminent Harvard professor, had unwittingly placed himself at the center of a furious controversy by writing an ambitious textbook on animal behavior. The 1975 publication of *Sociobiology: The New Synthesis* almost overnight transformed Wilson into a notorious and, to some people, repugnant political figure.

Wilson became the target of angry public attacks by both scientists and nonscientists that left many people with the impression that sociobiology was promoting racism, sexism, classism, and even Nazism. **Sociobiology,** the study of behavior from an evolutionary perspective, got such a bad name, in fact, that biologists finally renamed it **behavioral ecology.**

The firestorm of debate that Wilson ignited was fueled by several separate but intermingling currents of thought, each with its own long history. These included the ongoing political debate over the role of genetics in shaping human behavior, the related "nature-nurture" debate, and a simmering feud between two groups of scientists.

Until about 1972, two separate disciplines for the study of animal behavior existed—the European science of **ethology,** the systematic study of animal behavior from a biological point of view, and the primarily American science of **experimental psychology.** Both ethologists and psychologists were interested in animal behavior, but their approaches, emphases, and philosophies differed dramatically.

Experimental psychologists focused on how animals learn. For example, the Russian physiologist Ivan Pavlov discovered that he could teach a dog to salivate at the sound of a bell. First, Pavlov would ring a bell each time he fed the dog. When the dog saw the food, it salivated. In time, the bell alone would cause the dog to salivate, a behavior Pavlov termed a **conditioned reflex.**

Experimental psychologists were primarily interested in human behavior, and animals were only substitutes for humans. These researchers thought of behavior as being mostly learned. The adaptive value of behavior was of little interest to them, so all animals were considered more or less equivalent. As the famous experimental psychologist B.F. Skinner wrote in 1959, "Pigeon, rat, monkey, which is which? It doesn't matter."

Ethologists, by contrast, thought of behavior—however much of it was learned—as an adaptation that affected the survival of each individual. Early ethologists preferred to study the normal behavior of animals living in the wild, with special emphasis on behavior that was inherited, or "innate." For example, the tendency of newborn babies to suck milk from a nipple is considered an innate behavior—even though practice improves a baby's ability to suck.

Experimental psychologists were suspicious of ethology for two reasons. First, they believed that because of humans' tremendous ability to learn, genetics and evolution were irrelevant. Second, ethologists' emphasized the idea that behavior can be inherited, which reminded the psychologists of the 19th-century idea that whites were intrinsically (genetically) superior to

Oxford Scientific Films/Animals Animals

- All behavior develops under the influence of both genetic inheritance and environmental experience.
- Virtually all animals are capable of learning.
- The ecology of an animal determines the adaptive value of its behavior.
- Living in groups has both advantages and disadvantages.

all other races. In the 20th century, this idea was used to justify a variety of evils, including Hitler's human breeding program and the Nazis' mass murder of Jews, Gypsies, and others during World War II.

When E.O. Wilson's book came out in 1975, it brought these conflicts to a head. In his book, Wilson argued that by the year 2000 both ethology and experimental psychology would wither away, to be replaced by sociobiology and neuroscience, the study of the nervous system. Psychologists couldn't help but be offended. For them, acknowledging the value of the evolutionary viewpoint was one thing. Retiring meekly to the sidelines while neuroscientists and ethologists (turned sociobiologists) sorted out the problems of behavior was quite another.

Worse, Wilson speculated freely about the possible genetic basis of various aspects of human social behavior. Many people felt that Wilson's book provided ammunition that could be used to discriminate against the poor, the uneducated, minorities, and women.

A handful of biologists unfairly accused Wilson of being a "genetic determinist," someone who believes that all human behavior is determined by genes. But no biologist takes such an extreme view, and naturally Wilson denied the accusation. Today, the debate has lost much of its heat. Biologists do not argue about whether behavior is controlled by genes or by environment; all agree that behavior is always influenced by both.

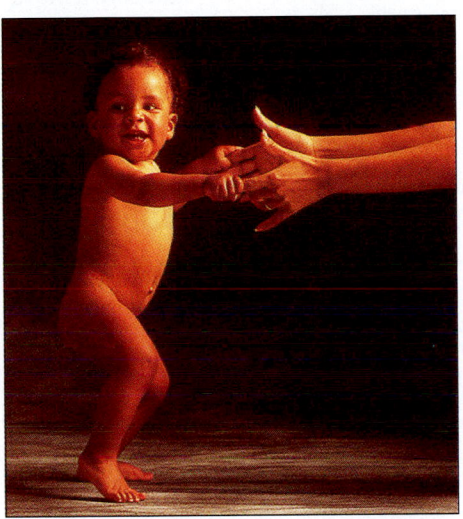

Figure 29-1 Programmed behavior. Humans all begin walking at about the same age. However, the style of walking, whether short steps or long, for example, is learned. *(Andy Cox/Tony Stone Images)*

HOW IS BEHAVIOR ACQUIRED?

An animal's behavior contributes to its ability to survive and reproduce in the same way that its anatomical, cellular, and molecular characteristics do. Since genes can shape behavior, natural selection can operate on genes that influence behavior, just

as it does on genes that influence structure. The adaptiveness of a given behavior depends to a large extent on ecology—the relationship of an individual to its environment. In this chapter, we will examine the ecological significance of behavior, as well as the role of natural selection in shaping behavior.

What Is Innate Behavior?

The human ability to walk is innate. Provided that we are physically able, we all begin walking at about 1 year. Even toddlers whose legs have been in casts during infancy, still begin to walk on schedule (Figure 29-1). **Innate** behavior is behavior that an animal engages in regardless of previous experience. The tendency for a dog to bark at strangers is innate. Most breeds bark at least occasionally. A dog's upbringing and psychological well-being may influence how much it barks and at whom, so that environment plays a role as well, but the dog's tendency to bark is inborn.

By contrast, a dog *learns* who is a stranger and who is familiar. **Learning** is a change in an animal's behavior in response to a specific previous experience. Mammals may be the most sophisticated learners, but essentially all animals—from flatworms to fish—can learn.

Different species of animals are able to learn different things. A grizzly bear is good at learning how to catch salmon in a stream, while a killer whale is good at learning how to catch salmon in the ocean. Every animal has innate tendencies to learn different things. We can see, then, that learned and innate behavior are not completely separable.

One interesting example comes from two races of garter snakes (*Thamnopis elegans*). Garter snakes that live along the coast of California are known to relish slugs, while inland snakes will not touch them. To see if this difference in behavior was innate, researchers hatched baby snakes in the laboratory so the snakes would have no opportunity to learn about

Figure 29-2 A modal action pattern. A greylag goose rolls a wayward egg back to her nest. If the egg is removed in the middle of this procedure, the goose continues the tucking movements until she is back on the nest.

slugs, then offered the young snakes a menu that included slugs. Despite their inexperience, snakes from the coastal population readily ate the slugs, while those from inland populations refused them. This result suggested that genes might determine the snakes' taste for slugs. In a separate experiment, the researchers crossbred the two kinds of snakes. Among the hybrids, more snakes refused slugs than accepted them. The researchers suggested that a gene influencing slug refusal might be dominant.

Any behavior is likely to include both innate and learned aspects.

What stimulates innate behavior?

A highly stereotyped innate behavior is called a **modal action pattern.** Examples of modal action patterns abound. All dogs scratch their ears in the same way, lifting a rear leg above a foreleg. This behavior is so stereotyped that if a human scratches a dog's ear, the dog's hind leg often will rise above the foreleg and move up and down in midair as if scratching. Parrots and many other birds also scratch their heads by bringing a foot over the wing. This mode of head scratching is not necessarily required by their anatomy, however. A parrot is per-

fectly capable of bringing a foot under the wing, and does so to clean its bill.

A modal action pattern can be partly fixed and partly flexible. For example, when an egg rolls out of the nest of a greylag goose, the goose can roll the wayward egg back into her nest by pushing it with the underside of her bill (Figure 29-2). This bill-tucking movement allows her to pull a loose egg back toward her, but she also must adjust to the sideways wobble of the egg by moving her head from side to side. If the goose is tucking the egg back into its nest and an experimenter suddenly removes the egg, the goose will continue tucking. But the bird's adjustments to the side-to-side movements of the egg will stop. The tucking movement is fixed, but the sideways adjustments are flexible.

Modal action patterns serve a purpose only when performed at the proper time and place. Some signal in the environment must stimulate a modal action pattern. In human infants, a pair of eyes stimulates a smile (Figure 29-3). A loose egg stimulates a greylag goose's chin tucking. A fly provokes the characteristic flick of a frog's tongue. In each case, the animal's senses detect a stimulus and the nervous system issues the order to act. The stimulus for a modal action pattern is called a **sign stimulus.**

Innate behavior often has a learned component. For example, the chick of a herring gull will peck at its mother's beak

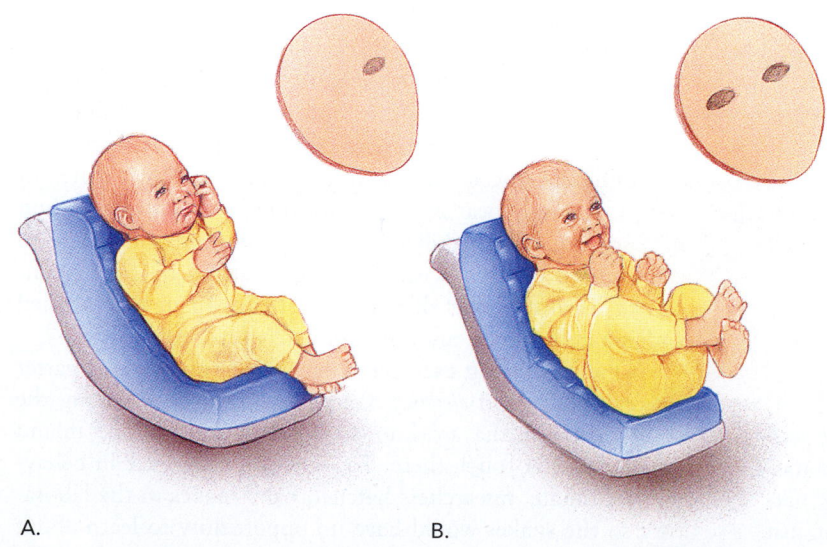

A.   B.

Figure 29-3 Releasing a smile. A flat, face-sized mask with one eye spot will not make a young baby smile. The same mask with two eye spots triggers a happy smile. The two eyes constitute a kind of sign stimulus. Older babies learn to prefer human faces.

Figure 29-4 A herring gull chick begins begging for food soon after it hatches, a behavior that is partly innate and partly learned. *(Thomas W. Martin/Photo Researchers, Inc.)*

Figure 29-5 **Tick sign stimuli.** A female tick needs a blood meal to produce her eggs, but she must wait for a meal to come to her. For as long as 18 years, she sits unmoving at the end of a twig. At the moment a warm animal passes beneath her twig, she awakes, lets go of the twig, and drops. If she lands on a suitable animal, she inserts her feeding organ into the skin and sucks herself full of blood. If she lands on the ground, she climbs up onto a nearby plant and waits even longer. A tick responds to three separate sign stimuli. In response to light, a tick moves to the end of the twig. In response to either carbon dioxide (which animals exhale) or butyric acid (found in sweat), a tick lets go of the twig. Finally, in response to any warm surface—even a balloon filled with warm water—a tick will begin feeding. Light, carbon dioxide, butyric acid, and warmth all act as sign stimuli for a tick's behavior.

to beg for food the first time it sees its parent (Figure 29-4). Pecking is therefore regarded as innate. The pioneering ethologist Niko Tinbergen reasoned that since a herring gull chick always pecks at a red spot on its parent's bill, the red spot may be what provokes the pecking behavior. To test this theory, Tinbergen presented chicks with models of gull bills with different colored spots. He found that the red spot elicited the pecking behavior better than any other colored spot, regardless of the shape of the model.

But even this behavior is partly learned. Later research by biologist Jack Hailman showed that a newly hatched gull chick pecks at anything that looks remotely like a beak, including the beaks of its siblings. Over the course of several days, however, the chick *learns* to recognize both its parents and food, so that it pecks in response to a much more limited range of stimuli.

Not all sign stimuli are visual. Touch, sound, and smell can also trigger stereotyped behavior patterns (Figure 29-5). A female silk moth emits a potent chemical, called bombykol, whose odor can attract males from over a mile away. When the male detects as few as 200 molecules of bombykol, he flies upwind until he loses the scent, flies in a zigzag pattern until he detects the bombykol again, and flies upwind again. He continues this pattern until he finds the female. Chemicals like bombykol that influence the behavior of another individual of the same species are called **pheromones.** Pheromones can activate modal action patterns.

Most sign stimuli are quite simple. Even with our own extraordinary brains and sensory systems, we simplify the complexity of the world. As a result, we can easily recognize public figures from simple line drawing caricatures (Figure 29-6). Animals with less developed nervous systems than our own also respond to caricatures. When Tinbergen presented gull chicks with a stick with three red spots, instead of just one, the chicks pecked at it even more than if it had been a real gull's bill. For a herring gull chick, three red spots constitute a **supernormal stimulus,** a stimulus even more stimulating than the normal stimulus.

Figure 29-6 A caricature is a kind of supernormal stimulus. A caricature of Oprah Winfrey exaggerates her features. The drawing doesn't really look like her, but we instantly recognize who it is supposed to be. *(Barbara Ericson)*

> *A modal action pattern is a stereotyped behavior triggered by a sign stimulus. The sign stimulus can be seen, heard, felt, or smelled.*

How Much Can Animals Learn?

In an early episode of the television show *Sesame Street*, Ernie is astonished that Bert has been able to teach Bernice the pigeon to play checkers. "Why, a pigeon that can play checkers!" Ernie exclaims. "That must be the smartest pigeon in the whole world!" "She's really not that smart," demurs Bert. "In the ten games we've played? She's only beaten me twice."

Pigeons may or may not be able to beat Muppets at checkers, but they are surprisingly capable learners. Pigeons and chickens can learn to distinguish between an underwater scene with fish and one without. They can even understand abstract ideas. Pigeons can learn to choose one of two disks according to whether it matches the color of a third disk.

Every animal can learn. Pigeons can beat humans at tic tac toe. Rats can learn not only to run a maze accurately and consistently, but to run a mirror image of the maze, switching back and forth on cue. Even planarians, among the simplest of all animals, can learn (Figure 29-7).

Animals pay close attention to one another's behavior. Experiments have shown that blue jays prefer to eat what they have seen other jays eating. Crows, upon seeing another crow sicken or die after eating poisoned bait, will shun the bait. One of the most striking examples of learning in natural populations of animals has occurred in at least 11 species of small English birds, including blue tits, titmice, and others. Starting in around 1920, a few birds learned to open the caps of milk bottles delivered to the front steps of English houses. Beneath the caps was a rich meal of pure cream. The cream was so attractive that the birds attacked the bottles as soon as the delivery person was gone, and, in some areas, flocks of birds followed the milk trucks down the streets. If the bottles were capped with cardboard, the birds pulled the whole cap off or

peeled it off in layers of paper. If the bottle was capped with foil, the birds poked a hole in the foil, then peeled the foil off in strips. This behavior first appeared in a few limited areas, then, over a period of years, gradually spread, suggesting that the birds learned from one another.

> *Animals can learn a great deal from their environment and from each other. Even behavior strongly influenced by genes is still open to modification through learning.*

Can old dogs learn new tricks?

Just as humans learn languages most easily when young, many birds learn to sing their species song only when they are young. For many kinds of learning, timing is critical. A chaffinch raised in isolation, so that it cannot hear adult birds singing, never learns to sing an adult song. If a chaffinch hears the song as an adult, he cannot learn it. If, however, the chaffinch hears the correct song during the first few months of life, he can learn to sing a song that resembles the chaffinch song—even if he cannot hear the song when he is actually practicing. The period of time during which an animal can learn a particular behavior pattern is called the **sensitive period.**

Sensitive periods exist for a variety of different kinds of behavior. For example, children born blinded by cataracts can learn to see if the cataracts are surgically removed before about age 10. If the cataracts are removed later in life, however, these people see only a random patchwork of colored shapes, which they cannot interpret. As infants we learn to interpret complex patterns of shading and color as a three-dimensional world. But for newly sighted people, the patterns may be meaningless.

The process by which an animal learns such behavior during the sensitive period is called **imprinting.** The ethologist Konrad Lorenz provided the most famous example. Geese learn to recognize their mother by "imprinting" on the first moving object they see after hatching (Figure 29-8). Usually this

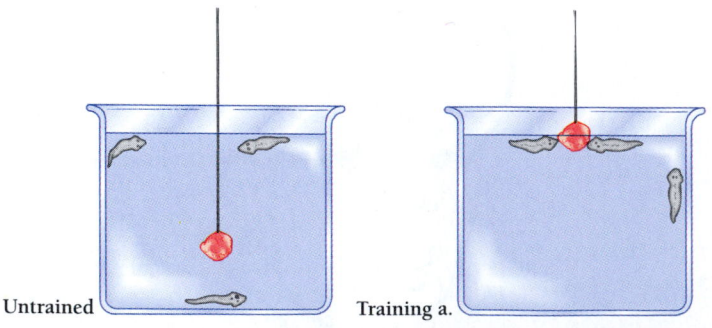
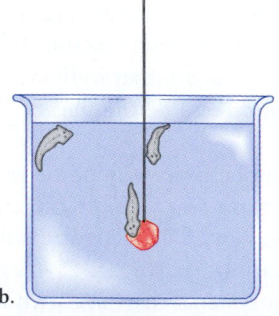
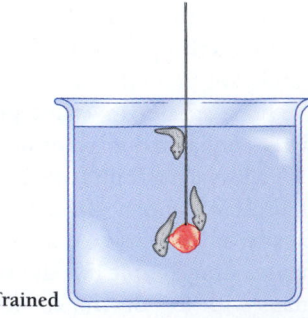

Untrained   Training a.   b.   Trained

Figure 29-7   Learning in flatworms. The simple planarian, with barely a brain to call its own, learns to find food suspended from a string. Untrained worms search the surface or the bottom of a container for food and miss the meat hanging in the middle. During training, a piece of meat is suspended at the surface and then gradually lowered into the water. The flatworms learn to locate the string at the surface and then to follow it down to the meat. Trained worms quickly find food suspended in the water.

Figure 29-8 Imprinting. Cygnets learn to recognize their mother and follow her soon after they hatch. (*Gerard Lacz/Peter Arnold, Inc.*)

object is the mother goose. But newborn geese will also follow a matchbox on a string, a rubber ball, a flashing light, or even a person.

Just as the development of normal vision depends on visual experience during a critical period, the development of normal social behavior depends on experience at sensitive periods. A young monkey isolated from other monkeys for six months any time during the first year and a half of life does not develop social interactions with other monkeys when later given the chance. Instead, the deprived monkey crouches in the corner of its cage, rocking back and forth, much like a severely disturbed human child. An adult monkey subjected to the same isolation returns to its friends without such serious behavioral problems.

Harry and Margaret Harlow, psychologists at the University of Wisconsin, attempted to find out what aspects of a young monkey's social life normally prevented such behavioral disorders. The researchers isolated young monkeys with and without a cloth-covered wooden dummy. The baby monkeys that had a wooden dummy, or "surrogate mother," showed fewer symptoms of isolation than the one with no dummy.

The Harlows could completely prevent the symptoms of isolation by allowing the young monkeys to spend just 15 minutes a day with a normally socialized monkey that spent the rest of its time with the colony. Such limited social experience usually produced full socialization only if it occurred during the sensitive period in the first year and a half of life. In a few cases, however, monkeys reared in isolation could be partly "cured" by later exposure to persistently friendly monkeys. In monkeys, at least, positive experiences later in life can sometimes partly make up for a horrific early life.

For some complex behavior, then, the sensitive period is not the only time an animal can learn. We can look at humans for another example. Although children learn new languages faster and better than adults do, adults can learn new languages rather well if they are willing to work at it.

Many forms of learning occur only at particular times, called sensitive periods. Other forms of learning take place most easily during sensitive periods, but can occur later as well.

How Do Genes Influence Behavior?

Behavior is the result of both genetic influences and experience (Figure 29-9). The nesting behavior of two small African parrots—the peach-faced lovebird and Fischer's lovebird—offers a perfect example of this. In both species of lovebird the female cuts long strips of bark or leaves for nest building. A peach-faced lovebird can carry up to six strips at a time by tucking them into her rump feathers, while a Fischer's lovebird carries these strips to the nest one at a time in her beak.

A hybrid of the two species tries both methods. When she builds her first nest, she starts out by tucking strips into her feathers, like her peach-faced parent. But she is inept, and the strips fall out before she can get to the nest. With time, she learns to carry just one strip at a time in her beak. After about 3 years, she gives up tucking strips into her feathers. However, just before she flies off with a strip in her beak, she still makes little sideways movements of her head, as if she were about to tuck a strip into her rump feathers.

Still, the exact role of genes in influencing behavior is hard to pin down. In rare cases, a single gene can affect behavior. For example, in humans, a rare mutation in one gene causes Lesch-Nyhan syndrome, in which people compulsively mutilate themselves. The vast majority of physical and behavioral traits, however, are "polygenic," influenced by many genes.

Figure 29-9 Behavior is partly genetic. Specific types of behavior, such as smiling, are often shared by groups of related species. (*Tim Davis/Photo Researchers, Inc.*)

Indeed, biologists have found that only about 2 percent of all traits are under the main influence of just one gene. Two or more—possibly hundreds—of genes influence the expression of almost every trait.

The complex genetics of all phenotypic traits—including behavioral traits—means that biologists cannot easily determine the exact degree to which a given behavior pattern is genetically or environmentally influenced. For example, despite years of research on the heritability of IQ scores—just one aspect of human intelligence— scientists can say only that IQ is somewhere between 30 and 70 percent inherited.

The balance is environmental. Nutrition, health, and environmental stimulation—in the womb, in infancy, and in early childhood—all affect intelligence to some degree. One study found that premature baby boys fed breast milk later had several more IQ points than boys who had been fed formula. Yet, in premature girls, the researchers found no difference.

HOW DOES ECOLOGY INFLUENCE BEHAVIOR?

Economists have debated whether people make rational economic decisions. For example, if food costs twice as much at one store as another, most people go out of their way to shop at the cheaper store. On the other hand, if travel costs are high, they will not spend more money getting to the cheaper store than they would save on the groceries. To a remarkable degree, then, people behave rationally. Animals, with far less intellectual capacity than humans, also behave surprisingly rationally about such decisions.

How Do Animals Make the Best Choices?

Crows open and eat thick-shelled sea snails, called whelks, by picking up the largest one they can find, carrying it to a height of 5 meters, and dropping it on the rocky shore. If the whelk breaks, the crow feasts. If the whelk stays intact, the crow carries it aloft and drops it again, repeating the procedure until the whelk's shell breaks.

Ecologists studying this tedious feeding strategy wondered if this was the best strategy to get the most food for the least energy expended. To see if the crows' strategy was "optimal," researchers themselves dropped variously sized whelks from different heights. The biologists soon found that the largest whelks were the most likely to break and that none of the whelks broke consistently unless dropped from a height of 5 meters or more. The crows' strategy of choosing only the largest whelks and carrying them to about 5 meters was indeed optimal. Further experiments revealed that if a whelk didn't crack open on the first try, persistence with a single whelk was more likely to succeed than starting over with a new whelk. So, a crow's decision to drop the same whelk over and over proved equally sensible. Overall, the crows' feeding behavior is an ef-

ficient way of spending energy to get more energy. Such energy-efficient feeding is called **optimal foraging.**

When animals forage, they may efficiently maximize the amount of food they get for the energy expended. Ecologists call this process optimal foraging.

How do competition and predation influence foraging behavior?

Maximizing energy intake is not the only consideration that guides feeding behavior, however. When predators are present, an animal must adjust its behavior to increase its own chances of surviving. One study of chickadee feeding, for example, revealed that the chickadees would eat seeds at a researcher's feeding tray only if the tray seemed safe. When researchers mounted a model of a hawk nearby, the chickadees took the seeds away to a safer place. Taking the seeds away cost the birds time and energy, but given the apparent presence of a hawk, moving the seeds was a good way to both eat and survive.

The presence of competing species can also change feeding behavior. For example, North American mink introduced into Sweden in the 1920s competed directly with European otters for food and caused otter populations to decline. However, the two animals continued to coexist. In America, the mink normally eat almost anything that moves, including crayfish, fish, muskrats, rabbits, birds, and small burrowing mammals. The European otters have a more specialized niche than the mink, feeding on fish, crayfish, crabs, and other aquatic invertebrates only. In Europe, the larger otters defend this narrow niche from the mink aggressively, sometimes killing mink that poach fish and crayfish. As a result, in areas where the otters persist, mink eat less fish and crayfish than those in America.

Foraging behavior depends, in part, on how an individual interacts with members of other species.

How do animals express aggressive feelings?

Reproduction and survival necessitate competition as well as cooperation. Among the most dramatic competitive kinds of behavior are the ritualized fights between pairs of bighorn sheep. The outcome of such combat determines which male will get to mate with a female. The clash of horns is mostly symbolic: usually the fighters do not seriously injure each other. Other species use other symbolic contests—who can roar the loudest, stand the tallest, or stare the longest.

These contests are examples of **agonistic behavior** [Greek, *agonistes* = champion]—all the aspects of competitive behavior within a species, including aggression (outright attacks and fighting), aggressive displays, appeasement, and retreat (Figure 29-10). Ethologists use the term "agonistic behavior"

Figure 29-10 Agonistic behavior in rabbits. Fighting between males is not confined to large animals such as deer or lions. Here, two rabbits fight to establish dominance. Agonistic behavior also includes aggressive behavior and appeasement displays. *(Art Wolfe/Tony Stone Images)*

instead of the word "aggression" to acknowledge that animals never exhibit pure aggression or pure fear. What most animals (humans included) show is a mix of fearful behavior and aggression, sometimes more of one, sometimes more of the other.

Some animals use agonistic interactions to establish a **dominance hierarchy,** a ranking of individuals that fixes which animals have first access to resources or mates. Dominance hierarchies establish rules that allow animals to live in the same territory with limited resources. The classic example of a dominance hierarchy occurs in a flock of hens. Such a flock quickly establishes a "pecking order," which specifies who pecks whom without being pecked in return.

One way that animals compete with others of their own species for mates, resources, and rank is through agonistic encounters.

Territoriality

In many species, an animal engages in competitive behavior to establish and defend a **territory,** an area occupied by an individual (or group of individuals), from which other individuals of that species are excluded. Some species, a few birds and fish, for example, use territories exclusively for mating. More commonly, species use territories for foraging or for raising young as well.

Establishing and defending a territory necessarily involves agonistic behavior. Animals use a wide variety of signals to mark their territories, including chemical scents, brightly colored plumage, or patrolling behavior.

Defending a territory requires energy expenditure. Ecologists have studied the "economics" of territoriality and have shown that some bird species defend smaller territories when food is abundant than when food is scarce. When food is abundant, less territory is necessary to raise the same number of offspring. So, to conserve energy, birds defend a smaller territory—an example of optimal foraging. A few bird species even adjust the size of their territories on a daily basis.

Animals defend territories for raising young, mating, or foraging—as long as defending the territory will increase their reproductive potential.

How does female choice influence the evolution of males?

The long-term result of competition for mates is **sexual selection,** the differential ability of individuals with different genotypes to acquire mates (Chapter 16). Some biologists distinguish sexual selection from other forms of natural selection, arguing that natural selection always results in the accumulation of adaptive structures and behavior, while sexual selection often results in "maladaptive" structures and behavior. But all adaptations to the various demands of survival and reproduction are a constant balancing act for organisms.

Natural selection can act both on inherited variations in structures that are important in mating behavior, such as a buck's antlers or a peacock's tail, and on inherited mating behavior. For example, in the tungara frog (*Physalaemus*), which lives in Panama, female frogs consistently prefer larger males to smaller ones. The females choose the larger males, however, not by comparing male physiques but by listening to the male frogs' mating calls. The song consists of a whine, followed by one or more low-pitched "chucks." Bigger frogs have lower "chucks," so female frogs prefer frogs with deep voices (Figure 29-11A). The lower the "chuck," the more the female frog is attracted, even when the "chuck" comes from a tape recorder.

Unfortunately, the male frog's deep "chucks" also attract the attentions of a frog-eating bat, which, like the female frog, prefers a big frog to a little one (Figure 29-11B). The bat presents the male frog with a serious dilemma. If he chucks very low to attract females, he risks his life by attracting bats. If he does not chuck low enough, he will not be able to reproduce. To maximize his chances of reproducing while minimizing his chances of being eaten, the frog must optimize the number and pitch of his chucks. To whatever extent that variation in pitch is under genetic influence, sexual selection will tend to favor the evolution of low-pitched chucks—but maybe not too low.

Competition for mates over evolutionary time results in sexual selection, the differential ability of individuals with different genotypes to acquire mates.

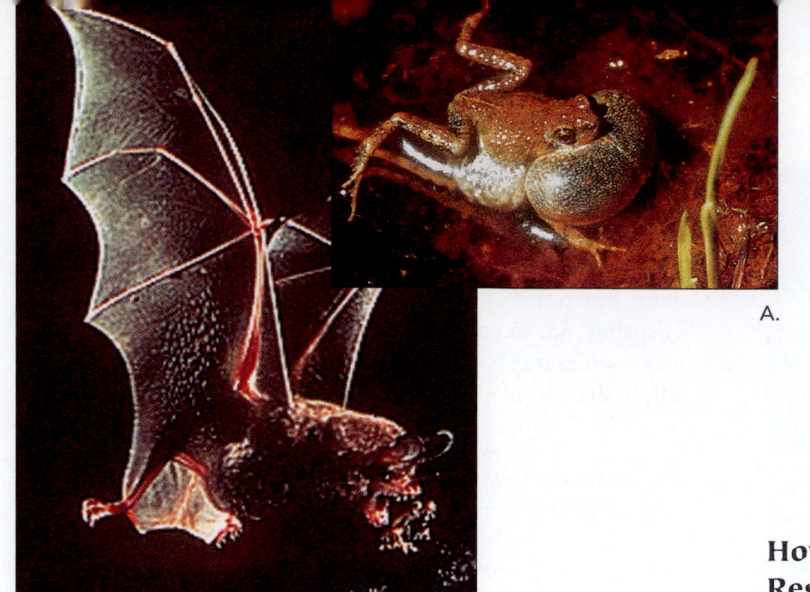

Figure 29-11 Cost-benefit analysis by frogs. The female tungara frog of Panama prefers a male that makes a deep chuck when he calls. Larger males make deeper chucks. Males with deepest voices have the best chance of attracting a female and, unfortunately, the best chance of being eaten by a frog-eating bat. A. A male tungara frog calling while floating in his pond. B. A frog-eating bat swoops away with a frog. (A, M.J. Ryan, University of Texas; B, Merlin D. Tuttle, Bat Conservation International/Photo Researchers, Inc.)

How Do Animals Compete for Resources or Mates?

Animals of the same species may cooperate or compete with one another, according to the demands of their current environment and the constraints of their evolutionary histories. For example, in many species, reproduction requires close cooperation between the male and the female. Such animals exhibit stylized, species-specific courtship behavior, including the well-known mating rituals of birds and fish (Figure 29-12).

Mating rituals allow females to assess the health and vigor of potential mates. Also, because mating season is often a time of heightened aggression, mating rituals allow both sexes to minimize the risk of being injured or killed by their partner. We can see that when such behavior patterns are influenced by the genes, natural selection must operate powerfully—increasing the frequency of those behavioral types that reproduce most successfully.

SOME ANIMALS LIVE IN SOCIAL GROUPS

A tiger lives a solitary life. It meets other tigers rarely and for not much longer than it takes to mate. Yet, the tiger's close relative the lion lives in a tightly knit group called a pride (Figure 29-13A and B). The members of a pride of lions are never far apart. They sleep together, hunt together, and raise their

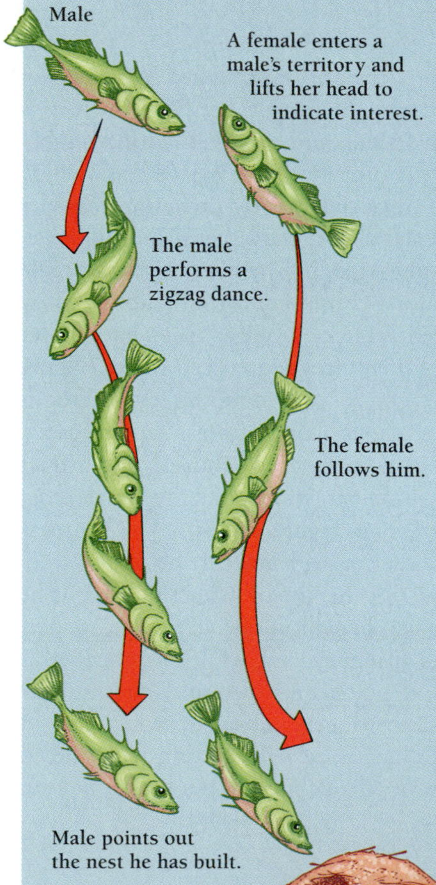

Male

A female enters a male's territory and lifts her head to indicate interest.

The male performs a zigzag dance.

The female follows him.

Male points out the nest he has built.

The female enters the nest, and the male touches her tail with trembling motions.

The female lays her eggs.

Female leaves the nest.

Male enters nest and fertilizes eggs.

Figure 29-12 Courtship in the stickleback. Mating rituals allow animals to recognize and evaluate one another.

A.

B.

Figure 29-13 Loners and social animals. A. Tigers hunt alone and kill smaller animals. B. Lions, which hunt in groups, can kill animals much larger than themselves. *(A, Tom Brakefield/DRK Photo; B, Anup & Manoj Shah/DRK Photo)*

young together. Why do some animals live socially, while others live apart?

How Is Social Behavior Advantageous?

Living in groups affords distinct advantages. The greatest advantage to group living is that large groups of animals can more quickly spot predators and other dangers than can solitary animals. Researchers have found, for example, that the more pigeons that are in a flock, the sooner the pigeons will notice a hawk and the less likely that the hawk will catch one of the pigeons.

Groups of animals can also better defend themselves. When threatened, a herd of musk oxen forms a circle with their young in the middle and their horns pointed outward. And the thousands of honeybees in a hive can drive off a predator or an animal bent on stealing their honey where a lone bee would be helpless. Finally, group living can be an advantage in finding food. Predators that hunt cooperatively, such as lions and wolves, can kill prey far larger than themselves (Figure 29-13B).

Social living has disadvantages as well, however. Animals living in groups must compete with each other for food, mates, and breeding areas. Individuals waste time and energy fighting or threatening one another and searching for food in the same places. Groups of animals also transmit diseases to one another and attract predators.

Whether animals live in groups depends on the ecological costs and benefits. Many songbirds stake out territories in the spring and actively repel all but their own mates and offspring. Yet when autumn arrives, and food becomes harder to find, the same birds abandon their territories and roam the countryside in large communal flocks.

Sociality allows animals to spot both prey and predators more easily. But groups of animals are themselves more visible to predators, and individuals in groups must constantly compete for food, mates, and breeding areas.

Honeybees Are an Extraordinary Example of Social Behavior

The coordination of individuals within a colony of social insects rivals that of the cells within an individual. We can view an insect colony as a superorganism, in which thousands or even millions of individuals function as a single unit. Social insects form enormous colonies. An African termite mound, for example, may contain 2 million workers. But the number of individuals alone is not what makes the social insects special. It is social organization that allows social species to take advantage of environmental resources unavailable to more solitary animals.

The familiar honeybee provides an excellent example. Humans long ago domesticated the honeybee to harvest the bees' useful honey and wax. Scientists since Aristotle have studied the workings of beehives and published tens of thousands of papers on bees. The fascination with bees has extended beyond the practical to the almost mystical. Humans admire the fact that the "busy bee" works so hard and loyally and ferociously defends its queen (long thought to be a king). One writer even suggested that honeybees have "voluntarily escaped from the Garden of Eden with poor fallen man for the purpose of sweetening his bitter lot."

A honeybee hive consists of double-layered wax combs that contain thousands of hexagonal cells. In these wax cells, the bees store honey and pollen or house developing larvae. Each bee grows from an egg deposited in a separate cell. The egg develops into a larva, which then develops into a pupa and finally into an adult.

A hive may contain 20,000 to 80,000 workers, each engaged in a succession of tasks. At the beginning of her life, a worker feeds nectar and pollen to the larvae. Later, she makes wax to enlarge the comb, removes the dead and the dying, guards the hive against intruders, and scouts the surrounding territory for food and nesting sites. Finally, toward the end of her short, 6-week life, she undertakes the most hazardous job of all, foraging for nectar and pollen among the flowers.

All the workers in the hive are female, and all are sisters. None reproduce, however. Only a single female—the queen—lays eggs. Most of these eggs develop into sterile workers, but a few may become queens. Workers can create queens by feeding a few, select larvae a special rich diet. Normally, the hive's queen produces a "queen substance," a pheromone that prevents the workers from building special wax cells to raise new queens. In the spring, however, the queen's production of this chemical falls off, and the workers build the royal cells. Workers soon push the old queen out of the hive, and she flies off with a large swarm of workers to start a new hive (Figure 29-14).

About 8 days after the old queen leaves the hive, the new queens begin hatching. Each one can form a swarm and fly off to start a new colony. If a queen elects to stay in the old hive, however, she first seeks and destroys rival queen larvae. She identifies these competitors by making quacking sounds outside the queen cells. If another queen responds to this challenge, a fight to the death ensues. There can be only one queen in each colony.

Once a queen has disposed of her rivals and established her queenship, she mates repeatedly in a series of "nuptial" flights. She leaves the hive, approaches a group of males, and releases small amounts of queen substance. The pheromone attracts males, one of whom will mate with her. The male releases his sperm into the queen's genital chamber and quickly dies. The queen may mate with as many as 17 males over a 4-day period. In this way, she obtains enough sperm for a lifetime. She stores the sperm and uses it to fertilize her eggs, which she deposits one at a time into separate wax cells. A queen may live and lay eggs for more than 5 years.

All of the social insects, including termites, ants, wasps, and honeybees, have three traits in common: (1) they cooperate in raising their young; (2) only a few individuals reproduce, while sterile workers assist and defend the fertile queen; and (3) generations overlap, so that offspring help their mother.

Honeybees display extraordinarily complex social structure.

Is Altruism Adaptive?

In Chapter 16 we learned that natural selection favors traits that increase the likelihood that an organism's offspring will survive. Why, then, we must ask, do the sterile workers in a beehive persist in a behavior that can have no adaptive value for them individually? The workers increase the reproductive output of the queen, but at the expense of their own reproduction. Biologists call such an arrangement **altruism,** behavior that benefits others at the expense of the animal that performs the behavior.

People sometimes jump into frozen lakes to rescue others. We know that this behavior is risky, because many of these would-be rescuers themselves die. Such behavior is clearly altruistic. Similarly, ground squirrels risk their own lives to warn one another of an approaching predator. And the young of many species of birds help their parents rear the next generation. The social insects, however, demonstrate the most extreme example of altruism.

If altruistic behavior has a genetic basis, then biologists must ask how it could possibly evolve. We would expect natural selection to favor sacrifice by parents to help their offspring but to act against the sacrifice of offspring to help their parents. Why do worker bees work so hard to help their mothers reproduce?

Does altruism arise by natural selection?

How could evolution produce workers, however industrious, if they leave no offspring? For a while, Darwin himself thought that this paradox was fatal to his whole theory of evolution by natural selection. To save his theory, Darwin argued that a family (the queen and her offspring, for example) must be the unit

Figure 29-14   Swarming honeybees. A swarm of bees will stay in one place for days at a time while scouts look for a suitable site for a new hive. The scouts return to the swarm and dance the location of a hole in a tree or cave. Each scout, however, is dancing a different location. When all the scouts agree, the swarm sets off to build a new hive. *(Gilbert Grant/Photo Researchers, Inc.)*

of selection, rather than the individual. By their self-sacrifice, the workers perpetuate genes that they share with their sisters, one of whom will become a queen. What is selected is the ability to be altruistic. The altruistic family is successful in passing on its genes.

In the 1960s, the British biologist W.D. Hamilton extended Darwin's idea, suggesting the theory of **kin selection.** Hamilton reasoned that an individual increases its reproductive output by helping relatives, which share its genes, to reproduce. Sisters (and brothers) are just as related to one another as they are to their parents. They share half of their genes. Thus, a worker bee helps ensure the continuation of her genes within her sisters and their descendants (recall that some of the sisters become queens).

Hamilton's work established the concept of "inclusive fitness." Fitness is a measure of selective advantage defined as the contribution to the next generation of one genotype in a population relative to the contribution of other genotypes. **Inclusive fitness** is the sum of an individual's genetic fitness (which includes that of its own direct descendants) plus all its influence on the fitness of its other relatives. Inclusive fitness, then, is the total success in passing alleles to the next generation, either by the efforts of a parent or by the altruistic acts of a relative.

One explanation for the adaptive value of altruism is inclusive fitness, which includes the concept of kin selection.

Why might intricate societies have evolved among certain insects?

Intricate societies evolved at least 11 separate times among the bees, ants, and other members of the order Hymenoptera, yet only once (in the termites) among all other insects. Why the difference? One reason may be that the bees and wasps have an unusual kind of sex determination, called **haplodiploidy.** In haplodiploidy, males develop from unfertilized eggs and are therefore haploid, while females are diploid. Other than providing sperm in a single mating event, males contribute nothing to the economy of the hive. Haplodiploidy dramatically alters the usual genetic relationships among parents and siblings. It means that a worker's inclusive fitness is greater if she devotes her energies to her siblings than if she reproduces herself. Let's see how this works.

First consider the usual case, say in humans. Half your alleles come from your mother, half from your father. Your brothers and sisters also share half of their alleles with your mother.

What, then, is the chance of sharing an allele with your sister? Half of all your alleles came from your mother, and the chance that you and your sister inherited the same allele is 0.5. So the chance that you each inherited a given allele from your mother is $0.5 \times 0.5 = 0.25$. Similarly, the chance that you each inherited the same allele from your father is 0.25. So

the chance that you and your sister share any allele is $0.25 + 0.25 = 0.5$, exactly the same as the chance that you share an allele with your mother or father. A similar calculation shows that you share one-eighth of your alleles with your first cousins. The British evolutionary biologist J.B.S. Haldane summarized these relationships by joking, "I would gladly lay down my life for two of my brothers or eight of my first cousins."

As in the case of ordinary sex determination, the queen passes half her alleles to her female offspring. However, since males are haploid, a male passes on all his alleles. That is, his daughters inherit all of his genes. As a result, the (female) workers share 75 percent (rather than 50 percent) of their alleles with one another. Because the queen mates with many different males, not all of her offspring share the same father. However, the average relatedness will be greater than 50 percent.

If the workers were to have their own offspring, they would pass on 50 percent of their alleles. They can therefore pass on more of their genes by putting their energies into making sisters than they can by reproducing and making daughters. To put it another way, the workers more efficiently perpetuate their genes by helping their mother, the queen, produce more offspring (including new queens) than by producing eggs themselves.

Haplodiploidy in some social insects probably facilitates kin selection.

Does inclusive fitness always explain altruistic behavior?

The haplodiploid social insects are not the only animals that help one another. Grazing animals, birds, and other animals that live in groups commonly help protect one another. In as many as 300 species of birds, "helpers" may assist a breeding pair feed and protect the nestlings. Whether such behavior can be explained by kin selection, however, depends on how closely the altruists are related to one another.

A few studies suggest kin selection as an explanation for altruism. For example, when a pair of Florida scrub jays breeds, as many as six other jays may help feed and protect the nestlings. These helpers are usually the breeding jays' offspring from previous clutches, offspring that have yet to mate themselves. Because territories suitable for breeding are in short supply, these helpers may not be able to breed effectively. Like the honeybee workers, the young jays may be better off helping their siblings than trying to breed themselves.

An even more persuasive example comes from Belding ground squirrel populations high in the Sierra Nevada, in California. In the summer, female squirrels defend territories where they raise their offspring. The two biggest dangers to Belding ground squirrels, besides winter snows, are predators and other squirrels. Unrelated or distantly related squirrels regularly kill other ground squirrels, killing up to 8 percent of all pups. At the approach of a coyote or strange ground squirrel, some female squirrels stand up tall and sound a shrill alarm

BOX 29.1

Tit for Tat and Prisoner's Dilemma

The classic game theory analysis of the risks and benefits of cooperation and altruism is presented in the "game" called prisoner's dilemma. Imagine that you and a friend are arrested for committing a crime, and you are held in separate jail cells, unable to communicate. The police immediately commence interrogating the two of you.

Suppose you claim innocence, but your friend confesses and incriminates you. You'll be in prison for a long time, but your friend will get off scot-free. On the other hand, you could get off yourself if you incriminate your friend and your friend pleads innocent. If you each confess (and incriminate each other), the state's job of convicting you will be easy, and your sentences will be light. But if you both plead innocent, the state will have little evidence against you, and your sentences will be even lighter.

What to do? Regardless of what your friend does, you are better off confessing, which is true for your friend, too. On the other hand, if both of you confess, you will be worse off than if both of you pleaded innocent. The safest solution is to confess. Even though pleading innocent might get you off scot-free, it's the riskiest solution to the problem.

The game of prisoner's dilemma seems to show that selfish behavior is safer than cooperative behavior. In 1981, however, a political scientist named Robert Axelrod and a biologist named W.D. Hamilton combined forces to explore what happens when the game is played over and over. They reasoned that in real life, people and animals interact repeatedly. Perhaps cooperation might pay off in the long run.

To make the endeavor more interesting, Axelrod invited game theorists from sociology, political science, economics, mathematics, and biology to design strategies to compete in an ongoing computer tournament. Predictably, players that al-ways cooperated were taken advantage of. But strategies of constant betrayal also did badly.

The winning long-term scheme was a simple strategy called tit for tat. In tit for tat, the player always cooperates (pleads innocent) in the first round. After that, the player does whatever the opponent did in the last round. In short, you reward your friend for cooperating and punish your friend for ratting on you. Over time, tit for tat always won.

In the early 1990s, an even more successful strategy emerged from Axelrod's tournament. "Modified tit for tat" occasionally forgives other players for a betrayal, just often enough to nudge the game in the direction of cooperation and reciprocity. Although the prisoner's dilemma is a simple game, it suggests that reciprocity is a possible explanation for altruism in some situations.

---

call that alerts other nearby squirrels (Figure 29-15). Other individuals who first sight a coyote or a stranger make no alarm and simply dive into the nearest burrow silently, leaving the other squirrels to fend for themselves. The squirrels that sound the alarm increase their own chance of being attacked, so their behavior is altruistic.

Paul Sherman, a biologist now at Cornell University, asked why these squirrels risk their own lives to help others, and why some squirrels behave altruistically, while others do not. Sherman discovered that most of the squirrels that sound the alarm are yearling and older females, who are warning their female relatives—sisters, mothers, and daughters, as well as juvenile sons. However, males and childless females rarely warned other squirrels. The older female incurs a risk that is counterbalanced by the chance that she will save the life of a close relative. Her altruistic behavior may contribute to her inclusive fitness.

In the past 20 years, many biologists, anthropologists, and other enthusiasts have tried to extend the concept of kin selection to virtually all altruistic behavior. But not all examples of altruistic behavior conform to the kin selection model. Dwarf mongooses, for example, help breeding pairs care for their young, but not all of the helpers are relatives. Kin selection does not account for such altruism.

There are other explanations for why animals behave altruistically, but one of the most interesting was first suggested by ethologist Robert Trivers. Trivers argued that animals capable of recognizing one another might engage in altruism on the understanding that the recipient would return the favor sometime in the future. Trivers called this idea **reciprocal altruism,** since the recipient reciprocates. Such a system at first seems too susceptible to cheaters to work. Biologists wondered, How can an animal ensure that the recipient of an altruistic act will return the favor? But research in the area of "game theory," a branch of mathematical logic applied to business, military, sociobiological, and other problems, suggests that altruism can pay off in spite of cheaters (Box 29.1).

Kin selection may explain altruism in other animals besides the social insects. But other explanations, such as reciprocal altruism, are possible, as well.

A.

B.

Figure 29-15 How kinship affects behavior. A. At Tioga Pass, California, students watch individual Belding ground squirrels marked with numbers written with hair dye. These ground squirrels live in colonies consisting of related females and their offspring. B. Older females sound a loud alarm when predators approach. Males take no such chances and scurry silently away. *(A and B, George D. Lepp)*

Study Outline with Key Terms

The systematic study of natural (wild) behavior is called **ethology.** In contrast, the laboratory study of behavior common to many animals is called **experimental psychology.** Social and other forms of behavior viewed in the context of evolution and ecology is the concern of the field of **sociobiology,** now called **behavioral ecology.**

Animal behavior is partly influenced by genes, partly by environment. Humans and other animals learn from one another as well as from their environment. Discussions of animal behavior can quickly become politicized.

Biologists are very interested in the relative importance of genes and environment in a variety of animal behavior. If genes help to shape behavior, and natural selection operates on behavioral variants, then biologists must ask, What ecological factors determine the adaptiveness of various kinds of behavior?

Innate behavior includes a baby's first smile and the salivating of a dog over his dinner bowl (a **conditioned reflex**), but all behavior in all animals can be modified through **learning.** A **modal action pattern** is highly stereotyped innate behavior performed in response to a **sign stimulus.** An exaggerated sign stimulus is a **supernormal stimulus.** Stimuli may be sights, sounds, tactile experiences, or smells in the form of **pheromones.** Some forms of learning, called **imprinting,** occur only at special times in development, called **sensitive periods.**

Animals seem to accurately weigh the costs and benefits of different foraging strategies in a process called **optimal foraging.** When animals compete for mates or resources, they engage in **agonistic behavior,** which includes aggression, appeasement, and retreat. Animals such as chickens, dogs, and primates use agonistic behavior to establish **dominance hierarchies.** Competition for mates results in **sexual selection.** Many animals defend a **territory,** which may encompass space and resources for mating, foraging, or raising young.

Social groups allow animals to spot both prey and predators more easily. However, groups of animals are themselves more visible to predators than are solitary animals. Also, the individuals in a group must constantly compete with one another for food, mates, and breeding areas.

The social insects display complex social structure and **altruism.** One explanation for the adaptive value of altruism is **inclusive fitness,** which includes the concept of **kin selection.** All of the social insects except the termites are **haplodiploid,** which means that sisters are more closely related to one another than to their own offspring. Such extreme relatedness facilitates kin selection and altruism. Another explanation for altruism is **reciprocal altruism.**

Review and Thought Questions

Review Questions

1. In the past, what distinguished ethology from experimental psychology?
2. Define a modal action pattern and give an example.
3. What is a pheromone?
4. Give an example of a supernormal stimulus in advertising you have seen.
5. Give an example of a sensitive period.
6. Define and give an example of optimal foraging.
7. Distinguish between aggression and agonistic behavior.
8. Give an example of territorial behavior in dogs.
9. Why, according to the theory of inclusive fitness, are honeybee workers better off not reproducing?

10. Give an example of an innate response modified by learning in an adult human.

Thought Questions

11. Three pizza parlors offer specials on identical giant pizzas. One parlor will deliver the giant pizza for $25. The other two don't deliver. One is just around the corner, and its pizza is $20. The other is 8 miles away and charges $12. It costs you 50¢ a mile to drive your 1980 Buick Skylark, and the car needs a new battery. If you break down in the neighborhood where the $12 pizza is, you might get mugged. What is the optimal solution to this foraging problem?
12. If humans were haplodiploid, how would human societies be different?

About the Chapter-Opening Image

English birds called tits learned from each other how to uncap milk bottles and sip the cream from bottles left on the front steps of English houses.

 On-line materials relating to this chapter are on the World Wide Web at **http://www.harcourtcollege.com/lifesci/aal2/**

part VI

Structural and Physiological Adaptations of Flowering Plants

Chapter 30
Structural and Chemical Adaptations of Plants

Chapter 31
What Moves Water and Sugars?

Chapter 32
Growth and Development of Flowering Plants

Chapter 33
How Do Plants Regulate Growth and Development?

Scanning electron micrograph of sundew trichomes.
(Fred Hossler/Visuals Unlimited)

30 *Structural and Chemical Adaptations of Plants*

Should We Eat Our Vegetables?

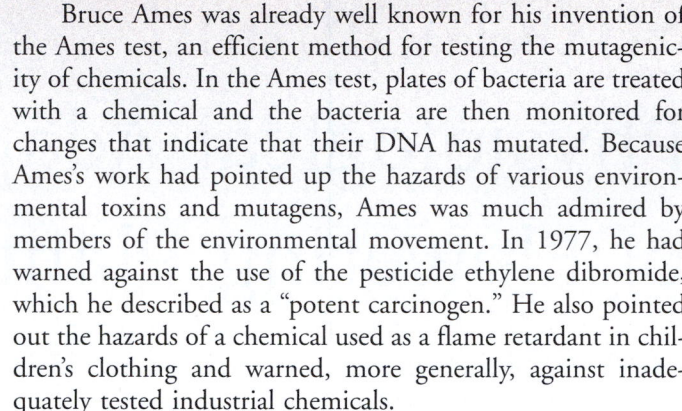

In 1983, a biochemist at the University of California, Berkeley, caused a stir in both the scientific community and in the health food industry when he asserted that enormous numbers of ordinary vegetables were loaded with mutagens, chemicals that cause DNA to mutate. Under the right circumstances, mutagens are capable of causing cancer, and natural chemicals, the biochemist argued, were as carcinogenic as anything manufactured by American industry.

Bruce Ames was already well known for his invention of the Ames test, an efficient method for testing the mutagenicity of chemicals. In the Ames test, plates of bacteria are treated with a chemical and the bacteria are then monitored for changes that indicate that their DNA has mutated. Because Ames's work had pointed up the hazards of various environmental toxins and mutagens, Ames was much admired by members of the environmental movement. In 1977, he had warned against the use of the pesticide ethylene dibromide, which he described as a "potent carcinogen." He also pointed out the hazards of a chemical used as a flame retardant in children's clothing and warned, more generally, against inadequately tested industrial chemicals.

But many people saw his 1983 paper as an about-face, in which he seemed to assert that pesticides, air pollution, and other environmental hazards were nothing compared to the toxins found in ordinary vegetables and meat. Indeed, Ames reviewed the scientific literature on mutagenic and toxic compounds in food and concluded, "The human dietary intake of 'nature's pesticides' is likely to be several grams per day—probably at least 10,000 times higher than the dietary intake of man-made pesticides." Ames listed scores of toxic chemicals found naturally in ordinary foods, including burned or browned meats and other foods, most fats—especially burned or rancid fats—and great numbers of compounds made naturally by plants.

Herbal teas made from plants in the family Euphorbiaceae, for example, contain potent promoters of carcinogenesis. Wilted carrots contain a mutagen. Alfalfa sprouts contain a highly toxic analog of the amino acid arginine, called canavanine, which causes a lupuslike immune disease in monkeys fed alfalfa sprouts. The metabolism of the alcohol in beer, wine, and spirits generates a compound that is both a mutagen and a teratogen, a compound that causes birth defects. As we will see later in this chapter, many plants have evolved the capacity to defend themselves chemically. In addition, many plants respond to any kind of damage by secreting high levels of toxins, some of which are mutagenic, teratogenic, or carcinogenic. Ames's list of plant and other food toxins went on for four pages in the journal *Science*.

Every plant that we eat has something in it that is bad for us, argued Ames. If we add all these compounds together, they far outweigh the traces of pesticides typically found on fruits

Bushnell/Soifer/Tony Stone Images

- Most flowering plants rely on underground roots and aboveground shoots to obtain water, minerals, carbon dioxide, and sunlight.

- A plant's vascular system consists of xylem, which carries water and minerals absorbed by the roots, and phloem, which carries sugars and other nutrients produced by photosynthesis.

- Plants produce secondary plant compounds that protect plants against herbivores, pathogens, and other plants; provide structural support; or attract animal pollinators and seed dispersers.

and vegetables or in drinking water. This is not to say that pesticides and other toxins are not dangerous at high levels. People who regularly expose themselves to toxic chemicals—field workers who spray pesticides, for example—are at great risk. But for the average person, Ames argued, the amounts of industrial chemicals we ingest are minuscule compared to what we get naturally from plants.

Ames's fellow researchers were appalled, accusing Ames of ignoring the well-known hazards of smoking, of backing off from his stance against both the use of dangerous pesticides and the chemical pollution of air and water. They even accused Ames of aiding corporations that manufacture and use dangerous chemicals. If nothing else, they asserted, he was confusing people about what constituted a real hazard and what did not.

Ames reacted defensively, arguing that he still believed in the strict testing of industrial chemicals. He insisted that the natural components of a diet could well play a role in the incidence of both heart disease and certain kinds of cancer, particularly colon cancer and breast cancer, the two most common forms of cancer after lung cancer, which is caused by smoking. But that role can be both positive and negative. For example, one study showed that smokers who did not eat fruit regularly had a 30 percent higher incidence of lung cancer than those who ate fruit.

A large part of Ames's 1983 paper had been devoted to natural plant compounds, such as vitamin E, vitamin C, and β-carotene, that seem to prevent cancer. These compounds, called antioxidants, work by preventing the formation of free radicals, compounds that damage DNA. Free radicals are formed during normal metabolic processes in both animals and plants. Because chlorophyll-mediated photosynthesis tends to generate destructive free radicals, all plants manufacture β-carotene or other similar antioxidants to protect their DNA and other important molecules.

In subsequent years, Ames increasingly focused on the production of free radicals during normal metabolism as the ultimate cause for all the degenerative diseases associated with aging—including heart disease, cataracts, and cancer. Lack

of antioxidant-containing fruits and vegetables in the diet, argued Ames, was a major cause for these degenerative diseases.

We can summarize Ames's argument by saying that if we don't smoke or drink too much, stay away from obvious hazards such as asbestos mines, and eat our fruits and vegetables, we need not worry very much about trace amounts of pesticides or other contaminants in our food and drinking water. Nutritionists now regularly advise Americans to eat more fruits and vegetables, not only for the vitamins, but also because of the high levels of antioxidants (Table 30-1).

On the other hand, the high levels of toxins that many workers are exposed to—in farming and in the processing of metals, for example—pose serious risks. Likewise, the high levels of contaminants found in toxic waste dumps pose distinct hazards to people and other organisms. Indeed, water and air pollution are frequently more threatening to other organisms than to us.

In the rest of this chapter we discuss plant structure and the chemical compounds found in plants. Most of our discussion will concern the angiosperms (flowering plants). Despite their diversity, flowering plants have evolved common organizational features in response to common environmental demands.

TABLE 30-1 FOODS THAT ARE GOOD SOURCES OF THREE ANTIOXIDANTS

| Food | % RDA | | |
	Vitamin A	Vitamin C	Vitamin E
Apples (1)	—	10	—
Apricots (3)	50	—	—
Blueberries (1 cup)	—	33	—
Strawberries (1/2 cup)	—	100	—
Cantaloupe (1/2)	> 100	> 100	—
Orange (1)	—	> 100	—
Broccoli	90	200	—
Carrot (1)	400	—	—
Red pepper (1)	84	250	—
Winter squash (1 cup)	150	—	—
Sweet potato	500	—	—
Almonds (1/2 cup)	—	—	170
Mayonnaise (1 T)	—	—	80
Peanuts (1/2 cup)	—	—	60
Safflower oil (1 T)	—	—	45

RDA = Recommended daily allowance.

WHAT ARE THE PHYSICAL CONSTRAINTS ON PLANTS?

Plants are photosynthetic. With a few exceptions, plants live by turning carbon dioxide into carbohydrates. To do this, plants need ample supplies of carbon dioxide, water, and sunlight. Recall from our discussion of photosynthesis in Chapter 7 that plants combine the hydrogen from water molecules (H_2O) with carbon dioxide (CO_2) to build the simple sugar glucose (Figure 30-1). The energy to accomplish this transformation comes from sunlight.

Until about 500 million years ago, photosynthetic organisms were largely aquatic, as was most life on Earth. Then, more than 450 million years ago, during the Ordovician Period, the first photosynthetic organisms invaded the land. Paleontologists have found traces of these little-known organisms in the form of spores, bits of cuticle, and tubes similar to the vessels that carry water in some modern vascular plants.

Biologists suspect these early plants may have been descended from green algae, or chlorophytes, which can live in fresh water and even on land. Modern chlorophytes associate with fungi as land-dwelling lichens. Some researchers have suggested that plants evolved from "inside-out" lichens, in which the photosynthetic chlorophyte became the dominant element. The fungal partner could have provided a way of getting water and minerals from the soil. (Such associations are so valuable that—as we discussed in Chapter 21—four-fifths of modern plants have fungal partners, called *mycorrhizae,* on their roots.)

From some such primitive form, recognizable plants evolved in the Silurian and diversified throughout the 100 million years of the Devonian and Carboniferous. Plants have become the dominant feature of land life. Especially successful are the flowering plants, with more than 275,000 species now populating our planet. They are the major primary producers of terrestrial ecosystems and the ultimate source of food for humans and all other terrestrial organisms. In Chapter 22, we discuss the origins and organization of plants.

How Did Terrestrial Plants Evolve?

For the aquatic ancestors of terrestrial plants, the land offered distinct advantages. Aquatic plants are limited to shallow water, where nutrients and sunlight combine in a narrow plane at the surface. In deeper waters, little light is available.

The land offered plenty of light and nutrients and a higher concentration of carbon dioxide than was available in the sea. Compared to marine and freshwater environments, however, the land could provide little water, and the carbon dioxide was in a gaseous state, not dissolved in water. In addition, whereas water tends to buoy plants and other organisms, gravity exerts a crushing force on these same organisms in a terrestrial environment.

Photosynthetic organisms therefore needed new mechanisms for obtaining water and carbon dioxide and for supporting their bodies against the pull of gravity. Whereas aquatic plants absorb water and dissolved carbon dioxide from the water that completely surrounds them, terrestrial plants have evolved roots that pull water from moist soil, as well as tiny pores in their leaves that collect carbon dioxide and release water. These pores, called **stomata** [Greek, *stoma* = mouth], open and close like tiny mouths, also providing a path for the movement of water through the plant, as we discuss in Chapter 31.

In addition, terrestrial plants harvest energy from sunlight using their leaves. Leaves are solar collectors: they are often broad and thin, which maximizes their exposure to the sun, and they are loaded with chlorophyll, the green pigment that plants use to capture the energy of sunlight. (There are, in fact, nonphotosynthetic plants, such as dodder, a close relative of morning glory; dodder lives as a parasite on other plants.)

Besides roots and leaves, most terrestrial plants rely on stems. Stems not only support leaves, flowers, and fruits but also provide an avenue for the transport of materials between the ground and the air. The leaves use water from the roots to support photosynthesis. The roots, in turn, use sugars manufactured in the leaves to supply energy to the ionic pumps that pull water from the surrounding soil.

Not all terrestrial plants have stems and transport tissues. **Vascular plants,** which have specialized conducting systems,

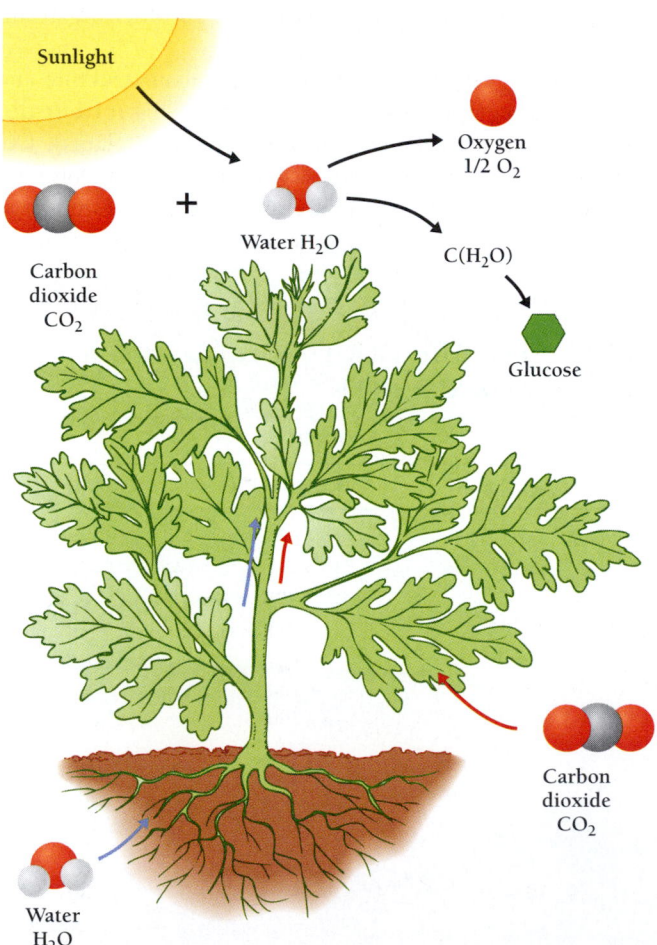

Figure 30-1   Photosynthesis and water flow through a flowering plant.

did not evolve until about 435 million years ago. And some 18,000 species of mosses and other nonvascular plants (*bryophytes*) flourish today, despite their lack of roots, stems, leaves, and vascular tissues (Chapter 22). The vast majority of terrestrial plants, however, are vascular plants, with roots, stems, and leaves.

Despite differences in detail, the basic form of vascular plants has remained unchanged for hundreds of millions of years. This form, which solves all of the major problems of terrestrial life, is highly successful. Vascular plants live virtually everywhere on the planet where there is sunlight, carbon dioxide, and any amount of water at all. Only the polar ice caps lack vascular plants (although one species of grass grows in the dry valleys of Antarctica).

Terrestrial plants have evolved structures and mechanisms to perform photosynthesis, to obtain water and carbon dioxide, to support themselves mechanically, and to transport minerals and nutrients throughout the plant.

What Are the Roles of Shoots and Roots?

While the defining characteristic of plants is that they are multicellular photosynthesizers, plants are not mere bags of photosynthetic protoplasm. Like animals and fungi, plants are highly structured multicellular organisms. Their forms represent solutions to easily defined problems.

The anatomy of a flowering plant reflects its need to derive sustenance from both ground and air. Most plants consist of an aboveground **shoot system**—stems, photosynthetic leaves, and flowers and other organs of reproduction—and an underground **root system,** which anchors the plant and absorbs water and minerals from the soil. Shoots and roots depend on one another utterly: roots could not exist without the energy provided by photosynthesis in the shoot, and shoots could not exist without the water and minerals absorbed by the roots.

Materials travel between shoot and root in the plant's **vascular system** [Latin, *vas* = vessel], the network of conducting tubes through which fluids move. The vascular system consists of two tissues, the **xylem,** which transports water and minerals, and the **phloem,** which transports the energy-rich products of photosynthesis. The phloem transports sugars from leaves to growing regions, reproductive structures, and root and also from storage areas (such as those of sugar beets, potatoes, or carrots) to the rest of the plant. Xylem, which consists mostly of dead cells, and phloem, which consists mostly of living cells, associate together in vascular bundles.

Terrestrial plants must collect water, carbon dioxide, and light, but they must also support their own bodies against the considerable pull of gravity and, finally, they must reproduce. Every plant is a compromise between design constraints imposed by the different functions of the plant. The traits that are best for intercepting light, for example, may conflict with those needed for conserving water.

The more plants spread their leaves and branches to maximize their exposure to sunlight, for example, the stronger the stems and branches have to be to support the weight. If you take this textbook in your hand, you will notice that it is harder to hold it straight out to the side or in front of you than to hold it down at your side or even straight up above your head. Plants that spread their leaves in the sun face the same problem. Their ability to do so depended on the evolution of materials that are both enormously strong and remarkably lightweight.

The single plant structure that is most important to supporting a plant is the cell wall. Like us, plants are eukaryotes, whose cells have a nucleus, ribosomes, and other organelles. Unlike our cells, however, plant cells also have a **cell wall,** which consists of **cellulose,** other carbohydrates, and specialized proteins. For its weight, cellulose is the strongest material known. The nutshell of the Australian macadamia nut is stronger than commercial-grade aluminum, concrete, glass, or brick; yet the density of the nutshell is less than half that of these other materials.

All plant cells have a **primary cell wall** outside the plasma membrane. Some cells make a thicker **secondary cell wall** between the cell membrane and the primary cell wall (Figure 30-2). Cells with a secondary cell wall are frequently cells specialized to provide the plant with extra mechanical support. A polymer called **lignin** gives additional strength to the secondary wall, and in some cases to the primary wall, as well. Lignin has some of the same qualities as a hard, stiff plastic.

The cytoplasm of most plant cells fills only part of the space within the cell wall, often as little as 10 percent. The rest of the internal space consists of a large central vacuole, filled with a watery fluid called the cell sap. Water pressure in the central vacuole pushes a cell's contents hard against the resisting cell wall. The pressure of the fluid contents of the vacuole against the hard cell wall, called turgor pressure, helps support plants against the pull of gravity. We have all seen that plants under water stress wilt. This is because they have lost turgor pressure.

Some plants, called *herbs* (or herbaceous plants), are entirely soft and green and depend mainly on turgor pressure for support. Herbs have no woody aboveground parts and are highly susceptible to wilting. Perennial herbs may have woody roots. (To a botanist, the term "herb" does not mean familiar kitchen herbs such as rosemary or thyme, some of which are, in any case, woody.) Woody plants, such as trees, shrubs, and other perennial plants, have woody parts that provide support independent of turgor pressure.

Terrestrial plants include a shoot, with stems and leaves, and a root. A plant is supported against the pull of gravity by turgor pressure within its cells and by structural materials, such as cellulose and lignin, within the cell walls of support cells.

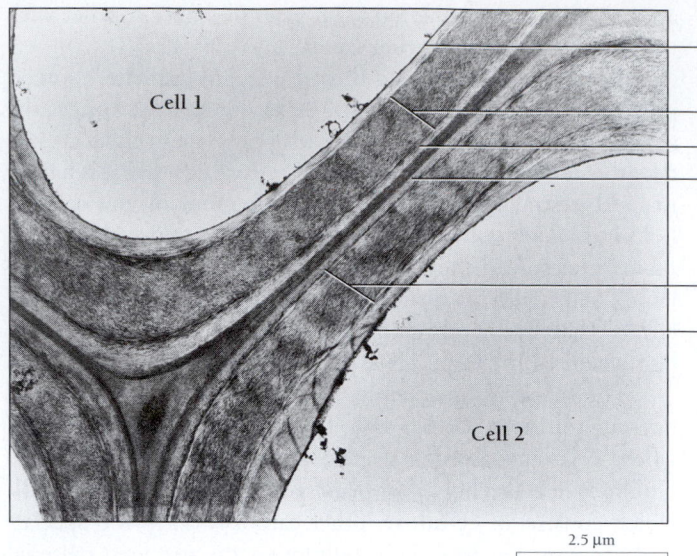

Plasma membrane of cell 1

Secondary cell wall of cell 1
Primary cell wall of cell 1
Primary cell wall of cell 2

Secondary cell wall of cell 2

Plasma membrane of cell 2

2.5 µm

Figure 30-2   Primary and secondary cell walls. All plant cells have a primary cell wall, which is composed of cellulose, hemicellulose, and pectin (a gluelike substance used to gel jellies). Some cells deposit a secondary cell wall on the inner face of the primary cell wall. The secondary wall is made of cellulose and lignin, a hardening agent. *(Biophoto Associates)*

HOW ARE FLOWERING PLANTS STRUCTURED?

The cells of plants are organized into tissues, which are organized into tissue systems. Four tissue systems occur in all organs of a plant. These are the dermal, vascular, and ground tissue systems, and the undifferentiated meristem tissue, from which new cells arise. The dermal tissue system makes up the outer, protective covering of the plant, the equivalent of our skin. The vascular tissue system comprises all of the tissues responsible for conducting. The vascular tissue system is embedded in the ground tissue system. The important differences in structure in root, stem, and leaf derive from differences in the way the vascular tissue system is distributed in the ground tissue.

The **dermal tissue system** is the protective covering of the plant. It consists of the **epidermis,** which is primary tissue, and the **periderm,** which is secondary tissue (Chapter 32). The epidermis of the shoot secretes a **cuticle,** a waxy covering that keeps the aboveground parts of the plants from losing water. At first glance, the epidermis appears to be a single layer of identical cells. Closer examination, however, reveals that the epidermis is a *complex tissue,* meaning that it consists of more than one cell type, each with a distinct functional role. (Tissues that consist of one cell type only are called *simple tissues.*) For example, the epidermis includes generalized epidermal cells, as well as the guard cells and the subsidiary cells of stomata. In addition, different plants may have glands, trichomes (hairs), secretory cells, and other specialized cells.

The **ground tissue system** consists of three main types of tissue—**parenchyma,** collenchyma, and sclerenchyma. Parenchyma and collenchyma are both simple tissues made of just one kind of cell, distinguished by the character of their cell walls. The cells of parenchyma [Greek, *para* = beside] are living cells that have only a thin primary cell wall. They are

usually soft and succulent, with large vacuoles. Parenchyma cells make up most of the softer tissues of plants and they frequently contain starch (Figure 30-3A).

The cells of **collenchyma** [Greek, *kolla* = glue] have a thick primary cell wall and often provide mechanical support, usually (but not only) in the growing regions of stems and leaves. Collenchyma cells, which are alive at maturity, often form bands or sheets just inside the epidermis. In celery, these bands are easy to find; they are the long strings just under the surface. In some collenchyma cells, lignin accumulates irregularly in the secondary cell walls, particularly in the corners (Figure 30-3A).

Sclerenchyma [Greek, *skleros* = hard] consists of one of two cell types, fibers or sclereids. Fibers and sclereids, each of which exist in several varieties, have both primary and secondary walls and furnish mechanical support to the plant. Fibers and sclereids (unlike collenchyma cells) are usually not alive at maturity. Fibers consist of long, thin cells that join together into strong plates or cords, while sclereids occur as individual cells or in small groups (Figure 30-3B and C). Linen, which is woven from strands of sclerenchyma from the flax plant, illustrates the strength of sclerenchyma fibers.

The **vascular tissue system** consists of the xylem, phloem, and associated tissues. Both xylem and phloem are themselves complex tissues. Xylem transports water and dissolved minerals and also provides mechanical support. Phloem transports dissolved sugars and other organic substances and provides some mechanical support.

All three specialized tissue systems develop from regions of actively dividing cells called **meristems** [Greek, *meristos* = divided]. Meristems at the tips of roots and shoots are called **apical meristems** [Latin, *apex* = top]. Growth from apical meristem tissue is termed **primary growth.** Many plants also possess **lateral meristems,** which are cylinders of actively dividing cells within the roots and stems. Growth arising from

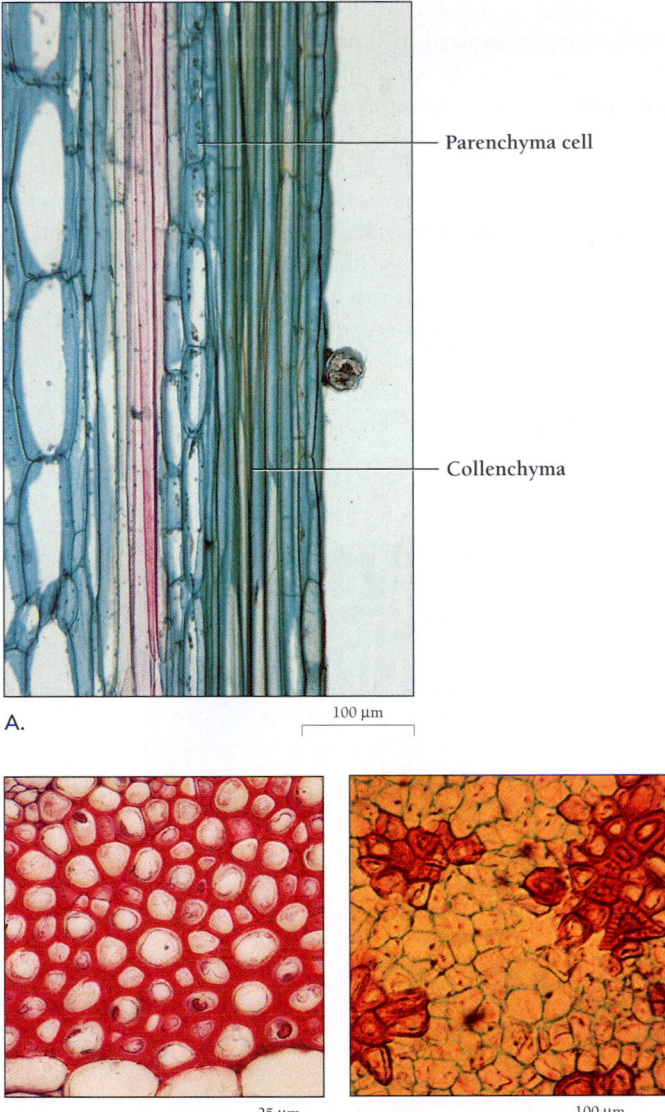

A.

100 µm

Parenchyma cell

Collenchyma

B.   25 µm

C.   100 µm

Figure 30-3   Cell types in plants. Most parts of a plant are made of just a few cell types. A. Delicate parenchyma cells make up most ground tissue, and tough-walled collenchyma cells strengthen stems and leaves. B. Fibrous sclerenchyma is tougher than collenchyma. C. Hard sclereids give pears their gritty texture. The ripening of fruits often requires the breakdown of the lignin in sclereids. *(A, Ray Evert; B, Biophoto Associates/Science Source/Photo Researchers, Inc.; C, Larry Mellichamp/Visuals Unlimited)*

How Is the Vascular System Organized?

Xylem carries water and its dissolved substances in two kinds of cells, tracheids and vessel elements. Both kinds of cells have lignin-reinforced secondary cell walls inside of primary cell walls. Although tracheids and vessel elements are the cells that conduct fluids, the complex xylem tissue also includes parenchyma cells, fibers, and secretory cells.

Xylem carries fluids in two kinds of nonliving cells

Tracheids are long, thin, spindle-shaped cells. Water and dissolved substances move from cell to cell through **pits,** regions that lack secondary cell walls and watertight lignin (Figure 30-4). Pits may occur anywhere along a tracheid, but most form at the tapered ends of the tracheids. Pits form as pairs between adjacent cells, and water flows smoothly through the pits, from one hollow tracheid to the next.

In a **simple pit pair,** a region of secondary wall is absent (Figure 30-5A). While the tracheid is alive, **plasmodesmata** connect adjacent cells through the pit. As the tracheids mature, they lose their nucleus and cytoplasm. The primary walls remain, however, as illustrated in Figure 30-5.

Because simple pits lack secondary cell walls, their presence in the xylem reduces its mechanical strength. Plants therefore also connect tracheids with another kind of pit pair, the **bordered pit pair.** In these, adjacent secondary wall overarches the pit membrane and reinforces the wall of the tracheid (Figure 30-5B). In conifers,

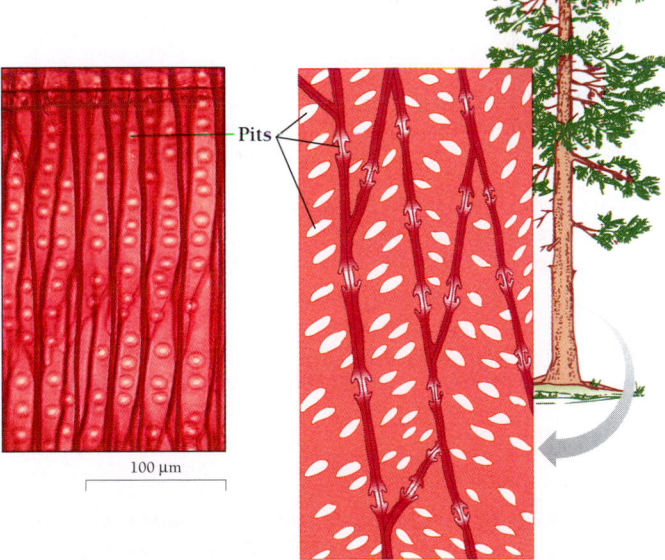

Pits

100 µm

Figure 30-4   Tracheid pit connections in a conifer. While water flows smoothly from one vessel element to another, water in the tracheids must pass from one tracheid to another by way of pits, which are specialized gaps in the secondary cell wall (see Figure 30-5). Flowering plants have both tracheids and vessel members, but gymnosperms such as this pine have tracheids only. *(Photo, John D. Cunningham/Visuals Unlimited)*

the lateral meristems, called **secondary growth,** increases the thickness of a shoot or a root. Secondary growth is commonly (although not always) accompanied by the development of wood, the familiar hard, fibrous substance, which actually is just secondary xylem. Secondary growth can also produce cork (part of the outer bark), which replaces the epidermis as a protective covering. We discuss secondary growth (and wood) in more detail in Chapter 32.

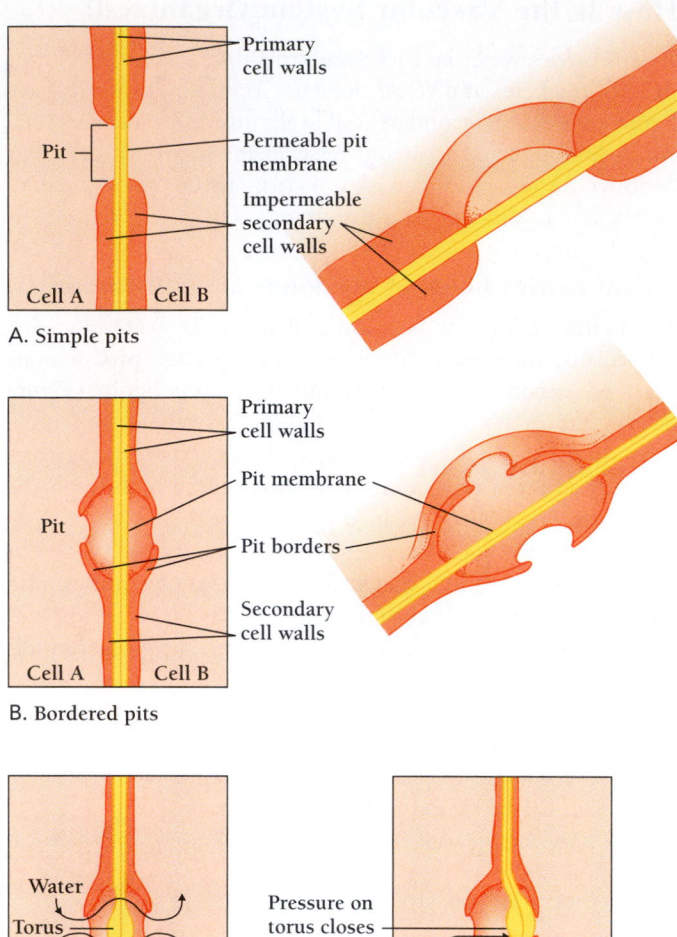

A. Simple pits

- Primary cell walls
- Permeable pit membrane
- Impermeable secondary cell walls

Pit

Cell A   Cell B

B. Bordered pits

- Primary cell walls
- Pit membrane
- Pit borders
- Secondary cell walls

Pit

Cell A   Cell B

C. Conifer bordered pit: Water pressure equal in cells A and B.

Water
Torus
Water

Cell A   Cell B

Pressure on torus closes pit

H_2O

Cell A   Cell B

Pressure higher in cell A than in cell B; torus closes pit.

Figure 30-5 Pit pairs in living plant cells are simply openings in the secondary cell wall. A. Two pits in adjacent cells form a pair of simple pits, also called a simple pit pair. B. In tracheids, a bordered pit forms, which is stronger than a simple pit pair. C. In conifers, a bordered pit functions as a valve. When more water is flowing on one side of the valve than on the other, a buttonlike structure called the torus shuts the pit, preventing water loss.

bordered pits have an additional feature that allows each pit to function as a valve. A thickened region of the cell wall, called the *torus,* forms a buttonlike thickening in the middle of the pit (Figure 30-5C). If the hydrostatic pressure from one cell is greater than that on the other side of the pit, the torus presses against the pit opening and prevents water flow. During the growth of a shoot or root, these valves reduce water movement away from the growing regions. In addition, the valves guide water away from tracheids that have been blocked by gas bubbles.

A larger, more efficient kind of conduction cell, called a vessel element, evolved in flowering plants. (Tracheids are the only conducting cells in conifers and most other nonflowering plants.) **Vessel elements** are shorter, broader, and less tapered than tracheids. Vessel elements, which are open at each end, connect end to end to form long open channels (Figure 30-6). They are found almost exclusively in flowering plants. The end wall of each vessel element, called a **perforation plate,** contains one or more large holes, through which water flows more easily than through the much smaller pits. While a tracheid pit lacks only the secondary wall, a vessel perforation lacks both primary and secondary walls. Water flows far more rapidly through the larger holes of the perforation plate than it can through the much smaller pits.

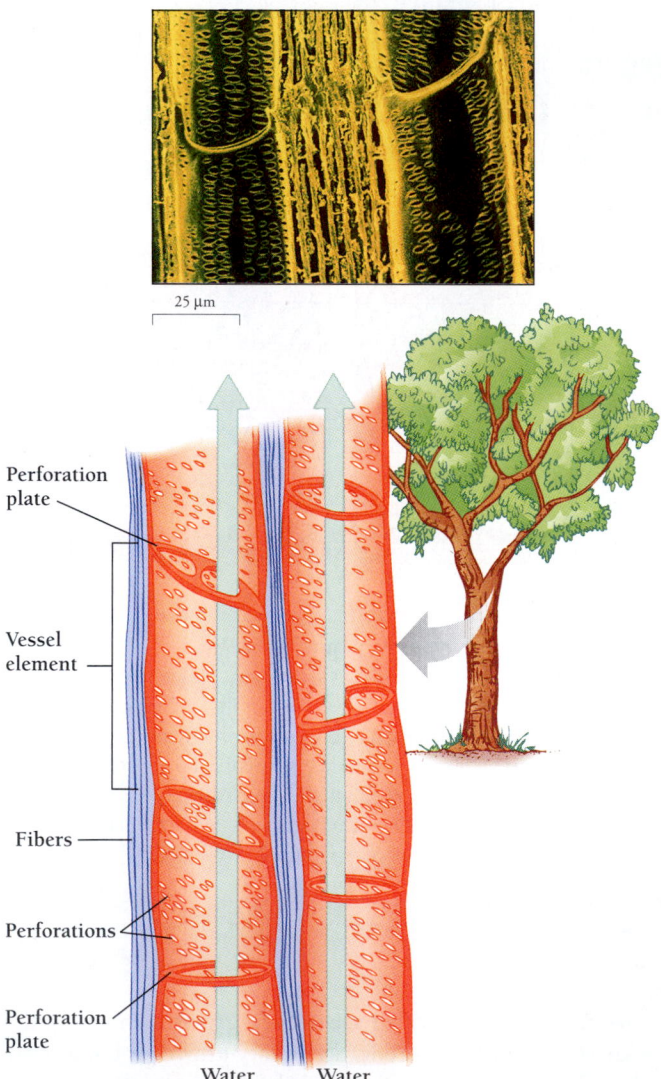

25 μm

- Perforation plate
- Vessel element
- Fibers
- Perforations
- Perforation plate

Water   Water

Figure 30-6 Vessel elements. Water passes directly through the perforation plate by way of holes that lack both primary and secondary cell walls. Vessel elements conduct water more efficiently than tracheids, but do not prevent the passage of bubbles of water vapor. *(Photo, G. Shih–R. Kessel/Visuals Unlimited)*

The conducting elements of phloem are living cells

Unlike water transport in the xylem, the movement of materials through the phloem depends on the action of living cells. The cells that actually conduct fluid are the **sieve tube members,** so named because they have sievelike plates in their end walls. Stacks of them form the pipelike channels called **sieve tubes** (Figure 30-7). The solution within a sieve tube contains up to 20 percent sucrose, as well as other nutrients produced in photosynthetic tissue. The complex phloem tissue also includes companion cells, parenchyma cells, sclerenchyma fibers, sclereids, and other cell types, depending on species. The cell walls of the sieve tube members do not contain lignin.

The end wall between adjoining sieve tube members is called a **sieve plate.** A sieve plate has more and bigger pores than the sides of the sieve tubes, permitting more fluid flow through a tube than in and out of it. Even the pores in the sieve plate, however, resist water flow more than the much larger perforations in the vessel elements of the xylem, since (as living cells) sieve tube members are surrounded by a plasma membrane. In addition, the sieve tubes are much narrower than the conducting tubes of the xylem. The sieve plate contains many sieve pores, open channels that develop from modified plasmodesmata.

Although sieve tube members are living cells, they are highly specialized and lack nuclei, ribosomes, and vacuoles. Without nuclei and ribosomes, they are incapable of synthesizing the proteins they need to live. Next to each sieve tube member is a **companion cell**—a long, nucleated, fully functional cell that supplies the sieve tube member with both proteins and energy-rich molecules. The companion cell is connected to the sieve tube member by plasmodesmata, through which materials flow to the sieve tube member.

Xylem consists of two kinds of nonliving cells: narrow tracheids and wider vessel elements. Phloem is made of living cells called sieve tube members, which accommodate the flow of materials through the phloem.

How Do Monocots and Dicots Differ?

Almost all modern plants are flowering plants (angiosperms), as we discussed in Chapter 22. About 500 species of modern plants, however, are *gymnosperms,* plants that produce seeds, but no fruits or flowers. Gymnosperms include pine trees and other conifers, as well as palmlike cycads. In addition, about 18,000 species of modern plants are *bryophytes,* with no vascular systems at all.

Botanists divide the 275,000 species of angiosperms into two groups, based on their early development. When a seed first germinates, it has a root and a shoot. The very first leaves are quite simple in form and are called **cotyledons,** or seed leaves (Figure 30-8A). All flowering plants are classified into one of two categories according to whether their seeds have one cotyledon or two.

Monocotyledons, or **monocots,** have one cotyledon, while **dicotyledons,** or **dicots,** have two. Monocots consist of about 65,000 species, including all of the grasses, lilies, palms, and orchids. Most monocots are herbaceous. Dicots include almost all trees and shrubs (but not conifers and other gymnosperms, which are not flowering plants), as well as many herbs.

Other traits that distinguish monocots from dicots are the number of flower parts, the arrangement of vascular bundles, and the structure of the root system. The flower parts of a dicot usually come in multiples of four or five, while those of a monocot come in multiples of three (Figure 30-8C). The leaves of monocots and dicots have characteristic patterns of veins (Figure 30-8B): in monocots the veins generally run parallel to one another (as the leaves of corn and other grasses), while those of a dicot usually form a complex netlike pattern (as in a tomato leaf or a maple leaf).

Inside a dicot's stem, the vascular bundles lie in the form of a ring, while those of a monocot are scattered throughout the stem (Figure 30-8C). Inside the vascular cylinder of a dicot's root, xylem and phloem often form a star shape (in cross section) that runs the length of the root. In many monocot roots, however, the vascular system (called the *stele*) has a central nonconducting core, made of ground tissue called **pith.** Bundles of xylem and phloem lie around the outside of the central pith. Monocot and dicot roots are not consistently different, however. Some dicots have a central pith, and some monocot roots have central xylem.

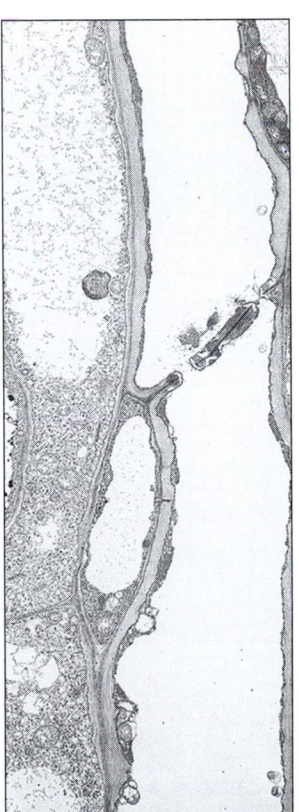

5 μm

Figure 30-7   **The living cells of phloem.** Phloem differs dramatically from xylem in consisting of living cells. Like xylem cells, phloem cells come in two types. Sieve cells are long and narrow with pointed ends, like tracheids. Sieve tube members are short and wide with flat ends, like vessel members. Similarly, only flowering plants have sieve tube members. Modified plasmodesmata between adjacent sieve tube members allow sugar to pass from cell to cell. Nearby companion cells load sugars in and out of the sieve tube members and probably supply the sieve tube members, which lack nuclei, with energy and useful molecules. *(Ray Evert)*

MONOCOT
(One cotyledon)

DICOT
(Two cotyledons)

A. Embryos

Parallel

Netlike

B. Leaf venation

Stem

Stem

Vascular
bundles
(xylem and
phloem)

1.0 mm

0.5 mm

0.5 mm

Vascular bundles
(xylem and phloem)

1.0 mm

C. Stems and roots   Root

Root

D. Flowers

Fibrous

Taproot

E. Roots

Figure 30-8 Characteristics of monocots and dicots.
A. The monocot embryo (corn) has a single cotyledon, while the dicot (bean) has two cotyledons. B. Monocots have parallel veins in their leaves, while dicots have a more netlike arrangement of veins. C. Cross sections through the stem and root of a corn plant (a monocot) and a buttercup (a dicot). In monocots, vascular bundles in the stem are scattered throughout the ground tissue of the stem, but form a tight cylinder in the roots. In dicots, the vascular bundles cluster near the outer surface of the stem, but crowd to the very center in the roots. D. The flower parts of monocots, such as this iris, as well as lilies, corn, and other grasses, come in groups of three. The flower parts of dicots, such as this buttercup, more often come in groups of four or five. E. Monocots have masses of small roots. A dicot may have a single, large taproot, with smaller branch roots. (*A, left, Barry L. Runk/Grant Heilman Photography, right, Runk/Schoenberger from Grant Heilman; C, monocot stem, © Dwight Kuhn; others, Runk/Schoenberger from Grant Heilman; D, left, John Gerlach/ Visuals Unlimited, right, John Gerlach/Tom Stack & Associates*)

Monocots and dicots differ also in the structure of their root systems (Figure 30-8E). Dicot roots have lateral meristems and secondary growth, while monocots have no lateral meristems and are incapable of secondary growth. For this reason, monocots generally have a delicate system of many small roots branching from the stem. These tiny, branching roots often form a thin, fibrous mat, familiar to anyone who has tried to dig in a grassy lawn or field. In contrast, most dicots produce one or several large roots, with much smaller roots branching off. In many dicots, the entire root system connects to a single vertical root, called a taproot. A carrot, for example, is a taproot that has enlarged to serve as a storage organ.

Flowering plants fall into two classes that have characteristic differences in their embryos, flowers, leaves, roots, and vascular systems. These two classes are monocots and dicots.

HOW DO TERRESTRIAL PLANTS OBTAIN WATER AND MINERALS?

Water and dissolved nutrients flow from the soil into the vascular system of the roots. They then move up through the stem to the plant's leaves and other organs. Some of the water contributes hydrogen atoms for the synthesis of sugars and other molecules. Ultimately, however, more than 90 percent of the water that enters a plant through the roots subsequently departs by way of the stomata in the leaves as water vapor, which drifts away into the atmosphere.

Plants need other materials besides water from the soil and carbon dioxide from the air. To make amino acids and nucleotides, for example, plants need large amounts of nitrogen, sulfur, and phosphorus. Plants also need fairly high levels (about 1 milligram per gram of soil) of such elements as magnesium, calcium, and potassium. Finally, plants need small amounts (less than 100 micrograms per gram of soil) of micronutrients such as molybdenum, copper, zinc, manganese, boron, iron, and chlorine. Plants obtain these nutrients almost entirely as ions dissolved in the water absorbed by their roots, so roots are the major suppliers of essential nutrients as well as of water.

Roots serve other functions as well: they anchor the plant to the ground, store carbohydrates, and produce several hormones that regulate growth. Finally, roots are important in *vegetative reproduction,* the propagation of plants without fertilization or seed production.

How Do Roots Carry Water?

Roots vary greatly in their external form, but they have a common internal organization. A cross section of a primary root shows three concentric rings: the epidermis on the outside, the large cells of the **cortex** beneath the epidermis, and the vascular system in the center.

Most of the water that enters a root comes through **root hairs,** each of which is the extension of a single epidermal cell. Numerous root hairs form just behind the root tips. Root hairs greatly increase the surface area of the epidermis. A single rye plant, for example, may have 10 billion root hairs and have a surface area of more than 400 square meters, as much as the walls of a small house. The root hairs extend the root's absorbing surface far to the sides of the root's axis, reaching water that would otherwise be missed. The epidermis and the root hairs are coated with a slimy substance called **mucigel,** which lubricates the root tips as they force their way between soil particles.

Most of a root's cortex consists of parenchyma cells, large cells with large vacuoles. In many plants, these cells contain stored starch, most dramatically in sweet potatoes and carrots. The innermost layer of the cortex differs from the rest of the cortex and is called the **endodermis** [Greek, *endon* = within + *derma* = skin].

Water flows from the soil to the vascular system by two paths (Figure 30-9). In both cases, the water flows from the epidermis through the cortex to the endodermis, where the two paths converge. The first path, the **symplast,** runs through the cytoplasm of the parenchyma cells of the cortex. Water and minerals pass from cell to cell via plasmodesmata. The second path, called the **apoplast,** lies in the material of the cell walls of the cortex. Water and dissolved minerals soak into the cell wall material of epidermis and cortex much as they soak into the cellulose fibers of paper towels.

Most minerals arrive at the endodermis via the symplast rather than apoplast. Active transporters in root hairs pump ions from the soil into the cytoplasm. The dissolved ions then move (along with water) through plasmodesmata into the endodermis.

The endodermis includes a structure that is key to roots' ability both to absorb water and to retain minerals. Each cell of the endodermis secretes into its walls a waxy, fatty substance called *suberin* that is impermeable to water (Figure 30-9). Suberin fills spaces between cellulose molecules—spaces that would otherwise be filled with water—making those cell walls as waterproof as waxed paper. The boxy endodermal cells deposit suberin on only four of their six surfaces. The other two surfaces have normal cell walls and membranes that regulate the flow of water and minerals into the cells of the endodermis. But the presence of this waxy wall, called the **Casparian strip,** prevents minerals and water from moving between the cells. For water to move from the cortex to the vascular system, it must pass through the membrane and the cytoplasm of the cells of the endodermis. The membrane of the endodermis thus regulates the kinds of solutes and the amount of water that enter the vascular system.

The endodermis is a selective barrier to the loss of solutes from the xylem to the soil and to the movement of undesirable solutes from the soil into the xylem. Like other cell mem-

BOX 30.1

The Distinctive Anatomy of C₄ Plants

In Chapter 7, we described the C₄ pathway for the assimilation of carbon dioxide, especially in grasses. Recall that in C₄ plants, the first reaction of carbon dioxide produces a 4-carbon compound, oxaloacetate, instead of the 3-carbon molecule phosphoglycerate. Oxaloacetate is converted to another compound that moves into the cells that make carbohydrates. There it breaks down to release carbon dioxide (Figure A).

In addition to a distinctive chemistry, C₄ plants also have a distinctive leaf organization. Surrounding each vascular bundle in the leaf is a sheath of cells, one or two layers thick, with especially large numbers of chloroplasts (Figure B). These bundle sheath cells are the principal sites of glucose formation. Outside the bundle sheaths are mesophyll cells, which are primarily responsible for the uptake of carbon dioxide and its conversion into oxaloacetate. The mesophyll cells deliver the absorbed carbon dioxide, in the form of an organic acid, to the bundle sheath cells, which regenerate and use the carbon dioxide. Only the bundle sheath cells contain

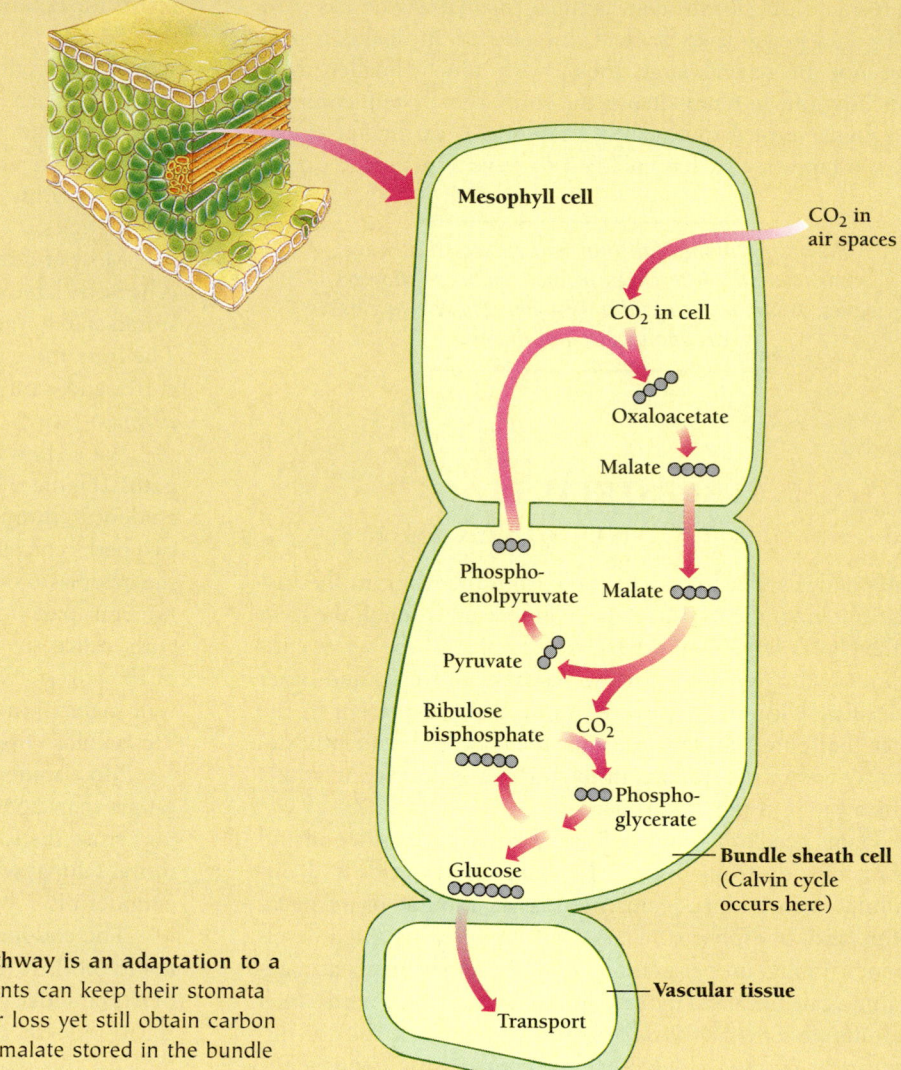

Figure A The C₄ pathway is an adaptation to a dry environment. C₄ plants can keep their stomata closed to minimize water loss yet still obtain carbon for photosynthesis from malate stored in the bundle sheath cells.

branes, the membranes of endodermis cells block the passage of some substances, allow other substances to diffuse passively into and out of the cell, and provide special transport mechanisms for still others. The endodermis thus serves as a gatekeeper, regulating the contents of the fluid that enter and leave the xylem. Were it not for the Casparian strip of the endodermis, most of the actively transported minerals would move back into the soil via the apoplastic path.

Within the cylinder formed by the endodermis is the **stele**, which contains the root's vascular system (Figure 30-9). At the center of the stele are the xylem and the phloem. Surround-

ing them is a sheath of meristem cells, called the **pericycle** [Greek, *peri* = around + *kykos* = circle]. The pericycle gives rise to branch roots and also thickens older roots by adding secondary xylem and phloem.

Water from the soil enters the root by way of root hairs, travels through the loose parenchyma cells of the cortex, and then moves through the endodermis into the vascular tissue.

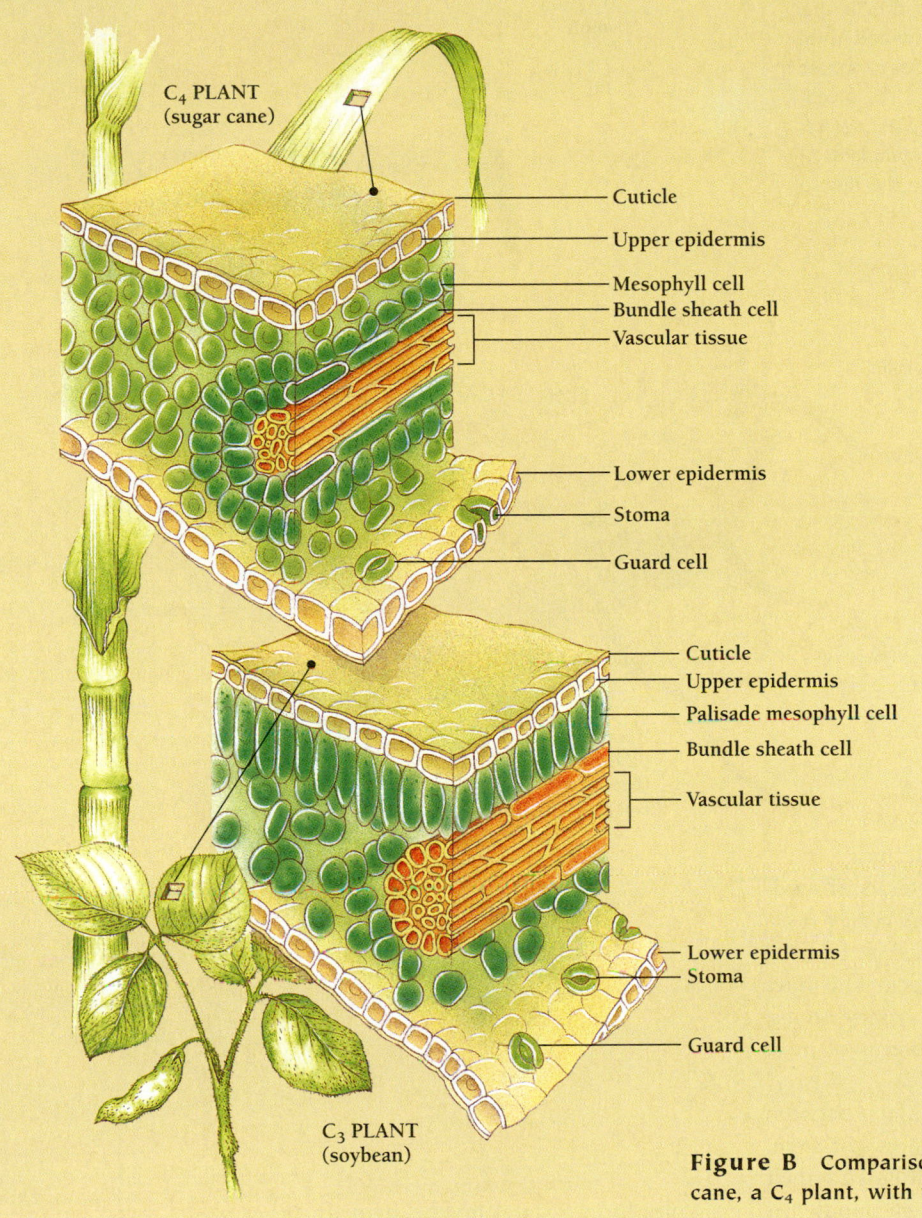

C₄ PLANT
(sugar cane)

Cuticle
Upper epidermis
Mesophyll cell
Bundle sheath cell
Vascular tissue

Lower epidermis
Stoma
Guard cell

Cuticle
Upper epidermis
Palisade mesophyll cell
Bundle sheath cell
Vascular tissue

Lower epidermis
Stoma

Guard cell

C₃ PLANT
(soybean)

ribulose bisphosphate carboxylase (Rubisco), the first enzyme in the pathway of sugar synthesis. The distinctive anatomy that almost always accompanies the C₄ pathway is called kranz anatomy [German, *Kranz* = wreath].

The C₄ pathway and kranz anatomy concentrate carbon dioxide in the bundle sheath cells. These cells thus have a much higher concentration of carbon dioxide than the corresponding cells of C₃ plants. The carbon dioxide effectively displaces oxygen from the Rubisco, so the plant does not waste accumulated energy in photorespiration. Despite the energy cost of transporting C₄ compounds to the bundle sheath (which occurs by active transport), C₄ plants are about twice as productive as C₃ plants when temperatures and light intensity are high.

Altogether, at least 3 monocot and 16 dicot families contain members with C₄ photosynthesis. All of these families also include other members that use the C₃ pathway. Biologists conclude from this that the C₄ pathway and kranz anatomy have evolved independently many times.

Figure B Comparison of the anatomy of sugar cane, a C₄ plant, with that of a soybean, a C₃ plant.

Stems Carry Water and Minerals and Nutrients Between Roots and Leaves

We usually think of stems as long and thin, with various arrangements of attached leaves. But stems may have many forms and functions. Stems produce leaves, flowers, seeds, and fruits; elevate these organs toward the sun, pollinators, and seed dispersers; store nutrients and water; and preserve perennial plants through the stress of winter or summer. Some stems are photosynthetic.

The region of the stem to which a leaf attaches is a **node**, and the region between nodes is an **internode**. Plants are modular in their organization, so that a leaf-node-internode unit can be repeated over and over to create a very large plant. In Chapter 32, we will discuss the developmental roles of apical meristems of both stems and roots. Our discussion here, however, focuses on the stem's role in transporting water and nutrients.

Like primary roots, primary stems consist of dermal, ground, and vascular tissues. As in the root, the dermal tissue

Figure 30-9   The two paths of water through the roots. The *symplast* runs through the cytoplasm and plasmodesmata of the cells of the cortex, while the *apoplast* consists of the extracellular spaces within the cortex. The endodermis is a layer of cells that surrounds the vascular cylinder. The walls between endodermal cells contain waterproof *suberin,* which defines the Casparian strip. The Casparian strip prevents movement of water and dissolved minerals between the vascular cylinder and the apoplast. Water and minerals can only cross the endodermis in the symplast, allowing the membranes of the endodermis to regulate the flow of solutes and water.

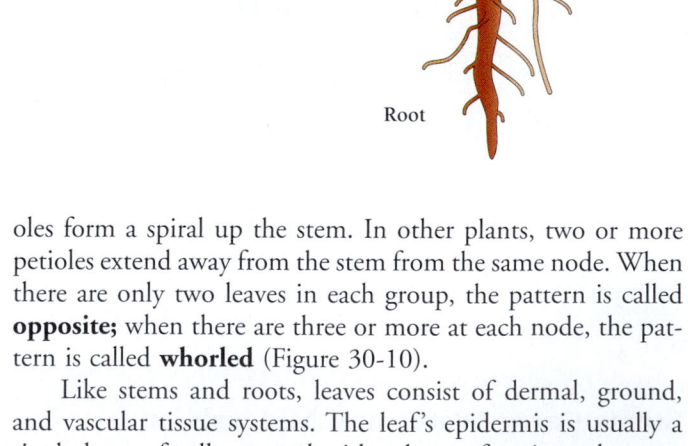

(epidermis) is on the outside with the vascular system embedded within the ground tissue.

The stem's epidermis differs from that of a root in three respects—the absence of root hairs, and the presence of stomata and a thick cuticle. The cuticle, which encases the stem and prevents water loss, is a waterproof layer made of wax and *cutin,* a polyester similar to that used in clothing.

Inside the stem's epidermis is a ring of cortex, surrounding a core of vascular tissue and pith. The stem's cortex is often thin, sometimes no more than a few millimeters. In some cacti, however, the cortex can be 20 to 30 cm thick. The outer parenchyma cells of the cortex, particularly of young stems, are sometimes green and photosynthetic. The stems of most plants do not have an endodermis or a pericycle.

The arrangement of vascular bundles differs between monocots and dicots. In a cross section of a monocot stem, the vascular bundles appear to be scattered. By following their courses, however, we see a defined pattern for each bundle. As a bundle ascends a stem, it gradually moves toward the center. It periodically branches outward to connect to other bundles or to leaves.

What Do Leaves Do?

Most leaves have a thin, flat part, called the **blade,** and a stalk, called a **petiole,** which connects the leaf to the stem (Figure 30-10). The arrangement of leaves on their stems, as well as the sizes and shapes of leaves, influences their ability to capture light. Many plants position their leaves by movements of the petioles, so that leaf blades move out of the shade of other leaves.

The arrangement of leaves on a stem is characteristic of each species, with three principal patterns. In the most common arrangement, the **alternate** or spiral, the bases of the peti-

oles form a spiral up the stem. In other plants, two or more petioles extend away from the stem from the same node. When there are only two leaves in each group, the pattern is called **opposite;** when there are three or more at each node, the pattern is called **whorled** (Figure 30-10).

Like stems and roots, leaves consist of dermal, ground, and vascular tissue systems. The leaf's epidermis is usually a single layer of cells covered with a layer of cutin and waxes. Many leaves have epidermal extensions, called leaf hairs, or trichomes. Leaf hairs have a variety of forms and functions, including shading, retardation of water loss, and defense against herbivores.

Both the upper and lower surfaces of a leaf contain specialized pores, called stomata [singular, stoma; Greek, = mouth], through which carbon dioxide, oxygen, and water vapor enter and exit the interior of the leaf (Figure 30-11). While

Figure 30-10

Figure 30-10 Leaf arrangements. Taxonomists have invented dozens of ways to describe the arrangements, shapes, and textures of leaves. Here are just three arrangements (from left to right): alternate, whorled, and opposite. *(Paraskevas Photography)*

the lower surface of a leaf may contain tens of thousands of stomata per square centimeter, the upper surface in dicots usually has few or none. In Chapter 31, we discuss the role of stomata in regulating the flow of water from leaves to the air (*transpiration*) and the consequent flow of water through the plant itself.

Surrounding each stoma are two guard cells, specialized epidermal cells that serve as a valve, opening and closing the stoma according to environmental conditions inside and outside the leaf. Stomata open, for example, when the plant lacks carbon dioxide and yet can afford to lose water vapor through its leaves. Stomata close when the plant cannot afford to lose water, and in the dark, when they do not consume CO_2. In many species, guard cells have chloroplasts and respond to light. Other epidermal cells, however, generally do not contain chloroplasts.

Stomata can also regulate a leaf's temperature. Just as an animal can cool itself by losing heat in evaporating sweat, an overheated leaf can cool itself by opening its stomata more widely, allowing more water to evaporate.

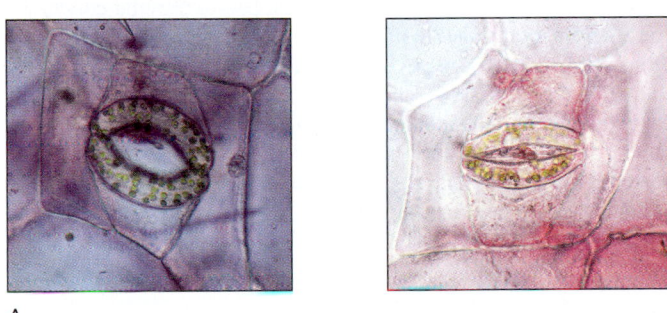

A.

Figure 30-11 The path of water in the leaf. Leaves consist of upper and lower epidermis enclosing a green tissue that carries out photosynthesis and a central layer of mesophyll, through which run the vascular bundles, or veins. A. Stomata in the surface of the leaf open and close to regulate the flow of water vapor and carbon dioxide in and out of the leaf. B. Inside the leaf, water passes from the xylem into the parenchyma and then to the stomata. Carbon dioxide in the atmosphere enters the plant through the stomata and is taken up by photosynthetic parenchyma cells. *(A, Ray Simons/Photo Researchers, Inc.)*

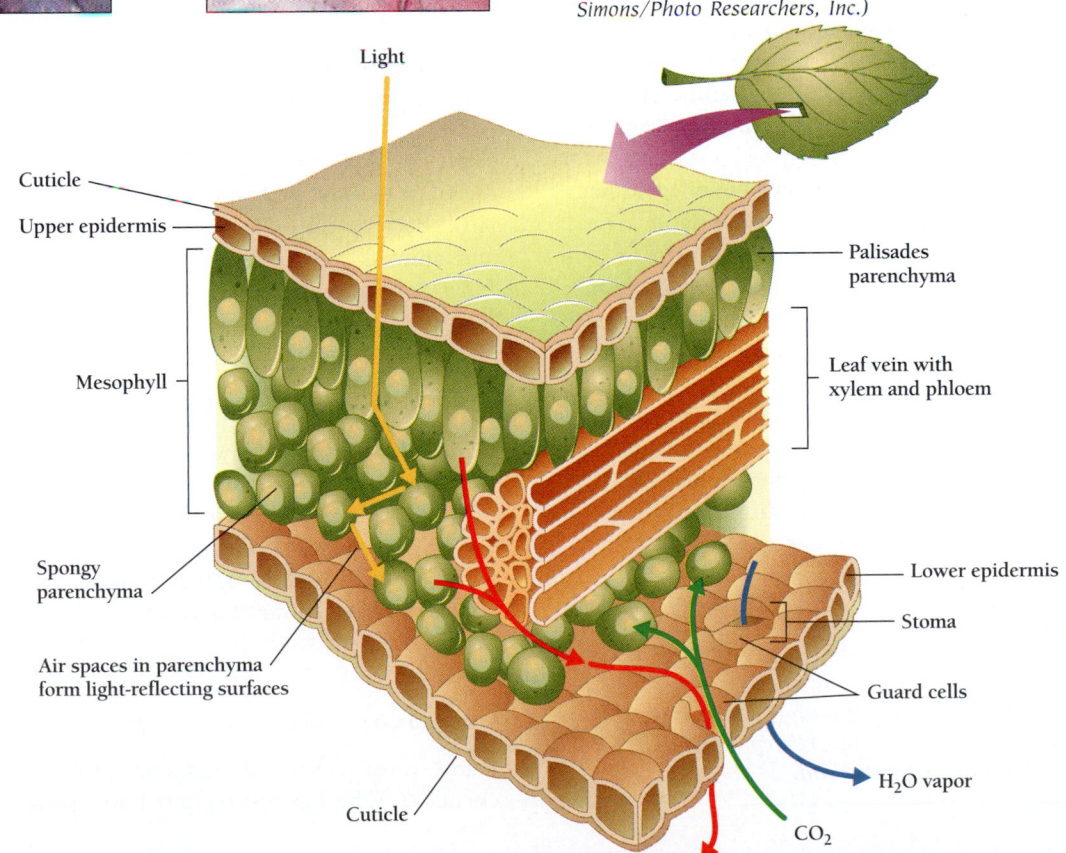

B.

The ground tissue of the leaf consists of **mesophyll** [Greek, *mesos* = middle + *phyllon* = leaf], green parenchyma cells that are responsible for most of a plant's photosynthesis. Most leaves have two kinds of mesophyll—**palisade parenchyma** and **spongy parenchyma** (Figure 30-11). The cells of the palisade parenchyma, which lie just under the leaf's upper epidermis, are elongated and packed with chloroplasts. Below them are the cells of the spongy parenchyma, which are irregular but rounded in shape and separated by numerous air spaces. These also contain numerous chloroplasts.

Like the air-filled pockets in our lungs, the air spaces in the spongy parenchyma provide a route by which oxygen and carbon dioxide can diffuse to and from the outside air, via the stomata. Water vapor follows the same route out of the plant (Figure 30-11). The internal air spaces also create many bright, reflecting surfaces within the leaf. Sunlight bounces back and forth within this tiny hall of mirrors, providing the photosynthetic pigments with many opportunities to absorb a photon.

Because cells near the top of a leaf shade those near the bottom, the lower cells have less light for photosynthesis. To minimize this effect, most leaves are thin, so that light can penetrate to the bottom-most parenchyma cells. In general, the tops of leaves are more specialized for light absorption and the bottoms for the exchange of gases and water.

One of the most obvious structures in a leaf is its vascular system. With the unaided eye, we can see the leaf's veins, which are its larger vascular bundles. The vascular bundles of a vein are continuous with those of the petiole and with the leaf traces of the stem. Inside the leaf, the veins branch into finer and finer vascular bundles. These may join to one another to form a continuous circular path, or they may end in special cells called terminal tracheids. The veins bring the leaf water to use in photosynthesis (or for cooling) and carry sugars back to the rest of the plant. In addition, they keep the blade extended.

In monocots, leaf veins usually run parallel to one another. In dicots, the leaf usually has only one or a few large veins. But these branch into fine networks, so that every part of the leaf is no more than a few cells away from the vascular system. If the vascular bundles of a single dicot leaf were placed end to end, they would extend over two city blocks. The pattern of veins varies greatly among species.

Besides their major role in photosynthesis, leaves may have a variety of other functions—protecting buds and flowers and storing food for the embryo and young plant (Figure 30-12). Leaves also help move water from the roots through the xylem, as we will see in Chapter 31.

> Leaves, the primary photosynthetic organs of plants, consist of an outer layer of epidermis and a central region of mesophyll through which run the vascular bundles, or veins. The epidermis is dotted with stomata.

Figure 30-12 A diversity of leaves. Leaves may be cotyledons, sepals, bud scales, or floral bracts. As organs of photosynthesis, they may be large or small, flat or round, waxy or hairy. Shown here are the flowerlike leaves of the poinsettia, the camouflaged leaves of the stoneflower, pine needles, and the bud scales, sepals, and new leaves of a horse chestnut. (*Poinsettia leaves, Lefever/Grushow from Grant Heilman; lithops (stone plant), Kjell B. Sandved/Photo Researchers, Inc.; conifer needles, David Sieren/Visuals Unlimited; horse chestnut, W. Ormerod/Visuals Unlimited*)

HOW DO PLANTS DEFEND THEMSELVES AGAINST HERBIVORES?

Plants are good to eat and they can't get away. Since the evolution of the first herbivores, plants have been fighting to stay alive. Bacteria eat plants, protists eat plants, fungi eat plants, and animals eat plants. Even other plants parasitize plants. How do plants defend themselves against this onslaught of hungry organisms?

Mechanical Defenses

Plants have numerous ways of preventing other organisms from taking a bite. One defense is the plant's waxy cuticle, which excludes both bacteria and fungi fairly well. Hairy leaves or very sticky leaves discourage caterpillars and other tiny herbivores (Figure 30-13). Spines (which are modified leaves), thorns (which are branches), and prickles fend off larger mammalian herbivores.

Chemical Defenses

Plants' most effective defenses are probably not physical but chemical. Who has not regretted an afternoon jaunt through

Figure 30-13 **Mechanical defenses.** Leaf hairs protect delicate plant epidermis from excessive sun and hungry insects, as well as from drying out and other dangers. Some leaf hairs secrete tarlike substances that make the leaf too sticky for caterpillars to walk on and too bitter for larger herbivores to eat. *(© 1981 by Darwin Dale/Photo Researchers, Inc.)*

a patch of poison ivy or poison oak? Over hundreds of millions of years plants have evolved a pharmacopoeia of chemical defenses against herbivores. Most of these compounds are **secondary plant compounds. Primary compounds** include glucose, amino acids, ATP, and DNA, compounds that are essential for plants' day-to-day functioning. In contrast, secondary plant compounds are diverse chemicals that function in very specific ways. Many are particular to small groups of plants or even to a single species.

Just as plants have evolved thousands of ways of defending themselves against herbivores and pathogens, herbivores and pathogens have evolved as many ways of overcoming these defenses. For every shield there is a stronger sword, and for every bulletproof vest, a more piercing bullet. The coevolution of plants and their attackers has led to a balanced ecosystem in which plants have so far managed to stay ahead. Cutworm larvae raised on an artificial diet, for example, grow six times as large as those raised on any natural diet of leaves to which they are adapted. Without plant defenses, the world would be inundated with insects and other herbivores.

Plants manufacture thousands of these compounds, but most secondary plant compounds fall into three groups: terpenes, phenolic compounds, and alkaloids. **Terpenes** are lipids and the largest class of secondary compounds. Many are actually primary compounds that play important roles in plant metabolism. But vast numbers of terpenes seem to function solely to drive away or poison herbivores. Pyrethroids from *Chrysanthemum* leaves, for example, are insecticides. Peppermint, lemon, basil, and sage all derive their aromas from mixtures of terpenes called **essential oils,** which repel or poison insects. Many plants store these repellent and toxic oils in glandular hairs on the surface of the epidermis, where the oils advertise the plant's toxicity to arriving insects.

Some terpenes, such as those found in sunflowers and sagebrush, repel mammals as well as insects and taste bitter to us. Certain varieties of alfalfa contain saponins, which kill cattle and other grazing animals by rupturing the membranes of red blood cells. Cotton plants derive much of their insect, bacterial, and fungal resistance from the terpene **gossypol,** which also happens to work as a contraceptive in human males.

Phenolic compounds are aromatic substances that play a variety of roles in plants. Aromatic chemicals are those whose molecular structure includes a flat carbon ring (Chapter 3). Some phenolics repel herbivores and pathogens, and some attract pollinators or fruit dispersers.

Others, such as lignin, both repel herbivores and provide mechanical support. Lignin is a major constituent of the walls of cells specialized to provide mechanical support or transport water. But lignin also makes plants relatively indigestible to both animals and microscopic pathogens. Surprisingly, lignin also has been isolated from the cell walls of algae, where it is believed to protect the algae from attack by bacteria and other microbes. Biologists now believe that lignin is an ancient compound, whose original function, to repel pathogens, has expanded to include mechanical support for terrestrial plants.

Some plants secrete phenolic compounds that poison any plants growing nearby. Desert plants, especially, resort to this unneighborly practice to eliminate competition for precious water. Other phenolics secreted into the soil regulate gene expression in nitrogen-fixing bacteria. Still others rapidly accumulate in the plant in response to bacterial or fungal infections. These phenolics form in a matter of hours around the infection site and kill a broad spectrum of fungal and bacterial plant pathogens. **Tannins,** commonly found in woody plants, are, like lignin, phenolic polymers that reduce growth and survivorship in many herbivores. Cattle, deer, humans, and apes avoid eating unripe fruits and other plant parts with high levels of tannins because of the tannins' sharp, astringent taste. This effect is caused by tannins' binding to enzymes in the saliva. Yet, humans like a bit of tannin in their tea, apples, blackberries, and red wine. Finally, the pigments responsible for most of the red, pink, purple, and blue colors of flowers, fruits, leaves, and other plant parts, called **anthocyanins,** are also phenolic compounds.

Alkaloids are nitrogen-containing secondary plant compounds found in 20 to 30 percent of all vascular plants. They include many of the most powerful drugs known to humans, including nicotine (the highly addictive active ingredient in tobacco), atropine, cocaine, morphine (from which heroin is made), the hallucinogen psilocybin, and the toxin strychnine. The alkaloids in lupines, larkspur, poison hemlock, and other plants are highly toxic to herbivores, killing livestock and humans ignorant enough to eat them. (The Greek philosopher Socrates carried out his own execution with a tea made from poison hemlock.) These compounds can also be highly teratogenic. Milk from cattle and goats that have grazed on lupine can induce high rates of birth defects in pregnant humans or other animals who drink the contaminated milk. At lower doses, alkaloids such as caffeine, nicotine, cocaine, and morphine are used widely as stimulants and sedatives, for both medical and nonmedical purposes.

Other secondary plant compounds besides the terpenes, phenolics, and alkaloids include the **mustard oil glycosides.** These compounds give cabbage, broccoli, radishes, and other plants in the mustard family their characteristic pungent odor and flavor. Many insects and mammals are repelled by mustard oil glycosides. But others have learned to cope with these compounds, and some insect herbivores specialize on plants in the mustard family (Figure 30-14). For these insects, the mustard oil glycosides act as an attractant. Herbivores that specialize on noxious plants can often avoid competition with other herbivores. Many compounds that originally evolved as repellents to herbivores now act as attractants for certain groups of animals.

Secondary plant compounds are chemicals not needed for the major metabolic pathways of a plant. Many secondary plant compounds repel or poison insects and other herbivores.

Figure 30-14 Attractive insect repellent. Most insects avoid plants in the mustard family because the leaves contain unpleasant compounds. The cabbage butterfly, however, prefers cabbages and other mustards. *(Holt Studios International/Photo Researchers, Inc.)*

In this chapter we have seen that the standard root-and-shoot form of a terrestrial plant allows the plant to collect water and minerals from the soil and to collect solar energy and carbon dioxide from the air. We have briefly discussed how the vascular system moves water and nutrients from one organ to another and how plants use secondary plant compounds to strengthen their bodies against the pull of gravity and to protect themselves from herbivores and pathogens. In the next chapter, we discuss in more detail how plants raise water from the roots to the tops of trees that may be hundreds of feet tall.

Study Outline with Key Terms

Despite the enormous diversity of flowering plants, they share characteristic organizational features. Most consist of an underground **root system** and an aboveground **shoot system.** The shoot usually contains a stem, leaves, and flowers.

Materials travel between shoot and root through the plant's **vascular system,** a network of two kinds of transport tissues. The **xylem** transports water and minerals from root to shoot, and the **phloem** transports the energy-rich products of photosynthesis from leaves or storage areas to the rest of the plant.

Flowering plants fall into two classes—the **monocotyledons,** or **monocots,** and the **dicotyledons,** or **dicots**—that have characteristic differences in their embryos, flowers, leaves, roots, and vascular systems. The embryos of monocots have one **cotyledon** and the embryos of dicots have two. In both classes of plants, roots, stems, and leaves consist of three types of tissue systems—dermal, vascular, and ground. The **dermal tissue system** (or **epidermis** and **periderm**) is the protective outer covering, which includes the **cuticle.** The **vascular tissue system** lies inside, surrounded by **ground tissues.** Roots, stems, and leaves differ in the arrangement of vascular system and ground tissue. Tissues may consist of **parenchyma, collenchyma, sclerenchyma,** and other cell types.

The **primary growth,** or lengthening, of a plant occurs only in specialized regions of dividing cells, called **apical meristems,** at the tips of roots and shoots. **Secondary growth** is an increase in diameter of shoot or root that occurs in a cylinder of dividing cells called a **lateral meristem.**

Plant cells have **cell walls** made of **cellulose,** other carbohydrates, and proteins. All plant cells have a **primary cell wall** outside the plasma membrane. Some have a thicker **secondary cell wall** between the primary wall and the membrane. The polymer **lignin** strengthens many plant cell walls. Channels called **plasmodesmata** often extend through the cell walls and connect the cytoplasms of adjacent cells.

Stems carry water and minerals from roots to leaves and nutrients from leaves to roots. Like roots, stems consist of dermal, ground, and vascular tissue systems. In dicots, the ground tissue of a stem consists of two regions, the **cortex** and the **pith.** Xylem and phloem lie in **vascular bundles,** which lie between the pith and the cortex. The xylem lies toward the center and the phloem toward the outside of each bundle. The arrangement of vascular bundles differs between monocots and dicots.

Leaves are the major organs of photosynthesis. Each leaf consists of a (usually) flat **blade** and a stalk, called a **petiole,** that attaches the leaf to the stem. The petiole attaches to a **node** on the stem. In between each pair of nodes is the **internode.** The arrangement of leaves on a stem is characteristic of each species, with three principal patterns: **alternate, opposite,** and **whorled.** The surfaces of a leaf usually contain specialized pores, called **stomata,** through which carbon dioxide, oxygen, and water vapor enter and leave. The interior of a leaf is composed of **mesophyll,** which is made of **palisade parenchyma** and **spongy parenchyma.**

Xylem consists of two types of conducting cells—long, tapered cells, called **tracheids,** and shorter, flat-ended cells, called **vessel elements.** The xylem functions only after the cells die, leaving behind empty cell walls. Water (as well as dissolved substances) moves through the tracheids from dead cell to dead cell by way of pairs of **pits,** either **simple pit pairs** or **bordered pit pairs.** Water moves through vessel elements through **perforation plates** at either end.

The **sieve tubes** of phloem are made of living **sieve tube members** and their **companion cells.** Each end of a sieve tube member is a perforated **sieve plate** that allows materials to pass from cell to cell.

Water moves from soil to air by way of the xylem. Absorption and transport begin in the roots. Roots consist of three concentric cylinders of tissue. At the center is the vascular tissue **(stele),** which is surrounded by cortex (ground tissue), which is surrounded by epidermis. Water flows from the soil to the vascular system by way of the **mucigel**-covered **root hairs** and cortex. The innermost layer of the cortex is the **en-dodermis.** Within the endodermis is a layer of waxy, waterproof cell walls called the **Casparian strip,** which prevents water from flowing *between* the endodermal cells into the **peri-cycle** and stele. Water and minerals can move through the root by two pathways, the **symplast** and the **apoplast.** The Casparian strip prevents movement of water and dissolved minerals between the vascular cylinder and the apoplast. Water must flow instead via the symplast, *through* the living cells of the endodermis. The membranes regulate the movement of both ions and water into the stele.

Secondary plant compounds are chemicals not needed for the major metabolic pathways of a plant. Many secondary plant compounds repel or poison insects and other herbivores. **Primary compounds,** which include glucose, amino acids, ATP, and DNA, are compounds that are essential for plants' day-to-day functioning. Plants manufacture thousands of secondary plant compounds, but most fall into three groups: **terpenes, phenolic compounds,** and **alkaloids.** The terpenes, the largest class, include the pyrethroid insecticides, and the **essential oils** in peppermint, lemon, basil, and other strong-smelling plants. The terpene **gossypol** gives cotton plants much of their insect, bacterial, and fungal resistance.

Phenolic compounds include lignin, **tannins,** and **anthocyanins.** Alkaloids, such as nicotine and cocaine, are nitrogen-containing secondary plant compounds found in 20 to 30 percent of all vascular plants. Other secondary plant compounds besides terpenes, phenolics, and alkaloids include the **mustard oil glycosides.**

Review and Thought Questions

Review Questions

1. What adaptations do terrestrial plants have that aquatic plants do not?
2. How is the root necessary for photosynthesis? How is the shoot necessary for photosynthesis?
3. How do monocots differ from dicots?
4. Is bread made from plants that are monocots or dicots? Name six other plants you know that are monocots and six that are dicots.
5. Name three general roles that secondary plant compounds can play in the life of a plant.
6. What is the difference between primary growth and secondary growth?
7. What does the word "meristem" mean literally and how does that meaning relate to what it is and does?
8. Draw a diagram of the endodermis, including the Casparian strip, and explain how these structures regulate the flow of water and minerals into the root xylem.

Thought Questions

9. Certain "prostrate" plants spread out over the ground without growing very tall. These plants can maximize their exposure to sunlight. Why don't all plants conserve energy and adopt this form? In other words, what are the advantages of growing tall?
10. When Bruce Ames first published his papers on the carcinogenicity of secondary plant compounds, many people felt that he was behaving irresponsibly. They worried that the public would misunderstand and (1) not bother to eat fruits and vegetables, and (2) not bother to avoid toxins that are genuinely dangerous. Some of his critics thought that Ames was endangering the lives of innocent lay people. Do you think Ames's behavior was irresponsible? Are scientists more obligated to publish what they know is true or more obligated to limit what they publish, according to how the public will likely respond to such information?

About the Chapter-Opening Image

This opium poppy symbolizes the beauty of plant structure and the power of chemical compounds manufactured by plants.

 On-line materials relating to this chapter are on the World Wide Web at **http://www.harcourtcollege.com/lifesci/aal2/**

The Devastating Dry Leaf Creature

I n the late 1980s, disaster threatened the burgeoning California wine industry. For years, ignoring the advice of European grape experts, growers had been planting a kind of grapevine that had only weak resistance to a devastating insect pest. By the early 1980s, fifty thousand acres of fertile farmland had been planted with these vines—all of them vulnerable to *Daktulosphaira vitifoliae,* a tiny sucking insect that feeds on roots and leaves. The insect's common name, phylloxera, means "dry leaf" in Greek [*phyllos* = leaf + *xeros* = dry].

It was only a matter of time before these tiny relatives of aphids developed a means to overcome the vines' feeble re-

sistance. Indeed, by the late 1980s, phylloxera were destroying whole vineyards in California. The pattern was always the same. The leaves of a few plants would turn red and drop from the vines in midsummer. Within months, the branches and roots were dead. Soon nearby plants began to lose their leaves as well. The phylloxera epidemic caught California experts off guard.

It wasn't as if California grape growers hadn't heard of phylloxera. In the 19th century, the infamous pest had devastated the European wine industry. In France alone, the insect utterly destroyed some two and a half million acres of vineyards—a national disaster. Nor was the invasion confined to Europe. The tiny pest infested vineyards all over the world, including those in California. Indeed, in 1880, the State Legislature directed the new University of California to create a department of viticulture ["grape growing"]—specifically to study phylloxera.

The cure for the great wine blight of the 1880s, as it came to be called, came from a combination of American science and French technology. The American entomologist C.V. Riley worked out the complex life cycle of phylloxera, showing that it lived on wild grapes of eastern North America. The American grapes had adapted to the pest and flourished despite the phylloxera feeding on their roots. By the turn of the century, French viticulturists had put two and two together, solving the phylloxera problem in Europe by grafting high-quality French wine grapes onto wild American roots, or "rootstock." The French hailed Riley as a savior for helping to save their wine industry (Figure 31-1).

How do phylloxera kill grapevines? The insects cluster on the roots. There they use their tiny mouthparts to pierce the outer layers of the root and inject saliva containing enzymes and auxin, a plant hormone. Auxin induces the roots to form a tumorlike ball of tissue around the insects. Inside this protective gall, the insects safely feed, digesting the tissues of the roots.

Phylloxera themselves do not usually kill the vine, for they do not take enough energy and materials from the vine to hurt it much. Instead, the thousands of insects cutting into the roots make the plant susceptible to infections by fungi and bacteria. These "secondary infections" invade and destroy the vascular

Tom Benoit/Tony Stone Images

Key Concepts

- In plants, water moves from soil to air through the vascular system.

- The main driving force for the ascent of water is transpiration, with upward transport depending on water's cohesive and adhesive properties.

- Stomata regulate the rate of transpiration by opening and closing.

- The bulk flow of water drives the transport of sugars in the phloem.

system in the roots. Unable to transport water and nutrients, the vine dies.

Nineteenth-century grape growers developed and tested thousands of different rootstocks. Some did well in the cool, wet climate of northern Europe. Some grew better in warm, dry soil. A few resisted phylloxera. Those rootstocks were found to secrete tannins and other substances that wall off the root galls and so prevent infections. One rootstock that had only moderate resistance to phylloxera but produced heavy loads of fruit was named AXR1. It was tempting to use it.

Initially, the phylloxera left AXR1 alone. But first in Sicily, and then in France and Africa, grapes grown on AXR1 flourished for a decade or two, then began to wither and die. French scientists suspected that phylloxera had adapted physiologically to overwhelm AXR1's resistance. In the 1920s the scientists began advising their growers against using the rootstock.

By then, however, Americans were in the midst of Prohibition, the 1919 law that prohibited the manufacture and sale of all alcoholic beverages—including wine. Phylloxera was the least of California grape growers' problems. They dug up their sprawling vineyards and planted apricots, peaches, and plums, or sold the fertile land for housing developments. Phylloxera all but disappeared both from the land itself and from people's minds. When Prohibition ended, in 1933, and Californians began replanting vineyards, they knew little about phylloxera.

In the 1950s, a researcher at the Davis campus of the University of California tested 18 rootstocks. Three of them yielded large crops of fruit and had other desirable traits for California vineyards. For simplicity, the university chose to recommend just one rootstock, AXR1. Although its resistance to phylloxera was "not high," the researcher wrote, he thought phylloxera would not be a serious problem in vigorous vineyards planted in deep, fertile soils with good irrigation.

In the 1970s, American wine lovers recognized for the first time that California wines were surprisingly good and far cheaper than French wines. The California wine industry

Figure 31-1 Memorial at the School of Agriculture, Montpellier, France (1911). A sculptor has represented the rescuing of the French wine industry with American rootstock as a young American rescuing the ailing, old French vine. *(Don Ellis)*

boomed. Soon everyone, from retired dentists to multinational corporations, was planting vineyards. Growers around Napa and Sonoma counties, the finest wine-producing area in the United States, planted nearly three-quarters of their vines on AXR1. If California's phylloxera adapted to AXR1, or if the European strain of AXR1-munching phylloxera found their way into the state, the insects would eventually feast on nearly 50,000 acres of vines.

In the summer of 1983, the first sign of trouble appeared. Austin Goheen, a plant pathologist (an expert on plant diseases) from the University of California, Davis, who had been called in to inspect some mysteriously dying grapevines in Napa County, recognized the telltale yellow colonies of the tiny phylloxera insects on the roots of the stricken plants. John Baritelle, the vineyard's owner, was horrified.

But although Goheen recognized the phylloxera, he couldn't be sure they were killing the vines. He knew that small populations of the insects still existed in many regions of California and almost every kind of rootstock harbored at least a few of the troublesome pests. Was this a new virulent strain of phylloxera that was sapping the life out of the roots, or was something else doing the damage?

It would take 2 more years to find out. Baritelle claimed his vineyard was planted on AXR1 rootstock, but Goheen couldn't be sure. In those days before DNA fingerprinting, the only way to tell one rootstock variety from another was to carefully examine the leaves for such characteristics as leaf color and the shape of leaf hairs. But the only leaves that grew on Baritelle's vines came from the wine grape variety grafted onto the rootstock. For two summers, biologists searched the vineyard for shoots growing from the rootstocks. What they found was a confusing mixture of AXR1 and other rootstocks.

Not until late in 1985 could researchers state with certainty that John Baritelle's vines harbored a new strain of phylloxera, one that could attack and kill AXR1 vines. But many questions still remained unanswered. How fast would these new, virulent creatures spread? Should growers who lived miles from an infection site tear out their vines now, or wait until the insects moved closer? Would this new insect ravage vines in other grape-growing regions of California where soils and climate were different?

The researchers didn't know. Should they recommend that growers stop using AXR1? Tearing out the vines and planting new ones would cost millions of dollars. The researchers felt they needed more information before issuing such an extreme warning. After all, they had only found the insects in a few vineyards so far. They studied and watched the phylloxera for another 3 years. Still, the researchers were not sure. Meanwhile, panicked growers clamored for guidance, as the insects spread relentlessly from site to site.

Still uncertain, the researchers finally issued a press release through the university in 1989 warning growers to stop using AXR1. Many growers immediately began tearing out their AXR1 vines and replacing them. But no one knew which rootstock to use instead. Some growers used four or five different rootstocks in a single vineyard, hoping some of it would withstand phylloxera. The change was costing the wine industry a fortune.

Many growers were furious with the university, blaming the researchers for recommending AXR1 in the first place. And why, they demanded, hadn't the university made its announcement earlier? Nurseries were stuck with millions of dollars worth of AXR1 they wouldn't be able to sell. Some nurseries threatened lawsuits if the researchers proved wrong.

Despite all the finger pointing, everyone shared some blame in the disaster. Growers and researchers alike knew that other nations didn't use AXR1. But as long as the rootstock had continued to do well in California, no one wanted to think about it or study it.

Another problem was finding funding for research for phylloxera. The wine industry provides funds for most research in viticulture, and it decides what kind of research to fund. Until phylloxera appeared on Baritelle's vines, no one was worried about the insect. As Jim Wolpert, University of California's chief viticulturist, said, "If I'd put in a proposal in 1982 to the winegrape industry for funding to work on AXR1 and phylloxera, they would have said I was crazy." Not until phylloxera started killing vines in California did these minute creatures garner the attention they deserved. And by then scientists lagged far behind in their race to save the state's vineyards.

Phylloxera continue to infect northern California's vineyards, but newer rootstocks are able to wall off the tissue-piercing insects, and water and minerals from the soil flow smoothly up to the leaves and grapes in the hot sun above. The flow of water is as necessary to a plant's life as the flow of blood is to each of us. The flow of water both sustains photosynthesis and carries minerals and nutrients from the roots to leaves and from leaves to roots. In the first part of this chapter we will see what forces drive water and minerals from the roots upward. Later, we will see how sugars manufactured in the leaves are transported down to the roots and other parts of the plant.

WHAT DRIVES WATER UP?

We know that water moves up from roots to leaves through the tracheids and vessels of the xylem: if we place a plant cutting in water that contains a dye, we can see the path the dyed water takes through the plant. If the stems were transparent, we would be able to see the water move, for during the day it ascends rapidly, at about 10 to 100 cm/minute.

What Is the Route of Water from Soil to Air?

As we saw in the last chapter, water enters the roots via the root hairs, moves through the cortex, crosses the endodermis, and enters the xylem. Once in the xylem, water ascends from the roots to the leaves through parallel and usually interconnected xylem vessels in the stem. These vessels branch into leaf traces, and water flows through petioles into the leaves.

The water then enters cells of the mesophyll and evaporates at the cell surfaces. The water vapor diffuses through the air spaces of the mesophyll and exits the leaf by way of the stomata. The passage of water vapor from leaf to air is called **transpiration.**

The rate of transpiration may be extraordinary: a leaf can transpire its own weight in water in less than an hour. A tree may transpire 50 gallons of water in an hour, up to 30,000 gallons during a growing season; a 40-acre cornfield uses 15 million gallons before harvest. In a plant's lifetime, it typically loses a hundred grams of water through transpiration for every gram of dry material it accumulates.

These tremendous quantities of water must sometimes move great distances. The tallest trees in the world are the California redwoods, *Sequoia sempervirens.* These trees are so large that entire churches have been built from a single tree. One tree, with a diameter of 20 feet and a height of over 200 feet, contained enough lumber to build 22 houses. The tallest redwood of all is 367 feet, the same height as a 29-story building, and nearly a third the height of the Empire State Building.

To pump water to such a height would require a pump exerting 150 pounds per square inch (psi) of pressure. But plants have no central pump. The cells of the xylem are themselves dead, so we can assume that these empty hulls exert no force. Plant scientists early in this century proposed three possible mechanisms for water transport: (1) water is pushed up from the roots; (2) capillary action in the xylem pulls water in the same way that a paper towel pulls water out of a glass; or (3) evaporation at the leaves creates a suction that pulls water up the xylem. But which mechanism was the right one? Many generations of plant scientists puzzled over this question before arriving at the currently accepted view.

Root Pressure: Can the Roots Push Water to the Tops of Tall Trees?

Recall from Chapter 4 that water tends to move from areas of purer water to areas with high concentrations of solutes. Thus, sugary raisins and cytoplasm-filled red blood cells both fill with water when placed in pure water. The tendency for water to move across a membrane in response to a difference in concentration of solutes is called **osmosis** [Greek, *osmos* = push, thrust].

When a plant cell is in a plain water (no solutes), water flows into the cell (which is full of solutes), causing the cell to swell. Most of the water accumulates in a central vacuole. As the vacuole swells, it exerts physical pressure against the cell wall. The pressure against the cell wall resulting from osmosis is called **turgor pressure.** The rigid cell wall itself resists this pressure, pushing back with "wall pressure."

The xylem in the roots usually contains higher concentrations of solutes than the soil; water from the soil therefore flows into the root xylem by different paths. The pressure that develops in the roots as a result of the inflow of water is called **root pressure.** As water flows into the xylem, the water pushes water already in the xylem up the roots and into the stem. In small plants, root pressure can push water all the way up the stem and out of tiny holes at the margins of the leaves, in a process called **guttation** [Latin, *gutta* = a drop].

Early in the morning gardeners sometimes see the leaves of grass, tomatoes, strawberries, and other plants rimmed with tiny drops of water (Figure 31-2). These drops, which are not dew, are the result of root pressure. Guttation occurs on cool, humid mornings, when the air is too saturated for water to evaporate from the leaves.

How far can root pressure push a column of water? Researchers can directly measure root pressure by attaching a pressure-measuring device to a cut stem (Figure 31-3). Root pressures measured in this way are usually less than 1 atmosphere, which can support a column of water about 10 meters

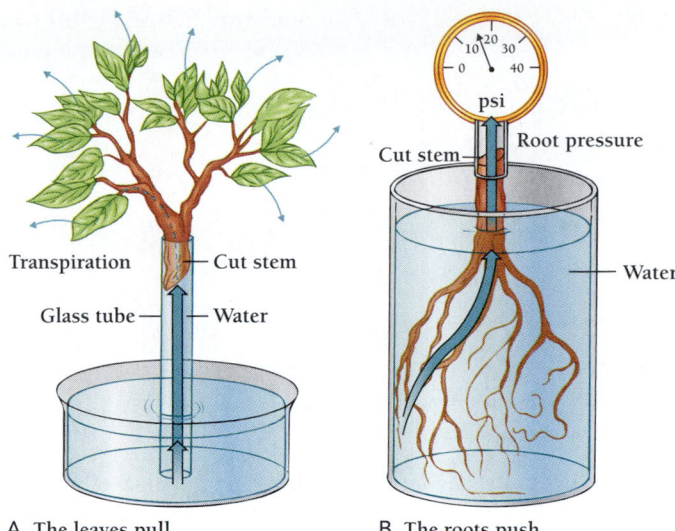

A. The leaves pull   **B. The roots push**

Figure 31-3 **Root pressure.** A. We've all seen cut flowers pull water up through their stems. B. Roots can also push water up through a cut stem. The pressure generated by the roots is a physical pressure that can be measured with a pressure gauge similar to that used to measure tire pressure.

(30 feet) high—impressive, but not nearly enough to move water to the top of a redwood or eucalyptus tree. In any case, many plants (including redwoods and other conifers) do not develop any root pressure at all.

Another clue that showed plant physiologists that root pressure cannot drive water to the tops of tall trees is that the water in the xylem is not usually under pressure. If we poke the xylem of a stem with a needle, water does not come spewing out, as if under pressure. Instead, air is sucked up into the xylem, indicating that the water in the xylem is being sucked up, as if by a vacuum.

Finally, the rate of water flow due to root pressure is too slow to account for the rapid transport of materials from roots to leaves. The driving force of water transport must therefore come from some source other than root pressure.

The most pressure that roots can generate is enough to raise water some 30 feet, not the 300 feet or more of the tallest trees.

Can Capillary Action Raise Water to the Tops of Tall Trees?

The movement of water also depends on its adhesive and cohesive properties. Because water is **cohesive,** it sticks to itself very well, and columns of water do not break as they move up narrow tubes, or **capillaries.** Indeed, water easily moves up the inside of most capillaries, including glass tubes and xylem (Figure 31-4). Could such capillary action drive water to the tops of tall trees? As it turns out, ordinary capillary action can drive

Figure 31-2 **Guttation demonstrates the action of root pressure.** The droplets at the margins of these strawberry leaves are not condensation, or dew, but water forced from tiny holes. (*Angelina Lax/Photo Researchers, Inc.*)

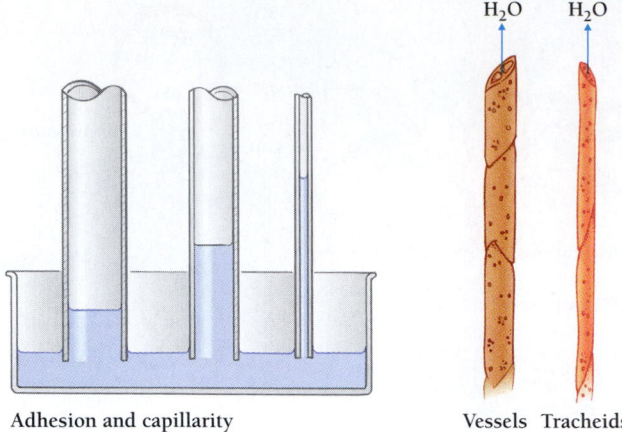

H₂O   H₂O

Adhesion and capillarity   Vessels Tracheids

Figure 31-4 Is capillarity enough? Water rises higher in narrower tubes than in wide ones. In xylem vessels and tracheids, water rises by capillarity about 0.5 meters, not nearly enough to get it to the top of a tall tree.

water up only about half a meter (less than 20 inches), not nearly enough.

But water is also highly **adhesive,** having a tendency to cling (by hydrogen bonds) to the surfaces of carbohydrates and other polar substances. Water's adhesiveness is what causes it to soak into the cellulose of paper towels or cotton towels. Molecules of water also stick tightly to the cell walls along which water travels.

Although root pressure and capillary action are not enough to drive water to the tops of tall trees, both effects are nonetheless important to water movement in plants, as we will now see.

Water's adhesiveness and cohesiveness enable it to move up capillaries, but not 300 feet up.

The Driving Force for the Ascent of Most Water Is Transpiration

If neither root pressure nor adhesion of water to the inside of the xylem is enough to drive water up, then what does? The now-accepted explanation for the ascent of water is called the **cohesion-tension theory,** first propounded in 1914 by the Irish botanist Henry Dixon. The cohesion-tension theory states that transpiration in the leaves pulls water up the stem in continuous columns (Figure 31-5A). For decades after Dixon's proposal, many plant physiologists doubted that the theory was correct. Acceptance of Dixon's hypothesis rested on answering two questions. First, did transpiration generate enough tension to pull water up hundreds of feet from the soil? And, second, were long columns of water strong enough to be pulled up a

tree like pieces of rope? Wouldn't the narrow columns of water in the xylem break under such tension?

Henry Dixon's cohesion-tension theory raised two questions—one about the amount of tension generated by transpiration and one about the tensile strength of water.

How Can Transpiration Draw Water to the Tops of Tall Trees?

We've seen that, through osmosis, water tends to move from areas of low solute concentration into areas of high solute concentration. But water also tends to move into areas with very few water molecules. Water molecules move into dry air as readily as they move into sugary raisins or cells. This is because the concentration of water molecules in air is low compared to the concentration of water molecules in a cell.

The tendency of water molecules to move along gradients of concentration and pressure is called **water potential.** Water potential is always relative to something else. So pure water outside of a cell full of solutes has a "high water potential" because the pure water has a strong tendency to flow into the cell. Likewise, liquid water in the presence of dry air has a high water potential because the water has a strong tendency to evaporate into the air.

Inside the mesophyll of a leaf, water is evaporating from virtually every cell surface, saturating the air spaces in the mesophyll with water vapor. If the air spaces remained saturated, water would stop moving out of the mesophyll cells and transpiration would cease. Outside the stomata, however, is the whole atmosphere, which is almost never saturated.

Air can be dry, as in the Mojave Desert, or very humid, as in much of the eastern United States. Even at 98 percent relative humidity, however, the tendency of water to evaporate is very high. The water potential that results from the difference in water concentration between the humid air and the liquid water inside a plant generates enough pull to support a 300-meter column of water, nearly three times the height of the tallest trees. In principle, then, the transpiration of water from the air spaces of the leaves to the atmosphere could provide the driving force for the ascent of water.

Further, as water moves from the air spaces in the mesophyll to the atmosphere, more water moves into the air spaces by evaporation from the mesophyll cells. As water leaves these cells, water flows into them from the xylem. The result is the movement of water from soil to roots to xylem to leaves to air. The driving force for all this movement is water potential, the tendency for liquid water to evaporate into the air.

Water potential in humid air can raise water nearly three times as high as the tallest trees. Water potential in dry air is even greater.

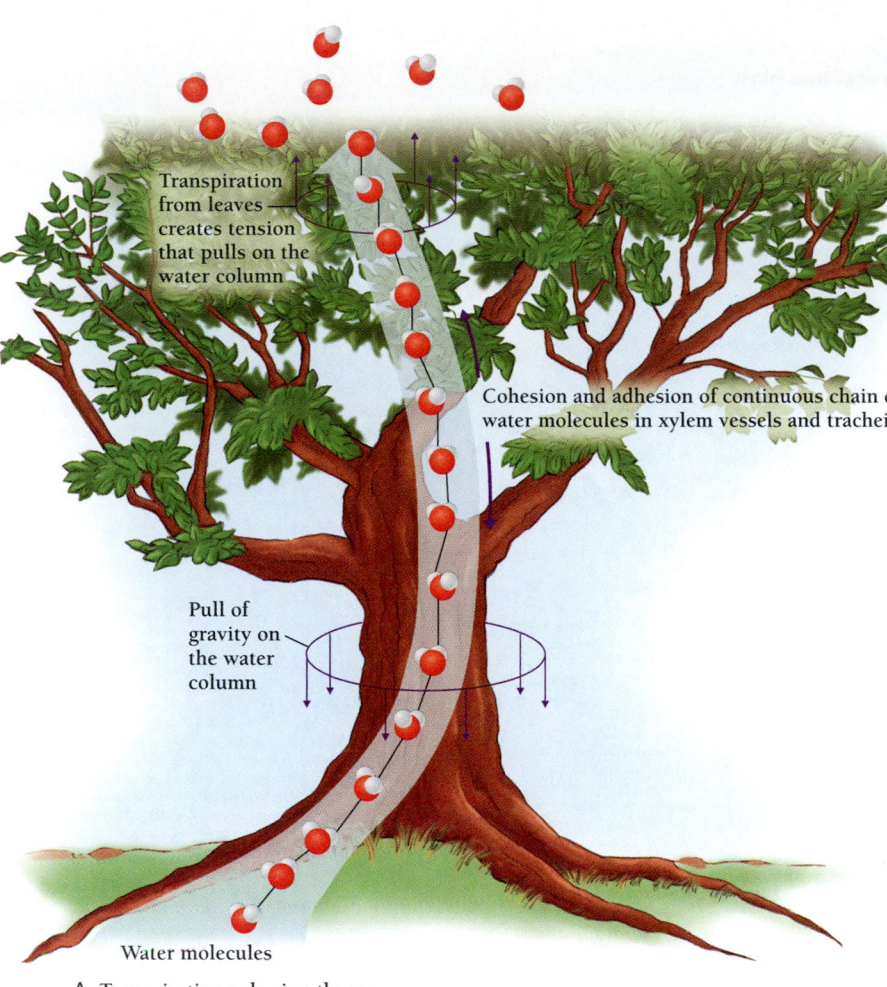

Transpiration from leaves creates tension that pulls on the water column

Cohesion and adhesion of continuous chain of water molecules in xylem vessels and tracheids

Pull of gravity on the water column

Water molecules

A. Transpiration-cohesion theory

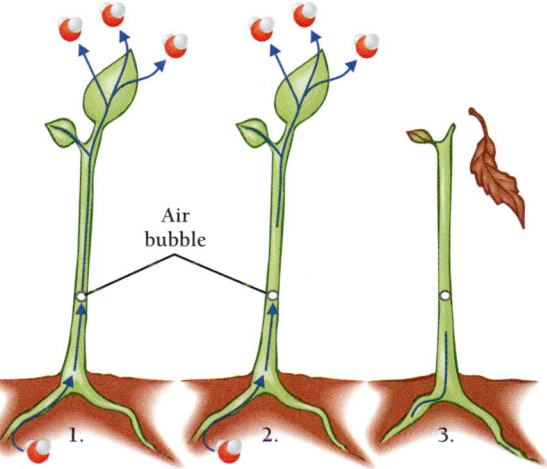

Air bubble

1.   2.   3.

B. Breaking water column

Figure 31-5   Cohesion-tension theory. A. The evaporation of water molecules from the leaf cells creates a vacuum that pulls water from xylem cells in a continuous column from roots to leaves. Only the enormous tensile strength of water (a result of hydrogen bonds between the molecules) keeps the column from breaking. B. If a bubble should break a water column, the water inside will stop ascending to the leaves.

Is the Flow of Water Through the Stem Coupled to Transpiration?

It is one thing for a process to be theoretically possible and another for it to actually occur and contribute to the workings of an organism. Plant scientists therefore asked whether water flow in the xylem increases as transpiration increases and decreases as transpiration decreases. In other words, are the two processes really linked? To address this question, researchers heated small amounts of water in the stem and then noted the time required for the heated water to arrive at a higher point. They found that fluid moves more quickly as transpiration increases during the day.

Is Water Cohesive Enough To Sustain the Transpiration Pull in a Tall Plant?

The cohesion-tension theory depends on the continuity of the water columns in the xylem. Opposing forces pull the column apart—transpiration pulling upward as gravity pulls downward. This tug-of-war extends throughout the plant, with each water molecule pulling at its neighbors via hydrogen bonds. If

the columns should break apart, the top part of the column will be pulled up, but the bottom section will be left behind (Figure 31-5B). But the columns usually do not break. The great cohesiveness of water allows columns to hold together under tremendous pull.

If an air bubble should interrupt a water column, however, the column breaks. In fact, air bubbles frequently arise in the xylem. In winter, when the air and the plant are cold, gases stay in solution and bubbles do not form easily. With the warmer temperatures of spring, however, gases in the xylem fluid form bubbles, disrupting the flow of water and minerals when it is most needed.

Woody plants have evolved a solution to this springtime hazard. Each spring, a burst of secondary growth creates new xylem routes. In many trees, most transport occurs in the newly formed xylem, in the outermost growth ring. The result is the familiar light and dark annual rings.

The cohesive strength of water is sufficient to keep it from breaking under the tension generated by transpiration.

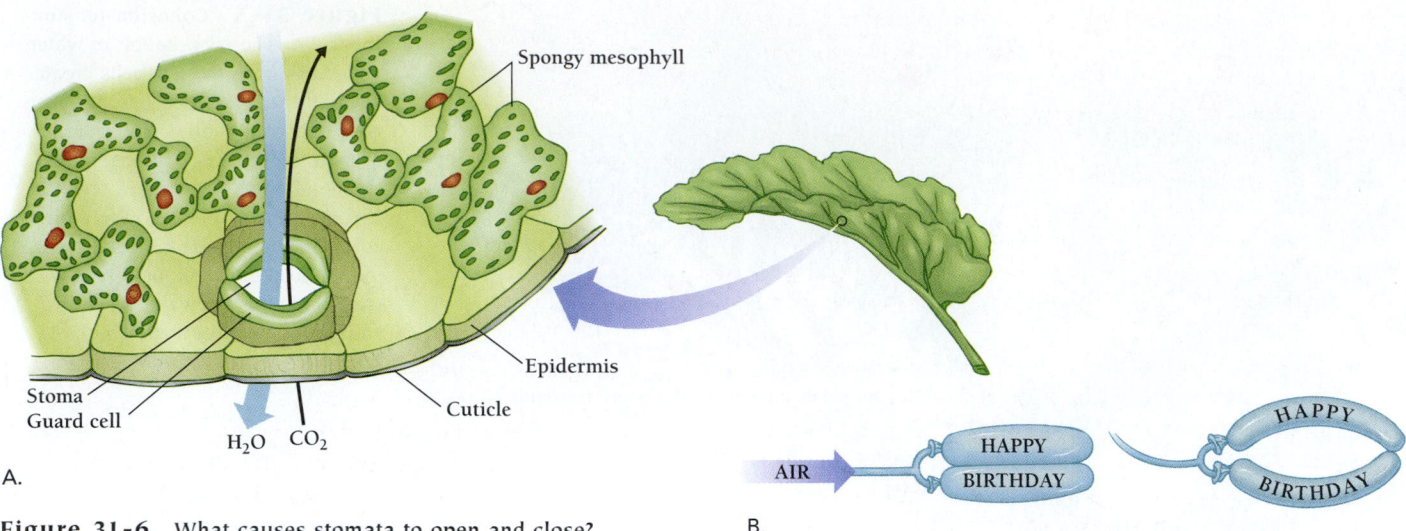

A.

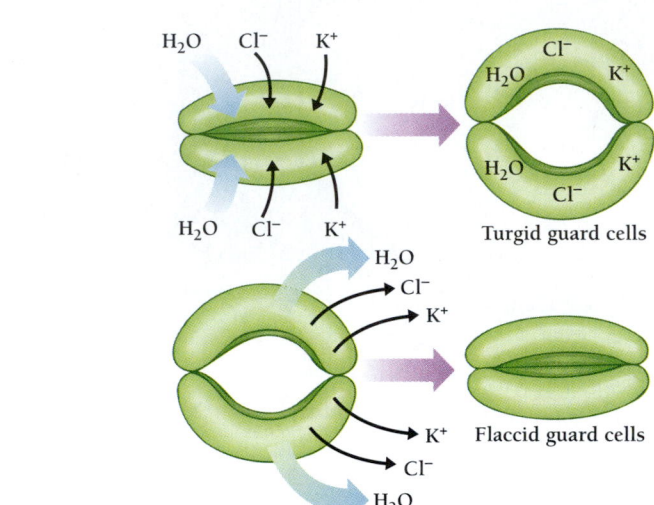

B.

C. Increased concentration of abscisic acid

Figure 31-6   What causes stomata to open and close?
A. When the stomata are open, water molecules from the inside of the leaf diffuse into the surrounding atmosphere and carbon dioxide gas diffuses into the interior of the leaf. B. Like a pair of balloons being inflated, stomata fill with water and bend away from one another. C. Each stoma has a pair of guard cells, which actively pump in potassium ions. Chloride ions, then water molecules follow passively, filling the guard cells with water and causing the stoma to open. When abscisic acid causes potassium ions to flow out of the guard cells, the chloride ions and water molecules soon follow, closing the stoma.

Is Water in the Xylem Really Under Tension?

The greater the rate of transpiration, the greater the tension in the xylem. The tension in the xylem of a big tree is strong enough to actually contract the walls of the xylem—something like when you suck in your stomach. Careful measurements show that during times of maximum transpiration tree trunks actually become smaller in diameter. When transpiration slows, the trunk relaxes to a larger diameter.

STOMATA REGULATE TRANSPIRATION BY OPENING AND CLOSING

Stomata [Greek, *stoma* = mouth] are the gates through which carbon dioxide enters and water and oxygen leave. Water often escapes faster, however, than carbon dioxide enters. In many environments, plants can lose more water through transpiration than is available in the soil to replace the water. Plants therefore have various structural adaptations for preventing water loss.

All plants closely regulate water loss by opening and closing their stomata. Closing the stomata reduces water loss, but, of course, it also prevents the plant from collecting carbon dioxide and therefore limits photosynthesis. The tradeoff between saving water and maintaining photosynthetic

productivity is called the **transpiration-photosynthesis compromise.** The role of the stomata is to "provide food while preventing thirst."

To balance these needs, stomata usually open in response to light and to a decrease in levels of carbon dioxide. When the air is hot and dry, and excessive transpiration threatens to wilt the plant, the stomata close. Without supplies of carbon dioxide from the outside air, however, most plants drastically reduce the rate of photosynthesis. This is one reason that crop yields are low during a drought.

How Do Stomata Open and Close?

Although numerous, stomata are so tiny that they occupy only about 1 percent of the total surface of a leaf. Each stoma lies over an internal air space, which in turn connects to the rambling air spaces within the mesophyll. The size and distribution of stomata permit the efficient diffusion of carbon dioxide into the air spaces and of water vapor from the air spaces into the atmosphere (Figure 31-6A).

Two guard cells surround each stoma. These cells can change shape, opening and closing the stoma. As the guard cells gain water, they bulge with water and bend outward, opening the stoma (Figure 31-6B). When water exits the guard cells, they become flaccid and close the gap between them.

Guard cells regulate the flow of water by actively pumping potassium ions (K^+) from adjacent cells of the epidermis (Figure 31-6C). As potassium ions flow into the guard cells, chloride ions follow them inwards. Because of osmosis, water follows rapidly on the heels of these ions, the cells swell, and the stoma opens.

Several factors control the pumping of potassium ions and the opening of stomata, including light levels and water availability. For example, the stomata close in response to water stress as a result of the action of the plant hormone **abscisic acid.** As water becomes less available, the concentration of abscisic acid increases. Abscisic acid inhibits the flow of potassium ions into the guard cells and, therefore, closes the stomata.

Guard cells open in response to light and low carbon dioxide by pumping potassium ions from adjacent cells, which draws water into the guard cells and opens the stoma. In the presence of abscisic acid, guard cells lose potassium ions, water leaves the cells, and they relax, closing the stoma.

How Can Plants Save Water Without Limiting Photosynthesis?

The transpiration-photosynthesis compromise prevents most plants from living in the driest deserts. In such places, most ordinary plants would not be able to open their stomata long enough to collect the carbon they need for photosynthesis. They would wilt before they got enough carbon dioxide to fuel photosynthesis.

Plants such as the cacti, however, have evolved a biochemical adaptation that gets them around this dilemma. Unlike other plants, **crassulacean acid metabolism (CAM)** plants can collect the carbon dioxide they need by opening their stomata only at night, when temperatures are cool and transpiration is low (Figure 31-7).

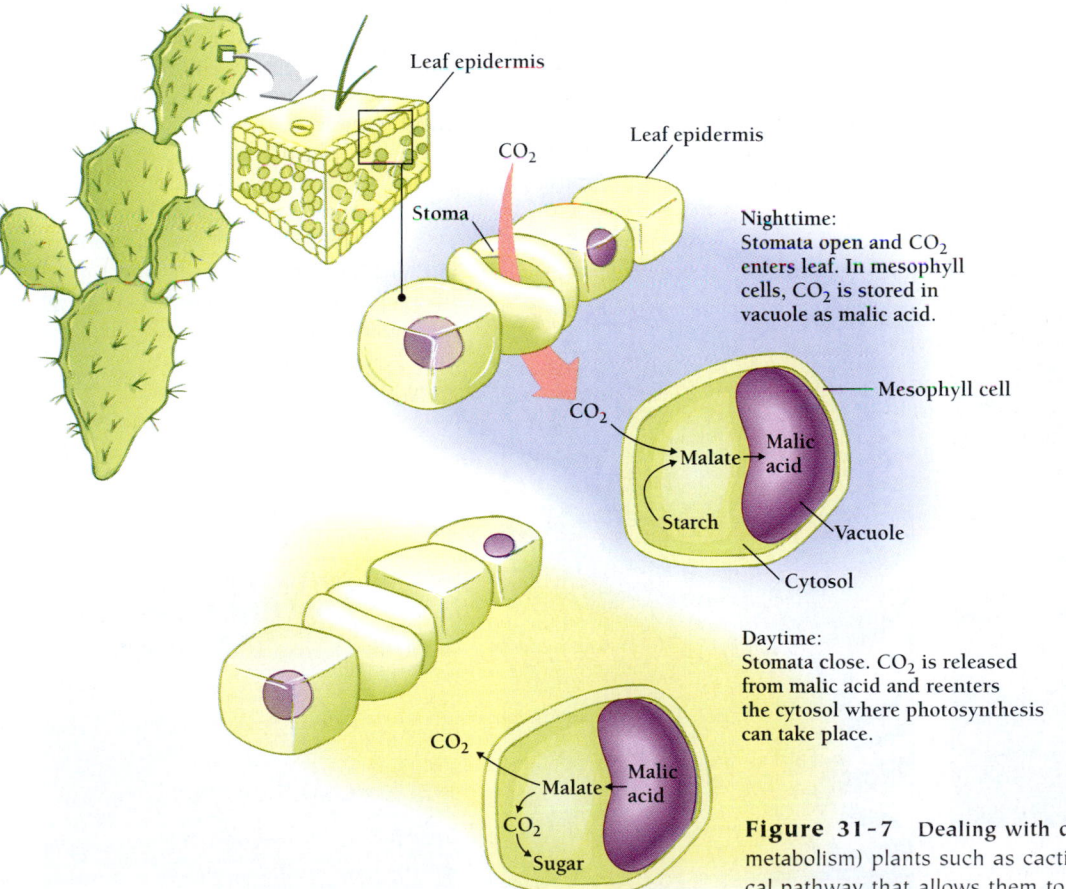

Nighttime:
Stomata open and CO_2 enters leaf. In mesophyll cells, CO_2 is stored in vacuole as malic acid.

Daytime:
Stomata close. CO_2 is released from malic acid and reenters the cytosol where photosynthesis can take place.

Figure 31-7 **Dealing with dryness.** CAM (crassulacean acid metabolism) plants such as cacti have evolved a unique biochemical pathway that allows them to store carbon dioxide. These desert-adapted plants open their stomata to collect carbon dioxide at night, when temperatures are cool and water loss is minimal. CAM plants store the carbon dioxide as malic acid until morning. During the day, CAM plants close their stomata (thus saving water) and use the carbon dioxide in malic acid to photosynthesize.

BOX 31.1

Adaptations to Limited Water Supply

All plants need ample supplies of water, so dry climates require special adaptations. Biologists sometimes divide plant adaptations to dry environments into four strategies—escape, resist, avoid, and endure. Desert annuals, for example, escape from drought (Figure A). Their seeds lie dormant, indifferent to any drought, until enough rain falls to wet the soil. Then, in a matter of days, they germinate, grow to maturity, and produce new seeds. Such seeds can survive a single dry season or 20 years of drought.

Cacti and other succulents resist drought by storing water (Figure B). CAM photosynthesis allows them to save water by keeping their stomata closed during the day. Shallow root systems quickly absorb water from the soil after each rainfall, and these plants store this water for future use.

Other plants such as desert palms and mesquite avoid water stress by growing only where rare water supplies exist. Palms grow only at oases, where their roots have access to relatively large amounts of groundwater. Mesquite bushes thrive by growing roots deep into the ground to water tables as much as 50 meters down.

Finally, some plants survive by enduring tremendous water losses. For example, while most plants die after losing 25 to 50 percent of their weight in water, the creosote bush can lose 70 percent of its weight in water and still survive (Figure C). How these plants survive such punishment is still unknown.

Even plants that grow in regions that ordinarily supply plenty of water have adaptations that allow them to survive water stress. The most common responses to drought are the slowing of cell growth and the reduction of transpiration by closing the stomata. Most plants can recover from temporary dry spells, although they may not make up for the lost period of growth.

In temperate climates, freezing temperatures make water unavailable for part of the year, and many species of plants cannot survive prolonged frosts. Surprisingly, the danger from frost turns out not to be from the ice itself. Ice crystals form only outside of cells and do not necessarily damage cells. Instead the presence of ice reduces the water potential, causing cells to lose water and therefore turgor.

Many plants can tolerate freezing temperatures, in some cases −25°C or lower. Winter rye and winter wheat, crocuses, tulips, and daffodils all grow slowly but steadily during the winter months. These frost-resistant plants prevent water loss by producing their own "antifreeze," which in some cases just consists of a concentrated solution of an amino acid. When spring comes, these plants are able to grow more rapidly, taking advantage of unfiltered sunlight before trees and taller plants leaf out.

Plants that are resistant to water stress—winter wheat, for example—are said to be acclimated or hardy. As plant biologists breed hardier plants, they greatly increase world food production. For example, if winter wheat and winter rye could withstand temperatures just 2°C colder than they now can, they could replace large tracts of spring wheat and rye in the United States, Canada, and Russia. Because winter crops make better use of the abundant spring rains, such a replacement could increase grain harvests by an astounding 25 to 40 percent.

Figure A Desert annuals mostly exist as dormant seeds. When rain comes, the seeds germinate, grow into mature plants, bloom, and go to seed—all in a period of days or weeks. *(S. Krasemann/ Photo Researchers, Inc.)*

Figure B Cacti and other succulents resist drought by storing water. *(M.C. Chamberlain/DRK Photo)*

Figure C The creosote bush endures punishing dehydration. It can lose up to 70 percent of its water and still live. *(M. Patterson/Photo Researchers, Inc.)*

TABLE 31-1   NUTRIENTS FROM THE SOIL

Nutrient	Concentration*	Symptoms of deficiency
Nitrogen	1–4%	Uniform loss of color in leaves, first on oldest leaves
Potassium	0.5–6.0%	Yellowing at margins of leaves
Calcium	0.2–3.5%	Terminal bud dies, new leaves hooked at tip. Tips and margins withered
Phosphorus	0.1–0.8%	Plants stunted, leaves abnormally dark green
Magnesium	0.1–0.8%	Yellow between veins of leaf
Sulfur	0.05–1.0%	Leaves pale green with dead spots
Chlorine	100–10,000 ppm	
Iron	25–300 ppm	In young leaves, large veins, green; rest of leaf yellow
Manganese	15–800 ppm	Dead spots scattered over leaf surface; only veins and veinlets remain green
Zinc	14–100 ppm	
Copper	4–30 ppm	
Boron	5–75 ppm	Petioles and stems brittle; bases of young leaves break down
Molybdenum	0.1–5.0 ppm	

*Typical concentrations in healthy plants in % weight or parts per million (ppm)

But photosynthesis cannot occur in the dark. The problem, then, is to store carbon dioxide until daylight. CAM plants have evolved a biochemical trick for storing carbon dioxide collected at night for use in photosynthesis during the day. In CAM plants, carbon dioxide enters the stomata at night and combines with a breakdown product of starch or another carbohydrate to form an organic acid (Figure 31-7). During the day, this acid, which is stored in the vacuole, gradually diffuses back into the cytoplasm, where it releases carbon dioxide for photosynthesis. One can actually taste this process: the leaves are sour in the early morning, from the accumulated acid, and sweet at the end of the day, after the acid disappears.

Crassulacean acid metabolism gives some plants another way of saving water.

Water in the Xylem Carries Minerals from the Soil

Plant scientists have determined the elements needed for plant growth by growing plants with their roots in water instead of soil, a technique called hydroponic culture. By varying the minerals added to the water, researchers can determine which elements, and how much of each, are required for normal growth. These experiments have led to the identification of the elements essential for growth and completion of the life cycle. The same experiments have revealed the symptoms that result from deficiencies of particular nutrients (Table 31-1).

Most Plants Obtain Both Water and Essential Minerals from the Soil

Both water and minerals usually come from the soil surrounding a plant's roots. Soils themselves are the product of both living and nonliving processes. They consist of particles of weathered rocks and partially decayed organic matter.

The best soils for plant growth hold enough water to provide a continuous supply after a rainfall, but not so much that the roots cannot get the oxygen they need for respiration. The water capacity of soils depends in part on the sizes of the particles of weathered rock. Soils consisting only of small particles (less than 2 μm) are called **clays;** soils of medium sized particles (from 2 to 20 μm) are called **silts;** and soils of large particles (from 20 to 200 μm) are called **sands. Loams** consist of mixtures of all three particle sizes. The ideal loams contain about 20 percent clay, 40 percent silt, and 40 percent sand.

In most soils, earthworms, insects, bacteria, and fungi quickly convert dead organisms into large, organic molecules. The residue of such decay is a black or brown material, called **humus,** that decays much more slowly. Good soil has a high capacity for positively charged ions, which bind to negatively charged particles of clay and humus. Humus does not provide the plant with organic matter (as was once thought), but it does provide a good reservoir of water and bound ions while also permitting oxygen to diffuse easily.

Most soils are composed of some combination of clay, silt, sand, and humus.

Why Do Roots Need Oxygen from the Soil?

Groundwater is full of dissolved minerals, usually in the form of ions. The concentrations of these ions in groundwater, however, are much less than their concentrations in plants. This is because plants actively transport ions into their root cells and concentrate them there, especially in the xylem. Because such active transport requires energy, root cells use large amounts of ATP. They obtain ATP by oxidizing the products of photosynthesis. Thus, roots depend on the presence of oxygen in the soil. Most plants drown in waterlogged soils.

> *Because root cells actively transport ions*
> *from the soil, root cells need oxygen to generate*
> *ATP by means of respiration.*

Why Do Some Plants Obtain Minerals from Animals?

Many plants live in areas where minerals, especially nitrogen, are in short supply. In many swamps and wetlands, for example, highly acidic water interferes with plants' ability to take up dissolved nitrates. Plants such as the Venus flytrap solve this problem by trapping and digesting insects for their nitrogen compounds, phosphates, and other ions (Figure 31-8).

Figure 31-8   Hungry plants. Some plants, such as this Venus flytrap, live in environments that have insufficient supplies of nitrogen. Such plants may evolve techniques for trapping and digesting insects and other small animals. *(William E. Ferguson)*

HOW DO PLANTS TRANSPORT THE PRODUCTS OF PHOTOSYNTHESIS?

We have seen that roots depend on the products of photosynthesis, but cannot themselves photosynthesize. How do roots and other plant parts obtain the energy and materials for building and maintaining cells and their myriad processes? To find out, modern biologists have tracked the movement of the products of photosynthesis.

Which Way Does Sugar Move?

By exposing leaves to carbon dioxide containing radioactive ^{14}C, researchers can study the distribution of the radioactive carbon. Such studies show that the products of photosynthesis move entirely within the sieve tubes of the phloem.

In woody stems, functional sieve tubes are confined to the inner bark (Figure 31-9A). Removing a strip of bark around a tree trunk, a technique called **girdling,** interrupts the phloem and blocks the transport of photosynthetic products (Figure 31-9B). Girdling always kills the tree. Early colonists in North America girdled trees in the spring and felled them in the fall to harvest partially dried firewood.

In the early 18th century, two distinguished scientists—the Italian anatomist Marcello Malpighi and the English physiologist Stephen Hales—studied how girdling kills trees. They noted that while the bark below the girdle shriveled and died almost immediately, the bark above the girdle remained healthy at first. But a girdled tree usually fails to leaf out in the following spring, and before long, the whole tree dies. What causes this pattern of death? The answer is that girdling first prevents sugars from flowing from the leaves to the roots. The leafy top of the tree remains alive for a while, but once the roots die, they no longer deliver water and minerals to the leaves, which then die.

Sugars are said to move from **source,** the site of production, to **sink,** the site of storage or consumption. A source may be either a site of photosynthesis, such as the leaves, or a storage site (such as the root) that releases sugars by breaking down starch. Sinks include the growing tips of roots and shoots; young leaves that are not yet active producers; fruits; and storage organs that are storing rather than releasing energy-rich molecules. The same organ—a potato tuber (an underground stem), for example—may be a sink in the summer, as it accumulates carbohydrates, and a source during the following spring, when it provides the plant with energy for the new growing season.

> *Sugars in plants move from sources, usually leaves or*
> *storage areas, to sinks, usually roots or storage areas.*

Figure 31-9   Sugar on the outside. A. In woody plants, working sieve tube elements are confined to the inside of the bark. B. Stripping the bark from a tree will kill it if the strip goes all the way around. Such girdling destroys the sieve tubes and prevents the products of photosynthesis from reaching the roots.

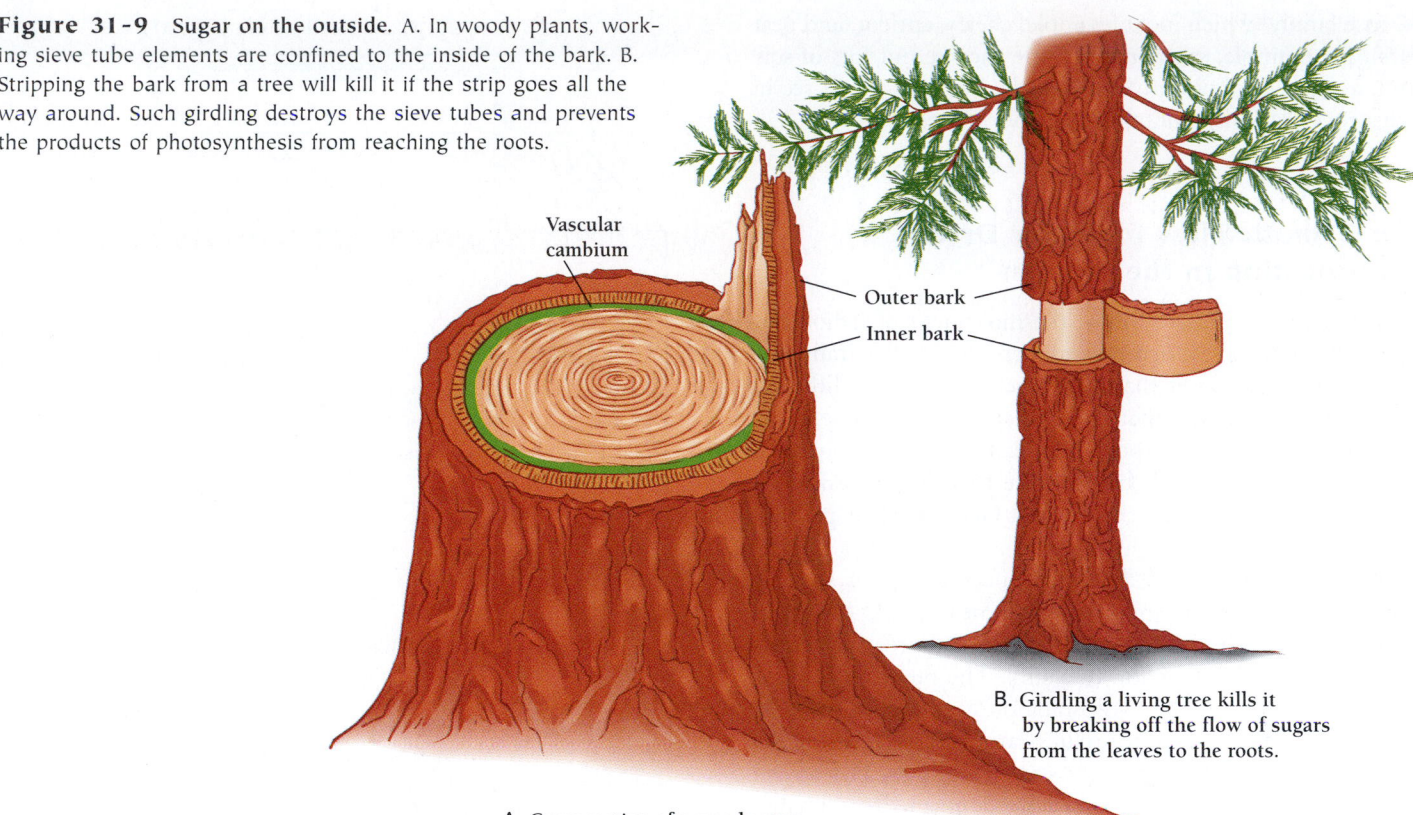

Vascular cambium

Outer bark

Inner bark

B. Girdling a living tree kills it by breaking off the flow of sugars from the leaves to the roots.

A. Cross section of a woody stem

What Is Inside a Sieve Tube?

Analyzing the contents of phloem was long a difficult problem for plant physiologists because the sieve tubes are small and the sap inside quickly seals any holes made in the sieve tubes. Amazingly, the most useful method for analyzing phloem contents came from a 1953 suggestion of two insect physiologists.

These researchers were studying the nutrition of aphids, which derive their food from sap, the fluid contained in phloem (Figure 31-10). An aphid has a specialized mouth part, called a *stylet*, that pierces a sieve tube. Sap then flows from the phloem, through the stylet, directly into the aphid's body. The aphid extracts what it can, and the rest appears as "honeydew" at its rear end.

By cutting off the aphid and leaving the creature's stylet embedded in the sieve tube, plant physiologists have been able to sample and to study the contents of the phloem. Botanists have found that the phloem contains a concentrated solution of sugars. Sucrose (table sugar) may make up as much as 30 percent of the phloem contents. Phloem sap also contains a few minerals, especially potassium ions, and carbohydrate derivatives, which vary considerably from plant to plant. Within

Figure 31-10   An aphid in action. Aphids unintentionally help researchers study the contents of phloem. An aphid passes its stylet into the phloem for a drink of sugar water. To collect the contents of the phloem, a researcher removes the aphid from its stylet. *(Jerome Wexler/National Audubon Society/Photo Researchers, Inc.)*

the rose family (which includes apple, cherry, apricot, and pear trees), for example, sap contains little sucrose but lots of sorbitol, a sugar alcohol derived from glucose. As we will see in Chapter 33, sap also contains plant hormones.

The Osmotic Flow of Water Drives Translocation in the Phloem

By using aphid stylets to follow the movement of radioactive sugars through the phloem, researchers found that translocation occurs much faster than could be explained by diffusion alone. They concluded that the contents of the phloem undergo some kind of transport. The explanation for how this transport occurs is called the **pressure flow hypothesis,** first advanced in 1926 by Ernst Münch, a German plant physiologist and artist.

Münch's insight was that the concentration of sugar is higher in the phloem near a source than near a sink. More water therefore flows into the sieve tubes near a source and pushes the contents in the direction of a sink. The pressure flow hypothesis predicts that the hydrostatic pressure within the sieve tubes is greater near a source than near a sink. Measuring the pressure within phloem tubes has been difficult, but several experiments have succeeded and have confirmed the prediction (Figure 31-11).

Ernst Münch suggested the pressure flow hypothesis to explain how sugars move once they are in the phloem.

What Draws Water and Sugar into the Phloem?

Once the concentrated sugar is in the phloem it makes sense that it should flow away from the areas of highest concentration. But what prevents the sugar (made in the mesophyll) from flowing back into the mesophyll? And what forces act to concentrate the sugar in the phloem? The answer to both questions is, once again, active transport.

The cells of the mesophyll are connected to one another by plasmodesmata, and sugars flow freely from cell to cell. There are few plasmodesmata between the mesophyll cells and the cells of the phloem, however, and sugar must therefore pass through two cell membranes, that of a mesophyll cell and that of a phloem cell (Figure 31-12A). Both passages require energy and are selective. The two cell membranes transport some molecules (such as sucrose) but not others (such as glucose). The sucrose concentration in phloem may be several times that in the mesophyll.

Sugar does not, however, flow directly from the mesophyll cells into the sieve tube members. Instead, sugar often flows into **companion cells,** specialized cells adjacent to the sieve elements (Figure 31-12B). Unlike sieve tube members, com-

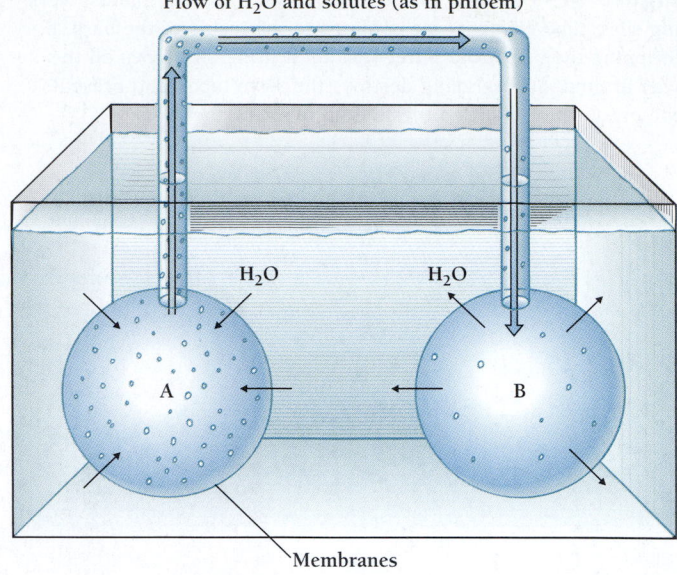

Flow of H₂O and solutes (as in phloem)

H₂O   H₂O

A   B

Membranes

PRESSURE FLOW HYPOTHESIS

Figure 31-11   Münch's model for the pressure flow hypothesis. To test his hypothesis, Münch built a laboratory model that consisted of two selectively permeable membranes connected to each other by a tube. In this simplified diagram, the first membrane (labeled A) represents the source, and the second membrane (labeled B) represents the sink. Inside the first membrane, A, is a solution more concentrated than the solution outside of it. Water consequently flows into A and then up into the connecting tube. The solution inside of B is less concentrated than the solution outside. Osmosis therefore moves water from the connecting tube into B. In accordance with Münch's hypothesis, the flow of water carries along any dissolved solutes at the same rate.

panion cells possess nuclei and ribosomes and can make polypeptides.

The active pumping of sucrose is coupled to a pH gradient across the cell membrane (Figure 31-12C). The cell creates this gradient by spending ATP to pump out hydrogen ions, as in the case of the pH gradients across chloroplast and mitochondrial membranes. Hydrogen ions flow back into the companion cell via a transport protein that also carries sucrose.

Once the sieve tube contains a high concentration of sucrose, osmosis drives water into the phloem from the surrounding cells, greatly increasing the pressure and pushing the sugar away. The pressure in the phloem is more than ten times that in a car tire. Sap moves through the sieve tubes with a velocity of about 1 meter per hour, about the speed of the tip of the minute hand on a large classroom clock. (In contrast, water flows much more rapidly through the xylem, at a rate near that of the clock's second hand.) At the sink end of a phloem vessel, the phloem unloads its sucrose (aided by companion

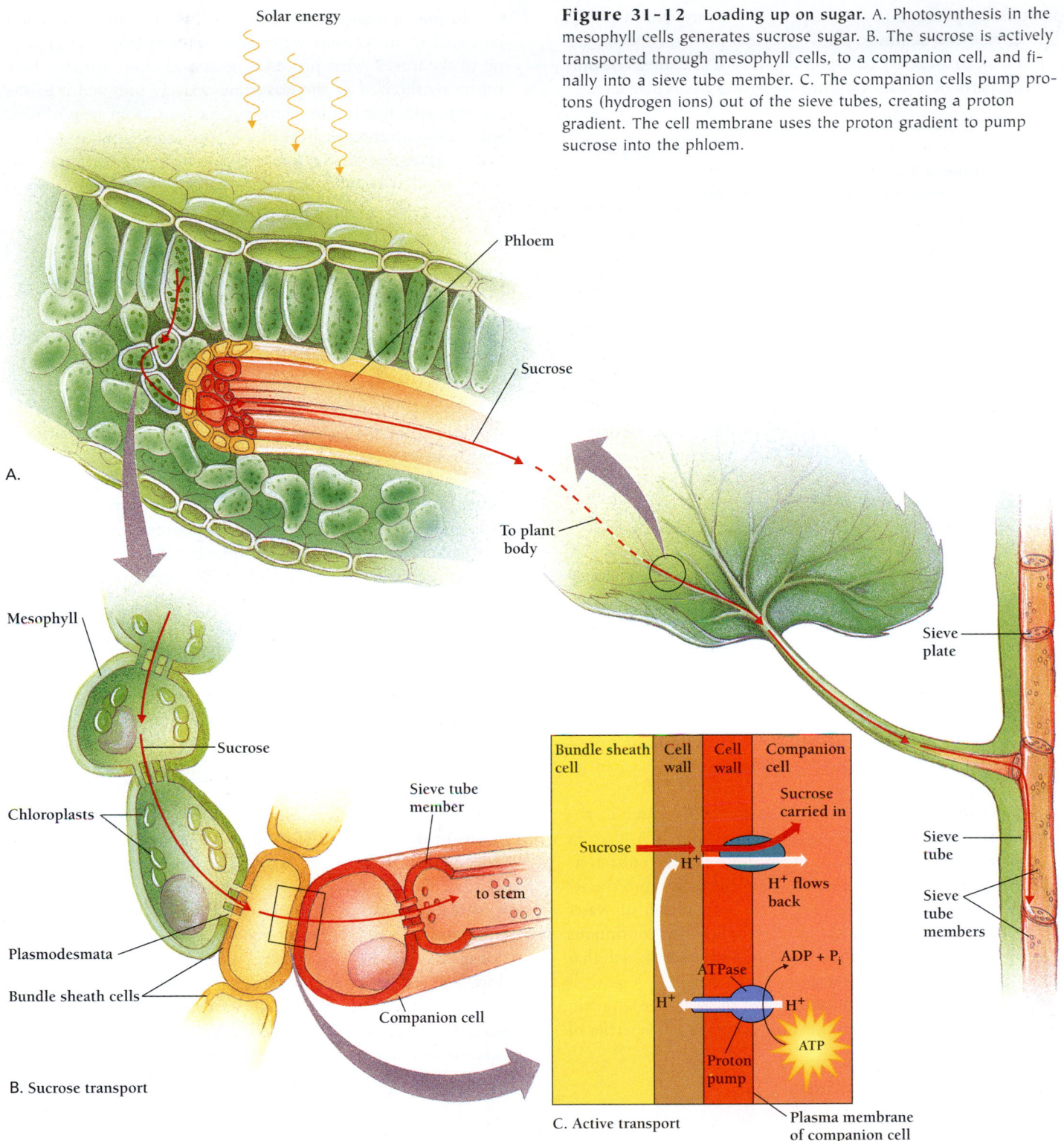

Figure 31-12 **Loading up on sugar.** A. Photosynthesis in the mesophyll cells generates sucrose sugar. B. The sucrose is actively transported through mesophyll cells, to a companion cell, and finally into a sieve tube member. C. The companion cells pump protons (hydrogen ions) out of the sieve tubes, creating a proton gradient. The cell membrane uses the proton gradient to pump sucrose into the phloem.

cells), and the water then enters nearby xylem and flows back up to the mesophyll.

The pressure flow model for phloem transport works in a single direction—from a source, with a high concentration of sucrose, to a sink, with a low concentration. One of the rea-

sons that plant physiologists were slow to accept this hypothesis is that phloem transport can often occur both up and down the stem. Most researchers now think, however, that the flow within a single sieve tube goes in a single direction at a given time. Unlike the circulation of blood, then, the "circulation" of fluids in plants does not depend on a mechanical pump. The movement of fluids, both in xylem and in phloem, depends instead on active transport and turgor pressure. The only moving parts of a plant's pumps are at the molecular level.

Cells in the mesophyll pump sugars by active transport from cell to cell and then into the phloem.

In this chapter we have seen how plants exploit the water potential in air to draw water and minerals from the ground up to the leaves; what processes open and close stomata; how sugars synthesized in the leaves are actively pumped into the phloem; and, finally, how those sugars move from areas of high sugar concentration to areas of low sugar concentration. In the next chapter, we discuss how developing plant embryos generate a body plan and form the elements of root and shoot.

Study Outline with Key Terms

Water moves into and out of living cells by **osmosis,** the flow of water through a selectively permeable membrane in response to differences in the concentrations of solutes. Water moves from areas of low solute concentration to areas of high solute concentration, and from areas of high water concentration to areas of low water concentration. The tendency of water molecules to move along gradients of concentration and pressure is called **water potential.**

In plant cells, the mechanical pressure of the cell wall, or "wall pressure," opposes the inward pressure of water from osmosis, or **turgor pressure.** Turgor pressure provides mechanical support to nonwoody plants. Solutes in the xylem cause water to flow into the roots, resulting in **root pressure.** Root pressure is beautifully demonstrated in **guttation,** the expression of droplets of water from the edges of leaves on cool, damp mornings.

Because of water's **adhesiveness** and **cohesiveness,** it tends to move up narrow tubes, or **capillaries.** But neither root pressure nor capillarity is enough to account for the movement of water to the tops of tall trees.

The major driving force for the ascent of water in the xylem is **transpiration.** And the driving force for transpiration is low water potential in air. According to the **cohesion-tension theory,** water vapor passes from the leaves into the atmosphere through open stomata. The evaporation of water from the leaves pulls water up the xylem in continuous columns. Experimental measurements have established that water in the xylem is under tension, like a dangling rope with a weight at the bottom. The integrity of the water columns in the xylem is critical and depends on the cohesive properties of water.

Plants regulate transpiration and photosynthetic activity by opening and closing their stomata, a process mediated by the hormone **abscisic acid.** During hot, dry periods, plants must balance the rate of transpiration against the rate of photosynthesis, a trade-off called the **transpiration-photosynthesis compromise.**

Many plants have special biochemical and anatomical adaptations that reduce water loss while permitting the entry of enough carbon dioxide to sustain photosynthesis. **Crassulacean acid metabolism (CAM)** plants collect carbon dioxide by opening their stomata at night, when temperatures are cool and transpiration is low.

Most plants obtain minerals as well as water from the soil. Roots actively transport ions from the soil through cells of the cortex into the xylem. **Loams** are different combinations of variously sized particles of **sand, silt,** and **clay.** Decayed organic matter in soils is called **humus.** Insectivorous plants obtain nutrients from animals.

Many parts of a plant do not perform photosynthesis and require energy-rich molecules transported from the leaves. The movement of such molecules, mostly in the form of sucrose, occurs in the phloem. **Girdling** a tree or shrub interrupts the phloem and blocks the transport of photosynthetic products.

The osmotic flow of water, described in the **pressure flow hypothesis,** drives the flow of sugar water through the phloem. **Companion cells,** adjacent to the sieve elements, actively transport sugars into the phloem, and water follows by osmotic flow. The resulting pressure carries the contents of the phloem to other parts of the plants. Sugars move from a **source** to a **sink.** The circulation of nutrients in the phloem, like the circulation of water and minerals in the xylem, is best understood in terms of differences in water potential.

Review and Thought Questions

Review Questions

1. Define "water potential." In what different ways do plants use water potential to move materials from place to place?
2. Which is higher—the water potential of a plant cell in air or the water potential of the same cell in a lake? In which direction would water move in each case?
3. Describe root pressure and guttation.
4. Explain why water's natural cohesiveness is an essential element of the cohesion-tension theory.
5. How would you test the cohesion-tension theory?
6. What problem do CAM plants solve?
7. Name the three main particle sizes in soil. Which is biggest and which is smallest? The most fertile soils, or loams, are generally made up of what combination of these three kinds of particles?

8. Describe the pressure flow hypothesis. Is a potato a source or a sink?

Thought Questions

9. Why should xylem be composed of dead cells while phloem must be composed of living cells? What differences in the way these two parts of the vascular system function could explain this difference?
10. Do you feel that any of the characters, whether scientists or grape growers, described in the story about phylloxera did anything wrong or unethical? Explain why or why not.

About the Chapter-Opening Image

Racks of wine barrels resemble the stacked vessel elements that carry water minerals through the xylem of plants. This photo has been turned sideways to enhance the resemblance to vessel elements.

 On-line materials relating to this chapter are on the World Wide Web at **http://www.harcourtcollege.com/lifesci/aal2/**

32 Growth and Development of Flowering Plants

The Origin of Corn

When Columbus left Spain in 1492, he was searching for a shortcut to China, Japan, and the rest of Asia. There he hoped to trade European goods for spices and silk. But when he reached the New World, one of his most valuable discoveries was not spices or silk but corn. Corn, also called maize, was good to eat, easy to grow, and bountiful.

Indeed, when Portuguese sailors finally reached the eastern coast of Asia in 1516, they brought, not European delicacies, but American corn. Within 50 years of Columbus's arrival in the West Indies, corn grew in every corner of the world. Wherever corn was introduced, it replaced native crops as an indispensable part of the diet. Corn became so thoroughly adopted in every tiny village where it was introduced that peoples in Africa, India, China, and Turkey all came to believe that maize was native to their region.

Where did corn come from? Certainly it is American. Archaeologists have found corn (*Zea mays*) that is thousands of years

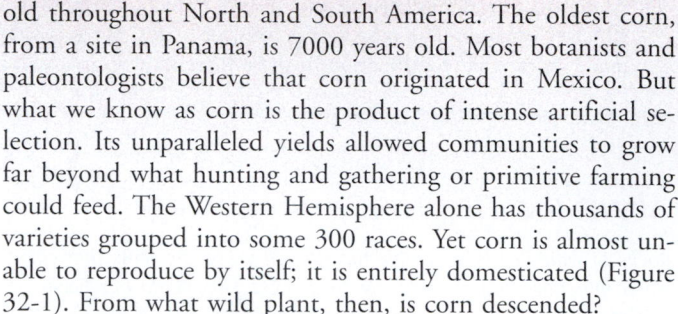

old throughout North and South America. The oldest corn, from a site in Panama, is 7000 years old. Most botanists and paleontologists believe that corn originated in Mexico. But what we know as corn is the product of intense artificial selection. Its unparalleled yields allowed communities to grow far beyond what hunting and gathering or primitive farming could feed. The Western Hemisphere alone has thousands of varieties grouped into some 300 races. Yet corn is almost unable to reproduce by itself; it is entirely domesticated (Figure 32-1). From what wild plant, then, is corn descended?

In the 1930s, the Harvard botanist Paul Manglesdorf backcrossed various corn hybrids to produce what he suggested was a model for wild-type corn. This approach would be similar to crossing several different breeds of dog to produce something that looked like a wolf. Manglesdorf found that as he backcrossed the corn, the cobs became smaller and the number of kernels inside the husk decreased until each corn kernel, or seed, was wrapped in its own husk, in the same manner as wheat. The ancestor of corn, he predicted, would be found to be an extinct, pod-bearing popcorn, which he called *pod corn.*

In the summer of 1948, two Harvard graduate students, in anthropology and botany, took a bus to Bat Cave, New Mexico. The cave had been occupied 3000 years earlier by people living on the shore of an ancient lake. As the two young men dug, they found hundreds of pieces of corn. The deeper they dug, the smaller and more primitive the corncobs became. When they reached the bottom, they found tiny cobs of popcorn in which each kernel was enclosed in its own husk—in other words, Manglesdorf's pod corn.

Manglesdorf, who had given the two students $150 for bus fare and sleeping bags, was delighted. He later wrote, "Seldom in Harvard's history has so small an investment paid so large a return." Manglesdorf took a few kernels of the ancient Bat Cave corn, dropped them into a pan of hot oil, and they popped!

But while Manglesdorf's own students and other archaeologists found the pod corn hypothesis persuasive, geneticists had their own ideas. The geneticist George Beadle had proposed, in 1939, that corn was descended from a Mexican grass called teosinte (tay-o-SIN-tay). Nineteenth-century naturalists had first suggested that teosinte might be the ancestor of corn, and Beadle found that kernels of teosinte popped in hot oil were "indistinguishable from popped corn." Teosinte, however,

Paraskevas Photography

Key Concepts

- In developing seeds, plant embryos establish a body plan soon after fertilization and then pause after the seed matures.

- The primary growth and development of plants comes from apical meristems. Secondary growth comes from lateral meristems.

- Specific genes control the pattern of development in flowers.

has 40 chromosomes whereas corn has only 20, a difference that would prevent the two plants from interbreeding. Without being able to crossbreed the two plants, classical geneticists could tell little about them.

Manglesdorf dismissed Beadle's teosinte hypothesis with a counter theory. The ancestor of corn is pod corn, he argued, and teosinte is a hybrid of ancient pod corn and another grass called *Tripsacum.* Modern corn, he suggested, is a mix of these three related grasses. *Tripsacum* did not have 20 chromosomes either, however, and the question of corn's ancestry remained unresolved. For 40 more years, Manglesdorf and Beadle, as well as their intellectual descendants, picked at each other's theories. By the 1970s, however, many botanists thought that an extinct teosinte would prove to be the ancestor of corn.

In 1978, Hugh Iltis, a professor of botany at the University of Wisconsin, sent a Christmas card to a colleague at the University of Guadalajara, in Mexico. The card bore a drawing of a fanciful extinct teosinte, something that might resemble the ancestor of corn. Amused, the Mexican botanist challenged her students to find such a plant in the wild. Young Rafael Guzman took the challenge and spent his Christmas vacation searching for the hypothetical plant. In the mountains near Jalisco, Guzman found a new species of teosinte, seemingly the model for Iltis's drawing.

Guzman sent Iltis some seeds, and Iltis found, to his delight, that the plants had half the number of chromosomes that teosinte has. Like corn, *Zea diploperennis* has 20 chromosomes. The search for corn's ancestor, Iltis declared, was over. Indeed, in the 1980s, genetic research at the University of Minnesota showed that approximately five genes control the traits that distinguish corn from *Zea diploperennis.* Once these genes are cloned, researchers hope to investigate how they are expressed.

Is the corn debate over? Most people now agree that corn is about 10,000 years old and probably descended from some form of teosinte from central Mexico. A

likely ancestor for corn seems close at hand, yet paleobotanists and geneticists continue their comfortable bickering. As Richard Schultes of Harvard's Botanical Museum has said, "The origin of corn is a mess."

For most of us, it is a mess we can safely overlook. In comparison, the changes in corn's morphology under thousands of years of selective pressure are dramatic and revealing. Corn kernels, each of which is actually a fruit, are so tightly wrapped in the husk that the seeds can never disperse. Even if the whole husk is buried in moist earth, the resulting seedlings are so crowded that most of them die. Corn's artificially induced anatomy has made it entirely dependent on humans for reproduction. It is nearly alone in this odd defect. Nearly every other species of plant on Earth can reproduce and grow with no help at all.

In this chapter we will see how plants reproduce, how the seed forms, and how the embryo in the seed develops into a mature plant.

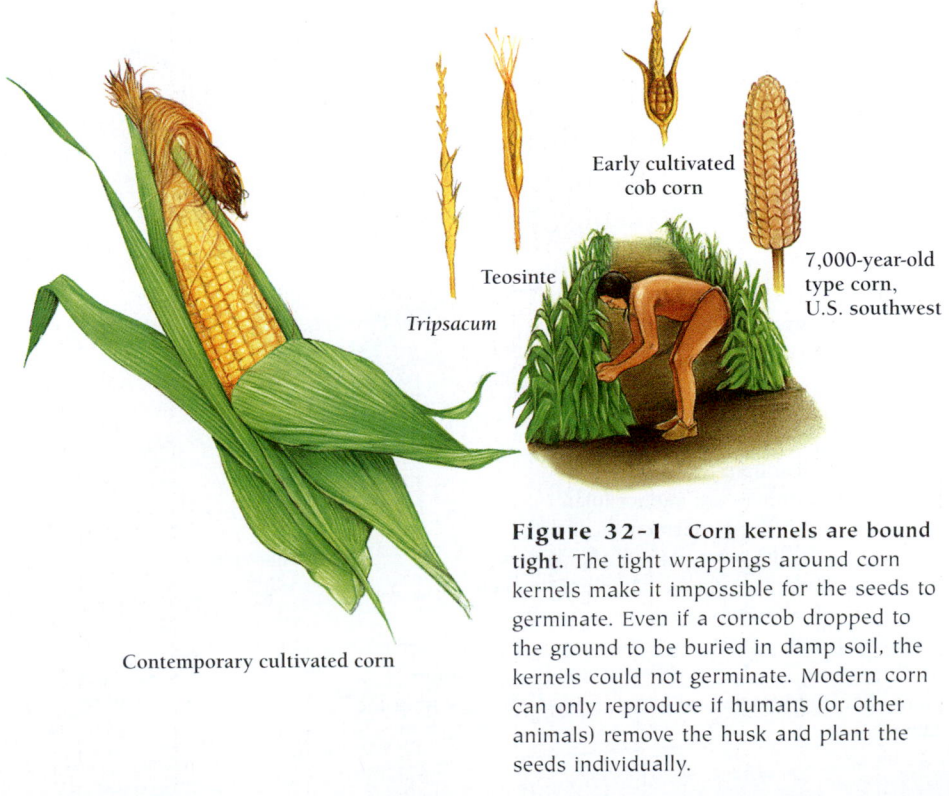

Figure 32-1 Corn kernels are bound tight. The tight wrappings around corn kernels make it impossible for the seeds to germinate. Even if a corncob dropped to the ground to be buried in damp soil, the kernels could not germinate. Modern corn can only reproduce if humans (or other animals) remove the husk and plant the seeds individually.

677

HOW DO PLANT EMBRYOS ESTABLISH A BODY PLAN?

Most plants develop continuously throughout life, modifying their development in response to environmental cues. Because a plant cannot move to a new environment and because it continually faces environmental hazards ranging from hungry herbivores to harsh weather, its ability to modify its development is crucial to its survival. In Chapter 33, we will see how plants sense environmental cues and coordinate their responses. In this chapter, we focus on how a flowering plant produces gametes, accomplishes fertilization, and begins its development. We will also see how apical meristems give rise to shoots and roots and how lateral meristems produce secondary growth.

The life cycle of a flowering plant consists of two phases: the diploid **sporophyte,** which produces haploid spores by meiosis; and the haploid **gametophyte,** which produces haploid gametes by mitosis (Chapter 22). (By comparison, the haploid phase of animals—including humans—is extremely limited, since, in animals, neither sperm nor eggs divide mitotically.)

The cycling of the multicellular sporophyte and gametophyte phases is called **alternation of generations** (Figure 32-2). Most of our discussion here will concern the diploid, sporophyte generation, the dominant phase of the life of every flowering plant. (In bryophytes such as liverworts and mosses, the gametophyte is the dominant phase in the life cycle, as we discussed in Chapter 22, but we do not discuss these plants at all in this chapter.) We begin by describing the development of the gametophytes, the production of gametes, and their union to produce a diploid zygote.

How Do Plants Make Sperm?

In flowering plants, sperm are made inside tiny gametophytes called **pollen grains,** which are released from the flower to travel to other plants. Pollen is made in flower parts called **stamens.**

The anther is the actual pollen-producing part of the stamen, attached to the flower's base with a filament (Figure 32-3A). Within the anther are four pollen sacs. Inside these sacs, diploid cells undergo meiosis to form haploid **microspores** [Greek, *mikros* = small], each of which develops into a male gametophyte.

Each microspore undergoes mitosis to produce two haploid cells, one *generative cell* and one *vegetative cell.* (A few flowering plants produce two generative cells and one vegetative cell.) The vegetative cell completely surrounds the generative cell. The two cells—generative and vegetative—are together enclosed within a common wall (that of the original microspore) to form a pollen grain. The outside of the pollen grain is often elaborately and distinctively sculpted (Figure 32-3B). Indeed, botanists often can recognize plants, extinct or living, by their pollen alone.

The anthers release the pollen, often immense quantities of it, and insects, wind, or other agents then carry some of it to the (female) stigma of a flower. Most pollen never reaches the stigma, however; it ends up, instead, in streams and lakes, on the ground, in the hives of bees, or even packed into small jars in health food stores.

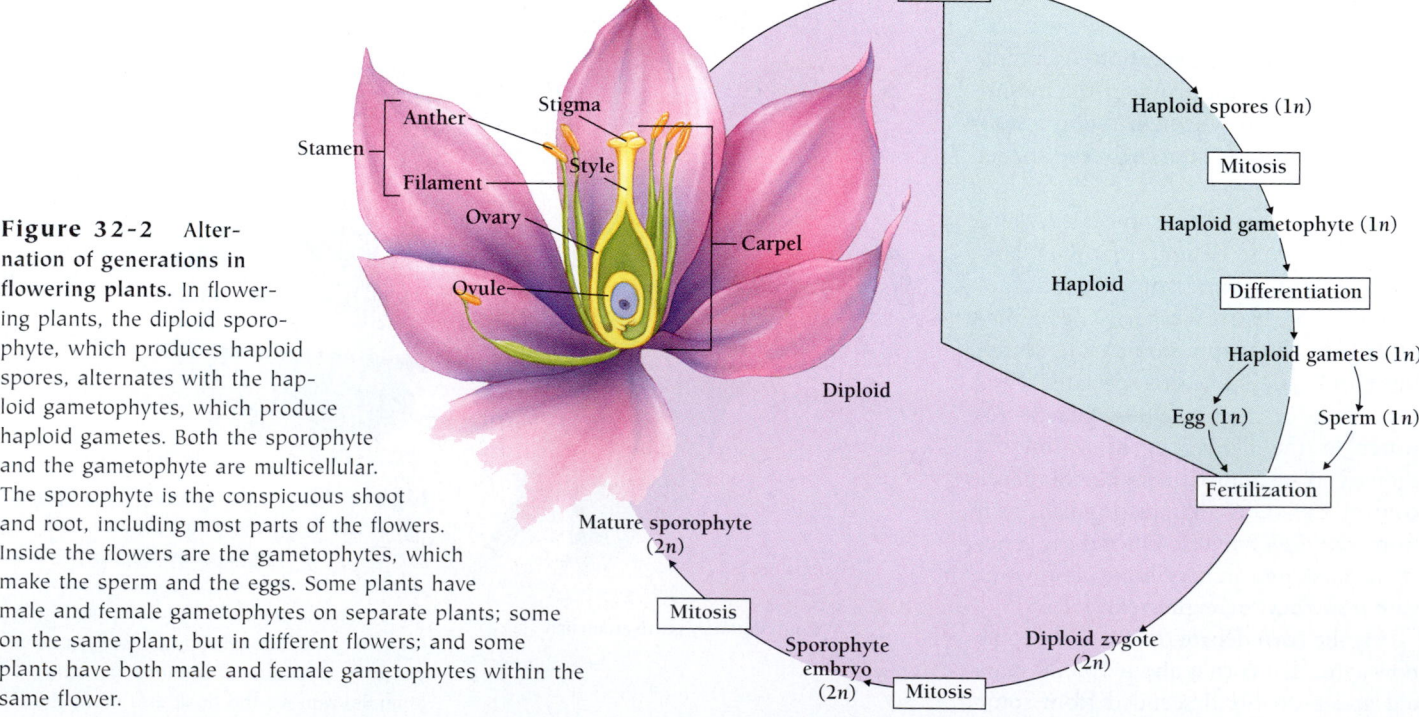

Figure 32-2 Alternation of generations in flowering plants. In flowering plants, the diploid sporophyte, which produces haploid spores, alternates with the haploid gametophytes, which produce haploid gametes. Both the sporophyte and the gametophyte are multicellular. The sporophyte is the conspicuous shoot and root, including most parts of the flowers. Inside the flowers are the gametophytes, which make the sperm and the eggs. Some plants have male and female gametophytes on separate plants; some on the same plant, but in different flowers; and some plants have both male and female gametophytes within the same flower.

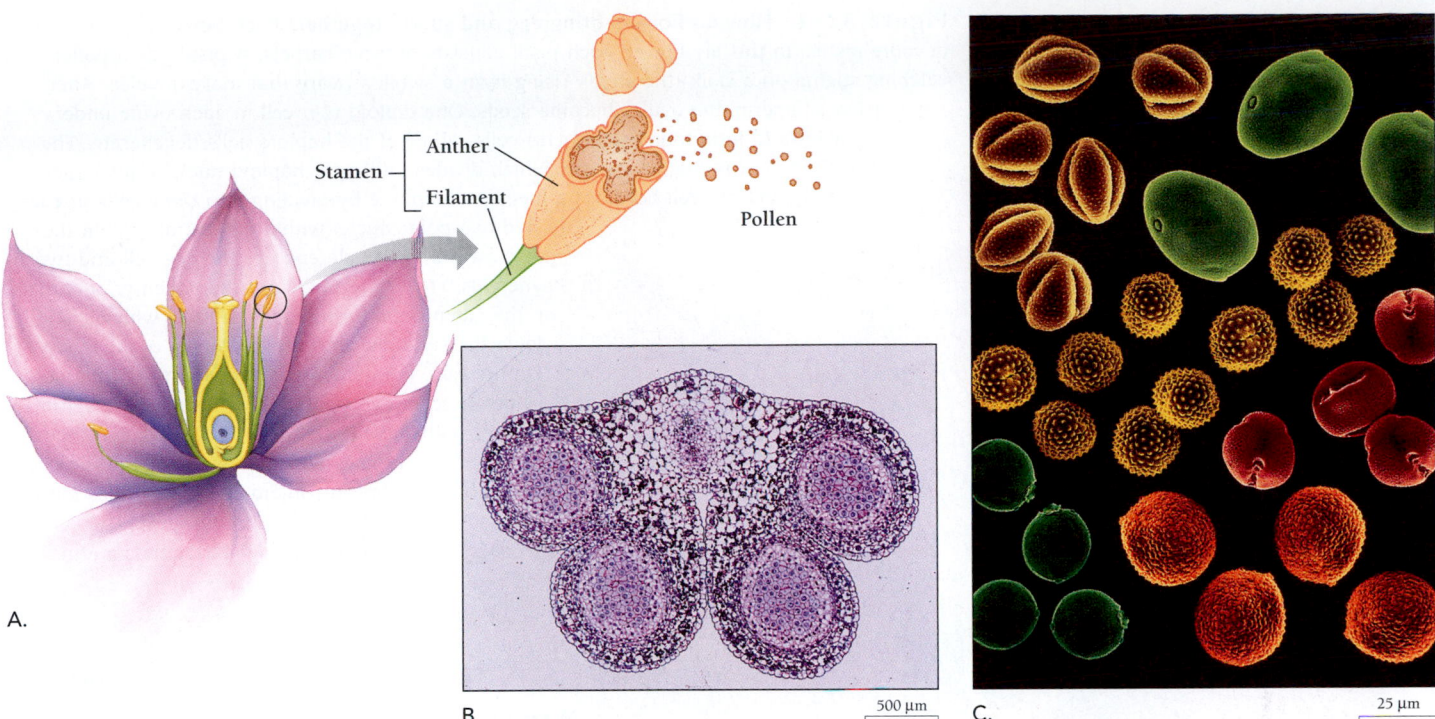

A.

B.

500 µm

C.

25 µm

Figure 32-3 **How do flowers make pollen?** Anther of a lily. A. The anther is the pollen-producing part of the stamen. B. Each anther includes four pollen sacs, inside of which haploid microspores are formed. Each microspore divides and develops into a male gametophyte, a two-celled pollen grain. C. Pollen grains have distinctive textures and patterns that experts recognize. Shown here are pollen from ragweed, timothy brush, alder, and poplar. (B, Cabisco/Visuals Unlimited; C, Dennis Kunkel/Phototake, NYC)

Stamens produce haploid microspores, which develop into immature male gametophytes known as pollen grains.

How Do Plants Make Egg Cells?

The female reproductive part of a flower is called a **pistil.** A flower may have one pistil or several, and some flowers lack them altogether. The pistil is the most central of the flower's parts and consists of one or more **carpels.** The swollen base of the pistil is the **ovary** [Latin, *ovum* = egg], which produces the female gametophytes (Figure 32-4A). When a pistil consists of several carpels, the carpels join to form a compound ovary. A projection, called the **style,** connects the ovary to the **stigma,** the structure that receives the pollen grains.

The wall of the ovary produces one or more **ovules,** structures that, after fertilization, form a seed. An ovule contains a fertile diploid cell within a jacket of cells called the **nucellus,** which in turn is enclosed in two more jackets of cells called **integuments** (Figure 32-4B and C). These jackets help to shel-

ter and nourish the developing embryo. The integuments eventually form the seed coat.

The path to egg cells begins when the fertile cell undergoes meiosis. One diploid cell in each ovule undergoes meiosis to form four haploid cells. One of the four cells, called the **megaspore** [Greek, *megas* = large], develops into the **embryo sac,** which is the mature female gametophyte, while the other three haploid cells degenerate. The megaspore nucleus undergoes three rounds of mitosis to form eight haploid nuclei. These eight nuclei form only seven cells, however, for one cell, the *central cell,* contains two nuclei. One of the other six cells of the embryo sac is the egg cell. The surrounding diploid cells of the ovule are also dividing as the haploid gametophyte develops, giving rise to the nucellus and the integuments.

Carpels produce haploid megaspores, which develop into female gametophytes called embryo sacs. Each embryo sac has one egg.

How Does Fertilization Occur in Plants?

After a pollen grain lands on the stigma, the vegetative cell of the pollen grain forms a **pollen tube,** into which the sperm nuclei pass. The pollen tube grows down the style and through the *micropyle,* the opening between integuments, to reach the embryo sac (Figure 32-4B). When the pollen tube starts to grow, the generative cell of the pollen grain fuses with the tube cell, and its nucleus divides mitotically to form two sperm

Figure 32-4   How do flowers bring egg and sperm together? Each flower can have one or more pistils. In this lily flower, each pistil consists of three carpels. A pistil has a pollen-catching stigma on a stalk (the style), rising from a swollen ovary that makes ovules. After the arrival of sperm, the ovules become seeds. One diploid (2n) cell in each ovule undergoes meiosis to form four haploid (1n) cells. Three of the haploid cells degenerate. The remaining cell is the megaspore, which divides into eight haploid nuclei. This large multinucleate cell becomes a megagametophyte by dividing into three cells at each end and two polar nuclei within a "central cell" in the middle. At the micropyle end are the egg cell and two "synergids." Before fertilization, the pollen grain lands on the stigma and forms a pollen tube, which grows down the style and into the micropyle of the ovule. On the way, the generative cell within the tube divides to form two haploid sperm cells. One sperm fuses with the egg cell to form the zygote. The other sperm fuses with the two nuclei of the central cell to form a triploid cell that divides mitotically to form the triploid endosperm. The developing seed includes embryo, endosperm, nucellus, and integument.

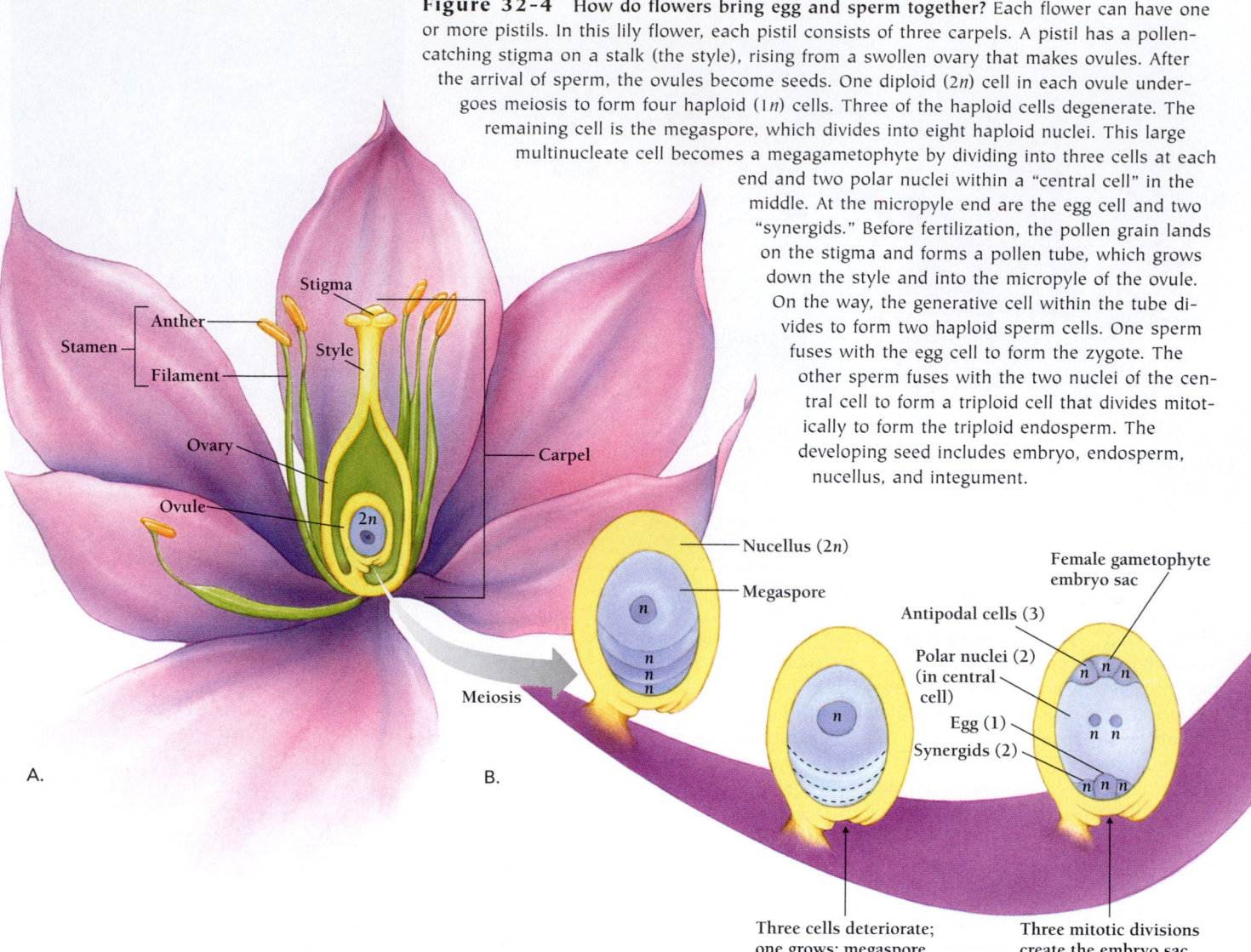

A.

B.

Nucellus (2n)

Megaspore

Meiosis

Female gametophyte embryo sac

Antipodal cells (3)

Polar nuclei (2) (in central cell)

Egg (1)

Synergids (2)

Three cells deteriorate; one grows: megaspore

Three mitotic divisions create the embryo sac

C.

nuclei. When the pollen tube arrives at the embryo sac within the ovule, the sperm cells are discharged.

Flowering plants differ from all other organisms in undergoing **double fertilization,** the simultaneous fusion of the two sperm nuclei with two nuclei in the embryo sac (Figure 32-4D). One sperm fuses with the egg to form the diploid zygote, which develops into an embryo and eventually into a plant. The other sperm fuses with the two central nuclei forming a triploid nucleus (with three sets of chromosomes). The triploid central cell divides mitotically to form the seed's **endosperm,** a triploid tissue that provides nourishment and hormones for the growing embryo.

Shortly after fertilization, both the zygote and the endosperm nucleus undergo repeated rounds of mitosis. The developing seed consists of an embryo, the endosperm, the nucellus, and the integuments. In most species, the endosperm is used up during development, and the mature, or ripe, seed has only the embryo, the nucellus, and the seed coat (derived from the integuments).

Flowering plants undergo double fertilization, the fusion of two sperm nuclei with two cells of the embryo sac to form (1) a diploid zygote and (2) a triploid endosperm.

How Does a Plant Form a Seed?

The pattern of cell divisions in the early embryo decides the positions of its cells and the overall shape of the embryo. A soybean embryo, for example, changes its shape from a sphere to a heart to a torpedo.

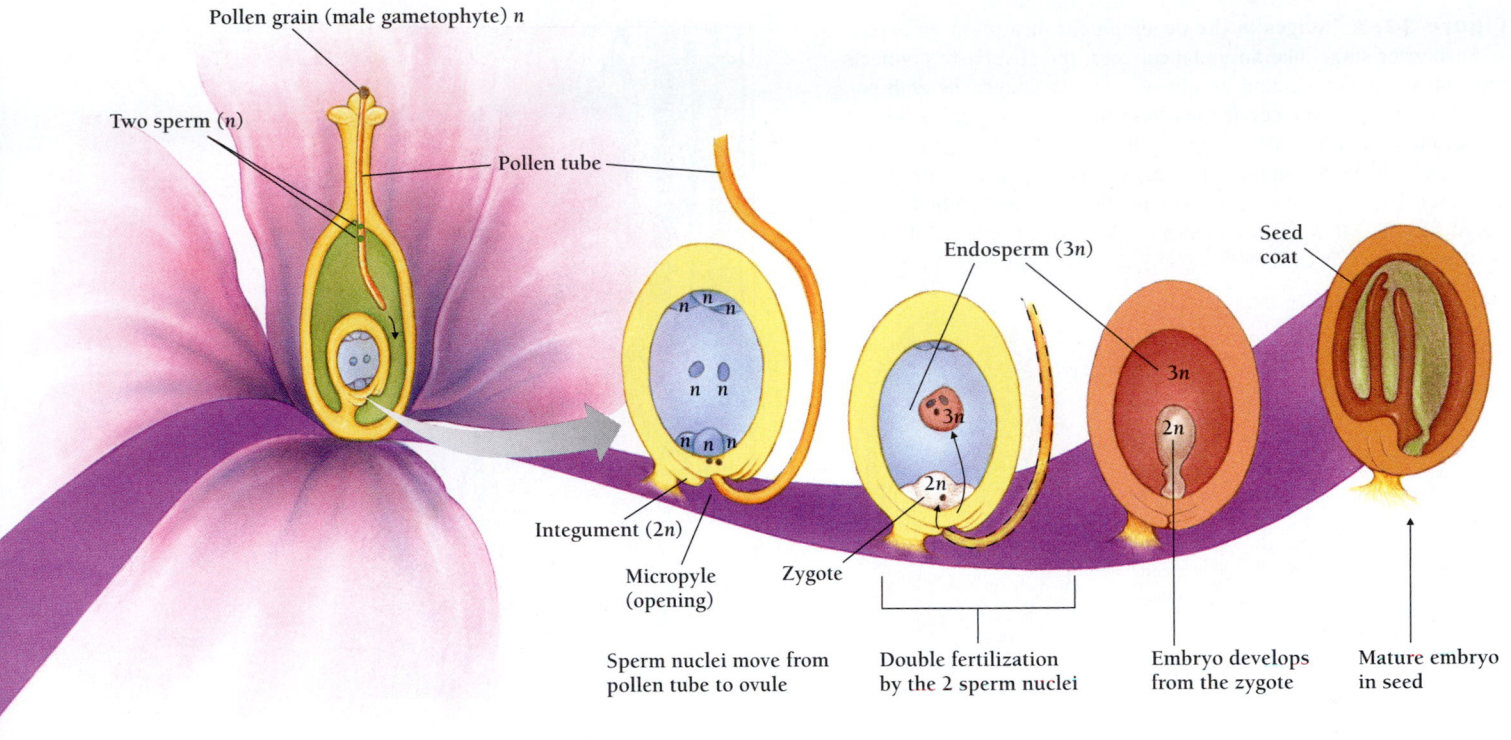

Pollen grain (male gametophyte) *n*

Two sperm (*n*)

Pollen tube

Endosperm (3*n*)

Seed coat

Integument (2*n*)

Micropyle (opening)

Zygote

Sperm nuclei move from pollen tube to ovule

Double fertilization by the 2 sperm nuclei

Embryo develops from the zygote

Mature embryo in seed

D.

E.

The first cell division in a soybean embryo divides the zygote into two cells of unequal sizes. The larger, basal cell divides several times to form a narrow column of cells. The *suspensor* consists of a large basal cell and the narrow stalk that, like an umbilical cord, attaches the embryo to the surrounding tissue and nourishes the embryo (Figure 32-5A).

The smaller cell divides to form a spherical embryo, one end of which forms a shoot apical meristem while the other end forms a root apical meristem (Figure 32-5B). The fundamental shoot-root polarity of the embryo is thus established early in plant development.

Continuing division of the spherical embryo leads to the formation of three regions: the tissues that later form the epidermis, the ground tissue system, and the vascular system.

Cells of the shoot apical meristem soon begin to grow and divide more rapidly, producing an embryo with a heartlike shape (Figure 32-5C). The continued division of cells in the shoot end produces "seed leaves," or **cotyledons.** The number of cotyledons divides flowering plants into two huge classes, the dicots and the monocots. **Dicots**—such as soybeans, roses, and most flowering trees—have two cotyledons, while **monocots**—such as palm trees, corn, and other grasses—have only one. The cotyledons absorb nutrients from the endosperm and later distribute them to the developing plant.

Above the attachment point of the cotyledons, the axis is called the *epicotyl* [Greek, *epi* = above], and below it is called the *hypocotyl* [Greek, *hypo* = under]. The first bud of the embryo, which includes the epicotyl and the first true leaves, is called the *plumule* [Latin, *plumula* = a small feather]. At the end of the hypocotyl is the **radicle,** a primordial root that contains the root's apical meristem. As the cotyledons and the radicle grow longer, they give the embryo a torpedo shape. Sometimes the cotyledons grow so long that they must curve backward to fit into the seed, forming a mature embryo, as in Figure 32-5D.

At this stage, most of the embryo's cells resemble the large, differentiated cells of a mature plant, except that their plastids do not yet contain chlorophyll. The meristem cells remain small and undifferentiated. The shoot apical meristem is nestled between the cotyledons, just above their points of attachment, while the root apical meristem lies near the point of attachment of the suspensor. In most plants, by the time the embryo is mature, the endosperm is fully consumed. Some plants retain the endosperm in the mature seed. In cereal grains, such as corn and wheat, most of what we eat is endosperm. In dicot plants such as peas, the developing plant derives nourishment from the cotyledons. In still other dicots, such as squash, the cotyledons nourish the seedling by photosynthesizing.

After the "torpedo" stage, the seed dries, and the seed coat hardens around the embryo. In most seeds, the embryo then enters a **dormant** [French, *dormir* = to sleep] stage, in which growth and development are suspended.

A seed contains a dormant embryo, with one or two cotyledons, surrounded by a tough seed coat.

Figure 32-5   Stages in the development of a plant embryo.
A. Suspensor stage: like an umbilical cord, the suspensor connects the embryo to surrounding tissues. B. Globular stage: the embryo as a sphere. C. As the cotyledons form the embryo's shape becomes more heartlike. D. As the cotyledons grow out, the embryo takes on a "torpedo" shape. The cotyledons may later grow so long that they must bend back to form the "mature" embryo, illustrated here. *(A, B, John D. Cunningham/Visuals Unlimited; C, D, Jack M. Bostrack/Visuals Unlimited)*

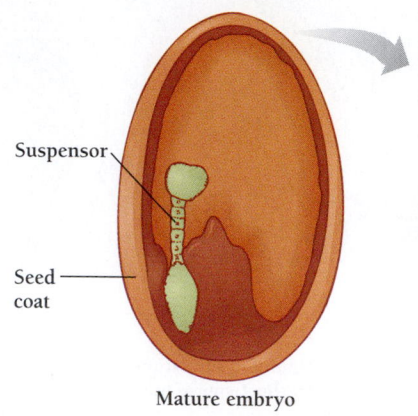

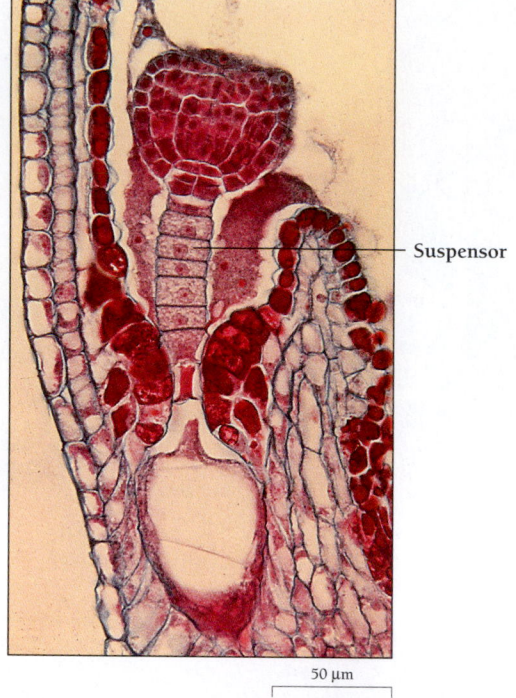

A.

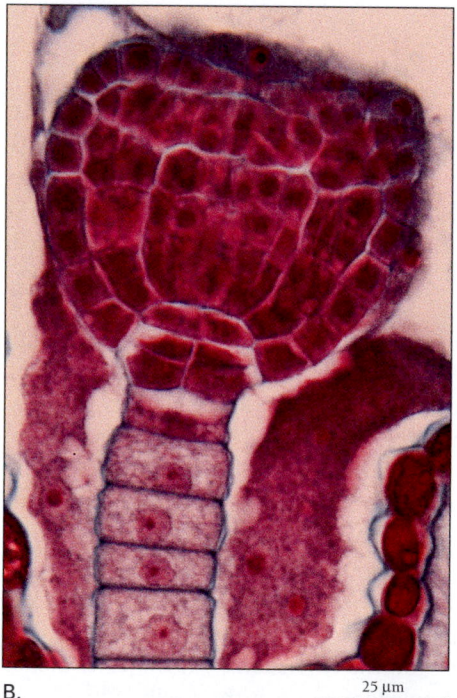

B.

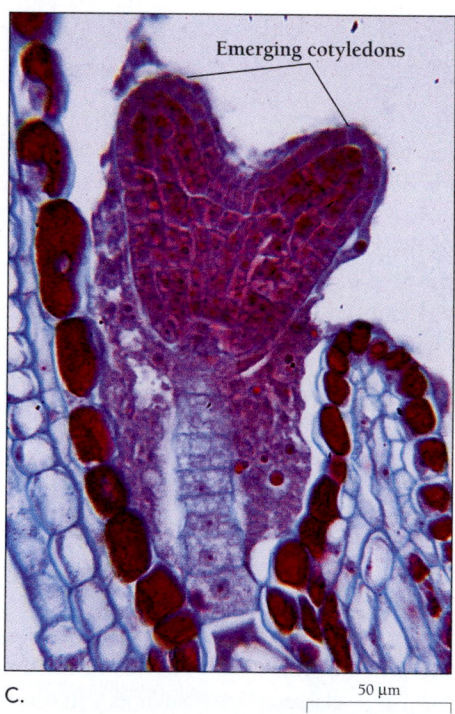

C.

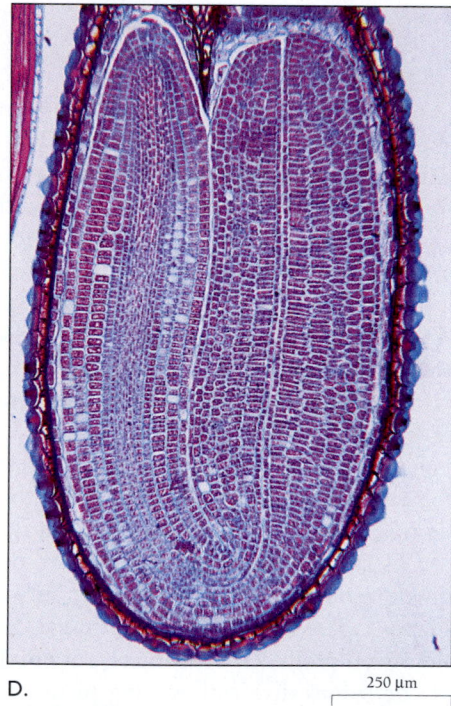

D.

How Does the Embryo Resume Development?

The resumption of growth is called **germination.** The first stage of germination is **imbibition** [Latin, *imbibere* = drink], the swelling of the seed with water. During imbibition, the seed may double or triple in size. Once the dormant embryo is hydrated, it awakes, with a burst of metabolic activity, cell division, and growth.

The first part of the embryo to break out of the seed coat is the radicle (Figure 32-6). The radicle immediately turns downward into the soil and forms a functioning root system that provides the growing plant with water and minerals. In the soybean and many other dicots, the top of the hypocotyl

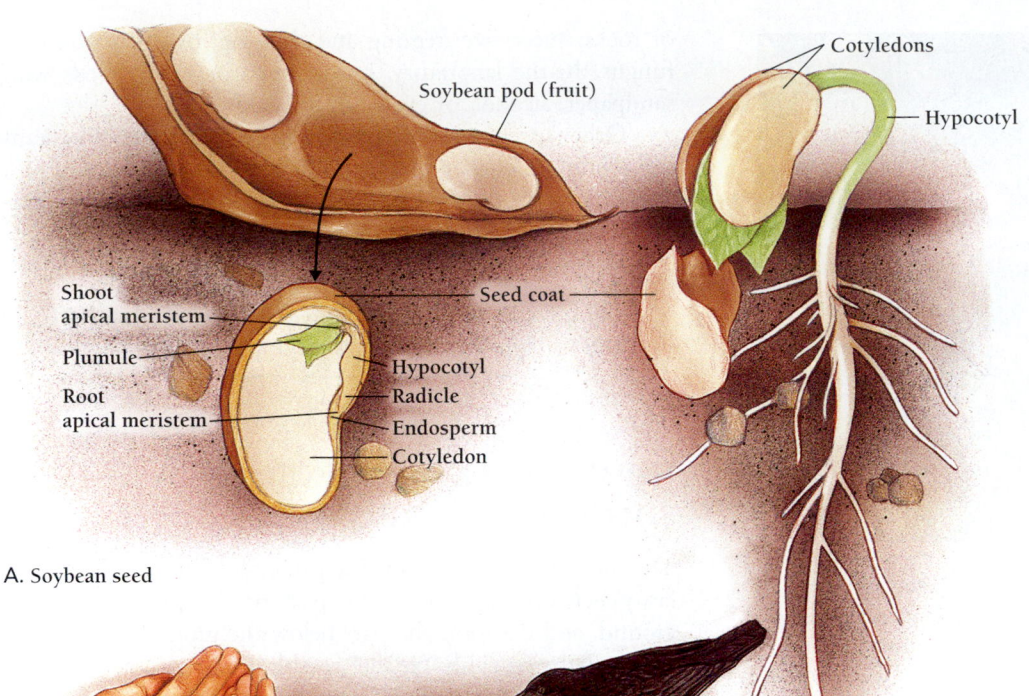

Soybean pod (fruit)

Cotyledons

Hypocotyl

Shoot apical meristem

Plumule

Root apical meristem

Seed coat

Hypocotyl

Radicle

Endosperm

Cotyledon

A. Soybean seed

Figure 32-6 Germination in a dicot and a monocot. A. When a soybean germinates, the radicle breaks out of the seed coat and turns down into the ground. When the shoot emerges into light, the bent hypocotyl straightens and the two cotyledons unfold. B. In corn, the shoot grows straight out of the kernel through the tube of the coleoptile and the root grows down through the tube of the coleorhiza.

Leaf (plumule)

Endosperm

Cotyledon (scutellum)

Coleoptile

Coleoptile

Shoot apical meristem

Coleorhiza

Coleorhiza

Radicle

Root apical meristem

B. Corn seed

How Does the Embryo Obtain Nourishment?

Cells in the dormant embryo are not dead; they merely maintain a low level of metabolism. Germination, however, triggers a huge increase in biochemical activity, with energy and building blocks derived from stored starches, oils, and proteins. The presence of these stores also means that seeds are rich foods for a wide variety of animals, including ourselves (Figure 32-7).

The breakdown of these stores requires the action of enzymes that are not present (or are present in much lower amounts) in the ungerminated seed. In cereals such as corn, barley, and wheat, these enzymes are secreted by the **aleurone layer,** an outer layer of cells that surrounds the endosperm. After germination begins, the aleurone layer secretes enzymes that digest the starches and other storage molecules of the endosperm. One of these enzymes is **α-amylase,** which helps to break down starches stored in the endosperm.

The embryo itself must stimulate the aleurone layer to make digestive enzymes. If researchers remove the embryo, the aleurone layer secretes no enzymes and the endosperm sits undigested. The chemical signal that causes the aleurone layer to make digestive enzymes is the plant hormone gibberellin. As little as 10^{-11} grams (one hundred-billionth of a gram) of gibberellin induces the expression of the gene that encodes α-amylase.

forms a hook that pushes up through the soil toward the sun, pulling the cotyledons and epicotyl behind (Figure 32-6). Once above ground, the arching hypocotyl hook straightens, the green cotyledons spread out in the sun, and the epicotyl points upward and resumes its growth. In other seeds, such as peas and corn, the hypocotyl doesn't grow, and the cotyledons stay underground. In these cases, the epicotyl grows to push the plumule and apical meristem through the soil.

At germination, the seed swells with water, and the radicle breaks out of the seed coat and grows down into the soil. Development resumes with a burst of metabolic activity.

At germination in cereals, the aleurone layer secretes enzymes such as α-amylase that release nutrients to the growing embryo.

Figure 32-7   The size of the endosperm and the resulting seed varies widely among species. Lettuce and tomato seeds are about 1 to 2 mm in diameter, while peas, beans and corn are about ten times larger. The smallest known seeds are the microscopic seeds of orchids, and the largest are the "double coconuts" of the Seychelles nut palm. In general, the smaller the seeds, the more of them a plant can produce. Plants that produce very large seeds have only energy enough to make a few. These few large seeds, however, generally have a good chance of surviving. *(D. Cavagnaro/Visuals Unlimited)*

What Environmental Cues Trigger Germination?

We can see the obvious effect of environment on plant growth by keeping a germinating dicot seedling in the dark. Instead of straightening, the hooked stem keeps its hook as it grows upward. The plant displays a condition called *etiolation* [French, *etioler* = to blanch or whiten], with a thin, spindly appearance, poor leaf development, and little or no chlorophyll production (Figure 32-8).

As a seed, a plant embryo can suspend growth for days, months, years, decades, even centuries, depending on the species. Radioactive dating of lotus seeds from a former swamp bed in Manchuria, China, showed that they were about 1000 years old. Yet they still germinated. Still more amazingly, researchers have succeeded in growing plants from lupine seeds that were found in 10,000-year-old frozen sediments in Alaska. Botanists have concluded that seeds do not have a built-in clock that determines the time of germination, but instead depend on cues from their environment.

Plants use a wide variety of mechanisms to prevent premature germination. In many legumes, for example, a hard seed coat prevents the entrance of water or oxygen needed for germination. In these and other species, germination often requires *scarification*, harsh treatment to break the seed coat barrier. In nature, scarification conditions include fire; acid and enzymes in an animal's digestive system; scraping against sand

or rocks; successive freezing and thawing; or the action of a fungus. In the laboratory, researchers can scarify seeds with sandpaper, alcohol, or even sulfuric acid.

Other species prevent premature germination with chemical inhibitors, such as sodium chloride, cyanide, mustard oils, and other organic compounds. Rainfall then stimulates germination by washing away the inhibitors.

Light, water, temperature changes, stomach acid, and other environmental cues can all trigger germination.

HOW DO PLANTS GENERATE A SHOOT AND A ROOT?

The most important organizing principle of a plant is the polarity between the shoot, the part of the plant above the ground, and the root, the part below the ground. The shoot-root polarity, roughly parallel to the head-tail polarity of an animal, is established early in the plant embryo.

Just behind the tip of both shoot and root are **apical meristems,** groups of *undifferentiated* cells that generate the specialized cells of the shoot and root (Figure 32-9). Undifferentiated cells are cells that resemble those in the embryo. Meristem cells can either divide, to give rise to more meristem cells, or differentiate.

An apical meristem can repeatedly generate a set of plant parts, which serves as a kind of construction module. Each module, similar to segments in animals, may be repeated any number of times, depending on the conditions of plant growth.

A shoot apical meristem can generate stem segments, leaves, and buds. (Later in life, it may also generate a flower.)

Figure 32-8   Etiolation. Bean plants grown in the dark are long and spindly compared to those grown in normal light. *(Blanche C. Haning)*

Figure 32-9   The root and shoot each have an apical meristem. The anatomy of a flowering plant demonstrates a regular repetition of meristem-derived modules. Each node, internode, attached bud, and leaf constitute a module.

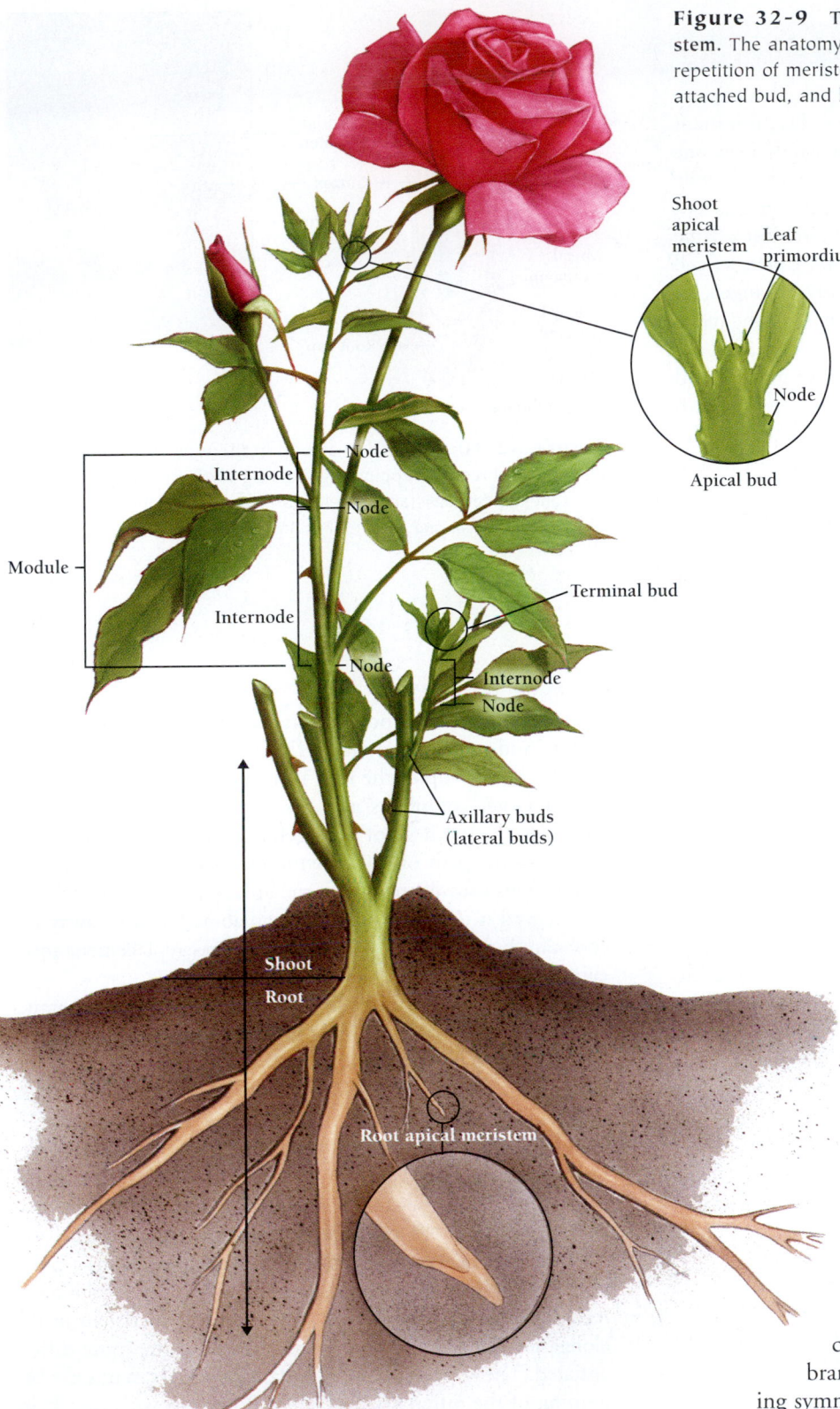

with its own apical meristem, and a few tiny leaves below the meristem. In both shoot and root, the action of the apical meristems is responsible for the plant's primary growth.

Each species of plant has a characteristic pattern of modular repetition. Modular organization, fixed repetition rules, and controlled outgrowth of buds underlie the branching patterns of stems, as well as the pleasing symmetry of many flowers and fruits. The modular organization of plants allows plants great flexibility in responding to environmental changes.

The region of the stem to which the leaf attaches is called a **node,** and the region between nodes is an **internode** (Figure 32-9). If we look from node to node, we see repeating modules, each consisting of a segment of stem (the internode), a leaf, a node, and a bud. A **bud** contains another shoot tip,

Plants repeatedly generate modular structures at the apical meristems at the ends of the shoot and root.

How Do Meristem Cells Elongate and Differentiate?

In both shoot and root modules, undifferentiated meristem cells themselves are relatively small, with thin walls, prominent nuclei, and small vacuoles. When a meristem cell divides, one daughter cell stays in the meristem and keeps dividing, while the other cell is pushed out of the meristem into an adjacent region. There it can grow—mainly by taking water into a large central vacuole—and differentiate into any of a variety of cell types. Elongated differentiating cells, already arranged in columns, may be up to 50 times larger than those of the meristem.

Unlike animal cells, plant cells cannot move around and form new associations during development. After division ceases, they can only expand and sometimes bend. Plant development therefore depends heavily on orderly cell division and cell expansion. Every cell has its place. As cells divide within the meristem, the planes of cell division and rates of expansion establish future spatial relationships.

Cell expansion depends mainly on the osmotic influx of water into the central vacuole. Cellulose in the cell walls is laid down in such a manner that the cell wall tends to be weaker in one direction than in others. As a result, as the vacuole swells, the cells elongate in that direction, rather than grow bigger in all directions. The exact orientation of cellulose fibers, which determines the direction of plant growth, depends on a number of plant hormones. Auxin and ethylene, for example, influence the direction of microtubules bound to the plasma membrane. The oriented microtubules in turn determine the pattern of cellulose deposition in the cell wall.

Because of their effects on the patterns of cell division and expansion, plant hormones can dramatically influence the growth of an entire plant. Ethylene, for example, causes a longitudinal orientation of the cellulose fibers. The parallel cellulose fibrils constrain cell expansion to the horizontal direction, leading to the production of short, fat seedlings. In contrast, auxin stimulates the transverse orientation of cellulose fibrils, leading to tall, thin shoots.

Although most of this discussion has centered on the growth and the differentiation of cells in the shoot, the same general considerations apply to the growth of the root. Nonetheless, the growth of roots and shoots have several distinctive characteristics.

The development of form in plants depends on oriented cell division and cell expansion.

Root Tips Grow and Mature in Three Zones

Primary growth of roots occurs almost exclusively near their tips. The growth and maturation of root tissues is complex. In general, however, the growth zone has four overlapping re-

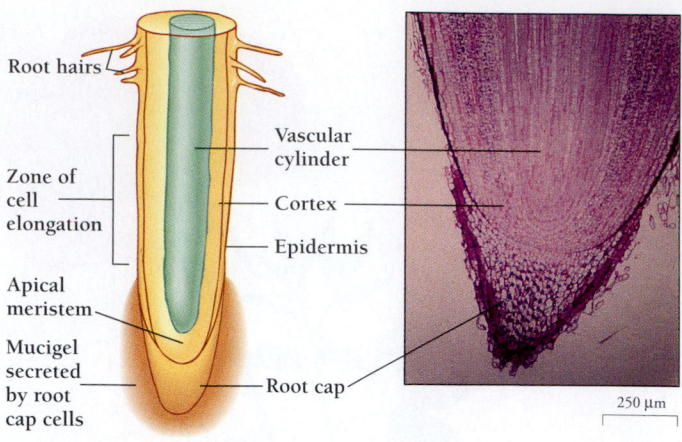

Longitudinal section of a root tip

Figure 32-10   Growth in a root. In the root, plant biologists distinguish three overlapping regions of growth—the root cap and the zones of cell division, cell elongation, and cell differentiation. *(Photo, Kevin and Betty Collins/Visuals Unlimited)*

gions—the root cap and the zones of cell division, cell elongation, and cell differentiation (Figure 32-10).

At the very tip of the root is the **root cap,** whose cells are full of Golgi complexes and vesicles. These organelles contribute to the synthesis of **mucigel,** a polysaccharide slime that lubricates the path of the growing root through the soil. The root cap also serves to protect the apical meristem, which lies just behind it. As the root grows, soil abrasion wears away the root cap, but the apical meristem produces replacement root-cap cells, also capable of secreting mucigel.

We can demonstrate that the apical meristem contains most of the dividing cells in the root. If we allow growing roots to take up a radioactive precursor of DNA (^{3}H-thymidine), almost all of the labeled DNA shows up in the meristem, where mitosis is occurring.

The cells within the meristem have a small, uniform size, but the pattern of cell division produces files of cells whose organization already resembles that of the mature root. Farther back along the root's axis is the **elongation zone.** The cells in each file are increasingly large. Cells farther from the tip are longer than those nearest the meristem, which have arisen more recently. Cells that are farther away from the meristem in the elongation zone are more mature and, therefore, more differentiated. The appearance of abundant root hairs marks the beginning of the differentiation zone. Root cells with root hairs are fully mature.

Growing root tips include three zones: the slimy root cap, the apical meristem, and the elongation zone.

Primary Growth of Shoots Is More Complicated Than That of Roots

In the shoot, the apical meristem must produce stem, leaves, branches, and flowers, so primary growth is more complicated than in the root. The region of primary growth may extend back as much as 10 to 15 cm behind the shoot's tip and may include several nodes (Figure 32-9). As in the root, the stem apical meristem contains most of the dividing cells. Cells produced in the meristem form files of cells that grow longer, pushing the meristem upward. The cells that are farthest away from the meristem are the most mature and the most differentiated.

Each leaf originates as a **leaf primordium**, a tiny extension of the apical meristem. Each primordium contains meristem cells, which together divide, expand, and differentiate to form a leaf. A bud, as we said above, is a shoot tip with an apical meristem and several young shoot modules that it has produced. Each module contains a node, a very short internode, and one or more leaf primordia whose cells have not elongated (Figure 32-9).

In addition to the active terminal shoot tips, most plants have **axillary buds,** which form just above the points where leaves join the stem and which can develop into branches. Like the terminal shoot tips, the axillary buds contain an apical meristem.

Many species show **apical dominance,** meaning that the presence of terminal shoot tips keep nearby axillary buds dormant (inactive). The terminal shoot tips do this by releasing a plant hormone. Axillary buds near the tip of a plant are more inhibited than those farther away. That is why branches near the bottom of a conifer tree, for example, are much longer than those at the top. Removing the **terminal bud** allows the axillary buds to grow, and many gardeners routinely produce bushier plants by pinching off the terminal bud.

Each leaf starts as a tiny extension of the apical meristem. Terminal buds suppress the growth of axillary buds, an effect called apical dominance.

How Does a Flower Derive from a Specialized Bud?

Some of the most interesting questions being addressed by contemporary plant biologists concern the molecular and cellular bases of flower development. These questions are important for both intellectual and economic reasons. Plant biologists want to understand just how plants regulate these beautiful patterned organs. And farmers (and seed companies) want to know to manipulate seed production, for example, by producing flowers that are exclusively male or female. In the last decade, plant biologists—with the help of genetics—have made major advances.

To explore pattern formation in the flower, we need a clear picture of this organ's parts and developmental origins. Some buds are specialized to form flowers or a flowering shoot. A flower consists of up to four sets of parts arranged in **whorls,** concentric circles of modified leaves that develop into specialized flower parts (Figure 32-11A). The outer two whorls, #1 and #2 in Figure 32-11A, are the **sepals,** the usually green parts at the base of a flower, and the **petals,** the usually showy, colored parts of the flower. Whorl #3 includes the male stamens, and whorl #4, the innermost whorl, includes the female pistils, each containing one or more carpels. The two inner whorls are said to be fertile, because they ultimately produce gametes—egg cells in whorl #4 and sperm cells in whorl #3.

Botanists have long noted that not all flowers have all four whorls. A flower that contains all four whorls is said to be a *complete* flower, while an *incomplete* flower is missing one or more whorls. A flower that contains both carpels and stamens (and produces gametes of both sexes) is said to be *perfect*. A flower missing either stamens or carpels is said to be *imperfect*. The primroses in Figure 32-11A are complete and therefore perfect flowers.

In contrast, the corn flowers in Figure 32-11B and C are incomplete and imperfect. Numerous male flowers make up the tassel-like corn flowers at the top of the plant. These flowers have stamens but no carpels. Numerous female flowers, in the "ears," have carpels but no stamens. Corn is an example of a *monoecious* [Greek, *monos* = single + *oikos* = house] species, in which separate male and female flowers appear on the same plant. Willow trees and date palms are *dioecious* [Greek, *di* = two], with male and female flowers on different plants. (If we were to use this terminology for animals, we would say that humans are dioecious and snails are monoecious.) Any plant with perfect flowers (with both stamens and carpels) is neither monoecious nor dioecious.

The generalized flower develops from four circles, or whorls, of primordia. These form, from outside to inside, sepals, petals, stamens, and carpels. Many flowers are missing one or more of these parts.

How Do Genes Regulate the Development of Flowers?

In the last few years, major insights have come from the genetic analysis of plant mutants, particularly of the humble weed *Arabidopsis thaliana*. Just as the genetic analysis of *Drosophila* and of *Caenorhabditis elegans* has revolutionized the understanding of animal development, so, it seems, that similar studies of *Arabidopsis* will lead to a new view of plant development. Many of the cellular decisions in plant development surprisingly resemble corresponding decisions in animal development, depending, for example, on regulation of gene expression by specific transcription factors.

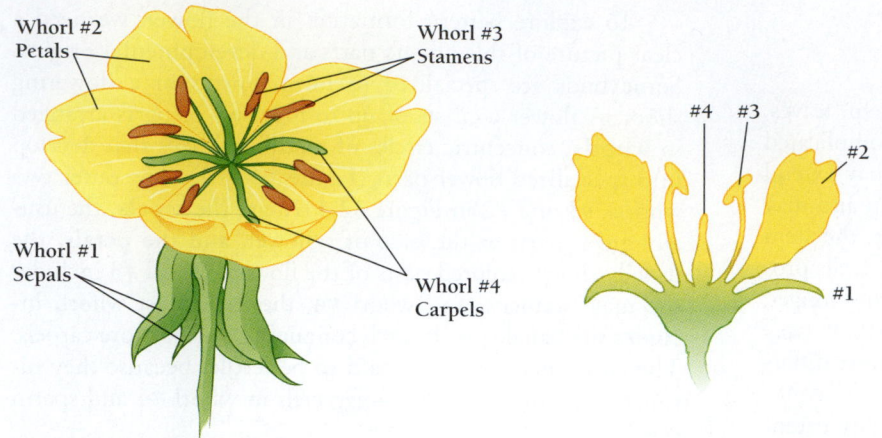

Whorl #2
Petals

Whorl #3
Stamens

Whorl #1
Sepals

Whorl #4
Carpels

#4 #3
#2
#1

Figure 32-11   The four whorls of flowers.
A. A flower consists of up to four whorls of modified leaves. These are the sepals, the petals, the stamens, and the carpels. A flower with both carpels and stamens is perfect. A flower with all four whorls is complete. The evening primrose in A is complete and also perfect. B and C. Corn flowers possess two kinds of flowers on each plant, neither of which is complete or perfect. One set of flowers, the tassels, lack carpels, and the other set, the silk, lack stamens. (*B, William E. Ferguson*)

A. The four whorls of a complete, perfect flower—the evening primrose

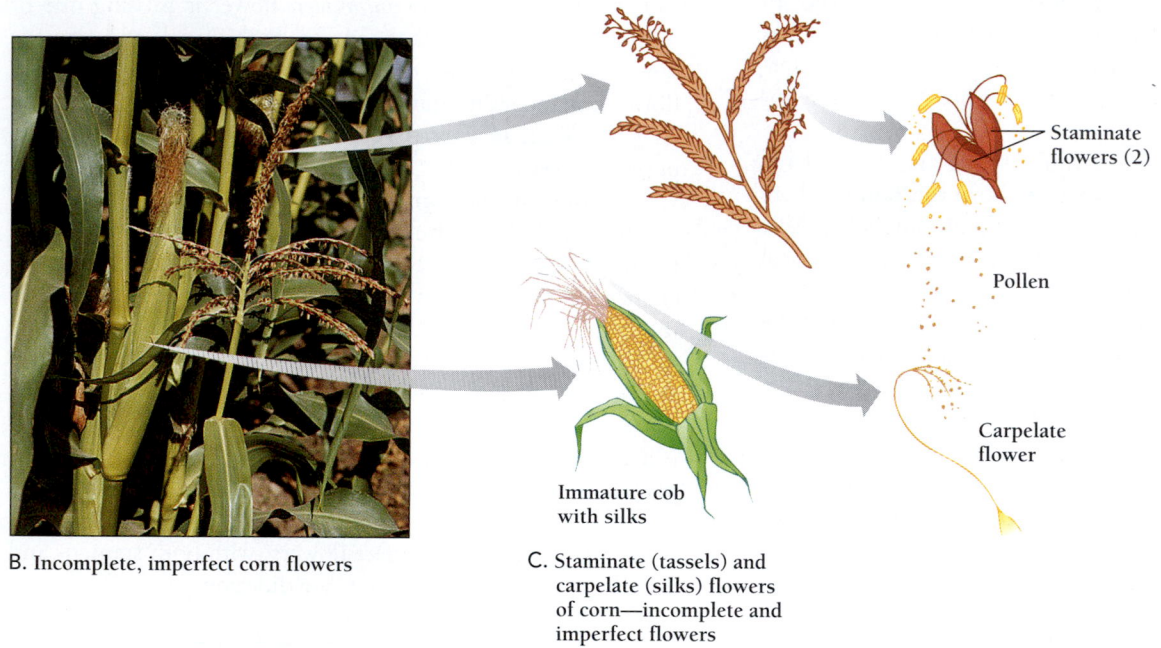

Staminate
flowers (2)

Pollen

Carpelate
flower

Immature cob
with silks

B. Incomplete, imperfect corn flowers

C. Staminate (tassels) and carpelate (silks) flowers of corn—incomplete and imperfect flowers

Box 32.1 shows an abnormal *Arabidopsis* flower that results from a homeotic mutation, a mutation in a gene that influences the location where organs form. **Homeotic selector genes** are striking because the same sequences of DNA seem to regulate the development of a great diversity of organisms. Homeotic selector genes in fruit flies, for example, also regulate the development of the body in vertebrates (Chapters 20 and 44). And genes with similar sequences and functions have been found in plants, vertebrates, and yeast cells. Despite great differences between plants and animals, both share common strategies and—incredibly—even common molecular designs for regulating the development of form.

In the *Arabidopsis* homeotic mutation *agamous,* the parts that would usually develop into stamens develop into petals, and carpels are completely missing. Similarly, the *apetala* 3 mutation results in flowers with carpels instead of stamens and

sepals instead of petals, and *apetala* 2 converts sepals into carpels and petals into stamens. Each of these homeotic selector genes changes the fate of one or more whorls.

Plant homeotic selector genes probably serve the same roles in all or most plants. Researchers have already found that similar genes serve the same roles in flower development in both *Arabidopsis* and snapdragons.

All of the homeotic selector genes for flower development discovered so far fall into one of three classes. To understand the action of these genes, researchers created a "triple mutant." This plant does not express any of the genes known to determine the identity of flower parts. The result is a "flower" made entirely of leaves. If any of the three homeotic selector genes is expressed, however, the whorls then develop into flower parts. Researchers interpret this result to mean that meristems produce leaves unless regulatory genes cause them to form

Why Is *Arabidopsis thaliana* Well Suited to Genetic Analysis?

Arabidopsis thaliana, a small, common weed, also called "wall cress," is so well suited to genetic studies that it has been called the "botanical *Drosophila*." The major advantage *Arabidopsis* has over most plants is that it can be grown from a seed to a mature plant in as little as 5 weeks. In addition, *Arabidopsis* can self-fertilize, making it much easier to isolate homozygous strains than in *Drosophila*. *Arabidopsis* is small, usually less than 30 cm tall, and spindly enough that many plants can be grown easily and inexpensively in a small area. Its seeds are so tiny that 35,000 seeds occupy about 1 milliliter.

Geneticists have already identified more than 100 mutations in *Arabidopsis*. Some of these mutations lead to specific enzyme deficiencies, others to alterations in response to plant hormones, and still others to alterations in pattern (Figure A). For example, one mutation leads to a missing shoot, another to a missing stem, another to a missing root. Mutations affect mature plants as well as seedlings. One mutant completely lacks a shoot meristem, both in the intact plant and in tissue culture. Other *Arabidopsis* mutations affect the formation of specific types of tissue, while still others prevent the proper establishment of the embryo shoot-root axis.

Once a mutation has been found, it is a relatively straightforward—if time-consuming—task to identify and sequence the gene in which the mutation lies. The function of the gene can then be studied by putting it into normal or mutant *Arabidopsis*. Researchers have performed such studies for a relatively small number of genes but have already learned much about *Arabidopsis* development.

One of the additional appeals of *Arabidopsis* is the small size of its genome, only 120 million nucleotides. Most of this DNA has already been sequenced, and the complete sequence is expected by the end of the year 2000. Knowledge of the sequence will improve the ability of plant biologists to understand not only how specific homeotic selector mutations give rise to altered flowers but also how flowering plants coordinate their responses to environmental cues.

(a)

(b)

Figure A Developmental mutations in the flower *Arabidopsis thaliana*. (a) A normal flower with all four whorls: sepals, petals, stamens, and carpels. (b) The *apetala 2* mutation converts sepals into carpels and petals into stamens. *(Leslie Sieburth, McGill University, Montreal)*

something else. They say that leaf development is the "default pathway" of meristem development.

The homeotic selector genes (also called "MADS" genes) provide striking examples of the influence of genes on plant development. Researchers have found that many of these genes specify proteins ("transcription factors") that regulate the expression of other genes (Chapter 11). This work is rapidly uncovering a cascade of events that regulate flower development.

Homeotic selector genes help regulate the development of flower parts in Arabidopsis.

Secondary Growth Depends on Two Kinds of Lateral Meristems

The tallest and arguably the most magnificent of Earth's organisms are the redwoods (which are not flowering plants but gymnosperms). These magnificent trees, as well as woody angiosperms, require both secondary and primary growth.

Secondary growth refers to the thickening, rather than the lengthening, of stems, branches, and roots. Secondary growth occurs not in the apical meristems but in two kinds of lateral meristems, or **cambia** [singular, cambium]. These are the **vascular cambium**, which produces the secondary xylem and phloem, and the **cork cambium**, which gives rise to *cork,*

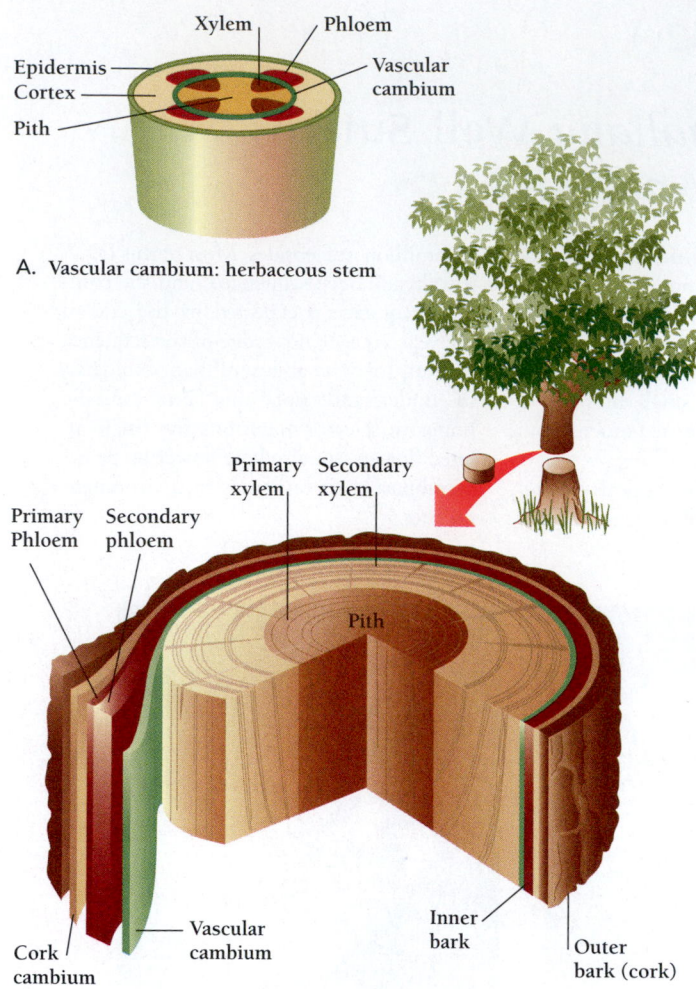

A. Vascular cambium: herbaceous stem

Xylem   Phloem
Epidermis
Cortex   Vascular cambium
Pith

Primary   Secondary
xylem   xylem
Primary   Secondary
Phloem   phloem

Pith

Cork   Vascular   Inner   Outer
cambium   cambium   bark   bark (cork)

B. Vascular and cork cambium: woody stem

These are cells and cell walls
of spring and summer growth rings.

250 μm

Ring 1   Winter   Ring 2   Winter

C.

Figure 32-12   Secondary growth. A. In a herbaceous stem, the cylinder of vascular cambium produces primary xylem (to the inside) and primary phloem (to the outside). B. In a woody stem, the vascular cambium produces secondary xylem (to the inside) and secondary phloem (to the outside). Everything outside of the vascular cambium is bark—including the secondary phloem, the cork cambium, and the periderm. C. Cross section through a woody stem showing two growth rings, each consisting of spring wood (big cells) and summer wood (small cells). The dark line at the outer edge of each ring indicates lack of growth during the winter. *(C, William E. Ferguson)*

the waterproof, insect-proof covering on woody stems and roots (Figure 32-12). The cells of both kinds of cambia lie in open-ended cylinders that extend almost the entire length of woody stems and roots. Only the tips of woody stems and branches that are still undergoing primary elongation lack cambia. **Bark** includes everything outside the vascular cambium, including phloem, cork, storage tissues, and the stem surface.

The vascular cambium lies between the xylem and the phloem. The vascular cambium adds secondary xylem to the outside of the primary xylem and secondary phloem to the inside of the primary phloem. In a woody stem, such as that of a tree, this process continues year after year. Each year the vascular cambium lays down new xylem outside the old and new phloem inside the old. The xylem accumulates from year to year as wood (Figure 32-12B). Among the most useful products of secondary growth are lumber, pulp for paper, cork, and root vegetables such as carrots, beets, turnips, and radishes.

As the trunk of a tree expands, it tears and destroys the mature epidermis. Just inside the epidermis, however, is the

cork cambium, a layer of meristem cells that produce rows of protective cork cells. These tissues grow both outward, toward the surface of the tree, and inward, toward the phloem. The cells toward the outside deposit waxy, fatty suberin and other secondary compounds in their cell walls. The cells then die and the resulting waterproof cell walls (cork) protect the tree from dehydration and attacks by herbivores and pathogens. As the tree grows, cork, cortex, and older phloem are torn and crushed, so the living tissue in the bark makes up only a thin layer, a few millimeters thick. New inner layers repeatedly provide a continuous sheath of living and dying cells. We can identify different tree species by the characteristic pattern of tears in the bark—including the smooth strips of the *Eucalyptus,* the rough bark of the oak, and the regular cracks in the bark of ponderosa pine.

Secondary growth changes from season to season. In temperate zones, the vascular cambium is most active in the spring, producing large tracheids and vessel elements. In the summer, activity slows, and tracheids and vessel elements are much smaller. As a result, summer wood is much denser and darker than spring wood. The difference in appearance between the dark summer wood and the lighter spring wood creates a distinct layer of added vascular tissue called a **growth ring.** Each ring represents one season, or year, and the number of rings in a tree trunk reveals the age of the tree.

Secondary growth, which depends on vascular cambium, changes from season to season, resulting in characteristic growth rings in trees and other woody perennials.

MERISTEMS MAY PRODUCE WHOLE PLANTS

The same meristems that produce primary and secondary growth can also produce whole new plants. In a process called **vegetative reproduction,** plants reproduce asexually, without fertilization or seeds. New strawberry plants form from runners, for example, and the houseplant *Kalanchoë diagremontiana* forms tiny new plants on its leaves.

Many people deliberately propagate their favorite houseplants by means of vegetative reproduction. In some species, a stem cut from the main plant (a *cutting*), placed in water, generates new roots. These roots are said to be *adventitious* [Latin, = not properly belonging to], because they do not develop from a root apical meristem.

Some species that usually do not form such adventitious roots do so in the presence of the plant hormone auxin. Amateur and professional gardeners often dip cuttings into a solution of auxin to enhance root formation. When the rooted cutting is planted, it grows into a clone—a mature plant that is genetically identical to the parent plant. In fact, farmers can use such cuttings to produce whole crops that are genetically identical.

One of the oldest and most commonly used means of artificial vegetative propagation is grafting. Horticulturists propagate several crops almost exclusively by grafting, including oranges, grapefruit, lemons, and limes; avocados, apples, and plums; roses; and grapes (Figure 32-13). In **grafting,** a horticulturist grafts a cutting of one woody plant onto the root of another plant. The stem cutting is called the *scion* and the root is called the *rootstock* (or simply the *stock*). Recall from Chapter 31 that French grape varieties are universally grafted onto American rootstocks. Rather than try to breed a single hybrid with the best grapes and the best roots, grape growers simply graft a variety that makes good grapes onto a rootstock that resists pests or supplies nutrients well.

Most commercial fruit trees and roses are the products of such grafts. The trick in grafting is to bind scion and stock so that their cambia come together. Secondary growth establishes continuity both within the xylem and the phloem of scion and

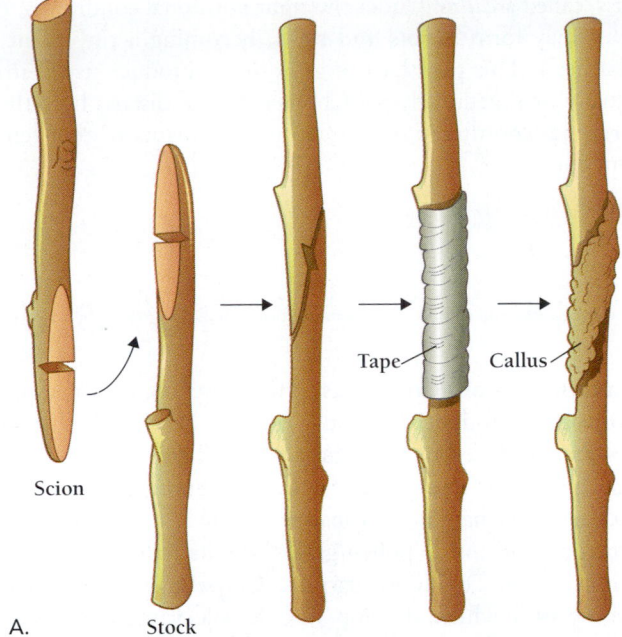

A. Scion Stock Tape Callus

B.

Figure 32-13 **Grafting allows a scion of one variety to grow on the stock of another.** A. The process of creating a graft. B. A tree with many citrus fruits growing on the same rootstock. *(B, A.B. Joyce/Photo Researchers, Inc.)*

stock. The buds, leaves, flowers, and fruits all derive from the scion, while the root system derives entirely from the stock.

Another method of vegetative propagation is through tissue culture. Tissue culture has become increasingly important with the advent of genetically engineered plants. Carrots, cotton, tomatoes, and Douglas fir trees have all been propagated this way.

The starting material for plant tissue culture is a block of tissue taken from a plant and placed on a sterile nutrient medium that contains sugar, minerals, and hormones. The cells grow and divide to make an unorganized, undifferentiated mass, called a *callus*. Under the right hormonal conditions, the callus may form shoots and roots, becoming a tiny plant (a "plantlet"). This plantlet can grow and reproduce itself either sexually or vegetatively. In Chapter 33, we discuss how these hormones coordinate other responses of plants to their environments.

Meristems allow plants to reproduce themselves asexually in a process called vegetative reproduction that does not involve fertilization or seed production. Cuttings, grafting, and tissue culture allow the artificial propagation of desirable plant varieties.

Study Outline with Key Terms

In **alternation of generations** in flowering plants, the diploid **sporophyte** produces haploid spores, and the haploid **gametophyte** produces haploid gametes. A flower contains both male and female reproductive structures. **Stamens** (male structures) produce haploid **microspores,** which develop into male gametophytes, inside pollen grains. **Pistils** (female structures) consist of one or more carpels. **Carpels** produce haploid **megaspores,** which develop into female gametophytes, the **embryo sacs.** The male and female gametophytes in turn produce male and female gametes. Each carpel contains a **style,** which connects the ovary to the **stigma.** Each **ovary** (made of one or more carpels) produces one or more **ovules,** which form the embryo sac, the **nucellus,** and the **integuments** of the developing seed.

Upon pollination, a **pollen grain** develops into a **pollen tube** with two sperm cells. Flowering plants undergo **double fertilization,** in which the two sperm nuclei fuse with two cells of the embryo sac within the carpel. One fertilization event produces the zygote, which grows into an embryo. The other gives rise to the **endosperm,** a triploid tissue that nourishes the growing embryo.

The zygote divides to form an embryo that first resembles a sphere, then a heart, then a torpedo. Development stops after the seed matures, when the embryo loses most of its water and becomes **dormant.** A seed consists of a dormant embryo and its adjacent endosperm, surrounded by a seed coat.

The resumption of growth is called **germination,** a process that begins with **imbibition,** the taking in of water. The first part of the embryo to emerge from the seed is the **radicle,** which contains the root's apical meristem and quickly forms a functioning root system. The shoot apical meristem lies above the point of attachment of the **cotyledons,** or seed leaves. Flowering plants are divided into two classes—the **dicots** and the **monocots**—based on the number of cotyledons.

A seed may maintain its dormant embryo for many years, until the proper environmental changes bring about germination. Germination triggers a huge increase in biochemical ac-

tivity, for which the growing plants derive energy from stored starches, oils, and proteins. In grains, **α-amylase** and other enzymes secreted by the **aleurone layer** break down starches in the endosperm.

The primary growth of both roots and shoots depends on **apical meristems.** The growing root tip has three distinct zones: the **root cap,** which secretes **mucigel;** the apical meristem; and an **elongation zone.** The growing tip of a stem, however, must be able to produce stem, leaves, branches, and flowers. Each leaf originates as an extension of the apical meristem, called the **leaf primordium.** At each **node,** one or more **buds** each contain several leaf primordia. In between the nodes are the **internodes.** A bud may lie at the tip of a shoot, in which case it is called a **terminal bud,** or just above the points where leaves join the stem, in which case it is called an **axillary bud.** Many species show **apical dominance,** meaning that terminal buds suppress the development of axillary buds.

The four **whorls** of a flower may form carpels, stamens, **sepals,** and **petals.** Mutations in several **homeotic selector genes** in *Arabidopsis* can result in interesting deformities such as extra petals, but no stamens. Similar genes are found in animals (including vertebrates) and fungi.

Secondary growth depends on two kinds of lateral meristems: the **vascular cambium,** which produces secondary xylem and phloem, and the **cork cambium,** which produces cork, the outer protective covering of woody stems and roots. **Bark** includes phloem, cork, and storage tissues. The rate of xylem growth changes from season to season, resulting in characteristic **growth rings.**

In **vegetative reproduction,** the same meristems that produce primary and secondary growth produce whole new plants. Many plants propagate in nature by producing adventitious roots or stems. Farmers, horticulturists, and researchers also propagate desirable varieties by stimulating adventitious root formation with auxin, by **grafting** a cutting of one plant together with the root or stem of another, or by allowing tiny plants to develop from undifferentiated tissue in artificial culture.

Review and Thought Questions

Review Questions

1. What are the two products of double fertilization?
2. Describe the relationship between the embryo, aleurone layer, and starchy endosperm in promoting growth and germination in cereals.
3. Compare the following terms:
   (a) meristem versus node versus internode
   (b) terminal bud versus axillary bud
   (c) epicotyl versus hypocotyl versus radicle
4. Describe the three zones of growth in a root.
5. Explain how a single gene mutation can dramatically alter flower morphology.
6. Why is leaf development considered the default pathway of meristem development? How was this demonstrated?

Thought Questions

7. What interesting roses might be produced by applying our knowledge of *Arabidopsis* regulatory genes?
8. Why would yeasts, plants, and vertebrates share similar genes for regulating development?

About the Chapter-Opening Image

Not only do modern corn kernels "pop," so do the kernels of corn's presumed ancestors.

 On-line materials relating to this chapter are on the World Wide Web at **http://www.harcourtcollege.com/lifesci/aal2/**

33 How Do Plants Regulate Growth and Development?

A Life Lived Full

In 1928, a young English chemist found himself on the job market in the middle of the worst economic depression in modern times. Unable to find an academic position in England, 25-year-old Kenneth Thimann landed a job as an instructor in biochemistry at the new California Institute of Technology, in southern California. In 1929, Thimann and his artist wife, Ann, sailed for America on their first wedding anniversary. It was the beginning of a new life, a life that would profoundly influence American agriculture and even the course of the Vietnam War.

Like many a biologist's career, Thimann's rests on an intriguing observation by Charles Darwin. In the 1880s, Charles Darwin and his son Francis performed a series of experiments designed to explain **phototropism**, the tendency of plants to bend toward light (Figure 33-1A). The Darwins showed that something in the tip of an oat seedling causes the seedling to bend toward light. When Charles and Francis Darwin masked the tips of seedlings with a light-proof foil, the seedlings no longer bent toward the light (Figure 33-1B). Later researchers found that the same mask applied in the growth region just below the tip had no effect on bending; the seedlings grew toward the light in a normal fashion.

Forty-five years later, the Dutch plant physiologist Frits Went and his father demonstrated that the "something" was a chemical substance in the tip of the plant. In 1926, the younger Went was serving his compulsory military service during the day and spending his evenings and nights working as a graduate student in his father's plant physiology laboratory at the University of Utrecht. Frits Went set out to isolate the growth-promoting substance from the tip of the oat seedling, using the strategy illustrated in Figure 33-2. On a tiny cube of agar, he placed as many plant tips as he could fit. After an hour, he removed the tips, reasoning that the hypothetical substance would have entered the cube of agar. He then placed the block on one side of a seedling stump that was growing in the dark and waited to see what would happen.

If the agar contained a substance that could alter growth, then the stumps should bend toward or away from the side with the cube of agar. At first nothing happened. By 3 a.m., however, the seedling had begun to curve away from the agar cube. Excitedly, Went ran home, burst into his parents' bedroom, and declared, "Father, come and see, I've got the growth substance!" Sleepily, his father suggested he repeat the experiment during the day. "If it is any good," the elder Went said, "it will work again, and then I can see it." Indeed, it worked again, and Frits Went named the mysterious growth substance **auxin** [Latin, *augmentum* = increase].

When Kenneth Thimann arrived in California three years later, one of the first people he encountered was fellow instructor Herman Dolk. Dolk had been at the University of Utrecht with Frits Went, and he recruited Thimann to help him identify the chemical nature of auxin. Thimann and Dolk worked doggedly for 2 years, struggling to isolate the growth substance and solidifying a friendship.

Thimann's life seemed to couple tragedy with triumph, however, for in 1931, Dolk was killed in a car accident. Thimann was left to finish their work by himself (Figure 33-3). "It was kind of a lost feeling," he recalled almost 60 years later. "When Herman was killed, I was so deep into it that there was no choice but to go on."

Thimann went on to isolate the first known plant hormone, indole-3-acetic acid (IAA), which he showed to be the compound responsible for all the actions of auxin. Then, in an inspired move, he determined that by substituting other

Helmut Gritscher/Peter Arnold, Inc.

Key Concepts

- Plants alter their activities and their development in response to environmental cues that include gravity and light.

- The internal signals in plants that help them regulate these activities include several kinds of hormones, the most prominent of which are the auxins, gibberellins, cytokinins, abscisic acid, and ethylene.

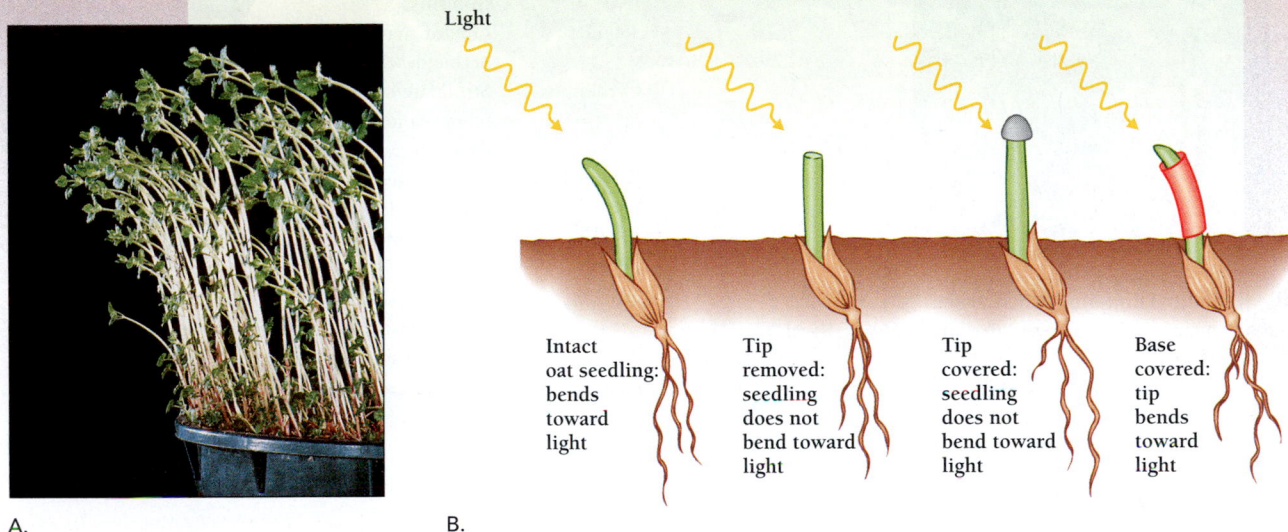

Figure 33-1 **What makes plants bend toward light?** A. Seedlings bending toward light. B. Charles and Francis Darwin showed that the part of the seedling that responds to light is the tip. When the tip is covered, the plant will not bend. Later research showed that covering the growth region, below the tip, has no effect on bending. *(A, Jerome Wexler/Photo Researchers, Inc.)*

chemical structures for the indole in natural auxin, he could make synthetic auxins with the same properties as the natural hormone (Table 33-1). But these synthetic auxins differed from the natural ones in one important respect. Plants did not possess enzymes for breaking down the synthetic versions. The synthetic auxins, therefore, persist in the plant much longer than natural ones.

Thimann and others began a series of experiments to see how auxin affects plant growth. Auxin, they found, stimulates rapid growth in the apical meristems where it is produced. Auxin also travels down the stem, stimulating cells behind the apical meristem to lengthen.

In 1934, Thimann and Frits Went, who had by then left Holland to join him at Caltech, discovered that auxin also

Figure 33-2 **Does a substance cause plants to bend toward light?** To answer this question, Frits Went placed the tips of seedlings on blocks of agar, in the hope that if a chemical substance were responsible, it would diffuse into the blocks of agar. He then balanced the blocks on the stumps of seedlings so that any substance in the agar would diffuse into only one side of the stem. The results confirmed his suspicions. Some substance causes the seedlings to bend. In addition, he learned that the seedling bends *away* from the substance.

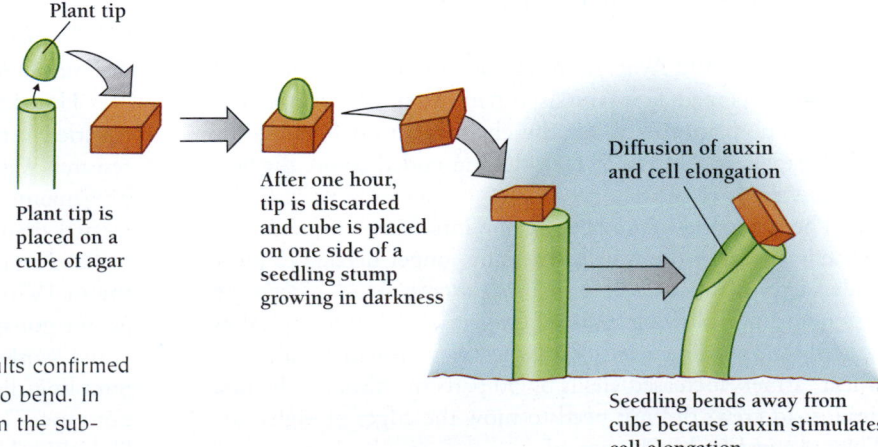

695

Figure 33-3   Kenneth Thimann. When asked for some advice to give to students, Thimann said, "Never give up." *(University of California Santa Cruz Photo Lab)*

TABLE 33-1	NATURAL AND SYNTHETIC PLANT HORMONES AND THEIR EFFECTS
Hormone	**Effects**
Auxins	Stimulate cell elongation and differentiation, apical dominance, development of fruit; act in tropisms
2,4-D	One of many synthetic auxins used as herbicides
Cytokinins	Stimulate cell division, promote bud initiation, relieve apical dominance, inhibit root initiation, delay abscission and senescence
Gibberellins	Promote seed germination, stem elongation, and fruit development
Abscisic acid	Closes stomata during water stress; causes dormancy in seeds
Ethylene	Triggers leaf and fruit abscission, fruit ripening, and defensive responses after injury; reorients cell expansion, alters gravitropic responses

stimulates plants to grow roots. It was a tremendously valuable discovery. Today, virtually every nursery in the United States and Europe uses artificial auxin to induce the formation of roots in cuttings.

In the same year, Thimann and Folke Skoog, at the University of Wisconsin, showed that auxin could inhibit growth as well as promote it. Auxin is responsible for apical dominance, the tendency of the apical meristem to suppress growth in lateral buds. In addition, auxin inhibits leaves, fruits, and flowers from falling. Today, many fruit growers spray their orchards with synthetic auxin to prevent the fruit from falling before all of the fruit in the orchard is ready for harvest.

But it was Thimann's next discovery that influenced agriculture most dramatically. Thimann had attracted the interest of Harvard University, which invited him to join its faculty. At Harvard, he noticed that although a little auxin may either promote growth or inhibit growth, large amounts of auxin are fatal to plants. Synthetic auxins are especially deadly because plants lack enzymes necessary to break down the synthetic versions.

During World War II, researchers at the University of Chicago developed Thimann's synthetic auxins for use as herbicides. The plan was to use the chemicals to defoliate areas hiding Japanese soldiers. In fact, the war ended before the herbicides were ever used.

The herbicides nonetheless came into heavy domestic use immediately after the war. Farmers no longer needed to plow under acres of weeds before planting a crop. Herbicides were easier, and not plowing reduced erosion of valuable topsoil as well. By suppressing weeds with selective herbicides, corn and wheat farmers increased yields by 30 percent. Also, for the first time, road crews did not need to mow the edges of highways. They sprayed herbicide instead.

The most famous herbicides are the synthetic auxins 2,4-D and 2,4,5-T (Table 33-1). In the 1960s, American military strategists revived the idea of using herbicides in the war against North Vietnam. American planes sprayed Vietnamese forests with heavy doses of herbicides to expose enemy positions. (They also destroyed crops with the same herbicides.) The trees dropped their leaves, revealing soldiers and others to American and South Vietnamese aircraft flying overhead.

The U.S. government experimented with a variety of herbicides, but the most commonly used herbicide was "Agent Orange," a mixture of 2,4-D and 2,4,5-T. Chemical companies, under pressure to supply huge amounts of these synthetic auxins to the government, produced bulk quantities that turned out to be heavily contaminated with the poison dioxin. In later years, American veterans who had been exposed to the dioxin attributed a variety of illnesses to Agent Orange. In the 1980s, 27,000 Vietnam veterans sued five chemical companies for illnesses they suffered after returning from Vietnam. The chemical companies settled out of court for $180 million.

Thimann took credit for the many uses to which his discoveries were put, but with reservations. He insisted that his research was always basic research, not applied research. "Improvements that are based on pure research are more fundamental than research based on finding applications," he said. He also argued that 2,4,5-T was not dangerous, pointing to a fire in 1976 at an herbicide factory in Seveso, Italy. "A lot of people got spattered with 2,4,5-T that contained dioxin." Exposed workers developed chloracne, a severe but temporary pimpling that is characteristic of dioxin exposure. "But that's not cancer," said Thimann. "It heals by itself after a time, just like any other acne. They've done studies galore, studied their babies and everything. In the years since, nobody's shown any serious effects, and researchers know the workers were really dosed."

Thimann spent nearly 30 years at Harvard University, devoting himself to teaching and mentoring hundreds of students. When he was 58, he left Harvard to return to California. At the brand new University of California, Santa Cruz, he assumed the task of building a first-class science faculty and creating an environment in which students could come into close contact with their professors. His secretary once recalled that the two of them sometimes worked until midnight to get all their work done, but Thimann never refused to see a student. Neither his research nor the work of building a new university campus ever prevented him from putting students first. Even after he retired from teaching at the age of 68, he continued to supervise undergraduate research for more than 15 years. In his 90s, he still went to work every day.

In 1983, the Italian government awarded Thimann the Balzan Foundation Prize, one of the highest awards in biology. Great Britain's Royal Society and France's Academy of Science each elected him as a member. When Thimann's wife, Ann, died in 1991, he moved to Pennsylvania to be near his grown daughters. The University of Pennsylvania immediately provided him with an office, and he continued his involvement in research and academic life until he died in January 1997. He spent his last years studying senescence in plants. He never lost his love of science or of plants, once remarking, "We look on nature as a book—every now and then we can turn a page. It's a great thrill."

HOW DO PLANTS DETECT ENVIRONMENTAL CHANGES, AND HOW DO THEY COORDINATE THEIR RESPONSES TO THOSE CHANGES?

Animals move to find new supplies of both energy and nutrients. To this end, they have evolved dozens of styles of locomotion, including swimming, running, and flying. In contrast, plants derive their energy from sunlight and their carbon from the air around them—an arrangement that makes large-scale movements unnecessary. Further, because most plants obtain water by means of roots that are permanently embedded in soil, they cannot move rapidly from place to place.

Plants are nonetheless active and responsive to their environments. Unlike animals, which have a permanent mature shape, plants change shape throughout their lives. We may compare the life of an animal to a play: it has a script with a beginning, a middle, and an end. In contrast, the life of a plant resembles an improvisation, whose form, content, and length depend on cues from the audience.

A plant shaded by other plants grows toward any available light. A low-growing plant may spread along the ground, rerooting itself as it goes. A vine may grow up toward light or out across the ground. The roots of a plant growing in dry soil grow downward toward water. A plant suddenly exposed to more light than it is used to grows new leaves that are less sensitive to light. A plant exposed to high winds grows stems or a trunk of increasing girth and strength.

Not all plant activity is slow or accomplished by means of gradual growth. Plants continuously adjust the position of their leaves or flowers to track the sun from hour to hour as it moves across the sky. Flowers open and close, according to the time of day and other variables. Some plants forcefully eject seeds from their fruits. Species of *Mimosa* (called "sensitive plants") rapidly fold their leaves in response to a light touch.

Despite such activity, plants are limited in their ability to seek desirable environments or avoid undesirable ones. They must make the most of what is available, and environmental cues dictate the pattern of growth. In response to light and gravity, for example, the shoot grows up and the root grows down into the earth. In response to changes in day length, temperature, or soil humidity, plants may drop their leaves, go dormant, or burst into flower. Depending on environmental cues, the same cells that give rise to leaves and stems may also give rise to flowers or roots.

Plants' ability to respond to changes in their environment seems remarkable, since their cells, which are enclosed in a cellulose wall, are more rigid and more tightly glued to one another than those of animals. When a leaf turns toward the morning sun, for example, cells on one side of the petiole must lengthen, while those on the other side remain the same size. For the leaf to grow toward the sun, cell expansion must be perfectly coordinated.

In the rest of this chapter, we see some of the molecular and cellular mechanisms for coordinating the responses of plants to environmental cues.

Like other living organisms, plants are highly responsive to changes in their environment. Over time, a plant may alter its shape, its structure, and even the positions of its leaves and flowers in ways that increase its ability to survive.

How Do Plants Respond to the Pull of Gravity?

Growing toward or away from a stimulus is called a *tropism* [Greek, *trope* = turning], so plants' response to gravity is called **gravitropism** (Figure 33-4A). In response to gravity, roots, especially primary roots, grow down. Primary shoots usually grow upward, but may later turn to grow horizontally; lateral branches generally grow at other angles. Gravitropism allows plants to respond vigorously to their environment; if we turn a young seedling on its side, the root turns and grows downward and the shoot turns and grows upward. How exactly does a plant "know" which way is up?

A.

B.

50 µm

Figure 33-4 How do plants know which way is up? A. A fallen tree often sends new shoots upward, away from the pull of gravity. B. Amyloplasts drop to the bottoms of cells and help plants detect gravity. *(A, John D. Cunningham/Visuals Unlimited; B, Ed Reschke/Peter Arnold, Inc.)*

A simple experiment demonstrated the location of a root's gravity detector: removal of the root cap abolishes the root's gravitropism. The response of the root must depend on the cells of the root cap. But which component of the root cap cells detects gravity?

Large cells in the root cap (and in other tissues throughout the plant) contain organelles called **amyloplasts**, each of which contains several starch grains (Figure 33-4B). The amyloplasts are denser than other organelles, and they consistently settle to the lowest part of the cell. Plant physiologists long suspected that the amyloplasts act as **statoliths** [Greek, *stato* = standing + *lithos* = stone], or gravity detectors.

To test this hypothesis, researchers treated roots with gibberellin and cytokinin, which induce the digestion of the starch grains. With no amyloplasts, the root no longer exhibited gravitropism. Later, the same roots made new starch grains and recovered their gravitropism.

But this was not the end of the story. More recent research has not supported the hypothesis that amyloplasts are plants' only gravity sensors. Timothy Caspar and his coworkers at Michigan State University genetically engineered *Arabidopsis thaliana* so that the plants could manufacture no starch and therefore no amyloplasts. These plants nonetheless demonstrated gravitropism. However, the response was not as strong as in plants with starch grains. Other research suggests that pressure between the cell membrane and cell wall stimulates gravitropism. Perhaps proteins between the cell membrane and cell wall detect such pressure.

Biologists do not yet know how plants sense, or detect, gravity, but they do understand how plants respond to gravity. As in phototropism, a change in the position of a shoot or root causes large differences in the rate of elongation of cells on the top and bottom. For example, the cells on the bottom of a horizontal stem may grow ten times faster than those on

the top. This difference is responsible for the upward bending of the stem.

Since this response is so similar to the phototropic response studied by the Darwins and the Wents, we should not be surprised to learn that auxin plays a role in gravitropism as well. Auxin is made in the tips of shoots and is transported downwards, creating a chemical difference that helps the plant define "up" and "down." Other signaling molecules also participate in gravitropism, including ethylene and abscisic acid, which we discuss later. The initial trigger, however, seems to be calcium ions.

When a root is oriented horizontally, calcium ions accumulate along the lower surface of the root and trigger the transport of auxin to the lower side, where it inhibits elongation. Meanwhile, the upper side, with less auxin, elongates more rapidly. As a result, the cells on the upper surface of the root grow more rapidly and force the root tip downward. When the root tip is oriented vertically once again, the asymmetrical distribution of calcium and auxin disappears, and the root grows straight.

The mechanism for gravitropism is an area of active research, as plant physiologists are finally close to answering a question that has vexed them for 50 years. The research even has a glamorous side, as interplanetary travel may ultimately depend on the ability to raise crops in the weightless environment of space.

Until recently, biologists believed that gravitropism was facilitated by starch grains called amyloplasts. Contrary evidence suggests that pressure between the cell membrane and the cell wall may stimulate the release of calcium ions and auxin. Auxin mediates the bending of roots and shoots in response to gravity.

Figure 33-5   Solar tracking. Flowers and leaves of some plants turn to face the sun as it moves across the sky. *(D. Newman/Visuals Unlimited)*

What Pigment Facilitates Phototropism?

A plant's survival depends on its ability to get enough light to support photosynthesis, and plants therefore respond to light (or its absence) in many ways. Among these are phototropism and solar tracking, the turning of leaves to follow the sun (Figure 33-5).

After much work and many confusing leads, researchers have recently made much progress in identifying the light detector responsible for these responses. The early approach to this question depended on studies of the effects of light of different wavelengths (which we perceive as different colors).

Every light-dependent process has a characteristic *action spectrum,* the effectiveness of different wavelengths in triggering a specific response (Chapter 7). Because the effect of light on a process depends on its absorption by some molecule, the action spectrum might be expected to match that molecule's *absorption spectrum,* the relative absorption of different wavelengths of light (Figure 7-14). If chlorophyll were the detector for phototropism and solar movements, then the action spectrum for these processes should correspond to the absorption spectrum of chlorophyll. But this is not the case. The light-absorbing molecule responsible for phototropism and solar movements is therefore not chlorophyll, but some other compound that absorbs blue, but not red or yellow light.

Researchers have recently found several types of blue-light detectors by analyzing *Arabidopsis* mutants whose seedlings cannot respond to blue light. They found two classes of proteins called *cryptochromes* and *phototropin,* both of which are surprisingly similar to proteins involved in cellular regulation in animals and bacteria. Somehow, these light-absorbing compounds respond to blue light by stimulating the transport of auxin from the illuminated to the shaded side of a growing shoot tip.

Phototropism depends on pigments other than chlorophyll.

How Do Plants Detect Changes in Season or Daylight?

Every gardener, farmer, or hiker knows that different plants flower in different seasons and at different times of day. The flowers of crocuses and daffodils appear in the spring, and those of carnations and black-eyed susans in the summer. Appropriate timing of flower and seed production determines how much sun and water will be available to new plants and which other organisms will be around to compete, to consume, or to pollinate, for example. In this section, we ask, How do plants detect and respond to seasonal changes?

In some species (such as cucumbers, peas, and tomatoes), the time of flowering depends only on maturity or size. Flowering may occur after a plant has grown for only a few weeks, or flowering may take months, or even years. In other species, flowering always happens around a certain date, even if the plant reaches maturity much earlier.

Understanding the seasonal control of flowering is a practical as well as a scientific issue. Soybean farmers, for example, once tried to stagger their harvests by planting fields at 2-week intervals. But they found that all the flowers appeared at once, late in the summer, and the whole harvest was ready at the same time regardless of the planting time.

For any species, the date of flowering varies with latitude and altitude. Spring-flowering plants in Massachusetts, for example, generally flower later than those in Georgia. From this observation, biologists conclude that flowering does not result from the operation of an internal annual clock. The flowering response turns out to be an example of **photoperiodism,** the response to the relative lengths of day and night as they change during the year. *Short-day plants* flower when days are short and nights are long, in the late summer, fall, or even winter. Examples include chrysanthemum, cocklebur, and corn (Figure 33-6).

In contrast, *long-day plants* require long days and short nights. These generally flower in the late spring or early summer and include spinach, sugar beets, and black-eyed susans. Finally, in *day-neutral plants,* such as tomatoes and dandelions, flowering does not depend on day length. Plants from tropical areas are often day neutral: lacking both night-length variations and temperature fluctuations during the year, tropical plants have been under little pressure to regulate the seasons of their flowering. By contrast, plants closer to the poles have evolved responses that take advantage of the night-length signals. This adaptation ensures that plants will flower early enough to set seeds before cold weather strikes.

One of the first questions that researchers asked was whether flowering depended on the length of the day or the length of the night. To find out, they grew short-day plants—cocklebur in one laboratory, soybeans in another—in artificial cycles of light and dark. The plants flowered whenever the "nights" were longer than a critical length. It did not matter how night lengths compared with "day" lengths.

In 1938, Karl Hamner and James Bonner, plant physiologists at the University of Chicago, tried a different experi-

Figure 33-6   **Short-day and long-day plants.** A. Short-day plants such as chrysanthemums flower when nights are longer than a critical length. Long-day plants such as black-eyed susans flower when nights are shorter than a critical length. B. The plant physiologists Hammer and Bonner asked if plants were sensitive to day length or night length. When they interrupted the day with a period of darkness, it made no difference in whether either kind of plant flowered. But when they interrupted the long nights with flashes of light, the short-day (long-night) chrysanthemum failed to flower, and the long-day (short-night) black-eyed susan flowered. The two researchers concluded that in both kinds of plants, it is the *length of the night* that determines flowering.

ment. They interrupted the light period with a period of darkness or the dark period with a period of light (Figure 33-6). For both short-day and long-day plants, interrupting the artificial day made no difference in flowering. For short-day plants, breaking up a night that was longer than critical prevented flowering, and for long-day plants, breaking up a night that was longer than critical promoted flowering. The interruption did not need to be large—an ordinary light bulb turned on for half a minute was enough to prevent flowering. The researchers concluded that both short-day and long-day plants measure the length of the night, not the day. Short-day plants should really be called "long-night plants," and long-day plants should really be called "short-night plants."

Other work on the basis of photoperiodism was going on at the U.S. Department of Agriculture Laboratory in Beltsville, Maryland, where photoperiodism had first been discovered in the 1920s. The early work had immediately helped agriculture: farmers were better able to control when their crops flowered and set seed, breeders could test new varieties under controlled light-dark conditions, and florists could provide year-round supplies of flowers that had once been seasonal. But still no one understood the chemical basis of photoperiodism.

In 1940, however, researchers at Beltsville realized that they could learn about darkness detection by determining the action spectrum of light that most effectively interrupted the long nights required for soybeans to flower. Determining the action spectrum for such a complex process, however, was no easy matter. The researchers first demonstrated that the light effect was in the leaves and that illuminating a single leaf was enough to suppress flowering in the whole plant.

They built a large instrument that would light individual leaves in separate groups of plants with light of different colors. This experiment showed that red light and violet light were most effective in promoting flowering. The researchers concluded that the light detector must be a pigment that absorbs red and violet light—in other words, a green or blue pigment, perhaps even chlorophyll. Further measurements revealed, however, that the pigment was a new one, which they named **phytochrome** [Greek, *phyton* = plant]. Phytochrome regulates a variety of processes that depend on the timing of dark and light, including flowering, germination, and leaf formation.

The Beltsville group discovered that they could cancel the effect of red light by immediately following the red flash with a light flash of longer wavelength. The most effective wavelength for the red-light effect was about 660 nm, and the most effective wavelength for the reversing flash was about 730 nm, in the "far-red" region of the spectrum. Sterling Hendricks, the plant physiologist who had built the instrument to determine the action spectrum, made a radical hypothesis to explain the far-red reversal. Phytochrome must exist in two interconvertible forms, one absorbing red light and one absorbing far-red light. He called the red-absorbing form P_r and the far-red–absorbing form P_{fr}. When P_r absorbs red light, it converts to P_{fr}, and far-red light converts P_{fr} back to P_r.

Hendricks and his colleagues proposed the following explanation for the effect of light and dark on flowering: (1) Sunlight contains more energy in the red than in the far-red part of the spectrum; (2) during the day, the red component in sunlight converts P_r to P_{fr}; and (3) at night, P_{fr} converts back to P_r. According to their hypothesis, P_{fr} promotes flowering of long-day plants and inhibits flowering of short-day plants. That is, in short-day plants, flowering occurs only when P_{fr} levels are sufficiently low (Figure 33-7). Red light supplied at night decreases P_r and increases P_{fr}, suppressing flowering. High levels of P_{fr} thus inhibit flowering in short-day plants and promote flowering in long-day plants. Other plant physiologists were initially skeptical about the existence of phytochrome. One doubter waggishly termed it "a pigment of the imagination."

Hendricks' hypothesis, however, made a strong prediction: purified phytochrome should be a single compound whose absorption spectrum can switch from one form to another. This prediction stimulated a biochemical search that was long and difficult for two reasons: (1) phytochrome acts catalytically, so plants do not contain great amounts; and (2) P_{fr} is unstable, so researchers had to work in the dark to try to keep phytochrome in the P_r form.

After many years of work, several research groups isolated pure phytochrome. It is a protein linked to a much smaller organic compound that absorbs light. Purified phytochrome, however, behaves like the pigment imagined by Hendricks and his colleagues almost 40 years earlier. Under natural sunlight, which contains both red and far-red light, both forms of phytochrome exist, with about 40 percent P_r and about 60 percent P_{fr}. In darkness, the P_{fr} level declines, either by reversion to P_r or by destruction by enzymes.

More recently plant biologists have learned much more about night timing, by analyzing the effect of mutations on phytochrome-initiated events. The current view is that plants, like animals (and even microorganisms), have an internal clock that cycles every day. Phytochrome starts the clock by detecting the end of a day (when P_{fr} declines), and restarts the clock at daybreak, when P_{fr} again increases.

The action spectrum for other light-regulated changes suggested that phytochrome is important in many developmental processes. These processes vary from species to species but include seed germination, the elongation of new seedlings, the beginning of chlorophyll synthesis, and the production of enzymes needed for photosynthesis. For example, in mustard seedlings that have been grown in the dark and transferred to the light, phytochrome controls the following changes:

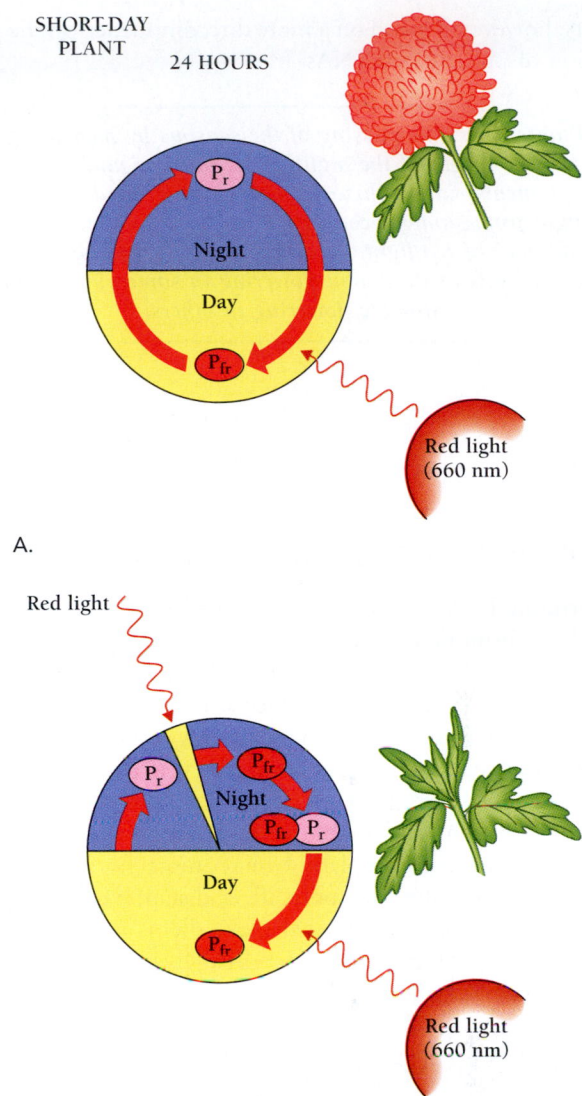

Figure 33-7 How long is the night? The flowering of long-day and short-day plants depends on the conversion of phytochrome from P_r to P_{fr} and back again. A. During the day, sunlight converts P_r into P_{fr}. During the night, some P_{fr} is destroyed or converted back to P_r. Phytochrome cycling between P_r and P_{fr} permits a biological clock to measure time. Short-day plants flower if the clock indicates that the nights are long. B. A flash of light in the night resets the clock, converting P_r to P_{fr}. The interruption inhibits flowering, since the clock registers a short night.

opening of the hypocotyl hook, development of primary leaves, differentiation of xylem, degradation of storage protein, chlorophyll synthesis, protein synthesis, and RNA synthesis.

In almost every case, the developmental change controlled by phytochrome depends on the accumulation of P_{fr}. Red light (or daylight) thus promotes these processes and far-red light inhibits them. Phytochrome permits the sun, or the plant physiologist, to turn a molecular switch on or off. Some of the effects of phytochrome may depend on induced changes in the synthesis, degradation, or transport of plant hormones. Some effects, however, depend on a more direct influence on the production of particular mRNAs.

Plants detect the passing of the seasons by measuring the length of the night by means of a pair of pigments, called phytochromes. During the day, red light from sunlight converts P_r to P_{fr}. At night, in the absence of red light, P_{fr} is destroyed or reverts to P_r. High levels of P_{fr} inhibit flowering in some plants and promote flowering in others.

HORMONES COORDINATE ACTIVITIES IN CELLS THROUGHOUT THE PLANT BOTH DURING DEVELOPMENT AND IN RESPONSE TO ENVIRONMENTAL CUES

A **hormone** is an organic compound that serves as a chemical signal. A hormone is produced in a tissue or organ and transported to another tissue or organ, called the **target**, where it produces one or more specific effects. Like animal hormones, plant hormones move from source to target both by diffusion from cell to cell and by transport in the vascular tissues. Plant hormones also exert their effects at very low concentrations (usually less than 10^{-6} moles per liter), in doses of as little as a millionth of a gram per plant. Unlike animal hormones, however, plant hormones not only affect distant target cells but also influence the very cells that make them.

Plant physiologists have identified five families of compounds that are traditionally viewed as hormones. The five families are the auxins, gibberellins, cytokinins, ethylene, and abscisic acid. There are three naturally occurring auxins, at least 60 gibberellins, and several cytokinins. Other related compounds can trigger the same biological responses, but are not made by plants. These include compounds isolated from animals, fungi, and bacteria, as well as the products of laboratory synthesis.

We may ask how a few types of hormones can control complex sequences of development and respond to an ever-changing environment. The answers are that:

1. The same hormones affect different tissues differently.
2. Hormones have different effects at different concentrations.
3. Hormones interact with one another in complex ways.

Auxin Coordinates Many Aspects of Plant Growth and Development

Frits Went's early experiments with auxin showed that auxin moves in a polar fashion, away from the tip of the coleoptile. To explain phototropism, he proposed that light coming from one side could divert the auxin flow, producing a lateral transport to the dark side. Since auxin stimulates elongation, this would produce bending toward the lighted side. In the early 1960s, Winslow Briggs, a plant physiologist at Stanford University, tested this prediction (Figure 33-8). Briggs first asked whether light affected the total amount of auxin in the growing tip of an oat seedling. After placing tips on blocks of agar, he illuminated some tips and kept others in the dark. He then used the degree of curvature to estimate the amount of auxin in each tip. Briggs found no difference in the amounts of bending or, therefore, auxin in the two sets of tips. Phototropism, he concluded, must depend not on the total amount of auxin in an illuminated tip, but on the distribution within the tip.

Briggs then looked for differences in auxin concentrations between the side of the tip nearer the light and the side away from the light. He inserted a thin piece of mica between the two sides. The mica completely separated the lighted side from the shaded side, and Briggs found no difference in the auxin levels of the two sides. When Briggs left a gap of half a millimeter near the very end of the tip, however, the auxin concentration on the shaded side rose to nearly twice that on the illuminated side. The light somehow caused the transport of auxin from the lighted to the shaded

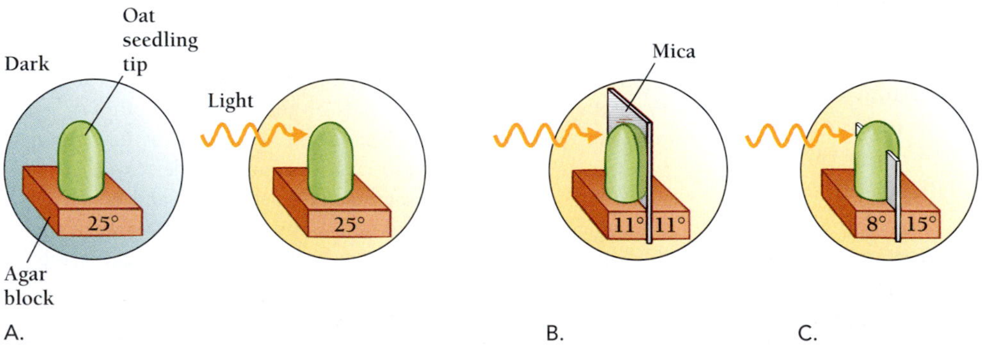

Figure 33-8   Does auxin move from one side of the seedling to the other? Or does the seedling synthesize extra auxin on one side? Winslow Briggs's experiment demonstrated that auxin is redistributed in response to light. Briggs put oat seedling tips on blocks of agar, then measured the auxin content of the blocks under different conditions. The numbers on the blocks of agar are the approximate degrees of bending the blocks caused in oat seedlings. A. The auxin concentration from a seedling in the dark is the same as the concentration when the seedling is exposed to light. Therefore, the seedling tip does not synthesize more auxin in response to light. B. The oat seedling tip is exposed to light on one side only and divided in half by an auxin-proof barrier. The resulting auxin concentrations are the same on each side. Therefore, extra auxin is not synthesized on the shaded side. C. If the barrier is incomplete, so that auxin can move from one side to the other, the auxin concentration becomes much higher on the shaded side. Therefore, auxin moves from the lighted side to the shaded side of the seedling tip.

BOX 33.1

What Is a Bioassay?

A fixed amount of auxin gives a reproducible amount of curvature. Went's experiment established a *bioassay,* a method that estimates the concentration of a substance by measuring its biological activity. Went's bioassay allowed plant physiologists to estimate auxin concentrations by measuring seedling curvature with a protractor (Figure A). Another, newer approach is to measure the concentration of auxin and other plant hormones using chromatography, which depends on chemical rather than biological properties. However, bioassays continue to be useful for the identification of hormones and other functionally important molecules not only in plants, but also in other types of organisms.

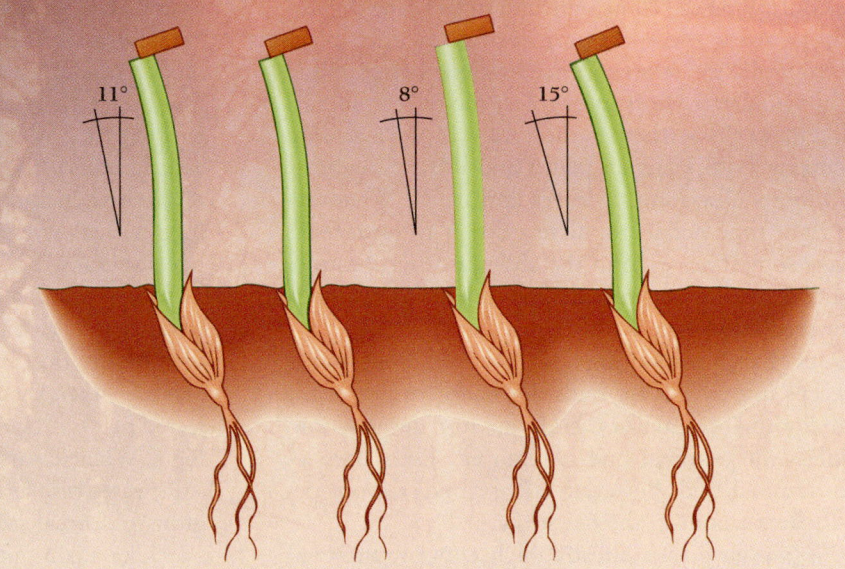

Figure A In Winslow Briggs's experiment on how auxin is redistributed, he used Went's oat-seedling bioassay as a measure of auxin concentration.

side of the tip. The higher concentration of auxin on the shaded side then caused that side to elongate more rapidly, leading to a curve in the direction of the light.

Briggs's experiments, like those of Went, showed that auxin stimulates cell elongation in both shoots and roots. In addition to its effects on cell elongation, auxin also affects the gene expression in target cells. Some of these changes occur rapidly, within 10 to 20 minutes after exposure to auxin.

One area of influence is probably the cell wall. Auxin stimulates cell expansion by stimulating biochemical events that loosen the cell wall. Turgor pressure exerted by the vacuole tends to stretch the wall, but the wall resists being stretched. Auxin stimulates the cell to secrete a protein, called *expansin,* that loosens the noncovalent bonds between the long polysaccharides (including cellulose) in the cell wall. As these bonds loosen, the turgor pressure pushes the cell's contents to expand, causing elongation.

Auxin differs from animal hormones, whose concentrations are the same throughout the body, because auxin's concentration differs from place to place. Auxin originates in shoot tips. Special transporter proteins transport auxin through the vascular system. One of these proteins actively transports auxin into the top (apex) of a cell adjacent to a vascular bundle. Another protein speeds the diffusion of auxin out of the base of the cell (Figure 33-9). The combined action of the two transporters establishes an auxin gradient, a graded change in con-

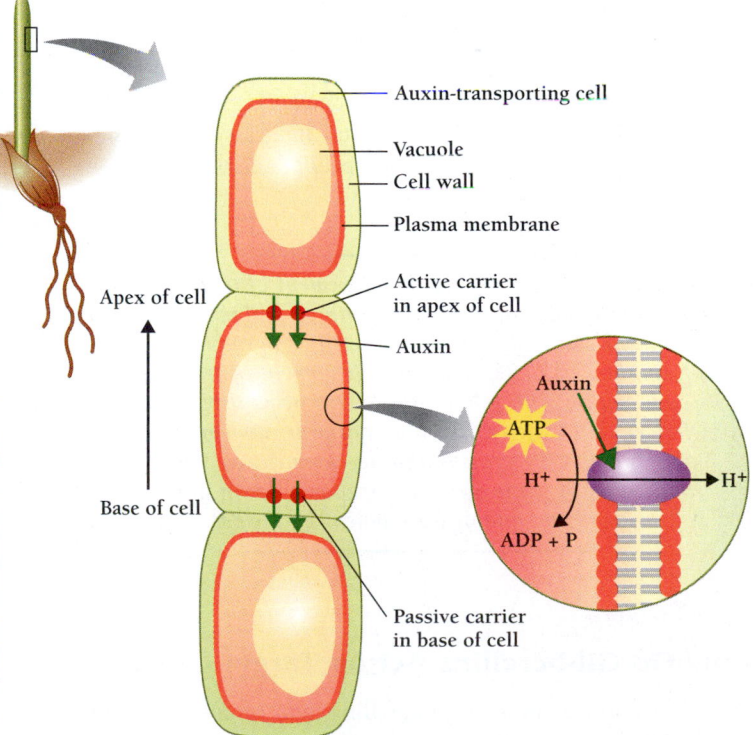

Figure 33-9 Auxin transport. One transport protein actively moves auxin into the top of a transporting cell; another speeds the diffusion of auxin out of the bottom of the cell.

703

centration from the top to the bottom of the shoot. In the root, these transporter proteins also establish an auxin gradient. Transporter mutations that prevent the root from creating an auxin gradient destroy the plant's ability to respond to gravity (gravitropism).

Besides controlling cell elongation and stimulating the formation of roots, auxin affects the growth of many other plant tissues. For example, the relative concentrations of auxin and gibberellin in vascular tissue help establish the developmental fate of individual cells—whether the cells will develop into xylem or phloem, both in intact plants and in tissue culture.

Auxin inhibits abscission of leaves and fruits. When applied at the right time it can stimulate a flower to form fruits without seeds, sometimes even in the absence of fertilization. Unfertilized tomato flowers treated with auxin, for example, develop into seedless tomatoes.

Besides its growth-promoting activity, auxin from an active shoot keeps nearby axillary buds dormant. In plants that show apical dominance, axillary buds do not develop into branches unless the terminal bud is removed. Even after removing the terminal bud, however, axillary buds do not develop if one places auxin on the cut surface.

Finally, as we have seen already, high concentrations of auxin are effective as herbicides. Synthetic auxins are especially effective because plants cannot break down these substances, and so they persist for a long time. Three synthetic auxins— 2,4-D (2,4-dichlorophenoxyacetic acid), 2,4,5-T (2,4,5-trichlorophenoxyacetic acid), and MCPA (methylchlorophenoxyacetic acid)—have been particularly popular because they are toxic to dicots and not to monocots. Grasses, corn, and wheat are monocots, so farmers can spray an entire field of corn or wheat, for example, killing all the dicot weeds without harming the crop. Similarly, ranchers can selectively kill sagebrush and mesquite, both dicots, in grassy pastures or rangelands, and lawnkeepers can kill every dandelion in a lawn of grass.

Phototropism results from the movement of auxin from the lighted side of the growing tip of a plant to the shaded side. Auxin stimulates cell elongation, the formation of roots, and the development of fruit. In conjunction with gibberellin, auxin helps determine whether cells differentiate into xylem or phloem. Auxin also inhibits the growth and development of axillary buds. Finally, in large concentrations, auxin and its synthetic analogs are potent herbicides.

How Do Gibberellins Act on Target Cells?

Plants have more kinds of **gibberellins** than of any other hormone. All the gibberellins are complicated organic molecules with 19 or 20 carbon atoms grouped into 4- or 5-ring structures (Table 33-1). Most plant species have several different kinds of gibberellin, some containing as many as 15. Most researchers think that most of the different forms are precursors or degradation products of a few active forms.

Gibberellins generally affect the overall growth of intact plants far more than the other plant hormones. In fact, their discovery arose from studies of a disease of rice plants caused by a fungus called *Gibberella fujikoroi*. The disease, first described in Japan in the 1890s, causes rice plants to grow so tall that they cannot support their own weight. The plants topple over into the water and rot. Japanese farmers called the disease *bakanae,* "foolish seedling." In the 1930s, two Japanese researchers identified the chemical compound made by the fungus that is responsible for the extravagant growth of the rice seedlings. The active compound is the gibberellin GA_3.

Later research in the 1950s showed that uninfected plants normally make gibberellins. Some plant mutants—of peas, corn, beans, rice, and other crops—lack gibberellins altogether and are much shorter than their normal counterparts. In infected plants, the fungus raises the gibberellin level abnormally high. When dwarf mutants are treated with gibberellins, in some cases with as little as a billionth of a gram per plant, they grow to normal size.

Like auxin, gibberellins seem to promote cell elongation by loosening cell walls, but with a more dramatic effect (Figure 33-10). If a segment of oat stem is treated with gibberellin and sucrose (to provide energy to the nonphotosynthesizing

Figure 33-10   The plant hormone gibberellin causes stems to lengthen dramatically. This plant's stem "bolts" when levels of gibberellic acid rise. (*Courtesy of B.O. Phinney, University of California, Los Angeles*)

tissue), the oat stem will grow 15 times as long as an untreated stem.

Gibberellins also stimulate the breakdown of starches and sucrose to monosaccharides (simple sugars). The increased glucose and fructose concentrations not only increase the availability of energy but also increase the solute concentration inside the cell. More water then flows into cells and contributes to their elongation.

In some cases, gibberellins can stimulate cell division—for example, in the apical meristem of the shoot. This effect is especially clear in biennial plants, which keep their stem very short for the first year, forming a rosette plant. In the second year, the stem rapidly elongates ("bolts"). Studies of these plants show that rosette plants are short because they lack cell divisions in the internodes. Gibberellin—produced in the plant or applied experimentally—causes bolting by stimulating cell division. Finally, gibberellins can act to change the pattern of expression of specific genes, as in the case of α-amylase induction in the germinating barley seed (Chapter 32).

Gibberellins appear to be important in the normal growth of shoots, the germination of seeds, the flowering of plants, and the mobilization of food reserves from the endosperm of cereals and cotyledons of other plants. Because gibberellins are present in most tissues of a plant and because they are active at such low concentrations, however, their precise role has been difficult to understand.

Gibberellins are complex organic molecules that promote cell elongation, mainly in stems; the breakdown of starches and sucrose; cell division; and gene expression.

Cytokinins Stimulate Growth and Differentiation

Since the early years of this century, researchers have known that some substances, now called **cytokinins,** stimulate cytokinesis (cell division) in plant tissue culture. Extracts of coconut milk, for example, can produce this stimulation. In all cases, however, these materials stimulated cell division only in the presence of appropriate levels of auxin. Thus, cytokinin and auxin *together* affect cell division. As we will see, such interactions are the rule in plants, not the exception.

In addition to stimulating cell division, cytokinin and auxin together help determine the developmental fate of plant tissue. The relative levels of cytokinin and auxin establish, for example, whether a callus culture will continue to divide without differentiation or will differentiate into shoots or roots.

Although cytokinin and auxin work together, they also work in opposition to one another. Auxin promotes roots initiation, while cytokinin opposes it. In addition, the two hormones have contrasting effects on the dormancy of existing axillary buds and the initiation of new buds. Auxin promotes bud dormancy and inhibits bud initiation; cytokinin relieves dormancy and promotes initiation of new buds. A plant bacterial infection called "witches' broom disease," provides an extreme example of the effects of cytokinin. The disease-causing bacteria secrete cytokinin, which stimulates the increased lateral branching. Although experimentally applied cytokinin and auxin affect the pattern of terminal and lateral branching, researchers are still not certain of the precise roles of these hormones in the normal regulation of lateral branching.

Auxin and cytokinin together coordinate the development of shoots and roots. The complementary actions of cytokinin and auxin depend not only on their different effects on cells but also on their differing distributions within the plant. As we noted earlier, auxin (which stimulates root formation) originates in shoot tips and moves downward. In contrast, cytokinin originates in roots and moves upward through the xylem. In a cutting, auxin therefore accumulates at the base and stimulates root formation. Conversely, cytokinin accumulates at the surface of a tree stump and stimulates bud formation.

When a leaf is removed from a plant, it usually begins to lose chlorophyll and to turn yellow. This yellowing is part of the process of **senescence,** the breakdown of cellular components leading to cell death. Application of cytokinin to a cut leaf, however, delays senescence.

During the normal life cycle of a plant, cytokinin transported from roots to leaves by the xylem appears to prevent senescence from occurring prematurely. Several fungi, and even two species of caterpillars, create "green islands" on otherwise yellowing leaves by secreting cytokinins.

Naturally occurring cytokinins are present at exceedingly low levels. The first cytokinin isolated from plant tissue, for example, required the extraction of 60 kg of corn kernels to obtain 1 mg of pure cytokinin. Researchers have found only a few cytokinins in plant tissues, but a number of other compounds, including natural products of several fungi and bacteria, have cytokinin activity. Modern techniques, however, have made it possible to determine the amounts of specific cytokinins at levels as low as a billionth of a gram, a nanogram (ng). This sensitivity has allowed the measurement of cytokinin concentrations in many plant tissues. By analyzing the distribution of and effects of cytokinins in wild-type and mutant plants, researchers have begun to understand how cytokinins act on plant cells. This work suggests that cytokinins act by regulating transcription and altering the pattern of gene expression.

Together with auxin, cytokinins stimulate cell division, delay senescence, and determine how cells differentiate. Unlike auxin, cytokinins stimulate cell division and growth in axillary buds. Cytokinins function at extremely low concentrations.

How Did Researchers Identify Ethylene as a Plant Hormone?

Ethylene is a much simpler molecule than the other plant hormones, consisting of just two carbon atoms and four hydrogen atoms (Table 33-1). Unlike the other plant hormones, ethylene is a gas and is not transported in the vascular system. And besides being made in plant cells, ethylene is produced in reactions that have nothing to do with plants—in smoke from fires, for example.

The most dramatic, delicious, and economically important effect of ethylene is on fruit ripening in fruits that have a distinct ripening phase. The ripening of a fleshy fruit (such as an apple or a tomato) is a complex process. Among the chemical changes that occur during ripening are the breaking down of chlorophyll, changing the fruit's color. Ripening also involves the breakdown of starches, increasing the fruit's sweetness, and the breakdown of cell walls, making it softer. Ethylene starts or speeds up these changes.

In fruits, ethylene also triggers the increased production of more ethylene, so that once ripening starts, it spreads rapidly, both within a single fruit and from fruit to fruit. This positive feedback acceleration of ripening has practical consequences for fruit lovers. One can hasten the ripening of green fruit by keeping them in an enclosed bag to trap the ethylene and by adding one piece of fruit that is particularly rich in ethylene (such as an apple or any fruit that is quite ripe).

Ethylene's effects also explain why "one rotten apple spoils the barrel." A rotten apple releases large amounts of ethylene and speeds up the ripening of those around it. These apples, in turn, release ethylene, and soon the whole barrel is filled with rapidly overripening apples.

Fruit growers and shippers knew how to hasten ripening long before they knew the chemical cause. The ancient Chinese ripened fruit in a room with burning incense, and Puerto Rican growers once built bonfires near their crops to stimulate flowering of their pineapples. Both these procedures work because the incense and wood fires produce ethylene. The identification of the plant hormone ethylene came from a practical problem first recognized in Germany in 1864. Before the invention of electric power, city streets were lit with gas lights. Gas lines carrying "illuminating gas" to each street light often developed leaks. When nearby shade trees began to lose their leaves, workers discovered that the gas was the cause. The demonstration that the active component of illuminating gas was ethylene came, however, only in 1901. A Russian graduate student named Dimitry Neljubov showed that illuminating gas stimulated pea plants to grow horizontally and that the active component of the gas was ethylene.

Ethylene is present almost everywhere in a plant. Under the right conditions, it causes plants to drop both leaves and fruit, a process called **abscission**. In addition, it participates in many other responses to environmental changes. For example, abnormally high levels of ethylene cause a "triple response" in the stems of tomato seedlings—stems thicken, their elonga-

A.

B.

Figure 33-11   **The triple response to ethylene.** In seedlings, the stem thickens, elongation slows, and the stem begins to grow horizontally. A. Triple response of normal tomato seedlings. B. Response of mutant tomato seedlings that are insensitive to ethylene. *(Yen, Lee, Tanksley, Klee, and Giovanoni,* Plant Physiology *[1995] 107:1343–1353)*

tion slows, and they begin to grow horizontally (Figure 33-11). Tomato seedlings grow in a similar way if they encounter a barrier as they emerge from the soil. The production of ethylene, primarily in the hook of the seedling's epicotyl, seems to coordinate the plant's ability to grow around objects. When a seedling finds its way to light, the amount of ethylene decreases (since there is no compact soil to retard its diffusion), and the seedling resumes its upward growth. Tomato mutants that are unresponsive to ethylene do not show any aspect of the triple response (Figure 33-11).

Ethylene also plays an important role in signaling cell injury and in stimulating the plant's defensive responses. Every injury stimulates ethylene production, and ethylene stimulates the formation of cork and the formation of lignin near a wound site. Ethylene also stimulates the production of salicylate, which (as discussed later) stimulates defensive responses in other parts of the plant.

Ethylene is a gas that stimulates fruit ripening and the release of more ethylene. Ethylene also triggers the abscission of fruit and leaves, the triple response in seedlings, and a plant's defensive responses.

Abscisic Acid Promotes Dormancy in Buds and Seeds

Abscisic acid [Latin, *abscissus* = to cut off] takes its name from the fact that it accumulates in abscising (dropping) of leaves and fruit (Figure 33-12). Despite its name, abscisic acid plays only a minor role in abscission, but it does play a major role in the suspension of bud and seed development, in signaling and preventing dehydration, and in regulating the opening and closing of stomata.

Unlike the other identified plant hormones, abscisic acid is primarily an inhibitor, not a stimulator. It antagonizes the effect of gibberellins. Abscisic acid slows the growth of oat seedlings, inhibits the germination of wheat seeds, and prevents the synthesis of α-amylase by the aleurone layer of barley seeds. At a cellular level, it inhibits the transcription of some genes and stimulates the transcription of others. Abscisic acid also coordinates plant responses to a variety of environmental stresses, including drought, excess salt, waterlogging, cold, and mineral deficiency.

Abscisic acid suspends development in buds and seeds and helps stimulate abscission, thus antagonizing the effects of gibberellins and auxins.

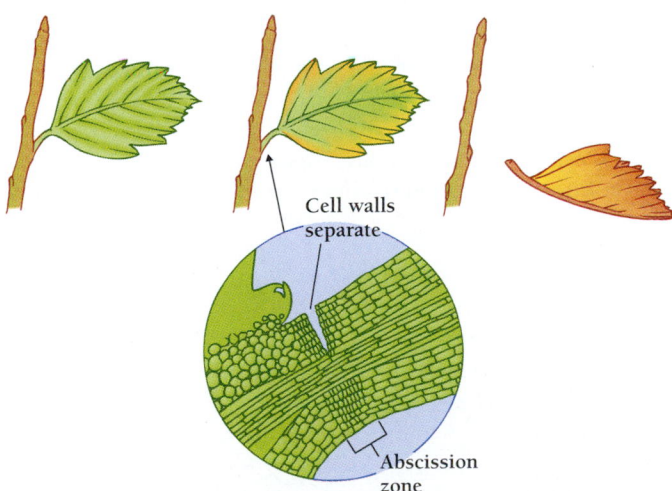

Figure 33-12 **Abscission.** Ethylene stimulates cells in the abscission zone at the base of a petiole to shrink and separate from one another. Finally, the attached fruit or leaf drops to the ground.

Cell walls separate

Abscission zone

Are There Other Plant Hormones?

Research in the past decade has shown that a number of other chemicals can coordinate cellular activities in different parts of a plant. For example, wounding a plant leaf can evoke both local and long-distance effects, suggesting the involvement of one or more plant hormones. When caterpillars chew on a leaf, the plant may produce *systemin,* a short polypeptide, 18 amino acids long. Systemin stimulates the synthesis of type of lipid called *jasmonic acid.* Jasmonic acid in turn stimulates the transcription of genes—called "systemic wound response genes"—that promote a set of defensive responses. One of these genes encodes a protein that inhibits the action of protein-digesting enzymes. The result is a bad case of indigestion for the caterpillar. Plants can also develop resistance to specific viruses, bacteria, or fungi, not only at the site of infection but also far away. Plant biologists have identified signaling molecules that may mediate both local and long-range effects.

In addition to systemin and jasmonic acid, plant researchers have identified other molecules that, somewhat unexpectedly, act as plant hormones. Some of these signals are steroids, which until recently had been thought not to be present in plants at all.

Another molecule recently identified as a plant hormone is *salicylic acid.* Salicylic acid derives its name from the willow tree [Latin = *salix*], whose bark was used for centuries by Native Americans and ancient Greeks to treat aches and fevers. Remarkably, salicylic acid, which we know as aspirin (acetylsalicylic acid), not only helps cure human ills, but helps prevent plant infections. In plants, salicylic acid sets off defensive processes and promotes flowering.

One of salicylic acid's most striking actions is the flowers of a number of heat-generating (thermogenic) plants, such as those of skunk cabbage and the voodoo lily. In these flowers, salicylic acid stimulates the production of a mitochondrial electron carrier that produces heat instead of ATP. The heat production is so dramatic that, on the day of flowering, the flower's temperature increases by 14°C (25°F). The high temperature stimulates the vaporization of foul-smelling chemicals that attract pollinating insects.

A plant hormone is an endogenous plant substance that acts at low concentration to affect physiological processes.

In this chapter of our section on plant anatomy and physiology, we have examined the ways that plants respond to their environments and the roles of plant hormones in plants' growth and development. We have reviewed both classic experiments and some of the latest research in these areas. In the last section of *Asking About Life,* we study the anatomy and physiology of animals, with an emphasis on vertebrates.

Study Outline with Key Terms

A plant's response to environmental cues depends on its ability to detect such cues. Plants employ different detection mechanisms for different kinds of stimuli. Large cells in the root cap called **amyloplasts** seem to detect gravity, an example of **gravitropism**. Amyloplasts function as **statoliths**. One kind of unidentified pigment detects light and brings about phototropism. Another kind of pigment, called **phytochrome**, serves as a molecular switch for the **photoperiodism** of many seasonal processes, including flowering and seed production.

Absorption of red light converts phytochrome into a form that no longer absorbs red light, but only far-red light. Similarly, absorption of far-red light switches phytochrome back to a form that absorbs red light. Most of the processes controlled by phytochrome depend on the accumulation of enough of the far-red–absorbing form.

Plant physiologists have identified five plant **hormones**. Each of these is an organic compound produced in one tissue or organ and transported to another tissue or organ, called the **target,** where it produces one or more specific effects.

The first plant hormone to be recognized was **auxin**, which (among other effects) coordinates **phototropism**, the growth of a plant toward light. Researchers can measure the amount of active hormones either with a *bioassay,* which estimates the amounts of hormones from their biological effects, or with chemical measurements. Illumination of one side of the growing tip of an oat seedling causes the transport of auxin away from the lighted side. The higher concentration in the shaded side causes it to grow more rapidly, leading the plant to curve toward the light.

Auxin triggers target cells to secrete hydrogen ions. This secretion makes the extracellular space more acidic and activates enzymes that break bonds within cell walls. Auxin also affects gene expression in target cells.

Other plant hormones have effects distinct from those of auxin. **Gibberellins** promote germination, trigger the mobilization of food reserves in germinating seeds, and stimulate growth in mature plants. **Cytokinins** stimulate cell division and growth and prevent leaf **senescence**. **Ethylene** promotes leaf senescence and fruit ripening and inhibits elongation of stems and roots. And **abscisic acid** promotes **abscission** and dormancy in buds and seeds.

Review and Thought Questions

Review Questions

1. What are the primary effects of auxin?
2. How is the process of senescence affected by cytokinins?
3. What is the triple response induced by ethylene? What is the experimental evidence suggesting that ethylene alters gene expression?
4. Why might you expect the level of abscisic acid to increase in a plant subjected to environmental stress?
5. Explain why botanists are beginning to doubt that amyloplasts help plants detect the pull of gravity.

6. Why can phytochrome be thought of as a molecular switch?

Thought Questions

7. Why do you think that some bacteria and fungi produce plant hormones? What advantage accrues to *Gibberella fujikoroi* in forcing seedlings to grow so fast?
8. What is a bioassay? Try to think of an example of a bioassay that could be used in animals.

About the Chapter-Opening Image

This slender seedling represents the tendency of a plant to bend toward light and the tendency of an engaged mind, such as Kenneth Thimann's, to bend toward a new question.

 On-line materials relating to this chapter are on the World Wide Web at **http://www.harcourtcollege.com/lifesci/aal2/**

part VII

Structural and Physiological Adaptations of Animals

Chapter 34
Form and Function in Animals

Chapter 35
How Do Animals Obtain Nourishment from Food?

Chapter 36
How Do Animals Coordinate the Actions of Cells and Organs?

Chapter 37
How Do Animals Move Blood Through Their Bodies?

Chapter 38
How Do Animals Breathe?

Chapter 39
How Do Animals Manage Water, Salts, and Wastes?

Chapter 40
Defense: Inflammation and Immunity

Chapter 41
The Cells of the Nervous System

Chapter 42
The Nervous System and the Sense Organs

Chapter 43
Sexual Reproduction

Chapter 44
How Do Organisms Become Complex?

Photo: Scanning electron micrograph of a human fallopian tube epithelium. *(Yorgos Nikas/Phototake)*

34

Tyrannosaurus Wrecks:
T. rex Treks, But Slowly

One of the high points of the movie classic *Jurassic Park* comes when an agile but stupid *Tyrannosaurus rex* pursues a Jeep at breakneck speed through a driving rain. Just when the *T. rex* seems ready to snap up actor Jeff Goldblum and down him like a dog biscuit, the gargantuan reptile runs out of steam and the three actors in the Jeep escape to (temporary) safety.

Could *T. rex* have run that fast? Zoologists and paleontologists say "no way." When an adult human runs, the force on each foot is hundreds of pounds— equal to the mass of the body times the acceleration of the foot against the ground. A 6.5-ton *T. rex* running at 45 miles per hour, say biologists, would exert so much force on its legs that they would snap like toothpicks.

Part of the drama of *Jurassic Park* stems from moviegoers' expectations that dinosaurs will be sluggish and clumsy. Indeed, the original reconstructions and paintings of dinosaurs from the early part of this century usually show dinosaurs moving very slowly or not at all. Early 20th-century paleontologists believed that dinosaurs were incapable of the kinds of sprints at which warm-blooded animals excel. Accordingly, dinosaurs were nearly always depicted with all four feet on the ground (Figure 34-1). Yet, the velociraptors and other predatory dinosaurs in *Jurassic Park* and its sequel *The Lost World* leap, run, whorl, cock their heads like birds, and attack, swiftly and often intelligently. How can we tell which image is correct?

Because dinosaur skeletons share many features with modern reptiles—similar jaws, for example—early paleontologists assumed that dinosaurs resembled modern reptiles in other ways as well. They assumed that dinosaurs laid eggs instead of bearing live young, that they were solitary rather than social, and that they were heterotherms— cold-blooded animals whose activities depended on air temperature. In the last 20 years, all of these assumptions have been called into question.

In the 1970s, the maverick American paleontologist Robert Bakker electrified the imaginations of paleontologists, scientific illustrators, and small children by suggesting that dinosaurs might have been warm-blooded animals, more like birds than modern reptiles. In particular, Bakker examined *T. rex*'s leg bones and concluded that they were long enough to have allowed the monster to run up to 45 miles per hour. But Bakker's quick calculation left plenty of room for further analysis.

British zoologist R. McNeill Alexander, an expert on the physics of animal move-

© John Sibbick

Key Concepts

- Form and function mutually determine one another.

- Increased surface area speeds the radiation of heat and the diffusion of molecules.

- Countercurrent systems are a way for organisms to establish useful gradients.

- Muscle contracts when actin and myosin fibers slide past one another.

- Animals have adaptations for both homeostasis and tolerance.

ment, jumped into the argument with a more thorough study of dinosaur leg bones. Alexander concluded that although many dinosaurs had the long legs that would have enabled them to run fast, their legs were not strong enough to carry them at high speeds. Alexander calculated the strength of a bone by comparing the size of a cross section of a bone with its length and the weight of the animal.

The femur (or thigh bone) of an African elephant, for example, has a fairly low "strength indicator" ratio of about 7. A running elephant moves at little more than 11 miles per hour. (Charging elephants only seem to be moving much faster.) For comparison, the femur of an African buffalo, which is capable of running 30 to 35 miles per hour, has a strength indicator of about 22. Alexander calculated that the femur of *T. rex* has a strength of about 9, a little stronger, perhaps, than that of an elephant, but nowhere near that of a buffalo. At best, Alexander estimated, a *T. rex* could run no more than about 18 miles per hour.

Alexander's analysis seemed conclusive until 1995, when a paleontologist and a physicist at Indiana University took the whole argument one heavy step further. What would happen, asked James Farlow and John Robinson, if a 6.5-ton *T. rex* pursuing a Jeep at 45 miles per hour happened to trip and fall? The answer, derived from simple physics, was decisive.

In a running position, *T. rex*'s head would be about 11 feet above the ground. At 45 miles per hour, the 5-foot-long head of a falling *T. rex*, with another 45 feet of its 6-ton body coming from behind, would hit the ground with a deceleration equal to 16 times the pull of Earth's gravity—about 40 tons—enough to pulverize the creature's skull and excavate a crater 8 inches deep. Then, the colossal corpse would have

Figure 34-1 **Could a dinosaur outrun a human being?** Early researchers assumed that because dinosaurs were so big, they were as slow moving as elephants. *(K. Perkins/J. Beckett, courtesy of the Department of Library Sciences, American Museum of Natural History)*

skidded some 50 feet until the mangled body finally overtook the head, breaking the monster's neck. Not a pretty sight.

Thus, Farlow and Robinson independently confirmed Alexander's calculations, estimating that *T. rex*'s maximum safe speed was probably 18 to 22 miles per hour. A Jeep would easily leave the dinosaur in the dust. As for Jeff Goldblum and the rest of us, it is a good thing the problem of outrunning *T. rex* and its 6-inch teeth is purely theoretical. It's a rare human who can run that fast.

How can biologists know the anatomy, physiology, and habits of long-dead animals? Based on zoologists' extensive knowledge of the anatomy and habits of living animals, biologists can draw detailed conclusions from the skeletons and teeth of extinct animals. Before R. McNeill Alexander estimated the movements of dinosaurs, he spent years studying how living animals run and jump.

As we ask how animals move in the rest of this chapter, we will focus on general ideas that account for why their skeletons and their muscles look and act the way they do. We will see that the shape of a structure determines how it works. Zoologists sum up this idea by saying, "Form follows function."

WHY ARE ANIMALS SHAPED THE WAY THEY ARE?

We can study form at many levels. We can look at the overall shape of an animal or we can look at the shape of the parts of an animal. For example, the outside of a fish is streamlined, which eases its passage through water. We can look at the heart to see how it pumps, or we can look at the microscopic structure of heart muscle to see how it differs from other kinds of muscle tissue.

Every organ of the body is shaped in a way that reveals its function. The shape of a leg bone can tell us whether its owner was a runner, a jumper, a walker, or a swimmer. The shape of a single tooth tells volumes about the owner's way of life. The large shearing teeth of lions reveal that they survive by tearing flesh from other animals. In contrast, the broad, flat grinding teeth of antelope tell us that they eat plants. The stomachs of lions and antelope are just as different (Figure 34-2). In cells also, shape reveals function. We can see how the branched shape of a nerve cell is related to how it interacts with other cells. Even the shapes of molecules can tell us something about how they work.

As we discuss the shapes of skeletons and muscles in this chapter, we need to remember that the forms of organisms re-

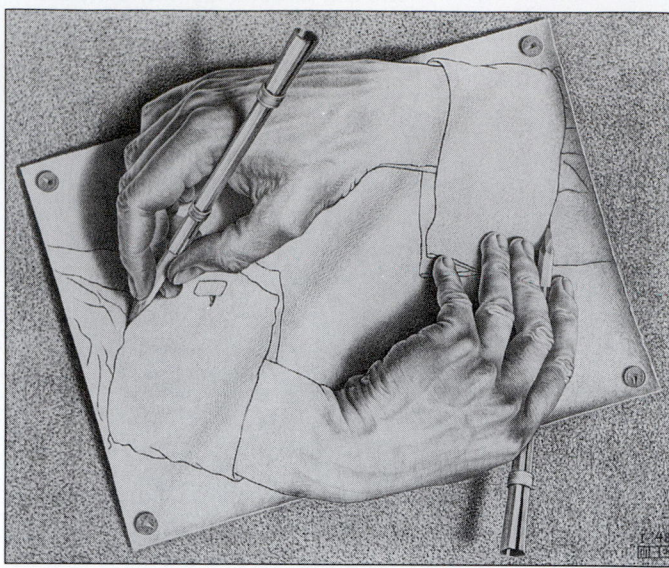

Figure 34-3 Form follows function, but function also follows form. (*Drawing Hands by M. C. Escher. © 2000 by Cordon Art-Baam-Holland. All rights reserved.*)

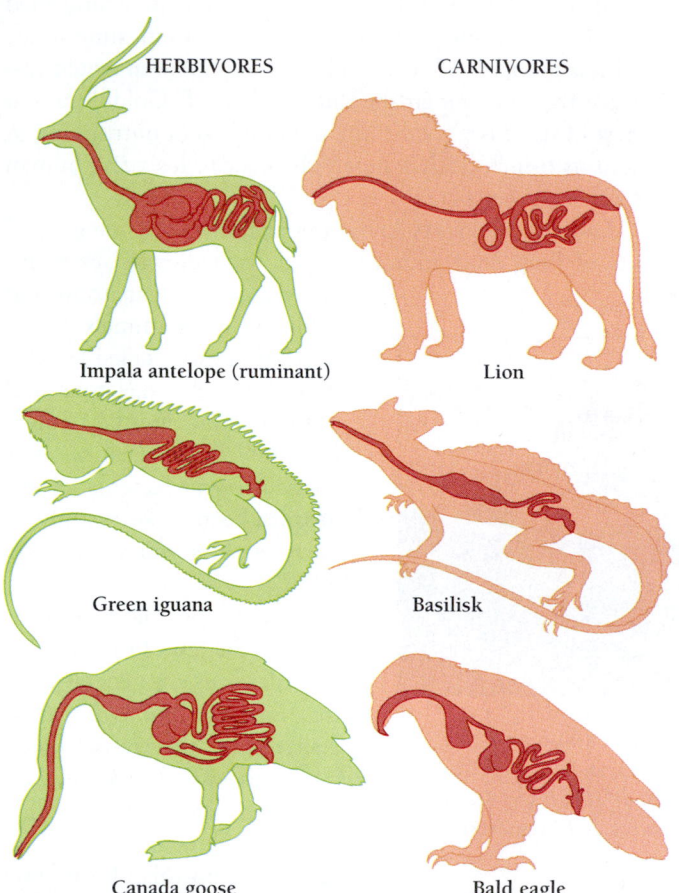

HERBIVORES   CARNIVORES

Impala antelope (ruminant)   Lion

Green iguana   Basilisk

Canada goose   Bald eagle

Figure 34-2   Grass eaters have big digestive tracts. Because plants, especially grasses, contain large amounts of cellulose and other indigestible compounds, a plant eater requires a larger digestive system than a meat eater. Herbivores must eat more material than carnivores to obtain the same quantity of nutrients. And, for full digestion to occur, the large amounts of low-quality food must remain in the digestive tract for a long time. Because meat is a rich source of protein and other nutrients, animals that eat meat generally need to process their meals very little. As a result, their digestive systems tend to be simple and small.

sult from successive adaptations over long periods of time. Our hands are adapted from paws, which evolved from the lobed fins of long-extinct fishes. Likewise, the bones in our ears evolved from the jawbones of fishes. Every organ bears within it clues both to its function and to its evolutionary history.

As an animal evolves, a bone may take on a new function. One that formerly supported a small amount of weight may have to support a greater weight. Natural selection will favor a heavier skeleton. In this respect, form follows function. That is, form evolves in response to the demands placed on the body part—whether it is a leg bone or a cell membrane protein.

But it is just as true that function follows form. The immediate function of an organ is limited by its shape. So, while in evolutionary terms, form follows function, in day-to-day terms, function follows form. We can think of this as like one of the drawings of M.C. Escher, where the hand draws a hand, which draws a hand (Figure 34-3).

What determines the form of an individual animal? Form can change over time. Most of us have had the experience of doing physical work for a while and developing big muscles, then losing them when we slack off.

But most form takes shape during the development of an animal from embryo to adult. Such form results from the interaction of genes handed down from parents with the environment of the developing embryo. The relationship between genes and environment is complex. The environment influences the expression of genes and the genes influence the environment, which continues to influence the expression of genes (Chapter 9). This complex interplay continues into adulthood.

How Are the Parts Laid Out?

When we go on a journey, we often look at a map first, to see where we'll be going. In Part VII of our book, on anatomy and physiology, it helps to keep in mind a map, the overall body plan, of an animal. Although we will discuss invertebrates occasionally for comparison, our emphasis is on vertebrates.

All vertebrates and their embryos share a common body plan—a tube inside a tube. The inner tube is the intestinal tract, which extends from mouth to anus. The intestines and associated organs (such as the liver and pancreas) lie within a cavity, the **coelom,** which separates the inner and outer tubes. In all vertebrates, from fish to birds, the coelom has two distinct parts: the **thoracic cavity,** which encloses the heart and lungs, and the **abdominal cavity,** which encloses most of the length of the intestinal tract, as well as the liver, pancreas, and kidneys.

Running along the back is the backbone, or spinal column. At the **anterior,** or front, end of the spinal column is the skull; at the **posterior** end is the tail. Just behind the skull, the *pectoral girdle* supports the forelimbs—fins in fishes, forelimbs in horses and humans. Behind the pectoral girdle are the ribs, also attached to the backbone, and finally, the *pelvic girdle,* which supports the pectoral fins in fish and the hind limbs of all other vertebrates. The bones of the tail are continuous with the bones of the backbone.

What Are Animal Bodies Made Of?

Animals are made up of different kinds of specialized cells. These can include nerve cells, skin cells, and hundreds of others. Groups of similar cells form **tissues** such as skin or bone that are units of both structure and function. Tissues include not only cells, but also any materials the cells secrete, called **matrix.** For example, bone cells secrete a matrix of proteins and minerals that give bone tissue strength and flexibility.

Biologists group most tissues into four classes: epithelial, connective, muscle, and nervous. **Epithelial tissues** tend to be flat sheets of tightly interconnected cells. Epithelial cells cover the outsides of our bodies, as the outermost layer of the skin, and line the inside of the digestive system, the lungs, and the reproductive system. Epithelial tissue may consist of several layers, as in the epidermis of the skin, or a single, thin layer, as in the lining of the intestines. Very similar to epithelial tissue is the *endo*thelial tissue, flat cells that line the heart and blood vessels. This tissue has a different name because it develops from a different group of cells in an early embryo.

Connective tissues such as cartilage and bone provide mechanical support to the rest of the body. Connective tissues often consist of individual cells dispersed in a flexible matrix of proteins and other materials. Connective tissue forms a layer beneath the skin, as well as fat, ligaments, tendons, and the thin membranes surrounding the heart, lungs, and gut.

Muscle tissue consists of all kinds of cells that can contract. Muscle tissue includes the **skeletal muscle** in our limbs, the **smooth muscle** that works our intestines and reproductive tracts, and the **cardiac muscle** of the heart.

Nervous tissue consists of two kinds of cells, the familiar branched **neurons,** which conduct electrochemical signals from one end of the body to the other, and **glial cells,** which support the neurons. Nervous tissue forms the brain, the spinal cord, and the nerves.

Organs are two or more kinds of tissues that together form distinct structures with specific functions. A kidney, for example, contains both epithelial and connective tissues. Groups of organs together form **systems** of functioning organs. A mammalian kidney is just one part of the "excretory system." The kidney directs urine into the ureters, which go to the bladder, which empties into the urethra. The reproductive system, the circulatory system, and the nervous system are other examples of systems, all of which we examine in more detail in later chapters.

How Are Size, Surface Area, and Shape Related?

When we look at a mouse and an elephant, the most obvious single difference is that one is big and one is small. An elephant's size greatly limits its shape. To support its tremendous weight, its legs must be thick and straight, its feet wide and well padded. In contrast, the legs of a mouse are slender and flexed, a shape typical of smaller animals.

But animals differ in other respects as well. Depending on where they live (their habitat) and how they live (their niche), they may have organs with large surface areas or small ones; they may have complex digestive systems or simple ones.

How does size affect shape?

A mouse could not be as large as a mammoth, and a polar bear could not be as small as a cat. For every type of animal, there is an appropriate range of sizes, and the evolution of a bigger or smaller size demands a change in form, as well.

One reason is gravity. As the linear dimensions of an animal (or any object) double, its area increases fourfold. In other words, both the surface area of an animal and the cross-sectional area of its bones increase four times. In contrast, the volume and weight increase eight times. The result is that the animal's bones and muscles bear a proportionately heavier load.

In general, we can say that the heavier the animal, the thicker its bones must be to support its weight. For the graceful little gazelle to become as big as a rhinoceros, for example, it would have to evolve thick, heavy legs, so that every pound of weight had the same cross-sectional area of bone to support it. A little gazelle cannot simply become a big gazelle. It must become something different. Its whole lifestyle must change, for it can no longer prance about and turn on a dime to escape a swift predator. Being big, however, the animal can defend itself against many predators if it is sufficiently aggressive, which a rhinoceros is.

Humans are no exception to this rule of size. Very tall humans, like very tall dogs, tend to suffer hip problems, a result of the enormous stresses generated by too much weight on too little bone. Humans, as a species, lack adaptations for a height of 7 or 8 feet.

Gravity affects the lives of large and small animals in other ways as well. Gravity will cause an ant (or any other object) dropped from an immense height to accelerate as it drops. Because force equals mass times acceleration, the force with which an object hits the ground is equal to its mass times the acceleration (which is 1 g, or gravity, on Earth). But an ant's mass is so small that the force with which it hits the floor is small—not enough to hurt it. As a result, an ant dropped from the top of a ten-story elevator shaft will likely walk away unscathed.

The same is true of mice, which can fall from great heights without injury. But larger animals are another matter. Occasionally, a small child miraculously survives a fall from a second-or third-story window. More often, the child is seriously hurt. An adult human could not possibly survive such a fall without broken bones or worse.

But small animals' lack of mass is only part of the reason they survive falls better than large animals. Ants and other small creatures are also saved from falls by their relatively large surface areas. Wind resistance, which is proportional to surface area, slows any falling object until the object reaches a "terminal velocity." Once an object reaches terminal velocity, it won't fall any faster. Like a parachute, an ant's relatively large surface area quickly slows the falling ant, and its terminal velocity is therefore low. The ant hits the ground lightly and slowly, rights itself, and walks away. In contrast, an elephant, whose relative surface area is small, would take a very long time to reach its terminal velocity—far longer than it takes to fall ten stories.

We can draw interesting conclusions from this kind of discussion. We can, for example, see that small animals are more suited to life in trees and other high places. We are not surprised to learn that while most primates live in trees, the larger primates, such as humans, gorillas, baboons, and chimpanzees, are more likely to be found safely on the ground. Likewise, large herbivores that eat the leaves of trees, such as elephants and giraffes, do not generally climb trees. Instead, they have evolved necks or trunks for reaching up into tall trees while keeping four feet firmly on the ground. As we saw at the beginning of this chapter, large animals must be careful to avoid not only heights, but even running too fast.

Because of gravity, large animals must have proportionately stronger and thicker bones than small animals. Large animals also fall much harder than small animals.

Why Does Metabolic Rate Depend on Size?

As gravity influences size, so size influences metabolic rate. Mice, shrews, hummingbirds, and other small, warm-blooded animals have proportionately greater metabolic rates than larger animals (Figure 34-4). Such small animals use both more calories per gram of body weight and, likewise, more oxygen to fuel this higher metabolism. To survive, a mouse must eat about one-quarter its own weight in food every day. Of course, a mouse doesn't eat nearly as many calories as an elephant, just more per gram of body weight.

Why large animals have proportionately slower metabolisms is not entirely understood. One factor may be that larger animals have proportionately smaller surface areas. As a result, large animals lose a smaller proportion of their metabolic heat. One physiologist calculated that a mouse with the low metabolic rate of a steer would need an 8-inch-thick fur coat to stay warm (Figure 34-5). Likewise, a steer with the high metabolic rate of a mouse would produce so much heat that it could only maintain a stable internal body temperature if its hide were well above the boiling point of water.

But even though metabolic rate is higher per gram in small animals, it's still greater overall in large animals. In fact, among all organisms, including "cold-blooded" animals (heterotherms), plants, and even one-celled organisms, the larger the organism the higher the metabolic rate. How and why size affects metabolic rate in sea stars and other heterotherms is one of the unanswered questions of biology.

Metabolic rate is partly a function of size, but biologists do not yet know why.

How Do Animals Use Expanded Surface Areas?

Animals of the same size can vary enormously in shape. The webbed flippers of a sea lion look nothing like our own hands and feet. And the broad, flat grinding teeth of a horse appear to have little in common with the sharp teeth of a lion. One

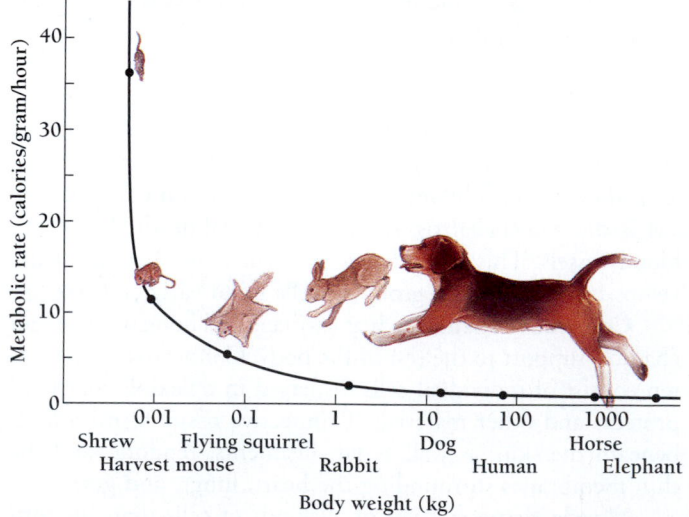

Figure 34-4 Small animals burn more fuel per gram than big animals.

of the many changes that can occur in an organ over the course of evolution is an increase or decrease in surface area. Animals may increase or decrease the surface area of limbs or teeth, as well as the lungs, the digestive tract, and tiny vessels of the circulatory system.

Webbed feet are a way to increase surface area without increasing overall size. Aquatic animals from frogs to ducks must push large volumes of water to propel themselves through the water. Likewise, animals that fly must push large volumes of air to keep themselves aloft. The wings of birds and bats are limbs whose surface area has increased enormously, whether by means of extended bones, long feathers, or wide membranes.

Animals also use large surface areas to increase the radiation of excess heat and to speed the diffusion of molecules across membranes (Figure 34-6). We diffuse oxygen (in the lungs), nutrients (in the digestive tract), and water and salts (in the kidneys). As we will see in the coming chapters, the lungs, the digestive tract, the kidneys, and the circulatory system all have enormous surface areas, whose function is to move nutrients and wastes in and out of the cells of the body.

The rate at which oxygen and other nutrients diffuse across a membrane depends on two factors. The first is how easily the material diffuses across the membrane, and the second is the difference between the amounts of material on the two sides. How easily oxygen moves from the lungs to the blood, for example, depends on how easily oxygen diffuses through the separating membranes, and on how much oxygen is on either side. If the membranes are thick, diffusion will be slow. If the membranes are thin, diffusion will be more rapid.

The rate of diffusion also depends on the gradient between the two sides. If there is a lot of oxygen on one side and little on the other, diffusion will occur rapidly. If, on the other hand, both sides have nearly the same amount of oxygen, then diffusion will occur more slowly. Just as much oxygen will move in one direction as in the other.

If a membrane is thin enough and a gradient is steep enough, diffusion rates can be quite high. But even so, some animals need to move more material even faster. Many evolve a larger diffusion surface. For example, mammals and birds,

Figure 34-6 **Big ears, cool head.** This blacktail jack rabbit rids itself of excess heat through its ears. Many animals cool themselves off by radiating heat through enormous ears or other expanded surface areas. *(Stan Osolinski/Dembinsky Photo Associates)*

which are warm blooded, use more oxygen per gram per minute than heterotherms such as frogs. Mammals and birds therefore have highly expanded lung surfaces. For example, a bit of frog lung the size of a farm blueberry has an internal surface area of about 20 square centimeters (about three lines of text on this page), compared to about 300 square centimeters (about two-thirds of this page) in the same volume of a human lung.

In the digestive system, thousands of folds in the lining of the intestines speed the absorption of nutrients. In the circulatory system, miles of tiny capillaries speed the distribution of those same nutrients, as well as oxygen from the lungs. In the organs of excretion—the kidneys in vertebrates—large surface areas increase the total rate at which water and salts are moved in and out of the blood.

Figure 34-5 **Small animals generate more heat per gram than big ones.** A mouse with the metabolic rate of a steer would need an 8-inch-thick fur coat to maintain a normal body temperature.

Many organs of the body rely on enormous surface areas to function. These include the webbed feet of aquatic animals and the lungs, intestines, and circulation.

Bulk Flow

In single cells and tiny animals such as the tartigrade, or water bear, diffusion is enough to move oxygen, nutrients, and wastes where they need to go (Figure 34-7). But diffusion works only over very short distances. Larger animals depend on **bulk flow** to move materials longer distances. Examples of bulk flow include the movements by which the digestive system pushes food into the stomach and through the intestines, the movement of blood through the circulation, the movement of air in and out of the lungs, the passage of water over the gills of a fish, and the movement of urine through the urethra.

In each case, some force pushes a fluid or substance from one place to another. In the digestive system, muscular movements of the esophagus, stomach, and intestines squeeze the food along. In the circulation, the heart pressurizes the blood, forcing it into lower-pressure capillaries and veins. In respiration, muscles in the wall of the chest and the diaphragm muscle together pump air in and out of the lungs.

How Do Animals Use Countercurrent Systems?

Bulk flow and diffusion often work together to move materials. The heart pumps blood in bulk to the lungs. From the lungs, oxygen diffuses into the blood. The circulation then carries the oxygen in bulk to the tissues of the body, where the oxygen then diffuses into individual cells.

Bulk flow and diffusion can work together in interesting arrangements called **countercurrent systems,** systems in which fluids or gases run past each other and exchange heat or materials. Essentially, countercurrent systems enable animals to establish steep gradients of temperature or concentration.

Animals use countercurrent systems in dozens of ways. Whales use countercurrent systems to keep their fins from radiating too much heat into cold Arctic water. Gazelles living on the hot plains of Africa use countercurrent systems to keep their heads from getting too hot. Fish use countercurrent systems to maximize the absorption of oxygen in the gills and a

similar system to keep their swim bladders pumped up with pressurized gases.

The structures of countercurrent systems are intricate and can involve any tissues in the body. Inside the noses of many vertebrates, including humans, are elaborately folded **turbinate** bones, which increase the surface area of the inside of the nose. Since the mucous membranes inside the nose are moist, evaporation from this large surface area tends to cool the inside of the nose.

Dogs depend on their cool noses to cool themselves off on a hot day. A dog inhales air through its wet nose, which cools the air before it enters the lungs. In the lungs, the cool air absorbs heat from the lungs and is then exhaled out through the mouth. As a result, the dog continually inhales cool air and exhales warm air. Because so much water evaporates inside the nose, it is important for dogs to drink lots of water on a hot day.

Desert animals, which conserve water by letting their body temperature rise on hot days, use their cool noses even more efficiently than dogs. The brain, unlike the other organs of the body, functions poorly at high temperatures. Many of us have experienced the delirium that sometimes accompanies a high fever, for example. In the desert, keeping the whole body cool would require more water than is usually available. The solution is to allow the body to heat up while keeping the head cool.

Many gazelles and antelopes keep a large volume of cool blood from the nose in an expanded blood vessel called a sinus. As hot blood from the heart enters the head, it separates into a network of hundreds of small arteries, which pass through the sinus filled with cool blood from the nose. In the oryx, such an arrangement keeps the brain nearly 3°C cooler than the central arteries of the body (Figure 34-8A).

Dolphins and whales use countercurrent systems to keep the cold blood in their flippers from chilling the rest of the body (Figure 34-8B). The body of a whale that swims in the freezing waters of the Arctic Ocean is well insulated against the cold. But its flukes and flippers, which are thin and flat, become extremely cold.

To prevent blood that has cooled in the flukes from chilling the body, the veins that carry the cold blood back to the body run adjacent to and surround the arteries that carry warm blood from the heart to the flippers. As a result, the warm arterial blood leaving the body heats the cold venous blood before it enters the body. Likewise, the cold venous blood leaving the fins cools the arterial blood before it enters the flippers.

This system, specifically called a *countercurrent heat exchanger,* occurs also in the legs of wading birds and in the testes of mammals. In most mammals, sperm develop best when the testes are cooler than the rest of the body. A countercurrent heat exchanger, along with other adaptations, keeps the testes cool but prevents the them from radiating too much heat and cooling off the body.

The end result of a countercurrent system is a steep gradient—often in temperature. But the same system can be used to create pressure gradients or concentration gradients. In Chapter 39, we'll see how such gradients in the kidneys control the balance of water and salts in the blood and urine.

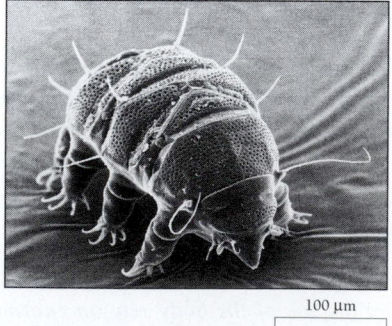

100 μm

Figure 34-7 Look Ma, no circulation! Tardigrades, such as this one, as well as rotifers and other microscopic animals, rely on diffusion alone to move nutrients and wastes. They need no circulatory system. (*Diane R. Nelson*)

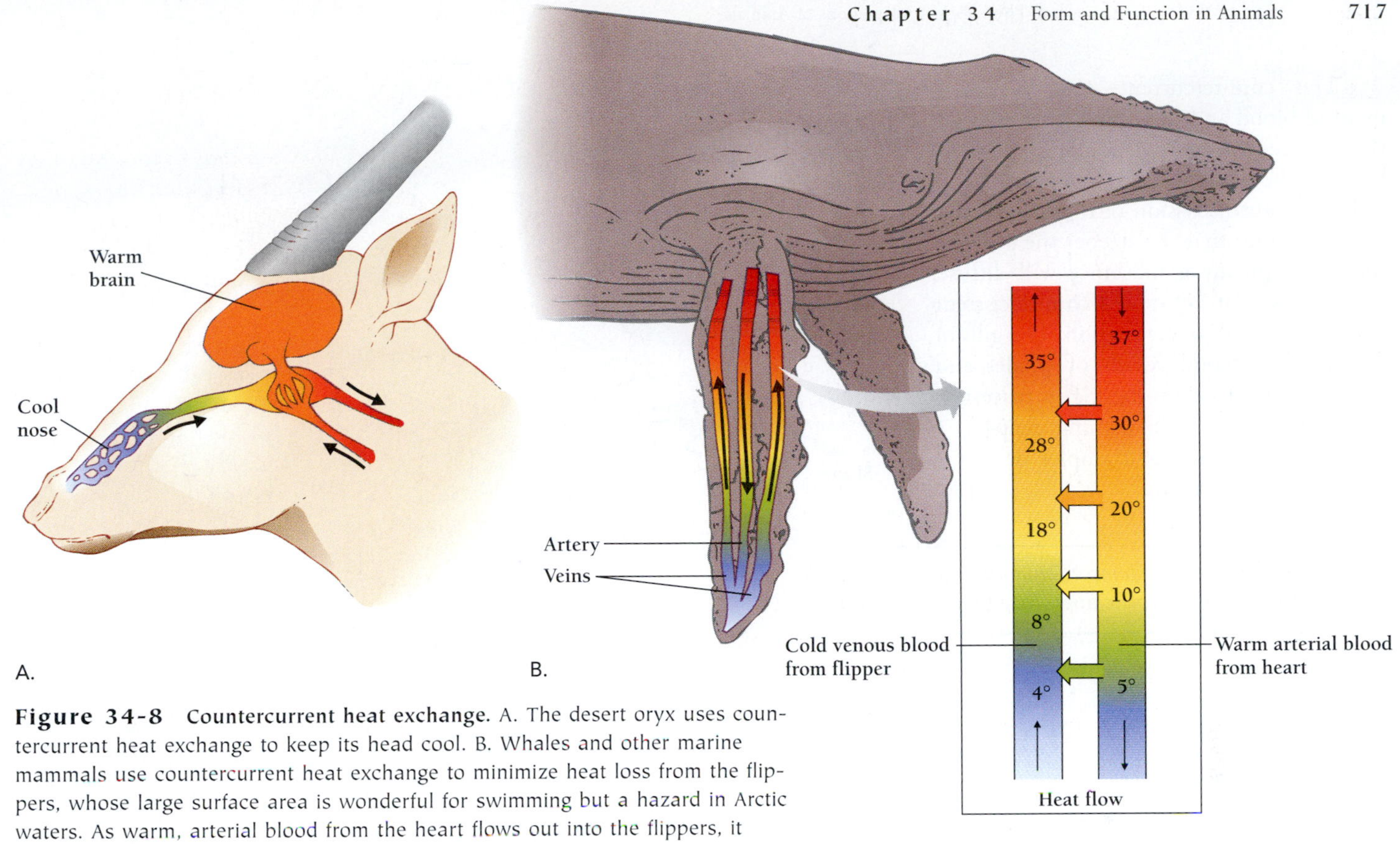

A.

B.

Figure 34-8   Countercurrent heat exchange. A. The desert oryx uses countercurrent heat exchange to keep its head cool. B. Whales and other marine mammals use countercurrent heat exchange to minimize heat loss from the flippers, whose large surface area is wonderful for swimming but a hazard in Arctic waters. As warm, arterial blood from the heart flows out into the flippers, it passes in close proximity to the returning venous blood, which is cold. The arterial blood becomes cooler and cooler as it enters the flippers, while the venous blood, warmed by the arterial blood, becomes warmer and warmer. In this way, body heat stays in the interior and the flippers remain cool.

Fish use a countercurrent system to extract oxygen from water flowing past their gills. The gills of fish consist of a huge surface folded into flat panels, or lamellae. Water enters a fish's mouth and flows out of the body by way of the gills (Figure 34-9). Blood vessels inside of the lamellae are arranged so that the blood flows in the opposite direction to that of the water. The water that first meets the blood in these vessels is rich in oxygen. But the blood coming the other way has already picked up oxygen from water it has already passed. So the blood is carrying a heavy load of oxygen already. Because the fresh water had even more oxygen, however, the blood picks up just a bit more. The water then moves past blood that is ever poorer in oxygen. At every point along its path, the water carries more oxygen than the blood and oxygen will diffuse into the blood.

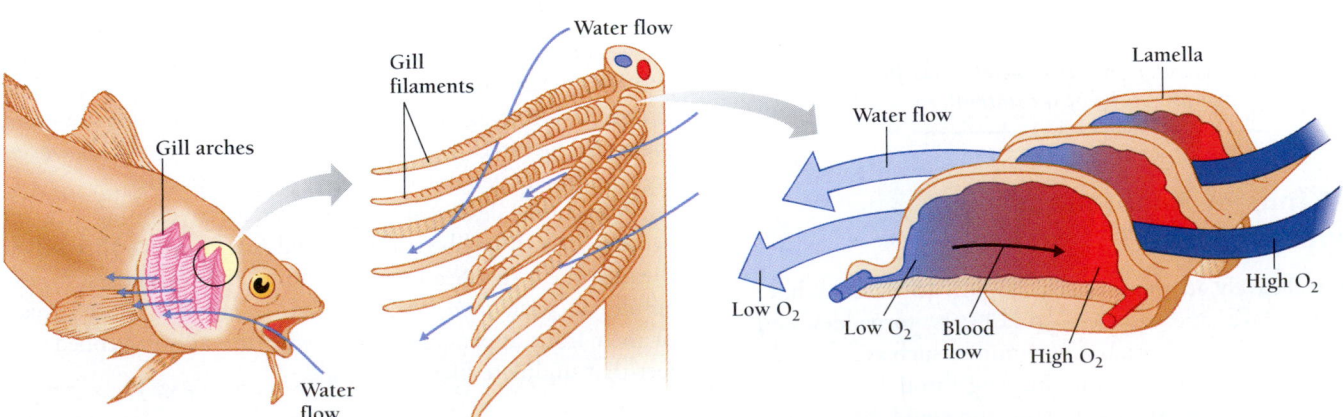

Figure 34-9   Countercurrent oxygen exchange. In fish gills, a countercurrent system ensures that the maximum amount of oxygen is transferred from water flowing past the gills to blood flowing in the opposite direction.

The countercurrent arrangement of blood and water maximizes the steepness of the oxygen gradient between the blood and the water. Recall from our discussion of diffusion earlier in this chapter that the steeper the gradient, the more quickly a substance will diffuse.

The exact structure of a countercurrent system is critical to its function. The gills of fish and the blood vessels of whales and oryxes must develop the right pattern to function effectively. In coming chapters, we will see other examples of countercurrent systems.

Seal

Fish

Shrimp

A.

Countercurrent systems enable animals to establish steep gradients of temperature or concentration.

HOW DOES STYLE OF LOCOMOTION INFLUENCE SHAPE?

An animal's style of locomotion determines shape just as size does. In order to propel themselves forward, animals must push against whatever is available. Aquatic animals push against water, terrestrial animals push against the ground or other objects, and flying animals push against the air. Both fast swimmers and flyers tend to be streamlined. Slower movers are less so. But whereas most fish and dolphins use their tails for propulsion and their appendages for guidance, birds use their appendages for propulsion and their tails for guidance.

Terrestrial animals can slither, crawl, walk, run, or jump. Each style of movement uses the limbs in different ways. A deer and a kangaroo are similar in size and habits: both are medium-sized herbivores that escape from predators by running. But the kangaroo is adapted to hop on two legs and the deer is adapted to bound on four. These different styles give the two kinds of animals very different anatomies.

Style of locomotion partly determines the overall shape of an animal.

What Kinds of Adaptations Do Swimmers Have?

Animals exclusively adapted to swim rapidly in water all have the same oblong shape, also seen in fast cars, torpedoes, and the fuselages of aircraft. Vertebrate swimmers such as fish, seals, and eels propel themselves by undulating through the water. Invertebrate swimmers, such as shrimp and squid, have a similar shape (Figure 34-10A). The strongest swimmers—dolphins, tuna, and mackerel, for example—all propel themselves

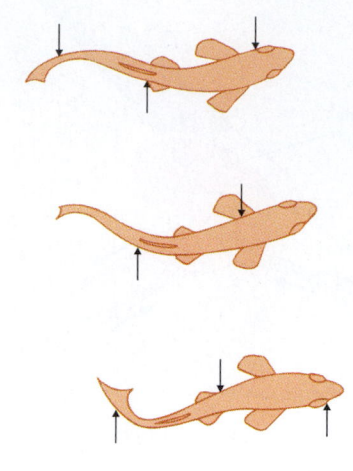

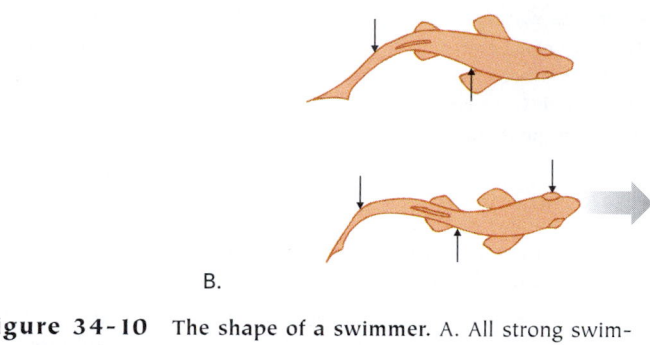

B.

Figure 34-10   The shape of a swimmer. A. All strong swimmers share the same streamlined shape, including shrimp, fish, and seals. B. When a fish swims, it pushes alternately from side to side. The sideways forces cancel out, but the slight push backward against the water pushes the fish forward with each wriggle.

by sweeping their tails back and forth. A swimmer's tail is broad and flat, a shape that forces water to flow against the tail. The tail exerts forces against the water toward both the sides and the rear. The forces to the side cancel each other out, however, and the net movement is forward (Figure 34-10B).

Vertebrates less well adapted for fast swimming rely on various sorts of webbed flippers and feet to move themselves through the water. Sea lions, penguins, and platypuses, for example, all push themselves through water with short strokes of their front appendages, often with the aid of webbed rear feet that undulate like a fish's tail.

Most swimmers undulate through the water.

How Do Snakes Crawl?

Snakes usually undulate across the ground in the same way that eels move through water. Each curve of a snake's body pushes against small objects and rough surfaces on the ground. As with the side-to-side sweep of a fish's tail, the lateral forces mostly cancel each other out, while the forces toward the rear push the snake forward (Figure 34-11). Snakes can use this motion to go straight up a small cliff.

Snakes on extremely smooth surfaces cannot easily move by undulating, as there is nothing for them to push against. In these circumstances, many snakes, especially short ones, resort to **rectilinear** movement, in which the snake progressively bunches up and relaxes its ventral (belly) scales, or scutes.

A very different kind of snake motion is **sidewinding,** in which the snake moves across loose or sandy soil by throwing successive coils sideways. The snake's body is in contact with the ground at two or three places at a time. The parts of its body in contact with the ground are momentarily stationary, whereas the loops that span the tracks are moving and are held clear of the ground. Most snakes are capable of this peculiar movement. Desert species such as the sidewinder rattlesnake, however, are especially good at sidewinding, which has the advantage of keeping much of the snake's body off the hot ground (Figure 34-12).

Snakes generally undulate like fishes, but many also move by rectilinear movement and sidewinding.

What Kinds of Adaptations Do Flyers and Gliders Have?

Flying offers enormous advantages to both invertebrates such as butterflies, locusts, and beetles and to vertebrates such as birds and bats. Flying offers a quick escape route from ground-dwelling predators. Flyers can wander great distances in search of mates, nest sites, and food; and flyers can migrate seasonally, so that they can enjoy summer and spring all year round. Flyers can feed from the flowers at the tops of trees, as well as from any kind of food that is dispersed over a wide area. Some flyers even feed on other flyers. Bats and many birds eat insects, while some hawks specialize on other birds. Finally, good flyers can disperse widely, so that one species can colonize whole continents or hemispheres.

Birds can fly enormous distances and achieve amazing speeds. Thousands of golden plovers fly nonstop from the Aleutian Islands to the Hawaiian Islands every fall, neither eating nor sleeping for days. The total distance is some 2500 miles. Each year the Arctic tern flies 12,000 miles from the Arctic Circle to Antarctica, and then back. One pilot estimated the speed of a flock of migrating sandpipers at 110 miles per hour. A peregrine falcon in a vertical dive, or stoop, was clocked at a speed of 125 miles per hour. And wind tunnel experiments with Lagger falcons suggest that this species may reach a terminal velocity of 225 miles per hour.

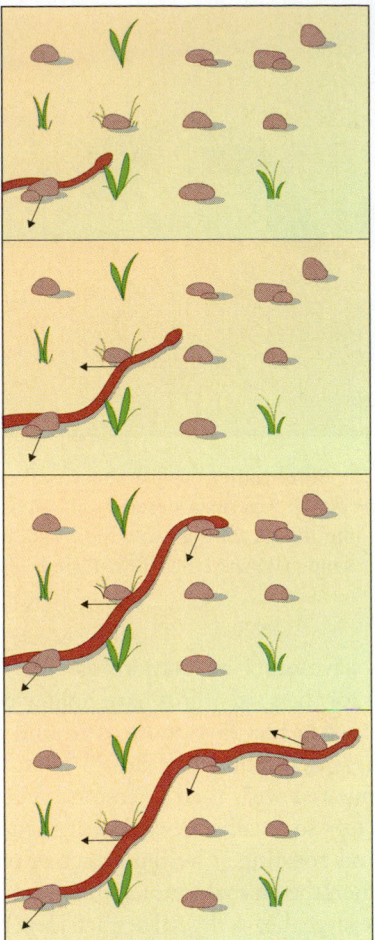

Figure 34-11   Swimming on land. When snakes, such as this blotched king snake, undulate from side to side through grass or across a bare rocky ground, they are doing the same thing as fish. The sideways forces cancel out, but the slight push backward against a rock or plant pushes the snake forward. (*Photo, E.R. Degginger/Dembinsky Photo Associates*)

Flying demands extreme specializations that can limit flyers when they are on the ground. The most obvious adaptation is wings. All vertebrate flyers, including bats and birds, have adapted the forelimbs for use as wings. Wings are virtually useless for anything but flying. As a result, most vertebrate flyers walk on two legs instead of four. Because birds have no forelimbs, they use their hind limbs for grasping food and twigs in the same way that most other animals use the forelimbs.

Figure 34-12   Sidewinding. Unlike most snakes, the sidewinder of the North American deserts rolls along, throwing successive coils into the air. In this way, the snake minimizes its contact with hot sand. *(Wayne Lynch/DRK Photo)*

But flyers have other adaptations as well. First, they are extremely light for their size. Birds have hollow bones and their skeletons are much lighter than those of ground dwellers. Birds also lack teeth, which tend to be heavy. Most flyers have reduced legs compared with their ancestors. A few flyers, such as swifts, have legs so small and weak that they use them only for perching and roosting. The reproductive organs of flyers often shrink when the animal is not actually breeding. Because herbivores have large, heavy digestive tracts, few flyers eat leaves or grass. Most birds eat insects, flesh, or seeds, which, like meat, are high in protein and fat and low in cellulose. The birds that do eat leaves—geese and quail, for example—are large-bodied birds that take flight reluctantly and with difficulty.

Another requirement of flying is that the body be compact and stiff. Bicyclists know that a bicycle frame must be extremely stiff, so that the force from the pedals is transmitted to the wheels and not lost in flexion of the frame. The skeleton of a flyer must be similarly rigid, so that the power in the breast muscles

Pelvic girdle ——

—— Sternum

—— Sternum keel

is transmitted to the wings. Much of a bird's skeleton is fused into a light, inflexible cage of bone (Figure 34-13). The sternum is enlarged into a keel, which serves as an attachment site for the huge breast muscles.

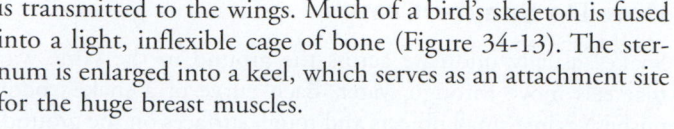

Flying animals have many specializations, including powerful wings; a light, rigid skeleton; hollow bones; reduced jaws and legs; and reproductive structures that shrink when not in use.

How Do Animals Walk and Run?

Most animals walk as if each leg were an upside-down pendulum. The foot is the pendulum's fulcrum, and the animal's center of mass is the hanging weight. At the beginning of each stride, an animal works against gravity, vaulting its center of mass up and over the foot/fulcrum. Gravity then takes over, and the body swings downward until the next leg hits the ground.

When an animal runs, however, the legs behave less like pendulums and more like pogo sticks. When the first leg hits the ground, the animal's body sinks down on the leg. (During walking, in contrast, the body rises up.) But tendons in the legs act like springs, which store and release energy, just like the spring in a pogo stick. These tendons propel the animal upward and forward. Amazingly, nearly all animals run this way, not only horses and cheetahs, but ants, crabs, and cockroaches, too.

It doesn't matter whether an animal runs on four legs, six legs, or two, the pogo stick model applies to all. For example, intensive studies of cockroaches have shown that, using pogo stick locomotion, an American cockroach can run 5 feet per second (more than 3 mph). At top speed, a roach rises up off its front legs so that it is sprinting on just its hindmost two legs. Wind resistance keeps its surfboardlike body from falling back down, so that it can run bipedally. Such bipedal (two-legged) running has evolved in many other lineages, including lizards, dinosaurs, kangaroos, rabbits, rodents, and apes.

What Kinds of Adaptations Do Runners Have?

The speed at which an animal runs depends on two factors—the length of each stride and the number of strides per unit of time. A galloping horse covers about 23 feet (7 meters) at a stride. A racehorse can manage well over 150 strides per minute, so that it can run up to 70 km per hour (45 miles per hour). The much smaller cheetah has the same stride length

◀ **Figure 34-13**   Built for flight. The bones of a flying bird are fused into a rigid frame—like a good bicycle frame—to allow for the efficient transfer of force from the breast muscles to the wings. The backbone is fused to the pelvis and the ribs are fused to the backbone and sternum. The sternum is enlarged into a keel to which the huge breast muscles attach.

as a horse, but it can sprint one and a half times as fast as a racehorse because it moves its legs one and a half times as fast. How can the cheetah have a stride as long as that of a racehorse? And how does the cheetah manage to move its legs so much faster than the horse?

Stride length

All fast-running animals attain great speed by increasing stride length and stride rate. One way that stride length increases over the course of evolution is for the leg to lengthen in proportion to other parts of the body. The parts of the leg farthest from the body usually lengthen the most (Figure 34-14A). One reason that cockroaches run on two legs when they are sprinting is that their hind legs are the longest, which gives them a longer stride. Another way is to make the shoulder highly flexible. Cheetahs and other fast-running mammals have a reduced shoulder girdle that allows them to move their forelimbs as far as possible. The clavicle, or collar bone, is a mere vestige. Effectively, this lengthens the leg still more. Cheetahs, dogs, weasels, horses, and other fast runners also extend their stride length by alternately flexing and extending their spines (Figure 34-14B).

Stride rate

One might think that animal runners would move their legs faster simply by contracting their muscles faster. But, in fact, large animals, which include most of the fastest runners, contract their muscles more slowly than small animals. Muscles that attach far from a joint, or lever, have more leverage than those that attach close to the joint. The armadillo uses its forelimbs to dig through soil with considerable force. As a result, its muscles tend to attach to the bone far from the joint, which gives it more leverage. Its forelimbs move much more slowly than those of a cheetah but with more force. In contrast, muscles that attach close to the joint move much faster, but can exert less force. Animal runners' muscles tend to attach very close to the joint.

Animal runners have other ways of increasing stride rate, but one important way is to keep the legs light. It is easier to move light legs quickly than heavy ones. Consequently, antelopes, cheetahs, and other fast runners are lightly built compared with slower-moving animals. A racehorse, for example, has longer, lighter legs than a draft horse. In runners, the bones of the outer parts of its limbs tend to be reduced and are often fused together for increased strength.

Animal runners increase overall speed by increasing stride length and stride rate. Long legs, specialized hips and shoulders, and a flexible spine all increase stride length. Light legs and muscle attachments close to the joint increase stride rate.

HOW DO ANIMALS USE SKELETON AND MUSCLE TO MOVE?

Up to now, we have focused on general principles of form and function, including how size, surface area, and style of locomotion influence structure in animals. Now we will see how muscles and bone work together to move the body and what molecular processes allow muscles to actually contract.

How Do Muscles and Skeletons Work Together?

All animals move by contracting muscles against a rigid framework called a skeleton. The skeleton can be made of bone, like ours, or out of other materials. It can be internal or external. An earthworm, for example, coordinates muscle action against a **hydrostatic skeleton**, a rigid, fluid-filled space inside the body called the coelom (Figure 34-15).

The majority of animals are arthropods—such as insects, spiders, and crustaceans—whose skeleton is external. We say they have an **exoskeleton** [Greek, *exo* = on the outside], which resembles a hollow cylindrical tube. In some ways, this design is superior to the **endoskeleton** of a vertebrate. A hollow tube can support much more weight than can a solid rod of the same

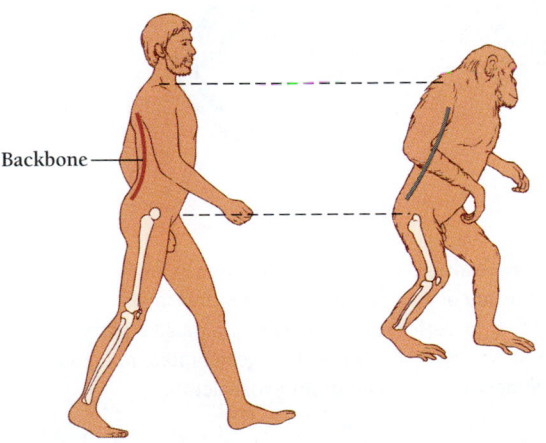

A. Proportionately longer legs

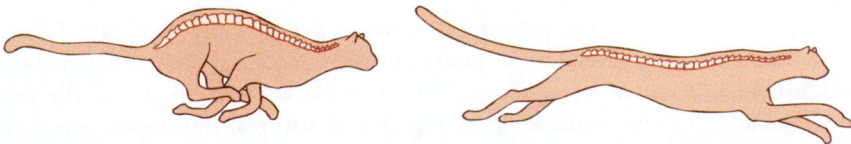

B. Flexible backbone

Backbone

◀ **Figure 34-14   How to increase stride length.** A. Humans are better adapted for running than chimps because, compared with the rest of our bodies, our legs are much longer than those of chimps. B. Runners also frequently have flexible spines, which increases the effective length of the stride.

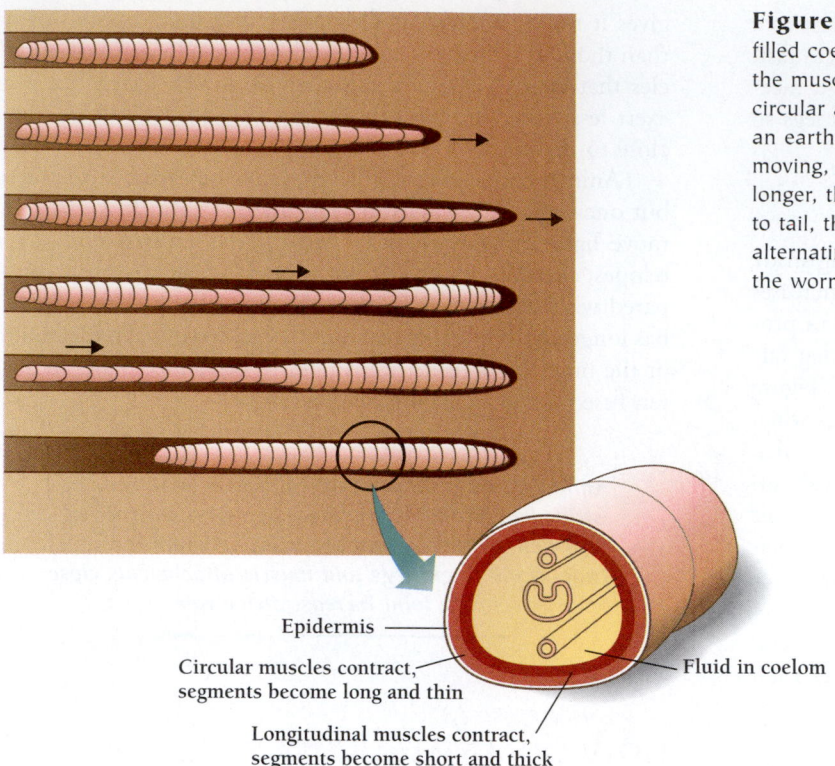

Epidermis

Circular muscles contract,
segments become long and thin

Fluid in coelom

Longitudinal muscles contract,
segments become short and thick

Figure 34-15 **Hydrostatic skeleton.** An earthworm's fluid-filled coelom provides a rigid hydrostatic skeleton, against which the muscles can push or pull. An earthworm uses a combination of circular and longitudinal muscles to inch through the soil. At rest, an earthworm is the same diameter from head to tail. To begin moving, circular muscles contract, squeezing the body into a longer, thinner cylinder. Longitudinal muscles, running from head to tail, then contract (shorten), pulling the body forward. Waves of alternating contractions of circular and longitudinal muscles move the worm through the soil.

weight. In fact, gram for gram, the exoskeleton of an insect can support more body weight than a vertebrate skeleton can.

In all animals, muscles come in pairs that counteract each other's movements. If a butterfly lifts one leg, for example, it can then lower it again. For every action, we need muscles to perform the opposite action.

The flexion and extension of an insect's leg involves a pair of **antagonist muscles.** Each muscle is attached to the exoskeleton in two places. When a muscle contracts, it pulls the two parts of the skeleton together. In the case of a leg, the two antagonistic muscles take turns pulling the leg one way or the other. Muscles always pull when they contract. They never push. Humans and other vertebrates also have antagonistic muscles. Insects' exoskeleton requires an arrangement of muscles and joints different from ours, but the same mechanical principles apply to both invertebrates and vertebrates (Figure 34-16).

Animals move by alternately contracting pairs of
antagonistic muscles, which pull against some
kind of skeleton.

Naming the Parts of the Vertebrate Endoskeleton

The defining characteristic of vertebrates is the **backbone,** a column of hollow bony segments called **vertebrae** [singular, vertebra; Latin, *vertebratus* = jointed]. The **skull,** the bony case that encloses and protects the brain, lies at the forward

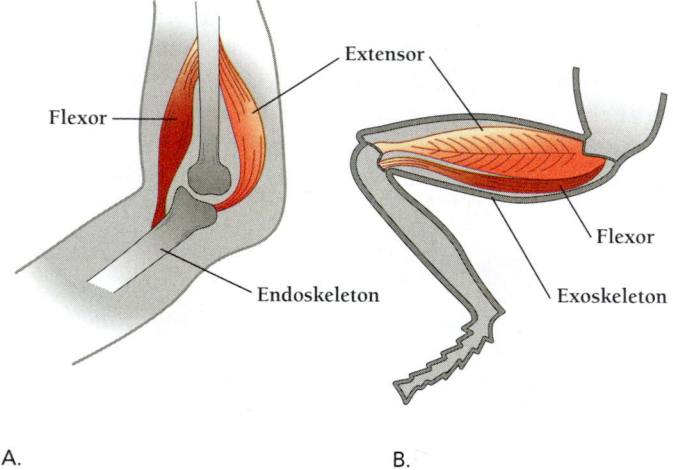

Extensor

Flexor

Endoskeleton

Flexor

Exoskeleton

A.   B.

Figure 34-16 **Antagonistic muscles.** All muscles come in pairs that work in tandem, pulling the elements of a skeleton one way or another. A. In vertebrates, antagonistic muscle pairs attach to the outside of an endoskeleton. B. In invertebrates, antagonistic muscle pairs attach to the inside of an exoskeleton.

(or, for humans, the top) end of a vertebrate's skeleton (Figure 34-17).

The point of attachment between any two bones is called a **joint.** Some joints are mere faces between different bones that never move. The joints between the bones of the skull, for example, grow together at **sutures,** immovable joints that fuse separate bones into a rigid and protective helmet for the brain (Figure 34-18A).

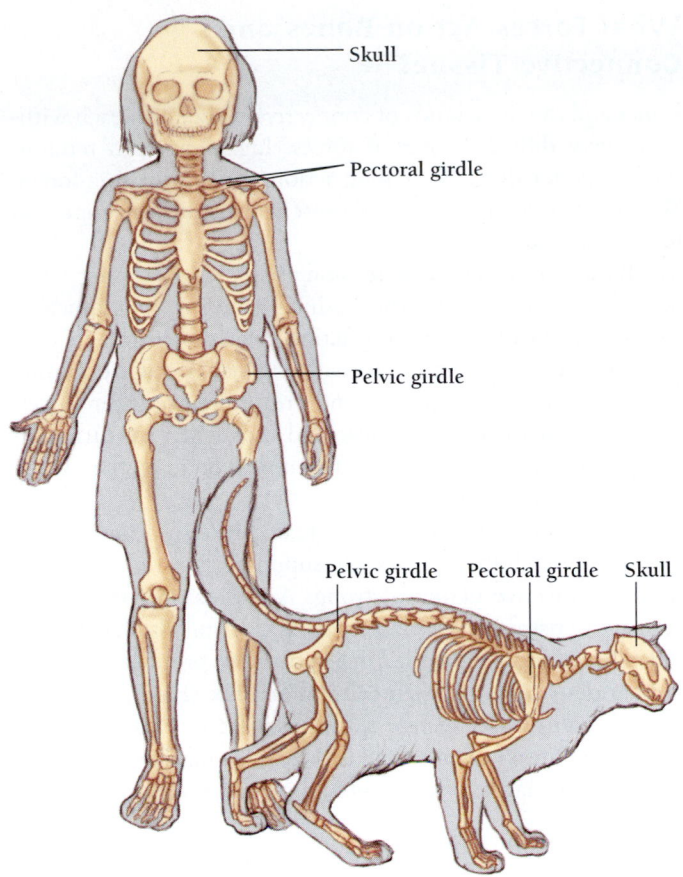

Figure 34-17 **The vertebrate skeleton.** In all vertebrates, the "axial skeleton" includes the skull, rib cage, and spinal column. The bones of the "appendicular system" include the shoulder, or pectoral girdle (clavicle and scapula), and arms, as well as the pelvic girdle and legs. In a few vertebrates such as snakes, many or all the bones of the appendicular system may be missing or vestigial.

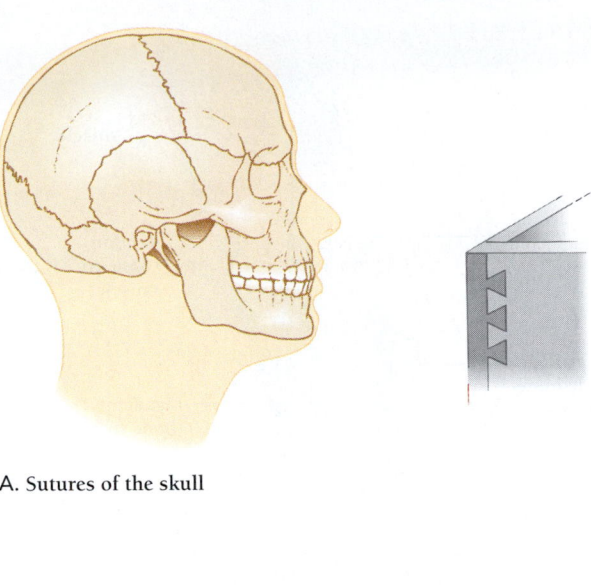

A. Sutures of the skull

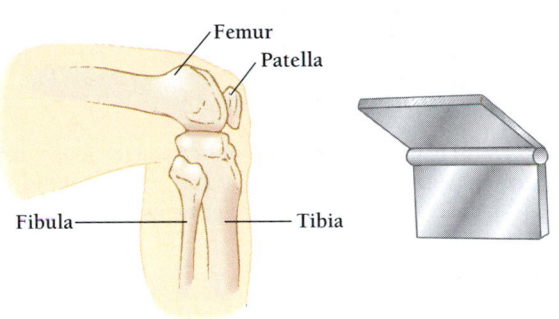

B. Knee: Hinge joint

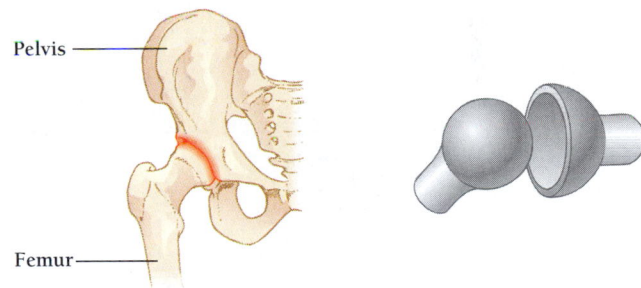

C. Hip: Ball-and-socket joint

Figure 34-18 **How do joints move?** A. Most of the bones of the head lock together in sutures. These bones cannot move relative to each other. B. Most joints allow some degree of motion. Like a hinge on a door, a knee moves in one plane only. An array of flexible tendons and ligaments limit sideways or twisting motions. C. The thigh bone (femur) attaches to the pelvic girdle by means of a ball-and-socket joint. Like a joystick, the femur moves freely both from side to side and up and down.

Most joints, however, allow bones to move relative to each other. The exact structure of a joint determines how the bones move. A knee joint, for example, allows the lower leg to swing back and forth, but not from side to side (Figure 34-18B). This joint is like the hinge on a door; it moves in two dimensions only. In contrast, the joint between the thigh and the hip allows the thigh to swing in wide arcs in any direction (Figure 34-18C).

Three kinds of connective tissue connect the bones and muscles of joints. Spongy connective tissue, called **cartilage,** cushions the joints at the ends of many long bones such as the tibia and femur of the leg. But we can also find cartilage on the tips of our noses and in our outer ears. The skeletons of sharks and rays are entirely cartilage, including no bone at all, as are the skeletons of fetal mammals, including humans. Wrapped around the different parts of a joint and binding them together are tough bands of connective tissue called **ligaments** that resemble rubber bands. Finally, **tendons** attach muscles to the bones (Figure 34-19).

Bone consists of three major tissue types. The outer layer of bone is hard, dense **compact bone tissue** (Figure 34-20). In mammals, the centers and ends of most bones are soft **spongy bone tissue.** Embedded in the spongy bone is an aggregation of **stem cells,** undifferentiated cells from which blood cells develop, called **red bone marrow.**

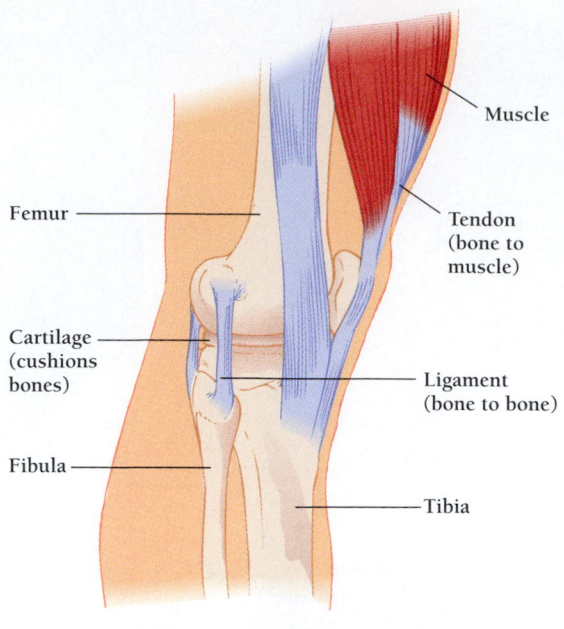

Connective tissues

Figure 34-19   Ligaments, tendons, and cartilage. Tough ligaments and tendons bind together the bones and muscles of the knee. Tendons connect muscle to muscle, ligaments connect bone to bone, and cartilage cushions the joint.

What Forces Act on Bones and Connective Tissue?

Bones and the three kinds of connective tissue must each withstand very different sorts of forces. Ligaments and tendons must resist **tension,** the pulling action of two opposing forces. When we play tug-o-war, we exert tension on the rope and tension on our arms.

Bones must also resist tension as well as two other kinds of stress—**compression,** the pushing action of two opposing forces, and **shear,** the twisting action created by forces that are not opposite one another. Tornadoes twist because opposing winds slide past one another and create intense shear forces. We feel shear forces in our legs and hips when we turn suddenly while running, and we feel compression in our legs when we jump up and down.

Bones and connective tissue are enormously strong: a tendon half an inch in diameter can support a two-ton car. What makes connective tissue so strong? A microscope reveals that connective tissue consists largely of flat, irregularly shaped cells called **fibroblasts** embedded in a network of proteins and polysaccharides called the **extracellular matrix** (Figure 34-21A).

The fibroblasts secrete a fibrous protein called collagen and other substances into the extracellular matrix. The tough extracellular matrix is responsible for the diverse properties of

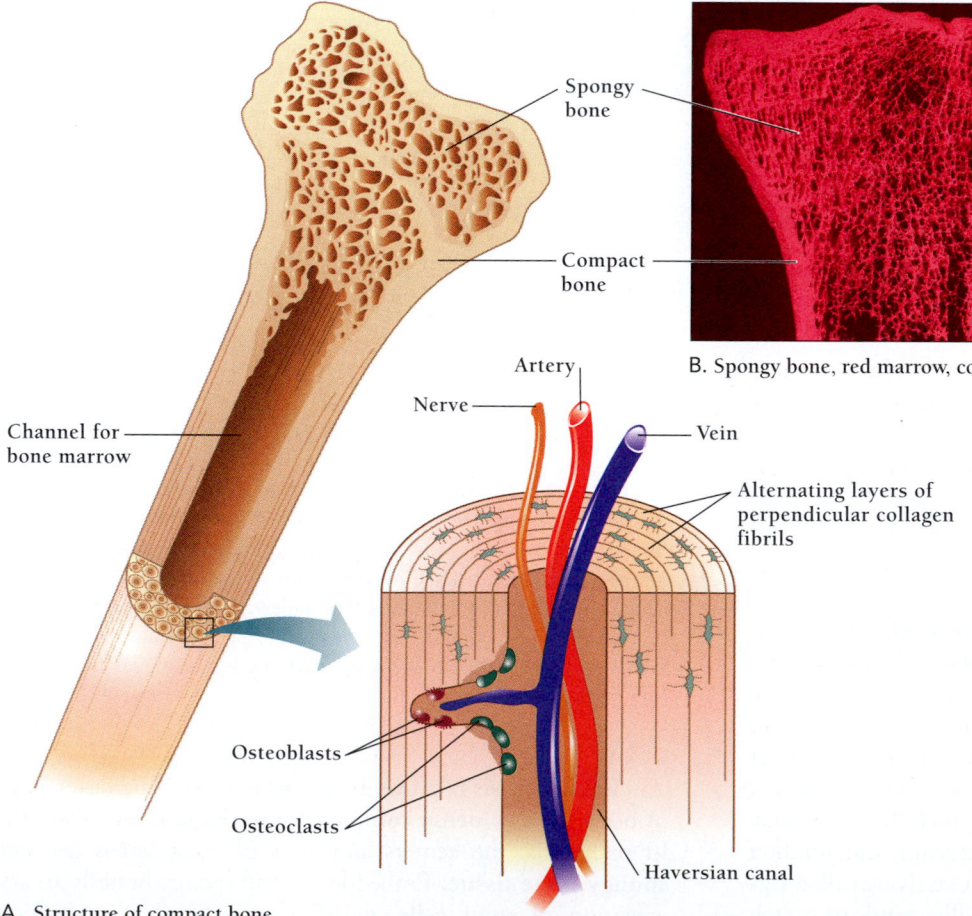

B. Spongy bone, red marrow, compact bone

A. Structure of compact bone

Figure 34-20   The structure of bone. A. Compact bone consists of dense layers of perpendicular fibers of collagen. Embedded in the collagen layers are crystals of calcium and phosphate, which give compact bone its hardness. Within the bone are Haversian canals, thin tubes that run parallel to the length of a bone and contain blood vessels and nerve cells. B. Spongy bone, which lies interior to the compact bone and near the ends of long bones, consists of a lattice of collagen and calcium phosphate, similar to that of compact bone but more open. At the very center of some of the larger bones of the body is bone marrow, where blood cells develop. *(B, Don W. Fawcett/Visuals Unlimited)*

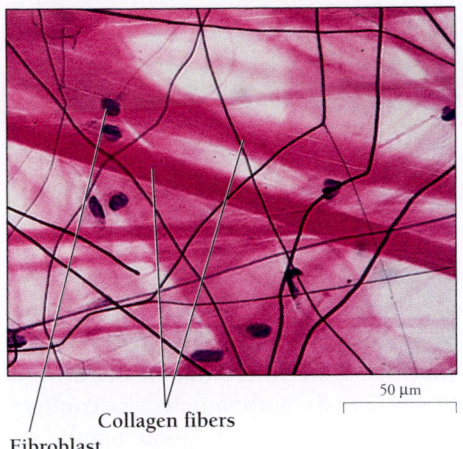

Collagen fibers

Fibroblast

50 µm

A.

Figure 34-21   Why is connective tissue so strong? A. The connective tissue of vertebrates is composed of fibroblasts and collagen fibers. B. A collagen fiber is composed, like a muscle fiber, of successive groups of fibers. Collagen molecules entwine to form collagen fibrils, bundles of which make up each collagen fiber. In connective tissue, these bundles often form perpendicular layers. Like plywood, which is made in the same way, these layers are together resistant to shear forces. *(A, Ed Reschke/Peter Arnold, Inc.)*

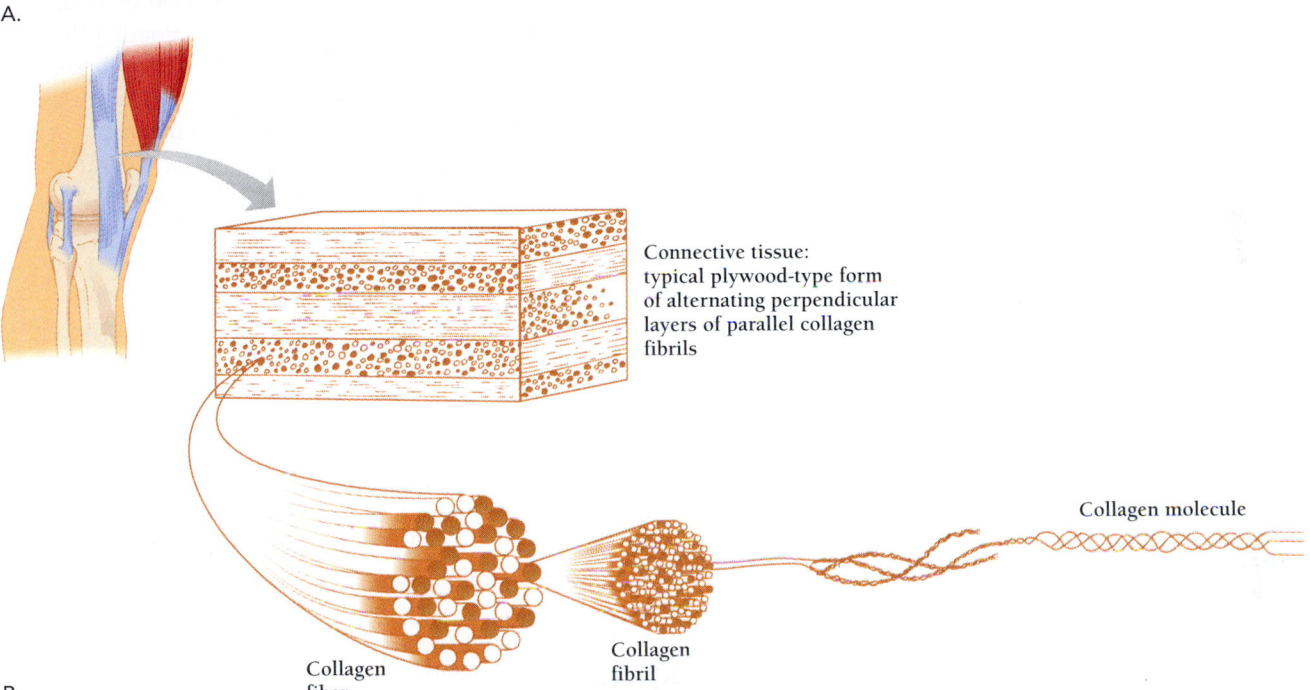

Connective tissue: typical plywood-type form of alternating perpendicular layers of parallel collagen fibrils

Collagen molecule

Collagen fiber

Collagen fibril

B.

connective tissues, which can vary from the soft transparency of the cornea of the eye to the rocklike properties of bone. **Collagen,** which strengthens cartilage, ligaments, tendons, and bones, is the single most abundant protein in the extracellular matrix and amounts to 25 percent or more of all animal protein. Collagen molecules form triple-stranded helices, which assemble into cross-linked cables, called **fibrils.** Fibrils in turn form large, interconnected **fibers.**

The organization of collagen helps connective tissues resist both tension and twisting shear forces. The fibrils resist tension by assembling in parallel along lines of stress, like the fibers of a rope. The fibrils resist shear forces by assembling in perpendicular layers, much like the alternating layers found in plywood (Figure 34-21B).

Bone's resistance to tension depends primarily on the strength of parallel collagen fibrils. But bone derives additional strength from needlelike crystals of calcium phosphate embedded in the extracellular matrix. The calcium phosphate

crystals enable bones to resist both compression and shear forces. If a bone should crack, for example, the calcium phosphate crystals usually keep the crack from spreading far.

Tough collagen fibers arranged in parallel, like the fibers of a rope, or in layers, like plywood, give bones and connective tissue much of their strength.

How Does Bone Respond to Use?

We tend to think of our skeletons as solid and unchanging. Certainly bones are solid, but unchanging they are not. Bones not only grow and repair themselves against continual wear and tear, they also change shape according to how they are used. The building and remodeling of bone is the task of two kinds of cells, the osteoblasts and the osteoclasts. The **osteoblasts** [Greek, *osteon* = bone + *blastos* = bud] are specialized cells that resemble fibroblasts and manufacture the bone matrix. The **osteoclasts** [Greek, *osteon* = bone + *klastos* = broken] are similar cells that digest collagen and bone matrix. In all living bone, the osteoclasts continually break down and absorb bone, while the osteoblasts continually form new bone.

Normally, the rate at which bone is created is equal to the rate at which it is absorbed. In bones that are growing, or in bones that are repairing damage, the osteoblasts produce more bone than the osteoclasts destroy.

Bones adjust their size and density according to the amount of compression they experience. The bones of people who exercise become thick and strong, while those of couch potatoes (and astronauts living at zero gravity) tend to become thin, brittle, and decalcified. If a person favors an injured leg for a long period, the injured leg becomes thin and decalcified, while the other leg bone becomes heavier and stronger. An unused bone can lose nearly a third of its mass. For this reason, physicians now encourage people with healing bone fractures to begin using the limb as soon as possible.

As we age, the osteoclasts outpace the osteoblasts and we gradually lose bone mass. Older people may have such weak bones that they break easily, a condition called **osteoporosis.** In addition, tiny fractures in the vertebrae cause the spinal column to bend, giving many older people a stooped posture.

People who start out with less bone in early adulthood are the most prone to osteoporosis. Women are especially likely to get osteoporosis, not only because they begin adulthood with less bone than men, but because women lose bone mass unusually rapidly while nursing babies and in the years after menopause.

Osteoclasts and osteoblasts continually shape and reshape bones as we grow, heal, and change our habits and physiology.

How Do Skeletal Muscles Move Bones?

The human body contains more than 600 skeletal muscles, all the muscles that are attached to bones (Figure 34-22). Skeletal muscles, together with the nerves that coordinate them, are responsible for our ability to move. Together, the skeletal muscles are the largest tissue in the vertebrate body, making up more than 40 percent of its weight. Some individual muscles are small, with only a few hundred muscle cells; others contain hundreds of thousands of muscle cells. The size, organization, and arrangement of our muscles establish which movements we can and cannot perform.

Like the muscles of invertebrates, our muscles can exert force in only one direction. Most body movement therefore depends on the arrangement of **antagonistic pairs** of muscles, which pull in opposite directions. The contraction of the biceps muscle, for example, flexes the forearm, while the contraction of its antagonist, the triceps muscle, extends the forearm.

In the human body, some 600 skeletal muscles, arranged in antagonistic pairs, work our bodies by pulling against bone and each other.

How Do Smooth and Cardiac Muscle Differ from Skeletal Muscle?

Vertebrates have other muscles besides those that move the skeleton. These include cardiac muscles, which pump blood through the heart, and smooth muscles, which line the walls of hollow internal organs such as the intestines, the uterus, and the blood vessels (Figure 34-23). Smooth muscles also contract and relax the iris of the eye, and they form "goose bumps" to make hair stand on end when we are frightened or cold.

Smooth muscle, as its name suggests, is smoother than either cardiac muscle or skeletal muscle, both of which are striated, or striped. The **striations** consist of cross stripes spaced every 2 to 3 μm and reflect the regular arrangement of the protein filaments that are responsible for muscle contraction.

Because humans and other animals have voluntary control over most of their body movements, skeletal muscles are often called **voluntary muscles.** In contrast, smooth muscle and cardiac muscle are **involuntary muscles.** Most people cannot consciously control the rate at which their intestines move or their hearts beat. In reality, however, some people can control the action of the heart and other smooth muscles.

The difference between voluntary and involuntary muscles lies not only in their structure but in the types of nerves that control them. Skeletal muscles are generally under the control of a system of nerves called the **motor axis.** In contrast, the smooth muscles and the cardiac muscles are controlled by a parallel system of nerves called the **autonomic nervous system.**

The autonomic nervous system consists of two distinct sets of nerves—the *sympathetic* and the *parasympathetic* nerves. The **sympathetic nervous system** brings about the responses of the "fight-or-flight" reaction, slowing movements of the digestive system and speeding up the heart in preparation for vigorous action, for example. In contrast, the **parasympathetic nervous system** acts to save resources, by slowing the heart and increasing the action of the intestines.

Smooth muscles and cardiac muscles are not attached to bones and are under the control of the autonomic nervous system, not the motor axis.

How Do the Two Kinds of Muscle Fibers Work?

Every skeletal muscle consists of many parallel **muscle fibers,** giant cells up to 4 cm long. Each muscle cell, or fiber, contains up to 100 nuclei, as well as the proteins responsible for muscle contraction.

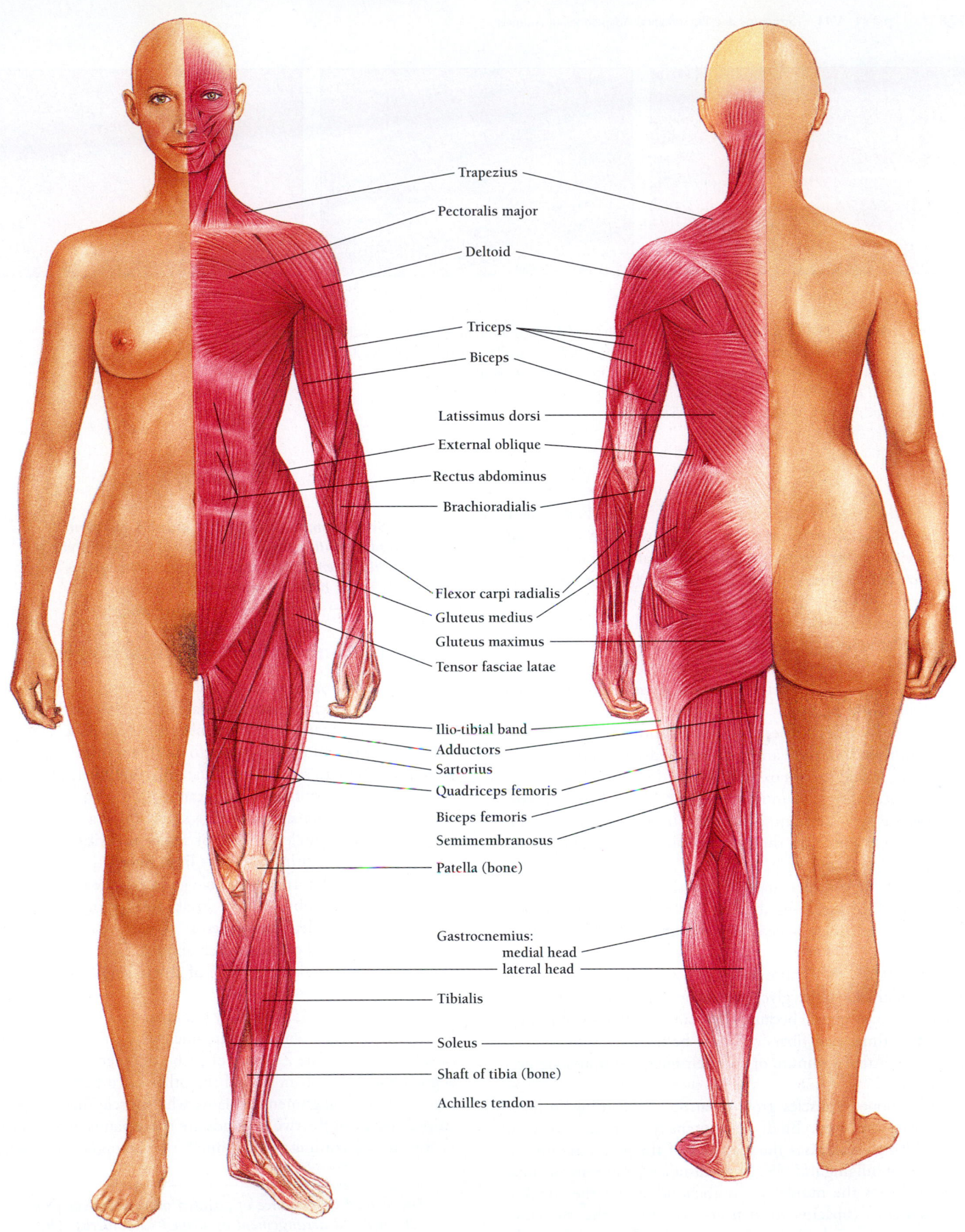

Figure 34-22 **Muscles of the human body.** Shown here are only some of the largest of the hundreds of muscles in the human body.

Trapezius

Pectoralis major

Deltoid

Triceps

Biceps

Latissimus dorsi

External oblique

Rectus abdominus

Brachioradialis

Flexor carpi radialis

Gluteus medius

Gluteus maximus

Tensor fasciae latae

Ilio-tibial band

Adductors

Sartorius

Quadriceps femoris

Biceps femoris

Semimembranosus

Patella (bone)

Gastrocnemius:
medial head
lateral head

Tibialis

Soleus

Shaft of tibia (bone)

Achilles tendon

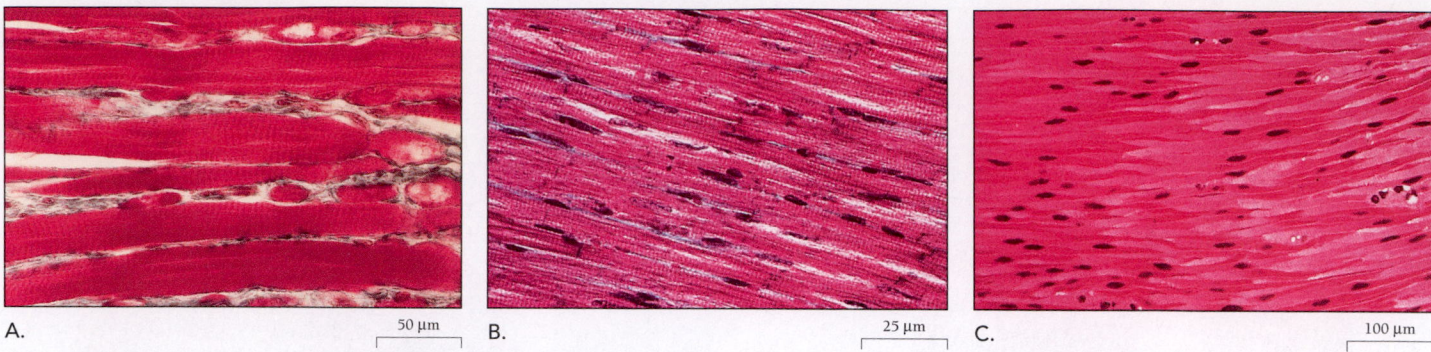

Figure 34-23   **The three kinds of muscle.** A. The striations that enable skeletal muscle to contract are distinctive in micrographs such as this. B. Cardiac muscle is also striated. C. Lacking striations, the muscles of the digestive tract and urogenital organs are called "smooth muscle." *(A, David M. Phillips/Visuals Unlimited; B, Biophoto Associates/Science Source/Photo Researchers, Inc.; C, M.I. Walker/Science Source/Photo Researchers, Inc.)*

Muscle contains two kinds of fibers. Thick fibers, called **glycolytic fibers,** derive most of their energy from glycolysis, the anaerobic breakdown of glucose sugar. Thinner fibers, called **oxidative fibers,** derive most of their energy from cell respiration, the aerobic breakdown of glucose by mitochondria, a process that depends on a good supply of oxygen. Oxidative fibers have many mitochondria, whereas glycolytic fibers have only a few. The thin oxidative fibers power muscles over long periods, whereas the thick glycolytic fibers power muscles in short bursts.

Oxidative fibers depend on a continuous supply of oxygen, and therefore they are well supplied with blood capillaries. The blood carries oxygen on the protein hemoglobin. The oxidative fibers of muscles contain high levels of **myoglobin,** an iron-containing protein similar to hemoglobin that pulls oxygen from the hemoglobin in the blood. Both myoglobin and the rich supply of blood give oxidative fibers, and the muscles that contain them, a distinctive red color. The "red meat" in chicken and turkey legs is red because these muscles have a large proportion of oxidative fibers.

In contrast, the "white meat" of a chicken breast consists mainly of thick, white glycolytic fibers. Glycolytic fibers are also called "fast fibers" because they can deliver a lot of power in a short time. Fast fibers can only be used for short bursts of activity. After a minute or so, their energy supplies are exhausted.

Like bones, muscles grow or atrophy according to how much they are used. Brief, high-intensity exercise—such as weightlifting—increases the diameter of the glycolytic fibers, resulting in bulging muscles. Long-distance swimming or running increases the number of mitochondria and the circulatory system's capacity to deliver oxygen to the muscles. Sprinters and weightlifters tend to develop big muscles with many glycolytic fibers, while long-distance runners tend to develop wiry muscles with many thin, oxidative fibers.

Skeletal muscles consist of parallel glycolytic and oxidative fibers. With heavy use, glycolytic fibers tend to enlarge, while oxidative fibers build up more mitochondria.

How Are Contractile Proteins Arranged?

When early biologists first examined skeletal muscle under the light microscope, its striated appearance suggested that muscle was highly organized at the molecular level. As a result, in the 1950s, when biologists began to use the transmission electron microscope to study subcellular structure, skeletal muscle was among the very first tissues examined.

These early microscopic studies showed that a muscle fiber from a skeletal muscle consists of many **myofibrils** [Greek, *myos* = muscle + Latin, *fibrilla* = little fiber], threads 1 to 2 μm in diameter that run the whole length of the fiber (Figure 34-24). Each myofibril thread itself consists of a string of smaller units, called **sarcomeres** [Greek, *sarx* = flesh + *meros* = part], small cylinders each about 2.5 μm long. A 2.5-cm-long myofibril, then, consists of about 10,000 sarcomeres laid end to end.

Each sarcomere is marked by light and dark stripes of varying width. At each end of a sarcomere is a thin, dark stripe called a **Z band.** The Z bands of each sarcomere are perfectly aligned with the Z bands of all the other sarcomeres within a myofibril. This alignment explains why muscle fibers appear striped. Between the two Z bands are more bands, which vary in thickness according to how much the sarcomere contracts.

The striped appearance of striated muscle derives from the ordered arrangement of striped sarcomeres. The light and dark bands of each sarcomere are aligned with those on adjacent myofibrils.

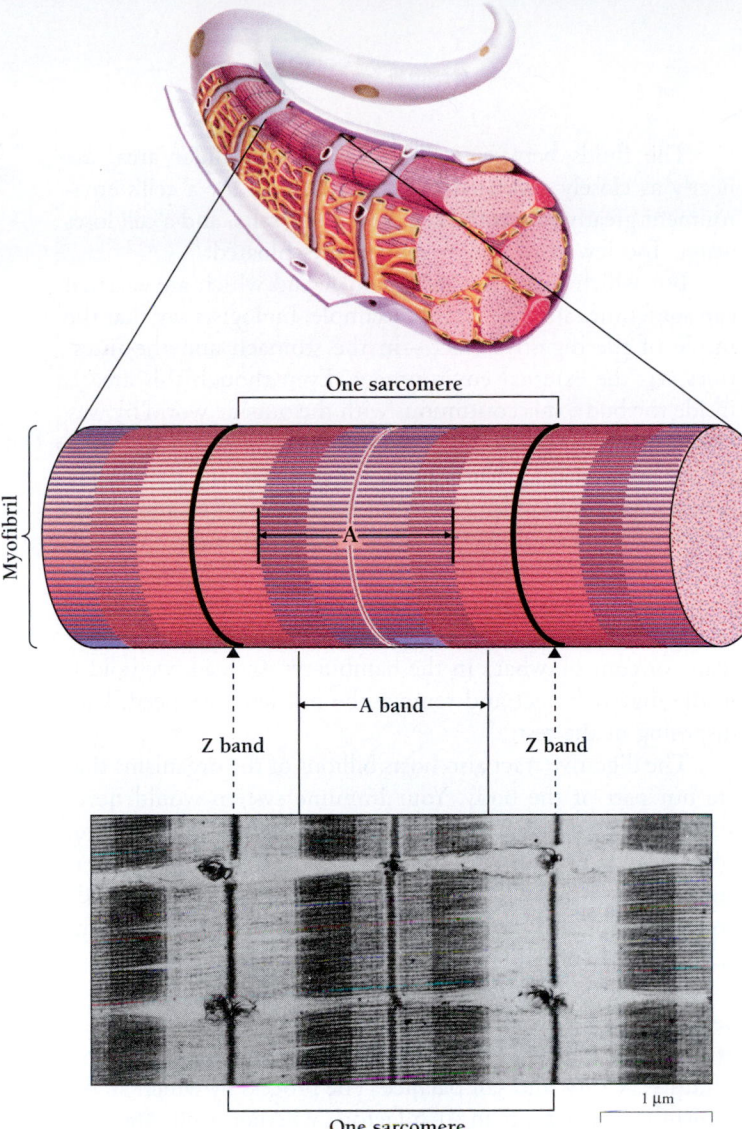

One sarcomere

Myofibril

A band

Z band   Z band

One sarcomere

1 μm

Figure 34-24   How does a muscle contract? Striated muscle consists of muscle fibers, which consist in turn of bundles of myofibrils. Each myofibril consists of bundles of thick and thin filaments. The functional unit of a myofibril is a sarcomere, a short cylinder of thick and thin filaments. At each end of a sarcomere is a thin, dark stripe called a Z band. Between the two Z bands is a band of thick filaments called the A band. *(Photo, Hugh Huxley, Brandeis University)*

How Do Striated Muscles Contract?

When a muscle fiber contracts, the sarcomeres contract. The distance between the Z bands changes in direct proportion to the length of a muscle. For example, if the muscle shortens by 20 percent, each sarcomere shortens by 20 percent. The question "How does a muscle contract?" therefore becomes "How does a sarcomere contract?"

An electron microscope reveals that each sarcomere consists of hundreds of tiny, parallel filaments. These filaments come in two sizes—**thin filaments** and **thick filaments.** Thin filaments are made of a protein called **actin.** Thick filaments are made of a protein called **myosin.**

The actin filaments come in pairs that extend from the Z bands inward toward each other (Figure 34-25). Parallel to the

pairs of thin actin filaments are thick myosin filaments. During contraction, each pair of actin filaments slides together past the myosin filaments. When a muscle relaxes, the pairs of actin filaments slide apart again. This model of muscle contraction is called the **sliding filament model.**

We can imagine how the filaments slide past each other during contraction, but what force pulls the actin molecules together and holds them together? And what role do the myosin filaments play? The answer came from particularly good electron microscope photographs. These showed that the thick and thin filaments were linked to each other at regular intervals by tiny **cross bridges.** The cross bridges are actually extensions of the myosin molecules that make up the thick filaments (Figure 34-25). Researchers guessed that the cross bridges perform the work of muscle contraction by pulling the thin and thick filaments over one another.

The hypothesis that the cross bridges provide the force for contraction suggested a testable prediction: the force generated by a muscle should depend on the degree of overlap between the thick and thin filaments. That is, if a muscle is sufficiently stretched, the thick and thin filaments have little overlap, and few cross bridges are available to move the filaments. Researchers tested this prediction by stretching muscles to different lengths and then measuring the amount of force the muscles could exert (by having the muscle pull against a spring). The results were exactly as predicted by the sliding filament model.

Relaxed sarcomere

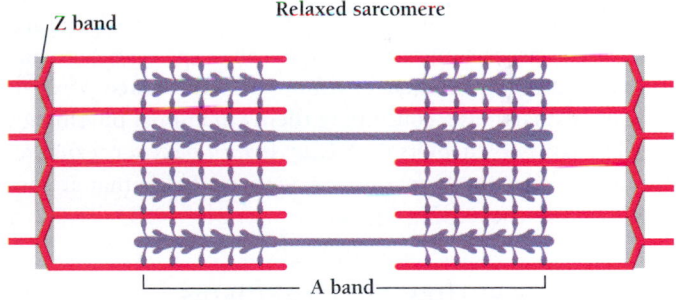

Z band

A band

Contracted sarcomere

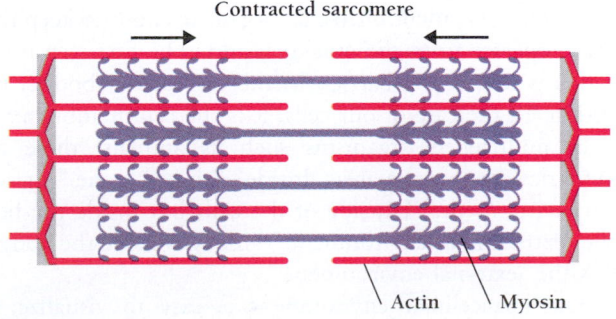

Actin   Myosin

Figure 34-25   What makes the filaments slide? Thin filaments attached to the Z bands and thick filaments (the A band) slide past one another during the contraction and relaxation of the muscle. Cross bridges between the thin filaments and the thick filaments provide the force for contraction.

Muscles contract when the cross bridges on thick filaments pull the thin filaments together, increasing the overlap between them.

INTRODUCTION TO PHYSIOLOGY: HOMEOSTASIS AND TOLERANCE

In the last 11 chapters of *Asking About Life,* we review some of the major ideas in animal anatomy and physiology. So far in this chapter, we have discussed some principles that help determine shape and we have briefly discussed how animals move. As you read the rest of this chapter, keep in mind that form and function are closely related. If you know what the heart is supposed to do (it pumps blood), that will help you remember how it is shaped (it is hollow). And if you can remember how something is shaped, that will help you remember what it is for.

The next five chapters in part VII cover the "housekeeping" functions of the body: digestion, hormonal regulation, circulation, respiration, and excretion. These systems all serve to keep the body operating. Next we cover defenses, which prevent the body from being overrun by other organisms, and the nervous system, which integrates all of the processes so far discussed. All of these topics deal with how the body is organized and how its parts work together, or **physiology.**

One theme that ties all of these topics together is the idea of **homeostasis,** the tendency of living organisms to maintain a constant internal environment. Mammals and birds, for example, usually maintain a nearly constant body temperature. But homeostasis is sometimes energetically expensive. So many animals change with their environment. **Tolerance** of environmental change is a continuing theme in animal physiology.

The last two chapters of *Asking About Life* cover sexual reproduction and development—two aspects of life that are not considered housekeeping functions.

Homeostasis: How Do Organisms Self-Regulate?

In Chapter 1, we mentioned that organisms need to keep their insides separate from the outside world. The first step in doing this is to create a barrier. In the case of our bodies, it is our skin. In the case of our cells, it is the cell membrane.

In multicelled organisms such as animals, these two boundaries create three areas. Inside each cell is the "intracellular environment." Outside of the cells but inside the body is the "extracellular environment." Finally, outside the body itself is the "external environment."

The intracellular environment is easy to visualize; it's everything inside the cell membrane (Figure 34-26). The contents of a cell tend to be closely regulated. If pH and salt concentration are not just right, the metabolic work of the cell cannot take place. Proteins lose their shape and function; metabolic processes come to a halt.

The fluids between cells, in the extracellular area, are nearly as closely regulated as inside the cell, for a cell's environment greatly affects the cell. Too many salts, and a cell loses water. Too few salts, and a cell becomes bloated.

But which spaces are extracellular and which are external can sometimes surprise us. For example, biologists say that the inside of the digestive tract—in the stomach and the intestines—is the external environment. Even though this area is inside the body, it is continuous with the outside world by way of the mouth and the anus.

When you eat a hamburger, for example, the hamburger can contain thousands of harmful microorganisms and an array of molecules completely unlike anything inside the body proper. The salt balance of a hamburger may be unlike that in your cells, the pH of the pickles too low, the temperature higher or lower than that of the body. Our bodies do not try to regulate, or control, what's in the hamburger. Instead, we hold it in the digestive tract and remove the nutrients we need, later disposing of the rest.

The digestive tract also hosts billions of the organisms that are not part of the body. Your immune system would never stand for billions of *E. coli* bacteria growing in the muscles of your legs, or in your abdominal cavity. Yet, your body tolerates billions of bacteria and yeast cells happily growing inside your digestive tract. Technically, then, the inside of your stomach is outside of your body.

Outside of the digestive tract is another story. The space between our cells and the blood in our veins is all part of the extracellular environment. Here, animals may regulate temperature, acidity, and salt balance. The process by which an organism resists changes in such factors, whether inside the cells or between the cells, is homeostasis [Greek, *homeo* = same + *stasis* = standing still]. Homeostasis always involves two stages:

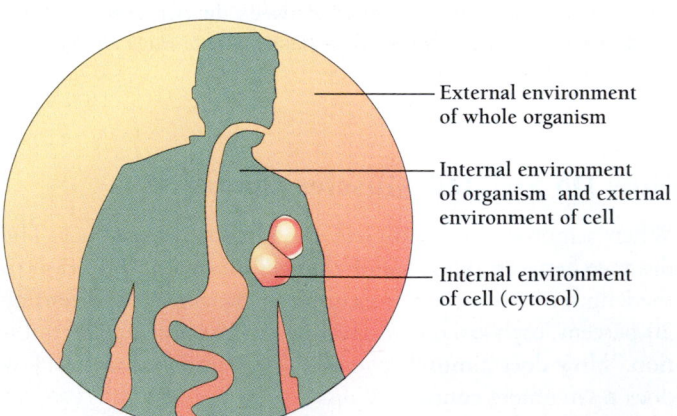

Figure 34-26   Inside out? Physiologists recognize three areas of the body—the external environment, the intercellular environment, and the intracellular environment. Animals regulate the insides of cells and the intercellular environment. But they do not regulate the external environment, which includes not only everything outside the skin, but also the contents of the gut.

External environment of whole organism

Internal environment of organism and external environment of cell

Internal environment of cell (cytosol)

(1) detecting change and (2) counteracting such change. Counteracting change is called **negative feedback.**

If you are driving a car, you use negative feedback to regulate the speed of the car. One way to do that is to compare your speed, as measured by the speedometer, with a rate set by a speed limit sign. If the sign says 55 mph and the speedometer says you are going 65, then you ease up on the gas and maybe even touch the brake to slow down.

In biology, the speed limit sign is called a **set point.** You may have heard that some physiologists believe that we have a set point for weight. Not everyone agrees with this idea, but we definitely have a set point for temperature. Normally, human body temperature ranges from 98° to 99° Fahrenheit (37°C). When we first get up in the morning, we are a bit cooler. Later in the day we get warmer. If we exercise strenuously we can become quite warm. But a healthy human rarely gets much warmer than 99°F or much cooler than 98°F. We can sit in a sauna with an air temperature approaching 200°F, then walk outside and lie in a snow bank, yet still maintain the same body temperature.

How does the body regulate temperature? In vertebrates a structure at the base of the brain called the hypothalamus detects changes in temperature. When the temperature is too low, the hypothalamus sends signals to other parts of the brain, which stimulate heat-seeking behavior, shivering, and increased heat production by the cells of the body. When we are cold, we also wrap our arms around our bodies and crouch, trying to reduce our surface area. Animals with fur contract tiny muscles at the base of each hair, which forces the hairs to stand up, forming a thick, insulating coat. Even nearly hairless humans get goose bumps, the visible contraction of those same muscles, and a relict from our furry past. Birds ruffle their feathers similarly.

When an animal is too hot, the hypothalamus detects the change and signals the brain to initiate a different set of responses. Some animals lay down feathers or fur. Some pant or sweat. Body heat is generated by all chemical reactions in the body, especially physical activity. Hot animals become quiet and move as little as possible to minimize the production of excess heat.

Most of us have learned that mammals and birds regulate temperature and most other animals do not. But this is an oversimplification. Many other animals regulate body temperature, although usually not to the same degree.

The body temperature of most aquatic animals is close to that of the water around them—cool. But tuna, sharks, and some other fast-swimming fish can generate more power in their swimming muscles if they keep them warm. These animals use a countercurrent heat exchanger like the one in the fins of whales to keep their muscles up to 14°C warmer than the water around them. Swordfish, marlins, sailfish, and mako sharks keep their brains and eyes warm using countercurrent heat exchangers, the better to detect their prey.

Some insects also heat parts of their bodies. Most insects are unable to fly in cold weather. We've all seen torpid flies on a windowsill on a cool, autumn morning. But many wasps, bumblebees, and large moths and butterflies, for example, heat their flight muscles and are able to fly in the cold. A bumble-

bee can fly in 5°C weather if it can get enough nectar (sugar) to keep its flight muscles at 30°C. The large, hummingbird-like sphynx moth can fly at night in temperatures of only 10°C and yet maintain a body temperature of 41°C, greater than that of a human (37°C). Biologist Bernd Heinrich says that many insects aren't just warm blooded, they are "hot blooded."

All organisms resist change through homeostasis and negative feedback mechanisms.

How Much Change Can Animals Tolerate?

As we examine the different systems of the body in the coming chapters, we will encounter many examples of homeostasis. It is important to remember, however, that animals often give in to environmental change instead of trying to resist it. For example, even as we humans regulate our core body temperature, we allow our legs and arms to warm up and cool down (Figure 34-27).

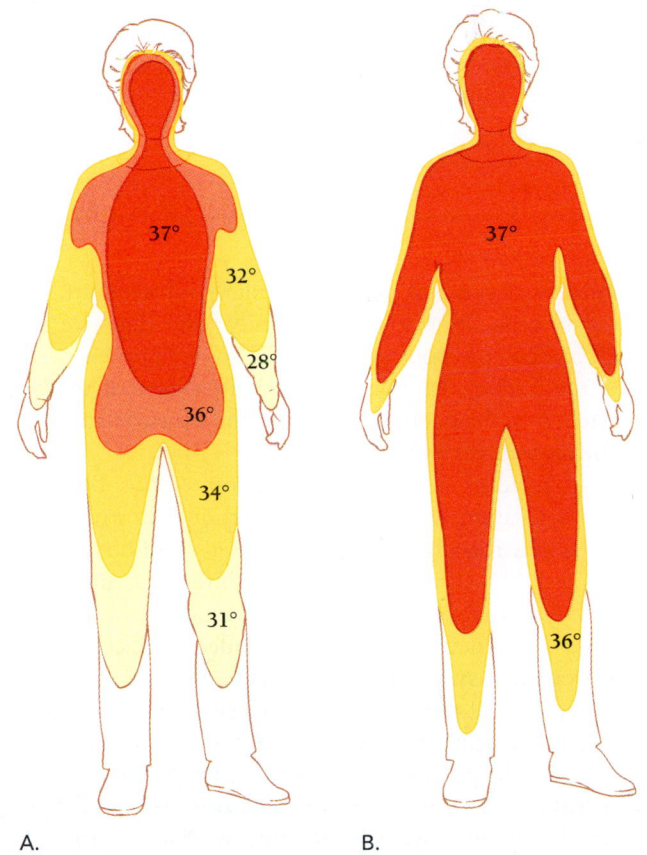

A.   B.

Figure 34-27 Tolerance in humans. A. In cold weather, human arms and legs become cold. The brain and the core of the body normally remain at 37°C, while the very center of the extremities drops to 31°C. The skin temperature is, of course, even lower, approaching that of the environment. B. When the temperature rises, we let our whole body warm up by increasing circulation to the arms and legs. If it's even hotter, we let our hands and feet get hot to help radiate excess heat.

Most animals will die at body temperatures above 50°C (122°F), and many die at much lower temperatures. Heat kills by changing the three-dimensional structure of proteins, and by altering the behavior of cell membranes. Yet regulating temperature is energetically expensive. For many animals, a better solution is to adapt or tolerate changes in temperature.

How much heat can an animal stand? The eggs of some shrimplike crustaceans can be boiled and yet still live and develop into healthy adults. Silver ants in the Sahara desert can behave normally at temperatures as high as 53°C—but only for a few minutes.

Other animals can tolerate extreme cold. Some organisms can be frozen to −296°C, close to the temperature of space, then thaw out and live a normal life. Some polar intertidal invertebrates may freeze and thaw twice a day. When the tide goes out, they freeze in the subzero air. When the tide comes back in, the warmer water thaws them out. As these animals freeze, the water between their cells freezes first. From inside the cells, more water diffuses outward, freezing in turn and sometimes severely distorting the muscles and other tissues. Worse, as the water freezes, it leaves a concentrated solution of salts and proteins inside each cell. So intertidal invertebrates that repeatedly freeze and thaw must tolerate not only ice crystals that deform their bodies, but high osmotic concentrations inside of their cells.

Many animals cannot tolerate ice in their tissues, yet cannot generate enough heat to stay warm when the temperature drops. One solution is to have antifreeze for blood. Certain fish and many insects contain glycerol or other substances that prevent their tissues from turning to ice when the temperature drops below freezing. Their body temperature drops below freezing, but no ice forms.

Many animals conform to their environment, a phenomenon called tolerance.

We have seen that the function of an organ determines its form and that the form of an organ determines its function. We have also seen what physical constraints affect form and physiology and how animals use skeleton and muscle to move. In the next chapter we will find out how animals derive nourishment from food. We will examine both digestion, the breakdown of food particles to their constituent building blocks, and absorption of those building blocks by the gut. In the remaining chapters of this last part of *Asking About Life,* we will see many more specific examples of how form follows function.

Study Outline with Key Terms

To understand function, we must understand structure. In all coelomate animals, the body is a tube (the digestive tract) within a tube (the body wall). In the space between the two tubes, the coelom, lie many of the main organs of the body. In vertebrates, the **coelom** consists of the **thoracic cavity** and the **abdominal cavity.**

The shape of an animal depends on its size. Because of gravity, large animals must have proportionately stronger and thicker bones than small animals. Size also influences metabolic rate. Webbed feet, the lungs, the intestines, and the circulatory system are all organs with expanded surface areas.

One important reason for expanded surface areas are **countercurrent systems,** which enable animals to establish steep gradients of temperature or concentration that help direct the movement of small molecules and temperature. For example, inside the noses of many vertebrates are elaborately folded **turbinate** bones with expanded surface areas that help cool the brain. To move large amounts of material from place to place, animals use **bulk flow.**

Style of locomotion also determines shape. Swimmers, which undulate through the water, have a streamlined shape. Snakes undulate like fish, but many also move by **rectilinear movement** and **sidewinding.** Flying animals have many specializations, including powerful wings, a light, rigid skeleton, hollow bones, reduced jaws and legs, and reproductive structures that shrink when not in use. Runners increase overall speed by increasing stride length and stride rate. Long legs,

specialized hips and shoulders, and a flexible spine all increase stride length. Light legs and muscle attachments close to the joint increase stride rate.

All animals have some kind of skeletal system. Earthworms and some other invertebrates coordinate muscle action using a rigid, fluid-filled space called a **hydrostatic skeleton.** Arthropods rely on an external **exoskeleton.** Vertebrates rely on an **endoskeleton.** The defining characteristics of vertebrates are the **backbone,** a column of hollow bony segments called **vertebrae,** and the **skull,** the bony case that encloses and protects the brain. The skull lies at the **anterior** end of the body, and the tail is at the **posterior** end. The point of attachment between any two bones is called a **joint.** Immovable joints among the bones in the head are called **sutures.**

All animals consist of tissues, including **epithelial tissue, muscle tissue, connective tissue,** and **nervous tissue.** Nervous tissue consists of **neurons** and **glial cells.** Tissues form **organs,** and groups of organs form **systems,** such as the digestive system and the nervous system.

Three kinds of connective tissue connect the bones and muscles of joints: **ligaments, tendons,** and **cartilage.** Ligaments and tendons must resist **tension.** Bones must resist tension and also two other kinds of stress—**compression,** the pushing action of two opposing forces, and **shear,** the twisting action created by forces that are not opposite one another.

Both bone and other kinds of connective tissue consist of flat, irregularly shaped cells called **fibroblasts** embedded in a network of proteins and polysaccharides called the extracellular **matrix.** The matrix consists of fibrous collagen and other substances secreted by the fibroblasts. **Collagen** molecules form triple-stranded helices, which assemble into cross-linked cables, called fibrils. Fibrils in turn form large, interconnected fibers. The fibers, which are arranged in parallel, like ropes, or in layers, like plywood, give bones and connective tissue much of their strength.

Adult bone consists of three major tissue types: **spongy bone tissue** (including **stem cells**), **red bone marrow,** and **compact bone tissue.** Specialized **osteoblasts,** similar to fibroblasts, manufacture the bone matrix. **Osteoclasts** continuously digest collagen and bone matrix. Osteoclasts and osteoblasts continuously shape and reshape bones as we grow, heal, and change our habits. In older people, the breakdown of bone by osteoclasts can lead to brittle bones, or **osteoporosis.**

In the human body, some 600 skeletal muscles—arranged in **antagonistic pairs**—move our bodies by pulling against bone and one another. Besides skeletal muscle, vertebrates also have **cardiac muscles,** which pump blood through the heart, and **smooth muscles,** which line the walls of hollow internal organs such as the intestines, the uterus, and the blood vessels. Smooth muscle and cardiac muscle are not attached to bones.

Because we have voluntary control over most skeletal muscles, they are often called **voluntary muscles.** In contrast, smooth muscle and cardiac muscle are **involuntary muscles.** Skeletal muscles are generally under the control of a system of nerves called the **motor axis,** while the smooth and cardiac muscles are controlled by a parallel system of nerves called the **autonomic nervous system.** The autonomic nervous system consists of the sympathetic nervous system, which brings about the "fight-or-flight" responses, and the parasympathetic nervous system, which slows the heart and increases the action of the intestines.

Unlike smooth muscle, cardiac and skeletal muscle are **striated,** or striped. The striations consist of cross stripes spaced every 2 to 3 μm and reflect the regular arrangement of the protein filaments that are responsible for muscle contraction. Every skeletal muscle consists of many parallel muscle fibers, giant cells up to 4 cm long. Each muscle cell, or fiber, contains up to 100 nuclei, as well as proteins responsible for muscle contraction.

Muscle fibers are of two kinds. The thick **glycolytic fibers** derive most of their energy from glycolysis and have few mitochondria. The thinner **oxidative fibers** derive most of their energy from respiration and have many mitochondria. Because oxidative fibers depend on a continuing supply of oxygen, they contain high levels of **myoglobin.**

Striated muscle consists of muscle fibers, which consist in turn of bundles of myofibrils. Each **myofibril** consists of bundles of **thick filaments** and **thin filaments.** The functional unit of a myofibril is a **sarcomere,** a short cylinder of thick filaments, which are made of **myosin,** and thin filaments, which are made of **actin.** At either end of each sarcomere is a thin, dark stripe called a **Z band.** According to the **sliding filament model,** thin filaments attach to the Z bands and thick filaments slide past during the contraction and relaxation of the muscle. **Cross bridges** between the thin filaments and the thick filaments provide the force for contraction.

In this last part of *Asking About Life,* we look at the **physiology** of the body—how it is organized and how its parts work together. Two overriding themes in all physiology are **homeostasis,** the tendency of living organisms to maintain their internal environment, and **tolerance** of environmental change. Most animals achieve homeostasis by means of **set points** and **negative feedback.**

Review and Thought Questions

Review Questions

1. Why are an elephant's legs thick and straight?
2. Explain how countercurrent exchange systems work. Give an example.
3. Why are fish, birds, and ships streamlined?
4. Fibroblasts exist in which tissues of the body? What function do they serve?
5. What are the three kinds of bone found in adults?
6. Name the two partners in an antagonistic pair of muscles. What is the value of these two muscles working against each other?
7. What is the difference between voluntary muscles and involuntary ones?

8. Explain why long-distance runners' muscles are smaller than those of body builders.

Thought Questions

9. Many people can consciously slow or speed up the rate at which their hearts beat. What mechanisms can you imagine that would allow voluntary control over this autonomic function?
10. Imagine a giant, 12 feet tall. For his legs to be as strong as a normal man's, how thick would his femurs need to be?

About the Chapter-Opening Image

Tyrannosaurus rex, featured in the lead story, shows how much we can learn about how animals work by looking at their bones and other structures.

 On-line materials relating to this chapter are on the World Wide Web at **http://www.harcourtcollege.com/lifesci/aal2/**

35

How Do Animals Obtain Nourishment from Food?

Another American Shot Heard Round the World

American physiology began with a "bang." In the summer of 1822, a shotgun accidentally discharged into the belly of a 19-year-old French Canadian named Alexis St. Martin. His misfortune added immeasurably to the scientific understanding of the process of digestion.

St. Martin worked for the American Fur Company as a "voyageur," transporting goods and supplies between the company's remote outposts. That summer, St. Martin was one of a crowd of soldiers, Indians, and trappers who had jammed into a trading post on Mackinac Island, between Lake Huron and Lake Michigan, to buy and sell pelts, coats, and moccasins and to trade stories after a long and lonely winter (Figure 35-1). St. Martin was less than 3 feet from the shotgun's muzzle, and the hole it made was bigger than the palm of his hand.

As the horrified crowd stood around the fallen young man, one of the store's clerks ran to fetch Dr. William Beaumont, the surgeon at nearby Fort Mackinac. When Beaumont arrived minutes later, he found a rib and part of a lung protruding from the wound, as well as a portion of St. Martin's stomach.

The protruding stomach had a puncture, which, Beaumont later wrote, was "large enough to receive my forefinger, and through which a portion of his food that he had taken for breakfast had come out and lodged among his apparel. . . . I considered any attempt to save his life entirely useless." Beaumont nonetheless cleaned and dressed the wound, "not believing it possible for him to survive twenty minutes."

In fact, St. Martin recovered amazingly. By December, St. Martin's health was "daily improving, his spirits good, his appetite regular, his sleep refreshing." But St. Martin was to live the rest of his life with a hole that led from the outside world directly into his stomach, and it would be more than a year before he fully recovered his health.

Beaumont stopped the leakage of St. Martin's stomach with a "plug of lint." But the young man was unable to return to his demanding job carrying supplies through the wilds of Canada, and he had no way to support himself. Beaumont took St. Martin into his own home and, in time, St. Martin regained his strength so that he could walk and take care of his personal needs. But he was still unable to return to work as a voyageur.

Then, one winter, almost 2 years later, Beaumont realized that St. Martin's protruding and perforated stomach provided "an excellent opportunity for experimenting upon the gastric fluids and the process of digestion." Beaumont began his first series of such experiments in May 1825, in Mackinac. A month later, the doctor was ordered to Fort Niagara, New York, and he took St. Martin with him in order to continue his experiments. Beaumont had come to realize that the unique experiments he could do with St. Martin would bring him great prestige in America's infant scientific community. In 1833, Beaumont published a full description of his experiments and his conclusions in a volume called *Experiments and Observations on the Gastric Juice and the Physiology of Digestion.*

Beaumont's work is a model of careful observation and experiment. In one study, for example, Beaumont showed that a piece of beef suspended into St. Martin's stomach on a string was digested over a period of several hours. Virtually the same visible changes occurred if a similar piece of beef was placed in a glass vial containing the "gastric juice" obtained from St. Martin's stomach with a glass tube. The gastric juice in the vial, Beaumont showed, worked much better at body temperature, however, than at room temperature.

Beaumont later showed that the gastric juice contained hydrochloric acid. Yet, hydrochloric acid alone would not digest the meat. We now understand Beaumont's results as demonstrating the action of the stomach enzyme *pepsin,* which hydrolyzes proteins into smaller fragments each containing a few amino acids. In the intestine, other enzymes break these fragments into individual amino acids. Pepsin catalyzes hydrolysis far more powerfully than hydrochloric acid. Pepsin works best in an acid environment, however, and much more rapidly at body temperature than at room temperature.

At the end of his volume, Beaumont listed 51 conclusions derived from his experiments and observations. This list was as important for the questions it raised as it was for the knowledge it imparted. For example, Beaumont noted that the secretion of gastric juice depended on the presence of food. From a 20th-century perspective, we want to know what chemical signals in food stimulate the secretion of gastric juice. How are the signals delivered? And what mechanisms trigger the production of hydrochloric acid and pepsin?

The questions are even more compelling if we ask them not only about humans but also about other animals. For example, Beaumont noted that how finely St. Martin chewed his food affected the way the food was digested in the stomach. We can then ask, How do birds, which lack teeth, and reptiles, which have teeth for biting but not grinding, digest their food efficiently?

WHY MUST ANIMALS EAT?

The task of the digestive system is to derive both energy and raw materials from food. Food comes in many forms, and diets vary enormously among species, but digestive systems perform the same basic task: they break down the complex molecules of food, especially macromolecules and lipids, into simpler molecules that can be absorbed and incorporated in the animal's own metabolism.

Simple molecules—such as glucose—undergo a series of enzyme-catalyzed conversions to still simpler molecules, such as carbon dioxide and water. These conversions—including the reactions of glycolysis, respiration, and oxidative phosphorylation—produce the ATP molecules that power most of an animal's activities.

Complex molecules—such as polysaccharides, proteins, fats, and oils—first undergo hydrolysis to building blocks or still simpler molecules. These small molecules can then enter these same energy-producing pathways (Chapter 6, Figure 6-11).

Although an animal can make many of the building blocks needed to produce its own large molecules, its food also provides a ready-made supply, including some that the animal cannot make itself. Most animals, for example, can make only half of the 20 amino acids needed to make proteins, but they must obtain the other half from their food. Animals need other kinds

Figure 35-1 **William Beaumont and Alexis St. Martin.** (*The Granger Collection, New York*)

of raw materials as well: minerals, inorganic elements, vitamins, and organic compounds that the animal cannot itself synthesize.

Individual species of animals derive their food from differing sources. **Herbivores** [Latin, *herba* = vegetation + *vorare* = to swallow], such as caterpillars and cows, eat only plants. **Carnivores** [Latin, *carn* = flesh], such as spiders and seals, eat only animals, with many species specializing on particular parts, such as the blood or the skin. Still other animals ("detritivores") consume only detritus, the remains of plants and animals, broken down into smaller fragments by the action of microorganisms. Humans are **omnivores** [Latin, *omnis* = all], and our diets range widely. Still, all animals digest the complex molecules and metabolize the small.

Animals eat to get energy and raw materials.
The strategy for processing food is always the same:
digest the complex molecules into small molecules
that can enter metabolism.

How Much Must an Animal Eat?

Food must provide energy for all of an animal's activities. Physiologists usually estimate energy use by measuring oxygen consumption in animals performing various activities. Each liter of oxygen used corresponds to the expenditure of just under 5 kilocalories of energy. (Recall that one kilocalorie is 1000 calories, where one calorie is the energy required to raise the temperature of one gram of water by 1°C; writings on diet and nutrition usually count "Calories," with a capital *C*, which are the same as kilocalories.)

Animals differ in their energy requirements, according to their size, species, and activity levels. Within groups of related species, energy consumption is roughly proportional to weight. For a moderately active college student, for example, daily energy use (in kcal) is about 15 to 20 times the weight in pounds. So a 120-pound woman may require 1800 to 2400 kcal per day, while a 160-pound man may require 2400 to 3200 kcal per day.

The consumption of energy depends on particular activities: running or swimming requires more energy than standing or floating. But every animal requires a certain amount of energy just to stay alive and awake. This minimal energy requirement is called the **basal metabolic rate.** For humans, the basal metabolic rate is about 1400 kcal per day for a 120-pound woman and about 1700 kcal per day for a 160-pound man. The same amount of energy would keep a 75-watt bulb burning all day and night or would fuel a one-mile drive in a small car. For humans and other mammals, the basal metabolic rate is usually slightly more than half the energy used in ordinary activities.

Individual animal species have widely varying basal metabolic rates, even after accounting for their different sizes. For example, the basal metabolic rate, per gram, of an adult human corresponds to about 1 calorie per hour, whereas the corresponding basal metabolic rate for a shrew is 35 times higher. A 5-gram shrew must therefore consume about 4 kcal per day just to stay alive and another 4 kcal per day (a total of 8 kcal/day) to lead a normal shrew life, searching for beetle larvae in underground tunnels. A diet of pure protein or carbohydrate contains about 4 kcal per gram, so the shrew could sustain itself with only 2 grams per day on a diet that consists only of protein and carbohydrate. But a wild shrew is unlikely to find so refined a diet, and it actually eats an amount each day corresponding to its own body weight. To minimize the amount of food that must pass through its gut, a shrew must seek a high-quality diet, with little indigestible material. The best diet would be rich in fat—insect larvae, for example—since fat contains 9 kcal per gram, compared with 4 kcal per gram in carbohydrates and proteins.

Food must provide enough energy to supply an
animal with its basic metabolic needs and to
fuel its various activities.

How Does an Animal Know How Much Food To Eat?

Most animals adjust their food intake to match their activities. For example, when researchers diluted ordinary rat food with nonnutritive materials, the rats increased the amount they ate so that they acquired the needed amount of food energy.

Damage to the hypothalamus, the part of the brain that controls appetite, however, can abolish the ability to adjust the diet. A rat with a damaged hypothalamus may feed without stopping, while another, with damage to another part of the hypothalamus, may starve itself to death.

The hypothalamus exerts its effects on other parts of the body through chemical signals called *hormones* (Chapter 36). One kind of hormone, called "orexin," triggers feeding behavior. On the other hand, fat cells in the body of a well-fed animal make another hormone called "leptin," which signals the hypothalamus to stop feeding behavior. Pharmaceutical companies are spending millions of dollars trying to find chemicals that will alter orexin and leptin levels, hoping that they will be able to provide chemical help for people who want to regulate their appetite.

What Are the Consequences of Too Much or Too Little Food?

If an animal takes in more food energy than it uses, the excess is stored as glycogen or as fat. This situation—called overnourishment—is common in human industrialized societies. Many people in the United States, for example, have stored more energy than they will ever need, causing strain on the heart and circulation.

If, on the other hand, an animal takes in less food energy than it needs, it must derive the extra energy from its own body. When stored energy runs out, animals begin to break down their muscles and other tissues. Undernourishment—taking in too few calories—occurs in modern human populations mostly in times of drought, war, or other catastrophe. In contrast, **malnourishment,** a deficiency in one or more essential nutrients, is unfortunately all too common: it affects at least 500 million people, 10 percent of the world's population. Malnourishment is an increasing problem even in wealthy countries such as the United States, where not everyone has equal access to a balanced diet or good nutritional advice.

Animals usually regulate their food intake to match their activities, but this regulation can go awry either because of an inadequate food supply or because of a failure in normal regulation.

What Are the Causes of Malnourishment?

Malnourishment may result from a deficiency in protein, in minerals, or in vitamins. Each type of deficiency leads to a recognizable set of symptoms, many of which were first described in sailors long at sea.

Protein deficiency, which affects some 100 million people worldwide, is the principal cause of human malnourishment. Protein deficiency disease is also called **kwashiorkor,** a term that originated in Ghana. Kwashiorkor is characterized by lethargy, severe anemia, change in hair color, inflammation of the skin, and a potbelly, caused by excessive water retention ("edema"). The cause of this edema is unknown, but probably depends on the failure to make proper blood proteins. Kwashiorkor is especially prevalent in children just after weaning, when their diet switches from milk to a single kind of starch—most often rice, corn, or cassava.

Understanding kwashiorkor also allows us to understand the causes of other, less severe protein malnourishment. The digestive system breaks down the proteins in food into individual amino acids, which are then transported to cells that make proteins. Although animals can themselves make ten or more amino acids from other molecules in food, they depend

on dietary protein to provide the **essential amino acids,** those that it cannot produce itself. Table 35-1 lists the essential amino acids in humans. Protein deficiency may result either from too little total protein or from too little of one or more essential amino acids.

Animals do not store amino acids between meals. (In contrast, we do store glucose in the form of glycogen.) Each meal must therefore provide the proper proportions of essential amino acids. For carnivores and for human populations that eat animal proteins such as those in milk, eggs, or meat, the balance of amino acids is usually about right. But for herbivores and human vegetarians, the problem of amino acid balance may dominate all other health issues.

Amino acid balance is especially critical in the many human societies that depend on a single crop, such as rice or corn. These crops grow well in the local climate, are resistant to local pests, and are therefore inexpensive. But the proportions of amino acids in any single plant species can never be the same as those in an animal's own proteins. Proper balance requires mixing protein sources (Figure 35-2). An appropriate mixture—one with *complementary proteins*—provides enough of each essential amino acid to sustain protein synthesis. Where enough food is available, traditional human societies have developed diets that contain complementary proteins—corn and beans, wheat and chickpeas, rice and lentils. Corn, for example, is low in lysine but high in tryptophan, while beans are low in tryptophan but high in lysine. In many parts of the world, however, only one protein source may be affordable, and protein deficiency disease is rampant.

Mineral deficiencies can also cause malnourishment. Animals require inorganic ions for a wide variety of structures and processes. Some—calcium, phosphorus, potassium, sulfur, sodium, chlorine, and magnesium—are required in relatively large amounts. Others—iron, manganese, and iodine—are needed in smaller amounts, while still others—copper, zinc, molybdenum, cobalt, and selenium—are needed only in trace amounts. Finally, some minerals—including vanadium, silicon, and nickel—are needed by some animals but not necessarily by humans. Altogether, 26 of the 92 naturally occurring

TABLE 35-1   ESSENTIAL AMINO ACIDS IN HUMANS

Amino acid	Poor sources	Good sources
Histidine	—	Essentially any good protein source: legumes, cereal grains, nuts, dairy, meat, eggs
Isoleucine	—	Essentially any good protein source
Leucine	—	Essentially any protein source
Lysine	Peanuts, corn, oats, rice, wheat	Soybeans, other beans, milk, eggs, meat
Methionine	Beans, lentils, peanuts	Cereal grains, milk, meat, eggs
Phenylalanine	—	Essentially any good protein source
Threonine	—	Essentially any good protein source
Tryptophan	Beans, peanuts, corn, almonds	Cereal grains other than corn, wheat germ, milk, meat, cheese, eggs
Valine	—	Essentially any good protein source

Figure 35-2   A nutritious diet must provide complementary proteins. Some foods are poor in lysine, while others are poor in tryptophan. *(Fish and shellfish, dairy products, Paraskevas Photography; beans and peas, grains and cereals, nuts and seeds, Amy Dunleavy)*

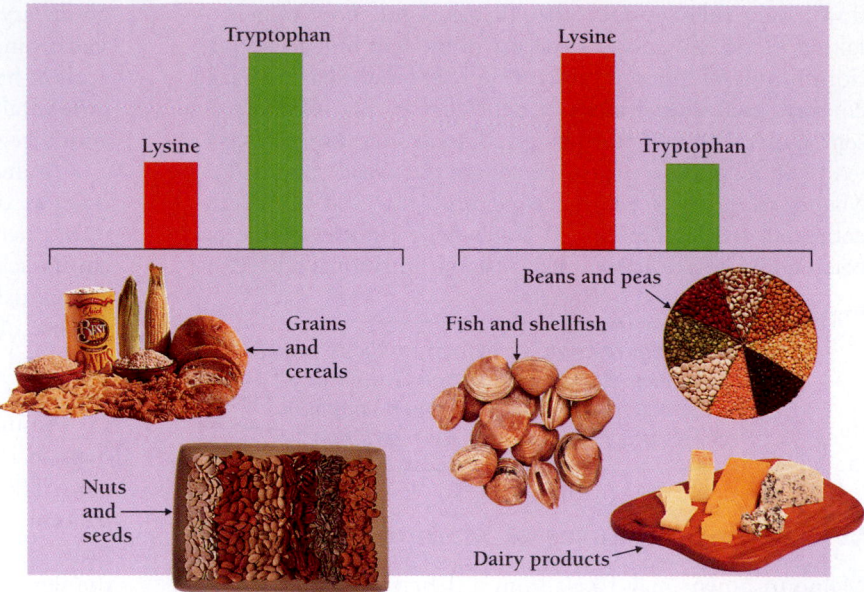

elements are known to be necessary for rats and chicks. Table 35-2 lists the essential minerals for humans, along with the symptoms of insufficiency for each mineral. Although humans need these 26 elements, however, some of them (such as lead or selenium) are harmful in large amounts. Dietary supplementation is not always safe!

As is the case for the consumption of energy-rich molecules and of amino acids, animals appear to have some mechanisms for ensuring that they take in the proper minerals. For example, herbivores seek salt licks to supplement the meager sodium present in plant tissues (Figure 35-3).

Vitamin deficiency also causes malnourishment. The first vitamin deficiency disease to be understood was *beriberi,* which afflicted colonial Dutch troops in Indonesia. The symptoms of this disease included the degeneration of muscles and nerves,

spasms or rigidity of the legs, and mental confusion. In the late 19th century—a time of great excitement about Pasteur's discoveries that bacteria cause disease—the Dutch government sent a team of researchers to find the cause of beriberi, suspecting that it resulted from a bacterial infection. The team worked for two years but they found no infectious agents.

Much of the troops' diet, however, consisted of "polished" rice, from which the husks had been removed to prolong storage. One of the researchers, Christian Eijkman, noticed that chickens that fed from the food from the kitchen and the mess hall developed symptoms like those of the stricken soldiers, whereas chickens fed elsewhere did not. Eijkman showed that he could cure both the sick chickens and the sick men by feeding them unpolished rice. He later showed that the antiberiberi agent was a water-soluble organic compound, vitamin B_1, or

TABLE 35-2   MAJOR MINERALS IMPORTANT IN THE HUMAN DIET

Mineral	Good sources	Symptoms of deficiency	Functions
Sodium	Normal diet; table salt	Muscle cramps	Water balance; operation of muscles and nerves
Potassium	Fruits, vegetables, grains	Irregular heartbeat, fatigue, muscle cramps	Operation of muscles and nerves; acid-base balance
Chlorine	Table salt	Unlikely	Acid formation in stomach; nerve signaling
Calcium	Dairy, bony fish, leafy green vegetables; dried legumes	Osteoporosis	Formation of bone and teeth; clotting; nerve signaling
Phosphorus	Dairy, meat, cereals	Bone loss, weakness, lack of appetite	Formation of bone and teeth; energy metabolism (e.g., ATP); membrane phospholipids
Magnesium	Nuts, greens, whole grains	Nausea, vomiting, weakness	Enzyme action; nerve signaling
Sulfur	Protein foods	None, when protein needs are met	Proteins

ATP = adenosine triphosphate.

Figure 35-3 Animals often obtain essential minerals at a salt lick. *(Frank S. Balthis Photography)*

thiamine, now known to be a coenzyme in carbohydrate metabolism. Beriberi is still a serious problem in places where rice is the major food: rice is low in thiamine, especially after "polishing," which removes the rice grain's outer layer.

The blocking of specific biochemical pathways because of a coenzyme deficiency also surrounds the discovery of vitamin C, or ascorbic acid. Ascorbic acid is the specific substance needed to cure *scurvy*, a disease characterized by bleeding gums, loosening teeth, and slow wound healing. Scurvy was common on sailing ships until the 18th century, when the British Navy began to supply lemons to their crews. When, in 1865, the Navy substituted limes for lemons, the word "limey" came to mean a British sailor.

The structure and biochemical function of ascorbic acid were not determined until the early 20th century, almost 200 years after its use to prevent scurvy. Ascorbic acid participates in the reactions that modify and strengthen collagen, the principal structural protein of connective tissue. Ascorbic acid also serves as an *antioxidant*, limiting the damage caused to cells by oxygen and its products.

Some other vitamins have been harder to identify, because they are present at high enough concentrations in the ordinary diet that deficiency diseases are rare. Despite these difficulties, we now know of 13 essential vitamins required in the human diet (Table 35-3). Nutritionists usually speak of the vitamins by their letter names (such as "vitamin C"). Now that their chemical identities are known, however, biologists prefer their chemical names (such as "ascorbic acid").

Chemically, the 13 vitamins fall into two broad groups: *water-soluble vitamins* and *fat-soluble vitamins*. The water-soluble group includes vitamin C (ascorbic acid) and the B-complex vitamins, which have a common designation only because they usually appear in the same foods. Each of the water-soluble vitamins is a precursor of a particular *coenzyme*, a small organic molecule that binds to an enzyme and plays a role in catalysis. The fat-soluble vitamins, including vitamins A, D, E, and K, play diverse roles: vitamin A, for example, is the precursor of retinal, the light-sensitive molecule in the visual pigments, whereas vitamin K participates in blood clotting.

The listed substances are essential for humans, but not for all animals. Vitamin C appears to be essential only for humans, monkeys, and guinea pigs; other species can produce it themselves. Similarly, the fat-soluble vitamins are essential only for

vertebrates. Nutritionists and physicians disagree, however, on how much of each vitamin a healthy diet should contain.

Malnourishment in humans usually results from an imbalance of amino acids in dietary protein, but it may also result from deficiencies in minerals or vitamins.

HOW DO SINGLE CELLS AND INVERTEBRATES DIGEST FOOD?

Most animals feed on complex fare—usually the whole, the parts, or the remains of other organisms. Their food usually consists of intact tissues, often in large chunks. The mechanical breaking of food into smaller bits is one of the first jobs in feeding. A human accomplishes this task with mouth, tongue, cheeks, and teeth; a bird uses its **gizzard**, a muscular region of the gut, specialized for the grinding of food.

The most nourishing parts of the food consist of carbohydrates, proteins, and lipids. In order to obtain this nourishment, however, an animal must (1) **digest** the food, that is, convert it into smaller molecules, such as glucose, amino acids, fatty acids, and glycerol; and (2) **absorb** the smaller molecules, that is, take them up into cells and into the circulation. Digestion and absorption are the business of the digestive system.

Digestion is a chemical process that splits the bonds that hold building blocks together in macromolecules and lipids—the bonds between sugars in polysaccharides, between amino acids in proteins, and between fatty acids and glycerol in lipids. Each splitting is a hydrolysis, accompanied by the addition of a water molecule.

None of these hydrolysis reactions requires energy: all are exergonic and occur spontaneously. For digestion to occur at a useful rate, however, specific enzymes must participate. *Proteases* accelerate the hydrolysis of proteins, *glycosidases* the hydrolysis of polysaccharides, *lipases* the hydrolysis of lipids, and *nucleases* the hydrolysis of DNA and RNA. But the enzymes and food molecules must first come into contact—no easy task for molecules hidden within chunks of food.

During digestion, enzymes hydrolyze macromolecules and lipids into their component building blocks.

Harsh Conditions Help Disrupt Tissue Interactions

Freeing food molecules to interact with enzymes requires harsh conditions. Humans help this process by cooking food at a high temperature or by marinating it in an acid solution such as vinegar, wine, or lemon juice. Like many other animals, we also break food into more accessible bits by using our teeth to bite, puncture, tear, and grind the food in our mouths. In our stomachs, we expose the food to strong acid and powerful enzymes, which begin to break up the large macromolecules—

TABLE 35-3   ESSENTIAL VITAMINS IN THE HUMAN DIET

Vitamin	Good sources	Functions
Fat-soluble		
A (retinol)	Leafy green and yellow vegetables, liver, eggs, fortified milk	Visual pigment; development of bone and teeth; development and maintenance of skin, gut, and other epithelia
D	Synthesized in skin exposed to sunlight; eggs, fortified milk	Bone growth; calcium absorption
E	Meat, milk, vegetable oils, whole grains	Prevents oxidative damage to cells; maintains levels of vitamin C
K	Green leafy vegetables; production by intestinal bacteria	Clotting; electron transport
Water-soluble		
B_1 (thiamin)	Meat, milk, eggs, grains, legumes	Coenzyme in production of carbohydrate and nucleic acids; development of connective tissues
B_2 (riboflavin)	Milk, meat, grains	Coenzyme in energy metabolism e.g., FAD)
Niacin	Meat, bread, potatoes	Coenzyme in energy metabolism and biosynthesis (e.g., NAD and NADP)
B_5 (pantothenic acid)	Milk, meat, eggs, yeast	Coenzyme in synthesis of fatty acids and steroids
B_6 (pyridoxine)	Meat, potatoes, spinach	Coenzyme in metabolism of amino acids
Folic acid (folate)	Vegetables, cereals, eggs, meat; also produced by bacteria in the gut	Coenzyme in metabolism of amino acids and nucleic acids
B_{12} (cobalamine)	Milk, meat	Coenzyme in metabolism of nucleic acids
Biotin	Legumes, nuts, eggs, liver	Coenzyme in metabolism of fats, glycogen, and amino acids
C (ascorbic acid)	Citrus fruits, potatoes, green leafy vegetables	Coenzyme in carbohydrate metabolism and in formation of connective tissue

carbohydrates, proteins, and nucleic acids. Then, the small intestine adds detergents (bile salts) that disrupt the nonpolar interactions of fats and oils.

The harsh conditions of digestion are a two-edged sword. They suitably prepare the consumable tissue, but they may also destroy the tissues of the consumer. To avoid this "self" digestion, animals perform most digestion in special compartments, either outside the cell entirely, or in a lysosome or peroxisome, where a membrane separates the digestive enzymes and the digesting food molecules from most of the cell's own proteins.

Animals protect their own cells from digestive enzymes by making digestion an extracellular process.

Digestion by Individual Cells

The simplest example of the isolation of the digestive process is the ingestion of liquid and dissolved molecules by a single cell. *Pinocytosis* is a form of *endocytosis,* the process by which a cell traps extracellular materials into a vesicle (an *endosome*) by folding of the plasma membrane inward (Figure 35-4A). The pouchlike endosome, typically about 0.1 μm in diameter, fuses with a *lysosome,* full of hydrolytic enzymes, and the resulting small molecules then move into the cytosol (Chapter 4).

Some individual cells can accomplish another type of feeding and digestion, called *phagocytosis.* Phagocytosis differs from pinocytosis in that phagocytosis involves larger particles and that the plasma membrane expands to surround the particle. Animals usually employ phagocytosis not for feeding but for removing cellular debris.

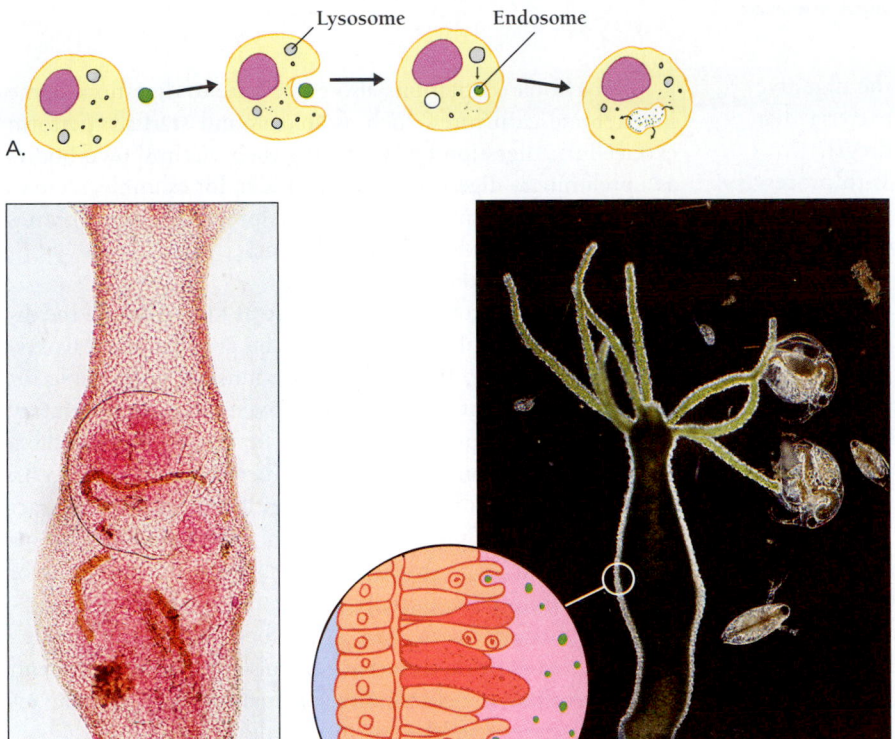

A.

B.

Food particle

Nutritive cell

Figure 35-4 Digestion in single cells and in simple organisms by endocytosis and phagocytosis. A. Endocytosis: a cell takes in extracellular materials into an endosome. The endosome fuses to a lysosome, which is filled with digestive enzymes. After digestion, small molecules are released into the cytosol. B. Phagocytosis: in a hydra, initial digestion occurs outside the cells of the primitive digestive tract. (*Left photo, © Wagner/Phototake NYC; right photo, Roland Birke/Okapia/Photo Researchers, Inc.*)

Digestion in Cavities with One Opening

How do animals deal with food particles too large to enter a single cell? Such food sources require an extracellular digestion chamber. The simplest digestion chambers, such as that of a hydra or a flatworm, have a single opening, which serves as both mouth and anus (Figure 35-4B). The hydra's arms move food into the chamber, and secreted enzymes then break down the food into smaller pieces, which then enter the surrounding cells by endocytosis. The hydra discharges the remnant of indigestible food through the opening. In flatworms, the digestive cavity ramifies through the body and also serves as a primitive circulatory system.

What Are the Advantages of a Two-Ended Digestive Tract?

Animals more complex than hydras and flatworms have digestive systems with two ends (Chapter 23). Food enters through the mouth and passes through an extended tube, variously called the **digestive tract,** the *gastrointestinal tract,* the *alimentary canal,* or the **gut** [Middle-English, *guttes* = entrails] (Figure 35-5). The undigested and unabsorbed material leaves

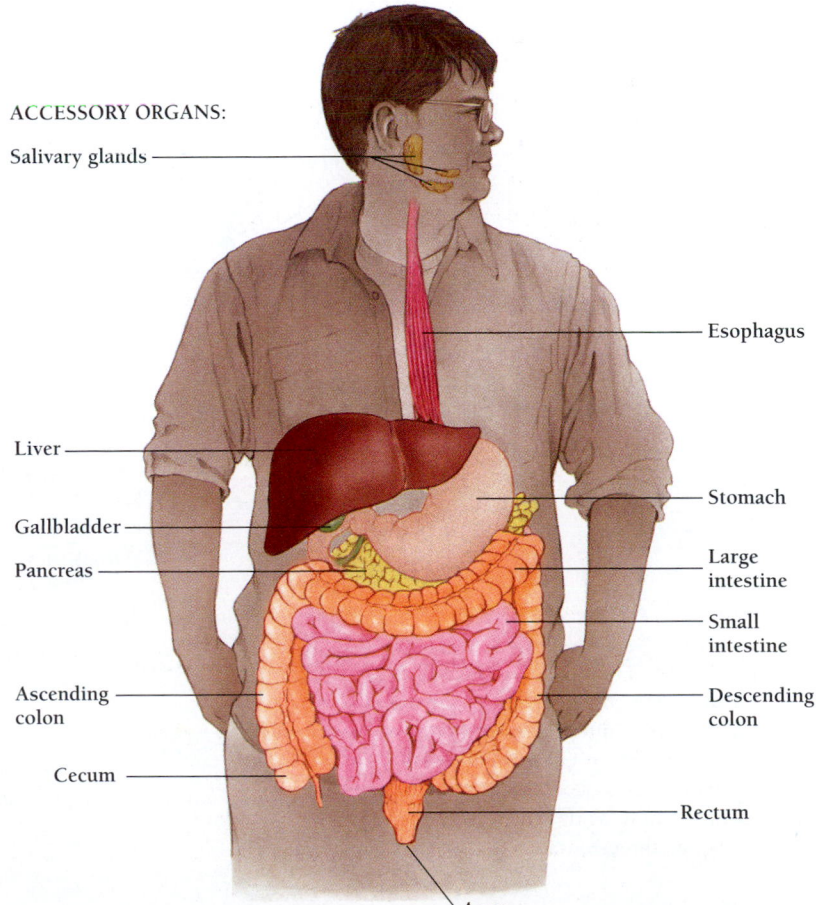

ACCESSORY ORGANS:

Salivary glands

Liver

Gallbladder

Pancreas

Ascending colon

Cecum

Esophagus

Stomach

Large intestine

Small intestine

Descending colon

Rectum

Anus

▶ **Figure 35-5** The human digestive tract, with accessory organs.

through the **anus,** the opening at the far end of the digestive tract. The remnants of the food, together with bacteria that inhabit the tract, form the **feces** [Latin, *faeces* = dregs].

A digestive tract with two ends allows animals to process food more efficiently than a one-ended digestion cavity. Since the food passes each point in the tract only once, sequential processing can occur. Distinct regions of the tract may become specialized for different processes, each of which may require a different set of environmental conditions and a different set of enzymes.

The evolution of the digestive tract anticipated, by about half a billion years, the advantages of an industrial assembly line, in which workers, machines, and equipment are arranged so that the product passes consecutively from operation to operation until it is completed. As food passes down the digestive tract, it undergoes a series of successive reactions that ultimately allow the animal to absorb a variety of valuable molecules, while leaving indigestible materials behind. Indeed, we may think of the digestive tract as a "disassembly line" (Figure 35-6).

Behavioral adaptations also contribute to digestion. Some invertebrate carnivores, such as spiders and starfish, perform extracellular digestion by employing their victims' own bodies as a preliminary digestion vessel. A spider, for example, secretes digestive enzymes into the body of its prey. After the enzymes have done their hydrolytic job, the spider ingests its now liquefied prey for further processing.

In many invertebrates, the passage of food through the digestive tract is highly controlled by the animal's nervous system. In the lobster, for example, the muscles responsible for digestive movements are striated muscles like the voluntary muscles discussed in Chapter 34. The contraction of these muscles depends on stimulation by networks of nerve cells, which cause four different sets of digestive muscles to contract with different rhythms: (1) a rhythm responsible for the ingestion of food; (2) a rhythm responsible for the emptying of the sac where food is initially stored; (3) a rhythm that controls the calcified "teeth" that grind food within the lobster's stomach; and (4) a rhythm that controls a system of valves and sieves that allow the elimination of small particles and the retention of larger particles for further digestion.

A two-ended digestive tract allows the specialization of different regions for specific processes that take place in a fixed order.

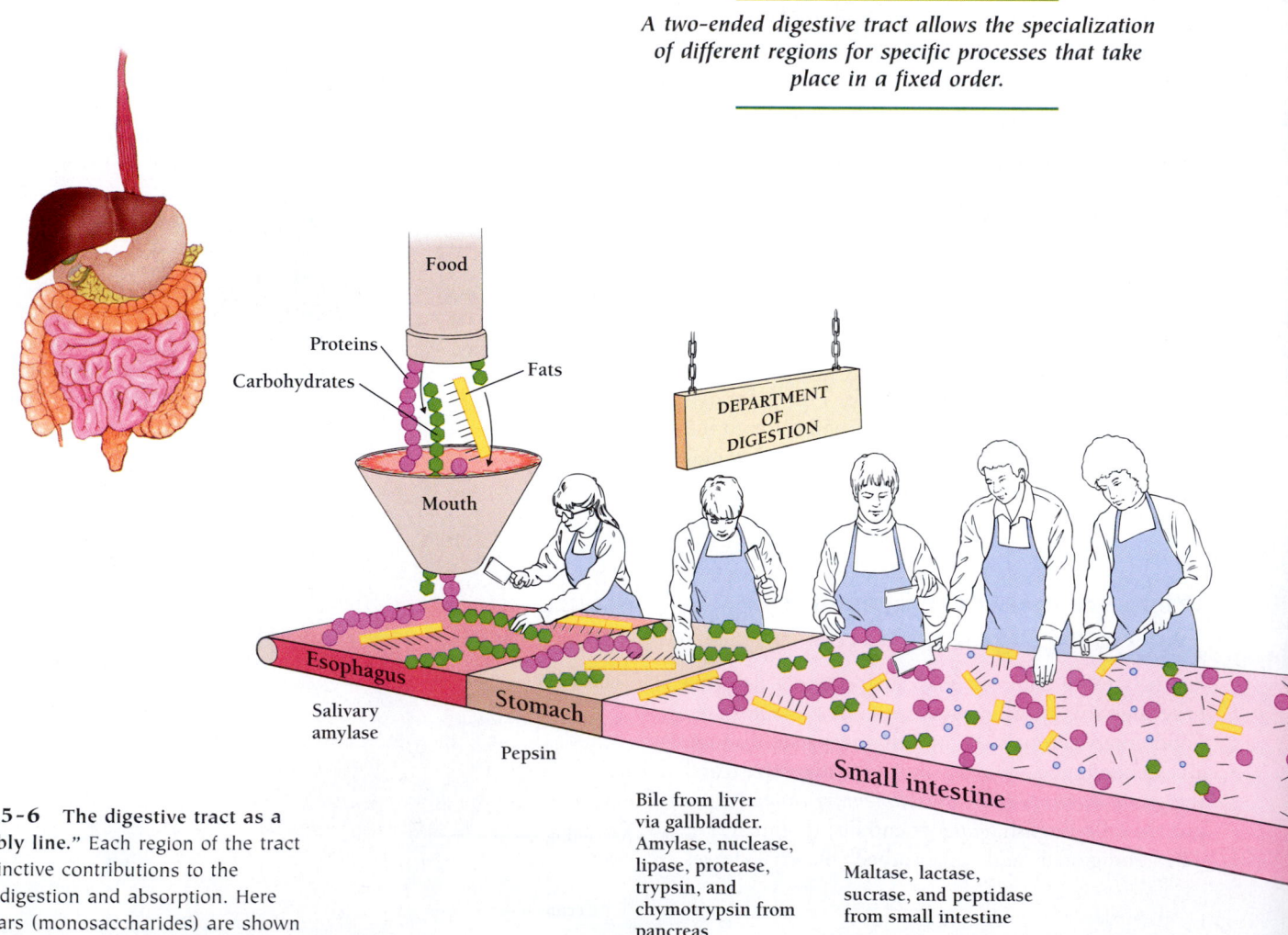

Figure 35-6   The digestive tract as a "disassembly line." Each region of the tract makes distinctive contributions to the process of digestion and absorption. Here simple sugars (monosaccharides) are shown as hexagons, amino acids as circles, and triglycerides as three-tooth combs.

HOW DOES THE VERTEBRATE DIGESTIVE TRACT FUNCTION AS A "DISASSEMBLY LINE"?

The digestive system actually performs five functions: (1) **movement** (or "motility")—agitating the food within specialized regions and pushing food through the system; (2) **secretion**—the production of lubricants, enzymes, and detergents; (3) **digestion**—the breaking down of large molecules to their component building blocks; (4) **absorption**—taking up small molecules into cells or into the circulation; and (5) **storage**—processing a meal a bit at a time, so animals do not have to feed continuously. To visualize how the disassembly process occurs, we will follow the progress of a hamburger and lettuce on a bun as it moves from mouth to anus. Before we do so, we will first describe the basic divisions of the tract.

How Is the Vertebrate Digestive Tract Organized?

Figures 35-5 and 35-6 show the general organization of a vertebrate digestive system into five specialized divisions: (1) mouth and throat *(headgut)*, (2) esophagus and stomach *(foregut)*, (3) small intestine *(midgut)*, (4) pancreas and biliary system, and (5) large intestine *(hindgut)*. The **lumen** [Latin, = opening, light] of the gut, the inner space within the tract, is continuous—from mouth to throat to esophagus to stomach to small and large intestines to anus. The pancreas and biliary system connect to the gut within the small intestine by means of separate ducts that empty into the gut.

The diameter of the gut's lumen varies widely from region to region, usually being widest in the stomach. Knowledge of the specialization of each division comes from anatomical studies. Microscopic studies have shown that in each division the lining of the gut lumen is an **epithelium**, a tightly connected sheet of cells. In each division, the cells of the epithelia are specialized for the particular functions of that division.

Researchers have followed the movement of materials through the digestive tract using a number of methods. Beaumont and others inserted tubes into surgically prepared openings to sample the processed food at each stage. More recently, x-ray studies have allowed researchers and physicians to observe both the movement of food and the movement of the tract itself, without actually disturbing the process.

The digestive system consists of five specialized divisions, each able to carry out distinctive parts of the digestive process.

The Human Digestive System and Some Vertebrate Variations

To visualize the progress of a meal through the human digestive tract, imagine Lexie St. Martin, the great-great-great-great-great-great-great-great-granddaughter of Alexis St. Martin, whom we discussed in the beginning of the chapter, is a student at the University of New Hampshire, only 200 miles from Alexis's Montreal. Lexie is about to enjoy one of the burgers she has prepared for a picnic.

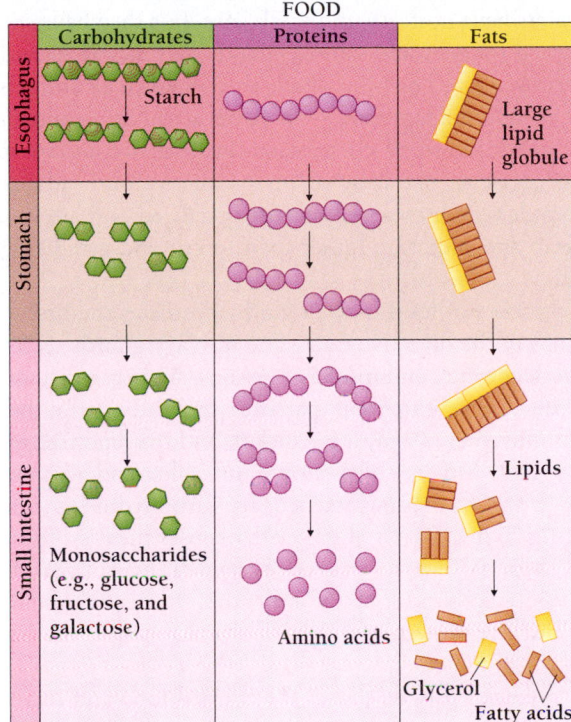

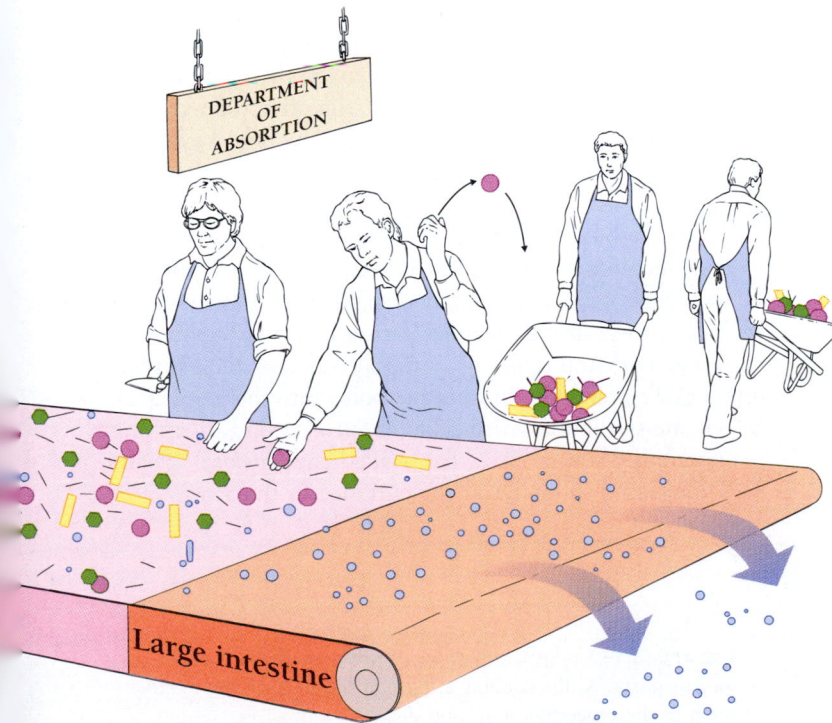

What does the headgut (mouth and throat) contribute to digestion?

Lexie's first bite contains some bun, some lettuce, and a chunk of meat and fat. As she chews, her teeth tear and grind the bite into little bits.

Food enters the digestive tract through the mouth and throat. The mouth captures the food and prepares it for entrance into the gut. All vertebrates other than lampreys and hagfish have jaws, and nearly all (except for birds, turtles, and a few other groups) have teeth. Jaws, beaks, teeth, tongue, and lips all participate in the grabbing and tearing of food.

Mammals are the only animals that can actually chew their food. Four sets of adaptations—jaws, tongues, cheeks, and teeth—allow mammals to pulverize food more finely than other vertebrates. Mammalian jaws and their attached muscles allow horizontal movement, needed for good grinding between the uneven surfaces of the large molars and premolars (Figure 35-7). Cheek and tongue muscles, which produce the suction

needed for nursing, also allow the manipulation of food for grinding. Mammalian teeth are highly specialized as well: human teeth include *incisors,* whose chisel-like edges cut; *canines,* whose pointed crowns tear; and *premolars* and *molars,* whose flat, ridged surfaces grind and crush (Figure 35-7A).

Mammals with different types of diets differ in the specialization of their teeth and other mouth parts. Carnivorous mammals have well-developed incisors and canine teeth, whereas herbivorous mammals may lack canines altogether and have incisors adapted for cutting vegetation (Figure 35-7C and D). Anteaters have no teeth and weak jaws, but they have specially adapted tongues and salivary glands that secrete quantities of viscous saliva. The blue whale, the largest of all animals, has another mouth adaptation—an elaborate filtering apparatus that catches the small crustaceans (krill) that compose its diet. In contrast to the array of mammalian mouth adaptations, a reptile's teeth appear to be unspecialized and are used only to grab and puncture rather than to cut and grind (Figure 35-7B).

The presence of food in the mouth stimulates the first secretions of the digestive tract. The **salivary glands** secrete both enzymes and **mucus,** a viscous, slippery substance that coats the food particles and lubricates their movements within the mouth and the digestive system. Mucus consists of water, salts, and *glycoproteins* (proteins attached to carbohydrates). The grinding action of the teeth exposes the food to the salivary enzyme **amylase,** which hydrolyzes the polysaccharides of the bun into shorter fragments. But the protein and lipid molecules within the bite are still intact.

Lexie's tongue shapes the chewed food and the lubricating mucus into a ball, called a **bolus.** Movement through the tract begins as the tongue and other muscles of the mouth push the bolus back into the throat, or **pharynx.** The pharynx is the common entryway both for food into the digestive tract and for air into the lungs. This being so, how does Lexie prevent the bolus from moving into her airways instead of her gut?

The bolus continues to move through the esophagus and into the stomach. Whether it will go into the esophagus or into the airway depends on the workings of *sphincters,* rings of muscles that surround a tube. Sphincters surround both the opening of the airway and the **esophagus.** Lexie's swallowing pushes the bolus into the esophagus. When she swallows, which takes about a second, breathing temporarily stops, and a flap, called the *epiglottis,* covers the closed sphincter at the entrance to the airway.

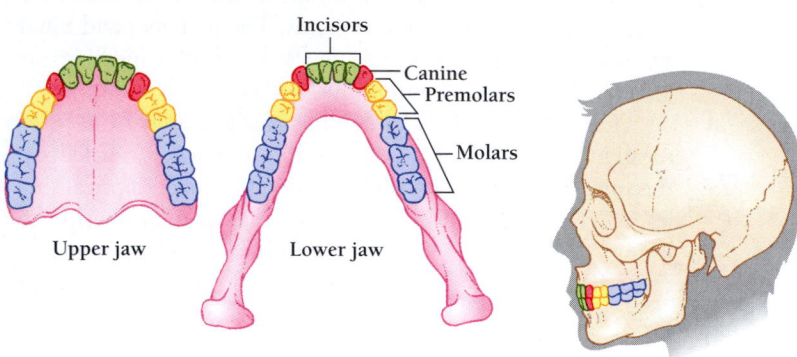

A. Human

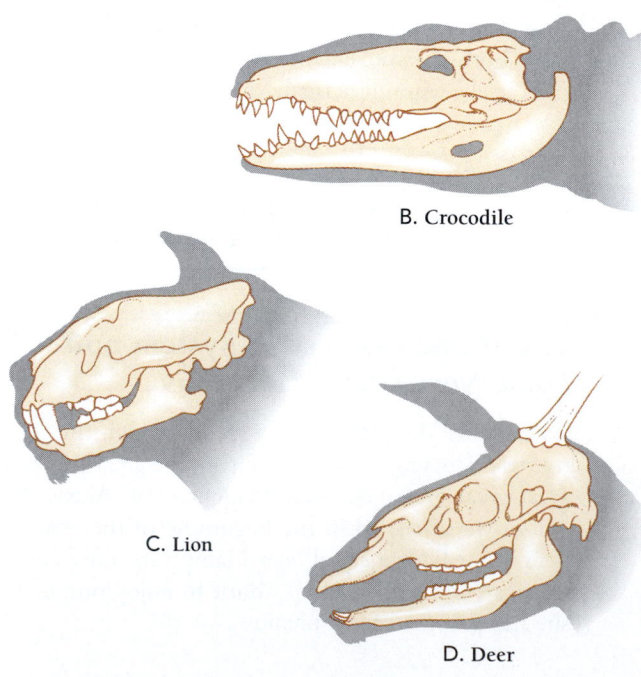

B. Crocodile

C. Lion

D. Deer

Figure 35-7 Specialization of jaws, teeth, and other mouth parts. A. Illustration of human teeth. B–D. Jaws and teeth of the crocodile, lion, and deer.

If a piece of food is too large to enter the esophagus, it may block the entrance to the *glottis,* through which air enters the lungs. The resultant choking can sometimes be fatal. The Heimlich maneuver can dislodge such a blockage by forcing the obstruction back into the mouth.

The initial movement of the bolus depends on swallowing. But how does Lexie control the complex movements of her swallowing? The answer is that her control is only partly conscious: Animals can voluntarily initiate chewing and swallowing, but both processes are also involuntary reflexes, automatic responses to the presence of food in the mouth. Pressure receptors sense food against the gums, palate, teeth, and tongue, and chemical receptors on the tongue detect food by its taste. Swallowing involves the coordinated response of both smooth and striated muscles in the pharynx, the sphincters, and esophagus, as well as the inhibition of muscles responsible for breathing. A swallowing center in the lower part of the brain (in the medulla) manages the operation through a network of nerves that connect to the involved muscles.

Digestion begins in the mouth with the division of the food, the start of enzyme hydrolysis, the formation of a bolus, and the movement of the bolus into the gut.

What does the foregut (esophagus and stomach) contribute to digestion?

Smooth muscles propel the bolus through the esophagus. **Peristaltic waves,** coordinated contractions of smooth muscles, move down the esophagus toward the stomach (Figure 35-8). During a wave, the muscles relax in front of the bolus and

contract behind it, pushing the bolus along toward the stomach. The path, about 25 cm long, goes straight down the chest cavity and through an opening in the diaphragm muscle, which is chiefly involved in breathing. Despite its downward course, gravity is not necessary for the movement of food into the stomach: Lexie could swallow her burger standing on her head, and astronauts can swallow when weightless.

Some animals have a saclike extension of the esophagus, called the **crop,** which can store food for later digestion. A bird, for example, can fill its crop with seeds or berries, which it can later regurgitate to feed its nestlings.

The esophagus empties directly into the **stomach** [Greek, *stoma* = mouth], the most dilated and most muscular section of the digestive tract (Figure 35-9). In humans, the empty stomach holds only about 50 ml (less than 2 ounces), but it can expand to contain an entire meal, which may occupy as much as 1.5 liters. As Beaumont's studies showed, the stomach accumulates food, disrupts it mechanically, and begins the process of protein hydrolysis. As Beaumont also observed, the stomach's lining has a velvetlike appearance, reflecting the folds of its epithelial surface.

In Lexie's case, bolus after bolus of Lexie's burger arrives in the stomach from the esophagus. After each bite, the sphincter at the stomach's entrance closes to prevent regurgitation of the contents back into the esophagus. Another sphincter at the stomach's exit prevents the contents from moving on prematurely.

The presence of the food in the stomach stimulates the secretion of hydrochloric acid and pepsinogen. In the acid environment of the stomach, the pepsinogen immediately unfolds. The acid cuts away a part of the polypeptide chain to

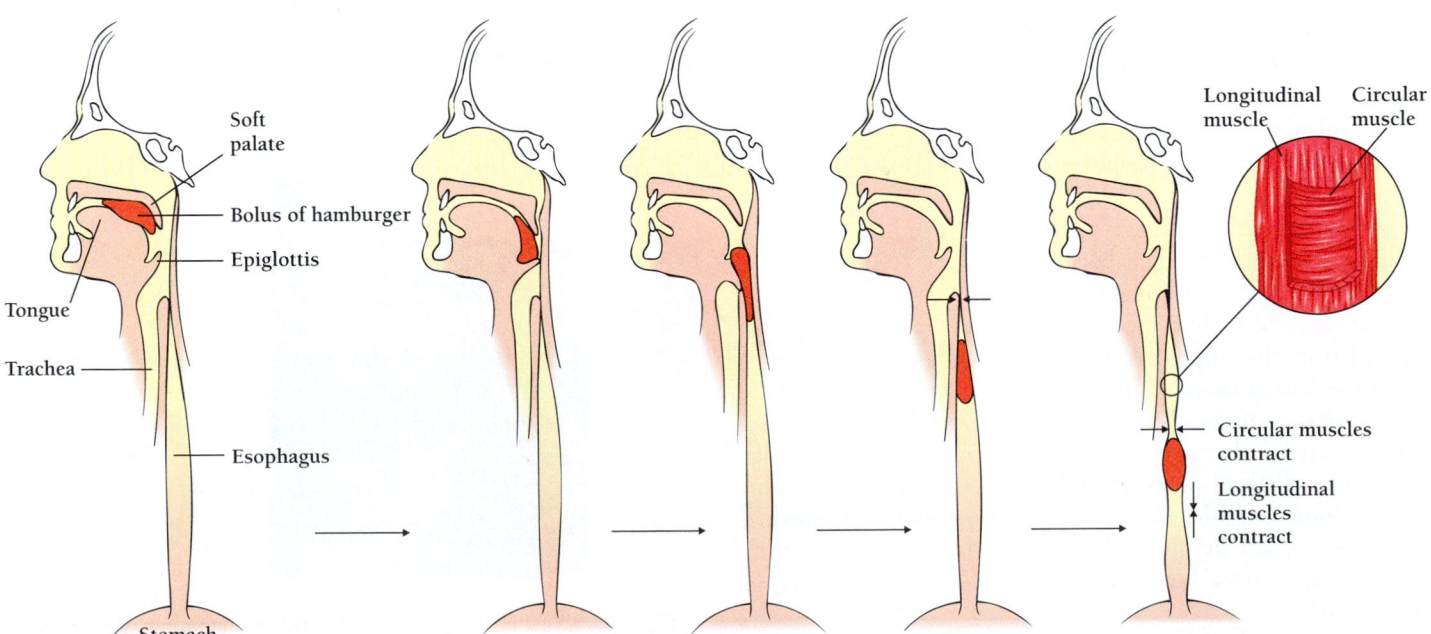

Figure 35-8 Swallowing and the beginning of a bolus. Peristaltic waves push the bolus down the esophagus and into the stomach.

form **pepsin,** the stomach's protease, which begins to hydrolyze proteins from Lexie's burger. The acid also terminates the action of amylase, which entered the stomach along with the processed burger.

Sometimes, particularly after a fatty meal or alcohol, this sphincter at the stomach's entrance fails to stay closed, and the acidic contents from the stomach bubble back up into the esophagus, producing an uncomfortable burning feeling in the upper chest, called "heartburn," though it has nothing to do with the heart at all.

The smooth muscles surrounding the stomach churn its contents, and the hydrochloric acid denatures the hamburger's proteins, increasing their susceptibility to pepsin digestion. The end result of the process is **chyme** [Greek, *khumos* = juice], a creamy, acidic liquid that passes into the small intestine. The fat from Lexie's burger remains completely undigested, and lipid molecules remain as suspended droplets within the chyme, giving the chyme the thick consistency of a fatty, burger-and-bun soup. The cellulose from the lettuce also remains undigested, and pieces of it remain dispersed within the chyme.

Besides causing proteins in the food to denature, the acid environment of the stomach also stimulates the activity of some digestive enzymes while destroying the activity of others. Enzymes are proteins, and their structures depend on pH, which deeply affects the formation of hydrogen bonds. The low (acid) pH of the stomach increases the activity of pepsin, but abolishes the activity of amylase (from the saliva). Similarly, the enzymes that will break down lipids in the small intestine cannot work in an acid environment.

Peristaltic waves move the chyme to the stomach's exit. The exit sphincter opens briefly, allowing a small spurt of chyme to pass at a time. Infections or actual physical blockage can interfere with peristalsis in the stomach and intestines, and chyme cannot move normally down the intestinal tract. The result is *vomiting,* the sometimes-violent upward movement of chyme back into the esophagus and mouth.

In birds, the contents of the esophagus are less finely divided than in mammals, for birds have no teeth with which to tear or grind their food. Food passes quickly from the esophagus (or crop) through an unspecialized stomach into the gizzard. The inner surfaces of the gizzard are coated with a horny material that serves as an abrasive, along with small stones that the bird picks up and swallows. Seed eaters, such as the domestic chicken, have the most muscular gizzards. Gizzards contain pepsin and hydrochloric acid, so that the chyme that leaves a bird's gizzard is ready for the next steps of digestion.

The stomachs of cows—and those of moose, buffalo, camels, giraffes, and other ungulates—have special adaptations that enable them to extract energy efficiently from the hard-to-digest grasses that make up most of their diets (Figure 35-10). These animals are all **ruminants**—hoofed, horned, or antlered herbivores that can regurgitate partly digested food, called *cud,* for further chewing. A ruminant's stomach contains four distinctive chambers, which in a large cow may hold as much as 200 liters. Bacteria and protists in these chambers produce enzymes that break down cellulose, a feat that no animal can do by itself. So, in contrast to the chyme from Lexie's meal, the chyme entering a cow's intestine contains no undigested cellulose, but only energy-rich molecules that result from the action of microbial enzymes.

The stomach mechanically disrupts food while hydrochloric acid and pepsin begin to hydrolyze proteins. Food leaves the stomach as semiliquid chyme.

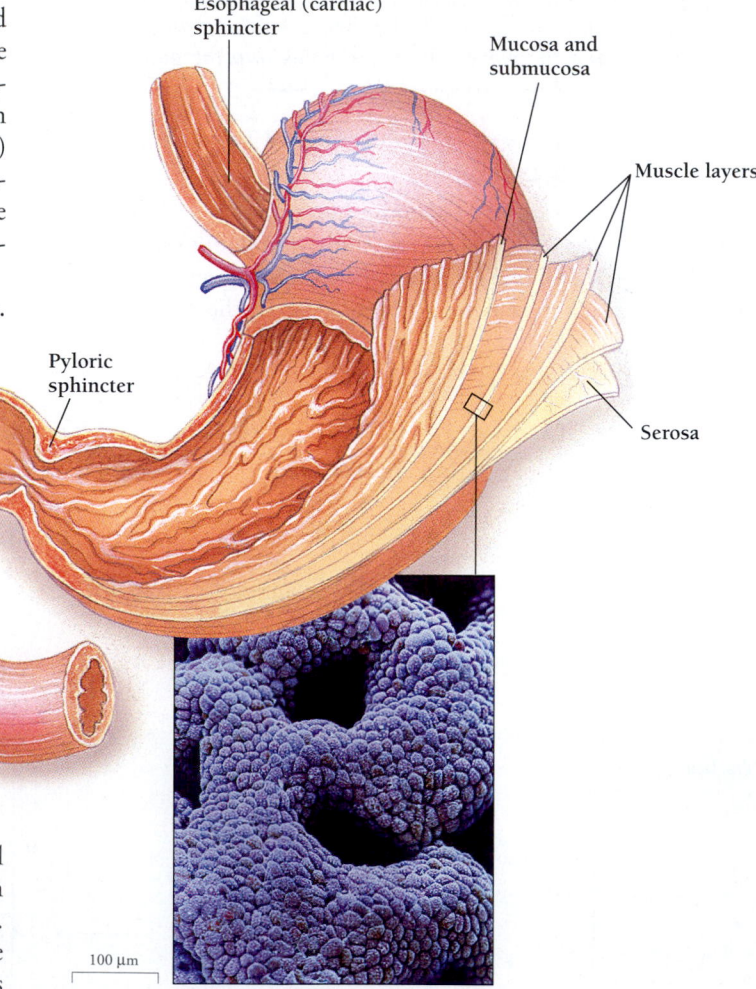

Figure 35-9   The tissues and cells of a stomach. The deep folds (villi) are the source of the velvetlike appearance that Beaumont saw in St. Martin's stomach. *(Photo, Quest/Science Photo Library/Photo Researchers, Inc.)*

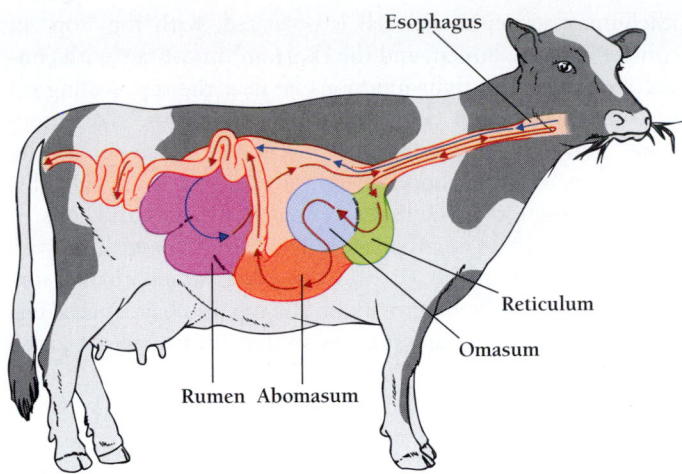

Figure 35-10 The four stomachs of a ruminant.

What does the midgut (small intestine) contribute to digestion?

Adaptations of the small intestine promote digestion and absorption. Chyme now enters the **small intestine,** the longest division of the digestive tract. The small intestine accomplishes two tasks: (1) it completes the digestion of macromolecules in the chyme and (2) it absorbs most of the useful products of this digestion.

Complete digestion requires a relatively long time for enzyme action, and absorption requires a large surface area. In humans the small intestine is typically 3 to 5 cm in diameter and about 6 m long. The substantial length of the small intestine allows time for digestion and absorption.

In herbivores other than ruminants, undigested cellulose interferes with digestion and absorption. The small intestine is therefore proportionally much longer in these herbivores than in humans or other omnivores. In contrast, strict carnivores have a proportionally shorter small intestine: the carnivorous adult frog, for example, has a much shorter small intestine than a tadpole, which lives on a vegetarian diet.

The organization of the small intestine also reflects its specialization (Figure 35-11). As in the rest of the tract, an epithelium lines the lumen. Surrounding the epithelium is a layer of connective tissue, which contains blood vessels, nerves, and lymphatic ducts, which drain extracellular fluid into the circulatory system. Outside the connective tissue is a thin layer of smooth muscle. Together, the epithelium, the connective tissue, and the thin muscle layer are called the *mucosa.* Although the mucosa of the small intestine roughly resembles that in the other regions of the digestive tract, it is much more highly folded, greatly increasing the effective surface of the epithelium. Each fingerlike fold is called a **villus** [Latin, = shaggy hair; plural, **villi**], and each extends about 1 mm into the lumen.

Surrounding the mucosa is another layer of connective tissue and then two additional layers of smooth muscle. These are the muscles responsible for moving the chyme. Contrac-

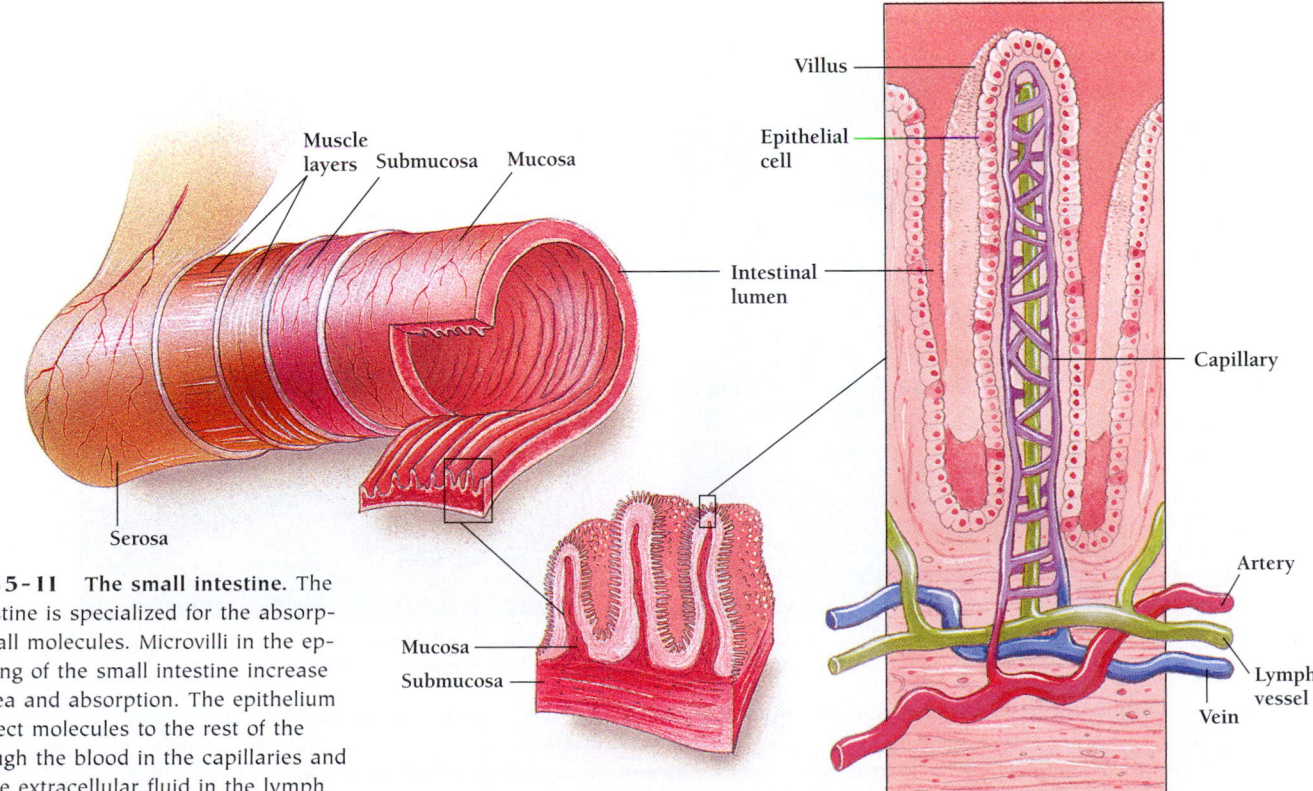

Figure 35-11 The small intestine. The small intestine is specialized for the absorption of small molecules. Microvilli in the epithelial lining of the small intestine increase surface area and absorption. The epithelium passes select molecules to the rest of the body through the blood in the capillaries and through the extracellular fluid in the lymph vessels.

tions of the inner, circular muscles squeeze the contents of the lumen; contractions of the outer, longitudinal muscles shorten the tube. Finally, another layer of connective tissue, the *serosa,* surrounds the entire tract, and the *mesenteries* connect the tract to the walls of the abdominal cavity.

Digestion within the **duodenum,** the first section of the intestine, and absorption within the rest of the small intestine depend on the continued mixing of the contents of the lumen. These movements depend on the action of the smooth muscles surrounding the small intestine, which are able to agitate the chyme in two distinct patterns. The first pattern, called *segmentation,* forces the contents back and forth, more or less in place. This action thoroughly mixes the chyme with the secretions of the liver and pancreas. It also allows prolonged absorption through the intestinal lining. Only after the intestine has absorbed most of the nutrients in a meal does the pattern of muscular contraction change. The second pattern of muscle contraction generates peristaltic waves. These push the remaining material toward the large intestine, the last region of the digestive tract.

The expansion of the epithelial surface extends to the cellular level, with the surface of each epithelial cell covered with **microvilli,** the membrane-covered cellular extensions (Figure 35-12). The epithelial cells are joined to each other through tight junctions, so no molecules or ions can leak from the lumen into tissues or blood without passing through the ep-

ithelium. Each epithelial cell is polarized, with the "top" in contact with the lumen, and the "bottom" in contact with connective tissue. The tight junctions lie near the top, sealing off the intestine's lumen (Figure 35-13B). Below the tight junctions, however, are adhering junctions, which physically strengthen the connections between adjacent epithelial cells, and gap junctions, which allow small molecules and ions to pass freely between adjacent cells. The gap junctions, however, do not allow materials to pass from the lumen to the surrounding tissue. The microvilli on the top also have special enzymes attached, for example, those that break down sucrose molecules (Figure 35-13B).

The polarized distribution of specific molecules also allows the epithelium to take up molecules, such as glucose, with great efficiency. The epithelial cells of the small intestine, for example, contain a special glucose transporter, which uses the differences in sodium ion concentration between the inside and outside of cells (high outside, low inside) to drive glucose into the cytosol of epithelial cells (Figure 35-13B). Another type of glucose transporter on the bottom surface of the each cell allows glucose to move passively out of the epithelium into the blood. The blood then carries glucose to all the cells in the body.

The rate of glucose absorption depends directly on the total surface area of the epithelium. Villi and microvilli greatly increase the total area of the small intestine. If the surface of

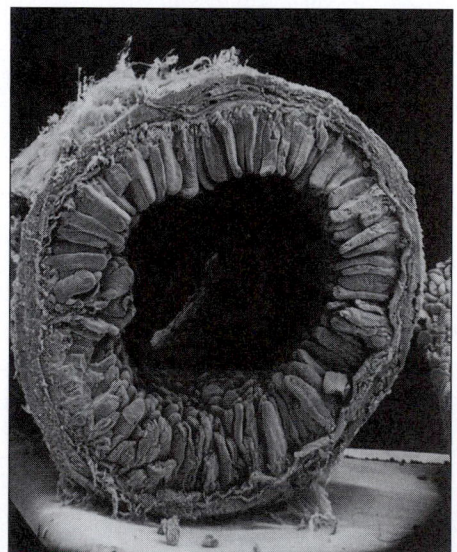

A. Cross section of the   500 µm
   small intestine

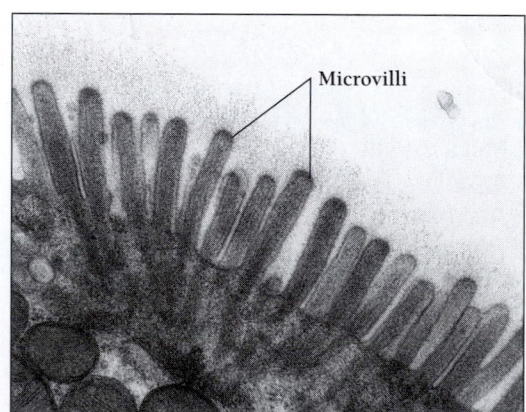

B. Microvilli in the lining of   1 µm
   the human intestine

Figure 35-12   Cell specialization of the small intestine. A. Cross section of the small intestine, showing how villi increase the surface area of the intestinal epithelium. B. An electron micrograph showing how microvilli increase the surface area of single epithelial cells. *(A, G. Shih-R. Kessel/Visuals Unlimited; B, David M. Phillips/Visuals Unlimited)*

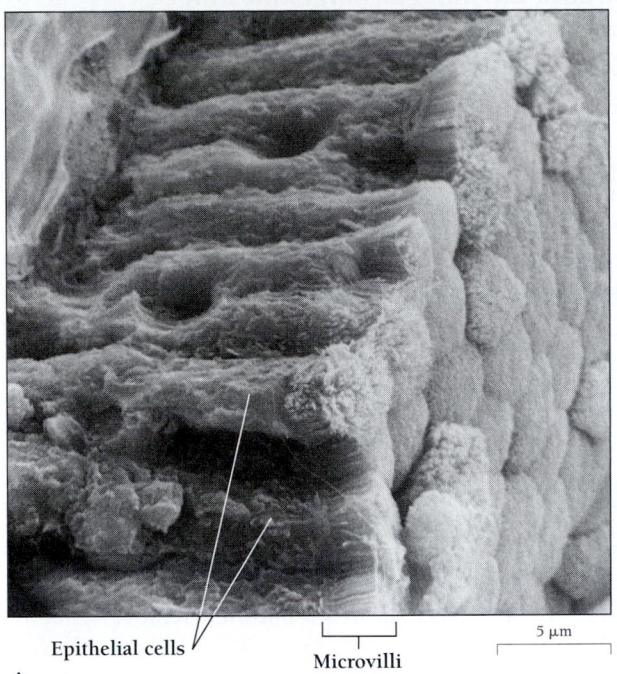

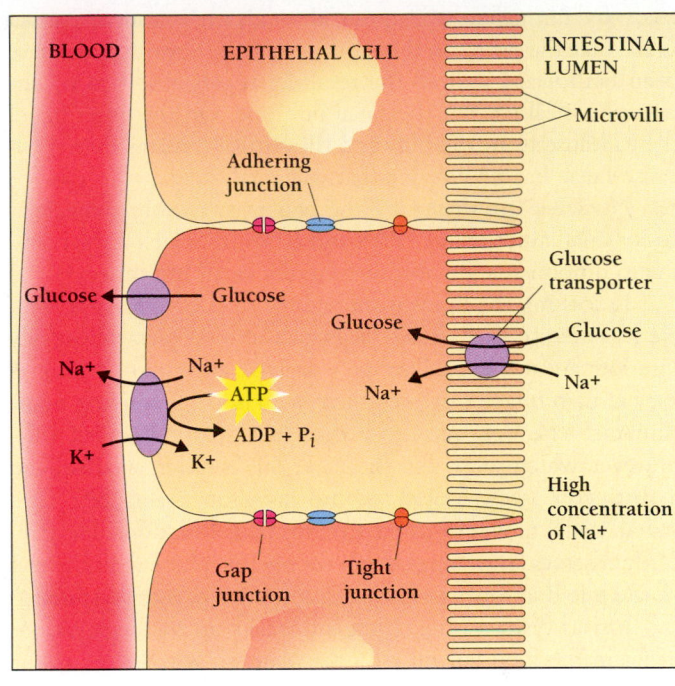

Figure 35-13 **Epithelial cells are polarized.** Tight junctions seal the lumen off from the underlying tissues and blood. Just below the tight junction are adhering junctions, which strengthen the epithelial sheet, and gap junctions, which allow molecules and ions to pass between cells. The top surface of each cell, which contacts the lumen, has different molecules from those of the basal surface. The sodium-driven glucose transporter is at the top surface, while the passive glucose transporter is on the basal surface. *(A, From R. G. Kessel and R. H. Kardon,* Tissues and Organs: A Text-Atlas of Scanning Microscopy, *1979, W. H. Freeman & Co.)*

the human small intestine were completely smooth, for example, its area would be about the size of a 28-inch television screen; instead, the total area through which it absorbs nutrients is almost 300 square meters, about the area of a tennis court.

The length of the small intestine allows sufficient time for digestion and absorption. Villi and microvilli increase the surface area of absorption. Smooth muscles agitate the contents of the lumen and propel them onward.

What do the pancreas, liver, and gallbladder contribute to digestion?

Digestion within the small intestine depends on secretions by the pancreas and liver. Most of the digestion within the intestine occurs in the duodenum, which, in humans, is about 25 cm long [about 12 "finger widths," hence its name, which comes from the Latin for "twelve"]. The rest of the small intestine is specialized for absorption.

A common duct carries materials from the pancreas, the liver, and the gallbladder. The **pancreas** provides a panoply of digestive enzymes, which enter the intestine through a special duct. The pancreatic enzymes, unlike pepsin, do not work at acidic pH, so the pancreas also secretes bicarbonate (HCO_3^-) ions, which neutralize the acidic chyme as it enters the duodenum.

Proteases from the pancreas carried into the lumen then chop the peptide fragments, produced by pepsin in the stomach, into lengths of a few amino acids. Other enzymes, called *peptidases,* which lie on the surface of the microvilli, then complete the hydrolysis of polypeptides into free amino acids, which can then be absorbed by the small intestine. By the time the remains of Lexie's meal have left the small intestine, essentially all the proteins have been hydrolyzed and all the resulting amino acids have been absorbed.

The digestion of polysaccharides (starch and glycogen) is also completed in the small intestine. Pancreatic amylase finishes the digestion of polysaccharides begun by salivary amylase. Maltose, sucrose, and other disaccharides accumulate in the small intestine. Again, enzymes on the surfaces of microvilli hydrolyze the disaccharides into simple sugars (such as glucose), ready for immediate absorption through the intestinal epithelium. Like the proteins, then, the starches in Lexie's chyme have been hydrolyzed and all the resulting sugars absorbed by the time the meal has left the small intestine. On the other hand, the cellulose from the bun and the lettuce remains intact, because Lexie has no enzyme that will catalyze its hydrolysis.

Nucleases produced by the pancreas hydrolyze the relatively small amounts of DNA and RNA in Lexie's burger, and their components are absorbed by the small intestine.

Processing the lipids is more difficult. Lipids are not soluble in water, so they usually aggregate into droplets whose oily interiors are not accessible to hydrolytic enzymes. As long as they remain intact, however, there is no hope of deriving nourishment. The liver secretes a detergent solution, called **bile,** however, which solves this problem. Bile is bitter, alkaline, and a distinctive green-yellow or brown-yellow color. Bile (also called "gall") moves from the liver to the **gallbladder.** The gallbladder serves as a storage depot for bile. A sphincter at the end of the bile duct opens only when food enters the intestine.

Just as dishwashing detergent breaks up the aggregations of fat in the food left on a dinner plate, so bile breaks up the fat droplets within the chyme. The movements of the small intestine mix the chyme with bile to form an *emulsion,* which contains tiny droplets of fat, about 1 μm in diameter, each coated with detergent molecules. Lipases, secreted by the pancreas into the duodenum, now hydrolyze the lipid molecules into fatty acids, glycerol, and other small molecules. All of these can be absorbed later by the small intestine.

Absorption of small molecules across the epithelium of the small intestine depends both on passive diffusion (for example, of free fatty acids across the plasma membrane) and on specific protein molecules on the plasma membrane. Some of these proteins are enzymes that finish the very last steps of digestion. Others are transport molecules. Some can carry molecules in either direction, but primarily they carry nutrient molecules from the lumen (where their concentration is high) into the cytosol (where their concentration is low). In some cases, the epithelial cells expend energy to ensure the effective transport of specific molecules, even when the concentration in the epithelium is higher than in the lumen.

The overall process of absorption is so efficient that, normally, by the time the chyme appears in the large intestine, no nourishing molecules remain—only water, dissolved ions, cellulose, and various unabsorbed molecules, together with mucus and other secretions of the digestive tract itself. On average, an adult human consumes about 2 liters of food and drink and secretes about 7 liters of mucus, acid, and other fluids each day. The small intestine absorbs about 95 percent of this total, so that the total amount of liquid entering Lexie's large intestine on a typical day is 450 to 500 ml (about a pint).

Digestion depends upon enzymes produced by the pancreas and bile produced by the liver and stored in the gallbladder.

What does the hindgut (large intestine) contribute to digestion?

The **large intestine** recovers water and dissolved ions. Most of the length of the **colon,** or large intestine, is specialized for the absorption of water and ions. Although the mucosa of the colon is not folded into villi like that of the small intestine, the epithelial cells that line the lumen have extensive microvilli. In humans, the junction between the small intestine and the colon is located near the bottom of the abdominal cavity, on the right side. The colon has three major sections: the *ascending colon,* which extends up the right side of the abdomen; the *transverse colon,* which crosses over to the left; and the *descending colon,* which extends downward on the left. The descending colon ends in a straight portion, called the **rectum** [Latin, = straight], in which feces are stored until their elimination through the anus.

The large intestine removes most of the water that enters with the remains of the chyme, so that feces typically contain only about 100 ml of water and 50 g of solids per day. Both the small intestine and the large intestine remove water by pumping ions across the epithelium. Water follows, moving from a solution in the lumen with low solute concentration to a solution with high solute concentration.

If the colon is irritated, its contents may move too fast for efficient absorption, resulting in **diarrhea,** or watery feces. On the other hand, if the intestine's contents move too slowly, the epithelium absorbs too much water, resulting in **constipation,** the inability to move the bowels. Undigested cellulose provides *roughage,* also called *fiber* or *bulk,* which stimulates peristalsis. Insufficient roughage also leads to constipation.

The colon contains large numbers of bacteria, which make up about half the dry weight of the feces. These bacteria live on unabsorbed energy-rich molecules within the colon. Some of the products of bacterial metabolism make their way through the epithelium into the blood—including many vitamins from the B complex and some small products of bacterial fermentation.

Some bacteria, however, can interfere with the normal function of the colon, leading to diarrhea. "Traveler's diarrhea" is a common result of infection with bacteria whose products prevent proper water absorption in the colon. Other infections can also cause diarrhea, for example, by protists such as *Giar-dia,* a common contaminant of once-pure mountain streams now frequented by increasing numbers of sheep, cattle, and campers. Diarrhea is the single greatest cause of illness and death on the planet. Most of its victims are young children.

Near the junction between the large and small intestine is the **cecum** [Latin, *caecus* = blind], a blind sac that diverts from the main route through the digestive tract. The human cecum is small and does not appear to serve any role in digestion or absorption. The *appendix,* a small, fingerlike projection at the top of the cecum, is now thought to have some role in the immune system. It is best known, however, for its tendency to become infected, usually prompting its surgical removal.

In contrast to the human cecum, the cecum of other mammals, especially herbivores, may be relatively large. In rabbits, for example, the cecum contains bacteria and protozoans that digest cellulose. Undigested material from the small intestine enters the cecum, where additional digestion can occur. The newly digested material, now containing energy-rich glucose molecules and other nutrients, leaves the cecum and travels down the large intestine. But the large intestine is not able to absorb these molecules, and they pass out in the feces. Rabbits, however, have evolved an effective (if unappetizing) recycling mechanism: they actually produce two types of feces, one of which consists of the energy-rich materials from the cecum. The rabbits then reingest these feces (but not the others) and extract the valuable molecules during the repassage through the gut.

The large intestine removes ions and water.
Bacteria grow in both the colon and the cecum;
in some species, bacteria in the cecum
digest cellulose.

HOW DOES AN ANIMAL COORDINATE PROCESSES IN INDIVIDUAL PARTS OF THE DIGESTIVE SYSTEM?

The digestive system must coordinate processes performed by cells that are far from one another. When food arrives in the stomach, for example, there must be acid and pepsin; and when chyme arrives in the intestine, there must be pancreatic enzymes, bile, and bicarbonate. How does this happen?

Hormones and Nerve Cells Coordinate the Actions of the Gut

In multicellular organisms, chemical signals are responsible for almost all intercellular communication and are especially important in coordinating the responses of different organs to changes in the external or internal environment. These signals contribute to **homeostasis,** the process of resisting changes in an organism's internal environment (Chapter 34).

Many important chemical signals are **hormones,** substances made and released by cells in a well-defined organ or structure, that move throughout the organism and exert specific effects on specific cells in other organs or structures (Chapter 36). Both animals and plants use hormones to coordinate the actions of physically distinct tissues and organs. The structures upon which hormones act are called **target organs.** Target organs respond to hormones by changing their metabolism, the activity of their enzymes, or the expression of specific genes. The regulation of appetite by the hormones orexin and leptin, for example, contributes to homeostasis—helping to maintain, within relatively defined limits, not only body weight but also water balance and nutrient supply.

Both nerve cells and hormones use chemical signals to coordinate muscle activity, secretion, digestion, and absorption. After the intestine has absorbed the products of digestion, glucose serves as the principal energy carrier to the whole body. Hormones coordinate the absorption of blood glucose into cells throughout the animal. In addition, part of the control of the digestive system depends on a complex network of nerve cells within the digestive tract itself. This network continuously receives information about the status of the tract from specialized receptors on the epithelial membranes. These receptors report on the contents of the lumen—how full it is, its total solute concentration, its pH, and the concentration of specific digestion products.

The nerve network processes this information and sends appropriate commands to cells within the tract. For example, when the small intestine is full (distended), the network increases the activity of its smooth muscles to increase mixing, digestion, and absorption. At the same time, other signals inhibit the smooth muscle activity in the stomach, preventing more material from entering the intestine.

The intestinal nerve network also receives information from the brain. Thus we respond to the sight, smell, and taste of food. The response includes activity of the salivary glands and the increased secretion of acid in the stomach. The Russian physiologist Pavlov discovered that the mere sounding of a bell would evoke acid secretion in dogs, provided that the dogs had learned to associate the bell's ringing with a forthcoming meal.

Both hormones and nerve networks coordinate the
actions of different parts of the digestive system, for
example by regulating the secretion of hydrochloric
acid, bicarbonate, bile, and digestive enzymes.

How Does the Gastrointestinal Tract Replenish Itself When Cells Die?

The movement of food particles and chyme subject the cells of the gastrointestinal tract to particularly harsh wear and tear. Maintaining the gut lining is yet another example of homeostasis, here at a cellular level. As cells die, new replacements move into place and take up their former roles. Because the epithelium contains many different types of cells, replacement requires more than simple cell division; the right types of cells must specialize and move into the proper place.

All the cells of the gut epithelium derive from common *stem cells.* These cells, which lie within the "crypts" between villi, undergo regulated cell division and specialization. Chemical signals, called *growth factors,* regulate the stem cells' processes of dividing and of acquiring specialized features, such as microvilli, gap junctions, and specific transporters. Exposure to radiation can destroy the ability of stem cells to divide. In many survivors of the 1945 atomic bomb attack on

Hiroshima and in people who have received large doses of radiation therapy for cancer, stem cells are less able to repopulate the gut, leading to gastrointestinal disorders (such as diarrhea) or, in extreme cases, to death.

Regulation of cell division can fail, however. Such failures are particularly common in the large intestine (colon), giving rise to *polyps,* small wartlike growths. Sometimes, cell division gets completely out of hand, resulting in cancer. Colon cancer is the second most common form of cancer in the United States, with more than 100,000 new cases a year. Some forms of colon cancer result from mutations in genes that regulate cell division.

The cells of the gut epithelium wear out and must be replaced. Radiation or cancer-causing mutations can interfere with the needed repair and replacement.

Study Outline with Key Terms

The digestive system allows animals to derive both energy and raw materials from food. Energy comes from the metabolism of small molecules, especially glucose. These molecules result from the action of hydrolytic enzymes on the macromolecules in food. The amount of food that an animal needs to consume depends on its **basal metabolic rate,** the minimum energy needed to stay alive and awake, on its size, and on its activities. Failure to obtain enough of particular raw materials—including the **essential amino acids,** minerals, and vitamins—leads to **malnourishment** and disease. Protein deficiency disease, **kwashiorkor,** is disturbingly prevalent in the world population.

Individual animal species differ in their diets; **herbivores** eat only plants, and **carnivores** eat only animals. Humans are **omnivores,** eating plants, animals, and other organisms. All organisms must both **digest** and **absorb** their food. Harsh conditions are necessary for hydrolytic enzymes to gain access to the molecules in food. The hydrolytic reactions of digestion always take place apart, away from the cytosol, in lysosomes, in a digestive cavity, or in a **gut,** or **digestive tract.**

The digestive tract of vertebrates consists of a continuous tube that leads from the mouth to the **anus.** Food enters through the mouth and undergoes a sequence of transformations as it passes through the **lumen** of the digestive tract. Enzyme action hydrolyzes macromolecules, and the resulting small molecules are absorbed in the small intestine. The remnants of food, together with secretions and bacteria from the large intestine, pass out of the anus as **feces.**

The digestive system has five specialized divisions: the mouth and **pharynx,** the **esophagus** and **stomach,** the **small intestine,** the **pancreas** and biliary system, and the **large intestine.** The digestive system accomplishes **movement, secretion, digestion, absorption,** and **storage.**

The mouth is specialized for the capture and initial preparation of food. **Mucus,** secreted by the **salivary glands,** lubricates the movement of the food in the mouth, and **amylase** begins to hydrolyze starches and glycogen. The mouth shapes the food and mucus into a **bolus,** to be swallowed by the pharynx. Smooth muscle contractions within the esophagus form **peristaltic waves,** which propel the bolus into the stomach. The stomach agitates the food with more mucus and secretes hydrochloric acid and **pepsin,** a protease that works best in the acid environment of the stomach. The result of the stomach's action is that the food is converted to **chyme,** a soupy liquid.

Individual groups of animals differ in the adaptations of their digestive tracts. Many birds have both a **crop,** an extension of the esophagus where food can be stored, and a **gizzard,** a specialized region beyond the stomach, where food can be ground against a hardened surface together with swallowed grit. **Ruminants** have highly specialized stomachs that allow the breakdown of cellulose, using enzymes produced by bacteria that live in the stomach.

In each region, the lumen is surrounded by an **epithelium.** The epithelium is specialized for different functions in different regions. In the small intestine, the epithelium and immediately surrounding tissue layers are folded into **villi,** and the epithelial cells themselves have extensively folded plasma membranes, or **microvilli.** Digestion occurs within the **duodenum,** using digestive enzymes from the pancreas and **bile** from the **gallbladder.** The resulting small molecules are absorbed by the rest of the small intestine, as the chyme continues its movement. The large intestine or **colon** removes water and ions as the remains of the food travel toward its last section, the **rectum.** Bacteria growing within the colon contribute about half the volume of the feces. Irritation of the colon can decrease absorption, leading to **diarrhea,** and insufficient bulk in the diet can lead to **constipation.** The **cecum** is a sac that extends just beyond the beginning of the colon and, in some species, is also the site of bacterial growth.

To achieve **homeostasis,** nerves and chemical signals coordinate muscle activity, secretion, digestion, and absorption within the digestive system. **Hormones** are chemical signals that are released into the circulation by specialized cells or organs. Hormones evoke distinctive effects on specific **target organs.**

Review and Thought Questions

Review Questions

1. Define the basal metabolic rate.
2. Why must vegetarians worry about balancing complementary proteins?
3. Give an example of a herbivore, a carnivore, and an omnivore. How do the teeth and digestive tracts of herbivores and carnivores differ?
4. Why is digestion always outside the cytosol? Describe the three major ways that organisms have for accomplishing this.
5. What is the function of the saliva?
6. How do hydrochloric acid and pepsin interact in the stomach?

7. Explain how peristaltic waves move food through the digestive tract. Draw a picture illustrating this process.
8. Explain what bile is, where it comes from, and how it functions in the digestive tract.

Thought Questions

9. Why must cows and other ruminants have such large stomachs?
10. If your pancreas no longer secreted its digestive enzymes, what food groups would you have to give up?

About the Chapter-Opening Image

A shotgun such as this one opened a hole in the stomach of Alexis St. Martin.

 On-line materials relating to this chapter are on the World Wide Web at **http://www.harcourtcollege.com/lifesci/aal2/**

36

How Do Animals Coordinate the Actions of Cells and Organs?

Environmental Effects on Sexual Development

In the early 1990s, researchers announced a trend that grabbed headlines and the attention of men everywhere. Western men of the 1990s, smirked TV newscasters, make half the number of sperm that men did in the 1930s. As one reproductive physiologist quipped at a Congressional hearing in 1995, "Every man in this room is half the man his grandfather was."

The news media had a field day. "Sperm counts down? Penises shriveled? Hey, . . . don't blame it on the feminists," declared *Newsweek* magazine. The culprits instead appeared to be chemicals in the environment that mimic or antagonize the powerful effects of natural hormones. Some researchers blame the downward trend in sperm counts on a class of environmental pollutants called environmental estrogens, or *estrogenics,* natural and synthetic substances that mimic the

physiological effects of the female sex hormone estrogen. Other researchers have found that other compounds—common solvents that make plastics flexible—interfere with the action of the male sex hormone testosterone.

Everyone agrees that some compounds in some environments are poisoning sexual development, but no one can say how widespread the problem or how important a particular chemical. Some scientists indict industrial and agricultural compounds that are known to interfere with sexual development, while others point to toxic compounds produced by plants (*phytoestrogenics*) or by tiny invertebrates.

The pesticides atrazine, chlordane, and DDT, for example, all have estrogenic effects. A 1980 spill of DDT and related compounds in a Florida lake, for example, appears to have reduced the birth rate of alligators there by 90 percent. And the young male alligators that grew up in the DDT-laden lake had smaller-than-average penises.

Are human penises also shriveling? Are men throughout the industrial world being doused in overwhelming quantities of estrogen? Will industrial pollution finally contribute to a fall in fertility? Well, maybe not, or at least not yet.

Certainly sperm counts are down. But men generally produce far more sperm than they need, and men's fertility—as measured by the number of children they conceive and by other measures—is fine. And while it's true that a growing number of both natural and synthetic chemicals are turning out to have effects similar to those of estrogen, no one is sure what effects those chemicals actually have on our health. The idea that sperm counts are down because of environmental pollutants is no more than an intriguing hypothesis.

Still, the topic of environmental estrogens has excited a lively debate among researchers. On one side are those who argue that the rise in environmental estrogens very likely accounts for a host of problems, including a sharp rise in the rate of testicular cancer and a more gradual rise in the rate of breast cancer, not to mention the decline in sperm counts. On the other side are those who think the whole problem

© Stan Osolinski 1993/FPG

has been greatly exaggerated. Above all, the debate has raised more questions than researchers know how to answer.

Some questions researchers are asking are, What kinds of effects can environmental estrogens exert? and How do environmental estrogens compare with the real hormones they are said to mimic?

Estrogens are a group of steroid hormones found in vertebrates that are responsible for the development of female secondary sexual characteristics and the functioning of the female reproductive organs. Androgens, such as testosterone, are also steroid hormones, responsible for both the initial development of male genitalia and for the later acquisition of secondary sexual characteristics.

Environmental estrogenics include hundreds of different compounds. They are present at low levels in our environment and we ingest them in food and water. Most can bind to estrogen receptors on cells throughout the body, turn on estrogen-sensitive genes, and cause certain kinds of cells to proliferate. Some estrogenics have effects similar to natural estrogens made by the body in that they activate estrogen receptors. Some compounds are antiestrogenic, meaning they counteract the effects of estrogen by preventing natural estrogens from binding to their receptors. Still other compounds are antiandrogenic, meaning they counteract the effects of testosterone and similar hormones, usually by blocking the receptors for testosterone.

Most environmental estrogenics are very weak in their effects compared with natural estrogens. Because estrogenics are so weak and because individually they exist in low concentrations in our environment, many researchers have dismissed the idea that they could produce any dramatic effects in humans. But some researchers have shown that combinations of different compounds can have much more powerful effects than any single compound. For example, PCBs (polychlorinated biphenyls) applied to turtle eggs individually had no effect on turtle development. When two different PCBs were combined, however, their estrogenic effects turned males into females.

In 1996, researchers at Tulane University reported studies that used yeast cells that had been genetically engineered to express the human estrogen receptor. This report claimed that combinations of different pesticides had up to 1600 times the estrogenic potency of any of the chemicals individually, a result that supported the conclusions of the studies of turtle egg development. But no one could replicate the yeast studies, and the authors retracted their original paper. So, for the moment at least, researchers do not know the molecular basis of the cooperative effects of diverse estrogenics.

How can minuscule amounts of synthetic chemicals evoke powerful biological responses? The answer is probably that industrial chemists have accidentally made molecules with sizes, shapes, and charge distributions similar to the hormones and other molecules that animals use for internal signaling. Ironically, then, industrial chemists have unwittingly accomplished a major goal of many pharmaceutical chemists, whose very purpose may be to produce drugs that mimic natural chemical signals that may be lacking in some people. Understanding the biological effects of environmental chemicals requires knowledge of natural signaling molecules, the subject of this chapter.

> *The young male alligators that grew up in the DDT-laden lake had smaller-than-average penises.*

HOW DO CHEMICAL SIGNALS COORDINATE RESPONSES IN MANY CELLS?

An animal depends on the coordinated activities of many cells and organs. Just maintaining the function of individual cells depends on the animal's ability to regulate its own **internal environment,** which means the external environment of the animal's cells (Figure 36-1). Most animal cells have only limited tolerance for environmental variation, and natural selection has produced remarkable adaptations that promote a stable chemical composition of the extracellular space. The properties of every animal organ system—from the digestive system to the circulatory system to the respiratory system—reflect the need for **homeostasis** [Greek, *homos* = same +

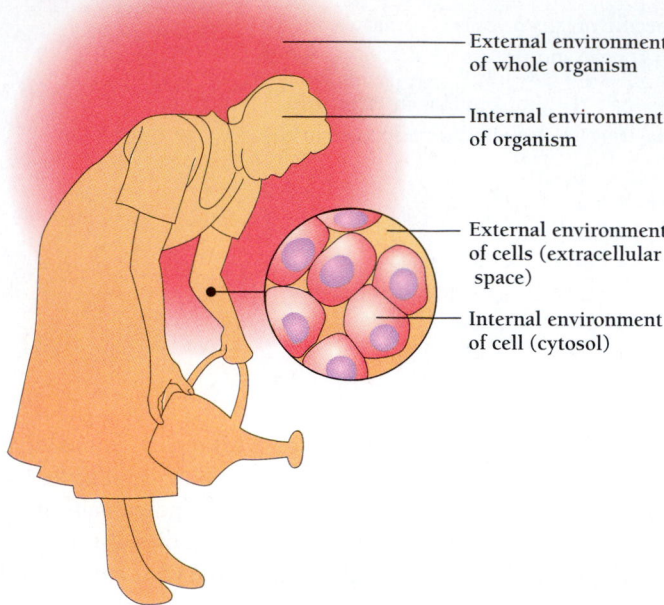

External environment of whole organism

Internal environment of organism

External environment of cells (extracellular space)

Internal environment of cell (cytosol)

Figure 36-1 An animal's "internal environment" is the extracellular environment of its individual cells.

stasis = standstill], the maintenance of a constant internal environment in the face of changing external conditions. In vertebrates, for example, extracellular fluids are constantly renewed by the blood, which carries nutrients and oxygen to individual cells and waste products to the kidneys and lungs.

Coordination of physically separated cells depends on chemical signals, compounds that influence the activities of individual cells, as well as of tissues and organs. An animal's circulatory system commonly carries chemical signals as well as nutrients and wastes. These signals mediate the responses to changes in the external or internal environment. In Chapter 35, for example, we saw how the presence of food in the stomach produced chemical signals—hormones—that stimulated the secretion of hydrochloric acid by the stomach, of hydrolytic enzymes by the pancreas, and of bile by the liver. Animals also use the electrical activity of nerve cells to coordinate events in physically separated organs.

Animals must also coordinate the actions of many cells and organs in response to special challenges, ranging from the development of sperm and eggs to embryonic development to birth to the stress of a final or a football game. Even the behavioral responses to seasonal changes—such as the migration of billions of birds each spring and fall—depend on chemical signaling among different organs.

To maintain a constant internal environment and respond to special challenges, animals must coordinate the activities of many cells and organs.

How Do Organisms Send Chemical Signals?

Biologists distinguish among three types of chemical signals (Figure 36-2):

1. **Pheromones** are substances secreted by one organism that influence the behavior or physiology of another organism of the same species. Examples of pheromones include compounds that allow ants to follow long trails and those that stimulate mating behavior.

2. **Hormones** are chemical signals, made and released by a well-defined organ or structure, that travel long distances through the circulation and exert specific effects on their **targets,** specific cells in other organs or structures that respond to a particular hormone. Examples of hormones are estrogen, which maintains female sexual function, insulin, which regulates blood glucose levels, and epinephrine, which coordinates an animal's "fight-or-flight" responses.

A. Pheromones

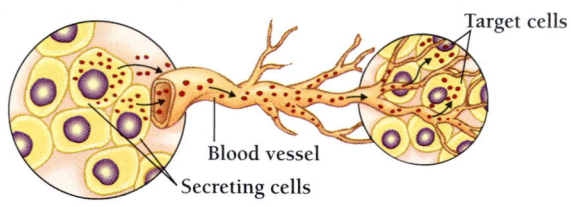

Target cells

Blood vessel

Secreting cells

B. Hormones

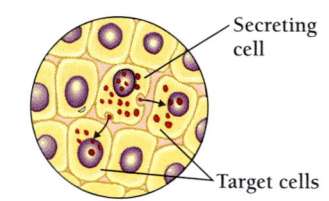

Secreting cell

Target cells

C. Paracrine signals

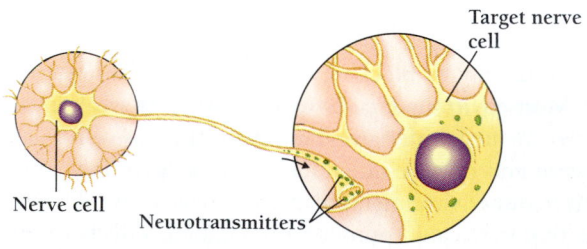

Target nerve cell

Nerve cell

Neurotransmitters

D. Neurotransmitters

Figure 36-2 **Types of chemical signaling.** A. Pheromones work between individuals. Ants lay down a trail of pheromones, which their nestmates follow and reenforce. B. Hormones travel long distances within the body. C. Paracrine signals travel short distances. D. Neurotransmitters are specialized types of paracrine signals.

How Do Researchers Identify Hormones and Glands?

Endocrinology, the study of hormones, has traditionally asked three questions about a suspected hormone: (1) What organ is responsible for its synthesis? (2) Does the suspected hormone move through the circulation? and (3) What is its chemical identity?

To answer the first question, a researcher would determine whether a product of one organ affected the function of another. For example, a researcher would destroy a suspected endocrine organ and note the resulting changes. An often-cited example is the effect of removing the testes of a juvenile animal. The result is the failure to develop the secondary sexual characteristics that appear in males at sexual maturity. In roosters, a large red comb is a secondary sexual characteristic, and removing the testes results in the withering of the comb.

The second question—whether the suspected hormone travels through the circulation—distinguishes a hormone from a secretory *(exocrine)* product or a short-range *paracrine* signal. Researchers have addressed this question by restoring the suspected endocrine organ to another place in the body. In the case of a castrated rooster, transplanting testes into the abdominal cavity restores the comb. This result shows that the hormone responsible for the secondary sexual characteristics moves through the circulation rather than through a particular duct.

The third question—the chemical identity of the signaling molecule—is the hardest to answer. Researchers begin by demonstrating that an extract of the endocrine gland (for example, of a rooster testis) can produce the same effect as the intact gland. The next task is to isolate and identify the active compound and to determine its chemical structure. Finally, researchers synthesize the suspected hormone in the laboratory and show that the synthetic substance has the same effect as the suspected hormone. Again, in the case of the rooster's comb, pure testosterone can restore the comb just as well as an extract of rooster testes. The ability of a pure chemical, made in the laboratory, to mimic a hormone's effect establishes that hormone's chemical identity. The action must depend on that single hormone rather than some unknown contaminant in the original extract or on two or more hormones.

This straightforward approach has not always worked. Endocrine organs sometimes make more than one hormone, so that researchers must replace a number of substances in order to reverse the effects of organ removal. The pancreas, for example, makes both insulin and glucagon. And the pituitary gland releases at least nine hormones.

In vitro cultures of target cells or target organs provide another way that contemporary endocrinologists identify and study hormones, particularly in complex systems that involve many different signals and targets. Using both the traditional surgical methods with whole organisms and the more recent approaches with cultured cells and organs, endocrinologists have identified more than 50 hormones and more than 15 hormone-producing structures in vertebrates and similar numbers in invertebrates.

3. **Paracrine signals** [Greek, *para* = beside] are chemical signals that affect only cells in the immediate vicinity of the signaling cells. Paracrine signals include **neurotransmitters,** which carry information between communicating nerve cells and from nerve cells to muscle cells; **prostaglandins,** which play an important role in blood clotting, inflammation, and uterine contractions; and **growth factors,** which stimulate cell division and promote cell survival.

Since the beginning of the 20th century, physiologists have distinguished between signaling in two physiological systems—the *nervous system* and the **endocrine system** [Greek, *endo* = within + *krinein* = to separate]. At the chemical level, however, the two systems are not that different. In most cases of nerve signaling, actual intercellular communication depends on signaling molecules that are identical or closely related to hormones.

Hormones are produced by endocrine cells or organs, which are specialized for secreting hormones into the general circulation. (In contrast, an *exocrine* [Greek, *exo* = outside] cell or organ makes products that are carried to specific targets by ducts, such as the duct that carries digestive enzymes into the small intestine.)

The major difference between hormonal and neural signaling is that hormones reach essentially all the tissues of the body, while nerves carry signals only to particular targets. A hormone signal resembles a broadcast message in that it may be received by any cell in the body, whereas a nerve signal can be received only by cells to which the nerve cell connects. (We will have more to say about neurotransmitters in Chapter 41.)

Another similarity of hormones and broadcast messages is that the same message can evoke different responses in different recipients. Consider, for example, the possible responses of people hearing a weather forecast that predicts a heavy snowfall: skiers will react with delight, while farmers may fall into a funk. Similarly, the same hormone, secretin, inhibits acid secretion in the stomach, stimulates bile secretion in the liver, and triggers the release of digestive enzymes in the pancreas.

Hormones are chemical signals that travel through the entire body. They evoke characteristic responses in individual cells and organs.

HOW DO HORMONES COORDINATE PHYSICALLY DISTANT ORGANS?

Among the most dramatic examples of hormone action is the emergence of a butterfly or moth from a wormlike caterpillar. During this transformation *(metamorphosis)*, the insect reorganizes every one of its organs.

How Do Hormones Control Insect Metamorphosis?

In a butterfly or moth, the fertilized egg develops first into a *larva* (a "caterpillar"), which is specialized for eating. As the larva grows, it undergoes several molts, each time casting off its outer skin and emerging as a slightly larger larva. The final larva undergoes a major reorganization, or metamorphosis, into a *pupa,* an apparently dormant stage, which is enclosed in a silken cocoon. Finally, the pupa dissolves its cocoon, and the adult butterfly or moth emerges. The adult has a strikingly different body from that of the larva or the pupa: it is specialized for reproduction and, in some species, cannot even eat (Figure 36-3).

Insects are particularly useful experimental subjects because their bodies are so hardy. Tying a loop of fine string around a larva, for example, results in pupa formation in just the front (anterior) half of the larva, while the rear (posterior) half stays a larva. Similar experiments allowed biologists to discover that molting and metamorphosis depend on hormones made in three distinct organs—two pairs of glands, called the *prothoracic glands* and the *corpora allata,* and the brain. At least three hormones are required: *juvenile hormone,* a small molecule, derived from a fatty acid and made in the corpora allata, that prevents molting and metamorphosis; *ecdysone,* a steroid made in the prothoracic glands, that promotes molting and metamorphosis; and *prothoracicotropin* (PTTH), a small protein made in the brain, that stimulates ecdysone production. Metamorphosis of larva into pupa occurs when juvenile hormone is low and ecdysone is high; metamorphosis of pupa into adult occurs when PTTH production—triggered by cold temperatures, extreme drought, or short days—stimulates ecdysone production in the absence of juvenile hormone.

Metamorphosis in amphibians also depends on hormone action, this time thyroid hormone. Similarly, the steroid sex hormones—testosterone and estrogen—trigger the metamorphosislike transitions of puberty in adolescent humans.

In both invertebrates (such as insects) and vertebrates, specific hormones regulate growth and development.

How Do Hormones Help Keep Blood Glucose at a Nearly Constant Level?

In mammals, the homeostatic regulation of blood glucose levels depends on separate biochemical events in the liver, the pancreas, and the small intestine, as well as in organs, such as muscle, that use glucose. Shortly after a meal, cells in many organs begin to take up needed fuel and building blocks. The body is said to be in an **absorptive state,** in which cells take up glucose, make glycogen, and increase the synthesis of fats and proteins. The result of these processes is a drop in the blood concentration of glucose and of other fuel molecules. The body then goes into a **postabsorptive state,** in which liver cells produce more blood glucose from the breakdown of glycogen. Other cells return to deriving energy from internal sources of glycogen, fats, and proteins.

Two major hormones—insulin and glucagon—coordinate the switch between the absorptive and the postabsorptive states. **Insulin** is a protein hormone produced by endocrine cells of the pancreas (Figure 36-4). When the insulin-producing cells detect high levels of glucose, they increase the production of insulin. Insulin, which has been called the "hormone of plenty," stimulates cells throughout the body to take up glucose, thereby decreasing the glucose concentration in the blood. Insulin also stimulates biochemical processes in the liver, the muscles, and adipose (fat) tissue. The responses in the target organs are characteristic of the absorptive state—the increased synthesis of fats, proteins, and glycogen.

When no glucose remains in the small intestine, blood glucose decreases, and insulin secretion stops. Other cells of the endocrine pancreas begin to secrete **glucagon,** a protein hormone that, together with the low insulin concentration, signals the postabsorptive state: cells reduce their uptake of glu-

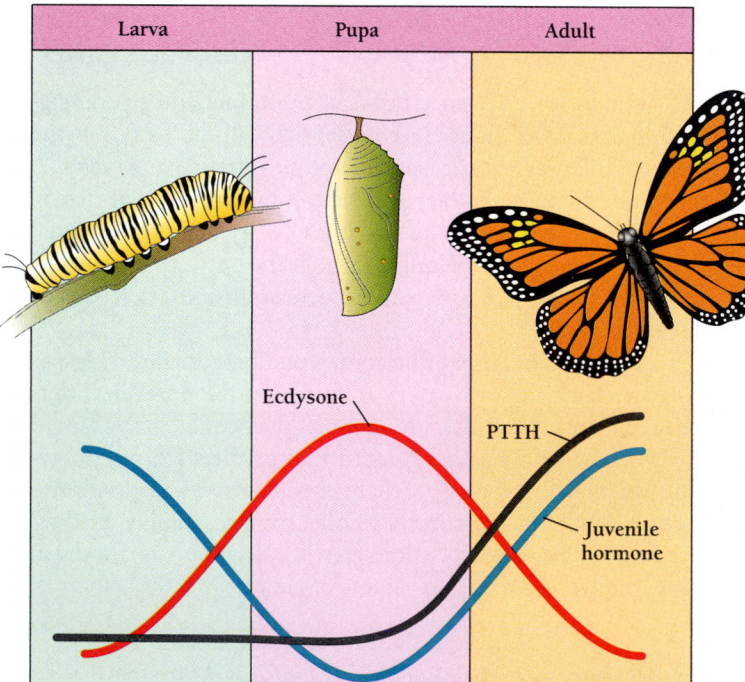

Figure 36-3 **Metamorphosis of a caterpillar into a butterfly.** At least three hormones regulate insect metamorphosis—juvenile hormone, ecdysone, and prothoracicotropin (PTTH).

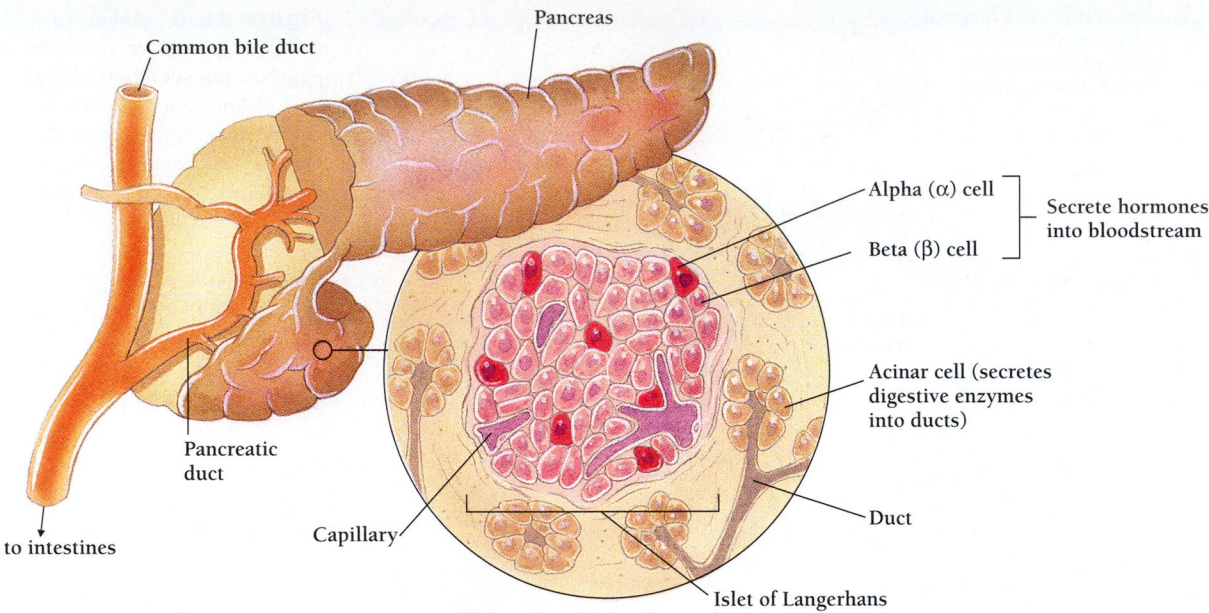

Figure 36-4   The pancreas. The pancreas produces hormones in the endocrine cells within the islets of Langerhans. It also produces digestive enzymes in its exocrine acinar cells.

cose and begin to break down glycogen, fats, and proteins. Whereas insulin promotes the synthesis of glycogen and inhibits its breakdown, glucagon inhibits glycogen synthesis and stimulates its breakdown into glucose (Figure 36-5).

Where are insulin and glucagon made?

The endocrine pancreas consists of 2 million clusters of cells that are physically distinct from the pancreas's exocrine cells (Figure 36-4). These clusters, called the *pancreatic islets* or *islets of Langerhans,* consist of at least three different cell types, each of which makes a different hormone. The α (alpha) *cells* produce glucagon, which stimulates the breakdown of glycogen to glucose in muscle cells; the β (beta) *cells* make insulin, which stimulates the uptake of glucose from the blood into muscle and other cells; and the δ (delta) *cells* make somatostatin, which decreases secretion, absorption, and smooth muscle action within the gut and inhibits the production of insulin and glucagon within the pancreatic islets themselves. Both α and β cells respond to the concentration of glucose in the blood: low glucose stimulates glucagon production by α cells, while high glucose stimulates insulin production by β cells.

Other parts of the digestive system also contain hormone-producing cells. Specialized cells of the stomach lining make *gastrin,* which stimulates hydrochloric acid production, and specialized cells in the small intestine make *secretin* and *cholecystokinin,* both of which inhibit hydrochloric acid production in the stomach (Chapter 35).

What happens when glucose regulation fails?

In the disease diabetes, the failure to produce insulin or to respond to insulin keeps the body in a permanent postabsorptive state. The result is an excessive concentration of glucose in the blood, which leads to its presence in the urine and an excess production of urine. This condition, relatively common, is called **diabetes mellitus** [Greek, *diabetes* = siphon + Latin, *mellitus* = sweet], meaning the overproduction of sweet urine. The production of sugar-containing urine was in fact responsible for the discovery that the pancreas was an endocrine organ. In 1889, an observant animal caretaker noticed flies gathering around the urine produced by a dog whose pancreas had been removed. Without a pancreas, we now realize, the dog was unable to produce insulin, and glucose accumulated in the blood and urine, as it does in the blood and urine of diabetics.

About 10 to 20 percent of people with diabetes suffer from *insulin-dependent diabetes mellitus* (IDDM), also called *type 1 diabetes* or *juvenile diabetes* because it generally begins early in life. People with IDDM fail to make any insulin because their immune systems have mistakenly attacked and destroyed the insulin-producing cells of the pancreas. Normal life depends on daily injections of insulin.

Most people with diabetes, however, do make insulin, but their target cells fail to respond to it. This condition is called *type 2 diabetes,* or *non-insulin-dependent diabetes mellitus* (NIDDM). NIDDM usually manifests itself in middle age, and it can usually be controlled by diet and by oral medicines. Type 2 diabetes has become more common, even among children, as more people indulge in rich diets and sedentary lifestyles.

Insulin, made when blood glucose is high, stimulates glucose absorption from the blood. Glucagon, made when blood glucose is low, promotes the production of more blood glucose from glycogen.

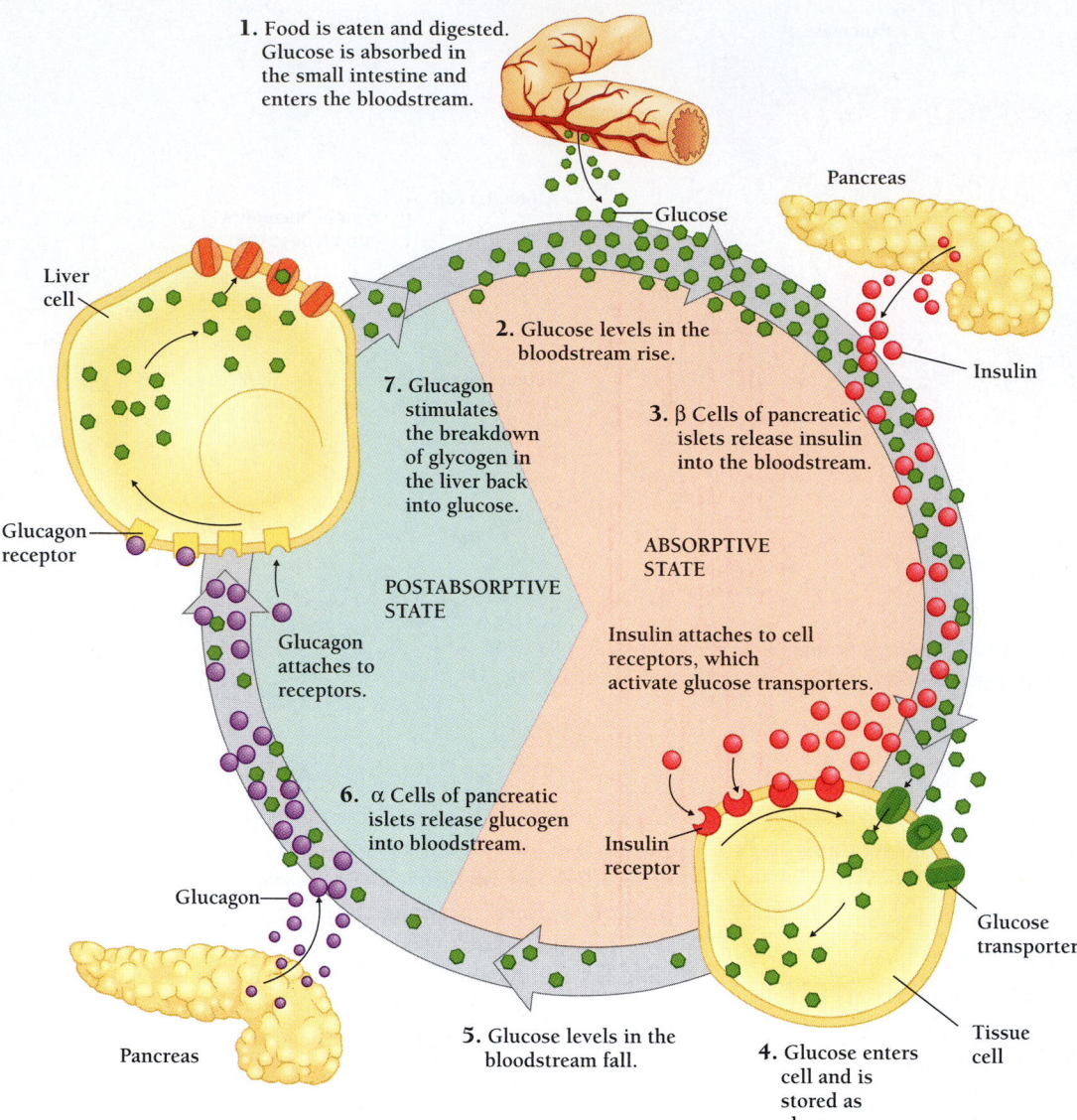

1. Food is eaten and digested. Glucose is absorbed in the small intestine and enters the bloodstream.

Glucose

Pancreas

Liver cell

Insulin

2. Glucose levels in the bloodstream rise.

7. Glucagon stimulates the breakdown of glycogen in the liver back into glucose.

3. β Cells of pancreatic islets release insulin into the bloodstream.

Glucagon receptor

ABSORPTIVE STATE

POSTABSORPTIVE STATE

Glucagon attaches to receptors.

Insulin attaches to cell receptors, which activate glucose transporters.

6. α Cells of pancreatic islets release glucogen into bloodstream.

Insulin receptor

Glucagon

Glucose transporter

Pancreas

5. Glucose levels in the bloodstream fall.

4. Glucose enters cell and is stored as glycogen.

Tissue cell

Figure 36-5 Absorptive and postabsorptive states depend on the secretion and action of hormones. Absorptive and postabsorptive states depend on the pancreatic hormones, insulin and glucagon.

How Do Hormones Coordinate the Response to Stress?

Stress—the response to a noxious or potentially noxious stimulus—is part of the life of any animal. Stress producers include physical trauma, intense heat or cold, infection, pain, and fright. Faced with any of these *stressors,* mammals regulate biochemical reactions in many organs to increase the availability of energy-rich molecules, especially glucose, in the blood. Muscles increase the breakdown of proteins into amino acids; fat increases the breakdown of triglycerides into fatty acids in the blood; the liver increases the rate of conversion of amino acids into glucose. All these changes depend on a steroid hormone called **cortisol.**

At the same time, **epinephrine** [Greek, *epi* = upon + *nephros* = kidney], a hormone derived from the amino acid tyrosine, coordinates a complementary set of responses. Epinephrine, also called *adrenaline* [because it is made by the adrenal gland], speeds the heart, dilates the blood vessels, and

increases the liver's production of glucose from glycogen. All these responses contribute to the "fight-or-flight" response that prepares an animal for immediate action and energy expenditure in the face of stress or danger (Figure 36-6).

Both cortisol and epinephrine are made in the adrenal glands [Latin, *ad* = to + *renes* = kidneys], which lie just next to the kidneys. Each adrenal consists of two parts—the outer **adrenal cortex** and the inner **adrenal medulla.** The adrenal cortex makes two hormones: cortisol and *aldosterone,* which helps regulate kidney function. The production of aldosterone depends on signals from the kidney (Chapter 39). The production of cortisol depends directly on the brain and its perception of a stress-inducing situation. Cortisol production in the adrenal cortex depends on the level of adrenocorticotropic hormone (ACTH), produced by the anterior pituitary. ACTH production in turn depends on the release of corticotropin-releasing hormone (CRH) from the hypothalamus, as we discuss later.

The adrenal medulla, on the other hand, receives neural signals from the peripheral nervous system. Its primary hor-

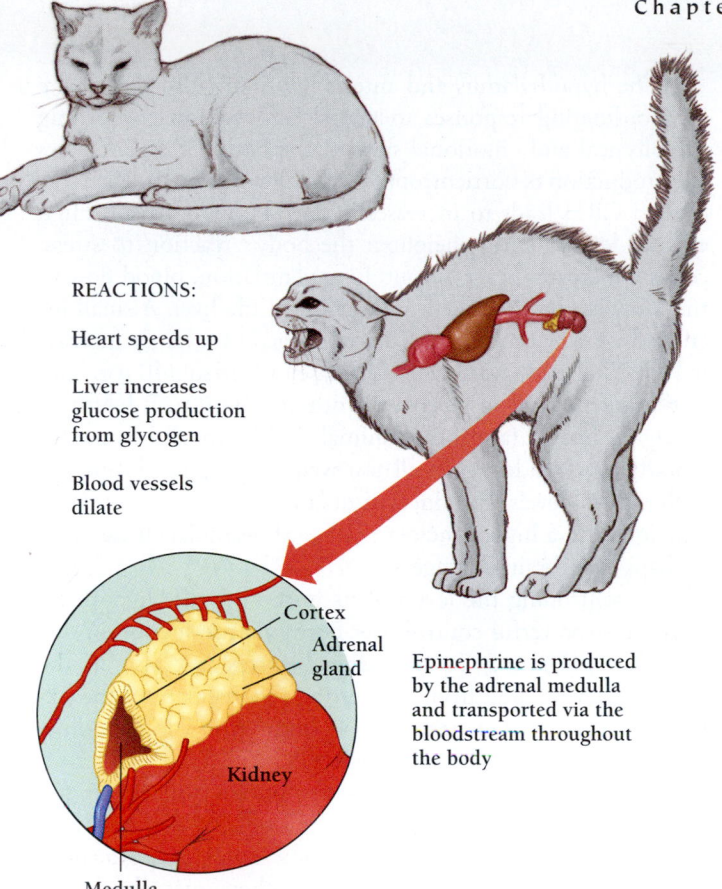

REACTIONS:

Heart speeds up

Liver increases glucose production from glycogen

Blood vessels dilate

Epinephrine is produced by the adrenal medulla and transported via the bloodstream throughout the body

Cortex
Adrenal gland
Kidney
Medulla

Figure 36-6 Fight or flight. Epinephrine (adrenaline) mediates the "fight-or-flight" response.

monal output is the hormone epinephrine, the mediator of the "fight-or-flight" response (Figure 36-6).

The release of epinephrine and cortisol into the blood readies the animal for energy expenditure and immediate action. Release depends on the perception of danger or stress. Many kinds of stimuli can trigger this perception; the perceived danger can range from an approaching bear to an overbearing teacher.

Transmitting a danger signal to the adrenal medulla involves both the central nervous system, which interprets the danger, and the peripheral nervous system, which transmits information to the adrenal medulla. The production of both epinephrine and cortisol depend on the perception of stress by the brain. Epinephrine release depends on nerve signals from the brain to the adrenal medulla, via part of the peripheral nervous system, which we will discuss in Chapter 42. The production of cortisol, on the other hand, depends on two other hormones, one from part of the brain (the hypothalamus) and the other from a nearby endocrine organ, the anterior pituitary gland.

The adrenal gland produces two hormones in response to stress-indicating hormone signals from the brain. The adrenal cortex produces cortisol, and the adrenal medulla produces epinephrine, the major mediator of the "fight-or-flight" response.

How Do the Pituitary Gland and Hypothalamus Regulate the Production of Hormones by Other Endocrine Organs?

The **pituitary gland** is a small structure, about the size of a pea, at the base of the brain (Figure 36-7). Its name [Latin, *pituita* = slime] came from the mistaken belief that it produced the mucus of the nose. In humans, the pituitary has two main parts, called the *posterior pituitary* and the *anterior pituitary.*

The **posterior pituitary** is actually part of the brain itself; it consists of the terminals of nerve cells whose cell bodies lie in a part of the brain called the **hypothalamus** [Greek, *hypo* = under + *thalmos* = inner chamber] (Figure 36-7). Some of the cells of the hypothalamus extend into the pituitary, and their terminals make up the posterior pituitary. These cells release two main hormones, vasopressin and oxytocin. *Vasopressin,* also called *antidiuretic hormone* (ADH), stimulates water reabsorption in the kidney (Chapter 39). *Oxytocin* acts at childbirth to stimulate the uterus to contract and the mammary glands to eject milk.

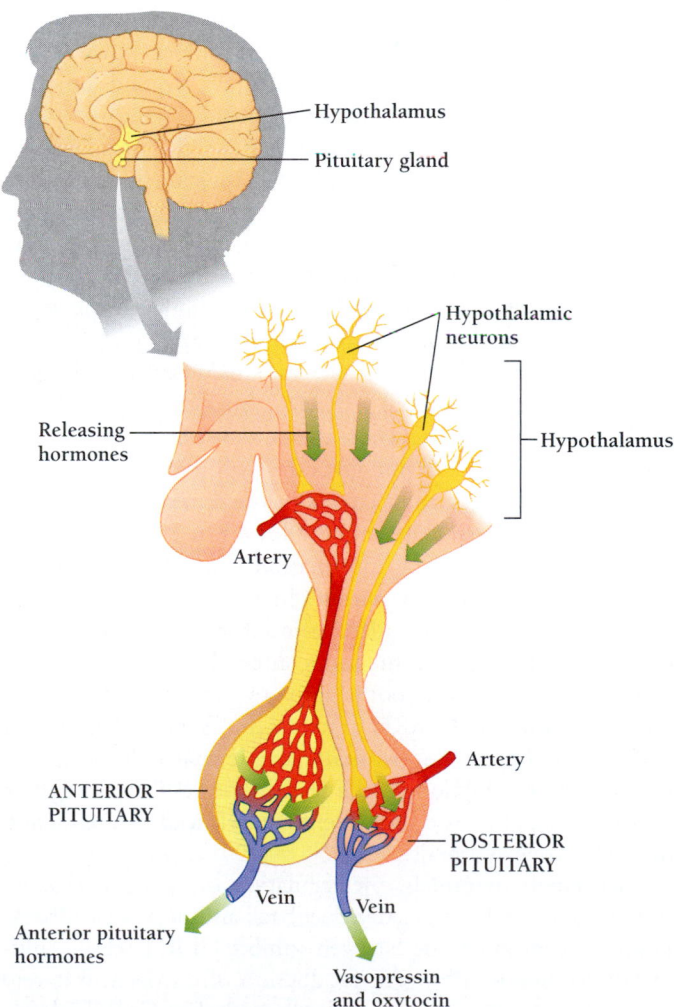

Hypothalamus
Pituitary gland

Hypothalamic neurons

Releasing hormones

Hypothalamus

Artery

ANTERIOR PITUITARY

Artery

POSTERIOR PITUITARY

Vein
Vein

Anterior pituitary hormones

Vasopressin and oxytocin

Figure 36-7 The hypothalamus and pituitary and their hormones. This diagram is stylized to emphasize the relationships of the hypothalamus and pituitary.

The **anterior pituitary** is not part of the brain, but a separate gland. The anterior pituitary produces and releases eight peptide hormones: *corticotropin* (also called adrenocorticotropic hormone, or ACTH), which stimulates the production of cortisol in the adrenal cortex; *endorphin,* a natural pain suppressor; *thyroid-stimulating hormone* (TSH), which stimulates the production of thyroxin in the thyroid gland; *growth hormone* (GH), which stimulates tissue and skeletal growth; *prolactin,* which stimulates milk production; *melanocyte-stimulating hormone* (MSH), which stimulates pigment production in specialized skin cells; *follicle-stimulating hormone* (FSH), which in females stimulates estrogen production by the ovaries and promotes the maturation of the follicle during the menstrual cycle; and *luteinizing hormone* (LH), which in females induces ovulation and stimulates estrogen production and in males increases testosterone production.

Distinct sets of cells within the anterior pituitary specialize in the production of each of these hormones. Some cells, for example, produce either growth hormone, ACTH, or TSH. Other cells produce FSH and LH, both of which act on gonadal tissue (ovaries and testes) and are therefore called *gonadotropins.*

Because the anterior pituitary stimulates so many target organs to produce other hormones, many endocrinologists have referred to it as the "master gland." In fact, however, the release of hormones by the anterior pituitary depends on other hormones produced by the hypothalamus.

The hypothalamus produces at least seven hormones that control the release of hormones by the anterior pituitary. Five of these are **releasing hormones,** each of which stimulates the release of a specific hormone, and two are **inhibiting hormones,** each of which inhibits hormone release. For example, *gonadotropin-releasing hormone* (GnRH) stimulates the secretion of FSH and LH by the anterior pituitary.

Hypothalamic hormones such as GnRH can enter the general circulation, but special veins flow from the hypothalamus to the anterior pituitary, where they divide and form a second capillary bed. This diversion gives the hypothalamic hormones more direct access to their targets in the anterior pituitary.

The regulation of hormone release often involves **negative feedback,** the use of information about the output of a system to reduce further output. In male mammals, for example, FSH and LH stimulate testosterone production in the testes. Testosterone from the testes enters the circulating blood and then acts on the hypothalamus to inhibit production of GnRH. The lowered GnRH in turn reduces the release of FSH and LH by the anterior pituitary, a state that reduces further testosterone production by the testes. The result is a negative feedback loop: increased testosterone levels lead to a reduction in testosterone synthesis.

In female mammals, the regulation of estrogen production depends on both negative feedback and positive feedback. High concentrations of estrogen inhibit GnRH release, ultimately leading to a lowered production of estrogen. But very high estrogen concentrations actually stimulate GnRH release, leading to a surge of LH. It is this surge of LH that triggers ovulation midway through the menstrual cycle.

The hypothalamus and anterior pituitary also participate in coordinating responses to neural information concerning the physical and emotional state of the body. Stress increases the production of corticotropin-releasing hormone (CRH). Increased CRH leads to increases in ACTH and cortisol. High cortisol levels in turn heighten the body's reaction to stress: cortisol increases heart output, lung ventilation, blood flow to the muscles, and glycogen production in the liver. A small increase in CRH production ($0.1~\mu g$) leads to a larger increase in ACTH ($1~\mu g$), a still larger increase in cortisol ($40~\mu g$), and a still larger effect on glycogen synthesis ($5600~\mu g$). Unfortunately, in people (and other animals) with stressful lives, this amplifying cascade works all too well, leading to chronically high cortisol levels, sustained stimulation of the heart and circulation, and a higher incidence of cardiovascular disease and perhaps even brain damage.

In controlling the levels of its hormones, the hypothalamus exerts powerful control over hormone production in distant sites—not only in the adrenal cortex but also in the gonads, thryoid, and liver. In exerting this regulation, the hypothalamus integrates information both from the endocrine system and from the nervous system.

Information from the endocrine system also affects the functioning of the brain. Hormonal changes can lead to altered behavior. For example, a female dog in heat will accept the mating behavior (mounting) of a male. At other times, the female may respond to the same male behavior by just walking away. Experimenters can change the dog's response by injecting appropriate hormones. So the communication between the hormone system and the nervous system runs in both directions.

The hypothalamus is the major mediator of information between the brain and the endocrine system. Hormones made by the hypothalamus regulate hormone production elsewhere, especially in the anterior pituitary, which also produces hormones that regulate hormone production elsewhere.

HOW DO HORMONES FIND THEIR TARGETS?

Studies of the chemical identities of hormones and their modes of action have shown that hormones fall into two classes—those that are **lipid-soluble signals** (such as estrogen, testosterone, and cortisol), whose nonpolarity allows them to pass directly through the plasma membrane of the target cell, and **water-soluble signals** (proteins such as insulin and glucagon, polypeptides such as CRH, and amines such as epinephrine), which cannot pass through the plasma membrane.

For both classes, action on a target cell begins when the signal molecule binds to a specific protein, called a **receptor.** Lipid-soluble signals bind to specific receptor molecules *inside* their target cells, while water-soluble signals bind to specific receptor molecules on the *surface* of their target cells (Figure 36-8).

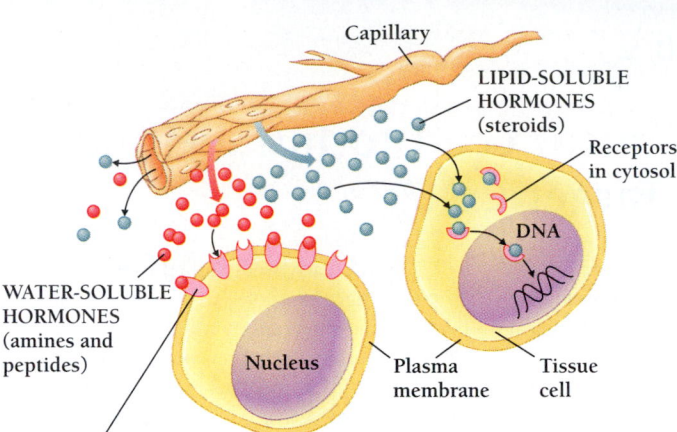

Figure 36-8 The differing actions of lipid-soluble and water-soluble hormones.

Among the most important lipid-soluble signaling molecules are the **steroids,** nonpolar molecules that derive from cholesterol. *Glucocorticoids* (such as cortisol) and *mineralocorticoids* (such as aldosterone) are made in the adrenal cortex, while the *sex steroids* (estrogen, progesterone, and testosterone) are made in the gonads. Chemically synthesized compounds that resemble estrogen and progesterone are widely used in birth control pills, and steroids that resemble testosterone have been abused by some athletes to build muscle mass. Use of these *androgens* [Greek, *andro* = man; the generic term for testosteronelike steroids] increases the risks of sterility, cancer, and psychiatric disorders.

Water-soluble chemical signals include most of the **amines,** the simplest chemicals that act as hormones. A water-soluble amine, such as epinephrine, cannot pass through the plasma membrane of its target cells and must act on the cell's surface. A few amines, however, such as the thyroid hormone thyroxin, are lipid soluble. Thyroxin passes directly through the plasma membrane and acts in the same way as steroids and other lipid-soluble hormones.

Peptides, which are water soluble, are by far the largest class of hormones. A peptide hormone may contain as few as 3 amino acids or well over 100. Among the shorter peptide hormones are those made by the hypothalamus, including the releasing factors, vasopressin, and oxytocin. The longer peptide hormones, which are classified as proteins, include insulin, glucagon, and the hormones of the anterior pituitary. Neither the small peptides nor the much larger proteins are lipid soluble, and they do not cross the plasma membrane of their target cells. The action of these hormones depends entirely on their interactions with receptors on the surface of their target cells. These receptors in turn stimulate the synthesis or release of intracellular "second messengers" that influence the inner workings of their target cells.

For all kinds of chemical signals, signaling between cells involves six steps: (1) a cell must make the signaling molecule; (2) the cell must release the signaling molecule; (3) the mole-

cule must travel to the target cells; (4) the target cells must detect the signal; (5) the target cells must respond to the signal; and, finally, (6) something must end the signaling process.

Hormones are chemically diverse, including steroids, amines, and peptides. Hormones are divided into those that are soluble in lipids and those that are soluble in water.

How Do Target Cells Detect Chemical Signals?

The response of a target cell to a particular signaling molecule depends on the presence of a specific receptor protein. The receptor binds to the signal and initiates the cell's response.

A receptor binds to a signaling molecule, such as a hormone, in much the same way that an enzyme binds to a substrate. Just as a substrate forms noncovalent bonds with amino acids at the *active site* of an enzyme, so a signaling molecule forms noncovalent bonds at the **binding site** of a receptor.

The affinity of a hormone receptor for a specific hormone is usually very high: The receptors on a cell may be fully saturated when hormone concentrations are as low as 10^{-9} moles per liter, one-millionth those at which most enzymes are saturated. The strong binding of hormones to their receptors helps explain why environmental contaminants that are present at extremely low concentrations can nonetheless interfere with natural signaling processes.

The response of a target cell to epinephrine can be mimicked by other synthetic compounds. Molecules that resemble a natural hormone can bind to the same receptor and initiate the same response. For example, the synthetic compound *isoproterenol* binds to the same receptors as epinephrine, and either compound can be used to open the airways and ease the congestion caused by allergies and asthma.

Other molecules that are chemically similar may bind to the receptor and thereby prevent the binding of a natural signal. For example, *propranolol,* another synthetic compound that resembles epinephrine, prevents the action of epinephrine on heart muscle. By antagonizing the effect of epinephrine, propranolol prevents stressful increases of heart rate and blood pressure. For this reason, propranolol is widely used to treat people who have high blood pressure.

Signaling molecules, such as epinephrine, play an important part in the coordination of distinct organ systems. Studies of the action of hormones such as epinephrine have led to new therapies, not only for asthma and high blood pressure but for many other conditions as well.

Hormone action begins with binding to a receptor. Molecules that resemble a natural hormone can bind to the same receptor and initiate or prevent the same response.

BOX 36.2

How Does an Extracellular Chemical Lead to Intracellular Responses in a Target Cell?

The binding of a chemical signal (such as a hormone) to a receptor triggers a sequence of events that changes the characteristics and the function of the target cell. This sequence of events, collectively called the *signal transduction pathway,* leads to changes either in the amounts or in the activities of particular proteins within the target cell.

Lipid-soluble signals—including the steroids and the thyroid hormones—easily cross the plasma membrane and bind to intracellular receptors. All known receptors for steroids and thyroid hormones are related to one another. All are proteins that can move from cytosol to nucleus, where they serve as *transcription factors (activators)* (Chapter 11). These receptors can bind both to a specific signal and to specific sequences in DNA, so that the hormone-receptor complex "enhances" the transcription of one or more genes in the target cells. (In some cases, however, the hormone-receptor complex decreases the rate of transcription of specific genes.) Steroids can have other effects as well, but the most dramatic and the best understood are those that change the production of messenger RNAs.

Water-soluble signaling molecules—including most amine and all peptide hormones—do not enter their target cells but instead bind to receptors on the outer surface. In some cases, such as the insulin receptor, the receptor is itself an enzyme that modifies other components of the plasma membrane. Such receptors are often **protein kinases,** which transfer a phosphate group from ATP to another protein. The addition of one or more phosphate groups can change the activity of the protein, altering the cell's metabolic or physiological activities.

Most receptors for water-soluble signals, however, are not enzymes themselves. Instead, the receptor plays the role of a baton in a relay race in which the hormone or signaling molecule is just the first runner, or "first messenger." When the hormone binds to the outer surface, the receptor changes shape and alters its interaction with another protein, closely coupled to the receptor in the plasma membrane. This receptor partner is called a *G protein* because it can bind GTP and GDP. When the receptor binds to a chemical signal, the coupled G protein exchanges a GDP molecule for a GTP molecule. This exchange activates the G protein, so that it now stimulates some other target molecule within the plasma membrane. This stimulation is brief, because the G protein also catalyzes the splitting of GTP into GDP.

It is the target of the activated G protein that initiates the intracellular events that follow hormone binding. In some cases, the activated G protein can interact with an ion channel, altering the cell's ionic balance. More commonly, the activated G protein stimulates the production or the destruction of a **second messenger,** which relays to the cell's interior the information that a signal has arrived at the cell's surface. Researchers now know of four second messengers whose production is stimulated by activated G proteins—cyclic AMP, calcium ions, inositol trisphosphate, and diacylglycerol.

A particularly well-studied part of the "fight-or-flight" response is an increase in the availability of energy, which is brought about by the increased breakdown of glycogen. Epinephrine increases the breakdown of glycogen by activating *glycogen phosphorylase,* the liver enzyme that hydrolyzes glycogen to glucose (Figure 36-9). Epinephrine itself acts only on the cell surface: the second messenger responsible for phosphorylase activation is **cyclic AMP (cAMP).** Addition of cAMP alone increases the activity of phosphorylase in liver extracts. Glucagon, another hormone that stimulates glycogen breakdown, also activates glycogen phosphorylase by means of cAMP.

The effects of cAMP (or of any second messenger) differ from cell to cell. For example, in the intestine, cAMP stimulates the pumping of sodium ions and water out of cells. People suffering from the infectious disease *cholera* have excess cAMP, which overstimulates the pumping of sodium ions and water, leading to diarrhea. The overproduction of cAMP results from the action of *cholera toxin,* a protein produced by *Vibrio cholerae* bacteria. Cholera toxin modifies the protein (called a G protein) that couples the receptor to the production of cAMP. The coupling protein then becomes locked into its active form and cannot be switched off.

In some cases the proteins affected by second messengers, such as cyclic AMP, are themselves transcription factors. The result is that water-soluble signals, like lipid-soluble signals, can also alter the pattern of gene expression and the total amounts (as

Hormones May Target More Than One Kind of Receptor

Isoproterenol and propranolol are only two among hundreds of synthetic epinephrinelike compounds that researchers have studied. As these studies proceeded, researchers became increasingly aware that epinephrine could not be binding to only one kind of receptor molecule, since different tissues responded differently to the synthetic compounds. For example, although isoproterenol mimics the effects of epinephrine on the heart, isoproterenol does not inhibit smooth muscle contractions in the gut, as does epinephrine. One explanation for this difference was that there were several types of epinephrine receptors that differed in their ability to bind other agonists and antagonists. This explanation turned out to be exactly correct.

well as the activities) of specific proteins. In general, however, changes in biological activity induced by second messengers occur more rapidly than those induced by lipid-soluble hormones.

The binding of many water-soluble signals such as epinephrine to certain membrane receptors stimulates an increase of free calcium ions (Ca^{2+}) within the cytoplasm. Free Ca^{2+} ions then function as a second messenger to change the activity of specific enzymes and other proteins (including transcription factors).

Finally, a more complex signal transduction pathway involves the simultaneous production of two second messengers, inositol trisphosphate (IP_3) and diacylglycerol (DAG) (Figure A). The production of these second messengers depends on the action of an enzyme called *phospholipase C*. This enzyme converts a membrane lipid, phosphatidylinositol bisphosphate (PIP_2) into IP_3 and DAG. IP_3 moves into the cytosol, where it releases intracellular stores of calcium ions, which—as described earlier—can activate enzymes and transcription factors. DAG, on the other hand, stays in the inner portion of the plasma membrane. There DAG can activate another protein kinase, called protein kinase C, which acts on specific membrane proteins, altering their activity by adding phosphate groups. DAG can also serve as a precursor of certain paracrine factors, called prostaglandins.

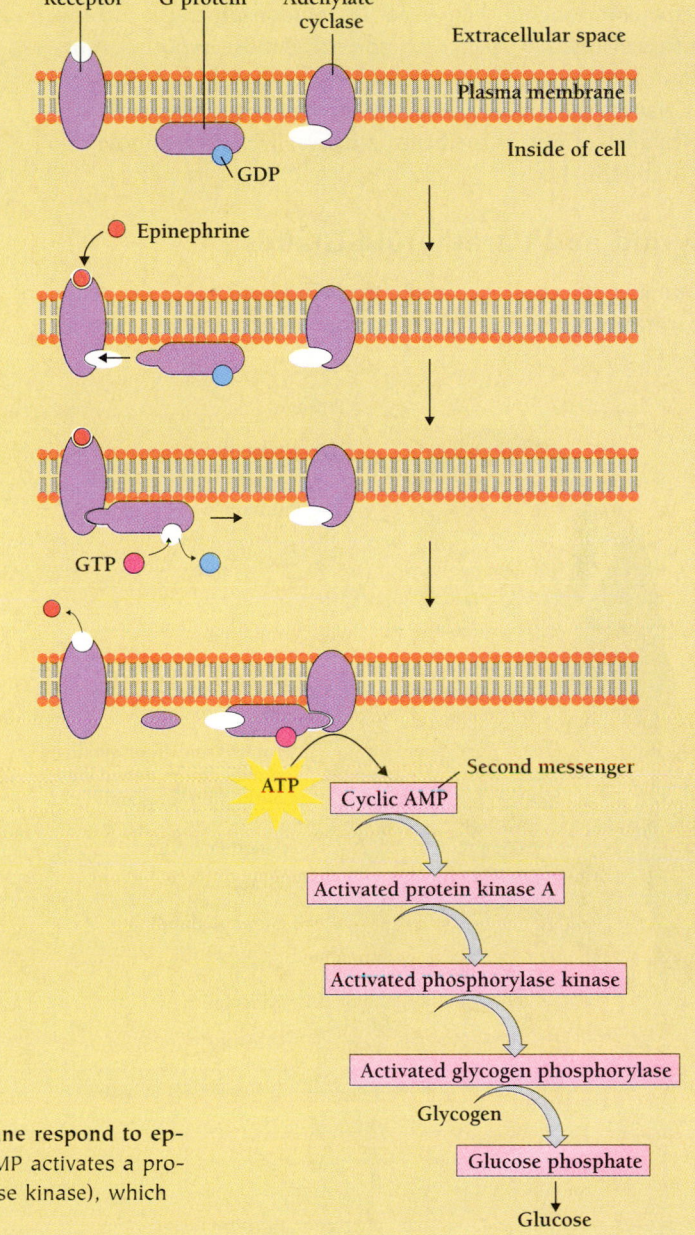

▶ **Figure A** Some adrenergic receptors in the plasma membrane respond to epinephrine by stimulating the synthesis of cyclic AMP. Cyclic AMP activates a protein kinase, which activates another protein kinase (phosphorylase kinase), which stimulates the enzymatic activity of glycogen phosphorylase.

Biochemists were able to isolate the receptor molecules that bind to epinephrine and other hormones. Using molecular biological techniques, researchers have been able to identify and sequence the genes that encode each of these receptors. They found, as expected, that different tissues had different forms of epinephrine receptors. In fact, for virtually every type of signaling molecule, researchers have found several types of receptors, which are deployed differently in different types of target cells. These findings begin to explain how different types of cells can respond so differently to the same signal.

Epinephrine binds to different receptor types in different tissues.

How Many Other Endocrine Organs in a Mammal?

Chemical signaling between organs is common, and an increasing number of organs are now considered to have endocrine components (Figure 36-9). In addition to the already mentioned organs—pancreas, adrenal glands, hypothalamus, and pituitary—we can list the gonads, the thyroid and parathyroid glands, the pineal gland, the intestines, the placenta, and the thymus, as well as the heart, kidneys, liver, and adipose tissue (Table 36-1).

Thyroid and Parathyroid Glands

In humans, the two lobes of the **thyroid gland** are in the neck, just in front of the windpipe. In other vertebrates, the thyroid

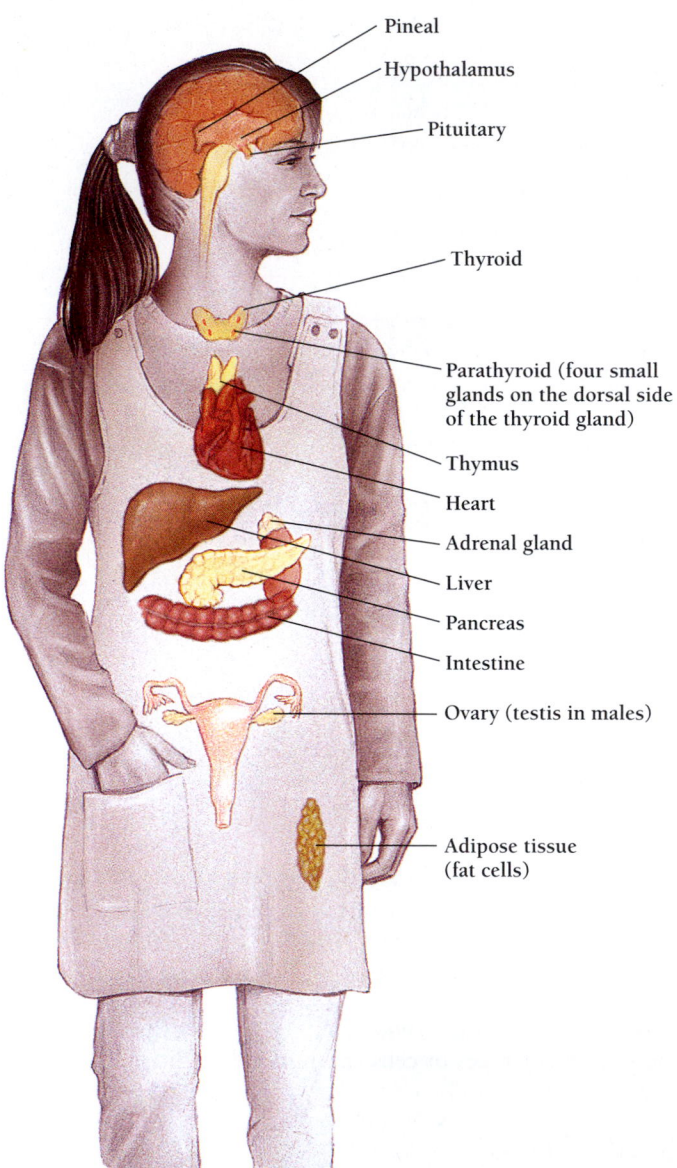

Figure 36-9 The major endocrine glands in a human.

Labels in figure:
Pineal
Hypothalamus
Pituitary
Thyroid
Parathyroid (four small glands on the dorsal side of the thyroid gland)
Thymus
Heart
Adrenal gland
Liver
Pancreas
Intestine
Ovary (testis in males)
Adipose tissue (fat cells)

lies just below the windpipe. The thyroid produces *thyroid hormone,* which regulates metabolism and growth. Overproduction of thyroid hormone, called *hyperthyroidism,* results in uncontrolled, rapid metabolism, whereas underproduction, called *hypothyroidism,* may cause such symptoms as lethargy, weight gain, and intolerance of cold. Hypothyroidism during fetal life may result in abnormal development, mental retardation, and low metabolic rate. The thyroid gland also produces *calcitonin,* which stimulates tissues to remove calcium ions from the blood.

Adjacent to the thyroid gland are the **parathyroid glands.** These glands produce *parathyroid hormone,* whose effect is to increase the concentration of calcium ions in the blood—exactly the opposite effect from that of calcitonin. The antagonistic actions of parathyroid hormone and calcitonin are responsible for calcium homeostasis: between them, they keep blood calcium relatively constant. Low blood calcium triggers the parathyroid to release parathyroid hormone, and high blood calcium triggers the thyroid to release calcitonin.

Hormones from the thyroid gland regulate metabolism, growth, and calcium levels in the blood. Parathyroid hormone also contributes to calcium regulation.

Testes and Ovaries

The gonads (testes and ovaries) produce the sex steroids: the **testes** produce *testosterone,* and the **ovaries** produce *estrogen* and *progesterone.* The sex steroids are responsible for the development and the maintenance of the secondary sexual characteristics as well as for the proper development of germ cells (Figure 36-10). Progesterone is especially important in maintaining pregnancy. We will have more to say about the role of the sex steroids in reproduction in Chapter 43.

The gonads secrete sex hormones.

PARACRINE SIGNALS ACT OVER SHORT DISTANCES

Some chemical signals between cells do not travel through the blood, but act at much shorter distances. Such compounds are called *paracrine* signals (in contrast to *endocrine* signals, or hormones). Paracrine signals affect only cells in the immediate vicinity of the signaling cell. Paracrine signals include *neurotransmitters* (signals by which nerve cells communicate), *prostaglandins* (signals by which other cells communicate, all derivatives of a 20-carbon unsaturated fatty acid), and *growth factors* (which stimulate cell division).

TABLE 36-1   MAJOR ENDOCRINE ORGANS IN HUMANS

Organ	Hormones (examples)	Target organ (example)	Effect (example)
Pancreas			
α cells	Glucagon	Muscle	Stimulates glycogen breakdown
β cells	Insulin	Many targets	Stimulates glucose uptake
δ cells	Somatostatin	Gut	Decreases secretion and absorption
Adrenal			
Cortex	Cortisol	Muscle	Increases protein breakdown
	Aldosterone	Kidney	Increases ion secretion
Medulla	Epinephrine	Heart	Increases heart rate
Hypothalamus	Corticotropin-releasing hormone (CRH)	Anterior pituitary	Stimulates release of corticotropin
	Thyrotropin-releasing hormone (TRH)	Anterior pituitary	Stimulates release of TSH
	Growth-hormone–releasing hormone (GHRH)	Anterior pituitary	Stimulates release of growth hormone
	Gonadotropin-releasing hormone (GnRH)	Anterior pituitary	Stimulates release of LH and FSH
	Dopamine	Anterior pituitary	Inhibits release of prolactin
	Orexin	Brain	Triggers feeding behavior
Pituitary			
Posterior	Oxytocin	Uterus	Stimulates contractions
	Vasopressin (antidiuretic hormone, ADH)	Kidney	Prevents water loss
Anterior	Corticotropin (adrenocorticotropic hormone, ACTH)	Adrenal cortex	Stimulates cortisol production
	Endorphin	Brain	Suppresses pain
	Thyroid-stimulating hormone (thyrotropin, TSH)	Thyroid	Stimulates thyroxin production
	Growth hormone (GH)	Bone	Stimulates skeletal growth
	Melanocyte-stimulating hormone (MSH)	Pigment-producing cells	Stimulates pigment production
	Follicle-stimulating hormone (FSH)	Gonad	Promotes gamete production
	Luteinizing hormone (LH)	Gonad	Stimulates ovulation, gamete production
	Prolactin	Breast	Stimulates milk production
Gonads			
Ovary	Estrogen	Anterior pituitary	Regulates secretion of FSH and LH
	Progesterone	Uterus	Stimulates proliferation of uterine lining
	Inhibin	Anterior pituitary	Regulates secretion of FSH
	Testosterone	Brain	Stimulates sex drive
Testis	Inhibin	Anterior pituitary	Regulates secretion of FSH
Thyroid	Thyroxin	Muscle	Stimulates growth and metabolism
	Calcitonin	Bone	Regulates calcium uptake
Parathyroid	Parathyroid hormone	Bone	Stimulates movement of calcium into blood
Pineal	Melatonin	Brain	Regulates daily rhythm
Intestinal tract	Gastrin	Stomach	Stimulates production of hydrochloric acid
	Secretin	Pancreas	Stimulates bicarbonate secretion
	Cholecystokinin (CCK)	Stomach	Inhibits production of hydrochloric acid
Placenta	Chorionic gonadotropin (CG)	Corpus luteum	Maintains the corpus luteum during pregnancy
	Estrogen	Uterus	Maintains uterus during pregnancy
	Progesterone	Uterus	Maintains uterus during pregnancy
Thymus	Thymopoeietin	T lymphocytes	Regulates T-lymphocyte function
Heart	Atrial natriuretic factor (ANF)	Kidney	Regulates sodium excretion and blood pressure
Kidney	Angiotensin	Kidney	Regulates water retention
Liver	Insulinlike growth factor (IGF)	Bone	Regulates growth
Adipose tissue	Leptin	Hypothalamus	Reduces orexin secretion, stops feeding behavior

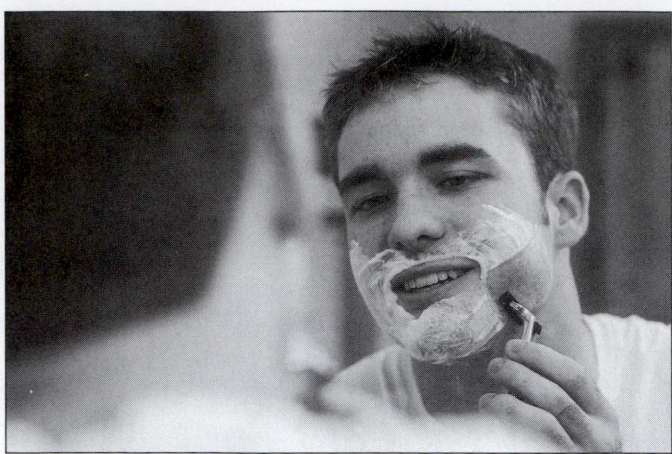

Figure 36-10 A single hormone may have different effects on different target tissues. Increased testosterone levels at puberty, illustrated here, stimulate increased growth in muscles, growth of the larynx (thereby deepening the voice), growth of facial hair, and the development of male sexual behavior. (*Roy Morsch/ The Stock Market*)

Some prostaglandins play an important role in the clotting of blood, others help bring about the inflammatory response, and still others induce smooth muscles to contract. For example, prostaglandins promote uterine contractions. The name "prostaglandin" (referring to the prostate gland in the male reproductive system) came from a practice, previously not understood, of using semen (which contains prostaglandins) to induce labor in pregnant women who had come to term.

A series of enzymes converts arachidonic acid, which is a minor component of membrane lipids, into various prostaglandins. Aspirin, ibuprofen, and acetaminophen all inhibit these pathways, so they act as both anti-inflammatory agents and anticlotting agents. By reducing the formation of blood clots that can block the blood vessels that supply the heart, small amounts of aspirin significantly diminish the risk of a heart attack.

Prostaglandins are paracrine factors that exert their efforts through second messengers.

Study Outline with Key Terms

To maintain a constant **internal environment,** an animal must coordinate the activities of many cells and organs. **Hormones,** produced in the organs of the **endocrine system,** contribute to **homeostasis** by evoking characteristic responses within their **target cells** and **target organs.** Hormones are chemical signals that are carried by the circulation and act at long distances. **Paracrine signals** act only at short distances, while **neurotransmitters** act only over the short distances between nerve cells. **Pheromones,** on the other hand, influence the activities of other organisms of the same species.

Insulin and **glucagon** contribute to the regulation of blood glucose. Insulin establishes an **absorptive state,** in which cells take up glucose from the blood and make glycogen. Glucagon establishes a **postabsorptive state,** in which the liver produces more glucose and breaks down glycogen. The failure of insulin production or insulin action leads to excessive blood glucose, a condition called **diabetes mellitus. Cortisol** and **epinephrine** contribute to the stress response.

Endocrinologists have used surgical experimentation, chemical synthesis, and tissue culture to identify the hormones and organs that contribute to the endocrine system. Among the endocrine organs are the pancreas, stomach, small intestine, adrenal glands, **thyroid, parathyroid,** and gonads (**testes** and **ovaries**). The **adrenal cortex** produces hormones that are distinct from those of the **adrenal medulla.** Similarly, the **anterior pituitary** is distinct from the **posterior pituitary,** which is actually part of the **hypothalamus,** a part of the brain.

The hypothalamus and **pituitary gland** regulate the production of hormones by other endocrine organs. The hypothalamus produces **releasing hormones** and **inhibiting hormones,** which regulate the release of hormones by the anterior pituitary. The anterior pituitary produces hormones that regulate hormone production elsewhere in the body, in a manner often subject to **negative feedback.**

Hormones are chemically diverse and include **steroids, amines,** and **peptides.** Hormones are either soluble in lipids or soluble in water.

The binding of a hormone to a **receptor** initiates the response of the target cell. The hormone attaches to the receptor's **binding site.** Some compounds evoke the same response as the hormone itself, whereas other compounds prevent hormone binding and the subsequent target cell responses.

The receptors for **lipid-soluble signals** are usually intracellular proteins that regulate transcription. The receptors for **water-soluble signals** are usually membrane proteins with seven transmembrane segments. Signals can often bind to a family of slightly different receptors, which may lie on different target cells and which may evoke different cellular responses.

Most water-soluble signals (such as **epinephrine**) act by stimulating the production of a **second messenger,** which triggers the cell's responses. The most common second messenger is **cyclic AMP (cAMP).** Cyclic AMP binds to a **protein kinase** and stimulates the phosphorylation of specific protein targets. Several types of epinephrine receptor stimulate cAMP synthesis through a coupling factor. In the case of glycogen breakdown, cAMP amplifies the relatively small signal of epinephrine into the formation of thousands of glucose molecules.

Receptors may also use other second messengers, such as calcium ions. Like cAMP, calcium ions stimulate the phosphorylation of specific proteins. Some water-soluble hormones do not work through second messengers but directly stimulate protein phosphorylation by membrane-bound kinases.

Prostaglandins and **growth factors** are paracrine factors. Prostaglandins regulate blood clotting and smooth muscle contractions.

Review and Thought Questions

Review Questions

1. How do hormones, paracrine signals, and pheromones differ from one another?
2. What is the difference between the nervous system and the endocrine system? What do these systems share in common?
3. What are the two major classes of hormones? How do they differ from each other?
4. What is a second messenger? Give an example of one.
5. How does diabetes mellitus (type 1 diabetes) differ from type 2 diabetes?
6. How is the anterior pituitary different from the posterior pituitary?
7. Why is the anterior pituitary not "the master gland"?
8. Name the eight peptide hormones released by the anterior pituitary and tell what each one does.

Thought Questions

9. How can different cell types respond differently to the same signal?
10. Suggest a mechanism by which DES, the synthetic estrogen given to pregnant women in the 1950s and 1960s, might cause cancers of the reproductive organs in the grown daughters of these women. How would you test your hypothesis?

About the Chapter-Opening Image

Some industrial and agricultural pollutants mimic the molecular structure of sex hormones. These compounds can interfere with the sexual development of alligators, and, many people fear, of humans as well.

 On-line materials relating to this chapter are on the World Wide Web at **http://www.harcourtcollege.com/lifesci/aal2/**

37

How Do Animals Move Blood Through Their Bodies?

Charles Drew and the Battle of Britain

On the night of March 31, 1950, four African-American physicians started an all-night drive from Washington, DC, to Tuskegee, Alabama. One of the four in the car, Charles Drew, was—at 45—one of the most famous doctors in America: his work on blood banking had saved thousands of lives during World War II (Figure 37-1). Drew and three friends were on their way to an annual free clinic where black physicians from all over the United States converged each year to treat people who had no other access to medical care.

But Drew never arrived in Tuskegee. He had just spent a busy day in Washington, performing operations and attending meetings at Howard University's College of Medicine, where he taught surgery. After taking the wheel just south of Richmond, Virginia, he fell asleep and drove off the road. The car rolled over several times, pushing Drew against the steering wheel, crushing his chest, breaking his neck, and severing the major vein that carries blood back to the heart.

Drew's friends called an ambulance, which hurried Drew to the emergency room of the Almanance General Hospital in Burlington, North Carolina. (Like other hospitals in the South in 1950, the Almanance did not treat blacks, except in its emergency room.) Nonetheless, the physician on duty immediately recognized Drew and did all he could to stop the bleeding and to replace the lost blood. Despite the doctor's efforts, Drew bled to death in less than 2 hours.

Ironically, Charles Drew's scientific work had saved thousands of other people from bleeding to death. During World War II German bombs devastated London in 1940, leaving tens of thousands of people with crushed limbs, broken bones, and horrendous burns. Although doctors could treat these injuries, their patients often died suddenly from *shock*, a life-threatening collapse of the circulatory system. Blood transfusions saved many victims, but the blood supply was limited. No one could keep blood cells from dying and spoiling after about a week. Despite the best efforts of British medical workers, a steady supply of freshly donated blood was impossible to maintain. Thousands of bomb victims died not from their wounds but from shock.

Fortunately, the liquid part of the blood, the plasma, can prevent shock almost as well as whole blood. And plasma can be stored for weeks rather than days, so plasma collected in the United States could be shipped both to Europe and to the Pacific. American plasma saved the lives of people wounded in the Battle of Britain, as well as in later battles. In 1940, the United States had not yet joined in the effort to defeat Hitler's armies, but American plasma made an invaluable contribution to Britain's ability to withstand the nightly air attacks.

From 1938 to 1940, Drew was a fellow at New York City's Presbyterian Hospital, associated with Columbia University. Drew had written a thesis, "Banked Blood," that established him as a leading authority on blood storage and processing. In 1940, Drew was chosen to lead an effort to develop ways of collecting, storing, and shipping plasma to Britain. Although plasma lasts longer than blood cells, it is chock full of glucose, proteins, and other nutrients, making it a superb growth medium for bacteria. In order to store and ship plasma, Drew had to establish procedures for preventing bacterial growth and for detecting such growth if it did occur. Drew worked out a foolproof system that allowed no bacterial contamination as blood moved from a donor's arm into a collecting vessel, from the collecting vessel into a centrifugation system that removed the blood cells and produced plasma, and from the centrifuge into a shipping bottle. To separate cells from plasma, Drew adapted a dairy apparatus that separated cream from milk. This invention allowed him to increase greatly the amount of blood processed each day.

Impressed by the success of Drew's program and procedures, the Red Cross set up a new blood bank at Presbyterian Hospital, this time for the U.S. Army and Navy, with Drew as its Assistant Director. At the same time, however, the Armed Forces and the Red Cross decided not to accept blood from African-American donors as part of the war effort despite the certain knowledge that this policy had no scientific basis. (The stated reason for this decision was the fear that some white soldiers would be afraid of transfusions if they might come from black donors.) In the face of this offensive policy, Drew resigned from his important position and returned to teaching surgery to black medical students at Howard University.

Drew's greatest contribution to the country after the war was his work in educating the next generation of African-American physicians. His goal was to change the image of African-American doctors as "just country practitioners . . . not particularly interested in advancing medicine." In Drew's day, for example, the American College of Surgeons would not admit Drew or any other African American of his generation to its membership, and many chapters of the American Medical Association also barred the membership of black physicians.

Drew remains a hero to this day. Amherst College, from which Drew graduated and where he ran track and played football, has established the Charles Drew House to promote the study of African and African-American culture. The Martin Luther King–Charles Drew Medical Center (associated with UCLA) in Los Angeles also honors Drew's contributions to medicine and his inspiration to generations of African-American physicians and medical students.

Figure 37-1   Charles Drew. (*National Portrait Gallery*)

WHY DO ANIMALS NEED BLOOD?

The "internal environment" of an animal is actually the *external* environment of its individual cells, specifically the **extracellular fluid** that surrounds individual cells (Figure 37-2). The constancy of this internal environment is a necessary condition for life. Both individual cells and organ systems contribute to homeostasis, keeping internal conditions constant in the face of a changing external environment. Homeostasis requires a constant exchange of materials between the extracellular fluid and an animal's external environment.

Blood is the indispensable intermediary in this exchange. Blood carries nutrients from the outside world to the extracellular fluid and cellular wastes to organs that can dispose of them. In mammals, for example, blood carries oxygen from the lungs and basic building blocks from the gut to the extracellular fluid all over the body. Blood simultaneously moves carbon dioxide to the lungs and nitrogenous wastes to the kidneys. Blood also serves several other functions: it distributes hormones, it transports the molecules and the cells of the immune system, and it conducts heat to all parts of the body. Figure 37-3 shows the heart and the major blood vessels in the human circulation.

What Is the Connection Between Blood and the Extracellular Fluid?

The extracellular fluid contains many substances dissolved in water. Some, such as bicarbonate and phosphate ions, act as *buffers,* which prevent the fluid from becoming too acidic or too basic. These substances establish a stable environment. In addition, blood carries energy-rich molecules, building blocks, and waste products. Blood ferries glucose, amino acids, nucleosides, and other small molecules from the gut to the extracellular fluid, oxygen from the lungs to respiring cells all

771

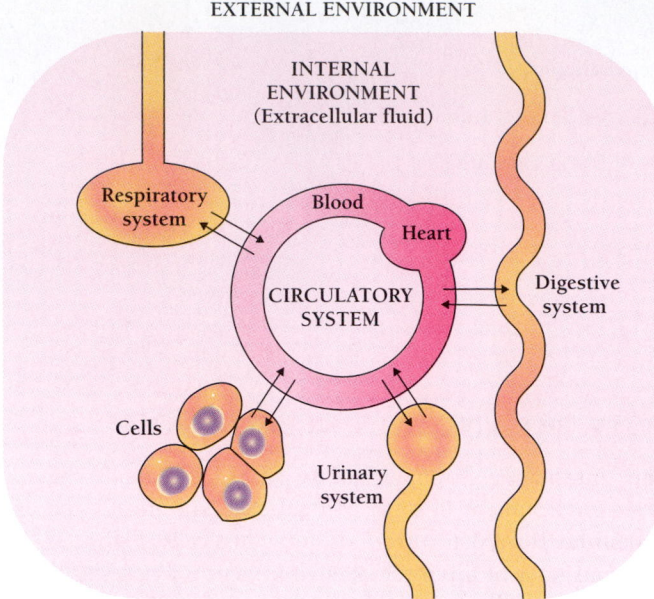

Figure 37-2 The extracellular fluid, which surrounds individual cells, is constantly renewed by the blood.

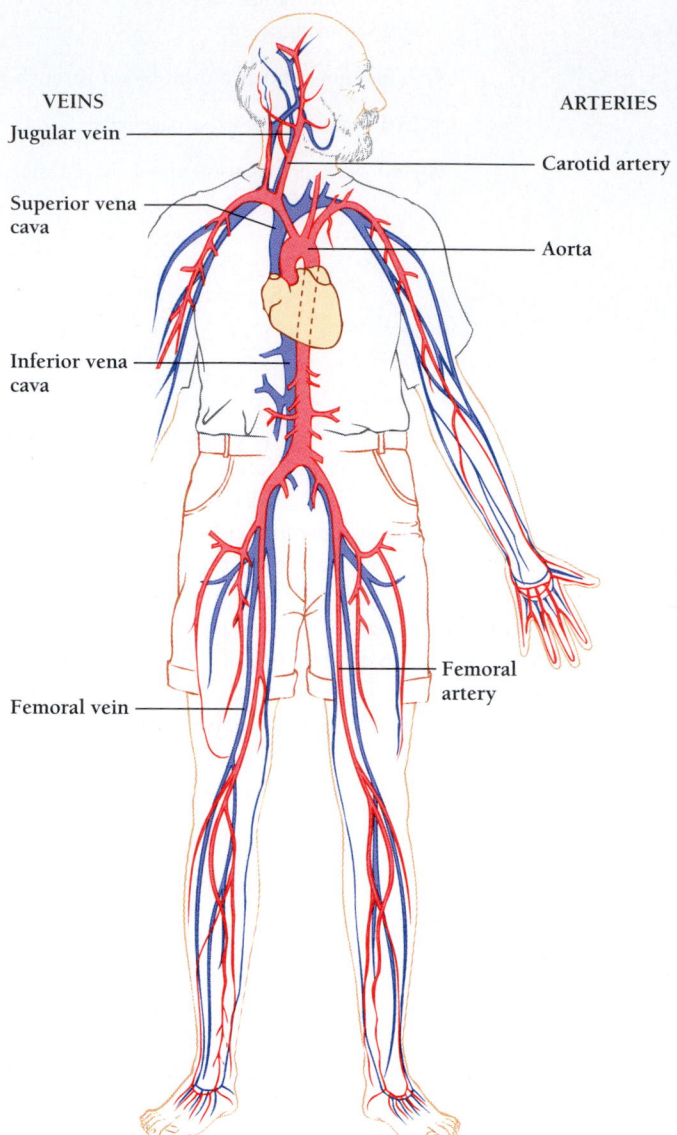

Figure 37-3 The major vessels of the human circulatory system. In addition to the heart, the circulation includes arteries, arterioles, and capillaries, which carry oxygenated blood from the heart, and venules and veins, which carry deoxygenated blood (and carbon dioxide) back to the heart. The *aorta* is the major artery through which blood leaves the heart and travels to the rest of the body. The *venae cavae* are the major veins that carry blood back to the heart.

over the body, and carbon dioxide from other tissues to the lungs.

All the cells of the body are bathed in the extracellular fluid. The cell membranes serve as active gatekeepers that regulate the internal environments of the cells. The membranes usher in some components of the extracellular fluid, denying entry to others. The extracellular fluid would quickly become a cesspool of cell wastes if it were not for the blood, which carries these wastes to the liver for detoxification or to the kidneys for excretion.

Circulating blood maintains the extracellular fluid by removing wastes and delivering nutrients.

What Does Blood Contain?

Blood carries all the substances that enter or leave the extracellular space. The fluid part of blood is called **plasma** [Greek, form or mold]. The chemical composition of plasma closely resembles that of the extracellular space. The major difference is that plasma contains about 7 percent protein, while the extracellular space contains only about 2 percent protein.

In vertebrates, blood contains both plasma and vast numbers of cells. The blood carries three kinds of cells—red cells, or **erythrocytes** [Greek, *erythros* = red + *kytos* = receptacle]; white cells, or **leukocytes** [Greek, *leukos* = white]; and **platelets** [little plates] (Figure 37-4). By far the

most numerous cells in mammalian blood are red cells. Each milliliter of human blood contains about 5 billion erythrocytes, and a human adult contains some 25 trillion erythrocytes.

The most distinctive characteristic of vertebrate blood, its bold red color, comes from the red blood cells themselves. And what makes red blood cells red is the oxygen-binding protein **hemoglobin** [Greek, *haima* = blood + Latin, *globus* = ball].

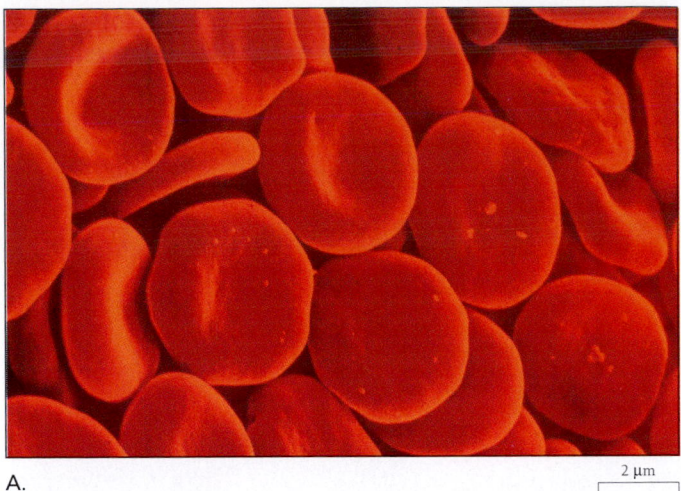

A.

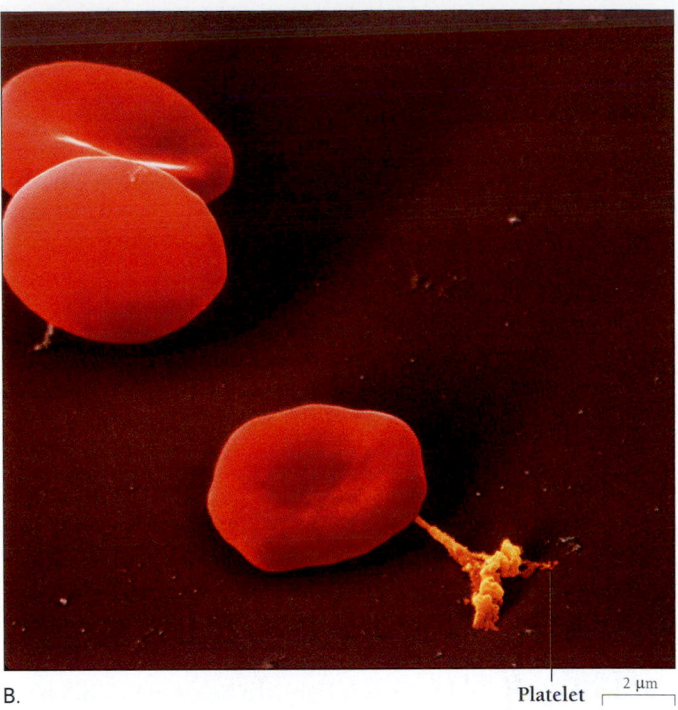

B.   Platelet

Figure 37-4   Blood cells. One milliliter of human blood normally contains about 5 billion erythrocytes, as well as numerous leukocytes and platelets. A. In mammals, red blood cells lose their nuclei during the process of maturation. B. Platelets (such as the one attached to the red blood cell on the lower right) are actually cytoplasmic fragments of larger precursor cells, and they also lack nuclei. (*A, David M. Phillips/Visuals Unlimited; B, Meckes/Ottawa/Photo Researchers, Inc.*)

Each erythrocyte contains about 30 million hemoglobin molecules. The bold red color of hemoglobin comes from the iron-containing heme groups that help this protein bind oxygen in the lungs and then release it to other tissues. In contrast, hemocyanin, the oxygen-carrying protein of many invertebrates, contains copper rather than iron and can be a beautiful blue color.

In invertebrates, blood has fewer cells, and oxygen-binding proteins (such as hemoglobin or hemocyanin) are extracellular rather than intracellular. Invertebrate blood cells help defend against infection, parasites, and blood loss.

White cells, or leukocytes, include several types of cells that contribute to defenses against infection and tumors. The platelets are small cellular fragments in the blood that help blood coagulate. Because platelets are essential to prevent blood loss, we will discuss them with other aspects of the body's defense system in Chapter 40.

Red cells, white cells, and platelets begin as true cells, with nuclei. In mammals, however, both the platelets and the red blood cells lose their nuclei, so that only the leukocytes are true cells. As the erythrocytes mature, they lose their nuclei—in mammals, although not in other vertebrates. The loss of the nucleus probably allows mammalian erythrocytes to pass more freely through the narrow capillaries of the circulatory system.

Blood consists of liquid plasma and different types of blood cells with specialized functions.

HOW DID SCIENTISTS DISCOVER THAT THE BLOOD CIRCULATES?

Ancient scientists understood that both blood and air were crucial for life, but they did not understand how the heart and blood vessels actually worked. The greatest contributor to the understanding of the relationship between blood and air was the Greek physician Galen (129–199 A.D.). At the age of 28, Galen became chief physician to the gladiators in the city of Pergamum. His close examination of their horrendous wounds revealed much about the movement of blood in the body. Later, he performed dissections on living animals for the education and entertainment of onlookers. These dissections led him to conclude, for example, that arteries carry blood and that nerves control movements.

Galen's extensive writings were preserved by medieval scholars and passed down to the anatomists of the 16th century. Fourteen hundred years after his death, the Catholic Church endorsed Galen's views, which became so influential that challengers were treated not just as scientific rebels but as religious heretics. Like some of the astronomers who speculated on the circling of the planets, some of the anatomists who speculated on the circulation of the blood paid for their speculations with their lives.

But Galen's understanding was only partly accurate. Galen thought that the movement of blood derived from the pulsing of the arteries. Galen viewed the heart itself not as a pump that pushed the blood through the vessels of the body but as a sort of furnace that heated the blood. Galen thought that blood originated in the liver and flowed to the right side of the heart, where it mixed with air from the lungs. The blood was then further refined, he believed, as it passed through a fine filter between the right and left sides of the heart.

In many respects, Galen's understanding was accurate. For example, the blood does release toxic fumes into the lungs. These fumes are carbon dioxide, the waste product of cellular respiration throughout the body. But Galen postulated microscopic holes in the heart that allowed the blood to flow from one side of the heart to the other. These holes do not exist, and later anatomists looked for them in vain. Most importantly, neither Galen nor any one else before the 17th century had any idea that the actual route between the left and right halves of the heart was a circuit from heart to arteries to veins and back to the heart.

The discovery of this route came in 1628 from the work of William Harvey, an otherwise traditional English physician (Figure 37-5). Harvey's analysis depended on both a handful of radical ideas from different sources and on meticulous observation and experiment.

Unlike modern biologists, Harvey worked in comparative isolation. He could not afford to discuss his ideas with others until he was certain that he was right. To be wrong would have been disastrous for his medical practice. As it was, many of his patients abandoned him when he published his classic work *de Motu Cordis* [*On the Movement of the Heart and Blood*]. Seventeenth-century physicians looked on new ideas with suspicion. Harvey probably could propose such an extraordinary reinterpretation of human physiology because he was so established and otherwise so orthodox. Because he was a widely respected doctor and a close friend of the king of England, Harvey could not only say what he thought but he could also publish his ideas.

In addition, the intellectual climate of the 17th century had fostered an enthusiasm for learning from direct experience. Among the factors contributing to this atmosphere were a decline in respect for political and religious authorities and an accelerating pace of technical invention. Among the new inventions were water pumps used by miners and firefighters. Some scholars have suggested that the water pump inspired Harvey to conceive of the heart as a pump.

Harvey recognized early that the heart functioned as a pump, but its movements were so rapid that he initially despaired of ever understanding them. Today a scientist who wants to study a rapid process videotapes the process, then watches it in slow motion. Long before slow-motion cinematography, however, Harvey found a convenient way to slow down the motion of the heart. Instead of trying to observe the rapidly beating hearts of "warm-blooded" mammals and birds, he began by studying the hearts of toads and snakes. By keeping these animals cool, he could slow down their hearts enough to observe the separate beats of the different parts of the heart.

With this understanding, Harvey performed dissections on living dogs and other animals so that he could see how the heart beat and where the blood went. As the animals died, their hearts beat more slowly, and Harvey could see the movements of each beat of the heart.

Figure 37-5   William Harvey. The English physician William Harvey constructed the first accurate description of the circulatory system. (*The Granger Collection, New York*)

The heart of a mammal consists of four chambers—two ventricles and two atria. The atria fill with blood, which flows down into the more muscular ventricles through one-way valves. The valves shut behind the blood, and the ventricles then squeeze the blood forcefully out.

By cutting the tip off the left ventricle, Harvey was able to show that the muscular contraction of the ventricle squeezes the blood out of the heart. In contrast, earlier European

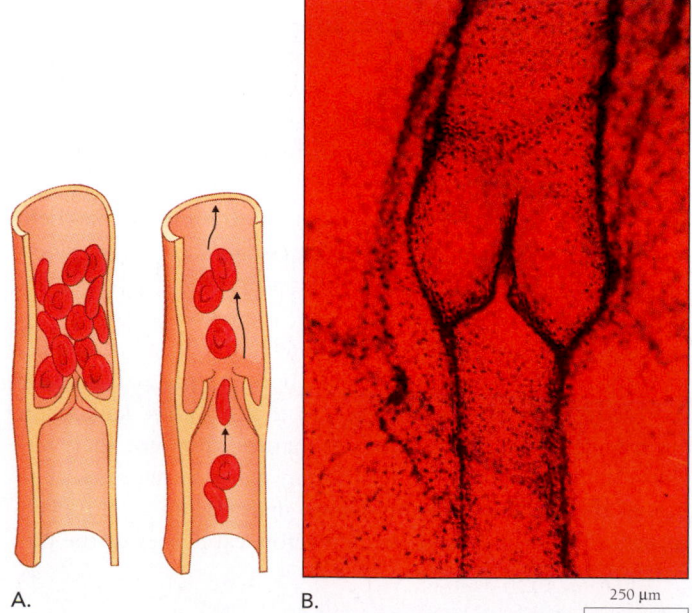

A.   B.   250 μm

Figure 37-6   Veins contain one-way valves. A. Tiny flaps inside of veins prevent blood from flowing back toward the capillaries but allow blood to move toward the heart. B. A closed valve. (*B, John D. Cunningham/Visuals Unlimited*)

anatomists had thought that the heart worked by sucking blood in. Harvey could see, however, that the heart pushed but did not pull. But if the heart cannot pull, he wondered, What forces the blood from the veins into the heart?

Harvey knew from his anatomical studies that the veins are filled with valves that allow blood to move toward the heart but not away from it (Figure 37-6). Like the valves in the heart itself, these flaps permit the passage of the blood in only one direction. The undeniable one-way flow of blood from the veins to the heart refuted Galen's hypothesis that blood moves back and forth like the tides. Harvey concluded that the veins deliver blood to the heart and the heart pumps the blood into the arteries. He was able to show that the pulse of the arteries was due to shots of blood pushed into them by the heart and was not something that the arteries did automatically.

How the blood got from the arteries back to the veins was a question he could not answer. But that the blood circulated—from heart to arteries to veins and back to heart—he was certain.

In addition, he now saw how the pulmonary circulation explained the complexity of the heart. The blood from the veins emptied into the right side of the heart, which pumped it to the lungs. The blood returning from the lungs entered the left side of the heart to be pumped into the arteries. In essence, the circulation consisted of two loops, one through the lungs and one through the body (Figure 37-7). Both loops joined at the heart, like the two parts of a figure 8.

Harvey bolstered his idea that the blood circulates with one final argument. He measured the amount of blood pumped with each heartbeat and showed that the amount of blood pumped each hour far exceeds the total amount of blood in the whole body. This demonstrated that the same blood was passing through the heart over and over again. Somehow the blood was being recycled.

Harvey's model was not complete, however, because he could not show the connections between the outgoing arteries and the incoming veins. He simply had no way of seeing the capillaries. As Galen had hypothesized microscopic pores between the two halves of the heart, Harvey guessed that the arterial blood passed through the tissues and organs and back to the veins. Unlike Galen's hypothesis, however, Harvey's was confirmed. Just 40 years later, the great Italian microscopist Marcello Malpighi discovered the minute capillaries that complete the circulation.

Harvey's achievements built upon the work of those who came before him. His anatomical training in Italy, his exposure to Italian standards of observation, and his knowledge of the pulmonary circulation and the valves in the veins all prepared him to appreciate better than anyone the workings of the heart and circulation. Finally, his position in society and

his relative safety in England enabled him to publish his theory without fear.

Harvey's work established that the blood circulates.

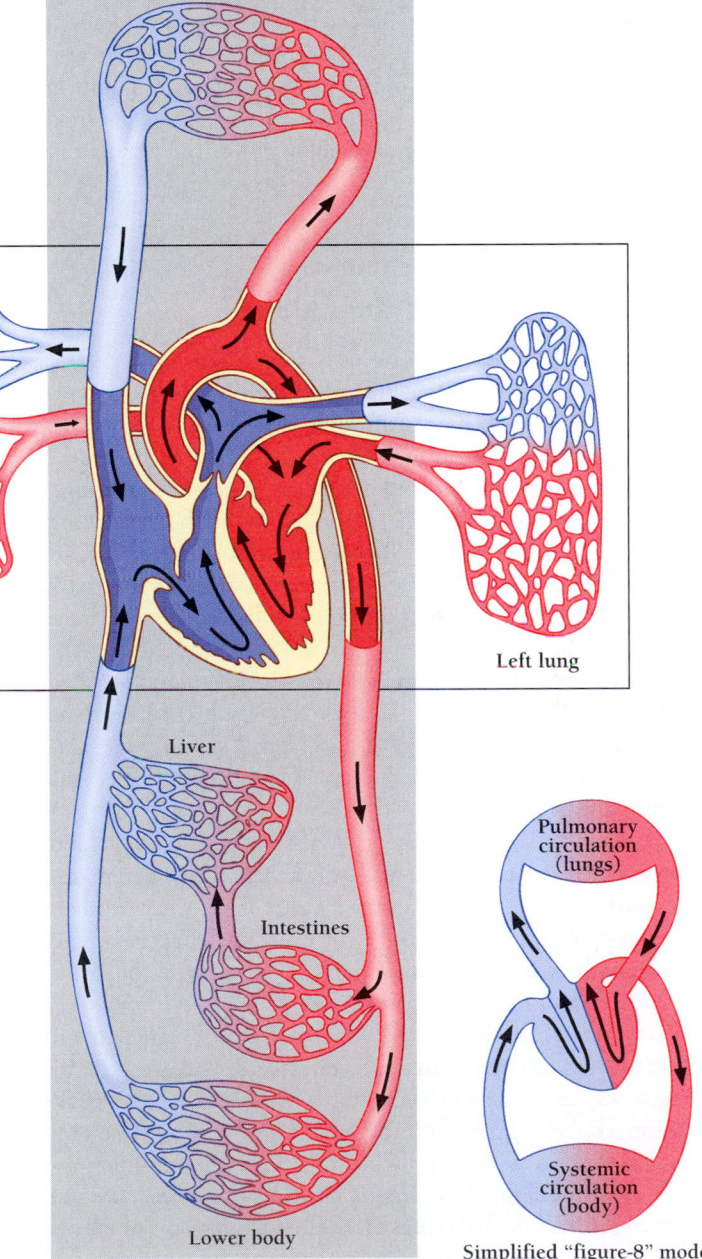

Figure 37-7 Harvey's figure-8 model of the circulatory system. The blood from the veins empties into the right side of the heart, which pumps it to the lungs. The blood returns from the lungs and then enters the left side of the heart, which pumps the oxygenated blood to the arteries.

Why Do Many Animals Need a Blood Pump?

Animals less than 1 mm in diameter get by without a circulatory system. These animals distribute materials by diffusion alone—nutrients, oxygen, wastes, and salts. But diffusion is a slow, unregulated process, and all larger animals have a circulatory system with some kind of pump.

In mammals, the rhythmic contractions of the heart accomplish most of the pumping, with blood leaving the heart through arteries and returning to the heart through veins. Other animals have hearts that are usually much less elaborate. In an insect, for example, the heart consists of the enlargement of a blood vessel that runs just above the gut. In other animals, the distribution of the blood depends on contractions in the blood vessels themselves or on the squeezing of the blood vessels during body movements. Valves within the circulatory system ensure that blood flows in a single direction.

How Does the Heart Pump Blood Through the Body?

As Harvey showed, humans and other mammals have a double circulation: blood flows through two adjoining circuits, similar to a figure 8 (Figure 37-7). In the first circuit, deoxygenated blood from the veins passes from the heart to the lungs, where oxygen attaches to the hemoglobin in the red blood cells. The newly oxygenated blood then returns to the heart and passes into the second circuit, which delivers the oxygenated blood from the lungs to all the tissues of the body.

The blood, the heart, and the blood vessels together make up the **cardiovascular system.** The route of the blood is called the **circulation** [Latin, *circus* = circle]. The route through the lungs is called the **pulmonary** [Latin, *pulmo* = lung] **circulation,** and the route through the rest of the body is called the **systemic circulation.**

The muscular organ responsible for pushing the blood through the circulatory system is the **heart.** Birds and mammals have a single four-chambered heart, but other animals may have hearts with two, three, or even five chambers, and some animals have more than one heart. Some types of earthworm, for example, have ten hearts.

In humans, the heart is about the size of a clenched fist and is located in the chest just beneath the breastbone, or **sternum** [Greek, *sternon* = chest]. A fibrous sac, the **pericardium** [Greek, *peri* = around + *kardia* = heart], encloses the heart within a watery lubricating fluid. Four interconnected rings of connective tissue provide the frame for the heart's organization. Two rings form the openings between the atria and the ventricles, and two rings form the exit from the ventricles to the circulatory system. The rings are covered by the valves, which determine the direction of blood flow. The muscular walls of the heart—the **myocardium** [Greek, *myos* = muscle]—attach to this fibrous skeleton to form the four chambers.

Several large arteries and veins attach directly to the heart, and it is sometimes hard to see where these vessels end and the heart itself begins (Figure 37-8). The largest vessel is the **aorta,** the artery that carries blood from the left ventricle to the rest of the body. The aorta arches over the top of the heart. The first branch of the aorta sends blood to the head (Figure 37-2). The two largest veins, which run up through the center of the body and carry deoxygenated blood from the body to the right atrium, are the superior and inferior **venae cavae.** The **pulmonary artery** carries deoxygenated blood from the heart to the lungs, and the **pulmonary veins** carry oxygenated blood from the lungs to the heart.

The mammalian heart consists of two separate halves. The right half pumps oxygen-poor blood to the lungs; the left half pumps oxygen-rich blood to the head and the rest of the body. Each half of the heart contains two muscular sections: (1) a relatively thin-walled entrance chamber called an **atrium** [Latin, courtyard or entry] and (2) a thick-walled pumping chamber called a **ventricle** (Figure 37-8). The mammalian heart, then, consists of four chambers: the right and left ventricles and the right and left atria.

To review, oxygen-loaded blood from the lungs enters the left atrium and pours down into the left ventricle, the most muscular of the four chambers. The left ventricle then pumps the blood through the arteries to the capillaries, which supply and wash all the tissues of the body. The oxygen-depleted blood then returns from the capillaries to the veins, which converge and enter the right atrium. The right atrium pumps the blood into the right ventricle, which pumps the blood into the lungs.

In mammals, the heart pumps blood through two different circuits. One circuit passes through the lungs, while the other passes through all the other tissues of the body.

How Do Birds and Mammals Prevent Oxygen-Rich Blood from Mixing with Oxygen-Poor Blood?

In many animals, the freshly oxygenated blood mixes with deoxygenated blood from the tissues. This means that the tissues never receive blood with the highest possible oxygen content. For animals that sustain movements and metabolism for long times, that arrangement isn't good enough. Birds and mammals need to maintain a constant body temperature, highly developed brains, and quick muscular responses, all of which require large supplies of concentrated oxygen.

The four-chambered hearts of mammals and birds prevent the oxygen-poor blood from the tissues from returning to the tissues before it has acquired oxygen in the lungs, always keeping the two kinds of blood separate. Because mammals and birds maintain a constant body temperature (that is, they are "homeotherms"), they use more energy and require more oxygen than do heterotherms. The organization of their circulatory systems maximizes the ability of the blood to deliver oxygen.

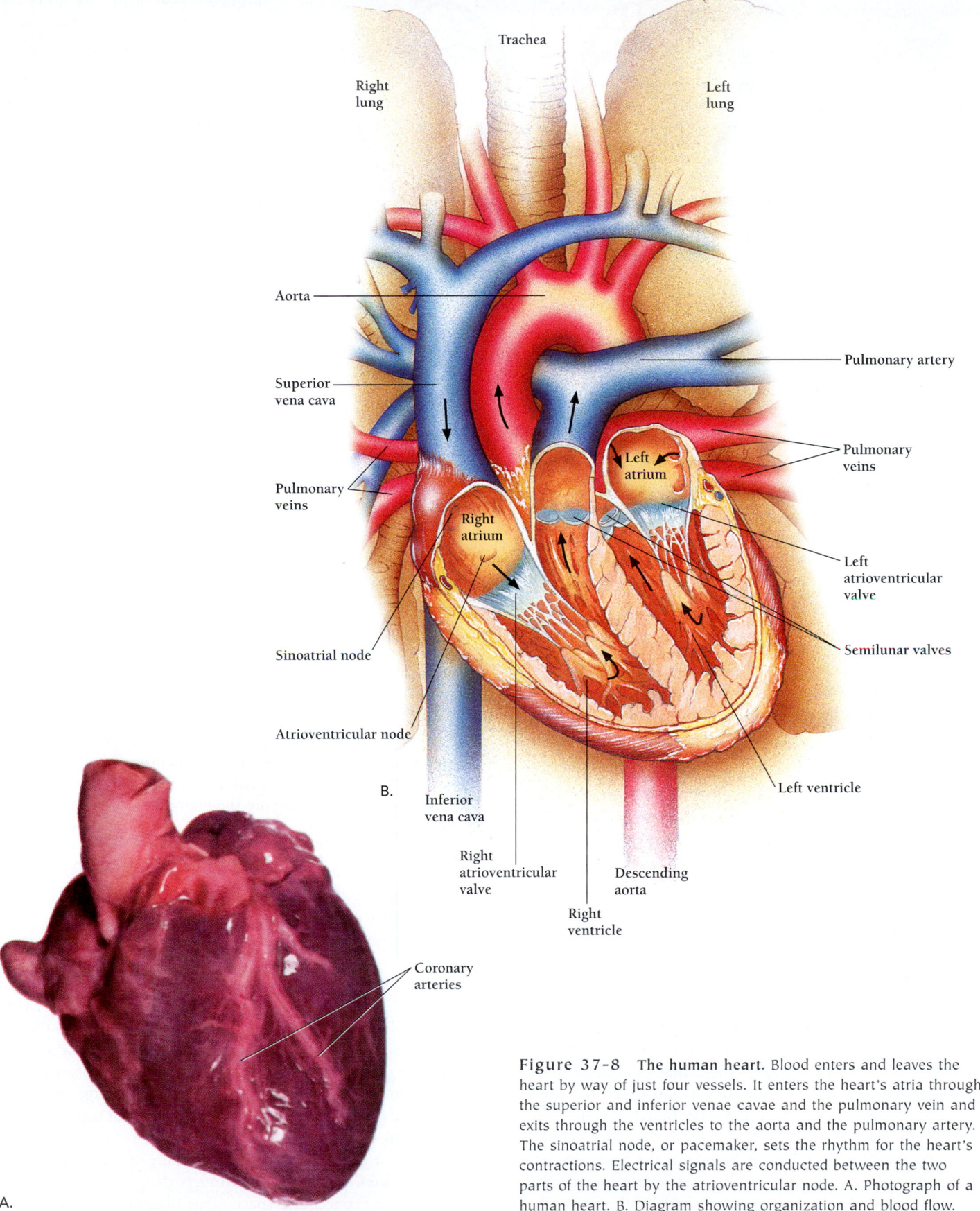

Trachea

Right
lung

Left
lung

Aorta

Superior
vena cava

Pulmonary
veins

Right
atrium

Sinoatrial node

Atrioventricular node

B.

Inferior
vena cava

Right
atrioventricular
valve

Right
ventricle

Descending
aorta

Pulmonary artery

Pulmonary
veins

Left
atrium

Left
atrioventricular
valve

Semilunar valves

Left ventricle

Coronary
arteries

A.

Figure 37-8   The human heart. Blood enters and leaves the heart by way of just four vessels. It enters the heart's atria through the superior and inferior venae cavae and the pulmonary vein and exits through the ventricles to the aorta and the pulmonary artery. The sinoatrial node, or pacemaker, sets the rhythm for the heart's contractions. Electrical signals are conducted between the two parts of the heart by the atrioventricular node. A. Photograph of a human heart. B. Diagram showing organization and blood flow. *(Photo, Morris Huberland/Science Source/Photo Researchers, Inc.)*

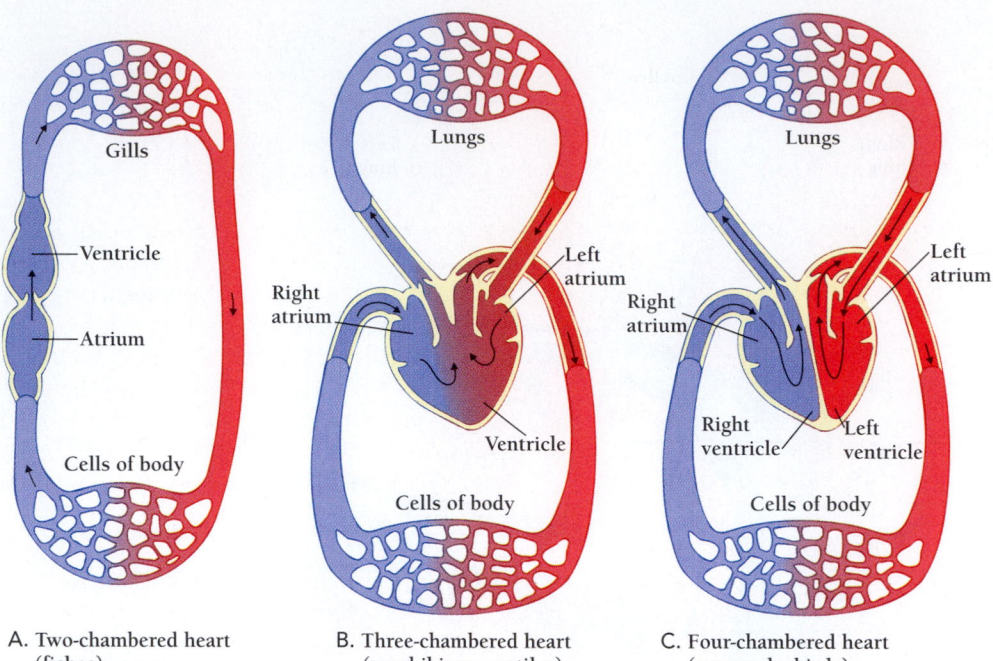

Figure 37-9 Evolution of the heart. A. Most fish have a two-chambered heart, which pumps a single stream of unoxygenated blood forward into the arteries that feed the gills. B. Three-chambered hearts, intermediate between those of fishes and mammals, are found in lungfish, amphibians, and reptiles. These hearts are not inefficient versions of the four-chambered heart, but more flexible versions. Blood passing through a turtle's three-chambered heart, for example, can bypass the lungs when a turtle is diving and oxygen supplies are low. C. The four-chambered hearts of mammals and birds pump two streams of blood, one oxygenated and one deoxygenated.

A. Two-chambered heart
(fishes)

B. Three-chambered heart
(amphibians, reptiles)

C. Four-chambered heart
(mammals, birds)

In contrast, the hearts of reptiles and amphibians contain only three chambers—two atria and one ventricle (Figure 37-9). Consequently, some mixing of oxygen-rich and oxygen-poor blood occurs. The ventricle's structure minimizes this mixing, however, so that the system works better than one might at first suspect.

Bony fishes get by with only two chambers, consisting of one atrium and one ventricle. The ventricle pumps blood to the gills, where it obtains oxygen in the gill capillaries. The oxygen-rich blood then flows directly from the gills to the tissue capillaries, where it provides oxygen and other nutrients.

The four-chambered hearts of birds and mammals keep deoxygenated blood from mixing with oxygenated blood. Such an arrangement helps guarantee that oxygen-rich blood flows to all of the tissues of the body.

HOW DO THE BLOOD AND THE EXTRACELLULAR FLUID EXCHANGE SMALL MOLECULES AND IONS?

Physiologists distinguish between open and closed circulation. In an **open circulation,** blood flows from the heart, through an artery, and into a large open space that occupies as much as 40 percent of the body's volume. As illustrated for a mollusk and a bumblebee in Figure 37-10, blood directly bathes all of an animal's tissues. An open circulation does not offer much resistance to the flow of blood, so the heart does not have to generate much pressure; a mollusk heart generates pressures of about 0.15 pound per square inch (psi) compared to about 35 psi in a car tire.

In a **closed circulation,** blood moves through vessels in a continuous circuit—from heart to arteries to veins and back to the heart. The route from arteries to veins lies within many tiny capillaries. The total blood volume in a closed circulation is only about 5 to 10 percent of the total body volume, much less than in an open circulation. The capillaries' small diameter allows close contact with cells throughout the body but requires that the heart generate significant pressure. Each time it beats, for example, a human heart generates a pressure of about 3 pounds per square inch.

A closed circulation allows animals more powerful homeostatic possibilities than an open circulation. By changing the action of the heart, an animal can regulate the flow through the whole system. By varying the diameter of blood vessels, an animal can regulate the transport of oxygen, nutrients, wastes, and heat to specific organs.

The disadvantage of a closed circulation is that the blood, contained in vessels, does not directly bathe the body's cells. To come into contact with the tissues, molecules and ions must pass through vessel walls. Most of the contact comes within rich beds of fine capillaries. The capillary beds are so extensive that each cell in a vertebrate's body is no more than two or three cell diameters away from the circulating blood. Many small molecules and ions move from the capillaries to the tissues by simple diffusion. Except in the brain, the cells that make up the capillaries are rather loosely joined, so small molecules can diffuse between adjacent cells. Even large molecules can move through the gaps between the vessels' cells, but less easily than small molecules and ions.

Because of the relatively high pressure of the blood within the capillaries, plasma oozes through these gaps to form the fluid of the extracellular space. This plasmalike fluid collects in the lymphatic system and eventually rejoins the blood. Even

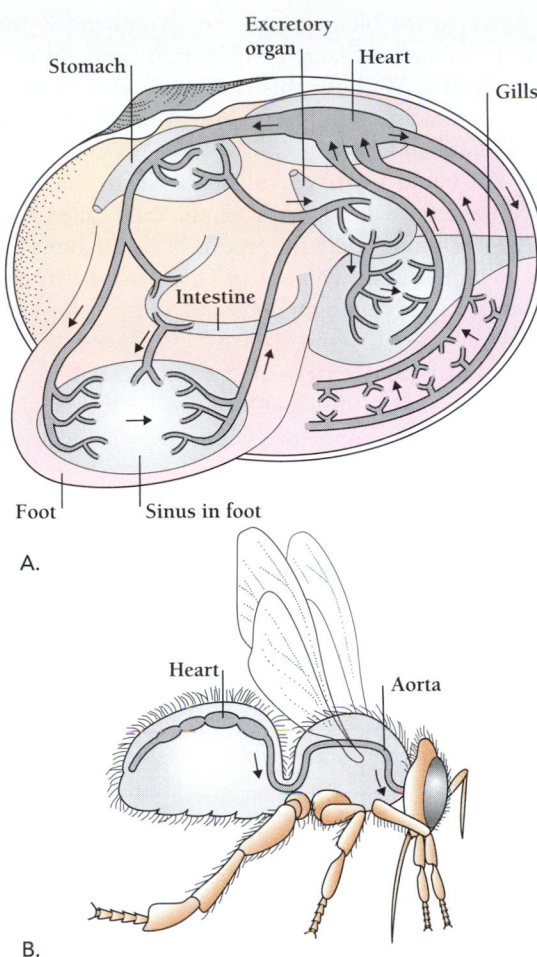

A.

B.

Figure 37-10   Open circulatory systems: A. In a mollusk (such as a clam), blood flows from the heart through closed arteries to contained open spaces (such as the sinus in the foot) and to the fluid (hemolymph) that bathes all the internal organs; blood returns to the heart through closed veins. B. In an insect (such as a bee), oxygen and carbon dioxide move by dissolving in the blood but through tubes called tracheae (Chapter 38); the heart pumps blood from the posterior toward the anterior end of a continuous tube that runs along the top (dorsal) of the bee's body; blood flows into the open hemolymph and ultimately returns to the dorsal tube through small openings.

though they are leaky, the capillary walls prevent almost all of the blood's cells from moving into the tissues. The sievelike walls also filter some of the larger molecules, so that the lymph contains considerably less protein than does the plasma.

In insects and other arthropods, and in most mollusks, the blood travels part of its circuit through open spaces, called **sinuses.** In an animal with an open circulatory system, the blood mixes freely with extracellular fluids and bathes the organs of the body. In an animal with a closed circulatory system, however, blood runs only within enclosed vessels. How then do materials move to and from the extracellular space? To answer this question, we must look at the structure of the finest blood vessels, the capillaries.

Arteries, Veins, and Capillaries

In most animals, blood leaves and returns to the heart via different routes, with different kinds of blood vessels. In vertebrates, the vessels that carry blood away from the heart are called **arteries,** and those that carry blood back to the heart are called **veins.**

As the arteries leave the heart, they branch many times, and blood flows into smaller vessels called **arterioles.** The arterial blood then enters still finer vessels called **capillaries.** The network of microscopic capillaries is so extensive that no cell of the body is more than three or four cells away from a capillary (Figure 37-11). If all of the blood vessels in your body were placed end to end, they would extend 2½ times around Earth's equator.

The capillaries are the immediate source of the extracellular fluid. Although arteries and veins do not normally leak at all, capillaries leak constantly, seeping nutrients and oxygen into the surrounding space. The thin, leaky walls of the capillaries also allow them to absorb cell wastes. The absorbed fluid, together with the suspended blood cells and the dissolved

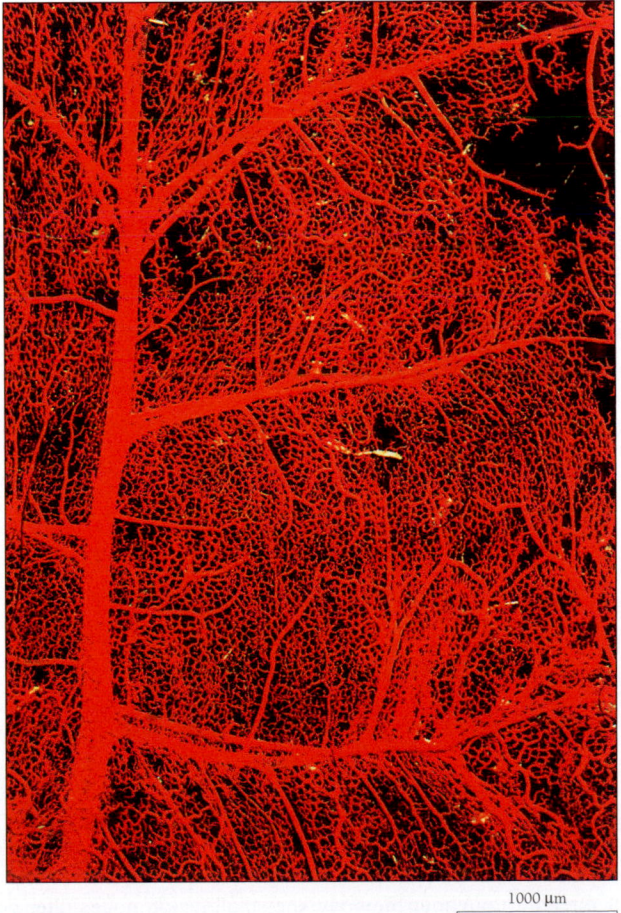

1000 μm

Figure 37-11   Capillaries. Arterioles branch into capillaries, so fine and so numerous that no cell is more than four cells away from a capillary. (*Biophoto Associates*)

molecules and ions, flows into the smallest veins, called **venules.** The venules, in turn, converge to form larger veins for the final return to the heart.

Arteries carrying blood from the heart branch to form smaller arterioles, which branch to form extensive capillary beds. Capillaries converge to form venules, which converge into veins, which pour blood back into the heart.

The Lymphatic System

Not all of the extracellular fluid reenters the circulatory system through the veins. Some fluid reenters through the vessels of the **lymphatic system** (Figure 37-12). The lymphatic system provides a secondary route for fluids from the extra-cellular space to the bloodstream. The lymphatic system also carries proteins and any large particles that cannot directly enter the capillaries. Without this system, we would die within 24 hours. Mammals, birds, reptiles, amphibians, and many fish all have a lymphatic system.

Like the blood in the circulatory system, the lymphatic vessels contain specialized white blood cells called **lymphocytes.** These immune cells are produced in the bone marrow and the thymus. As we will see in Chapter 40, lymphocytes respond to foreign proteins and kill microorganisms tagged by antibodies.

Extracellular fluid enters the lymphatic system through lymphatic veins and is funneled into two large thoracic ducts, which drain into the circulatory system by way of the veins of

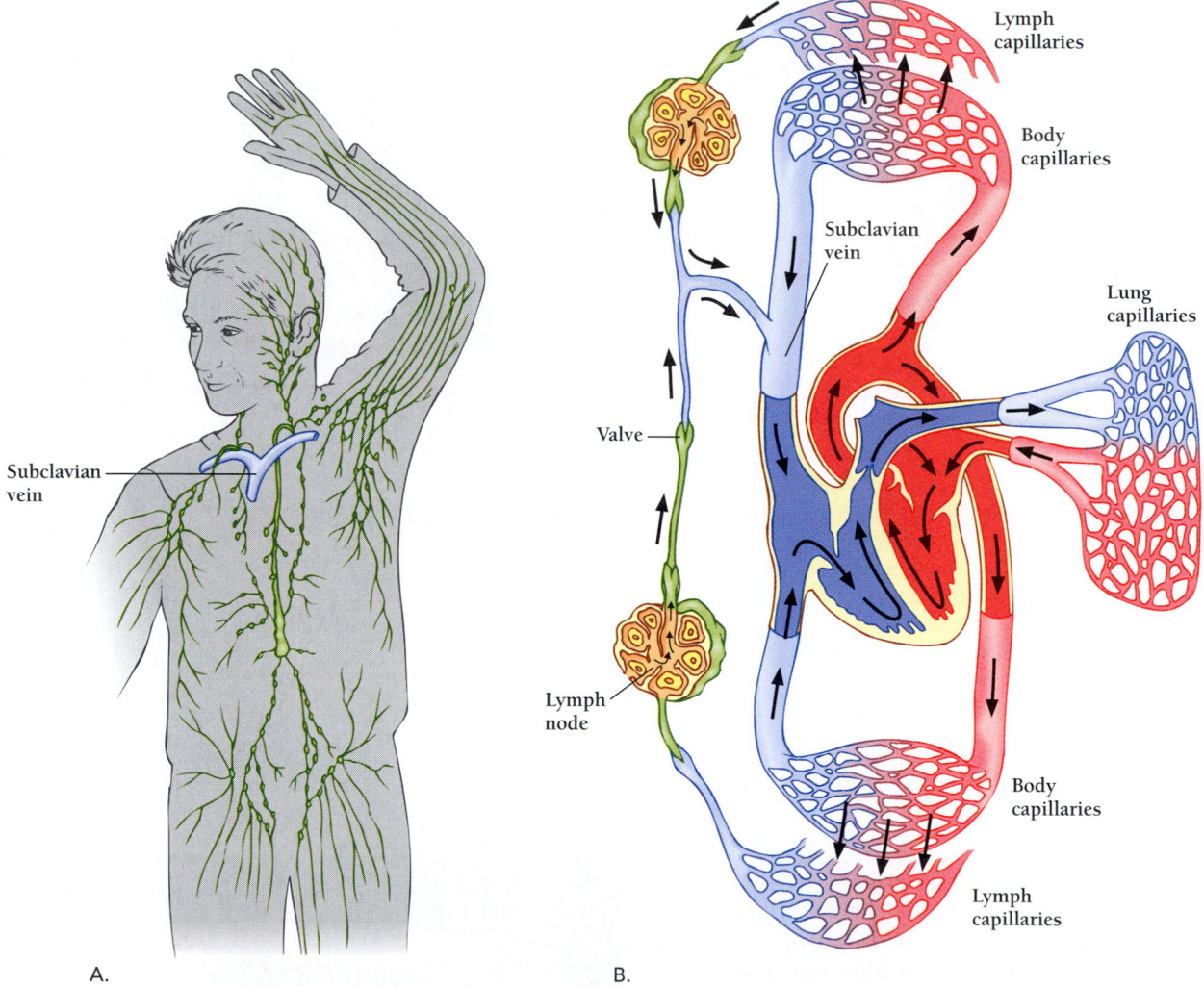

Lymph capillaries

Body capillaries

Subclavian vein

Lung capillaries

Valve

Lymph node

Body capillaries

Lymph capillaries

Subclavian vein

A.

B.

Figure 37-12 **The lymphatic system.** The lymphatic system carries the small amount of extracellular fluid that does not make its way back into the vascular capillaries. More importantly, the lymphatic system carries wastes from the extracellular fluid to the lymph nodes. A. Scattered throughout the body, the small lymph nodes filter dead cells, foreign cells, and other organic waste from the lymph. Lymphocytes ingest and digest the waste particles and the clean lymph exits the node. One-way valves keep the lymph draining in one direction. B. All the clean lymph drains into the "subclavian" veins (at the base of the neck) and reenters the circulatory system.

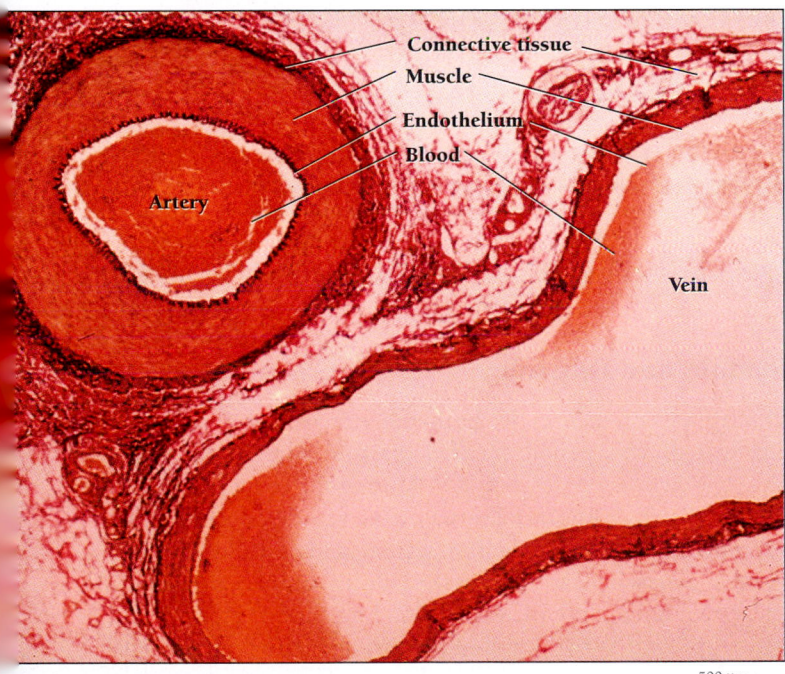

A.

ARTERIOLE

Connective tissue

Muscle fibers

ARTERY

Smooth muscle
Connective tissue
Endothelium

Elastic connective tissue

Endothelium

Connective tissue

Valve

Endothelium

Smooth muscle

CAPILLARIES

VEIN

VENULE

Connective tissue

Figure 37-13 Blood vessel.
A. Veins, arteries, and capillaries all include a thin layer of endothelial cells. Veins and arteries are further surrounded by connective tissue and smooth muscle, which contracts and relaxes to change the diameter of the vessel. Inside the lumen of a vein are one-way valves that force the blood back toward the heart. B. The walls of arteries are especially thick, the better to withstand the high blood pressures generated by the heart. (B, Biophoto Associates)

Connective tissue
Muscle
Endothelium
Blood

Artery

Vein

B.

500 µm

the neck. Fluid, dead or foreign cells, and proteins form the **lymph,** which moves passively through the lymphatic system. The normal movements of the body squeeze the lymph past one-way valves in the lymph vessels.

Lymphatic veins contain **lymph nodes,** regions where filterlike tissue separates cells and other detritus from the lymph. Inside the lymph nodes are phagocytic cells, which engulf material trapped in the lymph nodes. Dense concentrations of immune cells in the lymph nodes also monitor the lymph for signs of infection.

Dead and foreign cells and proteins in the extracellular fluid flow into the lymphatic system. In the lymph nodes, wastes are filtered from the lymph and ingested by phagocytic white blood cells.

The Structure of Blood Vessels

The cellular organization of the blood vessels—arteries, capillaries, and veins—reflects their differing roles in the circulatory system. All three types of blood vessels are hollow tubes. Surrounding the space inside, called the **lumen,** is a thin layer of cells called the **endothelium** (Figure 37-13).

Unlike the epithelial cells of the skin or gut, which are tough and tightly bound together, the endothelium of the capillaries is only one layer thick. The thin walls of capillaries permit the easy passage of substances from the lumen to the surrounding extracellular space.

The linings of arteries and veins are thick and relatively impermeable. In addition to their endothelial layers, the walls of arteries and veins contain layers of elastic fibers, collagen, and smooth muscles. Veins generally have thinner and less muscular walls than do arteries. The elastic walls of arteries and large veins allow these vessels to narrow or expand—in response to changes in blood pressure or in response to signals from the autonomic nervous system to the smooth muscles.

But the vessels—especially the arteries in which blood pressure is highest—must not be too elastic. If they stretch too much, they can balloon out and flatten the capillaries in the surrounding tissues. To prevent such ballooning, collagen

781

fibers, like those found in connective tissue and bone, strengthen and stiffen the vessel walls. If the collagen sheath of an artery fails and ballooning occurs, the result is an **aneurism.** An aneurism in the brain can cause tremendous damage. Sometimes an aneurism ruptures a major artery, and death soon follows.

The endothelium of a capillary is only one cell thick. The walls of larger vessels are thicker and surrounded by elastic fibers, collagen, and smooth muscle.

How Does the Blood Acquire Oxygen?

Oxygen is one substance delivered by the blood that the body needs continuously. In the United States the single most common cause of death is the failure of the circulatory system to deliver oxygen to either the brain or the heart. A failure of the blood supply to the heart is called a **heart attack,** and a failure of the blood supply to the brain is called a **stroke.** A failure can result from damage to the blood vessels either by blockage or breakage.

In air-breathing vertebrates such as ourselves, oxygen enters the body and then the blood by way of the lungs. Blood from the heart flows into the lungs through the pulmonary arteries. As in other tissues, the arteries in the lungs divide into arterioles and then into a large network of capillaries.

In the lungs, oxygen moves from lung tissue *into* the capillaries and binds to the hemoglobin molecules in the red cells (Figure 37-14). Elsewhere in the body, oxygen moves *out of* the capillaries to the body's tissues. The blood in the lungs also releases carbon dioxide—the waste from cellular respiration—that has accumulated during the blood's passage through the rest of the body.

The refreshed blood—low in carbon dioxide and high in oxygen—then returns to the heart by way of the pulmonary veins. The pulmonary veins are the only veins in the body that carry red, oxygen-rich blood. From the heart, the oxygen-rich blood moves through the other loop of the circulatory system, releasing oxygen to respiring cells.

Oxygen in the lungs moves into the blood in the capillaries, where it is transported to the heart in the pulmonary veins and then from the heart to the rest of the body.

WHAT MAKES THE HEART BEAT?

A healthy heart strictly coordinates the contractions of the atria and ventricles. In a resting human adult, the heart beats about 70 times per minute. Each beat consists of a cycle of contractions by the different chambers of the heart: each contraction begins in the right atrium, with the left atrium contracting almost simultaneously; the two ventricles contract in concert, shortly thereafter.

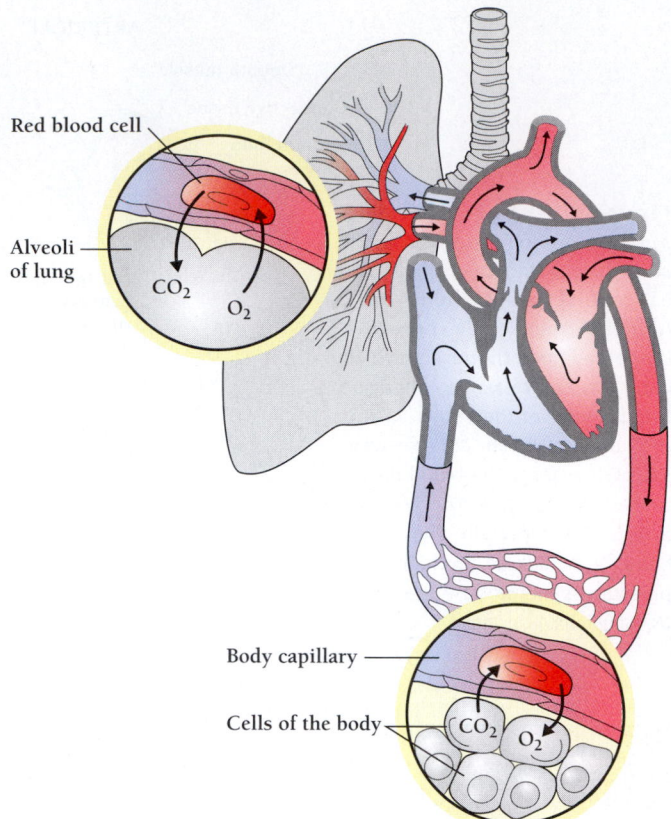

Figure 37-14   Movement of oxygen and carbon dioxide. In the lungs, oxygen moves from the tissues of the lung into the capillaries. Elsewhere, oxygen moves from the capillaries into the tissues of the body. Simultaneously, carbon dioxide moves from the tissues into the capillaries and from the capillaries into the lungs.

Systole and Diastole

During half of each cycle, both the atria and the ventricles are relaxed. This period is called **diastole** [Greek, *dia* = between + *systellein* = to contract]. During the other half of the cycle, the atria contract, followed by the ventricles. The period of contraction lasts about 0.3 second and is called **systole** [Greek, *systellein* = to contract]. During systole, the blood pressure momentarily increases. Blood pressure is therefore always expressed as two numbers, referring to the pressure generated during systole and during diastole. The units of pressure are mm Hg—the height (in millimeters) of a column of mercury (Hg) that could be supported by that pressure. (Atmospheric pressure, approximately 15 psi, can support a column of mercury that is 760 mm in height.) A normal blood-pressure reading for a person at rest is 120/80 mm Hg (read as "120 over 80"), corresponding to 2.4 psi and 1.6 psi. The higher number is the systolic blood pressure and the lower number is the diastolic blood pressure.

Within the heart, a system of valves ensures that the blood flows in one direction only. The valves between the atria and ventricles are called **atrioventricular (AV) valves;** those between the ventricles and the arteries are called **semilunar** [Latin, halfmoon] **valves.** Each valve consists of flaps of con-

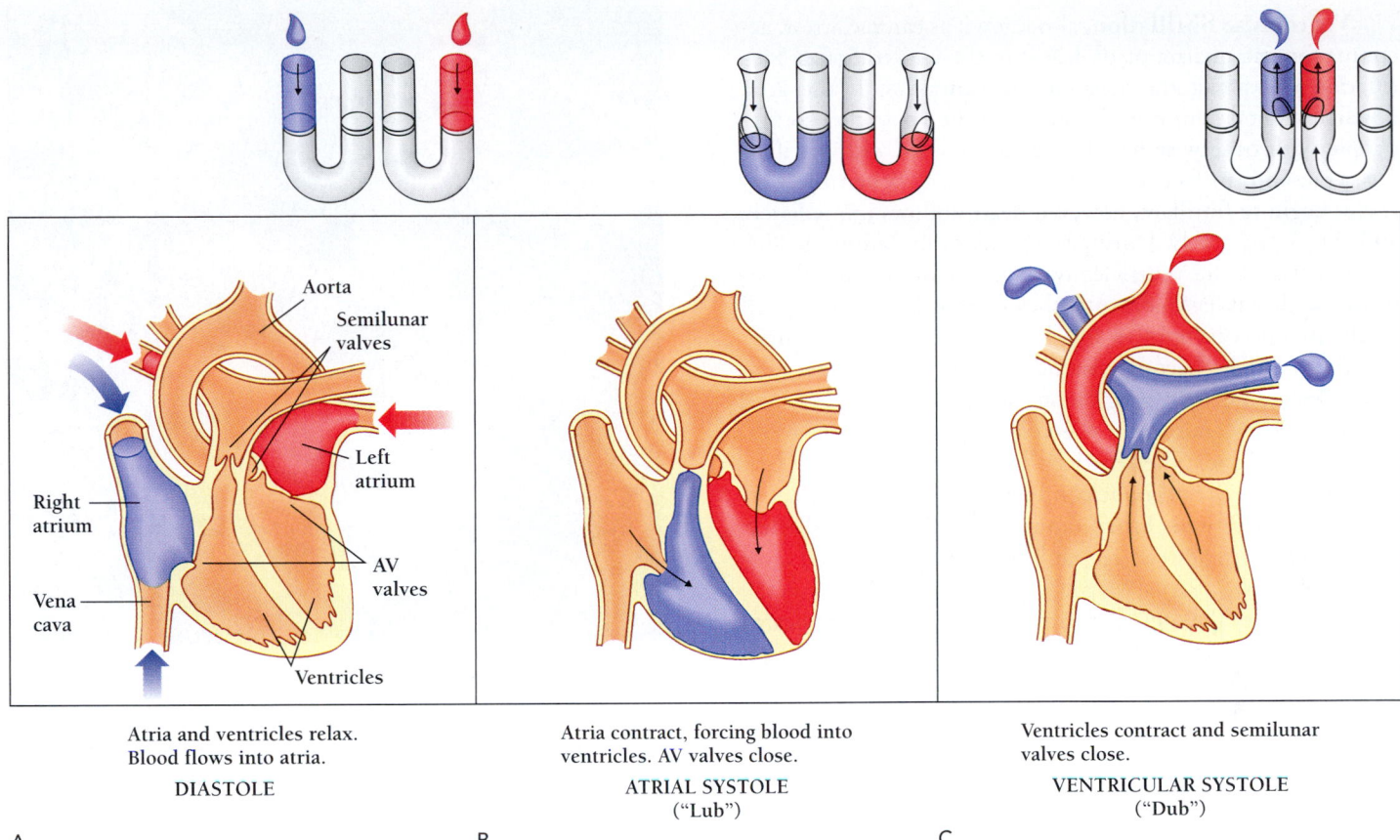

Atria and ventricles relax.
Blood flows into atria.
DIASTOLE

Atria contract, forcing blood into
ventricles. AV valves close.
ATRIAL SYSTOLE
("Lub")

Ventricles contract and semilunar
valves close.
VENTRICULAR SYSTOLE
("Dub")

A.   B.   C.

Figure 37-15   What's that sound? The heart beats in three parts. A. During diastole, the atria and ventricles relax and blood fills the atria. The semilunar canals are still closed from the last beat. B. *Lub (atrial systole)*. The atria contract, forcing the blood into the ventricles, and the atrioventricular valves close behind the blood, making a low-pitched "lub." C. *Dub (ventricular systole)*. During ventricular systole, the ventricles contract, forcing the blood into the aorta and the pulmonary artery. The semilunar valves close noisily behind the blood, producing the "dub."

nective tissue, which allow fluid to move in one direction but not the other. Blood from the ventricles cannot reenter the atria, and blood from the arteries cannot reenter the ventricles (Figure 37-15). Smaller valves in the veins ensure that blood flows in only one direction through the circulatory system.

You can hear the heart working if you listen to your own heart with a stethoscope or to a friend's heart by putting your ear on his or her chest. You will hear that each heartbeat consists of two separate sounds—a low-pitched "lub" and a higher-pitched "dub." The first sound is that of the two ventricles contracting and the atrioventricular valves closing (Figure 37-15). The second is the closing of the semilunar valves behind the blood that the ventricles have expelled.

A trained person can detect some heart abnormalities just by listening. A defective valve, for example, can produce a "heart murmur," a low, hissing sound in addition to the "lub-dub." This sound comes from the movement of blood in the wrong direction, for example, from a ventricle back into the atrium. Such a defect may result from abnormal development of the valve or from an infection such as rheumatic fever.

The period when the muscles of the heart contract is called systole and the period in between, when they relax, is called diastole. One-way valves allow blood to flow from the atria to the ventricles but not back, and from the ventricles to the arteries but not back.

What Coordinates the Beating of the Heart Muscle?

The action of the ventricles in pumping blood to the lungs and the rest of the body is essential to life. To pump blood, the muscle fibers of the ventricles must all work together—contracting together, relaxing together, and pausing together. Occasionally, the muscles of the heart begin working against one another, so that many of the fibers are contracting when they should be relaxed. The result is **fibrillation,** continuous, disorganized contractions.

Ventricular fibrillation, also known as cardiac arrest, accounts for one-quarter of all deaths in the United States. Fibrillation can affect the atria or the ventricles. If the atria fibrillate, blood continues to flow into the ventricles, so that people with otherwise healthy hearts can live with fibrillating atria, and, in fact, it may go unnoticed. When the ventricles begin to fibrillate, however, death follows immediately and almost inevitably. During ventricular fibrillation, the different parts of the ventricles no longer contract simultaneously. Within a few moments, they cannot pump any blood at all. Blood continues to fill the ventricles from the atria, and the ventricles become distended with blood. Because the ventricles are not pumping blood, however, their own blood supply, which comes from the coronary arteries, is cut off. Within 90 seconds, the ventricles become so weak from lack of oxygen that they cannot contract at all. A strong electric shock can sometimes bring about *defibrillation:* fibrillation stops, and ventricles can reestablish normal, organized contractions.

The cells of the muscular myocardium perform the heart's work. Like skeletal muscles, heart muscles are striated, and contraction depends on sliding protein filaments, actin and myosin (Chapter 34). The organization of heart muscle differs, however, from that of skeletal muscle. Whereas skeletal muscle fibers consist of single cells with multiple nuclei, heart muscle fibers consist of many cells, each with its own nucleus. Heart muscle cells are nonetheless intimately connected with one another, through specialized structures called *intercalated discs* (Figure 37-16), which join two cells end to end. An intercalated disc contains both *desmosomes,* which reinforce the muscle against mechanical stress, and *gap junctions,* which allow ions and small molecules to pass freely, so that the cells behave nearly as one cell (Chapter 4).

All the cells of the atria are interconnected in this way, as are all of the cells of the ventricles. If one muscle fiber in the ventricles is stimulated to contract, all of the other cells in the ventricles also contract, in a wave starting from the first fiber. The same is true of the atrial cells. But the atrial cells are physically separated from the ventricle cells by fibrous tissue, so stimulating the atrial cells to contract does not necessarily cause the ventricles to contract. The two groups of cells are, however, connected by a special tissue called the **atrioventricular (AV) node,** which conducts electrical signals from one part of the heart to the other.

As in the case of skeletal muscle, the immediate stimulus for contraction in cardiac muscle is an increase in calcium ion concentration. This increase depends on electrical changes across the cell membrane. In skeletal muscle, these electrical changes depend on stimulation by nerves. In cardiac muscle, however, the cells contract spontaneously.

About 1 percent of the cells of the myocardium are capable of spontaneous, rhythmic electrical activity, made up of patterned flows of ions across their plasma membranes. These cells do not contract themselves, but they stimulate contractions in the cells to which they connect. One group of these cells forms a structure called the **pacemaker,** whose frequency

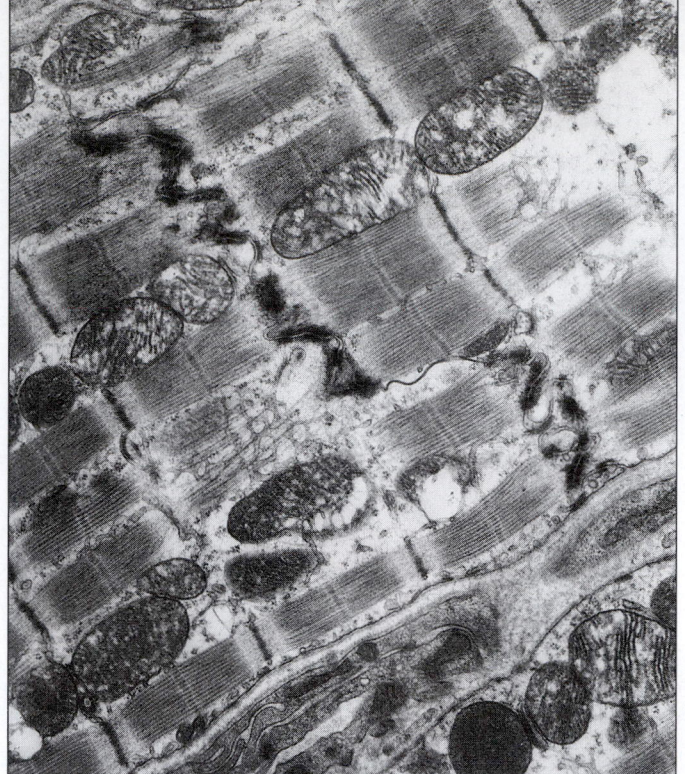

0.5 µm

Figure 37-16 Intercalated discs. Heart muscle cells form close contacts through gap junctions at these specialized structures, ensuring that the connected cells contract simultaneously. Desmosomes strengthen the connections, allowing them to withstand repeated mechanical stresses. *(Biophoto Associates/Photo Researchers, Inc.)*

of electrical activity is slightly higher. By initiating contraction before other cells in the heart, the pacemaker cells set the pace of contraction for the rest of the heart.

In the mammalian heart, the pacemaker lies near the top of the right atrium and is called the **sinoatrial (SA) node.** Electrical signals travel from the SA node to the left atrium and to the AV node. The AV node acts as a relay station and triggers contraction of the ventricles, transmitting its electrical activity throughout the ventricles through specialized gap junctions. The pacemaker thus determines the rhythm of the heart and coordinates the action of atria and ventricles.

Ordinarily, the SA node admits pulses of calcium ion at a frequency of 70 times per minute, thereby setting the resting heart rate. If the SA node is damaged, however, the AV node can take over as a pacemaker, but at a slower rate—setting the heart rate to about 50 beats per minute. If both nodes are damaged, then other conducting cells set the frequency at only 30 beats per minute, too slow to support oxygen delivery for anything like a normal life. Fortunately, *artificial pacemakers* can stimulate more rapid contractions, allowing otherwise incapacitated people to lead active lives.

The cells of the atria contract almost as one, as do the cells of the ventricles. Electrical signals for such contractions are conducted between the two parts of the heart by the atrioventricular (AV) node. A pacemaker, called the sinoatrial (SA) node, sets the rhythm for the heart's contractions.

Blood Pressure Depends on the Action of the Heart and the Properties of Blood Vessels

As William Harvey deduced, the pumping of the heart generates pressure that drives blood through the circulatory system. This pressure in the arteries changes from moment to moment as the heart contracts and relaxes. We can feel the fluctuations of pressure in the larger arteries, such as in the one that crosses the wrist, where we can feel our pulse. Each beat of the pulse corresponds to an expulsion of blood from the heart (systole). Medical workers usually measure blood pressure in the arteries of the upper arm, at the same level as the heart.

The pressure within the vessels decreases as the arteries divide into finer and finer tubes. In the capillaries, the pressure barely varies, and capillary blood flows smoothly. The capillary blood is still under pressure, however. This pressure drives some of the plasma from the capillaries into the surrounding extracellular space. Blood in the veins is under still less pressure and fluctuates hardly at all (Figure 37-17).

NERVES, GLANDS, AND OTHER TISSUES REGULATE THE FLOW OF BLOOD

The amount of blood pumped by the heart (the cardiac output) depends on two factors: (1) the **heart rate,** the number of beats per minute, and (2) the **stroke volume,** the volume of blood delivered by each ventricle. Because the blood travels in a circuit, the volume that passes through the left ventricle must equal the volume that passes through the right ventricle. In a resting adult, the heart beats about 70 times a minute with a stroke volume of about 70 ml. So the volume of blood pumped each minute is about 70 × 70 ml = 4900 ml, or about 5 liters. This is approximately the total volume of blood. Each of us, while sitting still, pumps nearly all of our blood through the circulatory system once a minute.

When we exercise, both heart rate and stroke volume increase, and the amount of blood pumped by the heart may go as high as 35 liters (about 9 gallons) per minute. Our physical and emotional states affect both the heart rate and the stroke volume. Running up stairs will dramatically increase both heart rate and stroke volume. But subtler influences, such as a disturbing thought, can also influence the heart. Both nerves and hormones regulate the beating of the heart.

The heart varies the amount of blood it pumps depending on our activities and moods. The amount of blood pumped is a function of how fast the heart beats and how much blood it moves with each beat.

Regulation by Nerves and Hormones

Two systems of nerves influence heart rate—the **sympathetic** [Greek, *sym* = together + *pathos* = suffering] **nervous system** and the **parasympathetic** [Greek, *para* = beside, next to + sympathetic] **nervous system.** Sympathetic nerves mobilize

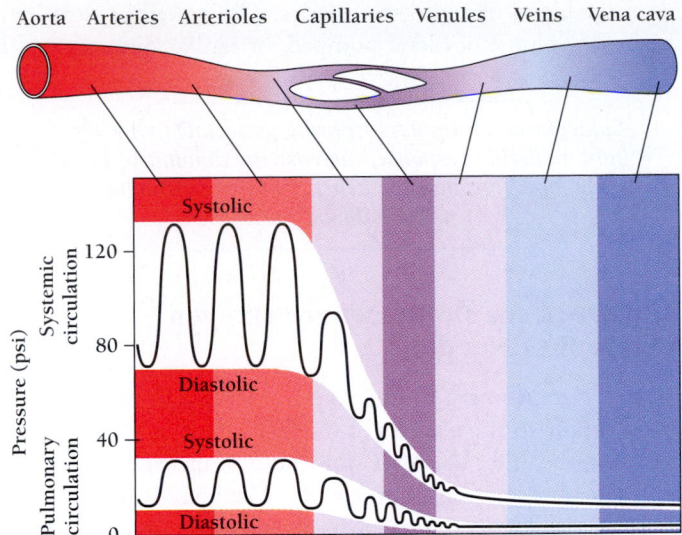

Figure 37-17   Pressure in the blood vessels. Blood pressure is highest in the aorta as the blood leaves the ventricles. In the arterioles and capillaries, the mean pressure drops from 100 to about 20 pounds per square inch. Finally, as the blood flows into the veins, it is under virtually no pressure at all.

the body in times of stress, speeding up the heart and narrowing capillaries. A sympathetic nerve acts on the heart's pacemaker by secreting a small molecule called **norepinephrine.** Parasympathetic nerves slow the heart by secreting a different molecule, **acetylcholine.**

Both norepinephrine and acetylcholine are **neurotransmitters,** molecules that transmit signals from nerve cells either to other nerve cells or to muscles or glands. The rate at which the heart beats depends on the action of norepinephrine and acetylcholine. In the absence of either neurotransmitter, the heart beats about 100 times per minute. Under the influence of acetylcholine, as in a person at rest, the heart slows to about 70 beats per minute. Under the influence of norepinephrine, the heart may beat more than 100 times a minute.

A third molecule, **epinephrine,** which is closely related to norepinephrine, also speeds the heart. Epinephrine (also known as adrenaline because it is produced by the adrenal glands) is a hormone, carried to its targets by the blood (Chapter 36). Epinephrine is responsible for the "fight or flight" response, which, in vertebrates, characterizes an animal's behavior in times of danger.

Both epinephrine and norepinephrine increase stroke volume. They do so directly and also indirectly—by speeding the contraction of the ventricles. The two molecules both increase stroke volume by causing the smooth muscles of the veins to contract. As the veins narrow, the blood flowing through them comes under pressure, increasing blood pressure. The blood gushes into the heart with much greater force, filling the heart more thoroughly than before. The heart becomes distended, which stretches the muscle fibers and increases the strength of their contraction. Because of the increase in blood pressure, then, more blood flows into the heart, which reacts by pumping more blood out into the arteries. The net effect is to increase the volume of blood pumped, or cardiac output.

Epinephrine and norepinephrine, associated with the "fight or flight" response, increase the amount of blood pumped through the heart, while acetylcholine reduces blood flow.

How Does the Body Deliver Oxygen Where It Is Needed?

When we are running to catch a bus, we need to increase the blood supply to the muscles of the legs as well as to the heart itself. On the other hand, we don't need to increase the blood supply to the brain or the intestines. The body is remarkably good at delivering oxygen-rich blood where it is needed. Changes in the diameters of specific arterioles route the blood to the specific organs or muscles that are most active at a given time. In fact, at any moment, only 5 to 10 percent of all the capillaries in the body are fully open to the passage of blood. The local constrictions of arterioles restrict flow into the other 90 to 95 percent of the capillaries. How does the body regulate the opening and closing of arterioles?

One form of control is highly localized and depends on the concentrations of oxygen, carbon dioxide, hydrogen ions, and other molecules that reflect metabolic activity. When the oxygen level in a tissue begins to drop, arterioles in that tissue dilate. Ordinarily, the smooth muscles that surround the arterioles maintain a certain "tone," or normal level of contraction. Decreased oxygen in the surrounding tissue reduces the contraction of the arteriole's smooth muscle. The result is **vasodilation,** the opening of the arterioles, which allows more blood to flow and more oxygen to arrive. Conversely, high oxygen levels stimulate **vasoconstriction,** the contraction of the arteriole's smooth muscle and the reduction of blood flow and oxygen delivery.

Another signal that causes vasodilation comes from the endothelium, the single layer of cells, under the smooth muscle, that lines the blood vessels. This signal, originally called "endothelium-derived relaxation factor" ("EDRF"), is now known to be the gas nitric oxide (NO), which we discussed in Chapter 2.

Blood flow is also regulated by the nervous system and by hormones and other substances. The sympathetic nervous system, for example, can rapidly override local control of arterioles by releasing norepinephrine, which causes vasoconstriction.

The effects of the sympathetic nerves vary from place to place in the body, according to physical and emotional states. For example, fear increases norepinephrine release in the area of the skin arterioles, causing the arterioles to contract and the skin to pale. A fever, or just an embarrassing moment, can decrease the activity of the same sympathetic nerves, causing vasodilation. In that case, blood flow increases and the skin reddens. If the temperature of the body rises, the sympathetic nervous system dilates the arterioles in the skin, allowing them to carry more blood to the surface of the body. Here the blood loses some of its heat and the body cools off.

Many substances influence the expansion and contraction of the vessels of the circulatory system. These include several that we will encounter in later chapters, including prostaglandins, bradykinins, angiotensin, and vasopressin. Anyone who has ever suffered from an allergy, an insect bite, or even poison ivy may be familiar with one of those substances—**histamine,** a small molecule released by cells in damaged tissues, which dilates capillaries and increases their tendency to leak fluid. Histamine increases blood flow into the damaged region, but it also causes the swelling that we associate with a stuffy nose, a mosquito bite, or hives. The increased blood supply provides nutrients that sustain the process of tissue repair as well as white blood cells that combat infection. Antihistamines, available in any drug store, provide relief by preventing histamine from acting.

The body regulates blood flow to different parts of the body (1) by means of local control over vasodilation and vasoconstriction, which is mediated by local fluctuations in the concentration of oxygen, carbon dioxide, and other molecules, and (2) by means of central control mediated by the nervous and endocrine systems.

In this chapter we have discussed the importance of blood in helping to provide and maintain a constant internal environment. We have seen how William Harvey correctly deduced the mechanical workings of the human heart and circulatory system, and we have touched on some of the factors that control the rate of blood delivery. In the next chapter, we learn how the lungs take up oxygen.

Study Outline with Key Terms

The blood, the **heart,** and the blood vessels together make up the **cardiovascular system.** The route of the blood is called the **circulation.** The route through the lungs is called the **pulmonary circulation,** and the route through the rest of the body is called the **systemic circulation.**

In humans, the heart is about the size of a fist and is located inside a fibrous sac called the **pericardium** in the chest, just beneath the **sternum.** The muscular walls of the heart—the **myocardium**—attach to the pericardium to form four chambers.

Several large arteries and veins attach directly to the heart—the **aorta,** the superior and inferior **venae cavae,** the **pulmonary artery,** and the **pulmonary veins.** Each half of the heart contains two muscular sections—a thin-walled **atrium** and a thick-walled **ventricle.**

The blood of invertebrates lacks cells, but the blood **plasma** of vertebrates contains large numbers of **erythrocytes, leukocytes,** and **platelets.** The **hemoglobin** in vertebrate red blood cells enables these cells to take up oxygen efficiently and also gives blood its bold red color.

Arthropods and mollusks have **open circulation,** and the blood travels part of its circuit through open **sinuses.** Vertebrates, as well as earthworms and many other invertebrates, have **closed circulation,** in which blood runs only within enclosed vessels.

Arteries carrying blood from the heart branch to form smaller **arterioles,** which branch to form extensive **capillary** beds, which converge to form **venules,** which converge into **veins,** which pour blood back into the heart.

Dead and foreign cells and proteins in the **extracellular fluid** flow into the **lymphatic system.** In the **lymph nodes,** wastes are filtered from the **lymph** and ingested by **lymphocytes.**

Each blood vessel consists of a thin layer of cells called the **endothelium,** which surrounds the hollow **lumen.** The endothelia of **capillaries** are only one cell thick, while the walls of the larger vessels are thicker and surrounded by elastic fibers, collagen, and smooth muscle. If the collagen sheath of an artery fails and ballooning occurs, the result is an **aneurism.**

The period when the muscles of the heart are contracted is called **systole,** and the period in between, when they are relaxed, is called **diastole.** One-way valves, called the **atrioventricular (AV) valves,** allow blood to flow from the atria to the ventricles but not back. The **semilunar valves** allow blood to flow from the ventricles to the arteries but not back.

The cells of the atria contract almost as one, as do the cells of the ventricles. Electrical signals for such contractions are conducted between the two parts of the heart by the **atrioventricular (AV) node.** A **pacemaker,** called the **sinoatrial (SA) node,** sets the rhythm for the heart's contractions.

A failure of the blood supply to the heart is called a **heart attack,** and a failure of the blood supply to the brain is called a **stroke. Fibrillation** is the disorganized contraction of the muscles of the heart. **Ventricular fibrillation,** also known as cardiac arrest, kills one-quarter of all persons in the United States.

The amount of blood that the heart pumps varies depending on activities and moods. The amount of blood pumped is a function of how fast the heart beats and how much blood it moves with each beat. The amount of blood pumped by the heart (the cardiac output) depends on two factors: (1) the **heart rate,** the number of beats per minute, and (2) the **stroke volume,** the volume of blood delivered by each ventricle.

The **sympathetic** and **parasympathetic nervous systems** both influence heart rate. Sympathetic nerves mobilize the body in times of stress, speeding up the heart—by secreting **norepinephrine** around the cells of the heart's pacemaker. Norepinephrine also causes **vasoconstriction**—the contraction of the smooth muscles surrounding the arterioles. Heat, illness, or embarrassment can all decrease the activity of the sympathetic nerves, allowing **vasodilation.** Parasympathetic nerves slow the heart by secreting **acetylcholine.**

Norepinephrine and acetylcholine are **neurotransmitters,** while **epinephrine,** which acts like norepinephrine, is a hormone. **Histamine** dilates capillaries and increases their tendency to leak fluid, increasing both blood flow and swelling.

Review and Thought Questions

Review Questions

1. What advantages do you think an open circulatory system has over a closed system?
2. What advantages does a closed circulatory system have over an open one? What are the advantages of double circulation?
3. How would you expect your blood pressure to vary
   a. if it is measured in your leg rather than in your arm?
   b. after strenuous exercise?
   c. after severe blood loss?
   d. after you have been frightened?
   e. after a big meal?
   f. after you stand up following a nap?
4. Compare the action of acetylcholine on muscular contraction in your arm and in your heart.
5. Why does an isolated heart keep beating, but an isolated arm muscle does not contract by itself?
6. How do the structures of the walls of arteries, veins, and capillaries relate to the role of each in the circulation?
7. Why does the heart need separate vessels to provide it with oxygen and nutrients?
8. Why does blood flow in one direction only?
9. What determines the direction of blood flow?

Thought Questions

10. Suppose you cut a vessel in your arm. How can you tell whether you have severed a vein or an artery?
11. Do you think William Harvey's conservative tendencies sped the acceptance of his ideas on circulation? Or did his conservative ways hinder their acceptance? Why?

About the Chapter-Opening Image

A blood poster from World War II. Charles Drew developed methods for shipping blood plasma overseas, saving thousands of lives during the war.

 On-line materials relating to this chapter are on the World Wide Web at **http://www.harcourtcollege.com/lifesci/aal2/**

38

How Do Animals Breathe?

Stanton Glantz and the Tobacco Industry

In 1995, a committee of the U.S. Congress singled out a scientist from among tens of thousands of others and canceled his research grant, astonishing medical researchers all over the United States. For Congress to assail a particular scientist in this way was virtually unheard of.

Normally, Congress gives the different government-funded research agencies a budget to distribute at their discretion. These agencies include the National Science Foundation, NASA, the National Institutes of Health (NIH), and several others. The National Cancer Institute (NCI), by far the largest of the institutes at NIH, had awarded a 3-year grant to Stanton Glantz, a professor of medicine at the University of California, San Francisco, to study the effects of public policy on tobacco use.

In particular, he had focused on how the tobacco industry works to keep Americans smoking. He then published a series of scientific papers in *JAMA,* the prestigious *Journal of the American Medical Association,* describing how the tobacco industry had for 30 years carefully concealed its knowledge that tobacco products are both deadly and addictive. Glantz's research also showed that elected officials who receive campaign contributions from tobacco companies were statistically more likely to vote in ways that helped the tobacco industry, a finding not likely to please some elected officials.

Skull with cigarette, by Vincent van Gogh, 1885/Art Resource, New York

The Appropriations Committee of the U.S. House of Representatives argued that the NCI had no right to fund Glantz's work because it involved social science and political science. But Glantz argued that the tobacco industry was much like the mosquitoes that carry the disease malaria from person to person. Researchers say that a mosquito is a "vector" for malaria, and Glantz insisted, "You need to understand the vector of a disease in order to control it. The tobacco industry is the vector for lung cancer and heart disease."

For Glantz, the cancellation of his grant was only the latest in a series of conflicts with the tobacco industry. Much of the turmoil had begun on May 12, 1994, when a mysterious box arrived at Glantz's office. The box contained 4000 pages of secret internal memos and research reports from a major tobacco company. The documents, Glantz later learned, had been stolen from the company by a paralegal who was astonished by what he was reading. The secret papers proved that the company, Brown and Williamson, had known since the 1960s that smoking is both deadly and addictive.

Brown and Williamson immediately went to court for the return of the documents, demanded the names of any people who had read them, and, allegedly, ordered a stakeout of the university library, where the documents were stored.

But it was too late. Copies of the documents had also arrived at several major newspapers and television networks, and the story became national news. A judge eventually ruled that the papers were in the public domain. In July 1995, the University of California at San Francisco posted the Brown and Williamson documents on the Internet, and Glantz's scientific papers analyzing the contents of the documents began appearing in *JAMA*. Within weeks of the first paper's publication, the House Appropriations Committee canceled Glantz's grant.

Why Was the Tobacco Industry Fighting So Hard?

If Brown and Williamson fought hard for the return of their papers, they had good reason. For the makers of Kools, Viceroys, and other popular cigarettes, the documents were acutely embarrassing. But more important, the documents could (and would) be used to attack the whole tobacco industry. A multibillion-dollar American industry, one that remains central to the economies of a half-dozen states, was in serious danger.

Until the late 19th century, tobacco smoking was viewed as a dirty habit, and only a few people smoked, nearly all of them men. In the 1880s, however, tobacco manufacturers discovered a way to process tobacco that made it possible to inhale without coughing. This allowed smokers to inhale deeply,

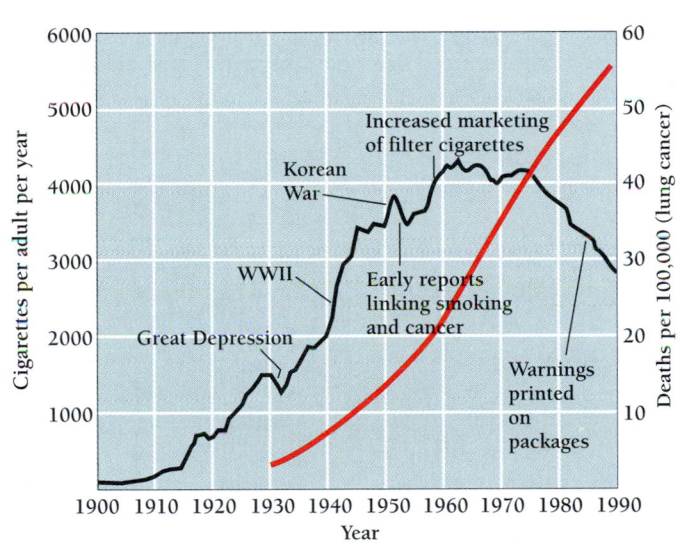

Figure 38-1 A rising death rate. As cigarette sales have increased, so has the death rate from lung cancer. In 1930 the lung cancer death rate was about three cases per 100,000 people. Today, it is nearly 20 times greater.

increasing their exposure to nicotine, which is addictive. For many people, what had once been an occasional cigarette became a regular habit. During World War I, soldiers were given cigarettes to calm their nerves, and smoking and smoking addiction increased among men.

Starting in the 1920s, even women began smoking, encouraged by sophisticated advertising campaigns. In the 1920s, the American Tobacco Company marketed Lucky Strikes cigarettes to women as a "healthy" cigarette that wouldn't "irritate" the throat. Other ads represented Lucky Strikes as symbolizing freedom for women. New York's 1929 Easter Parade featured beautiful, marching models, each brandishing a Lucky Strike cigarette—a "torch of liberty" (Box 38.1).

During World War II, cigarettes were once again distributed to GIs, along with K rations and letters from home. Hollywood movies featured virile actors and alluring actresses lighting one another's cigarettes. Smoking was glamorous, seductive, romantic. More than just accepted, smoking became expected.

By the 1960s, cigarette sales in the United States had skyrocketed from 2 billion cigarettes per year in the 1880s to over 600 billion. And as cigarette sales skyrocketed so, eventually, did lung cancer death rates (Figure 38-1).

As early as the 1880s, a physician published a report suggesting that smoking caused cancer. Back then, when so few people smoked, doctors noticed that patients with cancer of the mouth, throat, or lungs were nearly always smokers. Re-

BOX 38.1

The Green Ball

In the 1930s, the makers of Lucky Strikes found to their dismay that women were passing up the company's cigarettes. The company suspected that the problem might be the cigarette's new green packaging, which clashed with most women's clothes. Green just wasn't popular. What to do?

The American Tobacco Company turned to Edward Bernays, the world's most famous public-relations man. Bernays's solution was to make green the most popular color in the United States. He held a "Green Ball" for well-connected women, and he enticed European fashion designers to create a collection of fashions, all in green. He held a Green Fashions Luncheon for fashion editors from major newspapers and magazines. He invented a Color Fashion Bureau that sent out press releases announcing that green was the color of the season. In only 6 months, Bernays succeeded in making green the top fashion color of 1934. Women no longer hesitated to buy Lucky Strikes because of the color, and sales increased.

Figure A Actress Jean Harlow in 1932—wearing green and smoking Lucky Strikes cigarettes. When this advertisement and others like it failed to attract female smokers, American Tobacco hired promoter Edward Bernays to convince American women that green was the fashion color (and Lucky Strikes, the fashion cigarette). (*The Granger Collection, New York*)

---

searchers now know that smoking causes 90 percent of all cases of lung cancer. In societies that do not smoke, lung cancer is rare.

But the first big surge in lung cancer deaths among American men didn't begin until the 1930s. Because lung cancer takes years to develop, a 20-year lag appears between an increase in smoking and increase in lung cancer deaths. Because women started smoking later in the century than men, they began dying of lung cancer much later, in the 1960s and 1970s (Figure 38-2A).

In fact, the lung cancer death rate of the 1960s and 1970s was a public health disaster far greater in magnitude than that of the AIDS epidemic of the 1980s and 1990s. Doctors were horrified. Lung cancer is among the deadliest of all cancers, accounting for one-third of all cancer deaths (Figure 38-2B). Untreated lung cancer patients live an average of 8 months. Fewer than 15 percent of all lung cancer victims survive even 5 years. For comparison, 80 percent of breast cancer victims survive 5 years, and often much longer.

In 1964, the Surgeon General of the United States issued a report stating that cigarette smoking was hazardous to health. For the first time in 75 years, cigarette sales leveled off. Two years later Congress passed a bill requiring cigarette manufacturers to label their product as hazardous. For the first time ever, cigarette sales actually dropped.

It was a wakeup call for tobacco companies. Desperately, the tobacco industry began trying to develop a "safe" cigarette.

In the meantime, they did what they could to keep people smoking. Billions of dollars in sales and thousands of jobs were at stake.

Publicly, tobacco companies insisted that the scientific evidence for the cigarette-cancer link was weak. They blitzed television viewers with advertisements showing healthy young people smoking cigarettes against scenic backdrops—green meadows, waterfalls, and soaring mountains. The Brown and Williamson documents show, however, that the company's own research confirmed that cigarettes were both addictive and carcinogenic. Indeed, the quality of the tobacco industry's research was superior to that done by university researchers of the time.

In the late 1960s, the Federal Communications Commission (FCC) ruled that broadcasters must air one public service warning of tobacco's hazards for every four cigarette commercials. Viewers were suddenly besieged with shocking ads from the American Cancer Society, and cigarette sales dropped precipitously. Finally, in 1970, to get rid of the antismoking ads on television, cigarette companies pulled their own ads off the air. But by then millions of people had quit smoking. Between 1965 and 1979 alone, the percentage of smokers dropped from 42 percent of all adults to just 32 percent. Today, the numbers are even lower.

In 1988, the Surgeon General issued another report, this time stating that smoking was not only hazardous but also highly addictive, like heroin or cocaine. The tobacco industry vigorously attacked the report, arguing that the scientific evi-

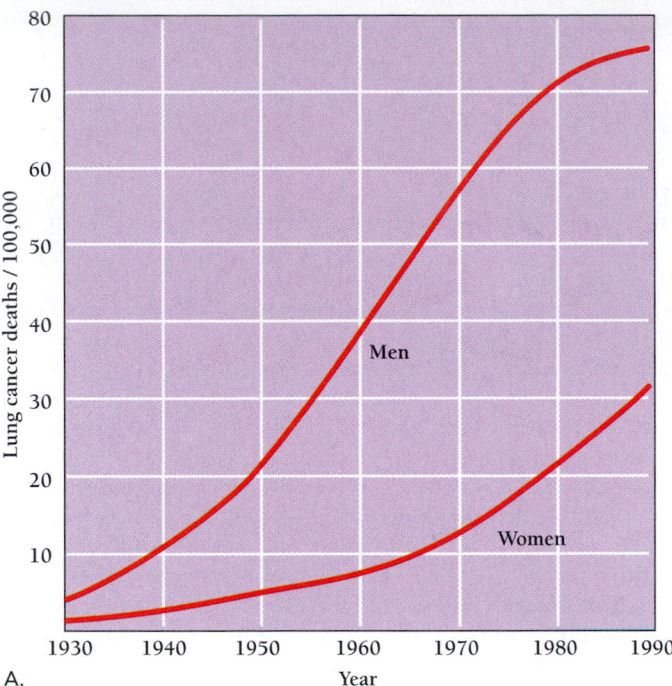

A.

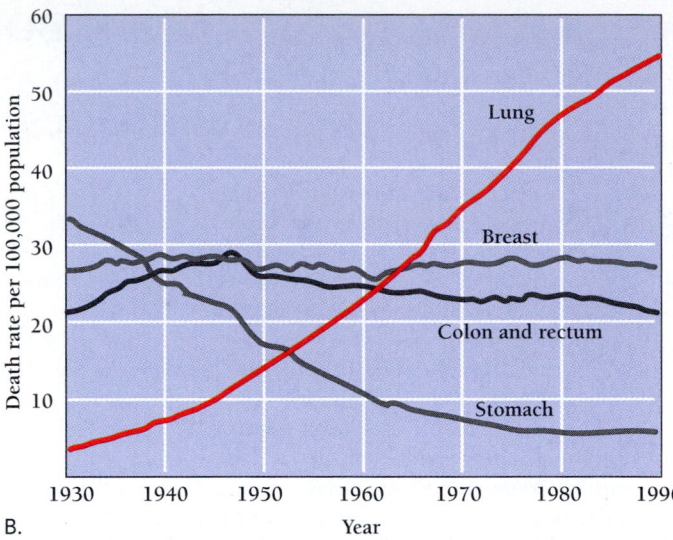

B.

Figure 38-2   Lung cancer's separate path. A. Because men began smoking before women, they also began dying of lung cancer and other diseases sooner. B. Nearly every form of cancer remained stable or declined during the 20th century. Among both men and women, lung cancer alone increased dramatically, as shown in this graph of death rates from different kinds of cancer.

dence for such a conclusion was scanty or nonexistent. In fact, the Brown and Williamson documents show that the company had fully understood the addictive nature of tobacco for at least 25 years.

The "safe" cigarette the industry had sought never materialized. Filters, low-tar, and low-nicotine cigarettes only force smokers to inhale more deeply or to smoke more cigarettes to get the same amount of nicotine. All tobacco smoke interferes with the healthy functioning of both the lungs and the cardiovascular system. In this chapter we will see how healthy lungs (and gills) work to absorb oxygen and rid the body of carbon dioxide and how the circulatory system moves oxygen and carbon dioxide between the lungs and the cells of the body. Along the way, we'll see how the lungs, heart, and blood vessels are all interconnected.

Meanwhile, the U.S. military—whose officers once ordered soldiers to take a rest with the words, "Rest! Light 'em up!"—has instituted a massive program to eliminate all tobacco use from the armed forces (Box 38.2).

HOW DO ANIMALS EXCHANGE CARBON DIOXIDE AND OXYGEN?

Animals usually get most of their energy from cellular respiration, a biochemical process that requires oxygen. Because cellular respiration takes place in almost every cell of the body, animal cells require the means both to acquire oxygen and to distribute it to the cells of the body. In addition, cell respiration generates carbon dioxide, a toxic waste product.

So animals must also be able to rid themselves of carbon dioxide.

Aquatic animals such as fish and many insect and amphibian larvae exchange oxygen and carbon dioxide across the surfaces of gills. Terrestrial animals such as birds, mammals, and reptiles all use lungs. In both gills and lungs, oxygen diffuses passively into the blood. The circulatory system then carries the oxygen to the different tissues of the body. The champion oxygen extractors are birds, some of which can fly at altitudes where human mountain climbers can barely walk.

Terrestrial insects use neither lungs nor gills. Instead, they have a network of open passages called trachea that carry oxygen directly to the tissues and carbon dioxide back out of the body. Insects have blood, which carries glucose and other materials, but, unlike ours, it does not transport oxygen or carbon dioxide.

Lungs and gills take up oxygen and release carbon dioxide.

Can Animals Take Oxygen Directly from the Air?

Oxygen is far more plentiful in the air than in water. In air approximately 1 of every 5 molecules (or 21 percent) is oxygen, whereas in water, oxygen molecules make up fewer than 1 in 170,000 molecules (or 0.0006 percent).

Land animals—particularly insects and land vertebrates—have flourished so well, in part, because they are able to take oxygen directly from the atmosphere. For oxygen to be useful, however, it must first dissolve in water, since only dissolved oxygen can participate in the biochemical reactions of a cell. A land animal's first task, then, is to create a thin, wet surface on which the oxygen can dissolve. We begin, then, with a discussion of the way that oxygen and other gases dissolve in water. Aquatic animals take oxygen that is already dissolved in the water around them. But that oxygen comes from air, so aquatic animals depend on atmospheric oxygen dissolving in water as much as terrestrial animals do.

Oxygen is far more plentiful in air than in water. But both land animals and aquatic animals need oxygen that has dissolved in water.

How Do Gases Exert Pressure?

In a gas such as air, the average distance between molecules is more than ten times the distance between the molecules in a liquid. Because the molecules in a gas are so far apart, they interact very little with one another. Instead, each molecule moves about at random, with an average speed that depends only on the temperature: the higher the temperature, the faster each molecule moves.

When a molecule in a gas strikes the surface of a liquid, it will usually bounce off into the space above the liquid (Fig-

ure 38-3). Each time a gas molecule strikes a surface, it exerts a tiny impulse. All the gas molecules together exert a pressure on the surface of the liquid. The total pressure depends on the number of gas molecules that strike the surface and on the speed with which each molecule strikes the surface (that is, the temperature).

We can measure this pressure using a barometer, a device that measures air pressure. Air pressure varies from day to day and place to place. For example, air pressure is lower at high altitudes than at low altitudes. During a storm air pressure can drop suddenly, hence the phrase, "the barometer is falling."

How much pressure can air exert? The original barometer, invented in the 17th century by Evangelista Torricelli, consisted of a column of mercury (Hg). At sea level, air exerts a total pressure that is enough to support a column of mercury about 760 mm (30 inches) high (Figure 38-3). We say that the pressure is 760 mm Hg, or "760 torr," after Torricelli. In this book, we use mm Hg as our unit of measure.

Each molecule contributes independently to the total pressure. But air is made up of different kinds of molecules. About 21 percent of the molecules are oxygen (O_2), about 0.03 percent are CO_2, and almost all the rest are nitrogen (N_2). Each kind of gas exerts a separate pressure, depending on the concentration and temperature of the molecules. Physiologists describe the pressure exerted by a type of molecule in a gas as the **partial pressure,** the pressure exerted by that one type of molecule (Figure 38-4). The total pressure is the sum of all the partial pressures. In air, the partial pressure of O_2 is about 160 mm Hg, that of N_2 about 600 mm Hg, for a total of about

760 mm Hg. CO_2 makes up such a small part of our atmosphere that its partial pressure is only about 0.2 mm. Even the nicotine in a puff of cigarette smoke has a measurable partial pressure (Box 38.3).

Each kind of molecule in a gas has a different partial pressure.

How Much Dissolved Gas Will a Solution Hold?

We said earlier that when a gas molecule strikes the surface of a liquid, it usually bounces off. Sometimes, however, the gas molecule penetrates the surface and becomes part of the liquid. We say the gas "dissolves" in the liquid. Such a dissolved gas molecule can escape from the solution and become a gas molecule again. When the number of gas molecules entering and leaving the solution is the same, we say the molecules have reached **equilibrium.**

The concentration of a molecule in a solution at equilibrium depends on the partial pressure of the gas and the tem-

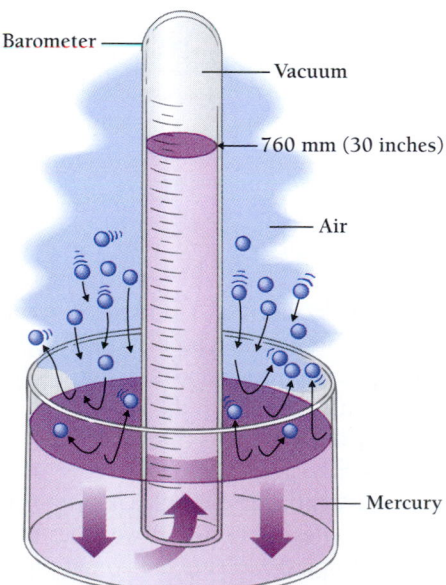

Figure 38-3 A mercury barometer. At sea level, the pressure from the atmosphere can support a column of mercury about 760 mm (30 inches) high. Barometric pressure results from the bouncing of gas molecules on the mercury.

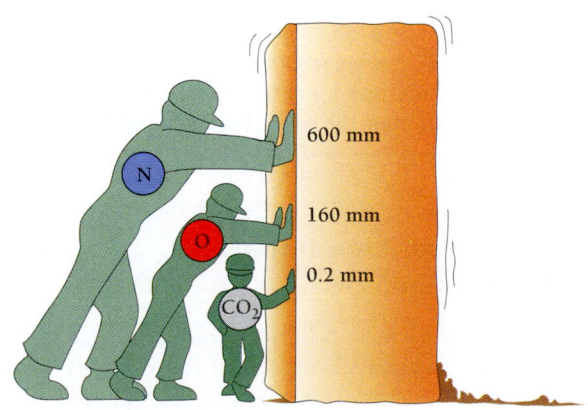

Figure 38-4 Partial pressure. The pressures of the different gases in the atmosphere add up to about 760 mm. Because nitrogen gas contributes about 600 mm Hg of pressure, we say it has a partial pressure of 600 mm Hg. Oxygen in the atmosphere has a partial pressure of about 160 mm. Carbon dioxide gas barely pushes at all, with a partial pressure of just 0.2 mm.

▶ **Figure 38-5 Gills and lungs.** Animals need a high surface-to-volume ratio to speed the uptake of oxygen. A. Organisms smaller than about 1 mm start with high surface-to-volume ratios and need not expand their surface area. B. In larger aquatic animals, gills increase the surface through which oxygen can diffuse. In a sea star, gills result from the evagination of the animal's outer surface. C. The extensive gills of fish are capable of extracting a lot of oxygen in a short time. D. Spiders develop internal gill-like structures, called "book lungs." E. The lungs of amphibians and other terrestrial vertebrates arise by invagination (infolding) rather than evagination (outfolding).

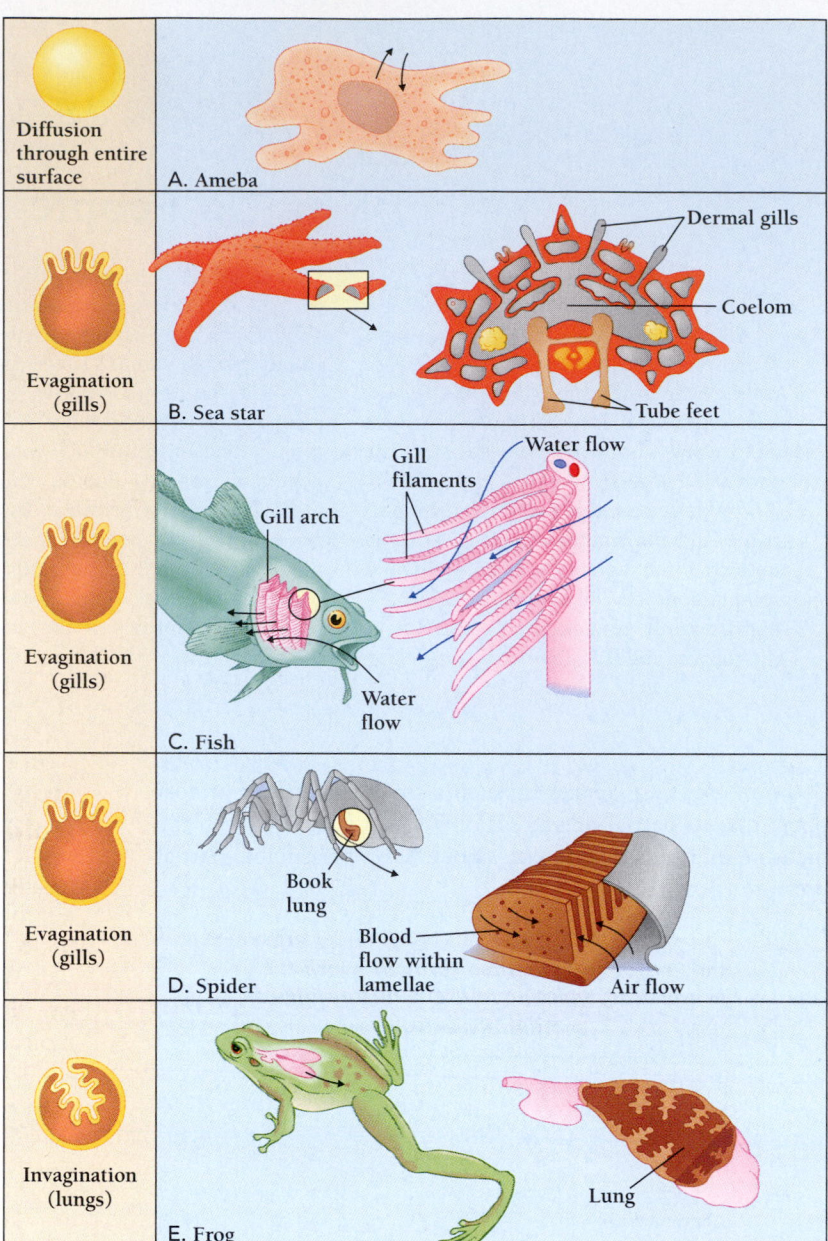

perature of the air and water. Not surprisingly, the higher the partial pressure of oxygen, the higher the concentration of oxygen in the water. In other words, the harder and faster the molecules hit the water, the more of them enter and dissolve. More formally, we say that a gas moves from an area of high partial pressure to an area of low partial pressure.

Less obvious is the effect of temperature. Cooks know that most solids dissolve better at higher temperatures. For example, more sugar will dissolve in hot water than in cold water. But gas molecules are *less* soluble at higher temperatures. So as you heat water, gases come out of solution, forming bubbles. A carbonated drink does the same thing. A cold soda contains dissolved carbon dioxide under pressure that is two to three times greater than atmospheric pressure. When you open a cold bottle of soda, a little of the CO_2 comes out of solution and forms tiny bubbles. If you open a warm bottle, however, much more CO_2 comes out of solution. If the bottle is hot enough, so many big bubbles form that the whole foamy mess overflows the bottle.

Most molecules of a gas bounce off the surface of a solution, but some enter and dissolve.
The amount of dissolved gas in a solution depends on the temperature and the partial pressure of the gas.

Why Is Surface Area Important in Gas Exchange?

If the partial pressure of atmospheric oxygen is high enough, oxygen molecules will pass from the air in our lungs into our blood. But what if the oxygen moves into the blood too slowly to provide all the oxygen we need? How fast does it have to move and what determines how fast it moves into the blood?

If we know temperature and partial pressure, we can estimate the eventual concentration of oxygen in a solution such as blood. What we don't know is how long it will take for the oxygen to reach that concentration. We can restate our ques-

tion and ask, What determines how fast oxygen in solution reaches equilibrium with the gas phase?

The rate at which molecules of a gas can enter a solution depends on three factors: temperature, partial pressure, and surface area. Both the molecules of a gas and the molecules dissolved in a solution move randomly in a process called *diffusion* (Chapter 4). In the lungs, oxygen spontaneously diffuses from a region of high partial pressure (the air[1] in the lungs) to a region of low partial pressure (the blood).

The amount of oxygen that moves across a boundary in a given time (per second, for example) is proportional to the difference in partial pressures on the two sides of the boundary. That is why hospital patients are sometimes given high concentrations of oxygen to breathe; at high concentrations,

[1]The composition of the gas in the lungs is not actually the same as that of air, as there is more water vapor and 40 times more carbon dioxide than in ordinary air.

the partial pressure of oxygen is higher and the oxygen molecules more quickly dissolve in the fluid covering the lungs.

Finally, large surface areas also speed the exchange of molecules. We understand intuitively that water in an open pan will evaporate more quickly than the same amount of water in a narrow-necked bottle. Similarly, oxygen exposed to water in an open pan will dissolve and equilibrate much more rapidly than oxygen exposed to the same amount of water in a narrow-necked bottle.

The same holds true for cells. Being small increases the ratio between surface area and volume, a cell's **surface-to-volume ratio.** That is, small cells have proportionately more surface than big cells. Because of their higher surface-to-volume ratio, small cells can more rapidly absorb all the oxygen they need than big cells can. For really big cells, diffusion occurs too slowly to provide enough oxygen. As a result, most animal cells are no more than 20 to 30 μm in diameter.

Small multicelled animals such as rotifers or nematodes—up to about 1 mm in diameter—can obtain enough oxygen by diffusion alone (Figure 38-5A). But larger animals need special surfaces through which oxygen can diffuse. All animals increase surface area for gas exchange in one of two ways. They either fold a surface outward (evagination) or fold a surface inward (invagination) (Figure 38-5). Evaginated breathing structures are called **gills,** and invaginated breathing structures are called **lungs.**

The rate of gas movement into a cell or an organism depends on the surface area through which the gas can move. Evaginated breathing structures are called gills, and invaginated breathing structures are called lungs.

HOW DO AQUATIC ORGANISMS OBTAIN DISSOLVED OXYGEN?

Aquatic animals larger than about 1 mm obtain dissolved oxygen through gills. A sea star, for example, has tiny protuberances all over its surface, and a sea scallop has specialized gills within its shell (Figure 38-5B). In each case, oxygen diffuses from the surrounding water across the expanded surface area into the animal's internal environment.

As oxygen diffuses from the water into the surface of the gills, however, the surrounding water loses oxygen. To exchange the oxygen-poor water for new, oxygen-rich water, aquatic animals either move their gills or move the water. Moving the gills through the water works well only if the gills are small. Most organisms therefore move water over the gills. In the scallop, for example, cilia on the gill surface move the water past the gill surfaces. Lobsters, crayfish, and crabs all use large internal paddles to move water over their gills.

Dissolved oxygen can diffuse into an organism by way of gills. Many animals replace oxygen-depleted water with oxygen-rich water by pumping water across the gills.

How Do Gills Obtain Enough Oxygen To Power a Fast-Swimming Fish?

The oxygen requirements of a fast-swimming fish are far greater than those of most invertebrates and slower-moving fish. Because fish need more oxygen per minute than, say, a sponge filter-feeding on the ocean floor, fast-swimming fishes have enormous surface areas on their gills. The fast-swimming mackerel, for example, has about 50 times as much gill area as a slow-moving bottom feeder such as a flounder. Like scallops and lobsters, fish move water continuously over their gills. Some fish use muscles and valves in the mouth to pump water over the gills. Others just hold their mouths partly open and swim constantly. Water flows into a fish's mouth and out past the gills.

Fish adjust their behavior according to how much oxygen they need. In one study, researchers kept mackerel in moving water that forced the fish to swim at a constant speed. When the researchers reduced the oxygen content of the flowing water, each fish opened its mouth wider, allowing more water to pass through the gills and more oxygen to be extracted.

The gills of fish are highly organized. On each side of the head lie several gill arches, all covered by a protective bony flap called the "operculum." Each arch carries two rows of gill filaments, and each filament carries many rows of parallel, plate-like lamellae (Figure 38-5C). Water passes through the lamellae in a single direction, and a dense network of capillaries exposes the circulating blood to the oxygen in the water.

A fish moves water over its gills by pumping or by swimming with its mouth wide open.

Why Can't Land Animals Breathe with Gills?

Why do fish die when they are taken from the water? Why can't gills extract oxygen from the air? Actually, some fish survive quite well out of water. An eel, for example, can breathe if it stays cool and moist. Eels can even crawl over land—usually at night and in moist grass—from one body of water to another. But, in air, most of an eel's oxygen comes through its skin, not through its gills.

The problem with gills out of water is that they collapse and flatten. Because gills lack mechanical rigidity, their surfaces fold and stick together when out of water, so that they cannot provide an expanded surface area for the diffusion of gases. Most gills also dry out and lose the water into which oxygen must dissolve.

One solution to this problem is to keep the gills inside the body. Spiders, for example, have stacks of internal plates called a *book lung* (Figure 38-5D). Book "lungs" are not actually lungs, however. They evaginate outward, like the pages of a book, not inward, like a lung. Anatomical comparisons suggest that they evolved from gills. Supporting bars keep the plates from collapsing.

Another solution is to breathe through the skin or some other broad surface. Some kinds of lungless salamanders, for example, breathe entirely through their skin. And some amphibians absorb air through the skin inside their mouths. But most land animals, including most amphibians, have inward-folded lungs (Figure 38-5E).

In air, gills need special support and protection to prevent them from collapsing and drying out.

HOW DO AIR-BREATHING ORGANISMS OBTAIN OXYGEN?

Aside from spiders, air-breathing organisms nearly all obtain oxygen from infolded (invaginated) surfaces that do not collapse easily. Land snails, many amphibians, and all reptiles, mammals, and birds depend on lungs, localized organs of gas exchange that are always associated with the circulation. Insects, on the other hand, depend on *tracheae,* a complex set of tubes, which also arise by invagination, that carry air throughout the animal's body.

Lungs and tracheae are both extensions of tubes within an animal's body. They lie within the body cavity, where they are more protected than gills and less likely to collapse.

Almost all air-breathing vertebrates depend on moist lungs to acquire oxygen.

What Qualities Enable the Lung To Exchange Gases?

We have seen that the surface area of gills limits the oxygen and therefore the energy available to fish and other aquatic animals. In the same way, the surface area of the lungs limits the oxygen and therefore the energy available to terrestrial animals. Birds and mammals, which are endotherms, burn more calories than other animals. Birds and mammals therefore require more energy and more oxygen and have lungs with huge surface areas. Indeed, the surface area of a human lung is about the size of a tennis court.

A lung's ability to absorb oxygen depends not only on its surface area but also on the blood supply that carries oxygen away from the lungs. The more blood that flows through the lung, the more rapidly oxygen diffuses into the blood.

In addition, the membrane between the blood and the gases must be extremely thin. If it is thick, it can block the diffusion of oxygen. In humans, for example, the thickness of the membrane between the air in the lungs and blood is only about 0.5 μm, just 1/15th the diameter of a red blood cell.

Rapid oxygen uptake requires a large, thin surface area and a rich blood supply.

By What Path Does Air Enter the Lungs?

Lungs do not function by themselves. They are one part of a **respiratory system,** which consists of all the structures responsible for the exchange of gases between the blood and the external environment. In vertebrates, the respiratory system consists of the lungs, the airways, and the muscles that move the air (Figure 38-6).

Air enters the body through the mouth and nose. These two openings lead to the throat, or **pharynx,** a common passage for food and air. Food and air soon diverge into two separate branches, however. The *esophagus* leads to the stomach, while the *glottis* leads to the lungs. A flap called the *epiglottis* prevents food from entering the air tube during swallowing.

The glottis leads directly into the **larynx.** The larynx contains the *vocal cords,* folds of membrane that vibrate as air passes over them, producing sound. The larynx leads directly into a long tube, called the **trachea,** or windpipe. The trachea enters the chest, where it forks into two smaller tubes, the **bronchi,** which lead to the two lungs. Inside the lungs, the bronchi branch into smaller and smaller extensions, with thinner walls. The smallest of the bronchi are the **bronchioles,** of which there are millions.

No gas exchange takes place in the larynx, the bronchi, or the bronchioles: they are only passages to the working surfaces of the lungs. The surfaces on which oxygen dissolves are called the **alveoli** [singular, alveolus; from Latin, *alveus* = a hollow], hollow sacs that are richly supplied with blood. Here in the alveoli, hemoglobin molecules in the passing blood pick up oxygen and carry it to the rest of the body.

Each lung of a healthy young adult contains about 300 million alveoli. The fine branching of the air passages and alveoli greatly increases the surface over which oxygen molecules come into contact with cell surfaces. If the inside of each lung were like a smooth, hollow sphere instead of spongy mass, the surface area of the lungs would be about 0.1 square meter (1 square foot). Instead, as mentioned earlier, the lungs have the surface area the same as a standard tennis court, more than 80 times the outer surface area of the entire body.

The mammalian respiratory system consists of finely divided airways leading to hundreds of millions of blind sacs (alveoli) where oxygen is absorbed.

How Do We Breathe?

In mammals, the alveoli are dead ends. Air leaves each alveolus the way it arrived, through a bronchiole. Only by actively breathing can we move air in and out of the alveoli. At rest, before you inhale, your lungs contain about 1500 milliliters (ml) of air. After you exhale, they contain about 2000 ml (2 liters). The amount of air drawn in and then expelled in a single breath is called the **tidal volume.** For a healthy adult

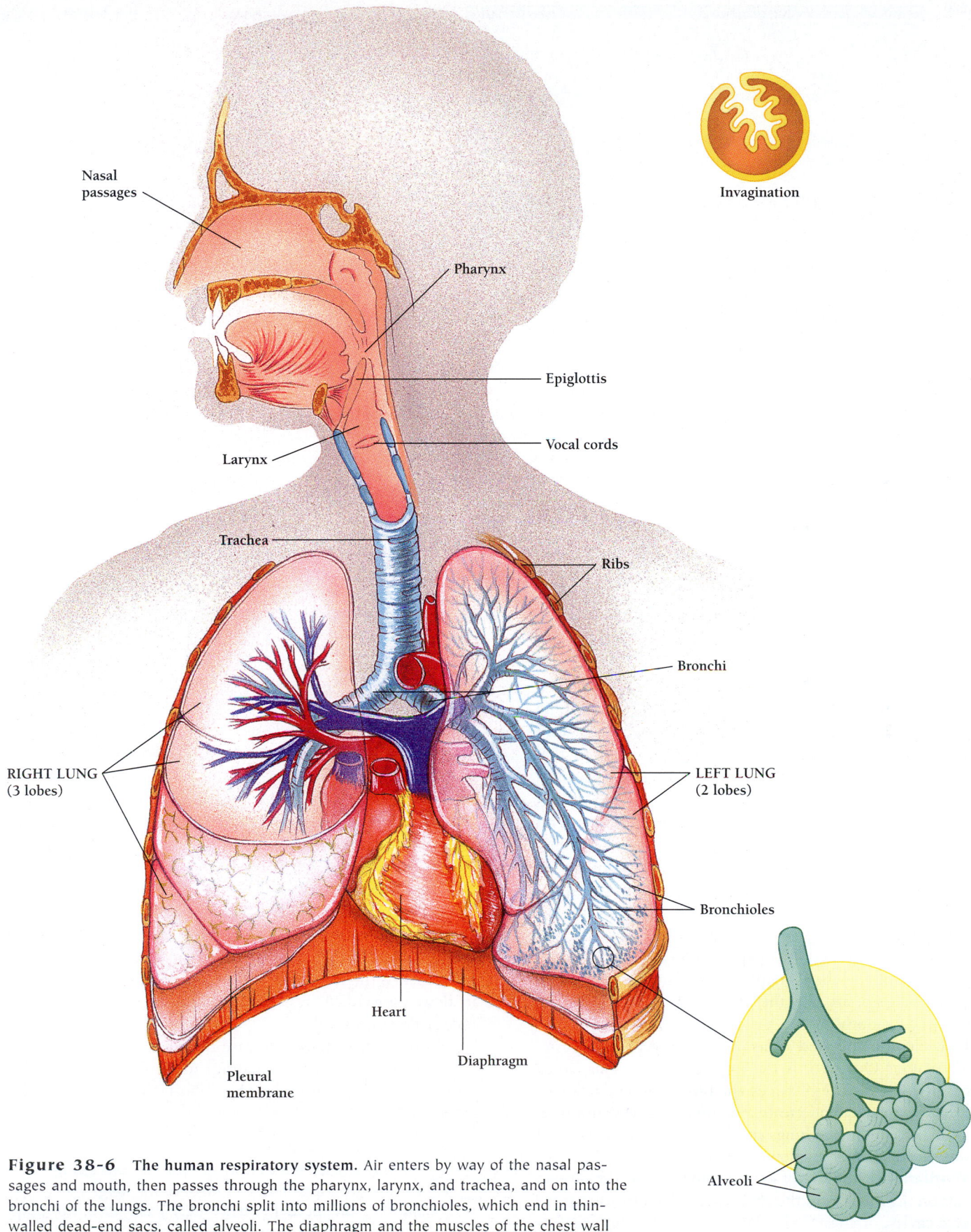

Invagination

Figure 38-6   The human respiratory system. Air enters by way of the nasal passages and mouth, then passes through the pharynx, larynx, and trachea, and on into the bronchi of the lungs. The bronchi split into millions of bronchioles, which end in thin-walled dead-end sacs, called alveoli. The diaphragm and the muscles of the chest wall pull the air into the lungs.

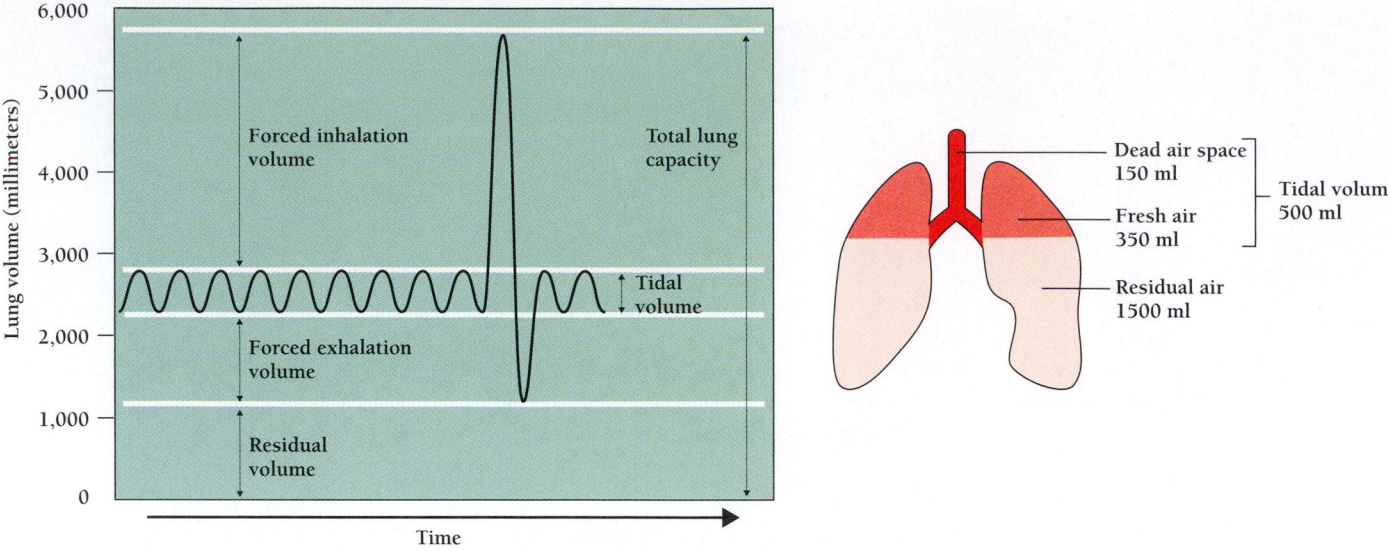

Figure 38-7 **Tidal volume.** At rest, we take small breaths, and the tidal volume of air entering and exiting the lungs is small, no more than 500 ml. Of this, 150 ml of dead air never enters the alveoli, but stays in the trachea, bronchi, and bronchioles. In addition, 1000 to 1500 ml of air in the lungs is residual air, air that is always in the lungs and cannot be emptied out. The harder we exercise, the larger the volume of air we take in and expel. A person in good condition can take in as much as 3000 ml more. All the different kinds of air in the lungs mix together during each breath. The separate layers shown here are for illustration only.

human, the tidal volume at rest is about half a liter (500 ml). During vigorous exercise, the tidal volume can increase to 3 liters or even more (Figure 38-7).

Not all of the 500 ml of air enters the alveoli. The trachea, bronchi, and bronchioles together contain about 150 ml of air. The air in these conducting tubes represents the **dead space** of the airways, that is, the volume of inspired air that does not make it into the alveoli. In a resting person, then, each breath brings into the lungs about 350 ml of fresh air (500 ml tidal volume − 150 ml dead space). This fresh air mixes with the **residual** air remaining within the alveoli after the last breath. The resting lungs can hold 2000 ml of air. Yet after a resting breath, less than 20 percent (350 ml) of that air is fresh air. The other 80 percent is air that has already lost some oxygen.

Mammals, birds, and most reptiles pull air into the lungs using muscles of the chest cavity. Amphibians and some reptiles push, or gulp, air into the lungs using the muscles and valves of their mouths.

How do humans and other mammals generate the required suction? Our trick is to pump it in using our closed chest cavity. The muscles between the ribs of our chest can expand the chest cavity and create a vacuum. The expansion reduces the pressure in the lungs compared with the pressure of the outside air, so that air flows passively inward.

Ventilation, the flow of air into and out of the alveoli, depends on the muscles of the chest cage, which surrounds the **thoracic cavity,** or chest cavity. Surrounding the thoracic cavity are the spinal column, 12 pairs of ribs, and the sternum, or breastbone. Muscles and connective tissue separate the top

of the cavity from the head. The **diaphragm,** a sheet of muscle beneath the lungs, separates the thoracic cavity from the abdomen (Figure 38-6). Muscles and elastic connective tissue also run between the ribs and form the sides of the cavity. The lungs themselves lie within a fluid-filled sac attached to these connective tissues of the chest.

During **inspiration,** or inhalation, the diaphragm contracts and moves downward (Figure 38-7A). This expands the chest cavity and decreases the pressure inside the lungs. At the same time, the rib muscles contract, which expands the chest. Since the pressure of the outside air now exceeds that inside the alveoli, air flows inward.

During **expiration,** or exhalation, the diaphragm relaxes upward into the chest cavity. When it is relaxed, it has a steep dome shape (Figure 38-7B). The ribs resume their relaxed positions, which further contributes to a decrease in the size of the cavity. As the volume decreases, the pressure increases, and air flows outward. At rest, we have only to relax our muscles and air flows outward.

During exercise, we increase the tidal volume by forcefully expelling (and then inhaling) more air. We do this by contracting special muscles that boost the lungs' internal pressure.

Breathing causes tides of air to move in and out of the alveoli. Mammals pull air into the lungs by expanding the thoracic cavity. They push air out of the lungs by relaxing the muscles of the thoracic cavity.

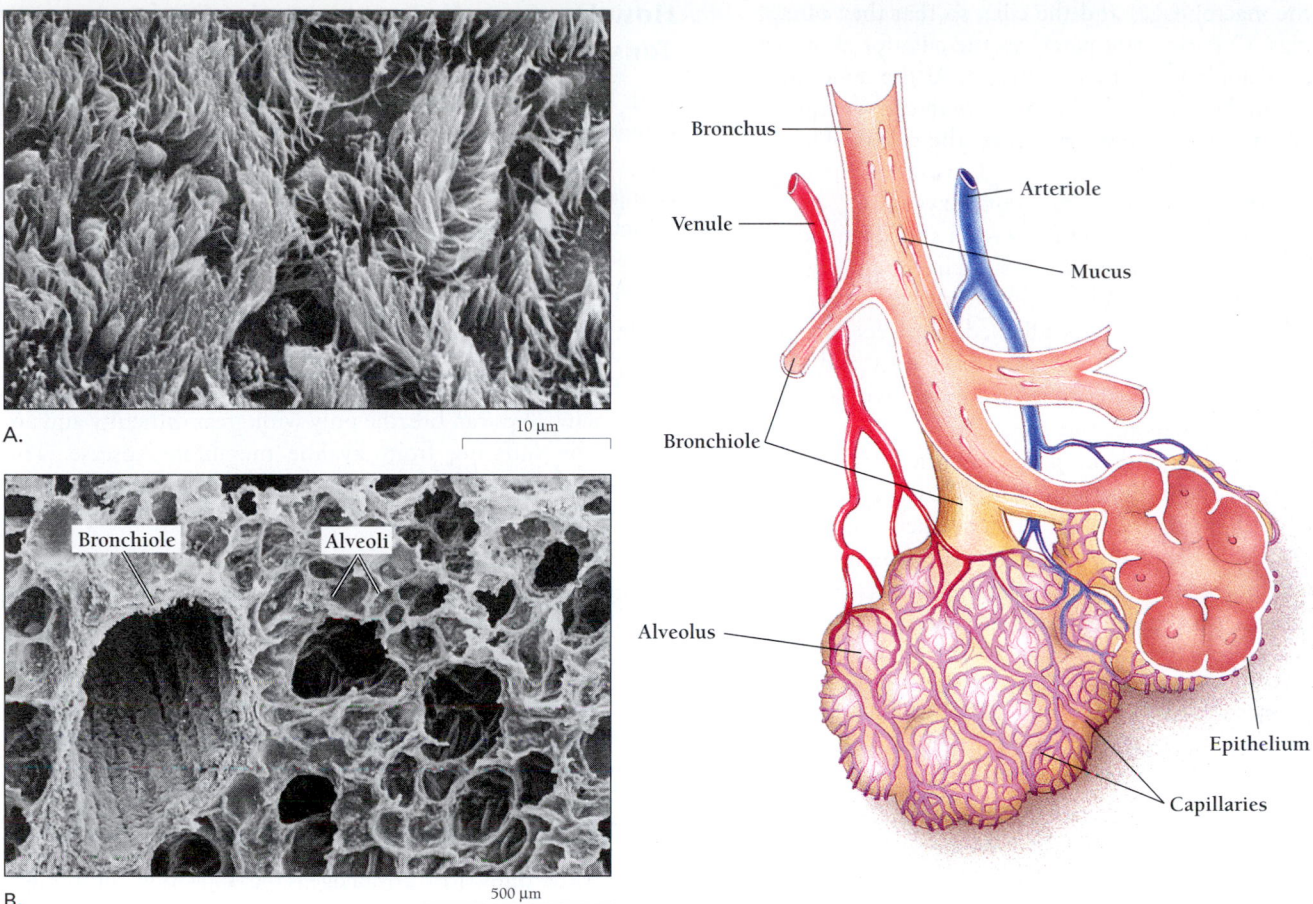

Figure 38-8   Epithelia in the respiratory tract. The epithelium that lines the airways does not absorb oxygen, whereas the epithelium that lines the alveoli does absorb oxygen. A. SEM of the inside of the trachea. B. SEM of a section of lung tissue, showing part of a bronchiole surrounded by spongy alveoli. *(Biophoto Associates)*

What Keeps the Airways Moist and Clean?

Tightly connected epithelial cells line both the nonabsorbing surfaces of the airways and the absorbing surfaces of the alveoli. As we might expect from their differing roles, the cells of the airways differ from those of the alveoli (Figure 38-8).

Most of the cells that line the trachea, bronchi, and bronchioles (but not those of the alveoli) have fringes of constantly beating **cilia** that carry away dirt, bacteria, viruses, and mucus. Many other cells in the airway linings secrete **mucus,** a solution of glycoproteins like that secreted by the membranes of the nose. Mucus keeps the surface of the airways wet, which in turn keeps the air within the airways moist and prevents the surfaces of the alveoli from drying out.

Mucus also helps protect the lungs from becoming clogged with dust particles or infected with bacteria. Dust particles, often carrying bacterial passengers, stick to the mucus, and the beating cilia drive the dust-laden mucus out of the airways into the throat. Swallowing then pushes the mucus into the digestive tract. The ciliated epithelium of the airways serves as an escalator for particles that would otherwise accumulate in the blind alveoli,

where they could interfere with gas exchange and cause infection. The airways also contain scavenger white blood cells called *macrophages* that engulf the accumulated debris in the airways.

The mucus in the respiratory tract traps dust and bacteria, which cilia propel towards the mouth. Macrophages also engulf debris in the airways.

How Does Tobacco Smoke Damage the Lungs?

The three most dangerous substances in tobacco smoke are nicotine, carbon monoxide, and tar. Tar, a brown, oily substance like that lining a chimney or a car's exhaust pipe, damages the lungs. Nicotine and carbon monoxide work together to damage the heart and arteries.

Lungs need to be clean in order to collect oxygen effectively. So tar interferes with lung function immediately. Within minutes of the first puff on a cigarette, tobacco smoke paralyzes both the macrophages and the cilia, so that they cannot clear the lungs. One cigarette paralyzes the cilia for about an hour. Regular smoking kills them outright. At the same time, the irritating smoke causes the lungs to secrete extra mucus, which accumulates in the lungs and clogs the alveoli. The tar in the smoke settles throughout the lungs, irritating the lungs and impairing the lungs' ability to absorb oxygen.

The only way to clear the lungs is for a smoker to cough constantly, the classic "smoker's cough." The lungs, clogged with mucus and tar and unable to rid themselves of bacteria and viruses, become infected and inflamed. Smokers have frequent respiratory infections, especially bronchitis and even pneumonia. These chronic infections, combined with continual coughing, further damage the lungs.

The constant coughing of chronic bronchitis actually breaks the walls of the individual alveoli. Nitric acid and sulfuric acid in the burning tobacco further weaken the walls of the alveoli. As the walls of the alveoli break, the many small, bubblelike chambers become a few larger chambers, and the lung gradually loses its enormous surface area.

In the disease *emphysema,* so many alveoli are lost that the lungs can no longer absorb enough oxygen to support life. In the last weeks or months of life, a person with emphysema must breathe from an oxygen tank. Finally, even that is not enough.

Tobacco smoke tar also causes cancer, the unregulated growth of the body's own cells. Tobacco smoke contains at least 50 different known *carcinogens,* chemicals that cause cancer. Because tobacco smoke ruins the lungs' ability to cleanse themselves, these carcinogens permanently coat the lungs. Some of them damage genes that control cell division, so that the cells divide without restraint. Others damage enzymes that help regulate cell division. Still others enhance the effect of the other carcinogens. In time, small tumors develop, which eventually metastasize, or spread, to other areas of the body. Invasion of nearby nerves, for example, may cause partial paralysis.

How Do Alveoli Overcome the Surface Tension That Resists Expansion?

Each inhaled breath pulls air into the alveoli, expanding the volume of each of these tiny sacs. The small size of each alveolus enables the lungs to have a huge area available for gas exchange. But the small size of each sac also increases the difficulty of expanding each alveolus, just as small balloons are harder to blow up than giant ones.

A liquid's resistance to an increase in surface area is called surface tension. To overcome this resistance, the alveoli produce a **surfactant,** a detergent that reduces surface tension and allows expansion. Infants born prematurely often lack surfactant. Such babies can breathe only with great difficulty and are said to be suffering from hyaline membrane disease. The prospects for survival are excellent, however, if the baby's doctor supplies a substitute surfactant.

Surfactants within alveoli reduce surface tension and allow the lungs to expand more freely.

How Do Birds Breathe So Well?

Among all the vertebrates, birds are the best at absorbing oxygen from the air. Many birds routinely fly long distances at altitudes above 6000 meters (nearly 4 miles), far above where most mammals can live. Birds' great advantage is that they move air through their lungs in one direction. Mammals move air in and out, and much of the air in the lungs is residual air that is not fresh.

When a bird takes a breath, the air passes into a group of sacs called the *posterior air sacs,* which, like our own bronchi, do not allow gas exchange (Figure 38-9). When the bird exhales, the contents of the posterior air sacs move forward into the lungs. On the next breath, air moves out of the lungs into another set of sacs, the anterior air sacs. Finally, on the second exhalation,

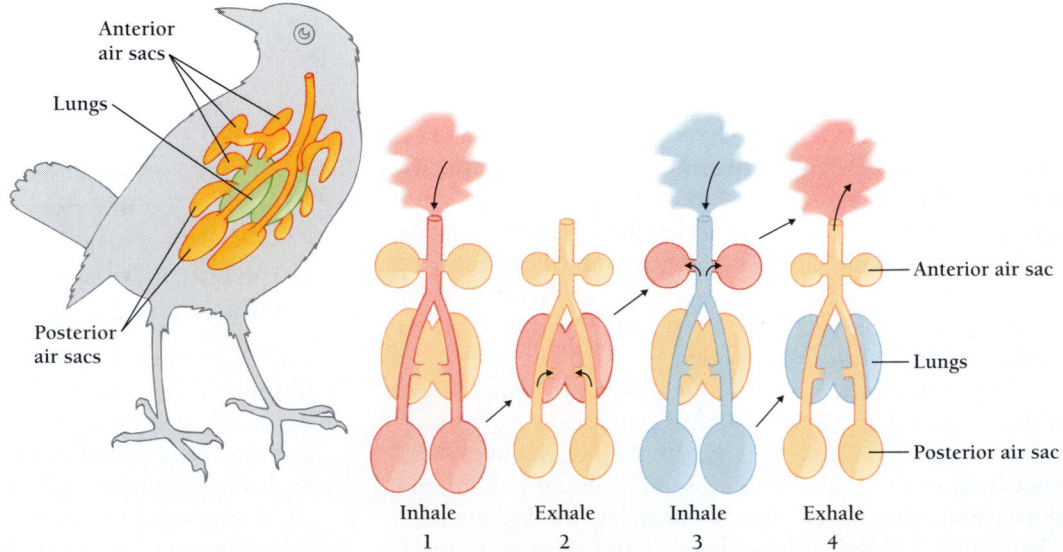

Figure 38-9   Bird breathing. The two-cycle respiration of a bird is more efficient in extracting oxygen than the one-cycle respiration of a mammal.

the contents pass back out into the atmosphere. Air moves in a single direction through the lungs, and the air that enters the lungs is fully oxygenated. In mammals, the residual air that makes up the bulk of the air in the lungs has lost some oxygen.

In a bird's lungs, air flows in a single direction, allowing birds to take up oxygen more rapidly than mammals.

How Do Insects Distribute Oxygen?

In all vertebrates and most invertebrates, the circulatory system distributes dissolved oxygen. Even in animals, such as salamanders, that absorb oxygen through the skin, blood vessels just under the skin take up dissolved oxygen and carry it to the rest of the body. In most invertebrates and in fish, the gills pass oxygen into the circulating blood. And in most terrestrial vertebrates, lungs pass oxygen to the circulation.

Insects, however, do not distribute dissolved oxygen at all. Instead, they distribute air itself through a system of tiny ventilation pipes called **tracheae** (Figure 38-10). Air enters the body through openings called **spiracles,** which lead directly to the tracheae. The advantage of this system is that it allows oxygen to diffuse as a gas, saving the energy of pumping it through the circulation. This separate system for distributing oxygen more than compensates for the limitations of insects' open circulatory systems.

In insects, oxygen diffuses to tissues through air tubes called tracheae.

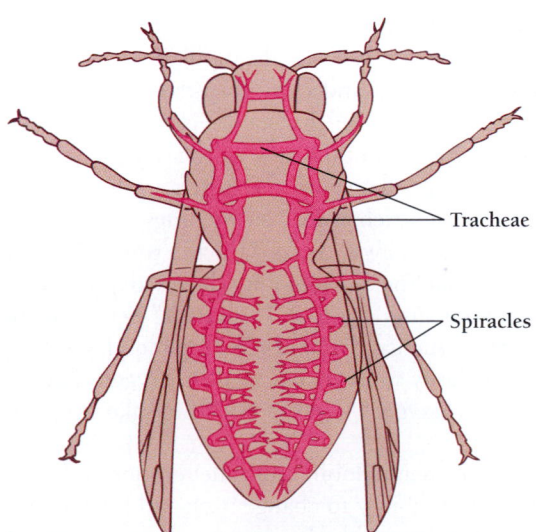

Figure 38-10   Insect breathing. The respiratory system of an insect. Instead of distributing dissolved oxygen, insects distribute air itself through an extensive system of tracheae.

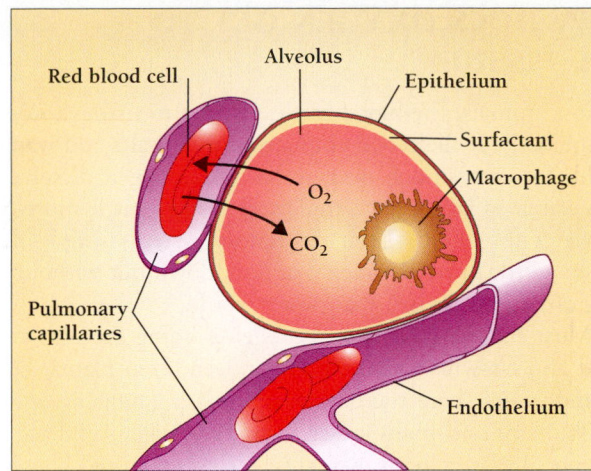

Figure 38-11   Gas exchange in the alveoli. Oxygen molecules move from the inside of the alveolus, through the epithelium of the alveolus and the endothelium of the pulmonary capillary, into the oxygen-carrying red blood cell. CO_2 moves in the opposite direction—from the blood cells and plasma through the endothelium and the epithelium into the airway.

How Do the Lungs Foster Gas Exchange?

The structure of the cells of the alveoli speeds the diffusion of oxygen and carbon dioxide to and from the blood. The epithelial cells lining the alveoli are so thin that only a spongy network of connective tissue keeps them from collapsing. Surrounding the alveoli is a dense network of tiny capillaries. These capillaries are so narrow that the red blood cells must squeeze through one at a time, their passage lubricated by a thin layer of plasma. The plasma also leaks out of the capillaries and into the alveoli, helping to wet the inside surface of the lungs.

Oxygen from the air in the alveoli dissolves in the liquid that coats the epithelium and diffuses across the thin epithelium of the alveoli. The oxygen then crosses the *endothelial* layer that makes up the walls of the capillaries and into the blood plasma (Figure 38-11). Once in the plasma, the oxygen molecules diffuse into the red blood cells and bind to hemoglobin. The total distance diffused is about 1 μm, about 1/100th the thickness of a hair.

The blood carries the oxygenated hemoglobin through the pulmonary veins to the heart for distribution to the rest of the body. Carbon dioxide follows the reverse path, from red blood cells to plasma to alveoli to exhaled air.

The cellular structure of the alveoli promotes the diffusion of oxygen and carbon dioxide to and from the blood.

HOW DOES OXYGEN GET TO THE TISSUES?

In most animals, the blood carries oxygen to tissues throughout the body. If the blood consisted only of salts and water, it would not carry much oxygen, for little oxygen dissolves in plain water. In fact, a few simple animals have little more than salt water for blood. Their circulatory systems simply move oxygen a little faster (by means of bulk flow), but without carrying any more oxygen than water would.

Most animals have special oxygen-binding proteins in the blood and elsewhere in the body. Indeed, even plants, bacteria, and yeasts share the oxygen-binding pigment myoglobin with humans and many other animals. Oxygen-binding proteins are always colored pigments: for example, red hemoglobin in vertebrates and some worms, mollusks, and crustaceans; green chlorocruorin in some annelid worms; and blue hemocyanin in many mollusks and crustaceans. The presence of these proteins can increase the oxygen-carrying capacity of blood by up to 50 times.

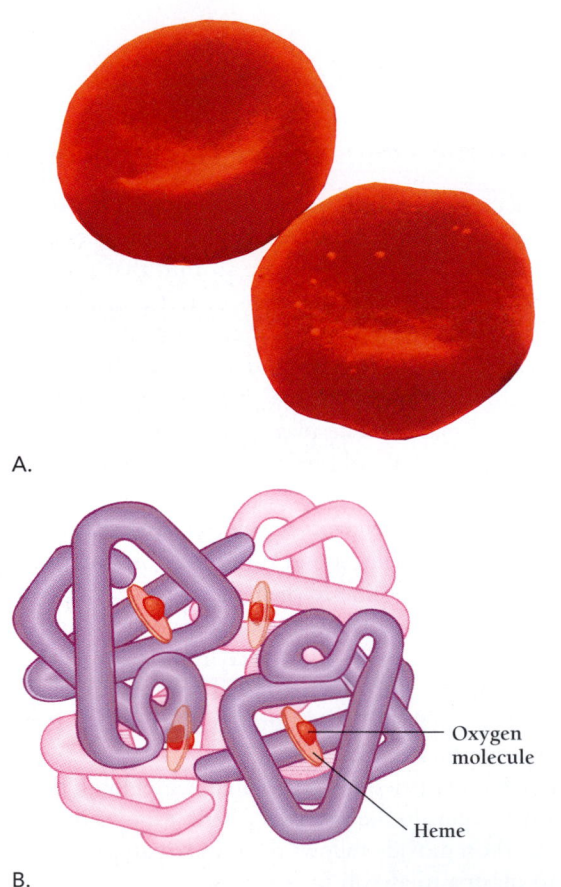

A.

B.

Figure 38-12   The oxygen ship. A. Red blood cells are giant transport ships that carry millions of hemoglobin molecules. B. Each molecule of hemoglobin can carry four oxygen molecules. *(A, David M. Phillips/Visuals Unlimited)*

Oxygen molecule

Heme

How Do Red Blood Cells Transport Oxygen?

Hemoglobin and the other pigments carry oxygen molecules in the blood much as a raft transports passengers down a flowing stream. In invertebrates, the oxygen-carrying pigments are usually dissolved directly in the blood. In vertebrates, however, hemoglobin molecules are packed into red blood cells.

The red cells are only containers for hemoglobin. Each red cell is like a giant transport ship, with no life of its own. It lives only a few weeks and lacks mitochondria and, in mammals, even a nucleus (Figure 38-12A). But a red cell may hold tens of millions of hemoglobin molecules.

Each molecule of hemoglobin consists of four polypeptide chains folded together. Each polypeptide enfolds a small molecule, called a **heme,** at the center of which is a single atom of iron (Figure 38-12B). One oxygen molecule can bind to each heme, so that each hemoglobin molecule can bind up to four oxygen molecules. Hemoglobin with one or more bound oxygens is called **oxyhemoglobin.** Hemoglobin with no oxygen is called deoxyhemoglobin, or simply hemoglobin.

Oxygen-binding proteins greatly increase the oxygen-carrying capacity of blood.

How Does Hemoglobin Load and Unload Oxygen?

The partial pressure of oxygen in the alveoli is about 100 mm Hg, while the partial pressure of the deoxygenated blood coming into the lungs is only 40 mm. As a result, the oxygen molecules in the alveoli passively diffuse into the blood (Figure 38-13). Because the surface area of the lungs is large, the oxygen molecules quickly reach equilibrium, and the partial pressure of oxygen in blood leaving the lungs is as high as that in the alveoli—100 mm.

Once the oxygen is inside the capillaries, hemoglobin molecules quickly bind it and carry it to the heart to be pumped to the different tissues of the body. Tissue cells usually have a partial pressure of about 46 mm, so the oxygenated blood diffuses down the gradient into the tissues.

In the oxygen-rich tissues of the lungs, hemoglobin binds oxygen and takes it away. Indeed, by the time blood leaves the lungs, nearly all the hemoglobin carries oxygen on all four hemes. Physiologists say it is "saturated." Yet, in tissues with little oxygen—such as the hard-working muscles of a runner—hemoglobin easily unloads its cargo of oxygen. For hemoglobin to deliver oxygen, it must not only take up oxygen but also let it go.

How can a hemoglobin molecule respond so flexibly? The answer lies in its ability to change back and forth between two alternate three-dimensional structures. In one form, the four polypeptide chains lie a certain distance from each other and hemoglobin is only moderately likely to bind an oxygen. In

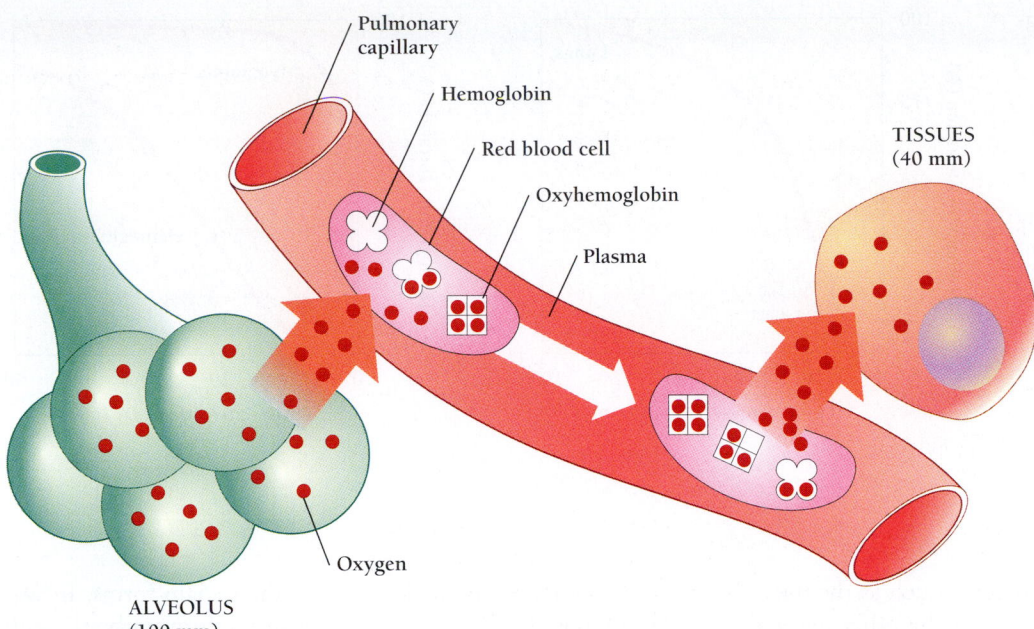

Figure 38-13 How does hemoglobin carry oxygen? Hemoglobin molecules associate with oxygen in the lungs, where the partial pressure of oxygen is high (100 mm). Oxyhemoglobin, in red blood cells, moves through the circulation. In oxygen-consuming tissues, such as muscles, the partial pressure of oxygen is lower (40 mm) and oxyhemoglobin dissociates into deoxyhemoglobin and oxygen. The released oxygen diffuses into the surrounding tissues.

the other form, two of the four chains move closer together and the whole molecule becomes much more likely to bind oxygen.

Physiologists say that hemoglobin changes its "affinity" (or fondness) for oxygen. Each time a hemoglobin molecule binds an oxygen molecule, the hemoglobin becomes more "fond" of oxygen. In contrast, as hemoglobin unloads its oxygen molecules, it loses its affinity for oxygen.

Hemoglobin changes its structure according to the partial pressure of oxygen around it (Figure 38-13). In the blood in the lungs, the partial pressure of oxygen is high. So the hemoglobin molecules bind oxygen and also increase their affinity for oxygen. Because of their high affinity for oxygen, the hemoglobin molecules do not let any of their oxygen go.

In the tissues of the body, the partial pressure of oxygen is much lower, around 40 mm. During vigorous activity, the partial pressure of oxygen in a muscle may drop as low as 10 mm. Even though hemoglobin's affinity for oxygen is still high at first, a few of the many hemoglobin molecules will each lose one or two oxygens. These are quickly taken up by nearby cells. Meanwhile, the hemoglobin molecules lose their affinity for oxygen and become even more likely to lose oxygen.

Hemoglobin has two alternate structures. One has a high affinity for oxygen and the other has a lower affinity for oxygen. In lungs, hemoglobin assumes the high-affinity state. In the tissues, hemoglobin assumes the low-affinity state.

What Determines Hemoglobin's Affinity for Oxygen?

We can see that hemoglobin absorbs oxygen from the lungs and drops oxygen where it is needed most. But what factors determine hemoglobin's changing affinity for oxygen?

Researchers can easily study hemoglobin's oxygen binding because oxyhemoglobin is redder than deoxyhemoglobin. The oxygen-loaded blood of the arteries is bright red, while the oxygen-depleted blood of the veins is slightly blue-red. Using a spectrophotometer, an instrument that measures color, researchers can tell how much oxygen is bound to hemoglobin at different partial pressures of oxygen. The amount of oxygen bound to the hemoglobin graphed against oxygen partial pressure is called an oxygen dissociation curve (Figure 38-14).

In adult humans, the percent of oxyhemoglobin increases from a partial pressure of about 10 mm Hg to about 60 mm. Above 60 mm, the percent of oxyhemoglobin does not increase much: the hemoglobin is effectively saturated with oxygen. The shape of the whole curve vaguely resembles the letter *S*.

Tissues that use oxygen have lower oxygen partial pressures than the lungs. The oxygen in a resting muscle, for example, has a partial pressure of about 40 mm. At this level, the hemoglobin holds only about 75 percent as much oxygen as in the lungs. As the blood moves through the circulation from lungs to heart to resting muscle, it gives up about 25 percent of its oxygen.

After leaving the lungs, saturated hemoglobin easily loses its fourth oxygen. But under most circumstances, the other three oxygens stay bound to hemoglobin. They form a sort of emergency supply of oxygen. Only if the blood passes through

Figure 38-14 Hemoglobin's changing affinity for oxygen. A. Hemoglobin. At low partial pressures of oxygen, like those found in active muscles, hemoglobin readily releases oxygen molecules to the tissues. At high partial pressures, like those found in the lungs, hemoglobin binds oxygen tightly and releases little. B. Myoglobin. Closely related to hemoglobin is an oxygen-binding protein of muscle cells called myoglobin. Like hemoglobin, myoglobin is fully saturated at 100 mm. But myoglobin is also fully saturated at 40 mm, and nearly so at 20 mm. Myoglobin could never work well as an oxygen carrier in the blood because it does not unload oxygen easily.

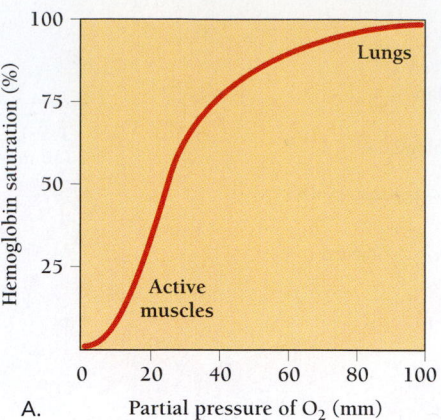

A.

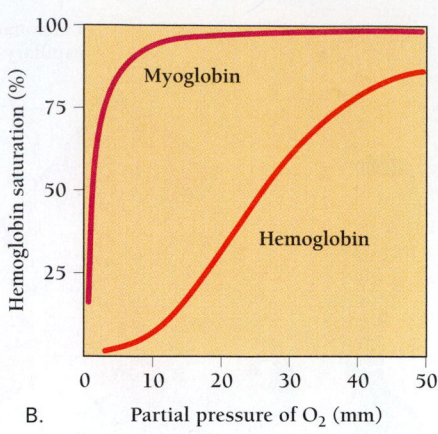

B.

tissues that are severely depleted of oxygen—such as the muscles of a runner—does hemoglobin unload the other oxygen molecules—and then very quickly (Figure 38-14).

Cells that are especially active often exceed their ordinary supply of oxygen. In that case, they depend (temporarily) on glycolysis for ATP. The result is the accumulation of lactic acid, which decreases pH. The lowered pH decreases hemoglobin's affinity for oxygen, increasing the release of oxygen.

Cells that are especially active also release lots of carbon dioxide as they burn sugar (combine oxygen and glucose). When hemoglobin binds the waste carbon dioxide, its affinity for oxygen decreases. So hemoglobin in active cells releases its oxygen more easily.

Still another adaptation of hemoglobin helps increase oxygen delivery. In many mammals, hemoglobin binds to an abundant three-carbon molecule called BPG (2,3-bisphosphoglycerate). When BPG binds to hemoglobin, it decreases hemoglobin's affinity for oxygen, allowing the hemoglobin to release more oxygen in the tissues. The molecular properties of the hemoglobin molecule favor the unloading of oxygen in more active tissues (Box 38.4).

Hemoglobin's affinity for oxygen changes according to the partial pressure of oxygen, pH, and the amount of CO_2 and BPG.

How Does Blood Carry Carbon Dioxide?

In addition to absorbing oxygen, the respiratory system also helps dispose of carbon dioxide (CO_2). As oxygen enters the blood, CO_2 leaves. Just as oxygen in the blood comes to equilibrium with oxygen in the alveoli, CO_2 in the blood comes into equilibrium with CO_2 in the alveoli. The difference between the CO_2 partial pressure in the blood and the alveoli is small compared to difference for oxygen—46 mm Hg in the blood and 40 mm in the lungs. Despite this small difference, however, the CO_2 readily diffuses into the alveoli.

CO_2 moves about the body in three main forms. In water or blood, CO_2 combines with water to form carbonic acid:

$$CO_2 + H_2O \longrightarrow H_2CO_3$$

Carbonic acid rapidly breaks down into a hydrogen (H^+) ion and a bicarbonate (HCO_3^-) ion:

$$H_2CO_3 \longrightarrow H^+ + HCO_3^-$$

As a result, about two-thirds of the CO_2 in the blood is in the form of bicarbonate—some of it inside red blood cells and some in the plasma. About one-third of the CO_2, however, combines with hemoglobin inside the red blood cells. CO_2 attaches to hemoglobin at a different place than oxygen, however, so the two molecules do not compete with each for space on the hemoglobin raft. Finally, a tiny amount of CO_2, about 5 percent, dissolves directly into the blood.

CO_2 readily diffuses from the blood into the alveoli. The blood carries most carbon dioxide in the form of bicarbonate ions.

WHAT FACTORS CONTRIBUTE TO OXYGEN HOMEOSTASIS?

Climb a tall mountain and you'll find yourself gasping for breath. Sprint to catch a bus or plane and you may wish you could transport oxygen a little faster. Happily, the body has adaptations that maintain oxygen delivery under most conditions.

How Can Blood Deliver Oxygen at High Altitudes?

At high altitudes, the difference in the oxygen partial pressures in the lungs and muscles is not as great as at sea level. When

BOX 38.4

How Does a Fetus Obtain Oxygen?

A fetus has a special problem in obtaining oxygen. Since it cannot get oxygen from its own lungs, it must get oxygen from its mother's blood. Maternal blood comes into close contact with fetal blood only in the *placenta*, a structure in the womb that conveys nutrients from the mother to the fetus. In the placenta, the maternal and fetal blood pass in close proximity without actually flowing together. Still, oxygen and other nutrients can pass out of the mother's blood and wastes can pass from the fetus to the mother.

In order for the fetus to sop up oxygen from the maternal blood, the fetus's hemoglobin must have a higher affinity for oxygen than the mother's hemoglobin. And, indeed, a fetus makes a different kind of hemoglobin, called hemoglobin F.

Amazingly, however, when researchers studied the oxygen-binding characteristics of hemoglobin F, they found that it bound oxygen with exactly the same affinity as adult hemoglobin, hemoglobin A. For decades, this lack of difference in oxygen affinity baffled researchers: how did fetal blood do its job?

The answer came from the discovery of the action of BPG. In humans, the major difference between hemoglobins A and F is that hemoglobin F does not bind BPG. When BPG binds to hemoglobin A, it decreases the molecule's affinity for oxygen. As a result, the hemoglobin can more easily give up oxygen to tissues that need it. Because hemoglobin F doesn't bind BPG, its affinity for oxygen is high, and the fetal hemoglobin absorbs oxygen from maternal blood. The hemoglobin F can still unload its oxygen in the fetal tissues because they have a lower oxygen partial pressure and lower pH than does the blood in the placenta.

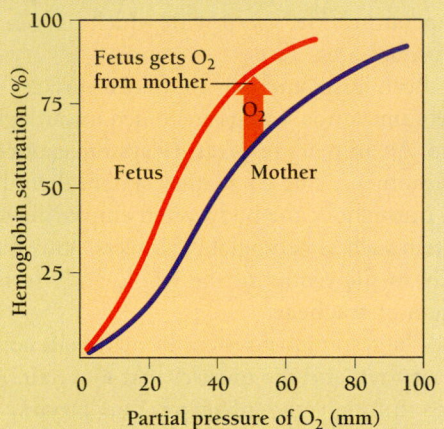

Figure A

Figure A Oxygen dissociation curves for fetal and maternal hemoglobin. Because fetal hemoglobin has a higher affinity for oxygen than does maternal hemoglobin, the maternal hemoglobin gives up oxygen to the fetal hemoglobin.

---

a person climbs a mountain, one of the first changes in the blood is an increase in the level of BPG, which binds to hemoglobin and decreases its affinity for oxygen. As a result, hemoglobin more easily delivers its cargo of oxygen to the brain and other tissues.

The body also adjusts to high altitude by building up a bigger supply of hemoglobin-rich red blood cells. In response to low oxygen levels, the kidney releases a protein hormone, called erythropoietin, that stimulates the production of new red blood cells in the bone marrow.

In humans, low oxygen in the tissues triggers an increase in BPG, which increases oxygen delivery by hemoglobin.

How Do Animals Regulate Breathing in Response to Changes in Oxygen and Carbon Dioxide Delivery?

The lungs and the blood, working together, accomplish both oxygen delivery and CO_2 removal. But how does an animal respond to lung failure, blood loss, or increased muscle activity, all of which may challenge the blood's ability to deliver sufficient oxygen?

As tissues use oxygen for respiration, they also produce CO_2 as a waste product. Increased respiration always leads to an increase in CO_2. In mammals, an increase in the partial pressure of CO_2 in the blood signals the need for deeper or more rapid breathing. Physiologists do not yet know exactly how this signal acts, but do know that the coordination of breathing depends on a nerve complex in the brain's medulla, called the **breathing center,** which regulates the rate of breathing.

The firing of nerves in the breathing center triggers the contraction of the diaphragm and other muscles needed for inspiration. As the lungs expand, stretch receptors report back to the breathing center, the contraction signals to these muscles cease, and expiration occurs. High CO_2 stimulates the breathing center to coordinate faster and deeper breathing. The medulla also controls the circulatory system, changing the heart output according to the needs of the body. The maintenance of a constant internal environment means coordinating the actions of heart, blood vessels, and lungs to control the levels of oxygen and CO_2 as the status of the tissues changes.

Study Outline with Key Terms

For oxygen to participate in biochemical reactions, it must first dissolve in water. The final concentration of dissolved oxygen at **equilibrium** depends on its **partial pressure** and on the temperature. The speed with which oxygen dissolves depends on surface area.

Both lungs and gills are folded surfaces with large surface-to-volume ratios. **Lungs** are folded inward, **gills** are folded outward. An animal's **respiratory system** extracts oxygen from the environment. The circulatory system then distributes the oxygen throughout the body, often employing an oxygen-binding protein such as hemoglobin. Insects, however, distribute air directly to the tissues of the whole body from **spiracles** into a system of **tracheae.**

The respiratory system in mammals consists of the lungs, the airways, and the muscles that move air. Air moves into the lungs through the **pharynx,** the **larynx,** the **trachea,** the **bronchi,** and the **bronchioles.** The bronchioles lead to the **alveoli,** the sites of gas exchange.

Contractions of the **diaphragm** and the muscles of the chest are responsible for **ventilation,** the **inspiration** and **expiration** of air. Muscle contraction expands the **thoracic cavity,** pulling air into the lungs. Muscle relaxation expels air. Each breath draws in and expels a volume of air called the **tidal volume.** Part of the new air stays within the **dead space** of the airways, while most of it enters the lungs. The fresh air mixes with the **residual** air remaining within the alveoli after the last breath. Most air in the lungs is residual.

The cells of the respiratory tract are specialized for several different functions. Some cells secrete **mucus,** which traps dust, bacteria, viruses, and other impurities. Other cells secrete **surfactants,** which allow the lungs to expand smoothly and freely. Most cells of the upper tract have **cilia** that push dust-laden mucus out toward the pharynx. The cells of the alveoli form a specialized epithelium, well adapted for gas exchange. Oxygen and carbon dioxide diffuse through the epithelium of the alveoli and the endothelium of the capillaries into the blood plasma.

Hemoglobin is made of four polypeptide chains and four **heme** groups. Each heme can bind one oxygen molecule, for a total of four oxygens per hemoglobin. Oxygenated hemoglobin is called **oxyhemoglobin.** Hemoglobin binds and releases oxygen according to the partial pressure of oxygen in the surrounding tissues. More active muscles have lower oxygen partial pressures, causing a greater release of oxygen. The low pH and CO_2 released by active muscle also favor the release of oxygen.

The binding of oxygen changes the three-dimensional structure of hemoglobin, increasing its affinity for oxygen. Conversely, as the hemoglobin loses oxygen, it becomes more likely to lose oxygen. This arrangement is at the heart of hemoglobin's ability to bind oxygen in the lungs and release it in the tissues of the body.

The respiratory system also serves to dispose of CO_2. CO_2 in the tissues combines with water to form bicarbonate ions. Some CO_2 also binds directly to hemoglobin molecules. In the lungs, CO_2 passes through the alveoli and exits from the body. Accumulated CO_2 in the tissues stimulates the brain's **breathing center** to command an increase in breathing rate.

Review and Thought Questions

Review Questions

1. How does the partial pressure of a gas influence how much of it dissolves in a fluid?
2. Do gases dissolve better at high temperatures or at low temperatures?
3. Which is larger, the surface-to-volume ratio of a pumpkin or that of a peach?
4. In what ways does a lung differ from a gill?
5. What characteristics do gills and lungs have that maximize the rate of oxygen diffusion?
6. Since atmospheric air is richer in oxygen than water, why do fish die when out of water?
7. Imagine air passing through the respiratory system from the nose to the alveoli. Name the parts along the way. Which part enables you to speak?
8. Define tidal volume, dead space, and residual volume. Which is the largest?
9. How do the lungs rid themselves of dust and microbes?
10. How does the diaphragm muscle help expand and contract the chest cavity? Draw a diagram.
11. Name the parts along the path that oxygen takes from the alveoli to the hemoglobin inside a red blood cell.
12. What change increases or decreases hemoglobin's affinity for oxygen?
13. Name four factors that cause this change in affinity.
14. How does the body cope with the low partial pressure of oxygen at high altitudes?
15. What is the partial pressure of oxygen in the lungs? In the tissues, approximately?
16. In what three forms does carbon dioxide exist in the blood? Which is most common?

Thought Questions

17. Why are scientific discoveries sometimes not immediately accepted by the general public?
18. Considering what you know about both the circulatory system and the respiratory system, explain why a drug that is inhaled reaches the brain twice as fast as one that is injected.

19. Explain what a vector is. What did Stanton Glantz mean when he said that the tobacco industry is the vector for heart disease and lung cancer?

About the Chapter-Opening Image

In 1885, Vincent van Gogh painted this portrait of a skeleton smoking. Physicians of the 19th century already recognized that smoking was dangerous to health.

 On-line materials relating to this chapter are on the World Wide Web at **http://www.harcourtcollege.com/lifesci/aal2/**

39

How Do Animals Manage Water, Salts, and Wastes?

Will Corn Chips Raise Your Blood Pressure?

In the spring of 1996, a team of researchers from the University of Toronto published a paper in the *Journal of the American Medical Association*. The paper's conclusions ran counter to decades of authoritative nutritional advice. Julian Midgley, Andrew Matthew, and their colleagues wrote that for most people an extremely low-salt diet would not lower blood pressure and in some people might actually cause a heart attack. At a time when salt consumption in the United States had fallen by 30 percent and supermarket aisles were jammed with low-salt crackers, low-salt chips, and low-salt soups, the researchers' stance dramatically contradicted accepted wisdom.

Americans love corn chips, potato chips, salty buttered popcorn, hot dogs, pizza, and greasy hamburgers. The more salt and fat, the better these foods seem to taste to us. Indeed, nutritionists uniformly agree that Americans consume far too much fat and far too much salt. Fat, especially animal and other highly saturated fat, contributes to both obesity and atherosclerosis, or hardening of the arteries. Both of these are major risk factors for heart attacks and strokes. Salt, we have been told, causes high blood pressure, a major risk factor for heart attacks, strokes, and kidney failure (Figure 39-1).

Even with recent declines in salt consumption, the average American consumes twice the recommended daily intake of salt. The minimum sodium requirement for normal day-to-day health is 115 milligrams (mg). (For comparison, a teaspoon of salt has about 2000 mg of sodium.) Yet most of us take in nearly 4000 mg of sodium each day. Government nutritionists, in an effort to recommend something reasonable, suggest a diet containing no more than 2400 mg per day.

Most researchers suggest limiting salt intake because they fear that salt causes the body to retain water. More salt means more water intake, and more water means increased blood pressure. In the short term, individuals can get rid of excess sodium, mostly in the urine, but physicians have long suspected that, long term, increased sodium increases blood pressure, leading to higher risk of heart attacks and stroke.

The strongest evidence that increased sodium actually leads to increased blood pressure has come from comparisons of high blood pressure in different countries. In countries where sodium consumption is high, blood pressure is also high, while in countries where sodium consumption is low, average blood pressure is also lower. One study of 32 countries found a consistent relationship between sodium consumption and high blood pressure, or *hypertension.*

But, according to Midgley and Matthew, a host of papers in the last ten years show that the connection between sodium and hypertension is indirect. To begin with, some researchers argue that a closer examination of people's behavior usually shows that those who consume a lot of sodium also drink more alcohol and tend to be overweight. Alcohol and obesity each cause hypertension, and these factors, rather than the salt itself, may be the cause of hypertension.

Researchers have long known that only a small percentage of people with high blood pressure clearly benefit from a low-salt diet. In about half of all people with hypertension (just 5 to 10 percent of all Americans) the kidneys are unable to compensate for large amounts of sodium. In these "sodium-sensitive" individuals, a high-salt diet causes blood pressure to rise, while a low-salt diet causes blood pressure to drop. Sodium sensitivity gradually increases in those older than 45 or 50 years of age. Apparently, as we age, our kidneys lose some of their ability to regulate salt and fluid levels

Key Concepts

- Animals must regulate water and solutes both inside each cell and in the extracellular fluid.

- Animals excrete nitrogenous wastes, usually diluted with water, by means of kidneys or other excretory organs.

- In mammals, water and small molecules from the blood flow into the kidney, which then "puts back" essential ions and molecules.

- Hormones coordinate salt and water excretion by altering blood flow and water permeability.

in the blood (Table 39-1 and Figure 39-1).

In most people, however, the kidneys compensate almost magically. Thanks to kidneys, a typical 20-year-old can maintain normal blood pressure even after eating, over the course of a day, five cafeteria pancakes, with a total of 4000 milligrams of sodium; two slices of pizza, with 2000 milligrams of sodium; and a can of spaghetti and meatballs, with 1000 milligrams of sodium. Nonetheless, considering the international data showing the strong association between sodium consumption and blood pressure, it was reasonable for health experts to recommend that Americans cut back on salt.

In January 1996, Midgley and Matthew decided to take a second look at all of the

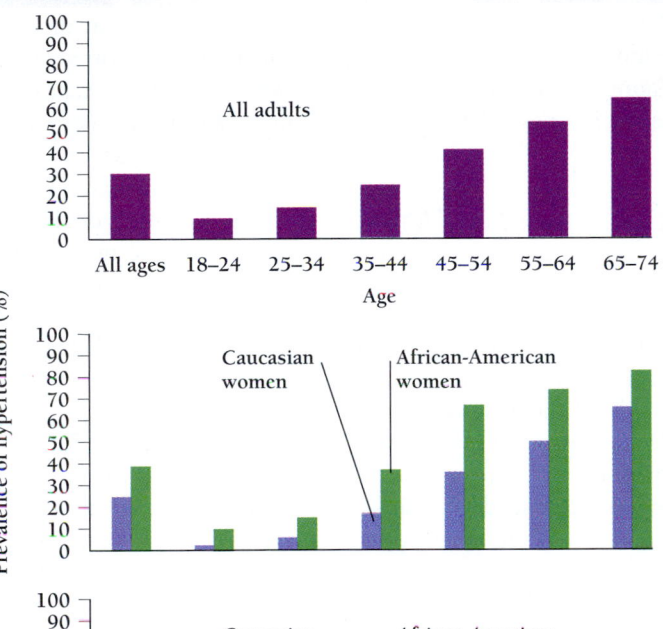

Figure 39-1 Prevalence of hypertension.

studies on blood pressure and sodium consumption in the previous several years. After discarding dozens of studies that they thought were poorly designed, they reanalyzed data from 56 clinical trials involving more than 3000 people. Midgley, Matthew, and their colleagues concluded that a low-sodium diet lowers blood pressure only in people older than 45 years of age who have high blood pressure. Even then, blood pressure drops only modestly. And a low-sodium diet has virtually no effect in people with normal blood pressure or young people with high blood pressure.

The researchers also questioned a basic assumption that any effect in older hypertensives would be beneficial. They argued that even in patients who had lowered their blood pressure, there is no direct

TABLE 39-1 PREVALENCE OF HYPERTENSION (≥ 140/90) IN U.S. ADULTS, AGES 18–74

	White men (%)	Black men (%)	White women (%)	Black women (%)	All adults (%)
All ages	33	38	25	39	30
18–24	16	11	2	10	9
25–34	21	23	6	15	14
35–44	26	44	17	37	24
45–54	43	55	36	67	41
55–64	51	66	50	74	53
65–74	59	67	66	83	64

evidence that blood pressure lowered by means of a low-sodium diet actually prevents heart attacks and strokes or saves lives. Indeed, a super-low-sodium diet may even be dangerous. In one study, for example, men with high blood pressure on extremely low-sodium diets had four times the risk of a heart attack as men consuming more sodium.

Should we eat as much salt as we can? Almost certainly not. Even for those of us whose kidneys are not sensitive to salt, too much salt burdens the kidneys and may possibly contribute to osteoporosis later in life. Will a high-salt diet raise blood pressure? It is beginning to appear that in most people, it does not.

We will undoubtedly be hearing more about the long-term effect of salt in the diet, and we cannot resolve the controversy here. What we can do is equip ourselves with enough knowledge to follow the debate as it unfolds. In particular, it will be useful to understand how kidneys remove salt from the blood, how different hormones regulate kidney function, and how sodium influences the release of such hormones.

WHY DO ANIMALS NEED TO REGULATE WATER AND SALT?

Water is the major constituent of organisms: in humans, for example, it makes up more than half of the body's weight. It is the medium in which cells are constantly bathed and is the major component of cells themselves. Most biochemical reactions occur in water, and water itself participates in many biochemical reactions. Cells and organisms therefore closely regulate the compositions of the solutions both inside and outside of cells. Even small changes in the concentrations of specific ions can dramatically change protein structures, reaction rates, and other cellular processes.

Organisms face different types of challenges to proper water balance—from deserts to oceans to ponds—and a variety of adaptations contribute to keeping them from gaining or losing too much water (Figure 39-2). In terrestrial environments, keeping enough water in the body and its cells is a constant challenge. In freshwater environments, keeping water *out* is the challenge, since the fresh water tends to move into more concentrated solutions within cells.

Surprisingly, the marine environment is more like a terrestrial one than a freshwater one. Because the ocean is salty, marine fish tend to lose water through osmosis. Marine fish must expend energy to keep from losing water.

Having the right amount of water, however, is not enough. Individual cells also need to have the right concentrations of ions (for example, sodium and chloride) both inside and outside. To maintain the correct concentrations of salts and other materials, organisms must somehow detect deviations from the norm. Then, when the deviation is large enough, the organism must take corrective action. Maintaining salt concentration is another example of homeostasis, the capacity of organisms to maintain a stable internal environment.

In vertebrates, the principal guardians of the body's internal environment are the **kidneys.** Kidneys regulate the volume and composition of the blood and the extracellular fluid. Invertebrates face many of the same problems, and often solve them using the same strategies, but they use simpler organs. These organs also maintain constant salt concentrations in body fluids and regulate the disposal of water, wastes, and toxins.

Which Way Does Water Flow?

To understand the flow of water between an organism and its environment, we can (for a moment) think of an organism as a bag of watery fluid. For a single-celled organism, this description is literally true: the plasma membrane surrounds the cytoplasm, which contains a dilute solution of salts and organic molecules. The external environment of such an organism is always aqueous, with more or less salt.

A.

B.

C.

Figure 39-2   Desert, sea, or fresh water. Healthy kidneys maintain correct blood pressure and water and salt balance regardless of the external environment. Thanks to adaptable kidneys, turtles and tortoises have evolved to live in environments in which the amount of salt and water vary enormously. A. An endangered desert tortoise (Mojave Desert). B. A green sea turtle (Hawaii). C. An eastern chicken turtle (North Carolina). *(A, Robert Winslow/Tom Stack & Associates; B, William E. Ferguson; C, Jack Dermid/Photo Researchers, Inc.)*

Multicellular organisms, however, contain another "compartment," the *extracellular fluid,* which lies between the external environment and the membranes of individual cells. For organisms with circulatory systems, the extracellular fluid is continuous with the liquid part of the blood (the *plasma*). The extracellular fluid contains a dilute solution of salts and organic molecules. The concentrations of individual ions and molecules differ from those in cytoplasm, but the total concentration of dissolved substances *(solutes)* is the same inside and outside the cell (Chapter 4). In a human, the cytoplasm makes up about 40 percent of the body's volume, the extracellular fluid about 20 percent, and the blood plasma about 4 percent.

Water can often flow freely between the extracellular fluid and the external environment and between the extracellular fluid and the cytoplasm. Which way, then, does water flow? In Chapter 4, we described *osmosis* [Greek, *osmos* = impulse, thrust], the flow of water from dilute solutions into more concentrated solutions (Figure 4-20). Net water flow between two solutions stops when the total concentrations, or *osmolarity,* of the two solutions become equal.

Water flows from a solution with lower osmolarity to a solution with higher osmolarity. The water flow itself does not require energy: water flows passively. Creating regions of high and low osmolarity, however, may require lots of energy for active transport, especially of sodium ions.

Between the external environment and the insides of the cells of multicellular organisms lies the extracellular fluid. The movement of water, ions, and molecules among the outside environment, the extracellular fluid, and the insides of cells depends on the osmolarity of each fluid.

Water and Salt Balance in Aquatic Animals

Water itself moves only passively, in the direction that leads to equalizing the osmolarity on the two sides of a membrane. When the osmolarity of two solutions is the same, they are said to be *isotonic.* We can say, for example, that the body fluids of most marine invertebrates are isotonic with seawater. As a result, there is no net flow of water between these animals and their environment. These animals must still regulate their internal environment, however, for the concentrations of specific ions and molecules differ from those in seawater.

In contrast, freshwater animals are saltier than their environments, and their body fluids are *hypertonic* (Figure 39-3). A hypertonic cell in a solution of fresh water takes in water, sometimes to the bursting point. Freshwater animals such as trout can somewhat reduce the relentless inflow of water with water-resistant skins. But having an impermeable skin limits a water-dweller's ability to obtain oxygen, which must diffuse from the water into the gills. Consequently, freshwater animals must tolerate considerable water influx and dispose of the excess in their urine.

The kidneys of a freshwater animal excrete great volumes of urine that is more dilute than the body fluids. Freshwater animals produce such dilute, or *hypotonic,* urine by pumping ions out of the urine and back into the body. Producing a lot of urine, however, cannot by itself solve all the problems of water and salt balance. Urine derives from the fluids of the extracellular space, and it contains dissolved salts and organic molecules. Excreting large amounts of urine, then, would result in the loss of essential molecules and ions if the animal had no way of "reabsorbing" what it needed.

Salt ions are removed from urine and returned to body

Salt water

Large volume of hypotonic (dilute) urine is excreted

Small volume of hypertonic (salty) urine is excreted

Salt ions are excreted

Water and salt ions enter through gills

Saltwater fishes: body fluids are less salty than surrounding water

Freshwater fishes: body fluids are saltier than surrounding water

Figure 39-3   Salt water and fresh water present different problems. The bodies of saltwater fishes are less salty than the surrounding seawater, so they tend to lose water. These fishes must work to retain water while excreting salts and other ions and molecules. In contrast, the bodies of freshwater fishes are saltier than the surrounding water, and they tend to gain water. Freshwater fishes must therefore excrete large amounts of water, while retaining precious salts. At the same time, the kidneys of both kinds of fishes need to excrete nitrogenous wastes.

How Do Saltwater Fish Hold on to Water?

Nearly all vertebrates have body fluids whose salt concentration is only about one-third that of seawater. (The jawless hagfish, among the most primitive of vertebrates, is the single exception: its body fluids have the same salinity as seawater.) Saltwater animals must therefore counteract the tendency of water to flow outwards (Figure 39-3). So, whereas a freshwater fish pumps ions out of its urine to concentrate salts in its body fluids, a saltwater fish actively pumps salt ions from its body fluids back into the sea.

To compensate for water loss, a marine fish drinks lots of seawater. (In contrast, a freshwater fish does not drink at all.) This nearly constant drinking takes in salt as well as water, however, and marine fish must remove the extra salt, mostly through its gills. Sea turtles and seabirds, which share this problem, also have special organs that remove excess salt (Figure 39-4).

Marine mammals generally do not drink seawater, but instead get most of their water from the organisms they eat. In addition, marine mammals can derive water from the breakdown of glucose or fat by cell respiration. (Recall from Chapter 6 that the breakdown of one molecule of glucose produces six molecules of water.)

In some cases, however, a marine animal may "burn" fat or carbohydrate just for the water needed to get rid of the excess salt in its diet. For example, when a seal eats marine invertebrates, whose bodies are isotonic with seawater, it has a greater need for water than when it eats bony fish, whose bodies are hypotonic to seawater. A seal will therefore grow fat on fish but get thin on invertebrates.

Freshwater animals rid themselves of excess water by excreting dilute urine. Marine and terrestrial animals tend to excrete salts and other wastes in minimal amounts of water.

How Do Land Animals Balance Water and Salt?

The major advantage of life on land is a rich supply of oxygen and a rich supply of plants to eat. The major problem, however, is getting and keeping water. Like marine animals, land animals derive water from three sources: drinking, eating, and metabolism.

Some land animals—such as frogs, salamanders, and earthworms—absorb water through their skins. Certain frogs, for example, have special abdominal patches that allow them to take up water directly from a puddle, with the skin using the same kind of salt-pumping mechanism as a kidney. Certain insects can also absorb moisture directly from the air. But most familiar animals obtain water by drinking or by eating.

The average adult human, for example, drinks about 1200 ml of fluids a day (a bit more than a quart) and obtains another 1000 ml in food. Metabolism yields another 350 ml of water, for a total input of about 2.5 liters per day.

Normally, the body loses the same amount of water it takes in. Water leaves the bodies of land animals by three main routes: urine, feces, and evaporation. Evaporation takes place both from the outer surface of the body and from the surfaces of the lungs or, in invertebrates, other respiratory organs.

In humans, most water leaves as urine. An additional 100 ml per day is lost in the feces, and about 50 ml in sweat. Most of the rest leaves as evaporation from the lungs (Figure 39-5).

Species differ in the total amount of water that they process each day. Desert animals, which have access to almost no water, conserve water meticulously. The kangaroo rat, a common desert rodent in the American Southwest, lives entirely on water derived from the metabolic breakdown of macromolecules in the seeds that it eats.

The kangaroo rat is much stingier with its water than nondesert animals. It lacks sweat glands entirely, its urine is highly concentrated, and it spends the daylight hours in cool underground burrows whose air is moister than that outside. It loses water mostly through evaporation from the lungs. Even this water loss, however, is less than it would be in other animals. The kangaroo rat has long nasal passages that

◀ **Figure 39-4   Salty tears.** Like many sea animals, this tortoise excretes excess salt in glands located near the eye. (*M.P. Kahl/Photo Researchers, Inc.*)

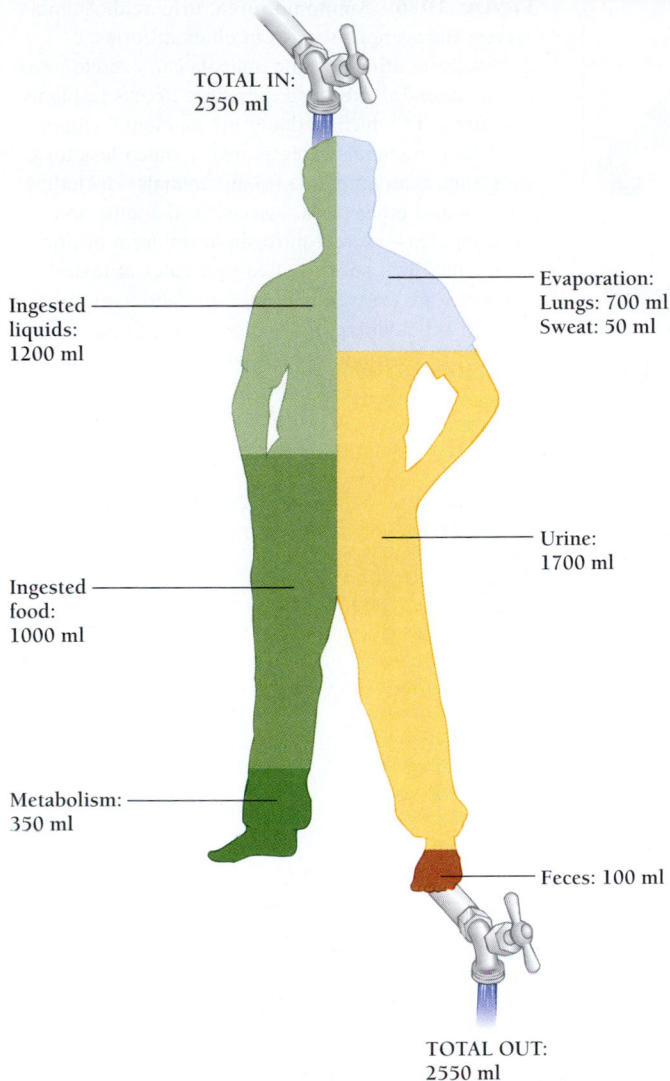

TOTAL IN:
2550 ml

Ingested
liquids:
1200 ml

Evaporation:
Lungs: 700 ml
Sweat: 50 ml

Ingested
food:
1000 ml

Urine:
1700 ml

Metabolism:
350 ml

Feces: 100 ml

TOTAL OUT:
2550 ml

Figure 39-5   Water in, water out. An average human ingests and excretes 2550 ml of water each day. Water comes into the body as food and drink and leaves in the urine, feces, and sweat and in the form of vapor from the lungs and other moist surfaces.

cool and condense the water in exhaled air. The condensed water is then reabsorbed before the air leaves the kangaroo rat's body.

In all animals, water intake and loss may vary greatly from day to day. A human, for example, may drink more one day than the next, sweat profusely during vigorous exercise, or lose too much water because of diarrhea or vomiting. Humans and other animals must therefore regulate water flow. We do so mostly by varying our drinking and urine production. We are so successful that, under most circumstances, our water content varies less than 1 percent. The failure of such regulation leads either to **dehydration** (in which too much water flows out of the body) or to *water intoxication* (in which too much

water flows into the body) and **edema** [Greek, = swelling] (in which water accumulates in the tissues).

Land animals acquire water from drinking, food, and metabolism. They lose water in urine, feces, and evaporation from lungs, skin, and other moist surfaces.

HOW DO ANIMALS DISPOSE OF NITROGEN-CONTAINING WASTES?

In land animals, water and salt regulation is tightly connected to the disposal of a variety of wastes. Chief among these are the **nitrogenous** (nitrogen-containing) **wastes,** which are the breakdown products of proteins, nucleic acids, and other nitrogen-containing compounds. Nitrogenous wastes come not only from food, but also from the normal turnover of cell components.

Cells and organisms rid themselves of nitrogenous wastes by means of **excretion,** a disposal process by which cells pass materials across a cell membrane. In contrast, the disposal of the remains of digested food is called **elimination.** Many animals, including mammals, excrete nitrogenous wastes in urine and eliminate digestive wastes in feces. Some animals, including birds and insects, mix the products of excretion with those of elimination.

Fish and most aquatic animals convert excess nitrogen to **ammonia** (NH_3). Ammonia-loaded urine flows rapidly into the surrounding water, where it provides usable nitrogen to a variety of aquatic microorganisms and plants.

Ammonia, however, reacts rapidly with organic molecules and is highly toxic to animals. Aquatic animals can dilute the ammonia to harmless concentrations, but land animals cannot. Land mammals solve this problem by using liver enzymes to convert ammonia into **urea,** a relatively nontoxic organic compound (Figure 39-6). Urea is less toxic, but it dissolves only in a large amount of water. Animals must therefore secrete a minimum amount of water each day—in humans, at least 500 ml.

Insects, land snails, most reptiles, and birds require even less water than mammals. Instead of disposing of nitrogenous wastes as urea, they produce a nearly insoluble organic compound called **uric acid** (Figure 39-6). Some insects, as well as the embryos of reptiles and birds, go one step further. They do not excrete the uric acid at all. They squeeze out as much water as they can and then deposit dry uric acid in their own bodies.

Organisms rid themselves of nitrogenous wastes by excretion of ammonia, urea, and uric acid.

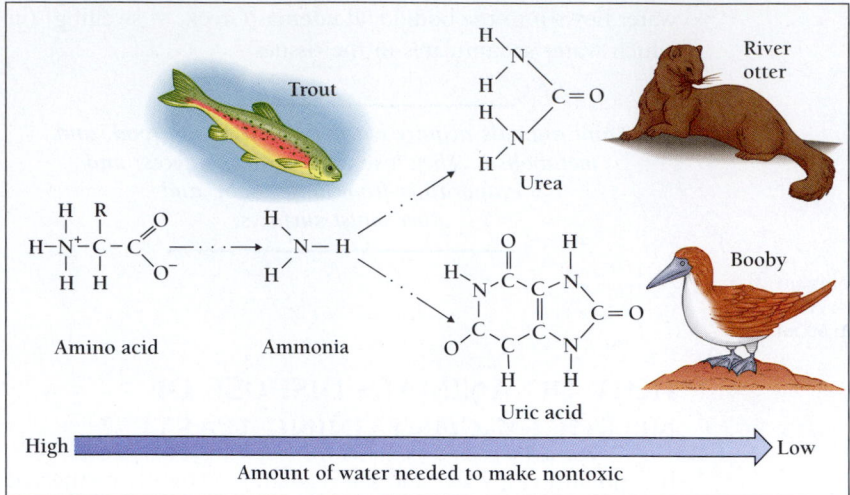

A.

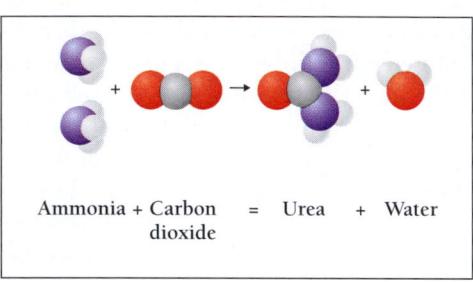

Ammonia + Carbon   =   Urea   +   Water
                dioxide

B.

Figure 39-6   Ammonia, urea, uric acid. Animals excrete nitrogenous wastes in different forms. A. Trout and other aquatic animals can excrete nitrogen as ammonia because they have access to plenty of water with which to dilute the ammonia. Otters and other mammals excrete urea, a much less toxic substance than ammonia. Many animals—including boobies and other birds, insects, land snails, and most reptiles—excrete nitrogen in the form of uric acid, which is a solid. B. Two molecules of toxic ammonia are transformed into urea and water through the addition of a molecule of carbon dioxide by a set of reactions carried out in the liver.

HOW DO KIDNEYS WORK?

In vertebrates, kidneys are the major organs of excretion. In addition, kidneys regulate water balance, as well as the concentrations of salts and various organic molecules, including wastes and toxins.

Kidneys are paired structures that lie against the lower back (Figure 39-7). Blood, carrying wastes from all over the body, arrives at each kidney by a **renal artery** [Latin, *renes* = kidney] and leaves by a **renal vein.** The two kidneys take wastes and water from the blood and produce urine that flows down into the **bladder** through a pair of tubes called the **ureters** [Greek, *ouron* = urine]. The bladder in turn drains to the outside of the body through the **urethra.**

In female mammals, the urethra empties directly to the outside. In male mammals, the urethra joins the reproductive tract and passes through the penis, a common exit for both sperm and urine. In most birds, reptiles, and amphibians, the ureter runs directly into a **cloaca,** a common exit chamber for the digestive, excretory, and reproductive systems.

How Does a Kidney Filter the Blood?

The bland exterior shape of a kidney reveals little about how it works. The interior of a kidney, however, is intriguingly intricate (Figure 39-8). A section through a kidney shows an

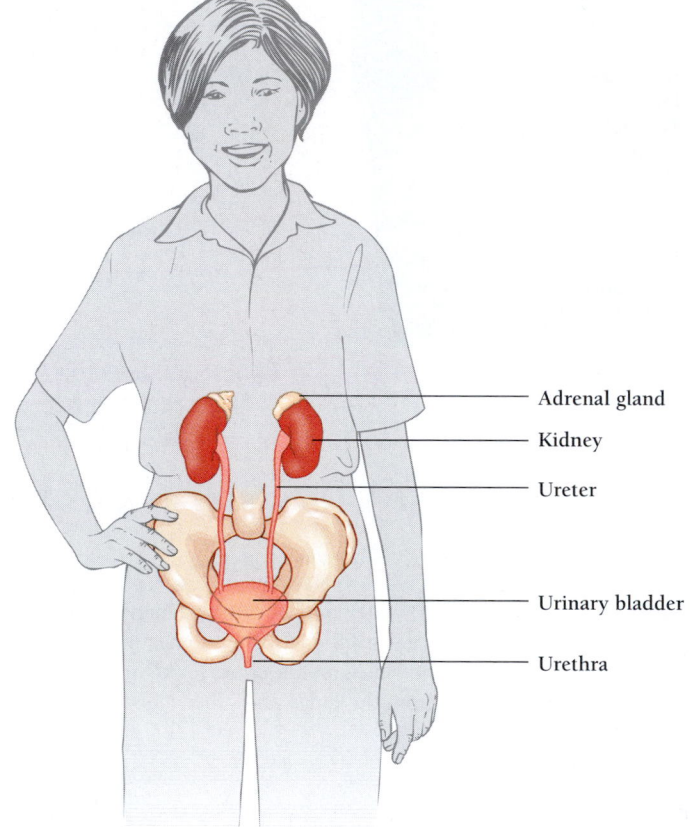

Adrenal gland

Kidney

Ureter

Urinary bladder

Urethra

Figure 39-7   Two kidneys, two ureters, one bladder, one urethra. The two kidneys lie just below the diaphragm close to the ribs of the back. Each kidney drains into a ureter, which drains into the bladder.

outer region, called the **cortex,** and an inner region, called the **medulla.**

The renal arteries enter the kidneys and then branch into fine arterioles in the cortex. Each arteriole forms a tangled network of capillaries, called a **glomerulus** [Latin, little ball; plural, *glomeruli*]. A single human kidney contains about a million glomeruli (Figure 39-9).

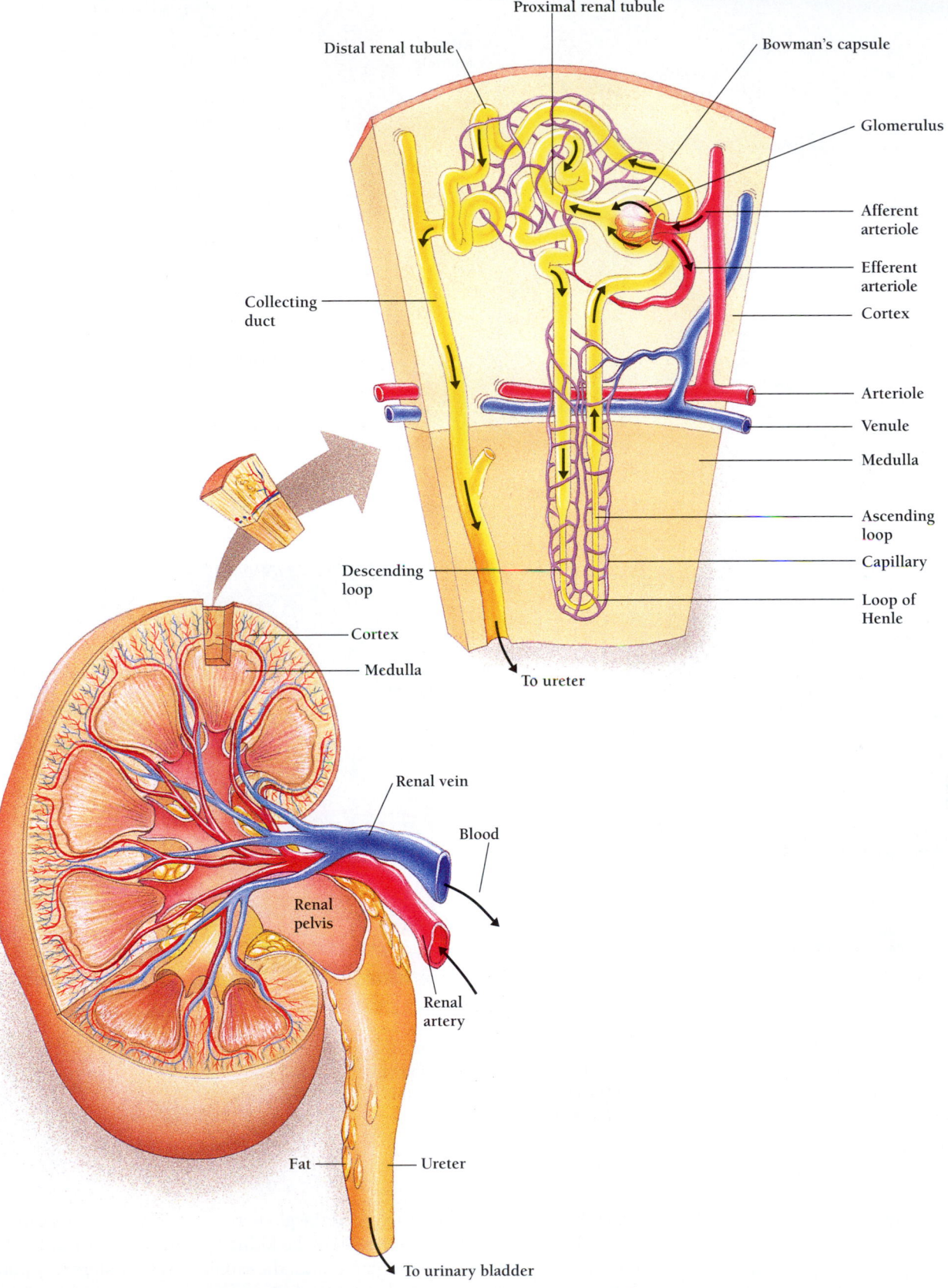

Figure 39-8   Cross section through a human kidney. The outer cortex contains the renal tubules and Bowman's capsule for a million nephrons. The inner medulla contains all the loops of Henle.

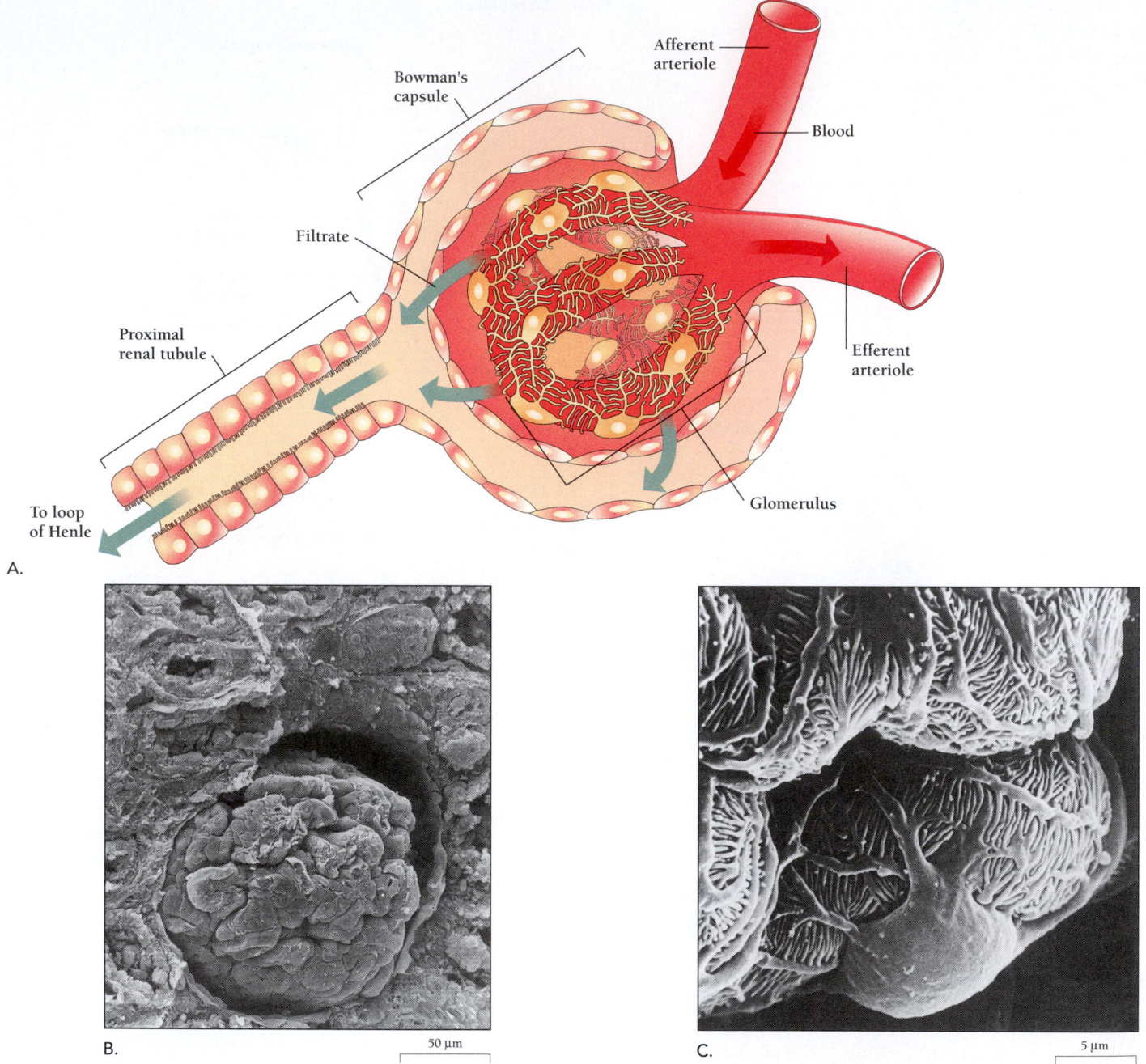

Figure 39-9   A glomerulus. A mammalian nephron includes a network of capillaries called the glomerulus. Blood pressure forces fluid out of the capillaries of the glomerulus and into an enclosed space called Bowman's capsule. The fluid in Bowman's capsule then flows into the proximal renal tubule for processing by the kidney. *(B, © Fred Hossler/Visuals Unlimited; C, F. Spinelli, D. Fawcett/Visuals Unlimited)*

Surrounding each glomerulus is a bulb called **Bowman's capsule,** which leads to a long, narrow tube, called the **renal tubule.** The glomerulus, Bowman's capsule, and the tubule together form a **nephron** [Greek, *nephros* = kidney], the functional unit of the kidney. If we can understand the workings of a single nephron, we can understand the workings of the whole kidney.

The structure of the glomerulus and of Bowman's capsule provides a clue to how the kidney performs its first task—filtration. In the glomerulus, the capillary walls contain tiny pores through which the blood plasma can filter, leaving blood cells and macromolecules behind. The filtered plasma then immediately moves through sievelike *filtration slits,* formed from cells of the inner surface of Bowman's capsule.

Blood from a renal artery enters each glomerulus and filters through the pores and filtration slits into the interior (lumen) of the tubule. The driving force for this filtration is the blood pressure within the glomerulus's arterioles. Water, salt ions, urea, and other small molecules in the blood freely pass into the lumen to form the **filtrate,** which moves into the narrow renal tubule. The initial filtrate is similar in composition to the blood's plasma, but without proteins.

The glomerulus and Bowman's capsule filter the blood plasma, so that cells and protein molecules do not enter the interior of the renal tubule.

How Do the Tubules Recycle Essential Molecules and Ions and Recapture Water?

In humans, the total volume of filtrate is almost 200 liters per day (enough to fill a very large bathtub). But the nephrons must retain enough water to maintain the volume of the blood, as well as all the essential ions and molecules that form an indispensable part of the internal environment. The nephrons typically recover more than 99 percent of the 200 liters of water and proteins. As a result, only 1.5 liters per day actually leaves the body as urine.

After the glomerulus and Bowman's capsule remove nearly everything from the blood, the filtrate flows down the length of the tubule. The cells of the tubule then **reabsorb** salt, water, glucose, and other ions and molecules and return them to the blood (Figure 39-10). Reabsorption is so efficient that, in a healthy person, no glucose at all escapes in the urine. The overall strategy of the kidney is the same as many people use to clean out a cluttered drawer—"empty it all out and put back what you really want to keep."

As the tubule pumps ions and molecules out (and back into the plasma and extracellular fluid), water follows. At any point in the passage through the proximal tubule, the osmolarity is the same within the lumen and in the surrounding capillaries.

The capillaries of the glomerulus are particularly leaky, so that, in humans, they filter all the body's plasma about 60 times per day. The blood pressure within Bowman's capsule is only about one-quarter that of the blood, but it is high enough to drive the fluid from the capsule down into the rest of the tubule. At the end of the tubule, the remaining filtrate exits via a collecting duct and empties into the ureter.

The renal tubules reabsorb ions and molecules from the filtrate. Water follows passively.

What does the structure of a tubule suggest about its workings?

Each tubule has a characteristic hairpin shape, with three main regions: (1) the wide **proximal tubule,** which lies just next to Bowman's capsule, in the kidney's outer cortex; (2) the narrow

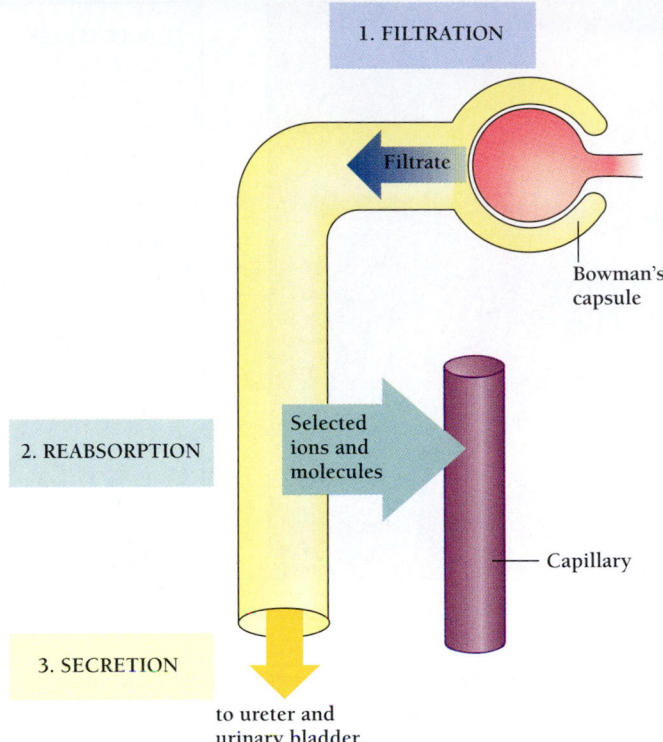

Figure 39-10 **Filtration, reabsorption, and secretion.** Each nephron *filters* cells and large molecules from the blood, *reabsorbs* salts and water, and *secretes* urea and water and waste materials.

loop of Henle (pronounced HEN-lee), which first descends into the inner medulla and then ascends back into the cortex; and (3) the wide **distal tubule,** which twists and turns in the cortex. The tubule's contents then flow into a collecting duct that again descends through the kidney's medulla, where many collecting ducts converge (Figure 39-8).

We can get a better idea about how a tubule works by examining the microscopic structure of its **epithelium,** the sheet of cells that lines the lumen of the tubule. The proximal tubule has great numbers of fingerlike *microvilli,* which form fuzzy-looking *brush borders* (Figure 39-11A). These microvilli greatly increase the surface area of the epithelium, and sodium pumps in the proximal tubule remove almost three-quarters of the sodium ions from the lumen. Chloride ions follow the sodium ions across the tubule wall, maintaining electrical neutrality; and water also follows, maintaining osmolarity (Figure 39-11B). The proximal tubule also has dense concentrations of mitochondria, which provide ATP to power the sodium pumps.

In contrast, the descending limb of the loop of Henle and the deepest part of the ascending limb have neither brush borders nor many mitochondria. We can guess from this difference that these parts of the loop of Henle do not actively pump materials out of the lumen.

Further along, in the upper portion of the ascending loop of Henle, in the distal tubule, and in the upper part of the collecting duct, we find that the epithelium again has brush

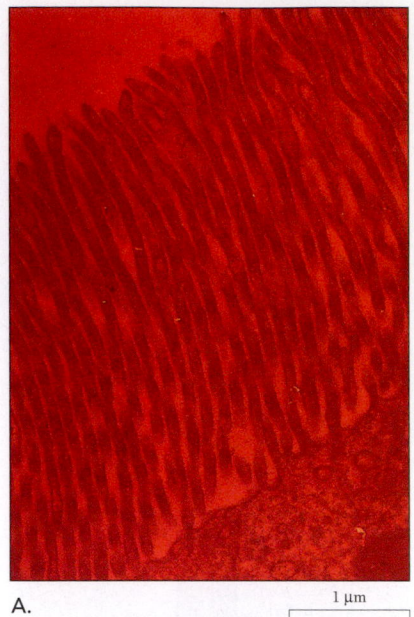

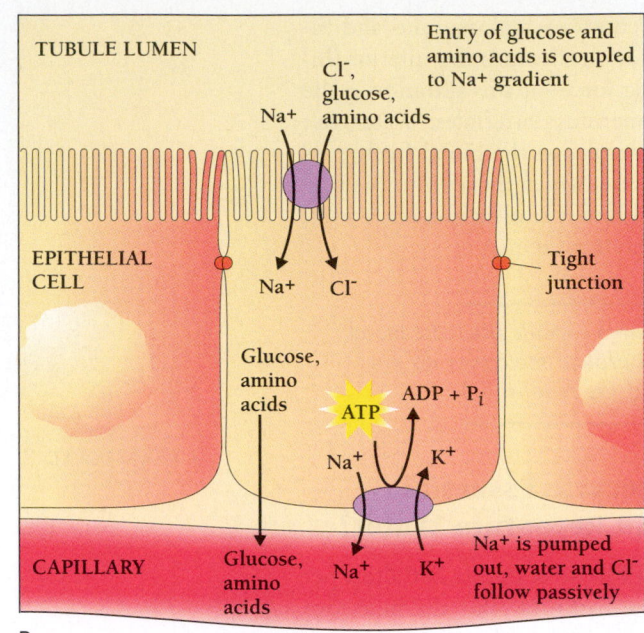

Figure 39-11   Brush borders. A. Microvilli, or brush borders, in the proximal tubules increase surface area and so promote the movement of ions. B. In the proximal tubules, brush border cells provide a greater area for the absorption of Na$^+$ ions from the filtrate. Water then follows the ions out of the tubule. *(A, Fred Hossler/Visuals Unlimited)*

borders and great numbers of mitochondria. These segments also pump sodium ions out of the lumen.

> *The three parts of a renal tubule differ in their structure and function. The proximal tubule is lined with microvilli, which increase the area through which sodium ions move across the epithelium. The descending limb of the loop of Henle is smooth and does not actively pump ions. The upper part of the ascending limb, the distal tubule, and the upper part of the collecting duct are lined with microvilli and also pump sodium ions.*

How do the tubules regulate the ion content of the lumen?

Researchers have been able to study the workings of the nephron by taking samples of the contents of different parts of the tubules with fine pipettes. By analyzing these samples, researchers found that the concentrations of specific ions and molecules change along the length of the tubule. These concentration changes result both from the active pumping of ions across the membranes and from osmosis.

The epithelium of the proximal tubule actively pumps sodium ions out of the filtrate in the lumen. Water follows the salt out of the lumen and into nearby capillaries. So, because the epithelium is freely permeable to water, the ion concentration in the lumen does not change.

By the time the filtrate in the lumen enters the loop of Henle, about 75 percent of the water has moved out of the filtrate and back into the extracellular fluid and the blood. The remaining filtrate still has the same salt concentration as blood plasma and the initial filtrate, and it still has many small molecules, including urea.

As we guessed from the absence of mitochondria and a brush border, neither the descending limb of the loop of Henle nor the deep part of the ascending limb actively transports ions (Figure 39-12). The deep part of the medulla, at the bottom of the loop of Henle, is very salty, however, as we will discuss shortly. Because of the difference in osmolarity, water in the lumen moves out through the walls of the tubule by osmosis, concentrating the filtrate as it moves to the bottom of the loop.

Beyond the turn of the loop, the epithelium is impermeable to water and no more water escapes the lumen. The upper part of the ascending limb, however, actively pumps salt out of the lumen, and so does the upper section of the collecting duct (Figure 39-12). As a result, by the time the fluid arrives in the collecting duct, the salt concentration in the lumen is lower than that in the fluid of the surrounding cortex. No water can escape from the lumen, however, until the lumen's contents reach the collecting duct. Water then flows out of the lumen, raising the total osmolarity in the lumen to that of the surrounding cortex.

BOX 39.1

The Artificial Kidney

Infections, poisons, genetic diseases, tumors, and physical trauma can all severely damage the kidneys. In fact, more than 25,000 people die of kidney failure in the United States each year. When a person's kidneys begin to fail, urea and other substances build up in the blood. This condition is called uremia ("urine in the blood").

Many patients with kidney failure stay alive with the help of a device called a kidney dialysis machine. The patient's blood runs into a chamber containing a membrane that allows the passage of salt ions, as well as urea and other small molecules (Figure A). On the other side of the membrane is a large volume of a solution whose composition is the same as that of normal blood plasma. Sodium, chloride, and potassium ions, as well as glucose and other molecules,

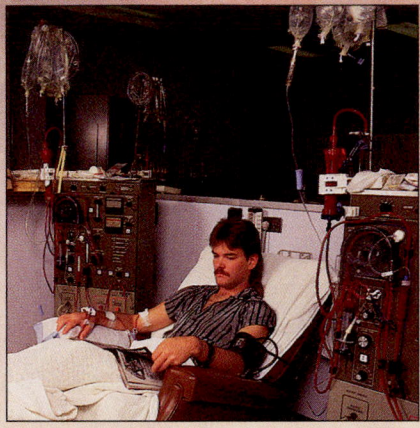

Figure A   Kidney dialysis. A young man with damaged kidneys waits as a dialysis machine cleans urea and other small waste molecules from his blood. *(SIU/Photo Researchers, Inc.)*

move passively across the artificial membrane and come to the same concentrations on each side. Urea and other small waste molecules diffuse across the membrane into the large volume, so that their concentration falls nearly to zero. The kidney dialysis machine thus restores the blood plasma to a nearly normal composition.

Patients who depend on dialysis machines cannot regulate their water balance with hormones, as the rest of us do. Instead, they must carefully balance their intake of food and drink to match losses from evaporation and elimination. Depending on the remaining abilities of their damaged kidneys, these patients produce little or no urine. Even the best dialysis machine, then, cannot approach the feats of the simplest kidney.

Active pumping of ions by the cells in the wall of the renal tubules creates osmotic gradients that pull water from the filtrate.

What happens to the urea?

Like salt, urea enters the tubule in the original filtrate. But unlike salt, urea does not move out of the lumen of the tubule. As the filtrate moves along its path, the salt and water leave the lumen, leaving behind an increasingly concentrated solution of urea. Only when the urea-loaded filtrate reaches the far end of the collecting duct can some urea diffuse out. As the concentrated urea exits the tubules, it raises the osmolarity of the fluid surrounding the deep part of the loop of Henle. The high urea content of this fluid, as well as its saltiness, is the driving force that pulls water and salt passively out of the lumen—in the deepest part of the descending limb and in the collecting duct (Figure 39-12).

Urea accumulates in the renal tubule until the filtrate reaches the far end of the collecting duct. There, urea diffuses into the surrounding tissues and creates the concentration difference that draws water from the remaining fluid in the tubule.

How Does the Kidney Form Concentrated Urine?

The filtrate leaving the distal tubule is still dilute and not yet ready to be excreted. In the collecting ducts, the water content of the filtrate can remain dilute or become highly concentrated, depending on variables that we discuss shortly.

The collecting ducts pass through the deepest part of the medulla, where the concentrations of salt and urea are the highest. There, water and urea passively move out of the lumen. The urine that leaves the duct can therefore attain the same concentrations of sodium, chloride, and urea as the surrounding extracellular fluid in the deepest region of the loop of Henle. For humans, this means a final maximum salt concentration almost four times that in blood plasma. In the kangaroo rat, the loop of Henle is even more effective. The salt concentration in the urine of a kangaroo rat can be 14 times that in the animal's blood plasma.

In addition to managing the flow of salt, urea, and water, the nephron also controls the concentrations of other substances in the blood and the urine. For example, the tubule pulls ("reabsorbs") ions, glucose, and many small molecules back into the blood. It also moves specific substances into the urine in a process called **secretion.** The kidney can secrete potassium and hydrogen ions, ammonia, and organic acids and bases. In addition, the kidney secretes foreign substances such as antibiotics and toxins, which are chemically tagged by the liver for secretion.

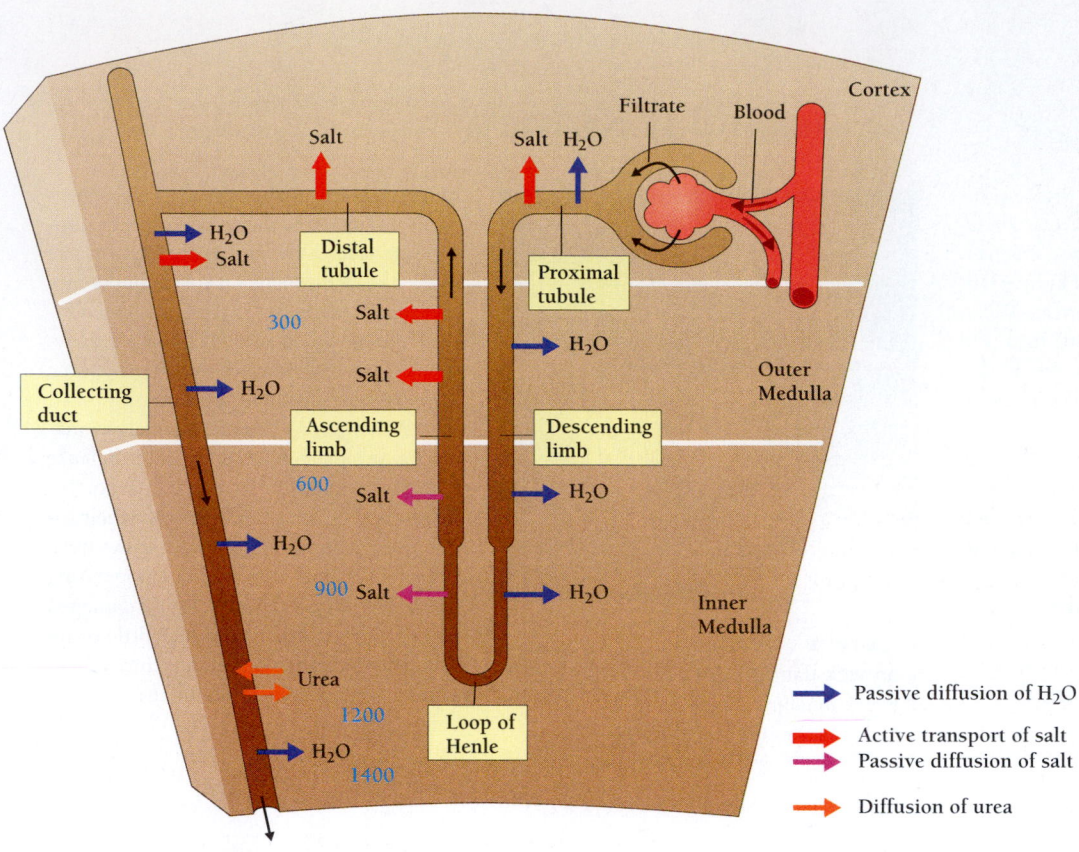

Figure 39-12   Countercurrent concentration of solutes in a nephron. A nephron is a countercurrent system that filters the blood and concentrates certain ions and molecules to form the urine. The epithelium at the beginning of the renal tubule actively pumps sodium ions out of the filtrate. The filtrate, containing water, urea, and other molecules, descends into the loop of Henle, which is surrounded by the increasingly salty medulla. Water passively diffuses from the tubule into a network of capillaries surrounding the loop of Henle. Beyond the turn of the loop, the epithelium is impermeable and water remains in the tubule. While the water is trapped inside, however, the same region of the tubule actively transports sodium ions out. The fluid entering the collecting duct is consequently dilute. As this dilute fluid descends into the salty medulla, water can flow out passively, and the urea and salts left in the duct become concentrated. Finally, as the water and urea pass through the saltiest portion of the medulla, urea flows out passively, further increasing the concentration of ions and molecules in the medulla. The remaining fluid, the urine, now flows from the collecting duct to the ureters and the bladder.

A kidney nephron generates urine that can contain high concentrations of urea, as well as various ions, ammonia, organic acids and bases, and foreign molecules. The nephron normally reabsorbs glucose and other nutrients and ions.

How Do the Kidneys of Nonmammalian Vertebrates Meet the Challenges of Different Environments?

In general the structure and function of nephrons are similar in all vertebrates. Not surprisingly, however, there are signifi-

cant differences. The kidneys of marine bony fish, for example, have nephrons without glomeruli and Bowman's capsules. Bony fish adjust the composition of the fluid in the tubules by secretion and reabsorption only. In contrast, freshwater bony fish have highly developed glomeruli, the better to rid the fish of excess water.

Both freshwater and saltwater fish have special cells in their gills that can pump chloride and other ions. The direction of the pumping depends on the environment. Freshwater fish pump salt ions into the body, whereas saltwater fish pump them outwards. Salmon and other fish that migrate from fresh water to salt water and back change the organization of these cells and the direction of salt pumping according to whether they are in fresh water or salt water.

The kidneys of most birds resemble those of mammals. Marine birds, which live on marine fish and drink salt water, however, have another means of disposing of salt. Like the sea tortoise of Figure 39-4, they possess special salt glands just above their eyes, which secrete a salty solution with twice the concentration of seawater. The salty tears produced by these salt glands appear to help these animals rid themselves of excess salt.

HOW DO ANIMALS ADJUST WATER AND SALT IN RESPONSE TO CHANGING CONDITIONS?

Humans and other animals consume different amounts of water and salt from day to day, and they eliminate different amounts in their feces, sweat, and urine. Maintaining a constant level of water and salt concentration in the blood means that the kidneys must constantly adjust urine concentration and flow.

Such homeostasis depends on a variety of complex mechanisms, including the control of fluid intake and the regulation of blood flow through the kidneys. Three hormones contribute to this regulation—aldosterone, antidiuretic hormone (ADH, also called vasopressin), and atrial natriuretic factor (ANF).

How Does the Brain Control Fluid Intake?

We animals control water balance in the body by varying how much water we drink. In mammals, the hypothalamus of the brain includes a "thirst center" that has *osmoreceptors,* receptors that detect changes in salt concentration. An increase in salt concentration in the extracellular fluid stimulates these receptors, and we experience the sensation of thirst, which lasts until we drink some water.

Once we drink water, we do not feel thirsty for about 15 minutes (30 minutes on a full stomach). If the salt concentration in the blood has not normalized, however, we get thirsty again, and we drink until the proper concentration is achieved. A thirsty animal almost never drinks more water than the amount needed to relieve dehydration. We always drink almost exactly the right amount.

When salt concentrations in the blood increase, the brain's hypothalamus detects the increase and triggers a sensation of thirst.

The Rate of Filtration Depends on Renal Blood Flow

Blood flow to the renal arteries of the kidneys is generally much higher and more constant than blood flow to other areas of the body. Nonetheless, the arterioles entering and leaving the glomerulus respond to changes in salt concentration in the blood. When the concentration of salt (or of nitrogenous wastes) rises, blood flow to the kidney increases. The kidneys can then filter more and remove more salt, for example, after we have eaten a bag of corn chips or a few slices of pizza.

An increase in salt concentration in the blood causes an increase in the blood flow to the kidneys.

Angiotensin and Renin

Angiotensin, a polypeptide hormone, is the most powerful constrictor of blood vessels known. One ten-millionth of a gram (0.1 μg) of angiotensin can cause the blood vessels to constrict and blood pressure to rise. Angiotensin is derived from a precursor polypeptide made in the liver. **Renin,** an enzyme secreted by the kidneys, produces angiotensin by cutting the precursor molecule.

Either a decrease in blood pressure or a decrease in the concentration of sodium ions in the blood causes the kidneys to secrete renin, thereby speeding the production of angiotensin. Angiotensin constricts the arterioles that lead to the glomerulus, thereby increasing the filtration rate and the rate of movement of the filtrate through the nephron. The result is increased retention of both water and salt. This salt and water retention increases total blood volume. This increased volume of blood, confined in a circulatory system of limited size, increases blood pressure. Angiotensin compensates for a decrease in sodium ions or blood pressure with a homeostatic buildup of body fluid and sodium ions (Figure 39-13A).

A decrease in blood pressure or in sodium ions causes the kidneys to secrete renin, which increases angiotensin production. Angiotensin constricts the arterioles, increasing blood pressure and causing the kidneys (and therefore the body) to retain more water and salt.

Aldosterone

The steroid hormone **aldosterone** comes from the cortex of the **adrenal glands** [Latin, *ad* = towards + *renes* = kidneys], which lie on top of the kidneys. Aldosterone helps keep the body from losing salt by stimulating sodium and potassium reabsorption in the distal tubules and collecting ducts of the kidneys, as well as in the sweat glands and in the intestines (Figure 39-13A). When the concentration of sodium in the blood decreases, the adrenal glands secrete more aldosterone.

Aldosterone regulates water loss as well as salt loss. People who cannot secrete aldosterone because of malfunctioning adrenal glands lose large amounts of salt in the urine. Along with the salt goes a tremendous quantity of water. Most people can compensate by consuming salty foods and lots of water. If not, however, a person lacking aldosterone can become severely

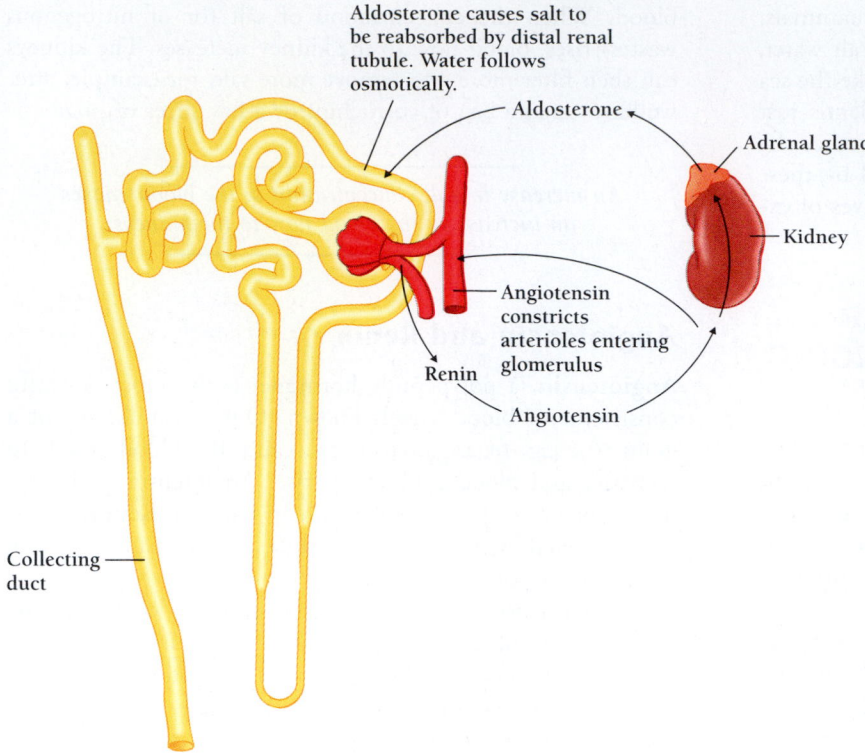

Aldosterone causes salt to be reabsorbed by distal renal tubule. Water follows osmotically.

Aldosterone

Adrenal gland

Kidney

Angiotensin constricts arterioles entering glomerulus

Renin

Angiotensin

Collecting duct

A.

Figure 39-13 Hormones and blood pressure.
A. When blood pressure or sodium concentration drops, the kidneys secrete the enzyme renin, which acts to form angiotensin in the blood. Angiotensin counteracts a drop in blood pressure by constricting the arterioles and veins, which increases blood pressure and causes the kidneys to retain both water and salt. The hormone aldosterone also increases blood pressure by specifically stimulating reabsorption of salt (and, therefore, water) in the distal tubules, thus increasing total blood volume.
B. When the concentration of salt in the blood rises, the posterior pituitary gland secretes ADH (vasopressin), which increases water permeability in the collecting ducts of the kidneys. By causing the body to retain water, ADH can raise blood pressure considerably. The heart limits such increases in blood pressure by secreting the hormone ANF. ANF inhibits the secretion of renin (and therefore angiotensin), aldosterone, and ADH (vasopressin). ANF also lowers blood pressure by relaxing smooth muscles, reducing thirst, and increasing the kidney's elimination of Na^+ and water by closing channels in the collecting ducts.

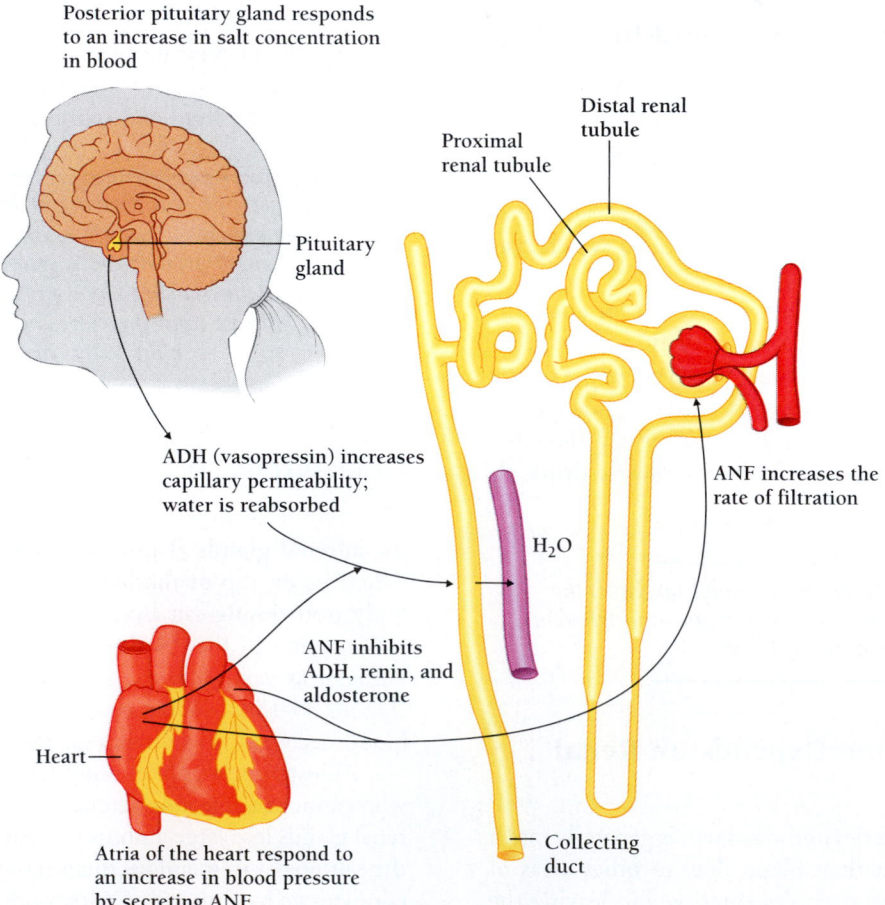

Posterior pituitary gland responds to an increase in salt concentration in blood

Proximal renal tubule

Distal renal tubule

Pituitary gland

ADH (vasopressin) increases capillary permeability; water is reabsorbed

ANF increases the rate of filtration

H_2O

ANF inhibits ADH, renin, and aldosterone

Heart

Atria of the heart respond to an increase in blood pressure by secreting ANF

Collecting duct

B.

dehydrated and salt depleted. In that case, the volume of blood is so low that the heart cannot supply blood to the tissues. The result is severe shock and, often, death.

Aldosterone secretion depends not only on the adrenal glands themselves but also on chemical signals from the liver and electrical signals from the nervous system. This complex control system allows regulation of sodium ion reabsorption in response to changes in blood pressure, salt concentration, water loss, and disease.

Aldosterone prevents water loss and salt loss by stimulating the reabsorption of sodium and potassium ions in the kidneys, the sweat glands, and the intestines.

Antidiuretic Hormone (ADH), or Vasopressin

Like aldosterone, **antidiuretic hormone (ADH)** prevents water loss. ADH, also called **vasopressin,** is a polypeptide hormone made in the hypothalamus and released from the posterior pituitary (Chapter 36). Antidiuretic hormone, as its name suggests, is the opposite of a *diuretic,* a substance, such as caffeine or alcohol, that stimulates water loss. ADH causes the body to retain water by increasing water permeability in the collecting ducts of the kidneys. Water then returns to the blood through the walls of the collecting ducts and the urine becomes more concentrated (Figure 39-13B). By this means ADH can raise blood pressure even more than angiotensin.

The cells of the posterior pituitary respond to changes in the salt concentration of the blood. After we eat a bag of corn chips or a slice of pizza, the concentration of salt in the blood rises and the posterior pituitary responds by secreting ADH into the blood. When the ADH reaches the kidneys, it increases water uptake in the collecting ducts. As a result, the urine becomes more concentrated, the blood becomes more dilute, and the salt concentration of the blood returns to normal. The total volume of the blood increases, however, and the result is an increase in blood pressure.

Other factors also influence ADH production and secretion. Alcohol, for example, is a well-known diuretic. Alcohol inhibits the release of ADH, increasing urine flow. But alcohol also stimulates an appetite for salt. A beer drinker reaches for a bowl of salty pretzels, then for another beer. But while the salt stimulates the release of ADH, the alcohol inhibits its release. Pain, fear, cold, and stress can all decrease ADH levels and increase urine production.

The same hormone systems that mediate changes in kidney operations also affect brain centers to stimulate thirst and appetite for salt. The body thus controls the composition of the blood and other fluids by regulating the transport of specific substances and altering drinking and eating behavior.

When the concentration of salt in the blood rises, the posterior pituitary secretes ADH, which increases water permeability in the collecting ducts of the kidneys. By causing the body to retain water, ADH can raise blood pressure.

Atrial Natriuretic Factor (ANF)

As we saw earlier, kidney function depends on blood pressure to provide the force for filtration and flow of the filtrate through the tubules. We have seen that an increase in ADH lowers urine volume and raises the volume of blood in the vessels. The result is an increase in blood pressure.

The heart itself helps to limit such increases. In response to excessive blood pressure, cells in the smaller chambers of the heart (the atria) produce a hormone called **atrial natriuretic factor** [Latin, *atrium* = vestibule, referring to the heart chamber, + *natrium* = sodium + Greek, *ouron* = urine]. ANF inhibits the secretion of renin, aldosterone, and vasopressin, relaxes smooth muscles, reduces thirst, and increases the kidneys' elimination of sodium ions and water by closing channels in the collecting ducts. ANF thus increases water loss and lowers blood pressure (Figure 39-13B).

The heart produces ANF, a hormone that increases water loss and lowers blood pressure.

HOW DO INVERTEBRATES SOLVE PROBLEMS OF WATER, SALTS, AND WASTES?

Animals in different environments must solve different problems of salt and water balance. Their solutions differ in ways that at least partly reflect their distinct evolutionary histories.

Flatworms

The simplest excretory organs are those of flatworms. These organs mainly excrete water, for most nitrogenous wastes are eliminated through the one-ended gut. The excretory organs of flatworms consist of two or more tubes that run the length of their bodies and empty through tiny pores. Leading into the tubes are many bell-shaped chambers, each formed from a single cell. Cilia, which line the interiors of these bell-shaped cells, propel liquids into the tubules. Under the microscope, the cilia appear to be flickering, an illusion that gives these cells their curious name, **flame cells.**

Earthworms

Although we think of the earthworm as a land animal, it spends most of its life in damp tunnels in the soil. Its body fluids therefore have a higher concentration of ions and molecules than the surrounding water, and the earthworm has the same problem as a freshwater animal—getting rid of excess water.

Like mammalian kidneys, earthworm **nephridia** (singular, nephridium) both filter and reabsorb. Each body segment contains a pair of nephridia. Fluid in the earthworm's body cavity, or coelom, is under pressure. The pressurized fluid pushes through an opening in the tubular part of the nephridium. Inside the nephridium, ions and molecules reenter the circulatory system. The dilute urine left behind passes from each nephridium through a pore in the body wall.

Marine Invertebrates

Crabs and most other marine invertebrates have body fluids with about the same concentration of salts as seawater. Many of these animals allow the concentrations of ions and molecules in their body fluids to fluctuate with that of their environment. In general, the disposal of excess water is not a problem for such animals.

Still, marine invertebrates must get rid of nitrogenous wastes and adjust the concentrations of specific ions. They accomplish these tasks with kidneys that, like those of vertebrates, work by filtration, absorption, and secretion. Like other aquatic animals, they eliminate excess nitrogen in the form of ammonia. Some invertebrates have arrangements of excretory organs very different from ours. Lobsters and other crustaceans, for example, perform excretion in their heads, through *green glands,* which function much like other invertebrate kidneys. And the abalone's two kidneys differ in function—the left kidney specializes in reabsorption and the right in secretion.

Insects

The excretory system of insects differs from all the other systems we have discussed. Insect excretory organs, called Malpighian tubules, are blind outpocketings of the gut. These tubules may number from two to several hundred. Their blind ends are bathed in the insect's body fluids (Figure 39-14).

Fluid flows into the tubules, but not because of a pressure difference (as in earthworms and marine invertebrates). Instead, the tubules actively pump potassium ions. Water, with

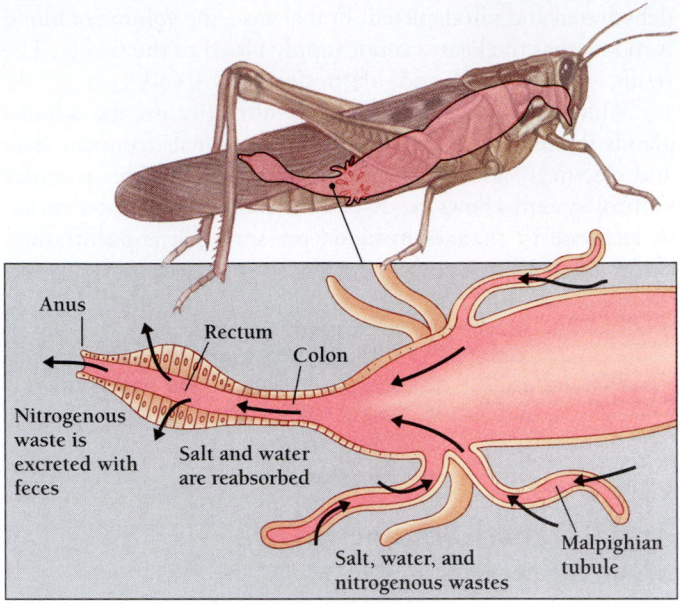

Figure 39-14   Malpighian tubules. In insects, the excretory organs are blind outpocketings of the gut.

dissolved nitrogen wastes, passively flows into the tubules by osmosis. The contents then flow into the gut, where the residues of digestion (the feces) mix with the products of excretion (the urine).

The insect's hindgut and rectum are especially efficient in removing water, so that the combination of feces and urine can be extremely dry—depending on the insect's diet. An insect that lives on fresh leaves has abundant water intake and excretes lots of liquid urine, while a termite, which dines on wood, excretes dry little pellets with the consistency of the compacted sawdust "logs" that city people buy at the supermarket for their fireplaces.

In this chapter we have seen why animals need kidneys and how they function to excrete water, salts, and nitrogenous wastes. In the next chapter we will see how animals defend themselves against infection by microorganisms, promote healing, and minimize blood loss—all important functions of the immune system.

Study Outline with Key Terms

The "internal environment" of an animal is the external environment of individual cells. Animals must regulate the composition of the fluids in this extracellular space.

An animal's major problem in regulating its internal environment depends on the total salt concentration in its external environment. Freshwater animals tend to gain water and lose salts, while marine animals tend to lose water and gain salts. The major problem of land animals is to get and con-

serve water. Every animal has characteristic adaptations for regulating salt and water, for preventing **dehydration** or **edema.**

Animals must rid themselves of **nitrogenous wastes** from the proteins in their food. Marine invertebrates and fish excrete excess nitrogen in the form of **ammonia.** Mammals instead convert the waste nitrogen to **urea,** which is less toxic than ammonia. Birds, insects, and other animals excrete **uric acid. Excretion** and **elimination** are separate processes. Flatworms excrete nitrogenous wastes using **flame cells,** while earthworms use more complex **nephridia.**

In vertebrates, the **kidney** is the major organ of excretion and of salt and water balance. The kidney regulates the internal environment by regulating the composition of the urine. This regulation depends on three processes—**filtration, reabsorption,** and **secretion.** Blood carries wastes to the kidneys through paired **renal arteries. Renal veins** carry the cleansed blood away from the kidneys. Urine leaves the kidneys in paired **ureters,** fills the **bladder,** and exits the body through a **urethra** or **cloaca.**

The **cortex** and **medulla** of the kidney consist of many separate functional units, called **nephrons.** Each nephron consists of a filtering unit—the **glomerulus** and **Bowman's capsule**—and **renal tubule** that ultimately drains into the ureter. As the **filtrate** passes through the tube, the cells of the sur-rounding **epithelium** adjust the concentrations of specific ions and molecules.

In mammals, the wide **proximal tubule** of the nephron bends into the narrow **loop of Henle** and exits the kidney through the wide **distal tubule.** This lies close to the end of the corresponding collecting duct. This arrangement allows the mammalian kidney to produce a concentrated urine.

Several hormones regulate the operation of the kidney. **Angiotensin,** whose production depends on **renin,** causes the kidneys to retain water. **Aldosterone,** from the **adrenal glands,** controls the reabsorption of salt in the nephrons. **Antidiuretic hormone (ADH),** or **vasopressin,** and **atrial natriuretic factor** (ANF) regulate water reabsorption in the collecting ducts. ADH is secreted by the posterior pituitary, and ANF is secreted by the heart.

Review and Thought Questions

Review Questions

1. Why do marine fish drink continuously, but freshwater fish not at all?
2. Sketch a nephron and collecting duct. How do the cells of the epithelium vary from segment to segment? How does the transport of ions vary from segment to segment? How does the epithelium's permeability to water vary?
3. Explain how differences in the transport of salt ions and in the permeability to water change the concentration of the filtrate as it passes down the tubule.
4. How does the cellular organization of the glomerulus and Bowman's capsule account for the passage of urea into the nephron? How does urea get to be more concentrated in the urine than in the blood? How does glucose get to be less concentrated in the urine than in the blood?
5. What are the water balance problems of flatworms, earthworms, lobsters, and insects? What organs does each use to solve these problems?
6. Why are people with kidney problems also likely to have high blood pressure?

7. Suppose you have an exceptionally spicy meal and drink two extra glasses of water at dinner. What effects do you expect the extra water to have on the secretion of hormones by the hypothalamus, adrenal glands, and heart?
8. Suppose you eat an exceptionally salty meal, without drinking any extra water. What effects do you expect the extra salt to have on the secretion of hormones by the hypothalamus, adrenal glands, and heart?

Thought Questions

9. Why do seabirds and sea turtles have salt glands? Why can a seagull survive by drinking seawater, but a shipwrecked sailor cannot?
10. Whales and seals do not have salt glands, but live well without fresh water. How do you think they might manage this?
11. What purpose might the highly salty tears of humans serve?
12. Why does alcohol act as a diuretic?

About the Chapter-Opening Image

The salt box symbolizes the huge amount of salt that Americans consume, forcing their kidneys to work overtime removing it from the blood.

 On-line materials relating to this chapter are on the World Wide Web at **http://www.harcourtcollege.com/lifesci/aal2/**

40

Defense: Inflammation and Immunity

The Danger Model

In the spring of 1996, several groups of researchers simultaneously published scientific papers that together challenged basic ideas in immunology that have stood for 50 years. In the 1940s, a series of crucial experiments persuaded biologists that the immune system protects the body by distinguishing self from nonself and learns which molecules are self (and which are not) around the time of birth. This idea is called the *self-nonself model.*

Specifically, the experiments done in the 1940s seemed to show that the immune systems of newborn mice worked very differently from those of adult mice. Newborn mice would tolerate injections of foreign cells that adult mice would reject. Not only that, the young mice tolerated the same foreign cells when they grew up, a sign that their immune systems had adopted these foreign cells as self.

But research by Polly C.E. Matzinger, a biologist at the National Institutes of Health, and that of two other teams of researchers suggests that newborn immune systems differ little from those of adults. Furthermore, according to Matzinger, the entire self-nonself model is wrong (Figure 40-1). The immune system does not distinguish between self and nonself at all, she says, but between cells and molecules that are dangerous and those that are not.

This new theory, which Matzinger calls the *danger model,* has provoked strong feelings in her fellow researchers. A few are delighted with the danger model. Transplant biologists and surgeons, in particular, see great practical value in the danger model. If the model is correct, surgeons may be able to perform organ transplants far more safely and successfully than ever before.

The majority of biologists, however, dismiss the danger model as a nice idea with little supporting evidence. For Matzinger, the most frustrating criticisms come from a few biologists who insist that she isn't saying anything new at all, that her theory is just a different way of saying the same thing. Says Matzinger, "They haven't really jumped off the cliff with me."

Jumping Off the Cliff

When Polly Matzinger was in her early twenties, she could hardly have anticipated that her scientific work would be featured on the cover of one of America's best-known scientific journals. Although she had studied biology in school, she abandoned science to try dog training, waitressing, and carpentry.

It was while serving drinks to two biologists at a restaurant in Davis, near the University of California, that she rediscovered science. The two biologists were arguing about an experiment, when Matzinger joined the conversation, pointing out some errors in their thinking. Astonished, they welcomed her into the conversation, and one of them ultimately persuaded her to return to school.

Today, Matzinger heads an immunology laboratory at the National Institutes of Health, near Washington, DC (and still trains border collies). Her perpetual restlessness now expresses itself, however, in her intellectual outlook. Matzinger has never been one to assume that those who came before her necessarily knew best.

Early studies in immunology focused on the transplanting of organs or skin from one individual to another. Normally, such transplanted organs are rejected: the cells of the immune system literally attack and destroy the cells of the transplanted organ. Most heart, liver, and other transplants today are therefore done with the aid of drugs that suppress the immune system.

In the 1940s, researchers noticed that cattle tolerated skin grafts from a twin sister or brother. In identical twins, this might not have been surprising since such individuals share the same "cell-surface" molecules that identify the cells of the

Key Concepts

- The skin is the first barrier against invading bacteria, viruses, fungi, and other pathogens.

- Inside the body, inflammation, complement, interleukins, and interferon provide nonspecific internal defenses against infection.

- Two key components of the immune system recognize and eliminate foreign molecules and cells—circulating antibodies and cytotoxic cells.

- The immune response demonstrates specificity, memory, diversity, and the ability to distinguish self from nonself.

body. But even fraternal twins, which have different cell-surface markers, could accept skin grafts from each other—as long as they had shared a blood supply inside their mother. Researchers speculated that the twins learned to recognize each other's cells as self while they were in the womb together.

To test this theory, English biologist Peter Medawar showed that he could "teach" newborn mice to tolerate foreign cells. Medawar injected cells from a white mouse into genetically distinct, brown fetuses while they were still developing inside their mother. When the brown mice grew up, he grafted onto them patches of skin from the same white mouse. All of the brown mice that had been exposed to white mouse cells in the womb accepted these grafts and sported white patches of healthy fur. In contrast, mice that had not been exposed to the white mouse's cells rejected skin grafts from that mouse.

Medawar's work confirmed the prediction of an Australian virologist named Frank MacFarlane Burnet, the researcher who had first formulated the self-nonself model of immunology. Specifically, Burnet had proposed that the immune system's main task is to distinguish self from nonself and that it learns this distinction just once—during fetal development. In 1960, Medawar and Burnet shared a Nobel prize for their trailblazing work.

But the self-nonself model has always had a few problems, problems that had bothered Matzinger when she was just beginning to study immunology. The self-nonself model seemed to work, however, and she put her doubts aside.

Then, in 1989, Matzinger met a younger scientist named Ephraim Fuchs who had similar doubts. How can it be, asked Fuchs, that the immune system only learns to recognize self before birth when that self keeps changing throughout life? At puberty, for example, our bodies go through major changes

Figure 40-1 Biologist Polly C.E. Matzinger. Matzinger argues for a model of immune function called the danger model. *(Myriam I. Rosado)*

and develop new kinds of cells. A pregnant woman hosts fetal cells, which because they reflect the genetic makeup of the father cannot be recognized by the body as self. Yet, mothers' immune systems only rarely attack their fetuses. How could the immune system of the mother possibly learn to recognize the cells of the father, who she will not meet for years?

In addition, many foreign cells and viruses take up residence in our bodies long after the immune system is supposed to have learned what self is. After birth, our intestines become loaded with bacteria, which our immune systems tolerate. And that is essential, for many of these bacteria are necessary to life. Some help us digest milk; some make vitamin K, which helps our blood clot; and others protect us from invasion by less friendly bacteria. The immune system tolerates bacteria in our mouths, throats, and noses. None of these organisms is self. How can the immune system learn to recognize all these foreign organisms long before it encounters them?

Matzinger recollected all her old doubts. She and Fuchs agreed that it was unlikely that the immune system actually recognized self. There were too many exceptions. But, in that case, what then did the immune system do?

Matzinger's answer came from her knowledge of the way cells die. A normal cell that has lived a full life dies quietly and then summons **macrophages,** big, cell-eating cells. The macrophages arrive and then neatly consume the dead cell, leaving no trace. This quiet death is called *apoptosis* [Greek, = falling away]. In contrast, a cell under attack by viruses eventually bursts open, releasing not only virus particles but also the cell's own cytoplasm and organelles. It's a mess guaranteed to attract the attention of the immune system.

In fact, this scenario perfectly explains the behavior of **T cells,** special immune cells that recognize and destroy infected cells, cancerous cells, and also the cells of transplanted tissues. T cells recognize the body's own cells by means of a set of highly individualized cell-surface molecules, called MHC (major histocompatibility complex) proteins.

According to the self-nonself theory—as it has evolved over the years—T cells that bind to the body's own healthy cells automatically die unless a mysterious *second signal* indicates that the cells are infected or foreign. Cells that die naturally during apoptosis do not trigger this second signal. But infected or damaged cells do, which triggers an immune response. If T cells detect the second signal, they become activated and migrate to a lymph node, where they activate other immune cells.

According to the self-nonself model, T cells are not activated by this second signal during fetal development. As a result, all the T cells that respond to anything in the fetal environment die or become inactivated. Medawar's mice tolerated the injection of foreign cells because the T cells that recognized them had died or become inactive. There were therefore no T cells to recognize and destroy the injected cells.

No one knows, however, what the "second signal" is nor why it would fail to operate during fetal development. One theory is that the second signal comes from "dendritic cells." Like T cells, dendritic cells are scattered throughout the tissues of the body, and, like T cells, they migrate to the lymph nodes when activated. Dendritic cells may collect molecules from foreign (or damaged) cells and display them to the T cells. These molecules, displayed as surface molecules, tell the T cells to multiply and launch an attack against cells bearing the telltale molecules.

But there is one hitch. Unlike T cells, dendritic cells cannot distinguish self from nonself. If dendritic cells carry the mysterious second signal and they cannot themselves distinguish self from nonself, then how do they avoid triggering constant immune attacks against the body's own tissues?

According to Matzinger and Fuchs, the dendritic cells are activated by the cytoplasm-spilling deaths of cells—regardless of whether the cells belong to the body or not. If cells are dying messily, that's a danger signal the dendritic cells can understand. If a virus-infected skin cell bursts open, spilling skin cell proteins and virus proteins, a nearby dendritic cell registers the disaster, drinks up the cell's contents, and heads for the nearest lymph node (where the T cells congregate) to sound the alarm (Figure 40-2). There, the dendritic cell displays some of these proteins on its cell surface, activating nearby T cells and marking which cells to attack.

While the self-nonself model says that the immune system learns self from nonself during the fetal period only, the danger model says that the immune system constantly redefines what is dangerous. If virus-infected skin cells are dangerous today, they may not be tomorrow. If no cells are being damaged, then the dendritic cells stop giving the second signal, and T cells that bind to skin cells die. As a result, the immune system can "quiet down," and become unreactive. In Matzinger's view we are constantly acquiring tolerance to our own proteins, to those of the bacteria that live with us, and to those of our own offspring. The self is redefined in a continuous process of adjustment.

Matzinger's theory explains why tumors and warts caused by viruses are often left alone by the immune system. The cells in these tissues are healthy and are not dying unnatural, messy deaths, so no second signal is given. After a while—without a second signal—the T cells that might react to them die off.

Matzinger speculates that organ transplants are rejected not because they are nonself but because they are full of dendritic cells activated by the trauma of being cut from the donor. As soon as the donated organ is transplanted, the donor's dendritic cells head for the lymph nodes of their new host and sound an alarm.

When Matzinger and her colleagues repeated Peter Medawar's mouse experiment, they were able to show that newborn mice are perfectly capable of rejecting foreign cells, so long as the cells are injected along with activated dendritic cells from the donor. The dendritic cells alert the T cells in the newborn mice.

Other researchers likewise found that infant mice, long thought to be incapable of fending off a certain virus, could mount a vigorous immune response, provided the dose of virus was lowered a thousandfold from what other researchers had been giving the mice. They too concluded that the immune response of infant mice works in the same way as that of adult mice. The researchers suggested that in previous experiments the young mice, which have fewer T cells than adults, were simply overwhelmed by too many viruses.

Does the immune system recognize self versus nonself, or does it recognize safe versus dangerous? The research of Matzinger and others seems to pose a challenge to the currently accepted self-nonself model of immune response and offers some possible answers to longstanding difficulties with that model. The danger theory may eventually displace the current model. For now, the danger model is raising new questions, encouraging biologists to reexamine long-accepted assumptions, and taking research in new directions. Nonetheless, in

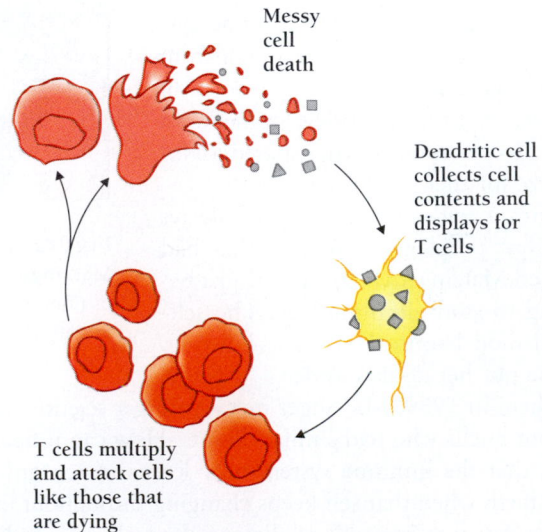

Messy cell death

Dendritic cell collects cell contents and displays for T cells

T cells multiply and attack cells like those that are dying

Figure 40-2   A simplified view of the danger model.

the rest of this chapter, we discuss the facts of immunology in the context of the self-nonself model, which most biologists still embrace.

THE CAST OF CHARACTERS

Our biosphere teems with potentially harmful viruses, bacteria, fungi, protozoans, and parasitic worms. Among vertebrates, at least, evolutionary success has depended heavily on our ability to prevent and combat invasions of these tiny pathogens. In this chapter we review three lines of defense used by mammals to protect themselves against pathogens.

The first line of defense is a barrier, the skin, that keeps out pathogens as effectively as a castle wall keeps out human invaders. However, any opening in such a barrier—whether by the mouth and eyes or a cut or puncture wound—allows pathogens free access to the body.

The next line of defense consists of **blood clots,** which plug any holes, and **inflammation,** which mobilizes white blood cells that "phagocytose" (engulf and consume) bacteria, viruses, debris, and other elements.

The last line of defense is the **immune system,** a highly complex system involving roughly eight kinds of white blood cells. Some of these cells phagocytose pathogens, some punch holes in pathogens, and some secrete **antibodies**—molecular flags that mark certain cells for destruction by other white cells. The immune system differs from the other two lines of defense in its *specificity.* It can distinguish among individual pathogens, responding more vigorously to some than others and tolerating the body's own cells in some cases and not in others.

All of the cells engaged in the body's defense are white blood cells, or **leukocytes** [Greek, *leukos* = clear, white], which come from "stem cells" in the bone marrow. (Stem cells are cells that can either differentiate into different mature, specialized cells or divide to produce more identical stem cells.) Four kinds of white blood cells—neutrophils, eosinophils, basophils, and mast cells—participate mostly in the inflammation response. Two kinds of **lymphocytes**—B cells and T cells—participate in the immune responses. In addition, both **natural killer cells (NK cells)** and macrophages, the giant cell-eating white cells, can participate in immune responses (Table 40-1).

The body defends itself against invading pathogens with three lines of defense: the skin, blood clots and inflammation, and the immune system.

HOW DOES THE SKIN KEEP PATHOGENS OUT?

The skin is a flexible, stretchable barrier that repels the great majority of potential invaders. In humans, the skin makes up

TABLE 40-1   TYPES OF WHITE BLOOD CELLS (LEUKOCYTES)

NONSPECIFIC (INFLAMMATION RESPONSE)
Neutrophils—phagocytic on bacteria and larger parasites
Eosinophils—attack bacteria and larger parasites
Basophils—reside in the bloodstream and release histamines, which trigger inflammation
Mast cells—reside in the tissues and release histamines, which trigger inflammation and attract neutrophils (IgG antibodies stimulate mast cells to release histamines.)
Natural killer cells (NK cells)—kill microorganisms and debris that are marked with antibodies
Macrophages—phagocytic on microorganisms and debris that are marked with antibodies and release four kinds of cytokines, or intercellular signals
SPECIFIC
T cells—recognize and kill cells that are infected or "altered self"
B cells—make memory cells and plasma cells, which make antibodies

---

about 15 percent of the body's weight. Like the linings of the gut, mouth, and other internal surfaces of the body, the skin consists of epithelial cells. Unlike the thin, internal epithelia, however, the skin epithelium consists of a thick pile of cell layers, which together make up the **epidermis** [Greek, *epi* = on + *derma* = skin] (Figure 40-3).

The outermost cells of the epidermis are nondividing cells, many of which are dead or dying, that protect the dividing cells beneath. These outer cells are continually exposed to scraping, drying, infection, and other injuries. On average,

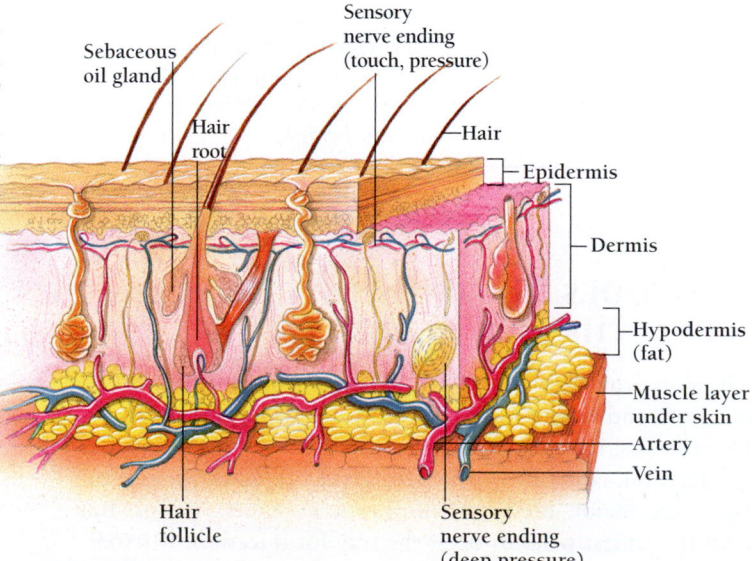

Figure 40-3   The epidermis is the first layer of defense against pathogenic microorganisms.

these cells live only about a month. As the outer cells die, still-dividing cells from the lower layers replace them.

Beneath the epidermis are the **dermis,** a thicker layer of living cells, and a layer of **subcutaneous tissue.** Embedded within these layers are hairs, nerve endings, muscles, sweat glands, oil glands, blood vessels, and sensors for heat, cold, and pressure.

At the top of the epidermal layers, the oil and sweat secreted by glands in the skin create a slightly acid environment, which is hostile to the growth of many microorganisms. The skin nonetheless swarms with microorganisms. Most are benign residents that cause little damage. Indeed, many actively compete with pathogenic organisms, reducing the chance that pathogens can enter the body.

The skin and its community of organisms are so effective a barrier that most microorganisms cannot enter the body unless they do so by way of the mouth, nose, or eyes. Even those that do enter on particles of food or dust usually die, because few survive exposure to **lysozyme,** an enzyme, found in saliva and tears, that digests the cell walls of many microorganisms.

A few microorganisms nonetheless reach the digestive tract or respiratory tract. In the respiratory tract, cilia sweep mucus and trapped microorganisms back toward the esophagus. This movement flushes most dust-borne microorganisms into the digestive tract, where they join a host of other microorganisms swallowed with food. These pass from the esophagus into the stomach.

The stomach's acid environment is enough to kill most microorganisms. Some—*Salmonella,* for example—survive and pass into the digestive tract. Vast numbers of microorganisms flourish in the digestive tract. Indeed, more individual bacterial cells inhabit the gut than there are cells in the body. As in the case of the microorganisms of the skin, however, most are benign and usually compete with pathogenic bacteria for space and resources. As a result, under normal conditions, foreign microorganisms cannot survive in the gut.

The first barrier to invasion by microorganisms is the skin. Lysozyme in saliva and tears kills many foreign microorganisms that enter the eyes, mouth, or nose. Microorganisms that manage to enter the intestinal tract must compete with millions of others—almost all benign.

HOW DOES THE BODY DEFEND ITSELF WHEN THE SKIN IS BROKEN?

The skin provides the first line of defense against infection or invasion, and relatively few microorganisms breach its barrier. Nonetheless, injuries, even minor ones, expose the inner body to infection. Animals continually suffer breaks in their skins. Some are major, life-threatening injuries, others minor. But even the tiniest wounds open the way for infection. Microorganisms that penetrate the skin, however, encounter a series of active defenses.

Any significant wound damages blood vessels as well as other tissues. Damaged blood vessels pose two dangers beyond the immediate infection of local tissues: blood loss and the rapid dispersal of microorganisms throughout the body by way of the circulating blood.

How Does the Body Control Blood Loss and Keep Pathogens Out of the Circulatory System?

Mammals have evolved two adaptations, each of which limits both bleeding and infection. In **wound healing,** normal cells—in both the epidermis and the dermis—divide when they lose contact with other cells and stop dividing once they have filled the gap. But wound healing takes time—days or even weeks. A short-term response that can occur in seconds is **hemostasis** [Greek, *haima* = blood + *stasis* = standing], the several mechanisms that hinder both the flow of blood

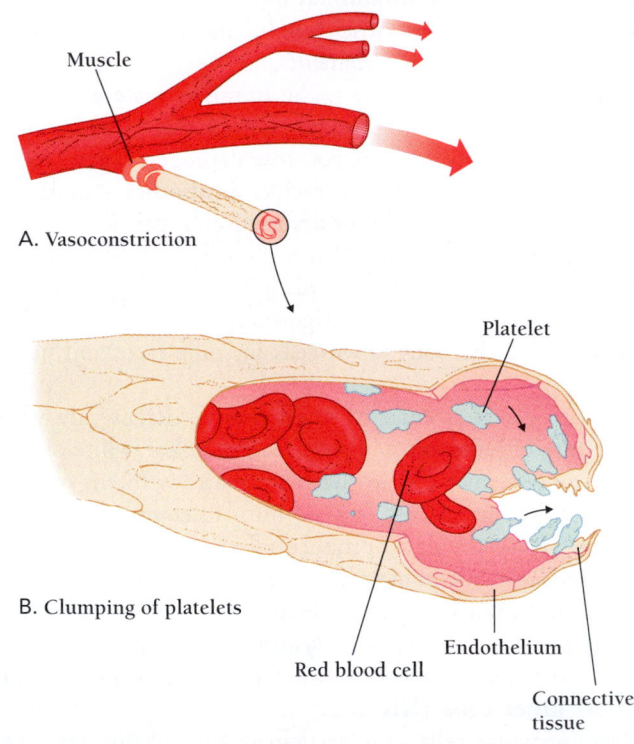

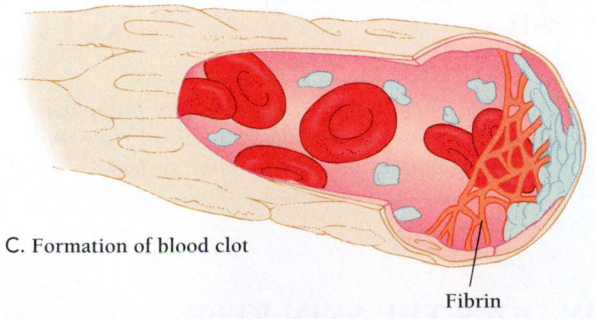

Figure 40-4   Hemostasis. The circulatory system has three ways of preventing blood loss. A. Vasoconstriction. B. Clumping of platelets. C. Formation of blood clots.

out of the body and the passage of microorganisms into the body.

Hemostasis involves at least three interconnected mechanisms (Figure 40-4): (1) the contraction of smooth muscles in the damaged vessels, called **vasoconstriction;** (2) the clumping of small cellular elements in the blood, called **platelets,** at the site of injury; and (3) the formation of a blood clot, a mass of coagulated blood.

Injury immediately triggers the contraction of nearby blood vessels, which sometimes seals the leaky vessel completely but at least reduces the loss of blood. Injury to the wall of a blood vessel also exposes collagen, the major structural protein of connective tissue. Platelets in the blood adhere first to the collagen and then to each other, forming clumps of platelets that can completely plug tiny openings in the wall of the blood vessel. Such minute ruptures occur hundreds of times every day, only to be sealed by platelets. In addition, the platelets stimulate further vasoconstriction and initiate clotting, the mechanism by which the blood forms a gel-like lump that blocks blood flow.

Although platelets can help initiate clotting, clotting is strictly a property of the **plasma,** the liquid, noncellular part of the blood. Clotting can occur in the absence of any blood cells, even in a tube of plasma outside the body.

The body closes breaks in the skin by means of wound healing and breaks in the circulatory system by means of vasoconstriction, the aggregation of platelets, and blood clots. All of these processes prevent excessive blood loss and help prevent microorganisms from entering the circulatory system.

How Does the Body Prevent Damaging Blood Clots?

Although blood clots can keep an animal from bleeding to death, blood clots are themselves extremely dangerous. A single clot can block an arteriole in the brain (stroke), heart (heart attack), or lungs (thrombosis), causing permanent damage or death. As a result, clots form only after a series of molecular checks (Figure 40-5A).

Biochemists have studied the process of clotting intensively and identified several proteins responsible for clot formation. The clot itself consists of a network of fibers made of the protein **fibrin.** Fortunately, fibrin is not present in un-clotted blood—otherwise clots would form constantly.

Fibrin derives from a precursor protein called **fibrinogen,** which becomes fibrin only when the enzyme **thrombin** digests away part of the fibrinogen polypeptide. Thrombin itself is not normally in the blood either. Otherwise it too would cause the blood to clot all the time.

Instead of thrombin, blood plasma contains **prothrombin,** a precursor of thrombin. At the site of tissue damage, still another enzyme, called *Factor X* ("factor ten"), converts inactive prothrombin to active thrombin, initiating the clot. Fac-

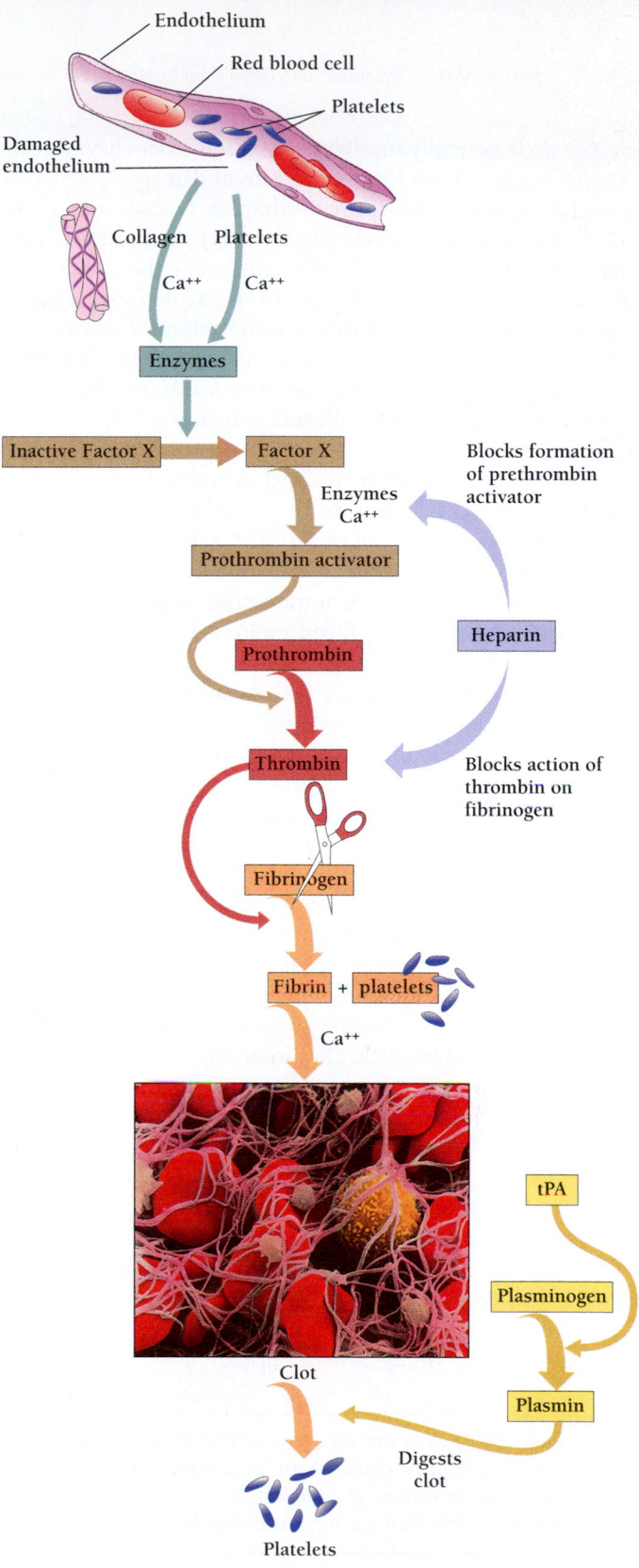

Figure 40-5 Making and unmaking blood clots. Exposed collagen or blood platelets trigger the first steps in the "clotting cascade." Thin arrows indicate the work of enzymes that facilitate transformation of one molecule into another. Fat arrows indicate the transformation. For example, the enzyme thrombin (scissors) cuts off a piece of fibrinogen, leaving fibrin, which combines with platelets to form a clot. Heparin prevents formation of clots, while the enzyme tPA (tissue plasminogen activator) triggers breakdown of clots. (Photo, CNRI/Science Photo Library/Photo Researchers, Inc.)

tor X is itself normally inactive, but several other enzymes can convert it to an active form at the site of damage.

The various steps form a complex cascade of events that lead to clot formation. The first step that initiates this cascade is an ordinary molecule, collagen—the major protein of connective tissue (Figure 40-4). At each subsequent step, a single molecule of one enzyme activates many molecules for the next step. The result is a tremendous multiplication of the initial effect, so that a minor injury that exposes just a little bit of collagen results in a large amount of fibrin.

The clotting system is so dangerous that it requires not only a series of checks but also a powerful antagonistic system that destroys fibrin (Figure 40-5B). Like the clotting system, the anticlotting system consists of a multiplying cascade of enzymes. The last enzyme in the cascade is **plasmin,** which digests the bonds between fibrin molecules and dissolves the clot.

Plasmin, in turn, derives from **plasminogen,** a precursor protein in the plasma. The activation of plasminogen depends either on the action of plasma enzymes or on **tissue plasminogen activator** (tPA), enzymes that activate the anticlotting system. Some tissues have more tPAs than others. The uterus, for example, contains high levels of a tPA that prevents excessive clotting of menstrual blood.

Physicians treat heart attack victims with synthetic tPA, which, when given intravenously, stimulates the production of plasmin. The plasmin breaks up the clots that are blocking the arteries, and blood flow to the heart muscle resumes.

Besides tPAs, mammalian blood also contains various **anticoagulants,** substances that interfere with the action of the clotting cascade. One such anticoagulant is **heparin,** which is made in specialized cells of the liver, lung, and gut. Physicians use heparin to prevent blood clots in the heart, brain, or lungs, and researchers often add heparin to blood samples to prevent them from clotting. Warfarin, a chemically synthesized anticoagulant, is used both as a drug, in people, and as a poison, in rats and mice, which die of internal bleeding. Blood-sucking mosquitoes and leeches also use anticoagulants to keep the blood of their victims flowing freely.

Because blood clots are dangerous, the synthesis of the active ingredient in clots, fibrin, is carefully regulated. In addition, a system of anticlotting enzymes ensures that any clots that do form are rapidly dissolved.

How Do Plasma Proteins Contribute to Nonspecific Defense?

We have seen that a wound can trigger vasoconstriction, tiny platelet plugs, and blood clots. A fourth kind of defense is the **complement,** a set of blood proteins that attack microbiological invaders. Some of these proteins kill foreign organisms by boring holes through their membranes. Other complement proteins coat the microbes' surfaces to identify them as targets

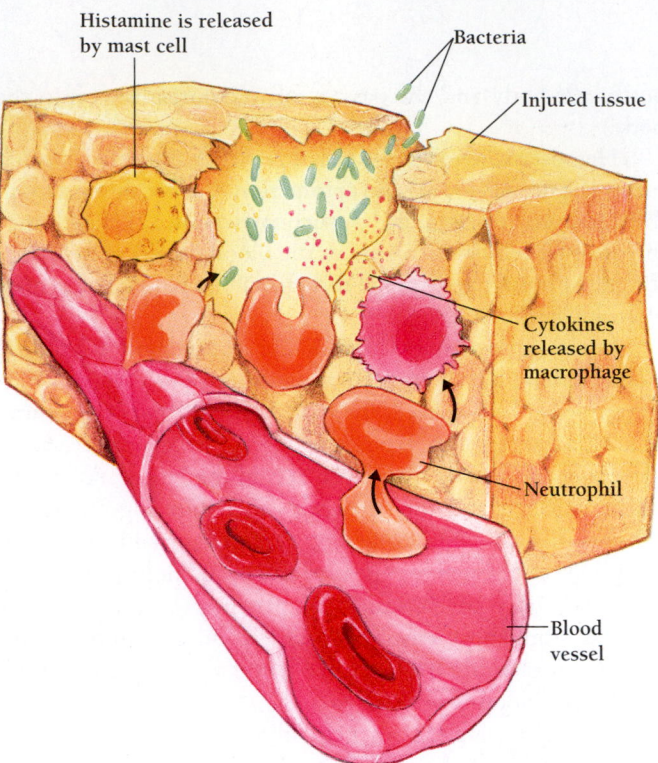

Figure 40-6   Tissue damage such as a cut in the skin stimulates mast cells. Tissue damage stimulates the granules in a mast cell to release histamine. Mast cells also release cytokines, which attract migrating neutrophils and help initiate the (slower) immune response.

for phagocytic cells. Still other proteins stimulate special white blood cells called *mast cells.* Mast cells help trigger the next line of defense, which is **inflammation,** the redness, heat, swelling, and pain that result from local injuries.

Mast cells are large, round cells filled with small vesicles, or granules. Mast cells are distributed throughout the body's connective tissues (Figure 40-6). Any kind of damage to a tissue stimulates the mast cells to release their granules, which are loaded with **histamine,** the major stimulus for the inflammatory response. Allergy sufferers are especially familiar with histamine's effects, which are swelling and itchiness in the eyes, nose, throat, and other tissues. *Antihistamines* are drugs that block the effects of histamines.

Complement proteins and mast cells react nonspecifically to damage to body tissues. Mast cells release histamines, which help promote the inflammation response.

How Does Inflammation Help the Body Resist Infection?

Microscopic observations reveal a sequence of cellular events that contribute to the inflammatory response: (1) Capillaries open, or *dilate,* in the injured region. (2) Special white blood cells, called **neutrophils,** attracted by chemicals released by the mast cells, attach to the capillary walls. (3) The neutrophils move through the capillary wall into the damaged tissue. (4) The neutrophils phagocytose (engulf and digest) microorganisms in the damaged tissue (Figure 4-12).

The open blood vessels increase the delivery of cells and plasma proteins, which contribute to the inflammatory and the immune responses. Many cells arrive at the inflamed region, among them, macrophages, which counterattack invading organisms.

Macrophages [Greek, *makros* = large + *phagein* = to eat] are enormous, phagocytic cells that are widely distributed throughout the body (Figure 40-7). Immature macrophages (monocytes) make up about 5 percent of the cells in the blood. Mature macrophages are found in connective tissue in the liver, brain, spleen, lungs, and lymph nodes. Macrophages consume microorganisms and cellular debris. They also play a major role in initiating inflammation.

During inflammation, macrophages produce three small proteins called **cytokines.** Cytokines regulate cell division, protein synthesis, and migration in other white blood cells involved in both inflammation and the immune response. The three cytokines produced early in the inflammatory response are **tumor necrosis factor-α** (TNF-α), **interleukin-1** (IL-1), and **interleukin-6** (IL-6). The two interleukins derive their name from the fact that they act as messenger molecules between different kinds of white blood cells (leukocytes). By binding to specific receptors and initiating specific biochemical responses, interleukins initiate and regulate both inflammation and specific immune responses.

Interleukin-6 stimulates stem cells in the bone marrow to produce more macrophages. TNF-α acts more locally, stimulating macrophages and neutrophils to increase their phagocytic activity.

IL-1 stimulates responses in leukocytes and in various organs distant from the site of infection. IL-1 attracts more phagocytic cells to the site of injury and stimulates T and B cells. IL-1 can cause a fever by stimulating cells in the *hypo-*

thalamus, a region of the brain that regulates body temperature. A fever of two or three degrees slows the growth of bacteria, improves the performance of complement proteins, and causes the T cells and B cells of the immune system to multiply faster.

Viral infection stimulates the infected cells to produce still another set of cytokines, called **interferons.** Interferons interfere with the replication of viruses by limiting protein synthesis in virus-infected cells. Even though the shutdown of protein synthesis may kill the cell, it prevents the release of more viruses. Interferons also stimulate phagocytosis and the immune response.

Inflammation helps the body resist infection by opening blood vessels, thus increasing the delivery of white blood cells, such as neutrophils and macrophages, as well as plasma proteins. Cytokines released by the macrophages regulate the inflammation response, stimulate the immune response, and send chemical signals to other parts of the body.

HOW DOES THE IMMUNE SYSTEM RECOGNIZE MOLECULES?

The most powerful and the most specific defenses against infection are those of the immune system. The immune system produces two distinct responses to invading organisms. The first is the production of antibodies, proteins that identify and label foreign molecules, whether they are free floating or attached to foreign cells and viruses. The second immune response is the production of cells that recognize and attack any cells of the body that have become infected by viruses or microorganisms.

What Distinguishes the Immune System from Other Defenses?

Most of us have had contact with someone infected with chickenpox, a common childhood disease. If we have not had chickenpox, or been vaccinated against it, we are susceptible to infection. Once we have suffered its itchy ravages, however, we are no longer susceptible; we then say that we are immune to chickenpox. Four attributes characterize immunity: *specificity, memory, diversity,* and *self-nonself recognition.*

Immunity is **specific:** cells of the immune system recognize specific invaders and not others. If we are immune to chickenpox, we may still be susceptible to measles or other diseases. The specificity of the immune response distinguishes it from the **nonspecific** responses of clotting, complement proteins, inflammation, and interferons.

The immune system has **memory:** once the immune system has developed defenses against a particular invader, it can attack that molecule, organism, or virus repeatedly, whenever it appears—even decades later. Furthermore, with experience,

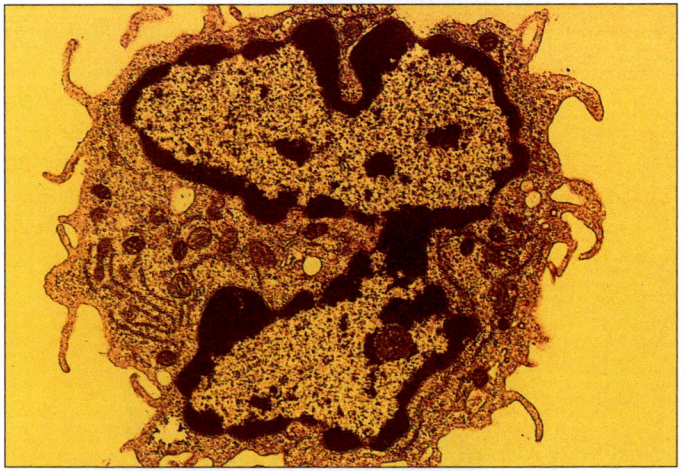

Figure 40-7 A macrophage. Notice at left the endoplasmic reticulum and Golgi, which produce and secrete proteins such as interleukins. *(Don Fawcett/Visuals Unlimited)*

2.5 μm

Figure 40-8 Cells involved in the immune response. Lymphocytes first develop in bone marrow. Those that migrate to the thymus mature into T cells (which recognize infected or cancerous body cells). Those that stay in the bone marrow mature into B cells (which make antibodies).

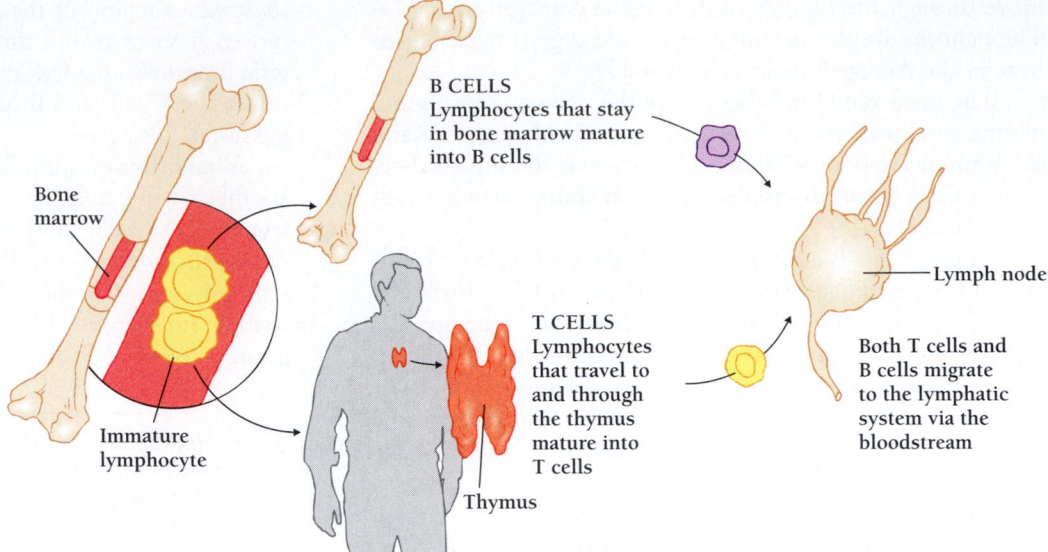

Bone marrow

Immature lymphocyte

B CELLS Lymphocytes that stay in bone marrow mature into B cells

T CELLS Lymphocytes that travel to and through the thymus mature into T cells

Thymus

Lymph node

Both T cells and B cells migrate to the lymphatic system via the bloodstream

the immune system removes pathogens more quickly. The reappearance of a pathogen stimulates a **secondary immune response,** which is usually faster and more efficient than the first.

The immune system is capable of recognizing and attacking an incredible **diversity** of foreign invaders, including parasitic worms, fungi, bacteria, viruses, and pollen, as well as purified proteins and other molecules. Because the immune system responds powerfully to so many different kinds of cells and molecules, its greatest challenge is to refrain from attacking the cells of its own body. The immune system **tolerates** the other cells in the body, ignoring them unless they are infected by a pathogen. The immune system's apparent ability to distinguish the body's own components from others is called **self-nonself recognition,** in accordance with the model currently accepted by most biologists.

The immune response demonstrates specificity, memory, diversity, and the ability to distinguish self from nonself.

What Gives the Immune System Its Specificity?

Until the 1950s, explaining these four properties of the immune response—specificity, memory, diversity, and self-nonself recognition—seemed an almost impossible challenge to biologists. Perhaps the most demanding aspect of the problem was the tremendous range of the immune response. The mammalian immune system can respond not only to proteins and other molecules created by any organism or virus, but even to molecules that are not natural products, ones that have been synthesized by organic chemists. How can an animal produce proteins and cells that specifically recognize substances that neither the animal itself nor any of its ancestors could ever have encountered in all of evolutionary history? And how can an animal have specific responses to so many different challenges?

The answer to all these questions was antibodies. The immune system produces a huge array of proteins called antibodies that selectively bind to specific molecules that have a particular shape and charge. Specifically, an antibody binds to an **antigen** [an *anti*body *gen*erator],

Figure 40-9 A macrophage consuming an illegally parked car. In the same way that an expired parking meter marks an illegally parked car, an antibody marks a cell for destruction. It's a good thing that meter maids don't operate like macrophages.

Antigen

any molecule that triggers an immune response. The array of antibodies is huge, but not infinite. Each antibody recognizes a characteristic shape and charge distribution rather than a specific molecule. While the number of possible atomic arrangements of foreign molecules is impossibly large, there are, fortunately, a limited number of shapes: different arrangements of atoms may look the same to the antibodies.

*The immune system's specificity begins
with antibodies.*

Which Cells Mediate the Immune Response?

Immune responses depend on cells. The immune cells are a special class of white blood cells called lymphocytes (T and B cells) that develop within the lymphoid tissues (including lymph nodes, spleen, thymus, and tonsils). A healthy human immune system contains about a trillion (10^{12}) lymphocytes. Under a light microscope, lymphocytes all look similar, but they vary enormously in what they do.

Lymphocytes fall into three main classes: B lymphocytes, T lymphocytes, and natural killer cells. **B lymphocytes,** or **B cells,** make antibodies, which recognize foreign cells and molecules (Figure 40-8). In particular, antibodies recognize and bind to bacteria, fungi, and protists, as well as any other cells and molecules that do not belong.

Although antibodies are essential parts of the immune system, they are themselves mere markers and cannot kill bacteria or other cells. Antibodies are like the chalk marks the meter reader puts on the tires of cars parked in front of a parking meter. Chalk marks indicate when a car should be ticketed, or even towed. In the same way, antibodies mark cells for destruction by macrophages (Figure 40-9).

Antibodies have another important limitation. They do not recognize the body's own cells even when those cells harbor viruses or other pathogens. Because so many infections, such as AIDS, are intracellular (inside the cell), the body has a second set of lymphocytes, the **T lymphocytes,** or T cells, that recognize and destroy body cells that have become infected or damaged (Figure 40-8). Helping the T cells are **natural killer cells,** or **NK cells,** which attack tumor cells and also cells infected by a pathogen.

*B cells, T cells, and natural killer cells are responsible
for the immune response.*

How Do Antibodies Recognize Cells and Molecules?

Although all the B cells look alike, they are not alike. Different cell lines produce different antibodies. In cell cultures, for

example, each line of B cells makes just one kind of antibody. A normal animal has millions of different kinds of B cells and therefore millions of different kinds of antibodies.

The specificity of the immune system depends on each antibody's capacity to bind to an antigen with a particular shape and charge. Among the most biologically important antigens are the coat proteins of viruses and the surface molecules (proteins, carbohydrates, and glycoproteins) of bacteria.

Every antigen contains one or more **epitopes** [Greek, *epi* = upon + *topos* = place], the specific shapes and charge distributions recognized by antibodies (Figure 40-10). Epitopes are often only small parts of molecules. A single epitope on a protein antigen, for example, typically consists of just five to eight amino acids. It is impossible to count the total number of possible epitopes, but biologists estimate that the immune system can make antibodies that recognize more than 100 million different epitopes. Because antibodies recognize shapes and charges rather than exact atomic structure, an antibody may bind epitopes that are chemically different.

All antibodies are globular proteins called **immunoglobulins.** There are five kinds of immunoglobulins—IgG, IgM, IgA, IgD, and IgE. These letters stand for "immunoglobulin G," "immunoglobulin M," and so on. All the immunoglobulins contain antigen-recognition sites made from two kinds of polypeptide chains, **light chains** and **heavy chains.** Immunoglobulins use thousands of different kinds of light and heavy chains. The five kinds of immunoglobulins differ in their heavy chains (Figure 40-11). But all of them use the same kind of light chains. The heavy chains are either 440 or 550 amino acids, while the light chains are about 220 amino acids long.

The most abundant class of immunoglobulins, **IgG,** consists of four polypeptide chains—two identical light chains and two identical heavy chains. Bonds between side chains connect the heavy chains to the light chains and the two heavy chains to each other. The resulting structure is shaped like a Y, although some people have compared it to the head of the

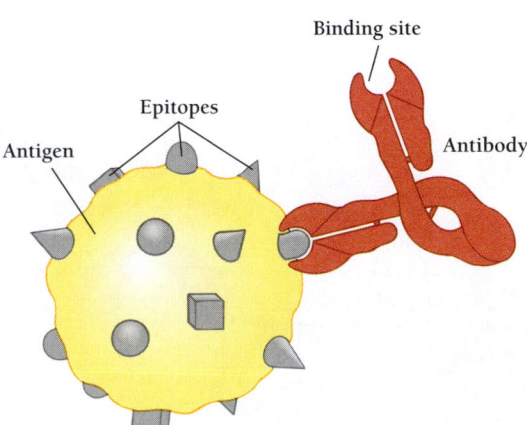

Figure 40-10   An antibody molecule binding to the epitope of a cell covered with different antigens.

mooselike cartoon character Bullwinkle. Either way, antigens bind at the top of each of the two arms of the antibody, so that each IgG molecule contains two identical binding sites for antigens (Figure 40-11).

While most of the immunoglobulins have just two binding sites, IgA has four and IgM has ten. The IgM immunoglobulins, for example, have five identical IgG-like subunits linked together in a shape like a snowflake.

The presence of two or more binding sites on each antibody explains the effectiveness of antibodies in combating infection. The surface of a microorganism may contain thousands of identical protein or carbohydrate molecules. An antibody with two binding sites can bind to the same site on two different bacteria. Because hundreds of antibodies can bind to different pairs of bacteria, the bacteria become linked together into a tangled mass. Masses of antibody-covered bacteria are easy targets for phagocytic cells.

Each of the five classes of antibodies plays a different role in the immune response. IgG and IgM, for example, activate complement proteins, which then destroy any cells marked with IgG or IgM. IgM molecules appear after the first expo-sure to antigen, while IgG molecules appear in greater amounts after a second exposure.

The remaining antibodies, IgA, IgD, and IgE, do not activate complement proteins. IgA protects all of the body's surfaces and is found in the intestinal, respiratory, and urogenital tracts. IgA antibodies are also abundant in tears, saliva, and milk.

IgE antibodies most likely evolved to defend the body against parasitic worms. Many people, however, mount an elaborate IgE defense in response to antigens that are not dangerous at all. Such a response to a harmless antigen is called an *allergy*. Harmless antigens that provoke this response include substances in the feces of dust mites, cat dander (dandruff), pollen, certain drugs, as well as foods such as nuts and shellfish.

When one of these antigens enters the respiratory passages of a sensitized person, for example, IgE molecules attach themselves to mast cells, which then release histamine. Histamine triggers the inflammatory response, increased mucus secretion, and contraction of the airways. Typical allergy symptoms include congestion, sneezing, a runny nose, and difficulty in breathing. The most extreme response can result in death.

In some cases, an allergic response involves only the antibodies, macrophages, and other cells mobilized by inflammation and happens rapidly. In other cases—commonly with antigens in the leaves of poison ivy or poison oak or even certain kinds of metal jewelry—T cells attack tissues days after exposure (Figure 40-12).

Antibody molecules recognize antigens by means of the specific shapes of their epitopes.

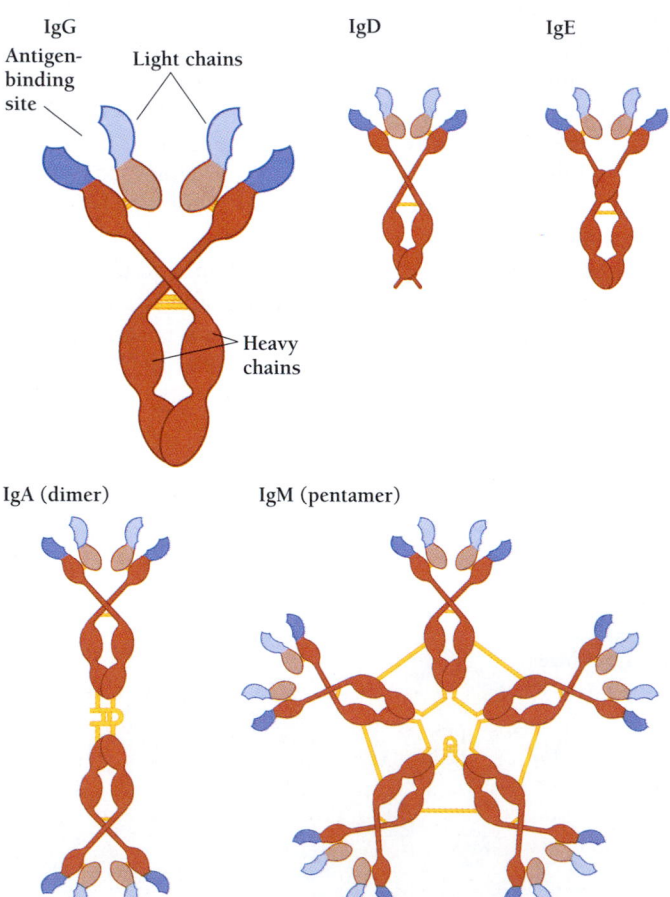

Figure 40-11 **The five classes of antibodies.** The five kinds of immunoglobulins differ in their heavy chains. Notice that IgG, IgE, and IgD have two binding sites each, while IgA has four and IgM has ten.

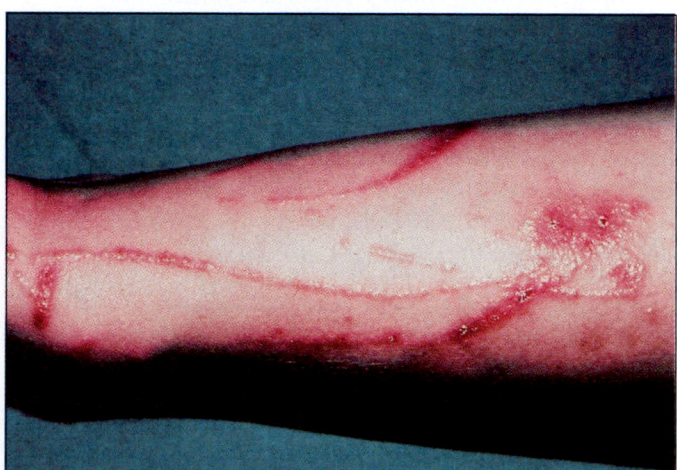

Figure 40-12 **Poison ivy and cytotoxic T cells.** Oils from poison ivy bind to proteins in skin cells, marking them as "foreign." Cytotoxic T cells then attack and destroy these otherwise healthy skin cells. Cortisone and other similar steroid creams can suppress the T_C cell response. *(Ken Greer/Visuals Unlimited)*

HOW DOES THE IMMUNE SYSTEM GENERATE SO MANY ANTIBODIES?

The diversity of antibodies depends on the immune system's ability to produce antibodies with many different three-dimensional structures. Each kind of antibody has a distinct amino acid sequence that results in a distinct three-dimensional structure able to bind to a distinct epitope. The immune system is able to field millions of kinds of antibodies—perhaps 100 million.

Each line of B cells that produces each antibody is a "clone," a group of identical cells. We can conclude, then, that the body generates up to 100 million kinds of B cells. From these, the immune system *selects* the B cells that produce antibodies needed at the moment. Immunologists call this the **clonal selection theory.**

If every different antibody amino acid sequence required a separate gene, the body would require 100 million genes. But that's more genes than most vertebrates have in all their DNA. The human genome, for example, contains fewer than 100,000 genes. And only a fraction of those are devoted to the immune system. How can a few thousand genes carry the information to make millions of antibodies?

How Can Thousands of Genes Produce Millions of Antibodies?

Two facts make this problem less perplexing:

1. As we have seen already, every antibody is made from two different polypeptides, so that different polypeptides can be combined in ways that multiply their differences.
2. Two separate genes can be spliced together to specify one polypeptide, with surprising results. This is the reverse of the idea that one gene may specify more than one polypeptide (Chapter 11).

If lymphocytes could combine 10,000 light chains with 10,000 heavy chains, the cells could generate 100 million antibodies. But that would require 20,000 antibody genes, and most biologists doubt that the immune system makes up even that large a share of our genome.

Our genome, however, is just the genetic information passed from parent to child. What if, after we inherited our few thousand antibody genes, we let the genes in different cells become different from each other? In other words, what if antibody diversity was not inherited at conception but arose during the development of the individual?

In 1965, two researchers argued just that, suggesting that separate gene segments responsible for encoding different regions of immunoglobulin polypeptides somehow recombine during the development of each individual. Just as we can combine five shirts and two pairs of jeans to make ten unique outfits, so lymphocyte cells could combine 300 variants of one polypeptide segment with four variants of another to make 1200 different completed polypeptides.

The idea was intriguing, but for a long time, biologists had no way of knowing if this could really happen. Then, in 1976, biologist Susumu Tonegawa and his colleagues discovered that lymphocytes have unique arrangements of antibody genes not found in other cells. Tonegawa showed that the novel gene arrangements in lymphocytes result from the joining of separate DNA fragments to make new genes. In a revolutionary discovery, he showed that two separate DNA sequences could join forces to code for a single polypeptide.

We now know that every light chain is formed from three different gene segments. One of the three segments is always the same, one has four variant forms, and the third has approximately 300 variants. These different light chains can combine in 3600 different ways. Heavy-chain polypeptides, which form from a combination of four different gene segments, are even more diverse, with about 130,000 types. The 130,000 heavy-chain polypeptides can be combined with the 3600 light-chain polypeptides to form almost half a billion different antibodies (Figure 40-13).

As if half a billion were not enough, immunoglobulin genes mutate at high rates during the production of lymphocytes, increasing antibody diversity by a factor of ten. It now appears that mammals can make more than a billion different kinds of antibody molecules.

Thousands of gene segments can produce millions of antibodies by (1) mixing thousands of polypeptides in different combinations and (2) splicing gene segments into thousands of combinations not originally specified by the genome.

The Selection Theory Explains Memory and Self-Nonself Recognition

The generation of variation in immunoglobulin genes helps explain both the *specificity* and *diversity* of the immune response. To understand *memory* and *self-nonself recognition*, however, we need to understand the behavior of whole cells. First, we will look at the development of B cells, which make antibodies to cells and other elements that are clearly nonself. Then, we will discuss the roles of T cells, which recognize and destroy any of the body's own cells that have become a threat to the body.

Every B lymphocyte contains just two rearranged genes—one that constitutes an active heavy-chain gene and one that constitutes an active light-chain gene. As a result, every B cell and its descendants make only one kind of antibody. The particular gene arrangements within the B cell determine the structure of that antibody.

During development, precursors of B cells form a **lymphocyte library,** with each lymphocyte in the library containing a unique random arrangement of immunoglobulin genes. Each member of the B-lymphocyte library—and all its descendants—can make only one kind of antigen recognition

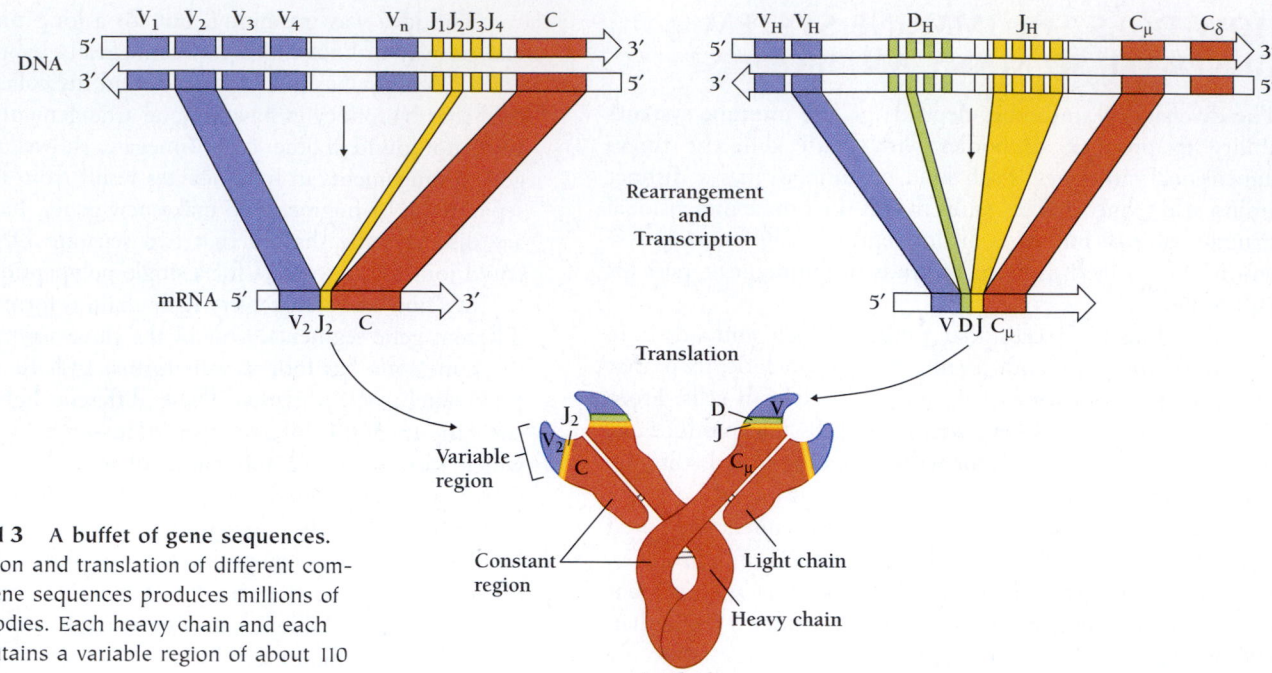

Figure 40-13   A buffet of gene sequences. The transcription and translation of different combinations of gene sequences produces millions of different antibodies. Each heavy chain and each light chain contains a variable region of about 110 amino acids.

site. But how do the right cells "know" when to spring into action, dividing and producing antibodies when their particular antigen appears?

B cells are able to respond to particular antigens because they have **B-cell receptors,** cell-surface proteins with the same antigen recognition sites as the antibody that they produce. In fact, B-cell receptors are just membrane-bound forms of the antibodies.

When an antigen appears in the blood and binds to the B-cell receptors, the B cells begin dividing and differentiating along two separate pathways. One set of cells become **plasma cells,** which actually make and secrete antibodies. The other set of B cells become **memory cells,** which can make more plasma cells if the antigen ever appears again Figure 40-14).

Memory cells are basically identical to unstimulated B cells. The only change is that there are many more of them

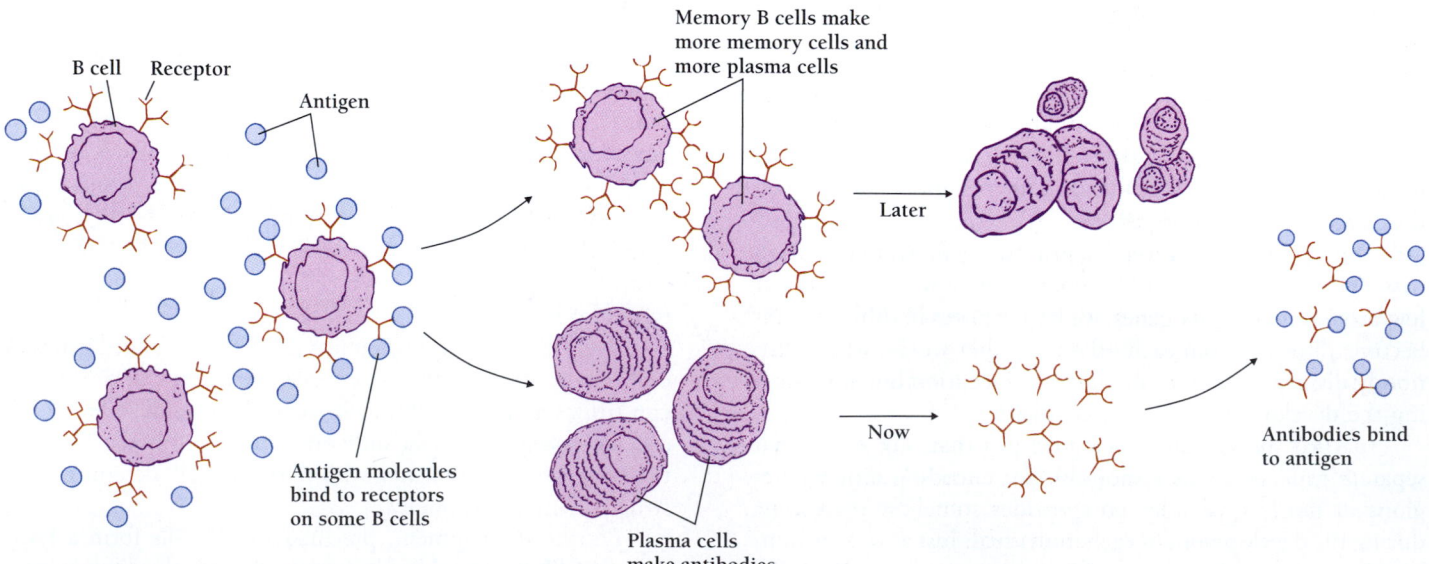

Figure 40-14   The making of memory. When B cells are exposed to antigen, they produce plasma cells and memory cells. If the same antigen appears again, the many memory cells quickly produce large numbers of plasma cells, mounting a much greater and faster immune response.

Acquired Immunodeficiency Virus

In 1996, the number of Americans with AIDS, or acquired immunodeficiency syndrome, declined for the first time since the disease became known in 1980. Researchers were delighted and heralded the six percent drop in new AIDS cases (from 60,620 cases to 56,730) as proof of the success of both preventive programs and treatments. Preventive programs have helped people avoid becoming infected with HIV (human immunodeficiency virus). Treatment programs have helped those infected with HIV to avoid becoming ill and those who are ill to survive longer.

In the United States, 50,000 people die of AIDS each year. The majority of AIDS patients contract the disease either through homosexual relations or from intravenously injecting illegal drugs using contaminated needles. Safe-sex education programs have dramatically reduced the transmission of AIDS among homosexual men, and similar education programs have reduced the infection rate among intravenous drug users.

About 13 percent of AIDS patients acquire the virus through heterosexual contact, and that number is increasing. Heterosexuals are much less likely to practice safe sex than homosexuals and, as a result, the percentage of AIDS cases contracted through heterosexual contact is rising.

HIV is a retrovirus, an RNA virus that relies on the enzyme reverse transcriptase to transcribe its RNA into DNA inside the cells the virus infects. As you might recall from Chapter 12, different infectious viruses attack different kinds of cells in the body. Influenza virus attacks cells in the respiratory tract, for example, and poliovirus infects cells of the intestinal tract and, rarely, the nervous system. HIV infects cells of the immune system, specifically helper T cells. Helper T cells send cytokine messages that activate both B cells and cytotoxic T cells.

Different helper T cells have different receptors. About 60 percent of T cells have *CD4* receptors and another 20 to 30 percent have *CD8* receptors. T cells with CD4 receptors are called "$CD4^+$" T cells, and T cells with CD8 receptors are called "$CD8^+$" T cells. Large numbers of $CD4^+$ T cells help activate other T and B lymphocytes. In contrast, $CD8^+$ T cells suppress the immune system.

Two of the main symptoms of AIDS are very low numbers of $CD4^+$ T cells and higher-than-average numbers of $CD8^+$ T cells. Healthy people have about 1000 $CD4^+$ T cells per ml of blood, while the sickest AIDS patients average less than 50 $CD4^+$ T cells per ml, with many individuals lacking these crucial immune cells altogether. The lack of helper T cells means that AIDS patients are unable to mount effective immune responses to infections or to certain kinds of cancers. As a result, they tend to die of pneumonia, fungal infections, and other diseases that most people's immune systems fend off.

Yet HIV continues to perplex medical researchers. As far as they can tell, HIV infects only one $CD4^+$ cell in every 400. Even the few cells that are infected seem relatively unaffected. And only one $CD4^+$ cell in 100,000 actually translates and assembles virus particles. Researchers wonder why the many uninfected cells don't multiply and protect AIDS patients from infections. Mysteriously, however, the $CD4^+$ cells just disappear.

Researchers suspect that the CD4 receptors are involved. Monocytes and macrophages also carry CD4 receptors on their surfaces, and they too are infected by HIV. They have fewer of these receptors, however, and perhaps for the same reason, fewer of them seem to die.

---

after exposure to antigen. These cells have the same genetic rearrangements as the parent B cell. And, like their parent cells, they do not themselves make antibodies. When they next encounter the antigen, however, the memory cells rapidly produce plasma cells. For this reason, the second exposure to an antigen often causes a much faster and stronger response than the first. The response may be so strong, in fact, that the body successfully fights off infection entirely. The multiplication of memory cells explains the lifelong immunity that children acquire after only one exposure to a disease such as chickenpox.

Despite the apparent specificity of antibodies, they are not perfectly specialized. Because lymphocytes respond to epitopes, which are very small parts of molecules, rather than to whole molecules, antigens whose epitopes are identical can precipitate an immune response for each other. In other words, antigens that mimic disease-causing bacteria or viruses are as effective at producing immunity as the pathogens themselves. A heat-killed polio virus, for example, produces immunity against polio, and an infection of cowpox, a mild disease, produces immunity against deadly smallpox.

> *When an antigen binds to the B-cell receptors, the B cells divide and differentiate into plasma cells, which actually make and secrete antibodies, or memory cells, which make more plasma cells if the antigen ever appears again.*

HOW DOES THE BODY DEFEND AGAINST GOOD CELLS GONE BAD?

The B cells produce antibodies that can recognize bacteria and bacterial products and target them for destruction. But because B cells do not respond to the body's own cells, B cells cannot

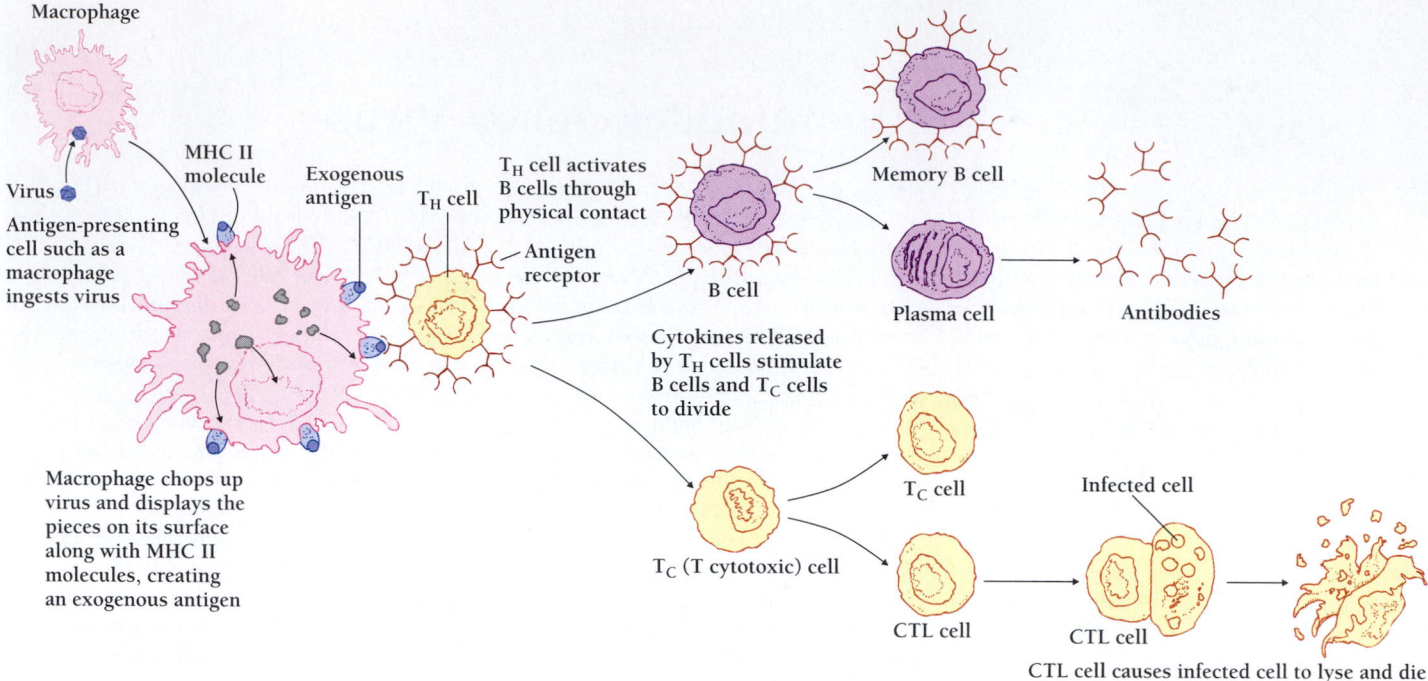

Figure 40-15 T_H cells stimulate antibody production or the production of cytotoxic T lymphocytes. The cells of the body display unique combinations of MHC proteins that identify the cells as belonging to the body. Phagocytic immune cells combine the MHC proteins of a dying cell with antigens from the virus or microorganism that infected the cell to form exogenous antigens. Exogenous antigens cause T_H cells to activate both B cells and T_C cells.

respond to body cells that have become infected by viruses or other pathogens. Because infected cells are actively producing viruses or bacteria, it is important for the immune system to destroy them. We can think of them as good cells gone bad. Biologists call such cells "altered self."

Cancer cells are also good cells gone bad. Recognizing body cells that have become infected or cells that are growing out of control is the task of the T cells (or T lymphocytes). T cells are also involved in rejecting organs and skin grafts surgically transplanted from other individuals. T cells not only recognize altered cells, they also kill them. T cells kill "bad" cells, a property called *cytotoxicity*, by boring holes in the cell membrane. Such T cells are called cytotoxic T cells.

We know the T cells are central to such defense because a child born without a thymus, where the T cells mature, lacks T cells and cannot mount an immune response to certain kinds of infections. Such a child has plenty of B cells and can withstand bacterial infections that multiply outside of cells. But the child is continually beset with infections by viruses, as well as by intracellular bacteria and fungi, all of which multiply inside of cells.

How Do T Cells Recognize Self?

T cells are of two kinds. **T cytotoxic cells (T_C cells)** kill altered cells of the body (whether they are infected by viruses or are cancerous), and **T helper cells (T_H cells)** alert nearby B

cells to make antibodies and stimulate T cytotoxic cells. Both kinds of T cells—T_C cells and T_H cells—recognize nonself antigens by means of **T-cell receptors,** cell-surface proteins that recognize and bind to antigens or altered self. But T_C cells and T_H cells serve different defensive needs (Figure 40-15).

T_H cells produce chemical signals (cytokines) that activate both T and B cells. While some cytokines (such as IL-1, mentioned previously) can act throughout the body, most cytokines are *paracrine signals,* molecules that activate only nearby cells. Some T_H cells promote the production of antibodies by B cells, while others promote T-cell immunity. T_H cells also bind the antigen and present it in a more effective form to the responding B cells.

B cells and T cells recognize different antigens. These two kinds of antigens differ from each other in two ways: (1) antibodies recognize foreign antigens in solution as well as on cell surfaces, but T-cell receptors bind foreign antigens only on the surfaces of cells that are recognizably self; and (2) antibodies recognize the three-dimensional shapes of proteins and carbohydrates, but T-cell receptors recognize only linear segments of polypeptide chains. These segments are short, usually less than 20 amino acids long.

T cells recognize the cells of the body by means of the proteins encoded in a set of 40 to 50 genes called the **major histocompatibility complex,** or **MHC.** MHC proteins appear on the surface of essentially every cell in the body other than sperm cells.

MHC genes are extremely variable, with each gene existing in as many as 100 different alleles, so (except for identical twins) every human's particular mix of MHC alleles is unique. It is on the basis of MHC proteins that transplant surgeons "match" donated organs. The MHCs of a donated organ must be similar to the MHCs of the recipient.

MHC proteins fall into three classes, two of which are important for our discussion. **MHC Class I** molecules are present on the surface of essentially every nucleated cell (except sperm). T_C cells recognize MHC Class I molecules coupled to **endogenous antigens**—proteins made by the body's own cells. In contrast, **MHC Class II** molecules occur on the surfaces of cells such as macrophages or dendritic cells that are specialized to present antigens, so-called "antigen-presenting cells." Recall from the beginning of this chapter that dendritic cells supply the mysterious "second signal" that marks infected or foreign cells. When an antigen-presenting cell phagocytoses a virus or microorganism, the cell couples antigens from the pathogen to MHC II molecules on its surface, creating **exogenous antigens**—proteins that alert the T_H cells to the presence of cell-infecting viruses or microorganisms (Figure 40-12).

> *T cells only recognize altered-self antigens, specifically, antigens that are bound to MHC proteins. Unlike B-cell receptors, which recognize the shapes of epitopes, T-cell receptors recognize linear segments of polypeptide chains. MHC Class I and II genes specify membrane proteins that bind antigens and present them to T cells.*

How Do T Cells Recognize Altered Self?

Unlike B cells, T cells do not recognize completely foreign cells. Human T cells would not, for example, recognize mosquito cells as invaders (although human B cells would). T cells recognize only those cells that are like others in the body but somehow different. Another way of stating this is to say that T cells recognize only nonself antigens in a self context, called **altered self.**

T cells recognize such altered-self antigens with the help of T-cell receptors—molecules on the surfaces of T cells that recognize foreign antigens as altered self. The genes for T-cell receptors are related to immunoglobulin genes. As in the B-cell immunoglobulin genes, rearrangements of T-cell receptor genes produce a T-lymphocyte library, each member of which contains one antigen recognition site. T-cell receptors are even more diverse than immunoglobulins.

> *T-cell receptors recognize altered self.*

How Does the Body Eliminate T Cells That Might Attack Healthy Cells?

Cells of the immune system undergo stringent selection as they develop. First, T cells that do not recognize self must be eliminated. During the maturation of T_C cells in the thymus, the main survivors are precursor cells that can recognize antigens within the individual's own MHC Class I molecules. Similarly, T_H precursor cells that can recognize antigens bound to the individual's own MHC Class II molecules survive.

Second, those T cells that do not distinguish between self and altered self must go. The T-cell precursors that respond to self antigens (combined with self MHC receptors) also disappear, leaving only T cells that can recognize nonself antigens within self MHC molecules. Altogether, 90 to 99 percent of T-cell precursors either die or become inactivated within the thymus. The T cells that leave the thymus are highly selected to recognize altered-self antigens. Researchers are intensively studying what causes death or inactivation and how cells distinguish between harmless self antigens and harmful nonself or altered-self antigens.

Individuals Have Distinctive Antigens on Blood Cell Surfaces

The diversity and individuality of MHC proteins also allow a definitive identification of cells from one person and the tracing of paternity in genetic studies and legal disputes. Other surface antigens besides the MHC proteins, however, can direct an immune response to transplanted cells. Antigens on the surfaces of blood cells are especially important, because of the common use of blood transfusions to treat patients who have lost blood in an injury or in surgery. Individuals differ genetically in the kinds of antigens they produce, and each individual may be classified according to a number of "blood groups."

The most important classification is that of the ABO blood groups. Individuals of blood group A have an antigen called A on the surfaces of their red blood cells; those of blood group B have another antigen, called B; those of group AB have both A and B; those of blood group O have neither A nor B. An individual who does not make A antigen (that is, someone with type O or B blood) has antibodies to A antigen. They have these antibodies even if they have never been exposed to type A blood, probably because similar epitopes are present on the surfaces of common gut bacteria. Similarly, individuals who do not express the B antigen (people with blood type O or A) have antibodies against B.

Before physicians understood the nature of the antigenic differences, blood transfusions were dangerous procedures. If a person with type A blood received a transfusion of type B blood, their antibodies attacked the red cells, linked them together in big clumps, and clogged the blood vessels. Such an occurrence could easily be fatal.

Another antigen on the surface of red cells of some individuals is called the Rh factor—after the rhesus monkeys in which the factor was first identified. Unlike the situation with the A and B antigens, individuals who lack the Rh antigen do not generally make antibodies against it unless they are first exposed to Rh-positive blood. This would happen, for example, after transfusions of Rh-positive blood into an

Rh-negative recipient—a situation that modern physicians avoid by testing the blood.

Another situation, far more common, is harder to control. Pregnancy frequently exposes Rh-negative women to Rh-positive blood. If a fetus receives the Rh factor gene from the father, the mother may develop antibodies against Rh factor. Such antibodies are not usually present at high concentrations during the first pregnancy with an Rh-positive fetus.

During the delivery of the baby, the mother is likely to be exposed to Rh antigens, triggering the creation of memory cells. Subsequent pregnancies with an Rh-positive fetus may then provoke a strong secondary response in the mother. The mother's antibodies can cross the placenta and attack the fetus's red cells. The fetus may die in utero or die at birth. Physicians can prevent this immune response by treating the mother

with anti-Rh antibodies, which destroy Rh-positive fetal cells, so the mother never makes anti-Rh antibodies herself.

MHC proteins and other cell-surface antigens make the cells of every individual unique.

In this chapter we have seen how the body defends itself against pathogenic viruses, bacteria, and other foreign invaders. The power of the immune system—its specificity, memory, diversity, and self-nonself recognition—comes from the ability of its cells to communicate with one another. In the next chapter we will find out how the cells of the nervous system communicate with one another and with the muscles and organs of the body.

Study Outline with Key Terms

The body defends itself against invading pathogens with three lines of defense: the skin and the populations of microorganisms on the skin and in the gut, **blood clots** and **inflammation,** and the **immune system.** The skin consists of the **epidermis,** the **dermis,** and the **subcutaneous tissue.** The saliva and tears contain **lysozyme,** an enzyme that attacks bacteria. Inflammation and immune responses result from the work of several kinds of white blood cells, or **leukocytes.**

The body closes breaks in the skin by means of **wound healing** and breaks in the circulatory system by means of **hemostasis,** which includes **vasoconstriction,** the aggregation of **platelets,** and blood clots. All of these processes prevent excessive blood loss and help prevent microorganisms from entering the circulatory system.

Because blood clots are dangerous, the synthesis of the **fibrin** in clots is carefully regulated. Fibrin derives from **fibrinogen** when the enzyme **thrombin** digests away part of the fibrinogen polypeptide. Blood **plasma** contains **prothrombin,** a precursor of thrombin. Another system of anticlotting enzymes destroys the fibrin in clots. **Plasminogen** in the blood plasma is turned into **plasmin,** which digests the bonds between fibrin molecules, dissolving the clot. **Tissue plasminogen activator** (tPA) activates the anticlotting system. Mammalian blood also contains **anticoagulants,** such as **heparin.**

Complement proteins and **mast cells** react nonspecifically to damage to body tissues. Mast cells release **histamines,** which help promote the inflammation response. Inflammation, a set of responses to local injury, helps an animal to resist infection by increasing circulation, attracting and mobilizing phagocytic cells, and inducing signaling molecules. During inflammation, **neutrophils,** attracted by chemicals released by mast cells, attach to the capillary walls, move through

the capillary wall into the damaged tissue, and phagocytose any microorganisms in the damaged tissue. **Macrophages** also consume microorganisms and cellular debris and play a major role in initiating inflammation by releasing three kinds of **cytokines: tumor necrosis factor-α** (TNF-α), **interleukin-1** (IL-1), and **interleukin-6** (IL-6). The interleukins act as messenger molecules between different kinds of white blood cells, initiating and regulating inflammation and immune responses. Interleukin-1 stimulates responses in phagocytic cells—in the endothelial cells of the capillaries and in the brain cells that control body temperature. Another set of cytokines, called **interferons,** interferes with the replication of viruses by limiting protein synthesis in virus-infected cells.

The immune system responds to invading organisms by producing **antibodies** and by producing lymphocytes—T cells and B cells—that recognize and attack cells of the body that have become infected by viruses or microorganisms. Four attributes characterize immunity: specificity, memory, diversity, and self-nonself recognition.

Cells of the immune system recognize **specific** invaders and not others. The immune system's antibodies recognize **antigens** by their **epitopes.** The specificity of the immune response distinguishes it from the **nonspecific** responses of clotting, complement proteins, inflammation, and interferons. The immune system also has **memory.** The reappearance of an invader stimulates a **secondary immune response,** usually even stronger than the first response.

The immune system can defend against a **diversity** of foreign invaders. But the immune system **tolerates** many cells in the body. In accordance with the model currently accepted by most biologists, the immune system's apparent ability to distinguish the body's own components from others is called **self-nonself recognition.**

Immune responses depend on three kinds of **lymphocytes,** which develop within the lymphoid tissues: **B lymphocytes,** or **B cells,** which make antibodies; **T lymphocytes,** or **T cells,** which recognize antigens on the surfaces of the body's own cells; and **natural killer cells,** or **NK cells,** which attack tumor cells and cells infected by a pathogen.

Antibodies are globular proteins called **immunoglobulins.** The simplest antibody molecules, called **immunoglobulin G** or **IgG,** consist of four polypeptide chains, two **heavy chains** and two **light chains.** Antibody diversity depends on the ability to combine many different light chains with many different heavy chains.

While the immune system can produce more than a billion antibodies, a single antibody-producing cell can make only one kind of antibody. Individual antibodies have different amino acid sequences. The immune response depends on an animal's ability to select among an existing repertoire of cells, each of which is able to make a single type of antibody.

Both light and heavy chains of immunoglobulin are encoded in genes that undergo rearrangement. To form a functional gene, a light-chain gene undergoes two rearrangements, and a heavy-chain gene undergoes three rearrangements. Somatic mutations can further increase the diversity of immunoglobulin sequences.

During embryonic life, precursors of B cells form a **lymphocyte library,** with each lymphocyte containing a unique random rearrangement of immunoglobulin genes. B cells are able to respond to particular antigens because they have **B-cell receptors,** cell-surface proteins with the same antigen recognition sites as the antibody that they produce. B-cell receptors are membrane-bound forms of antibodies. According to the **clonal selection theory,** an individual B lymphocyte proliferates when exposed to an antigen that binds to its particular antibody. Its descendants include **plasma cells,** which produce and export antibodies, and **memory cells,** which can later develop into plasma cells. The clonal selection theory also suggests a mechanism for self-nonself recognition and immunological tolerance. Cells in the lymphocyte library that recognize self antigens are eliminated or suppressed.

T cells are of two kinds: **T cytotoxic (T_C)** cells, which kill cells recognized as **altered self,** and **T helper (T_H)** cells, which send messages to nearby B cells and to T cytotoxic cells. Antigen recognition by T cells depends on **T-cell receptors** on the surface of T cells. T-cell receptors are even more diverse than antibodies.

T-cell receptors do not recognize free antigens, but only short polypeptides bound by the proteins of the **major histocompatibility complex (MHC).** Two classes of MHC proteins can present antigens to T cells: **MHC Class I** molecules, which are on the surface of all cells in the body, present **endogenous antigens** to T cytotoxic cells, and **MHC Class II** molecules, which present **exogenous antigens** on the surfaces of specialized antigen-presenting cells that interact with T helper cells.

Review and Thought Questions

Review Questions

1. Name the body's three lines of defense against pathogens.
2. What is hemostasis?
3. What two systems prevent all of the blood from clotting at once? How do these two systems work?
4. Where do interleukins come from and what do they do?
5. How does the immune system make enough antibodies?
6. Lymphocytes come in three types. What are they and what function does each one serve?
7. Explain what a lymphocyte library is and why it is important to immune function.

Thought Questions

8. Why do you think that sperm lack MHC Class I molecules on their cell surfaces? What might happen if sperm had these identifying markers?
9. Which facts presented in this chapter support the danger model and which support the self-nonself model?

About the Chapter-Opening Image

A dendrite cell looks something like a neuron, but it is an immune cell that may help other immune cells distinguish damaged cells from healthy ones.

 On-line materials relating to this chapter are on the World Wide Web at **http://www.harcourtcollege.com/lifesci/aal2/**

41

The Cells of the Nervous System

Rita Levi-Montalcini

When Rita Levi-Montalcini (1909–) was a 21-year-old medical student, she asked her professor to suggest a suitable research project. Her professor, a neuroanatomist at the Turin School of Medicine, in Italy, suggested that she determine how the convolutions of the human brain were formed during development.

"It was a really stupid question, which I couldn't solve and no one could solve," Levi-Montalcini recalled at age 82. When the professor called her in a few months later to see how she was doing, he pronounced her attempts "real trash" and said she was not cut out for scientific research. Neither of them could guess that she would eventually win a Nobel prize and become one of the most influential neuroscientists of the century. Nor could either of them know that her subsequent work would suggest new therapies for spinal cord injury and Alzheimer's disease.

Yet the 22-year-old didn't let her professor discourage her. After her initial failure, Levi-Montalcini quickly found a more manageable problem and never looked back. Levi-Montalcini gradually taught the professor to respect her. In the end, he proved a faithful friend and mentor until his death

many years later. All her life, Levi-Montalcini has skillfully navigated obstacles that would stop most people. Even in her 80s, she continued to inspire and charm new generations of biologists.

Born into an intensely patriarchal Italian-Jewish family, Levi-Montalcini made up her mind when she was still a child that she would not live under a husband's thumb, as her own mother did. She rebelled against her parents' assumption that she and her twin sister Paola would marry straight out of high school, and she begged, in vain, to be sent to a high school that would prepare her for college (Figure 41-1). Instead, she graduated from a high school for girls that emphasized literature but not math or science. Though she read widely throughout her adolescence, she had no particular interests. She buried herself in books in which she was only mildly interested. She had ambition but neither goals nor occupation.

When Levi-Montalcini was just 20, however, her beloved governess, Giovanna, died of stomach cancer, and Levi-Montalcini resolved to become a doctor. She knew that she had not had the education that she needed to go to medical school, so she decided to find a tutor to prepare her for the entrance exams. To make the endeavor more fun, she persuaded her cousin Eugenia to join her. Eight months later they both passed the entrance exams with flying colors. In 1930, they entered the Medical School at Turin, Italy.

By 1936, both young women had earned degrees in medicine with top honors and had apprenticed themselves to a flamboyant Jewish neuroanatomist at Turin. After the false start with the human brain, Rita began work that fascinated her. "For the first time," she wrote in her autobiography, "I became passionate about research. . . ."

It was a passion that would be severely tested. In 1924, Mussolini had become dictator of Italy, and beginning in the 1930s his alliance with Hitler obliged him to take action against Italy's Jews. By the fall of 1938, a series of state decrees forbade all Jews from pursuing any kind of profession, including academic research and teaching. In 1939, Levi-Montalcini left the university at Turin, where she could no longer work without endangering both herself and her non-Jewish colleagues.

- The nervous system consists of neurons, which carry electrical signals, and glia, which provide metabolic and structural support to the neurons.

- Neurons can carry electrical signals, called action potentials, over long distances.

- Connections between neurons may be either electrical, mediated by gap junctions, or chemical, mediated by the release of signaling molecules called neurotransmitters.

- Neurotransmitters act by binding to receptors in the plasma membrane of the target neurons.

For a year, she lived quietly with her mother, her sister Paola, and her brother Gino. Her father had died several years earlier. She and Eugenia spent much of their time treating Jews who had fallen ill but could not be treated by non-Jewish doctors.

Then one day she received a visit from an old friend from medical school who demanded to know what sort of research she was doing. Under the wartime conditions, it hadn't occurred to her that she could do any research, and she said nothing. Distressed by her apathy, he scolded her, "One doesn't lose heart in the face of the first difficulties. Set up a small laboratory and take up your interrupted research. Remember Ramón y Cajal, who in a poorly equipped institute, in the sleepy city that Valencia must have been in the middle of the last century, did the fundamental work that established the basis of all we know about the nervous system of vertebrates."

It was all the impetus Levi-Montalcini needed. She quickly set up a laboratory in her bedroom and decided to study chick embryos, as chicken eggs were relatively easy to obtain. Partly inspired by the thought of the famous Spanish anatomist Ramón y Cajal, she chose to study the development of the nervous system.

From 1941 until 1943, Levi-Montalcini studied the development of the nervous system in her bedroom. Always practical, she would carefully remove the tiny chick embryos from the eggs after she had finished an experiment and turn the eggs into omelets in the kitchen downstairs. Her brother Gino, after watching this operation once, categorically refused to eat any more of her omelets, although he had loved them until then.

Her research into how embryonic nerve tissue differentiates into specialized cells went splendidly. She described her fascination beautifully:

Figure 41-1 As girls, Rita Levi-Montalcini (shown here) and her sisters, Paola and Nina, hoped to emulate the Brontë sisters, the famous 19th-century trio of English writers. As they grew older, Rita became a Nobel prize-winning neuroscientist and Paola a famous painter. (*Washington University Archives*)

Now the nervous system appeared to me in a different light from its description in textbooks of neuroanatomy, where its structure is described as rigid and unchangeable. Only by following from hour to hour in different specimens, as in a cinematographic sequence, the development of nerve centers and circuits, did I come to realize how dynamic these processes are; how individual cells behave in a way similar to that of living beings; how plastic and malleable is the entire nervous system.

During development, the nerves of the peripheral nervous system appear to grow out from the spinal cord toward the limbs. The biologist Viktor Hamburger, at Washington University in St. Louis, had shown that embryos whose limb buds have been amputated do not develop nerves to supply the limb. He proposed that the limbs produce some special substance that tells the unspecialized nerve cells in the spinal cord what kind of cells they should turn into, as well as in what direction they should grow.

Levi-Montalcini's research suggested an alternative. She thought that the nerves in the embryonic spinal cord do specialize, or differentiate, but that they soon die without some special substance from the limb buds to sustain their growth. Her discovery of this substance revolutionized the understanding of neural development.

But Levi-Montalcini's work was again interrupted and her very life endangered. In 1943, German tanks approached Turin, and Levi-Montalcini and her family fled to Florence, abandoning the makeshift lab, their home, and their friends. In Florence, they took up lodging with a tolerant Catholic woman—assuring her that they were not Jewish. But Florence, too, was taken by the Germans, who mined the streets, blew up the bridges, and killed civilians. Finally, in September 1944, when the British liberated Florence, Levi-Montalcini went to work for the Red Cross as a practicing physician. Not until 1945 did she return to her research.

After the war, Hamburger, who had read her papers, invited her to come to Washington University, in St. Louis, to work with him. Her hypothesis that some substance sustains developing nerve cells turned out to be correct. In the 1950s, she and biochemist Stanley Cohen isolated this substance, which they called *nerve growth factor.* Levi-Montalcini earned a place on the faculty at Washington University, and in 1986, she and Cohen shared the Nobel Prize for Medicine and Physiology.

Their discovery led not only to a new understanding of neural development but also to the discovery of a whole set of related growth factors. These proteins stimulate the growth of neurons during early development and help sustain them during adult life. Growth factors, many researchers believe, may prevent cell death—and even promote regeneration—in Alzheimer's disease, Huntington's disease, Parkinson's disease, Lou Gehrig's disease, and spinal cord injuries.

Levi-Montalcini, now in her 90s, lives in an apartment in Rome that she shares with her twin sister Paola, a well-known artist. As a Nobel laureate, Levi-Montalcini serves on numerous scientific committees, a role that allows her to influence the direction of science both in Italy and internationally. She and her sister run a foundation to help teenagers find a path in life. Every week she talks to teenagers about their interests and helps them find work that interests them.

"You never know what is good, what is bad in life," Levi-Montalcini told the magazine *Scientific American* of the Nobel prize-winning research done in her bedroom. "I mean, in my case, [working in partial isolation] was my good chance."

HOW ARE NERVE CELLS SPECIAL?

Even a generation after the acceptance of the cell theory, biologists were still not certain that the nervous system contained conventional cells. Until the late 19th century, anatomists studied the brain in the same ways that they studied other tissues, cutting, staining, and examining the tissues under a microscope. But when they examined the tissues of the brain, they did not see distinct cells, only dots, blurs, and dark spots that resembled nuclei (Figure 41-2A). The problem, we now know, was that the brain contains so many intertwined cells that the only way to see an individual cell is to stain one cell, while leaving its neighbors unstained. In the 1880s, Camillo Golgi developed just such a method, now called the Golgi method, for staining only a few cells in each sample (Figure 41-2B).

Shortly afterward, the great Spanish neuroanatomist, Santiago Ramón y Cajal, used Golgi's method to study the nervous system of a variety of species. Ramón y Cajal's careful work provided neuroscience with many deep insights into the organization and development of the nervous system. His most important contribution, however, was clear evidence that the nervous system consists of cells. Ironically, Golgi did not accept Ramón y Cajal's conclusions. The two men shared the 1906 Nobel Prize, and their formal acceptance speeches contradicted one another.

Nerve cells have exactly the same kinds of organelles as other cells (Figure 41-3). Nerve cells come in a variety of shapes

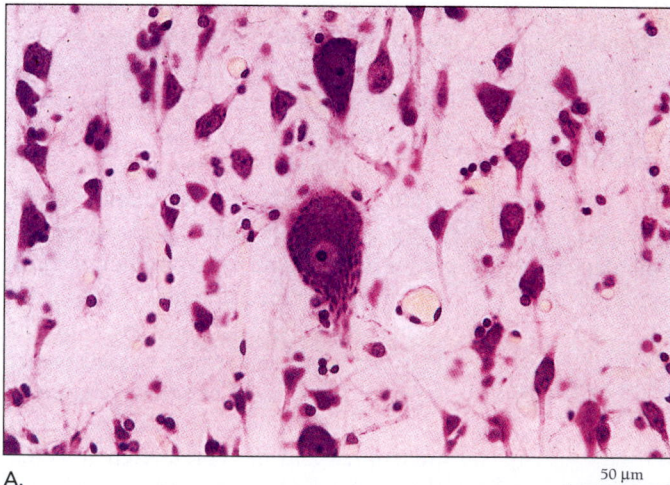

A.   50 µm

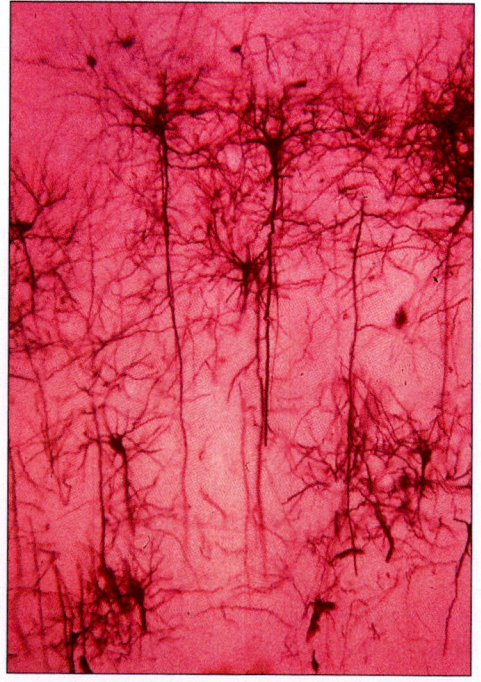

B.   100 µm

Figure 41-2   In the late 19th and early 20th century, anatomists debated whether the brain actually contains individual cells. A. With standard staining techniques, it is impossible to tell where one cell ends and another begins. B. The silver stain developed by Camillo Golgi allowed researchers to see the extended nature of individual cells. *(John D. Cunningham/Visuals Unlimited)*

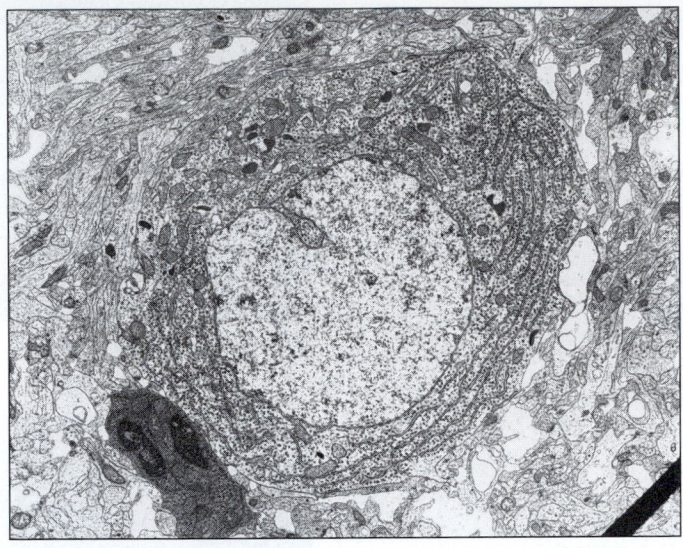

0.5 μm

Figure 41-3 Electron microscopy reveals that nerve cells have exactly the same kinds of organelles as others cells—including mitochondria, ribosomes, and endoplasmic reticulum. *(Dennis Kunkel/Phototake)*

and sizes, but all have a distinctive look (Figure 41-4A). The most obvious difference between nerve cells and other cells of the body is their size. They can be immense. Nerve cells can extend several meters, for example, in the legs of a giraffe.

Nerve cells that carry information in the form of electrical and chemical signals are called **neurons.** But the nervous system also includes **glial cells,** which provide metabolic and structural support for the neurons. Glial cells do not themselves transmit electrical and chemical signals. Some glial cells serve as guides and scaffolding during neural development. Others, called *oligodendrocytes* and *Schwann cells,* wrap around nerve cell extensions in a *myelin sheath* (Figure 41-4B). As we will discuss later, the myelin sheath speeds the transmission of signals through the nerves.

Neurons exhibit many variations on a basic plan (Figure 41-4). Most have a distinct center, called the *cell body* (or *soma*), which is about the same size as most other cells. All neurons have long extensions from the cell body. Numerous short extensions, called **dendrites,** usually relay signals from other cells to the cell body. Longer, thicker extensions, called **axons,** usually carry signals away from the cell body and connect to other cells. Axons may be quite long, up to a meter or more. They may branch at their ends to make contact with target cells, but the number of axon branches is trivial compared with that in some dendrites. One type of cell in the brain's cerebellum, for example, has enough dendrites to relay information from more than 50,000 other cells. Between the axons and dendrites are highly specialized contacts, called **synapses.**

All of the specialized structures in neurons help them to carry information over long distances and transmit it to specific other neurons. Neurons also have particularly large num-

bers of mitochondria and ribosomes, suggesting that they are active users of energy and synthesizers of proteins. Studies of brain metabolism support this conclusion: the human brain, which makes up only 2 percent of the body's weight, uses nearly 20 percent of all the energy the body uses.

How Do Nerve Cells Carry Information?

Nerve cells carry innumerable different kinds of information. Whether we cut ourselves or catch sight of a loved one, the information about the event is carried to the brain by nerves, and the brain's response is transmitted to our muscles by nerves. How do nerves do it?

Scientists have known since the late 18th century that nerves respond to electrical impulses. A small electric shock applied to the neurons that control a frog's leg, for example, will cause the leg to kick. The electrochemical pulses carried by nerves are entirely different, however, from the electric current that flows in the wires of your house. Household electric current, carried by electrons jumping from atom to atom in a metal wire, flows much more rapidly than nerve impulses, carried by ions moving back and forth across cell membranes. More importantly, however, electric current gradually loses energy as it flows passively through a wire. In contrast, the electric pulse that passes through the body's nerves is constantly regenerated as it goes. The energy of the pulse at the end of a nerve is the same as that at the beginning.

Another important difference between electric current and nerve impulses is the standard size of a nerve impulse. Electricity can flow at a great rate or at a trickle, in millions of volts or in tiny millivolts. A nerve cell, however, always carries the same voltage. If a stimulus, such as a mild electric shock, is sufficient to trigger a nerve impulse, the impulse will be the same size no matter how strong or weak the stimulus is. On the other hand, if the stimulus is too weak to trigger an impulse, the nerve cell makes no response at all. The nerve cell has an *all-or-none*, or **threshold,** response. Any stimulus above a certain level triggers the same response.

We can ask ourselves, if the neuron has an all-or-none response, how can we detect subtle differences in temperature, light intensity, and other variable stimuli? Nerves can convey differences in intensity in two ways. First, as the intensity of a stimulus increases, the number and the frequency of pulses generated and transmitted also increase. Second, neighboring nerve cells have different thresholds, so that as the intensity of a stimulus increases, the total number of responding cells increases (Figure 41-5). The result of these two processes is that as the intensity of a stimulus—say the temperature of your bath water—increases, the number of impulses reaching your brain each second increases. In addition, the impulses come from a greater number of receptor cells if the bath water is very hot than if it is just warm.

If nerve impulses are not simple electric currents, what are they? In the rest of this chapter we will try to understand how neurons generate and transmit impulses and how these impulses are then passed from one neuron to the next.

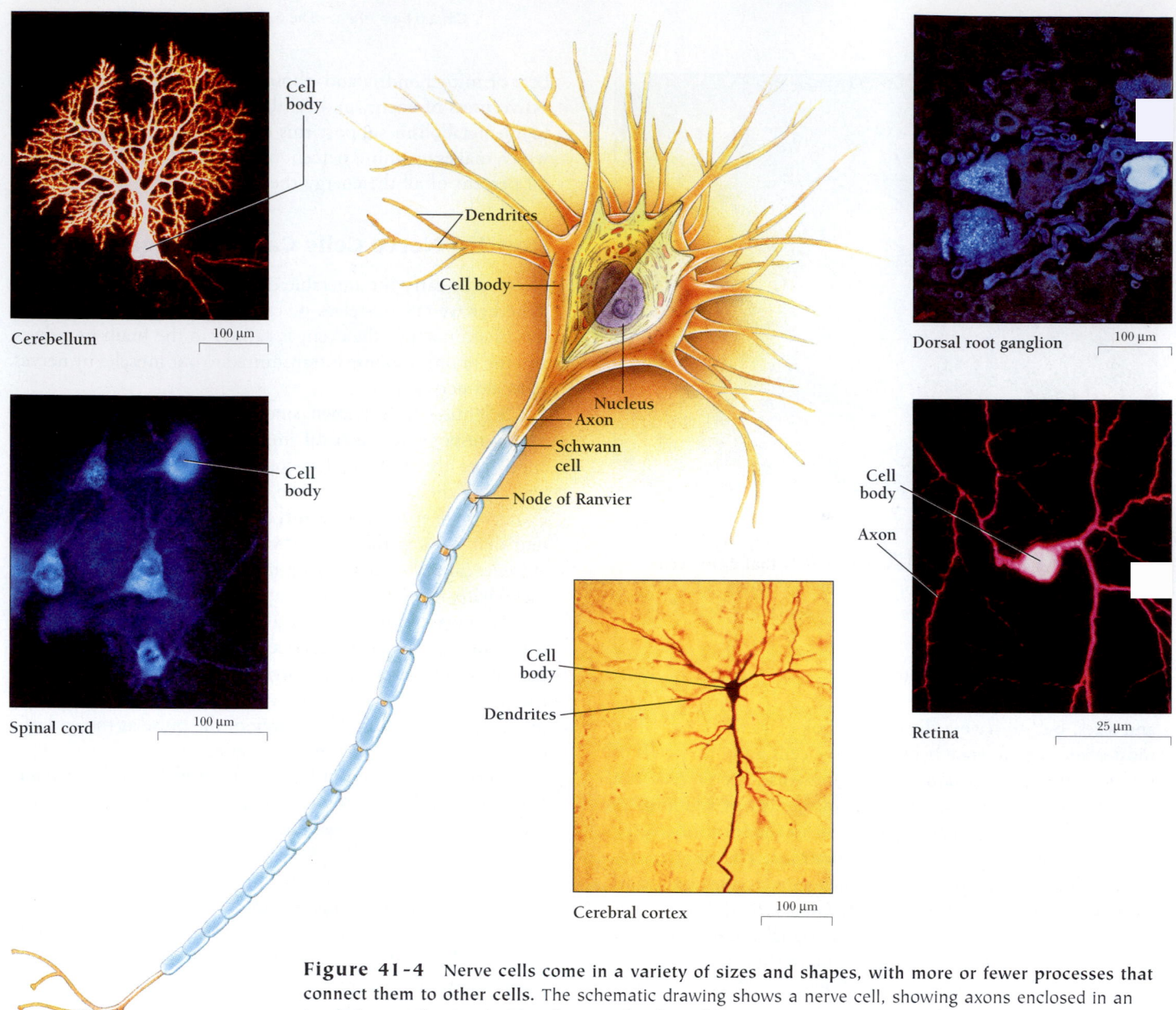

Figure 41-4 Nerve cells come in a variety of sizes and shapes, with more or fewer processes that connect them to other cells. The schematic drawing shows a nerve cell, showing axons enclosed in an insulating myelin sheath. The photographs show different types of nerve cells, including cells from the cerebellum, the dorsal root ganglion, the cerebral cortex, the retina, and the spinal cord. *(Photos, dorsal root ganglion and spinal cord, courtesy of Dr. Niranjali Tillakaratne, UCLA; cerebellum and retina, David Becker/Tony Stone Images; cerebral cortex, Carolina Biological/Phototake, NYC)*

Unlike electrical lines, nerves carry energy pulses that are always the same size and that do not dissipate during transmission. A nerve cell transmits an impulse only when stimulated by a signal that exceeds a minimum threshold.

How Does a Neuron Generate an Electrical Pulse?

Every cell has a chemical composition that is different from its surroundings. This difference results mostly from the activities of the cell membrane. As we saw in Chapter 4, cell membranes allow some molecules and ions to pass freely, while serving as impermeable boundaries to others. Membrane proteins that serve as **channels** allow the passage of specific molecules and ions. Some of these channel proteins have **gates,** which open and close the channels in response to environmental signals. Gates regulate the passive movement of ions or molecules through the membrane. Other membrane proteins actively **pump** ions or molecules through the membrane using the energy from ATP.

As a result of the operation of these channels, gates, and pumps, the inside of a cell has different concentrations of certain ions and molecules from the outside. A difference in the concentration of molecules and ions on the two sides of a cell

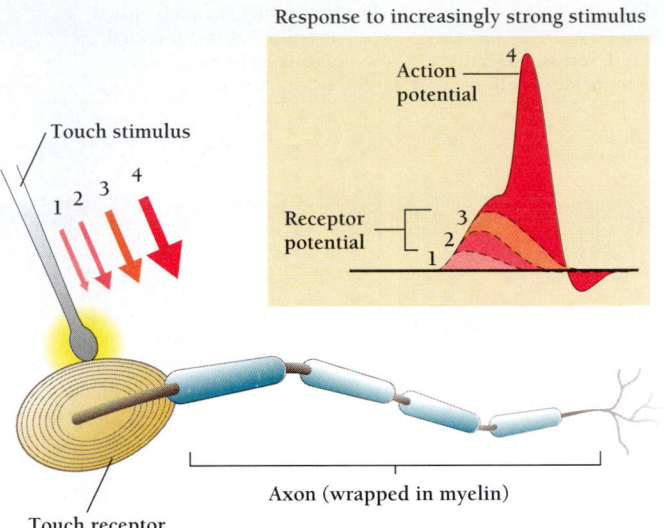

Response to increasingly strong stimulus

Touch stimulus

Action potential — 4

Receptor potential — 3
2
1

Axon (wrapped in myelin)

Touch receptor

Figure 41-5 Increased stimulation of a touch receptor leads to increased depolarization and an increased chance of firing. The stronger the touch, the greater the receptor potential. Once the threshold is reached, the cell fires an action potential.

membrane is a **concentration gradient,** which is a source of potential energy. If a concentration gradient also results in a net electrical charge on one side of the membrane relative to the other, then an electrical gradient also exists. Such a gradient is called **voltage,** or **electrical potential.**

Voltage is the driver of electric current. If we compare electricity to a waterfall, then the **current** is the amount of water, or the number of electrons, flowing per minute. The voltage is comparable to the height of the waterfall. The greater the voltage, the greater the work that can be performed.

In a living cell, the voltage across a cell's membrane results from the difference in charge between one side of the membrane and the other. Researchers can measure the voltage, and, as we discuss later, they can even study the dependence of the voltage on the concentrations of particular ions. At rest, cells maintain a potential (voltage) of about 70 millivolts across the plasma membrane, with the inside of the cell negative with respect to the outside (Figure 41-6A). During a nerve impulse, this net charge actually reverses itself for just a fraction of a second. This impulse (or "action potential") passes along the length of the nerve cell membrane like a wave.

The −70-millivolt potential comes from a difference in the distribution of both negative and positive ions on the two sides of the plasma membrane. The concentrations of negative ions, such as chloride (Cl^-) and various proteins, are constant during the generation of a nerve impulse. In contrast, the positive ions potassium (K^+) and sodium (Na^+) jump back and forth across the membrane like children playing hopscotch.

The efficient sodium-potassium pump in the cell membrane uses ATP to pump sodium ions out and potassium ions into cells. Thanks to this pump, the inside of a nerve cell contains a much lower concentration of sodium ions compared with that outside the cell. In contrast, the inside of a cell has more potassium ions than the outside. When a nerve cell is stimulated, sodium channels open and the cell membrane suddenly becomes permeable to sodium ions. As a result, sodium ions rush across the cell membrane to the sodium-impoverished cell interior, propelled by the difference of concentration on the two sides of the membrane and by the negative charge on the inside of the plasma membrane.

As sodium ions rush in, the inside of the cell gains a slight positive charge, at least in the vicinity of the open sodium channels (Figure 41-6B). A few milliseconds later, the membrane loses its permeability to sodium ions, but it can temporarily become even more permeable to potassium ions. Potassium ions now rush out of the cell, propelled by the concentration difference and repelled by the momentary positive charge inside the cell. Only a small fraction of the cell's sodium and potassium ions move during these events, and the voltage changes across the membrane center around the patch where the channels are opening and closing. Finally, the cell then regains its customary −70-millivolt potential and it is ready to transmit another pulse.

The −70-millivolt potential of a nerve cell at rest is called the **resting potential.** Like the coil in a car engine (which stores the charge for the spark plugs), a nerve cell's resting potential represents a certain amount of potential energy that can be used to pass an impulse, under the right circumstances. In contrast, the sudden electrical change that occurs during an impulse is called the **action potential.**

Nerve cells maintain differences in ion concentrations across the cell membranes, leading to a resting potential of −70 millivolts.

How Does a Neuron Maintain Its Resting Potential?

The main reasons that a resting potential exists are (1) the membrane's sodium-potassium pump and (2) the membrane's selective permeability to sodium (Na^+) and potassium (K^+) ions. The sodium-potassium pump in a cell membrane uses the energy of ATP to push sodium ions out and pull potassium ions inside. For every ATP spent, three sodium ions leave and two potassium ions enter so that sodium ions accumulate outside the cell and potassium ions inside. The membrane then maintains the resulting uneven concentration of ions by restricting the passage of sodium and potassium.

At rest, all sodium channels and most potassium channels are closed. The potassium channels allow potassium ions to move out of the cell, in a direction that would favor their equal concentration on both sides of the plasma membrane. As potassium ions move out, however, they leave a net negative charge that resists further movement of potassium ions. The open potassium channels allow potassium ions to come to equilibrium, with the tendency to move out (because of the difference in concentration) exactly balanced by the tendency to move in (because of the negative charge inside the cell). For every ion, the voltage at equilibrium, called the *equilibrium*

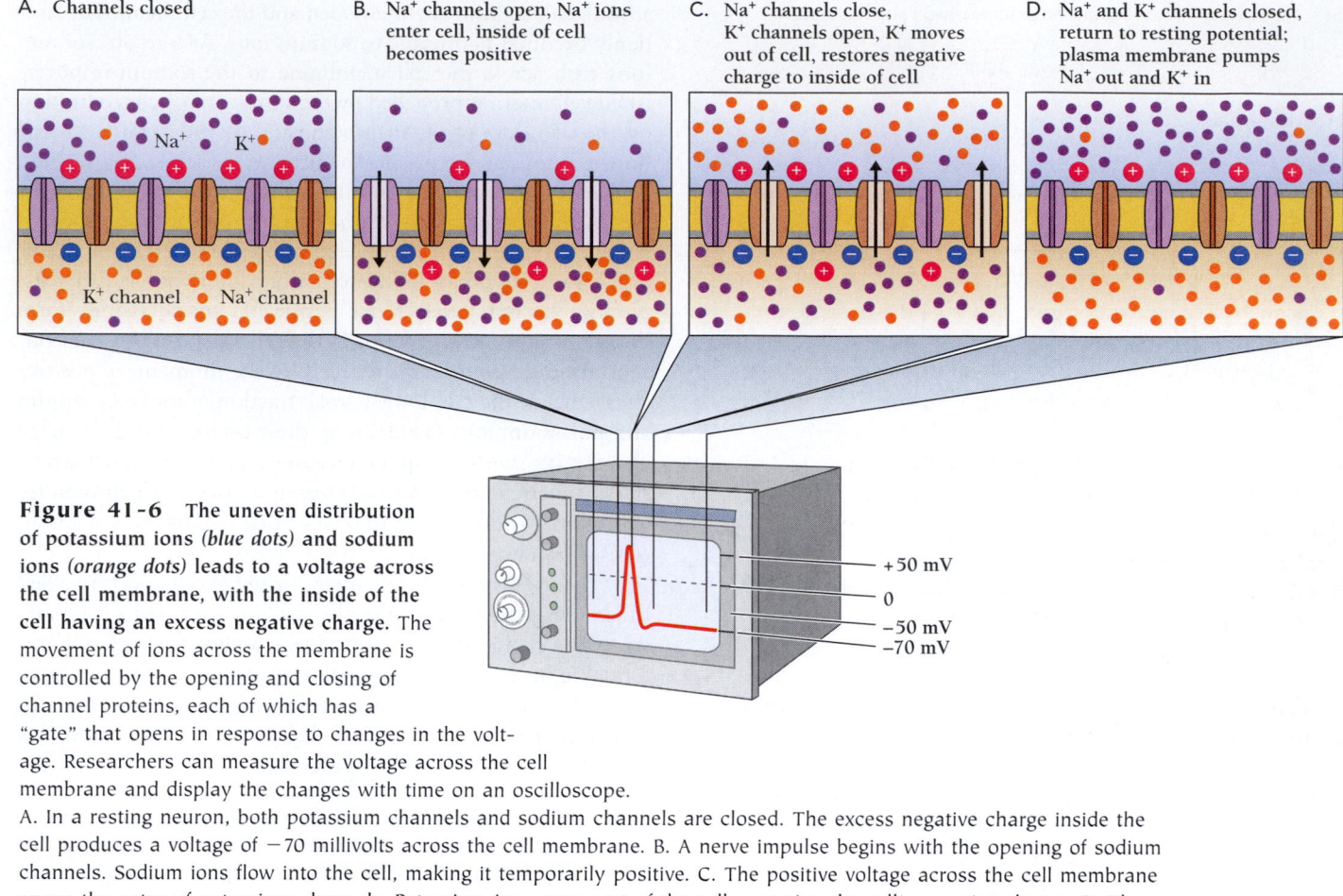

A. Channels closed

B. Na⁺ channels open, Na⁺ ions enter cell, inside of cell becomes positive

C. Na⁺ channels close, K⁺ channels open, K⁺ moves out of cell, restores negative charge to inside of cell

D. Na⁺ and K⁺ channels closed, return to resting potential; plasma membrane pumps Na⁺ out and K⁺ in

Figure 41-6 The uneven distribution of potassium ions (*blue dots*) and sodium ions (*orange dots*) leads to a voltage across the cell membrane, with the inside of the cell having an excess negative charge. The movement of ions across the membrane is controlled by the opening and closing of channel proteins, each of which has a "gate" that opens in response to changes in the voltage. Researchers can measure the voltage across the cell membrane and display the changes with time on an oscilloscope. A. In a resting neuron, both potassium channels and sodium channels are closed. The excess negative charge inside the cell produces a voltage of −70 millivolts across the cell membrane. B. A nerve impulse begins with the opening of sodium channels. Sodium ions flow into the cell, making it temporarily positive. C. The positive voltage across the cell membrane opens the gates of potassium channels. Potassium ions move out of the cell, restoring the cell's negative charge. D. The sodium channels are again closed, and the cell remains at its resting potential, which depends just on potassium ions.

potential, depends upon the difference in that ion's concentration on the two sides of the membrane. For a neuron, the calculated resting potential is almost exactly the equilibrium potential of the potassium ion, indicating that the resting potential depends almost entirely on the distribution of potassium ions.

The resting potential of a neuron arises because of the selectivity of ion pumps, which concentrate sodium ions outside the cell and potassium ions inside, and because of the selectivity of channels within the cell membrane, which allow different ions to pass at different rates.

How Does a Neuron Generate an Action Potential?

Electricians measure voltage with a *voltmeter,* a device that determines the tendency of electrons to flow between two contacts, called *electrodes.* The voltmeter measures this tendency in units called *volts.*

The voltages across cell membranes are (fortunately for curious biologists) much smaller than those in wall sockets. Even the small batteries used for toys and household appliances have voltages about 20 times greater than those of a cell. A standard radio battery, for example, generates about 1.5 volts, while most cells have voltages measured in millivolts. One millivolt is 1/1000 of a volt, so a radio battery generates 1500 millivolts. Neurobiologists measure these tiny voltages with an *oscilloscope,* a televisionlike screen that displays variations in voltage over time (Figure 41-6). Instead of standard metal electrodes, biologists use *microelectrodes,* tiny glass tubes that contain salt solutions and are small enough to maintain contact with a single cell.

When a physiologist pokes a microelectrode into a nerve cell, the oscilloscope registers a negative voltage of about 70 millivolts, the resting potential. When stimulated, the nerve cell can suddenly and repeatedly change its electrical potential, an action potential that the electrodes also detect.

Our understanding of how nerve cells generate resting potentials and action potentials has come from experiments in which researchers measured and manipulated the internal voltages of cells. Such experiments were impossible for a long time, however, because nerve cells are so small and standard electrodes are so big. Researchers were simply unable to put electrodes into cells. Doing the proper experiments required either a very large nerve cell or a very small electrode. Ultimately, biologists found the first and invented the second.

Much of our current appreciation of how a nerve cell fires—how it creates an action potential—came from experiments done on giant neurons from the squid. These neurons are responsible for the signals the squid uses to produce a jet of water, by which it can rapidly propel itself. The axons of these nerve cells are immense, up to 1 mm in diameter. They are so big, in fact, that until 1936, biologists thought that these nerve cells were blood vessels. With the discovery of their real nature, physiologists began collecting squid for work on neuron action potentials.

The most penetrating and influential work on squid action potentials was that of the English physiologists Alan Hodgkin and Andrew Huxley. In the 1930s and 1940s, Hodgkin, Huxley, and their collaborators took advantage of the size of the squid axon to measure voltage changes in the axon as the researchers varied ion concentrations inside and outside the cell. They found that the concentration of potassium ions determined the level of the resting potential. They also found a way to measure the ion concentrations within an axon by using a rubber roller to force the contents of the axon out into a dish.

An action potential, unlike a resting potential, depends on the flow of sodium ions into the cell. In one experiment, for example, Hodgkin and Huxley removed all sodium ions from the extracellular fluid, replacing them with a larger ion, choline, that could not pass through a sodium channel in the membrane—even when the channel was open. The result was that the axon could no longer produce an action potential at all (though its resting potential was unchanged). This experiment underscored the conclusion that understanding the action potential requires an understanding of how the membrane temporarily changes its permeability to sodium ions.

The resting potential depends on the presence of channels that allow potassium ions to move across the cell membrane, whereas an action potential depends on the temporary opening of channels that allow sodium ions to move across the cell membrane. The difference in the concentration of potassium ions between the inside and outside of a cell determines the size of the resting potential, whereas the difference in the concentration of sodium ions determines the size of the action potential.

How Does a Membrane Change Its Permeability to Sodium Ions?

Hodgkin and Huxley discovered that a neuron opens its sodium channels after they developed a method that allowed them to control the voltage across a cell membrane. At the normal resting potential, they found, sodium channels are closed. When the cell is *depolarized,* or made less negative, however, the sodium channels open briefly and admit a small horde of sodium ions. Because the channel's permeability to sodium ions depends on the membrane potential, sodium channels are said to be **voltage gated,** meaning that they open or close according to the voltage across the membrane.

Hodgkin and Huxley were able to explain the generation of action potentials in terms of the voltage dependence of sodium channels. When a nerve cell is sufficiently depolarized, the membrane reaches a threshold voltage, at which point sodium channels open and sodium ions flow inward. The flow stops within about 2 milliseconds. The sodium channels close and become less likely to open than before. During this time, called the **refractory period,** potassium ions flow out of the cell and restore the original resting potential.

During each action potential, then, some sodium ions move into the cell and some potassium ions move out of the cell. The flow of potassium ions outward increases during the action potential because the potassium channels are also voltage gated. But the number of potassium ions lost is small compared with the total inside the cell, and the membrane's sodium-potassium pump soon restores the original distribution of both ions.

The time course of an action potential depends on the opening and closing of voltage-gated sodium and potassium channels.

How Do Voltage-Dependent Sodium Channels Open and Close?

Experiments such as those of Hodgkin and Huxley established the characteristics of the channels responsible for the resting potential and the action potential, but could not identify the responsible molecules. For almost 30 years, neuroscientists knew that the molecules—presumably transmembrane protein—must exist, but they had no idea of what they were. Our modern understanding of nerve communication, however, depends on new detailed knowledge of the structure of ion-channel proteins, whose isolation depended on knowledge of natural poisons.

Many poisons and stimulants affect the nervous system profoundly by binding to specific proteins important to neural function. For example, *tetrodotoxin,* which comes from the Japanese puffer fish, blocks the passage of sodium ions through the voltage-gated channels (Figure 41-7). Using tetrodotoxin and similar toxins, biochemists have isolated the protein that forms the voltage-gated sodium channel. Electric eels are a particularly rich source of the sodium-channel protein: their electric organs—which consist of cells stacked on top of one another like so many batteries in a long flashlight—generate big voltages in the same way as neurons generate small voltages.

Soon after biochemists isolated the channel protein, a group of Japanese scientists, led by Shosaku Numa, isolated a

Figure 41-7 In Japan, the puffer (fugu) is a delicacy, but chefs are careful to remove the highly toxic skin and entrails before preparing the fish for the table. The active agent in the toxin is tetrodotoxin, which blocks sodium channels. Researchers used tetrodotoxin to isolate the sodium channel protein. *(Jeffrey L. Rotman/Peter Arnold, Inc.)*

gene encoding the channel protein, which they then sequenced. Knowing the amino acid sequence of the sodium channel allowed researchers to guess how it might work. Subsequent research confirmed that four polypeptide segments assemble to form a pore through which sodium ions can flow, as illustrated in Figure 41-8. Researchers were able to study the behavior of individual channel molecules by capturing tiny patches of membranes in fine glass tubes (in "patch electrodes"). They could then study the effect of changing voltage or ion concentrations on the passage of sodium ions.

Sodium ions flow when the "gate" is open, and the gate opens only when the voltage across the membrane becomes positive, that is, when the cell is depolarized. Shortly after opening, however, the gate closes again, and no sodium ions can enter the channel for several milliseconds. This period of inactivation is, as we will see later, responsible for the inability of an action potential to reverse directions.

A sodium channel is a membrane protein that allows sodium ions to pass when the outside of the membrane has a positive charge.

How Does an Action Potential Move Down an Axon?

The properties of the voltage-gated sodium channel explain not only how action potentials occur but also how they can move for long distances along nerve cell membranes. We can list six consecutive events in the production of an action potential in a single spot in the membrane:

1. An alteration in ion distribution in the nerve cell membrane causes a local depolarization that is large enough to exceed the threshold.

2. Sodium channels open, and sodium ions flow into the cell.
3. As sodium ions flow inward, the inside of the membrane becomes locally positive.
4. The decreased polarization of the membrane causes more channels to open, increasing the positive charge on the inside of the membrane.
5. Finally (that is, after less than a millisecond), the sodium channels close (spontaneously).
6. Potassium ions then flow outward, restoring the membrane to the resting potential.

Notice that the pumping of sodium and potassium ions does not directly enter into the events of the action potential. This pumping serves only to maintain the distribution of ions responsible for the resting potential.

Now we can see how the action potential propagates itself. As the membrane's cytoplasmic face becomes more positive at one spot, positive ions diffuse, and the positive charge spreads to an adjacent patch. As an adjacent patch becomes depolarized, its sodium channels open, leading to further depolarization and further spreading. In this way the action potential propagates itself down the axon, regenerating itself as it goes.

Once an action potential begins to move, it moves in just one direction. Recall that once the sodium channels have

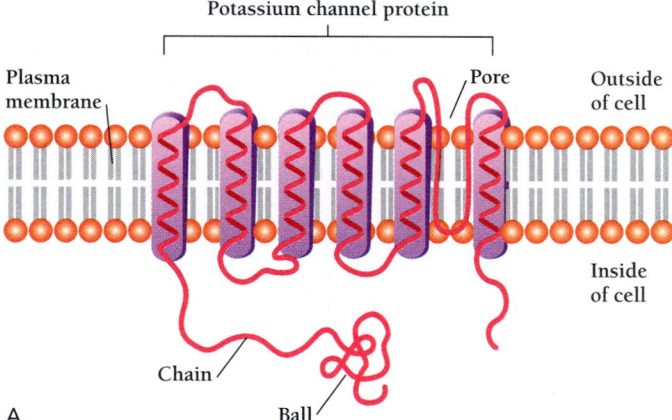

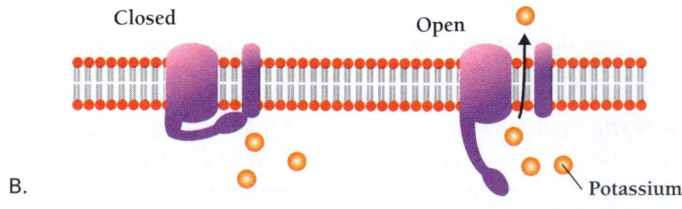

Figure 41-8 How does a potassium channel work? A. The channel is a transmembrane protein that crosses and recrosses the cell's membrane. Part of the polypeptide chain lines a transmembrane "pore," through which potassium ions can flow. Another part of the polypeptide chain forms a "chain" and a "ball" that blocks the pore. B. Depolarization of the cell causes the ball to move out of the way, allowing potassium ions to move out through the pore.

opened, they cannot immediately open again. As a result, an impulse cannot pass through again until after the refractory period. By that time the impulse has moved on. This prevents an impulse from moving backward. Each action potential passes through each membrane region only once.

An action potential propagates because a region of depolarization can move along the cell membrane, opening adjacent sodium channels and leading to further depolarization. Because each channel has a refractory period, the charge can move in only one direction.

How Do Nerves Speed the Transmission of Action Potentials?

Although the thick axons of the squid conduct impulses at speeds up to 10 meters per second, most nerve cells have much smaller axons that would normally conduct impulses much more slowly. Given the expected rate of conduction in the sensory nerves leading from our feet, for example, it would take a painfully long time for us to realize that we had stepped on a nail or come too close to a campfire. Most vertebrate nerves, however, are wrapped in **myelin,** a specialized, glistening sheath that allows a much more rapid conduction of nerve impulses.

The myelin sheath insulates most of the axon's membrane, preventing the passage of ions across the membrane (Figure 41-9). Ion flow is only possible in the **nodes of Ranvier,** gaps between adjacent myelin wrappings, spaced about every millimeter along the axon. As a result of this restriction, depolarization occurs only at the nodes and requires much less ion

movement. Instead of running down the axon continuously, then, the nerve impulse moves by **saltatory conduction,** jumping down a myelinated axon at rates up to 100 times faster than down an unmyelinated axon. An ordinary nerve conducts at about 1.2 meters per second, or about $2^{1}/_{2}$ miles per hour, the speed of a normal walk. In contrast, a myelinated nerve conducts at up to 120 meters per second, or about 250 miles per hour.

Myelin is present only in vertebrates, however. Invertebrates speed conduction by having axons of large diameters, such as those of the giant neurons of the squid.

Nerves wrapped in myelin conduct action potentials rapidly because nerve impulses jump from node to node.

HOW DO NEURONS COMMUNICATE WITH ONE ANOTHER?

Ramón y Cajal's studies demonstrated that the nervous system consists of discrete cells. Since then, studies with the electron microscope have revealed that there is a distinct connection point, the synapse, between most communicating neurons (Figure 41-10). The neuron sending information is the **presynaptic** neuron, while the receiving neuron is the **postsynaptic** neuron. We have already seen that a neuron can send information over long distances down an axon. But we still need to understand how a presynaptic neuron (lying "upstream" from the synapse) influences the electrical activity of a postsynaptic neuron (to which it connects through a synapse).

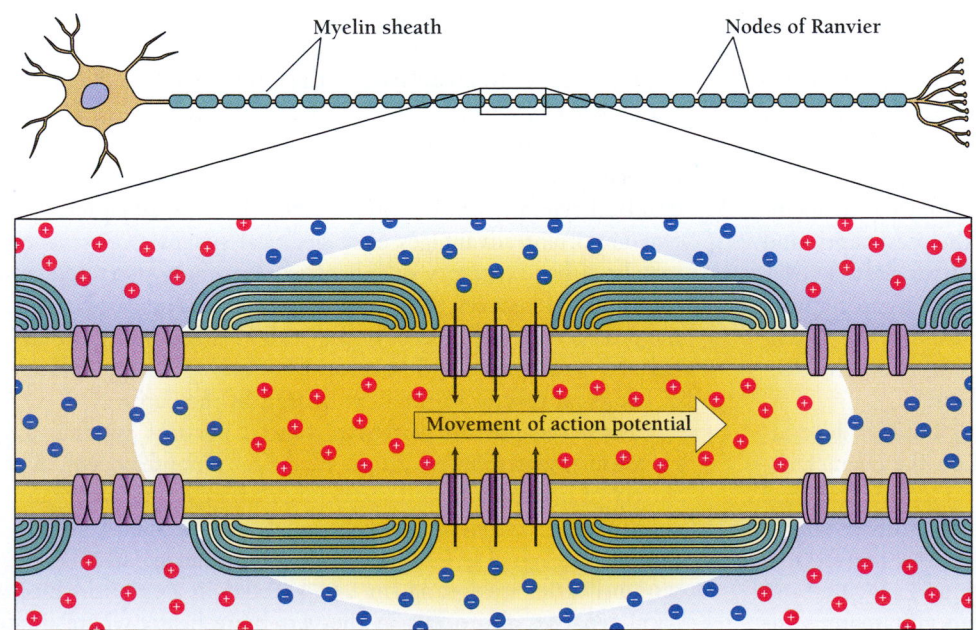

Myelin sheath   Nodes of Ranvier

Movement of action potential

Figure 41-9 A nerve impulse moves much more quickly down a myelinated axon because it jumps from node to node. Ion currents can only flow where the myelin sheath is interrupted, at the nodes of Ranvier.

Figure 41-10 Neurons communicate with each other (and motoneurons communicate with muscle cells) through specialized contacts called synapses. This drawing illustrates a chemical synapse that releases acetylcholine, the neurotransmitter that stimulates muscle contraction. In the brain, glutamate is the most commonly used excitatory neurotransmitter.

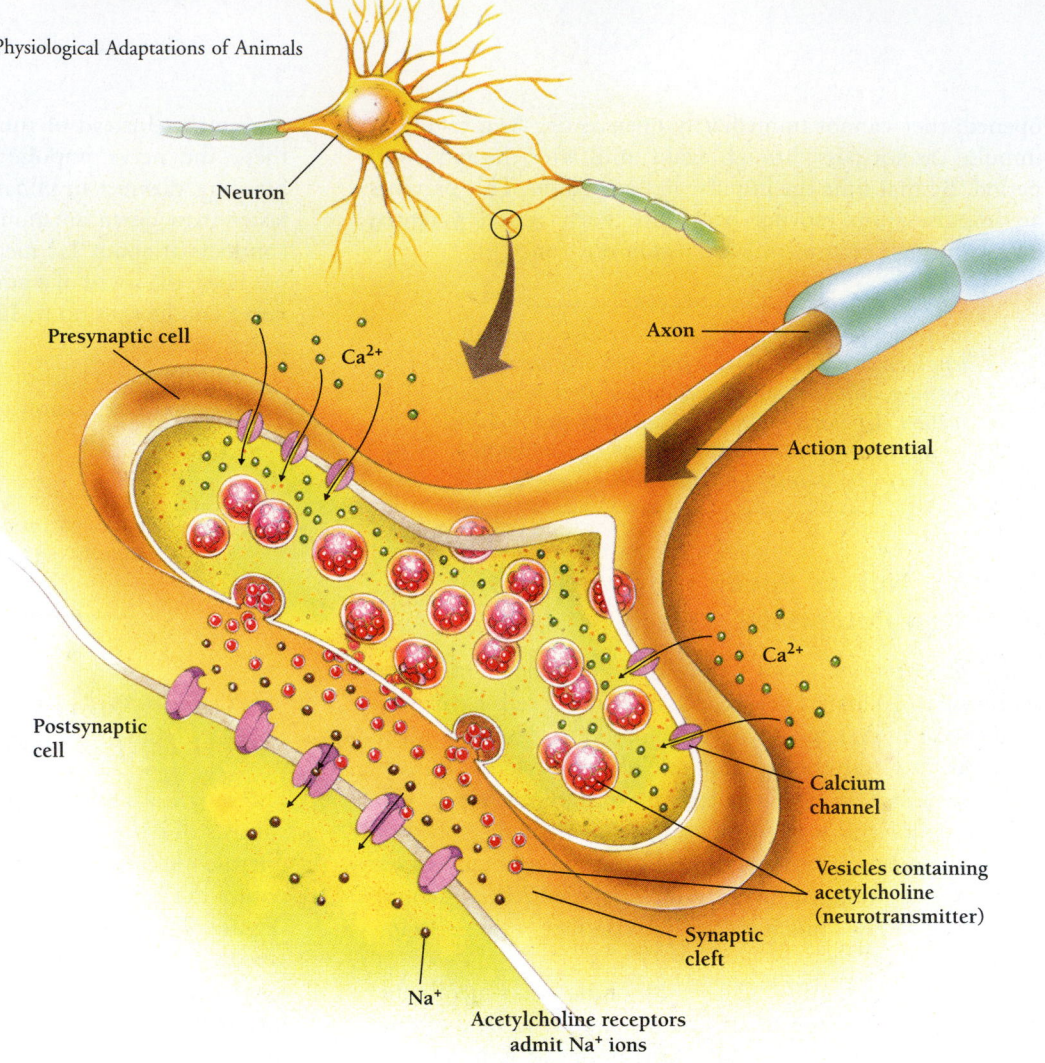

Neuron

Presynaptic cell

Ca^{2+}

Axon

Action potential

Ca^{2+}

Postsynaptic cell

Calcium channel

Vesicles containing acetylcholine (neurotransmitter)

Synaptic cleft

Na^+

Acetylcholine receptors admit Na^+ ions

The simplest synapse is the **electrical synapse,** which joins presynaptic and postsynaptic neurons through gap junctions, channels through the membranes of adjacent cells that allow ions and small molecules to pass freely from one cell to the next. An electrical synapse allows an action potential to continue traveling to the postsynaptic cell in the same manner and at about the same rate as it traveled down the presynaptic axon. All electrical synapses are **excitatory,** meaning that action potentials in the presynaptic cell stimulate action potentials in the postsynaptic cell.

The more common type of synapse, however, is the **chemical synapse,** in which presynaptic and postsynaptic membranes do not join (Figure 41-10). Instead, the presynaptic and postsynaptic cells are separated by the **synaptic cleft,** a space of about 20 nm. Communication across such a gap requires the diffusion of one or more **neurotransmitters,** small signaling molecules made in the presynaptic cell that affect the electrical charge of the postsynaptic cell membrane.

Chemical synapses may either excite or inhibit, depending on the particular neurotransmitter and on the characteristics of its receptors. So far, biologists have identified at least 20 different neurotransmitters, and there may still be dozens more. Among the small molecules that serve as neurotransmitters are several amino acids (glutamate and glycine) and amino acid derivatives (GABA, derived from glutamate; serotonin, derived from tryptophan; and dopamine, norepineph-

rine, and epinephrine, all derived from tyrosine). Other small-molecule neurotransmitters are acetylcholine and adenosine (Figure 41-11).

A nerve impulse in a presynaptic neuron triggers the release of a neurotransmitter, which then diffuses across the synaptic cleft and binds to a receptor. A neurotransmitter can either excite or inhibit the activity of the postsynaptic neuron, but the effect is always delayed by the time needed for the presynaptic cell's machinery (about 0.4 millisecond) to release the neurotransmitter and for the neurotransmitter to diffuse across the gap (about 0.1 millisecond). No such delay occurs in an electrical synapse. On the other hand, chemical synapses have two important advantages over electrical synapses. First, they may be either excitatory or **inhibitory** (meaning that the neurotransmitter reduces the probability of an action potential in the postsynaptic cell). Second, they can greatly amplify the signal of a small presynaptic neuron by releasing a large amount of neurotransmitter to the next neuron in the pathway.

Most synapses in vertebrate central nervous systems are chemical rather than electrical. Electrical synapses occur in a variety of invertebrate and vertebrate nerve circuits, where fast conduction between cells is advantageous and subtle modifications of a signal are unimportant. One such synapse is that responsible for mediating the escape reflex in a crayfish. Electrical synapses also occur in the vertebrate heart, where they coordinate the synchronous contraction of heart muscle cells.

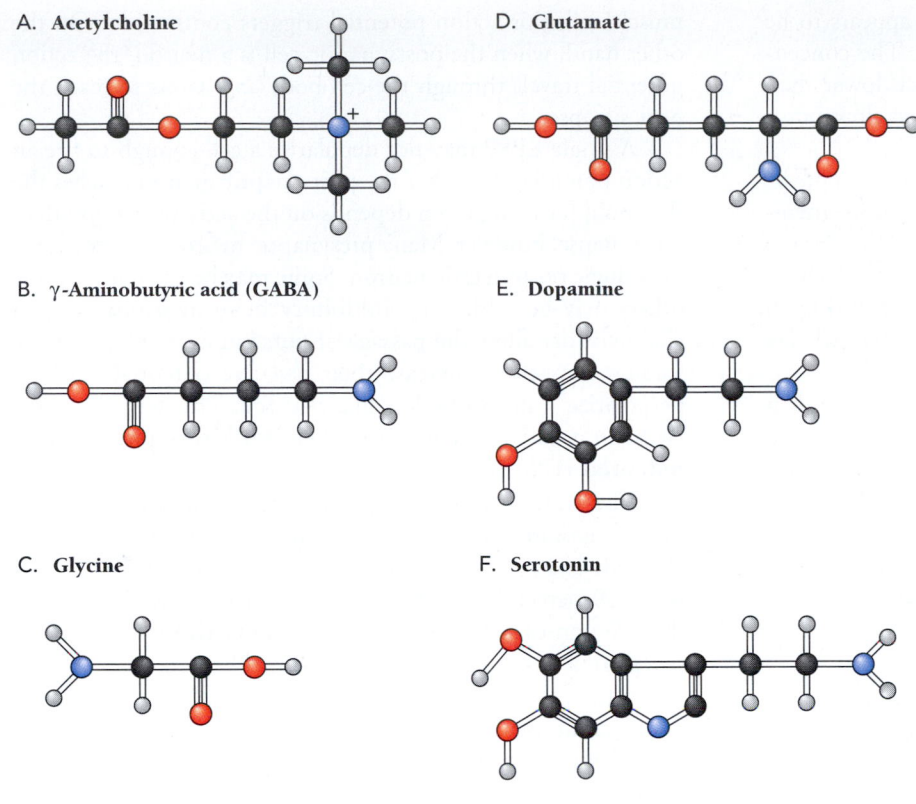

A. Acetylcholine

D. Glutamate

B. γ-Aminobutyric acid (GABA)

E. Dopamine

C. Glycine

F. Serotonin

Figure 41-11   The most commonly used neurotransmitters—acetylcholine, GABA, glycine, glutamate, dopamine, and serotonin. Carbon atoms are shown in black, oxygen in red, nitrogen in blue.

A synapse may be electrical or chemical. Electrical synapses are always excitatory, but chemical synapses may be excitatory or inhibitory.

How Does a Chemical Synapse Work?

Understanding how chemical synapses work is important not only to our understanding of the brain but also to medical and social problems. Essentially all the medications that affect the functioning of the brain act on synapses. Among these substances are sleeping pills, tranquilizers, antipsychotic medicines, and narcotics such as codeine, heroin, and cocaine. Each of these chemicals acts by mimicking or interfering with the production, release, or action of some neurotransmitter.

Let us look more closely at the structure of a chemical synapse. The best-understood chemical synapse is the vertebrate *neuromuscular junction,* the synapse between a motoneuron and a voluntary muscle cell (Figures 41-10 and 41-12). These synapses are similar to those between neurons, but they are larger and easier to study.

On the presynaptic side, the electron microscope reveals tens of thousands of tiny vesicles, each about 50 nm in diameter and enclosed by a membrane. These vesicles, which release their contents by exocytosis, are full of the neurotransmitter **acetylcholine.**

Knowing this much, we now want to know (1) what triggers the release of transmitter and (2) how the released neurotransmitter causes electrochemical changes in the postsynaptic cell, altering its likelihood of firing an action potential.

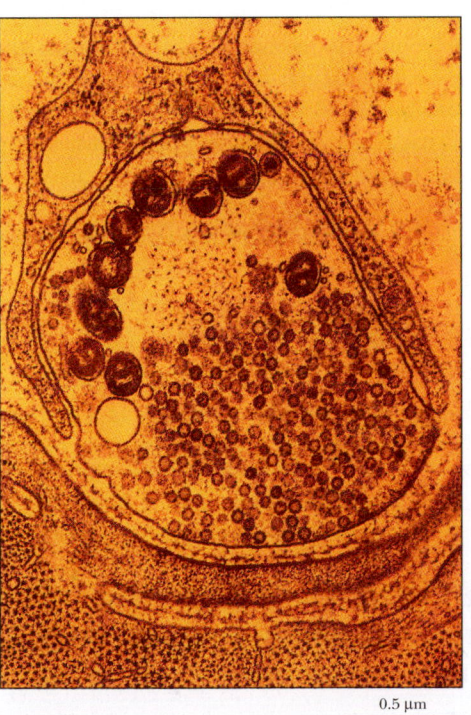

0.5 μm

Figure 41-12   The neuromuscular junction is a chemical synapse between a motoneuron and a muscle cell. In this electron micrograph, note the clustering of the small vesicles, which contain high concentrations of acetylcholine. Some of these vesicles are released by exocytosis when the motoneuron fires. *(T. Reese & D.W. Fawcett/Visuals Unlimited)*

The trigger for release of the acetylcholine appears to be an increase in intracellular calcium (Ca^{2+}) ions. The concentration of calcium ions in the cytoplasm is much lower than in the extracellular fluid. Calcium ions will always flow *into* a cell—if there is a route.

During an action potential, calcium ions flow into a presynaptic neuron, but they do so only locally—through the membrane next to the synaptic cleft and nowhere else (Figure 41-12). The calcium ions move through special calcium channels that open in response to the change in voltage during an action potential. Like sodium channels, calcium channels are voltage gated and close soon after they open.

Most neurons have many fewer calcium channels than sodium channels. There are so few calcium ions in the cytoplasm, however, that even a small flow is enough to cause a big change in the local concentration of calcium ions near the calcium channels. Since the machinery for releasing neurotransmitters is right next to the calcium channels, the increased local concentration can immediately trigger the release of acetylcholine.

At neuromuscular junctions, an increase in the local concentration of calcium ions triggers the release of the neurotransmitter acetylcholine into the synaptic cleft.

What Does the Neurotransmitter Acetylcholine Do?

Once acetylcholine is released, it diffuses across the synaptic cleft of the neuromuscular junction to the postsynaptic membrane, where it binds to a receptor molecule. This acetylcholine receptor is itself an ion channel that allows the passage of sodium and potassium ions. Like the voltage-gated sodium channels and the voltage-gated calcium channels, this ion channel is gated. But it is acetylcholine rather than voltage that opens the gate. A molecule that specifically binds to another molecule is called a ligand, so the acetylcholine receptor is an example of a **ligand-gated channel.**

The acetylcholine receptor in muscle is a ligand-gated ion channel. When acetylcholine binds to the channel protein, it opens a pore that allows the passage of sodium and potassium ions.

Neurotransmitters May Produce Either Excitatory or Inhibitory Effects in Postsynaptic Neurons

When sodium ions pass through the channel of the acetylcholine receptor, they depolarize the membrane, making the inside of the membrane less negative than before. This voltage change is called a **postsynaptic potential.** The postsynaptic potential triggered by acetylcholine excites the cell and is therefore called an **excitatory postsynaptic potential (EPSP).**

If an EPSP is sufficiently depolarizing, the postsynaptic cell fires an action potential. When the postsynaptic cell is a

muscle cell, this action potential triggers contraction. On the other hand, when the postsynaptic cell is a neuron, the action potential travels through the cell body and axons to reach the next synapse.

A single EPSP may not depolarize a cell enough to fire an action potential. Whether the postsynaptic neuron reaches the threshold for firing often depends on the activity of more than one synapse, however. Many presynaptic neurons can converge on a single postsynaptic neuron. Some may be excitatory, while others may be inhibitory. Inhibitory neurotransmitters open channels that allow the passage of potassium or chloride ions, leading the cell to increase their negative potential (to "hyperpolarize") and to be less likely to fire. This temporary increase in negative voltage is called an **inhibitory postsynaptic potential (IPSP).**

The most common inhibitory neurotransmitter in the brain is **gamma-aminobutyric acid (GABA)** (Figure 41-11). Virtually every neuron in the brain can respond to GABA, and some 30 percent of all brain neurons make GABA. Most of these responses depend on the binding of GABA to the **$GABA_A$ receptor**, a ligand-gated chloride channel like the acetylcholine receptor of the neuromuscular junction.

Binding of GABA to $GABA_A$ receptors opens a channel that admits chloride ions, making the inside of the cell more negative. The negative charge of these ions counterbalances any positive sodium ions flowing into the cell and decreases the chance that the postsynaptic cell will fire an action potential.

The activity of a postsynaptic cell depends on the summing of the effects of excitatory and inhibitory potentials. A single postsynaptic neuron may integrate inputs from as many as 100,000 different presynaptic cells. Each neuron effectively serves as a microcomputer that determines whether to fire or not to fire according to the final balance of depolarizing and hyperpolarizing responses.

The firing pattern of the postsynaptic cell depends on the summing of the responses to both excitatory and inhibitory inputs.

Neurons May Respond to the Same Neurotransmitters in Different Ways

By the mid-1970s, researchers had discovered and studied a handful of neurotransmitters. Some of these, like acetylcholine, were excitatory, while others, like GABA, were inhibitory. But neurobiologists soon realized that the same neurotransmitter could have different effects in different postsynaptic neurons.

These differences result from the existence of more than one type of receptor on each postsynaptic neuron. Different receptors have different responses to neurotransmitters and other compounds. The *nicotine* found in tobacco, for example, mimics the effects of acetylcholine in skeletal muscle and in some neurons but has no direct effect on heart muscle. On the other hand, the mushroom toxin *muscarine* stimulates heart muscle but not skeletal muscle. And muscarine stimulates only those neurons that respond to acetylcholine but not to nicotine.

Neuroscientists group acetylcholine receptors into two varieties, *nicotinic receptors* and *muscarinic receptors*. Skeletal muscle has nicotinic receptors, and acetylcholine triggers them to fire. Smooth muscle and heart muscle, on the other hand, have muscarinic receptors, and acetylcholine inhibits the production of an action potential. As a result, acetylcholine (or compounds that act like muscarine) reduces the heart rate and slows the peristaltic contractions of the smooth muscles in the intestines.

Neurotransmitters work in the same ways as hormones and other signaling molecules (Chapter 36). Postsynaptic receptors act in one of two ways. They are either *ionotropic* receptors or *metabotropic* receptors. An ionotropic receptor, also called a ligand-gated ion channel, changes its shape after binding to its neurotransmitter and admits specific types of ions through the cell membrane. In contrast, a metabotropic receptor is not an ion channel itself. When it binds its neurotransmitter, its altered shape changes its interaction with another membrane protein, such as one of the G proteins (Box 36.2). Many metabotropic receptors act by changing the concentration of a second messenger such as cyclic AMP.

Receptors for neurotransmitters respond either by opening ion channels or by triggering intracellular biochemical events, such as the production of cyclic AMP.

How Are Neurotransmitters Cleared from the Synapse?

We have seen how a neurotransmitter alters the postsynaptic neuron. But what terminates the action of the neurotransmitter?

Neurotransmitters must be cleared quickly from the synaptic cleft to make way for the next set of signals. Termination depends on diffusion of the transmitter away from the synaptic cleft. Termination may also involve specific **reuptake,** the active pumping of transmitter from the synaptic cleft either into the presynaptic neuron or into surrounding glial cells. The recovered transmitter is then either recycled into vesicles or degraded by specific enzymes. In the case of GABA, for example, an enzyme called GABA transaminase converts GABA into an inactive compound. In the case of acetylcholine, an extracellular enzyme called **acetylcholinesterase** destroys the acetylcholine soon after it is released into the synapse. Both GABA and acetylcholine therefore have only brief periods in which to act on the postsynaptic cell.

The time during which a neurotransmitter acts is limited by its diffusion and by its active removal by enzymes and transporters.

Many Psychoactive Drugs Act on Chemical Synapses

Not all the molecules that bind to neurotransmitter receptors are neurotransmitters. A variety of substances—including, for example, nicotine (from tobacco), curare (a frog toxin used in poison darts), sarin (a nerve gas, developed for chemical warfare), and α-bungarotoxin (from snake's venom)—all bind to the nicotinic acetylcholine receptor. While nicotine activates the receptor, however, the others block the receptor and prevent the initiation of muscle contraction. Similarly, the smooth muscle-relaxing drug atropine inhibits the muscarinic acetylcholine receptor. All of these drugs are said to be psychoactive, meaning they alter mood and perception. And all work by either mimicking or disrupting the action of a neurotransmitter.

Many compounds that bind to neurotransmitter receptors have proved to be useful medications. For 2000 years, for example, physicians have treated diarrhea with belladonna, an extract of a tall, bushy herb called deadly nightshade *(Atropa belladonna)*. We now know that the extract worked because *atropine* in the belladonna blocks the effects of acetylcholine in the smooth muscles of the intestines.

Other naturally occurring or chemically synthesized compounds affect other neurotransmitter receptors. We generally call a substance a medicine or a drug when it helps and a poison when it hurts. Some pesticides also affect the nervous system. Certain organic phosphates (malathion and parathion, for example) inhibit the breakdown of acetylcholine in insects, leading to the overstimulation of their muscles and ultimately to their death. *Benzodiazepines,* commonly prescribed antianxiety drugs, bind to $GABA_A$ receptors and increase the effectiveness of GABA in opening the chloride (Cl^-) channel.

Still other substances influence the release, reuptake, or inactivation of specific neurotransmitters. Some of these are effective as painkillers, others are tranquilizers, still others are stimulants. Some drugs, such as cocaine, bring about euphoria and anesthesia, possibly by blocking the uptake of the neurotransmitter dopamine. Among the most widely used medicines are the *selective serotonin reuptake inhibitors (SSRIs),* which include fluoxetine (Prozac) and sertraline (Zoloft); these drugs elevate mood and fight depression by blocking the reuptake of the neurotransmitter serotonin from certain synapses. Many drugs that act on the brain are highly addictive and, therefore, have enormous social consequences.

HOW DO NEURAL NETWORKS MEDIATE BEHAVIOR?

Just as the connections within the sensory pathways establish what we extract from our sense organs, the connections within neural networks determine how an animal will respond to a stimulus.

The connections are often as direct and mechanical as those in household wiring. When we flip a switch the light comes on. Similarly, if we poke a frog's right leg, the frog will withdraw the leg.

In one experiment, however, a researcher "rewired" a frog. He cut the sensory nerve that led from the right hind limb of a frog to the spinal cord. He then reconnected the nerve to the other side of the spinal cord. Unpleasant stimulation of the right leg then caused the frog to move its left leg.

How Do Neurons Mediate a Reflex?

Movements and behavior ultimately depend on coordinated signals from neurons in the spinal cord, called **motoneurons.** Motoneurons represent the final common pathway of the instructions that direct behavior.

The most complex behaviors—such as threading a needle, performing a laboratory experiment, or taking a midterm examination—originate in commands from the brain, particularly in the cerebral cortex. The least complex behaviors—such as withdrawing a hand from a flame—often do not depend on the brain at all but on simple and rapid signal processing in the spinal cord. Behavioral activity mediated by the spinal cord tends to be the most automatic, that is, the least influenced by thought or previous experience.

A **reflex** is the most automatic behavior pattern, with motor activity directly responding to a sensory stimulus. A reflex familiar to anyone who has had a medical examination is the jerking of the lower leg after a physician taps the knee with a little hammer. Let us see exactly how this happens.

The hammer hits a tendon, which stretches over the kneecap, or patella. This tendon connects a muscle in the thigh, the quadriceps, to a bone in the lower leg. The physician's hammer pulls the tendon and stretches the attached muscle. The muscle contains *stretch receptors,* modified muscle cells whose electrical output depends on the degree to which they are stretched. Information from these receptors (that they have been stretched) flows to **sensory neurons** in the dorsal part of the spinal cord, which in turn command the muscles to contract, which jerks the knee. The information passes through just one synapse, so the stretch reflex is said to be *monosynaptic.* The normal function of this reflex is to restore the position of the lower leg after a rapid movement in the other direction.

Information from the stretch receptors also travels to a second group of spinal cord neurons (Figure 41-13). These neurons inhibit the hamstrings muscles, which are "antagonistic" to the quadriceps muscles, that is, they pull the lower leg in the opposite direction. By inhibiting the hamstrings from inhibiting the quadriceps, this inhibitory neuron allows the knee to jerk. Not only does the signal from the stretch receptor trigger one set of motoneurons to fire but it also inhibits another set. Yet another set of neurons in the spinal cord carries information about the status of the leg muscle to brain areas that coordinate movements and posture.

The connections involved in the stretch reflex form the simplest known neural circuit in vertebrates. Other circuits are more complicated, involving at least one interneuron. An **interneuron** is a neuron that connects one or more neurons to other neurons with parallel functions (rather than to a sensory neuron or a motoneuron). An **inhibitory interneuron** prevents or slows the firing of the neuron with which it connects. Interneurons can also integrate information from many sources—from stretch receptors, from other somatic receptors, and from elsewhere in the nervous system, including the brain. Circuits in which such integration occurs are called *convergent.*

An interneuron can also participate in a *divergent* circuit, in which a single interneuron may coordinate the action of many motoneurons. Walking, for example, requires the cyclical activity of different groups of motoneurons. The needed coordination does not require the involvement of the brain, however; the interneurons of the spinal cord are enough. Input from the brain, however, can modify the activity of spinal cord motoneurons involved in walking.

The behavior that follows the activation of a neural circuit may be quite complex—walking and breathing are two clear examples. It seems likely that many behaviors—including the modal action patterns that we discussed in Chapter 29—depend upon such neural connections. Once the appropriate connections develop, the proper stimulation of a "trigger" neuron can evoke a standard pattern of behavior. The trigger for action is called a "releaser."

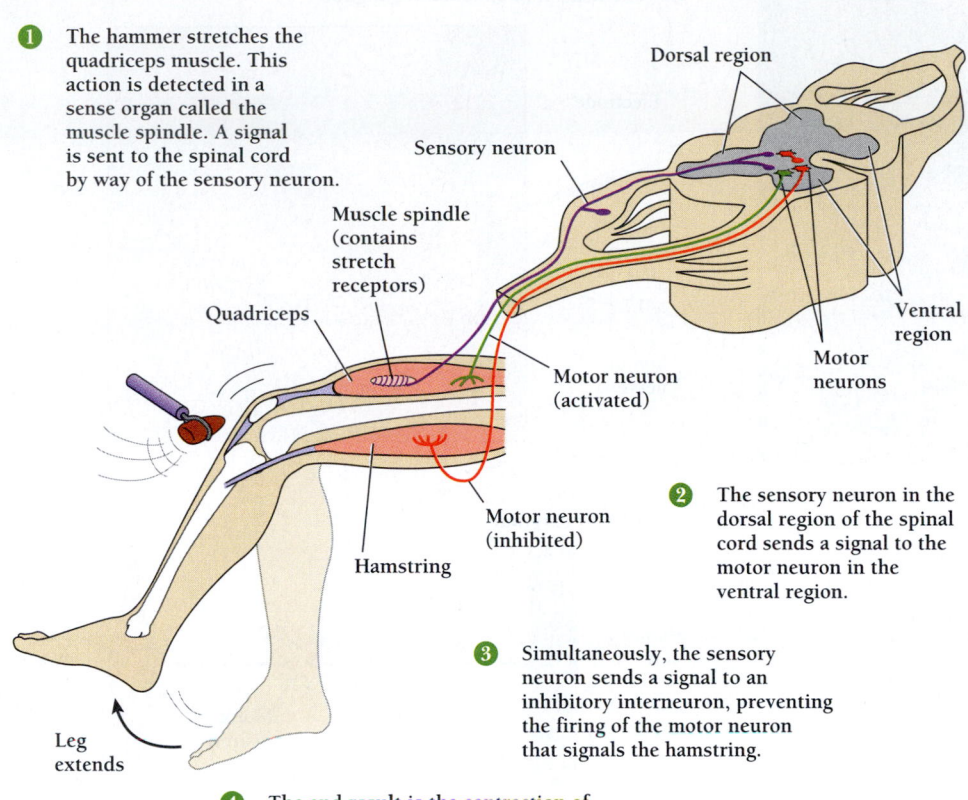

① The hammer stretches the quadriceps muscle. This action is detected in a sense organ called the muscle spindle. A signal is sent to the spinal cord by way of the sensory neuron.

Dorsal region

Sensory neuron

Muscle spindle (contains stretch receptors)

Quadriceps

Ventral region

Motor neurons

Motor neuron (activated)

Motor neuron (inhibited)

Hamstring

Leg extends

② The sensory neuron in the dorsal region of the spinal cord sends a signal to the motor neuron in the ventral region.

③ Simultaneously, the sensory neuron sends a signal to an inhibitory interneuron, preventing the firing of the motor neuron that signals the hamstring.

④ The end result is the contraction of the quadriceps and the inhibition of hamstring contraction.

Figure 41-13 The knee jerk is an example of a monosynaptic reflex (and of a two-synapse reflex). The physician's hammer stretches the quadriceps muscle. Stretch receptors in the quadriceps stimulate one set of motoneurons to fire, leading to the rapid contraction of the quadriceps. The stimulated stretch receptors also activate a set of inhibitory interneurons, which temporarily prevent the firing of the motoneurons that stimulate the hamstring.

In a simple reflex, input from a sensory receptor directly stimulates a motoneuron. Complex reflexes and more complicated neural circuits involve interneurons, which carry information among neurons with parallel functions.

HOW DOES EXPERIENCE MODIFY NEURAL NETWORKS?

Many neural networks are capable of **learning,** modification of neural activity and behavior as the result of experience. We humans are exquisitely aware of our learning, for example, as we master a new language.

Researchers distinguish between two types of learning— nonassociative learning and associative learning. In *nonassociative learning,* a person or other animal becomes more or less sensitive to a stimulus after repeated exposure. *Habituation,* for example, refers to the decrease in a behavioral response following repeated exposure to a stimulus: after we've lived in a new house for a while, we hardly notice the sound of the walls snapping at sunrise and the sounds of central heating going on and off. *Sensitization* refers to the increased behavioral response to a noxious stimulus: if a painfully loud sound is repeated, we are more likely to move away if we hear it again. The sound of the neighbor's dog barking is far more irritating after 5 hours than after 5 seconds.

Associative learning refers to the pairing of seemingly unrelated stimuli. The classic example of associative learning comes from the work of the Russian physiologist Pavlov, who trained dogs to salivate at the sound of a bell. Pavlov's trick was first to ring the bell (the *conditioned stimulus*) just before presenting the dog with dinner (the *unconditioned stimulus*). The dog would salivate at the sight and smell of the meat, which was accompanied by the ringing of the bell. Gradually the dog salivated at the sound of the bell alone. But the sound of a can opener works just as well.

Researchers classify learning into nonassociative learning (habituation and sensitization) and associative learning (the pairing of stimuli).

Experience Can Modify the Gill Withdrawal Reflex of the Sea Hare, *Aplysia*

Experimenters have studied both nonassociative and associative learning in *Aplysia,* a type of mollusk also known as the sea hare. *Aplysia's* simple nervous system consists of just eight paired clusters and one unpaired cluster of nerve cells called *ganglia* [singular, *ganglion*] (Figure 41-14). Each ganglion consists of fewer than 2000 neurons whose large sizes (up to 1 mm in diameter) make them easy to identify.

Neurophysiologists also like to work with *Aplysia* because it reflexively withdraws its gills in response to a range of stimuli, from a simple touch on the siphon to a sharp blow or shock to the head or tail. *Aplysia* can learn to modify this reflex.

The neurons responsible for the gill withdrawal reflex lie in the abdominal ganglion. A sensory neuron from the siphon makes a synaptic connection to the motoneuron that drives gill withdrawal, called L7. In addition, another neuron, called

A.

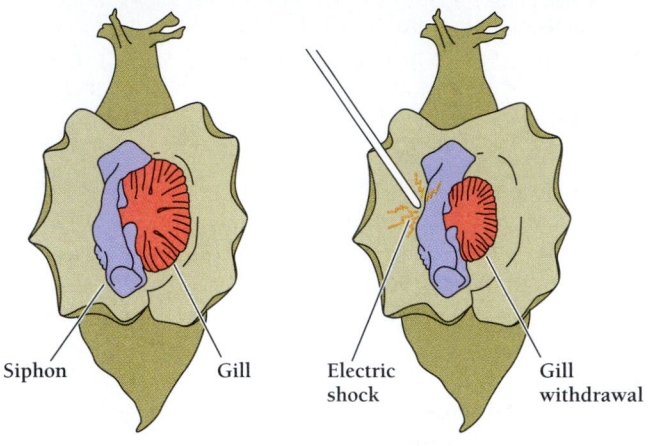

B. **Gill withdrawal reflex**

Siphon   Gill   Electric   Gill
shock   withdrawal

C. **L7 neuron in abdominal ganglion**

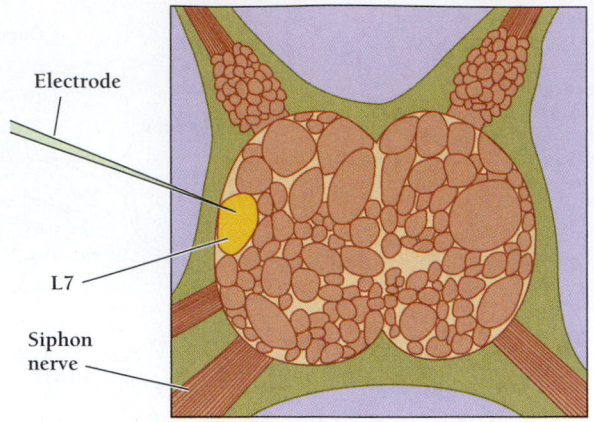

Electrode

L7

Siphon
nerve

D. **Circuit diagram**

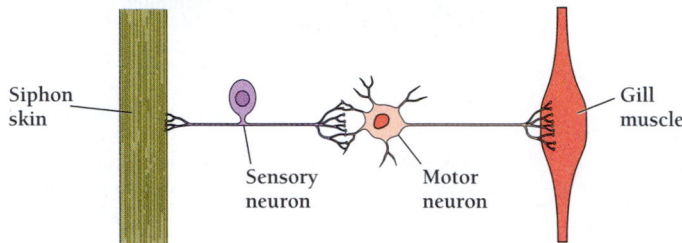

Siphon
skin

Sensory
neuron

Motor
neuron

Gill
muscle

Figure 41-14 The "sea hare" *Aplysia californica*. Despite its simple nervous system, this mollusk is still able to learn. A. *Aplysia*. B. Gill withdrawal, a reflex response to an electric shock. C. An L7 neuron in the abdominal ganglion. D. A circuit diagram of the gill withdrawal reflex. (*A, Dan Gotshall/Visuals Unlimited*)

a **facilitating interneuron,** synapses with the axon terminal of the sensory neuron.

A single stimulation of the sensory neuron produces an action potential in L7. Repeated stimulation of the sensory neuron, however, leads to a *decrease* in the size of the action potential in L7. This decrease, called *synaptic depression,* underlies habituation. Depression results from a decrease in the amount of a released neurotransmitter called **serotonin.**

How Does an *Aplysia* Learn?

When researchers stimulate the facilitating interneuron, the synaptic depression reverses and the L7 motoneuron is said to be "facilitated." This synaptic facilitation explains sensitization. At Columbia University, Eric Kandel and his colleagues have determined the mechanism for synaptic facilitation. When the facilitating interneuron is stimulated it releases the neurotransmitter serotonin. Serotonin acts on receptors on the presynaptic axon terminals of the sensory neuron, stimulating the synthesis of cyclic AMP, which indirectly inactivates a potassium channel that helps repolarize the sensory neuron after firing. Serotonin has the effect of lengthening the period during which the axon terminal is depolarized. During the extended depolarization, more Ca^{2+} ions enter the axon terminal, which leads to a release of more neurotransmitter than usual and, consequently, a stronger response by the postsynaptic motoneuron.

A single stimulus to the tail (or to the facilitating neuron) sensitizes the gill withdrawal reflex for several minutes, as described earlier. Repeated stimulation of the tail, however, brings about a much longer period of sensitization, lasting days or even weeks. This long-term sensitization is a form of associative learning, not unlike that in Pavlov's dogs. Like short-term sensitization, long-term sensitization depends on the inactivation of a repolarizing potassium current. But, unlike short-term sensitization, long-term sensitization results from a change in gene transcription and protein synthesis. When researchers inhibit protein synthesis during the training period, the sea hare cannot learn for the long term. Within a few hours, it "forgets."

The sea hare, Aplysia, *has provided neuroscientists with the opportunity to study the cellular bases of both associative and nonassociative learning. Learning depends upon changes in synaptic signaling between individual neurons.*

The insights from *Aplysia* have led to new approaches to studying the cellular basis of learning in other animals, including humans. In Chapter 42, we will see how the organization of the vertebrate brain—especially the human brain—allows behaviors far more complex than those of invertebrates.

Study Outline with Key Terms

The operation of the nervous system depends on the properties of individual nerve cells and the way in which they connect with one another. The cells of the nervous system include **neurons,** which carry information in the form of chemical and electrical signals, and **glial cells,** which provide mechanical and metabolic support. A neuron's **axon** transmits a signal, while the **dendrites** receive signals from other cells.

Like all other cells, neurons have different concentrations of molecules and ions inside than outside. The selective action of **channels, gates,** and **pumps** results in a greater concentration (a **concentration gradient**) of sodium ions outside the cell and of potassium ions inside, as well as an **electrical potential** (or **voltage**) of about -70 millivolts across the cell membrane. This **resting potential** can suddenly change when a neuron is stimulated, giving rise to an **action potential,** which can travel over large distances. Action potentials move more rapidly through axons that are surrounded by **myelin** or that have a large diameter.

The generation and propagation of an action potential depend largely on the **voltage-gated** sodium channels in the neuronal membrane. These channels open when the voltage across the plasma membrane exceeds a **threshold,** and the electrical **current** into the cells increases. The **refractory period** of the channels ensures that action potentials move in a single direction down an axon. In myelin-enclosed axons, ion flow occurs only at the **nodes of Ranvier,** giving rise to rapid **saltatory conduction.**

Nerve cells communicate with one another at **synapses,** which may be either **electrical** or **chemical.** Electrical synapses allow action potentials to propagate in much the same way as they travel down axons. At a chemical synapse, the **presynaptic** cell responds to the arrival of an action potential by releasing a store of **neurotransmitter** (such as **acetylcholine, glutamate, GABA,** or **serotonin**) contained in synaptic vesicles.

The released neurotransmitter diffuses across the **synaptic cleft** to act on receptors on the **postsynaptic** neuron where its action may be either **excitatory,** evoking an **excitatory postsynaptic potential (EPSP),** or **inhibitory,** evoking an **inhibitory postsynaptic potential (IPSP).** These receptors may be **ligand-gated** ion channels, as in the case of the nicotinic acetylcholine receptors in muscle cells and **GABA$_A$ receptors** in neurons. Other neurotransmitter receptors resemble the membrane receptors for adrenaline and bring about their postsynaptic effects by stimulating or inhibiting the synthesis of second messengers, such as cyclic AMP. The action of neurotransmitters is terminated by **reuptake** or by degradative enzymes, such as **acetylcholinesterase.** Many psychoactive drugs act by mimicking or inhibiting the production, release, degradation, or action of neurotransmitters.

The simplest neural circuit is a **reflex,** which involves only one synapse. More complex circuits involve **inhibitory interneurons.** Even simple nerve circuits can mediate **learning,** as in the case of the sea hare, *Aplysia.* In this animal, the characteristics of the three neurons—a **sensory neuron,** a **motoneuron,** and a **facilitating interneuron**—are responsible for the modification of the gill withdrawal reflex. Short-term learning in this system depends on the effects of serotonin on potassium channels, while long-term learning depends on changes in protein synthesis.

Review and Thought Questions

Review Questions

1. How are glia different from neurons? What role does each kind of cell play in the nervous system?
2. What is a resting potential? How does a neuron generate a resting potential?
3. What is an action potential? How does a neuron generate an action potential? Why does an action potential travel in only one direction down an axon?
4. How does an electrical synapse differ from a chemical synapse?
5. What is a neurotransmitter?
6. Give an example of an excitatory neurotransmitter and of an inhibitory neurotransmitter.
7. How does a chemical synapse work?

Thought Questions

8. What consequences—behavioral or otherwise—would you expect to see in an animal missing a gene that allows the synthesis of serotonin?
9. Eating carbohydrates causes an increase in blood sugar, which causes sleepiness. Suggest a hypothesis about neurotransmitters that would explain this phenomenon. How would you test your hypothesis?

About the Chapter-Opening Image

The freshly hatched chick reminds us of Rita Levi-Montalcini's studies of the development of chick embryo nervous systems, carried out in her bedroom during World War II.

 On-line materials relating to this chapter are on the World Wide Web at **http://www.harcourtcollege.com/lifesci/aal2/**

42 The Nervous System and the Sense Organs

Does a Falling Tree Make a Sound When No One Is There To Hear It?

Philosophers and philosophically minded students have argued about this question for centuries. We now know that the answer is "no." A falling tree makes the air vibrate, but it makes no sound. Sound is a **perception,** the recognition and interpretation of an outside stimulus. In the case of a falling tree, the stimulus is a series of pressure waves, or vibrations, which arrive at the ear and are interpreted by the brain.

As humans, the air vibrations to which we have the strongest and most varied reactions are patterns that we perceive as words. English speakers know that the vibration pattern we hear as "ma" conveys a particular meaning. Remarkably, people who speak many different languages also recognize slight variations of this particular vibration pattern as meaning "mother." But both the perception of the sound and the interpretation of its meaning depend on the experience of the hearer: a Chinese speaker, for example, hears tonal inflections in the vibration pattern that an English speaker might miss. To a Chinese speaker, "ma" can mean "mother" or "horse" or a fiber-producing plant, all depending on how the word is said.

We hear with our brains, not our ears. People with excellent hearing may be unable to understand language. Most English speakers do not understand Chinese language. And people who have suffered strokes on the left side of the brain may be unable to recognize or to form words.

For the last 150 years, researchers have been trying to understand the basis of word recognition and hearing in general. Early researchers knew enough to look beyond the ears to the brain itself, the site of perception and learning. But they could not agree whether the brain worked as an integrated whole or as a set of parts, more like the cogs of a machine.

The notion that the brain is subdivided into specialized parts first gained prominence in the early 19th century from the work of Franz Joseph Gall, a German physician and neuroanatomist who worked in Vienna. Gall tried to establish a correlation between the size of a brain area and a particular emotional or intellectual capacity.

Gall estimated these traits, moreover, not from the brain anatomy of people who had died but from the ridges and bumps on the skull of his living subjects. This practice came to be known as *phrenology* [Greek, *phren* = heart, mind], an exercise that captured the imagination of many 19th-century intellectuals, including the writers George Eliot and Charles Dickens. While phrenology was more of a social than a scientific endeavor, it helped stimulate interest in the physical basis of the mind.

In the early 1820s, a young scientist subjected Gall's ideas to actual experiments. Pierre Flourens, who had recently graduated from medical school, studied what happened when he damaged specific parts of the brains of experimental animals. Flourens found, for example, that destroying the *cerebellum,* a structure in the lower part of the brain, led to a loss of coordination. But when he removed slices of the *cerebral cortex,* the convoluted surface layer of the brain, he did not find any specific loss of particular traits, but rather a vague, general loss of mental function.

In humans, the cerebral cortex is far larger than in other an-

Key Concepts

- Nervous systems contribute to homeostasis by regulating internal organs and movements.

- Nervous systems are also responsible for sensation, cognition, learning, emotion, mood, and consciousness.

- Most sensations depend on a sense organ, with specialized receptor molecules, and on specific neural pathways within the brain.

imals. Indeed, like our upright posture and hairlessness, our immense cerebral cortex is one of the handful of traits that sets us apart from other primates. Flourens argued that his studies supported the idea that humans have a single, integrated mind and a single soul.

Though many scientists shared his opposition to the idea that the brain consisted of a collection of independent machines, other neurologists continued to try to associate specific functions—such as touching or hearing—with specific regions of the brain. For example, the French neurologist Pierre Paul Broca described a patient who could understand words but could not speak. After the patient's death, Broca found damage near the front, also on the left side of the cerebral cortex. Similarly, the British neurologist John Hughlings Jackson showed in 1870 that specific parts of the cerebral cortex were responsible for different types of sensation and movement.

In 1876, a young German neurologist named Carl Wernicke described a patient who was able to speak, but who could not understand what he himself was saying. After the patient died, Wernicke dissected his brain and found that the cerebral cortex had been damaged in a single place, probably by a stroke. Wernicke knew about the area Broca had described and its importance for language. He also knew that the cerebral cortex contained another area where damage led to a general failure of hearing.

Wernicke realized that word perception involved several different elementary processes, each of which occurred in a different region of the brain. His work became the basis for the now-accepted idea of *distributed processing,* in which different regions of the brain together perform specific components of a complex process such as word perception. Even today, Wernicke's model of how the brain processes language still influences the thinking of neuroscientists.

Today, neuroscientists can actually light up the different regions of the brain involved in word recognition and other thought processes (Figure 42-1). For example, researchers can compare the metabolic activity of different regions of the cerebral cortex as human subjects perform various tasks, as revealed by such techniques as *positron emission tomography (PET)* and functional magnetic resonance imaging (FMRI).

If a person has heard a single word read aloud, metabolism increases in the very areas predicted by Wernicke. When a person reads the same word silently, however, metabolism increases in a totally different region. When the person speaks the word, the researchers observed metabolic increases in still different regions, including Broca's area. And when the person is asked to merely think about the word, a new pattern emerges that includes not only Wernicke's and Broca's areas but also other areas of the cortex.

Contemporary neuroscientists accept the view that individual regions of the brain perform specific functions and that complex functions of the brain require cooperation and information flow between several regions. In this chapter, we will

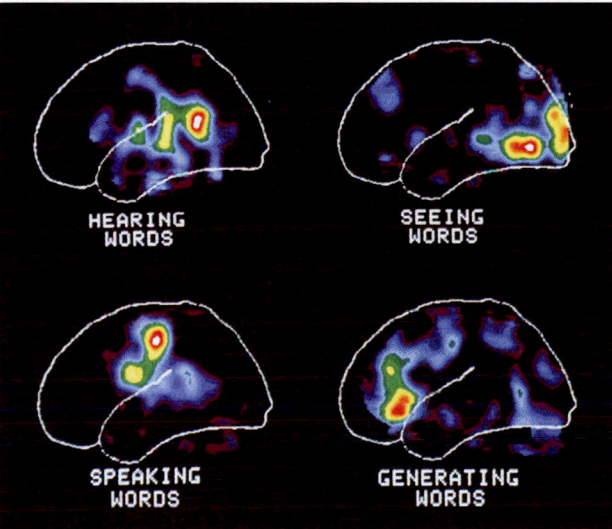

Figure 42-1   Different mental tasks involve different brain regions. Modern imaging techniques, such as these positron emission tomography (PET) scans, allow neuroscientists to observe the metabolic activity of individual regions of the brain. In these images, the hotter colors (red and yellow) show regions of high metabolic activity; cooler colors (blue and green) show regions of low metabolic activity. These images show metabolic activity just after a subject has heard words spoken, seen words written, spoken words, and thought words. In these images, we are looking at side views of the brain, with the front of the brain on the left and the rear on the right. (*Marcus Raichle, Department of Neurology, Washington University School of Medicine*)

focus on the central and peripheral nervous systems and on the roles of individual brain regions in **sensation**—the detection of external stimuli by vision, hearing, touch, taste, and smell. We will begin our discussion with an introduction to the nervous system.

HOW DOES THE NERVOUS SYSTEM RESPOND TO ITS ENVIRONMENT?

Nervous systems are major contributors to homeostasis. An animal's nervous system receives reports on the state of the exterior world, as determined by vision, touch, hearing, smell, and taste. At the same time, the nervous system also collects information about the interior world, as determined by receptors that monitor specific features of the internal environment, such as temperature, pH, oxygen supply, and levels of individual hormones.

The nervous system is not only the ultimate homeostatic organ, but it also coordinates simple and complex behaviors. Some behaviors contribute directly to homeostasis: shivering, for example, contributes to temperature regulation. Other behaviors may contribute to homeostasis only indirectly or not at all: knitting a sweater also contributes to temperature regulation, but playing the piano has no known homeostatic function. A pain, a smell, a sound, or a visual image can send a signal that directs a homeostatic response, such as withdrawing a hand from a hot stove or running away from a charging elephant.

For the nervous system to maintain homeostasis, it needs to be able to (1) *detect changes* in its internal and external environments, (2) *interpret* the meanings of those changes, and (3) *respond* by sending appropriate signals to the muscles and organs of the body. Our eyes, nose, and other sense organs detect environmental changes, while our brain and spinal cord interpret signals from the sense organs and command the muscles and endocrine systems.

Connecting the sense organs and the brain are the nerves. Nerves carry signals from the sense organs to the brain or spinal cord, and from the brain or spinal cord to the muscles, organs, and endocrine system. The intricate connections between nerve cells—in the brain, spinal cord, and nerves—determine how animals respond to a variety of stimuli.

An animal's nervous system detects changes in external and internal environment and coordinates homeostatic responses.

Do Invertebrate Animals Have Brains?

The simplest organisms that have nervous systems are jellyfish, hydra, and other cnidarians (Figure 42-2A and B). A hydra, for example, has a web of interconnecting neurons, called a **nerve net,** which carries information to all parts of its body. A stimulus anywhere on the hydra's body triggers reflexive

A. Hydra

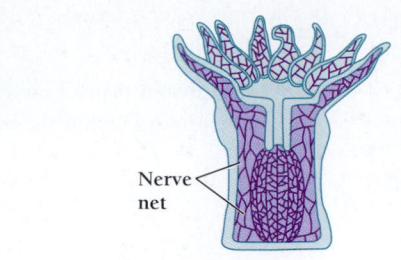

B. Jellyfish

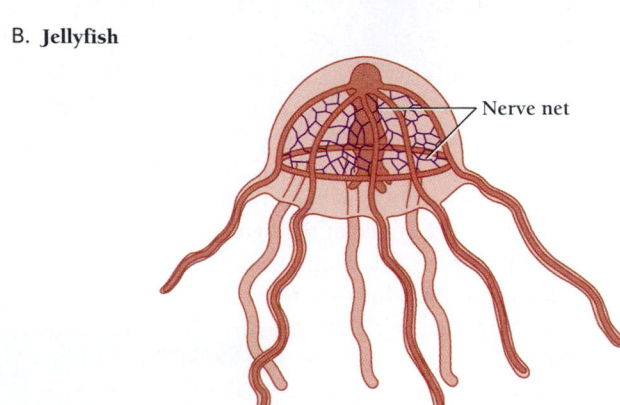

C. Planarian (flatworm)

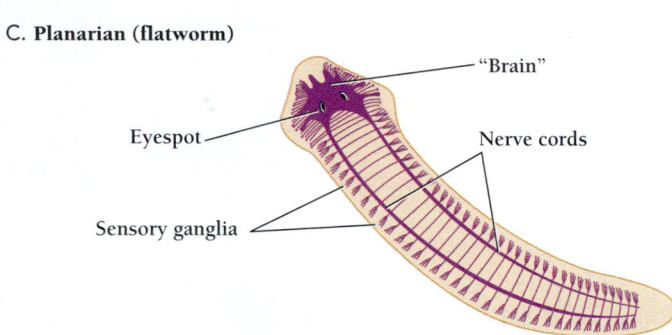

Figure 42-2   **Invertebrate nerve nets and nervous systems.** A. Nerve nets in a hydra. B. Nerve nets in a jellyfish. C. The nervous system of a flatworm, the simplest animal that is able to learn.

muscle contractions. In jellyfish, nerve nets coordinate complex swimming movements. But cnidarians have no brains, and, as far as anyone can tell, neither a hydra nor a jellyfish can learn. Their behavior does not appear to be influenced by experience.

The simplest animals capable of learning are the flatworms, and their nervous system resembles that of more complex animals (Figure 42-2C). The flatworm's nervous system consists of two parallel nerve cords, with nerves extending outward to the muscles. The two nerve cords converge near the front of the body in the simplest possible "brain." Like other bilaterally symmetrical animals, flatworms have a well-defined direction of movement. They have eyespots at the front end, and they process sensory information in clusters of nerve cells called *ganglia*.

In even a simple brain, the convergence of two nerves in a ganglion provides many opportunities to generate complex patterns of neural activity and behavior. A simple brain also allows animals to modify such patterns by learning through association. Flatworms can, for example, be taught to move in a variety of patterns in order to find food (Figure 29-8). More complex animals have increasing degrees of *cephalization,* the concentration of sense organs and ganglia in the front (anterior) end.

Flatworms are the simplest animals that can learn. They have parallel nerve cords that converge near the front of the body.

How Is the Vertebrate Central Nervous System Organized?

The nervous systems of vertebrates show the highest degree of cephalization and the most sophisticated sense organs. Most neurons are concentrated in the **central nervous system (CNS),** which consists of the brain and the spinal cord (Figure 42-3). In vertebrates, the CNS lies *dorsal* to (toward the back of) the body, whereas in invertebrates the CNS is usu-

ally *ventral* (toward the front of the body). (See Chapter 24 for a discussion of the surprising parallels between the dorsal nervous systems of vertebrates and the ventral nervous systems of invertebrates.)

In vertebrates, the central nervous system consists of the brain and spinal cord. The brain lies within a thick bony structure, called the *cranium,* and the spinal cord passes through a bony canal made by the stacks of vertebrae. Both the cranium and the vertebrae help protect the CNS from traumatic damage.

Also protecting the brain and spinal cord is a set of three membranes, which together are called the *meninges.* Bacterial or viral infection of the meninges—called *meningitis*—can be fatal. The central canal of the spinal cord and the connected *ventricles* of the brain are filled with a fluid, called the *cerebrospinal fluid (CSF).* The CSF also surrounds the brain and spinal cord, filling the space between the outer surface and the meninges. The CSF cushions the brain and spinal cord and protects them against injury.

The CSF derives from the liquid part of the blood and, in many respects, resembles blood plasma in its chemical composition. But the capillaries that supply blood to the spinal cord and the brain are different from those that supply the rest of the body. The cells of the capillary walls are tightly connected to one another so that the passage of molecules is highly restricted, and the capillary walls are said to form a *blood-brain barrier* (or in the case of the spinal cord, a *blood-nerve barrier*).

Except for molecules with special transporters, only small, nonpolar molecules pass through the blood-brain barrier. Among the molecules that easily traverse the barrier, however, are alcohol, nicotine, heroin, and cocaine—partially accounting for the potency of these drugs of abuse. Pharmaceuticals that act on the brain or spinal cord must also traverse the blood-brain barrier, a major consideration in the design of medicines for neurological and psychiatric disorders.

The vertebrate central nervous system consists of the brain and spinal cord. Both are surrounded by bones, membranes, and cerebrospinal fluid.

The spinal cord

The **spinal cord,** a long bundle of specialized nerve fibers that runs from the brain to the tail, receives *sensory information* regarding the position of the limbs and sends *motor instructions* to the muscles of the limbs and trunk. The spinal cord also controls or influences involuntary *autonomic functions,* such as the beating of the heart and the movements of the intestines.

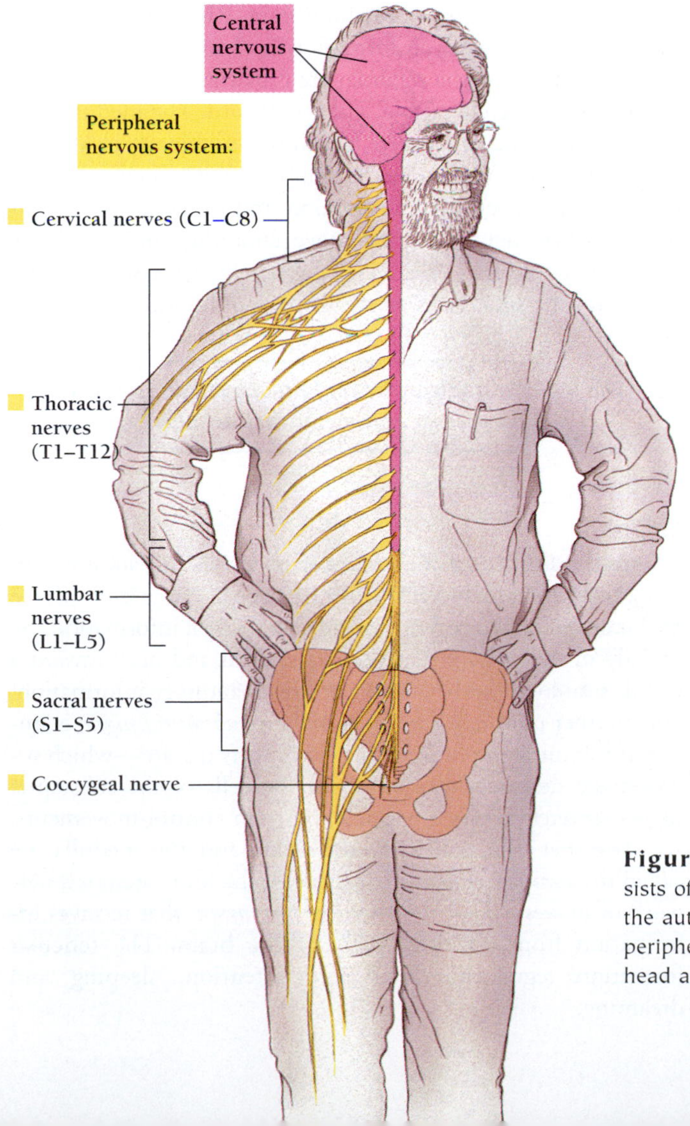

Central nervous system

Peripheral nervous system:

■ Cervical nerves (C1–C8)

■ Thoracic nerves (T1–T12)

■ Lumbar nerves (L1–L5)

■ Sacral nerves (S1–S5)

■ Coccygeal nerve

Figure 42-3 Nervous system of a human. The central nervous system consists of the brain and the spinal cord. The peripheral nervous system consists of the autonomic nervous system and the peripheral nerves. The cranial nerves are peripheral nerves that connect the brain to muscles, receptors, and glands in the head and in the thoracic and abdominal cavities.

The spinal cord contains the sensory neurons and motoneurons that are responsible for stretch reflexes, such as the knee jerk, as we discussed in Chapter 41.

Motoneurons in the spinal cord directly stimulate muscle contraction. The nerve fibers from the motor cortex cross from left to right (or from right to left) as they pass through the base of the brain, so that the left side of the brain controls muscles on the right side of the body.

In cross section, the inside part of the spinal cord resembles a grey butterfly (Figure 42-4). This *grey matter* contains nerve cells, but no myelin. The lower parts of the double wings, which are called the *ventral horns* (or, in upright humans, the anterior horns), contain the motoneurons. The upper parts of the double wings, which are called the *dorsal horns* (or, in humans, the posterior horns), receive inputs from sensory nerves—including the stretch receptors. (The cell bodies of the sensory nerves are adjacent to the spinal cord in the *dorsal root* ganglia.) In the center of the grey matter is the fluid-filled *central canal,* which is the remnant of the interior of the *neural tube,* from which the entire central nervous system originally developed (Chapter 44).

The exterior part of the spinal cord consists of *white matter,* whose glistening appearance results from myelin, which encloses the nerve tracts that run up and down the spinal cord. These spinal tracts carry information from the brain to the motoneurons and from the sensory neurons back up to the brain.

The spinal cord can itself direct relatively complex behaviors even without a connection to the brain: a person or animal with a severed spinal cord can learn to move his or her

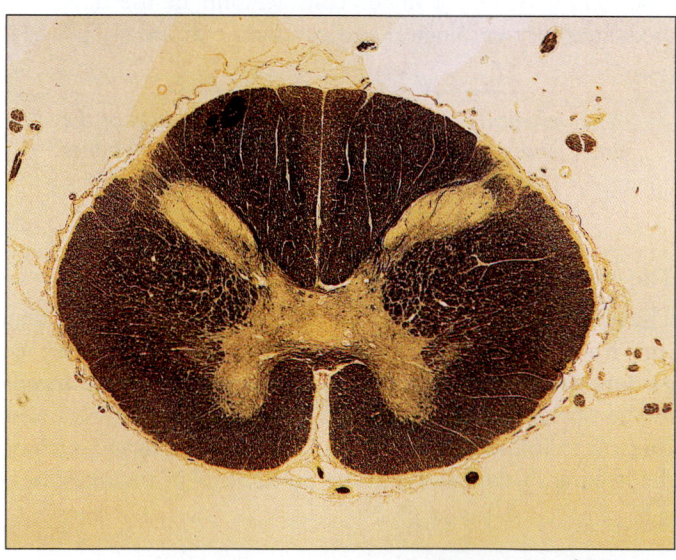

Figure 42-4   The organization of the human spinal cord. In this cross section, the white matter (darkly stained) lies outside the grey matter (lightly stained), which surrounds the central canal in a shape that resembles a butterfly. The ventral horn contains the motoneurons that command muscle contraction, and the dorsal horn contains the cell bodies of sensory neurons. *(J.C. Revy/Phototake)*

2.5 mm

legs in a coordinated rhythm and even to lift a leg over an obstacle. The ability of the spinal cord to direct coordinated movements without input from the brain is well known to people who have raised and slaughtered chickens: a beheaded chicken can run about a farmyard for some minutes after decapitation.

The spinal cord receives somatic sensory inputs concerning the position of body parts and sends motor instructions to voluntary muscles. It also coordinates many autonomic functions.

The vertebrate brain

Paleontologists' studies reveal that the brains of jawless fishes, almost half a billion years old, consisted of three major divisions: (1) a hindbrain, (2) a midbrain, and (3) a forebrain (Figure 42-5). The same divisions are apparent in today's fishes and in the embryos of contemporary vertebrates.

Hindbrain

The **hindbrain,** which forms a kind of cap on the spinal cord, consists of three parts, the **pons,** the **medulla oblongata,** and the **cerebellum.** Many of the nerves of the spinal cord form connections to neurons in the hindbrain, which integrates and processes both sensory information and motor instructions. Both the pons and the medulla are crossroads of information that flows between the brain and spinal cord. The medulla also contains centers that regulate breathing and blood circulation. The cerebellum, really an extension of the hindbrain, coordinates incoming sensory information and outgoing motor instructions so that the limbs move smoothly and the body maintains its upright posture. The cerebellum is also especially important for such tasks as eye-hand coordination.

The hindbrain integrates and processes both sensory information and motor instructions.

Midbrain

The **midbrain** is a major processor of visual and auditory information. In humans and other mammals, each side of the midbrain contains a region that processes visual information, especially in the reflexes that orient the head and neck toward a visual stimulus. Another region processes auditory information. Still another midbrain region, called the *substantia nigra,* is important in initiating movement. Parkinson's disease—which involves the destruction of many of the cells of the substantia nigra—severely limits a person's ability to control movements.

Together, the midbrain, the pons, and the medulla are called the *brain stem.* Running through the brain stem is a network of nerves, called the *reticular formation,* that receives information from all other parts of the brain. The reticular formation regulates arousal and attention, sleeping and dreaming.

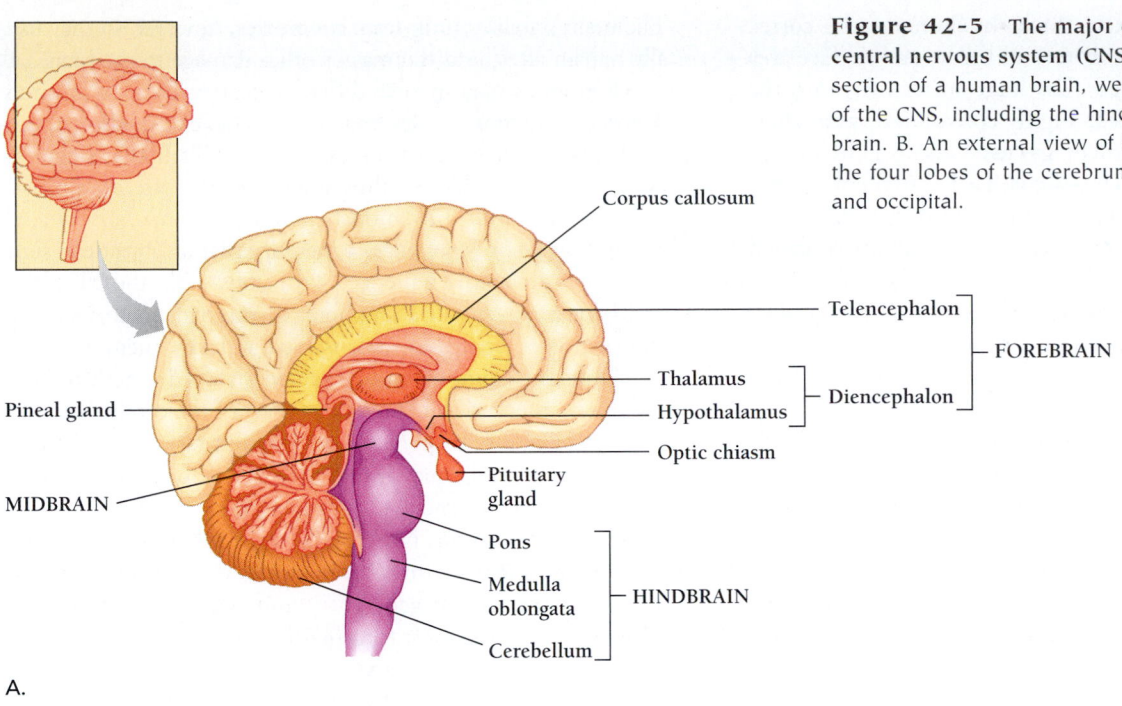

Figure 42-5   The major divisions of the mammalian central nervous system (CNS). A. In a midline (sagittal) section of a human brain, we can see the major divisions of the CNS, including the hindbrain, midbrain, and forebrain. B. An external view of the human brain illustrates the four lobes of the cerebrum—frontal, temporal, parietal, and occipital.

A.

B.

The midbrain relays sensory information, helps organize movements, and regulates attention.

Forebrain

In humans the **forebrain** is the locus of almost all conscious activity. It plays crucial roles in sensation, perception, movement, learning, memory, and mood. The forebrain consists of two divisions: the diencephalon and the cerebral hemispheres.

The **diencephalon** has two parts: the **thalamus,** the major integrator of sensory information about the *external* world, and the **hypothalamus,** the major integrator of information about the *internal* world. The hypothalamus is the master regulator of

homeostasis. It regulates body temperature, eating and drinking, growth, response to stress or threats, mating, and pleasure.

The **cerebral hemispheres** consist of a thin surface layer, called the **cerebral cortex,** and a number of underlying structures. The cortex's grey matter—neurons, glia, and unmyelinated neuron extensions—surrounds the white matter—the myelinated axons. In contrast, recall that in the spinal cord, the white matter is outside and the grey matter is inside.

In humans the cerebral cortex has become so large that it is packed around the rest of the brain like a wadded-up bed sheet. The cerebral cortex completely dominates the appearance of the human brain. It consists of just six layers of cells and is only a few millimeters thick. In humans, the cortex contains more than 10^{11} neurons and has a total surface area of about half a square meter. This huge surface fits into the brain case only because it is so highly convoluted, with characteristic folds (**gyri**) [singular, **gyrus**; Latin, = circle] and grooves, called **sulci** [singular, **sulcus**; Latin, = furrow] (Figure 42-5). The gyri and sulci provide landmarks for the division of the human cerebral cortex into four areas, called the *frontal, parietal, temporal,* and *occipital lobes.* The cerebral cortex, like the rest of the brain, consists of two almost symmetrical halves. The main direct connection between the two hemispheres is a thick bundle of nerve fibers called the *corpus callosum.*

Each of the four lobes of the cerebral cortex (on each side of the brain) shares the same cellular organization. The cortex consists of six cellular layers, with each layer containing neurons of characteristic shapes. Each lobe receives sensory signals (mostly via the thalamus), processes that information, and transmits instructions to some other part of the brain or spinal cord. The function of the rest of the cortex—called the **association cortex**—contributes to processing of both sensory input and motor output. More complex animals devote larger fractions of their cortex to association than do simpler animals. In a mouse, 95 percent of the cortex is sensory and motor and 5 percent is associative. In humans, more than half of the cortex is associative.

The frontal lobe deals principally with movement and smell, the parietal lobe with somatic sensation, the occipital lobe with vision, and the temporal lobe with hearing. The temporal lobe is particularly involved in the formation of memory, as is the underlying **hippocampus,** which is responsible for the first stages of memory storage.

Beneath the cerebral cortex lie other structures—the amygdala and a group of structures collectively called the basal ganglia. The **amygdala** coordinates autonomic (involuntary) responses to emotional states, especially anxiety. The **basal ganglia** help establish patterns of movement.

The forebrain plays important roles in sensation, perception, movement, learning, memory, and mood. The cerebral hemispheres consist of the cerebral cortex and underlying structures, which include the hippocampus, the amygdala, and the basal ganglia. In humans, the cerebral cortex is divided into four lobes, all of which share the same cellular organization.

How Do Humans Learn?

The association cortex, especially in the temporal lobe, is intimately involved in memory. **Memory** is the storage of knowledge about the world. So far, biologists have not been able to discover how or where memories form in the brain or how they are retained and recalled.

We do know that memories form in stages. **Short-term memories**—of experiences within the previous few seconds—can disappear after a blow to the head, electric shock, or drug treatment.

The brain stabilizes **long-term memories,** however, so they usually remain after head traumas or other damage to the brain.

Memories may involve different degrees of consciousness. A *reflexive* memory is one that has a reflexlike quality—automatic and not dependent on awareness. Skills (like riding a bicycle) and habits (like twirling a lock of hair) are examples of reflexive memories. In contrast, a *declarative* memory is a recalled thought or experience, a memory that a subject can summarize in a declarative sentence. Some skills—like those involved in driving a car—may be conscious, declarative memories at first and then later emerge as unconscious, reflexive memories.

Different brain structures may be involved in reflexive and declarative memories. Damage to a part of the cerebellum may destroy the ability to perform a particular task, while damage to the temporal lobe may interfere with the formation of long-term declarative memories.

The hippocampus has an important role in the formation of long-term memory from short-term memories and is also an important site in the generation and regulation of epileptic seizures. One of the most forceful illustrations of the role of the hippocampus came from a young man (called "H.M.") whose hippocampi were removed in an effort to stop persistent and otherwise untreatable seizures. After his operation, H.M. could still speak and understand language, he maintained memories from before his surgery, and he could remember for seconds or minutes. But he lost the capacity to form new long-term memories. Studies of both humans and experimental animals all point to the hippocampus and surrounding regions as a crossroads of short-term and long-term memory.

One of the most amazing discoveries about memories came from tests performed during neurosurgery. Wilder Penfield, a Montreal brain surgeon, stimulated the temporal lobes of several of his patients. These patients vividly experienced past events. For example, one heard a specific melody each time that Penfield stimulated a particular spot in her brain. Despite the incredible amount of information gathered by physicians, psychologists, and neuroscientists, however, we are still far from understanding the basis of memory. As in all scientific disciplines, the early stages are, of necessity, more descriptive than theoretical.

Memories form in stages, with short-term memories including experiences of the preceding few seconds and long-term memories including experience from many years before. The hippocampus has an important role in the formation of long-term memories.

How Is the Vertebrate Peripheral Nervous System Organized?

The **peripheral nervous system,** which consists of all the nerve cells that are outside the central nervous system, connects the central nervous system (CNS) to the sense organs, muscles, and other organs. **Afferent** nerves carry *sensory* information *to* the CNS. **Efferent** nerves carry instructions from the CNS to both muscles and other organs, including the endocrine glands (Figure 42-6).

The efferent nerves that carry information to the striated skeletal muscles make up the **somatic,** or voluntary, nervous system. Other efferent nerves, those of the **autonomic,** or involuntary, nervous system, carry instructions to the smooth muscles of the endocrine, digestive, reproductive, excretory, respiratory, and vascular systems, as well as to the cardiac muscles of the heart (Figure 42-6).

The autonomic nervous system consists of two distinct sets of nerves—the **sympathetic** and the **parasympathetic** nerves. These nerves use different neurotransmitters, which act on different receptors in the same target cells to bring about opposite effects. The sympathetic nervous system, which uses *norepinephrine* as a final neurotransmitter, helps coordinate the "fight-or-flight" reaction (together with the hormone *epinephrine*). This response involves the mobilization of energy stores as preparation for vigorous action. In contrast, the parasympathetic nervous, which uses *acetylcholine* as a neurotransmitter, helps conserve energy by slowing the heart and increasing intestinal absorption.

The peripheral nervous system connects the CNS to sense organs and muscles. It includes nerves of both the somatic nervous system, which carries information to voluntary muscles, and the autonomic nervous system, which carries instructions to endocrine organs, smooth muscles, and cardiac muscles.

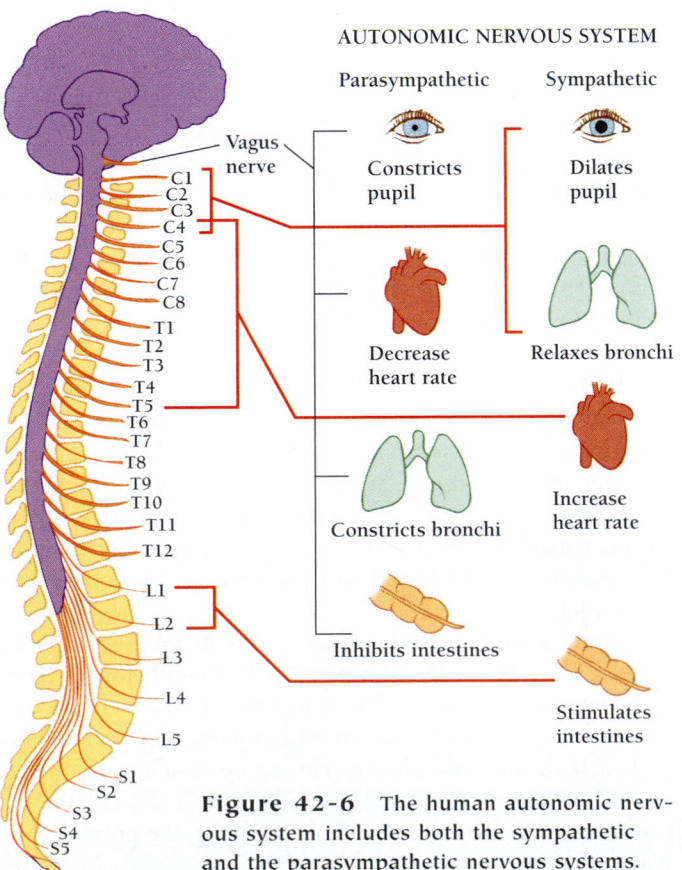

Figure 42-6 The human autonomic nervous system includes both the sympathetic and the parasympathetic nervous systems.

How Does Information Move Through the Nervous System?

The nerves of the peripheral nervous system are further classified according to what kind of information they carry. The **auditory system,** for example, contains all the cells responsible for hearing, and the **visual system** contains all the cells responsible for seeing.

The receptor cells' information is in the form of **receptor potentials,** changes in the electrical voltages across their plasma membranes. Information from receptor cells moves to **relay neurons,** which convert information from one or more receptor cells into a series of electrical discharges, or action potentials, which, as we saw in Chapter 41, consist of temporary changes in ion movements across the plasma membrane. The relay neurons carry information (in the form of action potentials) to particular sets of neurons in the brain.

Whereas action potentials are all of the same voltage, receptor potentials come in a range of sizes. Neurons can fire repeatedly, slowly or over and over in rapid succession. The graded receptor potentials alter the frequency of action potentials, rather than their sizes.

The electrical properties of nerve cells allow neurobiologists to probe individual cells and the interconnections of cells within the nervous system. Researchers can see how the brain responds to external stimuli by recording electrical currents from the surface of the brain, from within individual cells, and even across minuscule patches of nerve cell membranes.

Researchers can find out which regions of the brain control which parts of the body by electrically stimulating the brain and then recording what happens. Such tests on humans are common during neurosurgery, when the patient is usually awake and aware. Since the brain has no pain receptors, such a procedure is not painful (after the scalp receives local anesthesia). Researchers can study both the final response, such as the movement of a limb, and also the intermediate steps along a neural pathway. In this way, researchers have begun to define the pathways of information transfer within the nervous system.

Neuroscientists can also map neural pathways in experimental animals by using special dyes. Nerve cells transport these dyes in the same direction in which they transmit information, so that when such a dye is injected into the motor cortex, for example, researchers can watch the dye move through axons of the cortex neurons, through the medulla, and into the spinal cord.

Researchers have also found other dyes that move in the direction opposite to the flow of neural commands. These dyes allow neuroscientists to determine the sources of information processed by a particular set of nerve cells. Such studies show, for example, that cells of the motor cortex receive information from the spinal cord, from other regions of the cortex, and from brain structures that lie below the cortex. Each part of the nervous system and almost every nerve cell receives information from many sources. Ultimately, however, all external information comes from the sensory systems.

Receptors detect characteristics of the external world and generate graded receptor potentials. Relay neurons convert the information encoded in receptor potentials into patterns of action potentials. Relays of neurons in each sensory pathway convey information to the brain.

HOW DO ANIMALS SENSE THEIR ENVIRONMENT?

Many individual parts of the brain contribute to *sensation,* the detection of external stimuli, and many more to *perception,* the conscious recognition and interpretation of those stimuli. Animals—human or otherwise—can respond to sensation, however, in ways other than conscious perception. Sensory stimuli can help control movements, regulate internal organs, and maintain arousal, without our being conscious of making any changes. For example, changes in light intensity lead to changes in pupil dilation, of which we are generally unaware. Even more strikingly, women regularly exposed to men's perspiration have more regular menstrual cycles than women not exposed to this constellation of chemicals.

What Senses Contribute to Sensation and Perception?

Humans have long recognized five senses—hearing, vision, smell, taste, and touch. Neuroscientists include balance in the list of sensory **modalities** (classes of sensation): **audition,** or hearing; **vision,** or seeing; **gustation,** or tasting; **olfaction,** or smelling; **vestibular** sensation, or balance; and **somatic sensation.** Somatic sensation actually includes four independent modalities—touch (or pressure), pain, temperature, and sensing limb motion and position. Some animals detect additional qualities of the external world, such as electric fields (some fish), magnetic fields (some migrating birds), or ultraviolet light (bees and many birds).

Sensation begins in **sensory receptors,** specialized cells that detect a particular property of the external world. Each receptor contributes to only one sense modality: receptors in the eye respond to light, those in the ear to vibration, and those in the taste buds and nose to particular chemicals. The sensory receptors for hearing, vision, taste, and smell are in specialized **sense organs**—the ears, eyes, taste buds, and the olfactory epithelium of the nose. Somatic sensation, on the other hand, depends on sensory receptors, dispersed in skin and muscles, which report separately on pressure, pain, temperature, and limb position.

Every sensation except olfaction involves three stages:

1. **Transduction** translates the energy of a stimulus (such as light) into a change in the chemistry of the receptor cell.
2. **Transmission** carries a signal from the receptor cells to the central nervous system.
3. **Integration** combines the signals from many receptors to form a mental image of the outside world.

Each sensory system—including sensory receptors and the connections within the central nervous system—responds quantitatively to three characteristics of a stimulus—*intensity, duration,* and *location.* Sensory receptors within the ear, for example, produce larger receptor potentials in response to greater vibration intensities, which we perceive as louder sounds. The electrical activities of sensory receptors, relay neurons, and brain neurons indicate the beginnings and ends of each sound. Finally, nerve nets within the brain also allow an animal to coordinate information from both ears to locate a sound. Some species are much better than others at location: bats, dolphins, and even owls, for example, can precisely locate sounds that a human would be lucky to hear at all. On the other hand, a human can precisely locate visual stimuli, but a hawk or an eagle can do even better.

Sensory receptors also respond to other qualities of a stimulus. The ear, for example, distinguishes among different frequencies, which are perceived as high notes and low notes. Similarly, the eye distinguishes among different frequencies (or wavelengths) of light, which we perceive as different colors.

Sensation depends on signal transduction by the sensory receptor, the transmission of a signal to the central nervous system, and the integration of signals to form a mental image of the outside world.

What Path Does Sensory Information Take to the Brain?

Sensory receptors feed information to relay neurons. Some relay neurons simply transmit the information from a single sensory receptor to the next step of the neural chain. Other relay neurons do more: they integrate information from several sensory receptors or from several other relay neurons. When such integration occurs, the relay neuron extracts particular information from the combination of inputs. For example, some relay neurons in the eye are active when the center of a stimulus is brighter than the surrounding area, while other relay neurons are active when the center is darker than the surrounding ring.

Relay neurons usually send information first to the thalamus and then to particular regions of the cerebral cortex. Although all routes pass through the thalamus, they pass through different regions *(nuclei)* of the thalamus. And although all routes end in the cortex, they end in different regions of the cortex: visual information ends in the occipital lobe, auditory information in the temporal lobe, and somatic sensation in the parietal lobe.

Information from the olfactory system takes a different route, going first to the other parts of the brain before arriving at the thalamus and cortex. The uniqueness of the olfactory system probably reflects its ancient evolutionary origins and helps explain both our difficulty in naming odors and our intense emotional responses to certain smells such as the cologne or perfume of a parent or lover. Marcel Proust, the great French novelist of the early 20th century, provided an extraordinary account of his responses—including thoughts, memories, and fantasies—to the smells and tastes of tea and cake.

Each sensory receptor feeds information into a characteristic route. Each route is called a *labeled line* because the brain recognizes it as corresponding to a particular sense. Stimulation of a particular receptor, therefore, always evokes the same sensation, even when the stimulation is not entirely appropriate. For example, a person who suffers deafness because of a defect in the ear will "hear" a tone when a researcher electrically stimulates the relay neurons within the auditory pathway. (This ability to "hear" an electrical stimulus has allowed the development of electrical devices, called *cochlear implants,* that have allowed thousands of deaf people to hear.) Similarly, someone who receives a blow to the head may "see stars" because the blow mechanically stimulates receptors along the visual pathway.

After the neural response to a sensation first arrives in the cortex, the primary sensory region further processes information and sends it to another region of the cortex. In the case of the visual system, there are at least 32 areas of the cortex that have representations of the visual world. Researchers do not yet know the significance of all these representations, but each is thought to extract different features of the visual image. One region, for example, appears to be highly specialized for the recognition of faces: people with damage to that area cannot identify faces but have otherwise normal vision and memory.

The functional organization of sensory systems shows that the brain is an *active* agent in sensation; it is emphatically not a passive recipient of information from the sense organs. The brain extracts only certain features of the external world. The brain also communicates back to some sensory receptors and relay neurons, actively modifying even the very beginnings of sensation.

Sensory information from each sensory modality passes separately through the thalamus and converges separately in different regions of the cerebral cortex.

HOW DO ANIMALS HEAR?

Hearing, like the detection of light or touch, depends on the ability to translate, or **transduce,** energy into electrochemical events in neurons. To understand how this transduction takes place, we must first understand the physical nature of a stimulus.

What we perceive as sound is actually pulsations of the molecules of the air. If we pluck the string of a guitar, it vibrates with a characteristic frequency. As it moves, it alternately pushes molecules together and apart, so that the surrounding air contains waves of pressure.

We can describe a sound wave in terms of the *frequency* of vibration, which we ultimately perceive as **pitch,** and of the *amplitudes* of the pressure differences, which we perceive as **loudness.** While this simple description works well for simple tones, such as those produced by a vibrating string, we will also want to understand the perception of more complicated sounds, such as those of a word. Fortunately, the same description applies, for any sound can be regarded as the sum of many tones of different frequencies and amplitudes. If we can understand how the ear transduces simple vibrations, then we will also understand how it transduces speech or music.

How Does the Mammalian Ear Detect Frequency and Amplitude?

The mammalian ear consists of three parts, called the **outer ear,** the **middle ear,** and the **inner ear** (Figure 42-7). The outer ear collects sound the way a funnel collects liquid. It channels sound through the **auditory canal** to the **tympanic membrane,** or eardrum. The tympanic membrane vibrates in response to variations in air pressure.

Behind the tympanic membrane is the middle ear, which, like the outer ear, is filled with air. This air comes into the ears by way of two tubes (one on each side), called the *Eustachian tubes,* from the throat. The Eustachian tubes are normally closed, so that the air in the middle ear is isolated from the air outside. Because the tubes are usually closed, the air in the middle ear can become pressurized relative to the outside air. A person who rapidly gains altitude, in a plane or even on an elevator, sometimes experiences painful pressure on the eardrum. Fortunately, swallowing, chewing, or yawning can open the Eustachian tubes and allow the middle ear to attain the same pressure as that outside.

Inside the middle ear, three small bones—called the *malleus* (hammer), *incus* (anvil), and *stapes* (stirrup)—transmit and amplify the movement of the tympanic membrane to the fluid-filled inner ear, through a membrane-covered opening called the *oval window.* The middle ear bones amplify the vibrations of the tympanic membrane, so that the fluid in the inner ear vibrates with less amplitude but far more forcefully than if it were directly linked to the eardrum.

Inside the inner ear is an organ that translates, or transduces, vibration into signals that we understand as sounds. This organ, called the **organ of Corti,** lies within the **cochlea** [Latin, = snail], an enclosed tube that is coiled like a snail shell. When the fluid within the cochlea vibrates, the organ of Corti moves from side to side (Figure 42-8).

In the 1940s, Georg von Békésy devised a way to watch these movements inside the cochlea. Von Békésy removed a cochlea from a fresh human cadaver and sealed the lens of a microscope into the wall of the cochlea. Von Békésy observed that low sound frequencies moved the organ of Corti toward the end of the cochlea, while higher frequencies moved it toward the middle ear.

The inner ear interprets this movement as sound by means of **hair cells,** cells that have stiff bundles of actin filaments, called *stereocilia,* embedded in the tectorial membrane of the cochlea (Figures 42-7 and 42-8). When the basilar membrane of the organ of Corti moves, relative to the tectorial membrane, shear forces bend the hair cells. The bending of the hair cells alters the permeability of the membrane to ions.

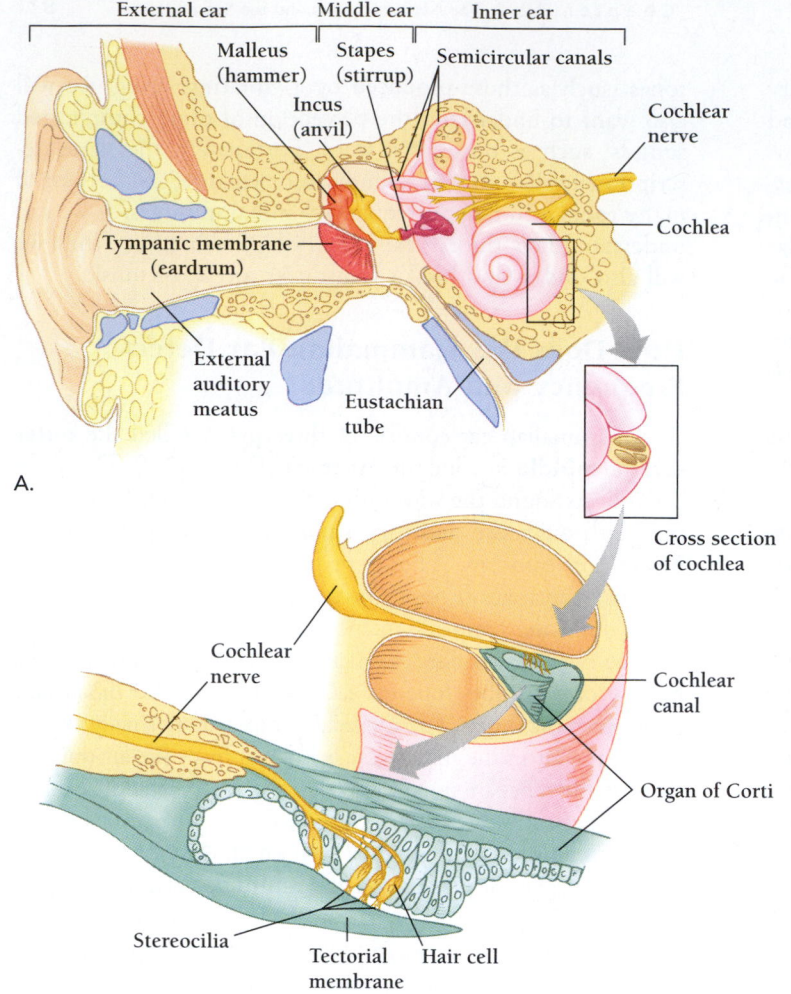

External ear | Middle ear | Inner ear

Malleus (hammer)
Stapes (stirrup)
Semicircular canals
Incus (anvil)
Cochlear nerve
Cochlea
Tympanic membrane (eardrum)
External auditory meatus
Eustachian tube

A.

Cross section of cochlea

Cochlear nerve
Cochlear canal
Organ of Corti
Stereocilia
Tectorial membrane
Hair cell

B.

Figure 42-7 **The human ear, showing the external ear, the middle ear, and the inner ear.** The inner ear contains the organ of Corti. A. The organization of the ear. B. Cross section of the organ of Corti, which detects vibrations in the fluid of the inner ear. C. Scanning electron micrograph, showing hair cells within the organ of Corti. (*Photo, G. Bredberg/Science Photo Library/Photo Researchers, Inc.*)

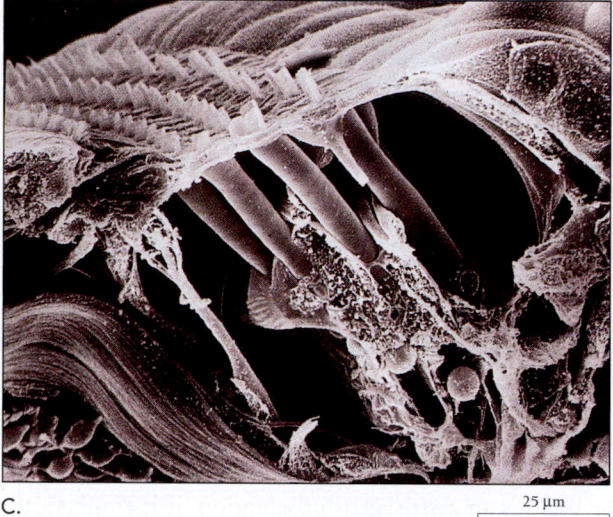

C.

25 μm

Each frequency stimulates a different set of hair cells, and the signals pass from the organ of Corti to relay neurons in the rear of the brain. Relay neurons connect, in turn, to several other regions of the brain. One pathway leads to a part of the thalamus, then to the *primary auditory cortex* in the temporal lobe of the cerebral cortex.

As in the visual and the somatosensory systems, the brain simultaneously processes several separate "images" of sound.

Each image provides different information. One set of cells, for example, appears to register differences in the timing of signals from the two ears; another set of cells registers differences in sound intensity. These two kinds of information are especially useful in locating the source of a sound.

Channels within the ear carry sound to the eardrum. The vibrations of the eardrum move the fluid in the inner ear, and hair cells in the organ of Corti detect the motion of the fluid. The brain interprets the patterns of signals from the organ of Corti as sounds.

How Does the Inner Ear Help Maintain Balance?

Hair cells are receptors that bend in response to the mechanical energy of air vibrations and transduce this energy into sound. But hair cells also bend in response to acceleration or gravity. Consequently, many animals, including humans, use hair cells to determine the movement and orientation of their heads.

Hairs

5 μm

◀ **Figure 42-8** **Sensory cilia ("hairs") of the pressure-sensitive hair cells in the organ of Corti.** (*G. Bredberg/Science Photo Library/Photo Researchers, Inc.*)

Fish and frogs also use hair cells to provide sensory information both about their own movements and those of potential predators and prey. In fish and frogs, the hair cells are part of the **lateral line system,** which consists of two canals that run just under the skin on each side of the body. The hair cells detect changes in water movements through the canals.

Within the mammalian inner ear are the **semicircular canals,** three fluid-filled tubes curved into half-circles that lie at right angles to one another. As an animal turns its head, the fluid in the canals tends to remain in place because of inertia, and the turning stimulates the movement of the fluid against the cilia of the hair cells in the canals. Other hair cells similarly detect movement against the force of gravity.

The pattern of electrical activity from the hair cells tells the brain's cerebellum and hindbrain how the animal's head is oriented and in what direction it is moving. The cerebellum then integrates this information with other sensory information to determine how the muscles should move to maintain proper posture or to execute a particular movement.

Hair cells, within the semicircular canals of mammals and within the lateral line system of fishes and frogs, help establish the position and the movement of the body.

HOW DO ANIMALS SEE?

Animals have evolved many different kinds of light detectors. In all cases, the primary event in vision is the action of light energy on a pigment molecule that changes from one form to another. This change takes place within a specialized light-detecting cell called a **photoreceptor.**

Different kinds of eyes and different visual systems use different strategies to deliver light to the receptor pigment and to present the resulting information to the rest of the nervous system. Planarians, a type of flatworm, have photoreceptors within a sense organ called an *eye-cup.* The receptors detect a shadow at the cup's edge, thereby identifying the direction of a light source. But the planarian's eye does not form any image of the outside world, and the flatworm lives in a world of light and shadow only.

In contrast, the eye of every vertebrate animal forms a distinct single image of the outer world. This image is projected against the back of the eye to a sheet of cells called the **retina** in much the same way that an image is projected against the film at the back of a camera. And just as photographic film records different frequencies and intensities of light from an image, photoreceptor cells in the retina contain pigment molecules that record the images before us.

In insects, the eye forms an image in a piecemeal fashion. The insect eye is really a collection of thousands of individual eyes, each of which forms a part of the full image. Together, the parts of an insect's *compound eye* compose an accurate image of the outside world. Although an insect eye sees less detail than the best vertebrate eye, insects are extraordinarily good at detecting movement—especially of large objects.

The primary event in the detection of a visual image is the capturing of light energy by a light-sensitive pigment within a photoreceptor cell.

How Does a Vertebrate's Eye Form an Image of the Outside World?

The vertebrate eye is roughly spherical (Figure 42-9). Encasing and protecting the eye is a tough connective tissue, called the *sclera.* Light enters the front of the eye through a transparent region of the sclera, called the **cornea.** Just inside the sclera is a pigmented layer, the *choroid.* Behind the cornea, the choroid forms a muscular, donut-shaped disk called the **iris** that gives the eye its color. The iris surrounds the **pupil,** the dark, central opening through which light reaches the retina. The iris controls the size of the pupil in the same way that the iris of a camera controls the aperture of the lens. In dim light, the iris opens the pupil to let in more light. In bright light, the iris closes the pupil, limiting the amount of light that reaches the retina and increasing the depth of field, the range of distances that will be in focus.

Just behind the pupil is a clear **lens,** which helps the cornea focus images on the rear surface of the retina. The problem in both a camera and an eye is that the point in space where an image comes into focus depends on how close or far away an object is. Faraway objects form images immediately behind the lens, while nearby objects form images some distance further. A photographer deals with this problem by "focusing," moving the lens farther from the film for nearby objects and closer to the film for faraway objects.

The eyes of fish and amphibians work like a camera, moving the lens relative to the retina to focus objects of different distances. The mammalian eye, however, deals with the problem quite differently: it actually changes the shape of the lens according to whether objects are nearby or far away. Because a thicker lens bends light more, it forms the image closer to

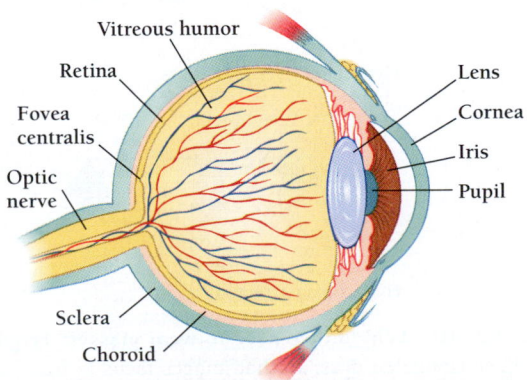

Figure 42-9 **The human eye.** The lens focuses images on the retina, which contains photoreceptor cells.

the lens. When focused on nearby objects, the mammalian lens thickens to form a closer image; when focused on faraway objects, the lens thins. The focusing of objects at different distances is called **accommodation.**

In a vertebrate eye, the lens focuses an image on the surface of the retina.

Why Must Many of Us Wear Glasses?

A person who is **nearsighted** has eyes that are too long from front to back. In spite of the thinning of the lens, the images of faraway objects come to a focus too soon in the space in front of the retina, so that the image on the retina is out of focus. A corrective concave lens spreads the light before it enters the eye, so that images converge on the retina (Figure 42-10).

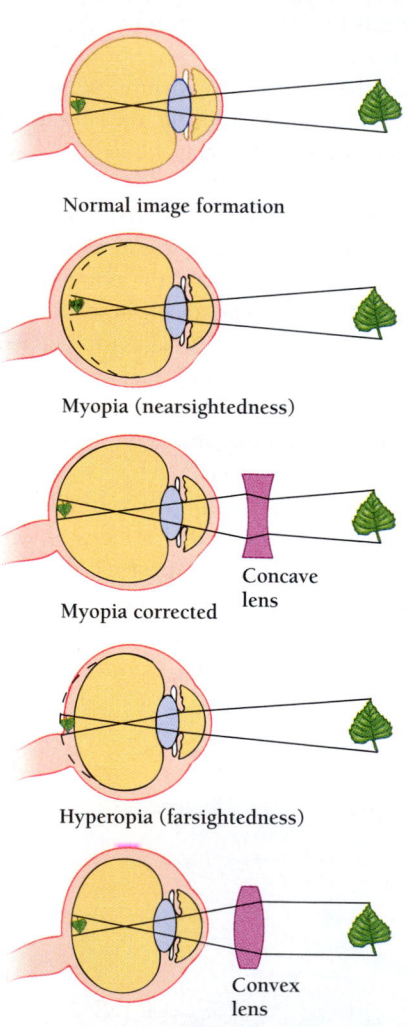

Normal image formation

Myopia (nearsightedness)

Myopia corrected   Concave lens

Hyperopia (farsightedness)

Hyperopia corrected   Convex lens

Figure 42-10   Why do we have to wear glasses? People with myopia have elongated eyes, so that images focus in front of the retina. People with hyperopia have shortened eyes, so that images focus behind the retina. In both cases, lenses can be used to correct vision.

A person who is **farsighted** has eyes that are too short. The result is that nearby images come to a focus beyond the retina (or would if they got that far), and images on the retina are out of focus. Even the thickening of the lens does not allow the formation of images of nearby objects. Correction of farsightedness requires a converging (convex) lens. Normally, people over 40 or 45 lose the ability to thicken their lenses and to focus on close objects, leading them to depend on "reading glasses."

The lens changes its thickness and converging properties by means of muscles that lie in a structure called the *ciliary body.* Sometimes parts of the lens become cloudy, leading to the obscuring of vision and even to blindness. Such cloudy spots are called **cataracts.** An eye surgeon can remove a lens with a cataract and replace it with an artificial lens.

The eye's ability to focus also depends on the fluids through which light passes before and after it travels through the lens. The large space between the lens and the retina contains a transparent, jellylike substance, called the *vitreous humor.* A more watery substance called the *aqueous humor,* produced by the ciliary body, fills the space between the iris and the cornea. Too much pressure within the aqueous humor—caused by blockage of the ducts that drain the cavity—is called *glaucoma,* a condition that damages the eye and causes blindness.

Defects in the focusing apparatus cause eye disorders.

How Does the Retina Detect an Image?

The retina is a thin sheet of cells behind the vitreous humor. It consists of five types of cells arranged in distinct layers. The most numerous of these cells are the photoreceptors, which actually lie in the bottom layer of the retina, farthest from the light entering the eye. Light from the outside world must pass through the cornea, the lens, and the four other cell layers, including overlying nerve cells, before it reaches the photoreceptors. It is the photoreceptors, however, that convert the image into electrical signals that the brain can interpret.

The retina contains two kinds of photoreceptor cells: cylindrical **rods** and cone-shaped **cones** (Figure 42-11). The rod cells are most dense slightly away from the edges of the retina and are phenomenally sensitive to light. When we are outside on a dark night and the pupil is opened up all the way, the 100 million rod cells in the human eye can detect even slight amounts of light and movement. The image that we see, however, is colorless and poorly defined. Because the rods are primarily around the periphery of the retina, we can actually detect faint objects more successfully by looking just to one side of the object, so that the image is projected not to the center of the retina, but to the edge. This phenomenon, useful to amateur astronomers, is called *averted vision.*

The cone cells are made for full-color vision in bright light. Unlike the rods, the 6 million cones of the human retina are concentrated in the central portion of the retina, called the

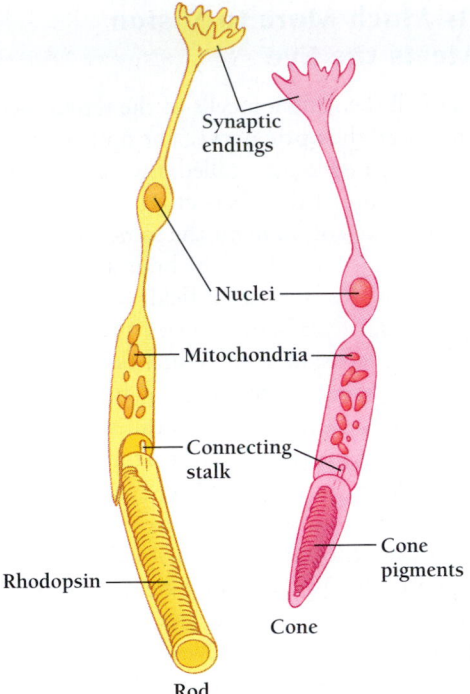

Figure 42-11 Rods and cones. The two types of receptors in the vertebrate eye. Rods have only a single type of light-detecting pigment molecule, but cones may have one of three different pigments.

fovea. As a result, even in bright light we see color most fully when we look directly at an object. We perceive the whole visual world as colored both because some cones also lie outside the fovea and because we constantly move our eyes so that different images impinge on the fovea.

Both rods and cones contain special pigments called **rhodopsins** that absorb visible light and transmit a signal. Rods and cones differ, however, in the wavelengths of light to which they respond. Rods contain a single kind of rhodopsin, which is most sensitive to blue-green light. Cones come in three varieties, each of which has a distinct type of rhodopsin; one of these is most sensitive to blue light, one to yellow, and one to green. The four pigments are distinct proteins, called **opsins,** each of which binds the same small molecule, called **retinal.**

Each cone has just one kind of opsin. The brain compares the images produced by the different kinds of cones and interprets the differences as colors. Color blindness results from the lack of one or more type of cone opsin. Because the genes for two of the cone opsins are on the X chromosome, men (who have only one X chromosome) are more likely to lack one of these genes than women (who have two X chromosomes). Consequently, color blindness is more common in men than in women. About 5 percent of men lack the green-sensitive opsin and are red-green color blind.

Photoreceptor cells contain pigments that absorb visible light. The absorption of light alters the cell's electrical properties and initiates a series of events that transmits visual information to the brain.

How Does Information Travel from the Retina to the Brain?

We know a lot about the path that carries visual information from the photoreceptors to the brain. This information comes largely from the systematic examination of the electrical signals from individual cells in the pathway. It is clearly impossible to study each of the hundreds of millions of cells, but neuroscientists have managed to study thousands of them. Researchers find the same patterns of organization repeated throughout the visual system. In addition, the principles of information processing in the visual system seem to apply to other sensory systems.

At the beginning of this chapter we said that every sensory system consists of three stages: transduction, transmission, and integration. Sometimes, however, these three stages are not completely separate. In the visual system, the process of integration actually begins during transduction and transmission.

We have seen that transduction occurs in the rods and cones. These cells are located in the inner nuclear layer of the retina, and their absorbing pigments lie in membrane-bound compartments still deeper than their nuclei (Figure 42-12). The photoreceptor cells make contact with two other kinds of cells. These cells are called the **bipolar cells** and the **horizontal cells.** The bipolar cells are the main conduits

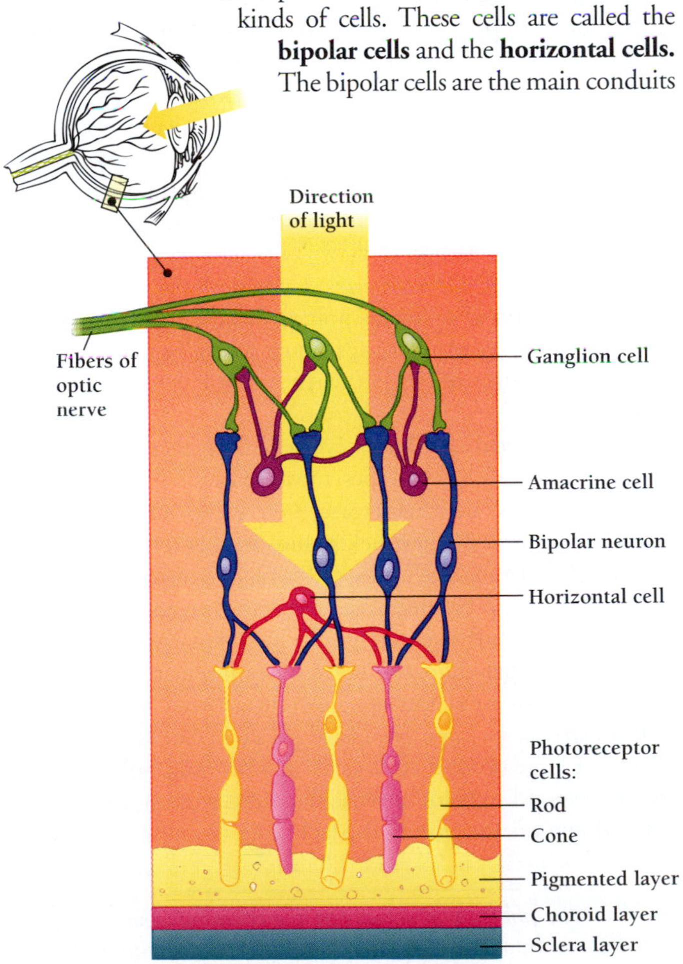

Figure 42-12 The organization of the retina. Light must pass through the outer layers of the retina before encountering the photoreceptor cells, the rods and cones. Information passes from the photoreceptor cells to the bipolar cells to the ganglion cells. Horizontal cells and amacrine cells are interneurons.

for information from the photoreceptors. They connect to **ganglion cells.** The ganglion cells in turn carry signals away from the retina to the brain itself. Also connected to the ganglion cells are **amacrine cells,** which lie in the same cell layer as the bipolar and horizontal cells.

The major pathway of transmission of visual information is from photoreceptors to bipolar cells to ganglion cells to the brain. There is not a point-to-point correspondence, however, as there is, for example, between the signal of a video camera and the image on a television screen. Instead, information from the more than 100 million receptors in the retina funnels into less than a million ganglion cells. This can happen because the ganglion cells receive a processed, or *integrated,* version of the visual world detected by the rods and cones.

The signals generated by each bipolar cell depend not only on the pattern of connections with photoreceptor cells but also on the action of the interconnecting horizontal cells. Similarly, each ganglion cell receives inputs both from bipolar cells and from amacrine cells. The horizontal and amacrine cells are examples of *interneurons,* which carry information between cells with parallel functions (Chapter 41). Interneurons often inhibit the action of neurons away from the immediate vicinity. In the retina, for example, the horizontal and amacrine cells inhibit nearby cells in a way that increases contrast in the image.

Because researchers can study the electrical activity of a single ganglion cell, they can discover what specific aspects of the external world a ganglion cell reports to the brain. Such studies have shown that every ganglion cell responds when light reaches a neighboring group of photoreceptors. Each ganglion then responds to a spot of light. Different ganglion cells respond to spots of different sizes. Cells near the center of the visual field respond to small spots and those near the side respond to larger spots. This difference partly explains why we can see with better resolution when we look directly at something.

These experiments have revealed more than 30 types of ganglion cells, with two types most prominent in primates: "on-center" cells, which increase their firing rate when light falls on the center of their fields, and "off-center" cells, which decrease their firing rate under the same circumstances. A ganglion cell seems to be specialized for detecting changes in light intensity between the center and the outside of its field of photoreceptors. A ganglion cell is relatively insensitive to the actual amount of light hitting the retina. Some respond with high sensitivity to large groups of photoreceptors and probably detect large objects, such as an approaching bull elephant. Others respond to color and the fine details of an image, such as the down on the skin of a peach.

The photoreceptor cells of the retina transmit information to the bipolar cells, and the bipolar cells transmit to the ganglion cells. Connections among the photoreceptor cells, the bipolar cells, and horizontal cells and among the ganglion cells and amacrine cells are responsible for the integration of different aspects of the visual world.

There Is Much More to Vision Than Meets the Eye

The axons of all the ganglion cells of the retina together form a thick stalk called the **optic nerve.** The optic nerves from each eye converge in a structure called the *optic chiasm* (Figure 42-13). There, some of the fibers cross to the opposite side of the brain and some continue on the same side. The fibers are arranged so that the left side of the brain receives information about the right half of the visual field, from both eyes. Similarly, the right side of the brain receives information from both retinas about the left side of the visual field.

Most of the optic nerve fibers in humans end in a region (or "nucleus") of the thalamus called the *lateral geniculate nucleus.* There the cells lie in distinct layers, each of which receives input from the axons of one eye. Each lateral geniculate neuron receives input from several ganglion cells, and each neuron can integrate the information from these different ganglion cells. Because these neurons receive information from the ganglion cells, which responds to patches of light and dark, the neurons likewise respond to small spots or bars, not to overall changes in light level. Neighboring cells in the lateral geniculate nucleus connect to neighboring ganglion cells, so the overall organization of the image is maintained.

The signals that enter the lateral geniculate nucleus continue on to the *primary visual cortex,* a special region at the back of the brain (Figure 42-13). The cells of the visual cor-

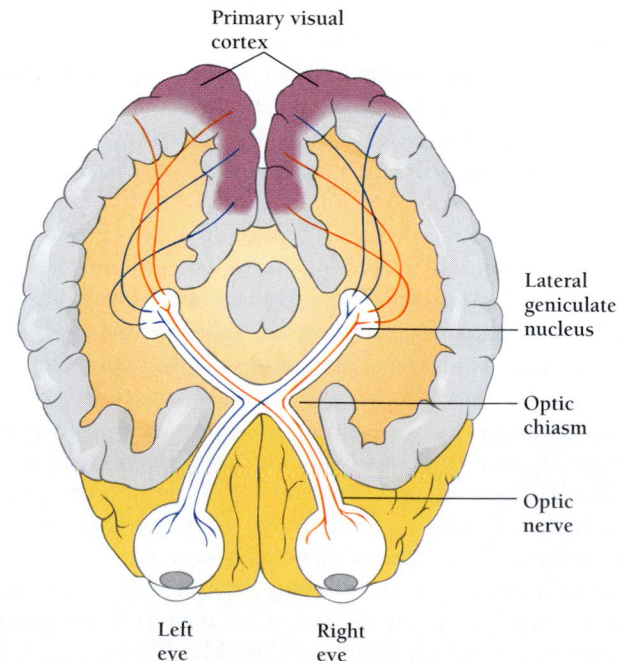

Figure 42-13   The pathway of visual information. Information from the retina passes by way of the optic nerve into the brain to the lateral geniculate nucleus of the thalamus. From there the information passes to the primary visual cortex on each side of the brain. Many other brain areas also receive visual information.

tex respond to still more integrated versions of the visual information gathered by the photoreceptors. Some cells respond to stationary bars or edges; others specifically respond to moving bars or edges. As in the case of the retina, connections between adjacent cells—both directly and through interneurons—increase the contrast in the image.

Information from the lateral geniculate and from the primary visual cortex also flows to other parts of the brain. In each case, the pattern of electrical activity in the neurons represents a "mental image" that reflects—although in distorted form—the spatial organization of the visual image. Each mental image represents the world slightly differently, emphasizing a particular aspect such as vertical bars, corners, color, three-dimensional form, or movement. We interpret these patterns of nerve activity as a visual representation of the world, even as we realize the limitations of our own perceptions. Ultimately, the interneural connections within the retina and brain establish not only what we see but also how we respond to it.

The axons of the ganglion cells form the optic nerve, which carries visual information to the brain. Optic nerve fibers end in the lateral geniculate nucleus of the thalamus, which serves both as an integrator of visual information and as a relay station.

OTHER SENSORY PATHWAYS

Our other sense organs present our brains with reports on other aspects of our external and internal worlds. In each case the corresponding mental image consists of a pattern of neural activity. Depending on which brain cells are active, we interpret these chemical changes and ion flows differently—as sound or smell, touch or taste.

Perception refers to the interpretation of these neural events as the conscious experience of objects and events in the external world. This definition raises a basic philosophical question: just who is doing the interpreting? Some people have envisioned the sensory pathways as television or telephone cables that carry electrical representations of pictures and sounds to some inner observer. The hard thing to understand about the brain is that there is no such inner observer sitting at a switchboard or video monitor. The brain does not convert the electrical impulses back into "real" images. Mental images are electrochemical patterns. The marvel is our ability to interpret these patterns and to respond to them. A particular mental image may thus stimulate a person to run, to talk, or just to feel good.

Somatic Sensations

The visual system processes information only from the external world. In contrast, the somatic sensory system reports on information from three sources—from the external world, as it impinges on the body's surface; from the position of body

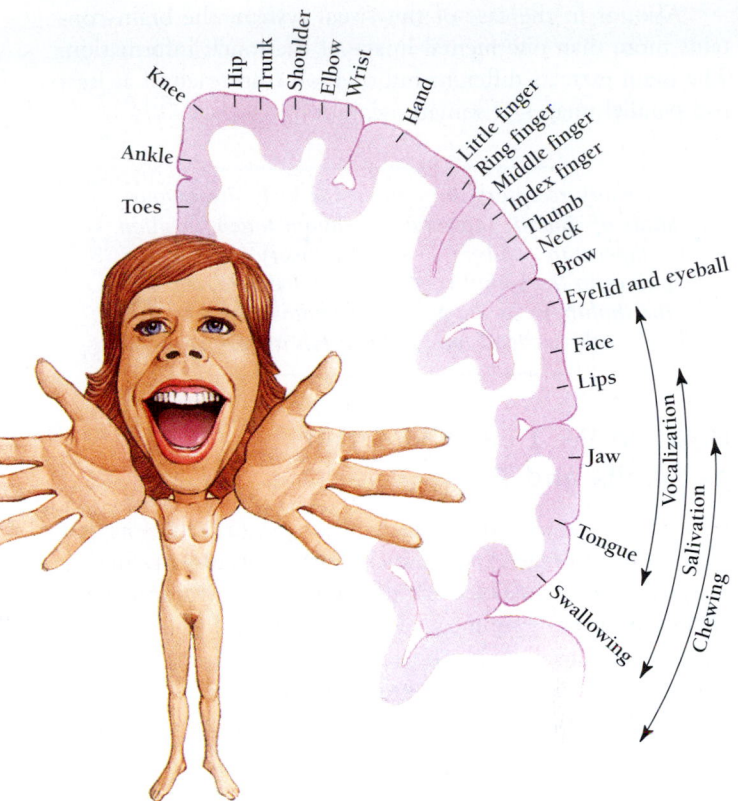

Figure 42-14 **Both sensory inputs from touch receptors to the cortex and motor outputs from the brain to individual body parts are highly organized.** The motor "homunculus," shown here, illustrates the relationships and the relative sizes of the areas devoted to each body part. Face and hands are particularly well represented compared, say, with the back and legs.

parts with respect to one another; and from the interior of the body. While the visual system responds to a single kind of stimulus, the somatic sensory system has four types of receptors—for touch (pressure), position, temperature, and pain.

Each kind of receptor converts information into electrical signals, which, like the signals of the visual system, travel to the central nervous system. For example, receptors in the skin transduce pressure into electrical activity. The receptors connect to neurons in the spinal cord or in the part of the brain called the medulla. The neurons in the spinal cord or medulla in turn connect to others in the thalamus, although in a different region from the lateral geniculate neurons that process visual information. Finally, the thalamus neurons connect to neurons of the cerebral cortex, in the region called the primary sensory area.

As in the visual system, the pathways of information roughly maintain the spatial organization of the receptors, so that the sensory cortex has a complete, although distorted, map of the whole body (Figure 42-14). Parallel pathways carry corresponding information from different kinds of receptors—for example, for touch or for position—to adjoining columns of cells within the sensory cortex.

Also, as in the case of the visual system, the brain contains more than one mental image of the touch information. The brain extracts different information from each of at least five parallel images of somatic sensation.

Sensory receptors throughout the body detect four kinds of somatic information—about touch, position, temperature, and pain. Somatosensory information passes to the spinal cord or to the medulla, through the thalamus, to the primary somatosensory area of the parietal lobe of the cerebral cortex.

How Do We Detect Molecules as Smells and Tastes?

Perception of smell and taste depends on sensory receptors that respond to molecules from the external environment. Similar receptors are responsible for determining the chemical status of the internal environment. These initial receptor molecules are proteins that bind to the small molecules responsible for smell or taste. Like the receptors for hormones and neurotransmitters, binding initiates a cascade of cellular events that ultimately result in a signal being sent to the brain (Chapters 36 and 41).

Taste sensations depend on taste buds on the tongue and on the lining of the top of the digestive tract—the mouth, pharynx, larynx, and upper esophagus. A single taste bud consists of about 50 cells. All are modified epithelial cells. Some cells respond to dissolved molecules, while some provide support (Figure 42-15).

Biologists classify tastes into just four qualities—sweet, sour, bitter, and salty. A single taste bud can detect all four qualities, but taste buds at different locations tend to be more sensitive to some tastes than others. The human tongue, for example, detects sweet and salt near the tip, bitterness at the back, and sourness at the sides. Information from the taste buds flows to the medulla. Like sensory information from the eyes, ears, and skin, the mouth's contributions funnel through a distinct part of the thalamus before reaching the cerebral cortex. The pattern of electrical activity in individual nerve fibers differs according to the presented taste, so that some fibers seem to carry more information about sweetness and others about saltiness.

Four tastes are not, of course, enough to convey the subtle flavors of food. What we usually call the "taste" of a food is actually both taste and smell. The aromas of chocolate, vanilla, garlic, and lemon all come to our brains by way of our noses, not our tongues. That is why a stuffy nose can impair our ability to taste what we are eating. Smelling, or olfaction, is far more sensitive than taste. Devoted wine tasters, for example, can distinguish more than 100 different flavors in wine alone. Altogether a well-trained human can distinguish about 10,000 odors. Yet this is nothing compared with the olfactory powers of dogs, which may have 200 million receptors to our 10 million. Dogs can distinguish far more odors and at much fainter concentrations.

Researchers now suspect that humans have from 100 to 1000 distinct types of odor receptors. Genes for about two dozen of these have been found.

The smell receptors lie in the olfactory epithelium of the nose. Smell receptors, unlike taste receptors, are true neurons that connect directly to the brain's **olfactory bulb,** which extends into the space next to the nose. Each receptor neuron contains a single type of olfactory receptor protein, which can bind a range of related molecules, thereby triggering neural activity. The pattern of electrochemical activity of individual receptor neurons in the olfactory bulb reflects the specific type

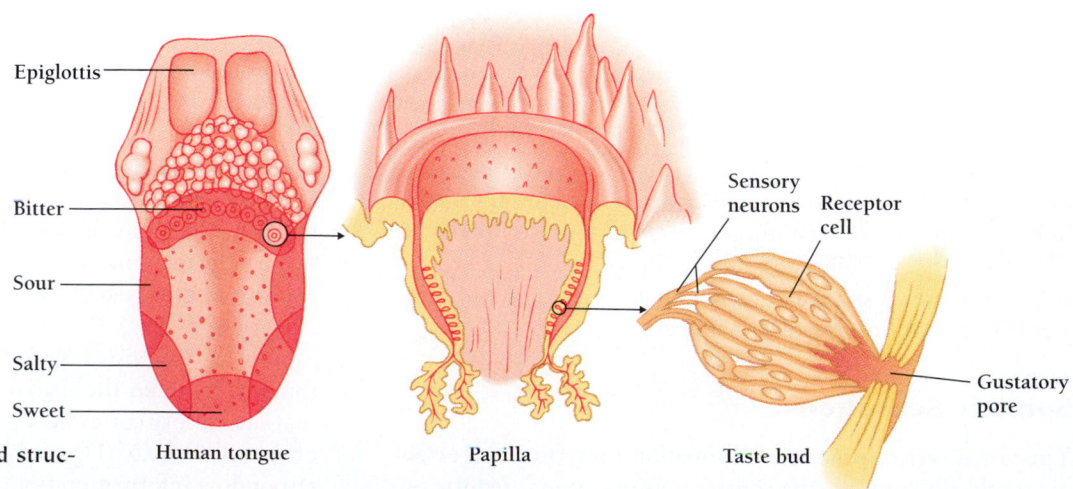

Figure 42-15 Distribution and structure of taste buds on the tongue.

Epiglottis

Bitter

Sour

Salty

Sweet

Human tongue

Papilla

Sensory neurons   Receptor cell

Gustatory pore

Taste bud

Figure 42-16   Olfactory receptors on the olfactory bulb.

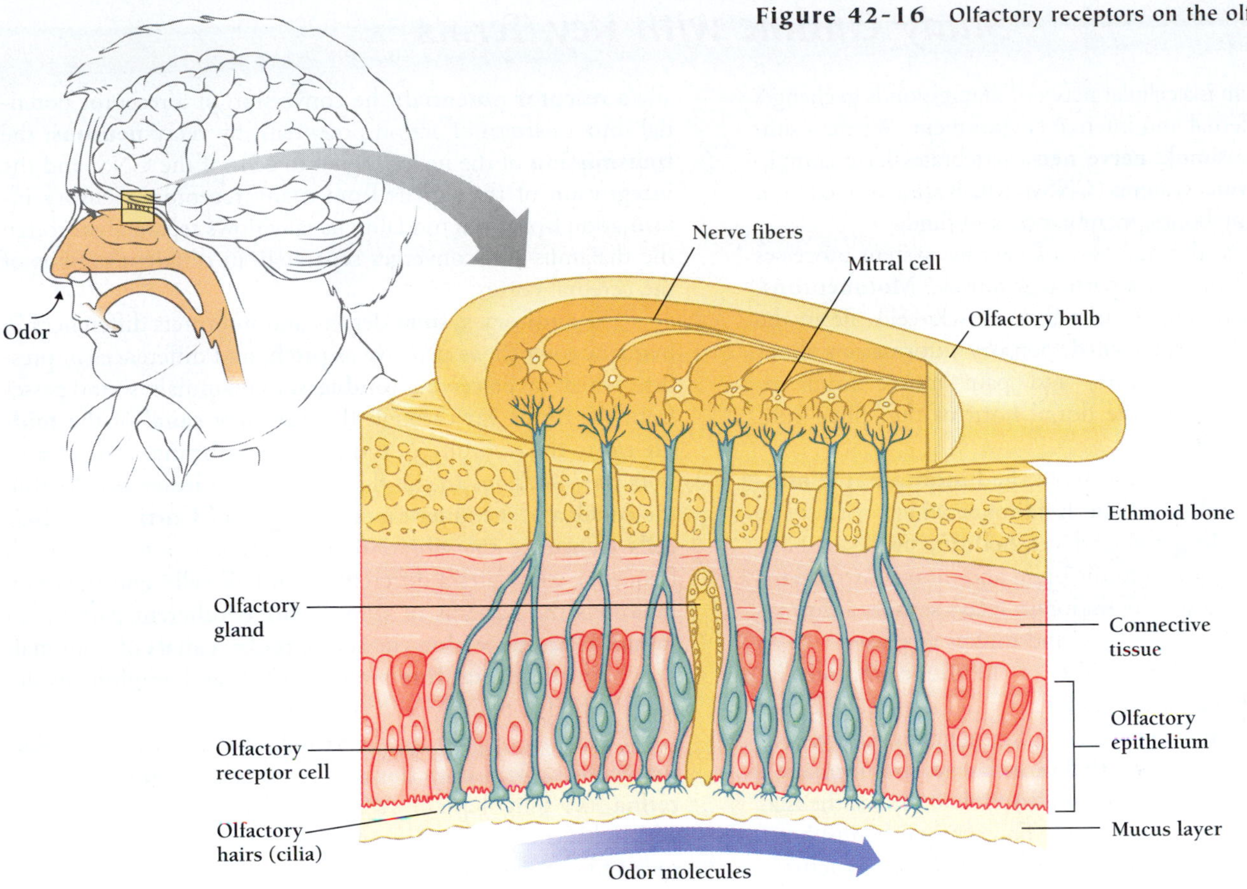

of odor (Figure 42-16). Unlike the relay stations for vision, hearing, and touch (all of which are in the thalamus), the olfactory bulb does not reflect the spatial organization of the outside world. As a result, our sense of smell provides no clues about the location of an odor.

Smell differs from the other senses in another way, as well. Electrical signals that represent smells do not reach the cerebral cortex directly, as do the signals for sound and sight. Instead, smell information flows to the parts of the brain that are most involved with emotions. These ancient regions of the brain—collectively called the **limbic system**—are thought to have developed early in vertebrate evolution and are sometimes half-jokingly called our "reptilian brain." The distinctive path that olfactory information takes suggests that odors have long played an important role in the responses of organisms to their environment.

Sir John Eccles, who won a Nobel prize for his studies of the synapse, once quipped, "The brain is so complicated that it staggers its own imagination." It seems unlikely that we will ever completely understand exactly how the brain accomplishes its uncounted activities. We cannot predict how much biology may reveal about the processing of sensory information, of the control of muscle movements, and of the modulation of mood, memory, personality, mind, and soul. One thing, however, is certain: intelligent, wondering women and men will continue to use these marvelous organs to ask questions about ourselves and our fellows creatures.

The primary event in taste and smell is the binding of a small molecule to a specific protein. This event triggers changes in nerve activity in the brain. Information from taste receptors passes to the medulla, through the thalamus, to the cerebral cortex. Olfactory receptors, however, themselves lie within a brain structure, the olfactory bulb, and the signals that result from olfaction do not directly reach the cerebral cortex.

Study Outline with Key Terms

The nervous system is a cellular network that responds to changes in an animal's external and internal environment. Whereas simpler animals have simple **nerve nets,** vertebrates have complicated **central nervous systems (CNSs),** which are protected from physical damage by bones, membranes, and fluid.

In many cases, the outcome of nervous system processes is a command to contract a particular muscle. **Motoneurons,** the nerve cells that directly stimulate muscle cells, lie in the ventral horn of the **spinal cord.** Sensory information about touch, position, temperature, and pain moves from somatosensory receptors to the dorsal horn of the spinal cord and then up to the brain.

The vertebrate brain consists of the **hindbrain,** the **midbrain,** and the **forebrain.** The hindbrain, which consists of **pons, medulla oblongata,** and **cerebellum,** processes both sensory information going to the brain and motor instructions coming from the brain. The midbrain relays sensory information, helps organize movements, and directs attention.

The forebrain, which consists of the **diencephalon** and **cerebral hemispheres,** is the site of activities that include sensation, perception, movement, learning, memory, and mood. Within the diencephalon, the **thalamus** integrates information about the external world, while the **hypothalamus** is the major regulator of the internal world. The cerebral hemispheres consist of the **cerebral cortex** and underlying structures, including the **hippocampus, amygdala,** and the **basal ganglia.** In humans, the cerebral cortex is highly folded, with many bumps **(gyri)** and grooves **(sulci).** The cortex receives sensory information and also processes that information, largely in the **association cortex,** which is intimately involved in **memory. Short-term memories** are less stable than **long-term memories,** whose formation depends on the workings of the hippocampus.

The **peripheral nervous system** consists of both **afferent** nerves, which carry sensory information to the CNS, and **efferent** nerves, which carry motor instructions from the CNS. Efferent nerves include both **somatic** nerves that carry instructions to the voluntary muscles and the **autonomic** nerves that carry instructions to internal organs. The autonomic nervous system contains both the **sympathetic** and the **parasympathetic** nerves.

Sensation refers to the detection of a stimulus in the external world, and **perception** refers to its conscious recognition and interpretation. Both sensation and perception depend on **sensory receptors** and on the central nervous system. The receptors for four of the five traditional sensory **modalities—audition** (hearing), **vision** (seeing), **gustation** (tasting), and **olfaction** (smelling)—are in specialized **sense organs.** The fifth sensory modality—touch, or **somatic sensation**—includes touch, position and motion, pain, and temperature and depends on receptors that lie throughout the body. A sixth modality is **vestibular** sensation (balance). Sensation depends on **transduction** of an external signal into a neural signal, usu-

ally a **receptor potential;** the conversion of a receptor potential into a pattern of action potentials, by **relay neurons;** the **transmission** of the neural signal to cells of the CNS; and the **integration** of the signals from many receptors. Sensory information from each modality usually flows separately through the thalamus and converges separately in different regions of the cerebral cortex.

The **auditory system** detects and interprets differences in frequency, which we perceive as **pitch,** and differences in pressure, which we perceive as **loudness.** In mammals, sound passes from the **outer ear** through the **auditory canal** of the **middle ear** to the eardrum, or **tympanic membrane,** which transmits the vibrations to the fluid within the **inner ear.** Within the **cochlea** of the inner ear is the **organ of Corti,** whose **hair cells transduce** the vibrations into receptor potentials. Each frequency stimulates a different set of hair cells, and the brain interprets the patterns of stimulation as different patterns of sound. Hair cells within the **semicircular canals** of mammals or within the **lateral line system** of fish and amphibians detect the body's position and motion.

Detection of light by the **visual system** starts with **photoreceptors,** light-detecting cells—**rods** and **cones**—in the **retina,** the thin sheet of cells at the back of the eye. Cones, which detect color, are concentrated in the central **fovea.** Both rods and cones contain **rhodopsins,** a light-absorbing pigment that consists of a protein **(opsin)** and **retinal.** Light enters the eye through the transparent **cornea** and passes through the **pupil,** whose opening depends on the action of muscles within the **iris.** The **lens** of the eye focuses images onto the retina. Misshapen lenses may result in **nearsightedness** or **farsightedness.** A cloudy spot on the lens is called a **cataract.** The ability to alter the shape of the lens to focus near or far is called **accommodation.**

The photoreceptor cells of the retina transmit information to the **bipolar cells,** and the bipolar cells transmit information to the **ganglion cells.** Connections among the photoreceptor cells, the bipolar cells, and **horizontal cells** and among the ganglion cells and **amacrine cells** are responsible for the integration of different aspects of the visual world. The axons of the ganglion cells form the **optic nerve,** which carries visual information to the brain. Optic nerve fibers end in the lateral geniculate nucleus of the thalamus, which serves both as an integrator of visual information and as a relay station.

The primary event in taste and smell is the binding of a small molecule to a specific protein. This event triggers changes in nerve activity in the brain. Information from taste receptors passes to the medulla, through the thalamus, and to the cerebral cortex. Olfactory receptors, however, themselves lie within a brain structure, the **olfactory bulb,** and the signals that result from olfaction do not directly reach the cerebral cortex. Olfactory information passes directly into the **limbic system,** the parts of the brain that are most involved with emotions.

Review and Thought Questions

Review Questions

1. In what way is a nervous system a homeostatic organ?
2. What kind of cell detects and transduces the signal received by the sense organs?
3. How do sense organs relay signals to the brain or the spinal cord?
4. How does the output of a sensory receptor differ from the output of a neuron?
5. How do relay neurons integrate signals?
6. How does olfaction differ from other senses?
7. Distinguish between afferent and efferent neurons.

8. Draw a picture of the cochlea of the inner ear, including the organ of Corti, and explain how this structure registers sound as an electrochemical signal.

Thought Question

9. Explain in what ways the brain is integrated (a single mind) and in what ways the brain is a group of separately operating units (a confederacy of minds).

About the Chapter-Opening Image

In the mid-19th century, scientists and the public began to realize that different parts of the human brain were responsible for different mental activities. Phrenologists imagined that this spatial specialization extended even into qualities of personality and spirituality. This idea, summarized in diagrams like this one, was soon abandoned by the scientific community.

 On-line materials relating to this chapter are on the World Wide Web at **http://www.harcourtcollege.com/lifesci/aal2/**

43 Sexual Reproduction

Why Did Sex Evolve?

Textbooks tend to say that the purpose of sex is to recombine genes in novel ways, thereby increasing genetic diversity in each new generation. In truth, however, no one knows why sex evolved, why it persists in the majority of multicelled organisms, and why some organisms nevertheless seem to do fine without it. In addition, biologists have asked themselves why most organisms have just two sexes. Why not 13 different sexes, as certain slime molds have? What is the purpose of the sex chromosomes and how have they evolved?

Nonscientists and scientists alike also wonder about some other aspects of sex. Why are men and women different from each other? Why are men bigger than women, while female hawks are bigger than their male mates? Which differences in behaviors and attitudes are "hardwired" during embryonic development? Which are the result of different levels of hormones such as estrogen and testosterone and might be changed with a few injections? Which differences are learned? What does it mean to learn to behave a certain way? Are women inherently better drivers? Are men inherently better navigators? If so, why? How do homosexual men and women become homosexual? What attributes do they share with one another or with members of the opposite sex? And why? Discussions of all these questions nearly always lead to heated arguments.

Amazingly, biologists cannot confidently answer a single one of these questions. Instead, they themselves continue to argue about the most fundamental aspects of sex, wrangling as fiercely as any group of nonscientists. Nonetheless, biologists have some fascinating insights into a few of these questions.

The idea that sex evolved to increase genetic diversity is not universally accepted. For one thing, such increased diversity is not necessarily a good thing. Most organisms live in environments similar to those of their parents. If their parents survived and reproduced successfully in that environment, why shake up the genome on the off chance that something better

will come up? Sex can erase adaptive traits as easily, probably a good deal more easily, than it can create them. If it ain't broke, why fix it?

One alternative theory comes from Richard E. Michod of the University of Arizona. Michod argues that sex evolved to repair damaged genes. Michod compares the genome to a vintage car that needs a steady supply of spare parts to keep it running. Any car collector knows that the best place to get parts is from another car of the same model. Chances are good that what is not broken or worn out in one car will be usable in the other. From two broken cars, one whole one can be reconstructed.

In the same way, organisms can reconstruct a damaged genome using spare DNA. If a single strand is damaged, the adjoining strand can be used as a template for repair. But what if both strands are damaged? In that case, the correct sequence must be obtained from somewhere else.

Michod's research showed that bacteria can survive DNA damage (from mutagens such as ultraviolet light or chemicals) by swapping their own DNA with that of dead bacteria of the same species. In fact, Michod has found that only bacteria with damaged DNA actively scavenge for spare DNA, while those with healthy DNA do not bother. Bacteria best at recombining were most likely to survive.

As multicellular eukaryotes evolved from bacteria, says Michod, they used the same kind of recombination when making sperm and eggs. Then when the egg and sperm join, the two chromosomes, one from each parent, line up, duplicate, and swap DNA. As a result, each chromosome is a mix of genes from both parents. Recombinations, adds Michod, are most likely to occur in gaps in the DNA, often a sign of damage.

Such an exchange costs an individual, however. Thanks to sex, each individual passes on only half of its genes to each of its offspring. A female, for example, who could reproduce asex-

Art Resource, New York

Key Concepts

- In mammals, male and female external genitalia develop from the same embryonic tissues, while internal genitalia develop from separate embryonic tissues.

- Oogenesis and spermatogenesis are parallel processes that produce haploid oocytes in females and sperm in males.

- The joining of a haploid sperm and a haploid egg join to form a diploid zygote is facilitated by sexual intercourse, a process that, in humans, is divided into four phases: excitement, plateau, orgasm, and resolution.

- A suite of related steroid hormones regulate the production of ova and the maintenance of pregnancy and lactation in females, as well as the production of sperm in males. The anterior pituitary gland makes and releases hormones that control the secretion of these steroid hormones in both sexes.

ually could pass on all of her genes to every offspring, doubling her genetic representation in the next generation. Some females do reproduce this way. Plants often reproduce asexually, but so do whiptail lizards, aphids, and a variety of other organisms.

But offspring inheriting both sets of chromosomes from one parent run the risk of getting a double dose of damaged DNA. Sex keeps harmful recessive mutations masked, says Michod. Still, he does not explain why Chihuahua whiptail lizards, all of which are female, manage so well without spare parts from males (Figure 43-1). Michod's ideas are well known, but they are not yet accepted.

Figure 43-1 Chihuahua whiptail lizard. This unusual animal reproduces by parthenogenesis, the process by which eggs become activated and develop without any paternal genetic contribution. All specimens are females, and males are unknown. (*M.P. Kahl/Photo Researchers, Inc.*)

HOW DO MAMMALS FORM GAMETES?

Few subjects occupy more attention, energy, and wonder than sexual reproduction. For species such as the mayfly, adult life consists only of mating and dying, sometimes without even the diversion of feeding. Like annual plants, such animals die before their young even appear. Humans and other mammals have a longer and more complex adult life, with considerable energy devoted to the care of the next generation.

In mammals, successful reproduction involves many steps—production of eggs or sperm, mating and fertilization, nurturing the embryo within the uterus, delivery, nursing and cleaning, and continued care even after nursing has stopped. In both males and females, many organs contribute to these processes. All are influenced by hormones, organic compounds produced in one tissue or organ that produce specific effects in other tissues or organs. Several of the hormones that coordinate reproductive processes also regulate the development of the reproductive structures and behaviors in fetuses and, later, in adolescents.

In this chapter we examine the specific tissues and organs that allow sexual reproduction, focusing on human reproduc-

tion. We begin with the **gonads,** the paired organs where the gametes form and mature. Humans and nearly all other sexually reproducing organisms form just two kinds of gametes, eggs and sperm. Eggs, or **ova** [singular, *ovum;* Latin, = egg], form in the **ovaries.** Sperm form in the **testes** [singular, *testis;* Latin]. In humans, the testes are called **testicles.**

Both eggs and sperm are haploid: they have just one set of chromosomes each. When an egg and a sperm come together at fertilization, or **syngamy,** the nuclei of the two gametes fuse and the resulting **zygote** is diploid: that is, it has two sets of chromosomes. In humans, fertilization, syngamy, and **conception** all refer to the same event.

If all goes well, the zygote begins to divide and makes its way down the **oviducts** [Latin, *ovum* = egg + *ductus* = duct], two long tubes to the **uterus** [Latin, = womb]. The multicelled embryo implants itself in the uterus (Figure 43-2). When the embryo has developed the basic features of the organism it is to become, the bare outlines of all its organs, limbs, and so on, it is a **fetus.** In humans, the embryo is called a fetus 9 weeks after fertilization.

How Do Female Mammals Produce Ova?

The female sex organs, or ovaries, are solid, almond-shaped organs, about 4 cm long. The ovaries lie on the side walls of the lower abdominal cavity, or pelvis, one on each side, where they produce ova, or egg cells, and also sex hormones. Much of the process of **oogenesis,** the production of eggs, begins before birth.

In both females and males, mature germ cells (eggs or sperm) are descendants of **primordial germ cells,** the cells that represent the germ line in the early embryo. During the development of a female fetus, the primordial germ cells migrate to the outer surfaces of the ovaries. These cells divide thousands

884

Figure 43-2 The female reproductive tract in humans. Two ovaries supply oocytes to the uterus. Oocytes (eggs) from the ovaries pass into the fallopian tubes. Sperm from the male pass into the vagina, through the uterus, and into the fallopian tubes. In the fallopian tubes, sperm may fuse with the oocyte. The resulting zygote later passes into the uterus and develops.

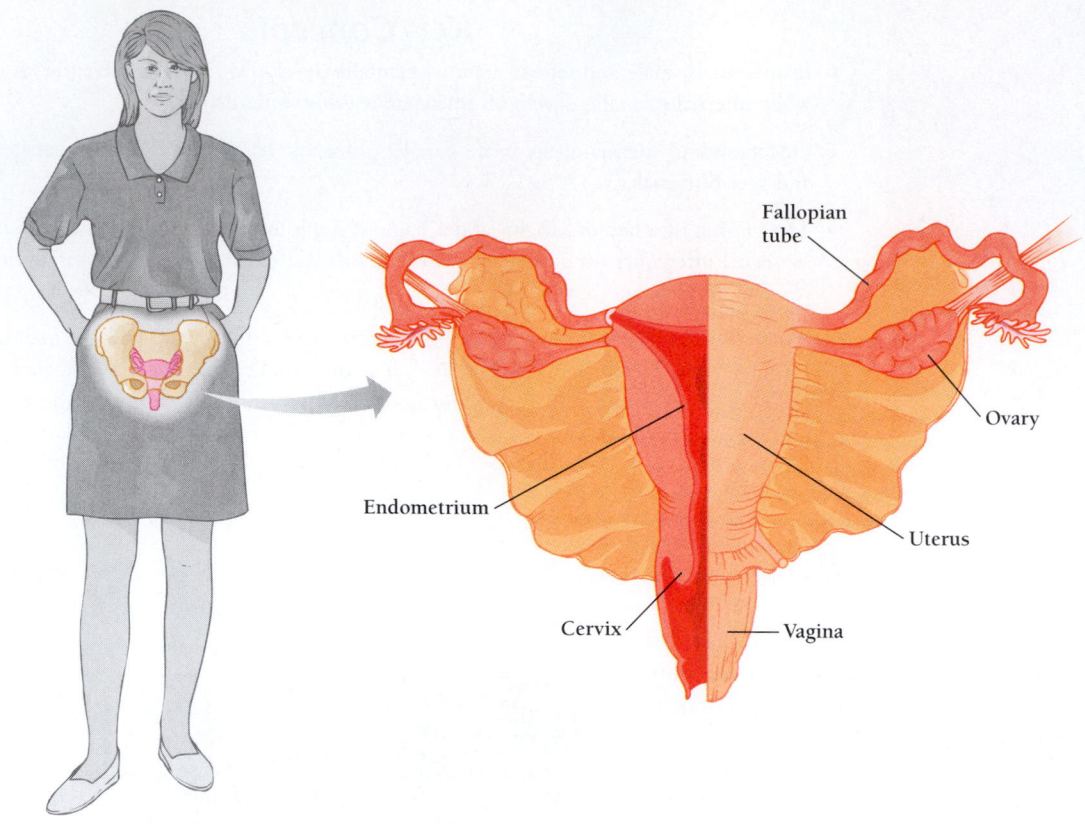

Fallopian tube

Ovary

Uterus

Vagina

Cervix

Endometrium

of times by mitosis to produce up to a million **oogonia,** diploid cells that, when they grow a little larger, are renamed **primary oocytes,** also diploid. The primary oocytes are each capable of dividing by meiosis to form a haploid ovum. They double their DNA and enter the first stage of meiosis (Figure 43-3). They will not complete meiosis, however, for 12 to 50 years.

A baby girl is born with an average of about 750,000 primary oocytes, each one partway through meiosis. During her childhood, many of these cells die, so that by the time she reaches puberty she has about 200,000 primary oocytes. Each primary oocyte, together with surrounding cells, called *granulosa* cells, forms a **follicle** [Latin, *folliculus* = small ball] (Figure 43-4). At puberty, the follicles begin developing, a few each month.

Several follicles fill with fluid each month, but usually only one of the several follicles in the two ovaries continues to ex-

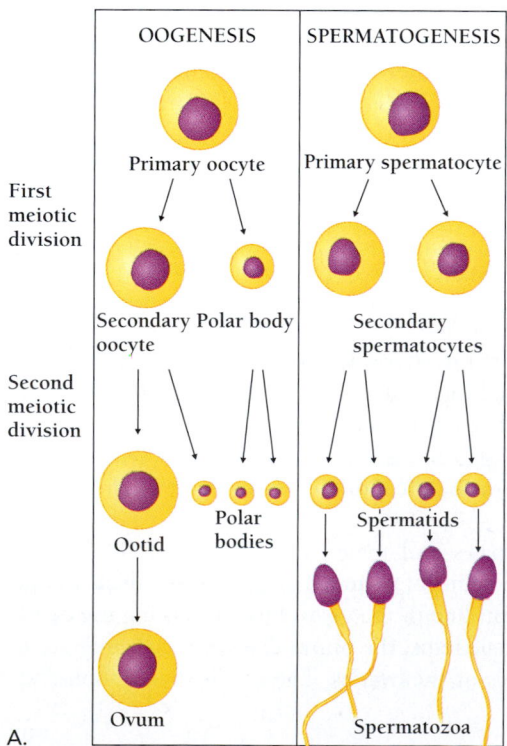

A.

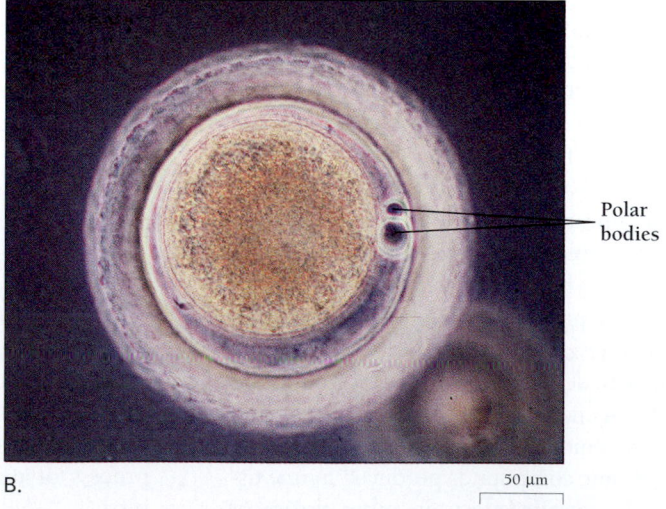

B.

Polar bodies

50 μm

Figure 43-3 **Oogenesis and spermatogenesis.** A. The diploid primary oocyte and spermatocyte divide by meiosis (which is two divisions) to produce four haploid cells—called ootids and spermatids. One of the ootids contains virtually all of the cytoplasm from the original oocyte and becomes the ovum, or egg. The other three cells, called polar bodies, fall away. In contrast, all of the spermatids mature into sperm, developing a head, a midsection, and a tail, and losing their cytoplasm. B. In this photo of a rabbit zygote, two polar bodies still cling to the surface at about 3 o'clock. (B, Biophoto Associates)

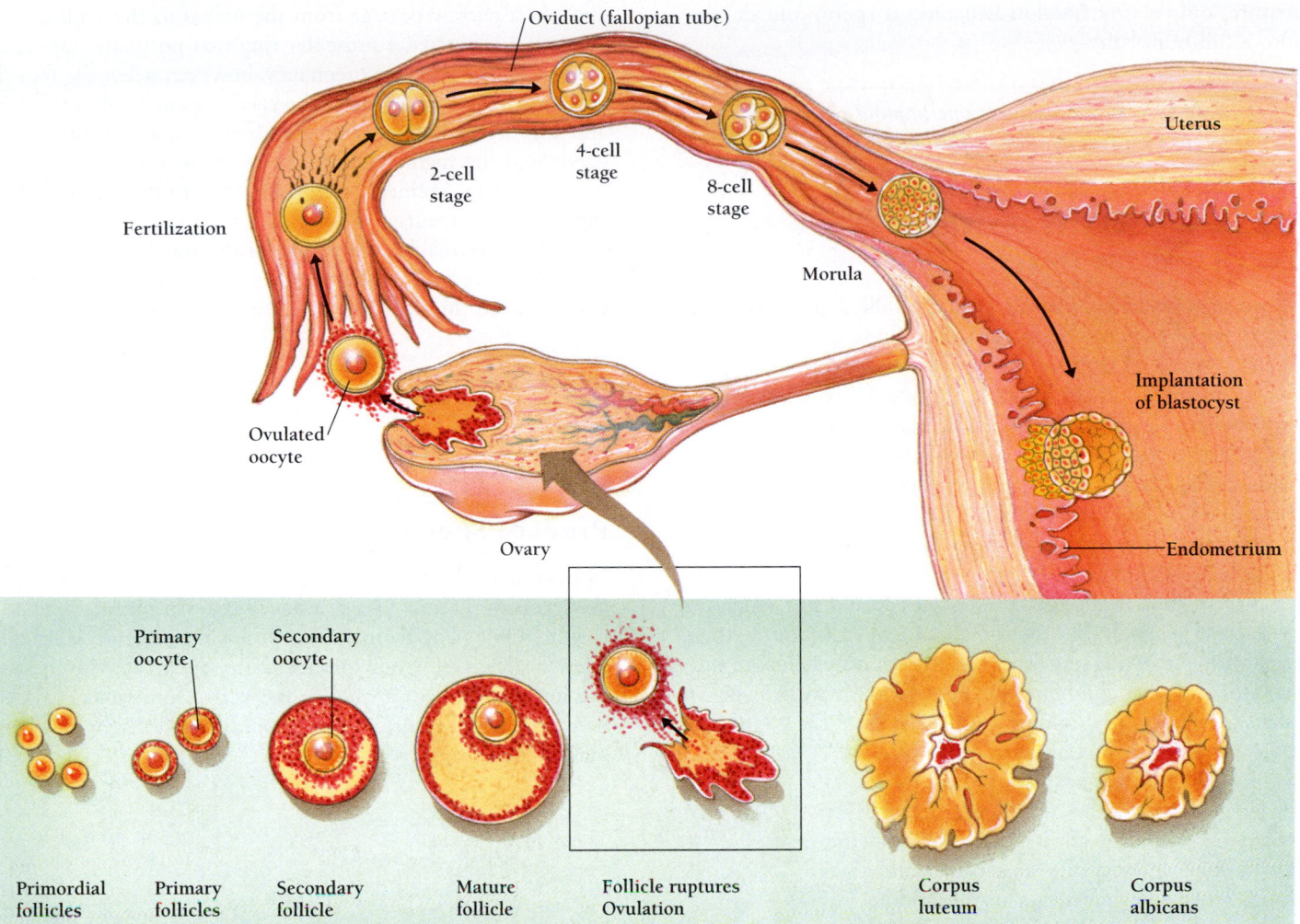

Oviduct (fallopian tube)

2-cell stage

4-cell stage

8-cell stage

Uterus

Fertilization

Morula

Implantation of blastocyst

Ovulated oocyte

Ovary

Endometrium

Primary oocyte

Secondary oocyte

| Primordial follicles | Primary follicles | Secondary follicle | Mature follicle | Follicle ruptures Ovulation | Corpus luteum | Corpus albicans |

Figure 43-4 **Development and maturation of primary oocytes and follicles.** Inside the ovary, the primordial follicles (*bottom left*) enlarge to form secondary follicles. While inside the follicles, the primary oocytes mature into secondary oocytes. After the follicle ruptures (ovulation), the secondary oocyte migrates into the fallopian tubes. If the oocyte fuses with a sperm, it finishes meiosis II and divides to form a mature oocyte and a third polar body. The zygote begins dividing. After about a week, the blastocyst embeds itself in the wall of the uterus. In the ovary, the ruptured follicle meanwhile matures into a corpus luteum, then corpus albicans.

pand. The single maturing follicle expands so much, however, that it occupies as much as an eighth of the ovary's volume and bulges from the side of the ovary (Figure 43-4). At **ovulation,** the follicle bursts open, releasing the oocyte.

Before ovulation, the primary oocyte completes meiosis I, producing two haploid cells—a large **secondary oocyte** and a small **polar body.** Polar bodies are tiny, spermlike cells that contain chromosomes but very little cytoplasm. Nearly all of the cytoplasm is reserved by the oocyte. The polar bodies are a mechanism for reducing the number of chromosomes, and they do not participate in fertilization. The polar body may divide, but it cannot develop into an ovum.

Just before ovulation, the oocyte (secondary) enters meiosis II. By the end of meiosis II, the oocyte will divide again to form the ovum and another polar body, but in humans this does not happen until after it unites with a sperm. The polar

bodies may remain attached to the ovum until the ovum fuses with a sperm (Figure 43-3B).

The ovum itself contains a single set of chromosomes tightly packed into its nucleus and a large amount of cytoplasm, including all of the organelles and structures that most cells contain. In addition to mitochondria, endoplasmic reticulum, Golgi complex, and microtubules, for example, the egg also contains an abundance of high-fat yolk particles, which supply energy to the developing embryo.

Over a woman's lifetime, approximately one primary oocyte will mature into an ovum every 28 days or so. Of the 200,000 oocytes initially present in the ovaries, then, only 400 or 500 will complete meiosis between the onset of ovulation, at about age 13, and the end of ovulation, at around age 50. All the rest of the oocytes degenerate, and women older than 50 have few if any oocytes. Of the 400 or so oocytes that mature, only a tiny fraction will meet a sperm and develop into a full-term baby.

A diploid oocyte produces one haploid ovum and two or three polar bodies.

Where Do Oocytes Go After They Leave the Ovary?

When the oocyte has erupted from the follicle in the ovary, it enters the abdominal cavity and then finds its way into the opening of one of the two oviducts that carry the egg into the uterus (Figure 43-4). In humans, the oviduct is called the **fallopian tube.** The end of each oviduct flares and partly encloses the adjacent ovary. Hairlike cilia around the opening of the oviduct draw fluid into the oviduct, sweeping the oocyte into the tube. The fallopian tubes do not fully enclose the ovary, however, and the oocyte often floats free in the abdominal cavity, at least for a time.

One might think that the oocytes would not enter the oviduct very reliably, but they do about 99 percent of the time. Even women missing one ovary and the opposite fallopian tube can still conceive children with ease. For this to happen, an oocyte from the functioning ovary must cross the abdominal cavity to the opposite fallopian tube.

Inside the fallopian tube, more cilia propel the oocyte toward the uterus, the chamber in which—if the oocyte meets a sperm—the oocyte may develop into an embryo and then into a fetus. The uterus lies in the middle of the pelvis, above and behind the bladder. Both oviducts open into the uterus's hollow center, the womb (Figure 43-2).

The uterus is about the size and shape of an upside-down pear. It has thick muscular walls, bordered on the inside surface with a specialized lining called the **endometrium** [Greek, *endon* = within + *metro* = mother].   A rich supply of blood in the endometrium carries nutrients to the early embryo.

The narrower, lower part of the uterus is called the **cervix** [Latin, = neck], which encloses a narrow passage from the uterus to the vagina. The cervix is a sphincter, a muscular ring that normally stays contracted. At the end of a pregnancy, however, when the fetus is **full term,** ready to be born, the cervix opens to about 10 cm and the uterus contracts to push the fetus out through the cervix and the vagina and into the wide world.

The vagina connects the genital tract to the outside. During sexual intercourse, the vagina encloses the penis and receives the ejaculated sperm. At birth, the baby passes out through the vagina, which stretches amazingly but may sometimes tear during this violent process.

The secondary oocyte moves from the ovary, through the oviduct, into the uterus.

How Do Male Mammals Produce Sperm?

In mammals, the male gonads, or testes, are contained within the **scrotum** [Latin, = bag], a pouch that lies outside the body (Figure 43-5). The temperature within the scrotum is usually a few degrees below body temperature. Sperm develop best at the lower temperature outside the body. Sometimes, in fact, hot baths or tight clothes can temporarily interfere with sperm maturation (although not reliably).

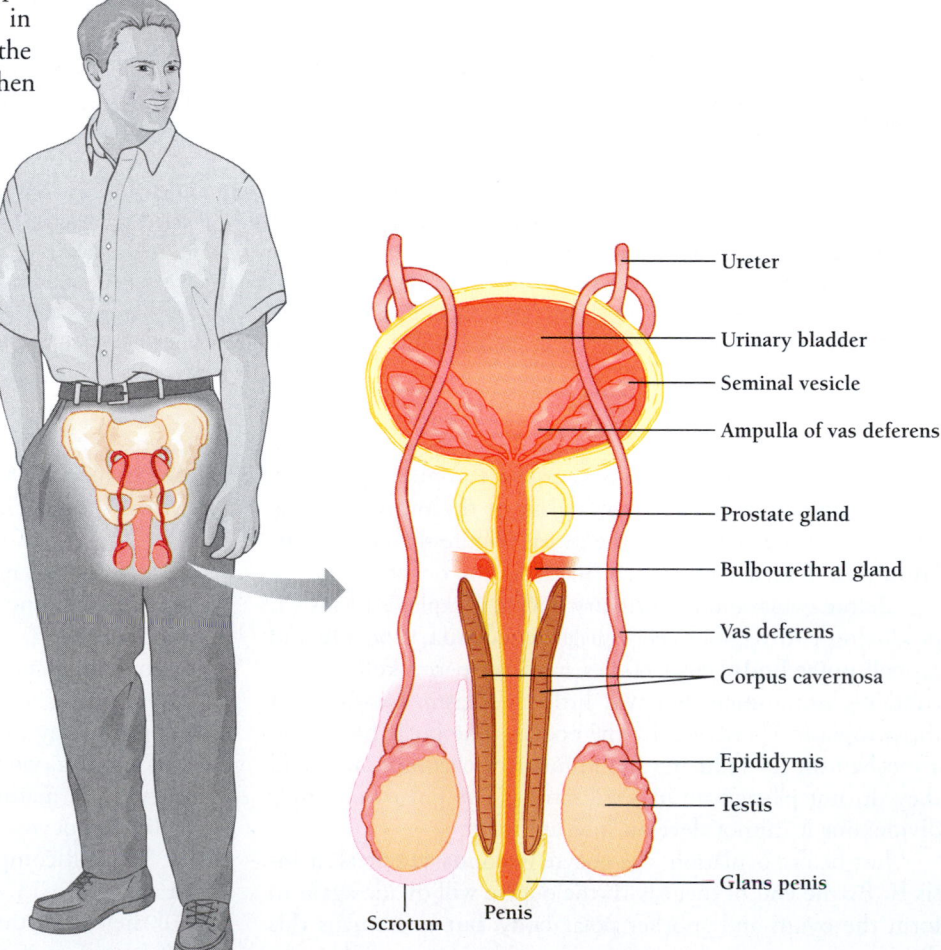

Figure 43-5   The male reproductive tract. Sperm form and mature in the testis, then pass into the epididymis until ejaculation, when they pass into the vas deferens and the penis. Fluids from the paired seminal vesicles and bulbourethral glands, as well as the single prostate gland, activate the sperm and speed them on their way.

Ureter

Urinary bladder

Seminal vesicle

Ampulla of vas deferens

Prostate gland

Bulbourethral gland

Vas deferens

Corpus cavernosa

Epididymis

Testis

Glans penis

Scrotum   Penis

Each egg-shaped testis is 4 to 5 cm long, about the size of a golf ball, although not so heavy. A testis consists of hundreds of separate chambers filled with tightly coiled ducts, called **seminiferous tubules** [Latin, *semen* = seed + *ferre* = to bear]. The sperm mature in the seminiferous tubules (Figure 43-6). Each sperm cell, or **spermatozoan,** consists of a head, a midpiece, and a tail. Sperm cells are specialized for rapid movement. The head contains a dense nucleus, with a tightly packed haploid set of chromosomes. The **midpiece** contains a microtubule organizing center and dense concentrations of mitochondria—sources of ATP for the journey to the egg. Finally, the tail consists of a long flagellum, which, powered by the mitochondria, whips about and keeps the sperm cell moving (Figure 43-7).

Sperm, like ova, are descendants of primordial germ cells. In males, primordial germ cells undergo several mitotic divisions in the testes before beginning meiosis. Sperm derive from the dividing diploid cells called **spermatogonia** [Greek, *sperma* = sperm + *gonos* = offspring], which lie on the inner walls of the seminiferous tubules. When a spermatogonium divides through meiosis, the resulting haploid cells move into the lumen of the seminiferous tubule. The two meiotic divisions (meiosis I and II) produce four haploid cells, each of which is called a **spermatid** (Figure 43-3). These two divisions are anal-ogous to the meiotic divisions that produce the ootid (ovum) and its three polar bodies. The spermatids are like polar bodies. Each spermatid then undergoes further development to form a mature sperm with midsection and tail (Figure 43-7).

Each sperm becomes highly elongated and the nucleus becomes highly condensed, occupying only a fraction of the length of the mature sperm. Most of the length is taken up by a **flagellum,** which propels the sperm forward (Figure 43-7). The sperm also contains a special structure at its front—the **acrosome,** a lysosome whose enzymes will eventually allow the sperm to enter the egg.

The process of meiosis and development takes more than two and a half months. During this time, the developing sperm are surrounded by specialized cells, called **Sertoli cells,** which create a continuous lining between the outside of the tubule and its lumen. The Sertoli cells nourish the developing sperm, regulate the passage of nutrients from the blood, and secrete a fluid that fills the lumen (Figure 43-6). Among the seminiferous tubules is a matrix of **interstitial cells,** which synthesize the male sex hormone testosterone. The interstitial cells make up about 20 percent of the mass of the testes.

Spermatogonia undergo mitosis to form more spermatogonia, so the tissue continually renews itself and provides a continuous supply of sperm from puberty on. Each hour about 200 million sperm mature, far more than can be used. Each ejaculation releases only about 200 million sperm, so unused sperm are absorbed by other cells in the testes.

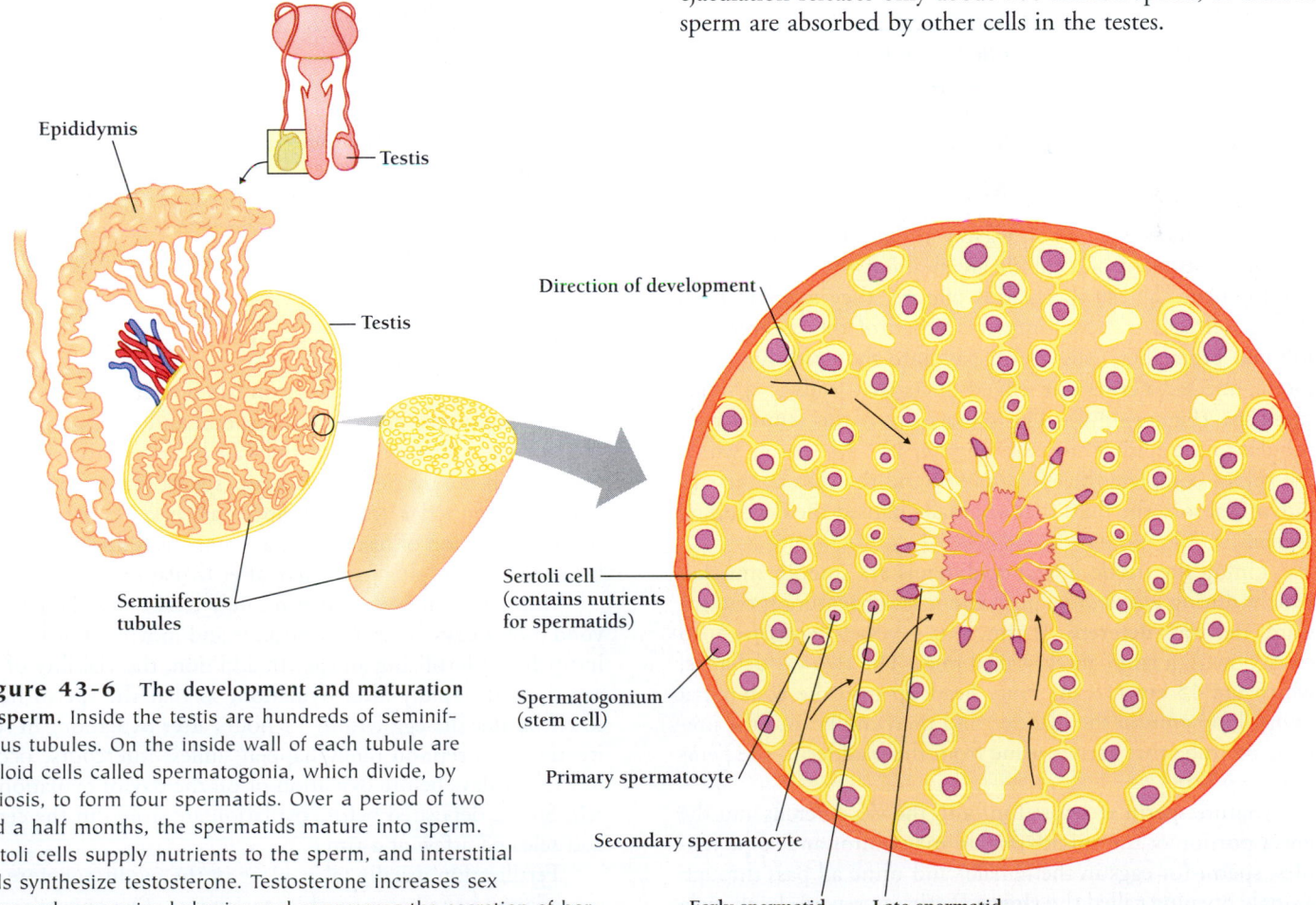

Figure 43-6 The development and maturation of sperm. Inside the testis are hundreds of seminiferous tubules. On the inside wall of each tubule are diploid cells called spermatogonia, which divide, by meiosis, to form four spermatids. Over a period of two and a half months, the spermatids mature into sperm. Sertoli cells supply nutrients to the sperm, and interstitial cells synthesize testosterone. Testosterone increases sex drive and aggressive behavior and suppresses the secretion of hormones that stimulate the interstitial cells to release testosterone.

Epididymis

Testis

Testis

Seminiferous tubules

Direction of development

Sertoli cell (contains nutrients for spermatids)

Spermatogonium (stem cell)

Primary spermatocyte

Secondary spermatocyte

Early spermatid   Late spermatid

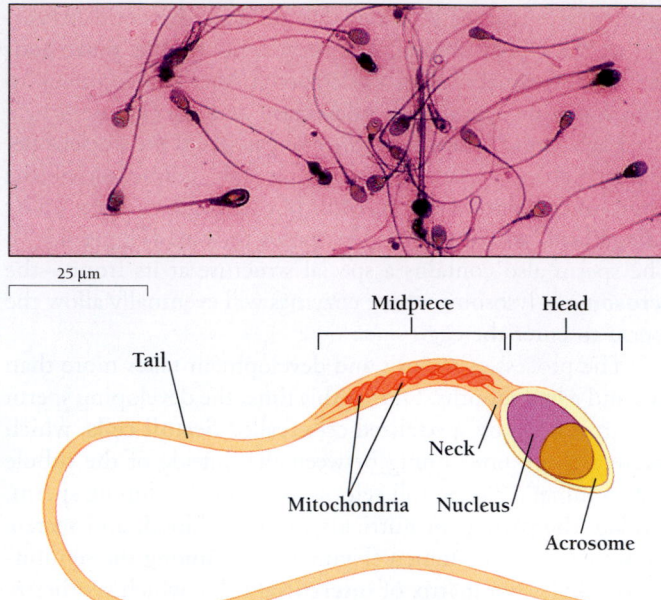

25 µm

Midpiece   Head

Tail

Neck

Mitochondria   Nucleus

Acrosome

Figure 43-7   Human sperm. A mature sperm, or spermatozoan, consists of a chromosome-packed head, a mitochondrion-rich midpiece, and a long flagellum, or tail. *(Photo, Ed Reschke/Peter Arnold, Inc.)*

Sperm cells form within the seminiferous tubules.

How Do Sperm Cells Travel from the Testes to the Penis?

Sperm maturation occurs in waves, so that all the spermatids in a given segment of a tubule are at about the same stage of development. As the spermatids mature, they separate from the Sertoli cells and pass into the lumen of the seminiferous tubules. From there, the sperm pass into the **epididymis,** an interconnected network of coiled ducts that leads to a single exit tube (Figure 43-6). The passage from the seminiferous tubules to the epididymis takes another 12 days, during which the sperm complete their maturation and the cells of the epididymis absorb the fluid that originally bathed the developing sperm.

Lining the inside of the epididymis are smooth muscles, which, during *ejaculation,* push the mass of mature sperm toward the **vas deferens** [Latin, *vas* = vessels + *deferre* = to carry down], a large, thick-walled duct. The two vas deferens, which are 45 cm (18 in) long, head up into the abdominal cavity, loop around the bladder, and then empty into the urethra, the tube that carries urine from the bladder to the penis (Figure 43-5).

Mature sperm are stored in both the vas deferens and the lower portion of the epididymis. In many birds and some reptiles, sperm (or eggs in the female) and urine all pass through a single opening called the **cloaca** [Latin, = sewer]. In all male

mammals except the monotremes, as well as in some birds, some reptiles, and a variety of invertebrates, a penis delivers sperm directly into the female genital tract during sexual intercourse.

Just before the vas deferens ducts empty into the urethra, they pass a junction with one of the two **seminal vesicles,** which secrete sugars and other nutrients around the sperm. The two vas deferens ducts then join as they pass through the **prostate gland,** which secretes a thin, milky alkaline fluid into the lumen of the urethra. Next, the **bulbourethral glands** inject a small amount of mucus into the mix (Figure 43-5). The sperm cells, together with the fluid from the seminal vesicles, the prostate gland, and the bulbourethral glands, make up the **semen** [Latin, = seed].

Sperm cells move from the seminiferous tubules, through the male genital tract, and out through the penis.

SEXUAL INTERCOURSE: HOW DO THE EGG AND SPERM RENDEZVOUS?

Fertilization in mammals (and many other animals) occurs within the body of the female. Reproduction therefore requires that the male deliver mature sperm into the female genital tract.

> Said an ovum one night to a sperm,
> "You're a very attractive young germ.
> Come join me, my sweet,
> Let our nuclei meet
> And in nine months we'll both come to term."
> —Isaac Asimov

When and Where Does Fertilization Occur?

For fertilization to occur, the secondary oocyte must encounter a sperm at a time when both the sperm and the oocyte are in their prime. Oocytes survive only 24 hours after they are released from the ovaries. In contrast, sperm can survive in the female reproductive tract for up to a week. As a result, the maximum period of fertility for a woman is 8 days, the 7 days before ovulation plus just 1 day after ovulation.

In practice, however, sperm rarely retain any viability beyond 3 or 4 days. They may be alive and moving, but they are incapable of fertilizing an egg. In addition, the viability of the egg declines rapidly after ovulation, so that the sperm needs to encounter the egg within 12 hours after ovulation. In reality, then, conception rarely happens unless intercourse occurs in the few days before ovulation or on the day of ovulation itself. Sperm deposited before ovulation are stored in the cervix and released a few at a time.

Fertilization usually takes place in the oviduct, before the secondary oocyte has arrived in the uterus. The oocyte moves

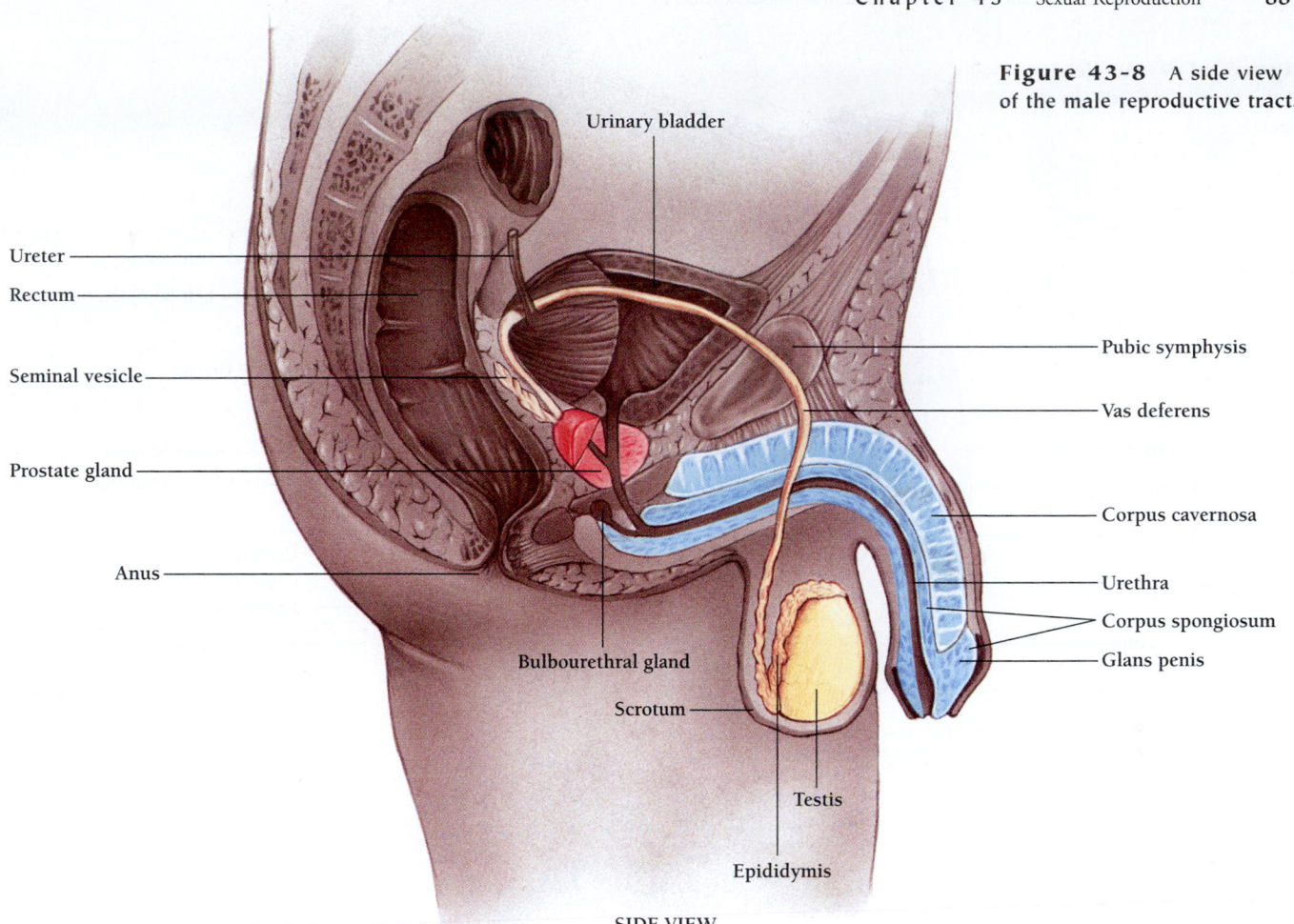

Figure 43-8   A side view of the male reproductive tract.

Urinary bladder

Ureter

Rectum

Seminal vesicle

Prostate gland

Anus

Bulbourethral gland

Scrotum

Testis

Epididymis

Pubic symphysis

Vas deferens

Corpus cavernosa

Urethra

Corpus spongiosum

Glans penis

SIDE VIEW

down the oviduct to meet the upward moving sperm. As we will see in Chapter 44, a single sperm fuses with the oocyte's cell membrane and ejects its haploid nucleus into the oocyte. Only then does the oocyte complete the second division of meiosis to form the mature ovum and a second (or third) polar body. The chromosomes of the egg and sperm then align, and the ovum becomes a zygote.

Usually, human sperm are viable for no more than 4 days after ejaculation, and eggs are viable for no more than 12 hours after ovulation. Consequently, conception rarely happens unless intercourse occurs four or fewer days before ovulation. Fertilization usually occurs in the oviduct.

How Do the External Genitalia Facilitate Sexual Intercourse?

The human penis contains a single exit tube, the **urethra,** which ends in a slitlike opening. Surrounding the urethra is a spongy cylindrical tissue, the **corpus spongiosum,** enlarged at each end (Figure 43-8). At the base of the penis, this tissue is attached to the body, below the pelvis. At the far end of the penis, the tissue forms a smooth cap, called the **glans.** The

glans is full of sensory nerves that make the penis especially sensitive to mechanical stimulation. Above this tissue lie two additional spongy cylinders, each called a **corpus cavernosum** (plural, **corpora cavernosa**), which run along the length of the penis. These cylinders fill with blood during erection, as we will discuss shortly.

Fibrous tissue surrounds and separates the three cylinders. Blood vessels and nerves enter the penis within the fibrous tissue. Around the outside of the penis is a loose layer of skin, which ends in a flap, called the **foreskin,** that folds over the glans. The foreskin is one of the most sensitive parts of the penis, being at least as well supplied with nerves as the glans. In many cultures, it is surgically removed in the first days of life. Such removal, called **circumcision,** remains a common practice.

The external genitalia of the female are collectively called the **vulva** [Latin, *volvere* = to wrap] (Figure 43-9). Two thin skin folds, called the **labia minora** [Latin, = small lips], surround the mouth of the vagina and the end of the urethra. These join at the front and form a hood over the **clitoris,** a diminutive penis. The clitoris includes two corpora cavernosa, a **bulb of the vestibule** that corresponds to the penis's corpus spongiosum and a glans that is especially sensitive to stimulation. On either side of the labia minora are two more skin folds, called the **labia majora** [Latin, = large lips], which

Figure 43-9 A side view of the female reproductive tract.

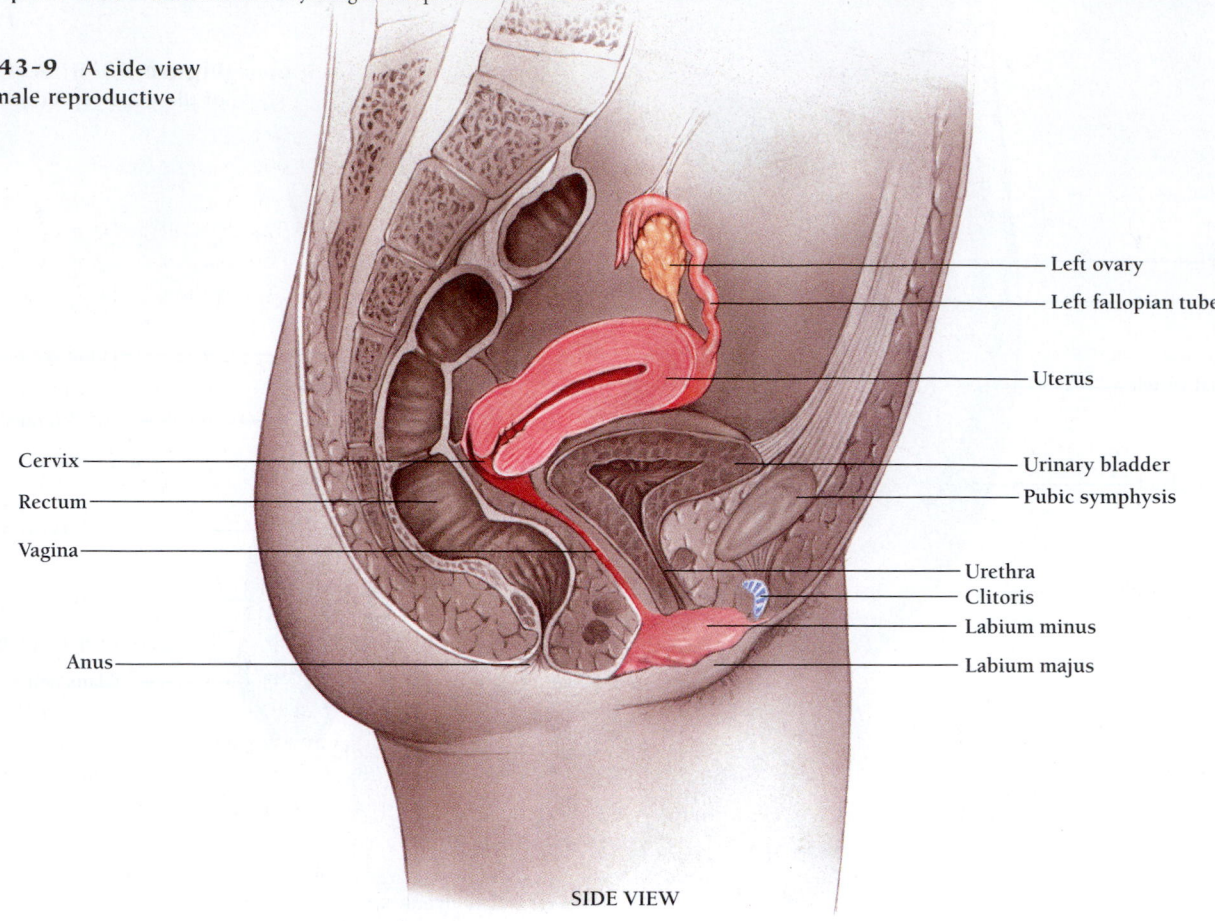

- Left ovary
- Left fallopian tube
- Uterus
- Urinary bladder
- Pubic symphysis
- Urethra
- Clitoris
- Labium minus
- Labium majus

Cervix
Rectum
Vagina
Anus

SIDE VIEW

enclose small amounts of fatty tissue. Bartholin's glands secrete mucus at the mouth of the vagina.

Just inside the mouth of the vagina is a thin membrane of irregular ragged shape, called the **hymen.** The hymen often impedes penetration by the penis, but only at first. The first sexual intercourse nearly always tears a hymen that closes the vagina, so an intact hymen was long regarded as the only proof of virginity. Other events can also break the hymen, however, including infection, a fall, or vigorous exercise.

How Are the Male and Female Genitalia Alike?

The external male genitalia are most simply described as a female genitalia that have grown together and fused at the midline, with the clitoris enlarged to include the urethra. Testes and ovaries develop from the same embryonic structures. The labia majora and scrotum develop from the same structures and are supplied with the same sets of nerves. The surface of the labia minora likewise corresponds to the ventral surface of the penis and is similarly sensitive (Figure 43-10).

The internal genital tracts of males and females develop from separate embryonic structures. All embryos develop two kinds of ducts, Wolffian ducts and Müllerian ducts. In the male, however, the vas deferens and its attachments develop from the paired Wolffian ducts. In the female, the oviduct, uterus, and vagina develop from the paired Müllerian ducts. The ducts not used by each sex degenerate.

Many parts of the reproductive system develop from the same embryonic precursors in males and females.

The Male and Female Sexual Responses Each Consist of Four Stages

Studies of the human sexual response reveal a stereotyped cycle of responses in both males and females. These responses involve changes in blood flow and in the contractions of smooth and skeletal muscle. Sexual physiologists divide the sexual response cycle into four phases: excitement, plateau, orgasm, and resolution.

The first phase, **excitement,** is marked by increased blood flow to the clitoris, the labia minora, and the breasts in the female and to the penis and testes in the male. In each sex, the affected tissues become engorged with blood, causing **erection** of both penis and clitoris.

Erection results directly from an increase of blood flow and may happen quickly, sometimes within 5 to 10 seconds. Blood fills the tiny spaces within the spongy tissues of the clitoris and penis, causing them to fill and assume a firm shape like that of a balloon filled with water or air. In females, erection simultaneously engorges the clitoris and tightens the tissues around the base of the vagina. During sexual intercourse, this tightening squeezes the penis.

The bulbourethral glands of the penis release a small amount of mucus, which lubricates the head of the penis and

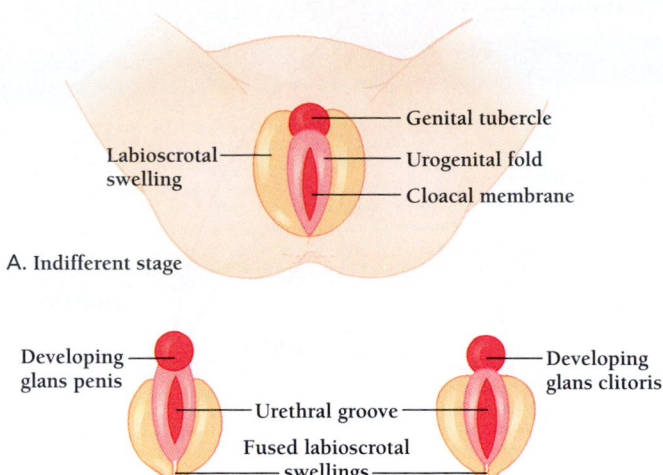

A. Indifferent stage

- Labioscrotal swelling
- Genital tubercle
- Urogenital fold
- Cloacal membrane

B. About 9 weeks development

- Developing glans penis
- Developing glans clitoris
- Urethral groove
- Fused labioscrotal swellings
- Anus

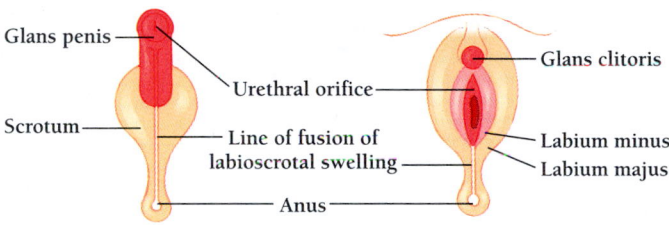

C. About 12 weeks development

- Glans penis
- Glans clitoris
- Urethral orifice
- Scrotum
- Line of fusion of labioscrotal swelling
- Labium minus
- Labium majus
- Anus

Figure 43-10   The common origin of male and female external genitalia. A. From 4 to 7 weeks' gestation, the genitals of male and female human embryos are indistinguishable. The "indifferent" genitals consist of a genital tubercle, a urogenital fold, and paired labioscrotal swellings on either side of the urogenital fold. B. By 9 weeks, the anus has formed and the labioscrotal swellings have begun to fuse along the midline at top and bottom. The structures of females and males begin to diverge, or differentiate. The labioscrotal swelling of males, for example, have fused more than those of females. C. By 12 weeks, the labioscrotal swellings have fused completely in males. In females, the genital tubercle has shrunk, and the labioscrotal folds have completed fusion at top and bottom and developed into the labia majora and labia minora.

eases entry into the now tightened entry of the vagina. Further lubrication comes from the female's **Bartholin's glands,** located beneath the labia minora and derived from the same structures as the bulbourethral glands in the male.

Erection in both females and males can occur as a result of physical stimulation of any part of the urogenital region, including both the sexual organs and the urinary tract. Stimulation of the glans, in particular, triggers a reflex response in which the blood vessels dilate.

Because nerve pathways from the brain control blood flow, visual stimuli, thoughts, or feelings can all cause erection and excitement even without physical stimulation. Just as the brain can stimulate erection and excitement, however, it can also inhibit excitement. Anxiety, fear, anger, or depression, for example, can all inhibit sexual response.

During the second, or **plateau,** phase, breathing and heart rate increase in response to continued stimulation. The end of the vagina dilates and the uterus pulls upward, creating a space near the cervix, where the semen can form a pool.

The third stage is **orgasm,** which refers to a whole complex of changes—smooth muscle contractions in the genital tract, skeletal muscle contractions throughout the body, and feelings of intense pleasure. Orgasm is typically brief, usually lasting 3 to 5 seconds.

In females orgasm increases the probability of conception, but it is not required. Conception can occur without orgasm. In males, however, full sperm delivery requires **ejaculation,** the propulsion of sperm out of the penis. During ejaculation, contractions of the genital ducts and muscles in the penis force sperm out of the male's genital tract into that of the female. Ejaculation consists of two phases: (1) smooth muscle contractions in the prostate, vas deferens, and seminal vesicles force semen into the urethra in the penis and (2) smooth muscle contractions in the urethra force the semen out of the urethra. Some physiologists refer to the first set of contractions as orgasm, and they reserve the term ejaculation for the second set of contractions. A human male typically releases about 3.5 ml of semen (a teaspoon), containing, on average, 200 million sperm (although these numbers are highly variable).

In the female, the smooth muscles in the uterus and the outer part of the vagina contract rhythmically and the cervix drops toward the pool of sperm. These contractions are, like those of the male, a reflex mediated by the spinal cord. Some research suggests that women who have orgasms less than a minute before the man, or up to 45 minutes after, retain most of the sperm in the reproductive tract, while those who have no orgasm, or have one more than a minute earlier, lose more sperm. It is also believed that orgasm-mediated contractions of the uterus and oviducts propel the sperm upward toward the descending ovum. In cows, the oviducts deliver sperm to the oocyte in 5 minutes, a rate of travel ten times as fast as that which the sperm can accomplish by beating their tails.

In the last phase, **resolution,** blood flow and muscle tension return to normal. In males, the penis ceases to be erect, and a second erection is not possible for a period of time that ranges from minutes to hours.

Sexual physiologists divide the sexual response cycle into four phases: excitement, plateau, orgasm, and resolution.

How Do the Sperm Reach the Egg and Penetrate Its Surface?

Once inside the vagina, the 200 million or so sperm will pile up against a mucous plug in the cervix, which keeps all but the healthiest, most active sperm out of the uterus. Those that cannot wiggle through the mucous plug end their days in the upper end of the vagina. The few hundred that make it through the cervix fan out into the uterus and up into the oviducts. In most instances of sexual intercourse, they find nary an oocyte. At the right time of the month, however, 400 to 500 sperm may arrive in one of the fallopian tubes and encounter a mature oocyte.

In many species of aquatic animals and other organisms the egg releases a chemical that attracts the sperm. Sea urchins, jellyfish, and seaweeds, for example, all use chemicals to attract sperm. Whether mammals use such mechanisms is unknown.

Even though the mammalian sperm may arrive near the oocyte within minutes of sexual intercourse, the sperm are incapable of fertilizing an egg until they have been in the reproductive tract for 5 or 6 hours. (Since the egg is only viable for 12 hours, intercourse must occur no later than 7 hours after ovulation for conception to result.) The head of each sperm is covered with a glycoprotein coat, which must be dissolved by enzymes in the female reproductive tract. Sperm not exposed to fluids from the female reproductive tract cannot fuse with an egg. After about 7 hours, the coating is dissolved, in a process called **capacitation.** The sperm become more motile and are now able to penetrate the outer layers of the egg.

Between the sperm's nucleus and the outer membrane is the acrosome, a lysosome packed with digestive enzymes. Under the influence of the female hormone progesterone, the outer membrane of the acrosome fuses with the membrane of the sperm head and releases the enzymes in the acrosome. These enzymes break down the outer layers of the egg, the corona radiata and the zona pellucida.

Once the sperm passes through the zona pellucida, this layer undergoes a reaction that makes it impenetrable to any other sperm. Fertilization by more than one sperm, or *polyspermy,* leads to an abnormal number of chromosomes, resulting in embryonic death or abnormal development. Eggs that are more than 12 hours old are unable to prevent polyspermy, and the resulting embryos soon die.

After enzymes in the female reproductive tract capacitate the sperm, it swims to the egg. Progesterone causes the sperm's acrosome to release enzymes, which break down the outer layers of the egg, the corona radiata and the zona pellucida. Once the sperm has passed through the zona pellucida, it swells to prevent the entry of other sperm.

HOW DO HORMONES CONTROL GAMETE PRODUCTION?

A collection of hormones regulate both the production of sperm in males and the production of ova in females. Both sexes use the same set of related hormones. What controls the production of the steroid hormones by the gonads? How does the body ensure, for example, that estrogen and progesterone are made in the right amounts and at the proper times?

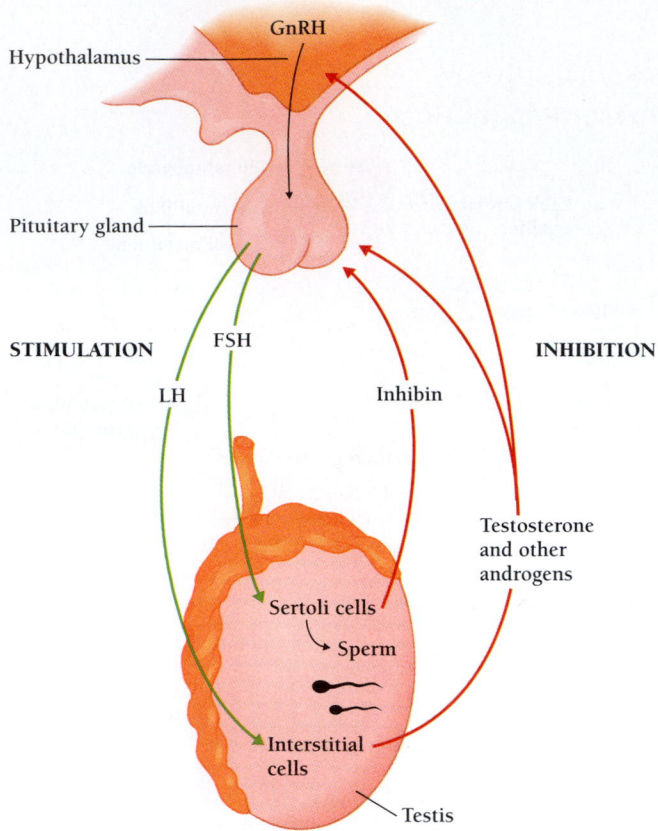

Figure 43-11   Hormonal regulation of sperm production in mammals. Here are just a few of the complex feedback systems that regulate male sperm production (and sexual behavior) in humans and other mammals. LH and FSH stimulate (*green arrows*) sperm production. Inhibin, testosterone, and other androgens inhibit (*red arrows*) sperm production.

The answer is that another endocrine organ, the **anterior pituitary gland,** makes and releases two hormones that together control the secretion of the sex steroids—estrogen and progesterone in females and testosterone in males. The pituitary gland, embedded in the base of the brain, consists of two lobes, the anterior and the posterior. The pituitary gland lies directly beneath the **hypothalamus** [Greek, *hypo* = under + *thalamos* = inner room], a part of the brain that regulates the expression of many hormone systems (Chapter 36).

The anterior pituitary fulfills all the definitions of an endocrine gland. Removal of the anterior pituitary affects several target tissues, including the gonads, which shut down the production not only of the sex steroids but also of a variety of other hormones. Extracts of the anterior pituitary can restore all these functions. This restoration, however, depends not on a single compound but on six distinct molecules, all of them small proteins or polypeptides.

Two of these polypeptides, called **gonadotropins,** are hormones that act on the gonads in both males and females. The names of the gonadotropins, **follicle-stimulating hormone (FSH)** and **luteinizing hormone (LH),** come from their major effects in females. In males the concentration of FSH and LH are relatively constant after puberty. In females, however, the concentrations of FSH and LH change in a monthly cycle.

How Do Hormones Control Sperm Production?

The testes secrete **testosterone,** the most potent of a class of androgens, or male hormones. Like the ovaries, however, the testes also secrete estrogen, about one-fifth as much as is secreted by an adult female. We do not know, however, if estrogen plays a central role in male reproduction.

Some researchers suspect that it does. Like testosterone, estrogen can stimulate sexual behavior in a variety of male mammals, including primates, even after individuals have been castrated. In addition, the male brain seems to convert testosterone to estrogen, and estrogen levels in the brain rise and fall in synchrony with blood testosterone levels.

The role of testosterone in mediating sperm production and sex drive is clear. Once testosterone is released into the blood it circulates for only 15 to 30 minutes before it is bound by tissue receptors or degraded. As a result, testosterone levels are constantly renewed and closely regulated.

Hormones regulate the production of both sperm and testosterone by means of negative feedback. In the brain, the hypothalamus secretes **gonadotropin-releasing hormone, GnRH,** which stimulates the pituitary gland, also in the brain, to produce two hormones that males share with females— luteinizing hormone (LH) and follicle-stimulating hormone (FSH). LH stimulates the interstitial cells of the testes to produce testosterone and other androgens, while FSH stimulates the Sertoli cells to produce more sperm (Figure 43-11).

But testosterone and other androgens released by the interstitial cells into the circulation inhibit the production of both GnRH in the hypothalamus and LH in the pituitary. When LH production increases, androgen levels increase, shutting down the production of LH and, therefore, of androgens. Similarly, the Sertoli cells secrete the hormone *inhibin,* which

enters the circulation and suppresses the secretion of FSH by the pituitary, shutting down sperm production.

Sperm production and testosterone production are thus intimately tied together. Not coincidentally, testosterone increases sex drive and aggressive behavior in general in both males and females (Figure 43-12). When sperm are ejaculated, the pituitary releases LH and FSH, which together stimulate renewed production of sperm and testosterone.

Human males do not begin substantial testosterone production until they are about 10 years old. By 13, they are usually sexually mature and testosterone production increases rapidly until about age 20, when it begins a steady decline (Figure 43-13). By the late forties or fifties, most men begin to experience a mild decrease in sexual function similar to that

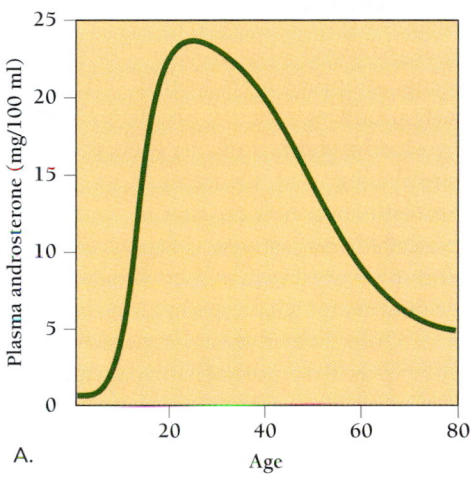

A.

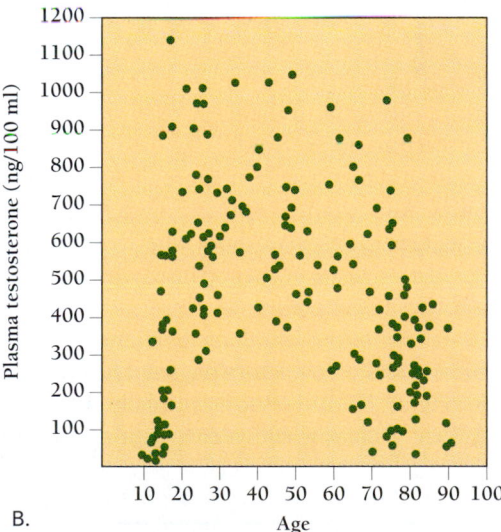

B.

Figure 43-13 Increase and decrease in testosterone as a function of age. A. The concentration of testosterone in the blood of human males increases dramatically at puberty, then gradually drops beginning at about age 20. B. The line in "A" looks smooth because the data are averaged together, but the actual data are scattered around. Men vary enormously in testosterone levels. A 90-year-old man can have the same testosterone levels as a 19-year-old.

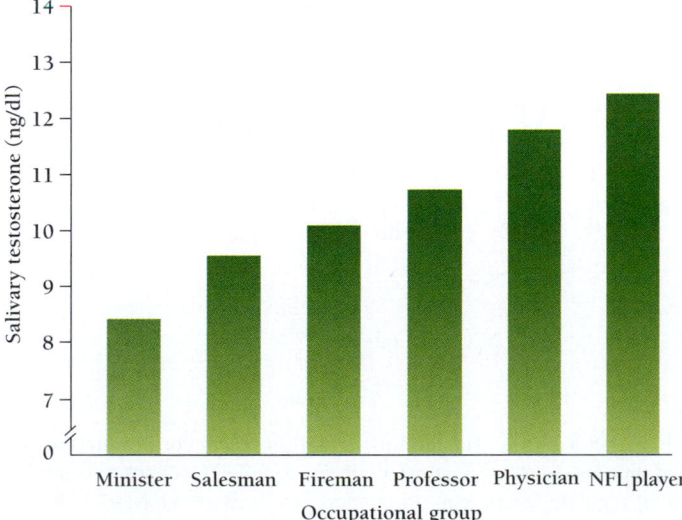

Figure 43-12 Testosterone hierarchy. Dominant and aggressive males tend to have higher testosterone levels than other males. The same effect occurs in females. (*Randy Nelson, Sinauer*)

which women experience during menopause. This decrease is more gradual than in women, but, in some cases, it can include the hot flashes that can characterize the transition into menopause. Treatment with either testosterone or estrogen can reduce such symptoms in men.

In men, the hypothalamus secretes GnRH, which stimulates the pituitary gland to produce LH and FSH. LH stimulates the production of testosterone and other androgens, while FSH stimulates the Sertoli cells in the testes to produce sperm.

What Are the Main Events of the Menstrual Cycle?

Beginning at puberty (between ages 10 and 17), human females undergo monthly reproductive cycles. Each cycle simultaneously produces a mature oocyte and prepares the wall of the uterus for **implantation,** the process in which a zygote burrows into the lining of the uterus. If a zygote does not implant, or implants but dies, the uterine lining is sloughed off in a process called *menstruation.* Each cycle is a highly coordinated series of events mediated by several hormones, many of the same ones that mediate sperm production in males. The cycle is divided into three phases: the menstrual phase, the preovulatory phase, and the postovulatory phase.

In 90 percent of healthy adult women, the entire menstrual cycle ranges from 21 to 38 days. The average is said to be about 28 days, sometimes called a *lunar month* because the phases of the moon also occur in a 28-day cycle. The menstrual cycles are *not,* however, coordinated with the phases of the moon.

Adolescent girls tend to have *irregular* cycles, which vary in length from one cycle to the next. As girls mature, their cycles tend to become more regular. Women who live together tend to ovulate and menstruate at approximately the same time of the month. And women who live with men are likely to have more regular cycles than those who live alone. Menstruation and ovulation cease at menopause, which usually occurs in the late forties or early fifties.

Most variation in the length of the menstrual cycle occurs in the menstrual and preovulatory phases. The menstrual phase can last as long as 12 days, but it is usually 3 to 7 days. The period between ovulation and menstruation is nearly always 14 days.

The menstrual cycle is divided into three phases, the menstrual phase, the preovulatory phase, and the postovulatory phase. The menstrual and preovulatory phases vary in length, but the postovulatory phase is nearly always 14 days.

How Do Hormones Regulate the Ovarian Cycle?

Hormones regulate the **ovarian cycle**—the production of the oocytes and the regulation of the state of the endometrium (Figure 43-14). As in the male, the hypothalamus secretes gonadotropin-releasing hormone, GnRH, which stimulates the pituitary to produce LH (luteinizing hormone) and FSH (follicle-stimulating hormone), which together stimulate the ovary to develop mature oocytes and to produce estrogen and other steroid hormones.

During the menstrual phase, an increase in FSH (from the pituitary) stimulates 5 to 12 follicles in the ovaries to grow and mature (Figure 43-15). FSH causes growth and development of the oocyte and, together with LH, causes the follicle cells to release increasing amounts of estrogen, which promotes the growth of the follicle. This positive feedback cycle causes a buildup of estrogen in the blood in the days preceding ovulation. All but one of the follicles degenerates, leaving just one (usually) to develop a mature oocyte. The remaining follicle continues to secrete estrogen (and also progesterone).

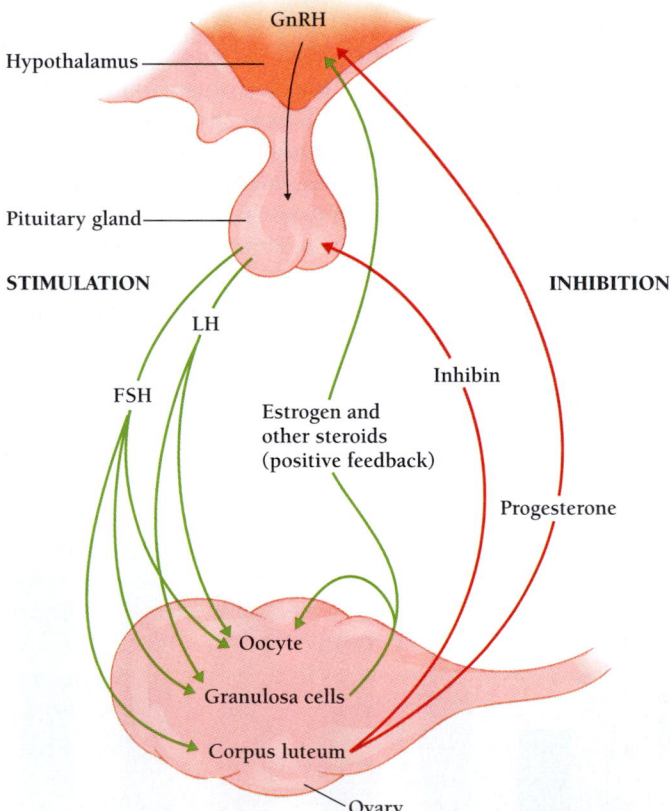

Figure 43-14   Hormonal regulation of oocyte production in mammals. Here are just a few of the complex feedback systems that regulate female egg production (and sexual behavior) in humans and other mammals. As in males, LH and FSH released by the pituitary stimulate (*green arrows*) oocyte production. Progesterone and inhibin inhibit (*red arrows*) the production of LH and FSH.

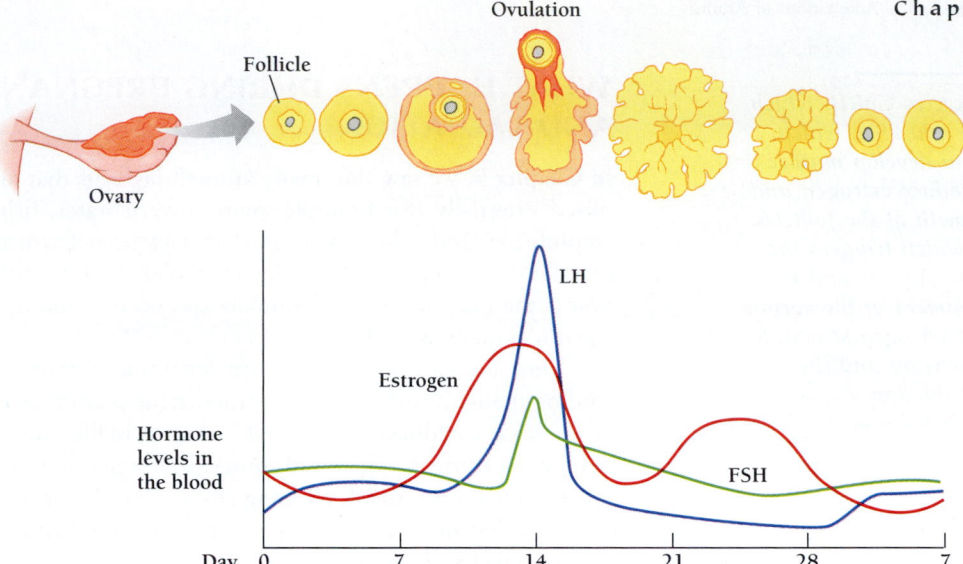

Figure 43-15 Hormonal regulation of the menstrual and ovarian cycle in humans. Estrogen, LH, and FSH dominate ovulation, while progesterone and inhibin dominate the luteal (premenstrual) phase.

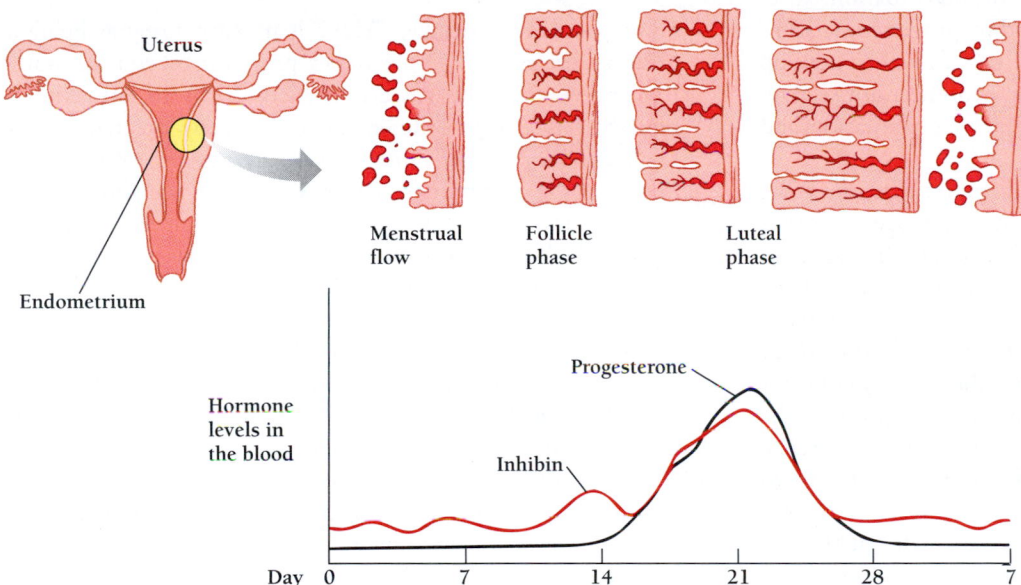

The increasing levels of estrogen trigger a second positive feedback cycle. Right around the time of ovulation, the high levels of estrogen stimulate the hypothalamus to release a surge of GnRH, which causes the release of a large surge of LH and FSH (Figures 43-14 and 43-15). The LH spike triggers ovulation. After the oocyte bursts from the follicle, the follicle collapses inside the ovary. There, stimulated by LH, it enlarges to form the **corpus luteum** [Latin, = yellow body], an important structure that secretes progesterone and a little estrogen. The progesterone stimulates the endometrium, the lining of the uterus, to thicken in preparation for implantation by the embryo (Chapter 44). The progesterone also signals the hypothalamus to stop producing GnRH. As a result, the pituitary stops producing LH and FSH. In addition, the corpus luteum produces inhibin, which also inhibits the pituitary.

If the egg is not fertilized, the corpus luteum degenerates in 8 to 10 days. Inhibin and progesterone levels drop, and the pituitary begins producing FSH and LH once more. If the egg *is* fertilized, the zygote produces *chorionic gonadotropin,* a hormone similar to LH that maintains the corpus luteum. The presence of chorionic gonadotropin, or **HCG (human chorionic gonadotropin),** in blood or urine is the basis of most pregnancy tests. During early pregnancy, the corpus luteum enlarges and produces even more progesterone until about the fourth month of pregnancy.

Estrogen, then, participates both in negative feedback and—once each cycle—in positive feedback. In contrast, progesterone only promotes negative feedback. Together these two feedback systems bring about the regular cycles of ovaries and uterus.

In women, the hypothalamus secretes GnRH, which stimulates the secretion of LH and FSH, which together stimulate the ovaries to develop mature oocytes (and follicles) and to produce estrogen and other steroid hormones. The growth of the follicles causes a buildup of estrogen, which triggers the release of more GnRH, and, therefore, LH and FSH. LH triggers ovulation and development of the corpus luteum, whose progesterone and estrogen stimulate the thickening of the endometrium and the cessation of GnRH production.

How Do Hormones Regulate the Menstrual Cycle?

The lining of the uterus undergoes changes that parallel the changes that occur in the ovaries. While the follicles are growing, the lining of the uterus, or endometrium, increases to two or three times its thickness at the end of menstruation. After ovulation, the corpus luteum forms and its progesterone stimulates the endometrium, the lining of the uterus, to develop further. Blood vessels increase, and the endometrium secretes glycogen and other materials that facilitate implantation and nourish the zygote.

Estrogen and progesterone have other effects besides those on the endometrium, however. For example, estrogen increases the activity of cilia in the oviduct, which propel the zygote to the uterus. In contrast, progesterone decreases cilial activity. Estrogen stimulates the cervix to secrete a clear, fluid mucus, which eases the passage of sperm into the uterus and oviduct. Progesterone, on the other hand, triggers the secretion of thick cervical mucus, which forms a mucous plug that prevents the entry of sperm or bacteria from the vagina into the uterus. Once conception has occurred, the plug protects the embryo from infection.

If no fertilization has occurred, however, the richly prepared lining of the uterus disintegrates. The blood vessels coil and constrict, and some 50 to 150 ml (about 3 to 10 tablespoons) of blood, mucus, vaginal secretions, and endometrial tissue washes out through the vagina. This bleeding, which may include bits of membranous tissue and blood clots, is called **menstruation** [Latin, *mens* = month]. Although light bleeding may occur after conception, several days of medium to heavy bleeding means that conception either has not occurred or it has occurred but the zygote has died.

After ovulation, the corpus luteum forms and its progesterone stimulates the endometrium, the lining of the uterus, to develop further. If no embryo implants, the endometrium breaks down and is washed away in a flow of blood, called menstruation.

WHAT HAPPENS DURING PREGNANCY AND LACTATION?

In Chapter 9, we saw that many animals lay eggs that are fertilized externally (for example, many invertebrates, fish, and amphibians) and others lay eggs that have been fertilized internally (birds and mammals, for example). Internal fertilization is the rule among land animals and occurs among many aquatic animals as well.

Most animals whose eggs are fertilized internally—including reptiles, birds, and even monotreme mammals (platypuses and echidnas)—lay "eggs" that actually contain a developing embryo, not a single unfertilized egg cell. But many other animals keep the developing embryo inside for up to 18 months (elephants) and then give birth. Giving birth to live young, called **viviparity,** occurs among sharks, snakes, and of course all marsupials and true mammals such as humans.

How Can a Woman Tell if She Is Pregnant?

Some people like to joke, "There is no such thing as being a little bit pregnant. You are either pregnant or you are not." But any woman who has ever carried a baby to term knows that there is a huge difference between early pregnancy and late pregnancy. In her ninth month, a pregnant woman is so huge that she may have difficulty walking up and down stairs, and the various discomforts make a good night's sleep nearly impossible. Yet, when she first becomes pregnant, the signs are so subtle that she may miss them all together. She is indeed only "a little bit" pregnant.

Pregnancy is the state in which a woman is carrying a fertilized egg (zygote), embryo, or fetus inside her body. It begins with the fusion of the two nuclei of the egg and the sperm. At that moment, two cells become one cell.

Confusingly, doctors say pregnancy begins 2 weeks *before* fertilization, at the beginning of the last menstrual period. Thus, a doctor will tell a woman carrying a 2-week-old embryo that she is "4 weeks" pregnant. A normal pregnancy lasts about 38 weeks, but doctors say that it is 40 weeks, because, by tradition, they count from the last menstrual cycle. In this book, we refer to the actual number of weeks, rather than adding the extra 2 weeks.

About 1 week after fertilization, the ball of cells that is the human embryo burrows into the lining of the mother's uterus ("implantation") and begins secreting human chorionic gonadotropin (HCG). HCG alters the physiology of the mother, causing the corpus luteum to double in size and produce enough progesterone to prevent menstruation.

The embryo quickly develops the same set of four membranes found in all "amniotes," animals that have amniotic eggs (Chapter 24). The amniotic eggs of reptiles, birds, and mammals contain an amnion, yolk sac, allantois, and chorion, which we discuss in Chapter 44. In addition, mammals develop a **placenta,** a flat organ rich in blood vessels that conduct nutrients from the mother's blood to the embryo and wastes from the embryo to the mother. Connecting the embryo to the placenta is a thick "umbilical cord."

As the blastocyst implants and begins releasing hormones, many women notice nothing at all. Others, especially those who have been pregnant before, may recognize the subtle signs immediately. At 1 week, the woman may feel a slight cramping from implantation. Not long after, other changes may begin, including sleepiness, dizziness, nausea, breast tenderness, bloating, and frequent urination.

The most obvious early sign of pregnancy, however, is the absence of a menstrual period. In the absence of pregnancy, menstruation occurs 2 weeks after ovulation. When a period is about a week late, most women immediately guess that they may be pregnant. To be sure, they may buy a home pregnancy test from a pharmacy or have a test done by their doctor.

But menstrual cycles can be irregular in teenagers, whose hormone cycles are not yet fully established. Irregular periods also occur in strenuous dieters, for low body fat can interfere with normal hormone cycles in women. In both cases, a young woman with subtle signs may not realize she is pregnant for months, particularly if pregnancy is not on her mind.

Modern pregnancy tests are extremely sensitive. A pregnancy test is a chemical test for the presence of HCG in either urine or blood. Lab tests can detect HCG in the blood as early as 8 days after conception, and even inexpensive drugstore versions can detect it in the urine as early as 14 days after conception. These tests are highly reliable, but not perfect. Occasionally, a pregnancy test can indicate that a woman is not pregnant when she is.

Early tests can lead to disappointment or needless worry. This is because many pregnancies end in **spontaneous abortions,** or miscarriages. A large fraction of such embryos die because of chromosomal abnormalities in the first week or two, usually before a woman even knows she is pregnant. A woman who is not eager for a baby may suffer needless anxiety, while a woman whose drugstore test tells her to anticipate a healthy baby soon may be disappointed if she later finds she is no longer pregnant. Fortunately, most spontaneous abortions are eventually followed by a successful pregnancy.

Secretion of HCG by the embryo after 1 week causes the corpus luteum to enlarge and secrete progesterone and estrogen, prevents menstruation, and makes early pregnancy tests possible.

What Happens Between Conception and Delivery?

Physicians divided the 38 weeks of pregnancy into three "trimesters," of about 3 months each. In the first trimester, a woman probably won't "show" much, but may gain a few pounds. Increasing levels of HCG stimulate the corpus luteum to secrete increasing amounts of estrogen and progesterone. Estrogen causes increases in the size of the breasts and the uterus, and sometimes intense nausea (morning sickness) and fatigue. By the end of 9 months, the breasts will be about 2 pounds heavier and the uterus will be 22 times its usual size. Progesterone prevents the uterus from contracting and, along with estrogen, stimulates the development of milk-producing cells in the breasts.

The fetal placenta also produces a **human chorionic somatomammotropin (HCS)** , which reduces the woman's sensitivity to insulin. As a result, glucose is redirected from the mother to the fetus, which needs large amounts of this sugar for growth. In compensation, HCS increases the metabolism of fat in the mother's body, so that she has access to energy-rich fatty acids instead of glucose.

In the second trimester, the corpus luteum degenerates and HCG levels decline. But the fetal placenta produces even more estrogen and progesterone, which sustains the pregnancy (Figure 43-16). The embryo becomes a fully formed fetus, and the woman feels its first kicks at about 4 months. Traditionally, this stage is called "quickening." The mother's abdomen swells noticeably, but her energy levels increase. Morning sickness usually goes away, and most women feel at their best during the middle trimester. The mother's appetite often increases.

By the third trimester, the mother's blood volume has increased by about 30 percent, or 1 or 2 liters. The growing fetus begins to impinge on the mother's internal organs. The weight of the fetus compresses the bladder, contributing to frequent urination and even incontinence. The pressure of the fetus on the blood vessels in the pelvic region can inhibit blood flow from the legs. Numbness, painful varicose veins, and leg cramps are all common, as are hemorrhoids. The mother's diaphragm no longer moves freely, so some mothers find breathing difficult. Pressure on the stomach and other physiological changes make heartburn common. Most women sleep badly and can easily become exhausted. Fortunately, these difficulties are often mitigated by happy anticipation.

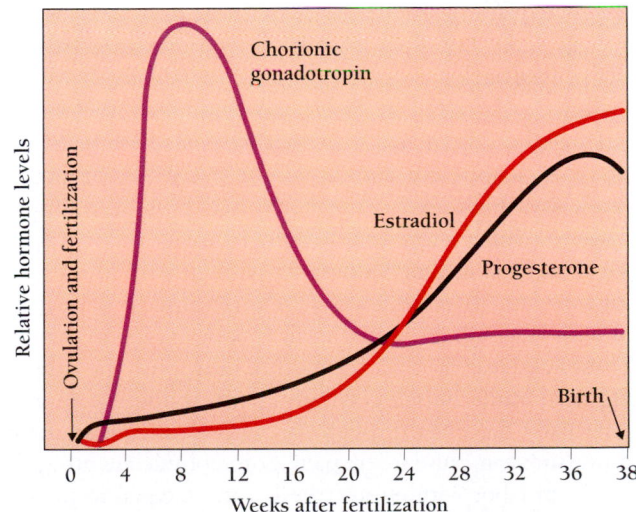

Figure 43-16 Hormone levels during pregnancy. Early in pregnancy, the embryo secretes human chorionic gonadotropin (HCG). HCG causes the corpus luteum to enlarge and secrete progesterone and estrogen (estradiol), which prevent the uterus from expelling the embryo. In the latter half of pregnancy, HCG levels decline, but the fetal placenta secretes ever higher levels of estrogen and progesterone until birth. The drop in these hormones at birth precipitates birth.

On average, pregnant women gain about 25 pounds, of which 7 pounds is the fetus and 4 pounds is amniotic fluid, placenta, and fetal membranes. In addition, blood and extracellular fluid add another 6 pounds, and the breasts and uterus each increase in weight by about 2 pounds. Beyond that, many women gain several pounds of fat, which they will need when they begin nursing the baby.

Between conception and delivery, a complex web of hormones help balance the needs of the fetus and mother.

Parturition: What Happens on D(elivery) Day?

Evidence now suggests that the baby helps decide the moment of birth, called **parturition,** by sending chemical signals to the mother. Late in pregnancy, the fetus begins to secrete **cortisol,** a steroid hormone associated with stress that increases the likelihood of labor. The mother's pituitary secretes the hormone **oxytocin,** which (1) stimulates the muscles of the uterus to contract and (2) stimulates the fetal membranes to secrete prostaglandins, which also increase contraction. Prostaglandins are chemical signals that increase inflammation and muscle contraction (Chapter 36). Aspirin and other pain relievers work by blocking the synthesis of prostaglandins.

The pressure of the baby on the inside of uterus, especially the cervix, also increases the tendency of the uterus to contract. Over a period of days, contractions may begin, then stop, repeatedly. But at some point, contractions begin to come regularly—every 20 minutes, for example. When contractions come every 3 to 5 minutes, labor has begun. During labor, contractions increase the pressure of the baby's head on the cervix, which stimulates the release of more oxytocin, and consequently more and more-intense contractions, an example of **positive feedback.**

Because of positive feedback, contractions come closer and closer together. By the time the mother is ready to deliver the baby, contractions are coming about every 90 seconds. In women with high levels of oxytocin and prostaglandins, the progression from a few contractions to many occurs rapidly, often in as little as 2 or 3 hours. In women with lower levels, labor may last 36 or 48 hours. The often intense pain associated with labor results at first from lack of oxygen in the uterine muscles. Near the end, the mother must push the baby out using these same muscles, as well as the abdominal muscles. As the baby descends through the vagina, intense stretching and pressure can cause severe pain. Some physicians attest that the pain of labor without anesthesia can be equal to the pain that would occur with amputation of the fingers. Fortunately, modern anesthesia can help reduce the pain.

At parturition, cortisol, oxytocin, and prostaglandins combine to drive the uterus into a positive feedback cycle of stronger and faster contractions that ultimately expel the baby.

Nursing the Baby

A few minutes after a woman delivers her baby, she also delivers the placenta. (She also feels ever so much better!) Someone must physically cut the umbilical cord, which connects the baby to the placenta. Then, within a few hours, a woman's body begins reverting to its former state. Excess fluid is excreted and the uterus begins to shrink. The 11 pounds that make up the baby, the placenta, and the amniotic fluid are lost immediately. Within the first week, the 6 pounds of excess blood and extracellular fluid vanish as well.

What do not normally revert to their former state are the breasts. In the first few days, the breasts secrete **colostrum,** a fluid similar to milk but with less fat. Colostrum also contains high levels of antibodies (mostly immunoglobulin A). As the baby suckles at each breast, the stimulation of the nipple causes the mother's pituitary to secrete the hormone **prolactin,** which stimulates the milk glands to produce milk. After about 3 days, the mother's milk usually "comes in." Human milk is better for babies than cow's milk. Human milk has more sugar and less fat and protein. It also contains immune cells and antibodies that help keep the baby from becoming ill.

The more often the baby sucks, the more prolactin is released and the more milk the breasts produce. A hungry baby may not be satisfied with the milk produced on one day. But if it is allowed to suck as long and as often as it likes, the next day will bring more milk. By the same token, an infant who leaves milk in the breasts all day will find less milk the next day.

Although the baby must suck hard to stimulate the nipples, a process that can be painful for the first few days, the milk ducts actually eject the milk into the baby's mouth with considerable force. Sucking stimulates the release of oxytocin (which also stimulated birth). Oxytocin stimulates tiny muscle cells surrounding the milk glands. A few seconds after an infant begins sucking, these muscles contract and shoot the milk into the hungry baby's mouth, a process called **milk ejection,** or the "let-down" reflex.

Nursing a baby places great metabolic demands on a mother and also increases her risk of bone loss. If her diet is low in calcium, the osteoclasts of the bones break down the mother's bone to supply calcium to the baby. Later in life, the mother is more subject to osteoporosis, porous and fragile bone. But if a nursing mother eats well and makes sure she gets plenty of rest, the result is healthful and satisfying for both mother and baby.

During the first few days after birth, babies get a low-fat version of human milk called colostrum. The baby's suckling stimulates the release of oxytocin, which causes the milk glands to eject milk, and prolactin, which ensures the synthesis of milk for subsequent meals.

HOW DO HUMANS CONTROL REPRODUCTION?

Humans, more than other animals, have sexual contact for reasons other than reproduction. Many students of human behavior have speculated on the evolutionary significance of sexual behavior that does not lead to new offspring. One hypothesis is that sex strengthens social bonding, increasing the chance that a couple will work together to nurture their young. In this way, sexuality may contribute to reproduction even when it does not lead to conception. Some biologists have speculated that this bonding is tied to a characteristic of our species unusual in other mammals: the human female is sexually receptive even at times when fertilization is impossible.

Humans do not necessarily want to bear as many children as possible, however, and the avoidance of conception is a major preoccupation for couples around the world (Figure 43-17). Although worldwide overpopulation is a real issue and of concern to millions of people, most individuals make decisions about how many children they want for personal, not ideological, reasons.

In societies where bottles, refrigeration, and formulas are unavailable, most mothers nurse their children until they are 3 years old. Because frequent nursing (at least eight times a day) inhibits ovulation, such mothers tend to bear children about 4 years apart. In developed countries, however, most women wean their babies much earlier. In the United States, for example, pediatricians regularly urge mothers to start giving babies solid food at 6 months, which often precipitates weaning. Many women nurse for only a few weeks, feeding their babies cow's milk, artificial milk (formula), and even solid food instead. In such cases, the mother begins to ovulate soon after giving birth and children may be spaced as little as 10 or 11 months apart.

In areas of the world where infant mortality is high, families often have many children in the hopes that a few will survive to adulthood. Farm families often have several children, because the children can work and increase the productivity of the farm. But tradition also plays a role. Most societies are now largely urban, but most city dwellers are the children or grandchildren of farmers, and the tradition of having many children continues.

Governments and religions also increase birth rates either by explicitly forbidding the use of birth control or by means of incentives (including tax deductions) that reward families with many children. In the past, societies ensured a steady supply of cheap labor and soldiers by discouraging the education of women and by banning information about contraception and other forms of birth control.

China, which has the largest population in the world, has dramatically lowered its birth rate over a period of decades by means of laws that punish people who have too many children. But such "top-down" policymaking doesn't always work. Italy has the lowest birth rate in the world, despite the fact that nearly all Italians are Catholic and the Catholic Church forbids both contraceptives and abortion. Some modern governments actively promote birth control.

In addition, in many cases, sexual activity precedes the beginning of reproductive life. Many societies, for example, expect couples to defer reproduction until long after sexual maturity, so that more young couples are sexually active than are reproductively active. **Birth control,** the conscious regula-

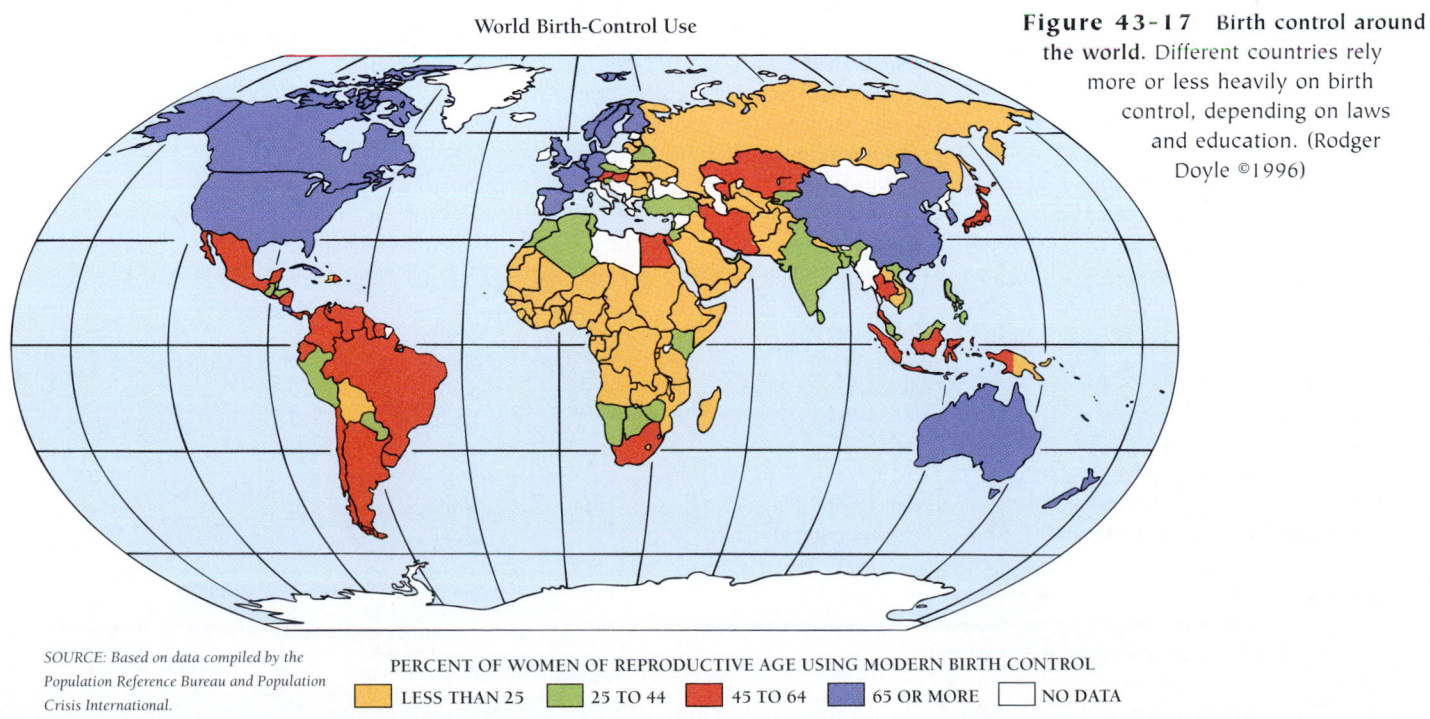

World Birth-Control Use

Figure 43-17   Birth control around the world. Different countries rely more or less heavily on birth control, depending on laws and education. (Rodger Doyle ©1996)

SOURCE: Based on data compiled by the Population Reference Bureau and Population Crisis International.

PERCENT OF WOMEN OF REPRODUCTIVE AGE USING MODERN BIRTH CONTROL

LESS THAN 25   25 TO 44   45 TO 64   65 OR MORE   NO DATA

tion of reproduction, is an important issue for many sexually mature humans. People have used four general approaches to birth control: abstinence, contraception, abortion, and **sterilization.** Each of these approaches is widely used and each has been the subject of intense social debate.

Abstinence

The most certain contraceptive method is to abstain from sexual intercourse altogether. **Abstinence** is common among adolescents. In the United States, about half of all teenagers younger than 19 abstain. Many adults in their twenties also continue to abstain, frequently until marriage. In addition many older adults abstain for a variety of reasons. Abstinence is favored, for example, by people who believe that the only reason for sexual intercourse is to reproduce (Table 43-1).

The **rhythm method** is a form of modified abstinence in which a couple refrains from sexual intercourse at times in the monthly cycle when conception is likely to occur. The Catholic Church, which opposes most forms of birth control, allows the use of the rhythm method.

As we have seen, the mature oocyte is only fertilizable for 12 hours after ovulation, while viable sperm can wait around in the female reproductive tract for 3 or 4 days for the egg to burst from the ovary. If a couple abstains from 4 days before ovulation to 1 day after ovulation, no conception should occur. But because the preovulatory phase can vary enormously, the exact date of ovulation is hard to predict, and in many women, its timing may vary from month to month. In addition, some evidence suggests that women's interest in intercourse increases in the few days prior to ovulation. It is not surprising then, that the rhythm method does not provide reliable birth control. Of every 100 women practicing the rhythm method, as many as 20 become pregnant in one year (Table 43-1). For comparison, among sexually active women who practice no birth control, 85 out of 100 become pregnant in a year.

Several factors can increase the success of the rhythm method. If a woman has regular and predictable cycles, she has a better chance of correctly estimating the time of ovulation. If she knows when she ovulates each month, she can estimate when it will occur the following month. One clue is a one-half to one-degree increase in body temperature on the day of ovulation. In the days before ovulation, the quality of the mucus in the vagina becomes clear and also thinner and stringier than usual. In addition, the rupture of the follicle at ovulation sometimes causes pain, usually minor. Pain dur-

TABLE 43-1   CONTRACEPTIVE METHODS AND THEIR EFFECTIVENESS

Method	Risk of pregnancy*	Health risks	Protection from STDs?	Mortality rate
Abstinence	0	None likely	Yes	0
Vasectomy	<1	Possible increase in prostate cancer, heart disease	No	1 in 300,000
Tubal ligation	<1	Infection, increase in heavy bleeding	No	1 in 67,000
Condom	3	None known	Latex, good	0
Birth control pill	<1	Heart disease, stroke (especially in smokers)	No	1 in 64,000†, 1 in 16,000‡
Depo-Provera	<1	Possible increase in osteoporosis	No	—
Norplant	<1	Infection at implantation site, abnormal vaginal bleeding	No	—
Morning-after pill	<1	Nausea, cramps	No	—
RU-486/Mifepristone	<1	Nausea, cramps	No	—
Intrauterine device (IUD)	<2	Pelvic inflammatory disease	No	1 in 100,000
Diaphragm with spermicide	6	None known	Some	0
Cervical cap	9	None known	Some	0
Contraceptive sponge	9	None known	Some	0
Female condom	5	None known	Yes, good	0
Spermicide alone	6	None known	Some	0
Rhythm method (partial abstinence)	6–10	None likely	None	0
No method	85	Pregnancy and childbirth	None	13 in 100,000
Legal abortion within 7 weeks after fertilization	<1	—	None	1 in 500,000

*When method is used correctly; expressed as percent per year.
†In nonsmokers.
‡In smokers.

ing ovulation is called **mittelschmerz** [German, *mittel* = middle + *schmerz* = pain, pain in the middle of the month]. Finally, a thorough knowledge of female reproductive cycles can help. Some women use a modified rhythm method, using contraception during the week before ovulation.

To some people abstinence and periodic abstinence are the only acceptable methods of birth control.

Withdrawal

Another technique is **coitus interruptus** [Latin, *coitus* = sexual intercourse], or withdrawal, in which a man attempts to avoid conception by removing his penis from the vagina before ejaculation. Although this method is widely practiced in some parts of the world, it is one of the least reliable forms of birth control. One reason is that the secretions of the bulbourethral gland sweep small numbers of sperm into the vagina before ejaculation. In addition, withdrawal before orgasm requires the man to exert considerable resistance to the stereotyped pattern of the sexual response. In other words, he does not always remember to withdraw.

Barrier Methods Prevent the Union of Sperm and Ovum

Couples may use one or more forms of contraception while still having sexual intercourse. The easiest of these methods to understand are those that physically prevent the union of sperm and ovum by imposing a **barrier** between them (Table 43-1).

Several contraceptive methods prevent the union of mature sperm and ova. A **condom** is a thin rubber sheath that covers the penis and prevents sperm from entering the vagina. By preventing actual physical contact between the male and female genital tracts, condoms prevent not only pregnancy but also sexually transmitted diseases (STDs), such as acquired immunodeficiency syndrome (AIDS), chlamydia, syphilis, and gonorrhea.

The major problems with condoms is that they break and sometimes fall off. In addition, if the condom is put on late, sperm may leak into the vagina before ejaculation, as in the withdrawal method. A condom must be put on as soon as the penis is erect, and it somewhat reduces the perceived stimulation of the penis and the vagina. It must be removed *immediately* after ejaculation, so that the sperm does not leak out from around the flaccid penis. Some couples prefer other methods of contraception, which, they feel, do not disrupt their sexual encounters.

A second method for preventing sperm from reaching the ovum is a **cervical cap,** a thimble-shaped rubber or plastic cap, about an inch in diameter, that fits tightly over the cervix. A **diaphragm** is larger than a cervical cap but works in the same way, serving as a barrier to the entrance of sperm. A **spermicide** [Latin, *cidere* = to kill], a cream or jelly that kills sperm, is an essential part of the effectiveness of both cervical caps

and diaphragms. A woman may insert a diaphragm up to several hours before intercourse. A cervical cap may stay in place for several days. However, additional doses of spermicide must be inserted before each instance of intercourse.

When properly used, condoms, cervical caps, and diaphragms are effective contraceptives, with only two pregnancies per year per hundred women. With all three methods, however, improper usage and breakage lead to actual pregnancy rates of 10 to 13 per year per hundred women (Table 43-1).

The contraceptive sponge, now available in the United States again, also provides a barrier to the entry of sperm into the uterus. Like the cervical cap and the diaphragm, the sponge fits over the cervix, but, unlike them, it requires no special fitting. Instead, the woman dips the sponge into water and inserts it over the opening of the cervix. The sponge, which may be left in place for as long as a day, not only blocks sperm entry but also releases a spermicide. Sponges by themselves give pregnancy rates of about 17 per hundred women per year and are therefore most effectively used in conjunction with another method, such as condoms.

Some couples attempt to prevent fertilization with spermicides alone. These include jellies, foams, creams, and suppositories. Each of these methods, by themselves, gives pregnancy rates of about 15 per one hundred women per year. Finally, some women attempt to prevent pregnancy with a **douche** [French, = wash], the washing of the vagina immediately after intercourse. This method is ineffective, with a pregnancy rate of about 40 per one hundred women per year.

Barrier methods such as condoms and diaphragms prevent the sperm from reaching the oocyte.

Hormonal Contraceptives and Intrauterine Devices

Researchers are now trying to develop chemical means of preventing sperm maturation. So far, however, no such "male pill" is yet available. In contrast, a great variety of *systemic* contraceptives, those which go into the bloodstream, exist for women.

Chemical birth control can prevent ovulation, fertilization, or implantation. One version of the female **birth control pill** consists of a combination of estrogen and progesterone. Together these prevent the anterior pituitary from secreting LH and thus stop ovulation. In one formulation, a woman takes the pills for 21 days and then stops for 7 days. The fall in steroid levels triggers menstrual flow, so that the result is a regular cycle without ovulation. Various versions of the *combination* birth control pill use different combinations of estrogen and progesterone (or similar compounds).

If used correctly, these formulations are almost entirely effective in preventing pregnancy, with pregnancy rates of less than 1 percent per year. The relatively high levels of hormone used, however, sometimes cause side effects, especially in the circulation, including high blood pressure and an increased risk of strokes and heart attacks. These risks are higher among smokers than among nonsmokers. The chances of fatal complications, however, are far lower than the mortality associated with pregnancy itself. For many couples the pill is the preferred method of birth control.

Another group of birth control pills contain only progesterone, with no estrogens. The pills (sometimes called "mini-pills") have few of the side effects of the combination pills, but they have a slightly higher pregnancy rate. They appear to work by inhibiting the movements of both ovum and sperm, as well as by interfering with implantation.

Another chemical method that interferes with ovulation involves the release of a substance from a capsule surgically placed under the skin. A single such implanted capsule, called Norplant, may last for up to 10 years.

Another chemical contraceptive is a progesterone derivative called DMPA, or **Depo-Provera.** DMPA is taken as an injection, which appears both to suppress ovulation and to inhibit implantation for 3 months at a time. The risks associated with DMPA, however, are still the subject of debate. Some people argue that it is safer than the available birth control pills, while others point to serious problems, including depression and abnormal menstruation.

The **morning-after pill** is an oral contraceptive taken in higher-than-normal doses within 72 hours of unprotected intercourse. Like standard doses of birth control pills, morning-after pills effectively interfere with both conception and implantation. Side effects, such as nausea and cramps, are unpleasant but transient.

Two general methods are currently available for preventing implantation of the embryo in the uterine lining. The first of these, the **intrauterine device, or IUD,** is a small piece of plastic that is placed into the uterus, where it interferes with implantation. The idea of such a method of birth control appears to have come from an ancient practice of camel drivers, who would put pebbles into the uterus of a female camel before a long trip. The pebbles kept the camel from becoming pregnant during the trip. In a similar way, an IUD in the uterus of a human female appears to prevent pregnancy, although researchers do not know exactly how.

More than 100 types of IUDs exist, some of which consist only of plastic, while others also contain copper or progesterone. One type of IUD that was sold in the United States between 1971 and 1974, the Dalkon Shield, produced serious side effects, including inflammations of the pelvis and spontaneous abortions, in a few cases resulting in death. Other types of IUDs are safer, although some women have infections, excessive bleeding, cramps, or other side effects. IUDs must be fitted and inserted (and eventually removed) by an experienced physician or other health care professional.

Hormonal contraceptives suppress ovulation and inhibit implantation. The IUD also inhibits implantation.

Sterilization Provides Effective But Generally Irreversible Birth Control

Quick and relatively inexpensive surgical methods of male and female sterilization are now widely available. Although sterilization can sometimes be reversed, it is usually a permanent change, so individuals choosing it as a method of birth control must carefully consider such a choice.

Vasectomy, the cutting and tying of the vas deferens, prevents sperm from entering the urethra (Figure 43-18). Vasectomies are safe, do not require hospitalization, and are performed in less than an hour with local anesthesia. The procedure is usually permanent, however, and is suitable for men who are certain that they do not want to reproduce in the future.

Tubal ligation cuts or blocks the oviducts, or fallopian tubes, thus preventing eggs from reaching the uterus after ovulation (Figure 43-18). It is a more complicated operation than vasectomy. Because the oviducts lie within the abdomen, a tubal ligation requires one or more surgical incisions into the abdomen and a hospital-like setting. A surgeon can, however, perform a tubal ligation with only two small incisions in the woman's navel. The surgeon then inserts a laparoscope, a thin tube containing a viewing instrument, in one incision and the surgical instruments in the other incision. While modern tech-

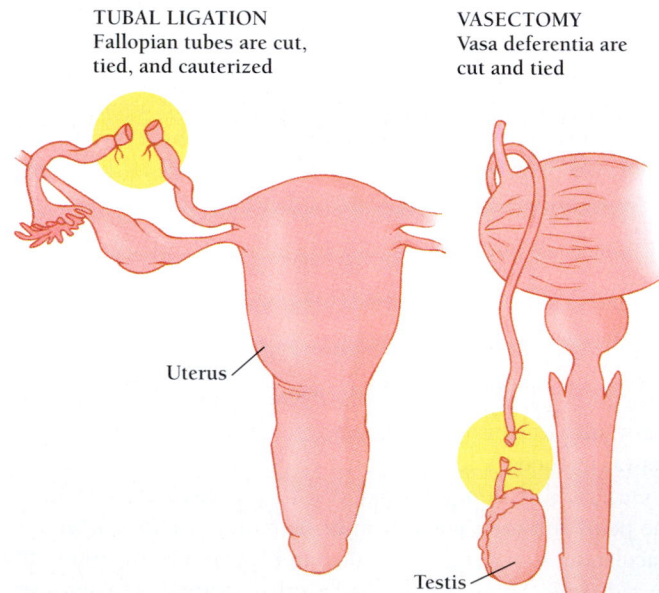

TUBAL LIGATION
Fallopian tubes are cut, tied, and cauterized

VASECTOMY
Vasa deferentia are cut and tied

Uterus

Testis

Figure 43-18   Tubal ligation and vasectomy. In tubal ligation, the fallopian tubes are either cut and tied, cauterized (burned), closed off with very tight rubber bands, which kill the tissue, or clamped with a tight clip. A tubal ligation requires abdominal surgery. Cutting and tying the vas deferens prevents sperm from reaching the penis. It is a much safer and simpler procedure than tubal ligation. Neither procedure should be considered reversible.

niques make this operation short and relatively uncomplicated, full recovery takes several weeks. Although reversals are occasionally possible, a tubal ligation should be considered permanent. Women should elect this option only if they are certain that they will never again want to conceive.

Sterilization—vasectomy and tubal ligation—is highly effective and usually irreversible.

Induced Abortions Terminate Pregnancies After Implantation

An **induced abortion** is the deliberate removal of an embryo from the uterus. The ease of the procedure decreases with the age of the embryo. In the first trimester (the first 3 months), the procedure is brief and can take place in a clinic or doctor's office. The embryo is usually removed from the uterus by gentle suction between 3 and 10 weeks after conception. Early abortions, between 3 and 5 weeks after fertilization, are called **menstrual extractions.** Abortions between 5 and 10 weeks after fertilization are done by dilation and extraction (**D&E).**

In several European countries, doctors use the drug **RU 486,** also called **mifepristone,** along with prostaglandins, to induce abortions within the first 9 weeks after conception. RU 486 blocks progesterone, thus allowing menstruation. A committee of the U.S. Food and Drug Administration (FDA) has recommended that RU 486 also be approved for use in the United States. But as of early 2000, this drug is unavailable in the United States.

Induced abortions in the second trimester (the second 3 months) require different methods, which are performed only in a hospital. The most common method is to induce uterine contractions by injecting a solution of salt or of prostaglandin. RU 486 is also given in conjunction with prostaglandins. Induced abortions in the third trimester are rare.

Until the late 19th century, abortion was legal in the United States before the "quickening" of the fetus, at 4 to 5 months. In 1873, however, President Ulysses S. Grant enacted a federal law prohibiting all contraception and abortion, as well as the publication or discussion of information about contraception. Similar laws were passed by individual states over the succeeding decades. In 1973, however, all of them were overturned and declared unconstitutional by the U.S. Supreme Court. The decision, called *Roe v. Wade,* declared that abortions in the early stages of pregnancy were legal in the United States, but that states could restrict abortions in the third trimester as long as the mother's health was not at risk. Today, abortion is legal in all parts of the United States, and remains deeply controversial.

An induced abortion is the removal and destruction of an embryo or fetus.

What Can Couples Do To Overcome Infertility?

About one couple in six in the United States is infertile, meaning that they do not conceive after a year of sexual relations without contraception. Many causes may underlie infertility, but about half of infertile couples are able to achieve a pregnancy. About a third of the time, infertility results from the man's failure to produce and deliver viable sperm; about a third of the time, the problem lies in the woman's inability to produce a fertilizable ovum or maintain a pregnancy once it has begun; and about a third of the time, combined factors are responsible.

A man's inability to produce mature sperm may result from infections (such as mumps), poor nutrition, or chemicals (including alcohol and other drugs) or radiation in the environment. Problems of sperm delivery may be due to blockage or scarring in the genital tract, which may result from sexually transmitted diseases, or from sexual problems such as impotence (inability to maintain an erection during intercourse) or premature ejaculation.

Infertility in a woman may result from infections of the genital tract (including sexually transmitted diseases), mechanical barriers (such as scarring) in the genital tract, scarring of the uterine wall or of the fallopian tubes, or from hormonal problems that interfere with ovulation or pregnancy.

Many infertile couples elect to adopt children. Others take advantage of an enormous industry of fertility specialists who sell sperm and eggs to infertile couples. If the woman is normal but the man cannot produce sperm, the woman can have a baby if someone places sperm from a donor, known or anonymous, in her vagina at the right time of the month. In rare cases, women who want to have a child without a male partner opt for the same procedure. In other cases of infertility, a man may produce normal sperm but not be able to deliver it in sexual intercourse. In that case, the sperm can be removed from the testes.

Women with normal ovaries and a normal uterus but blocked oviducts (fallopian tubes) are candidates for a technique called *in vitro fertilization* (IVF), the fertilization of an ovum in the laboratory, which is followed by the implantation of the embryo back into the uterus.

A physician can use the woman's own ova (which must be removed surgically at the right time of month) or the ova of an anonymous egg donor. To accomplish IVF, a physician treats a woman (either the prospective mother or the donor) with gonadotropins to increase the number of mature follicles. Then, using ultrasound imaging to tell whether a follicle is mature, a surgeon removes the ova from the ovaries. The mature ova are placed in a dish and exposed to sperm.

After fertilization and some subsequent development, the physician transfers the embryos back into the uterus, where they can then develop normally. Many fertility clinics implant several embryos because most or all are likely to die after transfer to the uterus. Occasionally several will survive. And the previously infertile couple goes home with triplets or even quintuplets. In such cases, premature birth is extremely likely. Prematurity impairs the health of the baby and hospital costs are frequently astronomical. In the 1990s, the cost of each *in vitro* baby averaged $60,000 to $100,000. A single premature baby can cost more than a half-million dollars.

In cases in which the woman is infertile, a couple may use the man's sperm to fertilize the ovum of a surrogate mother, a woman who will conceive and also carry the resulting embryo to term. Surrogate motherhood is a controversial procedure that has been the subject of much public debate because surrogate mothers sometimes ask for their children back. As a re-sult, fertility clinics are increasingly using egg donors, healthy young women willing to donate eggs, usually for $3000 to $5000. One fertility clinic recently advertised for an egg donor from Stanford University. For eggs from an intelligent, blonde athlete, the potential parents were offering $100,000.

Besides adoption, biotechnology offers a wealth of choices to infertile couples, from in vitro fertilization to surrogacy.

Study Outline with Key Terms

Mammals form haploid gametes (**ova** and **spermatazoa**) for sexual reproduction from **primordial germ cells** in the **gonads** —the **testes (testicles)** in males and the **ovaries** in females. During **oogenesis,** the ovaries form diploid **oogonia,** which grow into **primary oocytes.** Each month several primary oocytes develop, with surrounding cells in the ovaries, into **follicles,** which swell with fluid. Just before **ovulation,** a primary oocyte completes meiosis I and forms a **secondary oocyte** and a **polar body,** which enters meiosis II but does not complete it.

In the male's **scrotum** are the testes, within which are the **seminiferous tubules,** which form the sperm. The seminiferous tubules are surrounded by **Sertoli cells,** which nourish the developing sperm, and **interstitial cells,** which secrete **testosterone.** On the inside walls of the seminiferous tubules, diploid **spermatogonia** divide through meiosis to form four haploid **spermatids,** which mature into sperm, with a head, a **midpiece,** and a tail-like **flagellum.** The front of the head contains an enzyme-packed lysozome called the **acrosome.**

The fusion of the egg and sperm (ovum and spermatozoan) is called fertilization, **syngamy,** or **conception.** The sperm moves into the **oviducts,** or **fallopian tubes,** where it encounters the secondary oocyte and fuses with it. The secondary oocyte and the polar body divide once more, forming a mature ovum and two or three polar bodies. The nucleus of the sperm and egg fuse, and the resulting diploid **zygote** moves down one of the two oviducts, or fallopian tubes, to the **uterus.**

The developing embryo then implants in the **endometrium** of the uterus. At about 9 weeks, all the major organs have formed, at least in general outline, and the embryo is called a **fetus.** When the fetus has matured at 9 months, it is said to be **full term,** and is expelled by muscular contractions of the uterus.

During **ejaculation,** sperm leave the testes by way of the **epididymis,** the **vas deferens,** and the **urethra** of the penis. (In many birds and in some reptiles, the sperm leave by way of a **cloaca.**) Along the way, fluids from the **seminal vesicles,** the **prostate gland,** and the **bulbourethral glands** contribute to the formation of the **semen.**

The penis consists of the **corpus spongiosum,** paired **corpora cavernosa,** and the **glans,** all well supplied with blood vessels and nerves. A loose layer of skin, called the **foreskin,** covers the glans in males who have not been **circumcised.**

The female external genitalia consist of the **vulva** —including the **labia majora** and **labia minora** —and the **clitoris.** Like the penis, the clitoris includes a glans, paired corpora cavernosa, and a **bulb of the vestibule,** like the corpus spongiosum. Inside the entrance to the vagina is a membrane called the **hymen** that is usually damaged during first intercourse.

In humans, sexual response consists of four stages—**excitement, plateau, orgasm,** and **resolution.** During the excitement stage, increased blood flow to clitoris and penis results in **erection** and tightening of the vaginal walls. In addition, the **Bartholin's glands,** in the female, and the bulborurethral glands, in the male, contribute lubrication. Enzymes in the female reproductive tract break down a coating on the sperm, a process called **capacitation.**

Gonadotropin-releasing hormone (GnRH) secreted by the **hypothalamus** regulates the secretion of the two **gonadotropins luteinizing hormone (LH)** and **follicle-stimulating hormone (FSH)** by the **anterior pituitary gland.** In males, hormones regulate the production of sperm and testosterone by means of negative feedback. LH stimulates the production of testosterone and other androgens, while FSH stimulates the Sertoli cells in the testes to produce sperm. But the Sertoli cells also secrete inhibin, which suppresses the secretion of FSH, shutting down sperm production.

In women, LH and FSH regulate the menstrual cycle and the **ovarian cycle.** As in men, the hypothalamus secretes GnRH, which stimulates the secretion of LH and FSH. These stimulate the ovaries to develop mature oocytes (and follicles) and to produce estrogen and other steroid hormones. The growth of the follicles causes a buildup of estrogen in the blood in the days preceding ovulation, which triggers the release of more GnRH, and therefore LH and FSH. LH triggers ovulation and development of the **corpus luteum.** Progesterone and estrogen from the corpus luteum stimulate the cessation of GnRH production and the development of the endometrium, the lining of the uterus. If no embryo implants, the endometrium breaks down and is washed away by a flow of blood, called **menstruation.** If fertilization occurs, the resulting zygote secretes **human chorionic gonadotropin (HCG),** which, like LH, maintains the corpus luteum.

Like sharks and many snakes, humans and most other mammals are **viviparous.** The young develop inside the mother. After fertilization, an embryo's secretion of HCG doubles the size of the corpus luteum and increases the secretion of estrogen and progesterone, which helps prevent menstruation and sustains the pregnancy. Testing for HCG is the basis for pregnancy tests. The embryo constructs a set of four membranes and a **placenta.** The placenta secretes **human chorionic somatomammotropin (HCS),** which helps redirect glucose from the mother to the embryo.

Physicians divide pregnancy into three "trimesters." During the first trimester, the mother may experience fatigue and nausea. During the second trimester she will likely feel the first movements of the fetus, her nausea will probably go away, and she will feel energetic. In the third trimester, the increasing size of the fetus will impinge on her own organs and make her distinctly uncomfortable.

At **parturition, cortisol, oxytocin,** and prostaglandins combine to drive a **positive feedback** cycle of faster and faster uterine contractions. These contractions, combined with pushing by the mother, force the baby (and later the placenta) out through the **cervix** and vagina. After birth, the mother's breasts secrete colostrum for the nursing baby. **Colostrum,** a low-fat version of human milk, is replaced by milk after about 3 days. Suckling stimulates the release of oxytocin, which stimulates **milk ejection,** and **prolactin,** which stimulates the synthesis of more milk.

Birth control measures include **sterilization (vasectomy** and **tubal ligation), abstinence,** the **rhythm method, coitus interruptus, intrauterine devices (IUDs),** and a variety of **barrier** methods. The rhythm method requires that the woman be alert to signals that indicate the pace of the menstrual cycle and the day of ovulation. These include the consistency of vaginal mucus, body temperature, and **mittelschmerz.** Barrier methods include two kinds of **condoms,** one to be worn by the man and one to be worn by the woman, the **cervical cap,** and the **diaphragm.** The efficacy of the cervical cap and the diaphragm are much improved by the use of **spermicides.** (A **douche** is so ineffective that it cannot be considered a method of birth control at all.)

Systemic contraceptives are synthetic hormones that prevent ovulation, fertilization, or **implantation.** These include the **birth control pill, Depo-Provera,** and the **morning-after pill.** All of these chemical contraceptives are for women only.

When contraceptives fail, some women arrange to have a doctor perform an **induced abortion,** the intentional destruction and removal of the embryo or fetus. An induced abortion is distinct from a **spontaneous abortion,** the death and expulsion of an embryo due to a chromosomal abnormality or other damage. Abortions may be induced chemically, using **RU 486,** or **mifepristone,** for example, or physically. The most common two techniques are **menstrual extraction** and dilation and extraction **(D&E).** Both methods are performed in the first 3 months after fertilization.

Review and Thought Questions

Review Questions

1. Compare oogenesis with spermatogenesis. What is the main difference between these two processes?
2. What is the difference between a primary oocyte and a secondary oocyte? Which fuses with the sperm and what happens then?
3. Draw a diagram with labels showing a cross section of a seminiferous tubule that illustrates how the sperm form and develop.
4. Describe the acrosome, where it is, and what its function is.
5. Can nonmotile sperm reach the oocyte? Why or why not?
6. What glands contribute fluids that make up the semen?
7. What gland secretes mucus during sexual intercourse?
8. What are the four stages of the human sexual response?
9. How do GnRH, LH, and FSH regulate the formation of mature oocytes and sperm? Draw a diagram from memory showing where these hormones are made and how they interact with the gonads. What do the initials stand for?
10. If a woman has regular 28-day cycles and her period is 1 week late and a pregnancy test gives a positive result, how old is the embryo?

Thought Question

11. In all mammals except monotremes, the testes are external to the body cavity and sperm cannot form effectively inside the warmth of the body cavity. Why do you think this might be so? Did mammals evolve external testes and then evolve sperm that were temperature sensitive or did the temperature sensitivity evolve first? Why would either happen? Why are we different from birds and other vertebrates, which make healthy sperm inside of their bodies?

About the Chapter-Opening Image

This Faberge egg symbolizes the female's investment of energy and materials in the oocyte.

 On-line materials relating to this chapter are on the World Wide Web at **http://www.harcourtcollege.com/lifesci/aal2/**

44 *How Do Organisms Become Complex?*

How Do We Become Complex?

Most people know that reproduction requires two parents. But what exactly does each parent contribute? Until the 19th century, biologists were not sure. The Greek philosopher Aristotle argued that the female ovum provides formless substance, to which the male sperm gives shape, molding the egg into an individual as a potter shapes clay into a bowl. How sperm might accomplish such crafting, Aristotle did not say.

The question was—and remains—how does the complexity of a multicelled organism arise from the simplicity of a single egg? Eighteenth-century biologists answered this question by denying that simplicity ever existed. In this view, all the parts of the adult organism already exist at the earliest stages of life. The adult is *preformed* in the sperm or the egg.

The French biologist Charles Bonnet formalized this theory, called **preformation,** in 1745. He argued that each egg contains a complete embryo, and each embryo contains more complex embryos inside itself, and so on, like a set of Russian dolls. Many of Bonnet's contemporaries argued, however, that the complete embryo was in the sperm, not in the egg. The Dutch microscopist Antoine van Leewenhoek had already demonstrated the existence of sperm almost a hundred years before, and some microscopists even claimed to have seen a tiny creature, called a **homunculus,** curled up inside the sperm head, shown at left. Preformation implied that the whole organism resulted from the growth of a preformed miniature. At that time, no scientist had seen a mammalian egg or witnessed the coming together of an egg and sperm in either plants or animals.

The doctrine of preformation not only avoided the difficult problem of explaining development, it also provided a simple view of genetics: a child resembled its parents because it was already preformed in the ovum of the mother or the sperm of the father. The doctrine of preformation also presented a problem. If a child were already preformed—whether in egg or sperm—how would both parents contribute to the genetic makeup of the child?

Preformation left little room for evolution. Because all future generations must be contained within past generations, only a limited number of generations could be contained. Eventually, the tiniest embryos inside embryos would be too tiny to exist. Some biologists even tried to calculate how many embryos within embryos must have existed in Eve's ovaries.

Furthermore, preformation only worked as a theory as long as people believed that the Earth was only 4000 years old. As geologists discovered that life was millions of years old and chemists discovered that matter was not infinitely divisible, the underpinnings for the theory of preformation crumbled.

It was an Estonian embryologist, however, who provided the evidence that finally disproved preformation. In the 1820s, Karl Ernst von Baer described the gradual development of a mammalian zygote. He saw that a single-celled zygote divided and formed three layers of tissue, from which the embryonic organs and tissues then developed. It was a gradual process, like the one that Aristotle had suggested 2000 years before. The discovery and description of fertilization—the fusion of egg and sperm—in both plants and animals further undermined the theory of preformation.

By the middle of the 19th century, it was clear that Bonnet's theory of preformation was incorrect. Yet von Baer's work raised a new set of questions: if a single cell could divide and each descendant cell could specialize into different kinds of cells, how was this specialization accomplished? Did some cells get one set of genes while other cells got another set of genes? Or did every cell get a complete set of all the genes? In short, was each cell **totipotent,** capable of developing into any kind of cell? Or was each cell irreversibly specialized early on?

Nineteenth-century experimentalists were working at a disadvantage since they did not yet know that the genes were on the chromosomes or how genes were encoded in the DNA of the chromosomes. They only assumed that cells had genetic particles of some sort. In

The Granger Collection, New York

Key Concepts

- During development, an animal changes from a single cell to a complex organized multicelled organism by means of cell division, cell movement, cell specialization, and pattern formation.

- At the beginning of development, each cell contains all the genetic information necessary to produce an entire animal. As development proceeds, however, most cells become increasingly limited to a particular developmental pathway.

- The pattern of gene expression of individual cells and their participation in the formation of organs and body parts depend on chemical signals, many of which act by regulating transcription of specific genes.

- Virtually all animals appear to use the same mechanisms to achieve pattern formation and programmed cell death.

order to answer the question raised by von Baer's work, the German biologist Wilhelm Roux conceived an experiment that would tell whether or not embryonic cells were totipotent.

Roux reasoned that as the cells divided, each one got only some of the genes present in the zygote. If that were true, and if only half of the embryo were allowed to develop, the result would be an embryo with half its structures missing. In 1888, Roux took a two-celled frog embryo and destroyed one of the two cells by piercing it with a hot needle (Figure 44-1A). The remaining cell formed only half an embryo. Roux concluded that embryonic cells are not totipotent. Instead, he argued that at each cell division the daughter cells receive half of the complexity in the dividing cell. As the zygote divides into more and more cells, these cells receive smaller and smaller shares of the genes in the zygote's original nucleus.

Within 4 years, however, Roux was challenged by another German biologist, named Hans Driesch. Driesch performed a similar experiment, but with dramatically different results (Figure 44-1B). Driesch used sea urchin embryos, and, instead of killing one cell and leaving it in place, he shook the two-celled sea urchin embryos until the pairs of cells fell apart. As he watched, each of the two cells grew into a complete sea urchin. Driesch's result disproved the idea that each of the cells of an embryo had different information. An embryo, apparently, was not a mosaic of information. Instead, each cell may be totipotent, capable of forming all the structures of the adult.

Driesch's demonstration of totipotency in embryonic cells conclusively discredited the theory of preformation. For if each cell of an embryo could develop into a separate adult, then neither the egg nor the sperm could be said to exclusively harbor the individual.

But why had Roux's experiment suggested preformation? The German embryologist Hans Spemann wondered if Roux's embryos developed abnormally because the dead cell interfered with the development of the remaining embryo. Spemann repeated Roux's experiment, but instead of killing one cell, he used a fine hair (from a baby) to gently separate the cells of a two-cell salamander embryo (Figure 44-1C). Like Driesch, Spemann saw each cell develop into a whole, normal embryo. His result confirmed Driesch's view that each cell of a two-cell embryo can give rise to a whole adult.

The experiments of Driesch and of Spemann permanently laid to rest the theory of preformation and suggested that each cell inherits all of the genetic information contained in the zygote, not just part of it. But more questions remained. Where does complexity come from? How do cells become different from one another? How do organs and tissues form? How do the different parts of the embryo know how to arrange themselves?

As we will see later in this chapter, Spemann and his students went on to answer many of these questions, but each answer provoked still more questions. Near the end of his career, in 1936, Spemann gave a series of lectures at Yale, in which he despaired of ever understanding the mechanisms of development. In the last 10 years, however, developmental biologists—aided by genetics and molecular biology—have

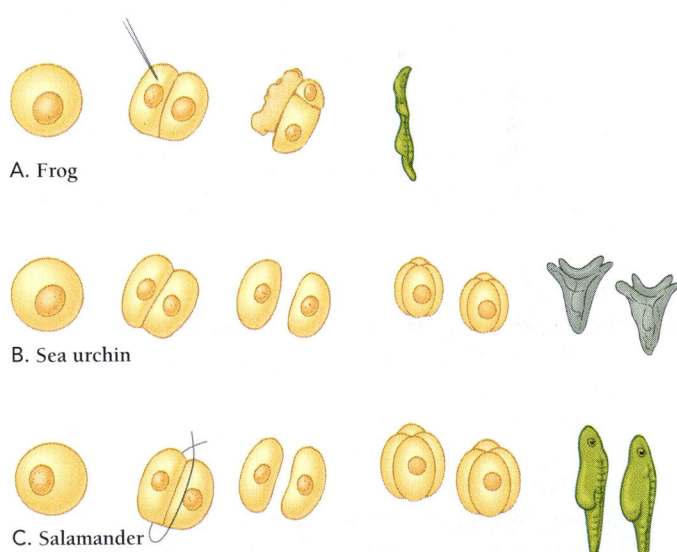

A. Frog

B. Sea urchin

C. Salamander

Figure 44-1 **Totipotency and preformation.** Is a single cell of an embryo as capable of forming a whole tadpole as all the cells in the embryo? A. Wilhelm Roux killed one cell of a two-cell frog embryo and got half a tadpole. B. Hans Driesch shook apart the two cells of a two-cell sea urchin embryo and got two complete sea urchins—the first known artificial clones of an animal. C. Hans Spemann repeated Roux's experiment. Instead of killing one cell, however, he separated the two cells of the salamander embryo with a fine hair. Both cells developed into normal salamander larvae.

begun to understand the molecular mechanisms responsible for complexity. Not only have biologists answered many of Spemann's perplexing puzzles, but they have also come to understand the basis of other strange phenomena. We can now say, for example, why the mutant fruit fly shown in Figure 9-26 has legs growing where its antennas should be. And, as we discussed in Chapter 24, we can also now understand the molecular mechanisms responsible for the upside-down parallel between vertebrate and invertebrate body plans, first noticed almost 200 years ago by Geoffroy St. Hilaire.

Of all the wonders of life, none is so amazing as development. Each of us has developed from a single cell, barely visible to the naked eye. The central question of this chapter is how this development can occur. Much of our focus will be on the development of vertebrates, as they are familiar and illustrate the general problems of animal development. Along the way, however, we will also encounter some of the powerful insights revealed by research on flies and other invertebrates.

HOW DOES FERTILIZATION INITIATE EMBRYONIC DEVELOPMENT?

Fertilization, or syngamy, joins the haploid ovum and sperm to form a diploid zygote. Fertilization brings together genetic information from the maternal and paternal genomes and begins the life of a new individual.

Within a minute after a sea urchin sperm binds to the ovum, the ovum increases its rate of oxygen consumption. Within 10 minutes, the zygote begins intense metabolic activity, using energy at a much higher rate. Protein synthesis also increases dramatically, and the embryo begins to divide.

At one time, researchers guessed that the initial burst of protein synthesis in the sea urchin's zygote resulted from transcription (DNA to RNA) in the newly formed nucleus. But this hypothesis was wrong. Embryos that are treated with a drug that suppresses transcription display the same burst of protein synthesis as untreated embryos. But protein synthesis can only occur if mRNA is present. Where was the mRNA coming from if not from the nucleus of the zygote? Researchers were forced to conclude that the mRNA that directs early protein synthesis is mRNA that is already present in the egg cell. Such mRNA, called **maternal mRNA,** is produced in the egg cell—transcribed during oogenesis (egg formation) from the genes of the mother, not those of the zygote.

The early development of a zygote is entirely controlled by maternal proteins. In sea urchins, the zygote's nucleus is not only inactive during early development, it is unnecessary. Embryos lacking active nuclei still divide and develop into multicelled embryos.

Fertilization is both a genetic and a developmental event. Protein synthesis and development in early embryos is directed by maternal mRNA—mRNA transcribed from the genes of the mother.

HOW DO CELLS OF THE EMBRYO GIVE RISE TO CELLS OF THE ADULT?

Developmental biologists view the body plan of a vertebrate as "a tube within a tube" (Figure 44-2). The outermost tube, or **ectoderm** [Greek, *ecto* = outside + *derma* = skin], consists of the part of the animal that is in contact with the outside world—the epidermis, or outer skin layer, the nervous system, and the sense organs. The innermost tube, or **endoderm** [Greek, *endo* = inside + *derma* = skin], is the gastrointestinal tract, together with associated organs such as the pancreas and liver.

Between the ectoderm and the endoderm is the **mesoderm** [Greek, *mesos* = middle + *derma* = skin], which consists of connective tissues, such as bones, muscles, and tendons, and the cells of the blood. The mesoderm also includes a number of organs including the heart and kidneys.

All three layers of the adult vertebrate arise from a single-celled zygote. To see how this basic organization is accomplished we can study the early events of development and ask when the three-layer body plan first appears.

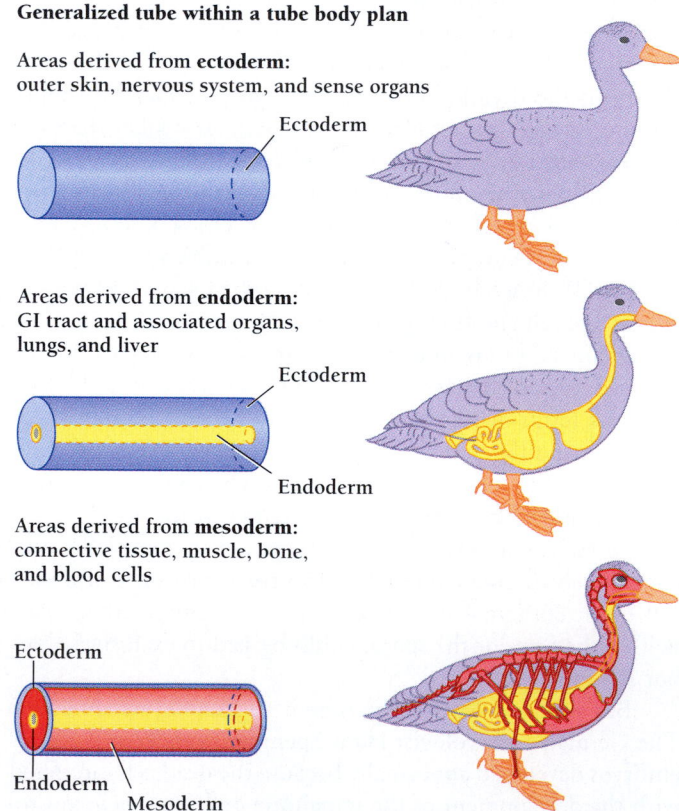

Generalized tube within a tube body plan

Areas derived from **ectoderm:**
outer skin, nervous system, and sense organs

Ectoderm

Areas derived from **endoderm:**
GI tract and associated organs, lungs, and liver

Ectoderm

Endoderm

Areas derived from **mesoderm:**
connective tissue, muscle, bone, and blood cells

Ectoderm

Endoderm

Mesoderm

Figure 44-2   A tube within a tube. An animal consists of a tube of endoderm (*yellow*) within a tube of ectoderm (*blue*). In between the two tubes the mesoderm (*red*) develops. In vertebrates such as this duck, the ectoderm forms the outer skin, nervous tissue, and sense organs; the mesoderm forms muscle, bone, blood cells, and connective tissue; and the endoderm forms the intestines, liver, and most other internal organs.

Biologists Separate Vertebrate Development into Eight Stages

Although the process of development is continuous, biologists often divide it into eight separate stages (Figure 44-3).

1. **Gamete formation**—the production of sperm and eggs, a process involving meiosis and cell specialization.
2. **Fertilization**—the fusion of the haploid egg and sperm to form a diploid zygote; fertilization initiates cleavage.
3. **Cleavage**—the division of the zygote into many smaller cells; in many vertebrates these cells form a hollow ball of cells called a **blastula;** in mammals the endpoint of cleavage is a **morula,** a solid ball of cells.
4. **Germ layer formation**—the movement of embryonic cells to form the three germ layers (ectoderm, mesoderm, and endoderm); the resulting structure is called a **gastrula** and the process is called **gastrulation.**
5. **Organ formation**—the movement and differentiation (specialization) of cells to form functioning organs such as the heart, kidneys, and nervous system.
6. **Growth**—the increase in size of an organism after the organs and body plan are established; growth involves both cell division and the production of extracellular materials

(such as bone, cartilage, and hair) and converts the collection of beginning organs into an adult.

7. **Metamorphosis**—a series of changes in form in which the larval form (for example, caterpillar or tadpole) changes into an adult form (for example, butterfly and frog), which has a different morphology and lifestyle; not all animals undergo metamorphosis.
8. **Aging**—further development, which inevitably leads to death; once the adult form is established, cells die, extracellular tissues change, and the efficiency of organs decreases. Aging occurs at different rates in different species and in different individuals within a single species.

Some developmental biologists divide the process of human development into four stages:

1. *Gamete formation*—takes place in the gonads of males and females (Chapter 43)
2. *Embryonic development*—the 8-week period between fertilization and organ formation
3. *Fetal development*—the 30-week period of growth and development through birth
4. *Postnatal development*—further growth and development, followed eventually by aging and death

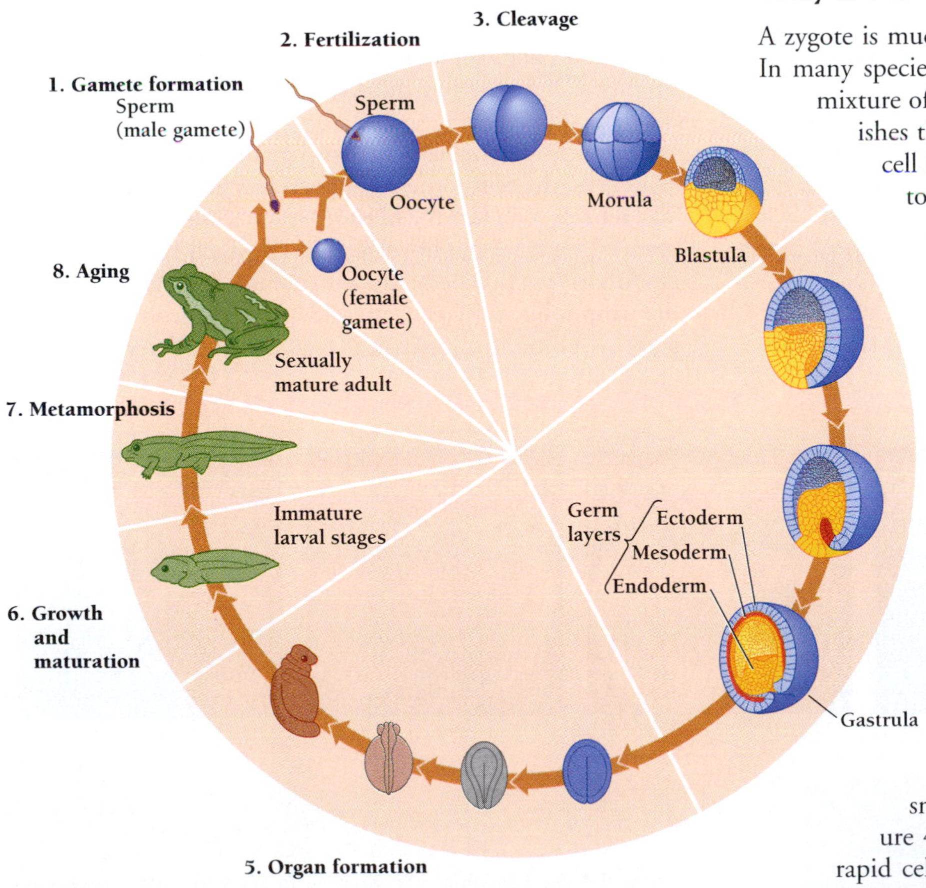

Why Does a Zygote Divide?

A zygote is much larger than the average body (somatic) cell. In many species, the zygote contains large stores of yolk—a mixture of proteins, lipids, and carbohydrates that nourishes the embryo until it can feed itself. Such a large cell has two problems—too little surface area and too little nucleus for the amount of cytoplasm it contains. Because of a zygote's small surface-to-volume ratio, its membrane can transport relatively fewer molecules—whether nutrients or wastes—than a smaller cell.

The zygote's nucleus presents a similar problem. The nucleus of a large cell and a small cell are the same size, containing the same amount of DNA and the same amounts of regulatory proteins. Yet the nucleus of a zygote has far more cytoplasm and membrane to regulate than a smaller cell. A single nucleus cannot meet the enormous demand for mRNA that an embryo has. In addition, the nucleus of a large cell cannot adequately control the distribution of mRNA (and, therefore, protein) in the cell.

The division of the zygote into many smaller cells solves both of these problems (Figure 44-4). A zygote undergoes cleavage, a series of rapid cell divisions. Because the cells do not grow between divisions, the resulting embryo is about the same size as the zygote. The subdivision of the zygote's cytoplasm into 100 or more cells increases both the surface area and the number of nuclei for the embryo's cytoplasm.

Figure 44-3 Stages of vertebrate development. 1. Gamete formation. 2. Fertilization. 3. Cleavage. 4. Germ layer formation. 5. Organ formation. 6. Growth and maturation. 7. Metamorphosis. 8. Aging.

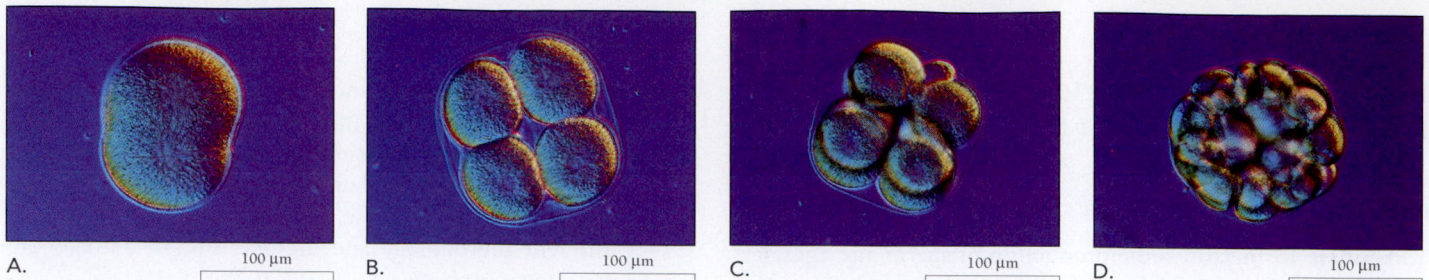

Figure 44-4 Cleavage in a sea urchin embryo. A. The beginning stages of the first cleavage. B. After two divisions, the embryo consists of four cells. C. The 8- to 12-cell stage. D. The 32-cell stage. *(David Fromson, California State University, Fullerton)*

The specialization of cells in a multicellular organism requires that different cell types produce different kinds of mRNA. It should not surprise us, then, that the first process undertaken by the embryo is rapid cell division. Cleavage not only increases surface area and potential mRNA production it also allows each nucleus (and its cell) to specialize.

Cleavage converts the embryo from a single large cell to many smaller cells that can operate independently.

How Do Zygotes Cleave?

Although the zygotes of all species of animals cleave, species differ dramatically in the pattern and pace of these divisions. The different cleavage patterns depend both on the orientation of the mitotic spindles in each division and on the distribution of yolk within the egg. Eggs that contain a lot of yolk either have relatively slow cleavage divisions or do not completely divide the yolk. Bird eggs and fish eggs, loaded with yolk, confine their cleavage divisions to a tiny disc of incompletely separated cells floating on the surface of the yolk (Figure 44-5).

Eggs with little yolk, such as those of the sea urchin, divide more or less equally and symmetrically. In the sea urchin,

each division takes less than an hour. After seven such divisions the sea urchin embryo becomes a blastula, a hollow, fluid-filled ball of 128 cells (Figure 44-6A). The blastula consists of a single layer of cells, which continue to divide, eventually reaching 1000 to 2000 cells, surrounding a cavity called a **blastocoel.**

During the cleavage stage of the amphibian embryo, cells near the top of the embryo, the **animal pole,** divide more rapidly than those near the bottom, the **vegetal pole.** The resulting blastula is several cell layers thick and more asymmetrical than the sea urchin blastula. The cells near the vegetal pole are quite large and laden with yolk.

Cleavage in mammals occurs slowly, each division taking 12 to 24 hours. The orientation of the cleavage divisions is also unique, and, while the cells of a sea urchin embryo divide simultaneously, the individual cells of a mammalian embryo divide at different times.

After a mammalian embryo has divided to form about eight cells, the cells form a compact structure called a morula. The morula soon forms a **blastocyst** [Greek, *blastos* = germ + *cystos* = cavity], a modified blastula in which the cells enclose an internal cavity. The blastocyst contains two types of cells—the **trophoblast** [Greek, *trephein* = to nourish + *blastos* = germ], a prominent outer cell layer, and the **inner cell mass,**

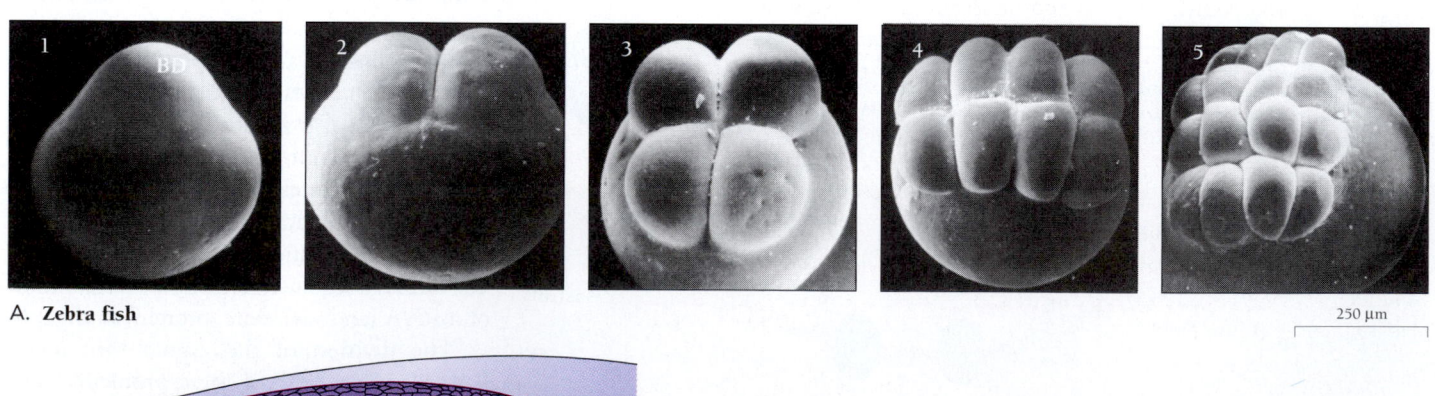

A. Zebra fish

250 μm

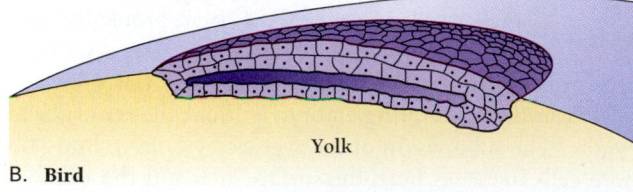

Yolk

B. Bird

Figure 44-5 Discoidal cleavage. A. In the yolky embryo of a zebra fish, cleavage is confined to a small disk on the outer surface of the yolk. B. The embryos of birds also divide by discoidal cleavage, which results in a flattened blastula. *(A, from Beams and Kessel, 1976)*

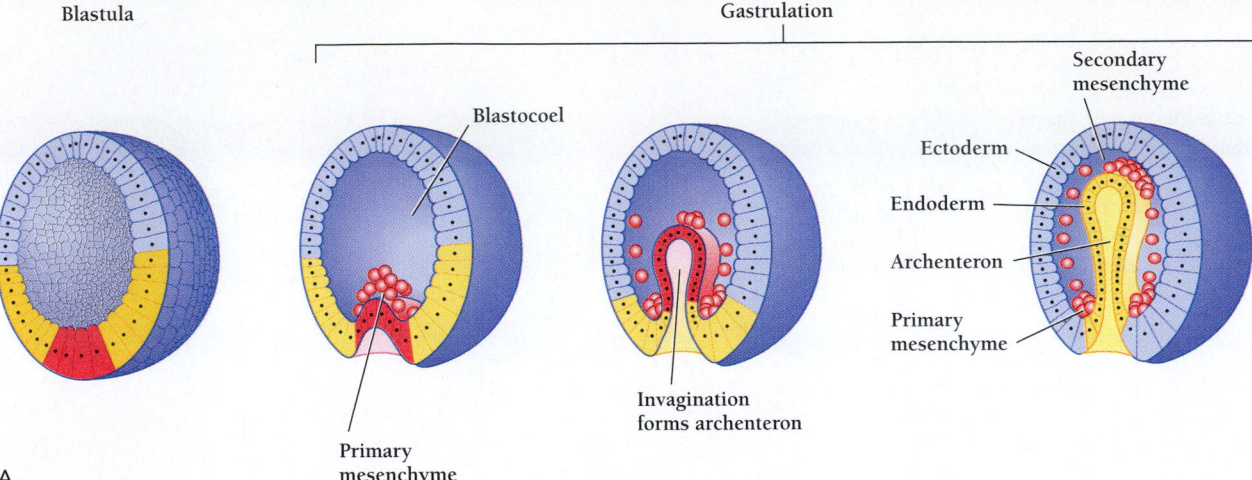

Blastula   Gastrulation

Blastocoel

Secondary
mesenchyme

Ectoderm

Endoderm

Archenteron

Primary
mesenchyme

Invagination
forms archenteron

Primary
mesenchyme

A.

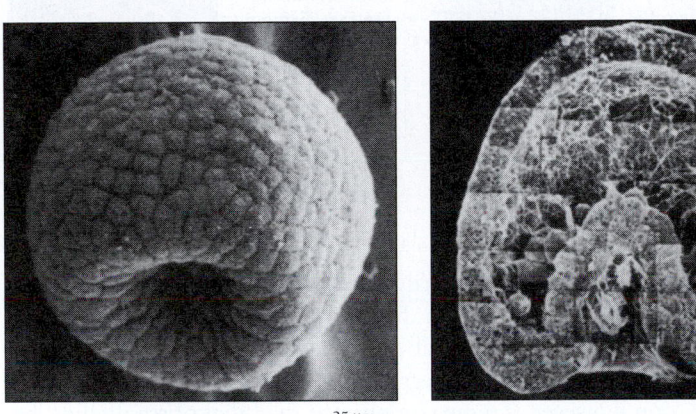

B.   25 µm   C.   25 µm

Figure 44-6 Gastrulation in a sea urchin. A. The sea urchin blastula is a hollow ball of cells. Cells from the bottom of the blastula, called primary mesenchyme, move into the blastocoel and become mesoderm. Endodermal cells invaginate smoothly in through the blastopore, to form the archenteron. Because sea urchins are deuterostomes, like vertebrates, the blastopore will eventually develop into the anus of the mature gut. As invagination proceeds, the cells at the tip of the archenteron detach to form the secondary mesenchyme. These cells form contacts with the wall of the blastocoel and actually pull the archenteron up to the wall. The cells of the secondary mesenchyme disperse into the blastocoel and proliferate to form most of the mesoderm. B. Scanning electron micrograph of early sea urchin gastrula C. Cross section of sea urchin gastrula. (B and C, from Morrill and Santos, 1985)

a small group of cells that will eventually grow into the embryo itself (and subsequently into the adult). The outermost layer of trophoblast cells, called the *chorion,* together with cells from the uterine lining *(endometerium),* form the **placenta,** the organ responsible for the connections between the embryo and the mother. The placenta is a highly vascularized organ—consisting of both maternal and fetal tissue—through which mother and fetus exchange nutrients and wastes.

Fingerlike projections of the chorion, called *chorionic villi* [singular, *villus*], extend into the endometrium and serve as a conduit for nutrients from the maternal blood. Later in development, cells of the inner cell mass proliferate to form a membranelike epithelium, called the *amnion,* which lines the space between the inner cell mass and the trophoblast. This space fills with *amniotic fluid,* which provides the growing embryo and fetus with a stable extracellular fluid and protects them from mechanical shocks. The chorionic villi consist of embryonic or fetal cells, and the amniotic fluid contains unattached embryonic or fetal cells. Both sets of cells can be sampled for DNA (Chapter 14).

Embryos divide at different rates and in different ways depending on species and the amount of yolk in the egg. Sea urchins and amphibians divide to form a blastula. Mammals divide to form a blastocyst, consisting of a trophoblast and an inner cell mass.

Germ-Layer Formation: How Does Gastrulation Set Up the Three-Layered Structure?

The embryo sets up the three germ layers and becomes "a tube within a tube" during the process of gastrulation. The most visible aspect of this process is the formation of the **archenteron** [Greek, *arche* = beginning + *enteron* = gut] or "primitive gut," the space inside the innermost tube. The archenteron is lined with endoderm and will become the digestive tract.

For an easy way to imagine the conversion of a blastula to a layered structure, think about pushing in one end of a tennis ball (Figure 44-6). Imagine pushing hard enough to create a sort of cup. The pushing-in process is called "invagination" and results in a two-layered cup. Now punch a hole in the closed end of the cup. (You will have to cut through two layers.) This exercise does not really explain how the layers form in real animals, but it does illustrate the geometry of gastrulation.

Gastrulation is driven by the changing properties of individual cells. These changing properties include alterations in the attachments of cells to one another and transformations in cell shape, brought about by the cytoskeleton (Figure 44-7).

Figure 44-7 Changes in cell shape underlie cell movement. During gastrulation in a sea urchin, the cytoskeletons of the cells at the bottom of the embryo elongate the cells and then constrict them at their bases. The result is an arch of cells that begins invagination.

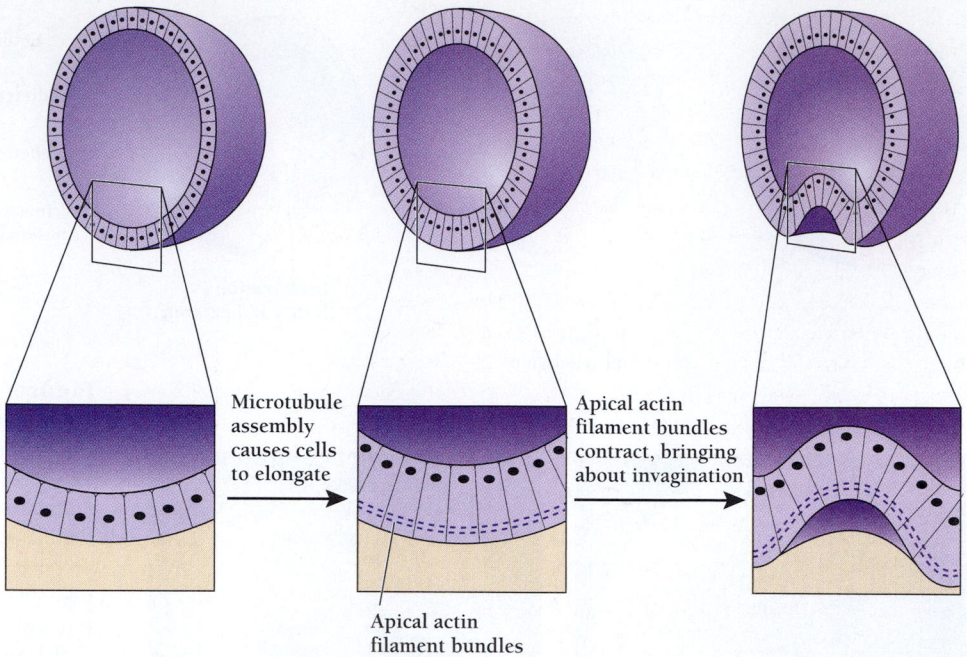

Microtubule assembly causes cells to elongate

Apical actin filament bundles contract, bringing about invagination

Apical actin filament bundles

During gastrulation, a blastula (or blastocyst) invaginates, forming a two-layered cup. The inside of the cup is the archenteron, the inside of the future gut.

How Do Amphibians Gastrulate?

Gastrulation in amphibian embryos depends on the coordinated behavior of sheets of cells and on the interactions of individual cells with extracellular matrices. Amphibians are favorite experimental animals for devel-

opmental biologists, partly because of the relatively large size of the egg (up to 1 mm in diameter). In addition, amphibians are the only vertebrates that can develop in a simple salt solution (rather than within an eggshell or uterus). Amphibian gastrulation, however, is more complicated than that of sea urchins, because of the presence of yolk.

Because the yolk-filled cells near the vegetal pole can move only sluggishly, amphibian cells gastrulate at the embryo's equator, rather than at the vegetal pole (Figure 44-8). Am-

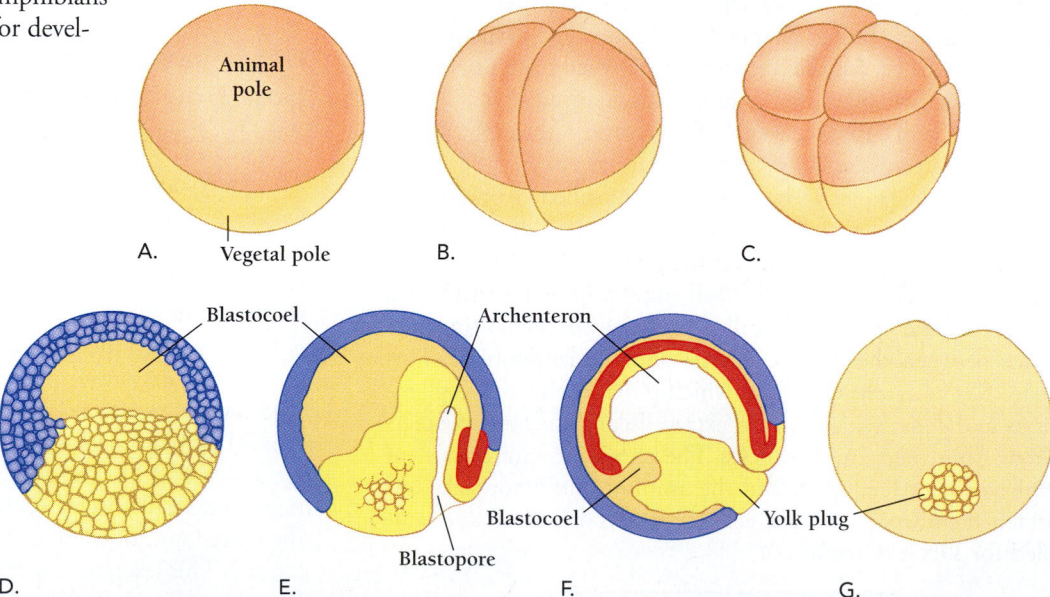

Figure 44-8 Early development in a frog. A. Zygote, showing animal pole and vegetal pole. B and C. Early cleavage stages. The cells at the animal pole divide into many smaller cells than those in the yolky vegetal pole. D. A longitudinal section of a blastula. E and F. Longitudinal sections through gastrulating embryos. As the endoderm (*yellow*) invaginates, a collection of mesodermal cells (*red*) forms at the blastopore (the future anus) and spreads across the embryo's back. This streak of red mesoderm will go on to form the notochord, which takes part in the formation of the spinal cord and spinal column. G. End view of late gastrula showing yolk plug and an indentation at the top that will become the neural fold.

phibian gastrulation begins when an indentation, called the **blastopore,** forms on the side of the blastula. Invagination at the blastopore depends on striking shape changes in cells at the blastopore lip, called bottle cells. Shortly after the initial invagination, adjacent cells follow the bottle cells into the invagination. The result is the formation of a hollow archenteron that is lined with endodermal (gut) cells from the surface of the blastula. As in the sea urchin, a few cells detach from the invaginating layers to form the mesoderm.

The cells of the future mesoderm help power gastrulation. These cells, which are among the first invaginating cells, crawl over the extracellular matrix on the roof of the blastocoel. Two proteins enable the mesodermal cells to attach to the blastocoel roof—*fibronectin,* in the extracellular matrix, and *integrin,* on the cell surfaces. Experiments have shown that when fibronectin and integrin are prevented from binding to each other the embryo cannot gastrulate.

Amphibians invaginate smoothly through the blastopore. The cells of amphibian gastrula can spontaneously organize themselves into three layers of tissue.

How Do Mammals Gastrulate?

The early development of a mammalian embryo follows a pattern more similar to that in birds and reptiles than that in sea urchins or amphibians. In reptiles and their descendants, the birds and mammals, germ layer formation depends on the separation of parallel sheets of cells followed by movements of individual cells.

Recall that a mammalian blastocyst consists of a trophoblast and an inner cell mass. The first separation of cells within the inner cell mass produces two layers, called the **hypoblast** and the **epiblast** (Figure 44-9). The cells of the hypoblast enclose what will become the primitive gut, but these cells are later replaced by endoderm cells, which derive from the epiblast. The hypoblast stays just below the epiblast (separated from it by the blastocoel) to form the blastodisc, the equivalent of a blastula. All the structures of the adult derive from the blastodisc.

The formation of the germ layers now takes place. Cells of the blastodisc move toward a central line, leading to the formation of the **primitive streak,** the mammal's equivalent of the blastopore (Figure 44-10). Individual cells now move through a groove in the streak, into the blastocoel. These cells advance independently into the blastocoel, where they later separate into endoderm and mesoderm.

The mammalian pattern of gastrulation, also found in birds and reptiles, is well suited to the yolky eggs of birds and reptiles, since only the relatively yolk-free cells of the epiblast move. This pattern appears to have evolved early in the history of the land-dwelling vertebrates, when the yolky, eggshell-encased eggs made reptiles and their descendants independent of water. The common pattern of embryonic development in

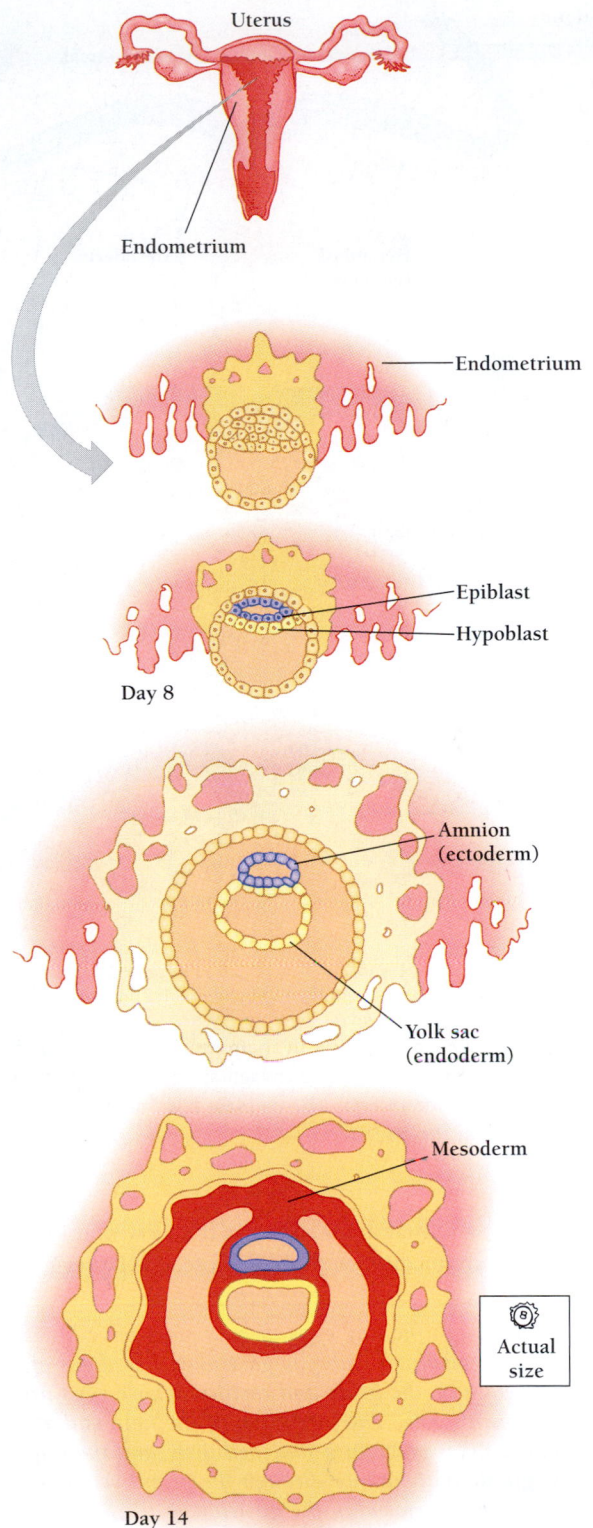

Figure 44-9 A developing human. In mammals, the inner cell mass splits into the epiblast and hypoblast. The hypoblast forms the blastodisc. These form the amnion and the embryonic disc.

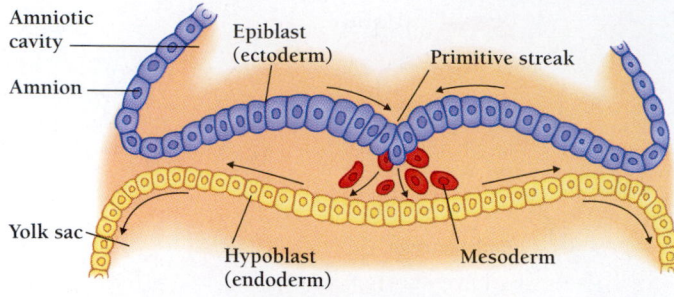

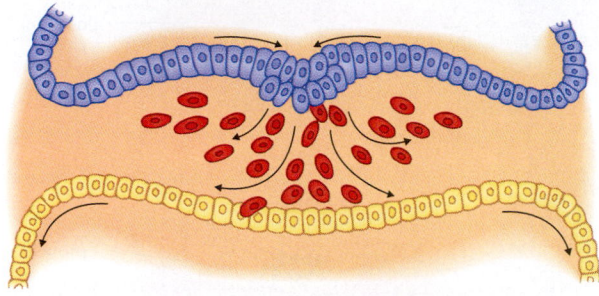

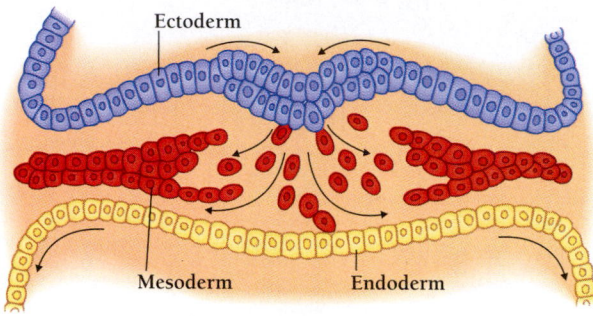

Figure 44-10 Gastrulation in mammals. In humans and other mammals, the epiblast cells invaginate along the primitive streak forming the mesoderm. The epiblast cells may also enter the hypoblast to form the endoderm.

reptiles, birds, and mammals supports the view that mammals evolved from an ancestor with an enclosed, yolky egg.

Even though the tiny eggs of mammals have little yolk, the resulting embryos gastrulate like those of their relatives, the birds and reptiles, whose eggs are heavy with yolk. The formation of the mammalian embryo takes place in a flattened disk, and the cells invaginate individually through the primitive streak, forming the gastrula.

Organ Formation: How Does the Nervous System Form?

Once an embryo has established three germ layers, it next begins to form organs. An organ may derive from a single germ layer or may come from two different germ layers. In each case, organ formation consists of two major processes: **morpho-**

genesis, the creation of form, and **differentiation,** the specialization of cells. The formation of the brain in higher vertebrates illustrates many of the main features of organ formation.

At the end of gastrulation, mesoderm has separated from ectoderm and endoderm. In vertebrates, part of the mesoderm forms the **notochord,** a supportive cord that runs from head to tail, beneath the dorsal (back) surface. The notochord serves both as an internal skeleton for all vertebrate embryos and as an organizer for further embryonic development.

Just above the notochord, the ectoderm rearranges to form the nervous system. The cells along the central line of the embryo thicken to form the **neural plate,** a flat plate above the notochord (Figure 44-11). Then, along the embryo's long axis, the neural plate curls up, as individual cells in the neural plate constrict at one end. The elongated invaginated structure then pinches away from the surface to form a hollow tube, called the **neural tube.** The neural tube is the precursor for the entire central nervous system.

The cells left on the surface, above the neural tube, become skin ectoderm (epidermis), while the cells that had connected the tube to the surface—called **neural crest cells**—migrate away. Descendants of the neural crest cells develop into different types of cells, depending on where they go after they leave the neural tube. Many of them form nerve tissues. But others form pigment cells in the skin, for example. The entire process of forming the neural tube and neural crest cells is called **neurulation.**

The neural tube itself now folds to form the beginnings of the different regions of the brain and spinal cord. The front of the tube enlarges and then bulges on each side to start the development of the cerebral hemispheres. Just behind, a pair of bulges appears. These are the optic vesicles, which will become the eyes.

The development of the eye, summarized in Box 44.1, illustrates several generalizations about organ formation: (1) Organ formation precedes differentiation: only after cells have taken their places do they start to specialize. (2) Organ formation often involves complex folding of sheets of cells. (3) Organ formation depends on interactions among cells that are brought together by movements of tissues and cells.

The notochord induces the formation of neural plate tissue, which folds into a hollow neural tube. The neural tube bends to form the different regions of the brain and the spinal cord. Neural crest cells from the surface of the neural tube migrate into distant parts of the embryo to form nerve and other tissues.

Programmed Cell Death Contributes to Normal Development

Growth and development usually mean both the increase in size of individual cells and the increase in cell numbers due to cell division. Paradoxically, however, the death of cells is also a major part of development. In the last decade, researchers have increasingly realized that development requires extensive "programmed cell death," or **apoptosis**—the death and removal of cells. Cells undergoing apoptosis are distinctive both visibly and biochemically: their nuclei condense, their DNA is systematically digested, and they are methodically engulfed by other cells.

Apoptosis occurs in many developmental pathways in both vertebrates and invertebrates. It is particularly common among cells of the vertebrate immune and nervous systems, for example. Ordinarily, when a cell dies, as a result of attack by microbes, for example, the cell bursts, or lyses, and releases toxic cell products that can disrupt other developmental processes. Lysis can also provoke the immune system to initiate an immune attack on healthy cells. In apoptosis, cells die quickly. They are engulfed by phagocytes and digested without leaving a trace.

One of the best-studied examples in vertebrates is the programmed death of cells between the digits of the embryonic limbs (Figure 44-12). In the early embryo, a thin tissue lies between the digits (fingers and toes in humans). This tissue persists as the familiar webbing of a duck's hind limbs, but it is destroyed during development in chickens and other non-aquatic vertebrates (including humans). The process of making a hand or a foot resembles the work of a sculptor, who chisels away marble to create a statue.

Apoptosis is also well studied in the adult nematode worm *Caenorhabditis elegans.* The adult worm, which looks like a 1-mm-long transparent tube, consists of only about 1000 cells. Its simplicity and transparency have enabled researchers to track the precise pathway of every cell that arises during development.

A number of cells that arise in a *C. elegans* embryo undergo apoptosis. For example, the sisters of cells that develop

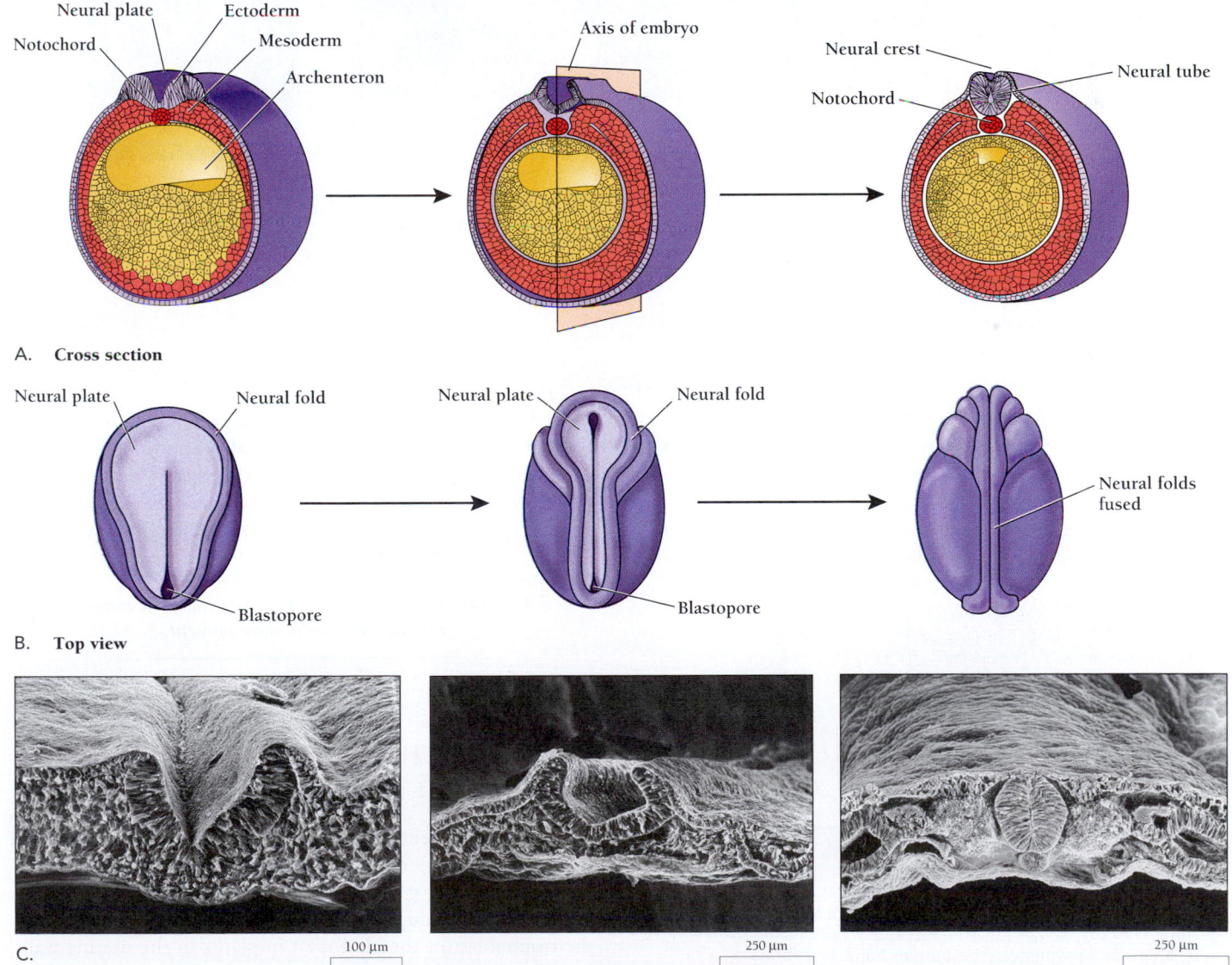

Figure 44-11   Formation of the neural tube. A. Cross section. B. Top view. C. Micrographs showing the neural folds coming together and fusing at the top. (*C, K.W. Tosney*)

BOX 44.1

How Does the Eye Form?

The eye forms as a result of a simple set of cell movements (Figure A). When the optic vesicle comes into contact with the overlying skin ectoderm, the ectoderm thickens to form a plate that will develop into the lens. Once the lens thickening occurs, the optic vesicle folds back on itself (or invaginates) to form a double-walled optic cup, and the lens plate curls and pinches away from the skin ec-

toderm to form the lens vesicle. The stalk that attaches the cup to the rest of the brain becomes the optic nerve, which will eventually carry visual information from the retina to the brain.

After this complex set of thickenings, foldings, pouchings, and pinchings, the cells of the lens, the retina, and the overlying skin ectoderm begin to differentiate. The cells of the lens make specialized pro-

teins, called crystallins, that give the lens its optical properties. The cells overlying the lens change character to become the cornea, the transparent covering of the eye. Finally, the cells of the retina develop into light-detecting rod and cone cells. Through these events a seemingly uniform layer of cells in the neural tube develops into an eye, our most prized contact with our surroundings.

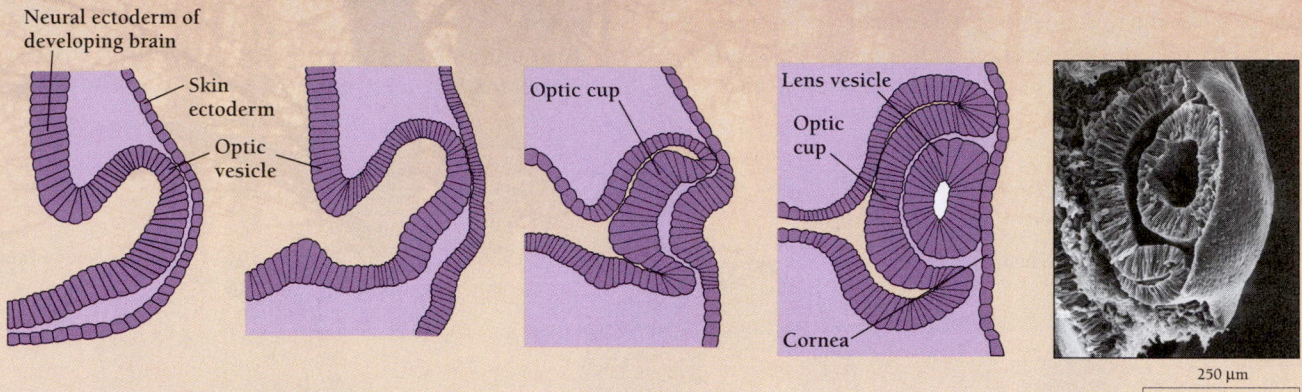

Figure A **Formation of the eye.** A sheet of neural ectoderm folds to form the optic vesicle and optic cup. The skin ectoderm responds by forming the lens, as shown in the micrograph on the right. *(Photo, courtesy of K.W. Tosney)*

---

into nerve cells of the adult worm normally undergo apoptosis. If a mutation prevents apoptosis, however, these nerve cells survive and grow and the worm's nervous system develops with too many neurons.

Apoptosis in vertebrates depends on gene action. Many of the genes that activate apoptosis are identical to genes that reg-

ulate the cell cycle. The product of one gene, called *bcl*-2, helps prevent apoptosis not only in mammals but also in *C. elegans*. Biologists can only conclude that the genes for cell death evolved at least 600 million years ago, when the common ancestor of mammals and nematodes first split into two lines.

Programmed cell death, apoptosis, plays a major role in embryonic development.

Figure 44-12 **Sculpture and development.** Like a sculptor chiseling stone, the programmed death of individual cells generates form in a developing animal. The human hand first forms as a fin-like paddle that lacks both fingers and thumb. As cells die in selected areas, however, the general shape of the digits emerges. Later, the specific tissues—bones, muscles, nerves, and blood vessels—will emerge within each digit.

HOW DOES HUMAN DEVELOPMENT PROCEED AFTER IMPLANTATION?

In mammals in general, and in humans specifically, fertilization occurs in the oviduct (fallopian tube), as described in Chapter 43. The zygote then begins its cleavage divisions, moving down the oviduct toward the uterus. After about 7 days, the trophoblast of the blastocyst implants in the uterine wall.

Between 9 days and 16 days after fertilization, the cells of the inner cell mass undergo gastrulation (germ layer formation) producing endoderm, ectoderm, and mesoderm. By 31 days, the placenta is fully functional, moving oxygen and nourishment to the embryo and carrying away carbon dioxide and other embryonic wastes.

The placenta can also transport less benign substances. Viruses, such as rubella ("German measles") and HIV (which causes AIDS), can move through the placenta. So can alcohol, legal and illegal drugs, and the toxic chemicals from tobacco smoke. Many of these substances are **teratogens**, chemical substances and infectious agents that cause abnormal development. In the late 1950s, for example, researchers discovered that thalidomide, a sleeping pill and tranquilizer widely prescribed by doctors in Germany and England, was a teratogen. Some 7000 pregnant women who took this drug gave birth to babies with abnormal stumps (resembling a seal's flippers) rather than normal arms and hands.

When development is not disturbed by teratogens, mutations, or other misadventures, the embryo develops into a fetus. An embryo is called a **fetus** when every organ appears as a recognizable **rudiment** [Latin, *rudimentum* = beginning], or initial stage, from which the final form will develop. In humans, all these organs are in their proper places by the end of 8 weeks. The embryo is then said to be a fetus.

But the fetus is far from fully formed. The final form of each organ requires extensive cell specialization and growth. Each human hand and arm, for example, consists of 43 muscles, 29 bones, and hundreds of nerve pathways. None of these are present in the rudiment of an arm. In fact, except for the heart, the organs of an early fetus are largely nonfunctional. In addition, a 9-week human fetus is just a little over an inch long. Only with further growth and development does an early fetus become a baby.

Different parts of the body grow at different rates. A 9-week fetus, for example, has a head nearly the same size as all of the rest of its body. Although the head will never again be as proportionately large, it continues to grow and develop in advance of the other parts of the body. The brain, eyes, jaws, lungs, stomach, intestines, and kidneys—all structures that the baby will need when it is born—grow and develop most rapidly, while the legs and feet lag behind. Between 9 and 12 weeks the fetus doubles in length, the external genitalia appear and begin to develop, and the kidneys excrete their first urine (Figure 44-13).

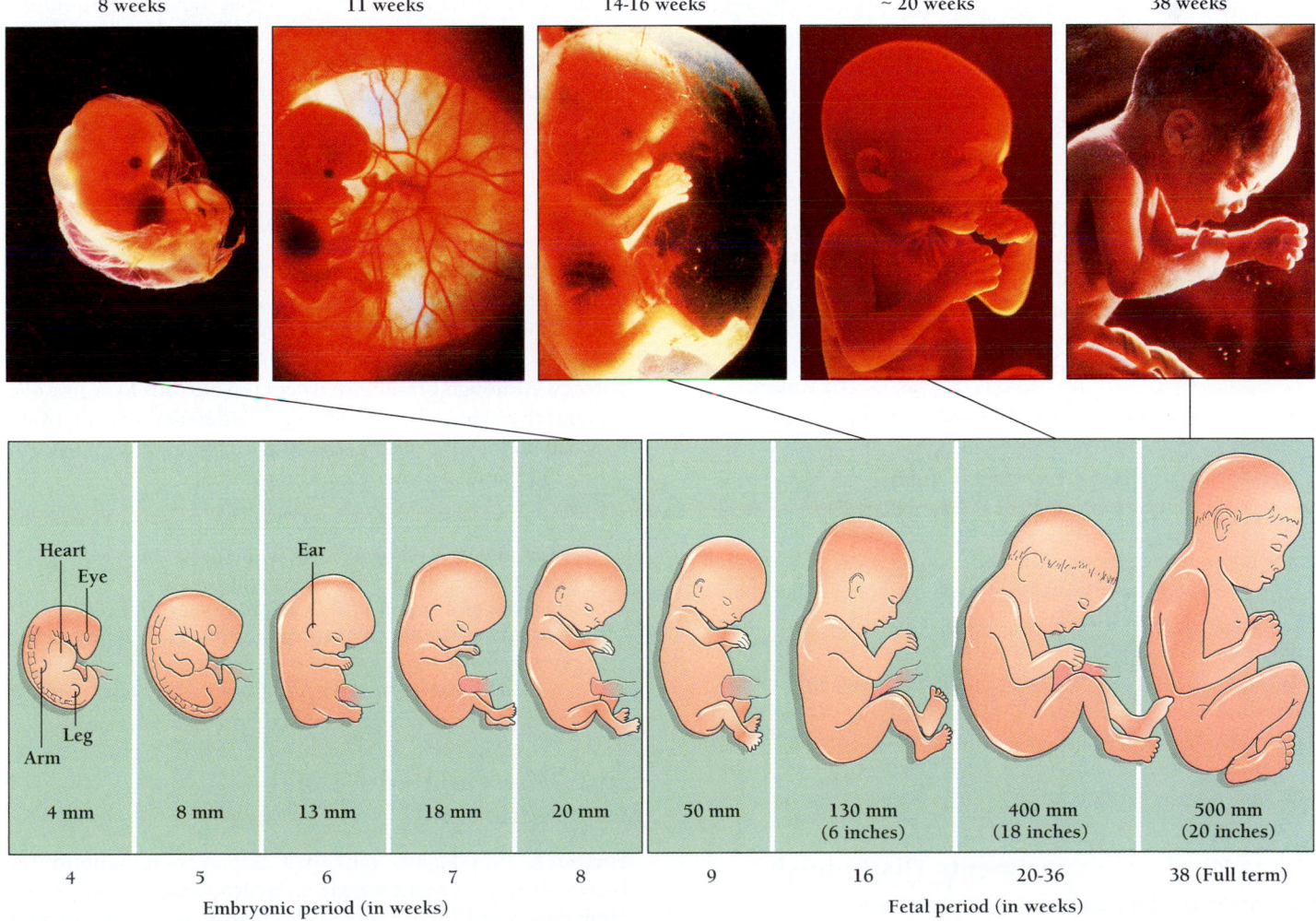

| 8 weeks | 11 weeks | 14-16 weeks | ~ 20 weeks | 38 weeks |

Heart
Eye
Ear
Leg
Arm

| 4 mm | 8 mm | 13 mm | 18 mm | 20 mm | 50 mm | 130 mm (6 inches) | 400 mm (18 inches) | 500 mm (20 inches) |

| 4 | 5 | 6 | 7 | 8 | 9 | 16 | 20-36 | 38 (Full term) |

Embryonic period (in weeks) | Fetal period (in weeks)

Figure 44-13   Stages of fetal development in humans. In humans, most organ formation occurs in the first 8 weeks, when the embryo is less than an inch long. At 9 weeks, the fetal period begins. Organs mature, and the fetus grows. (*Photos [from left to right], Science Pictures Ltd./SPL/Photo Researchers, Inc.; Petit Format/SPL/Photo Researchers, Inc.; Petit Format/Nestle/ Photo Researchers, Inc.; James Stevenson/SPL/Photo Researchers, Inc.; Petit Format/Nestle/Photo Researchers, Inc.*)

Between 12 and 16 weeks, a human fetus doubles its length again. By 16 weeks, the ovaries are differentiated and the primary follicles contain oogonia. Between 17 and 20 weeks, the mother feels the first fetal movements. By 20 weeks, the fetus is about 10 inches long, and, in males, the testes have begun to descend.

By 22 weeks, the lungs, intestines, and kidneys are sufficiently developed that it is sometimes possible—with intensive medical care—for the fetus to survive outside the womb. Between 20 weeks and 26 weeks, the fetus begins to fill out. At 24 weeks, the lungs begin to secrete a *surfactant,* a detergent that allows the lungs to inflate with air (Chapter 38).

After 26 weeks, a fetus born prematurely has a good chance of surviving—because its respiratory system is mature and functioning. Between 26 weeks and 36 weeks, when most babies are born, the fetus gains weight in the form of fat. By 36 weeks, the circumference of the baby's abdomen equals the circumference of its head.

Physicians usually describe the stages of human development in terms of **trimesters,** periods of about 3 months each. Most of the important developmental changes occur during the first trimester. The first 3 months are the time when there is the greatest chance of a **miscarriage,** or **spontaneous abortion.** Approximately 50 to 75 percent of embryos abort during the first trimester. Figure 44-13 summarizes the stages of human development. In Chapter 43, we briefly discuss the effects of pregnancy on the mother, as well as childbirth and lactation.

An embryo forms the bare outlines of all of its organs long before those organs are functional. At the end of 8 weeks, a human embryo possesses rudiments of all the major organs and is said to be a fetus, but only the heart is functional.

EMBRYONIC CELLS BECOME INCREASINGLY DIFFERENTIATED

Hans Driesch's experiment showed that each cell of a two-cell embryo has the ability to develop into a whole (rather than a half) sea urchin. With this and similar results in mind, embryologists have come to distinguish the **fate** of a cell—what it becomes during development—from its **potency**—what it could become if allowed to develop in another environment. In the case of the two-cell sea urchin embryos, the *fate* of each is normally to become the right or left half of the sea urchin, but each has the *potential* to become a whole organism.

What Kinds of Experiments Distinguish Potency and Fate?

At early stages of development and for relatively simple organisms, we may follow the fates of individual cells just by watching and photographing them. By observing the move-

ments of cells in developing sea urchins, for example, biologists have determined which cells give rise to endoderm, which to mesoderm, and which to ectoderm.

In a frog, a bird, or a mammal, however, gastrulation is too complicated to follow without marking individual cells—with dyes, radioactive tracers, or particles of carbon. By following the movements of marked cells, embryologists have been able to construct "fate maps" that show what happens to each cell during development.

Defining the potential of a cell is more difficult. One way of determining the development potential of embryonic cells is to transplant them from one embryonic environment to another. In 1918 the German embryologist Hans Spemann performed a now-famous transplant experiment in two species of newts. One species was darkly pigmented and the other was pale.

Spemann transplanted pieces of the dark, pigmented embryo into the pale, unpigmented "host" embryo. Because of the pigment, he could distinguish easily between structures formed from the host cells and those formed from the transplanted cells. Having studied the fate maps of the newt, Spemann knew which embryonic cells should form neural plate and which should form skin. Cells known to form a particular tissue are described as "presumptive," as in "presumptive ectoderm cells" or "presumptive skin cells."

Spemann transplanted presumptive neural plate (nerve tissue) into an area of presumptive skin cells. If he transplanted the presumptive neural cells at the early gastrula stage, they formed skin instead of a neural plate. That is, transplanted cells developed according to their new environment. In short, their potency was greater than their fate.

If, on the other hand, Spemann waited until the late gastrula stage, the presumptive neural cells developed into neural plate, even though they were in the wrong environment. Spemann concluded that later in development, cells become *committed* to their normal developmental fate, gradually losing potency. Although presumptive neural plate cells look just like presumptive skin cells, they are already **determined,** meaning they can no longer develop according to their environment. In short, their fate is greater than their potency.

The fate of a group of cells is the tissue they will become during the course of normal development. The potency of a group of cells is the tissues that they could become under varying circumstances in different environments. Hans Spemann devised experiments for distinguishing a cell's fate from its potency.

Are Differentiated Animal Cells Totipotent?

Spemann's experiments raise two important questions: (1) How does a cell's environment influence its development? (2) How does a cell become determined, so that its environment no longer influences its development?

Modern biologists think of determination as the process that establishes which genes will be expressed and which will

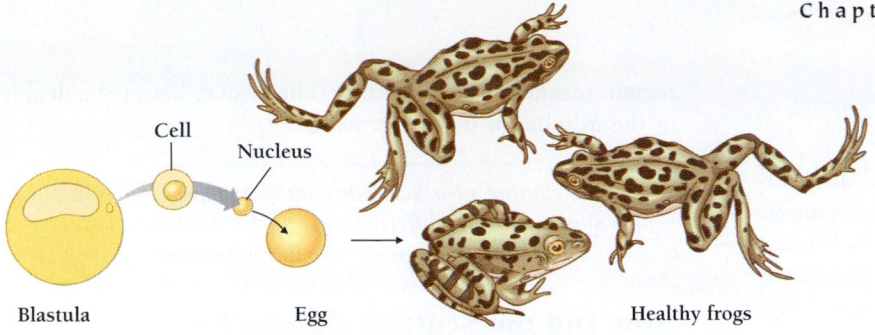

Cell

Nucleus

Blastula

Egg

Healthy frogs

Figure 44-14 Do the nuclei of developing embryos remain totipotent as they differentiate? The experiments of Briggs and King suggested that as cells differentiate, their nuclei become less able to direct complete development. A nucleus from a blastula cell could direct the development of an egg into a normal leopard frog; a nucleus from a gastrula could direct development in some cases; but a nucleus from a neurula cell could never direct complete development.

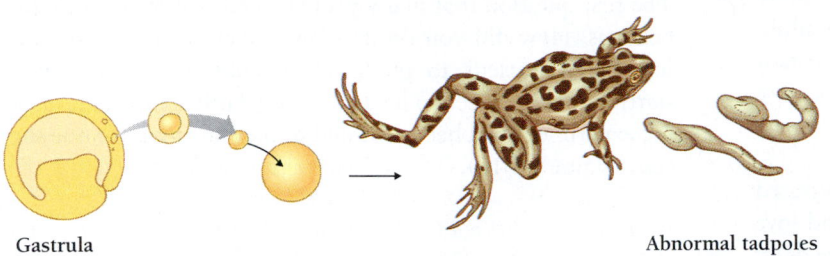

Gastrula

Abnormal tadpoles

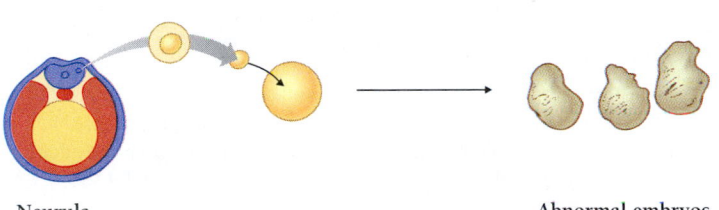

Neurula

Abnormal embryos

not, a process that is influenced by environmental signals. Spemann's transplantation experiment showed that determination in neural plate cells depends on environmental changes brought about by the movements of gastrulation. Further changes in the cellular environment bring about changes both in cell shape and in the pattern of gene expression that is responsible for neural development.

An easy explanation for determination would be that during development cells lose genes. For example, do presumptive neural cells lose the genes necessary to form skin proteins? In the 1950s, developmental biologists began to approach the

question experimentally. Robert Briggs and Thomas King devised a method for nuclear transplantation, in which they removed the nucleus of a leopard frog egg and replaced it with a nucleus from a cell from another frog (Figure 44-14). The egg developed under the control of the transplanted nucleus. Briggs and King soon discovered that nuclei from a blastula were totipotent: many of them could direct the development of the whole frog.

Yet Briggs and King never were able to produce a swimming tadpole from a nucleus that had developed beyond the neurula stage. The researchers concluded that, in leopard frogs at least, nuclei lose the potential to direct complete development.

Later researchers found that nuclei from the intestines of the South African clawed toad tadpoles *could* generate a normal adult (Figure 44-15). Were fully differentiated animal cells totipotent? Until 1997, this pressing question remained unanswered.

Early transplantation experiments did not make clear whether differentiated animal cells were totipotent.

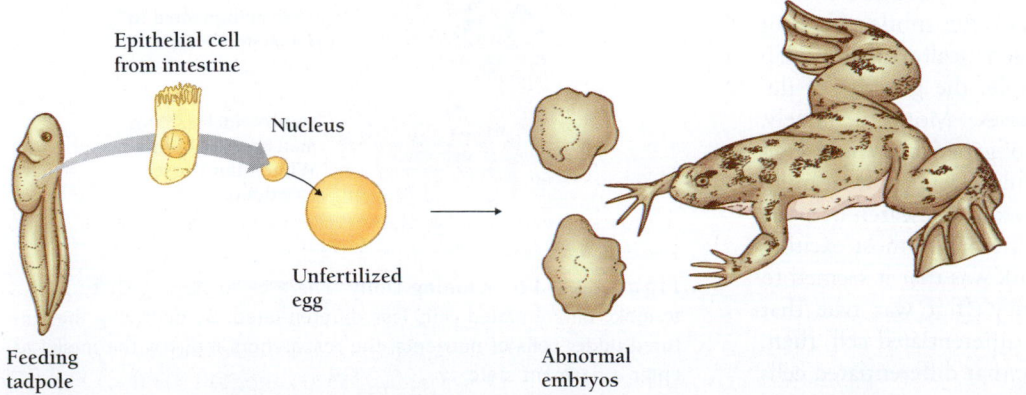

Epithelial cell
from intestine

Nucleus

Unfertilized
egg

Feeding
tadpole

Abnormal
embryos

Figure 44-15 Differentiated and totipotent? In the 1960s, John Gurdon repeated the experiments of Briggs and King in the South African clawed toad. Gurdon transplanted nuclei from the intestines of swimming tadpoles to egg cells. Although most of the resulting zygotes did not develop normally, about 1 percent developed into normal adult frogs whose cells contained nuclei identical to those of the donor tadpole.

Can a Differentiated Nucleus Direct Development from Egg to Adult?

A singular consequence of Briggs and King's nuclear transplantation experiments was that, for the first time, biologists could produce a group of genetically identical animals. Because the nuclei from the blastula cells were identical (all resulted from the cleavage of a single zygote), the resulting leopard frogs were clones of one another.

For the first time, zoologists could clone an animal. Gardeners had been cloning plants for centuries, for virtually every plant cell is totipotent and may develop into a complete adult. But no one had ever before cloned an animal. One important reason was that as animal cells differentiate, they seem to lose the ability to direct development. Why that should be so was not at all clear.

Animal breeders and biologists have long been motivated to clone animals. A farmer with a prize dairy cow would love to be able to clone such an animal. A whole herd of prize dairy cows would not only ensure a steady supply of milk for the life of the individual cows but it would so do indefinitely. More alluring still, a farmer with such a herd could sell the clones to other farmers with less productive animals.

Breeders would like to be able to clone adult animals with proven traits. No one can know an animal's worth until it is mature. But until recently, animal breeders never succeeded in cloning cells from adult animals. Biologists' efforts with mice had failed over and over. Animal breeders clone cells from embryos, but an embryo is basically an unknown: it is not a productive milk cow, a Derby-winning racehorse, or even a lab animal with a perfectly understood genotype and phenotype.

Then, in February of 1997, a team of biologists at the Roslin Institute, in Edinburgh, published a mind-bending result in the British journal *Nature*. The team, led by Ian Wilmut and Keith Campbell, had successfully reared a Scottish mountain sheep grown from an egg cell containing a nucleus transplanted from a mature cell from the udder of an adult sheep. In other words, the sheep, named Dolly, had been cloned from an adult animal. Dolly's impassive face appeared on the front pages of newspapers and magazines around the world. If biologists could clone sheep, commentators asked, why not people? Governments hastily passed laws banning the cloning of humans, although it was unclear if the cloning of humans would be possible.

Geneticists were equally excited. If they could clone lab animals, they realized, they could study the subtle effects of environment on whole colonies of genetically identical lab animals. They could study, for example, the genetics of development as well as genetic diseases. More lucratively, biotechnologists could generate herds of genetically engineered animals, such as goats that secrete useful proteins in their milk.

But developmental biologists viewed the research in a different light. From their perspective, one of the most exciting aspects of Wilmut and Campbell's work was that it seemed to answer the question about totipotency. If it was true that Dolly's nuclei had come from a fully differentiated cell, then, at least in sheep, it was possible to say that differentiated cells

remain totipotent. But do they? The answer lies in the details of the Edinburgh biologists' work.

The cloning of a Scottish mountain sheep in 1996 suggested that differentiated nuclei are totipotent.

How Did the Scottish Researchers Clone Dolly?

The first question that many people asked Wilmut and Campbell was, How did you do it? How had they prompted a differentiated nucleus to guide a full course of development starting in an egg cell? The answer was fairly simple, although the experiment had been repeated numerous times for one success (Figure 44-16).

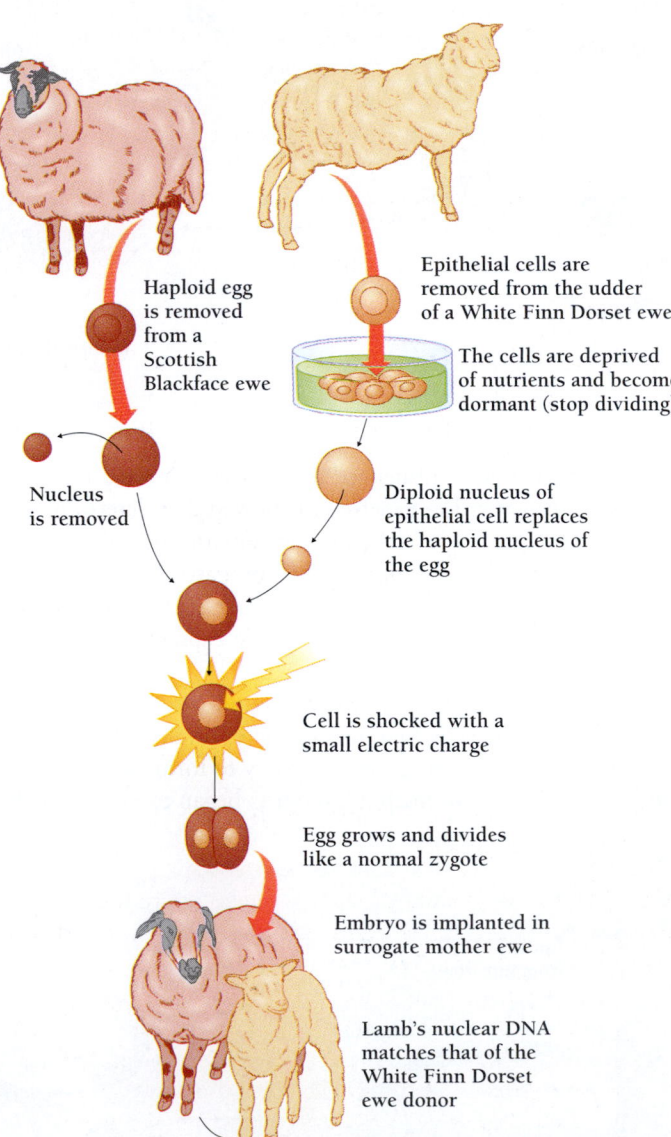

Haploid egg is removed from a Scottish Blackface ewe

Epithelial cells are removed from the udder of a White Finn Dorset ewe

The cells are deprived of nutrients and become dormant (stop dividing)

Nucleus is removed

Diploid nucleus of epithelial cell replaces the haploid nucleus of the egg

Cell is shocked with a small electric charge

Egg grows and divides like a normal zygote

Embryo is implanted in surrogate mother ewe

Lamb's nuclear DNA matches that of the White Finn Dorset ewe donor

Figure 44-16   Cloning Dolly. The trick to cloning Dolly was to make differentiated cells less differentiated. By depriving the cultured udder cells of nutrients, the researchers induced the nuclei to enter a dormant state.

The trick, Wilmut said, was to make the DNA of the donor cells behave more like the inactive DNA of an egg or sperm. They did this by growing the udder cells in culture and starving them of essential nutrients. Gradually, the cultured cells passed into a dormant state in which many genes shut down and replication was impossible. Once the DNA in the donor nuclei was dormant, the team of biologists began transferring the nuclei into egg cells. Painstakingly, they transplanted 277 nuclei, of which only one successfully developed.

During the first three divisions of the sheep embryo, the cells replicate their DNA without expressing any nuclear genes. All cell processes remain under the guidance of maternal proteins and RNA. The inactive DNA in the transplanted nucleus replicates but the cell transcribes no RNA from the DNA. Meanwhile, the nuclear DNA begins losing proteins that it brought from the udder cell, proteins that at first prevent the expression of the nuclear genes. By the third division, these proteins are replaced by proteins from the egg's cytosol. The maternal (egg) proteins then "reprogram" the nuclear DNA, and the embryo begins expressing its own genes.

Although maternal mRNA and maternal proteins are sufficient to guide cell division in sea urchins as far as the blastula stage, the nuclear DNA of most mammalian embryos comes into play much earlier. Mouse nuclei begin expressing their DNA after just one division. Human nuclei begin after two divisions.

Some biologists argue that the Scottish team succeeded where others had failed because the DNA of sheep takes so long to turn on. Three cleavage divisions perhaps allowed the egg cell proteins enough time to replace the udder proteins.

Wilmut, Campbell, and colleagues cloned Dolly the sheep by inducing cultured udder cells to enter a dormant state, in which gene expression was shut down. The nuclei from 277 such cells were then transplanted into sheep oocytes, which divided three times. Only one of the zygotes developed into an adult sheep.

Did Dolly Really Come from a Differentiated Cell?

Other biologists question whether Dolly is the product of a differentiated cell at all. Mammary cells, including cultured cells, typically contain not only differentiated cells for producing milk but also a variety of other cells. They are a mixed population of cells. Some of these cells are **stem cells,** cells that can either produce more cells like themselves by cell division or undergo differentiation to one or more specialized cell types.

Adult vertebrates contain many kinds of precursor stem cells. Any cells that must regenerate during adult life—such as the cells of the skin, the linings of intestines, and the blood—come from a supply of stem cells. But stem cells are less differentiated than other cells. The developmental fate of stem cells—whether they proliferate to make more stem cells

or differentiate to produce specialized cells—depends on cues from the environment.

Wilmut and Campbell do not know if the nucleus that gave rise to Dolly was a fully differentiated udder cell or an only partially differentiated stem cell. For the purposes of animal breeders, the difference is of minor importance. For developmental biologists, however, the question of whether a fully differentiated egg is totipotent is as interesting today as it was in 1950.

Because Dolly may have come from a stem cell nucleus, the totipotency of fully differentiated animal cell nuclei remains in question.

HOW DOES DEVELOPMENTAL FATE DEPEND ON CHEMICAL SIGNALING?

Since virtually every cell in an animal has the same genes, differences in cell fates must result from selective gene expression. Differences in the patterns of gene expression must be influenced by cellular environment. The question then is, How do cues from the embryo influence the developmental fates of individual cells?

Can the Extraordinary Actions of the Primary Organizer in Amphibian Development Be Explained in Terms of Cells and Molecules?

Arguably the most spectacular experiment in the history of developmental biology was one performed by Hans Spemann's student Hilde Mangold. Recall that Spemann had established that cells become determined during gastrulation. Mangold and Spemann asked *how* cells become determined.

The two researchers correctly guessed that determination of the ectoderm into neural tissue or into skin depended on contact between the ectoderm and the underlying mesoderm, which derives from the cells that invaginate during gastrulation (Figure 44-8). Mangold asked what would happen if the invaginating cells—the dorsal lip of the blastopore—were transplanted to another region of the embryo. Amazingly, the transplanted dorsal lip not only invaginated, it also induced the tissues around it to form a nearly complete second embryo (Figure 44-17).

Spemann termed the dorsal lip the **primary organizer,** meaning that it established the entire organization of the embryo. He was so astonished by Mangold's result that he compared the action of the primary organizer to the workings of the mind—suggesting that it could not really be understood in chemical and physical terms.

Subsequent work, however, has demonstrated that the primary organizer's mechanism is chemical. For example, chemical extracts of the dorsal lip induce differentiation of ectoderm.

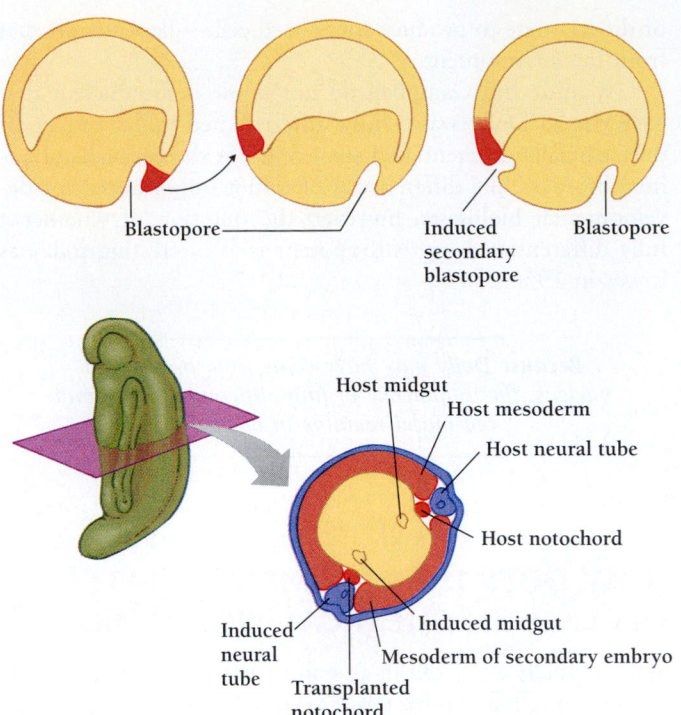

Figure 44-17   The primary organizer. Tissue from the lip of the blastopore, where gastrulation begins, can induce gastrulation and neurulation in an inappropriate place.

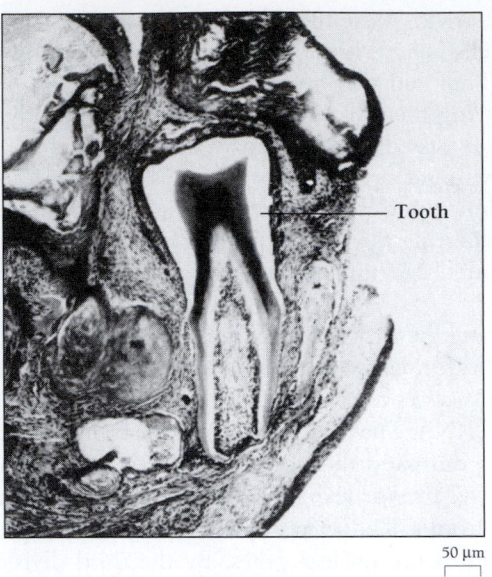

Figure 44-18   Waking sleeping genes. The power of tissues to induce differentiation in other tissues is enormous. Birds have not had teeth since they diverged from other reptiles millions of years ago. Yet when researchers transplanted mesoderm from the mouth of a mouse embryo under the epidermis of a chick embryo, the mouse cells induced the chick cells to make enamel in formations remarkably like teeth. *(From E. J. Kollar and C. Fisher, 1980. Science, 207:993-995. Copyright 1980 by the American Association for the Advancement of Science)*

Researchers do not yet fully understand the mechanism of embryonic induction. They have, however, identified a number of molecules that help to establish patterns of development within whole embryos and within individual organs.

In the last 10 years, developmental biologists have learned much about the molecules responsible for establishing such patterns. We now know, for example, that at least three proteins—called noggin, chordin, and follistatin—are responsible for inducing ectoderm to become neural tissue. These molecules all inhibit the action of another inducing protein—called bmp-4 [for "bone morphogenic protein"]—which prevents neural induction.

These same signaling molecules play important roles in the patterning of mesoderm. Noggin and chordin proteins can induce embryonic mesoderm cells that would otherwise form blood cells to form muscle and notochord, whereas bmp-4 prevents this induction. So the embryo uses the same molecular machinery to establish different patterns.

Still more surprisingly, invertebrate versions of the same proteins are responsible for pattern formation in *Drosophila.* As we discussed in Chapter 24, bmp-4 is the vertebrate version of a fly protein called dpp, and chordin is the vertebrate version of a fly protein called sog. Bmp-4 and chordin antagonize one another both in frogs and in flies. While bmp-4 promotes the development of ventral structures in vertebrates, however, its homolog dpp favors the development of dorsal structures in invertebrates. Frogs and flies use the same sig-

naling system, at least half a billion years old, to start neural development, but they use them in opposite ways.

Widely divergent species also use other common signaling molecules. Among these signals are many other proteins and a number of small molecules. One small molecule that is important in pattern formation is retinoic acid, which regulates the pattern of digit formation in the developing limb bud of the chick embryo and in regenerating limbs of newts. In the chick limb bud, added retinoic acid can induce the formation of extra digits, with the number of extra digits depending on the concentration of retinoic acid. In a regenerating newt limb, retinoic acid is present in a concentration gradient, with the highest concentration in the part of the limb farthest from the body.

Most of the proteins and small molecules that affect development affect the transcription of specific genes in their target cells. Retinoic acid, for example, binds directly to several proteins that regulate transcription directly. Many of the proteins that contribute to pattern formation also work by influencing transcription, but they do so less directly. Often a signaling protein (such as bmp-4) binds to a receptor on the surface of the target cell and stimulates a cascade of events that regulate the activity of a transcription factor. The altered transcription factor stimulates the production of new mRNAs and proteins, which in turn change the character of the target cell—

for example, stimulating an undifferentiated precursor cell to become a nerve cell.

Hilde Mangold's transplantation experiments showed that the primary organizer establishes the overall orientation (axis) of an embryo. Recent work has shown that the primary organizer contains many of the same pattern-forming proteins used by many species.

How Do Cells "Know" Where They Are and What To Do About It?

Development requires both that individual cells express appropriate genes and that cells be organized into functioning tissues and organs. The development of a limb, for example, requires the differentiation of the cells of the muscles, skin, and bone into a characteristic pattern, distinct from the organization of the same cell types in a foot. How does the developing organism achieve this pattern?

Part of the answer to pattern formation lies in the interactions between layers of cells. For example, the skin ectoderm of a chicken can give rise to several different kinds of structures—fully tufted feathers, partly tufted feathers, scales, and claws (Figure 44-18). The developmental fate of the ectoderm depends upon the source of the underlying mesoderm. Thus, if wing ectoderm is placed over foot mesoderm, claws will develop instead of feathers.

The overall pattern of a limb, however, does not depend on short-range interactions between adjacent cell layers. Instead, special mechanisms specify positional information, chemical cues that establish the position of a cell in a pattern. The concentrations of particular substances, called **mor-**

phogens, specify the contribution of each cell to a pattern, in the same way that a seat number can be used to specify the contribution of each member in a cheering section that shows the image of a team's mascot in a card display.

The formation of patterns during development depends on each cell's "knowing" *where* it is. It also depends on the interpretation of this information: each cell must also "know" *what* it is. At the tip of a developing leg bone, cells must develop into digits, while cells nearer the body must develop into the other bones, such as the femur and the tibia.

In vertebrates, one example of a morphogen is retinoic acid. In chicks, salamanders, and frogs, retinoic acid induces the formation of limbs—complete with humerus, radius, and ulna. Like a steroid, retinoic acid apparently directly influences gene expression by interacting with a transcription factor. Recall from Chapter 11 that steroid hormones such as progesterone and testosterone bind to receptor proteins inside a cell. The steroid-receptor complex then passes into the cell's nucleus and regulates the transcription of specific genes. Similar receptors bind to retinoic acid.

Which genes are expressed, however, differs from cell to cell. In the muscle cells of a male mammal, for example, testosterone stimulates cell division and growth, while in skin cells, testosterone stimulates the growth of hair. The cells' response to testosterone depends on already established differences among the cells.

Researchers have identified a number of genes responsible for pattern formation in *Drosophila.* The first of those identified were two similar genes that—when mutated—led to a couple of bizarre flies called bithorax, with two sets of identical wings rather than two different sets, and antennapedia, in which legs grew where antennae should have (Figure 44-19). Later studies revealed that the proteins encoded by bithorax and antennapedia act as transcription factors.

A.

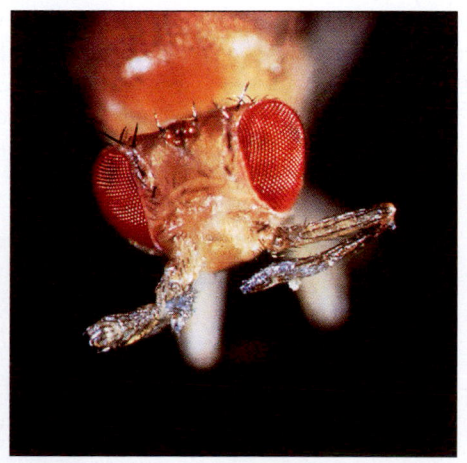
B.

C.

Figure 44-19 **Genes control pattern formation.** A. A normal fruit fly, *Drosophila melanogaster.* B. A fly with the *antennapedia* allele has legs where its antennae should be. C. A *bithorax* mutant has two identical pairs of wings instead of two distinct pairs of wings. *(A and B, Carolina Biological Supply Company/Phototake, NYC; C, David Scharf/Peter Arnold, Inc.)*

Bithorax and antennapedia are two of a family of **homeotic selector genes,** genes whose expression affects overall body plan.

Another example of a transcription factor that directs development is a protein called bicoid protein. Normal development of the early embryo requires bicoid. When the *bicoid* gene mutates, the resulting embryo has no anterior structures. Researchers can produce exactly the same phenotype in a normal fly by removing cytoplasm from the anterior end of the embryo.

Similarly, researchers could "rescue" *bicoid* mutants by injecting them with the mRNA for the protein bicoid. The injected embryos developed normal anterior structures. These experiments suggested that the bicoid protein, normally present in the cytoplasm near the anterior pole, regulates the expression of other genes during development.

Pattern formation seems to result from the action of morphogens such as retinoic acid. Retinoic acid influences the transcription of genes and therefore the differentiation of cells.

Mammals and Flies Use Many of the Same Transcription Factors To Establish Patterns During Development

Information about the structure and action of genes that regulate *Drosophila* development led immediately to questions about the role of similar genes in mammalian development. Researchers asked whether mammalian genomes had counterparts of the amazing homeotic genes in *Drosophila*.

Mammalian counterparts, called **Hox** genes, were easy to find, for their sequences are remarkably similar to those in *Drosophila*. The mouse gene *Hox 1.1*, for example, shares 59 out of 60 amino acids with the *Drosophila* gene *antennapedia*. Even more amazingly, when researchers inserted two vertebrate *Hox* genes into flies, the flies suffered the same mutation in phenotype as if they had the fly mutations antennapedia and deformed.

The biggest surprise in the study of the *Hox* genes, however, has been that corresponding genes in flies and mice are arranged on the animals' chromosomes in the same order. And the order of genes on the chromosome matches the pattern of action within the embryo. Genes that affect head development lie at one end of the cluster, while those that affect tail development lie at the other. The same pattern holds for the arrangement of the *Hox* genes on the mouse chromosome and their pattern of expression in the primitive nervous system of the mouse embryo.

The most recent common ancestor of *Drosophila* and mice lived some 600 million years ago. We can conclude that during the last 600 million years flies and mice have used—with very few modifications—a system of specifying positional information that had already evolved during or before the Cambrian explosion. Although researchers have little information about the genetics of development in other arthropods and vertebrates besides mice and flies, it's reasonable to assume that if flies and mice share these genes then probably all arthropods and vertebrates and many other groups share these homeotic genes.

Animals as different as mice and fruit flies use the same gene products to establish the position of the parts of the body.

POSTEMBRYONIC DEVELOPMENT

Metamorphosis Converts a Larva into an Adult

A larva, the first stage at which an animal has an independent life, may not at all resemble the sexually mature adult. A frog embryo, for example, develops into a larva called a tadpole, which has a fishlike life—propelling itself with a large tail fin, getting dissolved oxygen through its gills, and subsisting on a vegetarian diet. The tadpole **metamorphoses** into an adult frog, which, in contrast, jumps instead of swims, breathes air with its lungs, and consumes flies and other invertebrates with its long tongue and big mouth.

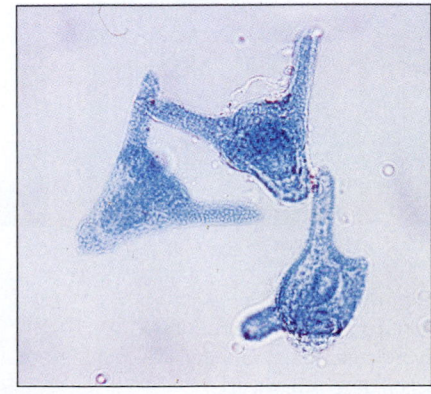

Figure 44-20 Metamorphosis. Larval sea urchins and adult purple sea urchins. (*Left, www.biodisc.com; right, Tammy Peluso/Tom Stack & Associates*)

100 µm

Figure 44-21 Metamorphosis in insects. Complete metamorphosis in the monarch butterfly. A caterpillar feeds on milkweed, then forms a pupa. Inside, the pupa metamorphoses into a butterfly. *(Photos, left to right, 1 and 3, Lior Rubin/Peter Arnold, Inc.; 2, Ed Reschke/Peter Arnold, Inc.; 4, Don Riepe/Peter Arnold, Inc.)*

Larvae are sometimes so different from the adult form that they appear to be different species. The axolotl, for example, inhabits lakes near Mexico City and has been considered a table delicacy since Mayan times. Only in 1920, when biologists induced metamorphosis in the laboratory, however, did they realize that the axolotl was actually a salamander.

Metamorphosis is more common among invertebrates than vertebrates. For example, the first independent form of a sea urchin is a larva, with an appearance and lifestyle totally different from the adult. The sea urchin larva, or **pluteus,** is mobile and elongated, moving easily on ocean currents (Figure 44-20). The adult sea urchin is spiny and more symmetrical, slowly grazing on the bottom of the sea.

The changing of a caterpillar into a butterfly is one of the most spectacular examples of metamorphosis. Many insects—including the beautiful butterflies, the more somber moths, and all the flies, beetles, and wasps—undergo **complete metamorphosis,** meaning that none of the tissues of the adult come from the larva (Figure 44-21). Instead, the adult, the *imago,* derives from 19 apparently unspecialized **imaginal discs,** groups of larval cells that are set aside for later development (Figure 44-22). In insects that undergo complete metamorphosis, the larval stages are specialized for feeding and growth, while the adults are adapted to dispersal and mating. An impressive example of this separation of tasks is the mayfly, whose adult form lives only 1 day. It flies about, mates, and dies.

During the development of insects that undergo complete metamorphosis, the embryo hatches into a wormlike larva (called a grub, a maggot, or a caterpillar). The larva may go through several stages of increasing size before entering an inactive phase, called a **pupa,** which may be enclosed in a cocoon. During the pupal stage, almost all the tissues of the larva are digested. The cells of the imaginal discs multiply, move about, and differentiate, to form the tissues and organs of the adult. Finally, the pupa digests away its cocoon and emerges as a sexually mature adult.

Not all metamorphoses are so complete, even in insects. Grasshoppers, cockroaches, and bugs, for example, undergo **gradual metamorphosis,** in which the embryo hatches into an immature form, called a **nymph,** which is more or less a miniature adult. The nymph then undergoes a series of molts,

each time getting bigger and looking more like the adult (Figure 23-25).

Do humans and other mammals undergo metamorphosis? Some scientists consider puberty a variation on the theme of metamorphosis. Before adolescence, mammals are dependent, sexually immature creatures, with size and body proportions different from those of adults. Adolescents undergo a burst of growth and develop a variety of sexual characteristics that enable them to reproduce. These changes, like those that occur during metamorphosis in insects and amphibians, are controlled by hormones.

Many amphibians and insects undergo complete metamorphosis. Some insects undergo gradual metamorphosis.

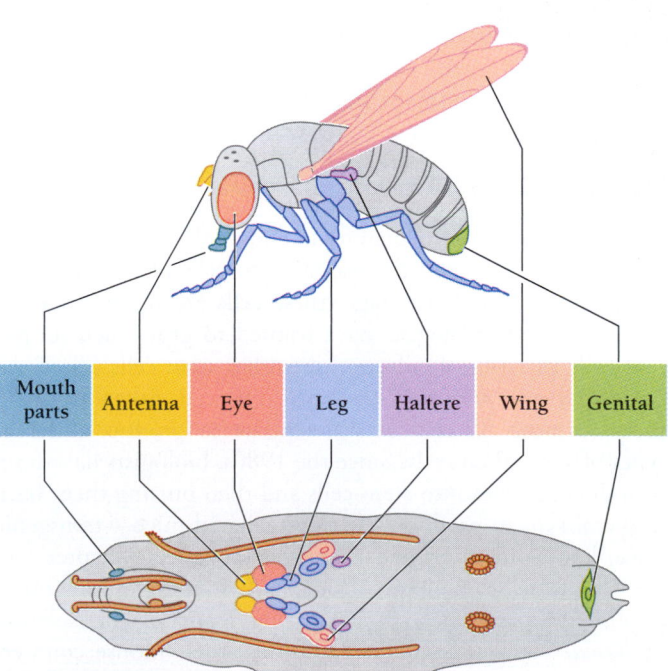

Figure 44-22 Imaginal discs. Imaginal discs in *Drosophila* are larval regions bound to develop into specific adult tissues.

925

Aging Depends on the Action of Both Genes and Environment

Multicellular organisms lose their ability to reproduce with age. Bacteria and other unicellular organisms, in contrast, cannot be said to age. They go on dividing as long as they survive.

From the point of view of natural selection, organisms that can no longer reproduce are dispensable. It is not surprising, then, that for many multicellular organisms, including humans, the chance of death increases with time. The relatively long life span of humans probably derives from our long period of fertility and from the importance of adults in caring for the next generation.

Human aging is a major focus of medical concern. Many measures of vitality decline with age, including fertility, cardiac output, brain weight, lung capacity, and muscle mass. In addition, extracellular proteins of connective tissues lose their earlier resilience. Arteries harden, skin gets less elastic, and joints stiffen.

Aging also occurs at the cellular level. Cells taken from a young organism and grown in tissue culture can divide more times than can cells taken from an older organism. Cells appear to lose their capacity for proliferation. The speed with which wounds heal also decreases with age.

Part of the aging process must depend on genes. For example, different strains of laboratory mice differ in their average life spans. Environmental factors, such as disease, predators, parasites, or inadequate nutrition, also influence how rapidly we age. And genetically determined differences in aging may involve differences in susceptibility to environmental factors. Like other aspects of development, aging depends on the interaction of genes and environment. Aging—the decline in function of the different tissues and organs of the body—results partly from the way our genes function and partly from environmental factors.

STEM CELLS: HOW CAN SOME CELLS RETAIN THE CAPACITY FOR REGENERATION?

All the cells of a mammalian fetus, and all the cells of an adult, ultimately derive from the dozen or so cells of the inner cell mass. At the blastocyst stage, these cells are interchangeable and totipotent. Biologists have learned to grow such totipotent cells in artificial culture. Like other stem cells (discussed earlier), these *embryonic stem (ES) cells,* can either divide to produce more ES cells or differentiate to produce different types of specialized cells. Since the 1980s, biologists have been introducing genes into stem cells and then putting them back into embryos to produce genetically altered mice—transgenic mice, knockout mice, and even knockin mice (Chapter 13).

In late 1998, biologists from the University of Wisconsin published a paper in *Science* magazine, reporting the isolation of *human* stem cells. These cells—like their mouse counterparts—not only proliferate in artificial culture but also differentiate into a variety of cell types, including cartilage, muscle, bone, and skin. Because of a U.S. ban on support of research on human fetal and embryonic tissue, the reported work was supported by private funds.

From a biological viewpoint, the *Science* paper represented a small milestone, but from a societal viewpoint, the paper seemed seismic. To many people, human embryonic stem cells were not merely a technical advance but the starting material for a revolutionary kind of replacement medicine. Some biologists think that stem cells will ultimately make it possible to replace human cells at will. Geneticists should, for example, be able to manipulate the stem cells so that they do not provoke an immune response after transplantation. So, after growing the cells under the proper conditions, physicians might be able to replace worn-out heart muscle or degenerated brain cells. Other biologists imagine that the human embryonic stem cells will even allow future biologists to engineer the human germ line.

The discovery provoked a storm of critical discussion, both among biologists and among lay people. Many people were upset that the embryonic stem cells came from a human embryo and they worried that the research was encouraging induced abortions. Some politicians called for a halt to all stem cell research, even as scientists and physicians were celebrating its potential. After a year of public discussion, the U.S. National Institutes of Health (NIH) established guidelines for human stem cell research. As of December 1999, the NIH ruled that public funding could support research on stem cell differentiation, but that it could not support the making of new stem cells. In January 2000, the University of Wisconsin announced that it would make their stem cells available to other researchers. The debate rages, another example of how strongly intertwined biology and society have become.

The prospects for cell replacement are not entirely focused on embryonic stem cells, however. In 1999, more than a dozen research laboratories also reported that adult tissues contain stem cells with broad developmental potential. Some findings completely amazed the researchers: stem cells from muscle could not only develop into muscle but into blood; stem cells from bone marrow could develop into liver cells.

Even the adult brain contains stem cells, contrary to previous expectations. Researchers are increasingly hopeful that such neuronal stem cells can replace neurons damaged in disease or injury.

AND SO WE SAY GOODBYE

In this final chapter of *Asking About Life,* we have surveyed the main events of developmental embryology. We have seen, for example, that although sea urchins, amphibians, and mammals may gastrulate differently, the result is the same—three germ layers that form skin, internal organs, and gut, respectively. We have also seen that the cells of early embryos are totipotent: each cell contains all the genetic information needed to construct a fully formed adult organism. The cloning of Dolly the sheep, along with other research, suggests that the nuclei of differentiated adult cells are very likely totipotent as well.

Does this mean that each cell in our bodies contains everything needed to make a new person? Could we scrape a cell from a woman's cheek and recreate her? The answer is no, for each of our cells and each of our nuclei is the product not only of the genes it carries and expresses but also of the environment in which it lives and develops.

The environment of a nucleus partly determines which genes are expressed and, therefore, the phenotype of the cell. The udder cell of a sheep may be driven into a state not unlike that of a stem cell, in which its nucleus can divide and redifferentiate to form the cloned sheep Dolly. But so far, this has occurred only when the nucleus has been transplanted into an egg cell. The cytoplasm of the original udder cell probably cannot sponsor embryological development.

In a similar way, we know the environment of a cell influences its fate. Ectoderm in contact with the notochord tends to become nervous tissue. Ectoderm in contact with the optic vesicle of the developing brain becomes lens tissue.

We can take another step back and see that the environment of a whole organism partly determines its phenotype as well. Wet soil produces a plant with a different phenotype than dry soil. Similarly, a calorie-rich, well-balanced diet produces an animal with a different phenotype than a diet deficient in either calories or specific nutrients.

Tongue in cheek, the British developmental and molecular biologist Sidney Brenner distinguished between two broad classes of developmental processes, those that occur by the "European Plan" and those that depend on the "American Plan." In the European Plan, Brenner argued, cells "do what their parents say," whereas in the American Plan, they "do what their neighbors say." In this chapter, we have learned that both plans operate—that development indeed requires genetic programs but that it also depends on environment and experience.

This lesson is a fitting place to end this book, since this conclusion leads to a slew of further questions: How does a set of genes specify a process that unfolds in time? Which features of the environment are important in affecting a particular process? How does a particular environmental trigger stimulate or inhibit the expression of the contributing genes? The answer to each question opens the door to still more questions.

Asking questions and seeking answers is one of the foremost ways of engaging with life. An unanswered question, like an unclimbed mountain or an unattained trophy, is always beckoning, calling us to do our best. We (the authors) hope that our book has provoked you (the readers) to ask your own questions about life.

Study Outline with Key Terms

Animal development begins with a zygote that appears simple and homogeneous. Each species undergoes a characteristic sequence of developmental events, but all vertebrates pass through the same stages: **gamete formation, fertilization, cleavage, germ layer formation, organ formation, growth, metamorphosis** (in many but not all species), and **aging.**

According to the theory of **preformation,** each zygote contains a complete embryo, within which are more complete embryos. The embryo is sometimes called a **homunculus.** In some versions, these embryos within embryos come from the sperm. In other versions of preformation, the embryos are in the egg. When it became clear that a zygote creates form and differentiated tissues and cells where none (or nearly none) existed before, embryologists asked if each cell of a developing embryo was **totipotent** or if each cell was fully **differentiated,** meaning it had a specific, unchangeable fate. We now know that the cells of early embryos tend to be totipotent and the cells of even adults can sometimes be induced to develop into a fully formed adult.

The early development of a zygote is influenced by proteins translated from **maternal mRNA** transcribed during oogenesis from the genes of the mother.

Embryonic life starts with fertilization, after which cleavage divisions lead to the formation of a hollow ball of cells. In amphibians, cleavage results in an asymmetrical ball of cells called a **blastula,** which has an **animal pole** that forms the embryo, and a yolky **vegetal pole.** Mammalian cleavage results in a solid ball of cells called a **morula,** which then forms a **blastocyst.** In mammals, the blastocyst consists of a **trophoblast,** which forms the tissues of the **placenta,** and an **inner cell mass,** which forms the embryo itself.

The hollow blastula or blastocyst then invaginates in a process called **gastrulation** that leads to the formation of a **gastrula,** with three germ layers—**ectoderm, mesoderm,** and **endoderm**—and a hollow space called the **archenteron,** which is the primitive gut.

In sea urchins, gastrulation begins when a few cells attach to the **blastocoel** wall and form a calcium-containing skeleton. Amphibian gastrulation begins when an indentation, called the **blastopore,** forms on the side of the blastula. Mammals gastrulate when the inner cell mass separates into two layers, the **hypoblast** and the **epiblast.** The hypoblast forms a blastodisc, which then invaginates along the **primitive streak** to form the three germ layers—ectoderm, mesoderm, and endoderm.

In vertebrates, part of the mesoderm forms the **notochord,** a supportive cord that runs from head to tail that also serves as an organizer for further embryonic development. In particular, the notochord induces **neurulation.** The notochord induces the formation of **neural plate** tissue, which folds into a hollow **neural tube.** The neural tube bends to form the different regions of the brain and the spinal cord. **Neural crest cells** from the surface of the neural tube migrate into distant parts of the embryo to form nerve, skin, and other tissues.

The optic vesicles of the neural tube induce the overlying ectoderm to form the retina, lens, and cornea of the eye. Organ formation of this sort depends both on differentiation, the acquisition of specific proteins and subcellular structures, and on **morphogenesis,** the creation of form. In 9-week-old human embryos, all of the major organs appear as recognizable, although nonfunctional, **rudiments.** The embryo is then called a **fetus.** Most **miscarriages,** or **spontaneous abortions,** occur in the first **trimester. Teratogens,** such as drugs and viruses, can cause malformations of the developing embryo. Programmed cell death, or **apoptosis,** contributes to normal development of the hands and feet and many other organs.

Embryologists distinguish a cell's **fate** from its **potency,** what it could become if it were given a different environment. Cells whose fate cannot be changed are said to be **determined.** Even in adults, **stem cells** remain totipotent and not determined.

In an embryo, cells differentiate according to their position in the embryo. Transplantation experiments have shown that, in amphibians, tissue from the dorsal lip of the blasto-pore, called the **primary organizer,** can establish the overall orientation of the embryo. Chemical signaling between different parts of the embryo provide positional information to individual cells, leading to the formation of harmonious patterns of organs and body parts. Substances whose concentration provides positional information are called **morphogens.** Retinoic acid, for example, is a transcription factor, whose presence influences the transcription of genes.

Studies of mutants that affect the early development of *Drosophila* have revealed that the establishment of pattern depends on the production and distinctive distribution of a cascade of transcription factors. The genes for these transcription factors are called **homeotic selector genes.** Many of the proteins responsible for pattern formation in *Drosophila* are virtually identical to mammalian proteins, whose genes are called *Hox* genes.

Many animals **metamorphose** from a larval form such as a **pluteus** into an adult form such as a sea urchin. In **complete metamorphosis,** tadpoles turn into frogs or salamanders, caterpillars develop into **pupae** and then into butterflies or moths, and maggots turn into flies. In flies, the adult form derives from **imaginal discs** in the larvae. In **gradual metamorphosis,** a succession of **nymphs** gradually change from a larval form to an adult form.

Review and Thought Questions

Review Questions

1. What are the eight stages of development in a vertebrate?
2. How did embryologists disprove preformation?
3. What does it mean for a cell to be totipotent? What does it mean for a cell to be determined?
4. In deuterostomes such as vertebrates and sea urchins, cleavage results in what basic form?
5. Draw a series of pictures of a mammalian embryo forming the blastocyst, trophoblast, and inner cell mass, and then gastrulating. Label as many tissues and spaces as you can.
6. Draw a picture of an embryo neurulating. What tissue induces neurulation?

7. How does the fate of a cell depend on morphogens? Give an example of a morphogen.

Thought Questions

8. If Dolly was cloned from a stem cell, rather than a fully differentiated udder cell, what are the consequences of that fact, both practical and theoretical?
9. Experiments in invertebrates suggest that the effects of maternal mRNA and other factors in the egg can influence phenotype for several generations. Design an experiment that would distinguish such maternal effects from those of the genotype. You may use any organism you like.

About the Chapter-Opening Image

If every individual were already preformed within a sperm (or within an egg) then development would not be such a mystery.

 On-line materials relating to this chapter are on the World Wide Web at **http://www.harcourtcollege.com/lifesci/aal2/**

Appendix A
Periodic Table of the Elements

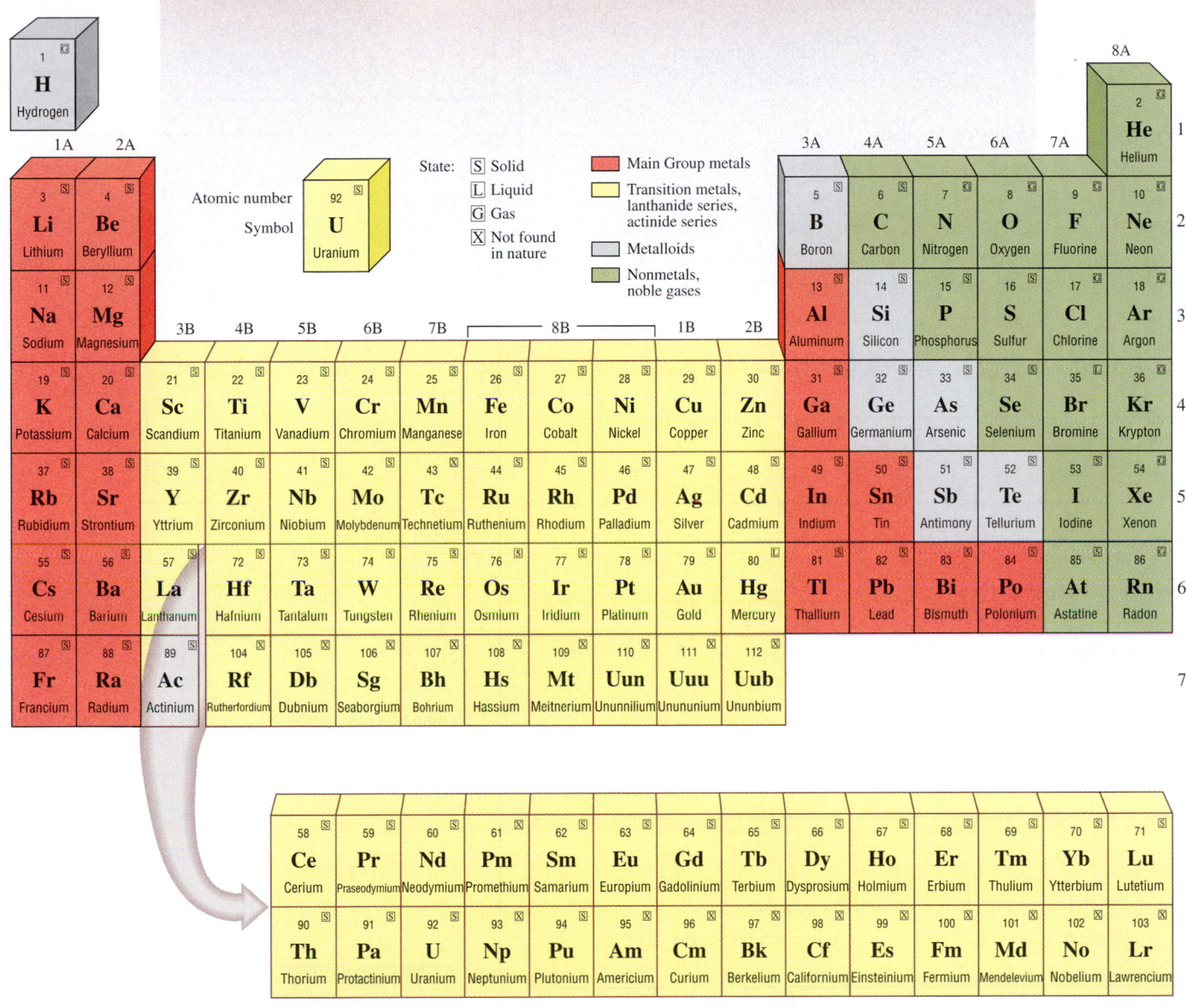

Appendix B
Units of Measure: The Metric System

Length

Standard Unit = Meter

1 meter (m) = 39.37 in

1 centimeter (cm) = 0.39 in   1 inch = 2.54 cm

1 kilometer (km) = 0.62 mi   1 mile = 1.61 km

Prefixes and Units of Length

Prefix	Meaning	Unit
kilo	thousand	kilometer (km) = 1000 m
centi	one-hundredth	centimeter (cm) = 0.01 m = 10^{-2} m
milli	one-thousandth	millimeter (mm) = 10^{-3} m
micro	one-millionth	micrometer (μm) = 10^{-6} m
nano	one-billionth	nanometer (nm) = 10^{-9} m

These prefixes are also used in units of volume and mass.

Volume

Standard Unit = Liter = 1000 cm^3

1 liter (l) = 1.06 qt   1 qt = 0.94 l

1 milliliter (ml) = 0.03 fluid oz   1 fluid oz = 30 ml

1 l = 0.26 gal   1 gal = 3.79

Mass

Standard Unit = Kilogram = 1000 grams

1 kilogram (kg) = 2.21 lb   1 lb = 453.6 grams (g) = 0.45 kg

Temperature

Standard Unit = Centigrade

°C = 5/9 (°F − 32)   °F = 9/5(°C) + 32

Interval equivalents: 1 °Centigrade (°C) = 1.8 °F

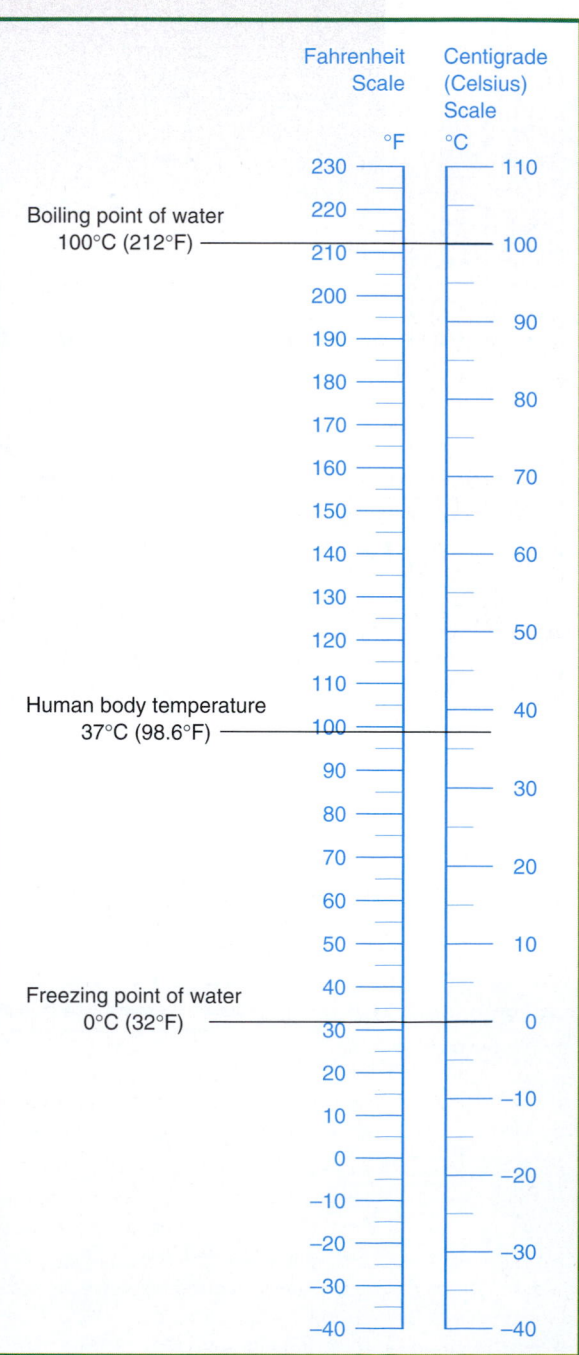

Boiling point of water 100°C (212°F)

Human body temperature 37°C (98.6°F)

Freezing point of water 0°C (32°F)

Glossary

3′ end The end of a polynucleotide at which the 3′ carbon of the nucleotide is not attached to another nucleotide but to a phosphate or a hydroxyl group.

5′ end The end of a polynucleotide at which the 5′ carbon of the nucleotide is not attached to another nucleotide but to a phosphate or a hydroxyl group.

9 + 2 pattern The pattern of microtubules in cilia and eukaryotic flagella; the pattern consists of nine doublet microtubules, each consisting of two microtubules fused along their length plus two singlet microtubules in the center.

α (alpha) carbon In an amino acid, the carbon atom to which the carboxyl group and the amino group are both attached.

α (alpha) helix A common secondary structure in proteins in which every carbonyl group of the polypeptide backbone is hydrogen bonded to the amino group four amino acids farther down the polypeptide chain.

A band (for *anisotropic,* meaning that it changes the direction of polarized light) In muscle, the region of a sarcomere that contains myosin thick filaments.

abdomen In a mammal, the part of the body between the chest (thorax) and the pelvis; the rear portion of an arthropod.

abdominal cavity The part of the coelom that encloses most of the intestinal tract, as well as the liver, pancreas, and kidneys.

abiotic Nonliving. Without life.

abscisic acid The plant hormone that causes the abscission of leaves, dormancy in buds and seeds, and the closing of the stomata (by stimulating the flow of potassium ions out of the guard cells).

absolute age The age in years of each of the layers of sedimentary rock.

absorb To take up into cells and into the circulation.

absorption spectrum The relative amounts of light of different wavelengths that a substance or solution absorbs.

absorptive state The state of an animal shortly after a meal, during which cells in many organs take up glucose and other small molecules to make glycogen, fats, and proteins.

abstinence Referring to birth control, refraining from sexual intercourse.

abyssal zone Ocean waters deeper than 1000 meters.

acetyl CoA A compound that consists of the two-carbon acetic acid (acetate) linked to a larger molecule called coenzyme A; acetyl CoA is the end point of glycolysis and a starting compound for the citric acid cycle.

acetylcholine A neurotransmitter used in the neuromuscular junction to stimulate muscle contraction; acetylcholine is also used by parasympathetic nerves to slow the heart.

acetylcholinesterase An extracellular enzyme that destroys acetylcholine soon after it is released into a synapse.

acid A molecule (or part of a molecule) that can give up a hydrogen ion.

acoelomates [Greek, *a* = without] The simplest animals, including the jellyfish and flatworms, that have no coelomic cavity.

acquired immunodeficiency syndrome (AIDS) A disease characterized by a lack of CD4 T_H cells; AIDS patients cannot mount effective immune responses to infections or to certain kinds of cancers.

acrasiomycota [Greek, *acrasia* = confusing two things + *mukes* = fungus] Cellular slime molds—peculiar organisms that resemble animals at some stages of their lives and plants at others.

acridine dye A chemical mutagen; it causes mutations by inserting itself into the backbone of the DNA double helix.

acrocentric Refers to a chromosome in which the centromere is near one end; such a chromosome looks like a bent capital L.

acrosome A specialized lysosome in a spermatozoan; it contains enzymes that allow the sperm to enter the egg.

actin The protein that makes up actin filaments in nonmuscle cells and the thin filaments in muscle cells.

actin filaments The most flexible elements of the cytoskeleton; formed by the association of actin molecules into long filaments about 7 nm in diameter.

action potential (also called a **nerve impulse**) A rapid, transient, and self-propagating change of voltage across the membrane of a neuron or a muscle cell; action potentials allow long-distance signaling in the nervous system.

action spectrum The relative effectiveness of different wavelengths in promoting a specific light-dependent process.

activation energy The minimum energy needed to initiate a chemical reaction or other process.

active site A groove or cleft on an enzyme's surface to which a substrate binds.

active transport Movement of a substance across a membrane in a manner that does not occur spontaneously but requires an expenditure of energy.

adaptation A genetically inherited structure or behavior acquired through the process of natural selection that contributes to the ability of an organism to live in its particular environment.

adaptive Contributing to the fitness of a species.

adaptive radiation The multiplication of a single species into many species, each with a separate ecological niche.

adenosine diphosphate (ADP) [Greek, *di* = two] A nucleotide that consists of adenosine, ribose, and two phosphate groups; ADP is the product of the hydrolysis of a single phosphate from ATP.

adenosine triphosphate (ATP) A nucleotide that consists of adenosine, ribose, and two phosphate groups; ATP is the universal energy currency, providing energy for many biochemical processes in all organisms.

adenylate cyclase The enzyme that catalyzes the conversion of ATP into cyclic AMP.

adhering junction A molecular assembly on the surface of an animal cell that connects the cell surface to actin filaments on the inner surface of the plasma membrane.

adhesiveness The tendency to cling to a surface.

adrenal glands [Latin, *ad* = towards + *renes* = kidneys] Endocrine glands that lie on top of the kidneys.

adrenal virilism The development of male secondary sexual characteristics in women with excessive adrenal gland activity.

adrenergic receptor [from Adrenaline (as in *adrenal* gland), the British trade name for **epinephrine**] One of a set of membrane proteins that binds to epinephrine and related molecules, thereby initiating a biological response.

aequorin   A protein, isolated from a luminescent jellyfish, that emits light when it binds Ca^{2+} ions.

aerobic   Requiring oxygen, usually for aerobic respiration.

afferent   Leading towards, as blood vessels entering an organ or nerve fibers carrying sensory information from peripheral nerves or sense organs to the central nervous system.

age structure   In a population, the fraction of individuals of various ages (or cohorts).

aging   Progressive developmental changes in an adult organism.

agonistic behavior   [Greek, *agonistes* = champion] All the aspects of competitive behavior within a species, including aggression, aggressive displays, appeasement, and retreat.

alchemy   An ancient study, one of whose aims was to turn "base" metals, such as lead or copper, into silver or gold.

aldosterone   A hormone, made by the adrenal glands, that stimulates sodium and potassium reabsorption in the distal tubules and collecting ducts of the kidneys.

aleurone layer   In a seed, a border of cells that surrounds the endosperm.

algae   Plantlike protists that perform photosynthesis using chlorophyll *a*.

alkaloid   A nitrogen-containing secondary plant compound.

allantois   One of the membranes that surrounds an amniotic embryo; the allantois functions in both respiration and excretion.

allele   One of several variant versions of the same gene.

allele frequency   The number of times the allele is present in the population divided by the total number of chromosomes on which the gene appears.

allergy   An inflammatory response to a harmless antigen; allergies involve the production of IgE antibodies and the release of histamine from mast cells.

allopatric speciation   [Greek, *allos* = other + *patra* = country] Species formation by geographical isolation.

allopolyploid   Description of the chromosome set that occurs when a hybrid doubles its chromosomes.

allosteric effector   (because it acts other than sterically)   A molecule or ion that changes the activity of an enzyme by binding to a site different from the active site.

altered self   In the immune response, nonself antigens presented in the context of an MHC molecule. These occur on the surface of a body cell when the cell is infected by a virus, bacterium, or other microorganism.

alternate   (also called **spiral**)   The arrangement of leaves on a stem in which the bases of the petioles form a spiral up the stem.

alternation of generations   A sexual life cycle in which haploid and diploid phases alternate.

altruism   Behavior that benefits others at the expense of the animal that performs the behavior.

alveoli   [singular, **alveolus**; Latin, *alveus* = a hollow] The expanded surfaces at the ends of the smallest passageways of the lungs; they provide the huge surface area for the exchange of oxygen and carbon dioxide.

Alzheimer's disease   A disease, associated with aging, in which patients suffer massive memory loss and disorientation.

ameboid movement   (also called **cell locomotion** or **cell crawling**) The movement of an individual cell over a surface by extension of pseudopodia.

amine   A chemical compound that contains an amino group; many chemical signals are amines, derived from amino acids by the removal of a carboxyl group.

amino acid   A small molecule that contains both amino and carboxyl groups; the building block of polypeptides and proteins.

amino group   A functional group that consists of a nitrogen and two hydrogen atoms (NH_2).

aminoacyl tRNA synthase   A "decoding enzyme" that links a transfer RNA to its corresponding amino acid to form a "charged" (aminoacyl) tRNA.

aminoacyl-tRNA   A transfer RNA molecule linked at its 3' end to the appropriate amino acid.

ammonia   A small molecule that consists of one nitrogen and three hydrogen atoms; it is secreted as a nitrogen-containing waste by many cells and organisms.

amniocentesis   [Greek, *centes* = puncture] Sampling of cells—usually with a syringe—in the amniotic fluid.

amnion   One of the membranes that surrounds a reptilian, avian, or mammalian embryo.

amniotic egg   An egg in which the embryo is surrounded by an amnion within a porous shell that allows the exchange of gases with the surrounding air.

amniotic fluid   [Greek, *amnion* = membrane around a fetus] The fluid that surrounds the fetus.

amphibians   Class of four-legged vertebrates that usually must reproduce in water.

amphipathic   [Greek, *amphi* = both + *pathos* = feeling] Having both a hydrophilic and hydrophobic region.

amylase   An enzyme found in saliva and pancreatic juice that hydrolyzes the polysaccharides of food into shorter fragments of maltose and glucose.

amyloplast   In a plant, a plastid that contains large granules of starches.

anabolism   [Greek, *ana* = up + *ballein* = to throw] The synthesis of complex molecules from smaller ones.

anaerobic   Without oxygen.

anaerobic photosynthetic bacteria   Photoautotrophic bacteria that live without oxygen.

anagenesis   (also called **phyletic evolution**)   Evolution within a single lineage.

analogous   Referring to structures that perform the same function though derived from different ancestral structures.

anaphase   [Greek, *ana* = up, again] The stage of mitosis during which sister chromatids separate.

anaphase A   The part of anaphase during which the chromosomes move.

anaphase B   The part of anaphase during which the poles of the spindle move apart.

analogy   See homoplasy.

anastral   Having neither centrioles nor asters; a characteristic of mitosis in vascular plants.

anchoring junctions   In animal cells, molecular assemblies that attach cells to each other or to extracellular matrix; anchoring junctions help provide strength to resist mechanical stress while not affecting the passage of molecules between them.

aneuploid   Having an abnormal number of chromosomes.

aneurism   Abnormal ballooning of the wall of a blood vessel.

angiosperms   [Greek, *angion* = vessel + *sperma* = seed] The flowering plants.

angiotensin   A hormone that stimulates the constriction of blood vessels.

Animalia   The kingdom of animals—multicellular, heterotrophic organisms that undergo embryonic development.

anion (so called because it will move toward a positively charged electrode, or anode)   A negatively charged ion.

annual rate of increase   The actual percentage by which a population increases each year.

annuals   Plants that complete their life cycles, from seed to mature plant, in a single growing season and then die.

antagonist   A molecule that prevents a chemical signal from binding to its receptor and triggering its characteristic response in target cells.

antagonistic pair   Muscles that pull in opposite directions.

antenna complex   In a chloroplast, an association of chlorophyll and carotenoids that traps light and transfers the energy to the chlorophyll molecules that actually participate in photosynthesis.

anterior   Toward the front of an animal.

anterior pituitary gland   An endocrine organ that makes and releases a number of peptide hormones.

anther   A thick pollen-bearing structure that is part of the stamen.

antheridium   In plants and protists, the structure that produces sperm.

anthocyanins   The phenolic pigments responsible for most of the red, pink, purple, and blue colors of flowers, fruits, leaves, and other plant parts.

Anthophyta   [Greek, *antho* = flower] Angiosperms.

Anthozoa   [Greek, *antho* = flower + *zoa* = animal] Sea anemones, corals, and related species.

anthropoid   A monkey, ape, or human.

antibiotic   A substance that kills (or interferes with the growth of) microorganisms.

antibody   A blood protein that forms complexes with molecules (antigens), such as those on the surfaces of microorganisms.

anticodon   A sequence of three nucleotides in transfer RNA that forms specific base pairs with the corresponding codon sequence in mRNA.

antidiuretic hormone (ADH) (also called **vasopressin**)   A hormone, released by the posterior pituitary gland, that prevents water loss in the kidneys and constricts blood vessels.

antigen processing   In the cellular immune response, degradation of an antigen into small peptides that associate with an MHC molecule.

antigen (an *anti*body *gen*erator)   A molecule that stimulates the production of an antibody or of another immune response; a molecule that binds to an antibody or other recognition molecule within an animal's immune system.

antigenic determinant (also called an **epitope**)   A specific shape and charge distribution to which antibodies or other recognition proteins can bind.

antioxidants   Natural plant compounds, such as vitamin E, vitamin C, and β-carotene, that prevent the damaging effects of oxygen.

anus   The opening at the far end of the digestive tract through which the undigested and unabsorbed material leaves.

aorta   The artery that carries blood from the left ventricle to the rest of the body.

apical dominance   In plant development, the suppression of lateral bud development by terminal buds.

apical meristem   [Latin, *apex* = top + Greek, *meristos* = divided] In a plant, a self-renewing group of undifferentiated cells, just behind the tip of the shoot or root, that generates differentiated structures.

apoptosis   [Greek, *apo* = away from + *ptosis* = fall] Programmed cell death.

aqueous   Watery; dissolved in water.

Archaea (also called **Archaebacteria**)   [Greek, *archein* = to begin] The kingdom (or "domain") of single-celled organisms that live under extreme environmental conditions and have distinctive biochemical features.

archegonium   In ferns and mosses, the female reproductive organ that produces and houses the ova.

archenteron   [Greek, *arche* = beginning + *enteron* = gut] The "primitive gut," the innermost tube of an animal embryo; it is lined with endoderm and will become the digestive tract.

aromatic   Having a benzene-like planar ring of atoms.

artery   One of the vessels that carry blood away from the heart.

arterioles   Smaller vessels that branch off from arteries.

artifact   A remnant or result of what the researcher does rather than of what exists in nature.

artificial selection   The process by which animal breeders and farmers create new lines of plants and animals by selecting the most desirable individuals for breeding.

Aschelminths   A group of eight phyla, all bilaterally symmetrical animals with either no coelom or a pseudocoel.

asci (singular, **ascus**)   The little sacs in which ascomycetes produce their sexual spores.

ascomycetes (also called **sac fungi**)   [Greek, *askos* = wineskin + *myketos* = fungus] So named because their sexual spores are always produced in little sacs.

asexual reproduction   A process that produces offspring with genes from a single parent.

associative learning   A type of learning in which the subject learns to respond to a stimulus (called the conditioned stimulus) that is not obviously related to the primary stimulus; the response to the conditioned stimulus arises because of its association with a stimulus that leads directly to a physiological response.

assortative mating   Nonrandom mating.

aster   [Latin, star] A starlike object visible in most dividing eukaryotic cells (other than those of vascular plants); the aster contains the microtubule organizing center.

astral   Having an aster and centrioles that participate in mitosis; characteristic of mitosis in animals and in nonvascular plants.

astral microtubules   Microtubules that extend from each pole of the mitotic spindle without attaching to any other visible structure.

asymmetric   A carbon atom surrounded by four different groupings of atoms.

atherosclerosis   Hardening of the arteries.

atom   The smallest unit of matter that still has the properties of an element.

ATPase   An enzyme that hydrolyzes ATP into ADP and phosphate.

ATP synthase   An enzyme responsible for making ATP.

atrial natriuretic factor   [Latin, *atrium* = vestibule, referring to the heart chamber, + *natrium* = sodium + Greek, *ouron* = urine] A hormone produced in the smaller chambers of the heart (the atria).

atrioventricular (AV) node   A special tissue that conducts electrical signals from one part of the heart to the other.

atrioventricular valves   The valves between the atria and ventricles of the heart.

atrium   [Latin, courtyard or entry] One of the smaller thin-walled entrance chambers of the heart, through which blood enters the adjacent ventricle.

australopithecine   [Latin, *australis* = south + Greek, *pithekos* = ape] An early hominid.

autonomic nervous system   Involuntary nervous system; it coordinates the responses of smooth muscles, cardiac muscles, and other effector organs—including those of the endocrine, digestive, excretory, respiratory, and cardiovascular systems.

autonomous replication   Propagation of a virus or plasmid in which the replicating DNA remains separate from the DNA of the host cell.

autoradiography   A method that detects radioactive compounds by their ability to expose a photographic film or a photographic emulsion.

autosomes   The chromosomes that do not differ between males and females.

autotroph   [Greek, *autos* = self + *trophos* = feeder] An organism that can make its own organic molecules from simple inorganic compounds (such as carbon dioxide, water, and ammonia).

axillary bud (also called **lateral bud**)   [Greek, *axilla* = armpit] A bud that forms just above the point where a leaf joins the stem and which can develop into a branch.

axon   A type of neuronal process that usually carries signals away from the cell body and connects to other cells.

β (beta) structure   A common secondary structure in proteins in which the amino and carbonyl groups of the polypeptide chain are hydrogen bonded to the other polypeptide chains or to distant regions of the same chain folded back on itself.

B lymphocyte   Lymphocytes that when stimulated divide to form antibody-producing effector cells or memory cells.

bacilli   [singular, **bacillus**; Latin, little rod] Rod-shaped prokaryotic cells.

backbone   A column of hollow bony segments called vertebrae.

bacterial flagellum   A long structure, about 10 to 20 nm thick and up to 10 μm long, whose rotation propels a bacterium through a liquid.

bacteriophage (or **phage**)   [Greek, *bakterion* = little rod + *phagein* = to eat] A virus that infects bacteria.

balanced polymorphism   A balance of different alleles in a population.

basal body   The base of an axoneme; a cylinder about 500 nm long that resembles a centriole; the microtubule organizing center of a cilium or a eukaryotic flagellum.

basal metabolic rate   The minimum amount of energy required just to stay alive and awake.

base   A molecule (or part of a molecule) that can accept a hydrogen ion.

base substitution   A type of point mutation in which one base (nucleotide) is replaced by another.

basidiocarp   [Latin, *carpus* = fruit] The fruiting body of a mushroom.

basidiomycetes, or club fungi   Include many of the most familiar fungi: the mushrooms, the bracket fungi, and the puffballs.

basidium   [plural, **basidia**; Latin, little pedestal] A tiny, club-shaped structure that produces spores.

Batesian mimicry   A deception in which one species mimics another's warning coloration for protection.

behavioral ecology (formerly called **sociobiology**)   The study of behavior from an evolutionary perspective.

behavioral isolation   Reproductive isolation that results from different mating behaviors.

benthic division   All the organisms that live on the ocean bottom.

biceps   The major muscle on the front side of the upper arm.

bilateral symmetry   Structural symmetry such that one cut through the axis will produce identical (or mirrored) halves.

bile   A detergent solution, secreted by the liver; bitter, alkaline, and an ugly green-yellow or brown-yellow color.

binary fission   The process of cell division (in prokaryotes) in which a cell pinches in two, distributing its materials and molecular machinery more or less evenly to the two daughter cells.

binding site   The region of a protein molecule to which a substrate or a chemical signal binds.

binomial   A two-part name that includes both the genus and species.

bioassay   A method that estimates the concentration of a substance by measuring its biological activity.

biochemistry (also called **biological chemistry**)   The study of the structures and reactions that actually occur in living organisms.

biogeochemical cycle   The movement of a substance, such as carbon or nitrogen, through many forms—both biotic and abiotic.

biogeography   The study of the past and present distribution of plant and animal species.

biological species   The largest unit of a population of similar organisms in which gene flow is possible.

Biological Species Concept   A definition of species that says species are groups of actually or potentially interbreeding populations, which are reproductively isolated from other such groups.

biology   [Greek, *bios* = life, *logo* = word] The science of life.

biomagnification   Toxic chemicals are concentrated at higher levels in the food chain.

biomass   Aggregate dry weight of all organisms in a community or ecosystem.

biome   A geographical region characterized by a distinctive landscape, climate, and community of plants and animals.

biosphere   The system of living things that covers the Earth.

biotic   Living.

bipedal   Consistently walking on two rather than four feet.

birth control   The conscious regulation of reproduction.

birth control pill   A formulation of estrogen and progesterone that prevents the anterior pituitary from secreting LH, thereby stopping ovulation and preventing conception.

birth rate   The annual number of births per 1000 individuals in a population.

bivalent (also called a **tetrad**)   A chromosome pair visible during meiosis I.

bivalves   [Greek, *bi* = two + *valva* = part of a folding door] Includes clams, oysters, scallops, mussels, and other mollusks with two shells—a right shell and a left shell.

bladder   Where urine is stored until its elimination.

blade   The thin, flat part of a leaf.

blastocoel   In an animal embryo, the interior of the blastula.

blastocyst   [Greek, *blastos* = sprout + *cystos* = cavity] A modified blastula in which the cells do not lie within a single layer, but do enclose an internal cavity.

blastopore   An opening at the end of the archenteron.

blastula   [Greek, little sprout] A stage of an animal embryo that consists of a sphere with cells on the surface and fluid inside.

blending inheritance   The pattern of inheritance in which offspring appear to have characteristics intermediate between those of their parents.

bolus   [Greek, *bolos* = lump of earth] In digestion, a ball of macerated food and lubricating mucus.

bone The hard connective tissue of a vertebrate; bone consists of fibers of collagen and crystals of calcium phosphate.

book lung Stacks of modified gills that provide a surface of gas exchange in some animals, including spiders.

bordered pit pair A region of the border between two plant cells in the xylem where the adjacent secondary wall overarches the pit membrane and reinforces the wall of a tracheid.

bottleneck The reduction of a large population to a few individuals.

bottleneck effect The restriction of genetic diversity of a population that results from the reduction of a large population to a few surviving individuals by a random disaster or harsh selection pressure.

Bowman's capsule In a kidney, the bulb that surrounds each glomerulus.

BPG (2,3-bisphosphoglycerate) A relatively abundant 3-carbon molecule in mammalian red blood cells. The binding of BPG to hemoglobin decreases oxygen affinity, allowing it to release more oxygen in the tissues than would hemoglobin without bound BPG.

branchial arches In fishes and amphibians, the tissues between the gill slits.

breathing center A nerve complex on which the coordination of breathing depends.

bronchi In the mammalian respiratory system, the two tubes that branch from the tracheae and carry air to and from the lungs.

bronchioles In the respiratory system, the smallest branches of the bronchi.

Brownian motion (after Robert Brown, a 19th-century British surgeon and botanist) Jerky movements of small particles, reflecting the random thermal motion of molecules.

bryophytes [Greek, *bryon* = moss] The mosses and their relatives.

bud In a plant, the precursor of a leaf or a flower, consisting of several leaf primordia, separated by internodes whose cells have not elongated.

budding yeast (*Saccharomyces cerevisiae;* also called baker's yeast and brewer's yeast) A widely used experimental organism that divides by budding rather than by binary fission.

buffer A molecule that easily converts between acidic and basic forms by donating or accepting one or more hydrogen ions.

bulb of the vestibule One of the corpora cavernosa of the clitoris that corresponds to the penis's corpus spongiosum.

bulbourethral glands The glands that inject a small amount of mucus into the semen.

callus Undifferentiated tissue that forms at the cut surface of a plant.

calorie A measure of energy; the amount of energy needed to raise the temperature of one gram of water 1°C.

calyx The fourth and outermost whorl of leaflike parts called sepals.

Cambrian explosion The burst of diversification—unmatched in the history of the world—in which all of the modern animal phyla that have fossilizable skeletons appeared.

camouflage A defense against being eaten in which organisms mimic materials in their environment in an attempt to be invisible.

canopy In a tropical rain forest, the continuous layer of trees 30 to 40 meters above the ground.

capacitation The process in which the glycoprotein coat that covers the head of each sperm is dissolved by enzymes in the female reproductive tract.

capillary One of the minute blood vessels that brings blood in closest contact with tissues; capillaries connect the finest arterioles with the finest venules.

capsid A coat of protein on the outside surface of a virus.

capsule A gelatinous layer of polysaccharides and proteins outside the cell wall; in disease-causing bacteria, the capsule protects from attack by the host's white blood cells, enabling the bacteria to overwhelm the body's immune system.

carbohydrate A compound that contains the equivalent of one water molecule (one oxygen atom and two hydrogen atoms) for every carbon atom; includes sugars and polysaccharides.

carbonyl A functional group that consists of a carbon atom attached to an oxygen atom by a double bond (C=O).

carboxyl A functional group that contains one carbon atom and two oxygen atoms; the carbon atom forms a double bond with one oxygen atom and a single bond with the other.

carcinogen A chemical that causes cancer.

cardiac muscle The muscles that pump blood through the heart.

cardiovascular system The blood, the heart, and the blood vessels together.

carnivore [Latin, *carn* = flesh] An animal that eats only other animals.

carotenoid A plant pigment, yellow or orange in color.

carpel (also called a **pistil**) In a flower, a female reproductive structure.

carrying capacity The plateau value of the S-shaped curve of population growth. The maximum number of individuals of a species that a habitat can support.

cartilage A type of connective tissue that cushions joints; cartilage consists of collagen, but without calcium phosphate.

Casparian strip In plants, a waxy wall that extends around the walls of endodermal root cells; it restricts the movement of water and dissolved solutes.

catabolism [Greek, *cata* = down + *ballein* = to throw] The reactions of metabolism that break down complex molecules, such as those in food.

catabolite activator protein (CAP) A positive regulator of transcription in *E. coli;* CAP binds to the promoter region of the *lac* operon and increases transcription whenever cyclic AMP is present at sufficiently high levels.

catalyst A substance that accelerates a chemical reaction and is not itself consumed in that reaction.

catarrhines [Greek, *kata* = down + *rina* = nose] Monkeys that have downward pointing nostrils and often no tail at all.

catastrophism The view that the discontinuities in the fossil record result from a series of catastrophes; some catastrophists argued that life arose anew after each catastrophe, rather than deriving from previously existing species.

cation (so-called because it will move toward a negatively charged electrode or cathode) A positively charged ion.

cecum [Latin, *caecus* = blind] A blind sac of the digestive tract, near the junction between the large and small intestine.

cell body The center of a neuron, including its nucleus, as distinct from its axon or dendrites.

cell center (also called the **centrosome**) The microtubule organizing center; a complex of proteins that in animal cells includes a pair of centrioles.

cell cortex A meshwork of actin filaments just below the surface of a eukaryotic cell.

cell crawling (also called **cell locomotion** or **ameboid movement**) The movement of cells over a surface.

cell cycle The orderly sequence of events that accomplish cell reproduction.

cell division   The process by which a parent cell gives rise to two daughter cells that carry the same genetic information as the parent cell.

cell locomotion (also called **cell crawling** or **ameboid movement**)   The movement of cells over a surface.

cell plate   In a dividing plant cell, the precursor of a new cell wall between two daughter cells.

cell theory   The summary of the cellular basis of organisms: (1) All organisms are composed of one or more cells; (2) cells, themselves alive, are the basic living unit of organization of all organisms; and (3) all cells come from other cells.

cell wall   An external rigid structure surrounding all plant cells and most prokaryotes.

cellular oncogene (also called a **proto-oncogene**)   The cellular counterpart of a viral oncogene; cellular oncogenes participate in the normal control of growth and differentiation.

cellulose   A polysaccharide made by plants; the major structural material of wood, cotton, and paper.

central cell   A cell within the ovule of a flower; it contains two haploid nuclei, derived from the megaspore and develops into endosperm after fertilization.

central dogma   The frequently violated principle that "DNA specifies RNA, which specifies proteins."

central nervous system (CNS)   The brain and spinal cord.

central vacuole   A large membrane-enclosed space within a plant cell.

centrifugation   A method for separating macromolecules and subcellular structures according to size (and shape) by subjecting them to high gravitational fields in a spinning tube.

centriole   A pair of small cylindrical structures, each about 0.2 μm in diameter and 0.4 μm long, that lie at right angles to one another; centrioles are present at each pole of the mitotic spindle in animal cells and in some other eukaryotes.

centromere   The point at which the two chromatids of a single chromosome are joined.

centrosome (also called the **cell center**)   A microtubule organizing center that (in many cells) also contains a distinctive organelle called the centriole; the centrosome plays an important role in cell division.

cephalization   The concentration of sense organs and ganglia in the anterior end of an animal.

cephalochordates (also called lancelets)   [Greek, *cephalo* = head] About 45 species of small, segmented, fishlike animals. Most belong to the single genus *Branchiostoma*, whose common name is amphioxus.

cervical cap   A birth control device; a thimble-shaped rubber or plastic cap, about an inch in diameter, that fits tightly over the cervix and prevents fertilization.

cervix   [Latin, neck] The narrower, lower part of the uterus.

cestoda   Tapeworms.

chaetognatha   Arrow worms.

channel   A membrane protein that allows the passage of specific molecules or ions.

chaparral   [Basque, *chabarro* = dwarf evergreen] The dense growth of shrubs that is characteristic of the American Southwest and other regions, all on the west coasts of continents, that share a "Mediterranean" climate, with mild, wet winters and hot, dry summers.

chaperone   One of a number of special proteins that prevent promiscuous interactions by binding to unfolded polypeptides and catalyzing correct folding.

character displacement   In evolution, a change in morphology, life history, or behavior that results from competition.

chelicerae   The first pair of appendages in spiders and other arachnids; they may serve as pincers or as fangs that are associated with poison glands.

chemical equilibrium   [Latin, *aequus* = equal + *libra* = balance] A balance between forward and reverse chemical reactions.

chemical reaction   A transformation in which different forms of matter combine or break down.

chemical synapse   A distinct boundary between two neurons through which neurotransmitters diffuse.

chemical work   The energy-consuming formation or breakage of chemical bonds.

chemiosmosis   [Greek, *osmos* = to push] The linking of chemical and transport processes.

chemistry   The study of the properties and the transformations of matter.

chemoautotroph   An autotroph that derives energy by oxidizing such inorganic substances as hydrogen sulfide (H_2S) or ammonia (NH_3).

chemoheterotroph   A heterotroph that uses no light, relying exclusively on organic molecules for both energy and carbon atoms.

chemotaxis   The movement of a cell toward a higher (or, in some cases, a lower) concentration of a particular chemical.

chiasma   [plural, **chiasmata**; Greek, cross] The sites of exchange of DNA between homologous chromosomes during meiosis; chiasma is visible during prophase of meiosis I.

chitin   A tough, nitrogen-containing polysaccharide in the cell walls of insects, arachnids, and crustaceans.

chlorophyll   The green pigment of plants and protists; chlorophyll is responsible for absorbing light for photosynthesis.

chlorophytes   [Greek, *chloros* = green] Green algae.

chloroplast   A large, green, membrane-enclosed organelle that performs photosynthesis.

chloroxybacteria   Certain bacteria that live as photoautotrophs.

cholesterol   A small molecule that consists of four interconnected rings of carbon atoms; a component of cell membranes and the starting compound in the synthesis of steroids.

chondrichthyes   [Greek, *chondros* = cartilage + *ichthys* = fish] Sharks and rays—fish with cartilaginous skeletons.

Chordata   All vertebrate animals—fish, amphibians, reptiles, birds, and mammals—and animals that have notochords, including sea squirts and lancelets.

chorion   [Greek, skin] One of the membranes that surrounds an embryo; it lies just beneath the shell within an amniotic egg.

chorionic villus sampling   A procedure for sampling cells of the early embryo that uses fetal cells present in the placenta.

chromatid   One of the two separate but connected bodies that make up a chromosome at the beginning of mitosis, when the chromosomes first become visible; a chromatid contains a single long molecule of DNA.

chromatin   [Greek, *chroma* = color] A diffuse material within the nucleus of a nondividing eukaryotic cell; chromatin consists of DNA and proteins.

chromatography   A method for separating molecules according to their relative affinities for a stationary support, called the stationary phase, and a moving solution, called the mobile phase.

chromoplast   A plastid that contains the pigments of fruits and flowers.

chromosomal mutation   A change in a relatively large region of a chromosome.

chromosome [Greek, *chroma* = color + *soma* = body; because it is stained by certain dyes] A discrete complex of DNA and proteins, visible with a light microscope within a dividing cell; originally used only for eukaryotic cells, but now also used to mean a single large molecule of DNA that contains the genes of a bacterium or a virus.

chromosome banding The distinctive pattern, visible in a light microscope, that results from the selective binding of certain dyes to individual chromosomes.

chrysophytes [Greek, *chrysos* = golden + *phyta* = plant] Golden algae and diatoms.

chyme [Greek, *khumos* = juice] A creamy, acidic liquid that passes into the small intestine.

ciliophora [plural, **ciliates**; Latin, *cilium* = eyelash] Some 8000 species of ciliated free-living, single-celled heterotrophs that live both in fresh water and salt water.

cilium [plural, **cilia**; Latin, eyelid, from the hairlike appearance of a cilium] A protein assembly, consisting of microtubules, that can move a cell through a liquid medium (or a liquid medium over a cellular surface); a single cell usually contains many cilia, often arranged in rows; cilia have the same organizational plan as eukaryotic flagella but cilia are much shorter.

circular For DNA, forming a continuous loop; examples of circular DNAs include those of bacteria, energy-producing organelles, and many viruses.

circulation (from "circle") The route of the blood throughout the cardiovascular system.

circumcision The surgical removal of foreskin.

citric acid cycle (also called the **Krebs cycle** or the **tricarboxylic acid cycle**) A set of reactions that converts the carbon atoms of acetyl CoA into carbon dioxide.

cladistics [Greek, *clados* = branch] An approach to taxonomy whose first goal is to describe the groupings of organisms in a way that shows their phylogeny.

cladogenesis Branching evolution.

cladogram A family tree based on shared derived characters. The cladogram itself, which represents the relations among species, is the basis for classifying organisms, not the characters.

class A taxonomic group that consists of one or more orders; a subdivision of a phylum (or, for plants, of a division).

clay A soil that consists only of small particles (less than 2 mm).

cleavage A series of rapid cell divisions following fertilization in early animal embryos; cleavage divides the embryo without increasing its mass.

cleavage furrow A groove formed from the cell membrane in a dividing cell as the contractile ring tightens.

climax community The long-lived community at the end of a succession.

clitoris A highly sensitive erectile tissue that is part of the female genitalia; it is homologous to the penis.

cloaca A common entrance and exit chamber for the digestive, urinary, and reproductive systems in many vertebrates.

clonal selection theory In immunology, a summary of the role of selective proliferation of responding lymphocytes, already programmed, into clones of cells that contain specific gene rearrangements and that recognize specific antigens.

clone [Greek, *klon* = twig] A population of genetically identical individuals or cells descended from a single ancestor.

closed circulatory system A circulatory system in which blood runs only within enclosed vessels.

cnidarians [Greek, *cnide* = nettle] One of two phyla of radially symmetrical acoelomates; cnidarians include three classes: Hydrozoa (hydroids), Scyphozoa (jellyfish), and Anthozoa (corals and sea anemones).

cnidocytes Specialized stinging cells that lie on the tentacles of cnidarians.

coacervate [Latin, *coacervatus* = heaped up] Discrete tiny droplet into which proteins and polysaccharides can spontaneously concentrate.

cocci [singular, **coccus**; Greek, *kokkos* = berry] Spherical-shaped prokaryotic cells.

codominant Two alleles that each contribute to the phenotype of a heterozygote.

codon A group of three nucleotides that specifies a single amino acid.

coelom [Greek, *koilos* = hollow] A cavity lined by a layer of mesoderm cells.

coelomates Animals with a coelom.

coenocytic [Greek, *koinos* = shared + *kytos* = hollow vessel] Fungi in which the nuclei all lie in a common cytoplasm rather than in separate membrane-bound cells.

coenzyme An organic molecule (but not a protein) that is a necessary participant in an enzyme reaction.

coevolution The mutual adaptation of two separate evolutionary lines.

cohesion An attraction between molecules of the same substance.

cohort The set of individuals that enter the population (or are born) at the same time.

coitus interruptus [Latin, *coitus* = sexual intercourse] A method of birth control, also called withdrawal, in which a man attempts to avoid conception by removing his penis from the vagina before ejaculation.

collagen A fibrous protein, secreted by fibroblasts, that provides mechanical strength in cartilage, ligaments, tendons, and bones; collagen is by far the most abundant protein in the extracellular matrix.

collagen helix A regular structure found principally in the structural protein collagen; it consists of three polypeptide chains wound around each other.

collenchyma [Greek, *kolla* = glue] Plant cells that have a thick primary cell wall and often provide mechanical support; collenchyma are usually in the growing regions of stems and leaves.

colon The lower end of the large intestine; it is specialized for the absorption of water and ions.

commensal The species that benefits from the protection of its host in a commensalistic relationship.

commensalism An association between individuals of two species in which one organism benefits without harming the other one.

communicating junctions Membrane-associated structures that allow small molecules to pass freely between two adjacent cells.

community An interacting group of species that inhabit a common area.

compact bone tissue The hard, dense bone that surrounds the spongy interior.

companion cell In plants, a long, nucleated, fully functional parenchyma cell that supplies a sieve tube member with proteins and energy-rich molecules.

comparative anatomy The study of morphological similarities and differences among organisms; it often provides clues to evolutionary relationships.

competition The interaction between organisms or species that depend on the same limited resource; competition occurs when one organism or species uses a resource in a way that limits the availability of that resource to others.

competitive exclusion principle   When two species compete directly for exactly the same limiting resources, the more efficient species will eliminate the other.

competitive inhibitor   A molecule whose inhibitory effects on an enzyme can be overcome by increased substrate concentration.

complement   A set of blood proteins that attack microbial invaders.

complete metamorphosis   The pattern of development in which there are distinct larval, pupal, and adult forms; this pattern is characteristic of many insects, including *Drosophila*, fleas, flies, beetles, wasps, moths, and butterflies.

completed family size   The average number of children that reach reproductive age born to each family.

complex tissue   Tissue consisting of more than one cell type, each with a distinct functional role.

complex transposon   A transposable element that contains other genes besides transposase, for example, a gene for antibiotic resistance.

compound   A substance produced, by a chemical reaction, from two or more different elements.

compound eye   An eye that consists of numerous simple light-detecting units; this type includes the eyes of most arthropod species.

compound microscope   A light microscope that contains several lenses.

compression   The pushing action of two opposing forces.

concave (or **type III**) **survivorship curve**   Description of a life history in which individuals have the greatest chances of dying early in life.

concentration   The number of molecules in a given volume.

concentration gradient (so named because the concentration changes in a graded manner)   A graded difference in the concentration of a substance.

conception   Fertilization or syngamy. The union of an egg and a sperm to form a zygote.

condensation (or **dehydration condensation**) **reaction**   The linking of two building blocks, accompanied by the removal of a water molecule.

conditional mutation   A change in DNA that alters a protein so that it can function under some conditions but not under others.

conditioned stimulus   In associative learning, the arbitrary stimulus that is paired with a stimulus (the "unconditioned stimulus") that directly evokes a physiological response.

condom   A thin rubber sheath that covers the penis and prevents sperm from entering the vagina; a condom is used both as a birth control device and to prevent the transmission of sexually transmitted diseases.

conformation   The three-dimensional arrangement of a polypeptide chain, the equivalent of its tertiary structure.

conjugation   A process in which two temporarily attached bacteria exchange genetic material.

connective tissue   A relatively sparse population of cells within a bed of extracellular matrix; includes bone and cartilage.

conservative   A possible pattern of DNA replication in which each parent molecule would remain intact and both strands of the descendant molecules would be newly assembled; actual DNA replication is not conservative, but semiconservative.

constant region   The part of an immunoglobulin light chain or heavy chain that is identical in all antibodies of a particular class.

consumer   An organism, usually an animal, that obtains energy by eating producers or other consumers.

contact inhibition of cell division   The ability of cells to stop dividing when neighboring cells touch each other.

contractile ring   A bundle of actin filaments that surrounds a dividing cell and pinches the cytoplasm in two during cytokinesis.

contrast   In microscopy, an object's ability to absorb more or less light than its surroundings.

control   In an experiment, the situation in which there has been no experimental manipulation; comparing measurements made in a control versus an experimental situation allows the evaluation of a hypothesis.

convergent evolution   The independent evolution of similar features in separate groups of organisms.

convex (or **type I**) **survivorship curve**   Description of a life history in which survivorship starts out high and decreases slowly with age until a certain point when survivorship begins to decrease more rapidly.

Coriolis effect   The twisting effect of rotation on movement, particularly of wind or of water.

corolla   In a flower, the whorl of petals.

corpus cavernosa   In the male reproductive system of mammals, two spongy cylinders that run along the length of the penis and fill with blood during erection.

corpus luteum   [Latin, yellow body] A structure within the ovary that develops from the ruptured follicle; it secretes progesterone and some estrogen.

corpus spongiosum   A spongy cylindrical tissue that surrounds the urethra.

corridors   Narrow strips of protected habitat designated for the purpose of allowing a species to travel from one preserve to another. Depending on the size of the corridor, the usual beneficiaries are small, slow moving animals such as frogs and salamanders or large predators such as mountain lions and wolves.

cortex   The outer region of a cell, a tissue, or an organ.

cortical microtubules   Microtubules that encircle a dividing plant cell; they determine the placement of a new cell wall.

corticotropin (also called **adrenocorticotrophic hormone** or **ACTH**)   A peptide hormone, made in the anterior pituitary, that stimulates the production of corticosteroids in the adrenal cortex.

cotransport   The coupled transport of two substances across a membrane; cotransport depends on specific transmembrane protein molecules.

cotransporter   A transmembrane protein that binds to two or more molecules or ions and transports them across a membrane.

cotyledon (also called **seed leaves**)   [Greek, *kotyle* = a deep cup] The very first leaves that sprout from a seed; present in the embryos of seed plants, they contain stores of nutrients for use after germination.

countercurrent system   A system in which heat, fluids, or gases run past each other in a manner that allows the exchange of heat or matter.

coupling factor (also called the F_0-F_1 **complex**)   A protein complex in the mitochondrial membrane that uses the flow of protons to drive the synthesis of ATP.

covalent bond   Shared arrangement of electrons that holds atoms together in molecules.

crassulacean acid metabolism (CAM)   A variant of the C_4 pathway of the light-independent reactions of photosynthesis; CAM is used by succulent plants such as cacti.

creationism   The view of the origin of life that says that the Earth was created very recently and that each species was created individually.

cristae   Elaborate folds of the inner membrane of a mitochondrion.

Cro-Magnon   An early species of *Homo sapiens*.

crop   In the digestive tract of birds, a saclike extension of the esophagus; the crop can store food for later digestion.

cross bridges   Extensions of myosin molecules that perform the work of muscle contraction by pulling the thin and thick filaments over one another.

crossbreeding (or **crossing**)   The interbreeding of two genetically distinct organisms.

crossing over   One type of genetic recombination; it involves the breakage and rejoining of single chromatids of homologous chromosomes.

crustacea   A subphylum of arthropods in which the first pair (or the first few pairs) of appendages are antennae, and the next pair are jaws, or mandibles.

crystal   A solid that is enclosed by geometrically regular faces; the starting material point for x-ray diffraction analysis.

ctenophores   [Greek, *ktenos* = comb + *phora* = motion] Comb jellies, a small phylum of about 90 living species.

cultural energy   Energy expended by humans in crop production.

current   The movement of electrical charges.

cuticle   A waxy covering that keeps the above-ground parts of plants from losing water.

cyanobacteria   Photosynthetic bacteria; previously called "blue-green algae"; the most ancient photosynthetic organisms known and the probable ancestors of chloroplasts.

cyclic AMP (cAMP)   Adenosine monophosphate in which the same phosphate group is linked to carbon atoms 3 and 5 of ribose; it is used as a "hunger" signal in bacteria and protists and as a second messenger in mammalian cells.

cyclic AMP phosphodiesterase   The enzyme that hydrolyzes cyclic AMP to AMP.

cyclic AMP-dependent protein kinase (also called **A-kinase**)   An enzyme that, when bound to cyclic AMP, transfers phosphate groups from ATP to sites in particular proteins.

cyclic photophosphorylation   The production of ATP from light energy by a series of electron transfers that regenerate the absorbing chlorophyll; in plants, cyclic photophosphorylation depends on the flow of electrons from excited P_{700} in a cycle that regenerates P_{700}, which is then able to absorb another photon of light.

cyst   An enclosed structure that contains reproductive cells.

cytochalasin   A poison that interferes with the growth of an actin filament.

cytochrome   [Greek, *kytos* = hollow vessel + *chroma* = color] One of a set of heme-containing electron carrier proteins that change color as they accept or donate electrons.

cytodifferentiation   The development of a specialized cell.

cytokine   One of a number of small proteins that regulate proliferation and protein synthesis in cells that participate in inflammation and in the immune response.

cytokinesis   [Greek, *kytos* = hollow vessel + *kinesis* = movement] The division of the cytoplasm and formation of two separate plasma membranes.

cytokinin   A plant hormone that influences the rate of division and differentiation.

cytoplasm   [Greek, *cyto* = a hollow container, i.e., a cell + Latin, *plasma* = a thing molded or formed] In eukaryotes, that part of the cell outside the nucleus but inside the plasma membrane.

cytoplasmic streaming   The movements of membrane-bounded organelles along actin filaments.

cytoskeleton   A network of protein fibers that runs through the cytosol of eukaryotic cells; it consists of microtubules, actin filaments, and intermediate filaments, along with other associated proteins.

cytosol   In eukaryotic cells, the part of the cytoplasm not contained in membrane-bounded organelles.

cytotoxic T lymphocyte (CTL)   An effector cell, derived from a T_C cell, that actually kills a target cell.

cytotoxicity   In the immune response, the ability of certain T lymphocytes to kill cells with altered surfaces.

daughter cells   The cells produced from a single parent cell after cell division.

dead space   The volume of air in the air passages that does not come into contact with the surfaces of the alveoli in the lungs.

death rate   The number of individuals per 1000 who die each year.

deciduous   Shedding leaves annually, as in many trees and shrubs.

decomposer   An organism, such as a bacterium or fungus, that lives on the energy in the complex molecules of dead organisms.

degenerate   Meaning that several codons are equivalent, that is, that they specify the same amino acid.

dehydration   A physiological state in which the body lacks sufficient water.

deletion   A type of mutation; the removal of one or more nucleotides.

demography   [Greek, *demos* = people + *graphos* = measurement] The statistical study of populations.

denatured   A protein that has lost its native, three-dimensional structure and its functional activity, while still having an unaltered primary structure.

dendrite   A type of neuronal process that usually carries signals to the cell body.

dendritic cell   In the immune response, a type of antigen-presenting cell; a highly extended cell present in many organs and tissues throughout the body.

denitrifying bacteria   Bacteria that convert nitrate to atmospheric nitrogen.

density gradient   A changing concentration of a dissolved substance—usually sucrose, glycerol, or cesium chloride—with the highest concentration at the bottom and the lowest at the top.

density-dependent   Referring to population growth, limited in proportion to population density—the denser the population, the more slowly it grows.

density-independent   Referring to population growth, limited by factors other than population density—such as fire, drought, and other natural disasters.

deoxyhemoglobin   Hemoglobin that has no bound oxygen.

deoxyribonucleic acid (DNA)   The long thin molecules in which organisms store and transmit genetic information.

Depo-Provera (DMPA, depo-medroxy progesterone acetate)   A birth-control pill; a progesterone derivative taken as an injection; it suppresses ovulation and inhibits implantation for three months at a time.

depolarizing   A voltage change across an excitable membrane that makes the inside of the cell less negative and more apt to produce an action potential.

depurination   The loss of a purine base from deoxyribose in DNA.

descent with modification   Evolution; change in organisms over time.

desert   A dry, relatively barren region, with lower productivity than most other biomes.

desmosome   An anchoring junction that consists of transmembrane proteins that attach to a cell's intermediate filaments.

detergent   An amphipathic molecule that interacts both with water and with hydrophobic molecules.

determined   In development, having a limited potency. A determined cell can no longer develop in accordance with new environmental signals.

detritus Particles of dead organic matter.

deuteromycetes Imperfect fungi.

deuterostomes [Greek, *deuteros* = second + *stoma* = mouth] Those animals in which the anus forms first, from the blastopore, and the mouth forms secondarily.

diabetes mellitus A disease characterized by the overproduction of sweet (sugar-containing) urine.

diagonal (or **type II**) **survivorship curve** Description of a life history in which survivorship decreases in proportion to age, so that the curve is a straight, declining line.

diaphragm (1) A sheet of muscle beneath the lungs that separates the thoracic cavity from the abdomen. (2) A birth control device. A membrane of latex that prevents sperm from entering the cervix.

diarrhea [Greek, *diarrhein* = to flow through] Watery feces.

diastole [Greek, *dia* = between + *systellein* = to contract] The half of the cycle of the heart in which both the atria and the ventricles are relaxed.

dicotyledons (also called **dicots**) Flowering plants that have two cotyledons; there are about 170,000 species, including almost all trees and shrubs, but not the conifers and other gymnosperms, as well as many herbs.

differential gene expression The production of differing amounts of individual proteins (and RNAs) in different types of cells and at different times of development.

differentiation The process by which tissues and cells become specialized and different from one another.

diffraction The scattering of electromagnetic waves by regular structures so that their interference produces a pattern of lines or spots; the diffraction of x rays by a crystal produces a pattern of spots that can be analyzed to reveal the structure of the molecules within the crystal.

diffusion [Latin, *diffundere* = to pour out] Random movements of molecules that lead to a uniform distribution of molecules both within a solution and on the two sides of a membrane.

digestion The process of hydrolyzing large molecules into smaller units such as glucose, amino acids, fatty acids, and glycerol.

digestive tract (also called the **gastrointestinal tract,** the **alimentary canal,** or the **gut**) The tube, extending from mouth to anus, in which animals accomplish digestion, absorption, and elimination.

dikaryotic Having two nuclei per cell, as in certain fungi.

dinoflagellates [Greek, *dinos* = whirling + *flagellum* = whip] Single-celled organisms that float freely as plankton in warm oceans.

diploid [Greek, *di* = double + *ploion* = vessel] Having two sets of chromosomes.

directional selection In evolution, selection that shifts the frequency of one or more traits in a particular direction.

disjunction In meiosis, the moving apart of two homologous chromosomes during anaphase I and of two sister chromatids in anaphase II.

dispersal The spread of a species or group of species through a large area.

disruptive selection In evolution, the opposite of stabilizing selection: it increases the frequency of extreme types in a population, at the expense of intermediate forms.

distal tubule In the kidney, the wide tubule from which the urine flows into the collecting duct and then to the bladder.

disulfide A covalent bond between two sulfhydryl groups; disulfides are a common cross link between two cysteine side groups within a single polypeptide or between two polypeptides.

diuretic A substance, such as caffeine or alcohol, that stimulates water loss.

divergent evolution The separation of one species of organisms into two (or more) species.

DNA polymerase The enzyme that strings together nucleotides into DNA.

dominance hierarchy A ranking of individuals that fixes (at least temporarily) who may dominate whom.

dominant Refers to an allele that alone determines the phenotype of a heterozygote.

dormant [French, *dormire* = to sleep] Referring to a stage in the development of a seed in which growth is suspended until restarted by environmental cues.

dorsal Refers to the top, or back, of an animal.

dorsal hollow nerve cord A flexible rod containing neurons and their processes that runs between the notochord and the surface of the back.

double circulation A pattern of circulation in which blood circulates separately between the heart and lungs and between the heart and the rest of the body.

double covalent bond A covalent bond in which two atoms share two pairs of electrons.

double fertilization In flowering plants, the simultaneous fusion of two sperm nuclei from the pollen tube with two cells of the embryo sac.

doubling time The time it takes a population to double.

Down syndrome (also called **Down's syndrome**) A disorder, resulting from a particular chromosomal abnormality, that leads to mental retardation and the abnormal development of the face, heart, and other parts of the body; it is almost always caused by trisomy of chromosome 21.

downstream In DNA or RNA, toward the 3′ end.

duodenum [Latin, twelve; referring to its length of 12 "fingers"] The first section of the intestine, where most of digestion occurs.

dynamically unstable Meaning that a microtubule or an actin filament loses tubulin or actin monomers and shrinks if it is not actively growing.

early cell plate In a plant cell, a small, flattened disc, formed in mitotic telophase, that is the beginning of a new cell wall.

Echinodermata "Spiny-skinned" animals such as sea stars.

ecological footprint The area of land needed to provide all of the resources a single person uses.

ecological isolation A barrier to reproduction that results from differences in the ways that species live.

ecology [Greek, *oikos* = home] The study of the interactions of organisms with one another and with their physical environments.

ecosystem A community of organisms together with the nonliving parts of the community's environment.

ectoderm [Greek, *ektos* = outside + *derma* = skin] The outermost cell layer of an embryo, including, in vertebrates, the future skin, sense organs, and central nervous system.

ectotherm [Greek, *ektos* = outside + *thermos* = heat] An animal that depends on external sources of heat.

edema [Greek, "swelling"] A physiological state in which water accumulates in the tissues.

edge effects A consequence of habitat fragmentation. Edge effects, such as incursions by light and predators, prevent a species from fully occupying a habitat island.

effector cell In the immune system, a cell that carries out a particular immune function.

efferent   Leading away, as blood vessels leaving an organ or nerve fibers carrying information from the central nervous system to the muscles and other organs.

egg (also called an **ovum**)   A female gamete.

ejaculation   The propulsion of sperm out of the penis.

electrical potential (also called **voltage**)   A measure of the potential energy that results from a charge separation.

electrical synapse   A connection between two neurons through gap junctions.

electrical work   Energy change caused by changing the separation of charges.

electrochemical gradient   A double gradient composed of a chemical gradient (the difference in hydrogen ion concentration, or pH) and an electrical gradient (the difference in charge).

electrogenic   Leading to the accumulation of charge on one side of a membrane.

electromagnetic radiation   A form of energy—including light—that is transmitted through space as periodically changing electrical and magnetic forces.

electron   A subatomic, negatively charged particle; its charge is exactly equal to that of a proton, but its mass is much smaller.

electron carrier   One of the molecules that carry electrons from high-energy, reduced compounds (NADH, NADPH, and $FADH_2$) to oxygen.

electron microscope   A powerful microscope that uses electrons instead of light waves to reveal structures; it has much higher resolution than a light microscope.

electron transport chain   The pathway of electrons in oxidative phosphorylation or photophosphorylation.

electronegativity   A measure of the tendency of an atom to gain electrons.

electrophoresis   A method for separating charged molecules (such as proteins and nucleic acids) according to their ability to move in an electric field.

electrostatic interaction   The attraction or repulsion of charges.

element   A substance that cannot be reduced to simpler substances by chemical means.

elimination   The disposal of the remains of digested food.

elongation   The adding of additional nucleotides or amino acids to a growing polynucleotide or polypeptide.

elongation zone   A region of the growth zone of a plant's root.

embryo   [Greek, *en* = in + *bryein* = to be full of] A set of early developmental stages in which a plant or animal differs from its mature form.

emergent property   A characteristic that arises only at complex levels of organization.

emergents   In tropical forests, trees with umbrella-shaped crowns extending to a height of 50 meters or more.

emigration   Movement out of the population.

emphysema   A disease of the lungs characterized by irreversible enlargement of the air spaces.

endemic   Found nowhere else in the world.

endergonic   Refers to a process in which free energy increases.

endocrine   [Greek, *endo* = within] Cells or organs that are specialized for secreting specific signaling molecules into the general circulation.

endocrine gland   An organ that is specialized for secretion of a hormone into the general circulation.

endocrinology   The study of hormones.

endocytosis   [Greek, *endon* = within] The process of taking in materials from outside a cell in vesicles that arise by the inward folding ("invagination") of the plasma membrane.

endoderm   [Greek, *endo* = inside + *derma* = skin] The innermost cell layer of an embryo, including the future gastrointestinal tract and associated organs such as the pancreas and liver.

endodermis   [Greek, *endon* = within + *derma* = skin] In vascular plants, the innermost layer of the cortex in roots and stems.

endometrium   [Greek, *endon* = within + *metro* = mother] The inner lining of the uterus.

endoplasmic reticulum (ER)   [Greek, *endon* = within + *plasmein* = to mold + Latin, *reticulum* = network] An extensive and convoluted network of membranes within a eukaryotic cell.

endorphin   One of three classes of polypeptides that act as natural pain suppressors.

endosperm   In flowering plants, a storage tissue that develops during double fertilization.

endosymbiosis   [Greek, *endon* = within + *syn* = together + *bios* = life] The close association of two organisms, one of which lives inside the other.

endosymbiotic theory   The generally accepted view that present-day energy organelles are descended from prokaryotes that once lived within the early eukaryotic cells.

endothelium   In the circulatory system, the epithelial cells that line the capillaries.

endotherm   [Greek, *endon* = within] An animal that warms its body by capturing the heat released by metabolism.

endotoxin   A toxic substance that is a component of a bacterial cell or other pathogen.

energy   The capacity to perform work, to move an object against an opposing force.

energy subsidy   The extra energy, usually provided by fossil fuels, that increases the productivity of a farm.

enhancer (because it stimulates the transcription of neighboring DNA, without itself serving as a promoter; also called a **response element** or *cis*-**regulatory element**)   A DNA sequence that regulates transcription by binding to a specific regulatory protein called a transcription factor.

entropy   A formal measure of disorder; entropy has a high value when objects are disordered or distributed at random and a low value when they are ordered.

enzyme   A large molecule, almost always a protein, that accelerates the rate of a specific chemical reaction.

enzyme-substrate complex   The association of enzyme and substrate that forms in the course of catalysis.

eon   One of several very long periods of time into which geologists and paleontologists have divided the history of life; eons are much longer than eras; a billion years or more.

epicotyl   [Greek, *epi* = above] The axis of a developing plant above the attachment point of the cotyledons.

epidemiologist   [Greek, *epidemia* = how widespread a disease is] A researcher who studies the incidence and transmission of diseases in populations.

epidermis   In animals, the outer layer of the skin; in plants, the outermost layer of the embryo and the plant.

epididymis   In the male reproductive tract of mammals, an interconnected network of coiled ducts in which sperm are stored.

epilimnion   The warm surface water layer of a lake or pond.

epinephrine (also called **Adrenaline**)   A hormone, made in the adrenal medulla, that speeds the heart, dilates the blood vessels, and increases the liver's production of glucose from glycogen.

epiphyte   A plant that grows entirely on other plants.

episome   A virus or other genetic system that can propagate either autonomously or as an integrated part of the host's chromosome.

epistatic gene   A type of modifier gene that limits the expression of another gene.

epithelial tissue   [Greek, *epi* = upon + *thele* = nipple] Cells tightly linked together to form a sheet with little extracellular matrix.

epithelium   A tissue that lines a surface (such as the outside of the body or the inside of the lungs).

epoch   A relatively short period of geologic time; one of the subdivisions of the Tertiary and Quaternary Periods of the Cenozoic Era.

equilibrium   In a chemical reaction, the point at which no further net conversion of reactants and products takes place.

era   A period of time, ranging in length from 65 million to several billion years, into which paleontologists usually divide the history of life; eras are shorter than eons.

erection   The rigid state of the penis or clitoris, caused by the blood filling the tiny spaces within spongy tissues.

erythrocyte   [Greek, *erythros* = red + *kytos* = receptacle] A red blood cell.

Escherichia coli (E. coli)   A common eubacterial resident of the human gut; a favorite experimental organism for thousands of research and industrial biologists.

esophagus   The part of the digestive tract that connects the mouth and pharynx with the stomach.

essential amino acids   The amino acids that an animal cannot itself produce.

essentialism   The view, originally argued by Plato, that individuals, whether chairs, daisies, or people, are only distorted shadows of an ideal, or essential, form.

estivate   To pass time in a sleeplike state (torpor) during the hottest and driest months of the summer.

estuary   [Latin, *aestus* = tide] A partly enclosed body of water where a freshwater stream or river meets the ocean.

ethology   The study of animal behavior.

ethylene   A plant hormone; a two-carbon molecule containing a double bond.

etiolated   [French, *etioler* = to blanch or whiten] Having a thin, spindly appearance, poor leaf development, and no chlorophyll production.

Eubacteria   [Greek, *eu* = true] The commonly occurring prokaryotes that live in water and soil or within larger organisms.

euglenophytes   [Greek, *eu* = true + *glene* = eyeball] Protists that have distinctive light-detecting eye spots.

Eukarya   The domain that encompasses the Animalia, Plantae, Fungi, and Protista kingdoms.

eukaryotic   [Greek, *eu* = true + *karyon* = nucleus] Cells that contain a central nucleus and other membrane-bounded organelles.

eukaryotic flagellum   [plural, **flagella**; Latin, whip] A protein assembly, consisting of microtubules, that can move a cell through a liquid medium (or a liquid medium over a cellular surface).

Eumetazoa   [Greek, *eu* = true + *meta* = middle + *zoa* = animal] The larger of two subkingdoms of animals, including all of the animals except the sponges and the Placozoa.

euploid   Having the correct number of chromosomes.

eutherians (also called **placental mammals**)   [Greek, *therion* = wild beast] Animals whose embryonic development takes place within the uterus of the mother.

eutrophic   A lake that has excessive minerals and organic matter and insufficient oxygen.

evergreen   Keeping leaves throughout the year.

evolution   The process by which species arise and change over time. The idea that all organisms have descended from common ancestors.

excitatory   In the nervous system, leading to the production of action potentials.

excitement   A phase of sexual arousal that is characterized by increased blood flow to the clitoris, the labia minora, and the breasts (in the female) and to the penis and the testes (in the male).

excretion   A disposal process by which cells pass materials across a cell membrane.

exergonic   Refers to a process in which free energy decreases.

exocrine   [Greek, *exo* = outside] A cell or organ that makes products that are carried to specific targets by ducts, such as the duct that carries digestive enzymes into the small intestine.

exocrine organ   An organ whose products are carried to specific targets by ducts.

exocytosis   The export of molecules from a cell by a process that is approximately the reverse of pinocytosis; molecules to be exported are surrounded by membranes that move to the cell surface

exon   A segment of a gene (or of a pre-mRNA) that is also present in mature mRNA; most exon sequences encode polypeptide segments.

exoskeleton   [Greek, *exo* = on the outside] External skeleton.

extinction vortex   The theory that inbreeding brings out genetic defects that hasten the extinction of a population.

exotoxin   A toxic substance secreted by bacteria.

exponential growth   A growth pattern in which a population repeatedly doubles in some constant period of time.

expiration   Exhalation.

extension   The unbending of a limb.

extracellular matrix   A network of proteins and polysaccharides found in connective tissue.

F-factor (for fertility factor)   A kind of episome in bacteria that can replicate either autonomously or in integrated form; it can move from one bacterium to another during conjugation.

F1 (or **first filial**) **generation**   [Latin, son] The initial progeny of a cross.

F2 (or **second filial**) **generation**   The progeny of the F1 generation.

facilitated diffusion   An increased rate of passive transport; it depends on the action of specific transporter molecules within the membrane.

facultative aerobe   A heterotroph that can grow either with or without oxygen.

facultative anaerobe   A microorganism that can live either anaerobically (by fermentation) or aerobically (using oxidative phosphorylation).

FAD (flavin adenine dinucleotide)   An electron acceptor in oxidative phosphorylation.

FADH$_2$ The reduced form of FAD.

fall turnover   In a lake, the annual mixing of the waters of epilimnion and hypolimnion.

fallopian tube   Human oviduct.

family   A taxonomic group that consists of one or more genera; a family is a subdivision of an order.

fat   A triacylglycerol that is solid at room temperature; the component fatty acids are usually saturated.

fate   What a cell or a tissue becomes during development.

fatty acid   A small molecule consisting of a hydrocarbon chain ending in a carboxyl group; a component of phospholipids and triacylglycerides.

feces   [Latin, "dregs"] Waste matter of digestion, discharged through the anus; feces consist of the remnants of food together with bacteria that inhabit the intestinal tract.

feedback inhibition   The inhibition of an enzyme reaction by the product of that reaction or of the end product of an entire pathway; a homeostatic mechanism.

female choice   A courtship pattern in which the female assesses and chooses the best male.

fermentation   [Latin, *fervere* = to boil] The anaerobic extraction of energy from organic compounds.

fermenting bacteria   Anaerobic heterotrophs that derive energy and carbon atoms from a variety of organic compounds in the absence of oxygen.

fertility   The number of offspring each individual in a cohort is likely to produce.

fertilization   The union of two haploid gametes to form a diploid cell or zygote.

fetal masculinization   The development of male characteristics in a fetus carried by a woman treated with male hormones during pregnancy.

fetus   In mammals, a stage of development in which all organs have formed.

fibril   A threadlike structure, made of smaller filaments; the term is used to refer to cross-linked cables of collagen molecules and to assemblies of actin and myosin filaments.

fibrillation   In the heart, continuous disorganized contractions.

fibroblast   A flat, irregularly shaped cell found in connective tissue.

fibronectin   A protein of the extracellular matrix.

fibrous protein   A protein with an elongated shape; fibrous proteins include most structural proteins.

filament   (1) A small protein fiber; (2) in plants, a thin stalk that connects the anther to the base of the flower.

filtrate   In the kidney, the water, urea, and other small molecules in the blood that freely pass into Bowman's capsule from the renal artery.

First Law of Thermodynamics   The statement that the total amount of energy stays constant in any process; that is, energy is neither lost nor gained—it only changes form.

fission yeast   *Schizosaccharomyces pombe,* a widely used experimental organism that divides by binary fission.

fitness   A measure of selective advantage; it is defined as the contribution to the next generation of one genotype in a population relative to the contribution of other genotypes.

fixation   A process that preserves cells for microscopic examination.

fixed nitrogen   Nitrogen in the form of ammonia (NH_3) or nitrate (NO_3).

flagellum   [plural, **flagella;** Latin, whip] In eukaryotes, a protein assembly, consisting of microtubules, that can move a cell through a liquid medium (or a liquid medium over a cellular surface); in prokaryotes, a protein assembly that moves like a propeller.

flexion   The bending of a limb.

fluid mosaic model   The accepted model of biological membrane structure; the model stresses that proteins and phospholipid molecules can move within each leaf of the lipid bilayer unless they are restricted by special interactions.

fluidity   A measure of the ability of substances to move within a membrane.

follicle   [Latin, *folliculus* = small ball] In the ovary, the granulosa cells surrounding a primary oocyte.

follicle-stimulating hormone (FSH)   A peptide hormone, made in the anterior pituitary that in females promotes the maturation of the follicle during the menstrual cycle and in males stimulates testosterone production.

food chain   The sequence in which consumers eat either producers or other consumers.

food web   The collection of all the interacting food chains of an ecosystem.

foot   In a mollusk, a muscular extension of the body that the mollusk uses for sensing, grabbing, creeping, digging, and holding on.

foraminifera (also called forams)   [Latin, *foramen* = little hole + *ferre* = to bear] Marine organisms whose tests are made of organic materials reinforced with grains of sand or minerals and full of tiny holes.

foreskin   A loose layer of skin around the outside of the penis, which ends in a flap and folds over the glans.

fossil   [Latin, *fossilis* = dug up] An object, usually found in the ground, that represents the remains or imprint of past life.

founder effect   The restriction of genetic diversity of a population that results from the founding of a new population by a small subset of a larger population.

frame-shift mutation   An insertion or deletion mutation that alters the groupings of nucleotides into codons.

free energy   A measure of available energy under the conditions of a biochemical reaction; the term is abbreviated G, after the American thermodynamicist J. Willard Gibbs.

free radical   A compound that is highly reactive because it contains an unpaired electron; it is biologically important because it can lead to DNA damage and to cell death; free radicals are formed during normal metabolic processes in both animals and plants.

free ribosomes   Ribosomes, present in the cytosol, that are responsible for the synthesis of soluble proteins.

fronds   Leaves of ferns that often have a lacy appearance.

fruiting body   A mushroomlike growth on cellular slime molds that produces spores.

functional group   A standard small grouping of atoms that contributes to the characteristics of an organic molecule.

fundamental niche   A species' potential ability to utilize resources.

Fungi   The kingdom that includes heterotrophic organisms, both multicellular and single-celled organisms.

G_0   The state of a cell that has withdrawn from the cell cycle.

G_1   The period of the cell cycle that represents the gap between the completion of mitosis and the beginning of DNA replication; it is also called the first growth phase.

G_2   The period of the cell cycle that represents the gap between the completion of DNA synthesis and the beginning of mitosis (of the next cell cycle).

gametangia   In plants, the enclosed reproductive organs in which gametes form.

gamete   [Greek, *gamos* = marriage] A specialized reproductive cell through which sexually reproducing parents pass chromosomes to their offspring; a sperm or an egg.

gamete formation   The production of sperm and eggs.

gametic isolation   Reproductive isolation that results from differences in the compatibility of gametes.

gametophyte   The haploid form of a life cycle characterized by alternation of generations.

gap junctions   Protein assemblies that form channels between adjacent animal cells.

gastrotricha   Microscopic, bottle-shaped animals that live in freshwater or marine environments.

gastrula   [Greek, little stomach] A stage of an animal embryo in which the three germ layers have just formed.

gastrulation   The process of forming a gastrula.

gate   The part of a channel protein that opens and closes the channel in response to environmental signals such as voltage or the binding of a neurotransmitter.

gene   [Greek, *gen* = to produce] The unit of inheritance. A DNA sequence that is transcribed as a single unit and encodes a single polypeptide, a set of closely related polypeptides, ribosomal RNA, or transfer RNA.

gene flow   The movement or flow of genes from one population to another by means of interbreeding.

gene pool   All the alleles of all the individuals in a population.

gene rearrangement   The cutting and splicing of segments of specific genes; it is known to occur within the immune system.

genetic code   The relationship between nucleotide sequence in mRNA and amino acid sequence in polypeptides.

genetic linkage   The tendency of two or more genes to segregate together.

genetic map (also called a **linkage map**)   A summary of the genetic distances between genes.

genetic recombination   Associations of genes that occur in offspring that did not exist in the parents.

genetics   The study of inheritance.

genome   The collection of all the DNA in an organism.

genotype   The genes present in a particular organism or cell.

genus   [plural, **genera**; Greek, *genos* = to be born] A group of species that share many morphological characteristics; a taxonomic grouping of one or more species; a subdivision of a family.

germ cells (or **germ line**)   Gametes and the cells from which they arise.

germ layer   One of the three tubes of the vertebrate embryo—ectoderm, mesoderm, and endoderm.

germ layer formation   The movement of embryonic cells to give the three germ layers (ectoderm, mesoderm, and endoderm); gastrulation.

germ line theory   The view that antibody diversity results from genetic information that is already present in the zygote.

germination   [Latin, *germinare* = to sprout] The resumption of growth by a seed.

gibberellin   A plant hormone.

gill slits   Openings that directly connect the throat to the outside.

gills   Evaginated breathing structures.

girdling   Removing a strip of bark around a tree trunk.

gizzard   A region of the intestinal tract of birds specialized for the grinding of food.

glans   The smooth cap formed by the tissue at the far end of the penis.

glial cell   A cell within the nervous system that does not itself transmit electrical and chemical signals, but which provides metabolic and structural support for neurons.

globular protein   A relatively compact protein that is roughly spherical in shape.

glomerulus   [plural, **glomeruli**; Latin, "little ball"] A tangled network of capillaries in the cortex of a kidney.

glucagon   A protein hormone produced in the pancreas; a signal for the postabsorptive state; glucagon inhibits glycogen synthesis and stimulates its breakdown into glucose.

glucocorticoid   A steroid hormone that stimulates the production of carbohydrate from protein.

glycerol   A polar three-carbon molecule with three hydroxyl groups; a starting compound for triacylglycerides and phospholipids.

glycogen   Polysaccharide used in animals for long-term energy storage.

glycogen phosphorylase   The enzyme that breaks down glycogen into glucose.

glycolipid   A molecule that consists of a lipid attached to carbohydrates.

glycolysis   [Greek, *glykys* = sweet (referring to sugar) + *lyein* = to loosen] A set of ten chemical reactions that is the first stage in the metabolism of glucose.

glycolytic fiber   A large muscle fiber that derives most of its energy from glycolysis and has few mitochondria.

glycoprotein   [Greek, *glykys* = sweet] A protein that contains covalently attached carbohydrates.

glycosidase   An enzyme that catalyzes the digestion of glycosidic bonds; it is used to digest starch or glycogen.

glycosidic bond   The covalent bond between two sugar molecules in a polysaccharide or oligosaccharide.

glycosylation   The process of adding sugars to a newly made protein; it takes place in the ER lumen and in the Golgi apparatus.

glyoxisome   A specialized peroxisome present in the seeds of some plants; glyoxisomes provide energy for the growing plant embryo by breaking down stored fats.

goblet cells   In the small intestine, specialized cells that produce mucus.

Golgi complex (also called the **Golgi apparatus**)   In eukaryotic cells, a set of flattened discs of membrane, usually near the nucleus, involved in the processing and export of proteins.

gonad   [Greek, *gonos* = seed] A gamete-producing organ; an ovary or testis.

gonadotrophin   One of two hormones (FSH and LH), made in the anterior pituitary, that act on gonadal tissue.

gonadotropin releasing hormone (GnRH)   A polypeptide hormone, made by the hypothalamus, that stimulates the secretion of FSH and LH by the anterior pituitary.

gossypol   A terpene, made by cotton plants, that is responsible for resistance to insects, bacteria, and fungi; it also works as a contraceptive in human males.

gradualism   The idea that species evolve gradually and continuously through the steady accumulation of changes, rather than through sudden changes.

granum   [plural, **grana**; Latin, grain] A stack of thylakoids within a chloroplast, bounded by the thylakoid membrane.

gravitropism   The response of a plant to gravity.

greenhouse effect   The warming of the Earth as the result of the absorption of heat by carbon dioxide and other gases.

ground tissue   In plants, the cells that occupy most of the interior of the embryo and the plant.

growth control   A regulatory mechanism that prevents cell division by allowing the cell cycle to proceed under some conditions and to stop under others.

growth factor   One of a number of protein paracrine signals that stimulate cell division and cell survival.

growth hormone (GH)   A peptide hormone, made in the anterior pituitary, that stimulates tissue and skeletal growth, milk production, and other processes in humans, cows, and other mammals.

gut   [Anglo-Saxon, "channel"] The gastrointestinal tract, the alimentary canal, or the digestive tract.

guttation   [Latin, *gutta* = a drop] The process by which root pressure can push water all the way up the stem and out of tiny holes at the margins of the leaves.

gymnosperms   [Greek, *gymnos* = naked + *sperma* = seed] Seed plants that do not have fruits or flowers, including, for example, pine trees and other conifers, as well as palmlike cycads.

habitat   The place in which an organism lives, along with the set of environmental conditions that characterize that place.

habitat destruction   The process by which humans make a habitat uninhabitable for a species. Habitat destruction can range from bulldozing and paving over the area to the elimination of a single, important (keystone) species.

habitat island   A fragment of habitat isolated from other similar fragments.

habituation   The decrease in a behavioral response following repeated exposure to a harmless stimulus.

half-life   The time required for half the atoms of a radioactive isotope to decay.

halophiles   The salt-loving bacteria, one of the three phyla of Archaea.

haplodiploidy   An unusual kind of sex determination in which males develop from unfertilized eggs and are therefore haploid, while females are diploid.

haploid   [Greek, *haploos* = single] Having a single set of chromosomes.

haplontic   Having the haploid phase dominate the life cycle; in a haplontic life cycle, the only diploid cell is the zygote, which itself undergoes meiosis to produce haploid spores.

Hardy-Weinberg equilibrium   A stable distribution of genotype frequencies maintained by a population from generation to generation.

heart   The muscular organ responsible for pushing the blood through the circulatory system.

heart attack   A failure of the blood supply to the heart.

heart rate   The number of contractions (beats) per minute.

heat   The form of kinetic energy contained in moving molecules.

heavy chain   One of the polypeptides present in an immunoglobulin molecule.

heme   An iron-containing organic molecule that gives hemoglobin and the cytochromes their red color; heme may donate or accept electrons, as its iron atom changes its charge between Fe^{2+} and Fe^{3+}.

Hemichordata   Acorn worms.

hemizygous   In male mammals, having one (instead of two) copies of a gene because the Y chromosome lacks the gene present on the X chromosome.

hemoglobin   [Greek, *haima* = blood + Latin, *globus* = ball] The oxygen-binding protein that makes red blood cells red.

herbaceous plant (also called an **herb**)   A plant that has no woody parts.

herbivore   [Latin, *herba* = vegetation + *vorare* = to swallow] An animal that eats only plants.

heterokaryon   [Greek, *heteros* = different + *karyon* = kernel, nucleus] A cell that has two nuclei from different sources.

heterosporous   Producing two kinds of gametophytes (male and female) from two kinds of spores on two kinds of sporangia.

heterotroph   [Greek, *heteros* = other + *trophe* = nourishment] An organism that cannot derive energy from sunlight or from inorganic chemicals but must obtain energy by degrading organic molecules.

heterozygous   Having two different alleles for a single gene (in a diploid organism).

hexose   [Greek, *hex* = six] A sugar that contains six carbon atoms; glucose is a hexose.

hierarchy   An arrangement in which larger groups include smaller groups, which include still smaller groups.

high-energy bond   A relatively unstable chemical bond that gives up energy as new, more stable, bonds form.

histamine   The amino acid histidine minus the carboxyl group, a major stimulus for the inflammatory response; it is released by cells in damaged tissues; it dilates capillaries and increases their tendency to leak fluid.

histone   [Greek, *histos* = web] One of a set of small, positively charged proteins that bind to DNA in eukaryotic cells.

homeostasis   [Greek, *homeo* = like, similar + *stasis* = standing] The tendency of organisms to maintain a stable internal environment in the presence of a changing external environment.

homeotic mutation   A mutation that causes the cells of an embryo to give rise to an inappropriate structure in the adult, for example, to legs instead of antennae.

homeotic selector gene   In a plant, a gene that establishes the fate of one or more whorls of a developing flower.

hominid   A member of the human family, Hominidae; the only living hominid is *Homo sapiens*.

hominoid   [Latin, *homo* = human] Hominids and apes; all have large skulls and long arms and tend to walk at least partially erect.

homologous   Arising from the same structures in a common ancestor.

homologous chromosomes   The two matching chromosomes that align during meiosis I.

homologous structures   Similar structures in two or more species. Homologus structures may perform different or similar functions.

homoplasy   Similar traits that have evolved separately. The possession by two species of a similar trait that is not derived from a common ancestor. "Homoplasy" has replaced the older word "analogy," which requires that two similar traits perform the same function. In contrast, similarities that are inherited from a common ancestor are said to be homologous.

homosporous   Producing just one kind of spore and one kind of gametophyte.

homozygous   Having two copies of the same allele (in a diploid organism).

hopeful monsters   Individuals with macromutations such as chromosome rearrangements whose deformities have adaptive value. A term coined by biologist Richard Goldschmidt to describe a mechanism for a saltationist theory of speciation.

hormone   [Greek, *horman,* = to urge on] A substance, made and released by cells in a well-defined organ or structure, that moves throughout the organism and exerts specific effects on specific cells in other organs or structures.

host range   For a virus, the set of hosts that a particular virus can infect.

Hox gene   Mammalian counterpart of a *Drosophila* homeo domain gene.

human chorionic gonadotropin (HCG)   A polypeptide hormone made after conception; the basis of the most common tests for pregnancy.

human immunodeficiency virus (HIV)   The retrovirus that causes AIDS; HIV infects and kills T_H lymphocytes by first binding to the CD4 protein on the cell surface.

humus   In soils, the residue of decayed dead organisms; a black or brown material that decays slowly.

hybrid   The progeny of a cross of two genetically distinct organisms.

hybrid breakdown   The weakness or sterility of the progeny of interspecies hybrids that are initially viable and fertile.

hybrid inviability   Chromosomal and genetic incompatibilities within hybrids.

hybrid sterility   The sterility of hybrids between two species.

hybridoma   A hybrid cell line derived from the fusion of a cancer cell (a lymphoma) to another cell, such as an antibody-producing cell.

hydrocarbon chain   A chain of connected carbon atoms, with hydrogen atoms sharing other available outer shell electrons.

hydrogen bond   A weak attraction between a hydrogen in one molecule that has a slight positive charge and a negatively charged atom in another molecule.

hydrogen ion (H^+)   The result of a dissociation of a water molecule; it is also called a proton, though it is actually a hydronium ion (H_3O^+).

hydrolysis   [Greek, *hydro* = water + *lysis* = breaking] Breaking the bond between two building blocks by adding a water molecule, reversing the dehydration-condensation reaction.

hydronium ion (H_3O^+)   A water molecule that has acquired an extra proton and a charge of +1; the proper name for a hydrogen ion.

hydrophilic   [Greek, *hydro* = water + *phili* = love] Water-loving; refers to a molecule (or a part of a molecule) that is soluble in water by virtue of its interactions with water molecules.

hydrophobic   [Greek, *hydro* = water + *phobos* = fear] Avoiding associations with water; nonpolar.

hydrophobic interaction   The association of nonpolar molecules.

hydrostatic skeleton   A rigid fluid-filled space (the coelom) that provides mechanical support in invertebrate coelomates.

hydroxide ion (OH^-)   A water molecule from which a hydrogen ion has dissociated; it has a charge of −1.

hydroxyl   The OH functional group, which allows molecules that contain it to form hydrogen bonds.

hymen   A thin membrane of irregular ragged shape just inside the mouth of the vagina.

hyperpolarizing   A voltage change (across the plasma membrane of a neuron) that makes the inside of the cell more negative and less apt to produce an action potential.

hypertension   High blood pressure

hypertonic   [Greek, *hyper* = above] Having a total concentration of solutes higher than that within a cell.

hypervariable region   The polypeptide segments, within the variable region of an immunoglobulin light chain or heavy chain, in which the most sequence variation occurs among antibodies; the hypervariable regions of an immunoglobulin molecule correspond exactly to the complementarity determining regions, where antigens bind.

hyphae   [singular, **hypha;** Greek, *hyphe* = web] The threadlike filaments of a fungus.

hypocotyl   [Greek, *hypo* = under] The axis of a developing plant below the attachment point of the cotyledons.

hypolimnion   In lakes and ponds, the layer of water beneath the epilimnion whose temperature stays about 4°C year-round.

hypothalamus   [Greek, *hypo* = under + *thalamos* = inner room] A part of the brain that regulates the expression of many hormone systems.

hypothesis   An informed guess—for example, about the way a process works or a structure is organized.

hypotonic   [Greek, *hypo* = under] Having a total concentration of solutes lower than that within a cell.

imaginal disc   In insect development, a group of larval cells from which adult structures later develop.

imago   Adult stage of an insect.

imbibition   The taking in of water at the time of seed germination.

immigration   Movement into a population.

immune response   The defense system by which animals resist microorganisms and other foreign tissues, including cancer.

immunoglobulin   An antibody; one of the members of a group of globular blood proteins called globulins.

impermeable junctions (also called **occluding junctions**)   Molecular assemblies that connect epithelial cells tightly together; they also prevent molecules from leaking between them.

implantation   In mammals, the process in which a zygote burrows into the wall of the uterus.

imprinting   The process by which an animal learns behavior during a sensitive period.

in vitro   [Latin, in glass] In a test tube.

in vivo   [Latin, in life] In a living organism.

inbreeding   Mating among close relatives, which greatly increases the number of homozygotes.

incipient species   Subspecies that do not interbreed.

inclusive fitness   The sum of an individual's genetic fitness (which includes that of its own direct descendants) plus all its influence on the fitness of its other relatives.

individualistic   The view that every species has an independent distribution, and that, in effect, every community is unique.

induced abortion   The deliberate removal of an embryo from the uterus.

induced fit   A change in the conformation of an enzyme brought about by the binding of the substrate.

induction   In embryonic development, the process by which one cell population influences the development of neighboring cells.

inflammation   A set of responses to local injury; characteristics of injured tissue include redness, heat, swelling, and pain.

inhibiting hormone   A peptide hormone, made by the hypothalamus, that inhibits the release of a specific hormone.

inhibitory   In the nervous system, tending to prevent the production of action potentials.

inhibitory interneuron   A neuron whose stimulation by one neuron prevents the firing of another neuron in the same area.

initiation   The start of synthesis.

initiation site   The first nucleotide actually transcribed from DNA into RNA.

innate   Behavior that an animal engages in regardless of previous experience.

inner cell mass   In a mammalian embryo, a small group of cells within a blastocyst that will eventually grow into the embryo itself and subsequently into the adult.

insertion   A type of mutation; the addition of one or more nucleotides.

insertion sequence (also called a **simple transposon**)   The simplest transposable element; a short length of DNA (up to a few thousand nucleotide pairs long) that can move from place to place in the genome.

inspiration   Inhalation.

insulin   A protein hormone, produced by the pancreas, that regulates glucose uptake; a signal for the absorptive state; it promotes the synthesis of glycogen and inhibits its breakdown.

integral protein (also called a **transmembrane protein**)   A membrane protein that spans the lipid bilayer.

integrated   The view that a community consists of characteristic assemblages of species that interact with each other in predictable ways.

integrated replication   A method for viral propagation in which the virus's DNA becomes integrated into the host cell's DNA and is replicated along with that of the host.

interferon   A cytokine that interferes nonspecifically with the reproduction of viruses.

interleukin-1 (IL-1)   A cytokine produced early during inflammation; it stimulates responses both in leukocytes and in organs distant from the site of infection.

intermediate filament   In eukaryotic cells, a component of the cytoskeleton that consists of filaments 8 to 10 nm in diameter, thinner than microtubules but thicker than actin filaments.

intermediate form   Fossils that grade from one form to another—for example, the fossils intermediate in both shape and in time from the tiny, horselike animal eohippus to the modern horse.

internode   In a plant, the region of a stem between nodes.

interphase   The part of the cell cycle in which the chromosomes are not condensed and the cytoplasm is not dividing.

interstitial cells   In the testes, the matrix amongst the seminiferous tubules that synthesizes the male sex hormone testosterone.

intertidal zone   The zone that lies between high tide and low tide.

intrauterine device (IUD)   A birth control device; a small piece of plastic or other material that is placed into the uterus, where it interferes with implantation.

intron (also called an **intervening sequence**)   A segment of a gene (or of a pre-mRNA) that is transcribed into RNA but excised before the primary transcript matures into functional mRNA.

invagination   The local folding of a cell layer to form an enclosed space with an opening to the outside.

inversion   A mutation in which a segment of a chromosome is turned 180° from its normal orientation.

involuntary muscle   A muscle that cannot be consciously controlled; smooth and cardiac muscles.

ion   An atom (or a molecule) with a net electrical charge, the result of a different number of electrons and protons.

ionic bond   A bond formed by ions with opposite charges.

iron-sulfur protein   One of at least six proteins that participate in the electron transport chain; each contains an iron-sulfur center with two to four iron atoms and an equal number of sulfur atoms.

isomers   Molecules that contain the same atoms arranged differently.

isotonic   [Greek, *isos* = equal + *tonos* = tension] Having a total concentration of solutes that is the same as a cell's interior.

isotopes   Forms of an element that have different numbers of neutrons.

J-shaped curve   A graph of the exponential growth of a population; the curve resembles the letter J.

joint   The point of attachment between two bones.

K-selected species   Species with adaptations that increase their ability to maintain populations as close to the carrying capacity (K) as possible; these species tend to live long lives, reproduce late in life, produce few and large offspring, and provide extended parental care to each offspring.

karyotype   [Greek, *karyon* = kernel or nucleus + *typos* = stamp] The chromosomal makeup of a cell.

keystone species   A predator or other organism whose presence promotes diversity or controls the structure of a community.

kidney   In vertebrates, the major organ of excretion and of salt and water balance; regulates the composition of the urine.

kin selection   The tendency of individuals in some species to increase their reproductive output by helping relatives, which share their genes, to reproduce.

kinetic energy   The energy of moving objects.

kinetics   [Greek, *kinetikos* = moving] The study of the rates of reactions.

kinetochore   A specialized disc-shaped structure that attaches the mitotic spindle to the centromere.

kinetochore microtubules   A subset of polar microtubules that run from a pole of the mitotic spindle to a kinetochore.

Koch's postulates   Rigorous criteria for identifying the pathogen for a given disease.

krummholz   A forest of stunted trees that grow near timberline on a mountain.

kwashiorkor   Protein deficiency disease.

labia majora   [Latin, larger lips] In the external genitalia of human females, the skin folds on either side of the labia minora that enclose small amounts of fatty tissue.

labia minora   [Latin, smaller lips] In the external genitalia of human females, the two thin skin folds that surround the mouth of the vagina and the end of the urethra.

***lac* repressor**   A protein that binds to the *lac* operator and prevents the expression of the *lac* operon.

lactose operon (or ***lac* operon**)   A region of *E. coli* DNA that encodes three proteins used to derive energy from lactose: the enzyme β-galactosidase (which splits lactose into galactose and glucose) and two other proteins, called permease and acetylase.

lagging strand   In replicating DNA, the newly made strand that is extended discontinuously.

larva   A feeding form of an animal distinct from the later adult.

larynx   The modified upper part of the trachea that contains the vocal cords, folds of membrane that vibrate as air passes over them, producing sound.

lateral bud (also called an **axillary bud**)   A bud that forms just above the point where a leaf joins the stem and which can develop into a branch.

lateral meristem   A cylinder of actively dividing cells within a root or stem.

leading strand   In replicating DNA, the newly made strand that extends continuously, with DNA polymerase adding nucleotides to its 3′ end.

leaf primordium   A tiny extension of the apical meristem; it grows into a leaf.

learning   Modification of neural activity and behavior as the result of experience.

leukocyte   [Greek, *leukos* = clear, white] White blood cell.

liana   A vine that is rooted in soil but climbs into the canopy.

ligament   A band of connective tissue by which joints may be joined together.

ligand   [Latin, *ligare* = to bind] A molecule that binds to a specific binding site in a protein.

light chain   One of the polypeptides present in an immunoglobulin molecule.

light source   The part of a microscope that illuminates a specimen.

lignin   A major constituent of the walls of cells specialized to provide mechanical support or transport water.

limiting resource   A resource that is in short supply.

limnology   The study of freshwater ecosystems—lakes, ponds, rivers, and streams.

linkage group   A set of genes that do not assort independently because they are physically close to one another on the same chromosome.

linkage map   A summary of the genetic distances between genes; a genetic map.

lipid   A compound that is less soluble in water than in nonpolar solvents.

lipid-soluble signal   A chemical signal that can enter a target cell by passing directly through the plasma membrane.

liposome   An artificially produced vesicle that is surrounded by a phospholipid bilayer.

loam   A soil that consists of a mixture of clay, silt, and sand.

locus   [plural, **loci**; Latin, place] The position of a gene on a chromosome.

logistic growth equation   An equation that describes the growth of a population.

loop of Henle   In a kidney, the hairpin-shaped portion of the renal tubule.

lophophore   In three phyla of marine invertebrates, a filtering apparatus that catches food.

low-density lipoprotein (LDL)   A carrier protein in the blood that binds to cholesterol.

low-energy bonds   Relatively stable chemical bonds.

lumen   [Latin, light, an opening] An enclosed space, bounded either by membranes (as in the ER lumen) or by an epithelium (as in the lumen of the gut).

lungs   Invaginated breathing structures; localized organs of gas exchange that are always associated with the circulation.

luteinizing hormone (LH)   A peptide hormone, made in the anterior pituitary, that in females induces ovulation and stimulates estrogen production and in males increases testosterone production.

lymph   Fluid containing dead or foreign cells and waste proteins that passively moves through the lymphatic system.

lymph nodes   Regions in the lymphatic veins where filterlike tissue separates cells and other detritus from the lymph.

lymphatic system   The network of lymphatic vessels and nodes that provides a secondary route for fluids from the extracellular space to the bloodstream.

lymphocyte   One of a class of white blood cells that develop within the lymphoid tissues (including lymph nodes, spleen, thymus, and tonsils); the cells responsible for the immune response.

lymphocyte library   The collection of B lymphocytes, each containing a unique random rearrangement of immunoglobulin genes.

lysis   [Greek, *lysis* = loosening] The breaking down of a plasma membrane.

lysogenic cycle   [Greek, *lysis* = loosen + *genos* = offspring, because it can generate the destructive lytic cycle] The reproduction of a virus along with its host without causing the host cell to lyse.

lysogenic virus   A virus that reproduces either in a lytic cycle, in which it destroys its host, or along with the host; the virus can lie in a dormant (lysogenic) state and can be activated to enter a lytic cycle.

lysosome   A small membrane-bounded organelle that contains hydrolytic enzymes that break down proteins, nucleic acids, sugars, lipids, and other complex molecules.

lytic cycle   Viral reproduction that destroys the host.

lytic virus   A virus that destroys its host cell by lysing it.

macroevolution   The processes by which species and higher groupings (taxa) of organisms originate, change, and go extinct.

macromolecules   [Greek, *macro* = large] Large molecules formed by the polymerization of smaller building blocks.

macrophage   [Greek, *makros* = large + *phagein* = to eat] A large phagocytic cell, widely distributed throughout the body.

magnification   The ratio of the size of an image to the size of the object itself.

major histocompatibility complex (MHC)   A set of 40 to 50 closely linked genes, some of which encode proteins on the surface of every somatic cell; MHC proteins present antigens to the immune system.

male competition   Competition among males, usually for territory or access to females.

malnourishment   A deficiency in one or more essential nutrients.

Malpighian tubules   Excretory organs in arthropods.

mammary glands   Milk-producing organs in the female that characterize the mammals.

mantle   In mollusks, a specialized tissue that secretes a shell onto the dorsal surface of the visceral mass.

mantle cavity   In mollusks, an enclosed space, formed by folds in the mantle, that contains the mollusk's breathing organs.

marsupials (also called **metatherians**)   Mammals whose young are born in a fetal stage; marsupials carry the young in a pouch until they can fend for themselves.

mass   A measure of the amount of matter.

mass number   The total number of protons and neutrons in an atom's nucleus.

mast cell   A large round cell; mast cells are distributed throughout the connective tissues and are filled with small histamine-containing vesicles, which are released at the sites of tissue damage.

maternal effect genes   Genes that are expressed in the mother during oogenesis.

maternal mRNA   An mRNA that is already present in the egg before fertilization.

matrix   The intercellular substance within a tissue, or the interior substance of a mitochondrion.

matter   Any substance.

mean   The average value of a set of numbers, also called the arithmetic mean.

mechanical isolation   A reproductive barrier that derives from incompatibilities between male and female reproductive organs.

mechanical work   Energy change resulting from the movement of an object against a force.

median eminence   A structure that carries hypothalamic hormones to the anterior pituitary.

medulla   A part of the brain that regulates the rate of breathing; the inner region of a kidney or adrenal gland.

medusa   The bell-like body plan that represents one of the two basic body plans of the cnidarians.

megaphyll   A large leaf containing several veins that form complex networks of transport tubes.

megaspore   [Greek, *megas* = large] In a flower, the haploid cell within an ovule that develops into the embryo sac, the mature female gametophyte.

meiosis   [Greek, *meioun* = to make smaller] The process by which haploid gametes arise from diploid cells; meiosis distributes chromosomes so that each of four daughter cells receives one chromosome from each homologous pair.

meiosis I   The first of the two divisions of meiosis, during which homologous chromosomes pair and are distributed into two daughter cells.

meiosis II   The second of the two divisions of meiosis, during which sister chromatids are distributed to daughter cells.

membrane envelope   A membrane that surrounds a virus particle.

membrane-bound ribosomes   Ribosomes associated with a cell's internal membranes; they are responsible for the synthesis of secreted proteins as well as many membrane-associated proteins.

memory cell   In the immune system, a cell that can later be stimulated to produce effector cells with particular antigen specificity.

menstruation   [Latin, *mens* = month] The monthly shedding of blood, mucus, vaginal secretions, and endometrial tissue out through the vagina.

meristem   [Greek, *meristos* = divided] In plants, a region of undifferentiated, actively dividing cells.

mesenchyme   [Greek, *mesos* = middle + *enchyma* = infusion] Loosely attached cells embedded within a jellylike substance.

mesoderm   [Greek, *mesos* = middle + *derma* = skin] The middle layer of a vertebrate embryo, including the future connective tissues (bones, muscles, and tendons) and the cells of the blood.

mesoglea   [Greek, *mesos* = middle + *glia* = glue] In cnidarians, the jellylike material that lies between the two cellular layers.

mesophyll   [Greek, *mesos* = middle + *phyllon* = leaf] Green parenchymal cells that are responsible for most of a plant's photosynthesis.

messenger RNA (mRNA)   The RNA molecules that carry information from DNA to the ribosomes, where the mRNA is translated into a polypeptide.

metabolism   [Greek, *metabole* = change] All the chemical reactions occurring within an organism.

metacentric   The term referring to a chromosome in which the centromere is at or near the middle; such a chromosome looks like a V.

metamere   One of the identical (or nearly identical) sections that make up an annelid's long, segmented body.

metamorphosis   In many animal species, a series of dramatic changes in form leading from a larva to an adult.

metaphase   The stage of mitosis or meiosis during which chromosomes move halfway between the two poles of the spindle, where they accumulate in the metaphase plate.

metaphase plate   A disc formed during metaphase in which all of a cell's chromosomes lie in a single plane at right angles to the spindle fibers.

metapopulation   A group of local populations, which may alternately go extinct and recolonize one another's vacant habitat islands.

metastasize   Referring to cancer cells, to spread to other parts of the body.

metatherians (also called **marsupials**)   [Latin, *marsupium* = pouch] Animals whose newborns are relatively undeveloped fetuses.

methanogen   A methane-producing bacteria, one of the three phyla of Archaea.

MHC restriction   In the immune response, the inability of T cells to recognize foreign antigens except in cells with the same MHC type as the responding animal.

micelle   A cluster of amphipathic molecules.

microelectrode   An electrode that is small enough to enter or maintain contact with a single cell.

microevolution   Changes in the frequencies of alleles of genes in a population.

micronutrient   A substance required in minute amounts for the life of an organism; examples include molybdenum, copper, zinc, manganese, boron, iron, and chlorine.

microphyll   A leaf with just one conducting vein.

microsomes   Vesicles derived from fragments of the endoplasmic reticulum after experimental disruption of eukaryotic cells.

microsurgery   The physical manipulation of subcellular structures.

microtubule   In eukaryotic cells, the largest elements of the cytoskeleton; they consist of tubulin molecules assembled into hollow rods, about 25 nm in diameter and of variable lengths, up to several μm.

microtubule-associated protein (MAP)   A protein that binds to microtubules and influences their organization.

microtubule organizing center (MTOC)   The region of a eukaryotic cell from which microtubules emanate.

microvillus (plural, **microvilli**)   A membrane-covered extension of an epithelial cell.

midbody   In a dividing plant cell, the thin connection between daughter cells that persists until the end of cytokinesis; the midbody is packed with microtubules from the spindle apparatus.

midpiece   In a spermatozoan, a section that contains a microtubule organizing center and dense concentrations of mitochondria—sources of ATP for the journey to the egg.

mimicry   Pretending to be something else.

mineral   An inorganic substance, especially one required for the life of an organism.

minipill   A birth control pill that contains only progesterone, with no estrogens.

missense mutation   A change in DNA that alters the codon for one amino acid into a codon for another amino acid.

mitochondrion   [plural, **mitochondria**; Greek, *mitos* = thread + *chondrion* = a grain] In eukaryotes, the major energy-producing organelle; in cells that do not directly harvest sunlight, mitochondria produce nearly all of the ATP the cell needs to power the chemical reactions of the cell.

mitochondrial matrix   The compartment surrounded by the inner mitochondrial membrane.

mitosis   [Greek, *mitos* = thread] The process of the equal distribution of chromosomes during cell division.

mitotic spindle   An elongated structure that develops outside the nucleus during early mitosis; contains the microtubular machinery that moves the chromatids apart.

mittelschmerz   [German, *mittel* = middle + *schmerz* = pain] Pain during ovulation.

mobile gene   A gene whose chromosomal address changes.

modal action pattern   A highly stereotyped innate behavior.

model   A simplified view of how the components of a structure operate.

modern synthesis   The marriage of genetics and evolutionary theory that elucidated the genetic basis of variation and natural selection.

modifier gene   A gene that regulates the expression of another, separate gene.

molarity   The concentration of a dissolved substance in moles per liter of solution.

mold   A protist or fungus that grows as a downy coating on animal or vegetable matter.

mole   The amount of a substance in grams equal to its molecular mass relative to a hydrogen atom.

molecular genetics   The branch of genetics that studies how DNA carries genetic instructions and how cells carry out these instructions.

molecular weight   The mass of a molecule relative to that of a hydrogen atom.

molecule   A specific combination of individual atoms held together by covalent bonds.

mollusk   [Latin, *molluscus* = soft] A soft invertebrate inside a shell.

molting   The shedding of feathers, fur, or exoskeleton and the secretion of new ones.

Monera   The kingdom that, until recently, was considered to include all prokaryotes, including both the eubacteria and the archaea; when used now, the term refers only to the eubacteria.

monoclonal antibody   An antibody produced from a single hybridoma clone; each monoclonal antibody has a defined chemical structure and has the ability to recognize a single epitope.

monocotyledon (or **monocot**)   One of a large group of flowering plants that have one cotyledon. Monocotyledons consist of about 65,000 species, including all of the grasses, lilies, palms, and orchids. Most are herbaceous.

monoculture   The raising of a single crop.

monomer   [Greek, *mono* = single] Each of the component parts of a polymer.

monophyletic   A taxon that includes an ancestral species and all its descendants.

monosaccharide   [Greek, *mono* = one + *saccharine* = sugar] A simple sugar with three to nine carbon atoms; the building block for polysaccharides.

monosomy   The presence of only a single copy of a chromosome in a diploid cell.

monotreme (also called **prototherian**)   [Greek, *monos* = single + *trema* = hole] An egg-laying mammal; like other mammals, monotremes have hair and mammary glands; their digestive and reproductive systems empty into a cloaca, hence their name

morning-after pill   A birth control method that is taken after unprotected intercourse.

morphogen   A substance that specifies the position of a cell within a pattern.

morphogenesis   The development of form.

mortality   The probability that an individual of a given age will die each year.

morula   [Latin, mulberry] A solid ball of cells in an early mammalian embryo.

motor cortex   A region of the brain responsible for movements, with each part of the body specified by a distinct area.

motor neuron (also called a **motoneuron**)   A neuron that directly connects with a muscle and commands muscle contraction; the final common pathway of instructions that direct movement.

mucigel   In plants, the slimy substance that the epidermis and the root hairs secrete that enhances the absorption of water and minerals.

mucus   A viscous, slippery substance that coats the food particles and lubricates their movements within the mouth and the digestive system.

Müllerian mimicry   The resemblance of two or more equally dangerous species to each other; the similarities in color or form represent similar dangers to common predators, which therefore avoid all the mimicking species.

multicellular   Made of many cells.

muscle fiber   A giant muscle cell that contains the proteins responsible for contraction.

mustard oil glycosides   Compounds that give cabbage, broccoli, radishes, and other plants in the mustard family their characteristic pungent odor and flavor.

mutagen   An agent that increases the rate of mutation; a mutagen can be a chemical or a form of radiation.

mutation   A change in the nucleotide sequences of DNA.

mutualism   An association between organisms of two species from which both organisms benefit.

mycelium   [Greek, *mukes* = fungus] In a fungus, the mass into which hyphae grow, branch, and intertwine.

mycorrhizae   [Greek, *mukes* = fungus + *rhiza* = root] Symbiotic association between the root of a plant with the mycelium of a fungus.

myelin   An insulating structure around a nerve fiber; myelin consists of extensions of the plasma membrane of a Schwann cell or an oligodendrocyte.

myoblast   A cell that is a precursor of a muscle fiber.

myocardium   [Greek, *myos* = muscle] The muscular walls of the heart.

myofibril   [Greek, *myos* = muscle + Latin, *fibrilla* = little fiber] A thread, consisting of actin and myosin and about 1 to 2 μm in diameter, that runs the length of a muscle fiber.

myoglobin   An iron-containing muscle protein that pulls oxygen from the blood.

myosin   A long, two-headed protein that interacts with actin and generates movement in an ATP-dependent manner.

myxomycota   [Greek, *muxa* = mucus + *mukes* = fungus] The plasmodial, or acellular, slime molds; they closely resemble the cellular slime molds in appearance and in life cycle.

NAD$^+$ (nicotinamide adenine dinucleotide)   The major electron acceptor in oxidative phosphorylation.

NADH   The reduced form of NAD$^+$.

Na$^+$-K$^+$ATPase (also called the **sodium-potassium pump**)   An important active transporter that simultaneously transports sodium ions (Na$^+$) out of cells and potassium ions (K$^+$) into cells, using the energy of ATP.

natural selection   The differential survival and reproduction of individuals with certain inherited traits; the major mechanism for evolution.

Neanderthal   One of the early species of *Homo sapiens*.

negative assortative mating   A process in which individuals choose mates who differ somewhat from themselves.

negative feedback   The process of neutralizing external changes.

negative regulator   A protein that reduces transcription of a particular gene or operon.

nematocyst   In cnidarians, a tiny barbed spear fired with water pressure by cnidocytes.

nematodes   Roundworms; slender, cylindrical, and unsegmented.

nematomorpha   Horsehair worms.

nemertea   Ribbon worms.

nephridia (singular, **nephridium**)   Tubular excretory organs that remove nitrogen-containing wastes from the coelomic fluid and regulate water and salt concentration.

nephron   [Greek, *nephros* = kidney] The functional unit of the kidney formed by the glomerulus, Bowman's capsule, and the renal tubule together.

neritic zone   The zone of the ocean that is out from shore, over the continental shelf.

nerve growth factor (NGF)   A protein paracrine signal necessary for the growth and differentiation of specific nerve cells.

neural crest   A set of embryonic cells derived from the roof of the neural tube; neural crest cells migrate to different locations in the embryo and develop into a variety of different cell types in the adult.

neural plate   A flat plate above the notochord in a vertebrate embryo; the future nervous system.

neural tube   A hollow tube that forms from the neural plate, above the notochord in a vertebrate embryo; the precursor of the central nervous system.

neuromuscular junction   The chemical synapse between a motor neuron and a voluntary muscle cell.

neuron   A nerve cell specialized for the conduction of electrical and chemical signals.

neurotransmitter   A signaling molecule that transmits signals from a nerve cell either to another nerve cell or to a muscle or a gland.

neurulation   The process of forming the neural tube and neural crest.

neutrons   Neutral particles without electric charge found in the nucleus of an atom.

niche   The way an organism uses its environment.

nicotine   A naturally occurring compound (especially in tobacco) that acts as a stimulant and narcotic; like heroin and cocaine, it is highly addictive.

nicotinic acetylcholine receptor   A ligand-gated cation channel that responds to acetylcholine and nicotine; such receptors are found in the neuromuscular junction and elsewhere.

nitrification   The process by which energy is obtained by converting ammonia to nitrite (NO_2) and nitrate (NO_3).

nitrifying bacteria   Bacteria specialized to convert atmospheric nitrogen (N_2) into ammonia or nitrate.

nitrogen fixation   The conversion of atmospheric nitrogen gas into ammonia, which makes nitrogen available to organisms; nitrogen fixation is carried out only by specialized prokaryotes, some of which live within the roots of some plants.

nitrogenous base (so-called because it contains nitrogen and can accept hydrogen ions)   A small molecule that contains one or two aromatic rings of carbon and nitrogen, with attached hydrogen atoms; a component of a nucleotide.

nitrogenous wastes   Nitrogen-containing products of protein breakdown.

node   The region of the stem to which a petiole attaches.

node of Ranvier   A gap between myelin wrappings along a nerve axon.

nonassociative learning   Learning in which the subject changes sensitivity to a stimulus after repeated exposure.

noncompetitive inhibitor   A molecule whose inhibitory effects on an enzyme cannot be overcome by increased substrate concentration; a noncompetitive inhibitor usually binds to an enzyme at a location other than the active site.

noncovalent interaction   Attraction (or repulsion) between separate molecules or ions; noncovalent interactions include electrostatic interactions, hydrogen bonds, van der Waals interactions, and hydrophobic interactions.

noncyclic photophosphorylation   The production of ATP from light energy by a series of electron transfers that do not directly regenerate the absorbing chlorophyll; in plants, noncyclic photophosphorylation occurs within photosystem II, with electron flow ultimately depending on both photosystems.

nondisjunction   The failure of homologous chromosomes or sister chromatids to move apart during meiosis; nondisjunction results in a gamete having too many or too few chromosomes.

noninducible   Not made at higher levels in the presence of an inducer that would normally increase production.

nonpolar   Having an approximately uniform charge distribution.

nonsense   Not specifying any amino acid; a nonsense codon is also called a "stop" codon.

nonsense mutation   A change in DNA that changes the codon for an amino acid into a nonsense or termination codon and thereby leads to a truncated polypeptide.

norepinephrine   A neurotransmitter derived from tyrosine; norepinephrine is used by the central nervous system and the sympathetic nervous system.

normal distribution   A distribution of numbers (data) where equal numbers of data points fall on each side of the mean and all fall within a certain distance of the mean, forming a bell shape. When data have a normal distribution, about 68 percent of the data will fall within one standard deviation of the mean, and about 95 percent will fall within two standard deviations.

notochord   A rod that runs from the front to the rear of a vertebrate embryo, beneath its back (dorsal) surface.

nucellus   [Latin, *nucella* = a small nut] In a flowering plant, the central portion of the ovule in which the embryo develops; the megasporangium.

nuclear envelope   The boundary of a nucleus; it consists of a double membrane separated by about 20 to 40 nm.

nuclear fission   The breakup of large nuclei, of uranium or plutonium, for example.

nuclear fusion   The fusion of two atomic nuclei to form a larger one—formation of a helium nucleus from two hydrogen nuclei, for example.

nuclear pore   An interruption of the nuclear envelope that forms channels between the contents of the nucleus and the cytosol.

nuclear reactions   The high-energy transformation of individual atoms; occurs in stars and in nuclear reactors.

nuclear transplantation   A technique for moving a nucleus from one cell to another.

nuclear winter   A world-wide darkening and cooling that is believed to be the devastating outcome of a large-scale nuclear war; the fine dust from nuclear bombs and from the fires they would ignite could create a dust cloud that would envelop the Earth and reduce sunlight (and photosynthesis) to a few percent of the present level; temperatures would plunge to below freezing in most parts of the world, triggering mass extinctions, similar to those of the Permian and Cretaceous Periods.

nuclease   An enzyme that catalyzes the hydrolysis of a nuclei acid.

nucleic acid   A macromolecule (DNA or RNA) formed by the polymerization of nucleotides.

nucleoid   In the prokaryotes, the restricted part of a cell that contains the cell's DNA; it is not surrounded by a membrane.

nucleolar organizer   A region of one or more chromosomes that contains the genes for ribosomal RNAs.

nucleolus   [plural, **nucleoli**; Latin, a small nucleus] A conspicuous structure within the nucleus, in which ribosomal RNAs are made.

nucleoplasm   The contents of the nucleus.

nucleoside   A nitrogenous base attached to a sugar by a covalent bond between a carbon atom of the sugar and a nitrogen atom of the base.

nucleosome   A DNA-histone complex, about 11 nm in diameter; each nucleosome contains a 146-nucleotide-long stretch of DNA and eight histone molecules (two each of H2A, H2B, H3, and H4).

nucleotide   A small molecule that consists of a nitrogen-containing aromatic ring compound, a sugar, and one or more phosphate groups.

nucleus   In eukaryotic cells, the membrane-enclosed structure that contains most of a cell's genetic information in the form of DNA.

objective   In a microscope, the lens nearest the sample.

obligate aerobes   Heterotrophs that require oxygen to live.

obligate anaerobe   A microorganism that can grow only in the absence of oxygen.

oceanic zone   The zone of the ocean that is beyond the continental shelf over the deepest water.

oceanography   The study of the seas and their ecosystems.

octet rule   The generalization that an atom is particularly stable and chemically unreactive when its outermost shell is full, meaning (usually) that it contains eight electrons.

ocular   The lens of a microscope nearest the eye.

oil   A triacylglycerol that is liquid at room temperature; the component fatty acids are usually unsaturated.

Okazaki fragment (named after its discoverer Reijii Okazaki)   In DNA synthesis, a stretch of DNA that is to be added to the lagging strand.

olfactory receptor   A protein, in the membrane of the olfactory epithelium, that binds to odorant molecules; binding is the first step in informing the brain about smell.

oligodendrocyte   A glial cell in the central nervous system; the oligodendrocyte's plasma membrane wraps around nerve cell extensions in a myelin sheet.

oligotrophic   Referring to a pond or a lake that lacks nutrients; oligotrophic lakes are clear and have no permanent algal blooms; they contain more oxygen and support a more diverse community of organisms than eutrophic lakes and ponds.

ommatidia (singular, **ommatidium**)   The simple light-detecting units that make up compound eyes.

omnibacteria   Bacteria that can perform aerobic respiration and can use nitrate (NO_3^-) as an electron acceptor; the most common organisms on Earth.

omnivores   [Latin, *omnis* = all + *vorus* = devouring] Animals that eat plants, herbivores, and other carnivores.

oncogene   [Greek, *onkos* = bulk, tumor] A gene whose product can change normal cells into cancerlike cells.

oocyte   A precursor of an egg cell.

oogenesis   The production of mature eggs.

oogonium   A precursor cell that is committed to forming female gametes (ova).

open circulatory systems   A circulatory system in which blood mixes freely with extracellular fluids and bathes the organs of the body.

open system   In thermodynamics, a region of space that exchanges both materials and heat with its surroundings.

operator   The DNA sequence in an operon to which a repressor protein binds.

operculum   In fish, a protective bony flap that covers several gill arches on each side of the head; in mollusks, a protective flap that covers the mantle cavity.

operon   In prokaryotes, a set of genes transcribed into a single mRNA.

opiates   Morphine, heroin, and related compounds.

opposite   The arrangement of leaves on a stem so that two petioles extend away from the stem from the same node.

optical isomers   Two molecules that differ only in the arrangement of four different groupings of atoms attached to a single carbon atom.

optimal foraging   Energy-efficient feeding.

orbital (so-called to distinguish from a planet's restricted "orbit")   A limited portion of an atom's space through which an electron moves.

order   A taxonomic group that consists of one or more families; a subdivision of a class.

organ   A structural unit with a distinctive function formed by two or more kinds of tissue.

organ formation   During embryonic development, the movement and specialization of tissues and cells to produce functioning organs such as heart, kidneys, and the nervous system.

organelle   A subcellular structure that performs a specialized task; in eukaryotic cells, many organelles are enclosed by membranes, which isolate the contents of the organelle from the rest of the cytoplasm.

organic   Carbon-containing; the term refers to all carbon-containing compounds, even when they have nothing to do with organisms.

organic chemistry   The study of the structures and reactions of carbon compounds.

organism   A living individual.

orgasm   A complex of changes that often accompany sexual intercourse—smooth muscle contractions in the genital tract, skeletal muscle contractions throughout the body, and feelings of intense pleasure.

origin of replication   A DNA sequence at which replication begins.

oscilloscope   A measuring instrument that displays voltages on a televisionlike screen, showing variations with time.

osmolarity   [Greek, *osmos* = push, thrust] The sum of the concentrations of all the ions and molecules in a solution.

osmosis   [Greek, *osmos* = push, thrust] The flow of water across a selectively permeable membrane as a result of concentration differences.

osmotic pressure   The pressure exerted by osmosis.

osteichthyes   [Greek, *osteon* = bone + *ichthys* = fish] Fish with bony skeletons.

osteoblasts   [Greek, *osteon* = bone + *blastos* = bud] Specialized cells that resemble fibroblasts and manufacture the bone matrix.

osteoclasts   [Greek, *osteon* = bone + *klastos* = broken] Specialized cells that digest collagen and bone matrix.

ostracoderms   [Greek, *ostrakon* = shell + *derma* = skin] Armor-plated jawless fishes, all of which are extinct.

outbreeders   Species with complex physical or behavioral adaptations that promote crossbreeding with individuals who are not closely related.

ova [singular, **ovum**; Latin, egg] Eggs.

ovarian cycle Refers to the events of the menstrual cycle within the ovaries—the production of the oocytes and the growth of the corpus luteum.

ovary [Latin, *ovum* = egg] The organ (in an animal or a flower) that produces female germ cells (or gametophytes); a female gonad.

overlapping genes A single stretch of DNA that contains the information for distinct polypeptides in different reading frames.

overshoot Population growth that exceeds the carrying capacity.

oviduct [Latin, *ovum* = egg + *ductus* = duct] One of two long tubes that lead to the uterus; in humans, it is called a fallopian tube.

ovulation The release of the oocyte from the follicle.

ovule [Latin, *ovulum* = little egg] The structure in a carpel that, after fertilization, forms a seed.

ovum [plural, **ova**; Latin, egg] A female gamete; an egg.

oxaloacetate A four-carbon compound that can combine with the acetyl group from acetyl CoA to produce citric acid, beginning the citric acid cycle.

oxidative fiber A thin muscle fiber that derives most of its energy from respiration.

oxidative phosphorylation The process that couples the oxidation of NADH and $FADH_2$ to the production of high-energy phosphate bonds in ATP.

oxidizing agent The electron acceptor in a redox reaction.

oxygen evolving complex In photosynthesis, a manganese-containing protein that directly participates in the electron transfers that convert coordinates of water to oxygen.

oxyhemoglobin Hemoglobin that is bound to oxygen.

oxytocin A peptide hormone, released by the posterior pituitary, that stimulates contractions of the uterus during childbirth.

ozone layer The layer of ozone gas 20 to 50 kilometers above the Earth; ozone is a highly reactive molecule made of three oxygen atoms, instead of the two in atmospheric oxygen.

pacemaker A group of the cells of the myocardium that are capable of rhythmic spontaneous contractions; they contract slightly more frequently than most myocardial cells; by initiating contraction before other cells in the heart, the pacemaker cells set the pace of contraction for the rest of the heart

paleontology [Greek, *palaios* = ancient + *on* = being + *logos* = discourse] The study of ancient life.

palisade parenchyma In plants, the mesophyll cells that lie just under the leaf's upper epidermis and are elongated and packed with chloroplasts.

paracrine signal A chemical signal that acts only in the immediate region of its production.

parapatric speciation A mechanism of speciation in which a species splits into two populations that are geographically separated, but still have some contact.

paraphyletic Refers to a taxon that contains some but not all the descendants of the ancestral species.

parapodia In polychaetes, a pair of leglike paddles in a single segment; parapodia are used in respiration and to swim, crawl, or burrow.

parasite An organism that consumes parts of a larger organism; it does not necessarily kill the host.

parasitism A symbiotic relationship in which one species benefits at another's expense.

parasympathetic nervous system [Greek, *para* = beside, next to + sympathetic] A division of the autonomic nervous system; it generally acts to husband resources, for example by slowing the heart and increasing intestinal absorption.

Parazoa The smaller of two subkingdoms of animals, which contains two phyla: the Placozoa and the Porifera (sponges); the parazoa differ greatly from all other animals in showing no symmetry and minimal organization and cell specialization—they possess only simple connective tissues and no organs.

parenchymal cells Plant cells with thin cell walls, chloroplasts, large vacuoles, and the machinery to perform photosynthesis.

parental generation (also called **P generation**) The original parents in a genetic cross.

passive transport Movement of a substance across a membrane that occurs spontaneously, without the expenditure of energy.

patch clamping A method that allows the measurement of currents across a tiny piece of membrane rather than across a whole cell membrane.

pathogens Agents that cause disease.

pattern formation The creation of a spatially organized structure during embryonic development.

pedigree A family tree; pedigrees are often used to show the inheritance of a disease within a family.

pelagic division The open waters of the ocean and the organisms within them.

pellet The part of a sample that, during centrifugation, moves to the bottom of the centrifuge tube.

penetrance The fraction of individuals with a particular genotype that show a corresponding phenotype.

pentose [Greek, *pente* = five] A sugar that contains five carbon atoms; ribose and deoxyribose, which are components of nucleotides, are pentoses.

pepsin The stomach enzyme that hydrolyzes proteins into smaller fragments each containing a few amino acids.

peptide A molecule that consists of amino acids linked by peptide bonds; some peptides are water-soluble signals.

peptide bond The covalent bond between two amino acid molecules in a polypeptide; these bonds are formed by the carboxyl group of one amino acid attaching to the amino group of another amino acid.

peptide bond formation During protein synthesis, the transfer of an amino acid from aminoacyl-tRNA to the growing polypeptide chain.

peptidoglycan The cell-wall material of many prokaryotes; peptidoglycan may consist of a single covalently linked molecule formed from polypeptides and polysaccharides.

perennial Refers to a plant that lives and produces seeds for two or more years.

perforation plate In the vascular system of a plant, the end wall of each vessel element; the perforation plate contains one or more holes through which water flows.

pericardium [Greek, *peri* = around + *kardia* = heart] A fibrous sac that encloses the heart itself within a watery lubricating fluid.

pericycle [Greek, *peri* = around + *kykos* = circle] In plant roots, a sheath of parenchyma cells just within the endodermis, but outside the vascular tissue.

periderm In plants, the outer, secondary tissue; periderm is primarily cork.

period In the geological time scale, a time interval of 30 to 75 million years that contains distinctive forms of life in its fossil record; a period is longer than an epoch, but shorter than an era.

peripheral nervous system (PNS) Nerve cells outside the central nervous system (CNS); the PNS carries information from the CNS to muscles and organs and to the CNS from sense organs.

peripheral proteins   Membrane proteins located on the outer and inner surfaces of the plasma membrane.

peristaltic waves   [Greek, *peristellein* = to wrap around] Coordinated contractions of smooth muscles.

permafrost   Permanently frozen ground less than a meter from the surface that underlies the soil during even the warmest summers.

peroxisome   In eukaryotic cells, a kind of membrane-bounded vesicle; peroxisomes contain enzymes that use oxygen to break down molecules by pathways that produce hydrogen peroxide (H_2O_2).

petal   One of the showy, usually colored, parts of a flower.

petiole   The stalk that connects the leaf to the stem.

pH   The logarithm (to the base of 10) of the molar hydrogen ion concentration.

pH scale   Measures the concentration of H^+ and OH^- ions in a solution.

phaeophytes   [Greek, *phaios* = dusky brown] Brown algae.

phagocytosis   [Greek, *phagein* = to eat + *kytos* = hollow vessel] A type of cellular ingestion in which the cell's membrane surrounds a relatively large solid particle, such as a microorganism or cell debris.

pharyngeal slits (also called **gill slits**)   Holes in the sides of the body that run from the inside of the pharynx to the outside surface of an animal.

pharynx   The throat; a common entryway for food into the digestive tract, for air into the lungs, and for water into the gills.

phenetics   A method of classification, based on all observable characteristics, that attempts to avoid the subjective choices of classical taxonomy.

phenolic compounds   Aromatic substances that play a variety of roles in plants; some phenolics repel herbivores and pathogens, some attract pollinators or fruit dispersers.

phenotype   [Greek, *phainein* = to show + *typos* = impression] The collection of all the properties of an individual organism.

phenotypic trait   A single aspect of phenotype in which individuals may vary.

phenylketonuria (PKU)   A human disease that results from the absence of the enzyme phenylalanine hydroxylase, which converts phenylalanine to tyrosine.

pheromone   A substance secreted by one organism that influences the behavior or physiology of another organism of the same species; a pheromone can also activate modal action patterns.

phloem   Conducting vessels that distribute the sugars and other organic molecules made in the leaves to the rest of the plant.

phosphate   The ion (PO_4^{3-}) formed by the dissociation of hydrogen ions from phosphoric acid (H_3PO_4).

phosphodiester bond   The links between nucleotides formed by the phosphate group of one nucleotide attaching to a carbon atom in the sugar component of another nucleotide.

phospholipid   An amphipathic derivative of glycerol in which two hydroxyl groups attach to fatty acids and the third to a phosphate ester; phospholipids are principal components of biological membranes.

phosphorylase kinase   An enzyme that transfers phosphate groups from ATP to glycogen phosphorylase.

photoautotroph   An organism that supplies itself with energy by means of photosynthesis.

photochemical reaction center   A complex of the chlorophyll molecules and proteins that convert captured light energy to chemical energy.

photoheterotroph   A heterotroph that uses light energy but also requires organic compounds.

photon   A package of energy; a light particle.

photoperiodism   The response of a plant to the relative lengths of day and night.

photorespiration   In plants, an oxygen-dependent process that does not produce ATP or NADPH; photorespiration converts ribulose bisphosphate into CO_2 and serine.

photosynthesis   [Greek, *photo* = light + *syntithenai* = to put together] Process in which sugars are synthesized within the chloroplast and temporarily stored there.

photosystem I   One of two distinct but interacting sets of electron transfer reactions responsible for storing light energy in high-energy chemical bonds; photosystem I best absorbs and uses light with wavelengths of about 700 nm.

photosystem II   One of two distinct but interacting sets of electron transfer reactions responsible for storing light energy in high-energy chemical bonds; photosystem II best absorbs and uses light with wavelengths of about 680 nm.

phototropism   The bending of a plant toward light.

phragmoplast   [Greek, *phragmos* = fence + *plasma* = mold, form] A set of microtubules that extends between two dividing plant cells at right angles to the cell plate.

phylogeny   [Greek, *phylon* = race, tribe + *geneia* = birth, origin] Evolutionary history.

physical anthropologist   [Greek, *anthropo* = human] Researchers who study the evolution, anatomy, and behavior of humans.

physiology   The study of how living organisms are organized and how their parts work together.

phytochrome   [Greek, *phyton* = plant] A plant pigment involved in many processes that depend on the timing of dark and light, including flowering, germination, and leaf formation.

pigment   A molecule that absorbs visible light and has color to human eyes.

pilus (plural, **pili**)   In bacteria, a long appendage that serves as the means of attachment of conjugating bacteria and a conduit for the transfer of DNA.

pinnae   [singular, **pinna;** Latin, feather] In plants, small leaflets that extend from a central stalk.

pinocytosis   [Greek, *pinein* = to drink] A type of endocytosis; the nonspecific uptake of bits of liquid and dissolved molecules.

pioneer community   The first community in a succession.

pistil (also called a **carpel**)   In a flower, a female reproductive structure.

pith   In plants, the unspecialized tissue within central nonconducting core of the vascular system.

pituitary gland   A pea-sized structure at the base of the brain that releases at least nine hormones.

placenta   [Latin, flat cake] Tissue that passes nourishment, oxygen, and wastes between the mother and the developing embryo in placental mammals.

placental mammals   Mammals whose embryonic development takes place entirely within the uterus of the mother.

placoderms   [Greek, *plak* = plate + *derma* = skin] Extinct armored fish with jaws.

plankton   [Greek, *planktos* = wandering] Organisms that float in the water, carried by currents.

Plantae   The kingdom that includes plants, autotrophic multicellular organisms that undergo embryonic development.

plasma   [Greek, form or mold] The fluid part of blood.

plasma cell   A cell specialized for the production of antibodies.

plasma membrane   The membrane that surrounds a cell.

plasmid   A circular DNA that can replicate autonomously.

plasmodesmata [singular, **plasmodesma;** Greek, *plassein* = to mold + *desmos* = to bond] Fine intercellular channels between plant cells, derived from vesicles trapped in the growing cell plate.

plasmodium   A mass of cytoplasm with many nuclei but no boundaries between cells.

plasmolysis [Greek, *plasma* = form + lysis = loosening] In a plant cell, the flow of water out of the vacuole in a hypertonic solution and the resulting separation of the plasma membrane from the cell wall.

plastid   A plant organelle surrounded by a double membrane; plastids include chloroplasts, chromoplasts, and amyloplasts.

platelet   A small, membrane-enclosed element that is a component of mammalian blood; it is formed as a cytoplasmic fragment of a precursor cell in the bone marrow; platelets contribute to clotting.

plate   In reference to the Earth's crust, a large block that moves with respect to other blocks at the rate of a few centimeters a year.

platyhelminthes   The phyla of the flatworms.

platyrrhines [Greek, *platyus* = flat + *rine* = nose] Monkeys whose flat noses have widely separated nostrils that point sideways.

pleiotropy [Greek, *pleios* = more + *trope* = turning] The capacity of a single gene to affect many aspects of phenotype.

point mutation (also called a **single-base mutation**)   A change in a single nucleotide pair of DNA; in some cases, a change in a few nucleotide pairs still qualifies as a point mutation.

polar   Having uneven distributions of electrical charge; having positive and negative ends (or poles).

polar body   The smaller daughter cell that results from the unequal meiotic division of an oocyte.

polar fronts   Regions of low pressure, where the westerlies end and polar easterlies begin.

polar microtubules   Microtubules that run from each pole of the mitotic spindle toward the equator.

pollen tube   An extension of a pollen grain, which carries the sperm nuclei into the ovule.

polyclonal   Referring to antibodies made in the descendants of many B cells, with each B-cell clone producing an antibody that recognizes a different epitope.

polygenic   Traits governed by many genes that vary smoothly and continuously within a population.

polymer [Greek, *poly* = many + *meros* = part] A large molecule that consists of smaller identical (or nearly identical) subunits, called monomers.

polymerase chain reaction (PCR)   A method that specifically and repetitively copies a segment of DNA between two defined nucleotide sequences.

polymerization [Greek, *polys* = many + *meros* = part] The assembly of many small molecules into a larger one.

polymorphic [Greek, *poly* = many + *morph* = form] Having two forms of a given gene; the term may refer to a gene or to a population.

polynucleotide   A chain of nucleotides held together by phosphodiester bonds; DNA and RNA are polynucleotides.

polyp   A body plan that resembles a cylinder; one of the two basic body plans of the cnidarians.

polypeptide   A chain of amino acids held together by peptide bonds.

polyphyletic   Refers to a taxon that includes descendants of more than one ancestor.

polyploid   Having two or more complete sets of chromosomes, as in some plants.

polyribosome (also called a **polysome**)   Several ribosomes attached to a single mRNA.

polysaccharide   A carbohydrate macromolecule formed by the polymerization of simple sugars (monosaccharides).

polyspermy   Fertilization by more than one sperm.

polytene chromosomes [Greek, *polys* = many + *tainia* = ribbon] Giant chromosomes, commonly studied in the larvae of *Drosophila* and other flies, that consist of about 1000 chromatids aligned in parallel.

polyunsaturated   Referring to a hydrocarbon chain (or fatty acid) with many double bonds.

Pongidae   Gorillas, chimpanzees, and orangutans.

population   A breeding group of individuals of the same species that inhabit a common area.

population genetics   The quantitative study of the processes by which variation is generated and passed on within populations.

population momentum   The continued growth of a population even though completed family size is below replacement level.

positional information   Chemical cues that establish the position of a cell in a pattern.

positioning   During protein synthesis, the binding of the next aminoacyl-tRNA to the A site in the ribosome-mRNA complex; the first step of elongation.

positive assortative mating   Choosing mates on the basis of similarities to the individual's own genotype.

positive regulator   A protein that increases the rate of transcription.

postabsorptive state   The state of an animal in which cells derive energy and building blocks by breaking down stored glycogen, fats, and proteins.

posterior   The back (tail-end) of an animal.

postsynaptic   Refers to the neuron that is "downstream" from a synapse; a postsynaptic neuron responds to the presynaptic neuron.

postsynaptic potential   A voltage change induced by the action of a neurotransmitter.

posttranscriptional processing   A series of chemical modifications that convert the primary transcript of a gene to a mature RNA.

posttranslational   Occurring after translation is completed.

postzygotic barrier   Barrier to gene flow which makes a zygote either inviable (certain to die) or sterile.

potency   What a cell or tissue could become during development if it were allowed to develop in another environment.

potential energy   A general term for energy that can ultimately be converted into kinetic energy.

pre-mRNA   The primary transcript of a protein-coding gene; the term is also used to refer to the partly modified, not yet mature mRNA precursor.

prebiotic evolution   The evolution of organisms from nonliving matter.

Precambrian   Referring to fossils that date from more than 590 million years ago, that is, before the beginning of the Cambrian Era.

predator   An organism that usually (but not always) kills its prey and consumes most of its prey's body.

preformation   The theory, now discredited, that all the parts of the adult organism already preexist at the earliest stages of life.

pressure flow hypothesis   The accepted explanation for the transport of sap through the phloem.

presumptive   Having a specified developmental fate.

presynaptic   Refers to the neuron that is "upstream" from a synapse; an action potential in the presynaptic neuron can trigger the release of neurotransmitter into the synapse.

prezygotic barrier   A barrier to gene flow that prevents the fusion of the sperm and egg to form a zygote (syngamy).

primary cell wall   The wall that lies outside the plasma membrane.

primary compounds   In plants, refers to compounds that are essential for day-to-day functioning; primary compounds include glucose, amino acids, ATP, and DNA.

primary growth   In a plant, growth from apical meristem tissue; the extension of length.

primary lysosome   A nearly spherical lysosome that has not yet fused with an endosome.

primary mesenchyme   [Greek, *mesos* = middle + *enchyma* = infusion] In an animal embryo, the first cells to move into the interior of a blastula.

primary oocyte   A precursor of an ovum; a diploid cell that is capable of dividing, by meiosis, to form a haploid ovum.

primary organizer   The term used by Spemann to describe the dorsal lip of the blastopore, meaning that its action established the organization of the entire early embryo.

primary productivity   The productivity of the first trophic level, the energy captured in the chemical bonds of new molecules each year for each square meter.

primary structure   The linear sequence of amino acids in each polypeptide chain of a protein.

primary succession   The invasion of a completely new environment such as a sandbar or new volcanic island.

primary transcript   The RNA first transcribed from a particular gene; the precursor of a mature RNA.

primates   The order of mammals that includes humans, apes, monkeys, and lemurs.

primer   An already existing polynucleotide to which additional nucleotides are added; all known DNA polymerases add nucleotides to a primer.

primitive characters   Traits shared by all the members of a lineage and also shared with more distant relatives. In members of the cat family, for example, long canine teeth are a primitive trait because they are shared with other members of the order Carnivora, such as canids (dogs) and bears.

primordial germ cells   The precursors of both sperm and eggs, which arise early in development.

Principle of Independent Assortment (also called **Mendel's Second Law**)   The generalization that the alleles for one gene segregate independently of the alleles of another gene.

Principle of Segregation (also called **Mendel's First Law**)   The generalization that a sexually reproducing organism has two "determinants" (or genes, in modern terms) for each characteristic, and these two copies segregate (or separate) during the production of gametes.

probability   The chance that a given event will happen; it is calculated as the number of times an event has actually occurred divided by the number of opportunities it could have occurred; in the most familiar example, the probability of a tossed coin turning up "heads" is 1 (the number of sides that are heads) divided by 2 (the total number of sides).

proboscis   A long, sensitive, retractable, and sometimes venomous, tubelike snout characteristic of ribbon worms; a long, flexible snout or trunk characteristic of elephants. Insects often drink using a proboscis.

producers   Organisms, such as plants, that harvest energy directly from sunlight or, rarely, from inorganic molecules.

product rule   In the calculation of probability, the statement that the probability of two independent events taking place together is the product of their probabilities.

proglottid   One of the repeated segments of a tapeworm; a proglottid contains both male and female reproductive organs.

programmed cell death   The death of specific cells as a normal part of development.

prokaryotic   [Greek, *pro* = before] Cells that contain neither a nucleus nor other membrane-bounded organelles.

prolactin   A peptide hormone, made in the anterior pituitary, that stimulates milk production.

prometaphase   [Greek, *meta* = middle] Prometaphase was previously called early metaphase, the stage of mitosis during which the nuclear membrane disappears and the chromosomes attach to the spindle fibers.

promoter   A DNA sequence that specifies the starting point and direction of transcription; it is where RNA polymerase first binds as it begins transcription.

prophage   [Greek, *pro* = before] The dormant form of a bacteriophage in which the phage DNA has integrated into that of the host cell.

prophase   [Greek, *pro* = before] The first phase of mitosis, when the diffusely stained chromatin resolves into discrete chromosomes, each consisting of two chromatids joined together at the centromere.

proplastid   Common precursor of all plastids.

prostaglandin   One of 16 paracrine signals derived from a 20-carbon fatty acid called arachidonic acid.

prostate gland   In the reproductive tract of a male mammal, a gland that secretes a thin milky alkaline fluid into the lumen of the urethra.

protease (also called a **peptidase**)   A digestive enzyme that catalyzes the hydrolysis of peptide bonds; used to digest proteins.

protein   A macromolecule consisting of one or more polypeptides.

protein kinase   An enzyme that transfers the phosphate group from an ATP to a protein.

protein kinase A   An enzyme that, when stimulated by cyclic AMP, transfers a phosphate group from ATP to specific proteins.

protein kinase C   An enzyme that, when bound to diacylglycerol, transfers phosphate groups from ATP to sites in particular proteins.

proteinoids   Polypeptides thought to have been the first to appear in prebiotic evolution; proteinoids are almost certainly without specific sequences of amino acids.

Protista   [Greek, *protos* = first] The kingdom that consists mostly of single-celled organisms but that also contains some related multicellular species; protists include algae, water molds, slime molds, and protozoa.

proto-oncogene (also called a **cellular oncogene**)   The cellular counterpart of a viral oncogene; proto-oncogenes participate in the normal control of growth and differentiation.

protobiont   A primitive cell—thought to represent a stage of prebiotic evolution—that could concentrate organic molecules and begin to evolve the first metabolic pathway.

proton   A positively charged particle found in the nucleus of an atom; the word proton is commonly but inaccurately used to refer to a hydrogen ion (H^+, actually a hydronium ion H_3O^+), which comes from the dissociation of a water molecule.

proton channel   The route that hydrogen ions follow as they flow down an electrochemical gradient.

proton gradient   The difference in H^+ concentration across a membrane.

protoplasm   [Greek, *proto* = first + Latin, *plasma* = a thing molded or formed] A term used by nonbiologists to refer to the living material of organisms.

protostomes [Greek, *protos* = first + *stoma* = mouth] Those animals in which the mouth develops first, from the blastopore; protostomes include mollusks, annelids, and arthropods.

prototherian (also called **monotreme**) Egg-laying mammal; like other mammals, prototherians have hair and mammary glands, but their digestive and reproductive systems empty into a cloaca.

protozoa [Greek, *protos* = first + *zoe* = life] The protists that most resemble little animals.

provirus [Greek, *pro* = before] The dormant form of a virus in which the viral DNA has integrated into that of the host cell.

proximal tubule The tubule that lies just next to Bowman's capsule, in the kidney's outer cortex.

pseudocoel [Greek, *pseudo* = false + *koilos* = hollow] A cavity in certain invertebrates that lacks the mesodermal lining of a true coelom.

pseudocoelomates [Greek, *pseudo* = false] Animals that have an organ-containing cavity that lacks the mesodermal lining of a true coelom.

Pseudomonads A group of bacteria that are typically straight or curved gram-negative rods with flagella; they seem to live everywhere—in soil, ponds, infected wounds, hot tubs, and even medicine bottles.

pseudopodia [Greek, *pseudos* = false + *pous* = foot] Temporary extensions of the cytoplasm.

pulmonary [Latin, *pulmo* = lung] Relating to the lungs.

pulmonary artery The artery that carries deoxygenated blood from the heart to the lungs.

pulmonary circulation Circulation from heart to lungs to heart.

pulmonary veins The veins that carry oxygenated blood from the lungs to the heart.

punctuated equilibrium The idea that species change very little most of the time (stasis), and that most anatomical or other evolutionary change in individual species occurs during a geologically brief period at the time of speciation.

pupa An inactive phase, following a larval stage, of insect development; a pupa may be enclosed in a cocoon.

purine A nitrogenous base that contains a particular nine-membered double ring, with five carbon and four nitrogen atoms; adenine and guanine are purines.

pyramid of biomass A graphic presentation of the total mass of organisms at each trophic level of an ecosystem.

pyramid of energy A graphic presentation of the amount of energy used at each trophic level in a specified time (usually a year).

pyramid of numbers A graphic presentation of the total numbers of organisms at each trophic level.

pyramid of productivity An energy pyramid, which shows how much sunlight energy enters the system how much is stored in the biomass of each trophic level, and how much flows between trophic levels.

pyrimidine A nitrogenous base that contains a particular six-membered ring, with four carbon and two nitrogen atoms; thymine, uracil, and cytosine are pyrimidines.

pyrophosphate A molecule that consists of two phosphate groups linked together in a high-energy bond.

pyruvate A three-carbon compound that is the end point of the first stage of glycolysis.

quantitative character A phenotypic trait that may be described numerically within a range of values rather than as clear-cut alternatives.

quaternary structure The relationship among separate polypeptide chains in a protein.

R group (also called a **side chain**) Each amino acid's characteristic group of atoms.

r-selected species A species that has adaptations that support high growth rate (r_{max}).

races (or **subspecies**) Morphologically distinct subpopulations that can interbreed.

radial symmetry Structural symmetry in which rotation along the central axis doesn't change the appearance.

radioactive isotope A radioactive (unstable) isotope of an element; examples include ^{14}C and ^{3}H.

radula In mollusks, a rasping tongue covered with teeth made from chitin.

rain shadow The area adjacent to a mountain range, away from the prevailing winds, where little rain falls.

random drift Changes in gene frequency not caused by selection, mutation, or immigration, but by random events.

random sample A sample chosen at random with respect to any differences among the members of a population.

reabsorption In the kidney, the transport of specific substances such as salt, water, and glucose from the filtrate back into the blood.

reading frame The grouping of nucleotides into codons that specify an amino acid sequence.

realized niche The resources that a species uses in a particular community.

receptor A protein to which a chemical signal first binds.

receptor-mediated endocytosis A cell's uptake of specific substances that are recognized by receptor proteins on the plasma membrane.

recessive The term referring to an allele that does not contribute to the phenotype of a heterozygote.

reciprocal altruism An explanation for apparently selfless behavior; the hypothesis that animals engage in altruism with the expectation that the recipient would return the favor some time in the future.

recombinant Containing a combination of genes not found in nature.

rectilinear Movement in a straight line, especially in snakes.

rectum [Latin, straight] The straight portion at the end of the descending colon in which feces are stored until their elimination through the anus.

reducing agent The electron donor in a redox reaction.

reductionism The effort to understand the whole in terms of the parts.

reflex The most automatic behavior pattern, with motor activity directly responding to a sensory stimulus.

refractory period In an axon, the period during which the further excitation does not result in channel opening and impulse propagation.

regulatory cascade A set of reactions that amplify a signal from a relatively small number of molecules into a response that affects many more molecules.

regulatory mutation A change in DNA that affects the amount (rather than the structure) of a polypeptide.

relative abundance A measure of species diversity that takes into account how common each species is.

releasing factor (also called **releasing hormone**) A peptide hormone, made by the hypothalamus, that regulates the release of specific hormones by the anterior pituitary.

renal artery [Latin, *renes* = kidney] The artery that carries blood to each kidney.

renal tubule In the kidney, a long narrow tube leading away from Bowman's capsule.

renal vein   The means by which blood leaves the kidneys.

renin   An enzyme secreted by the kidneys.

replacement reproduction   The family size at which each couple is replaced by just two descendants.

replication   [Latin, *replicare* = to fold back] Copying of a single DNA molecule (or a single set of DNA molecules) into two copies.

replication fork   A Y-shaped region of DNA where the two strands of the helix have come apart during DNA replication.

replication unit   In eukaryotic DNA replication cells, a group of 20 to 50 origins of replication that form replication forks at the same time.

reporter   A protein (or gene) whose activity serves as a measure of the ability of a test sequence to regulate expression in a particular type of cell.

reproductive unit   A species as the group of potential mates.

reproductively isolated   The term referring to populations that are unable to interbreed.

reptile   [Latin, *reptare* = to crawl] An air-breathing, egg-laying vertebrate with an external covering of scales or horny plates; reptiles include snakes, lizards, crocodiles, turtles, and dinosaurs.

resilience   One definition of ecosystem stability; the speed with which an ecosystem returns to a particular form following a major disturbance such as a fire.

resistance factor (also called an **R factor**)   A plasmid that contains a gene for an enzyme that inactivates an antibiotic.

resolution   (1) The minimum distance between two objects that allows them to form distinct images; (2) the part of the sexual response cycle in which blood flow and muscle tension return to normal after orgasm.

resource partitioning   Splitting an ecological niche.

respiration   The oxygen-dependent extraction of energy from food molecules.

respiratory chain   The pathway of electrons in oxidative phosphorylation.

respiratory system   All the structures responsible for the exchange of gases between the blood and the external environment.

response element (also called a ***cis*-regulatory element** or **enhancer**)   A DNA sequence that regulates transcription by binding to a specific protein, a transcription factor.

resting potential   The voltage across the cell membrane when a cell is not producing action potentials.

restriction enzyme   An enzyme that cuts DNA at a particular sequence.

retrovirus   An RNA virus that uses reverse transcriptase for viral replication.

reuptake   The pumping of transmitter from the synaptic cleft either into the presynaptic neuron or into surrounding glial cells.

reverse transcriptase   An enzyme that copies RNA into DNA; RNA-dependent DNA polymerase; reverse transcriptase is present in RNA tumor viruses and in HIV.

reversion   A mutation that restores the previous version of a gene sequence.

rhizoid   A thin, rootlike filament that enables mosses, liverworts, and fern gametophytes to anchor and to absorb nutrients.

rhizome   A horizontal stem that spreads on or below the ground from which fronds grow.

rhizopoda   Amebas.

rhodophytes   [Greek, *rodon* = rose] Red algae.

rhodopsin   The protein that detects light in the retina.

rhythm method   A birth control method; a form of modified abstinence, in which a couple refrains from sexual intercourse at times in the monthly cycle when conception is likely to occur.

ribbon model   A way of depicting the tertiary structure of a protein that stresses the secondary structures of a globular protein; in the ribbon model, an α helix is depicted as a coil, while β sheets are sets of arrows.

ribosomes   [Latin, *soma* = body + "*ribo*" because they contain *ribo*nucleic acid (RNA)] Complex assemblies of RNAs and proteins, about 15 to 30 nm in diameter, that are responsible for carrying out protein synthesis.

ribozyme   An artificial RNA enzyme.

RNA polymerase   The enzyme responsible for transcribing DNA into RNA.

root   The part of a plant below the ground.

root cap   A region of the growth zone of a plant's root; the root cap produces a polysaccharide slime that lubricates the path of the growing root through the soil and serves as a mechanical shield for the apical meristem, which lies just behind.

root hairs   Tiny projections, each of which is the extension of a single epidermal cell, through which most of the water that enters a root comes.

root pressure   Osmotic pressure generated by the difference in solute concentrations between root tissues and soil solution.

root system   An underground system that anchors the plant and absorbs water and minerals from the soil.

rotifers   Tiny aquatic aschelminths with saclike bodies.

rough ER   Endoplasmic reticulum that is associated with ribosomes.

RU 486   A drug used to induce abortions within the first nine weeks after conception.

rudiment   [Latin, *rudimentum* = beginning] Initial embryonic stage of an organ, from which the final form will develop.

ruminants   [Latin, *rumen* = throat] Hoofed, horned herbivores that can regurgitate partly digested food, called cud, for further chewing.

S (for DNA *s*ynthesis)   The period of the cell cycle during which DNA replicates.

salivary glands   Exocrine organs in the mouth that secrete both enzymes and mucus.

saltationism   The view, held by many 19th-century biologists, that evolution occurred by fits and starts.

saltatory conduction   The jumping movement of an action potential down a myelinated axon, from one node of Ranvier to the next; saltatory conduction occurs at rates up to 100 times faster than conduction down an unmyelinated axon.

sampling errors   Chance fluctuations that affect small samples but not large ones.

sands   Soils of large particles (from 20 mm to 200 mm).

saprophyte   [Greek, *sapro* = putrid + *phyte* = plant] A plant that decomposes dead material.

saprophytic   Deriving nourishment from dead organisms.

sarcomere   [Greek, *sarx* = flesh + *meros* = part] In muscle, a small repeating cylinder within a myofibril; each sarcomere is about 2.5 μm long; the contractile unit of a striated muscle.

sarcoplasmic reticulum   In a muscle fiber, a system of membrane-lined channels and sacs where Ca^{2+} ions are stored; a specialization of the endoplasmic reticulum.

saturated   A hydrocarbon chain (or fatty acid) in which all the bonds between carbon atoms are single bonds; the chain is called saturated because it contains the maximum number of hydrogen atoms.

savanna A tropical or subtropical grassland punctuated by solitary trees or small clumps of trees.

scanning electron microscope (SEM) A type of electron microscope in which electrons are reflected from the surface of the observed object.

scavenger An animal that feeds from whole carcasses.

Schwann cell A glial cell in the peripheral nervous system; its plasma membrane wraps around nerve cell extensions in a myelin sheet.

scientific method A formal manner of formulating, testing, and eliminating hypotheses.

sclerenchyma [Greek, *skleros* = hard] A plant tissue that consists of cells, with thick cell walls; may be fibers or sclereids.

scolex A specialized attachment organ at the anterior end of a tapeworm.

scrotum [Latin, bag] A pouch that lies outside the body that holds the testes.

Scyphozoa The class of animals containing the jellyfish.

Second Law of Thermodynamics The statement that, while the total energy in the universe does not change, less and less energy remains available to do work.

second messenger An intracellular molecule whose synthesis and degradation depends on the action of an extracellular signal and which stimulates changes inside a target cell.

secondary cell wall The wall that lies between the cell membrane and the primary cell wall.

secondary growth Growth at the lateral meristems that increases the thickness of a shoot or a root.

secondary immune response A robust immune response triggered by the reappearance of a previously encountered antigen or microbe.

secondary lysosome An irregular lysosome that results from the fusion of a primary lysosome and an endosome; a secondary lysosome contains material that is in the process of being digested.

secondary mesenchyme In an animal embryo, cells that become mesoderm as gastrulation proceeds.

secondary oocyte A precursor of the ovum; a haploid cell produced from the primary oocyte by meiosis.

secondary plant compounds Chemicals that are not essential to a plant's normal metabolism, but which often serve a defensive purpose.

secondary sexual characteristics Hallmarks of sexual differentiation that appear at sexual maturity; in humans these include an increase in body size, a deepening of the voice, hair growth, and the development of sexual drive.

secondary structure Regular local structures resulting from regular hydrogen bonding within adjacent stretches of a polypeptide backbone; secondary structures include α helices and β structures.

secondary succession The sequence of stages in a community that has suffered serious damage.

secretion In the kidney, transport into the filtrate; secreted substances include potassium and hydrogen ions, ammonia, and organic acids and bases.

secretory vesicles In eukaryotic cells, small vesicles formed on *trans* side of the Golgi apparatus; they contain glycoproteins that are to be secreted.

section A thin slice of tissue prepared for microscopy.

sedimentary rock Rock formed by pressure from succeeding layers turning layers of mud, sand, and other sediment into rock.

segmentation gene A gene that regulates the development of segments within the early embryo.

segregation The separation of two homologous chromosomes during meiosis.

selective theory The view that the immune response depends on an animal's ability to select the set of antibodies to make in response to the presence of a particular infection.

selectively permeable Allowing the passage of some ions and molecules (especially of water) much more rapidly than others.

self-splicing Removal of intervening sequences from an RNA without the participation of other molecules.

selfish gene A sequence of DNA whose only role is to reproduce itself.

semen [Latin, seed] The sperm cells, together with the fluid from the seminal vesicles, the prostate gland, and the bulbourethral glands.

semiconservative The pattern of DNA replication in which half of each parent molecule (one strand) is present in each daughter molecule.

seminal vesicles In the male reproductive tract, organs that secrete sugars and other nutrients into the semen.

seminiferous tubules [Latin, *semen* = seed + *ferre* = to bear] The tightly coiled ducts in the testis in which the sperm matures.

senescence Aging; deterioration; in plant cells, the breakdown of cellular components leading to cell death.

sensitive period The period of time during which an animal can learn a particular behavior pattern.

sensitization The increased behavioral response to a noxious stimulus.

sepal One of the parts at the base of a flower; sepals are usually small and green, but in some flowers they may be large and colorful.

septa In a fungus, the walls that separate the nuclei within a filament.

Sertoli cells In the testes, specialized cells that surround the developing sperm; they nourish the developing sperm, regulate the passage of nutrients from the blood, and secrete a fluid that fills the lumen.

sessile Permanently anchored to rocks, logs, or coral.

seta (plural, setae) In annelids, a stiff, bristlelike projection.

sex chromosomes The chromosomes that differ between males and females; in humans, the 23rd pair of chromosomes.

sex steroid A steroid hormone that stimulates, maintains, and regulates reproductive organs and secondary sexual characteristics.

sex-determining region In male mammals, a gene on the Y chromosome that encodes a regulatory protein that promotes the sexual differentiation of the embryonic gonads into testes.

sex-linked gene A gene that has a different pattern of inheritance in males and females because the gene lies on a sex chromosome (usually the X chromosome in mammals).

sexual dimorphism A phenotypic difference between the sexes.

sexual imprinting A special form of imprinting that helps animals recognize potential mates.

sexual reproduction A process that produces offspring that have inherited genetic information from two parents rather than one; because the genes from each parent are likely to differ, sexual reproduction provides new combinations of genes.

sexual selection The differential ability of individuals with different genotypes to acquire mates.

shared derived characters Traits shared by a lineage of organisms but not shared with more distant relatives. Among mammals, for example, fur is a shared derived character because other vertebrates do not have fur. In primates, fur is a primitive character, one that is shared with all other mammals. Cladists construct maps of relatedness, called cladograms, using shared derived characters.

shear   The twisting action created by forces that are not opposite one another.

shell   In an atom, a group of orbitals whose electrons have nearly equal energy.

shoot   The part of a plant above the ground.

shoot system   Stems, photosynthetic leaves, and flowers and other organs of reproduction.

sibling species   Two species that are extremely similar but reproductively isolated from one another.

sickle cell disease (also called **sickle cell anemia**)   A blood disorder that gets its name from the curled appearance of red blood cells in sickle cell patients; a genetic disease caused by a mutation in the gene encoding the β-polypeptide of hemoglobin.

sidewinding   Movement in which a snake moves across loose or sandy soil by throwing successive coils sideways.

sieve plate   In sieve tubes of a plant's phloem, the top and bottom end walls of each sieve tube member.

sieve tube members   In the phloem of a plant, the cells that actually conduct fluid.

sieve tubes   In the phloem of a plant, pipelike channels that carry organic matter from the leaves.

sign stimulus   The key aspect of an object that triggers a modal action pattern.

signal hypothesis   The statement that a hydrophobic signal peptide directs a growing polypeptide chain through the membrane of the endoplasmic reticulum into the ER lumen.

signal peptide   A polypeptide segment of 16 to 30 amino acids at the amino terminal end of the newly made protein; a signal peptide directs the newly made protein into the lumen of the endoplasmic reticulum.

silent mutation   A change in DNA that has no effect on phenotype.

silt   A soil of medium-sized particles (from 2 mm to 20 mm).

simple diffusion   [Latin, *diffundere* = to pour out] The random movement of like molecules or ions from an area of high concentration to an area of low concentration.

simple pit pair   A region between plant cells where the secondary wall is absent, allowing small molecules to pass.

simple transposon (also called an **insertion sequence**)   The simplest transposable element, a short length of DNA (up to a few thousand nucleotide pairs long) that can move from place to place in the genome.

sink   The site of storage or consumption of a substance.

sinoatrial (SA) node   The pacemaker, which lies near the top of the right atrium in the mammalian heart.

sinus   An open cavity; in insects and other arthropods and in most mollusks, it is a space through which blood travels.

sister chromatids   The two chromatids that make up a single chromosome; the sister chromatids are duplicate copies of the same genetic information.

site-directed mutagenesis   A technique that specifically alters a chosen nucleotide sequence.

skeletal muscles   The muscles that are attached to bones.

skull   The bony (or cartilaginous) case that encloses and protects the brain; it lies at the forward (or, for humans, the top) end of a vertebrate's skeleton.

sliding filament model   The accepted view of muscle movement, which holds that movement results from changing the relative positions of thick and thin filaments rather than from filament contraction.

slime layer   A tangled web of polysaccharides that, like a capsule, lies outside the cell wall.

smooth ER   Endoplasmic reticulum that is not associated with ribosomes.

smooth muscle   The involuntary muscles that line the walls of hollow internal organs such as intestines and blood vessels.

sociobiology (renamed **behavioral ecology**)   The study of behavior from an evolutionary perspective.

solute   A substance that dissolves within a solvent.

solvent   Any fluid in which other substances dissolve.

soma   The somatic cells of an organism, in contrast to germ cells or germ line.

somatic cells (also called the **soma**)   [Greek, *soma* = body] In a multicelled organism, the cells that do not give rise to germ cells or gametes.

somatic theory   The view that antibody diversity arises during development, rather than being inherited.

sori   [singular, **sorus;** Greek, *soros* = heap] In a fern, groups of sporangia clustered on the underside of the sporophylls.

source   The site of production (for example, of sugars in a plant).

speciation   The formation of a new species.

species   A biological species is a group of actually or potentially interbreeding populations that are reproductively isolated from other such groups. Species may also be defined differently, for example, according to phenetics or cladistics.

species diversity   A measure of diversity that takes into account how common individuals of each species are.

species richness   The number of species in an ecosystem.

sperm (also called a **spermatozoan**)   A male gamete.

spermatid   The four haploid cells produced during the two meiotic divisions (meiosis I and II).

spermatogenesis   The development of spermatogonia into mature sperm.

spermatogonium   [plural, **spermatogonia;** Greek, *sperma* = sperm + *gonos* = offspring] A diploid cell that is committed to forming sperm cells, but can still undergo mitosis.

spermatozoan (also called a **sperm**)   [Greek, *sperma* = seed + *zoos* = living] A fully differentiated male germ cell.

spermicide   [Latin, *cidere* = to kill] A cream or jelly that kills sperm; an essential part of the effectiveness of both cervical caps and diaphragms.

sphincters   Rings of muscles that surround a tube.

spiracle   In the insect respiratory system, an opening through which air enters the body; a spiracle leads directly to the tracheae.

spirilla   [singular, **spirillum;** Latin, *spira* = coil] Spiral-shaped prokaryotic cells.

spirochetes   Anaerobic heterotrophs that have a distinctive corkscrew shape.

spliceosome   A complex of pre-mRNA, proteins, and small nuclear RNAs that catalyzes the removal of introns and the splicing of exons.

spongy bone tissue   Tissue that generally lies at the ends and inside the bones.

spongy parenchyma   In a plant leaf, the mesophylls that lie just under the palisade parenchyma that are irregular but rounded in shape and separated by numerous air spaces.

spontaneous   A reaction that occurs without any external input of energy.

spontaneous abortion (also called **miscarriage**)   In mammalian development, embryonic or fetal death, usually resulting from congenital defects.

sporangium   [Greek, *spora* = seed + *angeion* = vessel] In plants and fungi, a structure within which cells undergo meiosis to produce the haploid spores.

spore   A cell that divides by mitosis to produce new individuals.

sporophyll   A leaf or leaflike structure that is specialized for the production of spores; in flowering plants, the term refers to carpels and stamens; in ferns sporophyll refers to a leaf that bears sporangia.

sporophyte   The diploid form of a life cycle characterized by alternation of generations; a sporophyte produces haploid spores that give rise to haploid gametophytes by meiosis.

sporozoans   Parasitic protists, including those that cause malaria; sporozoans undergo alternation of generations.

stabilizing selection   Selection that tends to act against extremes in the phenotype, so that the average is favored.

stage   On a microscope, a moveable platform that holds the observed specimen.

stain   A dye that binds differently to cell components and thereby increases contrast for microscopy.

stamen   In a flower, a male reproductive organ; the male sporophyll.

standard deviation   An estimate of the variation in a set of data calculated from how much the data deviate from the mean.

starch   Polysaccharides used in plants for long-term energy storage.

Start (also called the **restriction (R) point**)   The "point of no return" in the G_1 phase of the cell cycle; once a cell proceeds beyond Start, it proceeds through the rest of the cycle, including mitosis and cytokinesis.

stasis   Evolutionary stability over long periods of time.

statistics   The science of collecting and analyzing numerical data.

statolith   [Greek, *statos* = standing + *lithos* = stone] Gravity detector in a plant or animal.

stele   In roots and stems, a central cylinder through which the vascular bundles frequently run.

stem cell   A cell that can either produce more of itself by cell division or undergo differentiation to one or more specialized cell types.

stereoisomers   Molecules with the same atoms and functional groups; two stereoisomers differ only in the spatial arrangements of their atoms.

steric inhibition   Inhibition of an enzyme by a molecule whose shape resembles that of the substrate.

sternum   [Greek, *sternon* = chest] The breastbone.

steroid   A lipid-soluble molecule derived from cholesterol; many steroids are used as hormones.

stoma [plural, **stomata**; Greek, *stoma* = mouth] In a plant leaf, a tiny mouthlike pore that opens and closes to regulate the flow of carbon dioxide and other gases to the interior of the leaf.

stomach   [Greek, *stoma* = mouth] The most dilated and most muscular section of the digestive tract.

stop codon   A codon in mRNA that signals the end of a polypeptide chain.

striated   Striped.

striations   Cross stripes, 2 to 3 μm apart, that lie perpendicular to the long axis of a skeletal or cardiac muscle.

strobilus [plural, **strobili**; Greek, *strobilos* = cone]   A conelike structure (such as a pine cone) that consists of overlapping sporophylls grouped around a central axis.

stroke   Failure of the blood supply to the brain.

stroke volume   In the action of the heart, the volume of blood delivered by a ventricle.

stroma   [Latin, mattress] The region within a chloroplast bounded by the inner chloroplast membrane.

structural isomers   Molecules that contain the same atoms, grouped in different ways to produce different functional groups.

structural mutation   A change in DNA that leads to an alteration in the amino acid sequence of a polypeptide.

suberin   The waxy substance, which is impermeable to water, that surrounds each cell of the endodermis.

subsidize   Referring to the growth of crop plants, to provide additional energy.

substrate   A reacting molecule in an enzyme reaction; a substrate is usually (but not always) much smaller than an enzyme.

succession   Progressive, predictable change over time in kinds of species and numbers of species in a community.

successional communities   Intermediate stages of succession.

sucrose   Table sugar; a disaccharide consisting of two monosaccharides, glucose and fructose, linked together.

sugar   A simple carbohydrate; a molecule that has the equivalent of one molecule of water (that is two hydrogen atoms and one oxygen atom) for every atom of carbon.

superfecundity   A state of a population in which organisms are so fertile that their populations would increase exponentially if all offspring survived.

supernatant   The part of a sample that, during centrifugation, remains in suspension.

supernormal stimulus   A stimulus even more stimulating than anything normally encountered in nature.

superorganism   A group of individuals that cooperate more closely than the individuals of a colony and almost as closely as the cells of an individual.

suppressor mutation   A mutation in one gene that prevents the phenotypic changes caused by a mutation in another gene.

surface tension   The tendency of a substance to form a smooth round surface.

surface-to-volume ratio   The amount of surface area for each bit of volume.

survivorship curve   A graph that shows the fraction of a population that is alive at successive ages.

suture   An immovable joint that fuses separate bones.

swim bladder   An air-filled sac that helps fish control their buoyancy.

symbionts   [Greek, *syn* = together + *bios* = life] Organisms living in a symbiotic relationship with one another.

symbiosis   [Greek, *sumbios* = living together]   A close association between two organisms.

sympathetic nervous system   A division of the autonomic nervous system; the sympathetic nervous system initiates the "fight or flight" reaction.

sympatric speciation   [Greek, *syn* = together + *patra* = country] The splitting of one species into two without geographical isolation.

synapse   The distinct boundary between two communicating neurons.

synapsis   [Greek, union] The pairing of homologous chromosomes in prophase I.

synaptic cleft   A space of about 20 nm that separates the presynaptic and postsynaptic cells in a chemical synapse.

synchronous cell populations   Cells that are all at the same stage of the cell cycle.

syngamy   The coming together of an egg and a sperm at fertilization.

system   An assemblage of interacting parts or objects.

systematics   The scientific study of the kinds and diversity of organisms and of any and all relationships among them.

systemic circulation   The route of the blood through the body, minus pulmonary circulation.

systole   [Greek, *systellein* = to contract] The part of the cycle of the heart in which the atria contract, followed by the ventricles.

T cytotoxic cell (T_C cell)   A T lymphocyte that is responsible for the killing of cells recognized as nonself.

T helper cell (T_H cell)   A T lymphocyte that activates both the humoral and cellular immune responses.

T lymphocyte (because it matures in the thymus gland)   One of the cells responsible solely for cellular immunity or of the cells that participate in both humoral and cellular immunity.

T-cell receptor   A protein on the surface of a T lymphocyte that can bind to a complex of an antigen and MHC molecule.

tagmata (singular, **tagma**)   Fused segments.

taiga   The broad band of coniferous forest that extends across Canada, Alaska, Scandinavia, and Siberia.

tannin   A secondary compound commonly found in woody plants.

taproot   A single vertical root found in many dicots.

tar   An oily, viscous material, consisting of hydrocarbons, found in cigarette smoke and the insides of chimneys.

target organ   A structure upon which a hormone acts.

TATA box (after the first four nucleotides in the sequence)   A consensus sequence in eukaryotic promoters.

taxis   [Greek, to put in order] Directed movement.

taxol   A compound extracted from the bark of yew trees; taxol increases the stability of microtubules.

taxon   [plural, **taxa**; Greek, *taxis* = arrangement] A general term for any group of organisms at any level of the classification hierarchy.

taxonomist   A biologist who classifies organisms.

taxonomy   [Greek, *taxis* = arrangement] The naming and grouping of organisms.

telocentric   Refers to a chromosome in which the centromere is at one end.

telophase   [Greek, *telos* = end] The last phase of mitosis, during which the mitotic apparatus (including kinetochore, polar, and astral microtubules) disperses and the chromosomes lose their distinct identities.

temperate deciduous forests   Temperate forests that receive 80 to 140 centimeters of precipitation each year, composed of trees that lose their leaves in the winter.

temperate grasslands   Regions with well-defined seasons, with hot summers and cold winters; low annual rainfall keeps grasslands from turning to deserts but is not enough to sustain the growth of trees.

temperature-sensitive mutation   A change in DNA that alters a protein so that it functions normally at one temperature (called the permissive temperature, usually relatively low), but not at a second temperature (the restrictive temperature, generally higher than the permissive temperature).

template   A guide for the assembly of a complementary shape; in the context of DNA or RNA synthesis, one strand acts as a template for the assembly of a complementary sequence.

temporal isolation   Reproductive isolation that results from differences in the times at which two populations reproduce.

ten percent law   The generalization that the organisms of any trophic level provide the next higher trophic level with only 10 percent of the energy that they have assimilated from the lower trophic level.

tendon   A type of connective tissue that attaches the muscles to the bones.

tension   The pulling action of two opposing forces.

teratogen   A compound that causes birth defects.

terminal bud   A bud that lies at the tip of a shoot.

termination   The ending of chain growth.

termination signal   A DNA sequence that determines the end of transcription.

terpene   One of a class of unsaturated hydrocarbons that form the largest class of secondary compounds in plants.

terrestrial   Living on land.

territory   An area occupied by an individual (or group of individuals), from which other individuals of that species are excluded.

tertiary structure (or **conformation**)   The complete arrangement of all the atoms of a polypeptide.

testable   Refers to a hypothesis for which an experiment could be devised that would disprove the hypothesis if it were incorrect.

testcross   A cross between an individual that is homozygous for a recessive allele of a particular gene and an individual whose genotype (homozygote or heterozygote) for that gene is unknown; the phenotype of the progeny reveals the presence of the recessive allele in the parent.

testicles   Human testes.

testis (plural, **testes**)   A male gonad, which produces sperm.

tetrad, (also called a **bivalent**)   A united chromosome pair visible during meiosis I; a tetrad consists of four chromatids.

tetrapod   [Greek, *tetra* = four + *pous* = foot] A four-footed animal.

tetrodotoxin   A poison, isolated from a species of puffer fish, that blocks the passage of Na^+ ions through voltage-gated channels.

theory   A system of statements and ideas that explains a group of facts or phenomena.

theory of island biogeography   MacArthur and Wilson's theory that species diversity on an island is a function of the size of the island and its distance from a source of colonizing species, whether the mainland or other islands.

thermacidophiles   One of the three phyla of Archaea; the bacteria that inhabit hot sulfur springs such as those in Yellowstone National Park.

thermodynamics   [Greek, *thermo* = heat + *dynamis* = power, force] The study of the transformations and relationships among different forms of energy.

thick filament   In striated muscle, a myosin filament, which is 14 nm in diameter.

thin filament   In striated muscle, an actin filament, which is 7 nm in diameter.

thoracic cavity   Chest cavity; that part of the coelom that encloses the heart and lungs.

thorax   The middle portion of an arthropod.

thylakoid   [Greek, *thylakos* = sac + *oides* = like] A flattened disc surrounded by the innermost membranes of a chloroplast.

thylakoid membrane   A membrane, within the stroma of a chloroplast, that delineates stacked vesicles, called grana.

thymine dimers   Two adjacent T nucleotides linked together in the same DNA strand; they are formed as a result of exposure to ultraviolet light.

thyroid stimulating hormone (TSH)   A peptide hormone, made in the anterior pituitary, that stimulates the production of thyroxin in the thyroid gland.

tidal volume   The amount of air drawn in and then expelled in a single breath.

tight junction   The main kind of impermeable junction in vertebrates; they are formed by the fusion of the membranes of adjacent cells.

tissue   A group of similar cells and associated intercellular material that performs one or more specific functions; tissues are often integrated with other tissues to form an organ.

tolerance   In the immune system, a state of induced unresponsiveness to proteins and other molecules; tolerance prevents immune attacks on the body's own components.

torus   A thickened region of the cell wall between two plant cells; a buttonlike thickening in the middle of a pit.

totipotent   Able to develop into a whole organism.

trachea   [plural, **tracheae;** from Greek, *arteria trakheia* = rough artery] A tube, which arises by invagination, that carries air within an animal's body; in vertebrates, the windpipe.

tracheids   Long, thin, spindle-shaped cells in the xylem of plants.

tracheophytes   [Latin, *trachea* = windpipe] Vascular plants; tracheophytes include all the most familiar living plants.

tradewinds   Winds that blow from about 30° latitude, steadily toward the equator.

***trans*-acting factor** (also called a **transcription factor**)   [Latin, *trans* = across, because they regulate the expression of other molecules] A protein that binds to DNA and regulates transcription.

transcription   The production of RNA from DNA.

transcription factor (also called a ***trans*-acting factor**)   A protein that binds to DNA and regulates transcription.

transfer RNA (tRNA)   A small RNA molecule that serves as an adaptor in protein synthesis; each tRNA contains an anticodon that allows it to bind to a codon in mRNA and each becomes linked to a specific amino acid.

transformation   The transfer of one or more genes from one organism to another.

transgenic   Containing a particular recombinant DNA incorporated within the genetic material.

transition state   A distorted form of a substrate that is intermediate between the starting reactant and the final product.

translation   Protein synthesis; the conversion of information from an mRNA molecule into a polypeptide.

translational control   Alterations in the rate at which specific mRNAs are translated.

translocation   In chromosomes, a mutation in which part of one chromosome is moved to another chromosome; in protein synthesis, the movement of the growing polypeptide chain with its attached tRNA from the A site to the P site; the third step of the elongation cycle.

transmission electron microscope (TEM)   A type of electron microscope in which the electron beam passes through a specimen.

transmission genetics   The branch of genetics that deals with patterns of inheritance.

transpiration   The process by which plants pull water from the soil and release it as vapor through stomata in their leaves.

transpiration-cohesion theory   The accepted explanation for the movement of water from roots to leaves; transpiration in the leaves pulls water up the stem in continuous columns.

transpiration-photosynthesis compromise   The trade-off between saving water and maintaining photosynthetic productivity.

transposable element   A mobile gene that can replicate only in integrated form, as part of a host's chromosome.

transposase   An enzyme that catalyzes insertion of a transposable element into new sites.

tree line   The upper limit where subalpine trees grow.

trematoda   Flukes.

triacylglycerol (also called a **triglyceride**)   A nonpolar derivative of glycerol in which all three hydroxyl groups are attached to fatty acids.

triceps   The major muscle on the back side of the upper arm.

triple covalent bond   A covalent bond in which two atoms share three pairs of electrons.

trisomy   [Greek, *tri* = three + *soma* = body] The presence of three, rather than two, copies of a chromosome.

trophic level   Each level of a food chain.

trophoblast   [Greek, *trephein* = to nourish + *blastos* = germ] In a mammalian embryo, a prominent outer cell layer in a blastocyst.

tropical rain forest   A biome with relatively constant weather, with an average temperature of about 27°C (80°F) every month of the year and heavy rainfall, ranging from 2 to 4.5 meters per year.

tropism   [Greek, *trope* = turning] The bending or curving toward or away from a stimulus.

tropomyosin   In muscle, a rigid, rodlike protein, about 40 nm long, that stiffens the actin filament.

troponin   In muscle, a large protein that binds both to tropomyosin and to free calcium ions.

true-breeding   A breeding line in which offspring have the same phenotype, generation after generation.

tubal ligation   An irreversible method of birth control; cutting or blocking the oviducts, or fallopian tubes, preventing eggs from reaching the uterus after ovulation.

tubulin   One of two globular proteins, called α and β tubulin, from which microtubules are assembled.

tumor suppressor gene (also called an **antioncogene**)   A gene that antagonizes the action of an oncogene; this gene is intimately involved in the normal regulation of cell growth.

tumor virus   A virus that causes its host cell to lose its normal ability to regulate cell division.

tundra   [Finnish, *tunturi* = arctic hill] A vast, open land, with a cold, dry climate; it is dominated by grasses and low shrubs and covers more than a fifth of the Earth's surface.

turbellaria   The free-living flatworms; they reproduce either asexually or sexually.

turbinate   Elaborately folded bones inside the noses of many vertebrates.

turgor pressure   [Latin, *turgor* = swelling] In plant cells, osmotic pressure against the cell walls resulting from the higher concentration of dissolved molecules and ions in cytoplasm than in the surrounding fluid.

typological species concept   The view that each species consists of individuals that are variants of a fixed underlying plan.

ultracentrifuge   A high-speed centrifuge that can spin tubes at speeds up to 80,000 rpm, subjecting their contents to forces up to 500,000 × gravity.

ultrastructure   Subcellular structures visible only with an electron microscope.

unconditioned stimulus   In associative learning, the stimulus that directly leads to a physiological response.

unconventional gene   A gene without a stable cellular address.

uniformitarianism   The view, originated by Lyell, that the processes that now mold the Earth's surface—erosion, sedimentation, and upheaval—are the ones that have always molded it; geologic change is slow, gradual, and steady, not catastrophic.

Uniramia   A subphyla of arthropods in which the first pair (or the first few pairs) of appendages are antennae and the next pair are jaws, or mandibles.

universal   Used by all species.

unsaturated   A hydrocarbon chain (or fatty acid) that contains at least one double bond between two carbon atoms; the chain is called unsaturated because it could accept two more hydrogen atoms per double bond.

upstream   In a nucleic acid, toward the 5′ end.

upwelling   The upward movement of water that draws nutrients from the bottom of the ocean.

urea   The compound that mammals convert to ammonia to make it less harmful.

ureter   [Greek, *ouron* = urine] One of a pair of tubes through which urine flows from the kidneys to the bladder.

urethra   The means by which the bladder drains urine to the outside of the body; the single exit tube of the human urethra.

uric acid   A nearly insoluble organic compound produced by insects, land snails, most reptiles, and birds as nitrogenous waste.

uterus   [Latin, womb] The womb; in mammals, the portion of the female reproductive tract that is specialized to receive, nurture, and deliver a developing embryo.

vacuole   Within a eukaryotic cell, a membrane-bounded sac, without any obvious internal structure; a vacuole is larger than a vesicle.

van der Waals interactions   Weak attractive forces between molecules that are very close together.

variable region   In an immunoglobulin (antibody) molecule, the amino-terminal polypeptide segment, of about 110 amino acids, that contains the sequence variation responsible for antibody diversity.

vas deferens   [Latin, *vas* = vessel + *deferre* = to carry down] In the male reproductive tract, two large, thick-walled ducts that carry sperm from the testes.

vascular   Having a conducting system.

vascular system   [Latin, *vas* = vessel] In plants, the structures and cells that serve as a transport system; the vascular system forms the central core of the plant.

vasectomy   An irreversible method of birth control; the cutting and tying of the vas deferens, which prevents sperm from entering the urethra.

vasoconstriction   Contraction of the smooth muscles surrounding the arterioles.

vasodilation   A process causing arterioles to increase in size with consequent blood flow increase and reddening on the skin.

vasopressin (also called **antidiuretic hormone** or **ADH**)   A peptide hormone, released by the posterior pituitary, that stimulates water reabsorption by the kidneys.

vegetative reproduction   The process of producing a new plant without fertilization or seed production.

veins   In vertebrates, one of the vessels that carry blood back to the heart. In plants, the vascular tissue in the leaves.

velocity   In an enzymatic reaction, the rate of appearance of product.

vena cava (**superior** and **inferior**)   The two largest veins, which run up through the center of the body and carry deoxygenated blood from the body to the right atrium.

ventilation   The flow of air into and out of the alveoli.

ventral   The part of an animal facing the Earth.

ventricle   In the heart, a thick-walled pumping chamber.

ventricular fibrillation   Cardiac arrest.

venules   The smallest veins.

vertebrae   [singular, **vertebra**; Latin, *vertebratus* = jointed] The hollow, bony segments that make up the backbone.

vesicle   A membrane-bounded sac; some vesicles are present inside living eukaryotic cells and others form after a cell is broken open.

vessel elements   In flowering plants, structural elements of the phloem that are open at each end and connect end to end to form long open channels.

vestigial structure   [Latin, *vestigium* = footprint] A part of an organism, with little or no function, that reflects evolutionary history.

vicariance   The fragmentation of an already dispersed species or group of species.

villus   [plural, **villi**; Latin, shaggy hair] The highly folded mucosa of the small intestine that each extend about 1 mm into the lumen.

viral oncogene   A viral gene that affects growth control in a host cell.

virus   An assembly of nucleic acid (DNA or RNA) and proteins (and occasionally other components, such as lipids or carbohydrates) that can reproduce only within a living cell; viruses depend on cells to obtain energy and perform chemical reactions.

visceral mass   [Latin, *viscera* = internal organs] The main body of a mollusk; the visceral mass contains the intestinal tract, as well as the excretory and reproductive organs.

visible light   Electromagnetic radiation, with wavelengths from about 400 to about 750 nm, that can be perceived by human eyes and brains.

vitamin   An organic compound that an animal cannot itself synthesize but that is required in minute amounts for normal growth and metabolism; an essential component of the diet; many vitamins are cofactors in enzymatic reactions.

voltage clamping   A method that allows the control of the voltage across a cell membrane.

voltage (also called **electrical potential**)   A measure of the potential energy that results from a charge separation.

voltage-gated   Open or closed according to the voltage across the membrane.

voluntary muscles   Skeletal muscles; voluntary muscles enable voluntary control over body movements.

vulva   [Latin, *volvere* = to wrap] In animals, the collective name for the external genitalia of the female.

warning coloration   A bright, memorable design that helps the predator remember which prey to avoid.

water potential   The potential energy of water per gram (or kilogram); the difference between osmotic pressure and physical pressure on either side of a selectively permeable membrane.

water-soluble signal   A chemical signal that cannot pass through the plasma membrane and must act on its surface.

wavelength   The distance between the crests of two successive waves.

westerlies   Winds that come from the west caused by the rotation of the Earth, characteristic of latitudes between 30° and 60° in both hemispheres.

western blotting (after a similar blotting method, called "Southern blotting" because it was invented by Edward Southern)   A method for detecting a particular protein after electrophoresis by blotting the contents of the electrophoresis gel onto a paperlike support and incubating this "blot" with a specific antibody.

whorl   One of four concentric circles of flower parts at the end of a specialized stem.

whorled   The arrangement of leaves on a stem so that there are three or more petioles at each node.

wild type   An allele that is most highly represented in wild populations.

wobble   Nonstandard base pairing between the first base in the anticodon of transfer RNA and the third base of the codon in mRNA; wobble allows some tRNAs to recognize more than one codon for the same amino acid.

woody plants   Trees, shrubs, and other perennial plants that have woody parts that provide support independent of turgor pressure.

work   The movement of an object against a force, or the conversion of energy into electrical energy, chemical energy, or concentration energy.

x-ray crystallography   The analysis of the scattering, or "diffraction," of x rays by crystals.

xylem   [Greek, *xylon* = wood] Conducting vessels that carry water and minerals from roots to the photosynthesizing leaves.

yield   The amount of grain per acre.

yolk   In an amniotic egg, a mixture of proteins, lipids, and carbohydrates that nourishes an embryo until it can feed itself.

yolk sac   The membrane that surrounds the yolk in an amniotic egg.

Z band   In muscle, the boundary of a sarcomere.

zero population growth   The state of a population when its size reaches the carrying capacity and the environment cannot support any more growth; the state when a population reaches a stable size.

zoomastigina (also called **flagellates**)   [Greek, *mastix* = whip] A diverse group of single-celled heterotrophs.

zygomycetes   [Greek, *zygote* + *mukes* = fungus] Conjugation fungi.

zygosporangium   A structure that produces haploid spores from a diploid zygote.

zygote   [Greek, *zygon* = yoke] The first cell of an embryo, formed by the union of egg and sperm.

Index

Note: Page numbers in *italics* indicate a figure; page numbers followed by "t" indicate a table.

A

Abdomen, arthropod, 517, 521
Abdominal cavity, 713, 732
Abiotic factors, 546, 560
Abiotic synthesis, 404, 415
Abortion, induced, 903, 905
    spontaneous, 192–193, 205, 897, 905, 918, 928
    in U.S., history of, 904
Abscisic acid, 667, 674, 702, 707, 708
Abscission, 706, *707*, 708
Absolute age, 329
Absorption, 739, 750, 752
    by digestive tract, 743, 752
    in small intestine, 750
Absorption spectrum, 147, *148*, 159, 699
Absorptive state, 758, *760*, 768
Abstinence, sexual, 900–901, 905
Abyssal zone, 578, *579*, 582
Acacia, bull's horn, 590
Acacia cornigera, 590
Accommodation, 874, 880
Acetaminophen, 768
Acetone, 281
Acetyl-CoA, 127, *127*, 129
    generation of ATP from, 127–130
Acetylcholine, 7, 785, 787, 854, 855, *855*, 856, 861, 869
    functions of, 856
Acetylcholinesterase, 857, 861
Acid, definition of, 42, 45
Acid rain, in biomes, 575
Acoelomates, 501
    bilaterally symmetrical, 506–508, 520
        Nemertea, 508, *508*
        Platyhelminthes, 506–507, *508*
    radially symmetrical, 504–506, 520
        cnidarians, *504*, 504–505, *505*
        ctenophores, 505–506, *506*
Acorn worms, 527, 542
Acquired behavior, 627–632
Acquired characteristics, inheritance of, 347–348, 366
    theory of inheritance of, 325
Acquired immunodeficiency syndrome, 841
    design of drugs to combat, 271
    symptoms of, 841
Acrasiomycota, 464–465, 475
Acrosome, 887, 892, 904
Actin, 87, 729, 733
Actin filaments, 86–87
Action potential(s), 849, 850–851, 861
    generation by neurons, 849–850
    movement of, 852–853
    production of, 852
    transmission of, nerves and, 853, *853*
Action spectrum, 147–148, *148*, 159, 699, 700
Activation energy, 110–111, *111*, 117
Activators, 241–242, 764
Active site, 112, *113*, 117, 763
Active transport, 94, *95*, 101

Adaptations, 17–18, 20
Adaptive radiation, 376, *377*, 396
    plants and, 379
Adaptive traits, 321, 342
    accumulation of, 341
    in inheritance, 341
Adenine, 210, 233
Adenosine, 854
Adenosine diphosphate, as energy source, 105
Adenosine triphosphate, 61, *62*, 69
    as energy source, 105, 106, 116, 120
    generation of acetyl-CoA from, 127–130
    in glucose synthesis, 153
    light reactions and, 151
    phosphate bonds in, electrons and, 130–131
    photosystem II and, 151
    potential energy in, 104, *105*
    synthesis of, 135
        proton gradient and, 135–136, *137*
Adhering junctions, 98, *99*, 101
Adhesion, of water, 664, 674
Adrenal cortex, 760, 768
Adrenal glands, 760–761, 768, 821, 825
Adrenal medulla, 760–761, 768
Adrenaline, 760, 785
Adrenocorticotropic hormone, 762
ADT. *See* Adenosine diphosphate
Adventitious roots, 691
Aerobes, facultative, 446, 450
    obligate, 446, 450
Aerobic cells, 120, *120*
Aerobic heterotrophs, 446, *446*, 448
Afferent nerves, 868, 880
Agamous mutation, 688
Agar, 460
Agassiz, Louis, 324
Age, absolute, 329
Age distributions, life history and, *613*, 613–616, *615*
    male and female, 614
Age of Reptiles, 331
Age structure, 614, *615*, 625
Agent Orange, 698
Aging, genes and environment influencing, 926
    in vertebrates, 909, 928
Agnatha, 529–530, 542
Agonistic behavior, 632–633, *633*, 639
Agriculture, demand for food and, 619–620, *621*
    recombinant DNA in, 293
    slash-and-burn, 385, *385*
    subsidized, 619
Agrobacterium tumefaciens, 292
Agrostis tenuis, 376
AIDS. *See* Acquired immunodeficiency syndrome
Air, entering lungs, path of, 796
    plants and, 142, *142*
Air currents, 565, *566*
Air pollution, 622
Air sacs, posterior, 800, *800*
Airways, epithelial cells in, 799, *799*
Alchemist, by David Teniers the Younger, 25

Alcohol, as diuretic, 823
    metabolism of, 136
Aldosterone, 760, 821–823, 825
Aleurone layer, 683, 692
Alexander, R. McNeill, 710–711
Algae, 455, 457, 474
    brown, 460, *460*
    golden, 458
    green, 461, *461*
        in plant evolution, 481
    red, 460–461, 474
Alimentary canal, 741
Alkaloids, 657–658, 659
Alkaptonuria, 217, *218*, *219*, 220
Allantois, 536, 542
Allele(s), 194, 205, 350, 366
    dominant, 195–196, 205
    frequencies of, breeding between populations and, 358, *359*
        changing of, 355–366, *356*
        nonrandom mating and, 358
    hemizygous, 200
    multiple, genes with, 307, 307t
    recessive, 195–196, 205
Allergy, 836
Alligators, 534
    in DDT-laden lake, 754
Allopatric speciation, 374, *375*, 396
Allopolyploidy, 374, 396
Allosteric effector, 116
Allosteric inhibitors, *116*, 117
Alper, Tikvah, 46–47, 49
Alpha (α) carbon, 64, 69
Alternation of generations, in protists, 457, 474
Alternate leaves, 654, *655*, 658
Altitude, high, oxygen delivery at, 804
Altruism, 636
    inclusive fitness and, 637–638, 639, *639*
    natural selection and, 636–637
    reciprocal, 638, 639
Alvarez, Luis, 382
Alvarez, Walter, 382
Alveoli, 796
    gas exchange in, *801*
    surface tension and, 800
Amacrine cells, 876, 880
Amborella trichopoda, 490
Ambystoma tigrinum, 535
Amebas (Rhizopoda), 462, 475
Ames, Bruce, 642, 643
Ames test, 642
Amines, 763, 764
Amino acid balance, 737
Amino acids, 52, 54–55t, 68. *See also* Base(s); Nucleotides; Polypeptides
    definition of, 64
    essential, 737, 737t, 752
    linkage to form polypeptides, 63–66
    peptide bond joining, 64–65, *65*
    protein structure and. *See* Protein(s), structure of

Amino acids, *(continued)*
R groups of, 54–55t, 64
recognition of, by transfer RNA, 245, *246*
structure of, 64, *64*
transfer RNA and, *244*, 244–245, 255
Amino group, 51, 68
Aminoethane, 51
Ammonia, 813, *814*, 824
Ammonification, 558, 560
Ammonium ions, 557, 560
Ammospermophilus tereticaudus, 568
Amniocentesis, 312, *312*, 315
Amnion, 532, 538, 911
Amniotic egg(s), 428, 536, *537*, 542, 897
Amniotic fluid, 312, 315, 911
Amphibians, development of, primary organizer in, 921–923
disappearing, 535
first, 532, *532*
gastrulation in, *912*, 912–913
life cycle of, 532, 533, *534*
metamorphosis in, 758
orders of, 533
Amphipathic molecules, 40, *41*, 44–45
Amphotericin B, 460
Amygdala, 868, 880
Amylase, 744, 752
α-Amylase, 683, 692
Amyloplasts, 85, *86*, 100, 698, *698*, 708
Amystoma (Axolotl), 534
Anabolism, 129, *129*, 138
definition of, 114, 117
Anaerobes, 120, *120*
facultative, 126
obligate, 126, 446, 450
Anaerobic heterotrophs, 448
Anaphase, chromatids during, 172–173
in meiosis, 188
Anaximander, on evolution, 321
Androgen receptor, 255
Androgens, 763
Anemia, sickle cell. *See* Sickle cell anemia
Aneuploid cell, 191, 205
Aneuploidy, 231
Aneurism, 782, 787
Anfinsen, Christian, 68
Angiosperms, 490–493, 494
definition of, 481
evolution of, 481–482
fertilization in, 491
life cycle of, 491, *492*
reproduction by, 491
species of, 493
Angiotensin, 821, 825
Animal(s), agonistic behavior of, 632–633, *633*, 639
aquatic, dissolved oxygen for, 795
behavior of, ecology of, 626–640
body composition of, 713
body plan of, 713
body temperature regulation in, 731
breathing by, 788–806
carbon dioxide exchange for oxygen in, 791–785
cells and organs of, coordination of actions of, 754–769
characteristics of, 498–503
choices made by, 632
classification of, 499–503, *500*
developmental patterns and, 499–501, *501*
symmetry and, 502, *502*

cloning of, 920–921, *921*
defenses of, 588–589
of deserts, 568
water conservation by, 812
disposal of nitrogenous wastes by, 813, *814*, 824
eating by, 735–739
form of, 711–718
determinants of, 712
and function in, 710–733, *712*
grazing on savannas, 572
hearing by, 871–873
internal environment of, regulation of, 755, *756*
jaws, teeth, and mouth parts of, 744, *744*
land, water and salt regulation in, 812–813
learning by, 627, 630–631, 639
sensitive period for, 630
locomotion of, style of, as influence on shape, 718–721
movement of, skeleton and muscle in, 721–730
phenotype of, social environment and, 304, *304*
salt and water adjustment by, 821–823
sensing environment, 870–871
size of, gravity and, 714
as influence on shape of, 713–714
metabolic rate and, 714, *714, 715*
surface area of, and shape of, 713–714
in social groups, 634–638
sociality among, 635
structural and physiological adaptations of, 709–928
structure of, 499, *499*
surface areas of, use of, 714–715, *715*
of temperate deciduous forests, 573, *574*
vision of, 873–877
walking and running by, 720
water, salt, and waste management in, 808–825
water and salt regulation by, 810–813
Animal pole, 910, 928
Animalcules, 70
Animalia, 16, 20
Annelids, 514–517
Anomalocaris, *498*
Anopheles, 463
Anopheles maculipennis, 371
Antacids, 2
Antagonist muscles, 722, *722*
Anteaters, spiny, 541
Antenna complex, 148, 159
Antennapedia, 924
Anterior pituitary gland, 892, 904
Anther, 492, 493
Antheridium, 481, *482*
Anthocyanins, 657, 659
Anthophyta, 490, 493
Anthozoa, 504
Anthropologists, physical, 390, 397
Antibiotics, and bacteriophages, 265
broad-spectrum, 440
definition of, 436
functions of, 437
fungal infections and, 453
inhibition of elongation by, *250*
narrow-spectrum, 440
natural synthesis of, 436–437
and protein synthesis, 251, *251*

resistance to, 360, 436–438, *438*, 447
classification of, 437, *438*
and salmonella poisoning, 432–434, *434*
in ulcers, 4
Antibodies, 286, 295, 829
cells and molecules and, 835–836, *835, 836*, 842
immune system and, 837–839
production of, 837, *838*
Anticlotting agents, 768
Anticoagulants, 832, 843
Anticodon, 245, *246*, 256
Antidiuretic hormone, 761, 823, 825
Antigens, 834–835, 842
on blood cell surfaces, 841–842
endogenous, 841, 843
exogenous, 841, 843
Antihistamines, 832
Anti-inflammatory agents, 768
Antioxidants, 739
food sources of, 643t
Antisense strand, 240
Ants, and bull's horn acacia, 590
sex ratios in, 614
Anura (frogs and toads), 533, *533*
Anus, 741, 752
Aorta, *772*, 776, 786
Apes, classification of, 389
Aphids, 671, *671*
Apical meristems, 646
Aplacophora, 513
Aplysia, gill withdrawal in, 859, *860*
learning by, 860, 861
Apoplast, 651
Apoptosis, 827, 915–916, *916*, 928
Appendix, 751
Aquatic animals, dissolved oxygen for, 795
Aquatic ecosystems, 578–581
Aqueous environment, interactions in, 35, *35*, 44
Aqueous humor, 874
Arabidopsis thaliana, 687–688, 692, 698
genetic analysis of, 689
mutations of, 689, *689*
Arachidonic acid, 768
Archaea, 16, 20
classification of, 443, *444*, 450
Archegonium, 484, *484*
Archenteron, 499, 520, 911, 927
Archipelagos, 376
Arctic tern, *120*, 538, 719
Arendt, Detlev, 523
Arginine, 219, *220*
Aristotle, 141, 906
on classifications, 424
Aromatic compounds, 49, 50, 68
Arrow worms, 527, 542
Arterioles, 779, 787
Artery(ies), 775, 779, 787
pulmonary, 776, 786
renal, 814, 825
Arthropods, 522, 541
chelicerates, 518, 519
classification of, 518–520
crustacea, 518, 519–520, *520*
mouth parts of, 518, *519*
success of, 517–518
uniramia, 518, 520, 521
Artificial kidney, 818, *818*
Artificial selection, 340, 342
in pigeons, 340, *340*

Asaro, Frank, 382
Aschelminths, 509–510, 521
    nematodes, 510, 521
Asci, 469
Ascomycetes (sac fungi), 469, *470*, 475
Ascorbic acid, 739
Asexual reproduction, 185, *185*, 192, 204
Aspergillus, 468
Aspirin, 768
Association cortex, 868, 880
Associative learning, 859
Assortative mating, 358, 366
Assortment, independent, principle of, 198–200, *199*, 206
Aster, 166
Asteroids, 382, *382*
Atherosclerosis, *59*
Atmosphere, components of, 552, 552t
    evolution of, 405
    oxygen in, 791–792
    protective, 146, *147*
Atom(s), 23, *25*, 44
    composition of, 26
    electrons in, 30
    internal structure of, 27, *27*
    in ionic bonds, 33–34
    Ludwig Boltzmann and, 102–103
    masses of, 26
    nucleus of, 27, 44
    properties of, 26–30
    weight of, 26
Atomic force microscope, 26
ATP. *See* Adenosine triphosphate
ATP synthase, *134*, 134–135, *135*, 138
Atrazine, 156, 754
Atrial natriuretic factor, 823, 825
Atrioventricular node, 784, 787
Atrioventricular valves, 782–783, 787
Atrium, 776, 786
Atropa belladonna, 857
Atropine, 857
Audition, 870–872, 880
Auditory canal, 871
Auditory cortex, primary, 872
Auditory system, 869, 880
Australian marsupials, 380, *381*
Australopithecines, 390, 393, *393*, 397
Australopithecus, standing by, 390–392
Australopithecus afarensis, 391, 392, 393, *394*
Australopithecus anamensis, 390, 391–392
Autonomic functions, 865
Autonomic nervous system, 726, 733, 869, *869*
Autonomous replication, 264, 274
Autosomes, 193
Autotrophs, 119–120, 138, 412, 416, *444*, 448, 450, 548, 560
Auxin(s), 691, 694–696, 696t
    and cytokinin, 705
    discovery of, 694
    in gravitropism, 698
    as herbicides, 696, 704
    mechanisms of action of, 694–695, 696t
    naturally occurring, 702
    in phototropism, 694–696
    in plant growth and development, *702*, 702–704, 708
    synthetic, 695–696
    transport of, *703*, 703–704
Averted vision, 874
Avery, Oswald, 222, 224

Avogadro, Amadeo, 27
Avogadro's number, 27
Axelrod, Robert, 638
Axillary buds, 687, 692
Axolotl (*Amystoma*), 534, 926
Axons, 847, 861
AXR1 wine grapes, 661–662
Aysheaia, 498

B

B-cell receptors, 838, 843
B cells, *834*, 835, 843
B lymphocytes, *834*, 835, 843
Bacilli, 439, 450
Bacillus thuringiensis, 292
Backbone. *See* Spinal column
Backcross, 195, 197, *197*, 205
Bacteria, antibiotic resistance in, 437–438, *438*, 447
    cell walls of, 440–441, *440*
    colonial, 439, *440*
    as common ancestor of life, 448, *449*
    conjugation in, 265, *266*, 274
    fermenting, 448, 450
    functions of, 435
    gene cloning in, 284
    gene expression in, eukaryotic, 284
    genes in, 221
    genetically engineered, proteins from, 289
    gram negative, 439, *440*, 450
    gram positive, 439, *440*, 450
    kingdoms of, 16
    movement of, 441
    nitrifying, 558, 560
    nitrogen-fixing aerobic, 446, *446*, 448, 450
    normally in human body, 435–436, *436*
    as pathogens, 435, 450
    recombinant DNA and, 283–285
    two major groups of, 439
    in ulcers, 2
    viruses and, 260
Bacteriophage(s), 223, 232, 261, 274
    antibiotics and, 265
    to combat disease, 265
    life cycle of, 223, *224*, 268, *268*
    reproductive cycle of, 264, *264*
Bacteriorhodopsin, 135, 136
Badwater race, 562–564, *563*
Bakanae, 704
Bakker, Robert, 710
Balance, 870
    inner ear in, 872
Balanced polymorphism, *365*, 365–366
Balanus balanoides, 591, *592*
Baltimore, David, 269
Bark, 690, 692
Barnacles, 519–520, 591, 592
Barometer, mercury, 792, *793*
Barometric pressure, 792, *793*
Barracuda, *532*
Barrier methods, 901, 905
Bartholin's glands, 891, 904
Basal ganglia, 868, 880
Basal metabolic rate, 736, 752
Base(s), definition of, 42, 45
    substitution of, 231, 233
Base pairing, 215, *215*
Basidiocarp, *470*, 475
Basidiomycetes (club fungi), 469–470, *472*, 475

Basidium, 469, 475
Batesian mimicry, 589, 600
Bateson, William, 181, 200, 349
Beadle, George, 218, 219, 676
Beaumont, William, 734, *735*, 745
Becquerel, Henri, 29
Beef cattle, salmonella poisoning associated with, 432–434, 438
Beetles, grain, 591
    ladybird, 365
Behavior(s), acquisition of, 627–632
    agonistic, 632–633, *633*, 639
    ecology and, 632–634
    genes and, 627, *631*, 631–632
    homeostasis and, 865
    innate, *627*, 627–630, 639
    learned, 627, 628–629
    neural networks and, 857
    programmed, *627*
    social, 634, *635*
Behavioral ecology, 626, 639
Behavioral isolation, 373, 396
Beijerinck, Martinus, 260
Békésy, Georg von, 871
Belding ground squirrel, 637–638, *639*
Belladonna, 857
Benthic division, 578, 582
Benzene, *50*
Benzodiazepines, 857
Beriberi, 738–739
Bernays, Edward, 790
Bicoid gene, 924
Bile, 750
Binary fission, 167–168, *168*
Binding site, 763, 768
Binomial, 423, 431
Bioassay, 703, *703*, 708
Biobypass, 315
Biochemical processes, directions and rates of, 102–117
Biochemical reactions, 23
    water in, 40
Biodegradable pollutants, 621
Biodiversity, determination of, 593–596
Biogeochemical cycles, 552, 560
Biogeography, 338, 342
Biological molecules, 46–69
Biological species, 371
Biological species concept, 371, 396, 425, 431
Biologists, questions asked by, 4–13
Biology, conservation, 603, 624
    definition of, 13, 20
    molecular, comparative, 336–337
Biomagnification, 622, *623*, 625
Biomass, 549, 550
    inverted pyramid of, 550, *551*
    pyramid of, 550, *550*, 560
Biome(s), 546, 562, *563*, 564, 582
    definition of, 564
    determinants of type of, 564–565, *565*
    differences in, *567*, 567–578
    latitude and altitude effecting, 566, *567*
    major, temperature, precipitation, and productivity of, 567, 568t
    world distribution of, 567, *568*
Biosphere, 544, *545*, 547, *547*, 560
Biotechnology, 278, *280*, 294
Biotic factors, 546, 560
Bipedalism, 390, *390*, 397
Bipolar cells, 875, 880

Birds, *538*, 538–539
    adaptations of, 539, *539*
    breathing by, 800, *800*
    cladistic classification of, 427
    convergent evolution in, 333, *334*
    distances flown by, 719
    evolution of, 428
    mammals, and reptiles, cladogram for, 428, *429*
    marine, kidneys of, 821
    migration of, chemical signaling and, 756
    naming of, 423, *423*
    skeletal adaptations of, 720, *720*
    wings of, 719
Birth control. *See also* Contraception
    approaches to, 899–903
    worldwide, *899*
Birth control pill, 901–902, 905
Birth rate(s), 609, 624
    factors decreasing, 899
    factors increasing, 899
2,3-Bisphosphoglycerate, 804, 805
Biston betularia, *360*, 360–361
Bithorax, 924–925, *925*
Bivalves, *512*, 513, 521
Bladder, 814, 825
Blade, 654, 658
Blastocoel, 910, 927
Blastocyst, 910, 927
Blastopore, 501, *501*, 524, 913, 927
Blastula, 499, 520, 909, 927
Blending inheritance, 182, *182*
Blood, 770–776
    carbon dioxide in, 804, 805
    circulation of, 773–774, 775, *775*
    composition of, 773–774
    functions of, 771–772
    in oxygen delivery at high altitudes, 805
    oxygen delivery by, 782
    pH of, 42–43
    renal filtration of, 814–817, 825
Blood banking, 770
Blood-brain barrier, 865
Blood cells, antigens on surfaces of, 841–842
Blood clot(s), 831, 842
    prevention of, *830*, 831–832
Blood flow, regulation of, 785–786
    renal, filtration rate and, 821
Blood glucose, levels of, 758–759
Blood groups, human ABO, 307t
Blood loss, body control of, 830–831
Blood-nerve barrier, 865, 866
Blood pressure, 785, *785*
    high. *See* Hypertension
    hormones and, 821–823, *822*
    salt consumption and, 808–810
Blood pump, small animal, 776
Blood transfusions, 843
Blood types, variation of, among races, 302
    in Japan, 303, *303*
Blood vessels. *See also* Artery(ies); Veins
    major, *772*
    pressure in, 785, *785*
    structure of, *781*, 781–782
Blotting, 288, *288*
Body, human, three areas of, 730, *731*
Body defense, in broken skin, 830–833
Body plan, tube, *908*, 908–909
Bohr, Niels, 103
Boltzmann, Ludwig, 102–103, *103*, 105, 110

Bolus, 744, 752
Bombykol, 629
Bond(s), high-energy, 108, 116
Bone(s), 524, 538
    forces on, 724, 725
    responses to use, 725–726
    skeletal muscle and, 726
    structure of, 723, *724*
Bone marrow, red, 723, *724*, 733
Bone tissue, compact, 723, *724*, 733
    spongy, 723, *724*, 733
Bonner, James, 699–700
Bonnet, Charles, 906
Bonobos, 390
Book lung(s), 517, 521, *794*, 795
Bordered pit pair, 647–648, 658
Botstein, David, 296, *297*, 297–298
Bottleneck effect, 357, *357*, 366
Boveri, Theodor, 166, *167*, 182, 199
Bowman's capsule, 816, *816*, 825
BPG, 804, 805
Brachial arches, 334, *336*
Brachiostoma, 524
Bragg, Sir William Lawrence, 213, 214, 216
Brain, fluid intake and, 821
    in invertebrates, 864–865
    nerves, and sense organs, 865
    neurons of, 865
    regions in sensation, 863–864
    regions involved in various processes, 863, *863*
    sensory information and, 871–872
    of vertebrates, 866–868
Brain stem, 866
Brassica oleracea, 19, *19*
Brazilian rain forests, 385
Bread mold, black, 468, *469*
Breathing, 120
    by animals, 788–806
    by birds, 800, *800*
    by insects, 801, *801*
    mammalian, 796–798
Breathing center, 805, 806
Brenner, Sydney, 237
Bridges, Calvin, 200
Briggs, Derek, 496–497
Briggs, Robert, 919
Briggs, Winslow, 702–703
Broca, Pierre Paul, 863
Bromocriptine, 471
Bronchi, 796, 806
Bronchioles, 796, 806
Brown, Rachel, 452, 453
Brush borders, 817–819, *818*
Bryan, William Jennings, 318
Bryophytes, 479, 481, 483–484, 645, 649. *See also* Plant(s), nonvascular
    absorption of water by, 483, 494
    reproduction by, 483–484
Büchner, Eduard, 119
Büchner, Hans, 119
Bud(s), axillary, 687, 692
    flowering plant, 685
    terminal, 687, 692
Buffers, 43, 45, 771
Buffon, Count, 322–323, 325, 342
Bufo boreas, 535
Bufo bufo, 535
Bufo periglenes, 535
Building blocks, links of, 52–53
    molecular, 52, *53*, 54t

Bulb of the vestibule, 889, 904
Bulbourethral glands, 888, 904, 892
Bulk, 750
Bulk flow, in animals, 716, 732
Bullfrogs, 371, *371*, 531
Bult, Carol, 446
Burgess shale, 496, *498*
Burnet, Frank MacFarlane, 827
Butler, Samuel, 264
Butterfly, swallowtail, *363*, 363–364

C

C_3 plants, 158, 159
C_4 plants, 158, 159
    anatomy of, *652*, 652–653, *653*
Cabbage, sea, 19, *19*
Cacti, 158, 668, *668*
Cadaverine, 598
Caecilians (Gymnophiona), 533, *533*
Caenorhabditis elegans, 915–916
Cagniard de la Tour, Charles, 118
Calcitonin, 766
Calcium ions, in chemical synapse, 856
Callus, 690–691
Calorie(s), 106, 117
Calvin, Melvin, 153–155
Calvin cycle, 153–155, *155*, 159
Calyx, 489, 491
Cambium, cork, 689–690, *690*, 692
    vascular, 689, 690, *690*, 692
Cambrian explosion, 330–332, 380–383, 396
    feces in, 509
Cambrian Period, 330–331, 380, 381, 499
Camouflage, 588, *588*
Campbell, Keith, 920–921
Canavanine, 642
Cancer(s), cervical, and Pap smear, 164
    damage to DNA and, 231, *232*
    of lungs, *789*, 789–791, *791*
    patterns of gene expression in, *287*, 287–289
Candida albicans, 468
Canidae, hidden variation in, *332*
Canines, 744
Canis familiaris, 332
Canopy, 576, *577*, 582
Capacitation, 892, 904
Capillaries, 779, *779*, 787
    water movement in, 663–664, *664*
Capillary action, 39, 40, 663–664
Capsid, 261, 274
Capsule, 440, 447
Capybara, 424
Carbohydrates, 60, 69
Carbon, 553–554, 554t
    alpha (α), 64, 69
    capturing of, in Calvin cycle, 155, *155*
    isotopes of, 28
Carbon-14, 329
    half-life of, *28*, 28–29
Carbon atoms, structures formed by, 49–50
Carbon cycle, 554, *554*
    humans and, 555–557
Carbon dioxide, 553–554, 555
    atmospheric, temperature and, 555, *555*
    in blood, 804, 805
    in Calvin cycle, 155, *155*
    changing delivery of, breathing regulation in, 805
    exchange for oxygen, 791–795
    measuring levels of, for 40 years, 556, *556*
    movement of, 782, *782*

in photosynthesis, 155–157
to prevent photorespiration, 158
removal from atmosphere, 557
Carbon-fiber technology, 209
Carbon monoxide poisoning, 115
Carbonated beverages, 794
Carboniferous Period, 331, 481, 534, 536
Carboxyl group, 51, 68
Carcinogens, 800
Cardiac arrest, 784
Cardiac muscle, 713, 726, *728*, 733
Cardiac output, 786
Cardiovascular system, 776, 786
Carnivores, 548, 560, 736, 752
digestive tracts of, 712, *712*
β-Carotene, absorption spectra of, *148*
Carotenoids, 148, *148, 149*, 159
Carpels, 491, 494, 679, 692
Carrageenan, 460
Carrying capacity (K), 609
of Earth, 617–620, *618*, 624
factors influencing, 612–613
Cartilage, 528, 542
Cascades frog, 535
Caspar, Timothy, 698
Casparian strip, 651
Catabolism, 129, *129*, 138
definition of, 114–115, 117
Catabolite activator protein (CAP), *253*, 253–254, 256
Catalase, 84
Catalyst, 111, 116
Cataracts, 874, 880
Catarrhines, 389, *389*, 397
Catastrophism, 323–324, *324*, 341
Caviomorphs, 424
Cecum, 751, 752
Cell(s), 15
aerobic, 120, *120*
boundary, body, and genes of, 74–75
changes in, chemical and mechanical, 75
communication of, in multicellular organisms, 96–99, *98*
components of, 77–79, *78–79*
cytoplasm division in, 175
death of, programmed, 914–916, *916*, 928
depolarized, 851
determined, 919, 929
differentiated, totipotency of, 919–920, *920*, *921*
differentiation of, pattern formation and, *924*, 924–925
embryonic, differentiation of, 919–923
epithelial, of small intestine, 747–749, *749*
eukaryotic, 15, 16, 20
transmission electron microscope to assess, 77–79
fate of, 919, 929
features of, 77t
first, of new generation, chromosomes in, 186
functions of, 75
good gone bad, defense against, 839–843
growth control in, 177, *178*
loss and replacement of, by multicellular organism, 77
modern, evolution of, 412–414
of nervous system, 844–861
normal, division of, 176–177
determination of, 177
nucleus of, 74, *74*, 79–80, *80*, 100

organisms composed of, 70–101
organization of, 75–77
passage through cell cycle, 176–177
plant, 84–85
potency of, 919, 929
prokaryotic cells, 15, 16, 20
properties of, surface-to-volume ratio and, 76, *76*, 100
regulation of enzyme function by, 115, *115*
regulation of internal environment by, 75–76
reproduction of, 75
responses in, chemical signals and, 755–757
single, digestion in, 739
specialization by, 76
in small intestine, 747, *747, 748*
stem. *See* Stem cells
study of, using microscope, 72–73
use of genetic information by, 412
Cell biology, chemistry and, 21–160
Cell body, of neuron, 847, *848*
Cell cycle, passage of cells through, 176–177
stages of, chromosomes and, 168–172, *170*
Cell division, 163–167
cellular regulation of, 177, *178*
chromosomes in, discovery of, 165
daughter cells in, 167
determination in normal cells, 177
DNA and, 173
in eukaryotes, 168–172
mechanisms of, 163–165
of normal cells, 176–177
in plants, 686
in prokaryotes, 167–168, *168*
Cell expansion, in plants, 686, 703
Cell membranes, 772
Cell plate, early, 176, *176*
Cell reproduction, 162–179
cell division in, 163–167
chromosomes in, 165
mechanisms of, 163–165
coordination of cycles in, 177
process of, 167
Cell sap, 645
Cell senescence, 176–177
Cell theory, 74, 100
Cell wall(s), 93, 101, 645
primary, 645, *646*, 658
secondary, 645, *646*, 658
Cellulae, 70
Cellular respiration, 120
definition of, 120
energy capture in, 136, *137*
stages of, 121–122, *122*
Cellular slime molds, *464*, 464–465
Cellulose, 60, 69, 528, 645, 658
Cenozoic Era, 332
Central canal, 866
Central dogma, of molecular biology, 238, 239, *239*, 255
Central nervous system, mammalian, 866, *867*
in vertebrates, organization of, *865*, 865–868, 880
Centrifuge, 80–81, *81*
Centriole(s), 86, 100, 172
Centromere, 171
Centrosome, 86, 100
Cephalization, 865
Cephalochordates, 528
Cephalopods, *512–513*, 513–514, 521
eye of, 514, *515*

Cerebellum, 862, 866, 880
Cerebral cortex, 862–863, 867, 868, 880
Cerebral hemispheres, 867, 880
Cerebrospinal fluid, 865
Cervical cancer, and Pap smear, 164
Cervical cap, 901, 905
Cervical cells, normal, *164*
precancerous and cancerous, 164, *164*
Cervix, 886, *890*, 905
Cestoda, 506, 507, *508*, 521
Chaetognatha, 524, 541, 542
Chain, Ernst, 280
Chamberland, Charles, 260
Channel(s), ligand-gated, 856
sodium, voltage gated, 851, 852, 861
Channel proteins, of neurons, 848, 861
Chaos chaos, 460
Chaparral, 564, 571, *571*, 582
fires in, 571, *571, 572*
Chaperones, 68, 69
Character(s), primitive, 427, 431
shared derived, 427, 431
Character displacement, 592, *592*, 600
Chargaff, Erwin, 215, 217, 224
Chargaff's rules, 215–216, 233
Chase, Martha, 223–225, *225*
Cheetah, 721
genetic variability in, 353
Chelicerates, 518, 519, 521
Chelonia (turtles and tortoises), 534, *536*
Chemical bonds. *See* Bond(s)
Chemical defense(s), 588
plant, 656–657
Chemical equilibrium, 41, 45, 107–108, 115
Chemical reaction(s), 23
activation energy of, enzymes and, 113–114
coupled, 108, *109*
direction of, 104
prediction of, 104–106, 107, 116
rate of, determination of, 110–112
molecular motion and, 110, *111*
starting of, 110–112
stopping of, 110, *111, 112*
Chemical signaling, cellular responses to, 755–757
development and, 923–925
migration of birds and, 756
signals in, detection of, 763
six steps in, 763
types of, *756*, 756–757
Chemical synapses, 854, 861
action of psychoactive drugs on, 857
mechanism of action of, 855–856
structure of, 855, *855*
Chemical warfare, 588, *588*
Chemiosmosis, 131, 138
Chemiosmotic hypothesis, 131, 138
Chemistry, and cell biology, 21–160
definition of, 23, 44
of life, 24–26
Chemoautotrophs, 446, 450
Chemokines, 271
receptors for, AIDS and, 271
Chemotaxis, 442, 450
Chemotrophs, 119
Cherts, 401
Chewing, 744
Chiasma(ata), 189, *189*, 205
Chicxulib, 382
Chihuahua whiptail lizard, 883

Chimpanzees, 390, 392
    pygmy, 390
Chitin, 60, 69, 466, 475
Chlamydia, 435
Chlamydia trachomatis, 437, *446*
Chlamydomonas, 455, *457*, 461
Chloracne, 696
Chlordane, 754
Chlorella, 461
Chlorofluorocarbons and, 622, *624*
Chlorophyll(s), 85, 147, 159, 699, 701
    in chlorophytes, 461
    P$_{680}$, 149, 150, 151, 159
    P$_{700}$, 149, 150, 151, 159
Chlorophyll *a*, absorption spectrum of, 148
    photosynthesis and, *150*
Chlorophytes, 461, 474, 644
Chloroplast(s), 100
    anatomy of, 149, *150–151*
    DNA in, *272*, 273
    as endosymbionts, 413–414
    plant cell, 84–85, *86*
        functions of, 85
        structure of, 85
    requirements of photons for glucose produc-
        tion, 157
Cholecystokinin, 759
Cholera, cyclic AMP in, 764
Cholera toxin, 764
Cholesterol, biochemistry of, *58*, 58–59
Chondrichthyes, *530*, 530–531, 542
Chordata, 524, 531
    tissue layers of, *502*
Chordates, 522–542
    characteristics of, 527–528
Chordin, 923
Chordin gene, 523
Chorion, 536, 542, 911
Chorionic gonadotropin, human, 896, 897, 906
    pregnancy tests and, 898
Chorionic somatomammotropin, human, 898, 906
Chorionic villi, 911
Chorionic villus sampling, 312, *312*, 315
Choroid, 873
Chromatids, 188, 204–205
    during anaphase, 172–173
    separation of, 190, *190*
    sister, 171, 188, 190
Chromatin, 165, *165*, 166
    in cell, 79–80, *80*, 100
Chromatin-remodeling machines, 254
Chromatography, paper, 153–154, *154*
Chromoplasts, 85, *86*, 100
Chromosome(s), 100, 489, 490
    assortment of genes on, 201–202, *203*
    carrying of information by, 182–183
    definition of, 165
    discovery of, 165–166
    distribution by meiosis, to gametes, 188–192,
        191
    in first cell of new generation, 186
    function of, discovery of, 166–167
    homologous, 168, *169*, 185, 350, 366
        in meiosis, 188–190, 205
    inheritance from. *See* Inheritance, chromosomal
        theory of
    during metaphase, *172*
    movement of, during mitosis, 173
    mutations in, 231
        deficiencies, 231

    duplications, 231, 233
    inversions, 231
    monosomy, 231, 233
    translocations, 231
    trisomy, 231, 233
    number of, from generation to generation,
        185–188
    pairs of, 185, *186*
    recombination of two genes on, 204
    sex, in humans, 192–193, *194*
    of sexually reproducing eukaryotes, 168, *169*
    stages of cell cycle and, 168–172, *170*
    X, 192–193, 205
    Y, 192–193, 205
Chromosome banding, 175
Chrysemys scripta elegans, *120*, 125
Chrysophytes, 458–459, *459*, 474
Chthamalus stellatus, 591, *592*
Chyme, 746, 752
Cigarettes, addiction from, 791
    sales of, death rate and, *789*
        history of, 789
Cilia, 89, *89*, 100
    function of, 798–799, 806
Ciliary body, 874
Ciliates (ciliophora), 464, *464*, 475
Ciliophora (ciliates), 464, *464*, 475
Circuits, convergent, 858
    divergent, 858
Circulation, 776
    closed, 778–779, 787
    double, 532, 542
    open, 778, *779*, 787
    pulmonary, 776, 786
    systemic, 776
Circulatory system, 775, *775*
    pathogens sealed from, 830–831
Circumcision, 889, 904
Cirrhosis, 136
Citric acid cycle, 127–128, *128*, 138
    discovery of, 128–129
    other biochemical pathways and, 129–130
Citrulline, 219
Cladistics, 427–429, 431
Cladogram, 427, 431
    for birds, mammals, and reptiles, 428, *429*
    of primates, 427, *428*
Class, in hierarchy of classification, 423, 431
Classification, 418–431
    Aristotle on, 424
    difficulty in, 430
    kingdoms in, 430
    Linnaeus's system of, 422–423
    naming in, 422
    Plato on, 423–424
    red wolf controversy in, 418–421
    species grouping in, 421–430
        cladists and, 427–429
        classical evolutionary taxonomy and, 426
        disagreement over, 430
        DNA and protein sequences in, 429–430
        Foucault's Chinese encyclopedia and, 421
        guinea pigs and, 424
        pheneticists and, 426–427
        of reptiles, 428
Claviceps purpurea, 471
Clays, 669, 674
Cleavage, in vertebrates, 909, 910, 927
    in zygotes, *910*, 910–911, *911*
Cleavage furrow, 175, *176*

Clements, Frederick Edward, 585–587
Climate, of deserts, 564
    Earth, changing, 556
    sunlight, and movements of Earth, 564–566
    topography and, 566–567
Climax community(ies), 585, 597, 599, 600
Clitoris, 890, 905
Cloaca, 536–538, 542, 814, 825, 888, 904
Clonal selection theory, 837, 843
Clone, 185, 204, 285
Cloning, of animals, 920–921, *921*
    positional, *309*, 309–310, 315
Clonorchis sinensis, 506, *507*
Clostridium botulinum, 120, *120*, 126
Club fungi (Basidiomycetes), 469–470, *472*, 475
Club moss, life cycle of, *479*
Cnidarian(s), *504*, 504–505, *505*, 521
    tissue layers of, *502*
Cnidocytes, 504, *505*, 521
Coacervates, 408, *409*, 415
Cocaine, 857
Cocci, 439, 450
Coccidioides, 468
Cochlea, 873, *873*, 882
Cochlear implants, 871
Cockroaches, resistance in, 360, *360*
Codium magnum, 461
Codominance, 196, 205
Codon(s), definition of, 237, 255
    nonsense, 238, 255
    stop, 238, 255
Coelom, 501, 520, 713, 732
Coelomates, 501, 510–520, 524–531
    tissue layers of, *502*
Coenocytic fungi, 466
Coenzyme A, 127, 138
Coenzymes, 123, 138
Coevolution, 491, 590–591
Cohen, Fred, 48
Cohen, Mitchell, 432, 433
Cohen, Stanley, 846
Cohesion, water, in tall plant, 665
    of water, 663, 674
Cohesion-tension theory, 664, *665*, 674
Cohort, 614, 625
Coitus interruptus, 901, 905
Cold, tolerance by animals, 732
Collagen, 63, 725, 733
    helix in, 66, 67
    structure of, 65–66, *66*
Collagen fibers, 725, *725*
Collagen fibrils, 725, *725*
Collenchyma, 646, *647*, 658
Colon, 750, 752
Color blindness, 875
Coloration, warning, 589, *589*, 600
Colostrum, 898, 905
Comb jellies (ctenophores), 505–506, *506*, 521
Commensalism, 590, *590*, 600
Communication, between neurons, 853–857, *854*
Community(ies), 546, *547*, 560, 584–601
    changes in, 596–600
    climax, 585, 597, 599, 600
    individualistic, 586–587, *587*, 600
    integrated, 586, 600
    pioneer, 597, 597t, 600
    resilience of, 599, 600
    species in, interaction of, 587–593
    stability of, 599, 600
    successional, 597, 600

Compact bone tissue, 723, *724*, 733
Companion cells, 649, 672, *673*, 674
Comparative embryology, 334–336, *336*
Competition, 591–593, 600
    evolution and, 592–593
Competitive exclusion principle, 591, 600
Competitive inhibitor, 116, *116*
Complement, 832, 843
Complementary DNA (cDNA), 285, *285*, 286, 295
Complementary proteins, 737, *738*
Complex molecules, building of, 52
Compound eye(s), 518, 521, 873
Compound microscope, 72, *73*
Compound(s), 24, *25*, 44
Compression, 724, 732
Concentration, 91, 100
Concentration gradient, 848–849, 861
    in diffusion, 91–92, *92*, *100–101*
Conception, 883
    avoidance of, 900
Conditional mutation, 245
Conditioned reflex, 626, 639
Conditioned stimulus, 859
Condoms, 901, 905
Conduction, saltatory, 853, 861
Cones, 874, *875*, 880
Coniferous forests, 564, *564*
Conifers, 489
    tracheid pit connections in, *647*
Conjugation, in bacteria, 265, *266*, 274
Conjugation fungi (Zygomycetes), 468–469, *470*
Connective tissue, 499, *499*, 520, 713
    forces on, 724–725, *725*
Connell, Joseph H., 591, 592, 593
Conservation biology, 603, 624
Constipation, 750, 752
Constraints, developmental, 383, 396
Consumers, 548, 560, 588
Contact inhibition, 177, *178*
Contig, assembly of, 309, *309*
Continental drift, 339
Continent(s), history of, 339, *339*
    species diversification on, 379
Contraception, 899–902, 905
    abstinence as, 900–901, 905
    barrier methods as, 901, 905
    hormones as, 901–902, 905
    intrauterine devices as, 902, 905
    methods of, 900t
    sterilization as, 902–903, 905
    withdrawal as, 901, 905
Contraceptive sponge, 901
Contractile proteins, 728, *729*
Contractile ring, 175, *176*
Contrast, 72
Conventional genes, 259
Convergent circuits, 858
Convergent evolution, 376, 396
Coontz, Robert, 602
Copepods, 519
Copulation, adaptations to, *187*
Corals, 504
Corey, Robert, 214
Coriolis effect, 566, 582
Cork cambium, 689–690, *690*, 692
Corn, 687
    crossing over in, 203
    high-oil, 293
    jumping genes in, 258–260
    origin of, 676–677, *677*

pod, 676
prehistoric, *280*
Cornea, 873, 880
Corolla, 492, 494
Corpora allata, 758
Corpse, succession on, 598
Corpus callosum, 868
Corpus cavernosum, 889, 904
Corpus lutem, 895, 904
Corpus spongiosum, 889, 904
Correns, Karl Franz Joseph, 349
Corridors, in wildlife preserves, *607*, 607–608, 624
Cortex, 651, 658
    renal, 814–817, 825
Corticotropin, 762
Cortisol, 760, 768, 898, 905
Cotransport, 95, *95*
Cotton, transgenic, *293*
Cotyledons, 649, 658, 680, 692
Cough, "smoker's," 799
Countercurrent heat exchange, 716, *717*
Countercurrent heat exchanger, 716, 731
Countercurrent oxygen exchange, *717*, 717–718
Countercurrent systems, 716–718, 732
Coupling factor, 135, 138
Courtship, of stickleback, *634*
Covalent bond(s), 30, *32*, 44
    atoms in, 30
    double, 30, 44
    triple, 30
Cowles, Henry C., 585, 586, 600
Crab, horseshoe, *324*
Cranium, 865
Crassulacean acid metabolism, 158, 159, *667*, 667–669, 674
Creationism, 319, 341
Creighton, Harriet, 203–204, 259
Creosote bush(es), 567, *569*, 668, *668*
Cretaceous Period, 331–332, 339, 380, 382, 479, 536
Creutzfeldt-Jakob disease, 46
Crick, Francis, 208, 211, 212–213, *213*, 214, 216, 217, 225, 233, 234, 236–237, 239, 245, 269, 308, 411
    The Double Helix by, 237
Cristae, 84
Cro-Magnons, 395–396, 397
Crocodiles, 534
    Nile, *536*
Crocodilia (crocodiles and alligators), 534
Crop, 745, 752
Cross breeding, 194
Cross bridges, 729, 733
Crossing over, 188–189, *189*, 204
    in corn, 203
Crustacea, 519–520, *520*, 521
Cryptochromes, 699
Cryptococcus, 468
Crystallins, 916
Ctenophores (comb jellies), 505–506, *506*, 521
Cubozoans, 504
Cud, 746
Curie, Marie, 29, *29*, 329
Curie, Pierre, 29, 329
Current(s), ocean, 578–579
Cuticle, 478, 494
    of flowering plant, 646, 658
Cutin, 654
Cuvier, Baron Georges, 323, 325, 522, 523
Cyanobacteria, filamentous, 401, *401*, 402

Cycads, 489–490, *489*
Cyclic AMP, 764, 768
    in cholera, 764
    synthesis of, 765, *765*
Cyclins, 178
Cyclostomes, 529, 542
Cystic fibrosis, 305, 305–306, 310–311, 314
    transmembrane conductance regulator, 305
Cysts, in protists, 456, 474
Cytochrome *c*, phylogenetic tree of, 337, *338*
Cytochrome oxidase complex, 134
Cytochromes, 131, 138
Cytokines, 289, 833, 839, 843
Cytokinesis, 168, *170*, *171*, 175–176, *176*
    in plant cells, 176, *176*
Cytokinins, 696t, 702, 705, 708
Cytoplasm, 75, 100
Cytoplasm division, 175
Cytosine, 210
Cytoskeleton, 75, 86–87, *87*, 100
    function of, 86
Cytosol, 75, 80–81, 100
Cytotoxicity, 840

D

D & E, 903, 905
Daddy longlegs, 519
Daktulosphaira vitifoliae, 660
Dalldorf, Gilbert, 453
Dalton, John, 26
Dalton (unit), 26
Danger model, 826–829, *828*
Darrow, Clarence, 318
Darwin, Charles, 18–19, 319, *326*
    altruism and, 636–637
    artificial selection and, 321
    Beagle voyage of, 325–326, *370*
    blending of inheritance and, 346–347
    on conditions for production of living organism, 404
    on defining of species, 371
    distributions of species and, 338
    education of, 325
    evolution of species and, 325–327
    on Galápagos Islands, 368–369, 377–378
    gradualism and, 320
    natural selection and, 321, 338–341, 342
    phototropism and, 694, *695*
    speed of evolution of species and, 383–387
    theory of evolution of. *See* Evolution, Darwin's theory of
Darwin, Erasmus, 321, 325, 477
Darwin, Francis, 694
Darwin, Robert, 325
Daughter cells, 74, 165
    in cell division, 167
    in meiosis, 190
Daylight, plant responses to, 699–702, *700*, *701*
DDT, estrogenic effects of, 754
    poisoning from, 622
    resistance to, 360
de Vries, Hugo Marie, 349
Dead space, 796, 806
Dead zones, 559, *560*
Death rate, 609, 624
    smoking and, 789, *789*
Death Valley, 562
Decapods, 519
Decay, 329
Deciduous forests, 564, 582

Declarative memory, 868
Decomposers, 446, 450
Defenses, of animals, 588–589
    chemical, 588
    of plants, 588–589
Defibrillation, 784
Dehydration, 813, 824
Dehydration reaction, 52–53, 53, 68
Delbrook, Max, 223, 224, 224
Delphinium, 373, 373
Democritus, 26, 102
Demography, 616
Denaturing, of enzyme, 114, 117
Dendrites, 847, 861
Denitrification, 558, 560
Dennett, Daniel, 497–498
Density-dependent population growth, 611
Density-independent population growth, 611
Deoxyribonucleic acid. *See* DNA
Deoxyribose, 55t, 59, 69
Depo-Provera, 902, 905
Depolarized cell, 851
Dermal tissue system, 646, 658
Dermis, 830, 842
DeRobertis, Edward, 523
Desert(s), animals of, 568
        water conservation by, 812
    climate of, 564
    ecosystems of, 567–569
    plant water conservation in, 668
    plants of, 568
    soils of, 569
DeSilva, Ashanthi, *314*
Desmosomes, 98, *99*, 101, 784
Detergents, 40
Deterministic events, 604, 624
Deuteromycetes (imperfect fungi), 470, 471, *472*, 475
Deuterostomes, 501, 522–542, *525*, 541
    coelomates, 524–531
Development, chemical signaling and, 923–925
    vertebrate, stages of, 909
Developmental constraints, 383, 396
Devonian Period, 331, 479, 526, 527, 528
d'Herelle, Felix, 223
Diabetes mellitus, 759, 768
    insulin-dependent, 759
    juvenile, 759
    non-insulin-dependent, 759
    type 1, 759
    type 2, 759
Diacylglycerol, 765
Diaphragm, 798, 806, 901, 905
Diarrhea, 462, 750, 752
Diastole, 782–783, 787
Diatoms, 458–459, *459*
Dickens, Charles, 862
Dicots, 481, 649–651, *650*, 658, 680, 692
    germination in, 682–683, *683*
Dicotyledons, 649–651, *650*, 658
Didinium, 76, *76*
Diencephalon, 867, 880
Diet, low-salt, 808, 809–810
Differentiation, 914, 928
Diffusion, 91–92, 794
    facilitated, passive transport and, 94, 101
    simple, 91, 100
    in tiny animals, 716, *716*
Digestion, 121, *121*, 138, 739, 752
    behavioral adaptations and, 742
    in cavities with one opening, 741, *741*
    conditions of, 740

early studies in, 734–735
by individual cells, 740, *741*
in two-ended digestive tract, 741–742
Digestive tract, 752
    coordination of processes in, 751
    as "disassembly line," 741–742, *742–743*, 743
    human, 741, *741*, 743–751
    two-ended, 741–742
    vertebrate, organization of, 741–742, *742–743*, 743
Digits, programmed cell death and, 917, *917*
Dikaryotic fungi, 466
Dinkas, African, 352, *352*
Dinoflagellates, 457–458, *458*, 474
    test of, 457–458, 474
Dinomischus, 498
Dinosaurs, cloning of, 292
    speed of, 710–711, *711*
Dioxin, 696
Diphyllobothrium latum, 508
Diploid cells, 185, 204
Directional selection, *360*, 360–361, *362*, 362–363, 367
Disaccharides, 59, 69
Diseases, specific, location of genes for, 308–310
Dispersal, 379, 396
Disruptive selection, 363, 367
Distributed processing, 863
Distribution, normal, 12, *12*, 20
Diuretic, 823
Diuron, 156
Divergent circuits, 858
Divergent evolution, 376, 396
Diversity, 418–542
    abiotic factors influencing, 593–595
    biotic factors influencing, 593
    immune system recognition of, 834
    increasing, mass extinctions and, 383, *384*
    measurement of, 593
Dixon, Henry, 664
DNA, 16, 20
    alphabet, 239
    cell division and, 173
    as chain of nucleotides, 210, *211*
    in chloroplasts, *272*, 273
    chromosomal, tumor virus insertion into, 267, *268*
    circle of, 168, *168*
    cloning of, 922
    computer-generated model of, *218*
    damage to, consequences of, 231, *232*
        repair of, 231, *232*
    direction of DNA synthesis in, 225–229
    of energy organelles, 273
    in eukaryotes, 253
    fibers, 209, *210*
    folding of, 174
    in genes, 222, 223–225
    hybrid, 286, *286*, 295
    information in, 167
    loops of, 175, *175*
    in mitochondria, *272*, 272–273
    mitochondrial, of wolves and coyotes, compared, 419–420
    multiplication of, polymerase chain reaction and, 277, *278*, 286
    nucleotides of, 211, *213*
    in nucleus, 173–175, *174*
    phases of cell cycle and, 170, *170*
    and protein sequences, criteria for classification of, 429–430

protists and, 465
recombinant. *See* Recombinant DNA
replication of, 167, 225–227
    by dividing cells, 225–227
    origin of replication and, 228, *230*, 233
    polymerization in, 227, *228*, 233
    replication forks in, 227–228, *230*
    in 3′ and 5′ direction, 228
    Watson-Crick model for, 227
sequencing of, 226, *226*
structure of, 210–217
    discovery of, by Franklin, 211–216, *214*
    replication and repair of, 208–233
    3′ and 5′ ends in, 211, *213*
synthesis of, rules of, 226
    site of, 228, *230*
transcription of, 240
as transforming factor, 222
as twisted ladder, 210–211, *212*
viral, entry into host cell, 263, *264*
DNA array, 287, 295
DNA chip, 287, 295
DNA fingerprinting, 299, 315
    electrophoresis in, 300, *300*
    as evidence in U.S. courts, 299, 300–301
    in identification of kidnapped children, 302
    noncoding DNA and, 299
    reliability of, 299–302
    restriction enzymes in, 300, *300*
DNA ladder, *64*
DNA ligase, 228, 233, 283, 294
DNA polymerase, 227, *227*, 233
    in DNA assembly, 227, *229*
    in DNA sequencing, 226, *226*
DNA/RNA language, 237
DNA sequencing, and classification of prokaryotes, 448
DNA sequencing gel, 226
Dobzhansky, Theodosios, 338
Dog, cool nose of, *39*
Dolk, Herman, 694
Dolly, 292, 921, *921*
    cloning of, *920*, 920–921
Domains, 16, *17*, 20, 66
    in hierarchy of classification, 423, 431
Dominance, partial, 196, 205
Dominance hierarchy, 633, 639
Dominant alleles, 195–196, 205
Donohue, Jerry, 217
Dopamine, 854, *855*
Dormancy, plant embryo, 681, 692
Dorsal hollow nerve cord, 527, 542
Dorsal horns, 866
Dorsal root ganglia, 866
Dorsal structures, 523
Doubling time, 608, 624
Douche, 901, 905
Doudna, Jennifer A., 410
Down syndrome, 191, *191*, 193, 205, 312, 313
Drake, Frank, 403, *403*
Drew, Charles, 770–771, *771*
Driesch, Hans, 907, 918
Drosophila, imaginal discs in, 925, *925*
    mutant, 245
    pattern formation in, 922, 923–924, *924*, 928
Drosophila melanogaster, eye color gene in, 200, *201*, 202, *202*
    genetic research on, 352, 353
    life cycle of, 200, *200*
    mutant, *201*
    normal, *925*

sex-linked genes in, 201
wild type of, 201, *203*
Drosophila subobscura, courtship of, 364–365
Drugs, actions of, physical basis of, 858
Duck-billed platypus, 541, *541*
Duodenum, 747, 752
Dyneins, 86
Dynorphins, 858
Dysjunction, 190–191, 205, 190–191, 205

E
Ear, inner, 871, *872*, 872–873, 880
    and balance, 872–873
    mammalian, 872–874, *873*
    middle, 872, *873*, 882
    outer, 872, *873*, 882
    microorganisms inhabiting, 436
Eardrum, 871
Earth, carrying capacity of, 617–620, *618*, 624
    life of, origin of, 399–404, *402*
        theories of, 401–402
    movements of, sunlight, and climate, 564–566
Earthworms, 514, *516*
    salt and water balance in, 823–824
Easterlies, polar, 565, *566*, 582
Eccles, Sir John, 879
Ecdysone, 758
Echinodermata, 524, 541
Echinoderms, 522–542, 524–527, 541–542
Ecological footprint, 618, 620, *620*, 625
Ecological isolation, 373, 396
Ecological unit, species as, 425, 431
Ecology, behavior and, 632–634
    definition of, 546, 560
Ecosystem(s), 544–561, 546–547, *547*, 560
    aquatic, 578–581
    boundaries of, 546–547
    components of, 546
    of deserts, 567–569
    energy flow through, 548–552
    interactions of, 547
    nitrogen fertilization and, 559
    organisms cooperating in, 545, 546, *546*
    preservation of, 607–608
    properties of, 546–547
    recycling by, *552*, 552–559
    resilience in, 600
    terrestrial and aquatic, 562–583
Ectoderm, 499, 520, 908, *908*, 927
Ectotherms, 538, 542
Edema, 813, 824
Edge effects, 604, *605*, 624
Eels, 795
    electric, sodium-channel protein in, 851
Effector, allosteric, 116
    definition of, 114, 117
Efferent nerves, 868, 880
Egg(s), 186, 204
    amniotic, 428, 536, *537*, 542
    fertilization of, 897
    union with sperm, 186–188, *187*
Egg cells, growth of, 177
Ehrlich, Paul, 602
Eijkman, Christian, 738–739
Einstein, Albert, 4, 101
Ejaculation, 888, 891, 904
El Niño, 579
Elastin, 63
Eldredge, Niles, 387
Electrical potential, 849, 861
Electrical pulse, generation by neurons, 848–850

Electrical synapse, 854, 861, 854, 861
Electrochemical gradient, 133, 138
Electrochemical impulses, nerves and, 847
Electrodes, 850
Electron(s), 27, *27*, 44
    in atoms, 30
    in noncyclic electron flow, 150
    in photophosphorylation, 149
Electron carriers, 131, 138
Electron donors, 130
Electron microscope, 72–73, *73*
Electron orbitals, 30
Electron transport, 130–137
    energy of, harvesting of, 131
Electron transport chain, 131, *132*, *133*, 138
Electronegativity, 34, 44
Electrophoresis, 220, *221*, 226
    in DNA fingerprinting, 300, *300*
    gel, 288, 295
Elements, 23–24, 44
    of human body, *24*
    Periodic Table of, 30, 31, Appendix A
Elephants, unrestrained growth in, 608, *608*
Elimination, 813, 824–825
Eliot, George, 862
Ellis-van Creveld syndrome, 358
Elongation, 247, *248–249*
    inhibition of, *250*
Elongation zone, 686, 692
Embryo(s), cells of, and cells of adult, 908–916
    differentiation of, 919–923
    development of, fertilization and, 908, 927
    flowering plants, body plan of, 678–684
        development of, 680–683, *682, 683*
        nourishment of, 683, *684*
    formation of, historical views on, 906
    human, *336*
    mouse, *336*
    of vertebrates and invertebrates, compared, *524*
Embryo sac, 679, 692
Embryology, comparative, 334–336, *336*
Embryonic stem cells, 926
Emergent property, art as, 9, *9*
Emergents, 576, *577*, 582
Emigration, 610, 624
Emphysema, 800
Emulsion, 750
Encephalopathies, transmissible spongiform, 46
Endangered Species Act, 420–421
Endergonic reactions, 107, *107*, 116
    and exergonic reactions, coupling of, 108, 109, 116
Endocrine organs, 766, *766, 767*
Endocrine signals, 766
Endocrine system, 757
Endocrinology, 757
Endocytosis, 96, *97*, 101, 263, 741, *741*
    receptor-mediated, 96–97, 101
Endoderm, 499, 520, *908*, 908–909, 929
Endodermis, 651–652, 658, 659
Endogenous antigens, 840, 844
Endometrium, 886, 904, 911
Endoplasmic reticulum, 81–82, 100
    lumen of, 82, 100
    rough, 82, *82*, 100
    smooth, 82, *82*, 100
Endorphin(s), 762, 858
Endoskeleton, 721, 732
    vertebrate, parts of, 722–723, *723*
Endosome, 741
Endosperm, 491, 494, 680, 692

Endosymbionts, 413–414
Endosymbiosis, 414, *414*
Endosymbiotic theory, 273, 274
Endothelial tissue, 713
Endothelium, 22, 781, *781*, 787
    oxygen movement through, 801, *801*
Endotherms, 534, 538
Endotoxins, 435, 447
Energy, 30, 44
    activation, 110–111, *111*, 117
    extraction from glucose, by cells, 122–130
    extraction from macromolecules, by heterotrophs, 121
    flow through ecosystems, 548–552
    free, changes in, and direction of reaction, 107, 116
        concentration of molecules and, 109, *109*
        definition of, 107
        source of, during reaction, 108, *108*
    harvested from glucose, by cell, 136–137, *137*
    kinetic, from work, 104, 117
    loss of, during ecosystem recycling, 552, *552*
        as heat, 106, *106*
    potential, in work, 104, *105*, 117
    provision of, for biological processes, 108
    pyramid of, 551, *551*, 560
    subsidies of, 617–618, *619*, 625
    sugars for storage of, 60
    supply to organisms, 118–139
    transformation of, thermodynamics and, 103
Energy organelles, DNA and RNA of, 273
Energy requirements, of animals, 736
Energy supply, and energy consumption, 617–618
Enhancers, 255
Enkephalins, 858
Ensatina eschscholtzii, 375, *375*
Entropy, 109, *109*, 116, 544
Environment, adaptation to, 17–18
    animals sensing, 870–871
    conditions in, enzymatic reactions and, 114
    early life and, 412–413
    effects on sexual development, 754–755
    external, interaction of membranes with, 96–99
    as influence on aging, 926
    internal, of animals, regulation of, 755, *756*
    nervous system response to, 865–871
    nonmammalian vertebrate kidneys and, 820–821
    in phenotype, 183–185
    responses of plants to, 697–702
Enzyme(s), 108
    actions of, 112–114
    activation energy of chemical reaction and, 113–114
    active site of, 112, *113*, 117
    binding to reactant, 112–113, *113*
    as catalyst, 111, 112, 116
    denatured, 114, 117
    functions of, regulation of, 115, *115*
    genes and, 218–219, *220*
    influence of inhibitors on, 115–116, *116*
    to link nucleotides, discovery of, 234
    prevention from catalyzation of reaction, *116*
    proteins as, 63, 69
    reactions of, environmental conditions and, 114
        subreactions in, 114–115
    restriction, 270, 274, 283, 294
        in DNA fingerprinting, 300, *300*
        recombinant DNA and, *282*, 282, *283*, 294

Enzyme(s), *(continued)*
  specificity of, 112
  temporary binding of, 112, *113*
Eohippus, 334
Ephedra, 490
Ephedra viridas, 490
Epiblast, 913, 927
Epicotyl, 680
Epidemiologists, 432, 447
Epidermis, *829,* 829–830, 842
  of flowering plant, 646, 658
Epididymis, *887,* 888, 904
Epiglottis, 744, 796
Epinephrine, 760, 761, 763, 764, 765, 768, 785,
  787, 869
Epiphytes, 576, 582
Episomes, 265
  definition of, 268
  replication of, 268, 274
Epithelial cells, 97, 100
  in airways, 799, *799*
  of respiratory tract, 798, *799*
  of small intestine, 747–749, *749*
Epithelial tissue, 713
Epithelium, 98, 499, *499,* 817, 825
  in gut lumen, 743, 752
Epitopes, 835, *835,* 843
Epochs, 330, 342
Equatorial doldrums, 565
Equilibrium, chemical, 41, 45, 109–110, 117
  concentration and entropy affecting, 109–110
  factors influencing, 110, *111*
  of gas in solution, 793, 806
  punctuated, 387, *387,* 397
Equilibrium potential, 849–850
Equisetophytes, 487, *487*
Equisetum, 487
Eras, 330
Erection, 890–891, 904
Ergot, 471, *471*
Ergotamine, 471
Ergotism, 471
Erosion, soil, from chaparral fires, *572*
Erythrocytes, 772–773, *773,* 787
  "ghosts," 88
  membranes of, 87–88
  oxygen transport by, 802, *802*
  sickled, 220
  standard, *220*
Escherichia coli, 244, 252, 254, 255, 291, 361, 371–
  372, 435, 437, 439, *439,* 440, 441, 442, 446
  doubling time of, 608
  limitation of growth of, 609
  recombinant, 284
  restriction enzymes in, 270
  in study of DNA synthesis, 223, 228
Esophagus, 744, 752, 745–746, 752, 796
Esox lucius, 532
Essential amino acids, 737, 737t, 752
Essentialism, 321, 341
Estivate, 568, 582
Estrogen(s), 766, 894, 895, 896
  environmental, 754–755
  natural, 755
  during pregnancy, 897
  regulation of, 762
Estrogenics, 622, 754
Estuaries, 580, *580,* 582
Ethane, 50–51, *51*
Ethanol, 37, 51, *51*

in fermentation, 119
in genetic engineering of foods, 281
metabolism of, 136
properties of, 36t
water, and oleic acid, compared, 38t, 39
Ethology, 626, 639
Ethyl alcohol, 37
Ethylene, 30
  as plant growth hormone, 702, 706–707, 708
  in ripening, 706
  triple response to, 706, *706*
Ethylene dibromide, 642
Etiolation, 684, *684*
Eubacteria, 16, 20
  classification of, *444,* 445–448, 450
Eucalyptus trees, 490
Euglena, 89, 455, *455,* 457
Euglenophytes, 459–460, 474
Eukarya, 16, 20
Eukaryotes, 414, 430
  cell division in, 168–172
  DNA in, 253
  gene regulation in, 253–254
  nucleus of, *74*
  protein production and, 250
  in regulation of gene expression, 254
  sexually reproducing, of chromosomes, 168,
    169
  subcellular organelles of, *415*
  transcription and translation in, 241–242
Eukaryotic cells, 15, 16, 20
  transmission electron microscope to assess, 77–
    79
Eumetazoa, 499, *501,* 502, 520
Euphorbiaceae, 642
Euploid cell, 191, 205
Eustachian tube(s), 334, *336,* 871
Eutherians, 541, 542
Eutrophic lakes, 581, 582
Everglades, 580
Evolution, 16, 18–19, 317–416
  biotic, early, 409–412
  competition and, 592–593
  convergent, 376, 396
    in birds, 333, *334*
  before Darwin, 321–325
  Darwin's theory of, 18–19, *19,* 20, 319–321,
    326–327
  divergent, 376, 396
  evidence for, 318–367, 327–338
  fossil record of, 327–332, *328*
  genetic variation and, 353–354
  of horse, 333–334, *335,* 372, 380
  human, 388–396
  of males, female choice and, 633
  as opportunistic, *334*
  of organisms, 398–416
  preboitic, 404–409, 415
  process of, 320–327
  rate of, extinctions and, 383–388
Evolutionary taxonomy, classical, 426
Excitatory postsynaptic potential, 856, 861
Excitatory synapse, 854
Excitement, 890, 904
Excretion, 813, 824–825
Exergonic reaction(s), 108, *108,* 116
  and endergonic reactions, coupling of, 108,
    109, 116
Exhalation, 798, 806
Existence, struggle for, 341

Exocrine, 757
Exocytosis, 97, *97,* 101
Exogenous antigens, 840, 844
Exons, 242
Exoskeleton, 721–722, 732
  of arthropod, 517, 521
Exotoxins, 435, 447
Expansin, 703
Experiment(s), 6–7, 19
  "control," 7, 20
  design of, 7
  number of individuals in, 10
  statistics to plan and evaluate, 10
Experimental psychology, 626–627, 639
Expiration, 798, 806
Exponential growth, 608–609, 624
Expressivity, gene, 304, 315
Extinction(s), mass, 383, *384*
  and rate of evolution, 383–388
  and recolonization, *596*
Extinction vortex, 603, *603,* 604–606, 624
Extracellular fluid, 771–772, *772,* 811
Extracellular matrix, 96, 724–725, 732–733
Eye(s), 874–878, *875*
  brain information processing and, 876–878,
    878
  of cephalopods, 510, *511*
  compound, 518, 521, 873
  formation of, 916, *916,* 928
  of humans, *515, 873*
  microorganisms inhabiting, 436
  sight defects of, *875,* 875–876
Eye-cup, 873
Eyeglasses, 874, *874*
Eyepiece, 72

F

F_0-F_1 complex, *134,* 134–135, *135,* 138
F1 generation, 194, 205
F2 generation, 194, 205
  genotypes and phenotypes of, 194–195
Facilitated diffusion, passive transport and, 94, 101
Facilitating interneuron, 859–860, 861
Factor X, 831–832
Facultative aerobes, 446, 450
Facultative anaerobes, 126
FAD, 123, 138
$FADH_2$, 123, 138
Fairy rings, *467,* 470
Fall turnover, 581
Fallopian tubes, 886, 904
Family, in hierarchy of classification, 423, 431
Family size, completed, 616, 625
Farlow, James, 711
Farquhar, Marilyn, 82–83
Farsightedness, 874, 880
Fate maps, 919
Fats, acetyl-CoA and, 129
Fatty acids, functions of, *57*
  linking of, 56–59
  polyunsaturated, 56
  saturated and unsaturated, 53–56, *56*
Feces, 741, 752
  origin of, 509
Feedback, negative, 13, 730–731, 733, 762, 768
  positive, 898, 905
Female, choice of, evolution and, 633
Female choice, 364, *364,* 367
Fermentation, alcoholic, 118–119
  kinds of, 126, *126,* 138

Fern(s), 486–487
    life cycle of, *486*
    reproduction in, 486
Fertility, 614, 625
Fertilization, 166, 186, 204, 908, 909, 927
    double, 491, 494, 680, *680–681*, 692
    embryonic development and, 908
    internal, adaptations to, *187*
    in mammals, 888–889
    meiosis and, 188, *188*
    in plants, flowering, 679–680, *680–681*
Fertilizers, manufactured, 559
Fetus, 883, 917, 928
    full term, 886, 904
    oxygen for, 805
    stages of development of, 917–919, *918*, 929
Fiber(s), 750
    collagen, 725, *725*
Fibrillation, 783, 787
    ventricular, 784, 787
Fibrils, collagen, 725, *725*
Fibrin, 831, 842
Fibrinogen, 831, 842
Fibroblasts, 724, *725*, 732–733
Fibronectin, 913
Fibrous proteins, 65, 69
"Fight-or-flight" reactions, 760, *761*, 869
Filament, 492, 494
Filamentous cyanobacteria, 401, *401*, 402
Filtrate, 817, 825
Filtration rate, renal blood flow and, 821
Filtration slits, 816
Finch(es), adaptive radiation and, *377*, 377–378, *378*
    of Galápagos Islands, *592*, 592–593
    vampire, *378*
Fish, bony, body plan of, 531, *531*
    cladists and, 427
    countercurrent system in, *717*, 717–718
    freshwater, kidneys of, 820
        salt and water balance in, 811, *811*
    jawless, 529, *529*
    lobe-finned, 531, *532*
    ray-finned, 531, *532*
    saltwater, kidneys of, 820
        salt and water balance in, 811, *811*
        water retention by, 812
Fission, binary, 167–168, *168*
Fitness, inclusive, 637
    altruism and, 637–638, 639, *639*
Flagellates (zoomastigina), 464, 475
Flagellum(a), 100, 441, 450, 887, 904
    9+2 structure of, 89, *89*
Flame cells, 823
Flatworm(s), 506–507, 521
    excretory organs of, 823
    free-living, 506
    learning in, *630*
    nervous system of, 865, *865*
    tissue layers of, *502*
Flavor components, 280–281
Fleming, Alexander, 250, 280
Flemming, Walther, 165–166, 172, 182
Florey, Howard, 280
Flounder, *386*
Flowering, seasonal control of, 699
Flowering plants. *See* Plant(s), flowering
Flower(s), complete versus incomplete, 687
    development of, buds and, 687
        genes in, 687–689

functions of, 487–488, 678
    parts of, 492–493, 678, *678*, 679, *680–681*
    perfect versus imperfect, 687
    pollination of, 491
    whorls of, 687, *688*, 692
Fluid intake, brain control of, 821
Fluid mosaic model, 91, 100
Flukes, 506, *507*, 521
Fluorens, Pierre, 862–863
Fluoxetine, 857
Flyers, 719–720, *720*
Fly(ies), on corpse, 598
    fruit. *See* Fruit fly(ies)
Follicle-stimulating hormone, 762, 892, 894, *895*, 904
Follicles, 884–885, *885*, 904
Food(s), animal nourishment from, 734–753
    demand for, agriculture and, 619–620, *621*
Food chain, 548, *548*, 560
Food intake, animal, amount of, 736
    too much or too little, consequences of, 736–737
Food web, 548, *549*, 560
Foot (feet), of mollusk, 511
    webbed, 715
Foraging, competition and predation influencing, 632
    optimal, 632, 633, 639
Foraminifera, 462, *462*, 475
Forams, 462, *462*, 475
Forebrain, 867–868
Foregut, 743, 745–746
Foreskin, 889, 904
Forest(s), coniferous, 564, *564*
    decidous, 564, 582
    deciduous, temperate, 573–575, *574*
Fossil(s), *14*, 18, *18*, 20, 323, *330–331*, 342
    of Cambrian Period, 401
    formation of, *328*, 328–329
    of Precambrian Period, 401
Fossil fuels, 554, *555*
Fossil record, *327*
    of evolution, 327–332, *328*
    factors revealed by, 327–332
    for plants, evolution and, 481
Foucault's Chinese encyclopedia, 421
Founder effect, 357, 366
Fovea, 874–875, 880
Fox, Sidney, 409
Fraenkel-Conrat, Heinz, 235
Fragmentation, habitat destruction and, 604, 624
Frameshift mutation, 245
Franklin, Rosalind, 208, *209*, 209–210, 211–215, 233
    discovery of structure of DNA by, 211–216, *214*
Free energy (G). *See* Energy, free
Free radicals, 29, 156
Frog(s) (Anura), 533, *533*
    cascades, 535
    Chiricahua leopard, 535
    early development in, *912*
    gastric brooding, 535
    genes of, 523
    poison dart, 588
    South American, 534
    tarahumara, 535
    Yavapai leopard, 535
Fronds, 486
Fructose, *60*

Fruit(s), development of, 493
    functions of, 491
Fruit fly(ies), 352, 353, 364–365. *See also Drosophilia melanogaster.*
    advantages for genetic research, 200, *200*
    genes of, 523
Fruiting body, of cellular slime mold, 464–465
Fuchs, Ephraim, 827, 828
Functional groups, 68
    definition of, 50, *51*
    properties and structures of, 51, 52t
Fungal infections, 468
    antibiotics and, 453
    human diseases from, 452–453
Fungicidin, 453–454
Fungus(i), 16, 20, *466*
    characterization of, 466–467, *467*, 475
    classification of, 454, 467, *467*, 475
    ecological importance of, 465, 472–474
    infections from, 466
    multicellular, and protists, classification of, 452–475
    origin of, 474
    reproduction of, 467
Furchgott, Robert, 22–23
Fusion, membrane, 96–97, *97*

G

G protein, 764
GABA, 854, *855*
GABA transminase, 857
GABA$_A$ receptor, 856, 861
Gaia, 544–545, 585
Galápagos Islands, 368–369, 376–379, *377*
    finches of, *592*, 592–593
Galápagos tortoises, 368, *368*
Galen, 773–774
Galileo, 398
Gall, Franz Joseph, 862
Gallbladder, 750, 752
Gametangia, 478, 484, *484*, 494
Gametes, 185–186
    distribution of chromosomes to, by meiosis, 188–192, *191*
    formation of, 909, 927
        mammalian, 883–888, 904
    production of, control of, 892–896
Gametic isolation, 373, 396
Gametocytes, 463
Gametophyte(s), 474, 478, 483–484, 487–488, 678, 692
    in phaeophytes, 460
Gamma-aminobutyric acid (GABA), *855*, 856, 861
Ganglion(a), 859, 864–865
Ganglion cells, 876, 880
Gap junctions, 98, *99*, 101, 784
Garrod, Archibald, 217, 218
Garter snakes, 372, 627–628
Gas(es), in atmosphere, 405
    dissolution of, temperature and, 793–794
    dissolved in solution, 793–794
    greenhouse, reduction of, 556
    in lungs, 794
    pressure exerted by, 792, *793*
Gas exchange, in alveoli, *801*
    in lungs, 796, 801, *801*
    surface area and, 794–795
Gastrin, 759
Gastrointestinal tract, replenishment of, 751
Gastropods, 512–513, *513*, 521

Gastrotricha, 510
Gastrula, 499, 520, 909, 927
Gastrulation, 909, 927
    in amphibians, *912*, 912–913
    in mammals, 913–914, *914*
    process of, 912, *912*
    in sea urchin embryo, 910, *911*
Gates, of neurons, 848, 861
Gause, G.F., 591
Gel electrophoresis, 288, 295
Gene(s), in alkaptonuria, 217, *218, 219*, 220
    in bacteria, 221
    behavior and, 627, *631*, 631–632
    bicoid, 924
    cloning of, in bacteria and viruses, 284
    composition of, 221–225
    continuity and, 16
    conventional, 259
    definition of, 183, 217
    "designer." *See* Recombinant DNA
    Ds, 258
    effects pf, on biochemical processes, 217–219
    and enzymes, 218–219, *220*
    eukaryotic, 242
    expression of, 234–256
        and structure of, study of, 287–289
        by viruses, 269, *269*
    expressivity of, 304, 315
    in flower development, 687–689
    functions of, 183
    homeotic selector, 688, 689, 692, 924, 928
    Hox, 924, 928
    human, patenting of, 311
    as influence on aging, 926
    inheritance of, 16
        chromosomes and, 194
    jumping, 258–275
        in corn, 258–260
    mobile, 259–260, 274
        evolution of, 270–272
        replication by, 264–265, 267–269
    with multiple alleles, 307, 307t
    one gene-one enzyme, 218–219, 232
    one gene-one polypeptide, 219–221, 232
    penetrance of, 304–305, 315
    pleiotropy of, 306, 315
    polypeptide, 219–221
    on same chromosome, assortment of, 201–202, *203*
    sequenced, use of, 308–310
    sex-linked, in *Drosophila*, 201, *202*
    for specific diseases, location of, 308–310
    structure of, 16, 20
    traits expressed by, 305
    traits influenced by, *306*, 306–307
    transforming factor in, 221–222
Gene expression, cellular regulation of, 250–254
    eukaryotes in regulation of, 254
    prokaryotes in regulation of, 251–253
Gene flow, 358, *359*, 366, 376
Gene frequencies, 355–366
    chance events and, 356–358
    selection and, 358–360, *359*
Gene gun, 292, 295
Gene library, 286, 295
Gene pool, 354, 366
    draining of, extinction in, 604–606
Gene regulation, eukaryotes in, 253–254
Gene sequences, 39, *838*
Gene therapy, 313–314, *314*, 315

Generations, alternation of. *See* Alternation of generations
Genetic code, *237*
    description of, 237–238, 255
    discovery of, 234–237
    in mitochondria, 273
    of mitochondria, 413, *413*
    as redundant, 237
    as universal, 238, *238*, 255
Genetic counseling, 311
Genetic disease, prevention of, 310–315
Genetic diversity, in humans, 299–304
Genetic engineering, of animals, 291–293
    definition of, 278–281
    of grain, 278–280, *280*
    natural and induced variation in, 280
    of plants, 291–293, *293*
    and recombinant DNA, 276–297
    of tomatoes, 280–281, 293
Genetic equilibrium, in populations, 354–355
Genetic linkage, of *Drosophila*, 200–201, 202, *203*, 205
    Morgan's work on, 200–202, *202*
Genetic map, 204, 205
Genetic systems, evolution of, 409–412
    unconventional, 258–275
Genetic unit, species as, 425, 431
Genetic variation, 192, *193*, 350–351
    and evolution, 353–354
    from mutations, 229–231
    natural selection and, 353, 365–366
    in populations, 353
Genetics, 161
    definition of, 183
    human, 296–316
    Mendelian, rediscovery of, 349
    Mendel's work on, 183
    molecular, 183, 204
    rules of, 185
    transmission, 183, 204
Geniculate nucleus, lateral, 876, *876*
Genital warts, 164
Genitalia, external, female, 889–890, *890*
        male, 889, *889*
        male and female, common origin of, 890, *891*
        in sexual intercourse, 889–890
Genome, 14. *See also* Human Genome Project
    definition of, 183
    restriction enzymes cutting, 282
Genotype, 353, 366
    definition of, 183, 204
    and phenotype, relationship between, 304–307
    phenotype and, 183–185
Genus (genera), 332
    in hierarchy of classification, 422, 431
Geochelone elephantopus, 368, *370*
Geographic isolation, 376–378, *377*
    speciation and, 374
Geospiza difficilis, *378*
Geospiza fortis, 592, *592*
Geospiza fuliginosa, 592, *592*
Germ cells, 186, *187*, 204
    primordial, 883–884, 904
Germ layer, formation of, 909, 911–912, 927
Germ line, 186, *187*, 204
    continuity of, *187*
Germ plasm, 348, *348*, 366
Germination, 682–684, *683, 684*, 692
    environmental cues triggering, 684

Giardia, 465
Gibberella fujukoroi, 704
Gibberellins, 696t, 702, *704*, 704–705, 708
Gibbons, 389, *389*
Gibbs, J. Willard, 107
Gill(s), *337*, *794*, 795, 806
    fast-swimming fish and, 795
Gill slits, 527, 542
Gill withdrawal reflex, 859–860, *860*
Ginkgos, 490
Girdling, 670, *671*, 674
Gizzard, 739, 746, 752
Glands, identification of, 757
Glans, 889, 905
Glantz, Stanton, 788–789
Glaucoma, 874
Gleason, Henry Allan, 586–587
Glial cells, 713, 732, 847, 861
Globular proteins, 65, 67, 69
Glomerulus, 814, *816*, 825
Glossopteris, 339
Glossypol, 657, 659
Glottis, 745, 796
Glucagon, 758–759, 764, 768
    production of, 759
Glucocorticoids, 763
Glucose, *50*, 59, *60*, 69
    blood, levels of, 758–759
    in Calvin cycle, 154–155, *155*
    extraction of energy from, by cells, 122–130
    plant synthesis of, 153–157, 644, *644*
    during pregnancy, 898
    production from NADPH and ATP, 153
    regulation of, failure of, 759
    synthesis of, photons required for, 157
    uptake of, 864
Glucose transport, liposomes and, 94–95
Glutamate, 854, *855*
Glutamic acid, 281
Glyceraldehyde phosphate, 156
Glycerol, 56, 57, 69
Glycine, 854, *855*
Glycogen, 60, *69*
Glycogen phosphorylase, 764
Glycolipids, 88
Glycolysis, 122, *123*, 123–124, 125, 138
    anaerobic, 138
    regulation of rate of, 124, *124*
    without oxygen, 126–127
Glycolytic muscle fibers, 728, 733
Glycoproteins, 61, 69, 88
Glycosidases, 739
Glycoside bonds, 53
Glycosides, mustard oil, 658
Glycosidic bond, 59
Glyphosate, 156, 292
Gnetae, 490
Gnetophyte, 490
Gnetum, 490
Goheen, Austin, 661
Golden toads, 535, *535*
Goldschmidt, Richard, 386
Golgi, Camillo, 846
Golgi complex, functions of, 82, 100
    structure of, 82, *83*
Gonadotropin(s), 762, 892
    human chorionic, 895, 896, 897, 904
        pregnancy tests and, 896
Gonadotropin-releasing hormone, 762, 893, 894, 904

Gonads, 186, 204, 766, 883
Gondwanaland, 339, *339*
Goose (Geese), greylag, 628, *628*
    imprinting by, 630–631, *631*
Gosling, Raymond, 208, 209, 213, 214, 216
Gould, Stephen J., 387, 497
Gradualism, 386, 387, *387*, 396
    Darwin's theory of, 320, 342
Grafting, 691, *691*, 692
Grain, production of, *621*
Gram, Hans Christian, 439
Grana, 149, *150*
Granulosa cells, 884
Granum, 85, 100
Grapevines, phylloxera and, 660–662
Grass, 376
Grasshopper, lubber, 180, *181*
Grasslands, animals of, 570–571
    temperate, 570, *570*, 582
Gravitropism, 697–698, *698*, 708
Gravity, plant responses to, 697–698, *698*
    size of animals and, 714
Green algae, in plant evolution, 481
Green ball, 790
Green glands, 824
Green Revolution, 619
Greenberg, Bernard, 598
Greenhouse effect, 555
Greenhouse gases, reduction of, Martin, John H., 557
Grey matter, 866
Griffith, Frederick, 221
Griffith, J.S., 47
Griffith's experiment, 221, *222*
Ground squirrel, round-tailed, 568
Ground tissue system, 646, 658
Growth, and development, programmed cell death in, 916–917, *917*, 929
    exponential, 608–609, 624
    primary, 646, 658
    secondary, 647, 658
    in vertebrates, 909, 927
Growth factor(s), 289, 295, 752, 757, 766, 769, 846
Growth factor gene, 290, 307
Growth hormone, 289, 762
Growth rings, 691, 692
Grunberg-Manago, Marianne, 234, 235
Guanine, 210, 233
Guanosine diphosphate, in citric acid cycle, 127
Guanosine triphosphate, in citric acid cycle, 127
Guinea pigs, 424, *424*
Gurdon, John, 921
Gusella, James, 298, 299, 308–309
Gustation, 870, 880
Gut, 741, 752
    actions of, hormones and nerve cells in, 751
    primitive, 911
Guthrie, Woody, 298, *298*
Guttation, 663, *663*, 674
Guzman, Rafael, 677
Gymnophiona (caecilians), 533, *533*
Gymnosperms, 481, 489–490, 494, 649
Gyri, 868, 880

H
HA oxidase, 217
Habitat, 591, 600
    size of, and number of species, 594, *594*
    two species sharing, 591

Habitat destruction, 604, *605*, 624
    fragmentation and, 604, 624
Habitat islands, 595, 600
Habituation, 859
Hagfish, 525
Hailman, Jack, 629
Hair cells, 871, *872*, 880
Haldane, J.B.S., 361, 369, 404, 637
Hales, Stephen, 670
Half-life, 329
    of ^{14}C, *28*, 28–29
Hallucigenia, *498*
Halophiles, 443, *445*, 450
Hamburger, Viktor, 845, 846
Hamilton, W.D., 637, 638
Hamner, Karl, 699–700
Haplodiploidy, 637, 639
Haploid, 185–186, 204
Hardy, G., 354
Hardy-Weinberg equilibrium, 366
Hardy-Weinberg Principle, 354–355, 358, 359, 365–366
Harlow, Harry and Margaret, 631
Harlow, Jean, *790*
Harvey, William, *774*, 774–775
Hawaiian Islands, adaptive radiation in, 379, *379*
Hawaiian tarweeds, 379, *379*
Hawthorn fly, 376
Hazen, Elizabeth Lee, 452, 453
Headgut, 743, 744–745
Hearing, 870, 880
    mammalian, 871–873
    and perception of sound, 862
Heart, 771, 786
    anatomy of, 776, *777*
    chambers of, 774, 776, *777*, 778, *778*
    early studies of, 774–775
    evolution of, *778*
    functions of, 776
    muscles of, 784
Heart attack, 782, 787
Heart beat, 782–785
    coordination of, 783–784
    systole and diastole in, 782–783
    three parts of, 783, *783*
Heart murmur, 783
Heart rate, 785, 787
    parasympathetic nervous system and, 785, 787
    sympathetic nervous system and, 785, 787
Heat, capacity of water, 37, *39*
    loss of, during energy transformation, 106, *106*
    tolerance by animals, 732
Heat exchange, countercurrent, 716, *717*
Heat exchanger, countercurrent, 716, 731
Heinrich, Bernd, 731
Hela cells, 162–163
Helicase, 229, *229*
Helicobacter pylori, 2–3
Heme, 131, 802, 806
Hemichordata, 524, 541, 542
Hemizygous alleles, 200
Hemoglobin, 63, *65*, 66, 772–773
    affinity for oxygen, 804–805
    normal, 220
    as oxygen carrier, *802*, 803
    in oxygenation, 801–802, *802*
    of sickle cell patients, 220, 232
Hemoglobin F, 804
Hemostasis, *830*, 830–831, 842
Hendricks, Sterling, 701

Heparin, 832, 843
Herbaceous plants, 645
Herbal teas, 642
Herbicide(s), actions of, on plants, 156
    auxins as, 696, 704
Herbivores, 548, 560, 588, 736, 752
    digestive tracts of, 712, *712*
    plant defenses against, 656–658
Herbs, 645
Hereditary Disease Foundation, 296–297, 298–299
Heredity, 180, 341
Heron, blue, *538*
Herring gull chick, behavior of, 628–629, *629*
Hershey, Alfred, 223–225, *225*
Heterosporous plants, 487
Heterotrophs, 119–120, 138, 412, 416, *444*, 445–446, 450, 548, 560
    aerobic, 446, *446*, 448
    anaerobic, 448
    in energy extraction from macromolecules, 121
Heterozygous, 350, 366
Heterozygous alleles, 194, 205
Hierarchial classification, *422*, 423, 431
Hierarchy, 422
    dominance, 633, 639
High altitude, oxygen delivery at, 805
High-throughput screening, 310
Hilts, Philip, 793
Hindbrain, 866, 880
Hindgut, 743, 750–751
Hip, ball-and-socket joint of, 723, *723*
Hippocampus, 868, 880
Hirudinea, 512–513
Hirudo medicinalis, 512
Histamine, 786, 787, 832, 836, 842
Histidine, 236
Histone 4, 429–430
Histones, 174, 175, 254
Histoplasma, 468
HIV. *See* Human immunodeficiency virus
H.M.S. *Beagle*, *370*
Hodgkin, Alan, 851
Holmberg, Scott A., 432–434
Homeostasis, 13, 20, 751, 752, 755–756, 768, 771
    behaviors and, 865
    definition of, 730, 733
    oxygen, 805
    two stages of, 730–731, 733
Homeotherms, 776
Homeotic selector genes, 688, 689, 692, 924, 928
Hominidae, 322, 390, 397
Hominids, branching evolution of, 394, *395*
    earliest known, *393*, 393–394
Hominoids, *389*, 390, *390*, 397
    stance of, evolution of, 391–392
Homo, 322
Homo erectus, 394–395, 397
Homo ergaster, 394
Homo habilis, 394
Homo heidelbergensis, 395
Homo sapiens, 332, 390, 395, 397
Homolog, 168, 185, 205
Homologous characters, 426, 431
Homologous chromosomes, 168, *169*, 185, 350, 366
    in meiosis, 188–190, 205
Homologous structures, 333, 342
Homoplasty, 426, 431
Homosporous plants, 487
Homozygous, 350, 366

Homozygous alleles, 194, 205
Homunculus, *879*, 906, 927
Honeybees, hive of, 635–636
    social organization of, 635, 636
    swarming, 636, *636*
Hood, Leroy, 48
Hoof-and-mouth disease, 260–261
Hooke, Robert, 70–71
Hooker, J.D., 371
Horizontal cells, 875, 880
Hormones, 736
    blood flow regulation by, 785–786
    and blood pressure, 821–823, *822*
    as chemical signals, 756
    as contraception, 901–903, 906
    control of gamete production by, 892–896
    in coordination of distant organs, 758–762
    definition of, 702
    in digestion, 751, 752
    identification of, 757
    ovarian cycle and, 894–896, *895*
    plant responses to, 702–707
    of plants, 702–707, 708
    production of, by testes and ovaries, *892*, 893–
      894
    in regulation of menstrual cycle, 896
    targeting of, 762–766
Hornworts, 480
Horse(s), evolution of, 333–334, *335*, 372, 380
    thoroughbred, 353–354
Horsehair worms, 510
Horseshoe crab(s), *324*, 519
Horsetails, 487, *487*
Housman, David, 296, *297*, 298
Hox genes, 924, 928
Human(s), classification as primates, 319, *320*
    development of, *913*
      after implantation, 916–918
    digestive tract of, 741, *741*, 743–751
    eye of, *515*
    genetic diversity in, 299–304
    genetic variability in, 353
    learning by, 869
    modern, origin and migration of, 303
    respiratory system of, 796, *797*, 806
    stem cells in, 926
    tolerance in, 731, *731*
    water and salt balance in, 812, *813*
Human chorionic gonadotropin, 895, 896, 897,
  904–905
    pregnancy tests and, 896
Human chorionic somatomammotropin, 897, 905
Human evolution, 388–396
Human genetics, 296–316
Human Genome Project, 307–310, 315
    location of disease-specific genes in, 308
    mapping goals of, 308
    positional cloning in, *309*, 309–310
    sequenced gene applications in, 310
Human growth hormone, effects on dwarfism, *289*
    in mice, 290, *291*
Human immunodeficiency virus, 261, *262*
    drugs to combat, 271
    replication of, 271
Human papilloma virus, 164
Humboldt Current, 578–579
Hummingbird, blue-throated, *538*
Humus, 669, 674
Huntingtin polypeptide, 310
Huntington's disease, 294, 296, 297–298

gene expressivity and penetrance in, 304, 305
gene mapping of, 297
huntingtin gene and, 310
Milton Wexler and, 297
pedigree of family with, 308–309, *309*
Hutton, James, 323
Huxley, Andrew, 851
Huxley, Thomas, 346–347, 384
Hybrid breakdown, 374, 396
Hybrid DNA, 286, *286*, 295
Hybrid inviability, 374, 396
Hybrid sterility, 374, 396
Hybridization, 288
Hybridization probe, *286*, 286–287, 295
Hybrids, 194, 205
Hydra, 504, *505*
    nerve nets in, 865, *865*
Hydrocarbon chain(s), 49, *50*, 68
Hydrocarbons, polycyclic aromatic, *408*
Hydrochloric acid, in digestion, 734–735
Hydrogen bonding, in water, 39
Hydrogen bond(s), 35–36, *36*, 44
Hydrogen cyanide, 406, *407*
Hydrogen ion(s), 40, 45
    concentration of, 41–42
Hydroids, 504
Hydrolysis, 53, *53*
Hydronium ion, 40–41, 45
Hydrophobic interaction(s), 35, *35*, 44
    van der Waals attractions and, 33
Hydrostatic skeleton, 721, *722*, 732
Hydroxide ion, 40, 45
Hydroxyl group, 51, 68
Hydrozoa, 504, *505*
Hymen, 890, 904
Hyperopia, 875, *875*
Hypertension, prevalence of, in U.S., *809*, 809t
    sodium consumption and, 808
Hyperthyroidism, 766
Hypertonic solution, 93, *93*, 101, 811
Hyphae, fungal, 466, *466*
Hypoblast, 913, 927
Hypocotyl, 680
Hypothalamic hormones, 762
Hypothalamus, 731, 736, 761, *761*, 762, 768, 833,
  867, 880, 892
Hypothesis(es), definition of, 5, 6, 20
    disproving of, 6
    model of, *5*, 5–6
    related, 13
    testable, 6, 20
Hypothyroidism, 766
Hypotonic solution, 93, *93*, 101, 811
Hyracotherium, 333, *335*

I

Ibuprofen, 768
Ice, floating, 36, *36*
    winter, pond life under, 36, *37*
Icosahedron, 261, 274
IgA, 836, *836*
IgD, 836, *836*
IgE, 836, *836*
IgG, 835–836, *836*, 843
IgM, 836, *836*
Ignarro, Louis, 22–23
Iltis, Hugh, 677
Imaginal discs, in *Drosophila*, 925, *925*
Imago, 925, *925*
Immigration, 610, 624

Immune response, cells in, *834*, 835
    mediation of, 835
    secondary, 834
    structure of cells and, 73
Immune system, 829
    antibodies and, 837–839
    attack on body, 842
    memory of, 833–834
    molecules and, 833–836
    of newborn mice, 826
    recognition of diversity by, 834
    specificity of, 833, 834–835
    toleration by, 834
Immunity, attributes of, 833
    inflammation and, 826–844
    nonspecific, 833
    specific, 833
Immunoglobulins, 835–836, 843
Imperfect fungi (Deuteromycetes), 470, 471, *472*, 475
Implantation, 894
    human development after, 916–918
Imprinting, 630–631, *631*, 639
In vitro fertilization, 903–904
Inbreeding, 358, 366
Incisors, 744
Inclusive fitness, 637
    altruism and, 637–638, 639, *639*
Incus, 871
Independent assortment, principle of, 198–200,
  199, 206
Individual variation, 348
Individualistic communities, 586–587, *587*, 600
Induced abortion, 903, 905
Induced fit, 112, *113*, 117
Induction, 253, 256
Industrial melanism, 362
Infant, nursing of, 898
Infections, causing diarrhea, 750
    fungal. *See* Fungal infections
    resistance to, inflammation and, 832–833
Infertility, female, 903–904
    male, 903
    treatment of, 903–904
Inflammation, 829, 832, 842
    immunity and, 826–844
    resistance to infection and, 832–833
Influenza virus, 263, *264*
Ingram, Vernon, 220
Inhalation, 798, 806
Inheritance, of acquired characteristics, 347–348, 366
    adaptive traits in, 341
    blending model of, 182, *182*
    blending of, 346, *346*, 347, 366
    chromosomal theory of, 180–185
      acceptance of, 182–183
      discovery of, 180–181
      evidence for, 199–204
      Sutton-Boveri, 199, 204
    laws of, 185
    particulate, 346, *346*
    Sutton's six rules of, 200t
Inhibin, 893
Inhibiting hormones, 762, 768
Inhibition, 682, 692
Inhibitor(s), allosteric, *116*, 117
    competitive, 116, *116*
    definition of, 115
    noncompetitive, 116, *116*
    steric, 115–116, *116*, 117
Inhibitory postsynaptic potential, 856, 861

Initiation, 246, 256, *248–249*
Innate behavior, *627*, 627–630, 639
Inner cell mass, 910–911, 927
Inositol triphosphate, 765
Insect(s), 520
    breathing by, 801, *801*
    circulatory system of, *779*
    excretory system of, 824, *824*
    genetically engineered crops and, 294
    intricate societies of, 637
    metamorphosis in, 758, *758*, 925, *925*
    oxygen distribution by, 801
Insect repellent, mustard oil glycosides as, 658, *658*, 659
Inspiration, 798, 806
Institute for Genomic Research (TIGR), 311
Insulin, 99, 758, 768
    production of, 759
Insulin-dependent diabetes mellitus, 759
Integrated community, 586, 600
Integrated replication, 264–265, 274
Integration, 870, 880
Integrin, 913
Integuments, 679, 692
Intercalated discs, 784, *784*
Interferons, 833, 842
Interleukin-1, 833, 843
Interleukin-6, 833, 843
Intermediate filaments, 87, *87*
Intermediate forms, 334, 342
Interneuron(s), 858, 876
    facilitating, 859–860, 861
    inhibitory, 858, 861
Internode, 653, 658, 685
Interphase, 170
Interstitial cells, 887, 904
Intertidal zone, 578, *579*, 582
Intestine(s), microorganisms inhabiting, 436
    small, 747–749, 752
        absorption in, 750
        cell specialization in, 747, *747*, *748*
        epithelial cells of, 747–749, *749*
Intrauterine devices, 902, 905
Intrinsic rate of natural increase, 609, 624
Introns, 242, 254
Invagination, 912
Invertebrates, brain in, 864–865
    digestion in, 739
    marine, salt and water balance in, 819
    nervous system in, 866
    water, salts, and wastes in, 823–824
Inviability, hybrid, 374, 396
Involuntary muscle, 726, 733
Ion, 30
Ionic bond(s), 30, 44
    atoms in, 33–34
Ionizing radiation, 29
Ionotropic receptors, 857
Iridium anomaly, 382
Iris, 873, 880
Island biogeography, theory of, 594, 600
Islands, adaptive radiation on, 376
    diversity on, 595, *595*
        defaunation and, 594t
Islets of Langerhans, 759
Isolation, behavioral, 373, 396
    ecological, 373, 396
    gametic, 373, 396
    geographic, 376–378, *377*
        speciation and, 374

mechanical, 373, 396
reproductive, *373*, 373–374
    by polyploidy, 374, 396
    temporal, 373, 396
Isoproterenol, 764
Isotonic solution, 93, *93*, 101, 811
Isotope(s), 28, 44, 329
    decay of, 28, 29
Iwanowski, Dmitri, 260

J

J-shaped curve, *608–609*, 609, 610, *610*
Jack rabbit, blacktail, *715*
Jackson, John Hughlings, 863
Jacob, François, 251, 259
Japanese puffer fish, 851, *852*
Jasmonic acid, 707
Jawless fishes, 529, *529*
Jaws, mammalian, 744, *744*
Jellyfish, 504
    nerve nets in, 865, *865*
Jenkin, Fleeming, 347, 366
Jenks, Susan, 419
Joints, 722, 723, 732
Joliot, Frèdèric, 29
Joliot-Curie, Irène, 29
Judson, Horace, 215, 216
Junctions, adhering, 98, *99*, 101
    gap, 98, *99*, 101
    tight, 98, *99*, 101
Jurassic Park, 710, 711
Juvenile diabetes mellitus, 759
Juvenile hormone, 758

K

Kaback, Michael, 311
Kamen, Martin, 153
Kandel, Eric, 860
Kangaroo, red, 541, *541*
    western grey, *322*
Kangaroo rat, water balance in, 812–813
Keeling, C.D., 556
Keilin, David, 131
Kelps, 460, *460*
Kendrew, John, 208, 217
Keratin, 63, 87
Kettlewell, H.B.D., 361–362
Keystone species, 593, 600
Khorana, Gobind, 237, 238
Khrushchev, Nikita, 345
Kidney(s), adaptable, in turtles and tortoises, *810*
    anatomy of, 814, *814*, *815*, 824
    artificial, 818, *818*
    blood filtration by, 814–817, 825
    of freshwater fish, 820
    functions of, 810, 814–817, 825
    of nonmammalian vertebrates, environment and, 820–821
    of saltwater fish, 820
    sodium consumption and, 809
    urine concentration by, 819–820
Kidney dialysis, 818, *818*
Kin selection, 637, 639
Kinesins, 86
Kinetic energy, from work, 104, 117
Kinetics, application of, 103–104
    definition of, 103, 117
Kinetochore, 172, *173*

King, Mary Claire, 302, 303
King, Thomas, 919
Kingdom(s), 16
    in hierarchy of classification, 423, 431
Knee jerk reflex, 858, *859*
Knee joint, 723, *723*
Knockouts, 291, *291*
Koala, 541
Koch, Robert, 435
Koch's postulates, 435, *435*, 447
Kornberg, Arthur, 227
Krebs, Hans, 127, 129
Krebs cycle. *See* Citric acid cycle
Krill, *520*
K/T boundary, 382
Kuru, 46
Kwashiorkor, 737, 752
Kyoto Protocol, 556

L

La Niña, 579–580
Labeled line, 871
Labia majora, 889–890, 904
Labia minora, 889, 904
lac operon, 252–253, *252*, *253*, 256
lac repressor, 253, 256
Lacks, Henrietta, 162, 163, *163*
Lactate, in oxygen absence, 126–127
Lactation, 897–900
Lake Michigan, sand dune succession and, 585, *586*
Lake(s), 580–581
    temperature turnover in, 581, *581*
Lamarck, Jean Baptiste, 325, 342
Lamarckism, 348
Lampreys, 529
Land bridge, 380, *380*
Larkspur (Delphinium), *373*, *373*
Larva(ae), 758, 924–925
Larynx, 796, 806
Lateral geniculate nucleus, 876, *876*
Lateral line system, 873
Lateral meristems, 646, 658
Latimeria, 532
Latitude, temperature and, 565, *565*
Lavoisier, Antoine, 118, 142
Leaf(ves), arrangements of, 654–655, *655*
    autumn, *149*
    diversity of, 656, *656*
    functions of, 654–656
    water evaporation from, 664
    water path in, 654–655, *655*
    water transport in, 662
Leaf primordium, 687, 692
Leakey, Mary, 394
Learning, by animals, 627, 630–631, 639
    sensitive period for, 630
    associative, 859
    by humans, 869
    neural networks and, 859
    nonassociative, 859
    by sea hare, 860, 861
    sensitive period for, 630, 639
Leclerc, George-Louis, 322
Leder, Philip, 237
Lederberg, Joshua and Esther, 361
Leeches, 516–517
Legumes, scarification in, 684
Lemmings, population cycles of, 612
Length, measurement of, Metric System, Appendix B

Lens, 873, 880
Leptin, 736
Lesch-Nyhan syndrome, 631
"Let-down" reflex, 899
Leukocytes, 772, 773, *773*, 787, 829, 843
  types of, 829t
Levi-Montalcini, Rita, 844–846, *845*
Lewis, Sinclair, *Arrowsmith* by, 265
Lianas, 576, 582
Lichens, 573, 587
  description of, 473, *473*
  role of fungi in, 472–474, 475
Liebig, Justus von, 118
Life, chemical foundations of, 20–43
  continuity of, 161
  early, effects on environment, 412–413
  origin of, 399–404, *402*
    theories of, 401–402
  unity and diversity of, 2–20
Life history, survivorship and age distributions and,
  613, 613–616, *615*
Life history traits, 612, 625
Ligaments, 723, *724*, 732
Ligand-gated channel, 856
Ligand-gated ion channel, 857, 861
Ligase, DNA, 282–284, 295
Light, absorption of, 146–147, *147*
  definition of, 145
  photons in, 145, *145*
  plant productivity and, 157
  visible, 146, *147*, 159
  wavelength of, *145*, 145–146, *147*, 159
  waves of, 145, *145*
Light-independent reactions, *144*, 145, 159
Light microscope, 72, *73*
Lignin, 482, 645, 658
Limbic system, 879, 880
Limiting resource, 591, 600
Limnology, 580, 582
Linkage groups, 202
Linkage map, 204, 205
Linnaeus, Carolus, 332, 342, 422, 423, 476
  classification of plants by, 476–477, *477*, 493
  classification system of, 422–423
  evolution and, 322
Lions, 634, *635*
Lipases, 739
Lipid-soluble signals, 762, *763*, 768
Lipids, 52, 54–55t, 68
  processing of, 750
Liposomes, 409, 415
  glucose transport and, 94–95
Listeria monocytogenes, 273, *273*
Litter, forest, 573, *574*
Liver, cirrhosis of, 136
Liverworts, 483
Lizard(s) (Squamata), 534, *536*
  chihuahua whiptail, 883
  crested dragon, *536*
Loams, 669, 674
Lobe-finned fishes, 531, *532*
Lobster, digestion in, 742
Locomotion, animal, style of, as influence on shape,
  718–721
  in snakes, 719, *719*, *720*
Locus, 204
Locusts, plague of, *611*
Loewi, Otto, 7
Logan, Graham, 509
Logistic growth equation, 609, 624

Long-term memories, 869, 882
Loop of Henle, 817, 825
Lorenz, Konrad, 630
Loudness, 871, 880
Lovelock, James, 544, 545, 560, 585–586
Lubber grasshopper, 180, *181*
Lumen, 781, 787
  of endoplasmic reticulum, 82, 100
  of gut, 743, 752
  ion content of, tubules and, 819
Lunar month, 894
Lung(s), *794*, 795, 806
  air entering, path of, 796
  cancer of, *789*, 789–791, *791*
  damage due to smoking, 799–800
  gas exchange in, 796, 801, *801*
  gas in, 794
  of land animals, 795
  of mammals and birds, 715
Luria, Salvador, 223, 224, *224*
Luteinizing hormone, 762, 892, 893, 894, *894*,
  895, 904
Lycophytes, 487
Lycopods, 487
Lyell, Charles, 324, 325–326, 327
Lymph, 781, 787
Lymph nodes, 781, 787
Lymphatic system, 780, *780*, 787
Lymphocyte library, 837–838, 843
Lymphocytes, 780, 787, 829, 835
Lysenko, Trofim, 344–345
Lysergic acid diethylamide (LSD), 471
Lysis, 264, 274
Lysogenic cycle, 268, 274
Lysosomes, formation of, 83
  functions of, 83, *84*
Lysozyme, 830, 843
Lytic cycle, 268, *268*, 274
Lytic viruses, *263*, 264, 274

M

MacArthur, Robert H., 591, 593–594
MacArthur-Wilson theory of island biogeography,
  594
Mach, Ernst, 100
MacLeod, Colin, 222
Macrocystis, 460
Macroevolution, 350, 366, 368–397
  definition of, 368–397
Macromolecules, 49, 68
  extraction of energy from, by heterotrophs, 121
Macrophage(s), 799, 827, 833, *833*, *834*, 835
Mad cow disease, 47
Maggots, 398–399
Magnification, 72
Magnolia, 482
Maidenhair tree, 490
Major histocompatibility complex, Class I proteins
  of, 841, 843
    Class II proteins of, 841, 843
    T cells and, 840
Malaria, 463, *463*
  sickle cell disease and, *365*, 365–366
Male competition, 364, *364*, 367, 633
Malleus, 871
Malnourishment, 736–737, 752
  causes of, 737–739
Malpighi, Marcello, 670
Malpighian tubes, 517, 521, 824, *824*

Malthus, Thomas, 326
Mammalia, 322, *536*
Mammals, 539–541
  birds, and reptiles, cladogram for, 428, *429*
  breathing by, 796–798
  central nervous system of, 866, *867*
  gamete formation by, 883–889, 905
  gastrulation in, 913–914, *914*
  marine, water for, 812
  marsupial, 380
  placental, 380, 537, 538
  teeth of, 539–540
  transgenic, production of, *290*, 290–291
Mammary glands, 539, 540
Manglesdorf, Paul, 676
Mangold, Hilde, 921, 923
Mantle, 511, 521
Mantle cavity, 511, 521
Map unit, 204
Marathon, 1995 World women's, 125, *125*
Margulis, Lynn, 414, 545, 585
Marine birds, kidneys of, 821
Marine invertebrates, salt and water balance in, 819
Marine mammals, water for, 812
Markel, Michael, 557
Marrella, 498
Mars, life on, 544
Marsh, Charles, 423
Marshall, Barry, 2, *3*, 4
Marshes, 580
Marsupials, 537, 538
  Australian, 380, *381*
Martin, Gail, 307
  and growth factor gene, *311*, 307
Masiarz, Frank, 47
Mass, 26
  measurement of, Metric System, Appendix B
Mast cells, 832, *832*, 842
Mastophora, 519
Maternal mRNA, 908, 929
Mating, assortive, 358, 366
  nonrandom, and allele frequencies, 358
Mating rituals, 634, *634*
Matrix, 499, 713
Matter, definition of, 23
Matthei, Johann Heinrich, 234–236
  and Nirenberg, experiments by, 235–236, *236*
Matthew, Andrew, 808, 809
Matzinger, Polly C.E., 826, 827, 828
Mayer, Julius, 145
Mayr, Ernst, 370, 371, 386, 387
McCarty, Maclyn, 222
McClintock, Barbara, 203–204, 258–259, *259*,
  274, 278
McClung, Clarence, 180
Mean, 11, 20
  and standard deviation, 12
Measure, units of, Appendix B
Mech, David, 418–419
Mechanical defenses, plant, 656, *657*
Mechanical isolation, 373, 396
Medawar, Peter, 827, 828
Medulla, renal, 814–817, 825
Medulla oblongata, 866, 880
Medusa(ae), 504, *504*, 521
Megaphylls, 486, 494
Megaspores, 679, 692
Meiosis, 165, 183, 186
  distribution of chromosomes to gametes by,
    188–192, *191*

fertilization and, 188, *188*
homologous chromosomes in, 188–190, 205
Meiosis I, phases of, 188, *189*, 204
Meiosis II, phases of, 188, *189*, 204
Melanism, industrial, 362
Melanocyte-stimulating hormone, 762
Membrane envelope, 261, 274
Membrane fusion, 96–97, *97*
Membranes, cellular, functions of, 87
  interaction with external environment, 95–99
  plasma. *See* Plasma membrane(s)
  regulation of, space enclosed by, 91–96, *92*
  selectively permeable, 91, 100
  sodium ion permeability in, 851
  water moving across, 91–93
    direction of, 92–93
Memory, 868, 880
  declarative, 868
  of immune system, 833–834
  long-term, 868, 880
  reflexive, 868
  selection theory and, *838*, 838–839, *839*
  short-term, 868, 880
Memory cells, 838, 844
Mendel, Gregor, 19, 180, 183, 195, *195*, 217, 258, 348, 349
  genetics discoveries of, 195–199
    cross-pollinating and, 195, *196*, 197, *197*, 206
    principle of independent assortment and, 198–200, *199*, 206
    principle of segregation and, 197–198, *198*, 206
    traits of pea plants, 195–197, *196*
Mendelian genetics, rediscovery of, 349
Meninges, 865
Meningitis, 865
Menstrual cycle, 894
  hormonal regulation of, 896
  irregular, 894
Menstrual extractions, 903, 905
Menstruation, 894, 896, 904–905
Mercurialis annua, 488–489
Mercury barometer, 792, *793*
Meristem(s), 646
  apical, 646, 684, *685*, 692
  elongation and differentiation of, 686
  lateral, 646, 658
    secondary growth and, 689–691
  plant production by, 684, *685*, 692
Merozoite, 463
Meselson, Matthew, 236
Mesenchyme, in sponges, 503, 520
Mesenteries, 747
Mesoderm, 496, 520, 908, *908*, 927
Mesoglea, 504
Mesophyll, 656, 659, 672, *673*
  leaf, 664
Mesozoic Era, 331, 342, 534, 540
Messenger RNA. *See* RNA, messenger (mRNA)
Metabolic pathways, evolution of, 409, *410*
Metabolic rate, size of animals and, 714, *714, 715*
Metabolism, 120, 138
  of alcohol, 136
  of ethanol, 136
  food molecules entering, 129, *130*
  pathways of, 129, *129*
  regulation of, by cell or organism, 112–114
Metabotropic receptors, 857
Metallurgy, 24–25

Metameres, 514, 521
Metamorphosis, 517, *518*, 521, 533, *534*, 542, 909, *924*, 924–925, *925*, 928
  complete, 925, *925*, 928
  gradual, 925, 928
  in insects and amphibians, 758, *758*
Metaphase, chromosomes during, 172
  in meiosis, 188, 190
Metaphase plate, 172
Metapopulation, 595, *596*, 600
Metatherians, 541, 542
Meteorite, 403, *408*
Methane, 544
Methanococcus jannaschii, 447
Methanogens, 443, *445*, 450
Metric System, Appendix B
Mice, allopatric speciation of, *375*
  coat color in, genes and, *306*, 306–307
  human growth hormone in, 290, *291*
  knockin, 291
  knockout, 291, *291*
  newborn, immune system of, 826
  transgenic, 290, *291*
Michel, Helen, 382
Michod, Richard E., 882–883
Microelectrodes, 850
Microevolution, 344–367
  definition of, 349, 366
Microfilaments, 86–87
Microphylls, 486, 494
Micropyle, 679
Microsatellites, 420
Microscope(s), atomic force, 26
  compound, 72, *73*
  early history of, 70, 71
  electron, 72–73, *73*
  light, 72, *73*
  study of cells using, 72–73
Microspheres, proteinoid, 409, *409*, 415
Microspores, 678, 692
Microtektites, 382
Microtubule organizing center, 86, *88*, 100
Microtubules, 86, *87, 88*, 100, 172
Microvilli, 747, *748*, 752, 817–819, *818*
Midbrain, 866, 880
Midgley, Julian, 808, 809
Midgut, 743, 747–749
Midpiece, 887, 904
Mifepristone, 903, 905
Milk, human, 898
Milk ejection, 898, 905
Miller, Stanley, experiment by, 406, *406, 407*
Mimicry, 589, 600
Mineral deficiencies, 737–738
Mineralocorticoids, 762
Minerals, from animals, for plants, 670
  in plant growth, from soil, 669
Mini-pills, 903
Miscarriage, 898, 918, 928
Missense mutation, 245
Mitchell, Peter, 131
Mites, 519
Mitochondria, 84, 100, 127, 138
  DNA in, *272*, 272–273
  as endosymbiont, 413–414
  generation of proton gradient by, 133
  genetic code and, 238
  genetic code in, 273
  genetic code of, 413, *413*
  structure of, 85, *85*

Mitochondrial matrix, 133, 138
  pumping protons out of, *133*, 133–134
Mitochondria, transmission electron microscope to assess, 84, *85*
Mitosis, 165, 168, 170, 175, 188
  movement of chromosomes during, 173
  stages of, 171, *171*, 172–173
  triggering in, 177–178
Mitotic spindle, 172
Mittelschmerz, 900–901, 905
Mixture(s), 24, *25*, 44
Mobile genes. *See* Gene(s), mobile
Mockingbirds, 384
Modal action pattern, 628, *628*, 639
Model, 5, 20
Modern synthesis, 349, 366
Molars, 744
Molds, 468, 475
  bread, 468, *469*
  fungi as, 466
Mole, 42
Molecular biology, central dogma of, 238, 239, *239*, 254
  comparative, 336–337
Molecular clocks, 337
Molecular genetics, 183, 204
Molecular motors, 75, 100
Molecular weight, 26, 44
Molecule(s), amphipathic, 40, *41*, 44–45
  biological, 45–69
    organisms building, 49–53
    size of, 49
  bonding in, 30–35
  definition of, 24, *25*
  forces holding together, 35
  hydrophilic, 35
  hydrophobic, 35
  immune system and, 833–837
  in living organisms, 49
  motion of, rate of chemical reactions and, 110, *111*
  movement through plasma membrane, 93–95
  nonpolar, 34, 35
  organic, 25–26
  polar, 34, *34*
  shape of, 32, 32–33
  signaling, 254
Mollusks, 511–514, *512–513*, 521
  circulatory system of, *779*
Molting, 517, 521
Monarch butterfly, 589, *589*
Moncada, Salvador, 22–23
Monkey(s), 389, *389*
  socialization of, 631
Monocots, 479, 649–651, *650*, 658, 680, 692
  germination in, 682–683, *683*
Monocotyledons, 649–651, *650*, 658
Monod, Jacques, 251, 259
Monophyletic taxon, 427, 431
Monoplacophora, 513
Monosaccharides, 59, 69
Monosomy, 231, 233
Monotremes, 541, *541*, 542
Monsters, hopeful, 386, *386*, 396
Morgan, Thomas Hunt, 200–202, *202*, 203, 217
Morning-after pill, 902, 905
Morphogenesis, 914, 928
Morphogens, 923, 928
Morris, Simon Conway, 496–497
Morula, 909, 927

Mosquito, malaria-carrying, 371
Moss(es), 483
    club, life cycle of, *479*
    sphagnum, *483*
Moth, peppered, *360*, 360–361
Motile, 455, 474
Motoneuron(s), 858
    L7, 859–860, 861
    of spinal cord, 866, 867, 880
Motor axis, 726, 733
Motor instructions, 865
Motor neurons. *See* Motoneuron(s)
Mount St. Helens eruption, 584, *585*
    succession following, 585, *585*, 596, 597
Mouse, cactus, 568
Mouth, 744, 752
    microorganisms inhabiting, 436
Mouth parts, of arthropods, 518, *519*
Movement, in digestive tract, 743
Moyle, Jennifer, 131
Mt. Whitney, 562, 563, 564, 566
Mucigel, 651, 686, 692
Mucosa, of small intestine, 747
Mucus, 744, 752
    in respiratory system, 799, 806
Mudpuppy (*Necturus*), 534
Müllerian ducts, 890
Müllerian mimicry, 589, *589*, 600
Mullis, Kary, 276–278, *277*
Multicellular organism, loss and replacement of cells
    by, 77
Multicellular organisms, communication of cells in,
    97–99, *98*
Münch, Ernst, 672
Murad, Ferid, 22
Murex, 507
Murray, R.G.E., 447
Muscarine, 856
Muscarinic receptors, 857
Muscle(s), antagonist, 722, *722*
    antagonistic pairs of, 726, 733
    cardiac, 713, 726, *728*, 733, 784
    in movement of animals, 721–730
    reductionist study of, *8*, 8–9, *9*
    skeletal. *See* Skeletal muscle
    smooth, 713, 726, *728*, 733
    striated, contraction of, *729*, 729–730
Muscle fibers, glycolytic, 728, 733
    oxidative, 728, 733
Muscle tissue, 713
Mushrooms, 469–470
Mustard oil glycosides, 658
Mutagens, 230–231, 233
Mutation(s), allele frequency and, 355–356
    causes of, 230–231
    chromosomal, 231
    conditional, 245
    definition of, 233
    frameshift, 245
    frequency of, 231
    genetic variation from, 229–231
    missense, 245
    new, random drift and, 357
    nonsense, 245, *245*
    point, 231, 233
    rate of, 231, 233, 352
    reading of, by ribosomes, 245
    silent, 245
    spread through population, 352–353
    suppressor, 245

Mutation hot spots, 231, 233
Mutualism, 590, 600
Mycelium, 466, *467*, 475
Mycorrhizae, 473–474, *474*, 475
Myelin, 853, 861
Myelin sheath, 847, 853, 861
Myocardium, 776, 786
Myofibrils, 728, 733
Myoglobin, 728, 733
Myopia, 875, *875*
Myosin, 729, 733
Myxobacterium, *440*, 450
Myxomycota, 465, 475

N
NAD$^+$, 123, 138
NADH, 123, 138
NADPH, in glucose synthesis, 153
    light reactions and, 151
Naming, of organisms, 422
    of species, 423
National Aeronautics and Space Administration
    (NASA), 544
National Institutes of Health (NIH), 308
Natural killer cells, 835, 843
Natural selection, 360, *360*, 366, 633
    altruism and, 636–637
    analysis of, 341
    Darwin's theory of, 19, 20
    directional, *360*, 360–361, 366
    frequency of traits and, 358–362, *359, 360*
    genetic variation and, 353, 365–366
    preexisting variants and, 361
    pushing of populations by, 362–366
    stabilizing, *362, 363*, 366
Nautilus, 513, *514*
Neanderthals, 395–396, 397
    flute of, *395*
Nearsightedness, 874, 880
Necturus (mudpuppy), 534
Negative feedback, 730–731, 733, 762, 768
Negative regulators, 253, 256
Neljubov, Dimitry, 706
Nelson-Rees, Walter, 162
Nematocysts, 504, 521
Nematodes (roundworms), 510, 521
    apoptosis in, 917
Nematomorpha, 510
Nemertea (ribbon worms), 508, *508*, 521
Nephridia, 511, 514, 521, 824, 825
Nephron(s), 816, 825
    filtration, reabsorption, and secretion by, 817,
        817, 825
    solutes in, 819, *820*
Neritic zone, 578, 582
Nerve cells, *846*, 846–853, *847*
    in digestion, 751
    information transmission by, 847–848
    Levi-Montalcini's work on, 844–846
    organelles in, 846, *847*
    stimulation of, cellular response related to, 847,
        849
    threshold response of, 847
    types of, 846–847, *848*
    voltage of, sodium channels and, *849*, 849–
        850
Nerve cord, dorsal hollow, 527, 542
Nerve growth factor, discovery of, 846
    functions of, 846
Nerve nets, 864, *864*, 880

Nerves, afferent, 868, 880
    blood flow regulation by, 785–786
    efferent, 868, 880
    parasympathetic, 726, 869, 880
    sense organs, and brain, 865
    sympathetic, 726, 869, 880
Nervous system, autonomic, 726, 733, 869, *869*
    cells of, 844–861
    central, mammalian, 867, *868*
        in vertebrates, organization of, *865*, 865–
            868, 880
    formation of, 914–916
    information moving through, 869–870
    in invertebrates, 866
    parasympathetic, 726
        heart rate and, 785, 787
    peripheral, 868, 880
        organization of, in vertebrates, 870
    response to environment, 865–871
    and sense organs, 862–883
    signaling by, 757
    somatic, 869
    sympathetic, 726
        heart rate and, 785, 787
Nervous tissue, 713, 732
Neural crest cells, 914, 928
Neural networks, bahavior and, 857
    experience and, 859
    learning and, 859
Neural plate, 914, 928
Neural tube, 866
    formation of, 914, *915*, 927–928
Neurofibromatosis, 304, *305*
Neurons, 713, 732, 847, 861
    of brain, 864
    communication between, 853–857, *854*
    electrical pulse generation by, 848–850
    generation of action potential by, 850–851
    maintenance of resting potential by, 849–850
    motor. *See* Motoneuron(s)
    postsynaptic, 853, 861
    presynaptic, 853, 861
    reflex mediation by, 858–859
    relay, 869, 870, 880
    sensory, 858, 859, 861
    trigger, 858
Neuroscience, 627
Neurospora, 218, *219, 220*
Neurotransmitters, 757, 766, 768, 785, 787, 854,
    855
    receptors for, 856–857
    synapse clearing of, 857
Neurulation, 914, 928
Neutrons, 28, 44
Neutrophils, 832–833, 842
Newts, cellular potential experiments on, 919
Niche, 591
    fundamental, 591, 600
    realized, 591, 600
Nicholson, Garth, 91
Nicotine, 856
    addiction to, 793
    in tobacco smoke, 791, 792, 799
Nicotinic acetylcholine receptor, 857
Nicotinic receptors, 857
Nightshade, 857
Nirenberg, Marshall, *235*, 235–236, 238
    and Matthei, experiments by, 235–236, *236*
Nitric oxide, 22, 23, 786
Nitrification, 558, 560

Nitrifying bacteria, 558, 560
Nitrogen, fixed, 557–558
    sources of, 559, *559*
  supply of, 557–559
Nitrogen cycle, 557, *558*
  alteration by humans, 559
Nitrogen fixation, 448
Nitrogen-fixing aerobic bacteria, 446, *446*, 448, 450
Nitrogenous bases, 61, *62*, 69
Nitrogenous wastes, 813, *814*, 824
Nitroglycerin, 22
Nobel, Alfred, 22
Node, 653, 685
Nodes of Ranvier, 853, 861
Non-insulin-dependent diabetes mellitus, 759
Nonassociative learning, 859
Noncompetitive inhibitor, 116, *116*
Nonpolar compounds, small, 53–59
Nonsense mutation, 245, *245*
Norepinephrine, 785, 786, 787, 869
Norplant, 903
Northern blotting, 288, 295
Nose, microorganisms inhabiting, 436
Notochord, 527, 542, 914, 928
Nourishment, animal, from food, 734–753
Nowack, Ronald, 419, 421
Nübler-Jung, Katharina, 523
Nucellus, 487, 494, 679, 692
Nuclear envelope, 80, *80*, 100
Nuclear pores, 80, 100
Nuclear reactions, 24, 44
Nucleases, 739
Nucleic acid(s), linkage of nucleotides of, 61–62,
    63, 64
    linking of nucleotides in, 61–62
    proteins and, 52, 53
Nucleoid, 74, *74*, 100, 438, 447
Nucleosome, 174, *174*
Nucleotides, 52, 54–55t, 68
    DNA, 211, *213*
    linkage of, to form nucleic acids, 61–62
    of nucleic acid, linkage of, 61–62, *63, 64*
Nucleus, of atoms, 26, 44
    of cells, 74, *74*, 100, 79–80, *80*
    and characteristics of organisms, 166, *167*
    differentiated, development and, 920
    of eukaryotes, *74*
Numa, Shosaku, 851–852
Nursing, 898
    ovulation and, 900
Nutriceuticals, 291
Nutrients, from soil, 669t
Nymphs, 925, 928
Nystatin, 454, 468

O

Objective, 72
Obligate aerobes, 446, 450
Obligate anaerobes, 126, 446, 450
Observation, 4–5
  models of, 5, *5*
Ocean, ecosystems of, 578, 582
  fertilization of, 557
  zones of, 578, *579*, 582
Oceanic zone, 578, *579*, 582
Oceanography, 578, 582
Octane, *50*
Octet rule, 30
Odontogriphus, 498

Odum, H.T., 619
Oils, 35
Okazaki, Reiji, 228
Okazaki fragments, 228, 233
Oleic acid, properties of, 38t
    water, and ethanol, compared, 38t, 39
Olfaction, 870, 878–879, 880
Olfactory bulb, 878, *879*, 880
Olfactory receptors, 880, *881*, 882
Oligochaetes, 515, *516*
Oligodendrocytes, 847
Oligonucleotides, 285, 294
Oligosaccharides, 59, 69
Oligotrophic lakes, 581, 582
Olson, Carl, 598
Ommatidia, 518
Omnibacteria, *446*, 448, 450
Omnivores, 548–549, 560, 736, 752
Oncogenes, 267, 274
    proto-, 267, 274
    viral, 267, 274
Oocysts, 463
Oocytes, passage of, 885–886
    primary, 883–884, *885*, 904
    production of, 894, *894*
    secondary, 885, 888, 904
Oogenesis, 883, *884*, 904
Oogonia, 883–884
Oomycota, 465, 475
Opabinia, 498
Oparin, Alexander Ivanovich, 404, 408
Operator, 253, 256
Operculum, 511, 521
Operon, *lac*, 251, *252*, 255
Opiates, 858
    natural, 858
Opposite leaves, 654, *655*
Opsins, 875, 880
Optic chiasm, 876, *876*
Optic nerve, 876, *876*, 880
Optimal foraging, 632, 633, 639
Orbitals, 30, 44
Order, in hierarchy of classification, 423, 431
Ordovician Period, 472, 644
Orexin, 736
Organ of Corti, 871, *872*, 880
Organelles, 15, 74–75, 100
    energy, DNA and RNA of, 273
    in nerve cells, 846, *847*
Organic chemistry, 49, 68
Organic molecules, definition of, 49
Organism(s), air-breathing, oxygen for, 796–801
    cellular structure of, 15, *15*
    characteristics of, 13, *14*
    classification of, phenetics in, 426–427
    complex, 906–930
    composed of cells, 70–71
    differences in, 16–19
    domains of, 16, *17*
    energy supply to, 118–139
    evolution of, 398–416
    groups of, numbers of, 430
      study of, 7–8
    individual, as unique, 7–10, *10*
    naming of, 422
      and classification of, 421–425
    self-regulation by, 13–15, 730–731
    transformation in, 16
    useful, design of, 281
Organs, 499, 713

formation of, in vertebrates, 909, 927
  systems of, 713, 732
Orgasm, 891, 904
Ornithine, 219
Oscilloscope, 850, *850*
Osmolarity, 811
Osmoreceptors, 821
Osmosis, 92, 101, 663, 664, 674, 811
Osmotic pressure, 92
Osteichthyes, 530, 531, 542
Osteoblasts, 725–726, 733
Osteoclasts, 726, 733
Osteoporosis, 726, 733
Osterholm, Michael, 432–433
Ostracoderms, 529, 542
Ostrich, 392
Ottoia, 498
Outbreeders, 358, 366
Oval window, 871
Ovarian cycle, hormonal regulation of, *894*, 894–
  896, *895*
Ovary(ies), 186, 204, 489, 491, 766
    of flowering plants, 679
    in mammals, 883
Overnourishment, 736
Overpopulation, human, 621–624
Overton, William R., 319
Oviducts, 883, 904
Ovulation, 885, 904
    nursing and, 900
Ovules, 488, 494, 679, 692
Ovum(a), 186, 204, 883, 884
    production of, mammalian, 884–885
    sperm reaching, 892
Owens, Kelly, 303
Oxaloacetate, 652
Oxidation, 122–123
Oxidative muscle fibers, 728, 733
Oxidative phosphorylation, 122, 138
Oxidizing agent, 122, 138
Oxygen, 24
    for air-breathing organisms, 796–801
    in animals, source of, 791–792
    atmospheric, 791–792
    carbon dioxide exchange for, 791–795
    carried by hemoglobin, 802, 803
    carried to tissues, 801–802
    delivery by blood, 782, 802, *802*
    delivery of, 786
      changing, breathing regulation in, 805
      at high altitudes, 805
    dissolved, for aquatic animals, 795
    distribution of, by insects, 801
    hemoglobin affinity for, 804–805
    movement of, 715, 782, *782*
Oxygen debt, 125
Oxygen dissociation curve, *803*, 804, *804*
Oxygen exchange, countercurrent, *717*, 717–718
Oxygen homeostasis, 805
Oxygen saturation, 803
Oxygen ship, *802*
Oxyhemoglobin, 802, *802*, 806
Oxytocin, 761, 898, 905
Ozone layer, 622, *624*
    chlorofluorocarbons and, 622, *624*, 625

P

P_{680}, 149, 150, 151, 159
P_{700}, 149, 150, 151, 159

Pacemaker, 784
    artificial, 784
Palade, George, 82–83
Paleontology, 323
Paleozoic Era, 330–331, 342, 339, 481
Palisade parenchyma, 655, 656, 659
Pancreas, 749–750, 752, 758, 759
Pancreatic islets, 759
Pangaea, 339, 339
Pap smear, cervical cancer and, 164
Papanicolaou, George N., 164
Paper chromatography, 153–154, 154
Papilio dardanus, 363, 363–364
Paracrine signals, 757, 768, 766, 840
Paramecium, 73, 73, 76, 76, 89, 462
    reproduction by, 185
Paranthropus, 394, 397
Parapatric speciation, 376
Paraphylatic taxon, 427, 431
Parapodia, 515, 521
Paraquat, 156
Parasaurolophus, 18
Parasites, 588, 600
Parasitism, 590, 600
Parasympathetic nerves, 726, 869, 880
Parasympathetic nervous system, 726
    heart rate and, 785, 787
Parathyroid hormone, 766
Parazoa, 499, 503, 520
Parenchyma, 646, 647, 658
    palisade, 655, 656, 659
    spongy, 655, 656, 659
Parental (P) generation, 194, 205
Park, Thomas, 591
Parkinson's disease, 866
Parrots, African, 631
Parthenogenesis, 883
Partial pressure, 792, 793, 806
Partitioning, resource, 591, 600
Parturition, 898, 905
Passive transport, 93–94, 95
Pasteur, Louis, 118–119, 260, 399, 415, 434
    experiment of, 399, 400
Pasteur, Marie, 119
Pathogens, sealed from circulatory system, 830–831
    skin as barrier to, 829–830
Pauling, Linus, 34, 208, 209, 220
Pauling, Peter, 213, 217
Pavlov, Ivan, 859
Pea plants, backcross of, 195, 197, 197
    cross pollinating of, 196
    Mendel on, 195–196, 196
    testcross of, 197, 197
    two-factor cross of, 198, 198
Pectoral girdle, 713
Pedigree analysis, 308–309, 315
Pelagic division, 578, 579, 582
Pelvic girdle, 713
Penetrance, gene, 304–305, 315
Penfield, Wilder, 869
Penguin, emperor, 534
Penicillin, 250, 280, 281, 437, 440
Penicillinase, 234
Penicillium, 471
Pepsin, 734–735, 745–746, 752
Peptic ulcers, 2
Peptidases, 749
Peptide bonds, 53, 64, 69
    production of peptide bonds by, 247

Peptide hormones, 764
Peptidoglycan, 438–439, 447
Peptids, 763
Perception, 871, 881
    sensory pathways and, 878–879
    sound as, 862
Perforation plate, 648, 659
Pergolide, 471
Pericardium, 776, 786
Pericycle, 652
Periderm, of flowering plant, 646, 658
Period, geologic, 330
Periodic Table of Elements, 30, 31, Appendix A
Peripheral nervous system, 868–869, 880
    organization of, in vertebrates, 868–869
Peristaltic waves, 745, 745, 752
Permafrost, 569, 582
Permian Period, 331, 536
Peromyscus eremicus, 568
Peroxisomes, 84, 84, 100
Pesticides, 857
Petals, 492, 494, 687, 692
Petiole, 654, 658–659
pH, of blood, 42–43
    changes in, resistance to, 43–44
    of solution, 42–43
pH scale, 41–42, 43, 45
Phaeophytes, 460, 460, 474–475
Phage. See Bacteriophage
Phagocytosis, 83, 84, 101, 741, 741
Pharyngeal slits, 527, 542
Pharynx, 744, 752, 796, 806
Phenetics, 426–427, 431
Phenolic compounds, 657
Phenomones, 756, 768
Phenotype(s), 353, 366
    definition of, 183, 204
    environment in, 183–184
    environmental and genetic influences on, 184
    and genotype, 183–185
        relationship between, 304–307
Pheromes, 629
Pheromones, 636, 639
Phloem, 485, 485, 494, 645, 658
    conducting elements of, 649, 649
    contents of, 671–672
    osmotic flow in, 672, 672
    sugar and water in, 672–674, 673
    translocation of, 672, 672
Phosphates, organic, 857
Phosphatidylinositol biphosphate, 765
Phosphodiester bonds, 61
Phosphoglycerate, 652
    three-carbon, 156–157, 158
Phospholipase C, 765
Phospholipid(s), 57, 58, 69
Phosphorylation, oxidative, 122, 138
    substrate-level, 131
Photoautotrophs, 446, 450
    algae as, 457
Photochemical reaction center, 148, 159
Photoheterotrophs, 446, 450
Photolyase, 535
Photon(s), 145, 145, 146, 159
Photoperiodism, 699, 700, 708
Photophosphorylation, 144, 145, 159
    cyclic, 149, 152, 159
    noncyclic, 149, 151, 159
Photoreceptor(s), 873, 876, 880
Photorespiration, 158, 159

Photosynthesis, 84–85, 100, 140–160, 143, 144, 159, 667, 669
    atoms and, 143–145, 144
    carbon dioxide in, 155–157
    chlorophyll α and, 150
    evolution of, 412–413
    factors affecting, 157–158
    herbicide interference with, 156
    light reactions of, 150, 152–153
    molecules in, 146–148, 147
    by plant, 142
    products of, plant transport of, 670–674
        sorting of, 153–154, 154
Photosystem I, 149, 159
    electron transport in, 153
    noncyclic photophosphorylation in, 150, 152
Photosystem II, 149, 159
    electron transport in, 150–151, 152
Phototropin, 699
Phototropism, 694, 695, 699, 708
    auxin in, 694–696, 696t, 708
    pigments in, 699
Phrenology, 862
Phyllobates, 588
Phylloxera, 660–661
Phylogenetic tree, 337, 342
Phylogeny, 427, 431
    of primates, 388
Phylum, in hierarchy of classification, 423, 431
Physalaemus, 633, 634
Physalia, 505
Physiology, 730, 733
Phytochrome, 700–702, 708
Phytoestrogenics, 754
Phytophthora infestans, 465
Phytoplankton, 457
Pigeons, artificial selection in, 340, 340
    learning by, 630
Pigments, 147, 159
    in phototropism, 699
Pika, 596
Pikaia, 498
Pike, 532
Pilobolus, 470
Pine, life cycle of, 488
Pinnae, 486
Pinocytosis, 96, 101, 741
Pioneer community, 597, 597t, 600
Pioneer species, 598, 600
Pistil, 679, 692
Pit pair, bordered, 647–648, 658
    simple, 647, 648, 658
Pitch, 871, 880
Pith, 649, 658
Pitohui bird, 588, 589
Pits, tracheid, 647, 647
Pituitary gland, 761, 761, 768
    anterior, 761, 762, 768, 892, 904
    posterior, 761, 768
Placenta, 380, 541, 804, 896, 898, 905, 911, 917, 927
Placental mammals, 380
Placoderms, 530, 542
Placozoa, 503
Planaria, 506
Plankton, 455, 455, 474, 578, 582
Plant(s), actions of herbicides on, 156
    adaptation to dry land, 478–481
    adaptation to life on land, advantages and disadvantages of, 478

alternation of generations in, 478, *479*
   evolution in, 478, 480–483, *482*
   fossil record of, 481
adaptive radiation and, 379
and air, 142, *142*
bending toward light. *See* Phototropism
broad-leafed, 479
C$_3$, 158, 159
C$_4$, 158, 159
   anatomy of, *652*, 652–653, *653*
carbon and oxygen obtained by, 142–145, *143*
cells in, cytokinesis in, 176, *176*
classification of, 476–477
day-neutral, 699
defense against herbivores, 656–658
defenses of, 588–589
definition of, 478, 494
desert, water conservation by, 668
of deserts, 567, *569*
dioecious, 687
divisions of, 479, 480t, 481, *482*
embryos of. *See* Embryo(s), flowering plants
environmental responses of, 697–702
      to gravity, 697–698, *698*
      hormones and, 702–707
      to season and daylight, 699–702, *700, 701*
first, characteristics of, 480
flowering, alternation of generations in, 678, *678*, 692
      anatomy of, *685*
      egg cell production by, 679
      fertilization in, 679–680, *680–681*
      groups of, 481
      growth and development of, 676–708
      pollen production by, 678, *679*
      roots of, growth of, 686, *686*
      seed formation in, 680–681, *682*
      sperm production by, 678
      structure of, 646–651
genetic engineering of, *293*, 293–294
growth and development in, regulation of, 694–708
growth in, secondary, 689–691, *690*, 692
herbaceous, 645
herbicide-resistant, 156
hormones of, 702–707, *708*
lacking materials for photosynthesis, 157
long-day, 699, 700, *700, 701, 701*
minerals from animals for, 670
monoecious, 687
mycorrhizae and, 473–474, *474*
nonvascular, 479, 483–484
photosynthesis by, *142*
photosystems of, 148–149
physical constraints on, 644–645
prevention of photorespiration by, 158
production of, by meristems, 684, *685*, 692
productivity of, factors limiting, 157
      light and, 157
      water and, 157–158
seed, reproduction by, 487, 494
short-day, 699, 700, *700, 701, 701*
structural and chemical adaptations of, 642–659
during succession, 597, 597t
sun as energy source for, 145–153
terrestrial, evolution of, 644–645
      water and minerals for, 651–656
transgenic, *293*, 293–294

transport of products of photosynthesis by, 670–674
transport system of, 485
true-breeding, 194
turgor pressure and, 93, 101
vascular, 479, 485–490
      lacking seeds, 486–487
      reproduction and dispersion of, 485
vascular and nonvascular, compared, 479–480
vascular evolution of, 644–645
water conservation by, 667
Plant cell, chloroplasts of, 84–85, *86*
Plant compounds, primary, 657
      secondary, 657
Plantae, 16, 20
      major divisions of, *482*
Plaque, dental, bacteria causing, *441*
Plasma, 770–771, 772, 787, 811, 831, 842
Plasma cells, 838, 843
Plasma membrane(s), 74, 90, *90*, 100
      fluid layers of, 91
      molecules in, 87–90
      movement of molecules through, 93–96
      proteins in, 88, 91
      as selectively permeable, 91, 100
      structure of, and functional properties of, 90–91
Plasma proteins, in nonspecific defense, 832
Plasmid(s), 260, 274
      recombinant, 266, 274
      replication by, 265
      Ti, 292, 295
Plasmin, 832, 842
Plasminogen, 832, 842
Plasmodesmata, 98, *98*, 101, 647, 672, *673*
Plasmodial slime molds, 465
Plasmodium, 462–463, *463*
Plasmodium falciparum, 463
Plasmodium, definition of, 465, 475
Plasmolysis, 93, 101
Plastids, 84–85, 84, *86*, 100
Plateau, 891, 904
Platelets, 772, *773*, 831, 842
Plates, 339
Plato, 321
      on classifications, 423–424
Platyhelminthes (flatworms), 506–507, *502*, 521
Platypus, 541, *541*
Platyrrhines, 389, *389*, 397
Pleiotropy, 306
Plumule, 680
Pluteus, 925
Pneumococcus, 221
Poison ivy, and cytotoxic T cells, 836, *836*
Poisons, as pollutants, 621–622
Polar body, 885, 904
Polar easterlies, 565, *566*, 582
Polarity, 34, *34*
Poliovirus, 261, 263
Pollen grains, 678, *679*
Pollen tube, 679–680, 692
Pollutants, biodegradable, 621
      nondegradable, 621
Pollution, air, 622
      overpopulation and, 621–624
      thermal, 621
      water, 621–622
Polychaete worms, 515, *516*
Polygenic traits, *306*, 306–307, 315

Polymerase chain reaction (PCR), 277, 295
      DNA multiplication and, 277, *278*, 286
Polymerization, in DNA replication, 227, *228*, 233
Polymorphism(s), 350, *350*
      balanced, *365*, 365–366
      restriction fragment length, 297–298, 300, *300*, 315
      single-nucleotide, 298
Polypeptide(s), address tags of, 247–249
      cellular translation of, into messenger RNA, 243–249
      domains of, 66
      linkage to aminoacids to form, 63–66
      recombinant DNA and, 284–285
      synthesis of, stimulation of, 234
      termination of, 249–250
Polypeptide alphabet, 237, *237*
Polypeptide chains, 835–836, 837, 843
Polypeptide genes, 219–221
Polyphyletic taxon, 427, 431
Polyplacophora, 513
Polyploid, 493
Polyploidy, 493
      reproductive isolation by, 374, 396
Polyp(s), 500, *500*, 517, 752
Polysaccharides, 52, *53*, 59, 60, *61*
      digestion of, 750
      formation from sugars, 59–61, 69
      use by organisms, 60–61
Polysome, 247, *249*, 256
Polyspermy, 892
Pond life, under winter ice, 36, *37*
Pongidae, 390, 397
Pongids, 390
Pons, 866, 880
Popcorn, 676
Population(s), 20, 546, *547*, 560
      age structure of, 614, *615*
      annual rate of increase of, 616, *616*, 625
      breeding between, allele frequencies and, 358, *359*
      cycling up and down of, 612
      density-dependent factors and, 610, *611*
      density-independent factors and, 610–611
      differences in, checking for, 11
      evolution of, 344–367
      extinctions of, 604–608
            deterministic factors and, 604–606
            and explosions of, 602–625
      genetic equilibrium in, 354–355
      genetic variability in, 353
      growth of, 608–612
      interactions of, 595–596
      large, random drift and, 358
      measurement of, 616
      natural growth of, 612, *612*
      and overpopulation, 621–624
      pushing of, by natural selection, 362–366
      sampling of, 10–11, *11*
      size of, limitation of, 610
Population momentum, 616–617, *617*, 625
Porifera (sponges), 502, 503, *503*
Portuguese man-of-war, 504, *505*, 545
Positional cloning, *309*, 309–310, 315
Positive feedback, 898, 905
Positive regulators, 253–254, 256
Positron emission tomography, 863
Postabsorptive state, 758, *760*, 768
Posterior air sacs, 800, *800*

Postsynaptic neuron, 853, 861
Posttranscriptional control, 251, 256
Posttranslational control, 251, 256
Postzygotic barriers, 373, 396
Potassium channel, *852*
Potential energy, in work, 104, *105*, 117
Prairie chicken, extinction of, 605, 606, *606*
Prebiotic evolution, 404–409, 415
Prebiotic synthesis, 405–407, 416
Precambrian Era, 330, 342, 380, 381
Precipitation, of major biomes, 567, 568t
Predators, 588, 600
Preformation, 907, *907*, 927
Pregnancy, hormone levels during, 897, *897*
    and lactation, 896–898
    length of, 896
    and parturition, 898, 905
    Rh factor and, 843
    trimesters of, 897, 905, 918, 928
Pregnancy tests, 897
Premolars, 744
Prenatal testing, 311–313
    aborting of fetus and, 312–313
    amniocentesis in, 312, *312*, 315
    chorionic villus sampling in, 312, *312*, 315
    test-tube babies and, 313
Pressure, exerted by gases, 792, *793*
    partial, 792, *793*, 806
Pressure flow hypothesis, 672, *672*, 674
Presynaptic neuron, 853, 861
Prezent, Isai, 344
Prezygotic barriers, 373, 396
Priestley, Joseph, 142, *143*
Primary auditory cortex, 872
Primary cell wall, 645, *646*, 658
Primary growth, 646, 658
Primary oocytes, 883–884, *885*, 904
Primary organizer, 921, *922*, 928
Primary plant compounds, 657
Primary productivity, 550
Primary visual cortex, *876*, 876–877
Primase, 227, 233
Primates, 322, 388, 397
    cladogram of, 427, *428*
    phylogeny of, 388
Primer, 227, 233
Primitive characters, 427, 431
Primitive gut, 911
Primitive streak, 913, *914*, 927
Primordial germ cells, 883–884, 904
Principle of independent assortment, 198–200, *199*, 206
Principle of segregation, 197–198, *198*, 206
Prions, 46–49
    scrapie and, 48, *48*
Prisoner's dilemma, 638
Probe, *286*, 286–287, 295
Proboscis, 508, 521
Producers, 548, 560
Productivity, 549
    of major biomes, 567, 568t
    primary, 550
Products, free energy of, 105, 115
Progeny, 197
Progesterone, 766
    functions of, 897
    in menstrual cycle, 895–896
    during pregnancy, 898
    in pregnancy, 896
Proglottids, 507, *508*, 521

Progymnosperms, 481
Prokaryotes, 430, 432–451
    adaptations by, 442–443
    biochemical pathways of, 443
    cell division in, 167–168, *168*
    characteristics of, 438
    classification of, 443–449, *444*
        DNA sequencing and, 448
    growth of, 442
    metabolic diversity of, 442
    negative regulation of, 252–253
    nucleoid in, 438, 450
    outer surfaces of, 441
    and people, 435–438
    positive regulation of, *253*, 253–254
    in regulation of gene expression, 251–253
    replication of, 439
    shapes of, 439, *439*
Prokaryotic cells, 15, 16, 20
Prolactin, 762, 898, 905
Prometaphase, 172
Promiscuous interactions, 68
Promoters, 241, 253, 254, 255, 256
Prophage, 268, 274
Prophase, in meiosis, 188–190
    in mitosis, 172
Proplastids, 85
Propranolol, 763, 764
Prostaglandins, 757, 766, 768, 769, 899, 906
Prostate gland, 888, 904
Protease(s), 271, 739
Protease inhibitors, 271
Proteasomes, 81, *81*, 100
Protein(s), cellular direction of, 250
    complementary, 737, *738*
    conformation of, *65*, 66, 69
    construction of, 66–68
    contractile, 728, *729*
    of cytoskeleton, 86
    in endoplasmic reticulum and Golgi complex, 82–83, *83*
    as enzymes, 63, 69
    fibrous, 65, 69
    folding of, 48, 49, 66–67, 68
    globular, 65, 67, 69
    growth control in cells and, 177
    microtubule-associated, 86
    and nucleic acids, 52, *53*
    peripheral, 88
    in plasma membranes, 88, 91
    structure of, *65*, 65–66, 67–68
        α helix, 65, 67, 69
        β pleated sheet, 65, 67, 69
        primary, 65, *65*, 69
        quaternary, 66, 69
        secondary, 65, *65*, 69
        tertiary, 65, 66, 69
    synthesis of, antibiotics and, 251, *251*
        steps in, 247, *249*
    transmembrane, 88, *90*
Protein deficiency, 737
Protein-folding problem, 66, 68
Protein kinases, 764, 768
Protein sequences, DNA and, criteria for classification of, 429–430
Proteinoid microspheres, 409, *409*, 415
Proteinoids, 408, 415
Proterenol, 763
Prothoracic glands, 758
Prothoracicotropin, 758

Prothrombin, 831, 842
Protista, 16, 20, *456*
Protists, 427, 454
    algae, 455, 457
    characteristics of, 454–455, 474
    chlorophylls in, 459, 459t
    chlorophytes, 461, *461*
    dinoflagellates, 457–458, *458*
    diversity of, *454*, 455, *455*, *456*
    funguslike, 464–465
    and multicellular fungi, classification of, 452–475
    protozoa. *See* Protozoa
    reproduction in, 456
    rhodophytes, 460
    size of, 76
Proto-oncogenes, 267, 274
Protobionts, 408, 415
Proton gradient, 131, 138
    ATP synthesis and, 135–136, *137*
    generation by mitochondria, 133
Proton pumping, 135
Protons, 28, 40, *42*, 44
Protoplasm, 71, 74–75, 100
Protostomes, 496–521
    coelomates, 510–520
Protosun, 404–405
Prototherians, 537, *537*, 538
Protozoa, 462–464, 474, 475
    characteristics of, 462
    ciliophora, 464, *464*, 475
    foraminifera, 462, *462*, 475
    as protists, 462
    rhizopoda, 462, 475
    sporozoans, 462–463, 475
    zoomastigina, 464, 475
Proust, Marcel, 870–871
Provirus, 268, 274
Prusiner, Stanley, 46, *47*, 47–48, 66
Pseudocoelomates, 501
Pseudocoel, in aschelminths, 509, 520
Pseudomonads, 448, 450
Pseudopodia, 455, 475
Psilotophytes, 487
Psilotum, 487
Psychoactive drugs, action on chemical synapses, 857
Psychology, experimental, 626–627, 639
Pterophytes (ferns), 486–487
    life cycle of, *486*
    reproduction in, 486
Pterosaur, *322*
Puffer fish, 851, *852*
Pulmonary artery, 776, 786
Pulmonary circulation, 776, 786
Pulmonary veins, 776, 786
Pump, of neurons, 848, 861
Punctuated equilibrium, 387, *387*, 397
Punnett, Reginald, 195
Punnett square, 195, *195*, 205
Pupa(ae), 758, 925, 928
Pupil, 873, 880
Pure substances, 23
Purines, 61, 69, 210, *211*, 233
Pyramid, of biomass, 550, *550*, 560
    inverted, of biomass, 550, *551*
Pyramid of energy, 551, *551*, 560
Pyrimidines, 61, 69, 210, *211*
Pyruvate, 122, 138
    in presence of oxygen, 127, *127*

Q

QB, 156
QB protein, 156
Quartz, shocked, 382

R

R factors, 265, *266*, 274
r_{max}, 609, 624
Rabbit, digestion in, 751
Races, 302–304, 372, 396
  genetic variations within, *302*, 302–303
Radiation, adaptive, 376, *377*, 396
    plants and, 379
   benefits and dangers of, 29–30
   ionizing, 29
   from radioisotopes, 29
Radicle, 680, 682–683, 692
Radioactive dating, 329
Radioactive isotopes, 29, 44
Radioactivity, living organisms and, 29–30
Radioisotopes, 29, 44
Radula, 511, 521
Rain forests, tropical, 385, 575–578, *576, 577,* 582
Rain shadow, 566, *567,* 582
Ramón y Cajal, Santiago, 845, 846, 853
Rana cascadae, 535
Rana catesbeiana, 371, *371,* 535
Rana chiricahuaensis, 535
Rana muscosa, 535
Rana sylvatica, 371, *371*
Rana tarahumarae, 535
Rana yavapaiensis, 535
Randall, John, 209, 210
Random drift, 356, *356,* 357, 366
  founder and bottleneck effects in, 357, *357*
  large populations and, 358
  new mutations and, 357
Random sample, 10, 20
Rattlesnake, sidewinder, 719, *720*
Ray-finned fishes, 531, *532*
Rays, 530, *530,* 542
Reabsorption, renal tubular, 817, 825
Reactants, enzymes binding to, 112–113
  free energy of, 107, 117
Reactions, chemical. *See* Chemical reaction(s)
Reading frame, 244, *244,* 256
Receptor(s), 762, 768
  androgen, 255
  cell surface, 98, 101
  GABA$_A$, 856, 861
  ionotropic, 857
  metabotropic, 857
  muscarinic, 857
  for neurotransmitters, 856–857
  nicotinic, 857
  nicotinic acetylcholine, 857
  olfactory, 880, *881,* 882
  sensory, 870, 880
  stretch, 858, *859*
Receptor potentials, 869, 880
Recessive alleles, 195–196, 205
Reciprocal altruism, 638, 639
Recombinant DNA, 281, 285
  and bacteria, 283–285
  cloning of, 285, 294
  definition of, 281
  development of, 281, *282*
    research involved in, 283–284

DNA ligase and, 283
genetic alteration of animals and plants by, 290–295
genetic engineering and, 276–297
herbicide resistance and, 293
moral questions associated with, 294
protein products of, 290
reprogramming of cells by, 289, *289*
restriction enzymes and, 281, *282*
risks of, in agriculture, 293
uses of, 281–282
Recombinant plasmids, 266, 274
Recombination, of two genes on chromosome, 204
Rectilinear movement, 719, 732
Recycling, by ecosystems, *552,* 552–559
Red blood cells. *See* Erythrocytes
Red bone marrow, 723, *724,* 733
Red tide, 457, *458*
Red Wolf Reintroduction Program, 418, 420
Redi, Francesco, 398–399, 415
  experiment of, 398–399, *399*
Redox reactions, 122, 138
Reducing agent, 122, 138
Reduction-oxidation reactions, 122, 138
Reductionism, in study of muscles, 8, 8–9, *9*
Redwoods, 662
Reflex(es), conditioned, 626, 639
  definition of, 858
  "let-down," 899
  monosynaptic, 858, *859*
  neuronal mediation of, 858–859, 861
Reflexive memory, 868
Refractory period, 851, 861
Regulators, 254
  negative, 253, 256
  positive, 253–254, 256
Relative abundance, 593, 600
Relay neurons, 869, 870, 880
Release factor, 248–249, 249
Releaser, 858
Releasing hormones, 762, 768
Renal artery, 814, 825
Renal cortex, 814–817, 825
Renal medulla, 814–817, 825
Renal tubules. *See* Tubule(s), renal
Renal vein, 814, 825
Renin, 821, 825
Replacement reproduction, 616
Replication, autonomous, 264, 274
  of HIV virus, 271
  integrated, 264–265, 274
  of mobile genes, 267–269
  by plasmids, 265
  of transposons, 267, 272
  in viruses, 262, *263,* 264–265, 274
Replication fork, 228, *230,* 233
Replication units, 229, 233
Repressor, *lac,* 251, 255
Reproduction, asexual, 185, *185,* 192, 204
  human, control of, 899–903
  sexual, 185, 192, 204
    number of chromosomes and, 185
Reproductive isolation, *373,* 373–374
  by polyploidy, 374, 396
Reproductive tract, female, in humans, 884
  male, in humans, 887
Reproductive unit, species as, 425, 431
Reptiles, 530–534, *532, 533*
  birds, and mammals, cladogram for, 428, *429*
Reptilia, cladists and, 428

Residual air, 796, 806
Resilience, 599, 600
Resistance, to antibiotics, 360
  in cockroaches, 360, *360*
  to DDT, 360
Resistance factors, 265, *266,* 274
Resolution, 72, 891, 904
Resource, limiting, 591, 600
Resource partitioning, 591, 600
Respiration, 125, 138
  cellular. *See* Cellular respiration
Respiratory system, epithelial cells of, 798, *799*
  human, 796, *797,* 806
  mucus in, 799, 806
Response elements, 255
Resting potential(s), 849, 861
  maintenance by neurons, 849–850
Restriction enzymes, 270, 274, 283, 294
  in DNA fingerprinting, 300, *300*
  recombinant DNA and, *282,* 283, *283,* 294
Restriction fragment length polymorphisms, 297–298, 300, *300,* 315
Restriction fragments, 295
  joining of, 282–284, *284*
Restriction map, 282, 295
Restriction site(s), 283, 294
Reticular formation, 866
Retina, 873, 880
  image detection by, 874–875, *875*
  organization of, 875, *875*
Retinal, *875,* 880
Retroviruses, 269, 274
  reproduction of, 269–270, *270*
Reuptake, 857, 861
Reverse transcriptase, 269, 274, 284–285, 295
Reversion, 245
Rey, Jorge R., 595
Rh factor, 841–842
Rhagoletis pomonella, 376
Rheobatrachus silus, 535
Rhinoderma, 534
Rhizobium bacteria, 558
Rhizoids, 483
Rhizome, 486
Rhizopoda (amebas), 462, 475
Rhizopus stolonifer, 468, *469*
Rhodophytes, 460–461, 474
Rhodopsins, 875, 880
Rhythm method, 900–901, 905
Ribbon worms (Nemertea), 508, *508,* 521
Riboflavin, 281
Ribonucleic acid. *See* RNA
Ribose, 55t, 59, 69
Ribosomal RNA, 244, 411, *411*
Ribosome(s), 81, *81,* 100
  beginning of translation by, 246–247
  definition of, 243
  evolution of, 411, *411*
  functions of, *243,* 243–244, 256
  production of peptides bonds by, 247
  reading of mutations by, 245
  reading of single strand of mRNA by, 249
  structure of, 243
Ribozyme(s), 243, 410, 416
Ribulose bisphosphate, 155, *155,* 157
Riftia pachyptila, 442, 443
Riley, C.V., 660
RNA, alphabet, 239
  of energy organelles, 273
  messenger (mRNA), 240

RNA, alphabet, *(continued)*
cellular translation of, into polypeptide, 243–244
maternal, 908, 927
reading of ribosomes by strand of, 249
transcribed from DNA, cellular alteration of, 241
as original genetic material, 409–410
ribosomal (rRNA), 244, 411, *411*
in synthesis of polypeptides, 234–235
transcription of RNA polymerase into, by DNA, 240–243
transfer (tRNA), as adapter between mRNAs and amino acids, *244*, 244–246, 256
evolution of, 411–412
recognition of amino acid by, 246, *247*
viral, entry into host cell, 263, *264*
RNA editing, 242
RNA polymerase(s), 254
transcription by, beginning of, 240–241, *241*
of DNA into RNA, *241*, 241–243
RNA splicing, *242*, 242–243
RNA tumor viruses, 269–270
RNA world, 410, 416
Robins, *423*
Robinson, John, 711
Rodents, guinea pigs as, 424
Rods, 874, *875*, 880
Root cap, 686, 692, 698
Root hairs, 651
Root pressure, 662–663, *663*, 674
Root system, 645, 658
Roots, adventitious, 691
of flowering plants, growth of, 686, *686*
functions of, 645, 651–652
oxygen requirements from, 670
water flow through, 651, *654*
Rootstock, 691
Rotifers, 510, *510*
Roughage, 750
Roundup (glyphosate), 156, 292
Roundworms (nematodes), 510, 521
tissue layers of, *502*
Roux, Wilhelm, 166, 182, 907
RU 486, 903, 905
Ruben, Samuel, 153
Rubisco, 155, 158, 159
Rudiments, 917, 928
Ruminants, 746, 752
stomachs of, *747*
Runners, adaptations of, 720–721
stride length of, 721, *721*
stride rate of, 721
Running, by animals, 720
Rusts, 470
Rutherford, Ernest, 27, 30, 103
Rutherford's experiment, *26*

S

S-shaped curve, 609, *610*, 624
Sac fungi (Ascomycetes), 469, *470*, 475
Sagebrush, 567
Saint-Hilaire, Etienne Geoffroy, 522, 523
Salamander(s), 375, *375*, 532–533, *533*, 534, 926
North American tiger, 535
Salicylic acid, 707
Salivary glands, 744, 752
Salmonella, *439*, 830
Salmonella newport, 432, *434*

Salmonella poisoning, 432
antibiotics and, 432–434, *434*
Salt, 33, *33*
consumption of, blood pressure and, 808–810
in U.S., 808
and water, regulation of, in animals. *See* Water and salt regulation
Saltationism, 386, 396
Saltations, 383–384, 396
Saltatory conduction, 853, 861
Sampling, of population, 10–11, *11*
Sands, 669, 674
Sanger, Fred, 226
Saprophytes, 548, 560
Saprophytic oomycota, 465
Saquinavir, 271
Sarcomeres, 728, 729, 733
Savanna(s), 572–573, *573*, 582
animals in, 572
Savanna hypothesis, 390–392
Scallop(s), eye of, *511*
Scanning electron microscope, 73
Scarification, 684
Scavengers, 548, 560
Scheibinger, Londa, 476
Schleiden, Matthias, 71, 74
Schultes, Richard, 677
Schwann, Theodor, 71, 74, 118
Schwann cells, 847
Scientific method, 4, 20
Scion, 691
Sclera, 873
Sclerenchyma, 646, *647*, 658
Scolex, 507, 521
Scopes, John T., 318
Scopes trial, 318
Scorpions, 519
Scrapie, 46–49
prions and, 48, *48*
Scrotum, 886, *886*, 904
Scurvy, 739
Scyphozoa, 504
Sea anemones, 504
Sea hare, gill withdrawal in, 859, *860*
learning by, 860, 861
Sea lettuce, *461*
Sea stars, 524, *526*, 526–527
Sea urchin, larva of, *924*, 925
Sea urchin embryo, cleavage in, 910, *911*
gastrulation in, 910, *911*
Seabirds, excretion of salt by, 812
Search for Extraterrestrial Intelligence (SETI), 403
Second messengers, 98–99, 764, 765, 768
Secondary cell wall(s), 645, *646*, 658
Secondary growth, 647, 658
Secondary oocytes, 885, 889, 905
Secondary plant compounds, 588, 600, 657
Secretin, 759
Secretion, by digestive tract, 743
by tubule, 819
Sediments, 621
Seed(s), formation of, in flowering plants, 680–681, *682*
germination of, 682–683, 684
in plant fertilization, 487, 488
size of, *684*
vascular plants and, 485
Seed plants, reproduction by, 487, 494
without flowers, 487

Segmentation, 747
Segregation, principle of, 197–198, *198*, 206
Seizures, in epilepsy, 864
Selection, kin, 637, 639
natural, 633. *See* Natural selection
altruism and, 636–637
sexual, 633, *634*, 639
Selection theory, memory and, *838*, 838–839
self-nonself recognition and, 838–839, 844
Selective serotonin reuptake inhibitors, 857
Self-nonself model, 826, 828
Self-nonself recognition, 834, 837–838
selection theory and, 837–838, 843
Semen, 888, 904
Semicircular canals, 873, 880
Semilunar valves, 782–783, 787
Seminal vesicles, 888, 904
Seminiferous tubules, 887, *887*, 904
Senescence, 705, 708
Sensation(s), 870, 880
brain regions in, 863–864
classes of, 871
somatic, 879, *879*
Sense organs, 870, 880
nerves, and brain, 865
nervous system and, 862–883
Sense strand, 240
Sensitive period, for learning, 630, 639
Sensitization, 859
Sensory information, 865
brain and, 870–871
Sensory neurons, 858, 859, 861
Sensory pathways, perception and, 878–879
smells and tastes, *878*, 878–879
somatic, 877, *877*
Sensory receptors, 870, 880
Sepals, 492, 687, 692
Septa, fungal, 466
Sequoia forests, *40*, 599, *599*
Sequoia sempervirens, 662
Serosa, 747
Serotonin, 854, *855*, 860, 861
Serotonin reuptake inhibitors, selective, 857
Sertoli cells, 887, 904
Sertraline, 857
Sessile, 503, 520
Set point, 731, 733
Setae, 511, 517
Sex chromosomes, in humans, 192–193, *194*
Sex-linked genes, in *Drosophilia*, 201, *202*
Sex steroids, 763, 766, *768*
Sexual development, environmental effects of, 754–755
Sexual dimorphism, 364, 367
Sexual intercourse, external genitalia in, 889–890
Sexual reproduction, 185, 192, 204, 882–907
evolution of, 882
human control of, 900–905
number of chromosomes and, 185
variation and, 348
Sexual responses, stages of, 890–891
Sexual selection, 364–365, 367, 633, *634*, 639
Shared derived characters, 427, 431
Sharks, 526, 538
Shear, 724
Sheep, cloning of, *921*, 921–923, *922*
Shells, electron, 30, 44
Sherman, Paul, 638
Shock, 770
Shocked quartz, 382

Shoot system, 645, 658
Shoots, functions of, 645
    growth of, 687
Short-term memories, 868, 880
Sibling species, 371, 396
Sickle cell anemia, 66, 220, 305–306, 313
    malaria and, 365, 365–366
Sidewinding, 719, 720, 732
Sierra Nevada, 566, 566
Sieve plate, 649, 659
Sieve tube members, 649, 649
Sieve tubes, 649, 659, 670, 671–672
Sight, 871, 874–878, 882
    defects of, 875, 875–876
Sign stimulus, 628, 639
Signal peptide, 249, 255
Signal transduction pathway, 764
Signaling molecule, 254
Signals, chemical. See Chemical signaling
Silencers, 255
Silent mutations, 245
Silk, structure of, 67
Silk moth, behavior of, 629
Silts, 669, 674
Silurian Period, 519
Simberloff, Daniel, 594–595
Simple pit pairs, 647, 648, 658
Singer, Jonathan, 91
Single-nucleotide polymorphisms, 298
Sinoatrial node, 784, 787
Sinuses, 779, 787
Sister chromatids, 171, 188, 190
Skates, 530, 530
Skeletal muscle, 713, 726, 727, 728, 733
    and bones, 726
    neural control of, 726
Skeleton, of birds, 720, 720
    hydrostatic, 721, 722, 732
    in movement of animals, 721–730
    vertebrate, parts of, 722–723, 723
Skin, as barrier to pathogens, 829–830
    broken, body defense in, 830–833
    microorganisms inhabiting, 436
    transplantation between twins, 827
Skinner, B.F., 626
Skoog, Folke, 696
Skull, 722, 732
Skunk cabbage, 491
Sliding filament model, 729, 733
Slime layer, 441, 441, 450
Slime molds, cellular, 464, 464–465
    plasmodial, 465
Slug, 464, 464
Small intestine. See Intestine(s), small
Smell, 870, 878–879, 880
Smellie, William, 477
Smiling, by infants, 628, 628
Smith, William "Strata," 323
"Smoker's cough," 799
Smoking, causes of death related to, 792
    damage to lungs due to, 799–800
    history of, 789
    lung cancer and, 789, 789–791, 790
    reasons for, 793
Smooth muscle, 713, 726, 728, 733
Smuts, 470
Snakes, 534, 536
    garter, 627–628
    internal fertilization by, 187
    locomotion in, 719, 719, 720

Snapdragons, inheritance of flower color in, 194–195, 195
Social behavior, 634, 635
Social groups, animals in, 634–638
Sociobiology, 626, 639, 627
Socrates, 658
Sodium channels, voltage gated. See Voltage gated sodium channels
Sodium chloride. See Salt
Sodium ions, in membranes, 851
Sodium-potassium pump, 94, 101, 850
Sog gene, 523, 923
Soil(s), bacteria and fungi in, 435, 437
    desert, 569
    nutrients from, 669t
    of tropical rain forests, 576
Solar system, formation of, 404, 405
Solutes, 39, 811
    in nephron, 819, 820
Solution, pH of, 42–43
    water molecules in, 40–42
Solvent, water as, 39–40
Soma, of neuron, 847, 848
Somatic cells, 186, 204
Somatic sensations, 870, 880, 877, 877
Somatomammotropin, human chorionic, 897, 905
Soule, Michael, 602, 603, 604, 607
Sound, as perception, 862
Southern blotting, 288, 295
Spallazani, Lazzaro, 399, 415
    experiment of, 399, 400
Sparrow, Western song, 372
Speciation, allopatric, 374, 375, 396
    geographic isolation and, 374
    parapatric, 376
    sympatric, 374–376, 375
Species, 322
    biological, 371
        definition of, 425
    biological concept of, 425, 431
    in community, interaction of, 587–593
    competing, coexisting, 591–592
    definition of, 369–373
    density of, latitude affecting, 593, 594
    diversification of, on continents, 379
    diversity of, 369–371, 371
    as ecological unit, 425, 431
    evolution of, speed of, 383–388
    formation of, 373–376
        gene flow and, 376
        geographic isolation and, 374–376
    as genetic unit, 425, 431
    grouping of, 425–430
    incipient, 372, 396
    interactions among, 590t
    K-selected, 612–613
    migration of, 595, 597
    morphologically defined, 425
    multiplication of, 376–383
    naming of, 423
    r-selected, 612
    recognition of, 423–425
    reproductive isolation of, 373, 373–374, 396
    as reproductive unit, 425, 431
    sibling, 371, 396
    two, sharing habitat, 591
    typological, limitations of, 425
    typological concept of, 424, 431
Species richness, 593, 600
Specificity, of immune system, 834

Spectrum, absorption, 147, 148, 159
    action, 147–148, 148, 159
Spemann, Hans, 907–908, 918, 921
Sperm, 186, 204
    development and maturation of, 887, 888
    human, 887, 888
    passage of, mammalian, 888
    production of, 754
        hormones and, 892, 893–894
        infertility and, 903
        by mammals, 887–888
    reaching ova, 891–892
    union of egg with, 186–188, 187
    union with oocyte, 888
Spermatids, 887, 904
Spermatogenesis, 884
Spermatogonia, 887, 905
Spermatozoon. See Sperm
Spermicides, 901, 905
Sphagnum moss, 483
Sphincters, 744
Spiders, 519
    book lungs of, 795
    digestion in, 742
Spinal column, 722, 732
    anterior, 713
    posterior, 713
Spinal cord, 865, 865–866, 866, 880
Spiracles, 517, 521, 801, 806
Spirilla, 439, 450
Spirochetes, 448, 448, 450
Sponges (Porifera), 502, 503, 503
Spongy bone tissue, 723, 724, 733
Spongy parenchyma, 655, 656, 659
Spontaneous abortion, 192–193, 205, 897, 905, 918, 928
Sporangium(a), 468, 475, 484
Spores, 481, 490
    fungal, 467, 468
    in phaeophytes, 460, 474
Sporophores, 466, 475
Sporophylls, 486, 494
Sporophyte(s), 474, 484, 484, 494, 678, 692
    diploid, 484
    in phaeophytes, 460
Sporozoans, 462–463, 475
Sporozoite, 463
Sport utility vehicles, 555
Squamata (lizards and snakes), 534, 536
Squid, 514, 514
    neurological experiments on, 851
St. Anthony's fire, 471
St. Hilaire, Geoffrey, 908
St. Martin, Alexis, 734, 735
Stability, of community, 599, 600
Stabilizing selection, 362, 363, 367
Stalin, Joseph, 344
Stamens, 492, 494, 678, 692
Standard deviation(s), 11, 12, 13, 20
    calculation of, 12
    mean and, 12
Stanier, Roger, 447
Stanley, Wendel, 261
Stapes, 871
Starch, 69
    in plants, 60, 61
Starfish, 526, 526–527, 612, 612
Start, 177
Start site, 246, 256
Stasis, 386, 397

Statistical analysis, 20
Statistics, in estimation of value, 11
Statoliths, 698, 708
Stele, 649, 652, 654, 659
Stem cells, 723, 751–752, 921, 926, 928
      embryonic, 926
      human, 926
Stems, water flow through, 665
      water transport in, 653–654
Stereocilia, 871, 872
Steric inhibitor, 115–116, 116, 117
Sterility, hybrid, 374, 396
Sterilization, 900, 902–903, 905, 903–904, 906
Sterna paradisaea, 120
Sternum, 776, 786
Steroids, 58, 59, 763, 764, 768
Stickleback, courtship of, 634
Stigma, 679, 692
Stimulus(i), characteristics of, 870
      conditioned, 859
      sign, 628, 639
      supernormal, 629, 629, 639
      unconditioned, 859
Stochastic events, 604, 624
      extinction of species and, 606
Stomach, 745–746, 752
      of ruminant, 747
      tissues and cells of, 746
Stomata, 142, 142, 158, 159, 481, 494, 644, 655,
      659
      opening and closing of, 666, 666–667
      transpiration regulation by, 666–670
Storage, by digestive tract, 743
Streptococcus pneumoniae, 221, 222, 221, 438
Streptomyces, 445
Streptomycin, 361
Stress, response to, 760–761, 762
Stressors, 760
Stretch receptors, 858, 859
Striated muscles, contraction of, 729, 729–730
Striations, 726, 733
Stride length, 721, 721
Stride rate, 721
Strobili, 487, 494
Stroke, 782, 787
Stroke volume, 785, 787
Stroma, 149, 159
Stromatolites, 401, 401–402
Sturtevant, Alfred, 204
Style, 679, 692
Stylet, 671, 671
Subcutaneous tissue, 830, 842
Suberin, 651
Subspecies, 372, 396
Substantia nigra, 866
Substrates, 112, 113, 113, 117
Succession, 585, 600
      on corpse, 598
      plant traits during, 597, 597t
      primary, 596–597, 600
      reason for, 598
      secondary, 597, 600
Successional communities, 597, 600
Sucrose, 59, 69
Sugar(s), 52, 54–55t, 68
      formation of polysaccharides by, 59–61
      making of, in Calvin cycle, 155, 155
      movement of, 660–675
            in plants, 670, 674
      for storage of energy, 60

Sulci, 868, 880
Sulfolobus bacterium, 412
Sun, as energy source for plants, 145–153
Sunlight, 146
      evaporation of water by, 553
      as influence on ecology, 549–550
      movements of Earth, and climate, 564–566
Superfecunidity, 341, 342
Superorganism, 545, 560
Suppressor mutations, 245
Surface area, gas exchange and, 794–795
Surface tension, 38t, 39
      alveoli and, 800
Surface-to-volume ratio, 794–795
Surfactants, 622, 800, 806, 918
Surrogate motherhood, 905
Survivorship, life history and, 613, 613–616, 615
Survivorship curves, 613, 613–614, 615
Sutton, Walter S., 180–181, 181, 183, 199, 217
      six rules of inheritance of, 200t
Sutton-Boveri chromosomal theory of inheritance,
      199, 204
Sutures, 722, 723, 732
Swim bladder, 531, 542
Swimmers, adaptations of, 718
      shape of, 718, 718
Symbionts, 446, 450
Symbiosis, 461, 545, 560, 590
Symmetry, in animal classification, 502, 502
      bilateral, 502, 502
      radial, 502, 502
Sympathetic nerves, 726, 869, 880
Sympathetic nervous system, 726
      heart rate and, 785, 787
Sympatric speciation, 374–376, 375
Symplast, 651
Symplocarpus foetidus, 491
Synapse(s), 847
      chemical. See Chemical synapses
      electrical, 854, 861
      excitatory, 854
Synapsis, 188, 204
Synaptic cleft, 854
Synchronous cell populations, 177
Syngamy. See Conception; Fertilization
Synthesis, abiotic, 404, 415
      prebiotic, 405–407, 416
Syphillis spirochete, swimming, 448
Systematics, grouping of, 425, 431
Systemic circulation, 776
Systemin, 707
Systole, 782–783, 787
Szent-Györgyi, Albert, 128–129
Szostack, Jack, 410

T

T-cell receptors, 840, 843
T cells, 827–828, 829, 834, 835, 843
      altered, self-recognition of, 841
      cytotoxic, 840, 843
            poison ivy and, 836, 836
      elimination of, 841
      helper, 840, 840, 843
T lymphocytes. See T cells
Tagmata, 517, 521
Taiga, 564, 565, 572–573, 582
Tannins, 657, 659
Tapeworms, 506, 507, 508, 521
Tarahumara frog, 535

Tardigrade, 716
Target orders, 751, 752
Target tissues, 702, 708
Targets, 756, 768
Tarweeds, 379, 379
Taste, 870, 878, 880
      taste buds and, 878, 878
Taste buds, 878, 878
Tatum, Edward, 218, 219
Taxis, 441, 450
Taxon(s), 322, 423, 431
      monophyletic, 427, 431
      paraphyletic, 427, 431
      polyphyletic, 427, 431
Taxonomic categories, 372–373
Taxonomist, 421, 431
Taxonomy, 322, 332–333, 342, 425–426, 431
      classical evolutionary, 426
Tay-Sachs disease, 196, 311, 312, 359
Tears, salty, 812, 812
Teeth, human, 744
      mammalian, 539–540
Telophase, in meiosis, 188, 190
      mitotic apparatus during, 173
Temin, Howard, 269
Temperate deciduous forests, 573–575, 574
Temperature, atmospheric carbon dioxide and, 555,
      555
      body, regulation of, 731
            tolerance of, 731, 731
      changes in water, 36–37
      dissolution of gases and, 793–794
      equilibrium point of reaction and, 110, 112
      global, 555, 555
      latitude and, 565, 565
      of major biomes, 567, 568t
      measurement of, Metric System, Appendix B
      regulation of, in vertebrates, 538
Temperature turnover, in lakes, 581, 581
Template, 227, 233
Template strand, 240
Temporal isolation, 373, 396
Ten percent law, 551, 560
Tendons, 723, 724, 732
Teosinte, 676–677
Teratogens, 917, 928
Terminal buds, 687, 692
Termination, 248–249, 249
Termination signals, 255
Terminators, 241, 254
Tern, arctic, 120, 534, 719
Terpenes, 657, 659
Territoriality, 633, 639
Test, 457
Test-tube babies, 313
Testcross, 197, 197, 205
Testes, 186, 204, 766, 883, 887, 904
Testicles, 883, 904
Testicular feminization, 255
Testosterone, 98, 184, 755, 766, 893, 904
      aging and, 893, 893–894
      hierarchy of, 893, 893
      release of, 893
      role of, 893
Tetracycline, 441
      resistance to, 437
Tetrad, 188, 204
Tetrapods, 525, 538
Tetrodotoxin, 851, 852
Thalamus, 867, 880

Thalidomide, 917
Thamnopis elegans, 627–628
Theory, 13, 20
Theory of island biogeography, 594, 600
Thermal pollution, 621
Thermoacidophiles, 443, *445*, 450
Thermodynamics, 103, 104, 107
    First Law of, 104–105, *105*, 106, 116
        prediction of direction of reaction by, 104–106
    Second Law of, 105, *105*, 109, 116
Thermoplasma acidophilum, 414
Thiamine, 739
Thick filaments, 729, 733
Thimann, Kenneth, 694–697, *696*
Thin filaments, 729, 733
Thoracic cavity, 713, 732, 798, 806
Thorax, 517, 521
Threonine, 236
Threshold response, of nerve cells, 847
Throat, 744
Thrombin, 831, 842
Thylakoid space, 149, *152*
Thylakoids, 85, 100, 149, *150*
Thymine, 210
Thyroid gland, 766
Thyroid hormones, 764
Thyroid-stimulating hormone, 762
Tick, sign stimuli of, *629*
Tidal volume, 796, *798*, 806
Tiger(s), 634, *635*
Tight junction, 98, *99*, 101
Time, direction of, 104, *104*
Tinbergen, Niko, 629
Tissue(s), 499, *499*, 520
    connective. *See* Connective tissue
    endothelial, 713
    epithelial, 713
    muscle, 713
    oxygen carried to, 801–802
Tissue culture, 691–692
Tissue plasminogen activator, 832, 843
Tmesipteris, 487
Toad(s), 533
    boreal, 535
    clawed, nuclear transplantation in, 919, *919*
    golden, 535, *535*
    Western, 535
Tobacco. *See* Smoking
Tobacco industry, 788–791
Tobacco mosaic disease, 260
Tobacco mosaic virus, 260, *262*
    crystals of, 261, *261*
    filtration of, 260, *260*
Tobin, Allan, 297, 298
Tolerance, 730, 731–732, 733
    in humans, 731, *731*
Toleration, by immune system, 834
Tomato bushy stunt virus, *262*
Tomatoes, genetic engineering of, 280–281, 293
Tonegawa, Susumu, 837
Tongue, taste buds on, 880, *880*
Topography, climate and, 566–567
Torpor, 568
Torricelli, Evangelista, 792
Tortoise(s), 534, *536*
    desert, *810*
    excretion of salt by, 812, *812*
    Galápagos, 368, *368*
    giant, *536*
Torus, 648

Totipotency, differentiated animal cells and, 918–919, *919*
    in embryonic cells, 906–907, *907*, 927
Toxic waste, global problem of, first, 413
Trachea(e), 513, 517, 796, *797*, 801, *801*, 806
Tracheids, 647, 659
Tracheophytes, 479, 485–490. *See also* Plant(s), vascular
Tradewinds, 565, 582
Trait(s), adaptive, 321, 342
        accumulation of, 341
            in inheritance, 341
        expressed by genes, 305
        frequency of, selection and, 358–360, *359*
        influenced by genes, *306*, 306–307
        passing of, 19
        polygenic, *306*, 306–307, 315, 351, *351*
        sexual selection and, *364*, 364–365
Transcriptase, reverse, 269, 274, 284–285, 295
Transcription, 239–240, *239*, *241*, 241–242, 256
Transcription factors, 241–242, 256, 764
Transcriptional control, 250, 256
Transduction, 870, 871, 880
Transfer RNA. *See* RNA, transfer (tRNA)
Transgene(s), 290, 296
Transgenic organisms, 290, 295
Transition state, 113, *113*, 117
Translation, 238–239, *239*, *250*, 255–256
        evolution of, 411
        initiation in, 246, *248*
Translational control, 251, 256
Transmissible spongiform encephalopathies, 46
Transmission, 870
Transmission electron microscope, 72–73, *73*
        in assessment of mitochondria, 84, 85
        in investigation of eukaryotic cells, 77–79
Transmission genetics, 183, 204
Transpiration, 552, 560, 655, 662, 664–666, 674
        regulation of, by stomata, 666–670
Transpiration-photosynthesis compromise, 666, 667, 674
Transport, active, 94, *95*, 101
        passive, 93–94, *95*
Transposase, 266, 274
Transposon(s), 258, *266*, 267, 274
        complex, 267, 274
        replication of, 267, 272
        simple, 266–267, 274
Trees, carbon dioxide in, 557
        secondary growth in, 689–691, *690*, 692
        of tropical rain forests, 576
Trematoda, 506, *507*, 521
Treponema pallidum, 435, *448*
Triacylglycerol, 56, *57*, 69
Triassic Period, 331, 532, 536
Tribolium castaneum, 591
Tribolium confusion, 591
Trichoplax adhaerens, 503
Trigger neuron, 858
Tripsacum, 677
Trisomy, 231, 233
Trisomy 21, *191*, 191–192, *192*, 193, 205
        risk of, 193t
Trivers, Robert, 638
Trophic level, 548, 560
Trophoblast, 910, 927
Tropical rain forests, destruction of, 576–578
        productivity and diversity of, 575–576, *576*, *577*, 582
Tropics, 565, 582

Tropism, 697
True breeding, 194
Trypanosomes, 459–460, 474
Tschermak, Erich, 349
Tuatara, *536*
Tubal ligation, *902*, 902–903, 905
Tube body plan, *908*, 908–909, 912
Tube worm, *516*
Tuberculosis bacteria, resistant strains of, 437
Tuberculosis isolation wards, 437, *437*
Tubule(s), Malpighian, 824, *824*
        renal, 816, 825
            distal, 817, 825
                lumen ion content and, 819
            proximal, 817, *818*, 825
            reabsorption by, 817, 825
            recycling by, 817–819, 825
            secretion by, 819
            structure of, 817–818, *818*, 825
Tubulins, 86, 254
Tumor necrosis factor-α, 833, 843
Tundra, 564, *569*, 569–570
Tungara frog, 633, *634*
Tunicates, 528, *528*
Turbellaria, 506, 521
Turbinate bones, 716, 732
Turgor pressure, 93, 101, 663, 674
Turtle(s) (Chelonia), 534, *536*
        eastern chicken, *810*
        green sea, *810*
        red-eared, *120*, 125
        water and salt balance in, *810*
Twins, identical, 184, *184*
        transplantation of skin between, 827
Twort, Frederick, 261
Tympanic membrane, 871
Tympanuchus cupido, 605, 606, *606*
Typological species concept, 424, 431
Tyrannosaurus rex, 710

U

Ulcer(s), 2
        peptic, 2
Ulva, 461
Umbilical cord, 898
Unconditioned stimulus, 859
Uniformitarianism, 324, *324*, 341
Uniramia, 518, 520, 521
Unite de plan, 522, 523
Universality, 238
Updwelling, 578, 582
Uracil-DNA glycoside, 231
Urea, 558, 560, 813, *814*, 824
        flow of, 819, *820*
Uremia, 818
Ureters, 814, 825
Urethra, 814, 825, 889, 904
        microorganisms inhabiting, 436
Urey, Harold, 406
Uric acid, 558, 560, 813, *814*, 824
Urine concentration, by kidneys, 819–820
Urodela (salamanders), 532–533, *533*, 534
Ussher, Archbishop James, 321
Uterus, 883, 886, 904

V

Vacuoles, 79
Vagina, 886–887
        microorganisms inhabiting, 436
Value, estimation of, statistics in, 11

van der Waals attractions, 44
    and hydrophobic interactions, 35
van Helmont, Johann Baptista, 140–141
    plant experiment of, 141, *141*
van Leeuwenhoek, Antoine, 70–71, *71*, 399, 435, 906
van Niel, Cornelius, 143–144
Variant(s), creation and maintenance of, 346–347
    preexisting, natural selection and, 361
    in variation, 351–352, *352*
Variation, genetic. *See* Genetic variation
    individual, 341, 348
    natural and induced, in genetic engineering, 280
    sexual reproduction and, 348
    source of, evolution and, 350–355
    variants in, 351–352, *352*
Vas deferens, 888, 904
Vascular cambium, 689, 690, *690*, 692
Vascular system, 645, 658
    in plants, organization of, 647–649
    of sea stars, *526*, 526–527
Vascular tissue system, 646, 658
Vasectomy, 902, *902*, 905
Vasoconstriction, 786, 787, *830*, 831, 842
Vasodilation, 786, 787
Vasopressin, 761, 823, 825
Vauxia, 498
Vavilov, Nikolai, 344
Vector(s), 284, 295
Vegetal pole, 910, 928
Vegetative reproduction, 691, 692
Veins, 780, 787
    pulmonary, 776, 786
    renal, 814, 825
    valves of, *774*, 775, 779
Venae cavae, *772*, 776, 786
Venter, Craig, 308, 311
Ventilation, 798, 806
Ventral horns, 866
Ventral structures, 519
Ventricle(s), 776, 786, 865
Ventricular fibrillation, 784, 787
Venules, 779–780, 787
Venus flytrap, 670, *670*
Vertebrae, 722, 732
Vertebrates, adaptations evolved by, 532–541
    apoptosis in, 917
    back strength in, *529*
    brain of, 866–868
    central nervous system in, organization of, *865*, 865–868, 880
    development of, stages of, 909
    distinguished from other chordates, 528–529
    endoskeleton of, parts of, 722–723, *723*
    limbs of, homology in, 333, *333*
    nonmammalian, kidneys of, 820–821
    peripheral nervous system organization in, 878–869
    temperature regulation in, 538
Vesicles, 79
Vessel elements, 648, *648*, 659
Vestibular sensation, 870, 880
Vestigial structures, 334, *336*, 342
Vicariance, 379, 396
Villus (Villi), 747, 752
Virchow, Rudolf, 74
Virus(es), attachment of, to cell, 262–263, *263*
    and bacteria, 260

bacteriophages as, 223, 232
    definition of, 260–264, 274
    evolution of, 272
    expression of genes by, 269, *269*
    forms of, 261, *262*
    gene cloning in, 284
    life cycles of, *268*
    lytic, *263*, 264, 274
    multiplication cycles of, 271–272
    natural selection of, 270–272
    particles of, manufacture and release of, 264
    replication of, 262, *263*, 264–265, 274
    RNA tumor, 269–270
    tumor, insertion into DNA, 267, *268*
Visceral mass, 511, 521
Vision, 870, 873–877, 880
    of animals, 873–877
    averted, 874
    defects in, *875*, 875–876
Visual cortex, primary, *876*, 876–877
Visual illusions, 878, *878*
Visual system, 869, 880
Vitalism, 118, 138
Vitamin(s), as antioxidants, 643, *643*
    deficiency of, 738
    essential, in human diet, 739, 740t
    fat-soluble, 739
    water-soluble, 739
Vitamin A, 643
Vitamin B$_1$, 739
Vitamin C, 643, 739
Vitamin E, 643
Vitreous humor, 874
Viviparity, 897, 906
Vocal cords, 796
Volcano, 584
Voltage, 849, *850*
Voltage gated sodium channels, 851
    opening and closing of, 851–852, 861
    pumping of, 852
Voltmeter, 850
Volts, 850
Volume, measurement of, Metric System, Appendix B
Voluntary muscle. *See* Skeletal muscle
Volvox, 461
Vomiting, 746
von Baer, Karl Ernst, 906
Vulva, 889, 904

W

Wake, David, 535
Walcott, Charles Doolittle, 496, 497, *497*
Walcott, Helena, 492
Walking, 390, *391*
    by animals, 720
Wall cess, 689
Wallace, Alfred, 326, *326*, 348
Warfarin, 832
Warren, J. Robin, 2
Warts, genital, 164
Wasps, sex ratios in, 614
Wastes, nitrogenous, 813, *814*, 824
Water, in biochemical reactions, 40
    as coolant, *39*
    as essential, 36–44
    ethanol, and oleic acid, compared, 38t, 39
    heat capacity of, 37, *39*

hydrogen bonding in, 39
    movement of, 660–675
        roots in, 662–663
        from soil to air, 662
        upward, 662–663
    in plant growth, from soil, 669
    plant productivity and, 157–158
    polarity of, 34, *34*
    properties of, adhesion, 38t, 39
        cohesion, 38t, 39
        density, 38t
        polar, 40
    retention of, by saltwater fish, 812
    and salt, regulation of, in animals. *See* Water and salt regulation
    in solution, 40–42
    as solvent, 39–40
    temperature changes and, 36–37
Water and salt regulation, in aquatic animals, 811, *811*
    by kidneys, 814, 817, 819, 825
    in land animals, 812–813
    water flow in, 810–811
Water cycle, 552–553, *553*
Water intoxication, 813
Water pollution, 621–624
Water potential, 664, 674
Water-soluble signaling molecules, 764, 765
Water-soluble signals, 762, *763*, 768
Water supply, limited, adaptations to, 668
Watson, James, 234, 237, 308
Watson, James D., 208, 211, 212–214, *213*, 216, 217, 225, 233
Watson-Crick base pairs, 240
Wavelength, *145*, 145–146, *147*, 159
Wayne, Robert, 419
Webbed feet, 715
Weed, 156
Wegener, Alfred, 339
Weinberg, G., 354
Weismann, August, 166, 182–183, 347–348
Weissman, Charles, 46
Welwitschia, 490
Went, Frits, 694, 695–696, 702
Wernicke, Carl, 863
Westerlies, 565, 582
Western blotting, 288, 295
Wexler, Leonore, 298, *298*
Wexler, Milton, 296–297, 298, *298*
Wexler, Nancy, 298, *298*
Whales, countercurrent system in, 716, *717*
White blood cells. *See* Leukocytes
White matter, 866
Whittaker, Robert H., 430
Whittington, Harry, 496–497
Whorled leaves, 654, *655*
Whorls, 687, 688, 692
Wildlife preserves, 607–608
Wilkins, Maurice, 208–210, 212, 214, 216
Wilmut, Ian, 920–921
Wilson, E.B., 180
Wilson, E.O., 385, 574, 593–594, 626, 627
    Sociobiology: The New Synthesis by, 626–627
Wine blight, 660, *661*
Wings, 715
    of birds, 719
Withdrawal, 902
Wiwaxia, 498
Woese, Carl, 447, *447*

Wöhler, Friedrich, 120
Wolf, gray, disappearance of, 418, *419*
    red, as hybrid, 418, 420, *420*
        mitochondrial DNA of, 420
        origin of, 419, *419*
        species status of, 418–421
Wolfe, Ralph, 447
Wolffia, 490
Wolffian ducts, 890
Wolpert, Jim, 662
Woodpeckers, imitation, *378*, 378–379
Work, definition of, 104
    endergonic reactions and, 107, *107*
    kinetic energy from, 104
    potential energy in, 104, *105*, 116
Worm(s), acorn, 527, 542
    arrow, 527, 542
Wound healing, 830, 843

X

X chromosomes, 192–193, 205
Xylem, 485, *485*, 494, 645, 658
    function of, 647–648
    water in, 666, 669

Y

Y chromosome, 192–193, 205
Yeasts, in fermentation, 118, 126
Yellow fever, 260
Yolk, 536, 542
Yolk sac, 536, 542

Z

Z band, 728, 733
Zamia pumila, 487
Zea diploperennis, 677

Zea mays, 687
Zebra fish, discoidal cleavage in, 911
Zebroid, 346, *346*
Zero population growth, 609, 624
Zoomastigina (flagellates), 464, 475
Zygomycetes, 468–469, *470*, 475
Zygosporangium, 468
Zygote(s), 166, 186, *187*, 204, 290, 499, 520, 884, 889, 905
    cleavage of, 909–910, *910*
        mechanisms of, 910–912, *911*
    development of, 908
    division of, 909–910, *910*
Zymase, 119

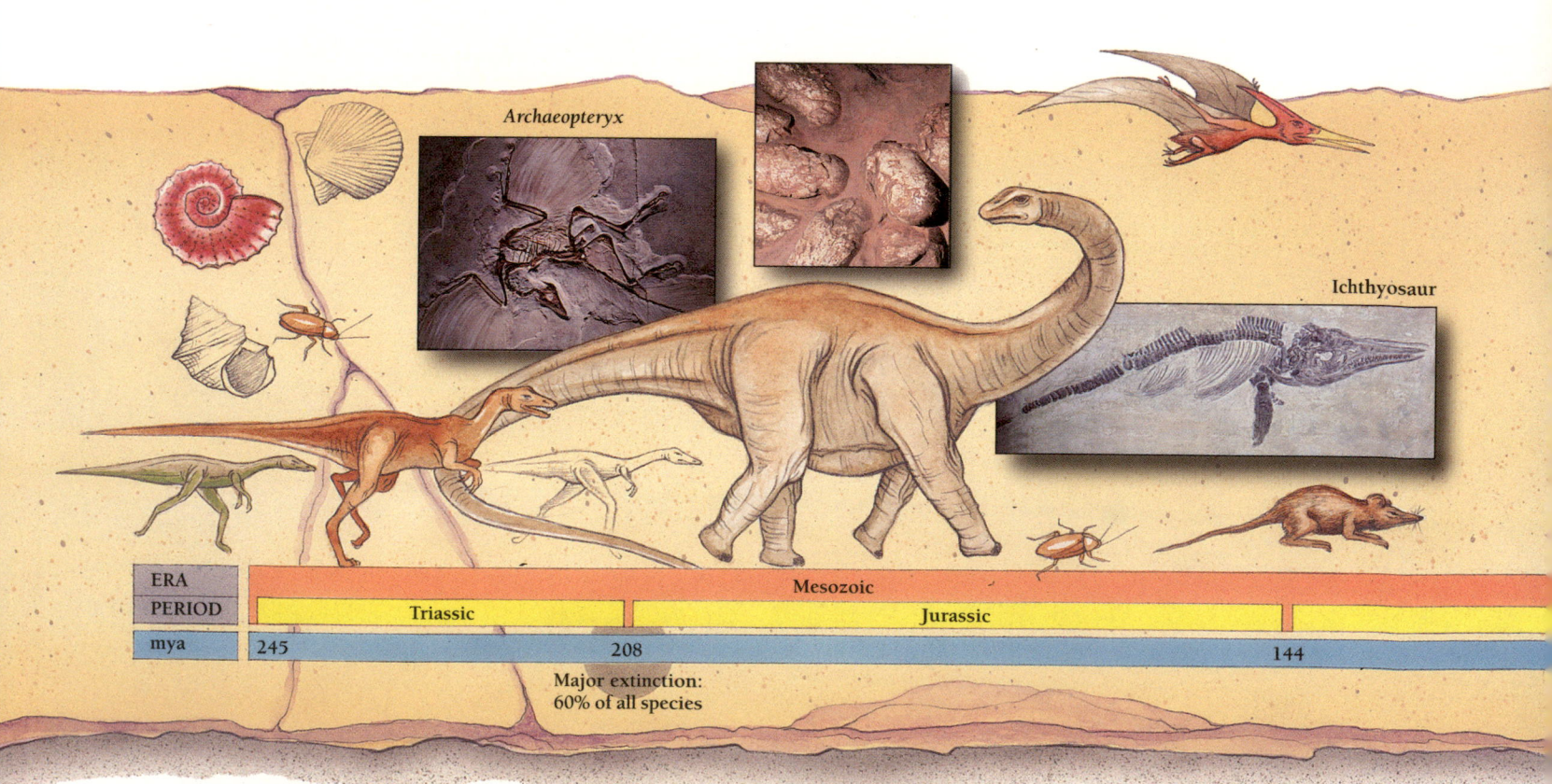

Archaeopteryx

Ichthyosaur

ERA		Mesozoic		
PERIOD	Triassic		Jurassic	
mya	245	208		144

Major extinction:
60% of all species